	Method (page no.)	Adult normal values
Creatinine	543(SU)	21–26 mg/kg/24 hours (♂ U)
	552(A)	16–22 mg/kg/24 hours (♀ U)
without Lloyd's reagent	544	0.9–1.4 mg/100 ml (♂ S)
		0.8–1.2 mg/100 ml (♀ S)
with Lloyd's reagent	546	0.7–1.2 mg/100 ml (♂ S)
		0.5–1.0 mg/100 ml (♀ S)
Creatinine clearance	1538	84–162 ml/min/1.73 m² (♂)
		82–146 ml/min/1.73 m² (♀)
Cystine, qual (U)	591	Negative
Cystine, quant (U)	592	10–100 mg/24 hours
Ferrohemoglobin solubility (B)	1177	88%–100%
Fetal hemoglobin by alkali denaturation (B)	1178	2% of total
Fetal hemoglobin by staining (E)	1180	
Fibrinogen (P)	458	0.20–0.40 g/100 ml
Folic acid	1406(S)	5–21 ng/ml (S)
	1408(E)	160–640 ng/ml (E)
N-Formiminoglu-tamic acid (U)	628	Less than 3 mg/24 hours
Galactosemia, screening (B)	910	
Galactose-1-phosphate uridyl transferase (E)	906	1.4–2.8 mol/ml/hour
Galactose tolerance test (B)	1009	Oral test, up to 60 mg/100 ml in 1 hour. IV test, up to 42 mg/100 ml in 1 hour
Gastric acid secretion	1561	Basal secretion test, 0–10 meq/hour. Augmented histamine test 1–40 meq/hour
Globulin, total (S)	453	2.4–3.5 g/100 ml
Glucose (o-toluidine)	1286(SP)	70–110 mg/100 ml (fasting SP)
	1287(B)	60–100 mg/100 ml (fasting B)
	1286(C)	40–80 mg/100 ml (lumbar C)
Glucose (oxidase method) (S)	1292	70–100 mg/100 ml (fasting)
Glucose (U)	1300	0–0.25 g/24 hours
Glucose-6-phosphate dehydrogenase (B)	849	140–280 units
Glucose-6-phosphate dehydrogenase, screening (B)	847	
Glucose tolerance test	1304	
Glutamic-oxalacetic transaminase (SC)	877	13–55 units (S). 7–49 units (C)

	Method (page no.)	Adult normal values
Glutamic-pyruvic transaminase	886(C)	12–53 units (♂ S)
	889 (A)	6–40 units (♀ S). None detectable (C)
Glutathione (B)	615	28–34 mg/100 ml
Glutathione reductase, screening (B)	855	
Guanase (S)	969	Up to 3 mU
Haptoglobin (S)	1202	21–195 mg/100 ml
α-Hydroxybutyr-ate dehydrogenase (S)	834	67–1⁀⁀
Hemoglobin (cyanmeth⁀ globir	1⁀⁀	(♂) (♀)
Hemog⁀ (iron a⁀		
Hemogl⁀		⁀nl
Hemoglo⁀ (B)		⁀tal
Hemoglob⁀ and E (B⁀		
Hemoglobin fractionation (B)	1182	Type AA
Hemoglobin H (B)	1190	
Hemoglobin S, dithionite-phosphate screening (B)	1175	Negative
Hexosamines (S)	471	75–120 mg/100 ml
Hippuric acid (U)	1007	Oral test, 3–3.5 g in 4 hours. IV test, 0.6–1.6 g in 1 hour
Homocystine, qual (U)	591	Negative
Homogentisic acid, screening (U)	624	Negative
Homogentisic acid (U)	625	Negative
Hydroxyproline (U)	609	Free, 0.2–1.8 mg/24 hours. Total, 25–77 mg/24 hours
Hyperlipopro-teinemias, phenotyping (S P)	1521	
Hyperlipopro-teinemia, phenotypes II, III, IV, indirect confirmation (S P)	1525	
Hyperlipopro-teinemia, phenotype, III direct con-firmation (S P)	1524	
Immunoglobulins (S)	475	IgG, 65–1600 mg/100 ml. IgA, 100–400 mg/100 ml. IgM, 40–160 mg/100 ml. IgD, 0–6 mg/100 ml
Inulin (SPU)	1314	
Iron	681(S). 686(A)	65–175 μg/100 ml

A	automated	P	plasma
B	blood	S	serum
C	CSF	U	urine
D	duodenal fluid	art	arterial
E	erythrocytes, RBC	cap	capillary
F	feces	ven	venous

For complete subject index, see p. 1617. Continued inside back cover.

Clinical Chemistry

PRINCIPLES AND TECHNICS

Clinical Chemistry

PRINCIPLES AND TECHNICS

Second Edition

Richard J. Henry, M.D.

Vice President, Bio-Science Enterprises,
Westwood, California

Donald C. Cannon, M.D., Ph.D.

Director, Bio-Science Laboratories,
Van Nuys, California

James W. Winkelman, M.D.

President, Bio-Science Laboratories,
Van Nuys, California

Medical Department

Harper & Row, Publishers
Hagerstown, Maryland
New York, Evanston, San Francisco, London

To the families
of those who write books.

Clinical Chemistry: Principles and Technics,
Second Edition.
Copyright © 1974 by Harper & Row,
Publishers, Inc.

Standard Book Number: 06-141181-7

Library of Congress Catalog Card Number:
73-20919

Contents

Contributors

CARL ALPER, Ph.D.,
Director, Bio-Science Laboratories, Philadelphia Branch, Philadelphia, Pennsylvania (Chapter 14)

DONALD C. CANNON, M.D., Ph.D.,
Director, Bio-Science Laboratories, Van Nuys, California (Chapters 16, 22, 30, 31)

JAMES A. DEMETRIOU, Ph.D.,
Assistant Director, Research Department, Bio-Science Laboratories, Van Nuys, California (Chapter 21, 27)

JOHN DI GIORGIO, Ph.D.,
formerly Senior Research Scientist, Bio-Science Laboratories, Van Nuys, California (Chapters 11, 17)

PATRICIA A. DREWES, Ph.D.,
Research Scientist, Bio-Science Laboratories, Van Nuys, California (Chapters 21, 26)

JERRY B. GIN, Ph.D.,
Assistant to the Director, Bio-Science Laboratories, New York Branch, Rockville Centre, New York (Chapters 20, 21)

RICHARD J. HENRY, M.D.,
Vice President, Bio-Science Enterprises, Westwood, California (Chapters 1, 12, 13, 15)

BENJAMIN N. HORWITT, Ph.D.,
Director, Department of Endocrinology, Bio-Science Laboratories, Van Nuys, California (Chapter 6)

FRANK A. IBBOTT, Ph.D.,
Director, Department of Chemistry, Bio-Science Laboratories, Van Nuys, California (Chapters 9, 18, 20)

JAMES A. INKPEN, Ph.D.,
Assistant Director, Department of Chemistry, Bio-Science Laboratories, Van Nuys, California (Chapters 16, 20, 28)

S. LAWRENCE JACOBS, Ph.D.,
Assistant to the Director, Bio-Science Laboratories, Van Nuys, California (Chapters 3, 22, 24)

THOMAS S. LA GANGA, Ph.D.,
Assistant Director, Department of Endrocrinology, Bio-Science Laboratories, Van Nuys, California (Chapters 20, 32)

NORMAN D. LEE, Ph.D.,
formerly Director, Century City/Beverly Hills Branches, Bio-Science Laboratories, Los Angeles, California (Chapter 8)

ANIS H. MAKAREM, Ph.D.,
formerly Assistant Director, Department of Chemistry, Bio-Science Laboratories, Van Nuys, California (Chapter 23)

WILLIAM B. MASON, Ph.D., M.D.,
Director, Affiliated Laboratories, Bio-Science Enterprises, Westwood, California (Chapters 1, 2)

IRVING OLITZKY, Ph.D.,
Director, Bio-Science Hospital and Clinic Laboratories, Los Angeles, California (Chapters 7, 16)

VINCENT J. PILEGGI, Ph.D.,
Assistant Director, Affiliated Laboratories, Bio-Science Enterprises, Westwood, California (Chapters 19, 25)

ALLEN H. REED, Ph.D.,
formerly Research Scientist, Bio-Science Laboratories, Van Nuys, California (Chapters 12, 13)

STANLEY M. REIMER, Ph.D.,
Director, Bio-Science Laboratories, New York Branch, Rockville Centre, New York (Chapter 4)

CHRISTOPHER P. SZUSTKIEWICZ, Ph.D.,
Assistant to the Director, Bio-Science Laboratories, Philadelphia Branch,
Philadelphia, Pennsylvania (Chapters 15, 25)

RALPH E. THIERS, Ph.D.,
Director Research, Bio-Science Laboratories, Van Nuys, California
(Chapter 10)

NORMAN WEISSMAN, Ph.D.,
Assistant Director, Department of Chemistry, Bio-Science Laboratories, Van
Nuys, California (Chapter 19)

JAMES W. WINKELMAN, M.D.,
President, Bio-Science Laboratories, Van Nuys, California (Chapters 22, 29)

DONALD R. WYBENGA, B.M.,
Assistant to the President, Bio-Science Laboratories, Van Nuys, California
(Chapters 5, 28, 29)

Preface to the Second Edition

The first edition of Clinical Chemistry was published in 1964. The cut-off date for literature review for that edition was September 1962, and of approximately 15,000 publications read and abstracted, some 7,000 were cited in the bibliography. The author of the first edition had more sense than to try to tackle the job of revision without assistance. This time he enlisted the aid of two co-editors, Drs. James W. Winkelman and Donald C. Cannon, President and Director, respectively, of Bio-Science Laboratories, Van Nuys, and many of the staff of Bio-Science Laboratories—hence the credit "by the Staff of Bio-Science Laboratories." The cut-off date for literature review for this second edition was June 1972. Approximately 10,000 more publications were reviewed, but the total number of references cited has been significantly curtailed. Frankly, this was dictated primarily by publication cost considerations. Many of the references in the first edition were cited for historical purposes; not infrequently in this revision the reader is referred back to the first edition for historical coverage of subject matter.

This new edition happily coincides with the twenty-fifth anniversary of Bio-Science Laboratories, an important milestone for our entire staff as well as for the still active founders of the laboratory. This second edition contains considerable new and revised material and some material unaltered or changed very little. The first edition covered clinical chemistry and thyroxine, but not other hormones or toxicology. Failure to cover these latter two areas was criticized in reviews, but the author had had very little personal experience in these fields; furthermore, near the end of the project it became clear that if these fields had to be included the book would never have appeared in print. This second edition is restricted to clinical chemistry only and does not include hormones (not even thyroxine this time) or toxicology. Again this will undoubtedly be criticized, but the alternative would have been no book at all.

The same general approach in the first edition has been retained in the second: (1) The book deals almost exclusively with analytical aspects of the tests. (2) Early chapters are concerned with general considerations of analytical procedures and with topics such as normal values and sample stability. (3) The remaining chapters deal with specific substances and include presentation of methods of analysis. Again, in regard to the choice of methods presented, no claim is made that those selected are the only good ones or are necessarily the best available. They are methods currently in use at Bio-Science Laboratories and, therefore, have withstood the examination of its Research Department and borne up under practical, everyday experience in a busy, quality conscious laboratory. It must be emphasized, however, that we are continually unearthing defects even in procedures that have been in use for long periods of time — this is hardly a static field. A table of normal values has been incorporated in the "Quick Index to Methods and Normal Values" which has been retained on the end papers of this volume. Also retained is the Appendix (revised), consisting of tables of information ordinarily widely scattered, e.g., buffers, pH reference standards, desiccants, etc. The abridged table of atomic weights and the table of "Useful Information about Concentrated Acids and Bases" have been included in the Appendix.

The editors especially wish to thank Dr. Sam Berkman, President of Bio-Science Enterprises, for his continued encouragement and support of the preparation of this second edition. They also wish to thank Dr. Orville J. Golub for helping in a number of ways in preparation of the revision. There are also various employees of Bio-Science Laboratories who were most helpful in the secretarial phases of preparation. In the latter category special thanks go to Marian B. Ellis. Finally the friendly and valuable assistance of the Medical Department of Harper and Row is gratefully acknowledged.

Because costs of publishing books have continued to rise, ways are always sought to conserve space. "Distilled water" is the reagent most commonly used in the methods presented. By shortening the term to just "water" probably several pages have been eliminated.

The preface to the first edition stated that simple statistical probability made it certain that the volume contained many errors and requested that errors be brought to the author's attention. The author was not disappointed — there was a steady stream of such communications over the ensuing years — and the corrections were made in subsequent printings of the first edition. We renew this request and will appreciate your comments and criticism.

R.J.H.
J.W.W.
D.C.C.

Van Nuys

Chapter 1

Photometry and Spectrophotometry

WILLIAM B. MASON, Ph.D., M.D.
RICHARD J. HENRY, M.D.

2

Photometric measurements are by far the most common terminal steps in quantitative determinations made in clinical chemistry. There are numerous reasons, the more important being speed and ease of measurements, adequate specificity and sensitivity, availability of relatively low cost instruments giving satisfactory accuracy and precision, and ready adaptability to automation. Because photometric measurements find such wide application, it is essential to have a clear understanding of theoretical and practical aspects as well as of instruments used. No instrument merits blind faith, and the analyst must constantly beware of certain pitfalls.

Photometry is an extremely broad subject and except for very brief references to turbidimetry and nephelometry, this chapter concerns itself only with absorption spectrophotometry. Simply stated, absorption spectrophotometry may be defined as measurement of the attenuation by a test material of incident radiation that is spectrally defined. In clinical chemistry the test material is usually a solution, and measurements are ordinarily made in the spectral range 220–800 nm. This range is subdivided into the visible region lying above 380 nm and the ultraviolet below 380 nm. The infrared region that extends above 800 nm finds little application in clinical laboratory work. References cited in the bibliography describe infrared technics and give additional details regarding topics included in the present chapter.

Spectrophotometry using visual comparisons has been practiced for well over a century, and comparisons by photoelectric technics have been in use for about 50 years. It is not surprising that such a long history has left a confusing vocabulary of terminology and symbols. Table 1-1 lists the terms and symbols currently preferred (35) as well as some older synonyms.

ABSORPTION PROCESS

In the absorption process an incident photon gives up its energy to a molecule (termed the *absorber*) resulting in excitation of the absorber to a higher energy level. This process is represented as

$$A + h\nu \longrightarrow A^* \longrightarrow A + heat$$

in which A is the absorber in a low (usually lowest) energy state, A^* is the absorber in an excited energy state, and $h\nu$ (h is Planck's constant; ν is the

3

TABLE 1-1. Definitions of Terms and Symbols Used in Photometry and Spectrophotometry

Preferred nomenclature		Definition	Other nomenclature	
Term	Symbol		Term	Symbol
Absorbance	A	$\log_{10}(1/T) = -\log_{10} T = 2 - \log_{10} \% T$	Optical density, density Extinction Absorbancy	OD, D, d E, ϵ A_s
Absorptimetry		Methods involving determination of absorptive capacity of a system for radiant energy	Colorimetry	
Absorptivity	a	Absorbance/unit concentration and thickness, i.e., specific absorbance $a = A/bc$, where b = internal cuvet length, c = concentration	Specific extinction Extinction coefficient Specific absorption Absorbance index Absorbancy index	K, K_{sp}, k K, ϵ, k a_s
Absorptivity, molar	ϵ	Absorptivity in units of liter/(mole cm.); concentration is in moles/liter, solution thickness in cm.; frequently designated as $A_{1\ cm}^M$	Molar or molecular extinction coefficient Molar absorbancy index	K_m, ϵ_{mol}, ϵ, K, E, A_c, a_M A.
Ångström	Å	Unit of length equal to $1/6438.4696$ of wavelength of red line of Cd; almost, but not exactly, 10^{-10} m.; difference is insignificant in applied spectroscopy		
Filter photometer		An instrument measuring relative light transmittance in which the spectral bandwidth is relatively broad, usually isolated by some type of transparent filter		
Frequency	f, v	Number of cycles of radiant energy/unit time; $\nu = v/\lambda$, where v = velocity of radiant energy, λ = wavelength		
Infrared Near Middle Far		Spectral region from ~ 0.78 to $300\ \mu$ $0.78-3.0\ \mu$ $3.0-30\ \mu$ $30-300\ \mu$		

4

Term	Symbol	Definition	Synonym	Symbol
Light		Radiant energy in spectral range visible to normal human eye ($\sim$ 380–780 nm)		
Micron	μ	Unit of length equal to 10^{-6} m.		
Nanometer	nm	Unit of length equal to 10^{-9} m.	Millimicron	$m\mu$
Photoelectric photometer		An instrument measuring relative radiant power in which radiant power is translated into electric energy by a sensitive element such as barrier layer cell or phototube, the electric energy in turn being measured	Photoelectric colorimeter	
Radiant energy		Energy transmitted as electromagnetic radiation		
Radiant power	P	Rate at which energy is transmitted in a beam of radiant energy	Radiant flux	I
Spectrophotometer		An instrument measuring relative radiant power as function of wavelength; filter photometer with a sufficient number of filters fulfills this definition, though it usually refers to an instrument permitting continuous adjustment of wavelength		
Spectroscope		An instrument which disperses light into a spectrum for visual observation		
Spectrum		Ordered arrangement of radiant energy according to wavelength, e.g., a rainbow is a "visible spectrum"		
Transmittance	T	Ratio of radiant power transmitted by sample (P) to radiant power incident on sample (P_0), both being measured at same spectral position and with same spectral bandwidth: $T = P/P_0$	Transmission Transmissivity Transmittancy	T T T_s
Percent transmittance	$\% T$	$\% T = 100\,T$	% transmission	$\% T, t$
Ultraviolet Far Near	uv	Region of spectrum from 10 to 380 nm 10–200 nm 200–380 nm		
Wavelength	λ	Distance, measured along line of propagation, between two points in phase on adjacent waves; expressed in nm or Å (angstroms) for uv and visible ranges, and in μ for infrared		
Wavenumber	σ, ν	Number of waves/unit length in a vacuum, reciprocal of λ_{vac}; usual unit of wavenumber is reciprocal centimeter, cm^{-1}		$\bar{\nu}$

frequency) is the energy of the incident photon having wavelength λ ($\nu \lambda = c$, where c is the velocity of light). A* is ordinarily unstable and quickly reverts to a lower energy state by loss of thermal energy. Only selected frequencies are absorbed, depending upon the structure of the absorber molecule.

LAWS OF RADIANT ENERGY ABSORPTION

The laws of absorption concern the relationship between the amount of absorber and the extent to which radiant energy is absorbed. Subject to certain limitations to be discussed later, for a solution of any particular absorber and wavelength, there are two variables affecting the degree of absorption. These are concentration of absorber and optical pathlength through the solution. The overall relationship is commonly given as

$$P = P_o\, 10^{-abc}$$

where: P = transmitted radiant power
P_o = incident radiant power, or a quantity proportional to it as measured with pure solvent in the cuvet
a = absorptivity, a constant characteristic of the absorber and the wavelength of incident radiation
b = length of lightpath through the absorber solution, usually in centimeters
c = concentration of absorber

Some authors prefer I_o and I for P_o and P, referring to *intensity* rather than *power* of incident and transmitted radiant energy.

The relationship $P = P_o\, 10^{-abc}$ is commonly referred to as Beer's law although Beer's only contribution was in showing the dependence of absorption upon the amount of absorber present. Bernard independently showed the proportionality to concentration in the same year, 1852. The relationship between absorption and path length was described much earlier by Lambert and still earlier by Bouguer. Historical aspects have been reviewed by Malinin and Yoe (24).

By transposition, Beer's law can be written in the form $P/P_o = 10^{-abc}$. In practice it is the ratio P/P_o, termed the *transmittance* (T) or multiplied by 100, the *percent transmittance* (% T) that is important, and it is convenient to take P_o as the transmitted energy when the absorber has zero concentration. This is accomplished by placing pure solvent or a "reagent blank" in the cuvet and measuring the transmitted radiant power. The solvent or blank is then replaced by the test solution and the measurement repeated in order to obtain the ratio P/P_o. Actually some energy is lost by reflections and scattering, and errors will result unless these losses are equivalent during the two measurements. Chances of error are less when the measurements are made on solutions free of turbidity and using a cuvet (or matched set) having parallel flat faces that are well polished.

Beer's law can also be written in the form $\log (P_O/P) = abc$. The term $\log (P_O/P)$ is defined as the *absorbance* (A). By substitution it is easily seen that absorbance and percent transmittance are related by the expression $A = \log (100/\%T) = 2 - \log \% T$. When measurements are made using a single cuvet (or matched set having constant light path) and optical effects due to the cuvet are reproducible, the term b in the absorbancy expression becomes a constant. Since a is also a constant characteristic of the absorber and wavelength of the incident radiation, absorbance is then directly proportional to concentration, i.e., $A = kc$ (where $k = ab$). The major advantage in using A rather than $\% T$ in quantitative work is that the relationship between concentration and absorbance is linear, whereas with percent transmittance it is not.

Since A is proportional to concentration:

$$A_1/A_2 = c_1/c_2$$

where: A_1 = absorbance of an unknown
A_2 = absorbance of a standard having known concentration
c_1 = concentration of the unknown
c_2 = concentration of the standard

then:

$$\text{concentration of unknown} = \frac{A \text{ of unknown}}{A \text{ of standard}} \times \frac{\text{concentration}}{\text{of standard}}$$

This is the most fundamental equation in analytical photometry.

Since $A = kc$, it is frequently recommended, for simplification of procedure, that k be determined. This is done by determining the absorbance for a known concentration; then:

$$k = A/c$$

This implies that once the k value is determined for a particular quantitative determination, it can be used henceforth in the calculation of unknowns, obviating the necessity of running concomitant standards. Such use of the k value is *not* recommended except in certain rare cases. In any event, the k value can be used only if the linearity between A and concentration has been established as holding for the entire range of concentrations over which it is to be used.

EFFECT OF WAVELENGTH ON ABSORBANCE MEASUREMENTS

The energy sources usually employed with photometers and spectrophotometers provide a continuous spectrum within certain broad limits. If the solution absorbed radiant energy more or less equally at all wavelengths, measurements of transmittance could be made without regard for wavelength. This situation does not exist and the solutions to be dealt with in photometry preferentially absorb energy in one or more regions of the spectrum and less or not at all in others. A hypothetical example is shown in Figure 1-1. The solution giving this absorption curve would be blue-green in color since it absorbs light in the red-orange-yellow region of the visible spectrum. If the beam incident on such a solution contained all visible wavelengths it is easily seen that only a small fraction of the total incident energy would be absorbed—only part of it above and none below 550 nm. Since absorbance depends upon the difference in transmittance between the solution and its reference blank, the use of white light in such an instance renders measurements very insensitive and thus practically useless. Naturally, this same principle applies in the invisible regions of the spectrum.

Let us now consider the other extreme and assume we are dealing with pure monochromatic light (only one wavelength present). Again referring to Figure 1-1, if the wavelength of this monochromatic beam were any value below 550 nm the transmittance would be 100%. The instrument could not differentiate between sample and solvent blank, i.e., there would be complete insensitivity. As the wavelength is increased above 550 nm the sensitivity would increase until a maximum sensitivity is attained at 650 nm, after which it would again decline. Generally speaking, absorbance measurements for quantitative purposes are made where the absorbance is maximal.

FIG 1-1. Hypothetical absorption curve. For explanation, see text.

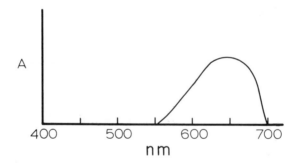

INSTRUMENTS

The basic components of all photometers and spectrophotometers are shown schematically in Figure 1-2 and described in the sections that follow. Included are a source of radiation, a means for isolating the desired wavelength or spectral band, a cuvet for holding the sample, and a means for

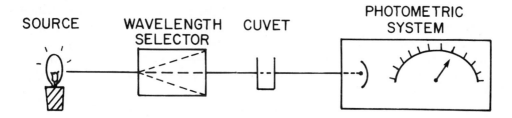

FIG 1-2. Basic components of a photometer and spectrophotometer.

measuring radiant energy and presenting data. In addition, there is an optical system that gathers radiation from the source, presents collimated (parallel) rays to the wavelength selector and cuvet, and focuses the transmitted radiation upon the detector. The overall instrument may be simple or complex depending upon its design and performance objectives.

SOURCES OF RADIANT ENERGY

A tungsten filament lamp is ordinarily used as the radiation source for visible and longer ultraviolet wavelengths, and a hydrogen or deuterium discharge lamp for wavelengths below about 360 nm. Actually, either lamp can be used over most of the range 320–400 nm depending upon specific conditions. As ordinarily operated, tungsten lamps emit most of their radiation in the infrared with only about 10% in the ultraviolet and visible range. By using a voltage slightly greater than the rated value, the lamp temperature is increased, which greatly increases output in the ultraviolet range. This permits operation at shorter wavelengths, but with appreciable reduction in lamp life. In any case, good voltage regulation is mandatory for a stable source since the output at wavelengths below 800 nm varies nearly as the fourth power of the applied voltage. This means, for example, that to reproduce a scale reading within 1% on a single beam instrument, the source voltage must be held within 0.25% or 15 mV when a 6 V lamp is used.

Electronic power supplies have been improved greatly within recent years and are now used almost exclusively as voltage supplies for tungsten lamps. Such power supplies are commonly included as an integral part of the instrument or the instrument manufacturer will recommend a suitable external unit. In former years, fluctuation of source voltage was often a serious problem and still may be on some occasions. In such cases, it may be advantageous to revert to use of a 6 V wet battery, which gives a smooth voltage although there is a slow drift during continuous operation due to battery discharge. The specific gravity of a wet battery should be maintained above 1.215 (half-charge), and the voltage should be kept between 5.9 and 6.2 V. After a storage battery has been fully charged (sp. gr. 1.280 at 20–25°C.), it must be run for about an hour before the output falls to 6.2 V, the region of maximum stability. Convenient charging of the battery can be accomplished by connecting into a trickle charger (battery booster) when not in use.

An excellent 6 V lamp power supply consists of a 5 A full-wave selenium rectifier battery charger receiving its power through a constant voltage transformer, the output of the charger being coupled in parallel to a wet storage battery that smooths out the direct current pulsations coming from the charger. Battery life is sufficiently prolonged to warrant the cost of the extra equipment. Terminal connections to the storage battery should be scraped occasionally to remove corrosion, since faulty connections are a frequent cause of instrument instability.

In most instruments the lamp and its socket are so designed that replacement of the lamp presents no problem since its filament, which is of standard shape and size, is automatically placed in the same position as the one removed. If such provision is not made, uncritical replacement of the lamp can lead to serious errors.

An incandescent lamp radiates considerable heat, and since the transmittances of filters and samples can be affected by temperature changes, it is necessary to filter out the major portion of the heat rays. In instruments employing barrier layer and photoconductive cells as the light sensitive element this is of further importance because their response varies markedly with temperature. This protection is usually afforded by a clear filter made of special heat-absorbing glass. A special type of interference filter that reflects infrared radiation can also be employed.

Aging of the lamp or accumulation of dirt can result in a change in absorbance readings on some instruments.

WAVELENGTH SELECTOR

Isolation of the required wavelength or spectral range can be accomplished by a filter or monochromator. Filters are much simpler, consisting only of a material that selectively transmits the desired band and absorbs all other wavelengths. In a monochromator, radiation from the source is dispersed by a prism or grating into a spectrum from which the desired band is isolated by mechanical slits. The term "monochromatic" is used rather loosely in photometric analysis. Strictly speaking, it means light limited to a single wavelength, as for example from lamps producing a discontinuous line spectrum suitable for wavelength calibration. Commonly, however, monochromatic refers to a narrow band of unspecified width. In such cases the term "narrow band" is preferable. Highly monochromatic light is not necessary in most spectrophotometric measurements for quantitative purposes.

FILTERS

There are two categories of filters, those with selective transmission characteristics, including *glass* and *Wratten filters*, and those based on the principle of interference, called *interference filters*.

Glass and Wratten Filters. These filters are available in a great variety of sizes, shapes, mountings, and transmission characteristics. The Wratten filter consists of colored gelatin cemented between clear glass plates, whereas the glass filters are composed of one or more layers of colored glass. In either case the filter transmits more light in some parts of the spectrum than in others. The color of the filter is attributable to this transmission selectivity, e.g., a red filter is red because it transmits a large part of the incident red light and absorbs a large part of the blue and green light. Glass filters, when available, usually are preferred to the gelatin type because they are more rugged and less susceptible to damage by heat and sunlight.

These filters can be classified further into three general types: the first is the *cutoff filter*, which produces a sharp cut in the spectrum, with almost complete absorption on one side and high transmittance on the other. This filter is used to eliminate unwanted "stray light," second order spectra, etc., and in composite filters as discussed below. The second type of filter is represented by the *neutral tints* or *grays* having relatively constant absorption over a wide spectral range. They have been used as absorbance standards to replace standard solutions. The third and most common type of filter is the *composite*, which transmits exclusively in one limited band. It is composed frequently of a combination of cutoff filters cemented together. The effect achieved is shown in Figure 1-3, where curves 1 and 2 are the

FIG 1-3. Typical composite filter. See text for explanation.

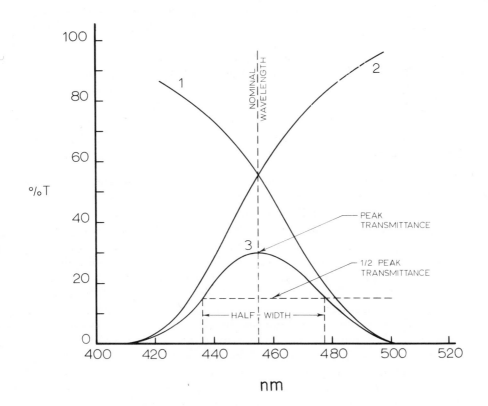

transmission curves of two hypothetical cutoff filters, and curve 3 is the transmission of the two combined. The transmission curves of composite filters should be smooth and sharp and have as complete cutoffs as possible on both sides. Some skewness is usually unavoidable.

It is customary to designate a filter by the wavelength at its *peak transmittance,* i.e., by its *nominal wavelength* (e.g., filter 54 or 540 has peak transmittance at 540 nm). The effectiveness of a filter with relatively symmetrical transmittance is commonly given in terms of its peak transmittance and its *half-width* or *half-bandwidth* (sometimes just called "bandwidth"), which is the wavelength interval between the two sides of the transmission curve at a transmittance equal to one-half the peak transmittance. This is depicted in Figure 1-3. Ideally, filters would be made with as narrow a half-width as possible, and up to a limit, this could be accomplished by combining cutoff filters whose cutoffs are closer and closer together, but this results in progressively less transmission of light. A glass composite with a half-width of 30 nm frequently passes only 10% of the incident light even at the peak.

Transmission characteristics of commercially available filters can be obtained from the manufacturers, e.g., Corning (37) and Alfa American Corporation, agents for Chance Brothers, for glass filters and Eastman Kodak for Wratten filters (38), from various publications of the National Bureau of Standards, and from the manufacturers of the photometers themselves.

Interference Filters. This type of filter is based on a different principle than the filters just discussed. The principle is the same as that underlying the play of colors from a soap film, namely, interference. When light strikes the thin film, some is reflected from the front surface while some of the light that penetrates the film is reflected by the surface on the other side. The latter rays of light energy have traveled farther than the former by a distance two times the film thickness. If the two reflected rays are in phase their resultant intensity is doubled, whereas if they are out of phase they destroy each other. Thus, when white light strikes the film some reflected wavelengths will be augmented and some destroyed, resulting in colors.

Interference filters are of two types:

a. One is designed after the Fabry-Perot interferometer. The interference effect is obtained as a result of multiple reflection from two reflecting but partially transparent parallel metallic layers placed very close together. The interference principle predicts that a series of spectral regions will be passed by such a filter at certain intervals. Thus, if such a filter has its "first order" peak transmittance at 700 nm, the "second order" and "third order" peaks will be at 350 and 233 nm, respectively (λ, $\lambda/2$, $\lambda/3$, etc.). Use of the higher order peaks for photometry offers the advantage of furnishing narrower bands and steeper curves, but the undesired adjacent bands are closer by. If these side bands fall within the response of the instrument they must be eliminated by cutoff filters, which are usually added directly to the main filter. Filters are available in the spectral range of 340—900 nm, with peak transmittance in the range of 25%—50% and with half-widths as narrow as about 10 nm. The transmission curves of these filters are symmetrical, very steep, and have very high rejection outside of about 50 nm from center. These filters can also be used in series to obtain even narrower bands and greater rejection outside the bands.

b. In the second type of interference filter the metallic layers are replaced by alternating layers of materials differing in their refractive indices, thus the name *multilayer interference filter*. The peak transmittance of this filter can be very high (90% or more), and the half-width can be as small as 4 nm. In this type of filter side bands of considerable width occur on both sides of the peak beginning about 50—100 nm from the band to be isolated. These side bands must be eliminated by auxiliary cutoff filters. The multilayer filters are usually more expensive than the metallic layer type. They are available with nominal wavelengths as low as 210 nm.

Interference filters permit passage of a light beam of large cross section, providing considerably more radiant energy for measurement than is obtainable from equivalent bandwidths obtained from a prism or grating, since in the latter cases the beam is limited to a small area by slits, as will be seen subsequently. Interference filters must be used with parallel light rays (collimated beam) to realize the half-widths claimed for the filters because their peak transmittance shifts with a change in angle of the incident beam.

Interference filters may be obtained from Photovolt, Baird-Atomic, Bausch & Lomb, and Fish-Schurman.

MONOCHROMATORS

Monochromators are capable of giving much narrower spectral bands than filters and have the additional advantage of being easily adjustable over a wide spectral range. The dispersing element may be a prism or a grating. In either case it is used with one or more slits, lenses, and mirrors in varying optical arrangements to provide an entrance slit through which white light (continuous mixture of all wavelengths) enters and an exit slit from which the isolated spectral band emerges. In recent years it has become common practice to use low dispersion grating monochromators rather than filters in low cost instruments intended primarily for repetitive measurements at fixed wavelengths.

Prisms. Glass prisms are satisfactory for the visible and near ultraviolet ranges, but quartz is required for the ultraviolet. Dispersion by a prism is nonlinear, becoming less at longer wavelengths where narrower slits are required to obtain the same spectral band width. Prism instruments, therefore, have adjustable slits that are usually curved. The entrance and exit slits are commonly controlled simultaneously by one adjustment and kept equal. Prisms give only one order of emerging spectrum, and thus provide higher optical efficiency since the entire incident energy is distributed over the single emerging spectrum. With both the entrance and exit slits of equal width, the radiant energy emerging from the exit slit is represented by an isosceles triangle, as shown in Figure 1-4. The *nominal wavelength* (value read on instrument scale) is the middle wavelength passing through the exit slit. This corresponds to the peak transmittance of filters when their trans-

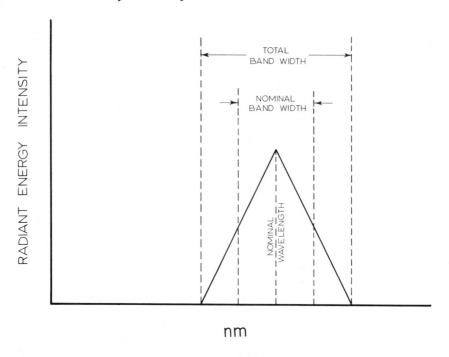

FIG 1-4. Energy distribution of beam emerging from slit in spectrophotometer.

mission curves are symmetrical. The *nominal bandwidth* corresponds to the half-width of filters and includes 75% of the radiant energy present in the emerging band. Actually, because of optical aberrations the emerging energy is not exactly represented by a perfectly symmetrical triangle, so that the *effective bandwidth* is slightly larger than the *nominal bandwidth* for any particular setting of the mechanical slits. Prism spectrophotometers are accompanied by graphs relating wavelength, effective bandwidth, and slit width setting. The *spectral* or *total effective bandwidth* is equal to twice the effective bandwidth. The Beckman Model DU Quartz Prism Spectrophotometer is still widely used in this country and permits isolation of effective bandwidths of approximately 0.5—1.5 nm within the range 220—950 nm.

Gratings. A grating consists of a large number of parallel, equally spaced lines ruled upon a glass or metal surface. An original grating is rather expensive, but many *replicas* can be made by coating the original with plastic and then stripping off the plastic. These are usually used in commercial instruments and are less expensive than prisms. They can be in the form of transmission gratings or reflecting gratings, the latter permitting use in the ultraviolet region since the radiant energy does not pass through the grating. When white light strikes a grating it is dispersed into a spectrum by the phenomenon of diffraction, similar to that described for interference filters. The energy is normally dispersed, so that for a given slit width the bandwidth is constant over the whole spectrum, thus permitting the use of fixed slits. The slits are straight. As with interference filters, several orders of spectra are produced, the first order generally being used. The energy distribution in the beam emergent from the exit slit again approximates a triangle, and the terminology used in its description is the same as that given previously for prisms.

CUVETS

The receptacle in which a sample is placed for photometric measurement is called a *cuvet*. There are many different sizes and shapes of cuvets and many commercial instruments are designed to employ more than one type. This is frequently desirable for two reasons: (a) the occasion arises when measurements are desired on a small sample volume, requiring a micro cuvet; (b) change of length of light path through the solution is a convenient method of bringing the absorbance values within the region of greatest photometric accuracy (see later section "Photometric Errors Originating in the Instrument").

Most cuvets are made of glass, which is satisfactory for the range 320–1000 nm, although plastic cuvets are also available for some instruments. For measurement below 320 nm it is necessary to use quartz (silica) cells. Such cells can of course be used at the higher wavelengths, but their considerably higher cost makes this impractical.

Two general types of cuvets are used, those having a square or rectangular cross section and those having a circular cross section, e.g., test tubes. Greater accuracy is achieved by the former with parallel faces made of optical glass, preferably fused rather than cemented. Aside from gross imperfections, there are at least two sources of error in cuvets having flat faces, neither of which, fortunately, is of serious consequence. Radiant energy that has already passed through the solution can be reflected back from the light sensitive element or a lens surface to the cuvet, where it can be reflected back again to the detector. The other source of error lies in the apparent nonhomogeneity of the receptor area of the light sensitive element (17). Thus, very slight differences in cuvets result in slight shifts of the area illuminated by the transmitted light. A "cuvet correction" can be determined for each cell to compensate for this error.

Cylindrical cuvets present further problems. They act as an astigmatic cylindrical lens, and the refractive index of a lens is an important factor in its focal length and hence the transmission of light to the detector (e.g., lower $\% T$ in an empty cylindrical cuvet than when filled with water). Loss of transmitted radiant energy also occurs as a result of reflection at the glass-solvent interface, the loss decreasing as the index of refraction of the solvent approaches that of the glass. Due to slight imperfections it is necessary to mark each cylindrical cuvet so that it can always be placed in the instrument in the same position. Another disadvantage of the cylindrical cuvet is the inability to calculate accurately the effective depth of solution for absorptivity values unless very narrow slit widths can be used, which is not usually the case. Fortunately, however, the use of cylindrical cuvets, even capillaries, does not significantly affect the obedience of solutions to Beer's law. Several instruments are designed for use with standard sizes of test tubes. This is very convenient and inexpensive but necessitates matching of cuvets by the user.

An important source of error is instability of the cuvet holder, both with respect to its positioning in the instrument if it is movable, and with respect to the precise reproducibility of positioning the cuvet in the holder. On the latter point several of the commercially available instruments are rather poorly engineered.

Micro Cuvets. Fused silica micro cuvets, designed by Lowry and Bessey (23), capable of being used with 50, 75, and 100 μl volumes with a 10 mm light path are available for the Beckman Model DU Spectrophotometer from Pyrocell. A special cell carrier and diaphragm attachment are required for their use. The 50 μl capacity cuvet is rather difficult to fill and empty.

Ordinarily, radiant energy traverses only a fraction of the total volume in a cuvet. This inefficiency is partially circumvented by the use of horizontal capillary cells (21), the bore of which need not be greater than the cross sectional area of the radiant energy beam. Capillary cells of 20 or 50 mm light path with bores of 0.7–8 mm and the required cuvet holders for the Beckman Model B and Model DU are available from Microchemical Specialties. Although sensitivity favors the capillary cell, they are difficult to fill without entrapment of air bubbles, and they require careful and reproducible alignment and positioning. Ease of filling and adjustment favors the Lowry-Bessey type of micro cuvet.

PHOTOMETRIC SYSTEM

Measurement of the radiant energy emerging from the cuvet includes several steps: detection, amplification of the signal when necessary, and final presentation as a galvanometer deflection, potentiometer reading, strip chart recording, etc.

DETECTORS

Two types of detectors commonly find use in the visible and ultraviolet regions, namely photovoltaic (barrier layer) semiconductive cells and photo-emissive phototubes. The choice depends largely on the radiant power available. Spectral isolation by filters and low dispersion monochromators gives sufficient energy so that barrier layer cells can be used. High resolution monochromators require the use of phototubes or photomultipliers.

Barrier layer (photovoltaic) cells. Barrier layer detectors consist essentially of a plate of copper or iron upon which a semiconducting layer of cuprous oxide or selenium is placed. The semiconductor is next covered by a light-transmitting layer of metal which serves as a collector electrode. Upon illumination through the transparent electrode an electron current flows if the circuit is closed. These cells have an output of about $120\,\mu\text{A}$/lumen (as much as 10 mA on exposure to direct sunlight), are very rugged, are relatively inexpensive, and are sensitive from the ultraviolet region up to about 1000–1200 nm. No external power supply is required; thus the term "self-regenerative" is frequently applied to barrier layer cells. The photocurrent produced is almost directly proportional to the radiant energy intensity for low values of external circuit resistance (100 ohms or less).

Barrier layer cells exhibit so-called "fatigue effects". Upon illumination the current rises rapidly to a value several percent above the apparent equilibrium value and then gradually decreases. It is therefore advisable to wait 30–60 sec before taking readings. There is complete recovery upon standing in the dark. Photosensitivity varies with wavelength, being maximum at about 550 nm for selenium cells and falling off to about 10% of the maximum at 250 and 750 nm.

Phototubes. A phototube resembles a radio tube in appearance but contains a cathode plate coated with a substance that emits electrons in proportion to the radiant energy striking it. The spectral response depends upon the cathode coating. The anode collects the electrons, which may amount to a current of 20–100 μA/lumen. The phototube requires an external source of voltage for operation, and since it has high internal resistance, its output is easily amplified. Some phototubes contain argon or other inert gas, but most are highly evacuated. Photomultiplier tubes, in addition to a cathode and anode, contain 9–14 dynodes that produce a final current that may be 10^8 times greater than the primary photocurrent. They are used with very low levels of illumination as occur with very narrow bandwidths or in reading unknowns against reagent blanks having high absorbance.

Phototubes when completely shielded from radiant energy emit a small current called the *dark current*. This current is not constant but fluctuates around an average value. The dark current (also referred to as "background noise") places a definite limitation on the sensitivity of the tube since it is amplified along with the photocurrent produced by the incident radiation.

Photoconductive cells. The photosensitive element in these cells consists of a material that is a very poor electrical conductor in the dark but undergoes marked change in resistance upon illumination. Their response is nearly linear since the conductivity increases virtually in direct proportion to the incident radiant power. The best known example is the so-called lead sulfide cell that is used extensively for measurements in the near infrared region, out to about 2500 nm. In the visible and near ultraviolet region, the sensitive material is usually cadmium sulfide or cadmium selenide. Photoconductive cells require an external voltage and provide a low impedance output.

SIGNAL AMPLIFICATION

The output from barrier layer cells is not easily amplified and such cells are used only in wide bandwidth instruments for the visible and near ultraviolet regions, where there is sufficient radiant power so that a sensitive galvanometer or ammeter can be coupled directly. This type of arrangement has been widely used in filter photometers. Another common arrangement employs two cells in a bridge circuit and uses a sensitive galvanometer as a null point indicator to signify when the outputs are balanced. Both cells are exposed to the same light source, and the sample is interposed between the source and one cell. The circuit is balanced on a blank and then again on the sample. By such an arrangement variations in light intensity are cancelled if the barrier layer cells are matched, and stabilization of the light source is less critical.

The current output of phototubes and photomultipliers requires further amplification, which is straightforward because these detectors have high internal resistance. Photoconductive cells have a low impedance output, and it is necessary to include a low impedance pre-amplifier. A consideration of amplifier circuits is beyond the scope of this chapter and would not be productive since amplifier technology is advancing rapidly. It cannot be overemphasized, however, that the amplifier is a very critical component and plays a major role in the stability, noise level, and final accuracy of the photometric measurement.

DATA PRESENTATION

The most useful form of output is in absorbance units rather than transmittance or % T since sample concentration is directly proportional to absorbance. Most instruments, however, still provide output that is linear in transmittance units because the associated electronic circuits are simpler. Conversion to absorbance units may be made by reading from a nonlinear scale alongside the linear transmittance scale. Output that is linear in absorbance units may be obtained by logarithmic amplification or by passing the linearly amplified transmittance signal into a logarithmic converter. Systems that measure transmittance and use logarithmic conversion have the apparent advantage of being able to display or record transmittance, but this is of little real value because data in transmittance units are rarely needed.

Digital indicating devices have come into use in recent years as a result of advancing electronic technology and are replacing meter displays. This permits greater precision in readings and eliminates the need for interpolation. Availability of data in digital form also simplifies linear scale changes (electronic scale expansion) and permits easy coupling to output printers. A notable advantage in recent years has been the incorporation of simple data handling capability into the photometer logic whereby calculations are greatly simplified and in some cases completely eliminated.

INSTRUMENT CLASSES

Photometers and spectrophotometers are commonly classified according to the spectral region in which they function, their spectral band pass, and characteristics of their electromechanical systems. Instruments are currently undergoing rapid evolution due primarily to new electronic technology, and the manufacturer's literature must be consulted for timely information.

Below about 300 nm, measurements usually require such narrow band-widths that a high quality monochromator is necessary. These instruments are referred to as *uv spectrophotometers* and commonly cover the range 380–210 nm.

Wide bandwidths are often satisfactory for measurements in the visible and near ultraviolet region and in this spectral range instruments are sub-classed according to whether their spectral band pass is greater or less than about 5 nm. Until about 20 years ago, instruments for wide band measurements almost always employed filters, hence their designation as *filter photometers*. More recently, however, simple grating monochromators giving low dispersion have been used because of the ease in making wavelength settings and to eliminate the cost and trouble of providing a large number of filters. These instruments are often referred to as uv-visible spectrophotometers which sometimes causes confusion with conventional spectrophotometers having much better resolution. For this reason the term "filter photometer" is often extended to include low dispersion spectrophotometers. Instruments providing spectral slit widths of 1 nm or less are classed as *high resolution spectrophotometers*. The significance of a 5 nm bandwidth in subdividing uv-visible instruments refers to the associated level of radiant power and the monochromator requirements. With spectral bands narrower than about 5 nm, the level of radiant power is so low that electronic amplification is required for the weak photocurrent that is produced. Also, the monochromator must be larger and more complex, with ad-justable slits. Overall there is an appreciable difference in cost for instruments in the two subclasses and the additional expense is seldom justified when repetitive quantitative measurements are the only appli-cation.

The electromechanical design of filter photometers and low dispersion spectrophotometers tends to be much simpler than for high resolution spectrophotometers. Wavelength selection is by manual means and photo-metric readout is usually by a meter or manually adjusted potentiometer, although automated digital readout has been introduced in recent years and strip chart recorders have long been available as accessories. Mechanisms for automatic wavelength scanning and data recording are usually provided with high resolution spectrophotometers. The manufacturer's literature should be consulted for description of the many conveniences and options available in newer instruments of all classes.

CHOICE OF INSTRUMENT

To some extent all instruments are compromises, no single instrument being ideal for all applications. Most laboratories performing a wide scope of clinical determinations find it necessary to have at least two instruments: (a) a spectrophotometer with manual or automatic wavelength scanning and fairly high resolution and (b) a simpler instrument such as a filter or low dispersion grating photometer for most routine work, as this is more rugged and simpler and faster to operate.

A few years ago, the "precalibrated" instrument gained wide popularity and acceptance. Photometers were supplied with tables or graphs relating concentration to photometer readings, standardization having been carried out at the factory. The implication usually made was that standardization of tests by the user was completely unnecessary. In the opinion of the authors, such "precalibrations" cannot be too severely condemned. Errors can occur as the result of changes in characteristics of the photometer or spectrophotometer itself (5, 12), making or buying new reagents, deterioration or contamination of old reagents, use of incorrect filters or wavelength settings, etc. Standardizing each test when a new instrument is obtained is not sufficient since the same errors can and do develop subsequent to standardization.

Among the factors to be considered in choosing an instrument for a clinical laboratory are the following:

(a) Intended Use. If the determinations to be run are the more common routine analyses, a filter photometer or low dispersion spectrophotometer with manually adjusted wavelength selector is usually best. In some cases, the use of broad bandwidths will give deviation from Beer's law and require a nonlinear calibration curve. If the determinations require measurements in the ultraviolet or measurements with very narrow bandwidths or with microcells, a more refined instrument is necessary. Some applications, especially for identification purposes as in toxicology, require an instrument capable of automatic wavelength scanning. Enzyme and other assays utilizing kinetic technics are most conveniently handled by instruments designed or modified specifically for this purpose. Measurements on electrophoretic strips also require special instruments.

(b) Performance. Basic operating parameters such as photometric range and linearity, stability, noise level, and resolution should be examined in comparing instruments, and most importantly, comparisons should be made for conditions that are realistic in terms of the intended use.

(c) Convenience in Operation. The instrument should be as simple and easy to use as is possible for the intended use. Greater complexity in operation usually means longer training time for the operator, greater time and trouble in operation, hence lower productivity, and increased likelihood for error especially when personnel are not highly trained.

(d) Stability. Many instruments are very annoying because of instability or drift that requires frequent resetting of controls or addition of a voltage regulator or perhaps an electronic power supply. The importance of circuits that provide stable performance cannot be overemphasized, especially when the instrument is to be operated from lines subject to wide voltage fluctuations, as may occur in a building containing heavy electrical equipment such as elevators.

(e) Reliability. Generally speaking the likelihood of instrument failure increases in proportion to the number of electronic and mechanical components present. With other factors being equal, the simplest design is to be preferred: Specifically, the fewest components necessary for the required function, the simplest mechanical structure, and fewest operating controls.

Gross malfunction is often common with newly designed instruments that have seen relatively little use under conditions existing in busy clinical laboratories. When reliability and trouble-free operation are important, it is advisable to choose an instrument that has been well proved by several years in the field.

(f) Cost. A high priced instrument does not necessarily guarantee better performance or greater accuracy than could be obtained at a lower cost. The expensive instrument may merely have features that are of no value for the intended use. Before selecting a specific instrument, contact users who have had experience with applications similar to those planned. Also, make certain the quoted price includes all necessary accessories.

PRACTICAL ASPECTS OF PHOTOMETRY AND SPECTROPHOTOMETRY

Each instrument should be checked periodically as to its working condition. This is especially true after replacement of a broken filter, lamp, phototube, etc. In the case of a spectrophotometer, both the wavelength calibration and the transmittance at various wavelengths should be checked. A filter photometer usually can be checked only for overall performance.

FACTORS AFFECTING NOMINAL WAVELENGTH

The effective wavelength of a filter photometer is determined largely, but not entirely, by transmission characteristics of the filter. Other factors are the relative distributions of source energy and of detector sensitivity over the spectral range transmitted by the filter. Both distributions vary with wavelength, quite steeply in some spectral regions. Aging of the source lamp or detector, or replacement of either, can produce a shift in effective wavelength, as can replacement of the filter itself since filters having the same designation are seldom identical. Wavelength changes occur with interference filters if there is a shift in angle between the incident radiation and the filter. All filters must receive proper care. They should be kept clean; a damp soft cloth should be sufficient for this purpose. Organic solvents should be avoided, but if used must be kept away from the edges where layers are cemented together. Sometimes the cement applied to the edges of filters develops cracks, admitting water vapor that causes changes in the separation between layers and results in a shift in nominal wavelength. Fortunately these effects are seldom significant since most applications of filter photometers are for measurements on solutions that have a relatively broad absorption maximum. Use of concurrent standards tends to correct for errors of this type.

Spectrophotometers having a direct reading wavelength scale that can be set to any value within the instrument's range are convenient but provide a greater opportunity for wavelength error than a filter photometer since the setting itself may be incorrect or there may be a faulty correlation between the scale reading and actual wavelength. With a manual spectrophotometer, it is important always to approach the wavelength setting from the same direction, usually from shorter wavelengths, to avoid hysteresis ("backlash") in the mechanical mechanism.

WAVELENGTH CALIBRATION

Special sources having strong emission lines are convenient for checking the wavelength calibration of a manual spectrophotometer and a mercury lamp is most commonly used. Care must be exercised to identify the lines correctly (34, 18). It is necessary either to substitute the calibration lamp for the regular source or to accomplish the equivalent by use of a mirror. The instrument manual that accompanies the instrument should be consulted for the exact procedure. A few points can also be checked easily using the isobestic (isosbestic) points of pH indicators (14, 30). Recording spectrophotometers that scan automatically are conveniently calibrated by recording the spectra for an absorber that has accurately known peaks. Didymium glass filters have been recommended but are being replaced by holmium oxide glass because the peaks are more reproducible. Interference packs (18) are also suitable with scanning spectrophotometers, and benzene vapor is used in the ultraviolet region.

TRANSMITTANCE STANDARDS

It appears not to be generally known that absorbance readings on spectrophotometers vary significantly between instruments even of the same make (4, 32). Beckman DU spectrophotometers showed differences up to about 20% in the ultraviolet (4). The reading on any given instrument can change significantly within a few weeks. Most spectrophotometers do not have adjustments for correcting their response. Such discrepancy is not a problem when results are calculated from the absorbance of an unknown relative to the absorbance of a standard. It is a serious problem, however, when calculations must be made from an established absorptivity because of the unavailability of a standard or the impracticability of running a standard, as in the determination of enzyme activity based upon change in absorbance at 340 nm.

The National Bureau of Standards has published transmission characteristics of solutions of copper sulfate, cobalt ammonium sulfate, and potassium chromate (9, 36). These solutions have not been completely satisfactory and other liquid standards are under development and evaluation (29). A series of colored glass filters (16) previously provided as reference standards by the National Bureau of Standards have now been replaced by three neutral glass filters designated SRM 930 (29). They are useful for checking the photometric scale in the region 400–700 nm. It is important to note that even with these "gray" filters, the observed transmittance depends to some extent on the beam geometry and spectral band pass of the instrument, and the positioning and surface condition of the filter. Wire mesh screens are also

useful for checking absorbance measurements (18) and are available commercially (Perkin Elmer).

What should one do when the reading on his spectrophotometer is significantly different from the accepted value for the reference standard? This
problem has not yet been completely resolved, but as a practical matter in
connection with enzyme assays based on change in absorbance at 340 nm, it
is recommended that the photometric measurements be standardized using a
solution of potassium dichromate, 50.0 mg/liter in 0.01 N H_2SO_4 (25). This
is accomplished by reading the dichromate solution at 350 nm and correcting the observed enzyme activity as follows:

$$\text{corrected} \, \Delta A_{340}/T_{min} = (\text{observed} \, \Delta A_{340}/T_{min})$$

$$\times \frac{0.535}{\text{observed} \, A_{350} \text{ for } K_2Cr_2O_7}$$

where T_{min} = time in minutes.

The term 0.535 is the accepted absorbance ($A_{1\,cm}$) at 350 nm for 50.0 mg
potassium dichromate/liter. The dichromate solution is read at 350 nm
rather than 340 nm to allow for a slight displacement of the peak absorbance
relative to NADH. The original paper should be consulted regarding
measurements made with cylindrical cuvets. We have also found it
convenient to utilize the potassium dichromate standard used for the icterus
index determination as an arbitrary standard for checking the constancy of
photometric readings since this solution is completely stable for many years.

CALIBRATION AND CARE OF CUVETS

Cuvets with flat windows are generally used with spectrophotometers, but
cylindrical cells (test tubes) are more common with filter photometers and
are sometimes used with spectrophotometers because they are less expensive
and easier to handle. Two types of correction must be considered with both
cuvet shapes. Flat-windowed cuvets are often sold in matched sets having
essentially zero correction between individual cells.

One correction, termed the "cuvet correction," relates to: (a) variations in
energy loss due to reflection and scattering of radiation at the cell surfaces,
and (b) differences in refraction by the cuvets which results in the transmitted radiation striking slightly different areas of the detector. Proceed as
follows: (a) Choose one cell, henceforth to be designated as the "reference
cell" and having no "cell correction." (b) Fill all cells with water. (c) With
the reference cell in position, adjust the instrument response arbitrarily to an
absorbance of 0.400. (d) Read all other cells in turn. (e) Differences in
readings from that of the "reference cell" constitute the "cell corrections,"
which must be added or subtracted as the case may be to all subsequent

measurements. Such "cuvet corrections" should be specifically determined for each wavelength (and slit, if variable) used. If the corrections are no greater than an absorbance of 0.002 they are of questionable value (5).

The second type of calibration is correction for variation in path length between cuvets. According to Beer's law, a 1% difference in this dimension results in a 1% change in the estimate of absorber concentration. To obtain this correction, a combination of a colored solution (e.g., aqueous $CuSO_4$) and a wavelength or filter that gives a system obeying Beer's law and an absorbance in the range of 0.2 to 0.7 (region of least relative analysis error per unit photometric error) is chosen. The absorbance of this solution in the cuvets is compared with the same solution in a reference cuvet. The ratio of the A values, corrected for the "cuvet corrections," is equal to the ratio of effective path lengths.

The procedure recommended for matching test tubes to be used as cuvets is as follows: (a) Decide on the tolerance to be permitted. (b) Set the instrument to 0 absorbance (100% T) with water in one tube. (c) Fill all tubes with a colored solution as described above, and make the measurements. If the solution absorbance is 0.50 and the tolerance decided upon is ±2%, then tubes with absorbances falling in a range of 0.49–0.51 (0.50±0.01) are accepted. Some of the rejected tubes may become acceptable at a certain position if they are rotated. Each tube must be carefully marked (usually the front) so it can always be placed in the same position. Contrary to the directions sometimes given, water cannot be used for matching test tubes.

When a series of solutions, including standards, are to be read in the same set of determinations the same cuvet can be used after rinsing out the cell with a little of the solution to be measured. For routine work, when successive solutions do not differ greatly in absorbance, omission of this rinsing procedure does not introduce a significant error.

Because highly volatile solvents introduce a possible source of error as a result of an increased concentration brought about by evaporation, rapid readings or stoppered cuvets are required.

All cuvets must receive proper care. They must be protected from any physical contact that might produce scratches and from chemicals that may etch them. Thus, hot concentrated sulfuric acid and strong alkali should not be used for cleaning. A mild detergent (not too alkaline) or organic solvent may be used provided the cuvet is not held together with cement that is attacked by the solvent. Optical surfaces should be cleaned with a soft dry cloth or lens paper and should not be touched during a series of measurements, especially when measurements are made in the ultraviolet.

REAGENT BLANKS

When a determination is made on an unknown, a similar determination is usually made substituting water for the unknown sample. This is referred to as the *reagent blank* and serves the purpose of determining the absorbance in the final solution derived *not* from the substances under test in the unknown, but rather from the reagents themselves. This absorbance can arise in two ways. First, it can represent contamination of the reagents with the substance under test, or the presence in the reagents of one or more substances capable of reacting in a similar manner. Second, the reagents themselves may be colored, and although the reaction with the substance to be determined may result in a second color, the two are superimposed, and correction for the reagent color must be made (assuming additivity of absorbances). An example of this is the biuret reaction for the determination of protein.

If the photometer or spectrophotometer is set at 0 absorbance (100% T) with water or other solvent, then the reading of the unknown is the A for the unknown plus the reagent blank. The reagent blank then is also read versus the solvent, and correction for reagent blank absorbance made by subtracting the reagent blank A from the unknown A. The resultant A then presumably represents absorbance due to the substance to be determined, provided of course that there are no interfering substances present in the unknown. When the reagent blank has a high absorbance, it is preferable to set the instrument at 0 absorbance on the reagent blank and then read the absorbance of the unknown. In this procedure the instrument is automatically subtracting the absorbance of the blank from that of the unknown.

Great caution must be exercised in instances where the reagent blank is very large (A greater than about 0.7 relative to pure water) whether or not the reagent blank is used to make the 0 absorbance setting, because small variations in the factors contributing to the large reagent blank may introduce large uncertainties in the net absorbance ascribed to the test substance. Also a large blank absorbance imposes the need for high spectral purity. A small amount of light having a wavelength different than the peak absorption always reaches the detector but ordinarily is negligible. When the blank is large, this heterochromatic component may give rise to a significant fraction of the total photocurrent developed by the detector. As a consequence, Beer's law will not be followed.

SPECTROPHOTOMETRY IN THE ULTRAVIOLET (uv) REGION

Many organic substances colorless to the eye have characteristic absorption curves in the uv region permitting direct determination after adequate isolation by such technics as extraction and chromatography. Specificity can usually be checked by taking measurements at two or more selected wavelengths and comparing the ratios obtained for the unknown with those obtained with known solutions. Curves in several different solvents or at different pH's are often helpful in distinguishing between two closely related compounds.

The choice of solvents is more limited than for the visual range. Table 1-2 lists the approximate "cutoffs" for some of the more commonly used solvents, the cutoff being arbitrarily taken as the wavelength below which the absorbance is greater than unity. The cutoff of most organic solvents is lowered with increasing purity and "spectrochemical grade" solvents are

TABLE 1-2. Ultraviolet Cutoffs of Common Solvents

Solvent	uv cutoff $(A_{1\,cm} = \sim 1.0)$ nm
Acetic acid, glacial	250
[a]Acetone	330
Aqueous HCl	< 200
Aqueous NaOH	< 220
[a]Benzene	280
[a]Carbon tetrachloride	265
Cellosolve	290
[a]Chloroform	245
[a]Cyclohexane	210
[a]1, 2-Dichloroethane	225
Dioxane	250
Ethanol	< 220
[a]Ethyl acetate	255
Ethyl ether	220
Glycerol	205
H_2SO_4	< 220
Heptane	210
[a]Hexane	210
[a]Isobutanol	230
Isooctane	210
[a]Isopropanol	210
[a]Methanol	< 210
[a]Methylene chloride	235
[a]N, N-Dimethylformamide	270
Octane	< 220
Pyridine	305
[a]Toluene	285
Water	< 200

[a] ACS spectrophotometric solvent

available from chemical supply houses. Among the buffers satisfactory for use in the uv range are acetic acid-acetate, borate, and phosphate solutions.

Care in the cleaning of cuvets and avoidance of finger marks on the optical surfaces are even more important in uv work than in the visible region. In storage of solvents and solutions and in preparative procedures prior to measurement, contact of solutions should be limited to glass (preferably Pyrex). Contact with grease lubricants is to be especially avoided. Separatory funnels must be used without grease lubrication of the stopcock, or funnels with Teflon valves must be employed. Studies in our laboratory have demonstrated that organic solvents in prolonged contact with polyethylene, Teflon, Lubriseal, and silicone stopcock lubricant may pick up material absorbing in the uv region. Even water stored in polyethylene bottles shows significant uv absorption (10). Whether significant contamination occurs in any specific procedure where such materials come into contact with the solutions used can be determined only by running appropriate blanks.

ABSORPTION CURVES

An *absorption curve* relates A or $\% T$ to wavelength and is useful in selecting the proper wavelength for quantitative measurements and for identifying the absorber. Measurements for identification purposes are usually made in the ultraviolet or infrared regions. The most common convention for presenting absorption data is to plot A versus wavelength, as in Figure 1-5. Identical curves are obtained by using semilogarithmic graph paper and plotting $\% T$ on the logarithmic axis (ordinate) as a decreasing function. Some workers prefer to plot log A versus wavelength as in Figure 1-6 since for different concentrations this results in a series of curves that are parallel at all wavelengths if Beer's law is obeyed. Alternatively this same

FIG 1-5. Oxyhemoglobin in 0.01% Na_2CO_3: (1) 3.8 mg/100 ml; (2) 7.6 mg/100 ml; (3) 15.2 mg/100 ml. Beckman Model DU Spectrophotometer, effective bandwidth about 1 nm.

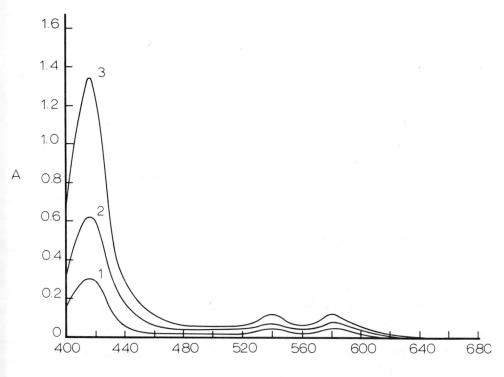

nm

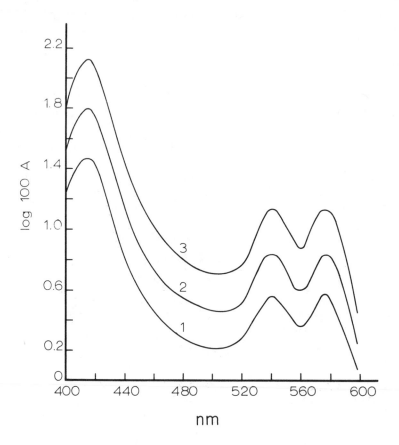

FIG 1-6. Oxyhemoglobin in 0.01% Na_2CO_3. Same data as in Fig. 1-5. Log 100 A used instead of log A, thus rendering all values positive.

result is obtained by plotting A versus wavelength on semilogarithmic graph paper.

CHOICE OF WAVELENGTH

As a general rule the nominal wavelength for quantitative work should be at an absorption peak of the absorber. If the peak falls below the spectral limit of the instrument or the peak absorbance is too intense to be measured (usually due to a very large blank), it may be necessary to work at a wavelength away from the peak. Obviously, if the absorbance of a solution is changing rapidly with wavelength (on the steep side of a peak), slight errors in the wavelength setting of the instrument will produce significant error in measurement. Conversely, if the rate of change of A with wavelength is small, as it usually is when working at the peak, a slight error in setting is of no great consequence. Other factors, however, can influence the choice of wavelength or filter: (a) In adapting methods to photometry that were originally designed for visual colorimetry, occasionally it has been found that if the general rule given above is followed the working range is quite restricted. To extend the working range to cover the clinically important values another wavelength may be chosen, thus sacrificing sensitivity and sometimes even photometric accuracy for expediency. An obvious alternative approach would be to start with a smaller sample. Dilution of the final color should never be done unless the validity of this procedure has been established by experimentation. (b) It is desirable to have a large absorption due to the substance being determined relative to the absorption due to other substances present. To achieve this it may be necessary to compromise on the general rule. For example, in measuring the turbidity in the thymol turbidity test of liver function, the general rule would dictate that measurements be made in the blue end of the spectrum. Both hemoglobin and bilirubin are frequently present in significant amounts in sera for this test, and since they have strong absorption peaks in the blue, sensitivity of thymol turbidity measurements is sacrificed by making measurements in the red in order to eliminate these undesirable interferences. The broad bands of filters frequently make it very difficult to eliminate interferences. (c) Other factors that may occasionally influence the choice of wavelength are the bandwidth obtainable at various wavelengths and instrumental error, which tend to be greater in the region where the photoreceptor element is least sensitive.

CALIBRATION CURVES

A *calibration curve* relates A or % T to concentration and is necessary in quantitative work where the amount of absorber has to be calculated. Where possible, construction of a calibration curve that covers the entire range of concentrations to be met in practice should be one of the first steps in setting up and standardizing a photometric procedure. The most conventional method is to plot A versus concentration as in Figure 1-7. Concentration can be expressed in any fashion (mg/100 ml, molarity, etc.), but the most convenient way is to use the same units that are to be employed in reporting the final result. Many photometers and spectrophotometers have scales reading directly in A values or in units proportional to A. Some instruments, however, do not have logarithmic scales and read only % T. In this case, the % T readings must be converted to A values either by using the equation $A = 2 - \log \% T$, by reference to special tabulations or logarithmic tables, or by use of a slide rule. Alternatively % T can be plotted versus concentration on semilogarithmic graph paper as shown in Figure 1-8. One cycle semilog paper is sufficient if the %T values all fall in the range 10—100.

It is usually *unsatisfactory* to plot % T versus concentration as shown in Figure 1-9. Six or more points are required to draw an accurate curve, and it is difficult to give proper experimental weight to each point. Also, it is virtually impossible from such a plot to decide whether Beer's law applies

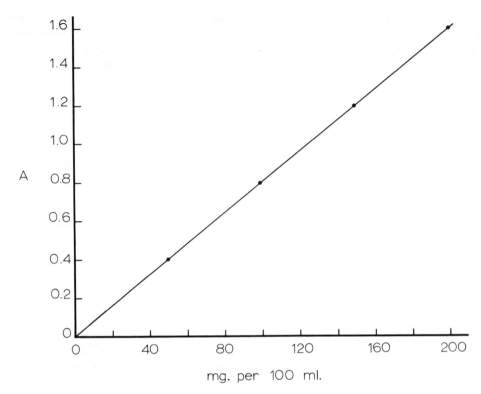

FIG 1-7. Calibration curve, A versus concentration. Oxyhemoglobin in 0.01% Na_2CO_3 at 580 nm. Beckman Model DU Spectrophotometer, effective slit width about 1 nm.

and results must be calculated by graphical reference to the curve itself. In the event there is a significant change in one or more of the standards used to check such a curve, it is necessary to reconstruct the entire curve in order to determine its proper shape. For these reasons a plot of A versus concentration is much preferable.

In order to determine whether the relationship between A and concentration is linear, a minimum of three points is required. With rare exceptions, A is 0 for 0 concentration, thus immediately locating one point. This means that, in addition to a reagent blank, a minimum of two other experimental points must be obtained. One point should represent the highest concentration of the range to be met in practice and the other should be one-half of the highest concentration. The experimental points should *not* be obtained by running the highest concentration and then making serial dilutions of the final colored solution. Each experimental point should be obtained independently. Failure to do this may, in some cases, lead to serious errors. It is preferable to run several replicates for each point and to employ greater care in technic than might be considered adequate for routine determi-

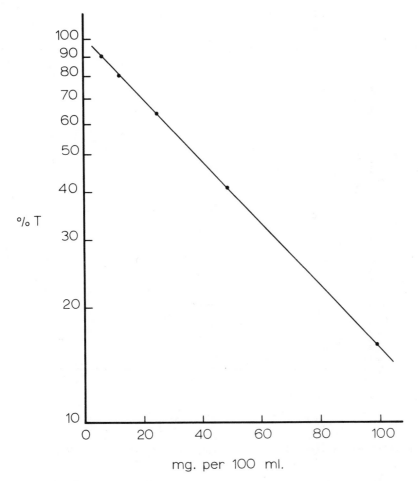

FIG 1-8. Calibration curve, % *T* versus concentration. Same data as in Fig. 1-7 plotted on one cycle semilog graph paper.

nations. Once the experimental points have been placed on the graph a transparent ruler is used to draw a straight line best representing all points. Since there is experimental error connected with each point determined it would be fortuitous if a straight line passes through all points, so the line is drawn which best represents the points, giving weight to each point (actually, the line decided upon may not pass through any of the points). From consideration of photometric errors, the points lying between A values of 0.1 and 1.0 (10–80%T) should be given greater weight in construction of this line. Statistical methods for drawing this line could be employed, but for routine work the added computations are not justified unless a programmed calculator is readily available.

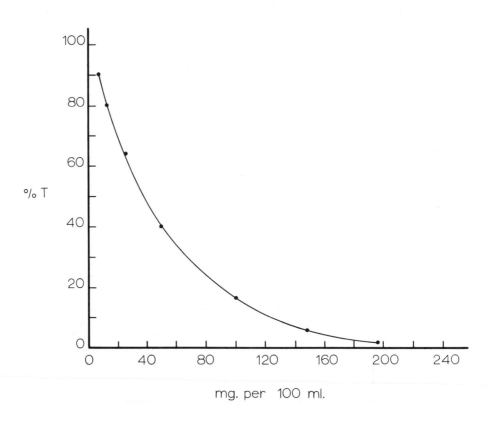

FIG. 1-9. Calibration curve, % *T* versus concentration. Same data as in Fig. 1-7.

The directions given above for drawing a straight line through the experimental points apply only to data for which a straight line would appear to be justified (Beer's law is obeyed), taking into consideration the degree of dispersion of replicate points and the degree of error involved if there is an apparent slight deviation from the straight line drawn. If it is obvious that a straight line will not represent the data (Beer's law is *not* obeyed) a curve must be drawn. This will frequently require more experimental points for its adequate description. Such curves are shown later in Figure 1-12. Concentrations of unknown can be read directly from the calibration curve and the curve checked by running one or two standards. For practical purposes a short segment of the curve can be considered to be a straight line. An unknown can be calculated from two standards falling near by if we assume Beer's law to be valid for this segment. The distance permissible between unknown and standards depends on the degree of curvature and the degree of error permissible in the result. In some cases the 0 standard can serve as a point.

DEVIATIONS FROM BEER'S LAW

Deviations from Beer's law are common. The law is valid only over a limited range under the most nearly ideal conditions. Causes for deviation may be either chemical or instrumental. Chemical causes include: (a) variable formation of absorber from the constituent to be determined; (b) interaction of absorber with itself (complex formation, polymerization); (c) interaction of absorber with the solvent (e.g., iodine in a nonpolar solvent such as carbon tetrachloride is purple in color, whereas in polar solvents the color is brownish); (d) interaction of absorber with some foreign material present; (e) change in refractive index with concentration; (f) displacement of equilibrium involving absorbers (e.g., changes in pH or ionic strength when absorbers are weak acids or weak bases). Instrument sources of deviation are discussed in a later section, but suffice it to say here that they may occur even in good spectrophotometers. Conformance to Beer's law should *never* be assumed but should be checked in each case under the conditions of the recommended or adopted procedure for the concentration range to be met in practice.

EFFECTS OF BANDWIDTH AND STRAY LIGHT

Pure monochromatic light is not realized in spectrophotometers, but with the best instruments bandwidths of the order of 1 nm or better can be obtained. If we assume exact reproducibility in setting the instrument to a given wavelength, a bandwidth of this magnitude will behave like a monochromatic beam. Difficulties can develop, however, as the bandwidth becomes wider and wider. Where the bandwidth occurs coincidental with a broad peak (absorption curve nearly flat in this interval) as in Figure 1-10 A, the situation is ideal, because the transmission of energy by the sample is relatively constant for all wavelengths present in the incident beam (represented by dotted line, the transmission curve of the filter). This latter condition is a prerequisite for Beer's law to be valid. It follows, therefore, that the narrower the peak, the narrower must be the band pass to satisfy this condition.

Filter photometers do not have the flexibility of a spectrophotometer and the situation can easily arise in which either the band pass of the filter is wider than the nearly flat segment of the absorption peak of the sample or the band pass of the filter is not coincidental with the absorption peak, as in B of Figure 1-10. In such instances, the transmission of energy by the sample is *not* the same at all wavelengths of the incident beam. In Figure 1-10, B, the light present in the beam up to 550 nm is nonabsorbable by the sample. The light above 550 nm is absorbable, but the light at 575 nm is relatively *less* absorbable than that at 600 nm. Such nonabsorbable energy (including "relatively" nonabsorbable energy) has been included by some authors in the term *stray energy*. Stray energy is perhaps more properly applied only to

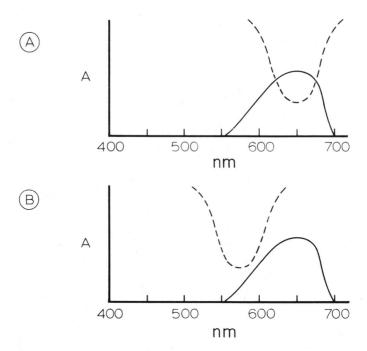

FIG 1-10. Hypothetical relationships between a colored solution and filters. Solid lines, absorption curve of colored solution. Broken lines, absorption curves of filters. For explanation, see text.

unwanted radiant energy arising from dust, scratches, or other optical imperfections on prisms, lenses, mirrors, slits, etc., and from light from the room or the light source of the instrument when the instrument itself is not "light tight." Regardless of source, the effect of stray energy is the same, and for simplicity nonabsorbable energy is included as stray energy in the present discussion.

The effect of stray energy on values of A is shown in Figure 1-11, where apparent A values have been calculated for true A values for circumstances in which 0%, 1%, 5%, and 10% of the incident energy is nonabsorbable. As the relative amount of stray energy increases, the curvature or deviation from a linear relationship increases. For the three conditions 1%, 5%, and 10% stray energy, as the true A approaches infinity (concentration approaches infinity), the apparent A values approach limits of 2, 1.3, and 1, respectively. Herein lies one of the most serious consequences of the presence of stray energy, namely, decreased accuracy at high concentrations. Simple mathematical technics are available by which, from values of A taken at several concentrations, correction for stray energy can be calculated for the entire curve relating A to concentration (20). Since the error produced by stray energy increases with absorbance, great errors may occur when a color is read against a reagent blank of high absorbance (15).

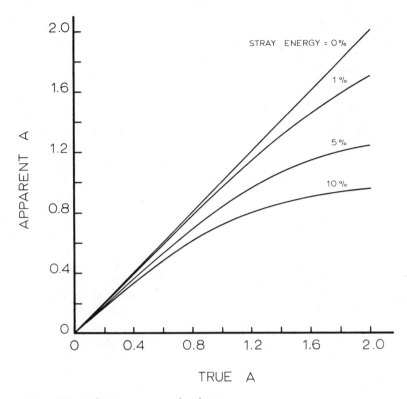

FIG 1-11. Effect of stray energy on absorbance.

As previously mentioned, when taking measurements at a narrow absorption peak of a substance, stray energy becomes quite significant when the band pass exceeds the flat portion of the peak, even when the nominal wavelength coincides with the center of the absorption peak of the substance. This is illustrated in Table 1-3, which gives the absorbance values

TABLE 1-3. Effect of Bandwidth on Absorbance of Oxyhemoglobin. (1:2500 dilution of 13.4 g hemoglobin per 100 ml 0.1% Na_2CO_3 versus a 0.1% Na_2CO_3 blank; Beckman Model B, at 414 nm)

Effective bandwidth, (nm)	A_{414}
0.56	0.390
1.2	0.390
2.2	0.390
4.0	0.365
7.4	0.352
14	0.332

obtained at various bandwidths for a solution of oxyhemoglobin, the nominal wavelength used being that of the 414 nm maximum of this substance. Thus, it is seen that an A value *per se* means very little without knowing the absorption curve and the band pass with which it was obtained.

Many filter photometers employ filters with rather broad half-bandwidths, and the stray energy factor is frequently significant. To illustrate this point, absorbance versus concentration curves were determined for oxyhemoglobin in 0.01% Na_2CO_3 at 415, 445, 490, 580, and 610 nm with the Beckman Model DU Spectrophotometer employing effective slit widths of 1 nm, with the Leitz filter photometer which employs filters of half-bandwidths of 30–75 nm, and with a Klett-Summerson filter photometer employing two filters and no filter at all (Klett readings are proportional to absorbance). The curves are shown in Figure 1-12, and should be interpreted in light of the absorption curve of this substance as shown in Figure 1-13. It is seen that Beer's law is followed for all measurements obtained with a 1 nm bandwidth regardless of whether or not the nominal wavelength of measurement was at the top or side of a peak. On the other hand, all measurements obtained with the filter photometers showed more or less deviation from Beer's law. The effect of bandwidth on A values is also illustrated in Figure 1-13, where the circles represent measurements made by the Leitz photometer. The loss of sensitivity by using a wide band filter when the absorption peak of the substance is relatively narrow, as illustrated by the relative A values for the two instruments at 415 nm, represents another type of restriction in the use of filters.

In instances where absorbance of a solution is measured against a blank having a high absorbance, deviation from Beer's law may occur even with a narrow half-bandwidth. This is because at high absorbance even a small amount of stray energy becomes a significant part of the total radiation available to the energy detector (33). By taking extreme precautions in eliminating "stray energy" of all types, obedience to Beer's law has been obtained up to absorbance values of 5 (27).

Fortunately, most solutions for photometric measurement in clinical analysis exhibit absorption bands sufficiently broad so that the average broad band filter is satisfactory, although certain exceptions occur. Of course, deviation from Beer's law *per se* does not mean the relationship is unusable, but there are disadvantages: (a) There is increased photometric error, and (b) a calibration curve must be constructed and referred to for each analysis, except for very limited concentration ranges. One of the disadvantages of filter photometers is that the number of filters routinely sold with most instruments is quite limited, although usually a wider selection is available at further cost. However, the inherent effect of stray energy in a filter photometer or a low dispersion spectrophotometer places a definite limitation on the use of such instruments especially for high absorbance measurements.

PHOTOMETRIC ERRORS ORIGINATING IN THE INSTRUMENT

Some instrumental errors are inconsequential for ordinary quantitative work because they are cancelled through use of calibration curves. Obviously,

FIG 1-12. Calibration curves. Oxyhemoglobin in 0.01% Na_2CO_3. Upper, Beckman Model DU Spectrophotometer, effective slit widths about 1 nm. Middle, Leitz photometer. Lower, Klett-Summerson photometer.

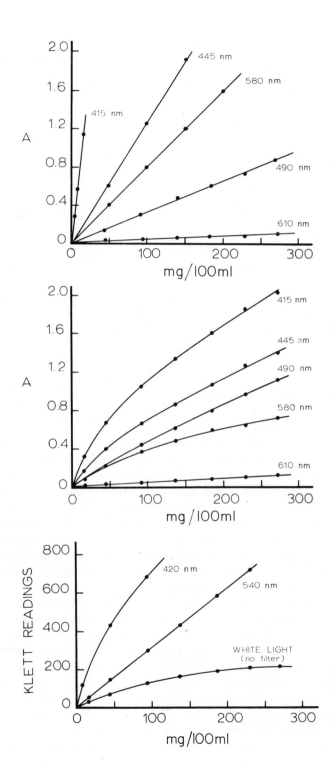

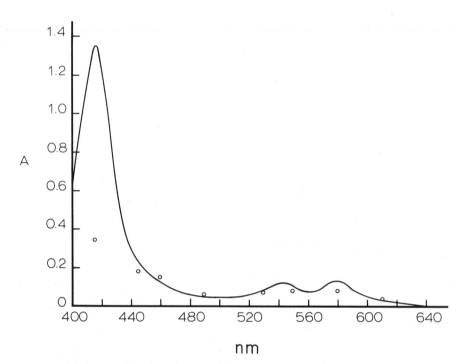

FIG 1-13. Oxyhemoglobin in 0.01% Na_2CO_3. Curve is that obtained on 15.2 mg/100 ml with Beckman DU Spectrophotometer, effective band width about 1 nm. Circles are readings made of same solution by Leitz photometer.

the curves must be checked regularly, preferably by including a blank and one or more standards along with each set of unknowns. Errors in this category include those due to nonparallel incident radiation and measurements with radiation that is not strictly monochromatic, use of cylindrical cuvets, reflections at cuvet surfaces and elsewhere along the optical path, and stray radiation reaching the detector. All such sources of error must be carefully controlled in determining molar absorptivities and other work relating concentration to true absorbance (8, 17). A technique for standardizing measurements of enzyme activity made at 340 nm has been described earlier (see "Transmittance Standards").

Another category of errors relates to technics employed in making the measurements. Errors due to mismatched, scratched, or dirty cuvets, to faulty positioning of cuvets, or to other grossly faulty technics can be avoided only by scrupulous attention to detail. The photometric measurement itself involves three steps or their equivalent: (a) setting the instrument at 0% T with no radiant energy striking the photosensitive element (adjusting for the dark current), (b) setting the instrument at 100% T with the reference blank, and (c) reading the radiant energy transmitted through the sample. An error in measuring the radiant power (P) is associated with each step. Assuming Beer's law applies, it is easily shown that the relative error in concentration measurement $(\Delta C/C)$ depends upon the error in measurement of radiant power (ΔP), the level of radiant power (P), and the observed absorbance (A), approximately as follows:

$$\frac{\Delta C}{C} = \frac{-\Delta P}{2.3\, AP}$$

In very simple instruments where the photocurrent is measured with a galvanometer or potentiometer having a linear scale, the uncertainty in reading the scale often becomes the limiting factor. In such cases, ΔP is independent of P and the photometric error is least at 37% T (0.43 A). The foregoing limiting condition is the basis for the socalled Twyman-Lothian error curve. In newer type instruments employing phototubes, electronic amplification and modern read-out techniques, the error in measuring radiant power tends to be proportional to the square root of the radiant power. Under such circumstances, photometric error is least at 0.87 A (13% T). The error curve has a broad minimum in both cases but increases steeply outside the range 0.5 M–2 M, where M is the absorbance corresponding to minimum error. For older type instruments the optimum range is about 20%–50% T (0.7–0.2 A). The range is broader with newer instruments (31), and one should be guided by specifications provided by the manufacturer. A systematic treatment of photometric error in terms of signal-to-noise theory is given in Ref. 26.

A common laboratory mistake is failure to recognize that progressively larger errors in concentration measurement occur as the absorbance readings deviate further and further from the optimum range. Ordinarily, measurements should be confined to the range 10%–80% T (1.0–0.1 A) unless the instrument is known to perform satisfactorily over a wider range. There are several possible ways to bring % T into the optimal range: (a) Adjustment of concentration by using a smaller or larger aliquot of sample. This approach is usually the simplest. (b) Use of cuvets of different light paths (short light path for low % T, long light path for high % T). This can be done only with certain instruments. (c) Use of a different wavelength or filter. This cannot always be done because of interfering substances. Even with single component systems a change in filter may result in marked deviation from Beer's law, which for practical purposes may be undesirable. Also, shifting to a wavelength where there is less absorption automatically means a decrease in slope of the calibration curve and a decrease in precision. In some cases this sacrifice in precision is overshadowed by the time saved, especially if the loss of precision is small. (d) Measurement of unknown versus a standard (transmittance-ratio technic), rather than unknown versus a blank. In this technic the photometer scale is set to 100% T on the solution of lesser absorbance, whether the unknown or the standard. A reading is then taken of the other solution. A calibration curve can be constructed in any of the usual ways. This technic not only permits extension of the working range but also results in a decrease in the relative analysis error. For further details see Refs. 6, 19.

A word of caution: It is usually unsatisfactory to simply dilute the final solution, except perhaps when the diluent has the same composition as the zero standard, because of chemical changes that occur in the reaction mixture. In any case, dilution should not be employed until it has been proved reliable by independent experiments.

PHOTOMETRIC ERRORS ORIGINATING IN THE SAMPLE

There are four possible sources of error originating in the sample itself: changes of absorption characteristics with temperature or time, turbidity, and fluorescence.

TEMPERATURE EFFECTS

Routine photometric and spectrophotometric measurements are usually made with the sample at "room temperature." This temperature, of course, is not precisely defined and, furthermore, after the instrument has been turned on for some time the cuvet compartment is usually at least several degrees above the prevailing room temperature. Variation in temperature may affect the measurement either by a change in solvent volume, which is about 0.1%–0.2% change per degree Centigrade, by physicochemical changes such as the degree of association or dissociation and solubility (1) or rarely, when working at high absorbance, by changing the molar absorptivity. Exposure to ultraviolet radiation, especially with organic solvents, may produce photochemical change in the solvent and thus change its absorption characteristics. The prime causes of significant temperature variation are excessive heating of the sample by the radiant energy source, which can occur in some photometers, failure to allow a cooled or heated sample to come to room temperature prior to measurement, and wide fluctuations of room temperature. Attachments (thermospacers) are available for some instruments for controlled temperature measurements and are mandatory for kinetic determination of enzyme activity and other assays based upon kinetic measurements.

THE "TIME" VARIABLE IN MEASUREMENTS

Many of the solutions met with in clinical chemistry upon which measurements are to be made have unstable color properties. The color intensity may increase or decrease, or the absorption peak may actually shift. Measurements on such systems must be made at some standard time interval after the color producing reaction is initiated, preferably when the rate of change in intensity or shift is minimal. Valid absorption spectra are very difficult to obtain on such solutions unless a stable plateau of relatively long duration exists. Whenever absorption curves are run on solutions of unknown stability it is necessary to establish their validity by rechecking intermittent points from time to time or when a recording spectrophotometer is used, by rescanning the entire spectrum after an appropriate interval.

TURBIDITY

One of the most common sources of error is the presence of turbidity in the test solutions. This fact cannot be overemphasized since it is a constant problem in many analyses. Particles in suspension, frequently colloidal in nature, result in decreased radiant energy transmission by a combination of true absorption and scattering. Since scattering is greater at shorter wavelengths, the presence of appreciable turbidity also causes an apparent shift in absorption peaks (22).

It should become routine practice for the analyst to examine solutions for turbidity prior to placing them in the instrument. One learns by experience in which tests this is likely to be a problem. It is not sufficient merely to hold up the solution to ordinary laboratory light or daylight for this will reveal only gross turbidity. By holding the solution directly in front of a strong spotlight such as a microscope lamp, much lesser degrees of turbidity can be easily detected. One should not look directly through the solution into the light, but rather through the solution against a dark background with the light entering the solution at right angles to the line of vision. The turbidity is thus visualized by its Tyndall effect. The photometer or spectrophotometer can detect degrees of turbidity not easily visible even by this technic.

Sometimes the solution can be cleared by centrifugation or by repeated filtration through very retentive filter paper. In certain analyses the situation is handled by special manipulations. Occasionally the situation may arise when the turbidity cannot be removed and yet a true reading of the desired absorption is necessary. In such cases it is possible to use the base-line technic of correcting the background absorption, using a plot of $\log A$ vs $\log \lambda$ since the relationship between $\log A$ and $\log \lambda$ is linear for noncolored particles (13).

FLUORESCENCE

In most photometers and spectrophotometers, the cuvet is placed after the filter or monochromator and just ahead of the detector. This is desirable to reduce heating and photochemical decomposition of the sample but gives rise to a photometric error when there is fluorescence of the sample or cuvet (28). For example, if a wavelength of 380 nm excites a fluorescence of any wavelength, the radiant energy of fluorescence is added to that transmitted through the sample at 380 nm. The magnitude of error depends on the ratio of fluorescent to transmitted energy and on the relative sensitivity of the detector to the fluorescent and transmitted energy. Fortunately, only a part of the fluorescent energy reaches the detector since the fluorescence is emitted equally in all directions. For this reason, the error produced by fluorescence is seldom very large.

In those instruments in which the cuvet is placed between the source and the filter or monochromator, the error caused by fluorescence is negligible.

MULTICOMPONENT SYSTEMS

In previous sections the concern has been with single component systems, i.e., solutions containing only one species of absorber at a given wavelength (it is assumed that the effect of other absorbers present in the reagents can be cancelled out by use of a reagent blank). Occasionally, situations are encountered in which a substance is to be determined in the presence of one or more absorbers having appreciable absorption at the same wavelengths. Two examples of such a situation are treated here.

TWO COMPONENT SYSTEM

The following conditions must apply to the system: (a) No interaction occurs between the two absorbers, i.e., the individual absorbances are additive. This should be proved experimentally by carrying known mixtures of the two components through the entire procedure and comparing the resulting absorbances with absorbances of the same concentrations of pure solutions of the individual components. (b) Both absorbers obey Beer's law. (c) At least one wavelength exists at which the absorptivities differ substantially.

The calculations involve the solution of simultaneous equations. Measurements of absorbance at two wavelengths are necessary. The ideal situation occurs when one component absorbs strongly at one wavelength while the other component has weak absorption at this wavelength and, at the other wavelength selected, the relation of absorbances of the two components is reversed. However, good results can be obtained even if both components absorb equally at one wavelength, provided that at the other wavelength selected their absorptivities differ markedly. Frequently, the optimal wavelengths are not clearly apparent but, for a two component system, they are easily determined graphically by plotting the ratio of the absorptivities of the two components versus wavelength. The maximum and minimum of this curve are the best two wavelengths. Spectrophotometers of fairly good resolution are required for this technic, unless the absorption maxima of the two components are far apart. Filter photometers ordinarily can be employed only if interference filters are used.

If the absorptivities of both components at the two wavelengths selected are known, the following equations are used for calculation of the unknown concentrations:

$$c_x = \frac{(a_{y\lambda_2})(A_{\lambda_1}) - (a_{y\lambda_1})(A_{\lambda_2})}{(a_{x\lambda_1})(a_{y\lambda_2}) - (a_{x\lambda_2})(a_{y\lambda_1})}$$

$$c_y = \frac{(a_{x\lambda_1})(A_{\lambda_2}) - (a_{x\lambda_2})(A_{\lambda_1})}{(a_{x\lambda_1})(a_{y\lambda_2}) - (a_{x\lambda_2})(a_{y\lambda_1})}$$

where c = concentration, x and y are the two components, λ_1 and λ_2 are the two wavelengths at which readings are made, A is absorbance, and a is absorptivity.

An *isobestic (isosbestic) point* is the wavelength at which the absorbances are identical for two substances that are interconvertible, as in acid-base, redox, or tautomeric equilibria.

DETERMINATION OF ONE COMPONENT IN THE PRESENCE OF "BACKGROUND ABSORPTION"

The occasion arises in spectrophotometric analysis when the absorption peaks of the desired component are superimposed on the absorption of undesired and frequently unknown components. The absorption due to these components is referred to as "background absorption." There are two approaches to this problem, both of which assume additivity of absorbances and obedience to Beer's law.

Analysis by Simultaneous Equations. The problem can be treated as a two component system, the desired component being the first and the background absorption considered as the second. For this solution the background absorption must be available without the component to be measured so that the ratio of its absorbances at the two wavelengths selected can be measured. The two wavelengths are selected as for any two component system.

The Allen "Base-Line" Technic (2, 3). This method is faster and simpler than the previously discussed technic. For its application, certain relationships must hold. It can be done graphically or mathematically. Because the graphic method illustrates the principle of the base-line technic very clearly it will be presented first.

The first step in the procedure is to run absorption curves on control samples substantially free of the substance to be measured. The regions of interest are those in the vicinity of the absorption peaks of the substance to be measured. Consider an absorption curve of "background" as shown in A of Figure 1-14 and let us assume that the absorption peak of the substance to be measured occurs at λ_x.

Now, let us run the absorption curve with concentration c of substance x added to a control sample as shown in C of Figure 1-14. Two wavelengths are selected, one on each side of λ_x, so that the A values at these three wavelengths for the background absorption lie on a straight line. The absorption curve for the background does not have to be linear between λ_1 and λ_2 but the three points at λ_1, λ_x, and λ_2 must lie on a straight line. A value called the *base-line absorbance,* which is proportional to the true absorbance (and concentration) of the unknown, is obtained graphically as follows: Draw a straight line between A values at λ_1 and λ_2 and measure the increment between e and f as shown in curve C of Figure 1-14. This "*base-line absorbance*" is proportional to concentration if Beer's law holds but is equal to the true absorbance of the unknown at λ_x only when the unknown has 0 absorbance at λ_1 and λ_2. Once standardized, only three measurements of A (λ_1, λ_x, λ_2) are required for graphic calculation of the base-line absorbance.

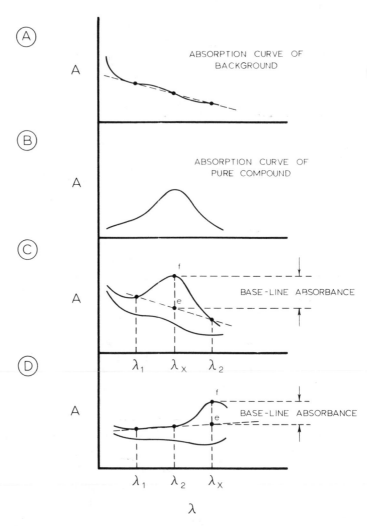

FIG 1-14. Graphic method of "base-line" technic. See text for explanation.

If two points on the absorption curve of the background cannot be found that lie on a straight line with the point at λ_x, a transformation of A, λ, or both might be found which brings the three points into line (e.g., the baseline technic applied to correction for turbidity discussed in an earlier section).

In example C of Figure 1-14, λ_1 and λ_2 are equidistant from λ_x, but this is not necessary. In some cases it is impossible to place λ_1 and λ_2 on opposite sides of λ_x. The method will work equally well if both λ_1 and λ_2 can be placed on the same side of λ_x, as shown in D of Figure 1-14.

When λ_1 and λ_2 are equidistant from λ_x, the base-line absorbance of substance x, A_x, can be calculated from the following equation:

$$A_x = A_{\lambda x} - \frac{A_{\lambda 1} + A_{\lambda 2}}{2}$$

When λ_1 and λ_2 are not equidistant from λ_x, the equation becomes:

$$A_x = A_{\lambda x} - A_{\lambda 2} + \frac{\lambda_x - \lambda_2}{\lambda_1 - \lambda_2} \left(A_{\lambda 2} - A_{\lambda 1} \right)$$

The base-line technic is especially advantageous in situations such as the determination of hormones in urine, where background absorption is almost invariably present. When using this technic one should be aware of its limitations: (a) Since the result is calculated from three measurements rather than one there is a compounding of error which increases markedly as the term $(A\lambda_1 + A\lambda_2)/2$ approaches $A\lambda_x$. (b) The validity of any one result rests on the *assumption* of a background absorption curve previously determined on another sample or samples; ordinarily, there is no way of proving this assumption to be correct for any particular unknown.

TURBIDIMETRY

Quantitation of turbidity by measurement of transmitted light is called *turbidimetry*. It is a great advantage in some cases to be able to quantitate a precipitate without separating it from the solution, especially when speed is desirable or no other practical method is available. There are several analytical procedures in clinical chemistry based on measuring the degree of turbidity produced in the test reaction, e.g., sulfosalicylic acid test for protein in urine or cerebrospinal fluid and the thymol turbidity test. The problem is to cause the desired constituent to precipitate in the form of finely divided particles of suitable size.

Measurements of turbidity can be made in any photometer or spectrophotometer. Values obtained are in terms of A or $\% T$, although loss of energy transmission through the sample is due to reflection and refraction of the energy rays by the particles as well as to absorption. According to Rayleigh's law, when the particle size is less than the wavelength of the incident radiant energy, the ratio of scattered to incident radiant energy is directly proportional to the cube of the particle diameter and inversely proportional to the fourth power of the wavelength. Reflection becomes a more important factor when the particle diameter is considerably larger than the wavelength of the incident radiant energy. In practice, the intensity of scattering by turbid solutions varies about as the inverse second power of wavelength. The relationship between $\log A$ and $\log \lambda$ is linear (13).

Since photometric sensitivity increases with decreasing wavelength, greatest sensitivity in turbidimetric methods is achieved with blue light, provided the suspending medium is colorless. On occasion, however, sacrifice of sensitivity is made to minimize the effect of interfering substances; e.g., in the thymol turbidity test, red light is used to minimize interference from bilirubin and hemoglobin present in the serum sample. Sensitivity is increased if the particles are opaque or highly colored. If the medium is colored, maximal sensitivity will usually be obtained at the wavelength of maximal transmittance of the medium.

The absorption curve of a suspension of colored particles is flattened, as compared with the curve of a solution of the pigment contained in the particles (11).

Within limits, the relationship between A and concentration of particles is usually linear for small sized particles. This correlation, however, may vary with wavelength (7). Any factor that produces a different distribution of particle size will alter the "absorbance" measurement. In fact, the most difficult aspect of turbidimetry is the preparation of reproducible and sufficiently stable suspensions. The use of protective colloids has been of value in some instances. All variables must be kept in careful control: the rate at which reagents are added, the manner of mixing or shaking, etc. Different cuvet shapes will give different "absorptivities." This must be kept in mind when comparing readings made between different instruments.

In general a turbidimetric method should be avoided if great accuracy and precision are required.

NEPHELOMETRY

Nephelometry differs from turbidimetry in that measurement is made of the light scattered at right angles to the incident light beam rather than the decrease in transmitted light. This arrangement results in greater sensitivity in measuring lower concentrations. The value of nephelometry lies in the ability to measure low concentrations that are beyond the sensitivity of turbidimetric measurements. The precautions necessary in nephelometry are the same as those discussed above under "Turbidimetry." Nephelometric attachments are available for some commercially available photometers and spectrophotometers.

SELECTED BIBLIOGRAPHY

BAUMAN RP: *Absorption Spectroscopy*. New York, Wiley, 1962

BRODE WR: *Chemical Spectroscopy*. New York, Wiley, 1943

EDISBURY JR: *Practical Hints on Absorption Spectrometry (ultraviolet and visible)*. New York, Plenum, 1968

GIBSON KS: Spectrophotometry (200 to 1000 millimicrons), *Natl. Bur. Std. Circ.* 484:1949

KOLTHOFF IM, ELVING PJ, SANDELL EB: *Treatise on Analytical Chemistry* Part I, Theory and Practice. Vol 5, New York, Interscience, 1964

LOTHIAN GF: *Absorption Spectrophotometry.* London, Hilger and Watts, 1958
MELLON MG: *Analytical Absorption Spectroscopy.* New York, Wiley, 1950

REFERENCES

1. ABBOTT R: *Soc Appl Spectroscopy Bull 4,* 1:1948
2. ALLEN E, RIEMAN W: *Anal Chem 25*:1325, 1953
3. ALLEN WM: *J Clin Endocrinol Metab 10*:71, 1950
4. BRODE WR, GOULD JH, WHITNEY JE, WYMAN GM: *J Opt Soc Am 43*:862, 1953
5. CASTER WO: *Anal Chem 23*:1229, 1951
6. CRAWFORD CM: *Anal Chem 31*:343, 1959
7. DAHLGREN SE: *Acta Chem Scand 14*:1279, 1960
8. DAVIES WG, PRUE JE: *Trans Faraday Soc. 51*:1045, 1955
9. DAVIS R, GIBSON KS: *Natl Bur Std (U.S.),* Misc. Publ. 114, 1931
10. DELHEZ R: *Chemist-Analyst 49*:20, 1960
11. DUYSENS LNM: *Biochim Biophys Acta 19*:1, 1956
12. EWING GW, PARSONS T Jr.: *Anal Chem 20*:423, 1948
13. FOG J: *Analyst 77*:454, 1952
14. FOG J: *Scand J Clin Lab Invest 14*:320, 1962
15. FRIDOVICH I, FARKAS W, SCHWERT GW Jr, HANDLER P: *Science 125*:1141, 1957
16. GIBSON KS, BELKNAP MA: *J Res Natl Bur Std (U.S.) 44*:463, 1950; *J Opt Soc Am 40*:435, 1950
17. GOLDRING LS, HAWES RC, HARE GH, BECKMAN AO, STICKNEY ME: *Anal Chem 25*:869, 1953
18. HEIDT LJ, BOSLEY DE: *J Opt Soc Am 43*:760, 1953
19. HISKEY CF, RABINOWITZ J, YOUNG IG: *Anal Chem 22*:1464, 1950
20. HOCH H, TURNER M, WILLIAMS RC: *Clin Chem 6*:345, 1960
21. KIRK PL, ROSENFELS RS, HANAHAN DJ: *Anal Chem 19*:355, 1947
22. LATIMER P: *Science 127*:29, 1958
23. LOWRY OH, BESSEY OA: *J Biol Chem 163*:633, 1946
24. MALININ DR, YOE JH: *J Chem Educ 38*:129, 1961
25. MARTINEK RG, JACOBS SL, HAMMER FE: *Clin Chim Acta 36*:75, 1972
26. McCARTHY WJ: *Advan Anal Chem Instrum:*9 493, 1971
27. MEHLER AH: *Science 120*:1043, 1954
28. MEHLER AH, BLOOM B, AHRENDT ME, STETTEN D Jr: *Science 126*:1285, 1957
29. MEINKE WW: *Anal Chem 43,* No. 6:28A, 1971
30. PARTHASARATHY NV, SANGHI L: *Nature 182*:44, 1958
31. SLAVIN W: *Appl Spectrosc 19*:32, 1965
32. VANDENBELT JM: *J Opt Soc Am 44*:641, 1954
33. VANDENBELT JM, HENRICH C, BASH SL: *Science 114*:576, 1951
34. WALKER IK, TODD HJ: *Anal Chem 31*:1603, 1959
35. *Anal Chem 43*:2038, 1971
36. *Natl Bur Std (U.S.),* Letter Circ LC-929, 1948; superseded by Letter Circular LC-1017, 1955
37. *Color Filter Glasses,* Corning Glass Works, Corning, N.Y., 1970
38. *Kodak Filters for Scientific and Technical Use,* Eastman Kodak, Rochester, N.Y. 1972

Chapter 2

Flame Photometry

W. B. MASON, Ph.D., M.D.

Flame photometry is widely used in clinical chemistry for determination of sodium, potassium, calcium, and magnesium and is becoming the method of choice for lead, mercury, copper, zinc, and other metals. Details of specific methodology are given in other chapters. Theoretical background and practical aspects having general application are considered here.

THEORY

The principles of flame photometry have been generally understood for many years and therefore are outlined only briefly. Comprehensive treatises that include references to an extensive literature are listed in the bibliography.

Flame photometry may be divided into two broad categories, *atomic emission*, synonymous with conventional flame photometry, and *atomic absorption*. A third category, *atomic fluorescence*, is just emerging but as yet finds virtually no application in clinical chemistry. It is potentially important, however, because of somewhat greater sensitivity. In all cases, the flame serves to convert the element to be measured into an atomic vapor. Thereafter the same physical phenomenon, namely reversible transformations between ground and excited electronic energy states, gives rise to the optical signal that is measured. This phenomenon may be represented as $A^* \rightleftharpoons A + h\nu$, where A^* represents an excited atom, A an atom having ground-state electronic energy, and $h\nu$ a photon. In conventional flame photometry (*atomic emission*), excited atoms are generated by heat and chemical reactions occurring within the flame, and the reaction that gives rise to the optical signal is from left to right. Instrumentation is designed to measure photons emitted, hence *atomic emission* flame photometry. In *atomic absorption*, the reaction is from right to left. A beam of monochromatic light of frequency ν is passed through the flame, and a measurement is made of the attenuation resulting from photons interacting with ground-state atoms to produce excited atoms. In *atomic fluorescence*, excited atoms are produced from ground-state atoms by illuminating the flame with an intense beam of light having frequency ν or greater, and the photons emitted when these atoms return to the ground state are measured. This measurement is ordinarily made at right angles to the exciting beam, as in conventional fluorometry. Since atomic fluorescence is still in an early developmental stage without significant application in clinical chemistry, it will not be mentioned further.

INSTRUMENTATION PRINCIPLES

Conceptually at least, the burner, optical system, and photometer are quite similar in atomic emission and atomic absorption instruments. In practice, however, there are major differences. Most atomic emission flame photometers used in clinical chemistry are limited purpose instruments designed specifically for determining Na and K in serum and urine and tend to employ very simple optical designs. Atomic absorption instruments are inherently more complex because a source of resonance radiation must be included. Over-all they resemble a spectrophotometer in which the sample cuvet has been replaced by a flame. In addition, atomic absorption flame photometers are multipurpose instruments capable of determining more than 65 elements. They were developed as general analytical instruments and only incidentally find use in clinical chemistry.

The two types of instruments are shown schematically in Figure 2-1 and are described briefly in the following paragraphs.

FIG 2-1. Schematic representations of atomic emission flame photometer (top) and atomic absorption flame photometer (bottom).

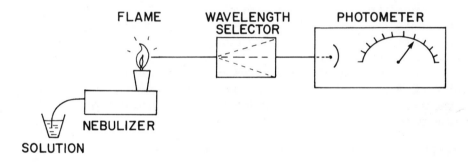

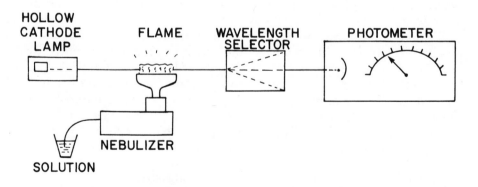

BURNER SYSTEM

In both emission and absorption instruments the solution must be converted into a fine spray or aerosol in the course of being introduced into the flame. This process is termed *nebulization*, and the nebulizer is ordinarily considered as being part of the burner. Within the flame, solvent evaporates from the aerosol leaving microscopic clinkers that disintegrate under the influence of heat and reactive combustion intermediates to yield atoms. Collectively these phenomena are termed *atomization*. Some atoms are in excited electronic states or even ionized, but most are in ground-state energy levels. In earliest flame photometers, aerosol was formed by spraying the solution into a chamber where it was thoroughly mixed with air and then drawn into the flame, usually a common Meker burner. Very little solution reached the flame because large droplets settled out in the nebulizer, but newer nebulizers are more efficient. This so-called "premix" design has found wide application and continues to be exceedingly popular and satisfactory for clinical laboratory instruments. Natural gas or propane are the common fuels for atomic emission, and acetylene for atomic absorption.

In another type of burner the solution is aspirated directly into the flame. The initial thought was that greater sensitivity would result by eliminating loss of solution during nebulization, hence the designation "total consumption" burner. It is now known, however, that considerable aerosol passes unchanged through the flame. Also, the micro clinkers are larger, and there is less efficient conversion into atoms. The flame is very turbulent, giving rise to audible noise that may be so intense as to be uncomfortable to the operator. This turbulence produces fluctuating concentrations of atoms, both excited and in the ground state, and the photometric signal is unsteady. At the present time total consumption burners find much less use in clinical chemistry than premix burners.

OPTICAL SYSTEMS FOR ATOMIC EMISSION INSTRUMENTS

Most emission flame photometers used in clinical chemistry employ very simple optical systems, often no more than a filter to select the desired spectral region and a single lens to focus an image of the flame on the photodetector. Such a design has been satisfactory because virtually all applications are for Na and K in serum and urine. The Na and K resonance lines at 589 and 766 nm are intense and widely separated from much weaker lines due to other elements present. The flame background itself is relatively weak. Thus, high spectral purity is not required. When better resolution is required, as in determination of calcium or magnesium, monochrometers similar to those used in spectrophotometers are employed. In fact, most commercial atomic emission instruments other than for Na and K are sold as attachments for general purpose spectrophotometers.

OPTICAL SYSTEMS FOR ATOMIC ABSORPTION INSTRUMENTS

An atomic absorption flame photometer is essentially a spectrophotometer having a hollow cathode lamp as source and a flame replacing the cuvet. Both single and double beam designs are used. An important difference from a spectrophotometer, however, is that the flame itself emits radiation, and provision must be made to correct for any photometric signal due to this radiation. Such correction is ordinarily made by modulating (chopping) the light from the hollow cathode lamp and using in the photometer an amplifier tuned to the same frequency. Except for small fluctuations at the modulation frequency, flame radiation is undetected. A monochromator is ordinarily used in atomic absorption instruments to isolate the desired wavelength. This is mainly for convenience and flexibility since atomic absorption instruments are expensive and are intended for multiple applications. Conversion from one element to another ordinarily entails changing only the hollow cathode lamp and making minor photometer adjustments, in addition to setting the monochromator to the new wavelength.

HOLLOW CATHODE LAMPS

In atomic absorption flame photometry, a beam of monochromatic light having the exact same wavelength as the characteristic radiation of the element to be determined must be passed through the flame. Attenuation of this beam by atoms present in the flame gives rise to the signal that is measured. Hollow cathode lamps are the most practical means for generating monochromatic light of the required spectral purity, and much of the progress in atomic absorption is due to the successful development and manufacture of high quality hollow cathode lamps. As the name implies, these lamps have a hollow or cuplike cathode that is constructed from the metal to be determined. Alternatively, the cathode may be lined with the pure metal or with an appropriate alloy. It is necessary to use a separate lamp for each element, except for a few instances where the cathode can be fashioned in such manner that a single lamp serves for two or perhaps three elements. The lamp is filled with an inert monatomic gas, usually neon or argon, at low pressure. In operation, metal atoms are continually sputtered from the inner service of the cathode by bombardment from positive inert ions, filling the cathode with an atomic vapor. Atoms in this vapor undergo electronic excitation by collisions with ions and electrons and the resulting excited atoms emit their characteristic radiation in returning to the ground-state electronic level. This creates highly monochromatic light having the correct wavelength for absorption by ground-state atoms in the flame.

PHOTOMETERS

In the simplest emission flame photometers (direct-reading), emitted light characteristic of the element being determined falls upon a photocell, and the resulting electronic signal, appropriately amplified, is shown directly on a sensitive galvanometer or other meter after appropriate amplification. Calculation of unknown concentration involves comparing this signal with that obtained for standard solutions run in series with the unknown. When an internal standard is used, a more complex photometer is required since it is necessary to find the ratio between the signals for unknown and internal standard, and to compare this ratio with that for standard solutions. In early instruments for Na and K, using Li as internal standard, this signal ratio was obtained by manually operating a potentiometer that was part of a simple null-balance circuit. A notable recent achievement has been the development of atomic emission instruments incorporating three separate detector channels, one each for Na, K, and Li, and the necessary analog circuits to present simultaneous digital readouts of Na and K concentrations, thereby avoiding all calculations by the operator.

Photometric readout for atomic absorption instruments differs from that for atomic emission in that no measurement is made of light originating in the flame itself. Instead, a measurement is made of the diminution that takes place when monochromatic light having wavelength equal to the resonance frequency of the element being determined is passed through the flame. Photomultiplier detectors are used almost exclusively in atomic absorption instruments and commonly in newer emission flame photometers. Amplification is ordinarily by ac technics, usually after modulation of the light giving rise to the signal, as this gives less noise. Output from the last stage of amplification is normally converted to dc and fed into a meter, potentiometric strip chart recorder, digital volt meter, or other readout device.

INTERNAL STANDARD TECHNIC

Variations in solution uptake, nebulizer performance, atomization efficiency, and flame parameters produce variations in photometer signal. Some degree of compensation can be achieved by the so-called internal standard technic in which an element foreign to the sample being tested but having flame characteristics similar to the element of interest is added in known concentration before the solution is fed into the flame photometer. For Na and K, the internal standard is usually Li. Both serum and urine require dilution as the first step, and Li is incorporated into the diluent. The calculation involves comparing the ratios of Na to Li and K to Li signals for the unknown with those for standards.

The element used as internal standard must behave in the flame in a manner similar to the element being determined. This requirement is easily met for Na and K, and the atomic emission instruments used for Na and K are sufficiently simple so that necessary optical and electronic components can be added without an unreasonable increase in cost. For these reasons, most atomic emission flame photometers used in clinical chemistry incorporate the internal standard principle. By contrast, atomic absorption flame photometers rarely use internal standards, due mostly to the large additional cost for providing internal standard capability. Atomic absorption already gives better accuracy and precision than is ordinarily obtained by chemical means for Ca, Mg, Pb, Cu, and similar elements. Furthermore, the accuracy and precision required for these elements is thought to be less than is required for Na. As better data accumulate regarding calcium levels, for example, it may become desirable to seek ways for improving atomic absorption flame photometers, and this may necessitate provision for an internal standard. At least one manufacturer (Instrumentation Laboratory, Inc.) already offers such an instrument.

FLAMELESS ATOMIC ABSORPTION

The flame functions only to convert the sample into an atomic vapor. It can be replaced by other atomization processes and two other technics find practical application. In one, the sample is dried on a carbon rod, tantalum ribbon, small boat, or other support and vaporized in an inert atmosphere by passing a high density electric current which creates an instantaneous temperature of about 2100°C. The flameless atomizer is positioned in the space normally occupied by the flame. This approach appears promising for Mg, Fe, Pb, Cu, Zn, and other metals. A second technic applicable to Hg uses wet chemical reactions to convert the metal into an atomic vapor. The sample is first decomposed by wet digestion, if necessary, and then treated with stannous chloride or other reducing agent to convert Hg into the elemental state. Thereafter a stream of gas is bubbled through the apparatus, sweeping Hg vapor into a cell with quartz windows mounted in the optical beam. Absorption measurements are made at the 2537 Å resonance wavelength.

CHOICE OF INSTRUMENT AND INSTALLATION

So many instruments are commercially available that it is obviously impossible to evaluate or even describe them all; since new models and components are being introduced every few months, to do so would be of only fleeting value.

The best practical guides in choosing a flame photometer are the recommendations of experienced chemists who have worked first-hand with the instrument. Lists of users should be available from the manufacturer or distributor of any instrument. Direct communication should be established with the person who is responsible for the instrument, and if possible it should be seen in operation under circumstances similar to those which are anticipated. Particular attention should be given to the needs of the laboratory since no one instrument is ideal for all applications. Where large numbers of serum Na and K determinations must be made, and reliability is a prime concern, the most satisfactory choice at present is a limited purpose atomic emission instrument designed specifically for Na and K and employing Li as internal standard. Instruments vary widely in sophistication and convenience of operation, including automatic sample handling, with a corresponding range in price. It is logical to do Li by atomic emission technic, and if use of Li as a therapeutic agent gains in popularity and requests for Li levels become more frequent, it is to be expected that special purpose Na and K instruments will be modified to include capability for Li determinations. This has already been done by at least one manufacturer (Instrumentation Laboratory, Inc.). Separate and more complex instruments, whether emission or absorption, should be obtained for Ca, Mg, and other metals. At present it appears that Ca and Mg are best determined by atomic absorption.

Suppliers of propane (LPG), acetylene, and other gases are ordinarily listed in the classified section of the local telephone directory. Alternatively, some instrument manufacturers supply small cylinders of special grade LPG for use with their flame photometers. It is advisable to consult the local fire department before bringing any bottled gas into the building. It is also prudent to advise the insurance carrier. Acetylene presents the greatest danger. Copper or copper alloy tubing must never be used with acetylene because of the possible formation of acetylide, which is explosive. Propane is heavier than air, and tends to collect at floor level in case of a leak and thus should never be used or stored in basements or other closed areas lacking adequate ventilation. Except for some Na and K instruments that employ very small flames, an efficient exhaust system, having its own blower, should be provided over the flame. This exhaust can discharge into a duct from the hood system or directly outdoors. For a steady supply of compressed air, the ballast tank should have a capacity of at least 200 liters and should be maintained at 70–140 psi.

GENERAL TECHNIC

Because details of procedure differ somewhat with each make of instrument, the manual accompanying the instrument should be consulted. The principles, however, are universal.

REAGENTS

It is becoming increasingly common for instrument manufacturers to offer the necessary standards and diluting fluids. In any case, the composition of specific reagents will be given in the brochure that accompanies the instrument. As a general rule stock solutions are prepared first and from them more dilute working solutions, both for standards and diluent. Certain basic principles apply to preparation of all these solutions.

Good analytical technic is essential, for the ultimate accuracy of the determinations cannot exceed that of the standards. Volumetric glassware must be accurately calibrated and appropriate pipets should be used. All solutions must be thoroughly mixed: a safe rule is to invert and shake each container 20 times after the solution has been diluted to final volume. Distilled water should be free of contaminants that would interfere in any way. If there is an appreciable blank, the water should be purified by ion exchange columns or by distillation or by both. Conductivity measurements are convenient for checking total ionic purity.

Storage bottles should be of polyethylene, Pyrex, No-Sol-Vit glass, or other material which will not introduce contamination. Prior to use, new Pyrex bottles should be soaked for several days in distilled water containing a little HCl, to remove the soluble alkali present on a new glass surface and then rinsed thoroughly with distilled water. Glass stoppers should not be given a "grinding twist" because of possible resultant contamination. Working standards are most likely to change with storage and must be checked periodically against fresh dilutions of stock standards. All standard bottles should be shaken immediately prior to use to recover the condensate that collects above the liquid level and insure a homogeneous solution.

Analytical grade salts are usually satisfactory, both in preparing standards and for diluting fluid, but various spectrochemically pure standards are available from chemical supply houses. The National Bureau of Standards offers selected standards. Chlorides are generally used for Na and K and carbonates for Ca and Li. Carbonates are converted to chlorides by HCl added during preparation of the solution. In any case, the material to be used as a standard should be of known composition and should meet ACS specifications or similar criteria. It should also be stable under conditions specified for drying. Concentration of stock solution should be selected so that the volumes required for preparing working standards can be measured accurately. Usually this means that not less than 5 ml of stock standard should be used for the weakest working standard, but this obviously is not a firm rule.

DILUTION

Serum and urine must be diluted prior to measurement to avoid clogging of the nebulizer and burner. Composition of the diluting fluid depends on the element being measured, the necessity of including substances to overcome interferences, the type of instrument (emission or absorption) and whether it is direct reading or uses an internal standard, and other factors such as fuel, oxidant, and flame temperature. The optimum concentration of internal standard is usually specified by the instrument manufacturer. For serum Na the dilution may be from 1:50 to 1:500, and for serum K from 1:10 to 1:100. It is convenient, of course, to be able to use one dilution for both elements, and 1:50 is commonly used. For Na and K in urine, one usually begins with at least two dilutions because of the much greater variation in concentration than occurs in serum. With internal standard instruments, it is good practice to dilute both the unknown and standards with the same diluent to avoid possible errors due to differences in concentration of the internal standard.

The dilution factor is determined from considerations of total solids present in the sample, sensitivity of the instrument, linearity of the relationship between signal and concentration, accuracy and precision attainable, and the amount of sample available. The degree of dilution *per se* does not affect the working range. For example, when both the unknown and standard are diluted 1:10 and the photometer is set to 100 on the diluted standard, the diluted unknown is found to read 50. If a second 1:20 dilution is made for the unknown and standard, and the photometer is set to 100 on the second dilution of standard, the second dilution of unknown will again read 50, assuming no interferences are present. Indeed, one of the best practical tests for interferences is to make a series of dilutions (preferably in the ratios 1:2:4· · ·) and determine the ratio between signals of unknown and standard at each dilution. The ratio will be constant in the absence of interferences. The likelihood of interferences decreases with increasing dilution. The greater the dilution, however, the weaker the photometric signal. This requires greater amplification and a concomitant increase in electronic "noise" in the readout. Choice of dilution factor therefore requires a careful compromise.

METHOD OF ANALYSIS

The flame photometer, whether atomic emission or atomic absorption, is capable only of comparing the signal produced by the unknown with that of standards. It is therefore essential that standards be correct, and that dilutions be accurate, both for standard and unknown. It is also essential that a proper procedure be followed in comparing unknowns and standards. Details for specific instruments will be found in the manufacturer's brochure.

Performance is usually more steady if the instrument is allowed to warm up for a few minutes before use. A solution closely approximating those to be run should be fed into the flame during warmup, and the electronics should be fully operational. There is usually some interaction between settings of the zero and full-scale photometer controls, so that successive adjustments are required to obtain constant readings. It is good practice to include a "zero standard" when making up the working standards, and to use it for the zero adjustment. In direct reading instruments the zero standard should be checked against pure solvent to determine the blank, and the entire series of standards replaced if the blank is excessive. With internal standard instruments it is necessary to use an independent method for determining the blank contributed by the internal standard itself. Stock internal standard diluted with pure solvent read against the zero standard provides a measure of blank contributed by ingredients other than the internal standard and solvent.

Probably the most unsettled question of technic is the number of standards to be run with each unknown or series of unknowns. Recommendations by various groups of workers include: run standards at each 10% or 20% of full scale with each set of determinations; check each unknown with the nearest standard; check each unknown with the next higher and next lower standard; run three standards spanning the range found in practice with each set of determinations. Because of the great accuracy required for serum sodium and potassium, at least three standards spanning the range found in practice should be run with each set of determinations no matter how small the set, and at least one of these standards should be interspaced between unknowns during the set. In general, a complete set of standards should be run immediately before any unknowns are read and again immediately after all unknowns are completed. Mid-range standards should be introduced between unknowns at frequent intervals, and the complete set of standards should be rerun whenever there appears to be significant drift.

A plot of standard readings against concentration gives a calibration curve from which unknown concentrations can be read, both with direct reading and internal standard instruments, whether atomic emission or atomic absorption. This is the preferred procedure since calibration curves are seldom perfectly linear. In some cases, however, there is sufficient linearity over the range of interest so that direct calculations can be made. Direct calculations are also used sometimes when the unknown falls so close to one of the standards that a negligible error results from assuming linearity.

ENHANCEMENT BY ORGANIC SOLVENTS AND BY EXTRACTION

Substitution of organic solvents for water commonly leads to greater sensitivity, both in emission and absorption modes of analysis. This is attributed to numerous factors including faster sample uptake, better nebulization and formation of finer particles, which makes atomization more effective, and higher flame temperature due to solvent combustion. The magnitude of the effect often depends on which portion of the flame is being used to generate the photometric signal; this undoubtedly has sometimes played a role in conflicting findings by different investigators. Sensitivities for Na, K, and Ca are such that organic solvents are unnecessary. Lead and other trace elements that may be extracted into organic solvents after formation of complexes with reagents such as ammonium pyrrolidine dithiocarbamate are conveniently handled by direct aspiration of the organic solvent.

SOURCES OF ERROR

The possible sources of error include those arising from the instrument itself and those from substances present in the sample or reagents, as well as those due to poor technic.

BURNER SYSTEM

The most critical component in a flame photometer is the burner and its associated nebulizer. A steady flame of constant thermal output is essential and cannot be obtained without good pressure regulation for both fuel and oxidant gases. This usually requires two-stage automatic reduction valves with a needle valve following the second stage. Gas flows should also be monitored by rotameters or other flow devices as a precaution against obstruction that may occur either in the nebulizer or burner. It is always necessary to filter compressed air to remove condensate and particulate matter. Glass wool or cotton filters are satisfactory. Other gases may also require filtering. City gas varies widely in composition and is difficult to regulate because pressures are low and variable. Cylinders containing propane as liquified gas (LPG) are commonly available, and propane finds wide use for sodium and potassium determinations by atomic emission. Commercial grade LPG contains appreciable butane and small amounts of other hydrocarbons that build up as the cylinder empties. This adversely affects the flame, so that the tank should be changed before it is completely exhausted. Acetylene tanks should also be changed before completely empty as the tail gas contains appreciable solvent.

In addition to steady pressures, it is important to have the correct pressures for both oxidant and fuel, and the proper ratio between the two. The pressures to be used, or the procedures to follow in establishing correct pressures, are given in the manufacturer's instruction manual. Optimal settings vary between instruments, sometimes even between instruments of the same model, and also depend on the element being determined and whether an emission or absorption technic is being used. Substantial deviation from optimum pressure results in decreased sensitivity and gives a signal that is unduly influenced by pressure fluctuations.

Assuming stable and optimum flame conditions, the next requirements are constant uptake and nebulization of the sample. Factors that may cause variable behavior from sample to sample or between unknown and standards include viscosity, density, vapor pressure, surface tension, temperature, solvent composition, salt content, and changing size of nebulizer bore due to buildup of protein or other foreign matter. In practice, variable nebulizer bore is the most troublesome. The others are mostly overcome by the large dilutions required with serum and urine. Including a wetting agent such as Sterox SE (Monsanto) in the diluting fluid is recommended to help reduce foreign deposits. A fine wire should always be at hand for use as a probe whenever partial obstruction is suspected. It is good practice to clean the nebulizer at regular intervals.

Sample uptake can be monitored by measuring the time required to consume a specified volume. Alternatively with pickup type nebulizers, the time interval between placing the capillary in the solution and first appearance of the flame signal can be measured. In instruments having gravity feed, readings should always be taken with the liquid level within narrow limits to minimize effects due to variable hydrostatic head.

MATRIX DISCREPANCIES

Variation between physical properties of test solutions and standards may give rise to errors. Factors already mentioned include: viscosity, density, vapor pressure, surface tension, temperature, solvent composition, and salt content. For good matrix matching, protein is sometimes included in standards when the serum dilution factor is small, as in determination of serum calcium. With Na and K, however, dilution is so large that inclusion of protein in standards is unnecessary. In actual practice, it is often difficult to obtain protein that is completely free of the element being determined, and this must be taken into account in preparing a proper blank.

CHEMICAL INTERFERENCE

With some elements, the presence of certain anions in the unknown results in formation of refractory compounds that are incompletely dissociated in the flame. Depression of photometric signal results. Calcium is the most common example in clinical chemistry. The effect is due principally to phosphate and occurs in both emission and absorption modes. The depressant effect of phosphate can be overcome by adding lanthanum or strontium; however, lanthanum is preferred since it forms a tighter complex with phosphate. Lanthanum phosphate is thought to precipitate first as the aerosol evaporates, leaving Ca to form a more easily dissociated compound. Chelating agents can also be used to control chemical interference, presumably by preventing reactions that lead to refractory compounds. EDTA, citric acid, glycerol, and a variety of other chelating agents are effective in relieving phosphate interference on Ca. Magnesium behaves similarly to Ca. Chemical interference may disappear at higher flame temperatures, such as are obtained in burning mixtures of nitrous oxide and acetylene. Internal standards ordinarily provide no relief from chemical interference.

SPECTRAL INTERFERENCES

In atomic emission flame photometers, spectral interferences result whenever the emission line of an extraneous element lies so close to the emission line being used for the measurement that the two are not resolved by the optical system, and both signals reach the detector. Similar interference may occur from molecular band emissions. Neither effect is serious in determining Na or K in serum or urine because of the large dilutions used and because other elements normally present give much weaker signals. Spectral interferences are generally much less serious in atomic absorption than in atomic emission and in actual practice are of no importance where atomic absorption finds application in clinical chemistry. Flame background can be considered a form of spectral interference but is ordinarily offset by use of a "zero standard."

IONIZATION INTERFERENCE

The equilibrium between atoms, ions and electrons that exists in a flame may be represented as $M^0 \rightleftharpoons M^+ + e$. The photometric signal is proportional to the number of free atoms (M^0), both in atomic emission and atomic absorption. Any increase in free electrons (e) will therefore *increase* the signal by shifting the equilibrium in favor of *free atoms*. For the effect to be significant, however, the element being determined must be appreciably ionized under conditions used for its determination. Ionization is strongly temperature dependent and with the low temperature flames commonly used in clinical chemistry amounts to about 3% for K and Ca and only 0.3% for Na. There is much variation depending upon the portions of the flame being examined. Variation in solution composition can produce only small changes in the free electron density in flames since free electrons arise mostly from chemical reactions that take place between the fuel and oxidant gases. In practice, ionization interference is very small for Na and K. With other elements, or with high temperature flames, the effect due to variable ionization may become significant. In such case it can be controlled by adding to the test solution an excess of an easily ionized element that stabilizes the flame by providing a high concentration of free electrons. Thus, in determining serum Ca by atomic absorption using a nitrous oxide-acetylene flame, it is good practice to include excess K (about 25 meq/liter) in both unknowns and standards to avoid a bias due to K enhancement in the unknowns. This is not necessary with an air-acetylene flame, which burns at a lower temperature.

ACCURACY AND PRECISION

For replicate measurements on the same solution, instrument manufacturers commonly claim coefficients of variation (C. V.) of 0.5% or better for Na and K by atomic emission and 1% for Ca by atomic absorption. In practice, under stringently controlled conditions, the within-laboratory reproducibility corresponds to a C. V. of about 1% for Na, 1.5% for K, and about 3% for Ca (2).

The error between laboratories is a better indication of accuracy obtained routinely and is of greater importance in evaluating the usefulness of results for diagnostic purposes. A recent survey (1) conducted by the U. S. Public Health Service, Center for Disease Control, Atlanta, Ga., covering nearly 500 laboratories participating under the Clinical Laboratory Improvement Act, showed C. V.'s of 2.8% for Na and 4.7% for K for 400 laboratories in the licensed and volunteer categories. Performance was somewhat better by another group of 86 laboratories selected in advance as "reference laboratories" for which C. V.'s of 1.8% for Na and 4.2% for K were reported. For the 400 licensed and volunteer laboratories, about 2.5% of the results for Na differed from the mean by more than 2.5 meq/liter. For K, about 3% of the results differed from the mean by more than 0.4 meq/liter. These limits are considered as defining the range required for medically significant results. In a similar survey in which 58 of the 472 participating laboratories reported Ca results by atomic absorption, the C. V. for these results averaged 3.8%. About 11% of the atomic absorption results differed from their mean by more than 0.25 meq/liter, which is considered the limit for medical significance for serum calcium.

SELECTED BIBLIOGRAPHY

CHRISTIAN GD, FELDMAN FJ: *Atomic Absorption Spectroscopy*, Applications in Agriculture, Biology and Medicine. New York, Wiley-Interscience, 1970

DEAN JA: *Flame Photometry*. New York, McGraw-Hill, 1960

DEAN JA, RAINS TC: *Flame Emission and Atomic Absorption Spectrometry*, vol 1. Theory; vol 2, Components and Techniques. New York, Marcel Dekker, 1969, 1971

SLAVIN W: *Atomic Absorption Spectroscopy*. New York, Interscience, 1968

REFERENCES

1. BAYSE D: Personal communication. Address: Chief, Clinical Chemistry Unit, Proficiency Testing Section, Laboratory Division, Center for Disease Control, Atlanta, Ga. 30333

2. YOUNG DS, HARRIS EK, COTLOVE E: *Clin Chem 17*:403, 1971

Chapter 3

Fluorometry

S. LAWRENCE JACOBS, Ph.D.

Fluorometry, or the determination of the characteristics and amount of luminescence produced by substances when examined under carefully controlled conditions, has been treated extensively and in detail in many review articles and texts. The reader is directed to these for a more comprehensive view of the subject than can possibly be given here (8, 10, 11, 12, 13, 14, 19).

THEORY

As has been pointed out in Chap. 1, molecules of many substances have the property of absorbing energy at specific portions of the electromagnetic spectrum. This occurs from the x-ray to the infrared regions. In addition to absorbing radiation, many molecules have the property of re-emitting some of this absorbed radiation as luminescence of one kind or another. This, too, occurs throughout the spectrum. However, for practical purposes, it is the region between about 250 and 700 nm that is useful to the clinical chemist and to which, therefore, the following discussion will be limited. The fundamental way that luminescence spectroscopy and absorption spectroscopy differ is that in the latter we are interested in the effect that the sample has upon the light beam impinging upon it, whereas with the former the reverse is true. In luminescence spectroscopy we are interested in the effect of the impinging radiation, called primary, activating or exciting radiation, upon the sample: What type of and how much luminescence will the impinging radiation evoke from the sample? What exciting wavelengths are most effective in evoking luminescence?

Luminescence may be of two types: fluorescence (the name is derived from the fluorescent mineral fluorspar) and phosphorescence. In the phenomenon of fluorescence, a molecule absorbs radiant energy, and electrons in the ground state are raised to an excited state, called an excited singlet state, wherein an electron with its spin in one direction remains paired with an electron with a spin in the opposite direction. Some of these electrons must lose energy by nonfluorescent decay in order to reach the lowest vibrational level of the excited state prior to returning to their ground state. The energy emitted in returning to the ground state, therefore, is less than that absorbed. Since energy (E) is related to vibrational frequency (ν) by the equation $E = h\nu$, where h is Planck's constant, and frequency is related to wavelength (λ) according to the expression, $\nu = c/\lambda$, where c is the velocity of light, it follows that $E = hc/\lambda$, or E varies inversely as λ. Thus, the wavelength of the emitted radiation must be greater than that of the exciting radiation. The process of fluorescence is instantaneous, occurring in about 10^{-8} sec. To date, fluorescence is the type of luminescence that has proved most useful in clinical chemistry.

Phosphorescence differs from fluorescence in that the spin of one of the electrons in the pair that is in the excited state has been reversed due to some internal energy transition. When this occurs, the electrons are said to be in the triplet state and in the process of going from the singlet to the triplet state have lost additional energy. Since subsequent return from the triplet to the ground state involves less of an energy loss than does dropping from the excited singlet to the ground state, phosphorescence occurs at a longer wavelength than does fluorescence. The process of phosphorescence is usually of the order of milliseconds but may require several seconds. This difference in the time taken for the processes of fluorescence and phosphorescence makes it fairly easy to differentiate between the two by appropriate instrumental means. Although phosphorimetry has been used in the determination of certain drugs and pesticides in blood and urine (20), there is as yet insufficient application of phosphorescence analysis in clinical chemistry to warrant further discussion here.

Luminescence spectroscopy is a powerful tool, for one can qualitatively and quantitatively describe (a) the excitation spectrum, (b) the fluorescence spectrum, and (c) the phosphorescence spectrum of a compound. Then, by judicious selection from these spectra of wavelengths for quantitative work, far greater selectivity, specificity, and sensitivity can be obtained than from a single absorption spectrum in absorptimetry.

From the mechanism of fluorescence described above, it can be predicted that for a molecule to fluoresce it must possess a structure that provides excitable electrons. The so-called π electrons present in compounds that have a conjugated system of double bonds (–C=C–C=C–, etc.) are such electrons. Thus, aromatic structures are very likely to fluoresce, especially if they also contain atoms that tend to contribute to the delocalization of these π electrons. Substituents that withdraw π electrons from the aromatic ring, and thus reduce their freedom, will also reduce or destroy the fluorescence properties of the molecule. Compounds not themselves fluorescent, or only weakly so, may be converted to highly fluorescent derivatives by chemical reactions such as condensation, substitution, dehydration, or oxidation, or by simply altering the pH.

Fluorescence intensity is directly proportional to the concentration of the fluorescing substance in dilute solutions. From the Lambert-Bouguer-Beer law:

$$I / I_O = 10^{-acd}$$

where I = intensity of light transmitted

I_O = intensity of incident light

$$a = \text{absorptivity}$$

$$c = \text{concentration}$$

$$d = \text{cell thickness}$$

then $1 - I/I_o = 1 - 10^{-acd}$ = fraction of light absorbed

and $I_o - I = I_o(1 - 10^{-acd})$ = amount of light absorbed

The quantum efficiency, ϕ, is the ratio of total emitted light or F (fluorescence) to total absorbed light. Therefore,

$$F = \phi(I_o - I)$$

or $F = \phi I_o(1 - 10^{-acd})$

or $F = \phi I_o \left[2.3\, acd - \frac{(-2.3\, acd)^2}{2} - \frac{(-2.3\, acd)^3}{6} \cdots - \frac{(-2.3\, acd)^n}{n!} \right]$

and if c and, therefore, acd is small,

$$F = \phi I_o\, 2.3\, acd,$$

and by combining all constants in a particular system, $F = Kc$. In the above derivation, c must be sufficiently small that no more than about 5% of the incident radiation is absorbed.

In absorptimetry, the sensitivity of measurement is dependent upon the fractional absorption of radiant energy by a substance in solution, which in turn is a function of the substance, its concentration, and length of the solution path traversed by the beam of radiant energy. In fluorometry, however, sensitivity of measurement is also dependent upon the intensity of the incident radiant energy, as has been shown above, and also upon the means of detecting and measuring the fluorescence. Thus, increased sensitivity can be achieved by merely increasing the intensity of the exciting radiation and/or the sensitivity of the detection device. Fluorometric procedures can in this way be more sensitive than spectrophotometry by an order of magnitude of at least $10^3 - 10^4$.

While it may at first seem inconsistent to say that, on the one hand, fluorescesce is only a fraction (ϕ) of the total light energy absorbed, and on the other hand, that fluorescesce is orders of magnitude more sensitive than absorptimetry, a consideration of the instrumental factors will account for the apparent paradox. In absorptimetry, light transmitted by a solution is compared with the amount of incident light. In practice, this is done by comparing a solution containing the absorbing material to a blank solution or solvent devoid of this material. The blank is considered to have a transmission of 100%. The sample, which is of a very low concentration, may have a transmission of 98%. If each of these readings were subject to a ± 0.5% error, then the 2% T *difference* between them would be subject to an absolute error of ± 1.0% T, corresponding to an uncertainty of ± 50%. In fluorescence, however, the blank solution would read, or could be adjusted to read, zero. The same sample used in the example above could be set to read 100. Then an error in each of these readings of 0.5% of the scale would contribute a combined error of only 1.0% to the *difference*. This is, in fact, the accuracy with which fluorescence measurements can be made in the concentration range of micrograms per liter. To put this explanation into more practical terms: In absorptimetry, the detector is looking at small differences in a light beam emanating directly from the light source; in fluorescence, the detector is looking at the relatively large differences in the radiation generated in the sample itself against a virtually black background.

There are several factors other than instrumental ones that will affect fluorescence intensity: (a) The concentration of the fluorescent material may be sufficient to absorb so much of the exciting radiation that a gradient in the intensity of the exciting energy is produced within the solution itself (inner filter effect). The I_0 is thus effectively reduced in the solution and consequently F, too, is reduced. Concentrated solutions, even opaque suspensions, can be quantitated fluorometrically (21) using equipment that detects the fluorescence on the same face as that upon which the exciting light impinges. Here too, however, a point can be reached at which linearity of fluorescence is lost as all of the incident radiation is absorbed. (b) Variation in the concentration of fluorescent substance may affect the fluorescing properties of the molecules by dissociation, association, and solvation. (c) Fluorescence is quite sensitive to certain variables in the solution such as the nature of the solvent (usually shifting to shorter wavelengths as the dielectric constant is increased), pH, temperature, impurities, and the presence of ions. Anions such as CNS^-, I^-, Br^-, and Cl^- have a marked quenching effect on the fluorescence of certain compounds. (d) Foreign materials may interfere by fluorescing themselves, by absorbing the exciting energy, or by absorbing the emitted light. Dichromate, traces of which may be present from cleaning of glassware, is particularly troublesome in this way. (e) Intramolecular effects such as competing electronic transitions in the light absorbing molecules and intermolecular effects involving bimolecular collisions or compound formation may interfere with fluorescence efficiency. This may cause the excitation energy to be dissipated as heat. Diminution in measured emission intensity from the theoretical, regardless of cause, is commonly referred to as *quenching*.

Light may be scattered in several ways and, if not recognized and dealt with, could interfere with fluorescence methods, particularly at low levels of analyte, where high instrumental sensitivity is necessary (9). *Rayleigh scattering* is reemission of radiation of the *same* wavelength as the incident light produced when electrons in the solvent or solute drop to the ground state from high vibrational level resulting from absorption of radiation. *Raman scattering* arises from similar phenomena but differs in that additional vibrational energy is gained or lost (generally the latter), causing a shift in the wavelength of the emitted light, usually to longer wavelengths. The observed Raman shift is characteristic of the solvent employed and is independent of wavelength. Parker (7) reports this shift for water to be about $0.34 \mu^{-1}$ toward the red end of the spectrum. In this laboratory the Raman shift for distilled water has been observed to be as shown in Table 3-1. Scattering of incident light also occurs because of particles in the solution (*Tyndall scattering*) and by reflection from surfaces of the instrument.

TABLE 3-1 Raman Shift in Distilled Water[a]

Excitation wavelength (nm)	Observed Raman peak (nm)
280	322
300	345
320	370
365	425
405	473
436	518
450	540

[a]Data obtained on an Aminco-Bowman spectrophotofluorometer with a 1P28 photomultiplier tube. No corrections were made for lamp or detector characteristics.

SPECTROFLUOROMETRY

The difference between spectrofluorometry and fluorometry is analagous to that between spectrophotometry and photometry. This difference is the sophistication of the instrumentation with respect to the means of spectral isolation of radiant energy. Thus, just as photometric methods must be studied with a spectrophotometer to determine the spectrum of the color produced, fluorometric procedures require knowledge of the characteristics of the fluorescence or other emissions (phosphorescence, Raman, etc.) produced in the sample by the substance of interest or by the solvent, reagents, or interfering materials. Perhaps more important is the necessity of knowing the characteristics of the excitation spectrum which is the fluorescence produced over the range of excitation wavelengths in the region of interest. This is observed by scanning the excitation wavelengths while viewing the fluorescence that is produced at one wavelength or region of optimal fluorescence emission. The excitation spectrum is, in fact, a very close approximation of the absorption spectrum of the substance of interest and, indeed, may even be used for qualitative identification of the substance, using published absorption spectra for reference.

Figure 3-1 illustrates diagrammatically the essential features of a spectrofluorometer (also referred to as a spectrophotofluorometer or fluorescence

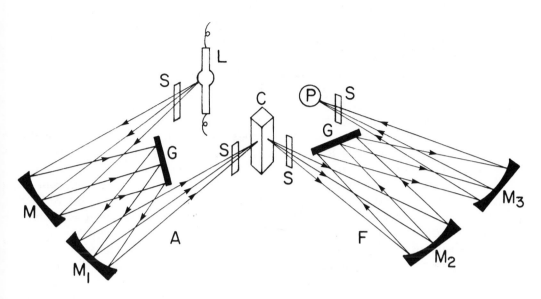

FIG 3-1. The spectrofluorometer. See text for identification of parts.

spectrophotometer). A xenon arc lamp (L) is usually the source of exciting radiation. This lamp produces a continuous emission from 200 to 700 nm. Mercury-xenon lamps are also sometimes used and can afford greater instrumental sensitivity at the wavelengths of the intense mercury lines. In some instruments, an ellipsoidal mirror is used to direct the radiation from the source into the primary monochromator. The various manufacturers of spectrofluorometers provide different slit (S) arrangements and mechanisms, but these all serve to define the spectral bandpass of the primary or activation (A) and secondary or fluorescence (F) monochromators. Spectral band passes and thus, the resolution of peaks of the order of 1 nm are obtainable. The monochromators usually consist of diffraction gratings (G) for dispersion of the exciting and fluorescent radiation used in conjunction with spherical mirrors (M) for collecting and directing the light as shown in the figure. Prism monochromators are also used but less commonly. Manual selection of wavelengths may be made on each monochromator, or automatic scanning may be used. The cell compartment (C) may be designed for many types of cuvets (square, round micro, flow-through, etc.). The most common configuration is as shown in Figure 3-1, where the fluorescent light is viewed at right angles to the exciting light. Other configurations are used for special applications; for example, the cuvet may be placed on an angle, and the fluorescence viewed at the front face of the cuvet so that the quenching effect of highly absorbing or concentrated samples is minimized. The photomultiplier tube (P) is usually a 1P21 or 1P28. The response characteristics of several photomultiplier tubes are shown in Figure 3-2. It is important to note how the response of these tubes decreases precipitously in the 600 nm region. This causes distorted fluorescence spectra where the peaks appear to be shifted to lower than the true wavelengths (2). Likewise, in the 300 nm region for the 1P21 tube, the peak is observed at too high a wavelength. This effect can be compensated for by using established correction factors (15) by using a photomultiplier tube (more expensive) such as the R446S (Fig. 3-2), or by using an instrument (very expensive) that makes the corrections electronically. For routine use, however, such corrections are not made.

The fluorescence readings may be made on a meter, oscilloscope, *X-Y* recorder, or strip-chart recorder or may be fed to a computer for further processing of the data. Readings may be taken at one wavelength setting for each monochromator or by automatic scanning with either the excitation or fluorescence monochromator. Figure 3-3 shows the excitation spectrum (E) of noradrenolutine, the highly fluorescent derivative of noradrenaline commonly used in its determination. Here the fluorescence monochromator was kept at 510 nm. The shape of this curve is dependent on the relative intensity of the light produced by the xenon lamp in the region scanned to produce the excitation spectrum. The output of the lamp falls off in the 300

FIG 3-2. Relative response of photomultiplier tubes used in spectrophotofluorometers throughout the uv and visible regions of the spectrum. Adapted from Aminco Laboratory News, vol 26, American Instrument Company, Silver Spring, Md., 1970.

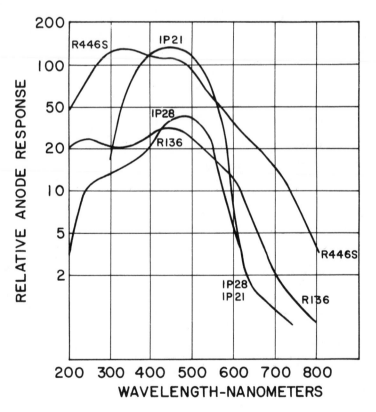

nm region (lower wavelengths). There are ways of compensating for this, and some instruments have the ability to do this so that a curve such as the broken one can be obtained. This type of correction is necessary if one wishes to report a "true" spectrum of the compound. For routine use, however, this is unnecessary. Thus, the excitation curve is far more characteristic of a fluorescing substance than is the fluorescence curve (6).

The fluorescence spectrum (F in Fig. 3-3) is obtained by keeping the excitation monochromator at a wavelength of maximum excitation, here 405 nm, and scanning with the fluorescence monochromator. Correction of this fluorescence spectrum for phototube response is discussed above. Rayleigh, Tyndall, and Raman scattering, also discussed above, are easily recognized in fluorescence spectra when high instrumental sensitivity settings are used.

There are several spectrofluorometers on the market, including those produced by the American Instrument Company, Baird-Atomic, Cary Instruments, the Farrand Optical Company, Perkin Elmer, the Schoeffel Instrument Corporation, G. K. Turner Associates, and Carl Zeiss, Incorporated.

FIG 3-3. Excitation and fluorescence spectra of noradrenolutine. Aminco–Bowman Spectrophotofluorometer.

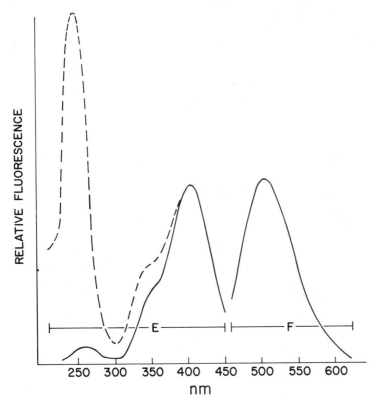

FLUOROMETRIC ANALYSIS USING FILTER FLUOROMETERS

Spectrofluorometers costing $5000 and up are beyond the reach of many laboratories that wish to use fluorescence analysis. Very good instrumentation in the form of fluorometers that sell for approximately $1000 are available. The usual configuration is shown in Figure 3-4, although complexity may vary in degrees from the simplest discussed here to a spectrofluorometer with price tags to correspond to the degree of sophistication.

In the fluorometer the exciting energy is usually produced by a mercury arc lamp or by a tungsten filament (L in Fig. 3-4). The mercury arc discharge lamp has emission lines at 313, 365, 405, and 436 nm. While these wavelengths may not be optimum for excitation of many substances, in most cases at least one of these lines is sufficiently close to an optimum wavelength to produce adequate excitation for assay purposes. An adjustable slit (S) is used to vary the amount of exciting energy directed at the sample in the cuvet (C) and thus acts as a sensitivity adjustment for the instrument. The wavelength band of the exciting energy is selected by the primary filter (PF), which may be a glass or gelatin (Wratten) filter or a combination of several of these, an interference filter, or a quartz interference wedge. Secondary filters (SF) are usually glass or Wratten filters and are chosen to exclude Rayleigh and Raman scattering as well as to select a wavelength region to optimize the detection of desired fluorescence and to minimize that from interfering substances. Semiquantitative estimation of fluorescence was at one time done visually. Current fluorometers, however, employ detectors (D) that may be barrier layer photocells, phototubes, or photomultiplier tubes. A shutter (SH) must be placed somewhere in the optical path to provide a means of zeroing the instrument by adjusting the dark current. In the author's opinion, any instrument that does not have adequate amplification of the signal will not have the sensitivity to be sufficiently versatile for the clinical chemist.

The actual mechanics of operating a fluorometer will vary somewhat from instrument to instrument and are described in detail in manuals accompanying them. The instrument is set at zero fluorescence on a reagent blank or with shutters closed so that no light reaches the detector and then some convenient reference point on the scale is adjusted to a reference standard, e.g., a reference point of 50 on the meter scale might be chosen for a standard equivalent to 50 μg/100 ml, the scale thus reading directly in μg/100 ml. In order quickly to adjust the instrument or recheck its setting, it is convenient to use a stable fluorescent solution such as quinine sulfate, sodium fluorescein, or a fluorescent glass block or rod that can be substituted for a cuvet.

It should be mentioned here that a filter fluorometer is perfectly satisfactory for most routine work *provided* that it has the required sensitivity at the desired wavelength bands. However, there is *often* no way of knowing whether or not the reading obtained is indeed derived from the substance for which the test is being run. That is, the specificity of the test is often suspect, particularly when patients are receiving a variety of drugs, many of which have fluorescent properties and which may come through the chemical isolation procedure. It has been shown in the analysis for urinary catecholamines by the trihydroxyindole fluorescence method that examination of the spectra of samples suspected of being elevated is essential (3). The initial elevated result obtained with a filter fluorometer is not adequate evidence of elevated catecholamines.

Fluorometers can be obtained from a number of companies including the American Instrument Company, Baird-Atomic, Beckman Instruments, Coleman, Farrand Optical, Jarrell-Ash, Klett, Photovolt, Technicon, and G. K. Turner (5). In purchasing a fluorometer, it is wise to make certain of the actual sensitivity obtainable under laboratory conditions for the tests to be performed (1).

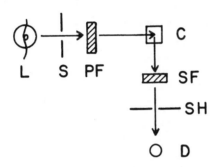

FIG 3-4. The fluorometer.

APPLICATIONS IN CLINICAL CHEMISTRY

Among the important precautions in fluorometric analyses in the clinical laboratory are the following:

a. The solvents used must be nonfluorescent and must not absorb radiation in the spectral regions used in the particular assay. With the most sensitive instruments, the lower limit of the working concentration range will be dependent on the "background" fluorescence of the solvent and on instrumental stability. Redistillation of water from bone black in an all glass still can be performed to reduce the trace fluorescence of impurities in water, although this is rarely necessary, and other solvents may be purified by suitable extraction and distillation procedures. "Fluorescent grade" solvents are now commercially available. It is occasionally necessary, when using the most sensitive methods, to specially purify all reagents used in an assay.

b. Pyrex or quartz cuvets should be used because of the slight fluorescence of soft glass. They should be kept free of scratches, which increase light scatter, and the windows should not be touched with fingers because of the likelihood of fluorescent or light-absorbing contamination. Cleaning of cuvets may be done with certain nonfluorescent detergents or with hot nitric acid. Traces of chromic acid may interfere by virtue of its absorption in the ultraviolet region.

c. Filters must be checked for contribution to blank readings because some of them fluoresce.

d. Contamination with stopcock grease is a serious problem in fluorometry. Wherever possible, the use of such grease should be avoided unless it has been shown that the lubricant does not interfere in the assay.

e. Fluorescent contaminants have been found in water or reagents which have been in contact with black rubber stoppers, polyethylene, or bakelite (4).

f. pH must be carefully controlled because of its great effect on fluorescence in the case of many substances.

g. Temperature must be fairly well controlled for two reasons: First, fluorescence usually decreases with increasing temperature. Second, temperature change during measurement may result in light scattering due to stratification, refractive index variation, and turbulence.

h. Solutions to be measured must be free of gas bubbles, suspended solids, and turbidity. To prevent formation of bubbles during measurement it may be necessary to boil and cool the solution immediately prior to measurement or to boil the distilled water and reagents before use. Excess scattering of light by particles or bubbles in a sample will cause low readings of the fluorescent radiation since some of the emitted light is scattered away from the detector. Suspended solids may be removed by centrifugation or by filtration through a sintered glass filter or a hardened filter paper. Removal of turbidity may not be possible, although centrifugation and filtration should be tried. Filter paper must be checked for fluorescent contamination of the solution being filtered.

i. The exciting radiant energy sometimes slowly changes the color or intensity of fluorescence. In such cases it is necessary to obtain the reading rapidly, to use an exciting energy of very low intensity, or to estimate the reading at zero time by extrapolation from several readings taken at intervals.

j. Loss frequently occurs in fluorescence analysis of submicrogram quantities of substances by adsorption on glassware, protein precipitates, etc., by photodecomposition of fluorescent derivatives on preparation and especially during the measurement, and by oxidation, as through the action of peroxides in certain solvents.

k. The presence of interfering substances that quench or alter the color of fluorescence can be detected by running an internal standard. Unusually low readings of this standard, which has been prepared in a portion of the sample extract, will be indicative of such interference. Comparison of the sample with the internal standard is often adequate to correct for the quenching of the sample.

In the clinical laboratory, fluorescence analysis has been used for: inorganic ions such as Mg, Ca, Cu, Fe, Zn, and Be; amino acids and related compounds such as phenylalanine, tyrosine, tryptophan, creatine, histidine, kynurenic and xanthurenic acids, catecholamines, metanephrines, serotonin, histamine, tryptamine, porphyrins, urobilin, and bilirubin; the nucleotides ATP, ADP, NADH, and NADPH; enzymes such as lactic dehydrogenase, transaminases, alcohol dehydrogenase, a-hydroxybutyrate dehydrogenase, isocitrate dehydrogenase, glucose-6-phosphate dehydrogenase, alkaline and acid phosphatases, creatine phosphokinase, galactose-1-phosphate uridyl transferase, lipase, cholinesterase and leucineaminopeptidase; steroids such as cortisol, aldosterone, progesterone, testosterone, estrogens, cholesterol, and bile acids; triglycerides; vitamins; and drugs including antibiotics, salicylates, sulfonamides, anticoagulants, barbiturates, quinidine, antimalarials, antihistamines, and alkaloids. References to these and other applications of fluorometry in the clinical laboratory may be found in recent reviews (11, 12, 13, 16, 17, 18).

REFERENCES

1. ANTON AH, FRIEDMAN AH: *Clin Chem 14*:187, 1968
2. HERCULES DM: *Science 125*:1242, 1957
3. JACOBS SL, SOBEL C, HENRY RJ: *J Clin Endicrinol Metab 21*:305, 1961
4. KORDAN HA: *Science 149*:1382, 1965
5. MARTINEK R: *Med Electronics and Data 1*:118, 1970
6. PARKER CA: *Nature 182*:1002, 1958
7. PARKER CA: *Analyst 84*:446, 1959
8. PARKER CA, REES WT *Analyst 87*:83, 1962
9. PRICE JM, KAIHARA M, HOWERTON HK: *Appl Optics 1*:521, 1962
10. ROTH M: *Methods of Biochemical Analysis*, edited by D Glick. New York, Interscience 1969, Vol 17, p 189
11. RUBIN M; *Advan Clin Chem 13*:161, 1970
12. UDENFRIEND S: *Fluorescence Assay in Biology and Medicine.* New York, Academic Press, 1962
13. UDENFRIEND S: *Fluorescence Assay in Biology and Medicine.* New York, Academic Press, Vol 2, 1969
14. WHITE CE, ARGAUER RJ: *Fluorescence Analysis, A Practical Approach.* New York, Marcel Dekker, 1970
15. WHITE CE, HO M, WEIMER EQ: *Anal Chem 32*:438, 1960
16. WHITE CE: *Anal Chem 40*:116R, 1968
17. WHITE CE: *Anal Chem 42*:57R, 1970
18. WHITE CE: *Anal Chem 44*:182R, 1972
19. WILLIAMS RT, BRIDGES JW: *J Clin Pathol 17*:371, 1964
20. WINEFORDNER JD, McCARTHY WJ, ST. JOHN PA: *Methods of Biochemical Analysis*, edited by D. Glick. New York, Interscience, 1967, Vol 15, p. 369
21. WINKELMAN J, GROSSMAN J: *Anal Chem 39*:1007, 1967

Chapter 4

Electrochemistry

STANLEY M. REIMER, Ph.D.

Electrochemistry is concerned with chemical transformations that involve proton or electron transfer and result in the flow of electricity. Electrochemical technics have been utilized in analytical chemistry for at least a century but until recently have had only limited and specialized applications to clinical chemistry except for the potentiometric measurement of blood pH, pCO_2, and pO_2. With the recent refinements of the glass electrode and the development of ion-selective electrodes, interest in electrometric titrations and other electroanalytical applications in clinical chemistry has gained added impetus. Also potentiometry, amperometry, and voltammetry (polarography) have now been introduced to the clinical chemistry laboratory in an effort to supplement the gravimetric, volumetric, and spectrophotometric technics of analytical chemistry presently in use. Coulometry, which involves measurement of the electron requirements necessary for the reduction or oxidation of an ionic species, is another technic that has recently attracted the attention of the clinical chemist as a method for precision analysis. These various methods will be discussed. More detailed and theoretical references in addition to those specifically noted can be consulted for further information (6, 16, 17, 19, 38, 61, 62, 65, 69).

POTENTIOMETRY

Two electrodes are required for potentiometric measurements. The working (indicator) electrode, e.g., glass anion or cation sensitive electrode, develops a change in potential in reference to a second electrode (reference electrode), such as calomel ($Hg-Hg_2 Cl_2$) or Ag-AgCl, which must exhibit a reproducible fixed potential with repetitive measurements. The change in potential elicited by the working electrode is directly related to the activity of the measured ion and therefore, indirectly related to the concentration of the ion species (32). The electrical behavior of an indicator electrode measuring one ionic species can be described by the Nernst equation (52):

$$E = E^o + 2.303 \, (RT/nF) \log a$$

where: E = measured half-cell potential, emf (electromotive force)
        E^o = constant (standard half-cell potential)
        R = gas constant

T = absolute temperature
n = number of electrons transferred (valence)
F = Faraday constant (96 500 C)
a = activity of ion being measured.

The most significant application of potentiometry in clinical chemistry today is the measurement of pH (negative log of H^+ ion activity) with the glass electrode. Since the Nernst equation applies to the glass electrode, pH is related linearly to the measured electrode potential except for a small junction potential at the reference electrode. The structure of the glass electrode consists basically of a hydrated glass outer layer in contact with the external solution, a dry glass middle layer constituting the major portion of the membrane thickness, and an inner hydrated glass layer in contact with an Ag-AgCl electrode in an internal solution of chloride ions at constant pH (19). The over-all potential of the glass electrode is the sum of the potentials across the various hydrated and dry glass layers.

In actual practice, all glasses respond to ionic species other than H^+ and the lack of observed linearity or "alkaline error" is due to Na^+ ion activity at high pH (low H^+ activity), e.g., pH>10. This is related to the chemical composition of the glass. Furthermore, errors occur when measurements are made at temperatures other than that at which the pH scale of the instrument has been standardized (66). Obviously, errors due to temperature effects are eliminated when electrodes, standard buffers, and test solutions are at the same temperature for pH measurements. Theoretically, although standardization of the indicator and reference electrodes can be facilitated by the use of a single standard buffer, any deficiency in electrode response due to aging or use can be corrected by using two standard buffers to bracket the unknown measurement. In general before a new electrode is put into routine use, it should be seasoned according to the instructions of the manufacturer. Usually this involves soaking the electrode for 1−2 hours or overnight in water or a neutral buffer (pH 7.0) if the electrode is to be used for pH measurement below 9 and in an alkaline buffer or borax solution if the electrode is to be used exclusively for more alkaline pH measurements. Failure to condition an electrode properly will require frequent standardization of the electrode assembly with standard buffer solutions inasmuch as the measured electrode potential will not be stable.

MEASUREMENTS WITH ION-SELECTIVE ELECTRODES

Changes in the chemical composition of the glass membrane induce selectivity for cations other than H^+, e.g., Na^+, K^+, NH_4^+, Ag^+, Cs^+, and Li^+. Glass consisting of 72.2% SiO_2, 6.4% CaO, 21.4% Na_2O is a typical glass for H^+ ion determinations. Glass composed of 71% SiO_2, 11% Na_2O, and 18% Al_2O_3 exhibits sensitivity to Na^+ ion that is 2800 times greater than to K^+ ion. Further reduction in the concentrations of SiO_2 and Al_2O_3 with a corresponding increase in Na_2O (68% SiO_2, 27% Na_2O, and 5% Al_2O_3) results in a K^+ : Na^+ ion selectivity of 20:1. It is now well established that all types of glass electrodes show both metal ion as well as pH response (50). Eisenman's theoretical and experimental investigations into the simultaneous response to two or more ionic species have provided the thrust for the development of selective anion and cation electrodes (18).

Major progress in analytical potentiometry during the last few years has been achieved through the development and introduction of various types of membrane electrodes that permit the measurement of ionic activities in solution with a high degree of selectivity. Electrodes formulated with glass, inorganic crystals, liquid, or heterogenous membranes are now available for the determination of a wide variety of cations and anions (9). Methods to study cation-protein interaction were developed by Carr (10), who prepared reference electrodes with collodion membranes to measure protein binding by ions. These electrodes may be calibrated for concentration measurements at constant ionic strength or as ion activity probes. Brand and Rechnitz (5) developed methodology for the direct differential potentiometric measurement of cations between two high impedance ion-selective electrodes. Glass and other membrane electrodes serve as reference electrodes without a liquid junction. The use of a membrane electrode as a reference for direct potentiometry and potentiometric titrations is advantageous inasmuch as problems (such as electrolyte leakage and unstable liquid junction potentials) that are associated with normal reference electrodes, e.g., calomel or Ag-AgCl, are eliminated. The relationship between the activity (a) of an ionic species and its concentration (c) was described by Jacobson (32). The formula for this relationship is $a_i = \gamma_i c_i$, where γ_i is the activity coefficient. To determine the concentration of an ion solely from its activity it is necessary to extrapolate ion-selective electrode measurements to infinite dilution to achieve an activity coefficient of 1. An alternative method is to maintain an essentially constant ionic strength of samples and reference standards with a buffer containing a moderate amount of an inert salt. This results in solutions with very nearly equal activity coefficients. It is thereby possible to compare directly the potentiometric measurements of samples with those of standards.

Membrane electrodes also obey the Nernst equation. Experimental values of the Nernst constant RT/nF which are significantly different from the theoretical value are often indicative of unreliable electrodes as a result of membrane failure. A given electrode may exhibit small variations with time from day to day. Although these are usually not significant frequent calibration is necessary for accuracy (10). Ion-selective electrodes are reported to exhibit Nerstian behavior down to the following limits of detection: Ca^{2+}, Zn^{2+}, Cl^-, F^-, 10^{-5} M; Cd^{3+}, 10^{-7} M; Cu^{2+}, Pb^{2+}, Ag^+; I^-, S^{2-}, 10^{-8} M (53).

The utility and practicality of selective cation and anion electrodes is yet to be fully realized. They have the advantages of relative low cost, portability, and ease of operation. New clinical applications of potentiometric methodology and ion-selective electrodes are continuously evolving: detection of cystic fibrosis by measuring sweat chloride (30, 35); adaptation of a digital pH stat for automated kinetic enzyme analysis (34); introduction of a redox electrode into a continuous flow system for glucose analysis by the ferricyanide reaction (68); new rapid enzymatic assays using enzyme electrodes for the analysis of urea, glucose, L-amino acids, D-amino acids, asparagine, and glutamine (29, 47).

Ion-selective electrodes are manufactured by: Beckman Instruments, Cambridge Instrument Company, Coleman Instruments Division of Perkin-Elmer, Corning Class Works, Materials Research Corporation, National Instrument Laboratories, Orion Research, and Photovolt Corporation.

SODIUM AND POTASSIUM MEASUREMENT

Portnoy and co-workers (59, 60) modified a routine glass electrode pH meter in order to determine sodium and potassium in blood, plasma, and cerebrospinal fluid. In these early investigations Na ion was 4−6 meq/liter and K was 0.3−0.5 meq/liter greater than that measured by flame photometry. Other investigators reported the development of a valinomycin K ion-selective electrode with high selectivity for K ions over Na ions (23, 57).

Recently, ion-selective electrode measurements of K and Na in whole blood, plasma, or serum have been shown to agree with flame photometric measurements of plasma K and Na to within 0.4 meq/liter and 4.0 meq/liter, respectively (44). The Na and K potentials were measured with ion-selective electrodes against an Ag-AgCl reference electrode. Neff *et al* (51) have compared plasma Na and K results obtained with a Beckman flame photometer, the AutoAnalyzer flame system, and ion-selective electrodes. Completely automated systems, computer assisted for data acquisition and processing, have been developed for Na^+, K^+, H^+, Cl^-, pH, pO_2, and pCO_2 (14, 51). Levy (37) reported on the development of a fully automated sodium and potassium discrete analyzer system that is self-standardizing and free of drift. Most recently, Na and K measurements by ion-selective electrodes have been adapted to continuous flow instrumentation (64). The satisfactory measurement of Na^+ in urine was reported by Annino (1) using a system composed of a glass indicator electrode, a calomel reference electrode, and a sensitive pH meter. Urine was diluted 2 parts urine with 18 parts Tris buffer at pH 8.0. Reimer (67) reported data for direct sodium and potassium measurements in urine that were comparable to flame analyses without necessary pretreatment of the sample. Abnormal electrode responses in the determination of Na^+ in the urine have been related to an H_2O_2 sensitive substance that lowers the electrode sensitivity (58).

CALCIUM MEASUREMENT

Ever since the pioneering efforts of McLean and Hastings (41) relating ionized calcium to the total serum protein and total serum calcium, accurate methods for the determination for ionized calcium have been actively sought. Ionic calcium values are an important aid in the diagnosis of hyperparathyroidism as it is this calcium fraction that is regulated by the parathyroid glands (4). The potentiometric method using ion-selective electrodes has commanded the research and development efforts of many investigators. This methodology is described in detail in Chap. 20.

FLUORIDE MEASUREMENT

Frant and Ross (22) reported an electrode for sensing fluoride ion activity in 1966. Specific F^- electrodes are constructed from single-crystal sections of rare earth F^- complexes. Crystals prepared from LaF_3, NdF_3, or PrF_3 exhibit a linear Nernstian response over more than five orders of magnitude (1 to > 10^{-5} M), provide a potential that depends linearly on the negative logarithm of F^- concentration and demonstrate a high selectivity over other common anions. Measurements of F^- activity have been reported in bone (3, 40, 74), serum (75, 78, 79), urine (76, 77, 82, 83), and saliva (24).

The potential of the electrode is developed against a reference electrode and calibrated with standard F^- solutions. The potential developed is strongly influenced by pOH (hydroxyl ions), temperature, ionic strength, and extraneous ions such as Ca^{2+} and Mg^{2+}. High concentrations of these ions, as might be present in bone ash extracts, would produce negative errors (74). Recovery experiments indicate 98.3% ± 3.86% (1s) recovery following the addition of 0.2 or 0.4 ppm F^- to urine (76). pH adjustments of urine in the range of 5.0–7.6 do not result in any change in the F^- activity, hence a more precise adjustment of pH is not necessary. Frant and Ross (21) developed a total ionic strength adjustment buffer for dilution of samples prior to measurement, which provided a constant ionic strength due to the increased ion concentration and buffering in a range which avoids OH^- interference.

Recent data provide evidence for the presence of two forms of fluoride in human serum and urine (79). Taves (78) reported that fluoride in serum is present in exchangeable and nonexchangeable forms. A significant amount of fluoride that is not ultrafiltrable seems to be associated with albumin. Measurements were made potentiometrically and with a fluorescein dye complex following serum ashing at 600°C for 3 hours.

VOLTAMMETRY

In a direct current system with metallic conductors the simple relationship of Ohm's law applies: $V = IR$, where V is the applied voltage or potential, I the current, and R the resistance. However, in electrolyte solutions there are extreme deviations from this law. The reason is that when ions are transported in a solution to an electrode, no oxidation or reduction will occur below a threshold voltage, and furthermore, the current is limited by diffusion. This, in turn, is a function of the concentration of the ionic species. The gradual application of increasing voltage and simultaneous measurement of current is the basic technic of polarography. Polarography is a specialized technic of voltammetry developed by Heyrovsky (31) in 1922 who received the Nobel Prize in 1959 for his achievement. Polarography is performed at controlled and varying voltage using a dropping Hg electrode. The Hg electrode provides a reproducible surface that is continuously renewed. Both the voltage and current are measured between the Hg electrode and a reference electrode, e.g., calomel or Ag-AgCl. At certain applied voltages, characterized by the half-wave potential (point of inflection of the current-voltage curve), a large current flow results. This potential is typical of the ionic species and serves for qualitative identification. The magnitude of the current flow resulting from this applied potential is quantitatively proportional to the concentration of the ion. Thus several ions may be determined both qualitatively and quantitatively in a serial sequence from a complex solution in a single run. The method is particularly suited for determining trace quantities of metals. It has also been used successfully for the determination of oxidizable and reducible organic compounds (68).

VOLTAMMETRY AT CONTROLLED VOLTAGE

The over-all rate at which electrode reactions proceed depends upon the mass transfer of ions that are being oxidized or reduced, the electrode surface areas, the transfer of electrons at the electrode surface, and the removal of toxic reaction products that can poison the electrode surface. Currents relating to ion transfer in an electrochemical cell are brought about by convection, migration, and diffusion forces. Convection currents usually result from the mechanical stirring of the solution whereas migration currents result from the internal movement of the cations and anions to the respective electrodes under the influence of an electric field. Inasmuch as the height of the current versus potential curve is proportional to the ions in solution it is most desirable to minimize the convection and migration currents so that the ions reach the electrode surface strictly by diffusion. This is accomplished without mechanical stirring in the presence of a concentrated solution of supporting electrolyte prepared from salt that is difficult to reduce (e.g., Na^+, K^+) because of the high voltage requirements. Thus, the high concentration of Na^+ or K^+ ions surrounding the cathode limit the migration of the ionic species being reduced so that they reach the cathode solely by diffusion (42).

VOLTAMMETRY AT CONTROLLED CURRENT (CHRONOPOTENTIOMETRY)

In this process, the potential is measured across the indicator and reference electrodes when a constant current is applied to the electroanalytical cell. The electrolysis proceeds and the concentration of the electroactive species at the electrode surface changes with time. The resultant potential-time curve can thus relate a change in concentration of a chemical species with a change in potential with time. Delahay and Mamantor (15) have reviewed the theory of this technic. Although chronopotentiometry has been known for many years only recently has it been applied to kinetic enzyme analyses of biochemical interest (63). Guilbault and collaborators (25) have used this technic for following rapid enzymatic hydrolytic reactions. Reports on the use of chronopotentiometry to study reactions involving cholinesterase and thiocholine esters (36), hydrolysis of organophosphorous compounds (26), glucose oxidase (27), and xanthine oxidase (28) attest to the increasing importance of this technic as an analytical tool.

ANODIC STRIPPING VOLTAMMETRY

In this specialized branch of electrochemistry, the electrolytic cell contains the unknown sample or standard solution in contact with a platinum reference electrode and a stationary electrode, e.g., a wax-impregnated, mercury-coated graphite rod (7). A variable negative (cathodic) potential is applied to the stationary electrode that results in plating onto the electrode surface of trace metals present in the sample or standard. This step serves to purify and concentrate the ions. By varying the voltage linearly in a positive (anodic) mode, the metals are then stripped from the electrode as a function of their unique standard electrode potentials. A chart readout provides both qualitative and quantitative analysis of the metals present as a result of peak location and peak height, respectively. As little as 50–100 μl of whole blood can be analyzed for trace metals such as Cu, Pb, Cd, Zn, etc., following digestion with perchloric acid.

PULSE POLAROGRAPHY

This electrochemical technic employs equipment similar to that described above for anodic stripping voltammetry (8). Both of these analytical methods provide reliable and rapid detection of trace toxic heavy metals at the levels of parts per billion and parts per trillion. This is in contrast to atomic absorption spectrophotometry that has sensitivity suitable for analyses in the parts per million concentration range. In contrast to normal polarography, wherein a slow moving potential is applied to the dropping mercury electrode, providing an essentially constant potential over the life of any one drop, in pulse polarography successive potential pulses of increasing amplitude are applied to the mercury drop only once during the drop life. The net effect of this technic is to increase the sensitivity of the measurement (e.g., ratio of the wave height obtained in pulse polarography to that obtained in normal polarography) due to decreased current decay during the measurement time. Consequently, the current is of much greater magnitude, making this mode of operation more sensitive than conventional polarography (54). Pulse polarography was developed by Barker (2), and details of this relatively new analytical technic have been published (55, 56).

COULOMETRY

Whereas volumetric titrations have reagents added to the reaction vessel from burets, in coulometric technics a reagent is generated *in situ* from an electrode surface.

The principles of coulometric titrations are based on Faraday's law, which relates the flow of electricity in Coulombs to the oxidation or reduction of gram-equivalents of a substance. Thus 1 gram-equivalent of a chemical species will be oxidized or reduced by 96 500 C of electricity (1 Faraday). Faraday's law is given by the formula

$$W = W_M it / 96500\, n$$

where W = the weight in grams of the substance transformed in electrolysis

W_M = atomic or molecular weight of the substance being oxidized or reduced

it = Coulombs of electricity (i = current in Amperes, t = time in seconds)

n = the number of electrons involved in the electrode process

Instrumentation for coulometric titrations is relatively simple and requires a timing device, a source of constant current, an ammeter, and a titration cell with an end-point detection system. Endpoints may be determined by color-producing chemical indicators, or by amperometric, potentiometric, or conductometric technics. Analysis is rapid and the titrant can be easily generated without being subject to evaporation or decomposition (43).

Detection of coulometric end points by amperometry is most advantageous. With this technic, two inert electrodes that are subjected to a small potential difference become polarized in a solution of oxidizing or reducing substances. Excess addition of the oxidizing or reducing titrant results in depolarization of the electrodes, thus allowing the end point to be detected by the initial current flow through the cell (63).

Coulometric titration technics have been used for the ultramicro determination of chloride (70), urea nitrogen in blood and urine (12), arsenic in urine (71), organic matter in urine (73), sodium seconal and sodium sandoptal in serum (46), uric acid in serum (80), and chromium in biologic materials (20).

Cotlove (13) was one of the first investigators to apply the coulometric titration technic to the determination of chloride. This automatic coulometric-amperometric titration utilizes silver ion titration in an electrolysis cell containing two pairs of silver electrodes in an acidic electrolyte solution. During the titration silver ions, generated at a constant rate by a constant current coulometric circuit, precipitate chloride as AgCl. By presetting a meter relay to automatically detect a limiting current which develops from the appearance of the first free silver ion in the vicinity, end points are reliably detected. This electrometric method is capable of high accuracy and precision. The titration time is linear with chloride concentrations over a thousandfold range. The lower limit of sensitivity is reportedly $10^{-5}\,M$. The precision obtained is routinely ± 0.5% (95% limits) and with optimum conditions can approach ± 0.1%.

AMPEROMETRY

The titration procedure in analytical chemistry is a well-established, precise, and accurate technic for the rapid analysis of an unknown quantity of a substance known to react to a specific end point with a known quantity of a known substance. The equivalence point is commonly determined by precipitation, complexation, or color change of an indicator. Amperometry refers to a titrimetric method wherein the equivalence point is determined by voltammetry at constant voltage. Specifically, a voltage is applied to the indicator and reference electrodes so that one of the constituents involved in the titration procedure can be either oxidized or reduced and the change in the current measured is a function of the concentration of the species involved in the electrode reaction. Equivalent points are determined graphically by plotting changes in the limiting currents measured versus the milliequivalents of titrant used. Amperometric endpoints have been used for the coulometric determination of maltose (49), amylase (48), glucose (72), oxygen consumption (33, 39), protein nitrogen of Kjeldahl digests (11), cholesterol (81), and barbiturates (81).

REFERENCES

1. ANNINO JS: *Clin Chem 13*:227, 1967
2. BARKER GC, GARDNER AWZ: *Anal Chem 173:*79, 1960
3. BARNES FW, RUNCIE J: *J Clin Pathol 21*:668, 1968
4. BERNSTEIN DS, ALIAPOULIOUS MA, HATTNER RW, WACHMAN A, ROSE B: *Endocrinology 85*:589, 1969
5. BRAND MJ, RECHNITZ GA: *Anal Chem 42*:616, 1970
6. BREZINA M, ZUMAN P: *Polarography in Medicine,* Biochemistry and Pharmacy, New York, Interscience, 1958
7. BULLETIN: *Environmental Science Associates, Inc.,* Burlington, Mass., 1972
8. BURGE DE: *J Chem Ed 47*:A81, 1970
9. CARR CW: *Ann N Y Acad Sci 148*:180, 1968
10. CARR CW: in *Electrochemistry in Biol and Med,* edited by Shedlevsky T, New York, Wiley, p 266, 1955
11. CHRISTIAN GD, KNOBLOCK EC, PURDY WC: *Clin Chem 11*:413, 1965
12. CHRISTIAN GD, KNOBLOCK EC, PURDY WC: *Clin Chem 11*:700, 1965
13. COTLOVE E: in *Standard Methods of Clinical Chemistry,* edited by Seligson D, New York, Academic 1961, vol 3, p 81
14. DAHMS H: *Clin Chem 13*:437, 1967
15. DELAHAY P, MAMANTOV G: *Anal Chem 27*:478, 1955
16. DELAHAY P: *New Instrumental Methods in Electrochemistry.* New York, Interscience, 1954
17. DURST RA: *Ion-Selective Electrodes, National Bureau of Standards.* Special Publication, 1969
18. EISENMAN G: *Ann N Y Acad Sci 148*:5, 1968
19. EISENMAN G: *Glass Electrodes for Hydrogen and Other Cations.* New York, Marcel Dekker, 1967
20. FELDMAN FJ, CHRISTIAN GD, PURDY WC: *Amer J Clin Pathol 49*:826, 1968

21. FRANT MS, ROSS JW: *Anal Chem 40*:1169, 1968
22. FRANT MS, ROSS JW: *Science 154:* 1553, 1966
23. FRANT MS, ROSS JW: *Science 167*:987, 1970
24. GRØN P, McCANN HG, BRUDEVOLD F: *Arch Oral Biol 13*:203, 1968
25. GUILBAULT GG, KRAMER DN, CANNON PL Jr: *Anal Biochem 5*:208, 1963
26. GUILBAULT GG, KRAMER DN, CANNON PL Jr: *Anal Chem 34*:1437, 1962
27. GUILBAULT GG, TYSON BC, KRAMER DN, CANNON PL Jr: *Anal Chem 35*:582, 1963
28. GUILBAULT GG, KRAMER DN, CANNON PL Jr: *Anal Chem 36*:606, 1964
29. GUILBAULT GG, SMITH RK, MONTALVO JG: *Anal Chem 41*:600, 1969
30. HANSEN L, BUECHELE M, KOROSCHEC J, WARWICK WJ: *Am J Clin Pathol 49*:834, 1968
31. HEYROVSKY J: *Chem Listy 16*:256, 1922
32. JACOBSON HJ: *Anal Chem 38*:1951, 1966
33. KADISH AH, LITLE RL, STERNBERG JC: *Clin Chem 14*:116, 1968
34. KERCHER RE, PARDUE HL: *Clin Chem 17*:214, 1971
35. KOPITO L, SCHWACHMAN H: *Pediatrics 43*:794, 1969
36. KRAMER DN, CANNON PL Jr, GUILBAULT GG: *Anal Chem 34*:842, 1962
37. LEVY GL: *Chim Chem 18*:696, 1972
38. LINGANE JJ: *Electroanalytical Chemistry*, 2nd ed. New York, Interscience, 1958
39. MAKIN HLJ, WARREN PJ: *Clin Chim Acta 29*:493, 1970
40. McCANN HG: *Arch Oral Biol 13*:475, 1968
41. McLEAN FC, HASTINGS AB: *J Biol Chem 107*:337, 1934
42. MELOAN CE: in *Instrumental Analyses Using Physical Properties.* Philadelphia, Lea and Febiger, p 104, 1968
43. MELOAN CE: Ref. 42, p 120
44. MIYADA DS, INAMI K, MATSUYAMA G: *Clin Chem 17*:27, 1971
45. MONFORTE JR, PURDY WC: *Anal Chim Acta 52*:25, 1970
46. MONFORTE JR, PURDY WC: *Anal Chim Acta 52*:433, 1970
47. MONTALVO JG: *Anal Chem 41*:2093, 1969
48. MOODY JR, PURDY WC: *Anal Chim Acta 53*:31, 1971
49. MOODY JR, PURDY WC: *Anal Chim Acta 53*:239, 1971
50. MOORE EW: *Ann NY Acad Sci 148*:93, 1968
51. NEFF GW, RADKE WA, SAMBUCETTI CJ, WIDDOWSON GM: *Clin Chem 16*:566, 1970
52. NERNST W: *Z. Physik Chem 4*:129, 1889
53. *Orion Information and Research Bulletin* 92-20
54. OSTERYOUNG JG, OSTERYOUNG RA: *Amer Lab 3(7)*:8, 1972
55. PARRY EP, OSTERYOUNG RA: *Anal Chem 36*:1366, 1964
56. PARRY EP, OSTERYOUNG RA: *Anal Chem 37*:1634, 1965
57. PIODA LAR, SIMON W: *Chimia 23*:72, 1963
58. POMMER AM, SIMON NS, CALCAGNO PL: *Clin Chem 13*:343, 1967
59. PORTNOY HD, GURDJIAN ES, HENRY B: *Am J Clin Pathol 45*:283, 1966
60. PORTNOY HD, GURDJIAN ES: *Clin Chim Acta 12*:249, 1965
61. PUNGOR E: *Anal Chem 39 (11)*:28A, 1967
62. PURDY WC: *Electroanalytical Methods in Biochemistry.* New York, McGraw-Hill, 1965
63. PURDY WC: in *Methods in Clinical Chemistry.* Basel-Munchen, Paris, New York, S Karger, p 82, 1970
64. RAO KJM, HOFSTETTER M, MORGENSTERN S: *Clin Chem 18*:696, 1972
65. RECHNITZ GA: *Anal Chem 37(1)*:29A, 1965

66. RECHNITZ GA: *Chem Eng News 45:*146, 1967
67. REIMER SM: *Applied Seminar on the Clinical Pathology of Respiratory Diseases,* edited by Sunderman FW, Institute for Clin Science Inc, Philadelphia, p 45, 1972
68. SAWYER R, FOREMAN JK: *Lab Pract 18:*35, 1969
69. SHEDLEVSKY T: *Electrochemistry in Biology and Medicine,* New York, Wiley, 1955
70. SIGGAARD-ANDERSON O: *Am J Clin Pathol 48:*444, 1967
71. SIMON RK, CHRISTIAN GD, PURDY WC: *Am J Clin Pathol 49:*207, 1968
72. SIMON RK, CHRISTIAN GD, PURDY WC: *Clin Chem 14:*463, 1968
73. SIMON RK, CHRISTIAN GD, PURDY WC: *Tech Bull: Reg of Med Tech 38:*733, 1968
74. SINGER L, ARMSTRONG WD: *Anal Chem 40:*613, 1968
75. SINGER L, ARMSTRONG WD: *Arch Oral Biol 14:*1343, 1969
76. SINGER L, ARMSTRONG WD, VOGEL, JJ: *J Lab Clin Med 74:*354, 1969
77. SUN MW: *J Am Ind Hyg Assoc 2:*133, 1969
78. TAVES DR: *Nature 217:*1050, 1968
79. TAVES, DR: *Talanta 15:*1015, 1968
80. TROY RJ, PURDY WC: *Biochem Med 3:*198, 1969
81. TROY RJ, PURDY WC: *Clin Chim Acta 26:*155, 1969
82. TUŠL J: *Clin Chim Acta 27:*216, 1970
83. ZIPKIN I, LEONE NC: *Am J Public Health 47:*847, 1957

Chapter 5

Electrophoresis

DONALD R. WYBENGA, B.M.

In general, an interfacial potential exists between two phases in contact. There are at least three ways this potential can arise: (a) by ionization of surface groups; for example, protein molecules owe their charge principally to ionization of their amine and carboxyl groups; (b) by preferential adsorption of anions or cations; (c) by orientation of adsorbed polar molecules. Helmholtz suggested in 1879 that an electrical double layer is formed at the interface, the first layer being the charged surface and the second layer consisting of a layer of oppositely charged ions. This has come to be known as the *Helmholtz double layer*. Modern theory, however, depicts the distribution of charges around a negatively charged surface as shown in Figure 5-1. The negatively charged surface is surrounded by an immobile layer of oppositely charged ions and this in turn is surrounded by a diffuse layer of ions of the same charge, which decreases in concentration and increases in mobility as the distance from the charged surface is increased. If the surface is positively charged, the surrounding ionic layers are negatively charged. The

FIG 5-1. Modern concept of the Helmholtz double layer.

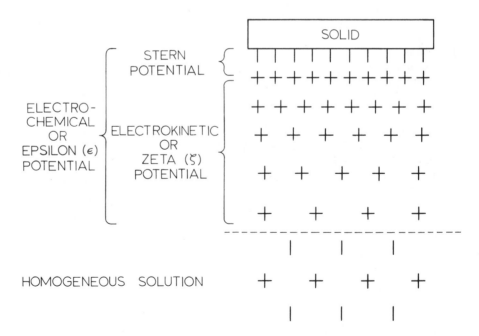

potential between the surface and the homogeneous body of solution is called the *electrochemical* or *epsilon* (ϵ) *potential,* which is partitioned into the sharp potential, between the surface and the immobile layer, called the *Stern potential,* and the diffuse gradient existing between the immobile layer and the body of solution, called the *electrokinetic* or *zeta* (ζ) *potential.*

The zeta potential is responsible for four "electrokinetic" phenomena:

(a) Streaming or flow potential. Water placed in a glass capillary has a positive charge with respect to the glass wall, probably a result of adsorption of hydroxyl ions. If the water is caused to flow by mechanical pressure, an electric potential develops between the ends of the capillary, as indicated in Figure 5-2. The excess hydrogen ions, responsible for the positive charge of the water and being mobile, are swept along with the water, resulting in a concentration gradient over the length of the capillary and hence a potential gradient. Thus a streaming potential results from the movement being caused by an external mechanical force.

(b) Electroendosmosis or electro-osmosis. Since water in a glass capillary is positively charged, it will flow toward the cathode of an impressed potential, as shown in Figure 5-3. It has been suggested (2, 14) that this transport of water, rather than being caused by the Helmholtz double layer, results from the cations being more highly hydrated than anions—hence more water is carried to the cathode than to the anode. In any event, electroendosmosis is the movement of a liquid relative to a solid under the influence of an externally applied electrical potential. This is the opposite of the streaming potential; and it is obvious that the effect of the potential developed by a mechanically induced flow is to oppose the flow.

(c) Dorn effect. Movement of a solid through a liquid under the influence of a mechanical force, usually gravity. This imparts a potential gradient through the liquid.

(d) Electrophoresis (formerly called *cataphoresis*), the movement of a solid phase with respect to a liquid under the influence of an externally applied electrical potential.

Historically, the *moving boundary technic* of electrophoresis dates back to 1886, but it was brought to fruition by A. Tiselius in 1930 and has since been improved by such workers as Tiselius himself, Philpot, Svensson, and

FIG 5-2. Streaming potential.

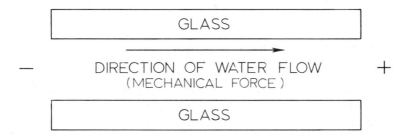

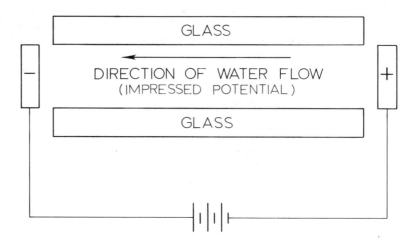

FIG 5-3. Electroendosmosis.

Longsworth. Tiselius' apparatus, and specific features of the technic he employed, are described in detail in the first edition of this book.

The "classical" Tiselius technic was expensive and its use was complicated and time consuming. Simpler methods of electrophoresis have now been developed. These methods generally employ a solid matrix to eliminate the problems caused by convection currents during electrophoresis and to stabilize the separation of boundaries. Filter paper was the first such matrix explored. The simplest form of electrophoresis was carried out on a strip of paper, which was moistened with buffer, blotted to remove excess liquid, and placed horizontally on a glass sheet so that each end dipped into a separate container with buffer. The sample was applied to the paper, which was then covered with another glass plate to prevent evaporation. A potential was applied across the strip and after a suitable time the strip was removed, dried in an oven, and treated with a reagent to locate the separated substances. The development of other supporting matrices like cellulose acetate, agar gel, starch gel, and acrylamide gel, and the improvement in electrophoretic equipment have resulted in very sophisticated electrophoretic technics. Certain general principles, however, are applicable to all technics.

PRINCIPLES OF ELECTROPHORESIS

In electrophoresis, the mobility of a charged particle is a function of the magnitude of the charge, which in turn varies with pH. For example, in the case of ampholytes such as proteins, at the isoelectric point, pI, mobility is zero. At any pH below pI the ampholyte has a net positive charge and moves toward the cathode; above pI the ampholyte has a net negative charge and migrates toward the anode. "Mobility" is the distance in centimeters a particle moves in unit time per unit field strength expressed as voltage drop per centimeter (mobility = μ = cm/sec/V/cm = cm^2/V sec).

The main functions of the buffer are to carry the current and to maintain a constant pH, thus ensuring that each component will possess a constant charge throughout the experiment. Buffers covering the whole pH range are available.

Besides the pH of the buffer its ionic strength is also of importance. The ionic strength (μ) of a buffer is calculated using the following formula:

$$\mu = \tfrac{1}{2}\sum Mc^2$$

where M is the molarity and c is the charge on the ion.

Current is carried by the ions present, and therefore, the more concentrated the buffer the greater the proportion of current carried by the buffer ions relative to other ions and the slower will those other ionic compounds move. Furthermore, the movement of ions surrounded by ions of opposite charge is retarded by the attraction of these ions, and the result will be a reduction in migration rates. Thus an increase in buffer concentration reduces migration by these two effects.

Resistance to current flow is a function of the medium, the buffer, and its concentration, and will determine the current and consequently, the heat generated by a given voltage. Since the resistance of the medium is fixed, it follows that the voltage must be raised if an increase in current is desired. With a constant voltage the current will increase during the electrophoresis period, because as heat increases and distillation of the solvent from the medium occurs, the resistance of the medium drops slowly. The decrease in resistance will result in an increase in migration rate. With low voltages this distillation effect is less and the rate of migration will not vary greatly. If the current is kept constant, the voltage will drop as the resistance of the medium decreases. To obtain the same degree of separation of migrating substances, either a constant voltage must be used irrespective of the number of strips of support medium per electrophoresis cell or the desired current must be multiplied by the number of strips.

A general question is whether the current or voltage should be kept constant during an electrophoretic run. With paper or cellulose acetate electrophoresis the heat generated is small and either constant voltage or constant current works satisfactorily for separation of most substances. This is not the case with gels and blocks, the thickness of which may vary from 0.3 to 1 cm causing heat dissipation to become a problem. Constant current is chosen in these procedures as well as for mobility measurements in paper or cellulose acetate electrophoresis.

Electrophoresis through a matrix such as paper is complicated by at least four factors not present in "free" (Tiselius) electrophoresis:

(a) *Electroendosmosis.* Filter paper is essentially a maze of capillaries and when filled with water the capillary surfaces of ordinary filter paper are negatively charged, with the contained water positively charged ("positive" and "neutral" papers have also been prepared). When the ends of a paper strip saturated with buffer are immersed in reservoirs of the same buffer and a potential is applied across the two reservoirs, water flows from the anode reservoir through the strip and into the cathode reservoir until the rate of water flow toward the anode resulting from differences in hydrostatic pressures in the two reservoirs exactly balances the electroendosmotic flow in the opposite direction. The magnitude of electroendosmosis is dependent on the paper, applied potential, ionic strength of the electrolyte, and its pH, addition of hydrogen ions reducing the charge of the negative pore walls and addition of hydroxyl ions having the reverse effect. Salts also affect the rate by selective adsorption of the ions onto the pore walls. The resultant mobility of a component on a paper strip is the algebraic sum of its electromigration velocity and the electroendosmotic velocity of the solution in the paper. For example, in routine paper electrophoresis of serum proteins, the γ-globulin may actually migrate toward the cathode although the molecules possess a net negative charge. This results from the electroendosmotic velocity toward the cathode exceeding the electromigration velocity of the molecules toward the anode. Theoretically, it would seem possible to correct mobilities for electroendosmotic flow by measuring the migration of uncharged particles, and attempts in this direction have been made with various substances, the most common being dextran (15). It appears, however, that this correction is valid only if the molecular volume of the "electroendosmotic indicator" is close to that of the particle under study (2).

(b) Streaming potential. A potential is developed by movement of the liquid through the paper that is directed against the applied potential.

(c) The "barrier effect." It was first proposed (19) that the path a particle must take in traversing a paper strip is tortuous and longer than the physical length of the strip and that the potential gradient across the migration path cannot be obtained merely by dividing the applied voltage by the length of the strip. This interpretation has been contested and the "barrier theory" introduced as an alternative (19). The paper fibers are viewed as barriers into which the migrating particles bump, momentarily slowing them down. The magnitude of this slowing down is a function of spacing and diameter of the fibers and the molecular volume of the migrant. The migration path is considered not to be increased nor the potential gradient decreased. According to the barrier theory then, the correction which must be applied to mobility data is a function of both paper and migrant. The presence of paper is believed to cause a decrease in the thermodynamic activity of the migrant—the greater the paper concentration, the greater the decrease in thermodynamic activity of the migrant and hence the lower the mobility (19). In the present author's opinion, however, it is difficult to see how a migrant can go around barriers without an increase in migration path. Increased migration path would seem to be responsible for the fact that mobilities determined by paper electrophoresis and corrected for electroendosmosis are still somewhat less than those measured by Tiselius electrophoresis (1).

(d) Chromatography. Chromatography through a filter paper matrix presumably occurs primarily as a result of partitioning of solute between two solvent phases, one mobile and the other bound to the cellulose fibers, and secondarily through physical adsorption and ion exchange. A certain element of chromatography is inherent in paper electrophoresis, and adsorption of components onto the paper is not uncommon (13, 14). In mobility determinations it is necessary to keep chromatographic interference at a minimum. This is accomplished by minimizing buffer flow through the paper by: (1) minimal evaporation of solvent from the paper, and (2) maintaining constant and equal levels of buffer in the two reservoirs.

Diffusion is another factor inherent in paper electrophoresis as well as in classical Tiselius electrophoresis. Thus, a spot or stripe applied to paper will increase in diameter and width, respectively, as electrophoresis progresses. Diffusion increases with concentration, time, and temperature, but is inversely related to viscosity of solvent. Unfortunately, the factors affecting the rate of diffusion also affect the rate of electromigration in the same way, so there is no means of decreasing the former's magnitude. On the other hand, electromigration rates far exceed diffusion rates, making electrophoretic separation of components possible.

PAPER ELECTROPHORESIS

The general procedure for an electrophorectic run on paper can be presented in six steps:

1. Placing the paper strip. The strip is wetted with buffer before or after setting it in place in the apparatus. The paper should not be too wet since this results in smearing the pattern (23). The strip is then placed in taut suspension or between glass or plastic plates depending on the type of apparatus.

2. Equilibration period. A period of about 0.5−1.5 hour with the potential applied is allowed for attaining equilibrium, during which time the paper may gain or lose buffer.

3. Application of sample. The current is turned off, and the sample is applied to the paper as a spot by a micropipet or as a stripe across the full width of the paper. The Beckman applicator, which holds 10 μl sample between two closely spaced parallel wires, is very convenient for applying the sample. The sample should not contain particulate matter since this may foul the run (15). The quantity and concentration of sample to be applied is dependent on width and thickness of the paper strip and the nature of the components. The quantities usually employed range from about 5 to 100 μl.

4. Electrophoresis

5. Identification of components. As soon as the strips have been removed from the cell they should be dried quickly in the horizontal position, or the bands will shift and diffuse. The components on the paper can then be identified by coloring the bands by dye adsorption, chemical reactions, ultraviolet light absorption, fluorescence, or radioautography.

6. *Quantitation.* The colored components can be quantitated by two methods: (a) the colored bands are cut out, and the color is eluted and then read photometrically, (b) the strip with the colored bands is scanned in a densitometer. The integrated areas under the curves are some function of the absolute quantities of the respective components. It is essential that the densitometer is standardized for each component to be scanned. If densitometer values are to be used in quantitation the values must first be converted to some function giving linear correlation with component concentration. The strip may be scanned dry, or the transparency of the paper can be increased with mineral oil (1, 3), methylsalicylate (16), benzyl alcohol (12), or 10:1 (refractive index 1.51, close to that of cellulose) (9, 10, 11, 17) or 1:1 mixtures of mineral oil and α-bromonaphthalene (refractive index 1.55) (5). That profoundly different results can be obtained between the dry and "oiled" strips (5) was confirmed in our laboratory.

CELLULOSE ACETATE ELECTROPHORESIS (CAE)

The principles of paper electrophoresis apply also to cellulose acetate electrophoresis. Cellulose acetate, however, has the following advantages over paper: (a) cellulose acetate strips can be used directly without any special preparation; (b) wet cellulose acetate is stronger than any other supporting medium and, therefore, easy to handle; (c) very small quantities of sample can be conveniently applied and are sufficient; (d) only a short time for electrophoresis is required; (e) better resolution of bands of components is achieved; (f) cellulose acetate strips can be easily cleared after staining; and (g) good transparency of the strips facilitates quantitation by densitometry.

CELLULOSE ACETATE STRIPS

Cellulose acetate strips are made by acetylation of cellulose, and the membranes generally do not vary much in pore size, density, thickness, and other physical properties. These membranes are stable in methanol, ethanol, propanol, butanol, benzene, toluene, and petroleum ether, but will dissolve in glacial acetic acid, chloroform, acetone, ethyl acetate, and methylene chloride. The thickness of the membranes is approximately 120μ, and the pores have a diameter of approximately 0.4μ. Cellulose acetate is also available with a Mylar backing from several commercial sources. The various types and brands of cellulose acetate are as follows:

Titan	Helena Laboratories
Celagram	Shandon Scientific Company, Incorporated
Sepraphore III	Gelman Instrument Company
Sartorius Membrane Filters	Brinkmann Instruments
Celotate	Millipore Corporation

Oxoid	Oxo, Limited
S and S 2500 Cellulose Acetate Membrane	Schleicher and Schuell

These membranes vary from each other by the degree of acetylation, thickness, and pore size. The decision as to which strip to use will depend on the procedure, the components to be separated, and the electrophoretic chambers to be used. The selection of strip is empirical, but it must be determined in such a manner that optimum results are obtained.

APPLICATION OF SAMPLE

Before applying the sample to a cellulose acetate strip, the position of application should be marked. Marking should be performed when the strip is dry with a marker that does not diffuse in either acid or alkaline solutions, adheres to cellulose acetate, and does not migrate in an electrical field. Grease pencils, ball point pens, and Visipoint pens can be used for this purpose.

The strips should be impregnated with buffer after visual inspection of the strip insures that imperfections such as spots and ridges are not present. If these faults are observed the strip should not be used since they will affect the electrophoresis of the components and the resolution of the bands. For impregnation with buffer the membranes should be allowed to float on the surface of the buffer in order to absorb the buffer from underneath. Then the strips are immersed completely in the buffer. After impregnation, strips are inspected for trapped air, which appears as white spots. The strips are then blotted between filter paper to remove excess buffer, but they should be kept moist and not allowed to develop dry areas.

The positioning of the strips in the electrophoresis cell or chamber depends on the type of chamber. In some chambers the strips are held in a horizontal position by magnetic bars; in others the moist ends are pressed against the edges of the supporting rack. The membranes with Mylar backing are positioned with the active surface on top and the Mylar backing on the bottom. After the strip is positioned the sample is applied to the strip in the form of a fine, uniform streak along the marked line. This can be carried out by using a capillary tube, micropipet or a wire applicator with parallel wires that are very close to each other. The sample size is generally $1-2 \, \mu l$ for proteins, $5-10 \, \mu l$ for lipoproteins and approximately $10 \, \mu l$ for lactic acid dehydrogenase isoenzymes. The samples are applied at the anodal or cathodal end of the strip depending on the charges of the substances to be separated and the pH of the buffer.

When samples are applied as stripes, "edge effects" are commonly seen. This is a convexity of the component bands in the direction of migration possibly due to increased evaporation of solvent at the edges, which results in increased electrolyte concentration, cooler strip temperature, and increased capillary flow from the reservoir, each of which would cause a slowing down of migration. This phenomenon becomes less pronounced as strip width is decreased. Every effort must be made to minimize this distortion, for it makes it difficult to cut out the bands for elution and introduces "stray radiant energy" in densitometry.

BUFFERS

The buffers used in cellulose acetate electrophoresis are generally more dilute than those used with paper. The ionic strength of the buffer will determine the migration distance; the lower the ionic strength the greater the distance. It is also true that diffusion of separated components is inversely related to the ionic strength of the buffer. The resolution of bands, however, gets poorer with increasing migration distance since the width of the bands will increase because of diffusion. Some of the commonly employed buffers are the barbital buffers with concentration that vary between 0.05 and 0.1 M and the Tris-EDTA-borate buffers at pH 8.4, 8.9, and 9.6. Buffer solutions should be changed frequently, especially at high ambient temperatures since the buffer will concentrate during use due to evaporation. Since the number of electrophoretic determinations will vary from laboratory to laboratory, it is recommended that the buffer be changed after a certain number of runs depending on the procedure.

INSTRUMENTS

Many electrophoretic chambers are available for the electrophoretic separation of proteins. The ones most commonly used are the R101 Microzone, Shandon, Photovolt, and Gelman chambers. These are all horizontal chambers with the exception of the Photovolt chamber, in which the strips with Mylar backing are not placed horizontally but rather form a bow type bridge. The densitometers used in most laboratories are the Beckman Analytrol, Photovolt Densicord, and the Microzone Densitometer. The main requirement for a power supply is that it can maintain either a constant voltage or a constant current. Since the resistance will change in an electrophoresis run, the current will change accordingly and a constant power stabilizer should therefore be used.

ELECTROPHORESIS

The optimum conditions for an electrophoresis run have to be determined for each technic. The factors affecting the electrophoretic separation of components are: (a) the type of supporting media, (b) composition and ionic strength of buffer, (c) temperature of buffer, (d) current or voltage applied, and (e) time of electrophoresis. Generally a protein electrophoresis will be completed in approximately 30 min when a current of 1.5 mA per strip is applied. If the ammeter reads offscale in the direction of high amperage, a "shorting" of the circuit as a result of spillage of buffer is the most likely cause. The direction of the current should be changed after each run to prevent the gradual change in buffer composition between the two reservoirs that would occur over many successive runs. Reference dyes are often used to monitor the rate of migration. For example, albumin stained with bromphenol blue is frequently used as internal control in electrophoresis (18).

STAINING

The staining solutions used for cellulose acetate electrophoresis are generally less concentrated than those used for paper electrophoresis. Aqueous as well as alcoholic staining solutions are employed by different workers, but the former ones are generally preferred. If alcoholic stains are used, additional washings through an aqueous buffer are required to prevent shrinkage of the strips. Proteins stain with the anionic dyes Ponceau S or Nigrosin by chemical reaction with the amino groups. Other chemical mechanisms that occur in staining include the oxidation or reduction of the dye. Sometimes the separated components are identified after their chemical conversion to fluorophors.

Strips stained with aqueous staining solutions are commonly washed with a 5% acetic acid solution until the background is white or until the washing solution is no longer colored with the dye. This rinse solution does not affect cellulose acetate. The stained strips are then dried by draining off excess rinse solution followed by air drying. Then they are placed in methanol, followed by a clearing solution containing glacial acetic acid or dioxan to make them transparent and suitable for densitometry. Cellulose acetate strips can also be cleared by oil impregnation. This procedure of clearing is not permanent, and the strips can be returned to their original state by washing them in a fat solvent such as ether.

QUANTITATION

Quantitative determinations can be carried out by cutting out the stained bands and eluting the dye that is then measured photometrically. The dye in the cut up band can also be dissolved completely in organic solvents prior to reading the color in a photometer or spectrophotometer.

A more convenient and rapid procedure for quantitation of components is continuous staining densitometry or fluorometry. The advantage of quantitation in this manner is that the cleared strip with the stained bands can be used without any further preparation. The responses of the photometer can be recorded and the resultant curves can be evaluated by a variety of manual or automated technics. An automatic calculation of densitometer scans of electrophoretic strips utilizing a disk integrator, automatic digital readout, and programmable desk-top computer was developed in our laboratory (24). Other refinements have been implemented, including replacement of the programmable desk-top computer with an "on-line" computer, which computes the percent of total protein, grams per 100 ml of each fraction, and the A/G ratio. The first part of this configuration is now available from Photovolt as the Densicord Model 552. An "on-line" computer approach (22) that requires human intervention for the selection of peak boundaries has also been reported. Either system requires the "off-line" entry of total proteins, if results are to be reported in grams per 100 ml. In another system an oscilloscope is employed for the display of the electrophoretic pattern (Phoroscope TM densitometer, Millipore Corporation), but a manual manipulation sets a cursor to the interpeak sites according to the judgment of the technician. The system used in our laboratory referred to above, has the advantages of good accuracy, reproducibility, and elimination of human error.

THIN-LAYER ELECTROPHORESIS

Thin-layer electrophoresis requires smaller amounts of sample than is needed for paper electrophoresis. Furthermore, the procedures are simpler and require less elapsed time for analysis.

Thin-layer supports. Kieselguhr, silica gel, cellulose, and alumina are usually used as supports. If the surface is cellulose the plates cannot be sprayed with strong reagents, and if alumina is employed, tailing of the dyes can be expected (4). In general, the best results have been obtained with Kieselguhr and silica gel (4). Prepared plates with most of the common supports are now available from commercial sources.

Buffers. Buffers containing volatile solvents such as butanol, acetone, pyridine, etc., are recommended since they can be eliminated easily when the plates are being dried and therefore do not interfere with the spraying reagent.

Application of sample and electrophoresis. The plate must be dry when the sample is applied in the form of a small spot with a micropipet. When the plate is sprayed with buffer care should be taken not to use too much. The plates are then subjected to electrophoresis for a period generally not longer than 1 hour. After electrophoresis the plates are dried at approximately 100°C and colors from the separated compounds are developed with a suitable reagent.

STARCH GEL ELECTROPHORESIS

In paper and cellulose acetate electrophoresis the medium is inert, but in starch gel electrophoresis this is not the case. The pore structure of the gel acts as a molecular sieve to separate components of a solution on the basis of molecular size. The difference between paper and starch is apparent when globulins are separated, since one band on paper is resolved into several bands on gel. Starch that is commonly available is too highly polymerized to form gels and must therefore be first hydrolyzed. Hydrolyzed starch for gel electrophoresis is now available from British Drug Houses.

Buffers. Most workers use gels within the range of pH 8–10 for separation of serum proteins by electrophoresis. A discontinuous buffer system in which the gel buffer was a Tris-citrate buffer at pH 8.7 and the reservoir buffer a sodium borate buffer at approximately pH 8.5 was described by Poulik (21). In this discontinuous buffer system the proteins will start migrating in the gel Tris-citrate buffer and will complete their migration in the borate buffer. The borate ions of the reservoir buffer have a faster migration rate than the protein ions in the gel and therefore overrun and pass them. The borate-Tris interface appears as a moving yellow line. This procedure results in a sharpening of the rear margin of each band. The use of lithium hydroxide instead of sodium hydroxide in the reservoir buffer resulted in even sharper bands (8).

Electrophoresis. Gel is prepared and poured into the holder unit that contains the gel plate on the bottom. The excess gel is squeezed out of the holder unit with a siliconed plastic sheet that is slowly flattened down, starting from one side so that no air bubbles are trapped. The gel is then cooled for 2 hours at the same temperature as it will subsequently be used. The gel should be covered at all times to prevent evaporation. When the gel is ready for use, the top plate is gently removed without tearing the gel. The gel plates can be used in horizontal or vertical position. When the gel is used in horizontal position, the sample is applied on a narrow strip of thick filter paper or as a slurried suspension in starch grains. With the vertical method the sample is applied directly into a slot cut into the gel. The advantage of this application is that the sample cannot adsorb on the paper or starch grains. If the sample is first applied on paper, the paper strip is inserted into the gel with the aid of a razor blade as follows: The blade is first placed in the gel and the paper strip rested against it. The paper is inserted into the gel by being pushed inwards while the blade is slowly withdrawn. After the paper is in place the surface of the gel is pressed together again without trapping any air. In case starch is used for application, the serum-starch mixture is deposited in the slot in the gel.

The current used in starch electrophoresis is generally 2 mA/cm width of gel for 5–6 hour or 0.7 mA/cm for overnight runs. Whatever current is used, the gel should not get warmer than 35°C. After electrophoresis the gel is stained. Extreme care must be taken since the gel plate is very fragile. Washing and staining should be done while the gel is supported by a plastic sheet.

Quantitation Before densitometric scanning the gel should be washed free of background stain and made transparent by immersing it in a solution containing 15% glycerol and 3% acetic acid in water for 60 min. A strong thin gel is obtained that can be preserved indefinitely. The gel patterns can then be scanned in a densitometer followed by quantitation using the integrated areas under the curves (24).

DISK ELECTROPHORESIS USING ACRYLAMIDE GEL

Disk gel electrophoresis was introduced by Ornstein (20) and Davis (7). This technic of electrophoresis is so named because the bands stack up as concentrated disks. The increase in resolution of gel disks over paper electrophoresis is evident from the fact that paper electrophoresis of human serum commonly resolves 5 protein bands while disk gel electrophoresis resolves more than 20 bands.

The advantages of disk over starch gel electrophoresis include greater simplicity of technic and much shorter running times. Most of the advantages of disk electrophoresis stem from the ability to make the gel to an exact concentration. An increase in gel concentration results in a decrease in pore size. Since the pore size of the gel can be determined by its concentration, separations are achieved by virtue of its property as a molecular sieve. A small pore gel containing a mixture of 20%—40% acrylamide in water, methylenebisacrylamide, tetramethylethylene diamine, ammonium persulfate and Tris-HCl buffer, is placed in a vertical tube (approximately 70 x 5 mm) that is closed on the bottom with a removable plastic cap. The spacer or larger pore gel, containing approximately 3% acrylamide in water, is placed on top of the small pore gel. The sample gel, which is a mixture of a few microliters serum and the large pore gel, is poured on top of the spacer gel. After removing the plastic cap on the bottom the gel tube is placed vertically in a Tris-glycine buffer of pH 8.3 that also is poured on top of the gel tube. The electrodes are connected with the buffer in the bottom reservoir and the buffer on top of the gel tube. A 5 mA current is passed through the tube for approximately 30 min. After electrophoresis the gel is removed from the tube and stained and excess stain removed. Alternatively, stain can be incorporated in the gel, obviating staining as a separate step. Gels will demonstrate horizontal bands or disks of separated components. The formation of bands is explained as follows: The glycinate ions in the buffer are placed on top of the chloride ions in the gel buffer. When a current is passed through the gel tube the protein ions will migrate down the wide pore gel with a mobility intermediate between that of the glycinate and chloride ions. Since the glycinate ions have a mobility less than the protein ions in the wide pore gel, on passage of a current, a sharp and fixed boundary is maintained at the interface of these ions. The protein ions, as they migrate through the spacer gel, will form discrete disks, which will form a stack at the interface of the small pore gel. When the protein ions migrate into the small pore gel their mobility decreases, since the pore size of the small pore gel that is within the range of dimensions of serum proteins will cause a sieving action. The glycinate ions will start to overtake the protein ions and form a sharp boundary at the chloride ion interface. The tris-HCl gel buffer is transformed to a Tris-glycine gel buffer of pH 9.5, and it is at this pH that the proteins will continue to migrate as a series of disks.

In summary, the following conditions have to be satisfied by a successful technic: (a) the ion in the running gel buffer must have the highest mobility of all ions in the direction of migration and must be that of a strong acid so that it is not affected by pH. (b) The ion that overruns the proteins must be from a weak acid so that its mobility will increase when the pH of the running gel buffer changes. (c) The formation of a sample gel to assure the layering of sample. (d) The presence of a spacer gel to allow stacking of protein disks.

SCOPE OF THE METHOD

Electrophoresis can separate *only* substances with *different* mobilities. Furthermore, the separation of a "component" is not *per se* an absolute criterion of its purity or homogeneity.

Electrophoresis has been applied to serum proteins, urine proteins, cerebrospinal fluid proteins, serum lipoproteins, serum protein-bound carbohydrates, the distribution of iodine, iron, and cholesterol in serum, hemoglobins, antibodies, nucleic acids, nucleotides, enzymes, antibiotics, toxins, viruses, polypeptides, amino acids, hormones, mucopolysaccharides, inorganic ions, and many other substances of interest to the biological chemist.

Conditions for electrophoresis must be worked out for each technic and the selection of supporting medium, pH and ionic strength of the buffer, duration of electrophoresis, and other parameters must be determined primarily empirically. Publications of electrophoretic technics are, therefore, only meaningful for setting up the published method if all conditions are disclosed.

REFERENCES

1. BLOCK RJ, DURRUM EL, and ZWEIG G: *A Manual of Paper Chromatography and Paper Electrophoresis.* New York, Academic, 1955
2. COLLET LH: *J Chim Phys 49:*C65, 1952
3. COOPER GR, MANDEL EE: *J Lab Clin Med 44:*636, 1954
4. CRIDDLE WJ, MOODY GJ, THOMAS JDR: *J Chromatog 16:*350, 1964
5. CROOK EM, HARRIS H, JASSAN F, WARREN FL: *Biochem J 56:*434, 1954
6. DARMOIS E: *Compt Rend 227:*339, 1948
7. DAVIS B: *J Ann N. Y. Acad Sci 121:*404, 1964
8. FERGUSON KA, WALLACE ALC: *Nature 190:*629, 1961
9. GRASSMANN W, HANNIG K: *Hoppe-Seylers Z Physiol Chem 290:*1, 1952
10. GRASSMANN W, HANNIG K: *Naturwissenschaften 37:*496, 1950
11. GRASSMANN W, HANNIG K, KNEDEL M: *Deut Med Wochschr 76:*333, 1951
12. KAWERAU E: *Analyst 79:*681, 1954
13. KRAUS KA, SMITH GW: *J Am Chem Soc 72:*4329, 1950
14. KUNKEL HG: *Methods Biochem Analy 1:*141, 1954
15. KUNKEL HG, TISELIUS A: *J Gen Physiol 35:*89, 1951
16. LATNER AL: *Biochem J 52:*xxix, 1952
17. MACKAY IR, VOLWILER W, GOLDSWORTHY PD: *J Clin Invest 33:*855, 1954
18. McDONALD HF, FORESMAN JL, BARNES EW Jr:*Proc Soc Exp Biol Med 94:*493, 1957
19. McDONALD HJ, LAPPE RJ, MARBACH EP, SPITZER RH, URBIN MC: *Ionography, Electrophoresis in Stabilized Media.* Chicago, Year Book, 1955
20. ORNSTEIN L: *Ann NY Acad Sci 121:*321, 1964
21. POULIK MD: *Methods Biochem Analy 14:*455, 1966
22. PRIBOR HC, KIRKHAM WR, FELLOWS GE: *Am J Clin Pathol 50:*67, 1968
23. SLATER RJ, KUNKEL HG: *J Lab Clin Med 41:*619, 1953
24. WINKELMAN JW, WYBENGA DR: *Clin Chem 15:*708, 1969

Chapter 6

Chromatography

BENJAMIN N. HORWITT, Ph.D.

INTRODUCTION

Since the beginning of the science of chemistry, chemists have faced several problems. One is the determination of the purity of a given preparation and the number of components in a given system. Another is the ability to resolve the system into its component parts so that they can be characterized and quantitated. The earlier chemists usually resolved these problems by fractional crystallization and distillation. Since the 1940s, the extensive use of chromatography has been instrumental in major advances in all fields of chemistry.

DEFINITION

By common practice a variety of properties are included in the term "chromatography," making a rigid definition difficult. Chromatography has been defined (25) as a general technic utilizing "those properties which allow the resolution of mixtures by effecting separation of some or all of their components in concentration zones on or in phases different from those in which they are originally present, irrespective of the nature of the force or forces causing the substances to move from one phase to another." Stated more simply, the technic separates mixtures of substances. This fractionation procedure may be considered analogous to distillation.

HISTORICAL

One might trace chromatographic technic back to Pliny the Elder. He reportedly described (1) a method for the detection of iron by applying a drop of solution to papyrus impregnated with an extract of oak-gall.

In the late 19th century Day, an American petroleum chemist, described the use of columns of Fuller's earth to separate crude petroleum mixtures (24). At approximately the same time a Russian botanist, Tswett, was interested in the separation of pigments of plants which, because of their similarity in properties and lability, presented peculiar analytic difficulties. Tswett was able to separate plant pigments dissolved in petroleum ether by pouring the extract over finely divided calcium carbonate contained in a vertical glass tube. He observed that the pigments were adsorbed near the top of the column. With the addition of more solution, the region of pigmentation became broader, and there was a separation of colors, becoming more marked when the column was washed with pure petroleum ether. The different pigments were being separated by adsorption at different levels on the column. By prolonged washing, the various colored bands became separated from each other by colorless regions (23).

Tswett named the colored column a chromatogram and the process chromatography. This was, perhaps, a poor choice since the method is far more universally applicable to colorless than to colored substances, a point recognized by Tswett in his paper. A better term might have been adsorption analysis, but unfortunately, the older name is now well established and it is unlikely that any generally accepted change will be made.

Tswett's discovery lay neglected for about 25 years until Kuhn *et al* (12,13) virtually rediscovered it in 1931, applying it to the resolution of plant carotenes into components. The impact of chromatography was tremendous. It is doubtful that the rapid progress in the fields of vitamins, hormones, and numerous other natural products would have been so rapid and so successful had it not been for this technic.

SCOPE OF CHAPTER

The revival of chromatography as an analytic tool has resulted in an abundance of papers on this subject. The scope of this section is to introduce the reader to the principles and various common forms of chromatography. Specific methods of analysis will not be discussed here but will be found throughout the book in specific chapters dealing with analytical procedures for various compounds.

CLASSIFICATION OF CHROMATOGRAPHIC METHODS

The common feature of all chromatographic methods is that there are two phases, one stationary and the other mobile. The separation of substances depends on the movement of the mobile phase relative to the stationary phase and the distribution of the substances to be separated between the two phases.

A classification can be made depending upon whether the stationary phase is solid or liquid. If it is solid, the method is termed *adsorption chromatography*; if it is liquid the method is *partition chromatography*

MAIN TYPES OF CHROMATOGRAPHY

In general, there are four main types which can be classified as follows:

1. Liquid-solid
   "Classical" adsorption chromatography (Tswett column)
   Ion-exchange chromatography

2. Gas-solid chromatography
3. Liquid-liquid
   "Classical" partition chromatography
   Paper chromatography
4. Gas-liquid
   Gas-liquid chromatography (GLC)

This classification is presented in a schematic form in Figure 6-1.

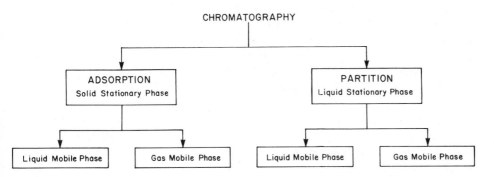

FIG. 6-1. Classification of chromatography.

DEFINITION OF TERMS

In the process of describing types of chromatography, we have already used some terms commonly used in this field of analysis, but there are a few more that need to be defined. The solid material that serves as the stationary phase in adsorption chromatography is called the *adsorbent*. The solid base which supports the liquid film in partition chromatography is called the *support*. When the mobile phase is caused to flow over the adsorbent or support the process is termed *development*. After the substances on the chromatogram have been separated by development, they are *detected* or *visualized*. If the substances on the chromatogram are actually washed off the adsorbent, they have been *eluted*. The substances being separated are usually called *solutes* or, as a group, the *sample*.

ADSORPTION CHROMATOGRAPHY

Adsorption column chromatography is the oldest form of chromatography. Whether two or more substances of a mixture can be separated by adsorption chromatography depends on a number of factors. Most important is the strength with which each component of the mixture is adsorbed and its solubility in the solvent used for elution. The degree to which a particular substance is adsorbed depends on the type of bonds which can be formed between the solute molecules and the surface of the adsorbent.

POLARITY AND SOLVENTS

Polarity, a frequently used term, basically means the possession by a molecule of separate positive and negative charge centers arising from the atoms involved and their arrangement or configuration. A molecule may have a dipole moment and may be attracted to other molecules that also possess polarity. The number of charges and extent of separation between the centers or poles will determine the degree of polarity and, therefore, the degree of attraction. In chromatography polarity is essentially a relative term and may be applied to solvents, solutes, and adsorbents. Water is a very polar solvent, and oxygenated organic compounds such as alcohols, ketones, esters, and ethers are less polar. Hydrocarbons are least polar of all. Trappe (22) was among the first to establish an eluotropic series. Table 6-1 presents a series of common solvents arranged according to relative polarities. The least polar (smallest dielectric constant) heads the list with the most polar (greatest dielectric constant) at the bottom. The order of eluting power of solvent observed in relation to a particular adsorbent is not necessarily the same when another adsorbent is substituted. In general, one can say "like dissolves like," i.e., polar solvents tend to dissolve polar solutes more readily than nonpolar solutes.

Generally speaking, the strongest adsorption occurs from nonpolar solvents such as petroleum ether and the weakest from highly polar ones such as alcohol. Solvents high in the list in Table 6-1 are appropriate for preparing the original solution to put on the adsorption column and those toward the bottom of the list are appropriate for eluting the component.

TABLE 6-1. Eluotropic Series of Solvents

Eluting power increases in numerical order.
1. Light petroleum (petroleum ether, hexane, heptane, etc.)
2. Cycolohexane
3. Carbon tetrachloride
4. Trichloroethylene
5. Toluene
6. Benzene
7. Dichloroethylene
8. Chloroform
9. Ether
10. Ethyl acetate
11. Acetone
12. N-Propyl alcohol
13. Ethyl alcohol
14. Methyl alcohol
15. Water

ADSORBENTS

Many different solids have been used as adsorbents, and there are several properties to be considered in choosing an adsorbent. The adsorbent must be insoluble in the solvents used. It must neither react with the substance to be separated nor catalyze the decomposition of the substances to be separated. If colored substances are to be separated and visualized, it is desirable for the adsorbent to be colorless. The properties of the adsorbent should be approximately the same regardless of the source from which it is obtained. This is sometimes a condition that is hard to realize.

Since adsorption is a surface phenomenon, it is important that the powder used for chromatography should be sufficiently small in particle size to give a large surface area. On the other hand, the rate of percolation through a column of very fine material may be exceedingly slow. A compromise between the ideal and the practical is necessary.

COMMON ADSORBENTS

Alumina, silica gel, magnesium oxide, and charcoal are among a great number of substances that have been used as adsorbents, with alumina being, perhaps, the most commonly used of all. Table 6-2 lists some of the common adsorbents and the types of compounds separated by them. There is little reference in the table to inorganic separations, which are usually more efficiently carried out with the aid of ion-exchange resins.

TABLE 6-2. Adsorbents

Solid	Separates
Alumina	Steroids, vitamins, esters, alkaloids
Silica gel	Steroids, amino acids
Carbon	Peptides, carbohydrates, amino acids
Magnesia	Similar to alumina
Magnesium carbonate	Porphyrins
Magnesium silicate (florisil)	Steroids, esters, glycerides, alkaloids
Calcium carbonate	Carotenoids, xanthophylls
Calcium phosphate	Enzymes, proteins, polynucleotides
Aluminum silicate	Steroids
Starch	Enzymes
Sugar	Chlorophyll, xanthophyll

STRENGTH OF ADSORBENTS

Adsorbents are sometimes classified according to the strength with which they adsorb substances. This is usually determined by measuring the rate at which a zone travels in an elution experiment; the greater the rate, the weaker the adsorbent. This strength would, of course, be influenced by the solvent employed. Surface area is another parameter: The greater the surface area the greater the amount of solute capable of being adsorbed. The strong adsorbents have relatively large surface areas. Any classification according to strength of adsorption must be considered only as a rough guide because of the many exceptions. The following is such a rough tabulation:

Weak adsorbents: Sucrose, inulin, starch, talc
Intermediate adsorbents: Calcium carbonate, calcium phosphate, magnesia
Strong adsorbents: Alumina (low water content), charcoal, Fuller's earth, silica gel (low water content)

ACTIVATION

The term "activity" often denotes the strength of the adsorbent. Many adsorbents such as alumina, silica gel, active carbon, and magnesium silicate can be obtained commercially, but they often require activation before use. Activation is achieved by heating and there is usually an optimum temperature for activation; for alumina, e.g., it is about 400°C. The period of heating is also important because prolonged heating will cause loss of activity. In the absence of good information, 200°C for 4 hours is safe for most solids. Part of this "activation" process is probably merely drying the adsorbent, but other factors may be involved.

Following activation it may be desirable to reduce activity by the controlled addition of water, and the subsequent activity is related to the amount of water added. Brockman and Schodder (2) established five grades of alumina. Grade I is the most active and is simply alumina heated at about 350°C for several hours. Grade II has about 2%–3% water added, Grade III 5%–7%, Grade IV 9%–11%, Grade V (least active) about 15%. This grading of activity was based on the position occupied on the column by each of a series of Azo dyes separated under standard conditions. One of the principal problems in adsorbent preparation is the difficulty of obtaining batches with reproducible properties. Even in the same batch adsorbent properties may vary. They may also "age" or deactivate with time.

REACTIONS WITH ADSORBENTS

Adsorption chromatography is a relatively mild fractionation process. Chemical reactions do occasionally occur, however, in the presence of adsorbents. Lists of such reactions have been compiled (3, 14). A few examples are the following: isomerization and polymerization of olefines with silica gel; deacetylation of acetylated sugars; oxidization of fatty acids; polymerization of acetone to diacetone alcohol; rearrangement of β, γ to α, β unsaturated ketones with alumina; aminolysis and oxidation of amino acids with charcoal. The active solids used for packing chromatographic columns may be good catalysts. Alkaline alumina can cause the condensation of aldehydes and ketones. Acetone is not recommended as a solvent when using alumina columns. Decomposition of the components to be resolved by the usual adsorbents is rare and generally only partial. Some irreversible adsorption may take place on columns, resulting often in incomplete recovery of the solutes.

PARTITION CHROMATOGRAPHY

A major breakthrough was achieved with the discovery of *partition chromatography*, so named by Martin and Synge (15, 16). They wished to separate the acetyl amino acids in a hydrolyzate of wool by counter current technics. The procedures available in 1941 were unsatisfactory for their needs and they decided to attempt to hold one of the liquid phases stationary and pass the other through it. Silica gel was saturated with water and packed into a column. The acetyl amino acids in chloroform were poured into the column, which was then eluted with additional chloroform. The acetyl amino acids were distributed between the stationary water phase on the silica gel and the mobile chloroform phase. They traveled down the column at different rates and were separated in different fractions of eluates.

PRINCIPLE

Whereas in adsorption chromatography the affinity of the components of a mixture for the solid surface and the solubility in the flowing solvent result in resolution of the mixture, in partition chromatography the solid adsorbent is replaced by a stationary liquid that is normally only partially miscible with the flowing liquid. Movement of the substances in the chromatogram depends on their relative concentrations in the stationary and the moving phases. The dissolved substances cross and recross the interfacial boundary from the moving to the stationary phase without becoming concentrated near the surface, as in the case in adsorption chromatography.

The rate at which a solute moves in such a system will depend upon its relative solubilities in the two phases. This is expressed as a partition coefficient (K), defined as the concentration of the solute present in one phase divided by the concentration of the solute in the second phase after equilibrium has been established. This is a constant at a given temperature when neither phase is saturated with respect to the dissolved substance. By definition, in chromatographic systems the solute concentration in the mobile phase is the numerator and in the stationary phase is the denominator. Where s is the solute concentration, the partition coefficient K is

$$K = \frac{s_{mobile}}{s_{stationary}}$$

SUPPORTS

In adsorption chromatography the solid material that serves as the stationary phase is termed the *adsorbent*. In partition chromatography the solid material acts as *support* for the liquid film. Where filter paper is used as the stationary phase the process is identified as *paper chromatography*. When the material to be separated is distributed between a moving gas phase and a liquid phase immobilized on a solid support, the process is known as *gas-liquid chromatography (GLC)*. Both of these systems have become very important methods of analysis.

It is possible to eliminate the support and allow one solvent to flow through the other in a dropwise manner. This common extraction technic is the basis of the Craig (6) countercurrent technic, which amounts to a series of conventional two phase solvent extractions carried out in succession. This, however, is not a separation in the chromatographic sense.

The solid support should be as inert as possible to the substances to be separated, adsorb and retain the stationary phase, expose a large surface to the mobile phase, and be mechanically stable and not impede solvent flow. No one support has all these properties. The greatest difficulty is the incursion of adsorption effects. As complete coverage of the surface may not be easy to achieve, adsorption may be a major influence on separation.

The solid supports most commonly used in partition column chromatography are silica gel, cellulose powder, and diatomaceous earths. Others such as starch or glass beads are used less extensively.

SILICA GEL

Silica gel is usually graded to particle size limits, which is a determining factor in the flow rate of the mobile phase. The most commonly used sizes are within the range 28–200 mesh. Within this range, silica gel is capable of retaining 50%–75% of its weight of water. Silica gel is almost always used with water or a buffered aqueous solution as the stationary phase.

Commercial silica gel usually contains water and may also contain impurities. It is usually necessary to clean and purify the silica gel before use. Purification is usually carried out by washing with two volumes of a mixture of equal parts HCl and water and stirring to form a homogenous slurry. After a few hours, the liquid is decanted and the silica gel thoroughly washed first with water and finally with methanol or ethanol. The gel is then dried at 110°C and stored in a closed container.

Silica gel possesses adsorptive properties and it is generally accepted that adsorption plays a part in all separations employing silica gel. The same kinds of chemical reactions can occur in partition systems as in adsorption systems, although they appear less marked since the influence of the surface of the solid support is diminished by the stationary liquid coating the surface.

DIATOMACEOUS EARTHS

Most commonly used diatomaceous earths consist of very fine particles and usually contain soluble impurities. If necessary they can be purified in the manner described for silica gel. They are commercially available as kieselguhr or Celite. When marketed as analytical grade quality they can be used without prior treatment.

CELLULOSE

Ashless cellulose powder only should be used. It is usually supplied ready for use, requiring no prior treatment. Use of cellulose powder in columns is an alternative to the use of cellulose in the form of sheets as in paper chromatography and it permits the separation of larger quantities of materials.

SOLVENT SYSTEM

Separation of substances by partition methods is dependent on the partition coefficients of the stationary and mobile phases. The simplest solvent system is an equilibrated pair of virtually immiscible liquids, one the stationary and the other the mobile phase. To avoid changes in phase composition during chromatography, the solvent should be equilibrated before use by thoroughly mixing the solvents in a separatory funnel. Each phase then consists principally of one component plus a little of the other. The individual phases are used to prepare and develop the chromatogram. The support is equilibrated with the stationary phase. In the case of column chromatography, this is achieved by mixing the support with the equilibrated stationary phase. In normal partition chromatography this is the more polar phase. Where the stationary phase is nonpolar and the mobile polar, the process is termed *reverse phase chromatography*. The mobile phase is the eluting solvent. Transferring the solute to the support depends on the chromatographic technic employed and is discussed in the appropriate section. The solute mixture distributes itself between the mobile and stationary phases, and the relative position of any one solute is determined largely by its relative solubility in the two phases. Those substances having a greater affinity for the stationary phase move more slowly than those having a greater affinity for the mobile phase.

The solvents should be pure since impurities may cause significant changes in solute distribution. It is also important that the temperature of equilibration should be the same as that used for the chromatographic process to prevent changes in mutual solubility of the solvent pair.

MANIPULATION OF SOLVENTS

The partition system can be altered by varying the nature of the two liquid phases. The solvent system is selected according to the solubility characteristics of the solute mixture. The solute distribution between the mobile and the stationary phases is a function of the solubility in each. This solubility distribution may be manipulated by changing the composition of the biphasic system.

Not only can the individual components of the system be changed, but the phases may be altered by using mixtures of a number of components rather than two pure substances. If three components are used, one phase will consist principally of two components and the other of one component. With four components, each phase may consist of two components. By adjusting the relative amounts of the various components it is usually possible to bring the distribution of the solute within the desired range.

Solvents may be arranged according to their ability to form hydrogen bonds, forming a series analogous to the eluotropic series of Trappe for adsorption chromatography. At the head of this series are those solvents that can form intermolecular hydrogen bridges (hydrophilic or polar solvents); at the bottom are those that cannot (hydrophobic or nonpolar solvents). Between these two extremes, there is a continuous transition formed by solvents of varying polarity. Table 6-3 lists solvents arranged according to their ability to form intramolecular hydrogen bridges, ranging from the most polar, water, to the least polar, paraffin oil. Up to tertiary butanol the solvents are completely miscible with water; the other solvents form two layers with water.

In selecting solvents, the components to be separated must be sufficiently soluble so that solutions of the desired concentration can be handled. If this is not the case at room temperature, it may be possible to meet this requirement by carrying out the chromatography at an elevated temperature. Another desirable feature of the solvent is volatility, which facilitates its removal from the support medium (e.g., paper) or eluates (e.g., column).

TABLE 6-3. Mixotropic Solvent Series (see ref. 7)

Water (most polar)	N-Amyl alcohol
Formamide	Ethyl acetate
Methanol	Ether
Acetic acid	N-Butyl acetate
Ethanol	Chloroform
Isopropanol	Benzene
Acetone	Toluene
N-Propanol	Cyclohexane
Tert-butanol	Petroleum ether
Phenol	Petroleum
N-Butanol	Paraffin oil (least polar)

ADSORPTION VERSUS PARTITION

Separation of solutes whether by adsorption or partition is dependent primarily on the difference in polarity of the solutes. Polarity is the major factor governing adsorption and solubility. Although this is a common feature of the two procedures, there are sufficient differences to favor one over the other for particular separations. In general, the choice between adsorption and partition chromatography will be governed by types of compounds being separated, the relative experimental difficulty, and the purpose and goal of the separation.

In partition methods, solubility in two liquids (phases) is the governing factor, making this process sensitive to small differences in molecular weight. Partition systems, therefore, are generally better for separation of members of a homologous series, especially those with more than five carbons.

Adsorption systems are quite sensitive to steric or spatial differences of the solutes. Adsorption processes, therefore, would be the preferred method for separating similar molecules having slightly different stereochemistry.

One would expect that adsorption chromatography could resolve any mixture, it being necessary only to find the solvent of the proper polarity. Unfortunately, this is not the case. Polar solutes, which are highly adsorbed to a surface, require highly polar solvents to elute them. These highly polar solvents will tend to produce no separations because they completely override the small differences in adsorptivity between the solutes.

As a general rule, highly polar materials such as amino acids, nucleotides, and carbohydrates are separated by partition methods, and less polar compounds such as hydrocarbons and monofunctional multicarbon organic molecules are separated by adsorption technics.

Polarity increases with number of functional groups and decreases with increasing molecular weight and number of carbon atoms. One can therefore state generally that hydrocarbons (nonpolar) are always separated by adsorption methods, and amino acids (highly polar) are always separated by partition methods. Between these extremes, either process might be applicable.

Experimentally, adsorption is easier than partition chromatography. The adsorbent can be kept constant and the solvents varied over the eluotropic series to find the solvent with a polarity that will effect a separation. In partition chromatography, neither phase can be maintained at a constant polarity, making it difficult to carry out systematic experimentation. If no other information were available, it would be reasonable to try adsorption first and then, if difficulty is encountered, partition.

Adsorption is better than partition as a preparative method. It can be used for separating larger quantities and to separate mixtures of solutes with widely varying polarity and structure. Partition is preferable in separating mixtures of very similar solutes, has a greater resolving power than adsorption, and with proper conditions, is a better qualitative method.

CHROMATOGRAPHIC TECHNICS

The previous discussion has been on the general principles of the different major classifications of chromatography, rather than on individual technics employed in chromatography. This portion of the chapter will be directed to four of the major technics: column, paper, thin layer, and gas chromatography. These categories refer to the type of apparatus used;

whether the chromatography is adsorption or partition will depend on the individual conditions being employed (see Fig. 6-1).

COLUMN CHROMATOGRAPHY

In its simplest form column chromatography consists of a solid medium packed into a column. After the mixture of solutes has been introduced into the column, the solvent is passed over the column. The solutes are eluted as the solvent percolates through the column. Column methods are useful for the separation of pure compounds in quantity from mixtures.

COLUMNS

Chromatographic columns are usually made of glass, to permit observation of the settling of the solid phase and the solvent level, inspection for air bubbles and channeling, and when dealing with colored substances, the movement of the colored bands.

For laboratory use, the effective length of the column varies between 2 and 150 cm. There is also considerable variation in the diameter/length ratio, but usually the length is at least 10 times the internal diameter. The diameter/length ratio to use is generally dependent on the ease or difficulty of the separation and must be determined empirically. The size of the column and the amount of packing are largely determined by the amount of solute to be separated. Practical considerations, such as convenience, are also factors in determining dimensions.

Various methods have been used to support and retain the solid phase at the bottom of the column, including glass-wool plugs, filter paper, and sintered glass disks.

To accommodate large volumes of solvent, columns can be fitted with reservoirs connected to the top. With long columns or columns of finely divided material, sources of pressure can be attached to force the eluting solvent through the column to speed the flow.

The column must be packed uniformly. Improper packing may lead to irregular development. Columns may be prepared by either dry or wet packing technics.

In dry packing, small quantities of adsorbent are introduced into the column. After each addition, the adsorbent is thoroughly shaken down by tapping the column, applying vibration to the column until the level of the adsorbent becomes constant, or tamping the adsorbent with a tamping rod. This procedure is repeated until the desired column height is obtained. Placing a small plug of glass wool on top of the column prevents disturbing the surface of the adsorbent. The column is then thoroughly washed with the initial solvent to be used in the elution process.

In wet packing, the adsorbent and solvent are mixed to form a slurry. This is poured into the column, and the adsorbent allowed to settle and form a bed at the bottom. During the settling, excess solvent is drained and additional slurry is added, and the process is continued until the desired height is obtained. Wet packing may also be carried out by introducing the solvent into the tube and then adding the slurry or the dry adsorbent in a fine stream through a funnel. As the adsorbent settles, the tube is gently tapped. When the excess solvent has been removed, the top of the column is covered with a pledget of glass wool to prevent disruption during elution.

Whatever way the column has been prepared, it is essential that the surface be flat and that once the column has been wetted with solvent the surface does not dry until the whole process of adsorption and development is complete. If the column becomes dry it is likely to shrink away from the walls of the tube. Additional solvent will then run down into the spaces so formed, upsetting the whole process.

DEVELOPMENT, ELUTION, AND DETECTION OF SOLUTES

It is important to apply the sample to the top as evenly and in as concentrated solution as possible, with no disturbance of the packing material. When the solution with the sample has been transferred to the column, it is then followed by eluting solvents. The actual solvents used for elution as well as the packing material used in the column depend on the solutes to be separated in the type of chromatography employed.

Colored substances can be detected by direct observation. However, most substances that are resolved are colorless, and various physical properties such as fluorescence, optical absorption, pH, etc., have been used to monitor the elution fractions for solutes to be separated. Chemical methods involving a reagent that produces some readily discernible reaction such as a color may be applicable. Most of the methods used are qualitative, but it is possible to evaluate the fractions quantitatively by physical or chemical means. In the case of substances with physiologic activity biologic methods may be applicable.

PAPER CHROMATOGRAPHY

Paper chromatography, as we now understand it, was a development of the partition system introduced by Martin and Synge in 1941 (15, 16). One of the solids that can be used to support the stationary phase in partition chromatography is cellulose powder. Consden, Gordon, and Martin in 1944 (5), while trying to separate products in partial hydrolyzates of wool proteins, hit on the novel idea of replacing the cellulose powder column with a sheet of paper suspended in a vapor-tight vessel.

Their technic has not undergone any fundamental changes. Although a wide variety of commercial apparatus is available, elaborate or expensive equipment is not essential and good results can be obtained with fairly simple apparatus and material. In addition to its simplicity, paper chromatography is very sensitive, capable of detecting a few micrograms of substance under suitable conditions.

A major difference between the paper technic and column forms of chromatography is that the former has nothing corresponding to the column effluent of the usual gas or liquid systems. The separated substances are detected on the paper and permanent records can be obtained. If it is desired to remove the separated materials from the paper, the appropriate parts of the paper containing the substances are cut out and each eluted separately.

GENERAL PROCEDURE

A drop of a solution of the mixure to be separated is placed on a marked position near one edge of a strip or sheet of filter paper. When the spot is dry, the paper is put in a suitable closed container with one end immersed in the solvent chosen as the mobile phase but not covering the applied spot. The paper can be supported so that the solvent runs upward (ascending chromatography) or downward (descending chromatography). By means of capillary action, the solvent percolates through the paper and moves the components of the mixture to different extents in the direction of the flow. When the solvent front has moved a suitable distance, and before the solvent front has reached the far end of the strip or sheet, the paper is removed from the apparatus, and the sheet allowed to dry. In some situations the solvent is allowed to run off the paper and is collected. This procedure is followed when one wishes to elute the compound of interest off the paper. The subtances separated are detected by various physical or chemical means depending on the nature of the substances.

APPARATUS

The basic requirement is a chamber or "tank" with a well-fitting cover to prevent escape of the solvent vapors and maintain a solvent saturated atmosphere. There is a support stand to hold up the solvent troughs and paper for descending chromatography. For ascending chromatography, only the paper needs support. The end of the paper in the troughs is held down by a glass rod or strip to keep the paper from slipping out of the trough because of increase of weight when it is saturated with the solvent. The paper passes over a glass rod that is supported parallel to the edge of the trough. This acts as an antisiphon rod to prevent siphoning of the solvent down the paper due to capillary action. By making the solvent rise a few centimeters this effect is eliminated. Gravitational flow begins when the front has passed over the glass rod. Solvent is placed in the bottom of the chamber to saturate the chamber atmosphere. In ascending chromatography the solvent at the bottom also serves as the developing solvent.

When the solvent is allowed to run in one direction, developing the chromatogram in this direction, the process is called single dimensional. By placing the solution to be investigated at a point a few centimeters from one corner of the square or rectangular sheet, allowing sufficient space for first one and then the second edge of the paper to be placed in troughs, it is possible to carry out two-dimensional paper chromatography. A one-dimensional chromatogram is first formed along one edge. After the sheet has been removed and thoroughly dried, a second solvent is run through it in a direction at right angles to the first. Substances that, because of similar R_f values, overlap in the first dimension may be separated by developing the chromatogram in the second dimension. R_f is the ratio of the distance the solute moves from origin to the distance the solvent front moves from origin.

The conditions that must be fulfilled regardless of the specific apparatus are that there must be a reservoir for the solvent of sufficient capacity to complete the chromatogram without interruption; solvent and vapor phases must always be in equilibrium; the paper must be freely suspended; large temperature changes must be avoided.

PAPER

The basic material for the paper used is a-cellulose of high purity. The original work in paper chromatography was done with Whatman 1 paper, which is still widely used. Manufacturers now supply specially selected products whose chemical and physical properties are closely controlled in manufacture to give papers of high purity and uniformity.

Papers may be classified according to thickness, flow rate, wet strength, purity, absorbancy, and prior treatment. The choice depends on the specific requirements of the separation to be made. When spots are to be located by ultraviolet absorption, requiring transmittance of ultraviolet light, thick opaque paper cannot be used very well. Thicker papers are useful for separations on a larger scale since they will accommodate more material without increase in the area of the original spot. Table 6-4 gives the speed and thickness characteristics of some types of Whatman paper.

TABLE 6-4. Relative Flow Rates of Whatman Papers

	Rate of flow		
	Fast	**Medium**	**Slow**
Thin papers	No. 4	No. 7	No. 2
	No. 54	No. 1	No. 20
	No. 540		
Thick papers	No. 31	No. 3	
	No. 17	No. 3MM	

SOLVENTS

The choice of solvent for the mobile phase naturally depends on the separation to be effected. It is usually a mixture of one main organic component, water, and additional organic solvents in lesser concentrations. The principal requirements of a solvent are that it should be reasonably cheap since large amounts may be used; mixtures of isomers such as xylene or petroleum ether, which tend to be variable in composition, should be avoided; it should be obtainable in a high state of purity; it should not be too volatile because of the need for more meticulous equilibration. On the other hand, high volatility makes for quick drying of the sheet after the run.

Some solvents when mixed in the proportions generally recommended give a two-phase system. The solvents are usually mixed in a separatory funnel so that the aqueous phase (stationary) and the solvent phase (mobile) are prepared at the same time. This preparation should not be carried out at temperatures higher than that at which the chromatography is to be conducted, otherwise the mobile solvent may separate into two phases during the chromatography procedure. In cases where the mixing of solvents yields a two-phase system, one uses the aqueous phase to saturate the atmosphere by putting it in a suitable vessel in the bottom of the tank; this is simply an economy measure since either layer will be in equilibrium with vapor with the same composition. The organic layer (mobile), which must be absolutely homogeneous, is put in the solvent trough.

Maintenance of equilibrium between the liquid and the vapor phases is most important. The object is to prevent evaporation of the solvent from the paper.

The time taken for the atmosphere to reach equilibrium with the solvent depends on the size of the tank and the volatility of the solvent. If very volatile mixtures such as those containing lower alcohols, ethers, or ketones are used, evaporation from the paper will be more rapid and equilibration is very important. The dry paper is positioned in the tank with the solvent trough empty. Vessels containing the solvent are placed at the bottom. To achieve saturation quickly, the walls of the chamber may be lined with paper soaked in the solvent. Evaporation from this large surface in the sealed tank will bring the tank atmosphere to equilibrium rapidly. It is essential that the tank be air tight. Any leakage will result in evaporation of the solvent phase from the paper, upsetting the equilibrium, causing erratic results. The tank should have a well-fitting glass cover which can be sealed with vaseline or silicone grease, care being taken that none gets on the paper or in the solvent. After the desired period to establish equilibrium, solvent is introduced into the trough with a pipet through a small hole in the lid that is then closed with a cork.

DEVELOPMENT AND DRYING

The time necessary to develop the chromatogram varies. It must be determined for each combination of mixture to be separated, solvent system, and type of paper. Whether the solute of interest is to be eluted from the paper, and the runoff collected in a container, or it is to be left on the paper, is also a determining factor. The actual development time must be determined empirically for each set of conditions.

When the development has proceeded for the desired time, the paper is removed from the tank and allowed to dry, either by hanging the paper on a rack and allowing it to dry by passive evaporation, or by suspending the paper from racks in suitably designed ovens and drying it at elevated temperatures. In either case the papers must not touch each other if more than one is being dried simultaneously.

DETECTION

The method used to detect the spots on the paper after development and drying will depend on the solutes. Commonly used methods employ color reactions with reagents, the visible colors of the compounds themselves, radioactivity scanning, fluorescence, ultraviolet absorption, and infrared absorption. The last three methods may require special instrumentation or photographic technics.

THIN-LAYER CHROMATOGRAPHY (TLC)

The first attempt at using a thin layer fixed on an inert rigid support as a chromatographic adsorbent was made by Izmailov and Schraiber in 1938 (9). They applied the sample to thinly spread layers of loose, dry powdered adsorbent as a central spot and developed the chromatogram by dropping the solvent onto the central spot so that the components moved out in concentric rings at different rates. Because the loose layer of adsorbent was easily disturbed, difficulties were encountered with the development of the chromatograms and the detection of the separated components. Meinhard and Hall (17) mixed the adsorbent with starch, which acted as a binder to hold the adsorbent to the glass. It was not until the late 1950s, when Stahl devised convenient methods of preparing plates and showed that thin-layer chromatography could be applied to a wide variety of separations, that this became a popular technic (20, 21). It was he who called the method "Dunnschicht-Chromatographie" or "thin-layer chromatography" and this designation has been widely accepted. He introduced a measure of standardization in the methodology employed, in the adsorbents used, and in the apparatus utilized for preparation of layers. These became commercially available, and applications of the method mushroomed. At the present time, prepared plates using a variety of adsorbents and supports are available from commercial sources.

OUTLINE OF METHOD

The chromatographic plate is a thin film of adsorbent of even thickness on a firm inert support. Glass plates are commonly used but other inert material, e.g., various plastics, are alternatives. Slurries of the coating material, usually made with water, are applied as thin films to the plates by spreading, pouring, dipping, or spraying, spreading with a commercial applicator being probably the most common. The plates are allowed to dry, and if necessary, activated by heating. The sample, in a volatile solvent, is applied with a pipet or syringe. After the spot has dried, the plate is placed vertically or at least in a nearly upright position in a small volume of mobile solvent in a suitable sealed chamber, and an ascending chromatogram is developed. Sometimes, for quick qualitative results, the plate is not placed in a sealed chamber but this yields less than optimal results. At the end of the run, the solvent is allowed to evaporate and the separated spots are located.

COATINGS

The stationary phase in TLC is generally called the adsorbent, even when it functions as a support for solvents in partition systems. All the types of adsorbents used in chromatographic columns may be used to prepare thin layers, but they must be of much smaller particle size than is usually used in column chromatography. They usually have a grain size of $1-25\,\mu$ and generally contain a binder such as gypsum.

The most common coatings are silica gel, alumina, kieselguhr, and cellulose. Silica gel is used more often than any other coating material. Silica gel and alumina, when activated by heat, have a high adsorption capability and can be used to separate less polar organic molecules. The developing solvents are organic, and the chromatography is of the adsorption type. Silica gel, kieselguhr, and cellulose, when properly coated with a liquid film, serve as supports for partition chromatography systems.

SOLVENTS

The solvent is determined by the principle of chromatography to be employed. For adsorption chromatography on alumina or other activated coatings the single solvents listed in the "eluotropic series" of Table 6-1 may be used. Mixtures may also be used if there is proper equilibration. Purity of the solvents is of greater importance in thin layer than in most other forms of chromatography because of the small amounts of material involved.

DETECTION

In general, the methods used to locate substances in paper chromatography are applicable to TLC. Most of the locating reagents used on paper for specific compounds or groups of compounds may be used in a similar manner in thin-layer separations. In addition, one technic can be used that is not available with paper chromatography. Corrosive spray reagents can be used to char, in an oven, most nonvolatile organic substances on inorganic layers. As a rule, the methods of visualization on thin layers are $10-100$ times more sensitive than on paper.

TLC VERSUS PAPER CHROMATOGRAPHY

The similarity in principles and technics of TLC and paper chromatography leads to a comparison: TLC claims the following advantages: greater speed; better resolution (frequently, substances that appear to be homogeneous with paper chromatography will give two or more spots with TLC); greater sensitivity; more reagents and higher temperatures may be used to detect compounds; it can be used to separate hydrophobic substances such as lipids and hydrocarbons, which are difficult to handle on paper.

Minor disadvantages of TLC as compared with paper chromatography are the following: There is greater difficulty in recording and preserving thin layer chromatograms; reproducibility of R_f values is not as easily achieved with TLC; greater cost of plates versus paper.

GAS CHROMATOGRAPHY

The systems of chromatography discussed so far have all had liquids as the moving phase. When the moving phase is a gas, the technic is known as *gas chromatography*. It has been pointed out earlier that gas chromatography may be either adsorption or partition in nature (Fig. 6-1). When the column is packed with an adsorbent such as activated carbon, and components of the mixture distribute themselves between the gas phase and the surface of the solid, the technic is *gas-solid chromatography (GSC)*. When the column is packed with an inert solid coated with a thin layer of a nonvolatile liquid as the stationary phase and separation of components is due to distribution between the gas phase and the stationary liquid according to their partition coefficients, the technic is *gas-liquid chromatography (GLC)*.

Gas-liquid chromatography was predicted by Martin and Synge in 1941 (16). In 1952, James and Martin (10, 11) published the first account of gas-liquid chromatography. At that time gas-solid chromatography was well established (8, 4, 18). It became apparent, however, that GLC offered great potential, and the publications of James and Martin marked the beginning of the rapid development of gas chromatography. Since then, many thousands of papers have been published on GLC, and gas-solid technics have been all but abandoned.

APPARATUS

GLC is the most sophisticated and highly instrumented of the various chromatographic technics. A gas chromatography instrument consists of the following basic parts: carrier gas and means for accurately regulating its flow; a device for introduction of sample; the column; a detector which can respond to low concentrations of material carried by the effluent gas; a device capable of amplifying the detector signals and presenting them to a strip chart recorder; oven with accurate temperature control to maintain column temperature. Brief comment on these parts is made below.

OUTLINE OF METHOD

The principle of GLC is that of partition chromatography. A small amount of the mixture is injected onto the column quickly. It is immediately vaporized by the elevated temperature in the chamber and is swept along by the carrier gas through the column. The column is filled with the stationary phase on an inert support. The volatilized mixture is separated into its components by the stationary phase and the components move through the column at different rates. As they are eluted from the column they come in contact with the detector, which responds with changes in electrical signal. This signal is amplified and recorded on a strip chart. The length of time between the time of injection and the time for each peak to be reported is called the *retention time,* which is characteristic of a given substance for given conditions. One can even consider the retention time as being analogous to R_f. The retention time can be used to identify the substance, and the height or area of the peaks recorded on the strip chart may be used to quantitate the components.

CARRIER GAS

Movement of the carrier gas causes the substances to move through the column. The nature of the gas employed depends on the detector used. Nitrogen, helium, argon, and sometimes hydrogen are used since they are readily available pure and dry. Ionization detectors employing radioactive materials require argon or helium. Electron detectors require nitrogen or argon. Flame ionization detectors can be used with any inert gas although nitrogen is generally used. Compressed air cannot be used because the oxygen in the mixture will oxidize the liquid stationary phase, the detector, and possibly the materials undergoing separation.

SAMPLE INTRODUCTION DEVICES

Liquids are most simply injected through a self-sealing septum. Solids are introduced as solutions in volatile solvents. Gas samples may be introduced by means of a gas-tight syringe.

The basic objective is to put the solution containing the sample into the apparatus as quickly as possible. The injection port should be hot enough to vaporize the injected sample immediately. The sample should also be placed as close to the column as possible. It is advisable that the septum used in the injection port be made of silicone rubber in order to withstand elevated temperatures.

COLUMN

The tubing used for the column is usually glass, stainless steel, aluminum, or copper. These can all be bent into a shape that fits into the oven. The diameter of the column can vary from capillary size to inches. The length of the column is also quite variable. Longer columns yield better separations but the longer the column the greater the pressure drop (the difference between the inlet and outlet gas pressures). A large pressure drop makes optimal operation difficult. Most columns are around 6 ft long, with an internal diameter of 1/8–1/4 in., and contain 5–10 g of coated support.

DETECTOR

Detectors should be sensitive, capable of rapid response directly proportional to concentration, and have a high signal-to-noise ratio. A large variety is available. The more common ones are thermal conductivity, flame ionization, and electron capture detectors. Flame ionization detectors are probably the most widely used. Specific needs and uses may require special types.

SIGNAL RECORDING

The gas chromatograph incorporates an electrometer that amplifies the signal and feeds it to a recording device such as a sensitive strip recorder. The instrument is also equipped with an attenuator that is a variable resistance potentiometer capable of reducing the signals to known fractions of the original signal. This enables one to retain the output of the detector on the chart and is particularly useful when the amount of sample being detected is so large that the peak would run off the chart unless attenuated. The linearity of the attenuator is usually good enough to permit its use in quantitative work.

OVEN TEMPERATURE

The temperature for GLC separations is an important variable. In modern instruments the column is heated by air circulated by a fan through an oven of proper design to permit rapid temperature equilibration. Close attention must be paid to the control of temperature within the system from the time of injection to the time of detection.

In isothermal chromatography, the column temperature is kept constant. Many instruments have a programmer that can increase the temperature at a linear and constant rate. As the temperature changes, the components of the sample are selectively eluted from the column depending on their differences in volatility.

SOLID SUPPORT

As in other partition chromatography systems, the solid support functions to distribute the liquid phase evenly over a large surface area. The most commonly used solid supports are diatomaceous earths. Others include glass beads, glass powder, metal helices, and Teflon.

STATIONARY PHASE

The material used for the liquid phase must be nonvolatile under the conditions of operation. Maximal operating temperatures will be determined by the stationary phase. Solids with low melting points have also been used as the stationary phase. Where these are used, there will be a minimal operating temperature dictated by the melting point, as well as a maximal operating temperature.

Every newly prepared column will "bleed" (slowly release) some stationary phase. This requires conditioning of the column by heating the column for a minimum of 10 hours at $10°-20°C$ above the temperature at which the column will be used.

Many materials have been used as stationary phases and a large variety is available commercially as well as already prepared columns.

GLC VERSUS OTHER CHROMATOGRAPHIC TECHNICS

GLC, the most sensitive of the chromatographic technics, is an extremely powerful tool. The information obtainable with relative ease in a single step from a strip chart recording is much more than can be obtained from most other chromatographic technics. GLC can separate the mixture, identify the components from the retention times, and from the peak heights or areas it can quantitate them. GLC is amenable to automation. Instruments are presently being marketed that will introduce samples automatically in succession at predetermined intervals. The peaks traced on the strip chart are capable of computer resolution. A fully automatic instrument, including quantitative analysis of components, is a distinct possibility. Unfortunately, the sophisticated instrumentation required is costly; most other chromatographic technics can be carried out with simple, inexpensive equipment.

OTHER CHROMATOGRAPHIC TECHNICS

The major systems of chromatography presently employed have been presented. There are additional materials which have found wide use. Two of these which deserve mention, however brief, are *ion exchange resins* and *gel filtration.*

ION EXCHANGE

Ion exchange consists essentially of a reversible exchange of ions between solid and liquid phases without radical change in the structure of the solid phase. It has long been known that certain natural materials such as zeolites have ion exchange capabilities, which have been put to practical use in water softening.

A wide range of synthetic resins with ion exchange properties have become available. These synthetic ion exchange resins consist of an insoluble matrix of a high polymer with polar groups attached. Polar groups confer the ion exchange properties to the resins. The groupings are of two main types: *cationic exchange resins,* which have acidic groups so that the resins have properties of a solid insoluble acid, and *anion exchange resins,* which have basic groups on basic resins. These may be subdivided according to the strength of ionizable groups. Strongly acid cation exchangers have $-OH$ (phenolic), $-SO_2OH$, and $-CH_2SO_2OH$ groups and are effective in the pH range $1-14$. Weak acid cation exchangers have the $-COOH$ radical as the functional group and are effective above pH 7. Strongly basic anion exchangers have a quaternary ammonium group as the functional group and are active over the pH range $1-14$. Resins with the groupings $-OH$, $-NH_2$, $-NHR$, and $-NR_2$ are weak anion exchangers and effective below pH 7.

The technic of using these materials in chromatographic columns is, in general, similar to that of adsorption chromatography. They have been particularly useful in separating biological material with ionizable groups such as amino acids, purine and pyrimidine bases, and alkaloids.

GEL FILTRATION

With the introduction of a cross-linked dextran polymer known commercially as Sephadex, a significant advance became possible. This is a granular, hydrophilic, water insoluble material that is nonionic and swells considerably in the aqueous solutions in which it is normally used. The material can be considered as very porous and each granule as a complex system of holes of the same order of size as molecules. When such material is used as the stationary phase of a chromatographic system, the large molecules that cannot penetrate the pores can be separated from smaller molecules which can move in and out of the pores. Thus the smaller molecules move through the chromatographic system more slowly than the larger ones.

The term "gel filtration" was suggested by Porath and Flodin (19) to describe their method of separating large molecules by Sephadex. It is sometimes referred to as "molecular sieving."

The adsorption effects on the external surfaces of the particles are usually not great as the larger molecules are eluted from the column quickly while the smaller ones are retarded as determined by their ease of diffusion into the gel particles. Gel filtration thus resembles partition chromatography.

The porosity of the gel is determined by the degree of cross linking. Commercial gels are supplied in a wide range of porosities with specific limits, i.e., the lowest molecular weight substance which will be completely excluded from the gel. The particular gel to be used for a given separation is determined by the molecular weights of the materials involved. Because gel filtration can separate large molecular weight compounds, it has found considerable application in the biologic field. Proteins, peptides, enzymes, hormones, and other compounds of biologic interest have been purified by this technic.

REFERENCES

1. BAILY KC, *The Elder Pliny's Chapters on Chemical Subjects*, Arnold, London, 1929
2. BROCKMAN H, SCHODDER H, *Ber 74*:73, 1941
3. CASSIDY HG, *Fundamentals of Chromatography* (Technique of Organic Chemistry, Vol X), Interscience, New York-London, 1957
4. CLAESSON S, *Arkiv Kemi 23A*:1, 1946
5. CONSDEN R, GORDON AH, MARTIN AJP, *Biochem J 38*:224, 1944
6. CRAIG LC, CRAIG D, *Laboratory Extraction and Countercurrent Distribution* (Technique of Organic Chemistry, Vol VIII), Interscience, New York-London, 1956
7. HAIS IM, MACEK K, *Paper Chromatography*, 3rd ed, Academic Press, New York, 1963

8. HESSE H, *Ann Chem 546*:251, 1941
9. IZMAILOV NA, SCHRAIBER MS, *Farmatsiya (Sofia)* 3:1, 1938, (*Chem Abstract 34*:855, 1940)
10. JAMES AT, MARTIN AJP, *Biochem J 50*:679, 1952
11. JAMES AT, MARTIN AJP, *Analyst 77*:915, 1952
12. KUHN R, LEDERER E, *Ber 64*:1349, 1931
13. KUHN R, WINTERSTEIN A, LEDERER E, *Hoppe-Seylers Z Physiol Chem 197*:141, 1931
14. LEDERER E, LEDERER M, *Chromatography*, 2nd ed, Elsevier, New York, 1957
15. MARTIN AJP, SYNGE RLM, *Biochem J 35*:91, 1941
16. MARTIN AJP, SYNGE RLM, *Biochem J 35*:1358, 1941
17. MEINHARD JE, HALL NF, *Anal Chem 21*:185, 1949
18. PHILLIPS CSG, *Disc Faraday Soc No. 7*:241, 1949
19. PORATH J, FLODIN P, *Nature 183*:1657, 1959
20. STAHL E, *Pharmazie 11*:633, 1956
21. STAHL E, *Chemiker Ztg 82*:323, 1958
22. TRAPPE W, *Biochem Z 305*:150, 1940
23. TSWETT M, *Ber Deut Botan Ges 24*:384, 1906, Eng. Trans., *J Chem Educ 44*:238, 1967
24. WEIL H, WILLIAMS TI, *Nature 166*:1000, 1950
25. WILLIAMS TI, WEIL H, *Nature 170*:503, 1952

Chapter 7

Immunochemical Technics

IRVING OLITZKY, Ph.D.

> "It is less hazardous to effect immunoelectrophoresis without basic knowledge of electrophoresis proper than to perform the experiment without being familiar with the multiple aspects of immunochemistry."
>
> R. J. Wieme, 1965 (162)

The author of a chapter on immunochemical technics as applied to clinical chemistry is confronted by two major problems: the outpourings of technical papers leading to a crescendo of modifications, improvements, and new applications, and the fact that other authors faced with the same task have, in general, succeeded so well. In regard to the current chapter there is a certain element of security in the fact that most readers will have that "basic knowledge of electrophoresis proper" referred to above by Wieme, and included in Chap. 5 of this volume. Since immunochemistry deals mainly with combinations of technics, immunology with chromatography, immunology with radioisotope labeling, etc., other chapters in this volume will serve as reference material. There was no attempt to compile a complete bibliography, but rather the objective was to accumulate, sort, and select published material found useful in our own laboratories. Many readers will note the lack of emphasis on the exquisitely sensitive and specific radioimmunoassay (RIA) technics. Most applications of RIA technics have been in the detection and quantitation of peptide and nonpeptide hormones, important biological compounds outside the scope of this book. A review by Berson and Yalow (19) lists 17 nonhormonal substances that can be measured by competitive radioassay where antibody is the specific reactor and 5 more which can be measured by a nonantibody specific receptor (also known as competitive protein binding). (See Chap. 27, Vitamin B_{12}.)

In Table 7-1 recommended references are listed for five basic topics pertaining to immunochemical technics. This chapter will be devoted to the fourth and fifth topics—methodology and application.

IMMUNOCHEMICAL TECHNICS: METHODOLOGY

The use of reagent antibody in clinical chemistry imposes the term *antigen* on those substances of biological importance that are objects of search (qualitative) and estimation (quantitative). This chapter will consider reagent

TABLE 7-1. Sources of Basic Information for Immunochemical Technics Used in Clinical Chemistry

Subject	References
Terminology of immunochemistry: What are antigens? antibodies?	29, 35 (Glossary); 40, 73 (Chap 1)
Antibody as reagent: basic antigen–antibody reactions	33 (Chap 1); 73 (Chaps 2–5); 86
Production and purification of reagent antibody	16, 29, 33 (Chap 1); 40, 41, 48, 64, 67, 73 (Chap 51); 86 (Chap 13); 109 (List of commercial sources); 161 (Pt A, Chaps 1-4 and Appendix II); 162 (Chap 5)
Immunochemical technics: general theory and methodology	9, 14, 23 (Pt 1); 24, 25, 32, 33 (Chap 1); 35, 61, 63, 64, 65, 67, 68, 69, 73 (Chaps 2, 8, 33); 86, 91, 97, 103, 116, 118, 119, 120 (Chap 1); 139, 154, 161 (Sec 3, Chap 19); 162 (Chap 5); 166
Applications in clinical chemistry: general	14, 23 (Pts II and III); 25, 31, 35 (Sec 4); 68, 69, 86 (Chap IV); 109, 117 (Appendix III); 120 (Chaps 2 and 3); 129, 130 (Chap 13); 139, 154, 166

antibody in terms of detection, identification, and quantitation of those nonhormonal biological substances of importance to clinical chemists.

It is perhaps characteristic of the cyclic nature of scientific discovery and rediscovery that the oldest technic, immunoprecipitation in liquids (IPL), languished as an analytical tool only to emerge relatively recently as a special application, the radioimmunoassay (RIA). Even more recently IPL has been adapted to a fully automated system for detection and quantification of certain serum proteins.

Immunoprecipitation in gels, described at the beginning of this century, burgeoned as an analytical tool in the late 1950s and early 1960s. For example, Ouchterlony (117) surveyed "about twelve hundred papers" that appeared between 1956 and 1960.

IMMUNODIFFUSION

SINGLE IMMUNODIFFUSION IN ONE DIMENSION

Often referred to as the Oudin technic (118), single immunodiffusion (SID) has basically remained a qualitative method. Antibody (Ab), or less frequently antigen (A), is incorporated into a gel, usually agar or agarose, which is in direct contact with a solution containing the other. A or Ab, the material in the liquid, diffuses into the gel forming a precipitin line at optimal A and Ab concentrations as shown in Figure 7-1(a).

Measurement can be made of the distance of the precipitin line from the interface between liquid and gel. Under proper conditions the distance is related to the concentration of the diffusing substance, e.g., antigen. The basic technic and its modifications are generally performed in long narrow-base glass tubes as reaction vessels. Various modifications, including double immunodiffusion (DID) have been described by Kwapinski (86).

DOUBLE IMMUNODIFFUSION (DID)

DID involves migration of both antigen and antibody: Wells are cut in a gel layered on a glass or plastic surface, and one is filled with A and the other with Ab. Diffusion of both A and Ab results in a precipitin line at the zone of optimal proportions [see Fig. 7-1(b)].

FIG 7-1. (a) Schematic diagram of single immunodiffusion (SID). In this case the reaction is said to be occurring in one dimension. (b) Diagram illustrates one of the many uses of double immunodiffusion (DID), i.e., comparison of antigens for similarity or dissimilarity. Well A contains two antigenic components, x and y. Well B contains only antigen y. Well C contains antibodies specific for both x and y. (c) Diagram illustrates the use of DID to quantitate antigen (see text and Ref. 44).

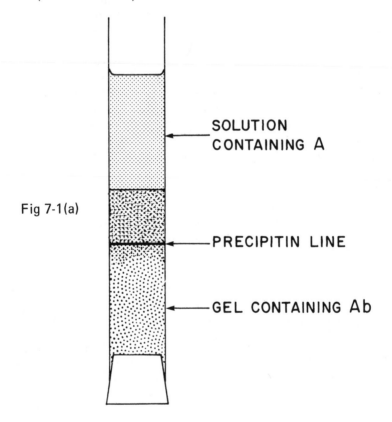

Fig 7-1(a)

SOLUTION CONTAINING A

PRECIPITIN LINE

GEL CONTAINING Ab

A useful micro modification of DID in one dimension employs a 50 mm length capillary tube with 1 or 2 mm inside dimension (44). A 10 mm plug of 1% agar serves as the reaction area into which *A* and *Ab* diffuse [see Fig. 7-1(c)]. The relative intensity of the precipitin band or the distance from the gel-liquid interface is compared with bands generated by solutions with known antigen content. Better quantitation is achieved by photographing the precipitin lines and scanning the negative with a densitometer.

A recently described DID method for the quantitation of serum proteins is capable of detecting albumin at a level of 4 μg/ml (163). Measurement of albumin was achieved in the range of 0.5–30 mg/ml with a between-run precision of 5% (1 C. V.). This order of precision is achieved with difficulty using the single radial immunodiffusion technics described below and in Chap. 16.

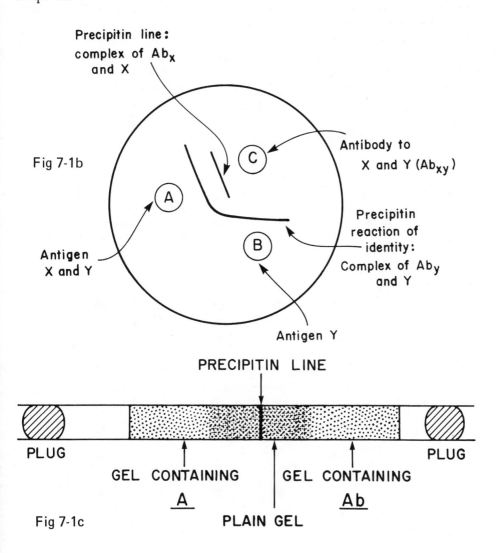

Precipitin line:
complex of Ab_x and X

Fig 7-1b

Antibody to X and Y (Ab_{xy})

Antigen X and Y

Precipitin reaction of identity: Complex of Ab_y and Y

Antigen Y

PRECIPITIN LINE

PLUG

PLUG

GEL CONTAINING A

GEL CONTAINING Ab

Fig 7-1c

PLAIN GEL

SINGLE RADIAL IMMUNODIFFUSION (SRID)

As of this writing SRID is the most widely used immunochemical technic for the estimation of proteins in body fluids. Small wells are made in antibody-containing gel and the wells filled with antigen (unknowns or standards). The antigens diffuse out into the gel to form circular precipitin lines. The diameter of the precipitin ring or the area of the circle within the ring is proportional to the concentration of antigen (see Chap. 16 for details and Fig. 7-2).

The popularity of the SRID method is directly related to the commercial availability of kits consisting of reaction plates covered with a 1–3 mm thick agar or agarose film purportedly containing optimum concentration of specific antibody. Antibody content is purposely reduced for low level or "pediatric" plates; however, the conditions should always be at slight antibody excess. Plates are available with 6, 24, or 36 well configurations. Wells are filled with patient's sera or standards that are supplied as a pooled serum with known concentrations of specific proteins. Standards should be used at several dilutions for the generation of the standard curve.

Specific procedures for SRID are based on either the Mancini method (98), in which diffusion and formation of precipitin lines are allowed to continue to completion, or the Fahey method (49), in which measurements are made in 4–24 hours. Kalff (75) and Arvan (10) have discussed the theoretical considerations and differences between the methods.

FIG 7-2. Schematic diagram of quantitation of serum proteins by SRID. Specific antibody, e.g., anti IgG, is incorporated into agar and wells are filled with standards or unknowns. Size of precipitin ring (diameter or area) is proportional to amount of IgG in the standards or unknowns. The four-point standard curve is generated from measurements of precipitin rings in wells 1, 2, 3, and 4.

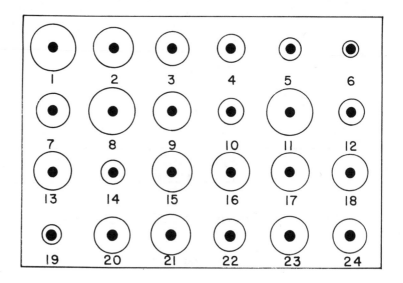

Modifications of the basic SRID method are numerous, and some are designed to minimize use of expensive reagents: (a) Use of rigid polyvinyl chloride masks with drilled holes performing as wells (51). (b) Incorporation of antigen or antiserum into small agar disks (101, 131) or on paper disks (159). (c) Antiserum sprayed (159) or layered (61), on the plain gel. (d) Use of cellulose acetate as reaction medium (110, 122, 159). (e) A radioactive SRID method described by Rowe (131, 132) is 60-fold more sensitive than conventional SRID and makes possible the estimation of IgE in normal serum (see Table 7-2 and Chap. 16).

Since the precipitin lines are so important in immunoprecipitation-in-gel technics, much effort has gone into methods by which the precipitin line is more easily visualized, measured, preserved, etc. One can very often at least double the sensitivity of a method by accentuating the lines. In Table 7-2 are listed some of the more useful technics for this purpose.

TABLE 7-2. **Aids in Visualizing and/or Preserving the Precipitin Lines in Immunoprecipitation-in-Gels Technics**

Principles	References
Dark field illumination with[a] or without photography	133
Antibody tagged with 1% haematoxylin	76
Acetic assay bath to intensify lines	7
Reverse staining: agar darkened by Nigrosin	88
Silver staining	77
Antibody labelled with fluorescein isothiocynate. Plates illuminated with wood lamp	57
Reaction chamber as negative in enlarger	144

[a]Principle of commercially available fixed focus enlargers using Polaroid film.

In spite of certain problems, SRID, when used judiciously, is a very valuable clinical and research tool for the estimation of antigen.

IMMUNOELECTROPHORESIS

QUALITATIVE IMMUNOELECTROPHORESIS

Immunoelectrophoresis (IEP) combines separation based on electrophoretic mobility and immunoprecipitation based on immunologic specificity. The macro method of Graber and Williams (62) and the Scheidegger micro method (136) have served as launching pads for myriads of modifications. As in simple electrophoresis and earlier immunodiffusion technics other supporting media were soon advocated: paper (152), cellulose acetate foil (108, 138), agar overlay on Mylar and Cronar (27, 28), acrylamide gels (38, 156), nitro cellulose membranes treated with Tween 60 (124), and stabilized gelatin (157). Membrane reaction strips conserve reagents and in some cases, result in sharper precipitin lines.

The micro IEP method is the most widely used. Microscope slides, prepared either with commercially available agar gel cutters or templates serve as reaction plates (36). It was claimed that by flooding the plate with antiserum the size, shape, position, and homogeneity of the antigen spot could be visualized (163). The usual method involves the diffusion of antibody out of a long rectangular trough. Systems suitable for clinical laboratories have been adapted to existing electrophoresis equipment. Purified agar can even be purchased in tablets. More than 40 distinct precipitin lines are seen when mammalian sera are subjected to IEP under ideal conditions. [See Fig. 7-3(a).]

It has been established that routine IEP is not a reliable *quantitative* procedure (89), but marked deficiencies of proteins can be established by methods such as described by Ironside (70).

QUANTITATIVE IMMUNOELECTROPHORESIS

A unique modification of IEP described by Ressler (127) served as the basis for many subsequent modifications and applications of IEP as a dependable quantitative method. These more complex technics basically consist of an initial electrophoresis followed by a second electrophoresis *into an antibody-containing agar* (replacing the double diffusion step of IEP). Many names have been given to the method designated by Laurell (89, 90, 90a) as *antigen-antibody crossed electrophoresis*. One name, *rocket electrophoresis*, refers to the shape of the precipitin curves, the heights of which are roughly proportional to the logarithm of antigen concentrations [Fig. 7-3(b)].

A recent modification of the Laurell method that utilizes a smaller reaction plate achieves significant savings of reagent antibody (39).

Afonso (2, 3, 4) described experiences in the analysis of serum proteins with a method that eliminates the antibody diffusion gradient inherent in IEP. Antibody was either incorporated into the agar gel before electrophoresis (as per Ressler), or was applied as a uniform liquid layer on the gel surface after electrophoresis. In subsequent reports (5, 6) Afonso defended his system against charges of nonlinearity. Not related to this controversy, Bednarik (17) substituted acrylamide gel into the Afonso technic and found it to be a useful modification.

Methods to identify and quantitate the many lines seen in IEP have been described (85). The maze of lines produced would preclude this methodology for routine use.

The Laurell method continues to be compared and contrasted with SRID primarily in the parameters of sensitivity and precision (30, 75, 96). Electroimmunodiffusion (EID) is emerging as a more acceptable descriptive term for the method whereby antigens previously separated by electrophoresis are subjected once again to electrophoresis into antibody-containing agar. The many variations of IEP, including EID, have been recently reviewed (52). Weeke (160), using EID as a quantitative tool, determined 18 individual serum proteins.

The quantitation of immunoglobulins (Ig) in dilute fluids such as cerebrospinal fluid is amenable to EID (104, 123). In another method applicable to such fluids carbamylation (potassium cyanate treatment) of the specimens resulted in improved accuracy and two-fold greater sensitivity than SRID (125). A 4 hour method for quantitative separation of 30 proteins with a sensitivity of 30 mg/ml was claimed by Firestone and Aronson (53). Because of its relatively low molecular weight and high negative charge, albumin is easily assayed in the EID system (134).

FIG 7-3(a). Photograph of a completed IEP run. Patients serum was placed in upper well and a normal serum in lower well. After electrophoresis the trough was filled with a limited spectrum antihuman antiserum produced in goats. (b) "Rocket" electrophoresis (EID). Characteristic shape of precipitin line results when solution containing protein is electrophoresed into a gel containing *Ab*.

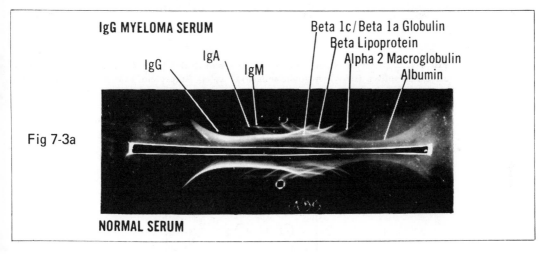

IgG MYELOMA SERUM

Beta 1c/Beta 1a Globulin
Beta Lipoprotein
Alpha 2 Macroglobulin
Albumin

IgG   IgA   IgM

Fig 7-3a

NORMAL SERUM

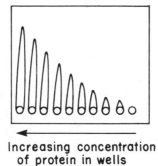

Fig 7-3b

Increasing concentration
of protein in wells

An interesting method to conserve reagent antibody was described by Gill, Fischer, and Holleman (58). Small, square reaction plates of Mylar-backed cellulose acetate membranes are submerged in diluted antiserum and allowed to hydrate. Commercially available or specially devised templates were used for application of specimen. Many of the proteins could be estimated at the end of 45 min with good precision (C. V. 5%).

COUNTER ELECTROPHORESIS

Used primarily as a qualitative procedure, counter electrophoresis (CEP) takes advantage of the fact that endosmotic flow in a negatively charged gel results in a net movement of the immunoglobulins in a direction opposite to the electrophoretic migration of proteins with a lower isoelectric point. Thus, during electrophoresis, the antigen is moving relatively fast toward the anode while the immunoglobulins move slowly toward the cathode (see Fig. 7-4). In the usual procedure rows of wells to be filled with sample are closer to the cathode than are the rows of wells to be filled with specific antiserum. The method has been used successfully for detection of alpha$_1$-fetoprotein (see Chap. 16).

FIG 7-4. Representation of counter-electrophoresis (CEP) plate for detection of hepatitis-associated antigen (HAA) (Australia antigen) or alpha$_1$ fetoprotein. Row A wells are filled with patients serum or quality control speciments. Row B wells are filled with specific antibody. Wells 1—8 contained the following: (1) positive control, (2) negative control, (3) strongly positive sample, (4) weakly positive sample, (5) negative sample, (6) positive sample, (7) nonspecific precipitate, and (8) moderately positive sample.

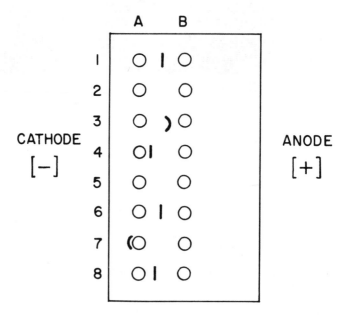

IMMUNOPRECIPITATION IN LIQUIDS (IPL)

QUALITATIVE IPL

As mentioned earlier, qualitative IPL was one of the first immunologic technics to be utilized outside the field of diagnostic or taxonomic microbiology. It has many uses in forensic medicine, e.g., identification of blood stains, as well as in public health entomology, whereby the host range of blood sucking insects is determined. Nevertheless, the IPL technic has been largely replaced by immunodiffusion in agar gel (20) since lines of identity can be used to verify the accuracy of identification.

QUANTITATIVE IPL

Broad applications of quantitative IPL in clinical chemistry awaited the development of the radioimmunoassay and automation. Clausen and Guttler (32) estimated albumin, β-lipoprotein, and IgG in spinal fluid by IPL. The amount of precipitate, which is related to the concentration of antigen, can be estimated by turbidimetric measurements (a basis for current automated methods) or by estimation of the total protein of the precipitate. Killingsworth and associates after working with automated IPL (see below) recently described their experiences with manual nephelometric methods (81, 135).

RADIOIMMUNOASSAY (RIA) AS A SPECIAL CASE OF IPL

RIA methods are 100–1000-fold more sensitive than SRID technics and thus are particularly useful in low protein body fluids such as saliva, spinal fluid, ocular fluids, etc. The literature of RIA and its applications is voluminous. A detailed and critical study of existing RIA methods citing 162 references was made by Aubert (11). The reader is also directed to a review by Grant and Butt (64) and the status report by Berson and Yalow (19) previously mentioned.

Illustrative of the wide range of applications of RIA in clinical chemistry is the estimation of immunoglobulins (Ig) (92, 99) and morphine (147). These same reports also illustrate the technic of insoluble antibody, frequently termed solid phase radioimmunoassay, in which reagent antibody globulin is chemically coupled to insoluble bromacetyl cellulose (99) or Sepharose 2B[3] (92). The RIA assay for morphine is typical of those situations in which the compound in question serves as a hapten and must be attached to a larger molecule for success in antibody production. Carboxymethylmorphine was coupled to bovine serum albumin in the presence of water soluble carbodiimide and injected with adjuvant into rabbits. Sensitivity of the method was such that 0.5 ng of morphine was enough to displace 20% of the tritiated dihydromorphine from its bond to reagent antibody.

AUTOMATION OF IPL

The estimation of serum proteins by continuous flow analysis has been reported (79, 80). The basic AutoAnalyzer system is used except for adaptation of a fluorimeter to function as a sensitive nephelometer. The ease by which serum proteins can be estimated suggests that routine screening may be in order for various proteins: IgA, IgG, IgM, C3, transferrin, alpha$_1$-antitrypsin, albumin, and haptoglobin. Although the sampling rate is 50 or 100/hour in the automated system, the analytical rate is half that because a sample blank is necessary to correct for turbidity. Preliminary estimates of reagent costs suggest substantial savings over the cost of SRID (using commercial plates). A coefficient of variation of 1.65% for C3 (79) and 5% for IgA, IgG, IgM (80) should do much to improve the diagnostic value of body fluid protein determinations by immunochemical means.

MISCELLANEOUS METHODS

CENTRIFUGATION OR CHROMATOGRAPHY AND ELECTRO-IMMUNODIFFUSION (EID)

To analyze serum Cordoba and associates (34) combined centrifugation in sucrose gradients with subsequent IEP of 25 consecutive fractions of 12 drops each. Electrophoresis was carried out on large gel-covered plates that were then flooded with antiserum. Characteristic patterns were related to various diseases.

High resolution was claimed with a method combining disk electrophoresis with immunodiffusion (95); the antiserum is incorporated into the gel in which disk electrophoresis is carried out. The migration of larger complexes of antigen and antibody is retarded and the noncombined antigen migrates to its usual position. Thus the presence of an antigen in the original body fluid can be proved by use of a specific antiserum.

Another method useful in body fluids with normally low protein concentration combines thin layer gel filtration through Sephadex G-200 (superfine) with immunological characterization (8). The proteins forming the precipitin bands are grouped according to sedimentation coefficients.

PARTICLES SENSITIZED WITH ANTIGEN OR ANTIBODY

The use of particles, e.g., bentonite, latex, or erythrocytes (RBC), coated with antigen provides an extremely sensitive, widely used method for detecting antibody. When RBC are used the direct procedure is called *hemagglutination*. A reverse procedure, *hemagglutination-inhibition (HI)*, has been used for the estimation of immunoglobulins (71): Formalized, tannic acid treated RBC are coated with antigen. When the sensitized cells are reacted with a specific antiserum hemagglutination occurs. Prior exposure of the antiserum to a test solution, e.g., serum containing antigen, will neutralize the agglutinating capacity of the antiserum and hemagglutination will not occur. Using varying dilutions of the test serum antigen can be estimated by comparing with a standard of known antigen concentration. Although much more sensitive than SRID the method is considerably less precise. Hemagglutination and HI lend themselves to continuous flow automation. The *automated passive hemagglutination-inhibition (APHI)* method described by Sturgeon and associates (150) reportedly measures insulin, ferritin, transferrin, ceruloplasmin, fibrinogen, and some of the immunoglobulins with a sensitivity of $0.1-0.4\,\mu g/ml$. Other applications to plasma proteins have also been described (114).

Fibrinogen antibody bound to latex particles is the basis for a commercial kit for measuring fibrinogen (26). Time for agglutination to occur is related to fibrinogen concentration by means of a standard curve.

COMPLEMENT FIXATION

In what will probably remain a research method, Barnett (12) reported on the use of complement fixation (CF) for the estimation of Ig and L chains. The sensitivity for IgA and IgM ranged from $1.5-3.5\,mg/100\,ml$, which is adequate for research studies on synovial fluid, spinal fluid, and tears.

ENZYME-LINKED ANTIGEN (ELA)

Engvall *et al* (45, 46) described a sophisticated procedure for measuring IgG which they called *enzyme-linked immunosorbent assay* (ELISA). This is essentially the same approach used by Van Weeman and Schurrs (158) to eliminate the problems of radionuclides as labelling agents in RIA. They linked the antigen human chorionic gonadotropin to the enzyme horse radish peroxidase. It is reasonable to expect that ELA technics will be adapted to diagnostic clinical chemistry.

BACTERIOPHAGE-LINKED ANTIGEN (BLA)

An ingenious use of a living reagent was described by Maron and Bonavida in their assay for human lysozyme in body fluids (100). Gluteraldehyde was used to conjugate purified lysozyme with T_4-bacteriophage. An antilysozyme serum produced in rabbits inhibits the phage activity of the antigen-phage reagent. Exogenous lysozyme can be estimated on the basis of the degree to which the inhibitory effect of antibody is neutralized.

IMMUNOCHEMICAL TECHNICS: APPLICATIONS

Some of the applications of immunochemical technics for clinical chemistry are shown in Table 7-3. The abbreviations used have in all cases been defined in the preceding text.

TABLE 7-3. A Summary of Applications of Immunochemical Technics in Clinical Chemistry.[a]

Biological compound	Body fluid	Immunochemical technic and references
Albumin	Serum	SRID (14, 149); EID (58, 123); AIP (method in Ref 80 applicable)
	CSF	EID (155); IPL (32)
	Urine	EID (123); RIA (105)
Alpha$_1$-antitrypsin	Serum	SRID (see Chap 16)
	CSF and urine	EID (125)
Alpha$_1$-glycoprotein (orosomucoid)	Serum	DID (82); SRID (15, 149); EID (30, 58)
Alpha$_1$-fetoprotein	Serum	See Chap 16
Alpha$_2$-HS-globulin	Serum	SRID (14, 149); EID (30, 58)
Alpha$_1$-lipoprotein	Serum	SRID (14, 149); EID (30, 58)
Alpha$_2$-macroglobulin	Serum	DID (82); SRID (1, 14, 149); EID (30, 58, 123)
	Urine	EID (123)
Barbiturates	Serum and urine	RIA (148)
Beta 1c/1a (C3)	Serum	SRID (149); EIP (107); EID (30, 58); AIP (79)
Beta-lipoprotein (Beta$_1$ lipoprotein)	Serum	SRID (14, 149); EID (30); IPL (18)
	CSF	IPL (32)
Beta$_2$ microglobulin	Serum, urine, colostrum, and saliva	RIA (47)
Blood stains		DID (20)
C$_1$ esterase inhibitor	Serum	SRID (15)
Carbonic anhydrases (B, C)	Blood	EID (112)
Carbonic anhydrases I, II	Blood	RIA (66)
Ceruloplasmin	Serum	DID (82); SRID (14, 149); APHI (114, 150); EID (30, 58); IPL (137)
Cholinesterase	Serum	SRID (14, 143, 149)
Chymotrypsinogen	Serum	RIA (153)
Digoxin	Serum and urine	RIA (115)
Esterase (acid stable)	Gastric juice	SLID (126)
Erythropoietin	Urine	RIA (54)
Fibrin degradation products	Serum and urine	HI (43)
Fibrinogen	Plasma	HI (55); DID (165); SRID (22)
	Serum	APHI (114, 150)
Fibrinopeptide A	Plasma	RIA (113)
Folate ("immunoreactive")	Serum	RIA (37)
Fructose 1, 6 diphosphatase	Tissue	RIA (84)
Haptoglobin	Serum	DID (82); SRID (14, 149); EID (30, 58); AIP (method in Ref 80 applicable); IPL (83)
	CSF	SRID (13)
Hemoglobin A$_2$	Blood	SRID (151)

(continued)

TABLE 7-3. A Summary of Applications of Immunochemical Technics in Clinical Chemistry.[a] (continued)

Biological compound	Body fluid	Immunochemical technic and references
IgA	Serum	SRID (see Chap 16); EID (58, 122); HI (71); AIP (80); RIA (50); APHI (114, 150); CF (12); IPL (81)
	CSF	SRID (13); CF (12); EID (125); IPL (135)
	Urine	EID (125)
	Saliva	EID (134); CF (12)
	Gastric fluid	EID (102)
	Tears	CF (12)
IgD	Serum	SRID (see Chap 16); HI (71); APHI (114)
IgE	Serum	RIA (59); APHI (114)
IgG	Serum	SRID (see Chap 16); EID (58, 122, 134); AIP (80); APHI (114, 150); HI (71); CF (12); IPL (81); ELIA (46)
	CSF	SRID (13); IPL (32, 135); EID (155); SLID (87); IPL (135)
	Gastric fluid	EID (102)
IgM	Serum	SRID (see Chap 16); EID (58, 122, 134); HI (71); APHI (114, 150); CF (12); AIP (80); IPL (81)
	CSF	SRID (13); IPL (135)
	Saliva	EID (134)
Lipoprotein (LP-X)	Serum	IEP (140)
Lysozyme (muramidase)	Serum	EID (72); BLA (100)
	Urine	EID (72)
	Tears and saliva	BLA·(100)
Morphine	Serum	RIA (147)
Myoglobin	Serum	SRID (128)
	Urine	SRID (128); HI (93)
	Muscle	SRID (74)
Ouabain	Plasma and urine	RIA (145)
Plasminogen (and plasmin)	Serum	SRID (14, 149); EID (58)
	Plasma	RIA (120)
Properdin	Serum	SRID (121)
Prostatic acid phosphatase	Prostatic fluid	EIP (121)
Prothrombin	Plasma	EID (56)
Retinal-binding protein	Plasma	RIA (146)
Transferrin	Serum	SRID (14, 42, 149); EID (30, 58, 123); APHI (114, 150); AIP (42)
Trypsin	Serum	SRID (141); RIA (153)
	Duodenal juice	RIA (153)
Thyroxin-binding globulin	Serum	EID (111); RIA (94)
Thrombosthenin	Platelets	SRID (21)

[a]Some references mentioned in text are not included in the table.

The broad range of applications shown in Table 7-3 may lead to the conclusion that immunochemical technics are capable of solving most of the analytical chemistry problems faced by the clinical chemist. This may very well be true but not until a reasonable solution is available for the problems associated with reagent antibody. It would be a disservice to the chemist if the immunologist did not expose these problems.

It may seem curious that a major area of investigation and applications with reagent antibody is the characterization and estimation of immunoglobulins since these are the antibodies of the patient. Certainly the facility with which one can produce reagent antibody against the immunoglobulins (relatively large proteins) is a factor. Production of antisera against human peptide hormones is much more difficult because of the low molecular weight of the antigen. Of course, the inherent lack of antigenicity of the substance in question is not an absolute barrier to the production of reagent antiserum. The haptenic qualities of the compound are an important consideration along with the availability of a suitable method for coupling the hapten to a protein with good antigenic characteristics. Obviously the coupling must not destroy those molecular arrangements that are associated with specificity.

Monospecific antisera are sometimes produced only with difficulty. If a pure antigen is not available for immunization the resulting antiserum may have to be absorbed. Furthermore, the definition of "pure" may need some revision for the clinical chemist. Purity of antigen may sometimes be achieved by dealing only with that portion of the molecule that confers absolute specificity. This is the procedure followed, for example, in producing specific anti-IgM antibody. Instead of immunizing with the entire IgM molecule only the heavy chain portion, which provides the specific differences among the five immunoglobulins, is used. More problems and solutions can be found in the references in Table 7-1.

Why should immunochemical technics be used rather than the chemical or physicochemical technics that have served the clinical chemist so well in the past? Many factors should be involved in this decision-making process:

Sensitivity. One can cite many examples where the chemical technics are not sensitive enough for certain biological compounds in body fluids or tissue. Myoglobin is an excellent example.

Specificity. It should be remembered that the immunological parameter is just another way of characterizing a biological substance. If separations based on electrophoretic mobility are sufficient for clinical purposes, no further value may be gained from immunoelectrophoresis. Myoglobin can again be used as an example, i.e., a potent, specific antimyoglobin antiserum can clearly distinguish between hemoglobin and myoglobin and estimate concentration of myoglobin *in the presence of hemoglobin*, whereas separation and identification cannot be achieved by electrophoresis unless myoglobin is present in large amounts.

Precision. Here the immunochemical technics do not fare as well, but again the critical question is the usefulness of the assay.

Cost. One should consider reagent cost and not neglect the question of availability. Some immunochemical technics need expensive instrumentation, e.g., the RIA. Labor cost analyses must certainly consider the kind of personnel needed, how much formal training?, experience?, etc.

Speed of availability of results. Clearly of importance to the clinical chemist in a hospital environment is the applicability of immunochemical technics to "stat" situations.

REFERENCES

1. ADHAM NF, DYCE B, HAVERBACK BJ: *Gastroenterology 62*:365, 1972
2. AFONSO E: *Clin Chim Acta 10*:114, 1964
3. AFONSO E: *Clin Chim Acta 14*:63, 1966
4. AFONSO E: *Clin Chim Acta 14*:567, 1966
5. AFONSO E: *Clin Chim Acta 17*:138, 1967
6. AFONSO E: *Clin Chim Acta 18*:95, 1967
7. AFONSO E: *Clin Chim Acta 10*:192, 1964
8. AGOSTONI A, VERGANI C, LOMANTO B: *J Lab Clin Med 69*:522, 1967
9. ALADJEM F, PALDINO RL, PERRIN R, CHANG F-W: *Immunochemistry 5*:217, 1968
10. ARVAN DA: *Laboratory Management 7*:21, Aug 1969
11. AUBERT ML: *J Nucl Biol Med 14*:4, 1970
12. BARNETT EV: *J Immunol 100*:1093, 1968
13. BAUER H, GOTTESLEBEN A: *Intern Arch Allergy Appl Immunol 36*:643, 1969
14. BECKER W, RAPP W, SCHWICK HG, STÖRIKO K: *Z Klin Chem Klin Biochem 6*:113, 1968
15. BECKER W, SCHWICK HG, STÖRIKO K: *Clin Chem 15*:649, 1969
16. BECKER W: *Clin Chim Acta 23*:509, 1969
17. BEDNARIK T: *Clin Chim Acta 17*:132, 1967
18. BERGQUIST LM, CARROLL VP Jr, SEARCY RL: *Lancet 1*:537, 1961
19. BERSON SA, YALOW RS: *Radioimmunoassay: A Status Report*, Chap. 30 in Immunobiology, edited by Good and Fisher. Sinauer Associates, Conn., 1971
20. BETTS A, SEWALL EL: *Am J Clin Pathol 43*:535, 1965
21. BOOYSE FM, ZSCHOCKE D, HOVEKE TP, RAFELSON ME: *J Lab Clin Med 79*:344, 1972
22. BRITTIN GM, RAFINIA H, RAVAL D, WERNER M, BROWN B: *Am J Clin Pathol 57*:89, 1972
23. BURTIN P: *Immuno-electrophoretic Analysis*, edited by Grabar and Burtin. New York, Elsevier, 1964, p. 94
24. CANNON DC: *Postgrad Med 46*:55, August 1969
25. CANNON DC: *Postgrad Med 46*:55, September 1969
26. CASTELAN DJ, HIRSH J, MARTIN M: *J Clin Pathol 21*:638, 1968
27. CAWLEY LP, SCHNEIDER D, EBERHARDT L, HARROUCH J, MILLSAP G: *Clin Chim Acta 12*:105, 1965
28. CAWLEY LP, EBERHARDT L, SCHNEIDER D: *J Lab Clin Med 65*:342, 1965
29. CHASE MW, BENEDICT AA: *Methods in Immunology and Immunochemistry*, edited by Williams CA, Chase MW. New York, Academic, 1967, Chap 2
30. CLARKE HGM, FREEMAN T: *Clin Sci 35*:403, 1968
31. CLAUSEN J: *Sci Tools 10*:29, 1963

32. CLAUSEN J, GÜTTLER F: *Scand J Clin Lab Invest 19:Suppl* 100:*12*, 1967
33. COOMBS RRA, GELL PGH: *Clinical Aspects of Immunology*, edited by Gell PGH Coombs RRA, 2nd ed. Philadelphia, F. A. Davis, 1968, Chap 1
34. CÓRDOBA F, GONZÁLEZ C, RIVERA P: Clin Chim Acta 13:611, 1966
35. CROWLE AJ: *Immunodiffusion.* New York, Academic, 1961
36. CROWLE AJ, LASKER DC: *Clin Med 59*:697, 1962
37. DA COSTA M, ROTHENBERG SP: *Brit J Haematol 21*:121, 1971
38. DARCY DA: *Clin Chim Acta 21*:161, 1968
39. DAVIES DR, SPURR ED, VERSEY JB: *Clin Sci 40*:411, 1971
40. DEUTSCH HF:in Methods in Immunology and Immunochemistry, edited by Williams CA, Chase MW. New York, Academic, 1967, Vol I, Chap 1, Sec 1A
41. DEUTSCH HF: in Ref. 40, Chap 3, Sec 3,A,2
42. ECKMAN I, ROBBINS JB, VAN den HAMER CJA, LENTZ J, SCHEINBERG IH: *Clin Chem 16*:558, 1970
43. EKERT H, BARRATT TM, CHANTLER C, TURNER MW: *Arch Disease Childhood 47*:90, 1972
44. EL-MARSOFY MK, ABDEL-GAWAD Z: *Experientia 18*:240, 1962
45. ENGVALL E, PERLMAN P: *Immunochemistry 8*:871, 1971
46. ENGVALL E, JONSSON K, PERLMAN P: *Protein Structure 251*:427, 1971
47. EVRIN PE, PETERSON PA, WIDE L, BERGGARD I: *Scand J Clin Lab Invest 28*:439, 1971
48. FAHEY JL: in Ref. 40, Chap 3, Sec 3,A,3
49. FAHEY JL, McKELVEY EM: *J Immunol 94*:84, 1965
50. FAULKNER W, BORELLA L: *J Immunol 105*:786, 1970
51. FEINBERG JG: *Nature 201*:631, 1964
52. FEINBERG JG, HILL CW: *Intern Arch Allergy Appl Immunol 33*:120, 1968
53. FIRESTONE HJ, ARONSON SB: *Am J Clin Pathol 52*:615, 1969
54. FISHER JW, THOMPSON JF, ESPADA J: *Israel J Med Sci 7*:873, 1971
55. FOX FJ Jr, WIDE L, KILLANDER J, GEMZELL C: *Scand J Clin Lab Invest 17*:341, 1965
56. GANROT PO, NILEHN JE: *Scand J Clin Lab Invest 21*:238, 1968
57. GHETIE V: *Immunochemistry 4*:467, 1967
58. GILL CW, FISCHER CL, HOLLEMAN CL: *Clin Chem 17*:501, 1971
59. GLEICH GJ, AVERBECK AK, SWEDLUND HA: *J Lab Clin Med 77*:690, 1971
60. GOLDER S, LOPEZ V, KELLER H: *Z Klin Chem Klin Biochem 7*:448, 1969
61. GRABAR P: *Methods Biochem Analy 7*:1, 1959
62. GRABAR P, WILLIAMS CA Jr, *Biochim Biophys Acta 10:*193, 1953
63. GRANT G: *Clin Chim Acta 22*:31, 1968
64. GRANT GH, BUTT WR: *Advan Clin Chem 13*:383, 1970
65. GRANT GH: *J Clin Pathol 24*:89, 1971
66. HEADINGS VE, TASHIAN RE: *Biochem Genetics 4*:285, 1970
67. HIRSCHFELD J: *Sci Tools 7*:18, 1960
68. HIRSCHFELD J: *Sci Tools 8*:17, 1962
69. HITZIG WH: *Protides of the Biological Fluids*, 8th Colloquim 1960, edited by Peeters H. New York, Elsevier, 1961, p 83
70. IRONSIDE P: *J Clin Pathol 22*:242, 1969
71. JOHANSSON SGO, HOGMAN CF, KILLANDER J: *Acta Pathol Microbiol Scand 74*:519, 1968
72. JOHANSSON BG, MALMQUIST J: *Scand J Clin Lab Invest 27*:255, 1971
73. KABAT EA, MAYER MM: *Experimental Immunochemistry*, 2nd ed. Springfield, Ill., C. C. Thomas, 1967

74. KAGEN LJ, CHRISTIAN CL: *Am J Physiol 211*:656, 1966
75. KALFF MW: *Clin Biochem 3*:91, 1970
76. KATSH S, MATCHAEL J: *Nature 194*:1186, 1962
77. KERENYI L, GALLYAS F: *Clin Chim Acta 38*:465, 1972
78. KILLINGSWORTH LM, SAVORY J: *Clin Chem 17*:936, 1971
79. KILLINGSWORTH LM, SAVORY J, TEAGUE PO: *Clin Chem 17:374, 1971*
80. KILLINGSWORTH LM, SAVORY J: *Clin Chem 17*:936, 1971
81. KILLINGSWORTH LM, SAVORY J: *Clin Chem 18*:335, 1972
82. KIM CY: *Lab Invest 23*:79, 1970
83. KLUTHE R, FAUL J, HEIMPEL H: *Nature 205*:93, 1965
84. KOLB HJ, GRODSKY GM: *Biochemistry 9*:4900, 1970
85. KRÖLL J: *Scand J Clin Lab Invest 24*:55, 1969
86. KWAPINSKI JB: *Methods of Serological Research.* New York, Wiley, 1965
87. LAFFIN RJ: *J Lab Clin Med 76*:816, 1970
88. LANGE CF: *Clin Chim Acta 18*:91, 1967
89. LAURELL CB: *Anal Biochem 10*:358, 1965
90. LAURELL CB: *Anal Biochem 15*:45, 1966
90a. LAURELL CB: *Scand J Clin Lab Invest 29 Suppl 124*:21, 1972
91. LAWRENCE M: *Am J Med Technol 30*:209, 1964
92. LEVIN AS, PIPKINS MO, FUDENBERG HH: *Vox Sanguinis 18*:459, 1970
93. LEVINE RS, ALTERMAN M, GUBNER RS, ADAMS EC Jr: *Am J Med Sci 262*:179, 1971
94. LEVY RP, MARSHALL JS, VELAYO NL: *J Clin Endocrinol Metab 32*:372, 1971
95. LOUIS-FERDINAND R, BLATT WF: *Clin Chim Acta 16*:259, 1967
96. LOPEZ M, TSU T, HYSLOP NE Jr: *Immunochemistry 6*:513, 1969
97. MACY NE, O'SULLIVAN MB, GLEICH GJ: *Proc Soc Exp Biol Med 128*:1098, 1968
98. MANCINI G, VAERMAN JP, CARBONARA AO, HEREMANS JF: *Protides of the Biological Fluids, 11th Colloquim* 1963, edited by Peeters, H. New York, Elsevier, 1963, p 370
99. MANN D, GRANGER H, FAHEY JL: *J Immumol 102*:618, 1969
100. MARON E, BONAVIDA B: *Biochim Biophys Acta 229*:273, 1971
101. MASSAYEFF RF, ZISSWILLER MC: *Anal Biochem 30*:180, 1969
102. McCLELLAND DBL, FINLAYSON NDC, SAMSON RR, NAIRN IM, SHEARMAN DJC: *Gastroenterology 60*:509, 1971
103. McKAY GG: *Practical Application of Immunoelectrophoresis in Lab Synopsis.* Behring Diagnostics, 1969, Vol 2, p 1
104. MERRILL D, HARTLEY TF, CLAMAN HN: *J Lab Clin Med 69*:151, 1967
105. MILES DW, MOGENSEN CE, GUNDERSEN HJG: *Scand J Clin Lab Invest 26*:5, 1970
107. MIYASATO F, POLLACK VE, BARCELO R: *Arthristis Rheumat 9*:308, 1966
108. NELSON TL, STROUP G, WEDDELL R: *Am J Clin Pathol 42*:237, 1964
109. NERENBERG ST: *CRC Crit Rev Clin Lab Sci 1*:303, 1970
110. NERENBERG ST: *J Lab Clin Med 79*:673, 1972
111. NIELSEN HG, BUUS O, WEEKE B: *Clin Chim Acta 36*:133, 1972
112. NÖRGAARD-PEDERSEN B, MONDRUP M: *Scand J Clin Lab Invest 27*:169, 1971
113. NOSSEL HL, YOUNGER LR, WILNER GD, PROCUPEZ T, CANFIELD RE, BUTLER VP Jr: *Proc Nat Acad Sci 68*:2350, 1971
114. NUSBACHER J, BERKMAN EM, WONG KY, KOCHWA S, ROSENFIELD RE: *J Immunol 108*:893, 1972
115. OLIVER GC, PARKER BM, PARKER CW: *Am J Med 51:186, 1971*

116. OUCHTERLONY Ö: *Acta Pathol Microbiol Scand 26*:507, 1949
117. OUCHTERLONY Ö: *Handbook of Immunodiffusion and Immunoelectrophoresis,* Ann Arbor, Mich, Ann Arbor Science Publishers, 1968
118. OUDIN J: *Ann Inst Pasteur 75*:30, 1948
119. OUDIN J, WILLIAMS CA, FINGER I, MUNOZ J, KOHN J, BECKER EL, POLSON A, RUSSELL B, ALLISON AC, DARBY, DA, PREER Jr, POULEK MD, CLARKE HGM, URIAL J, CHASE MW, EASTY GC, THORBECKE GJ, HOCHWALD GM, CROWLE AJ, HJERTEN S, SMITH H, MILGROM F: *Precipitation Analysis by Diffusion in Gels,* Chap 14 in Methods in Immunology and Immunochemistry, edited by Williams CA, Chase NM. New York, Academic, 1971
120. PEETOOM F: *Agar Precipitation Technique,* Springfield, Ill., Charles C. Thomas, 1963
121. PIETRUSKA Z, DROZD J, BOGDANIKOWA B: *Clin Chim Acta 21*:139, 1968
122. PIZZOLATO MCB, PIZZOLATO MA, AGOSTONI A: *Clin Chem 18*:237, 1972
123. PIZZOLATO MCB, DEL CAMPO GB, PIZZOLATO MA, VERGANI C: *Clin Chem 18*:203, 1972
124. PRĬSTOUPIL TL, HRUBÁ A: *Clin Chim Acta 14*:502, 1966
125. RAISYS VA, ARVAN DA: *Clin Chem 17*:745, 1971
126. RAPP W: *Clin Chim Acta 36:5, 1972*
127. RESSLER N: *Clin Chim Acta 5*:795, 1960
128. RICHTER RW, CHALLENOR YB, PEARSON J, KAGEN LJ, HAMILTON LL, RAMSEY WH: *J Am Med Assoc 216*:1172, 1971
129. ROSE NR: *Clin Chem 17*:573, 1971
130. ROWE DS: *Immunoglobulins,* Chap 13 in Clinical Aspects of Immunology, 2nd ed., edited by Gell PGH, Coombs RRA, Philadelphia, F. A. Davis, 1968
131. ROWE DS: *Bull World Health Organ 40*:613, 1969
132. ROWE DS: *Lancet 1*:1340, 1970
133. RUDGE JM: *J Clin Pathol 13*:530, 1960
134. SALVAGGIO JE, ARQUEMBOURG PC, SYLVESTER GA: *J Allergy 46*:326, 1970
135. SAVORY J, HEINTGES MG, KILLINGSWORTH LM, POTTER JM: *Clin Chem 18*:37, 1972
136. SCHEIDEGGER JJ: *Intern Arch Allergy Appl Immunol 7*:103, 1955
137. SCHEINBERG IH, GITLIN D: *Science 116*:484, 1952
138. SCHWARTZ HG: *Am J Clin Pathol 57*:326, 1972
139. SCHWICK HG, STÖRIKO K, BECKER W: in *Lab Synopsis,* Vol 1, edited by Hoaf E. Behring Diagnostics, 1969, p 1
140. SEIDEL D: *Clin Chim Acta 31*:225, 1971
141. SHAPIRA E, ARNON R, RUSSELL A: *J Lab Clin Med 77*:877, 1971
142. SHULMAN S, MAMROD L, GONDER MJ, SOANES WA: *J Immunol 93*:474, 1964
143. SIMMONS P: *Clin Chim Acta 35*:53, 1971
144. SMITH DB: *Anal Biochem 22*:543, 1968
145. SMITH TW, SELDEN R, FINDLEY W: *Clin Res 19*:356, 1971
146. SMITH FR, RAZ A, GOODMAN DDS: *J Clin Invest 49*:1754, 1970
147. SPECTOR S, PARKER CW: *Science 168*:1347, 1970
148. SPECTOR S, FLYNN EJ: *Science 174*:1036, 1971
149. STÖRIKO K: *Blut XVI*:200, 1968
150. STURGEON P, HILL MK, KWAK KS: *Immunochemistry 6*:689, 1969
151. SUHRLAND LG, ARMENTROUT SA, DANIEL TM: *J Lab Clin Med 71*:1021, 1968

152. TATARINOV YS: *Lab Delo* 6:37, 1960
153. TEMLER RS, FELBER J-P: *Biochim Biophys Acta* 236:78, 1971
154. TERRY WD, FAHEY JL: *Serum Proteins and the Dysproteinemias*, edited by Sunderman and Sunderman. Philadelphia, Lippincott, 1964, p 182
155. TOURTELLOTTE WW, TAVOLATO B, PARKER JA, COMISO P: *Arch Neurol* 25:345, 1971
156. VAN ORDEN DE: *Immunochemistry* 5:497, 1968
157. VAN ORDEN DE: *Immunochemistry* 8:869, 1971
158. VAN WEEMAN BK, SCHURRS AHM: *FEBS Letters* 15:232, 1971
159. VERGANI C, STABILINI R, AGOSTONI A: *Immunochemistry* 4:233, 1967
160. WEEKE B: *Arztl Lab* 18:12, 1972
161. WIER D, editor, *Handbook of Experimental Immunology*. Philadelphia, F. A. Davis, 1967
162. WIEME RJ: *Agar Gel Electrophoresis*. Amsterdam/London/New York, Elsevier, 1965
163. WIEME RJ, VEYS EM: *Clin Chim Acta* 27:77, 1970
164. WILSON AT: *J Immunol* 92:431, 1964
165. WOLF P, WALTON KW: *Immunology* 8:6, 1965
166. WUNDERLY C: *Advances in Clinical Chemistry*, edited by Sobotka H, Stewart CP. New York, Academic, Vol 4, 1961, p 207

Chapter 8

In Vitro Radioisotope Technics

NORMAN D. LEE, Ph.D

Radiosotopes have been used increasingly in routine analytical chemistry in general, and in routine clinical chemistry in particular, because they provide sensitivity, specificity, and procedural simplicity generally unattainable by other technics. Furthermore, the instruments used for quantitation are easily available, reliable, and simple to operate. Consequently, in 1972 there were 21 procedures in routine use at our laboratory that employed radioisotopes to identify and quantitate such clinically important substances in biological specimens as vitamins, enzymes, proteins, and hormones (e.g., vitamin B_{12}, ornithine carbamyl transferase, thyroxine-binding globulin, thyroxine, etc.).

This chapter will provide an overview of the fundamentals of radio-isotopes and their application to clinical chemistry. Only *in vitro* procedures will be presented; dynamic tests of function (e.g., lipid absorption, circulation time, red cell production and destruction, etc.) and imaging procedures (e.g., organ scans) are considered to be nuclear medicine rather than clinical chemistry. The interest of the clinical chemist lies in the use of radioactive materials as reagents. The emissions of radioisotopes are methodologically analogous to the properties of other substances to absorb photons of specific wavelength (i.e., photometry) or to emit photons (i.e., fluorometry). Therefore, the physical properties of radioactive materials as well as the methods for detecting and quantitating their emitted radiations deserve coverage in this chapter. A more detailed treatment of these subjects can be obtained by referring to the cited texts.

ATOMIC STRUCTURE AND THE INTERACTION OF RADIATIONS WITH MATTER

Elements commonly occur in nature in two or more isotopic forms. Each element has a unique and equal number of extranuclear electrons and intranuclear protons but exists in two or more atomic forms due to different numbers of neutrons within the nucleus. The total number of neutrons and protons determine the atomic weight, and the electrons determine the element's chemical behavior. Isotopes of a given element vary in atomic weight, which is shown in a superscript numeral to the chemical symbol (the current practice is to put this superscript before the chemical symbol rather than after it). For example, the four isotopes of carbon all possess six protons in the nucleus and six electrons, but ^{11}C has five neutrons in its nucleus, ^{12}C has six, ^{13}C has seven, and ^{14}C has eight.

The physical relationships and forces existing among the nuclear particles need not be discussed here. Suffice that in some nuclei these relationships and forces cause instability, which results in rearrangements and the emission of radiations, formation of a different nucleus, and thus, a different element. Such unstable atoms are termed radioistopes of the element. Thus, ^{11}C and ^{14}C are *radioisotopes* of carbon and ^{12}C and ^{13}C are *stable isotopes*. The occurrence of such nuclear rearrangement with its emission of radiations is called *radioactive decay*.

Emitted radiations are of three specific kinds: α particles, β particles, and γ rays. Table 8-1 shows their physical characteristics.

TABLE 8-1. Physical Characteristics of Nuclear Radiations

Name		Symbol	Mass, atomic units	Charge
Alpha particle		α	4	2+
Beta particle				
	Electron	β$^-$	5.5×10^{-4}	1−
	Positive	β$^+$	5.5×10^{-4}	1+
Gamma ray		γ	0	0

In addition to mass and charge, emitted radiations possess energy, conventionally expressed in terms of millions of electron volts (MeV). The electron volt is defined as the kinetic energy gained by an electron after passing through a potential difference of 1 V.

Radioactive decay is a complex process and possesses a number of features important to the applied use of radioisotopes. First, the time a given atom will decay is completely unpredictable and in a given population it is seen to occur randomly. When large numbers of unstable atoms are observed, decay can be seen to occur at an exponentially decreasing rate that approaches zero asymptotically. Although the life span of a radioactive atom cannot be determined, one can express the time required for the decay rate of a population to achieve half its reference value, and this is the "half-life" of the radioisotope and is as characteristic as its atomic number. Half-life is symbolically shown as $T_{1/2}$ and can vary from a fraction of a second to years. For example, the $T_{1/2}$ value for ^{11}C is 20.5 min and for ^{14}C it is 5568 years.

Second, the path of decay from parent to daughter is a characteristic of each radioisotope. Some decay is simply emission of only one electron or positron to form a stable daughter, e.g., ^{14}C and ^{11}C decay to ^{14}N and ^{11}B, respectively. Others decay in a complex manner. For example, ^{59}Fe releases 3 different β particles, each with a different energy range, to ^{59}Co. The daughter ^{59}Co exists briefly in an excited state and falls to the ground state by the emission of several γ rays, each with a different associated energy.

Last, emitted radiations have specific quantitative energy characteristics. Simple β decay, as occurs with ^{14}C or ^{32}P, results in the release of electrons with a range of associated energy values up to a characteristic maximum value. This value, designated as E_{max}, is 0.154 MeV for ^{14}C, and for ^{32}P, 1.707 MeV. Another group of radioisotopes exhibits a mixture of β and γ radiations. The β particles, again, cover ranges of energy and have E_{max} values. The γ radiations are, however, monoenergetic; each has a single energy value. For example, ^{59}Fe decays to stable ^{59}Co with the release of three populations of β particles. Of these 46% have an E_{max} of 0.271 MeV, 54% have an E_{max} = 0.462 MeV, and 0.3% have an E_{max} = 1.560 MeV. The excited daughter ^{59}Co decays by emitting three different γ rays, each with a single energy value of 0.191, 1.099, and 1.290 MeV, respectively.

Beta particles lose energy by interacting with matter in two ways. First, collision with and transfer of sufficient energy to an orbital electron results in the formation of a free electron and a positively charged atom; ionization has occurred with the formation of an ion pair. This interaction decreases the energy of the incident particle by about 35 eV. The electron continues to traverse the absorbing material, losing 35 eV/collision and forming ion pairs, until its energy is completely dissipated. The second occurs when the incident particle passes close to the nucleus of the absorbing atom. As it leaves the coulombic field of the nucleus, it slows down and loses energy by exciting orbital electrons. These later fall back to their ground state by releasing a photon, or γ ray, of characteristic energy. Such γ photons are called "brehmsstrahlung," the German term meaning "braking radiation."

Positrons, or positively charged β particles, interact with matter and lose kinetic energy by combining with an electron, both particles being annihilated by transformation to two γ rays. These γ photons each have 0.551 MeV of energy and are given off at a 180° angle to each other. Such photons are called "annihilation radiations."

Electromagnetic radiations, commonly termed γ rays or γ photons, have a much lower probability of interacting with matter because of their lack of charge and mass. Typical relative probability values for the interaction of α, β, and γ radiations with the matter are 2500, 100, and 1, respectively. Gamma photons, therefore, are highly penetrating radiations. Energy is lost by γ photons by three different types of interaction. The first, the "photoelectric effect", is important for photons of energy less than 0.5 MeV upon interaction with matter of high atomic number. Here the entire energy of the incident photon is transferred to the absorbing atom with the ejection of an orbital electron; ionization has occurred and an ion pair produced. A second process, the Compton effect, becomes quantitatively more important as the photon energy ranges from 0.5 to 1.0 MeV and/or the absorbing material is of lower atomic number. In this case energy transfer results in the photon having less energy, a "Compton photon." In addition, the excited atom returns to ground state by losing an orbital electron with the resultant production of an ion pair. The third process occurs only with γ photons having an energy of at least 1.02 MeV. When such photons are in the vicinity of the nucleus of an absorbing atom, the formation of a positron and an electron occurs, called "pair production." These particles then behave as described above, namely they interact through ionization and annihilation.

Radiations are detected and quantitated through their interaction with matter. Fundamentally different kinds of detecting devices are needed to measure the different radiations, depending on their mass, charge, and energy characteristics. For example, an end window Geiger Mueller counter easily measures ^{32}P (E_{max} = 1.707 MeV) whereas ^{3}H (E_{max} = 0.012 MeV) is undetectable. Iodine 125, a weak γ emitter, requires a sodium iodide crystal scintillation counter. For the clinical chemist, brehmsstrahlung is largely of academic interest. However, since its production increases with the atomic number of shielding material, it is customary to use shielding of low atomic number (plastics) for certain applications of β-emitting isotopes.

In practice one does not measure γ photons as monoenergetic radiations in spite of the fact they are so produced. This is because interaction with matter, including the specimen container and the detector itself, results in the production of Compton photons covering a whole range of energies. Examination of the γ radiations from isotopic decay will reveal continuous and multipeaked spectra of photon energies that are characteristic for each γ-emitting radioisotope. To display the monoenergetic photons requires the use of highly specialized instruments and technics that are irrelevant to the interests of the clinical chemist.

Most of the γ-emitting radioisotopes used by clinical chemists will exhibit both photoelectric and Compton interactions with matter. This situation is illustrated by the absorption of γ rays in aluminum. Figure 8-1 shows the relative contribution of these processes when radiations of different energy values are plotted as a function of absorption coefficient.

FIG. 8-1. The interaction of gamma rays with aluminum. From OVERMAN RT, CLARK HM: *Radioisotope Techniques.* New York, McGraw-Hill, 1960

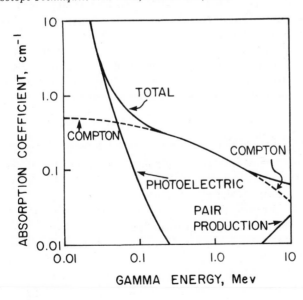

DEFINITIONS AND UNITS

Listed below are definitions for a number of nuclear terms. For a more complete listing refer to the manual by Chase and Rabinowitz (16). Conventional abbreviations are given in parentheses after the terms.

Absorption. A general term for the processes by which all or part of the energy of a radiation is transferred to the atoms and molecules of the material through which it passes.

Absorption coefficient. Quantitative expression of the decrease in intensity of a beam of radiation as it passes through matter. It is a function of the particular matter and the type and the energy of radiation.

Activity. The strength of a radioactive source expressed as disintegrations occurring, or radiations detected, per unit time.

Alpha particle (α). A helium nucleus; a unit of two protons and two neutrons with a charge of +2.

Background. The apparent strength of detected radiation in the absence of a sample; background radiation.

Beta Particle (β). An electron with charge −1 or +1, negatively charged betas (β) are sometimes called *negatrons* while positively charged betas (β+) are called *positrons*.

Brehmsstrahlung. Secondary γ radiation produced by the deceleration of a charged particle passing through matter.

Carrier. The nonradioactive counterpart of radioisotopic atoms or labeled molecules present in, or added to, a given preparation.

Carrier Free. A radionuclide preparation essentially free of its stable isotope.

Compton Effect. Gamma ray absorption by interaction with an orbital electron to produce a recoil electron and a scattered photon of lesser energy.

Count (c) (noun). The current pulse produced and detected by the action of a radiation on a detector; (verb) to register the frequency of detectable interactions between radiations and a detector.

Counts per Minute (cpm). The basic unit in practice for expressing the number of radiations detected per unit time; the relative unit of activity.

Curie (Ci). The quantity of a nuclide necessary to produce 3.7×10^{10} disintegrations per second; an absolute unit of activity.

Millicurie (mCi) = 3.7×10^7 disintegrations per second (dps)

Microcurie (μCi) = 3.7×10^4 dps

Dead Time. The period immediately following detection of a radiation in which the detector remains insensitive and unable to detect another radiation; recovery time; resolving time.

Decay Constant. Fractional value for number of radioactive nuclides disintegrating per unit time; disintegration constant.

Disintegration. Spontaneous transformation of a nucleus accompanied by the emission of radiation.

Electron Volt (eV). The kinetic energy gained by an electron in passing through a potential difference of 1 V, ordinarily expressed for radiations as thousands (KeV) or millions (MeV) of electron volts.

Excitation. The addition of energy to an atomic, molecular or crystal system thereby transferring it from its ground state to an excited state.

Excited State. The state of nuclei, atoms or molecules resulting from the energy addition of incident radiation. Also, that state immediately following nuclear decay and preceding photon emission.

Gamma Ray (γ). Electromagnetic radiation having its origin in nuclear decay or interactions of radiations with matter; the term "photon" is frequently used synonomously.

Geiger Counter. An ionization chamber operating in the Geiger-Mueller region.

Geiger-Mueller Region. The range of voltage applied to an ionization chamber in which the charge produced by an incident radiation is independent of its energy.

Ion Pair. A positive ion and electron produced by the action of ionizing radiations on matter.

Ionization Chamber. A detector that measures the amount of ionization produced in a given volume of gas by a radiation. It consists of two electrodes between which an electric field is maintained. The electrodes collect the ions produced by the radiation.

Isotope. One of several nuclides having the same number of protons but different numbers of neutrons in the nucleus.

Isotope Effect. The effects caused by differences in mass on the physicochemical behavior of isotopes of an element and of compounds of the element.

Neutron (n). Elementary nuclear particle having no charge and a mass of 1.

Nuclide. Synonymous with isotope. Current usage generally favors "nuclide."

Pair Production. An absorption process confined to the interaction of photons of energy exceeding 1.02 MeV with matter, resulting in production of an electron and positron pair.

Photoelectric Effect. An absorption process whereby all the energy of an incident photon is transferred to an orbital election, causing its ejection with the formation of an ion pair.

Plateau. That range of applied voltage in which a detector exhibits negligible changes in count rates.

Proton. A positively charged elementary particle of mass number 1; a hydrogen nucleus.

Pulse Amplifier. An electronic device for the amplification of electric impulses generated by radiation detectors.

Scaler. An electronic device that registers current pulses received in a given time interval.

Binary Scaler. A scaler with a scaling factor of 2 per stage.

Decade Scaler. A scaler with a scaling factor of 10 per stage.

Scatter. The change in direction of radiation produced by its interaction with matter.

Scintillation. The light flash produced in a phosphor by a radiation.

Scintillation Counter. Radiation detector system employing a phosphor, photomultiplier tube, and associated electronic circuitry.

Specific Activity (S; SA). The activity or decay rate of a radioisotope per unit of mass.

Absolute Units. mCi per mg; mCi per mmol; disintegration per second per mg.

Relative units. Counts per minute per mg.

Spectrum. A plot of the frequency of radiations emitted by a radionuclide as a function of their energy values.

Spectrometer. A system consisting of a detector, pulse amplifier, and circuitry designed to measure radiation across a range of energy values; β ray and γ ray spectrometers are the two kinds in use.

Standard, Radioactive. Radioactive sample in which the numbers and types of nuclides are known.

Statistical Error. Counting errors resulting from the random nature of radioactive decay.

Voltage, Operating. Voltage applied to a detector, or detector system, required for proper detection of radiations.

THE DETECTION OF RADIATIONS

DETECTORS

Since the interaction of radiations with matter commonly results in ion formation, ion collection is exploited in both detection and quantitation of radioactivity. A generalized device for ion collection is shown in Figure 8-2. It consists of a wire, or electrode, in the center of a chamber, the walls of which act as a second electrode, the two being separated by an efficient insulator. The electrodes are connected through suitable resistance to permit a potential difference to be applied. The chamber is filled with a gas mixture and when a radioactive source is brought into the vicinity of this arrangement, radiations passing through the gas cause ions to be produced; such ions directly arising from the action of incident radiations are termed "primary ions." By charging the central wire positively with respect to the other electrode, it collects the electrons produced, and positive ions migrate to the walls. Thus, a charge will be collected on the central wire and a current will pass through the external circuit to the other electrode. This means that detection and quantitation of radiations is a problem in the measurement of small currents. This problem has two facets: the behavior of ions in the collection chamber as a function of collection voltage and the detection of very small currents by the external system.

The relationship between ion production and applied voltage is shown in Figure 8-3 for both α and β particles; γ rays are omitted since ion collection is ordinarily not used because of the poor ability to cause ionization. Six distinctly different kinds of phenomena can be described for specific voltage ranges.

Range I: At low voltage, radiations cause relatively few ions to be detected because the weak electrostatic field allows some to recombine before collection on the electrodes.

Range II: Current produced is independent of applied voltage because the electrostatic field is strong enough to prevent recombination, yet

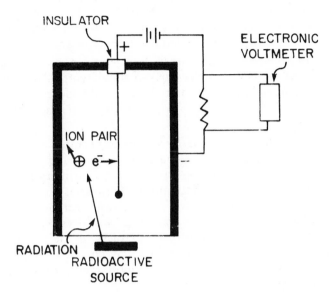

FIG 8-2. Schematic Diagram of an ion collection device. From OVERMAN RT, CLARK HM: *Radioisotope Techniques.* New York, McGraw-Hill, 1960.

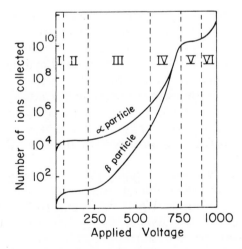

FIG 8-3. The relation between voltage applied to an ion collection detector and the ions produced per radiation detected. From OVERMAN RT, CLARK HM: *Radioisotope Techniques.* New York, McGraw-Hill, 1960.

insufficient to accelerate "primary" ions to become ionizing particles themselves (i.e., produce "secondary" ionizations). This range is termed the region of saturation voltages.

Range III is called the *proportional region*. Here the electrostatic field is of such strength that primary ions are accelerated and produce secondary ionizations. The number of secondary ions is directly proportional to the number of primary ions produced, regardless of the nature of the initial radiation and is an exponential function of voltage increase. The process of secondary ion production is called *gas amplification.*

Range IV is imprecisely defined at its lower limit and is largely related to the geometric relationships of the parts of the detector. Here the extent of gas amplification is limited by the shape of the detector chamber and current collection departs from its proportional relationship to applied voltage.

Range V is called the *Geiger-Mueller region* and is most commonly used in ion-collection counting. Here the charge collected is dependent neither on type of incident radiation nor on the number of primary ionizations. The intensity of the field around the central electrode is so high that ions of primary and secondary origin are further accelerated to produce additional ionization. In principle, any radiation produces an avalanche having the same number of final ionizations regardless of the number of ionizations initially produced. The relationship of charge collected to applied voltage over the entire Geiger-Mueller region shows only slight increase and is termed the *Geiger-Mueller plateau.*

Range VI is called the zone of continuous discharge, which results from the spontaneous ionization of the gas in the detector chamber because of the very high field intensity.

For the clinical chemist, Geiger counting is the most common form of ion collection counting. Generally a cylindrical chamber with a thin film over the end is employed. Radiations pass through this "window" on their way to create the primary ionizations. Such tubes are used mainly on survey instruments. Their principal limitation is relatively low efficiency and difficulties in sample preparation. Another type of detector permits a solid sample to be introduced into the chamber, which increases counting efficiency to about 50%. Another group of detectors requires conversion of the sample into gaseous form that is introduced directly into the counting chamber; here efficiency is close to 100%.

In all cases ionization occurs in a filling gas. Commonly this is a mixture of an inert gas, such as argon or helium, and a polyatomic compound, such as ethanol, butane, or chlorine. When the members of the ion pairs are neutralized at the cathode and anode, energy is released that can set off another train of ionizations. Unless stopped, this release of energy would keep the detector in a state of continuous discharge. The polyatomic molecules in the gas mixture "quench" this release of energy by absorbing it. This energy capture, however, results in molecular dissociation, and this places a limitation on the life of organically quenched Geiger-Mueller tubes. This is not true, on the other hand, for halogen quenched tubes since the dissociated atoms can recombine to form the diatomic halogen molecule.

The production of a current pulse arising from the detected ionizing radiation takes time. The ion pairs formed by the primary radiation move under the influence of the electric field toward the anode and the cathode causing secondary ionizations. These in turn cause additional ionizations until the tube is fully discharged. The electrons are quickly collected on the central wire, whereas the positively charged ions move more slowly toward the cathode because of their greater size and mass. Thus, a significant time interval, of the order of 50–200 μsec, elapses, during which the detector is insensitive to the presence of another ionizing radiation. This interval is termed the "dead time" or "resolving time" of the detector.

Another type of radiation detection takes advantage of a different physical process, one based on the transfer of energy from electrons when passing through certain materials known as "phosphors." Such electrons arise directly during β decay (e.g., ^{3}H or ^{14}C) or indirectly from the interaction of γ rays (e.g., ^{125}I and ^{131}I) with matter. The energy of these electrons is transferred to the outer orbitals of the phosphor molecules, raising them to an excited state. Such excited atoms then fall back to a ground state with the emission of energy as a photon with wave length in the ultraviolet or blue end of the spectrum. Detection of this emission provides a measure of the radiation. The process is called "scintillation counting."

Scintillation counting possesses several advantageous properties: The period between excitation and return to the ground state is very short, 1 μsec or less, a much faster process than ion collection; solid phosphors or liquid solutions of phosphors are denser than gases so that the probability of interaction with a γ photon is greater than with a gas; the number of photons generated in the phosphor is roughly proportional to the energy lost by the γ ray. The amount of light, therefore, can be related to the energy of the incident radiation. All these advantages are obtained in scintillation counting regardless of the type of phosphor used, be it an organic solution, as used in liquid scintillation counting for ^{3}H and ^{14}C, or a sodium iodide crystal, as used in crystal counting for ^{125}I and ^{131}I. With both, the ease of use and counting efficiency are considerably greater than generally obtained in ion collection.

Although sample conversion to a gas permits counting efficiencies approximating 100%, the associated technical problems of sample conversion and chamber filling and flushing are considerable. Hence this method of β counting is largely confined to special cases and has been largely supplanted by liquid scintillation counting.

Scintillation counting requires that a special phototube be coupled to the phosphor in order to detect the minute amounts of light produced. This is a photomultiplier tube with a photocathode surface that emits electrons when struck by the light scintilla. The number of electrons emitted is proportional to the number of photons striking the photocathode and thus to the energy of the radiation acting upon the phosphor. The total number of electrons produced per scintillation is, however, too small to be measured directly.

This problem is handled by a sequence of multiplier stages (dynodes) in the tube, each being maintained at a higher voltage than the preceding one. Electrons ejected from the photocathode are thereby accelerated and focused. The primary electrons cause the release of specific number of secondary electrons by striking the surface of the first dynode. These in turn are accelerated and strike the second dynode producing the same multiplication, and the process is repeated through a total of 10–14 dynodes. The result is the almost instantaneous production of $10^5 - 10^6$ electrons for each originally emitted from the photocathode. Under fixed conditions the total number of electrons in the output pulse is proportional to the number of scintilla created in the phosphor and thus to the energy transferred to the phosphor by the incident radiation.

Photomultiplier tubes have two properties of particular importance. First, it is possible for electrons to be spontaneously ejected from the photocathode and undergo multiplication just as if they had been ejected consequent to the action of a radiation on the phosphor. These are called "thermal electrons" because their rate of emission depends on the temperature of the photocathode. Although normally only few thermals are formed, they can be troublesome when one attempts to detect and count weak radiations. Photomultiplier tubes are, therefore, housed in a low temperature environment when used for counting a weak β emitter such as ^{3}H ($E_{max} = 0.018$ MeV).

Second, a change in applied voltage will change the size of the output signal from the photomultiplier tube. This means that variations in tube voltage will result in either the detection or escape of weak radiations. Furthermore, as tube voltage increases, thermal electrons will be increasingly detected as output signals. For this reason the high voltage supply for a photomultiplier tube, ordinarily 1000–2000 V, must be exceptionally stable.

A third type of radiation detection device employs semiconductors. At present they are of no practical use in clinical chemistry, but two intrinsic advantages and their rapid development suggest future significance. First, very small amounts of energy are needed to produce an "electron hole" pair as compared to that required for ion pair formation in a gas and photoelectron generation in a phosphor. Second, output pulse size is proportional to incident radiation energy. As a result, semiconductors have orders of magnitude greater resolving power in detecting radiations of different energy characteristics. Used in conjunction with suitable pulse amplifiers, such resolutions can be exploited in detecting low levels of radioactive materials and in increasing discrimination among different isotopes in the sample preparation.

ASSOCIATED ELECTRONIC EQUIPMENT

A number of items of associated electronic equipment are necessary for detection of output signals, selection according to magnitude, and recording of results. Commonly, these are obtained and used as a single piece of equipment. Figure 8-4 presents a schematic diagram of these associated electronic components.

FIG 8-4. Schematic diagram of a pulse counter. From OVERMAN RT, CLARK HM: *Radioisotope Techniques.* New York, McGraw-Hill, 1960.

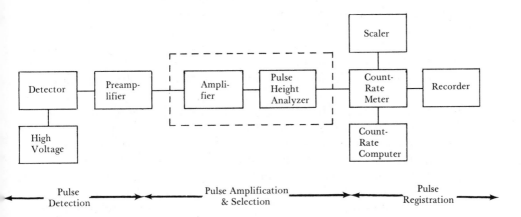

All radiation detectors require a stable high voltage source, ordinarily consisting of a high-voltage transformer, a rectifier, and one or more voltage regulation tubes. These will operate satisfactorily on line voltage between 90 and 135 V and provide output voltages ranging from 500 to 2500 V regulated to ±0.01%. Voltage regulation for Geiger-Mueller detectors is less demanding than for scintillation detectors because the former exhibit maximal signal output independent of detector voltage, whereas the magnitude of the output signal of scintillation detectors is directly related to the voltage applied to the photomultiplier tube.

Although Geiger-Mueller detectors may produce current pulses large enough to be accepted by a data registration unit, scintillation systems do not. For these instruments, pulse amplification in two or more steps is necessary. This is accomplished by the immediate preamplification of the output pulse of the photomultiplier tube by a factor of 20–30. Such circuits must have a constant amplification factor across the entire energy spectrum. Further amplification is accomplished by any of a variety of amplifier circuits compatible with the data registration stage of the counter. For scintillation detectors, the rigid control of pulse amplification can be achieved by a linear amplifier, an instrument that accomplishes pulse amplification by a fixed factor so that the relationship among pulse sizes and shapes is retained.

All electrical systems have random noise signals that must be differentiated from true pulses. All pulses, regardless of origin, are characterized by a voltage on the basis of which they can be accepted or rejected from the data registration part of the system. This is accomplished by a *pulse height selector*, which is a rather simple electronic switch. Selectors may be fixed or variable, the latter type usually being referred to as a *discriminator*. The selector can be adjusted to accept only those pulses having voltage greater than a given value, which becomes the input sensitivity limit for that system. Commonly, fixed selectors operate at 0.25 V input. In practical terms this means that true pulses from a detector must be amplified to exceed 0.25 V, whereas random signals, of less than that voltage, will be eliminated.

Sometimes it is necessary to determine pulse frequency as a function of pulse height, i.e., establish the energy or pulse height spectrum and to count radiations selected on the basis of energy value. This is accomplished in the selection stages of pulse handling by using two variable discriminators; the device to do this is called a *pulse height analyzer*. In addition to the discriminator circuits the analyzer has an *anticoincidence circuit*; this circuit prevents registration of those pulses that are simultaneously transmitted by both discriminators. Figure 8-5 shows a pulse height analyzer in block diagram.

Discriminator I is the baseline discriminator and accepts only pulses exceeding a set value, H. Discriminator II is the window discriminator set for pulses at a fixed incremental value $(H + \Delta H)$ with respect to Discriminator I. They are linked through an anticoincidence circuit that permits transmittal to the registration unit of only those pulses that exceed H and do not exceed $H + \Delta H$. When H and $H + \Delta H$ are both exceeded, the discriminators coincide in transmitting a pulse that is rejected by the anticoincidence circuit. Of course, pulses too small to exceed H will not exceed $H + \Delta H$ and escape registration. Thus, pulses within a given height range (ΔH) are recorded. By having Discriminator II linked to Discriminator I and having both variable, it is possible to establish the pulse height or energy spectrum of the material being radioassayed. Furthermore, based on this knowledge the conditions for radioassay can be selected to minimize random noise signals, environmental noise signals (contamination in the environment), and signals arising from radioisotopes in the preparation other than those of specific interest.

Data registration is accomplished through the use of a device called a *scaler*. These may operate either in a binary mode, data expression thus being limited to some power of 2, or in a decade mode, data being collected in a decimal system. The former is mainly of historical interest only since scalers in current use collect data with decade units. The use of such secondary devices as data printers is common and for practical reasons, necessary.

FIG 8-5. Block diagram of a differential pulse height analyzer. From OVERMAN RT, CLARK HM: *Radioisotope Techniques.* New York, McGraw-Hill, 1960.

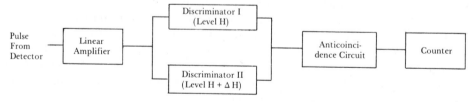

THE MEASUREMENT OF RADIATION

PHYSICAL CONSIDERATIONS

The basic objective in radioassay is identification of the signal caused by a radiation discharged from the specimen under study. Operationally, this means the quantitation of radiations from the sample versus those from the background. One source of background is the random noise of the electronic system. Another is the γ radiation secondary to the action of cosmic radiation on the Earth's atmosphere. A third is the natural radioactivity of the Earth's crust. Glass of specimen-holding test tubes, Geiger-Mueller tubes, and photomultiplier tubes will contain appreciable amounts of ^{40}K, a mixed β and γ emitter. Background count rates for a given system will be constant, showing only statistical fluctuations. However, there may be two variable additions to background, radioactive sources in the work area and radioisotopically "unclean" materials and equipment.

In radioassay the background value is subtracted from the specimen value to yield that due to the sample itself. This is satisfactory for Geiger counters, but further refinements are necessary when systems are used where pulse height is related to operating voltage.

Since background pulse height is variable, operating voltages are selected that maximize detector sensitivity to isotope radiations relative to background. A simple criterion is that the best operating voltage is the lowest that maximizes the ratio of the square of sample activity to background activity. Figures 8-6 and 8-7 show, respectively, a plot of isotopic and

FIG 8-6. Counting rate for an isotopic source and background as a function of voltage; NaI scintillation counter. From OVERMAN RT, CLARK HM: *Radioisotope Techniques.* New York, McGraw-Hill, 1960.

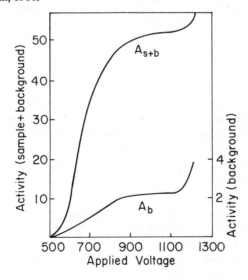

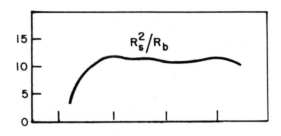

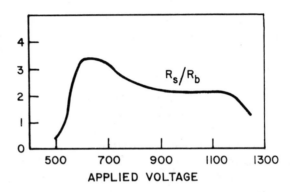

FIG 8-7. Criteria for selecting operating voltage; based on Fig. 8-6. From OVERMAN RT, CLARK HM: *Radioisotope Techniques.* New York, McGraw-Hill, 1960.

background activity as a function of operating voltage and of the suggested recalculation; in this case the voltage of choice would be 600. The voltage maximizing this ratio will depend on the energy spectrum of the isotope but in practice does not show appreciable differences from one isotope to the next.

Resolving time is another feature of radioassay systems. It was discussed previously in relation to ion collection. The initial radioactive interaction is always followed by a series of events ultimately leading to its registration. These take place during a discrete time interval during which the system is insensitive to another initiating event. The loss of counts caused by failure to distinguish between successive events is called coincidence loss, and its magnitude is a function of count rate and counter characteristics. Coincidence loss can be evaluated and corrected by empirical graphic means or by mathematical treatment. In the former case, a plot of detected versus expected activity of increasing known amounts of radioisotope will show a departure from linearity. Extrapolation of the early linear part of this curve, where coincidence loss is not apparent, will permit correction of count rates characteristic of the latter part, where coincidence loss becomes apparent and increasingly significant. Mathematically, coincidence corrections can be made using the expression:

$$N = \frac{n}{1-nT}$$

where N = true count

n = observed count

T = resolving time constant

Determination of T for Geiger-Mueller tubes can be done by the method of paired sources. Here the activities of two sources are measured individually (n_1, n_2) and together $(n_{1,2})$ and T calculated for the expression

$$T = \frac{n_1 + n_2 - n_1 n_2}{2 n_1 n_2}$$

As pointed out earlier, scintillation detectors have very short resolving times as compared to ion collection detectors, and hence coincidence correction is ordinarily not a serious consideration using these devices.

The energy of incident radiations determines the registered activity by assay systems because more ionizations are produced by more energetic β particles and more light quanta are produced in a crystal phosphor by more energetic γ rays. In many applications, this is simply handled by adjusting input sensitivity of the scaler to a fixed level and recording only interactions that exceed this level. This arrangement requires, however, that the signal size and shape be exactly controlled, that the activity be relatively great, and that the radiations be produced by a single isotope. When the latter two conditions are met, energy differentiation requiring a spectrometer is indicated.

Energy differentiation by counting radiations having a known and selected range of associated energy values improves radioassay precision by rejecting all background radiations not in the same range. This increases the difference between sample count rate and background count rate.

Energy differentiation is mandatory when methods employing two or more radioisotopes are used. Common examples are procedures employing ^{3}H and ^{14}C or ^{131}I and ^{125}I. In these cases energy differentiation allows selection and measurement of radiations of specific energy values associated with a specific isotope. Precise measurement can thus be made of isotopes individually although they are mixed with others in the same sample.

Both the characteristics of the specimen and the nature of the detector will influence the energy spectrum of the radiations counted. This can best be seen by two examples drawn from experiments in counting ^{131}I. Figure 8-8 shows the energy spectrum of ^{131}I as a point source in air and as a point source immersed in water. Immersion in water reduces the number of radiations having an energy of 0.364 MeV and increases those having lesser

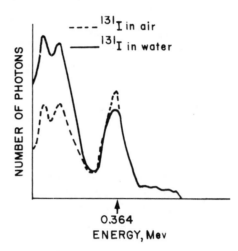

FIG 8-8. Energy spectrum of ^{131}I in air and in water. From OVERMAN RT, CLARK HM: *Radioisotope Techniques.* New York, McGraw-Hill, 1960.

energy, due to the Compton effect. Thus, the practical importance of immersion of the specimen under water is that the count of the emissions which are actually measured is increased. Figure 8-9 shows spectra obtained from an ^{131}I point source in air as detected by NaI crystals of different sizes; all spectra have been normalized on the 0.364 MeV peak. It is apparent that the proportion of low energy γ rays detected is reduced relative to the 0.364 MeV energy rays as crystal detector size increases. This is because the efficiency of the crystal for capturing the energy in the 0.364 MeV radiation increases with size. Fewer of these more energetic radiations escape from the crystal before being detected, whereas they fail to be detected by the smaller crystals that do register the lower energy γ photons.

The same type of phenomenon occurs with β particles. The interaction of electrons with the mass of the specimen results in loss of energy, sometimes resulting in undetectability because of complete absorption within the specimen. This is conditioned both by β particle energy and by the characteristics of the absorbing medium. However, unlike γ ray interaction, there is no concomitant production of additional lower energy radiations; i.e., the total count rate of the specimen will not increase.

Interaction of radiations with the mass of the specimen and with the specimen holder also results in radiation scatter, which can go undetected because it is deflected away from the sensitive part of the detector. This feature is also conditioned by the shape of the specimen. Figure 8-10 shows that with a NaI scintillation well crystal, the count rate of a fixed amount of ^{131}I diminishes as the sample volume changes. As long as sample size and shape are held constant throughout the set of determinations, geometric considerations are relatively unimportant. When variation occurs however, comparison of values becomes meaningless.

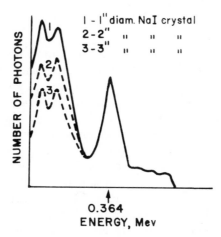

FIG 8-9. Effect of NaI crystal size on the energy spectrum of [131]I. From OVERMAN RT, CLARK HM: *Radioisotope Techniques.* New York, McGraw-Hill, 1960.

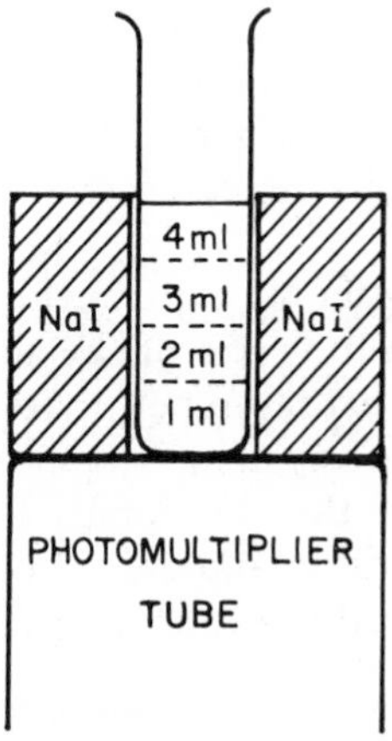

SPECIMEN VOLUME, ml	RELATIVE COUNT RATE
4 ml	84
3 ml	92
2 ml	94
1 ml	100

FIG 8-10. The effect of sample size on detector efficiency for ^{131}I, using a NaI well scintillation crystal.

STANDARDIZATION

It is sometimes convenient or necessary to be able to relate measurements to an absolute standard. The concept of a primary standard is not applicable to radioisotopic measurements as it is in analytical chemistry because all radioisotopes change in activity due to decay. An alternative is the concept of an absolutely standardized material, one that has its disintegration rate determined very accurately without reference to another standard of activity. Absolute standardization requires the use of highly complex equipment and expert interpretation. Such materials are normally not necessary for the purposes of the clinical chemist.

Reference materials have their activity measured by comparison, directly or indirectly, to absolutely standardized materials. They can be obtained from a number of commercial suppliers as well as from national standardizing laboratories (e.g., National Bureau of Standards, Washington, D.C.). Reference materials can be obtained as solutions, as evaporated specimens on a solid backing, as pure materials, and as sources incorporated into a solid rod. Each has its uses depending on the measuring system being calibrated and the type of specimen to which reference is being made. In all cases reference materials should be accompanied by a certificate stating: (a) measurement method, (b) measurement result, in absolute terms, (c) reference time and date, (d) recommended half-life, (e) description of radioactive material used in the source, and (f) assessment of precision of the quoted measurement.

The principal use of reference materials is calibration of measuring instruments, which in turn will permit determination in absolute units of laboratory specimen activities. Perhaps more important, such materials are useful in monitoring counting instruments with respect to variable performance. This can be accomplished by using the reference material to establish the calibration of the system and relating it concomitantly to the system's response to an otherwise uncalibrated source of longer half-life. Repeated measurements with the latter source then permits repeated checking of the calibration.

Calibration with a reference solution can be used to express the efficiency of counter performance. Efficiency is defined as the ratio of the actual count rate to the absolute count rate. Thus, when the reference material is prepared as the expected specimen and the detector system is operating under standard conditions, it is possible to determine the fractional efficiency of detection for both standards and specimens. Under comparable conditions in our laboratory, the same NaI well counter showed the following efficiencies: $^{131}I - 43.9\%$, $^{125}I - 37.6\%$, $^{57}Co - 89.5\%$

STATISTICAL CONSIDERATIONS

Since radioactive decay is a random event, the accuracy of its measurement becomes a statistical problem. The events detected are discretely recorded over fixed sampling intervals. For example, given the background and specimen count rates, the total number of nuclear disintegrations that must be detected in order to reduce the error due to imprecision to any particular acceptable level, can be specified. Figure 8-11, derived from statistical theory, shows these relationships graphically. It behooves the analyst, therefore, to be informed of the statistical considerations he can systematically control and evaluate the error of his measurements and efficiently use his instruments.

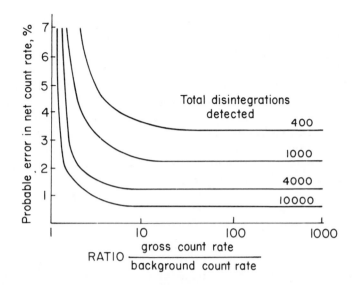

FIG 8-11. The relation between count rate and total counts to error. From *Radioisotope Training Manual. Part I Theory*, White Plains, N.Y. Picker X-Ray Corporation, 1960, p. 30.

ION COLLECTION COUNTING

Although γ rays can be counted using ion collection detectors, the process is inefficient and indirect. With the emergence of sodium iodide crystal counters, ion collection is presently limited to the detection and quantitation of β particles. From the practical standpoint, a decision to use ion collection will ordinarily be determined by the nature of the radiations to be counted and on the funds available to purchase instrumentation. Radioassay of β particles with energies equal to or greater than that of ^{14}C is practical using ion collection. To count weaker radiations, such as those emitted by ^{3}H, involves considerable effort in sample preparation due to the critical role of absorption and scatter.

Ion collection for β's almost always involves counting in the Geiger region. Two types of detectors are practical for routine work: the end-window Geiger tube and the windowless flow counter. The construction of an end-window tube is schematically shown in Figure 8-2, where the window is a thin sheet of mica having a density of a few milligrams per square centimeter. The position and distance of the source in relation to the window must be carefully controlled for reproducible results. With windowless flow counters, the sample is mechanically placed within the sensitive volume of the detector, and it is ready for counting after the tube has been purged with gas to remove introduced air. Windowless flow counters have the added advantage of being more efficient geometrically and more sensitive due to elimination of the mica window.

In addition to the geometric factors of sample positioning are those factors concerned with the sample itself. Ordinarily, the sample is prepared by forming a deposit on a flat metal surface, a *planchet*. If the deposit is a thin, effectively weightless film, no corrections are necessary for radiation absorption. This is not commonly the case, and the analyst must determine the mass on the planchet in some manner. By so doing he can relate the observed activity to previously developed curves that relate sample density (mg/cm^2) to fractional count rate loss. For practical reasons such curves are empiric since counting preparations are different from one analytic procedure to the next; e.g., one procedure may use benzidine precipitates of ^{35}S-sulfate and another barium precipitates of ^{14}C-carbonate.

CRYSTAL COUNTING

Crystal counting generally employs a thallium activated sodium iodide crystal. It is a 1 in. or more in diameter cylinder with a central well. Liquid samples in test tubes are inserted into the well for counting.

The advantages of crystal counting are considerable: Sample preparation is virtually nonexistent, accurate specimen volume measurement being all that is required; true count rates can be obtained by reducing background with a pulse height analyzer; relatively high efficiencies can be obtained. A limitation of crystal counting is its inefficiency for β particles.

LIQUID SCINTILLATION COUNTING

Although liquid scintillation counting can be used to count all three types of nuclear radiations, its principal contribution to clinical chemistry lies in its sensitivity to low energy β radiations. Thus, tritium ($E_{max} = 0.018$ MeV) can be counted with efficiencies approximating 25%. The principal limitations are two: First, sample preparation involves either solution or suspension in a liquid phosphor, which may present problems depending on the nature of the specimen. Second, the phosphor introduces a new cost, although only a few cents per specimen. The advantages, on the other hand, are considerable: sensitivity to low energy radiations; background reduction by detector energy selection, capability for radioassay of two or more different isotopes simultaneously.

For double isotope counting to be successful, the β energy spectra of the two isotopes must be sufficiently different to be separated by pulse height analysis. Many such pairs exist, the most common in clinical chemistry being the ^{3}H/^{14}C pair. Figure 8-12 is a simplified version of the energy spectra of these nuclides as revealed by progressive changes in discriminator settings of the pulse height analyzer. The significant observation is that ^{14}C can be counted at higher discriminator settings that completely reject the β energies of ^{3}H, but ^{3}H counting necessarily involves counting ^{14}C. Correction for ^{14}C in the count at the lower discriminator setting is done by subtracting

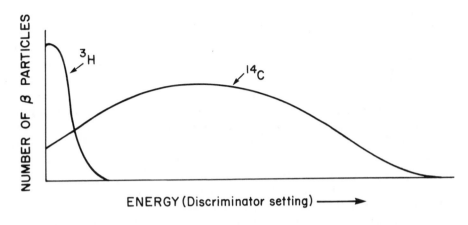

FIG 8-12. The β particle energy spectra for ^{3}H and ^{14}C.

the product of the higher setting ^{14}C count and the ratio of the ^{14}C efficiencies at both discriminator settings.

Using this feature of liquid scintillation counting and differential pulse height analysis, the analysis of biologic fluids for trace levels of hormones has been accomplished on a routine basis with the double isotope derivative assay technic. In this method, samples must be highly purified to avoid problems of quenching.

ISOTOPIC MATERIALS

The radioisotopic material used by the clinical chemist are the free radio-isotopic elements and compounds containing the radioisotopic element as part of their structure; these latter are referred to as *labeled compounds.* Examples of the former are ^{59}Fe used in the determination of the unsaturated iron-binding capacity of plasma and ^{45}Ca in determining the dialyzable calcium of plasma. Labeled compounds are more commonly employed, and these may be of two types: One type is that in which the radioactive isotope is substituted for the stable isotope of the same element; thyroxine - ^{131}I, aldosterone -^{3}H, and vitamin B_{12} -^{57}Co are three such examples. The second type involves substitution reactions with production of a labeled compound that is different from the native substance. Insulin-^{131}I is an example in which a hydrogen atom on a tyrosyl residue is replaced with a radioactive iodine atom.

Purity is at best a vague concept and frequently defined in terms of the use to which materials are put. In the case of radiochemicals there are two additional aspects to this concept. The first is "radionuclidic purity" and refers to the proportion of the total radioactivity in the form of the stated radioelement (radionuclide). For example, all solutions of ^{140}Ba are radionuclidically impure since it decays to its daughter ^{140}La, and both parent and daughter are mixed β and γ emitters. Radionuclides are ordinarily marketed in "pure" form, and radionuclidic purity is rarely of concern to the clinical chemist. However, an important facet of radionuclidic purity is the isotopic abundance of particular radioelements, i.e., the ratio of the radioisotopic to the stable forms of the element. This is important because of the need for high specific activity of labeled compounds. Thus, labeling protein hormones with ^{131}I or ^{125}I is usually done with preparations approaching abundances of 100%, i.e., almost no ^{127}I or any other form of iodine is present. Such preparations are termed carrier free, carrier in this case being the nonradioactive form of the element.

The proportion of the total radioactivity in the stated chemical form is another aspect of "radiochemical purity." This is important since it is usually assumed that in the preparation of radiolabeled compounds, no other radiolabeled compounds are present. This may not be the case since solutions of radiochemicals are particularly unstable because of radiation effects on the solvent. Furthermore, dilution analysis (discussed below) depends on isolation of product materials that are radiochemically pure. Thus, the methods of storage and use of radiochemicals is conditioned by the need to obtain acceptable levels of radiochemical purity, and specific analytic procedures must be employed to obtain products of similarly acceptable levels of radiochemical purity.

Another aspect of "purity" is a consequence of the fact that the physicochemical properties of radioactive tracers are not always identical to those of their stable congeners. Ion exchange chromatography of ^{14}C-labeled amino acids (52), absorption chromatography of cholesterol and cholesterol acetate labeled with ^{3}H and ^{14}C (38), and enzymic hydrolysis of ^{14}C urea (54) have all revealed the effects attributable to the presence of the isotopic element in the system studied. Another relevant consideration is that labeling by substitution, as with insulin-^{131}I, results in a mixture of labeled products (discussed below).

Should the chemist wish to establish compliance with some criterion of radionuclidic purity, the simplest ways are either through examination of the energy spectrum of the product for the isotope in question or by measuring its half-life. The former, although requiring complex instrumentation and careful technic, is fairly simple. The latter is not practical since it involves loss of a significant amount of the isotope originally present and is certainly limited to isotopes of short half-lives. The testing of radiochemical purity, however, is commonly conducted by other technics in routine laboratory use. One is chromatography in several solvent systems with demonstration of a single radioactive locus in all cases. Another is repurification to constant specific activity using a number of means such as recrystallization, derivatization, and chromatographic and solvent partition. Both approaches have inherent weakness since additional similar operations might reveal an impurity. Furthermore, the appearance of artifacts can be misleading. For example, impurities may form upon application of a labeled compound to a chromatographic strip, or a volatile impurity may be lost upon drying chromatographic strips. With highly labile compounds, such as iodinated proteins, the only practical approach is demonstration that the labeled protein and the unlabeled material behave identically in the test system.

Radiochemical purity is important since the effects of impurities may be deceptive because of the sensitivity of radioisotopic measurements. For example, if an impurity with greater specific activity than the labeled substance of interest is present in chemically undetectable concentrations, a reaction product may be chemically pure by conventional criteria while highly impure radiochemically. Furthermore, impurities may resist elimination through many separation steps. This is particularly true of labeled derivatizing agents that are used at the highest possible specific activity. Unless storage conditions are carefully controlled and the materials used only for limited periods of time prior to repurification, radiochemical impurities may be formed that invalidate the use of the material (33).

Probably the most common radiochemical impurities are those which develop as the result of self-radiation. Labeled compounds decompose because of absorption of the radiation energy by the compound itself or other components of the solution. When the former occurs the resultant excited molecules disintegrate in some manner; when the latter occurs the radiation energy produces free radicals and other reactive species that can destroy the labeled molecules.

The modes of decomposition of labeled compounds are presented in Table 8-2 (4). The primary (internal) mode may result in the production of an impurity due to the disintegration of an unstable nucleus in the structure of the compound. The fragments will be radioactive only if the parent molecule originally contained two or more radioactive atoms. This mode of decomposition is ordinarily not quantitatively important. However, it can be significant in dealing with protein molecules labeled with two or more radioactive iodine atoms. For obvious reasons the *primary* (external) mode is quantitatively even less important.

TABLE 8-2. Modes of Decomposition of Radioisotopically Labeled Compounds

Mode	Cause	Method of control
Primary (internal)	Natural isotopic decay	None
Primary (external)	Direct interaction of the radioactive emission (α, β or γ) with the molecules of the compound.	Dispersal of the labeled molecules.
Secondary	Interaction of excited environmental products with the molecules of the compounds.	Dispersal of radioactive molecules, cooling to low temperatures; free radical scavenging.
Chemical	Thermodynamic instability of compounds and poor choice of environment.	Cooling to low temperatures; removal of harmful agents.

Secondary decomposition is, however, the most damaging, troublesome to control, and most susceptible to minor variations in environmental composition and conditions. In aqueous solution ionizing radiations produce a multiplicity of "reactive species" (hydrogen radicals, hydroxyl radicals, and peroxides). Reaction between these reactive species and the labeled compound results in the formation of labeled impurities. Secondary effects are especially important with the weak β emitters since all the β ray energy may be absorbed by the environment.

Control of self-radiolysis from secondary decomposition is largely limited to reduction of the environmental temperature and the use of "scavengers" that react preferentially with reactive species. In general, storage of aqueous solutions at temperatures below $-20°C$ produces little additional benefit. Anomalously, tritiated compounds between $0°C$ and $-100°C$ often undergo accelerated self-radiolysis. Tritiated compounds in aqueous solutions should be stored, therefore, either just above the freezing point of the preparation or at temperatures below $-196°C$ (liquid nitrogen).

Many different compounds can, at relatively low concentrations, act as "scavengers" for "reactive species" formed in aqueous solution. Some of these are ethanol, sodium formate, benzyl alcohol, cysteamine, and ascorbic acid.

The mechanism of decomposition of labeled organic compounds in non-aqueous solution is poorly understood, and the transfer of energy and production of "reactive species" is quite different from that observed with aqueous solutions. The best precautions are the use of low temperatures and of highly purified solvents. Use of the "scavenger" principle does not seem to produce the beneficial results obtained in aqueous environments. Where stability is a significant problem for the clinical chemist it would be advisable to consult the comprehensive study and collection of data provided by Bagly and Evans (5, 6).

Radionuclidic compounds are no more susceptible to chemical damage than their inert congeners. Damage and the formation of other labeled species can result in erroneous interpretation of experiments employing them as tracers, in which only a single tagged species is assumed to be present.

CONDITIONS FOR USE

The United States Atomic Energy Commission has authority by Federal statute to regulate the production, use, transfer, and disposal of radioactive materials. With the exception of "generally licensed quantities" (microcurie quantities of nonhazardous materials), the user must be licensed by the Commission to obtain radioactive compounds. The manner in which he uses and disposes of these compounds is specified by the rules and regulations of the Commission. Licensure is obtained by submitting AEC Form 313, entitled "Application for Byproduct Material License" to the Commission for approval. In a number of instances the Commission has delegated licensing and regulating authority to the States. There this authority is commonly administered by a division of the State Department of Health; in these cases applications can be obtained from, and must be submitted to, the State Department of Health. The form will request such information as the type and quantity of isotopes, training and experience of the user, and facilities and radiation monitoring instruments available.

The license will specify the isotopes that the user may possess and will limit the quantities that may be possessed. Such limits are absolute and include isotopes in storage prior to use, in actual use, and in storage for disposal. The license may further specify conditions for the monitoring of personnel and facilities.

The pertinent regulations are available from the Commission and from the State Departments of Health and compliance is a condition of licensure. The regulations include requirements for personnel and area monitoring, the use of warning signs and labels, the kinds and amounts of isotopes that may be discharged into sewerage and into the atmosphere, and the maintenance of receipt, possession, and disposal records. A user is advised to be familiar with a series of bulletins issued by the National Bureau of Standards that are available from the Superintendent of Documents, Washington, D.C. These cover such subjects as safe handling of radioisotopes, control and removal of contamination from the laboratory, recommendations for waste disposal, maximum permissible concentrations of radionuclides in air and water, and radiological monitoring and instrumentation.

There are only three permissible methods of disposing of radionuclides: into sewerage, into the atmosphere and by transfer to a licensed receiver. Regulations for all three methods are found in the information provided by the Commission and by its delegated State Department of Health. The user of generally licensed quantities ordinarily has no problem with disposal. When quantities exceed permissible levels, it may be necessary to store wastes so they can decay to levels permissible for discharge into sewerage or incinerated and discharged into the atmosphere. If this is not practical the only other action is to transfer wastes to a licensed commercial disposal firm. There are many that operate on both a national and local basis.

Regulatory agencies also possess the authority to inspect facilities, procedures, and records. Inspections will include the location and design of the laboratory space, finishes of bench tops, understructures, and floor and wall surfaces, the design of hoods, the type and use of personnel and facility radiation monitoring devices, and the selection of storage areas.

The clinical chemist interested in initiating the use of radioisotopes should familiarize himself with the regulations and supporting information so that proper systems of records, monitoring, usage, labeling, storage, and disposal are established at the outset. In addition, consultation from a certified health physicist should be considered at the start up and for periodic review of operations. The local availability of such assistance can be determined by addressing an inquiry to the American Board of Health Physics, P.O. Box 156, East Weymouth, Mass., 02189.

RESOURCES

The following list is partial and by no means constitutes an exclusive recommendation; it merely reflects limitation in the experience of the author.

I. Information

a. Excellent bibliographies of general and specific nature can be obtained from:

1. Abbott Laboratories, Radio-Pharmaceutical Products Division
2. Nuclear-Chicago Corporation
3. Packard Instrument Company, Incorporated
4. Amersham-Searle
5. Picker Corporation, Nuclear Division
6. New England Nuclear

b. Films. An excellent compilation of films useful for training purposes has been published by the Society of Nuclear Medical Technologists.

c. Commercial publications. All of the above organizations that provide bibliographic material also provide interested persons with monographs on a variety of subjects, ranging from general manuals for the training of novices to subjects of specific interest to the sophisticated user. To these should be added the excellent monographs on liquid scintillation counting published by Research Products International Corporation.

II. Instruments and Radiochemicals

The six suppliers listed above will provide the chemist with their instrument and radiochemical catalogs upon request. In addition, many of the scientific and technical journals familiar to clinical chemists (e.g., Science, Analytical Chemistry, Lab World, etc.) publish comprehensive annual listings of suppliers of services, instruments, isotopes and radiochemicals.

APPLICATIONS TO CLINICAL CHEMISTRY

The basic premise in the use of radioisotopic materials is that the labeled substance is physically and biologically indistinguishable from its unlabeled congener. This is only approximately correct since the masses of nuclei vary, and according to kinetic theory, the rates of movement differ for atoms of different masses. In physiochemical systems, relatively large differences in behavior are observed between ^{3}H and ^{1}H, lesser differences between ^{14}C and ^{12}C, and still smaller differences between ^{127}I and ^{131}I. For example, ^{14}C-labeled amino acids (52) and ^{14}C- and ^{3}H-labeled cholesterol acetate have been shown to behave differently in chromatographic systems (39), and ^{14}C-urea is more slowly hydrolyzed by urease than is ^{12}C-urea (54).

Nevertheless, many useful applications in clinical chemistry have been found for radioisotopic elements and radioisotopically labeled materials. These can be grouped into three broad categories: (a) tracer analysis, (b) dilution analysis, and (c) saturation analysis.

TRACER ANALYSIS

Tracer analysis is probably the most straightforward application. A labeled substance is added to the analytic system and its radioactive emission is used to follow and quantify the fate of unlabeled material. The partition of ^{131}I-triiodothyronine between plasma binding proteins and a nonspecific adsorber (T3 test) and the amounts of precipitation of ^{131}I-insulin by antibodies in serum are examples of proportionate tracer analyses. In these cases no relationship to chemical or activity units can be made, only the proportion participating in the behaviour of the system can be expressed. This can be quite useful, however, in detecting and quantifying deviations from the normal state.

Tracer analysis can be conducted in systems where results can be quantitatively expressed. For example one can measure the binding capacity of the thyroxine-binding proteins of plasma in units of thyroxine per unit of plasma (22, 49), enzyme activities in units of substrate used per unit of plasma [e.g., ornithine carbamyl transferase (55), arginase (37), cholinesterase (53)], and the concentration of free, unbound thryoxine in serum (41).

DILUTION ANALYSIS

Classical analytical chemistry requires that substances be quantitatively isolated in pure form or have some specific measurable characteristic that remains unobscured by other substances present. In biologic systems, this frequently constitutes insurmountable problems because the substances may be structurally complex, occur in minute concentration, and be present in an extremely heterogeneous system (blood or urine). The substances may be too delicate and/or in too low concentration to permit quantitative isolation and may have measurable characteristics that are not uniquely or even easily distinguishable from other substances present. Dilution analysis is well suited for such substances since it eliminates the need for quantitative recovery.

The basic principle of radioactive isotope dilution analysis is that reduction in specific activity occurs when a known small amount of a radioactive substance is mixed with the corresponding unlabeled compound. It follows that if the new specific activity is measured, the amount of nonradioactive diluting material can be calculated. One needs only to isolate some of the diluted substance in pure form and determine its specific activity. Quantitative recovery is unnecessary.

The theory of isotope dilution analysis has been examined by Gest *et al* (25). Mathematically, radioisotope dilution can be developed as follows:

Consider a radioactive compound having an activity A_0 (cpm) and a mass W_0 (ng or μg); its specific activity is:

$$(1) \qquad S_0 = \frac{A_0}{W_0}$$

The inactive form of this compound has no radioactivity but has a mass W_u of unknown magnitude. If the active and inactive forms of the compound are mixed, the total activity is still A_0, but the mass is now $W_0 + W_u$, and the new specific activity is:

$$(2) \qquad S_1 = \frac{A_0}{W_0 + W_u}$$

Since A_0 and W_0 are known and S_1 can be determined on a purified aliquot recovered by the selected separation process, then W_u can be calculated from the equation:

$$(3) \qquad W_u = W_0 \left(\frac{S_0}{S_1} - 1 \right)$$

Inspection of Eq. (2) reveals the principal limitations of this technic from the standpoint of the clinical chemist. Commonly, substances of interest are present in minute quantity, of the order of μg/liter or ng/liter. This means that S_0 must be extremely high to maximize the role of W_u in diluting A_0. Inspection of Eq. (3) shows that the less A_0 is diluted, the closer S_1 approaches S_0, and the imprecision of the expression $(S_0/S_1 - 1)$ increases. Hence, the requirement for high specific activity for the radioactive tracer limits the utility of this approach.

A second problem is the need for the compound of interest to be available in a labeled form. In many instances this is not feasible commercially, and the analyst, therefore, cannot benefit from the power of dilution analysis.

These objections may be overcome by the use of derivative dilution analysis (51). In this procedure the compound of interest is derivatized with a radioactive reagent of known specific activity, and the resultant product is separated from the reaction mixture in pure form. Since the amount of radioactivity in the pure product A_1 is equivalent to the amount of radioactive reagent it contains and since the stoichiometry of the reaction between reagent and compound is known, calculation of the amount of original compound present in the final sample is relatively simple. Mathematically, this procedure rests on the following equation:

$$(4) \qquad W_u = \frac{A_1}{S_0}$$

There are two fundamental requirements in derivative dilution analysis. First is the necessity to establish conditions under which compound and reagent react either quantitatively or with a reproducibly high yield. Since derivatizing agents are nonspecific, i.e., reaction is with certain structural groups in the compound, and since the reaction must be carried out on an extract of a biologic material, achievement of quantitative reaction conditions largely depends on the amount and nature of extraneous materials present. Second, it is necessary to separate the derivatized product completely from all other radioactive materials present, including excess quantities of the reagent and its breakdown products. Separation can be conceivably accomplished in three ways:

(a) Direct quantitative separation of the radioactive reaction product is practical only if the product is easily separable from the reaction mixture and was originally present in relatively large quantity. Obviously, this approach is very limited in clinical chemistry.

(b) Separation of the radioactive derivative is simplified by the prior addition of large amounts of the nonradioactive derivative to the reaction mixture. Once a pure product is obtained it need only be radioassayed and its mass chemically or physically measured to permit calculation of the amount of radioactive derivative originally present. Mathematically, this is a variant of equation (4):

(5)
$$W_u = \frac{A_1}{S_0} \times \frac{\text{amount of carrier added}}{\text{amount of carrier recovered}}$$

The problem with this approach is the need for carrier quantities of the derivatized compound. Carriers may be difficult to obtain and/or may be quite expensive; this restriction is particularly significant in the assay of blood for steroid hormones.

(c) In (b) above the carrier not only serves to make purification simpler but also acts as a measure of loss during the recovery procedure. The need for carrier can be eliminated and the estimation of recovery loss can be achieved by the addition of small quantities of the same compound derivatized with a different radioisotope. This is called a *loss label*. This is the double isotope derivative assay and is broadly applicable to the measurement in biologic materials of nonprotein compounds present in trace amounts. Mathematically, this is a variant of equation (5):

$$W_u = \frac{A_1}{S_0} \times \frac{\text{activity of derivative added}}{\text{activity of derivative recovered}}$$

The principal analytical feature is the radioassay of the final preparation for two different isotopes, a subject discussed previously. Commonly, the isotope pairs used will be ^{14}C and ^3H, and the differences in their energy spectra make these isotopes easily discriminated and counted with a liquid scintillation spectrometer. Acetic anhydride labeled with ^{14}C or ^{3}H is frequently used as the labeled derivatizing reagent. The analysis of human urine for aldosterone is an excellent example of this approach (39). Both derivatizing agents are easily available in high specific activity. The isotope pairs $^{131}I + ^{35}S$ and $^{125}I + ^{35}S$ have been used in labeling p-iodophenylsulfonyl chloride and p-iodophenylsulfonic anhydride. From a radiometric point of view, these are assayed more simply and easily than the $^{3}H/^{14}C$ pair. However, since they are not easily available commercially and are prepared and stored with difficulty, their utility in clinical chemistry has been extremely restricted.

A few points of caution are worthwhile noting when considering the use of dilution analysis and its powerful variant, double isotope derivative assay. Addition of the loss label should be achieved as early as possible in the sequence of analytical steps. Preferably, the compound of interest in labeled form is added to the specimen before any processing occurs; this affords a more complete control of loss during the subsequent analytic steps. Both the added labeled compound and the endogenous unlabeled compound are derivatized by the reagent labeled with a different isotope, the constituent label. An example is the use of ^{3}H-labeled testosterone added to plasma and derivatization of the plasma extract containing the ^{3}H-testosterone + endogenous testosterone with ^{35}S-thiosemicarbazide (57).

Chemical and radiochemical purity of the derivatizing agents are essential and must be as high as possible. Chemical impurity or trace radiochemical impurity will result in erroneous value for S_0, and thus erroneous values in W_u. The existence of radiochemical impurities can be determined by procedures used to verify specific activity. Carrier quantities of a suitable compound can be derivatized and purification to constant specific activity will reveal this source of error.

When the loss label is added to either the specimen or the reaction mixture, it is important that mixing be complete. Ordinarily this is a minor problem easily solved by careful attention to technic.

If the loss label is added to an extract of the specimen, it is necessary to have an independent estimate of the loss of endogenous compound during the processing up to this point. This can be accomplished in model experiments employing specimens to which trace quantities of the labeled compound have been added. This means, however, that careful attention must be paid to standardize the extraction procedure when performing routine analyses with specimens.

The double isotope derivative assay procedure has been used in both research and routine analysis of human body fluids with great success. The preceding discussion is intended only to present general principles. The procedures are comprised of many operations involving chemical extraction, derivatization, and chromatographic separation. It is important in setting up a procedure to review carefully the literature in order to make the proper choice among the many modifications of a particular analysis. Table 8-3 lists a few applications of the double isotope procedure.

TABLE 8-3. Some Applications of the Double Isotope Derivative Assay Principle to Hormone Analysis

Compound	Specimen	Loss label	Constituent label	Sensitivity ng	Reference
Total catecholamines	plasma urine	^{3}H-norepinephrine	S-adenosylmethionine-methyl-^{14}C	1	23
Thyroxine	serum	^{131}I-thyroxine	^{3}H-acetic anhydride	100	29
Triiodothyronine	serum	^{131}I-triiodothyronine	^{3}H-acetic anhydride	2	30
Aldosterone	urine	Aldosterone-diacetate-^{14}C	^{3}H-acetic anhydride	10	38
	plasma	^{14}C-aldosterone	^{3}H-acetic anhydride	0.25	9
Hydrocortisone	urine	^{14}C-hydrocortisone	^{3}H-acetic anhydride	50	60
Estrone	plasma	^{3}H-estrone	^{35}S-p-iodobenzene-sulphonyl chloride	0.4	2
Estradiol	plasma	^{3}H-estradiol	^{35}S-p-iodobenzene-sulphonyl chloride	0.4	2
Testosterone	plasma	^{3}H-testosterone	^{35}S-thiosemicarbazide	3	57
Androstenedione	plasma	^{3}H-androstenedione	^{35}S-thiosemicarbazide	0.5	35
Progesterone	plasma	^{3}H-progesterone	^{35}S-thiosemicarbazide	0.5	56
Cortisol	plasma	Cortisol ester of ^{131}I-p-iodophenylsulfonic acid anhydride	^{35}S-p-iodophenylsulfonic acid anhydride	200	8

SATURATION ANALYSIS

Although dilution analysis, particularly the double isotope derivative assay principle, provides the means for increasing both analytical sensitivity and specificity, it also possesses a number of weaknesses.

The absolute requirement for sample purity is satisfied operationally. Commonly, a number of different chromatographic steps, with additional changes in derivative structure, are employed. This approach assumes that purity has been demonstrated if two or more successive operations produce no change in specific activity. To the extent that different compounds of similar structure partition or react similarly, this assumption is incorrect.

Purification of protein molecules, especially those present in dilute solution (see Table 8-4), is extremely difficult because of the lability of secondary and tertiary structural properties. In addition, the reactive groups of the plasma proteins are present in concentrations 6–8 orders of magnitude greater than those of the protein hormones, thus further complicating separations based on chemical technics.

TABLE 8-4. The Approximate Concentrations of Some Hormones in Serum

Hormone	Plasma concentration, molar
Thyroxine	1×10^{-7}
Triiodothyronine	2×10^{-9}
Cortisol	2×10^{-7}
Catecholamines	2×10^{-9}
Insulin	1×10^{-10}
Glucagon	1×10^{-10}
Adrenocorticotropin	1×10^{-12}
Growth hormone	6×10^{-10}

In practice, dilution analysis has been applied to substances in plasma existing at concentrations of $10^{-9}M$ or lower. At such dilutions precise radioassay requires great care since raw counting values begin to approach background and the influence of loss label radiations becomes significant. These factors prohibit satisfactory radioassay when dealing with compounds ranging from 1 – 5 orders of magnitude less concentrated (Table 8-4).

On the other hand, dilution analysis for thyroidal and steroidal hormones is possible because such substances possess distinctive chemical groupings and solubilities not possessed by protein and peptide hormones.

The *in vitro* analysis for protein hormones at physiologic concentrations has been made possible by the development of radioimmunoassay methods. This is best explained through consideration of the following equation:

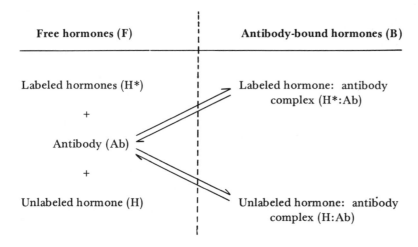

Labeled hormone (H*) binds to specific antibody (Ab) to form a labeled antigen-antibody complex (H*-Ab). Unlabeled hormone (H) in plasma, or other solution, competes with labeled hormone for antibody and thus inhibits or displaces the labeled hormone. As a result, the ratio of antibody-bound (B) to free (F) labeled hormone, denoted B/F, is diminished as the concentration of unlabeled hormone is increased. The concentration in an unknown sample can then be obtained by comparing the B/F ratio determined for the unknown with those obtained with a series of standard solutions containing known amounts of hormone (Fig. 8-13).

FIG 8-13. Idealized plot of B/F ratio as a function of unlabeled hormone added to the system.

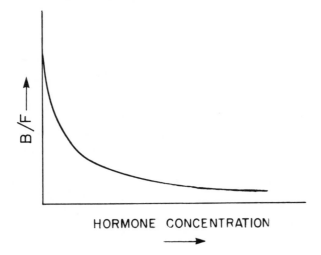

This formulation can be generalized to the assay of non-hormonal substances and nonimmune systems.

$$S* + R \rightleftharpoons S*R$$
$$+$$
$$S$$
$$\updownarrow$$
$$SR$$

where S is any substance to be measured, S* its radioactive form, and R is a specific reactor. The system thyroxine:thyroxine-binding globulin is one example where S=thyroxine, S* = ^{131}I-thyroxine and R = thyroxine-binding globulin.

Since the mid 1950s, when this approach to analysis first appeared (7), it has been the subject of intense reasearch. The basic principle was extended to the analysis of nonantigenic materials (20, 47) in which naturally occurring binding proteins were used in place of antibodies. More recently, nonantigenic materials have been coupled with proteins (13) and synthetic peptides (28) in order to function as antigens in the production of antibodies, extending this approach even further. Saturation analysis has now been extended to some 45 protein and nonprotein substances using antibody systems and to 8 substances using binding proteins other than antibodies.

Saturation analysis has been extensively reviewed and the clinical chemist should be thoroughly familiar with these reports before attempting to set up methods (1, 18, 21, 31, 63). The principles involved and the problems encountered are multifarious and require familiarity with the fields of protein chemistry, organic chemistry, radiochemistry, immunology, radioassay technic, and statistical analysis. In addition, many aspects of particular procedures are empiric. The remainder of this section is concerned, therefore, with discussion of issues central to saturation analysis from the standpoint of radioisotope application. Discussions of the principles involved in antigen purification, antibody production, or the conduct of quantitative antigen-antibody analytical methods will not be presented.

Before proceeding further, the name of this approach to analysis deserves consideration. Historically, several different names have been employed for different systems, e.g., radioimmunoassay, saturation analysis, competitive protein binding analysis, displacement analysis and radiostereoassay. None fits all applications of the principles described and some objection can be made to any one. "Saturation analysis" is preferred as a general term since it best fits most applications of current interest to the clinical chemist. All these applications include an immune or nonimmune reactor whose binding sites become saturated with both radioactive and nonradioactive forms of the ligand of interest.

The specific radioactive reagent used in saturation analysis is commonly a protein substituted on one of its tyrosyl residues with ^{131}I or [125]I. The chemistry of such substitutions has been reviewed by Hughes and others (18). The goal of iodination is the production of a labeled protein of high specific activity that will function as a tracer of the unlabeled protein in the analytical system. The labeled protein must exhibit behavior identical to the unlabeled protein in the system regardless of similarities and dissimilarities in other systems.

The attainment of high specific activity is determined by the specific activity of the ^{131}I or ^{125}I used and the number of tyrosyl residues available for substitution. Much research has shown that more heavily iodinated proteins lose immunoreactivity, and for this reason, a substitution level averaging less than one I atom/protein molecule is generally accepted. This means that the specific activity of the labeled protein will, therefore, be a direct function of the specific activity of the radioiodine used. The term "averaging less than 1 iodine atom/molecule" does not mean that all iodinated molecules are monosubstituted. Rather, such a preparation will consist of unsubstituted molecules, monosubstituted and multisubstituted molecules. Multisubstituted molecules will be of several types; those containing only monoiodotyrosine, and those containing mixtures of mono- and diiodotyrosine. Preparations will be substituted variably at the different tyrosyl residues where more than one exists per protein molecule. The degree and location of substitution will affect the immunoreactivity of the molecule. Hence, a labeled protein preparation must be regarded as a collection of molecules of various degrees of immunoreactivity. For these reasons, only "carrier free" iodine designated "for iodination" must be used since it will not contain the various reducing agents present in material for clinical use. Even then, the degree of labeling of protein under standard conditions is highly variable and unpredictable. Yalow and Berson have attributed this variability, in part at least, to the variable isotope abundance of ^{131}I in different lots of carrier-free material (range 5%−44%) (31).

Several different technics have been developed for iodinating microgram quantities of proteins with carrier-free iodine. A few are the use of I_2 (7, 42), the use of ICl (42) and the use of chloramine-T and ^{131}I-NaI (26). The chloramine-T technic appears to have been generally accepted as the method of choice although its chemistry is poorly understood, if at all. Hughes (18) has suggested the following reactions:

$$CH_3 \langle \rangle SO_2NHCl + 2\,I^- \rightarrow CH_3 \langle \rangle SO_2NH^- + Cl^- + I_2$$

$$I_2 + R \langle \rangle O^- \rightarrow + R \langle \rangle O^- + I^- + H^+$$

On the other hand, Hunter (36) postulates that chloramine-T forms an active intermediate in which the iodine carries a positive charge. The principal virtues of this method lie in the direct use of carrier-free iodine in ordinary reaction vessels and the fact that reaction is practically instantaneous.

After iodination the product must be quickly separated from the reaction mixture for two principal reasons. The assay method requires all radioactivity to be associated only with the labeled protein, and the labeled protein becomes "damaged" (immuno-nonreactive labeled protein is formed) during the process of labeling. This is termed "preparation damage," and is determined by several factors: the purity and chemical stability of the protein itself, the number of substituted tyrosyl groups/molecule, the presence of noxious substances in the radioiodine solution, and the effect of exposure to ionizing radiations. The ability of undamaged hormone to adsorb to cellulose is exploited for purification of ^{131}I-labeled peptides by adding a small amount of plasma to the reaction mixture before placing it on a cellulose column. Radioiodine and damaged fragments bound to plasma protein are unable to bind to cellulose and therefore, pass through the column; undamaged peptides bind to the cellulose. After washing the column with suitable buffers, the undamaged peptides can be eluted with larger amounts of plasma or other solvents, and the product is free from all other forms of radioactive material (62).

Immunochemical reactivity can be tested by incubating the labeled hormone with excess antibody and examining the mixture by paper chromatoelectrophoresis (64). Suitable preparations show no significant radioactivity migrating with normal serum proteins and all radioactivity completely localized in the γ globulin region, i.e., antibody bound. This is not an absolute proof since slight impairment of immunoreactivity may still be demonstrated in very dilute solutions of antibody. Additionally, different hormones behave differently under apparently identical labeling and purification conditions, hence the conditions must be specifically worked out for each hormone.

Another form of damage arises from exposure of the purified labeled hormone to the reactants of the assay system. This is termed "incubation damage" and seems to be related to the proteolytic factors in plasma. Incubation damage can be quantitatively variable among assay specimens and thus be a considerable source of inaccuracy. Various enzyme inhibitors and reducing agents have been found to be effective in controlling this source of difficulty (19, 64).

Another problem in saturation analysis lies in the separation of the "bound" (B) and "free" (F) forms of the labeled hormone. The ultimate measurement is the count rate of either the "bound" or the "free" fractions and its numerical value is directly related to the amount of unlabeled hormone in the specimen. Thus, the accuracy and precision of the result will be determined by the amount of contamination of one form by the other in the counting sample.

The first method used for B/F separation was employed in the radioimmunoassay of plasma insulin (7). This technic, termed paper chromatoelectrophoresis, depends on the property of insulin and ^{131}I-insulin to bind strongly to paper whereas the insulin-antibody complex does not. Separation is achieved by a combination of chromatographic partition on paper strips followed by electrophoretic migration. The B and F fractions can be then identified and quantitated by radioscanning of the paper strip. This technic provides virtually complete separation of B and F. Furthermore, radioactive iodide that arises in variable quantities from incubation damage is completely eliminated as a contaminant. The principal disadvantages of paper chromatoelectrophoresis are that it can be used only for those proteins which bind to paper, only a small aliquot of the incubation mixture (<0.5 ml) can be practically applied to the paper strip, and mass production is space consuming. Also this approach is relatively cumbersome and not easily adaptable to automated procedures and instrumentation.

Three other approaches to B/F separation are commonly employed and have found successful application for many forms of saturation analysis. One involves the ability of the labeled material and its unlabeled counterpart to adsorb to an inert solid with the antibody-antigen complex not being so adsorbed. Sephadex (24), ion-exchange resins (40), Florisil (3), charcoal coated with dextran (34), talcum, and precipitated silica (58) have been used as the "solid phase" in such systems. Rigid standardization of all details of the conditions is vital to obtain reproducible results, and one must be alert to the possibility of differences in adsorptive properties of the binder from one manufacturer's lot to the next.

Another approach employs the prior binding of the antibody, or non-immune receptor, to Sephadex (43, 61), to a copolymer of styrene and tetrafluoroethylene in the form of powders (12, 10) or disks (14), and to polypropylene or polystyrene surfaces, such as disks (17) and test tubes (11, 15). This technic has several significant advantages: The antibody-adsorbant preparation can be made in advance and stored for use; once the antigen-antibody reaction has come to equilibrium, relatively little handling is necessary prior to isotope counting (a few rinses in a suitable solvent suffice); incubation damage to the labeled antigen from serum factors is virtually eliminated since the serum is removed prior to adding the labeled antigen.

Probably the most widely used approach to B/F separation is the double antibody technic (46, 59). In this approach, labeled antigen, unknown specimen, and antibody are incubated, usually at 4°C for several days. To this mixture is added nonimmune serum obtained from the same species as the antibody, followed by the addition of an antibody preparation against the serum of this species prepared in a different species. For example, the first antibody is prepared in the guinea pig, and the second, an antibody against guinea pig serum, is prepared in the rabbit. The significance is that the second antibody precipitates the original antigen-antibody complex along with the added guinea pig γ-globulin. Counting the precipitate provides an estimate of the B fraction. The most significant problem with this technic is incomplete precipitation since the amount of γ-globulin in the reaction mixture is variable, being the sum of that in the specimen and that in the nonimmune guinea pig serum.

A variety of other technics has been used but only infrequently. These include the use of glutathione-activated ficin to degrade insulin into fragments nonprecipitable with trichloroacetic acid (44, 45) and of Na_2SO_4 (27) and alcohol-saline mixtures to precipitate the insulin-antibody complex (32, 48).

SELECTED BIBLIOGRAPHY

BIRKS JB: The Theory and Practice of Scintillation Counting. London, Pergamon, 1964

BRANSOME ED, editor: The Current Status of Liquid Scintillation Counting. New York, Grune Stratton, 1970

GLASSTONE S: Sourcebook of Atomic Energy, 3rd ed. New York, Van Nostrand, Reinhold, 1967

HINE GJ editor: Instrumentation in Nuclear Medicine. New York, Academic, 1967

LAPP RE, ANDREWS HL: Nuclear Radiation Physics, 3rd ed. Englewood Cliffs, N.J., Prentice-Hall, 1963

Liquid Scintillation Counting, 2nd ed. Des Plaines, Ill., Nuclear-Chicago Corporation 1966

OVERMAN RT, CLARK HM: Radioisotope Techniques. New York, McGraw-Hill, 1960

SHARPE J: Nuclear Radiation Detectors, 2nd ed. London, Methuen, 1964

U.S. Department of Commerce, National Bureau of Standards, A manual of radioactivity procedures. Recommendations of the National Committee on Radiation Protection & Measurement. (Handbook 80; N.C.R. Report #28) Washington, D.C., 1961

REFERENCES

1. AUBERT ML: *J Nucl Biol Med 14*:1, 1970
2. BAIRD DT: *J Clin Endocrinol Metab 28*:244, 1968
3. BAYARD F, BEITINS IZ, KOWARSKI A, MIGEON CJ: *J Clin Endocrinol Metab 31*:1, 1970
4. BAYLY RJ, WEIGL H: *Nature 188*:384, 1960
5. BAYLY RJ, EVANS EA: *J Labelled Compounds 2*:1, 1966
6. BAYLY RJ, EVANS EA: *J Labelled Compounds 3 (Suppl 1):349*, 1967
7. BERSON SA, YALOW RS, BAUMAN A, ROTHSCHILD MA, NEWERLY K: *J Clin Invest 35*:170, 1956
8. BOJESEN E: *Scand J Clin Lab Invest 8*:55, 1956
9. BRODIE AH, SHIMIZU N, TAIT SA, TAIT JF: *J Clin Endocrinol Metab 27*:997, 1967
10. CATT K, NIALL HD, TREGAR GW: *Nature 213*:825, 1967
11. CATT K, TREGAR GW: *Science 158*:1570, 1967
12. CATT K, NIALL HD, TREGAR GW: *Biochem J 100*:31c, 1966
13. CATT KJ, CAIN MC, COGHLAN JP: *Lancet 2*:1005, 1969
14. CATT KJ, NIALL HD, TREGAR GW, BURGER HG: *J Clin Endocrinol Metab 28*:121, 1968
15. CESKA M, GROSSMILLER F, LUNDQUIST U: *Acta Endocrinol 64*:111, 1970
16. CHASE GD: *Principles of Radioisotope Methodology.* Minneapolis, Burgess 1959
17. CROSIGNANI PG, NAKAMURA RM, HOVLAND DN, MISHELL DR Jr: *J Clin Endocrinol Metab 30*:153, 1970
18. DONATO L, MILHAND G, SERCHIS J Editors: *Conference on Problems Connected with the Preparation and Use of Labelled Proteins in Tracer Studies,* Pisa, January 17-19, 1966, Euratom, Brussels
19. EISENTRAUT AM, WHISSEN N, UNGER RH: *Am J Med Sci 255*:137, 1968
20. EKINS RP: *Clin Chim Acta 5*:453, 1960
21. EKINS RP, NEWMAN GB, O'RIORDAN JLH: *Saturation Analysis* in *Statistics in Endocrinology,* edited by JW McArthur, T Colton. Cambridge, Mass. MIT, 1970, p 345

22. ELZINGA KE, CARR EA, BEIERWALTES WH: *Am J Clin Pathol 36*:125, 1961
23. ENGELMAN K, PORTNAY B, LOVENBERG W: *Am J Med Sci 255*:269, 1968
24. GENUTH S, FROHMAN LA, LEBOVITZ HE: *J Clin Endocrinol Metab 25*:1043, 1965
25. GEST H, KAMEN MD, REINER JM: *Arch Biochem 12*:273, 1947
26. GREENWOOD FC, HUNTER WM, GLOVER JE: *Biochem J 89*:114, 1963
27. GRODSKY GM, FORSHAM PH: *J Clin Invest 30*:1070, 1960
28. HABER E, KOERNER T, PAGE LB, KLIMAN B, PURNODE A: *J Clin Endocrinol Metab 29*:1349, 1969
29. HAGEN GA, KLIMAN B, STANBURY JB: *Clin Res 15*:31, 1967
30. HAGEN GA, DIUGUID LI, KLIMAN B, STANBURY JB: *Clin Res 18*:602, 1970
31. HAYES RL, GOSWITZ FA, MURPHY BEP Editors: *Proceedings of a Symposium on Radioisotopes in Medicine*, Oak Ridge, November 13-16, 1967. U.S. Atomic Energy Commission, Div of Technical Information, Oak Ridge, Tenn, 1968
32. HEDING LG: *A Simplified Insulin Radioimmunoassay Method in Conference on Problems Connected with the Preparations and Use of Labeled Proteins in Tracer Studies*, edited by L Donato, G Milhaud, J Sirchis. European Atomic Energy Community, 1966, p 345
33. HENDERSON HH, CROWLEY F, GAUDETTE LE: *Advances in Tracer Methodology*, edited by S Rothchild. New York, Plenum, 1965, Vol 2, p 83
34. HERBERT V, LAU KS, GOTTLIEB CW, BLEICHER SJ: *J Clin Endocrinol Metab 25*:1375, 1965
35. HORTON R: *J Clin Endocrinol Metab 25*:1237, 1965
36. HUNTER WM: *Iodination of Protein Compounds in Proceedings of a Symposium on Radioactive Pharmaceuticals*, Oak Ridge, November 1-4, 1965 edited by GA Andrews, RM Knisely, HN Wagner Jr. U.S. Atomic Energy Commission, Div of Technical Information, Oak Ridge, Tenn, 1966, p 245
37. KAIHARA S, CARULLI N, WAGNER HN Jr: *J Nucl Med 9*:329, 1968
38. KLEIN PD: in Ref. 33, p 145
39. KLIMAN B, PETERSON RE: *J Biol Chem 235*:1639, 1960
40. LAZARUS L, YOUNG JD: *Austral J Biol Med Sci 46*:791, 1968
41. LEE ND, PILEGGI VJ: *Clin Chem 17*:166, 1971
42. McFARLANE AS: *Nature 182*:53, 1958
43. MILES LEM, HALES CN: *Biochem J 108*:611, 1968
44. MITCHELL ML, BYRON J: *Diabetes 16*:656, 1967
45. MITCHELL ML, COLLINS S, BYRON J: *J Clin Endocrinol Metab 29*:257, 1969
46. MORGAN CR, LAZAROW A: *Diabetes 12*:115, 1963
47. MURPHY BEP: *Nature 201*:679, 1964
48. ODELL WD, WILBER JF, PAUL WE: *J Clin Endocrinol Metab 25*:1179, 1965
49. OPPENHEIMER JH, MARTINEZ M, BERNSTEIN G: *J Lab Clin Med 67*:500, 1966
50. PENNISI F, ROSA U: *J Nucl Biol Med 13*:64, 1969
51. PETERSON RE: in *Advances in Tracer Methodology*, edited by S Rothchild. New York, Plenum, 1963, Vol 1. p 265
52. PIEZ HA, EAGLE H: *Science 122*:968, 1955
53. POTTER LT: *J Pharmacol Exp Therap 156*:500, 1967
54. RABINOWITZ JL, SALL T, BIERLY JN Jr, OLEKSYSHYN O: *Arch Biochem Biophys 63*:437, 1956
55. REICHARD H: *J Lab Clin Med 63*:1061, 1964
56. RIONDEL A, TAIT JF, TAIT SAS, GUT M, LITTLE B: *J Clin Endocrinol Metab 25*:229, 1965
57. RIONDEL A, TAIT JF, GUT M, TAIT SAS, JOACHIM E, LITTLE B: *J Clin Endocrinol Metab 23*:620, 1963
58. ROSSELIN G, ASSAN R, YALOW RS, BERSON SA: *Nature 212*:355, 1966

59. SOELDNER JS, SLONE D: *Diabetes 14*:771, 1965
60. VAGNUCCI AI, HESSER ME, KOZAK GP, PAUK GL, LAULER DP, THORN GW: *J Clin Endocrinol Metab 25*:1331, 1965
61. WEIDE L, AXEN R, PORATH J: *Immunochemistry 4*:381, 1967
62. YALOW RS, BERSON SA: *J Clin Invest 39*:1157, 1960
63. YALOW RS, BERSON SA: *Radioimmunoassays in Statistics in Endocrinology*, edited by JW McArthur, T Calton. Cambridge, Mass., MIT, 1970, p. 327
64. YALOW RS, BERSON SA: *Preparation of High Specific Activity Iodine-^{131}I-labeled Hormones, Use in Radioimmunoassay of Hormones in Plasma* in Proceedings of a Symposium on Radioactive Pharmaceuticals, Oak Ridge, Nov 1-4, 1965, edited by GA Andrew, RM Kinsely, HN Wagner Jr. U.S. Atomic Energy Commission, Div of Technical Information, Oak Ridge, Tenn, 1966, p 265

Chapter 9

Micro and Ultramicro Technic and Equipment

FRANK A. IBBOTT, Ph.D.

Micro and ultramicro technics are needed not only where there is an insufficient sample for the usual "macro" technics, especially in the case of infants when the sample size for blood determinations is necessarily limited, but also in an ever increasing number of manipulations associated with thin layer and gas chromatography. These latter technics are now firmly established in the clinical laboratory and their applications appear to be increasing.

There is still not complete agreement upon the use of the terms macro, semimicro, micro, and ultramicro. A recommendation has been made that when volumes of original sample are the order of 1 ml, this be referred to as macro (1), and volumes between 0.1 ml and 0.5 ml be termed semimicro (11). When 100 μl or less of sample is required, the term microliter method should be used (10). According to Meites (11), those microliter methods requiring 25 μl or more are "micro" methods, whereas Mattenheimer (10) would reserve this term for 10 μl. Where yet smaller volumes are used, the term ultramicro is appropriate.

Most macro procedures are uneconomical with respect to sample size and can be "scaled down" rather easily to the micro level when necessary. For example, in many procedures a 1:10 protein-free filtrate is made, starting with 1 ml of blood or serum and then only 1 or 2 ml of the filtrate is used. The proper way to scale down a procedure is to start at the end and work backward: If the final measurement is photometric or spectrophotometric, the limiting factor is the minimal volume required in the smallest cuvets available and the lowest concentration that will still give reasonable photometric accuracy (cf. section on photometric error in Chap. 1). Micro cuvets are available for many photometers. Even when 5 ml is required, however, most determinations can be performed on a 100 or 200 μl sample. For such modified procedures the only alteration in technic usually required is that the protein-free solution be obtained by centrifugation rather than by filtration because of the greater loss of solution in the latter method.

Tests on the ultramicro scale, however, require the use of micro apparatus, which in turn requires manipulative skill acquired only by practice. The procedures are more time consuming and are less susceptible to mass production than are macro or micro methods. When the necessary skill has been acquired, the micro method precision is at least equal to the macro method from which it was derived (10). Ultramicro technics are ordinarily employed when there is no other choice, or when the entire laboratory has been converted to this operating scale, such as a laboratory dealing only with pediatric patients.

When working with very small volumes, certain factors relatively insignificant in macro procedures assume importance. Evaporation (especially of the sample but also of solvents) can produce errors that render results unacceptable, particularly in low humidity regions.

COLLECTION OF BLOOD SAMPLES

Ultramicro procedures are most frequently used in pediatric practice, where venipuncture is difficult and to be avoided if possible. The alternative is blood collection by skin puncture using a disposable lancet, a scalpel blade, such as a Bard-Parker 11, or a spring lance. Sterile disposable lancets are recommended because of the danger of transmission of infectious hepatitis through incompletely sterilized scalpel blades or reusable lancets. Skin puncture may be performed in the adult on the finger tip or ear lobe; the heel or big toe are the sites used in infants. According to Josephson, (8) because of the danger of osteomyelitis heel puncture should be abandoned in favor of the big toe. The site of puncture is thoroughly cleaned with 70% alcohol, which may be followed by ether and then is dried with sterile cotton or gauze. Squeezing of the surrounding tissue is to be avoided; the first drop of blood is always wiped away.

A single puncture may give up to 1 ml of blood, and by multiple puncture over a limited area, several milliliters blood can usually be obtained.

The following method can be employed for blood collection (12). Prepare capillary tubes from Pyrex tubing (bore 1.5−2.7 mm; o.d. 2.8−4 mm). Heat the tube in a burner and draw out to a tip having a bore of about 0.7−1 mm. Break at the tip and 11 cm from the tip, after scoring with an ampul file. Fire polish the ends, clean the tubes, and dry. If whole blood or plasma rather than serum is desired, just before use draw up an anticoagulant solution (potassium oxalate, trisodium citrate, or heparin) in the tube and then expel it. Fill the tube to a height of 5−7 cm (approximately 100−300 μl) by capillarity by touching the tip to the drop of blood collected at the site of skin puncture and holding the tube in a nearly horizontal position. Then close the wide end of the tube in one of the following ways: Either use a small vaccine vial stopper (1) (No. 68MS orange V-3 stopper, West, fits over the end of a capillary tube having an outside diameter of 4 mm), or hold the tube at an angle of about 30° with the wide end down and warm the wide end in a flame along with DeKhotinsky cement or sealing wax, press the soft cement against the tube, and while it is still soft press the end of the tube vertically on a flat surface to force cement into the base of the tube.

Separation of cells and plasma or serum can be accomplished by placing the tube in an ordinary test tube or conical centrifuge tube and then centrifuging. After centrifugation the tube can be broken off just above the serum or plasma level after scoring with an ampul file, and serum or plasma removed by a micro pipet. For storage purposes, the tube can be broken off just above the junction of cells and supernate, and both ends of the tube segment containing serum or plasma can be capped by the small vaccine vial stoppers. Caraway blood collecting tubes are also convenient, in which case a special capillary sealant should be used to close the bottom of the tube prior to centrifugation.

Another alternative is that described by O'Brien *et al* (13) using small, rimless Pyrex tubes, which hold approximately 1 ml total volume; they are prepared with heparin and a layer of mineral oil to prevent evaporation of the sample. They are readily capped and centrifuged and are convenient in that they are of suitable dimensions to accept Folin pipets.

A further alternative has been described in which blood drawn into a plastic tuberculin syringe is transferred to 0.125 in. i.d. Tygon tubing, one end of which is then heat sealed. The device then may be placed in a special container made from the barrel of a plastic syringe and centrifuged (2).

The method of blood collection on filter paper has found popularity, especially for phenylalanine determinations (6), and its use for other analyses has been examined (7).

WEIGHING

Few gravimetric procedures are employed in clinical chemistry at the micro or ultramicro level. Examples that do occur are the determinations of sweat and nail electrolytes. A good four-place analytic balance, such as is found in most clinical laboratories, should suffice for virtually all requirements.

PIPETTING

Micro and ultramicro pipets are calibrated either to contain (TC) or to deliver (TD) a specified volume. Because of the relatively large ratio of glass surface to contained volume in capillary pipets, the drainage error in TD pipets may be quite large if the pipet is not perfectly clean or if the surface tension or viscosity of the solution pipetted differs significantly from that of water, for which the pipet is calibrated. The TC or "washout" type of pipet is capable of significantly greater accuracy.

Designs for micro pipets embodying new concepts have recently appeared. The following is intended as a representative rather than an exhaustive catalog:

The Kirk transfer pipet (TC), a modification of the Pregl pipet, is available for quantities from 1 μl to 1 ml from Microchemical Specialties, and other laboratory supply houses (see Fig. 9-1A). Filling this kind of pipet can be accomplished by mouth suction with the aid of a rubber tubing attached to the pipet and a mouthpiece (as is customary for drawing up blood in a Sahli hemoglobin pipet), but this method requires practice. More accurate control can be obtained by use of a *pipet control,* which is, essentially, a syringe with a gasket on the tip.

The plunger of the control should be lubricated, using a little mineral oil if it is snug fitting or petroleum jelly or silicone grease if it is loose fitting.

After the control end is first moistened the pipet is fitted tightly into this gasket. This can be done using a wick inserted in a small narrow mouth bottle containing distilled water and a small amount of detergent. If the pipet remains in the control too long the water will dry out, possibly leading to freezing of the pipet in the control. The pipet control can be operated by one hand by grasping the plunger end between thumb and forefinger and the barrel between the other fingers and the palm. The tip of the pipet is inserted into the fluid and the pipet filled with a slow, rotary movement of the plunger. The liquid is drawn exactly to, or *slightly* above, the mark. Taking the fluid above the mark and subsequently washing out the pipet above the meniscus may result in an error as great as that given with TD pipets of the same capacity. The pipet is then removed from the liquid and the tip wiped off in a quick stroke with cotton or absorbent tissue. If necessary, the meniscus can be brought exactly to the mark by touching the tip of the pipet to a fingertip (not if analysis is for Na^+ or Cl^-), to a clean cover glass, or to a piece of filter paper. The contents of the pipet are then delivered by reverse manipulation of the plunger of the pipet control. If the procedure is one of dilution, the contents are delivered slowly into the diluent when the sample will frequently sink, thus leaving clean diluent with which to rinse the pipet two or more times. If the procedure is one of delivery of the pipet contents, the pipet tip is washed off with solvent following the expelling of its contents and the pipet is then rinsed out at least two or three times with solvent, the washings being added to the first delivery.

The *self-filling transfer pipet* (TC) fills automatically by capillary action when the tip is dipped into fluid (Fig. 9-1B). It is available in capacities from 0.25 to 10 µl from the same companies as market the Kirk transfer pipets. They are emptied by pressure exerted either by a pipet control or a rubber bulb or by mouth through a rubber tube. For quantities greater than 10 µl a bulb is placed in the capillary, and the pipet becomes a *self-adjusting* or *overflow transfer pipet* (TC or TD). These pipets require suction for filling, the fluid being permitted to overflow the capillary orifice. Capillary force holds the liquid at this point. Overflow pipets are available commercially up to capacities of 500 µl or can be constructed by the analyst (12).

The *Lang-Levy delivery pipet* (TD) (Fig. 9-1C) is available in capacities of 5–500 µl from the same companies that market the Kirk transfer pipet. Fluid is sucked up just above the constriction, and with the pipet tip still below the liquid surface, the suction is released; the fluid automatically stops at the constriction. The pipet is withdrawn from the liquid, the bent tip is wiped, then placed against the side of the receiving tube, and the fluid ejected by gentle pressure from the mouth or a pipet control.

The *Grunbaum pipet* (TD) has been described as a self-contained analytic instrument in that it is capable of successive sampling with self-adjustment and self-cleaning, dispensing, and reagent storage (see Figs. 9-1D—9-1F). Carryover in the first rinse was found to be approximately 1% for the 500 and 100 µl sizes and about 0.2% for the 10 µl pipet. When the pipets were coated with a water-soluble silicone, carryover became negligible (5). This design of micro pipet is available in sizes from 1 to 500 µl. Reproducibility with aqueous solutions is claimed to be ±0.1% with the 25 μl and larger pipets, ±0.5% with the 10 and 20 μl sizes, and ±2% with the 1 μl pipet.

The *Sahli pipet* (TC), calibrated at 20 μl, is still commonly used in hemoglobin determinations. It is being displaced, however, by disposable micro pipets that are available in denominations from 5 to 100 μl and are made from precision bore capillary tubing and calibrated to within ±1%.

The *Drummond Microcap* (TC) disposable micro pipet is also manufactured from precision bore glass capillary tubing and is calibrated to contain specific volumes ranging from 0.5 μl to 500 μl with a tolerance no greater than ±1%. It may be used with or without a special holder-bulb assembly for which a plunger device is also available.

Beckman (Palo Alto) sells small reagent bottles for ultramicro work. Polyethylene pipets precalibrated for microliter volumes are fitted to polyethylene squeeze bottles. The first squeeze of the bottle fills the pipet and a second squeeze delivers the entire calibrated volume.

Several micro pipetting devices operating on the syringe principle are now available. They are of two types: in the first the liquid to be measured

FIG 9-1. Ultramicro pipets: (A) Kirk transfer pipet (TC), (B) self-filling transfer pipet (TC), (C) Lang-Levy pipet (TD). Functional modes of the Grunbaum pipet: (D) sampling, (E) dispensing, and (F) reagent storage. (From Grunbaum BW: *Microchem J 15*:680, 1970)

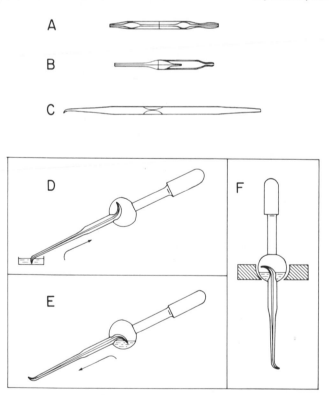

A

B

C

D

F

E

contacts only a disposable plastic tip, and in the second contact is made with the syringe plunger. In the first group, examples of which are manufactured by Oxford Laboratories, Kimble, Medical Laboratory Automation, Inc., and Eppendorf, the sizes available range from 5 μl to 1000 μl. Advantages claimed for this design are the use of disposable, nonwettable tips, no cross contamination, and high accuracy and precision combined with simplicity in use. An unpublished study of this type with equipment from four manufacturers showed deviations as great as ±11.6% from the mean delivery value using serum in 100 μl pipets (17). It was concluded that the pipets, although semiautomatic and appearing straightforward to use, require careful training of operators for acceptable results. Literature from the Eppendorf Company emphasizes the importance of carrying out the filling operation twice in order to coat the polypropylene tip with a film of serum. Only then does the thickness of the film remain constant so that the correct volume of sample can be dispensed.

The second group is represented by precision graduated syringes capable of the measurement of selected microliter volumes. Examples are those supplied by Hamilton, Precision Sampling Corporation, and Unimetrics. In the very small syringes, the solution being measured is restricted to the needle, displacement being indicated in the graduated syringe barrel. The Drummond Microtrol and Ultramicrotrol are syringe pipets that can be rapidly assembled and utilize Teflon tipped stainless steel plungers and precision bore glass tubes, both of which are disposable. The precision bore tube is mounted within a graduated, reusable barrel.

The pipets developed by the Manostat Corporation and marketed under the name *Digi-Pet* employ a digital readout device. The ultramicro pipet, which is available in 10.0, 1.0, and 0.1 ml sizes, uses a glass plunger that displaces titrant directly. Each division on the readout scale of the 0.1 ml pipet is equal to 0.02 μl.

The *Micromedic Automatic Pipet* has been designed for the measurement and dilution of samples in the volume range 2 μl to 1000 μl. It is a precision, two-pump device in which the rate of piston travel is sinusoidal, preventing shock and turbulence in the liquids being measured and thus contributing to a high standard of performance. The machine can be activated manually or by a foot pedal. Accuracy is claimed to be within ±2% and repeatability to ±0.1% over the entire range of the instrument.

Glass pipets must be clean and dry for use, except when the analyst is sure that prerinsing with the solution to be measured introduces no significant contamination (e.g., pipet wet with water to be filled from a stock reagent bottle containing a relatively large quantity of reagent). Nondisposable pipets are most easily cleaned in the same way as hematologic pipets: Attach the pipet to a water aspirator pump and suck through, in order, water, acetone, and air to dry. Cleaning should be performed as soon as feasible following use, or at least the pipets should be submerged in water until cleaning time. If necessary, pipets may be soaked in dichromate sulfuric acid cleaning solution or a detergent solution. Pipets soaked in the former must be rinsed profusely with water.

TITRATION

Micro burets fall into two general classes. The first is the *Rehberg Buret* (14) and its modifications, in which the fluid is contained in a graduated glass capillary and a screw-operated piston displaces mercury, which in turn forces solution out the other end of the glass capillary. The second type uses the displacement principle and dispenses with the graduations on the glass capillary. The volume delivered is measured by a calibrated micrometer screw or gauge.

One of the modifications of the Rehberg-type buret is that of Kirk, the latest model (16) of which is available from Microchemical Specialties. The buret consists of a capillary mounted on a rotating metal back, which is supported on a swivel clamp that also permits vertical adjustment on an iron stand. The capillary, available in capacities from 0.1 to 0.035 ml, lies on a scale graduated from 0 to 35 in large scale divisions, each of which is subdivided into tenths. The mercury contained in the buret does not contact the titrating liquid, being separated from it by a column of air. The height of the mercury column above the control plunger and gasket is so small that the danger of leakage past the gasket resulting from mercury pressure is negligible.

The *Scholander Micro Buret* (15) is available in capacities of 100, 250, and 500 μl from Microchemical Specialties, and in 800 μl capacity from Scientific Industries. In this micrometer screw type of buret there is contact between the mercury and the titrating fluid. With the 800 μl model, 10 μl volumes can be dispensed within $\pm 1\%$.

The *Gilmont Micrometer Buret* (3) is a modification of the Scholander buret in which a Teflon piston displaces fluid from a precision glass bore. Readings are made with a micrometer, one division of which is equivalent to 0.2 μl on the 200 μl buret and to 2 μl on the 2 ml buret.

The *Gilmont Ultra-Precision Micrometer Buret* provides a considerable increase in precision and accuracy over the buret described above. The plunger is of glass and is precision ground. The Swiss-made micrometer uses a combination digital readout and vernier so that one division is equivalent to 0.01 μl on the 250 μl buret while the 2.5 ml size can be read to 0.1 μl.

The buret-type *Digi-Pet* produced by the Manostat Corporation has a steel plunger which displaces mercury, thus indirectly delivering titrant. This buret, which has an optional synchronous motor drive, is available in 1.0, 0.1, and 0.01 ml capacities. The 0.01 ml buret has 0.002 μl scale divisions. The precision in delivering 10 μl from the 0.1 ml buret is $\pm 0.7\%$ (4).

The *SB-2 Syringe Microburet* marketed by Micrometric Instrument Company has a micrometer screw with a dial gauge that advances the plunger of a glass syringe to which is attached a special buret tip by means of a standard taper joint. Five sizes of syringes are available with volume deliveries ranging from 0.2 to 5 μl per dial division. This buret has the advantage that the syringes are easily interchangeable and a separate syringe can be kept filled with each reagent routinely used.

The *Model 153 Microtitrator*, marketed by Beckman (Palo Alto), delivers titrant by turning a micrometer screw readable to 0.005 μl. Titrations are carried out in disposable molded plastic cups with cone shaped inner surface. An adjustable vibrator stirrer with Teflon tip is built into the titrator.

The *Micro Syringe Burette*, marketed by California Laboratory Equipment, consists of a precision bore glass tube into which fits a Teflon plunger. Its capacity is 250 μl. A steel rod extends from the plunger to a micrometer control graduated to 0.1 μl. The delivery end consists of a Teflon plug into which a hypodermic needle or a glass buret tip is placed. The micrometer is equipped with a lock nut so as to give repeated delivery of the same volume.

The *Ultra-Buret*, marketed by Scientific Industries, has a capacity of 7 ml delivered by mercury drive from a hand wheel calibrated to 1 μl. A reagent reservoir permits easy refilling.

Micro burets are most accurately calibrated by weighing volumes delivered. Water can be delivered into small glass stoppered weighing bottles, or when mercury is the propellant and there is a micrometer screw or gauge, advantage can be taken of its greater density and the mercury can be delivered and weighed.

Before use, micro burets should be rinsed out several times with the titrating fluid; after use they should be kept filled with water.

Titration of micro and ultramicro quantities of a substance theoretically can be performed in two ways: (*a*) The titration can be carried out on a macro scale with a relatively large volume of a very dilute titrating solution. This is not feasible, however, because of the large error resulting from indefinite end points. (*b*) The second method is titration with volumes sufficiently small and concentrated (usually no less than 0.01 *N*) to give sharp end points. The very small volumes involved require a greater concentration of indicator to yield a visible color change at the end point. The *drop error*, the portion of titrating solution needed to change the indicator rather than titrate the sample, may thus be significantly larger in micro than in macro titrations. Drop errors are corrected for by standardizing the amount of indicator added and running proper blank and control titrations with standards (9).

Titrations of solutions less than 0.5 ml may be satisfactorily carried out in small, shallow, white porcelain dishes, end points being observed against the white background. Small test tubes also can be used when the progress of the titration is watched from above against a white surface. Specially designed vessels such as those supplied with the Beckman *Model 153 Microtitrator* are very satisfactory providing the volume to be titrated is suitable.

The tip of the micro buret is inserted into the solution during the titration. Stirring is accomplished by one of several means, including manual stirring with a thin rod, mechanical mixing by magnetic or vibration stirrers, or by use of a gas stream.

CENTRIFUGATION

Although technics and apparatus are available for micro filtration, separation of precipitates from small volumes of fluid is most easily accomplished by centrifugation. Several manufacturers build centrifuges designed for handling tubes of small capacity. The International *Microchemical Centrifuge* has a four-place head accommodating two tubes of 0.5, 1, 2, or 3 ml capacity together with two 5 ml tubes. The Fisher *Safety Centrifuge* can be supplied with a micro head that takes six 2 or 3 ml tubes. The Precision *Centricone Semimicro Centrifuge* has an eight-place head for 0.5, 1, 2, 3, or 5 ml tubes. These various centrifuges have top speeds ranging up to about 1725 rpm producing relative centrifugal forces up to about 400.

Microchemical Specialties Company markets a centrifuge specifically designed for ultramicro work. The *"Misco" Electric Micro Centrifuge* has a top speed of 19 000 rpm giving a relative centrifugal force of 21 500. Speed is controlled by a 7.5 Amp variable transformer. It holds eight tubes of 1 ml capacity but bushings are available for adapting to 100 and 200 μl tubes.

Beckman (Palo Alto) markets the *Model 152 Microfuge*, having a capacity of twenty 400 μl disposable plastic tubes, and a nominal speed of 15 000 rpm. It will sediment blood in 15 sec.

The Dade Division American Hospital Supply Corporation markets the Minifuge, a precision centrifuge designed for small samples. Twelve polycarbonate tube shields with rubber inserts accommodate samples in Caraway tubes. The centrifuge is a single speed machine developing 5500–6000 rpm equivalent to a relative centrifugal force of 3500 g.

For ultramicro work in clinical chemistry centrifuge tubes of 0.5 or 1 ml capacity are adequate. The only advantage of very high speeds is a more rapid centrifugation.

MISCELLANEOUS APPARATUS AND TECHNICS

The types of micro cuvets available and their use in photometry and spectrophotometry are considered in the section on cuvets in Chapter 1.

Whenever it is necessary to dilute a solution or sample to some fixed volume the container should be so designed that the mark is placed in a narrow stem. The test tube type of micro volumetric flask is undesirable in capacities less than 1 ml because the error in setting the meniscus corresponds to several percent of the total volume (4). Flasks down to 0.1 ml capacity and shaped similar to the conventional macro volumetric flasks are available commercially from Microchemical Specialties. Deproteinization centrifuge tubes are also available in calibrated capacities from 0.1 to 1.0 ml. These are used when it is desired to dilute to a fixed volume in the process of protein precipitation.

ULTRAMICRO ANALYTICAL SYSTEMS

In recent years a multitude of microliter analytical systems have been developed and marketed, covering the entire spectrum in sophistication from manual systems, supplied with appropriately designed instrumentation and bottles of liquid reagents, to those with premeasured or tableted reagents, to complex and computer-controlled machines operating virtually independently of human mediation. This last type of equipment emanates primarily from commercial organizations and critical evaluation has not yet appeared in the literature. Those desiring more information concerning these systems should obtain specifications from the manufacturer's literature and demonstrations.

REFERENCES

1. CARAWAY WT, FANGER H: *Am J Clin Pathol 25*:317, 1955
2. DECUIR M, ALEXANDER N: *J Med Lab Technol 27*:392, 1970
3. GILMONT R: *Anal Chem 25*:1135, 1953
4. GILMONT R: *Anal Chem 20*:1109, 1948
5. GRUNBAUM BW: *Microchem J 15*:680, 1970
6. GUTHRIE R, SUSI A: *Pediatrics 32*:338, 1963
7. HILL JB, PALMER P: *Clin Chem 15*:381, 1969
8. JOSEPHSON B: *Advan Clin Chem 1*:301, eds. Sobotka and Stewart, Academic Press, Inc, New York, 1958
9. KIRK PL: *Quantitative Ultramicroanalysis*, JOHN WILEY and SONS, Inc, New York, 1950
10. MATTENHEIMER H: *Micromethods for the Clinical and Biochemical Laboratory*, Ann Arbor Science, Publishers, Inc, Ann Arbor, Michigan, 1970
11. MEITES S, FAULKNER WR: *Manual of Practical Micro and General Procedures in Clinical Chemistry*, Thomas, Springfield, Illinois, 1962.
12. NATELSON S: *Am J Clin Pathol 21*:1153, 1951
13. O'BRIEN D, IBBOTT FA, RODGERSON DO: *Laboratory Manual of Pediatric Micro-Biochemical Techniques*, 4th Ed, Hoeber Med Div, Harper and Row, Publishers, Inc, New York, 1968
14. REHBERG PB: *Biochem J 19*:270, 1925
15. SCHOLANDER PF, EDWARDS GA, IRVING L: *J Biol Chem 148*:495, 1943
16. SISCO RC, CUNNINGHAM B, KIRK PL: *J Biol Chem 139*:1, 1941
17. Personal communication from the Diagnostest Division of The Dow Chemical Co, Life Sciences Group

Chapter 10

Automation and On-Line Computers

RALPH E. THIERS, Ph.D.

In clinical chemistry the word automation has a meaning more akin to "mechanization" than to its more general meaning, which implies a process under feedback control (37). It entered the vocabulary of the clinical chemist when the capability of mechanization jumped from one or two steps of some procedures to almost all steps (96). In its current practical usage all but one or two steps should be wholly mechanized before the word is applied. However, its application seems to be broadening rapidly with time.

As the field has developed, a large number of automated devices has been offered commercially with widely varying degrees of success (20). From a descriptive point of view these devices share certain properties of published procedures: (a) Not all of them perform as the printed page would lead one to expect; (b) barring first hand experience, one can judge them best by their degree of acceptance by the scientific community; (c) some that show promise and are much discussed today may have vanished from the scene next year; (d) an overall discussion of them, to be most useful, should focus on general principles common to various approaches rather than on the details of any one device.

Approaches to automation in clinical chemistry can be classified by employing three pairs of independent contrasting properties. This provides eight basic classes, into one of which almost every available device will fall. These three pairs are described by the adjectives *complete* vs *partial, serial* vs *parallel,* and *discrete* vs *continuous flow.* Any given device should be classified according to these criteria as a necessary preliminary to understanding it or to judging its applicability in any given situation.

In the logical extreme, *complete* automation would be represented by a device attached to the patient in such a way as to obtain its own specimen, provide an analytical result, and produce some automatic action on the patient, based on the result. Although such devices have been built, in the serial continuous flow mode (60, 61), the first and last steps are usually omitted as requirements for *complete* automation, with serum or some other fluid provided as the starting material and with the results being numerical concentration values or precursors of them. If any procedural steps are omitted, other than those mentioned, the term *partial* automation (or *semiautomation*) is applied. Automated photometry of solutions, in which the preceding steps had been performed manually, fits this classification. The measurement of Folate in Chap. 27, is one example.

In *serial* automation, the instrument works on each sample in sequence, and no two specimens occupy precisely the same stage in the procedure. Continuous flow analyzers such as the Technicon AutoAnalyzer, are examples of this class (96). In *parallel* automation the device analyzes sev specimens passing through the procedure simultaneously. Centrifugal lyzers, such as the Electronucleonics GEMSAEC, and the Union Carl Centrifichem, exemplify this class (7). In them, a rotor is loaded v reagents and with many specimens and rotated to carry out one determ tion on each.

In *discrete* methods, whether automated or not, each specimen witl accompanying reagents occupies a separate container. Although the sp men may change containers, it never shares a common container with o specimens. *Continuous flow* analyzers, on the other hand, send all specin through one greatly elongated container, and several specimens may re together in different portions of the container in physical contact with t adjacent specimens. In addition, convention has restricted this class devices involving a steady state (33).

In general, each automated device falls rather clearly into one of the ab classes. However, the "flow through" photometer, a continuous flow dev is employed as the last stage of many discrete analyzers and used discrete fashion. Instruments can also be conveniently classified on a fo basis depending upon how many different determinations can be perfor simultaneously on one specimen. Each determination requires one so-ca "channel." Single channel devices are most common, but multichan instruments having from two to twenty or more channels are made, example the Technicon SMA series, the AGA Autochemist, the Hycel M X, or the Vickers Multichannel 300.

One can also classify instruments that are designed to measure rate reaction differently from those designed to measure the equilibrium posi or end point of the reaction. Some devices are highly specialized for former task, for example ERA of Photovolt, ZYMAT of Bausch and Lo and products of Gilford, Coleman, and other companies. Other devices measure either equilibrium or rate reactions, sometimes intermixed, ACA of DuPont. The difference in instruments that measure rate ve those measuring end point are found in the data processing rather than sample handling portions of the instruments. These will be discusse length under "On-Line Computers" below.

This chapter will not attempt the fruitless task of describing all of automated instruments and procedures available, or of describing any instrument in detail. It is the nature of such devices to change with time rate that is rapid compared to book publication, as problems are found solved. This statement applies particularly to new devices, and most of th on the market fit this description. Instead a discussion of the underly principles will be presented to provide a basis on which the clinical chen can judge instrumental performance, choose devices best suited to his ne and optimize the output of his equipment with respect to the variable accuracy, speed, cost, etc.

The degree of utilization of continuous flow methods far exceeds tha discrete methods in automated clinical chemical analysis. The number of former devices in actual daily use probably amounts to between 10 and times the number of the latter, and the routine workload for the for between 100 and 1000 times the latter. Although the discussion in this chapter may seem weighted towards continuous flow methods, it will not reflect nearly so unbalanced a picture as actually currently exists.

UNIT OPERATIONS OF AUTOMATION

The common unit operations of clinical chemistry include: obtaining the specimen, centrifugation, obtaining a measured sample, dilution, reagent addition, mixing, splitting, transferring, separation of protein, separation of desired constituent, reaction to form the compound to be measured, incubation, measurement of some property of the desired compound, calibration, quantification, presentation of results, and calculation of the final data.

Obviously, any given procedure may not involve all of the above steps, or adhere to the above order. All of the steps have been subjected to partial automation, including the first (53, 61). Most of them form a step in one or another completely automated procedure. Therefore, discussion of these unit operations provides an appropriate unified approach to an otherwise fragmented subject.

CENTRIFUGATION

Automatic centrifuges have not been generally used in clinical chemistry either for separating cells from serum or for separating precipitated protein from the supernatant fluid. However, an innovative example of parallel discrete automation—the GEMSAEC device (7)—has the possibility of mechanizing this step effectively and promises to make it part of automated procedures (5).

PRESENTATION OF SAMPLE

In almost all automated devices the samples of serum, plasma, or other fluids must be presented to the instrument in some mechanical fashion that is highly specific to the device. The technician usually places the fluid in small disposable cups that are placed in turn on the sample presentation module. Manufacturing considerations seem to have dictated a disk turntable with cups around its periphery as the most popular form for this module (96). A fine tubular probe moves into and out of each specimen in turn, and aspirates sample, standard, wash fluid, or other materials submitted for analysis.

In many instruments for discrete automation secondary or tertiary circles of containers are placed on the same disk, and used to receive the diluted sample, or to act as reaction chambers. The now defunct Robot Chemist (81) was an early example of an instrument that employed this principle. In some cases the entire procedure, including photometry, is carried out in such a disk, for example the Abbott Bichromatic Analyzer. Usually the disk itself is moved by some indexing device that places each cup in turn under the probe or probes (101).

Frequently the probe retreats to a location between samples where it can pick up washing fluid. The exact relationship between sample container and probe, and the precise details of probe movement can be surprisingly important to overall accuracy of results (108, 109, 128).

Other means of presentation include devices that require the samples to be placed in cups on a rectangular tray, or a chain of either fixed or variable length, or variations on these. In the DuPont ACA instrument the individual sample cups, each attached to a patient information card, are moved down a channel in a line (85).

Devices that can present serum samples to an automated analyzer without transfer from the original drawing tube have obvious advantages (109), for example the Bio-Logics Incorporated Sample Presentation module.

In any such sampling system, contamination of any particular specimen with its predecessor must be minimized. Fluid can collect on the inside or outside of probes and be carried on to the following specimen. For this reason the probes are invariably small in diameter, and are often made of nonwetting material. If the fluid inside a probe of nonwettable material flows slowly and smoothly enough not to break the meniscus, almost no residue is left in the lumen. Similarly, if a fine tube pulls slowly out of a serumlike fluid, the drop that remains on the outside will be negligible when dealing with sample sizes over about 0.1 ml. As sample size decreases these problems increase, and greater attention to sample presentation is required. In some devices such as the Micromedic, the outside of the probe is automatically wiped by a clean portion of a roll of absorbent material.

The problem may often be present but unnoticed in automated analyzers because the drop on the outside of a probe tends to be washed off at the surface of the next sample, and the probe tip, nearer the bottom of the sample, fails to pick up surface contamination. Obviously this situation will reduce accuracy if an excess of sample is not placed in each cup because the probe will then pick up the contaminated surface fluid. Even if a threefold excess of sample is provided in each cup this problem frequently displays itself as inability to get precise duplicates when running a given tray of samples a second time, instead of taking new aliquots. Proper practice prohibits reusing trays of samples.

The sample presentation device can produce errors when its sample cups are open to the atmosphere, as is usually the case. Even where covers are provided they seldom make a complete seal over each sample, and far too often they are left off completely by operators who underestimate their importance. Loss of CO_2 can, of course cause low results in bicarbonate results (109). Evaporation of fluid has been observed to be significant by this writer, who has repeatedly relearned the importance of not placing sampler modules, sample cups or sample trays where a draft exists, however gentle. This problem frequently fails to be recognized in air conditioned rooms.

MEASUREMENT OF SAMPLE

In sample measurement the essential difference between discrete and continuous flow methods manifests itself most clearly. Discrete technics almost invariably involve some reciprocating device that aspirates a measured absolute amount of sample and stops, then reverses direction to deliver the measured volume. One can consider the syringe, in one or another variant, as a component common to discrete measurements. By contrast, in continuous flow measurement the driving pressures always move the measured fluid in the same direction. Rate of flow is held constant as the measured parameter, with the "peristaltic action pump" as the driving force (30, 31, 96).

DISCRETE MEASUREMENT

At least three modes of discrete measurement can be distinguished on the basis of how each one minimizes intersample contamination. The most common technic uses suction to aspirate the sample into a probe of known volume. A valve then moves to another position and pressure pushes a measured volume of diluent through the sample probe so that sample and diluent are delivered sequentially into some receiving vessel. Provided that the diluent washes the sample totally out of the probe, the device can be used immediately for the next sample. The valve simply returns to the aspiration position, and the probe moves into the new sample to repeat the cycle.

Devices using this technic find their ancestry in the Seligson pipet (77, 94) Figure 10-1. Mechanization of this basic principle formed one of the earliest examples of partial automation to gain wide application in the clinical laboratory (77), and dozens of different versions can be purchased today. In one example made by Micromedic Systems Incorporated, sample volumes from 2 μl to 5 ml can be measured from tubes placed in a sample presentation device, and diluted into tubes held in a receiver tray at the rate of 8 sec/sample.

The principle of the Seligson pipet finds application in almost every discrete automated analyzer and in fact makes most of them possible. It was published, interestingly enough, in the same journal as the basic article on continuous flow analysis, and just over 100 pages away. Both of these papers remain interesting, informative, and quite valuable reading today (94, 96).

In a second type, a measured volume of the sample is aspirated into a probe large enough to wholly contain it, then pushed into some receiving vessel by air pressure. The probe then moves to some other container and is washed with water or some other cleaning fluid. In the DuPont ACA, the probe is a fine needle that penetrates a rubber membrane of the receiving vessel, and after delivering the sample, it delivers computer chosen aliquots of a variety of solutions (85).

A third type of discrete measuring device employs disposable sample probes. This vastly decreases the problem of intersample contamination (82, 123) but increases mechanical complexity.

The simplest of the three, the first, requires the least mechanical complication—a factor that has proven to be of primary importance in practical use of automated equipment. Although washing the pipet-probe combination between uses, as in the second case, imitates familiar manual practice most closely it requires more complicated valving and associated moving parts. In both cases only a restricted number of materials can be used as reagent diluent or wash, because of possible attack on the materials of the device or because of formation of precipitate. These instruments are very sensitive to clogging by particles of solid in the fluid being handled or to errors caused by solid clinging to a probe wall.

Measurement and Dilution

The first of the above three types of measurement involves dilution as an integral step. This has been widely applied as the initial step in discrete automation whether complete or partial. When combined with modules for sample presentation, reception of sample and diluent, reagent addition, incubation and measurement, it forms the basis for most discrete analyzers.

If the measured sample is to be diluted with less than about a fivefold excess of diluent, great care must be taken to guarantee complete washing of the sample out of the pipet by diluent. One can check completeness of

FIG. 10-1. Seligson pipet. (A) three-way stopcock, and sample probe and pipet; (B) calibrated buret; (C) waste receiver. From SELIGSON D: *Am J Clin Pathol* 28:200, 1957.

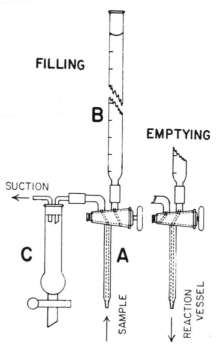

washing by using a concentrated solution of dye as the sample. The most convenient dyes to use are food colorings, from both the standpoint of availability and ease of removal from clothing, etc. One notes immediately, on doing such an experiment, that clean lines of hydrodynamic flow, and absence of any protrusions, sharp bends, or alcoves are essential to good washing with minimal diluent. Under ideal conditions one can get below a 5 to 1 ratio of diluent to sample and down to 2 or 3 to 1 but special care is required. Devices that place an air bubble between the sample and the diluent can do better in this respect for reasons discussed below.

CONTINUOUS FLOW MEASUREMENT

Without the peristaltic pump (30), and without specific major improvement to it (96), continuous flow measurement would probably be unknown to the clinical chemist. The diagrammatic sketch in Figure 10-2 illustrates its *modus operandi*. Rollers moving along a resilient tube of chosen diameter cause an occlusion of the lumen to move progressively to the left. Provided that at least one such occlusion exists at all times, fluid is forced down the tube at a linear rate set by the roller speed and at a volume rate approximately equal to the linear rate multiplied by the cross sectional area of the tube. A group of parallel tubes, as in Figure 10-2, is called a "manifold." By choosing appropriate inner diameters of tubes one can aspirate fluids at a wide variety of chosen rates simultaneously, at constant roller travel rate. Table 10-1 illustrates a few of these rates and their range of variation from tube to tube as specified by the manufacturer. Many sizes

FIG. 10-2. Peristaltic action pump moving fluid down several tubes at different but constant rates (diagrammatic). 1.–4. Elastic "manifold-tubes"; 5. sample pickup tube and sample cup; 6. downstream end of section of manifold tubing showing rollers and pumping action; 7. addition point for reagents from tubes 1 and 2; 8. addition point for sample from tube 3; 9. light source; 10. lens; 11. color filter; 12. flow through cuvet; 13. photoelectric cell.

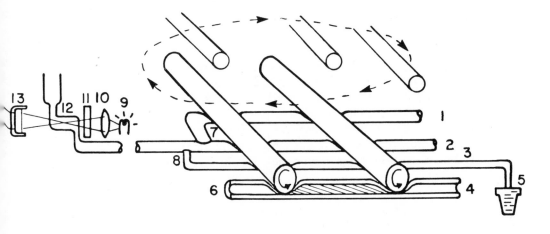

other than those in Table 10-1 are available, as are many tubes of other materials.

TABLE 10-1. Flow Rates of Manifold Tubes

Tube size i.d. (in.)	Volume pumped (ml/min) Mean	Range	Manufacturer's color code on tube ends
0.005	0.015	± 0.010	Orange black
0.010	0.050	± 0.018	Orange blue
0.020	0.16	± 0.03	Orange yellow
0.040	0.60	± 0.06	White white
0.081	2.50	± 0.13	Purple purple
0.110	3.90	± 0.15	Purple white

The peristaltic pump for continuous flow analysis has developed through several major and many minor changes. So fundamental a module is the pump that as problems such as inconstancy of pumping speed and wearing effect on manifold tubes have been solved, large improvements in precision, accuracy, reliability, speed, and other characteristics of methods have been realized. Similarly, improvements in tubing characteristics and increased variety in size and material have markedly advanced the field.

With a few exceptions (e.g., 80, 90) the rollers of peristaltic pumps run at constant speed, driven by high quality synchronous motors, and desired flow rates are achieved by choice of tubing diameter. Thus remarkable constancy of flow rates can be achieved over long periods of time. Table 10-2 gives representative data showing this constancy. Over short periods of time variations in the amount pumped during any given hour or day are too small to detect. Although the amounts pumped do not necessarily correspond to the nominal values, the precision of pumping remains remarkably constant. As days of elapsed pumping time pass the rates initially decline, but at a surprisingly slow pace. A good pump can maintain this degree of constancy in many tubes simultaneously. As pumps become worn this ability decreases.

Because flow rate should be absolutely constant, variations in downstream resistance should ideally have no effect on the flow rate in continuous flow systems. Such variations do occur and are quite significant. One can be quite surprised for example by the pressure required to move a bubbled stream down a tube of nonwettable material (23). If the tube holds no film of liquid on the wall between slugs of liquid, the pressure must overcome the sum of the meniscus tensions of all liquid gas interfaces. Figure 10-3 shows the relationship between flow rate and pressure for an AutoAnalyzer Pump II. If the flow rate were not so remarkably independent of downstream resistance, instability of flow rates would occur, which would produce unacceptable performance, due to inevitable variations in pressure in a driven bubbled stream.

Upstream pressure variations are a different matter, because the degree to which a resilient tube springs back to its resting size depends upon pressure

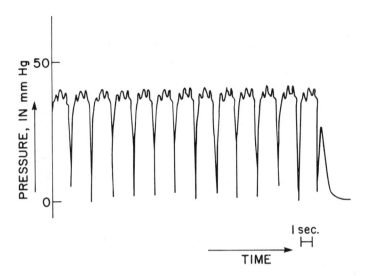

FIG. 10-4. Peristaltic pump pulsation.

CENTRIFUGAL MEASUREMENT

To date, sample measurement has been the unautomated step in centrifugal methods, although proposals exist for its integral automation (4). Centrifugal instruments available on the market utilize standard discrete components for measurement and transfer of sample (22). However, it seems likely that the unique features of centrifugal automation will produce more convenient technics.

MEASURED ADDITION OF DILUENT OR REAGENTS

Once the aliquot of specimen has been measured into its container, addition of diluent or reagent is a relatively simple step. It involves, in principle, merely an extension of previously described technics. In discrete methods another syringe pipet delivers another reagent. In continuous flow methods another manifold tube across the pump measures another reagent through a new connector into the flowing stream.

DISCRETE ADDITION

The mechanized pipets that can deliver successive aliquots of one reagent vary tremendously in design but have fewer operational problems than devices that measure samples, because of the uniformity of the reagent. In any given discrete instrument one must be able to adjust the amount of reagent delivered and the timing and sequence of delivery. These methodologic details have been expressed as "sequence diagrams" or "probe position and volume tables" (81), and in some devices like the American Monitor Programachem or the Instrumentation Laboratories/Harleco Clinicard, they are punched into cards. However, no single means of expressing discrete methodology has evolved to compare with the "flow diagram" of continuous flow (see below). Temperature changes, reagent action, or wear during operation can cause variations in the exact amount delivered by an automatic pipet. Normally these variations are small and randomly distributed about the desired mean. However, sticking of mechanical parts can occasionally cause large nonrandom errors. Slow change with time of the mean amount delivered can cause a phenomenon called "drift" wherein the value obtained for identical samples slowly changes with time. Means of observing and controlling or correcting both of these *must* be employed in automated analysis.

In discrete devices aliquots can be added as solids as well as liquids, which presents real advantages for reagent addition. By fabricating "pills" of reagent, the necessity of measuring anything but water or buffer can be removed, at least in theory. Substances that are unstable in solution are handled neatly by this technic, which forms a key feature in the DuPont ACA.

CONTINUOUS FLOW ADDITION

In continuous flow methodology a standard technic exists for expressing diagrammatically the relationships of the various flowing streams propelled by the pump. Such diagrams, called "flow diagrams," express a wealth of information to the careful reader. They can convey almost all procedural details except reagent composition.

This is illustrated in Figure 10-5, the flow diagram for a bilirubin method. These diagrams, by convention, show flow from right to left and therefore tend to read backwards. In the upper right the round symbol indicates a sampler module. The sampling rate (60 samples/hour) and the relative time the probe spends in and out of the sample (2:1 or 40 sec in and 20 sec out) are shown below the sampler as well as details of sample tube and probe. The numbered circles in a vertical column in middiagram indicate the positions of manifold tubes on the peristaltic pump. Horizontal lines through these circles represent tubes through which fluids flow from right to left. Above each line the nominal rate of flow and tubing diameter are listed, and to the right appears the name assigned to the fluid in the list of reagents.

One can deduce that the top tube in Figure 10-5 aspirates sample from each cup at the rate of 0.32 ml/min, and does so for 40 sec resulting in consumption of about 0.2 ml of each specimen. (One must remember that these numbers are approximate, see Table 10-2).

The tube second from the top pumps the reagent designated either "caffeine" for total bilirubin or "HCl" for direct bilirubin from its container (not shown) at 2.5 ml/min. Air, pumped by the tube third from the top, enters the stream at 1.2 ml/min. This must result in a bubbled stream with liquid slugs about twice the length of the bubbles.

BILIRUBIN, TOTAL OR DIRECT

FIG. 10-5. Flow diagram.

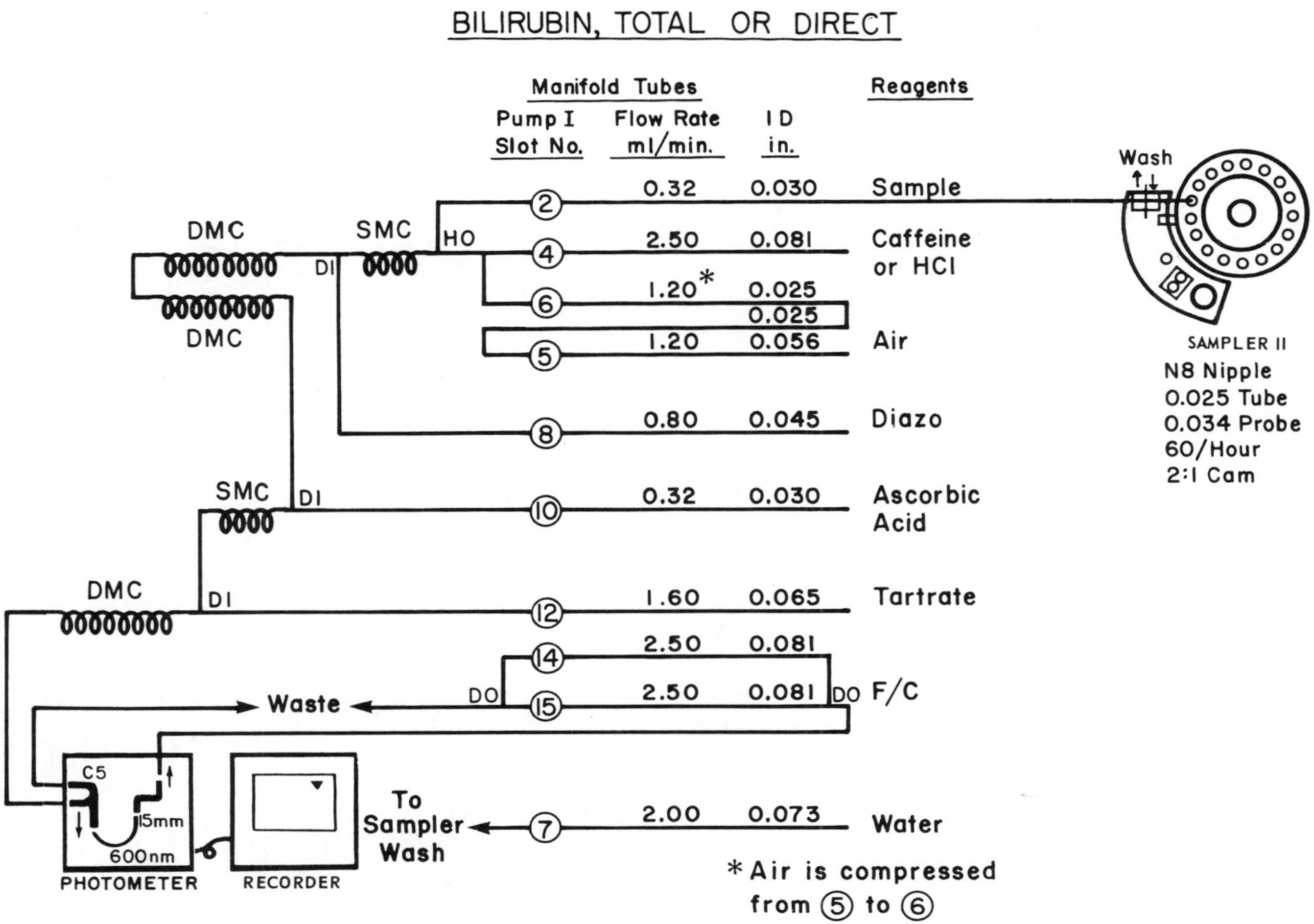

Immediately after the addition of the air the sample is added to the bubbled stream via a microplumbing fitting cataloged as H0 by the Technicon Company. This and some other fittings are illustrated in Figure 10-6.

FIG. 10-6. Some fittings and connectors for continuous flow analysis. These fittings are labeled with the designation of their manufacturer, Technicon Instruments Corporation. In general the A, D, G, H, and K series are for joining two streams of fluid; the B series for separating gas from liquid, the C series for removing bubbles from streams, the N series for joining tubes together, the PT series for splitting liquid streams. The N11 nipple is made of platinum, the N13 of stainless steel and the other nipples of Teflon, Kel F, or polypropylene. SMC stands for "single mixing coil" 3.0 ml in vol and 14 turns, DMC for "double mixing coil" 6.0 ml in vol and 28 turns, JSMC for "jacketed single mixing coil," SMC 3.0 mm i.d. for a 14 turn SMC 6.0 ml in vol and DMC 3.0 mm i.d. for a 28 turn DMC 12 ml in vol.

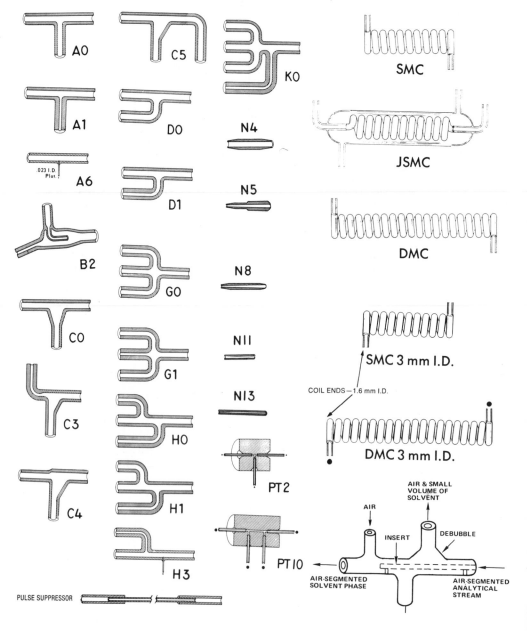

With sample added, the liquid slugs must be slightly longer than before. The rate of flow can now be calculated downstream from the H0, in two useful forms, rate and time. The rate must be the sum of the entering three, viz. 4.0 ml/min.

Time for a given portion of liquid to move from the H0 to the D1 fitting can be calculated from that rate plus the known standard volume of a single mixing coil (SMC), 3.0 ml, to give 45 sec. One can thus deduce that after the sample has been in contact with the caffeine or HC1 reagent for 45 sec a "Diazo" reagent is added at the rate of 0.8 ml/min. This of course increases the rate of flow to 4.8 ml/min. Flow through the two double mixing coils (DMC), which hold 2 x 6.0 ml, therefore takes 160 sec more.

This introduces the concept of "dwell time." Every drop of sample-reagent mixture passes the first D1 fitting after an identical dwell time, viz 45 sec. The dwell time from the H0 to the second D1 fitting reckons at 205 sec. In this case the double mixing coil clearly exists to provide dwell time in which the reaction between sample and reagents can proceed. At this point 0.32 ml/min of ascorbic acid reagent enters the stream and increases the overall flow rate to 5.1 ml/min.

The dwell time to the third D1 fitting reckons at 240 sec. After this point the fluid flows at 6.7 ml/min, of which 5.5 is liquid and 1.2 air. The dwell time to the C5 fitting is therefore 294 sec. With connections and allowances for tube size tolerance, the overall dwell time should measure at 5 ± 0.5 min. If one adds the time taken for the sample to move from the cup to the H0 fitting, an overall sample-to-peak time of about 6 min should be observed. If a different time is measured something is wrong. Analysts who ignore dwell times waste an essential disgnostic tool.

The dwell time can be calculated as exemplified above for any component at room temperature and pressure. In modules such as heating baths, dialyzers, long delay coils, and the like, thermal expansion or pressure contraction of the air bubbles produce dwell times that are less easy to calculate but are precisely reproducible. When a new method is first developed, or an established method is first reproduced in any laboratory, these hard-to-calculate dwell times should be measured and recorded in order to permit later trouble shooting. One unusual example of such a situation can be seen in the serum iron method of Chap. 19 (Fig. 19-4). A fairly viscous fluid is being pumped into two dialyzers in series at 3.6 ml/min. In order to keep the hydrostatic pressure within reasonable limits and to produce uniform flow the fluid is also pumped *out* of the dialyzer at 3.8 ml/min. Dwell time in the dialyzer is hard to predict under these conditions. However the rate of dialysis of iron is low enough that the dwell time has a large direct effect on sensitivity. Hence knowledge of the desired dwell time and observance of it can avoid some otherwise very confusing effects.

Depending on dwell time, the reaction between sample and reagents in the flowing stream may be complete or incomplete as the mixture passes any given point, such as the second D1 in Figure 10-5 or the photometer. If incomplete, it must be so to exactly the same degree for all portions of each sample, since all have the same dwell time. And unless the rate of the reaction is affected by the concentration of the constituent sought in a nonlinear fashion, then the degree of completeness of reaction must be identical for all samples as well as for all portions of each. For any given continuously fed sample, a sensor stationed at the second Dl or at the photometer would, under these conditions, measure no change in reactant concentration with time. It could only detect a difference by moving to a new position—at which location it could again see no change with time.

This situation has been described at length because it illustrates the key characteristic of continuous flow analysis, the "steady state" condition. This condition holds at any given point in the stream whenever the sample concentration is not changing with time at that point. Between samples one has the "transition state" from one steady state to another.

The steady state principle applies throughout the system, including at the photometer and at the strip chart recorder, where a steady state condition produces a smooth horizontal line. Deviations from steady state that give this line a ragged appearance are called "noise."

Temperature change, reagent action, wear during operation, or changes in tubing characteristics with repeated flexing can cause variations in the exact rates of delivery for manifold tubes. Slow changes in proportions of fluids can result. Even at constant proportions slow changes in dwell time can occur (29). All of these can result in drift, wherein the steady state value for zero sample concentration (or that for higher sample concentrations, or for both at once) changes slowly with time. Means of observing and controlling or correcting these drifts *must* be employed in continuous flow analysis (14).

Accurate addition, or to use a better term, accurate proportioning, is absolutely crucial to continuous flow analysis. So far only the coarse picture has been presented. Average flow rates have been considered, and average compositions assumed to result, throughout the stream. The more detailed picture of flow rate and composition has a vital bearing on a characteristic that the clinical chemist finds crucial—the rate at which samples may be run. Figure 10-4 demonstrates the regular pulses in the flow from a peristaltic pump. Other cyclical phenomena also exist in the continuous-flow device, e.g., bubble formation in the reagent stream and intersample bubble formation at the sampler. On earlier models of the pump produced by Technicon, the phasing of these phenomena was not controlled. As a result phase shifts between different cycle times and fluctuation in cycle times caused variations in proportioning that appeared as noise. In more recent models these problems are eliminated by phase control of air bubbles involving automatic time introduction of an air bubble at every negative pressure pulse (100). A simple technic for accomplishing this on older models is also available (48). As equipment improves and sampling frequency increases, it will presumably be necessary to control the relative phasing of intersample bubble formation, roller position, and stream bubble formation. When phasing is properly controlled, the composition at steady state of each individual slug of liquid in the flowing stream matches every other one exactly.

Elimination of the pulses has been attempted by pump design (59) and by resistance-capacitance pulse suppression (Fig. 10-7) with success in specific applications. A control valve for precise pulse free delivery of reagents is also available (57).

MIXING

Designers of automated equipment seem to have erred by underestimating the difficulty of adequate mixing very frequently. In several analyzers supplementary mixing devices were added after the first instruments had been marketed. In discrete analyzers inadequate mixing shows up as poor precision, which results from nonuniformity of reaction and measured product. In continuous flow it shows up as noise. In either case it *must* be monitored, quantified, and controlled in any acceptable procedure (55).

DISCRETE MIXING

In discrete analyzers the methods of mixing tend to be as varied as the devices themselves. Rotating or vibrating paddles have been used. Usually the motion of the paddle is designed not only to mix the solution in question but also to shake off all excess liquid before moving to the next. Vessels of special shape that cause turbulence on rotation have been used. Air bubbles rising from the bottom of the reaction vessel are employed effectively (6). In the DuPont ACA, kneading of plastic vessels causes mixing (85), and in centrifugal analyzers rapid acceleration or deceleration—usually the bane of good separation—is used to produce swirling and mixing. Mixing can also be accomplished by rotating magnets (124).

CONTINUOUS FLOW MIXING

In continuous flow methods successive samples occupy the same vessel. Mixing of one sample with another must be avoided by all possible means, but adequate mixing of each sample with itself is essential. Hence mixing and nonmixing demand more attention and understanding than in discrete analysis. The topic may not be as simple in theory as it is in discrete analysis, but it is simpler in practice. The answer lies in excellent proportioning more than in clever mixing.

If sample and reagent proportioning are slightly imperfect, and small variations in composition exist along the flowing stream they can be eliminated by small controlled amounts of mixing. But such mixing *along* the stream, which may be useful in the middle of the steady state for each sample, proves very undesirable between samples, because it lengthens the transition state and slows the rate of attainment of the new steady state.

Perfect sample and reagent proportioning, complete mixing *across* the axis of flow, zero mixing *along* the axis of flow—these are the ideals of continuous flow analysis. But the nature of streams forces them to behave in quite the opposite fashion. Normal laminar flow produces extensive longitudinal mixing with little radial motion (107). Figure 10-7 shows the cross section of different types of streams each with an abrupt longitudinal concentrational change—before any flow and after a very short passage down a tube. This figure illustrates all by itself the qualitative change in efficacy that the invention of the bubbled stream provided to continuous flow analysis (96). Bubbles inhibit longitudinal mixing by orders of magnitude and promote cross sectional mixing. A more complete description of the behavior of concentrational boundaries of a bubbled stream will be given below under the section "Transient State."

An additional effective means of mixing bubbled streams was invented at the same time as the bubble (96)—the mixing coil, a simple helical tube with its axis horizontal (Fig. 10-6). When a slug of liquid runs around such a helix it inverts itself twice per 360° turn. The slightest difference in density of an inhomogeneous solution causes rapid mixing under these conditions. Thus a single mixing coil of 14 turns (Fig. 10-6) proves more than adequate to mix thoroughly any properly proportioned stream. Therefore any flow diagram

FIG. 10-7. Plug, laminar, and bubbled flow in tubes. Cross sections are represented. The area of high concentration is cross hatched. Dotted lines indicate bubble starting positions. The thickness of the liquid film on the wall around the bubbles is much exaggerated for clarity. Arrows indicate intraslug flowpaths that produce intraslug mixing.

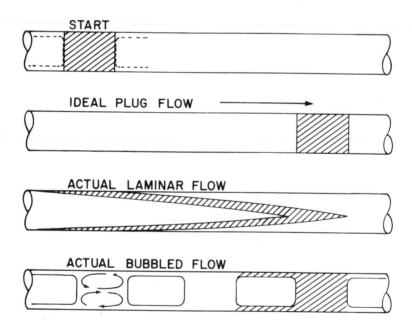

contains (without so designating) two fundamentally different kinds of streams, one bubbled and containing successive samples where mixing and nonmixing are crucial factors and the other unbubbled containing reagent of constant composition where mixing is complete. There are obvious exceptions to this rule. For example, in Figure 10-5, just before the stream passes through the photometer, a C5 fitting is used to pull 2.5 ml/min of the reacted stream, *without bubbles*, through the flow cell. (The bubbles plus leftover liquid go to waste.) Other exceptions are discussed below.

It is also possible to produce cross sectional mixing in an unbubbled stream, by passing it through a jet mixer (98) that produces intense turbulence between fine orifices and small mixing chambers.

SPLITTING AND TRANSFER

Quantitive automatic splitting of one solution or sample into two or more makes multichannel devices possible. In discrete analysis resampling of a diluted sample can provide measured aliquots for several different determinations, each on its own channel. Such splitting is best performed just after the last procedural step that is common to all channels. Of course, if no step is common, resampling of the original specimen is a valid form of splitting. Thus some discrete devices preclude splitting onto parallel channels but permit different determinations on sequential aliquots of each sample, e.g., the DuPont ACA. In either case the pipetting modules simply take new samples for each channel.

In continuous flow analysis, unbubbled streams of fluid can be split with accuracy into two streams of known flow rate, simply by pumping off part of a stream at some chosen rate, which reduces the first stream to a new known rate (98). Bubbled streams cannot split with high accuracy because of the compressibility of the bubbles. Therefore debubbling, as in the C5 fitting of Figure 10-6, precedes stream splitting as a rule. Multichannel analyzers having 6, 8, 12, or more channels make efficient and extensive use of automatic sample splitting.

Centrifugal analyzers also promise technics of splitting and transfer (8) and prototypes have been developed.

SEPARATION OF PROTEIN

Separation of desired constituents from protein, which may be the single most common step in clinical chemistry, has proved an almost insoluble problem for most discrete analyzers. The Beckman DSA 564 includes precipitation and filtration devices. Column chromatographic separation is used in the DuPont ACA. However, the commonest approach has been to avoid the separation by whatever possible means, usually increased chemical specificity. The difficulty of this step by other than continuous flow methods seems in fact to have stimulated much research on methods which do not require protein separation.

Dialysis

Continuous flow dialysis is one of the several inventions that made continuous flow analysis a practical reality (96). The steady state condition, by eliminating the need for completeness of reaction, made it possible to dialyze simply by running two bubbled streams parallel to each other separated by a semipermeable membrane. Dialysis of nonprotein molecules across the membrane, although far from complete, is reproducible, and separation from protein of those that do cross the membrane is total.

Complete separation of protein from smaller molecules does not occur in standard *con*current dialysis methods. However, continuous flow *counter*-current dialysis will provide complete separation where needed, as in determinations of protein in urine or cerebrospinal fluid (66).

Dialysis proceeds slowly compared to other reaction rates in analytical chemistry and becomes slower as molecular size increases. Surprisingly few data exist in the literature on the degree of completion of the dialysis reaction, but it is small (44). Apparently, reliance on steady state facilitates an empirical approach to development of methods without this information. If only 1% or 2% of the available constituent crosses the membrane in the dialyzer then a more sensitive color reaction must be used than if dialysis reaches equilibrium with about half on each side. But to a first approximation sensitivity is the only parameter affected by degree of dialysis, because of the steady state principle. The closer to *completion* any reaction reaches the less the dependence on identity of reaction *rate* for validity of the steady state concept. Thus dialysis proves to be the step at which the steady state assumption must be examined most carefully and matrix effects sought. Dialyzer pathlengths have been cut drastically in recent devices. With the longer dialyzers, intersample rate variation was an occasional problem; with the shorter ones, it must be a common one (12, 72).

Two streams exit from the dialyzer only one of which contains protein. Multichannel analysis often utilizes both of them for different determinations, debubbling and resampling if necessary.

SEPARATION OF DESIRED CONSTITUENTS

Separations, although they add a step to any method, solve many problems and errors due to interferences, blanks, matrix effects, etc. Because they add methodologic complexity, they are avoided where possible by chemical selectivity, specificity, or masking. As illustrations, (a) the *o*-toluidine method for glucose is specific enough that under controlled conditions protein separation is unnecessary (see Fig. 25-2), and (b) the effect of magnesium on some calcium methods is masked by adding 8-hydroxyquinoline. Nevertheless, separation by any of a wide variety of methods is an important step in many procedures.

SEPARATION BY PHASE CHANGE

Separation of a gas from the liquid phase has been used for determining CO_2 in both discrete (103) and continuous flow (97) methods. Carbon dioxide can be separated by its selective penetration of a silicone rubber membrane barrier between two liquid phases. This technic also forms the basis for electrodes sensitive to CO_2 and O_2 and will almost surely find increasing use in automation.

Separations by precipitation and filtration or precipitation and decantation are available as modules for continuous flow analysis but are not widely used in clinical chemistry, as distinct from hematology or industrial analytical chemistry (43, 105). Centrifugation of precipitates holds promise, but little experience exists to date. Separation of volatile solvents is possible to leave desired constituents behind (36).

COLUMN CHROMATOGRAPHY

This technic is routinely used in the DuPont ACA discrete analyzer. In continuous flow analysis chromatographic separations are not used as part of the steady state analytical procedure (76, 119). However continuous flow analysis is applied extensively to the effluents of column chromatography. These statements apply equally well to ion exchange column chromatography and gel filtration (21, 79).

SOLVENT EXTRACTION

Currently rare in discrete analysis, this technic is finding rapidly increasing continuous flow application. Mixing the phases is easy but until recently no really reliable fittings for separating organic and aqueous liquids had been devised, although some had come close (116). Simple modern fittings enable phase separation (15) and even washing and reseparation (91) (see Fig. 10-6). Problems remain in many cases where no known material for resilient tubing can tolerate the required solvent and where phases separate reluctantly.

WET ASHING OF ORGANIC MATTER

Organic matter may be separated from constituents with which it interferes by continuous flow wet ashing. This procedure employs an ingenious rotating horizontal glass tube with a helical groove (41). Portions of sample plus ashing reagents rest in the bottom of each turn of the groove like individual cups and are passed through areas of increasing then decreasing temperature during rotation of the tube. As in an Archimedes' screw they move along and are dumped into an annular container built into the tube at the end of the helix. The liquid is picked up from there, by a pumped line, for further processing. This has found extensive application for Kjeldahl nitrogen and protein-bound iodine measurements (41, 69).

REACTION

The reaction by which the compound desired for measurement is formed has as its key parameters time, temperature, and degree of completion. The real technical strength of automation lies in its superior control over these conditions, resulting not only in greater precision but also in greater flexibility of methodology. In manual methods, where one cannot guarantee well timed sequential performances, time dependence of reactions causes serious problems, as does extreme temperature dependence. Thus automatic control of these factors greatly widens the scope of chemistry available to the methodologist.

The analyst is interested in the *amount* of product formed by some reaction in equilibrium methods of analysis, and in the *rate* at which they form in enzymatic methods or in catalytic methods such as the ceric-arsenite measurement of iodine. The two requirements differ fundamentally, but prior to the last few years, and particularly before automation became widespread, more attention was paid to circumventing than to admitting this difference (114). For example, all four enzyme determinations in Vol. I of *Standard Methods of Clinical Chemistry* measure *amounts* of substance before and after a timed interval (102), as do most continuous flow automated enzyme methods (see Figs. 21-3 and 21-6 but also note Refs. 90 and 114). Although this procedure provides an average value for the rate over a time interval, information on variation in rate is not obtained. In some cases variations can occur that are large enough to make the average invalid. Happily the number of discrete automated analyzers capable of direct *rate* measurements appears to be increasing rapidly in recent years. Examples of these include the Bausch and Lomb Zymat, DuPont ACA, LKB Reaction Rate Analyzer, Photovolt Enzyme Rate Analyzer, and others (99). A direct comparison of results between an automated rate measuring and an automated amount measuring device has been presented (62, 125). A variety of other approaches exists to the problem of measuring a rate that may not be perfectly constant, and expressing it as a single value (27, 34, 56, 58, 68, 73, 113, 114).

In discrete analyzers reaction times are set by a wide variety of mechanisms that simply delay, for any chosen time, the occurrence of the next event (such as photometry) for each vessel in turn in sequential devices, or for all vessels together in parallel ones. For rate determinations, the fluid in question can rest in the photometer during the measurement, or some readings-separated-in-time technic can be used.

In continuous flow analysis dwell time is arranged to provide the desired reaction times. The steady state phenomenon provides direct measurement of amount but is not easily consistent with reaction rate measurements. Approaches to the latter have included single point measurements (see above), variation in pump speed to alter dwell time as a function of time (80, 90), and multipoint determinations by clever splitting of streams to provide simultaneous multiple dwell times that are read sequentially (35). In each case something is sacrificed in the compromise—in the first, reliability, in the last two, speed. However, the last technic utilizes continuous flow principles in a very ingenious manner and makes an excellent study for the interested reader.

MEASUREMENT

Automated instruments have, by now, utilized as sensors almost every quantitative transducer known to man and some qualitative ones, but the visible light photometer occupies first place in usage by a vast spread. In fact, all others might be considered exceptions to the rule, including flame photometers (3, 47, 71), fluorometers (87), spectrophotometers (2, 67), atomic absorptiometers (64), potentiometric electrodes (28, 92, 103), polarographic electrodes (60, 104), isotope counters (118), infrared spectrophotometers (87), polarimeters (49), particle counters (83), thermistors (39), and patterns on filters (65, 78).

MEASURING CHAMBERS

Obviously, photometer cuvets play a key role in automation. Whether in a discrete or a continuous flow apparatus the cuvet volume sets a series of parameters of the entire system including sample and reagent volume, rate of analysis, and overall machine size. Cuvet shape affects the same parameters—path length affecting sensitivity, and hydrodynamic properties such as ease of washing affecting the others. Since measuring chambers tend not to vary as much with type of sensor as do other components, these statements apply in areas other than photometry.

Flowthrough Cuvets

These cuvets can be continuous or intermittent in their flow depending on the device to which they are attached. They are very widely used in automation and semiautomation, probably being second only to variants of the Seligson pipet as the commonest mechanized unit. Designs are legion (9, 49, 54, 67, 93, 96). In all cases they possess the common problem that goes by many names, sample separation, wash, contamination, and in continuous flow analysis, carryover or interaction. When flowthrough cuvets are used alone, or in discrete analyzers, they may be emptied between samples and even washed to avoid this problem (6). In continuous flow analysis one sample must be washed out by the incoming next sample.

In any case air bubbles in the light path must be considered to be the nemesis of flowthrough cuvets. Bubbles can be extremely hard to control and sometimes to detect, particularly with discrete analysis with or without intermittent flow, where their only manifestation may be an incorrect but steady and valid-looking reading. The absolutely rigid habit pattern, common in manual photometry, of visually inspecting every cuvet for bubbles before placing it into a photometer bears mute testimony to the problem. Only scrupulous cleanliness of cuvet and solutions, with an assist from surfactants, protects against the problem.

In continuous flow cuvets, bubble trouble comes in two forms. The commoner is rapid passage of a bubble through the cuvet, carried by the stream. This produces a very characteristic sharp spike on the strip chart record towards higher absorbance. The other occurs when very tiny bubbles remain in the flow cell, causing small excursions of the recorder pen at the pump pulse frequency. The solution to the former is improved uniformity of bubble pattern in the flowing stream, best accomplished by bubble phase control. The latter should not occur in a clean system.

The need for a bubble-free stream at the photometer accounts for the "debubbler" fitting (C5 in Fig. 10-6) before the flow cell. This has been considered so important a factor that, in the cuvet, the laminar flow of a nonbubbled stream is tolerated, with its attendant longitudinal mixing (47). Successive generations of flow cells have decreased the length and effect of this "mixing chamber" formed by the debubbler and flow cell cuvet (100), but not debubbling at all eliminates the chamber and the problem (48). Figure 10-8 demonstrates the great importance of this phenomenon to the rate of transition between steady states. Increasing the fraction of the flowing stream that moves through the cuvet improves the transition rate markedly, but passing bubbles and the entire stream is best by far.

Several of the manifold diagrams for methods shown elsewhere in this book have been modified from their original source by two simple changes that greatly improve performance. The first of these is to increase the flow through the cuvet, the second is to control bubble timing with a "compressor" line, which not only makes the first change possible, but decreases steady state noise.

FIG. 10-8. Effect of flow rate through cuvet on rise and fall curves. From HABIG, RL, SCHLEIN BW, WALTERS L, *et al: Clin Chem* 15: 1045, 1969.

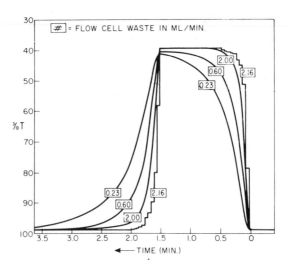

No—Flow Cuvets

A variety of cuvets exists in which no flow occurs. The DuPont ACA uses the transparent walls of a plastic reaction container as its cuvet ends (85). Several discrete devices employ unit cuvets or racks of cuvets that are all washed together or sometimes discarded as in the Abbott Bichromatic Analyzer. In any of these the bubble problem must be monitored and controlled carefully.

CALIBRATION

Automation in general, and multichannel continuous flow automation in particular, have changed the basic nature of calibration and calibration standards in clinical chemistry. The change has not been in itself for the better but has been a "trade off" for other advantages. Earlier methods depended for calibration and standardization on simple approaches. One was self-calibration in terms of fundamental constants, for example the volume and pressure of released CO_2 in the Van Slyke machine or the reagent weighed into the titrating solution for the Clark Collip method for calcium. Another was calibration against a simple solution of pure standard material such as glucose, urea, uric acid, 0.1 N HCl, or the like. These technics enabled the clinical chemist to get reasonable estimates of the accuracy of most of his methods and thus to control both accuracy and precision.

Automation calibrates by comparison only (10) and depends for accuracy on the correctness of the comparison standards and on their ability to behave in a fashion identical to the unknowns, throughout the system in question. This qualification covers many problems and pitfalls, especially when multiple standard constituents are to be combined. Matrix effects on rates of reaction, including dialysis rates (12, 72), cause differences in behavior between standards and samples. Therefore pure substances in aqueous solution are often unsatisfactory as standards. Even where they are, many pure compounds may not go into simple solution. In the early days of multichannel analysis, attempts by this writer to combine into 1 solution the 10 or 11 most commonly measured constituents of clinical chemistry developed a healthy respect for the problem but only a limited solution (109). An industry has developed for providing reference serums, originally for quality control but now utilized also for calibration in automation. If one uses one product (or even different products of one company) to calibrate his equipment and simultaneously for quality control specimens, subtle but serious errors can occur because of the complexity of this problem.

Actual technics of calibration are not changed by automation, only added to, because new parameters require calibration. A sufficient number of standards must still be run to produce a valid calibration curve. With a method in which Beer's law is obeyed exactly, and with a device that gives results proportional to concentration (e.g., by conversion of % T to A), this number may be as low as 2. Additional standards must be employed at reasonable intervals to measure drift, interaction between samples, and noise, and for quality control. The proper intervals are dictated by experience. An excessive number of standards is no real advantage and may be a disadvantage where it provides unwarranted confidence.

The term "drift" in automation refers to slow changes in calibration. It may be resolved into two independent types, zero drift and sensitivity drift (14). In the former the entire calibration curve moves without changing its slope. In the latter the curve rotates, changing its slope but not its zero intercept. Both can occur together, and of course do. Standards having zero concentration and some value high on the curve should be run with every batch or every day's work, depending upon experience, to measure these drifts. They must be controlled by proper maintenance of machines and methods. Under certain conditions mathematical or graphical technics of drift correction are appropriate (see below).

Standards for calibrating interaction between samples are discussed below. This, like drift, is simply an additional calibration required by automation. Both of these calibration technics are too frequently omitted by chemists experienced primarily in discrete manual methods.

Noise in continuous flow analyzers must be checked periodically by running standards, or preferably specimens, of high and zero concentration for prolonged periods, and observing the steady state recording obtained.

SCHEDULES

Automation deals with high workload situations and requires more standards of different types than are necessary when manual methods are employed in low volume situations. Therefore, the sequence in which the cups of fluid are presented to the machine (the work schedule) becomes important in terms of volumes of samples and expensive standards. This is complicated by interaction between samples, in that not only samples but also standards or controls following a high standard or sample may be unusable without correction (19). Therefore if one wishes to use the absolute minimum number of standards, calibration must include correction of standards for interaction, which may be complex (111). Attitudes far too simplistic seem to prevail on this point. This writer has visited many laboratories where the unwritten lore does not permit recognition of interaction but does dictate that calibration standards be sequenced in ascending order of concentration to hide it. Similar sequencing habits exist for multichannel continuous flow instruments, where a sample of roughly normal concentration may be placed before every standard, or two standards may always be run, with only the second being used for calibration.

No set rules can be provided for scheduling; the judgment of the analyst is involved. Nevertheless it is an important practical aspect of automation. One example may make this statement more concrete, even though the quantitative aspects of "carryover" or "interaction" that are involved in the example must be referred to in subsequent rather than preceding sections. Let us assume that cholesterol is being assayed on a sequential analyzer and that 5% of the value obtained for any given sample appears as part of the following sample. This actual figure is not uncommon either on certain discrete or continuous flow instruments. If one runs a 0 mg/100 ml standard, followed by two 200 mg/ml standards, the second high standard will read 10 mg/100 ml higher than the first. Which is correct? In fact only the first can be correct, since the second is affected by the first, and the first is unaffected by its 0 mg/100 ml predecessor. The proper solution to this problem is to precede every standard and sample by a 0 mg/100 ml standard. But a different schedule and a new assumption permits operation *without* these intervening standards, and therefore at twice the production rate, as follows. Because most of the specimens analyzed fall between 150 and 350 mg/100 ml one can count on most samples being preceded by another with a value of 250 ± 100. Let us therefore assume that any standard or sample is correct only when it follows a 250 mg/100 ml standard (and therefore contains a bias of + 12 mg/100 ml). On the basis of a conservative usage of Tonk's criterion (see Chap. 12), let us agree that an error of 7.5 mg/100 ml in a given result is an acceptable price to pay for doubled production rate. Then any given sample will provide results that can be accepted without correction unless preceded by a sample of concentration outside the range 250 ± 150, or 100 - 400 mg/100 ml. This includes most samples and requires identifying and rerunning or correcting only a few.

The schedule in this example would be as shown in Table 10-3 for a direct method that obeys Beer's law.

TABLE 10-3. Example of Schedule

Tray position	Cup contents	Comments
	(numbers are mg/100 ml)	
1	250 standard	Used only as "Primer"
2	250 standard	High standard
3	0 standard	Zero standard
4	0 standard	The difference between cups 3 and 4 permit calculation of interaction. Note that cup 4 will produce a "below zero" result
5	250 mg standard	Primer
6	Sample	
7	Sample	
8	Sample	
9	Sample	
10	Quality Control Sample	
11	Sample	
12–38	Samples	
39	250 standard	Sensitivity drift control
40	0 standard	Zero drift control. Note that cup 40 will normally give a reading above baseline.

PRESENTATION AND CALCULATION OF RESULTS

The strip chart recorder fits the needs of continuous flow analysis for result presentation exactly and is used universally. Discrete analyzers employ a variety of devices including printers, digital displays, teletypewriters, cathode ray oscilloscopes, etc. The mechanics of the discrete analyzer cause a reading to be taken from the sensor at precisely the right time for each sample. The signals are then conditioned for presentation—% T changed to A, and A to concentration (from preset calibrations) or the like. The results are then displayed or printed.

Among early users in continuous flow analysis, a biased approach to sample presentation developed that requires discussion because its effects still persist in many laboratories. As discussed above, and clearly stated by the inventor (96), the sample steady state forms the basis for readings of results. This point might have been communicated more successfully had long air bubbles not been placed between specimens in the original samplers. But these long air bubbles separated specimen signals into spikelike peaks. Steady state was not noticeable and was invariably approached from the same side as shown in Figure 10-9. Rates of analysis, in samples per hour, were increased by recognizing that even the reaction by which steady state is reached for each sample need not be complete in continuous flow devices (117). Then dwell time was almost totally forgotten or ignored, and the peaks took on some inherent significance of their own. This writer has repeatedly seen cases where analysts accepted as valid peaks that were obviously formed at a time totally inconsistent with the known dwell time of the method. Dwell time must be absolutely constant for valid operation, and if it is constant all peaks must be uniformly spaced and separated from each other by the known time interval between samples set by the sampler module. In contrast to Figure 10-9, Figure 10-10 shows the chart that results from a closer approach to steady state, and with minimal intersample times.

Beginning analysts who do not recognize these fundamental rules have on occasion failed to note a totally missing peak or a masked peak and have been known to ascribe all subsequent peaks to the wrong sample. Without scrupulous scheduling, the resulting incorrect total number of peaks can even be missed, producing the nightmare of automation—high speed production of incorrect data. The manufacturer and others have made valiant efforts to emphasize steady state and de-emphasize the peaks, and this problem is on the wane although still very much present.

In the second generation AutoAnalyzer, much closer approach to steady state was achieved in any given time. There is nevertheless the same tendency to increase production rates by decreasing the percent of steady state reached. This can be done only with knowledge and accompanying vigilance, or else precision suffers (33, 42), especially if sampler timing is inexact (79).

Whether or not steady state is reached, time forms the only valid basis for deducing where on the strip chart any given sample should be read. If all is well the peaks fall at regular intervals and indicate time, but extreme vigilance must be exercised to watch for those peaks that show timing error. For example, insufficient sample in the cup can produce a valid looking peak that is grossly but undetectably low, but it will invariably appear too early. Otherwise, dwell time, the most valuable quality assurance technic of continuous flow analysis, is ignored. On the other hand, if time is used as the criterion for when to read, accurate data can be obtained surprisingly far from steady state, without loss of this quality assurance factor (33).

Given a set of results from calibration standards, samples and controls, standard technics of deducing concentration can be employed, except for drift correction or correction for interaction, discussed below (see also the section "AutoAnalyzer Procedural Guidelines" below).

FIG. 10-9. AutoAnalyzer peaks. $t_{bs} = 60$ sec, $t_{in} = 40$ sec, $t_{out} = 20$ sec

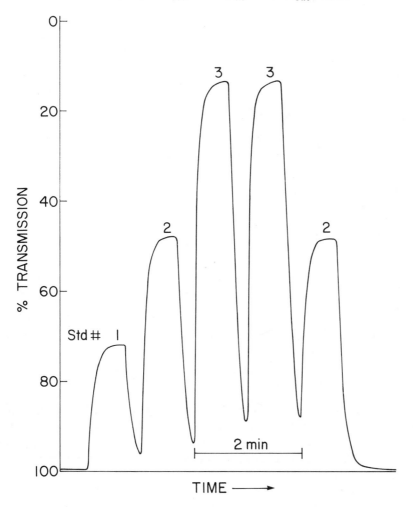

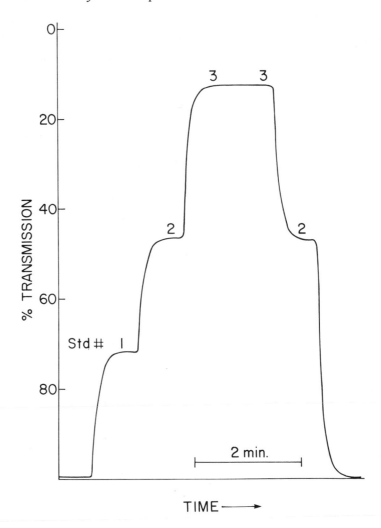

FIG. 10-10. Nonpeak records from an AutoAnalyzer. t_{bs} = 50 sec, t_{in} = 49 sec, t_{out} = 1 sec

SEQUENTIAL MULTICHANNEL RESULT PRESENTATION

The above paragraph implies that in a well functioning continuous flow analyzer only a small fraction of the strip chart record is really needed for each sample,—the peak or steady state. This fact has been exploited to permit multichannel presentation on one strip chart (98). By controlling dwell times with delay coils of chosen volume so that peaks occur sequentially, and by phased switching from one photometer to another, data from up to 12 channels can routinely be presented on one record. Figure 10-11 shows the "time-sharing" technic. Signals are conditioned to present

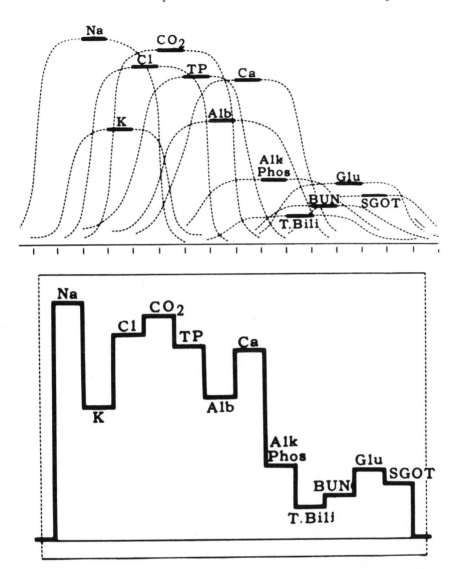

FIG. 10-11. Time-sharing in sequential multiple result presentation. From WHITEHEAD EC: in Technicon Symposium, *Advances in Automated Analysis,* edited by L. T. Skeggs, Jr., White Plains, New York, Mediad, 1966, p 439.

concentrations, and the instrument is calibrated "on the fly" to match preprinted paper, producing the very effective and familiar result presentation of Figure 10-12.

This approach demands a higher level of competence from the operator and greater understanding of continuous flow analysis than single channel operation, because problems present themselves more subtly, or not at all, in this time-shared output.

FIG. 10-12. Sequential multiple analysis report.

BIO-SCIENCE LABORATORIES

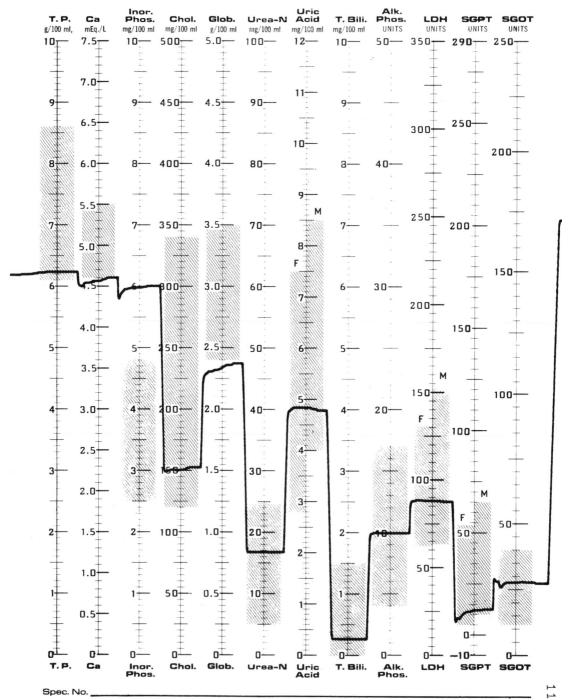

Spec. No. _____

Pos. No. _____

SPECIMEN IDENTIFICATION

Assignment of an analytical result to the wrong patient is the *bête noire* of the laboratory. However, a documented description of the problems associated with specimen identification and the magnitude of their consequences has not been developed. The topic is frequently addressed, although more by systems analysts and computer personnel than by laboratorians. Actually, little of practical utility has emerged from these considerations. Discussion centers around the possibility of this type of error in manual technics. The common current method involves simply keeping careful track of whose specimen goes where on the sample presentation tray. For example, serial numbers attached to incoming request slips and specimens are carefully recorded on the schedule for the automated device. Then data are recorded by serial number for each patient. Errors occur occasionally. Those that interchange the results for two patients are probably the most frequent and have little to do with automation per se except that each analyst handles more automated samples per unit time. However automation introduces the possibility of the "out of phase" error, e.g., where everyone gets the data of his predecessor or follower. However, in many years of work in highly active laboratories, this writer has seen only two or three cases where this error occurred in final reports, with only a few data involved in each case.

The schedule must be maintained scrupulously in spite of intrusion of stat samples, blanks, controls, repeated samples, and the like. A properly run laboratory seems to make so few errors of this type that mechanization can be justified only on the basis of work savings.

No solution to automated specimen or patient identification has yet been widely accepted. Most proposals solve only part of the problem or shift the position of the error-prone step in the sequence. Perhaps the most complete approach is that of the DuPont ACA (85) that automatically prints its results on a photocopy of its original request slip. Several devices, for example the Vickers 300, carry a digital code on each specimen tube which is machine readable. The report is then printed in association with this number. In some cases, such as the Bio-Logics S-Tab, the coded device can be attached to the tube of blood at the patient's bedside immediately after it is drawn, and the patient's name can be written on the same device. A sampler module is available that reads this device and stamps its number on the strip chart over the proper part of the record.

SPECIAL PARAMETERS OF DATA FROM CONTINUOUS FLOW ANALYSIS

As mentioned earlier continuous flow analysis has unique parameters that have no exact counterpart in either manual or discrete analysis. They result from the kinetic nature of continuous flow and are concerned with temporal relations in its flowing streams. These parameters divide themselves into those relating to the steady state condition and those relating to transition between steady states.

STEADY STATE

Considerations relating to steady state include proportioning, degree of completion of chemical and physical reactions, rates of such reactions, time, temperature, hydrostatic pressure, dwell time, noise, drift, sensitivity, specificity, and range. These have either been covered above or do not differ from discrete analysis. An excellent series of simple teaching exercises illustrating these properties has been developed (38).

Increased productivity was the original motivation for automation in clinical chemistry (96). Continuous flow analysis is founded on the ability to produce clean, noise free, steady states that relate reliably to concentration, and remain unchanged as time passes. Productivity, on the other hand, dictates that as little as possible of such steady states should be produced, since they take time without producing further information. Furthermore, any given steady state is approached along an exponential curve, (110, 133)—a situation that makes it impossible to state qualitatively when true steady state has been reached. In addition, all of the steady state information except noise level has appeared in less simple form long before steady state is reached (112, 120, 121). Therefore steady state is a goal to be striven for, a condition that must be understood and studied in every method, and a device for frequent evaluation of noise in the system. But understanding of the transient state provides the key to improvement in instruments, methods, and productivity in continuous flow analysis. This has been demonstrated by successive generations of AutoAnalyzers, wherein steady state has received much deserved attention and in regard to which user education has much improved, but wherein the true instrumental advances have been made by attention to the transient state. The improvements in transient state have been achieved by identifying those modules that showed up most poorly in regard to the parameters described in the following paragraph and redesigning them.

THE TRANSIENT STATE

This state must be discussed in terms that are unfamiliar to conventional analysts, such as rise curve, fall curve, degree of attainment of steady state (%SS), rate of attainment of new steady state, half-wash time ($W_{1/2}$), lag factor (L), interaction (I), time between samples (t_{bs}), and ratio of the time the aspirator is in the sample (t_{in}) to its time out of sample in air or wash or both (t_{out}), (110, 112, 121). These terms are illustrated in Figure 10-13.

Transition between steady states takes place along "rise curves" or "fall curves" (110) (Fig. 10-13). If these are reproducible, and their characteristics are quantitatively predictable, one need not wait for steady state to get desired information on sample concentration. Therefore the study of the

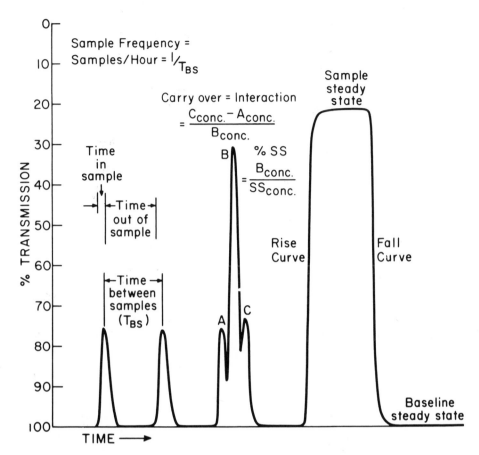

FIG. 10-13. Terms of continuous flow analysis. From THIERS RE, COLE RR, KIRSCH WJ: *Clin Chem* 13:451, 1967.

characteristics of these transitions and the origin of these characteristics are of basic importance to continuous flow analysis.

Separation of Samples

Transition between steady states occurs "between samples." When intersample contamination in the sample probe is negligible, the transition at the sample presentation device is a square wave as shown in Figure 10-14A. At the exit of the sample manifold tube, just as diluent or reagent is added, the laws of stream flow have begun to exert their rule and the square wave has changed, depending on apparatus, as shown in Figure 10-14B. The bubble or bubbles of air that separate samples are very important from the very beginning (96), as Figure 10-7 shows. The remainder of the flow system degrades the original square wave even further, until the actual output of a typical method resembles the curves of Figure 10-14C.

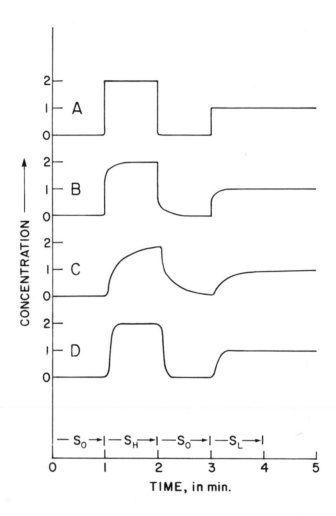

FIG. 10-14. Effect of exponential and lag factors on concentration changes in continuous flow analyzers. S_o is a zero concentration standard, S_H contains 2.0 g/100 ml globulin, S_L 1.0 g/100 ml. Curve A shows concentration changes at the sample probe, B just after the reagent is added, C and D at the photometer. Curve D has been derived from Curve C by Eq. (10-1).

The Exponential Factor

In generations of AutoAnalyzers up to the current one the most important contribution to deformation of the square wave rise or fall curve follows an exponential equation. The existence of this exponential factor was demonstrated quite early and was shown to cause the observed "interaction" between samples wherein apparently a constant fraction of the concentration of any given sample is carried over and appears added to the following one (108, 121). One form of the equation in question is as follows:

$$C_{td} = (b \ dA_t/dt + A_t) \ K \tag{1}$$

where C_{td} stands for the concentration of the constituent of interest at the sample probe of a continuous flow analyzer at time td (where td = time t minus a constant d, the dwell time from the probe to the photometer), b stands for the measured exponential factor of the system, A_t stands for the absorbance measured at the photometer at time t, and K stands for the constant for changing absorbance to concentration. In Figure 10-14D an actual stripchart absorbance record for a globulin method (Fig. 10-14C) has been recalculated by Eq. 1 to give concentration. The expected square wave regenerates itself to a remarkable extent, illustrating the quantitative importance of the exponential factor.

From a practical point of view the entire curve need not be regenerated. One need only perform a measurement of percent interaction according to the formula of Fig. 10-13. Then one can apply this correction to successive peak height concentration values in a series of samples to remove the interaction error (106, 108, 111). When performed by computer this correction provides striking improvements in accuracy or speed, or marked improvements in both together (111).

The exponential factor, whether expressed as b above, $W_{1/2}$ (half-wash time), or % I, also has value as a figure of merit that can be applied to any module, method or overall system of continuous flow analysis (110). The lower the $W_{1/2}$ the lower is the interaction, the better is the system in this regard, and the faster is the possible sample rate at given interaction error.

The exponential factor has been identified as a property of unbubbled streams in continuous flow analysis (112, 122). It results from the "mixing chamber" effect of such streams, especially where some turbulence may exist as in debubblers and photometer cuvets. Its decrease may be achieved by minimizing such components of stream flow, as in the SMA 12/60 (100) or by removing them completely (48).

The Lag Factor

The reconstruction of square wave transitions in Figure 10-14D is not perfect. A secondary factor perturbs the transition in the flowing stream— the lag factor. Plotting the data from the fall curve of Figure 10-14C on semilogarithmic coordinates gives a pattern familiar to biologists, particularly microbiologists, a lag phase followed by exponential change as shown in Figure 10-15 (110). The lag phase, measured in seconds as shown, forms a second figure of merit for continuous flow. Although currently secondary to the exponential factor in impact it may, as the latter is attacked and eliminated, become a primary factor, and the most characteristic one for continuous flow, because its origin proves to be the bubbled stream (112, 120). It originates in the film of liquid left at any given point on the wall of the transmission tube after each slug of liquid has passed that point and an air bubble is in the process of passing. This was long considered to be a

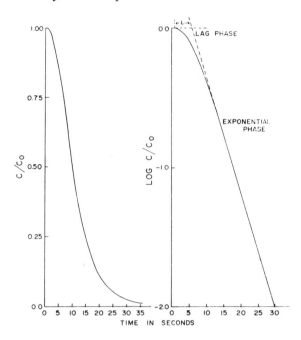

FIG. 10-15. Separation of a typical transition curve in continuous flow analysis into lag and exponential phases. From THIERS RE, REED AH, DELANDER K: *Clin Chem* 17:42, 1971.

negligible phenomenon because the tubes are made of nonwettable material and therefore should not hold such a film. In fact the film is really necessary for smooth stable flow of bubbled streams (23). Without it extremely high and variable resistance to flow occurs, causing instability. Detergents are routinely added to reagents to lower surface tension and increase wettability and to produce smooth flow. Therefore an interslug film exists that acts as a leakage pathway between slugs of liquid (23) and causes changes in the composition of any given slug as it moves down the tube, if its predecessor has a different composition.

The leakage rate is surprisingly high, our experiments indicating that in 1.6 mm i.d. glass coils with liquid slugs 9 cm long, and air bubbles 3 cm long travelling at 6 cm/sec, each slug contributed to its following one over 1% of its volume per second! This has two consequences, illustrated diagrammatically in Figure 10-16: First, that in a bubbled stream a sharp concentration boundary degrades to a distribution of concentration that initially matches a cumulative Poisson distribution (112) and finally a cumulative normal distribution (120), and second, that the boundary as a whole falls backward relative to the overall stream. Both of these effects are also seen in Figure 10-14D.

A general equation of continuous flow analysis has been derived using the above factors, which fits experimental data (120).

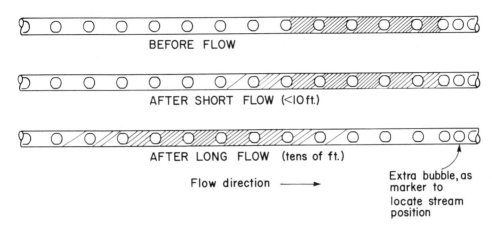

FIG. 10-16. Effect of bubbled stream flow on concentration interfaces. The figure illustrates one section of a stream which starts out with a small well defined section of different concentration. Concentration is expressed by degree of cross hatching. After flow of a short distance (one or two feet in a glass coil) the sample has spread out. After longer flow the sample has spread out further and moved back relative to the stream.

Rate of Analysis

Mathematical corrections for the above phenomena are possible and are very useful to owners of existing instruments because they enable operation at increased rates of analysis. However, improvements in instrument design based on knowledge of the above phenomena have proved a more universal pathway to increased rates and accuracy, without precluding use of the corrections to improve even these. Initially AutoAnalyzers were operated at 20 specimens/hour. Today rates 3-6 times this great are possible by using third generation devices directly, or second generation ones with arithmetic corrections. Promises of rates up to 10 times the initial ones exist in presently developing hardware and computer software (48, 111), without loss of precision. At such rates, factors in clinical chemistry other than instrumental begin to limit production. This is highly desirable because in the field of automation at present a compromise is necessary between economic and technical aspects, where production rate, volume of reagents, and technician time on one hand have to be balanced against degree of attainment of steady state, freedom from sampler timing problems, high sample consumption, and freedom from interaction on the other. When such compromise is no longer necessary the technical side will obviously benefit.

Although the above properties bear on the rate-accuracy relationship in analysis, it is well to seek out analogous problems in all automated systems. For example, now that the sources of interaction between samples is understood this problem is becoming a very minor one in continuous flow devices. But its analogous problem is still surprisingly high in some discrete automated analyzers (70).

PROCEDURAL GUIDELINES FOR AUTOANALYZERS

The Technicon AutoAnalyzer I has been used long enough and widely enough to appear in specific methods in various chapters of this book. Details that form the "Procedure" section in manual methods are automated in these cases and appear as flow diagrams. However, certain maneuvers are characteristic of such automation and are common to many methods. These will be outlined here as a generalized procedure, much of which applies equally to AutoAnalyzer version II and multichannel AutoAnalyzers.

GENERAL PROCEDURE

1. Stretch the manifold tubes across the pump and start the pump by engaging the rollers and platen.
2. Place all of the reagent lines in water which contains a detergent suitable to the determination.
3. Examine the entire manifold for leaks or wear. Look for worn "snaking" pump tubes. Replace leaking fittings and worn tubing.
4. Inspect the bubble pattern. If it is being disturbed or broken at any point, correct the problem by replacement of tubing, nipples, or fittings.
5. Place filters of the appropriate wavelength on the photometer or fluorometer. Set the apertures of these modules correctly and adjust the stripchart recorder to zero with no light and 100% with full light. (With the fluorometer, in direct methods, the 100% adjustment requires running reagents and a high standard.) Set the gain of the recorder to a maximum value without pen chatter.
6. Place the reagent lines in the appropriate reagents and the sample probe in the sampler wash reservoir.
7. Note the change in the 100% reading. If it does not correspond to that previously noted for the method in question the reagent lines should be checked for correctness of connection and the reagents for proper composition.
8. Note the noise level of the reagent baseline. If it exceeds the prestated tolerance for the method in question the source of the problem should be sought and corrected.
9. So called "inverse methods" involve a decrease in absorbance, fluorescence, or other measured parameter with increase in concentration of the substance being determined. (In "direct methods," the changes have a common direction.) In inverse methods the position of the reagent baseline is a particularly labile and important parameter, as is its noise level, and these should be noted and recorded with particular care.
10. At intervals that depend on the method, but always with a new or reactivated method, run a standard of concentration in the upper half of the calibration range. Measure the dwell time for the overall apparatus. If it does not correspond reasonably with the dwell time observed previously, or calculated from the flow diagram for the method, seek the

cause of the discrepancy and correct it. (This step can save much time and frustration by catching problems before they ordinarily show up.)

11. Note the reading of the standard. If it does not agree with expected or previously observed readings for its concentration do not proceed with the run (and probably waste patient samples thereby)—find and correct the problem.

In this writer's experience one invariably saves time by searching for such problems in a definite sequence. The first thing to come under suspicion should be the standard itself and the standards in general. If the standards seem correct then reagents should be examined next. In the vast majority of cases the problem will by this time have been located and solved. If not, the next item for investigation should be the apparatus. Much time is frequently lost by working on it before standards or reagents. At this point it is more important to take a careful history of the problem from the operator, than to do a physical examination on the instrument. The intuition of the operator is usually as good concerning apparatus as it is bad concerning standards and reagents. If by this time the problem remains unsolved then suspicion should fall on people, on the mode and sequence of operations, etc.

12. Allow the standard to run for several minutes and observe the rate of approach to steady state and the noise level of the steady state. If they are inconsistent with what is expected and acceptable for the method, seek and correct the problem.

13. On new or nonroutine methods run an interaction pattern (108) and correct any conditions that may be producing unacceptable high interaction.

14. Fill sample cups with standards and samples and carefully place them in a prenumbered clearly identified sampler tray in exact agreement with the written schedule. Use a standard format for the schedule, preprinted if possible. *Do not depart from the standard schedule.* If substitutions or changes are necessary (e.g., stats) adopt a standard system for making them—a system that retains all information about the original schedule, the change itself, and the new schedule.

15. Make measurement of interaction, standards drift, baseline (i.e., zero standard) drift, and control samples a routine part of the schedule.

16. Fill all cups to about the same height. Do not fill up to or beyond a level meniscus surface, to avoid physical contact of the specimen with any external surface or other materials.

17. *Cover the tray of filled cups.* Do not fill too many cups so that they sit and wait for long periods. For very long waits use individual cup caps. For waits while another tray is run use a complete tray cover such as a plastic pie holder. While trays are on the sampler *use the cover provided.* Avoid drafts. Never use the same cup of sample for more than one sampling operation.

18. Turn on the sampler and run the tray of specimens. Note the stripchart record for the first few standards and samples. Check that the standards

appear in the order expected and with the readings expected. Stop the run if the record shows discrepancies, is noisy, or is otherwise unacceptable.

19. After the run is completed label the stripchart record, before removing it from the recorder, with the identity of the tray of specimens. Date, operator, method, and other data should be written on it immediately after removal, preferably by filling in a stamped form to prevent omissions.

20. Inspect the stripchart record and the peaks obtained for noise, drift, badly formed peaks, missing peaks, or mislocated peaks. A simple and essential aid to this operation can be made from a strip of white plastic by marking it off like a ruler with the proper interpeak distances and numbering it to correspond to the schedule in use.

21. Identify those samples that must be repeated or diluted and repeated. If interaction correction is being applied to samples that follow high or low peaks between certain limits, identify the affected samples.

22. Use the standards to plot a calibration curve, in any of a variety of ways. For a photometer, linear paper or a Technicon reading board can be used provided the number of standards is great enough to accurately plot the curvature of the calibration line. For methods that follow Beer's law semilogarithmic paper can be used. For a fluorometer, flame, or other linear transducer, linear paper or the Technicon board can be used. These statements apply to both direct and inverse methods. One must be quite careful in changing from method to method, because not only may the direction of calibration change, but the apparent direction of time may change, and if one uses more than one type of recorder the two may change independently.

23. Determine the concentrations of the samples by interpolation into the calibration curve.

24. Calculate the interaction. Apply a correction to the indicated samples.

25. Check the values obtained for quality control samples and record them in a standard manner. React to problems of quality control as recommended in Chap. 12.

26. Dilute and repeat the indicated samples. Repeats due to interaction or drift can be avoided to a limited extent by technics discussed elsewhere (109).

27. Place the reagent and sample lines into water. Continue pumping until all traces of reagent have been removed from the system. Disengage the pump and unstretch the tubes.

28. Turn off nonthermostatted modules. Discard the cups of sample and clean the sample tray of spills—and erase the identity number.

29. Place a small square of nonwettable film between the pen and the paper of the stripchart recorder to prevent ink leakage by capillary action.

30. Store the schedule, stripchart, calibration curve, results, and all calculations in a standard systematic fashion.

31. If sample blanks must also be run (as with triglycerides, Chap. 28, Fig. 28-7) they are handled exactly as above and the values subtracted appropriately. Reagent blanks are of course handled by the original adjustment of the recorder scale.

ON-LINE COMPUTERS

Automation drastically altered the workload ratio of clinical chemistry. Prior to its advent so large a portion of the time was spent in manipulating samples and apparatus that calculation of final values from raw data was a negligible part of the whole. Now, in a highly automated laboratory, the analyst is swamped with so much calculation that it has become the repetitive, onerous, and error-prone portion of the overall task. At this same time computers have been evolving toward smaller, cheaper, and more powerful versions. The marriage of the need on the one hand, and the capability on tbe other, has not proved as easy as was confidently predicted a decade ago and has, in some cases, provided expensive lessons in the hidden complexities of laboratory operation. The computer is at its best in mass production situations, and promptly showed its value in the large laboratory (16, 17, 32); however, even today with over 75 on-line applications on record in clinical laboratories in this country (24), few standard technics exist, and there is an element of experimentation in every new installation. However, one common report comes from computer users, namely, that after the original problems of installation are over, performance of instruments, personnel, and equipment is upgraded by the computer. Perhaps this can be simply attributed to the fact that the computer is a tireless, ruthless inspector, and in routine situations "people tend to do not what is *ex*pected, but what is *in*spected".

The full role of the computer in the clinical laboratory has yet to be defined. We will consider here primarily its role in on-line processing of data from automated systems, that is computers physically connected to instruments (84) although off-line systems are also valuable (50, 63, 86). Nor will specific computer systems be described in detail. As with automation in general the field is changing rapidly. In the computer field technical descriptions have generally suffered peculiar difficulty in separating accomplished fact from untested plans. Therefore discussion of established principles can serve the clinical chemist best. Detailed descriptions and surveys can be found elsewhere (11, 13, 24, 26, 115).

ON-LINE HARDWARE

The automated instrument communicates with the computer by an electrical signal. In addition to passing information on absorbance or some other concentration-related parameter this signal must enable the computer to deduce when a series of analyses begins and ends, exactly when sample readings are to be taken, the type of sample being read, i.e., standard, sample, control, etc., and even which instrument is sending the data. In order to perform these deductions and complete its job the computer must store *all* of the detailed routine information, which the analyst takes for granted, about the determination in question and use it to interpret incoming signals. Nevertheless on-line hardware can be as simple as a single potentiometer and switch connected to the computer by two wires, (32, 45) or conversely, as complicated as amplifiers and multiwire devices setting up, passing, and decoding parallel, analog, or digital signals (18, 46, 95).

RETRANSMITTING POTENTIOMETER

The majority of automated instruments in use today present results on stripchart recorders. The majority of stripchart recorder-computer interconnections consist of so-called "retransmitting potentiometers." This provides d.c. voltage in proportion to pen position, from a single turn 360° potentiometer (such as Beckman model 5311R5KL.5 mounted on the shaft that controls the pen). With an appropriate power supply voltages from 0 to any chosen value can be obtained from the potentiometer. For example in one common system (32) the voltage is 0 with the switch off, and when the pen is at one end of the recorder travel the voltage can be chosen to be 0.1–0.3 V, at the other end 1.9–2.2 V. The sense of the voltage-position relationship can be reversed by a switch so that voltage always increases with increasing concentration.

INTEGRAL SIGNAL GENERATORS

Most modern measuring devices include a jack from which is supplied a built-in signal for transmission to a computer or include instructions on how to obtain a suitable signal from indicated circuit points. This statement applies across the broad spectrum of instruments from an inexpensive Turner spectrophotometer to a complex Technicon multichannel automated analyzer. In the latter case a variety of other signals is also available that indicate instrument status, sample timing, and other data.

Built-in signal sources tend to be very specific to the particular instrument, for example they may be d.c., interrupted d.c., a.c., serial digital, or parallel digital, and may work over a variety of voltage or current ranges. No general description is possible.

TRANSMISSION OF SIGNALS

On-line computers receive their signals via so-called hard-wired circuits. Errors in transmission, such as those that occur in noisy or poorly shielded cables, are intolerable and are unnecessary in passing on-line instrumental data to computers. In general the transmission distances are short (less than a mile), and quite standard cables and technics are involved. Experience in our laboratory has shown that where problems occur the cables are seldom at fault although faulty "open" wires have been found in new cables. For example, inexpensive well shielded single pair or multiple pair cables have sufficed for years in carrying d.c. signals from 0–2.0 V at 0.2% precision, from retransmitting potentiometers, over hundreds of feet through crawl-spaces crowded with the usual electrical gear of a laboratory. Line lengths have, in some cases, been so great that d.c. resistive line losses have been high enough to require compensatory adjustments at the power supply but with no other added problems.

Absolutely meticulous installation and connection of cables pays off in reliable signal transmission. Money spent on high quality junction boxes and scrupulously systematic wiring proves to be a worthwhile investment in accuracy and reliability. Short cuts cause problems that are often hard to diagnose, locate, and rectify.

Wiring for transmission of digital signals requires more expensive cable and connectors than does analog d.c. circuitry. Again the variety of equipment in this case makes detailed discussion of limited value.

ANALOG TO DIGITAL CONVERSION

Digital computers require binary signals. Concentration is a continuously varying, analog function. Therefore, an "analog to digital" (A/D) connector is an essential component somewhere between the sensor of the automated instrument and the numerical registers of the computer. Some small computers contain A/D converters. Others provide them as separate components. They operate by a variety of technics. In any case two key parameters are pertinent, accuracy and speed. Here accuracy means closeness of approach to the true value, in terms of significant figures, as distinct from freedom from error, and it is a function of the number of bits in the binary number to which the analog voltage is converted. Thus a "nine bit" A/D converter, which is free of error to the extent of ± 1 bit, can convert d.c. voltage to a nine bit binary number ± 1 in the rightmost, or least significant digit. The d.c. voltage that can have any of an infinite number of values must therefore be recorded as one of 2^9, i.e., 512 possible binary numbers. Thus the accuracy of a nine bit converter is ± 1 part in 512 or about $\pm 0.2\%$ of full scale. Although this degree of resolution is perfectly adequate for stripchart recorders, which generally are accurate to $\pm 0.5\%$ of full scale and precise to about $\pm 0.1\%$, and although it has proved so in use in several laboratories for years, one might feel more comfortable with a 10 bit converter at $\pm 0.1\%$ of full scale, or 11 bits or $\pm 0.05\%$. For clinical chemical purposes the further resolution of 12 or more bits seems a needless expense.

The speed of the A/D converter is stated as the number of microseconds it requires to calculate and store each binary number, and be ready for the next. In fact the combined speed of the converter and multiplexer (see below) make up the quantity which is pertinent to the user. This should be a small number of microseconds, e.g., 20 in a typical older machine. Otherwise this component cannot match the speed of the rest of the system, and may even limit the number of possible instruments that can be connected.

MULTIPLEXING

Computers receive incoming signals from more than one automated device. Eight, sixteen, or thirty-two are common independent input-channel capacities. The multiplexer, whether integral or separate from the basic computer, switches rapidly from one signal line to the next, pausing only long enough for the A/D converter to get its reading. Not only does the multiplexer set the number of possible incoming signal channels, but also the rate at which they are read. This rate combines with software to examine the signal from each automated device at 1 sec intervals, or 0.5 sec intervals, or whatever the capabilities of the equipment permit and the characteristics of the automated device demand. The choices made at this point tend to govern the size of computer files and even of the computer, in somewhat the same way that choice of serum sample size affects analyzer characteristics.

ON-LINE SOFTWARE

Computer software is so specific to its given task, equipment, and environment that only general principles warrant discussion here. The task under consideration involves taking incoming numbers from an automated device and translating them not only into concentrations but also into status signals, interruption indicators, times, specifications, and even requirements for comments. The goal of this section is to present to the reader those features of on-line computer utilization that have proved useful through experience and that have general application (1). Because of relative total laboratory hours of use and newness of other applications, discussion of on-line continuous flow analyzers must occupy the bulk of this section.

SIGNALS FROM CONTINUOUS FLOW ANALYZERS

Very slowly changing d.c. signals come from these devices. Readings are usually taken every second or half-second. But any program that stores all of these readings must have been written by one who fails to understand continuous flow analysis and wastes storage space on a grand scale. Only that part of the informational whole need be taken that gives data on the location and quality of the steady state for each standard or sample.

Originally the lore relating to peaks in AutoAnalyzers was so strong that every peak was sought and identified by technics such as continuous differentiation. When the differential of the stripchart curve passed from positive through zero to negative a peak was accepted. Realization of the fundamental importance of dwell time to continuous flow analysis, and its relationship to proper performance brought the much simpler and more elegant and diagnostically valuable approach of reading the signal when the peak or steady state *should* be present (32, 33, 128). In this technic a signal to the computer is turned on before the first peak forms. An initializing program monitors the signal and carefully identifies the time t_1 at which the first peak or steady state appears. If the sampler is operating at 90 samples/hour, i.e., the time between samples is 40 sec, the computer takes a set of readings 40 sec after t_1, and in a properly operating device the peak or steady state for the next sample will be there. If it is not, the computer can notify the operator that this problem exists (see below). In a commonly used system (32), between two and seven readings are taken at sample peak time, 1 sec apart, to gain further information about performance. In addition, information is stored about the zero reading, and if the signal falls below that level, or abruptly to zero volts the computer program recognizes a problem condition or a halt in analyzer performance.

DISCRETE SIGNALS

Discrete analyzers produce discrete signals, for timing and for absorbance or other concentration-related values. The variety of the analyzers and the specificity of their task is such that many of them incorporate their own special purpose computer (126).

RATE SIGNALS

One can find, in many elementary books on electronics, simple circuits for differentiating a signal to determine its rate of change. One can also connect such a circuit to the absorbance output signal of a photometer and thus learn that the measurement of reaction rates is not so straightforward. To do this successfully one must be able to measure slow reactions that change over only a small absorbance interval, and that show short term variations, or noise. A great deal of work has been done on this problem, with perhaps the best method being that of Malmstadt (27, 56, 58, 75). It seems likely that rate measurements will increase in importance to clinical chemistry as a direct result of improved hardware and software in this area.

TABULAR STORAGE OF REFERENCE DATA

In order to produce useful reports, even of a limited nature, the computer must store a large amount of reference data about the method, the results and the work schedule. The amount is surprising to the uninitiated because details that a trained technologist infers, or carries over from one method to the next, must be specified in exact detail for every case for the computer. The principle of storage of such reference data in easily changed tabular form rather than as part of any program has provided a vital advantage to certain systems for clinical laboratories. As an example, the reference data required for the AutoAnalyzer determination of triglyceride is presented in Table 10-4. The triglyceride procedure involves measurement of the difference between an assay and a blank, which can be run on two channels simultaneously as in the example, or on one, sequentially. The information divides itself into six basic parts: (a) data on the electrical signals from the analyzer, (b) data on the method itself, (c) tolerances acceptable for a variety of quality control parameters, (d) data on standards of various types, (e) data on the work schedule, and (f) specifications for results.

TABLE 10-4. LABCOM V — Table of Stored Information for Triglyceride Determination

Item No.	Required Information	Information stored	
		Assay channel	Blank channel
	A. Signal		
1.	Print reports on Teletype no.	6	6
2.	No. of input lines involved	2	2
3.	Computer input line no.	31	32
4.	No. of incoming signals, per cup	1[a]	1[a]
5.	Are signals multiplexed, as in SMA	No	No
6.	Time between samples, in sec.	60	60
7.	No. of readings retained per peak	7	7
8.	No. of cups per run	40	40
9.	No. of cups lead or lag	40[b]	40[b]
10.	Automatic sensing of 1st peak	Yes	Yes
11.	Code name for channel	TR2A	TR2B
	B. Method		
12.	Name	Triglyceride Assay	Triglyceride Blank
13.	Automatic drift correction	Yes	Yes
14.	Automatic interaction correction	Yes	Yes
15.	Are special calculations required?	Yes	Yes
16.	Code name of calculation program	DIF1[c]	DIF1[c]
17.	Change %T to A	No[d]	No[d]
18.	Units	mg/100 ml	mg/100 ml
	C. Tolerances		
19.	Maximum baseline drift[e]	±0.7	±0.5
20.	Maximum standards drift[e]	±0.8	±0.8
21.	Peak diagnostics tolerance[e]	±0.7	±0.7
22.	Maximum %I	10	10
	D. Standards		
23.	List of calibration standards, by number and concentration	1. 0	0
		2. 400	150
24.	List of controls, by number, 4 character code, concentration, and allowable range	1. SM-7[f] 147 ±14	SM-7 147 ±14
		2. LC-1[g] 120 ±10	LC-1 120 ±10
25.	Concentration of interaction standard	0	0
	E. Work schedule		
26.	Work schedule, by cup no.	1. Marker[h]	Standard 2[i]
		2. Marker	Interaction Standard
		3. Standard 1	Standard 1
		4. Standard 2[i]	Marker
		5. Interaction Standard	Marker
		6. Control 1	Control 1

TABLE 10-4. Continued

Item No.	Required Information		Information stored	
			Assay channel	Blank channel
		7–38.	Samples	Samples
		39.	Control 2	Control 2
		40.	Drift standard j	Drift standard j
	Between cups 40 and	41.	Calculate and print	Calculate and print

Item No.	F. Results		Assay	Blank	Net
27.	Column headings for results		TR-A	TR-B	TRIG
28.	No. of decimal places		0	0	0
29.	Multiple results by factor		1	1	1
30.	Extrapolate above highest standard?		Yes	Yes	—
31.	Extrapolate below lowest standard?		Yes	Yes	—
32.	Print full diagnostic numbers?		No k	No	No
33.	Identify out-of-normals		No	Yes	Yes
34.	Upper limit of normals		—	150	134
35.	Lower limit of normals		—	0	29
36.	Label results, by number,				
	as 1 STD, 2 CTL	1.	MRK l	STD l	NOP
	3 SPL m 4 MRK	2.	MRK	SPL	NOP
	5 NOP 6 MAN	3.	STD	STD	NOP
	7 CAL m	4.	STD	MRK	NOP
		5.	SPL n	MRK	NOP
		6.	CTL	CTL	CTL
		7–39.	—	—	CAL
		40.	SPL	SPL	NOP

a On a SMA-12 this number would be 12.

b This is the number of cups which one channel is permitted to get ahead of another in a parallel multichannel, or sequential correlated channel situation, such as the assay blank case here. If more than 40 data (here) of either channel appear, the computer prints out the partial information and notifies the operator that expected data are missing.

c DIF1 is a program that subtracts the results from the nth cup on one channel from the same cup on another channel, as when blanks are run.

d Triglycerides uses a fluorometer, not a photometer.

e In % of full scale.

These tolerances are given in the legends of figures showing strip chart records for AutoAnalyzers wherever they appear in this book, if the procedure in question is used routinely in our laboratory.

f Control for overall procedure, including saponification, etc.

g Control for automated measurement portion of procedure.

h A "Marker" is a cup of material that produces a special peak that is correct in position and shape but is unrelated quantitatively to the rest of the data. For example, the standards for the blank channel inevitably produce "Marker"-type peaks on the assay channel.

i The actual concentrations for the standards are given further down in the table and may be different on different channels for standards of the same name.

j These standards are for baseline drift. Drift of high standards is monitored from one calibration standard to the next.

k Only English interpretations and code symbols will be printed in case of noise, etc.

l The information in this column is redundant and is not stored twice but is included here for clarity.

m STD identifies standards; CTL, controls; SPL, special cups to be omitted from the DIF1 calculation; MRK, marker peaks; NOP, cups that are to have no results printed; MAN, a result which has been entered manually; and CAL, results from the DIF1 calculation.

n The %I is printed instead.

These data represent the interface between the analyst and the method. They must be alterable by the analyst, without the help of any programmer, using familiar noncomputer terms, and without reference to any programming conventions. This is accomplished by setting up tables of information like Table 10-4 in computer memory, in a standard form, so that all working programs get their operating data by reference to them, and by providing other programs that enable the analyst to change the tabulated data, preferably by engaging in a question and answer conversation with the computer program in simple English. For example, the user should be able to ask the computer what concentrations it has for calibration standards and to change one or more of these to the new values he plans to use in the next run, easily and quickly. Programs designed for this purpose are called user control programs, and a very effective way to evaluate any proferred computer system is to examine its user control programs.

THE COMPUTER ON LINE TO CONTINUOUS FLOW ANALYZERS

The broad tasks performed by a computer that already has the tables of reference data include calibration, quantification, error handling, and quality control. From the analyzer the computer has received and stored a series of numbers, in sequence. These must be identified by reference to the schedule. A calibration curve or its equivalent must be formed from the standards. Each sample must be quantified. Errors due to interaction and drift must be measured and corrected. A variety of unacceptable conditions, such as noise or incorrect dwell time, must be sought and identified by markers if found, and finally the data must be calculated and printed in final form.

CALIBRATION AND QUANTITATION

Some sensors provide a signal that changes linearly with concentration of the measured constituent, e.g., fluorometry. In such cases a minimum of two calibration standards suffices, although additional standards are of course required for drift interaction and quality control (see Tables 10-3, 10-4). The computer program uses these two standards to set up a linear equation relating concentration to signal and calculates unknowns from their signals. The limits between which such a calibration line is valid must be stored and used in the program to notify the operator of samples that must be diluted and rerun. Ideally the program should identify all specimens that lie outside the normal values for the determination.

The most common sensor, the photometer, provides a signal that can be made a linear function of concentration by a negative logarithmic transformation. This is not fast and easy for a small computer, at least compared to the above calculation. A simpler, and quite effective approach, makes a two dimensional table of standard concentrations and stripchart readings. Readings for unknowns are then interpolated linearly between standard readings and concentrations obtained by proportion. This is equivalent to a technologist drawing straight lines between calibration points in a curve, an often acceptable but not ideal technic, since it demands a number of standards large enough to make errors due to curvature negligible.

If Beer's law holds for the method and photometer in question, and if a negative logarithmic conversion is performed by the computer, changing % *T* to absorbance, *A*, then a straight line relationship holds for the relationship between the absorbance (*A*) and concentration. As with fluorescence, the computer then measures the constants m and n in the equation

$$C = m + nA \tag{2}$$

from the calibration standards and uses them in turn to calculate concentration for unknown samples from their absorbance. The minimum number of possible standards one can use becomes two, and this is frequently also the actual number, often called the "HiCal" and "LoCal" standards. If more are used the computer can calculate a regression line through all values and provide confidence limits or some other measure of their precision (14). Some programs evaluate each standard in terms of the others and discard out of limits standards. Transformation of other known relationships between standard readings and concentration is easy with a computer, for example the logit (88, 89). A very useful source of the most efficient algorithms for a wide variety of operations exists (25) and should not be overlooked.

DETECTION, CORRECTION, AND INDICATION OF ERRORS

As noted throughout this chapter certain types of errors form a predictable, systematic, and inherent part of automation. These can be measured, for a given machine or method or run, and corrections applied that effectively neutralize them. Other errors can occur that are random. The computer can be an unflagging monitor for these errors and can set up markers where they occur or in some cases print specific complaints about quality. Here the computer is most useful as a totally consistent inspector. The system most familiar to this writer is called LABCOM V. It is widespread enough in use, and straightforward enough in design, for discussion and example (32, 51, 52). It originated in the Clinical Laboratory of the University of Wisconsin, and is now offered commercially in expanded versions by Laboratory Computing, Incorporated.

LABCOM V involves a computer programming technic that is thought of by programmers almost as a physical device, and which is called a "window." The program is arranged so that the computer records in its memory only those signals which are "in" the window. In the case in point, the window can be thought of as a rectangle on the stripchart, with its horizontal dimension being time and its vertical dimension, percent of full scale. In the LABCOM system this window can be 1,2,3,4,5, or 6 sec long, corresponding to 2,3,4,5,6, or 7 readings, 1 sec apart. This dimension is set by the analyst, as item 7 in Table 10-4. Its height, which can be any fraction of recorder full scale but is properly less than ±1%, is set as the "noise tolerance" of item 21 Table 10-4. The horizontal location of successive windows is set by the sample frequency, item 6 Table 10-4. Thus a window exists in every place where a sample peak should exist. The vertical location of the top of any given window is set by the highest concentration reading appearing "in" the window time. The bottom of the window is set by the height listed in item 21 of Table 10-4. For linear transducers like fluorometers the window height is the same at any scale reading. For photometers the tabular constant for window height applies at high values of % *T* and is decreased at lower values so as to correspond better to constant absorbance difference. The window is used as a monitoring or diagnostic device to produce signals describing malfunction ("diagnostics") as described in the following sections on noise and timing errors. The other types of diagnostic messages use other detection technics also described below.

Noise

If several successive readings are taken of a signal which should be stable, such as a steady state or peak in a flow-through cuvet, or signal from a nonflow cuvet, the computer can be programmed to detect instability of the signal noise, and to distinguish between two different kinds of stability. This is illustrated in Figure 10-17 that shows various problems of AutoAnalyzer peaks in relationship to their windows. The top of the window is set by the highest of the seven incoming signals, after they have been processed so as to be proportional to concentration. Across the top of each rectangular window is a seven bit binary number, calculated by the computer. Each bit of the number is a one if the 1 sec reading that corresponds to that bit is "inside" the window and a zero if it falls "outside." Thus the binary number for a perfect signal is 1111111 (Fig. 10-17A). This is called the "diagnostic number." If an air bubble passes through the flow cell a noise spike is produced, and as illustrated in Figure 10-17E, the number for that sample is 0001000 or the like depending on the location of the spike. For methods with peaks in the opposite sense, where air spikes go downscale, it is 1101111 or the like. The other type of noise results from steady state variations and is illustrated by Figure 10-17B. It produces diagnostic numbers like 0011010, or similar mixed ones and zeros.

A typical tolerance for such diagnostic windows in practice is ±0.5% of recorder full scale. Hence, either of these types of noise can be detected with great sensitivity. Markers, such as an asterisk, are placed next to every reported noisy result if any are found, and a comment is printed at the end of the report to force attention to them.

The computer is programmed to distinguish, from the pattern of the diagnostic number, the two types of noise and to print an English language comment about any sample suffering from noise, such as "CUP #7 NOISE SPIKE" or "CUP #18 NOISE." The program also causes the computer to print out "ACTION" messages. In the former example the message would be "ACTION—PEAK MUST BE INSPECTED AND HAND CALCULATED—

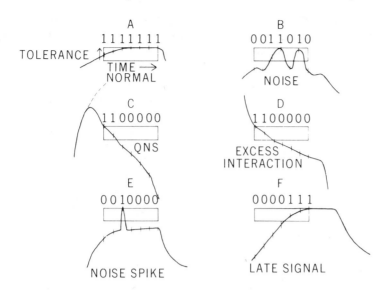

FIG. 10-17. Labcom "diagnostics"—strip chart peaks and windows. See text for description of origin of the "1" and "0" notations. From EVENSON MA, HICKS GP, THIERS RE: *Clin Chem* 16:606, 1970.

NOTIFY SUPERVISOR OF MAINTENANCE PROBLEM," in the latter, "ACTION–EXAMINE BUBBLE PATTERN AND MANIFOLD–NOTIFY SUPERVISOR." The diagnostic number can also be printed or not, as desired, Table 10-4 item 32.

Timing Errors

These are characteristic of automated devices as distinct from noise, the detection of which is useful even for a simple photometer connected to a computer. One sets the number of readings taken (window length) and tolerance (window height) for each method so that all cases of timing error serious enough to be detected correspond to unacceptable changes in dwell time. Insufficient sample will invariably cut short the formation of a peak or steady state so that it will appear too small and too soon (Fig. 10-17C). If the sample probe catches on the rim of a cup, then falls into the sample the peak will appear too small and too late (Fig. 10-17F). Surging flow in a system will cause peaks to be alternately too early and too late. This can be detected in patterns of successive diagnostic numbers more sensitively than by any other means (Table 10-5). System leaks, rapid deterioration of a pump tube or one type of sampler module malfunction will cause peaks to appear successively later (Table 10-5A). Overwhelming interaction from a preceding specimen will cause a peak to appear too soon and too high, or it may prevent peak formation altogether (Fig. 10-17D). These errors are also indicated by markers. Timing variation in a sampler will also be detected (Table 10-5 C and D).

TABLE 10-5. Patterns of Labcom Diagnostics.[a] The diagnostic numbers shown are associated with sequential or closely successive samples. See text for description of origin of the "1" and "0" notation.

A	B
0111111	1100000
0011111	0001111
0000111	1110000
0000001	0001111
Timing variation in sampler	Partial clot, or small dialyzer leak
C	**D**
0000011	1000000
0000001	1100000
0000111	1000000
0000011	1110000
Improper synchronization, or sampler interruption	Improper synchronization, or cam jump-ahead

[a]From EVENSON MA, HICKS GP, THIERS RE: *Clin Chem* 16:606, 1970.

Again the computer is programmed to deduce from the diagnostic numbers whether a peak is early or late, and by how many seconds for small deviations. Thus a typical quality control comment might read

CUP #12 3 S. EARLY

and the action message

ACTION – CHECK CUP FOR QNS – NOTIFY THE SUPERVISOR?

Drift

As indicated above, standards for the measurement of drift should be run at a frequency dictated by experience. Drift of zero level "zero drift" can be observed as slow shift in baseline and standards alike, and sensitivity drift as systematic changes in standard reading but not in baseline. The former seems most common and is certainly most easily corrected by hand or computer. A constant is simply added to every signal in a linearly changing fashion from one drift standard to the next to compensate for baseline drift. The inherent assumption in this treatment—that drift is not abrupt, but smooth and linear—seems to be borne out by experience. But correction should be employed only for small and consistent drifts.

Correction of sensitivity drift is a more serious problem and requires reconstruction of the calibration curve with each set of standards, and interpolation between successive curves. It should be undertaken only where drift is small and consistent, where all other means of controlling it have failed, and where the results of the correction technic have been tested. A better alternative has proved to be to monitor sensitivity drift from one run to the next and to have the computer mark and comment on any standard that exceeds the preselected tolerance (Table 10-4 item 20).

Suppose that standard 2 of the triglyceride assay channel in Tables 10-3 and 10-4 differed from the standard 2 most recently run by more than ±0.8% (Table 10-4, item 20). The quality control comment appearing on the report in this case would read

TR-A STANDARD DRIFT − CUP 4

and the action message

ACTION − CHECK CHARTS − STD DRIFT − IS RUN ACCEPTABLE?

Similar printouts result if excessive baseline drift is noted.

Standards

Because the computer stores concentrations of standards as tabular reference data, it can easily be programmed to detect if the standards are out of order, and to notify the operator of this condition.

In such a case the quality control message would be

TR-A RESULTS INVALID−STANDARDS OUT OF ORDER

and the action message

ACTION−REPEAT OR HAND CALCULATE

Interaction

Computer correction of interaction has proved very easy, accurate, and useful, because of the simple arithmetic rule followed by this error (108). Under proper conditions only one standard need be added to the list of calibration standards (111) to permit interaction measurement and correction. This totally eliminates the dependence of the calibration curve on the order of standards (which has always been a signal that all was not perfect in the world of continuous flow). Not only should the computer be programmed to measure and correct for interaction, it should show a marker and provide a comment whenever the interaction exceeds some chosen upper limit for each method.

In this case the typical quality control comment might be

RUN 3 EXCESSIVE INTERACTION 11%

and the action message

NOTIFY SUPERVISOR

Check Samples

It is customary, with every automated batch, to run at least one specimen of known concentration. Frequently two or more are run at different concentration levels or in different matrices, e.g., pooled serum and reconstituted lyophilized plasma. Because the expected values and the acceptable ranges for each of these are stored in the reference tables (Table 10-4, item 24), one can program the computer to show a marker and comment on any which fail to fall within the proper range.

In this case the quality control comment reads

CONTROLS OUT OF TOLERANCE

CUP #6 SM-7 147 ± 14

CUP #39 LC-1 120 ± 10

and the action message

NOTIFY SUPERVISOR

The supervisor can then go to the report for the results on cups 6 and 39 and note how far out of limits they are. He can then examine previous quality control data and take informed action.

Quality Control

In our laboratory all of these error detection programs are linked together into one set, which produce all appropriate markers, and which deduce, from the types of error observed, either possible sources of error, or appropriate action to be taken by the operator, or both. Such deductions are printed as a series of comments and action messages, as illustrated above, immediately after the data in each report for each run, or printed part of a run. The available list of comments includes "ALL QC OK." The obvious goal of the operator is to produce all of his work with only this comment attached, and no diagnostic numbers or other markers of any kind. Any given report that appears with this comment has run a formidable gamut of tests and checks, to a degree that simply cannot be reproduced in number, thoroughness, or consistency by manual methods.

Two comments are in order concerning the quality control comments described above: (a) As with any quality control system one of the unexpected practical supervisory problems is that of preventing operators from totally ignoring a perfectly clear and unequivocal instruction from the computer, such as "EXCESSIVE INTERACTION—NOTIFY YOUR SUPERVISOR." (b) For clarity the quality control comments and action messages printed here are longer and more verbose than those actually in use.

REPORTS

Table 10-6 shows a report from the on-line system used in our laboratory. It is an actual report taken from a typical run, for triglycerides, for which

TABLE 10-6. On-line Computer Report on Triglyceride Run. The on-line computer produces all but the second column which is the output of a separate off-line computer. I STD, in cups 2 and 5 stands for Interaction Standard, Bl for Blank. CTL and STD stand for control standard and calibration standard, respectively.

2018 HRS 11/18/72
TR2A, TR2B RUN 1

Cup #	ACC#	TR-A	TR-B	TRIG
1	Blank STD 2	MP	202V	
2	Bl I STD	MP	4.0%	
3	STD 1	118V	127V	
4	Assay STD 2	256V	MP	
5	Assay I STD	2.2%	MP	
6	CTL 1 SM-7	195	50	145
7	606248	158	20	138
8	606394	129	17	112
9	606423	81	15	66
10	606453	126	13	113
11	606539	148	17	131
12	606556	240	39	201
13	606566	120	16	104
14	606581	61	13	48
15	606611	367	25	342
16	606631	285	31	254
17	606647	126	17	109
18	606696	139	17	122
19	606705	64	23	41
20	606713	79	23	56
21	X45452	328	29	299
22	606868	152	27	125
23	606871	210	33	177
24	606876	123	17	106
25	X45418	148	23	125
26	606965	211	27	184
27	606997	166	25	141
28	607005	57	21	36
29	Y39974	164	23	141
30	607042	564	51	513
31	607049	570	40	530
32	607058	149	28	121
33	607072	396	35	361
34	607123	92	31	61
35	607125	85	27	58
36	607190	65	27	38
37	605981	250	37	213
38	606051	287	32	255
39	CTL 2 LC-1	126	7	119
40	STD 1	6	10	− 4

QC COMMENTS ON TR-A		QC COMMENTS ON TR-B	
2.2% I	ALL QC OK	4.0% I	ALL QC OK

the stored tabulated data are shown in Table 10-4. Its specific format has no general or inherent significance. Formats are, or should be, the most easily changed feature of computer programs. For example, in column 2 of this report the on-line computer prints the information in spaces 1–6, 39, and 40, and leaves a long series of gaps in spaces 7-38, over which is pasted the work list of specimen numbers from an off-line computer.

On-line systems of LABCOM V other than ours have been programmed to print the following additional reports, which assume that entry of patient information, undescribed here, has been carried out, either manually or from punch or mark sense cards.

Patient Directory

A complete list of all patients on the computer file can be printed at any time. The patients are sorted and the directory printed by any one of the following criteria on request.

(a) Name in alphabetical order.
(b) Hospital number in ascending order.
(c) Nursing station, room number.
(d) Physician code, nursing station, room number.
(e) Sex, race, date of birth.
(f) Record number.

With these six ways of sorting patients on file to produce a directory, patients can be rapidly categorized for many different purposes. Table 10-7 is an example of a few lines from the patient directory, which may include up to 2000 patients with 52 patients per page. Names used in all examples are fictitious.

TABLE 10-7. Patient Directory

PATIENT DIRECTORY				8:38 AM 9/26/72				PAGE 1
NAME	HOSP NO	NS	ROOM	PHYS	S	R	BIRTHDATE	RECORD
ABBER ET	424111	5E	522	LEE	M	W	8/18/1927	4342
BARNET PT	339585	CU2	202	PYE	M	W	5/30/1964	3380
BIZEL AC	603442	5C	502	LI	M	D	4/23/1922	6304
BYROK GEORGE G	121377	13	321	LOO	M	A	10/10/1931	1184
DART VIRGINIA A	433041	6PED	606	TOP	F	S	2/18/1943	4403
DICKINSON E	803112	3E	309	OTT	M	D	7/09/1950	8401
DULLER JOHN R	213400	5C	526	VOW	M	W	3/04/1929	2176
7 PATIENTS								

Laboratory Worklogs

A worklog is basically a list of test requisitions for any one day including patient identification and indicating the status of each requisition as complete, incomplete, or partially complete. Such information can be produced at any time according to several different options. A Master Log including all patients on file printed alphabetically or printed by accession number and including only incompleted work can be produced. Logs can be produced including only patients with work in defined laboratory sections, for example, automation section or special chemistry section. A log for an individual patient can be produced on entry of a patient name, hospital number, or accession number. For example, the following conversation at a terminal would produce an individual log by accession number. The user enters the information after the asterisks.

OPTION * A

ACC # * 1123

1123 2/15 SMITH ROBERT A [GLUC BUN %BGAS

The computer response tells that the patient's glucose is complete, BUN is not done, and the blood gas (BGAS) is partially completed. This is of obvious value to the laboratory which must respond to telephone inquiries concerning the status of work.

Tables 10-8 and 10-9 give examples of lines from a printed log for all patients and a log including only patients with incomplete work, respectively. The square bracket ([) indicates the test as complete and percentage sign (%) as partially complete. When all patients are requested as in Table 10-8, the date in parentheses indicates the date of the last requisition for a patient when there is no requested work on the date for which the log is produced. No entries at all means the patient has had no work requested on any day.

TABLE 10-8. Master Log–Alphabetic Order[a]

MASTER LOG FOR 9/25, 12:04 PM, 9/25/1971						PAGE 1
NAME	HOSP #	ACC #	REQUISITIONS			
ABBER ET	424111	450	%ELYT	SGOT	SAMY	SEPH
		677	[MA12			
BARNET PT	339585	455	%ELYT	SAMY	LDH	
BIZEL AC	603442	666	%ELYT	SAMY		
BYROK GEORGE G	121377	1000	[MA12			
DART VIRGINIA A	433041	1001	[MA12			
DICKINSON E	803112	1003	[MA12			
DULLER JOHN R	213400	(9/24)				
7 PATIENTS						

[a] From Laboratory Computing, Incorporated, Madison, Wis.

TABLE 10-9. Master Log—Accession Number Order

WORK IN PROGRESS LOG 9/26/72							PAGE 1
ACC #	DATE	NAME	HOSP #	REQUISITIONS			
450	9/25	ABBER ET	424111	%ELYT	SGOT	SAMY	SEPH
455	9/25	BARNET PT	339585	%ELYT	SAMY	LDH	
666	9/25	BIZEL AC	603442	%ELYT	SAMY		
END OF LOG							

Blood Drawing Lists

A blood drawing list with self-adhesive labels can be printed at any time on request. The list may be printed for selected nursing stations or all nursing stations in the hospital. The list is in order of ascending nursing station sorted by room number within each station. Each label contains the patient identification and location, accession number, and tests requested.

Figure 10-18 shows the label design. A new pair of labels is printed for each different tube type required, for example, RED, AMBR, etc.

Notice in the example that the first label on the right side of the list has four small transfer labels attached that contain the accession number. These transfer labels can be used as needed to identify split tubes, sample cups, etc.

Test Worksheets

The function of the worksheets is to provide an organized list of work to be done with all of the available information to assist in performing that work.

LABCOM produces two kinds of worksheets: (a) manual and (b) trayloading. A manual worksheet may be produced for any tests and/or batteries of tests to be performed off line. A trayloading worksheet may be generated for tests set up on line. A manual worksheet may be printed for tests even though the tests are to be run on an instrument on line. Whenever a worksheet is requested for on-line tests the laboratory personnel decide whether the loading sequence is to be determined by the computer. The manual and trayloaded worksheets may be used and intermixed as desired by the laboratory.

Table 10-10 is an example of a few lines from a manual worksheet. If a test is requested, the worksheet indicates how it was requested. For example, a potassium may have been requested alone as a urine potassium (UK) or as urine electrolytes (UELY). When results are already available they are automatically included in the worksheet, e.g., UNA for BIZEL is 205. Sometimes related results are produced in different areas of the laboratory. However, the results from one section may be needed to perform the work in another section. Urine electrolytes is an example. The urine volume and collection period are often measured in the laboratory receiving area. The volume and period are needed to calculate dilutions when the electrolytes

FIG. 10-18. Labels for sample containers.

```
ICU4 420A      # 468          H# 177518    # 468          H# 177518    # 468
4/01                          LANDOR DENNIS S             LANDOR DENNIS S
LANDOR DENNIS S    RED        RED : SAMY K    LDH         RED : SAMY K    LDH
H# 177518                     4/01 SGOT BSP               4/01 SGOT BSP
                                                    468    468
                                                    468    468

ICU4 420A      # 468          H# 177518    # 468          H# 177518    # 468
4/01                          LANDOR DENNIS S             LANDOR DENNIS S
LANDOR DENNIS S    RED        RED : BSP                   RED : BSP
H# 177518                     4/01                        4/01
                                                    468    468
                                                    468    468

ICU4 421B      #8109          H# 185596    #8109          H# 185596    #8109
4/01                          BURGESS JAMES G             BURGESS JAMES G
BURGESS JAMES G    RED        RED : PBI CRYO              RED : PBI CRYO
H# 185596                     4/01                        4/01
                                                   8109   8109
                                                   8109   8109

ICU4 421B      #2133          H# 185596    #2133          H# 185596    #2133
4/01                          BURGESS JAMES G             BURGESS JAMES G
BURGESS JAMES G    AMBR       AMBR: FE                    AMBR: FE
H# 185596                     4/01                        4/01
                                                   2133   2133
                                                   2133   2133

INF3 309       #1015          H# 91616    #1015           H# 91616    #1015
4/01                          HENSLOWE ROBERTA            HENSLOWE ROBERTA
HENSLOWE ROBERTA   RED        RED : ELYT                  RED : ELYT
H# 91616                      4/01                        4/01
                                                   1015   1015
                                                   1015   1015
```

271

TABLE 10-10. Worksheet for Manual Determination[a]

MANUAL WORKSHEET, ALL INC. 11:06 PM, 9/25/1971   PAGE 1

REF #	ACC #	NAME/DATE	HOSP #	UNA	UK	UCL	VOL	HR
1	37P	BIZEL AC		205	UELY	UELY	100	RAND
		9/24 5C	603442					
2	277	BYROK GEORGE G			UK		1280	24.0
		9/25 13	121377					
3	324	DULLER JOHN R		UELY	UELY	UELY	4280	24.0
		9/25 6E	213400					

END OF WORKSHEET

---

[a]From Laboratory Computing, Incorporated Madison, Wis.

are determined, usually in a different section. Worksheets should bring together such information to minimize the problems of communication between laboratory sections.

Accession numbers followed by a P have appeared on previous worksheets. RAND means random sample. "REF #" on a manual worksheet is simply a reference number for the patient on that particular worksheet.

Table 10-11 is a trayloading worksheet for an on-line instrument indicating where to put standards, controls, and each patient's specimen. The section on schedules describes the flexibility available to the user to develop loading sequences for trays.

On-Line Multichannel Reports

Table 10-12 is an example of an on-line report generated automatically by the computer for the technologist. This is not a report that leaves the laboratory but is provided for use by the person running the instrument. The on-line report provides several kinds of information useful to the technologist.

In Table 10-12 cups 101—109 are standards (STD) and marker peaks (MP), or both. Marker peaks are used to "fill in" cups where results are not to be produced. For example, in Figure 10-3, no standards are run in cups 101-104 for K, while standards are run for the other channels. Numbers followed by a V are voltage readings of standards rather than concentrations (the technologist knows the standard concentrations) so that a manual calibration curve can be made if ever necessary. A V followed by an L indicates that the standard was outside of the acceptable linearity tolerance for the calibration curve. Any value for a cup followed by a # has a reading "diagnostic" (see above). The column labelled BAL gives the electrolyte balance:

$$mEq\ Na + K - Cl - CO_2 - 12$$

TABLE 10-11. Worksheet for Automated Determination[a]

CUP #	ACC #	NAME/DATE	NS	HOSP #	NA	K	CL	CO2
LOADING WORKSHEET FOR NA 10:42 AM 9/25/1971								**PAGE 1**
1-MP								
2-STD								
4-STD								
5-STD								
6-STD								
7-STD								
8-STD								
9-CTL								
10	1	SUNCLAIR VB			NA			
		9/25 CUI		338555				
11	296	KABKA D			ELYT	ELYT	ELYT	ELYT
		9/25 6C		432234				
12	300	SHULTZ NW			ELYT	ELYT	ELYT	ELYT
		9/24 6C		221007				
13	400	BRENT CAROL			ELYT	ELYT	ELYT	ELYT
		9/24 6E		667834				
14	450	ABBER ET			ELYT	ELYT	ELYT	ELYT
		9/25 5E		424111				
15	455	BARNET PT			ELYT	ELYT	ELYT	ELYT
		9/25 CU2		339585				
16	467	KABKA FD			ELYT	ELYT	ELYT	ELYT
		9/24 6C		432234				
17	489	GRAHAM J			ELYT	ELYT	ELYT	ELYT
		9/25 2W		666262				
18	500	OLIVA LK			NA	K	CL	
		9/25 4B		551005				
19	530	SEGOYA A			ELYT	ELYT	ELYT	ELYT
		9/25 4W		600427				
20-CTL								
21	655	BRENT CAROL			ELYT	ELYT	ELYT	ELYT
		9/24 6E		667834				
22	666	BIZEL AD			ELYT	ELYT	ELYT	ELYT
		9/25 5C		603442				
23	670	GANDRY ML			ELYT	ELYT	ELYT	ELYT
		9/25 5B		626396				
24	780	THATCHER WE			ELYT	ELYT	ELYT	ELYT
		9/25 6E		164417				
25	800	DULLER JR			ELYT	ELYT	ELYT	ELYT
		9/24 5C		213400				
END OF WORKSHEET								

[a]From Laboratory Computing, Incorporated, Madison, Wis.

TABLE 10-12. Operator's Multichannel Report[a]

1144 HRS 9/24/70						PAGE 1
NA, K, CL, CO2						
CUP #	ACC #	NA	K	CL	CO2	BAL
101		90V	MP	325V	MP	
102		105VL	MP	346V	MP	
103		169V	MP	365V	159V	
104		235V	MP	343V	208VL	
105	STD	303V	167V	318V	281VL	
106	STD	367V	239VL	291V	368V	
107	STD	439V	323V	262VL	456V	
108		MP	413V	176V	495V	
109	MP					
110	CTL 1	156	3.3	111	25.4	
111	....	138	3.6	98	25.9	6
112	....	137	4.5	100	31.7	−3
113	....	141	4.6	101	26.7	6
114	....	135	4.5	93	32.7	1
115	....	125	4.5	93	20.5	4
116	....	133	4.4	95	30.5	0
117	....	136	4.1	101	27.7	−1
118	....	133	6.3	98	23.1	6
119	CTL 2	146	4.8	101	2.7	
120	MP					

[a]From Laboratory Computing, Incorporated, Madison, Wis.

Any similar calculation can be used.

Interim Patient Report

The interim report prints all of the data for each of any chosen set of patients on any one day. On request, interim reports may be printed for or automatically sorted by any one of the following:

    patient name
    nursing station
    hospital number
    medical record number
    doctor code

An example of an interim report is given in Table 10-13. The accession number, normal ranges, and the technologist's code are included in interim reports and abnormal results are flagged when desired.

Cumulative Patient Report

These vary tremendously in design. Table 10-14 gives one example that has the following features: The normal ranges are stratified by age and sex;

TABLE 10-13. Interim Report for One Patient Day

ONE DAY	INTERIM REPORT		TESTS FOR 10/01	PAGE 1
JONES ROBERT A	H # 80042 3C 357-A		PRINTED 11:21 AM 10-1/1971	
	R # 30427 OP			
			NORMAL	
ACC #			LO HI	TECH
55	HEM. SURVEY			
HEMATOCRIT	32%		40 54	13
HEMOGLOBIN	11.3 GMS %		13 18	13
RED CELL COUNT	3.5 M/CU.MM.		4.5 6.0	13
WHITE CELL COUNT	26.1 K/CU. MM.		5 10	13
55	CELL DIFFERENTIAL			
NEUTROPHILS	52%		50 70	13
BANDS	8%		0 5	13
EOSINOPHILS	1%		0 1	13
LYMPHOCYTES	28%		20 40	13
MONOCYTES	6%		1 6	13
PLT. EVALUATION	NORMAL			13
55	RBC MORPHOLOGY			
ANISOCYTOSIS	NONE			13
POIKILOCYTOSIS	NONE			13
POLYCHROMASIA	NONE			13
MACROCYT.TEND.	NONE			13
MICROCYT. TEND.	NONE			13
55				
SPHEROCYTES	MODERATE			13
55				
NUCLEATED RBC SM	1/100 WBC			13
1009	URINALYSIS			
SPECIFIC GRAV.	1.012		1.010 1.030	33
PH	7.5		4.8 7.8	33
PROTEIN-U. QUAL	0			33
GLUCLOSE-U. QUAL	0			33
KETONE BODIES	0			33
HEMASTIX	0			33
WHITE CELLS	1/HPF, CENT.			33
RED CELLS	45/HPF, CENT.			33
EPITHELIAL CELLS	MODERATE			33
BACTERIA	FEW			33
MUCUS	NONE			33
CRYSTALS	NONE			33
1009	URINE CASTS			
TYPE	HYALINE			33
QUANTITY	6/LPF, CENT.			33
3000	CHEMISTRY SURVEY			
	PENDING			

all abnormal results are flagged by an asterisk; a series of six inverted triangles (▽▽▽▽▽▽) indicates any day column with new results since the last report; repeats on the same day are sorted by increasing time and placed side

TABLE 10-14. Cumulative Report for One Patient

NAME: DOE JOHN H   NS: 6W   ROOM: 611

H# : 66778   DR: EXAM   AGE: 32 YR 8 MO   10:21 AM 1/10/71

R# : 87766

CUMULATIVE PATIENT REPORT   PAGE 1

	NORMALS	1/07	1/08	1/08	1/09	1/09	1/10	1/11
HEM. SURVEY			9:00A	9:00A	8:00A	12:00P	10:00A	▽▽▽▽▽ / ▽▽▽▽▽▽
HEMATOCRIT %	33– 42	36	34 / STAT		CLOT	35 / NLAB	37	
HEMOGLOBIN GMS %	11.0– 14.0	12.4	12.0 / STAT		CLOT	12.2 / NLAB	12.3	
RED CELL COUNT M/CU.MM.	3.4– 4.3	4.1	3.9 / STAT		CLOT	4.2 / NLAB	4.2	
WHITE CELL CT. K/CU.MM.	7.0– 10.0	9.6	10.5* / STAT		CLOT	10.4* / NLAB	9.8	
SODIUM MEQ/LITER	135– 148	129*	132*	9:00A 131*	8:00A 133*		10:00A 135	8:00A PENDING
POTASSIUM MEQ/LITER	3.5– 5.3	4.4					10:00A 4.6	8:00A PENDING
CHLORIDE MEQ/LITER	98– 108	89*	93*	92*	8:00A 95*		10:00A 97*	8:00A PENDING
CO$_2$ CONTENT MEQ/LITER	21– 28	28					10:00A 27	8:00A PENDING

				PENDING
CALCIUM MG/100 ML				
CREAT. CLEARANCE CLEARANCE RATE ML/MIN	87– 127	133*	129*	
CREATININE MG/100 ML	5– 1.2	.7	.6	
URINE CREAT. CREATININE MG/100 ML		40	39	
VOL. RECEIVED CC		1680	2021	
COLLECT. PERIOD HRS		24	24	
CHLORIDE MEQ/LITER				3:00P 125
SPECIMEN TYPE				CSF

NLAB = SPEC. NOT LABELED
CLOT = CLOTTED
DOE JOHN H
END OF REPORT

STAT = STAT
CSF = SPINAL FLUID

by side in columns; the formatted 8 ½ x 11 in. paper permits direct insertion into most charts without tearing and turning; the laboratory can change printing order and definition of groupings on the report any time without reprogramming; the pages are logically numbered, retaining consistent format, making it easy to find related results on different pages; the patient identification appears at the bottom of each page to make use easy when inserted in the chart.

Cumulative reports can also be sorted and selected in the variety of ways previously described for interim reports. For example, one can ask the computer to print all of the cumulative reports for a particular doctor. User control programs are employed to change these features at will and a wide variety of features and formats is available under user control.

Special Reports

Such an on-line computer system that also contains patient information can also be programmed to produce a variety of other reports, including bills for individual patients or for groups of patients, laboratory census, and lists of test information stored in tabular form in memory. The possible variety is endless and the need varies so greatly from laboratory to laboratory that only those reports of general application have been listed here.

OTHER SYSTEMS

Throughout this chapter the LABCOM V system has been used as an example because this author has experience with it in several different laboratories and can provide examples that have proved valuable in practice rather than simply listing information from the "technical literature" of manufacturers. However, many systems exist (24), and in the list of manufacturers in the appendix one can find names and addresses of manufacturers of both automated equipment and on-line computer systems. These lists cannot be complete or up to date but are provided for their utility as a guide in an area of clinical chemistry that is none the less important for being complex and commercialized.

REFERENCES

1. ABERNATHY MH, BENTLEY GT, GARTELMANN D, et al: *Clin Chim Acta* 30:463, 1970
2. AMADOR E, URBAN J: *Clin Chem 18*:601, 1972
3. AMADOR R, CECHNER RL, BARKLOW JJ: *Clin Chem 18*:668, 1972
4. ANDERSON NG: *Anal Biochem 23*:207, 1968
5. ANDERSON NG: *Anal Biochem 31*:272, 1969
6. ANDERSON NG: *Anal Biochem 32*:59, 1969
7. ANDERSON NG: *Clin Chim Acta 25*:321, 1969
8. ANDERSON NG: *Am J Clin Pathol 53*:778, 1970
9. ANNINO JS: *Clin Chem 14*:70, 1968
10. ANNINO JS, WILLIAMS LA: *Clin Chem 18*:488, 1972
11. Automation and Data Processing in Pathology, edited by TP Whitehead. London B.M.A. House, 1969

12. BABSON AL, KLEINMAN NM: *Clin Chem 13*:163, 1967
13. BALL MJ: *J Assoc Adv Med Instrum 6*:28, 1972
14. BENNET A, GARTELMANN D, MASON JE, et al: *Clin Chim Acta 29*:161, 1970
15. BLACKMORE DJ, CURRY AS, HAYES TS, et al: *Clin Chem 17*:896, 1971
16. BLAIVAS MA: in *Technicon Symposium*, 1965, edited by LT Skeggs Jr. New York, Mediad, 1966, p 452
17. BLAIVAS MA, MENCZ AH: in *Technicon Symposia*, 1967, edited by NB Scova. New York, Mediad, 1968, Vol. 1, p 133
18. BRECHER G, LOKEN H: *Am J Clin Pathol 55*:527, 1971
19. BROUGHTON PMG, ANNAR W: *Clin Chim Acta 32*:433, 1971
20. BROUGHTON PMG: *A Guide to Automation in Clinical Chemistry*, Association of Clinical Biochemists, Technical Bulletin 16, May 1969
21. BROWN HH, RHINDRESS MC, GRISWOLD RE: *Clin Chem 17*:92, 1971
22. BURTIS CA, JOHNSON WF, MAILEN JC, ATTRILL JE: *Clin Chem 18*:433, 1972
23. CHANEY AL: in *Technicon Symposia*, 1967, edited by NB Scova. New York, Mediad, 1968, Vol 1, p 115
24. Clinical Laboratory Computer Systems, J. Lloyd Johnson Associates, Northbrook, Ill., 1971
25. Collected Algorithms from Communications ACM, edited by JG Herriot. New York, Association for Computing Machinery
26. Computer applications in health services: *Calif Med 114*:106, 1971
27. CORDOS EM, CROUCH SR, MALMSTADT HV: *Anal Chem 40*:1812, 1968
28. COTLOVE E, NISHI HH: *Clin Chem 7*:285, 1961
29. DAVIS HA, STERLING RE, WILCOX AA, WATERS WE: *Clin Chem 12*:428, 1966
30. DEBAKEY M: *New Orleans Med Surg J 87*:386, 1934
31. EARLE H: *Device for Removing Ore Slimes from Settling Tanks*, U.S. Patent 829,516, 1906
32. EVENSON MA, HICKS GP, KEENAN JA, et al: in *Technicon Symposia*, 1967, edited by NB Scova. New York, Mediad, 1968, Vol 1, p 137
33. EVENSON MA, HICKS GP, THIERS RE: *Clin Chem 16*:606, 1970
34. FABINY DL, ERTINGSHAUSEN G: *Clin Chem 17*:696, 1971
35. FASCE CF Jr, REJ R: *Clin Chem 16*:972, 1970
36. FELLER B, BOYD WA, DiDARIO BE, et al: in *Technicon Symposia*, 1966, edited by NB Scova. New York, Mediad, 1966, Vol 1, p 206
37. FENNELL TRFW: *Proc Soc Anal Chem 1*:130, 1964
38. FISHER J: *Am J Med Technol 38*: 255, 1972
39. FORMAN DT, CHANGUS GC: *Clin Chem 14*:38, 1968
40. FRIEDMAN HS: *Clin Chem 16*:619, 1970
41. GAMBINO SR, SCHREIBER H, COVOLO G: in *Technicon Symposium*, 1965, edited by LT Skeggs Jr. New York, Mediad, 1965, p 363
42. GARDANIER SA, SPOONER GH: *Am J Clin Pathol 54*:341, 1970
43. GODSE DD, LYALL WAL, STANCER HC: *Clin Chim Acta 26*:89, 1969
44. GRADY HJ, LAMAR MA: *Clin Chem 5*:542, 1959
45. GRAY P, OWEN JA: *Clin Chim Acta 24*:389, 1969
46. GRIFFITHS PD, CARTER NW: *J Clin Pathol 22*:609, 1969
47. HABIG RL, WILLIAMSON WR: *Clin Chem 16*:251, 1970
48. HABIG RL, SCHLEIN BW, WALTERS L, et al: *Clin Chem 15*:1045, 1969
49. HELBING AR, DUIFS R: *Clin Chim Acta 31*:492, 1971
50. HERMANN G III, COHEN EL, SUGIURA HT: *Am J Clin Pathol 54*:226, 1970
51. HICKS GP, EVENSON MA, GIESCHEN MM, et al: *Computers in Biomedical Research*, edited by RW Stacey, BD Waxman, New York, Academic Press, 1969, p 16

52. HICKS GP, GIESCHEN MM, SLACK WV, et al: *J Am Med Assoc 196*:973, 1966
53. HOOVER PL, TEUBNER EJ, CURTIS FK, SCRIBNER BH: in *Technicon Symposium*, 1965, edited by LT Skeggs Jr. New York, Mediad, 1966, p 482
54. HUEMER RP, LEE KD: *Anal Biochem 37*:149, 1970
55. Information for authors (automated methods): *Clin Chem 18*:4, 1972
56. INGLE JD Jr, CROUCH SR: *Anal Chem 42*:1055, 1970
57. ISREELI J: in *Technicon International Congress*, 1970, edited by EC Barton. New York, Thurman Assoc., 1971, Vol 1, p 81
58. JAMES GE, PARDUE HL: *Anal Chem 40*:796, 1968
59. JANSEN AP, PETERS KA, ZELDERS T: *Clin Chim Acta 27*:125, 1970
60. KADISH AH, HALL DA: *Clin Chem 11*:869, 1965
61. KADISH AH: *Instrumentation Methods for Predictive Medicine*, edited by TB Weber, and J Poyer. Pittsburgh Pa., Instrument Society of America, 1966
62. KENNY MA, CHENG MA: *Clin Chem 18*:352, 1972
63. KLASS CS: *Am J Clin Pathol 55*:475, 1971
64 KLEIN B, KAUFMAN JH, MORGENSTERN S: in *Technicon Symposia*, 1966, edited by NB Scova. Mediad, New York, 1967, Vol 1, p 10
65. KLIMAN A, SMITH EP: in Technicon Symposia, 1966, edited by NB Scova. New York, Mediad, 1967, Vol 1, p 106
66. KLOSSE JA, DeBREE PK, WADMAN SK: *Clin Chim Acta 32*:321, 1971
67. KRIEG AF, HUTCHINSON BA: *Clin Chem 16*:443, 1970
68. KRIEG AF, HUTCHINSON BA: *Am J Clin Pathol 55*:141, 1971
69. KUTZIM H: in *Technicon Symposium*, 1965, edited by LT Skeggs Jr. New York, Mediad, 1966, p 393
70. LAESSIG RH: Wisconsin State Proficiency Testing Service, Personal Communication, June 1972
71. LEVINE JB, LARRABEE EW: in *Technicon Symposium*, 1966, edited by NB Scova. New York, Mediad, 1967, Vol 1, p 15
72. LOTT JA, HERMAN TS: *Clin Chem 17*:614, 1971
73. LUMENG J, BLICKENSTAFF L, MILLER J: *Am J Clin Pathol 55*:471, 1971
74. MADDOX WL, KELLEY MT: *Chem Instrum 1*:105, 1968
75. MALMSTADT HV, CORDOS EA, DELANEY CJ: *Anal Chem 44*:No. 12:26A, 1972
76. MARRACK D (Chairman): *Session on automated chromatography*, in Technicon Symposia, 1967, edited by NB Scova. New York, Mediad, 1968, Vol 1
77. MATHER A: *Am J Clin Pathol 33*:186, 1960
78. McGREW BE, DuCROS MJF, STOUT GW, et al: *Am J Clin Pathol 50*:52, 1968
79. McKENZIE JM, FOWLER PR, FIOICA V: in *Technicon Symposia*, 1966, edited by NB Scova. New York, Mediad, 1967, Vol 1, p 45
80. MORGENSTERN S, CHAPARIAN L, VLASTELICA D, KIEDERER A: in *Technicon International Congress*, 1970, edited by EC Barton. New York, Thurman Assoc., 1971, Vol 1, p 85
81. MORGENSTERN S, KAUFMAN JH, KLEIN B: *Clin Chem 13*:270, 1967
82. NATELSON S: *Microchemical J 13*:433, 1968
83. NEELEY WE, CECHNER RL, MARTIN RL, MARSHALL G: *Am J Clin Pathol 56*:493, 1971
84. PERONE SP: *J Chromatog Sci 7*:714, 1969
85. PERRY BW, HOSTY TA, COKER JG et al: *A Field Evaluation of the DuPont Automatic Clinical Analyzer*. Birmingham, Ala., University of Alabama Medical Center, 1970
86. PRIBOR HC, KIRKHAM WR, FELLOWS GE: *Am J Clin Pathol 50*:67, 1968
87. ROBBINS RC: *Clin Chem 15*:56, 1969
88. RODBARD D, BRIDSON W, RAYFORD PL: *J Lab Clin Med 74*:770, 1969

89. RODBARD D, LEWALD JE: in *Karolinska Symposium on Research Methods in Reproductive Endocrinology*, 2nd Symposium, edited by E Diczfalusy. Copenhagen, Bogtrykkeriet Forum, 1970, p 79
90. ROODYN DB: *Biochem J 119*:823, 1970
91. RUTTER ER: *Clin Chem 18*:616, 1972
92. RUZICKA J, TJELL JC: *Anal Chim Acta 47*:475, 1969
93. SEIBERT RA, BUSCH H: *Clin Chem 12*:439, 1966
94. SELIGSON D: *Am J Clin Pathol 28*:200, 1957
95. SIMPSON D, SIMS GE, HARRISON MI, et al: *J Clin Pathol 24*:170, 1971
96. SKEGGS LT, Jr: *Am J Clin Pathol 28*:311, 1957
97. SKEGGS LT, Jr.: *Am J Clin Pathol 33*:181, 1960
98. SKEGGS LT, Jr., HOCHSTRASSER H: *Clin Chem 10*:918, 1964
99. SMITH AF, BROWN SS, TAYLOR R: *Clin Chim Acta 30*:105, 1970
100. SMYTHE WJ, SHAMOS MH, MORGENSTERN S, et al: in *Technicon Symposia*, 1966, edited by NB Scova. New York, Mediad, 1967, Vol 1, p 105
101. SOBOCINSKI PZ, McDEVITT RP: *Clin Chem 18*:487, 1972
102. Standard Methods of Clinical Chemistry edited by M Reiner. New York, Academic 1953
103. STERLING RE, FLORES OR: *Clin Chem 18*:544, 1972
104. STEVENS JF: *Clin Chim Acta 32*:199, 1971
105. STRICKLER HS, SAIER EL, GRAUER RC: in *Technicon Symposium*, 1965 edited by LT Skeggs Jr. New York, Mediad, 1966, p 368
106. STRICKLER HS, STANCHAK PJ, MAYDAK JJ: *Anal Chem 42*:1576, 1970
107. TAYLOR GI: *Roy Soc London Proc Series A, 223*:186, 1953
108. THIERS RE, OGLESBY KM: *Clin Chem 10*:246, 1964
109. THIERS RE, BRYAN J, OGLESBY K: *Clin Chem 12*:120, 1966
110. THIERS RE, COLE RR, KIRSCH WJ: *Clin Chem 13*:451, 1967
111. THIERS RE, MEYN J, WILDERMANN RF: *Clin Chem 16*:832, 1970
112. THIERS RE, REED AH, DELANDER K: *Clin Chem 17*:42, 1971
113. TOREN EC, Jr, EGGERT AA, SHERRY AE, et al: *Clin Chem 16*:215, 1970
114. TRAYSER KA, SELIGSON D: *Clin Chem 15*:452, 1969
115. U.S. Department of Health, Education and Welfare Publication No. (HSM) 72-3004, A study of automated clinical laboratory systems
116. VALENTINI L: in *Automation in Analytical Chemistry*, 1966, edited by E Kaiverau et al. New York, Mediad, 1967, p 27
    Mediad, 1967, p 27
117. VARLEY JA, BAKER KF: *Analyst 96*:734, 1971
118. VERGANI C, STABILINI R, AGOSTINI A: *Clin Chem 15*:216, 1969
119. WADMAN SK, DeBREE PK, VAN DER HEIDEN C, et al: *Clin Chim Acta 31*:215, 1971
120. WALKER WHC: *Clin Chim Acta 32*:305, 1971
121. WALKER WHC, PENNOCK CA, McGOWAN GK: *Clin Chim Acta 27*:421, 1970
122. WALKER WHC, SHEPHERSON JC, McGOWAN GK: *Clin Chim Acta 35*:455, 1971
123. WALTER AR, GERARDE HW: *Clin Chem 10*:509, 1964
124. WEICHSELBAUM TE, PLUMPE WH Jr, ADAMS RE, et al: *Anal Chem 41*:725, 1969
125. WESTGARD JO, LAHMEYER BL: *Clin Chem 18*:340, 1972
126. WESTLAKE G, McKAY DK, SURH P, SELIGSON D: *Clin Chem 15*:600, 1969
127. WHITE WL, ERICKSON MM, STEVENS SC: *Practical Automation for the Clinical Laboratory*. St. Louis, Mo., Mosby, 1968
128. YOUNG DS, MONTAGUE RM, SNIDER RR: *Clin Chem 14*:993, 1968

Chapter 11

Commercial Kits and Reagent Sets

JOHN DI GIORGIO, Ph.D.

The U.S. Department of Health, Education, and Welfare published in July 1971 a booklet entitled "List of Test Kits for Clinical Laboratories," the purpose of which was to provide a reference to the increasing number of commercial test products available for use in clinical laboratories. The index lists the names of kits or reagent sets manufactured by at least 78 companies in the United States, for the analysis of about 105 serum constituents, from barbiturates to vanilmandelic acid. This illustrates the rapidity of growth and increasing importance of kits in the clinical laboratory field.

Clinical chemistry kits have been available for many years, either directly from manufacturers or through laboratory supply houses. Unfortunately, laboratory workers often have had no idea as to a given kit's accuracy and precision. Even though many clinical chemists and pathologists have evaluated these products, relatively few publications have appeared to inform and alert others. A search of the literature revealed only 26 such reports since 1964.

The exact definition of a "kit" or "reagent set" is not yet precise. There are two practical answers, however, as to their purpose: to fulfill a need in the clinical laboratory for convenient, rapid, and reliable laboratory determinations, and to be used in the physician's office or hospital laboratory for an infrequently performed method, as a "STAT" procedure, or as a backup method for an automated technic.

Manufacturers usually sell the chemicals in a kit or reagent set in one or a combination of three physical forms: (a) a freeze-dried mixture of reagents, (b) individual solutions or solids that must be mixed together before use, or (c) ready-to-use premeasured volumes of solutions. In addition, the manufacturer may or may not sell all the materials required for the particular method. For example, if reagents are provided in bulk, the company may or may not provide the vials or cuvets needed to carry out the visible or ultraviolet photometric method. If the method involves a separation step, the manufacturer usually sells the chromatographic plates or columns prepacked with the adsorbent or resin. Likewise, pipets that may be needed are packaged either with the reagent set or sold separately. Manufacturers of more sophisticated reagent sets usually also sell the accessories and instrumentation needed to perform the determination, e.g., test-tube racks, centrifuges, photometers, or spectrophotometers.

Naturally the vendor sells these chemicals, materials, and instruments in as attractive a package possible, but often neglects to provide pertinent product information to the user.

The American Association of Clinical Chemists (1) has published certain recommendations for the " · · · maintenance and improvement of analytical methods, products, and instrumentation in the field of clinical chemistry." Also, an attempt was made " · · · to define and to encourage high professional standards among commercial manufacturers and suppliers of reagents, reagent sets, and instruments." The policy states that companies should provide sufficient information to the prospective purchaser so he will have a preliminary idea of what to expect. The following information, if previously compiled (as it should have been) by the researcher during method development, should be made available to the customer: the chemical principle upon which the quantitative method is based, the exact reagent composition, the reagent stability with recommended storage temperature, the accuracy of the method and how determined, the precision of the method with confidence limits and how determined, the normal range obtained with the method, and the pertinent literature references used to develop the reagent set.

Since the above information is or should be included in a paper appearing in a scientific journal, the information should likewise be contained in the reagent set literature. In addition, the company should be explicit concerning how to calculate results, how to use the required instrumentation, and give precautions on technic in order to perform the method properly. Again, the purchaser should expect to receive all of the above information either in the company's package literature or in their journal and direct-mail advertising. Obviously, to do this the manufacturer must make a substantial financial investment.

Regardless of the research dollars spent and the quality of the research data proving the efficacy of the method, the reagent set sold is only as good as the final manufactured product. All reagent sets, therefore, must be manufactured under controlled conditions and subjected to a proper statistical quality control program. It is not enough, as we do in our laboratories when we prepare one liter of reagent, to buy the chemical, weigh or volumetrically measure a portion, dissolve, dilute to volume, and then mix. A reputable manufacturer of reagent sets must first establish specifications for all chemicals needed in a kit to insure reagent purity from lot to lot. Having determined by analysis that all raw materials meet these specifications, the manufacturer should then formulate the reagents under controlled manufacturing conditions of temperature, water purity, freedom from environmental contamination, or contamination from chemical carry-over from a previously prepared batch of a *different* reagent.

The manufacture of reagent sets should be further controlled as follows: the quality control laboratory must verify the accuracy of the reagent-set standard; glassware, whether vials or bottles, should be inspected before filling with reagents or standards; and the stability during storage at various temperatures should be determined for *final* manufactured reagents and standards.

The manufacturer should, whenever possible, chemically assay the standard immediately after manufacture to verify that its concentration is within reasonable tolerance limits, e.g., ±1% of stated label value. In addition to this chemical assay, quality control personnel should perform an "in-use" test with the exact method described in the package literature of the reagent set. The minimum requirement should be the obtainment of values within the acceptable range stated for a control serum with concentrations of constituents within the normal physiologic range and at least one control serum with constituents within the abnormal physiologic range.

A number of manufacturing variables, if not controlled, will affect these "in-use" test results as they would the results obtained in the purchaser's laboratory. For example, the lip or thread of a container should not be chipped or else the closure will not fit properly and reagent leakage may occur. Also, all glassware and plasticware should be free of carton dust, chipped glass, or glass stuck to the surfaces of the vial, especially if the vial is to be used as a cuvet. If the vial is to be used as the cuvet there should be no defect in the area through which light transmission is to occur. Equally important, every cuvet must have the proper outside dimension in the light-transmission area to avoid a loss in method precision.

Reagents and standards may deteriorate when stored at unusual temperatures. It is realistic to assume that once the reagent sets are shipped from the producer, the reagents will be subjected to temperature cycling from extremes of below 0° C to as high as 45° C, temperatures occurring during winter and summer months if the reagent sets are stored in shipping warehouses or on loading docks. The effects of temperature cycling on reagent stability should be determined by the manufacturer and revealed to the purchaser. Furthermore, temperature effects on reagent stability should be observed using the same package materials and types of containers intended for the finished product. They should be studied by performing chemical assays on standards and reagents, "in-use" tests with control sera, inspections of the physical appearance of the reagents, and noting whether there are changes in the amount of light absorption by the reagent blank as it is carried along in the analytic method.

The types of containers in which reagents are supplied is often quite important since the container material may not always be inert. There are Types I, II, and III glass, and the type chosen should be the one least likely to be affected by the acidity or alkalinity of the reagent. Similarly, various plastics react differently to acids, bases, and oxidizing or reducing substances, nor are plastic materials necessarily inert when used as cap liners for the vial, cuvet, or bottle. Especially important is the adhesive used to bind the liner to the cap since the possibility exists that certain reagents and solvents can leach adhesive ingredients and thereby contaminate the container contents.

The extent to which the manufacturer's quality control laboratory is involved in a statistical method for final product inspection determines whether the final product will perform accurately and precisely after its release to the customer's laboratory. Furthermore, the quality control laboratory should verify the accuracy of all volume measuring devices either included in the reagent set or recommended by the company for use with the method. Lastly, quality control personnel should maintain records of "in-use" test results for every lot of reagent sets prior to their release. This test should include not only the reagents and standards, but also the pipets and instrumentation intended for use by the laboratory personnel who will eventually purchase the reagent set.

Every clinical chemist should be aware of the manufacturing variables discussed. He should also be cautious of communication gaps between the research department, the manufacturing department, and the marketing department of any company. Consequently, the chemist should be able to ask the company for authentication of advertising method claims and even to obtain the raw data upon request.

Regrettably, not all kits or reagent sets can be recommended for use. For example, some do not provide a standard, several even relying on precalibration of the procedure, an approach that cannot be too severely condemned. In some kits or reagent sets the identity of the active ingredients is not revealed. No clinical chemist should ever agree to employ "secret" reagents.

Because of the recent rapid growth of the kit business and the failure of some manufacturers voluntarily to provide the user with the best reagent set possible from the state of the art, the U.S. Department of Health, Education, and Welfare has granted authority to the Food and Drug Administration to regulate this expanding field of medical diagnostic equipment. The patient should be the ultimate benefactor of this move.

REFERENCE

1. American Association of Clinical Chemists Committee on Standards and Controls. AACC Policy Regarding Reagent Sets and Kits. *Clin Chem 12:*43, 1966

Chapter 12

Accuracy, Precision, Quality Control, and Miscellaneous Statistics

ALLEN H. REED, Ph.D.
RICHARD J. HENRY, M.D.

Every analyst would admit that there is a degree of uncertainty in each result he obtains. For example, if a blood glucose result is 100 mg/100 ml, it is obvious that it probably is not 100.00 mg/100 ml. Furthermore, if the same analyst repeats the test, or another analyst performs it, the result may deviate somewhat from that obtained on first analysis. It is essential that the analyst has some conception of the subject of variability of results and the many factors that contribute to it. A result is worthless unless there is some knowledge of the degree of uncertainty associated with it. These problems cannot be approached quantitatively without the use of some simple statistical concepts and tools.

INTRODUCTORY STATISTICS

CLASSIFICATION OF OBSERVATIONS

All measured quantities are variables. Variables are of two types: The first is the discontinuous variable that varies only in discrete units, e.g., the number of animal deaths in an experiment; the second is the continuous variable that can take on any value within practical limits although it is reported in discrete units, e.g., the blood glucose level in normal people.

FREQUENCY DISTRIBUTIONS

If blood glucose determinations were performed on a large number of normal people, it would be expected that the greatest density of values obtained would occur at, or close to, the average of all values obtained, and the density would decrease steadily as we move away from the average in either direction. We would expect a similar pattern of results if many replicate determinations were made on the same blood specimen. The scatter of results in either case can be depicted graphically by what is called a *frequency curve*. Most of the statistical technics are based on the assumption that the distribution of values follows the symmetrical, bell shaped, so-called "normal" or Gaussian curve, shown in Figure 12-1. If the total area under the curve is equivalent to all the values plotted, then the area from $-s$ to $+s$ includes about 68% of the values, from $-2s$ to $+2s$ includes about 95.5% of the values, and from $-3s$ to $+3s$ includes about 99.7% of the values. The

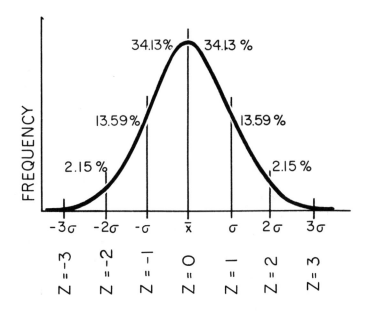

FIG. 12-1. Normal frequency curve.

value s is called the *standard deviation* and is sometimes written as the Greek letter σ. Standard deviation is a measure of the scatter of the values and can be calculated easily for any set of observations as follows:

$$s = \sqrt{\frac{\Sigma\,(x - \bar{x})^2}{(N-1)}}$$

where Σ = "sum of"
  $\bar{x}$ = mean or arithmetic average = $\Sigma x/N$
  x = observed values
  $x - \bar{x}$ = deviation of a value from the mean, $\bar{x}$
  N = number of observations

Often the mathematical labor is simplified by the use of another form of this equation:

$$s = \sqrt{\frac{\Sigma\,(x^2) - [(\Sigma x)^2 /N]}{N-1}}$$

A short cut has also been presented (18) to determine s from a small number of observations (20 or less):

$$s = K_w \times R$$

where R = range, i.e., the difference between the greatest and least values
in the series
K_w = a factor found in Table 12-1

TABLE 12-1. Table of K_w and Q Values (23, 24)[a]

No. of observations	K_w	Q
2	0.89	—
3	0.59	0.94
4	0.49	0.76
5	0.43	0.64
6	0.40	0.56
7	0.37	0.51
8	0.35	0.47
9	0.34	0.44
10	0.33	0.41
11	0.32	0.39
12	0.31	0.38
13	0.30	0.36
14	0.29	0.35
15	0.29	0.34
16	0.28	0.33
17	0.28	0.32
18	0.28	0.31
19	0.27	0.31
20	0.27	0.30
21		0.30
22		0.29
23		0.29
24		0.28
25		0.28
26		0.27
27		0.27
28		0.27
29		0.26
30		0.26

[a] Adapted from Introduction to Statistical Analysis by Dixon & Massey. Copyright (c) 1969 by McGraw-Hill, Inc. Used with permission of McGraw-Hill Book Company and from Dixon WJ: *Ann Mathematical Statistics* 22:68, 1951

This estimate of *s* is solely dependent on the outside values of the range, and occasionally one of these can be spurious, i.e., not representative of the population about which inference is desired. Such spurious values are called outliers. The treatment of outliers is discussed in the next chapter. Here, we give a general method for screening for outliers in a sample of 30 or less. This method in particular may be applied when calculating s by the short method using K_w. Calculate the distance between the doubtful value and its nearest neighbor and divide this distance by the range, R. If the result exceeds the tabulated value for Q in Table 12-1, the doubtful value should be discarded as an outlier.

Let us assume that we have a large set of observations and the values $\bar{x}$ and s have been obtained. If we assume that the distribution is a Gaussian one, we can now say that about 95% of past and future observations made in the same way on the same thing or things will lie within $\pm 2s$ from $\bar{x}$. Similarly, about 99.7% will lie within $\pm 3s$ of $\bar{x}$. Or, about only 1 in 20 and 1 in 300 observations will fall outside these limits, respectively.

The values $\bar{x}$ and s are the two most useful and fundamental indices in statistical work. Unfortunately, however, valid use of them in the simpler ways is predicated on the normal, Gaussian frequency distribution, which in many instances does not prevail.

ACCURACY

Accuracy is defined as the extent to which the mean measurement is close to the true value. Inaccuracy, then, is the extent to which a measurement will differ more or less constantly and in the same direction from what it should be and is frequently referred to as the "error of bias." Determination of accuracy consists of comparing observed results with the actual true value. The difficulty, however, is in the determination of the "true" value, and in clinical chemistry this is frequently impossible. One can, however, compare measurements made by the method in question with those obtained by an acceptable reference method. The suggestion (78) that the best means of assessment of accuracy is by recovery studies is invalid. This approach is a valuable part of a study of accuracy but contributes nothing to the aspect of specificity that is essential to accuracy.

The accuracy of a chemical analysis is a characteristic occasionally beyond the control of the analyst. Various methods for determining the same substance may give different results due to differences in specificity. For example, in the Folin-Wu method for determination of blood glucose, reducing substances other than glucose normally present in blood are determined as glucose. Such a test has poor accuracy, but this does not necessarily negate the value of the test, especially if the "interfering" substances do not vary markedly from individual to individual since the clinician becomes accustomed to the definite set of normal values given by the test.

PRECISION

Precision is the random error, the variation of results obtained by a method when the same sample is run repeatedly—in other words, the reproducibility of what is observed. The less variation observed, the greater is the precision. Neither accuracy nor precision is dependent one on the other; an inaccurate test may be very precise. Obviously, the ideal test possesses both great accuracy and precision.

Although most workers in the field use the terms precision, reproducibility, and repeatability interchangeably, some have chosen to make a distinction. For example, one group (78) feels that precision includes the two elements of reproducibility and repeatability. They define reproducibility as the deviation occurring with completely blind samples over a period of weeks or months, perhaps by different analysts, and define repeatability as the deviation obtained by one analyst without change of apparatus used where the apparatus may be a significant factor.

Precision is commonly expressed in terms of the standard deviation s and can be determined for every test procedure. When desirous of determining precision, a decision must be made first as to the precision of what. Thus, the "within-run" precision can be obtained by running about 30 replicate determinations on a single sample and calculating $\bar{x}$ and s utilizing the formulas given previously in the section "Frequency Distributions." This approach to determination of "within-run" precision is not very practicable in the routine operation of most clinical laboratories. A simpler approach requires running routine unknowns in duplicate until about 30 duplicate analyses are accumulated. Duplication must be complete, i.e., through the entire procedure, and the values used must be fairly close to the range at which knowledge of the precision is desired. Then (82)

$$s = \sqrt{(\Sigma d^2)/N}$$

where d = difference between duplicates and N = total number of determinations. For this approach to the determination of within-run precision to be valid the duplicates must be randomly distributed in a run rather than analyzed successively in the run. Otherwise there will be an unpredictably low bias in the estimate of s.

The precision of a test is usually given as "95% limits." These limits are $\bar{x}$ $\pm ts$, t being obtained from Table 12-2. The "degrees of freedom" (d.f.) are $N/2$ for the above formula. The value for t can be taken as 2 when d.f. > 30.

We have just discussed means of determining "within-run" precision of a procedure, but ordinarily, this is not what is really wanted in the laboratory. Since the physician may send a series of the same test on the same patient over a period of time or send in a number of the same tests on different patients over a period of time, he needs to know the "between-run" or "between-day" precision. As might be expected, this precision is usually poorer than "within-run" precision. This will be discussed further in a subsequent section ("The 'Error' of a Test"). The two most common methods of obtaining an estimate of "between-run" precision are (a) resubmitting patient samples, which have been already analzyed once, for reanalysis, completely blind, and calculating s from the paired results as

TABLE 12-2. Table of t Values at 5% Significance Level. This Is an Abbreviated Table. Complete Tables Are Given in Standard Texts on Statistics.

d.f	t	d.f	t	d.f	t
1	12.71	11	2.20	21	2.08
2	4.30	12	2.18	22	2.07
3	3.18	13	2.16	23	2.07
4	2.78	14	2.14	24	2.06
5	2.57	15	2.13	25	2.06
6	2.45	16	2.12	26	2.06
7	2.36	17	2.11	27	2.05
8	2.31	18	2.10	28	2.05
9	2.26	19	2.09	29	2.04
10	2.23	20	2.09	30	2.04

discussed above for "within-run" precision; (b) the quality control chart estimate of s obtained with serum or urine controls ordinarily is a reasonably valid estimate. Published reports (1, 32, 34, 77) are contradictory as to whether or not precision is estimated as poorer with blind control samples than with nonblind control samples. Extensive experience in our laboratory indicates that it frequently is poorer with blind samples.

The standard deviation may vary with the level of the substance being determined. The precision, therefore, of the test for abnormally low, normal range, and abnormally high values must be independently measured.

The standard deviation is in the same units in which measurements were expressed; e.g., if blood glucose values were expressed in milligrams per 100 milliliters, s would be in the same units. For practical interpretation it is more convenient to express s in terms of percent of the average of analyses used in the calculation of s. This is called the *coefficient of variation* (C. V.), and is calculated as follows:

$$C.\ V. = 100 \times (s/\bar{x})\%$$

The "95% limits" are now $\pm t$ (C. V.)%.

If the precision of a particular method is found to be inadequate, three courses of action are available to the analyst:

(a) First, the technic of the method might be improved, and greater reproducibility thus obtained. Each step and operation in an analysis is a continuous variable and, therefore, contributes its share of variation or error to the over-all variation of the analysis. Partitioning of the over-all variation in several common clinical chemical analyses has been worked out (37) and the following conclusions reached: (a) Except in cases in which a few operations contribute a large share of the over-all variation, material reduction in the over-all variation can be accomplished only by careful attention to every step along the way, e.g., rigid control of all variables such as temperature, meticulous care in pipetting with the most accurate type of pipet available for the particular operation, and use of a high caliber photometer or spectrophotometer; (b) with perhaps few exceptions, the contribution of pipetting errors and photometer errors to the over-all variation can be regarded as insignificant.

(b) The second method of attaining greater precision is through replication. The increase of precision, as the mean is taken of a greater and greater number of replicates, is given theoretically by the following formula:

$$\text{precision of mean of } N \text{ replicates} = \frac{\text{precision of a single analysis}}{\sqrt{N}}$$

If the 95% limits of a single analysis are ±10%, then the 95% limits of the averages of 2, 3, 4, and 10 replicates would be ±7.1, ±5.8, ±5.0, and ±3.2%, respectively. Replicates ordinarily are run in the same run so that the "between-run" precision, which is what should be known and used by the laboratory routinely, cannot be substituted in the above formula to obtain an estimate of the "between-run" precision of the mean of N replicates run in the same run. For this formula to be valid the replicates must be in different runs and in addition, completely blind. Gambino (32) found that the range for blind duplicates was three times greater than for nonblind duplicates. The latter showed severe bias towards zero difference. Thus, the increase in precision resulting from running samples in duplicate or triplicate in the usual manner is less than predicted by the formula. Although this is not a very efficient method of increasing the precision of a result, it occasionally may be the only approach available. The magnitude of increased precision varies with the step in a procedure at which replication is begun. For example, it is a common practice in blood analyses to begin replication after a protein-free filtrate is prepared. If replication is not started from the very first step, however, much of the efficiency of this device in increasing precision is lost. Certainly, replication greater than duplicates, or perhaps rarely triplicates, is totally impracticable in the clinical laboratory.

(c) If neither of the two approaches yields the required precision or if they are impracticable, one is left with only the third alternative: to discard the method and search for one that will give the desired precision.

The practice of some analysts to run a determination in triplicate, averaging the two closest results and discarding the third, is entirely unjustified, tends to overestimate greatly the precision of the procedure, and actually results in a less accurate determination. In fact, acceptance of the median of the three observations is better than the "best two out of three" (18).

RELIABILITY

The reliability of a test is the ability to maintain its accuracy and precision into the future. If a test has maintained a steady state of these attributes over a long period of time, then one can predict these characteristics for the future with assurance and the test is said to be reliable.

A concept that parallels and probably overlaps reliability is that of the "ruggedness" of a test (81). A procedure is "rugged" if results are immune to modest variations in technic, reagent concentrations, etc. An example of uncontrolled variation in a procedure is the effect of variable loading of a water bath on the time required for the tubes placed therein to reach the desired temperature (31). Lack of "ruggedness" can also account for failure of different laboratories to obtain the same results with a given procedure. This property of a method can be investigated by deliberately introducing minor, but reasonable, variations in the variables of the method and observing the results. Youden (81) presents an experimental design by which maximal information is obtained from minimal work.

THE ERROR OF LABORATORY RESULTS

THE "ERROR" OF A TEST

The term "error" is rather ambiguous. It is necessary whenever using it that it be quite clearly understood exactly what it is meant to include. Whitby *et al* (78) chose to define it as the difference between the observed value and the true value, i.e., the combination of the error of bias and the error due to lack of precision. Although this is a perfectly legitimate definition, the result obtained by one test cannot be compared with another result obtained by a test with entirely different specificity. Thus, in the discussion here, the factor of accuracy is excluded, and the term "error" is used only in the comparison of results obtained by the same method of analysis or by methods of very similar specificity.

Of primary concern to the analyst is how close the result he has obtained by a particular method on a certain sample is to the "correct" result by the method he is using. If we are concerned about only one particular method, there are several component variables that contribute to this total error. First, there is the variation between replicates in the same analysis (P). This is the "within-run" precision discussed previously. The second component is the variation between days for the same analyst in the same laboratory on the same sample (D). The third component is the variation between analysts in the same laboratory (A). The fourth is the variation between laboratories (L). According to the "addition theorem" of probability, the total error is equal to the square root of the sum of the squares of the component errors as follows:

$$\text{total error} = \sqrt{P^2 + D^2 + A^2 + L^2}$$

This theorem is valid if the four variables are independent of each other. If they are independent there is no correlation between any of them. This would seem to be a reasonable assumption, but for the purpose of discussion here it does not matter if it is not. The existence of these components of the total error has been verified experimentally (10, 34).

Any exact expression by an analyst of the "correctness" of his results would require his participation in complete statistical analysis as outlined above for each type of analysis made. This is obviously an impossibility. It is possible and desirable, however, to include the first three error components (P, D, A) in estimates of the precision of a laboratory's analyses. Approaches by which these estimates can be made were discussed in the section on precision. If the total error is distributed according to the normal Gaussian curve, limits of $\pm 2s$ or ± 2 C. V. and $\pm 3s$ or ± 3 C. V. placed on a single determination would give the assurance that there is only about 1 chance in 20 and 300, respectively, that the "correct" result exceeds these limits. Although the error distributions of chemical determinations frequently are not Gaussian (17, 66), it has become customary to consider $2s$ or 2 C. V. limits to be the "95% limits." If this assumption is correct, the analyst then can state that he has confidence the "correct" result has 19 chances out of 20 of lying somewhere within these limits; these limits may be referred to as "95% confidence limits." It must be definitely understood, however, that the chemist's guarantee can be no better than the state of control of the test. Interlaboratory accuracy surveys have revealed that in the vast majority of cases the state of control of most tests in many laboratories is very poor. Under the best of circumstances, however, it is to be expected that the interlaboratory error component is larger than any other. There is no easy solution to this problem of an analyst's guarantee of his results, and study of the problem is still in its infancy.

At the beginning of this section it was stated that the term "error" is rather ambiguous, and it is perhaps unfortunate that we are saddled with this term because to most people (physicians also) the word error implies a mistake. In the statistical sense considered here, however, this "error" is predictable and unavoidable and hence is not a "mistake." Sometimes it can be made smaller but it can *never* be eliminated. An incorrect result caused by adding the wrong reagent, spilling part of the tube's contents, incorrect calculation or transfer of the result, is an error of a different type, a *mistake*.

What is the "state of the art" of clinical chemical analyses? "Errors" and "mistakes" in laboratory results can be divided into two groups, one resulting from the analysis itself, and the other clerical (e.g., mixup of samples, incorrect keypunch, incorrect transfer of results, etc.). Although some studies (47) have revealed a "mistake" rate of nearly 3% due to the clerical category alone, careful review of many thousands of samples in our laboratory clearly shows that it is possible with carefully designed clerical procedures, including verifications at critical steps, to reduce this to a much lower level and in fact, to such a low level that it becomes rather insignificant compared to the error rate from the analysis itself. Assuming routine, feasible analytic methodology, since cost to the patient is a factor that must be considered in selecting methodology, we have arrived at the following estimates of the best that can be done by a "good" laboratory at the present "state of the art" (opinions were derived from published reports and our own experience with quality control in the clinical laboratory for a period of 25 years): (a) The "between-run" precision approaches a lower limit of about 5% for most tests, somewhat less than this for a few tests, and more than this (as much as 30%) for a number of tests. (b) If one includes the "between-laboratory" component of error, the precision becomes worse and this can be rather variable. (c) If we define a result to be "mistaken" if it deviates from the correct answer by more than one-fourth the normal range (Tonk's criterion), about 5% of the reports issued by a "good" laboratory will be "mistaken." The most frustrating thing of all in our entire clinical chemical laboratory experience is the failure to substantially improve on this level of performance, regardless of the amount of sincere effort expended.

PRACTICAL ASPECTS OF LABORATORY ERROR

In the first edition of this book the position was taken that it was probably wasteful to make procedures give greater precision than is actually required for medical diagnostic purposes. We now recognize this to be basically unsound. There is a reason to make the error of a test as small as possible—the smaller the error the greater the diagnostic power of the test (this will be discussed in greater depth in the next chapter). Nevertheless, in designing the routine procedures of a laboratory one must still stay within the bounds of practicability. Precision requirements vary considerably as might be indicated by the following consideration. If one accepts 4.1—5.6 meq/liter as the normal range of serum potassium, then ±16% from the middle spans the entire range. If one accepts 131–150 meq/liter as the normal range of serum sodium, then ±7% from the middle spans the range. It would seem reasonable that the precision requirement should be more stringent for sodium than for potassium. Nevertheless, even this rationalization may be misleading, for the normal range of any given individual may be considerably less than the range for the whole population (cf. Chap. 13). It is useful to have a rule of thumb for precision requirements and the one proposed by Tonks in 1963 (70) has become widely used and seems as good as any other proposed:

$$\text{allowable limits of error in \%} = \frac{\pm \frac{1}{4} \text{ of the normal range}}{\overline{x} \text{ of the normal range}} \times 100$$

Tonks originally set a maximum of ±10% for the allowable limits but subsequently (71) extended them to ±20%. With some few procedures even this is too restrictive.

One problem that is encountered is the comparison of two or more observations made on the same patient, e.g., two hemoglobin values obtained successively while the patient is on iron therapy, or two successive blood glucose values in a glucose tolerance test. This problem resolves itself into the question, how much must two determinations differ to be considered significantly different? Approximately 95% assuredness of a significant difference exists if the two results differ by at least 2.8 s, where s is the "between-run" standard deviation of the method used (54). At present, an estimate of s is not routinely provided by the laboratory with its lab report.

SIGNIFICANT FIGURES

A discussion of laboratory error would not be complete without pointing out a frequent abuse of mathematics made in reporting the results of chemical determinations. A result reported to be 16.5 mg/100 ml without qualification implies that the correct result by the method employed is between 16.45 and 16.55 mg/100 ml and that the maximal error, therefore, is about 0.3%. The implication is that all figures cited are significant; otherwise why cite them? Table 12-3 gives an example of a report with varying numbers of significant figures, the implied maximal limits and implied maximal errors.

TABLE 12-3. Significant Figures

Reported result	Number of significant figures	Implied limits	Implied maximal error, ±%
1	1	0.5–1.5	50
1.1	2	1.05–1.15	5
1.11	3	1.105–1.115	0.5
1.111	4	1.1105–1.1115	0.05
etc.			

Zeros to the right of a decimal point are significant, whereas zeros to the left of the decimal point may or may not be. From Table 12-3 it is obvious that, if the method employed has an error of the order of magnitude of 10%, it is rather misleading to give the report to three significant figures, and certainly absurd to give four significant figures. On the other hand, one is justified in retaining the second significant figure since the precision of the result is significantly better than that implied by only one significant figure. In routine work common sense should dictate the number of significant figures to retain in the reported value.

When collecting research data, however, a more rigorous criterion should be employed in the decision of how many digits should be retained. The loss of information in a figure is less than 1% if the group interval implied by the last retained digit does not exceed one-fourth of the standard deviation of the measurement itself (27). Thus, if s for the measurement of blood glucose at a level of 100 mg/100 ml is found to be 2.7, $s/4 = 0.68$. In an observation of 100.81, the fifth, fourth, and third significant figures have group intervals of 0.01, 0.1, and 1, respectively. Since a group interval of 1 is closest to $s/4$, three significant figures would be justified and according to the rounding rules, the value becomes 101. Actually, extra digits should be retained if possible when calculations for quality control are made, but unfortunately this is not always feasible in the routine work of a laboratory.

The dropping of digits is called "rounding," and the following rules for rounding are generally followed. If the digit to be dropped is less than 5, the last digit retained is left unchanged (called "rounding down"). For example, 23.4 becomes 23. If the digit to be dropped is greater than 5, or a 5 followed by digits greater than 0, the last digit retained is increased by 1 (called "rounding up"). For example, both 23.6 and 23.52 become 24. If the digit to be dropped is a 5 alone, or a 5 followed by zeros only, the last digit retained is rounded to the nearest even number. For example, both 23.5 and 24.5 become 24.

A significant observation (51) is that some analysts exhibit terminal digit preferences, e.g., preference for even numbers or preference for 130 and 140 meq/liter on AutoAnalyzer charts for serum Na.

CONTROL OF LABORATORY ERROR

In the previous sections it has been established that some degree of error is connected with every chemical determination, and the problem of the determination of the accuracy and precision of a method was discussed. Of equal importance is the control of error, i.e., the day to day assurance that the error is maintained within permissible limits and does not suddenly go out of bounds without the analyst being aware of it. That there is need for error control was more than adequately illustrated by interlaboratory accuracy surveys, beginning with the Belk and Sunderman survey of 1947. As late as 1966 a study of six "reputable" laboratories revealed only two obtaining acceptable results (28).

There are at least four independent forces existing today that bear on the control of laboratory error: (a) Complaints from the practicing physician that the results he received are not those he expected. This, of course, is a very valuable source, but frankly, it usually is very insensitive. The statement made as late as 1965 (74) that "the best method presently available for detecting errors is comparison of reported laboratory values with the patients' clinical states" is indeed regrettable and in our opinion totally incorrect. (b) Quality control. A quality control program in the laboratory with standards and control reference samples is the most important of the four sources for day-to-day control of error. A quality control program by itself does not appear, however, to guarantee improvement in a laboratory's performance (63). (c) State and Federal laws and regulations. These set standards for personnel and general procedures and may dictate a certain minimal quality control program as well as requiring proficiency testing. (d) "Proficiency testing," i.e., analysis of unknown samples submitted by some central source, is a valuable control since it usually monitors both accuracy and precision.

For over a decade utilization of quality control technics has been widespread. The federal government has laws and regulations for laboratories engaged in interstate commerce or reimbursed by Medicare. Many states have their own laws and regulations. In addition, a great many laboratories are participating in proficiency testing either voluntarily or because of legal requirement. Despite all of these, the problem still exists of inadequate control of laboratory error. We are convinced that the situation is better than it was 25 years ago, but we are equally convinced that it still is not as good as it must be. The ultimate in government control would be dictation of specific methodology, which perhaps could come about if all else fails. This has precedence and under certain circumstances (where the alternative is chaos), is clearly justified, e.g., in syphilis serology. It is truly hoped that clinical chemistry will not come to this because we believe nearly all clinical chemists fear that a stifling of research and innovation would result. Clearly, the only alternative is a persistent and continuing upgrading of the quality of laboratory analyses.

STANDARDS

DEFINITION OF TERMS

Radin (57) has proposed definitions for the various standards used in clinical chemistry that have become widely accepted:

Primary standard

A pure chemical substance used for assaying a volumetric solution of unknown strength or for preparation of a solution of known concentration. Although it is not always possible to select a substance for a primary standard with all the desired characteristics, it is generally stated that the primary standard should be selected with the following criteria in mind:

1. It must be a stable substance of definite composition.
2. It must be a substance that can be dried in the course of preparation, preferably at 105–110° C, without change in composition.
3. It should have a high equivalent weight in order that weighing errors may have a relatively small effect.
4. It must be a substance that can be accurately analyzed.
5. Desired reactions should occur according to single, well-defined, rapid, and essentially complete processes.

6. The purity of a primary standard must be assured through well-defined qualitative tests of known sensitivity, or through preparation by a method that has been demonstrated to yield consistently a pure product, and by storage under conditions in which the product is entirely stable.

A number of primary inorganic standards are available from the National Bureau of Standards (NBS). Some of these, and others, such as the ACS Reagent Grade chemicals, may be obtained from chemical manufacturers for use as primary standards. In addition, as of 1971, the NBS had available the following as "standard reference materials" (SRM) at a purity of 99.4% or better: glucose, cholesterol, urea, uric acid, creatinine, and $CaCO_3$. Their bilirubin (SRM 916) is provisionally certified at 99.0% purity.

Primary volumetric standard

A primary volumetric standard is usually prepared by dissolving and diluting an accurately weighed primary standard to volume in a calibrated volumetric flask. Both mass and volume, which are covered by the NBS and ACS specifications, are measurable with a high degree of precision.

Secondary volumetric standard

A reagent is dissolved and diluted to approximately the desired concentration. The concentration of the solution may be ascertained by titrating the solution (a) against a solution containing a known weight of a primary standard substance, or (b) against a measured volume of a primary standard solution. Alternatively, the solution to be standardized may be analyzed for the reagent substance by any analytical technic sufficiently accurate for this purpose.

Clinical primary standard

The classical characterization of a primary standard may not apply in the reference frame of the clinical chemistry laboratory. Many of the constituents found in the body fluids are not easily purified, and many are not stable. A clinical primary standard is being defined as a body fluid chemical substance that can be prepared within purity limits that would be acceptable for measurement of the quantity of the molecular entity by weight. The term "pure" must be used in an operational sense. It can be stated that a system of molecules is a pure compound if an exhaustive series of fractionations fails to produce fractions with different properties. What one calls a pure compound thus changes as new methods become available for separating material into fractions, or for measuring the properties of the fractions more accurately.

Clinical primary standard solution

A clinical primary standard solution is one in which an accurately known weight of a clinical primary standard substance is dissolved and diluted to an accurately known volume of solution. The solvent or solution for dissolving and diluting the standard must be defined.

Clinical secondary standard solution

This is a solution that is prepared by dissolving and diluting a weighed amount of a chemical substance to a known volume with a defined solvent or solution. If the weighed material is not a primary standard, the concentration of the chemical substance is then determined by chemical analysis. The secondary clinical chemistry standard may be defined as a body fluid chemical substance that cannot be prepared within purity limits that would be acceptable for the measurement of the quantity of the molecular entity by weight. When this secondary standard is used for the measurement of the desired constituent it becomes necessary to analyze a solution of it in terms of a primary standard.

Reference sample

A "reference sample" or "control material" is a sample in which the chemical composition and the physical characteristics simulate the specimen analyzed. Thus, a reference sample can be a serum or reference pool, kept in the original state, perhaps frozen or freeze-dried, or can be a synthetic mixture. Bovine serum (12) can be used for many determinations and has the advantages of low price and absence of risk of transmitting hepatitis, but it cannot be used for all analyses, e.g., protein by biuret, and probably should not be used for any of the serum enzymes. For preparation of reference sera with low concentrations of components, many components can be easily removed by dialysis (69). Various commercial normal and abnormal level reference samples are now available and they are supplied either unassayed or with assigned values arrived at usually by a consensus of a number of reference laboratories. Commercial reference samples with abnormally elevated levels of enzymes are usually prepared by adding enzymes of animal origin. As has been demonstrated by Bowers et al (14), the behavior of such animal enzymes can be entirely different than those occurring in human serum. Hence, reliance on a commercial serum enzyme control material as an enzyme "standard" is quite hazardous.

The between-bottle variation of commercial reference sample preparations should be and can be less than the analytic error of the methods used (53). We have seen unpublished data, however, that indicates that this may not always be the case. Such terms as "standard in serum," "internal standard control," "a primary serum standard," etc., are found in the literature concerning these commercial preparations. It has even been stated or implied by some commercial producers that reference samples can be used as standards. Although it is undoubtedly correct that standardization of a test with a reference sample is better than use of precalibrated tests without any standardization by the user, it is certain that reference samples should not be used for this purpose when a relatively pure form of the substance in question is available and a reasonably stable solution can be prepared (25,

46). A special problem still exists with enzyme determinations (48) since there are no stable pure reference standards available.

The National Committee for Clinical Laboratory Standards has under consideration a "Proposed Standard" (56) for the preparation of reference samples (originally proposed by an Ad Hoc Advisory Committee of the Center for Disease Control, Atlanta, and submitted in an earlier version to 50 clinical chemists and manufacturers of such materials before publication). Such materials are "designed for use in an analytical system or program to afford a means both for estimating precision and for detecting systematic analytical deviations that may arise due to reagent or instrumental defects. They find use also as samples for proficiency testing and interlaboratory surveys. If constituent levels are established with sufficient accuracy and precision, they may be used as calibration reference materials. Three classes of material were defined:

Calibration reference materials

Suitable for use as a calibration reference, and statistical confidence in the assigned value shall be such that the over-all uncertainty interval (assigned value ±2 standard errors) does not exceed 8% of the 95% normal range for the constituent. For values above the normal range, the acceptable uncertainty interval shall be increased in proportion to the ratio of the assigned value to the midpoint of the normal range. For values below the normal range, the permissible length of the uncertainty interval shall be the same as for values in the normal range.

Control materials with assigned values

For use as controls. Confidence limits for assigned value should be such that (a) the over-all uncertainty interval does not exceed 20% of the 95% normal range, or (b) the over-all uncertainty interval is not greater than the s of single measurements made on the same day in the same laboratory, using the best methods in common use. The manufacturer may select either alternative, but must designate which he has selected.

Control materials without assigned values

RUNNING OF STANDARDS

Whenever a laboratory sets up a method of determination it is inconceivable that there would ever be any justification for not standardizing the test wherever possible. As previously stated, the use of "precalibrated" test procedures is to be severely condemned. By relying on precalibrations, errors can occur as a result of making or buying new reagents, deterioration or contamination of old reagents, use of incorrect light filters, changes in the characteristics of the photometer, etc. Standardizing each test when it is first set up is not enough since the same errors can and do develop subsequent to standardization. Similar possibilities exist with all types of analytical methods other than photometric. Assuming the necessity of running standards, there are now several questions which must be considered.

(a) *What should be run and how often?* Let us state at the outset that as yet we cannot in every case give unequivocal answers to the question "what should be run?" The choice is principally between a *pure standard*, i.e., primary standard or clinical primary standard and a *reference sample* such as pooled serum or urine. Although it is true that ultimately any standard other than a pure standard must first be standardized against a pure standard, there is always the danger with a pure standard that substances present in the unknown that may interfere with or augment the measurement (color, titration, etc.) are not present in the pure standard. Furthermore, pure standards frequently are not run through the entire procedure. With some exceptions, there is no reason why they cannot be run through the whole procedure and, where possible, it is usually advisable. The danger of omitting steps is obvious. For example, treatment of aqueous standards as if they were protein-free filtrates of blood bypasses the reagents employed in preparing the protein-free filtrates of the unknowns. Any contamination of these reagents leading to alteration of results in the unknowns would not be impressed on the results of the standard. The advantage of the reference sample is that it is a sample similar to the unknown in most respects and must be run through the whole procedure. Assayed or unassayed commercial reference sample material can be purchased, or the laboratory can take a pooled serum or urine, determine many constituents with great care, split the pool up into small aliquots, freeze them, and then use them as desired. Many constituents of serum are stable for at least six months when stored at $-10°C$ (11). The component levels can in many instances be increased or decreased (38, 69). Urine pools can be prepared that are even stable for many components at room temperature (75). On the other hand, the stability of urine pools even when frozen can be rather shortlived and quite unpredictable and therefore, quite bothersome. In the previous section it was stated that the use of commercial reference samples for initial standardization of a test should be avoided, at least at the present time, because of lack of adequate confidence in the assigned values. A third type of standard, the *internal standard*, wherein a known amount of pure standard is added to the unknown finds little use in routine analysis. Its primary uses are in checking whether substances may be present that interfere with the analysis and in determining the percent recovery of methods in which steps such as extraction, chromatography, etc., are involved.

Since a reference sample is similar to the unknown, it is our opinion that for the purpose of quality control a reference sample should be run with every set of determinations when feasible. This applies even if there is only one unknown in the run. Running a reference sample or a pure standard less frequently, say once or twice a week, certainly is far better than not running them at all, but a definite risk is taken because several days may elapse before serious error is detected. Whether or not a pure standard should also be run along with a reference sample with each set of determinations is a debatable matter. If not with each set, they certainly should be run at intervals. Whitby *et al* (78) have claimed that the *same* sample cannot be used for both standardization and quality control. We are not certain we agree with this since once a procedure is standardized by a pure standard, the control chart for a reference sample should indicate the state of control of the test and at the same time, under certain circumstances that will be discussed later in greater detail in the section "Choice of Standard Value for Calculations," the reference sample value can and should be used for calculations. Some clinical chemists prefer running a pure standard in each run and running a reference sample only occasionally. Our experience indicates that this is in general less preferable. Thus, in decreasing order of preference: (a) run both pure standard and reference sample each run, (b) run reference sample each run and pure standard less frequently, and (c) run pure standard each run and reference sample less frequently. Admittedly, this conclusion is a guess devoid of proof. It is hoped that this important area will be subjected to statistical research in the future.

(b) *Is there need for standards or reference samples at more than one level?* If we are dealing with a photometric analysis and it does not obey Beer's law, a sufficient spread of standards must be run to characterize the standard curve over the range of unknowns in the run. On the other hand, if Beer's law is obeyed it is questionable that more than a reagent blank (a zero concentration standard), and one standard is required. It is probably best that the standard or reference sample be in the middle of the normal range except in those few instances where this happens to be in the very low area of absorbance where photometric error and analysis error is high (e.g., bilirubin determination). Some clinical chemists have taken the position that standards and reference samples should each be run at three levels—abnormal low level, normal range, and abnormal high level—but it seems unlikely that this is warranted with most procedures.

(c) *Where should standards be placed in a run?* It has always been a widespread practice in manual chemical analyses to place the reagent blank first, the standard or standards next, followed by the unknowns. Such a protocol does not detect a drift if it occurs, and it can occur for various reasons, e.g., variation in time of heating in such a series or variation in time elapsed before photometric reading. There are two approaches to this problem: (a) Randomize the positions that the blank and standard are placed in runs (4). Over a period of time, such difficulties may be detected if one is alert for them, but this approach is not specifically designed to detect a drift. It does insure, however, that the interrun precision estimate from a quality control chart of the standard will not be biased low. (b) Space blanks and standards at intervals throughout the run. For example, if there are 10 unknowns, place a blank and standard at the beginning, middle, and end. This approach is more costly than (a), especially for manual chemistries, but it has already become the preferred way with automated procedures. The simplest variation of this approach is presented later in the quality control section, "Duplicate Reference Sample Charts," wherein one standard is run at the beginning and one at the end.

REAGENT BLANKS

Running of reagent blanks on the materials used in a determination is nearly as important as the standard. In most types of determinations there is a significant blank value, and there are two purposes in determining its magnitude. First, measurements of the unknown and standard must be corrected for it. Just as for unknowns and standards, there is error connected with measurement of the blank. Frequently in photometric methods the analyst is advised to set the photometer at 100% transmission on the reagent blank, thus achieving an automatic correction for the blank in the readings of the unknown and standard (cf. Chap. 1). Much of the error contributed by variation of the blank can be eliminated by reading the standards, unknowns, and blanks against distilled water (or the solvent used), which is a fixed reference of absorbance, and then correcting the readings for the blank, using the average of many blanks run over a period of time as the "best estimate" of what the blank should be. This approach is feasible only when the absorbance of the reagent blank against water or solvent is fairly low. Consideration of how frequently blanks should be run leads us to the second purpose of blanks, which is to police the reagents for contamination. As for standards, the ideal frequency would be with every set of determinations, and the same considerations apply here as for standards. Reagent contamination is not self-evident when standards are read against reagent blanks.

QUALITY CONTROL

Whitby *et al* (78) have outlined the following stages in a program for "quality control":

(a) The appreciation and assessment of the sources of error inherent in the collection, transport, and reception of specimens for analysis and any initial processing steps, such as separation of plasma.

(b) The validation of new procedures introduced into the analytical repertoire of the laboratory with respect to accuracy, repeatability, and sensitivity by appropriate comparison and recovery experiments.

(c) The establishment of the reproducibility for the analytical methods in regular use, and the definition of acceptable limits of variation for control samples.

(d) The calculation of means and standard deviations and the preparation of quality control charts for displaying the constancy or variability of control data.

(e) The continuous independent monitoring of analytical performance by means of appropriate fictitious patient controls, carry-over samples, etc.

(f) When the program has been established, its value will be lost completely if the need for action indicated by its results is ignored.

This section deals principally with (d) as it applies to chemical analyses, but it should not be overlooked that a complete quality control program of a laboratory must be concerned with other things. For example, equipment such as clocks, water baths, timers, and thermometers (13) can perform other than represented, and this can result in significant errors.

QUALITY CONTROL CHARTS

The simplest way to keep track of standards, reference samples, and reagent blanks as they are run is merely to record them in a notebook reserved for the purpose. As expected, the values will vary from run to run, and if one value is suddenly greatly out of line suspicion will be aroused. There is a simple way, however, to treat these check data as they accumulate: by use of the *quality control chart* (13, 36). As we shall see, use of such a chart tells the analyst at a glance how much the checks can vary before suspicion is warranted and action indicated; in fact, trouble may be predicted before it actually occurs.

There are many ways in which control charts can be set up, but only a few will be presented. They are among the simplest but are quite adequate for control of chemical analyses.

Single standard and single reference sample control charts

It is very desirable from a management standpoint to put such matters as quality control programs into the form of clearly written directions for ready reference by all laboratory personnel. Such documents can become official "laboratory policies". The following might be typical directions for construction and use of control charts for tests for which the decision has been made to run one blank, one standard, and one reference sample in each run.

ABC LABORATORY

Laboratory policy for quality control program with single blank, standard and reference sample.

I. FORM

All control charts must be prepared using Dietzgen 341-M graph paper or other graph paper approved by the Laboratory Director. All charts are to be identified in the upper right hand corner with the name of the test and filed in a looseleaf notebook alphabetically. Each control chart is to have an "Out-of-Limits Log Sheet" (Exhibit A attached) facing it in the notebook so that both documents can be inspected at a glance. Only current charts and logs should be kept in the notebook. Old and outdated charts should be removed and kept on file.

The chart is divided horizontally into three sections, the bottom one for blank, the middle for standard, and the top for reference sample. The units on the vertical axis in each case are chosen according to the data to be entered. For photometric methods this scale for blank and standard is usually in terms of absorbance and the scale for the reference sample in terms of the final result, e.g., mg/100 ml. Ordinarily, before the control program is begun some idea is already available as to the values for these three so the required extent of these scales can be estimated fairly well. The abscissas represent successive batches with dates noted.

II. PREPARATION

At this point the control chart should look something like Figure 12-2 without any data entered or the horizontal solid or dashed control lines. The next step cannot be taken until at least 20 points for each have been plotted, so it is advisable to run standards frequently at least until this number is reached (indicated on Fig. 12-2 by a vertical line). The identification of the serum pool or urine should be clearly indicated on the control chart (as in Fig. 12-2).

The arithmetic average or mean, $\bar{x}$, is then calculated for the standard and reference sample and solid horizontal lines drawn on the graph at these points and extended to the right where future points will be plotted. The $\bar{x}$ for the reference sample represents the best estimate of its true value. Values for s for the standard and reference sample are now calculated. The decision as to the acceptability of the magnitude of these s values is to be made by the Supervisor (it is assumed that the precision of the test was adjudged adequate when the test was originally set up and evaluated; see section on "Control of Laboratory Error"). Values for s can be calculated in one of two ways:

(1) Use the standard formula (easiest form for use with calculators).

$$\sqrt{\left(\frac{\Sigma(x^2) - (\Sigma x)^2/N}{N-1}\right)}$$

(2) Treat successive determinations as pairs, i.e., x_1 and x_2 as a pair, x_2 and x_3 as a pair, etc. The absolute values of differences between pairs, R, and their average, $\bar{R}$, are calculated. Limits of $2s$ and $3s$ are then calculated by $1.77\bar{R}$ and $2.65\bar{R}$, respectively. This method usually yields a close approximation to s calculated by the standard formula.

Limits are not placed on the chart for the blank.

The next step is to establish control limits for the plotted points. There is a choice between $2s$ and $3s$ as "action limits" and this choice must be made by the Laboratory Director. If the $3s$ limits are chosen as the "action limits," lines drawn in red are placed 2*s* and 3*s* on each side of the *x*'s for standard and reference sample. These are "warning limits" and "action limits," respectively.

III. USE

A. Whenever a specimen or "batch" of specimens is analyzed, a blank, standard, and reference sample are included in the run. A "batch" or "run" is defined as the maximum number of samples that can be handled with assurance that no variable has changed that can reasonably be expected to change the result.

B. The reference sample result is calculated in the usual way from the blank and standard obtained in the run and then this result and the readings obtained on the blank and standard entered on the control chart in line with the date indicated. This is to be done immediately upon completion of the run and before results of unknowns are calculated.

C. If the blank reading appears to be in the right range and the points for the standard and reference sample within the "action limits," proceed with the calculations of the unknowns.

D. If the blank appears unusually large or if *either* the standard or reference sample value falls outside the "action" limits, the samples are not to be calculated, and this is to be brought immediately to the Supervisor. Similarly, if two consecutive values fall outside the "warning" limits but within the "action" limits, the Supervisor must be notified before proceeding. The Supervisor determines the action to be taken, e.g., report out, rerun, etc., and records his action on the "Out-of-Limits Log Sheet" (Exhibit A). In addition, the cause, if known, along with corrective action taken is recorded.

E. Whenever a new reference sample is introduced into the control chart, it must be indicated by designation and date on the chart. Introduction of a new standard or reagent must likewise be noted. In instances where the number of such introductions of reagents is large, which would result in the quality control chart being unreadable, the following is an acceptable alternate procedure for designation of introduction of new reagents (standards and reference samples must *always* be designated on the control chart itself):

   (1) A change in reagent is indicated on the chart by an asterisk.
   (2) Complete information on the reagent change must be entered on the "New Reagents Log Sheet," Exhibit B, attached.

Table 12-4 shows the calculations for the mean and for the limits of the standard on Figure 12-2 by both proposed methods. The mean and limits for the reference sample are calculated in an identical manner.

A number of comments are in order concerning the instructions given for construction and use of this type of control chart and Figure 12-2:

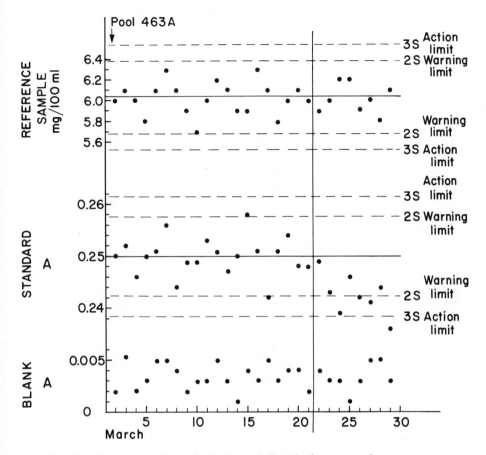

FIG. 12-2. Control chart for single blank, standard, and reference sample.

FIG. 12-3. Illustration of Gaussian distribution of control chart points around the Y axis.

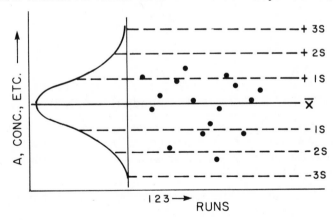

TABLE 12-4. Calculation of Mean and Control Limits of the Single Standard Control Chart Data of Figure 12-2

x	x^2	R
0.250	0.062500	—
0.252	0.063504	0.002
0.246	0.060516	0.006
0.250	0.062500	0.004
0.251	0.063001	0.001
0.256	0.065536	0.005
0.244	0.059536	0.002
0.249	0.062001	0.005
0.249	0.062001	0.000
0.253	0.064009	0.004
0.251	0.063001	0.002
0.247	0.061009	0.004
0.250	0.062500	0.003
0.258	0.066564	0.008
0.251	0.063001	0.007
0.242	0.058564	0.009
0.251	0.063001	0.009
0.254	0.064516	0.003
0.248	0.061504	0.006
0.248	0.061504	0.000

Method 1: $N = 20$

$$\bar{x} = \Sigma x / N = 5.000/20 = 0.250$$

$$s = \sqrt{\frac{\Sigma(x^2) - (\Sigma x)^2/N}{N - 1}} = \sqrt{\frac{1.250268 - 25/20}{20 - 1}} = 0.0038$$

Method 2: $\bar{R} = \Sigma R/N = 0.09/20 = 0.0045$
"2s" limits: $1.77\,\bar{R} = 0.00797$ (round off to 0.008)
"3s" limits: $2.65\,\bar{R} = 0.119$ (round off to 0.012)

    a.   In a graph of the usual frequency curve, dispersion of observations is plotted on the x axis and the frequency is plotted on the y axis. In a control chart the dispersion is turned around sideways, i.e., on the y axis, as indicated in Figure 12-3.

    b.   The absorbance values for the standard which are plotted can be corrected for the blank or not, whichever is preferred.

    c.   It is not necessary to have both "warning limits" (2s) and "action limits" ($3s$) on the chart. Either $2s$ or $3s$ limits alone can be used, in which case the limits must be "action limits", whichever is used. (See also later section, "Setting Control Chart Limits.")

    d.   The lines for the $\bar{x}$'s can be omitted. A cluttered control chart is difficult to use, and the more lines, limits, etc., the more it is cluttered.

    e.   The control charts in the notebook can be placed within what is called a "Loose Leaf Protector," which is a clear plastic overlay on which the limits can be drawn in red wax pencil. The advantage of this is that the red lines can be wiped off easily when new lines are desired.

    f.   There must be periodic review of control charts to decide whether there should be recalculation of $\bar{x}$'s and limits.

    g.   There is no easy way to calculate an upper limit for the blank. Either no line is used or the Supervisor can, after visual inspection, arbitrarily determine a horizontal "action limit" that is placed on the chart. As usual, if the blank exceeds this limit, the Supervisor must be notified for his decision and action. For example, in Figure 12-2, such a line might be placed at an absorbance value of 0.006.

What is the course of action to be taken when a value falls outside the limits? First of all, it must be remembered that approximately one value in every twenty should fall outside the $2s$ limits, but not far outside. But when two consecutive values fall between the $2s$ and $3s$ limits, or when one value is outside the $3s$ limits, the procedure should be checked carefully for an assignable cause, e.g., new reagent used, new standard, possible break in technic, etc. If none of these is obvious and the possible error involved, if it is an error, is not critical, one may wait to see what happens the next time. If the same thing happens next time, then there is cause for alarm since there is now a good indication that there is trouble. Actually, many times trouble can be predicted before it becomes significant. In Figure 12-2 it is noted that, after 21 March, there is a gradual trend downward in the points, although not until 29 March is there a value falling outside the lower action limit. Obviously, trouble began about 21 March, but not until 29 March is the test out of control if one is guided solely by whether or not a point is outside the "action" limit. This type of drift occurs when a reagent or the standard gradually deteriorates. Obviously, all bench analysts and super-visorial personnel must be carefully instructed to look continually for *more* than just points out of limits. Figure 12-4 shows two examples of charts where the tests clearly are *not* in control and yet action limits are not exceeded.

In Figure 12-2 it is noted that although the standard appears to begin to slip out of control about 21 March or soon after, there is no evidence of the reference sample going out of control even when the standard goes outside the $3s$ limits. This can occur when a reagent begins to deteriorate and pro-duces less chromogen. As long as the amount of color produced in the stan-dard and reference sample decreases proportionately, the results obtained

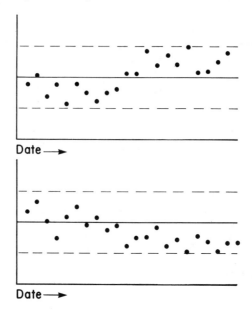

FIG. 12-4. Examples of tests "out of control" although action limits not exceeded.

on the reference sample, and presumably unknowns, remains valid. In such a case, although the control chart in Figure 12-2 signifies that it is time to make up a new reagent, the results obtained on unknowns may be perfectly valid and reportable.

Amador (2) studied this type of control chart with limits of $2s$, utilizing routine data as well as synthetic data and found that errors had to be about 1.5 times the C.V. before the control chart revealed the error. Furthermore, the sensitivity to errors was found to be a function of the magnitude of the C.V. When the method is precise, the control chart is more sensitive.

Youden has suggested a system of quality control based on analyzing mixtures of equal parts of patients' samples (mix equal parts of A and B, value for the mix then becomes A+B/2). This has been found to yield results about the same as a system using the usual reference sample pools and has been adjudged as not a satisfactory substitute (7).

Duplicate reference sample charts

More information can be obtained by running reference samples in duplicate. As before, calculations are made after about 20 pairs are accumulated. The mean, $\bar{x}$, of each pair and the absolute value of the difference between each pair (the range, $\bar{R}$) are plotted as shown in Figure 12-5. $2s$ and $3s$ limits for the $\bar{x}$ are $\pm 1.26\bar{R}$ and $\pm 1.88\bar{R}$, respectively, where $\bar{R}$ is the average of the R values. The upper $2s$ and $3s$ control limits for range R are $2.51\bar{R}$ and $3.27\bar{R}$, respectively. The lower limit is 0 in either case.

The main advantage of this method is that it is possible to tell whether the variation between runs is significantly greater than the variation in any one run. Such a situation would be indicated when the range values are apparently in control (within the control limits) but the averages of the duplicates are out of control. With many chemical procedures there is a definite tendency for this to be so, signifying the presence of insufficiently controlled variables, such as temperature.

Whitby *et al* (78) have recommended that duplicate controls be run "whenever possible." Whether the added labor and cost in running duplicates is worthwhile, however, depends on the individual situation. One can go even further with this type of control chart and with the aid of statistical analysis

FIG. 12-5. Duplicate reference sample control chart.

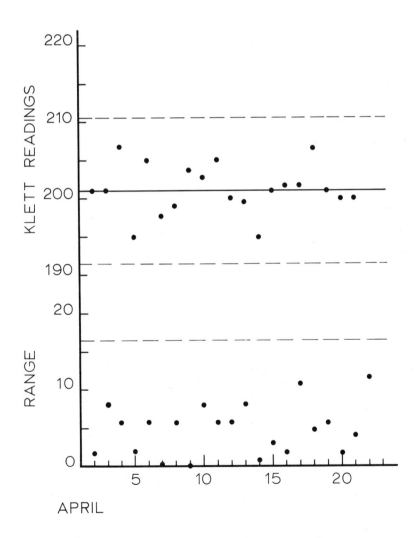

of variance sort out other contributions to the overall variability such as the effect of position in the run (4, 33, 62).

For detection of drift within a run, one control may be placed at the beginning and one at the end of the run. A third control chart, for differences, may then be constructed by computing the difference, including sign, of values obtained at the beginning and end of each run. The control chart is drawn by applying the same procedure to the differences as for single standard control charts. Since the process is in control when differences have a mean of zero the central point is taken as zero rather than the computed mean difference. An example is given in Table 12-5. Each line represents one run. x_1 is the control standard at the beginning and x_2 the value at the end of the run. The differences $x_1 - x_2$ are plotted on the control chart (Fig. 12-6).

In Figure 12-6 the first 20 plotted points represent values used to establish control lines (limits were calculated by using method 2 presented in the preceding section, "Single Standard and Single Reference Sample Control Charts.") Subsequent points show a rising trend indicating that a control

FIG. 12-6. Control chart for within-run drift.

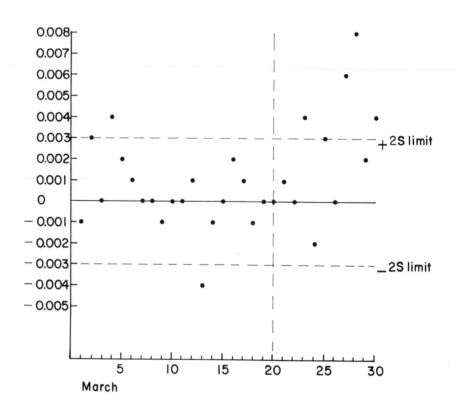

standard placed at the end of the run tends to be lower than the corresponding standard at the beginning. The calculations for method 2 are

$$\overline{R} = 0.033$$

$$\overline{R} = \Sigma R/\text{no. of } R \text{ values} = 0.0165$$

$$2s \text{ limits: } \pm 1.77\,\overline{R} = \pm 0.00292 \text{ (rounded to } \pm 0.003)$$

$$3s \text{ limits: } \pm 2.65\,\overline{R} = \pm 0.00437 \text{ (rounded to } \pm 0.004)$$

TABLE 12-5. Data for Within-Run Control Chart

x_1	x_2	$x_1 - x_2$	R
0.250	0.251	−0.001	—
0.252	0.249	0.003	0.004
0.246	0.246	0.0	0.003
0.250	0.246	0.004	0.004
0.251	0.249	0.002	0.002
0.256	0.255	0.001	0.001
0.244	0.244	0.0	0.001
0.249	0.249	0.0	0.0
0.249	0.250	−0.001	0.001
0.253	0.253	0.0	0.001
0.251	0.251	0.0	0.0
0.247	0.246	0.001	0.001
0.250	0.254	−0.004	0.005
0.258	0.259	−0.001	0.003
0.251	0.251	0.0	0.001
0.242	0.240	0.002	0.002
0.251	0.250	0.001	0.001
0.254	0.255	−0.001	0.002
0.248	0.248	0.0	0.001
0.248	0.248	0.0	0.0
0.250	0.249	0.001	
0.246	0.246	0.0	
0.251	0.247	0.004	
0.244	0.246	−0.002	
0.249	0.246	0.003	
0.251	0.251	0.0	
0.250	0.244	0.006	
0.251	0.243	0.008	
0.251	0.249	0.002	
0.248	0.244	0.004	

Cumulative sum method

Cumulative sum quality control charts, also called "cusum" charts, have the advantages over the charts previously discussed in that they more vividly display drift and they indicate a reliable estimate of the current $\overline{x}$ (21, 35,

78, 79). As of yet they have not been as widely used as the other types of charts but they may well become more popular.

A cusum chart is constructed as follows: Calculate the $\bar{x}$ of the last 10 or more standards or reference samples; this is the k value. Subtract k from each succeeding quality control point and accumulate the differences between each successive value and k. The successive accumulated differences are the cusums, and these are plotted. Thus, if s_1, s_2, etc. are points to be plotted and x_1, x_2, etc., are successive x's, then

$$s_1 = x_1 - k$$

$$s_2 = x_1 - k + x_2 - k$$

$$= s_1 + x_2 - k$$

$$s_n = s_{n-1} + x_n - k$$

$$= \sum_{i=1}^{n} x_i - nk$$

A horizontal line at 0 is drawn on the chart, but no control limits are drawn on this type of chart. It is possible to calculate control limits, but their derivation is complicated.

Using the data for standards from Figure 12-2 and Table 12-4, the k value obtained for the first 10 standards is 0.250. The points to be plotted for the next 18 standards are calculated as in Table 12-6.

TABLE 12-6. Cusum Calculations

x_n		s_n
0.251	$s_1 = x_1 - k$	+0.001
0.247	$s_2 = s_1 + (x_2 - k)$	−0.002
0.250	$s_3 = s_2 + (x_3 - k)$	−0.002
0.258	etc.	+0.006
0.251		+0.007
0.242		−0.001
0.251		0.000
0.254		+0.004
0.248		+0.002
0.248		0.000
0.249		−0.001
0.243		−0.008
0.239		−0.019
0.246		−0.023
0.242		−0.031
0.241		−0.040
0.244		−0.046
0.236		−0.060

The calculations are actually very simple—the difference between the last observation and the k value is added algebraically to the last point on the chart. Thus if the last point on the chart is +0.007 and the new value obtained is 0.242 and the k value is 0.250, −0.008 is added to +0.007 and −0.001 is obtained and plotted.

Values for s from Table 12-6 are plotted in Figure 12-7. It is seen in Figure 12-7 that the trouble that began shortly after 21 March and that can be seen in Figure 12-2 is harder to miss in Figure 12-7.

If k is an accurate estimate of the value obtained for the standard or reference sample and the test remains in good control after the cusum chart is started, the plotted points wander around the line at 0 more or less randomly, and difficulty is signalled by the points veering off upwards or downwards. If k is not an accurate estimate, the dotted points will wander around a line which is not horizontal. If the slope of this line is too steep it may make it more difficult to recognize a change in slope and the cusums are likely to go off the chart limits inconveniently often. Figure 12-8 shows the same data plotted with the k value taken as 0.260 and although there is a slope change after 21 March it is not nearly as detectable as the break in line on Figure 12-7.

Table 12-7 shows the average number of control values needed to detect an out-of-control condition such as a shift in mean value of magnitude 0.5σ, σ, or 1.5σ. The top entry of each pair refers to the usual control chart with horizontal control lines. The second entry refers to the cusum chart with appropriate cusum control lines. For this table "detection" occurs as soon as the first plotted point is outside control lines. The first entry under 1.5σ has

FIG. 12-7. Cusum control chart of data from Table 12-6.

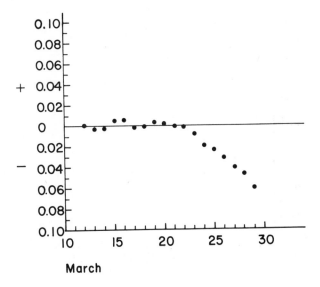

March

Table 12-7. Average Number of Points on Quality Control Chart before Shift of a Specific Magnitude is Detected.[a]

Type of chart	Average number of points			Significance level
	Amount of shift in mean			
	0.5 σ	1.0 σ	1.5 σ	
Ordinary	13.9	5.9	3.1	0.05
Cusum	29.5	7.4	3.3	
Ordinary	52.7	17.4	7.1	0.01
Cusum	42.4	10.6	4.7	
Ordinary	161.0	44.0	15.0	0.00270
Cusum	52.9	13.2	5.7	

[a]From JOHNSON NL, LEONE FC: *Statistics and Experimental Design,* New York, Wiley, 1964, Vol 1, p 342.

FIG. 12-8. Cusum control chart of same data as in Fig. 12-7 except with *k* value of 0.260.

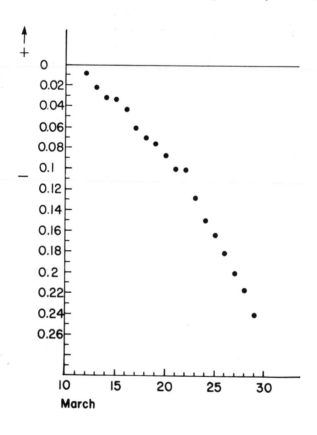

the following meaning. If a shift in mean in the amount 1.5σ occurs without any change in precision, then with the ordinary control chart, on the average slightly more than three consecutive values (i.e., 3.1 values) would fall inside the $2s$ lines before the first value occurs outside. The entry immediately below shows that the average number of points required by the cusum control chart is 3.3. It may be seen that the cusum control chart will detect an out-of-limits condition more readily than the standard chart when strict control limits, $3s$ limits, are used. On the other hand, the ordinary control chart is more sensitive for $2s$ limits.

We have experimented with cusum charts in our laboratory and have concluded that it is unlikely they will be used very much unless computerization or a statistician is available. This is because the control chart limits, which of course are needed, are significantly more difficult to calculate than the limits of the usual control charts.

Use of patient data for quality control

Waid and Hoffman in 1955 (76) proposed the use of patient data for quality control. In this first method they proposed they plotted the daily averages of all analyses performed that day of a given test. This method was replaced by these authors in 1963 (39) by their "number plus" method and two years later (38, 40) by their "average of normals" method. In this latter technic, at the end of each day the average was taken of all test values falling in the normal range, and these averages were plotted on a quality control chart. Although some (22, 45) have reported that this technic applied to some test procedures is equally or more powerful than the conventional sample method, considerable research by many other investigators (3, 30, 43, 52, 59, 61, 74) has shown that these technics are not as sensitive in signalling test difficulty. Reed (59) has described a mathematical model of the average-of-normals method. For the model eleven parameters are involved when considering the situation of three overlapping populations, the clinically normal population, the diseased population with low results, and the diseased population with high results. Three of the parameters are the relative proportions of these three groups occurring within the adopted normal range. Success of this method, at least as originally proposed, is dependent on stability of these proportions, an assumption that may not always hold. Reed proposed two modifications: Plotted points should consist of the same number of test values, and computation of control lines should be made without reference to the normal range. He concluded that, although this resulted in an increase in sensitivity of the method, more statistical research and further modifications are needed before the average-of-normals method will be useful for quality control in the clinical laboratory. Begtrup *et al* (9) and Lewis and Dixon (45) have made good starts in this direction. It is our opinion that such research is warranted since such an approach is quite attractive, especially when a laboratory has automatic data handling capabilities, and even if it cannot completely supplant quality control by the reference sample it may prove to be a useful adjunct. One of the main attractions of this approach is that it is not subject to analyst bias.

Setting control chart limits

There are always two dangers when using control charts: (a) looking for trouble that does not exist, and (b) not looking for trouble that does exist. When limits are made wider, danger (a) decreases and danger (b) increases. When limits are made narrower, the reverse occurs. Obviously, there must be a compromise. Historically, most statistical decisions are made on the "5%" decision point, i.e., if we have 19 chances out of 20 of being right or only one chance out of 20 being wrong we feel fairly comfortable—these odds are regarded as pretty good. So it is with quality control charts and control charts limits of $\pm 2s$ are limits within which 19 times out of 20 the standard value or the reference sample should fall, provided, of course, that no factors are operating other than those during the period in which the data were collected from which the $\bar{x}$ and s were calculated. For $2s$ limits to represent the 95% limits the error distribution must be Gaussian and as previously mentioned (in the section "The 'Error' of a Test"), this is often not so. If the deviation from Gaussian distribution is significant, more than one out of twenty values may fall outside the $2s$ limits. It is for this reason that $2s$ limits are often adopted as "warning limits" and $3s$ limits as "action limits." This is perhaps satisfactory if the precision of the test is good and comfortably greater than that required clinically. With many test procedures this is not the case, and therefore, $2s$ limits should be adopted as the "action limits," recognizing that we may "look for trouble that does not exist." This is a price that must be paid for the patient's benefit.

Choice of Standard Value for Calculations

If there is little or no correlation between the errors of standards and those of unknowns the control chart mean absorbance, $\bar{x}$, should be used in calculating unknowns. On the other hand if there is substantial correlation the control standard value x, obtained in the same run with the unknown, should be used so that small variations due to time and temperature are accounted for. In addition a control pool can also be used as a standard (i.e., it has been assigned a "true value" by some means which has been adjudged acceptable). Thus there are four possibilities: The choice between x and $\bar{x}$ applies both to a pure standard and to a control pool used as a standard. The problem of which absorbance to use in any given situation has been examined (60). Let us first consider the case where we wish to choose the more precise of two estimators: the one in which the standard used for calculation is the control pool average absorbance, and the other in which the control pool absorbance obtained in the current run is used as the standard for calculation. For tests that satisfy Beer's law, Reed showed that, over the range where a test has constant C. V., the choice of $\bar{x}$ or x depends only on the C. V. of the test and the correlation between control pool standard and unknown.

To explain this in greater detail we must show how to obtain an estimate of this correlation. Suppose two control pools x_1 and x_2 are run in each batch. Let x_{1i} be the value of x_1 determined from the ith run and let x_{2i} be the corresponding value of x_2 determined in the ith run. If the values are correlated we would expect x_{1i} and x_{2i} to be high or low together since they were analyzed in the same run. After $n = 30$ batches have been run, 30 pairs of determinations are available, $(x_{11}, x_{21}), (x_{12}, x_{22}), (x_{1n}, x_{2n})$. The estimate of correlation between x_1 and x_2 is

$$r = \frac{\sum\limits_{i=1}^{n} (x_{1i} - \bar{x}_1)(x_{2i} - \bar{x}_2)}{\sqrt{\sum\limits_{i=1}^{n} (x_{1i} - \bar{x}_1)^2} \sqrt{\sum\limits_{i=1}^{n} (x_{2i} - \bar{x}_2)^2}}$$

Note that r can be as small as -1 or as large as $+1$. Ordinarily in the laboratory we would expect r to be close to zero if there is little correlation between determinations analyzed in the same batch. If there is strong correlation r would be close to 1.

Reed showed that if $r < 0.5$, $\bar{x}$, the average of all pool standards, should be used in calculations. If $r > 0.6$, the current pool standard should always be used, whereas if $0.5 < r < 0.6$ the choice of x or $\bar{x}$ depends on r and the C. V. of the test. In the latter case the choice should be made by referring to Figure 12-9. In Figure 12-9 locate the point on the graph corresponding to the r and C. V. of the test. If this point is to the right of the curve in Figure 12-9 calculate concentration of unknowns using $\bar{x}$. If it is to the left, x, the current control pool determination, should be used.

In the general case, where there are four possible choices of standard, the problem is more complicated. A mathematical means for choosing the most precise of the four estimators is given in Ref. 60 based on the C. V. of the test and certain correlation coefficients between absorbances. However, because of complexity it is not suitable for everyday laboratory use. Instead, we will outline an experiment, which may be carried out during routine operation, in which the optimal estimator is determined empirically.

The experiment requires that a pure standard determination and two separate control pool determinations be made in the same run for a minimum of 20 runs. During this period it is important that neither the control pools nor the pure standard should be changed. Therefore, we recommend that the procedure begin with fresh pools and pure standard. After the data are obtained, one of the pools may be treated as if it is an unknown and its concentration calculated relative to the other control pool absorbance determined in the same run and also relative to the pure standard absorbance determined in that run. Concentration may also be calculated relative to average absorbances of the control pool and pure standard.

The data below were obtained from such an experiment. Data are shown for a test of glucose by the *o*-toluidine method. The pure standard absorbance in each of nine runs is denoted A_S and shown in column 1. Columns 2 and 3 are absorbances of two control pools analyzed in the same run as the pure standard, for nine runs. (They are actually aliquots from the same pool but placed in different positions of each run with the second pool always several samples removed from the first pool.) Although data are shown only for nine runs, this is for reasons of conserving space. In implementation of the method it is advisable that a larger number of runs be made, at least 20. Columns 4 through 7 show the four ways of calculating concentration of pool 1: relative to A_S (column 4), relative to the average, $\overline{A}_S$ (column 5), relative to A_2 (column 6), and relative to the average $\overline{A}_2$ (column 7). For the latter two columns, the average concentration for pool 2, C_p = 151.1, was used. Note that the s at the bottom of column 6 is less than 1/2 the s's in columns 5 and 7 and almost 1/4 as much as the s in column 4. Thus the precision of the test shows a fourfold difference between the best and worst cases.

FIG. 12-9. Chart for determining which standard value to use in computations.

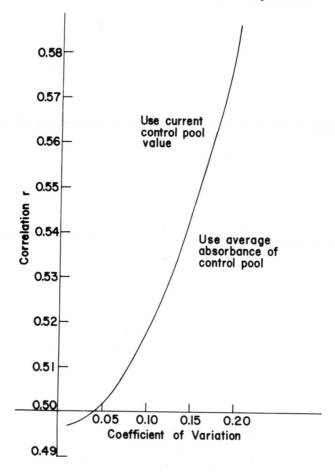

	(1)	(2)	(3)	(4)	(5)	(6)	(7)
	$A_s{}^*$	A_1	A_2	$100\dfrac{A_1}{A_s}$	$100\dfrac{A_1}{\overline{A}_s}$	$C_p\dfrac{A_1}{A_2}$	$C_p\dfrac{A_1}{\overline{A}_2}$
Run 1	0.296	0.452	0.452	152.7	152.2	151.1	152.1
2	0.294	0.449	0.452	152.7	151.2	150.1	151.1
3	0.303	0.448	0.452	147.9	150.8	149.8	150.8
4	0.292	0.453	0.459	155.1	152.5	149.1	152.5
5	0.288	0.441	0.442	153.1	148.5	150.8	148.4
6	0.304	0.432	0.435	142.1	145.5	150.1	145.4
7	0.300	0.442	0.436	147.3	148.8	153.2	148.7
8	0.299	0.460	0.461	153.9	154.9	150.8	154.8
9	0.301	0.341	0.450	149.8	151.9	151.4	151.8
Mean	0.297	0.448	0.449	150.5	150.7	150.7	150.6
s	0.00529	0.00817	0.00926	4.14	2.75	1.17	2.75

*concentration of standard was 100 mg/100 ml

This amount of reduction in method error is unusual and is due to the high correlation, $r = 0.940$, between pool 1 and pool 2 absorbances and the lack of correlation between the pure standard and pool 1 absorbance. (In this example the calculated correlation between the pure standard absorbance and the pool 1 absorbance is not significantly different from zero.) In our experience, however, a 50% improvement in method error is not uncommon.

The technic described can be applied to most photometric tests to select the type of standard that gives the most precise determination of unknowns. To do so, we recommend that a pure standard and two control pools be analyzed in the same run for a minimum of 20 runs. It is not necessary that the pools have different values. They can be aliquots from the same pool placed in different positions of the run, e.g., pool 1 might be at the beginning and pool 2 could be several samples removed (it is recommended that they not be adjacent). It is important, however, that pools or standards should not be changed during the data collection period. It is equally important that absorbance data for A_s and A_2 be free of outliers. When all runs are completed an outlier test such as the Q test should be applied. After excluding grossly deviant absorbances s's should be computed. Any value which exceeds the mean $\pm 3s$ should also be excluded. When an outlier is excluded, the remaining data for the run should not be included in the calculations.

After 20 runs have been made and the calculations have been completed, the additional pool and/or standard determinations can be discontinued. It is necessary, however, to repeat the experiment fairly often since the correlations can change, and therefore, the relative magnitudes of the four s's can change. It is thus important that they be monitored.

At this writing a tentative approach logically might be to repeat the entire procedure each time a new pool or pure standard is introduced for a minimum of three times. If the choice of standard is the same each time it probably is not necessary to repeat the procedure each time a new pool is used. Probably every third time will suffice. As more experience with the technic is accumulated alterations of this protocol are likely to be made.

Just how feasible the above approach will be in routine laboratory operation remains to be seen. We expect to employ it in at least a limited way in our laboratory over the next year. It is clear that many unanswered questions remain.

THE QUALITY CONTROL PROGRAM

What does it monitor? The "control of the quality" of a laboratory test must be concerned both with accuracy and precision. The control of accuracy can be effected by proper standardization procedures, comparison with reference methods, submission of samples with known values, and partaking in proficiency testing. Our laboratory has had a program for years in which blind known samples, some synthetic, some with internal standards, are submitted as part of our quality control program. The control of precision is effected by running standards and reference samples in one of the recommended ways previously discussed. The quality control chart does *not* control accuracy, only a *change* in accuracy.

How much quality control is enough? This is an important question because of cost. A recent survey in our laboratory showed that one out of every 5.7 samples run is a quality control sample. Is this enough or too much? We wish we knew. A good statistical study of "how much is enough" has not yet been made. The control system can be extended to cover entire ranges of tests and to each analyst individually in the laboratory (64).

Practical problems in the quality control program. We have had about 25 years of experience with quality control and have learned that the greatest single problem in an effective program is that the analysts learn to ignore the charts even though they are actually keeping them up to date. Examples such as shown in Figure 12-4 are surprisingly common. An example of this syndrome is illustrated by this story. We thought it would be a valuable addition to our quality control program to install a device on all water baths that would ring a bell and flash a red light when the temperature deviated from a narrow specified range. These devices were expensive, but the protection was adjudged worth it. The first thing that happened was many of the bells were disconnected. Next, some people learned to ignore the red light. Thus, it is a constant struggle, but it is one which must not be given up.

The use of a control chart assumes a reasonably steady state with the test over a period of time. Van Pennen *et al* (73) stated that in their experience the precision may vary as much as sevenfold. The extent of this variation has been debated (5, 72), and we feel that, although there is variation, it should not nearly reach this degree. Another interesting report (44) is that of "the 4 o'clock phenomenon." In that laboratory the precision fell sixfold as the day wore on, and the analysts either were tired or anxious to go home.

MISCELLANEOUS STATISTICAL TESTS

This section cannot possibly teach one how to use statistics. Statistics is a specialty field of mathematics just like clinical chemistry is of chemistry. For in-depth study of this specialty we can refer the reader to other sources (e.g., 16, 27, 67, 80, 81, 82). We will, however, present some statistical test procedures for which the clinical chemist may have frequent need and hope that with some precautionary notes the clinical chemist will not get into too much difficulty with self-delusion. Now as to the "precautionary notes." Everyone in the field of clinical chemistry who intends to use any statistical technic at any time is urged to read the paper by Schoolman *et al* (65). They point out that good answers come from good questions, not from esoteric statistical analysis. Statistical analysis is *not* an information generating process, it is rather an aid to data presentation and interpretation. Actually, the important function of statistical methods is in the planning of experiments. Thus, the time to consult a statistician is in the beginning of a study, not after the data are collected. They also concluded from a review of current literature that current practices have created a large and indeterminate risk to the reader if he accepts an author's conclusions based on statistical tests of significance.

There are two common misconceptions concerning statistical analysis (80). The first has to do with the 5% level of significance. There is nothing magical or fundamental about 5%, and it is only by convention that differences that occur with $P < 0.05$ (probability less than 5%) are called "statistically significant." The second misconception has to do with the interpretation of a test of significance. If a large value of P is obtained the conclusion should properly be that there is not sufficient evidence to reject the hypothesized relationship, H. It should be noted that this is quite different from concluding that H is true. In order to be able to assert with some confidence that the hypothesized relationship H is true, it is necessary to preset at suitably low levels both the significance probability and the probability of "type II error." Type II error is the error of failing to reject H when it is in fact false. It is possible to determine mathematically how many observations are needed in order that the significance level should be 5%, say, and the probability of type II error should also be suitably small. However, this is frequently not done. Consequently, it is easy for the test to fail to reject, not because the hypothesis is true, but rather because the probability of type II error is high, i.e., the number of samples is too few for the statistical analysis to have any value. In this case, the failure to reject H does not imply that H is true.

COMPARISON OF SAMPLES

Tests are needed to answer the following two types of question: (1) Is there a significant difference in results yielded by methods A and B? (2) Are the values in this set of data significantly higher than the values in the other set? For example, is the normal range for males higher than the normal range for females?

t TEST FOR COMPARISON OF TWO METHODS

The t test for paired data has been widely used for comparison of two methods utilizing patient samples. At the end of this section we shall indicate certain shortcomings of this test.

For paired observations (e.g., methods A and B run on a series of samples):

$$t = \sqrt{n}\ \frac{\bar{d}}{s_d} \qquad (\text{"degrees of freedom"} = n - 1)$$

where n = number of pairs of observation
  $\bar{d}$ = mean of the differences
  s_d = standard deviation of the differences

$$= \sqrt{\left(\frac{\Sigma\ (d)^2 - (\Sigma d)^2/n}{n - 1}\right)} \qquad \begin{array}{l}\text{(in calculation of } \Sigma d \text{ the signs}\\ \text{of } d \text{ are not ignored)}\end{array}$$

Critical values of t for several levels of significance can be found in statistical tables. If the calculated value of t exceeds the tabular value, the difference is significant at the probability chosen. The limits commonly used are 95% limits (see Table 12-2). For paired observations, the table is entered with $n - 1$ "degrees of freedom" (d.f.). If the t value obtained from the data exceeds the tabular value, the mean difference, $\bar{d}$, is significantly greater than 0, i.e., there is a significant difference.

When comparing two methods it is best that 40 or more samples be run with as equal distribution as possible in the low, medium, and high ranges.

Example: 10 patients' samples were run by methods A and B with the following results:

Method A	Method B	d	d^2
101	98	−3	9
116	119	+3	9
163	170	+7	49
86	96	+10	100
258	263	+15	25
143	139	−4	16
207	209	+2	4
322	329	+7	49
112	112	0	0
374	376	+2	4

$\Sigma d = 64$ $\Sigma d^2 = 265$

$$(\Sigma d)^2 = (29)^2 = 841$$
$$\bar{d} = 29/10 = 2.9$$

$$s_d = \sqrt{\left(\frac{265 - 841/10}{9}\right)} = \sqrt{(20.1)} = 4.5$$

$$t = (2.9/4.5)\ \sqrt{(10)}$$
$$= 2.02$$

Reference to Table 12-2 shows a value of t of 2.26 for nine degrees of freedom (Table 12-2 gives t values for P = 0.05, i.e., at the 5% level of significance). Thus, in this case d does not significantly exceed 0. Only 10 samples were compared, and the t value came reasonably close to 2.26. It would seem quite probable that if 20 or more samples were run we would achieve a significant result.

It is also very useful when comparing two methods to make an x vs y plot of the results.

The shortcoming of this test to which we previously alluded is that it gives equal weight to all differences without regard to the magnitudes of the original numbers. The $d = 7$ on the third line of the previous example, is actually a larger percentage of the original numbers, 163 or 170, than the $d = 7$ on the eighth line is of 322 or 329. The t test does not take this into account.

Barnett and Youden (8) recommend that, if inspection indicates that differences are dissimilar in magnitude at different levels, the data should be subdivided into 2 or 3 levels and separate t tests should be performed at each level. This, however, does not completely solve the problem. Subdivision of the data into smaller groups will reduce the power of the t tests (see previous section) and make it unlikely that statistical significance will be found unless the two methods are very different.

A subcommittee of the National Committee for Clinical Laboratory Standards is currently considering this problem. As of this writing the direction appears to be to use statistical regression technics for comparison of two methods rather than t tests.

TESTS FOR COMPARISONS OF NORMAL RANGES

In the first edition of this book it was recommended that the t test for unpaired observations, i.e., the t test for comparison of means, be used for comparison of normal ranges. This test is not appropriate for this purpose since it assumes Gaussian distributions of the two samples, and this is not a safe assumption in the case of normal ranges (see Chap. 13).

Until recently, the comparison of several groups of data was not very complicated. For comparison of two groups, the t test was applied while for comparing several groups the technic known as analysis of variance was used.

With continued statistical research the appropriateness of the t test has been revealed as unsatisfactory in several cases. Strictly speaking, the t test is applicable only when data are Gaussian distributed. The t test has been shown to be "robust" (i.e., not invalid) if data exhibit moderate departure from the Gaussian distribution (58). However, it is not valid for extremely non-Gaussian data. In addition, the above statements apply only to comparing groups for identical mean value. They do not apply to comparing two groups of non-Gaussian populations for identical normal range endpoints, i.e., identical 2.5 and 97.5 percentiles. In general, the t test is not appropriate for comparing extreme percentiles of two groups unless the two groups exactly follow the Gaussian distribution. Several studies have shown that distributions of chemical test results are not Gaussian distributed even for carefully selected healthy samples.

In recent years, statistical methods have been developed without regard to the form of the statistical frequency function that data obey. These methods, known as nonparametric methods, are valid whether or not data follow a Gaussian distribution. A nonparametric test for comparison of normal range endpoints is available. It is known as the Westenberg-Mood two-sample percentile test (15) applied to the 2.5 and 97.5 percentiles of the two groups. However, it is tricky to apply. Depending on the numbers of samples in each group, different mathematical approximations are needed. The aid of a statistician should be solicited if one wishes to use this test.

An alternative nonparametric test may be used. However, it not only tests for identical normal range endpoints but also tests for identity in all percentiles. This test, known as the Kolmogorov-Smirnov two-sample test, is fairly easy to apply (16). It compares the relative frequencies of data from two groups for each value of the independent variable. Of course, random sampling fluctuation is likely to produce a difference in relative frequencies even though the groups do not differ, but a very large discrepancy is not likely to be due to chance. The Kolmogorov-Smirnov test provides a criterion for determining how large this difference can be. Let M and N be the number of subjects in each group (it is necessary that $M > 40$ and $N > 40$). According to the Kolmogorov-Smirnov test, if there is no difference in the two groups the summed relative frequencies should not differ by more than

$$1.36 \sqrt{\frac{1}{m} + \frac{1}{n}} \qquad \text{(5\% significance level)}$$

or

$$1.63 \sqrt{\frac{1}{m} + \frac{1}{n}} \qquad \text{(1\% significance level)}$$

for each value of the independent variable.

An example is shown in Table 12-8 for data consisting of albumin/globulin (A/G) values from 464 normal women aged 20-49. The groups consisted of 341 women considered to have normal weight according to an actuarial criterion and 123 women considered to be overweight. After ordering the values in each group the cumulative relative frequencies are compared for

TABLE 12-8. A/G Values for 464 Normal Women Aged 20-49

A/G values	Frequency f	Cumulative frequency Σf	Cumulative relative frequency $\frac{1}{m}\Sigma f$	Frequency g	Cumulative frequency Σg	Cumulative relative frequency $\frac{1}{n}\Sigma g$	Unsigned difference of relative frequencies
0.7	1	1	0.0029	0	0	0.0	0.0029
0.8	0	1	0.0029	0	0	0.0	0.0029
0.9	2	3	0.0088	1	1	0.0081	0.0007
1.0	1	4	0.0117	1	2	0.0163	0.0046
1.1	17	21	0.0616	17	19	0.1545	0.0929
1.2	39	60	0.1760	23	42	0.3415	0.1655
1.3	54	114	0.3343	28	70	0.5691	0.2348[a]
1.4	90	204	0.5982	28	98	0.7967	0.1985
1.5	61	265	0.7771	15	113	0.9187	0.1416
1.6	45	310	0.9091	4	117	0.9512	0.0421
1.7	20	330	0.9677	3	120	0.9756	0.0079
1.8	6	336	0.9853	2	122	0.9919	0.0066
1.9	3	339	0.9941	1	123	1.0	0.0059
2.0	2	341	1.0000	0	123	1.0	0.0

[a]Largest unsigned difference.

each value of A/G. The maximum deviation occurs at the value of 1.3. This difference exceeds the 1% critical value:

$$1.63 \sqrt{\frac{1}{341} + \frac{1}{123}}$$

Thus we conclude at the 1% level of significance that the two groups were not obtained from identical populations of A/G values.

COMPARISON OF PRECISIONS

Tests are available to determine whether a significant difference in precision exists between two different methods, two analysts using the same method, etc.

F TEST

Let s_1^2 and s_2^2 be the sample variances for the two methods calculated from the data. Assume $s_2^2 > s_1^2$. Let $\sigma_1{}^2$ and $\sigma_2{}^2$ be the corresponding "population" variances.
Then

$$F = s_2^2 / s_1^2$$

is the criterion to use when testing the hypothesis that $\sigma_1{}^2 = \sigma_2{}^2$. The larger of the two sample variances is placed in the numerator except in the situation in which *a priori* knowledge dictates that one variance, $\sigma_1{}^2$, cannot be less than the other, $\sigma_2{}^2$. In this situation, $s_1{}^2$ is placed in the numerator even if it is smaller than $s_2{}^2$.

To obtain the limiting value for F for any particular level of significance, one enters a table of F, taking the "degrees of freedom" (d.f.) for the numerator and denominator as one less than the number in the respective sample. Table 12-9 lists F values for 95% probability. If the F value obtained exceeds the tabular value, the precision of the data corresponding to the numerator is significantly poorer than that corresponding to the denominator.

If the variances are calculated from duplicate analyses on different samples, the "degrees of freedom" equals the number of pairs of duplicates in each series.

TABLE 12-9. Critical Values of F, 95% Probability. This is an Abbreviated Table. Complete Tables are Given in Standard Texts on Statistics

								d. f. of numerator						
	1	2	3	4	5	6	7	8	9	10	15	20	30
1	161	200	216	225	230	234	237	239	241	242	246	248	250
2	18.5	19.0	19.2	19.2	19.3	19.3	19.4	19.4	19.4	19.4	19.4	19.4	19.5
3	10.1	9.55	9.28	9.12	9.01	8.94	8.89	8.85	8.81	8.79	8.70	8.66	8.62
4	7.71	6.94	6.59	6.39	6.26	6.16	6.09	6.04	6.00	5.96	5.86	5.80	5.75
5	6.61	5.79	5.41	5.19	5.05	4.95	4.88	4.82	4.77	4.74	4.62	4.56	4.50
6	5.99	5.14	4.76	4.53	4.39	4.28	4.21	4.15	4.10	4.06	3.94	3.87	3.81
7	5.59	4.74	4.35	4.12	3.97	3.87	3.79	3.73	3.68	3.64	3.51	3.44	3.38
8	5.32	4.46	4.07	3.84	3.69	3.58	3.50	3.44	3.39	3.35	3.22	3.15	3.08
9	5.12	4.26	3.86	3.63	3.48	3.37	3.29	3.23	3.18	3.14	3.01	2.94	2.86
10	4.96	4.10	3.71	3.48	3.33	3.22	3.14	3.07	3.02	2.98	2.85	2.77	2.70
15	4.54	3.68	3.29	3.06	2.90	2.79	2.71	2.64	2.59	2.54	2.40	2.33	2.25
20	4.35	3.49	3.10	2.87	2.71	2.60	2.51	2.45	2.39	2.35	2.20	2.12	2.04
30	4.17	3.32	2.92	2.69	2.53	2.42	2.33	2.27	2.21	2.16	2.01	1.93	1.84

d. f. of denominator

BARTLETT'S TEST (67)

This test will test for homogeneity of variances between more than two samples.

$$M = 2.3026 \left[\left(\Sigma f_i \right) \log \bar{s}^2 - \Sigma f_i \log s_i^2 \right]$$

where f_i = "degrees of freedom" in each sample = $n-1$

$$\bar{s}^2 = \Sigma f_i \, s_i^2 / \, \Sigma f_i$$

s_i = standard deviation

$$C = 1 + \frac{1}{3(a-1)} \left[\Sigma \frac{1}{f_i} - \frac{1}{\Sigma f_i} \right]$$

where a = number of s_i^2

$$\chi^2 = M/C$$

Example (simplest case, 2 samples):

Given 2 samples with 20 and 18 observations and standard deviations of 3 and 4, respectively, i.e.,

$$n_1 = 20 \qquad s_1 = 3$$
$$n_2 = 18 \qquad s_2 = 4$$

Sample no.	f_i	s_i^2	$f_i \, s_i^2$	$\log s_i^2$	$f_i \log s_i^2$	$1/f_i$
1	19	9	171	0.9542	18.1298	0.05263158
2	17	16	272	1.2041	20.4697	0.05882353

$\Sigma f_i = 36 \qquad \Sigma f_i s_i^2 = 443 \qquad \Sigma f_i \log s_i^2 = 38.5995 \qquad \Sigma 1/f_i = 0.11145511$

$$\bar{s}^2 = 443/36 = 12.30556$$

$$\left(\Sigma f_i \right) \log \bar{s}^2 = 36 \times 1.0902 = 39.2472$$

$$M = 2.3026 \, (39.2472 - 38.5995)$$

$$= 1.491394$$

$$C = 1 + \frac{1}{3(2-1)} \, (0.11145511 - 0.02777778)$$

$$= 1.027892$$

$$\chi^2 = M/C = 1.451$$

The χ^2 table (Table 12-10) is entered with "degrees of freedom" as one less than the number of samples. In this example, 2-1 = 1 degree of freedom. The χ^2 values for probabilities of 0.05 and 0.01 are found to be 3.841 and 6.635, respectively. Since the value of 1.451 is less than 3.841, the "null hypothesis" in this case is not disproved at the 5% level.

TABLE 12-10. χ^2. This is an Abbreviated Table. Complete Tables are Given in Standard Texts on Statistics.

Degrees of freedom	Probability	
	0.05	0.01
1	3.841	6.635
2	5.991	9.210
3	7.815	11.345
4	9.488	13.277
5	11.070	15.086

QUALITATIVE CHEMICAL TESTS

Qualitative tests are those that are either positive or negative, although frequently the positivity can be reported in several crude degrees, e.g., 1+, 2+, 3+. To the extent that these varying degrees can be consistently correlated with a quantitative measurement or the severity of a disease, the test can be considered semiquantitative. Since qualitative tests are usually very simple and relatively inexpensive, they can fulfill a definite and useful purpose. There are two main problems with such tests.

a. The first is that of *sensitivity*. Maximum sensitivity is not necessarily the ultimate goal. For example, a qualitative test for occult blood in feces may be too sensitive if minimal bleeding resulting from brushing one's teeth is sufficient to give a positive finding. The sensitivity should be such that the test is negative in healthy subjects and positivity is correlated to a high degree with the disease or pathology in question. If the sensitivity is too great, there will be false positive tests; if it is not sufficiently sensitive, there will be false negative tests.

b. The second problem is that of *specificity*. Poor specificity also results in false positive tests. The general problem here is actually the same as in a quantitative test giving results as a continuous variable when a critical point is chosen, above which the result is considered positive or abnormal, and below which it is considered negative or normal (or *vice versa*). It is obvious that, if this point is moved in one direction, false positives increase and false negatives decrease; conversely, when the point is moved in the opposite direction, false positives decrease and the false negatives increase. This is inescapable, and it is not necessarily best to select a point balancing these two kinds of errors. For example, in a screening test for cancer, it might be decided by diagnosticians that it is preferable to increase the number of false positives in order to obtain a decrease in the false negatives. A false positive can be eventually ruled out by further study of the patient, but a false negative may result in ignoring the patient until it is too late. This subject has been thoroughly considered by Dunn and Greenhouse (26), who have suggested that a screening test should correctly classify at least 90% of the diseased while it incorrectly classifies no more than 5% of the well. Such a decision for any particular test should be made by diagnosticians, not by the chemist.

There is the question of quality control of qualitative tests. Perhaps ideally a positive and negative control should be run each time or each day the test is run. Whether or not it is necessary to go to such extremes with qualitative tests depends on the test, stability of reagents, etc.

ACCURACY AND PRECISION OF VOLUMETRIC GLASSWARE

Standard tolerances for volumetric glassware have been established by the National Bureau of Standards (50) and have been incorporated into various Federal Specifications. Two grades are defined, grade A and grade B, the latter grade having twice the tolerances of the former. Pyrex and Kimble "class A" glassware conform to grade A tolerances and ordinary Pyrex and Kimax glassware conform to grade B tolerances. Some of the grade A tolerances are shown in Table 12-11.

TABLE 12-11. Federal Specifications for Tolerances for Class A Volumetric Glassware

Volumetric flasks (to contain)		Volumetric (transfer) pipets	
Capacity (ml)	Tolerance of total capacity (ml)[a]	Capacity (ml)	Tolerance of total capacity (ml)[b]
25	0.03	1	0.006
50	0.05	5	0.006
100	0.08	10	0.02
500	0.20		
1000	0.30		

Measuring (serologic or Mohr) pipets		Burets	
Capacity (ml)	Tolerance of total capacity (ml)[c]	Capacity (ml)	Tolerance of total capacity (ml)[d]
		5	0.01
1	0.01	10	0.02
5	0.02	50	0.05
10	0.03	100	0.10

[a] Interim Federal Specification NNN-F-289a

[b] Interim Federal Specification NNN-P-395c

[c] Interim Federal Specification NNN-P-350b

[d] Interim Federal Specification NNN-B-789a

In order to realize the greatest accuracy possible from volumetric glassware it is necessary to use extreme care and to employ the technic followed in its calibration (50). For example, with a volumetric pipet, it should be filled to about 2 cm above the zero, the setting to the zero mark should be made by slowly emptying from this point, excess fluid should then be removed from the tip, and then the contained liquid is delivered without restriction with the pipet in the vertical position; when the fluid reaches the upper end of the delivery tip, the tip of the pipet is touched to the side of the receiving vessel and contact maintained until emptying is complete, the small amount of fluid remaining in the tip not being blown out. Both pipets and burets are calibrated for unrestricted flow and are to be read without any waiting period. Delivery of partial volumes from either burets or pipets with the calibrated accuracy is thus an impossibility. This also means that the accuracy of delivery of partial volumes increases as the unrestricted delivery time increases, but many laboratory workers lack the patience for slow-delivery pipets and burets. Repair of delivery tips of pipets and burets nearly always changes the delivery rate and thus the calibration. Pipets are either calibrated "to contain" (marked TC) or "to deliver" (marked TD). Pipets calibrated for blow out of the small amount of fluid remaining in the tip after unrestricted draining have one or two etched bands about $\frac{1}{8}$ in. wide near or at the top. If the need arises, the analyst can calibrate his own volumetric glassware by weighing the water delivered in a stoppered weighing bottle. For pipets, the mean weight of five deliveries is taken and for volumetric flasks the weight of one filling is sufficient.

Following some of the above technics in the routine use of glassware in the clinical laboratory would be completely impractical. Table 12-12 gives the total errors of pipetting water from various Kimax pipets using ordinary laboratory technic, even to the holding of the pipets at about an angle of 30°-45° from vertical. These errors may appear to be sizable but they actually contribute little to the over-all error of the chemical test itself (37). It would seem that, except in certain rare cases, no great advantage would be obtained in using volumetric (transfer) pipets rather than the more convenient serologic type. The present authors, therefore, disagree with those who recommend transfer pipets for routine use in the clinical laboratory. In the use of the serologic type it should be remembered, however, that the greatest error occurs when the volume to be pipetted includes the fluid in the tip.

The temperature at which glassware is calibrated is usually etched on the glass, and the standard temperature employed in the United States is 20°C. Generally, however, this can be ignored since the difference in volume due to temperature is usually much less than the tolerance of calibration. For example, at ordinary laboratory temperatures, a 5°C change of temperature of the glassware results in a change of volume of about 0.005 to 0.015%, and a 5°C change in the temperature of water results in about a 0.1% change of its volume.

Volumetric glassware is calibrated for water; whenever a pipet or buret is used to deliver fluids having surface tension and viscosity differing from water, the calibrations will be in error. When whole blood is delivered from a

TABLE 12-12. Total Errors of Pipetting Water from Various Kimble Kimax Pipets (Manufacturer's Tolerances Twice that of National Bureau of Standards, Class A, i.e., Class B., see Ref. 37)

Pipet	Theoretical quantity pipetted (ml)	% total error of pipetting (95% confidence limits from nominal volume)
Serologic:		
1 ml	1.0	±1.5
	0.5	±3.0
	0.1	±4.6
5 ml	5.0	±1.3
	1.0	±2.8
10 ml	10.0	±0.76
	5.0	±0.77
	1.0	±2.8
Volumetric, 1 ml	1.0	±2.6

1 ml serologic pipet, the amount delivered is up to 3.5% short of 1 ml. With the Ostwald type of pipet, where the ratio of glass surface to contained volume is lower, the error is 1% or less. With serum both these figures are smaller. The question remains, however, whether for practical purposes in the clinical laboratory the difference between these two types of pipets is critical. For many analyses an additional error of 3% or 4% would not be serious, but for some analyses, such as serum calcium or sodium, it would place a strain on precision and accuracy requirements that are already difficult to meet.

In addition to the delivery error of burets there is the manipulation error, which has two components. The first is the error of reading the meniscus, which ordinarily is read at its lowest point. The reading can be facilitated by placing a shade of some dark material immediately behind and below the meniscus. For mercury, the meniscus is inverted and the top of the convex surface is employed for reading. Parallax can be avoided by using burets with lines extending completely around the buret barrel. The second component is called the "end point emergence" error and depends on the size of the drop or fractions of drops used. When filling a buret from the top, sufficient time must be allowed for drainage before setting or reading the meniscus.

The question can properly be raised as to whether or not a laboratory should check the calibration of purchased glassware. There certainly is no doubt that the responsibility lies squarely on the user, but if one purchases from a reputable company it is perhaps doubtful that continual checking of calibration is warranted. Experience in our laboratory indicates that it is not.

In conclusion it might be said that, with reasonable care, and assuming conformance to manufacturers' tolerances, glassware errors are usually unimportant in routine clinical chemical analyses. There is at least one possible source of error that might invalidate this conclusion. It has been termed the "chemical error" (37). For example, slight variations in the amount of sulfuric acid in thepreparation of a Folin-Wu protein-free filtrate might affect the final concentration of the substance under test in the filtrate in some way other than the dilution error involved.

REFERENCES

1. ALLAN JR, EARP R, FARRELL EC Jr., GRUMER HD: *Clin Chem 15*:1039, 1969
2. AMADOR E: *Am J Clin Pathol 50*:360, 1968
3. AMADOR E, HSI BP, MASSOD MF: *Am J Clin Pathol 50*:369, 1968
4. AMENTA JS: *Am J Clin Pathol 49*:842, 1968
5. AMENTA JS: *Am J Clin Pathol 51*:264, 1969
6. ANASTASSIADIS PA: *Can J Biochem Physiol 38*:1223, 1960
7. BARNETT RN, PINTO CL: *Am J Clin Pathol 48*:243, 1967
8. BARNETT RN, YOUDEN WJ: *Am J Clin Pathol 54*:454, 1970
9. BEGTRUP H, LEROY S, THYREGOD P, WALLOE-HANSEN P: *Scand J Clin Lab Invest 27*:247, 1971
10. BERRY RE: *Am J Clin Pathol 47*:337, 1967
11. BERRY RE: *Am J Clin Pathol 50*:720, 1968
12. BORNER K, FABRICIUS W, MAROWSKI B: *Z Klin Chem Klin Biochem 8*:170, 1970
13. BOUTWELL JH: Quality Control—Clinical Chemistry. Atlanta, Ga, U.S. Dept. of Health, Education and Welfare
14. BOWERS GN Jr, KELLEY ML, McCOMB RB: *Clin Chem 13*:595, 1967
15. BRADLEY JV,: Distribution-Free Statistical Tests. Englewood Cliffs, N.J., Prentice-Hall, 1968, p 203
16. CAMPBELL RC: Statistics for Biologists. Cambridge, Eng., Cambridge U.P., 1967
17. CLANCEY VJ: *Nature 159*:339, 1947
18. DEAN RB, DIXON WJ: *Anal Chem 23*:636, 1951
19. DEAN GA, HERRINGSHAW JF: *Analyst 86*:434, 1961
20. DEAN GA, HERRINGSHAW JF: *Analyst 86*:440, 1961
21. VAN DOBBEN DE BRUYN CS: Cumulative Sum Tests: Theory and Practice, No. 24, Griffins Statistical Monographs Courses. New York, Hafner, 1968
22. DIXON K, NORTHAM BE: *Clin Chim Acta 30*:453, 1970
23. DIXON WJ: *Ann Mathematical Statistics 22*:68, 1951
24. DIXON WJ, MASSEY FJ Jr: Introduction to Statistical Analysis. New York, McGraw-Hill, 3rd ed., 1969
25. DOBROW DA, AMADOR E: *Am J Clin Pathol 53*:60, 1970
26. DUNN JE, GREENHOUSE SW: Federal Security Agency, Public Health Serv Publ No. 9, 1950
27. FISHER RA: Statistical Methods for Research Workers. Edinburgh, Scotland, Oliver and Boyd, 8th ed., 1941
28. FLOKSTRA JH, VARLEY AB, HAGANS JA: *Am J Med Sci 251*:646, 1966
29. FRANCIS M, SOBEL E: *Anal Chem 42*:314, 1970
30. FRANKEL S, AHLVIN RC: *Am J Clin Pathol 48*:248, 1967
31. FREAKE R: *Am J Med Technol 35*:345, 1969
32. GAMBINO SR: The Variability of Laboratory Test Results. Talk given to Medical Directors of Pharmaceutical Companies at a meeting in Tampa, Florida, 29 March 1965

33. GOOSZEN JAH: *Clin Chim Acta 5*:431, 1960
34. GOWENLOCK AH, BROUGHTON PMG: *Z Anal Chem 243*:774, 1968
35. GRIFFIN DF: *Am J Med Technol 35*:644, 1968
36. HENRY RJ: *Clin Chem 5*:309, 1959
37. HENRY RJ, BERKMAN S, GOLUB OJ, SEGALOVE M: *Am J Clin Pathol 23*:285, 1953
38. HOFFMAN RG, WAID ME: *Am J Clin Pathol 43*:134, 1965
39. HOFFMAN RG, WAID ME: *Am J Clin Pathol 40*:263, 1963
40. HOFFMAN RG: Establishing Quality Control and Normal Ranges in the Clinical Laboratory. New York, Exposition Press, 1971
41. JOHNSON NL, LEONE FC: Statistics and Experimental Design. New York, Wiley, 1964, Vol. 1, p 342
42. KANNER O: *Am J Clin Pathol 41*:626, 1964
43. KILGARIFF M, OWEN JA: *Clin Chim Acta 19*:175, 1968
44. Letters to the Editor, The Four O'clock Phenomenon, *Lancet*, 1 Oct. 1966
45. LEWIS PW, DIXON K: *Clin Chim Acta 35*:21, 1971
46. LOGAN JE, ALLEN RH: *Clin Chem 14*:437, 1968
47. McSWINEY RR, WOODROW DA: *J Med Lab Technol 26*:340, 1969
48. MOSS DW: *Clin Chem 16*:500, 1970
49. MURPHY EA: *J Chronic Diseases 23*:1, 1970
50. Natl. Bur. Std., Circ 602
51. OWEN JA: *Proc Assoc Clin Biochem 5*:62, 1968
52. OWEN JA, CAMPBELL DG, FENWICK D, McCLEARY PF: *Clin Chim Acta 20*:327, 1968
53. PADMORE GRA, GATT JA: *Clin Chem 16*:15, 1970
54. PEARSON ES: *Biometrika 32*:301, 1942
55. PLYM AJ: *Am J Med Technol 31*:95, 1965
56. Proposed Standard: Standards for Calibration Reference and Control Materials in Clinical Chemistry, 1972. National Committee for Clinical Laboratory Standards, 2525 W. 8th St., Los Angeles, Ca. 90057
57. RADIN N: *Clin Chem 13*:55, 1967
58. RATCLIFFE JF: *Appl Stat 17*:42, 1968
59. REED AH: *Clin Chem 16*:129, 1970
60. REED AH: The Combination of Prior and Current Standard Determinations in Laboratory Assays, unpublished
61. RESSLER N, WHITLOCK LS: *Clin Chem 13*:917, 1967
62. RIDDICK JH, FLORA R, VAN METER QL: *Clin Chem 18*:250, 1972
63. RUDY JL: *Clin Chem 14*:583, 1968
64. SAX SM, DORMAN L, LIBENSON DD, MOORE JJ: *Clin Chem 13*:825, 1967
65. SCHOOLMAN HM, BECKTEL JM, BEST WR, JOHNSON AF: *J Lab Clin Med 71*:357, 1968
66. SHORE A, THOMSON LC: Guy's Hospital Reports, Vol 98, Nos. 1 and 2, 1949
67. SNEDECOR GW, COCHRAN WG: Statistical Methods. Ames, Ia, Iowa State U.P., 6th ed., 1967
68. SPARAPANI A, BERRY RE: *Am J Clin Pathol 42*:129, 1964
69. TEASDALE PR, BEAUMONT D, PARKES J: *Clin Chim Acta 30*:535, 1970
70. TONKS DB: *Clin Chem 9*:217, 1963
71. TONKS DB: *Can J Med Technol 30*:38, 1968
72. VAN PEENEN HJ: *Am J Clin Pathol 51*:265, 1969
73. VAN PEENEN HJ, TOWNSEND JF, GAMMEL G, LINDBERG DAB: *Am J Clin Pathol 49*:731, 1968

74. VAN PEENEN HJ, LINDBERG DAB: *Am J Clin Pathol 44*:322, 1965
75. WACHETER H, SALLABERGER G: *Clin Chem 16*:618, 1970
76. WAID ME, HOFFMAN RG: *Am J Clin Pathol 25*:585, 1955
77. WEINBERG MS, BARNETT RN: *Am J Clin Pathol 38*:468, 1962
78. WHITBY LG, MITCHELL FL, MOSS DW: *Advan Clin Chem 10*:65, 1967
79. WOODWARD RH, GOLDSMITH PL: Cumulative Sum Techniques. Edinburgh and London, Oliver and Boyd, 1964
80. WORCESTER J: *N Engl J Med 274*:27, 1966
81. YOUDEN WJ: Statistical Techniques for Collaborative Tests. Washington, D.C., Association of Official Analytical Chemists, 1967
82. YOUDEN WJ: Statistical Methods for Chemists. New York, Wiley, 1951

Chapter 13

Normal Values and the Use of Laboratory Results for the Detection of Disease

RICHARD J. HENRY, M.D.
ALLEN H. REED, Ph.D.

Since the purpose of laboratory results is to aid the physician in making decisions in the diagnosis and treatment of patients, there can hardly be a more important aspect of laboratory data than its interpretation. Considerable thought has gone into this subject since the first edition of this book, and we believe that nearly all the presentation there on this subject is now obsolete.

CONCEPT OF NORMALITY AND THE NORMAL RANGE

In statistics the normal range describes the prevailing condition, the usual state (32, 47). In medicine the term normal refers to a healthy person; departure from the normal range is identified as ill health. In general, healthy persons can be defined as those who have values of selected attributes, including laboratory tests, not characteristic of those defined states that seem important for the immediate purposes of the physician making the classification (47).

Murphy (34) takes a rather extreme position when he says, "Normalcy is a vestigial concept left in medicine from its unscientific era. It is properly a subject for the philosopher to explore and not one to be settled by observation and experimentation, which are the methods of science. We cannot come up with anything like an absolute definition of the normal from a scientific viewpoint." Recognizing difficulty with the term "normal," others have attempted to avoid it by recommending substitution of "standard" (Albritton's book on normal values) or "frequent" (59) for "normal," "clinical limits" for "normal limits" (14), and "cutoff points" for "normal range" (29). However, mere substitution of terms really does not change anything as long as the physician is not misled and utilizes laboratory data properly in his decision making. Cochrane and Elwood (10) felt that, because the distributions of healthy and diseased persons usually overlap, the concept of a normal range is invalid and misleading. This point is developed in greater detail in a later section (The Bayesian Approach) where it will be seen that sometime in the future there may be little or no use for the normal range as such, but that time has not yet arrived. At the present time the term "normal range" is firmly embedded in the medical field, and there is still a need for it by the practicing physician so that, rather than calling it by another name, we feel it is more fruitful to examine its nature, its shortcomings and its use.

Above everything else, including the method used in its calculation, the normal range is affected obviously by how the sample of healthy persons is chosen for study. Pryce (38) has said that the term normal means average, not healthy, optimum, or ideal. He points out (39) that if a "healthy" person is defined as one free from dental caries, alopecia, sinusitis, etc., everyone over the age of 20 is ruled out. Clearly, in picking a sample for determination of a normal range we could not accept such a limitation. Pryce believes, however, that standards should be derived from the population under study, a view apparently shared by Barnett (4). Others (16, 32) disagree and believe that the normal range that is sought is the one "ideal" for health. For example, some values change with age, and this may be due to inadvertent inclusion of abnormal values amongst the normal values. Many apparently healthy older persons are probably suffering from subclinical forms of degenerative disease. For example, the upper "normal" limits for serum urea and uric acid increase with advancing age when apparently healthy individuals are chosen for the normal sample. It is quite possible or even probable that these apparently healthy individuals who are responsible for the increase in the normal range actually have subclinical beginnings of disease. Acceptance of the increased upper normal limit thus decreases the sensitivity of case identification. Similarly, it has been shown (11) that normal ranges become more narrow for individuals whose health has been more reliably established with the passage of time and successive examinations. More thorough and longitudinal studies should be carried out to elucidate this problem.

It has also been pointed out (18) that the "healthy" hospital patient who has been lying in bed for one or more days is not the same as the "healthy" ambulatory patient.

PURPOSE AND USE OF LABORATORY TESTS

A laboratory test is run at the request of a physician to provide objective data that he hopes will aid in diagnosis and treatment. In the diagnostic area the test result should help him in his classification of the patient as healthy or diseased, and if the latter, to aid in subclassification as to what disease.

THE "NORMAL RANGE"

The earliest clinical laboratory tests were qualitative tests which were read as negative or positive, i.e., normal or abnormal. Although it was recognized that borderline results occurred occasionally and that different degrees of positivity were obtained with some test procedures, in general qualitative tests were viewed at least by most physicians as yielding either normal or abnormal results. As discussed in the last chapter (*Qualitative Chemical Tests*), however, interpretation of these tests is seldom quite that simple. Unfortunately, many physicians have come to expect a clear distinction even with quantitative tests. Most physicians have become aware that the normal range provided them by a laboratory has usually been 95% limits and that approximately 1 in 20 normal individuals will have a value outside the range. Most of these physicians, therefore, will consider a result normal if it is inside the limits, suspicious if it is somewhat outside, and abnormal if it is considerably outside. This, by and large, represents the general state of the art today of interpreting laboratory results.

The fact that, when using 95% limits, 1 out of approximately 20 results on normal individuals will be outside the limits is apt to be forgotten by the physician, and he may not be prepared, therefore, for the "abnormal" results frequently encountered when running test panels on apparently healthy individuals. The probability of obtaining one or more results outside the normal ranges in multitest profiles on healthy persons has been calculated by several workers (26, 41, 48, 51). Reed (41) has calculated that 95% of healthy persons may exceed no more than m normal ranges when undergoing n tests:

n	m
2 – 7	1
8 – 16	2
17 – 28	3
29 – 40	4
41 – 52	5

When using a normal range, the physician must know what percentage of the normal population the range is intended to include. The 95% limits have been so widely used that it is doubtful that any other have much chance of acceptance. This probably developed because of association with the 5% probability so commonly accepted in tests of significance as the point at which a difference is generally accepted as significant. The 95% limits cut off 2.5% at each end. It would have made more sense perhaps if 90% limits had been adopted so that 5% would have been cut off at each end. Using 95% limits, 1 normal individual in 40 is expected to exceed the top limit and 1 in 40 is expected to be below the lower limit.

THE BAYESIAN APPROACH

When we are dealing with quantitative data from two populations, the healthy and the diseased, there are three possibilities as shown in Figure 13-1. In A of this figure there is no overlap between the two populations so that the discriminatory or diagnostic power of the test is 100%. One could use the 95% normal limits, but obviously it would be better to select an upper limit that would be between the two populations. Unfortunately, this situation rarely, if ever, exists. In B we see the situation where the distributions for the two populations occur at essentially the same place on the x axis. In such a situation the test has 0% discriminatory power. This situation obviously is the most common of all. For example, a serum uric acid is of no help in diagnosing a coronary infarct. In C we see the most common situation with tests that have a high discriminatory power: The test has a sufficiently high discriminatory power but it is less than 100%. The two distributions overlap and the greater the overlap, the less the discriminatory power.

Let us assume distributions as in C of Figure 13-1 and a 95% normal range of 4.0–8.0. If we have a result of 8.0 on a patient, what does it mean? This value is at the 97.5 percentile but it does *not* follow that the chance of the result being normal is 1 in 40. This would be true only if the distributions did not overlap. Nor does it mean that there is a 40 to 1 chance of it being a pathological value. The probability of its being a pathological value depends not only on the value obtained but also on the number of diseased persons in the total population being sampled who have values of 8.0. To summarize, the probability that any given value is normal or abnormal depends on two things: (a) the relative proportions of the two populations and (b) the degree of their overlap.

FIG. 13-1. Three possible relationships between normal and abnormal populations. Solid line = normal population. Dashed line = abnormal population.

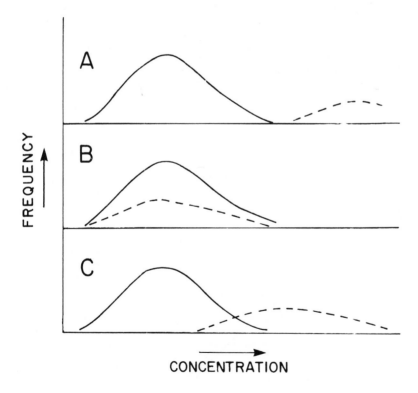

Thus, values found in normal subjects do not provide us with a probability that values in patients are, or are not, pathological values. Normal values simply enable us to estimate the percent of false positives that we would have if we adopt certain cutoff points and apply them to other subjects like those from whom we established the normal range (29).

Let us now examine the overlapping distributions in greater detail. In Figure 13-2 values to the left of vertical line 1 for practical purposes can always be considered normal, and those to the right of vertical line 2 always abnormal, but between lines 1 and 2 there is no way of telling from the value *per se* to which distribution it belongs. If we include the intermediate area with the normal range (taking line 2 as our upper limit of the normal range), a considerable number of abnormals will be judged normal. This error of missing a diagnosis is commonly referred to as a false negative result. With this limit, the test's *specificity* is maximized at the expense of *sensitivity*. If, on the other hand, we include the intermediate area with the abnormal distribution (taking line 1 as our upper limit of the normal range), a considerable number of normals will be adjudged abnormal. This error, calling a normal person diseased, is commonly referred to as a false positive result. Here *sensitivity* is maximized at the expense of *specificity*.

The proper place to put the limit in Figure 13-2, i.e., the proper balance between sensitivity and specificity, depends on the disease under consideration and requires careful evaluation of benefit in terms of morbidity or mortality, inconvenience, and distress caused to subjects by further investigation and treatment, and the costs of making the wrong decision (10, 27, 35). Murphy and Abbey (35) said that it becomes apparent that the normal range provides information that is mostly irrelevant and that there can and should be different ranges used for the same test in different

FIG. 13-2. Maximum sensitivity (vertical line 1) and maximum specificity (vertical line 2) points when distributions overlap. Solid line = normal population. Dashed line = abnormal population.

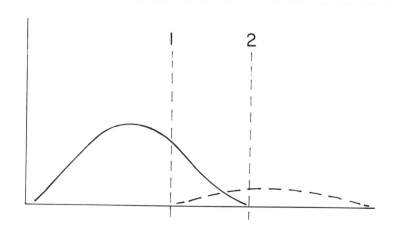

diseases. In fact, depending on the circumstances, there should be different ranges for the same test for the same disease. For example, a proper limit for the situation where 50% of the suspected sample has the disease is different than that for screening populations where only a few are diseased (27). Murphy and Abbey (35) proposed taking as the dividing line the intersection of the two distribution curves. At this point both types of error are minimized equally and not one at the expense of the other, i.e., the position is being taken that the two types of error are equally costly. The locations of the limits are shown in Figure 13-3 for the situations where 50% of the sampled population is diseased and where 5% of the sampled population is diseased. The two distributions are identical in these two cases, only their ratio being changed. It is seen that as the ratio of normal to diseased increases, the limit is shifted to higher values. Furthermore, whether or not the 97.5 percentile of the normal population and the point of intersection of the two distributions coincide is pure chance. It now becomes clear why the proper limit for use when performing multitest screenings where essentially normal people are being tested will be different than when preselection occurs—primarily only sick people go to doctors for diagnosis and treatment.

FIG. 13-3. Overlapping distributions where 50% of the sampled population is diseased (A) and where 5% of the sampled population is diseased (B). Solid lines = normal. Dashed lines = diseased. Vertical lines placed at intersections of normal and diseased distributions.

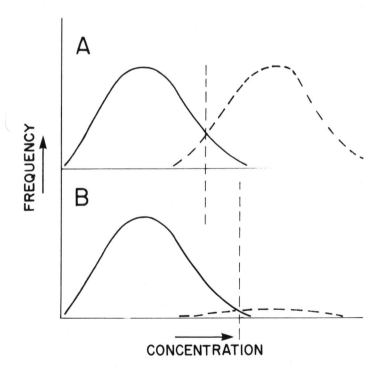

A and B of Figure 13-3 show the dividing lines for different ratios of normal to diseased populations when the two types of error possible are regarded as equally costly. If failing to identify a diseased person is regarded as threefold more costly than when a normal person is classed as diseased, the proper dividing line can be determined by amplifying the curve of the diseased population threefold (35). This is shown in Figure 13-4, where it is seen that this results in moving the intersection to a lower value. This of course is predicted since we are increasing the *sensitivity* at the expense of *specificity*.

The physician attempts, by use of the laboratory result, to place a patient in one of two categories—healthy or diseased. Bayes' theorem (23) deals with the probability that two events will occur concurrently, e.g., the probability that when a certain test is positive, a certain disease is also present, or conversely, the disease is absent when the test is negative. Vecchio (56) used the Bayesian approach to establish the predictive value of a qualitative test. This requires that the sensitivity and specificity of the test and the prevalence of the disease in the population being sampled be known.

FIG. 13-4. Overlapping distributions. The effect of regarding failure to identify a diseased person three-fold more costly than when a normal person is classed as diseased. Graph A = equally costly. Graph B = three-fold more costly. Solid lines = normal. Dashed lines = diseased. Vertical lines = dividing lines drawn at intercepts.

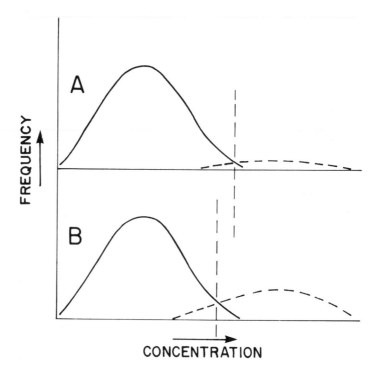

Vecchio's analysis is as follows: If *sensitivity* is the ability of a test to give a positive finding when the person tested truly has the disease under study, then

$$\text{sensitivity} = \frac{\text{number of diseased persons with a positive test}}{\text{all diseased subjects tested}} \times 100.$$

If *specificity* is the ability of a test to give a negative finding when the person tested is free of the disease under study, then

$$\text{specificity} = \frac{\text{number of nondiseased persons with a negative test}}{\text{all nondiseased subjects tested}} \times 100.$$

The total population consists of those who have the disease (p) and those who do not have the disease $(1-p)$. Then, the proportion of subjects in the total population yielding a positive test is

$$p\,a + (1-p)\,(1-b),$$

where

> a = proportion of subjects with the disease yielding positive tests (sensitivity). This must be estimated in a preliminary study.
> b = proportion of normal control subjects yielding negative tests (specificity). This must be estimated in a preliminary study.

The proportion of subjects in the total population yielding a negative test is

$$p\,(1-a) + (1-p)\,b.$$

To summarize,

Class	Total population	Yielding positive test	Yielding negative test
Diseased	p	$p\,a$	$p(1-a)$
Nondiseased	$(1-p)$	$(1-p)(1-b)$	$(1-p)\,b$
Total	$p + (1-p)$	$p\,a + (1-p)(1-b)$	$p\,(1-a) + (1-p)\,b$

The "predictive value of a positive test," i.e., the percent probability that the disease is present when the test is positive, is

$$\frac{\text{number (or proportion) of diseased persons with a positive test}}{\text{total number (or proportion) of persons with a positive test}} \times 100 = \frac{p\,a}{p\,a + (1-p)(1-b)} \times 100.$$

The "predictive value of a negative test," i.e., the percent probability that the disease is absent when the test is negative, is

$$\frac{\text{number (or proportion) of non-diseased persons with a negative test}}{\text{total number (or proportion) of persons with a negative test}} \times 100 = \frac{(1-p)\,b}{(1-p)\,b + p\,(1-a)} \times 100.$$

Vecchio (56) presented the following table for the situation where sensitivity and specificity are both 95%:

Actual disease prevalence in %	Predictive value of positive test	Predictive value of negative test
1	16.1	99.9
2	27.9	99.9
5	50	99.7
10	67.9	99.4
20	82.6	98.7
50	95	95
75	98.3	83.7
100	100	—

In addition, Vecchio gave tables of the predictive values of positive and negative tests over a range of sensitivities and specificities when the actual disease prevalence is 2%. He concluded that errors may be minimized by increasing specificity, even at the cost of decreased sensitivity, and by preselection of subjects to produce a higher disease prevalence in the population to be tested. Other tests should be performed in the positive group, when available, to attempt confirmation of the diagnosis and elimination of false-positive errors. Preselection of the population by the first test will greatly increase the predictive value of a second test.

Sunderman and Van Soestbergen (53), using Vecchio's approach slightly modified, give tables of predictive values of positive and negative tests for disease incidences ranging from 0.01% to 50% for sensitivity values of 90%, 95%, and 99% and specificity values of 90%, 95%, 99%, 99.5%, and 99.9%.

So far we have dealt with qualitative tests that simply give positive or negative results. With quantitative tests, clinical interpretation depends on the *degree* of positivity or negativity. Hall (23) developed the Bayesian approach to this situation, showing that the probability that "Y" disease is present when the result of the test is equal to any value "R" may be computed by the same reasoning that Vecchio applied to qualitative results. Utilizing probability notation,

Probability of result R in presence of Y disease $= P \ [\ R \ | \ Y \ ]$
(this reads "probability of R given Y")

Probability of result R in absence of Y disease $= P \ [\ R \ | \ \overline{Y} \ ]$
(this reads "probability of R in the absence of Y").

If

$$In \ = \text{incidence of } Y \text{ disease}$$

$$= \frac{\text{persons with } Y \text{ disease}}{\text{all persons in tested population}},$$

then,

Probability of Y disease, given result R

$$= \frac{In \times P \ [\ R \ | \ Y \ ]}{(1-In) \times P \ [\ R | \ \overline{Y} \ ] + In \times P \ [\ R \ | \ Y \ ]}.$$

This equation avoids categorization of test results as positive or negative and facilitates computer use for calculations of diagnostic probabilities of test results. Sunderman and Soestbergen (53) gave the following example of the use of this equation: Let us suppose that a fasting serum glucose of 150 mg/100 ml represents the 99th percentile of values in the healthy population and the 50th percentile of values in patients with untreated diabetes mellitus and that the approximate incidence of diabetes in the total population is 5%. For the 99th percentile of values in the healthy population $P \ [R \ | \ \overline{Y}]$ becomes 0.01 and for the 50th percentile of values in patients $P \ [R \ | \ Y]$ becomes 0.5. The incidence is 0.05. Then

Probability of diabetes for the result 150 mg/100 ml

$$= \frac{(0.05)(0.5)}{(0.95)(0.01) + (0.05)(0.5)}$$

$$= 0.725$$

$$= 72.5\%.$$

Test results in normal and diseased populations seldom conform to a Gaussian distribution (see "Determination of Normal Range") so that percentiles of results in the distributions should be determined by nonparametric means.

Certainly such an approach provides the physician precisely what he wants to know—what is the probability of a given disease actually existing in a patient on whom we have just obtained a certain value for a test. This requires knowledge of the distribution of results in the diseased as well as normal individuals in the population being tested and the incidence of the disease in the population being tested. Except in a few rather rare situations, these are facts not yet known. Perhaps sometime in the future, with the help of the computer, such an approach will become feasible. Interestingly, Mainland (30) has stated that he shares the skepticism voiced by Feinstein (15) regarding machine diagnosis with the Bayesian approach. Sunderman (52) has found the approach quite useful for those situations where he had knowledge of the distributions, except that where there was serious overlap between the healthy and diseased he found it impractical.

As stated earlier, Murphy and Abbey (35) hold that the normal range provides mostly irrelevant information and is largely ignored by the experienced clinician. Without knowledge of relative distributions, however, there is only the normal range to fall back on. Perhaps the experienced clinician intuitively shades the limits one way or the other depending on the circumstances.

MULTIVARIATE DISCRIMINANT ANALYSIS

Discriminant analysis is a statistical method that aims at allocating unclassified individuals to the correct class or population. Let us suppose that the unclassified individuals belong to one of K classes and that there are observations on samples of each of them known to be one of K classes. Furthermore, for each member, p variables are measured. A function of the variables, based on the observations, is produced so as to allocate the classified members to the correct class with minimal error. This function is called the discriminant function and is then used for classification of unclassified individuals.

This statistical approach permits the individual probabilities, obtained from different tests done on an individual patient, to be combined into a single, resultant probability, from which the most likely diagnosis can thus be determined (40, 44, 57). The mathematical complexity of this tool requires use of a computer. History and physical findings can be added to laboratory results in this type of analysis, and it is certain that some form of this approach will find increasing use in the future.

EFFECT OF PRECISION ON THE DISCRIMINATORY POWER OF A TEST

The first edition of this book contains the following statement (p. 128): "In designing the routine procedures of a laboratory one must stay within the bounds of practicability and still be assured of the accuracy and precision desired. Valuable time in the laboratory is wasted by complicating procedures to give greater precision than is actually required." Barnett (5) has likewise said that, "accuracy and precision of a degree greater than is useful clinically should not be required if extra time or expense is thereby made necessary." He employed the term "medically significant coefficient of variation" to indicate the precision desirable for diagnostic purposes. Similarly, Cotlove *et al.* (12) proposed the term "tolerable analytic variability," equal to $0.5S_B$, where S_B is the composite of personal and intergroup variability of the normal population.

Ressler and Whitlock (43) have developed the following line of reasoning that indicates that the above statements may be misleading. Referring back to Figure 13-2, the probability of a particular value in the overlap region being abnormal can easily be plotted from the ratio of the height of the dashed line at any given point to the sum of the heights at that point of the dashed and solid lines. This probability is plotted in Figure 13-5.

FIG. 13-5. Probability of test results in the overlap zone of Fig. 13-2 corresponding to an abnormal individual.

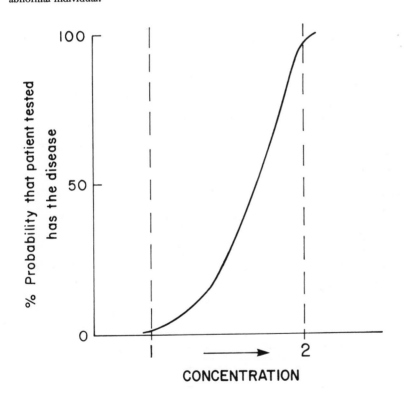

Figure 13-6 basically repeats Figure 13-5 but also illustrates how lack of precision decreases the diagnostic specificity of a test when the normal and abnormal frequency distribution curves overlap. If a concentration of b is obtained on an unknown sample, the true concentration will be within the range of $\pm 2s$, i.e., $a - c$, in 95% of cases. True concentrations of a, b, and c correspond to probabilities of 25%, 50%, and 80%, respectively, that the observed value is from a patient who has the disease. Since the imprecision reduces the discriminatory power of the test, whenever overlap is present, the smallest possible standard deviation is needed. Even if the "true" distribution curves do not overlap, if the deviations are greater than the concentration differences between the adjacent "ends" of the normal and abnormal curves, an artificial area of overlap may be created.

The imprecision of the observed result, which is always with us to a greater or lesser degree, must be added to the Bayesian approach discussed in an earlier section and becomes a complicating factor. Likewise, it complicates the use of the normal range. Obviously, the true value of an observed result close to one of the limits of a normal range may actually be on the

FIG. 13-6. Effect of imprecision on the percent probability that the patient tested has the disease. b = observed value, a = b − 2s, c = b + 2s.

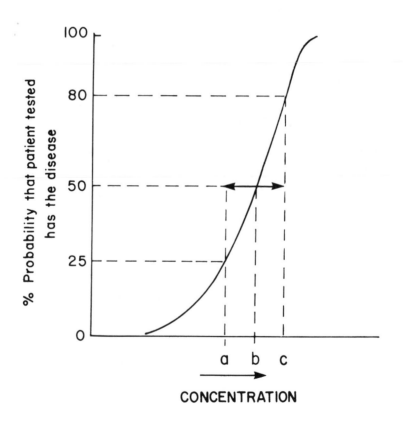

other side of the limit depending on the precision of the method. Acland and Lipton.(1) have treated this problem as follows:

Let μ_1 = true value,

μ_2 = midpoint of normal range,

σ_1 = standard deviation of the method,

σ_2 = standard deviation of the normal population,

K = $(\mu_2-\mu_1)/\sigma_2$ This is the "standardized result," i.e., the distance μ_1 from μ_2 in terms of σ_2 (e.g., K = 2 at the 95% limits)

r = σ_1/σ_2.

Table 13-1, from Acland and Lipton, gives the percentage of clinical mistakes when the 95% normal range is used.

To illustrate use of this table, take the protein-bound iodine determination. The 95% normal range is 4.0–8.0 μg/100 ml, so that μ_2 = 6 and σ_2 = 1. Assume a method standard deviation for this method, σ_1, of 0.2. If a serum sample that has a true value of 3.8 is analyzed, then

$$K = 6 - 3.8/1 = 2.2,$$

$$r = 0.2/1 = 0.2.$$

Entering the table at K = 2.2 and r = 1/5, we find that there is a 12% probability of getting a "normal result," i.e., >4.0, from this abnormal specimen. Similarly, for a true value of 4.2, K = 1.8, and we find that there is a 21% probability that an abnormal result, i.e., <4.0, will be obtained on this normal specimen. It must be noted that these calculations by Acland and Lipton assume Gaussian distributions that as we shall see later probably seldom occur. Nevertheless, the principle is valid, and the estimates should be reasonably close if the departure from Gaussian is not severe.

In a more recent paper (2) Acland and Lipton go a step further. Again using Gaussian assumptions, they derive a table that is useful for comparing diagnostic efficiencies of methods. This table gives, for all members of a normal population whose true analytic values are in normal limits, the proportions of observations that would lie outside normal limits as a result of method error.

DETERMINATION OF NORMAL RANGE

Barnett (4) holds it essential that every laboratory develop its own normal values for every test commonly used. Certainly published data (43) as well as

TABLE 13-1. Percentage of Clinical Mistakes due to Imprecision of Test When Using 95% Normal Range. (From Acland JD, Lipton S: *J Clin Pathol* 20:780, 1967). Figures are for Positive Values of $(\mu_2 - \mu_1)/\alpha_2$. For Negative Values of $(\mu_2 - \mu_1)/\alpha_2$, the Minus Sign is Disregarded.

K r	"Normal" result from abnormal specimen											"Abnormal" result from normal specimen									
	4.0	3.8	3.6	3.4	3.2	3.0	2.8	2.6	2.4	2.2	2.0	1.8	1.6	1.4	1.2	1.0	0.8	0.6	0.4	0.2	0.0
1	2	3	5	7	11	15	20	26	33	41	48	44	36	29	22	17	13	9	7	5	5
$\frac{3}{4}$	—	1	1	3	5	8	13	20	28	37	48	42	32	23	16	10	6	4	2	1	1
$\frac{1}{2}$	—	—	—	—	1	2	5	10	19	32	47	37	24	13	6	3	1	—	—	—	—
$\frac{1}{3}$	—	—	—	—	—	—	1	3	9	24	45	32	14	5	1	—	—	—	—	—	—
$\frac{1}{5}$	—	—	—	—	—	—	—	—	1	12	42	21	4	—	—	—	—	—	—	—	—
$\frac{1}{10}$	—	—	—	—	—	—	—	—	—	1	34	5	—	—	—	—	—	—	—	—	—

the results of proficiency test surveys would tend to support this viewpoint. There can be two reasons for the need of this: First, the normal range is best determined for the same population from which patients will be taken. This may be because of the effects of geography, race, etc., on values. If the laboratory receives samples from afar, however, this would seem to negate or at least complicate this reason. Second, the laboratory has a bias on the test in question from the values that should be rendered by the test. It seems that it is difficult to justify this as a reason for a laboratory to determine its own normal range since the problem could also be solved by removing the bias from the test. This seems a far safer approach since it cannot be assumed that the error of bias that may be caused by improper standardization, bad reagent, etc., will remain constant over a long period of time. Every laboratory today should be involved in interlaboratory proficiency testing so that such a bias will come to light and corrective steps taken. As we shall see later, a thorough establishment of a normal range is not an easy matter. This would not be true if derivation of a valid normal range could be made from patient data, but as shall be seen in the following section on parametric methods applied to patient results, we cannot recommend this approach. It would seem, therefore, that except in very special situations, it is impractical to expect each laboratory to establish its own normal range.

PARAMETRIC METHODS

PARAMETRIC METHODS APPLIED TO RESULTS FROM NORMAL INDIVIDUALS

In the earliest days of the clinical laboratory the normal state was often conveyed to the clinician by a single value, presumably the average normal. The next step was to give a normal range that was the total range observed in a series of clinically normal individuals. Frequently, the series was very small, sometimes only 6 or 10 people. For many years and occasionally even today, normal ranges are based on such totally inadequate data. When it became the vogue to apply statistical technics to the field of medicine, and admittedly medicine was in dire need of them, it was only natural that a simple statistical concept be applied to the normal range. This concept was the Gaussian curve, which, in statistical parlance, was often referred to as the "normal" curve. This use of the term has caused considerable confusion since it has no relation whatsoever to a "normal" person or a "normal" range (in this as well as Chap. 12 we have used the term "Gaussian" to avoid this confusion). The Gaussian curve is seen in Figure 12-1 (Chap. 12), and in a general way, it describes values found in samples of normal people, i.e., the greatest density of values is somewhere near the middle of the range and the density decreases steadily as we move away from the middle. If samples from a normal population actually conformed to this distribution, then the $\bar{x} \pm 2s$ that is derived easily from all the observed values represents the 95% limits of the normal range. For many years this was the thing to do and the $\bar{x} \pm 2s$

was applied blindly even to the point of absurdity, with frequent examples in the literature such as 160 ± 180. It was recognized that the cause of these absurdities was a skewing of the curve, and since the point was often made that biologic data frequently fall in distributions based on the geometric rather than the arithmetic mean, it was only a natural consequence that it was believed that a logarithmic transformation of normal values would result in conformance with the Gaussian distribution. This approach had two advantages: first, it was the only easy transformation available, and second, negative estimated normal limits could no longer occur. Graphical tests of goodness of fit to Gaussian or log-Gaussian distributions, such as the "normal equivalent deviate" (N.E.D.) method, were frequently used to justify this approach. It is now known, however, that such graphical methods are neither conclusive nor very sensitive.

In the first edition of this book it was pointed out that 95% of the normals could be calculated as $\bar{x} \pm 2s$ if the values for these two parameters were truly those of the population, but actually one was obtaining estimates of these from a relatively small sample. Accordingly, it was recommended that "tolerance limits" be calculated from the expression $\bar{x} \pm Ks$ and tables of K factors were provided. Reed *et al.* (42) have shown that this "K factor method" results in normal range estimates that are too wide unless the number of samples is very large (>1000). Furthermore, this method is dependent on the adequacy of the Gaussian assumption, i.e., the data or some transformation thereof must conform to a Gaussian distribution.

As Mainland (29) has pointed out, the Gaussian curve was invented by Gauss to represent the variation between different measurements of the same thing, hence the reason for another term used for this curve, the "error curve." It appears that Gauss had no intention that it be used to describe samples from a normal population. There does not appear to be any logical reason why it should. In fact, many statisticians claim that any defined population will *not* be Gaussian distributed if a sufficiently large number of samples is obtained. Pryce (39), on the other hand, believes that the Gaussian curve should apply not only to repeated measurements of the same thing but to normal values too, and that any deviation from a Gaussian distribution is due to failure to disentangle major sources of variation such as age, sex, season, etc. Whether these contentions are correct or not (and we do not believe they are), the crucial point is really whether or not the distribution we have, for which we wish to calculate a normal range, is or is not Gaussian distributed. The fact of the matter is that the evidence and the majority opinion of statisticians today are that many or most distributions of normal individuals are neither Gaussian nor log-Gaussian (3, 9, 14, 26, 29, 42, 52, 58). The experience of one of us (R.J.H.) in the application of N.E.D. plots for the goodness of fit of normal values to the Gaussian or log-Gaussian distribution, has been that over 100 instances over many years, many of the tests fit neither distribution adequately. Harris and DeMets (25) have recommended further research into procedures for transforming data to Gaussian form since nonparametric procedures require large sample sizes and are extremely vulnerable to the impact of outliers. It remains to be proved, however, whether such an approach would be feasible in estimating normal range endpoints. Since nonparametric methods are nearly as good as parametric if there is conformance with Gaussian or log-Gaussian distribution and far superior when there is not, we cannot recommend the parametric method at this time.

PARAMETRIC METHODS APPLIED TO PATIENT RESULTS

Hoffman (28) proposed a very interesting method for estimating the normal range from consecutive patient data. Assuming such accumulated data to be a mixture of individuals from both the normal and the diseased populations, the normal population being Gaussian distributed, the method involves a graphical dissection of the one from the other. More sophisticated statistical technics are available (18) for this dissection but they also assume Gaussian distribution. Although this approach has been defended by some (6, 7, 37, 38), rather convincing evidence and arguments have been presented (3, 14, 36, 42, 43) discounting its validity. Certainly it is invalid for those many instances involving non-Gaussian distributions (see previous section), which is unfortunate since this would be a far easier and cheaper way to estimate the normal range.

NONPARAMETRIC METHODS

A nonparametric estimate of the normal range makes no assumption at all as to the distribution form. The approach is valid whether the distribution is Gaussian, log-Gaussian, or neither. We and others (9, 29, 42, 52) believe this is the preferred approach for routine estimation of the normal range. Thompson's publication in 1938 (54) is the earliest statistical paper we have found on nonparametric normal range estimation but it took over two decades for the advantage of this method to be recognized.

Reed *et al* (42) studied two nonparametric methods, tolerance intervals and percentile estimates, with actual as well as artificial data; they found that such estimates are, for practical purposes, as accurate as those that assume the distribution to be Gaussian or log-Gaussian when the distribution is true. On the other hand, when the data are not distributed as assumed, nonparametric estimates are more accurate. They recommended the use of percentile estimates together with a nonparametric confidence interval for the true percentile. Furthermore, they recommended that the minimal number of samples needed to estimate the normal range with accuracy is 120, which sample is the smallest permitting 90% confidence intervals for the endpoints. Nonparametric 70% confidence intervals permit a sample size as low as 75.

A detailed treatment of the percentile estimates method with an example of its application follows:

1. Determination of 95% normal limits. Rank the test results in order of magnitude. An estimate of the 2.5 percentile of the frequency distribution of the normal population is the 2.5 percentile of the observed sample frequency distribution. The sample 2.5 percentile is the rth ordered sample from the low end value, where $r = 0.025 (n + 1)$, and n = number of observed values, i.e., the sample size. The corresponding estimate of the 97.5 percentile is the rth ordered sample value from the high end. For most values of n, r is not a whole number; one must then interpolate between the two ordered sample values whose ranks are nearest and on each side of r.

Example: Serum lactic dehydrogenase determinations were performed on 130 normal males aged 20-29. For the group, the lowest 10 values ranked in order were 47, 55, 57, 61, 61, 61, 62, 64, 65, and 66. The highest 10 ranked in order were 114, 114, 114, 118, 122, 126, 128, 130, 131, and 153. $r = 0.025 (130 + 1) = 3.28$. The 2.5 percentile is the 3.28th value, i.e., 28% of the distance from 57 to 61, which is 1.12, so that the lower normal limit is 58.12, which is rounded off to 58 since results are reported only in whole units. Similarly, the 97.5 percentile is the 3.28th value from the other end, i.e., 28% of the distance from 130 to 128, which is 0.56, so that the upper normal limit is 129.44, rounded off to 129.

2. Determination of Confidence Intervals for Each of the Normal Limits. A confidence interval is a continuous interval that covers the true value of the normal range limit with specified probability, say 0.90 (90%). Tables 13-2 and 13-3 list the ranks for 90% and 70% confidence intervals, respectively, for the 2.5 percentile. In each case, the ath and bth ranked test

TABLE 13-2. Nonparametric 90% Confidence Intervals for Normal Limits[a,b]

No. of samples, n		Rank	
From	To	a	b
120	131	1	7
132	159	1	8
160	187	1	9
188	189	1	10
190	216	2	10
217	246	2	11
247	251	2	12
252	276	3	12
277	307	3	13
308	310	3	14
311	338	4	14
339	366	4	15
367	369	5	15

[a]From REED AH, HENRY RJ, MASON WB: *Clin Chem 17*::275, 1971.

[b]ath lowest sample value = lower limit of 90% confidence interval for 2.5 percentile in target population. bth lowest sample value = upper limit of 90% confidence interval for 2.5 percentile in target population. To obtain ranks corresponding to a 90% confidence interval for the 97.5 percentile, subtract the values given for a and b from $n + 1$.

TABLE 13-3. Nonparametric 70% Confidence Intervals for Normal Limits[a,b]

No. of samples, n		Rank	
From	To	a	b
75	82	1	4
83	111	1	5
112	134	1	6
135	142	2	6
143	173	2	7
174	188	2	8
189	206	3	8

[a]From REED AH, HENRY RJ, MASON WB: *Clin Chem* 17:275, 1971.
[b]ath lowest sample value = lower limit of 70% confidence interval for 2.5 percentile in target population. bth lowest sample value = upper limit of 70% confidence interval for 2.5 percentile in target population. To obtain ranks corresponding to a 70% confidence interval for the 97.5 percentile, subtract the values given for a and b from $n + 1$.

values comprise the confidence intervals for the 2.5 percentile and for the values of n indicated. Thus, the probability is less than 0.1 or 10% that the upper limit of the confidence interval is less than the true 2.5 percentile or that the lower limit of the confidence interval is greater than the true 2.5 percentile. To obtain the ranks corresponding to the 90% or 70% confidence intervals for the 97.5 percentile, the values for a and b in the tables are subtracted from $n + 1$.

To obtain an indication of the precision of percentile estimates, consider limits of the population percentage excluded by use of the 2.5 percentile. Optimally, the 2.5 percentile will have 2.5% of the population to the left of its value. If so, the 2.5 percentile would exclude exactly 2.5% of the population from the estimated normal range. However, the percentile is obtained from a limited amount of data and it is unlikely to exclude this exact percentage. For example, if $n = 119$, the rank of the 2.5 percentile is $r = 0.025(120) = 3$. When $n = 119$, Table 13-4 indicates that $x_{(3)}$, the third smallest value, may exclude as little as 0.7% or as much as 5.3% of the population. Table 13-4 was constructed so that the above statement is true with 90% probability.

Although these limits narrow as n increases, they are still quite wide when $n = 400$. Thus one should strive to obtain as many values as possible.

As a working rule, we recommend that $n = 120$ should be taken as the smallest sample number from which to calculate normal range estimates and confidence intervals. It should be recognized that sometimes this number will be inadequate. We also recognize that sometimes it is impossible to obtain even 75 samples, the smallest number permitting calculation of 70% confidence limits. If $n = 39$, the 2.5 percentile has rank $r = 1$, and the normal limits are estimated by the lowest and highest values obtained.

TABLE 13-4. Limits of Population Percentage Excluded by Use of the 2.5 Percentile Estimate (Joint Probability Level = 0.90)[a,b]

No. samples, n	Rank of PE	Percentage of population excluded	
		Min.	Max.
39	1	0.1	7.5
119	3	0.7	5.3
199	5	1.0	4.6
279	7	1.1	4.3
399	10	1.2	4.0
∞	—	2.5	2.5

[a]From REED AH, HENRY RJ, MASON WB: *Clin Chem* 17:275, 1971.
[b]A percentile estimate (PE) of the 2.5 percentile will exclude from the normal range (values to the left of the 2.5 PE are excluded) a percentage of the population that is unlikely to be exactly 2.5%. This table shows the minimum and maximum percentages that are likely to be excluded as a function of the number of samples, n. Both limits hold simultaneously with 90% probability.

Example: Using the same enzyme data used to calculate the 2.5 and 97.5 percentile limits, we can calculate 90% confidence intervals. In Table 13-2 we see that for n of 120–131 the 90% confidence interval for the 2.5 percentile is obtained from values having ranks 1 and 7, i.e., 47–62. Similarly, the interval for the 97.5 percentile is obtained from values having ranks 131–1 and 131–7, i.e., ranks 130 and 124. The confidence interval then is 118–153.

This means that a lactic dehydrogenase value less than 47 or greater than 153 has less than 5% probability of being within the true normal range on the basis of these confidence intervals, and similarly, one greater than 62 and less than 118 has less than 5% probability of being outside the true normal range.

3. *The problem of outliers.* Unless the number of samples is extremely large, normal range estimation by nonparametric methods depends almost entirely on the one or two lowest and highest values. Thus, one or two persons who happen to be included in the sample but who are subclinically ill, and therefore do not properly belong to the population for which we wish to estimate the normal range, could have a considerable influence on the final normal range estimate. Also, there is always the danger that an extreme value may be the result of technical or clerical errors. On the other hand, it should be emphasized that nonparametric methods based on the extremes of ranked data are quite sensitive to overzealous discarding of unrepresentative cases. It is difficult to choose a middle ground.

The middle ground criterion we suggest is one modified from Dixon: Reject the largest value, $x_{(n)}$ if the ratio (r) between it and $x_{(n-1)}$ is more than one-third the range – i.e., reject $x_{(n)}$ if

$$r = \frac{x_{(n)} - x_{(n-1)}}{x_{(n)} - x_{(1)}} > \frac{1}{3}$$

This same criterion can determine if $x_{(1)}$ can reasonably be called an outlier. For this purpose the criterion is to reject $x_{(1)}$ if

$$r = \frac{x_{(2)} - x_{(1)}}{x_{(n)} - x_{(1)}} > \frac{1}{3}$$

Example: Again using the serum lactic dehydrogenase values, we test to see whether the lowest value, 47, is an outlier, i.e., not representative of the population about which inference is to be made. The second lowest value was 55 and the highest value was 153, thus

$$r = (55-47)/(153-47) = 8/106 = 0.075$$

is considerably less than $\frac{1}{3}$; therefore 47 cannot be considered an outlier by this criterion and should be retained. If the value were $> \frac{1}{3}$, the value should be discarded, and the 2.5 percentile recalculated.

A word of caution with respect to outliers: In currently available nonparametric methods extreme values play a much greater role in calculation of normal range estimates than do intermediate values. As a result nonparametric percentile estimates are vulnerable to distortion if outliers are present. In arguing that nonparametric are generally preferable to Gaussian or log-Gaussian estimates, our approach has assumed subjects were obtained from a homogeneous population and were characteristic of that population. In the more realistic case, where mixed populations are involved, where an individual's deviation from the intended population is not clear cut, but he may be in an early stage of disease, the problem of outliers clearly cannot be resolved by a mathematical formula such as the r criterion. The r criterion is, at best, a rough screening device, no substitute for careful specification of a method for obtaining subjects from the population about which normal range estimates are desired.

To guard against outliers arising from technical error, it is advisable to obtain a sufficiently large volume of each specimen so that repeat determinations can be made for the three highest and three lowest values. This assumes the constituent being measured is known to be stable for the period between analyses. The purpose of repeating the analysis is to ensure that those values that most influence normal range estimates truly reflect biologic variation and are not due to gross laboratory error. If a repeat value is substantially different from the original we would discard both values prior to calculating percentiles. We are not suggesting that values be averaged; this is statistically objectionable. Unless it is obvious that an original value is technically incorrect it should be retained unchanged.

A sufficiently large number of samples will reduce the effect of the outlier problem in normal range estimation considerably.

EFFECT OF PRECISION ON NORMAL RANGE ESTIMATES

The estimate of the normal range made by the parametric method is directly affected by the precision of measurements in the following way:

$$s_{\text{apparent normal range}} = \sqrt{s^2 \text{ method} + s^2 \text{ true normal range}}$$

Gowenlock and Broughton (20) graphed the percent error in the normal range versus coefficient of variation (CV) of the analytic method for different values of the true CV of the normal population. Barnett (5) stated that if the s of the method is $< \frac{1}{12}$ to $\frac{1}{20}$ the s of the normal range, the normal range is not materially increased. The actual increase in these two instances is 5.4% and 2%, respectively. It has even been claimed (55) that the fluctuations in precision within a laboratory make it difficult to set accurate normal ranges.

The estimate of the normal range made by nonparametric methods is affected by the precision of measurements in a similar way, but no rigorous statistical treatment of the subject has been made. Since the ratio of the variation due to the true normal range to the variation due to the method is usually of the order of 3:1 or greater, we believe that a reasonable approximation to correction of nonparametric percentile estimates for method error can be made by assuming that the percentile estimates were, in fact, determined parametrically.

Example: The 2.5 and 97.5 nonparametric percentiles for the serum lactic dehydrogenase data were previously calculated as 58 and 129. Let us assume a method error of $s = 5$. If we assumed the 58–129 distribution to have been Gaussian, the s for the normal range would be $(129 - 58)/4 = 18$. Using the same formula as for the parametric situation,

$$18 = \sqrt{5^2 + s^2 \text{ true normal range}}$$

$$s_{\text{true normal range}} = 17.3$$

The midpoint of the range 58–129 is 93.5, so the corrected normal range would be $93.5 \pm 2(17.3)$ or 59 to 128.

Although it has not been the custom to give an estimate of the normal range corrected for method error when publishing, future consideration should be given to this approach. If this is done, the laboratory adopting such a published estimate would have to add its own estimate of method precision back in. The fact that the precisions may be different in the two laboratories is, of course, the reason why this approach seems preferable.

FACTORS AFFECTING THE NORMAL RANGE

Establishment of normals is further complicated by many variables, including sex, age, race, weight, climate, geography, season, diet, time of day that the specimen was obtained, day to day variation, menstrual cycle, and whether or not the individual is in the postabsorptive state. In nearly all previous normal range estimates, the only variables sorted out have been sex and age, the latter usually being limited to designating infant, child, or adult.

Since a test's diagnostic sensitivity, its power, increases as the normal range is narrowed, it is best to have normal ranges for each subgroup differing significantly from any other subgroup. The reason is simple: As each significant variable is added into the group, the normal range increases. For example, the normal range for blood hemoglobin is higher for men than for women; thus physicians have for a long time used different normal ranges when interpreting a blood hemoglobin. Obviously, the power of this test would decrease if one range were used for both sexes. If data were available this could be extended much further. Thus, the normal ranges for blood hemoglobin in a 74-year old 200-lb. black male living in the hot climate of Mississippi (etc., etc.) may be narrower and different than for an 18-year old 110-lb. American Indian living in the colder climate of North Dakota.

INTRAINDIVIDUAL VERSUS INTERINDIVIDUAL VARIATION

When a normal range is determined from single observations on a group of normal individuals, the observed normal range includes "between-individual" and "within-individual" variation. Each individual has his own average and variability around this average. Thus, an individual has his own normal range, and it would be expected that this normal range is smaller than the normal range for all people with him included. The diagnostic power of a test would then be greatest if a result could be compared with the specific individual's own normal range. That "intraindividual" variation is indeed smaller in general than "interindividual" variation has been confirmed (12, 19, 24, 46, 47): There is less intraindividual variation with those properties essential to normal functioning of cells, e.g., osmolarity, pH, Na, K, Ca, and Cl (46). Cotlove *et al* (12) did not find a significant difference between "intra-" and "interindividual" variations for Cl, CO_2, and K. Harris *et al* (24) found that 85%–95% of within-individual observations were Gaussian distributed.

The ideal situation is to establish a data bank for the individual when healthy so that his own normal range can be established. The beginnings of this are already being seen in a small way. Complete blood counts and 12-test chemistry panels are being run on some patients at yearly intervals. If the ratios of "intra-" to "interindividual" variations are > 0.9, however, at the present state of the art such tests contribute nothing to a "normal personal profile." In some cases this is a result of a relative large method error.

The values for each individual may fluctuate in either a rhythmic or arrhythmic, oscillatory pattern (31, 49), further complicating the interpretation of laboratory tests. A single determination may be misleading or meaningless unless, of course, it is considerably outside the normal range. Mefferd and Pokorny (33) concluded that at least three determinations over a period of a few weeks are required before one can determine a trend or secure a reliable estimate of an individual's own mean.

REPORTING OF RESULTS

In reporting laboratory results to a physician, he must be given the information needed for proper data interpretation in a form that he can understand. The following are four ways of communication of data, listed in *decreasing* order of preference:

(a) Probability of a specific disease or diseases from the result or results obtained. This general approach was discussed in an earlier section ("Purpose and Use of Laboratory Tests: The Bayesian Approach") and could even be synthesized with the aid of computers along with pertinent facts from the history and physical examination of the patient, which is exactly what the physican tries to do mentally in any case. However, it is completely impossible for him either to store all the necessary data, even if it were available to him, or to perform all the necessary complex calculations. Except in a very few situations, this manner of communication to the physician is not yet available. A very large amount of data would have to be accumulated to make it feasible in any general way. Frankly, the problems involved are so complex that it is perhaps questionable whether this approach will be fruitful except in restricted areas. Note that this technic does not employ the normal range.

(b) Probability that the true result is outside the 95% limits of the normal range. Conspicuous by its absence perhaps is reporting the probability that a result is normal or abnormal. Actually, this is a form of the first way presented (a) and cannot be calculated from knowledge of the distribution of normal individuals alone. On the other hand it is possible from knowledge of the distribution of normals and the method error to give a probability that the true result of the report in question is inside or outside a 95% normal limit.

In developing this calculation we have assumed that the frequency distribution of method error is Gaussian and the distribution of patient values is uniform. In the latter case this is equivalent to assuming "no prior knowledge" for the distribution, i.e., all values are equally likely. The two frequency distributions are combined by Bayes formula in the following way:

Let X = the value obtained,

A = the lower limit of the stated normal range for the test,

B = the upper limit of the stated normal range for the test,

s = standard deviation of the test (precision),

Φ = values of the standard normal probability integral (found in many standard statistical texts, e.g., Ref 13) for the values of z where $z = (B-X)/s$ and $(A-X)/s$ in the next formula;

then,

Percent probability that the "true" result is outside the stated normal range when the value obtained is X $= 100 \left[1 - \Phi \left(\frac{B-X}{s} \right) + \Phi \left(\frac{A-X}{s} \right) \right]$.

Example: If we take the normal range for the serum protein–iodine determination as 4–8 μg/100 ml and have an observed value of 4.1 and a precision of $s = 0.25$, what is the probability that the "true" result is outside the stated normal range?

$$(B-X)/s = (8-4.1)/0.25 = 15.6.$$

Referral to a table of standard normal probability intervals reveals that for a z of 15.6 the "area" or Φ is >0.999 (areas for z values greater than 3.25 ordinarily are not listed) and therefore can be taken as 1.

$$(A-X)/s = (4-4.1)/0.25 = -0.4.$$

For a z of -0.4, Φ is 0.3446.

Probability that the "true" result is outside the stated normal range $= 100 \left[1-1 + 0.3446 \right]$

$= 34\%$

(c) Result and the standard deviation or coefficient of variation. A result by itself is totally meaningless to a physician. Unfortunately, most physicians do not even seem to be aware of this, nor do they seem to be aware that *all* tests have an unavoidable error, or if they do they appear to think it is always very small or at least should be. A major advance in the laboratory field would be to provide for the physician with each test result its possible error. A major but worthwhile educational campaign would be required to teach the physician how to use the information. The reason that (b) above was given preference to the present (c) is simply that it avoids the necessity of teaching the physician how to use the standard deviation or coefficient of variation. The approach outlined in (b) has this work done for

him. An interesting alternative to the customary method of reporting is that of giving results in terms of standard deviation units from the mean (21, 22, 45). The advantages are that the physician does not have to learn any normal ranges, and results from laboratories with different normal ranges would be directly comparable. The error of the method could be reported along with such results. This method of reporting, however, requires that the distribution of normal individuals be Gaussian or log-Gaussian, and as we have stated previously, this is the exception rather than the rule.

(d) The result by itself. This is the method, used today by nearly all laboratories, and it is not satisfactory since the physician *must* have some notion of method error before he can properly use the data.

REFERENCES

1. ACLAND JD, LIPTON S: *J Clin Pathol 20*:780, 1967
2. ACLAND JD, LIPTON S: *J Clin Pathol 24*:369, 1971
3. AMADOR E, HSI BP: *Am J Clin Pathol 52*:538, 1969
4. BARNETT RN: *Clinical Laboratory Statistics.* Boston, Little, Brown, 1971
5. BARNETT RN: *AM J Clin Pathol 50*:671, 1968
6. BECKTEL JM: *Clin Chim Acta 28*:119, 1970
7. BEST WR, MASON CC, BARRON SS, SHEPHERD HG: *Clin Chim Acta 28*:127, 1970
8. BOONSTRA CE, JACKSON CE: *Am J Clin Pathol 55*:523, 1971
9. BRUNDEN MN, CLARK JJ, SUTTER ML: *Am J Clin Pathol 53*:332, 1970
10. COCHRANE AL, ELWOOD PC: *Lancet I*:420, 1969
11. COLLEN MF, SIEGALAUB AB, CUTLER JL, GOLDBERG R: *Ann NY Acad Sci 161*:572, 1969
12. COTLOVE E, HARRIS EK, WILLIAMS GZ: *Clin Chem 16*:1028, 1970
13. DIXON WJ, MASSEY FJ Jr: *Introduction to Statistical Analysis,* 3rd ed. New York, McGraw-Hill, 1969 p.462
14. ELVEBACK LR, GUILLIER CL, KEATING FR Jr: *J Am Med Assoc 211*:69, 1970
15. FEINSTEIN AB: *Ann Intern Med 66*:789, 1967
16. FILES JB, VAN PEENEN HJ, LINDBERG DAB: *J Am Med Assoc 205*:684, 1968
17. GAMBINO SR: *J Am Med Assoc 212*:883, 1970
18. GINDLER EM: *Clin Chem 16*:124, 1970
19. GLASSY FJ, McBLUMENFELD C: *Calif Med 108*:172, 1968
20. GOWENLOCK AH, BROUGHTON PMG: *Z Analy Chem 243*:774, 1968
21. GRASBECK R, FELLMAN J: *Scand J Clin Lab Invest 21*:193, 1968
22. HALL GH: *Lancet 1*:1329, 1967
23. HALL GH: *Lancet 2*:555, 1967
24. HARRIS EK, KANOFSKY P, SHAKARJU G, COTLOVE E: *Clin Chem 16*:1022, 1970
25. HARRIS EK, DeMETS DL: *Clin Chem 18*:244, 1972
26. HEALY MJR: *Bull Acad Roy Med Belg 9*:703, 1969
27. HEMS G: *Lancet 1*:267, 1969
28. HOFFMAN RG: *Establishing Quality Control and Normal Ranges in the Clinical Laboratory.* New York, Exposition Press, 1971
29. MAINLAND D: *Clin Chem 17*:267, 1971
30. MAINLAND D: *Ann NY Acad Sci 161*:527, 1969

31. MARGOLESE S, GOLUB OJ: *J Clin Endocrin Metab 17*:849, 1957
32. McCALL MG: *J Chronic Diseases 19*:1127, 1966
33. MEFFERD RB, POKORNY AD: *Am J Clin Pathol 48*:325, 1967
34. MURPHY EA: *Perspectives Biol Med 9*:333, 1966
35. MURPHY EA, ABBEY H: *J Chronic Diseases 20*:79, 1967
36. O'HALLORAN MW, STUDLEY-RUXON J, WELLBY ML: *Clin Chim Acta 27*:35, 1970
37. PRYCE JD: *Lancet 2*:333, 1960
38. PRYCE JD: *Postgrad Med 36*:A-56, 1964
39. PRYCE JD: *J Am Med Assoc 212*:883, 1970
40. RAMSOE K, TYGSTRUP N, WINKEL P: *Scand J Clin Lab Invest 26*:307, 1970
41. REED AH: *Am J Clin Pathol 54*:774, 1970
42. REED AH, HENRY RJ, MASON WB: *Clin Chem 17*:275, 1971
43. RESSLER N, WHITLOCK LS: *Clin Chem 13*:917, 1967
44. RESSLER N, WHITLOCK LS: *Clin Chem 13*:931, 1967
45. RUSHNER RF: *J Am Med Assoc 206*:836, 1968
46. SARGENT F, WEINMAN KP: *Ann NY Acad Sci 134*:696, 1966
47. SCHNEIDER AJ: *Pediatrics 26*:973, 1960
48. SCHOENBERG BS: *Postgrad Med 47*:151, 1970
49. SOLLBERGER A: *Ann NY Acad Sci 161*:602, 1969
50. SPARAPANI A, BERRY RE: *Am J Clin Pathol 42*:133, 1964
51. SUNDERMAN FW Jr: *Am J Clin Pathol 53*:288, 1970
52. SUNDERMAN FW Jr: *Ann NY Acad Sci 161*:549, 1969
53. SUNDERMAN FW Jr, VAN SOESTBERGEN AA: *Am J Clin Pathol 55*:105, 1971
54. THOMPSON WR: *Ann Math Stat 9*:281, 1938
55. VAN PEENEN HJ, TOWNSEND JF, GAMMEL G, LINDBERG DAB: *Am J Clin Pathol 49*:731, 1968
56. VECCHIO TJ: *N Engl J Med 274*:1171, 1966
57. WEINER JM, MARMORSTON J: *Ann NY Acad Sci 161*:641, 1969
58. WERNER M, YOUNG DS, HEILBRON DC, DIXON WJ: *Clin Chem 16*:809, 1970
59. ZENDER R: *Ann Biol Clin 28*:15, 1970

Chapter 14

Specimen Collection and Preservation

CARL ALPER, Ph.D.

Modern clinical chemistry emphasizes skill in the use of instrumentation ranging from a simple photometer to gas chromatography and radioisotope technics, the application of automation such as continuous flow analysis, and the adaptation of the computer to analytic instrumentation and information retrieval. While such instrumentation has made a significant impact on the practice of clinical chemistry in terms of accuracy, precision, and analysis time, the final result is only as good as the specimen collection, the first step in all clinical chemical analyses, and the technics of specimen handling between the time of collection and the time of assay. The specimen must be representative and homogeneous to be biologically meaningful. This means that the technics of sampling, preservation, and sample transport from the bedside of the patient to the hospital laboratory or to the reference laboratory must take into account problems of patient preparation for laboratory studies, stability of metabolites being assayed in the laboratory, interferences in the assay arising from sampling and storage technics, and the interrelationship of the preceding to specimen transport. No amount of care or technical skill in the analytic procedure will yield a correct result if a source of error has been introduced by the technic of specimen collection or if a critical biochemical change occurs in the sample between the time it is collected and the time the analysis is performed.

The science of the collection of a representative biological sample requires: (a) a knowledge of the distribution and concentration of each metabolite in each of the body fluids (11, 15) as illustrated in Table 14-1; (b) a knowledge of the biological form of the metabolite being studied (i.e., free or bound); (c) a knowledge of the production, utilization, and homeostatic regulation of each metabolite to account for biological variability due to biological rhythmicity, stress, and feedback control of the blood level of the metabolite.

SPECIMEN COLLECTION

Since precautions specific to each determination are cited under the directions for the assay, only general concepts will be discussed here. Chemical cleanliness of all equipment used in specimen collection is a prime requisite. This includes syringe or evacuated blood collecting tube (e.g., Vacutainer) and the needle and the specimen container. The equipment must be washed with detergent and/or acid and then rinsed thoroughly with deionized or distilled water. Since the availability of the evacuated blood collecting tube, the problem has been substantially reduced. However,

TABLE 14-1. Concentration of Substances in Erythrocytes and Plasma[a]

Substance	Erythrocytes	Plasma
Glucose mg/100 ml	74.0	90.0
Nonprotein nitrogen mg/100 ml	40.0	8.0
Uric acid mg/100 ml	2.5	4.6
Total cholesterol mg/100 ml	139	194
Cholesterol esters mg/100 ml	0	129
Na^+ meq/liter	16	140
K^+ meq/liter	100	4.4
Cl^- meq/liter	52	104
Ca^{2+} meq/liter	0.5	5.0
HCO_3^- mmol/liter	19	26
LDH units	58000	360
Transaminase, glutamic-oxaloacetic, units	500	25

[a] Adapted from CARAWAY WT: *Am J Clin Pathol* 37:445, 1962

Pragay *et al* (59) demonstrated the presence of some calcium contamination in Vacutainer tubes and micropipets used for blood specimen collection, which was not a significant problem if the tube was properly filled with the required volume of blood. Special care is always required when assaying for substances normally present in the sample in only trace amounts.

When either serum or plasma is to be obtained from blood, hemolysis must be avoided because of the differing concentrations of some metabolites between red cell fluid and serum or plasma (11, 15). This can be minimized by using a dry syringe and slowly ejecting the blood from the syringe with a 25-gauge needle (55). It has been reported that greater hemolysis attributable to turbulence during forceful ejection occurs with wider bore needles than with smaller bore needles, contrary to popular belief and conventional practice. The conditions of this reported experiment may not have authentically represented those encountered in real life. This unexpected finding has not yet been confirmed by others. Except for selected instances the syringe and needle has been largely replaced routinely by the evacuated blood collecting tube. The tubes have been designed with a gum rubber stopper so that when used with the plastic needle holder unit, the stem of the double pointed needle punctures the vacuum tube stopper and blood is drawn directly into the collecting tube at a rate of about 1 ml/sec. The heavy walled evacuated tube, available in many sizes from 2 to 50 ml is used as the collecting and storage container. The specimen collected in this manner has less chance of loss due to contamination, spilling, clotting, or hemolysis. Hemolysis may result when a Vacutainer tube with an oxalate anticoagulant is shaken too vigorously. The rubber stopper is self-sealing, which will prevent problems due to evaporation and contamination, and is color coded to indicate the anticoagulant or special treatment for special analytic requirements. The blood specimen may be centrifuged directly in the Vacutainer tube.

Disposable blood lancets and blood collecting pipets, which may be heparinized when necessary, are routinely used by the laboratory today to obtain specimens for micro and ultramicro chemical analytic technics. Capillary blood is taken from the finger, heel, ear lobe (5), or scalp (6). Kaplan *et al* (43) showed no detectable difference in the commonly determined metabolites with the exception of glucose, which is somewhat lower in venous blood than capillary blood.

Recently, special technics have been developed to obtain repeated blood specimens from an arterial catheter for blood gas and acid-base studies (54), in critically ill patients undergoing prolonged surgical procedures, in diabetics to study the response to blood glucose regulating drugs (74), and in other studies employing pharmacodynamic agents (67). When specimens are taken from indwelling catheters, artifacts attributable to dilution from intravenous fluids and changes in the concentration of the constituent of interest from medications must be avoided (3).

PHYSIOLOGIC CONCERNS

Blood specimens of hospitalized patients are usually drawn in the morning, frequently before the patient has arisen and usually before breakfast. Thus the specimen drawn in the early morning is a fasting specimen (i.e., 10–14 hours after the last meal), and the patient has been in a recumbent position since the previous evening, though ambulatory and not confined to bed. If the patient is a new hospital admission, is an outpatient, or otherwise ambulatory, changes in many parameters may be the result of physical activity, recent eating, or simply the upright position versus recumbency. Furthermore, variations in levels of glucose and glucose tolerance (40), triglycerides (70), PBI (1), vanilmandelic acid (34), and other diagnostically important blood constituents (65) occur as diurnal variations and in circadian rhythms (20). Another factor of some consequence is the effect of stasis and the choice of anticoagulants (46) on plasma protein concentration (also enzyme activity) and on red cell volume. The randomly drawn specimen reveals the biochemical or hematologic profile of the patient at the time the specimen was drawn. To assess organ function or metabolic function, specimens may have to be obtained at specific time intervals following the ingestion of a metabolic load or after a specific physiological procedure. Time course studies are utilized to assess renal function, hepatic function, endocrinologic function, carbohydrate, lipid, and protein metabolism, and the metabolism of foreign organic (toxic) substances. If the patient is not in the fasting state one can observe significant differences on the circulating pool and consequently, concentration of glucose, lipids, uric acid, electrolytes, and inorganic phosphate (2, 81). Successful metabolic function studies require proper patient preparation (e.g., adequate carbohydrate diet for three days prior to the glucose tolerance determination).

Finally, certain medications can produce significant effects on some biochemical assays (18, 47, 64, 71, 72). If the drug is suspected of contributing to a false elevation or decline in the concentration of the metabolite under study, its administration should be stopped if possible for several days to allow for the establishment of the correct metabolite value for the patient.

Urine specimens may be either a randomly collected specimen (i.e., all or part of a random single voiding) or a timed specimen (i.e., all of the urine voided over a specified period of time. Analyses performed on a timed specimen yield excretion rates that are of value in the interpretation of metabolic processes, renal function, and renal clearance of intoxicants. It is most common to collect the urine for a 24-hour period, and this is, of course, critical for those substances that are excreted with a diurnal variation. If the substance is uniformly excreted by the kidney throughout a 24-hour period, then a much shorter time period of collection may be adequate for chemical studies. In every instance the bladder is voided and the specimen discarded at the onset of the collection period, and the final voided specimen is retained at the end of the collection period. A random urine specimen with infrequent exceptions is of value only for qualitative studies. To make certain that a proper specimen is collected the patient should be given a clean collection bottle and detailed printed instructions on how to collect a 24-hour or other timed specimen (Fig. 14-1).

From the analyst's point of view, the only thing to guard against in obtaining a sample of cerebrospinal fluid is a "bloody tap," i.e., contamination with blood caused by introduction of the needle itself, which would yield significant errors in protein and electrolyte assays and cell count.

SPECIMEN TRANSPORT

Transport of specimens to the laboratory in a hospital is always a problem. Specimens are drawn by a team of technicians or nurses who carry the specimens directly to the laboratory. During the day after morning rounds, specimens may come to the laboratory in haphazard fashion. The specimen vial or tube and the request slip must be carefully marked so that specimens are not lost. To speed up transport of a specimen to the laboratory, pneumatic tube transport systems have been developed in some large hospitals (50, 66). Some problems have been encountered with hemolysis of whole blood samples. Such a system, however, could conceivably save much transport time.

Of great concern in the hospital is the transport of specimens for blood ammonia and blood gases or other unstable substances. Such assay requests

PLEASE READ CAREFULLY

Instructions For Collection of 24-Hour Urine Sample

For proper evaluation of tests on a 24-hour urine sample it is important that a complete and accurate collection be made.

Drink less liquids during the collection period than you usually do, unless instructed otherwise by your physician. Do not drink any alcoholic beverages.

1. Empty your bladder when you get up in the morning.

   (.............A.M., ...
      (time)   (date)
   and discard this urine).

2. From then on collect in a clean bottle all urine you pass during the day and night.

3. Make your final collection when you empty your bladder the next morning

   at the same hour (........A.M., ...
            (time)   (date)

4. Keep the collected urine refrigerated, if possible, and bring it to the office or laboratory as soon as possible after the 24-hour collection is complete.

FIG. 14-1. Printed instructions on how to collect 24 hour urine.

should be directed to the laboratory by telephone, and the specimen drawn by a technician who then returns immediately to the laboratory for performance of the assay without further delay.

Another set of problems exists with the transport of specimens from a physician's office to a laboratory or from a laboratory to a reference laboratory. Knowledge of stability and preservation of specimens is required, and these will be discussed later. Availability of polyfoam containers and bus and air freight allow us to transport frozen specimens long distances in a satisfactory state. Samples can be delivered across the country or overseas within a 24-hour period.

COLLECTION OF SERUM

To obtain serum, blood is drawn into a plain collecting tube and allowed to stand until it has clotted spontaneously. This process normally requires

5-15 min at room temperature and considerably longer at refrigerator temperature. The clot that forms may adhere to the walls of the tube. "Ringing" or "rimming" by a wooden applicator stick or a thin glass rod in a single sweep around the periphery of the tube is necessary to release the clot prior to centrifuging down the clot. Ringing must be carried out with care to avoid undesired hemolysis. If the clot is then allowed to retract before centrifugation a larger volume of serum is obtained with less likelihood of hemolysis. Clot retraction occurs more quickly and to a greater extent at room temperature than at refrigerator temperature (14). The decision to obtain the maximal amount of serum by permitting clot retraction must be governed by knowledge of what changes may occur due to metabolite shifts from serum to cells and *vice versa*. Ordinarily, it is a rule of thumb that serum should be separated from the clot within 1 hour, although there are many exceptions. The stopper should not be removed from the blood collecting tube until the serum is ready to be used. This precaution will prevent evaporation (52) and contamination. After centrifugation, the serum is drawn off by a pipet with suction bulb or by mouth control through an attached rubber tube.

Most chemical analysis are carried out on serum rather than plasma or whole blood. Theoretically, the only difference between serum and plasma would be the presence of fibrinogen in the latter. Actually, however, certain other substances, e.g., lipemia "clearing factor" (26), appear to be coprecipitated or destroyed in the clotting process and LDH and aldolase are liberated from thrombocytes during clotting (19). Serum is usually optically clearer than the corresponding plasma.

During the clotting process CO_2 is lost from the serum unless collected anaerobically, e.g., in Vacutainers. The resultant effect is a shift of both water and electrolytes from erythrocytes to serum. However, as judged by the measurement of total protein content (17), the resultant dilution is no more than 1.5% and therefore negligible for routine work.

COLLECTION OF WHOLE BLOOD AND PLASMA

Many of the components of whole blood are present in unequal concentrations in the erythrocytes and in the extracellular plasma phase. In some instances, e.g., K^+ and hemoglobin, the substances are present predominantly or exclusively within the erythrocyte. Thus the determination of hemoglobin requires that the analysis be performed on whole blood, whereas the determination of serum K^+ is subject to serious error when the serum is hemolyzed. Any result obtained on whole blood in such instances obviously is dependent on the relative intracellular volume, i.e., the hematocrit. This fact must not be overlooked in the interpretation of such results, for an abnormally high hematocrit can cause an actually normal level of a substance to appear to be elevated.

Seldom is plasma called for in a chemical analysis with the exception of the determination of fibrinogen. Two advantages of plasma over serum are that there is no delay necessary in obtaining plasma by centrifugation of the whole blood and that usually a greater volume of plasma than serum is obtainable from a given volume of whole blood (57).

Delay of clotting by nonwettable surfaces

Clotting of whole blood can be delayed by collection in siliconized glassware, or in Teflon or polyethylene containers. Separation of cells from plasma can be achieved by immediate centrifugation. This technic is used principally in procedures that require that plasma be obtained without a chemical anticoagulant.

Anticoagulants

When whole blood or plasma is required, chemical anticoagulants are routinely used.

Heparin. Heparin may be considered to be the "natural" anticoagulant because it is present in blood, but in concentrations less than that required to prevent clotting in freshly drawn blood. Heparin is a sulfated mucopolysaccharide in which the repeating unit is proposed to be α-D-glucuronic acid in 1,4 linkage with α-D-glucosamine N-sulfate with an ester sulfate probably at the 2-carbon of glucuronic acid and an additional ester sulfate of the 6-carbon of glucosamine N-sulfate (75). In the presence of a protein cofactor in serum, heparin acts as an antiprothrombin and antithrombin thus interfering in the conversion of prothrombin to thrombin, the coagulation of fibrinogen by thrombin, and preserving platelets and preventing their agglutination (29).

Heparin is available as the Na, K, or Li salt. USP heparin sodium contains not less than 110 USP heparin units per milligram. Ordinarily, this anticoagulant is used in a concentration of about 0.2 mg/ml of blood, and it is usually added as a concentrated solution rather than in the dry form. The dry powder, which is hygroscopic, can be used, or the solutions can be dried in tubes at temperatures below 100°C. The higher the drying temperature, however, the more insoluble is the residue. It has been well established by many groups of workers that heparin, even in concentrations considerably higher than that required to prevent clotting (16), produces no change in erythrocyte volume. Commercial preparations appear to be contaminated with phosphate (7, 51), thus preventing their use if phosphate is to be determined. The rather high cost of heparin has prevented a more widespread adoption. Vacutainer tubes containing heparin are available in a number of sizes. Other contaminants in liquid heparin have cytotoxic effects on lymphocytes; therefore specimens collected for cytogenetic studies or lymphocyte transformation tests must be anticoagulated with crystalline heparin (76). The 7 ml tube contains 143 USP units of Li or Na heparin, which are sufficient to prevent coagulation.

Salts of Ethylenediaminetetraacetic Acid. Ethylenediaminetetraacetic acid, usually abbreviated EDTA, is an anticoagulant by virtue of removing calcium ions by chelation. It is available as the free acid or as the di-, tri-, or tetrasodium salt (EDTA-Na$_2$, etc.). A concentration of 1 mg of the disodium salt per milliliter of blood is sufficient to prevent coagulation. Concentrations even greater than this produce no detectable change in erythrocyte volume (12, 39) when stored in the refrigerator. It has been recommended (33) that the dipotassium salt be used in preference to the disodium salt because of its much greater solubility.

Vials are prepared by introduction of the proper volume of a 0.1% aqueous solution and drying at room temperature. EDTA-Na$_2$ or EDTA-K$_2$ does not interfere with determinations for glucose, urea, or creatinine (32, 33), but a prolongation of prothrombin times has been reported (82). EDTA contains nitrogen, and therefore, its use should, theoretically, yield slightly high values for nonprotein nitrogen (NPN), but the reports have been contradictory (32, 33). Vacutainer tubes containing EDTA are available in several sizes. The 7 ml tube contains 9 mg EDTA or 0.04 ml of a 15% solution.

Oxalates. Lithium, sodium, and potassium oxalates act as anticoagulants by removing calcium ions by precipitation as insoluble calcium oxalate. Potassium oxalate ($K_2C_2O_4 \cdot H_2O$) is still the most common anticoagulant; 1–2 mg/ml of blood is used. A customary procedure employs 0.01 ml of an aqueous 20% solution per milliliter of blood. This is 1 drop per 5 ml of blood. The instruction frequently given to neutralize the oxalate solution before use is unnecessary with the reagent grade of the salt. If the oxalate is to be dried before use it should be dried in a thin film on the walls of the vial to minimize hemolysis and the temperature of drying should not exceed 100°C lest the oxalate be decomposed to carbonate. Drying of the anticoagulant solution actually is unnecessary since the dilution error is only 1%, and there is less tendency for hemolysis with the solution than with the dry salt (16). The danger of hemolysis increases with increasing oxalate concentration and is almost unavoidable at concentrations greater than 3 mg/ml. Vacutainer tubes containing oxalate are available in several sizes with a variety of oxalates or combinations such as Li oxalate (12 mg), K oxalate (200 mg), K oxalate (14 mg) and NaF (17.5 mg), potassium oxalate (6 mg) and ammonium oxalate (4 mg).

The most serious and unavoidable objection to the use of oxalates as anticoagulants lies in the alteration of concentrations of plasma components (46). The hematocrit using potassium oxalate is as much as 8%–13% less than that obtained with heparin (48, 80). This shrinkage of erythrocytes results from a water shift from the erthyrocytes to the plasma caused by the addition of the salt to the plasma phase. Obviously, this shift will increase with increasing anticoagulant concentration (16), and if used in the same concentration on a weight basis, all anticoagulants will have this effect inversely proportional to their molecular weight. A decrease in hematocrit from 40 to 36 (10% decrease) results in a dilution error of plasma constituents of 5%. Measurements of certain components of plasma have shown that the use of oxalate rapidly causes a decreased concentration of 3%–10% (10), which may revert back to the original concentration in about 48 hours (10). Aside from the water shift there may also be alteration of erythrocyte permeability (7), which may explain the varied and inconsistent effects of oxalate and other salt anticoagulants on certain plasma constituents (7, 25).

The claim is frequently made that lithium oxalate is in general superior to the other oxalates. This apparently originates from the early observation that the presence of potassium ions produced turbidity in the determination of uric acid by the phosphotungstate method. The author, however, knows of no other instance in which the lithium salt is clearly preferable.

Because of the difficulty at the time in obtaining a satisfactory preparation of heparin commercially, Heller and Paul (36) introduced, in 1934, a "balanced" oxalate mixture for use in hematocrit and sedimentation rate, which contains three parts ammonium oxalate to cause swelling of the erythrocytes, balanced by two parts of potassium oxalate, which causes shrinkage, 2 mg of mixture being used per milliliter of blood. Data on the effect of this combination on the hematocrit are actually contradictory (80). Blood containing this anticoagulant combination obviously cannot be used for analysis for NPN or urea nitrogen by the urease method.

Miscellaneous Anticoagulants. Sodium fluoride is an effective anticoagulant because it removes Ca^{2+} but in view of its preservative action will be discussed in the next section. Because of the relatively high concentration required it produces significant water shifts and is also quite likely to produce hemolysis.

Sodium citrate is an anticoagulant at a concentration of 5 mg/ml blood by removal of Ca^{2+} but is seldom used in clinical chemistry because of the rather large water shift produced.

Sodium polyanetholsulfonate, marketed as Liquoid by Hoffmann-La Roche, has been used in Europe as an anticoagulant (4, 16). It is employed in concentrations of about 1–2.5 mg/ml blood, which is without effect on erythrocyte volume. It has been reported (60) to interfere with the determination of protein. It has recently become easily available in the U.S.

Vacutainer tubes are also available with each of these anticoagulants.

COLLECTION OF FECES

The primary objectives in fecal collection are avoidance of contamination with urine or blood and the obtaining of a correctly timed specimen. A special problem occurs with female patients when menstruating since blood may contaminate the fecal specimen. The can be avoided by the use of a vaginal tampon. Feces are collected in metabolic balance studies for periods of 24–72 hours. The specimen for study should be collected using two ingested markers such as carmine red or charcoal. The collection should include the specimen containing the first appearance of the first marker and all specimens up to but not including that showing appearance of the second marker. The specimen should be collected in a preweighed container so that the net weight of feces is easily ascertained.

SPECIMEN PRESERVATION

Relatively early in the history of clinical chemistry considerable investigation was made into the problem of preservation of glucose and various nonprotein nitrogenous constituents of blood. Since that time, however, there has been a surprising paucity of data on the stability and preservation of the large number of constituents of blood, urine, and cerebrospinal fluid of interest to the clinical chemist.

Preservation of a biological specimen is usually not a significant problem in the hospital clinical laboratory for the usually requested metabolites. The reporting time is usually 4–8 hours and, in a few instances, 24 hours on the commonly requested substances such as glucose, urea, Na^+, K^+, Cl^-, HCO_3^-, Ca^{2+}, PO_4^{3-}, uric acid, creatinine, bilirubin, alkaline phosphatase, LDH, GOT, GPT, and cholesterol. Those analyses, however, that are not performed daily or are sent to a reference laboratory may require the use of preservation technics to maintain a valid biological specimen. Most frequently, storage of a closed vial at a reduced temperature (35, 37) is the method of choice. Commonly employed reduced temperatures include $4°C$ (refrigeration), -5 to $-20°C$ (freezer), $-70°C$ (dry ice) (13, 21), and $-142°C$ (liquid nitrogen) (38, 45).

There are additional reasons for storing specimens: (a) Each sample should be retained for a period of time after analysis to permit a repeat analysis if deemed advisable, and (b) if they are time sequence specimens in a long term study they should be retained to be assayed all at the same time.

Alteration in the concentration of a constituent in a stored specimen can result from various processes, such as adsorption on to the glass or plastic walls of the blood collecting tube, partial denaturation of protein during formation of a monomolecular film, evaporation of volatile constituents or the aqueous phase of serum or blood, changes in erythrocyte permeability, continuing metabolic activities of the erythrocytes and leukocytes, including O_2 consumption and CO_2 production, altered pH and redox potential, hydrolytic (e.g., release of NH_3 from glutamine), proteolytic (e.g., increase in $\alpha-NH_2$ N), or phosphorolytic activities (e.g., phosphate release from organic phosphates), glycolysis (e.g., the metabolic degradation of glucose), and autodegradation (e.g., loss of enzyme activity). These have been reviewed and tabulated by Winsten (77, 78). Stability studies that test these parameters can be performed with a few dozen normal and abnormal samples.

Instability due to microbial growth is erratic but frequently severe. It cannot be studied with the simple protocol used for the causes of instability discussed above.

Changes in concentration of volatile substances such as O_2 and CO_2 are prevented, or at least hindered, by collection and storing of the specimen under anaerobic conditions. Absolute anaerobic conditions are difficult to maintain and are rather cumbersome. When mineral oil is used, it is almost impossible to prevent the admixture of some oil with the aliquot removed from the sample for analysis.

The problem of microbial growth enters whenever the sample is to be stored for longer than one day either at room or refrigerator temperature. This can be solved by four alternative courses of action: (a) collection and storage under sterile conditions (not always possible and certainly impractical as a routine procedure for all determinations); (b) freezing of the sample; (c) extreme alteration of pH; and (d) addition of an antibacterial agent (22, 23).

Lyophilized samples are stable with respect to many constituents for periods at least as long as ten years (13, 69). In the author's experience, components such as glucose can materially decrease in a period of one year if the lyophilization has not been complete.

The problem of stability has taken on particular significance because samples are being sent through the mail to reference laboratories better equipped to perform many of the more esoteric analyses required today in clinical chemistry. Many substances are stable at room temperature for at least several days, others can be stabilized by preservatives, and others as yet cannot be stabilized except by freezing.

TEMPERATURE EFFECT ON STABILITY

Three temperatures are usually readily available for the storage of specimens: "room temperature," which usually falls in the range of 18–30°C, refrigerator temperature (about 4°C), and the frozen state (– 5°C or lower) as low as dry ice temperature. With few exceptions the lower the temperature the greater the stability. Furthermore, microbial growth is considerably less at refrigerator temperature than at room temperature, and is completely inhibited in the frozen state. Even in the frozen state, however, some components of plasma deteriorate (27, 31, 42, 45, 53).

In the shipment of frozen specimens to a reference laboratory, the factors to be noted in maintaining proper specimens in transit are as follows: (a) Use a styrofoam container that will hold enough crushed dry ice to keep specimens frozen for 48–72 hours; (b) Mail or ship (by bus or air freight) the specimens early in the week so that they will not be in transit over the weekend; (c) Whole blood specimens should not be deposited if possible in outdoor mail boxes. In the winter, the intracellular fluid of the red cell will freeze and microcrystals of ice will damage the cell membrane and cause hemolysis; in the summer the hot sun will damage the cell membrane and cause hemolysis.

pH EFFECT ON STABILITY

There is no general rule, but certain components are considered stabilized by increasing the acidity or alkalinity of the sample by trapping the stable salt form or free acid form (e.g., catecholamines and citrate in urine). The stability of acid phosphatase can be prolonged by adjustment of the pH to 6.2 with citrate buffer (24). A very low or high pH will also prevent microbial growth in urine because such systems are incompatible with living systems.

CHEMICAL PRESERVATIVES

Chemical preservatives may be classified into two categories according to their function. The first category includes enzyme inhibitors that prevent chemical changes such as glycolysis (e.g., F^-). The second category includes bacteriostatic agents that interfere with and prevent microbial growth (58, 62).

Major (49) in 1923 introduced potassium fluoride as a preservative for blood. Because of its anticoagulant activity (formation of insoluble CaF_2), it is unnecessary to use another anticoagulant. Fluorides are preservatives through their inhibitory actions on certain enzymes such as those involved in glycolysis. Glucose, urea, nonprotein nitrogen, creatinine, cholesterol, and uric acid were found to be stable for as long as ten days at room temperature if the blood was sterile (49, 61), but fluoride is not sufficiently antibacterial to prevent microbial growth if the specimen is contaminated. Sander (63), in 1923, introduced the combination of 10 mg sodium fluoride + 1 mg thymol per milliliter of blood. The presence of the thymol effectively controlled microbial growth so that nonsterile specimens were stable for all of the above determinations (except nonprotein nitrogen) for at least two weeks. Other workers confirmed these findings, although not with equal success with respect to the period of preservation (9). Such high concentrations of fluoride cause significant water shift and tend to produce hemolysis, but it would appear that lower concentrations can be used (41). Vacutainer collecting tubes containing thymol and F^- are available.

An interesting possibility is the use of antibiotics to prevent bacterial growth. For example, 1 mg streptomycin base per 10 ml blood has been used for preservation of blood for hemoglobin and urea determinations (4).

Recently bactericidal agents, such as derivatives of mercuri-thiosalicylate (22, 23), have been used to prevent microbial growth in serum stored in the refrigerator. The metabolites studied were stable at $3-23°C$ for $3-4$ weeks.

Among the common preservatives for urine specimens are formaldehyde, thymol, toluene, and chloroform. All of these act primarily as antimicrobial agents, and the last two are added in sufficient quantity (a few milliliters) to saturate the urine. Thymol can be conveniently added to urine as a 10% (w/v) solution is isopropanol, $5-10$ ml being sufficient for a 24 hour collection (56). Our laboratory has made a comparison of the relative antibacterial action of boric acid, chloroform, formaldehyde, and toluene in urines seeded with bacteria. Boric acid at a concentration of 0.8% (one-fourth saturated) proved superior to the other substances and worked quite well. The efficacy of boric acid as a preservative was reported previously (30). Several preservative tablets are commercially available that are widely used by life insurance companies for the preservation of urine specimens. "Standard" Urokeep tablets are marketed by Standard Reagents, their composition being unpublished. Cargille Urinary Preservative Tablets are marketed by Cargille; these tablets buffer the urine to a pH of about 6 and owe their preservative action to a mixture of benzoic acid, formaldehyde (produced from Urotropin), and mercuric oxide. Apparently, they are made from the formula proposed by Kingsbury (44):

KH_2PO_4	0.100 g/tablet
Sodium benzoate	0.050 g/tablet
Benzoic acid	0.065 g/tablet
Urotropin (hexamethylenetetramine)	0.050 g/tablet
$NaHCO_3$	0.010 g/tablet
HgO, red	0.001 g/tablet

One tablet is used per ounce of urine and preserves sugar in the urine for five days at room temperature.

In general, the interest in preservation of biochemical metabolites has increased to the point that investigators are developing technics to observe sensitive changes in complex molecules such as the changes in visible and uv absorbance and changes in electrophoretic mobility in human serum lipoproteins on alteration (5) or denaturation.

REFERENCES

1. ACLAND JD: *J Clin Pathol 24*:187, 1971
2. ANNINO JS, RELMAN AS: *Am J Clin Pathol 31*:155, 1959
3. AUSTIN WH: *Am J Clin Pathol 53*:288, 1970
4. BAYLISS RIS, WOOTON IDP: *Brit Med J 1*:1151, 1954
5. BERMES EW Jr, McDONALD HJ: *Ann Clin Lab Sci 2*:226, 1972
6. BETKE K: *Deut Med Wochschr 97*:96, 1972
7. BLITSTEIN I: *Rev Belg Sci Med 7*:69, 1935
8. BOWIE EJW, TAUXE WN, SJOBERG WE Jr, YAMAGUCHI MY: *Am J Clin Pathol 40*:491, 1963
9. BOWMAN WM, ENTERLINE PE: *Public Health Reports 69*:246, 1954
10. BOYD EM, MURRAY RB: *J Biol Chem 117*:629, 1937
11. BOYD MJ, ALPER C: The Encyclopedia of Biochemistry, edited by Williams RJ, Lansford EM Jr. New York, Reinhold, 1967, p 147
12. BRITTIN GM, BRECHER G, JOHNSON CA, ELASHOFF RM: *Am J Clin Pathol 52*:690, 1969
13. BROJER B, MOSS DW: *Clin Chim Acta 35*:511, 1971
14. BUDTZ-OLSEN OE: Clot Retraction. Springfield, Ill., C. C. Thomas, 1951
15. CARAWAY WT: *Am J Clin Pathol 37*:445, 1962
16. CHORINE V: *Ann Inst Pasteur 63*:213, 1939
17. CHORINE V: *Ann Inst Pasteur 63*:367, 1939
18. CHRISTIAN DG: *Am J Clin Pathol 54*:118, 1970
19. COHEN L, LARSEN L: *New Engl J Med 275*:465, 1966
20. CONROY RTWL, MILLS JN: Human Circadian Rhythms. London, J. and A. Churchill, 1970
21. DAVIES DF: *Federation Proc. 24: Suppl. 15*:S249, 1965
22. de TRAVERSE PM, HENROTTE JG, DEPRAITERE R: *Ann Biol Clin 27*:253, 1969
23. de TRAVERSE PM, HENROTTE JG, DEPRAITERE R: *Sem Hosp Paris 45*:1907, 1969
24. DOE RP, MELLINGER GT, SEAL US: *Clin Chem 11*: 943, 1965
25. EISENMAN AJ: *J Biol Chem 71*:587, 1927
26. ENGLEBERG H: *Am J Physiol 181*:309, 1955
27. FAIRWEATHER DV, LAYTON R: *J Clin Pathol 20*:665, 1967

28. FAWCETT JK, WYNN V: *J Clin Pathol 13*:304, 1960
29. FERGUSON JH: Blood Cells and Plasma Proteins, edited by Tullis JL. New York, Academic, 1953, p 117
30. FROST DV, RICHARDS RK: *J Lab Clin Med 30*:138, 1945
31. GARRATTY G: *Am J Clin Pathol 54*:531, 1970
32. HADLEY GG, LARSON NL: *Am J Clin Pathol 23*:613, 1953
33. HADLEY GG, WEISS SP: *Am J Clin Pathol 25*:1090, 1955
34. HAKULINEN A: *Acta Pediatrica Scand Suppl 212*:1, 1971
35. HANOK A, KUO J: *Clin Chem 14*:58, 1968
36. HELLER VG, PAUL H: *J Lab Clin Med 19*:777, 1934
37. HILL HB, LOW EMY, SHOEN I: *Am J Clin Pathol 53*:918, 1970
38. HUNTSMAN RG, HURN BAL, IKIN EW, LEHMANS H, LIDDELL J: *Brit Med J 4*:458, 1967.
39. JACOBSON K: *Scand J Clin Lab Invest Suppl 14*:7, 1955
40. JARRETT RJ: *Brit Med J 4*:334, 1970
41. JOHN HJ: *Arch Pathol Lab Med 1*:227, 1926
42. JUUL P: *Clin Chem 13*:416, 1967
43. KAPLAN SA, YUCEOGLU AM, STRAUSS J: *Pediatrics 24*:270, 1959
44. KINGSBURY FB: *Proc Assoc Life Ins Med Directors America 13*:104, 1925
45. LAESSIG RH, PAULS FP, SCHWARTZ TA: *Health Lab Sci 9*:16, 1972
46. LANGE HF: *Acta Med Scand Suppl 176*:1, 1946
47. LUBRAN M:*Med Clin N Am 53*:211, 1969
48. MAGATH TB, HURN M: *Am J Clin Pathol 5*:548, 1935
49. MAJOR RH: *J Am Med Assoc 81*:1952, 1923
50. McCLELLAN EK, NAKAMURA RM, HAAS W, MOYER DL, KUNITAKE GM: *Am J Clin Pathol 42*:152, 1964
51. McGEOWN MG, MARTIN E, NEILL DW: *J Clin Pathol 8*:247, 1955
52. MOORE JL, WIDISH JR, SUDDUTH NC: *Am J Med Technol 36*:519, 1970
53. MORRISON G, DURHAM WF: *J Am Med Assoc 216*:298, 1971
54. MOSELEY RV, DOTY DB: *Surgery 67*:455, 1970
55. MOSS G, STAUNTON C: *New Engl J Med 282*:967, 1970
56. NAFTALIN L, MITCHELL LR: *Clin Chim Acta 3*:197, 1958
57. NISHI HH: *Clin Chim Acta 11*:290, 1965
58. PORTER IA, BRODIE J: *Brit Med J 2*:353, 1969
59. PRAGAY DA, HOWARD SF, CHILCOTE ME: *Clin Chem 17*: 350, 1971
60. RAPAPORT M, RUBIN MI, CHAFFE D: *J Clin Invest 22*: 487, 1943
61. ROE JH, IRISH OJ, BOYD JI: *J Biol Chem 75*:685, 1927
62. RYAN WL, MOLLS RD: *Am J Med Technol 29*:175, 1963
63. SANDER FV:*J Biol Chem 58*:1, 1923
64. SINGH HP, HEBERT MA, GAULT MH: *Clin Chem 18*:137, 1972
65. STAMM D: *Verh Deut Ges Inn Med 73*:982, 1967
66. STEIGE H, JONES JD: *Clin Chem 17*:1160, 1971
67. STEIN HH, DARBY TD: *Ann N Y Acad Sci 153*:559, 1968
68. STEPHENS SR, CLARK RB, BEARD AG, BROWN WE: *Am J Obstet Gynecol 105*:616, 1969
69. STRUMIA MM, McGRAW JJ, HEGGESTAD GE: *Am J Clin Pathol 22*:313, 1952
70. STUDLAR M, HAMMERL H, NEBOSIS G, PICHLER O: *Klin Wochschr 48*:238, 1970
71. SUNDERMAN FW Jr: *C R C Crit Rev Lab Sci 1*:427, 1970
72. VAN PEENAN HJ, FILES JB: *Am J Clin Pathol 52*:666, 1969
73. WALFORD RL, SOWA M, DALEY D: *Am J Clin Pathol 26*:376, 1956
74. WELLER C, LINDER M: *Metabolism 10*:669, 1961

75. WHITE A, HANDLER P, SMITH EL: Principles of Biochemistry, 4th ed. New York, Blakiston-McGraw-Hill, 1968, p 53
76. WINKELMAN J, MELNYK J, REED A, OLITZKY I: *Health Lab Sci 8:*11, 1971
77. WINSTEN S: Handbook of Clinical Laboratory Data, 2nd ed., edited by Faulkner WR, King JW, Damm HC. Cleveland, The Chemical Rubber Co., 1968, p 80
78. WINSTEN S: Standard Methods of Clin Chem, edited by Meites S. New York, Academic, vol 5 p 1, 1965
79. WIRTH WA, THOMPSON RL: *Am J Clin Pathol 43:*579, 1965
80. WITTGENSTEIN AM: *Am J Med Technol 19:*59, 1953
81. YOUNG DS, EPLEY JA, GOLDMAN P: *Clin Chem 17:*765, 1971
82. ZUCKER MB: *Am J Clin Pathol 24:*39, 1954

Chapter 15

The Preparation of Protein-Free Filtrates

RICHARD J. HENRY, M.D.
CHRISTOPHER P. SZUSTKIEWICZ, Ph.D.

Preparation of protein-free solutions is still of particular importance to the clinical chemist. Numerous analytical methods for determination of various components of blood, urine, and cerebrospinal fluid require preliminary treatment of the fluid to render it free of protein, so as to prevent foaming, turbidity, precipitate formation, or direct interference with certain color reactions. Any of these consequences may lead to errors in clinical determinations. The term "protein-free," however, is essentially a relative one, for some technics leave behind residual proteins ranging in concentration from 1 to 5 mg/100 ml.

METHODS OF DEPROTEINIZATION

Proteins carry both positive and negative charges and the effectiveness of the precipitation process depends strongly upon the final pH of the solution. In strongly acidic solutions, the proteins are positively charged whereas they are negatively charged at strongly alkaline pH. Attraction of solvent molecules for protein molecules, which favors solution, is opposed by the attraction of protein molecules for each other, which tends to prevent solution. Attraction of protein molecules for each other is maximal at the isoelectric point, i.e., the pH at which they are in the Zwitterion form with a net zero electrostatic charge. At this pH most proteins are least stable and most readily precipitated. Away from the isoelectric point the molecules acquire a net positive or negative charge that increases their solubility. The pH, therefore, should be carefully controlled in every procedure containing a step involving precipitation of proteins.

Protein can be effectively precipitated from a physiological fluid by either physical or chemical means. The former includes sorption onto adsorbents such as kaolin, ultrafiltration, microdiffusion, and heat denaturation. The optimal pH's for precipitation of proteins from serum or plasma and whole blood by heat denaturation are 5.3 and 6.4, respectively, 1 vol of serum or plasma and whole blood requiring 1 vol and 0.8 vol of 0.05 N acetic acid, respectively, to achieve the desired pH (29). In most instances, however, deproteinization is accomplished by chemical precipitation of the protein, followed by filtration or centrifugation. Such technics include antigen–antibody reaction, protein dehydration, and insoluble salt formation. The general nature of most of these reactions appears fairly certain although some of the details remain obscure. Plasma proteins vary widely, not only in their relative solubilities at the isoelectric point but also in the pH region in which they are isoelectric. For example, γ-globulins have their isoelectric point above pH 7 while albumins and some of the α-globulins are isoelectric below pH 5. Thus, isoelectric precipitation alone has relatively limited application; it is, therefore, frequently used in combination with other technics such as organic solvent precipitation or salting out.

PRECIPITATION BY ADDITION OF MISCIBLE SOLVENTS

It is well known that proteins can be precipitated from aqueous solution by the addition of organic solvents miscible with water. Solvents such as methanol, ethanol, acetone, and ethyl ether lower the solubilities of the proteins and precipitate them from highly polar solutions. It seems plausible that two factors are involved in this phenomenon: First, molecules of the nonaqueous solvent form hydrates, thus competing with proteins for water molecules. If competition is successful, the protein molecules become dehydrated. Second, the dielectric constant of the aqueous solution is lowered by addition of a miscible, nonaqueous solvent. Since the forces of electrostatic attraction and repulsion between solute molecules are inversely proportional to the dielectric constant, the force of repulsion between charged protein molecules is decreased. Expectedly, the less polar the solvent, the greater its precipitating action. Such treatment generally leads to denaturation of the protein. To avoid or minimize this denaturation of proteins precipitated by the addition of miscible solvents, it is necessary to work at a temperature of 0°C or lower. Denaturation is of no consequence if the purpose of protein precipitation is merely to obtain a protein-free solution. It is practically impossible to obtain complete protein precipitation (clear filtrates) unless cold solvent is used or unless the mixture is heated to coagulate the precipitate. Numerous factors such as pH, temperature, ionic strength, and organic solvent concentration influence precipitation of proteins by organic solvents. A systematic variation of all these parameters led to the well-known low temperature ethanol fractionation scheme for plasma proteins worked out by Cohn and co-workers (9).

Successful precipitation of proteins from serum or whole blood can be accomplished with a final concentration (v/v) of 63% cold acetone [1 vol sample + 9 vol 70% (v/v) acetone]. Similarly, 72% (v/v) ethanol [1 vol sample + 9 vol 80% (v/v) ethanol], and 76% and 81% (v/v) methanol are required for whole blood and serum, respectively, [1 vol whole blood + 9 vol 85% (v/v) methanol; 1 vol serum + 9 vol 90% (v/v) methanol]. If simultaneous extraction of lipid is desired, the protein solution must be added to the solvent slowly with constant mixing.

PRECIPITATION BY "SALTING OUT"

Addition of neutral salts decreases the attraction of protein molecules for each other, increasing their solubility. This is the "salting in" phenomenon.

As the salt concentration is increased, a point is reached at which, because of the successful competition with the protein molecules for the water molecules, the protein becomes dehydrated with a resultant decreased solubility. This precipitation is called "salting out" (not to be confused with protein precipitation by insoluble salt formation, which is discussed in the next section).

Salting out technics are relatively simple, although temperature and pH are significant variables in these systems. They are essentially preparative and have limited application in clinical chemistry.

Salts containing multivalent anions are the most efficient precipitating agents. Commonly used salts are sodium sulfate, ammonium sulfate, magnesium salts, phosphates, and citrates. Proteins precipitated by salting out are not denatured and usually redissolve when the salt concentration is lowered. The concentration of salt required for precipitation depends on the particular protein and the pH, and with few exceptions, minimal solubility occurs at the isoelectric point. Fractional precipitation of plasma proteins is considered in greater detail in the section on the determination of plasma proteins ("Salting Out Technics of Fractionation" in Chap. 16.)

PRECIPITATIONS DEPENDENT ON INSOLUBLE SALT FORMATION

This method is the most frequently used in clinical chemistry. Precipitation may be accomplished by addition of (a) anionic precipitants such as tungstic acid, trichloroacetic acid, and perchloric acid, or (b) cationic precipitants such as Zn^{2+}, Hg^{2+}, Fe^{3+}, and Pb^{2+}. Earlier thinking explained these precipitations in the following way: Acid precipitants form insoluble protein salts with the cationic form of the protein molecules, a prerequisite thus being that the pH be below the isoelectric point, pI, of the protein; heavy metal cations combine with the anionic form of the protein molecules to form insoluble proteinates. In the latter case the pH of the solution must be higher than the pI of the protein. The implication, of course, is that saltlike bonding occurs through positively charged amino groups in the case of anionic precipitants and through the negatively charged carboxyl groups in the case of heavy metals.

The above explanation may be true for anionic precipitants. The isoelectric points of reduced hemoglobin, oxyhemoglobin, serum albumin, α_1-globulin, β-globulin, γ_1-globulin (sometimes called β_2), γ_2-globulin, and fibrinogen are 6.8, 6.6, 4.9, 5.2, 5.4, 6.3, 7.3, and < 5.3, respectively (9, 30). Actually, these values are averages because of the heterogeneity of "pure" fractions. This is especially true of the γ_2-globulins, for it has been reported (2) that the pI's of this fraction form a normal curve ranging from 6.3 to 8.4; furthermore, the γ-globulins present in two different fractions from the low temperature ethanol fractionation scheme of Cohn *et al* have pI's of 7.3 and 6.85 (30). The pI's of whole blood, serum, and plasma thus fall in the range 4.9–7.3. Although a careful study does not appear to have been made with all the commonly used acid precipitants, both tungstic acid (5) and metaphosphoric acid (6) precipitate blood proteins below a pH of about 5.1.

The evidence does not support the view that heavy metals are precipitants as a result of combination with negatively charged carboxyl groups. Mercuric ions (51) and other heavy metal ions do precipitate proteins at pH's well below their pI's, although it has been reported (9) that the solubility minima of metal salts of proteins are somewhat alkaline to their pI's. It has been shown (48, 49) that Cu, Zn, Cd, and Pb ions bind to bovine serum albumin principally through the imidazole groups of the histidine residues of protein molecules and only weakly at the carboxyl groups.

Regardless of precipitant and mechanism there are two critical variables concerned with maximal precipitations: (a) pH of the reaction mixture and (b) concentration of precipitant. There must be sufficient reagent to combine with the protein present; ordinarily, a safe excess is used.

If desired, protein precipitates usually can be put back into solution, by acidification in the case of metal proteinates and by rendering the mixture strongly alkaline in the case of protein salts.

ANIONIC PRECIPITANTS

Among the acid precipitants that have been used in clinical chemistry are picric, molybdic, phosphotungstic, tungstomolybdic, sulfosalicylic, tungstic, trichloroacetic, metaphosphoric, and perchloric. The last four are still widely used and will be considered in detail. Krautman's reagent, a mixture of tungstic and metaphosphoric acids, will also be discussed.

Tungstic Acid

This precipitant was introduced by Folin and Wu (13) in 1919. One vol blood is mixed with 7 vol water, followed by 1 vol 10% sodium tungstate ($Na_2WO_4 \cdot 2H_2O$) and 1 vol $\frac{2}{3} N H_2SO_4$, giving a 1:10 dilution of the sample. The mixture is shaken and allowed to stand before filtration or centrifugation until the color changes to a chocolate brown (approximately 15 to 30 min). Besides the color change, signifying conversion of hemoglobin to methemoglobin, complete precipitation is indicated by absence of foaming and the presence of a clicking sound against a stopper when the mixture is shaken. Failure to reach the brown color can be remedied by adding 10% H_2SO_4 one drop at a time and shaking vigorously after each drop until the change is complete. Because of the lower concentration of protein in serum or plasma (about one-third that of whole blood), Folin (14) recommended using 0.5 vol of each precipitating reagent and increasing the water to 8 vol.

Haden (22) modified the Folin-Wu method by reversing the order of the addition of sodium tungstate and sulfuric acid, adding the acid first. The chief advantage of this modification, which is used almost universally today, is that the reaction is practically immediate and there is no necessity for delaying filtration. Haden also claimed more rapid filtration and a larger volume of filtrate. Other suggested modifications have been concerned chiefly with decreasing the number of solutions to be added. Haden actually employed 8 vol of $N/12$ H_2SO_4 (rather than 7 vol water) + 1 vol 10%

sodium tungstate, thus eliminating one addition. Combining of the water, sodium tungstate, and acid to give a single reagent of tungstic acid was suggested (50), but this reagent is stable for only two weeks and requires a waiting period for the precipitate to turn chocolate brown. Modifications have been proposed in which the tungstic acid solution is buffered so that tungstic acid does not precipitate, thereby stabilizing the reagent. Abrahamson (1) gives the following directions: Prepare the reagent by adding 11.11 g sodium tungstate and 5 g sodium citrate to about 700 ml water. Dissolve 13.6 g fused $NaHSO_4$ or 15.6 g $NaHSO_4.H_2O$ in about 200 ml water. Mix the two solutions and dilute to 1 liter with water. Let stand one week and filter. If mold grows in the reagent, 2 g benzoic acid may be added per liter of reagent. Filtrates containing the benzoic acid cannot be used for the urease determination of urea because of inhibition of the enzyme. Caraway (8) used orthophosphoric acid instead of citrate as the buffer and claimed a stability of at least four months. Goldenberg (16) developed a single tungstic acid reagent that is stable for 2 years at room temperature. The following directions are given: Dissolve 1.0 g polyvinyl alcohol (Elvanol 70–05, Dupont) in about 100 ml water with gentle warming (do *not* boil). Cool, and transfer to a 1 liter volumetric flask containing 11.1 g sodium tungstate ($Na_2WO_4.2H_2O$), reagent grade, previously dissolved in about 100 ml water. Mix by swirling. In a separate vessel add 2.1 ml concentrated sulfuric acid, reagent grade, to about 300 ml water, mix, and then add to the tungstate solution contained in the volumetric flask. Mix and dilute to 1 liter with water.

The $\frac{2}{3}$ N H_2SO_4 is prepared by weighing 35 g reagent grade acid to the nearest 0.1 g in a weighed beaker and diluting to 1 liter in a volumetric flask with distilled water, rinsing out the beaker. Pipetting the acid is unsatisfactory because of its high viscosity (aside from the danger involved). One need not titrate the acidity of the solution provided one is sure of the quality of the concentrated acid and the weighing procedure. The 10% sodium tungstate is prepared from the reagent grade salt. The original directions given by Folin and Wu included testing for and neutralizing the extraneous alkali (Na_2CO_3) present in the commercially available products of that time. Present day specifications for reagent grade sodium tungstate allow a maximum of 0.4 ml $0.1N$ acid to neutralize to pH 8 the extraneous alkalinity present in 2 g salt. Such salts give solutions having a pH ranging from about 9 to 9.5; it is unnecessary in routine work to neutralize the small amount of Na_2CO_3 present. The sodium tungstate solution is stable indefinitely, but eventually, especially if stored in a soft glass container, a slight amount of sediment accumulates.

When equal quantities of $\frac{2}{3}$ N H_2SO_4 and 10% sodium tungstate are mixed, there is a 10% excess of H_2SO_4 present. This excess, plus that tungstic acid not combined with protein, causes filtrates to be acidic. There has been a misunderstanding of a statement in Folin and Wu's original paper that "filtrates are nearly neutral." This statement did not refer to pH—no pH determinations were made—but rather to the fact that the titratable acidity should be very small (10 ml of filtrate should require only about 0.2 ml 0.1 N NaOH to neutralize with phenolphthalein as indicator). A study of the tungstic acid precipitation of blood protein (5) has disclosed the following: (a) the average pH of whole blood and serum filtrates is about 4.1 with a range of about 3.2 to 4.7; (b) the same filtrate pH is obtained whether acid or tungstate is added first; (c) maximal protein precipitation occurs only if the filtrate pH is 5.1 or less; (d) when the pH is above 5.1 the filtrate is turbid, either grossly, or demonstrable by a Tyndall effect; (e) the pH of the filtrate increases as the protein content of the blood increases, but protein precipitation is maximal (protein concentration in filtrate less than 2 mg/100 ml), and the filtrates clear over the range of protein likely to be met with clinically (up to 27 g/100 ml); (f) if the acid is added first to whole blood and the pH is below 3, the filtrates occasionally are slightly turbid and yellow and give a positive benzidine test, probably a result of the formation of acid hematin at these low pH values, as suggested by Haden (22); (g) contrary to the claims of Folin (14), the presence of up to 10 times the usually recommended amount of potassium oxalate as anticoagulant has no significant effect on filtrate pH.

Tungstic acid precipitation of protein in low concentration in fluids such as cerebrospinal fluid can be carried out by combining 2 vol of sample + 7.5 vol of water + 0.25 vol of each of the precipitating reagents (46).

By no means the least advantage of this means of protein precipitation is the crystal-clear filtrates almost invariably obtained.

Trichloroacetic acid (TCA)

This reagent was first used in 1915 by Greenwald (20). The original method called for 1 vol sample + 9 vol 2.5% acid, but due to occasional incompleteness of protein precipitation it was later suggested that the acid concentration be increased to 5 g/100 ml (20). The mixture should stand for 20 min prior to filtration or centrifugation; the pH of filtrates is about 1.0 (26). Filtration is quicker and the filtrate more voluminous than with tungstic acid (26). Often the protein-free filtrates from trichloroacetic acid precipitation are not optically clear. If the component to be determined can withstand heating, clarity can usually be obtained by heating the mixture at 90–95°C for about 15 min prior to filtration.

When boiling the filtrate at an acid pH has no ill effect on the desired constituents, the trichloroacetic acid may be decomposed by such treatment to chloroform and CO_2. Only about 70% of the acid, however, is removed by this treatment (35). An alternative (where feasible), extraction of the acid with ether, is difficult because of an unfavorable partition coefficient.

Binding of trichloroacetic acid to protein occurs at cationic side chain groups and at peptide bonds (21) as shown in Figure 15-1. Elimination of bound water and its replacement by the strong nonpolar $Cl_3C–C–$ radical destroys the water solubility of the protein and tends to make it soluble in nonpolar solvents such as benzene and chloroform.

Certain proteins are not precipitable by trichloroacetic acid, e.g., albumoses, peptones, and some amino acids (26) and acid seromucoid (38). Four percent to 20% of the urine protein present in proteinuria is not precipitated (4).

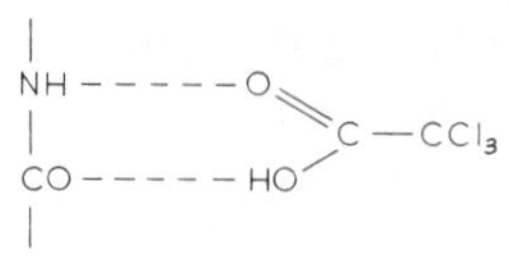

$$NH - - - - - O \diagdown \atop CO - - - - - HO \diagup C - CCl_3$$

FIG. 15-1. Binding of trichloroacetic acid to protein.

The reagent is stable.

Perchloric Acid

Deproteinization with perchloric acid was introduced in 1944 by Neuberg and co-workers (35, 36) as a substitute for trichloroacetic acid when the latter interferes in the reaction. Although considerably less appears to be required, a safe excess is provided by adding 1 vol blood, serum, or plasma to 9 vol 1.11 M $HClO_4$ to give a final concentration of 1 M (17). The mixture is shaken and allowed to stand for about 10 min. Filtration must be made through a highly retentive paper because a finely divided precipitate is obtained. In order to provide quicker and more complete precipitation, the reagent should be ice cold. The pH of filtrates is nearly 0.

1.11 M $HClO_4$ is prepared by diluting with water to 1 liter (a) 121 ml of the 60% acid (sp. gr. 1.53), or (b) 93 ml of the 71% acid (sp. gr. 1.68).

The excess perchloric acid can be removed from the filtrate by neutralization with KOH, K_2CO_3, or CH_3COOK and adding alcohol. $KClO_4$ precipitates and can be removed by filtration (35, 36).

Seromucoid (mucoprotein) (38), α_1-glycoprotein 3.5 S, haptoglobin, β_{1B}-globulin (MP$_3$ of Winzler) (18), peptones, and albumoses (35) are not precipitated by perchloric acid. A trace of albumin may also sometimes appear in serum filtrates (18).

Metaphosphoric Acid

The thorough review by Horvath (28) of deproteinization with this acid traces its use in Europe back as far as 1816. It was introduced in the clinical chemical literature in the United States by Folin and Denis in 1916 (15).

Metaphosphoric acid in water solution forms several polymers, $(HPO_3)_x$, and catalyzed by hydrogen ions, hydrolyzes to the ortho acid as follows: $HPO_3 + H_2O \longrightarrow H_3PO_4$. Since orthophosphoric acid does not precipitate protein, metaphosphoric acid solutions retain their activity for only about one week at refrigerator temperature (34).

Almost all of the studies on this protein precipitant have been performed with glacial metaphosphoric acid of variable composition. The glacial acid is marketed as pellets of a mixture of HPO_3 and $NaPO_3$, the latter added as a preservative. The glacial acid is actually a Na_2O–H_2O–P_2O_5 glass consisting of various long chain and branch chained phosphates (12). Even in the glacial form the surfaces of the pellets develop a white surface coating (ortho form?) and the product loses its protein-precipitating power. The American Chemical Society (37) has now established specifications for reagent grade metaphosphoric acid: 34.0–36.0% HPO_3 + 58.0–62.0% $NaPO_3$. According to specifications, the HPO_3 content is assayed as follows: weigh accurately about 5 g in a 250 ml glass-stoppered flask, dissolve in 100 ml water, and add 3 drops bromcresol green indicator; titrate with 1 N NaOH to first noticeable change in color from green to blue. 1 ml 1 N NaOH is equivalent to 0.07998 g HPO_3. Accuracy of this titration procedure is dependent on the almost complete absence of other phosphates. A scheme for the estimation of ortho-, pyro-, meta- and polyphosphates in the presence of one another has been presented (31). Titration using phenolphthalein as indicator, which has been proposed (34), will not give the concentrations of the meta acid. The pK_1 and pK_2 of H_3PO_4 are 2.2 and 7.2, respectively. Titration of H_3PO_4 with bromcresol green as indicator titrates the first H^+ and, since 1 mole HPO_3 converts to 1 mole H_3PO_4, does not distinguish between the two forms. It would appear that the only simple criterion of whether one has sufficient HPO_3 in solution is that of protein precipitation itself. It has been claimed (6) that the free acid present in the commercial product is in the ortho form whereas the sodium salt is in the meta form. If this is the case, solution of the glacial mixture would still yield PO_3^- for combination with protein.

A study (6) of blood protein precipitation by metaphosphate has revealed that, as the pH is decreased, precipitation begins at about pH 5.2. At pH's 2, 2.5, and 3, metaphosphate bound per gram of serum albumin is 0.28, 0.21, and 0.15 g, respectively. Calculation from various determinations (26, 34) for precipitation of blood or serum proteins reveals that the quantities of glacial metaphosphoric acid recommended provide approximately 0.6 g or more metaphosphate per gram of protein present, i.e., an amount in considerable excess over that required. Safe excess is provided by using 3 vol 5% glacial metaphosphoric acid for 1 vol whole blood, and 1 vol of the acid for 1 vol of serum or plasma. Water is also usually added. The mixture should be shaken for about 30 sec and allowed to stand about 15 min before filtration or centrifugation. The pH of filtrates thus prepared varies from about 2.1 to 2.6.

Since hydrogen ions catalyze the hydrolysis of the meta acid to the ortho form, it would be expected that by decreasing the hydrogen ion concentration the conversion rate would be decreased. It has been shown (7) that a solution of sodium metaphosphate adjusted to a pH of 8 to 9 by addition of Na_2CO_3 or NaOH is practically permanently stable at room temperature. The $NaPO_3$ can be prepared by heating $Na_2HPO_4 \cdot H_2O$ in a platinum crucible to 700°C for 1 hour (7). The melt, when cooled to the point at which it will just flow, is poured onto a platinum surface or removed from the crucible in drops on a platinum wire. To precipitate protein, an amount of solution is used equivalent to at least 0.2 g $NaPO_3$ per gram of protein, and sufficient acid such as hydrochloric or acetic is added to bring the pH to about 2.5–3.0.

Metaphosphoric acid precipitates no amino acids (7) or peptones and only a small part of the albumoses (52).

Even with fresh reagent prepared from a new bottle of metaphosphoric acid it is frequently difficult or impossible to obtain filtrates or supernates devoid of Tyndall effect. This reagent, therefore, should be avoided when possible in photometric analyses.

Krautman's Reagent (32, 33)

To prepare the stock solution, place 23.8 g sodium pyrophosphate and 50 ml glacial metaphosphoric acid solution (0.56 g/100 ml water) in a 1 liter volumetric flask. Mix and add slowly with shaking 29 ml concentrated H_2SO_4. Cool to about 40°C and then, while mixing, add 370 ml of 30% sodium tungstate. Heat and mix until clear. Cool to room temperature and add water to 1 liter. This stock solution, according to Krautman, is stable for at least five years if stored in a dark, rubber-stoppered bottle (cork or tinfoil will react with the reagent). The acidity of the stock solution can be checked by titration using alizarin indicator: 1 ml of stock reagent should require 0.39–0.5 ml 1 N NaOH.

The working solution is prepared by diluting the stock solution 1:10. Krautman claimed that the working solution is stable at least four years if stored in either a "white" or a dark bottle with rubber stopper. Deproteinization is effected by mixing 1 vol blood or other sample with 9 vol working reagent. The mixture should stand 5 min or in the case of blood, until brown. The pH's of serum and whole blood filtrates are about 2.3 and 3.0, respectively.

The purpose of the sodium pyrophosphate and metaphosphoric acid is to buffer the reagent so that tungstic acid does not precipitate, thereby achieving the goal of a single tungstic acid deproteinizing reagent. Certain difficulties, however, may be encountered with this reagent (1). There may be sufficient reducing substances in the chemicals or water used to produce in the reagent a yellow color that deepens with age. Perhaps the most serious objection to this reagent is the danger involved in use of filtrates with any reaction measured at an alkaline pH. Krautman's reagent is in effect a modification of the Folin-Denis uric acid reagent, which is a solution of phosphotungstic acid and gives a blue color in the presence of uric acid.

CATIONIC PRECIPITANTS

Among the cationic precipitants that have been used are Hg^{2+}, Cd^{2+}, Fe^{3+}, Cu^{2+}, and Zn^{2+} and also the combinations $Hg^{2+}–Ba^{2+}–Zn^{2+}$ and $Ba^{2+}–Zn^{2+}$. Only Zn^{2+}, $Ba^{2+}–Zn^{2+}$, and Cu^{2+} will be considered in detail here but references to use of the others may be found in the section, "Specificity of Methods for Glucose" in Chap. 19 of the first edition of this book (24).

Zinc

Deproteinization of blood proteins with zinc was introduced in 1923 by Hagedorn and Jensen (23)—blood was added to a mixture of zinc sulfate and sodium hydroxide. To obtain complete precipitation it was necessary to heat the mixture, a procedure deleterious to some substances. Somogyi (42) found that, if the zinc salt is added to laked blood and then the pH is brought to neutral or to slightly alkaline, precipitation is complete at room temperature.

The following methods involving zinc precipitation were introduced by Somogyi (42):

(a) 1 vol blood + 7 vol water + 1 vol 10% $ZnSO_4 \cdot 7H_2O$ + 1 vol 0.5 N NaOH. The blood must be laked before addition of the zinc sulfate and complete mixing performed before addition of the alkali. If the reagents are of proper strength, 10 ml of the zinc solution diluted to 50—70 ml with water will require 10.8—11.2 ml of the alkali to obtain a pink color with phenolphthalein, titrating slowly with constant shaking. For plasma or serum, Somogyi recommended mixing 1 vol sample + 8 vol water + 0.5 vol zinc sulfate solution + 0.5 vol sodium hydroxide solution.

(b) 1 vol blood + 8 vol acid-zinc sulfate reagent (dissolve 12.5 g $ZnSO_4 \cdot 7H_2O$ in water, add 125 ml 0.25 N H_2SO_4, dilute to 1 liter) + 1 vol 0.75 N NaOH. Titration of 50 ml of the acid-zinc sulfate reagent should require 6.70—6.80 ml of the alkali to obtain a pink color with phenolphthalein.

Method (b) gives more rapid filtration and a larger yield of filtrate than method (a).

Somogyi (42) also presented two methods for precipitation of small samples of blood:

(c) To 5.8 ml water add 0.2 ml blood and 1.0 ml 1.8% $ZnSO_4 \cdot 7H_2O$, mix and add 1.0 ml 0.1 N NaOH. 10 ml of the zinc reagent diluted with 60 ml water should require 12.0-12.2 ml of the alkali for titration to pink with phenolphthalein.

(d) To 6.8 ml zinc reagent (2.95 g $ZnSO_4 \cdot 7H_2O$ and 29.5 ml 0.25 N H_2SO_4 per liter) add 0.2 ml blood and 1.0 ml 0.15 N NaOH.

Calculations of the above titrations performed for checking the reagents reveal that the quantities of alkali fall short of the theoretical amounts required to reach the equivalance points in the titrations. Apparently zinc complexes are formed and the amounts of alkali used, therefore, are empirical.

The pH of whole blood filtrates [from method (a) above] has been found in our laboratory to be about 7.4 and their protein content by sulfosalicylic acid turbidity to be less than 2 mg/100 ml. The pH of serum filtrates obtained by using the suggested half amounts of precipitating reagents is about 7.4, but the filtrates are slightly turbid with a protein content of about 150 mg/100 ml. If the full amounts of reagents are used, however, the filtrates are usually clear, with a protein concentration ranging from 1 to 25 mg/100 ml and the pH still about 7.4.

In the precipitation of proteins by zinc there is complete coprecipitation of glutathione, ergothioneine, and uric acid, some loss of creatinine, but no loss of urea (43).

As noted above, zinc combines specifically and reversibly with the proteins of plasma. Knowledge of the fact that the zinc-protein complex is partially insoluble at 0.15 ionic strength and pH 7 has been applied to devise a system of protein fractionation (10, 47).

Zinc Plus Barium

Somogyi (44) noted that Zn^{2+} alone fails to give complete precipitation of proteins from serum or plasma and found that replacing the NaOH with $Ba(OH)_2$ results in complete precipitation, attributing this to the adsorptive capacity of the $BaSO_4$ formed. The precipitating reagent is prepared from a 5% solution of $ZnSO_4 \cdot 7H_2O$ and 0.3 N Ba(OH): [actually it is better to prepare a 0.34 N solution of $Ba(OH)_2$, i.e., 5.36 g $Ba(OH)_2 \cdot 8H_2O$ per 100 ml, to allow for impurities, including insoluble $BaCO_3$.] The accuracy of these concentrations is not too important, but the alkali must neutralize the zinc sulfate solution precisely, volume for volume, when titration is performed with phenolphthalein as indicator. Place 10 ml zinc sulfate solution in a flask, add about 100 ml water, and a few drops phenolphthalein indicator solution; then add alkali dropwise with continuous agitation until indicator turns pink and color persists for at least 1 min. Rapid titration gives a false end point. On the basis of the titration the solution that is more concentrated is diluted to match the other. The zinc sulfate solution is diluted 1:5 with water and the $Ba(OH)_2$ solution is diluted 1:5 with boiled water, and since contact with air must be minimal because of absorption of CO_2 with formation of $BaCO_3$, store under N_2 or in a bottle protected by soda lime from the CO_2 of air. If a visible precipitate forms the reagents must be restandardized. According to Somogyi, another advantage of this combination is that no salts from the precipitating reagent are introduced into the filtrate. Henry (24), however has analyzed Zn^{2+} and $Zn^{2+}-Ba^{2+}$ filtrates of a serum for Zn and found 176 and 60 μg/ml, respectively.

Copper

Deproteinization with copper is superior to zinc precipitation (45) in two ways: It gives more complete protein precipitation of serum or plasma, and the excess reagent is precipitated along with the protein. Protein can be precipitated by adding copper sulfate followed by sodium hydroxide, but in this case the excess copper appears in the filtrate. To eliminate this problem, sodium tungstate may be used in place of sodium hydroxide. Somogyi (45) presented the following method for copper precipitation: Lake 1 vol blood in 7 vol water. Add 1 vol 7% $CuSO_4 \cdot 5H_2O$ and mix. Add, with continuous shaking, 1 vol 10% $Na_2WO_4 \cdot 2H_2O$. Shake well and filter or centrifuge after a few minutes. For serum or plasma, use 5% $CuSO_4 \cdot 5H_2O$ and 6% $Na_2WO_4 \cdot 2H_2O$.

ULTRAFILTRATION

This method of filtration through a selective semipermeable membrane resembles dialysis in principle. Positive or negative pressure differential, however, must be applied to overcome the low permeability of the membrane and to separate the high and low molecular weight components of a physiologic fluid. The solvent and unbound low weight molecules pass through the membrane and form the ultrafiltrate, whereas the high molecular weight components remain behind. Obviously, the size of the molecules passing through the membrane depends on the pore size of the membrane.

Recently, size-selective membranes have been formulated into a cone configuration. Fabricated from inert polymeric materials, they are capable of retaining molecules with molecular weights in excess of 20, 000. Farese and Mager (11) have analyzed serum filtrates obtained by membrane cone ultrafiltration and filtrates prepared by conventional technics for glucose, urea nitrogen, creatinine, and uric acid. Their data showed no significant differences between the two methods. Ultrafiltration technics have also been applied in the determination of diffusible and proteinbound calcium and in the separation of carbohydrates by thin-layer chromatography.

The simplicity of the principle of ultrafiltration has a great appeal to a clinical chemist. The results are similar to those of dialysis, but ultrafiltration is quicker. In addition, there is no need for elaborate equipment or extensive manipulations.

Ultrafiltration procedure also presents several distinct advantages over deproteinization by protein precipitation (11): (a) no specific precipitating technics for individual procedures are required; (b) there is no significant change in the concentration of the serum solutes (conceivably there could be significant adsorption of certain solutes at trace levels); (c) the membranes are inert, thus no additional ions are added to the sample; (d) there is no coprecipitation of ultrafilterable serum components; (e) a very small sample volume may be used, giving a practical method for pediatric analyses.

PRECIPITATION BY SHAKING WITH CHLOROFORM

Protein can be removed from solution by adding 0.25 vol $CHCl_3$ + 0.1 vol of any foam-preventing substance, such as amyl alcohol, to the protein solution, shaking 15—60 min, followed by centrifugation (41). The mixture separates into two layers, the upper being the aqueous layer and the lower consisting of a $CHCl_3$-protein gel. If there is an excess of $CHCl_3$, a third layer of pure $CHCl_3$ appears at the bottom. Actually, this constitutes a very sensitive test for protein, for when the $CHCl_3$ layer is water clear and does not show a skinlike precipitate at the interface, the protein concentration of the original aqueous solution was less than about 3 mg/100 ml.

PRACTICAL PROBLEMS IN THE PREPARATION OF PROTEIN-FREE SOLUTIONS

The aims in the preparation of protein-free solution are to precipitate only protein and at the same time to precipitate it completely. Expectedly, however, certain components adsorb onto the precipitating protein or at least precipitate along with the protein, the nature of these components depending on the specific precipitating reagent and the pH. For example, acid reagents remove less nonprotein nitrogen than basic reagents. Thus, zinc filtrates contain up to 13 mg/100 ml less nitrogen than tungstic acid filtrates, the glutathione, ergothioneine, and uric acid being removed (43). Alcohol precipitation removes about one-third of the amino acids, whereas tungstic acid precipitates albumoses and peptones but not the amino acids. Trichloroacetic acid, on the other hand, passes all three substances into the filtrate (26). For complete recovery of calcium or phosphate in a filtrate a low pH is required to hold them in solution. In some instances, coprecipitation is exploited: For example, zinc precipitation of blood proteins removes substantially all of the nonglucose reducing substances present (42). Because of variation in protein concentration and unavoidable variation in delivery of reagents, the pH of filtrates from precipitating reagents other than those strongly acid may vary over a considerable range. It is quite possible that this may result in variation of coprecipitation with resultant analytical errors. This point has not been adequately investigated.

In order to ensure complete precipitation of protein there must be an excess of precipitating reagent. The excess of some reagents appears in the filtrate. In some cases the presence of the reagent in the filtrate may prohibit its use. For example, a perchloric acid filtrate cannot be used for glucose determination because of its oxidizing activity at elevated temperatures. Owing to variation in the protein concentration the amount of precipitating agent appearing in the filtrate also varies. It is conceivable that this could be a potential source of error in some determinations.

For practical purposes in nearly all determinations, protein precipitation can be considered to be complete if the filtrate is clear. Looking through the filtrate is *not* sufficient if the filtrate is to be subjected to photometric analysis. It must be examined for a Tyndall effect since slight turbidities otherwise missed can have a significant effect on results. The presence of foaming is also indicative of incomplete precipitation.

Removal of precipitated protein from the mixture can be effected either by filtration or by centrifugation. Centrifugation allows a greater recovery of protein-free solution, and the dangers of loss of solutes by adsorption onto filter paper or of introduction of new solutes from the paper are circumvented. With some precipitations, e.g., tungstic acid precipitation of whole blood, there is a small amount of scum that floats on top of the centrifugate, making removal of clear centrifugate difficult.

Preparation of deproteinized solutions by ultrafiltration presents a simple and rapid approach to the precipitation of proteins. The process can be carried out under conditions approximating a state of equilibrium yielding undiluted samples that are free of foreign ions.

A possible source of error in the determination of substances in protein-free filtrates that apparently is not commonly appreciated is the "volume displacement" error. This arises from the volume occupied by the protein precipitate itself, causing a somewhat increased concentration of solutes in the filtrate. An error of 2%–4% has been observed for the chloride (39), sodium (27), calcium (40), and total base (3). Thus, for these substances, a 1:10 filtrate would actually be about 1:9.7. Whether this effect occurs with all protein-precipitating methods and for all solutes has not been documented. The amount of filtrate or centrifugate recoverable varies significantly with various precipitants. Whether the "volume displacement" error would also vary is again unknown.

Protein-free filtrates in practically all instances appear colorless to the eye, and it has been tacitly assumed in photometric analyses that any nonspecific absorption that may be present in the filtrate is too small to contribute significantly to any specific absorption developed by a color reaction carried out on the filtrate. This is not a safe assumption, however, if the specific color produced is very weak or if the color reaction is carried out at an alkaline pH (25). The nonspecific "background" absorption generally appears to increase as the wavelength employed decreases and is greater with trichloracetic acid filtrates than with zinc or tungstic acid filtrates.

REFERENCES

1. ABRAHAMSON EM: *Am J Clin Pathol* Tech Suppl 4:75, 1940
2. ALBERTY RA, ANDERSON EA, WILLIAMS JW: *J. Phys Colloid Chem* 52:217, 1948
3. BALL EG, SADUSK JF Jr: *J Biol Chem 113*:661, 1936
4. BECKMAN WW, HILLER A, SHEDLOVSKY T, ARCHIBALD RM: *J Biol Chem 148*:247, 1943
5. BERKMAN S, HENRY RJ, GOLUB OJ, SEGALOVE M: *J Biol Chem 206*:937, 1954
6. BRIGGS DR: *J Biol Chem 134*:261, 1940
7. BRIGGS DR: *Proc Soc Exp Biol Med 37*:634, 1938
8. CARAWAY WT: *Am J Clin Pathol 32*:97, 1959
9. COHN EJ, GURD FR, SURGENOV DM *et al.*: *J Am Chem Soc 72*:465, 1950
10. COHN E: Blood Cells and Plasma Proteins. New York, Academic, 1953, p. 30
11. FARESE G, MAGER M: *Clin Chem 16*:280, 1970
12. FLASCHKA HA, WOLFRAM WE: *Chemist-Analyst 48*:65, 1959
13. FOLIN O, WU H: *J Biol Chem 38*:81, 1919
14. FOLIN O: Laboratory Manual of Biological Chemistry, 3rd ed. New York, Appleton, 1923
15. FOLIN O, DENIS W: *J Biol Chem 26*:491, 1916
16. GOLDENBERG H: *Anal Chem 28*:1003, 1956
17. GOTTFRIED SP, ERDMAN GL: *Am J Clin Pathol 21*:118, 1951
18. GRABAR P, DE VAUX ST. CYR C, CLEVE H: *Bull Soc Chim Biol 42*:853, 1960
19. GREENSPAN EM, DREILING DA: *Gastroenterology 32*:500, 1957
20. GREENWALD I: *J Biol Chem 34*:97, 1918
21. GRIMBLEBY FH, NTAILIANAS HA: *Nature 189*:835, 1961
22. HADEN RL: *J Biol Chem 56*:469, 1923
23. HAGEDORN HC, JENSEN BZ: *Biochem Z 135*:46, 1923
24. HENRY RJ: Clinical Chemistry Principles and Technics, 1st ed. New York, Harper and Row, 1964, p. 170
25. HENRY RJ, BERKMAN S: *Clin Chem 3*:711, 1957

26. HILLER A, VAN SLYKE DD: *J Biol Chem 53*:253, 1922
27. HOFFMAN WS, OSGOOD B: *J Biol Chem 124*:347, 1938
28. HORVATH AA: *Anal Chem 18*:229, 1946
29. HUNTER G: *J Clin Pathol 10*:161, 1957
30. JAGER BV, SMITH EL, NICKERSON M, BROWN DM: *J Biol Chem 176*:1177, 1948
31. JONES LT: *Anal Chem 14*:536, 1942
32. KRAUTMAN B: *Am J Clin Pathol 9*: Tech. Suppl. *3*:9, 1939
33. KRAUTMAN B: *Am J Clin Pathol 11*: Tech. Suppl. *5*:67, 1941
34. MacDONALD RP: *Am J Clin Pathol 25*:343, 1955
35. NEUBERG C, STRAUSS E, LIPKIN LE: *Arch Biochem 4*:101, 1944
36. NEUBERG C, STRAUSS E: *Exp Med Surg 3*:39, 1945
37. Reagent Chemicals, Am Chem Soc Specifications 1960, Appl. Pub., ACS, Washington, D.C., 1961, p. 343
38. ROBERT B, DE VAUX ST. CYR, Ch., ROBERT L, GRABAR P: *Clin Chim Acta 4*:828, 1959
39. SENDROY J Jr: *J Biol Chem 120*:335, 1937
40. SENDROY J Jr: *J Biol Chem 152*:539, 1944
41. SEVAG M, LACKMAN DB, SMOLENS J: *J Biol Chem 124*:425, 1938
42. SOMOGYI M: *J Biol Chem 86*:655, 1930
43. SOMOGYI M: *J Biol Chem 87*:339, 1930
44. SOMOGYI M: *J Biol Chem 160*:69, 1945
45. SOMOGYI M: *J Biol Chem 90*:725, 1931
46. SUNDERMAN FW, MacFATE RP, EVANS GT, FULLER JB: *Am J Clin Pathol 21*:901, 1951
47. SURGENOR D: *Quart Rev Internal Med Dermatol 9*:145, 1952
48. TANFORD C, WAGNER ML: *J Am Chem Soc 75*:434, 1953
49. TANFORD C: *J Am Chem Soc 74*:211, 1952
50. VAN SLYKE DD, HAWKINS JA: *J Biol Chem 79*:739, 1928
51. WEST ES, SCHALES FH, PETERSON VL: *J Biol Chem 82*:137, 1929
52. WOLFF E: *Ann Med 10*:185, 1921

Chapter 16

Proteins

DONALD C. CANNON, M.D., Ph.D.,
IRVING OLITZKY, Ph.D.,
JAMES A. INKPEN, Ph.D.

In addition to being major structural components of cells, proteins are involved to some extent in virtually all life processes—transport, enzymatic catalysis, homeostatic control, hormonal regulation, blood coagulation, immunity, growth and repair, and heredity. It is not surprising, therefore, that more has been written about proteins than any other subject in clinical chemistry and that this field continues to expand rapidly. Since the first edition of this book lipoproteins have achieved such prominence in clinical medicine that they now are relegated to a separate chapter. Immunochemical technics continue to provide an active stimulus to the identification and quantification of specific proteins. Consequently it is likely that both immunoglobulins and "miscellaneous proteins" will require separate mention in future editions of this book.

DETERMINATION OF TOTAL PROTEIN IN SERUM OR PLASMA

The concentration of protein in solution is universally expressed on the basis of weight per unit volume. Thus all methods ultimately must be standardized gravimetrically. Gravimetric methods are available (241, 540), but their complexity and the hygroscopic nature of dried proteins prevent their use as routine procedures. Even in the case of gravimetric determinations, however, the path to be taken is not clearly defined for it must be decided whether to include the lipid and carbohydrate moieties of protein conjugates in the protein weights determined. It has been proposed that the polypeptide content be established as the basis for the determination (452). This would be valid for some protein procedures but not for others as is discussed subsequently.

The classic approach to the determination of total protein has been the determination of protein N. The Dumas method in which the N is determined as N_2, is used to a very limited extent because it does not lend itself readily to the determination of total protein (294). In contrast, the Kjeldahl procedure has been widely accepted as the reference method for the determination of total protein in biologic material.

KJELDAHL DETERMINATION OF PROTEIN NITROGEN

The Kjeldahl determination of protein N is usually performed in the clinical laboratory only for evaluating or standardizing other methods (174).

In the Kjeldahl method all N-containing compounds are converted to NH_4^+ by oxidation in a digestion mixture of concentrated H_2SO_4, a catalyst, and a salt to increase the boiling point of the mixture. In some cases an oxidizing agent is added to complete the digestion and to clear the digestion mixture. By addition of alkali the NH_4^+ is converted to NH_3, which is usually steam-distilled into a receiver flask containing boric acid and then titrated with standard HCl (491). The NH_4^+ formed in the Kjeldahl digestion may also be determined by the Berthelot color reaction or by nesslerization (cf Chapter 17, *Determination of Nonprotein Nitrogen*), but these methods are generally regarded as less accurate and precise than the titrimetric approach (174). The NH_4^+ in the Kjeldahl digest can be converted to N_2 by the addition of hypobromite and the gas determined manometrically.

Kjeldahl in introducing his method in 1883 used a mixture of H_2SO_4 and H_3PO_4 for the digestion. Potassium permanganate was added to oxidize alkaloid compounds, which are resistant to oxidation. Many modifications of the original procedure have been proposed: (a) Addition of sodium or potassium sulfate to raise the boiling point and thus speed up digestion. Excessive amounts of salts, however, may increase the boiling point to a temperature where NH_3 is lost as a result of decomposition of NH_4HSO_4. (b) Addition of cupric or mercuric sulfate or metallic selenium as catalyst. Mercuric sulfate has the disadvantage that metallic mercury must be disposed of after the analysis is completed. The use of a copper salt makes the digest unsuitable for either direct nesslerization or gasometric determination of NH_3 (351). (c) Use of oxidizing agents such as potassium persulfate and H_2O_2.

For comparison of different Kjeldahl technics pure substances of known N content should be used. The most important factor in Kjeldahl technics is the digestion time. For serum samples most procedures call for continued heating for 0.5–12.0 hours after the mixture clears, with the usual time being 0.5–2.0 hours (134, 529).

The validity of the determination of protein based on a total N analysis rests on two assumptions: (a) complete or at least constant recovery of the protein N in the analysis, and (b) constancy of N content of the various proteins in the biologic sample. The first assumption is reasonably correct for serum samples even with the difficulty which occurs because of corrections for nonprotein N. Problems occur with the second assumption, however, since total N is converted to total protein by the use of a factor. Calculations generally are based on the assumption that proteins from biologic sources contain 16% N by weight, and the N content is multiplied by 6.25 to obtain the protein content. This factor has been in use for over 75 years since its obscure origin. Investigations of various protein fractions present in serum or plasma have shown that the factors actually extend over a range of 5.9–12.5 (13, 462). The higher factors belong to protein species that are lipid or carbohydrate conjugates and consequently contain less N. The usual premise is to disregard the nonprotein moieties and base all protein standardization strictly on the protein moieties. One study revealed that if the nonprotein moieties are disregarded the factors range from 5.69 to 6.52 for proteins in normal serum (350). The factor 6.54, however, has been recommended for converting N to total protein in normal serum (468). Unfortunately, it is a common but unwarranted assumption that the factor 6.25 is valid for proteins in all biologic fluids (162, 359). The use of this factor for calculation of total protein in urine or cerebrospinal fluid (CSF) is questionable since the protein distribution is significantly different from that of plasma or serum (162, 359, 406, 414).

DETERMINATION OF PROTEIN BY KJELDAHL ANALYSIS FOR NITROGEN

(method of Hiller et al, Ref 351, modified)

PRINCIPLE

The N of the protein and all other N-containing compounds is converted to NH_4^+ by oxidation in a digestion mixture of H_2SO_4, potassium sulfate, and copper sulfate. After alkalinization the NH_3 is distilled into a boric acid solution and titrated with standard acid. Protein N, calculated as the difference between nonprotein N and total N, is translated into serum total protein concentration using the factor 6.25.

REAGENTS

Copper Sulfate, Crystals, $CuSO_4 \cdot 5H_2O$.   Reagent grade.
Potassium Sulfate. Reagent grade.
H_2SO_4, conc. Reagent grade.
Boric Acid, 4%
NaOH, 40%
Methyl Purple Indicator. (Available in solution from Fleischer Chemical Co.)
HCl, 0.05 N
Standard N Solution. Dissolve 4.9520 g $(NH_4)_2SO_4$, reagent grade, in water and make up to 1 liter. 2 ml = 2.1 mg N.

PROCEDURE

1. Into a 100 ml digestion flask (with 24/40 joint) add the following:
   2 ml water + 0.2 ml serum (TC pipet) or 2 ml standard
   1.0 g K_2SO_4
   0.1 g $CuSO_4 \cdot 5H_2O$
   2 ml conc H_2SO_4. May use this acid to wash any crystals to the bottom of the flask. USE NO MORE THAN 2 ML CONC H_2SO_4. Add 2 or 3 glass beads.
2. Slowly digest mixture to SO_3 fumes. Be sure no material has climbed in the neck of the flask and evaded digestion. After the mixture turns green it should then be digested for 1.5—2.0 hours more by boiling gently. *Avoid* excess heat and consequent loss of N throughout the digestion.
3. Cool and add 5 ml water. (Flask may be stoppered and the determination interrupted at this point if necessary.)

4. Attach flask to semimicro steam distillation apparatus. When steam is being generated, add 10 ml 40% NaOH to sample in such a way as to preclude any loss of NH_3. Steam distill into 5 ml of boric acid containing 1 drop of methyl purple indicator. Tip of condenser must be immersed in boric acid. Distill for 10 min.

5. Titrate to indicator color change from green to purple with standardized 0.05 N HCl in a micro buret.

Calculations:

$$\text{total N in g/100 ml} = 14 \times \frac{100}{0.2} \times \frac{\text{ml acid}}{1000} \times \text{normality of acid}$$

$$= 7 \times \text{ml acid} \times \text{normality of acid}$$

$$\text{protein N} = \text{total N} - \text{NPN}$$

$$\text{serum protein, g/100 ml} = \text{protein N} \times 6.25$$

NOTES

1. Reagent blank. A blank must be run on all reagents by making a complete analysis and omitting the sample. With reagent grade chemicals and clean glassware this blank may be zero.

2. Recovery. Recovery runs can be made on accurately prepared standard solutions of $(NH_4)_2 SO_4$.

3. Determination of protein N in the presence of nonprotein N. When determining the protein content by nitrogen determination of a solution containing nonprotein nitrogen, three courses of action are available: (a) In the case of serum or plasma a normal NPN can be assumed and a correction made by subtracting 0.2% protein from the result. If the sample is a normal one, the error resulting from this assumption is negligible. (b) An NPN can be determined by nitrogen determination on a tungstic acid filtrate. (c) The Kjeldahl analysis can be run on a washed tungstic acid precipitate of the sample, dissolving the precipitate in the concentrated sulfuric acid and quantitatively transferring to the digestion flask by several small washes of water.

4. Determination of N in biologic material. The procedure as presented can also be used for the assay of total N in either biologic fluids such as urine or homogenates of feces or tissues. However, conversion of the total N value to a total protein value by the use of the same factor used for serum introduces an error which may be significant. In the procedure presented the upper limit for the amount of N which can be conveniently measured is about 10 mg. If samples containing larger amounts of N are to be assayed a smaller sample must be used or else the procedure must be scaled up proportionately.

5. Indicator. The methyl purple indicator is a mixed indicator solution, being green at pH 5.4, gray at pH 5.1, and purple at pH 4.8. The endpoints of this indicator are very sharp so that color standards are unnecessary. If at the beginning of distillation there is no color change, either there is no NH_3 or insufficient alkali was added to neutralize the sulfuric acid. A bromcresol green-methyl red indicator mixture (50 mg bromcresol green and 10 mg methyl red in 60 ml denatured ethanol) is a satisfactory substitute for methyl purple. This indicator mixture gives a blue-purple color in boric acid solution, changing to blue-green in the presence of a trace of NH_3 and to pink in the presence of a trace of mineral acid. Titration is continued until the blue disappears. An alternative is to titrate to pink before distillation and back to pink after distillation. Methyl orange, frequently employed in this titration, is not recommended because of its gradual color change.

6. Distillation. Distillation of the NH_3 into boric acid has three advantages over distillation into a measured amount of standard acid and titrating the excess unneutralized acid: (a) Only one standard solution is required. (b) There is no danger of spoiling a determination because of insufficient acid absorbent. (c) If part of the boric acid is accidentally drawn back into the distillation flask, distillation can be resumed.

ACCURACY AND PRECISION

This technic is highly accurate for the determination of N.

With the most careful technic the precision of this analysis is limited primarily by the measurement of the sample itself. For greatest accuracy, calibrated TC pipets should be used, in which case a precision of somewhat better than ±1% is obtainable.

NORMAL VALUES

See *Normal Total Protein Values in Serum.*

PROTEIN DETERMINATION BY THE BIURET REACTION

When urea is heated to about 180°C, it decomposes to give a product called "biuret," which in the presence of Cu^{2+} in alkaline solution forms a violet colored complex. The reaction is as follows:

UREA   BIURET   Cu COMPLEX

This color reaction is given by (a) substances containing two $-CONH_2$, $-CH_2NH_2$, $-C(NH)(NH_2)$, or $-CSNH_2$ groups joined either directly or through a C or an N; and (b) peptide structures containing at least two peptide linkages. This reaction also occurs with the peptide bonds of proteins (350) and is the basis for the biuret determination of protein in serum and other biologic fluids.

Early technics for the biuret determination of protein in serum used a considerable excess of copper salt. About 3% of the copper precipitated as $Cu(OH)_2$ when added to protein solutions at the alkalinity usually employed. This precipitate had to be removed by filtration or centrifugation. In 1942 the older biuret procedure was simplified by using a single reagent in which the copper salt at low concentration was stabilized in a 12% alkaline solution thereby avoiding precipitation of $Cu(OH)_2$ (233). This reagent had a limited stability period since the cupric salts are susceptible to autoreduction. General modifications have been proposed for stabilizing the biuret reagent and permitting the use of lower alkalinity without the formation of $Cu(OH)_2$ precipitates (94). Weichselbaum (500) used sodium potassium tartrate as a stabilizer and potassium iodide to prevent autoreduction; however, this reagent was found to be unstable on long storage (174). Gornall et al (153), after a thorough study of the biuret reagent, retained the tartrate as a stabilizer, finding that there must be a minimum ratio of 3:1 (w/w), tartrate to copper sulfate. Furthermore the potassium iodide was unnecessary since autoreduction occurred only when impure $CuSO_4$ was used.

Benedict's qualitative glucose reagent has also been used as a biuret reagent (145, 177, 197). This reagent contains citrate as a stabilizing agent and is stable at room temperature indefinitely because of its relatively low alkalinity (contains Na_2CO_3 rather than NaOH). The required amount of NaOH is introduced as a separate reagent. This procedure allows the use of a 3% NaOH solution, which is an advantage since maximum color is produced in the biuret reaction when the final concentration of NaOH is between 1.2% and 2.8% (153).

The absorption maximum of the protein-biuret complex in the visible range occurs at 545 nm when the absorbance is read against a biuret reagent blank (174). If the absorption curve is read against a water blank, however, the maximum occurs at 580 nm (see Fig. 16-1). The protein-biuret complex also has a maximum at 300 nm (145, 174). Measurement at 300 nm gives a fourfold increase in sensitivity, but interference by turbidity is also greatly magnified.

Absorptivities of various proteins compared on the basis of protein N concentration have been reported to be nearly identical, with few exceptions (177, 350). Results obtained with biuret reagents containing ethylene glycol appear to be an exception to this statement (308), but this may be due to interference by the ethylene glycol (530). The total protein in normal sera as determined by the biuret reaction agrees quite well with protein determined by the Kjeldahl method (174). The protein moieties of lipoproteins and glycoproteins have similar reactivity per unit weight (350). Their biuret equivalent varies from 95 to 116 with albumin arbitrarily set at 100 (350).

The biuret reaction has been adapted to the AutoAnalyzer. Earlier procedures used the biuret reagent described by Weichselbaum (500), which contains potassium iodide to prevent autoreduction (117, 537). Since potassium iodide has been shown to be unnecessary (153), the biuret reagent without potassium iodide has also been used in automated procedures (78). The manual biuret procedure as described by Henry *et al* (177) was adapted to the AutoAnalyzer by Rice (381).

DETERMINATION OF SERUM PROTEIN BY THE BIURET REACTION

(method of Henry et al, Ref 177)

PRINCIPLE

Protein forms a colored complex with Cu^{2+} in alkaline solution. The absorption curves of the biuret complex are shown in Figure 16-1.

REAGENTS

Biuret Reagent (Benedict's Qualitative Glucose Reagent). (a) Dissolve 17.3 g $CuSO_4 \cdot 5H_2O$ in about 100 ml hot water. (b) Dissolve 173 g sodium citrate and 100 g anhydrous Na_2CO_3 in about 800 ml water by heating.

FIG. 16-1. Absorption curve of protein-biuret: (a) reagent blank vs water, (b) protein-biuret vs water, (c) protein-biuret vs reagent blank. Beckman model DU spectrophotometer, half-bandwidth 1.3 nm at 545 nm (peak of curve c); 1.07 mg Armour's crystalline bovine albumin/ml.

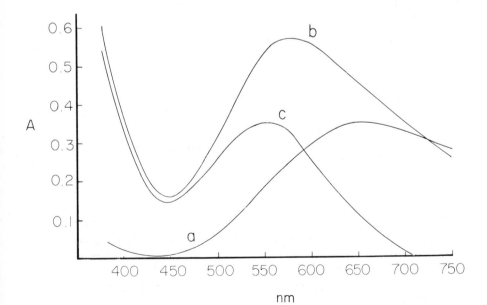

When cool, pour (a) into (b) while stirring and dilute to 1 liter. Stable indefinitely at room temperature.

NaOH, 3%

Protein Standards. There are two alternatives: (a) Pooled clear serum containing 6−8 g/100 ml standardized by Kjeldahl analysis, corrected for NPN, so that the actual concentration is known to within ±2%. Aliquot and store frozen; stable indefinitely. (b) Standardized and stabilized protein solution from commercial sources. The use of bovine serum albumin as a standard for human serum protein has been recommended (350) and is well established (177).

PROCEDURE

1. Set up in separate test tubes:
   Reagent blank: 5 ml 3% NaOH
   Standard: 0.1 ml protein standard (TC pipet) + 4.9 ml 3% NaOH
   Unknown: 0.1 ml unknown (TC pipet) + 4.9 ml 3% NaOH
2. Add 1.0 ml biuret reagent to each tube and mix. This step should follow step 1 without delay.
3. Wait at least 15 min, examine for Tyndall effect. If negative or only a trace, read absorbances of standard (A_s) and unknown (A_x) against the reagent blank at 545 nm or with a filter in this region. Color is stable for several hours although turbidity may develop. If unknown is significantly turbid, as it always is in the case of a lipemic serum, add about 3 ml ether, shake vigorously for 30 sec, and centrifuge. If clear, read against reagent blank and subtract 2% from calculated result (for routine work, ignore this correction). If turbidity is still present, try reextraction with fresh ether. If it is obvious that ether extraction will not work, start over and after development of biuret color read the absorbance, then add about 100 mg KCN and mix. After about 2 min read against a water blank and subtract this absorbance from the original biuret absorbance which included the turbidity.

Calculation:

$$\text{g protein}/100\ \text{ml} = \frac{A_x}{A_s} \times \text{conc of standard (g/100 ml)}$$

NOTES

1. Beer's law. Obedience to Beer's law should be established when the test is set up and standardized. In our experience there is no deviation up to a protein concentration of 15 g/100 ml even with a filter photometer. The protein-biuret color reaches a maximum within 15 min and is stable for several hours. With some biuret reagents it has been reported to be stable for several days (500).

2. Micro procedure. The total protein in 10 μl of serum can be determined by scaling down the above procedure tenfold.

3. Stability of serum and plasma. There is no change in the total protein of clear samples held for 1 week at room temperature. Serum is stable at refrigerator temperature for at least 1 month (174).

4. Interferences. (a) *Turbidity.* This interference arises most frequently with lipemic sera, and in most of these instances ether extraction clarifies the solution. Technics for removal of turbidity include extraction of the biuret reaction with ether; preliminary precipitation of the proteins with trichloroacetic acid; subtracting the absorbance due to turbidity after dispelling the copper color by addition of KCN and subtracting the absorbance of a blank containing the serum but devoid of the copper reagent. A reexamination of the turbidity problem led to the following conclusions (177): (a) Many biuret reactions which appear clear when examined in the usual manner exhibit a definite Tyndall effect that produces significant error. (b) The KCN technic usually overcorrects. (c) Preliminary precipitation with trichloroacetic acid or organic solvent mixtures occasionally fails. (d) Correction by a serum blank occasionally leads to a significant error. (e) Ether extraction is the most universally successful technic except in certain instances of icteric sera.

(b) *Bilirubin.* The presence of bilirubin in a concentration up to 29 mg/100 ml serum does not interfere (177). Occasionally, however, when the serum is both icteric and turbid, ether extraction does not successfully clarify the solution, in which case decolorization with KCN must be employed.

(c) *Salts.* Neither sodium sulfate nor sodium sulfite in the concentrations employed in salt fractionation alter significantly the readings made in square cuvets; this is not necessarily true for other biuret technics (177). If readings are made in round cuvets, however, a reagent blank containing the same salt concentration must be employed because of the change in refractive index caused by the salt (177). Large amounts of ammonium ions interfere by formation of a cupric ammonium complex, the copper then being unavailable for reaction with protein. The interference can be minimized by increasing the concentration of NaOH to about 10 g/100 ml. In the method presented, no significant error is introduced if the NH_3-nitrogen is no greater than five times the protein-nitrogen present (177).

(d) *Hemolysis.* Visible hemolysis in serum or plasma begins at a concentration of about 50 mg hemoglobin per 100 ml. The hemoglobin concentration may be determined by any method of sensitivity sufficient for the degree of hemolysis present. Thus for a hemoglobin level of 4 g/100 ml either the oxyhemoglobin or cyanmethemoglobin method is satisfactory. For low concentrations the benzidine method is required. The color produced in the biuret reaction by 1 mg hemoglobin is equivalent to 1.9 mg serum protein. Perform the biuret determination in the usual manner and subtract the protein equivalent of the hemoglobin present. This correction is valid up to a hemoglobin concentration of about 4 g/100 ml (177). Hemoglobin concentrations less than 100 mg/100 ml can be ignored.

(e) *Bromsulfophthalein.* The presence of this dye in serum causes a false elevation in results (109).

ACCURACY AND PRECISION

A blanket statement as to the accuracy cannot be made because of the various factors discussed previously. Practically, however, there seems to be no question but that the accuracy is always sufficient for clinical purposes provided there is no significant interference from turbidity.

The precision of the method is about ±3%.

NORMAL VALUES

See *Normal Total Protein Values in Serum,* this chapter.

AUTOMATED DETERMINATION OF SERUM PROTEIN BY THE BIURET REACTION

(method of Weichselbaum, Ref 500, adapted to AutoAnalyzer by Stevens, Ref 537)

PRINCIPLE

Serum and biuret reagent are mixed and the absorbance of the resultant color measured at 550 nm.

REAGENTS

Alkaline Iodide. Dissolve 8 g NaOH in about 800 ml water in a 1 liter volumetric flask. Add 5 g KI and stir until dissolved. Dilute to 1 liter and filter.

NaOH, approx 0.2 N

Stock Biuret Reagent. To 45 g of sodium potassium tartrate ($KNaC_4H_4O_6 \cdot 4H_2O$) in a 1 liter volumetric flask add 400 ml of 0.2 N NaOH. With stirring add 15 g $CuSO_4 \cdot 5H_2O$. After the copper sulfate is dissolved add 5 g KI and dilute to 1 liter with 0.2 N NaOH.

Working Biuret Solution. Dilute 200 ml stock biuret to 1 liter with alkaline iodide solution.

Blank Solution. Dissolve 9 g sodium potassium tartrate in 200 ml alkaline iodide solution and dilute to 1 liter with alkaline iodide solution.

Protein Standards. Crystalline bovine albumin with a known moisture and ash content. Store in refrigerator and protect from moisture absorption. Dissolve 10 g albumin corrected for nonprotein material in 80 ml 0.9% saline and dilute to 100 ml with saline. Prepare working standards to give protein concentrations of 2—8 g/100 ml by dilution of the stock with saline. Aliquot the various standards and store frozen until required.

PROCEDURE

Set up the manifold, other components, and operating conditions as shown in Figure 16-2. By following the procedural guidelines as described in Chapter 10, steady state interaction curves as shown in Figure 16-3 should be obtained and checked periodically. Place a set of working standards followed by the reagent blank and serum or plasma samples in the sample tray. Serum blanks are not normally run except for turbid sera (see *Note 2*). Results are calculated from the calibration curve in the standard manner for methods obeying Beer's law.

NOTES

1. See Notes 1, 3, and 4 of manual method.

2. Blank correction for turbid sera. Serum to be analyzed for total protein by this method should be clear and nonturbid. If turbidity is present a blank correction has to be made as follows: readjust the reagent base line with blank solution, run turbid serum sample with blank solution in place of the biuret reagent and subtract the resultant blank value from that obtained with the biuret reagent.

As noted under manual method (cf *Note 4*) the use of a serum blank to correct for turbidity may lead to a significant error, but with the automated method there is no other alternative.

ACCURACY AND PRECISION

Accuracy is sufficient for clinical purpose provided that correction is made for turbidity.

The precision (95% limits) is about ±3%.

NORMAL VALUES

See discussion of *Normal Total Protein Values in Serum,* which follows.

NORMAL TOTAL PROTEIN VALUES IN SERUM

There have been many studies on what constitutes the normal protein content of human serum. Nevertheless, because of the many variables the "normal" is difficult to define exactly. This is especially true for serum. Table 16-1 lists representative normal adult ranges obtained by various workers. It is obvious that since both the ranges and their averages differ significantly either the sampling was not representative of the whole

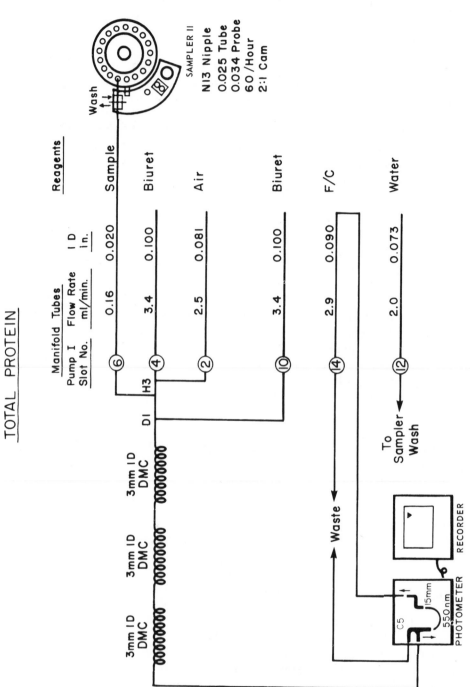

FIG. 16-2. Flow diagram for total protein in serum. Calibration range: 0.12 g/100 ml of serum. Conventions and symbols used in this

FIG. 16-3. Total protein strip chart record. S_H is a 7.1 g/100 ml control serum, S_L a 3.55 g/100 ml control serum. Peaks reach 96% of steady state; interaction between samples is less than 0.5%. $W_{1/2}$ and L are explained in Chapter 10.

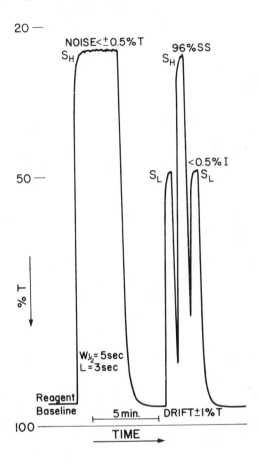

TABLE 16-1. Adult Normals for Total Serum Protein

Ref	Method used	Number in series	Total protein (g/100 ml) Mean	Range
37	Kjeldahl	87	6.9	5.6–8.2
376	Kjeldahl	80	7.23	6.1–8.3
377	Biuret	42	7.14	6.6–7.6
233	Biuret	40	7.50	6.6–8.4
175	Biuret	24	7.4	6.3–8.5
386	Biuret (automated)	400	—	6.0–7.4
372	Biuret (SMA 12/60)	1419	7.4	6.6–8.3
89	Biuret (SMA 12/60)	2400	6.91	6.0–7.9

population or the technics used were inadequately standardized. The biuret procedures, which are the most common methods for determining total serum protein, are influenced by the validity of the Kjeldahl which in many cases is used for standardization. Our laboratory determined the normal range to be 6.6–8.3 g/100 ml based on a study of 1419 clinically normal adults (372).

Variables which might affect the total serum protein concentration include the following:

(a) *Serum vs plasma.* Theoretically, the total protein content of serum should be less than that of plasma by an amount equal to the fibrinogen concentration, which ordinarily is about 0.3 g/100 ml. As a result of loss of CO_2 from the serum during clotting, water leaves the erythrocytes, thus diluting the serum proteins. If oxalate, citrate, or fluoride is used as an anticoagulant, water diffuses out of the erythrocytes to compensate for the increased salt concentration. This shift occurs within 15 min and is proportional to the concentration of the anticoagulant used. With the concentrations ordinarily employed the dilution that occurs yields a plasma protein concentration usually no higher and sometimes less than that of serum (73). If heparin is used as anticoagulant in a concentration of about 0.1 mg/ml of blood, no such water shift occurs (73).

(b) *Venous vs capillary serum.* There is no difference (149).

(c) *Vertical vs horizontal position.* Total serum protein is reported to be lower by 0.4–0.8 g/100 ml with the subject supine than when in the erect position. The minimum is obtained within 2–4 hours after assuming the horizontal position and the maximum within 2 hours after rising. This phenomenon is usually ascribed to hemoconcentration when in the erect position (254, 506).

(d) *Healthy ambulatory adults vs convalescents.* As might be expected, the former group has a higher average serum protein concentration than the latter (254). How much of this difference can be ascribed to the more frequent supine position to be expected in convalescents has not been investigated.

(e) *Exercise.* An increase of 6–12% occurs with vigorous exercise of short duration (254).

(f) *Age.* The normal total serum protein is about 3.6–6.0 g/100 ml for prematures and about 4.6–7.0 g/100 ml for full term infants (27, 174). The serum protein increases gradually, and adult levels are reached by the third year (27, 337). There is either no change or only a slight decrease in serum protein during old age (174, 372).

(g) *Meals.* There is no postprandial change (11).

(h) *Individual variation.* It has been reported that the individual variation in healthy young adults is about ±0.4 g/100 ml (525). Seasonal and diurnal variations are not observed (254, 525).

(i) *Sex.* A slight difference between sexes has been reported by various authors (258, 372) but is so small that it can be ignored. During pregnancy there is a decrease in total serum protein (327, 337).

(j) *Drugs*. Steroid contraceptive pills have been variously reported to increase or to decrease total serum protein (262, 263, 327). Estrogen therapy has been reported to decrease the total serum proteins (327).

MISCELLANEOUS METHODS FOR DETERMINATION OF TOTAL PROTEIN IN SERUM

Specific Gravity Measurements

In the method of Van Slyke *et al* (487) the total protein is computed by suspension of drops of serum or plasma in a series of copper sulfate solutions of known density. The method was designed mainly for field use and has been shown to be unreliable when the protein composition of the unknown is abnormal, when the levels of glucose or urea are grossly elevated, or when there is lipemia (33, 174).

Measurement of Refractive Index

This method is rapid, direct, and works well for clear sera (24, 174, 294). Materials such as glucose, bilirubin, cholesterol, and triglycerides, if present in abnormal amounts, can lead to erroneous results, as the determination of serum proteins by refractometry is dependent upon the nonprotein content of serum or plasma being constant and contributing only a small fraction to the refractivity (174, 498). The difference between the refractivity of equal weights of a lipoprotein and albumin is much less than the difference between the N contents. Therefore total protein determination of lipemic serum is higher by refractometry than by Kjeldahl or biuret (498).

Absorption of Radiant Energy

The fact that tyrosine, tryptophan, and phenylalanine of protein absorb in the 270–290 nm region has been used extensively for determining the concentration of individual proteins in solution. This approach has been used much less for the determination of total protein in serum or CSF (174, 294, 498). In fact, valid results for total protein cannot be obtained with this procedure since significant variations occur in tyrosine and tryptophan content among various serum protein fractions especially between albumin and the globulins (350, 463, 498), as well as variation between corresponding fractions from different sera (463). Many sera also contain nonprotein interfering materials—e.g., uric acid, bilirubin, free tyrosine, and tryptophan—which invalidates the use of this procedure (498).

The peptide bond absorbs energy at wavelengths between 200 and 225 nm. At 205 nm 70% of the absorption is attributable to the peptide bond with a specific absorption for most proteins 10- to 30-fold greater than that at 280 nm. The use of 215–225 nm has been recommended because of instrumental limitations (498). The absorption by proteins at these low wavelengths has been reported to obey Beer's law (498). Because of the low wavelengths used, exacting conditions and very high quality instrumentation are required. Another drawback is the interference from other materials such as tyrosine and tryptophan, which also absorb in this spectral region (344).

Turbidimetric Methods

This technic is regarded as less accurate than other rapid methods for quantitative determination of total protein in serum since it is difficult to control particle size (294). This approach has been used largely for quantitation of urine and CSF proteins and as such is considered clinically satisfactory. Turbidimetric procedures for urine and CSF are discussed in more detail under the determination of urine and CSF proteins.

Dye Binding ("Protein Error of Indicators")

Protein reacts with dyes to form complexes, some of which possess optical properties useful for quantitation. This approach has been shown to be useful for specific classes of protein; but because various proteins bind or react differently, this method is not recommended for determination of total protein in serum or plasma (24, 174, 294, 498). This approach is considered more fully later in this chapter for the determination of albumin.

DETERMINATION OF TOTAL PROTEIN IN CEREBROSPINAL FLUID

There are three major problems in determining the protein content of CSF: (a) the concentration of protein is low (normally 15—45 mg protein per 100 ml); (b) the concentration of nonprotein interfering substances is high relative to the protein concentration; and (c) the concentration of inorganic ions is high. Most of the methods for determination of protein in serum and plasma have been applied to CSF but have not generally proved satisfactory. The Kjeldahl digestion is frequently used as the reference procedure, but there are problems in obtaining an adequate volume of CSF (10—15 ml) and in correcting for the disproportionately large amount of nonprotein nitrogen present (379, 431). The biuret reaction has also been used (104, 145, 231), but the results obtained are generally high in comparison with other procedures (344, 380, 431). This discrepancy is believed to result from interference by proteoses, peptones, and polypeptides present in relatively high concentration. When a single biuret reagent is used, the final solution must be centrifuged before measuring the color to remove the white precipitate of $Ca(OH)_2$ and $Mg(OH)_2$ that forms. Precipitation can be prevented by the use of the disodium salt of ethylenediaminetetraacetic acid (397). Reading of the biuret color at 330 nm to increase sensitivity has also been suggested (145). Only one procedure using the biuret reaction has been reported to give results comparable with those obtained with micro-Kjeldahl and turbidimetric methods (380). In this procedure the CSF protein is precipitated with trichloroacetic acid, the biuret reaction is performed on the protein precipitate, and the resultant color is read at 330 nm. Direct spectrophotometric measurements of protein at 220 or 280 nm (344, 378, 480, 498) are subject to interference by ascorbic acid, tryptophan, nucleosides, peptides, and proteoses resulting frequently in falsely elevated results (344). Before using direct spectrophotometric methods, it has been recommended that CSF proteins be purified by gel filtration (344).

The two most satisfactory technics for the determination of CSF protein are the turbidimetric and Folin-Lowry methods.

TURBIDIMETRIC ESTIMATION OF PROTEINS

In 1914 Folin and Denis introduced the turbidimetric determination of protein in urine with sulfosalicylic acid. Subsequently, this technic was extended to protein determination in cerebrospinal fluid. The final concentration of sulfosalicylic acid usually employed is about 2.5%, and the final concentration range of protein determinable is about 1–25 mg/100 ml. A serious objection to the use of sulfosalicylic acid is the fact that the turbidity produced with CSF albumin is markedly greater (two- to fourfold) than that produced with CSF globulin (176, 315), so that results are dependent on the A/G ratio (431).

Mestrezat (314) in 1921 employed trichloroacetic acid for the turbidimetric determination of protein in spinal fluid, and others since have recommended its use in place of sulfosalicylic acid (176, 315). The discrepancy between turbidities produced with albumin and globulin is considerably less than with sulfosalicylic acid. In the studies performed in our laboratory γ-globulin produced about 20% higher turbidities than albumin (176). Several authors have used recrystallized bovine albumin as the standard for the trichloroacetic acid turbidimetric method (379, 380). In view of the difference in turbidity of protein fractions, the present authors recommend the use of a dilute human serum standard.

Turbidity has been reported to vary directly with temperature, but the variation is within ±5% over the range of normal room temperatures (20°–25°C) when using trichloroacetic acid (423). At temperatures above 25°C results obtained with the trichloroacetic acid procedure begin to resemble those obtained with sulfosalicylic acid, in that much more turbidity is found with albumin than with globulin (423)—indicating that the temperature has to be controlled to obtain reasonable accuracy with this procedure. As with sulfosalicylic acid, there is flocculation and settling of particles, and resuspension by shaking introduces troublesome air bubbles that do not easily float to the surface. The turbidities produced by trichloroacetic acid are roughly one-half those produced by sulfosalicylic acid (315).

The trichloroacetic and sulfosalicylic acid procedures agree reasonably well with biuret and Kjeldahl procedures performed on CSF protein precipitates (380, 431). Somewhat better agreement is obtained with trichloroacetic acid than with sulfosalicylic acid.

PROTEIN BY THE FOLIN-LOWRY METHOD

In 1912 Folin and Denis (129) found that a phosphotungstic-phosphomolybdic acid reagent gives a blue color with various substances containing a phenolic group. In 1922 Wu (521) utilized this reaction for colorimetric determination of plasma proteins; later Wu and Ling (522) extended the method to urine and spinal fluid. The difficulty of troublesome turbidities was largely overcome by using the reagent of Folin and Ciocalteu, which contains lithium salts instead of sodium salts (159). The turbidities were due to the formation of insoluble sodium salts, whereas the lithium salts are more soluble. Some turbidity, however, may develop even with this reagent (128).

Johnston and Gibson (211) developed the technic into the tyrosine equivalence method, which gives results comparable to the Kjeldahl method but is more difficult and time consuming than certain other procedures for the determination of CSF protein.

Pretreatment of the protein with an alkaline copper solution was found to increase the sensitivity of the phenol reaction 3- to 15-fold (181, 282). The color produced then results from reduction of the phosphotungstic and phosphomolybdic acids to molybdenum blue and tungsten blue by the Cu-protein complex and by the tryptophan and tyrosine of the protein (282). The latter two give color in the absence of Cu^{2+}, but the rest of the protein gives no color without Cu^{2+}. About 75% of the color is dependent on the presence of Cu^{2+} (282). Lowry and his colleagues (91, 282) utilized this information to devise a sensitive procedure for the determination of total protein which requires only 0.2 ml of CSF.

The main sources of error when determining CSF protein by the Folin-Lowry method are nonprotein substances present equivalent to about 6 ± 3 mg of protein per 100 ml CSF (470) and drugs such as salicylates, chlorpromazine, tetracyclines, and various sulfa drugs (109, 139, 533). To correct for these errors, the CSF protein can be calculated as the difference between simultaneous assays performed on the CSF directly and on a protein-free filtrate of the CSF. However, protein precipitation has inherent sources of error including both incompleteness of precipitation and co-precipitation of interfering substances (344). Comparison of the trichloroacetic acid turbidimetric technic with the Folin-Lowry method indicated that the results were in good agreement, although xanthochromic samples tended to give significantly higher values with the turbidimetric method (379). The sensitivity of the Folin-Lowry procedure makes it valuable for use with cerebrospinal fluid, in spite of problems with interfering substances.

Automated procedures for the phenol method have been developed, but their utilization for the determination of CSF protein has been limited (141, 192, 237).

DETERMINATION OF PROTEIN IN CEREBROSPINAL FLUID BY THE FOLIN-LOWRY METHOD

(method of Daughaday et al, Ref 91)

PRINCIPLE

Two reactions are involved: (a) an initial interaction of protein and Cu^{2+} in alkali (related to the biuret reaction); (b) a reduction of the phosphotung-stic and phosphomolybdic acids to molybdenum blue and tungsten blue (exact compositions unknown) both by the Cu-protein complex and by the tyrosine and tryptophan of the protein. The latter two give color in the absence of Cu^{2+}, but the rest of the protein gives no color without Cu^{2+}. About 75% of the color is dependent on the Cu^{2+}. The absorption curve of the colored products is given in Figure 16-4.

REAGENTS

Alkaline Tartrate Reagent. Dissolve 20 g Na_2CO_3 and 0.5 g Na or K tartrate in 1 liter 0.1 N NaOH.

FIG. 16-4. Absorption curve of color produced by Folin-Lowry reagent: (a) reagent blank vs water; (b) human albumin, 40 mg/100 ml, vs water; (c) human albumin, 40 mg/100 ml, vs reagent blank. Beckman DU spectrophotometer, 3.5 nm half-bandwidth at 755 nm peak.

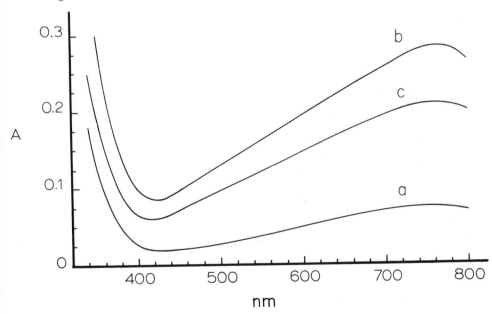

$CuSO_4 \cdot 5H_2O$, 0.1%

Working Alkaline Copper Reagent. Mix 45 ml alkaline tartrate reagent and 5 ml 0.1% $CuSO_4 \cdot 5H_2O$. Prepare fresh daily.

Folin and Ciocalteu Phenol Reagent. To a 2 liter flask, add 100 g sodium tungstate ($Na_2WO_4 \cdot 2H_2O$), 25 g sodium molybdate ($Na_2MoO_4 \cdot 2H_2O$), and 700 ml H_2O. When dissolved, add 50 ml 85% orthophosphoric acid and 100 ml conc HCl, and reflux gently for 10 hours (glassware used for the reflux apparatus must be equipped with ground glass joints). Add 150 g Li_2SO_4, 50 ml H_2O, and 2 or 3 drops of bromine. Boil the mixture for 15 min in a fume hood without condenser to remove excess bromine. (The bromine is added to reoxidize the slightly reduced reagent.) Cool, dilute to 1 liter, and filter if necessary. The reagent should be yellow without a green tint. Store in refrigerator. If reagent turns green during storage it can be restored by treatment with bromine again.

Dilute Phenol Reagent. Add 1.00 ml of concentrated reagent to 50 ml water and titrate to a phenolphthalein endpoint with standardized 0.1 N NaOH. This endpoint is not very sharp, and titration should be continued until the green color disappears and gives way to a pink-gray. On the basis of this titration, dilute the concentrated reagent so that 1 ml is equivalent to 9 ml 0.1 N NaOH (a dilution with H_2O to about one-half strength is usually required). The titration is checked with each batch of concentrated phenol reagent. It is usually recommended that the diluted reagent be made up fresh daily, but in our experience it is stable in the refrigerator for several months (discard if a green tint develops).

This standardization of the phenol reagent is required if the same readings are desired for each stock reagent prepared. If this is considered unessential, a routine dilution of 1:2 is satisfactory.

Standard. See *Note 2.*

PROCEDURE

1. Set up in test tubes:
   Reagent blank: 0.2 ml H_2O + 10 ml working alkaline copper reagent
   Standard: 0.2 ml protein standard containing 40 mg/100 ml (see *Note 1*) + 10 ml working alkaline copper reagent
   Unknown: 0.2 ml spinal fluid + 10 ml working alkaline copper reagent
   N.B. Greater accuracy is obtained if the 0.2 ml amounts are rinsed into the copper reagent from 0.2 ml TC pipets.

2. Mix tube contents and let stand 15 min.

3. Add rapidly to tubes, one at a time, 1 ml dilute phenol reagent. Mix *immediately* and thoroughly after addition (rapid mixing is necessary because of the instability of the reagent in alkaline solution).

4. Let stand at least 30 min (stable for several hours) and read absorbances at 750 nm (660 nm filter is satisfactory) against the reagent blank.

Calculation (see *Note 3*):

$$\text{mg protein/100 ml} = \left(\frac{A_x}{A_s} \times 0.08 \times \frac{100}{0.2} \right) - 6$$

$$= \left(\frac{A_x}{A_s} \times 40 \right) - 6$$

NOTES

1. Beer's law. A slight deviation exists even with a narrow half-band-width and does not increase materially with a broad-band filter. This suggests some kind of complexation or polymerization of the chromophore. By using a 40 mg/100 ml standard and assuming linearity, a maximal error of about 10% is incurred in the range of 20—60 mg/100 ml. For greater accuracy or for values in an extended range, a calibration curve must be prepared.

2. Standard. Since human γ-globulin yields about 20% more color than human albumin (174, 237, 498) the use of a purified standard containing only one of these proteins is inadvisable. Error due to difference in chromogenicity is minimized by using dilute serum as a standard. This can be prepared by proper dilution of the serum standard used in the biuret method.

3. Correction for nonprotein material. Color is obtained with trichloro-acetic acid filtrates of CSF, varying in protein equivalent from 3 to 9 mg/100 ml (91, 470). An average correction of 6 mg/100 ml is subtracted from all readings on unknowns.

ACCURACY AND PRECISION

In spite of the fact that proteins vary in chromogenicity, it has been claimed that the results are relatively unaffected by the A/G ratio (492). Results have also been reported to agree well with total protein determined by Kjeldahl analysis for normal CSF (108, 431) but are dependent on the validity of the correction for nonprotein material present (139, 379). Results can be spuriously high because of variability of the correction value for nonprotein substances present (e.g., tyrosine and tryptophan; drugs such as streptomycin, salicylates, acetylsalicylic acid, *p*-aminosalicylic acid, phenacetin, and chlorpromazine also give color) causing incorrect results (109). If results are above the normal range and interference by drugs is suspected, it is strongly recommended that CSF protein be calculated as the difference between simultaneous assays performed on the CSF directly and on a protein-free filtrate of the CSF.

The authors feel that in the routine use of this method one is running the slight risk of an occasional spuriously high result for a spinal fluid from a toxic patient with high circulating levels of endogenously produced phenolic substances.

The precision (95% limits) of the method is about ±10%.

NORMAL VALUES

The normal total protein for lumbar cerebrospinal fluid is 15–45 mg/100 ml for older children and adults (139, 174, 356, 379, 414). There is a tendency toward higher values in elderly adults (379). At birth the total protein levels are higher, usually less than 90 mg/100 ml (174, 379) but occasionally as high as 120 mg/100 ml (139). This upper limit gradually decreases to about 50 mg/100 ml during the first 5 years of life (139). In adults ventricular fluid protein is about 5–10 mg/100 ml and in cisternal fluid about 15–25 mg/100 ml (174, 379). In young children the cisternal fluid contains about the same protein concentration as lumbar fluid (174).

DETERMINATION OF TOTAL PROTEIN IN URINE

The quantitative determination of protein in urine presents problems which are similar to but more difficult than those for CSF. Interfering substances exist in greater quantity, inorganic ion content is higher, and protein levels are lower in urine than in CSF. Large discrepancies exist in the normal range established for urinary total protein by the various methods. This is partially explained by the presence in urine of relatively large amounts of low molecular weight proteins and polypeptides (48, 170, 406) and of mucoproteins (162, 412). Technics for determination of urinary protein apparently measure varying quantities of these materials in addition to the albumin and globulins that are present.

The lowest normal values for urine proteins are obtained using turbidimetric and dye-binding procedures, while higher values are obtained using the Biuret and Folin-Lowry methods.

The turbidimetric procedures have gained widespread acceptance for the determination of protein in urine because of their simplicity and sensitivity (cf section on *Turbidimetric Estimation of Proteins*, for discussion of factors affecting turbidimetric assays). Several precipitating agents have been used for turbidimetric assay of proteins: (a) Sulfosalicylic acid provides greater sensitivity of analysis but does not give the same quantity of turbidity with globulin as with albumin. Results are thus dependent upon the A/G ratio (176). This reagent also precipitates significant quantities of polypeptides from urine. (b) β-Naphthalene sulfonic acid precipitates fewer polypeptides from urine than does sulfosalicylic acid. The results obtained with this reagent correlate well with micro-Kjeldahl determination of dialyzed urine (172). (c) Trichloroacetic acid used in turbidimetric procedures has the advantage that albumin and globulin give similar turbidities, although there is some loss of sensitivity compared with sulfosalicylic acid procedures. Trichloroacetic acid does not precipitate mucoprotein at the concentration usually employed in turbidimetric procedures (412).

The biuret procedure has been used for the determination of total protein in urine. The procedures proposed usually require precipitation of the urine protein with trichloroacetic acid (174) or tungstic acid (174, 421) to remove protein from interfering material and to concentrate the protein. In one suggested method an equal volume of 10% trichloroacetic acid is added to an aliquot of urine containing 5–20 mg protein (183). The sample is then centrifuged, the supernate discarded, and the precipitate dissolved in 5 ml 3% NaOH. The analysis is concluded as in the manual procedure presented in the section *Protein Determination by the Biuret Reaction*. A correction for residual urine pigments may be made by measuring the absorbance of the protein precipitate dissolved in alkali versus water and subtracting this value from that obtained after forming the protein-biuret color. Savory *et al* (421) used 10% phosphotungstic acid in ethanol as the precipitating agent in which the protein in duplicate samples is precipitated at $0°C$ and the precipitate washed with cold absolute ethanol. One of the duplicates is dissolved in the biuret reagent, and the other is dissolved in an alkaline tartrate reagent to provide a sample blank. The normal range for adult urine protein by this method was found to be 40–150 mg/24 hours, with a sensitivity of 0.5 mg/100 ml and a 103% recovery ($1s = 3$) of added serum protein (421).

The Folin-Lowry procedure has also been used for the determination of urine protein. The urine protein is separated from interfering materials by precipitation, dialysis, or ultrafiltration (172, 214, 412). Saifer and Gerstenfeld (412) determined total urine protein by the Folin-Lowry procedure after precipitation and washing the protein with perchloric acid-acetone. The proteins determined with this precipitation and washing procedure include the mucoproteins (which can be up to 60% of the total protein in some urines) as well as albumin and the globulins. The mucoproteins can be removed by washing the precipitate with 5% trichloroacetic acid instead of perchloric acid-acetone, and the protein determined in the precipitate.

Since high mucoprotein levels can obscure changes in the albumin and globulin concentrations, it is frequently desirable to delete the mucoproteins from the determination of urine protein. This can be accomplished using the modification of the Folin-Lowry method described by Saifer and Gerstenfeld, or more simply by a turbidimetric procedure using trichloroacetic acid. Mucoproteins are soluble at the concentration of trichloroacetic acid usually employed in turbidimetric procedures.

DETERMINATION OF PROTEIN IN URINE BY TURBIDIMETRY WITH TRICHLOROACETIC ACID

(method of Henry et al, Ref 176)

PRINCIPLE

Addition of trichloroacetic acid to a solution of protein in low concentration precipitates the protein as a fine suspension which is quantitated photometrically.

REAGENTS

Trichloroacetic Acid, 12.5%. Reagent grade.
NaCl, 0.85%
Standard. A clear normal serum, the protein concentration of which is known, is diluted with 0.85% NaCl to a final protein concentration of 25 mg/100 ml. The protein standard employed in the biuret method can be suitably diluted. These dilutions should be prepared the day of use.

PROCEDURE

1. Set up in test tubes:
   Standard: 4.0 ml protein standard (1 mg protein)
   Unknown: 4.0 ml clear urine (centrifuge if turbid)
   Urine blank: 4.0 ml clear urine + 1.0 ml saline
2. To Standard and Unknown add 1.0 ml 12.5% trichloroacetic acid and mix immediately.
3. Between 5 and 10 min later measure absorbances of standard versus water and unknown versus urine blank. Before reading, remix, avoiding entrapment of air bubbles in the suspension.

Calculation:

$$\text{mg protein/100 ml} = \frac{A_x}{A_s} \times 1 \times \frac{100}{4}$$

$$= \frac{A_x}{A_s} \times 25$$

If A_x is above the range following Beer's law (see *Note 1*) start over using an appropriate dilution and make proper change in calculation.

NOTES

1. Beer's law. Beer's law is usually obeyed up to an absorbance of about 0.8–1.0 but should always be checked with the photometer used when first setting up the test.

2. Wavelength and bandwidth. Any wavelength in the short end of the visible spectrum can be used. The choice of 420 nm is entirely arbitrary. The half-bandwidth is not critical, straight line calibration curves usually being obtained even with glass filters.

3. Interfering colors. Urine color is usually corrected for by the use of the urine blank. If, however, the urine blank is obviously a different color from the urine to which the acid has been added, the method cannot be used. This may occur when certain drugs are being excreted.

ACCURACY AND PRECISION

Accuracy of the test for determination of protein on diluted normal and abnormal sera is about ±20%.

The precision is about ±8%.

NORMAL VALUES

The normal range for urine protein as determined in our laboratory is 25–75 mg/24 hours.

The normal protein content of urine varies greatly depending on the procedure used for its determination. Methods based on physical parameters give normal ranges for urine protein between 20–80 mg/24 hours (166, 320, 383, 476, 491). With biuret procedures the normal values are generally 40–150 mg/24 hours (52, 355, 421), and for Folin-Lowry methods the normal values are 40–400 mg/24 hours (172, 214, 412).

QUALITATIVE TESTS FOR URINE PROTEIN

Qualitative tests for urine protein are widely used to differentiate normal from abnormal concentrations. The success of such a test is dependent on the extent of the difference ordinarily encountered between the normal and abnormal concentrations and whether the sensitivity of the qualitative test is such that it will be negative for the former and positive for the latter.

Because of the possible significance of an elevated protein concentration in urine and the fact that this is one of the most common clinical laboratory tests requested, a qualitative screening test for "albuminuria" is most useful. None of the tests is specific for albumin, and it is more correct to refer to them as tests for protein rather than for albumin. Since the upper limit of normal for protein in urine is about 75 mg/24 hour output, the sensitivity of a qualitative test should not exceed about 5–10 mg/100 ml urine. Heller's nitric acid ring test in which concentrated HNO_3 is overlaid with urine has a sensitivity of about 10 mg protein per 100 ml urine but gives rings with mucin, proteoses, urea, uric acid, resinous bodies, and Bence-Jones protein. Robert's ring test, which uses a reagent of HNO_3 and $MgSO_4$, is superior to Heller's because there is less danger of nonspecific rings.

Formation of a protein precipitate by addition of sulfosalicylic acid or trichloroacetic acid can be used as a qualitative as well as a quantitative test. If performed for maximal sensitivity, these may detect as low as 0.25 mg protein per 100 ml urine (290). This of course is undesirable for a screening test for proteinuria, and a reasonable sensitivity is achieved either by proper dilution of urine in the test or by using the proper method for observing turbidity.

The authors recommend either the heat test or the sulfosalicylic acid test, one as a primary screening test and the other to be used as a confirmatory test.

Commercial materials are available for the qualitative testing of protein in urine. These tests are based on either the protein error of indicators or precipitation of urine protein. Some of the materials available as test reagent strips or as tablets include:

Albustix (Ames). The test reagent is the indicator bromphenol blue, buffered to a pH of about 3, impregnated in a paper reagent area attached to a plastic strip. At this pH the indicator is normally light yellow, but in the presence of protein at the same pH it is blue, the intensity of which increases with protein concentration. The sensitivity of the Albustix is about 20–30 mg/100 ml (19, 137). Other reagent strips have the protein test on the same plastic strip as reaction areas for other tests: *Bili-Labstix, Labstix, Hemo-Combistix, Combistix,* and *Uristix* (Ames).

Bumintest (Ames). This is a tablet of sulfosalicylic acid and an effervescent base used to prepare a 5% solution of sulfosalicylic acid. Equal parts of the solution and urine are mixed, and any resulting turbidity is graded from trace to 4+.

Protinur (Diagnostic Aids). This is a powdered mixture of sodium tetrametaphosphate and sulfamic acid which when added to urine precipitates all protein present except γ-globulin (368). The sensitivity of this test is about 5–10 mg/100 ml.

Quantitest Reagent (Quantitest). This is a precipitation test using β-naphthalenesulfonic acid as a protein precipitant (246).

The plastic strips with buffered indicators such as Albustix are the most widely used of the rapid qualitative tests for protein in urine. These "dip stix" tests are subject to considerable error and interferences of which the user should be aware.

The validity of results obtained with the reagent strip tests are dependent on the pH of the urine. In heavily buffered alkaline urine the buffer on the indicator stick is insufficient to achieve the pH required for adequate indicator function, thus leading to false positives (252, 477, 509). Deeply pigmented urines also tend to give false positives (252). The excretion of protein in the urine is not constant (7) and varies considerably over a 24 hour period with no particular pattern. Testing random samples therefore can be misleading.

Another problem which occurs with this type of rapid test is that the sensitivity of the indicators is not the same for all proteins, tending to be more sensitive for albumin than for globulins.

Comparison of the test strips with sulfosalicylic acid or Heller's nitric acid tests, indicate that the test strips are less sensitive (19, 256, 477).

HEAT TEST FOR URINE PROTEIN

PRINCIPLE

Urine is boiled at a pH (about 4–5) optimal for precipitation of heat coagulable proteins. pH control is obtained by the addition of acetic acid-sodium acetate buffer, which simultaneously provides adequate inorganic salt for protein flocculation in the case of a urine of low specific gravity.

REAGENT

Acetic Acid-Sodium Acetate Buffer, 5 M, pH 4.0. Dissolve 23.7 ml glacial acetic acid and 12.4 g $CH_3COONa \cdot 3H_2O$ (or 7.4 g of the anhydrous salt) in water and dilute to 100 ml.

PROCEDURE

1. Clarify urine by filtration or centrifugation, the latter method being convenient if a microscopic examination of the urine sediment is to be performed.
2. Place about 7.5 ml of the clear urine in a test tube having a diameter of about 16 mm, add 1 drop buffer for each 2.5 ml urine, and mix by inversion. If a precipitate of mucin occurs at this point, remove by filtration or centrifugation.
3. Heat *upper* half of urine column to boiling in a flame.
4. Compare heated portion with lower unheated portion of urine, using only ordinary room light. (Do *not* examine for Tyndall effect in a bright light, as this makes the test too sensitive.) If no difference is observed, report the test as "negative." If a barely perceptible turbidity is observed, report as "±" or "trace." Greater degrees of turbidity can be reported as 1+, 2+, 3+, and 4+.

INTERPRETATION

± barely visible turbidity	about 5 mg/100 ml
1+ distinct turbidity	10–30 mg/100 ml

2+ moderate turbidity	40–100 mg/100 ml
3+ heavy turbidity	200–500 mg/100 ml
4+ heavy flocculent precipitate	>500 mg/100 ml

SPECIFICITY

The test is positive with albumin, globulins, Bence-Jones protein, proteoses, and resinous acids from drugs such as benzoin. If clear urine becomes cloudy upon addition of the buffer and the turbidity lessens or disappears upon the application of heat, this indicates a condition of "pseudoalbuminuria" (discussed below under the sulfosalicylic acid test). If the turbidity increases when heated there undoubtedly is true proteinuria superimposed.

SULFOSALICYLIC ACID TEST FOR URINE PROTEIN

PRINCIPLE

Visual observation is made of the precipitation of protein by sulfosalicylic acid. Trichloroacetic acid can be used but has about one-half the sensitivity of sulfosalicylic acid. The test is performed essentially as a ring test so that comparison can be made of the turbidity produced with the overlying clear urine.

PROCEDURE

1. Clarify urine by filtration or centrifugation, the latter method being convenient if a microscopic examination of the urine sediment is to be performed.
2. Place about 3 ml of the clear urine in a test tube having a diameter of about 16 mm. Tilt the tube to about a 45° angle and add from a dropper or pipet 1 drop of 25% sulfosalicylic acid per milliliter of urine letting the acid run down the side of the tube. Being of greater specific gravity than the urine, the acid runs down to the bottom underneath the urine. Do *not* mix.
3. In a minute or so examine for turbidity in ordinary room light the urine near the bottom of the tube. (Do *not* examine for Tyndall effect in a bright light as this makes the test too sensitive.) If no turbidity is observed, report as negative. A barely perceptible turbidity is reported as "±" or "trace." Greater degrees of turbidity can be reported as 1+, 2+, 3+, and 4+.

INTERPRETATION

Same as for *Heat Test for Urine Protein*, above.

SPECIFICITY

The test is positive for albumin, globulins, proteoses, and Bence-Jones protein. The latter two substances go back into solution if heated to above 60°C. Uric acid may also precipitate but also redissolves on heating.

This test, in common with all qualitative tests involving acidification of the urine, may give false positive results in what has been termed "pseudoalbuminuria" (174). This occurs in about 10%–65% of cases receiving the organic iodine compounds Monophen, Telepaque, and Priodax in diagnostic x-ray procedures. It is most often seen during the first day following receipt of the compound but may occur as late as the third day. The heat test always differentiates a true proteinuria from a pseudoalbuminuria caused by organic iodine compounds. Pseudoalbuminuria can also occur because of the presence of the excretion product 1-butyl-3-p-carboxyphenylsulfonylurea of the drug tolbutamide. When this substance is present it is necessary to remove it by acidifying the urine to pH 2 with HCl and extracting with a mixture of ether or ethyl acetate and n-butyl alcohol.

FRACTIONATION OF PROTEINS

All biologic fluids contain a heterogeneous mixture of proteins, many of which are present in only trace amounts, e.g., enzymes and hormones. The chemical or physical separation of the protein components of such a mixture is referred to as "fractionation".

The earliest fractionation, dating to the last century, was achieved by salting out with ammonium sulfate. The fraction precipitated by 50% saturation was called *globulin*, and the fraction not precipitated by 50% but precipitated by complete saturation was called *albumin*. It was found that albumin is soluble in water in the absence of salts, whereas globulins require the presence of neutral salts for solubility. Subsequently, it was found that only a portion of the globulin is insoluble in pure water, this fraction being termed *euglobulin*, while the other freely soluble portion was called *pseudoglobulin*. Euglobulin has also been defined as that fraction precipitated from dilute solution by acidification with CO_2.

A new terminology was introduced with the advent of electrophoresis. The component with the greatest electrophoretic mobility was identified as albumin. The other three components observed were named arbitrarily α-*globulin*, β-*globulin*, and γ-*globulin*, in order of decreasing mobility. With the use of barbital buffers, a component appeared between albumin and α-globulin that had not separated out from albumin at the lower pH's used originally in electrophoresis. This new component was termed α_1-*globulin*, and the previously designated α-globulin became α_2-*globulin*. Technical modifications have resulted in improved resolution so that additional components may be identified in normal or abnormal sera. For example, a component sometimes appearing between the α_2 - and β-globulins is appropriately termed α_3-*globulin* (467). Similarly, a component appearing between β- and γ-globulins, which is accentuated in certain abnormal sera, has been called β_2 - or γ_1-*globulin* or simply the *T, M, ζ, or X* zone (179).

Various methods of fractionation have been applied to proteins in plasma and other body fluids. Many salts other than ammonium sulfate have been used in salting-out technics. Sodium sulfate was used as early as 1901 and was popularized by the fractionation scheme introduced by Howe (190) in 1921. Sodium sulfite was recommended by Campbell and Hanna (68) in 1937 and was incorporated with Span-ether in the scheme of Wolfson, Cohn, and co-workers (518). Other salts which have been advocated include magnesium sulfate, sodium hyposulfite, and sodium phosphate. The reader is referred to the first edition of this book for a detailed discussion of the various technics. Salting-out methods, being technically cumbersome and incapable of precise resolution, have almost disappeared from the clinical laboratory during the past decade.

Fractionation with organic solvents such as the ethanol fractionation of Cohn and co-workers (76), the methanol fractionation of Pillemer and Hutchinson (353), and the ether fractionation of Kekwick and Mackay (225), is adaptable to large batch processing and consequently has found important commercial application. The general approach is impractical for routine use in the clinical laboratory, however, because of the necessity for working at low temperatures (about 0°C).

Electrophoresis (cf Chapter 5) has now become firmly established as the method of choice for protein fractionation in the clinical laboratory to the virtual exclusion of other methods. Electrophoresis is also a simple replacement for salting-out technics in order to determine the albumin/globulin (A/G) ratio of serum (101), although determination of total protein and either albumin or globulin is also commonly used, as discussed later in this chapter. Since the first edition of this book, paper has been largely replaced as an electrophoretic medium by cellulose acetate, which has the advantages of speed, technical simplicity, and greater resolving power. Agar gel shares many of the advantages of cellulose acetate, particularly with Cronar, Mylar, or other plastics as a backing support instead of glass. Nevertheless agar gel remains more tedious than cellulose acetate, particularly in terms of preparation and storage of strips. Molecular sieving media such as starch gel and acrylamide gel have greatly augmented resolving power by effecting separation on the basis of molecular size as well as charge. For example, 20—30 protein bands are readily obtained with acrylamide disc electrophoresis of serum (92) in contrast to the 5—7 bands obtained with cellulose acetate, agar gel, or paper. The additional bands are generally difficult to interpret, however.

Other protein fractionation technics have limited and, at best, highly specialized applications to clinical chemistry. Chromatographic methods are discussed in Chapter 6. Accurate and precise resolution of plasma proteins is possible, for example, with DEAE-cellulose or DEAE-Sephadex columns (340). Chromatographic methods are technically demanding and slow, and generally require expensive equipment to collect and quantitate the fractions. By utilizing forces of 50 000–250 000*g*, ultracentrifugation can delineate two or more substances in solution which differ in molecular weight or density and sometimes only in molecular configuration (234). Although often referred to in publications, ultracentrifugation is not suitable for serum protein fractionation in routine clinical chemistry practice because of exceptionally high instrument cost and low productivity. When available and properly used at reference centers, the ultracentrifuge is a valuable adjunct in the study of lipoproteins (cf Chapter 29), serum macroglobulins, and immunoglobulin light chains or heavy chains.

ELECTROPHORESIS OF SERUM PROTEINS

Cellulose acetate was introduced as a medium for zone electrophoresis by Kohn (238, 239) in 1957. Cellulose acetate has many advantages over paper (220, 238, 443): Although brittle when dry, it has great tensile strength when wet. It is chemically uniform and almost pure. Uniform pore size allows improved resolution and sharpness of protein bands. Adsorption of proteins is minimal; therefore trailing of albumin is eliminated. A small sample suffices. The time required for adequate migration is greatly reduced. The background is easily washed clean after staining. With appropriate clearing, cellulose acetate can be rendered transparent to both visible and UV light of wavelength greater than 250 nm. Quantification is more accurate because protein is more evenly distributed in the fine membrane structure than in the irregular fibers of paper.

As with paper electrophoresis many modifications have been introduced, mostly for technical convenience or improved quantification. As discussed in Chapter 5, cellulose acetate is available from multiple manufacturers with some variation in quality and physical characteristics. Improved resolution has been reported with a special gelatinized cellulose acetate membrane, marketed as Cellogel, which does not rely on molecular sieving but purportedly separates 18–21 protein bands in serum (96). Various approaches use strips designed for either single or multiple samples varying in volume from approximately 0.1 to 10 μl. Recommended electrophoretic times are usually less than 1 hour. For example, the Microzone apparatus manufactured by Beckman is designed for eight samples per membrane and requires only 15–20 min for electrophoresis (164, 220).

The most commonly used buffer system is barbital-acetate at pH 8.6 and ionic strength (μ) varying from 0.025 to 0.075 (138, 242, 443). Brackenridge (54) investigated the effects of buffer pH and μ, and found the best electrophoretic separation at pH 8.66 and $\mu = 0.05$. Greater μ results in sharper zone resolution but a decreased migration rate and greater heat production. At lower μ, increased migration occurs concomitantly with some loss of zone resolution. Tris-barbital buffers, originally introduced for paper electrophoresis (369), have been used to advantage with cellulose acetate (510).

Another advantage of cellulose acetate over paper is that there is less urgency for drying and fixation upon completion of electrophoresis since air drying is usually rapid enough to prevent pattern distortion. Fixation of protein is usually achieved by chemical means, commonly using trichloroacetic acid or sulfosalicylic acid. Heating at $80°-90°C$ may also be used (443).

Visualization of the separated protein bands is achieved by staining with any suitable protein stain which is not dissolved in a cellulose acetate solvent such as dioxan (298). Ponceau S (Fast Ponceau 2B) is by far the most widely used dye (59, 101, 164, 220), and its characteristics have been reviewed by Luxton (285). Ponceau S is an indicator which is red at pH<10 and purple at pH> 12.5. The acid form of this dye has an almost symmetrical absorption peak at or about 523 nm. Other dyes employed in cellulose acetate electrophoresis include Nigrosin (238, 239), Amido Black 10B (Pontacyl Blue-Black Sx, Buffalo Black NBR, Naphthalene Black 10B, Amidoschwarz 10B) (138, 392, 443), azocarmine (54, 239, 443), bromophenol blue (443), Ponceau Red-fuchsin (239), and Procion Brilliant Blue M-RS (286).

Methods for quantitating the electrophoretically separated protein fractions such as N determination, the ninhydrin color reaction, or direct photometry in the far-ultraviolet range are rarely used in view of the much simpler technic of quantitating the bound dye. The dye employed would ideally have the following characteristics (285, 286): equal affinity for all protein fractions, uptake which is directly proportional to protein concentration, sufficient sensitivity to stain adequately all protein fractions, no affinity for cellulose acetate, stability of the dye solution and dye-protein complex, and simplicity of the staining procedure. Needless to say, the ideal dye has not been discovered (286). The most troublesome problem is differential affinity for various protein fractions. Most dyes have greater affinity for albumin than for globulin. Quantitative data are limited, but one study revealed that albumin binds 1.46 times more Ponceau S per unit concentration than does γ-globulin (285). This estimate would seem to be quite excessive, however, in view of the excellent quantitative correlation which has been reported between Ponceau S-stained electrophoresis strips and the moving boundary method of Tiselius (220). The relative affinity for various proteins varies with the particular dye employed. For example, β-globulins bind a greater percentage of bromophenol blue than Ponceau S, while the reverse is true for γ-globulins (242). Correction factors that have been recommended (138) can be criticized since they are based on normal sera and are not necessarily valid for abnormal sera (251, 392). The best solution to this problem would seem to be to establish a normal range for the particular dye and other electrophoretic conditions employed and to avoid affinity correction factors.

In general, dye binding is more nearly proportional to protein concentration for electrophoresis on cellulose acetate than on paper (1, 138). Ponceau S binding is linearly related to protein for concentrations well beyond those of normal serum (242, 285). A linear relationship has also been observed for other dyes, e.g., Amido Black 10B (138). Other dyes such as Lissamine Green deviate significantly from linearity above a critical protein concentration, and correction factors have been recommended for protein fractions which exceed this empirically determined limit (1).

As with paper electrophoresis there are two basic approaches to quantitating stained cellulose acetate electrophoretograms: densitometry and photometric determination of dye eluted from the segregated fractions. There is actually little difference in accuracy and precision of the two methods (59, 238). Resolution may be somewhat better with densitometry (251). Cutting out and eluting the stained fractions for photometric measurement is considerably more tedious than densitometry, however, particularly now that various reliable densitometers with automatic integrators are available. Various solutions are used for elution, depending upon the particular dye employed—e.g., 0.1 N or 0.4 N NaOH for Ponceau S (101, 242, 285), Na_2CO_3 in dilute methanol for azocarmine (443), or pH 6.0 phthalate buffer for Lissamine Green (54). Turbidity is a potential problem with Ponceau S if photometry is conducted at pH <8.0 (285). Instead of eluting the dye from the membrane, attempts have been made to dissolve away the cellulose acetate with suitable organic solvents, but insoluble residue is a serious problem (285).

Cellulose acetate electrophoretograms are usually cleared prior to densitometric quantitation, although some methods employ uncleared membranes in reflectance densitometry. Various oils of refractive index 1.470–1.478 have been used as clearing agents, e.g., Whitmore 120 oil, dodekahydronaphthalene (Dekalin) (138, 238). More commonly, mixtures of acetic acid with ethanol or methanol, often with added ethylene glycol, are employed. Clearing results in a collapse of the reticular structure and closure of the intermesh, thus imparting to the cellulose acetate membrane the optical and textural characteristics of cellophane (245). Kremes *et al* (245) emphasized that failure to obtain adherence to Beer's Law is related to the sieve-like character of uncleared media, whether paper or cellulose acetate. Threads of uncleared media are covered with stained protein while the intermesh completely transmits light, which has the effect of stray light from any other source in photometry. Special cams have been designed to compensate for nonlinearity in reflectance densitometry (2), but they are not ordinarily necessary in transmission densitometry of cellulose acetate electrophoretograms stained with Ponceau S.

Various ancillary technics have been proposed to aid in the identification of specific proteins on the electrophoretogram. In particular, interest has focused on identifying the specific immunoglobulins constituting the paraprotein (M-component, monoclonal gammopathy) bands commonly associated with lymphoreticular disease or plasma cell dyscrasias such as multiple myeloma or macroglobulinemia of Waldenstrom. Preincubation of serum with certain thiol compounds—e.g., mercaptoethanol (100) or penicillamine (40)—results in depolymerization of IgM so that paraprotein bands composed of this immunoglobulin are significantly altered. Alternatively Rivanol (6,9-diamino-2-ethoxyacridine lactate) selectively precipitates the IgM macroglobulins and most other proteins except IgG (99, 411). It is now known, however, that the paraprotein band may be composed of any one of the five immunoglobulins. Accurate identification of the constituent protein requires the specificity available only with immunochemical technics discussed in Chapter 7.

CELLULOSE ACETATE ELECTROPHORESIS OF SERUM PROTEINS

(method of Kohn, Refs 238, 239, modified)

PRINCIPLE

Serum proteins are separated by electrophoresis on cellulose acetate using a Tris-barbital buffer, pH 8.8. The electrophoretograms are stained with Ponceau S and quantified by transmission densitometry.

REAGENTS

Tris-Barbital Buffer, pH 8.8. Dissolve 5.75 g Tris(hydroxymethyl) aminomethane, 2.45 g barbital, and 9.81 g sodium barbital in approximately 250 ml warm water. Mix, allow to cool, and dilute to 1 liter. Adjust pH to 8.8 if necessary. Buffer is 0.0475 M with respect to both Tris and sodium barbital.

Ponceau S Solution. Dissolve 5 g in 1 liter 7.5% (w/v) trichloroacetic acid.

Acetic Acid, 5%

Methanol, Anhydrous

Clearing Solution. Mix 200 ml glacial acetic acid, 770 ml anhydrous methanol, and 30 ml ethylene glycol.

PROCEDURE

The procedure presented is for the Photovolt electrophoresis system which utilizes 12 x 75 mm cellulose acetate strips with Mylar backing and an integrating densitometer for quantification. This procedure is similar to those of several other commercially available systems. The manufacturer's instruction manuals should be consulted for technical details.

1. With a wax pencil label strips at one end of the Mylar side. It is important not to touch the acetate surface with bare fingers in this and all subsequent steps.
2. Place strips in a slide cradle and slowly immerse into a staining dish containing sufficient (approximately 250 ml) Tris-barbital buffer to cover strips. Soak for at least 2 min.
3. Add Tris-barbital buffer to appropriate levels in the electrophoresis reservoirs and tilt to equalize fluid levels.
4. Remove cradle from buffer solution and individually blot each strip between two blotters.
5. Place strips in electrophoresis chamber, acetate side up, by bowing across the holding rack.
6. Apply samples to cellulose acetate strips at the cathodic side of the chamber using the sample pickup tray, applicator, and applicator guide (see *Note 2*). Excess sample should be avoided by carefully blotting the under side of the applicator tip.
7. Place cover on electrophoresis chamber and allow electrophoresis to proceed for 20 min at a constant voltage of 110 volts.
8. Turn electrical current off, remove strips, and place in slide cradle.
9. Immerse in Ponceau S staining solution in staining dish for 3 min.
10. Drain for a few seconds and rinse for 3 min in each of three staining dishes containing 5% acetic acid.
11. Drain on absorbent paper for 10 sec. Immerse in absolute methanol for exactly 2 min.
12. Drain and immerse in clearing solution for 1 min.
13. Drain on absorbent paper for at least 3 min. Tilt cradle so that Mylar sides of the strips are against the cradle dividers and dry in a preheated oven (80°C for about 8 min). Allow strips to cool on a clean, flat surface with Mylar sides up.
14. Scan strips on an integrating densitometer using a 570 nm filter and a 0.1 x 6 mm slit. (*Note 1*).

Calculation:

Add up total integration counts. Divide counts for each fraction by the total. Multiply these figures by the total protein (cf *Determination of Total Protein in Serum or Plasma*) to obtain concentrations in g/100 ml for each fraction or multiply by 100 to obtain percentages.

NOTES

1. Relationship of densitometer reading to protein concentration. Densitometer readings are linearly related to concentration for pure γ-globulin and albumin up to a chart reading of 80% T with the system described.

Presumably this linearity also applies to other fractions as well. Consequently the densitometer scan should be adjusted to give a chart reading of 80% T with the most intense band (normally albumin). No correction is made for differences in relative dye affinity of the various fractions.

2. *Reuse of solutions.* The buffer solution in the staining dish as well as in the chamber should be changed daily. The polarity of the electrophoresis chamber should be reversed after each electrophoretic run. With the system described this is conveniently achieved by a toggle switch. The dye solution is suitable for 100 strips or 1 week. Rinse solutions should be serially replaced when coloration becomes excessive—i.e., discard the first rinse solution, move the second solution to the first position, the third to the second position, and place fresh solution in the third position. The absolute methanol and clearing solutions should be replaced daily.

3. *Sample stability.* It has been reported that β_2-globulin decreases and γ-globulin increases after 1 day at 20°C (264). On further storage at either 4°C or 20°C the β-lipoproteins tend to migrate more toward the α_2-globulins. In contrast, we have not found significant changes in the electrophoretic pattern of serum stored at 30°C for 4 days, 4°C for 2 weeks or −20°C for 6 months.

4. *Plasma vs serum.* Fibrinogen is represented by a narrow band in the β_2-globulin region, which may be mistaken for an M-component. It is therefore customary to use serum instead of plasma.

ACCURACY AND PRECISION

The Tiselius moving boundary method is usually considered to be the reference method. Kaplan and Savory (220) found reasonably close agreement between the results of cellulose acetate electrophoretograms stained with Ponceau S and the moving boundary method, although α-globulins were somewhat lower and β-globulins higher by the cellulose acetate method. Electrophoretic results with cellulose acetate can agree quite closely with those of paper (54).

The approximate precision (95% limits) of the method is as follows: albumin 7%, α_1-globulin 30%, α_2-globulin 25%, β-globulin 11%, γ-globulin 11%. These results are similar to those of other large studies (54, 100, 220). Results agree quite closely among laboratories using similar methods (101).

NORMAL VALUES

There are two methods for expressing the quantity of individual fractions in an electrophoretogram: (a) in relative terms as percentage of total protein, (b) in absolute terms of concentration, e.g., g/100 ml. There are arguments in favor of each method of expression. A disadvantage of expressing individual components as percentages of the total is that an increase or decrease in any one of the major fractions results in a corresponding decrease or increase in the percentages of all other fractions even though their absolute concentrations may be normal. On the other hand, in dehydration or water excess syndromes, the absolute concentrations may be increased or decreased, respectively, simply as a result of contraction or expansion of the intravascular fluid volume. An obvious solution would be to report the results in both ways.

Based on a study of 23 normal adults our laboratory established the following normal ranges for the various fractions (concentrations expressed as g/100 ml): albumin 3.5—5.0, α_1-globulin 0.1—0.4, α_2-globulin 0.5—1.1, β-globulin 0.6—1.2, and γ-globulin 0.5—1.5. Other reported normal ranges are in close agreement (54, 59, 220). Kaplan and Savory (220) found the following normal ranges expressed as percentages of total: albumin 52.2—67.0, α_1-globulin 2.4—4.6, α_2-globulin 6.6—13.6, β-globulin 9.1—14.7, γ-globulin 9.0—20.6

The effect of various factors is as follows:

Sex: There is no sex difference (203). Pregnant females at term have significantly decreased albumin and increased β-globulin and slightly increased α_1-globulin (336); α_2- and γ-globulins are not significantly altered during pregnancy.

Age: The effect of age has been investigated in detail by Oberman *et al* (336). Relative to adults, cord blood has decreased concentrations of total protein, albumin, and α_2- and β-globulins. α_1-Globulin is slightly increased, and γ-globulin is normal or increased. Jencks *et al* (203) reported similar results except for decreased α_1-globulin. Adult levels are reached for all fractions except γ-globulin by 3 months of age. γ-Globulin in cord blood is largely of maternal origin. Catabolism of the passively transferred protein results in a marked decrease in γ-globulin, which falls to a nadir at about 3 months. Adult levels are not reached until about 2—7 years. Albumin decreases and β-globulin increases after the age of 40 (274).

Race: Many studies have reported racial differences, particularly decreased albumin and increased γ-globulin in Negroes relative to Caucasians. It is uncertain whether such observations are true racial features or a reflection of socioeconomic factors (212, 236, 274). In one study both Negroes and American Indians had higher β- and γ-globulins and lower albumin than Caucasians (236). American Indians also had higher α_1-globulin and lower α_2-globulins than either Negroes or Caucasians.

Climate and season: There is apparently no significant seasonal variation in the United States, but this may not hold true for more extreme climates. Thus in the summer both Americans and Eskimos living in Greenland have increased concentrations of all protein fractions except albumin (287).

CEREBROSPINAL FLUID AND URINE PROTEINS

The principles of electrophoresis discussed for serum are applicable to other biologic fluids as well. However, most other fluids are so dilute that preliminary concentration is necessary to obtain adequate visualization of the stained electrophoretogram. In fact, a staining threshold has been observed for ponceau S—i.e., no staining occurs at protein concentrations below a critical limit (285). Cellulose acetate electrophoresis of unconcentrated CSF has been achieved in conjunction with nigrosin staining which yields greater sensitivity than Ponceau S (382, 434): α_1- and α_2-globulins were not clearly separated, however, and a prealbumin fraction was not regularly delineated. Since the normal protein concentration of CSF is only 15–45 mg/100 ml, the fluid should be concentrated about 100- to 200-fold to approach the protein concentration of serum. Even greater concentration of urine is necessary since the normal urinary excretion of protein, about 25–75 mg/24 hours in a urine volume of perhaps 1250 ml, is less than 6 mg/100 ml.

The concentration procedure should meet three criteria: speed, negligible loss of protein, and absence of protein denaturation. Unfortunately, many methods have gained wide acceptance without adequate documentation of the degree of denaturation or protein loss. Major methods which have been used for concentration of CSF include the following:

Pervaporation. CSF is placed in a closed cellophane bag which is exposed to moving air such as a fan or hair dryer (121, 147, 221). Preliminary dialysis against water is recommended to remove excess salts, which otherwise precipitate as concentration proceeds (221). Reported time requirements have varied from 6 to 7 hours for 100-fold concentration (221) to 5–7 days for 20-fold concentration (318). There is purportedly no distortion of proteins (121).

Concentration Dialysis. CSF is dialyzed through a semipermeable membrane against such polymers as PVP (121, 221, 318), Ficoll, Aquacides 1 or 2, Carbowax or Sephadex G 200 beads (221), dextran (147), or polyethylene glycol (81). Dialysis is usually conducted at refrigerator temperatures. Times vary somewhat but 1–20 hours is representative. These methods as a group tend to preserve albumin but may result in considerable loss or denaturation of γ-globulin (221).

Acetone Precipitation. Precipitation of proteins with cold acetone yields good recovery (388) but with some distortion of the subsequent electrophoretic pattern (121).

Freeze-drying. Freeze-drying requires preliminary dialysis to remove salts (486). The technic is relatively slow and requires expensive equipment.

Ultrafiltration. CSF is placed in a sac constructed of suitable semipermeable material—e.g., collodion—and filtered with the aid of pressure (compressed air or N_2) on the bag contents (302) or vacuum applied to the surrounding reservoir (65, 219, 221). In vacuum ultrafiltration the surrounding reservoir may be filled with dextran to aid in the concentration process (65), 0.9% NaCl (219), or an equal mixture of 0.9% NaCl with barbital buffer to avoid pH changes (221). Special commercially available apparatus is needed for the vacuum method. Using a vacuum of 500 mm Hg, Kaplan (219) achieved 100- to 200-fold concentration in 2—3 hours. Kaplan and Johnstone (221) found vacuum ultrafiltration to be superior to pervaporation and concentration dialysis in terms of protein preservation and electrophoretic reproducibility.

A commercially available (Amicon Corp.) ultrafiltration apparatus utilizes centrifugal force to accelerate the concentration process. CSF or other fluid to be concentrated is placed in a conical membrane supported by a polyethylene cone, which in turn is inserted in a plastic tube. With conventional laboratory centrifuges greater than 100-fold concentration can be achieved in a matter of minutes to an hour. Although the exact composition of the membranes has not been revealed, two different pore sizes are available to retain molecules greater than either 20 000 or 50 000 molecular weight. In addition to technical simplicity and speed, this method offers excellent preservation of proteins and electrophoretic reproducibility (510).

The stained electrophoretogram of CSF is similar to that of serum except for the prominent prealbumin band and decreased γ-globulin. A band occurring at the point of sample application between the β and γ fractions has been variously termed the β_2, ϕ or τ (tau) band. This band is the result of protein denaturation, and it is insignificant or absent when improved concentration methods are employed (221). With the greater resolution of polyacrylamide disc electrophoresis, 14—18 protein bands can be distinguished in CSF (87).

In contrast to serum or CSF, electrophoresis of urine proteins has relatively limited application to clinical diagnosis. Methods for concentrating CSF have also been applied to urine (158, 301, 519). Other concentration methods applied to urine include dialysis against powdered sucrose (300) and tannic acid precipitation (161).

Protein electrophoresis of normal urine reveals the same five major bands present in serum albeit in markedly different relative proportions. Most urinary proteins are immunochemically identical to plasma proteins, the major exception being the Tamm-Horsfall mucoprotein (cf *Glycoproteins*) (158). Berggard (32) found up to 20 different proteins in normal urine which are immunochemically identical to those in plasma. Specifically identified were prealbumin, albumin, α_1-lipoprotein, α_1-seromucoid, ceruloplasmin, haptoglobin, α_2-macroglobulin, transferrin, β_{1A}-globulin, β_{2A}-globulin, fibrinogen products, and γ-globulin.

CELLULOSE ACETATE ELECTROPHORESIS OF CEREBROSPINAL FLUID AND URINE PROTEINS

(method of Windisch and Bracken, Ref 510, modified)

PRINCIPLE

CSF or urine is concentrated by means of a commercially available membrane ultrafiltration system which utilizes centrifugation. Proteins are separated by electrophoresis on cellulose acetate using a Tris-barbital buffer, pH 8.8. The electrophoretograms are stained with Ponceau S and quantified by transmission densitometry.

REAGENTS

Same as for *Cellulose Acetate Electrophoresis of Serum Proteins.*

PROCEDURE

Concentration is achieved using the Centriflo Membrane Ultrafilter System of Amicon Corporation.

1. Soak a CF50A membrane cone in water for at least 1 hour.
2. Drop membrane cone into the polyethylene conical support. Rotate the support counterclockwise while pushing the cone gently downward until its tip protrudes through the bottom of the support thus locking the cone into place. (Caution: do not scratch or crease the interior of the membrane.)
3. Push the support into the polycarbonate tube and seat the flange on the tube.
4. Place CSF or urine specimen into cone (see *Note 3*).
5. Counterbalance centrifuge cups and centrifuge to achieve desired concentration. (Caution: do not exceed 1000*g* force. See *Note 4*.)
6. Withdraw the concentrate from the cone using any suitable micropipet or capillary tubing. Apply concentrated sample to electrophoresis applicator and proceed with electrophoresis as with serum.

NOTES

1. See Notes 1 and 2 of cellulose acetate electrophoresis of serum proteins.

2. Stability. There are no appreciable changes in CSF proteins following storage for 2 days at 30°C as determined in our laboratory or storage for 2 weeks at 4°C (434). CSF is stable indefinitely when frozen. We have found urine to be stable for 4 days at 30°C and indefinitely when frozen.

3. Sample volume. Do not exceed the following maximum volumes: 7 ml for horizontal (free-swing) centrifuge heads, 5 ml for 45° rotor, 3.5 ml for 60° rotor.

4. Centrifugation. We obtain good results using 2500 rpm in an International centrifuge (floor model with horizontal head) for times varying from 5 to 30 min depending on sample characteristics and desired concentration. In angle head centrifuges the seam of the membrane cone should be oriented toward the rotor shaft.

ACCURACY AND PRECISION

Electrophoretic results of CSF concentrated by centrifugal membrane ultrafiltration agree closely with those obtained following concentration by vacuum ultrafiltration and yield somewhat better precision (510). Windisch and Bracken (510) studied the effect of the concentrating process on precision using serum. Precision was determined for 30 replicate electrophoretic determinations of control serum and compared with 30 replicate electrophoretic determinations of the same serum concentrated 100-fold after a 100-fold dilution. Only slight deterioration in precision was attributable to the concentration procedure.

Data are not available for urine, but somewhat poorer precision would be anticipated as a result of loss of some of the lower molecular weight proteins, particularly in pathologic states, e.g., Bence-Jones protein. It is likely that this difficulty will be alleviated by a newer ultrafilter cone (CF20, Amicon Corp.) which is designed to retain molecules above 20 000 molecular weight.

NORMAL RANGE

The normal ranges of CSF determined in our laboratory are in agreement with those reported by Windisch and Bracken (510), expressed as a percentage of total protein: prealbumin 2.2–7.1, albumin 56.8–76.4, α_1-globulin 1.1–6.6, α_2-globulin 3.0–12.6, β-globulin 7.3–17.9, γ-globulin 3.0–13.0. These are in close agreement with other normal ranges using cellulose acetate electrophoresis after vacuum ultrafiltration (199, 221). There are no significant differences in the electrophoretic pattern between males and females (434) or in CSF obtained from ventricular, cisternal, or lumbar sites (182). The effect of various diseases has been reviewed by Goa and Tveten (147).

Data on urine are sparse. Utilizing concentration by pressure ultrafiltration followed by moving boundary electrophoresis, McGarry *et al* (301) found the following average distribution of protein (expressed as percentages of total protein) in urine from 13 normal subjects: albumin 37.9, α_1-globulin 27.3, α_2-globulin 19.5, β-globulin 8.8, γ-globulin 3.3.

DETERMINATION OF ALBUMIN AND TOTAL GLOBULINS

In addition to electrophoresis, the approaches which have been used for the determination of albumin and globulins are as follows: (a) separation of the albumin and globulins by precipitation technics, (b) turbidimetric assay for globulins, (c) determination of albumin by "protein error of indicators", (d) determination of total globulins based on tryptophan content. These technics have been applied mainly to serum or plasma and are of very limited value for CSF and urine.

DETERMINATION OF ALBUMIN AND TOTAL GLOBULINS BY PRECIPITATION TECHNICS

In this technic total protein is determined, usually by the biuret method, a salt is added to precipitate all the globulins, and the albumin is determined in the solution remaining. The salts most commonly used to precipitate the globulins are sodium sulfate and sodium sulfite, with ammonium sulfate and magnesium sulfate being used to a lesser extent (174).

Sodium sulfate was used for fractionation of proteins as early as 1901. In 1921 Howe (190, 191) described a scheme for protein fractionation with this salt, and his procedure was in wide use for many years in spite of the fact that it was well established that the salt concentrations used were not proper.

In the procedure of Majoor (291) sodium sulfate at a final concentration of 26% is used to precipitate the total globulins in serum. The final concentration of sodium sulfate was subsequently changed to 26.8% based on Tiselius electrophoresis (174). In order to remove the precipitated globulins, the salt-protein solution is shaken with ether 'to decrease the density of the globulins, which upon centrifugation collect as a compact mat at the interface of the two liquid phases. The albumin in the aqueous phase is then determined by a biuret procedure. The total globulins are determined by difference from the total protein determination of the original serum.

In 1937 Campbell and Hanna (68) introduced the use of sodium sulfite for fractionation of serum, and in 1948 Wolfson *et al* (517) presented a modified procedure which is still used to a considerable extent as a reference procedure. A final concentration of 26.9% sodium sulfite is used for precipitation of all globulins in serum (517). The precipitated globulins are removed from the salt-protein mixture by shaking with ether followed by centrifugation. This has the advantage that filtration is avoided. Filter paper absorbs protein from solution until it is saturated so the first half of the filtrate usually has to be discarded.

The advantages of the use of Na_2SO_3 over Na_2SO_4 are the buffering power of the former and the fact that the procedures can be carried out at room temperature without crystallization of the salt from solution. Owing to interaction between different protein anions and cations, fractionations should be carried out at a pH at which all protein molecules possess a net charge of the same sign. With sodium sulfite the pH is sufficiently high to be above the isoelectric points of all proteins present.

In a comparison of precipitation procedures with Tiselius electrophoresis, the 26.8% Na_2SO_4 gave better results than the 26.9% Na_2SO_3, but in both cases excessive shaking with ether denatured protein and increased the amount of protein precipitated, leading to erroneous results in both the albumin and globulin fractions (271). Na_2SO_3 solutions are unstable, as with time the sulfite oxidizes to sulfate and the amount of precipitation decreases. As a result of absorption of albumin on the precipitated globulins Na_2SO_4 fractionations give lower results for albumin than either a methyl orange procedure or electrophoresis (230).

A method for the precipitation of globulins by 2% trichloroacetic acid in ethanol has been used for the separation and determination of albumin and globulin in serum (390). In this method, serum is treated with 2% trichloroacetic acid in ethanol for 2 hours at room temperature to precipitate quantitatively the total globulins. The albumin in the supernate is then determined by a Kjeldahl analysis or biuret reaction. Hemoglobin is also soluble in 2% trichloroacetic acid in ethanol and when present results in an erroneously high value for albumin.

Although salt fractionation methods have been proposed for urine and cerebrospinal fluid (107), their use frequently leads to serious error (253, 414). The results obtained by salt fractionation methods usually are not comparable with those obtained by electrophoresis (414).

TURBIDIMETRIC ASSAY FOR TOTAL GLOBULINS

Because of their simplicity, turbidimetric determination of precipitates is attractive and methods have been proposed for the determination of total globulins in serum precipitated with 50% saturated ammonium sulfate (279), sodium sulfate (143), or sodium sulfite (174). Turbidimetric procedures using sodium sulfate (143) and sodium sulfite (371) have been automated. Turbidity can be affected by variables other than protein concentration (21), and although their careful control may yield satisfactory results the authors hesitate to recommend a turbidimetric technic when an alternative is available or when the greatest accuracy and precision are desired.

DETERMINATION OF ALBUMIN WITH DYES AND ORGANIC INDICATORS

Dye-binding technics for determination of albumin take advantage of the so-called "protein error of indicators." Addition of protein to a solution of certain dyes results in the formation of a dye-protein complex that exhibits optical properties different from the free dye. A few dyes are reasonably specific for albumin. The sites on protein at which dye is bound are also capable of reversibly binding other substances, such as bilirubin, free fatty acids, and drugs. When such substances are present in a protein solution, they interfere with dye-binding assays.

The use of methyl orange for the direct determination of albumin in serum was described in 1953 (53). Albumin in serum added to a methyl orange solution buffered at pH 3.5 binds and effectively removes some of the pink anion, so that a decrease in absorbance at 550 nm provides a measure of the albumin. The method was subsequently improved by the use of a serum blank (227) and by doubling the ionic strength of the dye (228). Although the methyl orange is purportedly bound only to albumin with little if any binding to γ-globulins (53, 152, 230, 499) this has been refuted by others who claim that β-lipoproteins as well as α_1- and α_2-globulins are capable of binding methyl orange (283, 395).

Quantification by the methyl orange method has been reported to give fictitiously high values for albumin especially at low concentrations (53, 85, 230, 283, 517). The methyl orange procedure has been automated (538) but has the same drawbacks as the manual method (283). Because of nonspecific binding the methyl orange method is not recommended.

Albumin can also be quantified from the increased absorbance when serum is added to a solution of 2-(4'-hydroxyazobenzene)benzoic acid (HABA) buffered at pH 6.2 (409). The original method was modified by using phosphate buffer at pH 6.2 instead of an acetate buffer to eliminate turbidity, by measuring the absorbance at 485 nm to increase sensitivity, and by carefully controlling temperature because of the thermolabile nature of the HABA-albumin interaction (330). The HABA assay for albumin has the advantage that the binding is specific for albumin, but in spite of modifications the sensitivity is low. Various materials such as salicylate, sulfonamides, penicillin, and conjugated bilirubin (but not free bilirubin) interfere with the binding (14). Heparin present in the usual anticoagulant amounts causes turbidity which interferes with the assay (14, 498). Another drawback of the procedure is that the albumin used as a standard in the procedure has to be either human albumin fraction V or human mercaptalbumin since heat treated or chemically stabilized albumin standards do not give complete binding of HABA. The HABA dye-binding procedure has been automated (539), but the basic shortcomings of the reaction have not been circumvented. The HABA dye procedure has been found to correlate poorly with cellulose acetate electrophoresis (14).

A bromcresol green (BCG) dye-binding procedure for the quantitative determination of albumin in serum was introduced in 1964 (25, 105, 389). Serum is diluted with buffered BCG at pH 7.0 and the decrease in absorbance read at 615 nm (390). The decrease in absorbance is linear with albumin concentration up to 5 g/100 ml, and hemoglobin and bilirubin do not interfere greatly at the wavelength used (106, 389). This method is more sensitive than the methyl orange or HABA procedures (498). Electrophoretically separated protein fractions, other than albumin, do not bind BCG. The manual BCG procedure has been automated (106, 180). Studies in which albumin levels were determined by BCG and electrophoresis show excellent correlation (106). The bromcresol green procedure is more specific for quantitation of albumin in serum and 20 times more sensitive than the

HABA method (317). The method using bromcresol green is much less affected by lipemia and high levels of hemoglobin and bilirubin. In clear sera with normal total protein levels but abnormal A/G ratios, the A/G ratios were found to be nearly identical using the HABA, bromcresol green, or the Wolfson-Cohn salt fractionation procedures (317).

Of the three dye-binding methods so far discussed for quantitation of albumin, bromcresol green appears to give the best correlation with electrophoresis and precipitation with 2% trichloroacetic acid in ethanol. The bromcresol green method is not affected by interfering materials as is the HABA procedure and bromcresol green does not bind to other serum proteins as does methyl orange.

Phenolsulfonephthalein (PSP) has been proposed for the determination of albumin in serum (498), but albumin levels below 1.5 g/100 ml give no appreciable binding (42) and plasmas from women more than 24 weeks pregnant are deficient in binding ability for PSP (316).

Bromcresol purple, a phthalein dye similar to bromcresol green, has been used to quantify albumin in urine and serum (69, 257). Salicylates, bilirubin, and lipids interfere with the analysis of albumin in serum (69).

Eosin has been reported to bind to albumin but should not be used for albumin assay in serum or plasma since free fatty acids compete for binding sites thus causing variable binding to albumin (498).

Methods based on the binding of biologic pigments have also been reported (498). In one such approach, the amount of methemalbumin produced after addition of serum to an alkaline hematin solution is measured; however, the binding affinity of hematin for albumin is low and is high for one or more of the β-globulins in plasma. In another method bilirubin in xylene is equilibrated with diluted plasma. The loss of bilirubin from the xylene correlates with the albumin concentration. Under experimental conditions the decreased absorbance of the bilirubin solution is proportional to albumin concentrations up to 4 g/100 ml, but in practice the results obtained are too high due to binding of bilirubin to β-lipoproteins.

Certain dyes which are nonfluorescent in aqueous solution but are fluorescent when absorbed or bound have been used for determining albumin. Examples of such dyes are 1-anilinonaphthalene-8-sulfonic acid and vasoflavine (498). These fluorescent dyes have been reported to bind only to albumin and not to other native proteins, and results are linear with albumin concentration. With 1-anilinonaphthalene-8-sulfonic acid, bilirubin at levels of 5 mg/100 ml or greater interferes by competition for binding sites (498). Procedures using fluorescent dyes have not yet been tested to any great extent.

Dye-binding procedures have been used to a very limited extent for determination of albumin in urine or CSF because of low protein concentration and high levels of interfering materials.

DETERMINATION OF TOTAL GLOBULINS BASED ON TRYPTOPHAN CONTENT

The tryptophan content of human serum albumin is 0.2% as compared with 2%–3% for the various serum globulins (41), and it has been reported that there is a direct correlation between total serum globulins and total serum tryptophan (55, 425). Increased protein tryptophan reported in some abnormal sera is presumably due to an increased globulin level rather than increased tryptophan incorporation in protein (55).

In a method proposed for the determination of total serum globulins, the serum proteins were precipitated with isopropanol and the tryptophan in the precipitate determined by the Tauber-Fischl reaction (416). This method gave excellent correlation with serum globulins determined by salt fractionation and electrophoresis. The procedure was subsequently modified by precipitating serum globulins with 2% trichloroacetic acid in ethanol and then assaying the precipitate for tryptophan (417). A serum of known albumin and globulin content was used as a standard, thus eliminating the need to correct for the small amount of tryptophan in albumin.

This approach for the assay of total serum globulins has been modified by Goldenberg and Drewes (149) who evaluated several variations of the Hopkins-Cole reaction for tryptophan and proposed a single reagent consisting of an aqueous mixture of glyoxylic acid, Cu^{2+}, H_2SO_4 and acetic acid to which serum is added. After a short heating period the reaction mixture is cooled and the absorbance of the resultant purple color measured at 540–560 nm. This globulin reagent is stable for 1 year or more at refrigerator temperatures. The method gives good reproducibility and conforms to Beer's law. The reaction with tryptophan in globulins is complete. Tryptophan as the free amino acid does not react with the reagent. Bilirubin at concentrations up to 20 mg/100 ml or moderately lipemic serum cause less than a 5% variation of the measured globulin levels. The results obtained with this method correlate well with those obtained by Na_2SO_4 fractionation and paper electrophoresis (230). Glucose, urea, creatinine, and cholesterol do not interfere with the reaction (133).

Savory *et al* (420) automated the Goldenberg and Drewes procedure and obtained values closely correlated with those obtained by electrophoresis but not with those obtained by difference using the HABA procedure for albumin. The recovery of γ-globulin added to serum was quantitative, and the coefficient of variation of the automated procedure was about 4.5%.

A photometric tryptophan procedure for determining total globulins in CSF has been proposed in which the globulins are precipitated with 10% trichloroacetic acid and reacted with glyoxylic acid reagent (414).

CHOICE OF METHOD

The methods for the determination of albumin in serum with dyes and organic indicators have not been compared to any great extent with accepted reference procedures (Tiselius electrophoresis and Wolfson-Cohn salt fractionation), and we feel it is inadvisable to recommend any dye-binding method in the absence of such comparison studies. We recommend that the determination of albumin and globulins in serum be performed by a direct determination of globulin based on tryptophan content and calculation of the albumin content by difference.

Good methods are not available for the determination of albumin and globulins in CSF and urine. Results obtained with salt fractionation technics and chemical determination of the resulting fractions do not correlate with immunochemical or electrophoretic technics (222, 414, 421).

DETERMINATION OF TOTAL GLOBULIN IN SERUM

(method of Goldenberg and Drewes, Ref 149)

PRINCIPLE

The total globulin in serum is determined by reaction with glyoxylic acid in an acid medium. The resultant purple color is measured photometrically at 540 nm (Fig. 16-5).

REAGENTS

Globulin Reagent. In a 1 liter volumetric flask dissolve 1.0 g $CuSO_4 \cdot 5H_2O$ in 90 ml water. Add 400 ml glacial acetic acid followed by 1.0 g 98% glyoxylic acid monohydrate (that from Matheson, Coleman and Bell is satisfactory). *Mix without delay.* Cautiously add with stirring 60 ml conc H_2SO_4. Cool to room temperature. Dilute to the mark with glacial acetic acid and mix. This reagent is stable at least a year refrigerated.

Standards. There are two alternatives: (a) a standard containing 7.5 g total protein per 100 ml consisting of 4.5 g human albumin/100 ml and 3.0 g human gamma globulin/100 ml; (b) a synthetic standard comprised of N-acetyl-DL-tryptophan equivalent to 3.0 g globulin/100 ml serum as determined by comparison with reference sera of known globulin content (see *Note 2*).

PROCEDURE

1. Set up in test tubes:
   Reagent blank: 5 ml globulin reagent
   Standard: 20 µl serum standard or 20 µl synthetic standard + 5 ml globulin reagent
   Unknown: 20 µl unknown serum (TC pipet) + 5 ml globulin reagent
2. Mix the contents of all tubes and place in boiling water bath for 5 min.
3. Cool tubes in tap water for 3 min and mix.
4. Read absorbances of standard (A_s) and unknown (A_x) against the reagent blank at 540 nm or with a filter in this region (e.g., Klett filter 54). The color is stable for at least 2 hours.
   Calculation:

$$\text{g total serum globulin/100 ml} = \frac{A_x}{A_s} \times \text{g globulin/100 ml standard}$$

NOTES

1. *Beer's law.* Beer's law is obeyed up to at least 7.5 g globulin per 100 ml serum. If globulin concentration is in excess of this value the sample may be diluted with an equal volume of 0.85% NaCl and the assay repeated.

2. *Sources of error.* The tryptophan in the albumin of a normal serum can contribute 7%–10% of the total absorbance. The use of a standard containing 4.5 g human albumin and 3.0 g human γ-globulin per 100 ml,

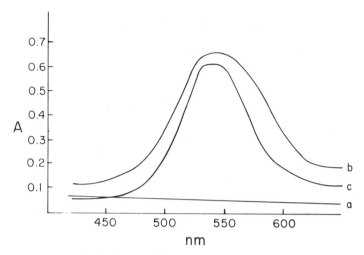

FIG. 16-5. Absorption curve of color produced by globulin reagent: (a) reagent blank vs water; (b) human globulin, 50 mg/ml, vs reagent blank; (c) human globulin, 50 mg/ml, vs reagent blank. Beckman DU spectrophotometer.

representing mean human albumin and globulin values, partially corrects for the albumin tryptophan contribution to the color reaction. Indole-3-acetic acid produces 64% as much color as acetyl tryptophan; however, normal levels in serum would contribute minimally to the total absorbance of the final solution (149). Bilirubin concentrations up to 20 mg/100 ml give less than 5% interference with the assay (133, 149). Moderately lipemic and hemolyzed sera contribute no appreciable interference (149, 420).

ACCURACY AND PRECISION

The method is relatively specific for tryptophan in the peptide linkage of proteins but is subject to inaccuracy as the tryptophan contribution of albumin in the sample varies (273). Also see *Note 2*, above.)

The precision (95% limits) of the method is about ±4%.

NORMAL VALUES

Table 16-2 lists representative normal adult ranges obtained by various workers. From a study of 1419 clinically normal adults, our laboratory obtained normal ranges of 2.4–3.5 g/100 ml for serum globulin and 3.7–5.1 g/100 ml for serum albumin calculated by difference. From a physiologic standpoint, it is the actual concentration of a protein, such as albumin and total globulin, not their ratio (A/G), that is of importance to the physician. A changed A/G ratio can result from increased or decreased albumin or globulin. The A/G ratio can also be normal, but the actual concentration of both albumin and globulin can be decreased or increased.

Variables which affect the normal ranges for albumin and total globulins are:

(a) *Age.* The normal serum albumin level is about 3–4 g/100 ml for premature infants (337) and about 3–5 g/100 ml for full term infants (27, 337). Adult normal levels are reached by the third month (336).

Table 16-2. Normal Adult Values of Albumin and Total Globulin in Serum

Ref	Method used	Number in Series	Protein (g/100 ml)		
			Albumin	Globulin	A/G ratio
516	Salt fractionation	15	3.5–5.0	1.7–4.2	1.0–2.2
416	Salt fractionation	26	3.5–4.9	2.7–4.5	0.7–1.7
416	"Globulin assay"	26	3.8–5.0	2.6–4.2	0.9–1.7
149	"Globulin assay"	20	–	2.5–3.6	–
377	Salt fractionation	42	3.8–4.6	2.4–3.5	1.1–1.7
386	Salt fractionation	400	3.6–4.7	–	–
497	Methyl orange	12	4.5–5.5	–	–
497	Bromcresol green	12	3.5–4.7	–	–
372	"Globulin assay"	1419	3.7–5.1	2.4–3.5	1.0–1.8
175	Paper electrophoresis	24	3.5–5.5	2.8–4.1	1.1–2.2
233	Cellulose acetate electrophoresis	40	3.7–5.2	1.8–4.2	0.9–2.0

A gradual decrease in serum albumin with age has been reported by several workers (341, 372, 386), the effect being more pronounced in males (372). The A/G ratio also decreases with age (372). The effect is small and for clinical purposes can be ignored.

(b) *Sex.* Using the method presented and calculating albumin by difference, our laboratory found no sex differences except for a slightly higher albumin level in males in the 20–39 year age group (372). Higher albumin in males was also found by salting-out (sodium sulfite) (386), but no sex difference was found for albumin and globulin fractions on paper electrophoresis (203). During pregnancy there is a significant decrease in serum albumin levels of about 15% (341).

The direct determination of total CSF globulins based on tryptophan content reveals a total globulin of 8–20 mg/100 ml, an albumin (by difference) of 10–31 mg/100 ml, and an A/G ratio of 0.95–1.95 (414).

FIBRINOGEN

Fibrinogen is a plasma glycoprotein of approximately 330 000 molecular weight (249) which is synthesized in the liver. The molecule is a dimer consisting of three pairs of polypeptide chains presumably linked by disulfide bonds (171). Of paramount importance in the blood coagulation mechanism, fibrinogen has been designated as Factor I by the International Committee for the Nomenclature of Blood Clotting Factors (520). Conversion of fibrinogen to a fibrin clot actually occurs in two phases (249). In the first or enzymatic phase thrombin, which is formed from the complex interaction of prothrombin with other coagulation factors in the presence of Ca^{2+}, cleaves four peptide bonds in the fibrinogen molecule to produce two pairs of small peptide chains, designated fibrinopeptides A and B, free carbohydrate, and the fibrin monomer. In the second or polymerization phase the fibrin monomers aggregate by end to end and side to side hydrogen bond linkages to form a three-dimensional insoluble mesh. In the presence of Ca^{2+}, the fibrin clot is stabilized by yet another plasma factor, fibrin stabilizing factor (Factor XIII).

Various methods for quantitating plasma fibrinogen are based on one or more of the following characteristics: enzymatic conversion to fibrin by thrombin, relative insolubility, immunochemical specificity, heat lability, or electrophoretic mobility. The classic method involving thrombin catalyzed conversion of fibrinogen to fibrin was introduced by Cullen and van Slyke (86) in 1920. In this method excess Ca as $CaCl_2$ is added to oxalated plasma, thus allowing intrinsic thromboplastin to convert prothrombin to thrombin, which in turn converts fibrinogen to fibrin. The clot is then washed and quantified by Kjeldahl N determination. The N content of fibrinogen has been reported as either 16.9% (56) or 16.6% (202). Many other thrombin conversion methods have been introduced. Exogenous thrombin is commonly used to assure rapid and complete conversion of fibrinogen, thus eliminating the effect of deficiencies of clotting factors necessary for the generation of thrombin. In addition to Kjeldahl determination, methods involve quantification of the fibrin clot by gravimetry (23, 156); photometrically by the biuret (494, 495), phenol-Cu (17, 370, 375), or ninhydrin (418) reactions; or by measuring the absorbance at 280 nm after dissolution of the clot with concentrated urea (202) or tryptic digestion (74). An isotope diultion method using ^{131}I-labeled fibrinogen has also been described (18).

A major source of error in chemical determinations of the fibrin clot is the occlusion of other serum proteins (202, 348). Whether specific protein occlusion occurs has not been determined with certainty. Morrison (321) found that although the amount of each extraneous protein in the clot was generally proportional to its plasma concentration, negligible amounts of albumin and large amounts of lipoproteins were occluded. Jacobsson (202) found no specific occlusion except for lipids and denatured lipoprotein. Manual compression of the clot (494), prolonged syneresis (202), dilution of plasma prior to clotting, and prolonged washing of the clot (86, 156, 321) help to minimize this source of error. Control of pH is also important—e.g., extraneous protein occlusion at pH 7.2 is two-to sixfold greater than at pH 6.3 (321). Occlusion is also decreased by the use of exogenous thrombin instead of simple recalcification of plasma (348). The other major source of error in chemical measurements of the fibrin clot is incomplete recovery of fibrin, which is particularly serious at low fibrinogen concentrations or low total protein concentration of the medium in which the clot is formed (119). In one of the best known modifications, Ratnoff and Menzie (370) agitated the clotting medium with crushed glass particles to facilitate complete harvesting of fibrin.

Equally important in quantification of the fibrin clot are various physical methods. For example, Losner et al (280) elaborated a method based on the fact that fibrinogen concentration is proportional to the increase in photometric absorbance which occurs upon clotting. This method is rapid, accurate, and free from interference by hemolysis or bilirubin, although elevated lipids do interfere (396). Another approach, introduced by Ellis and Stransky (110) and modified by others (49, 64), involves measurement of turbidity upon addition of thrombin to plasma diluted in barbitone buffered saline.

A rapid test which is useful for estimating clottable fibrinogen is based on the time required for fibrin clot formation after addition of thrombin to oxalated or citrated plasma (47, 433). Although commonly termed the "thrombin time", a more precise name for this test is "fibrin polymerization time" (488). The time required for clotting is proportional to fibrinogen concentration, but circulating heparin compounds or breakdown products of fibrinogen or fibrin interfere significantly (186). The specificity of the test is apparently improved by prior dilution of plasma (323). In one modification a micro sample of capillary blood is used instead of plasma (488). A semiautomated micro method has been described (323). Other coagulation procedures—plasma recalcification time, whole blood clotting time, one-stage prothrombin time, and partial thromboplastin time—also detect quantitative or qualitative deficiencies in fibrinogen but are relatively insensitive for this purpose and also are not specific since they are dependent upon many other coagulation factors.

Another rapid semiquantitative determination of fibrinogen is the titered dilution method. As introduced by Schneider (422) in 1952, serial dilutions of whole blood were prepared in Ringer's solution and the endpoint determined as the highest dilution in which clotting occurred. In later modifications thrombin is added to saline dilutions of oxalated or citrated plasma (142, 394, 433).

The second category of methods for fibrinogen determination—those based on relative insolubility of fibrinogen-involve salting-out followed by turbidimetric estimation or chemical or physical measurement of the precipitate. Howe (189) introduced salting-out in 1923 using chiefly 10.6% Na_2SO_4, although he noted that a variety of other salts—Li_2SO_4, $MgSO_4$, NaCl, and Na or K phosphates—gave essentially the same results. Perhaps the best known methods is that of Campbell and Hanna (67), which uses Na_2SO_3 in a final concentration of 11.9% followed by Kjeldahl N determination of the washed precipitate. A micromodification of the Na_2SO_3 method eliminates bilirubin interference (151). Claims have been made that either Na_2SO_3 (67) or glycine (157) is more specific for fibrinogen than are other salts. In general, the salt precipitation methods are closely correlated with clottable fibrinogen (67, 157, 189), although the fibrin clot methods are theoretically more physiologic. Higher results obtained with salt precipitation in the presence of *in vivo* fibrinolytic activity may be seriously misleading (352). The same criticisms should apply to turbidimetric methods, the best known of which is the $(NH_4)_2SO_4$ method of Parfentjev *et al* (343). Other turbidimetric methods employ $(NH_4)_2SO_4$ with citrate (195), NaCl (458), Na_2SO_4 (357), and phosphate buffer (299).

During the past decade various immunochemical technics with high specificity and sensitivity have been applied to fibrinogen quantification. In one rapid, semiquantitative method latex particles coated with anti-human fibrinogen are reacted directly with plasma (70). Reagents for this test, frequently termed the "Fi" test, are commercially available in kit form. Even more sensitive is a tanned red cell hemagglutination inhibition method which can detect fibrinogen at a level of 1 µg/ml (313).

Methods for quantifying fibrinogen on the basis of heat lability or electrophoretic mobility are largely obsolete. Fibrinogen is denatured by heating at 56°C for 10 min. Measurement of the precipitate is reported to correlate closely with the fibrin clot and Na_2SO_3 precipitation technics (150). In one technic applicable to capillary blood the volume of denatured precipitate is determined in a microhematocrit tube (281). Fibrinogen has an electrophoretic mobility intermediate between β and γ globulins, but in spite of method refinements (185) separation is incomplete.

Recently considerable interest has focused on proteolytic breakdown products of fibrinogen and fibrin which may appear in association with a variety of pathologic conditions (126). By interfering with the conversion of fibrinogen to fibrin by thrombin, the degradation products are also important in that they may cause artifactually decreased fibrinogen values in fibrin clot methods (186). Farrell and Wolf (120), however, found that under some circumstances fibrin degradation products may be incorporated into the fibrin clot thus resulting in an overestimation of physiologically active fibrinogen. Normally only a trace of degradation products is present in serum. Quantification of the degradation products requires the exquisite sensitivity of the various immunochemical technics—hemagglutination inhibition, coated latex particles, immunoelectrophoresis, or tube precipitin (70, 132, 312, 313). Cryofibrinogen is discussed in a subsequent section of this chapter. Also of clinical and research importance are various qualitative defects in fibrinogen, the dysfibrinogenemias, which are either hereditary or acquired. They are most often detected by the disparity between chemical measurements of fibrinogen, which are normal, and an abnormally prolonged fibrin polymerization time (thrombin clotting time) (292). The functional defect is in polymerization of fibrin monomers. Unlike congenital afibrinogenemia, which is inherited as an autosomal recessive condition, hereditary dysfibrinogenemias are autosomal dominants (292). A recent review cited eight familial cases (322). Following the precedent of hemoglobin nomenclature, the dysfibrinogenemias have been maned after the cities in which they were discovered, e.g., fibrinogens Bethesda (155), Cleveland (130), Baltimore (29), and Detroit (292). Fibrinogen Detroit has a decreased carbohydrate content and replacement of arginine by serine in the 19th position of the N-terminal disulfide knot (292). The specific molecular defects have not been established in all cases.

DETERMINATION OF FIBRINOGEN IN PLASMA

(method of Ware et al, Ref 495, modified by Ware, Ref 494)

PRINCIPLE

Plasma is diluted with buffered saline containing Ca^{2+}. Thrombin is added, causing rapid and quantitative deposition of fibrin. The fibrin clot is removed and washed, and the protein determined by the biuret reaction.

REAGENTS

Phosphate Buffer, M/5, pH 6.4. Dissolve 1.82 g KH_2PO_4 + 0.94 g Na_2HPO_4 and bring to 100 ml with water.

Thrombin Solution. To one vial (5000 units) of Thrombin-Topical (Parke-Davis) add 5 ml 0.85% NaCl. Dissolve and add 5 ml glycerol. This solution contains 500 units/ml and is stable indefinitely in the refrigerator.

$CaCl_2$ *1%*

NaOH, 3%

Biuret Reagent. See under earlier section, *Determination of Protein by the Biuret Reaction.*

PROCEDURE

1. Collect blood using citrate, oxalate, or disodium ethylenediaminetetra-acetate as anticoagulant (see *Note 1*).
2. Add 1.0 ml plasma to about 30 ml 0.85% or 0.90% NaCl in a flask containing 1 ml $M/5$ phosphate buffer and 0.5 ml 1% $CaCl_2$. Add 0.2 ml (100 units) thrombin solution.
3. Mix and let stand at room temperature for 15 min. Fibrin forms in less than 1 min and forms a solid gel if the fibrinogen concentration is normal.
4. Introduce a glass rod and wind the clot around the rod by pressing against the wall and rotating. Remove the rod with the attached clot, blot the clot carefully with filter paper, and then rinse with water. If difficulty is experienced with picking up the clot on a rod because of low fibrinogen concentration, transfer the entire contents of the flask to a 50 ml centrifuge tube, centrifuge, discard supernate, wash precipitate with a little water, recentrifuge, discard water, and proceed as below.
5. Place 5.0 ml 3% NaOH and 1.0 ml biuret reagent in a test tube and mix. Place glass rod with attached fibrin clot in biuret mixture, place in 37°C water bath, and stir until all of the fibrin is dissolved. If fibrin is collected by centrifugation, pour biuret mixture into the centrifuge tube and mix until fibrin is completely dissolved.
6. Allow 15 min for biuret color to develop and then examine for turbidity by looking for a Tyndall effect in a strong light. Turbidity caused by precipitation of $Ca(OH)_2$ can be removed by centrifugation.
7. Read as for total protein (cf *Protein Determination by the Biuret Reaction*) using the same reagent blank and standard.

Calculation:

$$\text{g fibrinogen/100 ml} = \frac{A_x}{A_s} \times \frac{100}{1} \times \text{g protein in standard}$$

PROCEDURE

The above technic can easily be scaled down tenfold. Wycoff (523) described such a micro adaptation.

NOTES

1. Anticoagulant. By interfering with the action of thrombin, heparin causes marked inhibition of fibrin formation. If the disodium salt of ethylenediaminetetraacetic acid (EDTA) is used as anticoagulant, the plasma must be obtained by centrifugation at 2000g to ensure removal of platelets, well preserved by this anticoagulant, which would otherwise be enmeshed in the fibrin clot and give fictitiously high results (259). Clots obtained in the presence of EDTA are friable with poor tensile strength (148).

2. Biuret Equivalent. The biuret color equivalent for fibrinogen is the same as for albumin and globulins (494).

3. Development of biuret color. Dissolving the fibrin clot in alkali prior to the addition of the biuret reagent yields lower results—breakdown of fibrinogen? (494).

4. Stability. It has been reported (110) and confirmed in our laboratory that plasma samples are stable up to 7 days at room temperature and up to 4 weeks in the refrigerator.

ACCURACY

The conditions employed in this method for conversion of fibrinogen to fibrin are those for greatest recovery with the least occlusion of other proteins. Nevertheless, other proteins are carried down, and although the calculation of error is difficult it has been estimated as varying from negligible amounts of albumin up to 10%—25% of lipoprotein (321). If the latter is correct, elevated values for fibrinogen obtained in the presence of high levels of lipoproteins should be regarded with suspicion. Jacobsson (202), however, maintains that lipoproteins are not occluded unless denatured.

A possible cause of low recovery is fibrinolysis of the clot before it is quantified. The fibrinolytic activity of different plasmas can vary considerably. This source of error can be minimized by avoiding delays in the analysis subsequent to clotting (202).

NORMAL VALUES

A compromise of normals obtained by various investigators using various methods is a range of 0.20—0.40 g/100 ml with a mean of about 0.28 g/100 ml (23, 168, 343, 349, 370, 375, 418). A slight tendency toward higher values in females has been reported (349, 481), but several other studies have found no significant sex differences (339, 370). Fibrinogen is increased during the first 3—5 days after birth (84, 481) and then decreases to a level which remains fairly constant until about age 20 (481). Progressive increases then occur in both males and females until at least age 50 (349). The level is relatively constant for any one individual (157) and is not significantly influenced by fasting, food, rest, or short, violent exercise (168). Values increase as much as 100 mg/100 ml by the last month of pregnancy (157).

GLYCOPROTEINS

Glycoproteins may be simply defined as proteins which have carbohydrate covalently linked to the peptide chains (453). All of the plasma proteins thus far isolated and characterized have covalently linked carbohydrate except for albumin (451). Their great diversity is further demonstrated by the fact that glycoproteins include enzymes, hormones, and important constituents of cell membranes, connective tissue, basement membrane, and mucous secretions.

Various classifications have been proposed for the carbohydrate-containing proteins. Winzler (514) proposed one of the better known classifications; it divides the carbohydrate-containing proteins into two major groups—*mucoproteins* and *glycoproteins*. In this classification mucoproteins are defined as proteins which are combined with mucopolysaccharides (polysaccharides which contain hexosamine and uronic or sulfuric acids) in polar or other labile linkage. Glycoproteins are defined as proteins which contain greater than 0.5% hexosamine firmly bound by stable bonds which can be split only by drastic treatment with acid, alkali, or enzymes. Winzler further subdivided glycoproteins into *glycoids,* which contain less than 4% hexosamine, and *mucoids,* which contain more than 4% hexosamine.

Glycoproteins were originally defined in 1908 by a joint committee of the American Physiological Society and the American Biochemical Society (540) as "compounds of the protein molecule with a substance or substances containing a carbohydrate other than nucleic acid." The current trend is to use the term *glycoprotein* as originally defined, irrespective of the carbohydrate content of the protein. This practice has a sound scientific basis, since structural studies have not revealed differences sufficient to warrant the use of separate terms for proteins of different carbohydrate content (450).

In addition to difficulties arising from attempts at classification, considerable confusion has been generated by imprecise terminology. Some terms have been defined differently by various investigators and at times by the same person. *Mucoprotein,* for example, has been used synonymously with *glycoprotein* although *mucoprotein* is usually restricted to proteins with a high carbohydrate content (449). In 1948 Winzler *et al* (515) referred to proteins soluble in dilute perchloric acid but precipitable by phosphotungstic acid as serum mucoproteins. Winzler (513) subsequently described this fraction as being of a mucoid nature since the carbohydrate can be cleared from the protein only by drastic treatment. In 1958 he classified mucoproteins as mucopolysaccharide-protein complexes with labile linkages (514). The true glycoproteins do not, however, contain hexuronic acid or sulfate esters, which are characteristic components of mucopolysaccharides (449). Mucoprotein must also be distinguished from *mucin,* which is a physiologic term referring to any viscous secretion. It seems desirable to follow Winzler's recommendation (513) that *serum mucoprotein* be replaced by *seromucoid,* a term introduced by earlier investigators to refer to a fraction obtained as a filtrate from serum deproteinized by heat coagulation and later found to be chemically identical to the perchloric acid-soluble, phosphotungstate-precipitable fraction of glycoproteins.

The clinical importance of serum glycoproteins has been discussed in recent reviews (241, 435, 449–452). An increase in serum glycoproteins occurs in association with various acute and chronic diseases, both localized and systemic—e.g., malignant neoplasms, collagen diseases, tuberculosis and other infections, diabetes mellitus, hepatic cirrhosis, psoriasis, and gout. The nonspecificity of increased serum glycoproteins has severely limited the usefulness of this determination in clinical diagnosis.

Although most studies from the standpoint of diagnosis have been concerned with serum or plasma, glycoproteins are normally excreted in urine (226, 265). Of paramount importance is the "mucoprotein" isolated and characterized by Tamm and Horsfall (472, 473) which is synthesized by the renal tubular epithelial cells (306). The function of this glycoprotein is unknown although it does constitute the matrix of hyaline casts (307). As with serum levels of glycoproteins, increased urinary excretion occurs in association with a wide variety of acute and chronic systemic diseases, stress, infection, and renal disease including nephrolithiasis (44, 232, 449). Various methods have been proposed for the isolation and quantification of urinary mucoproteins (9, 435). Analytic methods have also been adapted to cerebrospinal fluid and various tissues (435).

There are two general approaches to the measurement of glycoproteins for diagnostic purposes. In the chemical methods the glycoproteins are quantified by analysis for specific classes of carbohydrates or by measurement of either protein or carbohydrate after isolation by differential solubility in certain acids. The second approach involves electrophoresis. Other methods such as chromatographic separation have not been widely accepted for routine diagnostic use.

CHEMICAL METHODS OF ANALYSIS

Carbohydrate is bound to protein by firm covalent bonds and constitutes less than 1% to more than 80% of the weight of the molecule (449, 452). From one to as many as 800 individual carbohydrate units, varying from simple monosaccharides to polysaccharide chains with a molecular weight of several thousand, are present on a given molecule (452). Although several structurally different types of carbohydrate units may be present on the same molecule, most glycoproteins have a single type of unit. The carbohydrate is composed of neutral sugars, mostly hexoses, amino sugars or hexosamines, various derivatives of neuraminic acid (collectively termed sialic acids), and the methylpentose fucose.

Normal human serum contains an average protein-bound carbohydrate content of 273 mg/100 ml of which 121 mg is hexose, 83 mg hexosamine, 60 mg sialic acid, and 9 mg fucose (513). The carbohydrate composition varies widely from one specific type of plasma glycoprotein to another. Consequently the carbohydrate composition of the total serum glycoproteins may be significantly altered in disease since each of the various glycoproteins may respond differently. During adulthood there is no effect of aging on seromucoid or protein-bound hexoses, hexosamines, and fucose, but sialic acids do increase slightly with age (50). No sex differences have been found for seromucoid (15), protein-bound hexoses, hexosamines, or sialic acids (51).

HEXOSES

The major hexose constituents of the serum glycoproteins are mannose and galactose in about equal proportions. D-Glucose, D-xylose, and L-arabinose have been identified in glycoproteins from other tissues (453). Most analytic methods employ a preliminary precipitation of serum glycoproteins with 95% ethanol, although some rely on the difference between total and free hexose as a measure of bound hexose (265, 426). Treatment with strong H_2SO_4 hydrolyzes the glycoproteins and produces chromaphores which are reacted with cystine (265), tryptophane (22, 437), carbazole (426), orcinol (284, 513), anthrone (154, 436, 482), thymol (438), or resorcinol (432). The reaction with orcinol has been adapted to the AutoAnalyzer (215). The various methods vary rather widely in sensitivity and susceptibility to interfering substances. Although both galactose and mannose react with the above reagents, their relative color contributions vary from procedure to procedure. Fucose also reacts to a varying degree although its contribution is small. Hexosamines and sialic acids do not interfere with most reactions (435), although the relatively nonspecific cysteine reaction is an exception (265).

HEXOSAMINES

The hexosamine constituents of glycoproteins include 2-desoxy-2-amino-D-glucose (D-glucosamine or chitosamine) and 2-desoxy-2-amino-D-galactose (D-galactosamine or chondrosamine) present as the N-acetyl derivatives. The ratio of glucosamine to galactosamine in serum glycoproteins of normal individuals is approximately 10:1 (512). The methodology of hexosamine analysis has been reviewed by Gardell (140). Most methods are modifications of the reaction described in 1933 by Elson and Morgan, (111) which was an adaptation of a previously published method (346). Serum or plasma is subjected to acid hydrolysis followed by acetylation and conversion to cyclic oxazoles or pyrroles by treatment with alkali. Coupling with p-dimethyl-aminobenzaldehyde (Ehrlich's reagent) results in a stable violet color. Glucosamine and galactosamine have the same color equivalent and absorption curves in the Elson-Morgan reaction (140). The method is not specific for the 2-amino hexoses, since a mixture of neutral sugars with lysine or glycine reacts to a variable extent (140). Various methods have been proposed to increase the precision and specificity of the reaction by removing interfering chromogens (39, 140, 272). Modifications include preliminary separation of the hexosamines from the acid hydrolyzate by isolation on Dowex 50 ion exchange resin (43) or distillation of the volatile glucosamine-galactosamine derived chromogens into the p-dimethylamino-benzaldehyde reagent (72).

Other chemical approaches have been proposed. An ultramicro method suitable for determination of 5—20 ng amounts of hexosamines involves deamination with HNO_2, conversion to 2,5-hexose anhydrides with simultaneous Walden inversion at C-2, and reaction of the anhydrohexoses with indole in dilute HCl to form a color (102). A slightly different modification of this method involves preliminary separation of the hexosamines on ion exchange resin and reaction of the anhydrohexoses with pyrrole (114).

SIALIC ACIDS

Sialic acid is a collective term referring to various derivatives of neuraminic acid which chemically may be considered to be a nine-carbon aldol condensation product of pyruvic acid with either D-glucosamine or D-mannosamine (505). Although only the N-acetyl derivatives have been found in human plasma glycoproteins, N-glycolyl and diacetyl derivatives are present in glycoproteins from other sources (435). Sialic acids do not react in most of the reactions for hexoses since they are degraded into nonchromogenic products by the acid treatment (435). The usual terminal location of sialic acids on the carbohydrate units facilitates analysis following mild acid hydrolysis, although careful control of conditions is critical (83). Whitehouse and Zilliken (505) reviewed the various methods of isolation and measurement of sialic acids. Analytic methods include reactions with resorcinol-HCl (66, 469), thiobarbituric acid (83, 413, 435, 496), diphenylamine (415, 513), tryptophane-perchloric acid (427, 513), *p*-dimethylaminobenzaldehyde (Ehrlich's reagent) (265), and orcinol-$FeCl_3$ (Bial's reaction) (265). The thiobarbituric acid reaction has been adapted to the AutoAnalyzer (98). Both the thiobarbituric acid reaction (413) and the diphenylamine reaction (419) have been used to determine sialic acids in cerebrospinal fluid.

As with the determinations of hexoses and hexosamines, the normal range for sialic acids in serum varies somewhat depending upon the particular chemical method employed—e.g., a range of 63—92 mg/100 ml was reported for the diphenylamine reaction (415) and 39—81 mg/100 ml for the resorcinol reaction (66).

FUCOSE

Fucose is the only methylpentose present in plasma glycoproteins. Most analytic methods are based on the general reaction for methylpentoses described by Dische and Shettles (103): Serum glycoproteins are precipitated with 95% ethanol and heated with H_2SO_4. After cooling, color is developed by adding cysteine reagent. Specificity is increased by measuring absorbances at both 396 and ˙430 nm, which minimizes interferences from other sugars (513).

The fucose in serum glycoproteins of normal individuals is reportedly 8.9 ± 0.6 (1 *s*) mg/100 ml (513). This low level prevents the use of fucose determinations for the routine estimation of total glycoproteins in serum.

SEROMUCOID

Plasma seromucoid is a heterogeneous fraction of particularly soluble glycoproteins which can be separated by electrophoresis into five subfractions corresponding to the mobilities of the major plasma protein fractions (319). Additional heterogeneity has been demonstrated by immunoelectrophoretic (45) and chromatographic (361) methods. Specific glycoproteins in the seromucoid fraction include orosomucoid, haptoglobin, and α_2-glycoprotein (15). Orosomucoid, an α_1-glycoprotein of molecular weight 41 000, is the major component of seromucoid with a plasma concentration of 28—125 mg/100 ml (466). Although the function of orosomucoid is poorly understood, it is known to have genetically determined variants with codominant expression (209). The isolation and quantitation of orosomucoid has been effected by chromatographic (474), electrophoretic (466), chemical (503), and immunochemical (209) methods.

Although various methods have been used for isolation of the seromucoid fraction (513), the one most commonly employed involves precipitation of other plasma proteins with 0.6 N perchloric acid followed by precipitation of seromucoid with phosphotungstic acid (515). The isolated seromucoid has been quantified by measuring its content of protein (biuret or Kjeldahl analysis), tyrosine, hexose, hexosamine, or sialic acid (250, 513); by turbidimetry (93); and by complement fixation (475). A microchemical modification has also been reported (248).

Alterations in the seromucoid level as a consequence of disease tend to parallel those of total plasma glycoproteins as measured by hexose, hexosamine, sialic acid, or fucose analyses (435, 513). In most diseases the seromucoid hexose actually rises to 12%—20% of total protein-bound hexose from normal levels of about 10% (513). A significant exception is diffuse parenchymatous liver disease without infectious or neoplastic involvement (e.g., cirrhosis) in which the seromucoid hexose level may fall to 5%—6% of the total protein-bound hexose (513). The only other condition which results in a decreased seromucoid level is the nephrotic syndrome (513).

GLYCOPROTEIN ELECTROPHORESIS

Virtually all of the methods of zone electrophoresis have been employed in the study of glycoproteins: paper (241, 435), cellulose acetate (224), acrylamide gel (229), agar gel (485), polyvinyl chloride powder (326), and starch (332). The periodic acid-Schiff (PAS) reaction is the most common stain for glycoprotein electrophoretograms (224, 229, 241). Background staining in the PAS method is a serious problem with paper electrophoresis but not with cellulose acetate (224). A staining method successfully applied to agar gel involves oxidation by periodic acid, formation of aldehyde-aryl hydrazones, and conversion to formazan derivatives by coupling with a diazonium salt (485).

Variations exist in reported electrophoretic distributions of serum glycoproteins. Representative values for cellulose acetate electrophoresis as percentages of total serum glycoproteins are: albumin 10.9, α_1-globulin 18.6, α_2-globulin 29.6, β-globulin 21.6, and γ-globulin 19.3 (224). The glycoproteins of urine and cerebrospinal fluid are distributed in five major electrophoretic fractions corresponding to those of serum (435). Glycoprotein is not detected in the albumin fraction using electrophoretic methods with superior resolution, e.g., starch (332) or polyvinyl chloride powder (326). There are no sex differences in glycoprotein electrophoretic patterns (347). A reported increase in protein-bound hexose with age is associated with an increase in α_2- and β-glycoproteins (347). Changes in glycoproteins during pregnancy have been investigated by Mansfield and Shetlar (293). During normal pregnancy increased serum glycoprotein results from an absolute increase in all electrophoretic fractions. When expressed as percentages of protein in each electrophoretic fraction, however, glycoproteins associated with albumin and γ-globulin increase while those in the α_1-, α_2-, and β-globulin fractions decrease. During the puerperium glycoproteins associated with albumin and γ-globulin decrease while those in the α_1-, α_2-, and β-globulin fractions increase sharply. Alterations in the glycoprotein electrophoretic pattern vary widely with various diseases (224, 435), and no general comments can be given.

DETERMINATION OF SEROMUCOID IN SERUM

(method of Winzler et al, Ref 515, modified by Weimer and Moshin, Ref 502, and using the orcinol reaction adapted by Rimington, Ref 384)

PRINCIPLE

Proteins other than seromucoid are precipitated by perchloric acid. Seromucoid is precipitated from the filtrate by phosphotungstic acid and then dissolved in alkali. The oligosaccharides are cleared from the polypeptide chains by acid hydrolysis, and the hexoses converted to furfurals by prolonged heating (60). The color resulting from condensation with orcinol is read at the 520 nm peak rather than at 425 nm, the point of maximum absorption (Fig. 16-6), in order to minimize the difference in absorbances of the various hexoses and consequently minimize the error introduced if the hexose composition varies markedly from that of the galactose-mannose standard. Results are expressed arbitrarily as mg seromucoid per 100 ml in terms of the standard.

REAGENTS

Perchloric Acid, 1.8 M. Dilute 16.6 ml of 72% perchloric acid to 100 ml with water.

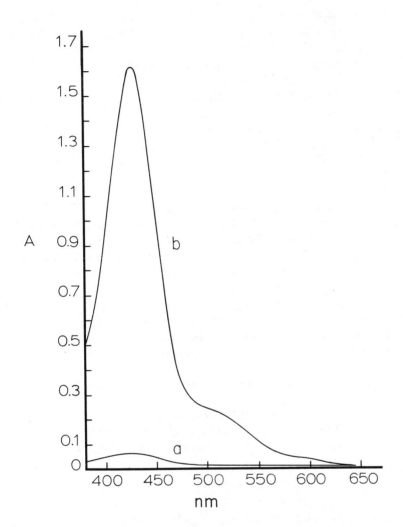

FIG. 16-6. Absorption curve of galactose-mannose standard and blank: (a) reagent blank vs water, (b) standard containing 0.05 mg galactose and 0.05 mg mannose vs water, Beckman model DU spectrophotometer.

Phosphotungstic Acid, 5% in 2 N HCl
NaCl, 0.85%
NaOH, 0.1 N
H_2SO_4, 60% (v/v)
Ethanol. 95%

Orcinol Reagent. 2% recrystallized orcinol in 30% H_2SO_4. To recrystallize, saturate about 25 ml of boiling water with orcinol, let cool, place in refrigerator overnight. Filter off crystals in a Büchner funnel, scrape off crystals into a dish, and dry in a desiccator. Original orcinol is lavender; final crystals should be off-white. The solution is usually stable for several months in the refrigerator. Discard if color (pink-brown tint) develops.

Stock Standard. Dissolve 100 mg galactose + 100 mg mannose in water and make to 100 ml. Saturate with benzoic acid and store in refrigerator. Stable indefinitely.

Working Standard. 1.0 ml stock standard + 9.0 ml water. Prepare fresh each day. 1 ml contains 0.2 mg galactose-mannose.

PROCEDURE

1. Add 0.5 ml serum to 4.5 ml 0.85% NaCl and mix.
2. Dropwise, add 2.5 ml 1.8 M $HClO_4$. Mix by inversion. At exactly 10 min, filter through Whatman 50 paper or its equivalent. To obtain a clear filtrate it may be necessary to refilter through the same filter paper.
3. To 5 ml filtrate in centrifuge tube, add 1.0 ml phosphotungstic acid reagent. Mix and let stand 10 min.
4. Centrifuge and discard supernate. Add 5 ml 95% ethanol, stir up precipitate, centrifuge, discard supernate, drain tube for 5 min.
5. Dissolve precipitate in 0.5 ml 0.1 N NaOH. This is the Unknown.
6. Set up Blank with 0.5 ml of water and Standard using 0.5 ml of working standard.
7. To Unknown, Blank, and Standard add 1.25 ml orcinol reagent and mix. Add 7.5 ml 60% $H_2 SO_4$ and mix.
8. Place in water bath at 80° ± 0.5°C for *exactly* 20 min.
9. Cool tubes in tap water and read against water at 520 nm within 15 min. Ignore any turbidity (positive Tyndall effect) in tubes.

Calculation:

$$\text{mg seromucoid/100 ml, expressed as galactose-mannose} = \frac{A_x - A_b}{A_s - A_b} \times 0.2 \times \frac{3}{2} \times 100$$

$$= \frac{A_x - A_b}{A_s - A_b} \times 30$$

NOTES

1. Beer's law. The standard suggested above is equivalent to a seromucoid value of 30 mg/100 ml, a value seldom exceeded in practice. Obedience to Beer's law can be assumed for any reading less than the standard, even when a Corning glass filter of 540 nm nominal wavelength is employed. At higher concentrations there may be deviation even with narrow bandwidths.

2. Standardization. A standard containing equal amounts of mannose and galactose is employed because plasma glycoproteins contain only these two hexoses in about equal amounts (512). From the absorption curves of the color produced by galactose and mannose independently at equal concentration, the point of equal absorbance was found in our laboratory to be at about 485 nm. With sera, however, no consistent difference was observed in results obtained by reading at 485 and 520 nm.

3. Temperature at the HClO₄ precipitation step (15). The temperature at this step should not exceed 26°C or spuriously low results will be obtained. If the room temperature exceeds 26°C, carry out this step in a bath of cool water.

4. Serum vs plasma. Results with plasma are as much as 30% lower than with serum and vary with the anticoagulant used (160).

5. Stability of serum. According to Greenspan (160), serum must be separated from the clot within 1–2 hours since results increase upon letting the serum stand at room temperature (no data presented). In experiments performed in our laboratory, however, serum was found to be stable at 30°C for periods of 2–7 days.

ACCURACY AND PRECISION

Hexosamine and sialic acid do not react with orcinol (265). On a weight basis, fucose results in about 80% of the absorbance of an equal mixture of galactose and mannose (265). The contribution of fucose is relatively slight, however, since its content in seromucoid is normally less than 10% of the hexoses.

Due to coprecipitation in the perchloric acid precipitation of the other proteins, recovery of seromucoid is only about 70% (515). The extent of coprecipitation is dependent on the dilution of the serum at the precipitation step, the method of mixing serum with the perchloric acid, the time and temperature of contact between the supernate and the precipitated proteins, and the NaCl concentration of the diluted serum (15, 513). It is essential, therefore, that the precipitation be carried out with scrupulous attention to constancy of technic. Another possible cause for low recovery, which has not been investigated in the procedure presented, is adsorption of seromucoid onto the filter paper in the filtration step (266). The empirical nature of the procedure is also attested to by the fact that results vary with variation in perchloric acid concentration. The accuracy of the test, however, is probably satisfactory for clinical use.

The precision of the test (95% limits) is of the order of ± 15%.

NORMAL VALUES

Expressed in terms of the galactose-mannose standard, serum seromucoid averages about 12 mg/100 ml with a range of about 9–15 mg/100 ml (250, 435, 515). There appears to be no significant difference between males and females (15, 93), although data to the contrary have been reported (250).

Studies of the effect of aging have been contradictory in that either no change has been found (51) or there is a trend toward higher levels with aging (424).

DETERMINATION OF PROTEIN-BOUND HEXOSE (NONGLUCOSAMINE POLYSACCHARIDE)

(method of Lustig and Langer, Ref 284, modified by Weimer and Moshin, Ref 502, and using the orcinol reaction as outlined by Rimington, Ref 384)

PRINCIPLE

The hexose moiety of protein-carbohydrate conjugates precipitated by 95% ethanol at room temperature (25°C) is determined by the orcinol reaction.

REAGENTS

Same as for seromucoid.

PROCEDURE

1. Add 0.1 ml serum to 5.0 ml 95% ethanol and mix.
2. Centrifuge for 15 min, decant, suspend, and wash precipitate with 5 ml 95% ethanol. Centrifuge and decant.
3. Proceed with step 5 under *Determination of Seromucoid in Serum*.

Calculation:

$$\text{mg protein-bound hexose/100 ml, expressed as galactose-mannose} = \frac{A_x - A_b}{A_s - A_b} \times 0.2 \times 5 \times 100$$

$$= \frac{A_x - A_b}{A_s - A_b} \times 100$$

NORMAL VALUES

The normal range by the orcinol method is reported to be 78–156 mg/100 ml (284). Somewhat different ranges have been reported for protein-bound hexose by other methods: 138–192 mg/100 ml (146) and 100–139 mg/100 ml (310) by the anthrone method and 73–131 mg/100 ml by the carbazole method (426).

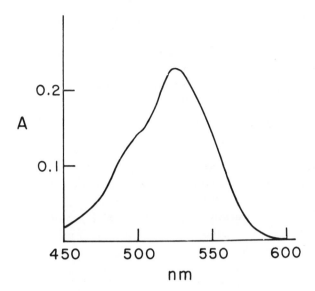

FIG. 16-7. Absorption curve of glucosamine standard vs reagent blank. The reagent blank has no absorbance when read vs water in the spectral range shown. Beckman DB-G spectrophotometer.

DETERMINATION OF HEXOSAMINES IN SERUM

(method of Elson and Morgan, Ref 111, modified from Winzler, Ref 513)

PRINCIPLE

Serum proteins are precipitated with alcohol, and the hexosamines are liberated from the glycoproteins by acid hydrolysis. Acetylation in alkaline medium cyclizes the hexosamines to pyrrole derivatives which are then coupled with p-dimethylaminobenzaldehyde (Ehrlich's reagent) to form a colored derivative which is measured photometrically at 530 nm (Fig. 16-7).

REAGENTS

Ethanol, 95%.   Ethanol denatured with methanol is satisfactory.
HCl, 3 N
Bromthymol Blue Solution.   Dissolve 0.1 g of bromthymol blue in 100 ml ethanol (ethanol denatured with methanol is satisfactory).
NaOH, 3 N
Acetylacetone Reagent.   Prepare 0.5 N Na_2CO_3 by dissolving 26.5 g of anhydrous Na_2CO_3 in water and dilute to 500 ml. Dilute 1 ml of acetylacetone (2,4-pentanedione) to 50 ml with freshly prepared 0.5 N Na_2CO_3.

Ehrlich's Reagent. Dissolve 0.8 g *p*-dimethylaminobenzaldehyde in a mixture of 30 ml of methanol and 30 ml of conc HCl.

Glucosamine Stock Standard. Dissolve 60 mg of glucosamine hydrochloride in water and dilute to 100 ml. Store in refrigerator.

Glucosamine Working Standard. 1 ml stock standard + 9 ml water. Prepare fresh daily. 1 ml = 0.05 mg.

PROCEDURE

1. To duplicate 0.2 ml portions of serum in test tubes or centrifuge tubes, calibrated at 10 ml, add 10 ml of 95% ethanol and mix.
2. Centrifuge for 15 min at 2000 rpm; discard the supernate by decantation. Resuspend the precipitate in 10 ml of 95% ethanol, recentrifuge, and decant as before. Allow tubes to drain for several minutes.
3. To the precipitated proteins, add 2 ml of 3 *N* HCl and hydrolyze by covering the tubes and placing in a boiling water bath for 4 hours. Cool at room temperature.
4. Neutralize by adding 1 drop of bromthymol blue solution and titrating to color change with 3 *N* NaOH. Dilute to exactly 10 ml with water.
5. To 1 ml aliquots of unknown and to 1 ml of water (blank) and working standard, add 1 ml of acetylacetone solution; mix.
6. Place tubes in boiling water bath for 15 min, capping the tubes tightly as soon as they are hot.
7. Cool tubes in tap water, add 7 ml of 95% ethanol, and mix.
8. Add 1 ml of Ehrlich's reagent; mix well. Let stand for 30 min.
9. Read absorbances of blank, unknown, and standard at 530 nm against water in 1 cm cuvet.

Calculation:

$$\text{mg hexosamine/100 ml, expressed as glucosamine} = \frac{A_x - A_b}{A_s - A_b} \times 0.05 \times \frac{100}{0.02}$$

$$= \frac{A_x - A_b}{A_s - A_b} \times 250$$

NOTES

1. Beer's law. Absorbances of standard solutions are linear up to 25 mg/100 ml (513).

2. Stability of color. Readings should be taken at 30 min. Beyond this time a gradual increase in absorbance has occasionally been observed in our laboratory.

3. Standardization. The absorbance of galactosamine is the same as that of glucosamine in this method so that the use of a glucosamine standard is justified (513).

4. Purity of acetylacetone. The reliability of the method is dependent upon the purity of the acetylacetone reagent. Yellow color or acidity of the acetylacetone indicates that it is impure. Purification of the commercial product by vacuum distillation may be necessary (140).

5. Stability of samples. As determined in our laboratory, hexosamines are stable in serum for 7 days at 30°C.

ACCURACY AND PRECISION

Recoveries of glucosamine added to precipitated serum proteins are satisfactory (513). Liberation of glucosamine from serum glycoproteins is essentially complete in 4 hours (513).

Precision of the method (95% limits) is about ± 5%.

NORMAL VALUES

The normal range is about 75−120 mg/100 ml (146, 501, 513).

IMMUNOGLOBULINS

Tremendous changes have occurred in the body of knowledge of "Abnormal Protein Fractions" as they were designated more than 9 years ago in the first edition of this book. Largely responsible for these changes are the developing applications of immunologic technics (cf Chapter 7). Those interested in normal or abnormal body fluid proteins now communicate in somewhat different language: "immunoglobulins" (Ig), "monoclonal and polyclonal gammopathies," "heavy and light chain" disease, "Fab and Fc Fragments," "kappa and lambda chains," etc. There is still some controversy over such terms as "paraprotein," "dysproteinemia," "M protein," "agam-maglobulinemia" (198), "plasma cell dyscrasia," etc. *Gammopathies* (more specifically, *immunoglobulinopathies*) are quantitative abnormalities of serum γ-globulins or immunoglobulins (Ig). *Monoclonal gammopathy* is the abnormal production of Ig (probably antibodies) by a clone (family) of antibody-forming cells which presumably descended from a single cell.

Reviews by Tomasi (479) and MacKenzie and Fudenberg (289) related the historical developments linking the electrophoretically characterized γ-globulin portion of the serum to serum protein fractions now called immunoglobulins (Ig). A system of nomenclature proposed by the World

Health Organization (535) originally dealt only with the immunoglobulins A, G, and M but was later extended to IgD and IgE (30a, 398). Refinements of the system were necessary when experimental work resolved the molecular structure of the Ig and defined those portions of the molecules associated with physicochemical and immunologic homogeneity and heterogeneity (58, 169, 398, 536).

Briefly, the Ig molecule is symmetrical and made up of two "heavy" and two "light" chains. The terms heavy and light refer to their relative molecular weights. It is now known that free light chains (L chains) of the Ig molecules are either identical to or very similar to Bence-Jones (BJ) proteins. Immunologic specificity is conferred by the heavy chain portions of the molecules, and the chains are designated α for IgA, γ for IgG, μ for IgM, ε for IgE, and δ for IgD. Light chains are predominantly of two immunologic types, designated kappa (K) and lambda (L); thus the K light chains of an IgM molecule are indistinguishable from the K light chains of an IgG molecule, etc. The two light chains comprising a single molecule are of the same type, e.g., K or L. A population survey of *normal* Ig reveals a ratio of K chain molecules to L chain molecules of roughly 2:1. Monoclonal Ig is generally composed of molecules of only one type of light chain; thus one might see a monoclonal immunoglobulinopathy described as "IgG, L type" (see "light chain typing" under *Bence-Jones Proteins,* below). An early definition of immunoglobulins (535)—"proteins of animal origin endowed with known antibody activity, and certain proteins related to them by chemical structure and hence antigenic specificity"—is still useful; however, IgD has no known antibody activity, and its biologic function is still under study (270). The relationship of the immunoglobulins to the myeloma proteins (M components) was recently reviewed (360) as were other physicochemical, immunologic, and clinical aspects (187, 295). That the monoclonal gammopathy state may be benign was recently revealed in a population survey involving almost 7000 subjects (20).

Particular attention has been directed to IgA because of its importance in local immunity. As the major Ig in body fluids and secretions, it is a dimer of normal 7S serum IgA combined with a transport component (T) more often referred to as the "secretory piece" (391, 444). The review by Collins-Williams *et al* (79) is helpful as it contains tables of IgA measurements in parotid fluid, saliva, and other fluids and secretions. The ratio of IgA to IgG in serum is about 1:5; in secretions total IgA often equals the combined concentrations of IgG and IgM. Quantitation of the secretory piece was reported by Brandtzaeg (57). Because the precipitates formed in cryoglobulinemia and pyroglobulinemia are usually intimately associated with Ig and monoclonal or polyclonal gammopathies, the discussion of these precipitable proteins are included in this section dealing with immunoglobulins.

DETERMINATION OF IMMUNOGLOBULINS IgA, IgD, IgG, AND IgM

(method of Fahey and McKelvey, Ref 116)

PRINCIPLE

The Ig's (except IgE) are measured by single radial immunodiffusion (SRID) (cf Chapter 7). Methodologies for the various proteins differ only in the specific antiserum incorporated in the diffusion medium and the particular concentration required to achieve adequate sensitivity and a useful working range.

REAGENTS

Immunodiffusion Plates Containing Antibody. These are available in various configurations from commercial sources. The 24-well plates are the most economical if relatively large numbers of specimens are to be tested. "Pediatric" or low-level plates contain less of the specific antibody to increase sensitivity. Plates are usually sealed in plastic bags to prevent water loss and if kept refrigerated have a shelf life of about 6 months. Some manufacturers arrange periodic production and shipment of plates containing the same lot and concentration of antiserum thus reducing between-run variation.

NaCl, 0.85%

Standard. A clear serum, with a specific protein content of known value, is diluted with 0.85% NaCl to three or more solutions of lower concentration. These serve as points for the standard curve. A typical four-point standard curve for IgG might include solutions with concentrations of about 2000, 1000, 500, and 250 mg/100 ml.

PROCEDURE

1. Immunodiffusion plates are removed from their plastic envelopes. Water of condensation should be wiped away and plates left at room temperature until the agar surface is free of moisture film.

2. Wells are filled with unknown sera or standard solutions using capillary pipets; over- or under-filling should be avoided. It is more convenient to place the standard solutions in order of increasing or decreasing concentrations. After wells are filled the reaction plates are put back into the plastic envelopes to maintain proper humidity. Because the method is quite responsive to changes in incubation time and temperatures, one or two control sera (of known Ig concentration) are included in each

reaction plate. Incubation times and temperatures vary with the system used; in our laboratory IgG, IgA, and IgD plates are incubated 16–18 hours, the IgG plates at 4°C and the others at room temperature. IgM plates are generally read at 24 hours after room temperature incubation.

3. The diameters of the precipitin rings are measured to the nearest 0.1 mm. Photographic technics are useful, and in our laboratory the Ig plates are photographed with an immunodiffusion camera (available from Cordis). Using a caliper, several diameters are measured for each ring in order to minimize errors due to occasionally noncircular rings.

Calculation:

The standard curve is plotted on semilog paper: concentration of Ig standards on the log axis, the corresponding mean diameters of the precipitin rings on the arithmetic axis (Fig. 16-8). A best straight line is drawn through the points and the concentrations of unknowns are read from the curve.

NOTES

1. Standard curve. Extension of the curve by extrapolation is hazardous. Antigen excess specimens (abnormally high concentrations of Ig) should be retested after appropriate dilutions with saline or with 4% ovalbumin. The curve does not start from zero but at a value slightly greater than the well diameter (Fig. 16-8).

2. Low-level specimens. Hypogammaglobulinemic or infants' sera should be tested with low-level plates, in which antibody concentration has been reduced. Increased sensitivity can also be achieved by intensification of the precipitin lines with tannin (439) and by other means (see Table 7-2).

FIG. 16-8. A typical standard curve for the SRID measurements of IgM.

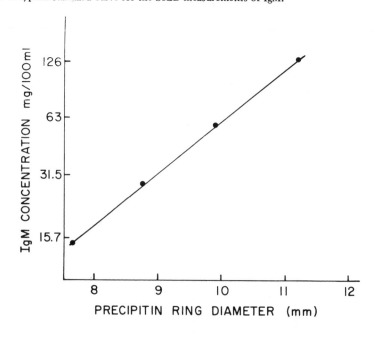

3. Specimen stability. Immunoglobulins A, D, G, and M are stable at 4°C and 30°C for at least 7 days. Long term storage requires the frozen state, preferably at −70°C or colder. Repeated freezing and thawing should be avoided (77).

4. Artifacts. Double rings which invalidate the test may be encountered if the patient has antibodies to serum protein(s) of the animal species in which the specific antiserum has been produced (8, 196).

5. Antisera preservatives. Sodium azide or Merthiolate is often incorporated into the diffusion media to increase shelf life by inhibiting microbial growth. These additives may, however, have a significant effect on test results; Merthiolate has been shown to produce larger than normal precipitin ring size (331).

ACCURACY AND PRECISION

Several factors affect the inherent accuracy of SRID, and at least two of these reflect reagent purity: (a) Specificity of the antisera used in the immunodiffusion plates. The best antisera are prepared against purified, specific heavy chains. Methods for evaluating and standardizing antisera are available (373, 374, 401). Antiserum specificity may be a problem in that, in general, the Ig used as antigens for antibody production are obtained from the sera of patients with monoclonal gammopathies, e.g., IgG myeloma (541). This plus the use of different standards are the main cause of interlaboratory differences. (b) The reliability of the estimate of Ig concentrations in the serum pool or reference preparations used for the standard curve. The use of myeloma Ig for reference primary standard or secondary standard results in extreme interlaboratory differences when the same samples are analyzed (493). A step toward solution of these problems was WHO's establishment of research standards for IgG, IgA, and IgM (399, 400); IgD (402); and IgE (404).

Since interlaboratory expression of Ig on a weight/volume basis varied excessively, the WHO expert committee recommended that measurements be expressed as International Units per milliliter, using comparisons with the WHO research standard (542).

The heterogeneity of IgM and IgA molecular populations is responsible for some of the inaccuracies and imprecision of Ig measurements. For example, more 9S IgA is needed to produce a precipitin ring of given diameter than is required by 7S IgA. Also, 7S IgM, found in some patients instead of 19S IgM, results in spuriously larger rings (213). The seriousness of this problem has been questioned, however (365). When the antigen is lacking some of the antigenic determinants, SRID measurements may be inaccurate (277). Other factors affecting accuracy have been discussed by Reimer (373).

Reported precision data for SRID measurements are difficult to assess since they are often obtained under the most favorable conditions. Reported expressions of precision are often not specified as to between-run or within-run. In our laboratory the coefficient of variation, calculated from the arithmetic mean of at least 30 separate between-run determinations on a serum pool, generally is about 7% for IgA and IgM and about 10% for IgG. The poorer precision of IgG measurements has also been found by others

(217). Somewhat better precision can be achieved with "identical" reaction plates, defined as plates prepared on the same day as the test (367).

NORMAL VALUES

It will be several years before Ig measurements are universally expressed in International Units. Complicating the lab-to-lab differences are the conflicting reports of the factors which influence the normal amount of Ig in individuals, in social and geographic groups, and in body fluids other than serum or plasma. As late as 1970 Kalff (218) stated: "It does not yet seem justified to speak of normal human immunoglobulin levels." Buckley and Dorsey (61) reported elevated IgM and/or IgA or low IgA, IgM, and IgG in a group of apparently healthy humans. In normal individuals the amount of any one Ig, at least for IgG, IgA, and IgM, appears to be independent of the concentrations of the others (3). Sophisticated statistical approaches for determining normal serum Ig levels have appeared (61, 62).

A selective compilation of normal ranges and means from data published since 1966 is in Table 16-3. In our laboratory the adult normal range in mg/100 ml is cited as: IgG 650–1600, IgA 100–400, IgM 40–160, IgD 0–6. For earlier studies of the normal range for Ig see Störiko (461) who compiled reports published from 1957 to 1966.

TABLE 16-3. Compilation of Recent Reports of Normal Values for IgA, IgG, IgM, and IgD (Adults, Male and Female Combined)

Reported range	Normal values (mg/100 ml)			
	IgG[a]	IgA[a]	IgM[a]	IgD[b,c]
Lower end of normal range	569–770	51–125	18–55	1.7
Means[d]	826–1381	158–321	84–159	4.0–15.2
Higher end of normal range	1401–2210	331–425	90–279	23.8

[a] References for IgG, IgA, IgM: 118, 165, 167, 207, 288, 334, 367, 460, 461, 464, 484.
[b] References for IgD: 207, 235, 403, 464.
[c] Data are sparse.
[d] Arithmetic and geometric means combined.

There is lack of agreement on sex differences (163, 393, 460) and seasonal differences (393). IgA, IgG, and IgM levels are clearly age-related (62, 207, 288, 367, 444, 457, 460, 461, 484). It is difficult to compile pediatric normals because of the overlap of age groups; however, some general observations can be made: Being the only Ig to cross the placental barrier, IgG levels at birth approximate maternal levels. Subsequent catabolism of the passively transferred IgG results in a rapid decline to a nadir, approximately one-third to one-half the level at birth, by 3 or 4 months of age. IgG then rapidly increases so that by age 2–7 years levels are nearly the same as in adults.

IgA concentrations are low the first year of life with a subsequent slow but constant increase to age 16, at which point some children have adult levels while others may have levels only about 60% that of adults.

IgM, although present at only 10%–50% of adult levels during the first month of life, approaches the adult level by the first year or certainly by the end of the second year. Between the ages of 5 and 8 years IgM may be higher than in adults but then decreases.

For further details regarding smaller age groups the following references should be consulted: 62, 63, 80, 113, 288, 342, 367, 448, 457, 460, 484. Fetal and maternal levels of Ig have been studied (167) as well as levels in normal and abnormal pregnancies and deliveries (464, 524).

Oral contraceptives do not seem to affect Ig levels.

IMMUNOGLOBULINS IN OTHER BODY FLUIDS

Cerebrospinal Fluid. The low concentrations of Ig in normal CSF requires prior concentration or use of a method more sensitive than SRID, e.g., counterelectrophoresis or electroimmunodiffusion (123, 311). IgA and IgM are usually not detectable in normal CSF; however, IgG appears in concentrations of 1.0–3.1 mg/100 ml (311). Delank and Wrede (95) using SRID found the ratio of CSF IgG/serum IgG to be 1:850 in normal CSF. By concentrating CSF five- to ten-fold prior to estimation by SRID, Bauer and Gottesleben (26) were able to detect these mean levels: IgA 0.226 mg/100 ml and IgG 1.76 mg/100 ml. IgM was not detected.

Urine. IgA and IgG were found in urine at mean concentrations of 0.045 mg/100 ml and 0.221 mg/100 ml, respectively, by Poortmans and Jeanloz (358). It was noted that the presence of light chains (Bence Jones), heavy chains, or Fab and Fc fragments might cause an overestimation of the IgA and IgG. The review by Pruzanski and Ogryzlo (362) is recommended, especially their Table 3. Urinary IgA was extensively studied by Bienenstock and Tomasi (36) who found that the average daily excretion of IgA in urine was 1.1 mg.

Amniotic Fluid. Although IgM and IgG are not detected in normal amniotic fluid, IgA appears at a concentration of about 0.79 mg/100 ml (88). In this study it was noted that the IgA antiserum used had been prepared against 7S IgA and that secretory IgA being dimeric has a sedimentation coefficient of 11S.

IMMUNOGLOBULIN E (IgE)

The discovery of a new myeloma protein which differed immunologically from IgA, IgG, IgM, and IgD was announced by Johansson and colleagues in 1967–1968 (31, 206). First called IgND (the initials of the patient), the protein was soon recognized as the fifth immunoglobulin, IgE (30a), and a research standard was set up by WHO (405). Problems related to standardization were discussed by Bazaral and Hamburger (28) who also studied stability.

The realization that IgE was somehow related to the atopic state generated considerable research interest and activity. The biologic characteristics of purified IgE myeloma protein were studied by Ogawa *et al* (338). Clinical aspects of IgE have also been reviewed (188, 345, 440, 459).

Compared with the other immunoglobulins, IgE is found in large amounts only in the few cases of IgE myeloma yet described (125). Even in the severe atopic state the levels of IgE are usually too low to detect with SRID unless a radioisotopic label is used (201). A simple, one-step SRID was recently described in which the IgG fraction of a specific anti-IgE antiserum was labeled with ^{125}I (12). Other sensitive methods include SRID with fluorescein-labeled or ^{131}I-labeled second antibody (71, 201).

Reported normal values by the radioimmunosorbent (RIST) test (205, 208), the radioactive single radial diffusion (RSRD) (12, 173), and the primary binding blocking test (PBBT) (455) are shown in Table 16-4. A recently introduced kit RIA system permits expression of IgE in International Units.

TABLE 16-4. Some Reported Normal Values for IgE

Method	Specimen	Normal values (ng/ml)	
		Range	Mean
RIST	Serum	660–1830	248
		105–1394	330
RSRD		Under 540	
		<20–8,850[a]	218
PBBT		41–697	177
	Cord serum	18–117	39

[a]"Random population samples"

BENCE-JONES (BJ) PROTEINS

Bence-Jones (BJ) protein first described in 1848 by Bence-Jones designates a group of urine proteins identified by the characteristic of precipitating at $45°–60°C$ and redissolving on boiling. Chemically and immunologically identical proteins may be found in serum. As with other serum and body fluid proteins the application of immunologic technics has greatly improved our knowledge of the BJ proteins; their physicochemical, biologic, and immunologic properties and their clinical significance are now fairly well established. The excretion of BJ protein is generally associated with malignant lymphoreticular disease, particularly multiple myeloma. A recent review relates BJ proteins to cases of abnormal proteinuria and describes light chain disease in which a clone of plasma cells is producing a large excess of free (not combined with heavy chains) light chains (362). One should be aware that sensitive and specific immunologic methods may detect proteins resembling BJ proteins in *normal* urines in very low concentrations. The realization that the BJ proteins are similar if not identical to the light chains or parts of the light chains of the normal immunoglobulins opened up vast new areas of research (34, 82, 115, 445). The following facts are relevant: (a) Bence-Jones proteinuria, free BJ in serum, or both situations may be found where ordinary serum electrophoresis reveals a hypoglobulinemic state. This can occur when there is either a suppression of heavy chain formation (112, 507) or in typical light chain disease (362, 447). (b) By appropriately sensitive immunologic methods, as little as 0.02 mg of BJ protein/ml can be detected. (c) BJ protein consists of either kappa (K) or lambda (L) light chains; however, there have been instances involving both types (38). A Kappa/Lambda incidence ratio of about 1.7:1 was found in a large study of light chain typing of paraproteinemia. (d) Light chain related materials (fragments of light chains?) have been detected, and although of different molecular weights they are antigenically similar in some cases (508) and dissimilar in others (446). (e) Bence-Jones proteins are a heterogeneous group. A "fingerprinting" study with over 100 specimens revealed that the two most similar BJ proteins still differed in at least five amino acids (75). The fragments found in urine and serum are thought by some to be building units rather than degradation products of Ig molecules or light chains (16, 478, 504). When urine is concentrated the light chains may be present as monomers or dimers, as well as submonomers—i.e., parts of the chain (30, 204). Molecular weights vary from 22 000 to 88 000; low molecular weight BJ proteins are in the 11 000—17 000 range. (f) Free light chains are increased even in benign monoclonal gammopathies (90, 362) and, if present in large amounts in serum, may simulate the double spike of biclonal gammopathy when electrophoresis is used for detection.

Immunochemical technics, because of their sensitivity and specificity, have advanced our knowledge of BJ proteins. It is now possible to detect BJ protein in urine from 70% of multiple myeloma patients. There are various reasons why BJ (light chain) detection and/or quantitation is clinically important (329). Kappa/Lambda typing of Ig myelomas by immunoelectrophoresis is illustrated in Figure 16-9. For detection of free light chains in the presence of intact Ig molecules one needs a specific antiserum. Absorption of a "raw" antiserum with human serum albumin and with Cohn

FIG. 16-9. Kappa/lambda typing of IgG monoclonal myeloma. Normal serum in top well. IgG myeloma (kappa type) serum in bottom well. Anti-kappa antiserum in trough reacts both with free kappa light chains and with the light chains on the intact IgG molecules, hence a reaction of identity. Bence-Jones proteins migrate ahead of IgG molecule. Fragments of light chains may migrate even further toward the anode (329).

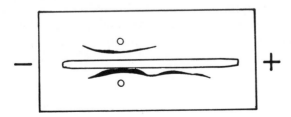

Fraction II IgG, itself devoid of free light chains, makes this possible (90, 527). The same kind of antiserum can then be used to quantitate free light chains by SRID — see section on Ig quantitation and the report by Zelkowitz and Yakulis (527). It has been suggested that the estimation of serum or urine light chains may be important for the early diagnosis of myelomas and Waldenström's macroglobulinemia (90).

A method of K/L typing which does not depend on previous isolation and purification of the light chain(s) was described by Rádl (366). Immuno-electrophoresis "selection" plates were prepared by incorporating anti-K or anti-L antisera into the agar. During electrophoresis entrapment of one kind of light chain would occur. In the subsequent immunodiffusion step specific antiserum in the trough would reveal the presence of the other light chain.

In spite of some problems with accuracy the test most widely used for the qualitative detection of Bence-Jones protein in *urine* is the heat test described below: Upon heating to 60°C a precipitate appears that disappears on boiling and reappears on cooling.

QUALITATIVE HEAT TEST FOR BENCE-JONES PROTEIN IN URINE

(method of Putnam et al, Ref 364)

The qualitative sulfosalicylic acid test for urine protein should be applied first; if negative, no detectable Bence-Jones protein can be present.

REAGENT

Acetate Buffer, pH 4.9, 2 M.   Place 17.5 g sodium acetate trihydrate in a 100 ml volumetric flask, add 4.1 ml glacial acetic acid, and add water to mark.

PROCEDURE

1. Place 4.0 ml clear urine in a test tube (centrifuge or filter urine if turbid). Add 1.0 ml acetate buffer and mix. Final pH should be 4.9 ± 0.1.

2. Heat for 15 min in a 56°C water bath. Any precipitation is indicative of Bence-Jones protein.

3. If there is a turbidity or precipitate, heat the same tube in a boiling water bath for 3 min and observe while in the bath. Any decrease in the amount of precipitate or decrease in turbidity is considered confirmatory for the presence of Bence-Jones protein. Many Bence-Jones proteins do not redisperse completely on boiling (see *Note 1*). An increase in turbidity or precipitate indicates the presence of albumin and globulin and the dissolving of Bence-Jones protein is obscured. Quickly filter the contents of the tube taken directly from the boiling water. Observe the filtrate closely. If at first it is clear, then cloudy as it cools, then clear again at room temperature, the test is considered positive.

NOTES

1. Stability of samples. Urine must be fresh or stored in a manner to prevent microbial growth. In the absence of bacterial digestion, Bence-Jones protein is stable for many years. The changes that may occur within several hours in urine kept at room temperature are: (a) ordinary heat-coagulable protein may precipitate at about 70°C but fail to dissolve at 100°; (b) non-BJ protein may decompose to substances behaving like Bence-Jones protein; and (c) Bence-Jones protein may decompose and fail to react in the usual way.

2. Excessive precipitate. If the precipitate forming at 60°C is too heavy to redissolve upon boiling, dilution of the urine with normal urine is recommended.

ACCURACY AND PRECISION

Some immunochemically defined BJ proteins do not precipitate partially or completely upon heating. Occasionally transferrin interferes when it is present by precipitating with the BJ proteins. Other proteins may also interfere with interpretation; there are different opinions on this matter (364).

NORMAL VALUES

With the heat test at least 0.5–0.8 mg/ml needs to be present to produce a visible precipitate (364); consequently BJ proteins are rarely or never detected in normal unconcentrated urine. Following concentration the more sensitive immunologic methods reveal levels of 0.05–0.47 mg/ml in normal urines (30). Abnormal amounts of light chains and IgG may be found even without clinical proteinuria as evidenced by one study utilizing 55-fold concentration of urine with dialysis and DEAE Sephadex (30).

IMMUNOGLOBULIN HEAVY-CHAINS

Since the first report by Franklin and associates (136) about 27 cases of heavy-chain disease (Franklin's disease) have been reported (35, 362, 428, 429, 483, 526, 532). The γ and α chain diseases predominate although one case of μ chain disease has been reported (131).

The typical laboratory finding in heavy-chain disease is proteinuria associated with abnormal proteins of lower molecular weight than intact Ig molecules. There is usually no evidence of light chain synthesis, although Bhoopalam *et al* (35) described a case of Waldenström's macroglobulinemia with circulating free light chains. They suggested an "unbalanced synthesis of IgM (heavy chains) with some unable to combine with light chains prior to secretion from cells." It is doubtful that the heavy chain is a catabolic product resulting from cleavage of the intact Ig molecules. Other *typical* laboratory findings are:

1. Electrophoresis. There is a characteristic β spike in serum or urine. *N.B.* Several cases in which pathologic proteins were not found have been described (429, 532).

2. Ultracentrifugation. Molecular weights of the abnormal components vary depending on the type of heavy chain and the tendency toward polymerization (428).

3. Immunoelectrophoresis. Abnormal immunoglobulin precipitin lines are seen with components closer to the anode. Usually the faster components form lines of identity with the heavy chains on the intact Ig molecules, but spurs, indicating some difference in antigenic structure, have also been noted (532).

CRYOGLOBULINS

A cryoglobulin is an abnormal globulin characterized by spontaneous but reversible gelling or precipitation upon cooling. Precipitation may even be in the form of needle-shaped crystals. This phenomenon of *cryoglobulinemia* was first noted in 1933 (511), and the name for the precipitate, *cryoglobulin,* was suggested by Lerner and Watson (268) in 1947. Where gelling rather than precipitation occurs the term *cryogelglobulinemia* has been used (442). Franklin (135) recently reviewed what he called the *cryoproteins,* a more general term.

Low levels of cryoglobulins are not usually associated with symptoms, and their presence may be an incidental finding when the patient's serum is stored at refrigerator temperature prior to or after other analytic procedures. Symptoms which do occur usually are of the cold intolerance type.

High levels of cryoglobulins are frequently associated with malignant lymphoreticular disease, especially macroglobulinemia or multiple myeloma. In a series of patients studied by Franklin (135), approximately one-half of the cryglobulinemias were associated with multiple myeloma (IgG type), one-fourth with macroglobulinemia (IgM type), and the remainder with the mixed type (both IgG and IgM).

Large amounts of cryoprotein may appear as homogeneous components on electrophoresis; the mixed cryoglobulins may present as normal or diffusely increased γ-globulin. Cryoglobulins may make up as much as 80% of the total serum protein and nearly all the γ-globulin.

A study of 21 patients with infectious mononucleosis revealed cryoprecipitates in sera from 20 (222). The precipitates contained IgM, IgG, and in 17 cases additional proteins with α_2 mobility. Precipitates consisting only of light chains have also been described (4).

Physicochemical studies of an IgG cryoprotein with lambda light chains led to the postulation that cryoprecipitation is the result of an inadequacy in water-protein interaction (410).

ANALYTIC METHODS

For the qualitative detection of cryoglobulins it usually suffices to leave serum devoid of precipitate in the refrigerator overnight. When only trace amounts (<6 mg/100 ml) are present, refrigeration for a period as long as 1 week may be required for precipitation to occur (267). Redissolving of any precipitate (discrete white particles) or gel by warming to 37°C constitutes conclusive evidence for the presence of cryoglobulin. For quantitation draw blood and allow to clot at 37°C. After removal of the clot, precipitate cryoglobulins in the refrigerator. Wash the precipitate in cold 0.9% NaCl, redissolve at 37°C in 0.9% saline, and reprecipitate in the refrigerator (269). The precipitate can be quantitated by N analysis or by the biuret reaction. It may be necessary to cool serum just short of freezing for complete precipitation (490). Repeated coolings and reheatings may lead to decreased ability to precipitate. Heating the serum to 56°C or diluting to 1:4 or greater results in loss of ability to precipitate.

Immunologic technics are employed for identification and are also used for quantitation. McIntosh and Grossman (303) analyzed serum from a patient with glomerulonephritis with a total cryoprecipitate of 12 mg/100 ml consisting of 9 mg of IgG, 1 mg of the third component of complement (C3), and 2 mg of fibrinogen and/or fibrinogen products. In another report from the same laboratory only IgG and C3 were found in the precipitate (304). The mixed cryoglobulins isolated from poststreptococcal glomerulonephritis were antigen-antibody complexes.

PYROGLOBULINS

Pyroglobulins are the antithesis of cryoglobulins since they are soluble in the cold but precipitated at temperatures greater than 56°C (296). They have been found in the serum and urine associated with various diseases and appear as an electrophoretically homogeneous fraction (193, 296, 297).

The presence of heat-precipitable protein is almost always discovered incidentally—e.g., as a precipitate occurring during the inactivation of serum specimens prior to VDRL, complement fixation, or other tests. Unsuspected myelomas have been uncovered in this manner (124, 278). In approximately 50% of cases pyroglobulins are associated with multiple myeloma (447). BJ protein is often found in the urine of these patients. Early case studies associated pyroglobulins only with IgG myeloma proteins. Stefanini *et al* (454), however, studied three cases of the IgM type discovered during the performance of VDRL tests. Two cases had kappa light chains, and the other had lambda light chains. In two of the patients ultracentrifugation revealed sharp peaks at 19S and 26S, while in the other patient only 19S IgM was found. Heating resulted in almost complete removal of IgM. A very recent report by Sugai (465) chronicled a case history involving an IgA kappa type multiple myeloma. Protein presenting as myeloma protein had a concentration of 10.3 g/100 ml while pyroglobulins represented 80% of a total protein of 14.0 g/100 ml. This illustrates that a protein can be classified in more than one way.

MISCELLANEOUS PROTEINS

CRYOFIBRINOGEN

Cryofibrinogen is a cold-precipitable protein found in plasma but not serum; it migrates electrophoretically like fibrinogen and has the clottable properties of fibrinogen. The precipitation is reversible on warming. It has been reported in association with coagulation disorders (216, 243, 305, 324, 385), phlebitis of pregnancy (144), neonatal infections (387), the use of oral contraceptives (354), and scleroderma (534).

In some cases immunologic studies of the cryoprecipitate reveals a mixture of proteins including fibrinogen (531). In one case the amount of fibrinogen precipitating as cryoprotein was about 65% of the total fibrinogen (354). Periodic sampling in another case revealed the fibrinogen present in the cryoprecipitate to range from 9% to 53% (324). Physicochemical studies also revealed a β_1-globulin, amounting to about 50% of the total protein, which was immunochemically similar to a normal serum protein with an 11S sedimentation coefficient. In this particular study no unclottable fibrinogen or fibrin breakdown products were found. In another detailed physicochemical study small amounts of Factor VIII were associated with the cryofibrinogen (325).

Analytic screening procedures are quite simple and similar to methods for the cryoglobulins, with the difference that *both* serum and plasma are refrigerated (354). The resulting precipitates are quantitated by N analysis or by the biuret reaction. The amount of protein in the plasma precipitate minus the amount in the serum precipitate is taken as cryofibrinogen.

α_1-ANTITRYPSIN

Laurell and associates discovered the relationship between decreased serum α_1-antitrypsin and emphysema (261), and recently discussed the validity of the premise that emphysema is a result of autodigestion (258). The clinical significance of this glycoprotein has been the subject of considerable research activity (46, 471).

Although trypsin-inhibitory capability (TIC) is present in other serum proteins with different electrophoretic mobility, 85%–90% of TIC is associated with α_1-antitrypsin (α_1-AT). With sensitive methods—e.g., electroimmunodiffusion (471) (cf Chapter 7)— α_1-AT may be detected in many other body fluids.

A recent review by Talamo (471) of the various methods did not include the fact that SRID is the most widely used technic for quantitation of α_1-AT in clinical laboratories. For genetic studies of α_1-AT deficiency more sophisticated technics are needed, such as measurement of functional activity. Lieberman *et al* (276) advocated protein electrophoresis on cellulose acetate membranes to screen for the homozygous or heterozygous state and compared results with a functional assay in which TIC is expressed in units (amount of trypsin inhibited by 1 ml of serum).

In most cases the homozygous state can be clearly defined with the absence or near absence of the α_1 peak in zone electrophoresis of serum. However, electrophoretic estimation of α_1-AT may lead to erroneous results especially when performed on paper. For most clinical applications the heterozygous state can be determined by measurement of the amount of α_1-AT in serum by the SRID technic. If there is some suspicion that the α_1-AT present is not completely active, the enzymatic assay should be used. The automated immunoprecipitin (AIP) technic (see Chapter 7, Table 7-3) and an electroimmunodiffusion technic recently described by Laurell (260) may be useful for mass screening.

DETERMINATION OF α_1-ANTITRYPSIN BY SINGLE RADIAL IMMUNODIFFUSION

(method for immunoglobulins, Fahey and McKelvey, Ref 116)

PRINCIPLE

Reaction plates are similar to those used for the Ig, except that a specific antiserum for α_1-AT is incorporated into the diffusion media.

NOTES

1. Standard curve. Three of four points covering the range 10–80 mg/100 ml are usually adequate if the patient's serum is diluted 1:10.

2. Specimen stability. Our data indicate stability for at least 7 days at 4° or 30°C. Serum is stable indefinitely when frozen, preferably at $-70°$C. Repeated freezing and thawing should be avoided.

ACCURACY AND PRECISION

Problems with accuracy and precision are similar to those encountered in the measurement of Ig by SRID.

NORMAL VALUES

The normal range determined with the use of SRID technics is 210–500 mg/100 ml (46). The arithmetic mean value for heterozygotes is 120 mg/100 ml as compared with a mean of 212 for normals and 25 ± 6 mg/100 ml for homozygotes (178). When over 1000 normal individuals were studied by electroimmunodiffusion the normal range was calculated as 200—400 mg/100 ml (210).

α_1-Antitrypsin is somewhat similar to an acute phase reactant in that it is elevated during infections, a situation which might mask the heterozygous state (333). Pregnancy and oral contraception can also produce falsely elevated values (275). However, there appear to be no significant changes during normal menstrual cycles (247).

FETO-PROTEINS

First described in calf serum and called fetuin the α_1 feto- (human) serum protein (AFP) has a molecular weight of 70 000 and contains about 4.3% carbohydrate (122, 407, 408). Normal AFP and that found in disease (see below) appear to be identical (407).

AFP was first thought to be associated normally only with neonatal and early postnatal states of mammalian development and was abnormally associated with primary hepatomas and testicular tumors (6, 194, 407, 441). Using an immunodiffusion method, AFP was found at a level of 120 mg/ml in a patient with prostatic carcinoma (309). AFP has also been found in cases of gastric carcinoma metastatic to the liver (5, 244) and in one-year-old Indian children with a childhood cirrhosis (328). With the development and application of more sensitive methods came the realization that "none detectable" findings in normal adults (10) were due to the insensitivity of the previous methodologies, i.e., electrophoresis, single radial immunodiffusion (SRID), or double linear immunodiffusion (DLID) (see Chapter 7). Recent findings using more sensitive methods are summarized in Table 16-5.

TABLE 16-5. A Comparison of Sensitivity of Methods for Detection and Quantitation of α_1- Fetoprotein

Method[a]	Ref	Sensitivity	Other findings
DLID	6	AFP not detectable in normal adults	
EP-PAG	255	60 000 ng/ml	
IEP	255	10 000 ng/ml	
SRID	127	2000 ng/ml	Range for babies at birth 2500—170 000 ng/ml
SRID	255		Method shows linear relationship 100 000 to 1 million ng/ml
CEP	240	AFP detected in fetal serum diluted 800-fold	
EID	335	200 ng/ml	Mean value in 28 newborns = 94 000 ng/ml
RIA	363	Accurate above 10 ng/ml	Mean value in Bantu miners = 11—13 ng/ml

TABLE 16-5 (cont'd)

Method[a]	Ref	Sensitivity	Other findings
RIA	430	0.25 ng/ml	Range in normal serum = 2–16 ng/ml Range in pregnancy = 18–550 ng/ml

[a]DLID (double linear immunodiffusion), EP-PAG (electrophoresis on polyacrylamide gel), IEP (immunoelectrophoresis), SRID (single radial immunodiffusion), CEP (counterelectrophoresis), EID (electroimmunodiffusion), RIA (radioimmunoassay). See Chapter 7 for method descriptions.

REFERENCES

1. ALBERT-RECHT F: *Clin Chim Acta 4*:627, 1959
2. ALBERT-RECHT F, OWEN JA: *Clin Chim Acta 10*:577, 1964
3. ALLANSMITH M, McCLELLAN B, BUTTERWORTH M: *Immunology 13*:483, 1967
4. ALPER CA: *Acta Med Scand Suppl 445*:200, 1966
5. ALPERT E, PINN VW, ISSELBACHER KJ: *New Engl J Med 285*:1058, 1971
6. ALPERT ME, URIEL J, de NECHAUD B: *New Engl J Med 278*:984, 1968
7. ALTMAN KA, STELLATE R: *Clin Chem 9*:63, 1963
8. AMMANN AJ, HONG R: *J Immunol 106*:567, 1971
9. ANDERSON AJ, MACLAGAN NF: *Biochem J 59*:638, 1955
10. ANDRIEU J, BREART G, RODIER B, ROBERT PE: *Presse Med 79*:1596, 1971
11. ANNINO JS, RELMAN AS: *Am J Clin Pathol 31*:155, 1959
12. ARBESMAN CE, ITO K, WYPYCH JI, WICHER K: *J Allergy Clin Immunol 49*:72, 1972
13. ARMSTRONG SH, Jr, BUDKA MJE, MORRISON KC, HASSON M: *J Am Chem Soc 69*:1747, 1947
14. ARVAN DA, RITZ A: *Clin Chim Acta 26*:505, 1969
15. ASHER TM, COOPER GR: *Clin Chem 6*:189, 1960
16. ASKONAS BA, WILLIAMSON AR: *Nature 211*:369, 1966
17. ASTRUP T, BRAKMAN P, NISSEN U: *Scand J Clin Lab Invest 17*:57, 1965
18. ATENCIO AC, BURDICK DC, REEVE EB: *J Lab Clin Med 66*:137, 1965
19. AUKLAND K, LYGREN T: *Scand J Clin Lab Invest 2*:172, 1959
20. AXELSSON U, HÄLLEN J: *Acta Med Scand 191*:111, 1972
21. BADIN J, SCHMITT F: *Ann Biol Clin (Paris) 15*:313, 1957
22. BADIN J, JACKSON C, SCHUBERT M: *Proc Soc Exp Biol Med 84*:288, 1953
23. BANG HO: *Scand J Clin Lab Invest 9*:205, 1957
24. BARTH WF: *Serum Proteins and the Dysproteinemias,* edited by Sunderman FW, Sunderman FW, Jr., Philadelphia, Lippincott, 1964, p 102
25. BARTHOLOMEW RJ, DELANEY A: *Proc Australian Assoc Clin Biochem 1*:64, 1964
26. BAUER H, GOTTESLEBEN A: *Intern Arch Allergy Appl Immunol 36 (Suppl)*: 643, 1969
27. BAUER S, DeVINO T: *Advances in Automated Analysis.* White Plains, NY, Mediad, Inc., 1970, Vol 3, p 31

28. BAZARAL M, HAMBURGER RN: *J Allergy Clin Immunol 49*:189, 1972
29. BECK EA, SHAINOFF JR, VOGEL A, JACKSON DP: *J Clin Invest 50*:1874, 1971
30. BELL CE Jr, CHAPLIN H Jr: *J Lab Clin Med 75*:636, 1970
30a. BENNICH HH, ISHIZAKA K, JOHANSSON SGO, ROWE DS, STANWORTH DR, TERRY WD: *Immunochemistry 5*:327, 1968
31. BENNICH H, JOHANSSON SGO: *Nobel Symposium, Gamma Globulins Structure and Control of Biosynthesis, Proceedings,* 3rd ed., edited by Killander J., New York, Interscience, 1967, p 199
32. BERGGÅRD I: *Clin Chim Acta 6*:413, 1961
33. BERNSTEIN RE: *South African J Med Sci 19*:131, 1954
34. BERROD J, LARRAT J, SANDOR G, SUREAU B: *Nature 202*:407, 1964
35. BHOOPALAM N, LEE BM, YAKULIS VJ, HELLER P: *Arch Internal Med 128*:437, 1971
36. BIENENSTOCK J, TOMASI TB Jr: *J Clin Invest 47*:1162, 1968
37. BING J, NAESER J, RASCH G, RØJEL K: *Acta Med Scand 126*:351, 1946
38. BLACKBURN R, CHAKERA TM, SOOTHILL JF: *Clin Chim Acta 14*:6, 1966
39. BLIX G: *Acta Chem Scand 2*:467, 1948
40. BLOCH HS, PRASAD A, ANASTASI A, BRIGGS DR: *J Lab Clin Med 56*:212, 1960
41. BLOCK RJ, WEISS KW, CARROLL DB: *Amino Acid Handbook,* edited by Block RJ, Weiss, KW, Springfield, Ill, Charles C. Thomas, 1956, p 252
42. BLONDHEIM SH: *J Lab Clin Med 45*:740, 1955
43. BOAS N: *J Biol Chem 204*:553, 1953
44. BOAS NF, BOLLET AJ, BUNIM JJ: *J Clin Invest 34*:782, 1955
45. BOGDANILOWA B, BERNACKA K, DROZD J: *Clin Chim Acta 13*:221, 1966
46. BOGGS PB, STEPHENS AL Jr, JENKINS DE: *Southern Med J 64*:426, 1971
47. BONSNES RW, SWEENEY WJ: *Am J Obstet Gynecol 70*:334, 1955
48. BOREL JP, CARANJOT JM, JAYLE MF: *Clin Chim Acta 16*:409, 1967
49. BOTTEMA JK, VERVOORT HJP: *Clin Chim Acta 9*:179, 1964
50. BÖTTIGER LE, CARLSON LA: *Clin Chim Acta 5*:664, 1960
51. BÖTTIGER LE, HOLMSTROM A: *J Lab Clin Med 63*:772, 1964
52. BOYCE WH, GARVEY FK, NORFLEET CM Jr: *J Clin Invest 33*:1287, 1954
53. BRACKEN JS, KLOTZ IM: *Am J Clin Pathol 23*:1055, 1953
54. BRACKENRIDGE CJ: *Anal Chem 32*:1357, 1960
55. BRACKENRIDGE CJ: *Clin Chim Acta 5*:539, 1960
56. BRAND E, KASSELL B, SAIDEL LJ: *J Clin Invest 23*:437, 1944
57. BRANDTZAEG P: *Acta Pathol Microbiol Scand (B) 79*:189, 1971
58. BRATCHER RL, FERGUSON WE, GOLDMAN AS, KOENIG VL: *Arch Biochem Biophys 148*:588, 1972
59. BRIERE RO, MULL JD: *Am J Clin Pathol 42*:547, 1964
60. BRUCKNER J: *Biochem J 60*:200, 1955
61. BUCKLEY CE III, DORSEY FC: *Ann Internal Med 75*:673, 1971
62. BUCKLEY CE III, DORSEY FC: *J Immunol 105*:964, 1970
63. BUCKLEY RH, DEES SC, O'FALLON WM: *Pediatrics 41*:600, 1968
64. BURMESTER HBC, AULTON K, HORSFIELD GI: *J Clin Pathol 23*:43, 1970
65. BURROWS S: *Clin Chem 11*:1068, 1965
66. CABEZAS JA, PORTO JV: *Clin Chem 10*:986, 1964
67. CAMPBELL WR, HANNA MI: *J Biol Chem 119*:15, 1937
68. CAMPBELL WR, HANNA MI: *J Biol Chem 119*:9, 1937
69. CARTER P: *Proceedings of the 7th International Congress of Clinical Chemistry,* edited by Roth, M, Basel, Karger, 1970, Vol 1, p 1
70. CASTELAN DJ, HIRSH J, MARTIN M: *J Clin Pathol 21*:638, 1968

71. CENTIFANTO YM, KAUFMAN HE: *J Immunol 107*:608, 1971
72. CESSI C, PILIEGO F: *Biochem J 77*:508, 1960
73. CHORINE V: *Ann Inst Pasteur 63*:213, 1939
74. CHRISTENSEN LK: *Scand J Clin Lab Invest 7*:246, 1955
75. CIOLI D, BAGLIONI C: *Studies on Bence-Jones proteins in man, Proceedings International Symposium on Gammapathies, Infections, Cancer and Immunity,* edited by Chini V, Bonomo L, Sirtori C. Milan, Carlo Erba Foundation, 1968, p 25
76. COHN EJ, GURD FRN, SURGENOR DM, BARNES BA, BROWN RK, DEROUAUX G, GILLESPIE JM, KAHNT FW, LEVER WF, LIU CH, MITTELMAN D, MOUTON RF, SCHMID K, UROMA E: *J Am Chem Soc 72*:465, 1950
77. COHNEN G, PAAR D: *Z Klin Chem Klin Biochem 7*:63, 1969
78. COLLIER M: *J Med Lab Technol 27*:86, 1970
79. COLLINS-WILLIAMS C, LAMENZA C, NIZAMI R: *Ann Allergy 27*:225, 1969
80. COLLINS-WILLIAMS C, TKACHYK SJ, TOFT B, MOSCARELLO M: *Int Arch Allergy Appl Immunol 31*:94, 1967
81. COLOVER J: *J Clin Pathol 14*:559, 1961
82. COOPER A, BLUESTONE R: *Ann Rheumatic Diseases 27*:537, 1968
83. COURTOIS JE, LABAT J, BARANCOURT N: *Ann Biol Clin (Paris) 21*:383, 1963
84. CRANE M, HEYWORTH N: *Am J Diseases Children 51*:99, 1936
85. CROWLEY LV: *Clin Chem 10*:114, 1964
86. CULLEN GE, van SLYKE DD: *J Biol Chem 41*:587, 1920
87. CUMINGS JN, SHORTMAN RC, TOOLEY M: *Clim Chim Acta 27*:29, 1970
88. CURL CW: *Am J Obstet Gynecol 109*:408, 1971
89. CUTLER JL, COLLEN MF, SIEGELAUB AB, FELDMAN R: *Advances in Automated Analysis,* White Plains, NY, Mediad, Inc., 1970, Vol 3, p 67
90. DAMMACCO F, WALDENSTRÖM J: *Clin Exp Immunol 3*:911, 1968
91. DAUGHADAY WH, LOWRY OH, ROSEBROUGH NJ, FIELDS WS: *J Lab Clin Med 39*:663, 1952
92. DAVIS BJ: *Ann NY Acad Sci 121*:404, 1964
93. DE LA HUERGA J, DUBIN A, KUSHNER DS, DYNIEWICZ HA, POPPER H: *J Lab Clin Med 47*:403, 1956
94. DE LA HUERGA J, SMETTERS GW, SHERRICK JC: *Serum Proteins and the Dysproteinemias,* edited by Sunderman FW, Sunderman FW, Jr, Philadelphia, Lippincott, 1964, p 52
95. DELANK HW, WREDE MT: *Klin Wochschr 47*:1270, 1969
96. DEL CAMPO GB: *Clin Chim Acta 22*:475, 1968
97. DEL CARMEN CHIARAVIGLIO E, WOLF AV, PRENTISS PG: *Am J Clin Pathol 39*:42, 1963
98. DELMOTTE P: *Z Klin Chem Klin Biochem 6*:46, 1968
99. DE ROSNAY D, DU PASQUIER P: *Ann Biol Clin (Paris) 16*:363, 1958
100. DEUTSCH HF, MORTON JI: *J Biol Chem 231*:1107, 1958
101. DIRSTINE PH, MacCALLUM DG, ANSON JH, MOHAMMED A: *Clin Chem 10*:853, 1964
102. DISCHE Z, BORENFREUND E: *J Biol Chem 184*:517, 1950
103. DISCHE Z, SHETTLES LB: *J Biol Chem 175*:595, 1948
104. DITTEBRAND TM: *Am J Clin Pathol 18*:439, 1948
105. DOUMAS BT, WATSON WA, BRIGGS HG: *Clin Chim Acta 31*:87, 1971
106. DOW D, PINTO PVC: *Clin Chem 15*:1006, 1969
107. EEG-OLOFSSON R: *Acta Psychiat Neurol Scand 7 (Suppl 50)*:1, 1948
108. EGGSTEIN M, KREUTZ FH: *Klin Wochschr 33*:879, 1955
109. ELKING MP, KABAT HF: *Am J Hosp Pharm 95*:485, 1968
110. ELLIS BC, STRANSKY A: *J Lab Clin Med 58*:477, 1961

111. ELSON LA, MORGAN WTJ: *Biochem J* 27:1824, 1933
112. ENGLE RL Jr, NACHMAN RL: *Blood* 27:74, 1966
113. EVANS HE, AKPATA SO, GLASS L: *Am J Clin Pathol* 56:416, 1971
114. EXLEY D: *Biochem J* 67:52, 1957
115. FAHEY JL: *J Immunol* 91:438, 1963
116. FAHEY JL, McKELVEY EM: *J Immunol* 94:84, 1965
117. FAILING JF Jr, BUCKLEY MF, ZAK B: *Am J Clin Pathol* 33:83, 1960
118. FALK GA, SISKIND GW, SMITH JP Jr: *J Immunol* 105:1559, 1970
119. FARRELL GW, WOLF P: *Med Lab Technol* 28:310, 1971
120. FARRELL GW, WOLF P: *Med Lab Technol* 28:328, 1971
121. FAULKNER WR, GARDNER M, LEWIS LA: *Clin Chem* 8:424, 1962
122. FETOPROTEINS (Anon). *Lancet* 1:397, 1970
123. FISCHER-WILLIAMS M, ROBERTS RC: *Arch Neurol* 25:526, 1971
124. FISHER G, SCHAER LR, MESSINGER S: *Am J Clin Pathol* 40:291, 1963
125. FISHKIN BG, ORLOFF N, SCADUTO LE, BORUCKI DT, SPIEGELBERG HL: *Blood* 39:361, 1972
126. FLETCHER AP, ALKJAERSIG N, SHERRY S: *J Clin Invest* 41:896, 1962
127. FOLI AK, SHERLOCK S, ADINOLFI M: *Lancet* 2:1267, 1969
128. FOLIN O, CIOCALTEAU V: *J Biol Chem* 73:627, 1927
129. FOLIN O, DENIS W: *J Biol Chem* 12:239, 1912
130. FORMAN WB, RATNOFF OD, BOYER MH: *J Lab Clin Med* 72:455, 1968
131. FORTE FA, PRELLI F, YOUNT W, KOCHWA S, FRANKLIN EC, KUNKEL H: *Blood* 34:831, 1969
132. FOX FJ Jr, WIDE L, KILLANDER J, GEMZELL C: *Scand J Clin Lab Invest* 17:341, 1965
133. FRANCO JA, SAVORY J: *Am J Clin Pathol* 56:538, 1971
134. FRANKEL S: *Gradwohl's Clinical Laboratory Methods and Diagnosis*, 7th ed., edited by Frankel S, Reitman S, Sonnenwirth AC, St. Louis, Mosby, 1970, Vol 1, p 38
135. FRANKLIN EC: *Am J Med Sci* 262:50, 1971
136. FRANKLIN EC, LOWENSTEIN J, BIGELOW B, MELTZER M: *Am J Med* 37:332, 1964
137. FREE AH, RUPE CO, METZLER I: *Clin Chem* 3:716, 1957
138. FRIEDMAN HS: *Clin Chim Acta* 6:775, 1961
139. FRIEDMAN HS: *Standard Methods of Clinical Chemistry*, edited by Meites S., New York, Academic Press, 1965, Vol 5, p 223
140. GARDELL S: *Methods of Biochemical Analysis*, edited by Glick D. New York, Interscience, 1958, Vol 6, p 289
141. GAUNCE AP, D'IORIA A: *Anal Biochem* 37:204, 1970
142. GIDDINGS JC, BLOOM AL: *J Clin Pathol* 24:467, 1971
143. GLENN JH: *J Clin Pathol* 18:131, 1965
144. GLUECK HI, HERRMANN LG: *Arch Internal Med* 113:748, 1964
145. GOA J: *Scand J Clin Lab Invest* 5:218, 1953
146. GOA J: *Scand J Clin Lab Invest* 7 (Suppl 22):1, 1955
147. GOA J, TVETEN L: *Scand J Clin Lab Invest* 15:152, 1963
148. GODAL HC: *Scand J Clin Lab Invest* 12 (Suppl 53):1, 1960
149. GOLDENBERG H, DREWES PA: *Clin Chem* 17:358, 1971
150. GOODWIN JF: *Clin Chem* 11:63, 1965
151. GOODWIN JF: *Clin Chem* 13:947, 1967
152. GOODWIN JF: *Clin Chem* 10:309, 1964
153. GORNALL AG, BARDAWILL CJ, DAVID MM: *J Biol Chem* 177:751, 1949

154. GRAFF MM, GREENSPAN EM, LEHMAN, IR, HOLECHEK JJ: *J Lab Clin Med* 37:736, 1951

155. GRALNICK HR, GIVELBER HM, SHAINOFF JR, FINLAYSON JS: *J Clin Invest* 50:1819, 1971

156. GRAM HC: *J Biol Chem* 49:279, 1921

157. GRANNIS GF: *Clin Chem* 16:486, 1970

158. GRANT GH: *J Clin Pathol* 10:360, 1957

159. GREENBERG DM: *J Biol Chem* 82:545, 1929

160. GREENSPAN EM: *Advances in Internal Medicine,* edited by Dock W, Snapper I, Chicago, Year Book Publishers, 1955, Vol 7, p 101

161. GREENWALT TJ, VAN OSS CJ, STEANE EA: *Am J Clin Pathol* 49:472, 1968

162. GRIEBLE HG, COURCON J, GRABAR P: *J Lab Clin Med* 66:216, 1965

163. GRUNDBACHER FJ: *Science* 176:311, 1972

164. GRUNBAUM BW, ZEC J, DURRUM EL: *Microchem J* 7:41, 1963

165. GUMPEL JM, HOBBS JR: *Ann Rheum Dis* 29:681, 1970

166. GUNTON R, BURTON AC: *J Clin Invest* 26:892, 1947

167. GUSDON JP Jr: *AM J Obstet Gynecol* 103:895, 1969

168. HAM TH, CURTIS FC: *Medicine* 17:413, 1938

169. HARKNESS DR: *Postgrad Med* 48:64, 1970

170. HARRISON JF, NORTHAM BE: *Clin Chim Acta* 14:679, 1966

171. HASCHEMEYER RH, NADEAU RE: *Biochem Biophys Res Commun* 11:217 1963

172. HEMMINGSEN L, SKOV F: *Clin Chim Acta* 19:81, 1968

173. HENDERSON LL, SWEDLUND HA, VAN DELLEN RG, MARCOUX JP, CARRYER HM, PETERS GA, GLEICH GJ: *J Allergy Clin Immunol* 48:361, 1971

174. HENRY RJ: *Clinical Chemistry: Principles and Technics,* New York, Harper & Row, 1964, p 173

175. HENRY RJ, GOLUB OJ, SOBEL C: *Clin Chem* 3:49, 1957

176. HENRY RJ, SOBEL C, SEGALOVE M: *Proc Soc Exp Biol Med* 92:748, 1956

177. HENRY RJ, SOBEL C, BERKMAN S: *Anal Chem* 92:1491, 1957

178. HEPPER NG, BLACK LF, GLEICH GJ, KUEPPERS F: *Proc Staff Meetings Mayo Clinic* 44:697, 1969

179. HEREMANS JF: *Clin Chim Acta* 4:639, 1959

180. HERNANDEZ O, MURRAY L, DOUMAS B: *Clin Chem* 13:701, 1967

181. HERRIOTT RM: *Proc Soc Exp Biol Med* 46:642 1941

182. HILL NC, McKENZIE BF, McGUCKLIN WF, GOLDSTEIN NP, SVIEN HJ: *Proc Staff Meetings Mayo Clinic* 33:686, 1958

183. HILLER A, GREIF RL, BECKMAN WW: *J Biol Chem* 176:1421, 1948

184. HILLER A, PLAZIN J, VAN SLYKE DD: *J Biol Chem* 176:1401, 1948

185. HIRSCH A, CATTANEO C: *Arch Biochem Biophys* 61:27, 1956

186. HIRSH J, FLETCHER AP, SHERRY S: *AM J Physiol* 209:415, 1965

187. HOBBS JR: *Advan Clin Chem* 14:219, 1971

188. HONG R, AMMANN AJ, CAIN WA, GOOD RA: *Am J Med Sci* 259:1, 1970

189. HOWE PE: *J Biol Chem* 57:235, 1923

190. HOWE PE: *J Biol Chem* 49:93, 1921

191. HOWE PE: *J Biol Chem* 49:109, 1921

192. HUEMER RP, LEE K: *Anal Biochem* 37:149, 1970

193. HUISMAN THJ, VAN DER WAL B, GROEN A, VAN DER SAR A: *Clin Chim Acta* 1:525, 1956

194. HULL EW, CARBONE PO, MOERTEL CG, O'CONOR GT: *Lancet* 1:779, 1970

195. HUNTER DT, ALLENSWORTH JL: *Am J Clin Pathol* 44:359, 1965

196. HUNTLEY CC, ROBBINS JB, LYERLY AD, BUCKLEY RH: *New Engl J Med* *284*:7, 1971

197. HUSSAIN QZ, SHAH NS, CHAUDHURI SN: *Clin Chim Acta 6*:447, 1961

198. HYPOGAMMAGLOBULINAEMIA (Anon). *Brit Med J 3*:66, 1971

199. IGOU PC: *Am J Med Technol 33*:501, 1967

201. ISHIZAKA K, NEWCOMB RW: *J Allergy 46:*197, 1970

202. JACOBSSON K: *Scand J Clin Lab Invest 7 (Suppl 14)*, 1955

203. JENCKS WP, SMITH ERB, DURRUM EL: *Am J Med 21*:387, 1956

204. JENSEN K: *Scand J Clin Lab Invest 26*:23, 1970

205. JOHANSSON SGO: *Lancet 2*:951, 1967

206. JOHANSSON SGO, BENNICH H, WIDE L: *Immunology 14*:265, 1968

207. JOHANSSON SGO, HÖGMAN CF, KILLANDER J: *Acta Pathol Microbiol Scand 74*:519, 1968

208. JOHANSSON SGO: *Intern Arch Allergy 34*:1, 1968

209. JOHNSON AM, SCHMID K, ALPER CA: *J Clin Invest 48*:2293, 1969

210. JOHNSON AM, ALPER CA: *Pediatrics 46*:921,1970

211. JOHNSON GW, GIBSON RB: *Am J Clin Pathol 8*:22, 1938

212. JOHNSON TF, WONG HYC: *Am J Med Sci 241*:488, 1961

213. JONES D: *Clin Chim Acta 29*:551, 1970

214. JØRGENSEN MB: *Acta Med Scand 181*:153, 1967

215. JUDD J, CLOUSE W, FORD J, VAN EYS J, CUNNINGHAM LW: *Anal Biochem 4*:512, 1962

216. KALBFLEISCH JM, BIRD RM: *New Engl J Med 263*:881, 1960

217. KALFF MW: *Clin Biochem 3*:91, 1970

218. KALFF MW: *Clin Chim Acta 28*:277, 1970

219. KAPLAN A: *Am J Med Sci 253*:549, 1967

220. KAPLAN A, SAVORY J: *Clin Chem 11*:937, 1965

221. KAPLAN A, JOHNSTONE M: *Clin Chem 12*:717, 1966

222. KAPLAN ME: *J Lab Clin Med 71*:754, 1968

223. KAPLAN A: *Standard Methods of Clinical Chemistry*, edited by MacDonald RP, New York, Academic Press, 1970, Vol 6, p 13

224. KELSEY RL, de GRAFFENRIED TP, DONALDSON RC: *Clin Chem 11*:1058, 1965

225. KEKWICK RA, MACKAY ME: *Separation of Protein Fractions from Human Plasma with Ether*, Special Report Series No. 286, London, Medical Research Council, 1954

226. KEUTEL HJ, KING JS Jr: *Clin Chim Acta 11*:341, 1965

227. KEYSER JW: *Clin Chim Acta 6*:445, 1961

228. KEYSER JW: *Clin Chim Acta 7*:299, 1962

229. KEYSER JW: *Anal Biochem 9*:249, 1964

230. KEYSER JW, STEPHENS BT: *Clin Chem 8*:526, 1962

231. KIBRICK AC: *Clin Chem 4*:232, 1958

232. KING JS Jr, WARNOCK NH: *Proc Soc Exp Biol Med 92*:369, 1956

233. KINSLEY GR: *J Lab Clin Med 27*:840, 1942

234. KIRSCHNER MW, SCHACHMAN HK: *Biochemistry 10*:1900, 1971

235. KLAPPER DG, MENDENHALL HW: *J Immunol 107*:912, 1971

236. KLEIN GC, CUMMINGS MM, HAMMARSTEN JF: *Proc Soc Exp Biol Med 111*:298, 1962

237. KLOSSE JA, DE BREE PK, WADMAN SK: *Clin Chim Acta 32*:321, 1971

238. KOHN J: *Clin Chim Acta 2*:297, 1957

239. KOHN J: *Clin Chim Acta 3*:450, 1958

240. KOHN J: *J Clin Pathol 23*:733, 1970

241. KÖIW E, GRÖNWALL A: *Scand J Clin Lab Invest 4*:244, 1952
242. KOROTZER JL, BERGQUIST LM, SEARCY RL: *Am J Med Technol 27*:197, 1961
243. KORST DR, KRATOCHVIL CH: *Blood 10*:945, 1955
244. KOZOWER M, FAWAZ KA, MILLER HM, KAPLAN MM: *New Engl J Med 285*:1059, 1971
245. KREMES B, BRIERE RO, BATSAKIS JG: *Am J Med Technol 33*:28, 1967
246. KROPP GV, McKEE RW: *Am J Clin Pathol 23*:403, 1953
247. KUEPPERS F, BRACKERTZ D, CZYGAN P J: *Clin Chim Acta 39*:131, 1972
248. KULENDA Z, RANDÁKOVÁ M: *Z Med Labortechn 12*:323, 1971
249. LAKI K, GLADNER JA: *Physiol Rev 44*:127, 1964
250. LANCHANTIN GF: *Standard Methods of Clinical Chemistry*, edited by MacDonald RP, New York, Academic Press, 1970, Vol 6, p 137
251. LANCHANTIN GF, NOTRICA SR, MEHL JW, WARE AG: *University of So Calif Med Bull 7*:4, 1955
252. LANE MK, PEARCE RH: *Can Med Assoc J 78*:843, 1958
253. LANGE C: *J Lab Clin Med 31*:552, 1946
254. LANGE HF: *Acta Med Scand (Suppl 176)* 1946
255. LARDINOIS R, ANAGNOSTAKIS D, ORTIZ MA, DELISLE M: *Clin Chim Acta 37*:81, 1972
256. LARSSON SO, THYSELL H: *Acta Med Scand 186*:313, 1969
257. LAUDERBOCK A, MEALEY EH, TAYLOR NA: *Clin Chem 14*:93, 1968
258. LAURELL CB: *Scand J Clin Lab Invest 28*:1, 1971
259. LAURELL CB: *Scand J Clin Lab Invest 7*:95, 1955
260. LAURELL CB: *Scand J Clin Lab Invest 29*:247, 1972
261. LAURELL CB, ERIKSSON S: *Scand J Clin Lab Invest 15*:132, 1963
262. LAURELL CB, KULLANDER S, THORELL J: *Clin Chim Acta 25*:294, 1969
263. LAURELL CB, KULLANDER S, THORELL J: *Scand J Clin Lab Invest 21*:337, 1968
264. LAURELL CB, LAURELL S, SKOOG N: *Clin Chem 2*:99, 1956
265. LAURENT B: *Scand J Clin Lab Invest 10 (Suppl 32)*:1, 1958
266. LEDVINA M, COUFALOVÁ S: *Clin Chim Acta 6*:16, 1961
267. LERNER AB, BARNUM CP, WATSON CJ: *Am J Med Sci 214*:416, 1947
268. LERNER AB, WATSON CJ: *Am J Med Sci 214*:410, 1947
269. LERNER AB, GREENBERG GR: *J Biol Chem 162*:429, 1946
270. LESLIE GA, SWATE TE: *J Immunol 109*:47, 1972
271. LEVIN B, OBERHOLZER VG, WHITEHEAD TP: *J Clin Pathol 3*:260, 1950
272. LEVVY G, McALLAN A: *Biochem J 73*:127, 1959
273. LEVY AL: *Clin Chem 17*:966, 1971
274. LICHTMAN MA, HAMES CG, McDONOUGH JR: *Am J Clin Nutr 16*:492, 1965
275. LIEBERMAN J, MITTMAN C, KENT JR: *J Am Med Assoc 217*:1198, 1971
276. LIEBERMAN J, MITTMAN C, SCHNEIDER AS: *J Am Med Assoc 210*:2055, 1969
277. LIETZE A, SINCLAIR C, ROWE A: *Clin Biochem 3*:335, 1970
278. LIPMAN IJ: *J Am Med Assoc 188*:1002, 1964
279. LOONEY JM, WALSH AI: *J Biol Chem 130*:635, 1939
280. LOSNER S, VOLK BW, JACOBI M, NEWHOUSE S: *J Lab Clin Med 38*:28, 1951
281. LOW EMY, HILL HB, SEARCY RL: *Am J Clin Pathol 47*:538, 1967
282. LOWRY OH, ROSEBROUGH NJ, FARR AL, RANDALL RJ: *J Biol Chem 193*:265, 1951
283. LUNDH B: *Scand J Clin Lab Invest 17*:503, 1965
284. LUSTIG B, LANGER A: *Biochem Z 242*:320, 1931
285. LUXTON GC: *Canad J Med Technol 30*:55, 1968

286. LUXTON GC: *Canad J Med Technol 30*:83, 1968
287. LUZZIO AJ: *J Appl Physiol 21*:685, 1966
288. LYNGBYE J, KRØLL J: *Clin Chem 17*:495, 1971
289. MacKENZIE MR, FUDENBERG HH: *Calif Med 103*:184, 1965
290. MAGATH TB: *The Kidney in Health and Disease*, Philadelphia, Lea & Febiger, 1935, p 440
291. MAJOOR CLH: *J Biol Chem 169*:583, 1947
292. MAMMEN EF, PRASAD AS, BARNHART MI: *J Clin Invest 48*:235, 1969
293. MANSFIELD RE, SHETLAR MR: *Proc Soc Exp Biol Med 112*:891, 1963
294. MARSH WH: *Protides of the Biological Fluids, 9th Colloquium*, 1961, edited by Peeters H. New York, Elsevier, 1962, p 11
295. MARTIN NH: *J Clin Pathol 22*:117, 1969
296. MARTIN WJ, MATHIESON DR: *Proc Staff Meetings Mayo Clinic 28*:545, 1953
297. MARTIN WJ, MATHIESON DR, EIGLER JOC: *Proc Staff Meetings Mayo Clinic 34*:95, 1959
298. MARTINEK RG: *Proc Assoc Clin Biochem III*:264, 1965
299. MARTINEK RG, BERRY RE: *Clin Chem 11*:10, 1965
300. McFARLANE H: *Clin Chim Acta 9*:376, 1964
301. McGARRY E, SEHON AH, ROSE B: *J Clin Invest 34*:832, 1955
302. McGUCKIN WF, McKENZIE BF: *Clin Chem 4*:476, 1958
303. McINTOSH RM, GROSSMAN B: *New Engl J Med 285*:1521, 1971
304. McINTOSH RM, KULVINSKAS C, KAUFMAN DB: *Intern Arch Allergy Appl Immunol 41*:700, 1971
305. McKEE PA, KALBFLEISCH JM, BIRD RM: *J Lab Clin Med 61*:203, 1963
306. McKENZIE JK, McQUEEN EG: *J Clin Pathol 22*:334, 1969
307. McQUEEN EG: *Lancet 1*:397, 1966
308. MEHL JW, PACOVSKA E, WINZLER RJ: *J Biol Chem 177*:13, 1949
309. MEHLMAN DJ, BULKLEY BH, WIERNIK PH: *New Engl J Med 285*:1060, 1971
310. MENINI E, FALHOLT W, LOUS P: *Acta Med Scand 160*:315, 1958
311. MERRILL D, HARTLEY TF, CLAMAN HN: *J Lab Clin Med 69*:151, 1967
312. MERSKEY C, KLEINER GJ, JOHNSON AJ: *Blood 28*:1, 1966
313. MERSKEY C, LALEZARI P, JOHNSON AJ: *Proc Soc Exp Biol Med 131*:871, 1969
314. MESTREZAT W: *Compt Rend Soc Biol 84*:382, 1921
315. MEULEMANS O: *Clin Chim Acta 5*:757, 1960
316. MILLER GH Jr, DAVIS ME, KING AG, HUGGINS CB: *J Lab Clin Med 37*:538, 1951
317. MIYADA DS, BAYSINGER V, NOTRICA S, NAKAMURA RM: *Clin Chem 18*:52, 1972
318. MIYASATO F, POLLAK VE: *J Lab Clin Med 67*:1036, 1966
319. MORAWIECKA B, MEJBAUM-KATZENELLENBOGEN W: *Clin Chim Acta 7*:722, 1962
320. MÖRNER KAH: *Skand Arch Physiol 6*:332, 1895
321. MORRISON PR: *J Am Chem Soc 69*:2723, 1947
322. MORSE EE: *Ann Clin Lab Sci 1*:155, 1971
323. MORSE EE, PANEK S, MENGA R: *Am J Clin Pathol 55*:671, 1971
324. MOSESSON MW, COLMAN RW, SHERRY S: *New Engl J Med 278*:815, 1968
325. MOSESSON MW, UMFLEET RA: *J Biol Chem 245*:5728, 1970
326. MÜLLER-EBERHARD HJ: *Klin Wochschr 34*:693, 1956
327. MUSA BU, DOE RP, SEAL US: *J Clin Endocrinol 27*:1463, 1967
328. NAYAK NC, CHAWLA V, MALAVIYA AN, CHANDRA RK: *Lancet 1*:68, 1972
329. NERENBERG ST: *CRC Crit Rev Clin Lab Sci 1*:303, 1970

330. NESS AT, DICKERSON HC, PASTEWKA JV: *Clin Chim Acta 12:*532, 1965
331. NEUBURG M, WETTER O: *Klin Wochschr 50:*119, 1972
332. NEUHAUS OW, LETZRING M: *J Lab Clin Med 50:*682, 1957
333. NOMIYAMA K, NOMIYAMA H, MATSUI H: *Bull World Health Organ 45:*253, 1971
334. NORBERG R: *Acta Med Scand 181:*485, 1967
335. NØRGAARD-PEDERSEN B: *Clin Chim Acta 38:*163, 1972
336. OBERMAN JW, GREGORY KO, BURKE FG, ROSS S, RICE EC: *New Engl J Med 255:*743, 1956
337. O'BRIEN D, IBBOTT FA, RODGERSON DO: *Laboratory Manual of Pediatric Micro-Biochemical Technics,* 4th ed. New York, Harper & Row, 1968, p 287
338. OGAWA M, McINTYRE OR, ISHIZAKA K, ISHIZAKA T, TERRY WD, WALDMANN TA: *Am J Med 51:*193, 1971
339. OGSTON CM, OGSTON D: *J Clin Pathol 19:*352, 1966
340. OH YH, SANDERS BE: *Anal Biochem 15:*232, 1966
341. O'KELL RT, ELLIOTT JR: *Clin Chem 16:*161, 1970
342. PANAYOTOU P, PAPADATOS C, PAPAEVANGELOU G, ALEXIOU D, SKARDOUTSOU A, KOUREA E: *Arch Disease Childhood 46:*671, 1971
343. PARFENTJEV IA, JOHNSON ML, CLIFFTON EE: *Arch Biochem 46:*460, 1953
344. PATRICK RL, THIERS RE: *Clin Chem 9:*283, 1963
345. PATTERSON R: *J Chronic Diseases 23:*521, 1971
346. PAULY H, LUDWIG E: *Z Physiol Chem 121:*170, 1922
347. PEARCE RH, WATSON EM, STODOLSKI R, MATHIESON JM, THEORET JJ: *Clin Chem 10:*1066, 1964
348. PERSSON I: *Scand J Clin Lab Invest 15:*353, 1963
349. PERSSON I: *Scand J Clin Lab Invest 7:*279, 1955
350. PETERS T Jr: *Clin Chem 14:*1147, 1968
351. PETERS JP, VAN SLYKE DD: *Quantitative Clinical Chemistry.* Baltimore, Williams & Wilkins, 1932, p 63
352. PHILLIPS LL, JENKINS EB, HARDAWAY M III: *Am J Clin Pathol 52:*114, 1969
353. PILLEMER L, HUTCHINSON MC: *J Biol Chem 158:*299, 1945
354. PINDYCK J, LICHTMAN HC, KOHL SG: *Lancet 1:*51, 1970
355. PISCATOR M: *Arch Environ Health 5:*325, 1962
356. PLUM CM, FOG T: *Acta Psychiat Neurol 34:*37, 1959
357. PODMORE DA: *Clin Chim Acta 4:*242, 1959
358. POORTMANS J, JEANLOZ RW: *J Clin Invest 47:*386, 1968
359. PORTER P: *J Comp Pathol Therap 74:*108, 1964
360. POTTER M: *New Engl J Med 284:*831, 1971
361. PRICE WH, MATANOSKI GM, MORRISON D, PREWER A, WAGNER G: *Bull Johns Hopkins Hosp 108:*227, 1961
362. PRUZANSKI W, OGRYZLO MA: *Advan Clin Chem 13:*335, 1970
363. PURVES LR, GEDDES EW: *Lancet 1:*47, 1972
364. PUTNAM FW, EASLEY CW, LYNN LT, RITCHIE AE, PHELPS RA: *Arch Biochem Biophys 83:*115, 1959
365. RADICHEVICH I, MORSE J, WERNER SC: *Proc Soc Exp Biol Med 139:*353, 1972
366. RÅDL J: *Immunology 19:*137, 1970
367. RÅDL J, SKVARIL F, MASOPUST J, KITHIER K: *J Hyg Epidemiol Microbiol Immunol 14:*488, 1970
368. RANE L, NEWHOUSER LR: *US Armed Forces Med J 5:*368, 1954
369. RAPP RD, MEMMINGER MM: *Am J Clin Pathol 31:*400, 1959
370. RATNOFF OD, MENZIE C: *J Lab Clin Med 37:*316, 1951
371. READ BAN: *Clin Chem 15:*1186, 1969

372. REED AH, CANNON DC, WINKLEMAN JW, BHASIN YP, HENRY RJ, PILEGGI VJ: *Clin Chem 18*:57, 1972
373. REIMER CB: *Health Laboratory Science 9*:178, 1972
374. REIMER CB, PHILLIPS DJ, MADDISON SE, SHORE SL: *J Lab Clin Med 76*:949, 1970
375. REINER M, CHEUNG HL: *Standard Methods of Clinical Chemistry*, edited by Seligson D. New York, Academic Press, 1961, Vol 3, p 114
376. REINER M, FENICHEL RL, STERN KG: *Acta Haematol 3*:202, 1950
377. REINHOLD JG: *Standard Methods of Clinical Chemistry*, edited by Reiner M. New York, Academic Press, 1953, Vol 1, p 88
378. RESSLER N, GOODWIN JF: *Am J Clin Pathol 38*:131, 1962
379. RICE EW: *Standard Methods of Clinical Chemistry*, edited by Meites S. New York, Academic Press, 1965, Vol 5, p 231
380. RICE EW, LOFTIS JW: *Clin Chem 8*:56, 1962
381. RICE JD Jr: *Am J Clin Pathol 45*:277, 1966
382. RICE JP, BLEAKNEY B: *Clin Chim Acta 12*:343, 1965
383. RIGAS DA, HELLER CG: *J Clin Invest 30*:853, 1951
384. RIMINGTON C: *Biochem J 34*:931, 1940
385. RITZMANN SE, HAMBY C, COOPER R, LEVIN WC: *Texas Rep Biol Med 21*:262, 1963
386. ROBERTS LB: *Clin Chim Acta 16*:69, 1968
387. ROBINSON MG, TROIANO G, COHEN H, FOADI M: *J Pediat 69*:35, 1966
388. ROBOZ E, HESS WC, TEMPLE DM: *J Lab Clin Med 43*:785, 1954
389. RODKEY FL: *Clin Chem 10*:606, 1964
390. RODKEY FL: *Clin Chem 11*:478, 1965
391. ROGERS AI: *Postgrad Med 48*:75, 1970
392. ROLLER E, BERG G, SCHEIFFARTH F: *Clin Chim Acta 5*:695, 1960
393. ROSE NR: *Clin Chem 17*:573, 1971
394. ROSENBERG AA: *Clin Chem 2*:331, 1956
395. ROSENBERG RM, LAVER WF, LYONS M: *J Am Chem Soc 77*:6502, 1955
396. ROSENFELD L: *J Lab Clin Med 72*:329, 1968
397. ROSENTHAL HL, CUNDIFF HI: *Clin Chem 2*:394, 1956
398. ROWE DS: *Nature 228*:509, 1970
399. ROWE DS, ANDERSON SG, GRAB B: *Bull World Health Organ 42*:535, 1970
400. ROWE DS, ANDERSON SG, FAHEY JL: *J Immunol 102*:792, 1969
401. ROWE DS, ANDERSON SG, SKEGG J: *Immunoglobulins*, edited by Merler E. Washington, DC, National Academy of Sciences, 1970, p 351
402. ROWE DS, ANDERSON SG, TACKETT L: *Bull World Health Organ 43*:607, 1970
403. ROWE DS, CRABBÉ PA, TURNER MW: *Clin Exp Immunol 3*:477, 1968
404. ROWE DS, TACKETT L, BENNICH H, ISHIZAKA K, JOHANSSON SGO, ANDERSON SG: *Bull World Health Organ 43*:609, 1970
405. ROWE DS, TACKETT L, BENNISH H, ISHIZAKA K, JOHANSSON SGO, ANDERSON SG: *Bull World Health Organ 43*:609, 1970
406. RUDMAN D, DEL RIO A, AKGUN S, FRUMIN E: *Am J Med 46*:174, 1969
407. RUOSLAHTI E, PIHKO H, SEPPÄLÄ M, VUOPIO P: *Scand J Clin Lab Invest 27 (Suppl 116)*:57, 1971
408. RUOSLAHTI E, SEPPÄLÄ M: *Int J Cancer 7*:218, 1971
409. RUTSTEIN DD, INGENITO EF, REYNOLDS WE: *J Clin Invest 33*:211, 1954
410. SAHA A, EDWARDS MA, SARGENT AU, ROSE B: *Immunochemistry 5*:341, 1968
411. SAIFER A: *J Lab Clin Med 63*:1054, 1964
412. SAIFER A, GERSTENFELD S: *Clin Chem 10*:321, 1964

413. SAIFER A, GERSTENFELD S: *Clin Chim Acta 7*:467, 1962
414. SAIFER A, GERSTENFELD S: *Clin Chem 8*:236, 1962
415. SAIFER A, GERSTENFELD S: *J Lab Clin Med 50*:17, 1957
416. SAIFER A, GERSTENFELD S, VECSLER F: *Clin Chem 7*:626, 1961
417. SAIFER A, MARVEN T: *Clin Chem 12*:414, 1966
418. SAIFER A, NEWHOUSE A: *J Biol Chem 208*:159, 1954
419. SAIFER A, SIEGEL HA: *J Lab Clin Med 53*:474, 1959
420. SAVORY J, HEINTGES MG, SOBEL RE: *Clin Chem 17*:301, 1971
421. SAVORY J, PU PH, SUNDERMAN FW Jr: *Clin Chem 14*:1160, 1968
422. SCHNEIDER CL: *Am J Obstet Gynecol 64*:141, 1952
423. SCHRIEVER H, GAMBINO SR: *Am J Clin Pathol 44*:667, 1965
424. SCHWARTZ CJ, GILMORE HR: *Circulation 18*:191, 1958
425. SEARCY RL: *Arch Biochem Biophys 81*:275, 1959
426. SEIBERT FB, ATNO J: *J Biol Chem 163*:511, 1946
427. SEIBERT FB, PFAFF ML, SEIBERT MV: *Arch Biochem 18*:279, 1948
428. SELIGMANN M, DANON F, HUREZ D, MIHAESCO E, PREUD'HOMME JL: *Science 162*:1396, 1968
429. SELIGMANN M, MIHAESCO E, HUREZ D, MIHAESCO C, PREUD'HOMME JL, RAMBAUD JC: *J Clin Invest 48*:2374, 1969
430. SEPPÄLÄ M, RUOSLAHTI E: *Am J Obstet Gynecol 112*:208, 1972
431. SETHNA S, TSAO MU: *Clin Chem 3*:249, 1957
432. SEXTON JS, AULL JC: *Am J Clin Pathol 42*:320, 1964
433. SHARP AA, HOWIE B, BIGGS R, METHUEN DT: *Lancet 2*:1309, 1958
434. SHERWIN RM, MOORE GH: *Am J Clin Pathol 55*:705, 1971
435. SHETLAR MR: *Progress in Clinical Pathology*, edited by Stefanini M. New York, Grune & Stratton, 1966, Vol 1, p 419
436. SHETLAR MR: *Anal Chem 24*:184, 1952
437. SHETLAR MR, FOSTER JV, EVERETT MR: *Proc Soc Exp Biol Med 67*:125, 1948
438. SHETLAR MR, MASTERS YF: *Anal Chem 29*:402, 1957
439. SIMMONS P: *Clin Chim Acta 35*:53, 1971
440. SIMONS MJ, HOSKING CS, HOGARTH-SCOTT RS: *Australian Ann Med 4*:366, 1970
441. SKOVRONSKY J: *Postgrad Med 49*:63, 1971
442. SLAVIN RG, SURIANO JR, DREESMAN G: *Intern Arch Allergy Appl Immunol 40*:739, 1971
443. SMITH DC, MURCHISON W: *J Med Lab Technol 16*:197, 1959
444. SMITH RT: *Pediatrics 43*:317, 1969
445. SOLOMON A, FAHEY JL: *Am J Med 37*:206, 1964
446. SOLOMON A, KILLANDER J, GREY HM, KUNKEL HG: *Science 151*:1237, 1966
447. SOLOMON J, STEINFELD JL: *Am J Med 38*:937, 1965
448. SOWARDS DL, MONIF GRG: *Am J Obstet Gynecol 112*:394, 1972
449. SPIRO RG: *New Engl J Med 269*:566, 1963
450. SPIRO RG: *New Engl J Med 269*:616, 1963
451. SPIRO RG: *New Engl J Med 281*:1043, 1969
452. SPIRO RG: *New Engl J Med 281*:991, 1969
453. SPIRO RG: *Ann Rev Biochem 39*:599, 1970
454. STEFANINI M, McDONNELL EE, ANDRACKI EG, SWANSBRO WJ, DURR P: *Am J Clin Pathol 54*:94, 1970
455. STEVENSON DD, ORGEL HA, HAMBURGER RN, REID RT: *J Allergy Clin Immunol 48*:61, 1971

456. STEVENSON GT: *J Clin Invest 39*:1192, 1960
457. STIEHM ER, FUDENBERG HH: *Pediatrics 37*:715, 1966
458. STIRLAND RM: *Lancet 1*:672, 1956
459. STITES DP, ISHIZAKA K, FUDENBERG HH: *Clin Exp Immunol 10*:391, 1972
460. STOOP JW, ZEGERS BJM, SANDER PC, BALLIEUX RE: *Clin Explt Immunology 4*:101, 1969
461. STÖRIKO K: *Blut 16*:200, 1968
462. STRICKLAND RD, MACK PA, GURULE FT, PODLESKI RR, SALOME O, CHILDS WA: *Anal Chem 31*:1410, 1958
463. STRICKLAND RD, MACK PA, PODLESKI TR, CHILDS WA: *Anal Chem 32*:199, 1960
464. STUDD JWW: *J Obstet Gynaecol Brit Commonwealth 78*:786, 1971
465. SUGAI S: *Blood 39*:224, 1972
466. SUNDBLAD L, WALLIN-NILSSON M: *Scand J Clin Lab Invest 14*:72, 1962
467. SUNDERMAN FW Jr, SUNDERMAN FW: *Ann Internal Med 51*:488, 1959
468. SUNDERMAN FW Jr, SUNDERMAN FW, FALVO EA, KALLICK CJ: *Am J Clin Pathol 30*:112, 1958
469. SVENNERHOLM L: *Biochim Biophys Acta 24*:604, 1957
470. SVENSMARK O: *Scand J Clin Lab Invest 10*:50, 1958
471. TALAMO RC: *J Allergy Clin Immunol 48*:240, 1971
472. TAMM I, BUGHER JC, HORSFALL FL Jr: *J Biol Chem 212*:125, 1955
473. TAMM I, HORSFALL FL Jr: *J Exp Med 95*:71, 1952
474. TANDON A, SAXENA KC: *Biochem Med 5*:552, 1971
475. TANDON A, SAXENA KC: *Biochem Med 5*:371, 1971
476. TÁRNOKY AL: *Biochem J 49*:205, 1951
477. THYSELL H: *Acta Med Scand 185*:401, 1969
478. TITANI K, WIKLER M, PUTNAM FW: *Science 155*:828, 1967
479. TOMASI TB Jr: *Blood 25*:382, 1965
480. TOMBS MP, SOUTER F, MACLAGEN NF: *Biochem J 71*:13p, 1959
481. TREVORROW V, KASER M, PATTERSON JP, HILL RM: *J Lab Clin Med 27*:471, 1942
482. TULLER EF, KEIDING NR: *Anal Chem 26*:875, 1954
483. TURNER MW, BERGGÅRD I: *Nature 224*:912, 1969
484. UFFELMAN JA, ENGELHARD WE, JOLLIFF CR: *Clin Chim Acta 28*:185, 1970
485. URIEL J, GRABAR P: *Anal Biochem 2*:80, 1961
486. VAN OSS CJ: *Advances in Immunogenetics*, edited by Greenwalt TJ. Philadelphia, Lippincott, 1967, p 1
487. VAN SLYKE DD, HILLER A, PHILLIPS RA, HAMILTON PB, DOLE VP, ARCHIBALD RM, EDER HA: *J Biol Chem 183*:331, 1950
488. VERMYLEN C, De VREKER RA, VERSTRAETE M: *Clin Chim Acta 8*:418, 1963
489. VINK CLJ: *Clin Chim Acta 15*:702, 1960
490. VOLPÉ R, BRUCE-ROBERTSON A, FLETCHER AA, CHARLES WB: *Am J Med 20*:533, 1956
491. WAGNER EC: *Ind Eng Chem 12*:771, 1940
492. WALDMAN RK, KRAUSE LA, BORMAN EK: *J Lab Clin Med 42*:489, 1953
493. WALLER M, HOFFMAN PF: *Am J Clin Pathol 56*:645, 1971
494. WARE AG: *Personal Communication*
495. WARE AG, GUEST MM, SEEGERS WH: *Arch Biochem 13*:231, 1947
496. WARREN L: *J Biol Chem 234*:1971, 1959
497. WATSON D: *Clin Chim Acta 15*:121, 1967
498. WATSON D: *Advances in Clinical Chemistry*, edited by Sobotka H, Stewart CP. New York, Academic Press, 1965, Vol 8, pp 237—282

499. WATSON D, NANKIVILLE DD: *Clin Chim Acta 9:*359, 1964
500. WEICHSELBAUM TE: *Am J Clin Pathol 16:*40, 1946
501. WEIDEN S: *J Clin Pathol 11:*177, 1958
502. WEIMER HE, MOSHIN JR: *Am Rev Tuberc Pulmonary Diseases 68:*594, 1952
503. WEIMER HE, MEHL JW, WINZLER RJ: *J Biol Chem 185:*561, 1950
504. WETTER O: *Klin Wochschr 48:*482, 1970
505. WHITEHOUSE MW, ZILLIKEN F: *Methods of Biochemical Analysis,* edited by Glick D. New York, Interscience, 1960, Vol 8, p 199
506. WHITEHEAD TP, PRIOR AP, BARROWCLIFF DF: *J Lab and Clin Med 24:*1265, 1954
507. WILLIAMS RC Jr, BRUNNING RD, WOLLHEIM FA: *Ann Internal Med 65:*471, 1966
508. WILLIAMS RC Jr, PINNELL SR, BRATT GT: *J Lab Clin Med 68:*81, 1966
509. WILLS MR, McGOWAN GK: *J Clin Path 16:*487, 1963
510. WINDISCH RM, BRACKEN MM: *Clin Chem 16:*416, 1970
511. WINTROBE MM, BUELL MV: *Bull Johns Hopkins Hosp 52:*156, 1933
512. WINZLER RJ: *The Plasma Proteins,* edited by Putnam FW. New York, Academic Press, 1960, Vol 1, p 309
513. WINZLER RJ: *Methods of Biochemical Analysis,* edited by Glick D. New York, Interscience, 1955, Vol 2, p 279
514. WINZLER RJ: *Ciba Foundation Symposium on the Chemistry and Biology of Mucopolysaccharides,* edited by Wolstenholme GEW, O'Connor M. Ciba Foundation Symposium, Boston, Little, Brown, 1958, p 245
515. WINZLER RJ, DEVOR AW, MEHL JW, SMYTH IM: *J Clin Invest 27:*609, 1948
516. WOLFSON WQ, COHN C: *Rapid Chemical Micromethods for the Analytical Fractionation of Serum Proteins.* Ann Arbor, Mich, Overbeck, 1951
517. WOLFSON WQ, COHN C, CALVARY E, ICHIBA F: *Am J Clin Pathol 18:*723, 1948
518. WOLFSON WQ, COHN C, CALVARY E, ICHIBA F: *Am J Clin Pathol 18:*723, 1948
519. WOLVINS D, VERSCHURE JCM: *J Clin Pathol 8:*140, 1955
520. WRIGHT IS: *J Am Med Assoc 180:*733, 1962
521. WU H: *J Biol Chem 51:*33, 1922
522. WU H, LING SM: *Chinese J Physiol 1:*161, 1927
523. WYCOFF HD: *Clin Chem 6:*429, 1960
524. YANG SL, KLEINMAN AM, ROSENBERG EB, WEI PY: *Am J Obstet Gynecol 109:*78, 1971
525. YOUNG DS, HARRIS EK, COTLOVE E: *Clin Chem 17:*403, 1971
526. ZAWADZKI ZA, BENEDEK TG, EIN D, EASTON JM: *Ann Internal Med 70:*335, 1969
527. ZELKOWITZ L, YAKULIS V: *J Lab Clin Med 76:*973, 1970
528. ZINNEMAN HH, SEAL US: *Arch Internal Med 124:*77, 1969
529. ZIPF RE, KATCHMAN BJ: *Serum Proteins and the Dysproteinemias,* edited by Sunderman FW, Sunderman FW Jr. Philadelphia, Lippincott, 1964, pp 40–46
530. ZISHKA MK, NISHIMURA JS: *Anal Biochem 34:*291, 1970
531. ZLOTNICK A, LANDAU S: *J Lab Clin Med 68:*70, 1966
532. ZLOTNICK A, LEVY M: *Arch Internal Med 128:*432, 1971
533. ZONDAG HA, VON BOETZELAER GL: *Clin Chim Acta 5:*155, 1960
534. ZVAIFLER NJ: *Arthritis Rheumat 15:*133, 1972
535. *Bull World Health Organ 30:*447, 1964
536. An extension of the nomenclature for immunoglobulins, *Immunochem 7:*497, 1970

537. *Autoanalyzer Methodology, Method File N-14b*. Tarrytown, NY, Technicon Corp.
538. *Autoanalyzer Methodology, Method File N-14C*. Tarrytown, NY, Technicon Corp.
539. *Autoanalyzer Methodology, Method File N-15C*. Tarrytown, NY, Technicon Corp.
540. Committee on Protein Nomenclature, American Physiological Society and American Biochemical Society. *J Bio Chem 4*:xlviii, 1908
541. Standards for immunoglobulins (Anon). *Lancet 2*:82, 1971
542. Measurements of concentrations of human serum immunoglobulins. *J Clin Pathol 25*:133, 1972

Chapter 17

Nonprotein Nitrogenous Constituents

JOHN DI GIORGIO, Ph.D.

There are about 15 nonprotein nitrogen (NPN) compounds known to be of clinical importance in serum, but only 6 of these are determined routinely in the clinical chemistry laboratory. The NPN fraction consists of urea nitrogen, which constitutes about 45% of the total NPN, followed in order of quantitative importance by amino acids, uric acid, creatinine, creatine, and ammonia. Amino acids are discussed in Chap. 18.

NONPROTEIN NITROGEN

The NPN is most commonly determined in serum. The NPN of whole blood is about 1.76 times higher than that of serum or plasma particularly because of the greater glutathione content of erythrocytes. The most satisfactory method for determining NPN is the Kjeldahl-Nessler method, various modifications of which have been introduced to facilitate complete sample digestion. The method is technically simple and involves three steps: (a) removal of serum proteins, (b) digestion of the NPN compounds with acid to convert all N into NH_4^+, and (c) color development to quantitate photometrically the amount of NH_4^+.

Two factors must be considered when choosing a protein precipitant: the amount of NPN constituents that coprecipitate with the proteins (101), and the ability to precipitate proteoses, peptones, and polypeptides. Folin and Wu (85) found that tungstic acid filtrates tend to give lower NPN values compared with trichloracetic acid filtrates. Hiller and Van Slyke (104) reported that when urea is initially degraded by urease, results for NPN are equivalent regardless of whether colloidal iron, trichloroacetic acid, tungstic acid, or metaphosphoric acid are used as precipitants (104). Alcoholic precipitants, however, definitely resulted in the lowest NPN values. Trichloroacetic acid has been recommended as a precipitant because, unlike tungstic acid, it does not appear to affect the nesslerization reaction (57). It has also been reported that tungstic acid, trichloroacetic acid, and metaphosphoric acid give equivalent NPN values while Somogyi's zinc filtrate results in NPN values that are about 10 mg/100 ml lower than other precipitants because of removal of glutathione, ergothioneine, uric acid, and some creatine (11). Peters and Van Slyke (178) and Henry (101) have reviewed the comparative results with various protein precipitants and conclude that tungstic acid is a better precipitant than trichloroacetic acid. Another method involves preliminary adjustment of the sample pH followed by heating to coagulate the protein. This results in a protein-free filtrate that contains about 25% more NPN than a tungstic acid filtrate (109).

Kjeldahl digestion transforms the N into NH_4^+, and the reaction is conveniently performed in an open test tube wherein a photometric reaction can be carried out. During digestion organic carbon is oxidized to CO_2, so that care must be taken to avoid oxidation of the NH_4^+ to N_2 (124). To prevent N loss and yet achieve an optimum rate of NH_4^+ formation, there is a choice of three ingredients in the digestion mixture: a catalyst to increase the digestion rate, a salt to elevate the digestion temperature, and an oxidizing agent to accelerate the conversion of carbon to CO_2.

Free mercury has been used as a catalyst (124, 181) as has mercuric sulfate (44). When mercury catalyst is used in the digestion step, zinc dust must be added to remove the free mercury before developing the color with Nessler's reagent (181). Selenium and copper salts as catalysts give low NPN results (139). A copper catalyst should not be used because the digest is not suitable for either direct nesslerization or the gasometric determination of ammonia (178). The decision whether or not to use a Kjeldahl catalyst will depend partly on the amount of nitrogen in the sample to be digested.

Several salts are used to elevate the digestion temperature. Potassium sulfate has the advantage that it avoids glassware etching (124). On the other hand, sodium sulfate in the digestion mixture can cause solidification upon cooling.

Even with the proper choice of a catalyst or salt, there can be a loss of N_2 gas during the digestion step if oxidizing agents are used other than potassium persulfate (235, 252) or H_2O_2 (124, 126, 154). H_2O_2 purportedly has the advantage of shortening the digestion time with no loss of ammonia during digestion even in the presence of chloride (126).

After digestion of the sample, the NH_4^+ is quantified by one of the following methods: (a) photometric measurement by addition of Nessler's reagent or Berthelot solution either to the original digestion mixture or to acid into which the NH_4^+ has been distilled, (b) manometric measurement of the N_2 released by addition of hypobromite to the digest, (c) acid-base titration of the acid solution into which NH_4^+ has been distilled after alkalinization of the digest, (d) coulometric titration of the digest with electrogenerated hypobromite.

Two color reactions can be carried out directly in the Kjeldahl digestion mixture: formation of colloidal amber-yellow with Nessler's reagent or formation of blue indophenol in the Berthelot reaction.

Direct nesslerization of the digestion mixture is thought to occur as follows:

$$(NH_4)HSO_4 + 2\,NaOH \longrightarrow Na_2SO_4 + 2\,NH_3 + 2\,H_2O$$

The final reaction of ammonia with the double iodide is not known, but it may be (227)

$$2\,HgI_2 \cdot 2KI + 2\,NH_3 \longrightarrow NH_2Hg_2I_3 + 4\,KI + NH_4I$$

Other formulas proposed for the Nessler complex include $Hg_2 NI \cdot H_2 O$ and HgOHNHHgI. The manner of adding the Nessler's reagent directly to the digest is important. During addition of the reagent, the digest must be constantly mixed to prevent uneven alkalinization and resultant turbidity. Folin and Wu (85) first digested tungstic acid filtrates in open test tubes with a sulfuric-perchloric acid mixture prior to the nesslerization step. The Folin-Wu procedure often results in turbid solutions, which can be prevented if the digest is cooled before adding the Nessler's reagent (11). Perchloric acid used instead of Folin's tungstic acid to make protein-free filtrates reportedly prevents turbidity by eliminating precipitation of Nessler's reagent by tungstate (94). The Folin-Wu method has been modified by: (a) digesting the tungstic acid filtrate with sulfuric acid, (b) substituting hydrogen peroxide for perchloric acid as the oxidizing agent, and (c) reformulating Nessler's reagent (126). For economic reasons, mercuric oxide (247) or mercuric iodide (227) can be substituted for metallic mercury in Nessler's reagent. Other modifications have been introduced to cope with the turbidity problem, e.g., adding gum ghatti and sodium citrate (57) or potassium cyanide (155) to the digest to prevent precipitation of Nessler's reagent. Potassium cyanide, however, results in decreased color intensity. Another approach to obviate precipitation problems involves adding acid to the sample, adjusting the pH to a suitable value, then heating to precipitate the proteins (109). When such solutions were digested and reacted with Nessler's reagent, the NPN values were higher than those obtained with tungstic acid filtrates.

The indophenol (Berthelot or phenol-hypochlorite) reaction (discussed in the section on ammonia) is the second method commonly used to estimate photometrically the NH_4^+ in the Kjeldahl digestion mixture. The sensitivity of this method allows measurement of from 1 to 15 μg of NH_4^+ N in a Kjeldahl digest (139). Since the indophenol reaction requires alkaline conditions for proper color development, tungstic acid, which results in a less acidic filtrate, has been recommended rather than trichloroacetic acid (70). The indophenol reaction applied to the digestion mixture has also been used for urine NPN (150). To avoid interference with this reaction all traces of $H_2 O_2$ used in a micro Kjeldahl digestion step must be removed (196) and a low mercury concentration maintained (207) if nitroprusside catalyst is used.

Rather than measure NH_4^+ photometrically in the Kjeldahl digest, Van Slyke (235) transferred the neutralized digestion mixture into the Van Slyke-Neill manometric apparatus and measured the N_2 gas released by reaction of ammonia with alkaline hypobromite. Alternatively, hypobromite was added to the protein-free solution before digestion. Variations of this methodologic approach include iodometric titration of the released bromine (184) or photometric measurement of the iodine liberated from potassium iodide added directly to the protein-free filtrate (75, 92). NPN values obtained with the hypobromite procedure may be 20% lower than Kjeldahl values (101).

Another approach is to distil the ammonia from the digestion mixture into acid and determine the ammonia concentration by titration (227), or isothermally diffuse microgram amounts of NH_3 from digested trichloro-acetic acid filtrates into boric acid and titrate the ammonium ion with a stronger acid (54).

NH_4^+ in the digest has also been titrated with electrogenerated hypobro-mite (44). The advantage of this coulometric method is that it is an absolute method requiring no standards.

DETERMINATION OF NONPROTEIN NITROGEN (NPN) BY NESSLERIZATION

(method of Koch and McMeekin, Ref. 126, modified)

PRINCIPLE

The nitrogen in a tungstic acid filtrate of whole blood or serum is subjected to oxidation with H_2SO_4 and H_2O_2 and the resultant NH_4^+ quantitated by nesslerization. Nessler's reagent is a double iodide of potassium and mercury, but its exact composition is as unknown as the reaction product with NH_4^+. The absorption curve of this reaction is shown in Figure 17-1.

REAGENTS

30% (v/v) H_2SO_4
30% H_2O_2 (Superoxol).   Store in refrigerator.
15% Sodium Citrate
Nessler's Reagent (according to Vanselow, Ref. 234). Dissolve 45.5 g mercuric iodide and 34.9 g potassium iodide in about 100 ml water. In a separate flask, dissolve 112 g KOH in about 500 ml water and cool to room temperature. Combine the two solutions and add water to 1 liter. Let the reagent age for about 2 or 3 days before use to permit any precipitate to settle out. Store in refrigerator in a Pyrex bottle with rubber stopper.
Standard, 0.1 mg N/ml. Dissolve 0.472 g $(NH_4)_2SO_4$ (reagent grade) in sufficient water in a 1 liter volumetric flask. Add a few drops of conc. H_2SO_4, then dilute to 1 liter with water. Stable indefinitely.

PROCEDURE

1. Prepare a protein-free tungstic acid filtrate as follows: Add 1.0 ml whole blood and 1.0 ml 2/3 N H_2SO_4 to 7.0 ml water and mix. Then add 1.0 ml 10% sodium tungstate and mix. Filter through Whatman 1 filter paper (see note 2). If the sample is serum, use half-quantities of precipitating reagents and make up the difference in volume with water.
2. Set up NPN digestion tubes graduated at 50 ml:
   Reagent Blank. 2.0 ml H_2O.
   Standard. 2.0 ml standard (0.2 mg N).
   Unknown. 5.0 ml of the 1:10 protein-free filtrate.

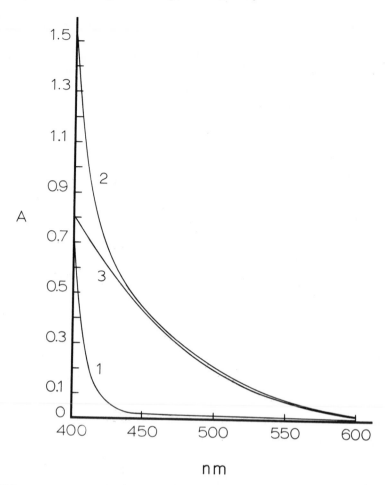

FIG. 17-1. Absorption curves of Nessler's reaction: (1) reagent blank vs water, (2) standard (20 µg N) vs water, (3) standard vs reagent blank. Beckman Model DU Spectrophotometer.

3. To each tube add 1.0 ml 30% (v/v) H_2SO_4 and a glass bead.
4. Heat over an open flame until water is driven off and dense white fumes of H_2SO_4 begin to fill the tube. Digestion must be carried out in a fume hood or the tubes must be inserted in a manifold connected to a water aspirator pump.
5. Remove tubes from the flame for about 30 sec and add 3 drops 30% H_2O_2 to each tube. Hold the tubes vertically so that the drops fall directly into the bottom of the tubes.
6. Resume digestion until H_2SO_4 fumes fill the tubes and continue digestion for 2 min. Discontinue heating (if solution is not colorless, repeat addition of H_2O_2) and allow the tubes to cool.
7. Add 2 ml 15% sodium citrate to each tube; dilute with water to the 50 ml mark and mix.
8. Transfer 2.5 ml of each digest to a large test tube (15–20 mm diam), add

2.5 ml water and mix. Then, with mixing, add 1.0 ml Nessler's reagent as rapidly as possible.

9. Between 1 and 30 min after adding Nessler's reagent, read absorbances of all tubes against water at 460 nm. If a filter photometer is used, employ a filter with a nominal wavelength of 420 nm and read standard and unknown against the reagent blank. All tubes should be examined for turbidity by Tyndall effect. If turbidity is a problem, read the colors *before* the turbidity appears.

Calculation: For spectrophotometer at 460 nm:

$$\text{mg NPN/100 ml} = \frac{A_x - A_b}{A_s - A_b} \times 0.01 \times \frac{100}{0.025}$$

$$= \frac{A_x - A_b}{A_s - A_b} \times 40$$

For filter photometer with a 420 nm filter:

$$\text{mg NPN/100 ml} = \frac{A_x}{A_s} \times 40$$

NOTES

1. Beer's law. The composition of the chromogen measured is unknown, but there seems to be little question that it is collodial in nature. According to Henry (101), it is impossible with the technic and reagents used to avoid a Tyndall effect when standards contain more than 60 μg NH_4^+-N in the reaction mixture. With the recommended Nessler's reagent, Beer's law is obeyed up to 60 μg NH_4^+-N (equivalent to an NPN of 240 mg/100 ml) at 460 nm in the spectrophotometer (A of about 1.1) or with a 420 nm glass filter, and up to 100 μg at 520 nm in the spectrophotometer or with a 540 nm glass filter. Some Nessler's reagents are more prone to form turbid solutions so that deviations from Beer's law at 520 nm or with a 540 nm glass filter are manifested in a calibration curve with *upward* concavity. This is presumably the result of turbidity. A calibration curve constructed from standards containing 10, 20, 40, and 80 μg NH_4^+-N in the final reaction mixtures should always be carried through with the Nessler's reagent and the instrument to be employed for this determination in order to verify the relationship between absorbance and concentration. If the method significantly deviates from Beer's law, results should be calculated from the curve rather than from the formula given.

In some methods, a wavelength longer than 520 nm is recommended to measure absorbances, but at such wavelengths the absorbance for a normal NPN is below the precise photometric range. Consequently, the shorter wavelength is recommended for routine use although NPN's above 200 μg NH^+_4-N give an absorbance above the precise photometric range. In these cases, nesslerize a smaller aliquot of the digest. An alternative is to standardize the test at both wavelengths.

2. *Error from filter paper.* With some filter papers ammonia is transferred to the protein-free solution during filtration. Acid-washed papers contain the largest amounts of ammonia. The amount found in Whatman 1 paper is insignificant for most purposes (final results may be high by 0.2 mg/100 ml). Filter paper can be checked by filtering 10 ml of water then adding 1.0 ml Nessler's reagent to 5 ml filtrate. Read and calculate the NH^+_4-N as for an unknown and divide by 20 to obtain mg N/100 ml in terms of an original blood sample.

3. *Importance of running reagent blanks.* Since gaseous ammonia is always a possible contaminant in the laboratory, reagent blanks should be set up frequently if not every time an analysis is performed. If a filter photometer is used with a filter having a nominal wavelength of 420 nm, a reagent blank must be used for setting the instrument to 0 absorbance. If a digested reagent blank is not used for this purpose, the following steps can be followed: mix 0.2 ml 15% sodium citrate, 0.1 ml 30% (v/v) H_2SO_4 and 4.7 ml water, then add 1.0 ml Nessler's reagent. The same amount of acid should always be employed with standards as with unknowns because the final color intensity depends on the titratable alkalinity (126). Furthermore, the acid frequently becomes contaminated with ammonia.

4. *Manner of addition of Nessler's reagent.* For accuracy and reproducibility of results there is probably no other single factor as important as *exact* reproducibility of the manner in which Nessler's reagent is added and mixed. Errors up to 20% have been reported from this factor alone (154). The easiest and most practical way to standardize the procedures is to add and mix the reagent as rapidly as possible. For this reason the use of wide test tubes is recommended, to facilitate rapid mixing while the reagent is being added. Henry (101) has also observed that immediate mixing with some Nessler's reagents decreased the probability of turbidity in the final color.

Another reason to reserve a special set of test tubes just for nesslerization is that mercury adsorbs to the glass and is not removed by ordinary washing procedures. This mercury can inhibit enzyme reactions subsequently carried out in these tubes. If the tubes cannot be reserved for this purpose, treat them after nesslerization with warm 20% HNO_3 to remove the mercury.

5. *Effect of temperature.* There is about a 10% increase in absorbance from 20 to 30°C (154).

6. *Anticoagulant.* If the analysis is to be performed on plasma, blood should be collected in a tube containing 2 mg potassium oxalate per milliliter. Obviously, an anticoagulant that introduces nitrogen into the sample must not be used.

7. *Preservative.* A sample preservative is not necessary if bacterial contamination is judiciously avoided. Sodium fluoride, 5 mg/ml, is a suitable bacteriostat.

8. Hemolysis. A slight amount of hemolysis in either serum or plasma will not significantly elevate the NPN.

ACCURACY AND PRECISION

The value for the NPN obtained is known to vary with the protein precipitant employed (109). The results obtained with this direct nessleriza-tion procedure (no preliminary isolation of the NH^+_4 from the digest) appear to agree with those obtained by the Kjeldahl-Gunning method (126).

The precision of the method is about ±5%.

NORMAL VALUES

Although the normal adult range for whole blood is frequently taken as 25–40 mg/100 ml (11, 219), there is justification for extending this to 25–50 (170). The values for males are slightly higher than those for females (170). The range for infants is somewhat higher than the normal adult range (101). After the age of 60, there appears to be a steady increase with age (170).

The normal adult range for serum NPN is about 20–35 mg/100 ml (219).

UREA

Urea nitrogen constitutes about 45% of the total NPN in serum or plasma. Many of the early studies of urea quantification were undertaken by biochemists who were interested in N balance, whereby all blood and urine nitrogenous constituents were added together in terms of their N content. Since then, clinical chemists have continued to report the concentration of urea as urea N, but this can be converted to milligrams urea by multiplying by 2.14.

Methods for determining urea are classified into three groups: (a) direct—condensation of urea with diacetyl to form a measurable chromogen; (b) indirect—determination of ammonia as a product of urease action on urea; (c) miscellaneous—procedures involving various photometric or physical principles of analysis.

The direct method using diacetyl monoxime has an important advantage: it does not determine $NH_4{}^+$. Fearon (74) was the first to show that urea and other compounds having the R_1 NH-CO-NHR$_2$ structure (when R_1 is H or a single aliphatic radical and R_2 is not an acyl radical) react with diacetyl monoxime in the presence of strong acid and an oxidizing agent to produce a chromogen. Ormsby (173) applied the Fearon reaction to measure blood and urine urea in a protein-free solution. He found that potassium persulfate enhances the color by oxidizing the hydroxylamine formed in the reaction between urea and diacetyl monoxime. Diacetyl has been used to eliminate hydroxylamine formation thereby obviating the need for an oxidizing reagent (163). The reaction mechanism is thought to proceed in two steps:

$$CH_3-\overset{\overset{\displaystyle O}{\|}}{C}-\overset{\overset{\displaystyle NOH}{\|}}{C}-CH_3 \quad + \quad H_2O \quad \xrightarrow{\ H^+\ }$$

DIACETYL MONOXIME

$$CH_3-\overset{\overset{\displaystyle O}{\|}}{C}-\overset{\overset{\displaystyle O}{\|}}{C}-CH_3 \quad + \quad HONH_2$$

DIACETYL   HYDROXYLAMINE

$$\begin{array}{c} CH_3 \\ | \\ C=O \\ | \\ C=O \\ | \\ CH_3 \end{array} + \begin{array}{c} NH_2 \\ \diagdown \\ \diagup \\ NH_2 \end{array}C=O \quad \xrightarrow{\ H^+\ } \quad \begin{array}{c} CH_3 \\ | \\ C=N \\ | \quad \diagdown \\ \quad \quad C=O + 2H_2O \\ C=N \diagup \\ | \\ CH_3 \end{array}$$

DIACETYL   UREA   DIAZINE DERIVATIVE
(YELLOW)

The following oxidants have been used to eliminate hydroxylamine: potassium persulfate (74, 140, 173), arsenic acid (4, 36, 87, 160, 243), cations (12, 141, 191, 253), perchloric acid (125), and phenazone (24, 39, 40, 41).

Phenylanthranilic acid (12, 246), thiosemicarbazide (56, 69, 141, 159, 177), or glucuronolactone (156) have been added to the reagent in order to intensify the color and minimize its photosensitivity. Various cations will also stabilize the final color (12, 253).

Rather than measure the color formed by the reaction between diacetyl and urea as a quantitative photometric method, its fluorescence can be measured at 415 nm (146).

The disadvantages of the diacetyl methods (101) include: (a) color develops rapidly and fades rapidly; (b) the color is photosensitive; (c) the color does not follow Beer's law with either a filter photometer or a spectrophotometer; (d) the unpleasant odor and irritant fumes of the reagents make it advisable to work in a fume hood; (e) with diacetyl monoxime the time of heating for maximal color development is dependent on the urea concentration; (f) the reaction is not completely specific, urease destroying all the chromogens in blood but only about 95% of those present in urine.

The indirect method uses the enzyme urease (urea amidohydrolase, E. C. 3.5.1.5) to decompose urea:

$$H_2N\text{-}CO\text{-}NH_2 + H_2O + 2H^+ \xrightarrow{\ UREASE\ } 2NH_4^+ + CO_2\uparrow$$

The NH_4^+ or CO_2 released from the ammonium carbonate is then quantitated. The sample must not be collected in a tube containing fluoride.

In 1913, Marshall (142) was the first person to use urease as a tool for the determination of urea in blood. Without prior removal of protein from the sample and after incubation with urease, ammonia was isolated by aeration and then quantitated by titration. The Van Slyke approach (236) of measuring manometrically the CO_2 released never became popular because it is easier to measure the NH_4^+.

After urease action, the NH_4^+ is measured either directly in the incubation mixture or by titration following its isolation by aeration (142) or diffusion (105). Nessler's reagent reacts with NH_4^+ (89, 121), but a serious restriction is imposed by the need for reading absorbances of the colored solutions at 420 nm within 1 min after addition of Nessler's reagent (102). This haste is dictated by the fact that substances are present other than NH_4^+ that develop a yellow color slowly. The NH_4^+ also reacts with salicylate-dichloroisocyanurate reagent, which is neither hazardous to laboratory personnel nor corrosive to automated instrumentation in contrast to the strongly acidic diacetyl monoxime reagent (201, 203).

A nonphotometric method for measuring urea N in blood or urine is to titrate the urease-liberated ammonia with coulometrically generated hypobromite ion (46).

Titration and nesslerization methods have been complemented by the indophenol reaction of Berthelot that is about 10 times more sensitive for ammonia than nesslerization (101). Serum urea is hydrolyzed with urease and the released ammonia then measured as indophenol (73). This method was cumbersome to perform until the four reagents were reduced to two, namely, the catalyst-phenol solution and the alkaline-hypochlorite solution (43). Weatherburn (244) studied the optimum concentrations for reagents and suggested changes in the NaOH and phenol concentrations in order to increase sensitivity. The indophenol reaction can be performed in either a protein-containing or a protein-free solution (120).

A study of buffer composition and pH effects on the rate and extent of urease hydrolysis of urea shows that a neutral dipotassium hydrogen phosphate buffer, rather than disodium hydrogen phosphate, will appreciably speed up the hydrolysis (242).

One of several catalysts is used in the indophenol reaction, but the most effective one is nitroprusside since it increases the speed of converting ammonia to indophenol (43, 244) and the results agree with a urease-aeration-nesslerization method (73). Fenton (77) has used either a combined acetone-nitroprusside catalyst or only nitroprusside.

There are a number of miscellaneous ways to determine urea N in serum or urine. Van Slyke and Kugel (237) used a manometric blood-gas apparatus to measure the N_2 released from a Somogyi filtrate after addition of alkaline hypobromite. Even though this reaction is rapid and requires no standard solutions, the hypobromite reaction is neither stoichiometric nor specific and requires an empiric correction by subtracting 1.2 mg urea N/100 ml (101). A simple and novel approach involves measurement of the height of precipitated dixanthylurea in microcapillary columns (202). Urea may be reacted with *p*-dimethylaminobenzaldehyde (Ehrlich's reagent) to yield a stable yellow color. The presence of the polysaccharide Dextran-40 in serum will interfere (128). A cadmium hydroxide filtrate (193) or a protein-free solution made from trichloroacetic acid containing activated charcoal (254) will eliminate interferences from other serum constituents. The activated charcoal will remove such interfering substances as sulfa drugs. Additional methods for determining urea N in blood or urine are the urease enzyme electrode (98), the N-bromosuccinimide gasometric method (10), the photometric nioxime method (210), and the photometric iso-nitrosopropiophenone method (59, 229).

This author has chosen for presentation both a manual and an automated method for determining serum and urine urea nitrogen. They yield equivalent results, require the same reagents, and the color developed is based on the same chemical principle. Also, the manual method is always available in emergency or "STAT" situations, or when the amount of work does not justify the use of the automated technic. In fact, these methods offer two additional advantages (69, 141): elimination of a protein precipitation step in the manual procedure and omission of the dialyzer in the automated procedure.

DIRECT MANUAL DETERMINATION OF UREA NITROGEN

(method of Wybenga et al, Ref. 253)

PRINCIPLE

Serum and urine urea react with acidic diacetyl monoxime during a short heating period. The presence of thiosemicarbazide in the reagent simultaneously intensifies the color of the reaction and decreases its photosensitivity. The absorption curve for the resultant purple-red solution is shown in Figure 17-2.

REAGENTS

Urea Solution, 2.6 mg/100 ml. Dissolve 26 mg urea/liter water.

Urea Nitrogen Reagent. Add 44 ml conc H_2SO_4 and 66 ml 85% orthophosphoric acid to about 100 ml of water contained in a 1 liter flask. Cool the solution to room temperature but do not use ice water as a cooling bath. Then add the following, dissolving each successively: (a) 50 mg thiosemicarbazide, (b) 2.0 g cadmium sulfate octahydrate, and (c) 10 ml urea solution. Mix and dilute to 1 liter with water. Transfer to an amber bottle. This reagent is stable for at least 6 months when refrigerated. The presence of a small amount of urea in the reagent improves the linearity of the calibration curve.

Diacetyl Monoxime, 2%. Add 20 g diacetyl monoxime (2, 3-butanedione monoxime; Eastman Kodak has been found to be satisfactory) to about 900 ml of water in a 1 liter volumetric flask, mix to dissolve, and dilute to volume. Transfer to an amber bottle. This reagent is stable for at least 6 months when refrigerated.

Urea Nitrogen Standards, 30 and 60 mg urea N/100 ml. Prepare the 60 mg urea N/100 ml standard by dissolving 128 mg urea in about 50 ml water, adding six drops of chloroform, and diluting to 100 ml. Prepare the 30 mg urea N/100 ml standard by diluting the 60 mg/100 ml standard 1:2 with

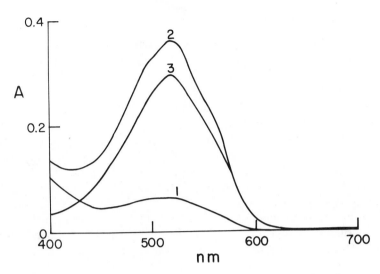

FIG. 17-2. Absorption curves of the diacetyl monoxime reaction: (1) reagent blank vs water, (2) standard (20 mg urea N/100 ml) vs water, (3) standard vs reagent blank. Beckman DB-G Grating Spectrophotometer.

water. Add several drops of chloroform for preservation. Transfer to amber bottles. These standards are stable for at least 6 months when refrigerated.

PROCEDURE FOR SERUM

1. To each of four test tubes labeled "Blank," "Standard 30," "Standard 60," and "Unknown" add 5.0 ml urea N reagent. Since the final color is developed in these tubes, cuvets may be used.
2. Using TC pipets, transfer 20 μl of each urea N standard and 20 μl serum to their respective test tubes and mix.
3. Add 0.5 ml diacetyl monoxime reagent to all test tubes, mix.
4. Place all test tubes into a 100°C heating block or boiling water bath.
5. Remove all test tubes after 12 min, immerse in cool tap water for 5 min, and mix.
6. Read A_s and A_x at 540 nm against the reagent blank.
7. Construct a two-point calibration curve and determine values of unknown samples (see note 1).

PROCEDURE FOR URINE

The method can be applied directly to urine and has sufficient sensitivity to permit analysis on micro volumes without deproteinization. The average urinary excretion of urea N is about 10 g/day, representing a concentration about 50 times that in serum. The test, therefore, is carried out with 20 μl of a 1:50 dilution of urine. The 30 mg/100 ml standard in this procedure is consequently equivalent to 1500 mg urea N/100 ml urine.

NOTES

1. Beer's law. Absorbance in the diacetyl reaction is not linear with concentration (4, 36, 160). The two-point calibration curve, however, is sufficiently accurate for routine use: The error introduced is less than 6% at a concentration of 15 mg urea N/100 and 2% at 50 mg/100 ml. For samples suspected of having a urea N value greater than 60 mg/100 ml, a 10 μl sample should be used initially.

2. Color stability. Thiosemicarbazide intensifies the color and promotes photostability of the colored complex (56). However, the final absorbance still decreases 7% in 30 min in the absence of cadmium ions. Increasing thiosemicarbazide does not improve photostability but rather results in decreased sensitivity (253). The use of Sb^{3+}, Cu^{2+}, Mn^{2+}, Fe^{3+}, and Cd^{2+} will stabilize the color formed in this reaction (12, 253). Manganese results in the lowest blanks. However, Cd^{2+}, (200 mg of $CdSO_4$ octahydrate/100 ml) is most effective (253) because the color decreases less than 5%/hour compared with a decrease of 10%/hour with Mn^{2+} (160 mg $MnCl_2$/100 ml) and 20%/hour with Fe^{3+} (33 mg $FeCl_3$/100 ml).

3. Dilution of final color. When the urea N concentration in urine gives an absorbance greater than 0.450, dilute the final colors of the unknown, standards, and reagent blank with an equal volume of water and immediately read again.

4. Sample stability. Urea in serum is stable for at least 1 day at room temperature, for several days in the refrigerator, and for 6 months when frozen. Fluoride (6 mg fluoride/ml blood) or fluoride plus thymol (10 mg fluoride plus 1 mg thymol/ml blood) are effective in preserving samples at room temperature for about 5 days.

Urea in urine is stable for several days in the refrigerator provided that the pH remains less than 4. In urine contaminated with bacteria, conversion of urea to ammonia may take place rapidly, resulting in alkalinizing the urine with subsequent loss of ammonia.

ACCURACY AND PRECISION

Results obtained by this method are not significantly different than those obtained by measuring the NH_4^+ produced by urease action (43). Citrulline and allantoin are physiologic compounds that also react with diacetyl to give a colored compound similar to that obtained with urea. The amounts of these substance in serum, however, are usually too small to contribute any measurable color. Elevated concentrations of serum bilirubin, as high as 10 mg/100 ml, or mild hemolysis do not interfere with the method.

The precision (95% limits) of the method is ±6% for urine and for serum with urea values between 10 and 80 mg/100 ml.

NORMAL VALUES

Serum and Plasma. In a study (186) performed by our laboratory of 1419 clinically normal adults from 8 geographical areas of the United States it was found that the upper normal limits are significantly higher for men than for women in the 20–39 year age groups and that age is accompanied by an increase in estimated normal limits for both sexes. These findings are in substantial agreement with previously published reports. Although significant statistically, no great error will occur, however, if the normal range is taken as 8–26 mg/100 for both men and women over the age span of 20 and 49. These are 95% limits determined by nonparametric statistical methods. Because urea freely diffuses through the erythrocyte membrane, investigators formerly believed the urea concentrations in erythrocyte water and plasma water were the same. The concentration in erythrocyte water is now known to be higher, the ratio being reported as 1.14:1 in the post absorptive state (183). As a result of the postprandial increase in urea level, this ratio may decrease to about 0.96:1 (183). The higher concentration in erythrocyte water is thought to result from hemoglobin binding of urea. On the other hand, because of the higher concentration of water in plasma than in erythrocytes, the net result is that the plasma urea N is only slightly higher than whole blood urea N (219).

Urine. In most cases urea N constitutes more than 80% of the total N in urine. When the protein intake is low, however, this value decreases. For adults, the normal range for 24 hour excretion of urea N is 6–17 g (114). Excretion is correlated with body surface area.

AUTOMATED DETERMINATION OF UREA NITROGEN

(method of Wybenga et al, Ref. 253, modified)

PRINCIPLE

Serum or urine reacts with acidic diacetyl monoxime during a short heating period in the presence of thiosemicarbazide, which intensifies the resulting color and decreases its photosensitivity.

REAGENTS

Urea Nitrogen Reagent. This reagent is the same as that used in the manual method except for the addition of 1.0 ml of "BRIJ-35" (Technicon Corporation) per liter.

Diacetyl Monoxime. This reagent is the same as that used in the manual method.

Urea Nitrogen Standards, 10, 40, and 80 mg/100 ml in 7% Protein Solution. Prepare a stock 100 mg urea N/100 ml standard by dissolving 214 mg urea in protein diluent and diluting to 100 ml. The diluent contains per 100 ml: 0.5 g phenol, 0.06 g sodium caprylate, and 7.0 g bovine albumin (Miles). Prepare the 10, 40, and 80 mg urea N/100 ml standards by diluting the 100 mg/100 ml standard 1:10, 4:10, and 8:10 with diluent. Store standards in amber bottles. These standards are stable for at least 6 months when refrigerated.

PROCEDURE

Set up manifold, other components, and operating conditions as shown in Figure 17-3. By following procedural guidelines as described in Chap. 10, steady state interaction curves as shown in Figure 17-4 should be obtained and should be checked periodically. Serum or urine previously diluted 1:50 with water is used for the determination. A blank is not necessary. Results are calculated from the calibration curve.

NOTES

1. Interaction. The carry over from one sample to another in this automated method is 2.0% at the rate of 60 determinations per hour. Moore (158) quoted a carry over of 0.7%, but his technic required dialysis.

2. Interferant. Bilirubin at a concentration of 10 mg/100 ml increases the urea N value by about 3 mg/100 ml of serum. This bilirubin effect is not observed for the manual procedure, and it may be due to the somewhat different heating and mixing conditions used in the two procedures.

3. Heating bath. Fluctuations may occur in the steady state which are caused by the on-off cycle of the heating bath. They can be smoothed by placing a variable transformer in the heating bath line and running it at about 90 V to even out bath temperature.

4. Flow diagram. This method is modified from that of reference 253 by controlling bubble timing to decrease noise.

ACCURACY AND PRECISION

Results obtained by this method are equivalent to those of the manual method (253).

Since dialysis is not performed, standards containing protein are required to obtain the needed precision. If aqueous urea nitrogen standards are used in the automated procedure, results will not agree with the manual method.

The precision (95% limits) is ±5.5% for urea N levels between 6 and 80 mg/100 ml of serum. This precision is comparable to that obtained with the manual method.

NORMAL VALUES

Normal values for serum and urine are identical to those for the manual method.

AMMONIA

There are basically two steps in the measurement of blood ammonia: alkaline conversion of NH_4^+ to NH_3 gas to effect separation and quantification of the ammonia.

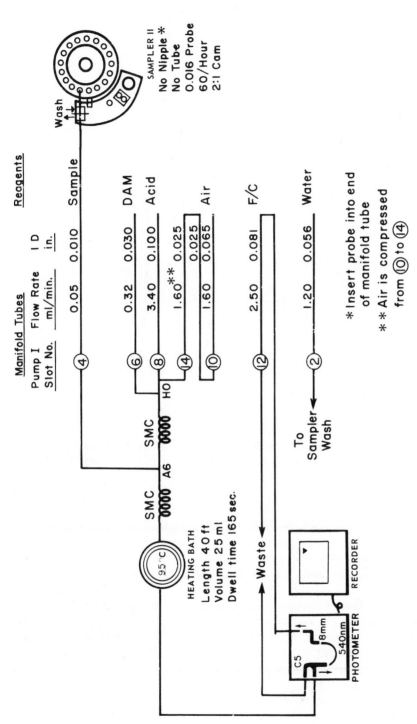

FIG. 17-3. Flow diagram for urea nitrogen in serum or urine. Calibration range: 0–80 mg/100 ml. Conventions and symbols used in this diagram are explained in Chap. 10.

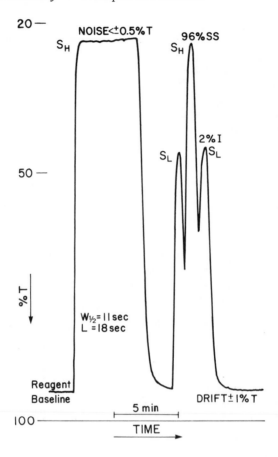

FIG. 17-4. Urea nitrogen strip chart record. S_H is a 80 mg/100 ml standard, S_L a 40 mg/100 ml standard. Peaks reach 96% of steady state; interaction between samples is 2%. $W_{1/2}$ and L are explained in Chap. 10.

Some methods (188, 206) do not require that proteins be separated prior to releasing NH_3 while others require a protein-free medium prepared variously as a trichloroacetic acid filtrate (91), by treatment with a mixture of mercuric chloride and lead acetate (88), or from a tungstic-sulfuric acid extract (147).

Blood or plasma ammonia can be measured using glutamic dehydrogenase in the presence of α-ketoglutaric acid (157, 171, 187). The decrease in absorbance at 340 nm, as NADH is oxidized to NAD, is proportional to the ammonia concentration. On the other hand, it is possible to measure the decrease in fluorescence of NADH when activated with 360 nm light and read at 460 nm (194).

Ammonia can be separated from plasma or serum by either a cation-exchange resin or isothermal diffusion. With a cation-exchange technic,

NH_4^+ is adsorbed to the ion-exchange resin, eluted, and then reacted with either Nessler's (60) or Berthelot's reagents (77, 78). Rather than separate plasma ammonia on a resin column, others (86, 153) have used a batch ion-exchange procedure followed by elution and quantification with the indophenol reaction.

Berthelot (17) discovered the reaction between ammonia and an alkaline solution of phenol-hypochlorite to give the blue chromogen, indophenol. The reaction proceeds in several steps (25): NH_3 and hypochlorite react to give chloramine, which reacts with phenol to form quinonechloramine. Quinonechloramine couples with another mole of phenol and forms the yellow associated indophenol, which then dissociates in alkali to give a blue color.

Lubochinsky and Zolta (138) introduced the catalyst sodium nitroprusside to increase the speed of the indophenol reaction. Horn and Squire (106) used an acetone-nitroprusside catalyst but reversed the usual order of reagents by adding the sodium hypochlorite before the sodium phenate. By adding reagents in this order, they claim an increased sensitivity and accuracy of the reaction. A year later they (107) published additional results that revealed their ability to further increase method sensitivity by omitting the acetone and substituting sodium ferrous nitritopentacyanide for sodium nitroprusside.

Probably the most popular method involves conversion of NH_4^+ in blood to NH_3 gas by alkalinization, isolation of the NH_3 by isothermal diffusion, and quantification by either of several methods: acidimetric titration (55, 72, 188), nesslerization (52, 206), photometric determination of the color produced with ninhydrin (164) or with the indophenol reaction using either nitroprusside or acetone and nitroprusside as a catalyst (30), or photometric determination of phenosafranin decolorization in the presence of hypobromite (216).

Once NH_3 has been separated from plasma or blood by isothermal diffusion, it can also be measured by coulometric titration with electrolytically produced hypobromite (45), reaction with acetylacetone to form a fluorescent compound (199), or formation of brilliant emerald green with alkaline salicylate-dichlorisocyanurate (185). The last method has been criticized because glycine, sarcosine, tryptamine, and lysine abolish the green color (205).

Leffler (132) prepared a protein-free solution with tungstic acid and quantified ammonia with the indophenol reaction after aeration into boric acid. Similarly, ammonia has been determined either titrimetrically or photometrically following aeration from a protein-free filtrate into a standard acid solution (88).

Methods for determination of ammonia in urine date back to 1850. Most technics which have been proposed begin with isolation of ammonia from the urine. Distillation with heat must be avoided due to the concomitant formation of ammonia from the urea present. More gentle approaches are required, and these have included (101): (a) distillation *in vacuo*, (b) aeration,

(c) adsorption onto Permutit followed by eltuion, and (d) isothermal diffusion. Direct nesslerization of ammonia in urine without preliminary isolation has been employed but is subject to positive error due to interference by creatinine and other substances unless they can be effectively removed or destroyed. Treatment with iodine reportedly eliminates such interference. For determination of urea in urine by direct nesslerization following urease treatment, correction for preformed ammonia should include the creatinine, etc. Hence, the value for preformed ammonia obtained by direct nesslerization is proper for this purpose. The indophenol reaction of Berthelot can also be used for determination of ammonia in urine.

A popular photometric approach is to nesslerize the ammonia isolated by the diffusion method of Seligson and Hirahara (190, 206). The author quite frankly does not know what method for blood ammonia is best and therefore, until order is made out of chaos, is presenting the method used at Bio-Science Laboratories.

DETERMINATION OF AMMONIA

(method of Seligson and Hirahara, Ref. 206, using the indophenol color reaction as modified by Chaney and Marbach, Ref. 42)

PRINCIPLE

A carbonate-bicarbonate buffer is added to the blood sample and the ammonia isothermally distilled into acid. The distillation is performed in vaccine bottles that are slowly and continuously rotated. The ammonia trapped by the acid is quantitated photometrically by the indophenol reaction as modified by Chaney and Marbach (42).

The ammonia content of urine is determined by applying the indophenol reaction directly to the sample.

SPECIAL APPARATUS

Distillation Bottles. Vaccine bottles, ~ 50–65 ml capacity, dimensions ~ 6.3 x 3.8 cm.

Rubber Stoppers. One set of solid stoppers to fit the vaccine bottles and a second set with one hole to fit the receiving rods.

Receiving Rods. 3 mm x 100 mm glass rod with 30 mm of one end ground so that when dipped in acid an acid film is retained over its surface. The rods are set in the one-hole rubber stoppers so that when placed in the bottle the ground end of the rod extends about halfway into the bottle.

Stainless Steel Rods. Approximately 7 mm in diameter and 40 mm long. Two are required per bottle (glass rods are not sufficiently heavy).

Rotator. A wheel rotating about 30 rpm with the axis horizontal. Spring clamps are attached to hold the flat bottom of the distillation bottles against the wheel and about 8 cm from the axis.

REAGENTS

Alkaline Buffer Mixture. Mix thoroughly—but do not powder—2 parts of $K_2CO_3 \cdot 1.5\ H_2O$ with 1 part anhydrous $KHCO_3$. Store this mixture in a closed container over aqueous saturated K_2CO_3.

H_2SO_4, *1 N.* Add 1 ml conc H_2SO_4 to 35 ml water.

Phenol Color Reagent. 50.0 g reagent grade phenol and 0.25 g reagent grade sodium nitroprusside per liter. Stable at least 2 months when kept cool and in an amber bottle.

Alkali-Hypochlorite Reagent. 25.0 g reagent grade NaOH and 2.1 g sodium hypochlorite per liter. If commercial bleach (Clorox), which contains 5.25 g NaOCl/100 ml, is to be used, then dissolve 25.0 g NaOH with 40 ml Clorox and dilute to 1 liter. Stable at least 2 months if kept cool and in an amber bottle.

Stock Ammonia Standard. 0.1 mg NH_3 N/ml. Dissolve 0.472 g $(NH_4)_2SO_4$ (reagent grade) in water in a 1 liter volumetric flask. Add conc. H_2SO_4 drop by drop to bring the solution to pH 2.0 (test with pH indicator paper), then dilute to 1 liter with water. Stable indefinitely.

Working Ammonia Standard. 2.5 μg NH_3 N/ml. Dilute 5.0 ml stock standard to 200 ml.

PROCEDURE FOR BLOOD

1. Add 3.0 g alkaline buffer mixture and 2 dry stainless steel rods to each distillation bottle required. Stopper and lay them on their sides until needed.
2. Set up bottles in duplicate in the following order:
   Blank. Add 2.0 ml water to bottle and immediately stopper.
   Standard. Add 2.0 ml working standard to bottle and immediately stopper.
   Unknown. Add 1 ml blood (collected with anticoagulant, see note 6) and 1 ml water to bottle and immediately stopper.
3. Mix contents of bottles on rotator for 1 min.
4. Working with one bottle at a time, wet the ground portion of a receiving rod in 1 N H_2SO_4, remove the solid rubber stopper from the bottle while it is lying on its side, and quickly insert the receiving rod with stopper, taking care not to touch the bottle neck as the rod is inserted.
5. Attach all bottles to the wheel and rotate at 30 rpm for 30 min.
6. Remove bottles from wheel and lay them gently on their sides. Remove receiving rods from the bottles one at a time, taking great care that the rods do not touch the necks of the bottles. Wash the acid off the receiving rod into a test tube with 1.0 ml phenol color reagent. Next, wash off the rod successively with 1.0 ml alkali-hypochlorite reagent and 1.0 ml water. Mix.
7. Cover tubes with Saran Wrap (Dow Chemical) and incubate in a water bath at 50°–60° for 3 min, 37° for 20 min, or 25° for 40 min.

8. Add equal volumes of water to all tubes to bring photometric readings into desirable absorbance range (preferably 0.2–0.8). For the Beckman DU Spectrophotometer with 1 cm curvets, add 7.0 ml water.
9. Read absorbances of blank (A_b), standard (A_s), and unknown (A_x) against water at 630 nm or use a filter with nominal wavelength in this region. If A_x is > 0.8, equally dilute the blank and unknown with water, read absorbances again, and make proper corrections in calculation.
Calculation:

$$\mu g \text{ ammonia N/100 ml blood} = \frac{A_x - A_b}{A_s - A_b} \times 5 \times \frac{100}{1}$$

$$= \frac{A_x - A_b}{A_s - A_b} \times 500$$

PROCEDURE FOR URINE

1. Measure total volume of the specimen (usually a 24 hour collection).
2. Set up the following in tubes:
   Blank. Add 0.2 ml water into 1.0 ml phenol reagent.
   Standard. Prepare 1:10 dilution of stock ammonia standard in water to yield 0.01 mg N/ml. Pipet 0.2 ml into 1.0 ml of phenol reagent.
   Unknown. Dilute urine 1:10 and 1:100 with water. For each dilution rinse 0.2 ml (TC pipet) into 1.0 ml of phenol reagent.
3. Add 1.0 ml alkali-hypochlorite reagent to each tube and mix.
4. Place in 37°C water bath for 20 min.
5. Cool, add 4.0 ml water. Mix.
6. Read at 630 nm vs water.
Calculation:

$$g \text{ NH}_3 - \text{N/24 hour} = \frac{A_x - A_b}{A_s - A_b} \times 0.002^* \times \frac{\text{TV (in liters)}}{0.02^{**}}$$

*0.002 for 0.01 mg/ml standard
**0.02 for 1:10 dilution of urine; use 0.002 for 1:100 dilution of the urine.

NOTES

1. *Beer's law.* Absorbance is linear with concentration of NH_4^+N on a Beckman DU to greater than 1.0. On a Klett photometer, Beer's law is obeyed with a Klett 54 filter only up to about 250 Klett units but not at all with a Klett 62 filter. The linearity of the method must be checked

when using filter photometers. If one chooses a nonlinear method, results must be read from a calibration curve.

2. *Color stability*. The color is stable for at least 24 hours.

3. *Stability*. There is contradictory evidence regarding the rate of ammonia formation from blood *in vitro*. Seligson and Hirahara (206) reported an increase of about 0.003 μg/ml blood/min. The rate in four normal samples studied in our laboratory was about 0.017 μg/ml blood/min at 25°. We have confirmed reports that freezing of blood at $-20°C$ prevents changes in blood ammonia for 24 hours (91, 137, 151, 172). If there is to be more than a few minutes delay in setting up the test, quick freeze a tube of the sample by immersing it in a mixture of dry ice and acetone at $-80°C$. Blood ammonia is stable for at least 7 days on dry ice. In a few experiments in our laboratory it was found that fluoride and/or mercuric chloride slowed down but did not stop the formation of ammonia. In contrast, Dienst (60) claims ammonia is stable in plasma for 1 hour at room temperature, and Gips and Wibbens-Alberts (91) stored heparinized blood collected anaerobically at room temperature for at least 2 hours without appreciable rise of ammonia content.

4. *Recovery*. Check the recovery of ammonia in the isothermal diffusion technic by applying the indophenol reaction directly to 2.0 ml aliquots of water (reagent blank) and working standard. Add 1.0 ml phenol color reagent, followed by 1.0 ml alkali-hypochlorite reagent and mix. Proceed with steps 7–9 but in step 8 dilute with 1.0 ml less water than used for standard and blank obtained by diffusion so that final volumes are the same.

$$\% \text{ recovery} = \frac{(A_s - A_b) \text{ by diffusion}}{(A_s - A_b) \text{ direct}} \times 100$$

The recovery should be 95%–100%.

5. *Absorbance of reagent blank*. A_b should not exceed 0.08 (40 Klett units with Klett 54 filter). If it does, the reagents either have become contaminated or are too old.

6. *Anticoagulant*. Blood may be collected with potassium oxalate, EDTA, or heparin. Different commercial brands of heparin, however, liberate varying amount of ammonia during the analysis thereby leading to spurious results (51). This extractable ammonia probably arises from contamination with ammonium compounds introduced during manufacture of the heparin. If contamination is not excessive a correction can be made by adding an equivalent amount of heparin to the blank and standard.

7. *Contamination by ammonia*. Because of the method sensitivity for ammonia, the method cannot be performed in a laboratory area where contamination is likely to occur, e.g., where a bottle of con NH_4OH will be opened to the air.

ACCURACY AND PRECISION

It is best to regard the results obtained with any method for blood ammonia as of questionable accuracy until several areas of confusion and contradiction already discussed are clarified.

The precision (95% limits) of the technic is about ±10%. Admittedly, however, it is difficult to maintain this level of precision.

NORMAL VALUES

Seligson and Hirahara (206) studied 29 normal adults and found the normal range for venous whole blood to be 75–196 μg ammonia N/100 ml, the normal range for plasma to be 56–122 μg/100 ml, and the normal range for erthrocytes to be 96–331 μg/100 ml. We analyzed seven normal whole bloods in our laboratory and the values all fell in the above range. Peripheral blood of normal newborn infants has a higher content of ammonia than that of adults (149).

The normal 24 hour urinary output of ammonia N is reported to be 0.14–1.5 g for adults and 0.56–2.9 g for infants (101).

Patients, according to Clarke *et al* (49), should fast at least 6 hours prior to the test to obtain basal blood levels.

URIC ACID

Research in uric acid methodology has largely been directed toward improving specificity. Three approaches have been used for preliminary separation of the uric acid from interfering substances present in serum: (a) isolation as the silver, magnesium, ammonium, cuprous or cupric salt, (b) isolation by an ion-exchange technic, and (c) precipitation of protein and analysis of the filtrate for its uric acid concentration.

An important early step in uric acid methodology was the preparation of a protein-free filtrate of blood and subsequent precipitation of uric acid from the filtrate as the silver salt (83). However, as a result of inadequate specificity in the salt precipitation step and of the appreciable solubility of the urate salt, significant negative errors were introduced (101). In a more sophisticated approach, an attempt has been made by Sambhi and Grollman (198) to increase specificity by preliminary isolation of uric acid by adsorption onto an ion-exchange resin followed by the arsenophosphotungstate color reaction (198) on the eluate. Their results were lower by 0.39 mg/100 ml than those obtained by the phosphotungstate-carbonate method of Henry (103) for the following reasons: (a) ascorbic acid was inactivated during its passage through the resin; and (b) sulfhydryl-containing compounds do not interfere because N-ethylmaleimide was added to serum prior to its application to the column. Shapiro *et al* (208) also used an ion-exchange technic but quantified the eluted uric acid by an ultraviolet method. These technics are time consuming and are not routinely used.

More convenient methods require only removal of proteins prior to the color development step. The most practical approach is to precipitate proteins using either tungstic acid (14, 80), trichloroacetic acid (21), or phosphotungstic acid (7, 38, 115, 131, 179). Heat coagulation of serum proteins with subsequent water extraction of the uric acid from the protein coagulum (144) or separation of uric acid from serum proteins by membrane ultrafiltration (71) has also been used. Once proteins are removed, the filtrate is analyzed for uric acid by one of the phosphotungstate methods. Most photometric methods are based on the reduction of phosphotungstate by uric acid in an alkaline solution to give a blue color (169). Folin and Denis (83) used this color reaction to quantify the uric acid previously separated as the silver salt from a protein-free filtrate of blood. The carbonate reagent used by Folin and Denis was later replaced with cyanide in order to increase method sensitivity, to prevent color fading, and to dissolve the silver urate (16). Folin subsequently published a series of papers (80, 81, 82) reporting the use of urea-cyanide as the alkaline reagent and directions for reagent preparation. The alkaline oxidation step can be carried out directly in a protein-free filtrate of serum (27) with phosphotungstic acid as the oxidizing agent in the presence of cyanide buffered with urea (162).

Other alkalis can be used to enhance the color: glycerin-silicate (119, 122); EDTA-sodium tungstate (176); sodium glycinate buffer (245); and carbonate-urea-triethanolamine (115). Henry *et al* (103) concluded that either sodium carbonate, sodium metasilicate, or sodium hydroxide may be used as the source of alkali in the development of color. Henry (101) has also discussed the reasons for turbidity and its prevention in the final colored solutions when determining uric acid by tungstic acid methods.

The analyst is not limited to using phosphotungstic acid as the oxidant. Others have used arsenotungstic acid (26, 165), arsenophosphotungstic acid (13), reduction of alkaline ferricyanide by uric acid before and after its destruction by uricase (32), or reduction of the alkaline cupric complex of 2, 9-dimethyl-1, 10-phenanthroline (21).

Uricase has long been used as a reagent to increase specificity by oxidizing uric acid to allantoin, which, unlike uric acid, does not have an absorption peak in the 290–293 nm region. Kalckar (117) and Praetorius and Poulsen (182) have determined uric acid concentration in plasma by measuring the absorbance change at 290 nm (differential ultraviolet spectrophotometry). Feichtmeir and Wrenn (76) measured absorbance differences at 293 nm for trichloroacetic acid filtrates of serum prepared before and after incubation of the sample with uricase. The carbonate-phosphotungstate method usually gives results that are about 11% higher than the Feichtmeir-Wrenn ultraviolet method. Troy and Purdy (232) published a coulometric method employing a uricase differential technic. The type of buffer used in these enzyme methods influences the uric acid oxidation products formed (33, 108, 133).

Alternatively, several methods are available wherein photometric endpoints in the visible region are used for quantification, e.g., oxidized *o*-dianisidine (61, 143) or an indamine dye (93) with a uricase-peroxidase system. Using the same enzyme system but with no deproteinization step, the Hantzch reaction has been applied to form the chromophore, 3,5-

diacetyl-1, 4-dihydrolutidine (116). The reduction of alkaline ferricyanide by protein-free filtrates of plasma before and after the destruction of uric acid by uricase is the basis for another analytic approach (32). Caraway and Marable (37) have proposed a uricase-carbonate-phosphotungstate method to determine the "true" uric acid concentration in serum.

Other approaches have been used to improve specificity without using uricase. Uric acid has been extracted from heat coagulated serum and then measured directly by its ultraviolet absorption (144). Alternatively, interfering reducing substances may be eliminated by: (a) alkaline incubation of serum with carbonate (7, 29) or trisodium phosphate (115); (b) alkaline incubation of protein-free solutions (21, 34, 67, 83); (c) storage of untreated serum in the refrigerator for three days (31, 65); or (d) incubation of untreated serum at elevated temperatures (65, 66, 110).

Three methods to determine serum uric acid have been included in this chapter: (a) a manual one-tube carbonate-phosphotungstate method using phosphotungstic acid reagent as both protein precipitant and oxidant with rapid color development at 100°C (179), (b) an automated method using cyanide-phosphotungstate reagent (28), and (c) an ultraviolet differential-spectrophotometric method employing uricase (103).

The manual carbonate-phosphotungstate method without cyanide has adequate sensitivity and also avoids the several disadvantages of cyanide when used with manual methods. These disadvantages include (a) appreciable blank absorbance which varies with age of the reagent and with the brand of cyanide used, (b) variable standard curves, and (c) its highly poisonous character. The automated method presented, however, needs the increased sensitivity provided by inclusion of cyanide, and the disadvantages listed for cyanide are less important with automated procedures. The third method, the uricase method, is recommended for research purposes and requires the use of an ultraviolet spectrophotometer.

DETERMINATION OF URIC ACID BY REACTION WITH ALKALINE PHOSPHOTUNGSTATE

(method of Pileggi et al, Ref. 179)

PRINCIPLE

Uric acid reduces phosphotungstate to tungsten blue, which is measured photometrically at 700 nm (Fig. 17-5). The absorbance of a reagent blank is less than 0.01 over the entire visible spectrum. The oxidation of uric acid in an alkaline medium proceeds as follows:

URIC ACID   ALLANTOIN

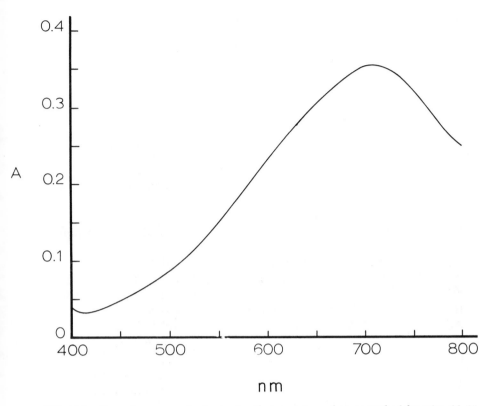

FIG. 17-5. Absorption curve of color in phosphotungstate–carbonate method for uric acid. 25 μg uric acid vs water. Beckman Model DU Spectrophotometer, 2 nm band width at peak.

REAGENTS

Phosphotungstic Acid Reagent. Dissolve 14 g molybdenum-free sodium tungstate dihydrate in about 50 ml of water in a 1 liter flask that can serve as a reflux reactor (see note 4). Add 11 ml 85% orthophosphoric acid and mix. Add several boiling chips, connect a reflux condenser, and reflux gently for 2 hours. Cool to room temperature, add about 50 ml water and 11 g $Li_2SO_4 \cdot H_2O$. Mix until dissolved. Dilute to 11 liter with water. The reagent is stable for at least 2 years when stored in the refrigerator.

10% Sodium Carbonate. Dissolve 10 g anhydrous Na_2CO_3 in water and dilute to 100 ml. Stable for at least 2 years when stored in a polyethylene bottle.

Uric Acid Standard, 62.5 mg/100 ml. This standard is equivalent to 5.0 mg uric acid/100 ml serum when used in the procedure as described. Dissolve 60 mg Li_2CO_3 in about 50 ml water in a 100 ml volumetric flask using gentle heat (do not allow temperature to exceed 50°C) and mixing. Add exactly 62.5 mg uric acid and continue to stir until completely dissolved. Add 2 ml formaldehyde solution and 0.15 ml glacial acetic acid. Mix and dilute to 100 ml with water. Stable for 2 years when stored in the refrigerator.

PROCEDURE FOR SERUM

N.B.: This method cannot be applied to plasma or whole blood containing potassium oxalate as the anticoagulant since insoluble potassium phosphotungstates are formed.

1. To each of three test tubes or cuvets labeled "blank," "standard," and "unknown" add 3.0 ml of phosphotungstic acid reagent.
2. Add 0.2 ml of serum or plasma to the tube marked "unknown." Cover or use stopper and mix contents of the tube by gently inverting 2 times. *Do not shake.*
3. Centrifuge the "unknown" tube for 5 min; do not decant the supernatant fluid.
4. Add 1.0 ml water and 20 μl uric acid standard (TC pipet) to the tube marked "standard" and 1.0 ml water to the "blank." (see note 5). Mix the contents of these tubes thoroughly for at least 5 sec. Do not proceed to the next step until step 3 is completed for all unknowns.
5. Add 1.0 ml 10% Na_2CO_3 to the "standard" and "blank" tubes and mix.
6. Add 1.0 ml 10% Na_2CO_3 to the "unknown" tube. Cover or use stopper and mix by inverting carefully one time in order not to break up the protein precipitate. If more than one unknown is analyzed, the 10% Na_2CO_3 should not be added to the subsequent tube until the previous tube has been inverted to evenly distribute the carbonate.
7. Insert all tubes into an appropriate heating block at 100°C for exactly 1 min.
8. Remove all tubes and immerse in cool tap water (must be 20°−25°C) for 3 min. Mix the contents of the blank and standard vials but do not mix the unknown vial. Centrifuge the unknown vial for 2 min to repack the protein precipitate.
9. Read absorbances of the standard (A_s) and unknown (A_x) against the blank at 700 nm or with a filter having a nominal wavelength between 640 and 720 nm. Readings should be made within 30 min (see note 9).
Calculation:

$$\text{mg uric acid/100 ml} = \frac{A_x}{A_s} \times 0.625 \times 0.020 \times \frac{4.0}{5.02} \times \frac{100}{0.20}$$

$$= \frac{A_x}{A_s} \times 5 \text{ (see note 6)}$$

PROCEDURE FOR URINE

1. If urine is cloudy, warm an aliquot to 60°C for 10 min to be certain that precipitated urates and uric acid are dissolved (see note 7). Centrifuge.

2. Make a 1:10 dilution of the urine with water. Begin at step 2 for serum, using 0.2 ml of the diluted urine. Since there is no denatured protein to remove, omit the centrifugation steps.

Calculation:

$$\text{mg uric acid/100 ml urine} = \frac{A_x}{A_s} \times 0.0625 \times 0.002 \times \frac{4.2}{5.02} \times \frac{100}{0.02}$$

$$= \frac{A_x}{A_s} \times 52$$

If the absorbance reading obtained with the 1:10 dilution is less than 0.2 or more than 0.8, repeat with an appropriate dilution of the urine and alter calculations accordingly.

NOTES

1. Beer's law. Beer's law usually is followed up to 20 mg uric acid/100 ml of serum using a glass filter with nominal wavelength of 660 nm. When a filter photometer is to be used, however, this linearity should be confirmed for the particular filter employed. If significant deviation is observed, a calibration curve must be employed for calculation of results.

2. Stability of samples. Dubbs *et al* (62) reported uric acid to be stable in serum at refrigerator temperatures for 3–5 days and in the frozen state for 6 months (240). Studies in our laboratory indicate that serum uric acid is stable for about 3 days at room temperature before its level begins to drop significantly. Fluoride or fluoride plus thymol preserves serum at room temperature for longer periods of time (101). Uricolysis can occur, however, when plasma is incubated with erythrocytes (19). Uric acid is usually stable in urine for about 3 days at room temperature although it may be rapidly destroyed by bacterial growth (167).

3. Effect of plasma pH. Using a phosphotungstic acid method, Talbot and Sherman (221) reported that the distribution of chromogen between erythrocytes and plasma is dependent on pH. A fall in pH is accompanied by a shift of plasma chromogen into the cells. The average decrease in plasma uric acid is 2.3 mg/100 ml for a pH drop of 1 unit (221). The presence of uric acid in erythrocytes has been confirmed by the differential spectrophotometric uricase method. The average cell/plasma ratio of uric acid is 0.55–0.56 (221). As might be expected, there is no significant change in uric acid concentration while serum is in contact with the clot (112) because of limited surface contact between serum and red cells.

4. Phosphotungstic acid reagent. Molybdenum is always present in tungsten ores from which sodium tungstate is derived, and unfortunately, molybdates are more sensitive than tungstates to reducing substances such as phenols, forming a molybdenum blue (84). Folin published a series of papers that culminated in a technic by which tungstate could be manufactured essentially free of this metal (82). Such a product is now available commercially ("sodium tungstate, according to Folin"). If there is doubt about a particular tungstate, apply the potassium xanthate test for molybdenum (84).

There are two forms of phospho-18-tungstic acid, and each gives a different color intensity on reaction with uric acid (101). In preparing uric acid reagent, it is necessary to minimize formation of phosphotungstic acid "A" because it is hypersensitive and behaves like a phenol reagent; also, it gives a high reagent blank. Lower blanks are obtained if most of the phosphotungstic acid is in the "B" form (82).

5. *Standard.* Since an aqueous standard is used, the phosphotungstic acid reagent must be diluted with 1.0 ml distilled water to simulate the phosphotungstic acid concentration in the "unknown" vial following protein precipitation.

6. *Volume displacement error.* The volume in the"unknown" tube is 4.0 ml and not 4.2 ml as predicted by summing the volumes of reagents and sample added to this tube. The 4.0 ml is the average measured volume of recovered centrifugate from 30 different sera. This difference between the recovered and theoretical volumes is due to the volume occupied by the protein precipitate. Failure to take this into account would have resulted in what has been referred to as the *volume displacement error* (101).

7. *Urine collection.* If a 24 hour urine is to be collected, add 10 ml 5% NaOH to the collection bottle to prevent urate precipitation (101).

8. *Interferences.* The effects of elevated serum bilirubin and hemoglobin were tested by adding solutions of purified bilirubin or hemoglobin to normal sera. Bilirubin did not interfere at concentrations up to 6.6 mg/100 ml and resulted in a positive interference of only 6% at a concentration of 8.5 mg/100 ml. Hemoglobin at levels up to about 200 mg/100 ml has a negligible positive interference of about 2–3%. The effect becomes progressively more serious, however, resulting in a +10.6% error at 424 mg/100 ml. Thus mildly hemolyzed sera can be used, but the analysis should not be carried out on moderately or severely hemolyzed specimens. It is possible, however, that the observed effect was not due to hemoglobin but to reducing substances present in the hemolysate used for the hemoglobin additions.

9. *Absorbance readings.* The absorbance of the unknown may be read in an optically acceptable tube that contains the protein precipitate provided that the precipitate is below the level of the light path for the instrument used. Instruments found satisfactory for making absorbance measurements over the precipitate include the Dow Photoelectric Colorimeter (660 nm filter), Klett Photoelectric Colorimeter, Leitz Model M Photrometer, Coleman Jr. Spectrophotometer, and Bausch and Lomb Spectronic 20.

ACCURACY AND PRECISION

Substances, other than uric acid, which reduce phosphotungstic acid include ergothioneine, glutathione, phenols, ascorbic acid, glucose, tyrosine, tryptophan, cystine, and cysteine (22). Since most of these nonspecific reducing substances are found inside erythrocytes, specificity is improved by analyzing serum or plasma rather than whole blood (35, 101). Substances that inhibit color formation have also been detected in protein-free filtrates (27, 32).

Substituted uric acids, i.e., the methyluric acids excreted in the urine following ingestion of caffeine or theophylline, are also determined by phosphotungstate methods. Salicylates taken by mouth cause spuriously high results in urine (255). Gentisic acid, a metabolic excretion product of salicylates, has been identified as the reactant. Salicylic acid and salicyluric acid are not chromogenic (255). The gentisic acid concentration of serum is insignificant (255), and our laboratory has confirmed that neither sodium salicylate nor acetylsalicylic acid added to serum in a concentration up to 25 mg/100 ml gives any interference.

The average of nonurate chromogens in serum after incubation of serum with uricase to destroy uric acid, is estimated at approximately 2% of the true uric acid value (37).

Feichtmeir and Wrenn (76) claimed very poor correlation between the results obtained by the alkaline-phosphotungstate method and those obtained by the differential spectrophotometric method with uricase. Henry *et al* (103) compared the carbonate-phosphotungstate method, a cyanide-phosphotungstate method, and the differential spectrophotometric method with uricase on a series of normal and abnormal sera and urines. The conclusions were as follows: (a) The two phosphotungstate methods yielded essentially identical values, and (b) there was a high degree of correlation between the phosphotungstate and uricase methods, but the methods are not equivalent.

A problem common to all uric acid methods employing protein-free filtrates of serum or plasma is the incomplete recovery of uric acid. Folin noted this in 1922 when he added the tungstate before the acid. When the order of adding reagents was reversed, there was less uric acid loss (32, 122); premixing the acid and tungstate before adding to serum was claimed to result in no uric acid loss (162).

These observations indicate that uric acid coprecipitates with proteins denatured by tungstic acid. In fact, Henry *et al* (103) have shown that recovery decreases significantly in filtrates with pH's below 3. With "half-quantities" of precipitating reagents, the pH of filtrates ranged from 3.0 to 4.3 and the recoveries were from 93% to 103% ($\bar{x}$ 98%). With "full quantities" of precipitating reagents, the filtrate pH's ranged from 2.4 to 2.7 and the recoveries were from 74% to 97% ($\bar{x}$ 87%). The results obtained with trichloroacetic acid precipitation are contradictory (76). Zinc hydroxide (Somogyi's reagent) cannot be used because it forms insoluble zinc urate.

Two types of recovery experiments have confirmed the validity of using a single phosphotungstic acid reagent as both precipitant and oxidant (179). In one series of experiments, serum was spiked with 2-^{14}C-uric acid. These sera were carried through the entire analysis and recovery determined by measuring the radioactivity present in the original serum and in the final solution following color development. Recoveries ranged from 92% to 94% with a mean of 93%. In the second experiment, the recovery of added nonradioactive uric acid at three different levels was also studied. The results were in excellent agreement; the mean recovery was 94%.

The precision (95% limits) of this method is ±5% for serum values ranging from 2.9 to 8.4 mg/100 ml.

NORMAL VALUES

Serum. Various investigators have reported different normal ranges of uric acid in serum for the phosphotungstic acid method. Whether this is a result of technical differences or the sampled population studied is difficult to say.

We have found in our laboratory that the range for men is higher than that for women. Sera from 1419 clinically normal adults were obtained over a period of 14 months from eight geographic areas of the United States. Normal range estimates (95% limits) by nonparametric statistical methods for uric acid were 4.0–8.5 and 2.8–7.5 mg/100 ml for men and women between the ages of 20 and 49, respectively (186). Values are somewhat higher for the newborn but by the age of one year they decrease to the adult range (47).

Urine. The normal 24 hour excretion of uric acid is about 250–750 mg (62, 114).

Effects of Variables. Additional data from our laboratory suggests a trend towards higher uric acid levels with aging for both sexes, but only the increase for females is statistically significant (186). Violent exercise may increase the serum uric acid level as much as 2.5 mg/100 ml and result in increased excretion for as long as 2 days following exercise. By contrast, 30 min of light muscular exercise results in an average increase of 0.16 mg/100 ml of serum. As a result of diurnal variation, uric acid levels are lower at night. There is no change following meals, but diets do affect uric acid levels, e.g., a purine-free diet causes a decrease in serum of up to 0.8 mg/100 ml, and a purine-rich diet increases the urinary excretion by as much as 1200 mg/day. Bishop and Talbot (20) also have reviewed thoroughly the effects of diet and other variables on uric acid levels.

AUTOMATED DETERMINATION OF SERUM AND URINE URIC ACID

(method of Brown, Ref. 27, adapted to AutoAnalyzer, Ref. 28)

PRINCIPLE

Serum or urine uric acid reduces alkaline phosphotungstate to tungsten blue, which is read at 660 nm.

REAGENTS

NaCl, 0.9%, for sample diluent. Add 9.0 g NaCl to about 500 ml water contained in a 1 liter volumetric flask. Add 0.5 ml Aerosol 22 (American

Cyanamid), then dilute to 1 liter with water. Mix, filter, and store in a polyethylene bottle.

NaCl, 0.9%, for sampler wash. Prepare as for sample diluent but omit the Aerosol 22.

Phosphotungstic Acid. Dissolve 40 g molybdenum-free sodium tungstate in about 300 ml water contained in a 1 liter reflux reactor. Add 32 ml 85% orthophosphoric acid and mix. After adding several boiling chips, connect the condenser and reflux gently for 2 hours. Cool to room temperature, add 32 g $Li_2SO_4 \cdot H_2O$ and mix until dissolved. Transfer solution to a 1 liter volumetric flask, rinse reflux reactor and transfer rinsings to the volumetric flask. Dilute to 1 liter with water and mix. Stored in an amber bottle in the refrigerator, the reagent is stable for at least 2 years.

Stock Sodium Cyanide, 10%. POISON—DO NOT PIPET BY MOUTH. Dissolve 100 g NaCN in about 950 ml water contained in a 1 liter volumetric flask. Add 2 ml conc NH_4OH then dilute to 1 liter. Mix, filter, and store in a polyethylene bottle.

Stock Urea Solution 20%. Dissolve 200 g urea in sufficient water and dilute to 1 liter. Mix, filter, and store in a polyethylene bottle.

Working Cyanide-Urea Solution. POISON—DO NOT PIPET BY MOUTH. When needed, mix one part of stock sodium cyanide solution with one part of stock urea solution.

Stock Uric Acid Standard, 0.1%. Dissolve 0.6 g Li_2CO_3 in about 150 ml water by shaking. Filter, heat the solution to 60°C, and transfer the warm solution to a 1 liter volumetric flask containing 1.0 g uric acid. Dissolve the uric acid within 5 min and then cool the flask contents under cold running water without delay. Add 20 ml 37% formalin, mix, and then add 400 ml water. Cautiously add, while shaking the flask, 25 ml 1 N H_2SO_4. Dilute to 1 liter with water, mix, and store in an amber bottle.

Working Uric Acid Standards. Dilute the following volumes of stock uric acid standard to 100 ml with water. These standards are stable for 2 years in the refrigerator.

Uric Acid (mg/100 ml)	Stock (ml)
2	2.0
4	4.0
6	6.0
8	8.0
10	10.0
12	12.0

PROCEDURE

Set up manifold, other components, and operating conditions as shown in Figure 17-6. By following procedural guidelines as described in Chap. 10, steady state interaction curves as shown in Figure 17-7 should be obtained and should be checked periodically. Clear nonhemolyzed serum, or urine previously diluted at least 1:10 with water, is used. A blank is not necessary. Results are calculated from the calibration curve in the standard manner for

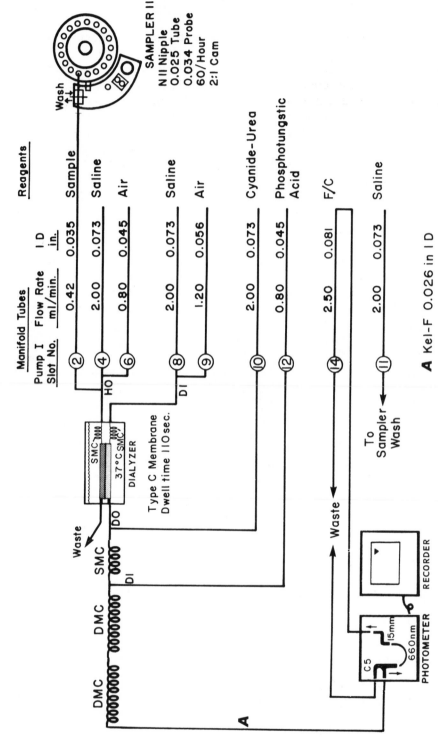

FIG. 17-6. Flow diagram for uric acid in serum and urine. Calibration range: 0–12 mg/100 ml. Conventions and symbols are used in this diagram are explained in Chap. 10.

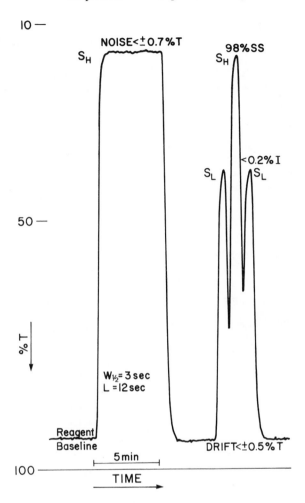

FIG. 17-7. Uric acid strip chart record. S_H is a 12 mg/100 ml standard, S_L a 6 mg/100 ml standard. Peaks reach 98% of steady state; interaction between samples is less than 0.2% T. $W_{1/2}$ and L are explained in Chap. 10. Computer tolerance for noise is set at ±0.7% T, and for standard drift at ±1% T.

methods obeying Beer's law. Specimens with uric acid values greater than the highest standard should be diluted with 0.9% NaCl and reanalyzed.

CAUTION: Flush the waste solution down the drain with cold water. If the solutions are collected, there should be a neutralizing powder such as trisodium phosphate in the container.

NOTES

1. Bubble pattern. In the event of excessive noise, check the bubble pattern through the manifold and out of the dialyzer. Aerosol 22 is purposely added to the saline to facilitate an optimal bubble pattern and low recorder-tracing noise.

2. *Color development.* Cyanide-urea solution prevents turbidity formation and enhances the color. If turbidity appears during an analysis, flush the manifold with water, then repeat with freshly prepared reagent.

3. *Method sensitivity.* A gradual decrease in method sensitivity may occur as the phosphotungstic or cyanide reagents age. Fresh reagent will restore sensitivity.

PRECISION

The precision (95% limits) of this method is ±5% at a concentration of 9.2 mg/100 ml serum.

NORMAL VALUES

Same as for manual method.

DETERMINATION OF SERUM AND URINE URIC ACID BY URICASE

(method of Henry et al, Ref. 103, modified)

PRINCIPLE

The absorption curve for uric acid has a peak between 290 and 293 nm (Fig. 17-8) whereas the degradation products after uricase action do not absorb light in this region. The decrease in absorbance is measured and is proportional to the uric acid concentration originally present.

REAGENTS

Acetic Acid, 0.05 N. Add 3 ml glacial acetic acid to a 1 liter volumetric flask and dilute to volume.

0.5 M Glycine Buffer, pH 9. Dissolve 37.5 g glycine in about 800 ml water and add 12.5 ml 10 N NaOH. Check pH with pH meter and if not within pH 8.9–9.1, adjust. Dilute to 1 liter with water and store in a Pyrex bottle in the refrigerator. Filter if necessary.

Uricase. Dissolve sufficient uricase in 0.5 M glycine buffer so that the reaction with a 20 μg uric acid standard goes to completion within about 15–30 min. Determine the required enzyme concentration by trial and error. Worthington's Uricase Powder (Worthington Biochemical Corporation) is satisfactory in a concentration of 1 mg/ml buffer. The solution retains adequate activity for 2 or 3 days in the refrigerator.

0.1% KCN. Use reagent grade salt. Stable for several days in refrigerator. The brand of reagent grade KCN used should be checked the first time the test is run for absorption at 290 nm. Add 0.2 ml of the solution to 3.8 ml

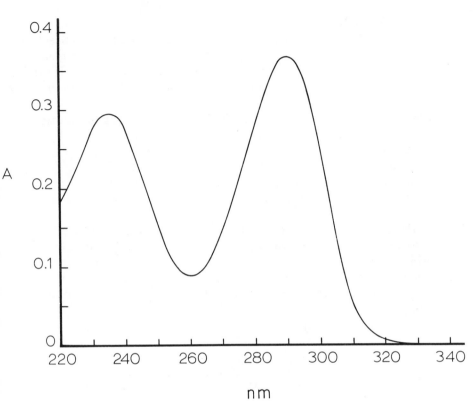

FIG. 17-8. Absorption curve of uric acid. 5 μg/ml water. Beckman Model DU Spectrophotometer, 1 nm band width at 290 nm peak.

water and read A_{290} against water. The absorbance must be less than 0.005. The 0.1% KCN solution may develop increased absorbance at 290 nm upon storage.

Uric Acid Stock Standard, 100 mg/100 ml. Place 100 mg uric acid and 60 mg Li_2CO_3 in about 50 ml water in a 100 ml volumetric flask. Warm to 60°C to dissolve, cool to room temperature, then dilute with water to 100 ml.

Uric Acid Working Standard, 0.10 mg/ml. Dilute the stock standard 1 to 10 with water.

PROCEDURE

1. Mix 2.0 ml serum with 2.0 ml 0.05 *N* acetic acid in a test tube. Stopper tube and place in boiling water for 4 min. Cool to room temperature and centrifuge. If any particulate matter persists in supernate, filter through a pledget of glass wool.

2. Set up blank and unknown:

Blank. Mix 1.4 ml buffer, 0.2 ml 0.1% KCN, 2.0 ml enzyme solution and 0.4 ml protein-free filtrate from serum (from step 1) or 0.4 ml of a 1:10 dilution of urine (do not acidify). If original urine is cloudy, warm to 60°C for 10 min, mix, then centrifuge. Stopper tubes.

Unknown. Mix 1.4 ml buffer, 0.2 ml water, 2.0 ml enzyme solution and 0.4 ml sample (as for blank). Stopper tubes.

3. Incubate at room temperature for 30 min and transfer to quartz cuvets (light path 1 cm). Zero the instrument on the blank at 290 nm (see note 1) with the absorbance set at 0.600 (with the Beckman DU, use the 0.1 scale unless a photomultiplier attachment is employed; band width of about 1 nm is desired).

4. Read the A of the unknown. Take a second reading about 5 or 10 min later to be sure that the uricase reaction has gone to completion. If the absorbance of the unknown reaches 0, the blank can be zeroed at an absorbance higher than 0.600, e.g., 0.8 or 0.9. If this is not sufficient and the A of the unknown is still less than 0, rerun the test with a dilution of the sample.

Calculation: (See note 1)

$$\text{mg uric acid}/100 \text{ ml} = \frac{A_b - A_x}{0.368} \times 0.02 \times \frac{100}{\text{ml sample used*}}$$

$$= \frac{A_b - A_x}{\text{ml sample used*}} \times 5.4$$

* ml sample used = 0.4 × dilution factor (ordinarily 1/10 for urine and 1/2 for serum).

NOTES

1. *Standardization.* Two things should be verified when the test is first set up: (a) The wavelength of maximum absorption by uric acid for the spectrophotometer to be employed (should be between 288 nm and 293 nm). This wavelength will be used for determinations. Mix 0.2 ml uric acid standard (0.1 mg/ml) with 3.8 ml water and read against a water blank. Absorbance at the peak should be 0.368. (b) A test should be carried out with the standard and uricase. Add 0.2 ml standard to 1.6 ml buffer. $A_b - A_s$ should be about 0.368.

2. *Sensitivity of uricase to cyanide.* We have noted as much as a 100-fold difference in sensitivity to cyanide for various lots of uricase. Incomplete inhibition of uricase would result in a low recovery for the standards (note 1). When using a high concentration of KCN (e.g., 10%) with a very sensitive enzyme preparation, sufficient cyanide can distil from the blank to the unknown in the cuvet compartment of a Model DU Spectro-

photometer within 10 min to stop the enzyme action. This distillation is prevented by covering each cuvet individually.

3. *Stability of serum samples.* See note 2 under "Determination of Uric Acid by Reaction with Alkaline Phosphotungstate."

ACCURACY AND PRECISION

As far as is known, the method is absolutely specific; uricase oxidizes only uric acid and no other purine compound.

In their original procedure, Henry *et al* (103) used serum rather than a protein-free filtrate. This technic worked well for a number of years until erratic results suddenly were encountered that were obviated by the removal of serum proteins. Recoveries by the technic presented have been 95% or better.

The precision of the method in the normal range is about ±8%, but it is better at elevated levels of serum uric acid.

NORMAL VALUES

The normal range of serum uric acid for the uricase method should be about 5%–15% lower than that for phosphotungstate methods (3, 32, 62, 76). If one deducts 10% from the normal ranges for the phosphotungstate method, the values for the uricase method are 3.6–7.7 and 2.5–6.8 mg/100 ml for men and women, respectively. The within-day and day-to-day variations in any one normal individual are reported to be up to an order of magnitude of about 1 mg/100 ml (118).

Urine. The normal 24 hour excretion of uric acid is about 250–750 mg (62).

CREATININE AND CREATINE

Most methods for determining creatinine or creatine are based on the Jaffe reaction described in 1886 (113). Creatinine, in this reaction, reacts with alkaline picrate to form an amber-yellow solution that is measured photometrically. Unfortunately, this simple method is not specific for creatinine since many pseudocreatinine substances in blood also react with alkaline picrate.

Folin and Wu (85) applied Jaffe's observation to the quantification of creatinine and creatine in a protein-free filtrate of blood. The nature of the substance formed was thought to be a salt of creatinine, picric acid, and sodium hydroxide (96). Improved specificity of the Jaffe reaction was claimed by measuring the difference in colors following addition of acid because the color from true creatinine is less resistant to acidification than the color from pseudocreatinine substances (94, 213). If the Jaffe reaction time for color development exceeds 15 min or if the temperature is elevated above 30°C, methylguanidine and presumably picramate are responsible for the increased color (8). Henry (101) has discussed the difficulties and effects of variables involved in the Jaffe reaction: picric acid, temperature, protein precipitation, and pH.

The specificity of the Jaffe reaction is improved if Lloyd's reagent (a purified fuller's earth, which is an aluminum silicate clay, as in kaolin) is used to adsorb and thereby separate the creatinine from other chromogens (101). Slot (213) claims his technic of adding acid after total color development and noting the difference between the two readings is more specific than this adsorption method. A heat-clot method of removing protein in conjunction with Lloyd's reagent purportedly avoids the creatinine loss that normally occurs when proteins are precipitated with tungstic acid (136).

Two other color reactions for creatinine were reported with claims that they are as specific as the Jaffe reaction using Lloyd's reagent: alkaline 3,5-dinitrobenzoic acid (15, 129) and potassium 1,4-naphthoquinone-2-sulfonate (218). A relatively nonspecific color reaction involves the reduction of potassium ferricyanide to ferrocyanide and reaction with ferric alum (238).

All of the above methods are based on color development for quantification of creatinine. There is, however, a physical method of analysis whereby creatinine is first adsorbed on an ion-exchange resin, eluted, and its concentration determined by ultraviolet absorption at 234.5 nm in alkaline solution (1, 2). Other methods for determining creatinine are as follows: (a) measurement of the picric acid color reaction before and after bacterial-enzyme destruction of creatinine (63, 152); and (b) oxidation of reducing substances by iodine followed by ether extraction of these noncreatinine chromogens (222, 224).

Rather than use Lloyd's reagent, a cation-exchange resin in a microcolumn has been substituted to adsorb the creatinine from acidified serum (225). Following elution, creatinine concentration in the eluate is determined using an automated technic with alkaline picrate reagent (180). Instead of a resin column, it is possible to use a resin-batch technic in a test tube to adsorb the creatinine which is then eluted with pH 12.4 phosphate buffer (192). Stoten (217) has a microprocedure for plasma or diluted urine whereby 5 mg of dry cation-exchange resin or 1 drop of 80% suspension in water adsorbs the creatinine that is then eluted into alkaline picrate.

There are two other methods for separating creatinine. After small volume electrophoresis, the electrophoretogram is sprayed to detect the creatinine and creatine that are then eluted and quantitated photometrically (79). Alternatively, after Sephadex gel filtration, the adsorbed creatinine is eluted and then measured at 235 nm (211). McEvoy-Bowe (148) using a similar technic but with DEAE-Sephadex, first separated urine creatinine, then determined its amount by absorption at 235 nm at pH 10.4. Their comparison study showed the Jaffe results to be higher by 12.9%.

Most analysts determine serum and urine creatine as creatinine by measuring the difference in colors produced with the Jaffe reaction before and after hydrolysis of the creatine to creatinine. There are, however, four other creatine methods which various investigators claim to be more specific:

(a) treatment of serum with iodine and potassium iodide to eliminate pyruvate, removal of proteins with zinc sulfate and barium hydroxide, and determination of serum creatine with creatine phosphokinase (123); (b) reaction of the serum or urine creatine with diacetyl-1-naphthol in alkaline solution (97, 127, 130, 166, 239, 252); (c) use of a coupled-enzyme system composed of creatine phosphokinase, pyruvate kinase, and lactic dehydrogenase. During the conversion of urine creatine to creatine phosphate, NADH is oxidized so that the absorbance change can be measured at 340 nm; (d) coupling the guanidine compounds with ninhydrin in alkaline solution and measurement of their highly fluorescent products (53). Henry (101) has discussed the difficulties encountered with these methods and their solutions.

MANUAL DETERMINATION OF CREATININE AND CREATINE

(modification of Owen et al, Ref. 175 and Taussky, Ref. 223)

PRINCIPLE

Creatinine is determined by quantitating the red pigment, alkaline creatinine picrate (Jaffe reaction). *Creatine* is determined by the difference in creatinine before and after conversion of the creatine to creatinine. This is brought about at the temperature of boiling water rather than in an autoclave for two reasons: (a) unless a large number of samples are to be run frequently, the use of an autoclave is inconvenient; (b) glucose does not interfere at this temperature. If conversion in an autoclave is desired the picric acid employed must be buffered. Two technics are presented for each, one without and one with adsorption onto Lloyd's reagent.

REAGENTS

Picric Acid, 0.036 M. 8.25 g anhydrous or 9.16 g reagent grade picric acid (containing 10%–12% added water), made up to 1 liter with water. Solution can be aided with heat. Recrystallization is not necessary. Exact standardization of the acid is not required. Solution is stable at room temperature but should be protected from sunlight.

Tungstic Acid. Dissolve 1 g polyvinyl alcohol (Elvanol 70-05, Dupont) with heat (do not boil) in 100 ml water. Cool to room temperature, then transfer solution to a 1 liter volumetric flask containing a solution of 11.1 g sodium tungstate dihydrate in 300 ml water. Mix. To another container add 2.1 ml conc H_2SO_4 to 300 ml water, mix, then add this solution to the tungstate solution. Dilute to 1 liter with water and mix. Stable for 24 months at room temperature. Do not refrigerate.

NaOH, 1.4 N. Dissolve 54 g sodium hydroxide in water and dilute to 1 liter with water. Store this reagent in a polyethylene bottle. Stable for 24 months at room temperature.

Oxalic Acid, Saturated Aqueous

Lloyd's Reagent. Approximately 100 mg aliquots of the powder are used in each adsorption. After weighing out such a sample the amount henceforth can be approximated on the tip of a spatula. An alternative is to use 1 ml of an aqueous 10% suspension. It has been claimed that not all brands of Lloyd's reagent will adsorb creatinine.

Creatinine Standard, 13.2 mg/100 ml. Dissolve 132 mg creatinine in 1 liter 0.1 N HCl. This standard is stable for at least 1 year at room temperature.

PROCEDURES WITHOUT LLOYD'S REAGENT

Serum Creatinine

1. Prepare a protein-free solution: Transfer to a test tube (preferably 16 mm x 100 mm) 4.0 ml tungstic acid, followed by 0.5 ml serum. Shake vigorously for 10 sec. Centrifuge for 10 min.
2. Set up in test tubes:
    Reagent blank. 3.0 ml H_2O.
    Standard. 3.0 ml H_2O + 0.05 ml standard (TC pipet).
    Unknown. 3.0 ml protein-free centrifugate.
3. To each tube add 1.0 ml picric acid. Mix thoroughly.
4. Add 0.5 ml 1.4 N NaOH to the first vial. Set a timer and mix vial contents. At 30 sec intervals, add NaOH to each of the remaining vials.
5. Exactly 15 min after adding NaOH read absorbances against reagent blank at 500 nm. Maintain 30 sec intervals between readings.
Calculation:

$$\text{mg creatinine/100 ml serum} = \frac{A_x}{A_s} \times 0.132 \times 0.05 \times \frac{100}{0.33} \times \frac{4.50}{4.55}$$

$$= \frac{A_x}{A_s} \times 2$$

Serum Creatine

1. Prepare a sufficient volume of protein-free solution for the creatinine and creatine determinations as follows: Transfer to a test tube (preferably 16 mm x 100 mm) 8.0 ml tungstic acid, followed by 1.0 ml serum. Shake vigorously for 10 sec. Centrifuge for 10 min.
2. Carefully pipet off the protein-free centrifugate into a clean, properly marked tube. Using 3.0 ml of the centrifugate, determine serum creatinine (see above procedure). Refrigerate remaining protein-free solution until needed.
3. Set up in graduated Pyrex centrifuge tubes:
    Reagent Blank. 6.0 ml H_2O.

Standard. 6.0 ml H_2O + 0.05 ml standard (TC pipet)

Known. 3.0 ml H_2O + 3.0 ml protein-free centrifugate.

4. Add 1.0 ml picric acid to each tube.
5. Immerse tubes in boiling water so that water level is above liquid level in tubes. Heat until volumes fall below 4.0 ml (requires about 1.5 to 2 hour). To prevent steam from condensing on tubes it is convenient to take a piece of sponge rubber mat, make holes with a cork borer slightly smaller than the diameter of the tubes to coincide with the holes in the top of the water bath, and then place the tubes in these holes.
6. Cool tubes to room temperature and dilute to 4.0 ml with H_2O.
7. Add 0.5 ml of 1.4 N NaOH to the first tube. Set a timer, stopper tube, and mix contents. At 30 sec intervals, add NaOH to the remaining tubes.
8. Exactly 15 min after adding NaOH, read absorbances against reagent blank at 500 nm. Maintain 30 sec intervals between readings.

Calculation:

$$\text{Total creatinine (mg preformed creatinine} + \text{creatine as creatinine}/100 \text{ ml serum)} = \frac{A_x}{A_s} \times 2$$

$$\text{mg creatine as creatinine}/100 \text{ ml serum} = \text{``total creatinine''} - \text{``preformed creatinine''}$$

Urine Creatinine

Same as for serum except:
1. For unknown, use 3.0 ml of a 1:200 dilution of the urine. If there is proteinuria, remove protein by making a 1:10 Folin-Wu protein-free filtrate in the manner ordinarily employed for blood. Make a further 1:20 dilution of this filtrate to yield the desired 1:200 dilution of the original urine.
2. For standard, use 0.1 ml standard (TC pipet) + 2.9 ml H_2O.

Calculation:

$$\text{mg creatinine/ml urine} = \frac{A_x}{A_s} \times 0.132 \times 0.1 \times \frac{1}{0.015}$$

$$= \frac{A_x}{A_s} \times 0.88$$

Urine Creatine

Same as for serum, except:
1. For unknown, use 3.0 ml of a 1:200 dilution + 3.0 ml H_2O. In case of proteinuria, remove as outlined in step 1 under "Urine Creatinine" above.
2. For standard, use 0.1 ml standard (TC pipet) + 5.9 ml H_2O.

3. Do *not* add 0.50 ml 1.4 *N* NaOH to unknown *before* heating.
   Calculation:

$$\text{Total creatinine (mg preformed creatinine + creatine as creatinine/ml urine)} = \frac{A_x}{A_s} \times 0.88$$

$$\text{mg creatine as creatinine/ml urine} = \text{"total creatinine"} - \text{"preformed creatinine"}$$

PROCEDURES WITH LLOYD'S REAGENT

Serum Creatinine

1. Prepare a protein-free solution: Transfer 4.0 ml tungstic acid to a test tube (preferably 16 mm x 100 mm) followed by 0.5 ml of serum. Shake vigorously for 10 sec. Centrifuge for 10 min.
2. Set up in test tubes:
   Reagent blank. 5.0 ml H_2O.
   Standard. 5.0 ml H_2O + 0.05 ml standard (TC pipet).
   Unknown. 2.0 ml H_2O + 3.0 ml protein-free centrifugate.
3. Add 0.5 ml saturated aqueous oxalic acid to each tube.
4. Add approximately 100 mg Lloyd's reagent or 1.0 ml 10% aqueous suspension to each tube.
5. Stopper tubes with rubber stoppers and shake intermittently for 10 min.
6. Centrifuge, decant and invert the tubes to drain well.
7. Add 3.0 ml H_2O, 1.0 ml picric acid and 0.5 ml 1.4 *N* NaOH to each tube.
8. Stopper and shake intermittently for 10 min. Centrifuge.
9. Transfer the centrifugates into cuvets and read at 500 nm against reagent blank at any time after 20 min after step 7.
   Calculation:

$$\text{mg creatinine/100 ml serum} = \frac{A_x}{A_s} \times 2$$

Serum Creatine

1. Prepare a sufficient volume of protein-free solution for the creatinine and creatine determinations as follows:
   Transfer 8.0 ml of tungstic acid, followed by 1.0 ml serum, to a test tube (preferably 16 mm x 100 mm). Shake vigorously for 10 sec. Centrifuge.
2. Carefully pipet off the protein-free solution into a clean, properly marked tube. Determine serum creatinine (see above procedure) using 3.0 ml protein-free centrifugate. Refrigerate remaining protein-free solution until needed.
3. Set up in graduated Pyrex centrifuge tubes:
   Reagent blank. 6.0 ml H_2O.

Standard. 6.0 ml H_2O + 0.05 ml standard (TC pipet).

Unknown. 3.0 ml H_2O + 3.0 ml protein-free centrifugate.

4. Add 1.0 ml picric acid reagent to each tube.
5. Immerse tubes in boiling water so that water level is above the liquid level in the tubes. Heat until volumes fall below 4.0 ml. (see step 5, "Serum Creatine, Procedures without Lloyd's Reagent.")
6. Cool tubes to room temperature and dilute to about 5.0 ml with water.
7. Add 0.5 ml saturated oxalic acid solution to each tube.
8. Add approximately 100 mg Lloyd's reagent or 1.0 ml 10% aqueous suspension to each tube.
9. Stopper each tube with rubber stoppers and shake intermittently for 10 min.
10. Centrifuge, decant centrifugate, and invert the tubes to drain well.
11. Add 3.0 ml H_2O, 1.0 ml picric acid, and 0.5 ml 1.4 N NaOH to each tube.
12. Stopper each tube with rubber stoppers and shake intermittently for 10 min. Centrifuge.
13. Transfer centrifugates to cuvets and read at 500 nm against reagent blank any time after 20 min after step 11.

Calculation:

$$\text{Total creatinine (mg preformed creatinine} + \text{creatine as creatinine/100 ml serum)} = \frac{A_x}{A_s} \times 2$$

$$\text{mg creatine as creatinine/100 ml serum} = \text{"total creatinine"} - \text{"preformed creatinine"}$$

Urine Creatinine

Same as for serum except:

1. For unknown, use 3.0 ml of a 1:200 dilution + 2.0 ml H_2O. In case of proteinuria, remove protein as outlined in step 1 of "Urine Creatine, Procedures without Lloyd's Reagent."
2. For standard, use 0.1 ml standard (TC pipet) + 4.9 ml H_2O.

Calculation:

$$\text{mg creatinine/ml urine} = \frac{A_x}{A_s} \times 0.88$$

Urine Creatine

Same as for serum, except:

1. For unknown, use 3.0 ml of a 1:200 dilution + 3.0 ml H_2O. In case of proteinuria, remove protein as outlined in step 1 of "Urine Creatinine, Procedures without Lloyd's Reagent."
2. For standard, use 0.1 ml standard (TC pipet) + 5.9 ml H_2O.

3. Do *not* add 0.50 ml 1.4 N NaOH to unknown *before* heating.
Calculation:

$$\text{Total creatinine (mg preformed creatinine} + \text{creatine as creatinine/ml urine)} = \frac{A_x}{A_s} \times 0.88$$

$$\text{mg creatine as creatinine/ml urine} = \text{``total creatinine''} - \text{``preformed creatinine''}$$

NOTES

1. Beer's law. With band widths of 20 nm or less the color obeys Beer's law from 485 nm to 520 nm; however, there is deviation with a wide band filter such as the Klett 54. Thus, if a filter photometer is employed, construct a calibration curve with at least two standards for each set of determinations. The wavelength of maximum absorption for alkaline creatinine picrate read against alkaline picrate is about 485 nm (see Fig. 17-9). For normal serum creatinine levels it would be better theoretically to make absorbance readings at 485 nm because at 500 nm the A readings are less than 0.4. Below 510 nm, however, absorption of light by the reagent blank rises steeply so that a slight volume error when adding the picric acid will have a great effect on results. Furthermore, using iodine oxidation for removal of interfering substances, Taussky (224) found better method specificity by reading absorbances at 520 nm because results for serum were 30% higher at 490 nm.

2. Recovery. When setting up the methods for creatine without Lloyd's reagent and for creatinine and creatine with Lloyd's reagent, internal standards should be used to check recovery. The recoveries should average between 98% and 100%.

3. Avoidance of heat while reading colors. Absorbances of the reagent blank and the alkaline creatinine picrate of unknowns and standards increase with temperature but not proportionately (175). Hence, if the instrument used is one that exposes cuvets to significant heat, it is important to make absorbance readings rapidly. This is especially true for the reagent blank since it is used repeatedly while reading a series of unknowns. The temperature effect is reversible. Color intensities of the blank, standards, and unknowns also increase upon exposure to light but, fortunately, they increase proportionately (175).

4. Interferences. Slight hemolysis does not affect the determination of creatinine in serum, but hemolysis can increase the creatine value 100%–200% (222). Serum or urine containing bromsulfophthalein or phenolsulfonephthalein (phenol red) cannot be assayed for creatinine without preliminary conversion of these dyes to their leuco form by treatment with granulated zinc (222, 224, 228).

5. Stability of samples. Aqueous solutions of creatinine at neutral pH eventually reach a point of equilibrium at which the ratio of creatine to

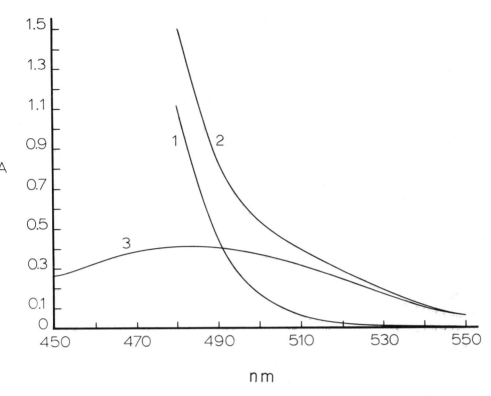

FIG. 17-9. Absorption curves of alkaline picrate and alkaline creatinine picrate: (1) blank vs water, (2) 20 μg creatinine vs water, (3) 20 μg creatinine vs blank. Beckman Model DU Spectrophotometer.

creatinine is about unity. Such a point is reached in urine at refrigerator temperature in about 1 year (161). Urine preserved with thymol or toluene, which do not interfere with the analysis (223), is stable for 24 hours at room temperature or for several days in the refrigerator with respect to creatinine (222). Taussky (224), however, has found thymol preservation to be unpredictable. Urine creatine and creatinine has been found in our laboratory to be stable at room temperature (30°C) for 4 hours or 4 days, respectively, without preservative. They are stable indefinitely when frozen. Clarke (50) calculated that a maximum of 10% is converted in 4 days at 37°C in a urine having a pH of 5. Neither acid nor alkali can be used to stabilize urine because of their catalytic conversion of creatine to creatinine (5, 224). Petroleum ether apparently preserves urine creatinine in the refrigerator for periods of 1–4 weeks. Creatine, ordinarily present only in a very small amount compared to the creatinine in urine, may increase significantly in 24 hours at room temperature (68). Both creatine and creatinine are reported (5) as stable for several months in the frozen state. Serum holds up fairly well at refrigerator temperature for a period of 24 hours (222). Fluoride or fluoride plus thymol preserves serum for about 5 days at room temperature. In our laboratory, serum creatine and creatinine have been found to be stable at room temperature (30°C) for 4 hours or 7 days without a preservative, respectively. They are stable indefinitely when frozen.

ACCURACY AND PRECISION

The method for serum creatinine without Lloyd's reagent is subject to at least two biases which are in opposite directions. As discussed, up to 20% of the chromogenic material in the Jaffe method is noncreatinine. On the other hand, since protein-free filtrates have a pH less than 2, whereas the standard is less acid, the unknown reads up to about 5% too low. Because of the larger uncertainty due to noncreatinine chromogens, there is no point in attempting to control the acidity. Nevertheless, the technic without Lloyd's reagent seems adequate for routine use.

The method employing the adsorption of creatinine onto Lloyd's reagent appears to be accurate for serum. Ordinarily there is no particular advantage in the use of Lloyd's reagent for urine analysis since there is usually little noncreatinine chromogen material present in urine and the problem of pH control does not arise.

The precision (95% limits) of the creatinine method in the normal range is about ±5%.

In the majority of cases, the accuracy of the creatine method is presumably the same as that for creatinine and subject to the same considerations. Sullivan and Irreverre (218), however, think there is little evidence of creatine in the urines of normal men and conclude that something other than creatine is formed during the heating with acid. In a study of about 300 urines from patients with muscular dystrophy, Linneweh and Linneweh (134) concluded the total creatinine after conversion of creatine was 10%—30% too high in about 10% of samples. They attributed this (as Sullivan and Irrevere did later) to the formation of nonspecific Jaffe-positive compounds during acid hydrolysis (135).

The precision of creatine determinations is less than for creatinine because results are calculated from the difference before and after hydrolysis of creatine. In urine this hydrolytic difference ordinarily is small compared with the total creatinine. Consider, for example, that a 24 hour urine with a creatinine content of 1500 mg and a creatinine + creatine content of 1550 mg has only 50 mg of creatine. If the reproducibility of these determinations is ±2%, a precision difficult to achieve, then the creatine, expressed as creatinine, could be anywhere from −11 mg to +111 mg. The implication that this has for the subject of normal urine creatine levels determined by such a technic is clear.

NORMAL VALUES

Serum Creatinine. The normal adult ranges for serum creatinine without Lloyd's reagent are 0.9 to 1.4 mg/100 ml and 0.8 to 1.2 mg/100 ml for men and women, respectively (230). With Lloyd's reagent, the ranges are 0.7–1.2 mg/100 ml and 0.5–1.0 mg/100 ml, respectively (64).

Evidence concerning the constancy of serum creatinine in an individual is contradictory. Sirota *et al* (212) claim there may be as much as a 10% change within a 4 hour period. However, Haugen and Blegen (100) found the daily variation to be less than 0.1 mg/100 ml. There appears to be no change in the serum creatinine level following meals (6).

The creatinine level for infants in the first week of life falls from an adult level (prematures start higher) to a low at 1 month; then there is a steady rise again to the adult level by puberty (99).

Serum Creatine. The normal adult ranges for serum or plasma creatine, determined by the Jaffe reaction without Lloyd's reagent, were reported by Tierney and Peters (227) to be 0.17–0.50 mg/100 ml and 0.35–0.93 mg/100 ml for males and females, respectively. More accurate data obtained by newer technics appear to be rather meager. Conversion of creatine to creatinine, followed by enzymatic destruction of the creatinine has indicated that the true serum creatine is about 70% of the total chromogenic material. The diacetyl-naphthol method gave normal plasma values of about 0.14–0.44 mg/100 ml (5). Taussky (224), employing iodine oxidation and ether extraction of interfering substances, reported a normal range of 0.16–0.40 mg/100 ml.

At Bio-Science Laboratories, the normal ranges employed for creatine (as creatinine) for males and females are 0.1–0.4 and 0.2–0.7 mg/100 ml, respectively.

Urine Creatinine. Since creatinine excretion is closely related to body weight, it can be reported as the *creatinine coefficient* in mg/kg/24 hours. Some prefer to express the output as *creatinine N coefficient*, where mg creatinine N = 0.372 x mg creatinine. Normal values for the creatinine coefficient are shown in Table 17-1. The normal creatinine coefficient is higher for men than women after the age of 12. Ryan *et al* (197) interpreted this to mean that creatinine excretion correlated more closely with muscle

TABLE 17-1 Normal Values of Creatinine Coefficient

Age	Creatinine coefficient (mg/kg/24 hours)	References
Newborn infant	7–12	148
1.5–22 months	5–15	134
2.5–3.5 years	12	148
4–4.5 years	15–20	148
9–9.5 years	18–25	148
Adult		
Male	20.7	18
	24–26	24
Female	15.9	18
	20–22	24

mass than with surface area or body weight. This is consistent with an earlier finding that obese, healthy individuals have lower coefficients than their lean counterparts (18). The output is little affected by diet (233), is independent of diuresis (241), and is somewhat increased by physical exercise (233).

At our laboratory the normal ranges employed for urine creatinine for males and females are 21–26 and 16–22 mg/kg/24 hours, respectively.

Urine Creatine. There is no agreement on the normal creatine output, which is not surprising in light of the previously discussed problems on the accuracy and precision of its determination. Furthermore, significant amounts of creatine in urine, estimated to be 10%–30% of the total (221), may originate from hydrolysis of creatinine during the course of urine accumulation in the bladder (68).

The above are some reasons that the reported normal creatine values are not in agreement. Clarke (50), using the Jaffe reaction, found the normal urinary excretion of creatine to be 0–800 mg (as creatinine) per 100 lb body weight per day with a mean of 275. The normal ranges obtained with creatine kinase (221), by Taussky's method employing iodine treatment (224), and with the naphthol-diacetyl reaction (90) are significantly lower.

The urinary creatine excretion in children is variable (48). At 3 years of age it is greater than creatinine but gradually decreases to the adult level at about age 18 (48).

Creatine excretion is reported to be higher during the day than at night (145) and to be increased by physical exercise (145) or by a few days of bed rest (200). Nevertheless, tentative ranges for creatine output, expressed as creatinine (Jaffe reaction), may be stated as up to 150 mg/24 hours for men and up to 250 mg/24 hours for women (241).

AUTOMATED DETERMINATION OF SERUM AND URINE CREATININE

(method of Folin and Wu, Ref. 85, adapted to AutoAnalyzer)

PRINCIPLE

Creatinine reacts with picrate under alkaline conditions (Jaffe reaction) to give a yellow-red solution which is measured photometrically at 505 nm.

REAGENTS

NaCl solution. Dissolve 9.0 g NaCl in about 500 ml water in a 1 liter volumetric flask. Add 0.5 ml Brij-35 then dilute to volume with water. Mix, filter, and store in a polyethylene bottle.

Sodium hydroxide, 0.5 N. Dissolve 20 g NaOH in sufficient water and dilute to 1 liter.

Picric acid, saturated. Add 13 g picric acid to about 500 ml water in a 1 liter volumetric flask. Shake and then dilute to volume. Continue to shake occasionally while allowing the excess picric acid to remain in contact with the saturated solution. Filter and store in a polyethylene bottle.

Stock creatinine standard, 1 mg/ml. Dissolve 1.00 g creatinine in a sufficient volume of 0.1 N HC1 and dilute to 1 liter with 0.1 N HC1.

Working creatinine standards. Dilute 5 and 10 ml stock creatinine standard to 100 ml with 0.02 N HCl to obtain 5 and 10 mg/100 ml, respectively.

Recipient reagent. Add 0.5 ml Brij-35 to 1 liter water and mix.

PROCEDURE

Set up manifold, other components, and operating conditions as shown in Figure 17-10. By following procedural guidelines as described in Chap. 10, steady state interaction curves as shown in Figure 17-11 should be obtained and should be checked periodically. The urine chart record is identical to that for serum except that the baseline and steady state noise levels are about twice as great for the former but still within the stated tolerances. Undiluted serum or urine is used. A blank for serum or urine is not necessary. Results are calculated from the calibration curve in the standard manner for methods obeying Beer's law.

NOTES

1. Noise. If the noise level is excessive, clean the picric-sodium hydroxide lines and coils as well as the flow cell with 10% acetic acid.

2. Temperature. The time-delay coil used for color development should be immersed in a container of water at room temperature to protect against rapid fluctuations in ambient temperature.

ACCURACY AND PRECISION

The precision (95% limits) is ±8% at a creatinine concentration of 2.9 mg/100 ml serum.

NORMAL VALUES

Same as for manual methods.

TOTAL FECAL NITROGEN

The determination of fecal N is primarily a test of pancreatic function. Enzymes contained in the pancreatic juice normally hydrolyze ingested proteins within the intestinal lumen so that the hydrolytic products can be absorbed. The N excreted, therefore, is composed of a small fraction of the ingested N that escapes absorption plus the N from bacteria, mucus, epithelial cells, and unabsorbed intestinal and digestive juices. If the

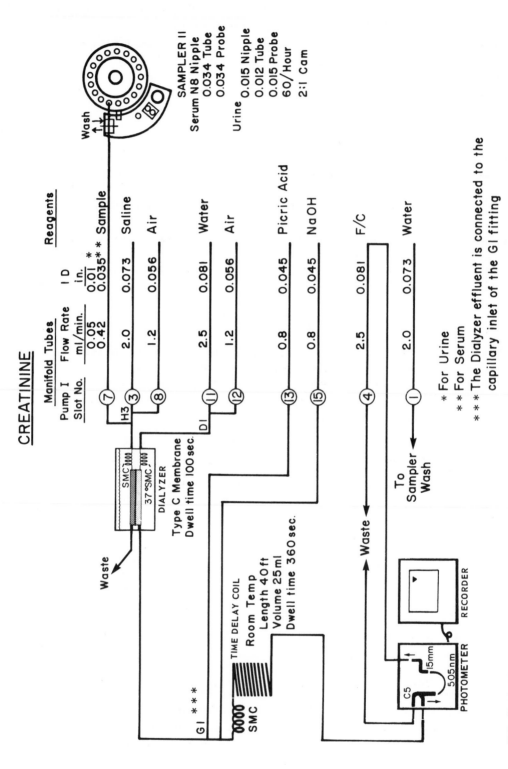

FIG. 17-10. Flow diagram for creatinine in serum or urine. Calibration range: 0–10 mg/100 ml for serum and 0–100 mg/100 ml for urine. Conventions and symbols used in this diagram are explained in Chap. 10.

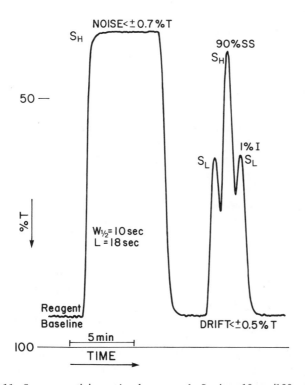

20 —

50 —

100 —

%T

S$_H$

NOISE<±0.7%T

90%SS
S$_H$

1%I
S$_L$

S$_L$

W$_{1/2}$= 10 sec
L = 18 sec

Reagent
Baseline

DRIFT<±0.5%T

5 min

TIME

FIG. 17-11. Serum creatinine strip chart record. S_H is a 10 mg/100 ml standard, S_L a 5 mg/100 ml standard. Peaks reach 90% of steady state; interaction between samples is 1%. $W_{1/2}$ and L are explained in Chap. 10. Computer tolerance for noise is set at ±0.7% T, for baseline drift at ±0.5% T, and for standard drift at ±1% T.

proteolytic enzymes are decreased or absent, a greater fraction of the ingested protein N passes through the gastrointestinal tract unabsorbed.

The method presented is for the determination of fecal N by Kjeldahl digestion and titration of the distilled ammonia. There are, in addition, three other methodological approaches: (a) Digestion of an aliquot of homogenized feces with the AutoAnalyzer digestor, followed by distillation and titration of the ammonia (111). This method for total fecal N is 7 times faster than the manual Kjeldahl method. One must analyze large numbers of samples, however, to warrant the financial investment in the digestor. (b) Digestion of a diluted aliquot of homogenized sample, removal of any catalysts used, followed by the indophenol reaction for ammonia (58). (c) Determination of fecal N by a micro-Dumas method in the Automated Coleman Nitrogen Analyzer (215). Stitcher *et al* found the Dumas results to be slightly higher than the Kjeldahl results.

DETERMINATION OF TOTAL NITROGEN IN FECES

PRINCIPLE

Fecal N is converted to ammonia by Kjeldahl digestion. The ammonia is then isolated by steam distillation and quantitated by titration.

REAGENTS

Boric Acid Solution, 4%
NaOH, 40% (10 N)
Indicator. Either (a) mixed indicator of bromcresol green and methyl red, prepared by dissolving 50 mg bromcresol green and 10 mg methyl red in 60 ml ethanol (ethanol denatured with methanol is satisfactory); or (b) Fleischers methyl purple, pH range 4.8–5.4 (available in ready-to-use solution from A.H. Thomas).
HCl, standardized at approx 0.1 N

COLLECTION OF SAMPLE

The normal range for fecal N is based on a 24 hour output that normally is dependent on the quantity and nature of the food intake. For the most accurate evaluation, it is necessary to place the subject on a standard diet such as the Nothman diet, and collect feces for several days. Details of the diet and method of fecal collection are given in the section "Fecal Lipids" in Chap. 28. Fortunately, the actual N content of standard diets is quite close to the calculated N content (189). The analysis of N in a sample from a single defecation, although admittedly more convenient, is only an approximation possessing significance when the results are markedly abnormal.

Specimens may be collected in cardboard ice cream containers, but if the stool is liquid, the containers may leak. Consequently, polyethylene containers with tight-fitting lids are preferable. Specimens must be stored at refrigerator temperature or, preferably, frozen until analyzed. To obtain the weight of feces collected, weigh the container before and after collection.

PROCEDURE

1. Record the weight of entire sample received. This sample, whether it is a single defecation or a 3 day collection, must be thoroughly mixed to assure representative sampling. A preliminary mixing is made by kneading with a spatula. In the case of a soft stool this may be all that is required. Approximately 1 g is weighed on a piece of filter paper to the nearest 0.01 g and placed in a 500 ml Kjeldahl flask. If the feces are very hard or very fluid, homogenize in a Waring Blendor (a measured volume of water is added to a weighed amount of feces if they are hard). An aliquot of the suspension, equivalent to 1.00 g feces, is then transferred to the Kjeldahl flask.
2. Add 10 g K_2SO_4, 0.5 g $CuSO_4 \cdot 5H_2O$, and 20 ml conc H_2SO_4.

3. Digest the sample and paper with very gentle heat in a fume hood, or connected to a Kjeldahl manifold, until the flask contents turn green. Then turn up heat and boil for 30 min.
4. Cool and rinse down the acid from around the neck of the flask with 150–200 ml water.
5. Set up a distillation assembly (complete with Hopkin's trap) and place the tip of the condenser below the surface of 50 ml boric acid solution contained in a 500 ml flask.
6. Gently and slowly pour 85–95 ml 40% NaOH down the side of the Kjeldahl flask. Do *not* mix. Immediately connect condenser to the Kjeldahl flask. Heat gently to boiling; then the heating can be increased.
7. Distil approximately two-thirds of the Kjeldahl-flask contents into the boric acid solution.
8. Add 5 drops indicator solution to the distillate then titrate the NH_4^+ with standardized HCl of about 0.1 N.

Calculation:

$$\text{g N/total weight feces} = \frac{\text{ml acid} \times \text{N of acid} \times 0.014 \times \text{g total feces}}{\text{g feces in aliquot used}}$$

NORMAL VALUES

Adults on an average diet excrete up to 2 g N/day (249). To illustrate the difficulty in interpretation of N excretion data without feeding a standard-ized diet, Rubner (195) reported that patients fed 1435 g meat/day excreted 1.2 g N per day or 2.5% of the N intake; whereas patients fed 1360 g black bread/day, excreted 4.26 g N per day or 32% of the N intake. Infants on an average diet excrete 0.11–0.52 g N/day (209).

The Schmidt diet (204) or the Nothman diet (168) contains about 110 g protein or 17.6 g N/day; the exact amount varies with the N content of the dietary components. Normally, less than 10% of the N ingested is excreted; thus, the N excreted per day is less than 1.8 g (23). However, on the diet of Wollaeger and co-workers (248), which contains about 118 g protein (18.9 g N) and is somewhat more complicated than Schmidt's or Nothman's, the normal N excretion varies from 0.8 to 2.5 g/day. Wolthius (250) subse-quently affirmed that the concept of estimating normal fecal N as 10% of intake is incorrect.

When N is determined on a single defecation, the best one can do for normals is to assume an average daily excretion of 100 g (wet weight) (174). On this basis, feces should contain less than 30 mg N/g wet weight.

REFERENCES

1. ADAMS WS, DAVIS FW, HANSEN LE: *Anal Chem* 34:854, 1962
2. ADAMS WS, DAVIS FW, HANSEN LE: *Anal Chem* 36:2209, 1964
3. ALPER C, SEITCHIK J: *Clin Chem* 3:95, 1957

4. ANDERSEN CJ, STRANGE B: *Scand J Clin Lab Invest 11*:122, 1959
5. ANDERSON DR, WILLIAMS CM, KRISE GM, DOWBEN RM: *Biochem J 67*:258, 1957
6. ANNINO JS, RELMAN AS: *Am J Clin Pathol 31*:155, 1959
7. ARCHIBALD RM: *Clin Chem 3*:102, 1957
8. ARCHIBALD RM: *J Biol Chem 237*:612, 1962
9. ARCHIBALD RM: *J Biol Chem 154*:643, 1944
10. BARAKAT MZ, SHEHAB SK, EL-SADR MM: *Arch Biochem Biophys 66*:444, 1957
11. BEACH EF: *Std Methds of Clin Chem 2*:100, 1958
12. BEALE RN, CROFT D: *J Clin Pathol 14*:418, 1961
13. BENEDICT SR: *J Biol Chem 51*:187, 1922
14. BENEDICT SR: *J Biol Chem 54*:233, 1922
15. BENEDICT SR, BEHRE JA: *J Biol Chem 114*:515, 1936
16. BENEDICT SR, HITCHCOCK EH: *J Biol Chem 20*:619, 1915
17. BERTHELOT MPE: *Repert Chim Appl* 282, 1859
18. BEST WR, KUHL WJ, CONSOLAZIO CF: *J Lab Clin Med 42*:784, 1953
19. BIEN EJ, ABBOT BH, ZUCKER M: *Proc Soc Exp Biol Med 93*:567, 1956
20. BISHOP C, TALBOT JH: *Pharm Rev 5*:231, 1953
21. BITTNER D, HALL S, McCLEARY M: *Am J Clin Pathol 40*:423, 1963
22. BLAUCH MB, KOCH FC: *J Biol Chem 130*:443, 1939
23. BODANSKY M, BODANSKY O: *Biochemistry of Disease.* New York, MacMillan, 1940, p 208
24. BOHUON C, DELARUE JC, COMOY E: *Clin Chim Acta 18*:417, 1967
25. BOLLETER WG, BUSHMAN CJ, TIDWELL PW: *Anal Chem 33*:592, 1961
26. BROWN H: *J Biol Chem 68*:123, 1926
27. BROWN H: *J Biol Chem 158*:601, 1945
28. BROWN H: adapted to AutoAnalyzer as described in "Uric Acid," Technicon Laboratory Method File No. N-13b, Technicon Instruments Corporation, Tarryton, New York as in Ref. 27
29. BROWN RA, FREIER EF: *Clin Biochem 1*:No. 2, 154, 1967
30. BROWN RH, DUDA GD, KORKES S, HANDLER P: *Arch Biochem Biophys 66*:301, 1957
31. BUCHANAN MJ, ISDALE IC, ROSE BS: *Ann Rheum Dis 24*:285, 1965
32. BULGER HA, JOHNS HE: *J Biol Chem 140*:427, 1941
33. CANELLAKIS ES, COHEN PP: *J Biol Chem 213*:385, 1955
34. CARAWAY WT: *Clin Chem 15*:720, 1969
35. CARAWAY WT: *Standard Methods of Clinical Chemistry 4*, edited by D Seligson. New York, Academic, 1963, p 239
36. CARAWAY WT, FANGER H: *Am J Clin Pathol 26*:1475, 1956
37. CARAWAY WT, MARABLE H: *Clin Chem 12*:18, 1966
38. CARROLL JJ, COBURN H, DOUGLASS R, BABSON AL: *Clin Chem 17*:158, 1971
39. CERIOTTI G: *Clin Chem 17*:400, 1971
40. CERIOTTI G, SPANDRIO L: *Clin Chim Acta 8*:295, 1963
41. CERIOTTI G, SPANDRIO L: *Clin Chim Acta 11*:519, 1965
42. CHANEY AL, MARBACH EP: *Clin Chem 8*:130, 1962
43. CHANEY AL, MARBACH EP: *Clin Chem 8*:131, 1962
44. CHRISTIAN GD: *Clin Chim Acta 14*:242, 1966
45. CHRISTIAN GD, FELDMAN FJ: *Clin Chim Acta 17*:87, 1967
46. CHRISTIAN GD, KNOBLOCK EC, PURDY WC: *Clin Chem 11*:700, 1965
47. CHRISTIANSSON G, JOSEPHSON B: *Acta Paediat 49*:633, 1960

48. CLARK LC Jr, THOMPSON HL, BECK EI, JACOBSON W: *Am J Diseases Childhood 81*:774, 1951

49. CLARKE JS, CRUZE K, McKESSOCK PK, OZERAN RS: *AMA Arch Surg 78*:836, 1959

50. CLARKE JT: *Clin Chem 7*:371, 1961

51. CONN HO: *New Engl J Med 262*:1103, 1960

52. CONN HO: *Std Methods of Clin Chem 5*:43, 1965

53. CONN RB Jr: *Clin Chem 6*:537, 1960

54. CONWAY EJ: *Microdiffusion Analysis and Volumetric Error.* Princeton, N.J., Van Nostrand, 1950

55. CONWAY EJ, BYRNE A: *Biochem J 27*:419, 1933

56. COULOMBE JJ, FAVREAU L: *Clin Chem 9*:102, 1963

57. DALY CA: *J Lab Clin Med 18*:1279, 1933

58. DAMBACHER M, GUBLER A, HAAS HG: *Clin Chem 14*:615, 1968

59. de NIEVES GV, PIKE RL: *Anal Biochem 2*:174, 1961

60. DIENST SG: *J Lab Clin Med 58*:149, 1961

61. DOMAGK GF, SCHLICKE HH: *Anal Biochem 22*:No. 2, 219, Feb. 1968

62. DUBBS CA, DAVIS FW, ADAMS WS: *J Biol Chem 218*:497, 1956

63. DUBOS R, MILLER BF: *J Biol Chem 121*:429, 1937

64. EDWARDS KDG, WHYTE HM: *Australasian Ann Med 8*:218, 1959

65. EICHHORN F, RUTENBERG A: *J Med Lab Technol 22*:233, 1965

66. EICHHORN F, RUTENBERG A: *New Istanbul Contrib Clin Sci 6*:87, 1963

67. EICHHORN F, ZELMANOWSKI S, LEW E, RUTENBERG A, FANIAS B: *J Clin Pathol 14*:450, 1961

68. ENNOR AH, STOCKEN LA: *Biochem J 55*:310, 1953

69. EVANS RT: *J Clin Pathol 21*:527, 1968

70. EXLEY D: *Biochem J 63*:496, 1956

71. FARESE G, MAGER M: *Clin Chem 16*:280, 1970

72. FAULKNER WR, BRITTON RC: *Cleveland Clinic Quart 27*:202, 1960

73. FAWCETT JK, SCOTT JE: *J Clin Pathol 13*:156, 1960

74. FEARON WR: *Biochem J 33*:902, 1939

75. FEE DA, CRUGER D, COLLIER HB: *J Lab and Clin Med 34*:873, 1949

76. FEICHTMEIR TV, WRENN HT: *Am J Clin Pathol 25*:833, 1955

77. FENTON JCB: *Clin Chim Acta 7*:163, 1962

78. FENTON JCB, WILLIAMS AH: *J Clin Pathol 21*:14, 1968

79. FISCHL J, SEGAL S, YULZARI Y: *Clin Chim Acta 10*:73, 1964

80. FOLIN O: *J Biol Chem 86*:179, 1930

81. FOLIN O: *J Biol Chem 101*:111, 1933

82. FOLIN O: *J Biol Chem 106*:311, 1934

83. FOLIN O, DENIS W: *J Biol Chem 13*:469, 1912-1913

84. FOLIN O, TRIMBLE H: *J Biol Chem 60*:473, 1924

85. FOLIN O, WU H: *J Biol Chem 38*:81, 1919

86. FORMAN DT: *Clin Chem 10*:497, 1964

87. FRIEDMAN HS: *Anal Chem 25*:662, 1953

88. GANGOLLI S, NICHOLSON TF: *Clin Chim Acta 14*:585, 1966

89. GENTZKOW CJ: *J Biol Chem 143*:531, 1942

90. GERBER GB, GERBER G, ALTMAN KI: *Anal Chem 33*:852, 1961

91. GIPS CH, WIBBENS-ALBERTS M: *Clin Chim Acta 22*:183, 1968

92. GOA J: *Scand J Clin Lab Invest 3*:311, 1951

93. GOCHMAN N, SCHMITZ JM: *Clin Chem 17*:1154, 1971

94. GOTTFRIED SP, ERDMAN GL: *Am J Clin Pathol 21*:118, 1951

95. GRAFNETTER D, JANOSOVA Z, CERVINKOVA I: *Clin Chim Acta 17*:493, 1967

96. GREENWALD I: *J Biol Chem 80*:103, 1928
97. GRIFFITHS WJ: *Clin Chim Acta 9*:210, 1964
98. GUILBAULT GG, HRABANKOVA E: *Anal Chim Acta 52*:287, 1970
99. HARE RS: *Proc Soc Exp Biol Med 74*:148, 1950
100. HAUGER HN, BLEGEN EM: *Ann Internal Med 43*:731, 1955
101. HENRY RJ: *Clinical Chemistry: Principles and Technics.* New York, Harper & Row, 1968
102. HENRY RJ, CHIAMORI N: *Am J Clin Pathol 29*:277, 1958
103. HENRY RJ, SOBEL C, KIM J: *Am J Clin Pathol 28*:152, 1957
104. HILLER A, VAN SLYKE DD: *J Biol Chem 53*:253, 1922
105. HOLM-JENSEN I:*Scand J Clin Lab Invest 13*:301, 1961
106. HORN DB, SQUIRE CR: *Clin Chim Acta 14*:185, 1966
107. HORN DB, SQUIRE CR: *Clin Chim Acta 17*:99, 1967
108. HÜBSCHER G, BAUM H, MAHLER HR: *Biochim Biophys Acta 23*:43, 1957
109. HUNTER G: *J Clin Pathol 10*:161, 1957
110. ISDALE I, BUCHANAN M, ROSE B: *Ann Rheum Dis 25*:184, 1966
111. JACOBS SC: *J Clin Pathol 21*:218, 1968
112. JACOBSON RM: *Ann Internal Med 11*:1277, 1938
113. JAFFE M: *Z Physiol Chem 10*:391, 1886
114. JELLINEK GM, LOONEY JM: *J Biol Chem 128*:621, 1939
115. JUNG DH, PAREKH AC: *Clin Chem 16*:247, 1970
116. KAGEYAMA N: *Clin Chim Acta 31*:421, 1971
117. KALCKAR HM: *J Biol Chem 167*:429, 1947
118. KANABROCKI EL, GRECO J, WILKOFF L, VEACH R: *Clin Chem 3*:156, 1957
119. KANTER SL: *Clin Chem 13*:No. 5, 406, May, 1967
120. KAPLAN A: *Standard Methods in Clinical Chemistry.* New York, Academic, 1965 Vol 5, p. 245
121. KARR WG: *J Lab Clin Med 9*:329, 1924
122. KERN A, STRANSKY E: *Biochem Z 390*:419, 1937
123. KIRBRICK AC, MILHORAT AT: *Clin Chim Acta 14*:201, 1966
124. KIRK PL: *Anal Chem 22*:354, 1950
125. KITAMURA M, INCHI I: *Clin Chim Acta 4*:701, 1959
126. KOCH FC, McMEEKIN TL: *J Am Chem Soc 46*:2066, 1924
127. KUROHARA SS: *Clin Chem 7*:284, 1961
128. LAMBOOY N, VAN AMSON G: *Lancet II*:1361, 1966
129. LANGLEY WD, EVANS M: *J Biol Chem 115*:333, 1936
130. LAUBER K: *Z Klin Chem 4*:119, 1966
131. LAVERY TD: *Clin Chim Acta 21*:415, 1968
132. LEFFLER HH: *Clinical Pathology of the Serum Electrolytes,* edited by FW Sunderman, FW Sunderman Jr. Chicago, Ill., Thomas, 1966, p 119
133. LIDDLE L, SEEGMILLER JE, LASTER L: *J Lab Clin Med 54*:No. 6, 903, 1959
134. LINNEWEH F, LINNEWEH W: *Klin Wochschr 13*:1581, 1934
135. LINNEWEH F, LINNEWEH W: *Klin Wochschr 13*:589, 1934
136. LONDON M, FREIBERGER IA, MARYMONT JH Jr: *Clin Chem 13*:970, 1967
137. LOWE WC: *Clin Chem 14*:1074, 1968
138. LUBOCHINSKY B, ZALTA JR: *Bull Soc Chim Biol 36*:1363, 1954
139. MANN LT Jr: *Anal Chem 35*:2179, 1963
140. MARSH WH, FINGERHUT B, KIRSCH E: *Am J Clin Pathol 28*:681, 1957
141. MARSH WH, FINGERHUT B, MILLER H: *Clin Chem 11*:624, 1965
142. MARSHALL EK Jr: *J Biol Chem 15*:487, 1913
143. MARYMONT JH Jr, LONDON M: *Am J Clin Pathol 42*:630, 1964
144. MARYMONT JH Jr, LONDON M: *Clin Chem 10*:937, 1964

145. MASES P, FALET F, MARTINOT: *Rev Pathol Gen Physiol Clin 56*:641, 1956
146. McCLESKEY JE: *Anal Chem 36*:1646, 1964
147. McCULLOUGH H: *Clin Chim Acta 17*:297, 1967
148. McEVOY-BOWE E: *Anal Biochem 16*:153, 1966
149. McGOVERN JJ, McDERMOTT WV Jr, McGOVERN MN, RUSSELL M, McGRATH E, HOLTZ A: *Pediatrics 23*:1160,1959.
150. MEIJERS CAM, RUTTEN JCJM: *Clin Chim Acta 24*:308, 1969
151. MERCHANT AC, GOLDBERGER R, BARKER HG: *J Lab Clin Med 55*:790, 1960
152. MILLER BF, DUBOS R: *J Biol Chem 121*:447, 1937
153. MILLER GE, RICE JD Jr: *Am J Clin Pathol 39*:97, 1963
154. MILLER GL, MILLER EE: *Anal Chem 20*:481, 1948
155. MINARI O, ZILVERSMIT DB: *Anal Biochem 6*:320, 1963
156. MOMOSE T, OHKURA Y, TOMITA J: *Clin Chem 11*:113, 1965
157. MONDZAC A, EHRLICH GE, SEEGMILLER JE: *J Lab Clin Med 66*:526, 1965
158. MOORE GR: *J Med Lab Technol 27*:139, 1970
159. MOORE JJ, SAX SM: *Clin Chim Acta 11*:475, 1965
160. MOORE ML, WEST CD: *Am J Med Technol 28*:309, 1962
161. MYERS VC, FINE MS: *J Biol Chem 21*:583, 1915
162. NATELSON S: *Standard Methods of Clin Chem*, edited by M. Reiner. New York, Academic, 1953, Vol 1, p 123
163. NATELSON S, SCORT ML, BEFFA C: *Am J Clin Pathol 21*:275, 1951
164. NATHAN DG, RODKEY FL: *J Lab Clin Med 49*:779, 1957
165. NEWTON EB: *J Biol Chem 120*:315, 1937
166. NIIYAMA Y, DAIGAKU OS: *Igaku Zasshi 10*:565, 1961
167. NORTON DR, PLUNKETT MA, RICHARDS FA: *Anal Chem 26*:454, 1954
168. NOTHMAN MM: *Ann Internal Med 34*:1358, 1951
169. OFFER TR: *Centr Physiol 8*:801, 1894
170. OLBRICH O: *Edinburgh Med J 55*:100, 1948
171. ORESKES I, HIRSCH C, KUPFER S: *Clin Chim Acta 26*:185, 1969
172. ORLOFF MJ, STEVENS CO: *Clin Chem 10*:991, 1964
173. ORMSBY AA: *J Biol Chem 146*:595, 1942
174. OSER BL: *Hawk's Physiological Chemistry, 14th ed.* New York, McGraw-Hill, 1965, p 530
175. OWEN JA, IGGO B, SCANDRETT FJ, STEWART CP: *Biochem J 58*:426, 1954
176. PATEL CP: *Clin Chem 14*:764, 1968
177. PELLERIN J: *Clin Chem 10*:374, 1964
178. PETERS JP, VAN SLYKE DD: *Qualitative Clinical Chemistry.* Baltimore, Md., Williams and Wilkins, Vol 2, 1932
179. PILEGGI VJ, DI GIORGIO J, WYBENGA DR: *Clin Chem Acta 37*:141, 1972
180. POLAR E, METCOFF J: *Clin Chem 11*:763, 1965
181. POLLEY JR: *Anal Chem 26*:1523, 1954
182. PRAETORIUS E, POULSEN H: *Scand J Clin Lab Invest 5*:273, 1953
183. RALLS JO: *J Biol Chem 151*:529, 1943
184. RAPPAPORT F, EICHHORN F: *J Lab Clin Med 32*:1034, 1947
185. REARDON J, FOREMAN JA, SEARCY RL: *Clin Chim Acta 14*:403, 1966
186. REED AH, CANNON DC, WINKELMAN JW, BHASIN YP, HENRY RJ, PILEGGI VJ: *Clin Chem 18*:57, 1972
187. REICHELT KL, KVAMME E, TVEIT B: *Scand J Clin Lab Invest 16*:433, 1964
188. REIF AE: *Anal Biochem 1*:351, 1960
189. REIFENSTEIN EC Jr, ABRIGHT F, WELLS SL: *J Clin Endocrinol 5*:367, 1945
190. REINHOLD JG, CHUNG CC: *Clin Chem 7*:54, 1961
191. RICHTER H, LAPOINTE Y: *Clin Chem 8*:335, 1962

192. ROCKERBIE RA, RASMUSSEN KL: *Clin Chim Acta 15*:475, 1967
193. ROIJERS AFM, TAS MM: *Clin Chim Acta 9*:197, 1964
194. RUBIN M, KNOTT L: *Clin Chim Acta 18*:409, 1967
195. RUBNER M: *Z Biol 15*:115, 1879
196. RUSSELL JA: *J Biol Chem 156*:467, 1944
197. RYAN RJ, WILLIAMS JD, ANSELL BM, BERNSTEIN LM: *Metab Clin Exptl 6*:365, 1957
198. SAMBHI MP, GROLLMAN A: *Clin Chem 5*:623, 1959
199. SARDESAI VM, PROVIDO HS: *Microchem J 14*:550, 1969
200. SCHØNKEYDER F, CHRISTENSEN PJ: *Scand J Clin Lab Invest 9*:107, 1957
201. SEARCY RL, FOREMAN JA, KETZ A, REARDON J: *Am J Clin Pathol 47*:677, 1967
202. SEARCY RL, KOROTZER JL, DOUGLASS GL, BERGQUIST LM: *Clin Chem 10*:128, 1964
203. SEARCY RL, REARDON JE, FOREMAN JA: *Am J Med Tech 33*:15, 1967
204. SCHMIDT A, translated by CD Aaron: *Test-Diet in Intestinal Diseases.* Philadelphia, Davis, 1909
205. SEELY JR, PETITCLERC JC, BENOITON L: *Clin Chim Acta 18*:85, 1967
206. SELIGSON D, HIRAHARA K: *J Lab Clin Med 49*:962, 1957
207. SHAHINIAN AH, REINHOLD JG: *Clin Chem 17*:1077, 1971
208. SHAPIRO B, SELIGSON D, JESSAR R: *Clin Chem 3*:169, 1957
209. SHOHL AT, BUTLER AM, BLACKFAN KD, MACHACHLAN E: *J Pediat 15*:469, 1939
210. SIEST G: *Clin Chim Acta 18*:155, 1967
211. SINHA SN, GABRIELI ER: *Clin Chem 15*:879, 1969
212. SIROTA JH, BALDWIN DS, VILLARREAL H: *J Clin Invest 29*:187, 1950
213. SLOT C: *Scand J Clin Lab Invest 17*:381, 1965
214. STEARNS G: *Infant Metabolism,* edited by IH Scheinberg. 1950 Proc. World Health Organization. New York, MacMillan, 1956, p 157
215. STITCHER JE, JOLLIFF CR, HILL RM: *Clin Chem 15*:248, 1969
216. STONE WE: *Proc Soc Exp Biol Med 93*:589, 1956
217. STOTEN A: *J Med Lab Technol 25*:240, 1968
218. SULLIVAN MX, IRREVERRE F: *J Biol Chem 233*:530, 1958
219. SUNDERMAN FW, BOERNER F: *Normal Values in Clinical Medicine.* Philadelphia, Saunders, 1949
220. TALBOT JH, SHERMAN JM: *J Biol Chem 115*:361, 1936
221. TANZER ML, GILVARG C: *J Biol Chem 234*:3201, 1959
222. TAUSSKY HH: *Clin Chim Acta 1*:210, 1956
223. TAUSSKY HH: *J Biol Chem 208*:853, 1954
224. TAUSSKY HH: *Standard Methods of Clinical Chemistry,* edited by D Seligson. New York, Academic, 1961, Vol 3, p 99
225. TEGER-NILSSON AC: *Scand J Clin Lab Invest 13*:326, 1961
226. THOMPSON JF, MORRISON GR: *Anal Chem 23*:1153, 1951
227. TIERNEY NA, PETERS JP: *J Clin Invest 22*:595, 1943
228. TILLSON EK, SCHUCKARDT GS: *J Lab Clin Med 41*:312, 1953
229. TIMMERMANS CJ: *Clin Chim Acta 7*:887, 1962
230. TOBIAS GJ, McLAUGHLIN RF Jr, HOPPER J Jr: *New Engl J Med 266*:317, 1962
231. TOLKACHEVSKAYA NF: *Clin Chim Acta 1*:501, 1956
232. TROY RJ, PURDY WC: *Clin Chim Acta 27*:401, 1970
233. VAN PILSUM JF, SELJESKOG EL: *Proc Soc Exp Biol Med 97*:270, 1958
234. VANSELOW AP: *Ind Eng Chem Anal Ed 12*:516, 1940
235. VAN SLYKE DD: *J Biol Chem 71*:235, 1927

236. VAN SLYKE DD: *J Biol Chem* 73:695, 1927
237. VAN SLYKE DD, KUGEL VH: *J Biol Chem* 102:489, 1933
238. VARMA SD, YADAVA IS, CHAND D: *Z Klin Chem U Klin Biochem* 6:111, 1968
239. VON GUNDLACH G, HOPPE-SEYLER GF, JOHANN H: *Z Klin Chem Klin Biochem* 6:415, 1968
240. WALFORD RL, SOWA M, DALEY D: *Am J Clin Pathol* 26:376, 1956
241. WANG E: *Acta Med Scand* Suppl *105*:1939
242. WATSON D: *Clin Chim Acta* 14:571, 1966
243. WEARNE JT: *J Clin Pathol* 11:367, 1958
244. WEATHERBURN MW: *Anal Chem 39*:971, 1967
245. WELLS MG: *Clin Chim Acta* 22:379, 1968
246. WHEATLEY VR: *Biochem J* 43:420, 1948
247. WICKS LF: *J Lab Clin Med* 27:118, 1941
248. WOLLAEGER EE, COMFORT MW, OSTERBERG AE: *Gastroenterology* 9:272, 1947
249. WOLLAEGER EE, COMFORT MW, WEIR JF, OSTERBERG AE: *Gastroenterology* 6:83, 1946
250. WOLTHIUS FH: *Acta Med Scand 171*, Suppl 373: 1962
251. WONG SY: *J Biol Chem* 55:431, 1923
252. WONG TAO: *Anal Biochem 40*:18, 1971
253. WYBENGA DR, DI GIORGIO J, PILEGGI VJ: *Clin Chem* 17:891, 1971
254. YATZIDIS H, GARIDI M, VASSILIKOS C, MAYOPOULOU D, AKILAS A: *J Clin Pathol 17*:163, 1964
255. YÜ TF, GUTMAN AB: *J Clin Invest 38*:1298, 1959

Chapter 18

Amino Acids and Related Substances

FRANK A. IBBOTT, Ph.D.

During the last decade, there has been a dramatic increase in efforts to elucidate the pathways of amino acid metabolism. This was stimulated by the realization that many of the genetically inherited aminoacidurias could be explained in terms of membrane transport derangements or of deficiencies in specific enzyme activities. With the development of analytical procedures for the identification and quantitation of amino acids and their metabolites by chromatographic and chemical means, characterization of the various biochemical lesions became possible thus facilitating logical approaches to diagnosis and to therapy.

Many, but not all, of the disturbances of metabolism in amino acid pathways lead to elevations of compounds in the plasma and/or urine that are characteristic of the specific disease entity. Table 18-1 lists 21 inborn errors of amino acid metabolism in which clinical laboratory procedures are helpful if not essential for accurate diagnosis as well as for control of therapy (214).

Laboratory methodology available for the investigation of the aminoacidopathies may be subdivided into the measurement of (a) total α-amino acid nitrogen, (b) one or more specific amino acids, and (c) metabolites of amino acids. A number of forms of liver disease including acute necrosis, congenital cirrhosis, Wilson's disease, and galactosemia, as well as renal tubular dystrophies, may be accompanied by excessive concentrations of a number of amino acids in the plasma or by a gross generalized aminoaciduria. A measure of the magnitude of this increase is obtained by determining the total α-amino acid nitrogen. On the other hand, diseases in which a single amino acid is affected are best monitored by the assay of that specific substance since, although the change in its concentration may be dramatic, the effect on the total amino nitrogen value may be insignificant.

α-AMINO ACID NITROGEN

Methods for the determination of total amino nitrogen involve the acceptance of a compromise. While all the α-amino acids possess certain characteristic structural groups, they are sufficiently diverse to react differently from one another in specific analytical systems. For an extensive review of methodology up to 1962, the reader is referred to the first edition of this book.

TABLE 18-1. Disorders of Amino Acid Metabolism

Amino acid disorder	Other names	Deficiency	Amino acids elevated in		Metabolites of interest
			Plasma	Urine	
β-Alaninemia		β-Alanine transaminase	BAIB[a], GABA[b],	BAIB[a], GABA[b], taurine	
Alcaptonuria	Ochronosis	Homogentisic acid oxidase (EC 1.13.1.5)	None	None	Homogentisic acid
Argininemia		Arginase (EC 3.5.3.1)	Arginine	Lysine, cystine, cysteine-homo-cysteine disulfide	Blood ammonia
Argininosuccinic aciduria		Argininosuccinase (EC 4.3.2.1)		Argininosuccinic acid	Blood ammonia
Citrullinuria		Argininosuccinic acid synthetase (ASAS) (EC 6.3.4.5)	Citrulline	Citrulline	Blood ammonia
Cystathioninuria		Cystathionase (EC 4.2.1.15)		Cystathionine	
Cystinuria		Transport		Cystine, arginine, lysine, ornithine	
Glycinemia		Glycine oxidase	Glycine	Glycine (during attacks: leucine, valine, threonine)	
Hartnup disease		Transport	Alanine, serine, threonine, asparagine, glutamine, valine, leucine, isoleucine, phenylalanine, tyrosine, tryptophan, histidine, citrulline		Indican, indoleacetic acid Indolylacryloyl-glycine
Histidinemia		Histadase (EC 4.3.13)	Histidine	Histidine	FIGLU[c]
Homocystinuria		Cystathionine synthetase (EC 4.2.1.13)	Methionine	Homocystine	

Hydroxyprolinemia Hypophosphatasia Lysinemia	Hydroxyproline oxidase ?	Hydroxyproline Phosphoethanolamine Lysine	Hydroxyproline Phosphoethanolamine Ornithine, GABA[b], ethanolamine
Maple syrup urine disease	Branched chain keto acid decarboxylase	Valine, leucine, isoleucine, alloisoleucine	Leucine, isoleucine, valine — Keto acids in urine
Phenylketonuria	Phenylalanine hydroxylase (EC 1.14.3.1)	Phenylalanine	Phenylalanine — Phenylpyruvic acid, Phenyllactic acid, Phenylacetylglutamine
PKU, Phenylpyruvic oligophrenia			
Prolinemia, type I	Proline oxidase (EC1.5.1.2)	Proline	Proline, hydroxyproline, glycine
Prolinemia, type II	Δ^1-Pyrroline-5-carboxylic acid dehydrogenase	Proline	Proline, hydroxyproline, glycine
Saccharopinuria	Lysine-ketoglutarate reductase	Saccharopine, homocitrulline, lysine, citrulline	Saccharopine, lysine, citrulline, histidine, homocitrulline, homoarginine, aminoadipic
Sarcosinemia	Sarcosine dehydrogenase ? (EC 1.5.1.3)	Sarcosine	Sarcosine
Valinemia	Valine transaminase	Valine	Valine

[a] BAIB β-Aminoisobutyric acid
[b] GABA γ-Aminobutyric acid
[c] FIGLU Formiminoglutamic acid

There is probably agreement between clinical chemists that the gasometric ninhydrin method of Van Slyke is still the ultimate reference for the quantitation of α-amino acids in urine (241), plasma (150), or blood filtrate (93). The general requirement that results be obtained rapidly and the virtually complete disappearance of the Van Slyke manometric apparatus from the clinical laboratory have combined to make this a method that is often preached but rarely practiced.

Many of the authors of modern approaches to the detection and quantitation of amino acids have employed the well-known color reaction with ninhydrin. A single drop of urine placed on base-impregnated filter paper and allowed to dry will be freed of the potential interference of ammonia and remain stable for several days. Subsequent dipping in a ninhydrin reagent followed by heating at 100°C causes color to appear, the intensity of which can be compared with similarly treated standards (111). An accurately measured volume of deproteinized plasma can be treated similarly and placed into a ninhydrin reagent at 100°C and the resultant color read in a photometer. Although the reagent recommended is nonaqueous, which has the advantage that urea does not react, the color yields of the various amino acids are more variable than in an aqueous reagent (235). Khachadurian *et al* (122), by using suboptimal conditions for the ninhydrin reaction, developed a method for urinary amino nitrogen in which urea, ammonia, and even small amounts of protein cause no interference. Removal of urea and ammonia can be effected by urease followed by drying in the presence of alkali (43), or by selective adsorption with Dowex-2 resin (153) prior to color development.

Ninhydrin reagents have been formulated in several ways. The preparation reduced with stannous chloride must be maintained at 4°C and under N_2 for stability (43), whereas that dissolved in ethylene glycol mono methyl ether (Methyl Cellosolve) and stored at 4°C includes cyanide (153) or phenol (63) in the reaction mixture.

Ninhydrin reactions have also been applied successfully to cerebrospinal fluid (247, 248).

The method first described by Pope and Stevens (190), in which amino acids chelate copper from its insoluble phosphate, has been modified by various workers with respect to quantitation of the solubilized copper. Whereas the original procedure concluded with an iodometric titration, photometric assays of the chelated copper have been described using tetraethylenepentamine (39) and 2,9 dimethyl-1,10-phenanthroline (Neocuproine) (245). Atomic absorption spectrophotometry is also not only suitable but is regarded as the most reliable step in the procedure (84).

In the presence of divalent cobalt ions, amino acids react with Folin and Ciocalteu reagents but show considerable variation in the intensity of color production and linearity of reaction (152).

Procedures using sodium β-naphthoquinone-4-sulfonate have been used successfully for several years (64, 222). Criticisms of this method lie mainly in the amount of technician time required, the critical nature of several of the timed steps and the tendency to produce a turbid final solution. The introduction of the reaction between amino acids and 1-fluoro-2,4-dinitrobenzene to form yellow dinitrophenyl (DNP) amino acid derivatives (206) has facilitated the development of reliable and rapid procedures for plasma (77, 193) and urine (76). An automated method has also been described using 2,4,6-trinitrobenzene sulfonate (176).

DETERMINATION OF AMINO ACID NITROGEN IN PLASMA AND URINE

(method of Goodwin, Refs. 76 and 77)

PRINCIPLE

Amino acid nitrogen is determined by photometric measurement of the yellow color produced when amino acids react with 1-fluoro-2,4-dinitrobenzene (FDNB) (see Fig. 18-1).

The absorption spectra given by different dinitrophenyl amino acids all show maxima in the vicinity of 356 nm. The imino acids are distinguished since their maxima occur at 370 nm. The absorbances of the DNP-amino and imino acids are similar at 420 nm.

The molar absorbances of the final products derived from the majority of the commonly found amino acids vary from 0.88 to 1.06 (glycine = 1.00). Exceptions are asparagine (0.86), lysine (1.97), ornithine (2.23), and tryptophan (1.09). For plasma, the reaction is applied to a protein-free filtrate using a hydrochloric acid-tungstate reagent.

REAGENTS

Phenolphthalein, 100 mg/100 ml ethanol. Ethanol denatured with methanol is satisfactory.

HCl, 1 N. Make a 1:12 dilution conc HCl with water.

NaOH, approx. 0.2 N. Dissolve 8 g NaOH in approximately 800 ml water and make up to 1 liter with water.

Stock Standard. Dissolve 1.05 g glutamic acid and 0.536 g glycine in approximately 100 ml water. Add 2.0 g sodium benzoate and 700 ml 1.0 N HCl and make up to 1 liter with water. Stable in refrigerator. 1 ml = 0.2 mg amino N.

Working Standard. Dilute 2.5 ml stock solution to 100 ml with water. Stable in refrigerator. 1 ml = 5 µg amino N.

Sodium tetraborate, 0.132 M. Dissolve 50 g $Na_2 B_4 O_7 . 10H_2 O$ in 1 liter water and allow to stand overnight before use. Stable at room temperature.

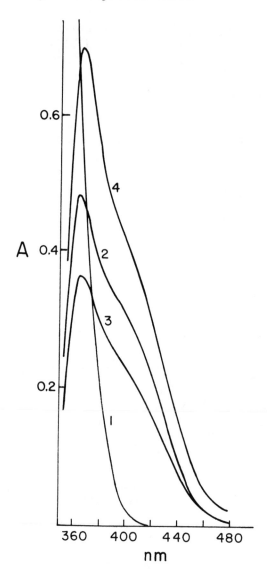

FIG. 18-1. Absorption spectra of dinitrophenylamino acids: (1) blank vs water, (2) mixed glycine–glutamic acid standard, 5 μg α-amino acid N vs reagent blank, (3) plasma vs reagent blank, (4) urine vs reagent blank. Beckman Model DBG recording spectrophotometer.

1-Fluoro-2,4,dinitrobenzene (DNFB). Eastman Kodak liquid preparation No. 6587.

Stock DNFB. Dissolve 0.65 ml in 50 ml acetone and store at 4°C. Stable for 2 months.

Working DNFB. Dilute stock solution 1:10 with $Na_2B_4O_7$ solution. Prepare immediately before use.

Acidified Dioxane. Add 10 ml conc HCl to 500 ml dioxane.

Sodium Tungstate, 13.3%. Dissolve 13.3 g sodium tungstate in approximately 80 ml water and make up to 100 ml with water.

HCl, 0.11 N. Dilute 110 ml 1.0 *N* HCl to 1 liter with water.
NaCl, 0.85%
Protein Precipitant. Mix 1 part sodium tungstate reagent with 9 parts
0.11 *N* HCl. Prepare fresh daily.

PROCEDURE FOR URINE

The determination should be performed on a 24 hour collection.

1. Measure the total urine volume, mix, and transfer 1 ml to a 100 ml beaker containing a few boiling beads.
2. Add 10 to 15 ml water, 2 to 4 drops phenolphthalein and sufficient 0.2 *N* NaOH (approximately 1 ml depending on pH of urine) to obtain a pink color.
3. Boil the diluted urine gently on a hot plate for at least 10 min, adding 0.2 *N* NaOH as necessary to retain pink color.
4. Cool the sample and bring volume to approximately 15 ml with water. Filter through Whatman 1 filter paper and make up to 20 ml with water. (If 24 hour urine volume is less than 500 ml make up to 40 ml with water).
5. Transfer 1 ml filtrate to a test tube marked "unknown." Transfer 1 ml water and 1 ml working standard to separate tubes marked "blank" and "standard," respectively.
6. Add 1 ml working DNFB solution to all tubes and heat in a 70°C water bath for 15 min.
7. Cool mixture and add 5 ml acidified dioxane to each tube. Mix thoroughly.
8. Read all absorbances versus reagent blank at 420 nm. If absorbance of the unknown is greater than 0.8, dilute the final color with reagent blank and multiply final result by dilution factor.

Calculation:

$$\text{mg } a\text{-amino acid N/24 hr} = \frac{A_x}{A_s} \times 5 \times 20 \times \frac{24 \text{ hour volume in ml}}{1000}$$

$$= \frac{A_x}{A_s} \times 0.1 \times 24 \text{ hour volume in ml}$$

If 24 hour urine volume is less than 500 ml and sample is diluted to 40 ml in step 4, multiply final result by 2.

PROCEDURE FOR PLASMA

1. Transfer 0.5 ml plasma and 0.5 ml saline to a test tube.

2. Add 4 ml protein precipitant and mix thoroughly. Centrifuge 15 min.
3. Filter through glass wool and proceed as for urine, following steps 5-8.

Calculation:

$$\text{mg } a\text{-amino acid N/100 ml} = \frac{A_x}{A_s} \times 5 \times 5 \times \frac{100}{0.5} \times \frac{1}{1000}$$

$$= \frac{A_x}{A_s} \times 5$$

NOTES

1. Beer's law. The color follows Beer's law up to at least 5 μg amino acid nitrogen per ml when read on narrow band spectrophotometers. Linearity is not obtained using filter and wide band instruments and a multipoint calibration curve should be processed simultaneously with the unknowns in these circumstances.

2. Stability of samples. Protein-free filtrates of plasma reportedly are stable for several days in the refrigerator (64). Plasma is stable for long periods of time if frozen (203) but at refrigerator temperature elevations of amino nitrogen concentration occur after 48 hours (77). It has been found in our laboratory that the instability occurring after 8 hours at 30°C may be overcome by the addition of 5 mg NaF per ml plasma so that satisfactory results may be obtained for 5 days.

Urine appears to be stable at room temperature for about 4 days without preservative (76, 222), although an additional precaution is to acidify to pH 2-3 with HCl. At refrigerator temperature unpreserved specimens are stable at least 5 weeks (222).

3. Plasma vs serum. It has been reported by several authors that serum yields higher values than plasma for amino nitrogen (63, 77, 93, 150) presumably due to the release of amino acids during the clotting process.

4. Hemolysis. While purified hemoglobin up to a concentration of 2 g/100 ml does not affect the determination of plasma amino nitrogen by the FDNB method, hemolysis liberates erythrocyte amino acids into plasma and causes false elevations of the level (77). For this reason hemolyzed samples should not be used.

5. Optimum conditions. The conditions for optimum color development are a pH of about 9.5, temperature 70°C, and heating time 15 min. The time and temperature do not appear to be critical.

6. Stability of final color. The final color is stable up to 5 hours.

ACCURACY AND PRECISION

The intensity of the final color varies with each amino acid and bears a relationship to the number of primary amino groups present. Lysine and ornithine produce 1.97 and 2.23 times as much color as glycine, respectively, whereas the terminal amino groups of arginine and citrulline do not appear to react. For this reason the method cannot be claimed to be accurate even though the dominant amino acids of plasma and urine produce quantities of color in the range 0.88–1.06 times that of glycine.

While NH_4^+ will react with FDNB if not removed, several important naturally occurring substances showed no interference even when present in relatively high concentrations. Examples are bilirubin (38 mg/100 ml), urea (4000 mg/100 ml), creatinine (800 mg/100 ml), and uric acid (400 mg/100 ml).

Reproducible results were not obtained using Folin-Wu precipitants (77), a fact also noted by other authors with respect to the ninhydrin method (63, 153).

Ascorbic acid in a concentration of 400 mg/100 ml produced color equivalent to 1.8 mg glycine per 100 ml. Albumin also interfered to the extent that 134 mg/100 ml and 368 mg/100 ml appeared as 3.7 and 5.0 mg amino nitrogen per 100 ml, respectively.

The precision (95% limits) is about ±7%.

NORMAL VALUES

The adult normal range of plasma amino nitrogen by the FDNB method is 3.6–7.0 mg amino nitrogen/100 ml based upon samples from 20 subjects aged 18–50 years (77). A group of normal children showed a range of 3.6–5.2 mg/100 ml plasma. Rapp (193) also found a range of 3.4–6.0 mg/100 ml.

For unknown reasons the mean value by the FDNB method is approximately 1 mg/100 ml higher than that given by a ninhydrin technic (77). The normal range given by the β-naphthoquinone method is 4-6 mg/100 ml plasma.

The FDNB procedure for urine correlates well with the β-naphthoquinone method (76), a fact that was confirmed in our laboratory. The adult range of normal is 50–200 mg/24 hour (99). Our laboratory observed a normal range of 0.5–2.8 mg/lb/24 hour for children aged 2.5–12 years. This is in close agreement with the values reported by Goodwin (76) of 1.20–5.60 mg/kg/24 hour for children aged 3–12 years. The normal output per unit body weight appears to be higher in infants, a range of 3.80–6.50 mg/kg/24 hour being reported (76).

AMINO ACID FRACTIONATION

Following the recognition that several of the inherited errors of amino acid metabolism result in severe sequelae including mental retardation and death, attempts were made to identify the affected families with a view to offering genetic counselling and anticipating the births of affected infants.

These studies became possible as clinical laboratories developed the methodology and skill to screen for these abnormalities and to confirm the findings by quantitative, specific procedures.

As a result, massive screening projects have been undertaken (33, 234, 252) with the majority of the States testing every newborn at least for phenylketonuria.

SCREENING TESTS FOR THE AMINOACIDOPATHIES

To be used on an extensive scale, screening tests for amino acid metabolic disturbances must be simple, cheap, and easy to perform. The samples required should be small, readily collected, and stable during shipment to the testing laboratory. The tests should be oversensitive with a tendency to false positive results with very few false negative findings, thus ensuring identification of nearly all abnormal samples (112).

The amino acid fractionation procedures in current use that are suitable for rapid screening purposes utilize chromatography or high voltage electrophoresis (HVE). High voltage electrophoresis is reported to be a rapid and satisfactory method for the separation and identification of amino acids. Plasma is extracted with ethanol and an appropriate volume applied to Whatman 3MM paper or to Whatman 1 paper. Following HVE, the papers are dried, stained and evaluated. A ninhydrin-copper nitrate reagent has been recommended as a general detection agent, but other reagents are available for the demonstration of certain specific amino acids. The amino acid bands may then be quantitated by use of densitometry. Precise control of development conditions, particularly low humidity, is required for good quantitation. Comparison of the color given by each band in the sample with the appropriate amino acid in a standard mixture is necessary (147, 170). A similar procedure has been recommended for urine using strips of Whatman 3MM paper (177).

High voltage electrophoresis followed by chromatography in the second dimension has also been recommended. Twenty to 40 μl volumes of urine may be applied in a band near one corner of a thin layer comprised of a mixture of cellulose and silica gel. Following HVE, water is allowed to ascend the layer for a short distance at 90° to the electrophoresis to condense the bands to spots. Then chromatography is carried out to separate further the amino acids (238). A similar procedure has been described using a 5 min HVE period on cellulose thin layers followed by a 15 min chromatography separation in the second direction (205).

Two-dimensional chromatography on paper has also been recommended. Plasma is passed through a Sephadex micro column to separate the amino acids from protein, the eluate desalted and lyophilised. An acid solution of the extract is then applied to the paper (155).

One dimensional paper chromatography systems have been described, especially for the detection of amino acid disorders in the newborn. Serum with or without preliminary precipitation of proteins may be used (36). This method is particularly useful for the detection of phenylalanine and tyrosine. A convenient method for the newborn consists of collecting blood on S and S 903 filter paper whereupon samples remain stable at room temperature for some time. Disks are punched out from each specimen and inserted into identically sized holes in a sheet of Whatman 3MM 20 x 20 cm chromatography paper. An isopropanol developing solvent elutes amino acids but not hemoglobin from the disks. The filter paper sheets are then dried and the blood collection disks removed. The sheets are then developed in the usual way for the amino acid separation (234).

A rapid procedure for screening newborns for phenylketonuria employs single dimension chromatography on paper and requires 2 hours to complete (219). Similar methodology has been applied to the detection of disorders of histidine metabolism by use of an appropriate solvent system and spray reagent (15). By densitometry and planimetry of the spots, it is possible to achieve quantitation that agrees well with column amino acid chromatography. In addition, the histidine metabolites, imidazole pyruvic acid and imidazole lactic acid, can be detected thus giving further information on a metabolic block in histidine metabolism.

Single-dimension thin layer chromatography on cellulose (45) or on Silica Gel G (197) also have been recommended as being more rapid and sensitive and giving better resolution than paper. The use of cyclohexylamine has been described as yielding improved amino acid separation on cellulose plates with extremely short development times (160, 161).

Two-dimensional thin layer chromatography for the quantitation of urinary amino acids has been reported. Urine is first passed through an ion-retardant resin column to remove salts and peptides. Taurine and urea are lost from the column, but the other amino acids are applied to a 20 x 20 cm thin layer of cellulose and separated. Following visualization with a ninhydrin-cadmium reagent, quantitation is achieved by densitometry (160).

The procedure described below is convenient and is sufficiently sensitive to show the gross changes in amino acid pattern associated with many of the inherited disorders of amino acid metabolism (97).

SCREENING METHOD FOR AMINO ACIDS IN PLASMA AND URINE

(method of Kraffcyzk et al, Ref. 129)

PRINCIPLE

Plasma or urine is chromatographed on precoated cellulose thin layer plates. The separated amino acids are made visible by reaction with ninhydrin. Elevated concentrations of amino acids are detected by comparison with the bands of a normal control on the chromatogram. Citrulline, homocitrulline, hydroxyproline, and proline are made visible with *p*-dimethylaminobenzaldehyde.

REAGENTS

Chromatography Solvents:
   A. *n*-Butanol:acetone, 1:1.
   B. Glacial acetic acid: water, 1:2.

Ninhydrin Reagent. Dissolve 7 g ninhydrin in 50 ml of solvent "A". Mix vigorously to dissolve. This reagent is stable at least 3 months in the refrigerator.

p-Dimethylaminobenzaldehyde Spray Reagent. Dissolve 10 g *p*-dimethylaminobenzaldehyde in 100 ml conc HCl. Mix 1 vol with 4 vol acetone and use immediately.

Cellulose Thin Layer Plates. 20 × 20 cm E. Merck #5537. Cut the cellulose coated aluminum sheets to 10 × 20 cm.

PROCEDURE

1. Mark each cellulose plate with a soft pencil and plastic ruler as follows (also see Fig. 18-2). The cellulose layer must not be touched with fingers.

   a. Lightly mark the plate 10 mm from the bottom to demarcate the sample 20 mm application lanes.

   b. Mark a finishing line (solvent front) 85 mm from the bottom.

2. Check each urine with pH paper. Samples must be between pH 4 and 7, otherwise adjust 1 ml aliquot to this range with either 1 N HCl or NaOH.

3. Apply 2 μl plasma or 5 μl urine along a 20 mm application line. Use a hair dryer to facilitate drying. To facilitate interpretation it is advisable not to mix plasma and urine samples on one plate. Include a normal serum or urine control, as appropriate, on each plate.

4. Write sample number above the finish line.

5. Mix 56 ml solvent "A" with 24 ml solvent "B" and add entire volume to chromatography tank No. 1. Cover tank with glass top.

6. For the first development, place cellulose plates into a holder and insert evenly into chromatography tank No. 1. Cover and place lead weights on top.

7. Allow solvent to migrate to finish line (approximately 45 min).

8. Raise holder, allow excess solvent to drain off, and place chromatograms in a 60°C forced draft oven for 10 min. Remove and allow to cool. Acetic acid odor should be faint or absent.

9. For the second development place chromatograms in chromatography tank No. 2 containing 56 ml solvent "A", 24 ml solvent "B", and 6 ml ninhydrin reagent. Allow solvent to migrate to the finish line.

10. Raise holders, drain and dry in 60°C forced draft oven for 10 min.

11. Place plates and holders in a 100°C forced draft oven for *exactly 5 min.*

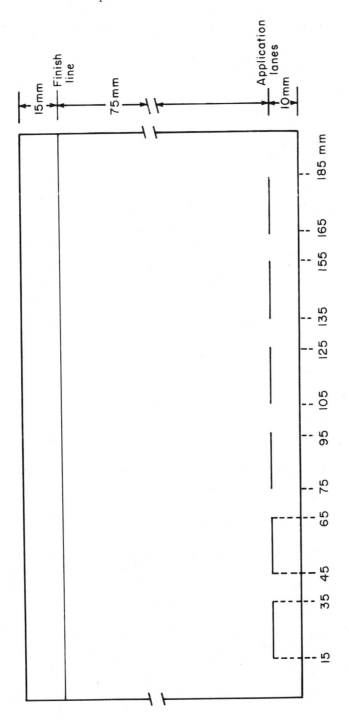

FIG. 18-2. Thin layer chromatography of amino acids. Layout of sample application sites.

12. Remove chromatograms from oven and allow to cool to room temperature.

13. The ninhydrin-development chromatogram plates are examined within 1 hour after color development. Record results (see note 1).

14. Mix *p*-dimethylaminobenzaldehyde with acetone (approximately 10 ml required per chromatogram) and use immediately to spray chromatograms.

15. Heat chromatograms 10 min at 100°C. Remove, allow to cool and examine for the following results:

Compound	Color	Migration distance mm
Citrulline	Red	16−20
Homocitrulline	Red	18−22
Hydroxyproline	Purple (intense)	20−22
Proline	Purple (weak)	33−35

16. Record results (see note 1).

NOTES

1. Evaluation. The absolute migration distances for the amino acids are reasonably reproducible but do show some variation, especially for urine samples. The sequence of the amino acids in the respective major bands and migration distances are shown in Figure 18-3.

2. Interferences. Normally, fresh plasma contains only small concentrations of glutamic acid. On standing, glutamic acid is formed by hydrolysis of glutamine (51) with a resultant increase in the intensity of the glutamic acid.

Urine samples acidified to pH 2 or below give anomalous migration results.

Since sweat contains amino acids, thin layer plates should be handled so that the chromatogram is not touched by the fingers (89). Plates should be held by the edges or the undersides.

Samples submitted for analysis on subjects receiving drugs and medications may produce anomalous findings. Fecal contamination of urine samples may increase the free amino acid content of the sample to the extent that the chromatogram suggests the presence of hyperaminoaciduria. Many amino acids will be increased including proline but hydroxyproline is not (141).

3. Stability. The amino acids of urine acidified to pH 3−5 with HCl and of heparinized plasma appear to be sufficiently stable at 30°C for 4 days to give useful information for screening purposes. If there is a likelihood that the sample will require more precise analysis, the appropriate precautions should be taken as described in the subsequent section on amino acid fractionation by column chromatography.

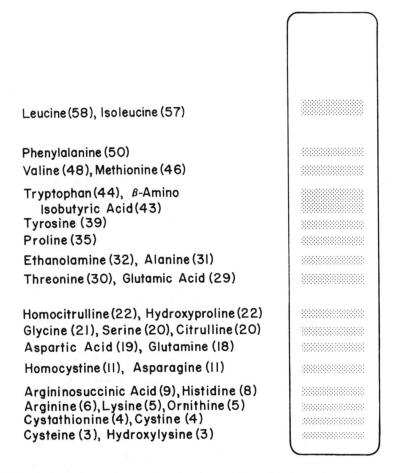

Leucine (58), Isoleucine (57)

Phenylalanine (50)
Valine (48), Methionine (46)
Tryptophan (44), β-Amino
    Isobutyric Acid (43)
Tyrosine (39)
Proline (35)
Ethanolamine (32), Alanine (31)
Threonine (30), Glutamic Acid (29)

Homocitrulline (22), Hydroxyproline (22)
Glycine (21), Serine (20), Citrulline (20)
Aspartic Acid (19), Glutamine (18)

Homocystine (11), Asparagine (11)

Argininosuccinic Acid (9), Histidine (8)
Arginine (6), Lysine (5), Ornithine (5)
Cystathionine (4), Cystine (4)
Cysteine (3), Hydroxylysine (3)

FIG. 18-3. Regions of migration of amino acids on thin layer chromatogram. (Migration distances expressed in mm in parentheses.)

INTERPRETATION

The appearance of one or more intense bands in comparison to the rest of the bands of the other samples on a chromatogram sheet is suggestive of a defect in amino acid metabolism. Abnormalities are more readily detected in plasma sample chromatograms because of the uniformity in the number and intensity of the bands. Normal urine samples give varying chromatographic patterns, but in most cases of hyperaminoaciduria, the abnormal bands are quite distinct.

Citrulline, homocitrulline, and hydroxyproline are easily detected with the p-dimethylaminobenzaldehyde spray. Proline is detectable only at high concentrations.

With few exceptions plasma offers greater sensitivity than urine for detection of primary aminoacidopathies, because urine reflects only secondarily the initial changes in plasma composition. The principal exceptions are argininosuccinicaciduria, homocystinuria, and cystathioninuria. In these cases the renal clearance of the substance is so high that urine is the preferable source for diagnosis.

All urine samples with suspected elevations in phenylalanine (band No. 2) should be tested for phenylpyruvic acid.

AMINO ACID FRACTIONATION BY COLUMN CHROMATOGRAPHY

The fractionation of amino acid mixtures on columns of Dowex-50 sulfonated polystyrene resin was reported by Moore and Stein in 1951 (168). Amino acids were eluted with a series of buffers, collected in 1 ml aliquots, and quantitated by the ninhydrin reaction (167):

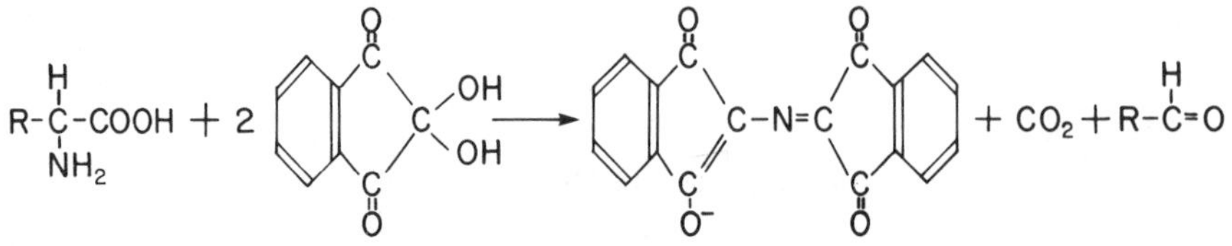

Since this first publication, a great deal of energy has been devoted to the investigation of the factors that influence the quality of the final chromatogram. These parameters include resin composition, degree of cross linking, particle size, shape, and uniformity; column length and diameter; single versus dual columns; buffer composition, pH gradient or abrupt increases of pH, pressure and flow rate, elution temperature programming; and reaction of ninhydrin with other chromogens. Efforts have been directed to achieve fully automated, accelerated systems that can be loaded with several samples, preferably of microliter volumes, to yield identified and completely calculated amino acid results on a computer printout in as short a time as possible.

In analytical column chromatography of solute mixtures, the hallmark of acceptability is usually characterized by obtaining a series of well separated, narrow peaks, one for each component in the minimum of time. Some of the factors that can affect such performance are discussed below.

No specific method is described as procedure will vary considerably with the particular equipment available. In general, manufacturer's instructions should be followed or a suitable recommendation selected from the literature. Nevertheless, some "notes" are given concerning factors common to all technics of ion-exchange chromatography of amino acids.

NOTES

Ion exchange resins. Columns of starch, as used for the original chromatographic separations of amino acids, had disadvantages that were quickly realized. Substitution of sulfonated polystyrene resin Dowex-50 eliminated these difficulties (168). Later, other cationic exchange materials with 7.5%–8.5% cross linkage were introduced for this purpose, examples of which are Amberlite IR 120 (Rohm and Haas) (92, 209) and Aminex A-7 (BioRad Laboratories) (90). While the earlier separations were carried out on columns packed with pulverized resin particles (92) there has been a transition to the use of homogeneous spherical resin particles of small mean diameter (13, 87, 90).

Effect of resin particle size. The range of resin particle diameters which is utilized in amino acid chromatography is up to 30 μ. The advantage of using smaller particle sizes has been shown mathematically and experimentally by Hamilton *et al* (87, 94). A decrease in particle diameter of 20% results in peaks which are about 13% narrower with a corresponding increase in height (91). Authors have tended with time to prefer smaller diameter resin particles, 40 μ (92), 20–40 μ (87), 17.5 ± 2 μ (88), 16 ± 6 μ (13), and 10 ± 1 μ (90), although this in turn creates a pressure increase within the system. When column length and diameter are defined, pressure varies with the reciprocal of the square of the resin particle diameter at a constant buffer input rate. Pressure is also proportional to column length when the other parameters are held steady so that small diameter particles in a short column can give the same pressure as larger particles in a longer column (91). Consequently, once the resin type and particle size have been selected to achieve acceptable resolution, equipment design must be directed to cope with the dimensions of pressure and flow rate.

Buffer systems. Moore and Stein (168) recommended a series of sodium citrate buffers of increasing pH values for elution of amino acids from Dowex 50. Succeeding workers appear to have found this system generally satisfactory, although myriad modifications have been made in the pH values of individual buffers and in the times that they are applied to the column.

An important change occurred when Peterson and Sober developed their pH gradient-producing device, the Varigrad (188). This was used successfully by Piez and Morris (189) and enabled them to depart from the two-column method of Spackman *et al* (225) to a one-column system that offers certain economies in operator time, a lower construction cost and simplified maintenance.

Another advance was made with the change from sodium to lithium citrate buffers. The latter were mentioned by Moore and Stein (168) in connection with an improved separation of certain amino acids at the expense of a less complete resolution of alanine and glycine. Benson *et al* (13) described a system using lithium citrate buffers in which glutamine could be separated from asparagine without loss of overall resolution (187). This separation is not possible with buffers containing sodium unless there is a concurrent sacrifice of resolution of other amino acids.

Sample preparation. Protein, in amounts greater than traces, interferes with the quantitative fractionation of free amino acids and several methods have been described for its removal. Picric acid was recommended as a suitable protein precipitant by Hamilton and Van Slyke (93) but, although it is an efficient precipitating agent, it is eluted in the amino acid chromatogram as a very large peak interfering with taurine and urea (51). Consequently, the picric acid should be removed from the sample by passage

TABLE 18-2. Normal Amino Acid Values in Plasma

Unit: μmol/100 ml

Age	Full term (Cord Blood)	2 days	5 days	9 days	16 days–4 mos	4 mos–2½ yrs	Pre-pubertal Children	Adults
References	70	70	70	70	30	18, 70	209	1, 51, 181, 183
Phosphoethanolamine	0–3.4	0–3.5	0–2.8	0.6–1.2		Tr	Tr–1.1	Tr–0.3
Taurine	6.0–27	3.1–24	0–18	0–18		1.0–12	5.7–12	3.5–14
Hydroxyproline							Tr–2.5	
Aspartic acid	0–5.1	0–4.6	0–4.1	0–3.2	1.4–2.3	0.7–1.9	0.4–2.0	1.1–5.4
Threonine	16–45	4.7–26	4.5–32	14	11–23[a]	5.0–18	4.2–9.5	7.5–25
Asparagine							5.7–47[b]	2.6–8.6
Glutamine	11–150	54–100	29–100	43–130		4.6–140		42–76
Serine	6.9–22	7.1–30	0–37	4.3–23	7.6–19	4.4–17	7.9–11	6.1–19
Glutamic acid	0–19	0–25	2.6–18	0–32		0–22	2.3–25	0–12
Proline	8.1–18	3.4–58	16–41	11–43	9.0–30	2.1–30	6.8–15	8.9–44
Citrulline	0–4.9	0.6–2.5	0.3–4.2	0–3.1		0.6–4.1	1.2–3.0	1.2–5.5
Glycine	12–34	20–44	0.7–46	6.0–35	14–28	10–31	12–22	13–49
Alanine	22–35	21–40	9.7–57	1.1–77	19–40	9.9–41	14–30	17–50
a-Amino-n-butyric acid	0–10	0–4.7	1.1–3.4	0.1–5.5		0.9–4.8	0.9–3.8	0.8–3.5
Valine	14–24	9.0–17	7.4–38	15–34	8.5–24	16–28	13–28	12–33
Cystine	2.9–3.7	2.8–6.8	1.8–6.0	1.2–5.0	2.4–5.9	0–4.5	4.5–7.7	3.1–14
Methionine	1.4–2.5	0.4–4.4	0.7–4.7	1.9–3.0	1.3–2.3	0–2.9	1.1–1.6	1.3–3.9
Isoleucine	3.3–7.9	1.5–8.1	2.1–12	5.0–10	2.3–5.6	2.4–13	2.8–8.4	3.5–10

Leucine	7.0–12	0.7–14	5.1–22	1.9–20	3.5–12	3.4–22	5.6–18	6.9–16
Tyrosine	0–17	10–17	5.3–25	1.7–24	1.2–9.6	1.1–12	3.1–7.1	3.2–8.7
Phenylalanine	3.3–7.9	0–16	0–18	2.9–11	3.3–7.2	0.6–10	2.6–6.1	3.4–12
β-Aminoisobutyric acid						0–2.2	0–Tr	
Ethanolamine								Tr–1.1
γ-Aminobutyric acid	0–5.3	0–9.4	0–5.3	0–3.8		Tr		
Ornithine	6.4–12	1.6–27	9.4–21	2.2–16	2.9–7.1	1.0–20	2.7–8.6	3.0–13
1-Methyl histidine							0–2.0	
Lysine	25–54	4.2–40	8.1–40	8.0–40	7.9–19	4.6–41	7.1–15	9.0–26
Histidine	0–26	1.0–17	0–21	3.8–13	5.0–11	2.4–13	2.4–8.5	5.6–12
3-Methyl histidine							0–Tr	
Arginine	0.6–21	0–14		0.4	4.3–8.2	0.5–15	2.3–8.6	4.6–15

[a] Includes asparagine.
[b] Includes glutamine.

TABLE 18-3. Normal Amino Acid Values

Sample	Urine	Urine	CSF	Aqueous fluid (eye)
Age	9 mos–2 yrs	Adults		
Refs	18	34, 47, 224, 228	182, 183	50
Units	μmol/24 hour	μmol/24 hour	μmol/100 ml (mean)	μmol/kg H_2O (1 case)
Phosphoethanolamine		26–101	0.44	
Taurine	18–110	63–2300	0.64	40
Aspartic acid				Trace
Threonine	26–74	85–440	3.2	160
Asparagine		270–700	0.74	180 (includes serine)
Glutamine	>12–>393	140–860	58	600
Serine	65–120	160–700	2.4	180 (includes asparagine)
Glutamic acid		55–270	0.17	20
Proline		<100		20
Citrulline			0.27	3
Glycine	140–560	160–4200	0.58	20
Alanine	64–160	60–800	3.3	230
α-Amino-n-butyric acid			0.38	20
Valine		14–51	2.1	380
Cystine		20–130		10
Methionine		20–95	0.25	50
Isoleucine		18–210	0.53	70
Leucine		21–200	1.5	170
Tyrosine		40–270	0.90	120
Phenylalanine		24–190	0.95	120
β-Aminoisobutyric acid	0–93	0–500		
Ethanolamine			1.2	
Ornithine			0.49	20
1-Methyl histidine	0–250	130–930		
Lysine	10–140	48–640	2.9	150
Tryptophan				40
Histidine	99–530	130–2100	1.2	70
3-Methyl histidine		180–520		
Arginine			2.2	110

through Dowex 2-X8 resin. Furthermore, its use results in considerable losses of tryptophan (181), citrulline, and homocitrulline (51).

Filtrates or supernates of serum obtained following the addition of solid sulfosalicylic acid (181) contain traces of protein (52) but nevertheless can be applied to the column directly since sulfosalicylic acid is not visualized by the ninhydrin reaction (181). Results obtained by picric acid and sulfo-salicylic acid precipitants show that both methods are similar in precision and accuracy (22). Tungstic acid was found to be unacceptable as a deproteinizing agent as its use is associated with amino acid losses, chiefly diamino acids, 30%–50% of which become firmly bound to the protein precipitate. The loss of neutral amino acids, with the exception of threonine, was insignificant (19). Trichloroacetic acid and acetone have been also investigated as protein precipitants (174). Ultrafiltration has been used with results in good agreement with the findings of others using chemical precipitants (244). This approach appears not to have been used extensively because a loss of cystine and hydrolysis of glutamine can occur (69), possibly due to the long time required (186). A method for the preparation of plasma for amino acid fractionation in which only a small portion of the protein is removed was described by Gerritson *et al* (69). Citrate buffer containing internal standards at pH 1.5 was added to plasma bringing the final pH between 2 and 2.5. After 30 min in the refrigerator the mixture was subjected to 18 000 g for 30 min and an aliquot of the supernate analyzed. The concentrations obtained by the proposed method were greater for most amino acids than by the chemical precipitation procedures, and recoveries were nearly complete.

Urine samples are examined for protein that, if present, should be precipitated with one-fifth volume sulfosalicylic acid (15 g/100 ml) and removed by centrifugation (90). The majority of the ammonia also may be removed by rendering the sample alkaline (pH 11–12) with NaOH and placing under reduced pressure for several hours. Water is then added to bring the sample back to the original volume (34).

Cerebrospinal fluid should be deproteinized by the addition of sulfosalicylic acid as described for urine and an appropriate volume placed on the analytical column (49). Aqueous fluid from the eye may be treated similarly (50).

For the amino acid analysis of feces, it is recommended that the sample be homogenized with dilute HCl to halt decomposition by enzymes and bacteria. Following precipitation with acetone the amino acids are isolated using a cation exchange resin (165, 210). Alternatively, amino acids can be extracted into isopropanol (10% v/v) acidified with HCl. A portion of the supernate is then deproteinized with picric acid (75).

Stability. Investigators studying the stability of amino acids in biologic samples have generally found that heparinized plasma is superior to serum (48). In addition, serum contains a greater level of amino nitrogen than plasma, probably due to the release of amino acids during the clotting process (63, 77, 93, 150). When plasma was maintained at 27°C for 24 hour, a 67% loss of cystine occurred (51), due presumably to the binding of cysteine by the plasma proteins (229). There was also a 52% increase in the concentration of glutamic acid probably resulting from the degradation of glutamine. Other experiments in which plasma was incubated at 37°C and at 6°C also showed very significant changes in the same amino acids. Following storage at −20°C, plasma cystine was reduced to 15% of the initial concentration in 7 days and to zero in 1 month (51), an observation similar to that made by Perry and Hansen (181). The latter authors also noted that homocystine and the mixed disulfide of cysteine and homocysteine were lost at this temperature. The aspartic acid level was constant for 1 week and then increased, a change which was approximately paralleled by glutamic acid. Methionine and tryptophan remained constant for 1 month. De Wolfe *et al* (48) also found storage at −20°C to be preferable to 6°C. When stored at −68°C, however, glutamic acid increased by 40% in 1 week, but no changes in any other amino acids were detected. The authors concluded that −68°C was a more satisfactory temperature than −20°C for the preservation of amino acids in plasma (51).

Recommendations for the stabilization of the urinary amino acids include the addition of 10 ml conc HCl per liter of urine (67), collection at 4°C under toluene and subsequent storage at −20°C (148). One urine sample showed no change in amino acid pattern when stored 10 days at 4°C or after freezing (123). The addition of 0.05 vol 0.1% merthiolate also has been used as a preservative (58).

Our laboratory has found that when urine samples are brought to pH2−3 with HCl, stability for one week at room temperature is achieved. Freezing preserves urinary amino acids for an indefinite period.

Interferences. The anticoagulants most often used for the preparation of plasma for amino acid analysis are heparin and ethylenediaminetetraacetic acid (EDTA). One batch of the latter material was found to contain ninhydrin-positive compounds that eluted in the chromatogram to yield spurious peaks (181).

Hemolyzed samples should not be analyzed since many artifacts have been shown to occur. These include large increases in the concentrations of taurine, phosphoethanolamine, glutamic acid, and aspartic acid. Reduced and oxidized glutathione, not normally detectable in plasma, also appear in large quantity. Argininine diminishes while ornithine increases possibly due to the release of erythrocyte arginase (181).

Contamination of the plasma with the contents of the buffy coat can cause increases in taurine, phosphoethanolamine, aspartic acid, and glutamic acid. For this reason, plasma should be removed before platelets rupture. It is also recommended that a 5 mm layer of plasma be left above the buffy coat when aspirating the plasma (181).

Since the amino acids of a single thumb print have been analyzed (89), it can be appreciated that inaccurate results may be caused by careless handling of samples.

ACCURACY AND PRECISION

Good resolution requires an analysis time that can be as long as 3 days (86). This approach however, regularly shows 130, and at times as many as 175, ninhydrin positive peaks in urine compared with 30−40 peaks in a 21 hour analysis. Similarly, up to 55 extra compounds may be revealed in plasma. Thus it can be seen that the degree of accuracy achieved is a direct result of the resolution obtained by the system. Peaks not separated obviously will elute simultaneously. Many of these unresolved compounds are present in low concentration but this is not a good reason for ignoring them (86).

Precision is about ±10% for most amino acids, larger peaks showing greater reproducibility.

Tables 18-2 and 18-3 list the ninhydrin positive compounds usually found by standard amino acid column chromatography in physiologic samples. By increasing the sensitivity of the system it is possible to demonstrate in urine as many as 80 (123) or even 175 constituents (90). The table has been assembled from several sources and the extreme values noted. Only references by those authors giving concentrations in mg/100 ml, or μmol/100 ml plasma, mg/24 hour, or μmol/24 hour urine were used. Other figures occur in the literature, but they are given in terms such as μmol/g creatinine or μmol/minute/1.73 m^2 body surface.

It should be emphasized that amino acid concentrations can vary even in a normal individual. Certain amino acids show circadian periodicity, and there are changes associated with the menstrual cycle and with pregnancy. Infants show significant differences in pattern and concentration from adults (59). Alterations in plasma amino acid levels occur in response to changes in the amount of dietary protein intake (221). Observations on young men in military service whose diet and living conditions were reasonably constant showed that there were seasonal variations in the excretion of several amino acids (83).

CYSTINE

The relationship between cystine and cysteine is as follows:

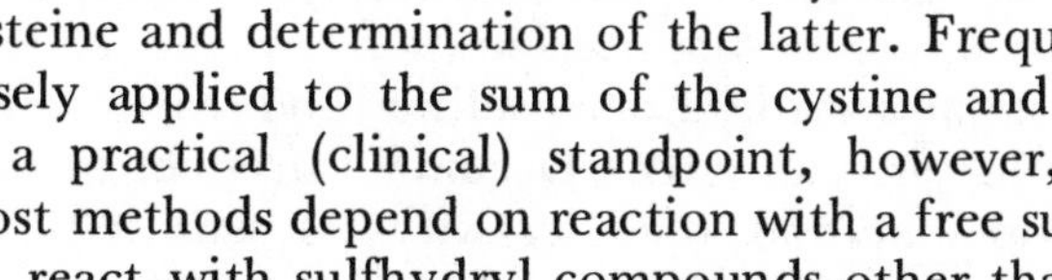

All chemical methods for the determination of "cystine" involve reduction of cystine to cysteine and determination of the latter. Frequently the term "cystine" is loosely applied to the sum of the cystine and the preformed cysteine. From a practical (clinical) standpoint, however, this is of no consequence. Most methods depend on reaction with a free sulfhydryl group and accordingly, react with sulfhydryl compounds other than cysteine and also determine reducible disulfides other than cystine.

The solubility of cystine is approximately ten times greater in urine than in water. In addition, the solubility of cystine in concentrated urine is considerably greater than in dilute urine. These factors may account for the observation that some urine samples which give strongly positive reactions for the presence of cystine do not show characteristic crystals in the urinary sediment when examined microscopically (56). The nitroprusside reaction for sulfhydryl groups following cyanide reduction of the cystine was introduced by Brand *et al* (27) as a simple qualitative screening test for cystinuria, there being insufficient sulfhydryl concentration in normal urine to produce a visible color. Homocystine also reacts. The original Brand test was compared with a simplified version using the commercially available Acetest tablets. It was found that while both tests gave some false positive and equivocal results, the tablet test also gave 3% false negatives (85). In an attempt to avoid the use of the toxic cyanide reagent, Middleton described a procedure in which reduction of cystine is carried out with H_2 formed from Zn and HCl. Cystine is detected by testing the sample with Ketostix before and after reduction (164). The cyanide-nitroprusside reaction with cystine is pH dependent and adjustment of sample pH to 7–8 has been recommended (68). A semi-automated version of the procedure using a triethanolamine buffer has been described (56). Sodium borohydride is a satisfactory substitute for cyanide as a reducing agent (200).

The Sullivan test using 1,2-naphthoquinone-4sulfonate (β-naphthoquinone-sulfonate) has formed the basis for several published procedures. These have the potential advantage that penicillamine (dimethylcysteine), a drug frequently used in the treatment of cystinuria and homocystinuria, does not react. The simple semiquantitative procedure described by MacDonald and Fellers (149) was found to require careful control of the incubation temperature and of each step of the reaction (46). Adsorption and subsequent elution of urinary cystine using Dowex 50-X12 resin followed by color development with 1,2-naphthoquinone-4-sulfonate has been recommended as a quantitative procedure (96). This method was found in our laboratory to have certain inherent disadvantages including the fact that calibration is not completely linear, and the absorption maximum is different for urine and a cystine standard. These findings are consistent with earlier work in our laboratory. At that time it was found that the color produced with standards by 1,2-naphthoquinone-4-sulfonate had an absorption peak at 505 nm, but the absorption curves produced with urines showed no evidence of such a peak. It was concluded, therefore, that any similarity between results obtained by the Sullivan-Hess (233) method and actual cystine content must be largely coincidental.

A highly specific and sensitive method for cystine proposed by Fernandez and Henry (60) involves its precipitation as the cuprous mercaptide, reduction to cysteine with hydrogen sulfide, and reaction with 2,6-dichloro-*p*-benzoquinone to yield an intensely colored product (60). While the linear calibration of this method together with its sensitivity, specificity, and precision make it attractive to the clinical chemist, the technician time involved together with the need for hydrogen sulfide are features that make it a less desirable procedure for the routine laboratory. This method also gave results in good agreement with that of Shinohara and Padis (217), which is given in detail below. Ellman's reagent, 5,5'-dithiobis-2-nitrobenzoic acid (DTNB), reacts with aliphatic thiols to yield reduced DTNB, which at alkaline pH is highly colored. Agents that reduce disulfides, such as sulfite and dithiothreitol, also react with DTNB. For this reason, a reducing agent that can be removed readily and completely is necessary. Thiolated Sephadex is a powerful reducing agent for disulfides that, since it is a particulate, can be filtered from the reaction mixture (198).

Advantage has been taken of the fact that cysteine will react with noradrenochrome converting the pink color to yellow. While the procedure will measure cysteine, cystine, and mixed disulfides containing cysteine, such as cysteine-homocysteine, it is not sufficiently sensitive to quantitate cystine in the concentrations found in normal urine but is suitable in cystinuria (201, 208).

The relatively straightforward method of Shinohara and Padis (217) is recommended for routine use since it gives results not significantly different from those of the highly specific but more complex procedure described by Fernandez and Henry (60).

QUALITATIVE TEST FOR CYSTINE AND HOMOCYSTINE IN URINE

(method of Rosenthal and Yaseen, Ref. 200)

PRINCIPLE

Cystine is reduced to cysteine with sodium borohydride. Ketones, which can interfere with the color reaction, are reduced to alcohols. Excess borohydride is then decomposed with acid and sulfhydryl groups detected with nitroprusside.

REAGENTS

Sodium Borohydride, Powder
HCl, conc
NaOH, 12.5 M. Dissolve 125 g NaOH pellets in water, cool, and make up to 250 ml.
NH₄Cl
Sodium Nitroprusside, 1%. Prepare fresh.
Comparison Standard in Normal Urine. Dissolve 5 mg cystine in 10 ml 0.1 N HCl. Make up to 100 ml with a fresh normal urine. Aliquot in 5 ml volumes and store frozen.

PROCEDURE

1. Transfer 5 ml urine specimen and 5 ml standard to 250 ml beakers.
2. While swirling the urine, add approximately 0.5 g sodium borohydride.
3. Allow to stand for 2 min.
4. Add conc HCl dropwise from a 1 ml graduated pipet until no more gas is

evolved. Usually, less than 1 ml is required but if more is used, note the volume.

5. Add 2 ml 12.5 N NaOH. (If more than 1 ml HCl was used in step 4, add the same volume of NaOH. The solution must be strongly basic at this point.)

6. Add NH_4Cl powder until the solution is saturated and some crystals remain undissolved after swirling.

7. Decant approximately 3 ml of the liquid into a test tube.

8. Add 5 drops sodium nitroprusside solution. Mix.

9. A distinct purple color indicates cystine or homocystine. Read the results at once as the color fades in a few minutes.

NOTES

1. Stability of samples. Urine is stable for this screening test for 1 week if acidified to pH 2–3 with HCl or for an indefinite period when frozen.

ACCURACY

While aqueous solutions of cystine give a positive reaction in this test down to a concentration of 5 mg/liter, urine obscures the color and an appreciable purple color requires about 50 mg/liter urine. Ketone bodies and acetone, which interfere with the original test (27), do not pose a problem with the modification presented here (200).

NORMAL VALUES

Normal urine is negative with this qualitative procedure. The normal urine cystine excretion is 10–100 mg/24 hour or a maximum concentration of 100 mg/liter. Results on samples are compared with the cystine standard which contains 50 mg cystine/liter. Hence weakly positive or doubtful results should be recorded as negative.

QUANTITATIVE DETERMINATION OF CYSTINE IN URINE

(method of Shinohara and Padis, Ref. 217, modified)

PRINCIPLE

Cystine is reduced to cysteine by bisulfite [one molecule of cystine yields one molecule of cysteine and one molecule of a salt of S-cysteinesulfonic acid (38)]. Other disulfide compounds that may be present are presumably also reduced. The sulfhydryl groups of cysteine and other sulfhydryl compounds present reduce phosphotungstic acid to a tungsten blue, which is determined photometrically. At the pH of the reaction, about 5, uric acid reduces phosphotungstic acid but slightly although other reductants present (e.g., ascorbic acid) react to a significant extent. A urine blank is run to which mercuric chloride is added prior to the phosphotungstic acid. The mercuric ions bind the sulfhydryl groups, rendering them nonreactive with the phosphotungstic acid, but they do not interfere with the reduction caused by other reductants such as uric and ascorbic acids. The difference in color intensity with and without the mercuric chloride thus represents the sulfhydryl compounds. The absorption curve of the blue color formed is shown in Figure 18-4.

REAGENTS

Sodium Acetate, 2 M. 82.0 g anhydrous sodium acetate or 136.0 g sodium acetate trihydrate per 500 ml.

Acetic Acid, 2 M. Dilute 11.4 ml glacial acetic acid to 100 ml with water.

Acetate Buffer. Add 2 parts 2 *M* acetic acid to 10 parts 2 *M* sodium acetate. Stable in refrigerator at least several days.

Sodium Sulfite Reagent. Add 12.6 g Na_2SO_3 and 0.2 g NaOH (or 5 ml 1 *N* NaOH) to 100 ml water. Stable about 1 month in refrigerator.

Mercuric Chloride, 0.1 M. 2.7 g $HgCl_2$/100 ml solution. Stable at room temperature.

FIG. 18-4. Absorption curves in cystine determination: (1) reagent blank vs water, (2) 0.3 mg cystine standard vs reagent blank. Beckman Model DU Spectrophotometer.

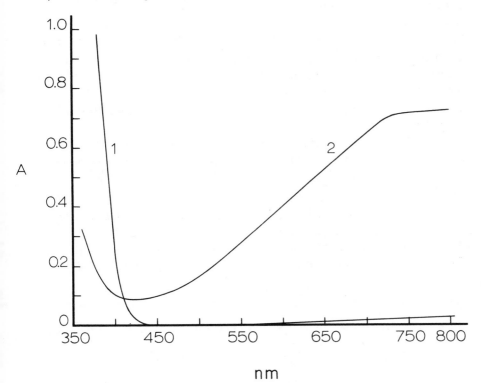

Phosphotungstic Acid Reagent. Dissolve 40 g molybdenum-free sodium tungstate in about 300 ml water. Add 32 ml 85% orthophosphoric acid. Reflux gently for 2 hours. Cool to room temperature and make to 1 liter with water. Mix. Dissolve 32 g $Li_2 SO_4 \cdot H_2 O$ in reagent. Stable indefinitely in refrigerator.

Cystine Standard, 0.4 mg/ml. Add 40 mg cystine to a 100 ml volumetric flask and 0.1 N HCl to volume. Stable in refrigerator.

PROCEDURE

1. If urine is not absolutely clear, centrifuge some urine and examine sediment for cystine crystals under a microscope. Cystine crystals are in the form of thin, colorless, hexagonal plates. If cystine crystals are found, centrifuge 10.0 ml of well mixed urine and pour supernate into another tube. Add 1.0 ml 1 N HCl to the precipitate and stir up. Heat in a 60°C water bath for 15 min with occasional mixing. Centrifuge and pour supernate into the original 10 ml of supernate. Mix. N.B.: This step is necessary only when the cystine concentration is so high as to result in formation of cystine crystals; hence it can be omitted if microscopic examination of sediment fails to reveal cystine crystals.

2. Set up the following in test tubes:
   Reagent blank. 1.0 ml water.
   Standard. 0.5 ml standard (0.20 mg cystine) + 0.5 ml water.
   Urine blank. 1.0 ml urine.
   Urine. 1.0 ml urine.

3. To each tube, add 1.0 ml acetate buffer and 0.3 ml sodium sulfite reagent.

4. To "reagent blank" and "urine blank" add 1.5 ml water and 0.2 ml mercuric chloride. Wait 2 min, add 1.0 ml phosphotungstic acid reagent, and mix.

5. To "standard" and "urine" add 1.7 ml water and 1.0 ml phosphotungstic acid reagent.

6. Within 15 min read "standard" and "urine" against their respective blanks at 600 nm or using a filter with nominal wavelength in this region. If absorbance of "urine" exceeds 0.8, start analysis over with a dilution of the urine from step 1.

Calculation: Concentration is read from a calibration curve (see note 1 below) and excretion/24 hour then calculated. The standard is run along with the unknowns as a check on the standard curve. If step 1 is *not* bypassed, correct results for the dilution in step 1 by multiplying by 11/10.

NOTES

1. Calibration curve. In our experience the color as read on a Beckman

DU Spectrophotometer deviates from Beer's law, showing a concavity *upward*. A standard curve is prepared by running standards containing 0.1, 0.2, 0.3, and 0.4 mg cystine. These are equivalent to 10, 20, 30, and 40 mg/100 ml urine. If it is desired to use cysteine to standardize the test, use half as much cysteine as cystine, since in the reduction of cystine to cysteine by sulfite only one molecule of cysteine is formed (38).

2. Stability of urine samples. Evidence has been presented for the presence of a cystine compound in urine that decomposes with the liberation of free cystine within 24 hours even at refrigerator temperature (6, 26, 27, 233). One group of workers (144) failed to observe such an increase, although, admittedly, they studied only one urine. It may be that the increases that have been observed simply represent hydrolysis of some of the conjugated cystine present, for approximately one-third of the total cystine excreted is in the conjugated form. For practical purposes, however, urines can be adequately preserved in the refrigerator with chloroform for periods up to 3 months (6). Urine is also stable in the frozen state.

ACCURACY AND PRECISION

The method probably gives a fairly accurate estimate of sulfhydryl compounds, but how accurately it reflects the actual cystine concentration is a different question. The maximal normal cysteine concentration in urine has been stated to be about 1 mg/100 ml (217). Determination of the normal 24 hour output of cystine plus cysteine by iodometric titration (242), the method of Sullivan and Hess (233), the phosphotungstate reaction (121, 217), the method of Medes and Padis (157, 158), polarographic determination (57, 195), and microbiologic assay (32, 65, 253)—all give a top limit of the order of magnitude of 100 mg. The top normal limit as determined by ion exchange chromatography, however, is about 20 mg/24 hour (29, 57, 228). This apparent disagreement has not been resolved. From the viewpoint of the clinician, however, it is of no great importance because several of the technics appear to be adequate for detection of cystinuria.

The reproducibility of the technic is about ±10%.

NORMAL VALUES

A normal 24 hour output of 10 to 100 mg was obtained in our laboratory in a study of 10 normal adult urines. This is in agreement with a compromise of the values reported in the references cited above under *Accuracy and Precision*. Output increases with the amount of protein in the diet (157). Reports on the normal range of total cystine obtained by acid hydrolysis (free + conjugated) are not in very close agreement (32, 195, 233), but the range appears to be about 50 to 150 mg/24 hour.

PHENYLALANINE

The determination of blood phenylalanine is now a test that is performed extensively since the majority of the States adopted a mandatory screening program requiring the examination of every newborn infant for phenylketonuria (PKU). The laboratory procedures that have met the criteria of acceptance for this purpose are the Guthrie inhibition assay, the quantitative fluorometric procedure, and amino acid chromatography (194). It has been recognized for several years that examination of urine for phenylpyruvic acid is not acceptable for the early diagnosis of PKU (191, 230).

La Du and Michael (132) described an enzymatic procedure for the quantitation of phenylalanine in blood. The amino acid is converted to its keto derivative by L-amino acid oxidase. In the presence of arsenate and borate the keto derivative is transformed to the enol-borate complex that exhibits a characteristic absorption in the ultraviolet region. Values for tyrosine and tryptophan can also be calculated from the absorbance readings since these amino acids also react producing complexes with different absorption spectra. The keto derivative can also be quantitated using 2,4-dinitrophenylhydrazine (117).

A procedure utilizing gas-liquid chromatography has also been described. In this approach, which is claimed to be rapid, specific, and precise, the amino acid is first converted to the volatile neopentylidene-phenylalanine methyl ester (113).

The method that probably has proved to be the most widely used is the fluorometric procedure of McCaman and Robins (154). This determination is based upon the reaction between phenylalanine, copper, and ninhydrin to give a fluorescent complex as product. This reaction is enhanced by several peptides including glycyl-DL-phenylalanine and L-leucyl-L-alanine. Sufficient sensitivity is obtained so that microliter volumes of blood are adequate for reliable measurements. The method, is, therefore, particularly suitable for use in pediatrics. Specificity is also of a high order, tyrosine giving 5% of the fluorescence of phenylalanine and other potential interfering substances giving approximately 1% or less (251). Studies have been made to establish optimum parameters for the reaction with regard to pH, reagent concentrations, removal of impurities, time, and temperature of heating (2, 4). Use of trichloroacetic acid as a deproteinizing agent appears to lead to erroneous results, but the error can be diminished by the addition of albumin to the standards. Alternatively, protein can be removed with ethanol (237).

In response to the demand for screening large populations, the procedure was automated using the Technicon AutoAnalyzer (3, 105). Further investigation showed that glycyl-L-leucine satisfactorily enhanced the excitation and emission spectra of the final fluorescent product and was considerably less expensive than L-leucyl-L-alanine. Also, substitution of a filter passing more light of a broader wave band increased sensitivity 2.5 times with no loss in specificity (104). Elimination of the dialyzer unit, use of a modified pump, and a special fluorometric cell enabled 120 specimens to be analyzed per hour with sufficient accuracy and specificity for screening purposes (25).

Great convenience has been achieved in sample collection, particularly from pediatric patients, by adoption of the device described by Guthrie (80) in which capillary blood is allowed to fill a printed circle on a filter paper form. Some authors recommend autoclaving the blood soaked filter paper (25, 212), while others have not found this necessary (103, 105).

MANUAL DETERMINATION OF PHENYLALANINE IN SERUM

(method of Hsia et al, Ref. 110)

PRINCIPLE

Serum is deproteinized and allowed to react with ninhydrin and copper to produce a fluorescent complex, the excitation and emission spectra of which are given in Figure 18-5. This reaction is enhanced by L-leucyl-L-alanine. Reagent and sample blanks are included so that in the event that the serum contains interfering fluorescent substances originating, for example, from drug therapy or diet, an appropriate correction can be made.

REAGENTS

N.B.: All reagents are prepared with deionized distilled water.
Succinic acid, 0.6 M, pH 5.88. Dissolve 70.9 g succinic acid in 600 ml

FIG. 18-5. Excitation and emission spectra of phenylalanine—copper—ninhydrin complex. For this excitation scan, emission wavelength was set at 490 nm, and for the emission scan, excitation was set at 365 nm. Aminco Bowman Spectrophotofluorometer.

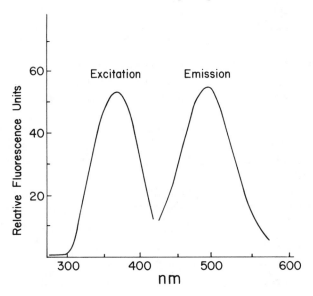

water. Adjust pH to 5.88 with 5 N NaOH. Dilute to 1 liter. Check pH at least once a week. Store at 4°C.

Ninhydrin solution 0.03 M. Dissolve 1.34 g ninhydrin in 250 ml water. Store at 4°C.

L-Leucyl-L-alanine, 0.005 M. Dissolve 253 mg leucyl-alanine in 250 ml water. Dispense in 1 ml volumes and store frozen.

Buffered Ninhydrin-Peptide Solution. Mix 5 vol succinate buffer, 2 vol ninhydrin solution, and 1 vol leucyl-alanine solution. Prepare fresh daily.

Buffered Ninhydrin-Water Solution. Mix 5 vol succinate buffer, 2 vol ninhydrin solution, and 1 vol water. Prepare fresh daily.

Sodium Carbonate, Potassium Sodium Tartrate Solution. Dissolve 1.33 g Na_2CO_3 and 55 mg $NaKC_4H_4O_6 \cdot 4H_2O$ in 500 ml water.

Copper Sulfate, 0.6 mM. Dissolve 150 mg $CuSO_4 \cdot 5H_2O$ in 1 liter water.

Copper Carbonate-Tartrate Solution. Mix 3 vol carbonate-tartrate solution with 2 vol copper sulfate solution. Prepare fresh each time.

Trichloroacetic Acid 0.6 N. Dissolve 98.0 g trichloroacetic acid in 1 liter water.

Albumin Solution, 7.5%. 7.5 g bovine albumin in 100 ml water.

Standards, 5 and 10 mg Phenylalanine/100 ml. Dissolve 10.0 mg L-phenylalanine in 100 ml albumin solution. For 5 mg/100 ml standard, dilute 10 mg/100 ml solution with an equal volume albumin solution. Pipet 1 ml aliquots of the standards into plastic vials and store frozen.

PROCEDURE

1. Set up the following in test tubes, using 0.2 ml TD pipets:
   Reagent Blank. 0.2 ml albumin solution.
   Standards. 0.2 ml 5 and 10 mg standards.
   Unknown. 0.2 ml sample.

2. Add 0.2 ml trichloroacetic acid solution to each tube. Mix thoroughly and allow to stand 10 min with occasional mixing. Centrifuge.

3. Set up the following in test tubes:
   Reagent Blank with Peptide. 0.3 ml buffered ninhydrin-peptide solution. Transfer 20 μl reagent blank supernate from step 2 using TC pipet.
   Standard Solutions. 0.3 ml buffered ninhydrin-peptide + 20 μl standards.
   Unknown. 0.3 ml buffered ninhydrin-peptide + 20 μl unknown supernate from Step 2.
   Unknown Blank. 0.3 ml buffered ninhydrin-water solution + 20 μl unknown supernate from step 2.
   Reagent Blank without Peptide. 0.3 ml buffered ninhydrin-water solution + 20 μl reagent blank supernate from step 2.

4. Mix thoroughly and incubate for 2 hour in a 60°C water bath.

5. Cool in cold tap water.

6. Add 5 ml copper carbonate-tartrate solution. Mix.

7. Read fluorescence of tubes in a fluorometer using an excitation

wavelength of 365 nm and an emission wavelength of 490 nm. Results should be read on the linear scale of the instrument.

Calculation: Select the reading of the standard nearest to that of the unknown and substitute in the following expression. See below for key to symbols.

$$\text{mg phenylalanine/100 ml} = \frac{F_x - F_b}{F_s - F_b} \quad \text{x} \quad \begin{array}{c}\text{concentration of standard}\\ \text{(mg/100 ml)}\end{array}$$

If unknown blank reads significantly higher than the reagent blank (with peptide), then use following equation:

$$\text{mg phenylalanine/100 ml} = \frac{F_x - F_b - F_y - F_c}{F_s - F_b} \quad \text{x} \quad \begin{array}{c}\text{concentration of standard}\\ \text{(mg/100 ml)}\end{array}$$

F_x = reading of unknown (with peptide)

F_y = reading of unknown (without peptide)

F_s = reading of standard

F_b = reading of reagent blank (with peptide)

F_c = reading of reagent blank (without peptide)

NOTES

1. Linearity of fluorescence. Linearity is obeyed up to a concentration of 10 mg/100 ml.

2. Fluorescence stability. Fluorescence of the final solution is stable for at least 1 hour.

3. Stability of serum samples. Most samples of serum will remain stable for about 4 days at room temperature and, if preserved with 5 mg sodium fluoride/ml, for at least 1 week.

ACCURACY AND PRECISION

It was reported by Wong *et al* (251) that compounds other than phenylalanine react under the conditions of the test to give a fluorescence intensity relative to phenylalanine as follows: tyrosine 5%, tryptophan 1.2%, leucine 1.1%, tyramine 0.6%, phenyllactic acid 0.3%, arginine 0.2%, phenylacetic acid 0%, phenylpyruvic acid 0%. Further lists of compounds of biologic importance and their potential influence on phenylalanine results are given by Ambrose (2). By summation of the interferences at the concentrations usually occurring in serum, it was concluded by Ambrose that phenylalanine levels in the newborn could be overestimated by 1.0–2.4 mg/100 ml and in adults by 1.0–2.2 mg/100 ml. 5-Hydroxytryptamine gave 20.9% the fluorescence of phenylalanine, but this compound appears only in low concentrations in the blood of full term and premature newborns. Due to the possible presence of substances in blood that are naturally fluorescent or react with ninhydrin to give a fluorescent product, serum blanks omitting dipeptide should be run.

The precision of the method is about ±7% at the 5 mg/100 ml level.

NORMAL VALUES

A level of 4 mg phenylalanine/100 ml is generally accepted as the division between normal individuals and those in which further investigation is indicated. In our laboratory, the normal range was found to be 1.6–3.0 mg/100 ml, which agrees with the values of 1.1–3.1 mg/100 ml reported by others (251) from the examination of 4000 newborns. Phenylalanine has a diurnal variation with highest values occurring around noon and the lowest between midnight and 6 am (40, 82). For testing newborn babies for elevated phenylalanine levels, it is recommended that blood be collected after the child has received milk for at least 24 hours. A further safety measure suggested is to retest all infants at 4–6 weeks of age (194). The age of the infant and the type of feeding, however, have a rather small effect on the phenylalanine level. As discussed subsequently in the section on tyrosine determination, the value of the phenylalanine:tyrosine ratio is of assistance as a test of heterozygosity for phenylketonuria (184, 185).

AUTOMATED DETERMINATION OF PHENYLALANINE IN BLOOD AND SERUM

(method of Hill et al, Ref. 104, 105)

PRINCIPLE

Blood is collected on filter paper and dried. Phenylalanine is eluted from the dried blood with water. The eluate is then dialyzed and phenylalanine is allowed to react with ninhydrin, L-leucyl-L-alanine, and copper reagent to produce a fluorescent compound with activation peak at 365 nm and an emission peak at 490 nm (Fig. 18-5).

REAGENTS

N.B.: All reagents are prepared with deionized distilled water.
Succinate Buffer, 0.04 M, pH 5.8. Dissolve 9.44 g succinic acid in approximately 1 liter water and then 5.50 g NaOH in about 600 ml water.

Slowly and with stirring, add the NaOH solution to the succinic acid solution. Transfer to a 2 liter volumetric flask and dilute to mark with water. Adjust the pH to 5.8 with 0.1 N NaOH or 0.1 N HCl. Add 1 ml of Brij 35, mix thoroughly, and store in refrigerator.

Ninhydrin, 0.53%. Dissolve 2.65 g ninhydrin in about 400 ml water. Stir with magnetic stirrer until material is completely dissolved. Add 0.2 ml Brij 35, dilute to 500 ml with water, and mix thoroughly.

Copper Reagent. Dissolve 16.0 g Na_2CO_3 in about 300 ml water. Dissolve 0.65 g potassium sodium tartrate in about 200 ml water. Transfer both solutions to a 1 liter volumetric flask and mix thoroughly. Dissolve 0.60 g $CuSO_4 \cdot 5H_2O$ in 200 ml water, add to the solution in the 1 liter flask, dilute to mark, and mix thoroughly.

Saline, 0.85%. Dissolve 8.5 g NaCl in a 1 liter volumetric flask containing about 600 ml water. Add 0.5 ml Brij 35, dilute to mark, and mix thoroughly.

L-Leucyl-L-Alanine, 80 mg/100 ml. Dissolve 800 mg L-leucyl-L-alanine in 1 liter water. Dispense in 100 ml quantities into plastic bottles and store in freezer until day of use.

Stock Standard, 8.0 mg/100 ml. Dissolve 40.0 mg L-phenylalanine in 500 ml water. Aliquot into 5 ml portions and store in freezer.

Working Standards. Prepare fresh daily as follows:

Stock Standard	Water	Chloroform	Equivalent to blood level of phenylalanine
1.0 ml	99.0 ml	0.1 ml	8 mg/100 ml
0.5 ml	99.5 ml	0.1 ml	4 mg/100 ml

PROCEDURE

Set up manifold, other components and operating conditions as shown in Figure 18-6.

By following procedural guidelines as described in Chap. 10 steady state interaction curves as shown in Figure 18-7 should be obtained and should be checked periodically. Blanks on blood and serum are usually low and sufficiently uniform that they may be ignored.

Preparation of Samples

A. Filter Paper Sample

Blood from a baby's heel or finger is made to flow into and fill a 3/8 in. circle printed on Schleicher and Schuell 903 filter paper. It is essential that the blood soak through to the reverse side of the paper. Allow the preparation to dry at room temperature.

1. Remove the central part of the dried blood spot into a test tube using a paper punch, diameter 6.3 mm.

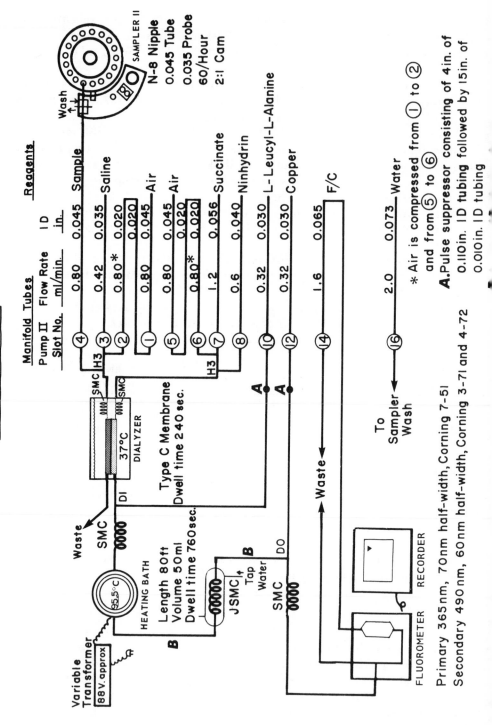

FIG. 18-6. Flow diagram for phenylalanine in serum and eluates of dried whole blood on filter paper disks. Calibration range: 0–8.0 mg/100 ml. Conventions and symbols used in this diagram are explained in Chap. 10.

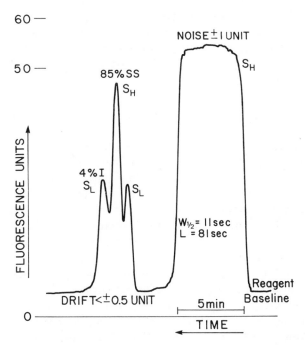

FIG. 18-7. Phenylalanine strip chart record. S_H is a 8.0 mg/100 ml standard, S_L a 4.0 mg/100 ml standard. Peaks reach 85% of steady state; interaction between samples is 4%. $W_{1/2}$ and L are explained in Chap. 10. Computer tolerance for noise is set at ±1 unit, for baseline drift at ±0.5 unit, and for standard drift at ±1 unit.

2. Add 1 ml water.
3. Let samples soak for 20 min with thorough mixing at the following time intervals: 0, 10, 15, and 20 min. Check to make sure that blood cells are completely lysed and that eluate appears homogeneous.
4. Centrifuge. Analyze supernate within 1 hour.

B. Serum Samples
1. Add 20 μl serum to 2.0 ml water with TC pipet. Mix.

Results from blood samples on filter paper are calculated from the calibration curve in the standard manner except that an allowance for fluorescence from the filter paper should be made. This is usually about 0.2 mg/100 ml. Results obtained for serum samples should be multiplied by 1.01 to correct for the volume of serum used.

NOTES

1. See notes 1 and 3 of "*Manual Method.*"

2. Flow diagram. This method is modified from that of Hill *et al* (105) by using a Technicon fluorometer, by increasing flow through the fluorometer cuvet to improve $W_{1/2}$ and by controlling bubble timing to decrease noise.

3. Stability of blood samples. Dried blood spots on filter paper have been reported to be stable as long as 11 weeks at room temperature (104). In our laboratory it was found that the eluate from the filter paper discs was stable in the freezer for at least 3 days. Diluted serum was stable in the refrigerator also for at least 3 days.

ACCURACY AND PRECISION

See Manual Method. Each new lot of Schleicher and Schuell 903 filter paper should be checked for fluorescence by processing 6.3 mm disks through the determination. The fluorescence obtained should be equivalent to no more than 0.2 mg phenylalanine/100 ml. The reliability of the final result obtained is dependent to a large extent upon the technic used in sample collection. It is imperative that the quantity of blood transferred to the filter paper be sufficient to fill an area of adequate size and to penetrate through the paper completely.

It has been shown that variations in hematocrit, irregularities in filter paper thickness and porosity, changes in atmospheric humidity (102), and differences in handling a paper punch all can contribute to the absolute volume of blood taken for analysis on the paper disc. Users should be aware of the shortcomings of the technic that, however, can give results of acceptable accuracy in a routine situation (103).

The precision (95% limits) of the automated method is about ±4% at a level of 6 mg/100 ml.

NORMAL VALUES

Same as Manual Method.

TYROSINE

Reaction with 1-nitroso-2-naphthol to yield a red compound has been used for many years as a test for tyrosine. The red color is unstable but can be converted to a stable yellow color by heating with HNO_3 and then separated from excess nitrosonaphthol (239). Fluorescence of the yellow nitrosonaphthol derivative of tyrosine allows greater sensitivity and specificity than the original photometric method (243). Ambrose *et al* (5) showed that extraction of the chromophore with ethylene dichloride is not necessary and that modification of the nitrosonaphthol reagent by addition of phosphoric acid prevents or minimizes instability. Modifications of these procedures requiring microliter volumes of sample suitable for pediatric use have been described (5, 251). Blood collected onto filter paper may be used. Trichloroacetic acid added to the disks precipitates protein on the filter paper thus eliminating the inconvenience of autoclaving to achieve the same purpose (213).

The procedure has been automated using AutoAnalyzer equipment, and this too is amenable to the use of blood spots collected on filter paper (107). Whereas in the manual version excess nitrosonaphthol is removed by shaking with ethylene dichloride, the same result is achieved in this approach by the addition of sodium metabisulfite.

Tyrosine also may be determined by the procedure of La Du and Michael (132) in which the amino acid is converted to its keto-derivative by L-amino acid oxidase. Subsequent formation of the enol-borate complex and measurement of absorbance at specific wavelengths in the ultraviolet region permit the determination of tyrosine, phenylalanine, and tryptophan.

DETERMINATION OF TYROSINE IN SERUM

(method of Wong et al, Ref. 251)

PRINCIPLE

Deproteinized serum is allowed to react with 1-nitroso-2-naphthol to form a fluorescent product. Excess reagent is removed by extraction with ethylene dichloride, and the resultant solution examined by fluorometry with excitation wavelength 470 nm and emission wavelength 545 nm. See Figure 18-8.

REAGENTS

Trichloroacetic Acid, 0.6 N.   Dissolve 98.0 g trichloroacetic acid in water and make up to 1 liter.

Trichloroacetic Acid, 0.3 N.   Mix equal volumes 0.6 N trichloroacetic acid and water.

Ethylene Dichloride.   Reagent grade.

1-Nitroso-2-Naphthol 0.2%.   Dissolve 200 mg 1-nitroso-2-naphthol in 95% ethanol and make up to 100 ml. Ethanol denatured with methanol is satisfactory. Store at 4°C.

HNO_3, 3 N.   Dilute 37.5 ml conc HNO_3 to 200 ml with water.

Sodium Nitrite, 0.10 N.   Dissolve 6.90 g $NaNO_2$ in water and make up to 1 liter. Store at 4°C.

Color Reagent.   Mix 2 vol 1-nitroso-2-naphthol reagent, 3 vol 3 N HNO_3, and 3 vol 0.1 N $NaNO_2$. Prepare just before use.

Stock Standard, 10 mg/100 ml.   Dissolve 10 mg L-tyrosine in 50 ml water contained in a 100 ml volumetric flask and make up to 100 ml. Freeze in 5 ml aliquots. Before use warm to 37°C to redissolve tyrosine and then bring to room temperature before pipetting.

Working Standard, 4 mg/100 ml.   Add 4 ml stock standard to a 10 ml volumetric flask containing 5 ml 0.6 N trichloroacetic acid and dilute to volume with water. Prepare fresh before use.

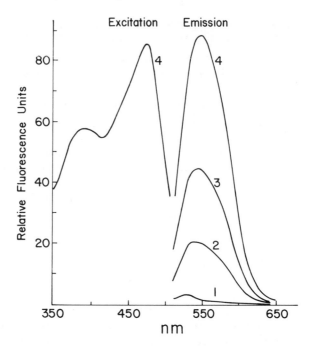

FIG. 18-8. Excitation and emission spectra of tyrosine–nitrosonaphthol chromogen. For the excitation scan, emission wavelength was set at 545 nm, and for the emission scan, excitation wavelength was set at 470 nm. (1) blank, (2), (3), and (4) standards containing 1, 2, and 4 μg tyrosine, respectively. Note shift in peak emission wavelength from 530 nm to 545 nm with increasing tyrosine concentration. Aminco Bowman Spectrophotofluorometer.

PROCEDURE

1. Pipet 0.5 ml serum and 0.5 ml 0.6 N trichloroacetic acid into a test tube. Mix thoroughly.
2. Allow to stand for 10 min. Centrifuge for 5 min.
3. Transfer 0.1 ml of the supernate into a 15 ml glass stoppered centrifuge tube. Also set up the following 15 ml glass stoppered centrifuge tubes:
   Blank: 0.1 ml 0.3 N trichloroacetic acid.
   Standard: 0.1 ml 4 mg/100 ml working standard.
4. To all tubes add 1 ml color reagent.
5. Mix and incubate at 32°C for 20 min.
6. Pipet 5 ml water into each tube. Mix.
7. Pipet 6 ml ethylene dichloride into each tube. Shake in a reciprocal shaker at medium speed for about 3 min.
8. Centrifuge. Transfer approximately 5 ml of the aqueous (upper) phase to a test tube.
9. Read fluorescence on a spectrophotofluorometer using activation wavelength of 470 nm and fluorescent wavelength of 545 nm. Reading should be taken within 30 min after centrifugation.

Calculation:

$$\text{mg tyrosine/100 ml} = \frac{F_x}{F_s} \times 0.004 \times \frac{100}{0.05}$$

$$= \frac{F_x}{F_s} \times 8$$

NOTES

1. *Linearity of fluorescence with concentration.* Linearity is obtained up to a value of at least 8 mg tyrosine/100 ml serum.
2. *Stability of samples.* Serum was found in our laboratory to be stable for tyrosine determination for 4 days at 30°C and for an indefinite period at −20°C. The addition of 3 mg NaF per ml serum prolonged the stability at 30°C to 7 days. Blood collected on filter paper and stored in closed glass containers at room temperature showed no deterioration in 1 month (213).

ACCURACY AND PRECISION

Studies of the manual fluorometric method have shown that of 50 compounds tested for interference, a considerable number of the amino acids, creatine, creatinine, and diodotyrosine gave no fluorescence (5). Parasubstituted phenols with no other substituents in the aromatic ring and the 5-hydroxyindoles gave appreciable fluorescence (20). The most interference was caused by *p*-tyramine, which gave the following different fluorescence yields relative to L-tyrosine: 193.8% and 161% on a weight basis in a modification of the manual method and automated method, respectively (5); 150% on an equimolar basis in the manual method (251). Interference due to *p*-tyramine is negligible probably because the kidney is very effective in clearing phenolic acids from the blood stream (20). A modification of the fluorescent tyrosine determination (5) gave results virtually identical to those using ion-exchange chromatography.

Precision is about ±4% at a level of 2.2 mg/100 ml.

NORMAL VALUES

The adult normal range (95% limits) found by out laboratory is 0.9–2.9 mg/100 ml based upon samples from 20 males and 20 females. There was no statistically significant difference between males and females. Other ranges (mean ± 2s) reported are 0.6–1.5 mg/100 ml (251) obtained from 88 normal fasting adults and 0.9–2.1 mg/100 ml (5) from 27 fasting individuals ranging in age from 5 days to 45 years. Ninety heterozygotes for PKU gave fasting values of 0.5—1.7 mg/100 ml while homozygous PKU patients showed results of 0.2—0.9 mg/100 ml (251).

Tyrosine is one of the amino acids that exhibits diurnal variation with a mid-morning peak value and a nadir occurring at 1:30—2:30 am. The level rises as much as 8% following vigorous exercise (254).

Neonatal tyrosinemia has been shown to occur in 10% of full term infants during the first week of life and in approximately 30% of premature infants (8) and is due to a deficiency of *p*-hydroxyphenylpyruvic acid oxidase (131). Another study showed that in Massachusetts approximately 1.8% of all infants experienced transient tyrosinemia within 6 weeks of birth (142).

Tyrosine levels may also be useful in combination with phenylalanine levels for the detection of PKU heterozygotes. A fasting serum phenylalanine level not less than 1.8 mg/100 ml together with a phenylalanine:tyrosine ratio not less than 1.6 is indicative of the carrier state. A fasting phenylalanine level of less than 1.8 mg/100 ml with a phenylalanine:tyrosine ratio less than 1.3 is consistent with the noncarrier. About 30% of the individuals tested could not be assigned to a group, but it was believed that the remainder were correctly identified (185).

HYDROXYPROLINE

The only mammalian protein that contains significant amounts of hydroxyproline is collagen. For this reason, the hydroxyproline found in biologic fluids is assumed to indicate collagen turnover. Of the total hydroxyproline found in normal urine, no more than 3% is free, and almost the whole of the remainder is bound in small peptides. Some is nondialyzable and presumably bound in large compounds (139). For this reason, the clinical laboratory is called upon for hydroxyproline analyses before and after sample hydrolysis.

Some workers prefer to improve the specificity of the determination by a preliminary chromatographic separation prior to chromophore formation and photometry. Hydroxyproline may be oxidized with peroxide or with chloramine-T. The latter is preferable since excess can be readily removed. The products of oxidation have been reported to be pyrrole and pyrrole-2-carboxylic acid. A chromophore that can be quantitated is then formed with *p*-dimethylaminobenzaldehyde (Ehrlich's reagent) either directly or following distillation or extraction. The chromophores produced with ninhydrin and isatin have also formed the basis of quantitative procedures (139). An enzymatic method also has been described that uses a crude preparation obtained from *Pseudomonas fluorescens A-312* adapted to utilize hydroxyproline as the sole source of carbon and nitrogen. Disappearance of color in an acid ninhydrin reaction is proportional to the hydroxyproline content (199).

Investigators are agreed that hydrolysates of urine contain substances, particularly pigments and salts, which interfere with the chloramine-T-Ehrlich's reagent photometric procedure. Purification of urine acid hydrolysate on Dowex 50-X8 gives better recoveries than when the chromatography step is omitted or when alkaline hydrolysis is substituted (61). Koevoet and Baars (126) examined three previously published methods and found all to be influenced by one or more of the following compounds: urea, ammonium chloride, glucose, and mannitol. They recommend that urine samples be purified by Dowex 50W-X2 resin prior to hydrolysis. Stegemann *et al* (227) advise that urine first be heated at 80°C to precipitate protein and carbonates and the supernate applied to a column of Amberlite IRA 410 (OH⁻) resin. Two eluates were obtained, the first containing collagen precursors or breakdown products and the second low molecular weight hydroxyproline-containing peptides and free hydroxyproline. Plasma was autoclaved to precipitate protein and the protein-free filtrate applied to the resin column. Hydrolysate was then prepared and used for photometric determination. Bergman and Loxley (14) recommend oxidation by chloramine-T since it is a more efficient process than peroxide oxidation, and the product, pyrrole, is more stable than the 1,2-dehydrohydroxyproline produced by peroxide. By carrying out the oxidation in high isopropanol concentration they were able to speed up the process and diminish the inhibitory effects of salt and excess *p*-dimethylaminobenzaldehyde. Automated versions of this method also have been described (109, 180) including one that recommends the use of an internal standard, running all samples in duplicate and using the second peak for calculation to diminish effects of the considerable carry over (118).

In 1960 a method was described in which specificity was obtained by removal of humin following hydrolysis and extraction of the chromophore into toluene. Later Kivirikko *et al* (124) considered removal of humin to be unnecessary for most urines and in addition, lengthened the time of the oxidation and color development steps. The product of hydroxyproline oxidation is volatile and can be distilled to separate the chromogen from potential interfering non-volatile products such as those yielded by tyrosine and tryptophan (166, 216). This procedure has been found to give very accurate results although special precautions must be taken to avoid losses of the highly volatile product (54).

The hydroxyproline of serum can be separated into five fractions on Sephadex G-200. The low molecular weight portion consists of monomolecular and peptide hydroxyproline (7) while the majority is precipitated with the serum proteins, a fraction termed hyproprotein (140).

DETERMINATION OF FREE AND TOTAL HYDROXYPROLINE IN URINE

(method of Kivirikko et al, Ref. 124)

PRINCIPLE

Urine is subjected to acid hydrolysis for the determination of total hydroxyproline. Following hydrolysis, heavy pigments are removed by centrifuging and hydroxyproline oxidized with chloramine-T in the presence of a measured excess of alanine. Oxidation is halted by the addition of thiosulfate and interfering substances extracted with toluene and discarded. The first products of hydroxyproline oxidation, Δ^1-pyrroline-4-hydroxy-2-carboxylic acid and pyrrole-2-carboxylic acid are not soluble in toluene. After heating, the pyrrole produced is extracted into toluene and the color reaction with *p*-dimethylaminobenzaldehyde carried out. The absorption spectrum of the reaction product is shown in Figure 18-9.

For the determination of free hydroxyproline, the acid hydrolysis step is omitted.

PROCEDURE FOR TOTAL HYDROXYPROLINE

REAGENTS

KOH, 12 N.   Dissolve 67.2 g KOH in water and dilute to 100 ml.
KOH, 10 N.   Dissolve 56.1 g KOH in water and dilute to 100 ml.
KOH, 1 N, 0.1 N and 0.05 N.   Dilute from 10 N KOH.
HCl, 6 N.   50 ml conc HCl diluted to 100 ml with water.

FIG. 18-9. Absorption spectrum of reaction product in determination of hydroxyproline. (1) reagent blank vs water, (2) 4 μg standard vs water, (3) urine vs water. Beckman Model DBG recording spectrophotometer.

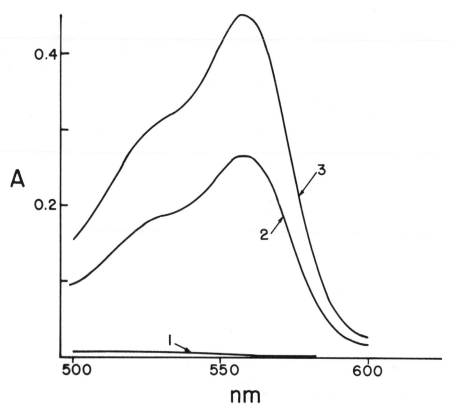

HCl, 1 N and 0.05 N. Dilute from 6 N HCl.

Phenolphthalein. Dissolve 1 g in 100 ml ethanol. Ethanol denatured within methanol is satisfactory.

Potassium Borate Buffer, pH 8.7. 61.8 g boric acid and 225 g KCl are dissolved in about 800 ml water. Adjust pH using a pH meter to 8.7 with 10 N and 1 N KOH. Dilute to 1 liter.

Alanine, 10%. Dissolve 10 g alanine in about 90 ml of water. Adjust the pH to 8.7 and then dilute to 100 ml.

Chloramine-T, 0.2 M. Dissolve 1.41 g chloramine-T in 25 ml peroxide-free methyl cellosolve. (Excessive effervescence indicates impure solvent.) Prepare just before use.

Sodium Thiosulfate, 3.6 M. Dissolve 89.3 g $Na_2S_2O_3 \cdot 5H_2O$ in 100 ml water. Solution can be kept for several weeks at room temperature when stored under toluene.

p-Dimethylaminobenzaldehyde (DMBA, Ehrlich's Reagent)

Solution A—27.4 ml conc H_2SO_4 is added slowly to 200 ml absolute ethanol in a beaker or Erlenmeyer flask. Ethanol denatured with methanol is satisfactory. Cool in running water while adding the acid.

Solution B—Add 200 ml absolute ethanol to 120 g p-dimethylamino-benzaldehyde. Ethanol denatured with methanol is satisfactory. Dissolve and cool. If DMBA does not dissolve, warm briefly in a 37°C water bath.

Solution C—Solution A is slowly added to Solution B in an ice bath. Solution is stable for several weeks.

Stock Standard, 36 mg/100 ml. Dissolve 18 mg hydroxyproline (Schwarz/Mann or Calbiochem is satisfactory) in water in a 50 ml volumetric flask. Add 0.2 ml conc. HCl and dilute to 50 ml with water. Prepare fresh on day to be used.

Working Standard, 36 μg/ml. Dilute 1.00 ml stock to 10.0 ml with water. Prepare fresh just before use.

PROCEDURE

1. Set up 16 x 150 mm screw capped (Teflon lined) culture tubes, graduated at 15 ml, for blank, standard and unknown.

2. Pipet 1 ml water, 1 ml working standard and 1 ml urine into the respective tubes.

3. To each tube add 1 ml of conc HCl.

4. Hydrolyze overnight (16-18 hours) by heating tubes in a heating block at 100°C.

5. Cool and add 1 drop phenolphthalein solution. Adjust pH to a pink color by addition of 0.95 ml 12 N KOH followed by 0.5 ml 1 N KOH. Make the final adjustment by the dropwise addition of 0.1 N KOH.

6. Dilute with water to the 15 ml mark.
7. Mix. If precipitated pigments occur in the sample tube, remove by centrifugation.
8. Pipet 3 ml aliquots from the three tubes from step 7 into other screw capped tubes.
9. Add 1 ml water to all tubes to make a total volume of 4 ml.
10. Add 1 drop phenolphthalein solution and readjust to a pale pink color with 0.05 *N* KOH or 0.05 *N* HCl.
11. Saturate the solution with about 3 g KCl.
12. Pipet 0.5 ml alanine reagent and 1.0 ml potassium borate buffer into all tubes. Mix well and stand at room temperature for 20 to 30 min with occasional mixing.
13. Add 1.0 ml 0.2 *M* chloramine-T solution and mix immediately. Let stand for 25 min, with occasional mixing.

    N.B.: The time of oxidation should be the same for all samples.
14. Add 3.0 ml 3.6 *M* sodium thiosulfate. Mix and add 5.0 ml toluene. Cap tightly.
15. Shake vigorously for 5 min in a mechanical shaker (horizontal type). Centrifuge, then aspirate the toluene layer (top layer) and discard. Replace and tighten caps.
16. Place in a boiling water bath for 30 min.
17. Cool tubes with running tap water and then add exactly 5.0 ml toluene. Recap, shake for 5 min and then centrifuge.
18. Pipet exactly 2.5 ml of the toluene extract into small tubes and add 1.0 ml of Ehrlich's reagent while mixing rapidly.

    N.B.: Do not transfer any of the aqueous phase since it will cause color to fade.
19. Let stand for 30 min and read absorbance against toluene as blank at 560 nm in a spectrophotometer. Absorbance should be between 0.1 and 0.6. If reading is too high, make a dilution of the toluene extract.

Calculation:

$$\text{mg total hydroxyproline/ml} = \frac{A_x - A_b}{A_s - A_b} \times 0.036$$

mg total hydroxyproline/24 hr = mg/ml $\times$ 24 hour volume in ml

PROCEDURE FOR FREE HYDROXYPROLINE

REAGENTS

All reagents as used for total hydroxyproline.
Working Standard, 3.6 µg/ml. Dilute 1 ml stock standard to 100 ml with water.

PROCEDURE

1. Set up culture tubes (as in method for total hydroxyproline) for blank, unknown, and standard.
2. Pipet water as follows: 4 ml into blank, 3 ml into standard and 1 ml into sample tubes.
3. Add 3 ml urine to tubes for unknown and 1 ml working standard into tubes for the standard.
4. Proceed from step 10 in the procedure for total hydroxyproline.

Calculation:

$$\text{mg free hydroxyproline/ml} = \frac{A_x - A_b}{A_s - A_b} \times \frac{0.0036}{3}$$

$$= \frac{A_x - A_b}{A_s - A_b} \times 0.0012$$

mg free hydroxyproline/24 hr = mg/ml × 24 hour volume in ml

NOTES

1. Beer's law. In our laboratory linearity is obtained up to a concentration of at least 7.2 μg hydroxyproline per ml urine, and according to Kivirikko *et al* (124) calibration is satisfactory up to 10 μg hydroxyproline. Absorbance of the final colored solution should lie between 0.1 and 0.6 for satisfactory linearity.

2. Stability of samples. Urine samples were found to be stable at 30°C for 5 days providing the pH had been lowered to 1–2 with HCl. Stability in the freezer was satisfactory for an indefinite period.

ACCURACY AND PRECISION

The addition of an excess of alanine eliminates the influence of other amino acids upon the yield of pyrrole (192). The method is highly specific for hydroxyproline, even in the presence of a 10 000-fold excess of other amino acids. Potentially interfering materials are removed by extraction with toluene prior to the formation of pyrrole in the heating step (124). With hydrolyzed samples the recovery of added hydroxyproline to urine was found to be 90% or greater by Kivirikko *et al* (124) and varied over the range 90%–104% in our laboratory.

Precision (95% limits) is approximately ±15%.

NORMAL VALUES

In our laboratory, normal excretion values (nonparametric 95% limits) for 39 individuals who observed no dietary precautions were found to be 0.2–

1.8 mg/24 hour for free hydroxyproline and 25—77 mg/24 hour for total hydroxyproline.

The daily excretion of hydroxyproline in individuals not suffering from diseases of bone or collagen tissues was found to rise significantly with age between 10 and 20 years and to fall between 70 and 90 years (204). Laitinen *et al* (135) examined 46 healthy volunteers who had not eaten gelatin, meat, or fish during the 24 hour period before urine collection and obtained ranges for total hydroxyproline of 20.3—55.1 and 15.1—42.8 mg/24 hours for the age groups 18—21 and 22—55 years, respectively (135). For free hydroxyproline, they found 0.36—1.12, 0.34—1.33, and 0.30—1.23 mg/24 hour for the age groups 18—21, 22—40, and 41—55 years, respectively.

GLUTATHIONE

Glutathione is the tripeptide of glycine, cysteine, and glutamic acid:

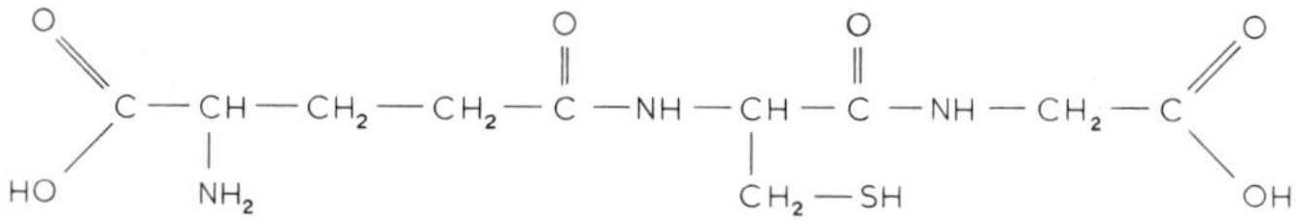

GLUTATHIONE

This is the reduced form of glutathione (GSH), the oxidized form being the disulfide (GSSG). Through reduced NADP and glutathione reductase, glutathione in the red cell is maintained in the reduced state. The functions of GSH seem to be to keep sulfhydryl groups in their active, reduced state and through glutathione peroxidase, to remove H_2O_2 (16).

Methods which have been proposed for the determination of erythrocyte reduced glutathione fall into several groups: (a) photometric, (b) enzymatic, (c) amperometric, (d) fluorometric, and (e) chromatographic.

A simple photometric procedure was described in 1963 using 5.5'-dithiobis-(2-nitrobenzoic acid) (DTNB). Only three stable reagents, precipitating solution, phosphate reagent, and DTNB, are required in the reaction postulated to be as follows (17):

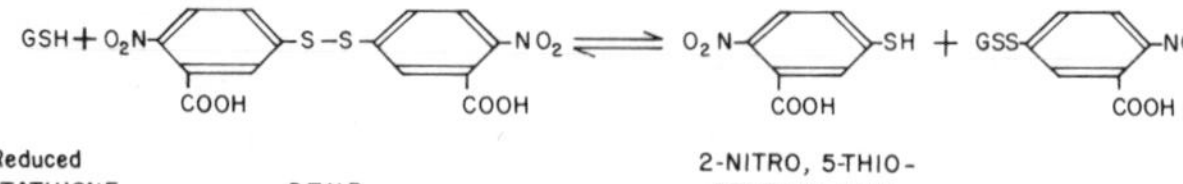

The first reaction product appearing in the equation is colored and can be measured at 412 nm (175). The method is not specific for glutathione since DTNB reacts with other sulhydryl groups. Glutathione is virtually the only nonprotein sulfhydryl compound present in significant levels in erythrocytes, ergothionine being present in relatively very small concentration, so the DTNB method gives useful information (17). If it is important to obtain higher specificity, the "alloxan 305" method of Patterson *et al* (179) may be used. The nitroprusside method has been re-examined and its drawbacks minimized, including an unstable color that must be read in not more than 1 minute and a final product concentration that is very sensitive to the temperature of reaction. Also, the nitroprusside reagent is unstable and solutions vary greatly in their activity and thus affect the method sensitivity (169). Methods involving enzymes also have been described. Glutathione functions as coenzyme for glyoxylase (115), formaldehyde dehydrogenase, maleylacetoacetate isomerase, DDT dehydrochlorinase, and indolylpyruvic acid enolketo tautomerase. In addition, glutathione is required for the activity of cis-trans maleylpyruvic acid isomerase which is the basis for a highly specific method (130).

Amperometric methods have been recommended in which the titration of the protein-free solution is carried out with silver salts (12, 79) to measure active sulfhydryl groups. A sensitive fluorometric method has been presented in which glutathione reacts with *o*-phthalaldehyde to yield a highly fluorescent product that is free of significant interference from a number of amino acids and sulfhydryl compounds (41). The method is sufficiently sensitive to be suitable for 5 μl blood samples. Recoveries range 91–103%.

Chromatographic methods also are available in which separation is carried out on paper that is then dipped in a ninhydrin-cadmium reagent. Glutathione concentrations down to 0.2 μg can be exactly estimated (246).

Although the DTNB method of Beutler *et al* (17) is very attractive for its simplicity in a routine laboratory, the classical "alloxan 305" method is retained here as a highly specific reference procedure.

DETERMINATION OF REDUCED GLUTATHIONE IN BLOOD

(method of Patterson and Lazarow, Refs. 178, 179)

PRINCIPLE

Glutathione reacts with an excess of alloxan to produce a substance with an absorption peak at 305 nm (see Fig. 18-10).

REAGENTS

Metaphosphoric Acid, Stock 25% Solution. Dissolve 25 g of the glacial pellets in water and dilute to 100 ml. Store in refrigerator.

Metaphosphoric Acid, Working 5% Solution. Dilute stock 25% solution 1:5 once each week. Store in refrigerator.

Alloxan Solution, 0.1 M. Dissolve 8.0 g of alloxan monohydrate (Eastman Kodak) in water and dilute to 500 ml. This reagent is stable for 1 week at refrigerator temperature or indefinitely if frozen. It is convenient to divide the reagent into 5 ml aliquots in glass vials and store in a freezer.

Equivalent NaOH Solution. Prepare a 0.5 N solution by dissolving 10.5 g NaOH in water and diluting to 500 ml. Standardize this solution so that 10 ml is equivalent to a mixture of 10 ml 5% metaphosphoric acid and 10 ml 0.1 M alloxan. Titrate the mixture with the NaOH to pH 7.5 with a pH

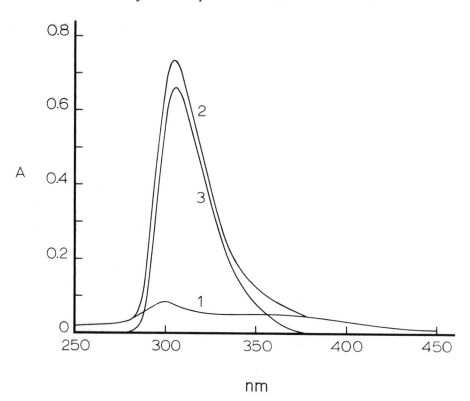

FIG. 18-10. Absorption curves in alloxan "305" method: (1) reagent blank vs water, (2) 50 µg standard vs water, (3) standard vs reagent blank. Beckman Model DU Spectrophotometer.

meter and dilute the NaOH solution appropriately. This titration should be made periodically since the reaction is pH sensitive.

Phosphate Buffer, 0.5 M, pH 7.5. Dissolve 59.7 g anhydrous Na_2HPO_4 (or 74.8g $Na_2HPO_4\cdot2H_2O$) and 10.88g KH_2PO_4 in water and dilute to a final volume of 1 liter. Check pH by meter on a 1:5 dilution.

NaOH, 1N (approx.). 40 g/liter.

Glutathione Standard, 50 µg/ml. Glutathione, "greater than 99% pure," is obtainable from Schwarz/Mann. Store in desiccator and preferably in a cool place. Dissolve 25.0 mg in 500 ml 5% metaphosphoric acid. Divide the standard in 1–2 ml aliquots in glass vials and store in deep freeze.

PROCEDURE

1. Mix 1.0 ml oxalated whole blood (not serum or plasma) and 7.0 ml water. Hemolysis must be complete. Add 2.0 ml 25% metaphosphoric acid, mix, and centrifuge.

2. Set up the following in test tubes:

Reagent blank. Set up 2 tubes:

b = 1.0 ml 5% metaphosphoric acid + 1.0 ml 0.1 M alloxan.

b_o = 1.0 ml 5% metaphosphoric acid + 1.0 ml H_2O.

Standard. Set up 2 tubes:

s = 1.0 ml standard (50 μg glutathione) + 1.0 ml 0.1 M alloxan.

s_o = 1.0 ml standard + 1.0 ml H_2O.

Unknown.   Set up 2 tubes for each unknown:

x = 1.0 ml filtrate + 1.0 ml 0.1 M alloxan.

x_o = 1.0 ml filtrate + 1.0 ml H_2O.

3. Line up all the tubes in order and add to the first tube 1.0 ml 0.5 M phosphate buffer, followed *immediately* by 1.0 ml of the "equivalent NaOH solution" and mix. Repeat this procedure with each successive tube at 30 sec intervals.
4. After exactly 6 min, add 1.0 ml 1 N NaOH to each tube (again adding at 30 sec intervals) and mix. This stops the reaction and stabilizes the "305" product for several hours.
5. At 305 nm, read b against b_o, s against s_o and x against x_o.

Calculation:

$$\text{mg glutathione}/100 \text{ ml} = \frac{A_x - A_b}{A_s - A_b} \times 0.05 \times \frac{100}{0.1}$$

$$= \frac{A_x - A_b}{A_s - A_b} \times 50$$

ULTRAMICRO PROCEDURE

The macro procedure can be scaled down as much as 20-fold.

NOTES

1. *Beer's law.* Beer's law is obeyed up to a concentration of at least 60 mg/100 ml.
2. *Stability of samples.* The reduced glutathione level in four samples of oxalated whole blood was followed in our laboratory for 24 hour at room temperature and little or no decrease in that period of time was observed. Samples stored at refrigerator temperature were more prone to hemolysis

which, in every case, produced a large decrease in the reduced glutathione level. Glutathione of shed human blood has been shown to be in a dynamic state since glycine is incorporated into the molecule. Glucose and low temperature are functional in keeping GSH of stored blood at its original level.

Concomitant with the disappearance of glucose in stored blood is the appearance of GSSG, and the addition of glucose restores the GSH level almost to the original level. Blood held for 22 hours at a temperature of 6°C showed a decline of 2.5% in GSH level (172).

ACCURACY AND PRECISION

When present in equivalent molar amounts, the absorbance of cysteine, cysteinylglycine, and ergothioneine at 305 nm is not more than 5% of that given by glutathione. The absorptivity of glutamylcysteine is 16% of that of glutathione. Ascorbic acid lowers the value of a glutathione determination 10%, 13%, and 20% when present in amounts equal to 0.5, 1, and 5 times the molar concentration of glutathione, respectively (178).

The reproducibility of the technic is about ±5%.

NORMAL VALUES

A range of 28–34 mg/100 ml blood was obtained in our laboratory in a series of 8 normal adults.

Glutathione levels in erythrocytes fall with the red cell age. Values fell from 74.9 mg/100 ml red cells for the young cells to 60.0 mg/100 ml red cells for the fraction containing the oldest cells. In G-6-PD deficient cells, the various red cell fractions showed lower values although the percent decrease was similar (207). It has been reported that the red cell GSH level is inversely related to the hematocrit level. There is no correlation with patient's age within the range 18–55 years (37).

METHYLMALONIC ACID IN URINE

Methylmalonyl-coenzyme A is converted to succinyl-coenzyme A by a cobamide linked isomerase. The excretion of excessive amounts of methylmalonic acid (MMA) is found in at least two conditions, vitamin B-12 deficiency (119), and the inherited disease congenital methylmalonic aciduria, which is associated with the lack of the isomerase activity (231).

The reaction between MMA and diazotized *p*-nitroaniline to give a green color has been utilized for the quantitative determination in urine. While a number of urinary constituents are found to react, passage of the urine through Dowex 3-X4 resin removes all interfering substances but malonate (71). Preliminary extraction of the sample with ether prior to passage through the resin yields a more sensitive procedure (249). Deacidite FF(IP), can be substituted for the Dowex resin (78). Organic acids including MMA have been separated from biologic fluids by column chromatography on silicic acid. Detection of the weak organic acids in the eluate was accomplished with *o*-nitrophenol and the identity of the eluted peaks confirmed by paper or gas chromatography (9). Preliminary isolation of MMA from urine has been achieved also by ion exchange chromatography on Dowex 2-X8 using water, 0.02 *M* cupric acetate and 2 *M* formic acid to elute low molecular weight fatty acids that would interfere in the subsequent gas chromatographic analysis (108). Gas chromatographic technics for analyzing solvent extracts of urine samples for organic acidurias have been described. After extraction with ether (44) or ethyl acetate followed by ether (95), the organic acids were methylated using diazomethane and separated on GLC using 10% Apiezon L (available from A. H. Thomas or Scientific Products) on Celite (44) or a 15% diethylene glycol succinate column (95).

Methylmalonic acid extracted from urine with ether has been isolated by thin layer chromatography on silica gel G, its migration being detected with methyl red indicator. Confirmation of identity and quantitation was achieved by removal of the spot from the TLC plate and analysis by GLC following methylation (73). This procedure was found to show peaks overlapping that of MMA, a disadvantage that could be avoided by preparing the di-trimethylsilyl esters directly from the ether extract. Then, a single uncontaminated symmetrical peak was obtained for MMA (226).

Simple screening procedures for methylmalonic aciduria have been reported using thin-layer chromatography on silica gel (53) or two-dimensional chromatography on Whatman 1 filter paper (10) with brom-cresol green as detection agent. A yet more rapid, specific, and sensitive method can be carried out on silica gel thin-layer chromatography followed by location of MMA with the stabilized diazotate of dianisidine, Fast Blue B (81). This method is described below.

SEMIQUANTITATIVE DETERMINATION OF METHYLMALONIC ACID IN URINE

(method of Gutteridge and Wright, Ref. 81)

PRINCIPLE

Methylmalonic acid is extracted from acidified urine and separated from other urine components by thin layer chromatography (TLC). The methylmalonic acid is visualized on the chromatogram using Fast Blue B and the amount present approximated by comparison with standards.

REAGENTS

Ethanol, absolute. Ethanol denatured with methanol is satisfactory.

HCl, conc

Ammonium Sulfate

Diethyl Ether

Solvent system. Combine just prior to use 90 ml amyl acetate, 30 ml glacial acetic acid, and 9 ml water.

Fast Blue B Reagent. Dissolve 500 mg Fast Blue B (Gurr, Ltd.) in 75% ethanol and make up to 100 ml. Ethanol denatured with methanol is satisfactory. Add 4 ml glacial acetic acid. Prepare just prior to use.

Stock Standard, 2 mg/ml. Dissolve 200 mg methylmalonic acid in 100 ml water. Store at 4°C.

Working Standard, 2 mg/100 ml. Dilute 1 ml stock to 100 ml with water. Prepare freshly before use.

TLC plate. 0.25 mm silica gel on 0.125'' glass. These can be purchased already prepared.

PROCEDURE

1. Transfer 1% of the 24 hr. urine volume to a 50 ml glass stoppered centrifuge tube and 10 ml working standard to a second tube.

   N.B. If a random sample is received, results should be reported on a per liter basis.

2. Acidify the sample and standard to pH 2 (pH paper) with conc HCl and mix.

3. Saturate the sample and standard with ammonium sulfate.

4. Extract the samples and standard three times with 15 ml aliquots of diethyl ether. Combine the ether extracts and evaporate to dryness under N_2 in a 37°C water bath.

   N.B. Emulsions which form may be broken by adding additional ether or allowing to stand for 5–10 min.

5. Dissolve the dried extracts in 250 μl ethanol.

6. Equilibrate a TLC tank with the developing solvent for 30 min.

7. Activate TLC plate by heating at 100°C for 30 min and allow to cool.

8. Apply 5 and 10 μl aliquots of the ethanol concentrate from the unknowns and standards as 1 cm streaks 1.5 cm from the bottom of the TLC plate.

9. Develop the plate, allowing the solvent to ascend 10 cm above the point of sample application.

10. Remove the plate from the TLC tank and dry in a stream of cool air.

11. Spray the plate with Fast Blue B reagent and dry in a stream of warm air.

12. Place the plate in a dark area and examine 30 min after spraying. Approximate the concentration of the methylmalonic acid in the urine unknowns by comparison with the two standard spots. The color of the MMA in the extract and standards after spraying is usually a dark purple color about 5 cm above the point of application. The purple color fades after about 4 hours.

Calculation: The 5 and 10 µl aliquots of standard contain 4 µg and 8 µg MMA, respectively, while the urine extract spots are derived from 1/5000 and 1/2500 the total volume of urine.

Estimate the micrograms of MMA in the sample spots by comparison with the two standard spots and calculate total excretion as follows:

$$\text{mg MMA/24 hour} = \mu\text{g MMA in sample spot} \times \frac{2500}{1000}$$

$$= \mu\text{g MMA in sample spot} \times 2.5$$

when 10 µl urine extract is compared. If 5 µl extract used,

$$\text{mg MMA/24 hour} = \mu\text{g MMA in sample spot} \times \frac{5000}{1000}$$

$$= \mu\text{g MMA in sample spot} \times 5$$

NOTE

1. Stability. Unpreserved urine samples are stable for at least 7 days at room temperature. When stored at $-20°$C samples are stable for at least 4 months (78).

ACCURACY AND PRECISION

The solvent system used for the thin-layer chromatography separates methymalonic acid from other compounds found in urine that react with Fast Blue B (81). This locating agent reacts intensely with methymalonic acid so that the concentrations found in the urine of normal individuals can be detected readily.

The test is semiquantitative, and no precision data are available.

NORMAL VALUES   m

Experience in our laboratory from running hundreds of normal samples indicates that the normal excretion of methylmalonic acid is less than 10 mg/24 hour in agreement with Gutteridge and Wright (81). Giorgio and Plaut (72), investigating the urine of 13 healthy persons by passing through Dowex 3 and carrying out the color reaction with diazotized *p*-nitroaniline, reported a range of 0–11.2 mg/24 hour. Gutteridge and Wright (81) conclude that few healthy individuals excrete more than 10 mg MMA/24 hour.

While Cox and White (44), who used a GLC procedure, found a normal excretion of up to 3.8 mg MMA/24 hour based on five normal individuals, Gompertz *et al* (74), employing a similar procedure, recommended a normal range of up to 9.0 mg/24 hour.

HOMOGENTISIC ACID

In 1859, Bödeker (23) named the disease *alcaptonuria* in which "alcapton" (also spelled "alkapton") appears in the urine. This rather rare disease is characterized by excretion of urine that upon standing gradually becomes darker in color and may finally turn black. It is now known to be an inborn error of phenylalanine and tyrosine metabolism transmitted as a recessive Mendelian trait (there is some evidence that it might be dominant in some cases). No method is available for detection of heterozygotes (202). The normal metabolism of these amino acids proceeds as follows:

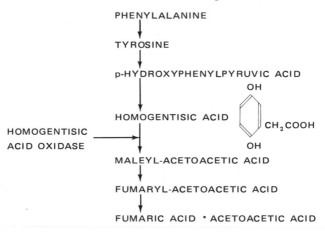

PHENYLALANINE

TYROSINE

p-HYDROXYPHENYLPYRUVIC ACID

HOMOGENTISIC ACID

HOMOGENTISIC ACID OXIDASE

MALEYL-ACETOACETIC ACID

FUMARYL-ACETOACETIC ACID

FUMARIC ACID · ACETOACETIC ACID

The defect in alcaptonuria is the absence of homogentisic acid oxidase, resulting in accumulation of homogentisic acid (2,5-dihydroxyphenylacetic acid) (134).

The qualitative aspects of homogentisic acid in urine include the following (114, 156): If the urine as passed is acid, its color is generally normal. When allowed to stand exposed to air, or if initially alkaline or made so, in the course of a few days the urine becomes dark brown to black. This process begins at the surface and gradually spreads throughout. Darkening of urine, however, may also result from gentisic acid (metabolite of salicylates), melanogen, indican, urobilinogen, porphyrins, and phenols. An orange precipitate forms upon boiling with Fehling's solution and a black solution with subsequent formation of a yellow precipitate is produced by boiling with Benedict's reagent. Benedict's reagent is also reduced in the cold. Clinitest (Ames) gives a positive test, but glucose oxidase tests are negative. There is no fermentation with yeast. A black precipitate is formed immediately with ammoniacal $AgNO_3$ in the cold (ascorbic and uric acids also reduce in the cold). Millon's test is positive. Addition of 10% ferric chloride drop by drop produces a deep blue color that persists for only a fraction of a second. This phenomenon can be repeated until oxidation is complete. The ferric chloride reaction is not absolutely specific.

A qualitative test for a single drop of urine has been proposed in which the homogentisic acid acts as a developer on photographic film (62). The test is positive if the film turns black upon addition of the urine and a solution of NaOH. The test can be performed in full daylight. A rough estimation of concentration can also be obtained, using hydroquinone in urine as a standard (223).

The reducing property of homogentisic acid may erroneously lead to a diagnosis of diabetes mellitus (156).

Normally, homogentisic acid cannot be detected in urine (173). The output in alcaptonuria is usually in the range of 3–5 g/day although it may go as high as 16 (62). Attempts to quantitate homogentisic acid in urine date back as early as 1891, when a method was proposed based on reduction of ammoniacal $AgNO_3$ (11, 250); this method, however, was not very specific (66). Reduction of Ag^+ at pH 4.4, precipitated in the form of a sol in the presence of a gold sol and gum arabic and read photometrically, constitutes a modification for which specificity is claimed (151). Reduction of phosphomolybdic acid with colorimetric measurement is another approach (28) with which there are difficulties, including nonspecificity (171, 211). Greater specificity has been obtained by prior extraction of homogentisic acid into ether from acidified urine, effecting separation from glutathione and ascorbic acid, which interfere (133). Measurement of the blue color produced by reduction of alkaline phosphotungstate constituted another approach (137). An iodometric method with back titration with thiosulfate has also been employed (21, 136, 145, 162, 163). The presence of catechol or dopa in significant amounts may cause misleading results (171).

Seegmiller *et al* pointed out that the error introduced in reduction methods by other substances with reducing properties similar to homogentisic acid is negligible at the high levels encountered in alcaptonuric urines, but that this is not true with the relatively low levels in alcaptonuric plasma (215).

Methods have been presented for the confirmation of homogentisic acid by paper chromatography (120, 133, 156, 240) and by thin-layer chromatography (202). Quantitation also has been achieved by comparing the stained areas given by unknown samples with reference spots (120), as well as by elution from the chromatogram with subsequent photometric measurement (42). A gas chromatographic procedure also has been described (24).

Skarżyński *et al* (220) described two cases in which urine showed two spots differing in R_f: One was homogentisic acid and ether soluble; the other was a glycine conjugate and water soluble. McKenzie *et al* (156) have expressed the opinion that paper chromatography should always be used for confirmation since no other test is completely reliable. This opinion was voiced prior to the introduction of a specific enzymatic method for plasma and urine by Seegmiller *et al* (215); the method is based on enzymatic conversion of homogentisic acid to maleylacetoacetic acid with purified homogentisic acid oxidase, and measurement of the absorption of the product at 330 nm. Using this method, which has a sensitivity of 1 $\mu g/ml$, normal plasma showed no detectable homogentisic acid.

Since homogentisic acid is not detectable in normal urine and since the problem of the clinical chemist is primarily one of establishing its presence or absence with a high degree of specificity, a quantitative procedure will not be presented here. If one or more of the qualitative tests previously mentioned is positive (darkening of urine upon alkalization, ferric chloride test, etc.), the presence of homogentisic acid should be confirmed by chromatography.

SCREENING TESTS FOR HOMOGENTISIC ACID IN URINE

(methods of Jensen, Ref. 114, and McKenzie et al, Ref. 156)

In order to select urine samples that may contain homogentisic acid, one or more of the following qualitative, screening tests should be performed.

REAGENTS

$FeCl_3 \cdot 6H_2O$, 10%
Benedict's Qualitative Solution. See Chap. 25 for preparation.
$AgNO_3$, 3%
NH_4OH, 10%

PROCEDURE (each step is a different test)

1. Add 2 drops 10% ferric chloride reagent to 2 ml urine. The appearance of a transient blue or green color is a tentative indication of homogentisic acid.
2. Add 0.5 ml urine to 5 ml Benedict's qualitative reagent. Mix and place in a boiling water bath for 3 min. Urine from an alkaptonuric individual causes the mixture to assume a green-brown or green-black color that upon standing shows a yellow precipitate of cuprous oxide. Benedict's solution is also reduced at room temperature by homogentisic acid.
3. Add 5 ml 3% silver nitrate solution to 0.5 ml urine and mix. Upon the addition of a few drops of 10% ammonium hydroxide the mixture turns black if homogentisic acid is present.

NOTE

1. Specificity. The tests described above are not specific for homogentisic acid. A green color with ferric chloride is given by phenylpyruvic acid, and a black reaction occurs when urine is examined with Clinitest (same reaction as Benedict's qualitative test) after the patient has received the x-ray contrast medium, sodium diatrizoate (Hypaque Winthrop) (138).

If one of the screening tests given above is positive, proceed to the chromatographic confirmation of homogentisic acid.

CONFIRMATION OF HOMOGENTISIC ACID IN URINE BY THIN-LAYER CHROMATOGRAPHY

(method of Roth and Felgenhauer, Ref. 202, modified)

PRINCIPLE

Urine is pretreated by passage through a column of cation exchange resin and an aliquot of the eluate subjected to thin-layer chromatography on silica gel. Homogentisic acid migrates to a characteristic R_f and is visualized with a molybdate spray reagent.

REAGENTS

Thin-layer Plates Coated with Silica Gel G. These can be purchased already prepared.

Solvent System. Hexane-ethyl acetate-acetic acid (50:50:2).

Cation Exchange Resin. Dowex 50W-X12, 200–400 mesh, H^+ form. Wash resin six times with water in a beaker. Allow to settle briefly and decant supernate with suspended fine resin particles.

Standard. Dissolve 5 mg homogentisic acid in 10 ml absolute ethanol (ethanol denatured with methanol is satisfactory).

Gentisic Acid Standard. Dissolve 5 mg gentisic acid in 10 ml absolute ethanol.

H_2SO_4, *5 N.* Add 15 ml conc H_2SO_4 to 93 ml water.

Spray Reagent. 5 g ammonium molybdate dissolved in 100 ml 5 N H_2SO_4.

PROCEDURE

1. Prepare a 2 x 1 cm resin bed in a glass column having a stopcock. Use glass wool to support the resin.
2. Transfer approximately 5 ml urine to the top of the column and allow to pass through the resin. Collect the eluate.
3. Add sufficient developing solvent mixture to a chromatography tank and allow the atmosphere to become saturated (requires not less than 30 min).
4. Apply 30 μl pretreated urine sample, undiluted and diluted 1:10 with water, and 30 μl of each standard to appropriate spaces 10 mm from one edge of a thin-layer chromatography plate.
5. Place the plate in the chromatography tank, opening the lid for the minimum time possible. Allow the solvent to migrate until the front has travelled 10 cm (approximately 1 hour).
6. Remove the plate and evaporate the solvent using a hair dryer.

7. Spray the plate with the molybdate reagent in a hood.
8. Direct a stream of hot air at the plate for at least 15 min or until the standard spots show up strongly.

INTERPRETATION

When urine from an alkaptonuric individual is examined, the chromatogram will show a bright blue spot of homogentisic acid parallel with the reference standard materials at R_f 0.17.

NOTES

1. Stability (215). Urine acidified and frozen is stable for at least 1 week. There is a 25% loss in frozen plasma in 1 week (test can also be performed on serum or plasma).

2. Specificity. Gentisic acid, if present, gives a blue color indistinguishable from that of homogentisic acid but at an R_f of approximately 0.29. It is, therefore, important to apply to every chromatogram a homogentisic acid reference solution.

N-FORMIMINOGLUTAMIC ACID (FIGLU)

A number of methods have been devised for the determination of urinary FIGLU. A sensitive spectrophotometric procedure which will detect 0.1 μg FIGLU/ml urine is based upon the enzymatic transfer of the formimino group of FIGLU to tetrahydrofolic acid (THF) to form 5-formimino-THF (236). Subsequently 5-formimino-THF is converted by the addition of acid to 5,10-methenyl-THF, which has an absorptivity of 25 000 at 350 nm. While the test appears straightforward, the preparation of FIGLU transferase from hog liver is laborious.

Microbiologic procedures have been described for the determination of fourteen amino acids by Henderson and Snell (98). Using *L. arabinosis,* which will respond to glutamic acid or to glutamine but not to FIGLU, it was possible to assay glutamate in urine that had been heated to remove glutamine and to convert FIGLU to glutamic acid (31). Silverman *et al* (218) described a method in which THF is formylated enzymatically by FIGLU to give 10-formyl-THF. This is converted nonenzymatically to 5-formyl-THF, which is assayed using *P. cerevisiae* as test organism. Urine samples are tested before and after autoclaving, which converts FIGLU to glutamate. The presence of FIGLU is assumed when the sample loses activity after heating. The THF-enzyme reagent that is extracted from chicken liver in a lengthy operation can be lyophilized and stored at $-20°$C for at least 6 months without loss of activity (35).

Several authors have reported methods based on electrophoretic separation followed by staining with ninhydrin or other detection agents. Using

Whatman 3MM paper Knowles *et al* (25) subjected 25 μl duplicate aliquots of urine to high voltage electrophoresis (6.0 kV, 60–80 A) in pyridine-acetate buffer. Samples containing known amounts of FIGLU were included in the run. Following electrophoresis, one paper was exposed to ammonia vapor to convert FIGLU to glutamic acid. Following neutralization of the ammonia, both papers were stained with ninhydrin. The strips were examined for the presence of a spot appearing only in the ammonia treated strip having mobility corresponding to known FIGLU. The method will detect 10 μg FIGLU/ml urine. Satisfactory separation of FIGLU from glutamic acid can be achieved on cellulose acetate at 400–500 V. Even this voltage is considered unnecessary since FIGLU can be demonstrated by comparison of the ammoniated with the nontreated strip at 200 V in spite of considerable overlap (128). This procedure can be made quantitative by cutting out the spots corresponding to the migration of FIGLU on both strips and eluting into chloroform:ethanol (9:1) or into an ethanol-copper sulfate reagent. The absorbance of the colored solutions is then measured. Glutamic acid may be used as standard since its equivalence to FIGLU is known (127). In another procedure glutamic acid was removed from urine by treatment with chloramine-T, following which the remaining amino acids and FIGLU were retained on a column of Zeocarb 225 (Permutit Company). After elution with ammonium hydroxide they were separated by electrophoresis on cellulose acetate and visualized with ninhydrin (143). When chromatographed on an automatic amino acid analyzer using a 50 cm resin column, FIGLU emerged between ammonia and lysine in a 6 hour run. The ninhydrin color factor for FIGLU was found to be 1.9 compared with 20 for alanine and 18.8 for glutamic acid. Recovery was 85% at both the 3 and 5 μM levels. (55). A photometric method using an alkaline ferricyanide-nitroprusside (AFN) reagent was examined by Tabor and Wyngarden (236) who found it ι less sensitive than their enzymatic method. Further work by Johnstone *et al* (116) employs the principle that the color developed in 30 sec is used as blank for the slowly appearing absorbance at 485 nm due to FIGLU. Urocanic and glutamic acids did not react with the reagent. A further refinement has been described in which urine is passed down a column of aluminum oxide (106). FIGLU is eluted with sodium tetraborate solution to which the AFN reagent is added and the color allowed to develop for 60 min before it is read at 485 nm. FIGLU in another aliquot of the eluate is hydrolyzed with alkali, and this tube is used as the blank. The procedure appears to be free from interference due to substances thought likely to react with the AFN reagent. Recoveries ranged from 70%–100%, and the reproducibility was about ±30%. Thin-layer chromatography on cellulose of 5 μl urine samples afforded a semiquantitative method for FIGLU (196). After development in *n*-butanol:glacial acetic acid:water, 114:38:60, the plate was exposed to ammonia vapor and sprayed with ninhydrin. FIGLU gave a discrete spot just ahead of glutamic acid that could be quantitated approximately by comparison with standards. The generalized aminoaciduria of patients with megaloblastic anemias did not interfere with the assessment of the FIGLU spot. A sensitive and quantitative method for FIGLU has been developed that uses stable and readily available reagents (159). Interfering substances are removed from the urine sample, and one portion subjected to alkaline hydrolysis. The oxidation of glutamic acid in hydrolyzed and nonhydrolyzed aliquots is effected enzymatically by glutamic acid dehydrogenase:

$$\text{glutamic acid} + NAD + H_2O \underset{\longleftarrow}{\overset{GAD}{\longrightarrow}} \alpha\text{-ketoglutarate} + NADH + NH_4^+ + H^+$$

The system includes NAD so that the reaction can be followed at 340 nm. This method is given in detail below.

DETERMINATION OF N-FORMIMINOGLUTAMIC ACID IN URINE

(method of Meiss et al, Ref. 159)

PRINCIPLE

Urine is brought to pH 7 and passed through a Permutit column to remove NH_4^+. The effluent is divided into two portions to one of which is added NaCl and activated charcoal. The charcoal removes substances that have strong absorbance at 340 nm and that cause nonspecific oxidation of NADH. NaCl blocks the adsorption of glutamic acid and FIGLU onto the charcoal. The second aliquot is brought to pH 12–13 and incubated at 37°C to convert FIGLU to glutamic acid. Following neutralization, NH_4^+ produced during the hydrolysis step is removed with Permutit and the solution treated with charcoal. Both solutions are then assayed for glutamic acid using glutamic acid dehydrogenase (GAD) and NAD. The increase in NADH is measured at 340 nm and compared with standards treated similarly.

REAGENTS

HCl, 6 N. Dilute conc HCl with an equal vol water.
HCl, 3 N. Dilute 1 vol conc HCl with 3 vol water.
HCl, 1.5 N. Dilute 1 vol conc HCl with 7 vol water.
HCl, 1.2 N. Dilute 1 vol conc HCl with 9 vol water.
NaCl, 0.3 M. Dissolve 17.6 g NaCl in 1 liter water.
NaOH, 6 N. Dissolve 48 g NaOH in 200 ml water.
NaOH, 3 N. Dilute 6 N NaOH with an equal vol. water.

Tris Buffer, 0.5 M, pH 8.1. Dissolve 12.1 g Tris [tris(hydroxymethyl) aminomethane] in 100 ml water. Adjust pH to 8.1 (pH meter) with conc HCl. Dilute to 200 ml with water.

Phosphate Buffer, 0.2 M, pH 8.0. Dissolve 5.68 g anhydrous Na_2HPO_4 in 100 ml water. Adjust pH to 8.0 (pH meter) with 1.5 N HCl. Dilute to 200 ml.

Phosphate Buffer, 0.2 M, pH 7.6. Dissolve 5.68 g anhydrous Na_2HPO_4 in 100 ml water. Adjust pH to 7.6 (pH meter) with 1.5 *N* HCl. Dilute to 200 ml.

Activated Charcoal (Norit A, Pfanstiehl). Add 1 liter 1.2 *N* HCl to 200 g Norit A. Bring mixture to a boil. Cool, filter, and wash well with water. Dry at 105°C.

NAD Solution. Dissolve 4 mg NAD/ml 0.5 *M* Tris [tris(hydroxymethyl) aminomethane] , pH 8.1. Prepare just before use.

Permutit (Ionac Company).

Glutamic Acid Dehydrogenase (GAD). Add 0.3 ml of commercially available GAD solution, 10 mg/ml (Boehringer, Type II in 50% glycerol), to 9.7 ml 0.2 *M* Na_2HPO_4 , pH 7.6 buffer. Make up once a week.

Glutamic Acid Stock Standard, 60.0 µg/ml. Dissolve 60.0 mg glutamic acid in 1 liter water. Store frozen in 2 ml aliquots.

Glutamic Acid Working Standard, 20 µg/ml. To 1 ml stock standard add 2 ml water. Prepare fresh and discard after use.

PROCEDURE

1. To a 20 x 150 mm test tube, add 0.1 ml 0.2 *M* phosphate buffer pH 8.0, 6.0 ml water, and 5.0 ml urine. Mix and chill in ice water.
2. Add a few drops of 6 *N* NaOH to bring pH to about 7 (pH paper). Let stand 10 min.
3. Centrifuge if any flocculation appears.
4. Prepare 1 x 3 cm Permutit columns (use glass columns with 1.5 mm bore ground glass stopcock). Wash columns thoroughly with water and allow water to drain completely prior to addition of samples.
5. Transfer the entire sample of diluted urine to a Permutit column. Adjust flow rate to about 10 ml/hour (4 drops/min).
6. Discard first 3 ml of eluate and collect remainder.
7. Transfer a 3 ml and a 4.5 ml aliquot of each eluate to 15 x 125 mm test tubes. The 3 ml set will be used for the glutamic acid and the 4.5 ml for the FIGLU assay.
8. Glutamic acid assay preparation:
   (a) To the 3.0 ml aliquots from each column add 1.0 ml 0.3 *M* NaCl and 150 mg Norit.
   (b) Shake vigorously for 1 min. Centrifuge and transfer supernates to clean tubes for use in step 10.
9. FIGLU assay preparation:
   (a) To the 4.5 ml eluate from each column, add 0.225 ml 3 *N* NaOH. pH should be 12–13. Place in 37°C water bath for 30 min.
   (b) Remove from water bath and add 0.225 ml 3 *N* HCl, 0.15 ml pH 8.0 phosphate buffer, and 1.5 ml 0.3 *M* NaCl; mix.

(c) Add 1.5 ml Permutit and shake for 2 min.

(d) Centrifuge and transfer supernates to 12 x 125 mm test tubes containing 225 mg Norit.

(e) Mix for 1 min and centrifuge. Transfer supernates to clean test tubes.

10. Assay Procedure:

(a) Set up the following in 13 x 100 mm test tubes:

	NAD Tris (ml)	0.2 M Na$_2$HPO$_4$ pH 7.6 (ml)	GAD (ml)	GA* Prep. (ml)	FIGLU Prep (ml)
GA* Blank	0.5	0.5	—	1.0	—
GA Test	0.5	—	0.5	1.0	—
FIGLU Blank	0.5	0.5	—	—	1.0
FIGLU Test	0.5	—	0.5	—	1.0

*GA = glutamic acid

(b) A standard curve is to be run with each batch. Set up the following in 13 x 100 mm test tubes:

µg/ml	Working std (ml)	Water (ml)	NAD-Tris (ml)	0.2 M pH 7.6 Na$_2$HPO$_4$ (ml)	GAD (ml)
0	0	1.0	0.5	0.5	0.5
2.5	0.125	0.875	0.5	0.5	0.5
5	0.25	0.75	0.5	0.5	0.5
10	0.5	0.5	0.5	0.5	0.5
20	1.0	0	0.5	0.5	0.5

(c) Mix. If a precipitate appears in the urine samples, centrifuge.

(d) Stopper and place in 37°C water bath.

(e) Between 15 and 30 min after placing tubes in water bath, read absorbances of all solutions against water at 340 nm in 2 ml cuvets with 1 cm light path. (See note 3.)

Calculation:

1. Plot calibration curve on graph paper and read concentration (C) of glutamic acid in each tube in µg/ml.

2. Substitute the values obtained in the appropriate expression below.

(a) μg endogenous GA/ml urine $= \dfrac{C \times 4 \times 11.1}{3 \times 5}$

$$= 2.96\ C$$

(b) μg total GA/ml urine $= \dfrac{C \times 6.6 \times 11.1}{4.5 \times 5}$

$$= 3.26\ C$$

(c) mg FIGLU/total volume urine

$$= \frac{174}{147}\ (\text{total} - \text{endogenous GA/ml}) \times$$

$$\frac{\text{total volume urine in ml}}{1000}$$

$$= 1.18\ (\text{total} - \text{endogenous GA/ml}) \times$$

$$\frac{\text{total volume urine in ml}}{1000}$$

where 174 = molecular weight of FIGLU

147 = molecular weight of GA

NOTES

1. *Calibration.* The plot of absorbance against glutamic acid concentration, which is quite reproducible, is nonlinear but does not give a pronounced curve until values above 15μg are reached.

2. *Stability of samples.* Formiminoglutamic acid is very unstable at alkaline pH with a half-life of 20 min at pH 11.5, 70 min at pH 10.8, and 40 hours at pH 9.2. It is relatively stable at pH 7 at 0°C and very stable in acid. At 25°C, no deterioration occurred after several months in 0.1 *N* acetic acid or in 0.1 *N* or 6 *N* HCl (236). Several authors recommend that the 24 hour urine sample be preserved by the addition of thymol crystals to diminish bacterial growth and conc HCl, 0.5 ml (125), 1 ml (101, 128), 2 ml (106) or 5 ml 5 *N*. The sample then will keep for 2 week at 28°C or 1 month at −20°C (106). Two ml conc HCl followed by storage at −20°C is also recommended (116) as is collection under toluene with the addition of 8 ml 8 *N* acetic acid (159) or sufficient glacial acetic acid to give a final concentration of 1 ml/100 ml urine (31).

3. *Stability of final test solutions.* Because the equilibrium level of NADH varies with temperature, solutions should remain in the water bath until they

are to be read. At that time they should be transferred quickly to cuvets and the absorbance measured at once (159). For this reason, a batch should consist of no more than six samples. It is also advisable to run all the total glutamic acid determinations first followed by those for endogenous glutamic acid. Solutions should be read in the sequence in which they were set up.

ACCURACY AND PRECISION

This enzymatic method (159) is about one-seventh as sensitive as that of Tabor and Wyngarden (236). Although reduction of NAD by norvaline and leucine in the presence of GAD has been reported, no such activity was noted by Meiss *et al* (159) with leucine concentrations up to 2 *mM* under these assay conditions. Glutamine at a concentration four times that found normally also did not interfere significantly.

Preformed and total glutamate can be quantitated with a precision of about ±6%. The values of FIGLU, derived from the difference between these fractions are considerably less precise, particularly in the normal range when concentrations are low.

NORMAL VALUES

In our experience normal excretion of FIGLU appears to be less than 3 mg/24 hours. Meiss *et al* (159), using the enzymatic method described above, found no FIGLU in the urine of 14 normal individuals. According to an electrophoretic method (128) the normal range was shown to be less than 2 mg/hour. Excretion rates up to 0.5μmol (86 μg)/hour were obtained (100) from 3 normal adults and 7 child and 16 adult patients who were non-leukemic and not receiving folate antagonists. Eleven healthy males gave values of up to 8.7 μg/ml by the alkaline ferricyanide-nitroprusside method (106).

By far the greatest use of this determination has been made following oral histidine loading. As originally described by Luhby *et al* (146) a load of 15 g L-histidine monohydrochloride daily was administered orally in three doses for two to three days prior to urine collection. Under these conditions, excretion was 0.5–30 mg FIGLU/24 hour in normal adults and 1.9–7.2 mg/24 hour in children. Seventeen normal individuals, following a single 15 g oral loading dose of L-histidine, excreted from less than 50–100 μg FIGLU/ml urine. The sample was collected over the period 3–8 hour following the loading dose (116). The excretion of FIGLU by normal individuals, after receiving 10 g L-histidine orally, was 15 μg/ml urine using a 5 hour collection period (101). A loading dose of 0.12 g L-histidine monohydrochloride monohydrate/pound body weight resulted in FIGLU excretions in 17 healthy males of up to 26 mg/8 hour (106). Roberts and Mohamed (196) found that following a single oral dose of 5–15 g histidine, the urinary FIGLU excretion was 0–10 mg/8 hour. Meiss *et al* using the procedure described above obtained FIGLU excretion values from 14 normal subjects ranging from 1.1 to 7.3 mg/8 hour following the ingestion of 15 g histidine.

REFERENCES

1. ACKERMANN PG, KHEIM T: *Clin Chem 10*:32, 1964
2. AMBROSE JA: *Clin Chem 15*:15, 1969
3. AMBROSE JA, ROSS C, WHITFIELD F: in *Technicon Symposia*, Vol 1, 1967, edited by NB Scova. New York, Mediad, 1968, p 13
4. AMBROSE JA, INGERSON A, GARRETTSON LG, CHUNG CW: *Clin Chim Acta 15*:493, 1967
5. AMBROSE JA, SULLIVAN P, INGERSON A, BROWN RL: *Clin Chem 15*:611, 1969
6. ANDREWS JC, RANDALL A: *J Clin Invest 14*:517, 1935
7. ARNOLD E, HVIDBERG E, RASMUSSEN S: *Scan J Clin Lab Invest 24*:231, 1969
8. AVERY ME, CLOW CL, MENKES JH, RAMOS A, SCRIVER CR, STERN L, WASSERMAN BP: *Pediatrics 39*:378, 1967
9. BARNESS LA, MORROW G, NOCHO RE, MARESCA RA: *Clin Chem 16*:20, 1970
10. BARNESS LA, YOUNG D, MELLMAN WJ, KAHN SB, WILLIAMS WJ: *New Eng J Med 268*:144, 1963
11. BAUMANN E: *Hoppe Seylers Z Physiol Chem 16*:268, 1892
12. BENESCH R, LARDY H, BENESCH R: *J Biol Chem 216*:663, 1955
13. BENSON JV Jr, GORDON MJ, PATTERSON JA: *Anal Biochem 18*:228, 1967
14. BERGMAN I, LOXLEY R: *Clin Chim Acta 27*:347, 1970
15. BERRY HK, PONCET IB: *Clin Chim Acta 29*:83, 1970
16. BEUTLER E: *"Red Cell Metabolism,"* in *A Manual of Biochemical Methods*. New York, London, Grune and Stratton, 1971
17. BEUTLER E, DURON O, KELLY BM: *J Lab Clin Med 61*:882, 1963
18. BIGWOOD EJ, SCHRAM E, SOUPART P, VIS H: *Exposés Ann Biochim Méd 23*:9, 1961
19. BITO LZ, DAWSON J: *Anal Biochem 28*:95, 1969
20. BLAU K, EDWARDS DJ: *Biochem Med 5*:333, 1971
21. BLIX G: Hoppe Seylers Z. *Physiol Chem 210*:87, 1932
22. BLOCK WD, MARKOVS ME, STEELE BF: *Proc Soc Exp Biol Med 122*:1089, 1966
23. BÖDEKER C: *Z. Rat Med 7*:130, 1859
24. BONDURANT RE, GREER M, WILLIAMS CM: *Anal Biochem 15*:364, 1966
25. BOURDILLON J, VANDERLINDE RE: *Pub Health Rep 81*:991, 1966
26. BRAND E, CAHILL GF, BLOCK RJ: *J Biol Chem 110*:399, 1935
27. BRAND E, HARRIS MM, BILOON S: *J Biol Chem 86*: 315, 1930
28. BRIGGS AP: *J Biol Chem 51*:453, 1922
29. BRIGHAM MP, STEIN WH, MOORE S: *J Clin Invest 39*:1633, 1960
30. BRODEHL J, GELLISSEN K: *Pediatrics 42*:395, 1968
31. BROQUIST HP, LUHBY AL: *Proc Soc Exp Biol Med 100*:349, 1959
32. CAMIEN MN, DUNN MS: *J Biol Chem 183*:561, 1950
33. CAMPBELL DJ: *New Engl J Med 274*:1213, 1966
34. CARVER MJ, PASKA R: *Clin Chim Acta 6*:721, 1961

35. CASEY HJ, WELLS DG: *Clin Chim Acta 19*:165, 1968
36. CAWLEY LP, GOODWIN WL, DIBBERN P: *Am J Clin Pathol 48*:405, 1967
37. CERNOCH M, MALINSKA J: *Clin Chim Acta 14*:335, 1966
38. CLARKE HT: *J Biol Chem 97*:235, 1932
39. CLAYTON CC, STEELE BF: *Clin Chem 13*:49, 1967
40. COBURN SP, SEIDENBERG M, FULLER RW: *Proc Soc Exp Biol Med 129*:338, 1968
41. COHN VH, LYLE J: *Anal Biochem 14*:434, 1966
42. CONSDEN R, FORBES HAW, GLYNN LE, STANIER WM: *Biochem J 50*:274, 1951
43. CONSTANTSAS NS, DANELATOU-ATHANASSIADOU C: *Clin Chim Acta 9*:1, 1964
44. COX EV, WHITE AM: *Lancet 2*:853, 1962
45. CULLEY WJ: *Clin Chem 15*:902, 1969
46. CURNOW DH, LYNCH WJ: *Am J Clin Pathol 51*:547, 1969
47. DANOVITCH SH, BAER PN, LASTER L: *New Eng J Med 278*:1253, 1968
48. DEWOLFE MS, BASKURT S, COCHRANE WA: *Clin Biochem 1*:75, 1967
49. DICKINSON JC, HAMILTON PB: *J Neurochem 13*:1179, 1966
50. DICKINSON JC, DURHAM DG, HAMILTON PB: *Invest Ophthalmol 7*:551, 1968
51. DICKINSON JC, ROSENBLUM H, HAMILTON PB: *Pediatrics 36*:2, 1965
52. DOLEŽALOVÁ V, BRADA Z: *Clin Chim Acta 10*:34, 1964
53. DREYFUS PM, DUBÉ VE: *Clin Chim Acta 15*:525, 1967
54. DUPONT A: *Clin Chim Acta 18*:59, 1967
55. EDOZIEN JC, MODIE JA: *Lancet 2*:1149, 1964
56. ETTINGER B, KOLB FO: *J Urol 106*:106, 1971
57. EVERED DF: *Biochem J 62*:416, 1956
58. FAHIE-WILSON MN: *J Med Lab Technol 26*:363, 1969
59. FEIGIN RD: *Am J Diseases Children 117*:24, 1969
60. FERNANDEZ AA, HENRY RJ: *Anal Biochem 11*:190, 1965
61. FIRSCHEIN HE, SHILL JP: *Anal Biochem 14*:296, 1966
62. FISCHBERG EH: *J Am Med Assoc 119*:882, 1942
63. FISHER LJ, BUNTING SL, ROSENBERG LE: *Clin Chem 9*:573, 1963
64. FRAME EG, RUSSELL JA, WILHELMI AE: *J Biol Chem 149*:255, 1943
65. FRANKL W, DUNN MS: *Arch Biochem 13*:93, 1947
66. GARROD AE, HURTLEY WH: *J Physiol 33*:206, 1905
67. GEER RP, HANTMAN RK, SWETT CV: *Clin Chem 14*:12, 1968
68. GEORGES RJ, POLITZER WM: *Clin Chim Acta 30*:737, 1970
69. GERRITSEN T, REHBERG ML, WAISMAN HA: *Anal Biochem, 11*:460, 1965
70. GHADIMI H, PECORA P: *Pediatrics 34*:182, 1964
71. GIORGIO AJ, LUHBY AL: *Am J Clin Pathol 52*:374, 1969
72. GIORGIO AJ, PLAUT GWE: *J Lab Clin Med 66*:667, 1965
73. GOMPERTZ D: *Clin Chim Acta 19*:477, 1968
74. GOMPERTZ D, JONES JH, KNOWLES JP: *Clin Chim Acta 18*:197, 1967
75. GOODMAN SI, McINTYRE CA, O'BRIEN D: *J Pediat 71*:246, 1967
76. GOODWIN JF: *Clin Chim Acta 21*:231, 1968
77. GOODWIN JF: *Clin Chem 14*:1080, 1968
78. GREEN A: *J Clin Pathol 21*:221, 1968
79. GRIMES AJ: *Nature 205*:94, 1965
80. GUTHRIE R: *J Am Med Assoc 178*:863, 1961
81. GUTTERIDGE JMC, WRIGHT EB: *Clin Chim Acta 27*:289, 1970
82. GÜTTLER F, OLESEN ES, WAMBERG E: *Am J Clin Nutr 22*:1568, 1969

83. HALE HB, ELLIS JP Jr, VAN FOSSAN DD: Air University, School of Aviation Medicine, USAF, Randolph AFB, Texas, June 1959
84. HALL FF, PEYTON GA, WILSON SD: *Am J Clin Pathol 51*:660, 1969
85. HAMBRAEUS L: *Scand J Clin Lab Invest 15*:657, 1963
86. HAMILTON PB: *Clin Chem 14*:535, 1968
87. HAMILTON PB: *Anal Chem 30*:914, 1958
88. HAMILTON PB: *Anal Chem 35*:2055, 1963
89. HAMILTON PB: *Nature, 205*:284, 1965
90. HAMILTON PB, in Ref. 3, p 317
91. HAMILTON PB: in *Technicon Symposium*, 1965, edited by LT Skeggs Jr., New York, Mediad, 1965, p 702
92. HAMILTON PB, ANDERSON RA: *Anal Chem 31*:1504, 1959
93. HAMILTON PB, VAN SLYKE DD: *J Biol Chem 150*:231, 1943
94. HAMILTON PB, BOGUE DC, ANDERSON RA: *Anal Chem 32*:1782, 1960
95. HAMMOND KB, GOODMAN SI: *Clin Chem 16*:212, 1970
96. HAUX P, NATELSON S: *Clin Chem 16*:366, 1970
97. HEATHCOTE JG, DAVIES DM, HAWORTH C, OLIVER RWA: *J Chromatog 55*:377, 1971
98. HENDERSON LM, SNELL EE: *J Biol Chem 172*:15, 1948
99. HENRY RJ: *Clinical Chemistry, Principles and Technics.* New York, Harper and Row, 1964
100. HIATT HH, GOLDSTEIN M, TABOR H: *J Clin Invest 37*:829, 1958
101. HIBBARD ED: *Lancet 2*:1146, 1964
102. HILL JB: *Biochem Med 2*:261, 1969
103. HILL JB, PALMER P: *Clin Chem 15*:381, 1969
104. HILL JB, SUMMER GK, HILL HD: *Clin Chem 13*:77, 1967
105. HILL JB, SUMMER GK, PENDER MW, ROSZEL NO: *Clin Chem 11*:541, 1965
106. HLA-PE U, AUNG-THAN-BATU: *Anal Biochem 20*:432, 1967
107. HOCHELLA NJ: *Anal Biochem 21*:227, 1967
108. HOFFMAN NE, BARBORIAK JJ: *Anal Biochem 18*:10, 1967
109. HOSLEY HF, OLSON KB, HORTON J, MICHELSEN P, ATKINS R: *Technicon International Congress 1, 105, 1969*
110. HSIA DYY, BERMAN JL, SLATIS HM: *J Am Med Assoc 188*:203, 1964
111. HYÁNEK J, CAFOURKOVÁ Z: *Clin Chim Acta 12*:599, 1965
112. JANEWAY CA: *New Eng J Med 284*:787, 1971
113. JELLUM E, CLOSE VA, PATTON W, PEREIRA W, HALPERN B: *Anal Biochem 31*:227, 1969
114. JENSEN B: *Acta Med Scand 153*:383, 1956
115. JESENOVEC N, FISER-HERMAN M: *Clin Chim Acta 14*:293, 1966
116. JOHNSTONE JM, KEMP JH, HIBBARD ED: *Clin Chim Acta 12*:440, 1965
117. JONES DD: *J Med Lab Tech, 24*:301, 1967
118. JONES JD, SKALET AC, BURNETT PC: *Anal Biochem 37*:194, 1970
119. KAHN SB, WILLIAMS WJ, BARNESS LA, YOUNG D, SHAFER B, VIVACQUA RJ, BEAUPRE EM: *J Lab Clin Med 66*:75, 1965
120. KANG ES, GERALD PS: *J Pediat 76*:939, 1970
121. KENNEDY GC, LEWIN DC, LUNN HF: *Guy's Hosp Rept 88*:34, 1938
122. KHACHADURIAN A, KNOX WE, CULLEN AM: *J Lab Clin Med, 56*:321, 1960
123. KING JS: *Clin Chim Acta, 9*:441, 1964
124. KIVIRIKKO KI, LAITINEN O, PROCKOP DJ: *Anal Biochem 19*:249, 1967
125. KNOWLES JP, PRANKHERD TAJ, WESTALL RG: *Lancet, 2*:347, 1960
126. KOEVOET AL, BAARS JD: *Clin Chim Acta 25*:39, 1969

127. KOHN J: *J Clin Pathol 17*:466, 1964
128. KOHN J, MOLLIN DL, ROSENBACH LM: *J Clin Pathol 14*:345, 1961
129. KRAFFCZYK F, HELGER R, LANGE H: *Clin Chim Acta 31*:489, 1971
130. LACK L, SMITH M: *Anal Biochem 8*:217, 1964
131. LA DU BN: *Am J Diseases Children 113*:54, 1967
132. LA DU BTN, MICHAEL J: *J Lab Clin Med 55*:491, 1960
133. LA DU BN, ZANNONI VG: *J Biol Chem 217*:777, 1955
134. LA DU BN, ZANNONI VG, LASTER L, SEEGMILLER JE: *J Biol Chem 230*:251, 1968
135. LAITINEN O, NIKKILÄ EA, KIVIRIKKO KI: *Acta Med Scand 179*:275, 1966
136. LANYAR F, LIEB H: *Hoppe Seylers Z. Physiol Chem 203*:141, 1931
137. LANYAR F, LIEB H: *Hoppe Seylers Z. Physiol Chem 203*: 135, 1931
138. LEE S, SCHOEN I: *New Eng J Med 275*:266, 1966
139. LeROY EC: *Advances in Clinical Chem 10*:213, 1967
140. LeROY EC, KAPLAN A, UNDENFRIEND S, SJOERDSMA A: *J Biol Chem 239*:3350, 1964
141. LEVY HL, MADIGAN PM, LUM A: *Am J Clin Pathol, 51*:765, 1969
142. LEVY HL, SHIH VE, MADIGAN PM, MacCREADY RA: *J Am Med Assoc 209*:249, 1969
143. LEWIS FJW, MOORE GR: *Lancet 1*:305, 1962
144. LEWIS HB, BROWN BH, WHITE FR: *J Biol Chem 114*:171, 1936
145. LIEB H, LANYAR F: *Hoppe Seylers Z Physiol Chem 181*:199, 1929
146. LUHBY AL, COOPERMAN JM, TELLER DN: *Proc Soc Exp Biol Med 101*:350, 1959
147. MABRY CC, KARAM EA: *Am J Clin Pathol 42*:421, 1964
148. MABRY C, TODD WR: *J Lab Clin Med 61*:146, 1963
149. MacDONALD WB, FELLERS FX: *Am J Clin Pathol 49*:123, 1968
150. MacFADYEN DA: *J Biol Chem 145*:387, 1942
151. MASES P, FALET F, MARTINOT: Rev Pathol Gen Physiol Clin 56:641, 1956
152. MATSUSHITA S, IWAMI N, NITTA Y: *Anal Biochem 16*:365, 1966
153. MATTHEWS DM, MUIR GG, BARON DN: *J Clin Pathol 17*:150, 1964
154. McCAMAN MW, ROBINS E: *J Lab Clin Med 59*:885, 1962
155. McEVOY-BOWE E, THEVI SS: *Clin Chem 12*:144, 1966
156. McKENZIE AW, OWEN JA, RAMSEY JHR: *Brit Med J 2*:794, 1957
157. MEDES G: *Biochem J 31*:12, 1937
158. MEDES G: *Biochem J 30*:1293, 1936
159. MEISS M, PEYSER DP, MILLER A: *J Lab Clin Med 64*:512, 1964
160. MELLON JP: *J Med Lab Technol 24*:146, 1967
161. MELLON JP, STIVEN AG: *J Med Lab Technol 23*:204, 1966
162. METZ E: *Biochem Z 190*:261, 1927
163. METZ E: *Hoppe Seylers Z Physiol Chem 193*:46, 1930
164. MIDDLETON JE: *J Clin Pathol 23*:90, 1970
165. MILNE MD, ASATOOR AM, EDWARDS KDG, LOUGHRIDGE LW: *Gut 2*:323, 1961
166. MITOMA C, SMITH TE, DAVIDSON JD, UNDENFRIEND S, DaCOSTA FM, SJOERDSMA A: *J Lab Clin Med 53*:970, 1959
167. MOORE S, STEIN WH: *J Biol Chem 176*:367, 1948
168. MOORE S, STEIN WH: *J Biol Chem 192*:663, 1951
169. MORTENSEN E: *Scand J Clin Lab Invest 16*:87, 1964
170. NAFTALIN L, STEPHENS A: *Clin Chim Acta 12*:365, 1965

171. NEUBERGER A: *Biochem J 41*:431, 1947
172. NIV J, HOCHBERG A, DIMANT E: *Biochim Biophys Acta 127*:26, 1966
173. O'BRIEN WM, LA DU BN, BUNIM JJ: *Am J Med 34*:813, 1963
174. OEPEN H, OEPEN I: *Klin Wochschr 41*:921, 1963
175. OWENS CWI, BELCHER RV: *Biochem J 94*:705, 1965
176. PALMER DW, PETERS T Jr: *Clin Chem 15*:891, 1969
177. PASIEKA AE, THOMAS ME, LOGAN JE, ALLEN RH: *Clin Biochem 2*:41, 1968
178. PATTERSON JW, LAZAROW A: *Methods of Biochemical Analysis* edited by D Glick. New York, Interscience, 1955, Vol 2, p 259
179. PATTERSON JW, LAZAROW A, LEVEY S: *J Biol Chem 177*:197, 1949
180. PENNOCK CA, MOORE GR, HOYLE MD: *J Med Lab Technol 27*:302, 1970
181. PERRY TL, HANSEN S: *Clin Chim Acta 25*:53, 1969
182. PERRY TL, JONES RT: *J Clin Invest 40*:1363, 1961
183. PERRY TL, HANSEN S, DIAMOND S, STEDMAN D: *Lancet 1*:806, 1969
184. PERRY TL, HANSEN S, TISCHLER B, BUNTING R: *Clin Chim Acta 18*:51, 1967
185. PERRY TL, TISCHLER B, HANSEN S, MacDOUGALL L: *Clin Chim Acta 15*:47, 1967
186. PETERS JH, BERRIDGE BJ Jr: *Chromatog Rev 12*:157, 1970
187. PETERS JH, LIN SC, BERRIDGE BJ Jr, CUMMINGS JG, CHAO WR: *Proc Soc Exp Biol Med 131*:281, 1969
188. PETERSON EA, SOBER HA: *Anal Chem 31*:857, 1959
189. PIEZ KA, MORRIS L: *Anal Biochem 1*:187, 1960
190. POPE CG, STEVENS MF: *Biochem J 33*:1070, 1939
191. Present Status of Different Mass Screening Procedures for Phenylketonuria. (Medical Research Council Working Party on Phenylketonuria). *Brit Med J 4*:7, 1968
192. PROCKOP DJ, UDENFRIEND S: *Anal Biochem 1*:228, 1960
193. RAPP RD: *Clin Chem 9*:27, 1963
194. Recommended Guidelines for PKU Programs for the Newborn. U.S. Department of Health, Education, and Welfare, 1971
195. REED G: *J Biol Chem 142*:61, 1942
196. ROBERTS M, MOHAMED SD: *J Clin Pathol 18*:214, 1965
197. ROKKONES T: *Scand J Clin Lab Invest 16*:149, 1964
198. ROOTWELT K: *Scand J Clin Lab Invest, 19*:325, 1967
199. ROSANO CL: *Anal Biochem 15*:341, 1966
200. ROSENTHAL AF, YASEEN A: *Clin Chim Acta 26*:363, 1969
201. ROSTON S: *Anal Biochem 6*:486, 1963
202. ROTH M, FELGENHAUER WR: *Enzymol Biol Clin 9*:53, 1968
203. RUBINSTEIN HM, PRYCE JD: *J Clin Pathol 12*:80, 1959
204. SALEH AEC, COENEGRACHT JM: *Clin Chim Acta 21*:445, 1968
205. SAMUELS S, WARD SS: *J Lab Clin Med 67*:669, 1966
206. SANGER F: *Science 129*:1340, 1959
207. SASS MD, CARUSO CJ, O'CONNELL DJ: *Clin Chim Acta 11*:334, 1965
208. SCHNEIDER JA, BRADLEY KH, SEEGMILLER JE: *J Lab Clin Med 71*:122, 1968
209. SCRIVER CR, DAVIES E: *Pediat 36*:592, 1965
210. SEAKINS JWT, ERSSER RS, GIBBONS IS: *Ann Clin Biochem 6*:38, 1969
211. SEALOCK RR, SILBERSTEIN HE: *J Biol Chem 135*:251, 1940
212. SEARLE B, MIJUSKOVIC MB, WIDELOCK D, DAVIDOW B: *Clin Chem 13*:621, 1967

213. SEARLE BG, LI M, BRIGGS J, SEGALL P, WIDELOCK D, DAVIDOW B: *Clin Chem 14*:623, 1968
214. SEEGMILLER JE: *Clin Chem 14*:412, 1968
215. SEEGMILLER JE, ZANNONI VG, LASTER L, LA DU BN: *J Biol Chem 236*:774, 1961
216. SERAFINI-CESSI F, CESSI C: *Anal Biochem 8*:527, 1964
217. SHINOHARA K, PADIS LE: *J Biol Chem 112*:709, 1935
218. SILVERMAN M, GARDINER RC, CONDIT PT: *J Nat Cancer Inst 20*:71, 1958
219. SINGER K: *Am J Clin Pathol 45*:647, 1966
220. SKARŻYŃSKI B, SARNECKA-KELLER M, FRENDO J: *Clin Chim Acta 7*:243, 1962
221. SNYDERMAN SE, HOLT LE Jr, NORTON PM, ROITMAN E, PHANSALKAR SV: *Pediat Res 2*:131, 1968
222. SOBEL C, HENRY RJ, CHIAMORI N, SEGALOVE M: *Proc Soc Exp Biol Med 95*:808, 1957
223. SOMMERFELT S Chr, WŸNSTROOT E: *Scand J Clin Lab Invest 9*:196, 1957
224. SOUPART P: *Clin Chim Acta 4*:265, 1959
225. SPACKMAN DH, STEIN WH, MOORE S: *Anal Chem 30*:1190, 1958
226. SPRINKLE TJ, PORTER AH, GREER M, WILLIAMS CM: *Clin Chim Acta 24*:476, 1969
227. STEGEMANN H, STALDER K: *Clin Chim Acta 18*:267, 1967
228. STEIN WH: *J Biol Chem 201*:45, 1953
229. STEIN WH, MOORE S: *J Biol Chem 211*:915, 1954
230. STEPHENSON JPB, McBEAN MS: *Brit Med J 3*:582, 1967
231. STOKKE O, ELDJARN L, NORUM KR, STEEN-JOHNSEN J, HALVORSEN S: *Scand J Clin Lab Invest 20*:313, 1967
232. STONER RE, BLIVAISS BB: *Clin Chem 11*:833, 1965
233. SULLIVAN MX, HESS WC: *J Biol Chem 116*:221, 1936
234. SZEINBERG A, SZEINBERG B, COHEN BE: *Clin Chim Acta 23*:93, 1969
235. SZENTIRMAI A, BRAUN P, HORVÁTH I, HAUK M: *Clin Chim Acta 7*:459, 1962
236. TABOR H, WYNGARDEN L: *J Clin Invest 37*:824, 1958
237. TERLINGEN JBA, VAN DREUMEL HJ: *Clin Chim Acta 22*:643, 1968
238. TROUGHTON WD, BROWN R ST CL, TURNER NA: *Am J Clin Pathol 46*:139, 1966
239. UDENFRIEND S, COOPER JR: *J Biol Chem 196*:227, 1952
240. VALMIKINATHAN K, RAO VNV, VERGHESE N: *J Chromatog 24*:283, 1966
241. VAN SLYKE DD, MacFADYEN DA, HAMILTON PB: *J Biol Chem 150*:251, 1943
242. VIRTUE RW, LEWIS HB: *J Biol Chem 104*:415, 1934
243. WAALKES TP, UDENFRIEND S: *J Lab Clin Med 50*:733, 1957
244. WALKER DG, PRASAD AS, SADRIEH J: *J Lab Clin Med 59*:110, 1962
245. WELLS MG: *Clin Chim Acta 25*:27, 1969
246. WERNZE H, KOCH W: *Klin Wochschr 43*:454, 1965
247. WILLIAMS EM, MATTHEWS DM: *J Clin Pathol 18*:771, 1965
248. WILLIAMS EM, DONALDSON D, MATTHEWS DM: *Clin Chim Acta 12*:468, 1965
249. WOLFF PR, PAYSANT P, NICOLAS JP, THENOT JP, SCHEMBERG M: *Ann Biol Clin 27*:749, 1969
250. WOLKOW M, BAUMANN E: *Hoppe Seylers Z. Physiol Chem 15*:228, 1891
251. WONG WK, O'FLYNN ME, INOUYE T: *Clin Chem 10*:1098, 1964
252. WOOLF LI: *Arch Disease Childhood 43*:137, 1968
253. WROBLEWSKI F: *Pure Appl Chem 3*:385, 1961
254. WURTMAN RJ, CHOU C, ROSE CM: *Science 158*:660, 1967

Chapter 19

Inorganic Ions

NORMAN WEISSMAN, Ph. D.
VINCENT J. PILEGGI, Ph. D.

SODIUM

Sodium has been determined in biologic specimens by the following technics: (a) chemical, (b) neutron activation (440), (c) flame photometry, (d) atomic absorption spectrophotometry (530), and (e) ion-selective electrode. Advances in flame photometer instrumentation, and more recently ion-selective electrode technology, have made chemical methods obsolete. Actually, the precision attainable by chemical methods is marginally satisfactory at best. The reader is referred to the first edition of this book for a complete review of chemical methods. Although atomic absorption spectrophotometry is the method of choice for many metal atoms, it is not ordinarily used for the determination of Na or K because simpler and readily available flame emission methodology is quite adequate. An automated flame photometry method for Na and K utilizing the AutoAnalyzer has been described (196) in which sample protein concentration, detergent concentration, room temperature, and the effects of sunlight intensity on the detection module were studied. Suggestions were made concerning ways to detect and to decrease random fluctuations in equipment response during routine Na and K determinations. An ion-selective electrode for Na employing the ternary glass system NAS_{11-18} (Na_2O 11; Al_2O_3 18; SiO_2 71 moles/100 ml) has been used to determine Na in body fluids (385). A computerized apparatus employing selective electrodes for Na, K, and chloride has even been described (120). The Na electrode measures activities, which must be related to concentration by calibration and conversion factors. Calculations show that the conversion factors vary significantly in disease states due to changes in protein content of sera (120). Two automated systems which claim to have solved the concentration and protein problems have been reported (see Chapter 4 and Refs 340, 463). If the claims advanced are borne out by trial and experience, the electrode method may supplant flame photometry. In any event, the method is promising, particularly for fluids low in protein such as sweat (407) and cerebrospinal fluid (448).

An ideal method for Na requires, in addition to accuracy and precision, speed and simplicity of performance. Flame photometry meets these requirements and is the method currently in routine use at our laboratory.

SODIUM DETERMINATION BY FLAME PHOTOMETRY

PROCEDURE

Since the details of analysis vary with the flame photometer employed, and these are given in the book of instructions accompanying each instrument, no attempt is made to present them here. Attention, however, is directed to references dealing with the flame photometric analysis of Na (55, 576) and to Chapter 2 on *Flame Photometry*. There are several suitable flame photometers available which have the following desirable characteristics: (a) high sensitivity through employment of photomultiplier cathode receptors and integrator measuring circuits; (b) simultaneous determination of Na and K on a single dilution of a micro sample; (c) digital or recorder readout with, in some cases, ability to interface with computers; (d) ability to measure Li separately; and (e) ready adaptation to automatic dilutors for sample preparation in conjunction with either manual or automated analyses.

The usual mode of analysis employs dilutions of unknown (usually 1:200) and standard solutions made with an internal standard of Li. These are aspirated into a propane-air flame where the excited metal atoms emit characteristic radiation. The 589 nm line of Na and the 766 nm line of K are isolated by suitable interference filters and transmitted to individual photoreceptors. The resulting photocurrent is amplified and read out on a digital display (61).

NOTES

1. Collection of samples. Scrupulous care is an absolute necessity in flame photometry. When normal serum Na and K are determined, the solutions measured contain 0.0007 and 0.000025 meq/ml (16 and 1 μg), respectively. It is obvious that the slightest contamination can drastically alter results. All glassware should be borosilicate (e.g., Pyrex or Kimble) and rigorously clean. Where possible, plastic containers should be employed to avoid ion contamination. If plasma is to be analyzed, either siliconized tubes, lithium heparin, or lithium oxalate anticoagulant may be used.

2. Stability of serum samples. Sodium in serum is stable for at least 2 weeks either at room or refrigerator temperature.

3. Determination of Na or K in food, feces, or tissues. This requires preliminary treatment of the sample. Samples can be dry ashed or wet ashed with H_2SO_4 and H_2O_2. Attempts have been made at extraction technics to avoid the necessity of ashing. In the method of Dean and Fishman (129) an extract of ashed tissue is prepared. This method has been in satisfactory use in our laboratory for a number of years for the determination of Na, K, and chloride in tissues.

ACCURACY AND PRECISION

Results obtained by flame photometry agree with chemical methods

(220), with spectrochemical analysis, with atomic absorption (608), and with selective-ion electrode technics (382).

Between-run precision (95% limits) is about ±3% for serum Na levels in the normal range.

NORMAL VALUES

Serum. The adult normal ranges for serum Na reported by nearly all workers are in fairly close agreement. They average about 135–155 meq/liter (18, 154, 534). Payne and Levell (421) redefined the normal range for serum Na based on a study of hospital inpatients, outpatients, and healthy staff and blood donors. It was suggested that two ranges are required, a normal range (137–147 meq/liter) to make assertions about alterations in specific diseases, and a range derived from patients likely to have no manifest disturbances of salt and water metabolism (135–144 meq/liter) to detect such disturbances. This type of use of multiple ranges is dealt with in detail in Chapter 13, *Normal Values and the Use of Laboratory Results for the Detection of Disease.* There appears to be no significant sex difference (171). There is a slight increase after meals (18). Studies on the normal range of premature and full-term infants disagree somewhat: some (416, 442) indicate that the normal is essentially the same as for adults, while others (540) indicate that the normal range is 5–10 meq/liter lower. Joseph and Bergstrom (282) ran 118 sets of duplicates on 59 subjects between the ages of 1 month and 16 years with very interesting results: (a) the total range was 138–161 meq/liter with a mean of 146. It was believed that the rather high mean may have been due to the high temperature mean prevalent during the study. (b) The standard deviation for intraindividual variation was 4 meq/liter, comprised of a laboratory error of ±2.5 meq/liter and a "true" intraindividual variation of ±3.2 meq/liter. (c) There was no significant difference between individuals. (d) Within 90% confidence limits, the maximal chance difference between two successive analyses on one individual whose mean fell within the normal range for means of two samples was 11 meq/liter. In another study of intraindividual variability performed on 20 normal adult males over a 4 week period, a wider range (±15.6 meq/liter) was reported (195).

Erythrocytes. The Na content of erythrocytes is about one-tenth that of serum. Funder and Wieth (178) carefully determined Na, K, and water in human erythrocytes and found a mean value for Na of 10.9±2.6 (±2s) meq/kg RBC.

Urine. The normal 24 hour output of Na in urine is quite variable, depending greatly on intake. On a "normal" diet it may range from 27 to 287 meq (78, 385). There is a large diurnal variation, the output while sleeping being as low as one-fifth the peak rate during daytime activity (542).

Feces. A novel method was proposed as a useful and convenient technic for measurement of the electrolyte content of human feces (623). Visking cellulose dialyzing bags filled with an inert colloidal solution were swallowed by normal individuals. The bag contents were analyzed 24–120 hours later when passed in the feces. The authors reported that the bag contents reached osmotic equilibrium with intestinal fluids within 3 hours. Potassium was the predominant cation followed by Mg, Ca, Na, and ammonium in decreasing amounts. In eight normal subjects the mean content of fecal dialysate for Na was 31.6 meq/liter (range 44–112) and for K 75 (range 29–147). The Na/K ratio was 0.42, a value in good agreement with that obtained in a study using conventional technics of fecal collection and analysis (132). The latter study reported Na excretion ranging from 0.24 to 7.86 meq/24 hours (0.2%–14% of daily intake), with an average Na/K ratio of 0.35.

Sweat. The normal Na content of sweat from adults is about 10–80 meq/liter (16, 371). Anderson and Freeman (16) found the upper limit to be 40 during the first year of life. McKendrick (371) observed an upper limit of 50 meq/liter from the age of 1 year to puberty, after which there was an abrupt rise to adult levels. *In vivo* measurements by an ultramicro capillary ion-selective electrode inserted into sweat ducts and glands—e.g., sweat gland of human nailbed—gave values of 19–42 meq/liter (407).

Cerebrospinal fluid. The limited data available for spinal fluid electrolytes indicate a normal range for Na of 123–154 meq/liter (385, 448).

POTASSIUM

Nine different technics for determining K in biologic fluids were reviewed in the previous edition of this book. These included chemical methods, neutron activation, and flame photometry. To these must now be added the ion-selective electrode for K (172, 340, 443, 463, 618). The electrode has a selectivity for K 5000 times greater than for Na and 18 000 times greater than for hydrogen. It is also highly selective for K over divalent cations. The electrode responds to K^+ concentrations from $1 M$ to $10^{-6} M$, with results in good agreement with those obtained by flame photometry (443).

The electrode holds promise as a nondestructive method and has possible ultramicro applications similar to the Na electrode mentioned previously (see *Chapter 4* and discussion under *Sodium*). Another ultramicro method for K in human tissue and serum is x-ray spectroscopy (355). As little as 0.1 μg K in 25 μg wet weight of tissue or in 0.5–1.0 μl serum can be measured with CVs equal to 1.4% and 2.6%, respectively.

For the same reasons cited in the discussion of Na technics, flame photometry and potentiometric measurement with the ion-selective electrode are the methods of choice. Potassium is usually determined in combination with Na as described under *Sodium Determination by Flame Photometry.*

POTASSIUM DETERMINATION BY FLAME PHOTOMETRY

PROCEDURE

Since the details of analysis vary with the flame photometer employed and these are given in the book of instructions accompanying each instrument, no attempt is made here to present them. Attention, however, is directed to references dealing with the flame photometric analysis of K (55, 61, 196, 220, 576) and to Chapter 2 on *Flame Photometry*.

For determination of K in feces, tissue, and foods, see the section on *Sodium Determination by Flame Photometry*.

NOTES

1. Specimen collection. The collection of blood specimens for K determinations requires special attention and technic. Venipuncture can be performed without removing the tourniquet, but it has been reported (157) that opening and closing the fist 10 times with a tourniquet applied results in an increase in the serum K of 10%–20%, persisting about 2 min. This combined effect of forearm exercise and restricted blood flow has been confirmed (528). Potassium can be determined either in serum or in plasma, although there is some disagreement as to comparative results (54, 178). It has been claimed that the concentration of K in plasma is less than the serum concentration by about 0.12 and 0.7 meq/liter (436, 437). The increase in serum has been attributed to a breakdown of platelets during coagulation, since their content of K is 10–20 times that of plasma (436). Others (54), however, have failed to observe this difference in man but did observe it in dogs. If plasma is used, obviously anticoagulants containing K cannot be employed. Since the concentration of K in erythrocytes is about 20 times that in serum or plasma (178), it is imperative that hemolysis be avoided. Minimal hemolysis occurs if the blood is drawn into a *dry* syringe, transferred to a *dry* tube slowly *after* removal of the needle from the syringe, allowed to clot, and then centrifuged. Do *not* centrifuge fresh unclotted blood since this produces hemolysis (597). Assuming a normal hemoglobin concentration and hematocrit, hemolysis sufficient to become just visible (i.e., to give a serum hemoglobin concentration of about 50 mg/100 ml) results in a serum K increase of about 0.15 meq/liter. If hemolysis occurs in a freshly drawn sample, the following correction can be made after a determination of hemoglobin is made (367):

$$\text{corrected serum K} = (\text{meq K/liter}) - (3.3 \times \text{g hemoglobin/100 ml})$$

The concentration of K in serum or plasma undergoes changes upon prolonged contact with erythrocytes. Counteracting the concentration gradient across the erythrocyte membrane is an active transport into the erythrocyte coupled with the absorption and phosphorylation of glucose (122). The direction of K shift under any particular circumstance is explainable on the basis of these two opposing forces. Thus K may enter the cells until all the glucose is used up and then there is a reversal of K shift (122, 259, 597). In circumstances under which phosphorylation is inhibited, such as at refrigerator temperature or in the presence of fluoride, the serum concentration of K increases steadily (200, 597). To illustrate the magnitude of these shifts, examples from one report might be quoted (122): at 37°C the serum K decreased 1.2 meq/liter in 4.5 hours and then increased 7.8 meq by 18 hours; at refrigerator temperature there was an increase of 0.9 and 3.8 meq in 4.5 and 18 hours, respectively. The necessity for separation of serum or plasma from erythrocytes within 0.5–1.0 hour after venipuncture thus becomes apparent.

2. *Stability of serum samples.* Potassium in serum is stable for at least 2 weeks either at room or refrigerator temperature.

ACCURACY AND PRECISION

Results obtained by flame photometry agree with spectrochemical analysis (534) and ion-selective electrode analysis (172, 340, 382, 443, 463, 618).

Between-run precision (95% limits) is about ±5% for serum K levels in the normal range.

NORMAL VALUES

A compromise of the reported normal adult ranges of serum K is 3.6–5.5 meq/liter (18, 154, 157, 534, 597, 625). The values in newborns are somewhat higher, and although there is general agreement that the lower limit of normal is about 4 meq/liter the upper limit has been variously reported as 6.8 (442), 7.1 (540), and 9 (153) meq/liter. There is no postprandial change (18). There is as much as a 30% increase upon transfer from a cold to a warm climate (239). Intraindividual variation has been reported to be ±1.3 meq/liter (95% confidence level) for samples obtained on different days (195).

Cerebrospinal fluid levels of K average 2.7 ± 0.4 (2s) meq/liter (448).

The normal 24 hour output in urine is quite dependent on intake, and may range from about 1 to 5 g (26–123 meq) (6, 78). There is a large diurnal variation, the output at night while sleeping being as much as fivefold less than at the peak rate during the day (542).

Fecal excretion of K has been studied (132, 623); see discussion under *Sodium.*

The normal concentration of K in sweat is about 5–17 meq/liter (51, 312).

CALCIUM

A bewildering array of technics is available for the determination of Ca. These include (a) reaction with an anion—e.g., oxalate, followed by quantification of the anion; (b) chelation with ethylenediaminetetraacetic acid (EDTA); (c) precipitation by chloranilic acid or naphthylhydroxamic acid followed by photometric determination of the anion; (d) flame photometry; (e) atomic absorption; (f) direct spectrophotometric measurement of colored reaction products formed with various indicators; (g) ion-selective electrode; (h) x-ray fluorescence; and (i) neutron activation. For a complete review of the earlier literature, the reader is referred to the first edition of this book.

Of the many methods published for quantification of the calcium oxalate precipitate, the redox titration of oxalate with potassium permanganate or ceric ion is the most popular (105, 291). In general, three technics have been employed in the precipitation of calcium oxalate: (a) preliminary ashing of the sample, (b) precipitation from a protein-free filtrate prepared with trichloroacetic acid, and (c) direct precipitation from the serum sample. Preliminary ashing of the sample is necessary for feces, but direct precipitation has proved satisfactory for the routine analysis of serum and urine. Nevertheless, preliminary ashing is regarded as a reference technic (517). Precipitation from a trichloroacetic acid filtrate is regarded as the least satisfactory technic (517). Furthermore, as a result of the volume occupied by the protein precipitate, the Ca concentration in the filtrate is too high by about 2.4%—the so-called "volume displacement error" (517). Direct precipitation of Ca from serum was introduced into routine clinical chemistry in 1921 by Kramer and Tisdall. Clark and Collip (105) simplified the test somewhat, and their modification is still widely used today.

The technics of Kramer-Tisdall and Clark-Collip employ titration of the oxalate in the precipitate with potassium permanganate. This reagent has the advantage of being its own indicator since an excess is signaled by its pink color. Its use has been considered disadvantageous by some workers because of the instability of permanganate solutions and because the endpoint is not as vivid as may be desired. Although both these points have some validity, the authors believe that they are not so strong as to warrant abandoning permanganate. The use of ceric ion as oxidant has gained popularity, however, because of greater stability and a sharp endpoint when used in conjunction with an external redox indicator. The indicator employed is ferrous *o*-phenanthroline (ferroin) or ferrous nitro-*o*-phenanthroline (nitro-ferroin). These indicators are orange (red if concentrated) when the iron is in the ferrous state, and pale blue or colorless when the iron is in the ferric state. The clarity of the endpoint is especially advantageous in micro titrations where the endpoint with permanganate is very poor (310). The various modifications employing ceric ions have included the following: (a) titration of the oxalate in sulfuric acid with ceric sulfate, using iodine monochloride as a catalyst so that titration can be carried out at $45°-50°C$, rather than at the higher temperature required without catalyst (291); (b) titration with perchloratoceric acid at room temperature; (c) addition of excess ceric ions, followed by KI and quantification of the free iodine liberated by photometric measurement of the color produced with starch (516); (d) addition of excess ceric ions, back-titrating the excess with ferrous ion (52); (e) addition of excess ceric sulfate and quantification of the excess by photometry—Ce^{4+} is yellow, Ce^{3+} colorless (602). The subject of cerate oxidimetry is thoroughly covered in the monograph by Smith (533).

The most successful methods for direct determination of Ca by chelation technics originate from the work of Schwarzenbach and co-workers (511, 512) on the formation of chelate complexes of alkaline earth metals with ethylenediaminetetraacetic acid (EDTA; trade names: Versene, Sequestrene). The analytic technics are based on the fact that color production between dissociated metal ions and certain indicators is prevented by the chelating agent (32, 474).

Ammonium purpurate (murexide) is a specific indicator for Ca, being orchid-purple in the absence of Ca^{2+} and pink in its presence. Eriochrome black T [1-(1-hydroxy-2-naphthylazo)-5-nitro-2-naphthol-4-sulfonic acid sodium salt] is an indicator which is red below pH 6.3 and blue above it. At elevated pHs, however, Ca^{2+} or Mg^{2+} cause it to be red. It has been claimed that the dye forms 1:1, 2:1, and 3:1 complexes with both Ca^{2+} and Mg^{2+}, the particular form or mixture of forms depending on the pH, but it has been shown (134) that such results stem from the study of an impure dye and that actually only 1:1 compounds are formed between azo dyes and Ca^{2+} or Mg^{2+}. The endpoint of ammonium purpurate is poor compared with that of eriochrome black T. Other indicators which have been used for Ca are: (a) Calcon or Cal-Red, the Na salt of 1-(2-hydroxy-1-naphthylazo)-2-naphthol-4-sulfonic acid (26, 245, 369); (b) Calcein, a fluorescein complexone more sensitive than ammonium purpurate, the endpoint being determined either by color change (14) or by fluorescence (19, 56); (c) Calcofast blue 2G and methylthymol blue (495); (d) Plasmocorinth B, a naphthylazobenzene (326); (e) hydroxynaphthol blue has been used for the direct measurement of Ca in serum, urine (95, 100), and feces (101); (f) o-cresolphthalein which is also used in a popular automated procedure (302); (g) glyoxal-bis-(2-hydroxyanil), which forms a red Ca complex in alkaline solution although it is unstable above 29°C (49); and (h) chlorophosphonazo III (260).

In a number of technics serum is titrated directly with EDTA without preliminary separation of the Ca. Thus Ca has been determined directly with ammonium purpurate as indicator (75, 266). When eriochrome black T is used as indicator, the results represent the sum of the Ca and Mg. Some workers have assumed a normal Mg content of 1.64 meq/liter and subtracted this from the total to obtain the Ca (535). This method cannot have high accuracy. Others have determined the Mg of serum (75, 605) and urine (613) by taking the difference in titrations with the two indicators.

Elevated bilirubin levels and the presence of hemolysis interfere with direct titration of serum, but the interference can be eliminated by the use of resins to effect a preliminary separation of Ca (100). Interfering fluorescence in urine can similarly be avoided by resin treatment (101, 582). A new chelating agent, DCTA (Na salt of 1,2 diaminocyclohexane-N-tetraacetic acid) has also been used with Calcein to determine Ca in hemolyzed or icteric sera (594). Metals such as Fe, Cu, and Zn also interfere with direct titration methods, although the metals can be complexed with cyanide (26, 75). At best, the endpoints for most indicators in direct titrations with sera are obscure (125). For this reason, modifications have been presented (303, 467, 605) employing photometric detection of the endpoints with ammonium purpurate and eriochrome black T. This requires a special instrument setup which, however, may be warranted when large numbers of samples are to be run.

An obvious approach involves preliminary separation of Ca from Mg by precipitating the Ca as the oxalate. This permits the determination of either or both with EDTA using the indicator with the sharper endpoint, eriochrome black T. In one proposed technic (174, 478) the sum of Ca and Mg is determined with eriochrome black T, the Mg is determined in the supernate with the same titration procedure after oxalate precipitation, and the Ca calculated by difference. Another technic determines both the Ca in the precipitate and the Mg, either with (614) or without (85, 250) separation of the Mg as $MgNH_4PO_4$, by titration with EDTA and eriochrome black T. Direct determination of Ca in the oxalate precipitate presents a problem because of its insolubility at the high pH required for the titration. One solution is preliminary destruction of the oxalate with perchloric acid (210) or with heat (543). A second approach consists of adding EDTA to the dissolved Ca oxalate which holds the Ca in solution during back-titration (467). Eriochrome black T is actually purple in the presence of Ca, whereas it is a purer red with Mg. Obviously, the endpoint is better when Mg is present (467, 613). Thus either the Ca has been titrated with EDTA after adding a known amount of Mg (250), or an excess of EDTA can be added and the excess back-titrated with either standard Mg (467), or standard Ca if Mg was introduced along with the EDTA (210).

Most workers (56, 85) have found the various technics employing EDTA to agree with the Clark-Collip method with permanganate titration, although in two reports (152, 605) the EDTA results were lower by 5%–7%.

Phosphate ions compete with EDTA for Ca^{2+}, but there is no significant interference in the case of serum because the samples are diluted and the phosphate concentration is then low enough so that the ion products are less than the solubility product of $CaHPO_4$. This may not be the case with urine, for which the P:Ca ratio may go as high as 250 (298). Addition of citrate appears to counteract this interference (26). Ethylene glycol bis(β-aminoethyl ether)-N,N'-tetraacetic acid has been used in the complexometric titration of urine Ca with the claim that phosphate does not interfere (627). Phosphate can also be removed by precipitation with morpholine nitrate and tungstate (35), or by adsorption onto an anion resin (169).

Fluorescence methods employing Calcein for the microdetermination of Ca in serum or urine have been described (276, 299, 587). In one method (276) 2–10 μg Ca was titrated in 0.25 N KOH with EDTA using Calcein W as indicator. The Calcein combines with Ca^{2+} to form a fluorescent complex, which is excited maximally at a wavelength of 490 nm and emits fluorescent radiation with a maximum at 520 nm. The rapid decrease in fluorescence which occurred when EDTA removed the Ca from the Calcein complex was followed on the fluorometer's galvanometer to mark the endpoint of the titration. Magnesium and phosphate did not interfere. Interference by Cu, Zn, Fe^{2+}, or Fe^{3+} was prevented by adding cyanide and triethanolamine. The authors claim that KOH does not fluoresce while NaOH does (587).

Another approach to determination of Ca is based on the fact that Ca is precipitated by chloranilic acid, a reagent that is red-purple. Teeri (560) first precipitated the Ca from serum as the oxalate, dissolved the oxalate with acid, then determined the Ca by photometrically measuring the decrease in color of chloranilate added in excess. Ferro and Ham (163) precipitated the Ca directly from serum as chloranilate, dissolved the precipitate in a solution of EDTA at high pH, and determined the color of the liberated chloranilate photometrically. Direct precipitation from lipemic sera leads to turbidity in the final colored solution that can be removed by ether extraction (102). The technic has been successfully adapted to the analysis of urine and feces and the ultramicro analysis of serum (102). There is some time saved, compared with the Clark-Collip method, when large numbers of samples are run. Webster (598) precipitated Ca as the chloranilate, then reacted the precipitated chloranilate with ferric chloride to form an intense and stable color which was measured photometrically. Calcium has also been precipitated as its naphthalhydroxamate, the precipitate dissolved in alkaline EDTA, and an orange-red color produced with acid ferric nitrate (570). A comparison of the Clark-Collip titrimetric and the Trinder hydroxamate methods by Wilkes (604) showed that the latter method was approximately twice as precise as the former.

The simplest chemical methods for determination of Ca are based on photometry of a colored reaction product of Ca and some reactant. Several such methods have been introduced: (a) Ca and Mg form colored complexes with alizarin which can then be extracted by n-octyl alcohol (393). (b) A color is formed in an alkaline solution of disodium 1-hydroxy-4-chloro-2,2-diazobenzene-1,8-hydroxynaphthalene-3,6-disulfonic acid (308, 626). Reportedly, total and free (not bound by protein) Ca and total and free Mg can be determined. One report (297) states that erroneous results are obtained in serum, presumably because of protein interference. (c) The color intensity formed with ammonium purpurate at pH 11.3−12 is proportional to the Ca concentration, and there is no interference by Mg at this pH (103). (d) Ca forms a colored complex with nuclear fast red (22). The foregoing direct methods are subject to interference by bilirubin and other serum or urine pigments, hemolysis, and lipemia. (e) Glyoxal-bis-(2-hydroxyanil) forms a red Ca complex in alkaline solution which can be extracted by chloroform (49) or measured in a mixture of methanol-acetone-water (359). The complex is unstable and has to be measured within 30 min. (f) o-Cresolphthalein complexone, Schwarzenbach's "metalphthalein" (512), has been used by several authors to determine Ca in various biologic specimens by automated procedures (108, 191, 302, 595). The dye forms a red Ca complex at pH 10−12. In the earlier methods it was necessary to use pH 12 in order to minimize interference by Mg (302). Subsequently, Mg interference was eliminated by adding 8-hydroxyquinoline to the reaction mixture (108, 191, 595), and the need for high pH was obviated. Iron interference, encountered in specimens prepared from ashed tissue, has been eliminated by prior precipitation with cupferron (191). Comparison of results for serum and urine analyses by o-cresolphthalein complexone versus Clark-Collip, flame photometry, and fluorometry showed good agreement with adequate reproducibility and recovery of added Ca (191, 302). Other automated photometric procedures have utilized alizarin (175), Calcein (165), plasmocorinth B (611), and glyoxal-bis-(2-hydroxyanil)-GBHA (77). The last

method employs a novel device to improve the accuracy of the spectrophotometric measurement. EDTA is added to remove part of the Ca present in
order to yield a Ca-GBHA complex with $A = 0.42$ for a serum containing 5
meq/liter.

Neutron activation analysis has been used for the determination of Ca in
serum and other biologic specimens (66, 232). Calcium may be measured
either as ^{49}Ca (half-life=8.8 min) or as its daughter ^{49}Sc (half-life=58 min).
The detection limit for Ca was 3 μg, equivalent to 0.03 μl serum. The
methods, however, require elaborate chemical preparation, very sophisticated instrumentation, and complex computer calculations employing many
correction factors, and are unsuited for any application other than very
specialized research. X-ray fluorescence analysis of ashed biologic samples
has been described (96). Four hours were required to prepare batches of
10–20 specimens of between 0.25 and 1 g, plus an additional 2 min per
specimen in the x-ray apparatus. Results agreed within 1% with those
obtained by an oxalate method. A polarographic method has also been
suggested (273).

A rapidly developing field is the use of ion-selective electrodes to measure
ionized Ca (this is discussed in the section on *Ionized Calcium*).

Calcium can be determined directly by flame photometry. In general,
however, more difficulty has been experienced with this cation than with
either Na or K. The flame spectrum of Ca has emissions at 422.7, 554, and
622 nm, the first being an arc line and the last two, molecular bands. Sodium
and K give positive interference (290, 389, 518); phosphate and sulfate
inhibit the calcium emission (30, 389), pyrophosphate and pyrosulfate
intermediates apparently being responsible (30). Interferences can be
removed by preliminary isolation of Ca as the oxalate (98, 309, 449) or the
phosphate (389), although if a flame photometer of sufficient sensitivity and
spectral resolution is employed the results obtained directly on the serum are
purportedly accurate (309). Calcium emission can be potentiated as much as
15-fold by use of organic solvents (264). When such solvents are used, a
non-ionic wetting agent such as Sterox must be added to prevent adsorption
by glass surfaces of the Ca from the organic solvents (491). The reported
precision of the flame photometric method for Ca has varied from less than
±1% to ±4% (491), and although the results obtained by some workers
appear to average a few percent higher than those given by the Clark-Collip
method (491, 617), most investigations have revealed no significant
difference between the flame photometric values and those obtained by
chemical technics involving preliminary precipitation of the Ca as oxalate
(263, 518). Because of the difficulties involved, the popularity of this
method for Ca determination has greatly diminished, particularly since the
introduction of atomic absorption spectrophotometry (AAS) and the
commercial availability of reliable instruments (see *Chapter 2*).

Winefordner *et al* (616) reported a comprehensive critical comparison of
atomic emission, atomic absorption, and atomic fluorescence flame spectrometry. The advantages of atomic absorption spectrophotometry that have
made it the preferred method for determining Ca are as follows: (a) It is
highly specific, there being almost no interferences between one metallic

element and another. (b) Sample preparation is minimal and for most routine analyses involves only dilution of body fluids. (c) Instrumentation is simple to operate, and little training is required. (d) Detection limits are low enough to permit ample dilution of small sample sizes. (e) The determination can be performed in less than 1 min per specimen (306).

Zettner and Seligson (643) were among the first to describe a practical procedure for the determination of Ca in serum by atomic absorption spectrophotometry. A study of interferences revealed that Mg, K, chloride, ammonia, and bicarbonate were without effect, while Na, phosphate, sulfate, oxalate, chelators, and protein depressed absorption. Effects of anions and chelators were diminished by the competitive cation technic, i.e., the addition of trivalent lanthanum. The effect of Na and protein were compensated for by adding these substances to reference solutions. With these modifications and the addition of butanol to enhance sensitivity, results obtained on diluted serum agreed with those obtained when Ca was first separated from other serum constituents by oxalate precipitation. A different approach to the problem of interferences was used in a method for the determination of both Ca and Mg on a single dilution of serum (552). Anion interference was overcome by adding strontium to samples and standards, while protein interference was avoided by deproteinizing with trichloroacetic acid.

A modification of the preceding method utilized a single dilution step to produce a protein-free supernate containing lanthanum (497). It was noted that trichloroacetic acid inhibited Ca atomic absorption by 10%, although that of Mg was slightly enhanced.

A method suitable for both serum and urine used a 1:50 dilution of serum, with 0.1% La as the diluent (571). Acidified urine was diluted 1:50 with 0.5% La to overcome the higher phosphate content of urine. Recovery of Ca added to Ca-free serum was 100%. Excellent agreement was found between the test method and each of two reference methods. The direct method also agreed with determinations performed on oxalate precipitates from both serum and urine. Atomic absorption analysis for Ca in serum has been automated (315). The serum sample, diluted with acidified lanthanum trichloride, was dialyzed against 0.1 N hydrochloric acid and a portion of the recipient solution pumped into the atomizer-burner of the atomic absorption spectrophotometer. A semilog plot of recorder readings versus Ca concentration was linear up to 7.0 meq/liter and curved between 7.5 and 10 meq/liter, being low by 5%. Recovery of Ca added to serum averaged 99%. The procedure consumed 0.6 ml serum at an analysis rate of 40/hour; for pediatric samples it was suggested that a rate of 60/hour (0.4 ml serum) would be practical, although this rate gave only 96% of the continuous sampling response. A study of the automated dialysis method led to the conclusion that an error is introduced when aqueous standards are used (351, 611). The fraction of Ca or Mg that dialyzes is proportional to the protein content of the solution. Incorporation of 7 g of decalcified albumin into 100 ml of standard yielded results by atomic absorption spectrophotometry that agreed with reference values (351). An automated simultaneous determination of serum Ca and Mg has been described (197). This method makes use of the availability of a dual channel, double beam instrument and performs 90 tests/hour while consuming only 30 μl serum. Calcium in cerebrospinal fluid and urine has been determined by automated atomic absorption spectrophotometry (317). An automated method for the simultaneous determination of serum Ca by atomic absorption spectrophotometry and phosphate by the molybdenum blue reaction has been reported (314).

A joint effort of the U.S. National Bureau of Standards and the National Institutes of Health has resulted in a proposal of the use of an isotope-dilution mass spectrometry method as a reference method (81a).

The foregoing discussion of the abundant variety of Ca methods, while far from complete, bears witness to the great advances in the field over the past decade. Although it seems likely that atomic absorption and ion-selective electrode methods will eventually be recognized as the primary methods for total and ionized Ca, many laboratories still do not possess the instrumentation to perform these methods. The same observation applies to improved chelometric technics that use measurement of fluorescence to determine Ca. These considerations have led to a choice of the following methods for presentation: (a) chelometric, (b) manual atomic absorption, and (c) selective-ion electrode (see section on *Ionized Calcium in Serum*). In addition, the Sulkowitch screening method for urine Ca is presented.

DETERMINATION OF CALCIUM BY EDTA TITRATION

(method of Bachra et al, Ref 26)

PRINCIPLE

Calcium is determined by titration with disodium ethylenediaminetetraacetate, using Cal-Red as indicator. Nonicteric and nonhemolyzed sera are titrated directly, but in the cases of urine and icteric or hemolyzed sera the Ca is first obtained as the oxalate.

REAGENTS

KOH, 1.25 N. Dissolve 70 g KOH pellets in 500 ml water and add water to 1 liter. Store in polyethylene bottle.

EDTA Solution. Dissolve 395 mg disodium ethylenediaminetetraacetate dihydrate in 500 ml water and add water to 1 liter.

Cal-Red Indicator Solution. Grind 1.0 g "Cal-Red Dilute" (Libecap Enterprises) in 10 ml water. Not all the material goes into solution. This suspension is usually stable in the refrigerator for about 2 weeks. It must be discarded if the endpoint of the titration is not sharp. Experience in our laboratory indicates that with some batches of the indicator the solution has to be prepared daily.

Sodium Citrate, 0.05 M. Dissolve 14.7 g of the dihydrate per liter water.

Caprylic Alcohol

Calcium Standard. Weigh out exactly 250 mg calcium carbonate and transfer to a 1 liter volumetric flask. Add 7 ml dilute HCl (1 part conc HCl +

9 parts water). Let stand until solid is dissolved, then add approximately 900 ml water. Adjust pH to 6.0 with 50% (w/v) ammonium acetate. Adjust volume to 1 liter with water and mix. 1 ml = 100 μg Ca.

HCl, 1.0 N. Dilute conc acid 1:12.

Ammonium Oxalate, 10%

PROCEDURE FOR SERUM

1. Add the following to clear, conical tipped 12 or 15 ml centrifuge tubes:

   Standard: 0.50 ml standard (50 μg Ca)

   Unknown: 0.50 ml serum + a very small drop of caprylic alcohol
2. Proceed with one tube at a time (if samples remain alkaline for more than 10 min, the endpoint is not sharp). Add 2.5 ml 1.25 N KOH, mix, and add 0.25 ml indicator solution (shake indicator well each time before pipeting).
3. Immediately titrate with EDTA solution until color changes from wine-red to blue. Observe the color change by looking through the solution against an incandescent light.

   Calculation:

$$\text{mg Ca/100 ml serum} = \frac{\text{ml unknown titration}}{\text{ml standard titration}} \times 0.05 \times \frac{100}{0.5}$$

$$= \frac{\text{ml unknown titration}}{\text{ml standard titration}} \times 10$$

$$\text{meq Ca/liter serum} = \frac{\text{ml unknown titration}}{\text{ml standard titration}} \times 5$$

PROCEDURE FOR ICTERIC OR HEMOLYZED SERA

1. To 0.50 ml serum in the conical tipped centrifuge tube add 0.63 ml water and 0.13 ml 10% ammonium oxalate and mix.
2. Place in 56°C water bath for 15 min.
3. Centrifuge for 10 min at 2000 rpm.
4. Decant supernate carefully, invert, and drain on absorbent paper.
5. Dissolve precipitate in 0.25 ml 1 N HCl and add 0.25 ml 0.05 M sodium citrate.
6. Proceed as in step 2 under *Procedure for Serum*, above. Also run a standard as set up in step 1 under *Procedure for Serum*.

PROCEDURE FOR URINE

1. If urine is clear and no precipitate is seen clinging to walls of container,

mix and remove approximately 10 ml. Acidify to pH 1 with conc HCl (wide range pH paper). If precipitate is present in original urine and if no other test is requested, then acidify entire urine specimen to pH 1. If other tests are requested, then shake and mix contents well, remove 10 ml aliquot, and acidify.

2. Warm acidified specimen to 60°C for at least 15 min, mixing occasionally.

3. Pipet 0.50 ml urine from step 2 into a conical tipped, 12 or 15 ml centrifuge tube, add 0.1 ml 10% ammonium oxalate, and mix. Add 1 drop methyl red indicator (0.1%) and then 5% NH_4OH dropwise to obtain orange color.

4. Place in boiling water bath for 20 min. Cool to room temperature, decant supernatant fluid carefully, invert, and drain on absorbent paper.

5. Dissolve precipitate in 0.25 ml 1 *N* HCl and add 0.25 ml 0.05 *M* sodium citrate.

6. Proceed as in step 2 under *Procedure for Serum*. Also run a standard as set up in step 1 under *Procedure for Serum*. If unknown requires less than 0.5 ml EDTA solution in titration, a larger aliquot of urine at step 3 should be precipitated.

Calculation:

$$\text{mg Ca/100 ml urine} = \frac{\text{ml unknown titration}}{\text{ml standard titration}} \times 10$$

$$\text{meq Ca/liter urine} = \frac{\text{ml unknown titration}}{\text{ml standard titration}} \times 5$$

$$\text{mg Ca/24 hours} = \frac{\text{ml unknown titration}}{\text{ml standard titration}} \times 0.05 \times \frac{1}{0.5}$$

$$\times \text{ml 24 hour volume}$$

$$= \frac{\text{ml unknown titration}}{\text{ml standard titration}} \times 0.1 \times \text{ml 24 hour volume}$$

$$\text{meq Ca/24 hours} = \frac{\text{ml unknown titration}}{\text{ml standard titration}} \times 0.0025 \times \frac{1}{0.5}$$

$$\times \text{ml 24 hour volume}$$

$$= \frac{\text{ml unknown titration}}{\text{ml standard titration}} \times 0.005 \times \text{ml 24 hour volume}$$

MICRO PROCEDURE

The original procedure of Bachra *et al* (26) was scaled down fivefold from the technics presented above.

NOTES

1. *Volume of Cal-Red indicator used in titration.* It has been the experience in our laboratory that slightly different volumes of EDTA solution are required for titration of the standard with different batches of Cal-Red indicator. If desired, this variation can be decreased or eliminated by altering the volume of indicator used in the test.

2. *Interference by heparin.* It has been reported that heparin *in vitro* or *in vivo* causes an apparent decrease in the serum Ca level of up to about 50% when determined by EDTA titration or flame photometry. This is difficult to understand, and the authors are unaware of any confirmation of this report.

3. *The endpoint.* Although the endpoint with Cal-Red is better than with some of the other indicators that have been used, not all analysts are successful in visualizing it with any confidence.

4. *Stability of samples.* Serum or plasma should not be allowed to remain in prolonged contact with the erythrocytes since with time the cells become permeable to Ca (430). This transfer is undoubtedly of less magnitude between serum and a clot than between plasma and suspended cells. If denatured protein develops in serum and precipitates on standing it can carry down some Ca (99). Fatty acids released from lecithin by the action of lecithinase B are said to combine with Ca to form a precipitate in aged serum (558). An analysis on the supernatant serum would, in such events, be erroneously low.

There appears to be no reason why Ca in feces, cerebrospinal fluid, and urine should not be stable, provided of course that in the case of urine proper precautions are taken to obtain solution of all precipitated Ca salts.

ACCURACY AND PRECISION

According to Bachra *et al* (26), phosphate at levels up to 30 mg/100 ml does not interfere. Magnesium does not interfere significantly, giving only 1% higher values at levels of 20 mg/100 ml. Results with serum correlate well with results obtained by precipitation of the oxalate and titration of the oxalate. Experiments in our laboratory, however, indicate that direct EDTA titration of urine yields results which are too high by about 35%. Isolation of the Ca as the oxalate, followed by EDTA titration of the Ca, yields results which average within 2% of values obtained by redox titration of the oxalate in the calcium oxalate precipitate.

The precision of the method is ±3%.

DETERMINATION OF CALCIUM BY ATOMIC ABSORPTION SPECTROPHOTOMETRY

(method of Trudeau and Freier, Ref 571, modified)

PRINCIPLE

Calcium is determined by atomic absorption in serum or urine diluted sufficiently with lanthanum chloride solution to avoid interference by protein or anions. The 422.7 nm line of a Ca hollow cathode lamp is used in the measurement.

REAGENTS

Stock Lanthanum Solution. Place 58.7 g of lanthanum oxide in a 1 liter volumetric flask and wet with water. Add 250 ml of conc HCl very slowly until the material is dissolved. Dilute to 1 liter with water. This provides a 5% lanthanum ion solution in 4 N HCl.

Working Lanthanum Solution (0.1% w/v). Dilute 10 ml of the La stock solution to 500 ml with water.

Working Lanthanum Solution (0.5% w/v). Dilute 50 ml of the La stock solution to 500 ml with water.

Stock Calcium Standard (25 meq/liter). Dissolve 124.9 mg reagent grade $CaCO_3$ in 3 ml 1 N HCl. Dilute to 100 ml with water.

Calcium Working Standards (5.0 and 7.5 meq/liter). Deliver 20 and 30 ml of stock Ca standard into 100 ml volumetric flasks. Dilute to volume with water.

PROCEDURE FOR SERUM

1. Serum assays should be performed on duplicate aliquots for required precision. Label 16 x 125 mm plastic tubes (or 10 ml polyethylene vials) for each unknown and for the 5.0 and 7.5 meq/liter standards.

2. Dilute 0.1 ml of each serum to 5.0 ml with 0.1% working La solution. This can be done by an automatic dilutor or 0.1 ml can be added to 4.9 ml with a TC pipet. Mix thoroughly.

3. Repeat step 2 with the 5.0 and 7.5 meq/liter working standards. Pour about 30 ml of the 0.1% La working solution into a 50 ml Erlenmyer flask for use as a blank.

4. Atomic absorption: No attempt is made to present the details of analysis here since they vary with the atomic absorption spectrophotometer employed and are given in the book of instructions accompanying each instrument.

PROCEDURE FOR URINE

The method for determining Ca in urine is the same as for serum with the following changes:

1. Follow instructions in steps 1 and 2 under *Determination of Calcium by EDTA Titration — Procedure for Urine* to prepare aliquots for analysis.
2. The 1:50 dilution is made with 0.5% La solution as diluent. This concentration of La is also used as the reference blank.
3. Beer's law is obeyed up to 20 meq/liter. Samples which have a greater concentration are diluted 1:100 and the assay repeated. The atomic absorption readings are multiplied by 2 to correct for the dilution.
   Calculation:

$$\text{meq Ca/24 hours} = \text{meq Ca/liter} \times \frac{\text{total urine volume (ml)}}{1000}$$

PROCEDURE FOR CEREBROSPINAL FLUID

The procedure for serum should be used. It may be advantageous to introduce a 2.5 meq/liter standard under some circumstances.

PROCEDURE FOR FECES

1. Calcium excretion in the feces is usually expressed on a 24 hour basis. Mix the 24 hour sample well and record the weight. Weigh approximately 1 g feces to the nearest 0.01 g into a platinum or fused silica crucible and record the weight.
2. Dry sample on steam bath.
3. Ash with a gentle flame for a few minutes until most of the carbon is burned off. Cool.
4. Add a few drops of water to the crucible. Break up residue with a stirring rod.
5. Dry on steam bath.
6. Ash again until no black spots (carbon) remain.
7. Cool. Add 3 drops 6 N HCl to dissolve the ash. Dry on steam bath.
8. Add 2 more drops 6 N HCl and transfer quantitatively with several small aliquots of hot water into any receptacle graduated at 25 ml (e.g., graduated cylinder, volumetric flask). Add water to a total volume of 25 ml and mix.
9. Proceed with atomic absorption analysis.
   Calculation:

$$\text{g Ca/24 hours} = \frac{\text{meq/liter}}{2000} \times \frac{\text{total weight}}{\text{weight of sample}}$$

$$\text{meq Ca/24 hours} = \frac{\text{meq/liter}}{40} \times \frac{\text{total weight}}{\text{weight of sample}}$$

NOTES

1. Interference by other substances. Use of a 1:50 dilution of serum in 0.1% La avoids the interference by Na, phosphate, and protein in combination. Similarly, use of 0.5% La with urine overcomes interference from Na, phosphate, oxalate, sulfate, and citrate (571). It has been reported that sera left overnight in plastic AutoAnalyzer sample cups may give low Ca values (221). Serum Ca values are falsely elevated in specimens collected in tubes with cork stoppers (324, 532).

ACCURACY AND PRECISION

Results obtained by atomic absorption agree with EDTA-Calcein chelometric methods, but average 3% lower than results obtained with the Clark-Collip method (643).

The precision (95% limits) of the method is about ±3.5% for serum Ca values in the normal range and ±7.5% for urine at a concentration of 14.4 meq/liter.

NORMAL VALUES

Serum or Plasma. A compromise of various published reports on adult normals is 4.5–5.7 meq/liter (458, 465). Two studies performed on large numbers of men and women at ages of 20 and above show significant differences for both age and sex (295, 465). Keating *et al* (295) studied 278 women and 298 men undergoing employment examinations and developed regression equations describing the relation of serum Ca to age:

$$\text{Women} \quad Y = 4.70 + 0.0005X$$
$$\text{Men} \quad Y = 4.89 - 0.00035X$$

where

$$Y = \text{meq Ca/liter}$$
$$X = \text{age in years}$$

The slope of the regression line for women was not significantly different from zero, although the sex difference was significant.

Our own study based on 1419 patients from eight geographic areas of the United States (465) gave normal range estimates for 603 males and 816 females as follows (values are given in meq/liter):

	All ages		*Ages 20 to 60+*	
			Proposed normal range for laboratory reporting	
Sex	*Normal Range*	*Median*	*Lower limit*	*Upper limit*
M	4.7–5.5	5.1	—	—
F	4.5–5.5	5.0	—	—
M + F	—	—	4.6	5.5

In this study significant differences between males and females were found only in the 20–29 and 30–39 age groups when the 10th and 90th percentiles were tested.

There is no appreciable variation throughout the day if exercise is avoided (593), and no difference between fasting and nonfasting levels (18, 549). The serum level appears to follow a sine-like pattern throughout the year (171). Data on newborns show that their values are lower than the adult range. A serum range of 2.8–4.6 meq/liter was found on the fifth day of life for 40 full-term infants, of whom 35 were breast-fed (540). Plasma Ca determined on small groups of bottle and breast-fed normal newborns during the first 7 days of life showed a lower Ca level in the bottle-fed on the sixth day, with a mean of 4.4 versus 5.0 meq/liter (229). Daily fluctuations were also noted with wide variations from 2.6 to 6.3 meq/liter. Values for serum Ca in infants at 3 months were 4.30–7.00 and at 6 months were 4.89–6.65 meq/liter (215).

There is no difference between venous and capillary serum (583).

Erythrocytes. The mean value for normal subjects is 0.634 μg/ml of packed erythrocytes (226). The Ca is mostly and perhaps exclusively located in the membrane, since after osmotic hemolysis the same amount is found in the ghost cells as was present in the erythrocytes from which they were prepared.

Cerebrospinal Fluid. The normal range is about 2.1–2.7 meq/liter (549).

Urine. The urinary output of Ca is dependent on endogenous factors (presumably endocrine), on dietary intake, and on skeletal weight (320). Adults on an average intake excrete about 2.5–20 meq/24 hours (320). A significant fall in Ca excretion in persons between 70 and 90 years of age has been attributed to normal aging of bone (496).

Feces. The average output is about 32 meq (0.64 g)/24 hours (268, 550).

ALBRIGHT-REIFENSTEIN CALCIUM BALANCE STUDY

Albright and Reifenstein (7) recommend the following balance study for instances in which knowledge of the urine output versus intake is desirable. Place the patient on the diet given below for 1 week and collect a 24 hour urine on the seventh day of the diet. Normally, the 24 hour urine contains less than 7.5 meq Ca; 7.5–10 meq is suggestive, and more than 10 meq definitely indicates a negative calcium balance. This diet contains approximately 6.8 meq Ca/day. Actual determinations of low Ca diets, however, have indicated that the Ca content may often be more than 25% less than that calculated (468). Others have studied the relation between dietary intake and urinary output of Ca in both normal individuals and patients (88).

Albright-Reifenstein Calcium Balance Diet

Breakfast

> Orange juice, 1 small glass
> Cooked farina or rice, 1/3 cup after cooking
> Uneeda biscuits, 4
> Oleomargarine
> Crisp bacon, 3 strips
> Coffee or tea
> Salt, sugar

Noon

> Lean meat, medium sized serving
> Potato, 1 medium sized
> White corn, ½ cup
> Uneeda biscuits, 4
> Oleomargarine
> Applesauce, ½ cup, or 1 medium sized apple
> Tea, salt, pepper, sugar

Night

> Chicken, 1 medium serving
> Macaroni, 1/3 cup (cooked)
> Canned tomato, ½ cup
> Uneeda biscuits, 4
> Oleomargarine
> Banana, 1 medium sized
> Tea or coffee
> Salt, pepper, sugar

Note: Use oleomargarine and sugar generously to keep up weight. Absolutely no butter, milk, cheese, or cream. Caution must be exercised to avoid cereals and oleomargarine which have been fortified with additional calcium.

SULKOWITCH DETERMINATION OF CALCIUM IN URINE

(method of Sulkowitch, Ref 33)

PRINCIPLE

Ca in urine is precipitated as the oxalate by the addition of a buffered oxalic acid reagent. The pH is such that phosphates do not precipitate. The degree of turbidity is noted visually. The results provide a rough estimate of the urine Ca level, and indirectly of the serum Ca level. Opinion is divided as to whether this test is sufficiently accurate for usefulness in clinical interpretation (8, 475). It has been recommended as a valuable ancillary aid to the early diagnosis of acute pancreatitis (435).

REAGENT

Sulkowitch Reagent. Dissolve 2.5 g oxalic acid, 2.5 g $(NH_4)_2 SO_4$, and 5 ml glacial acetic acid in water, and add water to a final volume of 150 ml.

PROCEDURE

1. To 5 ml urine in a test tube, add 5 ml Sulkowitch reagent and mix by inversion.
2. Observe turbidity after about 1–2 min. Grade the turbidity as 0, 1+, 2+, 3+, or 4+.

INTERPRETATION (8)

$$0 \text{ (no turbidity)} = \text{a serum Ca level} < 3.75 \text{ meq/liter}$$

$$3+ \text{ to } 4+ = \text{a serum Ca level} > 5.25 \text{ meq/liter}$$

ACCURACY

This a crude test, and interpretations must be qualified accordingly. Results are obviously affected by urine concentration and by Ca intake (8). Comparison of results of the Sulkowitch test with quantitative analysis of urine Ca has indicated that the urinary constituents greatly influence the degree of turbidity produced by any given Ca concentration (475).

IONIZED CALCIUM

The term "ionized" Ca (Ca^{2+}) is well entrenched in the literature as meaning the free Ca ions. This fraction of the total Ca is important because it is the physiologically active Ca. The studies of Ringer in 1882 revealed the basic role of Ca^{2+} by its effect on the contraction of frog heart ventricle (107). McClean and Hastings in a classic series of papers showed that serum Ca^{2+} could be determined by the use of isolated frog hearts. Other important physiologic processes which require Ca^{2+} are enzyme activation and blood clotting.

It is well recognized that serum total Ca represents the sum of three distinct fractions: (a) protein-bound calcium (CaPr), (b) diffusible calcium complexes (CaR), and (c) Ca^{2+} (384). Total ultrafilterable Ca, representing both Ca^{2+} and CaR, has been widely used as an index of Ca^{2+}. Moore (384) found the following average distribution of components in 27 normal sera: (a) nondiffusible (CaPr), 39.5%; (b) ultrafiltrable, 60.5%, the latter consisting of 46.9% Ca^{2+} and 13.6% CaR. The CaR fraction consists of complexes of calcium with citrate, phosphate, bicarbonate, and sulfate (107, 384, 589). Only 17% of a total serum Ca of 5 meq/liter is free, ionic, and effective (396).

McLean and Hastings (373) in 1935 represented the relationship between Ca and protein (Pr) in serum at 25°C and pH 7.35 by the following mass law equation:

$$\frac{[Ca^{2+}][Pr^{2-}]}{[CaPr]} = K = 10^{-2.22 \pm 0.07}$$

where K is a constant, 0.07 constitutes 1 standard deviation, and the brackets [ ] signify concentration. They recognized pH, temperature, and Mg and citrate concentrations as variables in this relationship but found their effects to be small. The exponent -2.22 was a weighted mean of the factors observed for albumin and globulin and an assumption of 1.8 as the normal albumin/globulin ratio. Subsequent work with various protein fractions (350) claimed that the Ca bound differs significantly for each protein, but, even more important, the binding for any particular protein can change considerably in certain disease states. Pedersen showed that many of the disparate results for the binding of Ca to albumin could be accounted for by an unrecognized high citrate content of the purified albumins used by previous workers (426). He established that in the pH region 6.5–8.5 Ca is bound reversibly to serum albumin by a single class of 12±1 identical binding sites per molecule (426), later identified as the imidazole groups of histidine (427). He also found that 2.4 calciums were bound per mol of purified γ-globulin, not via histidine. Studies of other serum proteins have shown that α- and β-globulins bind more Ca than does γ-globulin (350). A vitamin D-induced calcium-binding protein in the intestine of many species has been described (592). Moore (384) reported that [CaPr] appears to be linearly related to [Alb] by the function:

$$[CaPr]_{mmol/liter} = 0.11 + 0.019 [Alb]_{g/liter}$$

About 81% of the CaPr is CaAlb and the remainder CaGlob.

It is generally agreed that pH affects binding of Ca^{2+} to protein (384, 427). Thus at pH 6.8 an average of 54.3% of serum Ca was ionized, while at pH 7.8 it was 37.5% ionized. Serum $[Ca^{2+}]$ decreased about 4% for each 0.1 unit increase in pH, reflecting the pH dependency of Ca binding by serum proteins. $[Ca^{2+}]$ is about 5%–6% higher at 25°C when serum is ultrafiltered or ultracentrifuged than at 37°C (350, 425). By contrast, when $[Ca^{2+}]$ is measured at 10°C or 37°C in previously prepared ultrafiltrates the results are the same (425).

There are a number of ways to quantitate the Ca^{2+} of serum: (a) Determine the total Ca in cerebrospinal fluid, since it is the same as the Ca^{2+} in serum except for a small amount of bound but diffusible CaR. (b) Determine Ca^{2+} by dialysis or ultrafiltration across a semipermeable membrane (158, 384, 423, 424, 460, 476, 477, 612), a technic requiring special apparatus and careful control of pH and temperature. (c) Separate Ca^{2+} in the ultracentrifuge (70, 350). (d) Determine Ca^{2+} as the difference in Ca content of serum before and after removal of Ca^{2+} by an ion exchange resin (176, 506) or coated charcoal (72). (e) Determine the effect on coagulation of a normal decalcified plasma (537). (f) Measure *in vitro* calcification of rachitic rat cartilage (631). (g) Determine the effect on the isolated frog heart (373). (h) The concentration of metal ions can be determined with pM indicators by reading at two wavelengths (155, 249, 354). pM is the negative log of molar concentration of the free, solely hydrated metal ion (459). This approach is based on the additivity of absorbances of two forms of a metal indicator, free and bound—e.g., free murexide and calcium murexide—both forms being in a mass action equilibrium. (i) Use "ultrafiltration *in vivo*" (578). (j) Use gel filtration, a technic in which the unbound Ca is calculated from the difference in Ca concentration before and after equilibrium with dry dextran (Sephadex) gel (504). The method has been criticized on the grounds that the premises concerning gel-serum behavior are not valid for individual sera (456). (k) Determine the total Ca and total protein and calculate the "ionized" Ca from the nomogram of McLean and Hastings (375) or from the following formula (640):

$$mg\ Ca^{2+}/100\ ml = \frac{6\,[Ca] - (P/3)}{P + 6}$$

where [Ca] = mg total Ca/100 ml and P = g protein/100 ml. This is based on the original McLean-Hastings equation. Several studies have shown good agreement between values for $[Ca^{2+}]$ in normal sera obtained by direct measurement and values calculated by use of the McLean-Hastings nomogram (384), but there seems to be lack of agreement in results on sera from abnormal patients (452, 506). Neuman and Neuman (396) conclude that the McLean-Hastings formula has stood the test of time and that it does describe quite accurately the protein binding in most normal sera and in several disease states. It would appear that normally there is a direct correlation between the total Ca and total protein (373, 384). Efforts to show a similar correlation among [Ca], [Prot], and inorganic phosphorus have not succeeded (209). (l) Finally, the method which is preferred above all others at the present time (446), the Ca ion-selective electrode. For detailed information on this subject the reader is referred to *Chapter 4* and references dealing with the theory, development, and practice of ion-selective methodology (383, 464, 490). The first practical Ca electrode, developed by Ross (489), measured calcium ion *activity* in the presence of many common interfering ions. The electrode utilized a liquid ion exchanger membrane containing the Ca salt of a disubstituted phosphoric acid (0.1 M Ca didecylphosphoric acid) dissolved in di-n-octylphenyl phosphonate. The internal reference electrode contained 0.1 M $CaCl_2$ and an Ag-AgCl electrode. The lipophilic ion exchanger molecule (Ca didecylphosphoric acid) served as the mobile transport mechanism for Ca^{2+}. Over the range $10^{-1}M-10^{-4}M$, this electrode behaved in a completely Nernstian manner. From $[Ca^{2+}]$ $10^{-1}M$ to $10^{-4}M$, a plot of E_{cell} versus log $M_{\gamma\pm}$ (activity) had a slope of 87.5 mV. Theory predicted 87.9 mV. The empirical equation describing the behavior of an ion-selective electrode in mixed-ion media is (490):

$$E = \text{constant} + \frac{2.3\,RT}{z_A F} \log_{10}\left(A + \sum_i k_i B_i^{\,z_B/z_A}\right)$$

where R, T, and F are the gas constant, absolute temperature, and the Faraday constant, respectively; z is the charge of the ions A and B; A and B are activities of the sought-for ion and an interfering ion, respectively; $k_i = a$ selectivity constant, which is a measure of the degree to which the ion B interferes with the measurement of A (Ca^{2+} in this case). The value of k_i for Na^+, K^+, and $NH_4^+ = 10^{-4}$, indicating negligible effect on Ca^{2+} measurement; k_i for $Mg^{2+} = 0.014$. In serum the presence of 150 meq Na^+/liter and 1.0 meq Mg^{2+}/liter would result in about 2% enhancement of apparent Ca^{2+} concentration. The effect of Na^+ on serum measurements is eliminated by the incorporation of 0.15 mols NaCl/liter in Ca standard solutions. Above pH 6 the Ca^{2+} electrode does not "see" H^+. The electrode just described is called a "static" electrode and is no longer used because it had several disadvantages, the most serious of which was its slow response. A third generation "flow-through" electrode has now been perfected (Orion system 99-20, Fig. 19-1). This electrode has the following advantages (383): (a) Equilibration time is quite rapid, usually 30–60 sec. (b) It is more stable, with typical drift of 2–5 mV during the course of a day, as compared with 3–10 mV/day with static electrodes. (c) Less sample volume is required, several measurements being possible with a 1 ml sample. (d) The measurement is anaerobic, analagous to a blood pH measurement. The electrode is unresponsive to anions and chelates of Ca and is thus highly specific for Ca^{2+}. Below pH 5.5 the electrode no longer shows Ca^{2+} selectivity but behaves as a pH electrode. According to the Nernst equation, a tenfold change in calcium ion activity should yield a potential change of 30.8 mV at $37°C$ and 29.6 mV at $25°C$. These theoretical slopes have been observed occasionally; more typical values have been about 29 mV and 27 mV, respectively (383). Electrode response has been found to be linear over the Ca^{2+} range 2.0–20.0 meq/liter. We have developed a method, employing the flow-through calcium electrode, which determines Ca^{2+} in serum in the pH range 7.0–8.0, and by means of an empirically derived calibration curve corrects the value found to that at pH 7.4 (624).

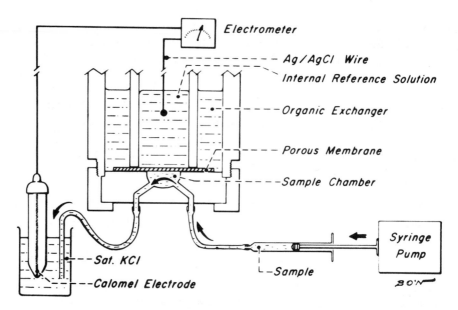

FIG. 19-1. Diagram of a calcium flow-through electrode. Courtesy of Orion Research, Inc.

DETERMINATION OF IONIZED CALCIUM IN SERUM

(method of Wybenga et al, Ref 624)

PRINCIPLE

Ionized Ca in serum is determined by an ion-selective electrode (Orion Research Inc.). Values are determined from a standard curve in which millivolts are plotted versus log $[Ca^{2+}]$. Electrode response is improved by conditioning the electrode with serum prior to standardization with aqueous standards. Since serum Ca ion activity is also a function of pH, results are corrected to pH 7.4.

REAGENTS

Triethanolamine (TEA), 1 M. Dissolve 14.9 g TEA in 100 ml water.
Trypsin, Crystalline
Stock Calcium Standard. Weigh out 500 mg reagent grade $CaCO_3$ and transfer to a 100 ml volumetric flask. Add 4 ml of 3 N HCl and let stand until dissolved. Add approximately 70 ml water and adjust pH to 6.0 with 50% (w/v) ammonium acetate. Adjust to volume and mix. 1 ml = 0.1 meq Ca^{2+}.
Stock Calcium/Sodium Chloride Standards.
1 meq Ca^{2+} and 150 meq Na^+/liter. Add 10 ml stock Ca standard to a 1 liter volumetric flask. Add 8.77 g reagent grade NaCl. Add approximately 900 ml water and dissolve NaCl. Adjust to volume and mix.

2 meq Ca²⁺ and 150 meq Na⁺/liter. Repeat above using 20 ml of stock Ca standard.

4 meq Ca²⁺ and 150 meq Na⁺/liter. Repeat above using 40 ml of stock Ca standard.

Working Standards. To 25 ml of each stock Ca^{2+}/Na^+ standard, add 0.075 ml TEA and 15 mg trypsin. Dissolve completely. Adjust pH to 7.4±0.05 with 0.4 *N* KOH using an expanded scale pH meter. Working standards are stable for 5 days refrigerated. *Bring to room temperature before use.*

PROCEDURE

Refer to *Note 1* for electrode assembly.

1. Equilibrate and calibrate the electrode as follows: Fill three 1 ml disposable capped syringes with working standards and a fourth syringe with any random serum which is clear. Cap tips until syringe content of Ca^{2+} is measured. Alternately pump through the electrode the 1 meq Ca^{2+} working standard, random serum, 1 meq Ca^{2+} working standard, random serum, etc., recording millivolt (mV) readings at exactly 2 min until the mV difference for successive standard readings is less than ±0.2 mV. Repeat for the other two working standards. Using two cycle, semilogarithmic graph paper, plot the potential (mV) developed in each standard (linear axis) against the concentration value of the standard (log axis). Draw a straight line between the 1 and 2 meq points and between the 2 and 4 meq points (see *Note 2*). A potential difference of at least 12 mV should be obtained between the 1 and 4 meq standards.

2. Using an expanded scale pH meter, determine and record serum pH to two decimal places at room temperature.

3. Immediately after the pH determination, draw up 1 ml of the serum into a 1 ml disposable plastic syringe, avoiding introduction of air bubbles. Immediately cap the syringe to minimize contact with air. The sample is stable in the capped syringe for at least 4 hours. If the sample cannot be assayed during this period, the pH must be redetermined.

4. Pump the unknown sample through the electrode and record the mV reading at exactly 2 min.

5. Pump the standards which give mV readings bracketing the unknown and record the mV reading of the standards at exactly 2 min. If these readings are different from the values obtained during electrode calibration at step 1, plot these new ones on semilog paper.

6. Repeat steps 3 and 4 until the mV variation for the unknown is less than ±0.2 mV.

7. Using the standard curve determine the Ca^{2+} in the unknown.

8. The pump lines are to be filled with any working standard after the samples are assayed. See also *Note 1, Electrode Assembly.*

Calculation: In this method, Ca^{2+} is determined at 23°C and at the pH of the serum (pH_X) as received in the laboratory. The value obtained may be corrected to any desired pH (pH_Y) by use of the following equation (see discussion under *Stability of Specimens*):

$$Ca^{2+}pH_Y \text{ meq/liter} = Ca^{2+}pH_X \text{ meq/liter} + [1.03 \, (pH_X - 0.17 - pH_Y)]$$

where

$Ca^{2+}pH_X$ = concentration of Ca^{2+} at 23°C and pH_X, (i.e., the concentration obtained in the above method.)

pH_X = pH of serum at 23°C at time of analysis

pH_Y = pH of serum at 37°C at time specimen was drawn

The factor 0.17 (14 x 0.012) is introduced to correct for the difference in temperatures—i.e., 37°C and 23°C—used for the two pH measurements. Assuming the original pH of the blood to be 7.4 at the time it was drawn can lead to errors as large as about 0.4 meq/liter at the extremes of pH which can be encountered physiologically. It is therefore imperative to know the original pH with accuracy in order to obtain accurate Ca^{2+} values.

NOTES

1. *Electrode assembly.* (See diagram in instruction manual for Orion 99-20 system.)
   a. Disconnect electrodes from meter.
   b. Disassemble and drain excessive fluids.
   c. Using tissue wipes, dry all parts as thoroughly as possible.
   d. Wash with petroleum ether 3 times and air-dry.
   e. Using plastic tweezers pick up one membrane, moisten the shiny side with ion exchanger, and immediately place shiny side down in the exact center of the membrane spacer.
   f. Replace flow-through bottom cap and retainer ring.
   g. Remove the top cap of the electrode and the flat washer.
   h. Using a glass 1 ml syringe, transfer 0.25 ml ion exchanger through the larger of the two outside filling holes. Replace the washer.
   i. Using a disposable 1 ml syringe, transfer sufficient internal filling solution to fill completely the center filling hole.
   j. Replace the top cap.
   k. Connect flow-through electrode to reference electrode and Luer-Lok with 0.025 in. I.D. catheters.
   l. Connect cables to digital readout pH meter. Set meter to read millivoltage.

m. Let the unit stand for 2 hours before carrying out the equilibrium procedure.

n. The capillary between the ion exchange electrode and the reference electrode should be disconnected, and the lines should contain any of the working standards when the unit is not in use.

2. *Calibration curve.* A point to point standard curve is drawn through the calibration points. Drawing a smooth curve or using a least squares method to obtain the best fit only makes the curve less precise because of Na error (383).

3. *Error of electrode.* Small errors in potential measurements may yield rather large errors in estimated Ca^{2+} in unknown solutions. Thus, with an electrode slope of 28 mV, an error of 1 mV in measured potential would yield about an 8% error in apparent Ca^{2+} (384, 490). The requirement for a reproducibility of ±0.2 mV in the present method brings the allowable error below 2%. This agrees with the reports of others using calcium-selective electrodes (342, 384, 490).

4. *Stability of specimens.* Wybenga *et al* (624) derived the equation used in the method described from a series of measurements on different sera adjusted to pHs from 7.0 to 8.0 These authors found that under the stated conditions, fresh sera, measured anaerobically and then allowed to be in contact with air for 24 hours, had the same Ca^{2+} content when the latter measurement was corrected for pH by use of the equation. It was further determined that sera stored at 4° or 30°C were stable up to 6 days, while sera frozen and held at −20°C gave reproducible results up to 8 days, Storage of serum at −18°C for 5 months had no effect on ultrafilterability and Ca^{2+} (424). Moore (384) reported that Ca^{2+} can be reliably determined in sera that have been frozen 72 hours, but that a decrease was frequently observed in frozen ultrafiltrates.

5. *Interferences.* Plasma Ca^{2+} is 0.06−0.12 meq/liter lower than serum, presumably due to heparin binding (342, 413, 494). Li *et al* found that Mg added both to the standard solutions and to serum depressed the potential readings (342). They concluded that the true concentration of Ca^{2+} in serum is likely to be higher than that measured with the calcium ion-selective electrode system. The same authors confirmed the finding by Hattner *et al* (230), that trypsin and triethanolamine (TEA) lowered the potential readings of standards by an amount equivalent to 0.10 meq Ca^{2+}/liter, thus raising the serum Ca^{2+} values. The addition of these two substances to standards stabilizes potential readings. It would seem that the net effect of Mg and trypsin-TEA is not significant.

ACCURACY AND PRECISION

The mean value obtained in our laboratory by the ion-selective technic for percent ionized calcium in 16 normal adults was 51.7% (624). Li *et al*, using the same system in a study of 231 normal adults found a mean of 53.4% (342). These values compare favorably with 53.7% found by Schatz (504) using dextran gel, 53.1% by Loken *et al* (350) using the ultracentrifuge, and 51% by Prasad and Flink (453) using the ultrafiltration method.

Precision of the ion-selective electrode method for normal sera is about ±2% (95% confidence limits).

NORMAL VALUES

The normal range for serum ionized Ca, using nonparametric statistics, for 76 normal, healthy adults in our laboratory is 2.42–2.78 meq/liter. Li *et al* (342), also using an ion-selective electrode technic, reported a normal range of 2.2–2.7 meq/liter; others (345, 384, 447) have reported similar ranges. Ionized Ca as determined by ultracentrifugation is 50%–58% of the total Ca (70, 350). The normal range for the unbound fraction as determined by the same technic is about 2.45–2.85 meq/liter. The McClean-Hastings formula method for calculating "ionized" Ca gives a normal range of 2.1–2.6 meq/liter (375) and is the same for adults, children, and infants (15, 566).

Approximately 22% of the Ca of normal urine is in the ionized form (403).

MAGNESIUM

Rapid changes in technology have affected the determination of Mg significantly (9). Among the methods available, the following are most useful: (a) atomic absorption, (b) flame photometry, (c) fluorescence spectrophotometry, (d) spectrophotometry of dye-lake combinations (titan yellow) or of other magnesium-indicator compounds, and (e) precipitation of magnesium ammonium phosphate with subsequent determination of phosphate by a variety of methods.

Magnesium, an alkaline earth element, is readily detectable by atomic absorption (551, 616). Many details discussed under *Calcium* apply equally to Mg, and that section should be consulted. Indeed, methods have been published for the simultaneous determination of Ca and Mg (351, 458, 497, 552). The 285.2 nm line of a Mg hollow cathode lamp is used. Interferences encountered with the atomic absorption determination of Ca in serum and urine do not occur with analysis for Mg (225), but in bone, tissue extracts, and erythrocytes there are ratios of Na^+, K^+, and phosphate which do affect Mg determinations (265, 356), and it is necessary to use lanthanum or strontium to overcome these interferences. The atomic absorption method for serum Mg has been automated (316). Atomic absorption for Mg is specific, has good detection limits, and is simple to perform. For these reasons, it is the method of choice.

Magnesium has a flame spectrum with a band emission with peaks at 370 and 383 nm and an arc line at 285.2 nm. Flame photometry has been used for determination of Mg but is not recommended when atomic absorption is

available (616). For a review of flame photometry technics, the reader is
referred to the first edition of this book.

Precipitation of Mg as magnesium 8-hydroxyquinoline has been the basis
for another method of determination. The hydroxyquinoline in the
precipitate has been quantitated photometrically with Folin's phenol reagent
(151) by the blue-green color formed with ferric ion in acid solution (248)
and by bromination followed by iodometric titration of the excess bromate
(208). Reaction with 8-hydroxyquinoline occurs as follows (208):

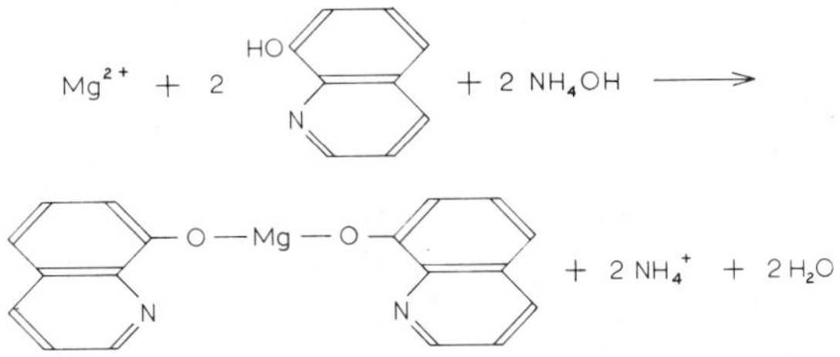

The reported disadvantages of this general method are that the precipitate is
very light, making it difficult to handle (523), and the technic is susceptible
to error from contamination with Zn (80). Another approach which avoids
precipitation of the Mg is fluorometric measurement of magnesium
8-hydroxy-5-quinolinesulfonate (498, 499). The fluorimetric method has
been tested as a manual procedure (563) and in several automated versions
(318, 408). Calcium interference has been eliminated by sequestration with
ethylene bis(oxyethylenenitrilo)tetraacetic acid (EGTA), which does not
affect fluorescence of the Mg complex (318). Several other fluorescence
methods have been described using other reagents: o,o'-dihydroxyazo-
benzene (135); calcein, using EGTA to mask Ca and 2,3-dimercapto-1-
propanol (BAL) to mask Zn in tissue extracts (641); and calcein plus the
chelating agent DCTA (1,2-diaminocyclohexanetetraacetic acid) (586).
Although the 8-hydroxyquinoline reagent has been recommended for urine
Mg determinations, other authors have warned of severe and frequent
quenching by unidentified constituents (563).

An attractive method for the direct determination of Mg stems from the
work of Kolthoff in 1927 (323). The method depends on the formation of a
red lake with the dye, titan yellow (Clayton yellow, thiazole yellow), in an
alkaline solution. The dye has been shown to be heterogeneous by thin layer
chromatography (222). A Mg-reactive component has been isolated by
chromatography on Sephadex (305). It is so reactive that it can be used at a
concentration as small as 0.008%, at which concentration the response is
linear between 20 and 150 μg Mg. The studies cited in the following
discussion were all carried out on the heterogeneous dye, and their
conclusions are subject to revision when and if they are repeated using purer
preparations.

The red lake consists of dye adsorbed on the surface of colloidal particles
of Mg (OH)$_2$ (183, 194). The earlier technics (246) did not call for

preliminary precipitation of protein, but this is a desirable precaution (180). There have been four problems with the titan yellow method: (a) The red lake formed has a tendency to precipitate out of solution. To stabilize the lake and give an optically clear solution, gum ghatti (180, 394), gum arabic (404), polyvinyl alcohol (233, 234, 338, 412, 538), a polyvinyl alcohol-sodium lauryl sulfate mixture (116), and a mixture of glycerol and starch (439) have been employed. (b) Some workers (329) experienced fading of the color and used hydroxylamine to prevent it, whereas others (394) found this to be of no help. It has also been reported that substituting LiOH for NaOH in the procedure produces a color that is stable for 24 hours when protected from light (529). (c) The color reaction is not particularly sensitive, some workers employing a 5 cm light path for measurement of absorbance (412). Fortunately, polyvinyl alcohol appears to potentiate the color produced about twofold (233). (d) Titan yellow methods, at least the one presented in this chapter, yield results on sera about 10%–15% higher than those obtained by the $MgNH_4PO_4$ method of Simonsen *et al* (523). This difference is seen also in normal values reported for the two methods. This discrepancy is explained if the data indicating that as much as 15% of the oxalate precipitate in the removal of Ca is Mg are correct. Aikawa and Rhoades (4), however, observed a maximum of 3% of the Mg in the calcium oxalate precipitate and obtained a final recovery of 95% of added ^{28}Mg. Dawson and Heaton (127) also found results by the $MgNH_4PO_4$ method to be only 1%–2% less than values obtained by atomic absorption spectrophotometry. There was a 2.2%–2.5% Mg loss almost balanced by a phosphate gain. Kolthoff (323) originally called attention to the fact that Ca intensifies the color of the Mg-titan yellow lake; and although it has been claimed (17, 34, 180, 293, 412, 538) that the Ca present in serum does not interfere, others (394) have observed intensification of about 15% with the Ca concentrations found in serum and recommend inclusion of optimal amounts of Ca in standards and unknowns alike. Kawashima and Ueda (293) added methylene blue to the final reaction mixture. A further discussion of titan yellow methods is given in ref 37.

A spectrophotometric method for the determination of serum Mg using Magon, 1-azo-2-hydroxy-3-(2,4-dimethylcarboxanilido)naphthalene-1-(2-hydroxybenzene), has been developed (76, 365). After acid digestion, interfering phosphate is removed as aluminum phosphate (76). The colored complex developed at pH 11.2 in an alcoholic medium is read at 505 nm. The dye is said to be eightfold more sensitive to Mg than is titan yellow (76), and the complex is stable for 3 days at room temperature. An ultramicro method using Magon requires only 40 μl of sample and is not affected by the presence of gluconate (29, 473). The method has also been adapted to the AutoAnalyzer (63).

The azo dye, chrome fast blue BG, has also been used in a photometric method for serum Mg (67).

Other dyes and indicators have been used to determine Mg complexometrically and have been discussed in some detail in the preceding section on

Calcium. One such method, automated and applicable to serum, plasma, urine, and tissue ash, determines Mg with eriochrome black T in the presence of strontium EGTA, which eliminates significant error from Ca and the necessity for separation of Ca from the Mg prior to analysis (192).

Ionized Mg has been determined by ultrafiltration (400) and ultracentrifugation (118, 400).

Minute amounts of Mg in tissue ash can be determined by neutron activation by first removing heavy metals with an extraction by sodium diethyldithiocarbamate. Magnesium is then transferred from the aqueous to an organic phase by use of TTA (thenoyltrifluoroacetone). Following neutron activation, ^{27}Mg is isolated for counting. The method is satisfactory only for research (218).

The Mg content of erythrocytes can be determined directly (143, 189, 341), or indirectly from whole blood and serum or plasma levels and the hematocrit (93, 207, 588).

We present a titan yellow method for those laboratories without atomic absorption equipment, and a preferred atomic absorption method we have found to be reliable.

DETERMINATION OF MAGNESIUM

(method of Garner, Ref 180, and Neill and Neely, Ref 394, modified)

PRINCIPLE

Magnesium is determined photometrically in urine or a tungstic acid filtrate of serum by forming a red lake with titan yellow in alkaline solution. Color formation is potentiated by polyvinyl alcohol. Figure 19-2 shows the absorption curve of the color formed.

REAGENTS

H_2SO_4, 2/3 N
Sodium Tungstate, 10%
NaOH, 4 N
Titan Yellow, 0.05%. Dissolve 50 mg in water to 100 ml. Store in the dark in an amber colored bottle. Stable about 1 week.

Polyvinyl Alcohol (PVA), 0.05%. Disperse 50 mg PVA (Elvanol 70-05, Du Pont) in 50 ml water. Dissolve by heating in a 65°C water bath and stirring. Dilute to 100 ml with water and filter. Store in refrigerator. Refilter if solution becomes cloudy.

Stock Magnesium Standard. Dissolve 50 mg Mg metal turnings, ribbon, or powder in 4–5 ml 1 N HCl in a 100 ml volumetric flask. Dilute to volume with water.

Working Magnesium Standard. Dilute stock 1:100 with water. 1 ml = 5 μg Mg.

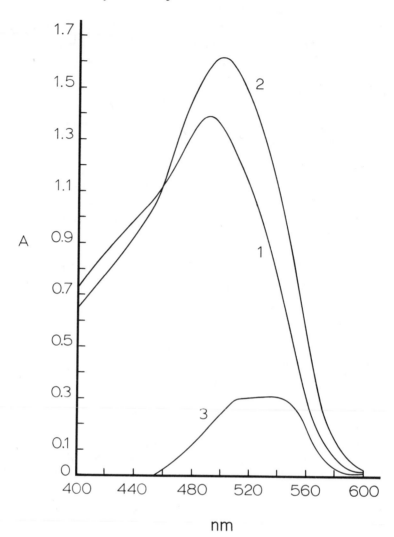

FIG. 19-2. Absorption curves in titan yellow method: (1) reagent blank vs water, (2) 8 μg standard vs water, (3) standard vs reagent blank. Perkin-Elmer recording spectrophotometer model 4000-A.

PROCEDURE FOR SERUM AND SPINAL FLUID

Serum must be separated from the clot as soon as possible, otherwise the serum Mg level increases significantly because of passage of Mg from the erythrocytes into the serum (266).

1. Set up the following in test tubes:
   Blank: 3.0 ml water
   Standard: 2.0 ml working standard + 1.0 ml water
   Unknown: 0.5 ml serum or cerebrospinal fluid + 2.5 ml water

2. Add 1.0 ml 10% sodium tungstate to each tube and mix. Add 1.0 ml 2/3 N H_2SO_4 to each tube and mix. Centrifuge the unknowns.
3. Transfer 2.5 ml from each tube to new tubes. Add 1.0 ml PVA to each tube and mix. Add 0.5 ml titan yellow to each tube and mix.
4. Add *slowly* 1.0 ml 4 N NaOH to each tube. Mix *gently*.
5. Read absorbance of standard (A_S) and unknown (A_X) versus blank at 540 nm (see *Note 1* below).

Calculation:

$$\text{mg Mg/100 ml} = \frac{A_x}{A_s} \times 0.005 \times \frac{100}{0.25}$$

$$= \frac{A_x}{A_s} \times 2$$

$$\text{meq Mg/liter} = \frac{A_x}{A_s} \times 1.64$$

$$\text{meq Mg/liter} = 0.82 \times \text{mg/100 ml}$$

PROCEDURE FOR URINE

Magnesium excretion in the urine is usually expressed on a 24 hour basis. Measure and record the total volume. If urine is clear and no precipitate is seen clinging to walls of container, mix and remove approximately 10 ml. Acidify to pH 1 with conc HCl (wide range pH paper). If precipitate is present in original urine and if no other test is requested, then acidify entire urine specimen to pH 1. If other tests are requested, then shake and mix contents well, remove 10 ml aliquot, and acidify. Warm acidified specimen to 60°C for at least 15 min, mixing occasionally. Proceed as for serum and correct A_x for urine color as follows: Run a Blank and Unknown through the procedure but substitute water for the titan yellow reagent. Read the "Unknown minus titan yellow" vs "Blank minus titan yellow," obtaining $A_{\text{urine blank}}$. Then:

$$\text{corrected } A_x = \text{observed } A_x - A_{\text{urine blank}}$$

NOTES

1. Beer's law. The color obeys Beer's law at 540 nm in a Beckman DU or B up to 8 μg Mg in the final tube, i.e., an A_{540} of about 0.31. This is equivalent to a sample concentration of 3.2 mg/100 ml. Unknowns reading above this point must be calculated from a calibration curve or rerun on a small aliquot. The color does not obey Beer's law with a Klett 52, 54, or 56 filter. It is quite probable, therefore, that a calibration curve must be employed with all photometers using glass filters.

2. *Color stability.* Colors are stable from 5 min after formation to at least 1 hour. Color intensity increases reversibly with increased temperature (548).

3. *Interference by calcium gluconate* (12, 31). The presence of calcium gluconate in serum, resulting from intravenous administration, causes results to be low by as much as 35%. No such interference occurs with $MgNH_4 PO_4$ methods.

4. *Interference by mercurial diuretics (31).* This can be removed by treatment with $H_2 S$.

5. *Interference by Shohl's citric acid-sodium citrate solution (234).* The titan yellow method underestimates Mg in urine of patients receiving this solution.

6. *Stability of serum samples.* Studies in our laboratory indicate that samples are stable for 1 week at room temperature.

7. *Determination of Mg in erythrocytes.* Our laboratory has had no experience with this analysis, but the titan yellow method has been applied to protein-free filtrates of washed, lysed erythrocytes (341).

ACCURACY AND PRECISION

When considering the accuracy of the titan yellow method for determination of Mg in serum, spinal fluid, and urine, the following are of concern: reports that (a) Ca does or does not intensify the color (previously discussed), (b) ammonia in large quantities (46) and iron (234, 412) depress the color, (c) phosphate may interfere (181, 439), and (d) the color varies directly with pH (329). These variables have been studied in our laboratory with the method presented with the following results: Calcium, ammonia, and phosphate were without effect up to concentrations of 20, 1000 (as N), and 200 (as P) mg/100 ml, respectively. The addition of Fe^{3+} to a protein-free filtrate, equivalent to an initial serum concentration of 200 μg/100 ml, resulted in a 31% depression of color. Addition of 500 μg of Fe^{3+} per 100 ml to the serum prior to tungstic acid precipitation was without effect, indicating that tungstic acid precipitation effectively removes this source of error. Variation of the amount of 4 N NaOH used in the test from 0.9 to 1.1 ml was without effect. Recoveries of Mg added to serum averaged 98%. Willis (609) believed that the titan yellow method may not be very accurate, basing his conclusion on variable recovery and the observation that results were about 5% higher after ashing samples (only two samples studied). Recovery may be as low as 50% if serum is not diluted prior to protein precipitation (538).

The precision of the method is about ±5% (17).

DETERMINATION OF MAGNESIUM BY ATOMIC ABSORPTION SPECTROPHOTOMETRY

(method of Hansen and Freier, Ref 225, and Perkin-Elmer Handbook of Analytical Methods for Atomic Absorption Spectrophotometry, Ref 428)

PRINCIPLE

Magnesium is aspirated at a suitable dilution into a three-slotted air-acetylene burner, and the absorbance of a hollow cathode Mg lamp output at 285.2 nm is recorded.

REAGENTS

Stock Lanthanum Solution. Place 58.76 g of lanthanum oxide in a 1 liter volumetric flask and wet with water. Add 250 ml of conc HCl very slowly until the material is dissolved. Dilute to 1 liter with water. This provides a 5% lanthanum ion solution in 4 N HCl. Store in refrigerator.

Working Lanthanum Solution, (0.2%, w/v). Dilute 20 ml of the stock solution to 500 ml with water.

Stock Magnesium Standard (40 meq/liter). Dissolve 48.6 mg of Mg metal turnings, ribbon, or powder, in 2 ml 1 N HCl in a 100 ml volumetric flask. Dilute to 100 ml with water. Store in refrigerator in a polyethylene bottle.

Working Magnesium Standards.

2.0 meq/liter: Dilute 5.0 ml stock to 100 ml with water.

3.0 meq/liter: Dilute 7.5 ml stock to 100 ml with water.

Store standards in plastic bottles in the refrigerator. Prepare fresh monthly.

PROCEDURE FOR SERUM, SPINAL FLUID, AND URINE

1. Label 16 x 125 mm tubes (or 10 ml polyethylene vials) for each unknown and for the 2.0 and 3.0 meq/liter standards.

2. Dilute 0.1 ml serum, cerebrospinal fluid, or urine (see *Procedure for Urine* under titan yellow method for Mg) to 5.0 ml with 0.2% working lanthanum solution. This can be done by an automatic dilutor, or 0.1 ml can be added to 4.9 ml with a TC pipet. Mix thoroughly.

3. Repeat step 2 with the 2.0 and 3.0 meq/liter working standards. Pour about 30 ml of the 0.2% working lanthanum solution into a 50 ml Erlenmeyer flask for use as a blank.

4. Atomic absorption: Since the details of analysis vary with the atomic absorption spectrophotometer employed and since these are given in the book of instructions accompanying each instrument, no attempt is made here to present them.

NOTES

1. Beer's law. Linearity exists up to 4 meq Mg/liter. For samples which assay greater than 4 meq/liter, prepare a twofold or more dilution of the original sample and take the diluted sample through the procedure. The result obtained should then be multiplied by the dilution factor.

2. Stability. Samples are stable at room temperature for at least 1 week.

3. Interferences. The use of the La diluent is of doubtful value unless Ca and Mg are to be measured simultaneously. Willis (610) claimed that protein causes a positive error in the determination of serum, which can be obviated by releasing Mg from the protein with Sr^{2+} or HCl. Hansen and Freier (225) could not confirm this claim. Iida *et al* (272) showed that serum diluted more than 1:50 did not require any diluent other than water. However, in cerebrospinal fluid and urine, where Mg and phosphate concentrations are greater than Ca, oxine was required to overcome phosphate interference (272).

ACCURACY AND PRECISION

Results yielded by atomic absorption spectrophotometry of serum are in good agreement with a standard ammonium phosphate precipitation method, and recovery of Mg added to serum is quantitative (9). Several investigators have compared four or more methods on the same sera and have found that the titan yellow technic gives results 0.2–0.3 meq/liter higher than those obtained by atomic absorption, with mean normal values of 2.0 and 1.7 meq/liter, respectively (457, 493).

Between-run precision (95% limits) of the method is about ±5% for serum values in the normal range.

NORMAL VALUES

Serum or Plasma. Mean normal values for plasma and serum Mg by a variety of methods have been summarized by Alcock (9, 10). Plasma values range from 1.57 to 2.05 meq/liter, serum from 1.67 to 2.05 meq/liter. The normal range obtained by titan yellow methods is about 1.5–2.4 meq/liter (17, 31, 293, 341). The mean plasma value for 610 subjects determined by atomic absorption, flame photometry, and $MgNH_4PO_4$ precipitation was 1.7 meq/liter (9). A reasonable compromise of all methods (95% limits) is a normal range of 1.3–2.1 meq/liter. There is no reported sex difference.

Of the Mg in normal human serum, 70% is unbound; the remainder sediments in the ultracentrifuge at the same rate as albumin, and little or none travels with β-lipoprotein or γ-globulin (118). Plasma from 23 normal adults was studied by ultrafiltration and ultracentrifugation at 20°C and pH 7.4 (400). The mean free Mg was 1.32 meq/kg plasma water, while the protein-bound Mg was 0.53 meq/kg plasma water. The free Mg thus equaled 71% of the total; of the bound, 77% was bound to albumin and 23% to globulins.

There is no appreciable variation throughout the day provided exercise is avoided (378, 593); there appears to be a sine-like pattern to the level of an individual throughout the year (633). Newborn levels are essentially the same as those for adults (13, 29, 544). Meals are without effect (207, 588). In women there appears to be a significant variation with the menstrual cycle, being highest at the menses (171).

Whole Blood and Erythrocytes. The levels in whole blood are higher because the erythrocytes contain about threefold more Mg than does plasma (39, 93, 207, 341). In contrast to Ca, which is firmly bound to the erythrocyte membrane, Mg is intracellular, and 96% is lost upon osmotic hemolysis (226). The normal content of erythrocytes as obtained by various analytic technics is about 4.4–6.0 meq/liter (93, 189, 207, 341, 588). One study of 200 individuals yielded a mean value of 4.6 meq/liter (189).

Cerebrospinal Fluid. Total Ca in the spinal fluid essentially equals the ionized Ca in serum. One might have expected a similar relationship with Mg, but there is more Mg in spinal fluid than ionized Mg in the corresponding serum (272). The normal range is about 2.0–2.7 meq/liter, with a mean value of 2.4 meq/liter (39, 130, 267, 272, 377).

Urine and Feces. Urinary excretion of Mg normally equals about one-third of the daily intake and averages 10.5 meq/day for men and 8.8 meq/day for women (156). Other values have been reported ranging from 1 to 24 meq (39, 40, 529). One study reported that a group of subjects with a daily intake of 61 meq Mg had an average 24 hour fecal excretion of 24.2 meq (39).

SERUM IRON

Human serum or plasma always contains some hemoglobin no matter how carefully the sample is obtained (cf *Plasma and Serum Hemoglobin*, Chapter 23). Irrefutable evidence has been obtained, however, that plasma or serum contains Fe in excess of that accountable to the hemoglobin present, although for many years there was controversy as to whether this fraction was an artifact derived from hemoglobin (579). This "serum iron" has been called by various workers nonhemoglobin iron, transport iron, acid-soluble iron, loosely bound iron, and ultrafilterable iron. Of the total 3.5–4.5 g of body Fe in a healthy 70 kg male, virtually all is bound to protein either as functional iron, such as in hemoglobin and the Fe-requiring enzymes catalase and cytochrome, or as storage or transport Fe (184). Almost all of the Fe in serum is bound to a β-globulin, transferrin, in the form of a Fe(III)-protein complex (258). The remainder is bound to free amino acids (454). Most of the technics for determination of serum Fe involve disruption of the Fe-protein complex by HCl; Barkan, who introduced this analytic approach, referred to the Fe so obtained as "iron that can be easily split off." The above names for "serum iron" do not necessarily refer to the same thing because of the different technics used to obtain the Fe. They have in common, however, the identity of nonhemoglobin iron.

The first step in the spectrophotometric analysis for serum Fe usually involves splitting the Fe from its protein combination by exposure to acid. One of three courses may then be followed: (a) removal of protein by precipitation with trichloroacetic acid, (b) arranging conditions so that proteins remain in solution without interfering with the subsequent analysis, and (c) in automated analyses, dialysis of the Fe from the protein.

It has been shown that Fe^{3+} is coprecipitated by trichloroacetic acid with serum proteins, whereas Fe^{2+} is not (492). It is essential for full recovery of Fe from serum to ensure that Fe dissociated from transferrin be reduced to Fe^{2+} prior to protein precipitation (461, 492). Caraway (82) incubated serum with ascorbic acid and HCl before deproteinizing with trichloroacetic acid and chloroform. This procedure yields a water-clear supernate with zero absorbance at 590 nm as a consequence of extracting interfering serum chromogens with the chloroform. In a similar procedure, thioglycolic acid has been used in place of ascorbic acid (433), but the use of either ascorbic or thioglycolic acid has been criticized by others who claimed that under some conditions those reducing agents reduce hemoglobin Fe and nearly double serum Fe values (395). These authors propose the use of another reducing agent, Nitroso-R-salt, which does not reduce hemoglobin Fe. Methods have been proposed (307, 395, 432) in which the Fe is split off the protein by the addition of acid, and the colored Fe complex extracted into isoamyl alcohol, thus avoiding protein precipitation and possibility of loss of Fe by coprecipitation. These technics encounter difficulties with icteric sera, however, because the bilirubin is extracted by the organic solvent and is oxidized to biliverdin, which leads to serious positive errors unless proper blank corrections are made (241, 487). Other approaches to the direct determination of Fe without protein precipitation include one employing a detergent, in the presence of which the proteins, the Fe reagent (bathophenanthroline), and its ferrous complex are all soluble (596), and another in which the Fe is dissociated by adjusting the pH to 2.8 and quantitated with bathophenanthroline (42). One of the most satisfactory technics appears to be precipitation of proteins with hot trichloroacetic acid (287, 569), a method that avoids losses of Fe and always yields an optically clear solution for photometry (241, 569). The protein-free filtrates from trichloroacetic acid contain a significant amount of "background absorbance" which can be corrected for by reading the absorbance before color development (241). A novel approach to splitting the Fe from protein is through displacement by the addition of tetravalent thorium (470).

Among the reagents used for the spectrophotometric quantitation of Fe are the following: thiocyanate, 2,2′ dipyridyl, 2,2′,2″-tripyridyl, 1,10-phenanthroline, and 4,7-diphenyl-1,10-phenanthroline (bathophenanthroline) (94, 136). The molar absorptivities of the Fe-chromogen complexes range from 7000 for thiocyanate to 22 140 for bathophenanthroline disulfonic acid (166). Organic compounds containing the "ferroin" group (=N–C–C–N=), which forms colored complexes with Fe^{2+}, are the most sensitive reagents for the determination of Fe in trace quantities. In addition to the phenanthrolines and di- or tripyridyls listed previously, the following new ferroin containing reagents have been introduced: (a) 2,4,6-tripyridyl-s-triazine (TPTZ), ϵ = 22 600; (b) 2,6-di-[pyridyl-(2)]-4-[p-methoxyphenyl] pyridine (DPMPP), ϵ = 26 900; (c) 3-(2-pyridyl)-5,6-bis(4-phenylsulfonic acid)-1,2,4-triazine (Ferrozine), ϵ = 27 900; (d) 2,6-bis(4-phenyl-2-2-pyridyl)-4-phenylpyridine (Terosite), ϵ = 30 200 (82, 89, 136, 507, 636).

Automated methods employing the ferroin reagents have been published (292, 507, 630, 634). Zak adapted an earlier manual procedure employing bathophenanthroline and found that Fe^{2+} dialyzes better and more reproducibly than Fe^{3+} (637). A modification of this method uses a citrate wash between samples to reduce background noise and carryover. Also, a more efficient dialysis was achieved by complexing the dialyzed Fe^{2+} ion in the recipient stream, enabling the use of a single dialysis instead of the original double dialysis (292). AutoAnalyzer methods require the addition of a neutral salt to the acid-reducing reagent to make the rate of dialysis of Fe^{2+} from protein solutions comparable to the aqueous standards (24, 25, 140).

Although one would have expected atomic absorption to be a popular method for serum Fe determination, this has not proved to be the case (162). Several methods have been published claiming good results (142, 410, 481, 557, 644). Direct measurement of serum Fe at low levels was found to give unreliable results because of limitations of sensitivity, matrix interferences, and the inability of atomic absorption to distinguish hemoglobin Fe from transport Fe (644). Extraction of chelated Fe into methyl isobutylketone gave improved results. Users of atomic absorption equipment have been aware of the fact that Fe hollow cathode lamps are more unstable and "noisy" than other commonly used lamps. Improved lamps have been manufactured, but at present the more sensitive spectrophotometric methods have a distinct advantage. Atomic absorption has been successful in the determination of Fe in urine, where higher levels are encountered, particularly for monitoring the course of Fe excretion following the therapeutic administration of chelators such as desferrioxamine (642).

Two methods for serum Fe are given, one manual (241) and the other automated (637). Both methods employ sulfonated bathophenanthroline and have given reliable results in our laboratory. It may be anticipated that the substitution of the newer ferroins, with at least 30% greater sensitivities, will lead to further improvements.

DETERMINATION OF SERUM IRON, MANUAL PROCEDURE

(method of Henry et al, Ref 241)

PRINCIPLE

Proteins are precipitated, and the Fe is freed simultaneously with hot trichloroacetic acid. The Fe in the supernate is reduced by hydrazine. The Fe^{2+} is quantitated photometrically by reaction with sulfonated bathophenanthroline. Correction for background absorption is made by reading the absorbance prior to color development. Figure 19-3 gives the absorption curve of the colored product.

REAGENTS

Trichloroacetic Acid, 20%.   Stable in refrigerator.

Ammonium Acetate, 50%

Hydrazine Sulfate, Saturated Aqueous.   Stable at room temperature in a brown bottle.

Sulfonated Bathophenanthroline, 0.05% Aqueous.   Obtainable as sodium salt from G.F. Smith. One can also prepare the sulfonated derivative from bathophenanthroline (4,7-diphenyl-1,10-phenanthroline, G.F. Smith) according to Trinder (569): Weigh 100 mg bathophenanthroline and place in a test tube; add carefully 0.5 ml chlorosulfonic acid. Boil gently for 0.5 min. Cool and cautiously add 10 ml distilled water. Mix and heat 5 min in boiling water bath to dissolve precipitate. Transfer to a 200 ml volumetric flask which contains about 150 ml water. Add 10%–20% NaOH dropwise to pH 4 (pH paper); then add water to 200 ml.

FIG. 19-3. Absorption curves in serum Fe determination: (1) reagent blank vs water; (2) 4 μg Fe standard vs water, (3) standard vs reagent blank. Beckman model DU spectrophotometer.

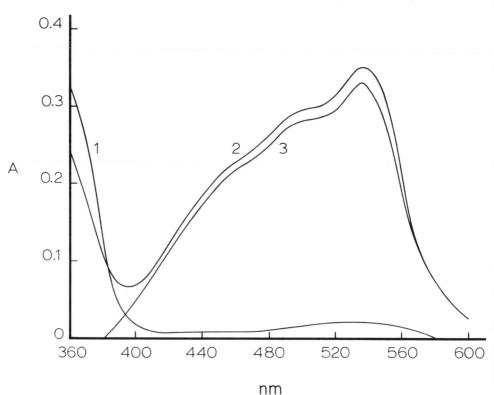

Iron Standard, 0.01 mg/ml. Clean a small piece of iron wire of known iron content (usually 99.8%). Weigh an amount of wire to contain 100 mg Fe (iron for standard can also be obtained from the National Bureau of Standards: ingot iron NBS 55e). Place in a 250 ml flask and add 10 ml conc HNO_3, reagent grade, and about 40 ml double distilled water. Heat to boiling and boil until Fe is completely dissolved. Cool and transfer quantitatively to a 100 ml volumetric flask. Dilute to volume and mix. Dilute this stock standard 1:100. 1 ml = 10 μg Fe. Stable indefinitely.

Glassware. All glassware used (reagent bottles, pipets, etc.) must be cleaned with hot HCl (concentrated acid diluted with an equal volume of water) and then rinsed with water.

Water. All water used in the preparation of reagents and in the test itself must be double distilled, the second distillation being from glass.

PROCEDURE

1. To 2.0 ml serum or plasma in a centrifuge tube, add 2.5 ml water and mix. Add 1.5 ml 20% trichloroacetic acid, drop by drop, with mixing.
2. Cover tube with an aluminum foil cap and place in water bath at $90°-95°C$ for 15 min, mixing tube contents at 5 min intervals. Cool tube to room temperature in tap water, mix contents again, and centrifuge.
3. Set up the following tubes:
   Blank: 3.0 ml water + 1.0 ml 20% trichloroacetic acid
   Standard: 2.0 ml diluted standard (1:5 dilution of the standard containing 10 μg Fe/ml) + 1 ml water + 1.0 ml 20% trichloroacetic acid
   Unknown: 4.0 ml supernate
4. To each tube add 0.35 ml ammonium acetate solution (resulting pH should be 4.5±0.5) and 0.3 ml hydrazine sulfate. Mix, transfer to cuvets, and read absorbances against water at 535 nm or use a filter with nominal wavelength in this region.
5. Return contents of each cuvet to its respective tube and add 0.4 ml 0.05% sulfonated bathophenanthroline. Mix, let stand 20 min, and read absorbances again. The color is stable for at least 2 hours. The absorbances of reagent blank A_b, standard A_s, and unknown A_x are calculated as the differences in readings before and after addition of the bathophenanthroline. (Once it has been established that the reagent blank and standard show zero absorbance before addition of bathophenanthroline, these readings can be omitted.)

Calculation:

$$\mu g\ Fe/100\ ml = \frac{A_x - A_b}{A_s - A_b} \times 4 \times \frac{100}{1.33}$$

$$= \frac{A_x - A_b}{A_s - A_b} \times 300$$

MICRO PROCEDURE

The determination can be successfully scaled down fourfold as follows: Use 0.5 ml serum, 0.6 ml water, and 0.4 ml trichloroacetic acid. For the blank, use 0.75 ml of a mixture containing 0.4 ml trichloroacetic acid and 1.1 ml water. For the standard, use 0.75 ml of a mixture containing 0.5 ml diluted standard (1:4 of the 10 μg/ml standard), 0.4 ml trichloroacetic acid, and 0.6 ml water. Use 0.75 ml of the unknown supernate. Add 0.1 ml ammonium acetate, 0.1 ml hydrazine sulfate, and 0.1 ml bathophenanthroline. Read absorbances in micro cuvets with a 1 cm light path before and after addition of the sulfonated bathophenanthroline. The standard is equivalent to a serum Fe concentration of 250 μg/100 ml.

NOTES

1. Beer's law. The color follows Beer's law even with a glass filter such as a Klett 54.

2. Stability of samples. By the method given, serum Fe is stable for at least 4 days at room temperature and 1 week at refrigerator temperature. Serum Fe has been reported to be unstable when determined by some methods (432).

least 4 days at room temperature and 1 week at refrigerator temperature. Serum Fe has been reported to be unstable when determined by some methods (432).

3. Interference by hemoglobin. The presence of hemolysis in serum is a possible source of error in the determination of serum Fe because 1 mg hemoglobin contains 3.4 μg Fe. Since in the determination of serum Fe the Fe is split off the transferrin by treatment with acid, there is the danger of simultaneous release of Fe from the hemoglobin. The level of hemoglobin concentration at which such interference becomes significant varies considerably with the intensity of acid·treatment (64, 241). With the technic presented, a positive error does not begin to appear until the serum hemoglobin concentration exceeds about 100 mg/100 ml (visible hemolysis begins at about 50 mg/100 ml) (241).

ACCURACY AND PRECISION

There is complete recovery of added Fe when using the hot trichloroacetic acid precipitation method (569). Complete recovery of internal standards, however, does not necessarily guarantee complete recovery of all the Fe originally present in the sample (492). Actually, the technic presented appears to determine about 80%–90% of the nonhemoglobin Fe present (241). Just what fraction or fractions of serum Fe are included is unknown.

The precision of the method is about ±5% (95% limits). It cannot be overemphasized that good precision is obtainable only with scrupulous technic. The determination is being made at the microgram level of a metal which is difficult to exclude as a contaminant. The effect of contamination increases as the size of the analyzed sample decreases. It is strongly recommended that, where feasible, determinations be run in duplicate.

NORMAL VALUES

Serum iron. The normal range for adults by the manual and automated methods presented is 65–175 µg/ml. A similar range has been reported for males only (166, 557) and for both sexes (231, 410, 422, 519, 521, 636). Normal ranges reported by others employing a variety of technics have varied within broader limits (83, 177, 334, 395). It is difficult to evaluate the influences of background absorption, turbidity, coprecipitation, reducing agent used, and instrumentation in most of these reports. Many investigators have found the range for males to be higher than that for females (177, 187, 231), but others have observed no such difference (83, 241, 410, 519, 521). Levels are high at birth, fall rapidly by the third or fourth day, rise by the ninth day, and then fall again to a minimum by the end of the second year. Other studies of newborns have shown a minimum level reached between 4 and 12 months (193, 636). Children between 4 and 12 years had a mean of 100 µg Fe (range 65–125). Adolescents between 11 and 22 years had a mean of 93 (range 10–189) with no difference between the sexes (514). The mean level of serum Fe was significantly higher in women taking oral contraceptives than in the "normal women" (366). No relation between age and serum Fe levels was found in two studies (83, 521), although others reported a decrease in the normal range in the elderly.

There is considerable diurnal variation in the Fe level—which is high in the early morning and low in the evening (83, 366, 539). There is considerable variation in the reported magnitudes of the diurnal swing, but its order of magnitude appears to be about 50 µg/100 ml (in one report a normal individual showed a fivefold swing). The observation that the diurnal change in serum Fe parallels the level of serum bilirubin prompted the suggestion that the phenomenon may be secondary to the metabolism of hemoglobin. The diurnal variation as described occurs when the individual is partaking of the usual three meals a day, but if the subject is placed on four meals a day spaced 6 hours apart the peak serum Fe level shifts to the afternoon or evening with the minimum occurring in the morning. The day to day and week to week variations are even larger than the within-day variation.

Urine iron. Excretion levels in 14 healthy children were 32–285 µg/24 hours (203). Other reports have given levels of 200 µg/liter and 200 µg/24 hours (346, 405).

Cerebrospinal fluid iron. Eighteen individuals with presumably normal CSF had levels of less than 1–2 µg Fe/100 ml (311). Transferrin levels in CSF, determined by immunodiffusion, correspond to an Fe content of 2.5 µg/100 ml. This is below the usual level of detection (173).

Feces iron. About 13 mg/day is excreted in feces, corresponding with the dietary intake (628).

Sweat iron. Twenty-eight normal adults (14 male, 14 female) were found to have a mean of 17 μg/100 ml of cell-free sweat. If cells are present, values are much higher (106).

AUTOMATED DETERMINATION OF SERUM IRON

(method of Zak, Ref 635, adapted to AutoAnalyzer by Zak and Epstein, Ref 637, modified)

PRINCIPLE

Using an AutoAnalyzer, Fe is separated from serum *transferrin* by acid treatment in the presence of ascorbic acid to convert the Fe to the ferrous state. The released Fe is dialyzed into a buffer and then reacted with 4,7-diphenyl-1,10-phenanthroline sulfonate to form a colored reaction product which is measured at 535 nm. Pepsin is added to prevent protein precipitation, particularly on the dialyzer membranes.

REAGENTS

Deionized distilled or double distilled water is used in the preparation of all reagents.

HCl, 1 N. Dilute 83 ml conc HCl to 1 liter.

Ascorbic Acid-Pepsin. This reagent must be prepared fresh daily. Dissolve 1 g ascorbic acid (USP grade) and 1 g pepsin (B grade, Calbiochem is satisfactory) in about 80 ml 1 N HCl. Dilute to 100 ml with 1 N HCl.

Brij 35. 30 g/100 ml water (McKesson Chemical Co.)

NaCl, 0.85%. Dissolve 8.5 g of NaCl in about 500 ml water. Dilute to 1 liter with water, add 1 ml Brij 35 solution, and mix thoroughly. Stable at room temperature.

Sodium Bicarbonate, 0.02 M. Dissolve 1.68 g of anhydrous $NaHCO_3$ in about 500 ml of water and dilute to 1 liter. Add 1 ml of Brij 35 and mix thoroughly. Stable at room temperature.

Stock Iron Standard, 1 mg/ml. Clean a piece of iron wire (usually 99.8% pure) using glass wool wetted with absolute ethyl alcohol. Weigh an amount of wire to contain 100 mg Fe and place in an acid-washed volumetric flask containing 15 ml water and 10 ml conc HCl. Dissolve wire completely, dilute to 100 ml with water, and mix thoroughly. Stable at room temperature.

Working Iron Standards, 1 and 2 µg/ml. Prepare two acid-washed volumetric flasks containing about 50 ml 0.85% NaCl each. To one add 0.100 ml and to the other 0.200 ml stock iron standard using TC micropipets. Dilute to 100 ml with the NaCl solution and mix thoroughly. Stable at room temperature.

Acetate Buffer 0.5 M, pH 4.65±0.05. Dissolve 544 g sodium acetate trihydrate in 1 liter water. To this add 230 ml glacial acetic acid, 8 ml Brij 35 solution, and dilute to 8 liters with water. Stable at room temperature.

Bathophenanthroline Sulfonate. Dissolve 25 mg of 4,7-diphenyl-1,

10-phenanthroline sulfonate (G.F. Smith) in 100 ml 0.5 M acetate buffer, pH 4.65. Stable at room temperature.

PROCEDURE

Set up manifold, other components, and operating conditions as shown in Figure 19-4. By following the procedural guidelines as described in Chapter 10, steady state interaction curves as shown in Figure 19-5 should be obtained and checked periodically. Results are calculated from the calibration curve in the standard manner for methods obeying Beer's law.

NOTES

1. *Beer's law.* The method is linear to 400 μg Fe/100 ml.
2. *See Notes 2 and 3 of manual method.*

ACCURACY AND PRECISION

Results obtained by the automated procedure are in excellent agreement with those obtained by the manual procedure (241, 635).
Precision (95% limits) is about ±10% for normal serum Fe values.

NORMAL RANGE

See manual method.

IRON-BINDING CAPACITY

In 1945 Holmberg and Laurell (252) observed that the addition of Fe^{2+} to serum resulted in the appearance of a red color. The following year Schade and Caroline (500) confirmed this phenomenon, recognizing it to be a result of combination of Fe with a protein found in Cohn's fraction IV-3,4. The protein, which has been isolated and crystallized (321), has been called siderophilin, transferrin, and β_1-metal-combining globulin. Its concentration in normal sera is about 0.21—0.36 g/100 ml (148, 201, 277). Evidence has been presented (420) that there is more than one transferrin and that, although there is only one present in the serum of the majority of individuals, sometimes two are present. Fourteen genetically controlled variants have been detected by starch gel electrophoresis. Transferrin C is the most prevalent type, found in high frequency in all populations (420). Transferrin contains four residues of sialic acid (21) and has a molecular weight of about 90 000 and an isoelectric point of about 5.9. It appears in the β_1-globulin fraction that is separated when Ca^{2+} is added to the buffer in paper electrophoresis or just ahead of β-globulin in starch gel electrophoresis.

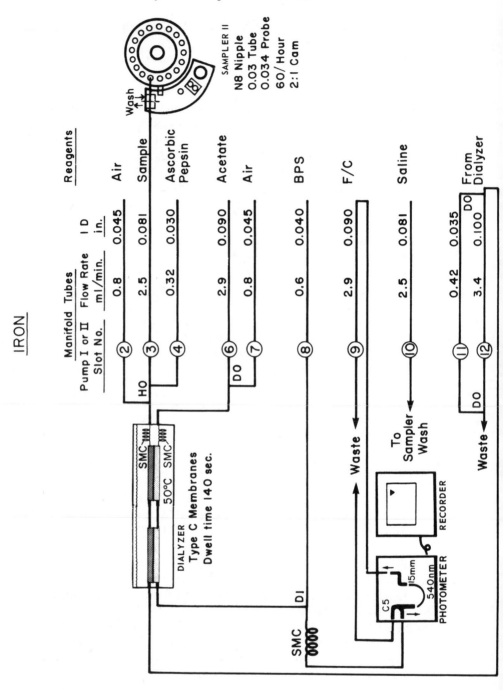

FIG. 19-4. Flow diagram for Fe in serum. Calibration range: 0–400 mg/100 ml. Conventions and symbols used in this diagram are explained in Chapter 10.

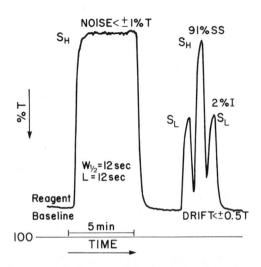

FIG. 19-5. Iron in serum strip chart record. S_H is a 200 mg/100 ml standard, S_L a 100 mg/100 ml standard. Peaks reach 91% of steady state; interaction between samples is 2%. $W_{1/2}$ and L are explained in Chapter 10. Computer tolerance for noise is set at ±0.7% T, for baseline drift at ±1% T, and standard drift at ±0.7% T.

The transferrin B variants—e.g., B_1, B_2—are the fastest moving, followed by C and D. Each molecule of the protein combines with two molecules of Fe^{3+} or Cu^{2+}, Fe more strongly than Cu (321, 554). The Fe in the complex is in the ferric state, the bond being ionic rather than covalent (1). It has been suggested that each chelating site involves two imidazole groups of the protein in addition to three tyrosyl residues (1, 554). Combination takes place when the pH is 6.5 or above, and 1 mg of protein combines with about 1.25 μg of Fe (74, 148). The color produced when combined with Fe^{3+} has been described as salmon-pink, with an absorption peak at 460–465 nm (501, 554). When Fe^{2+} is added to the protein, O_2 is required for development of the salmon-pink color, indicating autooxidation to the ferric state (321, 501). It has been claimed (501) that CO_2 ($HCO_3{}^-$) is essential for development of the color, but this has been denied (5, 170).

The concentration of transferrin in serum can be determined directly by immunochemical means (148, 201, 277), but this technic is not commonly used. A more feasible approach is to determine the protein concentration in terms of the Fe content after saturation. When all the transferrin present is saturated with iron, the Fe concentration of the serum is called the *iron-binding capacity (IBC)* or *total iron-binding capacity (TIBC)*. Normally, only part of the transferrin in serum is combined with Fe, and the difference between the serum Fe and total iron-binding capacity is called the *unsaturated iron-binding capacity (UIBC)*—also called the *latent iron-binding capacity*, or *LIBC* (364). The serum Fe expressed as a percentage of the *IBC* is the *% saturation*.

Many approaches have been made to the determination of *IBC* and *UIBC* (references to earlier methods which are little used today may be found in the first edition of this book): (a) In the method introduced by Schade and Caroline (500), an excess of Fe is added to the serum and the increase in the salmon-pink color is determined photometrically. The test measures the *UIBC* directly. (b) Fe is added in excess and the unbound excess is determined directly with dipyridyl, tripyridyl, bathophenthroline (65), or ferrozine (429). A modification of this method, carried out in one tube, employs 2,4,6-tripyridyl triazine (TPTZ) at pH 2.2 to measure serum Fe, and tripyridyl chromogen at pH 9 to measure excess Fe (606). In another procedure TPTZ is used to measure serum Fe and excess Fe at pH 2.2 and pH 9, respectively (411). (c) Radioactive Fe is added in excess and then the proteins are precipitated and the radioactivity in the precipitate determined (UIBC), or proteins are precipitated and the radioactivity in the supernate determined (radioactivity added—radioactivity in supernate = UIBC), or the excess Fe is removed by ion exchange resin and the radioactivity remaining in the serum measured (UIBC) (74, 322, 337). The resin may be used in batch form as free particles or in a column. A variation employs the resin embedded in a polyurethane sponge. The excess radioactive Fe is adsorbed by the sponge which is counted, where *UIBC* = Fe added—Fe in sponge (79, 322), or coated charcoal is used as an adsorbent in place of resin (242). (d) Fe is added in excess, the excess is removed by an ion exchange resin (110, 187, 241, 337) or magnesium carbonate (82, 166, 462), and Fe is determined in the supernate (IBC). (e) Excess Fe is added, the transferrin is precipitated by an alcohol-ether-water mixture which also serves as a wash reagent for removal of excess free Fe, and the bound Fe is removed from the precipitate by trichloroacetic acid and determined photometrically. (f) Excess Fe is added, the free Fe bound as a ferroin complex with *o*-phenanthroline, and fractionated on a Sephadex column. The *TIBC* in the effluent is measured with dipyridyl (399). (g) Excess Fe is added, an ultrafiltration technic removes protein-bound Fe, and the filtrate is analyzed for unbound Fe (337).

Several authors have compared the relative merits of resin versus $MgCO_3$ adsorption, using both photometric and radioisotope measurements (73, 79, 110, 322, 337). Whereas some workers have recommended $MgCO_3$ (73, 110), others have claimed that $MgCO_3$ not only binds all of the excess Fe but also removes some Fe from transferrin (79, 322, 337).

Ramsay (462) originally proposed a ratio of 100 mg $MgCO_3$ for every 5 μg of Fe^{3+} added and advised testing each batch of $MgCO_3$ for its adsorptive property, which varies from batch to batch. One of the critics cited above adds 5 μg Fe^{3+} and uses 400 mg $MgCO_3$; another gives no data on amounts of Fe^{3+} or $MgCO_3$, although mentioning a method which uses correct proportions (322, 337). Ample data have been published justifying the use of $MgCO_3$ as an adsorbent by comparison with reference methods and by clinical values obtained (73, 82, 166). It might also be mentioned that the resin method cannot be used in cases of Fe poisoning because the resin binds only the *anion* Fe^{3+}-citrate$^-$ and not the cations Fe^{2+} or Fe^{3+} which may be present in such patients (337).

A variety of AutoAnalyzer methods for *TIBC* have been published. In all but one, excess Fe is added, an adsorbent removes unbound Fe, and the *TIBC* is determined on the dialyzed supernate or filtrate as in the determination of Fe: (a) TPTZ and $MgCO_3$ adsorbent (634) using serum or heparinized plasma (but not EDTA); (b) bathophenanthroline sulfonate and $MgCO_3$ adsorbent (177, 328); (c) Ferrozine and $MgCO_3$ adsorbent (630); (d) TPTZ and a resin column (337); and (e) a continuous resin-loaded paper in a fully automated procedure with either bathophenthroline sulfonate or TPTZ as the chromogen (186). The last method also provides for a semiautomated procedure without use of the resin filter paper.

Two methods for *IBC* are presented, a manual method which appears to be accurate even with very lipemic and icteric samples (241), and an automated method based on the manual method of Caraway (82) for the preparation of the saturated transferrin combined with the automated Fe method described in the previous section.

DETERMINATION OF IRON-BINDING CAPACITY, MANUAL METHOD

(method of Peters et al, Ref 431, modified by Henry et al, Ref 241)

PRINCIPLE

Excess Fe is added as ferric ammonium citrate to saturate the transferrin. The excess Fe is removed by batchwise adsorption onto Amberlite IRA-401. This is an anionic exchanger which removes the Fe by adsorbing the citrate with which the Fe is associated. The bound Fe remaining in the supernate is the Fe-binding capacity and is determined by the same technic as used for serum Fe.

REAGENTS

Ferric Ammonium Citrate, 0.050 mg Fe/ml. Place 43 mg ferric ammonium sulfate in a 50 ml centrifuge tube, dissolve in 10 ml double distilled water (from glass), add 2 ml conc NH_4OH, mix, and centrifuge. Discard supernate, wash once with distilled water, centrifuge, and again discard supernate. Suspend the precipitate in about 15 ml distilled water, add a few crystals of citric acid, and warm to dissolve, adding more citric acid if necessary to dissolve. Adjust pH to 6.5–7.0 with dilute NH_4OH using narrow range pH paper. Transfer to a 100 ml volumetric flask and add water to 100 ml. This reagent is stable in the refrigerator for 1–2 months.

Barbital Buffer, pH 7.5. 6.4 g NaCl, 6.0 g barbital (Veronal), and 2.3 g barbital sodium (Veronal sodium) per liter.

Amberlite IRA-401 Resin. Reagent grade, Fisher Scientific. Prepare a stock of resin as follows: Suspend resin in approximately 3 *N* HCl overnight to wash resin free of Fe. Wash well with distilled water, discard supernate, suspend in barbital buffer, and bring to pH 7.5 (using narrow range pH paper) with dilute NaOH. Mix, allow resin to settle, discard supernate, and dry at 95°C.

Other Reagents. Same as for serum iron determination.

Glassware. All glassware employed must be rendered iron-free as for the serum Fe determination.

PROCEDURE

1. Place 1.0 ml serum or heparinized plasma (*not* citrated or oxalated) into a 15 ml centrifuge tube. Add 0.1 ml ferric ammonium citrate and mix. It is very important that the serum and ferric ammonium citrate be added directly near the bottom of the tubes and not be allowed to run down the sides. Let stand 10 min at room temperature.

2. Add 0.4 ml dry resin (previously measured out in a graduated centrifuge tube) to tube of step 1 and mix occasionally for 5–10 min by gentle "spanking," being sure that the resin reaches all the wet surface of the tube.

3. Add 5.0 ml barbital buffer and stir with a glass stirring rod for several minutes. Centrifuge.

4. Transfer 5.0 ml supernate to a test tube; add 1.0 ml 30% trichloroacetic acid and mix. Cover tube with an aluminum foil cap and place in water bath at 90°–95°C for 15 min, mixing tube contents at 5 min intervals. Cool tube to room temperature in tap water. Mix contents again and centrifuge.

5. Proceed with step 3 under *Determination of Serum Iron, Manual Procedure.*

Calculation:

iron-binding capacity (IBC), as μg Fe/100 ml

$$= \frac{A_x - A_b}{A_s - A_b} \times 4 \times \frac{100}{(5/6.1) \times (2/3)}$$

$$= \frac{A_x - A_b}{A_s - A_b} \times 732$$

unsaturated iron-binding capacity (UIBC), as μg Fe/100 ml serum $= \text{IBC} - \mu\text{g Fe/100 ml serum}$

$$\% \text{ saturation} = \frac{\text{serum Fe}}{\text{IBC}} \times 100$$

MICRO PROCEDURE

The procedure can be scaled down twofold with no great difficulty. Absorbances must be read in micro cuvets with a 1 cm light path.

NOTES

1. Stability of samples. Serum appears to be stable at room temperature for 4 days and at refrigerator temperature for at least 1 week when the method presented is employed (241). Frozen samples appear to be stable for many months (217).

ACCURACY AND PRECISION

The method of Peters *et al*, as modified herein, gives results which, on the average, agree quite closely with results obtained by the method of Schade and Caroline (241).

The reproducibility of the method is about ±5% (241, 431). The comments regarding contamination made earlier under *Accuracy and Precision* in *Determination of Serum Iron, Manual Procedure* are also pertinent here, and again it is strongly recommended that the determination be run in duplicate where feasible.

AUTOMATED DETERMINATION OF IRON-BINDING CAPACITY

(method of Caraway, Ref 82, and Zak and Epstein, Ref 637, modified)

PRINCIPLE

Excess Fe is added as ferric chloride to saturate the transferrin. Excess free Fe is removed with magnesium carbonate powder and the total Fe-binding capacity is determined by performing a total Fe determination on the supernate.

REAGENTS

Deionized-distilled or double distilled water is used to prepare all reagents and for the final rinse of all glassware prior to drying.

Reagents of the automated iron procedure.

Magnesium Carbonate, Powder, $4 MgCO_3 \cdot Mg(OH)_2 \cdot 4H_2O$. Mallinckrodt Chemical Co., supplies a suitable product.

Ferric Chloride Stock Solution A, 10%. Dissolve 10 g $FeCl_3 \cdot 6H_2O$ in about 60 ml 0.1 N HCl. Dilute to 100 ml with 0.1 N HCl and mix thoroughly. Stable at room temperature.

Ferric Chloride Stock Solution B, 50 mg/100 ml. Transfer 1 ml stock solution A to a 200 ml volumetric flask. Dilute to 200 ml with 0.1 N HCl and mix thoroughly. Stable at room temperature.

Ferric Chloride Working Solution, 5.2 µg Fe/ml. Transfer 5 ml stock solution B to a 100 ml volumetric flask. Dilute to 100 ml with water and mix thoroughly. Stable at room temperature.

PROCEDURE

N.B. An Fe determination must be run on each sample.

Set up manifold, other components, and operating conditions as shown in Figure 19-4. By following procedural guidelines described in Chapter 10, steady state interaction curves as shown in Figure 19-5 should be obtained and checked periodically. Results are calculated from the calibration curve in the standard manner for methods obeying Beer's law.

1. Using an automatic dilutor, aspirate 1 ml serum, then wash it down into a 15 x 125 mm test tube with 2.0 ml ferric chloride working solution. Mix vigorously for 10 sec and allow to stand for 5 min.
2. Add one scoop $MgCO_3$ (approximately 250 mg), mix vigorously and place on a rotator for 30 min at 30 rpm, then mix again vigorously.
3. Centrifuge at 2500 rpm for 15 min.
4. Pour the supernate into AutoAnalyzer cups (supernate must be clear with minimum volume of 1.7 ml/cup).
5. Proceed with the determination of Fe by the automated serum iron procedure given previously.

Calculation:

$$\text{iron binding capacity (IBC),} = \frac{A_x}{A_s} \times \text{ µg Fe in standard}/100 \text{ ml} \times 3$$
$$\text{as µg Fe}/100 \text{ ml}$$

$$\% \text{ saturation} = \frac{\text{serum Fe}}{\text{IBC}} \times 100$$

NOTES

1. Beer's law. Beer's law is obeyed up to 400 µg Fe/100 ml (637). If it is necessary to dilute any specimens, the dilutions should be made in test tubes using saline as the diluent.

2. Stability of samples. Serum appears to be stable at room temperature for 4 days and at refrigerator temperature for at least 1 week with the method presented (241). Frozen serum is stable for 90 days (186).

3. Interferences. AutoAnalyzer methods for Fe or IBC are not affected by turbidity, lipemia, or bilirubin (634). Interference by Cu is prevented by the acidity of the ascorbic acid reagent (186).

ACCURACY AND PRECISION

The automated method, as modified herein, gives results for serum Fe and IBC which agree quite closely with results obtained by the manual methods described previously.

The precision (95% limits) is about ±10% for serum Fe and IBC levels in the normal range.

NORMAL VALUES

The normal adult range of total serum Fe-binding capacity (*TIBC*) obtained by the methods presented is about 250–410 µg Fe/100 ml (241). Normal ranges obtained by use of other technics agree reasonably well (82, 83, 166, 231, 334, 410). The quantitative determination of transferrin in 101 normal sera by immunologic technics has given results of 220–372 mg transferrin/100 ml (equivalent to 265–465 µg Fe/100 ml) (469), while another study employing an automated technic for transferrin found a range of 139–325 mg/100 ml (equivalent to 170–400 µg Fe/100 ml) (148). It would seem that transferrin values should be the ultimate reference for *TIBC*, but there is evidence that atypical transferrins may occur (601). More data are needed to judge whether transferrin and serum Fe bear a simple relationship to each other (79, 387). There is no definite evidence of a sex difference (83, 410, 422, 521). The rather meager data available on children suggest that in early childhood the *IBC* is higher than in adulthood; data on children 4–12 years old and adolescents agree with the adult range and lack of sex difference (202, 514). The adult normal range for the unsaturated Fe-binding capacity *(UIBC)* is 140–280 µg/100 ml and that for "% saturation" is 20%–55% (65, 82, 166, 177, 231, 241).

In contrast to normal serum Fe levels, TIBC does not exhibit diurnal variation, while *UIBC* does (68, 539). *TIBC* is said to be higher in women taking oral contraceptives (366). Therapy with folate has led to misinterpretation of serum Fe and % saturation values (244). Obese adolescents have substantially higher *UIBC* and decidedly lower "% saturation" than nonobese subjects (513).

COPPER

The total body Cu content is estimated to be about 80 mg. The fraction circulating in the blood (less than 10% of the total) has hitherto received most attention. Recent findings of the fundamental role of Cu enzymes in the biosynthesis of collagen and elastin (84), and in the biogenesis of amines such as norepinephrine (199) have led to studies of Cu in other tissues.

Copper-64 given by mouth is bound by albumin immediately upon entering the bloodstream; after a few hours it shifts to ceruloplasmin, an α_2-globulin (44). Copper can also bind to transferrin (321, 554) but far less avidly than does Fe (253). Studies (92) have shown that from 0% to about 15% of the Cu in normal plasma reacts directly with diethyldithiocarbamate and is therefore termed the "direct-reacting fraction." This fraction is nondialyzable. A small fraction of serum Cu may also exist bound to amino acids and in equilibrium with albumin-bound Cu (397). The major form of Cu, representing that bound to ceruloplasmin and often termed the "indirect-reacting fraction," constitutes on the average about 93% of that normally present and reacts with diethyldithiocarbamate only after treatment with acid (92). The "direct-reacting fraction" is the Cu loosely bound to albumin and is concerned in the transportation of Cu from the gastrointestinal tract to the tissues and from one tissue to another (92).

Ceruloplasmin, the blue Cu-containing protein of serum, has the following properties: It has a Cu content of 0.32%, a molecular weight of about 151 000, eight atoms of Cu per molecule, an isoelectric point of about 4.4, and an absorption peak at 605 nm. Human ceruloplasmin isolated from Cohn Fraction IV-I was found to yield a single precipitin line on agar gel microimmunoelectrophoresis, and a single band that is positive to both the amido black and oxidase stains following starch gel electrophoresis (279). There is a direct proportion between serum Cu and the absorbance at 605 nm (254). The blue protein can be reduced to a colorless compound by hydrosulfite, vitamin C, hydroxylamine, cyanide, and thioglycollic acid. There are four Cu^+ and four Cu^{2+} in the molecule. Ceruloplasmin is a glycoprotein with 8 subunits (319) and 8−10 sialic acid-containing carbohydrate chains—transferrin also is a glycoprotein with different sialic acid chains (279, 280). The adult normal concentration of ceruloplasmin in serum, as determined by immunochemical analysis, is about 23−44 mg/100 ml (91, 92). This range is only slightly higher than that obtained by spectrophotometric determination. Ceruloplasmin has been demonstrated to be an oxidase with p-phenylenediamine as substrate (255, 445), but its physiologic function remained unclear until recently. It is now known to function as the enzyme which oxidizes Fe^{2+} to Fe^{3+}, helping to transport Fe across the intestinal wall, enabling the Fe to bind with apotransferrin, forming transferrin, and also helping to mobilize Fe from storage tissues (336, 414, 445). This discovery may explain the major importance of Cu deficiency in the pathogenesis of anemia. It has been proposed (445) that the name ceruloplasmin be replaced by serum ferroxidase (EC 1.12.3). The low levels of ceruloplasmin found in Wilson's disease may arise from synthesis of a defective molecule. Ceruloplasmin from which sialic acid has been removed disappears from the circulation in a few minutes and appears in the parenchymal cells of the liver. A reduced or absent sialyltransferase activity which normally completes the synthesis of ceruloplasmin might result in the release of the incompletely sialylated protein, which would not survive in the circulation (386). A nonceruloplasmin ferroxidase (ferroxidase II), which has no amine oxidase activity, has been isolated from human serum (568). Citrate has also been reported to contribute to the ferroxidase activity of serum (335). Ferroxidase II is said to account for 15% of the total ferroxidase activity in normal man and for nearly 100% of the activity in patients with Wilson's disease. The relationship of citrate to ferroxidase II is not clear.

In one study (92) the means for total serum Cu and ceruloplasmin were 114 µg/100 ml and 33 mg/100 ml, respectively. This same report gave a value of 7 µg/100 ml for direct-reacting Cu. Thus the amount of Cu in the ceruloplasmin was 33 mg × 0.32%, or 106 µg/100 ml. The two fractions account for the total serum Cu (92). Other Cu-proteins have been isolated and crystallized from human tissues. These are erythrocuprein from erthyrocytes, hepatocuprein from liver, and cerebrocuprein from brain (87, 228). All three have molecular weights of 33 600 and Cu contents of 0.38%, and have been found to be identical with respect to amino acid composition, immunochemical determinants, and ultracentrifugal behavior. The name "cytocuprein" has been suggested for these interesting compounds (87).

Copper may be determined in several ways: (a) atomic absorption spectrophotometry, (b) spectrophotometry, (c) neutron activation, and (d) radioisotope dilution. Atomic absorption spectrophotometry is undoubtedly the method of choice for the determination of Cu in biologic fluids, tissues, and foods. It offers the required sensitivity, specificity, reliability, speed, and ease of measurement that are needed. The resonance line at 324.7 nm has no nearby interferences and is in a region of the spectrum where photomultiplier detectors operate with ease at low gain settings, thereby eliminating "noise." Serum and plasma Cu have been determined by simple dilution and aspiration into the flame (126, 188, 271, 376, and 419). Suitable standards have to be prepared to compensate for viscosity effects of serum proteins (271, 419). Another approach involves the splitting of Cu from the proteins by treatment with acid, followed by precipitation of the protein and determination of Cu in the supernatant solution (409, 419, 553, 572). Plasma and erythrocyte Cu liberated by acid have been chelated with ammonium pyrrolidine dithiocarbamate (APDC) and extracted into an organic solvent for atomic absorption measurement (60). *Free* plasma Cu can be determined by the same technic minus the acid treatment. Copper is present in low amounts in urine, about 5 µg/100 ml, and has been determined by atomic absorption using APDC chelation following ashing (553), chelation without ashing (419), and direct aspiration of undiluted urine against standards containing a urine-salt matrix (126). In general, tissues, feces, and diets should be acid extracted (188, 419) or ashed and extracted (553).

One group of methods for determination of Cu in biologic fluids is based on photometric measurement of the color produced with diethyldithiocarbamate. One technic (149, 370, 398) involves wet digestion of the sample; another (251, 347, 398, 567) involves precipitation of proteins with trichloroacetic acid, with or without the prior addition of HCl, followed by development of the color with diethyldithiocarbamate and extraction of the color into amyl or isoamyl alcohol. Still another (479) involves extraction of the copper carbamate directly from serum with isoamyl alcohol. The latter technic cannot be employed with icteric sera since bilirubin and possibly other serum pigments are extracted by the organic solvent (241). Gubler *et al* (211) presented a modification of the technic employing trichloroacetic acid precipitation in which the copper carbamate color is read without extraction into amyl alcohol. The chief drawback of these methods employing diethyldithiocarbamate as color reagent is their very low sensitivity. Peterson and Bollier (434) used biscyclohexanoneoxalyldihydrazone as the color reagent. This reagent has about twice the sensitivity of diethyldithiocarbamate. Ressler and Zak used neocuproine (2,9-dimethyl-1,10-phenanthroline) as the color reagent, adding tetravalent thorium to the serum to displace the Cu from the protein. Zak and co-workers also used bathocuproine (2,9-dimethyl-4,7-diphenyl-1,10-phenanthroline) (635, 638) and its water-soluble, sulfonated derivative (331) as reagents. Other reagents employed for both serum and urine are diphenylcarbazone (333), oxalyldihydrazide (42, 472, 615), zinc dibenzyl dithiocarbamate (185), diphenylcarbazide in benzene (546), and 1,5-diphenylcarbohydrazide (380). The last mentioned is the most sensitive of all Cu reagents, with an $\epsilon =$ 158 000, forming a Cu:dye complex of 1:2. It can be used to determine Cu in 0.2 ml of serum, which makes it valuable for pediatric specimens.

Neutron activation employs the "n, γ" reaction: ^{63}Cu (n, γ)^{64}Cu. Biologic material such as liver biopsies, hair, nails, and other ultramicro specimens can be analyzed for as little as 10^{-9} g of Cu with a precision of ±10% (142, 160, 288, 289, 564). Copper-64 has been added to serum and an isotope dilution analysis carried out (59). A polarographic method for the determination of Cu in blood and urine has also been used (261).

Ceruloplasmin has been determined by immunochemical analysis (91) and by measuring the difference in absorbances at 610 nm before and after reduction (3, 133). The latter requires a cuvet of 5 cm light path. Ceruloplasmin can also be determined in terms of its oxidase activity. Consideration of ceruloplasmin as an oxidase and a photometric method for the determination of ceruloplasmin, using *p*-phenylenediamine as substrate, are presented in Chapter 21. In comparative studies (3, 91, 471) it has been shown that there is a linear correlation between serum Cu concentration, ceruloplasmin as determined spectrophotometrically, and oxidase activity with *p*-phenylenediamine as substrate. The relationship between serum Cu and ceruloplasmin in normal individuals has been given as follows (91):

$$\text{mg ceruloplasmin/100 ml} = \frac{\mu g \text{ Cu/100 ml} \times 0.94}{3.2}$$

For the determination of Cu, an atomic absorption method is presented.

DETERMINATION OF COPPER IN SERUM

(method of Piper and Higgins, Ref 444, modified)

PRINCIPLE

Serum is deproteinized with trichloroacetic acid and the supernate analyzed for Cu by atomic absorption spectrophotometry at a wavelength of 324.7 nm.

REAGENTS

All reagents and standards must be prepared with double deionized water or deionized-distilled water.

Glassware is soaked in 4 N HNO_3 overnight and prior to use is rinsed 3 times with double deionized or deionized-distilled water and dried in an oven at 100°C.

Trichloroacetic Acid, 25%. Dissolve 50 g trichloroacetic acid in water and dilute to 200 ml.

Stock Copper Standard, 1 mg Cu/ml. Dissolve 393.2 mg $CuSO_4 \cdot 5H_2O$ (reagent grade) in 100 ml water.

Dilute Copper Stock Standard, 100 µg Cu/ml. Dilute 100 ml of the stock standard to 100 ml with water. Prepare fresh weekly.

Working Copper Standards, 100 and 200 µg/100 ml. Dilute 1 and 2 ml of dilute stock to 100 ml with water. Prepare fresh shortly before using.

PROCEDURE

1. Label a set of acid-washed, 15 x 100 mm test tubes for the unknown sera.
2. Pipet 1.5 ml of serum into its corresponding tube. Add 1.5 ml of 25% trichloroacetic acid and mix immediately.
3. Let samples stand for 10–15 min with occasional mixing. Centrifuge all samples for 5 min or more at 2200 rpm. (See *Note 1* for handling of cloudy supernates.)
4. Transfer the supernates to clean, appropriately numbered tubes.
5. Prepare the following in 50 ml Erlenmyer flasks:
   100 *µg standard:* 10 ml working (100 µg) standard + 10 ml 25% trichloroacetic acid
   200 *µg standard:* 10 ml working (200 µg) standard + 10 ml 25% trichloroacetic acid
   Blank: 10 ml water + 10 ml 25% trichloroacetic acid
6. Atomic absorption: Since the details of analysis vary with the atomic absorption spectrophotometer employed, and since these are given in the book of instructions accompanying each instrument, no attempt is made here to present them.

NOTES

1. Cloudy supernates. When the supernatant fluid for a serum Cu assay is cloudy after centrifugation, the following procedure is to be used. Place tube containing precipitate and supernate in boiling water for 5 min. Stopper lightly with a polyethylene stopper or Parafilm-covered cork to prevent evaporation. Cool and centrifuge. Supernate should now be clear.

2. Beer's law. The 324.7 line of Cu is suitable for use over a wide range of values that encompass the levels likely to be found in serum (572).

3. Stability of samples. Copper is stable in serum up to 2 weeks at refrigerator temperature (3, 57, 211). Presumably, samples are also stable at room temperature.

4. Collection of serum samples. It has been reported (574) that the use of ordinary needles for blood collection may lead to significant Cu contamination because of the chrome-plated brass hub. This can be avoided by the use of all-stainless steel needles.

5. Interferences. Although the level of inorganic ions in urine is high enough to interfere with the determination of Cu, no such effect is encountered in serum assays (126). Organic components of serum produce no interference.

ACCURACY AND PRECISION

Results by atomic absorption are in good agreement with values obtained by spectrophotometric methods (126, 376, 444, 553). Previous treatment of plasma with HCl was found to be unnecessary for complete recovery of Cu when 20%–25% trichloroacetic acid was used in atomic absorption analysis (444). Other reports of satisfactory atomic absorption spectrophotometric analyses of serum treated with lesser amounts of trichloroacetic or HCl have been published (126, 188, 419, 553, 572). Addition of Cu to serum as an internal standard, however, does not constitute a valid test of recovery of serum Cu. Ceruloplasmin is completely saturated with Cu in normal serum (253), and Cu added to serum exists in combination with other proteins and is not as firmly bound as that originally present. The calculations of total, free, and ceruloplasmin Cu cited earlier (92) and the ashing experiments (126) are the best evidence for the true Cu content of serum by which to gauge other methods.

The precision of the method (95% limits) is ±8% for serum levels in the normal range.

DETERMINATION OF COPPER IN URINE

(method of Dawson et al, Ref 126, modified)

PRINCIPLE

Undiluted urine is acidified and aspirated directly into the atomic absorption flame. Copper is measured by absorption of the 324.7 nm resonance line. Standards are made up with a "urine matrix" salt mixture to compensate for the depressive effect of the salts on copper absorption.

REAGENTS

Refer to *Reagents* under *Determination of Copper in Serum* for instructions regarding water and glassware.

Urine Equivalent Solution (UES). Dissolve the following in approximately 800 ml of water: 0.67 ml conc H_2SO_4, 8.7 ml conc HCl, 0.312 g $CaCO_3$, 2.86 g KCl, 5.8 g NaCl, 0.418 g Mg $Cl_2 \cdot 6H_2O$, and 3.09 g $(NH_4)_2HPO_4$. Adjust volume to 1 liter.

Stock Copper Standard, 600 mg Cu/liter. Dissolve 0.350 g $Cu(ClO_4)_2 \cdot 6H_2O$ (G.F. Smith) in approximately 75 ml of UES and adjust to 100 ml with UES.

Dilute Stock Copper Standard, 6 mg Cu/liter. Dilute 1.00 ml stock standard to 100 ml with UES.

Working Standards:
600 µg Cu/liter: Dilute 10 ml dilute stock to 100 ml with UES.
300 µg Cu/liter: Dilute 50 ml of 600 µg standard to 100 ml with UES.
150 µg Cu/liter: Dilute 50 ml of 300 µg standard to 100 ml with UES.
75 µg Cu/liter: Dilute 50 ml of 150 µg standard to 100 ml with UES.
37.5 µg Cu/liter: Dilute 50 ml of 75 µg standard to 100 ml with UES.

PROCEDURE

1. Set up the following in test tubes:
   Unknown: Transfer approximately 5 ml urine specimen into a small test tube. Adjust to a pH of less than 2 with conc HCl (1–2 drops). pH paper is satisfactory to verify pH.
   Standards: Into 125 x 12.5 mm tubes place about 10 ml of the 37.5, 75, 150, 300, and 600 µg/liter standards.
   Blank: Pour about 25 ml UES solution into a 50 ml Erlenmyer flask since a larger volume is required for frequent calibration.
2. Atomic absorption. Since the details of analysis vary with the atomic absorption spectrophotometer employed, and since these are given in the book of instructions accompanying each instrument, no attempt is made here to present them.

Calculations: Read results from a standard curve. Multiply diluted samples by the dilution factor. Convert readings from µg/liter to µg/24 hours by multiplying result as follows:

$$\mu g \text{ Cu}/24 \text{ hours} = \frac{\mu g \text{ Cu}}{\text{liter}} \times \frac{\text{ml total 24 hour volume}}{1000}$$

NOTES

1. Beer's law. The standard curve is usually linear up to 300 µg/liter. The 600 µg point may be lower than expected.

2. Stability of specimens. 10 ml glacial acetic acid is added after collecting a 24 hour specimen. Samples are stable for at least 11 days at room temperature when so preserved.

3. Interferences. Suppression of up to 10% of the apparent Cu content can occur as a result of inorganic components of urine (126). This effect is overcome by the addition of inorganic salts (UES) to the standards.

ACCURACY AND PRECISION

The normal adult means and ranges for 24 hour urinary Cu excretion by atomic absorption (126, 376, 553) are about the same as those determined spectrophotometrically (81, 92, 185, 380, 391, 615) and by neutron activation (288). The excessive urinary excretion of Cu in Wilson's disease, where levels of 1000–2000 µg/24 hours are seen (42), permits the use of the direct atomic absorption method presented, which is simple and rapid. For greater precision and accuracy at low or normal Cu levels (15–50 µg/24 hours), it is recommended that a concentration procedure with chelation and organic solvent enhancement be used (444, 553).

Precision (95% limits) of the method is about ±11% at a level of 200 µg/liter.

NORMAL VALUES

Serum. Published reports on the normal range of total serum Cu vary somewhat (60, 92, 126, 376, 419, 444, 472, 553). This may be the result of differences in technic used. There is practically general agreement that there is a sex difference, with females showing higher levels than males. A compromise which agrees closely with the normal ranges obtained by atomic absorption is 70–140 µg/100 ml for adult males and 85–155 µg/100 ml for adult females. The normal range has been reported as 12–67 µg/100 ml in newborns, and as 27–153 µg/100 ml in children 3–10 years old (90, 505). The serum Cu concentration is not influenced by meals or by the menstrual cycle (92). Women on oral contraceptives have elevated serum Cu levels of over 200 µg/100 ml (60). The total serum Cu, direct-reacting Cu, and the concentration of ceruloplasmin are greatly increased during the third trimester of pregnancy, with mean values of 239 µg/100 ml, 29 µg/100 ml, and 84 mg/100 ml respectively (92). There is general agreement that there is some diurnal variation in serum Cu, but the degree and pattern are not well defined.

Erythrocytes contain about 90 µg/100 ml packed cells (92, 60). Red cell values show no sex difference, are not correlated with total serum Cu or ceruloplasmin concentration, and are not influenced by pregnancy or diseases associated with hypocupremia (92).

Urine. Urine normally contains 15–50 µg Cu/day (92, 185, 376, 380, 444, 553, 615). The excretion of Cu by the kidney probably represents the Cu which has been dissociated from the Cu-albumin complex during its passage through the kidneys. Ceruloplasmin is not found in the urine under normal circumstances (92). Some authors have reported normal Cu excretion to be as high as 120 μg (126, 160). The output is considerably higher if there is a concomitant proteinuria (391).

Cerebrospinal fluid. The mean Cu content of normal cerebrospinal fluid has been reported to be 27.8 μg/100 ml by neutron activation (289) and 16 μg/100 ml by atomic absorption (376).

Saliva. The Cu content of saliva in 30 normal subjects ranged from 5 to 76 μg/100 ml. This Cu is direct-reacting, and the saliva has zero oxidase activity (131).

Other tissues. Neutron activation analyses of hair have given results of 10–26 μg/g and 23 μg/g dry weight (160, 441). Nails had 18 μg/g (160). The Cu content of various organs has also been reported (92).

ZINC

Zinc is a trace element essential for the growth and propagation of cell cultures and the functioning of several enzymes; it also plays an unknown role in both insulin and porphyrin metabolism (123, 361, 575). Carbonic anhydrase was the first isolated Zn-protein shown to have a physiologic role (296). Other Zn metalloenzymes are carboxypeptidase A, alcohol dehydrogenase (liver and yeast), lactic dehydrogenase (muscle), and glutamic dehydrogenase. Zinc forms water-soluble complexes with serum albumin, α-lipoprotein, α-glycoprotein, alkaline phosphatase, serum esterase, and metal-combining globulin (575). In normal serum Zn exists in two fractions, firmly bound (34%) and loosely bound (66%). The firmly bound fraction (globulin) is a metalloprotein; the loosely bound fraction is a metal-protein complex (575). Radioisotope studies with ^{65}Zn have shown that Zn is bound to albumin, transferrin, and α_2-macroglobulin in normal serum (69), and to IgG in some sera (69). It has been suggested that transferrin and α_2-macroglobulin may have an important role in internal Zn exchange. Zinc also forms complexes with porphyrins. Zinc uroporphyrin is excreted in urine and feces in cases of intermittent acute porphyria. It is also found in the liver in this disease. Zinc coproporphyrin is excreted in lead poisoning and in the urine of patients with acute rheumatic fever (575).

Zinc in biologic materials is readily determined by atomic absorption spectrophotometry (123, 128, 179, 216, 444, 455). A method using simple dilution with buffer, or water and an air-hydrogen flame, gives a detection limit of 0.002 μg/ml. This is 10–100 times more sensitive than the dithizone method commonly used previously and requires only one-tenth as much time (179). Chelating agents (EDTA and o-phenanthroline) do not interfere, but high phosphate lowers the Zn level drastically. The presence of trichloroacetic acid also has a negative effect (179). Another report utilizing simple dilution of plasma revealed that standards made up in water result in an error of 33%. Standards made up in 3% dextran solution, however, give results similar to those obtained with trichloroacetic acid filtrates and water standards (216). In a method in which plasma is diluted 20-fold with 0.1 N HCl, whole blood 100-fold, and urine ten-fold (128), suppression of up to 15% of the apparent Zn content by inorganic components of the sample is overcome by addition of the appropriate amounts of those ions to the standard Zn solutions used in the determination (128). A method which utilizes both 2 N HCl and 10% trichloroacetic acid plus heating to determine plasma Zn has also been used for red blood cells and urine (455). It has been claimed that trichloroacetic acid gives an increased absorbance due to lowering of surface tension and formation of smaller droplets, thus allowing more specimen to reach the flame (278). Another technic of Zn extraction from plasma utilizes 3 N HCl, but yields and recovery of Zn are 10% lower than when trichloroacetic acid is used (343). Plasma has been diluted with an equal volume of 10% trichloroacetic acid and standards prepared in 5% trichloroacetic acid (123). Urine diluted 1:2 with water may be directly aspirated, but results are not satisfactory. Chelation with ammonium pyrrolidone dithiocarbamate (APDC) and extraction with methyl isobutyl ketone (MIBK) is a better procedure (444). Other methods for determining Zn are as follows: (a) spectrophotometric, with zincon (2-carboxy,-2'-hydroxy,5'-sulfoformazylbenzene) (607), with dithizone (235, 237, 284, 545), and with di-β-naphthylthiocarbazone (379); (b) fluorometric with 8-quinolinol, whose Zn complex fluoresces at 517 nm when excited at 375 nm; and (c) arc spectrographic (547).

We have chosen two different atomic absorption technics for serum and urine which proved suitable in our laboratory.

DETERMINATION OF ZINC IN SERUM

(method of Davies et al, Ref 123, modified)

PRINCIPLE

Serum is deproteinized with trichloroacetic acid, and the supernate is analyzed for Zn by atomic absorption at 214 nm.

REAGENTS

Water. Use double distilled or deionized-distilled water for the preparation of reagents and Zn standard solutions.

Trichloroacetic Acid, 25%. Dissolve 50 g trichloroacetic acid in water and dilute to 200 ml.

Stock Zinc Standard (200 μg Zn/ml). Dissolve 176.0 mg $ZnSO_4 \cdot 7H_2O$ in double distilled water, add 2 drops conc H_2SO_4, and dilute to 200 ml.

Working Zinc Standards (50–200 μg Zn/ml). To each of four 200 ml volumetric flasks add 0.5, 1.0, 1.5, and 2.0 ml of the stock Zn standard and dilute to the mark with double distilled water. This provides working Zn standards of 50, 100, 150, and 200 μg/100 ml. Store in plastic bottles. Solutions are stable for 2 weeks.

PROCEDURE

1. Set up the following:

   Blank: Add 10 ml water + 10 ml 25% trichloroacetic acid to a 50 ml Erlenmyer flask.

   Standards: Label four 50 ml Erlenmyer flasks 50, 100, 150, and 200. Add to each 10 ml of the respective working standard plus 10 ml of 25% trichloroacetic acid.

   Unknown: Into a 12.5 x 100 mm test tube, pipet 1.0 ml of serum and add with mixing 1 ml of 25% trichloroacetic acid. Let stand for 10 min with occasional mixing. Centrifuge for 5 min at high speed to obtain a clear, particle-free supernate. Decant the supernate into another small tube, taking care not to disturb the precipitate.

2. Atomic absorption: Since the details of analysis vary with the atomic absorption spectrophotometer employed and since these are given in the book of instructions accompanying each instrument, no attempt is made to present them here.

   N.B. Between each four serum specimens, check the readout by first aspirating the blank and then one of the standards. A low reading of a standard may indicate clogging of the nebulizer by protein particles. When this is the case, reverse the nebulizer flow until a strong flow of bubbles appears in the solution being aspirated. Then readjust the instrument.

DETERMINATION OF ZINC IN URINE

(method of Piper and Higgins, Ref 444, modified)

PRINCIPLE

Zinc is chelated by ammonium pyrrolidone dithiocarbamate (APDC) and extracted with methyl isobutyl ketone (MIBK). The organic phase is analyzed by atomic absorption at 214 nm.

REAGENTS

Methyl Isobutyl Ketone (MIBK), Saturated with Water. In a 1 liter bottle, shake 800 ml methyl isobutyl ketone with 100 ml water. Use the top phase.

Ammonium Pyrrolidine Dithiocarbamate Solution (APDC), 1%. Dissolve in water 1 g ammonium pyrrolidine dithiocarbamate, ground to a fine powder, and dilute to 100 ml. Filter if turbid.

Stock Zinc Standard (20 mg Zn/100 ml). Dissolve 176 mg $ZnSO_4 \cdot 7H_2O$ in double-distilled water, add 2 drops conc H_2SO_4, and dilute to 200 ml.

Working Zinc Standards. To two 200 ml volumetric flasks add 1.0 and 1.5 ml of the stock zinc standard and dilute to the marks with double distilled water. This provides working Zn standards of 1.00 and 1.50 mg/liter.

PROCEDURE

1. Check the pH of the urine specimens. If the pH is below 5, adjust with 1 N NaOH to between 5 and 6.
2. Set up the following in 35 ml glass-stoppered centrifuge tubes:
   Blank: 5 ml water + 1 ml APDC
   Standard (1.0 mg/liter): 5 ml 1.0 mg/liter working standard + 1 ml APDC
   Standard (1.5 mg/liter): 5 ml 1.5 mg/liter working standard ± 1 ml APDC
   Unknown: 5 ml urine + 1 ml APDC
3. Adjust the pH to between 5 and 6 with 1 N HCl or 1 N NaOH.
4. Add 10 ml MIBK to each tube and shake on a shaker for 10 min.
5. Centrifuge at 2000 rpm for 5 min. Do not separate phases since the top MIBK layer is to be aspirated directly for the analysis. Take care to aspirate only the MIBK layer into the atomic absorption burner.
6. Atomic absorption: Since the details of analysis vary with the atomic absorption spectrophotometer employed and since these are given in the book of instructions accompanying each instrument, no attempt is made here to present them.
   N.B. Between each 3 urine specimens, check the readout by aspirating MIBK followed by one of the standards.

Calculation: Read Zn concentration directly in mg/liter.

$$\text{mg Zn/24 hours} = \text{mg Zn/liter} \times \frac{\text{24 hour urine vol in ml}}{1000}$$

NOTES

1. Beer's law. A linear response is obtained over the range used for serum and up to 1.5 mg/liter for urine. For values above 1.5 mg/liter curve correction settings may be used on the digital readout using a 2.0 mg/liter standard. (See instrument manufacturer's manual for details of this procedure.)

2. Stability of samples. Plasma Zn values are stable for at least 14 days when stored at room temperature or 4°C (216). Serum has been reported to be stable for at least 2 months at 4°C (179).

3. Interferences. There are no interferences with the determination of Zn in urine when a MIBK extract is used. Reference has been made to the effect of inorganic salts when dilution methods were used for plasma and urine (128), and particularly the effect of high phosphate content in the medium analyzed (179).

ACCURACY AND PRECISION

Values by atomic absorption for the normal range of plasma and serum Zn agree with those found by spectrophotometry (179, 223, 235, 237, 343,

455, 545) and by fluorometry (360). Recovery of added Zn from plasma and urine ranges from 94% to 103%.

The precision (95% limits) of the method is about ±10% and ±13% for serum and urine, respectively.

NORMAL VALUES

Serum or Plasma. A compromise of various published reports on adult normals is 55–150 μg/100 ml (223, 235, 237, 360, 455, 545). There is little or no sex or age difference, nor have seasonal or diurnal variations been reported (223, 575). It has been claimed that the Zn content of serum is 16% higher than that of plasma due to platelet disintegration (223).

Red Blood Cells. A range of 8.6–16.1 μg Zn/10^{10} RBCs has been reported (488). Other reports give 1200–1300 μg/100 ml RBCs (575) and 1070–2850 μg/100 ml (226).

Leukocytes. A mean of 3.7×10^{-2} μg/10^6 cells has been reported for normal leukocytes (575). This value is 30 times higher than the RBC Zn content.

Urine. The normal 24 hour excretion has been reported to range from 150 to 1300 μg/24 hours (128, 237, 444).

Hair. Normal hair contains 103–120 μg Zn per gram (547). Hair analysis is claimed to be a reliable method of assessing body zinc stores.

Feces. About 5–10 mg/24 hours is excreted (575).

MANGANESE

The body of a 70 kg man contains 12–20 mg of Mn (573). The dietary Mn requirement for man has not been established, but Mn deficiency has been observed in the rat, chick, and pig (573, 603). Liver arginase contains Mn, and other enzymes activated by Mn include phosphoglucomutase, choline esterase, the oxidative β-ketodecarboxylases, certain peptidases, muscle adenosine triphosphatase, cytoplasmic isocitric acid dehydrogenase, "malic enzyme," and 6-phosphogluconic acid dehydrogenase.

Many lipid-handling pathways contain Mn-dependent steps. Among these are lipoprotein lipase (clearing factor), "palmitate-synthesizing system," acetyl CoA carboxylase, phosphatidyl inositol and phosphoglyceride synthesis, and dihydrosphingosine and ganglioside synthesis. Amino acid metabolism involving Mn includes leucine aminopeptidase, glycylglycine dipeptidase, and serine to glycine conversion.

Manganese toxicity in man is well known in the disease chronic manganism, which is accompanied by mental disorders and other neurologic manifestations (486). Although at present Mn toxicity in humans is confined to industrial exposure, the possible use of Mn as a catalyst in motor vehicle antismog devices may lead to increases in Mn in the general population.

Manganese is very difficult to determine in biologic fluids because of its low concentration. It is now recognized that previous estimates of blood and serum Mn were falsely elevated due to contamination with Mn of reagents and other materials used in the analyses (114, 161). Quantitative methods employed are as follows: (a) neutron activation (114, 159, 622); (b) atomic absorption spectrophotometry (362); (c) spectrography (256); and (d) spectrophotometry of Mn-catalyzed dye reactions (161, 541).

The neutron activation method employs the (n, γ) reaction: ^{55}Mn (n, γ) ^{56}Mn. It requires an atomic pile to generate a neutron flux and elaborate counting equipment. Although it is the most sensitive method and has good specificity, it is impractical for the majority of laboratories to use. Atomic absorption lacks the required sensitivity for biologic materials. It has been employed using the method of additions, which involves analyzing a series of serum aliquots to which increments of Mn are added, but the results for human serum were tenfold and more, higher than others have reported. Furthermore the method requires a 20 ml specimen of blood (362).

The method to be presented here depends on the Mn-catalyzed oxidation of the triphenylmethane dye, malachite green, to a colorless product (161). It is suited to analytic determinations in the range 0.2–3.0 ng/ml of final solution read.

DETERMINATION OF MANGANESE
(method of Fernandez et al, Ref 161, modified)

PRINCIPLE

Manganese catalyzes the oxidative decolorization of malachite green by periodate. The reaction is first order, the rate of decolorization of the green dye being dependent upon the concentration of Mn.

REAGENTS

All water used must be redistilled from all glass apparatus. Reagents are stored in acid-washed polyethylene bottles unless otherwise indicated. All glassware must be washed in hot HCl (1:2), rinsed in distilled water, and oven-dried using glass trays or beakers to hold the glassware. Do not allow glassware to be in contact with Fe. The glassware should then be stored in a special drawer and covered with a towel to prevent contamination by dust. The amount of Mn being determined is a few nanograms, and prevention of contamination is the decisive factor in obtaining satisfactory results.

NaOH, 1 N. Dissolve 4 g NaOH in 100 ml water.

Dithizone. 10 mg in 100 ml chloroform. Store in glass-stoppered bottle and keep in refrigerator.

Buffer. Transfer 27.6 g of reagent grade $NaH_2PO_4 \cdot H_2O$ to a 125 ml separatory funnel (Teflon stopcock, *no* grease) containing 60 ml water. Mix to dissolve. Add 1 ml dithizone, 15 ml chloroform, and shake for about 5 min. Allow chloroform phase to settle out and discard it. Wash the aqueous phase three times with 5 ml volumes of chloroform, discarding the chloroform phase each time. Allow the solution to stand for about 30 min so the chloroform settles out, and then drain off remaining chloroform. Transfer the aqueous solution to a 200 ml acid-washed beaker and heat on a hot plate *just* to boiling. Allow to simmer until the odor of chloroform is gone. Cool, add 12 ml glacial acetic acid and 30 ml water and mix. Add 40 ml 1 *N* NaOH, and mix. Transfer some of solution to a pH meter cup and measure pH, which should be 4.0–4.1. If too low, add more NaOH (0.5 ml at a time), mix, and measure pH. If pH is too high, add a drop of glacial acetic acid, mix, and measure pH. Store in an acid-washed polyethylene bottle.

Dilute Buffer. Dilute stock buffer 1:2 with water.

Malachite Green. (Matheson, Coleman and Bell.) Dissolve 10 mg in 100 ml water. Discard after 1 week.

HCl, approx 1 N. Dilute 21 ml conc acid to 250 ml (acid-washed volumetric flask) with water.

Sodium Periodate, 0.5%

Manganese, Stock Standard. Transfer 500 ml water to a 1 liter volumetric flask. Add 5.6 ml of conc H_2SO_4 and mix. Add 15.4 mg $MnSO_4 \cdot H_2O$ and mix to dissolve. Dilute to mark with water and mix. 1 ml = 5 μg Mn.

Manganese, Working Standard. Transfer 0.5 ml stock standard solution to a 250 ml volumetric flask (acid-washed) and dilute to mark with water. 1 ml = 0.01 μg = 10 ng Mn.

N.B. This dilute working standard is prepared just before needed and discarded after sampling. *Do not* store this working standard solution.

PROCEDURE FOR SERUM

Duplicate analysis should be performed because of the poor precision of the method in the normal range and because duplicates provide a means of detecting random contamination of glassware used in the analysis.

1. Pipet duplicate 3.0 ml aliquots of serum into 25 x 100 mm cylindrical quartz tubes (Cal-Glass).

2. Cover each tube loosely with aluminum foil pierced with holes, and place tubes in a heating block set at low temperature. Evaporate to dryness. Raise heat to carbonize contents.

3. Transfer the tubes in beakers to a muffle furnace at 540°C. Ash for 3 hours.

4. Remove beakers with tubes and allow to cool. Moisten ash with a few drops of redistilled water. With the help of a footed glass rod (which has been acid-washed), break up the ash. Rinse rod with water. Transfer beakers with tubes to steam bath and evaporate to dryness. Reposition beakers containing tubes in the muffle furnace and heat (540°C) for 1/2 hour. Repeat treatment with water, dry, and reheat in muffle furnace until ash is white.

5. To the cooled white ash, add 0.3 ml *N* HCl. Place tubes in hot water and agitate to dissolve ash. Cool.

6. Set up the following in square 10 x 100 mm glass cuvets equipped with glass covers (E. Leitz).
   Blank: 0.3 ml *N* HCl + 1.0 ml water
   Standard (2 ng): 0.3 ml *N* HCl + 0.2 ml working standard + 0.8 ml water
   Standard (5 ng): 0.3 ml *N* HCl + 0.5 ml working standard + 0.5 ml water.

7. To all tubes add 2.3 ml diluted buffer and mix.

8. To all tubes add 0.3 ml *N* NaOH reagent and mix.

9. To all tubes add 0.3 ml malachite green reagent and mix.

10. To all tubes add 0.3 ml of periodate reagent, mix, and record time.

11. Stopper cuvets and place in 26°C water bath.

12. Measure absorbance at 620 nm against water in Beckman DU equipped with light cover for 100 mm cuvet height at 1 hour and again 6 hours after addition of the last reagent in step 10. Return cuvets to constant temperature bath after each reading.

Calculation: Since this reaction is first order, a plot of log absorbance versus time yields a straight line, the slope of which is proportional to the amount of Mn in the solution.

$$\frac{\log A_O - \log A_x}{X} = \text{slope}$$

where

A_O = initial absorbance reading (1 hour)

A_x = final absorbance reading (6 hours)

X = number of hours elapsed between initial and final readings

Determine the slope for blank, standards, and unknowns, using the 5 hour interval.

The amount of Mn in the unknown is calculated as follows:

$$\text{ng Mn in aliquot of unknown serum} = \frac{\text{slope of unknown} - \text{slope of blank}}{\text{slope of standard (5 ng)} - \text{slope of blank}} \times 5$$

$$\mu\text{g Mn}/100 \text{ ml serum} = \frac{\text{ng Mn in unknown}}{\text{ml of unknown ashed}} \times \frac{100}{1000}$$

PROCEDURE FOR URINE

Normal urine samples and urine from patients not treated with EDTA contain so little Mn (less than 2 μg/24 hours) that both a 3 and 9 ml urine aliquot should be ashed. The procedure is then identical to that used for serum.

$$\mu\text{g Mn/24 hours} = \frac{K_x - b}{K_s - b} \times s \times \frac{24 \text{ hour volume (ml)}}{V}$$

where

K_x = slope of log A_x versus time for sample
K_s = slope of log A_s versus time for standard containing s μg Mn
b = slope of blank
V = volume of urine ashed

NOTES

1. Blood collection. Stainless steel silicone-coated needles with Vacu-tainer assembly should be used. Immediately after receipt of the specimen in the laboratory, centrifuge and transfer serum to an acid-washed plastic vial. Mild hemolysis does not interfere, but severely hemolyzed specimens cannot be used.
2. Linearity of method. The plot of log absorbance versus hours gives straight lines, with amounts of Mn added ranging from 0.02 to 0.10 μg (161).
3. Stability of samples. Serum samples are stable at room temperature and 4°C.
4. Interferences. Out of 22 cations and anions tested, only iodide gave positive interference at a level 300-fold greater than normal serum levels (161). At a tenfold greater than normal level, iodide had no effect.

ACCURACY AND PRECISION

Results obtained with the catalytic oxidation method presented are in good agreement with values obtained by neutron activation (114). Recovery of Mn added to serum was 76% (161).
The reproducibility of the test in the normal range is so poor that no significance can be attached to any normal result other than it is normal. In subjects suffering from manganism, serum values can be five times normal

(486), and the precision of the method is much improved (486) with a reproducibility of about ±25% (95% confidence limits).

Blood. The normal adult range for serum of both sexes, determined in our laboratory by the method given, is 0.08–0.26 µg/100 ml. A similar range has been obtained by neutron activation (114). Whole blood has been reported as containing 0.84µg/100 ml (114).

Urine. Very little Mn is excreted in normal urine (486). In our experience a small number of specimens have all given values less than 2 µg/24 hours.

Feces. The average output is about 4.0 mg/day (256).

Hair. Manganese is not detectable in the scalp hair of normal individuals, but concentrations as high as 4.7 µg/g have been reported in persons living in industrial areas where ambient air contains Mn (486).

CHLORIDE

Chloride is usually determined by one of a variety of technics after it has been bound to silver or mercury to form undissociated silver chloride or mercuric chloride. In the earliest method, chloride was precipitated by $AgNO_3$ and the excess Ag titrated with thiocyanate using ferric ion as indicator (584). Interfering organic matter could be removed by wet digestion or preparation of protein-free filtrates.

Chloride has also been determined by the displacement of the chromate ion from insoluble silver chromate with formation of silver chloride, or by addition of potassium chromate followed by titration with $AgNO_3$ (274). Other photometric procedures exchange chloride for the anion of silver dithizonate or mercuric chloranilate, resulting in the liberation of an equivalent amount of colored dithizone and chloranilic acid, respectively. The chloranilate method has been used in several micro and ultramicro methods for serum (23), biologic fluids (247), and serum, sweat, and urine (526).

In the iodometric determination of chloride, the Cl^- reacts with insoluble $AgIO_3$, added in excess to form insoluble AgCl and IO_3^-. The insoluble salts are filtered off and the IO_3^- in the filtrate determined gasometrically, titrimetrically, or photometrically (482).

Chloride ion reacts with mercuric thiocyanate to release SCN^-, which reacts with Fe^{3+} to yield a colored complex which can be measured spectrophotometrically at 480 nm (117, 224, 508, 536). This method is the one used in AutoAnalyzer systems (145, 190). It has been shown that in cases of brominism the error in chloride measurement is greater with the automated procedure than with the manual mercurimetric reference method (502). Sera which show a high chloride value and an apparent electrolyte imbalance should be checked for bromide (145). Manual SCN^- methods, however, gave good agreement with independent reference methods for both serum and urine (117, 508, 536).

Perhaps the most popular method for chloride in body fluids is the mercurimetric one (502). In this technic the Cl^- is titrated with a standard solution of Hg^{2+}, forming undissociated but soluble $HgCl_2$. The endpoint is signaled by a violet-blue color resulting when excess Hg^{2+} forms a complex with the indicator diphenylcarbazone or diphenylcarbazide. A more stable indicator, 2-(8-hydroxyquinolyl-5-azo)-benzoic acid in dimethylformamide solution, has also been described (182). A modification suitable for icteric sera employs H_2O_2 and a brief heat treatment, which eliminates the need for deproteinization and destroys interfering pigments such as bilirubin (164). A comprehensive study of each step of the mercurimetric method, pointing out sources of error, has been published (600).

Chloride has been determined electrometrically, conductimetrically, and by polarography (28, 97, 111, 363, 561). Automatic titrators are available commercially which measure Cl^- by titration with Ag^+ (ref 111 and Chapter 4). The method employs established principles of coulometric generation of titrant and of amperometric indication of the endpoint. The method is simple, rapid, sensitive, and accurate. It automatically delivers Ag^+, stops the titration at the endpoint, and registers the result. It is readily adaptable to a variety of biologic fluids and tissues and to a wide range of Cl^- concentrations (113). An ultramicro adaptation of the method requires only 8 μl of serum and allows the titration of sweat collected in a capillary tube (521).

Chloride has been determined by an isotope dilution method using ^{36}Cl (112). Flame photometry has been used to determine Cl^- indirectly. Excess $AgNO_3$ is added to the serum sample, the precipitated AgCl removed, and the silver remaining in solution determined by flame photometry at 337 or 338 nm (45, 332). Since the serum is diluted 400-fold, the method is free from interference. Atomic absorption has been employed in similar fashion, but here the Ag resonance line at 328 nm is used (36). Several types of ion-selective chloride electrodes have been described (ref 490 and Chapter 4). Their utility for clinical analysis remains to be determined. For the determination of Cl^- in tissues and food, several technics are available (113, 129).

Reference should be made to the first edition of this book and ref 502 for a detailed description of the mercurimetric method of Cl^- analysis. We present the electrometric method because of advantages cited earlier.

DETERMINATION OF CHLORIDE

(method of Cotlove, Ref 111, and Instruction Manual for Buchler-Cotlove Chloridometer, Model 4-2008)

PRINCIPLE

A coulometric (or generator) circuit generates silver ions which combine with Cl^-. An amperometric (or indicator) circuit indicates the first appearance of uncombined, free silver ions at the endpoint and triggers a relay circuit to interrupt the titration and stop a timer. The timer registers elapsed seconds, which measures the amount of silver ions generated at constant current and thereby at a constant rate. By comparison of the time required to titrate an unknown with that of a standard, the Cl^- concentration of the unknown can be calculated. Modern instruments also read directly in meq/liter. A gelatin solution is added to the test system to prevent reduction of AgCl and to promote uniform deposition of reduced silver ions on the indicator cathode.

REAGENTS

Use chloride-free water (distilled or deionized) for all reagent preparation.

Nitric-Acetic Acid Reagent (0.1 HNO_3 in 10% Glacial Acetic Acid). To 900 ml water add 6.4 ml conc HNO_3 and 100 ml glacial acetic acid. Mix thoroughly.

Gelatin Reagent. Prepare a mixture of the following dry pulverized chemicals in the weight ratio of 60:1:1—gelatin (Knox Unflavored Gelatin No. 1, Charles D. Knox Gelatin Co., Inc., or J.T. Baker Chemical Co.); thymol blue, water-soluble (MC/B Manufacturing Chemists); thymol, reagent grade crystals.

To 6.2 g of the dry mixture add approximately 1 liter of hot water and heat gently with continuous swirling until the solution is clear. Dispense into test tubes in portions sufficient for 1 day's analyses. The amount per tube is determined by the estimated daily workload; 0.2 ml is required per titration. The tubes are stable in the refrigerator for at least 6 months. Do not freeze, since this destroys the effectiveness of gelatin. A new tube of dissolved gelatin reagent should be used for each day's analyses. Immerse the tube in hot water to liquefy the gelatin.

NaCl Standard, 100 meq Cl^-/liter. Dissolve 5.85 g dried reagent grade NaCl in water and dilute to 1 liter in a volumetric flask.

NaCl Standard, 1 meq Cl^-/liter. To 10.0 ml of the 100 meq standard in a 1 liter volumetric flask add approximately 800 ml water, 6.4 ml conc HNO_3, and 100 ml glacial acetic acid. Dilute to 1 liter with water and mix.

NaCl Standard, 0.5 meq Cl^-/liter. To 5.0 ml of the 100 meq standard in a 1 liter volumetric flask add approximately 800 ml water, 6.4 ml conc HNO_3, and 100 ml glacial acetic acid. Dilute to 1 liter with water and mix.

PROCEDURE FOR SERUM AND CEREBROSPINAL FLUID

1. Set up the following in 10 ml polyethylene vials:

   Blank: Prepare 2 vials: 4 ml acid reagent + 0.1 ml H_2O.

   Standard: Prepare 3 vials: 4 ml acid reagent + 0.1 ml 100 meq/liter standard (TC pipet). These are needed for the calibration procedure carried out prior to analyzing each batch of unknowns.

   Unknowns: 4 ml acid reagent + 0.1 ml serum or cerebrospinal fluid (TC pipet).

An automatic dilutor is convenient for delivery of the acid reagent and standard or unknowns into the titration vials.

2. Immediately prior to carrying out the titration of each vial, add 4 drops of gelatin reagent. The pH indicator (thymol blue) in the gelatin mixture, which is red at pH 1.2 or lower, provides a visual check for the presence of gelatin and acid in each titration vial.

3. Titration procedure
   a. Place the vial to be titrated in its fully raised position in the sample holder.
   b. Set titration rate switch at desired position.
   c. Turn the "Titration" switch to ADJUST (position 1). The stirrer then starts and the indicator pointer falls within an interval of 10−30 sec to a stable value of less than 5 microamperes.
   d. Reset the timer to zero.
   e. Set the adjustable (red) pointer to 10 divisions (microamperes) above the indicator pointer (black).
   f. Turn the "Titration" switch to TITRATE (position 2). The timer then starts with the simultaneous generation of silver ions. Readings are in seconds when the "Direct Reader" selector switch is in the "Blank" position, and in meq/liter in the "Titrate" position. Avoid delay of more than 1 min between steps *b* and *e*.
   g. Remove the vial. Rinse the electrodes with water. The instrument is now ready for the next titration.

4. Calibration
   a. Titrate the two blank vials, setting the "Direct Reader" selector switch to "Blank" position and the titration rate switch at "Hi". The readings are given in seconds and are averaged.
   b. Set the average elapsed time of the blank on the "Time Delay" meter using the black pointer. When titrating unknowns, the blank time interval is automatically subtracted from the result and the instrument automatically resets itself.
   c. Turn the "Direct Reader" selector switch to "Titrate" and move the adjustable red pointer on the "Time Delay" meter until it coincides with the position of the black pointer. A distinct click is heard when this occurs. Reset the digital counter to zero.
   d. Titrate one of the "Standard" vials (100 meq/liter) as described in step 4. The number on the digital counter should read in the 99−101 range. If the number registered is less than 99 turn the "Blank Time" adjust screw ("Hi"), located above the digital counter, one or two divisions in a clockwise direction. If the number is greater than 101 turn the screw one or two divisions counterclockwise. Repeat procedure with additional standard vials until readings replicate between 99−100.

5. Titrate the unknowns using the same instrument settings used for the standard. Results are read directly from the digital counter in meq/liter.
   N.B. Optimum titration times are 40−60 sec at high rate, 60−100 sec at medium, and 100−150 sec at low.

PROCEDURE FOR URINE

1. The Cl⁻ concentration of urine specimens may range from under 1 meq/liter to over 300 meq/liter. Start the analysis by using the same procedure as for serum. If the titration time is less than 20 sec (estimate by use of watch or timer) repeat the titration at the medium rate, and for extremely low concentrations switch to the low rate with the following modifications of the serum procedure.

2. *Medium rate:*
   Blank: 4 ml acid reagent
   Standard: 4 ml of 1 meq/liter standard
   Unknown: 4 ml acid reagent + 0.1 ml urine
   Calibration: To calibrate the direct readout for the medium titration rate, turn the "Titration Rate" selector switch to "Med" position and repeat *Serum* calibration procedure. Adjust the digital counter to read 40 meq/liter.

3. *Low Rate:*
   Blank: 2 ml acid reagent
   Standard: 2 ml of 0.5 meq/liter standard
   Unknown: 1 ml or less (precise volume to be used should give titration time of between 60 and 150 sec) + sufficient nitric-acetic acid reagent to make a final volume of approximately 2 ml
   Gelatin reagent: Use 2 drops for low rate titration
   Calibration:
   To calibrate the direct readout for the low titration rate, turn the "Titration Rate" selector switch to "Lo" position and repeat *Serum* calibration procedure. Adjust the digital counter to read 1 meq/liter. The setting gives direct readings for 1 ml urine samples. If smaller volumes are used, calculate as follows:

$$\text{meq Cl}^-/\text{liter} = \text{reading in meq/liter} \times \frac{1.0}{\text{volume of specimen in ml}}$$

4. *High chloride concentrations.* If the high rate titration time is more than 100 sec, dilute the original specimen 1 + 1 with water and repeat the analysis. Apply a dilution correction to the result.

ULTRAMICRO PROCEDURE

1. Set up the following in 10 ml polyethylene vials:
   Blank: Prepare 3 vials: 2.5 ml acid reagent
   Standard: Prepare 3 vials: 2.5 ml acid reagent +10 μl 100 meq/liter standard
   Unknowns: 2.5 ml acid reagent +10 μl serum, urine, or cerebrospinal fluid. Use precision micropipets such as the Lang-Levy type made of glass, or the Sanz type made of polyethylene or Teflon. The volume delivered by

the latter type need not be known exactly if the same pipet is used for standard and unknown solutions.

2. Immediately prior to carrying out the titration of each vial, add 2 drops of gelatin reagent.

3. Proceed as in macro procedure, starting with step 3 but using low titration rate setting.

PROCEDURE FOR "SWEAT TEST"

The "sweat test," whereby sweat is analyzed for Cl^- (sometimes also Na^+ and K^+), was originally introduced as a measure of adrenocortical function. Analysis of sweat Cl^- is now used almost exclusively as a diagnostic aid for cystic fibrosis of the pancreas (139, 270, 348).

An adequate sample for analysis usually can be obtained after deliberate induction of sweating by one of several methods (269, 270). The patient may be placed in a plastic suit or bag and covered with a blanket. This thermal stimulus has been supplemented by hot water bottles or electric lights. However, methods which involve whole body heating are now discouraged, as they have resulted in hyperpyrexia and death (270). The current method of choice is to introduce by iontophoresis one of the parasympathomimetic agents, mecholyl or pilocarpine, through the intact skin of the volar surface of the arm, a procedure which can be concluded in a short time and usually yields an adequate volume of sweat (269). One technic for collecting sweat uses filter paper (269). Prior to the test, two 2.5 cm filter paper circles are placed in a capped plastic weighing bottle and weighed to the nearest 0.1 mg. The chosen skin site is iontophoresed for 5 min with pilocarpine to induce sweating. Following suitable preparation the test papers are placed on the sweating site, covered with Parafilm, and allowed to absorb sweat for $\frac{1}{2}$–1 hour. The papers are returned to the weighing bottle and reweighed. The difference in weight is the weight of sweat. About 100 mg of sweat should be collected.

To determine the Cl^- of sweat, proceed as follows (269): (a) Add 5.0 ml of the nitric-acetic acid reagent (see under *Determination of Chloride, Reagents*) to the plastic bottle containing the sweat sample and shake thoroughly. (b) Follow the procedure for serum Cl^-, employing a 4 ml aliquot of sweat extract, 4 ml of 0.5 meq/liter standard, and 4 ml of nitric-acetic reagent as blank. Titrate with the selector switch in the medium position. (c) Calculate the results using the following equation:

$$\text{meq } Cl^-/\text{liter} = 0.002 \times \frac{T - B}{S - B} \times \frac{5 + W}{4} \times \frac{1000}{W}$$

$$= \frac{T - B}{S - B} \times \frac{5 + W}{2W}$$

where

$$T =$$ titration of test

$$S =$$ titration of standard

$$B =$$ titration of blank

$$W =$$ sweat collected, in grams

Sweat collected by the gauze or sponge technic may be diluted with 4 vol of water and a 1 ml aliquot analyzed as for serum or urine.

Results of the sweat test are affected by the rate of sweating, skin temperature, intensity of heat stimulus, hormonal effects, salt intake, and the anatomic site of sweat collection (348). Chloride concentration is altered in women in a cyclical manner related to the salt and water retention associated with the menses, presumably the result of hormonal influences (270). The concentrations of Na^+, K^+, and Cl^- change with progressive sweating. It is also reported that the Cl^- concentration of sweat is higher in winter than in summer (348).

A number of attempts have been made to devise simple screening procedures in which finger or hand contact with a test material registers the skin Cl^- concentration in a qualitative or semiquantitative manner. Such testing is now deemed inadvisable (138, 270). The determination of Na^+ is also of value (325, 407).

NOTES

1. Precautions in use of instrument. Rinse the electrodes thoroughly following the last titration. If any of the following signs of impaired performance is noticed, polish the electrodes, using silver polish: (a) initial stabilization of indicator current is delayed more than 20–30 sec after the titration switch is turned to position 1; (b) the blank time is increased to above 2.0 sec at high current, 5.5 sec at medium current, or 11 sec at low current; (c) erratic variation of indicator current during titration.

2. Whole blood vs serum or plasma. It is generally conceded that the Cl^- level of serum or plasma is more meaningful than the level in whole blood. Approximately one-third of the Cl^- content of whole blood is in the erythrocytes, hence the whole blood concentration varies with the hematocrit (562). Changes in the pH of blood at the time of collection and before serum is separated cause a slight but measurable shift of Cl^- from the plasma into the erythrocytes. For the most accurate results, draw blood with minimal venous stasis and prevent exposure of drawn blood to air (loss of CO_2) before separating serum.

3. Stability. It has been claimed (502) that results obtained on refrigerated urine samples by the mercurimetric technic increase significantly within 24 hours. Our laboratory has been unable to confirm this and has found urine and serum samples to be completely stable at either room or refrigerator temperatures for 1 week. Although no data are available for spinal fluid, it would seem reasonable to assume a similar stability.

4. *Interferences.* As with any of the methods involving reaction of Cl^- with Ag^+ or Hg^{2+} (145), the electrometric method measures in addition to Cl^- other silver ion-combining substances such as Br^-, I^-, SCN^-, S^{2-}, or sulfydryl ions if these are present in the sample. The bromide ion is perhaps the most likely interfering ion encountered (145). As the serum Br^- increases, the serum Cl^- decreases meq for meq so that the total halide concentration remains fairly constant. The "true" Cl^- can be calculated from the "apparent" Cl^- found by analysis if an independent analysis for Br^- is made. Reagent solutions and specimens should be kept out of contact with rubber tubing or stoppers since aqueous solutions, especially if alkaline, extract sulfydryl groups from rubber. Metal surfaces should also be avoided. Metal ions with more than one ionic valence (such as Fe or Cu) can produce a positive amperometric current. The amounts of these ions in biologic fluids have no effect on the method.

ACCURACY AND PRECISION

The question of the accuracy of the determination of Cl^- in serum or plasma must be considered in two parts. First, the accuracy of determining the Cl^- present in a sample is quite satisfactory, Br^- being the only other ion commonly present which cannot be differentiated. Cotlove (111) has found that results obtained on serum and urine by the automatic titration method agree with those obtained with an isotope dilution method (112) within 0.5%.

This consideration is the only one pertinent for whole blood, cerebrospinal fluid, and urine. In the case of serum or plasma, however, there is a second consideration: The value desired is the Cl^- concentration in the circulating plasma, and several factors may affect the Cl^- concentration by the time the serum or plasma sample is available for analysis. If the blood sample is left exposed to the air, CO_2 is lost, resulting in a water shift from the erythrocytes into the serum or plasma, causing a dilution of about 1.5% (cf Chapter 14, *Specimen Collection and Preservation*). Heparin is a completely satisfactory anticoagulant (580), but other anticoagulants may significantly affect electrolyte distribution between erythrocytes and plasma. For example, potassium oxalate has a varied and inconsistent effect, but the plasma Cl^- concentration is changed by no more than about 3 meq/liter if the final concentration of potassium oxalate does not exceed 0.2%. When the usual procedure of removing the plasma from the erythrocytes after the sample has come to room temperature ("derived" plasma) is followed, there is about a 3% shift of the Cl^- from the plasma into the erythrocytes. In addition to these problems, the Cl^- concentration as determined in protein-free filtrates is about 2% too high because of the "volume displacement effect" (515).

The between-run precision of the method presented is about $\pm 2.6\%$ (95% confidence limits) at a level of 100 meq/liter.

NORMAL VALUES

The normal ranges for Cl^- in serum or plasma, whole blood, cerebrospinal fluid, urine, and sweat are given in Table 19-1. The ranges for serum or plasma and whole blood reported by various workers agree fairly well and the values given in the table are compromises of the references cited. Studies of the serum Cl^- levels in individuals have revealed (171) a sine-like pattern over the year. There is no postprandial change (18).

TABLE 19-1. Normal Values for Chloride

Specimen	Normal values	References
Serum or plasma	98—109 meq/liter	111,502,508,561,600
Whole blood	73—95 meq/liter	283,502,503,621
Erythrocytes	1555.5 meq/kg	113
Cerebrospinal fluid	122—132 meq/liter	111,502
Urine	170—250 meq/24 hours	78,502
Sweat[a] (normal children)	5—30 meq/liter	591
Sweat (normal children)	<30 meq/liter (96% of population) <60 meq/liter (4% of population)	270 270

[a]In cystic fibrosis a mean of 89 (range 55—130) meq/liter has been reported (591).

The ranges for serum or plasma (153) and whole blood (283) are slightly higher for women than men and a slight decrease with age has been reported (283). The range for serum Cl^- in the newborn is somewhat broader than the adult range, e.g., it has been reported as 94—118 (531) and 91—115 meq/liter (442). The range for the first year of life has also been reported to be rather broad, about 80—140 meq/liter (215).

The diurnal variation in urinary output of Cl^- is quite large, the output at night during sleep being up to fivefold less than the peak output during the day (325, 542).

The Cl^- content of normal sweat varies with age. The Na^+ concentration (and presumably Cl^- also) is low in children under one year of age and rises slowly to the adult level by about the 20th year (270). There is no sex difference, and the variation within subjects may be very large, e.g., one subject varied between 6 and 54 meq/liter (348).

PHOSPHATE

Phosphorus is present in blood chiefly as inorganic and organic phosphates, nearly all of the latter residing in the erythrocytes. The *total phosphorus* of blood is determined as inorganic phosphate after wet ashing with sulfuric acid and perchloric acid, nitric acid, chloric acid, or hydrogen peroxide (41, 47, 167, 236, 300, 306, 484). The phosphorus present in a trichloroacetic acid filtrate of whole blood is called the *acid-soluble phosphorus* and includes the inorganic phosphate ions as well as various organic phosphates. The total acid-soluble phosphorus is determined as phosphate after digestion of the trichloroacetic acid filtrate (48, 71, 167). The organic phosphate compounds in the trichloroacetic acid filtrate have been fractionated by adaptations of the method of fractional analysis of phosphate esters in which different esters are split in different periods of time of hydrolysis in hydrochloric acid at $100°C$ (47, 62, 212, 236, 300, 401). Another method of fractionation is based on solubility differences of the barium salts of the esters (150). Paper chromatographic fractionation of the acid-soluble phosphorus of normal blood has shown the following approximate distribution (357, 485): inorganic phosphate, 12%; 2,3-diphosphoglyceric acid, 50%; adenosine triphosphate, 22%—called "pyrophosphate" in early literature (300); fructose 1,6-diphosphate, 16%. Ordinarily, determination of inorganic phosphorus is the only analysis requested of the clinical laboratory; this is best performed on serum or plasma, since if performed on whole blood there is the danger of splitting organic esters in the course of the analysis. One solution to this problem is to use a citrate-arsenite buffer system which complexes the excess color reagent remaining after the inorganic phosphorus initially present has reacted. Then, if organic phosphates hydrolyze to release inorganic phosphate there is no chromogen available to react with it (27). The determination is usually referred to as *inorganic phosphorus* rather than inorganic phosphate, since the phosphate determined is usually reported in terms of phosphorus.

The technics of determination of phosphate ions can be grouped into several categories. The basis of the first category is that under suitable conditions molybdates react with phosphate (also arsenate and silicate) to form heteropoly compounds such as ammonium molybidiphosphate (also called ammonium phosphomolybdate) with a formula of $(NH_4)_3$ $[P(Mo_3O_{10})_4]$ (620). An ultramicro method has been described in which the ammonium molybdiphosphate formed is determined by acidimetric titration (344). Others have measured the absorbance at 340 nm of the unreduced phosphomolybdate formed in a centrifugal apparatus (121), or at 310 nm after solvent extraction with a mixture of xylene-isobutanol (65:35, v/v) (141). The latter method is quite sensitive, 1 µg P producing an absorbance of 0.93. Phosphomolybdic acid has been extracted by 2-octanol and Mo determined by atomic absorption, thus providing an indirect ultramicro phosphorus determination (639).

Most of the technics for the determination of phosphate, however, involve the photometric determination of the molybdenum blue formed by reduction of the molybdiphosphate under conditions which do not reduce the excess molybdate present. The formation of the heteropoly molybdenum blue is quite complex and undoubtedly heterogeneous (524). The various reduction technics that have been proposed differ chiefly in the reducing agent employed: (a) *Stannous chloride.* A chronology of the use of this reductant was given in the first edition. Of all the reductants employed, stannous chloride produces the most molybdenum blue from a given amount of molybdiphosphate, thus rendering the test more sensitive, but its use also has certain undesirable features, including unstable color, deviation from Beer's law, instability of the reducing agent itself, great sensitivity to change in acidity, and difficulty in reproduction of standards from day to day. Efforts have been made, however, to increase color stability. A mixture of stannous chloride and hydrazine, said to be a more stable reductant, has been used in the AutoAnalyzer (327). In our hands this method led to increased deposits in the mixing coils and a high noise level. Berenblum and Chain (50) introduced a modification in which the molybdiphosphate is extracted into isobutyl alcohol before reduction with stannous chloride. (b) *Phenylhydrazine (559), hydroquinone (48, 71, 484), ferrocyanide (565), aminonaphtholsulfonic acid (167, 306), hydrazine sulfate (205), and 2,4-diaminophenol · HCl (11)* have all been used. (c) *Ascorbic acid.* When this reagent is used at pH 4 the splitting of labile phosphate esters is 5% or less of that occurring with the technic using aminonaphtholsulfonic acid (20, 352). A "direct" method, not requiring protein removal, forms the phosphomolybdate complex in the presence of borate. The complex is reduced with ascorbic acid and the resulting suspension solubilized with carbonate (202). The method is sensitive and claimed to be free of interference. Also, if read at 320 nm instead of at 720, the sensitivity is increased fourfold. (d) *p-Methylaminophenolsulfate (Elon).* The color produced by this reagent has considerable acid tolerance, is quite stable, and has a sensitivity equal to that produced by aminonaphtholsulfonic acid (205, 390, 450). A "direct" method has been published which uses mono-ethanolamine as solubilizing agent. It is claimed that accuracy is unimpaired even in the presence of high serum protein (144). (e) *N-Phenyl-p-phenylenediamine.* Use of this reagent possesses the advantages of color stability and increased sensitivity. The sensitivity is higher, because in addition to the molybdenum blue formed the oxidized reagent itself is blue (146). (f) *Ferrous sulfate.* This technic has the advantages of color stability and of greater specificity when labile phosphate esters are present, because the reaction is carried out in the presence of less acidity than most of the other methods (257, 480, 556). A modification employing Fe^{2+} and thiourea has very stable reagents, a simplified technic requiring a minimum number of steps, a stable color with twice the sensitivity of the Elon and aminonaphtholsulfonic acid methods, and conforms to Beer's law over a wide range of phosphorus concentrations (198). The method has been successfully automated (629).

The second category of phosphate determinations involves the formation of molybdivanadophosphoric acid. This technic consists of photometric determination of the stable yellow color formed when an excess of molybdate is added to an acidified solution of an orthophosphate and a vanadate. The exact nature of the reaction is uncertain, but presumably a heteropoly complex such as $(NH_4)_3PO_4 \cdot NH_4VO_3 \cdot 16MoO_3$ is formed (620). This method has been applied to the determination of phosphate in blood and serum (522). In our experience results yielded on serum by this method run about 0.2–0.3 mg/100 ml higher than a molybdiphosphate method, possibly due to greater hydrolysis of organic phosphate esters present by the higher acidity employed in the molybdivanadophosphate method.

A third category for the determination of phosphate involves reaction of phosphomolybdate with the triphenylmethane dye malachite green (275). The complex shows a marked shift in absorption maximum from that of the dye alone. The method is the most sensitive known for phosphorus; 1 μg P gives an absorbance of 0.022 in the commonly used Fiske-Subbarow method compared with 0.642 in the malachite green assay, a ratio of 30:1. One drawback is the very high acidity (1 N) at which the complex is formed. Organic phosphates are hydrolyzed, but the reaction may be carried out at 5°C, at which temperature only creatine phosphate is labile in the acid medium. A micromethod for serum and urine which uses lesser amounts of malachite green and a different dispersing agent has also been reported (38). Other dyes react with phosphomolybdate, e.g., methyl green has been used in an automated method (577). With a short reaction time of 15 sec, it was claimed that hydrolysis of labile phosphates is negligible. Linearity and reproducibility are excellent for both the manual and automated methods.

An enzymatic method for inorganic phosphorus in serum and tissues is based on successive enzyme reactions catalyzed by glycogen phosphorylase, phosphoglucomutase, and glucose 6-phosphate dehydrogenase. NADPH is finally produced from NADP in proportion to the phosphorus present and is measured by its fluorescence or light absorption (509). The reactions take place at neutrality, thereby permitting measurement of inorganic phosphorus in the presence of unstable organic phosphates.

Although the dye-shift method offers the promise of extreme sensitivity coupled with practical utility, we have not had experience in its use. We have chosen to present a ferrous sulfate method whose sensitivity is adequate for the majority of clinical specimens.

DETERMINATION OF INORGANIC PHOSPHORUS

(method of Goldenberg and Fernandez, Ref 198, modified)

PRINCIPLE

The specimen is deproteinized by an iron-trichloroacetic acid reagent. The supernate is mixed with molybdic acid to form molybdiphosphate which is then reduced by Fe^{2+} to produce molybdenum blue (Fig. 19-6). Absorbance is measured at 660 nm.

REAGENTS

Iron-Trichloroacetic Acid, Stabilized. Transfer 50 g trichloroacetic acid to a 500 ml volumetric flask with the aid of 300 ml water. Add 5 g thiourea (Eastman Organic Chemicals No. 497) and 15 g Mohr's salt (ferrous ammonium sulfate hexahydrate), dissolve, and dilute to mark with water.

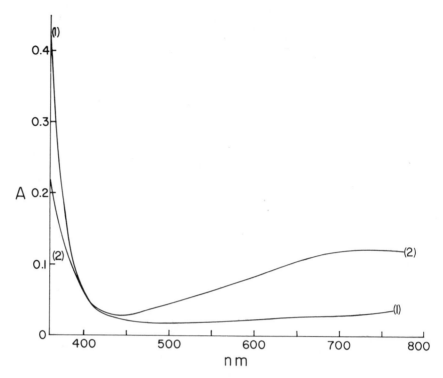

FIG. 19-6. Absorption curves in manual phosphorus determination: (1) reagent blank vs water, (2) 5 μg phosphorus standard vs reagent blank. Cary model 15 recording spectrophotometer.

Store in an amber bottle. A deposit of sulfur begins to form after a week but does not interfere with the analysis. The shelf life of this reagent is 6–12 months. Discard when the absorbance of the standard changes by more than 20% from its original value.

Ammonium Molybdate, 0.0355 M. Add 45 ml conc H_2SO_4 with cooling to a 500 ml volumetric flask containing 200 ml cold water. Dissolve 22 g $(NH_4)_6Mo_7O_{24} \cdot 4H_2O$ in 200 ml water and add to flask. Mix and dilute to mark with water. Reagent is stable for several years.

Phosphorus standard, 5 mg P/100 ml. Dissolve 0.220 g KH_2PO_4, dried for 1 hour at 110°C, in water in a 1 liter volumetric flask. Dilute to mark with water. Add a few drops of chloroform as preservative and store in a polyethylene bottle. Discard if there are signs of mold growth.

HCl, 6 N. Add 10 ml conc HCl to 10 ml water.

PROCEDURE FOR SERUM (or Plasma)

1. Deliver 0.2 ml serum into a 12 ml thick-walled centrifuge tube. Add 5 ml iron-trichloroacetic acid reagent with mixing. Let stand 10 min and centrifuge. The supernate must be clear.

2. Set up the following in 100 x 12.5 mm tubes:
   Blank: 5 ml iron-trichloroacetic acid + 0.2 ml water.
   Standard: 5 ml iron-trichloroacetic acid + 0.2 ml standard
   Unknown: Carefully and completely decant supernate from Step *1* into tube
3. To each tube, add 0.5 ml molybdate reagent and mix. The color develops rapidly and is measured after 20 min. The color is stable for at least 2 hours.
4. Read the standard and unknown against the blank at 660 nm in any suitable photometer.

Calculation:

$$\text{mg inorganic phosphate, as P/100 ml} = \frac{A_x}{A_s} \times 0.01 \times \frac{100}{0.2}$$

$$= \frac{A_x}{A_s} \times 5$$

$$\text{meq inorganic phosphate/liter} = \text{mg inorganic phosphate, as P/100 ml} \times 0.58 \text{ (see *Note 5* below)}$$

PROCEDURE FOR URINE

1. Mix urine and check pH with pH paper. If alkaline, acidify with conc HCl to about pH 6.0. Prepare a 1:10 dilution with water and mix.
2. Transfer 0.20 ml diluted urine to a 12 ml centrifuge tube, add 5 ml iron-trichloroacetic acid reagent and 0.5 ml molybdate reagent. Mix. Urine specimens containing protein develop turbidity upon addition of the iron-trichloroacetic acid reagent. If this happens, allow the protein to settle for 15 min before centrifuging. Decant the supernate into a clean 100 x 12.5 mm tube.
3. Proceed as for serum starting at *step 2,* setting up the blank and standard tubes.

Calculation:

$$\text{g inorganic phosphate, as P/24 hours} = \frac{A_x}{A_s} \times \frac{0.01}{1000} \times \frac{10}{0.2} \times \frac{24 \text{ hour urine}}{\text{volume (ml)}}$$

$$= \frac{A_x}{A_s} \times \frac{24 \text{ hour urine volume (ml)}}{2000}$$

NOTES

1. Beer's law. Beer's law is obeyed up to a concentration of at least 20 mg/100 ml.

2. Stability of samples. The problem of instability of samples lies chiefly with whole blood. Organic phosphates susceptible to hydrolysis exist principally inside the erythrocytes. The inorganic phosphorus of whole blood, held at 37°C, can double in a few hours (47, 357, 294). The inorganic phosphorus of the plasma in contact with the erythrocytes also increases as a result of leakage through the cell walls (294, 527). Phosphatases are responsible for at least part of the phosphatolysis that occurs (47). It is important to separate serum or plasma from the erythrocytes as soon as possible (527). Once the serum or plasma has been separated, the problem of instability, for most practical purposes, has been solved because although there still are some organic phosphates present the maximal increase in organic phosphate on standing even at room temperature is only a few tenths of a milligram per 100 ml (527). At refrigerator temperature serum is stable for at least 1 week (556).

Urine also contains some inorganic phosphates which decompose, even under aseptic conditions, within a few days to weeks at 37°C (368). When acidified with HCl, urine is stable for periods exceeding 6 months in the refrigerator (556).

3. Interferences. Results obtained with serum and heparinized plasma are identical (198). Plasma prepared with oxalate or citrate gives about 5% lower values than serum. Ingestion of a phosphate-containing drug, "Fleet phospho-soda," caused prompt elevation of serum phosphorus concentration (86). When digestion of tissues or phospholipids requires use of H_2O_2, some peroxide may remain even after 15 min of boiling. This gives a yellow color with molybdate but can be removed by treatment with permanganate (388). Increased values for serum phosphorus have been reported following the administration of alkaline antacids, vitamin D, cottonseed oil injection, heparin, Methicillin, Pituitrin, and tetracyclines (104). Decreased values were reported following the administration of aluminum hydroxide, epinephrine, ether anesthesia, insulin, and parathyroid injection (104).

4. Serum vs whole blood. Erythrocytes contain a negligible amount of inorganic phosphate (527). The inorganic phosphate of whole blood is thus affected by the hematocrit. This fact, plus the difficulties introduced by the large amounts of organic phosphates present in the erythrocytes, render the use of serum preferable.

5. Phosphate expressed in milliequivalents. Inorganic phosphate exists in serum as HPO_4^{2-} and $H_2PO_4^{-}$. At pH 7.4 about 80% is in the bivalent form. Accordingly, to convert mg/100 ml to meq/liter, the former is multiplied by $10(0.8 \times 2/31) + 10(0.2 \times 1/31)$ or 0.58. The ratio of the forms of phosphate ions varies with the blood pH. For example, at pH 7.1 the factor is 0.54. Conversion with the factor of 0.58 therefore assumes a normal blood pH. It is not wise to attempt conversion in the case of urine, where the pH can vary about 3 pH units. Furthermore, the pK_2' of phosphate in urine can vary between 6.56 and 6.98 (510).

ACCURACY AND PRECISION

Certainly one of the sources of error in all the methods based on reduction of molybdiphosphate is the hydrolysis of phosphate esters present during the running of the test since the reaction is carried out in acid (48, 168, 556). Furthermore, molybdate catalyzes acid hydrolysis of acid-labile P (599). The degree of acidity required for the color reaction is dependent on the reductant employed (198). Hydrolysis of phosphate esters can lead to significant inaccuracy with some types of samples, but with serum, urine, and cerebrospinal fluid this source of error is rather small. Trichloroacetic acid filtrates of serum possess a background absorption which introduces a positive analysis error of up to about 5% (240). It has been shown (109) that mannitol in urine interferes with color formation when using ferrous sulfate as the reducing agent. Whether interference also occurs with other reducing agents is not known. By the present method recovery of phosphorus, added to serum and urine, averages 97% and 98%, respectively.

The precision (95% limits) of the method is about ±5%.

AUTOMATED DETERMINATION OF INORGANIC PHOSPHORUS IN SERUM AND URINE

(method of Kuttner and Cohen, Ref 330, adapted to AutoAnalyzer by Kessler, Ref 301, modified)

PRINCIPLE

Molybdate reacts with phosphate to form molybdiphosphate. Stannous ion reduces the molybdiphosphate to heteropoly molybdenum blue, which is measured at 660 nm. For absorption curve see Figure 19-6.

REAGENTS

H_2SO_4, *1 N.* Dilute 28 ml conc H_2SO_4 to 1 liter. Add 0.5 ml wetting agent, Levor IV (Technicon Corp.), and mix well.

Ammonium Molybdate, 0.0121 M. Dissolve 15 g $(NH_4)_6Mo_7O_{24} \cdot 4H_2O$ in approximately 600 ml water. Cautiously add, with mixing, 55 ml conc H_2SO_4. Cool, dilute to 1 liter with water, and mix.

Stock Stannous Chloride. Dissolve 10 g $SnCl_2$ in 25 ml conc HCl. Store in glass-stoppered bottle in refrigerator.

Working Stannous Chloride. Dilute 1 ml stock to 1 liter with water. Prepare fresh daily.

Phosphorus Standard, 5 mg P/ml. See under *Reagents*, manual method described previously.

PROCEDURE

Set up manifold, other components, and operating conditions as shown in

Figure 19-7. By following procedural guidelines described in Chapter 10, steady state interaction curves as shown in Figure 19-8 should be obtained and checked periodically. Results are calculated from the calibration curve in the standard manner for methods obeying Beer's law.

N.B. Urine samples generally require a 1:10 dilution.

NOTES

1. Beer's law. Beer's law is obeyed up to 12 mg P per 100 ml.
2. See notes 2, 3, 4, and 5 of manual method.

ACCURACY AND PRECISION

The automated method yields results in excellent agreement with the manual procedure described previously.

The precision (95% limits) of the method is ±5%.

NORMAL VALUES

Published normals for serum inorganic phosphorus agree reasonably well, and a compromise for the normal range is 2.5–4.8 mg P/100 ml (1.45–2.76 meq/liter) (47, 236, 450, 522). There is no significant sex difference. All reports agree that the normal range for children is higher than that for adults. The approximate range for children in the first year of life is 4–7 mg/100 ml (2.32–4.06 meq/liter), the range gradually dropping to adult levels by the age of 20 (229). There is also evidence of decreasing values after middle age, the decrease apparently occurring only in males (206). Levels are significantly lower during the menstrual period than between menses. There is a slight decrease after meals (18).

The normal total phosphorus of whole blood is about 31–44 mg/100 ml (236). The normal acid-soluble phosphorus of whole blood is about 18–28 mg/100 ml (300).

The normal range for inorganic phosphorus in cerebrospinal fluid is 0.9–2.0 mg/100 ml (406).

The normal 24 hour output by adults of inorganic phosphate in urine is about 0.34–1 g, expressed as P (520). There is a diurnal variation in output, excretion being highest in the afternoon, and the hourly variation is as much as sixfold (353). In children, the output is about 0.53–0.84 g/day, the output increasing steadily with age (358).

SULFATE

The sulfur of blood, serum, and urine is present as the inorganic sulfate ion, as conjugated sulfate, and as a component of such compounds as glutathione and sulfur-containing amino acids. Determination of total S in biologic material involves oxidation of all S to sulfate followed by analysis

INORGANIC PHOSPHORUS

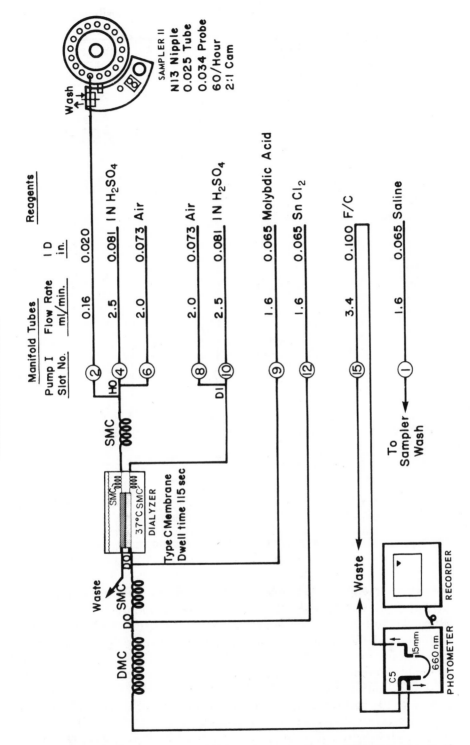

FIG. 19-7. Flow diagram for inorganic phosphorus in serum and urine. Calibration range: 0–10 mg/100 ml. Conventions and symbols used in this diagram are explained in Chapter 10.

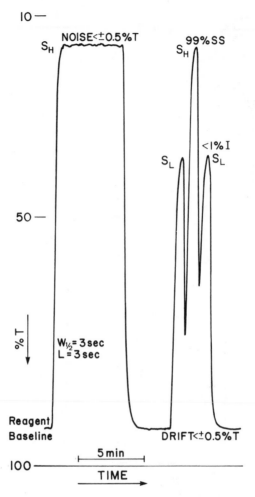

FIG. 19-8. Inorganic phosphorus strip chart record. S_H is a 10 mg/100 ml standard, S_L a 5 mg/100 ml standard. Peaks reach 99% of steady state; interaction between samples is less than 1%. $W_{1/2}$ and L are explained in Chapter 10. Computer tolerance for noise is set at ±0.5% T, for baseline drift at ±0.5% T, and for standard drift at ±1% T.

for sulfate. Oxidation is carried out by a mixture of H_2O_2 and HNO_3 or by oxidation in a Schöniger flask (483). "Total sulfate" is the sum of the inorganic sulfate and conjugated sulfate and is determined after acid hydrolysis of the protein-free sample which converts the conjugated sulfate into sulfate ions. The "conjugated" or "ethereal" sulfate therefore is the increment in inorganic sulfate produced by acid hydrolysis.

Serum sulfate has been determined by gravimetric determination of $BaSO_4$ (349) and by a technic in which the sulfate is precipitated as $BaSO_4$ by an excess of $BaCl_2$ and the excess Ba is precipitated as $BaCO_3$, which is then titrated with acid (286). Sulfate ions in urine (53, 392) and serum (53, 392) can also be determined by turbidimetric or nephelometric measurement of $BaSO_4$. These technics do not appear to possess high accuracy or precision (213). A turbidimetric method for inorganic sulfates in urine, based on the precipitation of barium sulfate, has been adapted to the AutoAnalyzer. Coating of the flowcell with $BaSO_4$ was avoided by an intersample air rinse (137). In another method the sulfate is precipitated by Ba containing some radioactive Ba, and the radioactivity remaining in the supernate is measured (381, 438). The accuracy claimed for this procedure is surprising in view of the tendency of $BaSO_4$ to form colloidal particles and to coprecipitate Ba. Gasometric (581) determination of sulfate involves addition of excess solid $BaIO_3$, which reacts with the sulfate to form $BaSO_4$, releasing an equivalent amount of iodate ions. The iodate is reacted with hydrazine to form nitrogen. Another gasometric method for water analysis involves reduction of sulfate to H_2S by a mixture of HI, acetic anhydride, and $NaHPO_2$. Although it has been suggested that it might be applicable to biologic fluids (123), the method does not seem suitable for clinical analyses. Another method proposed for determination of sulfate in serum involves adding excess insoluble barium chloranilate so that the sulfate exchanges for an equivalent amount of chloranilate. The $BaSO_4$ formed and the excess barium chloranilate are removed by centrifugation, and the chloranilate ion remaining in solution is determined photometrically. Unfortunately, in our experience this technic does not have sufficient sensitivity. Changes in ionic strength affect the absorbance of the barium chloranilate blank, leading to gross errors in the sulfate determination (555). A fluorometric method has been proposed which depends on the binding of thorium by sulfate, decreasing the fluorescence of a standard thorium-morin complex in an acid medium (214).

By far the most popular approach in clinical chemistry to the analysis for sulfate ion has been precipitation of the sulfate as benzidine sulfate. Among the various technics of quantitating the precipitate have been the following: (a) gravimetry (285); (b) alkalimetric titration (451); (c) oxidation with excess $K_2Cr_2O_7$ followed by iodometric back-titration of the excess (339); (d) photometric determination of the brown color produced with KI, I_2, and NH_4OH (632); (e) photometric determination of the yellow color produced with $FeCl_3$ and H_2O_2 (262); (f) photometric determination of the color produced by coupling phenol (285) or thymol (119) with the benzidine following diazotization; (g) photometric determination of the red-brown color produced with β-naphthoquinone-4-sulfonate (339). Quite early in the history of the technics based on benzidine sulfate precipitation, it was recognized that the presence of phosphate constituted a possible source of error because of benzidine phosphate precipitation (417). If the pH at precipitation is not sufficiently low, benzidine phosphate precipitates; if the pH is low enough to avoid precipitation of phosphate, the chloride, if present in high enough concentration, increases the solubility of benzidine sulfate. Attempts were made in the earlier methods to minimize the interference by phosphate by compromising on the pH for precipitation but this did not completely effect a separation of the two anions (339). Preliminary removal of the phosphate ion has been achieved in serum by uranyl acetate precipitation of protein, which also precipitates the phosphate. The necessity for removal of phosphate from either serum or urine has been contested by Kleeman *et al* (313).

DETERMINATION OF SULFATE
(method of Kleeman et al, Ref 313, modified)

PRINCIPLE

Serum is deproteinized with uranyl acetate, which simultaneously removes the phosphate. The sulfate in the filtrate is precipitated as benzidine sulfate, and the benzidine in the precipitate is then quantitated photometrically by reaction with sodium β-naphthoquinone-4-sulfonate. The absorption curve of the colored product is shown in Figure 19-9. The curves obtained are similar to those obtained in the β-naphthoquinone method for a-amino acid nitrogen, but the absorption peaks obtained are somewhat different.

The original method of Kleeman *et al* employed trichloroacetic acid for deproteinization, but since this acid is usually contaminated with phosphate and must be purified by distillation before use, and since there appears to be a question regarding interference by phophate ions, we have chosen to use uranyl acetate for deproteinization. This reagent is also used with urine, effectively removing about 98% of the phosphate ions present in urine.

REAGENTS

Benzidine, 1% in Ethanol. Prepare fresh each day; ethanol denatured with methanol is satisfactory. The benzidine should be recrystallized as follows: Dissolve in a minimum amount of acetone, add charcoal, warm for a few minutes in a 60°C water bath with occasional mixing. Cool, filter through a Büchner funnel. Cool in ice bath and add water until precipitate forms (preferably keep overnight in refrigerator). Filter off crystals on a Büchner funnel. Wash twice with 30 ml portions of cold water. Recrystallize again but do not use charcoal. Dry crystals in a vacuum desiccator.

Uranyl Acetate, 0.4%

Sodium Borate, 1% in 0.1 N NaOH. Store in polyethylene bottle.

Alcohol-Ether Wash Reagent. 2 vol ethanol (denatured with methanol is satisfactory) + 1 vol ether.

Sodium β-Naphthoquinone-4-Sulfonate. Dissolve 15 mg in 10 ml water. Prepare fresh daily.

Stock Sulfate Standard. 1.09 g K_2SO_4 /liter.

Working Sulfate Standard. 1:10 dilution of stock. 1 ml = $20 \mu g$ S.

Powdered Glass. Pulverize pieces of Pyrex glass in mortar with pestle. Wash with hot dilute HNO_3. Rinse thoroughly with water and dry.

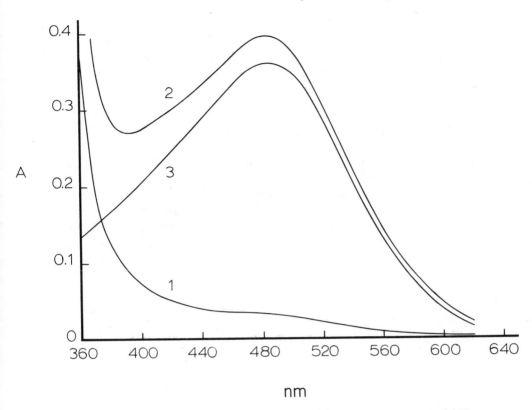

FIG. 19-9. Absorption curves of sulfate determination: (1) reagent blank vs water, (2) 15 μg sulfate (5 μg S) standard vs water, (3) standard vs reagent blank. Beckman model DU spectrophotometer.

PROCEDURE FOR SERUM

1. Add 1.0 ml serum to 3.0 ml 0.4% uranyl acetate, mix, and centrifuge.
2. Set up in 12 ml heavy walled centrifuge tubes:
   Blank: 0.5 ml water + 1.5 ml 0.4% uranyl acetate + 0.5 ml glacial acetic acid + 4.5 ml 1% benzidine + pinch of powdered glass (~5 mg)
   Standard: Same as blank except use 0.5 ml working standard (10 μg S) in place of 0.5 ml water
   Unknown: 2.0 ml supernate + 0.5 ml glacial acetic acid + 4.5 ml 1% benzidine + pinch of powdered glass
3. Stopper tubes and let sit in refrigerator at least 3 hours (overnight is satisfactory).
4. Centrifuge, preferably in a horizontal centrifuge, for 20 min. Siphon off supernate with reverse-tipped capillary. Invert and drain on absorbent tissue (e.g., Kleenex) for 3 min, (*no longer*, lest precipitate dry out and

fall). Wet 2 pieces of tissue with alcohol and carefully wipe mouths of tubes.

5. Rinse inside of tube with 1 ml alcohol without disturbing precipitate. Centrifuge. Siphon off supernate.

6. Add 10 ml alcohol-ether wash reagent. Stir up precipitate and powdered glass with glass stirring rod. Centrifuge and siphon off supernate.

7. Add 0.5 ml alcohol without disturbing precipitate. Centrifuge and siphon off supernate.

8. Add 0.5 ml sodium borate solution. Heat in 60°C water bath for 5–10 min. Cool. Add 5 ml water and 0.5 ml naphthoquinone reagent. Mix and allow to stand exactly 5 min. Add 1 ml acetone and mix. Centrifuge and carefully transfer supernate to test tube. Read colors against water at 485 nm or with filter having a nominal wavelength in this region (a Klett 42 or 50 filter is satisfactory). The color is stable for at least 2 hours.

Calculation:

$$\text{sulfate as mg S/100 ml} = \frac{A_x - A_b}{A_s - A_b} \times 0.01 \times \frac{100}{0.5}$$

$$= \frac{A_x - A_b}{A_s - A_b} \times 2$$

$$\text{mg sulfate/100 mg} = \frac{A_x - A_b}{A_s - A_b} \times 6$$

PROCEDURE FOR URINE

Since the concentration of inorganic sulfate in normal urine is approximately 1 mg sulfate as S per milliliter, the urine is diluted 1:100 with water. The diluted urine is then treated exactly as a serum. The uranyl acetate precipitates the phosphate and any protein which may be present.

Calculation:

$$\text{sulfate as mg S/24 hours} = \frac{A_x - A_b}{A_s - A_b} \times 2 \times 24 \text{ hour volume in ml}$$

$$\text{mg sulfate/24 hours} = \frac{A_x - A_b}{A_s - A_b} \times 6 \times 24 \text{ hour volume in ml}$$

NOTES

1. Beer's law. Beer's law is obeyed up to a concentration of 7.5 mg sulfate per 100 ml (2.5 mg S per 100 ml) with either a spectrophotometer or Corning glass filters with nominal wavelengths of 420, 490, or 540 nm. Above this concentration marked deviation occurs.

2. Stability of samples. It has been reported (313) that serum or urine stored at refrigerator temperature for 24 hours shows a significant increase in sulfate, presumably due to degradation of organic sulfates. In our experience, however, the increase occurring in 4 days at refrigerator or room temperature is too small to negate the values from a clinical standpoint. Frozen samples are more stable (313).

3. Washing the benzidine sulfate precipitate. Benzidine sulfate is somewhat difficult to centrifuge into a compact pellet in the centrifuge tip when acetone or ether-acetone is used as the wash fluid. The use of 80% acetone (445) was found in our laboratory to give very low recoveries. The use of powdered glass (313) is a definite help in successfully packing the precipitate. $BaSO_4$ has also been used to prevent loss of benzidine sulfate (219). Our laboratory also confirmed the claim that better recoveries are obtained with the use of an alcohol-ether wash fluid than with acetone alone as employed by Letonoff and Reinhold (339).

4. pH for precipitation of benzidine sulfate. It has been reported (372) that the optimal pH for precipitation is 3. The pH at precipitation for serum or urine in the method as presented is about 4.5. No significant differences in results were observed in our laboratory between precipitation at pH 4.5 or after acidification to pH 3.

5. Standards. In the benzidine method as it is usually presented, benzidine hydrochloride is run as the standard rather than sulfate itself. In the present authors' opinion, this is not advisable since all the steps in the procedure prior to color development are thus bypassed. When first setting up the test, however, it is useful to run a standard curve with benzidine to determine the recoveries obtained with sulfate standards. Place 0.1606 g benzidine hydrochloride in a 200 ml volumetric flask and add about 50 ml water. Dissolve by warming to 50°C. Cool and add water to mark. This stock standard should be stored in the refrigerator. 1 ml of a 1:10 dilution of this standard contains benzidine equivalent to 10 µg S. The benzidine standard curve is begun at step 8 under the procedure, substituting the benzidine standard for the water added, always keeping the total volume of standard and water at 5 ml. A reagent blank is also set up beginning at this step.

6. Possible carcinogenicity of benzidine. As discussed in Chapter 23 *(Occult Blood, Choice of Methods)*, there is a remote possibility that benzidine is carcinogenic, so that its use probably should be avoided when possible. Fortunately, the sulfate procedure is seldom requested.

ACCURACY AND PRECISION

It has been claimed that up to 50% of sulfate added to serum becomes bound and nonprecipitable by benzidine (339). This has been contested (313), and recoveries of 98%–100% were obtained in our laboratory with the technic as presented. The benzidine method is subject to error from occlusion of the excess benzidine and from incomplete precipitation of benzidine sulfate if the time of precipitation is too short, the concentration of sulfate or benzidine is too low, or the pH at precipitation is too far from the optimum of about pH 3 (372). The technic as presented, however, appears to minimize these errors.

The reproducibility of the technic is about ±5%.

NORMAL VALUES

There is fairly good agreement that the upper limit of normal for serum inorganic sulfate is about 3.6 mg/100 ml (1.2 mg S/100 ml). The lower limit has been reported to lie between 1.4 and 2.4 mg sulfate/100 ml (0.47–0.8 mg S/100 ml) (313, 339, 392, 466). There is no sex difference (381). There is very little sulfate in erythrocytes (119).

The "total sulfate" in serum has been reported to range from about 2.5 to 4.2 mg/100 ml (0.8–1.4 mg S/100 ml) (313, 585). The normal range reported for the conjugated or ethereal sulfate is not well defined, varying from 0 to 3.0 mg sulfate/100 ml (339, 451, 585). Since this fraction is determined by difference between inorganic and total sulfate, it is understandable that the precision of its measurement is usually poor.

Inorganic sulfate in spinal fluid has been reported to range from 1.5 to 3.6 mg/100 ml (0.5–1.2 mg S/100 ml) (466).

The 24 hour output of inorganic sulfate in urine is about 1.5–3.2 g (0.5–1.1 g as S). The total sulfate range is 1.8–4.1 g (0.6–1.4 g as S), the ethereal sulfate range is 0.18–0.36 g (0.06–0.12 g as S), and the neutral S—i.e., S of amino acids, thiosulfates, thiocyanates, taurocholic acid, urochrome—range is 0.18–0.54 g as sulfate (0.06–0.18 g as S) (227, 418). The outputs vary directly with protein intake (227).

OSMOLALITY

Determination of serum and urine osmolality is of considerable value in the understanding and clinical management of water and electrolyte disturbances (590). Colligative properties of a solution—i.e., those which depend upon the number of dissolved particles rather than upon their kind—are the freezing point, boiling point, vapor pressure, and osmotic pressure. The osmolality of a biologic fluid can be determined by measuring either freezing point, vapor pressure, or osmotic pressure. Freezing point measurement has been chosen for clinical work because of its simplicity.

The addition of 1 mol of nonelectrolyte to 1 kg of water (molal solution) depresses the freezing point by 1.86°C. This number has been called the *molal freezing point (depression) constant*. Electrolytes in solution dissociate to yield more particles per mol than nonelectrolytes, changing the colligative properties correspondingly more per mol than nondissociated solutes. The additive osmotic effect of individual particles is inherent in the fundamental equation for *osmolality* (*O*):

$$O = \frac{\Delta}{1.86}$$

where Δ = freezing point depression in °C. If a salt such as NaCl were completely dissociated, a 1.00 molal solution would have a Δ = 3.72° and O = 3.72/1.86 = 2.00. However, a 1 molal solution of NaCl actually freezes at −3.38°C, and its osmolality is 1.82 Osm/kg. The discrepancy between actual and theoretic freezing point depressions, Δ_a and Δ_t, is described by the *osmotic* coefficient, $\emptyset$, as follows:

$$\emptyset = \frac{\Delta_a}{\Delta_t}$$

A compound in solution can be assigned an appropriate number, *n*, such that ideal behavior is characterized by $n = 1$ for nonelectrolytes, $n = 2$ for univalent salts such as NaCl, $n = 3$ for $CaCl_2$, etc. Then, it can be shown (619) that

$$O = \frac{\Delta_a}{1.86} = \emptyset nc$$

where *c* = molality.

The *osmotic coefficient* ($\emptyset$) of NaCl is a function of concentration, and varies between 0.91 and 0.95 for the range of concentrations (0.05–1.1 molal) pertinent to clinical work (281). A 0.147 molal NaCl solution has a $\Delta_a = 0.51°$ and a $\emptyset$ of 0.93:

$$O = \emptyset nc = 0.93 \times 2 \times 0.147 = 0.273 \text{ Osm}$$

or 273 mOsm/kg water. A 0.273 molal solution of an ideal nonelectrolyte would have the same osmolality.

The freezing point of a solution may be defined as that temperature at which an infinitesimal amount of solid phase exists in equilibrium with the liquid phase at standard pressure (2). Water in biologic fluids does not start crystallizing at the freezing point, and it is necessary to supercool as well as to initiate crystallization by some means such as a vibrating rod. During crystallization, which is almost instantaneous once started, heat is liberated and the temperature rises until the freezing point of the now slightly more concentrated solution is reached (about 3% of the total water freezes for each 2°C of supercooling) (281). The different methods of correcting for the supercooling error and the rationale for the technic used in modern osmometry are discussed by Abele (2).

DETERMINATION OF OSMOLALITY OF SERUM AND URINE
(method of Johnson and Hock, Ref 281)

PRINCIPLE

The osmolality of serum or urine is determined by comparing its freezing point with that of a NaCl solution of known osmolality. The solution of super-cooled and crystallization initiated by a vibrating rod. During crystallization, heat is liberated, and the temperature rises until the freezing point is reached.

REAGENTS

NaCl Standards. Dry approximately 30 g reagent grade NaCl overnight between 200° and 500°C. Boil 3.5 liters of double distilled water in a 4 liter flask and then cool to room temperature. Fill a 1 liter volumetric flask to within 1 cm of the graduation mark with water and adjust the temperature to 20°C. Fill to the mark and add 1.8 ml more water before adding salt. This procedure avoids the necessity of calibrating a flask to contain 1 kg of water. Prepare the following standards:
NaCl standard, 100 mOsm/kg: 1 kg water + 3.094 g NaCl
NaCl standard, 300 mOsm/kg: 1 kg water + 9.476 g NaCl
NaCl standard, 500 mOsm/kg: 1 kg water + 15.93 g NaCl
Transfer the solutions to clean polyethylene bottles. Test the bottles for tight closure by inverting and applying pressure. Dissolved atmospheric gases alter the freezing points of standards. Pour sufficient standards for daily use into clean tubes (never pipet from the bottles).

PROCEDURE

Several commercial instruments are available which employ thermistor probes in a null meter circuit. The instruments consist of equipment to hold, cool, stir, and freeze a sample, and an electric thermometer (thermistor) to measure freezing temperature (2).

Most instruments are adapted to either 2 ml samples or micro samples, e.g., 0.25 ml. The specific instrument manual should be referred to for operational details.

NOTES

1. Instrument calibration. Since the normal range of urine osmolalities is broad, ranging up to 1200 mOsm, no predetermined setting is used for urine. Directions for preparing higher standards needed for concentrated urines have been reported (281). By following calibration instructions it is possible to get linear readings from 0 to 1000 mOsm.

2. Stability of samples. Serum osmolality has been reported to remain unchanged for 3 hours at room temperature and for 10 hours at 4°C (281). Experiments in our laboratory showed that serum specimens are stable for 3 days at 4°C and 30°C. Urine specimens are stable for 24 hours at 4°C or 30°C. It would seem best to freeze specimens of both serum and urine unless the determination is carried out within the time limits cited above. Frozen serum and urine specimens are stable for 1 week.

3. Interferences. Samples should be centrifuged to remove particulate matter which could start crystallization before the correct degree of supercooling is attained.

ACCURACY AND PRECISION

The accuracy of measurement of osmolality based on freezing point depression is ±1%.

Precision (95% limits) of the method is about ±1% at a 300 mOsm level.

NORMAL VALUES

Serum. The normal range for serum determined in our laboratory on 63 males and 70 females is 278–305 mOsm/kg serum water. An almost identical range was reported by others (590).

Urine. The total osmolal urinary output varies greatly with diet. The normal range established in our laboratory is 50–1200 mOsm/kg urine water with no significant sex difference. Separate normal ranges for males (767–1628 mOsm/24 hours) and females (433–1146 mOsm/24 hours) have been reported (590). After an overnight fluid fast, a normal person should concentrate his urine to 800–900 mOsm/kg urine water (590).

Other Fluids. Hendry (238) measured a series of body fluids and found that "true body fluids" had the same osmolality as a corresponding serum sample taken at the same time. "True body fluids" include: ascitic, cerebrospinal, hydrocele, edema, pericardial, pleural, spermatocele, and synovial fluids. Secretions not in osmotic equilibrium with the extracellular fluids of the body include gastric juice, saliva, and sweat.

REFERENCES

1. AASA R, MALMSTROM BG, SALTMAN P, VANNGARD T: *Biochim Biophys Acta 75*:203, 1963
2. ABELE JE: *Am J Medical Electronics 2*:32, 1963
3. ADELSTEIN SJ, COOMBS TL, VALLEE BL: *New Engl J Med 255*:105, 1956
4. AIKAWA JK, RHOADES EL: *Am J Clin Pathol 31*:314, 1959

5. AISEN P, AASA R, MALMSTROM BG, VANNGARD T: *J Biol Chem* 242:2484, 1967

6. ALBANESE AA, WAGNER DL: *J Lab Clin Med* 30:280, 1945

7. ALBRIGHT F, REIFENSTEIN EC Jr: *The Parathyroid Glands and Metabolic Bone Disease*. Baltimore, Williams & Wilkins, 1948, p 75

8. ALBRIGHT F, REIFENSTEIN EC Jr: *The Parathyroid Glands and Metabolic Bone Disease*. Baltimore, Williams & Wilkins, 1948, p 302

9. ALCOCK NW: *Ann NY Acad Sci* 162:707, 1969

10. ALCOCK NW, MacINTYRE I: *Methods Biochem Analy* 14:1, 1966

11. ALLEN RJL: *Biochem J* 34:858, 1940

12. ANAST CS: *Am J Diseases Children* 98:661, 1959

13. ANAST CS: *Pediat* 33:969, 1964

14. ANDERSCH MA: *J Lab Clin Med* 49:486, 1957

15. ANDERSCH MA, OBERST FW: *J Clin Invest* 15:131, 1936

16. ANDERSON CM, FREEMAN M: *Arch Disease Childhood* 35:581, 1960

17. ANDREASEN E: *Scand J Clin Lab Invest* 9:138, 1957

18. ANNINO JS, RELMAN AS: *Am J Clin Pathol* 31:155, 1959

19. APPLETON HD, WEST M, MANDEL M, SALA AM: *Clin Chem* 5:36, 1959

20. APRISON MH, HANSON KM: *Proc Soc Exp Biol Med* 100:643, 1959

21. ARENDS T, GALLANGO ML, PARKER CW, BEARN AG: *Nature* 196:477, 1962

22. BAAR S: *Clin Chim Acta* 2:567, 1958

23. BAAR S: *Clin Chim Acta* 7:642, 1962

24. BABSON AL, KLEINMAN NM: *Clin Chem* 13:163, 1967

25. BABSON AL, KLEINMAN NM: *Clin Chem* 13:814, 1967

26. BACHRA BN, DAUER A, SOBEL AE: *Clin Chem* 4:107, 1958

27. BAGINSKI ES, FOA PP, ZAK B: *Clin Chim Acta* 15:155, 1967

28. BAGINSKI ES, WILLIAMS LA, JARKOWSKI TL, ZAK B: *Am J Clin Pathol* 30:559, 1958

29. BAJPAI PC, SUGDEN D, RAMOS A, STERN L: *Arch Disease Childhood* 41:424, 1966

30. BAKER GL, JOHNSON LH: *Anal Chem* 26:465, 1954

31. BARKER ES, ELKINGTON JR, CLARK J: *J Clin Invest* 38:1733, 1959

32. BARNARD AJ Jr, BROAD WC, FLASCHKA H: *Chemist-Analyst* 46:46, 1957

33. BARNEY JD, SULKOWITCH HW: *J Urol* 37:746, 1937

34. BARON DN, BELL J: *J Clin Pathol* 10:280, 1957

35. BARON DN, BELL JL: *J Clin Pathol* 12:143, 1959

36. BARTELS H: *Atomic Absorption Newsletter* 6:132, 1967

37. BASINSKI DH: *Standard Methods in Clinical Chemistry*, edited by Meites S. New York, Academic Press, 1965, Vol 5, p 137

38. BASTIAANSE AJ, MEIJERS CA: *Z Klin Chem Klin Biochem* 6:48, 1968

39. BASTOS de JORGE F, CANELAS HM, ZANINI AC: *Rev Paulista Med* 65:95, 1964; *CA* 62:5647h, 1965

40. BATSAKIS JG, ORSINI FM, STILES D, BRIERE RO: *Am J Clin Pathol* 42:541, 1964

41. BAUMANN EJ: *J Biol Chem* 59:667, 1924

42. BEALE RN, BOSTROM JO, TAYLOR RF: *J Clin Path* 15:156, 1962

43. BEALE RN, CROFT D: *J Clin Pathol* 17:260, 1964

44. BEARN AG, KUNKEL HG: *Proc Soc Exp Biol Med* 85:44, 1954

45. BECHTLER G, GUTSCHE B, HERRMANN R, LANGE W, STAMM D: *Z Klin Chem Klin Biochem* 5:138, 1967

46. BECKA S: *Biochem Z* 233:118, 1931

47. BEHRENDT H: *Am J Diseases Children 64*:55, 1942
48. BELL RD, DOISY EA: *J Biol Chem 44*:55, 1920
49. BELLINGER JF. CAMPBELL RA: *Clin Chem 12*:90, 1966
50. BERENBLUM I, CHAIN E: *Biochem J 32*:295, 1938
51. BERENSON GS, BURCH GE: *J Lab Clin Med 42*:58, 1953
52. BERGER EY: *Clin Chem 1*:249, 1955
53. BERGLUND F, SØRBO B: *Scand J Clin Lab Invest 12*:147, 1960
54. BERLINER RW, KENNEDY RJ Jr, HILTON JG: *Am J Physiol 162*:348, 1950
55. BERRY JW, CHAPPELL DG, BARNES RB: *Anal Chem 18*:19, 1946
56. BETT IM, FRASER GP: *Clin Chim Acta 4*:346, 1959
57. BINNERTS WT, ACHTERPO T: *Tijdschr Diergeneesk 92*:639, 1967; *CA 68*:84871, 1968
58. BIRDSALL NJM, KOK D'A, WILD F: *J Clin Pathol 18*:453, 1965
59. BIZOLLON ChA: *Ann Pharm Franc 23*:461, 1965; *CA 64*:7032b, 1966
60. BLOMFIELD J, MacMAHON RA: *J Clin Pathol 22*:136, 1969
61. BOLING EA: *J Lab Clin Med 63*:501, 1964
62. BOMSKOV C: *Hoppe-Seylers Z Physiol Chem 210*:67, 1932
63. BOSSCHE HV, WIEME RJ: *Clin Chim Acta 14*:112, 1966
64. BOTHWELL TH, MALLETT B: *Biochem J 59*:599, 1955
65. BOUDA J: *Clin Chim Acta 23*:511, 1969
66. BOWEN HJM, CAWSE PA, DAGLISH M: *Analyst 89*:266, 1964
67. BOWEN WJ, MARTIN HL: *Proc Soc Exp Biol Med 101*:734, 1959
68. BOWIE EJW, TAUXE N, SJOBERG WE Jr, YAMAGUCHI MY: *Am J Clin Pathol 40*:491, 1963
69. BOYETT JD, SULLIVAN JF: *Metabolism 19*:148, 1970
70. BREEN M, FREEMAN S: *Clin Chim Acta 6*:181, 1961
71. BRIGGS AP: *J Biol Chem 59*:255, 1924
72. BRISCOE AM, RAGAN C: *J Lab Clin Med 69*:351, 1967
73. BROZOVICH B: *J Clin Pathol 21*:183, 1968
74. BROZOVICH B, COPESTAKE J: *J Clin Pathol 22*:605, 1969
75. BUCKLEY ES Jr, GIBSON JG II, BORTOLOTTI TR: *J Lab Clin Med 38*:751, 1951
76. BURCAR PJ, BOYLE AJ, MOSHER RE: *Clin Chem 10*:1028, 1964
77. BURR RG: *Clin Chem 15*:1191, 1969
78. BURRILL MW, FREEMAN S, IVY AC: *J Biol Chem 157*:297, 1945
79. BURROWS S: *Am J Clin Pathol 47*:326, 1967
80. BUTLER EJ, FIELD AC: *Analyst 81*:615, 1956
81. BUTLER EJ, NEWMAN GE: *J Clin Pathol 9*:157, 1956
81a. CALI JP, MANDEL J, MOORE L, YOUNG DS: NBS Spec Publ 260-36, 1972
82. CARAWAY WT: *Clin Chem 9*:188, 1963
83. CARD RT, BROWN GM, VALBERG LS: *Can Med Assoc J 90*:618, 1964
84. CARNES WH: *Federation Proc 30*:995, 1971
85. CARR MH, FRANK HA: *Am J Clin Pathol 26*:1157, 1956
86. CARRERA AE: *Am J Clin Pathol 49*:739, 1968
87. CARRICO RJ, DEUTSCH HF: *J Biol Chem 244*:6087, 1969
88. CARRUTHERS BM, COPP DH, McINTOSH HW: *J Lab Clin Med 63*:959, 1964
89. CARTER P: *Anal Biochem 40*:450, 1971
90. CARTWRIGHT GE, GUBLER CJ, WINTROBE MM: *J Clin Invest 33*:685, 1954
91. CARTWRIGHT GE, MARKOWITZ H, SHIELDS GS, WINTROBE MM: *Am J Med 28*:555, 1960
92. CARTWRIGHT GE, WINTROBE MM: *Am J Clin Nutr 14*:224, 1964

93. CARUBELLI R, SMITH WO, HAMMARSTEN JF: *J Lab Clin Med 51*:964, 1958
94. CASE FH: *A Review of Syntheses of Organic Compounds Containing the Ferröin Group.* Columbus, Ohio, G. Frederick Smith Chem. Co., 1960
95. CATLEDGE G, BIGGS HG: *Clin Chem 11*:521, 1965
96. CHAMPION KP, WHITTEM RN: *Analyst 92*:112, 1967
97. CHANIN M: *Science 119*:323, 1954
98. CHEN PS Jr, TORIBARA TY: *Anal Chem 25*:1642, 1953
99. CHEN PS Jr, TORIBARA TY: *Anal Chem 26*:1967, 1954
100. CHEN T, DOTTI LB: *Am J Clin Pathol 48*:136, 1967
101. CHEN T, DOTTI LB: *Clin Chim Acta 18*:453, 1967
102. CHIAMORI N, HENRY RJ: *Proc Soc Exp Biol Med 97*:817, 1958
103. CHILCOTE ME, WASSON RD: *Clin Chem 4*:200, 1958
104. CHRISTIAN DG: *Am J Clin Pathol 54*:118, 1970
105. CLARK EP, COLLIP JB: *J Biol Chem 63*:461, 1925
106. COLTMAN CA, ROWE NJ: *Am J Clin Nutr 18*:270, 1966
107. COMAR GL, BRONNER F: *Mineral Metabolism.* New York, Academic Press, 1964, Vol 2, Part A, p 368
108. CONNERTY HV, BRIGGS AR: *Am J Clin Pathol 45*:290, 1966
109. COOK BS, SIMMONS DH: *J Lab Clin Med 60*:160, 1962
110. COOK JD: *J Lab Clin Med 76*:497, 1970
111. COTLOVE E: *Standard Methods of Clinical Chemistry*, edited by Seligson D. New York, Academic Press, 1961, Vol 3, p 81
112. COTLOVE E: *Anal Chem 35*:95, 1963
113. COTLOVE E: *Anal Chem 35*:101, 1963
114. COTZIAS GC, MILLER ST, EDWARDS J: *J Lab Clin Med 67*:836, 1966
115. COX DW: *J Lab Clin Med 6*:893, 1966
116. CRAIG P, ZAK B, ISERI LT, BOYLE AJ, MYERS GB: *Am J Clin Path 21*:394, 1951
117. CROCKSON RA: *J Clin Path 16*:473, 1963
118. CUMMINGS NA, KUFF EL, SOBER HA: *Anal Biochem 22*:108, 1968
119. CUTHBERTSON DP, TOMPSETT SL: *Biochem J 25*:1237, 1931
120. DAHMS H, ROCK R, SELIGSON D: *Clin Chem 14*:859, 1968
121. DALY JA, ERTINGSHAUSEN G: *Clin Chem 18*:263, 1972
122. DANOWSKI TS: *J Biol Chem 139*:693, 1941
123. DAVIES IJT, MUSA M, DORMANDY TL: *J Clin Pathol 21*:359, 1968
124. DAVIS JB, LINDSTROM F: *Anal Chem 44*:524, 1972
125. DAVIS LR, SMITH MJH: *J Clin Pathol 6*:198, 1953
126. DAWSON JB, ELLIS DJ, NEWTON-JOHN H: *Clin Chim Acta 21*:33, 1968
127. DAWSON JB, HEATON FW: *Biochem J 80*:99, 1961
128. DAWSON JB, WALKER BE: *Clin Chim Acta 26*:465, 1969
129. DEAN RB, FISHMAN MM: *J Biol Chem 140*:807, 1941
130. DECKER CF, ARAS A, DECKER LE: *Anal Biochem 8*:344, 1964
131. de JORGE FB, CANELAS HM, DIAS JC, CURY L: *Clin Chim Acta 9*:148, 1964
132. DEMPSEY EF, CARROLL EL, ALBRIGHT F, HENNEMAN PH: *Metab Clin Exptl 7*:108, 1958
133. DEUTSCH HF: *Clin Chim Acta 5*:460, 1960
134. DIEHL H, ELLINGBOE J: *Anal Chem 32*:1120, 1960
135. DIEHL H, OLSEN R, SPIELHOLTZ GI, JENSEN R: *Anal Chem 35*:1144, 1963
136. DIEHL H, SMITH GF: *The Iron Reagents*, edited by McBride L, Cryberg R, Columbus, Ohio, G. Frederick Smith Chem. Co., 1965
137. DIEU JP: *Clin Chem 17*:1183, 1971
138. di SANT'AGNESE PA: *J Am Med Assoc 179*:388, 1962

139. di SANT'AGNESE PA, DARLING RC, PERERA GA, SHEA E: *Pediatrics 12*:549, 1953

140. DORCHE J, NYSSEN M: *Clin Chem 13*:813, 1967

141. DREISBACH RH: *Anal Biochem 10*:169, 1965

142. DREUX C, BOUCHET R, GIRARD ML: *Ann Biol Clin (Paris) 29*:251, 1971

143. DREUX C, GIRARD M: *Ann Biol Clin (Paris) 19*:627, 1961

144. DREWES PA: *Clin Chim Acta 39*:81, 1972

145. DRISCOLL JL, MARTIN HF: *Clin Chem 12*:314, 1966

146. DRYER RL, TAMMES AR, ROUTH JI: *J Biol Chem 225*:177, 1957

147. DUBINSKAYA NA, MIHELSONS HH, PELEKIS L: *Neitronoaktiv Anal Akad Nauk Latv SSR, Inst Fiz* 1966, p 109 (Russian); *CA Biochem Sect 68*:2646, (27471v) 1968

148. ECKMAN I, ROBBINS JB, VAN DEN HAMER CJA, LENTZ J, SCHEINBERG IH: *Clin Chem 16*:558, 1970

149. EDEN A, GREEN HH: *Biochem J 34*:1202, 1940

150. EGGLETON GP, EGGLETON P: *J Physiol 68*:193, 1929

151. EICHHOLTZ F, BERG R: *Biochem Z 225*:352, 1930

152. ELDJARN L, NYGAARD O, SVEINSON SL: *Scand J Clin Lab Invest 7*:92, 1955

153. ELKINGTON JR, DANOWSKI TS: *The Body Fluids.* Baltimore, Williams & Wilkins, 1955, p 120

154. ELLIOTT HC Jr, HOLLEY HL: *Am J Clin Pathol 21*:831, 1951

155. ETTORI J, SCOGGAN SM: *Clin Chim Acta 6*:861, 1961

156. EVANS RA, WATSON L: *Lancet 1*:522, 1966

157. FABER SJ, PELLEGRINO ED, CONAN NJ, EARLE DP: *Am J Med Sci 221*:678, 1951

158. FARESE G, MAGER M, BLATT WF: *Clin Chem 16*:226, 1970

159. FELDMAN MH, REVA RC, BATTISTONE GC: *J Nucl Med 7*:548, 1966

160. FELL GS, SMITH H, HOWIE RA: *J Clin Pathol 21*:8, 1968

161. FERNANDEZ AA, SOBEL C, JACOBS SL: *Anal Chem 35*:1721, 1963

162. FERNANDEZ FJ, KAHN HL: *Clinical Chemistry Newsletter.* Norwalk, Conn., Perkin-Elmer Corporation, Fall 1971, Vol 3, No. 2

163. FERRO PV, HAM AB: *Am J Clin Pathol 28*:689, 1957

164. FINGERHUT B, MARSH WH: *Clin Chem 9*:204, 1963

165. FINGERHUT B, POOCK A, MILLER H: *Clin Chem 15*:870, 1969

166. FISCHER DS, PRICE DC: *Clin Chem 10*:21, 1964

167. FISKE CH, SUBBAROW Y: *J Biol Chem 66*:375, 1925

168. FISKE CH, SUBBAROW Y: *J Biol Chem 81*:629, 1929

169. FOSS OP: *Scand J Clin Lab Invest 2*:211, 1959

170. FRAENKEL-CONRAT H: *Arch Biochem 28*:452, 1950

171. FRANK HA, CARR MH: *J Lab Clin Med 49*:246, 1957

172. FRANT MS, ROSS JW Jr: *Science 167*:987, 1969

173. FRICK E: *Klin Wochschr 41*:75, 1963

174. FRIEDMAN HS, RUBIN MA: *Clin Chem 1*:125, 1955

175. FRINGS CS, COHEN PS, FOSTER LB: *Clin Chem 16*:816, 1970

176. FRIZEL DE, MALLESON AG, MARKS V: *Clin Chim Acta 16*:45, 1967

177. FUHR J, STARY E: *Artzl Lab 16*:244, 1970

178. FUNDER J, WIETH JO: *Scand J Clin Lab Invest 18*:167, 1966

179. FUWA K, PULIDO P, McKAY R, VALLEE BL: *Anal Chem 36*:2407, 1964

180. GARNER RJ: *Biochem J 40*:828, 1946

181. GILLAM WS: *Anal Chem 13*:499, 1941

182. GINDLER EM: *Clin Chem 14*:1172, 1968

183. GINSBERG H: *Z Elektrochem 45*:829, 1940

184. GIORGIO AJ: *Med Clin N Am 54*:1399, 1970
185. GIORGIO AJ, CARTWRIGHT GE, WINTROBE MM: *Am J Clin Pathol 41*:22, 1964
186. GIOVANNIELLO TJ, DiBENEDETTO G, PALMER DW, PETERS T Jr: *J Lab Clin Med 71*:874, 1968
187. GIOVANNIELLO TJ, PETERS T Jr: *Standard Methods of Clinical Chemistry*, edited by Seligson D. New York, Academic Press, 1963, Vol 4, p 139
188. GIRARD ML: *Clin Chim Acta 20*:243, 1968
189. GIRARD ML, ROUSSELET F, DESCUBE J: *Compt Rend, Ser D 262*:2380, 1966; *CA 65*:7712h, 1966
190. GIRAUDET P, PRE J, CORNILLOT P: *Clin Chim Acta 28*:323, 1970
191. GITELMAN HJ: *Anal Biochem 18*:521, 1967
192. GITELMAN HJ, HURT C, LUTWAK L: *Anal Biochem 14*:106, 1966
193. GLADTKE E, RIND H: *Klin Wochschr 44*:88, 1966
194. GLEMSER O, DAUTZENBERG W: *Z Anal Chem 136*:254, 1952
195. GLENN WG, SHANNON IL: *Aerosp Med 37*:1008, 1966
196. GLICK MR: *Amer J Med Technol 33*:120, 1967
197. GOCHMAN N, GIVELBER H: *Clin Chem 16*:229, 1970
198. GOLDENBERG H, FERNANDEZ A: *Clin Chem 12*:871, 1966
199. GOLDSTEIN M: *The Biochemistry of Copper*, edited by Peisach J, Aisen P, Blumberg WE. New York, Academic Press, 1966, p 443
200. GOODMAN JR, VINCENT J, ROSEN I: *Am J Clin Pathol 24*:111, 1954
201. GOODMAN M, NEWMAN HA, RAMSEY DS: *J Lab Clin Med 51*:814, 1958
202. GOODWIN JF: *Clin Chem 16*:776, 1970
203. GOODWIN JF, MURPHY B: *Clin Chem 12*:58, 1966
204. GOODWIN JF, MURPHY B, GUILLMETTE M: *Clin Chem 12*:47, 1966
205. GOODWIN JF, THIBERT R, McCANN D, BOYLE AJ: *Anal Chem 30*:1097, 1958
206. GREENBERG BG, WINTERS RW, GRAHAM JB: *J Clin Endocrin Metab 20*:364, 1960
207. GREENBERG DM, LUCIA SP, MACKEY MA, TUFTS EV: *J Biol Chem 100*:139, 1933
208. GREENBERG DM, MACKEY MA: *J Biol Chem 96*:419, 1932
209. GREENWALD I: *J Biol Chem 93*:551, 1931
210. GRETTE K: *Scand J Clin Lab Invest 5*:151, 1953
211. GUBLER CJ, LAHEY ME, ASHENBRUCKER H, CARTWRIGHT GE, WINTROBE MM: *J Biol Chem 196*:209, 1952
212. GUEST GM, RAPAPORT S: *Physiol Rev 21*:410, 1941
213. GUILLAUMIN CO: *Bull Soc Chim Biol 22*:564, 1940
214. GUYON JC, LORAH EJ: *Anal Chem 38*:155, 1966
215. GYLLENSWARD C, JOSEPHSON B: *Scand J Clin Lab Invest 9*:21, 1957
216. HACKLEY BM, SMITH JC, HALSTED JA: *Clin Chem 14*:1, 1968
217. HAGBERG B: *Acta Paediat Scand 42 (Suppl 93)*, 1953
218. HAHN KJ, TUMA DJ, QUAIFE MA: *Anal Chem 39*:1169, 1967
219. HÄKKINEN IPT: *Nature 186*:232, 1960
220. HALD PM: *J Biol Chem 167*:499, 1947
221. HALL RA, WHITEHEAD TP: *J Clin Pathol 23*:323, 1970
222. HALL RJ, FLYNN LR: *Nature 208*:1202, 1965
223. HALSTED JA, SMITH JC Jr: *Lancet 1*:322, 1970
224. HAMILTON RH: *Clin Chem 12*:1, 1966
225. HANSEN JL, FREIER EF: *Am J Med Technol 33*:158, 1967
226. HARRISON DG, LONG C: *J Physiol 199*:367, 1968
227. HARROW B: *Textbook of Biochemistry*, 4th edition. Philadelphia, Saunders, 1946

228. HARTZ JW, DEUTSCH HF: *J Biol Chem 244*:4565, 1969
229. HARVEY DR, COOPER LV, STEVENS JF: *Arch Disease Childhood 45*:506, 1970
230. HATTNER RS, JOHNSON JW, BERNSTEIN DS, WACHMAN A, BRACKMAN J: *Clin Chim Acta 28*:67, 1970
231. HAUSMANN K, KUSE R, MEINECKE KH, BARTELS H, HEINRICH HC: *Klin Wochschr 49*:1164, 1971
232. HAVEN MC, HAVEN GT, DUNN AL: *Anal Chem 38*:141, 1966
233. HEAGY FC: *Can J Res Sect E 26*:295, 1948
234. HEATON FW: *J Clin Pathol, 13*:358, 1960
235. HELLWEGE HH, SCHMALFUSS H, GOSCHENHOFER D: *Z Klin Chem Klin Biochem 7*:56, 1969
236. HELVE O: *Acta Med Scand 125*:505, 1946
237. HELWIG HL, HOFFER EM, THIELEN WC, ALCOCER AE, HOTELLING DR, ROGERS WH: *Am J Clin Pathol 45*:160, 1966
238. HENDRY EB: *Clin Chem 8*:246, 1962
239. HENROTTE JG, KRISHNARAJ PS: *Nature 195*:184, 1962
240. HENRY RJ, BERKMAN S: *Clin Chem 3*:711, 1957
241. HENRY RJ, SOBEL C, CHIAMORI N: *Clin Chim Acta 3*:523, 1958
242. HERBERT V, GOTTLIEB CW, LAU K-S, FISHER M, GEVIRTZ NR, WASSERMAN LR: *J Lab Clin Med 67*:855, 1966
243. HERRING WB, LEAVELL BS, PAIXAO LM, YOE JH: *Am J Clin Nutr 8*:846, 1960
244. HILAI H, McCURDY PR: *Ann Internal Med 66*:983, 1967
245. HILDEBRAND GP, REILLEY CN: *Anal Chem 29*:258, 1957
246. HIRSCHFELDER AD, SERLES ER: *J Biol Chem 104*:635, 1934
247. HODGE CJ Jr, GERARDE HW: *Microchem J 7*:326, 1963
248. HOFFMAN WS: *J Biol Chem 118*:37, 1937
249. HOFMANN F, BUCHNER M: *Arztl Lab 9*:392, 1963
250. HOLASEK A, FLASCHKA H: *Z Physiol Chem 290*:57, 1952
251. HOLMBERG CG: *Acta Physiol Scand 2*:71, 1941
252. HOLMBERG CG, LAURELL CB: *Acta Physiol Scand 10*:307, 1945
253. HOLMBERG CG, LAURELL CB: *Acta Chem Scand 1*:944, 1947
254. HOLMBERG CG, LAURELL CB: *Acta Chem Scand 2*:550, 1948
255. HOLMBERG CG, LAURELL CB: *Acta Chem Scand 5*:476, 1951
256. HORIGUCHI K, HORIGUCHI S, TANAKA N, SHINAGAWA K, HAMAGUCHI T: *Osaka City Med J 12*:151, 1967; *CA 70*:11294q, 1969
257. HORWITT BN: *J Biol Chem 199*:537, 1952
258. HOSAIN F, FINCH CA: *J Lab Clin Med 64*:905, 1964
259. HOUTSMULLER AJ: *Clin Chim Acta 4*:606, 1959
260. HOWELL DS, PITA JC, MARQUEZ JF: *Anal Chem 38*:434, 1966
261. HUBBARD DM, SPETTEL EC: *Anal Chem 25*:1245, 1953
262. HUBBARD RS: *J Biol Chem 74*:v, 1927
263. HÜBENER HJ: *Hoppe-Seylers Z Physiol Chem 289*:188, 1952
264. HUMOLLER FL, WALSH JR: *J Lab Clin Med 48*:127, 1956
265. HUNT BJ: *Clin Chem 15*:979, 1969
266. HUNTER G: *Analyst 84*:24, 1959
267. HUNTER G, SMITH HV: *Nature 186*:161, 1960
268. HUSDAN H, RAPOPORT A: *Clin Chem 15*:669, 1969
269. IBBOTT FA: *Standard Methods of Clinical Chemistry*, edited by Meites, S. New York, Academic Press, 1965, Vol 5, p 101
270. IBBOTT FA: *Handbook of Clinical Laboratory Data*, 2nd edition. Cleveland, Ohio, Chemical Rubber Co., 1968, p 342

271. ICHIDA T, NOBUOKA M: *Clin Chim Acta 24*:299,1969
272. IIDA C, FUWA K, WACKER WEC: *Anal Biochem 18*:18, 1967
273. IRVING EA, WATTS PS: *Biochem J 79*:429, 1961
274. ISAACS ML: *J Biol Chem 53*:17, 1922
275. ITAYA K, UI M: *Clin Chim Acta 14*:361, 1966
276. JACKSON JE, BREEN M, CHEN C: *J Lab Clin Med 60*:700, 1962
277. JAGER B: *J Clin Invest 28*:792, 1949
278. JAMES BE, MACMAHON RA: *Clin Chim Acta 32*:307, 1971
279. JAMIESON GA: *J Biol Chem 240*:2019, 1965
280. JAMIESON GA: *J Biol Chem 240*:2914, 1965
281. JOHNSON RB Jr, HOCH H: *Standard Methods of Clinical Chemistry*, edited by Meites S. New York, Academic Press, 1965, Vol 5, p 159
282. JOSEPH JB, BERGSTROM WH: *Pediatrics 27*:597, 1961
283. JOSEPHSON B, DAHLBERG G: *Scand J Clin Lab Invest 4*:216, 1952
284. KAHN AM, HELWIG HL, REDEKER AG, REYNOLDS TB: *Am J Clin Pathol 44*:426, 1965
285. KAHN BS, LEIBOFF SL: *J Biol Chem 80*:623, 1928
286. KAHN M, POSTMONTIER RS: *J Lab Clin Med 10*:317, 1925
287. KALDOR I: *Austral J Exp Biol 31*:41, 1953
288. KANABROCKI EL, CASE LF, FIELDS T, GRAHAM L, MILLER EB, OESTER YT, KAPLAN E: *J Nucl Med 6*:780, 1965
289. KANABROCKI EL, FIELDS T, DECKER CF, CASE LF, MILLER EB, KAPLAN E, OESTER YT: *Intern J Appl Radiation Isotopes 15*:175, 1964
290. KAPUSCINSKI V, MOSS N, ZAK B, BOYLE AJ: *Am J Clin Pathol 22*:687, 1952
291. KATZMAN E, JACOBI M: *J Biol Chem 118*:539, 1937
292. KAUPPINEN V, GREF C-G: *Scand J Clin Lab Invest 20*:24, 1967
293. KAWASHIMA S, UEDA T: *Nara Igaku Zasshi 12*:1086, 1961; *CA 56*:14562b, 1962
294. KAY HD, ROBISON R: *Biochem J 18*:755, 1924
295. KEATING FR Jr, JONES JD, ELVEBACK LR, RANDALL RV: *J Lab Clin Med 73*:825, 1969
296. KEILIN D, MANN T: *Biochem J 34*:1163, 1940
297. KELLERMAN GM, DALE NE: *Med J Australia 471*:931, 1960; *CA 55*:663d, 1961
298. KENNY AD, TOVERUD SV: *Anal Chem 26*:1059, 1954
299. KEPNER BL, HERCULES DM: *Anal Chem 35*:1238, 1963
300. KERR SE, DAOUD L: *J Biol Chem 109*:301, 1935
301. KESSLER G: *Gradwohl's Clinical Laboratory Methods and Diagnosis*, 7th edition, edited by Frankel S, Reitman S, Sonnenwirth AC. St. Louis, Mosby, 1970, Vol 1, p 354
302. KESSLER G, WOLFMAN M: *Clin Chem 10*:686, 1964
303. KIBRICK AC, ROSS M, ROGERS HE: *Proc Soc Exp Biol Med 81*:353, 1952
304. KING EJ: *Biochem J 26*:292, 1932
305. KING HGC, PRUDEN G: *Analyst 92*:83, 1967
306. KING JS Jr, BUCHANAN R: *Clin Chem 15*:31, 1969
307. KINGSLEY GR, GETCHELL G: *Clin Chem 2*:175, 1956
308. KINGSLEY GR, ROBNETT O: *Am J Clin Pathol 29*:171, 1958
309. KINGSLEY GR, SCHAFFERT RR: *Anal Chem 25*:1738, 1953
310. KIRK PL, TOMPKINS PC: *Anal Chem 13*:277, 1941
311. KJELLIN KG: *J Neurochem 13*:413, 1966; *CA 65*:2608b, 1966
312. KLEEMAN CR, BASS DE, QUINN M: *J Clin Invest 32*:736, 1953
313. KLEEMAN CR, TABORSKY E, EPSTEIN FH: *Proc Soc Exp Biol Med 91*:480, 1956

314. KLEIN B, KAUFMAN JH: *Clin Chem 13*:1079, 1967
315. KLEIN B, KAUFMAN JH, MORGENSTERN S: *Clin Chem 13*:388, 1967
316. KLEIN B, KAUFMAN JH, OKLANDER M: *Clin Chem 13*:788, 1967
317. KLEIN B, KAUFMAN JH, OKLANDER M: *Clin Chem 13*:797, 1967
318. KLEIN B, OKLANDER M: *Clin Chem 13*:26, 1967
319. KLOTZ IM, LANGERMAN NR, DARNALL DW: *Ann Rev Biochem 39*:31, 1970
320. KNAPP EL: *J Clin Invest 26*:182, 1947
321. KOECHLIN B: *J Am Chem Soc 74*:2649, 1952
322. KOEPKE JA: *Am J Clin Pathol 44*:77, 1965
323. KOLTHOFF JM: *Biochem Z 185*:344, 1927
324. KOPITO L, SCHWACHMAN H: *N Engl J Med 273*:113, 1965
325. KOPITO L, SCHWACHMAN H: *Pediatrics 43*:794, 1969
326. KOVACS GS, TARNOKY KE: *J Clin Pathol 13*:160, 1960
327. KRAML M: *Clin Chim Acta 13*:442, 1966
328. KUNESH JP, SMALL LL: *Clin Chem 16*:148, 1970
329. KUNKEL HO, PEARSON PB, SCHWEIGERT BS: *J Lab Clin Med 32*:1027, 1947
330. KUTTNER T, COHEN HR: *J Biol Chem 75*:517, 1927
331. LANDERS JW, ZAK B: *Am J Clin Pathol 29*:590, 1958
332. LANG VW: *Z Klin Chem 3*:186, 1965
333. LAPIN LN: *Biokhimiya 22*:825, 1957; *CA 52*:4731, 1958
334. LAUBER VK: *Z Klin Chem 3*:96, 1965
335. LEE GR, NACHT S, CHRISTENSEN D, HANSEN SP, CARTWRIGHT GE: *Proc Soc Exp Biol Med 131*:918, 1969
336. LEE GR, NACHT S, LUKENS JN, CARTWRIGHT GE: *J Clin Invest 47*:2058, 1968
337. LEHMANN HP, KAPLAN A: *Clin Chem 17*:941, 1971
338. LEIFHEIT HC: *J Lab Clin Med 47*:623, 1956
339. LETONOFF TV, REINHOLD JG: *J Biol Chem 114*:147, 1936
340. LEVY GB: *Clin Chem 18*:696, 1972
341. LEWIS WHP: *J Med Lab Technol 17*:32, 1960
342. LI, TK, PIECHOCKI JT: *Clin Chem 17*:411, 1971
343. LINDEMAN RB, BOTTOMLEY RG, CORNELISON RL, JACOBS LA: *J Lab Clin Med 79*:452, 1972
344. LINDER R, KIRK PL: *Mikrochemie 22*:300, 1937
345. LINDGÄRTE F, ZETTERVALL O: *Israel J Med Sci 7*:510, 1971
346. LITTLEJOHN ESN, RAINE DN: *Clin Chim Acta 14*:793, 1966
347. LOCKE A, MAIN ER, ROSBASH DO: *J Clin Invest 11*:527, 1932
348. LOCKE W, TALBOT NB, JONES HS, WORCESTER J: *J Clin Invest 30*:325, 1951
349. LOEB RF, BENEDICT EM: *J Clin Invest 4*:33, 1927
350. LOKEN HF, HAVEL RJ, GORDAN GS, WHITINGTON SL: *J Biol Chem 235*:3654, 1960
351. LOTT JA, HERMAN TS: *Clin Chem 17*:614, 1971
352. LOWRY OH, LOPEZ JA: *J Biol Chem 162*:421, 1946
353. LUBELL D: *Helv Paediat Acta 12*:179, 1957; *CA 51*:16651, 1957
354. LUMB GA: *Clin Chim Acta 8*:33, 1963
355. LUND PK, MATHIES JC: *Am J Clin Pathol 44*:398, 1965
356. MacDONALD MA, WATSON L: *Clin Chim Acta 14*:233, 1966
357. MACHO L: *Nature 180*:1351, 1957
358. MACY IG: *Nutrition and Chemical Growth in Childhood*. Springfield, Ill., Charles C Thomas, 1942, Vol 1, p 161
359. MAGER M, FARESE G: *Clin Chem 12*:234, 1966
360. MAHANAND D, HOUCK JC: *Clin Chem 14*:6, 1968

361. MAHLER HR: *Mineral Metabolism*, edited by Comar CL, Bronner F. New York, Academic Press, 1961, Vol 1, Part B, p 783

362. MAHONEY JP, SARGENT K, GRELAND M, SMALL W: *Clin Chem 15*:312, 1969

363. MALMSTADT HV, WINEFORDNER JD: *Clin Chem 5*:284, 1959

364. MANDEL EE: *Clin Chem 5*:1, 1959

365. MANN CK, YOE JH: *Anal Chem 28*:202, 1956

366. MARDELL M, ZILVA JF: *Lancet 2*:1323, 1967

367. MATHER A, MACKIE NR: *Clin Chem 6*:223, 1960

368. MATHISON GC: *Biochem J 4*:233, 1909

369. McALLISTER HC Jr, YARBRO CL: *Clin Chem 6*:52, 1960

370. McFARLANE WD: *Biochem J 26*:1022, 1932

371. McKENDRICK T: *Lancet 1*:183, 1962

372. McKITTRICK DS, SCHMIDT CLA: *Arch Biochem 6*:411, 1945

373. McLEAN FC, HASTINGS AB: *J Biol Chem 107*:337, 1934

374. McLEAN FC, HASTINGS AB: *J Biol Chem 108*:285, 1935

375. McLEAN FC, HASTINGS AB: *Am J Med Sci 189*:601, 1935

376. MERET S, HENKIN RI: *Clin Chem 17*:369, 1971

377. MERRITT H, FREMONT-SMITH F: *Cerebrospinal Fluid*. Philadelphia, Saunders, 1937, p 12

378. METZ B, MOURS-LAROCHE MF: *Compt Rend Soc Biol 149*:579, 1955; *CA 50*:1153, 1956

379. MIKAC-DEVIC D: *Clin Chim Acta 23*:499, 1969

380. MIKAC-DEVIC D: *Clin Chim Acta 26*:127, 1969

381. MILLER E, HLAD CJ Jr, LEVINE S, HOLMES JH, ELRICK H: *J Lab Clin Med 58*:656, 1961

382. MIYADA DS, INAMI K, MATSUYAMA G: *Clin Chem 17*:27, 1971

383. MOORE EW: *Ion-Selective Electrodes*, edited by Durst RA. Washington, D.C., Nat Bur Stand Spec Pub 314, 1969, p 215

384. MOORE EW: *J Clin Invest 49*:318, 1970

385. MOORE EW, WILSON DW: *J Clin Invest 42*:293, 1963

386. MORELL AG, IRVINE RA, STERNLIEB IH, ASHWELL G: *J Biol Chem 243*:155, 1968

387. MORGAN EH: *J Lab Clin Med 75*:1006, 1970

388. MORTIMER JG, RAINE DN: *Anal Biochem 9*:492, 1964

389. MOSHER RE, ITANO M, BOYLE AJ, MYERS GB, ISERI LT: *Am J Clin Pathol 21*:75, 1951

390. MÜLLER E: *Z Kinderheilk 57*:243, 1934

391. MUNCH-PETERSON S: *Scand J Clin Lab Invest 2*:337, 1950

392. NALEFSKI LA, TAKANO F: *J Lab Clin Med 36*:468, 1950

393. NATELSON S, PENNIALL R: *Anal Chem 27*:434, 1955

394. NEILL DW, NEELY RA: *J Clin Pathol 9*:162, 1956

395. NESS AT, DICKERSON HC: *Clin Chim Acta 12*:579, 1965

396. NEUMAN WF, NEUMAN MW: *Chemical Dynamics of Bone Mineral*. Chicago, Univ. of Chicago Press, 1958

397. NEUMANN PZ, SASS-KORTSAK A: *J Clin Invest 46*:646, 1967

398. NIELSEN AL: *Acta Physiol Scand 7*:271, 1944

399. NIELSEN I: *Z Klin Chem Klin Biochem 6*:103, 1968

400. NIELSEN SP: *Scand J Clin Lab Invest 23*:219, 1969

401. NISSEN H: *Z Kinderheilk 57*:289, 1936

402. NOBLE S: *Proc Soc Exp Biol Med 95*:679, 1957

403. NORDIN BEC, TRIBEDI K: *Lancet 1*:409, 1962

404. NOYONS EC: *Rec Trav Chim 63*:248, 1944

405. NYSSEN M, DORCHE J: *Ann Biol Clin (Paris) 25*:137, 1967

406. ODESSKY L, ROSENBLATT P, BEDO AV, LANDAU L: *J Lab Clin Med 41*:745, 1953
407. OHARA K, NEWTON JL: *Jap J Physiol 18*:632, 1968
408. OKLANDER M, KLEIN B: *Clin Chem 12*:243, 1966
409. OLSON AD, HAMLIN WB: *At Absorption Newsletter 7*:69, 1968
410. OLSON AD, HAMLIN WB: *Clin Chem 15*:438, 1969
411. O'MALLEY JA, HASSAN A, SHILEY J, TRAYNOR H: *Clin Chem 16*:92, 1970
412. ORANGE M, RHEIN HC: *J Biol Chem 189*:379, 1951
413. ORESKES I, HIRSCH C, DOUGLAS KS, KUPFER S: *Clin Chim Acta 21*:303, 1968
414. OSAKI S, JOHNSON DA: *J Biol Chem 244*:5757, 1969
415. OSAKI S, JOHNSON DA, FRIEDEN E: *J Biol Chem 241*:2746, 1966
416. OVERMAN RD, ETTELDORF JN, BASS AC, HORN GB: *Pediatrics 7*:565, 1951
417. OWEN EC: *Biochem J 30*:352, 1936
418. PAPADOPOULOU DB: *Clin Chem 3*:257, 1957
419. PARKER MM, HUMOLLER FL, MAHLER DJ: *Clin Chem 13*:40, 1967
420. PARKER WC, BEARN AG: *Science 137*:854, 1962
421. PAYNE RB, LEVELL MJ: *Clin Chem 14*:172, 1968
422. PECHERY C: *Ann Biol Clin (Paris) 22*:357, 1964
423. PEDERSEN KO: *Scand J Clin Lab Invest 24*:69, 1969
424. PEDERSEN KO: *Scand J Clin Lab Invest 25*:199, 1970
425. PEDERSEN KO: *Scand J Clin Lab Invest 25*:223, 1970
426. PEDERSEN KO: *Scand J Clin Lab Invest 28*:459, 1971
427. PEDERSEN KO: *Scand J Clin Lab Invest 29*:75, 1972
428. PERKIN-ELMER, NORWALK, CONN: *Analytical Methods for Atomic Absorption Spectrophotometry*, March 1971
429. PERSIJN JP, VAN DER SLIK W, RIETHORST A: *Clin Chim Acta 35*:91, 1971
430. PETERS JP, VAN SLYKE DD: *Quantitative Clinical Chemistry*. Baltimore, Williams & Wilkins, 1932, Vol 2
431. PETERS T, GIOVANNIELLO TJ, APT L, ROSS JF: *J Lab Clin Med 48*:274, 1956
432. PETERS T, GIOVANNIELLO TJ, APT L, ROSS JF: *J Lab Clin Med 48*:280, 1956
433. PETERSON RE: *Anal Chem 25*:1337, 1954
434. PETERSON RE, BOLLIER ME: *Anal Chem 27*:1195, 1955
435. PFEFFER RB, COHEN T: *J Am Med Assoc 184*:422, 1963
436. PFLEIDERER Th: *Klin Wochschr 42*:640, 1964
437. PFLEIDERER T, OTTO P, HARDEGG W: *Klin Wochschr 37*:39, 1959
438. PICOU D, WATERLOW JC: *Nature 197*:1103, 1963
439. PIETERS HAJ, HANSEN WJ, GEURTS JJ: *Anal Chim Acta 2*:241, 1948
440. PIJCK J, HOSTE J: *Clin Chim Acta 7*:5, 1962
441. PIJCK J, RUTTINK J, CLAEYS A: *J Pharm Belg 16*:207, 1961; *CA 61*:9765e, 1964
442. PINCUS JB, GITTLEMAN IF, SAITO M, SOBEL AE: *Pediatrics 18*:39, 1956
443. PIODA LAR, SIMON W, BOSSHARD HR, CURTIUS HCh: *Clin Chim Acta 29*:289, 1970
444. PIPER KG, HIGGINS G: *Proc Assoc Clin Biochem 7*:190, 1967
445. PIRIE NW: *Biochem J 28*:1063, 1934
446. PITTINGER C: *CRC Crit Rev Clin Lab Sci 1*:351, 1970
447. PITTINGER C, CHANG PM, FAULKNER W: *Southern Med J 64*:1211, 1971
448. PORTNOY HD, GURDJIAN ES: *Clin Chim Acta 12*:429, 1965
449. POULOS PP, PITTS RF: *J Lab Clin Med 49*:300, 1957
450. POWER MH: *Standard Methods of Clinical Chemistry*, edited by Reiner M. New York, Academic Press, 1953, Vol 1, p 84
451. POWER MH, WAKEFIELD EG: *J Biol Chem 123*:665, 1938

452. PRASAD AS: *Arch Internal Med 105*:560, 1960
453. PRASAD AS, FLINK EB: *J Appl Physiol 10*:103, 1957
454. PRASAD AS, OBERLEAS D: *Proc Soc Exp Biol Med 138*:932, 1971
455. PRASAD AS, OBERLEAS D, HALSTED JA: *J Lab Clin Med 66*:508, 1965
456. PRUDEN EL, CREASON PL: *Clin Chim Acta 27*:19, 1970
457. PRUDEN EL, MEIER R, PLAUT D: *Clin Chem 12*:613, 1966
458. PYBUS J: *Clin Chim Acta 23*:309, 1969
459. RAAFLAUB J: *Methods of Biochemical Analysis*, edited by Glick D. New York Interscience, 1956, Vol 3, p 301
460. RAMAN A: *Clin Biochem 4*:141, 1971
461. RAMSAY WNM: *Biochem J 53*:227, 1953
462. RAMSAY WNM: *Clin Chim Acta 2*:221, 1957
463. RAO KJM, HOFSTETTER M, MORGENSTERN S: *Clin Chem 18*:696, 1972
464. RECHNITZ GA: *Ion-Selective Membrane Electrodes*. American Chemical Society Audio Course, 1971
465. REED AH, CANNON DC, WINKELMAN JW, BHASIN YP, HENRY RJ, PILEGGI VJ: *Clin Chem 18*:57, 1972
466. REED L, DENIS W: *J Biol Chem 73*:623, 1927
467. REHELL B: *Scand J Clin Lab Invest 6*:335, 1954
468. REIFENSTEIN EC, Jr, ALBRIGHT F, WELLS SL: *J Clin Endocrinol 5*:367, 1945
469. RENTSCH I: *Klin Wochschr 47*:433, 1969
470. RESSLER N, ZAK B: *Am J Clin Pathol 28*:549, 1957
471. RICE EW: *Clin Chim Acta 15*:632, 1960
472. RICE EW: *Standard Methods of Clinical Chemistry*, edited by Seligson D. New York, Academic Press, 1963, Vol 4, p 57
473. RICE EW, LAPARA CZ: *Clin Chim Acta 10*:360, 1964
474. RINGBOM A, VANNINEN E: *Anal Chim Acta 11*:153, 1954
475. RITTER S, SPENCER H, SAMACHSON J: *J Lab Clin Med 56*:314, 1960
476. ROBERTSON WG: *Clin Chim Acta 24*:149, 1969
477. ROBERTSON WG, PEACOCK M: *Clin Chim Acta 20*:315, 1968
478. ROBINSON HMC, RATHBURN JC: *Can J Biochem Physiol 37*:225, 1959
479. ROBINSON JC: *J Biol Chem 179*:1103, 1949
480. ROCKSTEIN M, HERRON PW: *Anal Chem 23*:1500, 1951
481. RODGERSON DO, HELFER RE: *Clin Chem 12*:338, 1966
482. RODKEY FL, SENDROY J Jr: *Clin Chem 9*:668, 1963
483. ROE DA, MILLER PS, LUTWAK L: *Anal Biochem 15*:313, 1966
484. ROE JH, IRISH OJ, BOYD JI: *J Biol Chem 67*:579, 1926
485. ROHDEWALD M, WEBER M: *Hoppe-Seyler's Z Physiol Chem 306*:90, 1956; *CA 51*:5235, 1957
486. ROSENSTOCK HA, SIMONS DG, MEYER JS: *J Am Med Assoc 217*:1354, 1971
487. ROSENTHAL HL, PFLUKE ML, JUD L: *Clin Chem 4*:290, 1958
488. ROSNER F, GORFIEN PC: *J Lab Clin Med 72*:213, 1968
489. ROSS JW: *Science 156*, 1378, 1967
490. ROSS JW JR: *Ion-Selective Electrodes*, edited by Durst RA. Washington D.C., Nat. Bur. Stand., Spec. Pub. 314, 1969, p 57
491. ROTHER CF, SAPIRSTEIN LA: *Am J Clin Pathol 25*:1076, 1955
492. RYALL R, FIELDING J: *Clin Chim Acta 28*:193, 1970
493. RYAN MP, HINGERTY D: *J Clin Pathol 21*:220, 1968
494. SACHS Ch, BOURDEAU AM, BALSAN S: *Ann Biol Clin (Paris) 27*:487, 1969
495. SADEK FS, REILLEY CN: *J Lab Clin Med 54*:621, 1959
496. SALEH AEC, COENEGRACHT JM: *Clin Chim Acta 21*:445, 1968
497. SAVORY J, WIGGINS JW, HEINTGES MG: *Am J Clin Pathol 51*:720, 1969

498. SCHACTER D: *J Lab Clin Med 54*:763, 1959
499. SCHACTER D: *J Lab Clin Med 58*:495, 1961
500. SCHADE AL, CAROLINE L: *Science 104*:340, 1946
501. SCHADE AL, REINHART RW, LEVY H: *Arch Biochem 20*:170, 1949
502. SCHALES O: *Standard Methods of Clinical Chemistry*, edited by Reiner M. New York, Academic Press, 1953, Vol 1, p 37
503. SCHALES O, SCHALES S: *J Biol Chem 140*:879, 1941
504. SCHATZ BC: *Determination of Protein-bound and Unbound Calcium in Serum using Dextran Gel*. Graduate School Thesis, University of Southern California, 1962
505. SCHEINBERG IH, COOK CD, MURPHY JA: *J Clin Invest 33*:963, 1954
506. SCHIRARDIN H, METAIS P: *Ann Biol Clin (Paris) 17*:465, 1959
507. SCHMIDT R, WEIS W, KLINGMULLER C, STAUDINGER HJ: *Z Klin Chem Klin Biochem 5*:304, 1967
508. SCHOENFELD RG, LEWELLEN CJ: *Clin Chem 10*:533, 1964
509. SCHULZ DW, PASSONNEAU JV, LOWRY OH: *Anal Biochem 19*:300, 1967
510. SCHWARTZ WB, BANK N, CUTLER RWP: *J Clin Invest 38*:347, 1959
511. SCHWARZENBACH G: *Anal Chim Acta 7*:141, 1952
512. SCHWARZENBACH G: *Analyst 80*:713, 1955
513. SELTZER CC, MAYER J: *Am J Clin Nutr 13*:354, 1963
514. SELTZER CC, WENZEL BJ, MAYER J: *Am J Clin Nutr 13*:343, 1963
515. SENDROY J Jr: *J Biol Chem 120*:335, 1937
516. SENDROY J Jr: *J Biol Chem 144*:243, 1942
517. SENDROY J Jr: *J Biol Chem 152*:539, 1944
518. SEVERINGHAUS JW, FERREBEE JW: *J Biol Chem 187*:621, 1950
519. SHARMA DC, SINGH PP, KHALSA JK: *Clin Biochem 2*:439, 1969
520. SHERMAN HC: *Chemistry of Food and Nutrition*. New York, Macmillan, 1941, p 254
521. SIGGAARD-ANDERSEN O: *Am J Clin Pathol 48*:444, 1967
522. SIMONSEN DG, WERTMAN M, WESTOVER LM, MEHL, JW: *J Biol Chem 166*:747, 1946
523. SIMONSEN DG, WESTOVER LM, WERTMAN M: *J Biol Chem 169*:39, 1947
524. SIMS RPA: *Analyst 86*:584, 1961
525. SINNIAH R, NEILL DW: *J Clin Pathol 21*:603, 1968
526. SITZMANN VFC: *Z Klin Chem 4*:290, 1966
527. SKAUG OE, NATVIG RA: *Scand J Clin Lab Invest 9*:39, 1957
528. SKINNER SL: *Lancet 1*:478, 1961
529. SKY-PECK HH: *Clin Chem 10*:391, 1964
530. SLAVIN W: *Atomic Absorption Spectroscopy*. New York, Interscience, 1968, p 195
531. SMITH BW, ROE JH: *J Biol Chem 179*:53, 1949
532. SMITH FE, REINSTEIN H, BRAVERMAN LE: *N Engl J Med 272*:787, 1965
533. SMITH GF: *Cerate Oxidimetry*. Columbus, Ohio, G. Frederick Smith Chem. Co., 1942
534. SMITH RG, CRAIG P, BIRD EJ, BOYLE AJ, ISERI LT, JACOBSON SD, MYERS GB: *Am J Clin Pathol 20*:263, 1950
535. SOBEL AE, HANOK A: *Proc Soc Exp Biol Med 77*:737, 1951
536. SOBEL C, FERNANDEZ A: *Proc Soc Exp Biol Med 113*:187, 1963
537. SOULIER JP, CROSNIER J: *Rev Franc Etudes Clin Biol 3*:157, 1958
538. SPARE PD: *Am J Clin Pathol 37*:232, 1962
539. SPECK B: *Helv Med Acta 34*:231, 1968; *CA Biochem Sections 69*:84814d, 1968
540. SPIVEK ML: *J Pediat 48*:581, 1956
541. SRIVASTAVA SP, PANDYA KP, ZAIDI SH: *Analyst 94*:823, 1969

542. STANBURY SW, THOMPSON AE: *Clin Sci 10*:267, 1951
543. STERN J, LEWIS WHP: *Clin Chim Acta 2*:576, 1958
544. STEVENS JF: *J Med Lab Technol 22*:47, 1965
545. STOJANOVSKI-BUBANJ A, KELER-BACOKA M: *Clin Chim Acta 25*:478, 1969
546. STONER RE, DASLER W: *Clin Chem 10*:845, 1964
547. STRAIN WH, STEADMAN LT, LANKAU CA, BERLINER WP, PORIES WJ: *J Lab Clin Med 68*:244, 1966
548. STROSS W: *Analyst 67*:317, 1942
549. STUTZMAN FL, AMATUZIO DS: *Arch Biochem Biophys 39*:271, 1952
550. SUNDERMAN FW, BOERNER F: *Normal Values in Clinical Medicine*. Philadelphia, Saunders, 1949, p 359
551. SUNDERMAN FW Jr: *Laboratory Medicine, The Bulletin of Pathology*. September 1969, p 287
552. SUNDERMAN FW Jr, CARROLL JE: *Am J Clin Pathol 43*:302, 1965
553. SUNDERMAN FW Jr, ROSZEL NO: *Am J Clin Pathol 48*:286, 1967
554. SURGENOR DM, KOECHLIN BA, STRONG LE: *J Clin Invest 28*:73, 1949
555. SUTHERLAND I: *Clin Chim Acta 14*:554, 1966
556. TAUSSKY HH, SHORR E: *J Biol Chem 202*:675, 1953
557. TAVENIER P, HELLENDOORN HBA: *Clin Chim Acta 23*:47, 1969
558. TAYEAU F, NIVET R, DUMAS M: *Bull Soc Pharm Bordeaux 95*:202, 1956; *CA 51*:7458, 1957
559. TAYLOR AE, MILLER CW: *J Biol Chem 18*:215, 1914
560. TEERI AE: *Chemist-Analyst 43*:18, 1954
561. TELOH HA: *Am J Clin Pathol 24*:1095, 1954
562. TELOH HA: *Am J Clin Path 26*:535, 1956
563. THIERS RE: *Standard Methods of Clinical Chemistry*, edited by Meites S. New York, Academic Press, 1965, Vol 5, p 131
564. THOMAS C Jr, TERCHO GP, SONDEL JA: *Nucl Appl 3*:53, 1967; *CA 66*:92354g, 1967
565. TISDALL FF: *J Biol Chem 50*:329, 1922
566. TODD WR: *J Biol Chem 140*:cxxxiii, 1941
567. TOMPSETT SL: *Biochem J 28*:1544, 1934
568. TOPHAM RW, FRIEDEN E: *J Biol Chem 245*:6698, 1970
569. TRINDER P: *J Clin Path 9*:170, 1956
570. TRINDER P: *Analyst 85*:889, 1960
571. TRUDEAU DL, FREIER EF: *Clin Chem 13*:101, 1967
572. ULLUCCI PA, MARTIN HF, GRIFFITHS WC: *Clin Biochem 3*:189, 1970
573. UNDERWOOD EJ: *Trace Elements in Human and Animal Nutrition*, 2nd edition. New York, Academic Press, 1962, p 188
574. VALLEE BL: *Metabolism 1*:420, 1952
575. VALLEE BL: *Mineral Metabolism*, edited by Comar CL, Bronner F. New York, Academic Press, 1962, Vol II, Part B, p 443
576. VALLEE BL, THIERS RE: *Flame Photometry in Treatise on Analytical Chemistry*, edited by Kolthoff IM and Elving PJ, New York, Interscience, 1965, Vol 6, p 3463
577. VAN BELLE H: *Anal Biochem 33*:132, 1970
578. van LEEUWEN AM, THOMASSE CM, KAPTEYN PC: *Clin Chim Acta 6*:550, 1961
579. VANNOTTI A, DELACHAUX A: *Iron Metabolism and its Clinical Significance*. London, Muller, 1949
580. VAN SLYKE DD, HILLER A: *J Biol Chem 167*:107, 1947
581. VAN SLYKE DD, HILLER A, BERTHELSEN KC: *J Biol Chem 74*:659, 1927
582. VEDSØ S, RUD C: *Scand J Clin Lab Invest 15*:395, 1963

583. VINK CLJ: *Clin Chim Acta* 5:702, 1960
584. VOLHARD J: *J Prakt Chem* 9:217, 1874
585. WAKEFIELD EG: *J Biol Chem* 81:713, 1929
586. WALLACH DFH, ESANDI MP: *Anal Biochem* 7:67, 1964
587. WALLACH DFH, STECK TL: *Anal Biochem* 6:176, 1963
588. WALLACH S, CAHILL LN, ROGAN RH, JONES HL: *J Lab Clin Med* 59:195, 1962
589. WALSER M: *J Clin Invest* 40:723, 1961
590. WARGHOL RM, EICHENHOLZ A, MULHAUSEN RO: *Arch Internal Med* 116:743, 1965
591. WARWICK WJ, HANSEN L: *Pediatrics* 36:261, 1965
592. WASSERMAN RH, CORRADINO RA: *Ann Rev Biochem* 40:423, 1971
593. WATCHORN E, McCANCE RA: *Biochem J* 26:54, 1932
594. WATSON D, ROGERS JA: *Clin Chim Acta* 8:168, 1963
595. WEATHERBURN MW, LOGAN JE, ALLEN RH: *Clin Biochem* 2:159, 1968
596. WEBSTER D: *J Clin Pathol* 13:246, 1960
597. WEBSTER JH, NEFF J, SCHIAFFINO SS, RICHMOND AM: *Am J Clin Pathol* 22:833, 1952
598. WEBSTER WW: *Am J Clin Pathol* 37:330, 1962
599. WEIL-MALHERBE H, GREEN RA: *Biochem J* 49:286, 1951
600. WENDLAND PE, WINSTEAD ME: *Am J Med Technol* 31:121, 1965
601. WESTERHAUSEN M, KELLER E, MAAS D, GERMANN HJ, KLEMENC-LIPPKAU G, SCHUBOTHE H: *Klin Wochschr* 47:1279, 1969
602. WEYBREW JA, MATRONE G, BAXLEY HM: *Anal Chem* 20:759, 1948
603. WHITE A, HANDLER P, SMITH EL: *Principles of Biochemistry*, 4th edition. New York, McGraw Hill, 1968
604. WILKES WC: *Am J Med Technol* 29:121, 1963
605. WILKINSON RH: *J Clin Pathol* 10:126, 1957
606. WILLIAMS HL, CONRAD ME: *J Lab Clin Med* 67:171, 1966
607. WILLIAMS LA, COHEN JS, ZAK B: *Clin Chem* 8:502, 1962
608. WILLIS JB: *Spectrochim Acta* 16:551, 1960
609. WILLIS JB: *Anal Chem* 33:556, 1961
610. WILLIS JB: *Methods Biochem Anal* 11:1, 1963
611. WILLS MR, GRAY BC: *J Clin Pathol* 17:687, 1964
612. WILLS MR, LEWIN MR: *J Clin Pathol* 24:856, 1971
613. WILSON AA: *J Comp Pathol Therap* 63:294, 1953
614. WILSON AA: *J Comp Pathol Therap* 65:285, 1955
615. WILSON JF, KLASSEN WH: *Clin Chim Acta* 13:766, 1966
616. WINEFORDNER JD, SVOBODA V, CLINE LJ: *CRC Crit Rev Anal Chem* 1:233, 1970
617. WINER AD, KUHNS DM: *Am J Clin Pathol* 23:1259, 1953
618. WISE WM, KUREY JJ, BAUM G: *Clin Chem* 16:103, 1970
619. WOLF AV: *Aqueous Solutions and Body Fluids.* New York, Harper & Row, 1966, p 9
620. WOODS JT, MELLON MG: *Anal Chem* 13:760, 1941
621. WOOTTON IDP, KING EJ: *Lancet* 1:470, 1953
622. WORWOOD M, TAYLOR DM: *Int J Appl Radiat Isotopes* 19:753, 1968
623. WRONG O, METCALFE-GIBSON A: *Proc Roy Soc Med* 58:1007, 1965
624. WYBENGA DR, CANNON DC, IBBOTT FA: *Clin Chem* 18:715, 1972
625. WYNN VS, MORRIS RJH, McDONALD IR, DENTON DA: *Med J Australia* 1:821, 1950
626. YANAGISAWA F: *J Biochem* 42:3, 1955

627. YARBRO CL, GOLBY RL: *Anal Chem 30*:504, 1958
628. YBEMA HJ, LEIJNSE B, WILTINK WF: *Clin Chim Acta 11*:178, 1965
629. YEE HY: *Clin Chem 14*:898, 1968
630. YEE HY, ZIN A: *Clin Chem 17*:950, 1971
631. YENDT ER, CONNOR TB, HOWARD JE: *Bull Johns Hopkins Hosp 96*:1, 1955
632. YOSHIMATSU S: *Tohuku J Exp Med 7*:119, 1926
633. YOUNG DS, HARRIS EK, COTLOVE E: *Clin Chem 17*:403, 1971
634. YOUNG DS, HICKS JM: *J Clin Pathol 18*:98, 1965
635. ZAK B: *Clin Chim Acta 3*:328, 1958
636. ZAK B, BAGINSKI ES, EPSTEIN E, WEINER LM: *Ann Clin Lab Sci 1*:14, 1971
637. ZAK B, EPSTEIN E: *Clin Chem 11*:641, 1965
638. ZAK B, RESSLER N: *Clin Chem 4*:43, 1958
639. ZAUGG WS, KNOX RJ: *Anal Biochem 20*:282, 1967
640. ZEISLER EB: *Am J Clin Pathol 24*:588, 1954
641. ZEPF S: *Clin Chim Acta 20*:473, 1968
642. ZETTNER A, MANSBACH L: *Am J Clin Pathol 44*:517, 1965
643. ZETTNER A, SELIGSON D: *Clin Chem 10*:869, 1964
644. ZETTNER A, SYLVIA LC, CAPACHO-DELGADO L: *Am J Clin Pathol 45*:533, 1966

Chapter 20

Blood pH, CO_2, and O_2

FRANK A. IBBOTT, Ph.D.
THOMAS S. LaGANGA, Ph.D.
JERRY B. GIN, Ph.D.
JAMES A. INKPEN, Ph.D.

Control of acid-base balance is maintained by renal and pulmonary mechanisms. Respiratory exchange includes not only CO_2 elimination but also O_2 uptake, both of which produce an effect upon blood pH. Thus levels of O_2 tension and saturation are closely linked to acid-base parameters. The subject of acid-base balance is quite intricate. Only certain basic aspects are presented here to elucidate what is being measured in the various analyses most commonly performed in the clinical laboratory.

BICARBONATE BUFFERING SYSTEM

Carbon dioxide dissolved in blood is in equilibrium between the plasma and the interior of erythrocytes, and undergoes the following reaction in both plasma and erythrocytes:

$$CO_2 + H_2O \overset{A}{\rightleftharpoons} H_2CO_3 \overset{B}{\rightleftharpoons} HCO_3^- + H^+$$

The reaction at A is about 1000-fold to the left and that at B is about 20-fold to the right. In plasma the hydration of CO_2 is slow and the formation of HCO_3^- from dissolved CO_2 in plasma negligible.

In the erythrocyte the dissolved CO_2 is hydrated to form H_2CO_3, which in turn dissociates to $H^+ + HCO_3^-$. The H^+ is taken up chiefly by the buffering action of the imidazole groups (anions) of the histidine residues of hemoglobin. The cations Na^+ and K^+ balance the imidazole anions, and after the H^+ neutralizes the imidazole anions the Na^+ and K^+ then balance the HCO_3^-. At the beginning of this process the HCO_3^- within the erythrocytes is in equilibrium with the HCO_3^- in the plasma. Formation of new HCO_3^- in the erythrocytes results in a higher concentration within the erythrocytes than in the plasma, and to restore equilibrium HCO_3^- diffuses out of the cells into the plasma. When HCO_3^- goes out, Cl^- goes in to take its place—this is the *chloride shift*. Since both H^+ and HCO_3^- are removed from the scene of action, the reaction

$$CO_2 + H_2O \rightleftharpoons H_2CO_3$$

proceeds to the right. Hydration and dehydration of CO_2 ordinarily occur at a relatively slow speed, but the enzyme carbonic anhydrase, present in erythrocytes, accelerates the reaction sufficiently so that the reverse of the above process—which must occur in the lungs in order for CO_2 to be released into alveolar air—can proceed within less than 1 second, which is the time spent by the erythrocytes in the lung.

Carbon dioxide in the form of carbamino CO_2 accounts for about 3% of the CO_2 combined with plasma proteins and about 2–10% of that combined with hemoglobin, depending on the degree of oxygenation; reduced hemoglobin carries somewhat more than oxyhemoglobin. The general reaction is as follows:

$$CO_2 \text{ dissolved } + R{-}NH_2 \rightleftharpoons R{-}NHCOO^- + H^+$$

The steady state equilibrium in the body is such that about 5% of the total CO_2 is in the form of the (physically) dissolved gas. The HCO_3^- constitutes about 90% and H_2CO_3 far less than 1% of the total CO_2.

The bicarbonate-carbonic acid system described above forms a major buffer system of the blood. The most important function of this system is maintenance of the blood pH. At the pH of blood this buffer is most effective in buffering H^+, which reacts with HCO_3^- to form H_2CO_3, a weak acid, the production of which has only a slight effect on the pH. It can also buffer OH^-, which reacts with H_2CO_3 to form HCO_3^- and H_2O leading to only a small pH change. The bicarbonate buffering system is effective owing to its high concentration in blood and because H_2CO_3 can be disposed of as CO_2 via the lungs or increased by retention of CO_2.

HENDERSON-HASSELBALCH EQUATION

The interrelationship between pH, pCO_2, CO_2, and HCO_3^- is expressed by the Henderson–Hasselbalch equation (91, 98), which is developed as follows:

$$H_2CO_3 \rightleftharpoons H^+ + HCO_3^-$$

According to the *law of mass action:*

$$\frac{[H^+]\,[HCO_3^-]}{[H_2CO_3]} = K$$

$$[H^+] = K\,\frac{[H_2CO_3]}{[HCO_3^-]}$$

As $-\log[H^+] = pH$, making $-\log K = pK$, the equation can be expressed logarithmically:

$$pH = pK + \log\frac{[HCO_3^-]}{[H_2CO_3]}$$

This is not a usable form of the equation because there is no way to measure $[H_2CO_3]$ directly. Since $[H_2CO_3]$ is proportional to the concentration of the dissolved CO_2, $[CO_2$ dissolved$]$ may be substituted, requiring a change in the numerical value of the constant pK (signified by changing pK to pK'):

$$pH = pK' + \log \frac{[HCO_3{}^-]}{[CO_2 \text{ dissolved}]}$$

The dissolved CO_2 (including the small amount of H_2CO_3) is proportional to the partial pressure of gaseous CO_2 in the alveolar air in the lungs with which the plasma is in equilibrium. Hence:

$$[CO_2 \text{ dissolved}] = a \cdot pCO_2$$

where "a" is a constant. Substituting:

$$pH = pK' + \log \frac{[HCO_3{}^-]}{a \cdot pCO_2}$$

As will be seen, $[HCO_3{}^-]$ can be determined by a titrimetric procedure, but the "total CO_2" is determined by gasometric procedures. Omitting the H_2CO_3, which is present in insignificant amounts:

$$\text{Total } CO_2 = HCO_3{}^- + CO_2 \text{ dissolved}$$

$$HCO_3{}^- = \text{total } CO_2 - a \cdot pCO_2$$

Substituting:

$$pH = pK' + \log \frac{\text{total } CO_2 - a \cdot pCO_2}{a \cdot pCO_2}$$

The constants "pK'" and "a" used in the equations are valid only for a given set of conditions. The pK' varies directly with temperature, increasing by approximately 0.004 per °C; pK' varies inversely with pH and ionic strength (11, 55, 92, 206, 253). The pK' is 6.103 at pH 7.4 when bicarbonate is determined titrimetrically, and 6.108 at pH 7.4 when bicarbonate is determined gasometrically (209). For serum and plasma at 37°–38°C, the pK' can be taken as 6.10, with an uncertainty no greater than 0.04 (55). The pK' is not a constant for whole blood, being different for cells and plasma and thus varying with hematocrit and degree of oxygenation (55,

78, 93, 206, 245). The pK' range for whole blood is 6.07 to 6.13 (18, 19), while for erythrocytes it is 6.18 (51). Conflicting reports on the alteration or lack of alteration of the pK' in various disease states have been made (245), with many of the variations reported being attributed to systematic errors in the actual determination, especially of pH and pCO_2, and variation in type of sample used (78).

The constant "a" is derived from the solubility coefficient (α) of CO_2 in serum or plasma. The α value is defined as the volume (in milliliters) of CO_2 that dissolves in 1 ml of serum or plasma at 37°C, corrected to standard conditions (0°C and 760 mm Hg). The value used for α is 0.510 at 38°C based on the water content of an average "normal plasma" of about 918 g/liter (198, 237, 245). The value α (hence "a") is not an absolute constant for serum or plasma since it may vary with water content and may increase in lipemic sera (134, 245). The constant "a," which changes the units in which CO_2 is expressed from mm of Hg to meq/liter, is obtained from α as follows:

$$CO_2 \text{ dissolved (meq/liter) at } 37°C = \frac{1000 \times \alpha \times pCO_2}{760 \times 22.26}$$

$$= 0.0303 \, pCO_2$$

$$= a \cdot pCO_2$$

The "a" value calculated using an α value of 0.510 is 0.0301 at 38°C and 0.0303 at 37°C, and can be used to calculate the amount of dissolved CO_2. Substituting the value for pK' and the appropriate value for "a" when total CO_2 is in meq/liter and pCO_2 in mm Hg:

$$pH = 6.10 + \log \frac{\text{total } CO_2 - 0.0303 \, pCO_2}{0.0303 \, pCO_2}$$

The pCO_2 can be calculated from the pH and total CO_2 by the following rearrangement of the Henderson-Hasselbalch equation:

$$pCO_2 \text{ (mm Hg)} = \frac{\text{total } CO_2}{0.0303 \, [\text{antilog}(pH - 6.1) + 1]}$$

The units mmol and meq are both used in the literature and for CO_2 are synonymous; however, meq will be used in this text.

Regardless of the mathematical form of the Henderson-Hasselbalch equation, there are three unknowns. For accurate knowledge of the status of acid-base balance of a patient, all three must be known. Although all unknowns except $H_2 CO_3$ can be determined directly, the relationship as described by the Henderson-Hasselbalch equation is sufficiently accurate for calculation of the third if any two of the values are known. For example, in calculation of pH from the other two known values it has been shown (246) that the total maximal error is 0.036 pH unit. To stay within such an error the other two quantities must be determined with an accuracy of about ±3%. Similarly, pH and $HCO_3{}^-$ or total CO_2 must be determined with accuracies of ±0.03 pH unit and ±3% to obtain an accuracy of ±5% in the calculated pCO_2. Calculation of values with the Henderson-Hasselbalch equation is tedious and time-consuming, and for this reason many authors have presented tables or nomograms relating pH, pCO_2, $HCO_3{}^-$, and CO_2 content (210, 211, 213, 216, 217).

An interesting fact is that throughout the literature on acid-base balance the temperature of blood in normal man has been variously taken as 37°, 37.5°, and 38°C. The normal mean rectal temperature in man is, however, frequently taken as 38°C with a diurnal swing around this value of 1°C or even more (232). Thus even in a normal individual an uncertainty is introduced from the effect of temperature on the various constants. In any event, selection of a "normal" reference temperature is arbitrary because the temperature differs in various parts of the body. Since pK' and the solubility coefficient of CO_2 in plasma vary with temperature, measurement of blood pH at 37° or 38°C, or use of the Henderson-Hasselbalch equation containing values for constants at these temperatures, would lead to error if any significant change in body temperature exists—e.g., a high fever. A nomogram for calculation of pCO_2 may be used when the body temperature differs from 37°C (130). Although the body temperature normally varies with time for any given individual, between individuals, and between various parts of the body, it is useful to have some commonly accepted reference "normal." The temperature which is recommended (80) and which is used throughout this chapter is 37°C. In the event of deviation of body temperature by more than 2°C from 37°C determined orally, measurements should be corrected to actual body temperature by use of the proper factors (17, 20, 80, 113, 162). Line charts are available for making this correction (Fig. 20-1).

pH-HCO₃⁻ DIAGRAM

The relationship between the three variables of the Henderson-Hasselbalch equation can be depicted graphically by substituting various values of $[HCO_3{}^-]$ in meq/1, and pCO_2 in mm Hg in the equation

$$pH = 6.10 + \log \frac{[HCO_3{}^-]}{0.0303 \, pCO_2}$$

solving for pH and then plotting $[HCO_3{}^-]$ versus pH. This is done for any

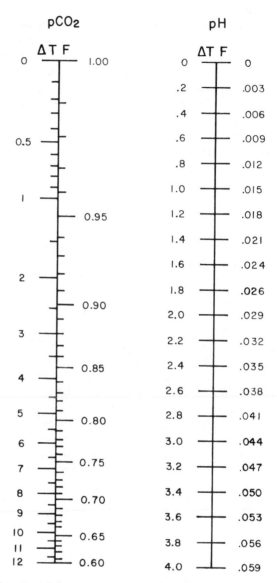

FIG. 20-1. Line chart for temperature correction of blood gas parameters. The ΔT is the difference between 37°C and the patient's temperature in $^{\circ}$C. For pCO_2: If the patient's temperature is less than 37°C, multiply the measured value by the factor appropriate to the particular ΔT. When the temperature is greater than 37°C, divide by the factor. For pH: If the patient's temperature is less than 37°C, add to the measured value the factor appropriate to the particular ΔT. When the temperature is greater than 37°C, subtract the factor.

desired values of pCO_2; in this way, pCO_2 isobars are obtained. Figure 20-2 shows this relationship with pCO_2 isobars of 20, 40, and 60 mm Hg.

If normal blood is drawn anaerobically, and the plasma separated and then equilibrated with gas mixtures of varying CO_2 content and analyzed for

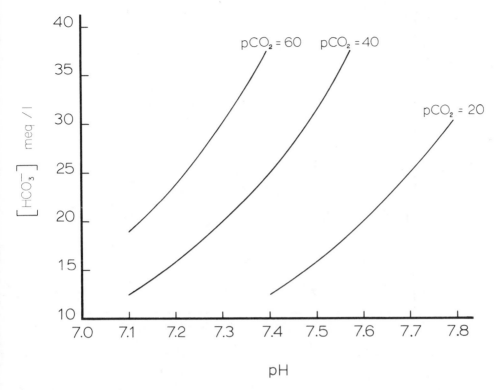

FIG. 20-2. pCO₂ isobars on pH-bicarbonate diagram. From DAVENPORT HW: *The ABC of Acid-Base Chemistry*, Fourth edition. University of Chicago Press, Chicago, 1958

HCO_3^-, the values fall on the *normal buffer line* of "separated plasma" (Fig. 20-3, line A). This must be contrasted with the results obtained with "true plasma." Plasma by itself is a poorer buffer than whole blood. If normal whole blood is equilibrated with gas mixtures of varying CO_2 content and *then* the plasma is separated and analyzed, the normal buffer line of "true plasma" is obtained which actually represents the buffering power of plasma plus erythrocytes. This of course is equivalent to the true situation *in vivo*. The buffer line of reduced blood is parallel but higher than the buffer line of oxygenated blood (Fig. 20-3, lines B and C), because oxygenated hemoglobin is more acidic than reduced hemoglobin. Note that the buffer lines of "separated plasma" and "oxygenated true plasma" cross approximately at the normal pCO₂ 40 isobar. The slope of these lines is directly related to the buffer capacity of the system; hence the slope obtained with "true plasma" increases with increasing hemoglobin concentration. These buffer lines were originally believed to be straight lines, but subsequent work revealed them actually to be slightly curvilinear (14, 165, 167). The pH-bicarbonate diagram was used quite successfully by Davenport in his monograph (48), *The ABC of Acid-Base Chemistry*, to depict the pathways of acid-base shifts (Figure 20-4).

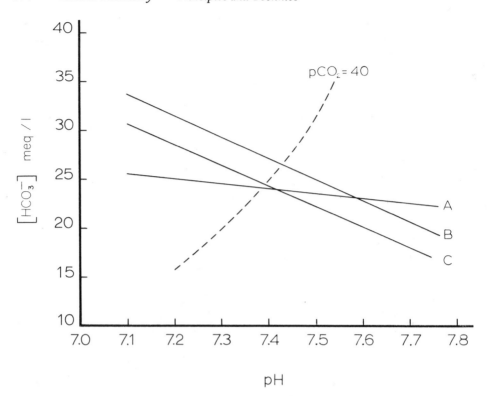

FIG. 20-3. Normal buffer lines on pH-bicarbonate diagram. (A) separated plasma, (B) true plasma from reduced blood, (C) true plasma from oxygenated blood. From DAVENPORT HW: *The ABC of Acid-Base Chemistry*, Fourth edition. University of Chicago Press, Chicago, 1958

As the temperature of whole blood drops, CO_2 diffuses into the erythrocytes and the pH of the blood rises (16). It is clear, therefore, that for plasma to reflect exactly the *in vivo* status of acid-base balance the plasma must also be separated at the temperature prevailing *in vivo*. Thus in the case of normal blood separated at room temperature, the pCO_2 of the separated plasma is too low by several millimeters, the total CO_2 is too low by about 0.5 meq/liter, and the pH at 38°C is too high by about 0.05 pH unit (14). Some workers observed no change in total CO_2, but this is regarded as a canceling out of two opposing effects (72, 134). Fortunately, there is no reason not to use the whole blood for pH measurements, and such an error involved in the determination of CO_2 content is not critical for clinical purposes. Most laboratories are not equipped to centrifuge blood at elevated temperatures.

RESPIRATORY AND METABOLIC COMPONENTS OF ACID-BASE BALANCE

The function of acid-base balance is primarily maintenance of a constant

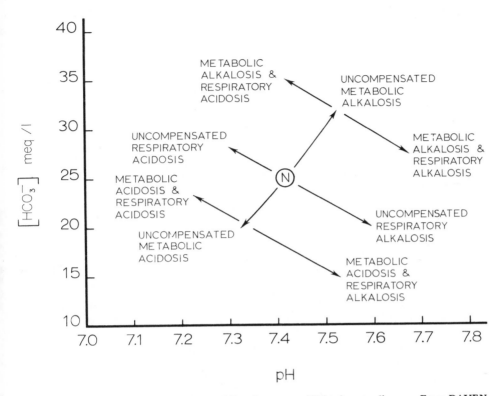

FIG. 20-4. Pathways for acid-base shifts shown on pH-bicarbonate diagram. From DAVEN-PORT HW: *The ABC of Acid-Base Chemistry*, Fourth edition. University of Chicago Press, Chicago, 1958

internal environment with regard to pH. Two systems interact to achieve a constant pH by adjusting the CO_2 and $HCO_3{}^-$ concentrations to meet the requirements defined by the Henderson-Hasselbalch equation. They are described as the respiratory and metabolic components. The respiratory component, by definition, is regulated via action of the lungs, and control is effected by alteration of the CO_2 concentration of blood. The mechanism for regulation of the metabolic component is a function of the kidneys and is reflected by changes in the $[HCO_3{}^-]$ term in the Henderson-Hasselbalch equation. Since the body is in a dynamic state with respect to acid-base balance, neither system functions alone but interrelates with the other in order to maintain a constant internal environment. However, for purposes of clinical analysis, it is necessary to be able to measure changes in one or both components, as well as pH, in order to provide adequate information as to the source of disturbances in the acid-base balance of the blood.

Excessive accumulation or loss of CO_2 comprises the respiratory disturbances of acid-base balance. A simple respiratory disturbance, such as hyperventilation, is characterized by excessive loss of CO_2 by the lungs, which leads to a transient increase in pH. The ensuing condition is described as a respiratory alkalosis. The increase in pH is readily predicted from the Henderson - Hasselbalch equation, since a decrease in the CO_2 dissolved increases the overall value of the ratio on the right side of the equation:

$$pH = pK' + \log \frac{[HCO_3^-]}{[CO_2 \text{ dissolved}]}$$

The opposite case—in which CO_2 excretion is diminished, as in emphysema—results in an increase in the dissolved CO_2. This leads to a transient decrease in pH or respiratory acidosis.

Derangements which give rise to changes in the concentration of $[HCO_3^-]$ are termed metabolic acidosis or alkalosis. Simple metabolic acidosis results when $[HCO_3^-]$ is decreased, while, conversely, metabolic alkalosis is associated with retention of $[HCO_3^-]$ and a concurrent increase in pH.

Clinically, such simple disturbances are experienced only on a transient basis, since the organism counters any change due to respiratory function by a corrective renal reaction and, conversely, a metabolic disorder is opposed by an adjustment in the rate of CO_2 removal by the lungs. The response of the two interrelated systems is directed toward preservation of a constant ratio between $[CO_2 \text{ dissolved}]$ and $[HCO_3^-]$, so that deviations of pH are restricted to the narrow confines of the normal range.

The most useful acid-base values are those which best express the respiratory and nonrespiratory disturbances in such a manner that each type of change can be isolated from the other. A primary respiratory disturbance is best characterized by the pH and pCO_2 in combination with a measure of the nonrespiratory component. The latter has been most commonly related to the total CO_2, the CO_2-combining power, standard bicarbonate, buffer base, and base excess.

CO_2-Combining Power

Historically, the most widely determined parameter of acid-base studies has been the CO_2-combining power, also called *CO_2-capacity* or *alkali reserve*. This technic was introduced in 1917 by Van Slyke and Cullen (238), but Peters and Van Slyke recommended in 1932 that the method be abandoned (169). Undoubtedly, the reasons for the continued widespread use of this technic are twofold: first, it has become firmly entrenched in most textbooks and manuals of clinical laboratory technics, and second, there is a definite attraction to a technic that does not require anaerobic storage and centrifugation of blood samples.

The CO_2-combining power is defined as the total CO_2 of plasma that has been equilibrated at a pCO_2 of 40 mm Hg at 25°C and is an index of the amount of CO_2 that can be bound. This value, which is determined gasometrically after correction for dissolved CO_2, is thus a measure of plasma bicarbonate. Nearly all the CO_2 taken up represents $BHCO_3$, where B represents the cation—hence the term "alkali reserve." According to the modern Brønsted concept, however, the buffer anion is the base, and cations such as Na^+ and K^+ are neither acids nor bases. As would be expected, the

CO_2-combining power agrees quite closely with the CO_2 content and also the direct determination of HCO_3^- except when the pCO_2 of the blood of the patient is abnormal—i.e., when there is respiratory acidosis or alkalosis. The CO_2-combining power thus is a measure of the metabolic component, but it has several serious defects: (a) The test is usually performed on blood which is not handled anaerobically so that any alteration in the state of oxygenation of the hemoglobin results in a change of the plasma HCO_3^-. (b) The test is performed on "separated plasma" so that the final HCO_3^- after equilibration is reached via the flat buffer line of "separated plasma" rather than via the steeper line of "true plasma," which would be the case if equilibration were carried out with whole blood (Fig. 20-3); the further away the pCO_2 of the whole blood sample is from 40 mm Hg prior to separation of the plasma, the greater is the error in the final result. If, in addition, the blood is not handled anaerobically, the more significant is the fall in pCO_2. (c) Cooling whole blood from 37°C to room temperature results in a slight decrease in the plasma bicarbonate level.

The laboratory procedure determines the dissolved CO_2 and H_2CO_3 as well as the HCO_3^-, and the result must be corrected by subtracting a value for dissolved CO_2. The solubility of CO_2 varies inversely with temperature, typical values being 1.3 meq/liter at 38°C and 1.8 meq/liter at 25°C.

CO_2 Content

The CO_2 content or total CO_2 is defined as the CO_2 derived from dissolved CO_2, H_2CO_3, HCO_3^-, and carbamino CO_2 in plasma from anaerobically drawn blood. The total CO_2 can be calculated from the pH and pCO_2 with the Henderson-Hasselbalch equation, or it can be determined directly. Total CO_2 is not a pure index of the metabolic or nonrespiratory component, as it varies with the pCO_2 and degree of O_2 saturation of the whole blood and does not directly give information on the accumulated amount of acid or base.

Bicarbonate

The bicarbonate value is defined as the concentration of HCO_3^- in plasma of anaerobically drawn blood and, as is discussed later, also includes carbamino CO_2. The HCO_3^- value is slightly lower that that for the total CO_2, and its use as a measure of nonrespiratory disturbances has the same drawbacks as the total CO_2 value. Useful clinical information is obtained with the use of the total CO_2 or the HCO_3^- value. If a sample of "true plasma" yields values of 15 meq HCO_3^-/liter and pH 7.3, and the reference point is placed on a pH-HCO_3^- diagram (such as Fig. 20-4), it is clear that the clinical state is one of primary metabolic acidosis partially compensated by a respiratory shift.

Standard Bicarbonate

Jørgensen and Astrup (108) introduced the standard bicarbonate value as a measure of the metabolic or nonrespiratory component, which is devoid of the defects of the CO_2-combining power previously discussed. It is defined as the bicarbonate concentration of the plasma of fully oxygenated blood at 38°C after equilibration of the whole blood at a pCO_2 of 40 mm Hg. The technic for its determination is rapid: after equilibration, the pH of the whole blood is measured and the standard bicarbonate calculated by the Henderson-Hasselbalch equation from the measured pH and the known pCO_2 40, or read directly from a nomogram (214).

Other advantages include the following: (a) There is no need for separation of plasma from the erythrocytes. (b) The sample need not be obtained anaerobically. (c) Slight hemolysis can be ignored. (d) The relationship to added strong acid or base is linear. (e) The amount of acid or base can be calculated in meq/liter provided the hemoglobin content is known. Table 20-1 presents data of Jørgensen and Astrup (108) illustrating that the standard bicarbonate is uninfluenced by variations in pCO_2 and oxygen saturation of the hemoglobin. As is true for all values calculated from the Henderson-Hasselbalch equation, however, results are uncertain to the extent that they are influenced by errors in assumed values for pK' and the solubility coefficient of CO_2 in plasma. For example, as previously discussed under *Henderson-Hasselbalch Equation*, the variation in pK' between different plasmas may be as much as ±0.04, yielding an uncertainty in the standard bicarbonate of about ±8%.

The standard bicarbonate value, as described, presumably eliminates the influence of respiration on the bicarbonate concentration and represents only metabolic changes in acid-base balance.

TABLE 20-1. Effects of Variation in pCO_2 and Oxygenation of Hemoglobin on Measurements of the Bicarbonate Buffer System[a]

Test	meq/liter		
	pCO_2 20	pCO_2 40	pCO_2 80
Fully oxygenated			
blood sample			
Total CO_2	16.8	22.2	30.0
HCO_3^-	16.2	21.0	27.6
CO_2-combining power	18.0	22.2	27.5
Standard HCO_3^-	21.0	21.0	21.0
Fully reduced			
blood sample			
Total CO_2	19.6	25.7	34.8
HCO_3^-	19.0	24.5	32.4
CO_2-combining power	21.0	25.7	31.9
Standard HCO_3^-	21.0	21.0	21.0

[a] From JØRGENSEN K, ASTRUP, P: *Scand J Clin Lab Invest* 9:122, 1957

Buffer Base

Values for bicarbonate, standard bicarbonate, or CO_2-combining power have the common drawback that they do not give directly the amount, in meq/liter, of buffer base in the whole blood. This is because the bicarbonate system is responsible for approximately 75% of the buffer capacity of blood when pCO_2 is kept constant. In 1948 Singer and Hastings (217) introduced the concept of *buffer base* (BB) as a parameter for the measurement of nonrespiratory disturbances of acid-base status. The *buffer base* is defined as the sum of the buffer anions (mainly HCO_3^- and proteinate ions), expressed in meq/liter at 38°C, in oxygenated blood or plasma derived from fully oxygenated blood equilibrated at a pCO_2 of 40 mm Hg. Buffer base is calculated with the Henderson-Hasselbalch equation after determination of the hemoglobin concentration and any two of the three parameters (pH, pCO_2, HCO_3^-). A nomogram (Fig. 20-10, below) simplifies the calculation (216, 217).

The *normal buffer base* (NBB) according to Singer and Hastings (217) is 40.8 meq/liter for plasma. The value for blood increases over that of plasma by 0.36 meq/liter for each gram of hemoglobin per 100 ml and can be calculated as follows:

$$\text{NBB (meq/liter)} = 40.8 + (0.36 \times \text{hemoglobin in g/100 ml})$$

The value 0.42 has also been used instead of 0.36 to calculate NBB of whole blood (210). The difference between the *buffer base* and the *normal buffer base* of a particular blood sample represents the amount of fixed acid or base causing the change and is a measure of the nonrespiratory or metabolic shift. This shift has been referred to as the ΔBB by Singer and Hastings, whereas Astrup and co-workers (15, 210, 213) call this quantity the *base excess* (signified by a positive value) or *base deficit* (signified by a negative value) depending on whether the buffer base is greater or less than the NBB value.

Base Excess

The base excess (BE) is defined as zero for fully oxygenated blood at pH 7.40 and pCO_2 40 mm Hg at 38°C. Positive values indicate an excess of base (or deficit of acid), while negative values indicate a deficit of base (or excess of acid). As indicated, the BE is the same as ΔBB. The calculation of the base excess is discussed in detail in the section *Calculation of Standard Bicarbonate, Buffer Base, and Base Excess.*

Note that the standard bicarbonate, buffer base, and base excess require the use of fully oxygenated blood at pCO_2 40 mm at 38°C. This eliminates the influence of the respiratory component, and any deviation of these parameters from the normal ranges represents the changes in acid-base balance caused by metabolic disturbances. This approach has been criticized because when used in isolation from other acid-base parameters (pH, pCO_2) it may lead to erroneous conclusions (190).

Acid-Base Ratio

The acid-base ratio is an approach to presenting acid-base data in a way that is easily comprehended (115, 233). The acid factor, A, of blood is defined as the ± deviation of the pCO_2 from the normal average pCO_2; the base factor, B, equals the ± deviation of base from its normal. The B factor is expressed directly by the base excess or ΔBB.

The concept of the A/B ratio is derived from the observation that the corrective mechanisms of the blood buffering system for overcoming an acid-base imbalance are designed to bring about compensation—i.e., a normal A/B ratio—and not necessarily to bring the A and B factors individually within their normal ranges. An A/B ratio of 2:1 is advantageous in that it is always associated with a normal pH (115). A nomogram has been developed which allows visualization of the A/B ratio in conjunction with pH (115, 233).

BLOOD COLLECTION

For a complete description of the state of acid-base balance of a patient it is preferable to use arterial rather than venous blood (80), because in the latter case the O_2 saturation may range from 35% to 75%, thus introducing an additional factor in the calculations of pCO_2 and whole blood buffer base (217, 218). The usual collection sites are the brachial, radial, or femoral arteries (79, 128). Although arterial puncture requires skill, may be painful, and can be followed by complications (192), well trained and supervised technicians frequently perform this procedure on adults (79). When several samples are to be obtained from the patient, a Cournand needle (Becton, Dickinson and Co.) or Seldinger catheter (Becton, Dickinson and Co.) may be used (79). Venous blood shows levels of pH, CO_2 content, and especially O_2 saturation and pO_2 that are different from arterial blood. Nevertheless, the differences in pH, pCO_2, and CO_2 content are relatively small when venous blood is collected carefully (75, 79, 80, 109, 192). Hence most routine determinations of these parameters are carried out on venous blood (76). For acceptable results to be derived from venous blood the patient should be warm and at rest in the supine position for 15 min prior to obtaining the sample (79). If a tourniquet is used, it should be left in place while the blood is drawn (80). The patient should be instructed not to clench and unclench his fist (75). "Arterialized" blood also may be used for the estimation of arterial blood acid-base values, provided the effects of local metabolism are minimized (80, 203). This is accomplished by ensuring that muscles are relaxed and by application of heat to produce "arterialization". Blood is drawn from the antecubital vein or a vein on the back of the hand after heating the hand and arm with electric pads for 15 min (33, 163) or soaking the arm in water at 46°–47°C for 10 min. A tourniquet is not usually required but, if so, it should be removed and a period of 30 sec allowed to elapse before drawing the blood (48). When drawn under ideal conditions, arterialized venous blood has pH and pCO_2 values agreeing with arterial blood to within 0.01 unit and 1 mm Hg, respectively (203). When arterial pO_2 is less than 90 mm Hg, agreement between the two sources is better than when the patient is breathing gas with a high O_2 content. Differences are too great, however, to permit accurate correction of venous to arterial pO_2 (69). The values obtained from capillary blood obtained by cutaneous puncture of hyperemic skin are virtually equivalent to those of arterial blood (75, 77, 123, 137, 214, 218). The earlobe, finger, thumb, or heel must be thoroughly warmed to prevent stagnation of the blood (80) and to make it unnecessary to resort to squeezing (17) if results are to be equivalent to those in arterial blood (80, 126, 224). Use of an electric heating pad reportedly causes an unpleasant tingling sensation in the earlobe (53). When coping with a struggling infant, better control is possible by using the child's thumb (224). Arterialized capillary blood yields values for most acid-base parameters that are in satisfactory agreement with those from arterial blood (53, 109, 119, 120). A striking exception is provided by pO_2 measurements which are found acceptable by some (53, 109, 119, 120) but not by others (109, 120). There is general agreement that it is essential for the earlobe or digit to be properly warmed (80, 126, 127), possibly by use of a special paste (120), and that the patient not be suffering from heart or lung diseases (259) and not have impairment of peripheral blood flow (119), circulatory collapse (119, 120, 126), or low cardiac output such as can occur in cardiac surgery (126).

Collection of blood for acid-base and respiratory gas analyses must be accomplished with an anaerobic technic. While this can be achieved using catheters (69, 119, 120, 192), particularly during cardiac catheterization, syringes are frequently used for this purpose. Glass syringes, preferably with matched barrels and plungers, should contain sterile anticoagulant and be lubricated with light mineral oil or vaseline to ensure a perfect seal (79, 80). Lubricants have been discouraged by some, however, because of the greater solubility of CO_2 in oil than water (203). Although plastic syringes have the advantage of possessing airtight seals and do not require greasing, plungers may stick at times (79). Their use has been criticized, however, because CO_2 and O_2 dissolve in the plastic (203). As the variety of plastic syringes available is considerable, it is necessary to check each brand for acceptability (79). After the blood is withdrawn, a few drops of mercury are taken into the syringe to facilitate complete mixing of the anticoagulant and blood (do *not* add the mercury to the syringe prior to venipuncture since cases of peripheral embolism from mercury have been reported from the use of this technic (124). The needle is removed and the syringe capped (Becton, Dickinson and Co., No. 425-A) or a cap can be made by removing the needle from its hub and applying a drop of solder over the hole in the hub. As a temporary expedient, the needle can be left on the syringe and the needle plunged into a rubber or cork stopper.

As a means of simplifying the collecting, storage, and handling of blood samples, the use of heparinized Vacutainers (Becton, Dickinson and Co.) has been recommended. Provided these tubes are filled completely, results for pH, CO_2 content, and O_2 saturation have been shown to be valid when compared with blood collection in plastic syringes (73). Other observations are that the tubes, if undisturbed, may remain unstoppered for up to 2 hours after centrifugation without significant effect upon plasma pH and CO_2 content; serum pH from blood collected into a plain Vacutainer is indistinguishable from heparinized plasma pH; other anticoagulants (Na oxalate, K oxalate, mixed K and NH_4 oxalates, citrate, and EDTA) all affect pH to some extent (73). While there appears to be satisfaction with the Vacutainer system for drawing venous samples, this is not true for the collection of arterial blood because of the short needle; a suitably modified design is now available (68) for this purpose. This collection set includes a 5-ml heparinized Vacutainer tube containing N_2 at a pressure of 152 mm Hg, and a 30-cm length of Tygon tubing to one end of which is attached a needle suitable for arterial puncture. At the other end is a Vacutainer puncture needle. During the 4-ml specimen collection, the N_2 is compressed to a pressure of 760 mm Hg.

The collection of capillary blood may be carried out using 1 x 100 mm heparinized glass tubes with a capacity of 70 μl. When the tube is filled, a 5-mm length of steel wire is introduced and the ends of the tube sealed with putty-like material. The sample is then mixed with the anticoagulant by stroking a magnet along the length of the tube, thus moving the wire inside (17, 79, 214). Alternatives are the larger 4 x 75 mm capillary with a capacity of 250 μl (Scientific Products) or polyethylene micro test tubes (79, 251) supplied by Beckman Instruments, Inc.

Although a thin, undisturbed layer of mineral oil overlaying a small area of sample inhibits diffusion of CO_2 from the blood to air and causes no adverse effects, large amounts of oil, agitation of the sample with oil (79), or drop by drop exposure *through* a layer of oil leads to significant loss of CO_2 (84).

BLOOD pH

The pH of blood can be determined directly or indirectly, although indirect methods are now rarely performed. The method of choice is the direct electrometric determination with a glass electrode. The design of electrode commonly used for this work is based on that of MacInnes and Dole (139, 140). Where the required equipment is not available, the pH can be determined colorimetrically with phenol red as indicator, either by visual comparison with standards (207), or preferably by photometric measurement of the color (185, 199, 218, 225, 239, 249). The pH thus determined has deviated in some studies from the value obtained by the glass electrode by only ±0.04 pH unit (218, 249). In another comparison, however, the photometric method yielded values averaging 0.13 pH unit higher (185).

The pH can also be calculated by means of the Henderson-Hasselbalch equation from the plasma bicarbonate concentration and the pCO_2 or from the total plasma CO_2 content and the pCO_2, with an error of the same order of magnitude as that for the indicator method (236, 246). Another indirect method involves the calculation of pH from the CO_2 content after equilibration with two gas mixtures of different known pCO_2 values (61, 110). The most convenient method for measuring blood pH employs the glass electrode. Cremer (43) recognized that a potential develops across a glass membrane separating two solutions of unequal H^+ concentration. Differences in potential are measured, as an absolute measure of electric potential cannot be made (252). The pH of blood is measured using a system consisting of glass and reference electrodes. Within the glass electrode is an Ag/AgCl wire immersed in an electrolyte of fixed pH. The reference electrode generally used is of the $Hg/Hg_2 Cl_2$ type. Several measuring systems for blood pH are commercially available and all have the following features in common (Fig. 20-5): (a) glass electrode enclosed within thermostated chamber, (b) capillary chamber, and (c) facility for aspiration of the specimen and cleansing solutions through the sample chamber. The pH meter has an expanded readout covering a range of approximately pH 6.0–8.0. Models are available from Beckman Instruments; Corning Scientific Instruments; and Radiometer-Copenhagen, The London Company.

FIG. 20-5. Thermostated capillary glass–calomel electrode system. From ASTRUP P, JØRGENSEN K, SIGGAARD ANDERSEN O, ENGEL K: *Lancet 1*:1035, 1960

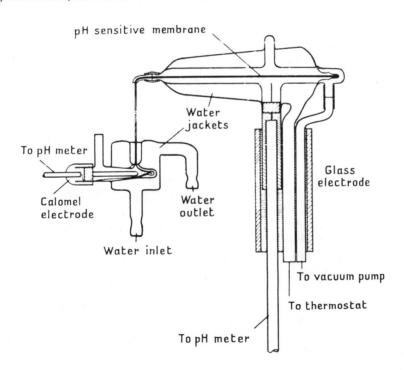

ELECTROMETRIC DETERMINATION OF BLOOD pH WITH THE GLASS ELECTRODE

SAMPLE

The pH values of plasma and whole blood at equilibrium theoretically are identical when measured at the same temperature as that existing at separation. Actually, it has been reported (205) that the pH of whole blood as measured by the glass electrode is about 0.01 pH unit lower than that of plasma. One explanation for this difference is that it is due to a "suspension effect" (205). The liquid junction potential at the KCl bridge operates in the direction of making the pH of suspensions appear lower than they are (103, 143). Another explanation is that the difference is due to alteration of the junction potential by precipitated and hemolyzed erythrocytes, not by intact erythrocytes (8). Gambino (72), on the other hand, observed no difference in pH of plasma and whole blood with the glass electrode. Certainly one distinct advantage in using whole blood is that less handling is involved. The pH of "separated" plasma at 37°C, i.e., plasma separated after the blood has come to room temperature, is higher by approximately 0.02—0.03 pH unit than the pH of the whole blood at 37°C (14, 16, 72, 150, 186). Although it is possible to make an approximate correction for this shift (see *Note 1*, below), this approach should be avoided if possible.

The electrode assemblies available require about 25 μl to 1 ml of sample. While heparin is usually regarded as the anticoagulant of choice, potassium oxalate also can be used (73, 262)—although addition of oxalate to serum or plasma causes a pH increase (e.g., 3 mmol/liter produces an increase of 0.07 pH unit). Presumably this is due to removal of Ca from protein leaving the protein molecule more negatively charged—i.e., in a more alkaline state. The pH change, however, is much less with whole blood because of the buffering capacity of erythrocytes. Ammonium oxalate, sodium citrate, and disodium ethylenediaminetetraacetate (EDTA; Sequestrene, Versene) cause a significant decrease in pH, whereas sodium oxalate causes an increase (73).

PROCEDURE

The anaerobic electrode assemblies commercially available are accompanied by specific instructions for use of the particular design. Use 0.85% saline for flushing out the electrode so that hemolysis and precipitation of euglobulins are avoided. It has been recommended (229) also that the electrodes be rinsed first with a neutral detergent for this purpose. Some models are designed only for use at room temperature, while others can be immersed in a constant temperature water bath permitting measurement at 37°C. The latter procedure is considered more accurate because it obviates the necessity of a temperature correction (see *Note 1*, below). The blood, serum, or plasma, however, should not be kept at 37°C for an extended period of time because apparently there is a slow upward drift in pH without change in CO_2 content (159).

Exact pH measurement is predicated upon accurate standardization of the electrodes against a buffer solution of known pH. The commercially available models usually employ a calibration system of two buffers having pH values of 6.841 and 7.384, respectively, at 37°C. Buffer solutions are available from the manufacturers for their particular instruments. Buffer solutions must be dispensed into a small vessel previously rinsed with the buffer, and then used to calibrate the instrument. Sampling directly from the stock bottle is improper technic since contamination of the buffer may occur. It is best to use a small plastic disposable container that is capped between samplings and then discarded at the end of the day.

The primary reference buffer should be close to pH 7.40 for four reasons (196): (a) Electrodes are more reproducible and stable over a narrow pH range. (b) Some electrode glasses do not exhibit complete theoretical potential response to be expected from the Nernst pH-potential relationship. (c) Errors from nonlinearity and meter scale are minimized. (d) Errors from changes in liquid junction potential are minimized. Mattock (143) recommended the following reference buffers: 0.025 M $KH_2PO_4 - 0.025$ M Na_2HPO_4 for setting the meter and 0.01 M borax and 0.05 M potassium acid phthalate for checking. At 38°C these buffers have pH values of 6.839, 9.085, and 4.025, respectively. The phosphate and phthalate buffers should be stable for at least several months. Spinner and Petersen (222) recommend the following buffer, which has a pH of 7.381 ± 0.002 at 38°C: 1.816 g KH_2PO_4 and 9.501 g $Na_2HPO_4 \cdot 2H_2O$ per 1000 g water. The National Bureau of Standards has proposed the following buffer for blood pH measurements (29): 1.179 g KH_2PO_4 and 4.303 g Na_2HPO_4 per liter at 25°C, using NH_3-free distilled water. The pH values of this buffer are as follows:

°C	pH
25	7.413
30	7.399
35	7.387
37	7.384
38	7.382
40	7.379

After the blood is in the electrode there may be a drift for about 1 min. Wait until the reading is stable, then use the mean of at least two readings. There is always a tendency to drift up, however, because of solution of electrode glass with discharge of base into solution (30).

NOTE

1. Temperature correction. The temperature correction factor for whole blood (pH to be subtracted per degree drop from 37°C) has been reported by various groups of workers to be 0.013 (158), 0.0149 (89), 0.015 (73, 94),

0.0147 (186), 0.0146 (188), and 0.017 (220); for serum and plasma it has been reported as 0.0118 (186) and 0.0110 (73, 188). The most commonly used correction factors were formerly those of Rosenthal (186). Rosenthal's $\Delta pH/\Delta T$ appeared to be independent of pH and remained constant when blood samples were diluted with plasma or saline. However, it has been found that the correction factor should include correction not only for temperature but also for pH and CO_2 content of the specimen (2, 35). The correction factor increases with an increase in pH and with a lowering in CO_2 content. Hematocrit values between 20% and 60% had no significant effect upon the $\Delta pH/\Delta T$; at hematocrits below 20% there was a decrease in $\Delta pH/\Delta T$, the value equaling that of plasma (0.0118/°C) at a hematocrit of 5% (2). Thus over the physiologic pH range, the $\Delta pH/°C$ can be expressed as a function of pH and CO_2 content by the following equation (2):

$$\Delta pH/°C = 0.0146 - [0.005 \, (7.40 - pH \, 38°C)]$$
$$+ \, 0.00005 \, (20 - CO_2 \text{ content in meq/liter})$$

where 0.0146 is the $\Delta pH/°C$ of a sample with plasma CO_2 content of 20 meq/liter at pH 7.4, and the coefficients 0.005 and 0.00005 are factors derived from the curve relating pH and CO_2 content to the $\Delta pH/°C$. The $\Delta pH/°C$ obtained above can then be substituted in the following equation to obtain the corrected pH:

$$\text{Blood pH } (38°C) = pH_t - \Delta pH/°C \, (38° - t)$$

where t = temperature of measurement.

It has been claimed that the maximal error for the whole blood correction is 0.05 pH unit (89,186). There is disagreement, however, as to whether the constants are independent of hemoglobin and plasma protein concentration (16, 186). Most disturbing is the report (158) that the temperature coefficient varies significantly with the particular pH meter employed. This may explain the variation in average values of the coefficient observed by different workers and indicates that the errors in calculated pH obtained by use of the coefficient can be kept minimal only by obtaining an average coefficient for the apparatus used, including electrodes.

2. *Stability of samples.* The pH of unpreserved blood begins to drop immediately after it is drawn. Havard and Kerridge (94) were the first to claim the existence of a "first acid change" in freshly shed blood, its magnitude being as much as 0.05 pH unit and occurring within 6 min at 38°C and 2 hours at about 22°C (94). Their observation was subsequently both branded as an artifact (170, 171) and confirmed (66, 125). More recent data indicate a linear decrease from zero time, the mean rate at 38°C reported to be 0.001 pH unit per minute (7, 161). At 20° and 5°C the rates of decrease are reported to be 0.0003–0.0004 (7, 137) and 0.0001 (137) pH unit per minute, respectively. That this phenomenon actually exists and

results from glycolysis is indicated by observations of the practically immediate increase in blood lactic acid in drawn blood (34, 64), although it is possible that glycolysis is not the only chemical change taking place (132, 161). By 4 hours at 37°C, the pH drops about 0.2 pH unit, but if the blood is immediately chilled to 4°C there is a drop of only about 0.03 pH unit in 5 hours. For practical purposes there is no significant change in 30-60 min at room temperature (186, 197, 229, 261).

The pH of fingertip blood starts to drift upward immediately, even when collected under oil, because of the release of carbonic anhydrase in greater concentration when the finger is punctured than when venipuncture is performed (159). Adding crushed erythrocytes to serum has the same effect (139). Hemolysis therefore should always be avoided. Capillary blood from the earlobe may be preferable to blood from the fingertip (234).

The use of 0.1 ml of 1% heparin-10% sodium fluoride per 2 ml of blood has been claimed to preserve the pH for 24 hours even at room temperature (89, 250) and to maintain the pH several days at refrigerator temperature (186). The ability of fluoride alone to inhibit glycolysis, however, has been contested (205); furthermore, its use has been condemned because it may alter pK ' (72). Nunn (161) studied the effect of fluoride and observed that in the presence of 25 mg NaF per 10 ml of blood the pH decreases as a quadratic function for the first 10 min, then increases. In some instances the pH becomes more alkaline than the original sample.

To summarize, if the determination is to be performed within 0.5 hour the specimen can be kept at room temperature, if within about 5 hours the specimen should be kept on ice or in the refrigerator.

ACCURACY AND PRECISION

The accuracy of commercial pH meters with calomel and glass electrodes ranges from 0.0025 to 0.1 pH unit. Obviously, this is a limiting factor, and an instrument should be used that is capable of an accuracy of at least 0.05 pH unit. Another limiting factor is the accuracy of the reference buffer on which the instrument is set. Since some handling of the blood is unavoidable, the overall accuracy must be somewhat less than the instrumental or standardization accuracy. The use of a temperature correction factor introduces further uncertainty. It is doubtful that routine blood pH determinations are usually more accurate than ±0.01 pH unit, which is the desired accuracy (260). The accuracy of blood measurement also depends on the assumption that the junction potential is the same for buffer and unknown. Semple (195), however, showed that values are about 0.11 pH unit lower with a 0.15 M NaCl bridge than with saturated KCl. Semple believed that the values obtained with saturated KCl are the correct ones.

NORMAL VALUES

The 95% limits of the normal range of arterial blood pH are generally taken as about 7.31–7.45 with a mean of about 7.40 (5, 76, 116, 142, 207,

217, 254). Some workers (9, 52, 85, 158, 205, 229) observed a significantly narrower range, about 7.37–7.42. It has been suggested (74) that the range is narrower if the subjects are at rest but wider if they are not. Values are the same for all ages and both sexes. Venous blood is, at most, about 0.03 pH unit less than arterial or cutaneous blood because of the fact that oxygenated hemoglobin is a stronger acid than reduced hemoglobin (5, 56, 72, 75, 166). There are significant differences in blood from different veins (161).

The intracellular erythrocyte pH is 0.15–0.23 ($\bar{x}$ 0.21) units lower than that of plasma.

CO_2 CONTENT

The "CO_2 content" or "total CO_2" of serum or plasma includes not only the dissolved CO_2 but also that in the forms of H_2CO_3, HCO_3^-, and carbamino-bound CO_2; and in whole blood it also includes the hemoglobin-bound CO_2. Normally the CO_2 content of whole blood is not measured because it does not directly give information on accumulated amounts of acid or base and the calculation requires determination of hematocrit and oxygen saturation (218). The classic technic for the determination of CO_2 content is gasometric and involves addition of acid, vacuum extraction of the CO_2, and measurement of the CO_2 volumetrically (247) or manometrically (155, 156, 241). Manometric technics have been developed which require 30–50 μl of plasma and are reported to be accurate and convenient (77, 155, 156, 242). The CO_2 released by acid can be absorbed into $Ba(OH)_2$ in a Conway or similar cell, precipitating as $BaCO_3$. The $Ba(OH)_2$-$BaCO_3$ mixture is then titrated with HCl using thymol blue, thymolphthalein, or a thymolphthalein-phenolphthalein mixture as indicator (44, 107, 144). This microdiffusion technic is reported to be too delicate and inconvenient for samples smaller than 100 μl (17).

An automated method utilizing the AutoAnalyzer is in widespread use in which the CO_2 released by acid is absorbed by an alkaline buffer containing phenolphthalein or cresol red (114, 219). The major disadvantage of the automated method is the loss of CO_2 while the specimen is in open cups waiting for sampling. The loss of CO_2 can be 1–3 meq/liter in 1–3 hours (71, 82). Cellulose acetate or plastic disks to cover most of the exposed sample surface have been used to minimize the loss of CO_2 (71). The disk must not be too thick or hard as it must turn to a vertical position to allow pickup of the sample when the probe strikes the surface of the disk. If the disk is too thin, it is likely to curl at the edges and occlude the tip of the sample probe.

The addition of one drop of 1 N NH_4OH per milliliter of plasma has been used as a method of preventing significant losses of CO_2 from plasma in open AutoAnalyzer cups (82). By bringing the pH of the sample to 8.6–8.8, the plasma pCO_2 approximates that of room air, so that no significant gain or loss of CO_2 occurs. Comparison of CO_2 content determined by the AutoAnalyzer method with those determined with the Natelson micro-gasometer, with the Astrup apparatus, or by calculation from pH and pCO_2 values reveals comparable results, provided the aliquot for testing is taken immediately after opening sealed samples or the plasma is made alkaline immediately upon being poured into AutoAnalyzer cups (82).

Another method for the determination of CO_2 content utilizes the pCO_2 electrode; here whole blood or plasma is diluted anaerobically with HCl converting HCO_3^- and carbamino-bound CO_2 to free CO_2 (182, 183, 200). The pCO_2 of this mixture is directly proportional to the CO_2 content of the original sample. The CO_2 content can also be determined by gas chromatography (60, 247), by calculation from pH and pCO_2 values with the Henderson-Hasselbalch equation, or with a nomogram such as that given in Figure 20-6.

FIG. 20-6. Nomogram for calculation of serum or plasma values by the Henderson-Hasselbalch equation. A straight line through given points on any two scales cuts the other two scales at points indicating simultaneously occurring values. From McLean (145).

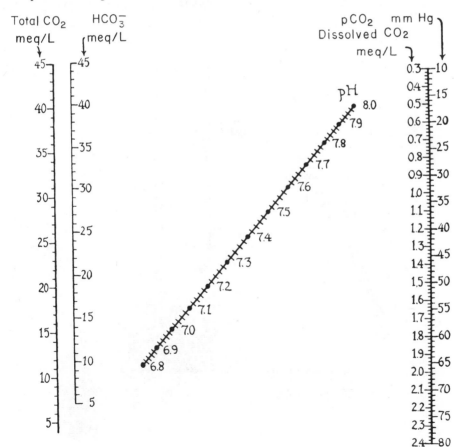

Method of Choice

The reference method for determination of CO_2 content is manometric, utilizing the Natelson (155) microgasometer (Fig. 20-7). This method is useful only for relatively small numbers of samples. The apparatus, available from Scientific Industries, Inc., adapts the Van Slyke-Neill (241) manometric gasometer to an ultramicro scale. A pipet calibrated at 0.01, 0.02, 0.03 and 0.1 ml is connected to the reaction chamber by a ball and socket joint. Reagents and sample are measured in this pipet while it is attached to the instrument; thus there is no intermediate transfer of liquids. Vacuum extraction is aided by agitation of the apparatus or by a magnetic stirrer in

FIG. 20-7. Natelson microgasometer for the manometric determination of CO_2. From Scientific Industries, Model 600, Natelson Microgasometer

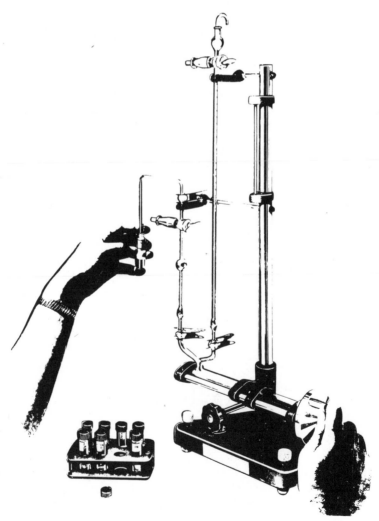

the reaction chamber. Pressure control over the mercury is afforded by a hand wheel that turns a screw in a mercury reservoir. The usual sample size used for CO_2 and O_2 determinations is $30 \, \mu l$.

The calculations for CO_2 using the microgasometer are minimal, as temperature is the only variable. A set of correction factors based on temperature variation is provided with the apparatus. Factors for gases besides CO_2 can be calculated using the Van Slyke-Neill equations (241, 243) as follows:

$$V_{0°;760} = a \times \frac{P}{760} \times \frac{1}{1 + 0.00384t} \times \left(1 + \frac{S\alpha'}{A - S}\right) i$$

where

$V_{0°;760}$ = volume of gas at $0°C$ and 760 mm Hg

a = fixed volume of gas (0.12 ml)

P = observed pressure (mm Hg)

t = temperature $(°C)$

S = total volume of solution from which gas is extracted (0.17 ml)

A = volume of chamber occupied by gas and solution (3 ml)

$A-S$ = volume of gas during extraction (2.83 ml)

α' = Henry distribution coefficient between gas and liquid phases (values for α' for different gases can be found in ref. 241)

i = correction factor for reabsorption of gas (1.017)

$\dfrac{1}{1 + 0.00384t}$ = correction factor for combined temperature effect on gas, mercury, and glass

$\left(1 + \dfrac{S\alpha'}{A-S}\right) i$ = correction for unextracted and reabsorbed gas

A micromanometric chamber for the manometric Van Slyke apparatus has been described utilizing a 50-μl sample. The accuracy is reported to be the same as, or higher than, that of the Natelson gasometer (242).

MANOMETRIC DETERMINATION OF CO_2 CONTENT WITH THE NATELSON MICROGASOMETER

(method of Natelson, Ref. 155, 156)

PRINCIPLE

The HCO_3^- of serum or plasma is converted to CO_2 by addition of acid; the dissolved CO_2, other gases, and H_2CO_3 are extracted *in vacuo*. The pressure of the gas mixture is measured at a fixed volume. Anaerobic addition of alkali causes selective reabsorption of CO_2, and the pressure of the residual gases is measured at the initial fixed volume. The difference in pressure is a measure of the CO_2 content.

REAGENTS

Lactic Acid, 1 N. Dilute 10 ml lactic acid to 100 ml with water.

NaOH, 3 N. Dissolve 12 g NaOH in approximately 70 ml water and dilute to 100 ml.

Anti-Foam Reagent. (Scientific Industries, Reagent No. 820): Use 10% solution as obtained. Shake bottle before each use.

Mercury

Working Solutions: For ease of operation and to avoid introduction of air into the apparatus, place the following into 10-ml vials (volumes are approximate):

2 ml mercury, 4 ml 1 *N* lactic acid
2 ml mercury, 4 ml Anti-Foam reagent (shake before each use)
2 ml mercury, 4 ml 3 *N* NaOH
1 *N* lactic acid for rinsing

PROCEDURE

Refer to Figure 20-7 for location of operational parts and to manufacturer's manual for detailed instructions.

1. Introduce 30 μl plasma into the gasometer pipet, followed by 10 μl mercury, 30 μl lactic acid, 10 μl mercury, 10 μl Anti-Foam, 10 μl mercury, 100 μl water, and mercury to 0.12-ml mark.

2. Close reaction chamber stopcock and draw mercury meniscus back to 3-ml mark.

3. Mix the reaction mixture for 1 min.

4. Advance aqueous meniscus to 0.12-ml mark and read manometer (P_1). Note temperature to the nearest degree centigrade.

5. Advance mercury just to top of manometer and open reaction chamber stopcock.

6. Add 0.1 ml 3 N NaOH, avoiding introduction of air, followed by mercury to 0.12-ml mark.

7. Close reaction chamber stopcock and draw mercury meniscus back to 3-ml mark.

8. Return the aqueous meniscus to 0.12-ml mark and read manometer (P_2).

9. Eject reaction mixture; rinse reaction chamber with water, then with lactic acid, then again with water.

10. Determine the blank value by following the procedure as described with the exception that in step 1 the 30 μl of plasma is not introduced into the apparatus. The difference between P_1 and P_2 is the "blank."

Calculation:

$$CO_2 \text{ content (meq } CO_2 \text{ /liter)} = (P_1 - P_2 - \text{blank}) \times \text{factor}$$

where the "factor" is found in Table 20-2.

TABLE 20-2. Factors for Calculating CO_2 Content with the Natelson Microgasometer

Temperature (°C)	Factor
20	0.237
21	0.236
22	0.235
23	0.234
24	0.233
25	0.232
26	0.231
27	0.230
28	0.229
29	0.228
30	0.227
31	0.225
32	0.224

NOTES

1. Factors. The factors listed in Table 20-2 include all variables, at each temperature (155, 156, 241).

2. Determination of CO_2-combining power (CO_2 capacity, alkali reserve). The CO_2-combining power can be determined with the microgasometer after equilibrating the serum or plasma with a gas mixture having a pCO_2 of 40 mm Hg and deducting 1.5 meq/liter from the CO_2 content to correct for dissolved CO_2. The recommended method for equilibration is to bubble a gas mixture containing 5.5% CO_2 and 94.5% N_2 gently through water, and then into 1.5 ml of serum or plasma to which a drop of an anti-foam agent (caprylic alcohol) has been added. Equilibration of the sample with expired alveolar air from the lungs of the analyst is not recommended since the assumption that expired alveolar air has a pCO_2 of 40 mm Hg is not necessarily valid, as the pCO_2 can vary from 33 to 48 mm Hg.

3. Standards. Technic can be checked by performing the analysis on a standard solution of $Na_2 CO_3$. A convenient standard is 25 mmol $Na_2 CO_3/$ liter (0.265 g $Na_2 CO_3$ per 100 ml). Caprylic alcohol must not be used with this standard in the CO_2-combining power procedure as the alcohol floats on top and CO_2 is more soluble in the alcohol (99). Lyophilized control serum is available (General Diagnostics Division, Warner-Chilcott) that approximates normal and abnormal acid-base conditions.

4. Anesthesia. Samples taken from patients under ether anesthesia may yield erroneously high results due to the fact that ether enters the gaseous phase during the first extraction (P_1), and after addition of alkali some of the ether is absorbed. There are technics for correcting this error (86, 94). Nitrous oxide used in anesthesia does not affect the CO_2 content (100).

5. Sample stability. The pH shift occurring in whole blood after it is drawn due to cooling to $25°C$ results in a drop of the plasma CO_2 content of about 5%. To avoid the pH shift, the sample should be separated at a temperature of $37°C$. Blood can be stored under anaerobic conditions for up to 4 hours in ice water and returned to room temperature or $37°C$ prior to separation. Once the plasma is separated from the erythrocytes, the CO_2 content is stable, provided loss of gaseous CO_2 is prevented.

6. Sources of error. The major source of error lies in the handling of samples: (a) If the sample is collected under oil a significant amount of CO_2 may be lost due to absorption of CO_2 by the oil (164, 249). (b) Plasma separated at room temperature has an erroneously low CO_2 content (the decrease for normal blood is about 0.5 meq/liter plasma) due to diffusion of CO_2 into the red cell (14). (c) Anticoagulants such as EDTA, citrate, oxalate, or sodium fluoride have a varied and inconsistent effect on the electrolyte distribution in blood and should not be used. Heparin (less than 100 units/ml blood) does not affect the CO_2 content (80, 255). The solubility of CO_2 is increased up to 8% in the presence of lipemia (245). These errors originate outside the actual analytic technic. The most common sources of technical error are leaks in the apparatus and introduction of air during aspiration of sample and reagents.

7. Whole blood versus serum or plasma. Plasma is preferred for total CO_2 determinations as the total CO_2 of whole blood also includes the carbamino-bound CO_2 (mainly bound to hemoglobin), which complicates the calculations of some of the factors and gives less accurate results. The hematocrit and O_2 saturation must be determined when whole blood is used for CO_2 content. The use of serum for CO_2 determination is not recommended because of errors caused by handling.

ACCURACY AND PRECISION

The technic is extremely accurate. Most errors are due to improper collection and storage of sample.

The precision during routine use is about $\pm 3\%$.

NORMAL VALUES

The normal values for CO_2 content are as follows:

Venous serum or plasma	24–34 meq/liter (74, 99)
Arterial serum or plasma	22–29 meq/liter (212, 227, 233, 242)
Venous blood	21–28 meq/liter (99)
Arterial blood	20–24 meq/liter (99)

There is a sex difference, with values for females being lower than for males by 2–3 meq/liter. The ranges for full term and premature infants appear to be lower than those for adults (223).

SERUM OR PLASMA BICARBONATE

While bicarbonate in the strictest meaning of the term is usually considered to be the HCO_3^- anion, in acid-base studies the true bicarbonate level cannot be determined directly but must be calculated, either from the pH and pCO_2 or the pH and total CO_2. In acid-base studies when bicarbonate is determined either titrimetrically or by precipitation the measurement includes not only the HCO_3^- but the CO_3^{2-} and carbamino CO_2, which together account for about 5% of the total (136).

The most common approach to the determination of bicarbonate is by titration, as introduced by Van Slyke and co-workers in 1919 (248); this involves addition of an excess of standard acid to plasma in order to release all the CO_2 from HCO_3^-, CO_3^{2-}, and carbamino CO_2, and then back-titrating the remainder of the acid. The back-titration should be to the original pH of the plasma at a temperature of 37°C to obtain the greatest accuracy (70), but for routine determination it usually suffices to titrate to the pH of normal plasma. The endpoint of the titration can be signaled by an indicator such as neutral red, phenol red, or diphenylcarbazone (28, 151, 191, 248); however, the use of a pH meter to determine the endpoint is more accurate and is recommended.

The titrimetric approach has been modified such that the proteins are first precipitated with methanol or ethanol and the supernate acidified and back-titrated to pH 5.0 using methyl red-methylene blue indicator (63). It has been claimed that there is no necessity to back-titrate to the original pH since the effect of protein is removed.

In another approach the protein is precipitated with ethanol, methyl red is added to the supernate, and the absorbance at 520 nm is determined. The HCO_3^- concentration in the plasma is calculated from the absorbance of HCO_3^- standard solutions (230). A criticism of the protein precipitation technic is that the CO_2 in the carbamino-CO_2 is not determined, and that normal levels of phosphate cause the results to be too high by 1 meq/liter (230).

The HCO_3^- can also be determined by precipitation with $Ba(OH)_2$, forming $BaCO_3$. The Ba in the precipitate can be determined by flame photometry and the HCO_3^- calculated (50). In another technic the mixture is ashed to remove organic material and boric acid added to form $Ba(H_2BO_3)_2$, which is then titrated with HCl to the pH of the original boric acid (221).

TITRIMETRIC DETERMINATION OF SERUM OR PLASMA BICARBONATE

(method of Segal, Ref. 193, modified)

PRINCIPLE

Acid is added in excess to serum or plasma. The CO_2 is released from HCO_3^-, an equivalent amount of H^+ being removed by the formation of H_2O. The excess of acid is then titrated with standard alkali.

REAGENTS

HCl or H_2SO_4, 0.050 N. Standardize to within 1%.

NaOH, 0.0100 N. Prepare 0.100 N from carbonate-free, saturated NaOH and standardize to within 1%. The 0.0100 N alkali is prepared by making a 1:10 dilution; if tightly stoppered, this is usable for several days.

PROCEDURE

1. Add 1.0 ml serum or plasma to 1.0 ml 0.050 N acid in a beaker of sufficient size to accommodate pH electrodes.
2. Swirl beaker contents for 1 min.
3. Add sufficient water to cover tips of pH electrodes (up to 30 ml is allowable).
4. Back-titrate with 0.010 N NaOH with stirring (electric stirring motor is convenient; titrate from 5-ml microburet). The technic required for handling samples and the pH to which the back-titration is carried are dependent on the accuracy desired. There are three alternatives, presented here in order of decreasing accuracy:
   a. Serum or plasma is obtained anaerobically and the pH determined at room temperature. Back-titration is carried to this pH. The original pH

can be determined with either aerobic or anaerobic electrodes; or, if the sample is limited, the serum or plasma can be diluted up to 30 times and the pH taken.

b. Blood is drawn and transferred to a tube, stoppered, allowed to clot at room temperature up to 2 hours, and centrifuged; the pH of the serum is determined as in (a). Back-titrate to 0.1 pH unit less. Experiments performed in our laboratory indicate an increase in pH ranging from 0.05 to 0.15 when serum is obtained by this technic.

c. If the history of the sample of serum or plasma is unknown or if it is known that the sample has had ample opportunity to lose CO_2, there is no choice but to back-titrate to pH 7.6 (see *Note 3*, below).

Calculation

$$\text{meq } HCO_3{}^-/\text{liter} = 1000 \ (\text{ml acid} \times N \text{ of acid}) - (\text{ml NaOH} \times N \text{ of NaOH})$$

MICRO PROCEDURE

The procedure can be scaled down as much as tenfold without difficulty, the limiting factor being the volume required by the electrodes used. This is no problem if an indicator is used to signal the endpoint. The macro technic described can be used with 0.5-ml samples and the results multiplied by 2.

NOTES

1. *Use of indicator rather than pH meter to signal endpoint.* Although this is not as accurate as electrometric titration, it is satisfactory for most routine purposes. A 0.1% aqueous solution of the sodium salt of phenol red is used. Use 1 drop/10 ml solution. If procedure (a) or (b) is employed, a blank is prepared by adding the indicator to a mixture of 1 ml serum or plasma and the amount of water employed in the analysis. Back-titration of the unknown is carried to a color match with this blank. If procedure (c) is employed, titrate to a definite red (not orange). The indicator begins to turn red at pH 7.4–7.5.

2. *Standardization with NaHCO₃.* A standard containing 25 meq $HCO_3{}^-$/liter can be prepared from $NaHCO_3$ (210 mg/100 ml). It is used as a serum unknown, and back-titration is carried to pH 7.0. If back-titration is carried to a higher pH, CO_2 is picked up rapidly from the air. Thus if phenol red indicator is used, the latter part of the titration must be carried out rapidly and stopped as soon as the solution becomes pink (fades rapidly). This occurs very slowly when serum is present because of the buffering action of the protein present.

3. *Magnitude of error when back-titrating to pH 7.6.* A serum or plasma pH of 7.6 at 25°C is approximately equivalent to a pH of 7.4 at 37°C, which is the mean normal pH of blood. The actual pH determined at 25°C can be converted to the pH at 37°C by use of Rosenthal's factor—decrease of 0.0147 pH units per degree increase in temperature—(186). In serum or

plasma with normal protein and phosphate content, the error involved if the pH of the serum or plasma at 37°C is actually 7.0 or 7.8 is about 3 meq/liter. If the protein content, however, is 12 g/100 ml, the possible error is approximately doubled.

4. *Vacuum extraction versus agitation to remove CO_2*. Vacuum extraction results in complete removal of CO_2 after acidification (44). Swirling in the manner outlined in the procedure was found in our laboratory to yield results identical with vacuum extraction.

5. *Determination of HCO_3^- in duodenal fluid*. This is performed the same as with serum except that titration should always be carried back to the original pH. The normal pH varies from about 6.4 to 8.0 (37, 59, 106), and the HCO_3^- content varies from about 1 to 40 meq/liter (3, 59). For greater accuracy, the pH should be taken immediately after specimen collection since CO_2 is lost and the pH increases rapidly (121).

6. *Stability of samples*. If the pH of the samples is determined immediately after the sample is obtained, the serum or plasma is stable at room temperature under aerobic conditions until bacterial decomposition begins. This is because—regardless of whether the sample loses CO_2 (pH increases) or gains CO_2 (pH decreases)—back-titration to the original pH is always the same. For stability of the sample pH, see under *Blood pH*. The bicarbonate concentration is not significantly influenced by the temperature at which plasma and cells are separated (136, 206).

ACCURACY AND PRECISION

The assay cannot be regarded as absolutely specific for HCO_3^- as CO_3^{2-} and carbamino CO_2 are also determined; however, these components are a small part of the total and are considered as HCO_3^-. The accuracy obtained also depends on whether the titration is carried to the specific pH of the fresh sample (see *Note 3*). Since titration is carried out at room temperature, there is a further inaccuracy of as much as ±1.5% due to variation in the pH temperature coefficient of serum or plasma (see *Blood pH*).

The precision utilizing a pH meter is about ±2%.

NORMAL VALUES

The normal range for venous serum or plasma HCO_3^- is 22–30 meq/liter, with a mean of about 25 meq/liter (4, 207, 217). Values are slightly higher in males than females, higher in venous than arterial plasma by about 1.3 meq/liter, and slightly lower in infants than in adults (4, 223). There is no postprandial change (10).

pCO_2

The dissolved CO_2 in plasma is proportional to the pCO_2 —i.e., the partial pressure of CO_2 in the gas phase with which the solution is in equilibrium.

The fact that the plasma is not in contact with a gas phase anywhere other than in the lungs should not be misleading, as the pCO_2 of a solution not in contact with a gas phase merely refers to the partial pressure of CO_2 in a hypothetical gas phase with which the dissolved CO_2 would be in equilibrium. The pCO_2—the partial pressure of CO_2—can be determined in several ways: (a) The alveolar CO_2 concentration is analyzed and the pCO_2 calculated from the prevailing barometric pressure corrected for the vapor pressure of water. The pCO_2 of alveolar air is virtually the same as the pCO_2 of arterial blood. (b) A bubble of gas is equilibrated with the sample and then analyzed for CO_2. (c) Calculation from blood or plasma pH and plasma total CO_2 content. (d) Interpolation from changes in pH after equilibration with gases of known pCO_2. (e) Direct determination of pCO_2. The first method is not pertinent and is not discussed in this text. The determination of pCO_2 by equilibration of a gas bubble with the sample at 37°C is a difficult technic to learn, tedious in its operation, and its precision is not high (23, 32, 38, 42, 181). The other methods for determining pCO_2 are discussed in greater detail below.

pCO_2 by Calculation

The pCO_2 can be calculated from the pH and CO_2 content by the Henderson-Hasselbalch equation as follows:

$$pCO_2 \text{ (mm Hg)} = \frac{CO_2 \text{ content}}{a\,[\text{antilog}(pH-pK')+1]}$$

The value for a at 38°C is 0.0301 and at 37°C is 0.0303 (80). When the CO_2 content is used in the equation, the pK' value is 6.108 for whole blood at pH 7.4 and for plasma is 6.098 (80, 209). Tables have been devised for facilitating this calculation (129, 147, 206).

The pCO_2 also can be calculated using the HCO_3^- value determined titrimetrically, in which case the pK' value for whole blood at pH 7.4 is 6.103 and for plasma is 6.093 (80, 209). For practical purposes the use of a pK' value of 6.10 is recommended. The nomogram of McLean (145), as shown in Figure 20-6, alleviates the necessity of tedious calculations with the Henderson-Hasselbalch equation.

pCO_2 by Interpolation

When pH of plasma or whole blood is plotted versus log pCO_2, a straight line is obtained (31, 79, 213). In reality, the lines for whole blood are very slightly convex to the right, but within the physiologic range they can be considered as straight lines (31, 213). Thus if the original pH of a sample is known and its pH is then determined after equilibration with gas at different known pCO_2 levels, a plot (Fig. 20-8) of pH versus log pCO_2 allows computation of the original pCO_2 of the sample (17, 24, 31, 79, 183, 213). In the assay, the pH of blood is measured and then aliquots of the blood are equilibrated with two different concentrations of CO_2 gas in O_2. After equilibration the pH of each aliquot is measured and the pH versus the log pCO_2 is plotted. A straight line is drawn between the two points obtained following equilibration; the point on a vertical line intersecting the actual pH indicates the pCO_2 (Fig. 20-8). This technic requires an accurate pH meter and equilibration with gases of accurately known CO_2 concentrations. The slope of the pH/log pCO_2 line varies with the protein and hemoglobin concentration, and the position of the curve varies with the nonrespiratory acid-base balance, thus requiring separate pH/log pCO_2 curves for each determination (13, 24).

A small error may be included in the interpolation technic, since all of the shift in pH is not due to changing the pCO_2; some is due to changing the original pO_2 of the sample. A correction factor has been recommended by some (112, 214), but is reported by others to be unnecessary (27, 175).

pCO_2 by Direct Measurement

The pCO_2 can be measured directly with a pCO_2 electrode as originally described by Stow *et al* (228) and later modified by Severinghaus and Bradley (204). This direct measurement is in reality a modified pH measurement. The electrode consists of a combination reference and glass electrode mounted behind a gas-permeable Teflon membrane (Fig. 20-9). The property of this Teflon membrane is such that small *uncharged* molecules (e.g., dissolved CO_2) readily traverse it, while on the other hand, charged particles (e.g., hydrogen ions) do not (78). The electrode is designed to allow pH measurements of a very thin film of bicarbonate solution that is in equilibrium with the CO_2 of the unknown sample. The dissolved CO_2 in a sample of blood, plasma, or serum diffuses across the membrane into the bicarbonate solution. The dissolved CO_2 in the bicarbonate solution reacts with water to form carbonic acid, and the change in pH caused by the carbonic acid is measured. The pCO_2 of the sample can be determined through use of a linear calibration curve relating log pCO_2 to pH or by direct reading with instruments which have a logarithmic scale for pCO_2. Instruments for direct pCO_2 determination are available from Corning Scientific, Instrumentation Laboratories, Beckman Instruments, and Radiometer. The actual pH of the sample has no effect on the pH of the bicarbonate solution because the membrane is impermeable to hydrogen ions (79).

DETERMINATION OF pCO_2 BY INTERPOLATION

(method of Astrup, Ref. 14, modified)

PRINCIPLE

The relationship of pH to log pCO_2 is linear for any particular blood sample. The values obtained by determining the pH of aliquots of blood

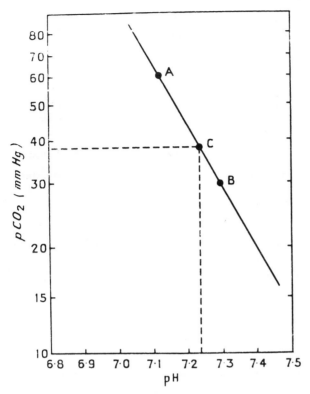

FIG. 20-8. pH/log pCO₂ line for a blood sample. Point A indicates the measured pH value after equilibration at pCO₂ 60 mm Hg. Point B indicates the measured pH value after equilibration at pCO₂ 30 mm Hg. If the pH of the anaerobically drawn blood, 7.23, the pCO₂ would be 38 mm Hg (point C). From ASTRUP P, JØRGENSEN K, SIGGAARD ANDERSEN O, ENGEL K: *Lancet* 1:1035, 1960

equilibrated with gas mixtures having different CO_2 concentrations define a line on a pH-log pCO_2 diagram (Fig. 20-8). By plotting the pH of the unequilibrated sample on this line, the pCO_2 of the sample can be read directly from the diagram.

REAGENTS

O_2-CO_2 Gas Mixtures. Two separate mixtures containing about 4% and 8% CO_2, respectively, in O_2. The exact percentage of CO_2 in each mixture must be accurately known (see *Note 3, pCO₂ of Equilibration Gases*). The pCO_2 of the gas mixtures can be determined as follows:

$$pCO_2 = \text{(barometric pressure to nearest 0.1 mm Hg} - 47.1)}$$
$$\times \text{ \% } CO_2 \text{ in gas mixture}$$

where 47.1 is the vapor pressure of water at 37°C.

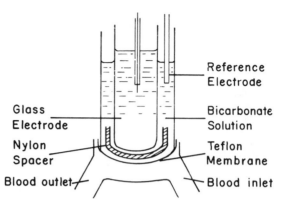

FIG. 20-9. pCO_2 electrode.

pH Reference Solutions, pH 6.841 and 7.384 at 37°C: Reference solutions should meet requirements of the National Bureau of Standards, Washington, DC (see *Electrometric Determination of Blood pH with Glass Electrode*).

NaCl, 0.85%

PROCEDURE

There are a number of good instruments designed to measure blood pH. A general outline of the procedure is presented, but for exact details the reader is referred to the manufacturer's instruction manual for the instrument being used.

1. Refer to the procedure for the *Determination of Blood pH* for general instructions on the calibration and use of the pH meter.
2. Transfer 1 ml of blood or plasma into each of two clean test tubes at 37°C.
3. Equilibrate tube 1 with gas humidified by bubbling through H_2O at 37°C from the low CO_2 cylinder for 10 min at a rate of 1 bubble/sec.
4. Equilibrate tube 2 with humidified gas at 37°C from the high CO_2 cylinder for 10 min at a rate of 1 bubble/sec.
5. Measure the pH of the unequilibrated blood maintained under anaerobic conditions and of the blood in tubes 1 and 2. Rinse the pH electrode with saline between each pH measurement.
6. Recheck the calibration of the pH meter with the pH 7.384 reference solution between each sample assayed.

Calculations:

Plot the pH values obtained on the equilibrated samples versus their respective pCO_2 on the pH-log pCO_2 diagram (Figure 20-8) and draw a straight line between the two points. Plot the pH of the unequilibrated sample on this line and read the pCO$_2$ value from the diagram.

NOTES

1. Sample type. This procedure is generally used only for whole blood, as the Astrup technic is based on fully oxygenated blood and for this reason the equilibration gases contain O$_2$ (see *Blood Collection*). The procedure can also be used to determine the pCO$_2$ of plasma or serum.

2. Temperature. For the most accurate results the instrument should maintain the sample chamber at 37° ± 0.1°C.

3. pCO$_2$ of equilibration gases. The pCO$_2$ of the equilibration gases must be accurately known; since it varies with the barometric pressure, the pCO$_2$ for the gas mixtures must be calculated prior to assay of samples. The 4% and 8% CO$_2$ gas mixtures can be obtained commercially as analyzed gases. The % CO$_2$ in unanalyzed gases can be determined with the Scholander-Roughton volumetric microgasometer or by gas chromatography (189).

4. Sample stability. The pCO$_2$ of drawn blood rises with time, with the actual change varying with temperature and white blood cell count. When the utmost accuracy is required, the pCO$_2$ should be measured within 5 min of sampling. For routine work samples under anaerobic conditions can be left at 25°C for 20 min. If the pCO$_2$ cannot be measured within this period of time, the sample should be stored in ice water where it is stable for 4 hours.

ACCURACY AND PRECISION

The error incurred in this test when the patient has a low hemoglobin concentration or undersaturation with O$_2$ is usually insignificant for clinical purposes, and correction of the results is reported to be unnecessary (27, 175).

The precision of the interpolated pCO$_2$ values is about ±2%.

NORMAL VALUES

See *Normal Values* under *Direct Determination of pCO$_2$*.

DIRECT DETERMINATION OF pCO$_2$

(method of Severinghaus and Bradley, Ref. 204, modified)

PRINCIPLE

The CO$_2$ of the sample diffuses rapidly across a membrane into a bicarbonate solution where H_2CO_3 is formed, causing a change in the pH of the solution. This alteration of pH, which is proportional to the pCO_2 of the sample, is measured using a pH-sensitive electrode system.

REAGENTS

O_2-CO_2 *Gas Mixtures.* Two separate mixtures are required containing about 4% and 8% CO_2, respectively. The percentage of CO_2 in each mixture must be accurately known (see *Note 3*). The pCO_2 of the gas mixtures can be calculated as follows:

$$pCO_2 = \text{(barometric pressure in mm Hg to the nearest 0.1 mm} - 47.1)$$
$$\times \text{ } \% \text{ } CO_2 \text{ in gas mixture}$$

where 47.1 is the vapor pressure of water at 37°C.
  NaCl, 0.85%

PROCEDURE

There are a number of instruments designed to measure pCO_2. A general outline of the procedure is presented, but for exact details the reader is referred to the manufacturer's instruction manual for the instrument being used.

1. Check membrane for leaks by measuring the resistance across the membrane.
2. Permit at least a 15-min warm-up after turning on the water bath.
3. Calibrate the pCO_2 electrode with two gases having known but different pCO_2 levels. The gases are to be humidified and warmed to 37°C just prior to entering the electrode chamber. The gases should always flow from the top stopcock down through the chamber and out the lower stopcock to help keep the membrane clean and free of water droplets.
4. Introduce the blood sample into the electrode chamber through the lower stopcock and inject until blood appears at the upper stopcock. Close the lower stopcock and leave the upper stopcock open to atmospheric pressure.
5. Read the pCO_2 when equilibrium is reached (usually 2–3 min).
6. Flush the electrode chamber with saline and water prewarmed to 37°C. Reflush the electrode chamber with one of the gases prior to measuring another blood sample.

NOTES

1. Sample type. The direct determination of pCO_2 can be performed on almost any biologic fluid. In acid-base studies the normal sample is whole blood or anaerobically separated plasma (see *Blood Collection*).
  2. Temperature. The pCO_2 is influenced by temperature, as the solubility coefficient for CO_2 varies with temperature and the electrode is

very sensitive to temperature changes. The temperature control should be within ±0.05°C. If cold blood is injected into the electrode chamber, the equilibration time must be longer.

3. *pCO₂ of equilibration gases.* The pCO_2 of the equilibration gases has to be accurately known; and since it varies with the barometric pressure, the pCO_2 for the gas mixtures must be calculated prior to assay of samples. The gas mixtures can be obtained commercially as analyzed gases of known composition, or unanalyzed gases can be analyzed in the laboratory with the Scholander-Roughton volumetric microgasometer or by gas chromatography (189).

4. *pCO₂ electrode.* The pCO_2 electrode is sluggish and takes several minutes to reach equilibrium after introduction of the sample. The equilibration time varies with sample temperature, membrane thickness, electrode buffer strength, electrode buffer volume, and the pCO_2. The Teflon membrane when stretched or damaged may develop pinholes permitting transfer of hydrogen ion from samples into the electrolyte, resulting in high pCO_2 readings. The resistance across the membrane should be checked before use according to the manufacturer's instructions to eliminate this source of error. Air bubbles which form in the electrode chamber can be removed by adding an antifoam compound to the saline wash.

ACCURACY AND PRECISION

This assay for the determination of pCO_2 is specific and accurate. The most common sources of error are improper calibration of the instrument and variation in temperature. A difference of 1°C causes about a 5% change in the result obtained (80).

The precision of the method is ±2.5%.

NORMAL VALUES

The normal range for the pCO_2 of arterial and capillary blood is about 33–48 mm Hg with an average of about 40 (13, 15, 79, 80, 88, 133, 208, 210, 213, 217, 233). The range for venous blood is about 38-52 mm Hg with an average of about 46 (74, 233). The pCO_2 of venous blood varies as much as 3 mm between different veins (5, 85, 161, 166). It is about 3 mm lower in infants, about 3 mm lower in females than in males, about 3 mm lower during erect posture than when recumbent, and about 8 mm higher during sleep (5). The pCO_2 also rises with exercise (22).

CALCULATION OF STANDARD BICARBONATE, BUFFER BASE, AND BASE EXCESS

As indicated in an earlier discussion, the standard bicarbonate and buffer base values cannot be determined directly but must be calculated from pH and pCO_2 or HCO_3^- values using the Henderson-Hasselbalch equation or a nomogram derived from the equation (214, 217). Base excess can be determined directly on plasma or whole blood; however, it is difficult to carry out the titration required without dilution of the sample, changes in ionic strength, or hemolysis, all of which alter the result. The base excess for plasma can be calculated from the pH and pCO_2, and for whole blood from the pH, pCO_2, and hemoglobin.

To facilitate calculation of these values Astrup *et al* (13, 15, 210, 213) developed a technic of determining all pertinent values of acid-base status on a nomogram (Fig. 20-10) containing the pH/log pCO_2 diagram (see *pCO_2 by Interpolation*). The nomogram originally developed by Siggaard Andersen and Engel (213), using whole blood containing heparin as anticoagulant and NaF as stabilizer, was subsequently considered incorrect by Siggaard Andersen (210) and was revised by him. The revisions involve maintenance of normal osmolar concentration and ionic strength, use of cooling rather than NaF to decrease the rate of glycolysis, and recalculation of the normal ranges to eliminate the significant effects caused by adding NaF to the samples (210). The revised nomogram (Fig. 20-10) can be used to calculate the pCO_2 (by interpolation technic), base excess, buffer base, and standard bicarbonate. The pH/log pCO_2 line from which the pCO_2 is determined by interpolation also intersects the base excess and buffer base curves in this figure. The point of intersection with the line for pCO_2 40 also intersects the standard bicarbonate line. A problem which arises in using this nomogram is that obtaining the data for the pH/log pCO_2 line requires equilibration of the sample at two pCO_2 levels as well as determination of three pH values. The flaw in this nomogram is that values for CO_2 content or bicarbonate cannot be used.

To overcome the limitations of the nomogram in Figure 20-10, Siggaard Andersen (211) developed an alignment nomogram (Fig. 20-11) from which values for buffer base and base excess can be read when any two of the following are known: pH, pCO_2, $HCO_3{}^-$, and CO_2 content. This new nomogram is a modification of one presented earlier by Van Slyke and Sendroy (244). Values may be obtained for plasma, and for blood if the hemoglobin concentration is known. The buffer base cannot be determined from the nomogram but must be calculated as follows:

Plasma buffer base = normal buffer base + base excess

Blood buffer base = normal buffer base + 0.36 (hemoglobin g/100ml)
         + base excess

The value for blood buffer base increases by 0.36 meq/liter for every gram of hemoglobin according to Singer and Hastings (217) and by 0.42 meq/liter according to Siggaard Andersen (210).

NORMAL VALUES

Standard Bicarbonate. The normal range for standard bicarbonate in

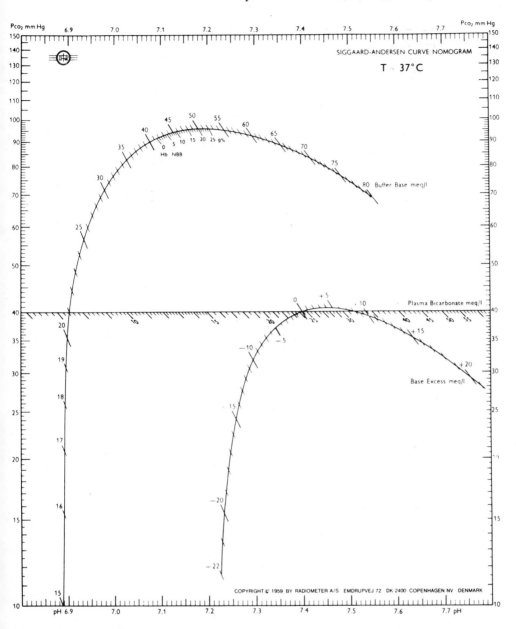

FIG. 20-10. Siggaard Andersen curve nomogram for the calculation of pCO$_2$, base excess, buffer base, and standard bicarbonate. From SIGGAARD ANDERSEN O, *Scand J Clin Lab Invest 14*:598, 1962

arterial plasma or capillary plasma is 21—25 meq/liter (15, 17, 190).

Normal Buffer Base. The normal buffer base for plasma as reported by Singer and Hastings (217) is 40.8 and as reported by Siggaard Andersen (210) is 41.7. The latter value is more commonly used today.

Base Excess. The normal range in capillary blood for base excess using

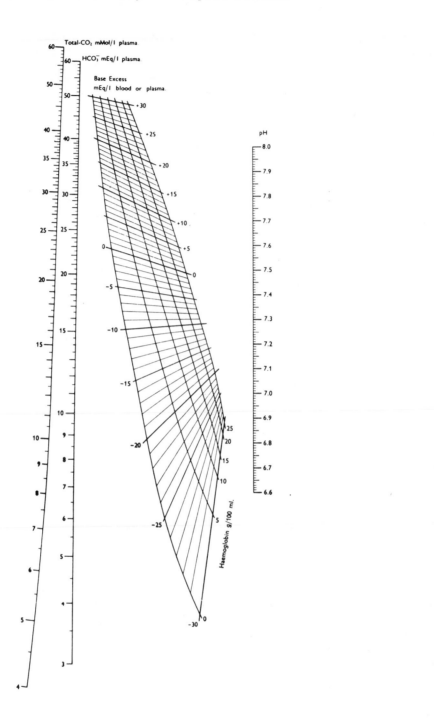

FIG. 20-11. Siggaard Andersen alignment nomogram for the determination of acid-base values of blood. From SIGGAARD ANDERSEN O: *Scand J Clin Lab Invest 15*:211, 1963

the revised nomogram is −2.4 to +2.3 for males and −3.3 to +1.2 for females (80). For venous plasma the range is 0.0 to +5.0 for males and −1.0 to +3.5 for females (233).

BLOOD O₂

The determination of O_2 carried in blood as oxyhemoglobin can be made by the classic gasometric technics of Peters and Van Slyke (volumetric) or Van Slyke and Neill (manometric) as discussed in the previous edition of this book (99) and in the original references (169, 241). The Natelson microgasometer, which also can be used, is a micro version of the Van Slyke apparatus. Other technics include spectrophotometry based on the fact that reduced hemoglobin and oxyhemoglobin possess different absorption spectra, polarography (160), electrochemical measurements (45, 46), gas chromatography (49, 132), microdiffusion (41), and mass spectrography (256).

Measurement of pO_2 by electrode has become the method of choice because of the speed of the determination as well as the small amount of sample necessary. The pO_2 of blood can be measured directly by use of the Clark electrode (36). This electrode (Fig. 20-12) consists of a half-cell constructed from a Pt wire cathode, $10–25\,\mu$ in diameter, sealed in glass with its tip exposed. The cathode and its glass support are covered with a polypropylene O_2-permeable membrane. Between the electrode and membrane is a film of buffered KCl. The other half-cell is a Ag/AgCl anode which also makes contact with the KCl solution. Oxygen molecules from samples or calibrating materials diffuse through the membrane to the negatively charged cathode where reduction occurs. Four electrons are required for the reduction of each O_2 molecule forming H_2O_2 and ultimately OH^-. This flow of electrons is interpreted by appropriate instrumentation and transformed to give a readout of pO_2 in mm Hg. The potential required for polarization of the electrode, −0.7 V, is supplied to the cathode from an external source, while the electrons necessary are generated from the Ag/AgCl half-cell.

FIG. 20-12. pO_2 electrode.

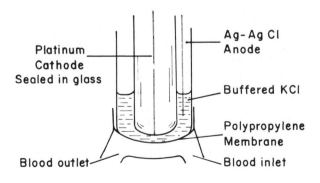

The Clark electrode consumes O_2, but by keeping the area of exposed Pt small and the membrane material somewhat restrictive to the passage of O_2 the fall in pO_2 of a blood sample can be limited to about 1.5% or less. For this reason it is satisfactory to calibrate the electrode with known mixtures rather than blood equilibrated with these gases. The current output of the electrode is linear within ±1% over the entire range of pO_2 values (203).

The membrane material preferred for its response and durability in blood pO_2 measurements is polypropylene. Although on occasion it becomes defective because of the presence of small holes, it is more uniform than polyethylene and more reliably maintains its fit over the electrode. Teflon is more permeable to O_2 than the other materials and gives shorter response times but also causes higher current flow and slightly greater differences between blood and gas measurements. Silastic silicone rubber material needs replacement daily and produces continuous drift in readings (157).

A necessary condition for assuring the linear relationship in electrode response to an increase in O_2 tension is that temperature must be held constant. The commercially available instruments for measuring O_2 tension accomplish this by enclosing both the specimen chamber and the O_2 electrode within the confines of a water jacket maintained at a constant temperature.

The determination of oxygen saturation is commonly carried out by a spectrophotometric procedure. Most of the various modifications of the spectrophotometric technic require special cuvets having a short light path ranging from 0.01 to 5.0 mm (57, 104, 138, 152, 154, 226). There is a greater error between duplicates of the same sample using Lucite cuvets than with glass cuvets (111, 177). This is thought to be due to Lucite's property of absorbing O_2 reversibly to an extent which is dependent upon the O_2 content of the sample previously in the cuvet (hysteresis effect). In addition, the transmittance characteristics among Lucite cuvets are more variable than glass cuvets. A survey of the literature shows a wide range of wavelengths used in the spectrophotometric procedure (47, 87, 138, 215, 235), the choice of wavelength being based on differences in extinction coefficients of oxyhemoglobin and reduced hemoglobin at the particular wavelength chosen and at that of their isosbestic point (65). Another factor to consider in choice of wavelength is the sensitivity of O_2 saturation to wavelength (148); thus the choice of 560 nm is not recommended because of low sensitivity. The choice of hemolytic agent is apparently important for greater accuracy and precision, with Triton X-100 being preferable to saponin (54, 104). Technics have been developed to correct for the presence of carboxy-hemoglobin (26, 97). The spectrophotometric approach is subject to error, especially at low hemoglobin concentrations; these errors arise from the optical characteristics of plasma and saponin and from the presence of cellular fragments. Results improve following high speed centrifugation (174). Abnormal blood pigments and plasma turbidity interfere with the method (153). The gasometric and spectrophotometric methods agree within about 3% (58, 102, 177), with a tendency for the spectrophotometric method to run higher (177). Hemoglobin O_2 saturation has also been determined by reflectometry (81, 176).

DETERMINATION OF pO₂ by O₂ ELECTRODE

PRINCIPLE

A pO_2 electrode consists of a Pt cathode, a Ag-AgCl anode, and an electrolyte solution consisting of buffered KCl. A constant potential of -0.7 V applied to the Pt results in the formation of a layer of H_2 at the cathode, which increases the resistance of the current. Oxygen from the sample, diffusing through a polypropylene membrane, is reduced and forms water at the cathode, which results in a decrease in resistance to the current. The subsequent increase in current is directly proportional to the pO_2 of the sample.

REAGENTS

NaCl, 0.85%
Calibration Gases. Two tanks of gas are needed, one containing no O_2 (such as CO_2/N_2) and one with a known proportion (mol%) of O_2 (such as $O_2/CO_2/N_2$).

PROCEDURE

There are a number of instruments (Radiometer, Corning, Instrumentation Laboratories) designed to measure blood pO_2 by O_2 electrode. A generalized procedure is presented. For specific details, the manufacturer's instructions should be consulted.

1. Calculate the pO_2 of the O_2 calibration gas by the following equation:

$$pO_2 = (\text{barometric pressure} - 47.1) \times \text{mol\% } O_2 \text{ in calibration gas}$$

   where 47.1 is the water vapor pressure at $37°C$, the temperature at which the instrument is set. The barometric pressure should be set to the nearest 0.1 mm Hg.
2. Inject water into the sample compartment to check for possible leaks in the membrane. A leak would be indicated by a full scale pO_2 reading.
3. After aspiration of the water, switch to the calibration gas containing no O_2. The gas is humidified by bubbling through water, the flow rate being about 1–2 bubbles/sec. After meter stabilization, adjust reading to 0mm Hg.
4. Switch to the O_2 calibration gas and, in a manner similar to step 3, adjust the reading to the calculated pO_2.
5. Repeat steps 3 and 4 several times to ensure that stable readings are being obtained.
6. It is essential to keep the membrane moist. If calibration is not completed within 10 min, inject a small amount of water through the sample chamber; then aspirate off the water.

7. Inject blood into and beyond the sample chamber, taking care that no air bubbles remain in the chamber.

8. Allow the pO_2 reading to stabilize, then record.

9. After each reading, flush the sample chamber and lines with 0.85% NaCl, followed by water. Do not flush out sample with air. After proper flushing, a duplicate reading may be obtained.

10. After the last reading, leave sample chamber filled with water.

NOTES

1. Stability of samples. Blood samples should be stored in ice (water ice, *not* Dry Ice) until ready for analysis (178–180). The stability of blood kept anaerobically in ice water varies with both temperature and initial pO_2 (131, 141, 258). It was observed that in ice pO_2 was unchanged for 5 hours at any initial pO_2 and for 1 day if the initial pO_2 was less than 100 mm Hg (258). If conditions are anaerobic, the blood may be kept for 2 hours at room temperature if the initial pO_2 is less than 80 mm Hg (258). Storage temperature affects pO_2, since utilization of O_2 is greater at 37°C than at 0°C (62, 146). Oxygen consumption is mainly by leukocytes (201), not mature erythrocytes (90). Cyanide and fluoride do not affect the rate of consumption (12, 105, 146).

2. Calibration of instrument. For practical purposes, calibration with gases which have been humidified by bubbling through water at low flow rates gives satisfactory results. Holmes *et al* (101) and Elridge and Fretwell (62) report no significant differences between calibration with gases and with blood or liquid equilibrated with the same gas. However, other authors (1, 131, 200, 201) report 2–5% higher readings when the electrode is calibrated with gas than with blood equilibrated with gas. At high pO_2 values (greater than 105 mm Hg) the difference in calibration between gas and blood equilibrated with gas ranges from 10% to 20% in one report (149) and up to 10% in another (131). A procedure (83) in which water equilibrated with room air at various temperatures—thus providing a range of pO_2 values—has been used for electrode calibration. Using this procedure, the resulting blood values are lower by 4% as compared to calibration with blood equilibrated with gas.

The nature of the electrode membrane affects calibration. The pO_2 of blood and of gas is about the same when a polypropylene membrane is used (149, 179). Teflon and polyethylene membranes gave nonlinear responses upon calibration (118, 149).

A longer time is required to obtain a zero reading following a high pO_2 than a low pO_2 (149). Moran *et al* (149) report that 1.5 min is required to obtain a zero reading following measurement of a sample with a pO_2 of 106 mm Hg, but 6 min after one with a pO_2 of 430 mm Hg. In addition, it requires more time to rezero the electrode the longer it is exposed to a high pO_2.

It has been shown that the electrode response using unstirred blood is not identical with the response obtained with flowing gas. This is due to formation of a diffusion gradient arising from the consumption of O_2 by the electrode, which results in a lowered electrode output when calibrated with blood rather than with gas. Stirred blood gives a pO_2 calibration value about the same as that obtained with flowing gas, whereas nonstirred blood gives a lower value. In order to determine the magnitude of the difference (and consequently the correction factor) between the pO_2 obtained by calibration with gas and with blood equilibrated with gas, two general technics have been used. The first technic involves tonometry of blood (1, 101, 201). Severinghaus (201) finds the tonometric procedure difficult and prefers a procedure involving stirring of blood to more closely simulate pO_2 values obtained by calibration with a flowing gas.

There are several factors which affect satisfactory calibration of the electrode. If the gas flow is too fast, the electrode membrane may dry out. Also, exposure to gas for too long a time results in cooling of the electrode (101). It is not necessary to prewarm blood from 0° to 37°C prior to analysis since the small volume of blood is rapidly brought to 37°C in the apparatus.

3. Miscellaneous factors and their effect on pO_2. Elevated numbers of reticulocytes and leukocytes result in an increased rate of decay of pO_2 owing to O_2 consumption by metabolic processes (131). Some factors that do not affect the pO_2 appreciably are the blood pH, hematocrit value, and the presence of heparin (179, 180). If electrode chambers are improperly flushed out, the presence of microorganisms growing mostly in deposits of old blood may result in reduced pO_2 (201). Sterilization of the chamber may be accomplished with benzalkonium chloride. If the electrode is equilibrated with a pO_2 quite different from that of the sample, diffusion of O_2 into electrode membranes, rubber gaskets, and other parts of the electrode may affect the pO_2 (201).

ACCURACY AND PRECISION

As described in each of the *Notes* of this section, many factors may affect the pO_2 reading. The accuracy of the pO_2 electrode can be no better than the accuracy of the calibration procedure. Unless special equipment such as the blood stirring device of Severinghaus (201) is used to improve the overall performance of the instrument, accuracy is of the order of 2–5%.

The reproducibility of the method for determination of pO_2 is about ±4%.

NORMAL VALUES

For normal resting adult males, the arterial pO_2 range is 80–104 mm Hg and the venous pO_2 range is 20–49 mm Hg (5a). Since transfer of

O_2 from blood to alveolar air does not take place as rapidly as the transfer of CO_2, there is a downward pO_2 gradient from alveolar air to arterial blood of about 9 mm at rest and about 16 mm during exercise (133).

CALCULATION OF O_2 CONTENT, CAPACITY, AND PERCENT SATURATION

(method of Severinghaus, Ref. 201)

PRINCIPLE

The position of the blood O_2 dissociation curve on the chart relating pO_2 to percent saturation is influenced by pH and temperature; the shape of the curve, however, remains unchanged (see *Note 4*). Thus for any pH and temperature, the observed pO_2 can be corrected to standard pO_2 (i.e. what the value would be at 37°C and pH 7.4) and the corresponding percent saturation obtained from the O_2 dissociation curve. Kelman and Nunn (113) compiled a useful line chart (Fig. 20-13) for obtaining percent saturation from the observed pH, temperature, base excess and pO_2. If the temperature is held at 37°C, there is no need for temperature correction. Similarly, if base excess is zero, there is no need to correct for this parameter. The O_2 capacity is calculated from the hemoglobin concentration. The O_2 content is then calculated based upon O_2 capacity and percent saturation.

PROCEDURE

Measure blood pO_2, pH (see *Determination of pO_2 by O_2 Electrode, Electrometric Determination of Blood pH with the Glass Electrode*, and *Calculation of Standard Bicarbonate, Buffer Base and Base Excess*), and hemoglobin concentration (*Determination of Blood Hemoglobin, Cyanmethemoglobin Method*, Chapter 23).

Calculations:

(a) % O_2 saturation:

Using Figure 20-13, lines A, B and C, read off the temperature factor (f_t), pH factor (f_{pH}) and base excess factor (f_{BE}) for correcting the observed pO_2 to pO_2 at 37°C pH 7.4, and a base excess of zero. Note that temperature in this line chart refers to the temperature of the blood during measurement of the pO_2 and not body temperature. Substitute the observed pO_2 and the correction factors in the following expression:

pO_2 at base excess of zero, 37°C, pH 7.4 = pO_2 observed

$$\times f_t \times f_{pH} \times f_{BE}$$

After making these corrections of the pO_2 read the corresponding % saturation from column D or E of the line chart.

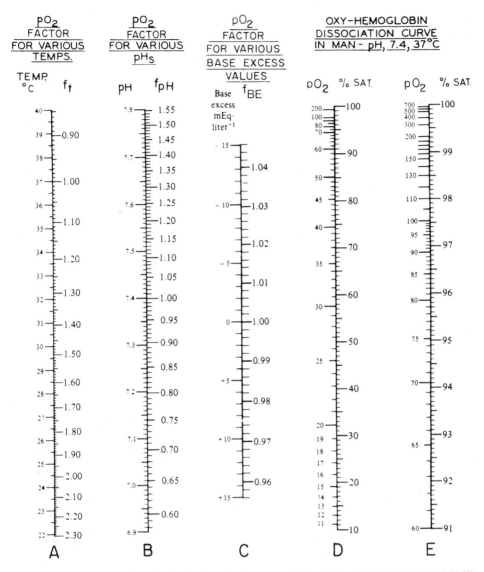

FIG. 20-13. Line chart for the O_2 dissociation curve of blood. From KELMAN GR, NUNN JF: *J Appl Physiol* 21:1484, 1966

(b) O_2 capacity (see *Note 1*):

$$O_2 \text{ capacity (ml } O_2/100 \text{ ml blood)} = \text{g hemoglobin}/100 \text{ ml} \times 1.36$$

where 1.36 is the factor based on the observation that 1.36 ml of O_2 fully saturates 1 g of hemoglobin.

(c)　O_2 content:

$$\frac{O_2 \text{ content}}{(\text{ml } O_2/100 \text{ ml blood})} = \% \ O_2 \text{ saturation} \times O_2 \text{ capacity} \times 10^{-2}$$

NOTES

1. O_2 capacity. The O_2 capacity of blood is its O_2 content when saturated with O_2—i.e., when all the available reduced hemoglobin is converted to oxyhemoglobin. Experimentally, it has been found that 1 g hemoglobin can bind with 1.36 ml O_2 (231). Thus O_2 capacity can be determined from hemoglobin concentration as described above under *Calculations.*

A spectrophotometric technic in which the difference in absorbance at a particular wavelength of a fully reduced and fully oxygenated sample is correlated with O_2 capacity has also been used (184). Gasometric technics following O_2 saturation of samples are the classic procedures used in determining O_2 capacity (99).

One occasionally sees the statement that the O_2 capacity of whole blood is unstable—that it decreases on standing due to formation of methemoglobin. Actually, the O_2 capacity may increase 1—4.5% upon standing at room temperature for 1—2 hours (187, 240).

2. O_2 saturation. The O_2 saturation may be expressed in one of two ways (39):

(a) % saturation $= 100 \times \dfrac{\text{combined } O_2 \text{ content (ml/100 ml)}}{\text{combined O2 capacity (ml/100 ml)}}$

$= 100 \times \dfrac{\text{oxyhemoglobin (g/100 ml)}}{\text{reduced hemoglobin (g/100 ml)} + \text{oxyhemoglobin (g/100 ml)}}$

(b) % saturation $= 100 \times \dfrac{\text{oxyhemoglobin (g/100 ml)}}{\text{total hemoglobin (g/100 ml)}}$

Total O_2 in blood is the sum of the dissolved O_2 and the O_2 from oxyhemoglobin. The term combined O_2 (content or capacity) refers to O_2 from oxyhemoglobin (i.e., total O_2 corrected for dissolved O_2). The two definitions above give the same value for % saturation when total hemoglobin is composed only of reduced and oxyhemoglobin. The Kelman and Nunn line chart is based on standard dissociation curves for normal human blood (113) with % saturation being defined according to definition (a).

Since the dissociation curve is an average curve, the resulting line chart is subject to some variation in normals and perhaps greater variation in the presence of disease. The O_2 saturation values obtained by spectrophotometry correlate with values obtained from polarographic measurement of pO_2 and subsequent determination of O_2 saturation from a dissociation curve (47).

If blood contains carboxyhemoglobin or "inactive" hemoglobin, definitions (a) and (b) do not give the same value for O_2 saturation. The term "inactive" hemoglobin is used to indicate hemoglobin derivatives that do not combine reversibly with O_2 or CO but can be converted into active hemoglobin by reducing agents that convert ferrihemoglobin into ferrohemoglobin. Various estimates of the top limit of "inactive" hemoglobin of freshly drawn blood from apparently healthy individuals have ranged from 2% to 6% of the total pigment (6, 187, 240). The blood of heavy smokers may contain up to 10% of the total hemoglobin as carboxyhemoglobin (97). The presence of carboxyhemoglobin increases the affinity of hemoglobin for oxygen and thereby affects O_2 saturation (97). The "inactive" hemoglobin decreases slowly after drawing until at 2 hours it equals the methemoglobin content, which is stable at room temperature for at least 4 hours and which normally is about 0.4% of the total hemoglobin (240). Thus if O_2 capacity were directly determined, it would increase as saturation with air is continued for 1 hour or more (173). As a consequence, the calculated "% saturation" of the original blood sample decreases. Whether it is more correct to base O_2 saturation on the O_2 capacity before or after this change is a moot point and is dependent on whether expression (a) or (b) above is the desired definition of percent O_2 saturation.

3. O_2 content. By the procedure presented, O_2 content of blood is calculated based on the percent saturation (as derived from the pO_2) reading and O_2 capacity (as derived from the hemoglobin concentration). An alternative procedure (21, 96, 117), using only the oxygen electrode, may be used to obtain O_2 content. The principle is based upon release of the O_2 that was combined with hemoglobin, after addition of carbon monoxide or ferricyanide, and then measuring the change in pO_2.

4. Effect of 1,3-diphosphoglycerate (DPG) on the O_2 dissociation curve. In addition to H^+ and CO_2, red blood cell DPG also enhances release of O_2 from hemoglobin by affecting the position of the O_2 dissociation curve (67). The DPG enters the core of the hemoglobin molecule between the β chains, and binds to each chain only when hemoglobin is in the deoxy form. The resulting stabilization of the deoxy shape of hemoglobin favors O_2 release. A term often used to describe how DPG (or H^+ or CO_2) affects the O_2 dissociation curve is P_{50}, defined as the pO_2 at which 50% of hemoglobin is saturated with O_2. A doubling of the normal red blood cell DPG from 15 to 30 $\mu mol/g$ hemoglobin would result in an increase in P_{50} from 26.5 to 36.8 mm Hg (67). Conditions which affect the DPG concentration—such as abnormalities in glycolytic enzymes resulting in decreased levels of DPG—would affect the binding of O_2 to hemoglobin. Abnormal hemoglobins which affect the affinity for O_2 could result in an abnormal O_2 dissociation curve.

ACCURACY AND PRECISION

The accuracy of measuring O_2 saturation and content by the procedure presented is dependent upon the accuracy of the pO_2 measurement as described in the previous section. It should be noted that when O_2 saturation is derived from the pO_2 value, it is calculated from an O_2 saturation-pO_2 dissociation curve based on data from several sources. Thus changes in DPG concentration or the existence of abnormal hemoglobin can affect the dissociation curve and result in a false O_2 saturation value.

NORMAL VALUES

A reasonable compromise on the reported normal limits for O_2 content is as follows: 15—23 vol% for arterial blood and 10—19 vol% for venous blood (4, 85, 95, 135).

The O_2 capacity of blood depends directly upon the hemoglobin content. The average % O_2 saturation of arterial blood is about 98% (58, 187, 257), with a range of about 95—100% (85, 177), and for venous blood about 55—71% (85).

Since the hemoglobin concentration of blood averages 6% lower in persons in the recumbent position than in those with erect posture, because of increases in plasma volume (122), the O_2 capacity and content also show a parallel change. Females average about 2 vol% lower in O_2 content than males, paralleling the sex difference in hemoglobin (5). There is also a fluctuation of these values in an individual throughout the day, ranging up to 2.3 vol% (144).

REFERENCES

1. ADAMS AP, MORGAN-HUGHES JO: *Br J Anaesth 39:*107,167.
2. ADAMSONS K Jr, SALHA SD, GILLIAN G, JAMES A: *J Appl Physiol 19:*897, 1964.
3. AGREN G, LAGERLOF H: *Acta Med Scand 90:*1, 1936
4. ALBRITTON EC: *Standard Values in Blood.* Philadelphia, Saunders, 1952, p 120
5. ALBRITTON EC: *Standard Values in Blood.* Philadelphia, Saunders, 1952, p 124
5a. ALTMAN PL, DITTNER DS (eds.): *Blood and Other Body Fluids,* Bethesda, Federation of American Societies for Experimental Biology, 1961, p 169
6. AMMUNDSEN E, TRIER M: *Acta Med Scand 101:*451, 1939
7. ANDERSEN OS: *Scand J Clin Lab Invest 13:*196, 1961
8. ANDERSEN OS: *Scand J Clin Lab Invest 13:*205, 1961
9. ANDERSEN OS, ENGEL K, JØRGENSEN K, ASTRUP P: *Scand J Clin Lab Invest 12:*172, 1960
10. ANNINO JS, RELMAN AS: *Am J Clin Pathol 31:*155, 1959
11. ARBUS GS, HERBERT LA, LEVESQUE PR, ETSTEN BE, SCHWARTZ WB:*N Engl J Med 280:*117, 1969
12. ASMUSSEN E, NEILSEN M: *Scand J Clin Lab Invest 13:*297, 1961
13. ASTRUP P: *Clin Chem 7:*1 1961

14. ASTRUP P: *Scand J Clin Invest 8*:33, 1956
15. ASTRUP P, JØRGENSEN K, SIGGAARD ANDERSEN O, ENGEL K: *Lancet 1*:1035, 1960
16. ASTRUP P, SCHRØDER S: *Scand J Clin Lab Invest 8*:30, 1956
17. ASTRUP P, SIGGAARD ANDERSEN O: *Adv Clin Chem 6*:1, 1963
18. AUSTIN WH, FERRANTE VV, ANDERSON C: *J Lab Clin Med 72*:129, 1968
19. AUSTIN WH, FERRANTE VV, RITCHIE RF: *Am J Clin Pathol 51*:799, 1969
20. AUSTIN WH, LITTLEFIELD SC: *J Lab Clin Med 67*:516, 1966
21. AWAD O, WINZLER RJ: *J Lab Clin Med 58*:489, 1961
22. BALDWIN EdeF, COURNAND A, RICHARDS DW Jr: *Medicine (Baltimore) 27*:243, 1948
23. BATES GD, OLIVER TK Jr: *J Appl Physiol 17*:743, 1962
24. BERGLUND E, MALMBERG R, STENHAGEN S: *Scand J Clin Lab Invest 16*:185, 1964
25. BERKENBOSCH A: *Acta Physiol Pharmacol Neerl 10*:101, 1961
26. BJURE J, NILSSON NJ: *Scand J Clin Lab Invest 17*:491, 1965
27. BONHAM TJ, SAMMONS HG: *Clin Chim Acta 29*:507, 1970
28. BOONE CW, FIELD JB: *Calif Med 79*:420, 1953
29. BOWER VE, PAABO M, BATES RG: *Clin Chem 7*:292, 1961
30. BREEN M, FREEMAN S: *Clin Chim Acta 6*:181, 1961
31. BREWIN EG, GOULD RP, NASHAT FS, NEIL E: *Guys Hosp Rep 104*:177, 1955
32. BRINKMAN GL, JOHNS CJ, DONOSO H, RILEY RL: *J Appl Physiol 7*:340, 1954
33. BROOKS D, WYNN V: *Lancet 1*:227, 1959
34. BUEDING E, GOLDFARB W: *J Biol Chem 141*:539, 1941
35. BURTON GW: *Br J Anaesth 37*:89, 1965
36. CLARK LC Jr: *Trans Am Soc Artif Intern Organs 2*:41,1956
37. COMFORT MW, OSTERBERG AE: *Arch Intern Med 66*:688, 1940
38. COMROE JH Jr: *Methods in Medical Research.* Vol 2, Chicago, Year Book Publishers, 1950, p 162
39. COMROE JH Jr: *Methods in Medical Research*, Vol 2, Chicago, Year Book Publishers, 1950, p 167
40. CONWAY EJ: *Microdiffusion Analysis and Volumetric Error.* Princeton, NJ, Van Nostrand, 1950
41. CONWAY EJ, MASTERSON BF: *J Lab Clin Med 68*:824, 1966
42. COTES JE, OLDHAM PD: *J Appl Physiol 14*:467, 1959
43. CREMER M: *Z Biol 47*:562, 1906
44. CULLEN GE: *J Biol Chem 30*:369, 1917
45. DAMASCHKE K, SALING E: *Klin Wochenschr 37*:826, 1959
46. DAMASCHKE K, SALING E: *Klin Wochenschr 39*:265, 1961
47. DANZER LA, COHN JE: *J Lab Clin Med 63*:355, 1964
48. DAVENPORT HW: *The ABC of Acid-Base Chemistry.* Fourth edition. Chicago, University of Chicago Press, 1958
49. DAVIES DD: *Br J Anaesth 42*:19, 1970
50. DAVIS S, SIMPSON TH Jr: *J Biol Chem 219*:885, 1956
51. DEANE N, SMITH HW: *J Biol Chem 227*:101, 1957
52. D'ELSEAUX FC, BLACKWOOD FC, PALMER LE, SLOMAN KG: *J Biol Chem 144*:529, 1942
53. DESAI SD, HOLLOWAY R, THAMBIRAN AK: *Lancet 2*:1126, 1965
54. DIEBLER GE, HOLMES MS, CAMPBELL PL, GANS J: *J Appl Physiol 14*:133, 1959
55. DILL DB, DALY C, FORBES WH: *J Biol Chem 117*:569, 1937
56. DILL DB, EDWARDS HT, CONSOLAZIO WV: *J Biol Chem 118*:635, 1937

57. DRABKIN DL, AUSTIN JH: *J Biol Chem 112*:105, 1935-1936
58. DRABKIN DL, SCHMIDT CF: *J Biol Chem 157*:69, 1945
59. DREILING DA, JANOWITZ HD: *Gastroenterology 30*:382, 1956
60. DRESSLER DP, MASTIO GJ, ALLBRITTEN FF Jr: *J Lab Clin Med 55*:144, 1960
61. EISENMAN AJ: *J Biol Chem 71*:611, 1926-1927
62. ELRIDGE F, FRETWELL LK: *J Appl Physiol 20*:790, 1965
63. ESCHENBACH C, RAUSCH-STROOMANN JG: *Klin Wochenschr 39*:693, 1961
64. EVANS CL: *J Physiol 56*:146, 1922
65. FALHOLT W: *Scand J Clin Lab Invest 15*:67, 1963
66. FERGUSON JH, DuBOIS D: *J Lab Clin Med 21*:663, 1936
67. FINCH CA, LENFANT C: *N Engl J Med 286*:407, 1972
68. FLEISHER M, SCHWARTZ MK: *Clin Chem 17*:610, 1971
69. FORSTER HV, DEMPSEY JA, THOMSON J, VIDRUK E, DoPICO GA: *J Appl Physiol 32*:134, 1972
70. FREIER EF, CLAYSON KJ, BENSON ES: *Clin Chim Acta 9*:348, 1964
71. FRIEDNER S, PHILIPSON A: *Scand J Clin Lab Invest 17*:185, 1965
72. GAMBINO SR: *Am J Clin Pathol 32*:270, 1959
73. GAMBINO SR: *Am J Clin Pathol 32*:285, 1959
74. GAMBINO SR: *Am J Clin Pathol 32*:294, 1959
75. GAMBINO SR: *Am J Clin Pathol 32*:298, 1959
76. GAMBINO SR: *Am J Clin Pathol 32*:301, 1959
77. GAMBINO SR: *Am J Clin Pathol 35*:175, 1961
78. GAMBINO SR: *Clin Chem 7*:236, 1961
79. GAMBINO SR: *Standard Methods of Clinical Chemistry.* Edited by S. Meites. New York, Academic Press, 1965, Vol 5, p 169
80. GAMBINO SR, ASTRUP P, BATES RG, CAMPBEL EJM, CHINARD FP, NAHAS GG, SIGGAARD ANDERSEN O, WINTERS R: *Am J Clin Pathol 46*:376, 1966
81. GAMBINO SR, GOLDBERG HE, POLANYI ML: *Am J Clin Pathol 42*:364, 1964
82. GAMBINO SR, SCHREIBER H: *Am J Clin Pathol 45*:406, 1966
83. GELDER RL, NEVILLE JF: *Am J Clin Pathol 55*:325, 1971
84. GEUBELLE F: *Clin Chim Acta 2*:442, 1957
85. GIBBS EL, LENNOX WG, NIMS LF, GIBBS FA: *J Biol Chem 144*:325, 1942
86. GOLDSTEIN F, GIBBON JH Jr, ALLBRITTEN FF Jr, STAYMAN JW Jr: *J Biol Chem 182*:815, 1950
87. GORDY E, DRABKIN DL: *J Biol Chem 227*:285, 1957
88. GRAHAM BD, KOEFF ST, TSAO MU, SLOAN C, WILSON JL: *Am J Dis Child 98*:593, 1959
89. GRAIG FA, LANGE K, OBERMAN J, CARSON S: *Arch Biochem Biophys 38*:357, 1952
90. HARROP GA Jr: *Arch Intern Med 23*:745, 1919
91. HASSELBALCH KA: *Biochem Z 78*:112, 1917
92. HASTINGS AB, SENDROY J Jr: *J Biol Chem 65*:445, 1925
93. HASTINGS AB, SENDROY J Jr, VAN SLYKE DD: *J Biol Chem 79*:183, 1928
94. HAVARD RE, KERRIDGE PT: *Biochem J 23*:600, 1929
95. HAWK PO, OSER BL, SUMMERSON WH: *Practical Physiological Chemistry.* Twelfth edition. Philadelphia, Blakiston, 1947, p 451
96. HEDEN M: *Br J Anaesth 42*:15, 1970
97. HELLUNG-LARSEN P, KJELDSEN K, MELLEMGAARD K, ASTRUP P: *Scand J Clin Lab Invest 18*:443, 1966
98. HENDERSON LJ: *Ergeb Physiol 8*:254, 1909
99. HENRY RJ: *Clinical Chemistry—Principles and Technics.* New York, Harper & Row, 1964, p 422

100. HOBSLEY M: *Clin Chim Acta 12*:493, 1965
101. HOLMES PL, GREEN HE, LOPEZ-MAJANO V: *Am J Clin Pathol 54*:566, 1970
102. HOLMGREN A, PERNOW B: *Scand J Clin Lab Invest 2*:143, 1959
103. JENNY H, NIELSON TR, COLEMAN NT, WILLIAMS DE: *Science 112*:164, 1950
104. JOHNSTON GW: *Standard Methods of Clinical Chemistry.* Edited by D Seligson. New York, Academic Press, 1963, p 183
105. JOHNSTONE JM: *J Clin Pathol 19*:357, 1966
106. JONES JA: *J Pediat 49*:672, 1956
107. JØRGENSEN K: *Scand J Clin Lab Invest 8*:168, 1956
108. JØRGENSEN K, ASTRUP P: *Scand J Clin Lab Invest 9*:122, 1957
109. JUNG RC, BALCHUM OJ, MASSEY FJ: *Am J Clin Pathol 45*:129, 1966
110. KAHN A: *J Lab Clin Med 46*:312, 1955
111. KARENDAL B, MICHAELSSON M, STROM G: *Scand J Clin Lab Invest 22*:57, 1968
112. KELMAN GR: *Clin Chim Acta 22*:277, 1968
113. KELMAN GR, NUNN JF: *J Appl Physiol 21*:1484, 1966
114. KENNY MA, CHANG MH: *Clin Chem 18*:352, 1972
115. KINTNER EP: *Am J Clin Pathol 47*:614, 1967
116. KINZLMEIER H: *Z Klin Med 154*:55, 1956
117. KLINGENMAIER CH, BEHAR MG, SMITH TC: *J Appl Physiol 26*:653, 1969
118. KOCH B: *Clin Biochem 2*:399, 1969
119. KOCH G: *Scand J Clin Lab Invest 17*:223, 1965
120. KOCH G: *Scand J Clin Lab Invest 21*:10, 1968
121. LAGERLOF HO: *Acta Med Scand [Suppl] 128*:1, 1942
122. LANGE HF: *Acta Med Scand [Suppl] 176*, 1946
123. LANGLANDS JHM, WALLACE WFM: *Lancet 1*:315, 1965
124. LATHAM W, LESSER GT, MESSINGER WJ, GOLDSTON M: *Arch Intern Med 93*:550, 1954
125. LAUG EP: *J Biol Chem 106*:161, 1934
126. LAUGHLIN DHE, McDONALD JS, BEDELL GN: *Fed Proc 22*:573, 1963
127. LAUGHLIN DHE, McDONALD JS, BEDELL GN: *J Lab Clin Med 64*:330, 1964
128. LEITNER MJ: *Am J Clin Pathol 40*:299, 1963
129. LEITNER MJ, THALER S: *Am J Clin Pathol 33*:362, 1960
130. LENFANT C: *J Appl Physiol 16*:909, 1961
131. LENFANT C, AUCUTT C: *J Appl Physiol 20*:503, 1963
132. LENFANT C, AUCUTT C: *Resp Physiol 1*:398, 1966
133. LILIENTHAL JL Jr, RILEY RL, PROEMMEL DD, FRANKE RE: *Am J Physiol 147*:199, 1946
134. LUBBROOK J: *A Symposium on pH and Blood Gas Measurement.* Edited by RF Woolmer. Boston, Little, Brown, 1959, p 34
135. LUNDSGAARD C: *J Biol Chem 33*:133, 1917
136. MAAS AHJ: *Clin Chim Acta 29*:567, 1970
137. MAAS AHJ, VAN HEIJST ANP: *Clin Chim Acta 6*:31, 1961
138. MAAS AHJ, ZUIJDGEEST LPWA, KREUKNIET J: *Clin Chim Acta 9*:236, 1964
139. MacINNES DA, DOLE M: *Industr Engin Chem Anal Ed 1*:57, 1929
140. MacINNES DA, DOLE M: *J Am Chem Soc 52*:29, 1930
141. MacINTYRE J, NORMAN JN, SMITH G: *Br Med J 5619*:640, 1968
142. MANFREDI F: *J Lab Clin Med 59*:128, 1962
143. MATTOCK G: *Symposium on pH and Blood Gas Measurement.* Edited by RF Woolmer. Boston, Little, Brown, 1959, p 19
144. McCARTHY EF, VAN SLYKE DD: *J Biol Chem 128*:567, 1939
145. McLEAN FC: *Physiol Rev 18*:495, 1938

146. MELLEMGAARD K: *Scand J Clin Lab Invest 18*:380, 1966
147. MILCH RA, BANE HN, ROBERTS KE: *J Appl Physiol 10*:151, 1957
148. MOOK GA, VAN ASSENDELFT OW, ZIJLSTRA WG: *Clin Chim Acta 26*:170, 1969
149. MORAN F, KETTEL LJ, CUGELL DW: *J Appl Physiol 21*:725, 1966
150. MURRAY JT: *Am J Clin Pathol 26*:83, 1956
151. NADEAU G: *Am J Clin Pathol 23*:710, 1953
152. NAERAA N: *Scand J Clin Lab Invest 16*:45, 1964
153. NAHAS GG: *J Physiol 47*:867, 1955
154. NAHAS GG: *Science 113*:723, 1951
155. NATELSON S: *J Clin Pathol 21*:1153, 1951
156. NATELSON S: *Microtechniques of Clinical Chemistry*.Second edition. Springfield, Ill, Charles C Thomas, 1961
157. NATELSON S: *Techniques of Clinical Chemistry*. Springfield, Ill, Charles C Thomas, 1971, p 536
158. NATELSON S, BARBOUR JH: *Am J Clin Pathol 22*:426, 1952
159. NATELSON S, TIETZ N: *Clin Chem 2*:320, 1956
160. NEVILLE JR: *J Appl Physiol 15*:717, 1960
161. NUNN JF: *Symposium on pH and Blood Gas Measurement*. Edited by RF Woolmer. Boston, Little, Brown, 1959, p 60
162. NUNN JF, BERGMAN NA, BUNATYAN A, COLEMAN AJ: *J Appl Physiol 20*:23, 1965
163. PAINE EG, BOUTWELL JH, SOLOFF LA: *Am J Med Sci 242*:431, 1961
164. PAULSEN L: *Scand J Clin Lab Invest 9*:402, 1957
165. PETERS JP: *J Biol Chem 56*:745, 1923
166. PETERS JP, BARR DP, RULE FD: *J Biol Chem 45*:489, 1920
167. PETERS JP, EISENMAN AJ, BULGER HA: *J Biol Chem 55*:709, 1923
169. PETERS JP, VAN SLYKE DD: *Quantitative Clinical Chemistry*. Baltimore, Williams & Wilkins, 1932, Vol 2
170. PLATT BS: *J Lab Clin Med 22*:1115, 1937
171. PLATT BS, DICKINSON S: *Biochem J 27*:1069, 1933
172. PURCELL MK, STILL GM, RODMAN T, CLOSE HP: *Clin Chem 7*:536, 1961
173. RAMSAY WNM: *Biochem J 40*:286, 1946
174. RAND PW, LACOMBE E, BARKER N: *J Lab Clin Med 69*:862, 1967
175. REDSTONE D, BEARD RW: *Clin Chim Acta 27*:317, 1970
176. REFSUM HE, HISDAL B: *Scand J Clin Lab Invest 10*:439, 1958
177. REFSUM HE, SVEINSSON SL: *Scand J Clin Lab Invest 8*:67, 1956
178. REPPETO NP: *Am J Med Technol 31*:425, 1965
179. RHODES PG, MOSER KM: *J Appl Physiol 21*:729, 1966
180. RHODES PG, MOSER KM, SCARBOROUGH WR: *Fed Proc 23*:568, 1964
181. RILEY RL, CAMPBELL EJM, SHEPARD RH: *J Appl Physiol 11*:245, 1957
182. RISPENS P, BRUNSTING JR, ZILJLSTRA WG, VAN KAMPEN EJ: *Clin Chim Acta 22*:261, 1968
183. RISPENS P, VAN ASSENDELFT OW, BRUNSTING JR, ZIJLSTRA WG: *Clin Chim Acta 14*:760, 1966
184. RODDIE IC, SHEPHERD JT, WHELAN RF: *J Clin Pathol 10*:115, 1957
185. RODKEY FL: *J Biol Chem 236*:1589, 1961
186. ROSENTHAL TB: *J Biol Chem 173*:25, 1948
187. ROUGHTON FJW, DARLING RC, ROOT WS: *Am J Physiol 142*:708, 1944
188. SAITO K, HONDO Y: *Nisshin Igaku 42*:167, 1955; *Chem Abst 49*:14084, 1955
189. SCHOLANDER PF, ROUGHTON FJW: *J Industr Hyg Toxicol 24*:218, 1942
190. SCHWARTZ WB, RELMAN AS: *N Engl J Med 268*:1382, 1963

191. SCRIBNER BH: *Mayo Clin Proc 25*:641, 1950
192. SEARCY RL, GORDON GF, SIMMS NM: *Lancet 2*:1232, 1963
193. SEGAL MA: *Am J Clin Pathol 25*:1212, 1955
194. SELIGSON D, SELIGSON H: *Anal Chem 23*:1877, 1951
195. SEMPLE SJG: *J Appl Physiol 16*:576, 1961
196. SEMPLE SJG, MATTOCK G, UNCLES R: *J Biol Chem 237*:963, 1962
197. SENDROY J Jr: *Am Soc Testing Materials Spec Techn Pub 190*:55, 1956
198. SENDROY J Jr, DILLON RT, VAN SLYKE DD: *J Biol Chem 105*:597, 1934
199. SENDROY J Jr, RODKEY FL: *Clin Chem 7*:646, 1961
200. SEVERINGHAUS JW: *Acta Anaesthesiol Scand 11*(Suppl): 207, 1962
201. SEVERINGHAUS JW: *Ann NY Acad Sci 148*:115, 1968
203. SEVERINGHAUS JW: *Handbook of Physiology.* Section 3, Respiration. Edited by W Fenn. Bethesda, Md. American Physiological Society, 1965, Vol 2, p 1475
204. SEVERINGHAUS JW, BRADLEY AF: *J Appl Physiol 13*:515, 1958
205. SEVERINGHAUS JW, STUPFEL M, BRADLEY AF: *J Appl Physiol 9*:189, 1956
206. SEVERINGHAUS JW, STUPFEL M, BRADLEY AF: *J Appl Physiol 9*:197, 1956
207. SHOCK NW, HASTINGS AB: *J Biol Chem 104*:585, 1934
208. SIGGAARD ANDERSEN O: *Scand J Clin Lab Invest 12*:311, 1960
209. SIGGAARD ANDERSEN O: *Scand J Clin Lab Invest 14*:587, 1962
210. SIGGAARD ANDERSEN O: *Scand J Clin Lab Invest 14*:598, 1962
211. SIGGAARD ANDERSEN O: *Scand J Clin Lab Invest 15*:211, 1963
212. SIGGAARD ANDERSEN O: *Scand J Clin Lab Invest 21*:289, 1968
213. SIGGAARD ANDERSEN O, ENGEL K: *Scand J Clin Lab Invest 12*:177, 1960
214. SIGGAARD ANDERSEN O, ENGEL K, JØRGENSEN K, ASTRUP P: *Scand J Clin Lab Invest 12*:172, 1960
215. SIGGAARD ANDERSEN O, JØRGENSEN K, NAERAA N: *Scand J Clin Lab Invest 14*:298, 1962
216. SINGER RB: *Am J Med Sci 221*:199, 1951
217. SINGER RB, HASTINGS AB: *Medicine (Baltimore) 27*:223, 1948
218. SINGER RB, SCHOHL J, BLUEMLE DB: *Clin Chem 1*:287, 1955
219. SKEGGS LT Jr: *Am J Clin Pathol 33*:181, 1960
220. SKOTNICKY J: *Z Physik Chem A191*:180, 1942
221. SOBEL AE, EICHEN S: *Proc Soc Exp Biol Med 79*:629, 1952
222. SPINNER MB, PETERSEN GK: *Scand J Clin Lab Invest 13*:1, 1961
223. SPIVEK ML: *J Pediat 48*:581, 1956
224. SPOCK A, HINTON ML, ALBERTSON TH: *J Pediat 68*:987, 1966
225. STANAGE WF, YANKTON SD, BROWN JM, OBERST BB: *J Pediat 47*:571, 1955
226. STEEL AE: *Proc Assoc Clin Biochem 4*:1, 1966
227. STILL G, RODMAN T: *Am J Clin Pathol 38*:435, 1962
228. STOW RW, BAER RF, RANDALL BF: *Arch Phys Med Rehabil 38*:646, 1957
229. STRAUMFJORD JV Jr: *Standard Methods of Clinical Chemistry.* Edited by D Seligson. New York, Academic Press, 1958, Vol 2, p 107
230. SUBBENAM TA: *Pharm Weekbl 103*:837, 1968
231. SUNDERMAN FW: *Hemoglobin, Its Precursors and Metabolites.* Edited by FW Sunderman, FW Sunderman Jr. Philadelphia, Lippincott, 1964, p 17
232. SUNDERMAN FW, BOERNER F: *Normal Values in Clinical Medicine.* Philadelphia, Saunders, 1949
233. *Todd-Sanford: Clinical Diagnosis by Laboratory Methods.* Fourteenth edition. Edited by I Davidsohn, JB Henry. Philadelphia, Saunders, 1962, p 653
234. TORJUSSEN W, NITTER-HAUGE S: *Scand J Clin Lab Invest 19*:79, 1967
235. TSAO MU, SETHNA SS, SLOAN CH, WYNGARDEN LJ: *J Biol Chem 217*:479, 1955

236. ULLIAN RB, KOGOS IG, BOLUB M, STEIN M: *N Engl J Med 265*:235, 1961
237. VAN SLYKE DD: *Personal communication*
238. VAN SLYKE DD, CULLEN GE: *J Biol Chem 30*:289, 1917
239. VAN SLYKE DD, HANKES LV, VITOLS JH: *Clin Chem 12*:849, 1966
240. VAN SLYKE DD, HILLER A, WEISIGER JR, CRUZ WO: *J Biol Chem 166*:121, 1946
241. VAN SLYKE DD, NEILL JM: *J Biol Chem 61*:523, 1924
242. VAN SLYKE DD, PLAZIN J: *Micrometric Analysis*. Baltimore, Williams & Wilkins, 1961
243. VAN SLYKE DD, SENDROY J Jr: *J Biol Chem 73*:127, 1927
244. VAN SLYKE DD, SENDROY J Jr: *J Biol Chem 79*:781, 1928
245. VAN SLYKE DD, SENDROY J Jr, HASTINGS AB, NEILL JM: *J Biol Chem 78*:765, 1928
246. VAN SLYKE DD, SENDROY J Jr, LIU SH: *J Biol Chem 95*:547, 1932
247. VAN SLYKE DD, STADIE WC: *J Biol Chem 49*:1, 1921
248. VAN SLYKE DD, STILLMAN E, CULLEN GE: *J Biol Chem 38*:167, 1919
249. VAN SLYKE DD, WEISIGER JR, VAN SLYKE KK: *J Biol Chem 179*:743, 1949
250. VINCENT D, MIGNON M: *Ann Biol Clin (Paris) 17*:104, 1959
251. WARE AG, NOWACK J, WESTOVER L: *Clin Chem 9*:340, 1963
252. WEBBER RB: *The Book of pH*. London, George Newnes Limited, 1957, p 70
253. WIESINGER P, ROSSIER H, SABOZ E, SAMPAOLO G: *Helv Physiol Pharmacol Acta 7*:28, 1949
254. WILSON RH: *J Lab Clin Med 37*:129, 1951
255. WILSON RH, JAY BE, CHAPMAN CB: *Fed Proc 20*:422, 1961
256. WOLDRING S, OWENS G, WOOLFORD DC: *Science 153*:885, 1966
257. WOOD EH: *J Appl Physiol 1*:567, 1949
258. WOOLF CR: *J Lab Clin Med 69*:853, 1967
259. WORTH G: *Lancet 2*:907, 1965
260. WRIGHT MP: *Symposium on pH and Blood Gas Measurement*. Edited by RW Woolmer. Boston, Little, Brown, 1959, p 5
261. WYNN V, LUDBROOK J: *Lancet 272*:1068, 1957
262. YOSHIMURA H: *J Biochem 22*:279, 1935

Chapter 21

Enzymes

JAMES A. DEMETRIOU, Ph.D.
PATRICIA A. DREWES, Ph.D.
JERRY B. GIN, Ph.D.

The concentration of an enzyme in a tissue or fluid can be determined by measuring the rate of a reaction catalyzed by the enzyme. Either the decrease in substrate concentration with time or the increase in some reaction product can be measured. In the past the concentration of an enzyme in solution was expressed in arbitrary units defined as the amount of transformation of a given quantity of substrate in a specified time under certain standard conditions. For example, the Somogyi unit of amylase is that amount of enzyme in 100 ml of serum which splits reductants equivalent to 1 mg glucose from starch substrate in 30 min at 40°C under certain other conditions specified in the test. The purity of an enzyme preparation is measured by the number of units of activity per gram or milligram of dry weight or per milligram of nitrogen. The Commission on Enzymes of the International Union of Biochemistry defined a unit of enzyme activity as 1 micromole (μmol) of substrate utilized per minute under specified conditions of pH and temperature control, and the specific activity as micromoles per minute per milligram of protein—i.e., units per milligram of protein (211). The commission further recommended that the concentrations of units be in terms of a milliliter, but units per liter is now considered to be more applicable in clinical chemistry (102). Where the nature or concentration of the substrate is in question, it was recommended that the units be in terms of microequivalents of the analyzable substances or groups—e.g., fatty acids, reducing groups, amino or carboxyl groups (102, 391). Although it is quite possible that such units will eventually supplant the older arbitrary units, the latter have become so firmly entrenched, at least with well established assay technics, that it is unlikely that this will occur easily or soon.

Many factors may affect enzyme activity, including substrate concentration, temperature, pH, and the concentrations of activators and inhibitors. All such variables can be and are controlled in a well designed enzyme assay, but activators and inhibitors present in biologic samples are usually unknown to the analyst and cannot be controlled. It follows, therefore, that measurement of enzyme activity is not necessarily a reliable measure of enzyme concentration.

Unfortunately, many methods available to the clinical chemist for determination of enzymes have been improperly designed. There have been instances, for example, in which variables such as pH or temperature were not controlled. In a well designed method, where possible, the reaction follows zero order kinetics—i.e., the reaction rate is constant and independent of substrate concentration (within wide limits). In such a situation results are directly proportional to enzyme concentration—e.g., use of half the amount of sample in the test yields half the results. To be sure, this is not always possible.

Since primary standards are not available, it is imperative that conditions of enzyme assays be duplicated exactly each time they are run and be exactly the same as the conditions used when normal values were obtained with the method. Inclusion of a pooled serum or lyophilized serum in each run is strongly recommended. This at least provides assurance of day to day constancy of technic.

For many enzyme tests that lack a suitable reference standard—especially those that use coenzymes as an indirect measure of substrate transformation per unit time—the absorbance reading is used to calculate enzyme units. The use of absorbance readings to measure dehydrogenases, transaminases, phosphotransferases, and lyases is widely practiced because of variations in purity, absorptivity, and stability of solutions of coenzymes as well as for simplicity and convenience. Henry (314), in pointing out the variation in absorbance readings with various spectrophotometers, brought to the attention of the clinical chemist the importance of using an arbitrary standard to check the constancy of photometers. The proposal to use potassium dichromate as an arbitrary standard (314) also was recommended for enzyme assays to guarantee "absolute" photometric absorbance measurements, eliminate differences between spectrophotometers, and ensure accuracy for those enzyme tests that require indirect standardization (481).

Regardless of the manifold problems concerned with the standardization of enzyme methods, efforts directed toward reducing the large variety of substrates, buffers, and conditions used for many of the routinely assayed enzymes are still continuing (102, 520, 585).

The order of appearance of enzymes in this chapter is according to the classification and numerical coding established by the Enzyme Commission. The trivial name of each enzyme is used, along with the generally accepted abbreviation, the systematic name, and code number—e.g., lactic dehydrogenase, LDH, L-lactate:NAD oxidoreductase, EC 1.1.1.27.

LACTIC DEHYDROGENASE

Lactic dehydrogenase (LDH, L-lactate:NAD oxidoreductase, EC 1.1.1.27) catalyzes the following reaction:

$$
\begin{array}{ccc}
\text{CH}_3 & & \text{CH}_3 \\
| & & | \\
\text{HCOH} \quad + \quad \text{NAD}^+ \;\rightleftharpoons\; & \text{C=O} \quad + \quad \text{NADH} \; + \; \text{H}^+ \\
| & & | \\
\text{COOH} & & \text{COOH} \\
\text{L-LACTIC} & & \text{PYRUVIC} \\
\text{ACID} & & \text{ACID}
\end{array}
$$

The reaction is reversible, with the equilibrium strongly favoring formation of lactic acid. The oxidation of lactic acid to pyruvate, commonly called the forward reaction, has an optimal pH in the alkaline range (8.8–9.8). The reduction of pyruvate, commonly called the backward reaction, has an optimal pH range of 7.2–7.8. The optimal pH is dependent upon a number of factors such as enzyme source (i.e., which isoenzyme is most abundant), temperature, buffer, and substrate concentration (248).

LDH is not specific for pyruvate or lactate. In the backward reaction, LDH reacts with α-keto and α, γ-diketo acids in general. In the forward reaction, LDH reacts with α-hydroxy and α-hydroxy-γ-ketoacids. The α-hydroxybutyric acid dehydrogenase activity is based on the nonspecificity of one particular LDH isoenzyme, LDH_1 (LDH_2 to a lesser extent), and its reaction with α-hydroxybutyric acid. LDH reacts only with the L anomer of lactate and only with the nicotinamide adenine dinucleotide (NAD or NADH) form of the coenzyme, not with nicotinamide adenine dinucleotide phosphate (NADP). NAD and NADP were formerly designated DPN and TPN, respectively.

LDH is a tetramer (four polypeptides) with a total molecular weight of about 135 000 (280, 432). The tetramer can be composed of two types of monomers, the H type (abundant in heart) or the M type (abundant in liver). Thus heart muscle is rich in the HHHH (LDH_1) form of LDH, and the liver is rich in the MMMM (LDH_5) form. A tetramer structure allows formation of five isoenzymes of LDH: H_4 or LDH_1, H_3M or LDH_2, H_2M_2 or LDH_3, HM_3 or LDH_4, and M_4 or LDH_5. The European system of nomenclature, in which LDH_1 moves fastest toward the anode and LDH_5 moves the slowest is used in this text. The American numbering system, not in popular vogue, numbers the isoenzymes in exactly the reverse direction. Thus the American LDH_1 is the European LDH_5.

Another commonly used nomenclature system, based on the electrophoretic mobility, considers LDH_1 as α_1, LDH_2 as α_2, LDH_3 as β, LDH_4 as γ_1, and LDH_5 as γ_2. A third form of LDH, LDH-C, is a tetramer composed of C (or X) subunits and is found in semen and postpubertal testes. Electrophoretically, LDH-C is located between LDH_3 and LDH_4 (432, 664, 780).

Sulfhydryls are apparently important for catalytic activity since mercuric chloride, p-chloromercuribenzoate, and N-ethylmaleimide inhibit LDH. Inhibition by p-chloromercuribenzoate is reversible with cysteine (432, 523). Oxalate inhibits LDH noncompetitively, the effect being about twice as great with LDH_5 as compared with LDH_1 (432, 780). Thus oxalate would be a poor choice of anticoagulant for studies of plasma LDH. Oxamate is inhibitory, inhibition being competitive with pyruvate (432, 780). Among anions, iodate, borate, and sulfite are inhibitory, with sulfite preferentially inhibiting LDH_1 and LDH_2 (523). Pyruvate in excess is inhibitory, resulting in product inhibition in the forward reaction (432). Inhibition by pyruvate is rather selective toward LDH_1, with much less inhibition toward LDH_5. Lactate is also inhibitory, being more inhibitory toward LDH_1 than LDH_5 (432). Use of selective inhibition of LDH isoenzymes with urea is the basis for a number of isoenzyme assays (664). At 2 M urea, LDH_1 is inhibited about 20% whereas LDH_5 is completely inhibited (432, 664).

Studies by Schmidt and Schmidt (647) show that LDH is located only in the cytoplasm of cells. In contrast, an enzyme such as glutamate dehydrogenase is found only in the mitochondria. Thus, after cellular damage (liver, heart), LDH is readily released from the cytoplasm. More serious damage must occur before release of a mitochondrial enzyme such as glutamate dehydrogenase. Friedel and Mattenheimer (239) concluded that the major portion of LDH in normal serum originates from erythrocytes and platelets released during the physiologic turnover of these cells.

A number of procedures have been developed for measuring total LDH activity. The most direct is the spectrophotometric assay based on following the rate of increase or decrease of absorbance at 340 nm, the absorption maximum for NADH; NAD^+ has no absorption at 340 nm (Fig. 21-1). Thus in the forward reaction the rate of increase of the absorbance at 340 nm (A_{340}) is measured (9), while in the backward reaction the rate of decrease is measured (319, 800). Using optimal conditions for both the forward and backward reaction and taking into account variations possible with the various isoenzymes, Gay *et al* (248) considered the assays in either direction equivalent. For the backward reaction the method of Wróblewski and LaDue (800), as modified by Henry *et al* (319) is quite reliable. The forward reaction as described by Amador *et al* (9), with slight modifications, has a number of advantages over the backward reaction. The forward reaction does not require preincubation, while the backward reaction requires a preincubation period for consumption of serum substrate by NADH-dependent enzymes which decreases the initial NADH concentration. Also, the reaction rate of the forward reaction is linear over a wider range of LDH activity as compared with the backward reaction. In the buffer system used in the LDH method presented in this chapter no spontaneous degradation is noticed with NAD after 3 hours (515) as compared with the slight but continuous degradation with NADH in the buffer system for the backward reaction (319).

Another procedure for the determination of LDH activity is based upon measuring the fluorescence of NADH (114). An alternative fluorometric procedure is based on the addition of methyl ethyl ketone to make NAD fluorescent (432).

A number of less direct visible range spectrophotometric procedures have been used to follow LDH activity. The principle for many of these is based upon the use of redox indicators, such as tetrazolium salts (483, 523) or 2,6-dichloroindophenol (483), which are colorless until reduced by NADH. The reduction of the indicator by NADH is indirect and involves an intermediate electron transfer from NADH to phenazine methosulfate. A disadvantage of this procedure is the fact that phenazine methosulfate is unstable and photosensitive. Also, when a tetrazolium salt is used the resulting formazan is sparingly soluble in an aqueous medium (515). Diaphorase has been used as an intermediate electron carrier instead of phenazine methosulfate, but it has the disadvantage of variable reproducibility (515). Another procedure involves measuring the pyruvate formed by the forward reaction by reacting pyruvate with 2,4-dinitrophenylhydrazine to form the colored hydrazone (515). This procedure has been criticized for

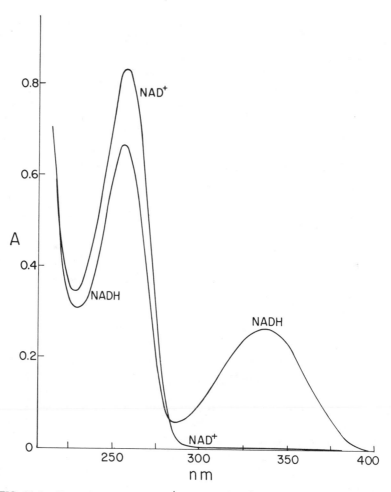

FIG. 21-1. Absorption curve of NAD$^+$ and NADH. Spectrum of NAD$^+$, 4.9 X 10^{-5} M in 0.1 M Tris, pH 7.5; and NADH, 4.1 X 10^{-5} M in 0.1 M Tris, pH 9.5. Beckman DB-G Grating Spectrophotometer.

lack of correlation with the spectrophotometric method (9, 485) and with the fluorometric method (114), as well as with clinical data (483). One procedure involves measurement of a cuprous-neocuproine complex formed by the coupled reaction with enzymatically generated NADH (515).

Many of the above procedures have been adapted to automated measurement of serum LDH: spectrophotometric, in the ultraviolet range (351, 406, 470, 708); fluorometric (216, 553); spectrophotometric (visible range) procedures based on phenazine methosulfate-tetrazolium salt (127), or cuprous-neocuproine complex (516). Automated spectrophotometric (visible range) procedures have also been devised for urine (327, 332).

Different investigators have used a variety of procedures to differentiate the various LDH isoenzymes. For obtaining a total isoenzyme pattern, electrophoresis appears to be the method of choice. The major advantage of electrophoresis is that it reveals the distribution of all five isoenzymes. Inhibition procedures generally inhibit one isoenzyme completely (either LDH$_1$ or LDH$_5$) and partially inhibit some of the structurally related isoenzymes. Thus one can see gross changes in the relative amounts of LDH$_1$ or LDH$_5$, but only a hazy indication as to the activities of the other three isoenzymes. A variety of support media have been used for electrophoresis of LDH isoenzymes. Paper and starch gel have been properly criticized from a clinical viewpoint because of the necessity for lengthy electrophoresis times of 14 and 18 hours, respectively (464). Disk electrophoresis with polyacrylamide gel has been used successfully under appropriate conditions (664), as has agar gel (432, 664, 810). There are a number of good reviews on the subject of LDH isoenzymes which describe technics as well as clinical significance (432, 664, 780, 810).

For agar gel electrophoresis—probably the most widely studied support medium—the type of agar used is apparently critical. Noble agar (Difco) is recommended (432). The inclusion of albumin (432) renders a series of different agars more suitable for detecting LDH$_5$, probably by improving its stability. With agar gel electrophoresis, Veronal buffer can inhibit the isoenzymes, but by the inclusion of citrate (activator, especially for LDH$_5$) the inhibition can be neutralized (352). Another precaution is the avoidance of a low buffer concentration, which can result in an abnormal isoenzyme pattern (432). For development of the isoenzyme bands, iodonitrotetrazolium (89) or tetranitroblue tetrazolium (702) appears to be the tetrazolium salt of choice. The optimal temperature for color development has been reported to be 37°–40°C (422). Cyanide, often used in the staining solution to trap the pyruvate formed, is detrimental to formazan formation; this effect can be offset by substituting pyrophosphate for cyanide (352). Fluorometric procedures have also been used for detection of isoenzymes (664).

Cellulose acetate has proved to be a useful support medium; it is commercially available in a convenient form (182) and thus avoids the tedium of gel preparation. In the procedure to be described (182), the same buffer, voltage, and electrophoretic apparatus can be used for both protein and LDH electrophoresis. The resulting cleared cellulose acetate strip is permanently stained and can thus be placed directly in the patient's chart.

The addition of just NAD to normal serum results in reduction of some of the NAD. This phenomenon is believed to be caused by the oxidation of endogenous serum lactate by LDH (502). However, after electrophoresis of the LDH isoenzymes in any of the common support media, and if the electrophoretic strips are stained in the presence of NAD and in the absence of substrate, some staining still occurs in bands corresponding to LDH (215). This phenomenon cannot be explained by oxidation of endogenous serum lactate and has been called the "nothing dehydrogenase." The apparent reason for the staining is the presence of groups similar to lactate which are insolubly bound to the various types of support media used in electrophoresis (215).

Heat stability (usually expressed as LDH activity after 15—60 min at 60°C as compared with activity prior to heat treatment) has been advocated as a procedure to estimate the amount of the more stable myocardial isoenzymes, primarily LDH_1 (432, 664, 780). The heat procedure, although useful as a screening test for detecting myocardial infarction and for use in laboratories without electrophoretic equipment, is generally considered to be significantly less informative than the electrophoretic method (146, 554, 681).

Urea at a concentration of about $1-2$ M inhibits LDH_5 to a much greater extent than LDH_1 and has thus been used as a procedure for determining whether a particular serum is richer in LDH_1 or LDH_5 (432, 664, 780). The fact that high lactate concentrations are more inhibitory to isoenzymes rich in H monomers (LDH_1 and LDH_2) than to isoenzymes rich in M monomers (LDH_4 and LDH_5) is the basis for another isoenzyme assay (432). A combination of high lactate concentration to inhibit LDH_1 and LDH_2 isoenzymes, which provides an estimate of LDH_5 and LDH_4 activity, and urea to inhibit LDH_4 and LDH_5 which provides an estimate of LDH_1 and LDH_2 activity, is the basis of Babson's assay (664). The resulting isoenzyme index—the ratio of LDH_1 activity to LDH_5 activity—derived from Babson's assay has been considered a useful index but not as reliable as electrophoresis for determining the relative amount of LDH_1 and LDH_5 (51, 181, 233, 516, 566).

Organic solvents have also been used in attempts to fractionate the LDH isoenzymes (664, 780). Column chromatography with DEAE cellulose or DEAE-Sephadex has been used but, again, is not considered as satisfactory as electrophoresis (146, 263, 269).

The fact that LDH_1 and, to a lesser extent, LDH_2 have a much higher reactivity with α-hydroxybutyrate (α-ketobutyrate by the backward reaction) than LDH_5 (44, 521) is the basis for the α-hydroxybutyrate dehydrogenase assay. The activity has been assayed spectrophotometrically (613, 614, 780), fluorometrically (67), by formation of the 2,4-dinitrophenylhydrazone of α-ketobutyrate (64, 571), and by an automated spectrophotometric procedure (196).

DETERMINATION OF LACTIC DEHYDROGENASE

(method of Amador et al, Ref 9, modified)

PRINCIPLE

Lactic dehydrogenase reversibly catalyzes the conversion of lactate to pyruvate, with the simultaneous reduction of NAD to NADH. Since NADH absorbs at 340 nm and NAD does not, the activity of LDH is assayed by following the increase in absorbance at 340 nm using lactate and NAD as substrates. The unit of activity is defined as the number of μmol NADH formed/liter serum/min at 32°C and under other specified conditions of the test.

REAGENTS

2-Amino-2-methyl-1-propanol

Lactic acid, 85%. Can be obtained commercially.

Working Lactic Acid Solution. To approximately 800 ml water add 63.9 ml 2-amino-2-methyl-1-propanol plus 10 ml 85% lactic acid. Mix and adjust to pH 9.0 ± 0.05 with 5N NaOH. Store in refrigerator.

Lactic Acid-NAD Reagent. For every 100 ml of working lactic acid solution to be used, add 400 mg NAD. Prepare fresh each day a sufficient volume for the number of specimens to be assayed. Keep solution in refrigerator or on ice at all times.

PROCEDURE FOR SERUM

The procedure is described for a Gilford 2400 recording spectrophotometer equipped with a temperature controlled cuvet compartment held at 32°C by a circulating constant temperature bath. Alternative approaches are discussed in *Note 1.*

1. Place 2.9 ml lactic acid-NAD reagent into a test tube and incubate in a water bath at 32°C ± 0.5°C for 4–5 min. During this period the cuvets are kept in the instrument's cuvet compartment so they reach temperature equilibrium (spectrophotometer and circulating water bath turned on at least 2 hours prior to use).

2. Add 0.1 ml serum by TC pipet, mix, and transfer as quickly as possible to the prewarmed cuvet, which is rapidly reinserted into the cuvet compartment. The cuvet compartment lid is left open for as short a time as possible. Immediately start automatic recording of the change in absorbance at 340 nm.

3. If the resulting recorded curve is linear for at least 6 min, proceed with calculations as described below. If the rate is too rapid to give a linear curve, rerun using the sample diluted with 0.85% saline (dilution required if result is over 300 units).

Calculation:

$$\text{units } (\mu\text{mol NADH/min/liter}) = \frac{\Delta A_{340} \text{ for } T \text{ min}}{T} \times \frac{1}{\substack{\text{ml sample} \\ \text{used in test}}} \times 483$$

The term 483 is used to convert A_{340}/min/ml to μmol NADH/min/liter and is based on 6.22×10^3 as the molar absorptivity of NADH, the use of a 3 ml assay volume, and correction to 1 liter of serum, i.e.:

$$\frac{A_{340}/\text{min}}{6.22 \times 10^3} \times 3 \times 10^3 \times 10^3 = 483$$

PROCEDURE FOR URINE

1. Centrifuge the urine, then place a 5 ml aliquot in a dialysis bag and weigh. Dialyze the aliquot against cold running water for 80 min, then reweigh. Recentrifuge if the urine is turbid.

2. Separately incubate 1 ml of dialyzed urine and 2 ml of lactic acid-NAD reagent at 32°C for 5 min. Transfer contents to a prewarmed cuvet, mix, and quickly insert into the spectrophotometer at 32°C. Follow absorbance at 340 nm.

Calculation:

The units (defined as nmol NADH formed/ml/min) per 8 hours volume of urine are calculated as follows:

$$\text{units/8 hr vol} = \frac{\Delta A_{340} \text{ for } T \text{ min}}{T} \times \frac{\text{wt after dialysis}}{\text{wt before dialysis}}$$

$$\times \text{ ml vol collected} \times \frac{8}{\text{number hr of collection}} \times 483$$

PROCEDURE FOR CEREBROSPINAL FLUID

The procedure and calculation for CSF are similar to those for serum except 0.5 ml CSF and 2.5 ml lactic acid-NAD reagent are used. Incubate the CSF and lactic acid-NAD reagent separately for 5 min at 32°C. Transfer contents to a prewarmed cuvet, mix, and insert in the spectrophotometer. The use of 0.2 ml of CSF and 2.8 ml of lactic acid-NAD reagent is possible with more sensitive instruments such as the Photovolt Enzyme Rate Analyzer.

NOTES

1. Temperature of reaction and spectrophotometer employed. The control of temperature or knowledge of the reaction temperature is crucial since there is an increase in activity of about 7% per °C or 70% increase for a 10°C rise in temperature (319). Schneider and Willis (650) and Henry *et al* (319) proposed a standard temperature of 32°C for the determination of LDH.

Temperature can be controlled with the Gilford or Beckman model DU equipped with thermospacers. If the reaction is run at a temperature other than 32°C, temperature corrections must be used. The following equation can be used to correct values to 32°C:

$$\text{Units}_{\text{corr to } 32°C} = f \times \text{units obtained at } t$$

where $t = °C$; factors (f) are listed in Table 21-1 (319).

TABLE 21-1. Factors for Correction of LDH to 32°C[a]

Reaction temperature, (°C)	Factor (f) to convert to 32°C	Reaction temperature, (°C)	Factor (f) to convert to 32°C
25	1.67	33	0.93
26	1.55	34	0.87
27	1.44	35	0.81
28	1.34	36	0.75
29	1.24	37	0.70
30	1.16	38	0.65
31	1.07	39	0.61
32	1.00	40	0.57

[a]From HENRY RJ, CHIAMORI N, GOLUB OJ, BERKMAN S: *Amer J Clin Pathol* 34:381, 1960

Amador *et al* (9) similarly calculate temperature coefficients, except values are calculated relative to 25°C instead of 32°C. The data in Table 21-1 were calculated based upon the backward reaction, whereas those of Amador *et al* (9) were calculated based upon the forward reaction. Both tables show a 1.67-fold difference in activity between 25° and 32°C. Thus within the temperature range of the two tables, the forward and backward reactions show the same ratio of activities at any two temperatures listed on the tables. The reaction should not be run above 40°C because of heat denaturation of the enzyme (319). For the Beckman model B or DU (without thermospacers), the temperature of the cuvet compartment parallels room temperature, being 7° higher in a model B and 5° higher in a model DU. For determination of temperature, a thermister or thermometer can be installed in the compartment of either model.

Considerable time is required for a spectrophotometer that is at room temperature (e.g., one that has been turned off overnight) to reach temperature equilibrium after it has been turned on. A model DU requires about 2 hours and a model B about 2.5 hours. To avoid such a delay when the thermospacers are not used, the instrument, including the lamp, must be left on at all times. When cuvets containing 3 ml of solution at room temperature are placed in the cuvet compartment, there is a time lag before the cuvet and its contents reach the compartment temperature. This lag is about 5 min for a model DU and about 20 min for a model B. Presumably because of the considerable air space around a cuvet in a model B and because of the large compartment lid, opening the compartment to add serum to start the reaction results in a temperature drop of the cuvet and its contents. About 15 min are required for reestablishment of temperature equilibrium. It would therefore appear that, short of rebuilding the cuvet compartment, the model B cannot be used for enzymatic analyses if greatest accuracy is required.

2. Dichromate blanks. If a recording spectrophotometer is not used, then dichromate blanks are necessary. A dichromate blank, prepared as described below, is used to adjust the spectrophotometer to zero absorbance so that the sample gives an initial absorbance at 340 nm of 0.2–0.3. Dichromate blanks are prepared as follows (689): A stock 0.001 M solution is prepared by dissolving 29 mg $K_2Cr_2O_7$ in 100 ml water and adding a few drops of conc H_2SO_4. A series of blanks are prepared from the stock. The following four concentrations suffice in nearly all cases: 0.00025, 0.000187, 0.000125, and 0.0000625 M.

3. Inhibition by reaction product. For high LDH activities, if the serum is not diluted inhibition by pyruvate can occur. Linearity between ΔA and time was obtained provided $\Delta A/\text{min}$ was less than 0.1/min over a 5 min period (9). Others (149, 668) claim that linearity can be obtained only at a lower LDH activity. With higher LDH activities, linearity can be obtained by diluting the serum or by a polynomial curve-fitting procedure (149). Alternatively, semicarbazide in the reaction mixture may be used to trap the pyruvate formed and thus reduce product inhibition (668).

4. Inhibitors in NAD preparations. It is well known that NADH preparations contain inhibitors of LDH activity (491, 709). Similarly, NAD preparations may contain LDH inhibitors which are not the same as those found in NADH preparations. Commercial preparations contain varying amounts of inhibitors (33). NAD can react with acetone, which may contaminate NAD preparations to form an inhibitor of LDH (188). Our laboratory has noticed no significant inhibition for routine LDH assays due to inhibitors in NAD preparations.

5. 2-Amino-2-methyl-1-propanol buffer. The buffer used in the present assay has a number of advantages, including absence of substrate inhibition at 500 mM lactate concentration as compared with other buffers and greater stability of NAD (no detectable deterioration after 3 hours at 25° or 37°C) compared to Tris buffer (515). Also Tris and phosphate buffers can adversely affect LDH activity (758).

6. Stability of specimens. There is an abundance of conflicting literature on the stability of serum LDH at various temperatures. Most data suggest LDH is reasonably stable at room temperature—e.g., up to 10 days (504); other authors (17) report about a 10% drop in activity after 24 hours at room temperature. A number of authors (504) suggest that LDH is stable at refrigerator temperatures for times ranging up to 4 weeks; others report loss of activity (504). Similar conflicts of data occur with freezing, reported stability ranging from 2 days to 4 weeks or 4 years with liquid nitrogen (175, 504). Other authors report loss of activity with freezing (17, 138). Quick freezing (acetone-dry ice) has been recommended (504), but Chilson *et al* (138) note that even a fast freeze and fast thaw (exposure to tap water with shaking) result in a 10% loss of activity (a slow thaw of 20–30 min at room temperature results in about a 40% loss of activity; a slow freeze and slow thaw results in a 70% loss of activity).

A possible explanation for some of the data can be seen in the study of Hoffmeister and Junge (333), which shows that LDH has a minimum stability at 0°C. At temperatures above or below 0°C there is greater stability. Thus freezing at very low temperatures or leaving the sample at room temperature is preferable to maintaining it at 0°C. Similarly, Amelung *et al* (17) showed greater stability of LDH at 20°C than at 4° or −8°C. Thus it appears that for serum it is best not to refrigerate or freeze but to hold the specimen at room temperature and assay it as soon as possible. In addition, it should be noted that LDH$_5$ is more heat- and cold-labile than the other isoenzymes (780), and that it is more stable in Tris than in phosphate buffer (758).

LDH is stable in CSF for 4−6 hours at room temperature or 2 weeks at 4°C (273). Most of the LDH activity in urine is lost in 5 days at 20°−25°C (640). About 50% of the activity is lost in 30 days at 2°−5°C. From our experience it is best to perform the assay on the same day the specimen is collected. After removal of inhibitors by dialysis, samples are stable at 4°C for 24 hours (191).

7. *Serum vs plasma.* It has been reported that heparin (88, 800) and EDTA (88) do not interfere with the test but that oxalate inhibits LDH (88, 800). For some unexplained reason, in one investigation (342) the LDH activity of both heparinized and oxalated plasma samples was about 50% higher than the activity in serum. It would seem advisable, therefore, to restrict the analysis to serum.

Leaving the serum with the clot is not advisable (253, 342, 368). Activity is claimed to rise 25% after leaving the serum with the clot for 1 hour (253, 342). Increased activity is presumably due to hemolysis from clotted erythrocytes.

8. *Hemolysis.* The concentration of LDH is about 100-fold greater in erythrocytes than in serum (800). Calculation shows that barely visible hemolysis–i.e., about 50 mg hemoglobin per 100 ml serum—results in an increase of about 15−20% activity in the serum. Thus any visible hemolysis cannot be tolerated.

9. *Turbid or icteric sera.* If a recording spectrophotometer is not used, for very turbid or icteric serum it may be necessary to run the test on a dilution of the serum so a dichromate blank may be used that allows the spectrophotometer to be set at zero absorbance, with an initial absorbance of the test mixture being about 0.5.

10. *Uremic sera.* LDH activity is reduced in uremic sera due to the presence of inhibitors—urea, oxalate, and other unidentified substances (781). Dialysis results in an increase in LDH activity (604).

11. *Urine.* Removal of LDH inhibitors in urine is necessary in order to determine LDH activity accurately. Dialysis (9, 191, 649), Sephadex (50, 327), and Bio-Rad AG 1−8X (332) have been used to remove the low molecular weight inhibitors. The dialysis procedure of Dorfman *et al* (191) is used in the urine procedure described. LDH values, using the above procedure, are approximately 1.9-fold larger than those described by Dorfman *et al* (191) and Amador *et al* (9). A contributing factor for this difference is probably the use of 2-amino-2-methyl-1-propanol (pH 9.0) as

buffer by the procedure described as compared with the use of sodium pyrophosphate (pH 8.8).

12. Units. A unit is defined as the number of μmol of NADH formed/min/liter of serum or CSF at 32°C. With urine, in order to avoid working with decimals, a unit is defined as the number of nmol of NADH formed/min/ml of urine; the number is then multiplied by the 8-hour urine volume to give units/8 hours volume. Much of the data in the literature is expressed in Wróblewski units which are defined as the amount of enzyme required to bring about a change in absorbance of 0.001/min/ml of sample. To convert Wróblewski units to units defined as μmol NADH/min/liter, multiply Wróblewski units by 0.483. However, because assay conditions by various authors are different, indiscriminate use of a conversion factor is not advisable. If one is comparing values by different authors, it is better to conduct an assay on a number of samples by the two methods and empirically obtain a conversion factor.

13. Spectrophotometer standardization. Since the enzyme units are based on absolute spectrophotometric absorbance units, the accuracy of the absorbance scale of the instrument must be checked. Mass-produced spectrophotometers rarely respond identically; and even if instruments were standardized by the manufacturer, gradual aging would make "absolute" absorbances unreliable. Furthermore, the absorbance scale of most spectrophotometers cannot be adjusted. Since the error from this source can be quite substantial, it is absolutely essential that a check be made periodically for this aspect of spectrophotometer performance.

Spectrophotometers used for assaying enzyme reactions involving NADH or NADPH consumption or production can be checked by the method of Martinek *et al* (481): Dry $K_2Cr_2O_7$ (National Bureau of Standards) at 100°–110°C for at least 4 hours and cool in a desiccator. Prepare a 40 mg/liter solution of $K_2Cr_2O_7$ by dissolving 160.0 mg in 4 liters of 0.01 N H_2SO_4. The absorbance at 350 nm (absorbance maximum) in a 1 cm cuvet, read against a water blank, should be 0.428. If possible, the instrument should be adjusted to obtain a reading of 0.428. Alternatively, the change in absorbance at 340 nm (absorbance maximum of NADH) can be corrected by the following equation:

$$\text{corrected } \Delta A_{340}/T_{min} = \text{observed} \Delta A_{340}/T_{min}$$

$$\times \frac{0.428}{\text{observed } A_{350} \text{ for } K_2Cr_2O_7 \text{ std}}$$

where T_{min} = time in minutes.

ACCURACY AND PRECISION

LDH follows zero order kinetics. However, for high LDH concentrations, errors can occur because of product inhibition by pyruvate unless precautions are taken (*Note 3*). The accuracy of the test is rather difficult to assess because the measurement of enzyme activity is not necessarily a reliable measure of enzyme concentration (735). In addition, there are five common isoenzymes of LDH, each isoenzyme having its own pH and substrate concentration optimum (248). Thus any single measurement of LDH activity under one set of conditions cannot measure optimal activity for all of the isoenzymes. When various pathologic conditions change the concentration of one or more isoenzymes, optimal conditions for that particular set of isoenzymes may very well differ from optimal conditions for a normal pattern of isoenzymes. The possible presence of inhibitors in NAD (see *Note 4*) would also affect accuracy and precision.

The precision (95% limits) for the test is about ±8%.

NORMAL VALUES

The normal ranges (95% limits) for serum of adult males and females, as determined in our laboratory, are 63–155 and 62–131 units, respectively. Unfortunately, at this time the determination of the normal range by the method presented has not been carried out for urine and spinal fluid. The values mentioned below are merely ranges calculated from normal ranges determined by the method of Henry *et al* (319). The correction factor for carrying out the calculation was determined by running both the method presented and the method of Henry *et al* (319) on a series of urine and spinal fluid specimens. The computed upper normal limit for urine is 3680 units/8-hour volume, with a unit being defined as the amount of enzyme which at 32°C brings about the conversion of 1 nmol of NAD/min. For CSF the computed normal range is 3–23 units, with units being defined as for serum LDH. The normal range for urine given by Dorfman *et al* (191) could not be used directly because of the difference in buffers used (pyrophosphate instead of 2-amino-2-methyl-1-propanol in the method presented). The present method gives about 1.84-fold higher values than those given by Dorfman *et al.*

In children serum values are generally reported to be higher than in adults (208, 283, 328), although one report (309) disagrees with this observation. In adults our studies indicate a difference between sexes; other studies indicate no such difference (145, 309). However, differences between sexes have been noted in isoenzyme studies (145), the difference disappearing after menopause. Conflicts as to whether LDH activity correlates with the menstrual cycle have been reported (189, 290). Diurnal variations in LDH activity ranging from 30% (800) to 200% (702) have been reported. Strenuous exercise produces a significant increase in LDH activity (291). From studies in our laboratory LDH values appear to be influenced by a number of factors. There is a seasonal variation, with LDH values being higher by 20% in the summer months as compared to the corresponding median value in winter months. The observation holds true for males as well as for females. For males and to a lesser extent for females values appear higher in the morning than in the afternoon. Finally, values appear to be higher in urban than in rural areas, the difference being more significant for males than females.

DETERMINATION OF LACTIC DEHYDROGENASE ISOENZYMES

(method of DiGiorgio, Ref 182)

PRINCIPLE

Lactic dehydrogenase isoenzymes are electrophoretically separated on cellulose acetate strips, visualized by the reduction of INT dye (2-*p*-iodophenyl-3-*p*-nitrophenyl-5-phenyltetrazolium chloride), followed by quantitation using an integrating densitometer.

REAGENTS

Buffer, 0.11 M, pH 8.8. Dissolve 2.45 g diethylbarbituric acid, 9.81 g sodium diethylbarbiturate, and 5.75 g tris(hydroxymethyl)aminomethane and dilute to 1 liter with water.

Acetic Acid, 0.36 N. Dilute 2 ml glacial acetic acid to 100 ml with water.

Clearing Solution. Mix 225 ml of *n*-propanol with 25 ml dimethylsulfoxide.

Sodium Lactate Working Solution. Add 1 ml sodium lactate syrup (5 *M* sodium lactate, 60 g/100 ml) to 3 ml buffer.

INT Solution. Dissolve 14 mg INT (2-*p*-iodophenyl-3-*p*-nitrophenyl-5-phenyltetrazolium chloride) in 3.5 ml buffer by stirring and warming to 37°C. Store refrigerated in the dark. Prepare daily.

PMS Solution. Dissolve 2 mg PMS (phenazine methosulfate) in 2 ml water. Store refrigerated in the dark. Prepare daily.

Substrate Medium. Prepare substrate medium during last 15 min of electrophoresis. Weigh 10 mg NAD and transfer to a suitable test tube. Add 1 ml buffer, 1 ml sodium lactate solution, and 3 ml INT solution. Mix, add 0.3 ml PMS solution, remix, then pour into a plastic container. Protect from light. The final molarity of lactate and NAD in the mixture is 0.31 and 0.028, respectively.

PROCEDURE

1. Presoak cellulose acetate strips (cellulose acetate on a plastic backing, approximately 2.5 x 7.5 cm), vertically supported in a glass slide cradle, in a dish of buffer for at least 5 min.

2. Fill electrophoresis chambers with buffer, making sure the buffer levels in both sides are equal.

3. Remove a cellulose acetate strip by holding it along its edges and, in order to remove excess buffer, press lightly between blotters. Using an appropriate serum applicator (such as a Beckman pipet to deliver 3.5 μl serum to a Gelman push-button serum applicator), apply about 10 μl (two applications) of serum to the acetate side of the strip.

4. Electrophorese for about 45 min at 115 V.

5. Prepare substrate medium during last 15 min of electrophoresis. (Caution: this medium is light-sensitive.) For each cellulose acetate strip being electrophoresed, saturate a dry cellulose acetate plate for at least 5 min in substrate medium.

6. Upon completion of electrophoresis, blot wet ends of strip and superimpose with a strip saturated with but drained of excess substrate medium. Avoid entrapment of air bubbles between strips by using a gradual, rolling technic. A slight change in relative position between the two strips during or after placement results in distortion of isoenzyme bands.

7. Place the two attached strips in a moist incubation chamber in total darkness at 37°C for 45 min.

8. After incubation, peel apart the strips and wash with three consecutive 1-min rinses of 0.36 *N* acetic acid. A slide cradle is useful for rinsing. Sample and substrate plates should be rinsed separately. If multiple strips are processed, make certain that the acetate (dull) sides all face the same direction in the cradle. Strips are stable in acetic acid and should be removed from the final rinse only when they can be conveniently carried through the remaining steps without delay.

9. Remove strips from acetic acid rinse, drain, and place in a dry cradle. Place in clearing solution for exactly 1 min. Agitate gently three times during this period.

10. Remove from cleaning solution, drain, and individually blot the long edge of each strip against blotting paper. Place on two layers of paper towels. This entire step should be completed within 2 min.

11. Dry strips in a gravity convection oven at 90°C for about 10–15 min.

12. Quantitate isoenzymes with an integrating densitometer. Add up total integration counts disregarding counts for the albumin band. Divide counts for each isoenzyme fraction by the total counts and multiply by 100 for relative percentages of each fraction. Alternatively, the percent of total lactic dehydrogenase activity for each fraction can be calculated as follows: read peak height of each fraction and divide by sum of all peak heights; multiply by 100.

NOTES

1. See *Notes for total LDH.* The notes there concerning stability, hemolyzed samples, and so on also apply for the isoenzyme procedure.

2. *Rinsing solutions.* A series of rinsing solutions were tried for their ability to diminish background color before the plates were cleared (182): (a) dilute HNO_3 3 ml plus 97 ml water, (b) 5% trichloroacetic acid, and (c) acetic acid, 5 ml or 2 ml diluted to 100 ml with water. Of these, the more dilute acetic acid was the most effective.

3. *Clearing solution.* Dimethylsulfoxide (10 ml) plus *n*-propanol (90 ml) was considered to be a better clearing solution than: (a) 20 ml acetic acid in

80 ml methanol; (b) acetic acid-methanol-water, 10:50:40, by volume; or (c) 30 ml dimethylsulfoxide in 70 ml methanol. Advantages of dimethylsulfoxide-propanol are its 1-min clearing time and the fact that it does not remove as much formazan from the protein isoenzyme bands as do the other clearing solutions (182).

ACCURACY AND PRECISION

Evaluating the accuracy of the procedure is difficult since there is no standard LDH serum available. However, as shown by the normal values obtained by the procedure presented, the values compare well with the results of many other laboratories.

NORMAL VALUES

The normal ranges for the LDH isoenzyme fractions are the same whether determined by integration or peak height measurements (182):

Isoenzyme	% of Total LDH activity
LDH_1	18-33
LDH_2	28-40
LDH_3	18-30
LDH_4	6-16
LDH_5	2-13

Comparable normal ranges have been obtained by other workers using cellulose acetate (182). Some differences have been reported in the distribution of isoenzymes in the serum of men and women (145). Values in women were about 7% higher for LDH_1 and 5% lower for LDH_2 as compared with values for men. No significant differences were observed for LDH_3 to LDH_5. These differences disappear after menopause. The distribution of LDH isoenzymes in tissues and various disease states has been reported (432, 664, 780, 810).

DETERMINATION OF α-HYDROXYBUTYRIC DEHYDROGENASE

(method of Rosalki and Wilkinson, Ref 613, 614, modified)

PRINCIPLE

α-Hydroxybutyric dehydrogenase reversibly catalyzes the reduction of α-ketobutyrate to α-hydroxybutyrate with the simultaneous oxidation of NADH to NAD. In this method the activity of the enzyme is measured by determining the rate of disappearance of NADH from the reaction mixture. A unit of activity is defined as the number of micromoles of NADH oxidized per liter of serum per minute at 32°C and under other specified conditions of the test.

REAGENTS

Tris buffer, 0.1 M, pH 7.4. Dissolve 12.1 g Tris base—tris(hydroxymethyl)aminomethane—in 800 ml of water, adjust to pH 7.4 ± 0.1 with conc HCl, then dilute to 1 liter with water.

NADH, 2.5 mg/ml Tris buffer. Stable for at least 1 month in the frozen state.

LDH Tubes. Make Tris buffer-NADH mixture in the ratio of 25 ml buffer to 2 ml NADH. Deliver 1.7 ml into each tube, label, stopper, and freeze. In order to ensure that the correct level of NADH is present in the LDH tube, dilute the 1.7 ml aliquot to 3.0 ml with water. The absorbance at 340 nm read against water should be 1.0—1.2. If below 1.0, add more NADH. If the spectrophotometer being used cannot read accurately above 1.0, read the absorbance of the NADH against a dichromate blank (see *Note 2*, LDH assay). The dichromate blank versus water should give an A_{340} of 0.500—0.600. Add the readings of the blank against water and the NADH-buffer against the blank; the A_{340} should total 1.0—1.2.

α-Ketobutyrate. Dissolve 12.4 mg sodium α-ketobutyrate per milliliter 0.1 *M* Tris buffer, pH 7.4. Prepare fresh daily.

PROCEDURE

The procedure is described for a Gilford 2400 recording spectrophotometer equipped with a temperature controlled cuvet compartment held at 32°C by a circulating constant temperature bath. *Notes 1* and *2* in the total LDH assay and *Note 1* in this section are also applicable as alternative approaches.

1. Pipet 0.1 ml serum by TC pipet and 1.0 ml water into an LDH tube. Mix and incubate 20 min in a 32°C water bath. During this period the cuvets are kept at 32°C (in the instrument's cuvet compartment or water bath).

2. Transfer contents of tube to a cuvet and place in the spectrophotometer. The cuvet compartment lid is left open as briefly as possible. Record the absorbance at 340 nm for several minutes at 1 min intervals to check for side reaction as evidenced by a drop in A_{340}. If no side reaction occurs or when it stops, proceed with the next step.

3. Pipet 0.2 ml α-ketobutyrate, which has been kept in a 32°C water bath, into the cuvet. Mix with a polyethylene stirring rod. Work rapidly so that the compartment lid is open as briefly as possible. Record A_{340}.

4. If the resulting curve is linear for at least 6 min, proceed with the calculations as described below. If the rate is nonlinear, rerun on an aliquot of sample diluted with buffer.

Calculation:

$$\text{Units} = \mu\text{moles NADH oxidized/min/liter } (32°\text{C})$$

$$= \frac{\Delta A_{340} \text{ for } T \text{ min}}{T} \times \frac{1}{\text{ml sample used in test}} \times 1000 \times 0.483$$

The term 0.483 is used to convert ΔA_{340} of 0.001/min/ml of sample to μmol of NADH oxidized/min/liter since 0.000483 μmol of NADH in a 3 ml volume gives an absorbance of 0.001 at 340 nm.

NOTES

With the exception of stability of serum and table of factors used to convert LDH activity at various temperatures to 32°C, the notes for LDH activity are also applicable for α-hydroxybutyric dehydrogenase.

1. Temperature correction factors. With spectrophotometers not having cuvet compartments controlled at 32°C, the factors given below are used to convert to units at 32°C using the equation (733):

$$\text{Units}_{\text{corr to } 32°\text{C}} = f \times \text{units obtained at } t$$

where t = temperature in degrees centigrade.

Cuvet temperature (°C)	Factor (f)
25	1.33
26	1.28
27	1.24
28	1.19
29	1.14
30	1.09
31	1.04
32	1.00
33	0.95
34	0.89
35	0.84

2. Stability of serum. The α-hydroxybutyric acid dehydrogenase activity is stable for at least 5 days at 30° and 4°C. The enzyme is reported to be stable at −20°C for 10 days (682).

3. Substrate. The sodium salt of α-ketobutyrate is more stable than the free acid when used as a substrate (782). α-Ketobutyric acid, when frozen, tends to form a cyclic decomposition product after prolonged storage.

4. Inhibitors in NADH. As mentioned in the note on the presence of LDH activity inhibitors in NAD preparations, inhibitors of LDH activity are also present in NADH preparations (491, 709). The inhibitor forms in frozen solutions of NADH, in NADH exposed to humid air, with repeated freeze-thawings, and especially in frozen alkaline solutions of NADH. The isolated inhibitor (709) behaves competitively toward NADH (709). Significant differences in amounts of inhibitor occur in different commercial preparations of NADH (491, 709). All four commercial samples in one study showed the presence of inhibitors (491). Generally, good batches of NADH are white, free-flowing, and have A_{260}/A_{340} ratios below 2.45 (491). Bad batches tend to be darkly colored. Tris buffer at neutral pH, such as 7.4, appears to be an excellent solvent for NADH, inhibitor formation being insignificant for short periods at 4°C (709). Keeping solutions of NADH frozen for long periods is not recommended.

5. Hemolysis. Since erythrocytes contain large amounts of LDH, the presence of any visible hemolysis produces invalid results (see *Note 7* under LDH assay).

6. Turbid or icteric sera. If a Gilford or a recording spectrophotometer with adjustable zero absorbance control knob is not being used, turbid or icteric sera may necessitate the use of a dichromate blank to set the zero absorbance and thus give the test mixture an initial absorbance of about 0.5.

7. Effect of temperature on pH optimum. The pH optimum is temperature-dependent: pH 7.1 at 37°C and pH 7.4 at 25°C (205).

ACCURACY AND PRECISION

The same statements regarding accuracy and precision for LDH are applicable to α-hydroxybutyric dehydrogenase.

NORMAL VALUES

The normal range for serum of adults is 67–169 units at 32°C as determined by our laboratory. Others report normal ranges of 73–186 units (202, 614) and 67–166 units (201) corrected to 32°C.

ISOCITRIC DEHYDROGENASE

Isocitric dehydrogenase (ICD, L_s-isocitrate:NADP oxidoreductase (decarboxylating), EC 1.1.1.42) in serum catalyzes the oxidative decarboxylation of isocitrate to α-ketoglutarate.

$$\begin{array}{c} COOH \\ | \\ HOCH \\ | \\ HC\text{-}COOH \\ | \\ CH_2 \\ | \\ COOH \end{array} + NADP^+ \rightleftharpoons \begin{array}{c} COOH \\ | \\ C\text{=}O \\ | \\ HC\text{-}COOH \\ | \\ CH_2 \\ | \\ COOH \end{array} + NADPH + H^+$$

L-ISOCITRATE   OXALOSUCCINATE

$$\begin{array}{c} COOH \\ | \\ C\text{=}O \\ | \\ HC\text{-}COOH \\ | \\ CH_2 \\ | \\ COOH \end{array} \rightleftharpoons \begin{array}{c} COOH \\ | \\ C\text{=}O \\ | \\ CH_2 \\ | \\ CH_2 \\ | \\ COOH \end{array} + CO_2$$

OXALOSUCCINATE   α-KETOGLUTARATE

Both reactions are catalyzed by the same enzyme with the decarboxylation step requiring Mn^{2+} (393). This enzyme is found in most cells in the soluble fraction with particularly high concentrations in the liver. In contrast, an NAD-dependent isocitrate dehydrogenase, which is found in the mitochondria but not in serum, is incapable of carrying out the decarboxylation of oxalosuccinate.

The pH optimum of serum isocitrate dehydrogenase is 7.2–7.8, with a mean of 7.5 (66). Studies with pig heart isocitrate dehydrogenase have shown molecular weights of 64 000 (393) and 58 000 (153). The enzyme appears to be a single chain polypeptide. Isocitrate dehydrogenase has been found in liver, heart, skeletal muscle, erythrocytes, brain, and kidney cells. Four isoenzymes have been observed (384, 393). In rat liver and heart different isoenzymes have been found in combined microsomal, mitochondrial, and nuclear fraction (66).

The preferred activator for the enzyme is Mn^{2+}; Mg^{2+} also activates, although less efficiently (522). Manganese appears to be essential for the decarboxylation step (393). Inhibitors include p-chloromercuribenzoate, iodoacetate, Ba^{2+}, Zn^{2+}, Ca^{2+}, CN^-, oxalate, and N-ethylmaleimide (522). Essential sulfhydryl (152, 277, 334) and methionine groups (151) appear to be involved in the catalytic function. Calcium chloride (98) and NaCl (206) can cause inhibition. With NaCl, 50% inhibition can occur with 0.3 M concentration. Buffers at concentrations of 0.3–1.0 M cause an increasing amount of inhibition.

The most direct procedure for assaying isocitric dehydrogenase is the spectrophotometric assay based on the increase in absorbance at 340 nm caused by the reduction of NADP to NADPH. The unit of activity is usually defined as the number of nmol of NADPH formed per ml of serum per hour at 25°C. The procedure of Wolfson and Williams-Ashman (793), as modified by Bowers (98), is widely used. One drawback involved in the assay of isocitrate dehydrogenase is the low activity of the enzyme in serum which necessitates the use of long assay times and large serum volumes in order to obtain a significant change in absorbance at 340 nm. A procedure developed to overcome this difficulty uses the Photovolt Enzyme Rate Analyzer, which offers greater sensitivity for measuring low enzyme rates. The unit describing the change in absorbance at 340 nm on the Enzyme Rate Analyzer is equivalent to a ΔA_{340} of 0.0001.

Ellis and Goldberg (206) developed a spectrophotometric procedure using triethanolamine buffer at pH 7.3 instead of Tris at pH 7.5 as had been recommended by Bowers (98) and Wolfson and Williams-Ashman (793). This method uses higher concentrations of NADP and isocitrate and avoids additional NaCl in the assay mixture.

Photometric procedures have been described (65, 730) which depend upon the reaction of 2,4-dinitrophenylhydrazine with the reaction product α-ketoglutarate. A photometric procedure based on the reduction of INT (2-*p*-indophenyl-3-*p*-nitrophenyl-5-phenyltetrazolium chloride) by NADPH, with PMS (phenazine methosulfate) as intermediate electron carrier, has also been described (522).

DETERMINATION OF ISOCITRIC DEHYDROGENASE

(method of Wolfson and Williams-Ashman, Ref 793, and Bowers, Ref 98, modified)

PRINCIPLE

Isocitric dehydrogenase catalyzes the oxidative decarboxylation of isocitrate to α-ketoglutarate, with the simultaneous reduction of NADP to NADPH. Since NADPH absorbs at 340 nm and NADP does not, the activity is assayed by following the increase in absorbance at 340 nm, using isocitrate and NADP as substrates. The unit of activity is defined as the number of nmol of NADPH formed/ml of serum/hour at 25°C.

REAGENTS

NADP, 0.00114 M. Prepare solution just prior to use. Dissolve 1.7 mg NADP, disodium salt, in 2 ml water. Keep NADP, disodium salt, in desiccator in refrigerator.

Tris buffer, 0.1 M, pH 7.5. Dissolve 12.1 g tris(hydroxymethyl) aminomethane in 800 ml water. Adjust pH to 7.5 with conc HCl and dilute to 1 liter with water.

Manganous Chloride, 0.01 M. Dissolve 200 mg $MnCl_2 \cdot 4H_2O$ in 100 ml water.

DL-Isocitrate, 0.1 M. Dissolve 26 mg isocitrate trisodium salt in 1 ml water.

PROCEDURE

The procedure described is particularly adaptable to the Photovolt Enzyme Rate Analyzer (ERA I) equipped with the print command module and printer.

1. Set the circulating water bath for the cuvet compartment at 32°C.
2. Put 1.0 ml Tris buffer, 0.4 ml NADP solution, 0.6 ml $MnCl_2$ solution, and 1.0 ml serum sample into a test tube.
3. Mix and filter three times through one piece of tightly packed glass wool.
4. Transfer above assay mixture to the cuvet and place cuvet in a 32°C water bath for 5 min.
5. Add 0.1 ml isocitrate solution (previously brought to 32°C). Mix.
6. After an initial lag (may be as long as 3–4 min), the rate of change in absorbance readings stabilizes. Record the increasing absorbance at 340 nm at 1 min intervals for at least 5 min after the lag phase (see *Note 7*).

Calculation:

$$\text{ICD units (nmol NADPH/ml/hour at 25°C)} = \frac{\Delta A_{340}}{T} \times 1.6613 \times 10^4$$

where ΔA_{340} is the absorbance change in T min (see *Note 1*).

NOTES

1. Derivation of formula. The factor 1.6613×10^4 is derived as follows:

$$1.6613 \times 10^4 = \frac{3.1 \times 60 \times 10^6}{6.22 \times 10^3 \times 1.0 \times 1.8}$$

where

$\quad$ 3.1 $\quad$ = assay volume (ml)
$\quad$ 60 $\quad\;\,$ = conversion from per minute to per hour
$\quad$ 10^6 $\quad\;$ = conversion of mmol NADPH to nmol NADPH
$\quad$ 6.22×10^3 = molar absorptivity of NADPH
$\quad$ 1.0 $\quad\;\,$ = ml serum used
$\quad$ 1.8 $\quad\;\,$ = temperature correction factor to convert values from 32° to 25°C (153)

2. Stability of specimens. Samples are stable for many days to weeks at 4°C (98, 793). The effect of freezing is somewhat equivocal. Although freezing the sample has been recommended if the analysis cannot be performed on the day of collection (730), another report indicates that freezing causes a 10–25% loss of activity (98). Data from our laboratory indicate no loss of activity after 7 days whether the sample is kept at 30°C or frozen.

3. Lag phase. A lag of up to about 3–4 min can occur before the reaction proceeds. The lag is shorter with higher enzyme activity.

4. Variation in rate with temperature. The temperature coefficient of ICD is larger than that for lactic dehydrogenase or glutamic oxalic transaminase (98). The reaction is run at 32°C, which allows a 1.8-fold greater activity than at 25°C, the temperature used for calculation of ICD units.

5. Hemolysis. Hemolyzed specimens are unsuitable for analysis because of the high levels of ICD in erythrocytes. Contact of serum with erythrocytes for as long as 5 hours at room temperature does not affect ICD activity so long as no hemolysis has occurred (793).

6. Effect of Mn^{2+}. ICD activity is negligible in the absence of Mn^{2+}. High concentrations of Mn^{2+} tend to shorten the lag phase (98, 793). Lesser degrees of activation can be achieved with Mg^{2+} and Co^{2+}

7. Spectrophotometer. With a spectrophotometer of less sensitivity than the Photovolt Rate Analyzer (1 absorbance unit equivalent to ΔA_{340} of 0.0001), it may be necessary to record the absorbance for considerably more than 5 min in order to obtain a satisfactory ΔA. For the Photovolt Rate Analyzer, a dichromate solution cannot be used to adjust the absorbance of the instrument. Instead, prepare an NADH solution in 0.1 M Tris buffer, pH 9.5, with an absorbance between 0.1–0.2 at 340 nm. Accurately measure the absorbance at 340 nm, reading against Tris buffer in a spectrophotometer calibrated as described in *Note 13* in the LDH Section. Then measure the absorbance of the NADH solution in the Photovolt Rate Analyzer according to the calibration procedure for the instrument, using the Tris buffer to zero the instrument. Correct any discrepancy by adjusting the check point control on the instrument.

8. Buffer and pH optimum. Maximum activity occurs over a broad pH range, 7.0–8.0 (206). Above pH 8.5 precipitation of $Mn(OH)_2$ becomes significant. Phosphate buffers yield a cloudy precipitate with Mn^{2+}; tricine (N-tris-hydroxymethyl-methylglycine) forms a brown-yellow color with Mn^{2+}; and collidine results in precipitation of serum proteins (206). Tris, cacodylate, and triethanolamine buffers are acceptable. The procedure described here and by Bowers (98) uses Tris, while Ellis and Goldberg (206) use triethanolamine. Buffers generally cause inhibition as concentrations increase from 0.3 to 1.0 M.

9. Inhibition with NaCl. Ellis and Goldberg (206) state that use of NaCl in the assay mixture should be avoided because of the inhibitory properties of NaCl (0.3 M NaCl can cause 50% inhibition). We found that the small amount of NaCl in the assay mixtures described by Bowers (98) and Wolfson and Williams-Ashman (793) causes no detectable inhibition.

ACCURACY AND PRECISION

Accuracy is affected by interfering Ca ions. Bowers (98) suggests that Ca in serum is already at a maximal inhibitory concentration since addition of Ca does not result in further inhibition. The temperature must be carefully controlled for greater accuracy, as a 10°C change in temperature causes a 2.4-fold difference in activity.

The precision (95% limits) for the test is about ±8%.

NORMAL VALUES

The normal range given by Wolfson and Williams-Ashman (793) is 50—260 units. The distribution curve is skewed in the direction of high values. Other normal ranges have been reported (98, 206), but the various authors agree that values below 195 units are normal and values above 300 units are clearly abnormal. There is no variation in ICD activity between males and females, or with age, race, or total protein concentration (793). Values for sera from term pregnancies were within the normal range. Newborn infants have elevated values, with a range of 123—487 units (793).

GLUCOSE-6-PHOSPHATE DEHYDROGENASE

Glucose-6-phosphate dehydrogenase (G6PD, D-glucose-6-phosphate: NADP oxidoreductase, EC 1.1.1.49), found in red blood cells, catalyzes the following reaction:

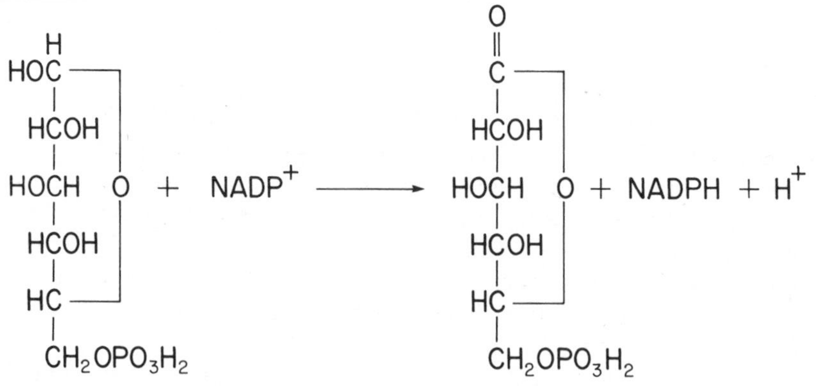

$$
\begin{array}{c}
\text{D-GLUCOSE-6-PHOSPHATE} + \text{NADP}^+ \longrightarrow \text{D-GLUCONOLACTONE-6-PHOSPHATE} + \text{NADPH} + \text{H}^+
\end{array}
$$

Also found in red blood cells are the enzymes gluconolactonase (EC 3.1.1.17) and 6-phosphogluconate dehydrogenase (6PGD, EC 1.1.1.44), which catalyze the following reactions:

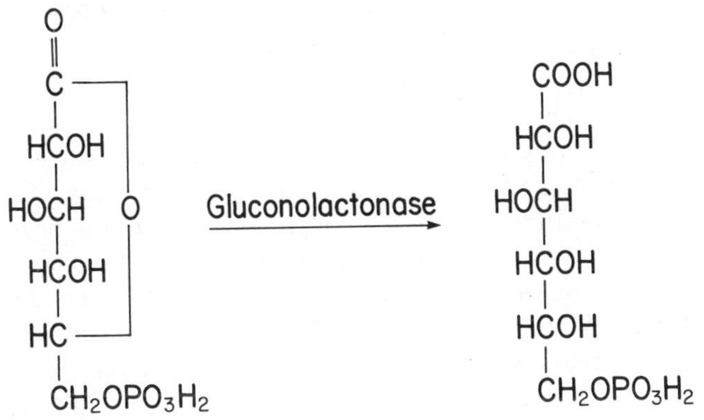

D-GLUCONOLACTONE-
6-PHOSPHATE

6-PHOSPHOGLUCONATE

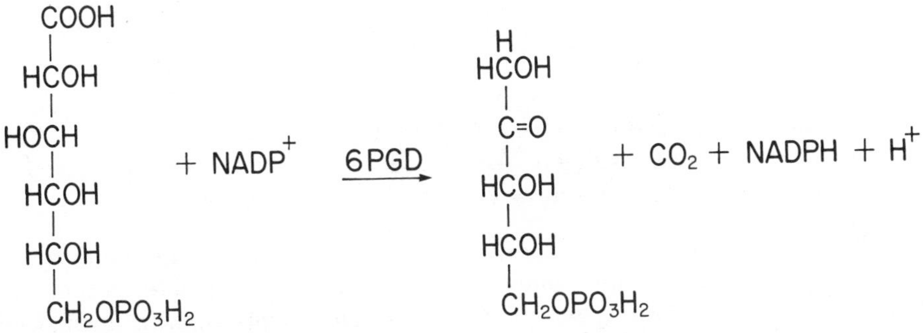

6-PHOSPHOGLUCONATE   RIBULOSE-5-PHOSPHATE

More than 1 mol of NADPH can be generated per mol of glucose-6-phosphate because of the presence of gluconolactonase and 6PGD.

G6PD is highly specific for the β-anomer of glucose-6-phosphate (685). There is slight reactivity for galactose-6-phosphate and 2-deoxyglucose-6-phosphate (387). G6PD requires NADP, the reaction being decreased more than 95% if only NAD is present.

The pH optimum range is 8–9 (387). The molecular weight of G6PD is variously stated to be 105 000–240 000. The apparent cause of this variability is the different degrees of aggregation, which are dependent upon ionic strength, pH, and protein concentration (147). It is believed that the dimer contains the essential requirements for catalytic function (subunit molecular weight of 53 000) and that the native protein is a tetramer (147).

For optimal G6PD activity, Mg^{2+} is required; Ca^{2+}, Mn^{2+}, and Ba^{2+} may also serve as activators but can cause inhibition at high concentrations (387). Phosphate buffer is inhibitory (90, 457), as are Ag^+, Hg^{2+}, Cu^{2+}, and p-chloromercuribenzoate (25, 581). The presence of NADP or NADPH tends to activate and stabilize the catalytically active G6PD dimer, whereas removal of NADP or NADPH results in dissociation to the monomer and inactivation (387).

There are more than 50 variants of G6PD, but only a few are of clinical significance (74, 387). Variants have been divided into three main groups (580): normal or slightly decreased G6PD activity, low G6PD activity associated with drug-induced hemolytic anemia, and uncommon variants with low G6PD activity associated with congenital nonspherocytic hemolytic anemia. Differences between normal and variant G6PD may be observed in electrophoretic mobility, G6PD activity, K_m for glucose-6-phosphate and NADP, ability to react with 2-deoxyglucose-6-phosphate, thermostability, and pH optima. A World Health Organization (WHO) Commission recommended a nomenclature system for the variants (794).

G6PD plays an important physiologic role as the initiator of the hexose monophosphate shunt, which results in production of 2 mols of NADPH per mol of glucose-6-phosphate. The resulting NADPH has a number of important functions (387) including methemoglobin reduction (evidence indicates that the NADH-dependent methemoglobin reductase is more important than NADPH-dependent methemoglobin reductase), maintenance of reduced glutathione via glutathione reductase, reduction of oxidant compounds and drugs, and involvement in lipid synthesis and thus possible maintenance of the red blood cell membrane.

The various assays available for G6PD activity have been reviewed (74, 214, 387) and include the following.

Heinz Body Test. Exposure of red blood cells to acetylphenylhydrazine results in a greater abundance of Heinz bodies (refractile granules of precipitated hemoglobin) within G6PD-deficient cells than in normal cells. The test is considered insufficiently accurate and specific for general use and is consequently not widely used.

Spectrophotometric Assays. These assays depend upon the formation of NADPH from NADP as detected by an increase in absorbance at 340 nm. Since 1 mol of glucose-6-phosphate results in formation of more than 1 mol of NADPH (6-phosphogluconate dehydrogenase activity), an error of 17–30% may result if correction is not made for 6PGD activity (387). Correction for 6PGD activity has been achieved by adding 6-phosphogluconate to the G6PD assay mixture and to a blank containing hemolysate (90, 259). The difference in the two activities is a measure of G6PD activity. Alternatively, 6PGD activity has been corrected by adding excess purified 6PGD to the G6PD assay mixture to ensure formation of 2 mol of NADPH per mol of glucose-6-phosphate (194). Thus the activity of G6PD is calculated as the number of mol of NADPH formed divided by two. The major disadvantage of the spectrophotometric procedures is the time-consuming nature of the assay which renders it impractical for use in large scale screening programs.

Fluorometric Assay. The fluorometric assay is also based upon the rate of formation of NADPH, but fluorescence of NADPH instead of UV absorption is the basis for quantitation. An automated fluorometric assay for G6PD has been described (726). Lowe *et al* (466) describe a manual fluorometric procedure which corrects for 6PGD activity by adding 6-phosphogluconate. The advantage of the fluorometric over the spectrophotometric assay is the greater sensitivity of the technic, which means a smaller sample requirement and thus less interference by hemoglobin. Another advantage is the elimination of time-consuming red cell washing procedures used in spectrophotometric procedures.

For both spectrophotometric and fluorometric procedures, enzyme activity is typically expressed as amounts of NADPH formed per unit of time per gram of hemoglobin, per 100 ml of blood, or per red blood cell (RBC). The least satisfactory unit is per volume of blood since this does not take into account differences in the hematocrit. Since G6PD is located within the RBC, the amount of NADPH formed per unit of time per RBC is the unit of choice. The use of hemoglobin as a standard reference would result in falsely high values of G6PD in patients with iron deficiency, hypochromic anemias, and microcytic anemias. The method of Lowe *et al* (466) uses nmol NADPH/min/10^9 RBC.

Glutathione Stability Test. This assay is an indirect measure of G6PD activity based on coupling the G6PD reaction with the glutathione reductase reaction via the NADPH formed. The procedure involves the assay for reduced glutathione by the nitroprusside method before and after incubation of red blood cells with an oxidant compound (acetylphenylhydrazine). The level of reduced glutathione is dependent upon the level of NADPH, which in turn is dependent upon the G6PD activity. There are several problems encountered in this procedure. (a) A protein-free filtrate must be prepared in order to assay for reduced glutathione. (b) Nitroprusside is not specific for reduced glutathione. (c) The procedure is not as sensitive as the spectrophotometric assay. (d) Nitroprusside and its colored complex with glutathione are unstable. (e) The procedure is too tedious for use in routine screening.

Methemoglobin Reduction Test. NADPH from the G6PD reaction reduces methemoglobin by coupling to NADPH-dependent methemoglobin reductase. Sodium nitrite is used to form methemoglobin. Methylene blue stimulates the hexose monophosphate shunt and methemoglobin reductase by serving as an electron carrier for the oxidation-reduction reactions. The color of the reaction mixture indicates the extent of reduction. The sensitivity of the quantitative form of the assay for detection of G6PD deficiency is comparable to the spectrophotometric assay and allows detection of 70–80% of the heterozygous females. The assay is time-consuming (about 3 hours) but inexpensive since there is no need to add glucose-6-phosphate and NADP although a supply of glucose must be present. If glucose-6-phosphate or NADP is exhausted, false answers can occur. Also, if there is an NADPH-dependent methemoglobin reductase deficiency, the result may be interpreted as a G6PD deficiency.

Methemoglobin Elution Test. This assay is a modification of the methemoglobin reduction test. Red blood cells are treated with $NaNO_2$ to produce methemoglobin and then incubated with Nile blue and glucose. Cells are treated with cyanide and smears made. In G6PD-deficient cells which contain mostly methemoglobin, staining by a methemoglobin elution technique results in unstained cells. Normal cells take the stain. The technic is very sensitive, being capable of detecting all heterozygotes, even when deficient cells constitute only 5% of the cell population. Since the technic requires considerable skill and time, it is used mainly as a research tool to study heterozygotes and mosaicism.

Dichloroindophenol (DCIP) Decolorization Test. The NADPH formed by the reaction of G6PD with NADP and glucose-6-phosphate is used to reduce the colored DCIP to a colorless form. Phenazine methosulfate is added as an intermediate electron carrier between NADPH and DCIP. The advantage of the test is that there is a visual endpoint, and thus the test is suitable for screening purposes. The disadvantages are the prolonged incubation required and the instability of phenazine methosulfate unless it is kept dark and dry.

Brilliant Cresyl Blue Test. The presence of NADPH is indicated by the presence of the redox indicator brilliant cresyl blue. The procedure is rapid and simple but lacks the sensitivity to detect most heterozygotes. Also, variable results are obtained with different lots of dye.

MTT Spot Test. The tetrazolium dye thiazolyl blue tetrazolium bromide forms the purple formazan upon reduction by NADPH in the presence of phenazine methosulfate as an electron carrier. The test is relatively simple.

G6PD Tetrazolium Cytochemical Method. Hemoglobin in red blood cells is converted to methemoglobin by incubation with sodium nitrite. Nile blue and glucose are added to stimulate the hexose monophosphate shunt. With normal cells methemoglobin is reduced to hemoglobin via NADPH-dependent methemoglobin reductase. The tetrazolium dye MTT is added and converted to the formazan by hemoglobin, resulting in purple-black granules within the cell. G6PD-deficient cells lack the colored granulation. The test is extremely sensitive and can detect practically all degrees of mosaicism. However, it is impractical for screening purposes since it requires several hours to perform.

Fluorescent G6PD Spot Test. After incubation of glucose-6-phosphate and NADP with G6PD from hemolysates of red blood cells, a portion of the reaction mixture is spotted on filter paper. The spot shows fluorescence under ultraviolet light if NADPH is present. For G6PD-deficient cells, no fluorescence is observed after a 30 min incubation. The method is quite good for screening purposes. The sensitivity of the test is such that consistently abnormal results are obtained in the presence of 60% abnormal cells (214).

Ascorbate Test. H_2O_2 is formed as a result of the oxidation of ascorbate in the presence of oxyhemoglobin. Cyanide is added to inhibit catalase, which would reduce H_2O_2. Cells with normal levels of G6PD produce sufficient NADPH to reduce H_2O_2 via NADPH-dependent glutathione peroxidase. In G6PD-deficient cells H_2O_2 results in peroxidative denaturation of hemoglobin as evidenced by a brown hemoglobin pigment. However, the test is not specific for G6PD deficiency since it also detects deficiencies in glutathione peroxidase, glutathione reductase, and pyruvate kinase. Such nonspecificity may be useful as a general purpose screening procedure to detect deficiencies in red blood cell enzymes.

In summary, all of the tests described detect males who are hemizygous for the enzyme defect. For detecting G6PD-deficient individuals the fluorescent spot test appears to be quite good for screening purposes. For heterozygotes, the methemoglobin elution and the tetrazolium cytochemical procedures appear to be the most sensitive since they are capable of detecting as few as 5% G6PD-deficient cells (214). The methemoglobin reduction test is capable of consistently detecting G6PD deficiency when 40% of the cells are G6PD-deficient (214). Both the spectrophotometric and fluorescent spot tests are capable of consistently detecting G6PD deficiency when 60% of the cells are G6PD-deficient (214). The brilliant cresyl blue test requires 80% G6PD-deficient cells in order to detect G6PD deficiency (214).

A person requires only about 30% of the normal G6PD activity in order to be resistant to drug-induced hemolysis (387). Consequently, for detecting persons with G6PD deficiency who are susceptible to drug-induced hemolytic anemia, tests with excessive sensitivity are not desirable for routine use. For a routine screening procedure the fluorescent spot test appears quite good. For quantitative results the fluorometric procedure is the method of choice (466).

DETERMINATION OF GLUCOSE-6-PHOSPHATE DEHYDROGENASE, SCREENING PROCEDURE

(method of Beutler and Mitchell, Ref 82)

PRINCIPLE

When G6PD is present in the hemolysate, glucose-6-phosphate is oxidized with resultant formation of NADPH from NADP. After spotting the reaction mixture on filter paper, the NADPH fluoresces when activated by ultraviolet light (360 nm).

REAGENTS

Glucose-6-phosphate, 0.01 M. Dissolve 0.306 g reagent grade glucose-6-phosphate in water and dilute to 100 ml. Dispense 5 ml aliquots into individual plastic vials and store frozen.

NADP, 0.0075 M. Dissolve 0.300 g in water and dilute to 50 ml. Dispense 5 ml aliquots into individual plastic vials and store frozen.

Saponin, 10%. Dissolve 1 g saponin in water and dilute to 100 ml. Dispense 5 ml aliquots into individual plastic vials and store frozen.

Tris-HCl Buffer, pH 7.8, 0.75 M. Dissolve 9.085 g of Tris in 85 ml water. Adjust pH to 7.8 with 8.5 ml of 6 *N* HCl before diluting with water to 100 ml. Dispense 5 ml aliquots into individual plastic vials and store frozen.

Glutathione, Oxidized (GSSG), 0.008 M. Dissolve 0.490 g in water and dilute to 100 ml. Dispense 5 ml aliquots into individual vials and store frozen.

Reaction Mixture. Mix the following solutions: water 0.2 ml, Tris buffer 0.3 ml, GSSG 0.1 ml, saponin 0.2 ml, glucose-6-phosphate 0.1 ml, NADP 0.1 ml. All reagents of the reaction mixture are stable for a month when stored frozen. The reaction mixture is stable at room temperature for 7 hours. One milliliter of the mixture is adequate for analysis of four samples.

Deficient Control. If a known or pretested deficient blood sample is not available, place a normal blood sample collected in glucose-EDTA (see *Note 1*) in a 56°C water bath for 6 hours. Store at 4°C in the refrigerator. It can be used for 1 month.

PROCEDURE

1. For each unknown, the deficient control, and the reaction mixture (to be used as reagent blank), draw two squares (about 2 x 2 cm) on Whatman 1 filter paper for spotting assay mixture at zero and 30 min.
2. For the deficient control and the unknown, dispense 0.2 ml reaction mixture into each of two 10 x 75 mm test tubes.
3. Add approximately 20 μl whole blood sample (Alsever's solution/blood, 1:1) and deficient control to reaction mixture in the appropriate tubes using a micro pipet (use 10 μl of blood if dried Alsever's mixture was used). Mix by shaking.
4. Using the same pipets as in step 3, immediately spot 10 μl of unknown and deficient control on the appropriate filter paper squares marked zero min. Also spot 10 μl reaction mixture as a reagent control.
5. Incubate samples at 45°C for 30 min and again spot 10 μl from each tube on designated squares.
6. Dry spots under hot air blower.
7. Examine under long wave UV light (Chromato-Vue, model CC-20, from Ultra-Violet Products, or an equivalent source of UV light).

INTERPRETATION OF RESULTS

The "0 min" spots should show no fluorescence. However, because of the rapidity of the reaction, some fluorescence may occur.

At 30 min, if normal G6PD activity is present the spot should fluoresce brightly. There should be no fluorescence in deficient samples or in the reagent and deficient cell controls. Because of the rate of the reaction, spotting after the 15 min incubation would be sufficient to distinguish the fluorescent spots from normal samples and the nonfluorescent spots from abnormal samples. However, 30 min is normally used to eliminate any possibility of ambiguous interpretation.

NOTES

1. Stability of specimens. G6PD is stable in erythrocytes preserved with Alsever's solution (1 ml Alsever's, either maintained as solution or taken to dryness, per ml of blood) for 3.5 weeks at 30°C. The composition of Alsever's solution is as follows: 2.05% glucose, 0.8% sodium citrate, 0.055% citric acid, and 0.42% NaCl. Alsever's solution is far superior to other anticoagulants such as heparin and glucose-EDTA. Whole blood collected in glucose-EDTA (1 mg EDTA and 5 mg glucose per ml of blood) is stable for 8 days at room temperature and 30°C. The literature data on stability tend to be contradictory with regard to stability of whole blood using ACD (acid-citrate-dextrose); some authors advocate the use of ACD (90, 494), while others feel ACD does not sufficiently stabilize G6PD in whole blood (598, 741). Heparin with tricresol is deleterious to G6PD (494). Hemolysates are not stable for more than a few hours at room temperature or 4°C (494, 741). NADP tends to stabilize the enzyme (494). With ammonium or potassium oxalate, the enzyme was found to be stable for 4 days at 4°C (90). Collection of blood in citrate, oxalate, heparin, thymol, and fluoride has been reported to be satisfactory (90).

2. Sensitivity. The lowest percentage of G6PD-deficient cells giving consistently abnormal test results is 60% (214). The highest percentage of G6PD-deficient cells giving consistently normal results is 10%.

3. Buffers. Tris buffer enhances the stability and initial intensity of fluorescence more than phosphate buffer (82).

4. Oxidized glutathione. In order to eliminate the slight fluorescence of NADPH sometimes observed in mild cases of G6PD deficiency, oxidized glutathione is added. This substance reduces small amounts of NADPH via glutathione reductase. Glutathione reductase activity is only about a quarter of the activity of the G6PD-6PGD system (6PGD, 6-phosphogluconic acid dehydrogenase). Thus one is actually measuring the difference between glutathione reductase activity and G6PD-6PGD activity. For this reason, samples with 10%–20% of the normal G6PD activity show essentially no fluorescence.

DETERMINATION OF GLUCOSE-6-PHOSPHATE DEHYDROGENASE AND 6-PHOSPHOGLUCONIC ACID DEHYDROGENASE

(method of Lowe et al, Ref 466)

PRINCIPLE

One molecule of NADPH is generated in the G6PD reaction and another in the 6PGD reaction. NADPH is quantitated fluorometrically. Measurement of G6PD is obtained by subtracting the amount of NADH formed with 6-phosphogluconic acid as substrate from the amount of NADH formed with glucose-6-phosphate and 6-phosphogluconic acid as combined substrate.

REAGENTS

Glucose-6-phosphate, 0.005 M. Dissolve 0.153 g of the disodium salt in water and bring to 50 ml. Dispense 1.5 ml aliquots into individual plastic vials and store frozen.

NADP, 1.5 mg/ml. Dissolve 0.075 g in water and bring to 50 ml. Dispense 2.5 ml aliquots into individual plastic vials and store frozen.

Tris-HCl, pH 7.8, 0.75 M. Dissolve 9.085 g Tris in 85 ml water. Adjust pH to 7.8 with conc HCl before diluting with water to 100 ml. Dispense 7 ml aliquots into individual plastic vials and store frozen.

Tris-HCl Buffer, pH 9.0, 0.75 M. Dissolve 9.085 g Tris in 85 ml water. Adjust pH to 9.0 with conc HCl before diluting with water to 100 ml. Stable at 4°C.

Digitonin, 1/3 Saturated. Place approximately 0.1 g digitonin powder in a 50 ml stoppered bottle and add 30 ml water to make a saturated solution. Place the stoppered bottle on a mechanical shaker at high speed and shake for 30 min. Filter, using Whatman 42 filter paper. Make a threefold dilution with water and store at 4°C.

6-Phosphogluconic Acid, 0.005 M. Dissolve 0.090 g of the trisodium monohydrate salt in water and bring to 50 ml. Dispense 2.5 ml aliquots into individual vials and store frozen.

Reaction Mixture I. Prepare fresh mixture by mixing the following solutions: 1 ml glucose-6-phosphate, 1 ml 6-phosphogluconate, 1 ml NADP, 3 ml Tris-HCl buffer, 4 ml water. This volume is adequate for analysis of three samples; it is stable at room temperature for 7 hours.

Reaction Mixture II. Prepare fresh mixture by mixing the following solutions: 1 ml 6-phosphogluconate, 1 ml NADP, 3 ml Tris-HCl buffer pH 7.8, 5 ml water. This is adequate for analysis of three samples and is stable at room temperature for 7 hours.

NADPH Stock Standard, 0.5 mM. Prepare a fresh NADPH solution slightly more concentrated than 0.5 mM by dissolving 3 mg NADPH (tetra sodium salt) in 5 ml water (solution A). Take 2 ml solution A and make a 2:25 dilution with 0.75 M Tris buffer (pH 9). Read the absorbance of this dilute solution (solution B) on a spectrophotometer at 340 nm (see *Note 13* in LDH section for accuracy of spectrophotometer used to measure NADPH). Since 6.22 = mM absorptivity for NADPH, the exact concentration of solution A can be calculated by the following formula:

$$\text{Solution A conc (m}M) = \frac{A_{340}}{6.22} \times \frac{25}{2}$$

Make correct dilution of solution A to give a final concentration of 0.5 m*M* of NADPH. Use the following formula:

Vol A x conc A = vol 0.5 m*M* NADPH x 0.5

PROCEDURE

For the following procedure a Turner fluorometer (model 111 with temperature stabilized sample holder and circulating water bath) and Corning filters 7-60 (primary) and 3-72 (secondary) are used in our laboratory. Other comparable equipment may be substituted.

1. Thaw reagents and prepare Reaction Mixtures I and II. Dispense 2.5 ml of each reaction mixture into separate 10 x 75 mm test tubes. Place in heating block or water bath at 30°C.

2. Adjust the circulating water bath of the fluorometer to 30°C.

3. Prepare a dilute hemolysate by adding exactly 50 μl of whole blood sample (micro pipet) to 0.45 ml of digitonin solution. Mix and let stand for 1 min.

G6PD and 6PGD Determination

4. Add 5 μl of the dilute hemolysate (micro pipet) to 2.5 ml of Reaction Mixture I. Mix. Incubate for 3 min at 30°C. For the Turner, the 10 x 75 mm test tubes are used as cuvets; other instruments may require transfer of the reaction mixture to cuvets. Bring the temperature to 30°C then record the fluorescence.

5. Wipe the tube clean and immediately place the tube in the cell compartment of the fluorometer at 3X scale. Adjust the blank knob so the fluorescence dial reads approximately 5, then record the increase in fluorescence for 2 min, taking readings once every 30 sec if an automatic recording device is not used. Calculate the increase in relative fluorescence units per minute and designate this figure as reaction rate I. (Relative fluorescence units is a term used to designate the arbitrary unit of fluorescence used with the Turner or other frequently employed fluorometers; or for automatic recording devices, it is the arbitrary unit used with the chart paper for the recording device. Since the arbitrary units and scales to expand the units on different fluorometers are different, the 3X scale and arbitrary unit of 5 is intended for the Turner. Settings for other instruments may be chosen using the 1 μM NADPH working standard as described in step 8.)

6PGD Determination

6. Add 5 μl dilute hemolysate to 2.5 ml Reaction Mixture II. Mix. Incubate for 3 min at 30°C.

7. Wipe the tube clean and immediately record the increase in fluorescence for 2 min, taking readings once every 30 sec if an automatic recording device is not used. Calculate the increase in relative fluorescence units per minute and designate as reaction rate II.

8. Dilute 3 ml of the pH 7.8 Tris-HCl buffer to 10 ml with water. Pipet 2.5 ml of the diluted Tris into a 10 x 75 mm test tube and use to zero the fluorometer. Add 5 μl of the freshly prepared NADPH solution to the 2.5

ml of diluted Tris buffer, giving a 1 μM working standard. Mix, then read and record the fluorescence of the standard.

9. Obtain the red cell count (RBC/cu mm) for the blood sample.

Calculation:

G6PD activity (nmol NADPH/min/10^9 RBC) =

$$\frac{\text{reaction rate I} - \text{reaction rate II}}{\text{fluorescence of 1 } \mu M \text{ standard}} \times 0.535 \times \frac{10^{10}}{\text{RBC count/cu mm}}$$

NOTES

1. Derivation of formulas and factors used in calculation. These calculations are necessary to understand the above equation.

G6PD + 6PGD activity (nmol NADPH/min/10^9 RBC) =

$$\frac{\text{reaction rate I}}{\text{fluorescence of standard}} \times C_{\text{standard}} \times 1.07$$

$$\times \frac{10^9}{[\text{RBC count/cu mm}] \, [1000 \times \frac{2.5}{5000}]}$$

$$= \frac{\text{reaction rate I}}{\text{fluorescence of standard}} \times 0.535 \times \frac{10^{10}}{\text{RBC count/cu mm}}$$

where

C_{standard} = nmol of NADPH standard in 2.5 ml = 2.5

1.07 = correction factor for the 7% quenching effect of the hemoglobin present

10^9 = factor to normalize observed RBC count to 10^9 RBCs

$\dfrac{[\text{RBC count/cu mm}]}{[1000 \times \frac{2.5}{5000}]}$ = factor converting RBC count/cu mm to RBC in 2.5 ml of reaction mixture where 1000 converts RBC/cu mm to RBC count/ml, 2.5 is the volume in ml, and 5000 is the dilution of the RBC in 2.5 ml of reaction mixture

Similarly, the activity for 6PGD is

$$\text{6PGD activity} = \frac{\text{reaction rate II}}{\text{fluorescence of standard}} \times 0.535 \times \frac{10^{10}}{\text{RBC count/cu mm}}$$

Since G6PD is the difference of G6PD + 6PGD activity and 6PGD activity:

$$\text{G6PD activity} = \frac{\text{reaction rate I} - \text{reaction rate II}}{\text{fluorescence of standard}} \times 0.535$$

$$\times \frac{10^{10}}{\text{RBC count/cu mm}}$$

2. *Stability of specimens.* Lowe *et al* (466) found that for this quantitative determination of G6PD activity in whole blood collected in Alsever's solution (1 vol blood plus 1 vol Alsever's) the enzyme was stable for 7 days at both 30° and 4°C. Similarly, G6PD activity was stable at 4° and 30°C with heparin as anticoagulant. For further discussion of G6PD stability, refer to *Note 1* in the preceding screening procedure for G6PD.

3. *Correction for quenching by hemoglobin.* Interference by hemoglobin has always been a problem in spectrophotometric methods. With the small sample size (50 μl) used in this method, interference by hemoglobin is minimal. Quenching caused by hemoglobin with the 50 μl sample size is 7% (466); hence a correction factor of 1.07 is employed in the calculation.

4. *Use of digitonin as lysing agent.* Digitonin and saponin were found to be better lysing agents than water or freeze-thawing, a conclusion in agreement with Bishop (90). Since saponin exhibits a certain degree of fluorescence resulting in a high blank, digitonin was the agent of choice for this method.

ACCURACY AND PRECISION

With no standard for comparison, the accuracy of the procedure is hard to assess. However, the G6PD normal range reported by the procedure of Lowe *et al* (466) is in agreement with the normal ranges of others (466).
Precision (95% limits) is about ±11%.

NORMAL VALUES

The normal adult range for G6PD by the method presented, as determined by nonparametric statistics (586), is 140-280 units, a unit being defined as a nmol NADPH/min/10^9 RBCs (466). There is no sex difference. Diurnal variation in G6PD activity is reported (303) but has been refuted (165). Higher activities are reported in newborns, the values decreasing to adult levels by the 10th to 12th months (707). For 6PGD, the normal range is 100—180 units (units as defined above).

GLUTATHIONE REDUCTASE

Glutathione reductase (NAD(P)H:oxidized glutathione oxidoreductase, EC 1.6.4.2) is a NAD(P)H and flavin adenine dinucleotide (FAD)-dependent enzyme found in red blood cells. Glutathione reductase catalyzes the oxidative cleavage of the disulfide bond of oxidized glutathione (GSSG) to form reduced glutathione (GSH):

$$GSSG + NADPH + H^+ \longrightarrow 2\,GSH + NAD^+$$

Glutathione reductase, purified 18 000-fold (657), appears not to be specific for NADPH but is activated to a lesser extent by NADH (375, 657). The enzyme also appears to have a weak dihydrolipoic dehydrogenase activity (657). The pH optimum is 6.75 for NADPH and 6.2 for NADH (657). From the FAD content of a 47 000-fold purified glutathione reductase, a minimum molecular weight of 56 000 was obtained; by Sephadex G-200 the molecular weight was 115 000 (694, 698). Thus the enzyme may be a dimer with each of its substituent monomers containing a mol of FAD. The FAD aids in stabilizing the enzyme (694). Flavin mononucleotide (FMN) does not serve as a coenzyme (695). Electrophoretic variants of glutathione reductase appear to exist (93, 375, 460).

One study indicated that glutathione reductase from normal individuals and from those with hemolytic anemia caused by glutathione reductase deficiency has differences in both pH optimum and in K_m for oxidized glutathione, the K_m being two to three times greater for the abnormal enzyme (762). Others repeated the same type of experiment and found no such differences (693). However, the observation was made that the FAD-binding ability of the glutathione reductase-deficient enzyme is decreased in comparison with the normal enzyme (693).

Using NADPH as coenzyme, chloride has an activating effect on glutathione reductase. With NADH as coenzyme, however, chloride causes a decrease in activity (84). Phosphate buffer enhances enzyme activity with NADPH but has no effect on activity when NADH is used as coenzyme (84). With NADPH as coenzyme, maximum velocity was found to be dependent upon Na^+ concentration, but this is not the case with NADH (697). NADH is a competitive inhibitor toward NADPH. NADPH and oxidized glutathione are both inhibitory to the reaction at high concentrations (697). FAD tends to activate glutathione reductase activity, the activation being inhibited by ATP (75).

The glutathione stability test (72, 809), as briefly described under assays for glucose-6-phosphate dehydrogenase, is a nonspecific, indirect measure of glutathione reductase activity. Since reduced glutathione is directly assayed in the stability test, a deficient level is indicative of either a defect in the NADPH-generating ability or a defect in glutathione reductase activity.

The spectrophotometric assay for glutathione reductase is based on the reaction of oxidized glutathione with NADPH and observing the decrease in absorbance at 340 nm caused by the conversion of NADPH to NADP (337, 578, 628). The normal range based on the study of 199 blood donors was 6.3–39.0 IU/10^9 erythrocytes (628). Age and sex have no effect on the normal range. This range may be in question, however, since addition of FAD to the reaction mixture augments glutathione reductase activity even at normal enzyme levels (75, 76). In order to correct for possible FAD deficiency, FAD should be added to the assay mixture (75, 76, 257). Assays for glutathione reductase have been conducted with and without FAD, thus obtaining an estimate of the corrected activity as well as testing for riboflavin deficiency (257, 578).

A convenient screening procedure was developed by Beutler (73). Glutathione reductase from lysed red blood cells is incubated with a reaction mixture containing oxidized glutathione and NADPH. After reaction for 1 hour at 45°C, the reaction mixture is spotted on filter paper and then examined under ultraviolet light. The disappearance of fluorescence resulting from conversion of NADPH to NADP is indicative of normal levels of glutathione reductase.

Red blood cell deficiency of glutathione reductase is one etiology for congenital nonspherocytic hemolytic anemia. Glutathione reductase deficiency has also been associated with various other clinical states including pancytopenia, thrombocytopenia, hypoplastic anemia, oligophrenia, Gaucher's disease, and α-thalassemia. However, with the recent observation by Beutler (73, 75, 76) that FAD deficiency may cause decreased glutathione reductase activity, the various conditions associated with decreased enzyme activity should be restudied to ascertain whether decreased activity is the result of deficient FAD or of abnormal or reduced levels of the enzyme.

DETERMINATION OF GLUTATHIONE REDUCTASE, SCREENING PROCEDURE

(method of Beutler, Ref 73)

PRINCIPLE

When glutathione reductase is present in a hemolysate, oxidized glutathione is reduced, with the concomitant oxidation of NADPH. Oxidation of NADPH is indicated by decreased fluorescence.

REAGENTS

Glutathione, Oxidized (GSSG), 0.03 M. Dissolve 1.84 g in water and dilute to 100 ml. Dispense 5 ml aliquots into individual plastic vials and store frozen.

NADPH, 0.015 M. Dissolve 1.33 g of the tetrasodium salt (grade A) in water and dilute to 100 ml. Dispense 5 ml aliquots into individual plastic vials and store frozen.

Phosphate Buffer, 0.25 M, pH 7.4. Mix 195 ml 0.25 M Na_2HPO_4 (7.10 g anhydrous Na_2HPO_4 in water to 200 ml) and 37 ml 0.25 M KH_2PO_4 (3.40 g KH_2PO_4 in water to 200 ml).

Saponin, 1%. Dissolve 1 g saponin in water and dilute to 100 ml. Dispense 5 ml aliquots into individual plastic vials and store frozen.

Reaction Mixture. All constituents of the reaction mixture are stable for at least 7 days when stored frozen. Thaw reagents and prepare 1 ml of reaction mixture (sufficient for three unknowns and one deficient control) according to the following proportions: phosphate buffer 0.6 ml, saponin 0.2 ml, oxidized glutathione 0.1 ml, NADPH 0.1 ml. The reaction mixture is stable at room temperature for 7 hours.

Deficient Control. If a known or pretested enzyme-deficient blood sample is not available, place a normal blood sample collected in glucose-EDTA (1 mg glucose + 1 mg EDTA per ml blood) in a 56°C water bath for 6 hours. Store at 4°C. Use control no longer than 1 month.

PROCEDURE

1. For each unknown, the deficient control, and the reaction mixture (to be used as a reagent blank), draw three squares (2 x 2 cm) on Whatman 1 filter paper for spotting assay mixture at 0, 30, and 60 min.
2. For the deficient control and the unknown, dispense 0.2 ml reaction mixture into 10 x 75 mm test tubes.
3. With a micro pipet add approximately 20 μl of blood sample (Alsever's solution/blood, 1:1) to reaction mixture (10 μl of blood sample if dried Alsever's mixture was used). Add 20 μl deficient control blood to control tube. Mix by shaking.
4. Using the same pipets as in step 3, immediately spot 10 μl unknown and deficient control on appropriate filter paper squares marked 0 min. Also spot 10 μl reaction mixture as a reagent control.
5. Incubate the samples at 45°C for 30 min and again spot 10 μl from each tube on designated squares. Return tube to heating block for continued incubation.
6. Dry spots under hot air blower.
7. Examine under long wave ultraviolet light (Chromato-Vue, model CC-20, from Ultra-Violet Products, or equivalent source of UV light). Continue incubation of all fluorescing samples for another 30 min at 45°C (*Note 5*).
8. After a 60 min incubation spot 10 μl on designated squares.
9. Dry spots under a hot air blower and again examine under long wave UV light.

INTERPRETATION OF RESULTS

a. All "0 min" spots should fluoresce from the presence of NADH.
b. At 30 min, no fluorescence of specimen spots indicates normal enzyme activity. Analysis is completed for such samples (*Note 5*).

c. Fluorescence at 30 min does not necessarily indicate enzyme deficiency. If fluorescence disappears by 60 min, the sample is normal.

d. Persistence of fluorescence at 60 min indicates enzyme deficiency.

e. For the test to be valid, fluorescence must persist at 60 min in both the reagent control and the deficient control.

NOTES

1. Stability of specimens. Data from our laboratory indicate that Alsever's solution is the anticoagulant of choice for the screening test. Glutathione reductase was stable for 25 days at 30°C when one part whole blood was mixed with one part Alsever's solution (or one part whole blood mixed with one part Alsever's solution taken to dryness; see *Note 1* in the G6PD section for composition of Alsever's solution). With heparin the enzyme was stable at −20°C for eight days and at 30°C for seven days. Beutler (75), found the enzyme still suitable for assay after 3−4 weeks at 4°C with 1.0−1.3 mg neutralized EDTA per ml of blood. ACD and heparin are also satisfactory (73).

2. Stability of reaction mixture. The reaction mixture is stable for at least 7 days frozen. It is stable for at least 3 months at 4°C if freeze-dried under nitrogen (73).

3. Sensitivity and specificity. According to Beutler (73), the rate of disappearance of fluorescence is proportional to the level of glutathione reductase activity as measured quantitatively. However, the assay has not been evaluated clinically because of the rarity of glutathione reductase deficiency. The assay appears specific for oxidized glutathione since there is failure of fluorescence to disappear from control mixtures lacking oxidized glutathione.

4. Riboflavin deficiency. Since glutathione reductase is FAD-dependent, deficiencies in riboflavin may give the same result in the test and be a "false positive" glutathione reductase deficiency (75, 76).

5. Incubation time. In our experience fluorescence disappears by 30 min in at least 95% of all normal samples.

CERULOPLASMIN

Ceruloplasmin (ferroxidase or ferro-O_2-oxidoreductase) is an α_2-glyco-protein (636). An EC designation of 1.12.3 has been proposed (547). Holmberg and Laurell (335) initially isolated this enzyme and named it *ceruloplasmin*. An x-ray crystallographic study (473) indicated that ceruloplasmin has a molecular weight of 132 000 with apparently six atoms of Cu per molecule. Earlier studies (394, 636), however, indicated molecular weights ranging from 143 000 to 160 000 with usually eight atoms of Cu per molecule. Studies using the ultracentrifuge (567) suggest that ceruloplasmin has an octamer structure consisting of two polypeptides of the same size but different charge ($\alpha_4 \beta_4$). Thus the size, structure, and number of Cu atoms has not been definitely established. In the crystalline state, 40% of the Cu is cupric (383). In solution, half of the Cu atoms per molecule are in the cupric and half in the cuprous state (249); other studies indicate a variable content of cupric ions (92). Half of the Cu atoms are not readily removable by the proteolytic enzyme chymotrypsin (167). A further observation is that half of the Cu atoms are exchangeable by the isotope ^{64}Cu in the presence of ascorbic acid (643). The role of ceruloplasmin as a ferroxidase has been studied (546, 547, 607). Its apparent biologic role (547, 607) is to increase the rate of saturation of transferrin with Fe and thus stimulate Fe utilization. It is believed that enzymatic oxidation of the ferrous state is essential in the formation of transferrin. Ferrous Fe is presented to the cell surface and oxidized to the ferric state, whereupon it is bound by apotransferrin.

Ceruloplasmin is also a general oxidase, being capable of oxidizing benzidine, *p*-phenylenediamine (PPD), N, N-dimethyl-*p*-phenylenediamine (DPP), and ascorbic acid. The pH optimum for oxidase activity is generally 5.0–6.0, depending on choice of buffer, substrate, and other assay conditions. One group of investigators (512) did not confirm the ascorbate oxidase activity, stating that this oxidase activity was caused by metals. Subsequent work (347, 548, 672) has shown that ceruloplasmin does show ascorbate oxidase activity. Ceruloplasmin also exhibits oxidase activity toward catechol, adrenaline, dopa, 5-hydroxytryptamine, 5-hydroxyindole-acetic acid, and dianisidine (394).

Ceruloplasmin oxidase activity is subject to inhibition by a variety of anions, EDTA, and a number of organic acids. Among the anions, azide and cyanide, with strong Cu binding properties, are very effective inhibitors of ceruloplasmin (169, 170). One mol of azide completely inhibits 1 mol of ceruloplasmin (169). Chloride, acetate, and EDTA apparently protect against azide inhibition. Other anions known to inhibit ceruloplasmin are thiocyanate, chloride, nitrite, bromide, fluoride, sulfate, and phosphate (636). The nature and ionic strength of buffers affect the activity of ceruloplasmin. For example, the activity of acetate is twice as great at 0.1 M as at 1 M; with phosphate, the activity at 0.1 M is about the same as 1 M acetate, and at 1 M phosphate no activity is detectable (154, 347). Organic acids such as maleic, citric, fumaric, salicylic, oxalic, ascorbic, and benzoic acid inhibit ceruloplasmin oxidase activity (113, 170, 549). Inhibition has been related to substances containing the

$$> C = CH - COOH$$

grouping (170). Activation of ceruloplasmin occurs with low ferrous ion concentrations (636). Bovine serum albumin, although previously reported to inhibit ceruloplasmin, apparently has no effect on ceruloplasmin activity (636). Also, hemoglobin and bilirubin do not interfere with the assay using PPD (715). Ultraviolet light at 253.7 nm inhibits by affecting Cu binding (27).

Immunochemical assays (394, 636) and spectrophotometric procedures (visible range) based upon the disappearance of ceruloplasmin's blue color (605 or 610 nm) under the influence of ascorbic or cyanide (394, 636) have been used to determine ceruloplasmin concentration. Such spectrophotometric procedures have the disadvantage that ceruloplasmin concentration in serum is low and thus a large volume of serum is needed for the assay. Also, slight turbidity in the serum significantly affects the result (636).

A number of assay methods are available utilizing the enzymatic properties of ceruloplasmin. The ferroxidase assay involves oxidation of ferrous to ferric ions in the presence of ceruloplasmin and apotransferrin (360, 546, 547). The transferrin formed is assayed by observing the increase in absorbance at 460 nm. The advantages of the ferroxidase assay over the more classic assays involving *p*-phenylenediamine (PPD) include: (a) smaller amounts of serum are required $(2-10 \mu l)$; (b) there is shorter reaction time; (c) hemolysis does not affect the rate; and (d) a single final product is formed instead of a series of intermediates, resulting in more linear rates (360). However, the PPD oxidase assay is considered the method of choice since it is more precise than procedures involving ferroxidase activity (715).

Oxidase activity was initially determined by following the rate of O_2 consumption (Warburg apparatus) with *p*-phenylenediamine as substrate (394). A more practical procedure for routine work is the spectrophotometric measurement of the colored products formed by oxidation of benzidine, N, N-dimethyl-*p*-phenylenediamine (DPD), or PPD (394, 636). A filter paper spot test for rapid screening (5) uses a solution of PPD in pH 5.7 acetate buffer which is dried on filter paper. An automated method using PPD has also been described (787).

The oxidation products of PPD and DPD appear to be complex. For DPD a semiquinone-type free radical (Wurster's red) has been suggested (6). A number of oxidation products with a possible dismutation equilibrium between them has been suggested (285, 563). For PPD "Bandrowski's base" (597) and a free radical of the semiquinone type (312) are possible products. Subsequent work suggests a complex of oxidation products (154, 558).

Ascorbic acid in serum produces a lag in the PPD assay for ceruloplasmin, indicating that PPD oxidized by ceruloplasmin is again reduced by ascorbic acid (320). Studies by Henry *et al* (320) demonstrate that up to 10 min from zero time is required to obtain a stable maximum rate. Thus the rate is not measured until after 10 min.

Oxidation of benzidine, PPD, or DPD by serum is catalyzed by the serum ceruloplasmin and by the Cu and Fe present in the serum as well as in the reagents used in the test (320). To determine the oxidation due to ceruloplasmin alone it is necessary either to correct for or to inhibit the catalytic oxidation by metals. Both approaches have been used. Ravin (583) used a serum blank in which the ceruloplasmin is inhibited by sodium azide. Ravin's original method contained a technical error in that serum was added to the unbuffered acidic PPD reagent prior to addition of buffer, thus leading to variable destruction of enzyme; Ravin modified this procedure in a subsequent publication (584). Nevertheless, this approach appears to be valid since azide in sufficient concentration completely inhibits oxidase activity without affecting metal catalysis (169, 320). In other methods (113, 346), catalytic oxidation of metals is inhibited by complexing them with EDTA. This approach, however, is questionable since Henry *et al* (320) observed that EDTA causes an inhibition of oxidase activity up to about 20%. Such inhibition could be the result of complex formation with ceruloplasmin through its Cu, removal of Cu from ceruloplasmin leading to inactivation, or nonspecific inhibition by a polyvalent anion. EDTA is capable of removing half of the Cu atoms from the ceruloplasmin molecule with a concomitant 50% reduction of its oxidase activity (347). The inhibition is noncompetitive (168).

DETERMINATION OF SERUM CERULOPLASMIN

(method of Henry et al, Ref 320)

PRINCIPLE

Ceruloplasmin concentration is determined from the rate of oxidation of *p*-phenylenediamine at 37°C and at pH 6.0. The rate of appearance of the purple oxidation product(s), which has an absorption peak at 520–530 nm, is measured spectrophotometrically or photometrically. Correction is made for oxidation catalyzed by the Cu and Fe present by running a serum blank in which the ceruloplasmin is inhibited by sodium azide. The unit of activity is arbitrarily defined as an increase of 0.001 in absorbance at 530 nm in 30 min under the conditions of the test. The test can also be performed on a Klett photometer using pontacyl violet 6R as an artificial standard. Figure 21-2 shows absorption curves of the colored product formed by oxidation of PPD and of pontacyl violet 6R.

REAGENTS

Acetate Buffer, 0.1 M, pH 6.0. Add 10 ml 0.1 *M* acetic acid (0.57 ml glacial acid diluted with water to 100 ml) to 200 ml 0.1 *M* sodium acetate (1.36 g $CH_3COONa \cdot 3H_2O$ per 100 ml). The pH must be 5.95–6.00.

Sodium Azide, 0.1% in 0.1 M Acetate Buffer. The pH must be 5.95–6.00. Store in refrigerator.

p-Phenylenediamine·2HCl, 0.25% in 0.1 M Acetate Buffer. Recrystallize commercial salt as follows: Dissolve in water, add Darco charcoal, warm in water bath at 60°C with occasional mixing, and filter. Add acetone to filtrate until turbidity appears, refrigerate for several hours, filter off the *p*-phenylenediamine·2HCl (PPD), and dry the crystals in the dark in a vacuum desiccator over anhydrous $CaCl_2$. Store in a brown bottle. To prepare the 0.25% reagent, dissolve 12.5 mg in 3.0 ml acetate buffer and, using narrow range pH paper, adjust the pH to 6.0 by adding 1 *N* NaOH dropwise from a 0.2 ml serologic pipet (approximately 0.1 ml required). Add acetate buffer to a final volume of 5 ml. This reagent can be used up to about 2 hours after preparation if kept in the dark.

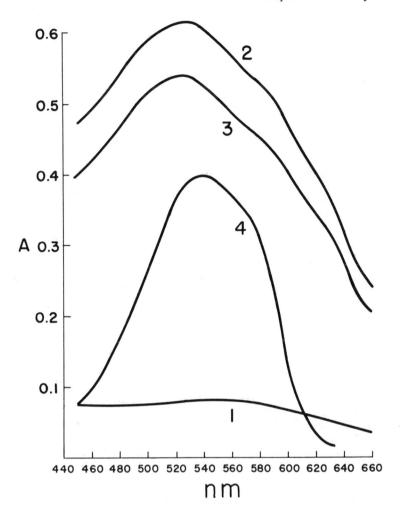

FIG. 21-2. Absorption curves in test for ceruloplasmin: (1) serum blank vs water at 30 min; (2) normal serum vs water at 30 min; (3) normal serum vs serum blank; (4) pontacyl violet 6R, 0.0159 mg/ml 0.5% acetic acid vs water. Perkin-Elmer Recording Spectrophotometer, model 4000-A. From HENRY RJ, CHIAMORI N, JACOBS SL, SEGALOVE M: *Proc Soc Exp Biol Med 104:*620, 1960

Artificial Standard for Use with Klett Photometric Procedure. Dissolve 15.9 mg pontacyl violet 6R (Du Pont) per liter of 0.5% acetic acid. This standard is equivalent to 400 units.

PROCEDURE WITH BECKMAN MODEL DU SPECTROPHOTOMETER EQUIPPED WITH THERMOSPACERS.

Keep the compartment temperature at 37°C.

1. Set up the following in cuvets with a 1 cm light path:
   Blank: 1.0 ml azide reagent + 1.0 ml buffer + 1 ml PPD reagent.
   Test: 2.0 ml buffer + 1 ml PPD reagent.

2. Place cuvets in compartment and allow 5 min for temperature equilibration.

3. Add 0.1 ml serum (heparinized or oxalated plasma is satisfactory) to each tube from a TC pipet, effecting mixing in the process.

4. Read absorbances of the Test against the Blank at 530 nm exactly 10 min and again 40 min after addition of serum.

Calculation:

$$\text{Ceruloplasmin units} = (A_{40\,min} - A_{10\,min}) \times 1000$$

PROCEDURE WITH KLETT FILTER PHOTOMETER

Follow these steps.

1. Set up the following in matched test tube cuvets:
   Blank: 2.0 ml buffer + 2.0 ml azide reagent + 0.2 ml serum (heparinized or oxalated plasma is satisfactory).
   Test: 4.0 ml buffer + 0.2 ml serum (TC pipet).

2. Incubate Blank and Test and PPD reagent in 37°C water bath for 5 min.

3. Add 2.0 ml PPD reagent to Blank and Test and mix. Cover water bath with heavy black cloth.

4. Read Test versus Blank at 10 min and again at 40 min with Klett 54 filter, leaving tubes exposed to light as briefly as possible.

Calculation:

$$\text{Ceruloplasmin units} = \frac{\text{Klett reading at 40 min} - \text{Klett reading at 10 min}}{\text{Klett reading of dye standard-vs water}} \times 40$$

NOTES

1. Artificial standard. Of a number of dyes studied, pontacyl violet 6R (Du Pont) possesses an absorption curve closest to that of PPD oxidized by ceruloplasmin. The curves are not identical, however, and the concentration of dye given as a 400 unit standard was established for a Klett 54 filter and may not be valid for other photometers. It is not valid for a spectrophotometer at 530 nm. Rice (597) suggested the use of Bandrowski's base (product of oxidation of PPD in aqueous ammoniacal solution by H_2O_2) as a standard.

2. Effect of light. Catalytic oxidation of PPD is increased by exposure to light. The test therefore must be carried out in the dark.

3. Variation in pH. In the technic presented the optimal pH for serum ceruloplasmin activity is fairly sharp at 6.0. At pH 6.0 and above, the rates after the lag phase are linear to 60 min. At pH 5.8 and lower, however, a decrease in rate occurs during the 30—40 min period. The cause for this is unknown. A pH range of 5.12—6.0 has been used by other investigators (715).

4. *Variation in temperature.* In the method presented Arrhenius plots of log of activity against reciprocals of absolute temperature give straight lines between 22° and 45°C with a slope of −3700. This gives an activation energy (μ) of 17 000 cal/mol and a temperature coefficient (Q_{10}) of 2.45 between 30° and 40°C. Sunderman and Nomoto (715) report a Q_{10} of 2.8 at 25°—58°C.

5. *Lag phase.* Evidence indicates that PPD oxidized by ceruloplasmin is again reduced by the ascorbic acid present in the serum until all the ascorbic acid is used up (6). A lag period greater than 10 min has never been observed by our laboratory with the method presented—i.e., a linear rate is established by 10 min.

6. *Hemolysis.* Hemoglobin at concentrations up to 200 mg/100 ml does not interfere with the method as observed in our laboratory. Others (715) similarly showed that the presence of hemoglobin and bilirubin does not interfere with the PPD assay for ceruloplasmin.

7. *Stability of samples.* Stability at room temperature is somewhat variable, some sera showing no degradation in 2 days, others decreasing up to 15% in 24 hours (320). Samples are stable at least 2 weeks at 4°C (2, 583, 584) or in the frozen state (320, 339, 584). Ultraviolet light of 253.7 nm wavelength inactivates the oxidase activity of ceruloplasmin by cleaving away the Cu (27).

8. *Calculations.* According to King (394), to convert the units described in the above procedure to mg/100 ml, multiply by the empirically determined factor 0.06.

ACCURACY AND PRECISION

The accuracy of any method for determination of the oxidase activity of ceruloplasmin is restricted by the absence of any reference ceruloplasmin standard. Oxidase activity can be plotted against ceruloplasmin concentration determined by immunochemical analysis (339), but this is not feasible for most laboratories. Crystalline human ceruloplasmin, unavailable to most laboratories, was used to standardize the method of Sunderman and Nomoto (715). In any event, the enzymatic activity observed in a method is governed not only by the enzyme concentration but also by the concentrations of various ions (394, 636). Such ions as nitrate, chloride, and acetate accelerate the activity at low concentrations and inhibit at high concentrations; others such as phosphate, citrate, oxalate, and sulfate appear to inhibit only. Ferrous ion, and to a lesser extent ferric and other cations at low concentration, enhance the oxidase activity of purified ceruloplasmin (636). Ferrous ions inhibit at higher concentrations. Albumin, previously reported to be inhibitory (441), appears to inhibit only nonenzymatic oxidation of PPD (584).

The precision (95% limits) of the test is about ±5%.

NORMAL VALUES

The 95% adult limits are 280–570 units. This range corresponds to 16.8–34.2 mg/100 ml if the 0.06 conversion factor of King (394) is used. No sex differences are observed (302, 509). Individual ceruloplasmin values are very constant throughout the day and between days (164, 750), and 1 hour of physical exertion does not affect the level (295). Age, however, does influence ceruloplasmin levels. Ceruloplasmin is significantly decreased in newborn infants (568, 642) but attains adult levels at the end of the first year (568). By age two, levels are slightly elevated and gradually return to adult normal levels by age 12 (164). Levels are not affected by body weight or menstrual cycle (164). Estrogen (oral contraceptives) may cause up to a twofold increase in ceruloplasmin (509). During pregnancy, levels up to three times higher than normal may occur (509).

NADH METHEMOGLOBIN REDUCTASE

The ferrous iron of hemoglobin readily undergoes spontaneous oxidation to the ferric state with the resultant conversion of hemoglobin to methemoglobin. Two enzymatic mechanisms are available for reduction of methemoglobin to hemoglobin: NADPH methemoglobin reductase and an NADH methemoglobin reductase (573). NADH methemoglobin reductase is synonomous with NADH (or DPNH) diaphorase or NADH ferrihemoglobin reductase. The NADH methemoglobin reductase is much more active than the NADPH-dependent enzyme (341, 573, 635). Thus deficiencies in NADPH methemoglobin reductase result in no clinical symptoms as long as the NADH methemoglobin reductase activity is normal (635). Deficiencies in NADH methemoglobin reductase are associated with certain types of congenital methemoglobinemia (307, 655, 658). Erythrocytic NADH methemoglobin reductase, purified about 1000-fold, has a pH optimum of 5.2 (306). NADH methemoglobin reductase is specific for NADH except in the presence of other electron acceptors such as methylene blue or cytochrome c, in which case there is some reaction with NADP (306). Purified NADH diaphorase (308) does not catalyze reduction of methemoglobin, although the ability to catalyze dichlorophenolindophenol is unaffected. Thus Hegesh et al (308) suggest that NADH diaphorase activity (reaction with dichlorophenolindophenol) be considered as separate from NADH methemoglobin activity (reaction with methemoglobin). An unknown factor might activate the methemoglobin reductase activity of the diaphorase. The molecular weight of the NADH-dependent enzyme is 30 000 and the NADPH-dependent enzyme 20 000 (308).

Electrophoresis of erythrocytic diaphorases (308) reveals five bands using NADH as electron donor and seven or eight bands using NADPH. One case of methemoglobinemia showed normal NADPH diaphorases but lacked three of the NADH diaphorase bands (308). Kaplan (376) found by electrophoresis only a single band of NADH diaphorase in erythrocytes from normal subjects; the NADPH diaphorase appeared as a double, not clearly resolved band. The NADH band split into three components after the sample aged. Schwartz *et al* (652) isolated a NADH methemoglobin reductase variant that showed decreased activity, different electrophoretic mobility, reduced affinity for NADH, and increased thermolability. Kaplan (376) found three categories of congenital NADH diaphorase deficiencies in electrophoretic studies: (a) The NADH diaphorase band was completely missing. (b) The NADH band was present with normal electrophoretic mobility but reduced activity. (c) The NADH band had a faster electrophoretic mobility. Hsieh *et al* (341) found six electrophoretic variants in cases of hereditary methemoglobinemia.

A number of tests have been developed for detecting NADH methemoglobin reductase. In the method of Scott (655, 658), as modified by Ross (615), NADH methemoglobin reductase from lysed erythrocytes catalyzes the reduction of 2,6-dichlorophenolindophenol, the reduction being measured by change in absorbance at 600 nm. An alternative procedure involves reduction of ferrihemoglobin by NADH methemoglobin reductase (305, 307). Ferrihemoglobin is formed by oxidation of hemolysates by ferricyanide. A resulting ferrihemoglobin-ferrocyanide complex is generated, which is believed to induce a change in the ferrihemoglobin structure. The complex allows NADH methemoglobin to approach the reactive site more readily and thus "activates" the reduction of ferrihemoglobin. The rate of reduction is measured by change in absorbance at 575 nm. A semiautomated procedure using the ferrihemoglobin-ferrocyanide complex has been described (699). A comparison of the two assay systems is given by Scott (656). Clinically, in the discrimination of phenotypes the results of the two assay systems gave identical results. The methemoglobin-ferrocyanide method appears somewhat more specific in measuring methemoglobin reductase than the dichlorophenolindophenol method. In practical terms preparing reagents for the methemoglobin-ferrocyanide method takes longer than for the dichlorophenolindophenol method, but the methemoglobin-ferrocyanide assay takes less time to perform because repeated washing of cells is not necessary.

A simple spot screening test for NADH methemoglobin reductase was described by Kaplan *et al* (378). Reduction of 2,6-dichlorophenolindophenol by NADH is catalyzed by NADH diaphorase. The disappearance of fluorescence of NADH is observed under ultraviolet light after spotting the reaction mixture on paper.

DETERMINATION OF NADH METHEMOGLOBIN REDUCTASE, SCREENING PROCEDURE

(method of Kaplan et al, Ref 378)

PRINCIPLE

In the presence of NADH methemoglobin reductase, the dye dichlorophenolindophenol is reduced by NADH. During the reaction, NADH, which fluoresces in the presence of long wave UV light, is oxidized to NAD which is not fluorescent. To prevent reduction of dichlorophenolindophenol by hemoglobin, blood is treated with sodium nitrite to oxidize the hemoglobin.

REAGENTS

2,6-Dichlorophenolindophenol (DCIP), 19 mM. Dissolve 0.0625 g of the sodium salt in water and bring to 100 ml in a volumetric flask. Dispense 5 ml aliquots into individual plastic vials and store frozen.

Tris Buffer, pH 7.6, 60 mM. Dissolve 0.7267 g Tris—tris(hydroxymethyl) aminomethane—in 85 ml water; adjust pH to 7.5 with conc HCl before diluting to 100 ml.

Ethylenediamine Tetraacetic Acid (EDTA), 0.27 mM. Dissolve 0.0112 g of the tetrasodium dihydrate salt in 60 mM Tris buffer and bring to 100 ml with the buffer. Dispense 5 ml aliquots into individual vials and store frozen.

NADH, Disodium Salt. Place 1 mg aliquots of NADH into vials. (Preweighed vials containing 1 mg NADH per vial are commercially available from P-L Biochemicals.) Refrigerate for prolonged storage. Refrigeration is not as important as keeping the vial perfectly dry and protected from light.

Sodium Nitrite, 1.24%. Add 0.62 g $NaNO_2$ to 50 ml of water. Prepare fresh daily.

Saponin, 1%. Dissolve 1 g saponin in 100 ml water.

Reaction Mixture. Add 1.33 ml EDTA solution and 60 µl 19 mM DCIP to the vial containing 1.0 mg NADH. Mix. The reaction mixture is stable at room temperature for 7 hours. The above volume is adequate for four samples.

Deficient Control. If a known or pretested deficient blood sample is not available, place a normal blood sample collected in glucose-EDTA (see *Note 1*) in a 56°C waterbath for 6 hours. Store at 4°C in the refrigerator. This is usable for 1 month.

PROCEDURE

1. For each unknown, the deficient control, and the reaction mixture (to be used as a reagent blank), draw three squares (2 x 2 cm) on Whatman 1 filter paper for spotting assay mixture at 0, 30, and 60 min.

2. To 10 x 75 mm test tubes labeled "unknown" and "control" add 2 µl freshly prepared 1.24% $NaNO_2$.

3. Add 20 µl whole blood sample (Alsever's solution/blood, 1:1) to the "unknown." (Use 10 µl of blood if dried Alsever'x mixture was used.) Add 20 µl of deficient control blood to the "control." Mix and allow to stand at room temperature for 30 min.

4. Add 0.04 ml 1% saponin to each tube. Mix.

5. Add 0.2 ml reaction mixture. Mix.

6. Using the same pipets as in step 3, immediately spot 10 µl unknown and deficient control on the appropriate filter paper squares marked 0 min. Also spot 10 µl reaction mixture as a reagent control.

7. Incubate the samples at 45°C for 30 min and again spot 10 µl from each tube on the designated square. Return the tube to the heating block or water bath for continued incubation.

8. Dry the spots under a hot air blower.

9. Examine under long wave UV light (Chromato-Vue, model CC-20, from Ultra-Violet Products, or equivalent source of UV light). Continue incubation of all fluorescing samples for another 30 min at 45°C (*Note 4*).

10. At the end of the 60 min incubation, spot 10 µl on the designated squares.

11. Dry spots under a hot air blower and examine under long wave UV light.

INTERPRETATION OF RESULTS

a. All "0 min" spots should fluoresce from the presence of NADH.

b. At 30 min, no fluorescence of specimen spot indicates normal enzyme activity. Analysis is completed for that sample (*Note 4*).

c. Fluorescence at 30 min does not necessarily indicate enzyme deficiency. If fluorescence disappears by 60 min, the sample is normal.

d. Persistence of fluorescence at 60 min indicates enzyme deficiency.

e. For the test to be valid, fluorescence must persist at 60 min in both the reagent control and the deficient cell control.

The test does not distinguish heterozygotes from normals.

NOTES

1. Stability of specimens. In quantitative assays methemoglobin reductase activity has been found stable at refrigerator temperatures for 3 days (378) to 2 weeks (655) when using ACD as preservative and anticoagulant for whole blood. Our studies indicate that methemoglobin reductase is stable in whole blood for 10 days at both 30° and −20°C using Alsever's solution (Alsever's solution or dried Alsever's solution/blood, 1:1); see *Note 1* in the G6PD section for composition of Alsever's solution. With glucose-EDTA (5 mg glucose plus 1 mg EDTA per ml blood), stability lasted 4 days at −20° and 30°C; with heparin, stability was 4 days at 30°C. Thus Alsever's solution is the anticoagulant of choice.

2. Correction for low hematocrit. Low hematocrits prolong the time required for disappearance of fluorescence. Thus when the hematocrit is lower than 35%, readjust the packed cell volume to normal by suspending the cells in an appropriate volume of saline or by removing some of the plasma (378).

3. Stability of reagents. The nitrite solution must be prepared fresh daily. All the other solutions are stable frozen for many months.

4. Incubation time. In our experience fluorescence disappears by 30 min in at least 95% of all normal samples.

ORNITHINE CARBAMOYL TRANSFERASE

Ornithine carbamoyl transferase or ornithine transcarbamylase (OCT, carbamoylphosphate: L-ornithine carbamoyltransferase, EC 2.1.3.3) is a liver mitochondrial enzyme which catalyzes the following reaction as part of the "urea cycle."

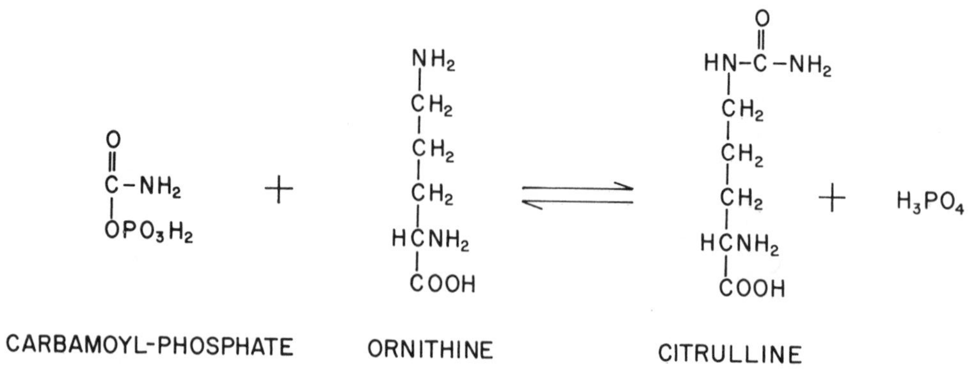

$$
\begin{array}{c}
O \\
\parallel \\
C-NH_2 \\
\mid \\
OPO_3H_2
\end{array}
\quad + \quad
\begin{array}{c}
NH_2 \\
\mid \\
CH_2 \\
\mid \\
CH_2 \\
\mid \\
CH_2 \\
\mid \\
HCNH_2 \\
\mid \\
COOH
\end{array}
\quad \rightleftharpoons \quad
\begin{array}{c}
O \\
\parallel \\
HN-C-NH_2 \\
\mid \\
CH_2 \\
\mid \\
CH_2 \\
\mid \\
CH_2 \\
\mid \\
HCNH_2 \\
\mid \\
COOH
\end{array}
\quad + \quad H_3PO_4
$$

CARBAMOYL-PHOSPHATE   ORNITHINE   CITRULLINE

Equilibrium favors the reaction proceeding to the right, as written.

The measurement of citrulline formation provides the basis for several OCT methods. Brown and Grisolia (115) decomposed any interfering urea with urease, then incubated the enzyme with carbamoyl phosphate and ornithine, stopped the reaction with perchloric acid, and determined citrulline in the supernate by the diacetylmonoxime reaction. In a similar procedure the enzyme reaction was stopped with phenylmercuric borate, and after precipitation with either perchloric or $BaSO_4$ the citrulline was determined photometrically with dimethylglyoxime (425). Filtration instead of centrifugation can be employed with citric acid precipitation in a modification of the Brown and Grisolia technic (444). Phenazone and ferric ion accelerate the photometric reaction with diacetylmonoxime (131) and are used in a more rapid ultramicro method for screening purposes (132). Improvements in this type of technic have involved the use of a noninhibitory buffer and corrections for nonenzymatic citrulline production (688).

The enzyme is capable of catalyzing the following reaction in the presence of arsenate:

$$\text{citrulline} \xrightarrow[\text{arsenate}]{\text{OCT}} CO_2 + NH_3 + \text{ornithine}$$

An intermediate, carbamoyl arsenate, forms and then breaks down non-enzymatically (588). Methods employing this arsenolysis reaction have been popular. Reichard and Reichard (589, 590) measured the $^{14}CO_2$ enzymatically released from ^{14}C-citrulline. Ammonia released during arsenolysis has been trapped in a Conway microdiffusion apparatus and then quantitated by nesslerization (560, 590). Alternatively, NH_3 may be measured by Berthelot's phenol hypochlorite reaction (513). A simple and rapid technic avoids the Conway apparatus and deproteinization by measuring NH_3 spectrophotometrically as indophenol (413, 414), but apparently lacks sensitivity, so that values below the upper normal range are not reliable (514). In other micro methods NH_3 is determined by an improved Berthelot reaction (511) or by the Seligson technic (765). Ammonia has even been estimated by the height of discoloration of a bromocresol green strip suspended in a test tube (579). The ornithine produced by arsenolysis has been measured by paper chromatography, and apparently there is agreement with the technic employing nesslerization of NH_3 (560). Correlation between NH_3 methods and the $^{14}CO_2$ procedure is good, although not stoichiometric because different substrate concentrations are used (590). The main advantages of the isotope method over the NH_3 methods, however, are greater accuracy in the low range of activity and more rapid performance.

One automated method for OCT has been devised along the lines of the Ceriotti and Gazzaniga diacetylmonoxime method (710). Another (256) uses the arsenolysis and Berthelot reactions while preventing the evolution of NH_3 from the serum by the use of EDTA.

DETERMINATION OF ORNITHINE CARBAMOYL TRANSFERASE

(method of Reichard, Ref 589)

PRINCIPLE

In the presence of arsenate, OCT catalyzes the breakdown of citrulline to CO_2, NH_3, and ornithine. With ^{14}C-labeled citrulline as the substrate, the amount of $^{14}CO_2$ released during the enzyme reaction is a measure of OCT activity.

REAGENTS

Arsenate Buffer, 0.5 M. Dissolve 7.8 g $Na_2HAsO_4 \cdot 7H_2O$ in about 30 ml deionized water. Adjust the pH to 7.1 with HCl and bring the total volume to 50 ml with deionized water.

L-Citrulline-ureido-^{14}C in 0.5 M Arsenate Buffer. Dissolve sufficient citrulline-ureido-^{14}C (New England Nuclear) and unlabeled L-citrulline (Calbiochem, A grade) in 0.5 M arsenate buffer, pH 7.1, so as to contain about 100 μmol citrulline and about 2.3 x 10^6 cpm/ml. For example, for material of specific activity 3.11 mCi/mmol, 15 mg of ^{14}C-citrulline and 528 mg L-citrulline per 30 ml solution would contain 103 μmol/ml.

The counts per minute representing 1 μmol of released $^{14}CO_2$ are experimentally determined. Dilute 0.2 ml L-citrulline-ureido-^{14}C in 0.5 M arsenate buffer to 10 ml with 10% KOH. Transfer 0.1 ml of this to a fluted filter paper wick (2 x 3 cm, Whatman 1) held by forceps; then place the wick in 10 ml scintillation fluid in a scintillation vial. Count in triplicate in scintillation counter with appropriate settings for maximum efficiency. Calculate the counts per minute per micromole of the original reagent for use in the test:

$$\frac{cpm}{\mu\text{mol citrulline/ml}} \times \frac{10}{0.1 \times 0.2}$$

In the above example, μmol citrulline/ml = 103.

H_2SO_4, 9 N. To 3 vol cold deionized water, add 1 vol conc H_2SO_4.

KOH 10%. Dissolve 1 g KOH in 10 ml CO_2-free water (boiled for 10 min or freshly deionized).

Ethanol-Toluene Solution. Combine 1 vol absolute ethanol (ethanol denatured with methanol is satisfactory) and 2 vol toluene.

Scintillation Fluid. Dissolve 2.0 g 2,5-diphenyloxazole (PPO) and 50 mg 1,4-bis-2(5-phenyl-oxazolyl)-benzene (POPOP) in 500 ml ethanol-toluene solution.

NaCl, 0.85%

PROCEDURE

1. Pipet into the outer well of 25 ml Warburg flasks: 0.5 ml serum or NaCl (blank) followed by 0.5 ml L-citrulline-ureido-^{14}C in 0.5 M arsenate buffer.
2. Add 0.3 ml 9 N H_2SO_4 to the sidearm of each flask and stopper the sidearm.
3. Make 2 x 3 cm fluted wicks of Whatman 1 filter paper. Moisten each with 0.1 ml 10% KOH, place the wick in the center well of the flask with the aid of forceps, and stopper immediately.
4. Incubate flasks in a Dubnoff metabolic shaking machine at 37°C, with gentle shaking, for exactly 2 hours.
5. Tilt each flask carefully so as to allow the H_2SO_4 to run down into the outer well. Shake flasks gently on the Dubnoff machine for 30 min at 37°C.
6. Using forceps, transfer the folded paper wicks to glass scintillation vials containing 10 ml of scintillation fluid. Cap and mix until solution is clear.
7. Place vials in the scintillation counter and allow to equilibrate for several hours or overnight. Count for an appropriate interval of time at the correct settings for maximum efficiency, corresponding to the conditions used for determination of specific activity of the labeled reagent.

Calculation:

$$\mu\text{mol } CO_2 \text{ formed/0.5 ml serum/2 hours} = \frac{cpm_{unknown} - cpm_{blank}}{cpm/\mu\text{mol}}$$

NOTES

1. Time curve. The formation of $^{14}CO_2$ during incubation with the enzyme is proportional to time up to 24 hours (589).

2. Filter papers. Filter papers should be large enough to absorb all of the KOH solution used for trapping the CO_2 without allowing excess liquid to drain and remain behind in the center well. In this way washing the center well is unnecessary. The position of the filter paper in the scintillation vials does not affect the counting rate (589).

3. Stability of samples. Ornithine transcarbamylase is a heat-stable enzyme with 45°C as the optimum temperature for activity (590) in the method described. Whole blood is stable for at least 1 day at room temperature (590). In serum the enzyme is stable for 1 week at 4°C and over 1 year in the frozen state (588). Repeated exposure to thawing and freezing, however, affects OCT in an unpredictable way (588), amounting to approximately 10% loss for one freeze-thaw cycle (688). In our laboratory serum samples have been found to maintain original levels of OCT after 3 days at 30°C.

4. Hemolysis. Slight hemolysis does not interfere with the determination (590, 710); in fact, massive hemolysis results in only small decreases in OCT activity (590).

5. Inhibitors. Phosphate and Tris buffers, excess ornithine, and mercuric salts are reported to inhibit the enzyme; the use of metal-free water is recommended for avoidance of the last type of inhibition (688). Heparin, oxalate, and citrate do not affect results (590).

ACCURACY AND PRECISION

The accuracy of the test is difficult to assess. Among merely technical parameters (e.g., substrate, pH, temperature, time) to be controlled in any enzyme test, the accuracy of the method depends also upon the determination of cpm/μmol tracer. This in turn presupposes absolute accuracy in specific activity as stated by the source.

The precision (95%) of the method is about ±20% at the top of the normal range.

NORMAL VALUES

The normal range for adults is below 0.005 μmol/0.5 ml/2 hours under the conditions of the test (590). There is no sex difference (425, 511, 587) or diurnal variation (587). Three groups of investigators noted no age difference (425, 511, 710), although Reichard (587) reported that values for those over 25 years of age tend to fall in the upper portion of the stated normal range and values for those under 25 lie mainly in the lower portion of the range. Hard physical work has no effect on enzyme levels (587). OCT may rise during delivery if chloroform is used as anesthetic (513). Chloroform dripped onto a gauze mask causes an increase in enzyme level, which peaks 3 days later (588). In contrast, a single dose of alcohol has no effect on the enzyme (587). Dietary conditions that increase the OCT level are associated with increased protein breakdown—i.e., switching to a high protein diet can raise the level two-fold and returning to a normal diet can raise the enzyme level three-fold (112).

γ-GLUTAMYL TRANSPEPTIDASE

γ-Glutamyl transpeptidase (GGTP, EC 2.3.2.−) is one of many peptidases cleaving terminal peptide bonds of proteins or peptides. It is a carboxypeptidase in that it attacks the terminal moiety with the free carboxyl group. Specifically, the C-terminal amino acid must be glutamic acid. Glutamic acid can be transferred to a variety of suitable acceptors including peptides, L-amino acids, and even water, but not D-amino acids (595). Specificity resides largely in the γ-glutamyl portion of the substrate; the adjoining portion of any substrate appears to influence only the rate of the reaction (545). Three different reactions have been attributed to this enzyme: (a) hydrolysis, in which the glutamyl residue is simply split off hydrolytically as free glutamic acid; (b) internal transpeptidation, in which the liberated glutamyl can be transferred back to the substrate itself—i.e., γ-glutamyl-γ-glutamyl-α-naphthylamide may result from γ-glutamyl-α-naphthylamide; and (c) external transpeptidation, in which the glutamyl is transferred to another suitable acceptor—e.g., amino acids or di- and tripeptides (718). Glycylglycine has been found to be the most avid of the acceptors. All three reactions may be shown to occur simultaneously in serum, and transpeptidase may be a common name for a group of enzymes in the microsomes and supernate of cells (718). The last of the three reactions, however, is stressed in recent investigations.

In the earliest description of the enzyme (293) paper chromatography was used for qualitatively isolating the products of GGTP action on glutathione or various synthetic γ-glutamyl dipeptides, with a variety of amino acids and peptides acting as acceptors. Methods for measuring the enzyme in serum are all based on the use of aromatic amines as substrate material. Goldbarg *et al* (261) employed γ-glutamylanilide as substrate; the aniline split off by the enzyme was measured by the Bratton-Marshall reaction ($NaNO_2$ and sulfamate) to form a blue azo dye. Glutathione is a competitive inhibitor and bromsulfophthalein a noncompetitive inhibitor in this reaction. With γ-L-glutamylnaphthylamide as substrate, the α-naphthylamine released by transpeptidation can also be assayed by the Bratton-Marshall reaction (545). This approach has been modified (717, 718) for estimation of enzyme activity in cerebrospinal fluid as well as in serum and tissue extracts. The incubation lasts 24 hours, but no protein precipitation or correction for product losses is necessary and small samples can be used. Without an acceptor, hydrolysis or internal transpeptidation occurs, but with glycylglycine there is external transpeptidation. The enzymatic activity can be estimated very simply by incubating for 12 min with γ-L-glutamyl-naphthylamide as substrate and glycylglycine as both acceptor and buffer, acidifying, and diazotizing to form a stable azo dye—fast blue (424). While this method is sensitive, there is a drawback in the fact that α-naphthylamine is carcinogenic. The peptide analog γ-glutamyl-*p*-nitroanilide can be used as both substrate and acceptor (543). Transpeptidation takes place between two of the substrate molecules, and the split product *p*-nitroaniline absorbs at 405 nm while the substrate itself is colorless. A kinetic method with γ-glutamyl-*p*-nitroanilide as substrate and glycylglycine as acceptor was developed by Szasz (721) for the rapid determination of a large number of samples using small sample volumes (0.1 ml). The enzyme can also be measured in urine by this method, although heat-stable, dialyzable inhibitors are present (722). A simple nonkinetic version of this method involves stopping the reaction with NaOH after 45 min of incubation (356). However, unless this method is kinetic, there is a disadvantage in measuring yellow color since the specimens of interest are likely to be icteric. To avoid color interference the Bratton-Marshall reaction can be substituted (524), resulting in the additional advantage of enhanced sensitivity. The method of Szasz would seem to be the most promising of the above procedures, particularly since the test materials are available commercially. Optimal substrate concentrations, sensitivity, speed, and ease of handling large number of specimens are features to recommend it.

The enzyme is not electrophoretically homogeneous. In separations by paper, activity has been found normally in the α_1 and α_2 fractions and also in the β- and γ-globulins in diseases of the liver, bile duct, and pancreas (411, 625). Seven fractions were demonstrated in hepatobiliary disease via starch gel electrophoresis (544). Diagnostic significance has not been established for the isoenzymes of GGTP, and further investigations are indicated (625).

GLUTAMIC-OXALACETIC TRANSAMINASE

The process of combined deamination and amination in which the amino group of an amino acid is reversibly transferred to an α-keto acid, the α-keto group in turn being transferred to the original amino acid, is called *transamination*. An enzyme catalyzing this reaction is called a *transaminase*. This type of reaction was of interest in cellular metabolism for many years but burst into importance in clinical medicine with the report by LaDue, Wróblewski, and Karmen in 1954 that serum glutamic-oxalacetic transaminase (GOT, aspartate aminotransferase, EC 2.6.1.1) is increased following myocardial infarction (427).

GOT catalyzes the following reaction:

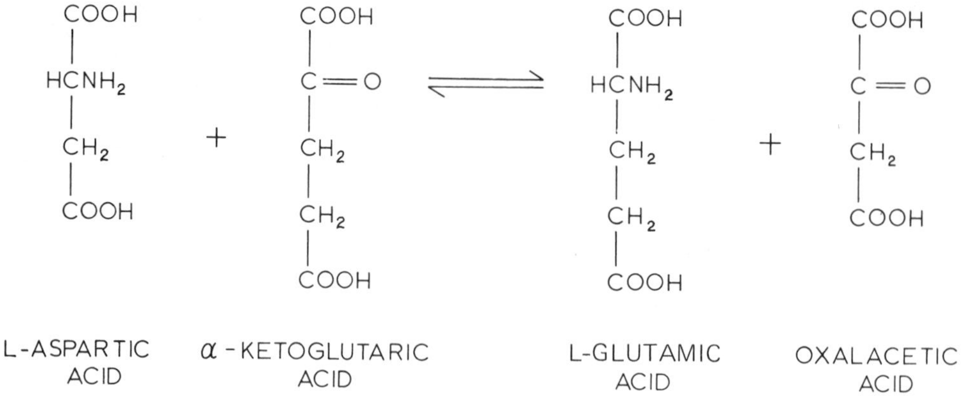

L-ASPARTIC ACID	α-KETOGLUTARIC ACID	L-GLUTAMIC ACID	OXALACETIC ACID

The rate of reaction is faster to the left than to the right, but there is no easy means of quantitating either aspartic or α-ketoglutaric acid. The methods proposed, therefore, for determining GOT employ aspartic and α-keto-glutaric acids as substrates and determine the rate of glutamic or oxalacetic acid formation.

The glutamate formed has been isolated and quantitated by paper chromatography (382), but the technic was employed only in research. Methods applicable to routine determinations in the clinical laboratory determine the oxalacetic acid directly or indirectly.

Measurement by ultraviolet spectrophotometry was introduced by Karmen (381, 427, 799) and is generally regarded as the reference method. Oxalacetate is determined by adding excess malic dehydrogenase (MDH) and NADH to produce malate and NAD:

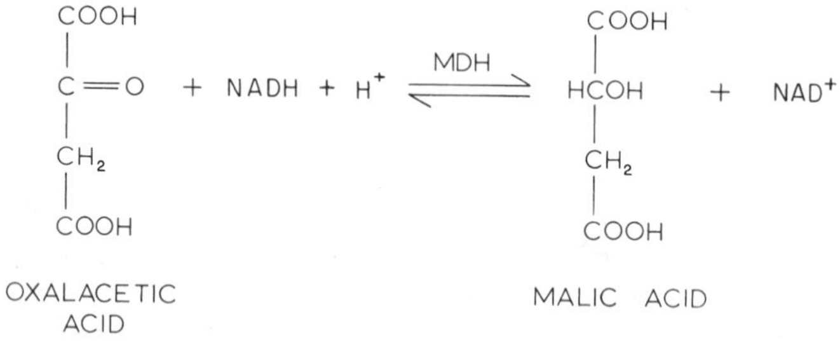

OXALACETIC ACID	MALIC ACID

The decrease in absorbance at 340 nm resulting from the oxidation of NADH per unit time is a measure of the rate of transamination. Later modifications of the method have stressed the importance of waiting until the reaction rate becomes linear (705), correcting all results to a standard temperature (704), maintaining a constant temperature in the spectro-photometer housing (650), and prechecking the activity of the MDH and NADH used (650). Another improvement was the use of a dichromate solution as a stable reference blank (689). Henry *et al* (319) combined the foregoing modifications with improvements in optimization of substrates to make a theoretically more sound technic. Amador and Wacker (15) made similar modifications to overcome inaccuracies in the original Karmen method. Devices such as timing schedules, graphic plotting, and nomograms increase speed and accuracy (555). The kinetic approach is usually considered superior, but the GOT reaction can be stopped with *p*-chloro-mercuribenzoate for simplicity of serial analyses (577). Oxalacetate has been converted to pyruvate by heat, followed by addition of lactic dehydrogenase and NADH in excess to form lactate and NAD. The decrease in absorbance at 340 nm is a measurement of transaminase activity (311).

It was perhaps natural to search for methods requiring a filter photometer so that measurements could be made in the visible spectrum. In a method for the determination of GOT in tissue, the oxalacetate formed is converted to pyruvate by treatment with aniline citrate, and the pyruvate then deter-mined photometrically after forming the colored dinitrophenylhydrazone and extracting into toluene (740). Cabaud *et al* (123) applied this method to the determination of serum GOT. Modifications of the Cabaud method have appeared (195, 345, 629), including ones in which protein precipitation and toluene extraction have been avoided (508, 804). Extraction by toluene apparently often results in the formation of a troublesome gelatinous mass (804).

In the method of Reitman and Frankel (592), oxalacetic acid is deter-mined directly as the dinitrophenylhydrazone. Both α-ketoglutaric acid and oxalacetic acids form hydrazones, but the absorptivity of the hydrazone of oxalacetic acid is much greater than that of α-ketoglutaric acid. With an increase in oxalacetic acid and a decrease in α-ketoglutaric acid, the resultant increase in absorbance of hydrazone color is proportional to the oxalacetic acid produced. Pyruvic acid, resulting from boiling the enzymatic-ally produced oxalacetate, has been measured by reaction with vanillin or β-hydroxybenzaldehyde in alkaline medium to form a yellow color (743). Icteric samples could pose a problem, however. Babson *et al* (41) introduced a photometric method in which the oxalacetate is coupled with a stabilized diazonium salt, azoene fast violet B, to yield a red color. After improvements were made in buffer and the use of serum in the standard curve (40), the method was adapted by others as a screening test in floating wells or impregnated paper disks (594, 774). Improvements regarding stabilization of the final color with ethoxylated tridecyl alcohol (36) or metabisulfite (243) followed, as did the optimization of substrates and coupling of another diazo dye at a pH capable of stopping the enzyme reaction (639). Formaldehyde was also used for the latter purpose (192). A revision of this original approach incorporated the best of the modifications and suggested the use of D-aspartate as a "blank" substrate (34). An ultramicro method involves coupling glutamate through the glutamic dehydrogenase reaction with NAD. Electrons are transferred via the electron carrier phenazine methosulfate, with subsequent reduction of iodonitro tetrazolium violet (INT) to its formazan with its characteristic color with absorption maximum at 520 nm (453). The same coupled system can be used to decolorize dichloro-indophenol (761). Another screening procedure couples the citrate syn-thetase system to the transaminase reaction through oxalacetate. The reduced coenzyme A produced then splits 5,5'-dithiobis-(2-nitrobenzoate) or DTNB to the TNB (yellow) form (354).

Technics not falling into the above categories include the manometric measurement of CO_2 released in the conversion of oxalacetate to pyruvate by aniline citrate (266), and determination of the fluorescence of the ketone condensation product of NAD formed in the Karmen-type reaction after an arbitrary period of time (437).

Among the foregoing manual methods, those of Karmen and of Reitman and Frankel have been the most popular. The advantages of the photometric hydrazone methods over the Karmen method are that (a) an ultraviolet spectrophotometer is not required, and (b) accurate temperature control of the enzyme reaction is easily obtained. The disturbing fact concerning the hydrazone methods is that oxalacetate, which inhibits the reaction (104, 106), is allowed to accumulate so that at elevated enzyme levels the hydrazone results may be half those of the spectrophotometric method (797) in which the oxalacetate is removed by the MDH-NADH system as fast as it is formed. Since the original Karmen and hydrazone methods have suboptimal substrate concentrations, the earlier literature is full of irreconcilable differences in units and results.

Many papers have appeared in which the Karmen reaction with NADH-NAD measurement at 340 nm has been adapted to automation (96, 406, 653). Another approach has been automation of the Reitman-Frankel dinitrophenylhydrazone method (640). Several groups have utilized the diazonium salt procedures (13, 85, 518, 640), although these methods suffer from interference by any metabolite or drug with a labile hydrogen atom between two carbonyl groups (121). Finally, a fluorometric version of the Henry *et al* method has been automated (447). The situation with regard to automated procedures may be considered unresolved at the present time. Although many of the procedures are routinely used in the clinical laboratory, their respective drawbacks preclude any particular recommendation.

Electrophoretic separation of serum GOT on paper usually shows enzyme activity to be in two fractions (226, 367). This may appear in α_1-globulin (209), in α_2-globulin (209, 313, 665, 670), between α_2- and β-globulin (670), or in β-globulin (209). In one study (497) activity was found in the albumin and γ-globulin zones as well. Separation on starch gel appears to yield two (108) to five (91) components. Agar gel (105, 106), cellogel (475), acrylamide gel (107), and cellulose polyacetate (609) have also been used for separation of isoenzymes. The main distinction appears to be between the supernatant or cytoplasmic enzyme and the mitochondrial enzyme (103, 104, 106, 107, 475). These differ in both chemical and physical properties. For example, their amino acid sequences around the lysine residue binding the cofactor (pyridoxal phosphate) are different (700). Antibody-binding technics have shown that a particular isoenzyme from different organs has the same characteristics; thus the enzymes are not organ-specific (430). Normally the liver mitochondrial enzyme is scarcely observable in serum (107, 353); however, the levels of both types rise in pathologic states (106).

DETERMINATION OF GLUTAMIC-OXALACETIC TRANSAMINASE

(method of Karmen, Ref 427, modified by Henry et al, Ref 319)

PRINCIPLE

L-Aspartic acid and α-ketoglutaric acid are incubated with the sample and the rate of formation of oxalacetic acid determined by the MDH-NADH system. Since MDH is added in excess, the concentration of GOT is rate-controlling and the rate of decrease in absorbance at 340 nm is taken as a measure of GOT activity. The unit of activity is defined as a decrease in absorbance of 0.001/min/ml of serum at 32°C and under other specified conditions of the test. It is important to note that the substrate concentrations employed in the modification of Henry *et al* are optimal concentrations resulting in reaction rates and, therefore, results which are 20% higher than results yielded by the method of Karmen when run at the same temperature.

REAGENTS

Phosphate Buffer, 1 M, pH 7.4. 136 g KH_2PO_4 plus 33 g NaOH per liter.

Phosphate Buffer, 0.1 M, pH 7.4. Dilute from 1 M buffer.

NADH Solution. 2.5 mg/ml 0.1 M phosphate buffer. Can be used for 1 week if kept frozen. Commercial preparations range from about 85% to 100% purity and are supplied as the tri- or tetrahydrate. The NADH concentration is not so critical that adjustment for purity or the one water molecule is necessary. (Also, see *Note 11*.)

α-Ketoglutarate Solution, 0.1 M, in 0.1 M Phosphate Buffer. To about 35 ml water in a beaker, add 5 ml 1 M phosphate buffer and 0.73 g α-ketoglutaric acid. Adjust to pH 7.4 ± 0.1 with 1 N NaOH (about 8.1 ml is required) using a pH meter. Dilute to 50 ml with water. Solution is stable in the refrigerator.

L-Aspartate, 0.375 M, in 0.1 M Phosphate Buffer. Dissolve 5.0 g L-aspartic acid in a 250-ml beaker containing about 50 ml water and 35 ml 1 N NaOH by mixing and warming on a steam bath (crushing large crystals helps speed solution). Cool to room temperature and add 10 ml 1 M phosphate buffer. Adjust to pH 7.4 ± 0.1 with 1 N NaOH using a pH meter. Dilute to 100 ml with water. Solution is stable in the refrigerator. DL-Aspartate cannot be used since D-aspartate not only is not acted upon by GOT but also actually inhibits the reaction.

Malic Dehydrogenase (MDH) Solution in 0.1 M Phosphate Buffer. Dilute the enzyme suspension (C. F. Boehringer) with 0.1 M phosphate buffer to give 10 000 units/ml. This dilute solution should be prepared daily. The stock solution is stable in the refrigerator for periods up to 4 months without significant loss of potency.

MDH solutions can be assayed by the following modification of the method of Siegel and Bing (678): To a cuvet add 2.6 ml phosphate buffer (0.1 M, pH 7.4), 0.2 ml NADH solution (2.5 mg/ml phosphate buffer), 0.1 ml MDH solution (if stock solution is about 50 000 units/ml, use a 1:100 dilution); mix. After incubation at 32°C for about 5 min, add 0.1 ml oxalacetate (1 mg oxalacetic acid per milliliter 0.1 M phosphate buffer, prepared immediately prior to use) and mix. Read absorbance (A) at 340 nm every minute for 5 min.

$$\frac{\text{MDH units/ml}}{\text{stock solution}} = \frac{A \text{ at 1 min} - A \text{ at 5 min}}{4} \times \frac{1}{\substack{\text{ml stock used} \\ \text{in test}}} \times 1000$$

Over the range of 25°–36°C the rate of increase in activity of MDH is about 4% per °C. If the assay is made at a temperature other than 32°C, an approximate correction to 32°C should be made. The "unit" employed here is the same as defined for GOT activity.

Commercial MDH preparations generally are sold with an assay given for GOT activity that is usually very small—of the order of 10 units/ml. The assay for GOT activity is made by setting up a blank determination for GOT, omitting serum. A number of such preparations have been checked in our laboratory, and no GOT activity was found when correction was made for spontaneous degradation of the NADH (see *Note 12*). Even if a preparation did contain this much GOT activity, it could be ignored.

It has been shown (808) that some batches of MDH contain an apotransaminase, causing falsely high results in GOT assays. MDH does not effect transamination in the absence of serum, however, unless the blank is run with pyridoxal 5-phosphate. The presence of apotransaminase can be ascertained by running a blank in which 0.2 ml solution containing 0.2 μg pyridoxal 5-phosphate replaces the 0.2 ml of serum.

Dichromate Blanks. A stock 0.001 M solution is prepared by dissolving 29 mg $K_2Cr_2O_4$ in 100 ml water and adding a few drops of conc H_2SO_4. A series of blanks are prepared from the stock. The following four concentrations suffice in nearly all cases: 0.00025, 0.000187, 0.000125, and 0.0000625 M.

PROCEDURE

The procedure is described for a Beckman DU spectrophotometer equipped with thermospacers and compartment temperature held at 32°C. Alternative approaches are discussed in *Note 2*.

1. To a cuvet with a 1 cm light path, add the following and mix: 1.3 ml 0.1 M phosphate buffer, 1.0 ml aspartate, 0.2 ml NADH solution, 0.1 ml MDH solution, and 0.2 ml serum or spinal fluid with a TC pipet.

2. Incubate for 30 min in the spectrophotometer (alternative procedure is to incubate mixture in a small test tube in water bath supplying the thermospacers).

3. Select a dichromate reference blank to set the spectrophotometer at zero absorbance so that the sample cuvet reads between 0.4 and 0.6 at 340 nm. The reading should be checked at about 2 min intervals until no further change occurs. Without removing the cuvet from the compartment, add 0.2 ml of α-ketoglutarate that has been kept at 32°C, mix quickly with a small polyethylene stirring rod, and quickly replace compartment lid. A timer should be started at once.

4. Read A every minute for at least 10 min. If ΔA/min is greater than 0.075, rerun the test on a dilution of the sample with buffer. The ΔA/min for each minute is calculated and the early readings in the lag phase (see *Note 1*) discarded. The later values appearing to be in a steady state within experimental error should be used in the calculation.

Calculation:

$$\text{GOT units} = \frac{\Delta A \text{ for } T \text{ min}}{T} \times \frac{1}{\text{ml sample used in test}} \times 1000$$

NOTES

1. Lag phase. Following initiation of the GOT reaction by addition of α-ketoglutarate, there is a delay of several minutes before a steady, maximal rate is reached. At zero time the oxalacetic acid concentration approaches zero; hence the MDH-NADH reaction also approaches zero. As oxalacetic acid accumulates as a result of the GOT reaction, the MDH rate increases. A point is reached at which the oxalacetate is removed by the MDH reaction as fast as it is formed by the GOT reaction, and a steady state occurs.

2. Temperature of reaction and spectrophotometer employed. In many of the earlier publications employing the spectrophotometric technic of determining GOT and GPT, the temperature was not controlled and the test was run at the equilibrium temperature of the cuvet compartment of the spectrophotometer. The temperature of the cuvet compartment of a Beckman model DU parallels room temperature but is 5°C higher, and room temperature can vary from about 22° to 32°C. Wroblewski and LaDue defined room temperature for their studies as 24°–30°C (799). Since there is an increase of about 7% per °C increase for both enzymes, it can be concluded that results could vary up to at least 70% because of uncontrolled temperature. Some investigators have controlled temperature, but no one standard temperature has been adopted. Schneider and Willis (650) and Henry *et al* (319) proposed a standard temperature of 32°C for the determination of GOT and GPT.

In running an enzyme analysis in a spectrophotometer there are two approaches: (a) Run it at a controlled temperature. This requires the use of thermospacers, which are available for a model DU. (b) Run it at a variable but known temperature. A thermister or thermometer can be installed in the compartment of a model DU. This method requires use of temperature corrections to correct values obtained at temperature t to 32°C as follows:

$$\text{Units}_{\text{corr. to } 32°} = f \times \text{units obtained at } t$$

Factors (f) are listed in Table 21-2. The reaction should not be run above a temperature of about 40°C because of heat denaturation of the enzyme (319).

TABLE 21-2. Factors for Correction of GOT and GPT to 32°C[a]

Reaction temperature (°C)	Factor (f) to convert to 32°C	
	GOT	GPT
25	1.57	1.59
26	1.47	1.48
27	1.38	1.39
28	1.29	1.30
29	1.21	1.22
30	1.13	1.14
31	1.06	1.07
32	1.00	1.00
33	0.94	0.94
34	0.88	0.88
35	0.83	0.82
36	0.78	0.77
37	0.73	0.73
38	0.69	0.68
39	0.65	0.64
40	0.61	0.60

[a] From HENRY RJ, CHIAMORI N, GOLUB OJ, BERKMAN S: *Amer J Clin Pathol* 34:381, 1960

Considerable time is required for a spectrophotometer that is at room temperature (e.g., one that has been turned off overnight) to come to temperature equilibrium after it has been turned on. A model DU requires about 2 hours. To avoid such a delay (when the thermospacers are not used), the instrument including the lamp must be left on at all times. When cuvets containing 3 ml of solution at room temperature are placed in the cuvet compartment, there is a time lag before the cuvet and its contents reach the compartment temperature. This lag is about 5 min for a model DU.

3. *Variation of rate with time.* Once the lag phase has been passed the decrease in A at 340 nm is constant down to zero absorbance. A total ΔA of 0.5 is equivalent to a decrease in NADH concentration of 0.08 mM from the initial concentration of 0.187 mM.

4. *Combination reagent.* If many GOT determinations are run per month, it is economical to prepare a combination reagent and freeze aliquots. A combination reagent in phosphate buffer is not very stable, but in Tris buffer it is stable at least 1 month (99). Dissolve 16.3 g Tris and 25.8 g L-aspartic acid in 1 liter of water, heating to effect solution. Cool to room temperature. Add 800 000 units malic dehydrogenase and 260 mg NADH. Adjust pH to 7.4 ± 0.1 with 40% NaOH, then add water to a final volume of 1344 ml. Dispense 2.6 ml aliquots into small test tubes, stopper, and freeze. The α-ketoglutarate is also prepared in 0.1 M Tris buffer: Dissolve 14.6 g α-ketoglutaric acid and 12.1 g Tris in 900 ml water, with heat if necessary. Cool and adjust to pH 7.4 ± 0.1 with 40% NaOH. Add water to 1 liter.

5. *Serum versus plasma.* Either can be used. Heparin, oxalate, EDTA, and citrate are satisfactory as anticoagulants and cause no enzyme inhibition (382, 704, 797).

6. *Hemolysis.* Two studies indicate that the concentration of GOT in erythrocytes is roughly ten-fold the normal serum level (382, 705). In another study (410) the distribution of GOT in whole blood was as follows: 80% in erythrocytes, 13% in platelets, 5% in leukocytes, and only 2% in the serum. Calculation shows that barely visible hemolysis—i.e., about 50 mg hemoglobin per 100 ml serum—results in an increase of only about 1–2% in the serum. The error from minimal hemolysis would seem to be negligible; however, others (398) report elevations of 10% in serum GOT at this level of hemolysis. It has been reported (274) that the presence of more than 400 erythrocytes or 200 leukocytes per cubic millimeter of cerebrospinal fluid results in a significant increase in GOT activity.

7. *Side reaction with NADH.* Oxidation of NADH by serum alone (step 2 of procedure) results from reduction of pyruvate present in the serum, catalyzed by lactic dehydrogenase also present in the serum (381). The more the pyruvate, or the less the lactic dehydrogenase, the longer it takes for this reaction to reach completion. Obviously, this reaction must reach exhaustion before proceeding with the analysis for GOT. Occasionally, if the pyruvate is very high the original supply of NADH may be exhausted, requiring the addition of more (650).

8. *MDH requirement.* In order for the GOT reaction to be rate-controlling, MDH must always be present in excess. As would be expected,

therefore, a linear relationship exists between GOT activity and that amount of MDH required to give optimal activity; 1000 MDH units in the 3 ml of test mixture is sufficient for GOT activity giving a $\Delta A/min$ of about 0.090. Serum itself contains some endogenous MDH (678), and at low GOT activity there may be sufficient endogenous MDH to provide optimal rate.

9. *Stability of samples.* The enzyme is stable in the freeze-dried state (382). Ebeling *et al* (199) added crystalline GOT to pooled human serum for use as a control serum and successfully stored aliquots at $-20°C$ for periods up to 8 months. Serum samples are stable frozen or at refrigerator temperature for at least 2–3 weeks (382, 427, 704, 705), but deterioration has been observed after 4 months in the refrigerator (704). Samples have been found by most workers to be stable at room temperature for 3–14 days (382, 427, 577), and one report (689) indicates stability for as long as 19 days even in the presence of bacterial growth. In another study (241), however, average decreases of 20–40% were found at 1 and 2 weeks, respectively. Even 10% loss in 24 hours at 5° or 20°C and in 2 weeks at $-15°C$ has been reported (437). Serum is significantly less stable when held at 37° (689). At temperatures between 37° and 58°C variable losses of activity are shown to be related to different isoenzyme forms (91, 806). The form present in normal serum is essentially thermostable (386). In our laboratory serum has been found to be stable for 4 days at 30°C and for 2 weeks frozen or refrigerated.

Spinal fluid GOT reportedly is stable only 6 hours at room temperature and 2 weeks at 4°C (626). Urinary GOT is asserted to be stable for 24 hours (179).

10. *Inhibitors.* Mercury and cyanide are reported to inhibit the enzyme (294). The cytoplasmic isoenzyme is sensitive to *p*-mercuribenzoate (700).

11. *Formation of inhibitor in NADH.* Fawcett *et al* (217) and Strandjord *et al* (711) reported the formation of an inhibitor of the LDH system in frozen solutions of NADH, the former workers finding it especially upon repeated freezing and thawing. It was not formed at room temperature, and no studies at refrigerator temperature were reported. There was no detectable alteration of the absorption spectrum, and it was believed to be a reduced nucleotide closely related to NADH. Kinetic studies indicated that it acted competitively with NADH. Fawcett *et al* also claimed that there was as much as a three-fold difference in the activity of different commercial preparations. A strong case is made for maintenance of a quality control chart on this test using a lyophilized serum standard.

12. *Spontaneous degradation of NADH.* The absorbance at 340 nm of the NADH in the test mixture decreases at the rate of about 0.0004/min at 32°C owing to spontaneous degradation. If a test is performed on 0.2 ml of a serum possessing 200 units of activity, $\Delta A/min$ would be 0.04. Spontaneous degradation of NADH in this case would be only 1% of the catalyzed oxidation—i.e., it would be insignificant. With 0.02 ml of the same serum, however, the catalyzed $\Delta A/min$ would be 0.004. The spontaneous degradation

would be added to this, giving an observed ΔA/min of 0.0044—i.e., a value 10% too high. For most purposes, however, spontaneous degradation can be ignored.

13. Turbid or icteric sera. In the case of a very turbid or icteric serum it may be necessary to run the test on a dilution of the serum; this is done so that a dichromate blank may be used that allows the spectrophotometer to be set at zero absorbance, giving an initial absorbance of the test mixture in the region of 0.5.

14. Spectrophotometer standardization. See *Note 13* under *Determination of Lactic Dehydrogenase.*

ACCURACY AND PRECISION

One would perhaps expect that the test is quite accurate if properly controlled with respect to such parameters as temperature. From voluminous reports of clinical studies it appears that the test does, in fact, possess creditable accuracy. There are, however, several disturbing points: (a) Pyridoxal phosphate and pyridoxamine phosphate serve as coenzymes for GOT. The addition of pyridoxal phosphate to serum *in vitro* has failed to show any effect on GOT, resulting in the conclusion that serum GOT is already fully activated (266, 382, 629). On the other hand, the addition of pyridoxine as a dietary supplement to humans results in an increase of serum GOT of 25–50% (258, 478). It has been suggested (508) it may be necessary to add coenzyme to serum in a case of severe vitamin B complex deficiency but so far this has not been demonstrated experimentally. (b) There is evidence of more than one GOT in serum. Optimal conditions for activity, temperature coefficients, etc, may not be the same for the different enzymes. Assays of the different electrophoretic components of serum GOT do not run parallel when different assay technics are used (367). (c) Instances of zero serum activity have been seen in our laboratory which raise the question of the possibility of enzyme inhibition by drugs administered to patients.

As is the case with other serum enzymes there is the possibility of variable levels of activators and inhibitors (735). The combination streptomycin-dihydrostreptomycin apparently in some way causes increased serum GOT activity (561). Heart muscle GOT is activated by Mg^{2+} (294); but if serum GOT is also activated by Mg^{2+}, the Mg^{2+} level of serum appears to be optimal (319).

As both Henry *et al* (319) and Amador and Wacker (15) pointed out, there is a decrease in absorbance in the test not due to the enzyme GOT (see *Note 12*) which, in the case of normal serum GOT levels, causes a positive error of about 10%. The normal values presented below have been corrected for this "blank" value, but for routine analyses it can be ignored.

The precision of the test at the top of the normal range is about ±8%, improving at higher values (319).

NORMAL VALUES

The adult normal range for serum is 13—55 units/ml, under the conditions of the test (585a). A slight sex difference has been reported, with levels in males appearing higher (434, 669), slightly higher (319, 680), or the same (10) as levels in females. Levels in cord blood serum may be as much as twice the adult values (258, 397), and during the first week or so of life these values (397, 445, 701) or even higher ones (418) are normal. Children by the age of about 1 year have levels similar to adult levels (208, 701). The elevation in plasma GOT of newborns seems related to higher erythrocytic enzyme levels, which also subside within a month (707). The day to day variation in any one individual appears to be quite small (426, 705). Some have observed no appreciable alteration of levels by meals (705, 797), while others report that some individuals may show significant rise in serum GOT after breakfast, particularly a protein-rich meal (as much as 27 units), followed by a decline to normal by 3 hours (763). Reports on the effect of moderately strenuous exercise are contradictory: no effect, particularly in those accustomed to such exercise (166); an increase, in most instances above normal (232, 593, 646), which returns to normal while moderate exercise is continued (593) or ceased (646); or slight decrease in the untrained (166) or 10 min after exertion (527). In one study (593) no correlation with physical fitness or muscle mass was seen. In another (434), levels were significantly dependent upon body weight, stature, and age. Most studies revealed no age difference (10, 22, 593, 669) unless associated with vitamin B_6 deficiency (292) of the aged. Serum GOT levels are not influenced by the menstrual cycle or menopause (189, 434). Pregnancy involves either no change (258, 416) or slight lowering (770) of enzyme level. In 48% of normal women a transitory elevation occurred after delivery (734). Hospitalized patients often have values 50% higher than healthy normals, in the absence of any condition suggesting true elevations (192). Drugs such as opiates (231) or erythromycin estolate (627) may cause an apparent rise, as do aspirin and sodium salicylate in 25—50% of children (476, 624).

The normal range (95% limits) for cerebrospinal fluid of adults is 7-49 units/ml (319). In one series (450) the levels of males were significantly greater than those of females. Apparently no age differences are seen among adults (158).

Normal urine has been reported to contain no GOT (110), less than 1 unit/ml (705), or less than about 9 units/ml (374).

GLUTAMIC-PYRUVIC TRANSAMINASE

Glutamic-pyruvic transaminase or alanine aminotransferase (GPT, L-alanine:2-oxoglutarate aminotransferase, EC 2.6.1.2) catalyzes the following reaction:

$$
\begin{array}{cccc}
\text{CH}_3 & \text{COOH} & \text{COOH} & \text{CH}_3 \\
| & | & | & | \\
\text{HCNH}_2 \quad + & \text{C}=\text{O} \quad \rightleftharpoons & \text{HCNH}_2 \quad + & \text{C}=\text{O} \\
| & | & | & | \\
\text{COOH} & \text{CH}_2 & \text{CH}_2 & \text{COOH} \\
 & | & | & \\
 & \text{CH}_2 & \text{CH}_2 & \\
 & | & | & \\
 & \text{COOH} & \text{COOH} &
\end{array}
$$

L – ALANINE   α - KETOGLUTARIC   L – GLUTAMIC   PYRUVIC ACID

               ACID   ACID

As reported in paper electrophoretic studies (313, 665) the enzyme is found in the β-globulin fraction. Some investigators (226, 430) have been unable to detect more than one component. In one study (497), however, activity was found in the albumin and in the α_1-, α_2-, β-, and γ-globulins. Furthermore, the distribution varies in different diseases.

Methods proposed for determination of the rate of this reaction can be classified as follows: (a) *Chromatographic method*: The glutamic acid formed can be isolated by paper chromatography and then quantitated (382). (b) *Spectrophotometric determination of the pyruvic acid by the LDH-NADH system*: Excess lactic dehydrogenase (LDH) and NADH are added to the system. As the pyruvate is formed it is reduced to lactic acid with concomitant oxidation of NADH to NAD. Since GPT is rate-controlling, the rate of decrease in absorbance at 340 nm is a measure of the GPT activity. The original method (799, 801) has been improved in such areas as substrate concentrations, precise temperature control, and allowance for the lag period, for optimum rate (69, 319). (c) *Determination of the pyruvic acid by fluorescence measurement of the NAD formed by the LDH-NADH system*: NAD has no native fluorescence but is made to fluoresce by the addition of methyl ethyl ketone (435, 438). (d) *Visible range photometric measurement of pyruvate*: Wróblewski and Cabaud (798) introduced a method in which pyruvate is converted to its hydrazone, which is then extracted into toluene for photometric measurement in alkaline medium. Xylene extraction has also been used (631). Modifications of this method by Mohun and Cook (508), Reitman and Frankel (592), and Yatzidis (804) eliminated the precipitation of serum proteins and toluene extraction. Extraction by toluene can lead to the formation of a troublesome gelatinous mass (804). The enzyme reaction has been coupled to glutamic dehydrogenase via glutamate, and the reduced NAD produced by this second reaction used to reduce a diazonium salt (INT) to its colored form (520 nm) in the presence of an electron carrier, phenazine methosulfate (452, 487). The same coupled system can be used to decolorize dichloroindophenol (761). (e) *Isotopic measurement*: Pollock *et al* (569) reported incubation with alanine-^{14}C, formation of the 2,4-dinitrophenyl-hydrazone of pyruvate-^{14}C, and extraction into toluene for counting the radioactivity.

Levine and Hill (448) adapted the LDH-NADH system methodology to automation, measuring the fluorescence of the NADH. This approach has been modified by others (535). Knight and Hunter (406) chose to automate the 340 nm spectrophotometric version of the LDH-NADH method.

Among the manual methods, the kinetic spectrophotometric methods based on a decrease in 340 nm absorbance of NADH are considered superior. Photometric procedures in which product is allowed to accumulate have often proved unreliable. Automated methods concentrate on the LDH-NADH system, utilizing either UV absorption or fluorescence measurement of NADH after a known interval of time during zero order kinetics.

SPECTROPHOTOMETRIC DETERMINATION OF GLUTAMIC-PYRUVIC TRANSAMINASE

(method of Henley and Pollard, Ref 311, modified by Henry et al, Ref 319)

PRINCIPLE

L-Alanine and α-ketoglutarate are incubated with the sample and the rate of formation of pyruvate determined by the LDH-NADH system. Since LDH is added in excess, the concentration of GPT is rate-controlling and the rate of decrease in absorbance at 340 nm is taken as a measure of GPT activity. The unit of activity is defined as a decrease in A of 0.001/min/ml of serum at 32°C and under other specified conditions of the test. It is important to note that the substrate concentrations employed in the modification of Henry *et al* are optimal concentrations resulting in reaction rates and therefore results that are 48% higher than those yielded by the method of Wróblewski and LaDue (801) and probably more than 48% higher than those obtained by the method of Henley and Pollard (311).

REAGENTS

Phosphate Buffer, 1 M and 0.1 M. As for GOT determination.
NADH Solution. As for GOT determination.
a-Ketoglutarate Solution. As for GOT determination.
DL-Alanine, 1 M, or L-Alanine, 0.5 M, in 0.1 M Phosphate Buffer. To about 75 ml water in a beaker, add 10 ml 1 *M* phosphate buffer and 8.9 g DL-alanine (or 4.45 g L-alanine). Adjust to pH 7.4 ± 0.1 with 1 *N* NaOH (about 0.8 ml is required). Dilute to 100 ml with water. Solution is stable in the refrigerator.
LDH Solution in 0.1 M Phosphate Buffer. This enzyme is obtainable from C.F. Boehringer. Dilute the commercial stock solution with buffer to 10 000 units/ml. The diluted solution can be used for 1 week if stored in the refrigerator.
Dichromate Blanks. As for GOT determination.

PROCEDURE

The procedure is described for a Beckman DU spectrophotometer equipped with thermospacers and compartment temperature held at 32°C. Alternative approaches are discussed in *Note 2* of *Determination of Glutamic-Oxalacetic Transaminase.*

1. To a cuvet with a 1 cm light path add the following and mix: 1.3 ml 0.1 *M* phosphate buffer, 1.0 ml alanine solution, 0.2 ml NADH, 0.1 ml LDH solution (1000 units), and 0.2 ml serum or spinal fluid (from a TC pipet).

2. Incubate for 20 min in the spectrophotometer. (Alternatively incubate mixture in a small test tube in water bath supplying the thermospacers.)

3. Select a dichromate reference blank to set the spectrophotometer at zero absorbance so that the sample cuvet reads between 0.4 and 0.6 at 340 nm. The reading should be checked at about 2 min intervals until no further change occurs. Exhaustion of endogenous keto and diketo acids is evidenced by a stable absorbance reading. Ordinarily, this occurs during the 20-min incubation period. Without removing the cuvet from the compartment, add 0.2 ml α-ketoglutarate solution, which has been kept at 32°C; mix quickly with a small polyethylene stirring rod and quickly replace compartment lid. A timer should be started at once.

4. Absorbance readings (*A*) should be made every minute for at least 10 min. If the change in absorbance per minute (Δ*A*/min) is greater than 0.05, the test should be rerun on a dilution of the sample with buffer. Δ*A*/min for each minute is calculated and early readings in the lag phase (see *Note 1*) discarded. The later values appearing to be in a steady state within experimental error should be used in the calculation.

Calculation:

$$\text{GPT units} = \frac{\Delta A \text{ for } T \text{ min}}{T} \times \frac{1}{\text{ml sample used in test}} \times 1000$$

NOTES

1. Lag phase. A lag of several minutes occurs before the rate of reaction reaches a stable maximum. In the preliminary side reaction the keto (e.g., pyruvic) and diketo acids are exhausted by the LDH-NADH reaction. The GPT reaction is then begun by adding α-ketoglutarate. At zero time of this latter reaction the pyruvate concentration approaches zero, at which concentration the LDH reaction approaches a rate of zero. As pyruvate accumulates as a result of the GPT reaction, the LDH rate also increases. A point is reached at which the pyruvate is removed by the LDH reaction as fast as it is supplied by the GPT reaction, and a steady state occurs.

2. Temperature of reaction and spectrophotometer employed. See *Note 2* of *Determination of Glutamic-Oxalacetic Transaminase* as it applies also to the GPT determination. Factors (*f*) for temperature corrections are given in Table 21-2.

3. *Variation of rate with time.* Once the lag phase has been passed, the decrease in absorbance at 340 nm is constant down to zero absorbance. A total ΔA of 0.5 is equivalent to a decrease in NADH concentration of 0.08 mM from the initial concentration of 0.187 mM.

4. *Combination reagent.* If many GPT determinations are run per month, it is economical to prepare a combination reagent and freeze aliquots. A combination reagent in phosphate buffer is stable only about 1 week in the frozen state, but in Tris buffer it is stable at least 1 month (99). Dissolve 30 g DL-alanine and 10.7 g Tris in 700 ml water. Add 510 000 units LDH and 170 mg NADH. Adjust pH to 7.4 ± 0.1 with conc HCl and add water to 880 ml. Dispense 2.6 ml aliquots into small test tubes, stopper, and freeze. The α-ketoglutarate is also prepared in 0.1 M Tris buffer: Dissolve 14.6 g α-ketoglutaric acid and 12.1 g Tris in 900 ml water with heat if necessary. Cool and adjust to pH 7.4 ± 0.1 with 40% NaOH. Add water to 1 liter.

5. *Hemolysis.* Erythrocytes contain about 3–5 times more GPT than serum (382, 410). Minimal hemolysis therefore can be tolerated.

6. *LDH requirement.* For the GPT reaction to be rate-controlling LDH must always be present in excess. As would be expected, therefore, a linear relationship exists between GPT activity and the amount of LDH required to give optimal activity; 1000 LDH units in the 3 ml of test mixture is sufficient for a GPT activity giving a ΔA/min of 0.50. Serum itself, of course, contains endogenous LDH, and at low ΔA/min (e.g., 0.01) there may be sufficient endogenous LDH to provide optimal rate. At higher GPT activities, however, exogenous LDH is required. A serum sample assaying 500 LDH units provides only 100 LDH units if 0.2 ml of sample is used in the test.

7. *Stability of samples.* There are conflicting data on stability of the enzyme in serum. Karmen *et al* (382) claimed that the sample is stable at room temperature for 4 days or refrigerated for up to 2 weeks, and that there was no change in activity on freezing or freeze-drying. Mosley and Goodwin (519), however, reported losses after 1 week of 15% at room temperature, 5% in the refrigerator, and 19% frozen, although samples could be kept for 99 days at −40° or −50°C. Our laboratory has found the enzyme to be stable in serum for 3 days at 30°C and 1 week refrigerated. Others have stated that there is some loss on frozen storage (95, 779), but Ebeling *et al* (199) have successfully kept a serum pool spiked with crystalline GPT (for use as a standard) at −20°C for 3 months.

8. *Serum versus plasma.* Oxalate, heparin, and citrate do not adversely affect the determination; therefore plasma as well as serum may be used (382).

9. *See Notes 12 and 13 under GOT determination.*

10. *Spectrophotometer standardization..* See *Note 13* under *Determination of Lactic Dehydrogenase.*

ACCURACY AND PRECISION

The accuracy of the test is very difficult to assess. The enzyme activity is being measured under a specific set of conditions, but whether the activity parallels enzyme concentration in any or all instances is unknown. There may well be activators and inhibitors present in varying concentrations. Also, there may be more than one GPT enzyme (497).

There is a decrease in absorbance in the test not due to the enzyme GPT (see *Note 12, Determination of Glutamic-Oxalacetic Transaminase*) which, in the case of normal serum GPT levels, causes a positive error of about 10%. The normal values presented below have been corrected for this "blank" value, but for routine analyses it can be ignored.

The precision of the test at the top of the normal range is about ±8%, improving at higher values (319).

NORMAL VALUES

The normal ranges (95% limits) for serum of adult males and females are 12–53 and 6–40 units, respectively (319). The top normal limit for infants is considerably higher. In investigations (418) employing the method of Wróblewski and LaDue (801) the top limit was about 85 units. Since the method presented yields results about 48% higher than the technic of Wróblewski and LaDue, the top normal limit for infants here should be about 125 units. After infancy there is no change with age (699). The menstrual cycle has no effect on serum levels (189). Moderately strenuous exercise results in only a slight increase (646). Normal spinal fluids contain no detectable GPT (319). Drugs such as iproniazid (807) and morphine (439) are reported to cause an increase in GPT. Salicylates produce an increased GPT in children but not in adults (624).

AUTOMATED DETERMINATION OF GLUTAMIC-PYRUVIC TRANSAMINASE

*(method of Henry et al, Ref 319, adapted to
AutoAnalyzer by Levine and Hill, Ref 448)*

PRINCIPLE

The method is based on the same coupled enzyme reaction as the manual method; however, NADH is excited at a wavelength of 340 nm and measured at a fluorescence wavelength of 470 nm. Enzyme activity is proportional to the decrease in fluorescence, since NAD does not fluoresce. Results are then converted to the same units as in the manual method. One unit is equivalent to a change in absorbance (due to oxidation of NADH) of 0.001/min/ml of serum at 32°C and under the other test conditions used. Values are therefore comparable.

REAGENTS

Tris Buffer, 0.1 M. Dissolve 12.1 g tris(hydroxymethyl)aminomethane in about 900 ml water. Add 1.0 ml Triton X-405 (Technicon) and mix. Adjust the pH to 7.6 and bring the total volume to 1 liter with water.

Substrate Solution. Dissolve 12.1 g tris(hydroxymethyl)aminomethane in about 900 ml water. Add 34.5 g DL-alanine and dissolve. Add 1.09 g α-ketoglutaric acid (A grade) and dissolve. Add 1.0 ml Triton X-405 and mix. Adjust the pH to 7.6 ± 0.05, then dilute to 1 liter with water. Store in the refrigerator. Warm to room temperature before use.

Blank Reagent. Dissolve 12.1 g tris(hydroxymethyl)aminomethane in about 900 ml water. Add 34.5 g DL -alanine and dissolve. Adjust pH to 7.6 ± 0.05 and dilute to 1 liter with water. Store in refrigerator. Warm to room temperature before use.

NADH Solution. Dissolve 20 mg NADH disodium salt in 100 ml cold Tris buffer (0.1 M, pH 7.6). Prepare fresh each day. While using, keep cold in an ice bath.

Lactic Dehydrogenase (LDH) Solution. Dilute 0.3 ml rabbit muscle LDH (C.F. Boehringer, 5 mg/ml, approximately 360 units/mg) to 100 ml with cold Tris buffer, pH 7.6. Prepare fresh each day. While using, keep cold in ice bath.

Pyruvic Acid Standard Solutions. Prepare a 10 mM stock standard solution by dissolving 110.1 mg sodium pyruvate (A grade) per 100 ml of 0.02 N HCl. Store in refrigerator. Make working standard solutions from this 10 mM stock by diluting the following volumes to 100 ml with 0.02 N HCl. Store in refrigerator.

Standard (μM)	Stock (ml)
200	2.0
400	4.0
600	6.0
800	8.0
1000	10.0

NaCl, 0.85%. Dissolve 8.5 g NaCl in about 900 ml water, add 1.0 ml Triton X-405 and bring to a total volume of 1 liter with water.

PROCEDURE

Set up manifold, other components, and operating conditions, as indicated in the flow diagram for the Technicon AutoAnalyzer (Fig. 21-3).

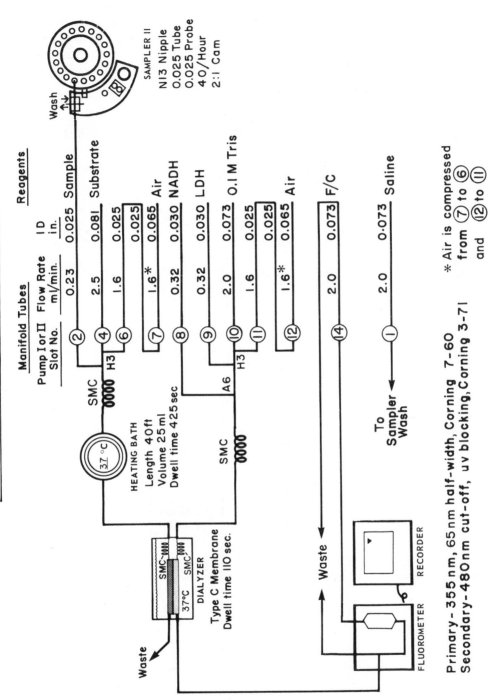

FIG. 21-3. Flow diagram for glutamic pyruvic transaminase (GPT) in serum. Calibration range: 0–100 IU/liter. Conventions and symbols used in this diagram are explained in Chap. 10.

By following procedural guidelines as described in Chapter 10, steady state interaction curves as shown in Figure 21-4 should be obtained and should be checked periodically. A parallel run is performed in which the substrate reagent is replaced by the blank reagent in order to correct for preexisting pyruvate. Sample splitter or double sampling device may be used with double manifold and two fluorometers.

Calculation:

Response is linear up to 200 μmol/liter. Obtain the μmol/liter for sample and blank in the standard manner (see *General Procedure for Automated Methods*, Chapter 10) and convert to spectrophotometric units at 32°C by the following formula:

$$\text{GPT units } (32°\text{C}) = \frac{\mu\,\text{mol/liter}_{\text{sample}} - \mu\,\text{mol/liter}_{\text{blank}}}{\text{time } (T) \text{ (see } Note\ 2)} \times \frac{0.73 \times 6.22 \times 10^3}{1000 \times 3}$$

$$= \frac{\mu\,\text{mol/liter}_{\text{sample}} - \mu\,\text{mol/liter}_{\text{blank}}}{\text{time } (T)} \times 1.51$$

FIG. 21-4. Glutamic pyruvic transaminase strip chart record. S_H is a 76.2 IU/liter control serum, S_L a 38.1 IU/liter control serum. Peaks reach 98% of steady state; interaction between samples is less than 0.2%. $W_{1/2}$ and L are explained in Chap. 10.

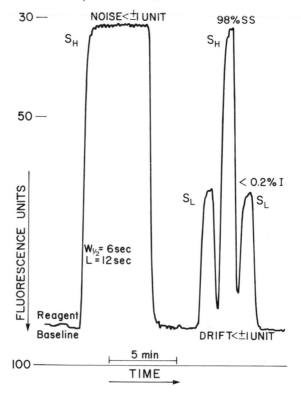

where 0.73 is the temperature correction factor from $37°$ to $32°C$ (see Table 21-2) and 6.22×10^3 is the molar absorptivity of NADH in 3 ml under the conditions of the spectrophotometric method (319).

NOTES

1. Flow diagram. This method is modified from that of Levine and Hill (448) by using a Technicon fluorometer, by increasing flow through the fluorometer cuvet to improve $W_{1/2}$, and by controlling bubble timing to decrease noise.

2. Measurement of incubation time. The incubation time is measured each day before starting the run. Pump colored solution (such as ink) through the system. Start timing at the point of sample addition to substrate; note times the reaction mixture enters and leaves the dialyzer plates. Calculate total incubation time as follows: the time (minutes, seconds) taken to reach the dialyzer plates plus half of the time (minutes, seconds) spent in the dialyzer plates. The total incubation time in minutes (to the nearest tenth) is substituted in the calculation.

3. Fluorometer settings. Maximum fluorescence is measured when the reagent base line is established. Increasing levels of enzyme decrease the amount of fluorescing NADH. Set recorder to 0% T with blank control with Tris buffer flowing in place of NADH, then replace NADH and choose the proper combination of slit and light aperture which gives a reagent base line of about 95% T.

4. Checking reagents for fluorescence. With all tubes pumping water, set the base line at 5% T. Then aspirate all the reagents except the NADH (replace with buffer) and check that the fluorescence is no greater than 10% T.

5. See Notes 5, 7, and 8 under manual method.

ACCURACY AND PRECISION

In our laboratory the automated method agrees with the manual method presented earlier. The same assessment of accuracy would apply to both.

The precision (95% limits) for GPT is about ±8% at the upper normal level.

NORMAL VALUES

Same as for manual method.

PYRUVATE KINASE

Pyruvate kinase (PK, ATP: pyruvate phosphotransferase, EC 2.7.1.40) catalyzes the following reaction:

$$\underset{\overset{||}{CH_2}}{\overset{COOH}{\underset{|}{C\text{-}OPO_3H_2}}} + ADP \xrightleftharpoons{Mg^{2+},K^+} \underset{\overset{|}{CH_3}}{\overset{COOH}{\underset{||}{C\text{=}O}}} + ATP$$

PHOSPHOENOLPYRUVATE   PYRUVATE

Red blood cell pyruvate kinase deficiency is second in frequency only to G6PD deficiency as a cause of congenital nonspherocytic hemolytic anemia (297).

The equilibrium strongly favors formation of pyruvate (119). The molecular weight of the enzyme is 205 000 (695). Pyruvate kinase from leukocytes is different from the erythrocytic enzyme with respect to genetic control and antigenic, kinetic, and chromatographic characteristics (410). Pyruvate kinase activity is dependent upon the presence of Mg^{2+} ions and is activated by the presence of a monovalent cation such as K^+, $NH_4{}^+$, or Rb^+ (40, 70, 370). Li^+ and Na^+, however, can counteract K^+ activation (370). Ca^{2+} is inhibitory, but this effect can be lessened by increasing the K^+ concentration. Copper is also inhibitory (122). It is not definitively established as to the number of forms of pyruvate kinase in erythrocytes since different authors, upon electrophoresis, report different numbers of bands (one to three) with pyruvate kinase activity (122, 377, 695, 696).

Pyruvate kinase deficiency may be caused by a number of factors ranging from quantitative enzyme deficiency (lack of one or more electrophoretic bands or decrease in intensities of the bands) to qualitative alterations of the enzyme as evidenced by kinetic studies (loss of allosteric properties, change in K_m for phosphoenolpyruvate, increased heat lability). Electrophoretic bands in PK deficiencies may be identical to those from normal erythrocytes, the intensities of the bands may be reversed, or bands may be missing (122, 377, 696). In certain cases PK is more heat-labile (696). The kinetics observed may be normal or may show a loss of allosteric properties (696). In other cases of PK deficiencies the K_m for phosphoenolpyruvate can be normal, increased, or decreased (377). Part of the results with differing K_m values for phosphoenolpyruvate may be explained by the fact the pyruvate kinase is allosteric. Adenosine triphosphate (ATP) is an allosteric inhibitor and fructose-1,6-diphosphate is an allosteric activator of PK. The enzyme responds sigmoidally to phosphoenolpyruvate at pH 7.2. The form showing allosteric kinetics has a large K_m for phosphoenolpyruvate, whereas the form showing normal Michaelis kinetics has a low K_m (377, 692, 695).

Pyruvate kinase may be assayed spectrophotometrically by coupling the pyruvate formed to lactic dehydrogenase. Lactic dehydrogenase then catalyzes the reduction of pyruvate to lactate with oxidation of NADH to NAD and, therefore, decreased absorbance at 340 nm. Conditions of the assay are usually adaptions of the general procedure presented by Bücher and Pfleiderer (119). Tanaka (727) reviewed the various assays with emphasis on the spectrophotometric method. Several modifications of the assay are available such as use of reduced glutathione to obtain more reproducible results (117); use of an imidazole-histidine buffer to increase the PK activity (150)—this claim has been contested (727); and use of a continuous flow procedure (507). Other procedures measure the formation of pyruvate by reacting pyruvate with 2,4-dinitrophenylhydrazine (507) or the shift in pH as the reaction proceeds (130).

The pH used for the assay is commonly 7.5. Collier *et al* (150) state that in their system the pH is critical and must be between 7.4 and 7.6. Beutler (77) uses a pH of 8.0. Since two enzymes and different buffers are involved in the assay, it would be expected that each assay system would have optimal concentrations of reagents and pH which apply to that system. Optimal ADP and phosphoenolpyruvate concentrations are said to be achieved at 2.0 and 3.0 mM, respectively (727).

Care must be taken to avoid leukocytic contamination in the assay since leukocytes have 300 times the PK activity of red blood cells (727). Blood is allowed to sediment, and the layer containing leukocytes plus platelets (buffy coat) is removed. After its removal, a cell count is made. It is not necessary to correct for leukocytic PK activity if the leukocyte count is below 1000/μl in the sedimented red blood cells (727). Another precaution in the spectrophotometric assay is the observation that dilute hemolysate is unstable and therefore must be used promptly (727).

A number of screening procedures are available. Tests based on auto-hemolysis (increased hemolysis of red blood cells after 48 hours of sterile incubation) have been made obsolete by the more specific and sensitive screening test of Beutler (73, 214, 282, 377, 663). The procedure of Brunetti and Nenci (116) or as modified by Tanaka (727) is based on the pH change involved in the pyruvate kinase reaction. In an unbuffered system the pH increases during the reaction. The use of *o*-cresol red as indicator results in a color change from yellow to violet red as the pH increases from 6.5 to 8.5. An advantage of this method is that whole blood from the finger is used directly in the assay mixture. However, variations in the hematocrit and white blood cell count may distort results. Homozygotes show no color change in this assay. Normal persons generally show a complete shift from yellow to violet-red. Heterozygotes show intermediate colors, although some normal subjects are indistinguishable from heterozygotes (727). The test is not consistently reliable (214, 727).

The fluorescence screening procedure of Beutler (73) is the method of choice. As with the spectrophotometric procedure, the formation of pyruvate is coupled to lactic dehydrogenase. The reduction of pyruvate to lactate results in oxidation of NADH to NAD. Since NAD is not fluorescent, application of a drop of reaction mixture to filter paper and observation of the loss of fluorescence under ultraviolet light is indicative of normal PK activity. As with the spectrophotometric assay, the buffy coat of the sedimented blood must be removed since PK activity may be normal in leukocytes but decreased in erythrocytes. This test readily detects homozygotes but sometimes cannot distinguish normals from heterozygotes (727).

DETERMINATION OF PYRUVATE KINASE, SCREENING PROCEDURE

(method of Beutler, Ref 73)

PRINCIPLE

When pyruvate kinase is present in the hemolysate, a phosphate group from phosphoenolpyruvate is transferred to ADP, forming pyruvate and ATP. Lactate dehydrogenase in the hemolysate catalyzes the reduction of pyruvate to lactate with the oxidation of NADH in the reaction mixture to NAD resulting in a loss of fluorescence.

REAGENTS

Phosphoenolpyruvate (PEP), 0.15 M. Dissolve 0.6984 g in 8.5 ml water and neutralize with 0.25 N NaOH (pH 7–8 with pH meter) before diluting to 10.0 ml. Dispense 0.5 ml aliquots into individual plastic vials and store frozen.

NADH, 0.015 M. Dissolve 0.1063 g in 8.5 ml water, neutralize with 0.25 N NaOH (pH 7–8 with pH meter), then dilute to 10.0 ml. Dispense 0.5 ml aliquots into individual vials and store frozen.

Adenosine-5'-diphosphate (ADP), 0.03 M. Dissolve 0.1508 g in 8.5 ml water, neutralize with 0.25 N NaOH (pH 7–8 with pH meter), then dilute to 10.0 ml. Dispense 0.5 ml aliquots into individual plastic vials and store frozen.

Potassium Phosphate Buffer, pH 7.4, 0.25 M. Dissolve 3.403 g KH_2PO_4 in 85 ml water. Adjust to pH 7.4, then dilute to 100 ml. Dispense 5 ml aliquots into individual vials and store frozen.

Magnesium Sulfate, 0.08 M. Dissolve 1.972 g $MgSO_4 \cdot 7H_2O$ in water and dilute to 100 ml. Dispense 5 ml aliquots into individual vials and store frozen.

Reaction Mixture. Thaw reagents (stable for a month if frozen) and prepare reaction mixture according to the following proportions: water 0.62 ml, KH_2PO_4 0.05 ml, $MgSO_4$ 0.1 ml, PEP 0.03 ml, ADP 0.1 ml, NADH 0.1 ml. Mix. One ml of the mixture is adequate for four analyses. The reaction mixture is stable at room temperature for 7 hours.

NaCl, 0.85%

Deficient Control. If a known or pretested deficient sample is not available, prepare a 20% cell suspension from a normal blood sample according to steps 3, 4, and 5 in the procedure. Place the 20% cell suspension in a 56°C water bath for 4 hours. Store at 4°C in the refrigerator. It can be used for 2 weeks.

PROCEDURE

1. For each unknown, the deficient control, and the reaction mixture (to be used as reagent blank), draw three squares (2 x 2 cm) on Whatman 1 filter paper for spotting assay mixture at 0, 30, and 60 min.
2. For the deficient control and the unknown, dispense 0.2 ml aliquots of

reaction mixture into two 10 x 75 mm test tubes.

3. Centrifuge 0.2 ml of blood sample in a 10 x 75 mm test tube.

4. Remove upper layers of plasma and buffy coat by aspiration.

5. Add approximately 0.2 ml of 0.85% NaCl to the remaining red blood cells in order to prepare a 20% cell suspension (4 vol saline to 1 vol red cells). Mix by lateral tapping of the tube.

6. Using a micro pipet add 20 μl of the unknown and deficient control 20% cell suspensions to reaction mixture in the appropriate tubes. Mix by shaking.

7. Using the same pipets as in step 6, immediately spot 10 μl of unknown and deficient control on appropriate filter paper squares marked 0 min. Also spot 10 μl reaction mixture as a reagent control.

8. Incubate samples at 45°C for 30 min and again spot 10 μl from each tube on designated squares. Return tubes to the heating block or water bath for continued incubation.

9. Dry the spots under a hot air blower.

10. Examine under long wave UV light (Chromato-Vue, model CC-20, from Ultra-Violet Products, or equivalent source of UV light). Continue incubation of all fluorescing samples for another 30 min at 45°C (*Note 4*).

11. At the end of the 60 min incubation, spot 10 μl on designated squares.

12. Dry spots under a hot air blower and again examine under long UV light.

INTERPRETATION OF RESULTS

a. All "0 min" spots should fluoresce from the presence of NADH.

b. At 30 min no fluorescence of specimen spots indicates normal enzyme activity. Analysis is completed for such samples (*Note 4*).

c. Fluorescence at 30 min does not necessarily indicate enzyme deficiency. If fluorescence disappears by 60 min, the sample is normal.

d. Persistence of fluorescence at 60 min indicates enzyme deficiency.

e. For the test to be valid, fluorescence must persist at 60 min in both the reagent control and the deficient cell control.

NOTES

1. Stability of specimens. Data from our laboratory indicate that Alsever's solution is the anticoagulant of choice for the screening test. Pyruvate kinase is stable for 3.5 weeks at 30°C when one part whole blood is mixed with one part Alsever's solution (or one part whole blood mixed with one part Alsever's solution taken to dryness). (See *Note 1* in the G6PD section for composition of Alsever's solution.) With heparin the enzyme is stable at 30°C for 8 days. Specimens cannot be frozen since the freeze-thaw process results in release of leukocytic pyruvate kinase. Beutler (77) found less than 10% loss of activity at 25°C for 5 days using ACD, EDTA, or heparin and at 4°C for 20 days using the same anticoagulants.

2. Stability of reaction mixture. The reaction mixture once prepared is stable at room temperature for 7 hours. When frozen it is stable for at least 4 but less than 10 days (73), and freeze-dried under N_2 for at least 3 months at 4° or 20°C (73).

3. Removal of white blood cells. Leukocytic pyruvate kinase activity is normal in subjects with hereditary PK deficiency. For this reason it is necessary to remove the buffy coat before screening for PK deficiency. In leukocytes the PK activity is 300 times as great as in red blood cells (370).

4. Incubation time. In our experience fluorescence disappears by 30 min in at least 95% of all normal samples.

CREATINE PHOSPHOKINASE

Creatine phosphokinase (CPK, adenosine triphosphate: creatine phospho-transferase, EC 2.7.3.2) catalyzes the following reaction:

$$\text{Creatine} + \text{ATP} \rightleftharpoons \text{creatine phosphate} + \text{ADP}$$

As written, the reaction to the right has been termed the "forward reaction"; the enzyme can also catalyze the "reverse reaction", in which ADP and creatine phosphate are converted to creatine and ATP, at a rate two to five times faster (159, 210, 618). Manual methods are easily categorized as those based on the measurement of the reverse and the forward reactions.

The creatine formed enzymatically from creatine phosphate and ADP (reverse reaction) has been measured photometrically by the diacetyl-*a*-naphthol reaction (210). Modifications such as stopping the enzyme reaction by zinc baryte precipitation (641) and the addition of cysteine as stabilizer (344) have given this approach more accuracy and sensitivity. Other authors have found residual CPK activity in Ba-Zn filtrates and have suggested the use of *p*-chloromercuribenzoate for stopping the enzyme reaction without protein precipitation (795). Enzymatically formed creatine has been measured fluorometrically by reacting with ninhydrin in strong alkali (638). Optimization of incubation conditions produced higher normal values (159), which allowed development of a micro method (409). A kinetic approach (541) employs the reverse reaction but is coupled to the hexokinase and glucose-6-phosphate dehydrogenase reactions to convert finally NADP to NADPH, which can be measured spectrophotometrically at 340 nm at a rate dependent upon the activity of the creatine phosphokinase:

$$\text{ATP} + \text{glucose} \xrightarrow{\text{hexokinase}} \text{ADP} + \text{glucose-6-phosphate}$$

$$\text{Glucose-6-phosphate} + \text{NADP} \xrightarrow{\text{G6PD}} \text{6-phosphogluconate} + \text{NADPH}$$

The introduction of bulk, freeze-dried substrate (612) has made this procedure extremely rapid and easy. In the forward reaction ADP has been coupled through the pyruvate kinase and LDH reactions to oxidize finally NADH to NAD in a kinetic spectrophotometric method (728). In a more sensitive modification of this approach, the pyruvate formed can be measured as the colored 2,4-dinitrophenylhydrazone (539). A spot screening method, based on the disappearance of NADH fluorescence by 40 min, is another variation of this approach (388). Phosphate released upon hydrolysis of creatine phosphate formed in the forward reaction has been photometrically determined (423). Although the substrates are less expensive for the forward reaction than for the reverse (618), more attention must be paid to the relative proportions of serum and substrate volumes as a result of heat-stable inhibitors (539).

The forward reaction has been automated by determination of phosphorus after hydrolysis of creatine phosphate (796). The reverse reaction has been automated by adaptation of the diacetyl-α-naphthol (223, 679) or ninhydrin fluorescence (785) methods.

At least three CPK isoenzymes have been distinguished by electrophoretic technics. Separations are accomplished on agar or agarose gel (268, 671, 752), paper (498), and cellulose acetate (742); visualization is by means of fluorescence (671) or by transference of electrons via phenazine methosulfate to the diazo dye NBT (268, 742, 752) in the absence of cysteine, which would immediately decolorize the dye (325).

The kinetic method involving the reverse reaction is considered to be the best operationally and also the most precise in the normal to moderately elevated region. Its popularity is increased by the availability of the substrate in compressed tablets or in lyophilized form.

DETERMINATION OF CREATINE PHOSPHOKINASE

(method of Rosalki, Ref 612, modified)

PRINCIPLE

In a series of linked reactions, ATP formed from creatine phosphate and ADP by the action of CPK reacts with glucose in the presence of hexokinase to form glucose-6-phosphate. Glucose-6-phosphate in turn is oxidized by NADP in the presence of glucose-6-phosphate dehydrogenase (G6PD) to form 6-phosphogluconate and NADPH. Conditions are such that the CPK reaction is rate-limiting. Consequently the rate of increase in absorbance at 340 nm due to the reduction of NADP to NADPH is a measure of the CPK activity. The unit of activity is defined as 1 μmol of creatine phosphate (equivalent to 1 μmol of NADP) converted per liter of serum per minute under the specified conditions of the test.

REAGENTS

Tris Buffer, 0.05 M. Dissolve 6.05 g tris(hydroxymethyl)aminomethane, analytic grade, in about 900 ml water. Adjust pH to 6.8 at 25°C by adding HCl; bring the total volume to 1 liter with deionized water.

ADP, 0.01 M. Dissolve 49.3 mg adenosine-5'-diphosphate trisodium salt in Tris buffer in a 10-ml volumetric flask, adjust pH to 6.8, and dilute to volume with buffer. Store frozen (stability unknown).

Creatine Phosphate, 0.1 M. Dissolve 363 mg creatine phosphate disodium hexahydrate in Tris buffer in a 10-ml volumetric flask, adjust the pH to 8.0 (some breakdown occurs at lower pH), and dilute to volume with buffer. Store frozen (stability unknown).

Glucose, 0.2 M. Dissolve 360 mg glucose in Tris buffer and dilute to 10 ml in a volumetric flask. Store frozen.

$MgCl_2$, 0.3 M. Dissolve 286 mg $MgCl_2$ in Tris buffer and dilute to 10 ml in a volumetric flask.

NADP, 0.008 M. Dissolve 60 mg NADP disodium salt in Tris buffer, adjust the pH to 6.8, and dilute to 10 ml in a volumetric flask. Store frozen.

AMP, 0.1 M. Dissolve 499 mg adenosine-5'-monophosphate disodium salt hexahydrate in Tris buffer, adjust the pH to 6.8, and dilute to 10 ml in a volumetric flask. Store frozen (stability unknown).

Cysteine, 0.05 M. Dissolve 0.88 g L-cysteine hydrochloride monohydrate in Tris buffer, adjust the pH to 6.8, and dilute to 10 ml in a volumetric flask. Prepare immediately before use to avoid deterioration.

Hexokinase, approximately 6 IU/ml. Dilute crystalline suspension of yeast hexokinase with Tris buffer so as to contain 6 IU/ml. For example, hexokinase with activity of approximately 140 IU/mg in 2 mg/ml suspension should be diluted 1:47 with Tris buffer. Store frozen (stability unknown).

Glucose-6-phosphate Dehydrogenase, approximately 3 IU/ml. Dilute crystalline suspension of yeast G6PD with Tris buffer so as to contain 3 IU/ml. For example, G6PD with activity of approximately 350 IU/mg in 1 mg/ml suspension should be diluted 1:117 with Tris buffer. Store frozen (stability unknown).

Substrate Mixture. On the day of use, bring all of the foregoing reagents to 4°C. Combine one part of each of the following: ADP, creatine phosphate, glucose, $MgCl_2$, NADP, AMP, cysteine, hexokinase, G6PD, and Tris buffer, keeping mixture at 4°C until immediately before use (can be used up to 4 hours after preparation). Check pH before use and if not 6.8, readjust. The mixture may be freeze-dried in aliquots of sufficient volume for 1 day's use and stored at refrigerator temperature. Reconstitute with appropriate volume of 4°C deionized distilled water for use, gently swirling and allowing to stand until solution is complete. Dried substrate powder is also available commercially in capsules or vials sufficient for single tests (Calbiochem, Boehringer). These are reconstituted with manufacturer's stated volume and diluent.

PROCEDURE

1. Pipet 2.9 ml (reconstituted) substrate mixture to a small (about 12 x 75 mm) test tube. See *Notes 2* and *10*.
2. Add 0.1 ml serum or plasma (see *Note 3*) and mix well.
3. Incubate for 6 min in a constant temperature water bath at any temperature between 25° and 37°C.

4. Transfer contents of tube to a cuvet (with a 1 cm light path) that has been kept in the cuvet holder of a temperature-regulated spectrophotometer capable of measuring absorbance at 340 nm. Temperature of cuvet well and water bath must be the same.

5. Record the absorbance at 1 min intervals over a period of at least 5 min against a reference cuvet containing water. Values appearing to be in a steady state within experimental error should be used in the calculation (see *Note 1*).

Calculation:

$$\text{CPK units} = \frac{\Delta A \text{ for } T \text{ min}}{T \times 2.07} \times \frac{1000}{\text{ml sample used in test}}$$

where 2.07 is the absorbance of 1 μmol NADPH in the 3.0 ml test volume.

CPK activity (IU/liter at 30°C) = CPK units × temperature factor

The temperature factor can be found in Table 21-3.

NOTES

1. Lag phase. When the substrate mixture is first reconstituted, consumption of endogenous contaminating substances may cause an increase in absorbance for about 5 min (612). When the serum containing CPK is added, the absorbance increases at an accelerated rate until about 6 min when the rate becomes constant. During the lag phase the myokinase reaction is occurring which converts two molecules of ADP to one each of ATP and AMP (612, 783). Myokinase is inhibited by AMP (612), which is added to the substrate for this purpose although an excess of AMP can also inhibit CPK itself (783). The lag phase has been reported to last longer than 6 min when CPK activity is low and is then followed by a longer linear phase than usual (324). The interval taken for ΔA must be within the linear phase of the enzyme reaction.

2. Stability of substrate mixture. It is recommended that the freeze-dried mixture be reconstituted immediately before use. The freeze-dried material is stable at 4°C for at least 6 months (612).

3. Plasma versus serum. Plasma collected in the presence of heparin or EDTA may be used instead of serum.

4. Stability of samples. The disagreement with respect to stability of specimens apparent in the literature may be related to the specific CPK method employed. Serum or plasma should be separated from the erythrocytes within 2 hours (795) and can be stored at −20°C for at least 2 weeks (159, 641, 795). There have been reports of 6% loss in CPK activity after 1 day at −16°C, increasing to 47% loss at 4°C and 74% loss at 20°C (617). Others have found a 90% loss after 48 hours at refrigerator temperature, which can be avoided by a quick-freeze technic (504).

TABLE 21-3. Temperature Correction Factors for CPK[ab]

Reaction temperature (°C)	Factor to convert to 30°C
25	1.47
26	1.35
27	1.25
28	1.16
29	1.08
30	1.00
31	0.92
32	0.85
33	0.79
34	0.73
35	0.68
36	0.63
37	0.58

[a] From ROSALKI SB: *J Lab Clin Med 69*:696, 1967

[b] To correct enzyme activity, multiply by factor appropriate to the reaction temperature.

According to Rosalki (612), serum is stable for 48 hours at room temperature, for 7 days at 4°C, and for at least 1 month at −18°C. CPK in serum from patients with muscular dystrophy may show very little loss of activity compared with CPK associated with myocardial infarction, which decreases to a variable extent overnight in the freezer (323). See *Note 10* for an explanation of the variability noted in temperature stability.

5. *Thiol compounds.* The reverse reaction is apparently enhanced by the presence of thiol compounds, whereas the forward reaction is affected to a negligible extent (159, 210). Thiol compounds have also been used as preservatives of stored CPK specimens (539, 612). The most popular compounds used have been cysteine, glutathione and dithiothreitol or Cleland's reagent (612). The last-named has the advantages of water solubility, little odor, and insensitivity to oxidation by air (143). The -SH group of these compounds is not so much an activator as a reactiva-

tor of the -SH groups of the enzyme (52). When thiols are added to the substrate mixture instead of the serum itself, they exert the same reactivating property (612) without altering specimen volume by the addition or the subsequent adjustment of pH.

6. *Lipemia.* Because turbidity can interfere with the absorbance (see *Note 9* under *Determination of Lactic Dehydrogenase*), lipemia should be avoided; however, fasting *per se* is unnecessary (612, 776). With instruments such as the Beckman DU spectrophotometer a dichromate blank can be used to bring the absorbance of a turbid sample into a readable range (see *Note 2* under *Determination of Lactic Dehydrogenase*). With instruments such as the Gilford spectrophotometer 2400 this is not necessary, since absorbance is linear to three. It is the responsibility of the analyst to make certain that accuracy and precision of measurements are satisfactory at such absorbance levels.

7. *Hemolysis.* Slight hemolysis can be tolerated by the test (612, 641) as red blood cells do not contain significant amounts of CPK (198). Gross hemolysis, however, liberates glucose-6-phosphate and ATP, which would interfere with the test (612).

8. *Temperature and rate of reaction.* The average temperature coefficient is about $7-8\%/°C$ over the range $25°-37°C$ (324, 612). The reaction may be run at any temperature in this range, but it should be standardized and any fluctuation kept within $±0.2°C$. Temperatures over $37°C$ inactivate the enzyme, particularly in stored or diluted specimens (612). Temperature correction factors shown in Table 21-3 are used to convert enzyme activity to the activity at the standard temperature of $30°C$ recommended by the International Union of Biochemistry.

9. *Dilution effects.* Dilution of the specimen with either distilled water or saline leads to the same discrepancy—i.e., erroneously high values in both the forward and reverse reactions (324). The factor responsible for this effect is a heat-stable inhibitor (272, 539), not related in any way to a particular disease (691). Specimens with high levels can be diluted with serum that has been heated to inactivate CPK (272), but preferably should be determined by taking the ΔA over a shorter interval of time (272, 612).

10. *Daylight.* CPK is inactivated by daylight, even at frozen temperatures. The blue portion of the spectrum is presumably responsible, since the effect has been seen to vary with the degree of amber color of the serum (736). Specimens should be stored in the dark and the assay performed in subdued light (736).

11. *Inhibitors.* Chloride and sulfate inhibit CPK (534); however, the sulfate contributed by the enzyme suspensions in the substrate is below the level that would result in inhibition. Traces of the divalent metals Zn, Mn, and Cu also affect the enzyme (739). It has been recommended that deionized, glass-distilled water and clean glassware uncontaminated by heavy metals be used (159).

12. *Spectrophotometer standardization.* See *Note 13* under *Determination of Lactic Dehydrogenase*.

ACCURACY AND PRECISION

It is difficult to assess the accuracy of the test, particularly in view of the known effects of various inhibitors and activators (see *Notes*). However, under the set conditions for optimal activity the test has been found to be definitely useful in clinical situations.

The precision (95% limits) of the method at the top of the normal range is about ±10%.

NORMAL VALUES

Normal serum CPK values (612) for men are 7—57 IU/liter and for women 6—35 IU/liter at 30°C. Other normal ranges have been quoted—e.g., 0—40 IU/liter at 30°C for men, women, and children (324). Males are generally reported to have higher normal values than females (776, 795) after age 10 (776), but this sex difference has been observed only in the reverse and not the forward reaction (159). That there is a sex difference if the reaction is run in one direction but not in the other is difficult to understand. Values for children under 6 years of age have been reported as no different from adults (324), lower than adults (776), or higher than adults (757), especially for those aged 6 months to 2 years (325, 757). Cord blood CPK has been estimated to be two to threefold higher than the adult level (776), or lower than the adult normal at first followed by a sharp rise in the 1 to 45 day group (757).

Vigorous physical activity or exercise normally elevates the serum CPK level (155, 281, 415, 538, 691). The effect may be negligible if exercise is on a short-term basis rather than prolonged (281), is less pronounced in physically conditioned persons (538), and subsides to normal after 48 hours (271). The muscular activity of labor and delivery produces an increase in CPK activity (155, 415). No rise is observed during the 26th to the 36th week of pregnancy (155). A correlation with body weight has been reported for CPK (795). No diurnal variation in enzyme levels occurs (538). Frequent intramuscular injections of penicillin cause a temporary (48 hour) rise in CPK, but injections of narcotics, barbiturates, and diuretics have no effect (322).

GALACTOSE-1-PHOSPHATE URIDYL TRANSFERASE

Galactosemia is an inborn error of metabolism characterized by the specific inability to convert galactose to glucose. Schwartz *et al* (654) first described the metabolic error in galactosemia and suggested the blocking of step (b), below, in erythrocytes, of the following transformations necessary for galactose to enter the mainstream of carbohydrate metabolism:

$$\text{Galactose} + \text{ATP} \xrightarrow{\text{galactokinase}} \text{galactose-1-phosphate} + \text{ADP} \qquad (a)$$

$$\text{Galactose-1-phosphate} + \text{uridine diphosphoglucose} \xrightarrow[\text{transferase}]{\text{gal-1-P uridyl}}$$

(b)

$$\text{glucose-1-phosphate} + \text{uridine diphosphogalactose}$$

Glucose-1-phosphate is converted either into glycogen by phosphorylase or into glucose-6-phosphate by phosphoglucomutase; uridine diphosphogalactose is recycled via epimerase to uridine diphosphoglucose. Kalckar *et al* (371) demonstrated that erythrocytes of galactosemics contain little or no galactose-1-phosphate uridyl transferase (UDP glucose: α-D-galactose-1-phosphate uridyltransferase, EC 2.7.7.12), while other enzymes concerned with galactose metabolism are present in normal amounts.

The UDPG consumption test introduced by Anderson *et al* (20) in 1957 measures enzyme activity by incubation in the presence of large excesses of galactose-1-phosphate (G-1-P) and uridine diphosphoglucose (UDPG), so that the rate of conversion of UDPG to uridine diphosphogalactose (UDPGal) is limited only by the enzyme under study. After 30 min at 37°C the reaction is stopped by heat inactivation and deproteinization. The amount of UDPG before and after incubation is determined by adding NAD and uridine diphosphoglucose dehydrogenase (UDPGDH) and measuring the increase in absorbance at 340 nm.

$$\text{Uridine diphosphoglucose} + \text{NAD} \xrightarrow[\text{dehydrogenase}]{\substack{\text{uridine} \\ \text{diphosphoglucose}}}$$

$$\text{uridine diphosphoglucuronic acid} + \text{NADH} + \text{H}^+$$

In a similar method the uridine diphosphogalactose produced during the enzyme reaction can be coupled to the epimerase reaction, which also involves an NAD-NADH change in absorbance at 340 nm (489):

$$\text{UDPGal} + \text{NAD} \xrightarrow{\text{epimerase}} \text{UDPG} + \text{NADH} + \text{H}^+$$

The original method cannot detect the below-normal enzyme levels found in heterozygotes without overlap (340), but with later modifications of the method this purportedly can be done (190). Thionicotinamide adenine dinucleotide (thio analog of NAD) has been used to bring the assay into the visible spectrophotometric range at 400 nm (536). The use of dithiothreitol for protection and reactivation of the enzyme results in an improved technic for both red and white blood cells (496). In the version of Beutler and Baluda (78) various corrections for linearity, turbidity, and nonspecific NAD reduction, in addition to increased substrate concentrations, render the technic reliable for detecting heterozygotes and the Duarte variant.

Galactose-1-phosphate uridyl transferase has been determined mano-metrically by estimating oxygen uptake of hemolysates incubated with galactose-1-phosphate (399).

Radioisotopic methods based on the measurement of $^{14}CO_2$ produced from incubation of leukocytes, whole blood, or hemolysate with galactose-1-^{14}C have been proposed (459, 768). The glucose-1-phosphate produced by the transferase is converted to glucose-6-phosphate, which then loses the first carbon as CO_2 upon entering the hexose monophosphate shunt. This is a screening method which may (768) or may not (459) detect heterozygous carriers. Upon modification of this technic for whole blood the enzyme level is found to be proportional to the genetically expected ratios of 2:1:0 for normal, heterozygous, and galactosemic individuals (530). An incomplete block was noted in several galactosemics. A more direct quantitative method by the same investigators (529, 531) utilizes the conversion of 1-^{14}C-galactose-1-phosphate to UDPGal-^{14}C. It is similar to the classic UDPG consumption test of Anderson *et al* but radioactive UDP-hexose is separated by charcoal (529) or more simply by DEAE cellulose strips (531) for counting. Clear distinction can then be seen between the three genetic groups, and correlation with the consumption test is good.

A methylene blue reduction method, in which blood is incubated anaerobically in order for a series of endogenous coupled enzyme reactions to decolorize the dye was introduced by Beutler *et al* (80) as a screening method. Severe anemia or deficiency in any one of several enzymes assumed to be present unfortunately affects the results. Another screening test devised by Beutler and Baluda (79) involves the same enzyme coupling systems monitored by the fluorescence of NADPH under long wave ultraviolet light after spotting the incubation mixtures on filter paper. In a more quantitative screening procedure (83), the spot test assay mixture is read in a photofluorometer for NADPH and in a photometer for hemo-globin, then results are converted to UDPG consumption units. Glucose-6-phosphate dehydrogenase must be added to the system in the case of Mediterranean type G6PD deficiency. A similar method measures fluo-rescence of NADPH but relates activity to packed red blood cell volume (160). Units are double the usual UDPG consumption test units because two NAD-NADPH reactions are involved.

Electrophoresis of the enzyme reveals faster mobility in the Duarte variant isoenzyme than in the normal (486), an unstable form of enzyme with slower mobility in the Indiana variant (135), and an abnormally slow-moving enzyme associated with the incomplete block in some galactosemics (667).

QUANTITATIVE DETERMINATION OF GALACTOSE-1-PHOSPHATE URIDYL TRANSFERASE

(method of Ng et al, Ref 529, modified)

PRINCIPLE

Whole blood hemolysates are incubated with radioactive galactose-1-

phosphate and uridine diphosphoglucose in order for the following reaction to occur:

$$^{14}\text{C-galactose-1-phosphate} + \text{UDPglucose} \xrightarrow{\text{transferase}} \text{glucose-1-phosphate}$$

$$+ \text{UDPgalactose-}^{14}\text{C}$$

The UDPgalactose-^{14}C is adsorbed to charcoal, eluted with ethanol-ammonia, and quantitated by counting the radioactivity. Unreacted ^{14}C-galactose-1-phosphate is effectively excluded by this process.

REAGENTS

NaCl, 0.9%

NaOH, 10 N. Carefully add 40 g NaOH pellets to water and stir until dissolved. Bring total volume to 1 liter with water.

Glycine Buffer, 1 M, pH 8.1. Dissolve 30 g glycine in about 200 ml water. Adjust the pH to 8.1 with 10 *N* NaOH. Bring the total volume to 400 ml with water.

^{14}C-Galactose-1-Phosphate, 2 mM. Dissolve sufficient ^{14}C-galactose-1-phosphate (the product from Calbiochem is satisfactory) and unlabeled α-D-galactose-1-phosphate, dipotassium dihydrate (Calbiochem, molecular weight 372) in water to produce a solution with about 2 μmol galactose-1-phosphate and 40 000–70 000 cpm/ml. For example, for material of specific activity 40 mCi/m*M* with 0.01 mCi contained in 17 μl, 5 μl of labeled material plus 7.4 mg nonradioactive material per 10 ml of solution would contain 2 μmol (0.52 mg) and 65 300 cpm/ml.

UDPG Solution, 2 mM. Dissolve 14.0 mg uridine diphosphoglucose in 10 ml water.

Trichloroacetic Acid (TCA), 5%. Dissolve 5 g TCA in water and bring to a total volume of 100 ml.

Activated Charcoal. Wash Merck's activated charcoal with 50% ethanol (ethanol denatured with methanol is satisfactory), then with water. Dry at 105°-110°C.

Ethanol-Ammonia Solution. Mix 50 ml water with 50 ml ethanol (ethanol denatured with methanol is satisfactory). Add 0.1 ml conc NH$_4$OH and mix.

Ethanol-Toluene Solution. Combine 1 vol ethanol (ethanol denatured with methanol is satisfactory) with 2 vol toluene.

Scintillation Fluid. Dissolve 2.0 g 2,5-diphenyloxazole (PPO) and 50 mg 1,4-bis-2(5-phenyl-oxazolyl)-benzene (POPOP) in 500 ml of ethanol-toluene solution.

PROCEDURE

1. Wash 4–5 ml heparinized blood sample three times by centrifuging at 4°C in plastic tubes. Remove supernate (including buffy coat), add an equal

vol saline, and resuspend cells. Final centrifugation should achieve firm packing of cells (see *Note 4*).

2. Hemolyze the packed cells by freezing in a dry-ice-ethanol bath and thawing. Repeat three times.

3. Just prior to incubation, prepare a 50% hemolysate by pipetting equal volumes of hemolyzed cells and water into a small test tube. Mix.

4. Set up incubation mixtures in small test tubes:
   Blank: 0.4 ml water + 0.4 ml glycine buffer + 0.1 ml ^{14}C-galactose -1-phosphate + 0.1 ml 50% hemolysate.
   Unknown: 0.3 ml water + 0.4 ml glycine buffer + 0.1 ml ^{14}C-galactose-1-phosphate + 0.1 ml 50% hemolysate + 0.1 ml UDPG.

5. Incubate the tubes at 37°C for exactly 30 min. Stop the reaction by adding 2 ml 5% TCA and mix.

6. Allow to stand for 10 min, then centrifuge at 12 000 g for 40 min (see *Note 4*).

7. Decant the TCA supernates into small test tubes containing 100 mg washed activated charcoal. Mix thoroughly and allow suspension to stand for 10 min.

8. Centrifuge for 10 min. Discard the supernates, which contain unreacted galactose-1-phosphate.

9. Wash the charcoal twice by suspending in 2 ml water, centrifuging, and discarding the supernate.

10. Elute the adsorbed UDP-hexose from the charcoal by adding 2 ml ethanol-ammonia solution. Centrifuge and decant the supernates into test tubes.

11. Elute once more with 2 ml ethanol-ammonia and combine each second supernate with the respective supernate from the first elution (F_1).

12. Repeat the entire elution procedure with two additional washes of ethanol-ammonia and combine these two supernates into a second set of small test tubes (F_2).

13. Transfer 1.0 ml of each eluate to scintillation vials. Evaporate to dryness in a 60°C water bath under a stream of N_2. Add 10 ml scintillation fluid; cap and shake the vials.

14. Allow vials to equilibrate in the scintillation counter (or in the dark at refrigerator temperature) overnight. With instrument set for maximum efficiency for ^{14}C, count for 10 min.

Calculation:

Enzyme activity is expressed as the total μmol galactose-1-phosphate transferred by 1 ml packed red blood cells in 1 hour. For both F_1 and F_2:

$$\mu\text{mol/ml/hour} = \frac{\text{cpm}_{\text{unknown}} - \text{cpm}_{\text{blank}}}{\text{cpm}/\mu\text{mol gal-1-P}} \times 160$$

where the dilution factor of 160 reflects the 20-fold correction to 1 ml packed RBCs, four-fold correction for the aliquot counted, and two-fold correction for incubation time.

$$\text{Total } \mu\text{mol/ml/hour} = F_1 + F_2$$

NOTES

1. *Linearity of reaction with time.* Consumption of galactose-1-phosphate is linear through 30 min of incubation under the conditions of the test (529).

2. *Nature of eluted UDP-hexose.* The radioactive UDP-hexose eluted from the charcoal has been confirmed by acid hydrolysis and chromatography of pooled eluates to be UDPgalactose (529). No UDPglucose-^{14}C would be expected to be present, since UDPgalactose-4'-epimerase is normally inactive under the conditions of the test (20). However, one exception to the above is a report that hemolysates from newborns exhibit epimerase activity which is independent of exogenous NAD (531).

3. *Stability of samples.* The enzyme is remarkably stable in intact erythrocytes. Sterile, heparinized blood may be stored for 14 days at room temperature without loss of activity, and at 4°C the enzyme is stable for 4 weeks in ACD-, heparin-, or perhaps EDTA-treated blood (78, 83). Blood has been reported to be stable for only about 1 week in the frozen state (529, 768). Red blood cell hemolysates are less stable than intact red blood cells; however, the enzyme activity can be restored by Cleland's reagent (496). Packed red blood cells can be stored for 7 hours in the refrigerator (160). In our laboratory washed red blood cells are stable for only 1 hour at 30°C and for 1 week frozen. Storage instability is one characteristic of the electrophoretically slow-moving Indiana variant (135).

4. *Centrifugation.* Adequate packing of cells can be achieved in step 1 by centrifugation at 1100g for 15 min (e.g., approximately 3000 rpm in a floor model International centrifuge with horizontal free-swing head). At step 6, centrifugation of TCA precipitates at 12 000g for 40 min (in polycarbonate tubes) is recommended (529); our laboratory has found that results average 9% lower if 1100g is used.

ACCURACY AND PRECISION

The accuracy of the method is comparable to the classic UDPG consumption test. The reliability of the labeled species is discussed in *Note 2*.

The precision (95% limits) of the test is approximately ±20% in the normal range.

NORMAL VALUES

Normal adults and children have enzyme activities of 1.4–2.8 μmol/ml/hour. In heterozygotes for galactosemia, the mean value is approximately half the mean value for normals.

SCREENING TEST FOR GALACTOSEMIA

(method of Beutler and Baluda, Ref 79)

PRINCIPLE

Whole blood is incubated in a buffer mixture containing galactose-1-phosphate, uridine diphosphoglucose, NADP, and a hemolyzing agent. If galactose-1-phosphate uridyl transferase is present, the following series of enzyme reactions proceeds:

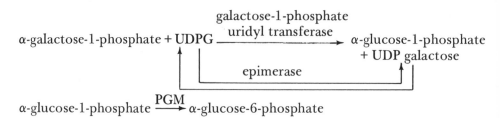

α-glucose-1-phosphate $\xrightarrow{\text{PGM}}$ α-glucose-6-phosphate

α-glucose-6-phosphate $\xrightarrow{\text{PHI}}$ β-glucose-6-phosphate

β-glucose-6-phosphate + NADP $\xrightarrow{\text{G6PD}}$ 6-phosphogluconic acid + NADPH

6-phosphogluconic acid + NADP $\xrightarrow{\text{6PGD}}$ ribulose-5-phosphate + CO_2 + NADPH

where PGM is phosphoglucomutase; PHI is phosphohexose isomerase; G6PD is glucose-6-phosphate dehydrogenase; and 6PGD is 6-phosphogluconic dehydrogenase. Transferase activity thus results in the generation of NADPH, which is detected by fluorescence of the reaction mixture spotted on paper.

REAGENTS

UDP-Glucose (UDPG) Solution, 9.5 × 10^{-3} M. Dissolve 59.7 mg UDPG · H_2O in water and bring to a total volume of 10 ml. Store at $-20°$C.

Galactose-1-Phosphate, 2.7 × 10^{-2} M. Dissolve 90.8 mg α-D-galactose-1-phosphate, dipotassium (A grade), in water and make up to 10 ml. Store at $-20°$C.

NADP Solution, 6.6 × 10^{-3} M. Dissolve 38.4 mg NADP, disodium salt, in water and make up to 10 ml. Store at $-20°$C.

EDTA Solution, 2.7 × 10^{-2} M. Dissolve 114.4 mg disodium salt of EDTA in water and bring to 10 ml. Stable at room temperature.

Digitonin, Saturated Solution. Add digitonin to water in an amount which leaves solids undissolved at $20°-30°$C after mixing well and allowing to stand for at least 24 hours. Stable at room temperature.

Tris-Acetate Buffer, 0.75 M. Dissolve 9 g tris(hydroxymethyl)amino-methane in about 60 ml water. Adjust to pH 8.0 with dropwise addition of glacial acetic acid. Bring total volume to 100 ml with water. Store in refrigerator.

Reaction Mixture. Mix the above reagents in the following proportions to make 6.0 ml of solution: 0.2 ml UDPG, 0.4 ml gal-l-P, 0.6 ml NADP, 2.0 ml Tris-acetate buffer, 0.8 ml digitonin, 0.03 ml EDTA, and 1.97 ml water. After mixing, 200-μl aliquots of the reagent may be conveniently dispensed into small stoppered test tubes and stored at $-20°$C for at least 8 weeks.

PROCEDURE FOR HEPARINIZED BLOOD

1. Thaw frozen reaction mixture to ice bath temperature and mix.
2. Add 20 μl heparinized blood specimen to 200 μl reaction mixture with a disposable microliter pipet. Mix, leave pipet in the reaction tube, stopper, and return to ice bath until all specimens are ready. Include a normal specimen with each run.
3. Incubate the tubes at 37°C for exactly 1 hour.
4. Using the same pipets as in step 3, immediately spot a few microliters of incubation mixtures onto Whatman 1 filter paper to make spots 4—10 mm in diameter. Allow spots to dry.
5. At the end of a 2 hour incubation, spot another portion of reaction mixtures as before. Allow spots to dry.
6. Examine spots under long wave ultraviolet light.

PROCEDURE FOR DRIED BLOOD ON FILTER PAPER

1. Using an ordinary paper punch, punch out a 1/4-inch disk from dried spot of blood specimen on filter paper. Include a disk from a normal blood specimen with each run.
2. Add the disks to tubes of incubation mixture in the ice bath.
3. Incubate the tubes at 37°C for exactly 2 hours.
4. Immediately spot a few microliters of each incubation mixture on filter paper. Allow to dry.
5. Examine spots under long wave ultraviolet light. Somewhat less fluorescence appears with dried blood spots than with heparinized blood.

INTERPRETATION OF RESULTS

Normals are characterized by the appearance of bright fluorescence in both 1 and 2 hour spots, assuming no other dependent enzyme is deficient. Congenital galactosemics show no fluorescence at 1 or 2 hours. Heterozygotes for galactosemia and homozygotes for the Duarte variant (heparinized specimens only) exhibit bright fluorescence in the 2 hour spot, but the 1 hour spot is considerably less bright (with dried blood specimen it is not possible to distinguish these from congenital galactosemics).

NOTES

1. Assumptions regarding endogenous enzymes. It is assumed that all of the auxiliary enzymes normally present in blood, particularly G6PD and PGM (see *Principle*, above), are actually present in the specimen in adequate amounts. Limiting amounts of auxiliary enzymes would result in less fluorescence, which would be interpreted erroneously as a deficiency in transferase.

2. Stability of samples. See *Note 3* under *Quantitative Determination of Galactose-1-Phosphate Uridyl Transferase.* Dried blood specimens are stable for 10 days even at room temperature(81). Blood collected in acid-citrate-dextrose (ACD) is unsatisfactory for the test (79). EDTA inhibits the reaction when present in excess amounts (over 0.1 mg/ml); blood collected with EDTA as anticoagulant (1 mg/ml blood) may be used in the test if EDTA is omitted from the incubation mixture (79). Heparinized samples are stable for 1 week at room temperature (79).

3. Stability of fluorescence. Fluorescence is relatively stable and fades only gradually over several days at room temperature. For a permanent record the filter paper is illuminated with ultraviolet light and photographed through an appropriate yellow filter (79).

4. Anemia. Anemia has little effect on results, since the decreased amount of hemoglobin exerts a proportionally lower quenching effect on the decreased amount of enzyme (79).

5. Stability of reaction mixture. The reaction mixture is stable at $-20°C$ for at least 8 weeks. It may also be freeze-dried, as long as the temperature does not exceed $20°C$ and the UDPG concentration is doubled to compensate for loss (79).

6. Glassware. Since it has been claimed (81) that contamination of glassware by unidentified substances may inhibit the chain of reactions and lead to false positives, it is preferable to use disposable glassware.

7. Simulation of heterozygotes. As guidelines for comparison, a galactosemic result can be simulated by withdrawing a portion of incubation mixture from a normal specimen at zero time, and a heterozygote for galactosemia can be simulated by withdrawing the mixture at 10–15 min.

ACCURACY AND PRECISION

The specificity of the test is such that no fluorescence appears if either galactose-1-phosphate or UDPG is omitted from the mixture. This applies to cord blood as well as adult blood. The dependence of the test upon the nonlimiting activities of the auxiliary enzymes has already been stated. Failure to develop fluorescence in the screening test is not necessarily diagnostic of galactosemia; the more specific quantitative test therefore should always be performed for verification.

LIPASE

Lipase (glycerol ester hydrolase, EC 3.1.1.3) catalyzes the hydrolysis of esters in the α positions of triglycerides to yield β-monoglycerides and fatty acids. An example of the reaction is the following:

$$CH_2-O-CO-(CH_2)_7-CH=CH-(CH_2)_7CH_3$$
$$CH-O-CO-(CH_2)_7-CH=CH-(CH_2)_7CH_3 \qquad \overset{\pm\ H_2O}{\underset{}{\rightleftharpoons}}$$
$$CH_2-O-CO-(CH_2)_7-CH=CH-(CH_2)_7CH_3$$

GLYCERYLTRIOLEATE (TRIOLEIN)

$$CH_2-OH$$
$$CH-O-CO-(CH_2)_7-CH=CH-(CH_2)_7CH_3 \quad + \quad 2\ HOOC-(CH_2)_7-CH=CH-(CH_2)_7CH_3$$
$$CH_2-OH$$

GLYCERYLMONOOLEATE   OLEIC ACID

The reaction products of pancreatic lipase on radioactively labeled trioleins have been identified as free fatty acids and mono- and diglycerides (140). Within the first 20 min of the reaction, the disappearance of triolein is proportional to the increase in fatty acids.

Of primary interest in clinical chemistry are the lipases in serum. At least three types of serum enzyme have been characterized by their action on different lipid substrates and by inhibition or activation with various compounds (49, 265, 551). Classification of these types of enzymes by Overbeek (551) as (a) aliesterases, (b) lipases, and (c) lipoprotein lipases was based on the hydrolysis of fats containing short or long chain fatty acids and on the hydrolysis of neutral fats with added taurocholate, quinine, atoxyl, or eserine. Others have used sodium cholate, sodium chloride, and intravenous heparin to identify the three fractions of pancreatic lipase in serum (49). Henry *et al* (321) showed that the lipolytic activity of serum from individuals with acute pancreatitis differs from the lipase of pancreatic extracts with reference to substrate specificity and response to activators.

With the introduction by Cherry and Crandall (137) of olive oil as a substrate for assay of serum lipolytic activity in cases of acute pancreatitis, a variety of modifications involving substrate emulsifiers, incubation times, indicators, pH optima, and types of buffer have appeared. Although the original procedure employed an olive oil-gum acacia emulsion (137), a study of seven gums, polyvinyl alcohol, polyvinylpyrrolidone, and propylethyleneglycol showed that gum arabic was the best emulsifier (599). Incubation times have varied from 24 to 16 (321), to 6 (737), and finally to 3 hours (472). Indicators used in the titration of the liberated fatty acids include phenolphthalein (137), thymolphthalein (321), and methyl red (484). The pH of the incubation medium has varied from 6.56 to 9.1 (137, 472, 729, 759), whereas the pH activity curve shows a plateau between pH 8.5 and 9.0 (599). Phosphate buffers were used in the methods conducted below pH 8 (137, 321, 729) and barbital or Tris buffers at 7.8 and above (472, 737, 759).

Other substrates reported for measurement of lipolytic activity include tributyrin (264), phenyl laurate (786), *p*-nitrophenylstearate (440), diolein (660), and triolein (606). Although a positive correlation of tributyrinase and cholinesterase levels in the sera of normal subjects has been demonstrated, no increase in serum tributyrinase was found in subjects with pancreatitis (120). Studies with triolein and olive oil (which consists of 50% triolein) as the substrate show excellent agreement between these two substrates on the sera of normal subjects and of patients with pancreatitis (606).

Fluorogenic fatty acid derivatives recently were studied for their specificity to pancreatic lipase (284, 355, 501). Sensitivity of 10^{-4} units during 2–3 min of incubation with an accuracy of about 1.5% has been reported for N-methyl indoxyl myristate (284). An automated technic based on the use of a fluorescein fatty acid ester requires only 0.1 ml serum (227).

Titrimetric methods suffer from difficulties in determining the endpoint. Rick (599) described a kinetic method by automatic titration at a constant pH. Another approach that circumvents titration is the reaction of fatty acids with Cu, extraction of the Cu soaps with an organic solvent, and then photometric measurement of the Cu (184, 440, 760). These methods require only 0.1 ml serum and a 10–30 min incubation during which period the reaction is zero order. The advantages of sensitivity, short incubation times, and zero order for the Cu salt methods are outweighed by the disadvantages of solvent extractions, centrifugations, and incomplete extraction of low concentrations of Cu salts. Therefore the simpler titrimetric procedure is preferred.

DETERMINATION OF SERUM "PANCREATITIS LIPASE" AND DUODENAL FLUID LIPASE

(method of Henry et al, Ref 321)

PRINCIPLE

Serum is incubated with an olive oil emulsion buffered with phosphate. The fatty acids liberated are titrated with standard alkali, employing either a pH meter or thymolphthalein.

The number of lipase units in the sample is the milliliters of 0.050 N NaOH required to neutralize the fatty acids released by 1 ml of sample under the conditions of the test.

REAGENTS

Olive Oil Substrate. To one part olive oil, USP grade or better (Old Monk

Olive Oil is excellent), add one part 5% gum acacia containing 0.2% sodium benzoate. Emulsify by either (a) passing through a hand homogenizer repeatedly until a smooth white emulsion is obtained (usually 5 to 12 times), or (b) mixing in a Waring Blendor for 15 min. Prepare an estimated 4 weeks' supply and store in the refrigerator. It is still usable for about 2 weeks after separation into a top emulsion phase and a bottom aqueous phase if shaken before use. Discard if olive oil separates out.

Phosphate Buffer, M/15, pH 7.0. 0.58 g anhydrous Na_2HPO_4 + 0.35 g KH_2PO_4, water to 100 ml.

NaOH, 0.050 N

Thymolphthalein Indicator. 1% in ethanol.

Ethanol. Ethanol denatured with methanol is satisfactory.

PROCEDURE

1. To a test tube, add 3.0 ml water, 0.50 ml buffer, 2.0 ml substrate, and 1.0 ml serum or 1.0 ml duodenal fluid diluted 1:10 with phosphate buffer. Mix and stopper.

2. Incubate overnight (about 16 hours) at 37°C.

3. Remix and pour into a 50-ml Erlenmeyer flask. Add 3 ml denatured alcohol to tube, shake vigorously, and add to flask.

4. Titrate with 0.050 *N* NaOH to pH 10.5 with pH meter or add 2 drops indicator and titrate to distinct blue color (color fades).

5. A Blank should be run with each set of determinations. To a test tube add 3.0 ml water, 0.50 ml buffer, and 2.0 ml substrate. Mix and stopper. Proceed as for an unknown, but immediately prior to titration add 1.0 ml serum or diluted duodenal fluid.

Calculation:

$$\text{Lipase units/ml serum} = \text{ml required for unknown} - \text{ml required for blank}$$

$$\text{Lipase units/ml duodenal fluid} = \text{ml required for unknown} - \text{ml required for blank} \times 10$$

MICRO PROCEDURE

By employing a micro buret the test can easily be scaled down fourfold, requiring 0.25 ml serum. Do not run the test with full amounts of reagents and reduced amount of serum and then multiply results by a factor to correct for reduced sample size, as the results obtained are not equivalent.

NOTES

1. Time of incubation. The original Cherry-Crandall technic called for a 24-hour incubation period. The shape of the curve relating units of activity and time of incubation deviates sufficiently from zero order that there is relatively little increase after 16 hours. Practically, it is more feasible to set up in late afternoon all samples received during the day, and then titrate them the next morning. If the answer is in great demand, the incubation can be shortened to 4 hours and the final result multiplied by 2 to obtain a result approximating the 16 hour result (372).

2. *Endpoint of titration.* The equivalence point for titration of the free fatty acids is at about pH 10.5. The original Cherry-Crandall method employed phenolphthalein as indicator which, under the conditions of the titration, changes color at pH 8.8. At this pH only about 70% of the fatty acids is titrated. Titration with a pH meter is the more accurate technic, but it is also more laborious and not necessary for routine work. Thymolphthalein changes to a distinct blue at about pH 10.5 under the conditions given. The endpoint is not very sharp, the solution becoming pale blue at pH 10.1, distinctly blue at pH 10.5, and deeply blue at pH 10.8. It is estimated, however, that an error of less than 5% of the total titration is involved in titrating to a "distinct" blue. With a highly icteric serum the endpoint is green.

3. *Blanks.* The blank titration includes the alkali required by the buffer, substrate emulsion, and serum to bring the pH to the endpoint. This remains practically constant for any particular olive oil, and the purpose of running a blank with each set of determinations is to check on technic and reagents. It is not necessary to run a blank for each serum; in fact, if one unknown is run and the quantity of sample is limited, a blank employing another serum can be used.

4. *Hemolysis.* Hemoglobin inhibits "pancreatitis lipase" (321, 729, 803). A concentration of 0.16 g/100 ml has no effect, but one of 0.5 g/100 ml produces about 50% inhibition. A normal result from a hemolyzed sample is therefore subject to question, and an elevated result, while significant, should have been still higher.

5. *Optimal temperature.* The optimal temperature of "pancreatitis lipase" is at about 40°C. The peak does not appear to be very sharp, however, and it is more convenient for most laboratories to incubate at 37°C.

6. *Optimal pH.* Tietz et al (737) reported that serum lipase has two distinct pH optima, 7.8–7.9 and 8.0–8.2. They introduced a modification of the olive oil method using a barbital buffer to bring the final pH to about 7.8. The presence of two optima so close together is rather difficult to prove. Henry et al (321) observed no greater activity at pH 8.4 than at pH 7.0. For the higher pH they used both NH_4OH-NH_4Cl and tris(hydroxymethyl) aminomethane (Tris) buffers. The pH of 8.4 is rather close to the second optimum observed by Tietz et al (737), and one would have expected, therefore, to find higher results at 8.4 than at 7.0. Rick found a broad pH activity curve extending from pH 8–9 and no indication of two optima (599).

7. *Stability of samples.* Serum is stable up to 1 week at room temperature and for many weeks at refrigerator temperature (321, 599). The addition of an equal volume of glycerin slows the deterioration of lipase activity in duodenal fluid (428).

ACCURACY AND PRECISION

The relationship between amount of serum used and results obtained is not linear. It can be assumed, therefore, that the results obtained are not necessarily linearly correlated with concentration of enzyme in the serum. The results thus are an arbitrary index of enzyme activity and are of sufficient accuracy to be useful as a diagnostic aid.

The precision of the test in the low normal range is poor. At a level of 3.5 units, the reproducibility is about ±10%.

NORMAL VALUES

The normal range of serum lipase is "up to 1.5 units" (321, 472). The serum level is not affected by meals (737). The normal range of lipase in duodenal fluid is 5–267 units (157).

CHOLINESTERASE

Esterases that hydrolyze choline esters at a higher rate than other esters have been designated as cholinesterases. There are two major types based on certain substrate specificities and susceptibility to inhibitors. First, the *acetylcholinesterases* (acetylcholine hydrolase, EC 3.1.1.7), also known as *red cell*, *true*, *specific*, or *type I cholinesterase*, hydrolyze acetylcholine and acetyl-β-methyl choline but not other types of choline esters. The second division of cholinesterases (acylcholine acyl hydrolase, EC 3.1.1.8), also known as *serum*, *nonspecific*, *pseudo*, or *type II cholinesterase*, acts on a variety of choline, phenyl, nitrophenyl, and other types of esters (30). The general reaction mediated by these enzymes is:

$$H_3C-\overset{\overset{O}{\|}}{C}-O-CH_2-CH_2-\overset{\overset{CH_3}{|}}{\underset{\underset{CH_3}{|}}{N}}-CH_3 \; \overset{\pm H_2O}{\rightleftharpoons} \; H_3\overset{\overset{O}{\|}}{C}-OH \; + \; HO-CH_2-CH_2-\overset{\overset{CH_3}{|}}{\underset{\underset{CH_3}{|}}{N}}-CH_3$$

ACETYLCHOLINE	ACETIC ACID	CHOLINE

Several comprehensive reviews of methods for assaying cholinesterases have been published (30, 31, 790). These include detailed discussion of the following principle methods of cholinesterase assay: gasometric, titrimetric, electrometric, photometric, and radiometric.

The number of reported isoenzymes varies from two to seven with molecular weights ranging from 80 000 to 260 000 (429). Genetic heterogeneity results in the biosynthesis of five types of cholinesterases which vary in their interaction with succinylcholine, dibucaine, fluoride ion, and other agents. The pH optima range from 7.5–8.5, but nonenzymatic hydrolysis of most of the common substrates increases markedly as the pH of the medium increases. The effects of various buffers, differential inhibitors, salts, and metallic ions on the enzyme activity of the different types of cholinesterases have been reported (246, 462).

One of the early practical methods for measuring either red cell or plasma cholinesterase activity was the electrometric technic of Michel (503). Measurement of the fall in pH, resulting from liberation of acetic acid by enzymic hydrolysis of acetylcholine was the basis for this method. Various modifications involving buffers, ionic strength, volume of serum, use of washed or unwashed erythrocytes, and substrate concentrations have been introduced (31). Elimination of the initial pH reading simplifies the procedure (791). The instability of acetylcholine chloride, because of its hygroscopic nature, has been circumvented by the use of the perchlorate salt (791). By coupling a recorder to the pH meter stable readings were obtained within 3 minutes (443). Automatic titrating and recording technics have been applied to the determination of enzyme activity (29, 364). Factors to convert pH-stat to Michel units have been reported (557).

A photometric method based on the reaction of acetylcholine and hydroxylamine to yield acetohydroxamic acid was described by Hestrin (326). This product reacts with ferric ions in an acidic solution to form a red-purple complex that is measured at 540 or 495 nm (772).

The introduction of acetylthiocholine as a substrate for cholinesterase provided two methods for the enzyme assay: (a) the disappearance of the absorption maximum at 229 nm; (b) the reduction of 2,6-dichlorophenol-indophenol by the liberated thiocholine at 600 nm (244). Subsequently this method was modified by reacting the liberated thiocholine with 5,5'-dithiobis-(2-nitrobenzoate) to produce 5-thio-2-nitrobenzoic acid (207). The resultant yellow color of this anion was kinetically measured at 412 nm. Propionyl and butyryl thiocholines have been investigated as substrates to differentiate red cell and plasma cholinesterases (677, 720). Slightly higher values of enzyme activity have been reported with hemolysates than with intact red cells (142). Furthermore, the inhibitory effect of borate and Tris buffers was not observed with phosphate or barbital buffers. A tenfold greater amount of fluoride is necessary for inhibition of the red cell enzyme than for the plasma enzyme (144). Serum cholinesterase assays with butyrylthiocholine have been adapted to an automated system (547).

Photometric methods based on the color change of various dyes have also been applied for both laboratory and field testing. Phenol red (128), bromthymol blue (87), and *m*-nitrophenol (582) have afforded convenient technics that are minimally affected by serum protein binding of the dye complexes. Automated systems have been described that employ phenol red (703) and phenophthalein (97).

Radiometric technics have been used for rapid testing of enzyme activity (675, 789). Samples as small as 0.01–1µl plasma or 0.01–1 mg tissue may be employed in this assay (675).

Two new approaches to cholinesterase assay are a fluorometric procedure (746) and an organic substrate selective electrode (53). Comparison of the latter technic with the Hestrin method (326) gave a correlation coefficient of 0.937.

DETERMINATION OF CHOLINESTERASE

(method of Michel, Ref 503, modified by Larson, Ref 431)

PRINCIPLE

The concentration of cholinesterase is determined by electrometrically measuring the decrease in pH (ΔpH) following incubation with acetylcholine bromide or acetylcholine chloride at 25°C for 1 hour. Correction is made for the drop in pH resulting from nonenzymatic hydrolysis occurring during the period of incubation.

REAGENTS

Red Cell Buffer:
(a) *Stock Solution:* Place 44.73 g KCl in a 250 ml graduated cylinder. Add hot water to about the 175 ml mark and dissolve the salt. Wash 4.124 g sodium barbital into a 200 ml volumetric flask with about 100 ml of the KCl solution and dissolve the sodium barbital by shaking. Dissolve 0.545 g KH_2PO_4 in about 50 ml of the KCl solution contained in a beaker. This solution and the remainder of the KCl solution are transferred to the volumetric flask. The graduate and beaker are rinsed with small amounts of water and rinsings added to the flask. The volume is then adjusted to 200 ml with water and the contents mixed. Other than for dissolving the KCl, heating should be avoided. This buffer can be stored at room temperature in a polyethylene bottle.
(b) *Working Solution.* To 20 ml stock buffer, add about 75 ml water and 2.5 ml of 0.1 N HCl. Adjust the pH to 8.10 with more acid, transfer to a 100-ml volumetric flask, and make to volume with water. The working buffer becomes more acidic on standing; when the pH has dropped 0.05 unit, fresh buffer should be prepared.
Serum or Plasma Buffer. Dilute 32 ml of red cell working buffer to 100 ml with distilled water. The pH should be about 8.00. The pH decreases slowly on standing and should be freshly prepared when the pH has dropped 0.05 pH unit.
Substrate. Acetylcholine bromide or acetylcholine chloride can be used. The commercial bromide salt must be purified by recrystallization as follows: Dissolve 50 g in 100 ml warm 95% ethanol. Precipitate by adding 100 ml ether, filter, and wash crystals with more ether. The crystals can be dried in an oven at 110°C for 5 min or in a vacuum desiccator overnight. Acetylcholine chloride is so hygroscopic that it is difficult to weigh.
(a) *Serum or Plasma Substrate.* Prepare fresh before use as follows: Dissolve 373 mg acetylcholine bromide in 10 ml water, or quickly weigh 100

mg of acetylcholine chloride into a small beaker or tube and dissolve the salt with 3.3 ml water.

(b) *Red Cell Substrate.* Add 1 vol water to 2 vol serum or plasma substrate.

Saponin Solution, 0.01%: Dissolve 10 mg in 100 ml water. Prepare fresh on day of use.

NaCl, 0.9% aqueous

MACRO PROCEDURE FOR SERUM OR PLASMA

1. Standardize the pH meter with phosphate buffer of about pH 7 and check operation with reference standards of lower pH—e.g., potassium acid phthalate or potassium acid tartrate. If electrodes have been used previously for alkaline solutions, soak in 0.1 N HCl for 10 min prior to use.

2. To 5 ml water in a 50-ml beaker, add 0.10 ml serum or plasma from a TC pipet, followed by 5 ml "serum or plasma buffer." Swirl to mix and place beaker in a 25°C water bath for about 10 min.

3. Take the pH of sample, rinsing the electrodes with water and wiping dry with tissue prior to each measurement. This is pH_1. Add 1.0 ml "serum or plasma substrate," mix, note time, and replace in water bath. Times of substrate addition to a series of samples should be staggered—e.g., at 1- or 2-min intervals.

4. After 55 min, restandardize pH meter. Take pH readings of samples exactly 1 hour after addition of substrate. This is pH_2.

Calculation:

$$\text{Cholinesterase units (}\Delta\text{pH/hour)} = pH_1 - pH_2 - b$$

The value for b is obtained from table 21-4.

MACRO PROCEDURE FOR RED CELLS

Same as for the macro procedure for serum and plasma with the following exceptions. Transfer 0.1 ml packed red cells with a TC pipet to a centrifuge tube containing 5 ml 0.9% saline. Mix, centrifuge, and discard supernate. Add 5 ml saponin solution to the red cells, mix, transfer to a 50-ml beaker, and proceed as for serum. "Red cell" buffer and substrate are of course used throughout.

MICRO PROCEDURE FOR SERUM OR PLASMA AND RED CELLS

The macro procedures are conveniently scaled down fivefold. A Sahli, 20-μl pipet can be used, as can 5-ml beakers. If a Beckman model G pH meter is employed, use the small cup and small internal electrodes.

TABLE 21-4. Cholinesterase Procedure: The Factor *b*

pH$_2$	*b* Red cell	*b* Serum or plasma
7.9	0.03	0.09
7.8	0.02	0.07
7.7	0.01	0.06
7.6	0.00	0.05
7.5		0.04
7.4		0.03
7.3		0.02
7.2		0.02
7.1		0.02
7.0		0.01
6.8		
6.6		
6.4		
6.2		
6.0		

NOTES

1. Serum versus plasma. Results on serum are identical to those obtained on plasma. If the red cell level is also desired, however, it is more convenient to use an anticoagulant, thus providing red cells and plasma in one sample. Citrate, oxalate, and fluoride must be avoided because of complexation of divalent ions, such as Ca^{2+}, which activate cholinesterase (790). Heparin is satisfactory as an anticoagulant. It is advisable to separate serum from the clot within 2 hours. A 20–25% increase in cholinesterase can occur over a 24-hour period if serum is left on the clot (677).

2. Temperature. Most of the studies with this method have used 25°C as the incubation temperature for the assay. King and Morgan contend that a true index of enzyme activity can be obtained only at 37°C, especially in inhibition studies to differentiate serum cholinesterase variants (396).

3. *Stability of samples.* Serum samples are reported to be stable for 17 days at 23° or 4°C and 3 months at −20°C (677). Plasma stored at various temperatures shows no significant differences in total enzyme activity, although changes in isoenzyme patterns do occur (369). The stability of red cell enzyme is still debatable, but Witter (790) reports that the activity is stable several weeks at 0° and 5°C.

ACCURACY AND PRECISION

Two conditions essential for the accurate measurement of cholinesterase activity are that: (a) the rate of the reaction is proportional to the amount of enzyme present, and (b) the enzyme measured under the conditions of the assay is a cholinesterase. This second condition is usually demonstrated by showing that low concentrations of an inhibitor (physostigmine) inhibits hydrolysis of the substrate. Other factors that affect accuracy are variations in pH, salt concentrations, and temperature which may occur. Besides these conditions and factors there are additional variables—e.g., activators, inhibitors, cholinesterase isoenzymes, and type of substrate—that could affect the accuracy of the measurement.

The electrometric method of Michel (503) is adequately precise, and results obtained parallel those obtained manometrically (30, 790). The sources of error in the Michel method are: (a) the activity of the enzyme decreases as the pH decreases; (b) the decrease in buffer capacity does not exactly parallel the decrease in enzyme activity; and (c) the measurement of pH at a fixed time interval is not an accurate determination of the rate of the reaction. Other errors such as nonenzymatic hydrolysis of the substrate and the deviation of the reaction rate from linearity during the first 5 min of incubation are corrected or contribute only a small error.

The precision (95% limits) of the method is about ±15% (600, 790).

NORMAL VALUES

There have been many studies of the normal ranges of serum and red cell cholinesterase by the method of Michel. Some workers conclude that adult male normal levels are 10−15% higher than female normal levels (304), while others observe no significant difference (773). If one pools the reported values for serum cholinesterase (approximately 95% limits) for 1636 adult males and females, a total range of 0.32−1.45 ΔpH units is obtained. Perhaps a more valid estimate of the true normal range is obtained by weighting the limits obtained in each study by the number of individuals comprising the study. Such an approach yields 95% limits of 0.42−1.32. The total range for red cell cholinesterase obtained by pooling values on 1123 adults is 0.44−1.09 ΔpH units (316). The weighted range is 0.57−0.98.

In one study the intraindividual variation of serum cholinesterase was found to be relatively low with a coefficient of variation of 8.4% (773). For red cell enzyme activity, Hecht and Stillger (304) reported a variation of 10−15% for each individual. Their study also indicates less variation of enzyme activity in the red cells than in plasma.

ALKALINE PHOSPHATASE

The nonspecific phosphatases that effect the hydrolysis of mono-phosphoric acid esters at an alkaline pH are designated as *alkaline phosphatases* (orthophosphoric monoester phosphohydrolases, EC 3.1.3.1). This class of enzymes acts on aliphatic, aromatic, and heterocyclic phosphoric acid esters in the following fashion:

$$\underset{\underset{OH}{|}}{\overset{\overset{O}{\|}}{R-O-P-OH}} \quad \underset{\pm H_2O}{\overset{[OH^-]}{\rightleftharpoons}} \quad R-OH + \underset{\underset{OH}{|}}{\overset{\overset{O}{\|}}{HO-P-OH}}$$

The optimal pH for measuring alkaline phosphatases varies with the type of substrate (392), the substrate concentration, incubation temperature, and the tissue source of the enzyme (100). It is not surprising, therefore, that reported pH optima range from 8.6 to 10.3.

Alkaline phosphatase activity in blood was demonstrated by Kay (385) in 1930 with β-glycerophosphate as the substrate. Assays with this substrate were modified and improved with the incorporation of diethylbarbiturate buffer (pH 9.3), incubation at 37°C, and reduction of incubation time to 1 hour (674). Phenyl phosphate as a substrate was introduced by King and Armstrong (390), with measurement of phenol released by the Folin-Ciocalteu phenol reagent. This procedure also went through a series of modifications with the eventual introduction of 4-aminoantipyrine as the phenol coupling reagent to yield a red quinone (389, 572). Another substrate for the assay of alkaline phosphatases, *p*-nitrophenyl phosphate, was introduced in 1937 (540) and another series of method improvements occurred (68, 71, 100). Three other commonly used substrates are phenolphthalein phosphate (32, 109), β-naphthyl phosphate (662), and thymolphthalein phosphate (148, 620). The alkaline phosphatase activity of human liver, bone, intestine, and bile extracts was measured with five different substrates: phenyl phosphate, β-glycerophosphate, *p*-nitrophenyl phosphate, phenolphthalein monophosphate, and β-naphthyl phosphate (792). Phenyl phosphate was the most rapidly hydrolyzed substrate and β-naphthyl phosphate was the slowest.

Kinetic methods have been described for both *p*-nitrophenyl phosphate (101, 234, 433) and phenolphthalein phosphate (35) as substrates. A 1- or 2-minute reading with 20 μl serum is sufficient for the assay.

Highly sensitive assays of alkaline phosphatase activity are also afforded by fluorescent substrates. Neuman (528) reported on the phosphoric acid esters of fluorescein, eosin, and 4-methyl-7-oxycoumarin. Others have used the phosphoric acid derivatives of 3-hydroxy-2-naphthanilides (naphthol AS derivatives) for quantification or for detection of isoenzymes in starch gel and thin film agarose electrophoresis (178, 361, 756).

Methods using phenyl phosphate (479, 775), β-glycerophosphate (738), and *p*-nitrophenyl phosphate (706) as substrates have been adapted to automated systems. The *p*-nitrophenyl system has been modified by substitution of 2-amino-2-methyl-1-propanol as the buffer to eliminate inhibition by the glycine buffer (517). Phenolphthalein diphosphate (156) reportedly eliminates the necessity of running serum blanks, but the monophosphate gives a lower blank than the diphosphate (242). The automated technic of Cornish *et al* (162) employs the fluorogenic substrate 4-methylumbelliferyl phosphate and requires only 10 μl serum.

Serum alkaline phosphatase activity is dependent on Mg^{2+} for activation but is inhibited by Ca^{2+} (385) and phosphate ions (101). Glucose and glycerol increase the activity of the enzyme by acting as phosphate acceptors (630). Diminution of activity with oxalate (60), citrate (213), and ethylenediaminetetraacetic acid (48) may be due to Mg^{2+} chelation by these agents. Ammonium hydroxide, borate, or glycine-based buffers (177, 467) have an inhibitory effect on the enzyme activity, whereas 2-amino-2-methyl-1-propanol buffer affords greater activity (467, 706). Two recent surveys of aminated alcohol buffers showed that some of these compounds increased enzyme activity two- to fourfold greater than other commonly used buffers (14, 490).

Serum alkaline phosphatase isoenzymes were recently reviewed (220). The concensus is that there are three isoenzymes in serum located in the α_1-, α_2- and β-globulin regions. A combination of electrophoretic, phenylalanine inhibition, and heat inactivation technics employed by Yong indicate that the source of serum enzyme in adults is primarily the liver, and in children bone (805). To differentiate the source of enzyme in normal individuals and those with certain diseases, ethanol precipitation (556); heat inactivation (570); phenylalanine, homoarginine, or tryptophan inhibition (221, 254); urea inhibition (338); and immunochemical (645, 716) procedures have been employed. The electrophoretic identification of the tissue source of serum isoenzymes in various clinical states has not been successful (788).

Alkaline phosphatase activity has been reported in urine (16). Dialysis is required to remove an inhibitor which may be inorganic phosphates (331). There is evidence of an additional nondialyzable inhibitor that becomes deactivated on storage of the dialyzed urine at 5°C (280). Fractionation of the urinary phosphatase isoenzymes on starch gel showed two major bands of enzyme activity in the α and $\alpha_2\beta$ fractions (482).

The following procedures are presented for the determination of alkaline phosphatase: (a) serum procedure—manual, (b) serum procedure—automated, and (c) urine procedure—manual.

DETERMINATION OF SERUM ALKALINE PHOSPHATASE

(method of Kind and King, Ref 389, modified)

PRINCIPLE

Serum is incubated with phenyl phosphate buffered at pH 10 for 15 min at 37°C. The hydrolysis product, phenol, is condensed with 4-aminoantipyrine and then oxidized with alkaline ferricyanide to give a red complex, which is measured photometrically at 500 nm. The absorption curve of the complex formed is shown in Figure 21-5.

REAGENTS

NaOH, 0.5 N. Dissolve 2 g in water and dilute to 100 ml.
Sodium Bicarbonate, 0.5 M. Dissolve 4.2 g in water and dilute to 100 ml.
Magnesium Chloride. Dissolve 41.8 g $MgCl_2 \cdot 6H_2O$ in water and dilute to 500 ml.
Carbonate-Bicarbonate Buffer, 0.1 M, pH 10. Dissolve 6.36 g anhydrous sodium carbonate plus 3.36 g sodium bicarbonate in water and dilute to 1 liter.

FIG. 21-5. Absorption curve of the color formed in the serum alkaline phosphatase determination: (1) blank vs water; (2) 10 μg phenol standard vs water; (3) standard vs blank. Cary Model 15 Recording Spectrophotometer.

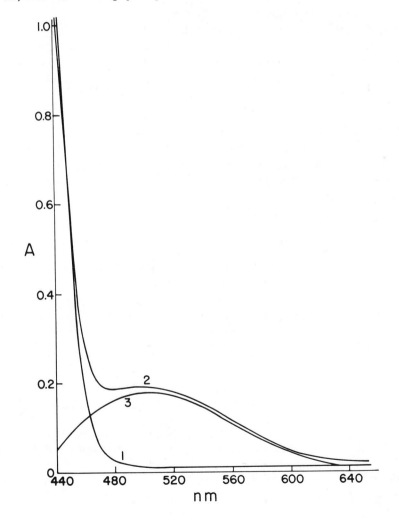

Phenyl Phosphate, 0.02 M.   Dissolve 436 mg disodium phenyl phosphate in water, add 0.4 ml magnesium chloride solution, and dilute to 100 ml. Add several drops of chloroform and store at 4°C. Solution is stable 3—4 weeks.

4-Aminoantipyrine, 0.6%.   Dissolve 600 mg in 100 ml water.

Potassium Ferricyanide, 4%.   Dissolve 4 g in water and dilute to 100 ml. Prepare this reagent fresh.

Stock Phenol Standard, 1 mg/ml.   Dissolve 100 mg crystalline phenol in 100 ml 0.1 N HCl.

Working Phenol Standard, 0.01 mg/ml.   Dilute 1.0 ml of stock to 100 ml with water. Prepare fresh.

PROCEDURE

1. To test tubes labeled "control" and "unknown", add 1 ml buffer plus 1 ml substrate, mix, and place in a 37°C water bath for at least 2 min.
2. Add 0.1 ml serum to the unknown by TC pipet, mix, and incubate exactly 15 min.
3. While the above is incubating, set up the following:
   Blank: 1.1 ml buffer + 1 ml water + 0.8 ml 0.5 N NaOH.
   Standard: 1.1 ml buffer + 1 ml working standard + 0.8 ml 0.5 N NaOH.
4. At the end of the 15 min incubation period, immediately add 0.8 ml 0.5 N NaOH to the unknown and mix. To the control add 0.8 ml 0.5 N NaOH plus 0.1 ml serum, mix, and set up for color development.
5. To all tubes add 1.2 ml 0.5 M NaHCO$_3$, 1 ml aminoantipyrine, and 1 ml ferricyanide; mix with each reagent added.
6. Read absorbance of the standard (A_s), control (A_c), and unknown (A_x) against the blank at 500 nm or with a filter having a nominal filter in this region (a 520 nm filter is satisfactory). The color is relatively stable, but it is preferable to read the samples within 20 min.

Calculation:

$$\text{Alkaline phosphatase (units/100 ml serum)} = \frac{A_x - A_c}{A_s} \times 0.01 \times \frac{100}{0.1}$$

$$= \frac{A_x - A_c}{A_s} \times 10$$

NOTES

1. Units of enzyme.   One phosphatase unit was originally defined by King and Armstrong as the amount of enzyme in 100 ml serum which liberates 1 mg phenol in 30 min at 37°C (479). In a subsequent publication that dealt with a new method for estimating the liberated phenol, Kind and King (389) reduced the incubation time to 15 min. The units of serum alkaline phosphatase for the manual method are expressed according to this shorter incubation period.

2. pH optimum. Roy (361) reported no appreciable change in enzyme activity over a pH range of 9.9—10.4. The concentration of buffer used is high enough to maintain the pH of the assays to within ±0.2.

3. Hemolysis. Minimal hemolysis does not appear to interfere significantly with the results for alkaline phosphatase since the concentration in red cells is only about six times that of serum with this substrate (385).

4. Serum versus plasma. Plasma is acceptable provided it is prepared from heparinized blood. Oxalate, citrate, and EDTA inhibit the enzyme (48, 60, 213).

5. Stability of samples. Enzyme activity is stable 7 days at 25° or 4°C (101).

ACCURACY AND PRECISION

The heterogeneous nature of alkaline phosphatase and the variations in proportions of the isoenzymes in serum and urine (220, 482) do not permit a definitive statement about the accuracy of this assay. Isoenzymes that contribute to the level of enzyme activity in serum can be derived from liver, bone, intestine, or placenta. Furthermore, these isoenzymes have different substrate affinities, so a change in substrate alters the apparent enzyme activity (792). The type of buffer, especially the aminated alcohols, can markedly affect the activity of the phosphatases (14, 490).

Precision (95% limits) for the test is ±5% except in the low normal range where the error can be significantly greater.

NORMAL VALUES

Normal serum level range is 4—17 units for adults and 17—33 units for children (713). There is no sex difference (390). Values tend to increase with increasing age for both sexes (400). Individuals with O or B blood group who are ABH secretors but also have Le(a—) factor (45) and individuals who are hyperthyroid show higher circulating levels of enzyme (251). Artificially elevated phosphatase levels have been observed 10—30 days after transfusion of albumin that has been prepared from human placenta (47).

AUTOMATED DETERMINATION OF SERUM ALKALINE PHOSPHATASE

(method of Kind and King, Ref 389, adapted to AutoAnalyzer by Marsh et al, Ref 479, modified)

PRINCIPLE

Serum is incubated with phenyl phosphate at pH 10.2 and 37°C. The hydrolysis product phenol is condensed with 4-aminoantipyrine and then oxidized with alkaline ferricyanide to give a red complex which is measured photometrically at 500 nm (Fig. 21-5).

REAGENTS

Buffer Solution, pH 10.2. Dissolve 1.58 g anhydrous Na_2CO_3 and 0.84 g $NaHCO_3$ in water and dilute to 1 liter.

Buffered Substrate Solution. Dissolve 2.0 g disodium phenyl phosphate, 1.58 g anhydrous Na_2CO_3, and 0.84 g $NaHCO_3$ in water and dilute to 1 liter. The pH should be 10.2. The solution should be stored in the refrigerator but warmed to room temperature before using.

4-Aminoantipyrine. Dissolve 1.0 g in water and dilute to 1 liter. Store in an amber bottle.

Potassium Ferricyanide. Dissolve 5.0 g in water and dilute to 1 liter. Store in an amber bottle.

Stock Phenol Standard. Dissolve 1.0 g reagent grade phenol in 0.1 N HCl in a 1-liter volumetric flask and dilute to volume. Store in an amber bottle.

Working Phenol Standard. Make dilutions to a total volume of 500 ml with 0.005 N HCl to prepare the following working standards, which are stable at least 2 months:

μg Phenol/ml	ml of Stock Standard
10	5
30	15
60	30
100	50
150	75
200	100

PROCEDURE

Set up manifold, other components, and operating conditions as shown in Figure 21-6. By following the procedural guidelines as described in Chapter 10, steady state interaction curves (Fig. 21-7) should be obtained and checked periodically. Samples are run initially with buffer solution *only*, to provide blank values for correction of phenolic substances that may be present in sera. Blank values are subtracted from the assay values, and the results are calculated from the calibration curve in a standard manner.

Calculation:

1 phosphatase unit is the amount of enzyme in 100 ml serum which liberates 1 mg phenol in 15 min at 37°C. Hence, where T is the incubation time in minutes:

$$\text{Alkaline phosphatase (units/100 ml)} = \mu g \text{ phenol} \times \frac{1}{1000} \times \frac{100}{1} \times \frac{15}{T}$$

$$= \mu g \text{ phenol} \times \frac{1.5}{T}$$

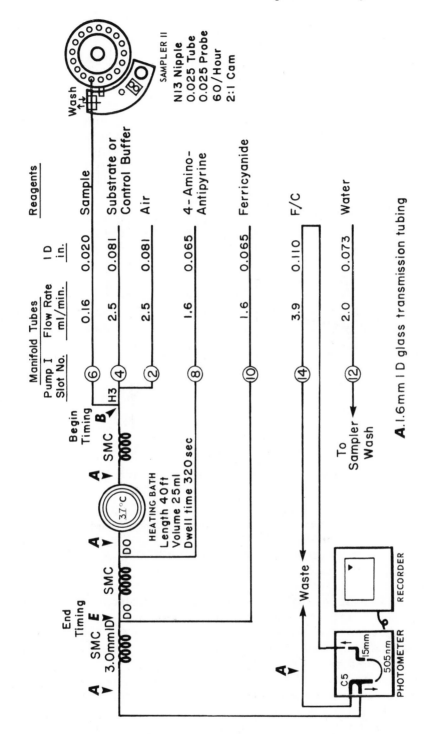

FIG. 21-6. Flow diagram for alkaline phosphatase in serum. Calibration range: 0–100 IU/liter. Conventions and symbols used in this diagram are explained in Chap. 10.

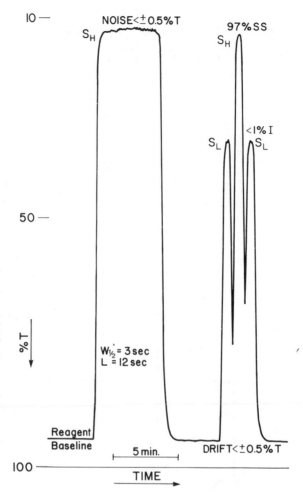

FIG. 21-7. Alkaline phosphatase strip chart record. S_H is a 64 IU/liter control serum, S_L a 32 IU/liter control serum. Peaks reach 97% of steady state; interaction between samples is less than 1%. $W_{1/2}$ and L are explained in Chap. 10.

NOTES

1. Enzyme units. The enzyme units calculated by the automated method are the same as the manual method because the difference in incubation time is corrected by the factor 15 min/T.

2. Incubation time. The incubation time is determined by pumping all reagents for 5 min, followed by inserting the buffer-substrate line into the ferricyanide reagent. When the yellow color reaches point B (Fig. 21-6), begin timing and continue until the color reaches point E. Note the elapsed time and express it to the nearest tenth of a minute (e.g., 6.5 min).

3. Phenol standards. The phenol standards used in this method show atypical behavior, giving inconstant higher values of $W_{1/2}$ than do serum samples; therefore such standards must be run with cups of water or saline between them, even though serum samples can be run one after the other. For the instrument from which Figure 21-6 was taken, phenol standards gave $W_{1/2}$ = 18 sec and I = 6% at 60/hour.

ACCURACY AND PRECISION

The same statements concerning accuracy apply for both the manual and automated methods.

Precision of serum alkaline phosphatase determinations by the automated method is ±5% for levels in the normal range.

NORMAL VALUES

Same range as for the manual method.

DETERMINATION OF URINARY ALKALINE PHOSPHATASE

(method of Shinowara et al, Ref 674, and Amador et al, Ref 16)

PRINCIPLE

Urine is dialyzed to remove enzyme inhibitors, and then an aliquot is incubated with β-glycerophosphate (pH 10.8) for 1 hour at 37°C. Urinary inorganic phosphate is determined before and after incubation, and the difference represents the phosphate released from glycerophosphate as a result of enzyme action. Phosphate ions are determined in a trichloroacetic acid supernate by formation of molybdophosphate, followed by reduction to molybdenum blue with *N*-phenyl-*p*-phenylenediamine (*p*-semidine). The molybdenum blue is determined photometrically. The absorption curve of the color formed is shown in Figure 21-8.

REAGENTS

Buffered Substrate, Stock Solution. Measure 3 ml petroleum ether and about 200 ml water into a 250 ml volumetric flask. Add 2.5 g sodium β-glycerophosphate and 4.25 g sodium diethylbarbiturate. Dissolve and dilute with water to bring the aqueous level to 250 ml. Mix. Stable in refrigerator about 3 months.

Working Buffered Substrate. To a 100 ml volumetric flask, add 3 ml petroleum ether, 50 ml stock substrate, and 2.9 ml 0.10 *N* NaOH. Dilute with water to bring the aqueous level to 100 ml and mix. Check pH with pH meter. (Solution is 0.06 *M* Na$^+$. For Beckman general purpose glass electrodes, the Na$^+$ correction is +0.027.) If the pH deviates more than 0.1 from pH 10.8, adjust with 0.1 *N* NaOH or HCl. The substrate slowly hydrolyzes in refrigerator but is generally usable for about 2 weeks.

Trichloroacetic Acid (TCA), 30%

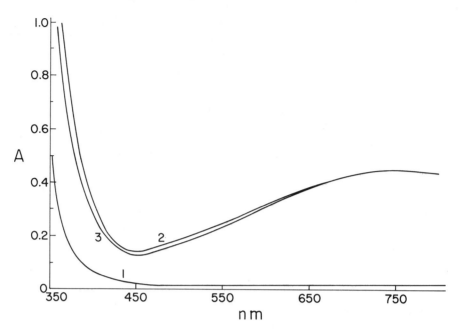

FIG. 21-8. Absorption curves in urinary alkaline phosphatase determination: (1) reagent blank vs water; (2) 20 μg phosphorus standard vs water; (3) standard vs reagent blank. Cary Model 15 Recording Spectrophotometer.

Stock Tris Buffer, 0.1 M, pH 7.4. Dissolve 12.1 g Tris—tris (hydroxymethyl)aminomethane—in 1 liter distilled water and adjust with 0.1 N NaOH or HCl to pH 7.4.

Working Tris Buffer, 0.001 M, pH 7.4. Dilute the stock buffer 1:100 with water.

Ammonium Molybdate, 0.0202 M. Dissolve 25 g $(NH_4)_6Mo_7O_{24} \cdot 4H_2O$ in approximately 700 ml of distilled water. Cautiously add, with mixing, 84 ml conc H_2SO_4. Cool, dilute to 1 liter with water, and mix. Store in a polyethylene bottle.

p-Semidine Reagent. Place 50 mg p-semidine (obtainable from Eastman Organic Chemicals as N-phenyl-p-phenylenediamine hydrochloride, No. 2043) in a flask, wet the solid salt with a few drops of 95% ethanol, and add 100 ml 1% $NaHSO_3$ with shaking. Filter off insoluble residue. Store in refrigerator. This is stable about 1 month. A slight discoloration may appear which does not impair the effectiveness of the reagent but does increase the absorbance of the reagent blank. After 1 month in the refrigerator, the absorbance of standards slowly decreases.

Stock Phosphorus Standard, 1 mg P/ml. Place 438 mg KH_2PO_4 in a 100 ml volumetric flask and add water to volume. Add a few drops of chloroform as preservative.

Working Phosphorus Standard, 10 μg P/ml. Transfer 1.0 ml stock standard to 100-ml volumetric flask. Add 16 ml 30% TCA, then water to volume.

PROCEDURE

1. Measure urine volume and adjust an aliquot to pH 7.4 with 1 *N* NaOH or 1 *N* HCl. Use pH meter and magnetic stirrer.

2. Centrifuge 15 ml urine at 2500 rpm for 10 min. If supernate is not clear, increase speed or time.

3. Transfer 10 ml urine to a cellulose dialysis bag made by tying the ends of a short length of cellulose tubing (Union Carbide). Weigh the bag containing the urine on an analytic balance.

4. Dialyze urine for 4 hours against 350 ml portions of 0.001 *M* Tris buffer, pH 7.4, changing the buffer solution at 30 minutes and 1 and 2 hours. Stir the buffer continuously.

5. After dialysis, dry the bag with a paper towel and reweigh.

6. Transfer dialyzed urine to a test tube and centrifuge at 2500 rpm for 5–10 min.

7. Place 4.0 ml working buffered substrate in a test tube and incubate in a 37°C water bath for about 2 min.

8. Add 1.0 ml urine, mix, stopper, and replace in water bath for exactly 1 hour.

9. While above is incubating, set up the following:
Substrate blank. 4.0 ml working buffered substrate + 1.0 ml water.
Unknown inorganic P. 4.0 ml water + 1.0 ml urine.

10. At the end of 1 hour of incubation, remove tubes from water bath; to all tubes from previous steps add 1.0 ml 30% TCA, mix, let stand 5 min, and centrifuge.

11. Set up and develop colors as follows:

Reagent blank: 0.33 ml 30% TCA + 1.67 ml water	(A_b).
Standard: 2.0 ml working phosphorus standard–20 μg P	(A_s).
Substrate blank: 2.0 ml of mixture from step 9	(A_{sb}).
Unknown inorganic P: 2.0 ml of supernate from step 10	(A_{ip}).
Phosphatase: 2.0 ml supernate from step 10	(A_x).

12. To each tube, add 0.4 ml molybdate reagent, mix, then add 4.0 ml *p*-semidine reagent and mix again.

13. After 10 min read absorbances of all tubes against water (see step 11 for designations) at 770 nm or with a filter having a nominal wavelength in this region (a 660 nm filter is satisfactory). Color is stable at least 1.5 hours.

Calculation:

Urine alkaline phosphatase units/8 hour =

$$\frac{A_x - A_{ip} - A_{sb} - A_b}{A_s - A_b} \times 0.2 \times \frac{\text{ml 8 hour vol}}{0.33} \times \frac{\text{weight postdialysis}}{\text{weight predialysis}}$$

N.B. If the volume of urine represents an excretion of less than 8 hours, calculate an 8 hour volume as follows:

$$\text{Vol collected} \times 8/\text{hours of collection}$$

NOTES

1. *Enzyme units.* The unit of enzyme activity in urine has been defined as the amount of enzyme in an 8 hour volume of urine that liberates 1 mg of phosphorus under the conditions of the assay (16).

2. *Stability of samples.* Studies in our laboratory have shown that 1 mg Omacide-2 (Olin Chemicals) per milliliter urine preserves samples for 4 days at room temperature.

ACCURACY AND PRECISION

Accuracy of urine enzyme determinations has the same qualifying factors as those in serum assays.

Precision for urine assays is ±10% at elevated levels.

NORMAL VALUES

The normal level for urinary alkaline phosphatase is less than 3.5 units/8 hours (16).

ACID PHOSPHATASE

The acid phosphatases, a class of nonspecific phosphomonoesterases (orthophosphoric monoester phosphohydrolase, EC 3.1.3.2), hydrolyze esters of orthophosphoric acid at acidic pH. Enzyme hydrolysis results in the splitting of the O-P bond by the following mechanism:

$$R-O-\overset{\overset{\displaystyle O}{\|}}{\underset{\underset{\displaystyle OH}{|}}{P}}-OH \underset{\pm H_2O}{\overset{[H^+]}{\rightleftharpoons}} R-OH + HO-\overset{\overset{\displaystyle O}{\|}}{\underset{\underset{\displaystyle OH}{|}}{P}}-OH$$

The acid phosphatases have a rather flat pH activity curve, 4–7, with the pH of most reactions being measured in the range of 4.8–6.0. The pH optimum varies with the source of the enzyme and the substrate employed.

Acid phosphatase activity is present in serum, red blood cells, platelets, leukocytes, prostate, liver, spleen, bone, kidney, and other tissues (1, 56, 287, 357, 621, 802). Most of these tissues contain two or more isoenzymes. At least six genetically determined isoenzymes have been reported in red cells (218, 219, 336, 673). Leukocytes normally have four bands of enzyme activity, but as many as seven bands are evident in some disease states (802). Roy *et al* (621), using DEAE-cellulose columns (686), found three enzyme fractions for liver and kidney. Prostate tissue contains two major acid phosphatases that have identical pH optima, substrate, and inhibitor specificities (686). Other studies have established the molecular weights, substrate specificities, and metallic ion interactions of the prostatic enzymes (461, 488).

Starch gel electrophoresis (212) and column chromatography (21) effect the separation of two major enzyme fractions in serum, one of which is identical to that of prostatic origin as characterized by tartrate inhibition studies. Three to five enzyme bands are found in the sera of normal children and adults with polyacrylamide gel electrophoresis (623). No sex difference was noted in the zymogram patterns.

Acid phosphatase activity in serum was first demonstrated in 1938 with the substrate phenyl phosphate (287), which subsequently was shown to have a two- to threefold greater rate of hydrolysis than β-glycerophosphate (288). *p*-Nitrophenyl phosphate was also successfully applied to the measurement of serum acid phosphatase (19). α-Naphthyl phosphate was introduced as a substrate specific for prostatic acid phosphatase and has the advantage that naphthol, which results from its hydrolysis, can be coupled with tetrazotized *o*-dianisidine to form an azo dye (38). Diazotized 5-nitro-*o*-anisidine, substituted as the coupling agent in the α-naphthyl reaction, affords greater sensitivity and color stability (37, 754). The liberated α-naphthol may also be measured directly with a spectrophotofluorometer (126). Adenosine-3′-monophosphate hydrolysis has been reexamined as a potential substrate for serum acid phosphatases (204). In a comprehensive study Roy *et al* (621) compared thymolphthalein monophosphate with five other substrates, the effect of L-tartrate and formaldehyde on the enzyme, and isoenzyme activities from prostate, liver, kidney, bone, erythrocytes, platelets, and urine with each substrate.

Automated methods for the determination of acid phosphatases have utilized either phenyl or α-naphthyl phosphate as a substrate (278, 402). Tartrate or copper ion has been incorporated into the systems to differentiate the prostatic and erythrocytic acid phosphatases (278, 401).

In the initial attempt to measure serum acid phosphatase, Gutman and Gutman (287) reported that fluoride inhibited enzyme activity. They subsequently noted that fluoride inhibited acid phosphatase of prostatic origin but not that from red cells (289). In a study of the effects of a large number of compounds and ions on erythrocytic and prostatic acid phosphatases, Abul-Fadl and King (1) found inhibition of the prostatic enzyme by tartrate, fluoride, and iron. Formaldehyde and copper ion specifically inhibited the red cell enzyme. Tartrate inhibition was later confirmed, and the list of inhibitory compounds was extended to include saccharate and oxalate, regardless of the substrate employed in the assay (532). Another important facet of this investigation was the effect of acetate, citrate, and Tris buffers on the pH optima of enzyme activity with different substrates. Varying degrees of inhibition of red cell isoenzymes were reported to occur with the addition of oxalate, fluoride, formaldehyde, ethanol, and calcium (250).

Activation of prostatic acid phosphatase observed with low concentrations of short chain alcohols suggests the occurrence of transphosphorylation (26). Reinvestigation of this phenomenon showed that simple glycols are more efficient phosphate acceptors than cyclic alcohols (533). Acid phosphatases from the prostate, red cells, and seminal fluid show significant activation with citrate and cyanide (1).

Following initial studies of the use of specific substrates and inhibitors to detect selectively the prostatic enzyme in serum (1, 287), efforts to improve substrate specificity have continued. Babson and associates (39) presented data to support the specificity of α-naphthyl phosphate in the discernment of prostatic acid phosphatase. However, other investigators cited the incidence of false positives (12% of a male population without prostatic cancer and in a similar percentage of females) as evidence of the lack of specificity of this substrate (11). Comparison of the relative activities of many types of human acid phosphatase isoenzymes toward various substrates—phenyl phosphate, phenolphthalein monophosphate, α-naphthyl phosphate, p-nitrophenyl phosphate, β-glycerophosphate, and thymolphthalein phosphate—indicates that the last named substrate is most specific for the prostatic enzyme (621). This specificity was confirmed by measurement of serum acid phosphatase activities before and after prostatectomy and the finding of values with thymolphthalein phosphate that are equivalent to tartrate-inhibited values with other substrates.

Urinary acid phosphatase activity has been measured in children and women (421). However, the major contribution of prostatic enzyme from adult males cannot be inhibited without concomitant suppression of the renal source of enzyme (420, 421).

DETERMINATION OF SERUM ACID PHOSPHATASE

(method of Roy et al, Ref 621, modified)

PRINCIPLE

Serum acid phosphatase hydrolyzes thymolphthalein monophosphate at pH 5.9 to yield thymolphthalein. The addition of alkali terminates enzymatic activity and simultaneously converts the colorless thymolphthalein to a blue chromogen that is measured photometrically (Fig. 21-9).

REAGENTS

Buffered Substrate. Dissolve 88.2 g sodium citrate, 16.4 g sodium acetate, and 1.5 g Brij-5 in about 800 ml water. Add 0.63 g magnesium thymolphthalein phosphate (Eastman Organic Chemicals) and dilute to 1

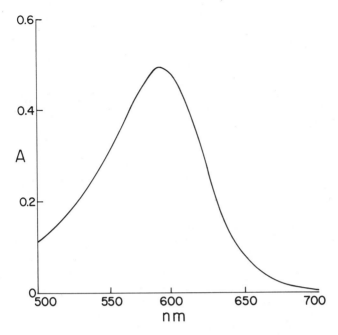

FIG. 21-9. Absorption curve for color in acid phosphatase determination: 21.5 μg thymolph-thalein standard vs water blank. Cary Model 15 Recording Spectrophotometer.

liter with water. Adjust to pH 7.0 ± 0.1 with glacial acid on a pH meter. Solution is stable 9 months at 4°C.

Acetic acid, 13% (v/v)

Alkali Reagent. Dissolve 20 g NaOH and 20 g Na_2CO_3 in water and dilute to 1 liter.

Thymolphthalein Standard, 150 μM. Dissolve 6.46 mg thymolphthalein in 75% *n*-propanol.

PROCEDURE

1. Add 1.9 ml buffered substrate to tubes labeled "control" and "unknown."
2. Add 0.1 ml acetic acid reagent to each tube.
3. Place both tubes in a water bath at 37°C and allow to stand 10 min.
4. Add 0.2 ml serum by TC pipet to the "unknown," mix, and incubate at 37°C.
5. After exactly 30 min add 1.0 ml alkali reagent to the "control" and "unknown" and mix. Then add 0.2 ml serum by TC pipet to the "control" and mix.
6. To a tube labeled "standard", add 0.2 ml thymolphthalein solution, 2.0 ml water, and 1.0 ml alkali reagent; mix.

7. Read absorbances of the standard (A_s), control (A_c), and unknown (A_x) against water at 590 nm.

Calculation:

$$\text{Acid phosphatase units} \atop (\mu\text{mol/liter/min})} = \frac{A_x - A_c}{A_s} \times 150 \times 0.2 \times \frac{1000}{0.2} \times \frac{1}{30}$$

$$= \frac{A_x - A_c}{A_s} \times 5$$

N.B. If absorbance of sample is greater than 0.8, make a 1:5 or 1:10 dilution of an aliquot of the blank and unknown with water. Do not dilute the standard. Read the absorbances of the blank and sample tubes, calculate as above, then multiply by the appropriate dilution factor. The standard contains 150 μmol thymolphthalein per liter. This is equivalent to 5 μmol or 5 IU/liter of serum per minute of incubation (divide 150 by 30 min incubation period).

NOTES

1. Definition of enzyme unit. The acid phosphatase unit is defined as the amount of enzyme which releases 1 μmol thymolphthalein per minute per liter of serum under the specified conditions of incubation. This is expressed as 1 International Unit (IU) per liter.

2. Color of solution. Although the dye complex of the standard is blue in alkaline solution, the final color of the sample may appear to be amber or green, depending on the level of enzyme in the sample. This is due to the yellow color of the substrate after addition of the alkali reagent and the relative amount of liberated thymolphthalein. The color does not fade measurably in 2 hours.

3. Reagent blank vs serum blank. Normally, little interference is expected by hemoglobin or bilirubin at 590 nm, but turbid, lipemic, heavily hemolyzed, or highly icteric sera can increase the reading of the sample. Therefore a serum blank is recommended for each sample (621).

4. Substrate stability. Brij is added to stabilize the buffered substrate solution and to prevent precipitation of some of the substrate after the addition of serum (621).

5. pH optimum. The enzyme activity is uniform over a pH range of 5.8–6.4; therefore small deviations in pH are without significant effect.

6. Serum vs plasma. Since this substrate is not hydrolyzed as readily as others by platelet enzymes released during clotting (621), serum and plasma are equally acceptable.

7. Hemolysis. As a substrate, thymolphthalein phosphate is 13- to

275-fold less sensitive that α-naphthyl phosphate to hydrolysis by the erythrocytic isoenzymes (621). Consequently even moderate hemolysis has a negligible effect on the assay.

   8. Stability of serum samples. Acidification of elevated acid phophatase sera with disodium citrate, 18 mg/ml, stabilizes the enzyme for 7—8 days at 25°, 5°, and −15°C (187). Similarly, 20% acetic acid, 10 μl/ml of serum, preserves the activity (204). At room temperature, however, serum which has been separated from the clot loses enzyme activity very rapidly, with considerable losses occurring within 1 hour. This deterioration results from an increase in the pH of serum due to loss of CO_2, and the magnitude of change is a function of the air space in the tube above the serum. If the serum is left on the clot, the pH of the serum does not increase and the acid phosphatase is stable at room temperature for as long as 24 hours, although erythrocyte phosphatase leaks out of the red blood cells even in the absence of hemolysis.

ACCURACY AND PRECISION

   In view of the relative specificity of thymolphthalein phosphate for the prostatic isoenzyme, the effect of nonprostatic phosphatases is minimal (621). Furthermore, in a study of 87 sera from four groups of patients (breast cancer, liver or biliary tract diseases, kidney diseases, and miscellaneous conditions) no abnormal results were obtained with thymolphthalein phosphate, whereas other substrates gave 0—48% abnormal results.

   Precision (95% limits) of this test is about ±5% in the normal range.

NORMAL VALUES

   Our laboratory has found a normal range of 0.15—0.31 IU/liter, while Roy *et al* (621) report a range of 0.11—0.60 IU/liter with this substrate. The difference in the upper limit of the normal range may be attributable to the age of the subjects surveyed, which was up to 63 years in our study and up to 84 years in the series of Roy *et al* (621). A slow, steady increase in acid phosphatase levels occurs with increasing age (666). There is no sex difference. A rise in serum enzyme levels has been noted in some cases following rectal palpation of the prostate (172, 621). A circadian rhythm of serum acid phosphatase levels has been observed with the highest levels between 9 AM and 3 PM (186).

5'-NUCLEOTIDASE

   The enzyme 5'-nucleotidase (5'-Nase, 5'-ribonucleotide phosphohydrolase, EC 3.1.3.5) specifically catalyzes the hydrolysis of 5'-nucleotides:

ADENOSINE-5′-PHOSPHATE   ADENOSINE

Two types of alkaline phosphatases, distinguished by their substrate specificity, are present in serum and tissues. The nonspecific alkaline phosphatases hydrolyze not only the phosphate groups of 5′-nucleotides but of many other biochemical compounds as well. In several studies of the pH optimum of 5′-Nase, a range of 7.5–7.9 was reported (4, 61, 591). However, a recent study reported a pH optimum of 6.8, which was attributed to the greater specificity of the method employed (330). A study of the effects of metallic ions on the activity of 5′-Nase showed activation by Mg^{2+} and Mn^{2+} and marked inhibition by Ni^{2+} and Zn^{2+} (4).

Most of the initial assay methods for this enzyme were based on the determination of liberated phosphate (4, 125, 185, 591). Since the nonspecific alkaline phosphatases present in sera or tissue also hydrolyze the adenosine-5′-phosphate (A-5′-P) substrate, the concomitant hydrolysis of phenyl phosphate or glycerophosphate was used as a measurement of nonspecific activity (61, 185, 591). Campbell (125) utilized nickel inhibition of 5′-Nase to correct for the hydrolysis of A-5′-P by alkaline phosphatase. An automated method based on the nickel inhibition technic compared well with the manual method (329). Selective inhibition of the alkaline phosphatases by a mixture of 2′- and 3′-adenosine monophosphates has been reported to be more suitable than EDTA or nickel inhibition for the subtraction of nonspecific phosphate hydrolysis (55). Another approach applied to reduce nonspecific alkaline phosphatase interference has been the incorporation of phenyl phosphate or glycerophosphate along with A-5′-P in the incubation mixture (63). This form of competitive inhibition or "enzyme diversion" is based on the high affinity of the nonspecific phosphatases for the secondary substrate, in turn resulting in a suppression of A-5′-P hydrolysis.

Three newer methods for the assay of 5′-Nase have been described that employ adenosine deaminase in a coupled enzyme reaction. One method utilizes the decrease in absorbance at 265 nm as a consequence of adenosine conversion to inosine by adenosine deaminase (62, 449). The second method measures ammonia, the other by-product of the reaction, by the Berthelot reaction (562, 751). A third method employs both adenosine deaminase and glutamic acid dehydrogenase to effect the reductive amination of α-ketoglutarate (203). The coenzyme NADH consumed in this reaction is measured by a decrease in absorbance at 340 nm. In a review (64) of the coupled enzyme methods for 5'-Nase, the major conclusions were: (a) the spectrophotometric assay at 265 nm is the most specific method, and (b) the photometric determination of ammonia liberated is the most sensitive method.

DETERMINATION OF 5'-NUCLEOTIDASE

(method of Dixon and Purdom, Ref 185)

PRINCIPLE

5'-Nucleotidase hydrolyzes adenosine-5'-phosphate to yield adenosine and inorganic phosphate. Nonspecific alkaline phosphatase activity is measured by incubating the sample with glycerophosphate to yield glycerol and inorganic phosphate. Phosphate ions are determined in a trichloroacetic acid supernate by formation of molybdophosphate, followed by reduction to molybdenum blue with aminonaphtholsulfonic acid. The absorbance of molybdenum blue is measured photometrically at 770 nm. (The shape of the absorption curves are similar to the curves shown in Figure 21-8.) 5'-Nucleotidase activity is calculated by subtracting the nonspecific alkaline phosphatase activity with glycerophosphate as the substrate from the total activity with adenosine-5-phosphate as the substrate.

REAGENTS

Buffered Adenosine-5'-phosphate Solution (A-5'-P). Dissolve 87 mg A-5'-P and 424 mg sodium barbital in 100 ml water. Adjust to pH 7.5 with 1 N HCl on a pH meter. Store at 4°C.

Buffered Glycerophosphate Solution (G-P). Dissolve 50 mg sodium β-glycerophosphate and 424 mg sodium barbital in 100 ml water. Adjust to pH 7.5 and store at 4°C.

Aminonaphtholsulfonic Acid Reagent. Dissolve 12 g sodium bisulfite and 400 mg sodium sulfite in 200 ml water. Add 200 mg 1,2,4-aminonaphtholsulfonic acid and shake to dissolve. Solution stable 4 weeks at 4°C.

Ammonium Molybdate, 0.0202 M. Dissolve 2.5 g $(NH_4)_6 Mo_7 O_{24} \cdot 4H_2 O$ in approximately 70 ml water. Cautiously add, with mixing, 8.4 ml conc $H_2 SO_4$. Cool, dilute to 100 ml with water, and mix. Store in polyethylene bottle.

Trichloroacetic Acid (TCA), 30%

Working Phosphorus Standard, 5 µg P/ml. Dissolve 22.0 mg anhydrous $KH_2 PO_4$ in water and dilute to 100 ml. Transfer 10 ml of this solution to a 100-ml volumetric flask, add 16 ml 30% TCA and dilute to mark with water.

PROCEDURE

1. Add 4.8 ml adenosine-5'-phosphate to test tubes labeled "A-5'-P unknown" and "A-5'-P control". Add 4.8 ml glycerophosphate to test tubes labeled "G-P unknown" and "G-P control". Place tubes in a 37°C water bath for 5 min.

2. Add 0.2 ml serum sample by TC pipet to both unknown tubes and incubate 2.5 hours at 37°C.

3. Remove tubes from the bath, add 0.2 ml serum to both "control" tubes followed by the immediate addition of 1.0 ml 30% TCA to all tubes. Mix contents of tubes thoroughly, allow to stand 10 min, and centrifuge.

4. Set up and develop colors as follows:
   Reagent Blank: 0.67 ml 30% TCA + 3.33 ml water (A_b).
   Standard: 4.0 ml working phosphate standard—20 μg P (A_s).
   A-5'-P Control: 4.0 ml supernate from step 3 (A_{ca}).
   A-5'-P Unknown: 4.0 ml supernate from step 3 (A_{xa}).
   G-P Control: 4.0 ml supernate from step 3 (A_{cg}).
   G-P Unknown: 4.0 ml supernate from step 3 (A_{xg}).

5. To each tube, add 0.5 ml molybdate reagent, mix, then add 0.5 ml aminonaphtholsulfonic acid reagent and mix again.

6. After 10 min read all tubes against water at 770 nm or with a filter having a nominal wavelength in this region (a 660 nm filter is satisfactory). Color is stable 1.5 hours.

Calculation:

$$5'\text{-Nucleotidase (units/100 ml)} = \frac{(A_{xa} - A_{ca}) - (A_{xg} - A_{cg})}{A_s - A_b} \times 6$$

NOTES

1. Beer's law. Beer's law is obeyed with a spectrophotometer or a Corning glass filter with nominal wavelength of 660 nm.

2. Stability of samples. Enzyme activity in serum is stable 4 days at room temperature or 4°C. Frozen samples are stable 4.5 months (62).

3. Serum vs plasma. Serum is assayed in most instances, but heparinized plasma is also suitable (751). Since the enzyme is activated by Mg^{2+} or Mn^{2+}, the metal-chelating anticoagulants should be avoided.

ACCURACY AND PRECISION

The hydrolysis of adenosine-5'-phosphate by nonspecific alkaline phosphatases, the use of inhibitors to suppress this hydrolysis, and the requirement of adequate amounts of Mg to activate the enzyme raises some question about the accuracy of the assay. However, the good agreement of results obtained with the nickel inhibition method of Campbell (125) and the adenosine deaminase method of Persijn *et al* (562) indicates a fair degree of accuracy (575). Furthermore, there is a general agreement of normal ranges regardless of the technic employed for the assay.

The precision (95% limits) of the method is ±7% for the assay of normal serum enzyme levels.

NORMAL RANGE

The normal range is 0—1.6 units (185). A significant correlation of increasing serum enzyme levels with increasing age was reported for females only (62).

AMYLASE

α-Amylase (α-1,4-glucan-4-glucanohydrolase, EC 3.2.1.1) catalyzes the hydrolysis of oligosaccharides, polysaccharides, starches, and glycogen. Hydrolysis of amylose yields primarily maltose and some simple polymers (4—12 glucose units). Amylase action on branched chained amylopectin (1,4 and 1,6 linkages) yields maltose and various polymers including limit dextrins, which are formed by cleavage of maltose units from each branch point until a 1,6 linkage prevents any further degradation.

The enzyme has a molecular weight of 45 000 and a pH optimum of 7. It is activated by Ca^{2+} and therefore can be classified as a metalloenzyme. Compounds that chelate Ca are powerful inhibitors. Halogen ions, especially chloride, also activate the enzyme.

The tissue sources of human serum amylase are chiefly the pancreas and parotids, possibly the liver, and to a minor extent other organs (778). Experiments on dogs with pancreatic ligations, fistulas, pancreatectomy, or hepatectomy point to the liver as a major source of serum amylase (537).

At present there is considerable evidence for at least two isoenzymes with electrophoretic mobilities of β- and γ-globulin (493). Three to five additional isoenzyme fractions have been detected, but the number varies with the electrophoretic technic and tissue source. Urine also contains at least two and possibly three to seven isoenzymes (493).

The historical background on amylase, development of the methodologies, investigation of appropriate substrates, and other pertinent aspects have been thoroughly covered by Henry (315), King (395), and Searcy *et al* (659). These reviews deal mainly with the following technics for the determination of amylase: (a) viscosimetric—decrease in viscosity of a starch solution; (b) turbidimetric—decrease in turbidity of a starch suspension; (c) iodometric (amyloclastic)—decrease in the starch-iodine color; and (d) reductometric (saccharogenic)—an increase in the reducing groups of a starch solution. Of the many methods and modifications published, the amyloclastic procedure of Street and Close (712) and the saccharogenic assay of Henry and Chiamori (318) continue to be used as reference methods. To some extent, Somogyi's modified method is also used (690).

Additional technics and procedures continue to appear. The photometric assay of reducing groups with the reduction of triphenyltetrazolium to red formazan has been reported (463). Coupling the products of the amylase reaction with α-glucosidase and glucose-6-phosphate dehydrogenase utilizes the increase in absorption of reduced coenzyme as a means of measurement (644). A coulometric method (510), radial diffusion assays (59, 133), and a technic that employs ^{14}C-labeled starch as the substrate (474) are other new methods.

Treatment of starch with sodium borohydride has reduced the blank values, improved the reliability, and extended the range of the assay (714). Solubilization of starch in dimethylsulfoxide purportedly results in a stable substrate solution (235). Reexamination of various starches demonstrated that the rate of hydrolysis of potato or corn starches differs with the biologic source or type of amylase (492).

Following the introduction of the dye-coupled starch remazol brilliant blue as a substrate for amylase (603), a plethora of dye-polysaccharide substrates have been synthesized and tested, including remazol yellow-GR or orange-3R (176), Cibachrom blue F-3GA amylose (403), Reactone-red-2B-amylopectin (42), blue starch polymer 51-A (134), and Procion brilliant red M-2BS-amylopectin (637).

Automated methods for amylase have also been examined with several technics and substrates—e.g., iodometric methods (296, 408), reductometric assays (237, 542), and the dye-polysaccharide complexes Amylose-azure (301) and starch-anthranilate (601).

Gambill and Mason (245) investigated serum and urine amylase and concluded that although the two levels increase concomitantly in pancreatitis, the level in urine is a more sensitive indicator and a better monitor of the course of this disease. Their recommendation of performing the assay on 1 hour urine samples has been reaffirmed (200). However, the recently described hyperamylasemia due to globulin-bound amylase (macroamylase) does not result in increased levels in the urine (777). A rapid detection method for macroamylase has been described that utilizes gel filtration to separate the two types of amylase (238).

Studies of the new dye-coupled substrates for amylase have been made in our laboratory. Comparison with the saccharogenic method of Henry and Chiamori (318) revealed widely varying results. Similar discrepancies have been reported by other investigators (403, 637). Until this variability in enzyme-dye substrate interaction is explained or resolved, the saccharogenic method remains the method of choice. Furthermore, the results obtained with the saccharogenic method show good correlation with the clinical status of patients, both with and without pancreatic disorders.

DETERMINATION OF AMYLASE IN SERUM, URINE, AND DUODENAL FLUID

(method of Somogyi, Ref 690, modified by Henry and Chiamori, Ref 318)

PRINCIPLE

The sample is incubated with a buffered starch solution at 37°C for 34 min. Reducing substances are then determined in a protein-free filtrate. The difference in reducing substances, expressed as glucose, in mg/100 ml of sample before and after incubation is taken arbitrarily as the units of amylase activity. The amylase activity of a timed urine sample can be expressed as units/hour, i.e., the calculated milligrams of reducing substances, expressed as glucose, produced by the volume of urine excreted per hour.

REAGENTS

Phosphate Buffer, 0.1 M, pH 7.0. Dissolve 4.55 g KH_2PO_4 and 9.35 g Na_2HPO_4 in water and make to 1 liter.

Buffered Starch Substrate. Add 1.14 g soluble Zulkowski starch (E. Merck AG) to 100 ml buffer. Mix gently. This modified starch is completely soluble in water in room temperature. Prepare fresh.

An alternative is to use 1.5 g starch (soluble potato starch, Lintner or ACS specifications; or corn starch) to 100 ml phosphate buffer and heat to boiling for about 3 min. Cool to room temperature and add phosphate buffer to a volume of 140 ml. The substrate is stable for at least 4 months at room temperature if kept sterile. Nearly all substrate preparations show some turbidity. Sodium benzoate in a concentration of 0.86% preserves starch substrates for 1 year at room temperature and has no effect on enzyme activity.

NaCl, 0.85%

H_2SO_4, 0.66 N. Dilute 5.5 ml acid to 300 ml with water.

Sodium Tungstate, 10%

Stock Glucose Standard, 1%. Use highest purity glucose. Saturate with benzoic acid. This is stable indefinitely in refrigerator. 1 ml = 10 mg.

Working Glucose Standard, 2 mg/ml. Prepare a 1:5 dilution with water.

Copper Reagent. Dissolve 40 g anhydrous Na_2CO_3 in about 400 ml water in a 1 liter volumetric flask. Add 7.5 g tartaric acid; when this has dissolved, add 4.5 g $CuSO_4 \cdot 5H_2O$. Mix and dilute to volume. This reagent is stable indefinitely at room temperature.

Phosphomolybdic Acid Reagent. To 70 g molybdic acid and 10 g sodium tungstate, add 400 ml 10% NaOH and 400 ml water. Boil for 20–40 min to drive off NH_3. Cool and add water back to about 700 ml. Add 250 ml conc orthophosphoric acid (85% H_3PO_4) and dilute to 1 liter with water.

PROCEDURE

1. The sample to be tested may be serum or plasma, a 1:2 dilution of urine, or a 1:25 dilution of duodenal fluid with saline as the diluent.
2. Place 0.5 ml sample into a test tube labeled "control" and 3.5 ml of buffered starch into another tube labeled "unknown".
3. Place both tubes in a 37°C water bath and after 5 min add 0.5 ml sample to "unknown".
4. After exactly 34 min, remove tubes from the bath. Immediately add 0.75 ml 0.66 N H_2SO_4 to each tube and mix; then add 0.25 ml 10% sodium tungstate and mix again. Addition of acid stops amylase action completely.
5. Add 3.5 ml buffered starch to "control" and mix.
6. A reagent blank and a glucose standard are also run by determining the reducing sugar in aliquots of the following mixtures:
   Blank: 3.5 ml buffered starch, 0.75 ml 0.66 N H_2SO_4, 0.25 ml 10% sodium tungstate, and 0.5 ml water; mix.
   Standard: 3.5 ml buffered starch, 0.75 ml 0.66 N H_2SO_4, 0.25 ml 10% sodium tungstate plus 0.5 ml working glucose standard; mix.
7. Filter or centrifuge and transfer 1.0 ml supernate to 15 x 125 mm tubes.
8. Add 1 ml copper reagent to each tube and mix.
9. Place tubes in boiling water for exactly 6 min, then cool to room temperature in a cold water bath.
10. Add 1.0 ml phosphomolybdic acid reagent to all tubes and place in boiling water bath exactly 5 min.
11. Transfer tubes to cold tap water bath. When cool, add 10.0 ml water and mix by inversion.
12. Read absorbances of "standard" (A_s), "control" (A_c), and "unknown" (A_x) against reagent blank at 420 nm (Figure 21-10).

Calculation:

For serum or plasma

$$\text{Amylase units/100 ml} = \frac{A_x - A_c}{A_s} \times 1 \times \frac{100}{0.5}$$

$$= \frac{A_x - A_c}{A_s} \times 200$$

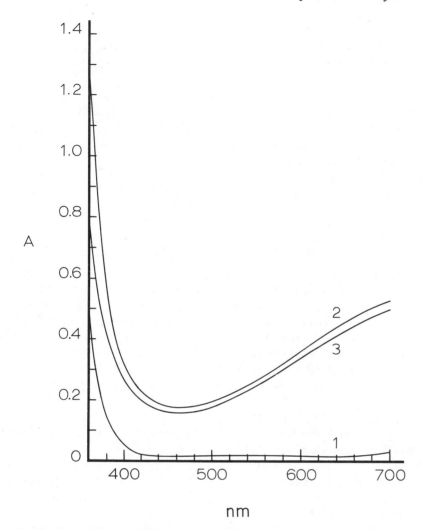

FIG. 21-10. Absorption curves in amylase determination by glucose reduction: (1) reagent blank vs water; (2) standard equivalent to 100 mg glucose/100 ml vs water; (3) standard vs reagent blank. Perkin-Elmer Model 4000-A Recording Spectrophotometer.

For duodenal fluid

$$\text{Amylase units/100 ml} = \frac{A_x - A_c}{A_s} \times 1 \times \frac{1}{0.02}$$

$$= \frac{A_x - A_c}{A_s} \times 50$$

For urine

$$\text{Urine amylase units/hour} \quad = \frac{A_x - A_c}{A_s} \times 1 \times \frac{\text{ml volume/hour}}{\text{ml sample used}}$$

N.B. If the amylase units calculate to be greater than 200 units with the volume of serum or duodenal fluid used, or greater than 400 units in the volume of urine used, repeat "unknown" using 0.5 ml supernate plus 0.5 ml water in step 7. If the values exceed 600 units for serum or duodenal fluid, or 1200 units for urine, then dilute the sample to an appropriate volume with 0.85% NaCl and rerun "unknown".

NOTES

1. Glucose standards (318). A glucose standard is run with buffered starch substrate and the protein-precipitating reagents because the high concentration of phosphate buffer affects the pH in the sugar test, altering its absorptivity. Thus in the Folin-Wu method a pure glucose standard has a pH of about 10.4 after addition of the alkaline copper reagent. Blanks, glucose standards, and urine or serum specimens containing the prescribed amount of buffered starch substrate and treated with the precipitating reagents had a final pH of 9.4—9.7 after addition of the alkaline copper reagent. This decrease in pH results in an increase in absorptivity of about 6%.

2. Anticoagulant. Heparinized plasma or serum yields identical results, but citrate or EDTA must be avoided because of the binding of Ca^{2+}, which is essential for enzyme activity (315).

3. Effect of sodium chloride. Chloride ion at a level of 0.01 M is required for optimal activity. If the final dilution in the test tube results in inadequate chloride concentration, the serum or urine sample must be diluted with 0.85% saline (318).

4. Effect of pH. The pH optimum for amylase activity in serum and urine is 6.9—7.2 (42, 318).

5. Stability of samples. Serum and urine samples, without bacterial contamination, are stable at room temperature 1 week and several months at 4°C (315). Duodenal fluid diluted 1:2 with glycerine is stable for several days at room temperature and for several months at 4°C (428).

ACCURACY AND PRECISION

The test appears to be accurate in the sense that it provides a reliable but arbitrary measurement of amylase activity present. The "amylase activity" is the net effect of enzyme concentration and the concentrations of activators and inhibitors which may be present. There has been some question about the reliability of amylase determinations in diabetes when the glucose concentration is elevated, but such an interference does not occur with the method presented even at a glucose level of 800 mg/100 ml (318).

Precision of the test varies according to the level of enzyme assayed—i.e., low levels ±10%, high levels ±3% (318).

There is a significant difference in the values for males (38–118 units) and females (46–141 units). An overall range of serum amylase in adults is 40–140 units/100 ml (318).

The normal range for urinary amylase by the Henry and Chiamori method (318) is 43–245 units/hour or 66–870 units/100 ml. Others have reported a similar range for timed urine collections (296). A study by Eberhagen *et al* (200) showed that either 2- or 24-hour collections can be used. The normal range for duodenal fluid amylase is 100–800 units/ml.

An International Unit of amylase has been proposed by Helger and Lang (310): 1 Street-Close unit = 3.1 Somogyi unit = 0.57 International Unit.

LYSOZYME (MURAMIDASE)

Lysozyme (mucopeptide N-acetylmuramylhydrolase, EC 3.2.1.17) acts upon mucopolymers of bacterial cell walls, which are composed of alternating N-acetylmuramic acid and N-acetylglucosamine residues with alternating β (1 $\rightarrow$ 4) and β (1 $\rightarrow$ 6) linkages, to liberate N-acetyl hexosamines. Bacterial cell wall polymers of this type are found in *Micrococcus lysodeikticus, Sarcina lutea,* and *Bacillus megatherium*, with *M. lysodeikticus* being the most labile to the action of this enzyme (363).

Lysozyme has a low molecular weight (14,500), is stable at acidic pH and unstable at alkaline pH, and has an isoelectric point between 10.5 and 11.0. Two recent reviews deal with the sources, physicochemical properties, crystallographic data, and interactions with synthetic substrates (141, 363). Lysozymal activity is primarily localized in the lysosomal or granular fraction of polymorphonuclear granulocytes and monocytes (111). Although the lysozymes from different organs or cell types have the same enzymic action, they possess different chemical structures and different reaction velocities with the bacterial cell wall substrate of *M. lysodeikticus* (455).

Since the initial discovery of the dissolution of *M. lysodeikticus* cell wall by lysozyme (228), different types of cellular suspensions of this organism have been used in various technics for measurement of the enzyme (94, 454, 576, 687). The method of Boasson (94) was modified by substituting freeze-dried cells and measuring the turbidity of the cell suspension at 540 nm (687). This basic technic was later modified, and the decrease in turbidity was measured over 6 min (454) or 3 min (576) at 645 nm. In the photometric methods the rate of lysis is not linear with time; however, a linear correlation is obtained by plotting 1/absorbance over a 6 min period (574). Other investigators continuously recorded the reaction velocity (552).

Another technic for the measurement of lysozyme is the agar diffusion assay, the lysoplate method (550). The enzyme in serum or urine is allowed to diffuse into an agar gel plate containing *M. lysodeikticus.* The diameter of the cleared zone is proportional to the log concentration of the enzyme. A comparative study of the lysoplate and spectrophotometric methods showed a correlation coefficient of 0.93 (812). More recently a quantitative immunochemical method was described (359) that utilizes electrophoresis of samples into antibody-containing agarose gel. The lower limit of detection, 0.3 mg/liter, allows quantitation of the levels in normal serum and of elevated levels in urine. A study of the relationship of lysozyme concentrations obtained by enzymatic and the immunochemical methods gave a correlation coefficient of 0.98. A sensitive immunoassay has been described that utilizes a human lysozyme-bacteriophage conjugate as the test system (477). Mucolytic activity measured by viscometry has also been used as an assay (471, 500).

Daniels *et al* (173) studied urinary lysozyme from the standpoints of pH optimum, temperature effect, temperature-pH dependence, and the effect of storage at different temperatures. Their method employs turbidimetric measurements (645 nm) at pH 6.3 for 3 min.

DETERMINATION OF SERUM OR URINE LYSOZYME

(method of Litwack, Ref 454, modified)

PRINCIPLE

Serum or urine is incubated with a suspension of *Micrococcus lysodeikticus.* The lysozyme present in the sample causes lysis of the cells. The decrease in turbidity is measured at 540 nm and is correlated with lysozyme concentration by means of a calibration curve.

REAGENTS

NaCl, 0.9%

Phosphate Buffer, M/15, pH 6.2. Mix 815 ml of KH_2PO_4 (9.08 g anhydrous/liter water) and 185 ml of Na_2HPO_4 (9.47 g anhydrous/liter water). Verify pH with a meter. Store in refrigerator.

Buffered Substrate. Dissolve 15 mg of dry *M. lysodeikticus* in 10 ml 0.9% NaCl and dilute to 100 ml with the phosphate buffer. Substrate should be aged 5–16 hours before use. The product from Worthington Biochemical Corp. was found to be suitable. Discard after use.

Stock Lysozyme Solution, 400 µg/ml. All enzyme solutions are prepared in siliconized glassware (Desicote, Beckman No. 18772). Dissolve 20 mg enzyme in 50 ml of 0.9% NaCl. Egg white lysozyme (recrystallized six times, salt-free, 99+% pure, 50,000 units/mg) is obtained from Miles Laboratories. The stock solution is stable in the refrigerator for 6 days.

Working Lysozyme Standard (20 µg/ml). Dilute 2.5 ml stock standard to 50 ml with 0.9% NaCl. Prepare fresh each time. From this solution, prepare a

4 μg/ml enzyme standard (2.0 ml of working standard + 8.0 ml 0.9% NaCl) and a 10 μg/ml standard (4.0 ml working standard + 4.0 ml 0.9% NaCl).

PROCEDURE

1. To tubes labeled S-4, S-10, and unknown add 5.0 ml substrate. To a blank tube add 5.0 ml 0.9% NaCl.
2. Equilibrate all tubes for at least 10 min at 37°C.
3. Zero the photometer or spectrophotometer (540 nm) with the blank and return tube to incubator.
4. Add 0.5 ml serum or urine to one of the unknowns, start the timer, mix, and continue incubation.
5. At exactly 30 sec, mix unknown, read and record first reading, then return tube to incubator.
6. Check zero of photometer at 2.5 min with blank.
7. At exactly 3 min, mix unknown, read, and record second reading.
8. Analyze 4 μg and 10 μg lysozyme standards, and the other samples in the same manner by taking readings at 30 sec and 3 min intervals after addition of the enzyme.

Calculation:

Calculate the change in absorbance (ΔA) for each sample:

$$\Delta A = A \text{ at 3 min} - A \text{ at 30 sec}$$

Using graph paper, plot ΔA versus lysozyme concentration for the 4 μg/ml and 10 μg/ml standards and draw a straight line through the two points. Determine the lysozyme values for serum or urine from the standard curve.

N.B. The 30 sec reading generally falls in the absorbance range of 0.11–0.16. If the suspension has already cleared at 30 sec, repeat the analysis by diluting the sample 1:10 with 0.9% NaCl. If any lysozyme value exceeds 15 μg/ml, then repeat the analysis by diluting with saline and making the appropriate dilution corrections for the final concentration.

NOTES

1. Serum vs plasma. Either can be used to measure the enzyme activity. EDTA is the preferred anticoagulant, for heparin interferes with the assay.

2. pH optimum. A pH of 6.0–6.5 is acceptable for measurement of the enzyme. A shift of the pH optimum to lower values occurs when incubation temperatures greater than 45°C are used (173).

3. Reaction temperatures. Although most investigators have used 25°C for the incubation, temperatures up to 52°C can be employed without deleterious effects (173).

4. Crystalline egg white lysozyme. Since the enzyme is used as a reference standard for the assay, it is important to note and use a preparation with 50 000 units/mg. Crystalline lysozyme preparations can be obtained from various suppliers that range in activity from 8 000 to 50 000 units/mg.

5. Wavelength of turbidimetric measurements. Wavelengths of 540, 640, or 645 nm yield comparable results (173, 454, 552, 687, 812).

6. Stability. Urinary enzyme is stable 4 days at room temperature at pH 4.5–6.3 (173). The higher the pH, the more rapidly the activity drops. At 4° and −20°C the enzyme is stable 8 and 18 days, respectively. Serum lysozyme activity is stable at −20°C.

ACCURACY AND PRECISION

Crystalline hen egg white lysozyme has been used widely as a reference standard for the conversion of enzyme activity to weight of enzyme present in a sample (173, 454, 576, 687). The general acceptance of egg white lysozyme as the reference standard has been mainly because of the stability of the preparation and the similarity of various properties of this enzyme with lysozyme from human sources (173). There is serious question as to the accuracy of lysozyme assays expressed in such absolute units because the reference enzyme standards used by various investigators have ranged from 8 000 to 50 000 units/mg. Johansson and Malmquist (359) reported the specific activity (enzyme activity/mg protein) of crystalline human urinary lysozyme to be three- to fourfold greater than that of egg lysozyme. A similar discrepancy in specific activity of human and egg lysozyme preparations has also been noted with the lysoplate method (550). The only significance of these facts is that attempts to express results of the enzyme assay in absolute units have failed so far. What is really important to the clinician is that elevated levels in serum and urine in renal disease (576), leukemias, and severe infections (812) are useful clinically.

The precision of the method varies from ±18% for values in the normal range to ±5% at a concentration of 25 μg/ml (173, 454). Precision in the normal range can be improved by performing the assays in duplicate.

NORMAL VALUES

The normal range is 2.8–8.0 μg/ml in serum and up to 2 μg/ml in urine. Prockop and Davidson (576) reported similar ranges for serum and urine in a survey of 100 adult subjects. Normal values obtained by an immunochemical technic were 1.2–4.4 μg/ml for serum and less than 0.3 μg/ml for urine (359).

LEUCINE AMINOPEPTIDASE

Leucine aminopeptidase (LAP, L-leucyl-peptide hydrolase, EC 3.4.1.1) is regarded as a typical aminopeptidase, not hydrolyzing acylated compounds such as benzoyl-L-leucylglycine but hydrolyzing leucylglycine, L-leucinamide, L-leucylglycylglycine, L-leucyl-L-leucylglycine, glycyl-L-leucinamide, L-glutamyl-L-leucinamide, glycylglycyl-DL-leucylglycine, and L-leucyl-L-glutamic acid (684). Hydrolysis occurs at the carboxyl end of the leucine residue, there being no endopeptidase activity. More than one LAP enzyme may be present in serum. The one normally present has a pH optimum of 8.3, whereas the one in hepatitis serum has a pH optimum of 8.7—8.8 (224).

L-Leucylglycine has been employed as a substrate for determination of LAP in serum, the extent of hydrolysis being determined photometrically by reaction of the liberated amino acid with ninhydrin (224). The method is adaptable to ultramicro measurements (225).

A number of synthetic substrates susceptible to enzymatic cleavage have provided the basis for measurement of LAP activity. L-Leucyl-β-naphthylamide is hydrolyzed enzymatically to produce β-naphthylamine, which has been quantitated by coupling to nuclear fast red B salt (5-nitro-2-aminomethoxybenzene), resulting in the formation of a red azo dye (634). An ultramicro modification (766) was adapted to the determination of LAP in cerebrospinal fluid (717). The β-naphthylamine was also coupled to diazo blue B—tetrazotized di-*o*-anisidine (276). Goldbarg and Rutenberg diazotized the β-naphthylamine by a Bratton-Marshall reaction and coupled it to N-naphthylethylenediamine hydrochloride (260, 262). This approach has been useful for tissues, urine, and bile as well as serum. It involves a 2 hour enzyme incubation in the presence of substrate and cofactors, followed by precipitation of protein and development of color in the supernate for 15—30 min. Units of activity were at first related arbitrarily to Klett units and later to μg β-naphthylamine released under standard conditions. This method has been adapted by others to cerebrospinal fluid (275) and for the handling of large numbers of specimens with greater simplicity and precision (480). Liberated β-naphthylamine has been coupled to diazotized 3-chloro-4-nitroaniline for photometric determination (499). It has been measured directly in serum, urine, and duodenal fluid by its fluorescence immediately after incubation (616), after stopping the enzyme reaction with EDTA in an alkaline medium (605), and after precipitation with ethanol to stop the reaction (747). Leucine hydrazide has been utilized as a substrate for the enzyme (300, 725). The hydrazine released is reacted with *p*-dimethylaminobenzaldehyde in acid to produce an orange-red color, which is measured photometrically.

A rapid kinetic method (719) based on an earlier nonkinetic approach (366, 525) measures directly at 405 nm the *p*-nitroaniline that is split off from leucine-*p*-nitroanilide by LAP. Ammonia released from leucinamide by the action of the enzyme has been determined by titration with acid after trapping in a Conway diffusion apparatus (299). Leucine itself can be determined by paper chromatography (299). Another kinetic approach employs the cleavage of L-leucylalanine by LAP to leucine and alanine, followed by conversion of the alanine to pyruvate by excess GPT and a-ketoglutarate, and finally conversion of the pyruvate to lactate by excess LDH and NADH. The change in absorbance at 340 nm due to oxidation of

NADH to NAD is a measure of the LAP activity (405). The substrate, unlike β-naphthylamine (480), is noncarcinogenic and is a more physiologic substrate. The kinetic approach of the *p*-nitroaniline and leucylalanine methods is considered intrinsically superior to stoichiometric technics.

Several isoenzymes for LAP have been demonstrated by starch gel electrophoresis (419), paper electrophoresis (683), and cellulose acetate electrophoresis (3). Normal serum exhibits one peak in the postalbumin or fast α_2 fraction. The serum of patients with liver disease shows additional peaks in the positions between the origin and slow α_2 and between the slow α_2 and β (419). Comparable positions were noted by others—normally in the α_1 but also in α_2, β, or albumin in various pathologic states or late pregnancy (683).

The serum "LAP" used in clinical diagnosis has been associated with the β-naphthylamine procedure for 15 years (719), although the classic LAP from pig kidney (specific substrates: L-leucinamide, leucine di- and tri-peptides) does not take any appreciable part in this reaction (525). The serum enzyme responsible for splitting L-leucine-β-naphthylamine is clearly not this LAP. With another artificial substrate, L-leucine-*p*-nitroanilide, the Michaelis constants for the pig kidney enzyme and the human serum enzyme are known to be $3.15 \times 10^{-3}M$ and $0.44 \times 10^{-3}M$, respectively, thus showing greater affinity of the serum enzyme for this artificial substrate also (719). Hydrolysis of both leucine-β-naphthylamide and leucine-*p*-nitroanilide by serum cannot be attributed necessarily to a single enzyme (719). It has been recommended that the enzyme(s) of clinical interest which split these artificial substrates be systematically named amino acid arylamidase instead of leucine aminopeptidase (719); but the old nomenclature continues to be generally used in medicine, in spite of the fact that no correlation between the two classes of enzyme has been found in normal or pathologic sera (525). In this evolving situation, we find ourselves describing a method for an enzyme properly called leucine arylamidase, under the general heading of the commonly misused name of leucine aminopeptidase (LAP).

DETERMINATION OF LEUCINE ARYLAMIDASE

(methods of Nagel et al, Ref 525, Jösch and Dubach, Ref 366, and Szasz, Ref 719, modified)

PRINCIPLE

Leucine arylamidase catalyzes the hydrolysis of L-leucine-*p*-nitroanilide to L-leucine and *p*-nitroaniline. The rate of formation of *p*-nitroaniline is determined by the rate of increase in absorbance at 405 nm (Fig. 21-11). Enzyme activity is expressed in terms of milliunits per milliliter (mU/ml), i.e., nmol product formed per min per ml specimen at 25°C under the conditions of the test.

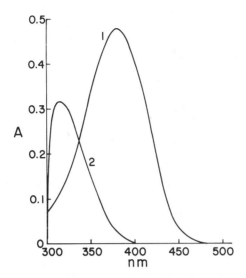

FIG. 21-11. Absorption curves for (1) *p*-nitroaniline and (2) leucine-*p*-nitroanilide; 0.1 ml of 1.25 m*M* solution in methanol added to 3 ml buffer and read vs methanol plus buffer. Beckman DB-G Grating Spectrophotometer and recorder.

REAGENTS

Phosphate Buffer, 0.1 M, pH 7.2. Dissolve 28.4 g Na_2HPO_4 and make up to 1 liter with water (solution A). Dissolve 27.6 g $NaH_2PO_4 \cdot H_2O$ and make up to 1 liter with water (solution B). Combine 72.0 ml solution A with 28.0 ml solution B, dilute to 200 ml with water, and check that the pH is 7.2. Stable for a year at 4°C.

L-Leucine-p-nitroanilide Reagent, 25 mM. Dissolve 63 mg L-leucine-*p*-nitroanilide in reagent grade methanol and bring to a total volume of 10 ml. Store in amber bottle. Stable for 6 months at 4°C.

PROCEDURE

The procedure is described for a Beckman DB-G spectrophotometer, with recorder and equipped for maintenance of compartment temperature at 32°C. A Haake (model FJ) temperature control unit is suitable (see *Note 1*, below).

1. To a cuvet with a 1-cm light path, add 3.0 ml 0.1 *M* phosphate buffer and 0.1 ml L-leucine-*p*-nitroanilide reagent. Mix; place cuvet in spectrophotometer for 5 min for contents to reach 32°C. Add water to the reference cuvet.
2. Add 0.1 ml serum or dialyzed urine (see *Note 6*) to the substrate mixture using a TC pipet. Mix and replace cuvet in compartment without delay.
3. Record the absorbance at 405 nm until a 5-min (at least) linear rate is obtained.

Calculation:

$$\text{mU LAP/ml} = \frac{\Delta A \text{ for } T \text{ min}}{T} \times \frac{10^6}{9.9 \times 10^3} \times \frac{3.2}{0.1} \times 0.63$$

$$= \frac{\Delta A \text{ (for } T \text{ min)}}{T} \times 2036$$

where 9.9×10^3 is the molar absorptivity of p-nitroaniline, 3.2 and 0.1 represent reaction volume and sample volume, respectively, and 10^6 is the nmol conversion factor for mol/liter to nmol/ml. The temperature factor of 0.63 is taken from Table 21-5 to convert the result obtained at 32° to 25°C.

TABLE 21-5. Temperature Correction Factors for LAP[a]

Reaction temperature (°C)	Factors to convert to 25°C
25	1.00
26	0.94
27	0.88
28	0.82
29	0.77
30	0.72
31	0.68
32	0.63
33	0.59
34	0.55
35	0.52
36	0.49
37	0.46

[a]Data are from Drewes (unpublished data)

NOTES

1. Temperature of reaction and spectrophotometer employed. The reaction may be run conveniently at any temperature between 25° and 37°C, but that temperature must be maintained constant (±0.2°C) during the entire test. The appropriate temperature correction factor is then applied to the calculation so as to express units at 25°C. If a recorder is not available the readings can be taken every minute, reproducible ΔA/min denoting the linear portion of the reaction. The test can also be performed with a filter with nominal wavelength between 400 and 420 nm. In this case the calculations cannot be based on the molar absorptivity of p-nitroaniline, but the system must be calibrated by running a standard curve of p-nitroaniline at the wavelength of choice.

2. Linearity of reaction rate. After the momentary disturbance of replacing the cuvet in the instrument, the reaction proceeds linearly with time for at least 30 min, unless the enzyme level of the specimen is extremely elevated. The reaction of elevated specimens is also initially linear, but the substrate is more rapidly consumed. A valid result is obtained by measuring the ΔA over a shorter time period in the linear part of the reaction. Alternatively, the specimen may be diluted with saline and rerun.

3. Blanks. New reagents or those that have been stored for some time should be checked for nonenzymatic hydrolysis of the substrate. Substitute 0.1 ml of water for the 0.1 ml of sample. Blanks are normally negligible, but any appreciable blank value should be subtracted from the samples.

4. Stability of samples. In our laboratory LAP has been found to be stable in serum for at least 7 days at 30°C and to be stable frozen. We also found urine specimens to be stable for 1 week at room temperature. Although serum samples were stable for at least 48 hours at room temperature in another study, losses of 5% and 10% were observed after 1 and 2 weeks, respectively, at both 4° and −20°C (719).

5. Hemolysis and icterus. Since the absorption curve of p-nitroaniline overlaps the curve of hemoglobin and bilirubin, endpoint determination of LAP cannot be done in hemolyzed or icteric specimens without error (366). With the kinetic method presented, however, minimal hemolysis and icterus may be tolerated since the error is constant throughout the period taken for ΔA; however, excessive color from the sample may raise the absorbance into a range rapidly exceeding the instrument scale.

6. Preparation of urine samples. An overnight urine specimen should be collected without preservatives. Dialyze 8–9 ml centrifuged urine for 90 min against running deionized water in cellulose casing of 2 cm diameter (e.g., Union Carbide). Continue with step 1 of the *Procedure for Serum*, substituting 0.1 ml clear dialyzed urine for 0.1 ml serum in step 2. Before and after dialysis the tubing is blotted and weighed to establish the dialysis ratio (Q).

$$Q = \frac{\text{weight postdialysis}}{\text{weight predialysis}}$$

The calculated results are then multiplied by Q.

7. Spectrophotometer standardization. Since the absolute absorbance of *p*-nitroaniline forms the basis of the calculation, the absorbance scale of the instrument must be checked. Mass-produced spectrophotometers rarely respond identically, and even if instruments were standardized by the manufacturer, gradual aging would make "absolute" absorbances unreliable. Prepare a 50 μM solution of *p*-nitroaniline in methanol (final concentration, 6.9 mg/liter). The absorbance at 405 nm in 1-cm cuvets, read against methanol, should be 0.495. If it is not, electrical adjustment of the instrument should be made if possible to obtain a reading of 0.495, or else sample absorbance at 405 nm can be corrected by the following equation:

$$\text{Corrected } A_{405\ nm} = \text{observed } A_{405\ nm} \times \frac{0.495}{\text{observed } A_{405\ nm} \text{ for } p\text{-nitroaniline}}$$

ACCURACY AND PRECISION

The accuracy of the method is difficult to assess. Mixing two sera (with high and low activity) in various ratios and comparing the measured activity with the calculated activity gives excellent correlation, indicating zero order kinetics (719).

The precision (95% limits) of the method is about ±10%.

NORMAL VALUES

The normal range of serum is 8–22 mU/ml at 25°C (525). A slightly higher level in males than females has been reported (719). The normal range in overnight urine samples is 0.6–4.7 mU/ml for males and 0.1–3.6 mU/ml for females (366).

PEPSIN, PEPSINOGEN, AND UROPEPSIN

Pepsin (EC 3.4.4.1) is secreted as the inactive zymogen *pepsinogen* by the chief cells of the gastric mucosa. Upon acidification, the proenzyme is broken down into two fragments—the active enzyme and a small polypeptide. Pepsin then acts autocatalytically to hydrolyze the proenzyme to the enzyme. At the pH optimum, 1.5–2.0, pepsin acts as an endopeptidase and preferentially attacks peptide bonds with adjacent aromatic amino acids (tyrosine or phenylalanine) or the bond between an aromatic amino acid and glutamic acid.

Although most of the pepsinogen produced by the gastric mucosa is secreted into the lumen of the stomach, a small amount diffuses into the circulatory system and is excreted in the urine. This form of the proenzyme, called *uropepsinogen*, is converted to *uropepsin* by incubation of the urine at pH 1.5–1.8.

Isoenzymes of pepsin and the proenzymes are known to occur (661, 731). At least two types of pepsinogens and three pepsins have been characterized in gastric fluid (661). In addition to confirming the presence of isoenzymes in serum and gastric fluid, electrophoresis has revealed the presence of four to six isoenzymes in urine (633).

Most pepsin and uropepsin determinations are based on incubation of the enzyme with hemoglobin as described by Anson and Mirsky (24), with measurement of the hydrolyzed phenolic acids by the Folin-Ciocalteu reaction. Mirsky and coworkers (505, 506) applied this method to clinical investigations. Other tests used to measure the activity of the enzyme include the increase in absorption at 280 nm of hemoglobin hydrolysates (139, 365), radial diffusion of the enzyme into acid-hemoglobin agar gel (632), or the fluorometric measurement of liberated tyrosyl groups coupled with 1-nitroso-2-naphthol (769). Casein (conversion to paracasein in homogenized milk) (267, 771), edestin (to edeston) (784), bovine serum albumin (252), and sulfanilamide-casein or nitrocasein (648) have been used as alternative substrates. Autolysis of the serum proteins by conversion of the serum pepsinogen to pepsin at pH 2 and measurement of the liberated tyrosine has been described (748). More recently albumin and hemoglobin labeled with radioactive iodine provided extremely sensitive assays of the proteolytic actions of pepsin and uropepsin (229, 404, 458, 745).

DETERMINATION OF PEPSINOGEN IN SERUM OR PLASMA

(method of Mirsky et al, Ref 506)

PRINCIPLE

Serum is incubated with acidified hemoglobin substrate, and the trichloroacetic acid-soluble tyrosyl groups are determined with phenol reagent of Folin and Ciocalteu. The absorption curve of this reaction is shown in Figure 21-12.

REAGENTS

Hemoglobin Substrate, 2.5%. Dissolve 2.5 g hemoglobin in 50 ml water. Adjust to pH 1.5 with 0.3 N HCl using a pH meter and constant stirring. Dilute to 100 ml. Store at 4°C.

HCl, 0.3 N. Dilute 2.5 ml acid to 100 ml with water.

HCl, 1.5 N. Dilute 12.5 ml acid to 100 ml with water.

Trichloroacetic Acid (TCA), 5%

Trichloroacetic Acid (TCA), 3%

NaOH, 0.5 N. Dissolve 2 g/100 ml water.

Stock Folin-Ciocalteu Phenol Reagent. As in *Determination of Proteins in Cerebrospinal Fluid by the Folin-Lowry Method, Chapter 16.*

Working Folin-Ciocalteu Solution. Dilute one part of the stock solution with two parts water.

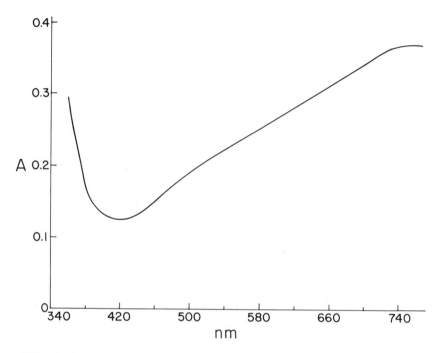

FIG. 21-12. Absorption curve of color produced by Folin-Ciocalteu reagent for the determination of pepsinogen: tyrosine (40 µg/ml) vs water. Cary Model 15 Recording Spectrophotometer.

Tyrosine Standard, 40 µg/ml. Dissolve 10 mg tyrosine in 50 ml of 3% TCA. Dilute 1 ml of solution to 5 ml with 3% TCA for use.

PROCEDURE

1. Add 1.0 ml serum to 5 ml hemoglobin substrate in a graduated test tube.
2. Adjust the pH to 1.5 with 0.3 N HCl using a pH meter (about 0.3 ml is required). Adjust the volume to 7.5 ml with water and mix.
3. Transfer a 3 ml aliquot to a tube containing 5 ml 5% TCA, mix, and allow to stand 10 min. Centrifuge, transfer 1 ml supernate to a tube labeled "control" and store in the refrigerator until needed.
4. Transfer another 3 ml of the pH 1.5 solution to a tube and incubate 24 hours at 37°C. Add 5 ml 5% TCA, mix, and allow to stand 10 min. Centrifuge and transfer 1 ml supernate to a tube labeled "unknown".
5. To 1.0 ml 3% TCA reagent blank, 1.0 ml tyrosine standard, and 1.0 ml supernates from steps 3 and 4, add 1.5 ml 5% TCA, 5 ml 0.5 N NaOH, and 1.5 ml working Folin-Ciocalteu solution. Mix thoroughly.
6. After 5 min read absorbances of reagent blank (A_b), tyrosine standard (A_s), "control" (A_c), and "unknown" (A_x) at 725 nm against water. Read within 30 min because color fades slowly.

Calculation:

$$\text{Units pepsinogen/ml} = \frac{A_x - A_c}{A_s - A_b} \times 40 \times \frac{1}{0.05}$$

$$= \frac{A_x - A_c}{A_s - A_b} \times 800$$

NOTES

1. Beer's law. There is slight deviation even with a narrow half-band width, which does not increase materially even with a broad band filter, signifying some kind of complexation or polymerization of the chromophore. By using a $40\,\mu g/ml$ standard and assuming linearity, a maximal error of about 10% is incurred in the range of 20–60 mg/100 ml. For greater accuracy or for values in an extended range, a calibration curve must be prepared.

2. Sample stability. The enzyme is stable 4 days at room temperature (506).

ACCURACY AND PRECISION

The accuracy of serum pepsinogen levels is difficult to assess because of the variety of substrates used to measure the enzyme. Low serum levels in patients with total gastrectomy or pernicious anemia support the specificity of the assay.

The precision of the method is ±15%.

NORMAL VALUES

The range for normal subjects is 350–750 units (506). Blood levels are fairly constant over 24 hour periods.

DETERMINATION OF UROPEPSINOGEN

(method of West et al, Ref 771)

PRINCIPLE

Following conversion of uropepsinogen to uropepsin by an initial incubation at an acidic pH, urine is incubated with buffered homogenized milk and casein is converted to insoluble paracasein. The time required for visible appearance of aggregates of paracasein particles on the wall of the test tube is taken as the endpoint of the reaction.

REAGENTS

Acetate Buffer, pH 4.9. Mix 42 ml 10% NaOH and 9.2 ml glacial acetic acid; then add water to 100 ml. Check pH with meter and, if necessary, adjust.

HCl, 2N. Dilute conc acid 2:13.

Methyl Orange Indicator, 0.2% Aqueous

Milk-Buffer Mixture. Mix equal parts of acetate buffer with fresh homogenized milk.

PROCEDURE

1. An accurately timed urine specimen is required. Measure and record total volume.
2. To 2.0 ml of urine, add 1 drop methyl orange indicator and 0.1 ml 2 *N* HCl. The pH should be 3 or less as indicated by a change in color to red. Another drop of 2 *N* HCl may be necessary.
3. "Activate" the urine by incubation for 1 hour or longer in a 37°C water bath.
4. In three separate tubes place several milliliters of water, acetate buffer, and milk-buffer mixture. Bring all reagents to 37°C in the water bath.
5. Transfer 0.20 ml of activated urine (of step 3) to a test tube, add 0.80 ml water and 1.0 ml acetate buffer (of step 4), and mix.
6. Start reaction by adding 0.5 ml milk-buffer mixture, mix at once, and start a stopwatch. Immediately place tube in 37°C water bath. Every 10–15 sec shake tube slightly and remove momentarily from water bath to facilitate detection of endpoint. The endpoint is a particulate precipitate observed in the thin film on the inner wall of the test tube above the surface of the mixture. A few seconds later a curdy precipitate appears. The determination should be done in duplicate, and duplicates should check within 10 sec. If the time is less than 80 or more than 400 sec, a suitable alteration in the amount of activated urine used and in the amount of water added should be made to bring the endpoint between these limits. The volume of activated urine plus water must be 1.0 ml.

Calculation:

Results are expressed in arbitrary units of uropepsin excreted per hour. One unit equals the number of milliliters of activated urine x 10 required to give the endpoint in 100 sec. This is equivalent to 0.26 μg of crystalline pepsin (771).

$$\text{Units/hour} = \frac{1}{10} \times \frac{V}{vh} \times \left(\frac{100}{T}\right)^{1.32}$$

where

V = total volume timed urine specimen in ml
h = hours during which V was excreted
v = volume in ml of activated urine used in test
T = seconds elapsed to reach endpoint

This formula can be used in calculation if a log-log slide rule is available. An alternative is to use the following form:

$$\text{Units/hour} = \text{antilog} \left[\log \frac{V}{vh} + 1.64 - 1.32 \log T \right]$$

Still another alternative takes advantage of the fact that v and T are log related with a slope y/x of -1.32. Set up the abscissas and ordinates of a sheet of one cycle, log graph paper as shown in Fig. 21-13. For any value of T obtained, draw a line of slope -1.32 back to the 100 sec abscissa. The value of v at this intercept is the milliliters of urine that will give the endpoint in 100 sec. With this value (v') the calculation then is:

$$\text{Units/hour} = \frac{V}{v' \times h \times 10}$$

For example, let us assume that a T value of 225 sec is obtained. A distance of 10 cm to the left of this point is measured off as in Figure 21-13, then a vertical distance of 13.2 cm. The two points are connected and the intercept of this line with the 100-sec abscissa is v'—in this case 0.29.

NOTE

1. *Stability of samples.* It has been reported that uropepsin is stable at room temperature for several days or more if there is no bacterial growth and that it is stable in the refrigerator (267, 505). The use of toluene as preservative has been recommended (404, 505). One group of workers, however, reported that even at refrigerator temperature with toluene there is a decrease in uropepsin activity of 10–30% in 4 days (267).

ACCURACY AND PRECISION

The chief possible source of inaccuracy appears to be lack of complete uniformity of the substrate used. A study of five different brands of homogenized milk (267) indicated a variation of ±8% resulting from use of the different milks. Six different brands of pasteurized, homogenized milk

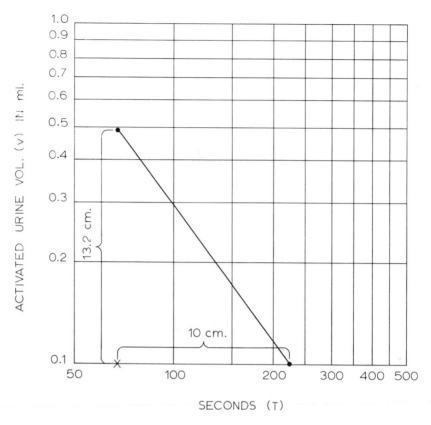

FIG. 21-13. Graphic calculation of uropepsin concentration (see text).

obtained in Los Angeles were compared in our laboratory, and the results were all within ±6% of the average. For most practical purposes, however, this degree of variation does not seem critical.

It is claimed that duplicate determinations should check within 10 sec. In terms of units per hour, this is about ±5—10%.

NORMAL VALUES

The normal output of uropepsin is about 15—50 units/hour (771). Normally, the output of males appears to be greater than that of females by about 60% (753). There is disagreement as to the constancy of output from hour to hour. Thus it has been reported that there is no diurnal variation, that the output is higher in the morning than in the evening, and that the output is higher at night than during the day (753). It may be that the blood level of pepsinogen is more stable.

Many workers who appraised the clinical usefulness of this test expressed the opinion that the wide day to day variation in output in individuals, the large variation between individuals, and the overlap between the output of normals and abnormals limit the usefulness of the test (267, 753).

TRYPSIN

Trypsin (EC 3.4.4.4) hydrolyzes the peptide bonds of proteins at the carboxyl groups of lysine or arginine. The variety of substrates used to measure the proteolytic activity of trypsin are denatured hemoglobin (23), serum (270), casein or modified casein (136, 236, 298, 362), numerous esters or amides of arginine and lysine (373, 526, 608, 610, 723), and hide powder dye complex (602). Methods for assessing the proteolytic action of trypsin include: titration of liberated carboxy groups, determination of phenolic substances formed, formol titration of liberated amino groups, nesslerization of nonprotein nitrogen formed, and viscosimetry (317).

Measurements of this enzyme have been performed on serum, duodenal, and fecal samples. Although the clinical correlations with enzyme levels in feces are useful (18, 46, 302), Babson (43) and others (86, 608) conclude that the trypsin assays are open to question because of the well established presence of trypsin inhibitors (86, 180).

A recently described radioimmunoassay for trypsin, chymotrypsin and chymotrypsinogen (732) may circumvent many of the problems of assays based on enzymatic activity.

Until a specific method can be devised to quantitate trypsin, the simple screening test in duodenal fluid and feces described below may suffice for the detection of deficient secretion of the enzyme.

DETERMINATION OF TRYPSIN IN DUODENAL FLUID AND FECES

(method of Johnstone, Ref 362, modified)

PRINCIPLE

A series of twofold dilutions of duodenal fluid or feces is made with barbital buffer, pH 8.0. Aliquots of these dilutions are placed on the gelatin coat of x-ray film and incubated at 37°C for 15 min. The endpoint is that concentration producing complete clearing of the x-ray film—i.e., complete digestion of the gelatin coat.

REAGENTS

Barbital Buffer, pH 8.0. Prepare 0.1 *M* sodium diethylbarbiturate by dissolving 20.6 g in water and make to 1 liter. Prepare 0.1 *N* HCl by diluting 8.3 ml conc HCl to 1 liter with water. Prepare pH 8.0 barbital buffer by mixing 7 vols 0.1 *M* sodium diethylbarbiturate and 3 vols 0.1 *N* HCl.

X-ray Film. Cut film into rectangles of about 1.5 x 2.5 in. Wet one side with warm water and scrape clean of gelatin coat. Make a series of five circles about 1 cm in diameter on unscraped side with a wax pencil. Label *outside* the circles as follows: C, 5, 10, 20, and 40. The film has a tendency to curl during a test. This can be prevented by stapling or clipping it to a piece of cardboard. Eastman Kodak film has been found to be suitable. If the "control" run in the test shows clearing, a film showing no such digestion must be found.

PROCEDURE

1. *Feces.* Mix entire sample thoroughly with a spatula. Weigh out 1.0 g feces into a 50-ml Erlenmeyer flask. *Duodenal fluid.* Pipet 1.0 ml into a test tube.
2. Add 4.0 ml barbital buffer and mix, making a 1:5 dilution of the sample. For a fecal sample, stir with a glass rod until a fine suspension is formed, then centrifuge and decant into a test tube.
3. Prepare the following twofold dilutions by serially diluting with barbital buffer: 1:10, 1:20, 1:40, and 1:80. Higher dilutions can also be run if an endpoint is desired in every case.
4. Place 1 drop of each dilution in corresponding circle of x-ray film and 1 drop of barbital buffer in circle marked C (control), using a pipet or dropper and beginning with buffer, then the highest dilution, etc. Place x-ray film in a petri dish and cover the dish.
5. Incubate at 37°C for 15 min.
6. Wash off film in a stream of cold tap water for 5 sec and remove from attached cardboard. Examine circles for evidence of digestion of gelatin film. Digestion can be graded 0, 1+, 2+, 3+, and 4+, with 4+ representing complete transparency. The "control" should show no evidence of digestion.

NOTES

1. pH optimum. Enzyme activity can vary up to 30-fold when water instead of buffers is used as the diluent (362). The pH optimum of 8.0 used by Johnstone (362) with gelatin as the substrate is applicable to other substrates as well (302, 602).

2. Temperature. Rate of digestion of the gelatin film varies directly with temperature. Results obtained at 30°C for 30 min and at 37°C for 15 min for both types of samples were found to be essentially the same in our laboratory. Since most laboratories have a 37°C incubator, the latter conditions were chosen.

3. Stability of samples. Enzyme levels in fecal samples are stable for 8 days at room temperature (18, 197), although it is advisable to store at 4°C (18). It is recommended that duodenal fluid be collected in chilled tubes, diluted with an equal volume of glycerine, and frozen (18). Repeated freezing and thawing or homogenization is deleterious to enzyme activity.

ACCURACY

False positives and false negatives can occur with this test, but false reports can be minimized if chymotrypsin is assayed and a 1:1 ratio of trypsin to chymotrypsin is obtained (18, 362). Bacterial gelatinase in feces yields a false-positive result (362).

NORMAL VALUES

4+ at 1:10 or higher dilutions. There is no relationship between age and the levels of fecal trypsin, with the exception of newborn infants where levels are lower (46).

ARGINASE

Arginase (L-arginine ureohydrolase, EC 3.5.3.1), first described in 1904 (417), catalyzes the following reaction in the urea cycle:

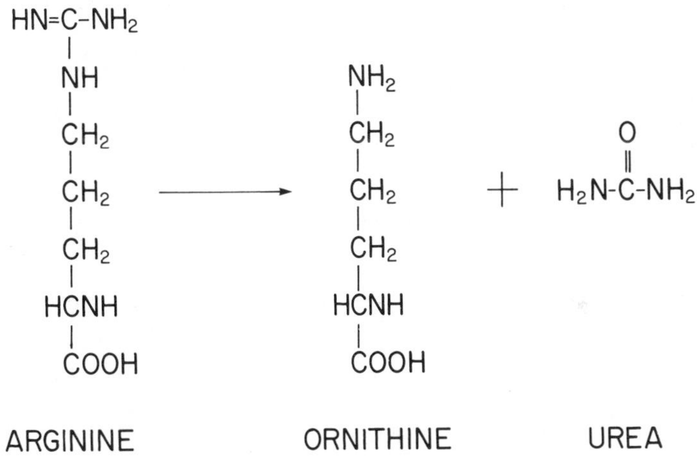

$$
\begin{array}{lll}
\text{HN=C-NH}_2 & & \\
\quad | & & \\
\text{NH} & \text{NH}_2 & \\
\quad | & \quad | & \\
\text{CH}_2 & \text{CH}_2 & \text{O} \\
\quad | & \quad | & \quad \| \\
\text{CH}_2 \longrightarrow & \text{CH}_2 \quad + & \text{H}_2\text{N-C-NH}_2 \\
\quad | & \quad | & \\
\text{CH}_2 & \text{CH}_2 & \\
\quad | & \quad | & \\
\text{HCNH} & \text{HCNH} & \\
\quad | & \quad | & \\
\text{COOH} & \text{COOH} & \\
\end{array}
$$

| ARGININE | ORNITHINE | UREA |

The enzyme is present in many tissues, but mainly in the liver. It is localized in the nuclei, mitochondria, and microsomes of the cells rather than the supernate (456). In normal blood, activity of the enzyme is related to hemoglobin content, thus indicating its presence in erythrocytes (358). The pH optimum lies above 9 (230), and the enzyme is activated by Mn^{2+} (358, 495) and Co^{2+} (349).

Early methods most often measured arginase in a three step procedure: hydrolysis of arginine, incubation with urease to decompose the urea to CO_2 and NH_3, and estimation of NH_3 by reaction with Nessler's reagent to produce an amber solution which was either read at 540 nm (349, 596) or titrated (174, 348). Alternatively, the CO_2 was determined manometrically (350, 755, 767). The ornithine produced by arginase can also be measured by the increase in alkali required to titrate to pH 12, at which point arginine is completely ionized and ornithine is not ionized (451). The arginine remaining after hydrolysis has also been measured photometrically by several modifications of the Sakaguchi reaction (255, 456).

Urea can also be measured directly, and this approach has led to less complex and more sensitive methods. The Archibald (28) reaction for urea involves α-isonitrosopropiophenone in the formation of a red complex, which, while photolabile, has been used for measurement of arginase in serum (161, 749) as has the xanthydrol color reaction for urea (279, 559). Another color procedure for urea employs diacetylmonoxime (240). A modification of this technic utilizes 2 atm pressure in an attempt to increase sensitivity for serum arginase (230). The carbamidodiacetyl reaction for citrulline was modified for measurement of urea by using thiosemicarbazide and diacetylmonoxime to yield augmented specificity and sensitivity (163). This produced a nonphotolabile complex with maximum absorption at 530–535 nm. Mellerup (495) adapted this photometric urea method to the measurement of arginase activity. This method appears promising for routine clinical laboratory use and purportedly gives greater accuracy in the normal range than older methods. Creatine, creatinine, uric acid, ammonia, and 20 amino acids do not interfere. Urea measurement by formation of a yellow color with 4-dimethylaminobenzaldehyde has also been utilized in an arginase method (358).

Because of arginase in red blood cells, hemolysis must be avoided (230). Serum is stable for months at $-20°C$ (495). The normal range appears to be higher in males (495).

GUANASE

Guanase (guanine aminohydrolase, EC 3.5.4.3) catalyzes the reaction:

GUANINE   XANTHINE

The pH optimum of the enzyme is about 7.5, and the range of activity is fairly broad, pH 6.45–8.10. Guanase activity is highest in liver, less in kidney and brain, and considerably less or not detectable in other tissues. Ordinarily, minimal amounts of this enzyme are present in serum and practically none in white or red blood cells (407). Therefore this test is considered to be a sensitive and reasonably specific indicator of hepatocellular damage.

Initially, guanase activity was quantified in tissues by measuring the decrease in absorbance at 245 nm resulting from the conversion of guanine to xanthine (619). This technic was subsequently applied to measure guanase activity in serum (343, 407). Caraway (129) developed a photometric method by employing the phenol-hypochlorite reaction to form a colored complex with the ammonia generated by the enzyme. A highly sensitive method has been described that utilizes radioactive guanine, xanthine oxidase to convert xanthine to uric acid, and an ion exchange column to extract the uric acid for radioactive measurement (7).

DETERMINATION OF SERUM GUANASE

(method of Knight et al, Ref 407, modified)

PRINCIPLE

Serum is incubated with buffered guanine for 1 hour at 37°C. The enzyme reaction is stopped by the addition of perchloric acid. The amount of guanine converted to xanthine is determined by measuring the decrease in absorbance at 245 nm. Figure 21-14 shows the absorption curves of guanine and xanthine.

FIG. 21-14. Absorption curves for guanase determination: (1) guanine (50 µg/ml) vs water; (2) xanthine (50 µg/ml) vs water. Cary Model 15 Recording Spectrophotometer.

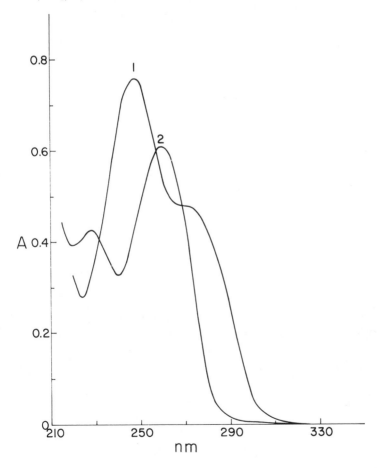

REAGENTS

NaOH, 0.1 N. Dissolve 1 g in 250 ml water.

Borate Buffer, pH 7.5. Dissolve 2.5 g boric acid in 200 ml water and adjust to pH 7.5 with 40% NaOH on a pH meter.

Perchloric Acid, 12%. Dilute 10.4 ml 70% perchloric acid (sp. gr. 1.68) to 100 ml with water.

Stock Guanine Standard, 1 mg/ml. Dissolve 50 mg guanine in 10 ml 0.1 N NaOH with gentle heating. Cool and dilute to 100 ml with water. Stable 1 month at 4°C.

Working Guanine Standard, 50 µg/ml. Dilute stock standard 1:20 with borate buffer. Prepare fresh.

Stock Xanthine Standard, 1 mg/ml. Dissolve 50 mg xanthine in 10 ml 0.1 N NaOH with gentle heating. Cool and dilute to 100 ml with water. Stable 1 month at 4°C.

Working Xanthine Standard, 50 µg/ml. Dilute stock standard 1:20 with borate buffer. Prepare fresh.

PROCEDURE

1. Set up the following solutions in 15 × 125 mm test tubes:

	Borate buffer	Guanine	Xanthine	Water	Serum
Blank	2.0	—	—	0.5	—
Guanine std	—	2.0	—	0.5	—
Xanthine std	—	—	2.0	0.5	—
Serum blank	2.0	—	—	—	0.5
Serum sample	—	2.0	—	—	0.5

2. Incubate all tubes containing serum 1 hour in a 37°C bath.
3. To all tubes add dropwise, with continuous mixing, 7.5 ml perchloric acid. Allow to stand 10 min and then centrifuge.
4. Transfer supernates to silica cuvets.
5. At 245 nm read the absorbance of guanine (A_g) and xanthine (A_{xant}) against the blank, and the serum sample (A_{test}) against the serum blank.

Calculation:

$$\text{Guanase, milliunits } (\mu\text{mol/liter/min}) = \frac{A_g - A_{test}}{A_g - A_{xant}} \times \frac{1}{0.5} \times 1000 \times \frac{1}{60} \times 0.66$$

$$= \frac{A_g - A_{test}}{A_g - A_{xant}} \times 2200$$

NOTES

1. Termination of reaction. The perchloric acid used in our method not only stops the enzyme reaction but also acts to deproteinize the solution and eliminate turbidity. Lowering the pH to 6.0 to stop the reaction as recommended by Knights *et al* (407) was found to be unsuitable because some sera formed turbid solutions.

2. Stability of samples. Serum guanase activity is stable 3 days at room temperature. The enzyme is reportedly stable 2 weeks at $4°C$ (7) and for 10 months at $-15°C$ (407).

ACCURACY AND PRECISION

The only criterion available for the accuracy of guanase activity is the good correlation of serum levels with the clinical state of individuals with hepatocellular damage.

The level of guanase activity in sera from normal subjects is very low, and therefore the precision in this range is poor. For this reason the volume of serum employed in the test was increased to 0.5 ml instead of the 0.2 ml used by Knights *et al* (407). Guanase activity rises tenfold or greater in certain disease states, and the precision at elevated levels is approximately ±10%.

NORMAL RANGE

The normal range for serum guanase is less than 3 mU (129, 407).

ALDOLASE

The enzyme aldolase (fructose-1,6-diphosphate: D-glyceraldehyde-3-phosphate lyase, EC 4.1.2.13), catalyzes the following reaction:

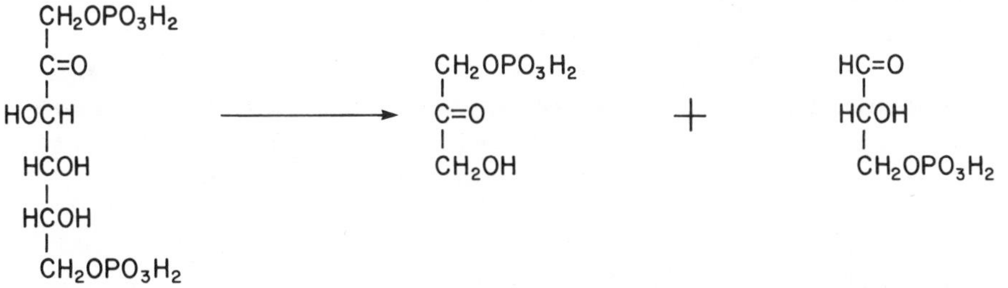

$$\begin{array}{l} CH_2OPO_3H_2 \\ | \\ C=O \\ | \\ HOCH \\ | \\ HCOH \\ | \\ HCOH \\ | \\ CH_2OPO_3H_2 \end{array} \longrightarrow \begin{array}{l} CH_2OPO_3H_2 \\ | \\ C=O \\ | \\ CH_2OH \end{array} \quad + \quad \begin{array}{l} HC=O \\ | \\ HCOH \\ | \\ CH_2OPO_3H_2 \end{array}$$

FRUCTOSE-1,6 -DIPHOSPHATE (FDP)	DIHYDROXYACETONE PHOSPHATE (DAP)	GLYCERALDEHYDE- 3-PHOSPHATE (GAP)

Aldolase is active in the pH range 5–9 with an optimum at about 6.8–7.2 (193). This reaction is catalyzed by at least four serum isoenzyme fractions, although in certain illnesses up to eight isoenzymes occur. Isoelectric focusing of human liver extract has resolved the enzyme activity into five distinct fractions (183). The aldolases have been characterized by their electrophoretic properties and by fructose diphosphate/fructose-1-phosphate activity ratios into three parental types: A, in most tissues; B, liver and kidney; and C, brain and other tissues (442).

A method for the measurement of aldolase activity devised by Warburg and Christian (764) consists of the following reactions involving triose-phosphate isomerase (TPI), glyceraldehyde phosphate dehydrogenase (GADP), and nicotinamide adenine dinucleotide (NAD):

$$FDP \xrightarrow{ALD} GAP + DAP \tag{a}$$

$$GAP \underset{}{\overset{TPI}{\rightleftharpoons}} DAP \tag{b}$$

$$GAP + NAD^+ \xrightarrow{GADP} PGA + NADH + H^+ \tag{c}$$

Reaction conditions are adjusted so that the formation of 3-phosphoglycerate (PGA) and NADH, which is measured at 340 nm, are a direct reflection of aldolase activity. Another enzymatic method utilizes reactions (a) and (b) above but substitutes glycerophosphate dehydrogenase (GPD) and reduced coenzyme to yield glycerol-1-phosphate (G-1-P) in the final reaction (58):

$$DAP + NADH + H^+ \xrightarrow{GPD} G-1-P + NAD^+$$

Aldolase activity is measured by the decrease in absorbance at 340 nm. Two additional studies were made of these two methods for aldolase involving variations in substrate concentrations, enzymes, coenzymes, and buffers (469, 564).

An application of the reaction of 2,4-dinitrophenylhydrazine with the triosephosphates and measurement of the resultant hydrazones at 540 nm for investigating the clinical significance of aldolase was reported by Sibley and Fleisher (676). The pH optimum, temperature optimum, inhibitors, trapping agents, and other factors of this reaction have been modified by other investigators (54, 118, 222, 565). An automated procedure (372) gave good correlation with the photometric method of Bruns (118).

DETERMINATION OF SERUM ALDOLASE

(method of Pinto et al, Ref 564, modified)

PRINCIPLE

Adolase converts fructose-1, 6-diphosphate to glyceraldehyde-3-phosphate and dihydroxyacetone phosphate. The addition of triosephosphate isomerase, glycerophosphate dehydrogenase, and NADH to the incubation mixture converts the dihydroxyacetone phosphate to glycerophosphate. The rate of the aldolase reaction is measured by the decrease in absorbance at 340 nm as a consequence of the conversion of NADH to NAD by the coupled enzyme reaction (Fig. 21-1). The reaction is measured at 32°C, and a temperature correction factor of 1.4 is used to calculate the units of activity at 37°C.

REAGENTS

Tris Buffer, 0.1 M, pH$_{25° C}$ *7.4.*   Dissolve 12.1 g tris (hydroxymethyl) aminomethane in 800 ml water and adjust with HCl to pH 7.4 ± 0.1 on a pH meter. Dilute with water to 1 liter.

Fructose-1, 6-diphosphate (FDP), 0.022 M. Dissolve 1.12 g FDP trisodium pentahydrate salt in 100 ml pH 7.4 buffer. Dispense 2 ml aliquots in vials and store at −15°C. Thaw daily as needed. Solution stable only 1 day at 4°C.

NADH, 0.9 mM. Dissolve 600 mg NADH in 100 ml pH 7.4 buffer and inspect to ensure complete solution of the powder. Dispense 0.5 ml aliquots in vials and store at −15°C. Thaw and use within 1 hour.

Glycerophosphate Dehydrogenase: Triosephosphate Isomerase (GD:TI), 10 mg/ml. Obtained as an ammonium sulfate suspension. Product from Boehringer or Calbiochem is satisfactory.

Buffered Enzyme Solution. Prepare the following solution immediately before use: 2.5 parts Tris buffer, 0.1 part NADH, and 0.05 part GD:TI.

PROCEDURE

1. Add 0.2 ml serum by TC pipet to 2.6 ml buffered enzyme solution in a cuvet, mix, and incubate 10 min at 32°C.
2. Place cuvet in a spectrophotometer with the sample compartment stabilized at 32°C and check whether absorbance at 340 nm is constant over a 2 min period. If absorbance is changing, continue incubation at 32°C an additional 10 min and check for constancy of readings.
3. Add 0.2 ml FDP solution warmed at 32°C, mix, and record absorbance at 340 nm at 1-min intervals over a 5-min period.

Calculation:

$$\text{Adolase units, } 37°C \, (\Delta A/\text{liter/min}) = \Delta A_{32°} \times \frac{1000}{0.2} \times \frac{1}{5} \times 1.4$$

$$= \Delta A_{32°} \times 1400$$

where 1.4 is the factor to convert from 32° to 37°C (see *Note 3*).

NOTES

1. Linearity of reaction. The reaction has been found to be linear during the 5 min period of measurement. If the change in absorbance exceeds 0.6, make a 1:5 dilution of the serum in buffer and repeat the test. Sera with very high levels give no change in absorbance because the NADH is completely consumed. In these cases a 1:10 or greater dilution is required.

2. Hemolysis. Care must be exercised to avoid hemolysis because red blood cells contain 10 times as much enzyme as does serum.

3. Temperature factor. A factor of 1.4 has been determined in our laboratory for the conversion of aldolase units from 32° to 37°C.

4. Stability of enzyme. Our laboratory studies show that the enzyme is stable 5 hours at room temperature, 5 days at 4°C, and 15 days at −15°C. These findings are in agreement with other reports (372, 676).

5. Plasma versus serum. Plasma is the more satisfactory specimen because platelets contain large amounts of the enzyme (171). If necessary, however, serum can be used because most aldolase released from the platelets during clotting is firmly bound in the fibrin clot.

6. Spectrophotometer standardization. See *Note 13* under *Determination of Lactic Dehydrogenase.*

ACCURACY AND PRECISION

Adolase measurements are considered to be fairly reliable based on the good agreement of enzyme activities in sera of normal subjects by photometric and enzyme coupled methods (469, 564).

Precision of this test varies from ±20% in the normal range to ±10% at elevated levels.

NORMAL RANGE

The normal ranges established by nonparametric statistics in our laboratory are 11−22 and 13−31 units for females (N=30) and males (N=31), respectively. Serum levels for children and infants are two and four times greater, respectively, than adult values. Pregnancy and nutritional status have no effect on enzyme levels (676).

ARGININOSUCCINATE LYASE

Argininosuccinate lyase (L-argininosuccinate arginine lyase, EC 4.3.2.1), an integral part of the urea cycle, catalyzes the following reaction:

```
        COOH
         |
HN=C-NH-CH              HN=C-NH₂
   |     |                 |
   NH    CH₂              NH                      COOH
   |     |                 |                       |
   CH₂   COOH             CH₂                      CH
   |                       |              +        ‖
   CH₂                    CH₂                      CH
   |                       |                       |
   CH₂                    CH₂                      COOH
   |                       |
   HCNH₂                  HCNH₂
   |                       |
   COOH                   COOH
```

ARGININOSUCCINATE   ARGININE   FUMARATE

Red blood cells contain large amounts of this enzyme. Plasma levels, which normally are very low, increase markedly in certain metabolic disturbances, especially those that involve the liver (124, 724). A genetic deficiency of this enzyme resulting in excretion of large amounts of argininosuccinic acid has been reported (8, 446). The pH optimum is 7.0 and has a narrow range (124). The method presented is the only one that has been applied to the measurement of argininosuccinate lyase activity in plasma.

DETERMINATION OF SERUM ARGININOSUCCINATE LYASE

(method of Takehara et al, Ref 724a)

PRINCIPLE

Serum is incubated with buffered argininosuccinate for 1 hour at 37°C and the reaction is stopped by the addition of trichloroacetic acid. Arginine in the trichloroacetic acid supernate is reacted with 2,4-dichloro-1-naphthol and the absorbance of the red colored complex measured at 515 nm. The absorption spectrum of the reaction product is shown in Figure 21-15.

REAGENTS

NaOH, 10%
Trichloroacetic Acid (TCA), 20%
NaOCl Solution. Dilute 12 ml of any commercial bleach solution (approximately 5.25% NaOCl) to 100 ml with water.
Phosphate Buffer, 0.2 M, pH 7.0. Dissolve 27 g $NaH_2PO_4 \cdot H_2O$ in 900 ml water and adjust to pH 7.0 with 10% NaOH using a pH meter. Dilute with water to 1 liter.

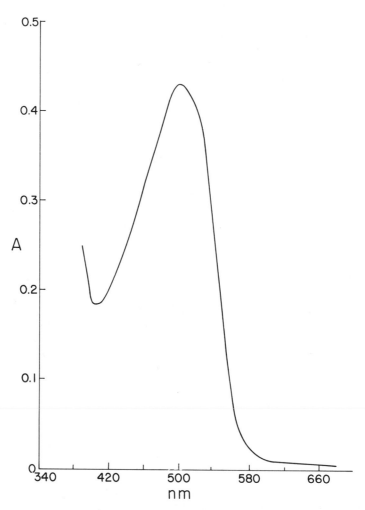

FIG. 21-15. Absorption curve of color formed in the Sakaguchi reaction for the determination of argininosuccinate lyase: arginine (211 µg/ml) vs water. Cary Model 15 Recording Spectrophotometer.

Ethylenediaminetetraacetic acid (EDTA), 0.2 M, pH 7.0. Suspend 5.84 g EDTA in 50 ml water and add 10% NaOH dropwise until the crystals have completely dissolved and the pH is 7.0. Dilute to 100 ml with water.

2,4-Dichloro-1-naphthol, 0.06%. Dissolve 60 mg of recrystallized dichloronaphthol in 100 ml 95% ethanol and store in refrigerator. Dichloronaphthol is recrystallized by dissolving 10 g in 40 ml benzene on a boiling water bath, adding 40 ml heptane, and cooling to 0°C. After standing 1 hour the suspension is filtered through a Büchner funnel, the crystals are allowed to air dry, and the recrystallized material is stored in a reagent bottle.

Thymine Solution. Dissolve 2 g thymine in 100 ml 10% NaOH. Store solution in refrigerator. Thymine from Calbiochem was found to be satisfactory.

Dichloronaphthol-Thymine Solution. Mix equal volumes of dichloro-naphthol and thymine on the day of the test.

Sodium Argininosuccinate, 20 mM. Dissolve 25 mg argininosuccinic acid, barium salt (Calbiochem, grade B was satisfactory), in 2.0 ml water. In another tube dissolve 25 mg anhydrous sodium sulfate in 0.5 ml water. Mix the two solutions thoroughly, allow to stand 30 min, then centrifuge and transfer the clear supernate of sodium argininosuccinate to a clean reagent bottle. Prepare fresh.

Buffered Substrate-EDTA Solution. Mix 2 ml sodium argininosuccinate, 40 ml phosphate buffer, and 20 ml EDTA solution. This solution is stable at 4°C for 3 weeks.

Arginine Standard, 1 mM. Dissolve 21.1 mg L(+)-arginine hydrochloride in 100 ml 0.1 N HCl and store at 4°C.

PROCEDURE

1. Set up the following in test tubes:
    Blank: 1.5 ml phosphate buffer + 0.5 ml water
    Standard: 1.5 ml phosphate buffer + 0.10 ml arginine standard + 0.4 ml water
    Control: 1.5 ml substrate
    Unknown: 1.5 ml substrate + 0.5 ml serum

2. Incubate the control and unknown tubes 1 hour at 37°C.

3. Add 0.5 ml serum to the control tube.

4. Add 1.0 ml TCA to all tubes, mix, and allow to stand 5 min.

5. Centrifuge the control and unknown tubes for 10 min.

6. Transfer 2.0-ml aliquots from each tube to clean test tubes and place in a 5°−10°C cold water bath.

7. To each tube add 0.4 ml 10% NaOH, 1.0 ml dichloronaphthol-thymine solution, and 0.4 ml NaOCl, mixing for 10 sec with the addition of each reagent.

8. Allow tubes to stand 15 min in cold water bath, then at 515 nm read the absorbances of the standard (A_s), control (A_c), and unknown (A_x) against water.

Calculation:

$$\text{Argininosuccinate lyase units} \atop (\mu\text{mol arginine}/100\text{ ml}/\text{hour}) = \frac{A_x - A_c}{A_s} \times 0.1 \times \frac{100}{0.5}$$

$$= \frac{A_x - A_c}{A_s} \times 20$$

If the absorbance of the unknown is greater than 1.0, dilute the sample fourfold with 0.85% NaCl, repeat the test, and make the proper correction in calculations.

NOTES

1. Beer's law. Beer's law is obeyed up to 80 units.

2. Hemolysis. The concentration of argininosuccinate lyase is 25- to 200-fold greater in erythrocytes than serum (724). Samples with slight hemolysis therefore are unacceptable.

3. Temperature of cold water bath. Campanini *et al* (124) performed the color reaction in an ice bath, but in our laboratory this condition resulted in crystallization of the contents of the tubes. With a cold water bath temperature of $5°-10°C$, no crystals were formed.

4. Color stability. The color of the reaction product is stable 30 min in the cold water bath.

5. Stability of samples. The enzyme is stable 6 days at $-15°C$. A 10% loss per day occurs on storage at $4°C$. It has been recommended that the assay be performed within 24 hours after sample collection (724).

ACCURACY AND PRECISION

No reference preparation is available to test the accuracy of the method. Reasonable accuracy can be assumed in view of the consistent levels of the enzyme over a 12-day period in a variety of patients with chronic liver disease, congestive heart failure, and other pathologic conditions (124).

Precision (95% limits) for normal levels is about ±25% but improves somewhat at elevated levels.

NORMAL RANGE

The normal range for adults is 0–4 units. There is no sex difference in enzyme levels. Liver damage is indicated if levels increase three- to four-fold. In parenchymal liver disease the serum argininosuccinate lyase levels usually exceed the upper limit of normal by three- to 100-fold (124).

TRIOSEPHOSPHATE ISOMERASE

Triosephosphate isomerase (TPI, D-glyceraldehyde-3-phosphate ketol-isomerase, EC 5.3.1.1) catalyzes the following reaction:

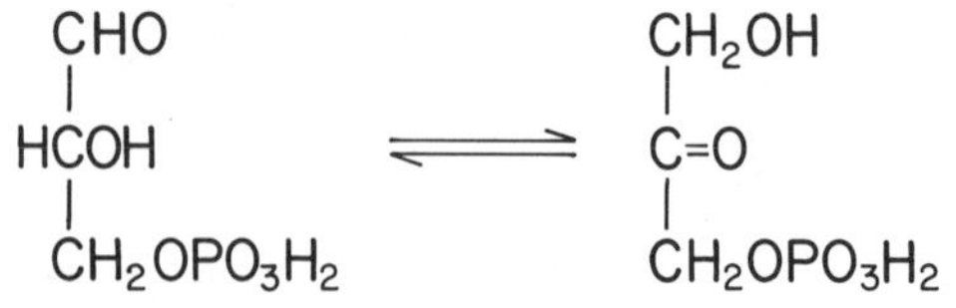

$$
\begin{array}{ccc}
\text{CHO} & & \text{CH}_2\text{OH} \\
| & & | \\
\text{HCOH} & \rightleftharpoons & \text{C=O} \\
| & & | \\
\text{CH}_2\text{OPO}_3\text{H}_2 & & \text{CH}_2\text{OPO}_3\text{H}_2
\end{array}
$$

D-GLYCERALDEHYDE-   DIHYDROXYACETONE
3-PHOSPHATE   PHOSPHATE

Equilibrium strongly favors formation of dihydroxyacetone phosphate (57). TPI deficiency in red blood cells is a rare inherited disorder associated with hemolytic anemia and marked neuromuscular dysfunction.

The enzyme has a molecular weight of 56 000 (622). Isoelectric focusing (622) or electrophoresis (380) reveals three bands. Chromatography on DEAE Sephadex reveals two peaks which may represent a monomer-polymer relationship (651). Ammonium sulfate inhibits the enzyme—e.g., 0.5 mM ammonium sulfate can cause 10% inhibition (77). Phosphate ions reportedly inhibit the calf muscle TPI; 0.5 M phosphate causes about a 25% reduction in activity (57). The pH optimum for calf muscle TPI is in the range 7–8 (57). pH of buffers that have been used for TPI assays include 7.5 (651), 8.0 (77, 379), and 7.9 (57).

Spectrophotometric assays are modifications of the procedure used by Beisenherz (57). The assay is based upon coupling the TPI reaction with the α-glycerophosphate dehydrogenase (α-GPD) reaction:

$$\text{Glyceraldehyde-3-phosphate} \overset{\text{TPI}}{\rightleftharpoons} \text{dihydroxyacetone phosphate}$$

$$\text{dihydroxyacetone phosphate} + \text{NADH} + \text{H}^+ \overset{\text{α-GPD}}{\longrightarrow} \text{α-Glycerophosphate} + \text{NAD}^+$$

The conversion of NADH to NAD results in decreased absorbance at 340 nm. The assay principle has been applied to TPI from red blood cells (77, 651). Excessive amounts of ammonium sulfate in commercial preparations of α-GPD can be avoided by using high activity preparations of the enzyme so that only minimal amounts are required (77). Since the natural isomer and substrate is D-glyceraldehyde-3-phosphate, the use of commercial DL-glyceraldehyde-3-phosphate results in an initial rapid reaction which should be measured in the spectrophotometric procedure while the subsequent slower reaction with the L isomer is ignored (77).

The principle of the screening procedure (379) is identical to that of the spectrophotometric procedure. Oxidation of NADH to NAD results in a loss of fluorescence. After reaction for 1 hour at 45°C, part of the reaction mixture is spotted on filter paper. The disappearance of fluorescence indicates normal levels of triosephosphate isomerase.

DETERMINATION OF TRIOSEPHOSPHATE ISOMERASE, SCREENING PROCEDURE

(method of Kaplan et al, Ref 379, modified by Lowe and Gin, Ref 465)

PRINCIPLE

Triosephosphate isomerase (TPI) in the hemolysate catalyzes the conversion of glyceraldehyde-3-phosphate to dihydroxyacetone phosphate.

α-Glycerophosphate dehydrogenase in the reaction mixture catalyzes the reduction of dihydroxyacetone phosphate to α-glycerophosphate. Concomitant oxidation of NADH to NAD is indicated by a loss of fluorescence.

REAGENTS

NADH, 5 mg/ml. Dissolve 0.050 g of the disodium salt in water and dilute to 10 ml. Dispense 0.5 ml aliquots into individual plastic vials and store frozen.

Triethanolamine-EDTA Solution. Dissolve 2.902 g triethanolamine and 0.650 g EDTA in approximately 100 ml of water, adjust pH to 8 with 4 N NaOH (approximately 0.7 ml). Dilute to 250 ml. Dispense 2 ml aliquots into individual vials and store frozen.

α-Glycerophosphate Dehydrogenase (GPD), 8 IU/ml. Dilute with water to give an enzyme concentration of 8 IU/ml. Dispense 0.1 ml aliquots into individual vials and store frozen.

Glyceraldehyde-3-phosphate (G3P), 6 mg/ml. Place 40 mg glyceraldehyde-3-phosphate diethyl acetal (barium salt) and 500 mg Dowex (50 W-X12 resin, ionic form H^+, 200–400 mesh) in a test tube. Add 3.0 ml of water and place in a boiling water bath for 3 min. Keep the resin in suspension by shaking the tube throughout the 3 min. Filter by suction. *N.B. The hydrolysate formed by this procedure is stable for 4 hours at room temperature and should be prepared fresh before each preparation of reaction mixture.*

Reaction Mixture. The volume here is sufficient to run the blank, control, and seven samples. Thaw reagents and mix in the following order: triethanolamine-EDTA solution 1.6 ml, GPD 0.006 ml, G3P (fresh hydrolysate) 0.18 ml, NADH 0.2 ml. *N.B. The reaction mixture is stable at room temperature for 30 min only. After preparation, proceed immediately with test.*

Deficient Control. If a known or pretested deficient blood sample is not available, place a normal blood sample collected in Alsever's solution in a 56°C water bath for 6 hours. Can be used for 1 month stored at 4°C.

PROCEDURE

1. For each unknown, the deficient control, and the reaction mixture (to be used as reagent blank), draw three squares (2 x 2 cm) on Whatman 1 filter paper for spotting assay mixture at 0, 30, and 60 min.

2. For the deficient control and the unknown, dispense 0.2 ml reaction mixture into 10 x 75 mm test tubes.

3. While reagents are being thawed, prepare hemolysates by adding 5 μl unknown blood sample to 1.0 ml water; mix by shaking. Similarly prepare a hemolysate of the deficient control. (If dried Alsever's mixture was used as an anticoagulant then add 5 μl blood to 2.0 ml water.)

4. Add approximately 20 µl of dilute hemolysates from step 3 to reaction mixture in appropriate tubes using micro pipet. Mix by shaking.

5. Using the same pipets as in step 4, immediately spot 10 µl of unknown and deficient control on the appropriate filter paper squares marked 0 min. Also spot 10 µl reaction mixture as a reagent control.

6. Incubate samples at 45°C for 30 min and again spot 10 µl from each tube on designated squares. Return the tube to the heating block for continued incubation.

7. Dry the spots under a hot air blower.

8. Examine under long wave UV light (Chromato-Vue, model CC-20, from Ultra-Violet Products, or an equivalent source of UV light). Continue incubation of all fluorescing samples for another 30 min at 45°C (*Note 6*).

9. After 60 min of incubation, spot 10 µl on designated squares.

10. Dry spots under a hot air blower and again examine under long wave UV light.

INTERPRETATION OF RESULTS

a. All "0 min spots" should fluoresce from the presence of NADH.

b. At 30 min, no fluorescence of specimen spot indicates normal enzyme activity. Analysis is completed for such samples (*Note 6*).

c. Fluorescence at 30 min does not necessarily indicate enzyme deficiency. If fluorescence disappears by 60 min, the sample is normal.

d. Persistence of fluorescence at 60 min indicates enzyme deficiency.

e. For the test to be valid, fluorescence must persist at 60 min in both the reagent control and the deficient cell control.

NOTES

1. Stability of specimens. Data from our laboratory indicate that Alsever's solution is the anticoagulant of choice for the screening test. TPI was stable for 3.5 weeks at 30°C when 1 part Alsever's solution was mixed with 1 part blood. (See *Note 1*, G6PD section, for the composition of Alsever's solution). According to Beutler (57), the enzyme shows less than a 10% loss of activity at 25°C for 2 days or at 4°C for 6 days with ACD, EDTA, or heparin.

2. Stability of reaction mixture. The reaction mixture is stable at room temperature for only 30 minutes. Once prepared, the assay should be carried out immediately. Although the procedure used in our laboratory calls for preparation of the reaction mixture immediately before use, Kaplan *et al* (379) report that the frozen reaction mixture is stable for at least 10 days.

3. Sensitivity of assay. The assay readily detects homozygotes. Its suitability for heterozygote detection has not been evaluated extensively, but based on the limited data it seems probable that heterozygotes could be presumptively identified by this technic (379). Heterozygotes are clinically unaffected (651).

4. Inhibition with ammonium sulfate. Ammonium sulfate can inhibit TPI (57). Thus if low-activity a-glycerophosphate dehydrogenase is used and the a-glycerophosphate dehydrogenase is preserved in ammonium sulfate, inhibition can occur. Ammonium sulfate (0.5 mM) can cause 10% inhibition.

5. Glyceraldehyde-3-phosphate. Glyceraldehyde-3-phosphate is prepared fresh for each day's use by acid hydrolysis of the stable glyceraldehyde-3-phosphate, diethyl acetal (barium salt). Dowex 50 serves as both the source of the acid for hydrolysis and as a convenient procedure for removal of the barium ion.

6. Incubation time. In our experience fluorescence disappears by 30 min in at least 95% of all normal samples.

REFERENCES

1. ABUL-FADL MAM, KING EJ: *Biochem J* 45:51, 1949
2. ADELSTEIN SJ, COOMBS TL, VALLEE BL: *N Engl J Med* 255:105, 1956
3. AGOSTONI A, TITOBELLO A: *Boll Soc Ital Biol Sper* 40:2073, 1964
4. AHMED Z, REIS JL: *Biochem J* 69:386, 1958
5. AISEN P, SCHORR JB, MORELL AG, GOLD RZ, SCHEINBERG IH: *Am J Med* 28:550, 1960
6. AKERFELDT S: *Science* 125:117, 1957
7. AL-KHALIDI UAS, AFTIMOS S, MUSHARRAFIEH S, KHURI NN: *Clin Chim Acta* 29:381, 1970
8. ALLAN JD, CUSWORTH DC, DENT CE, WILSON VK: *Lancet* 1:182, 1958
9. AMADOR E, DORFMAN LE, WACKER WEC: *Clin Chem* 9:391, 1963
10. AMADOR E, MASSOD MF, FRANEY RJ: *Am J Clin Pathol* 47:3, 1967
11. AMADOR E, PRICE JW, MARSHALL G: *Am J Clin Pathol* 51:202, 1969
12. AMADOR E, REINSTEIN H, BENOTTI N: *Am J Clin Pathol* 44:62, 1965
13. AMADOR E, SALVATORE AC: *Am J Clin Pathol* 55:686, 1971
14. AMADOR E, URBAN J: *Am J Clin Pathol* 57:167, 1972
15. AMADOR E, WACKER WEC: *Clin Chem* 8:343, 1962
16. AMADOR E, ZIMMERMAN TS, WACKER WEC: *JAMA* 185:953, 1963
17. AMELUNG VD, HOFFMANN L, OTTO L: *Dtsch Med Wochenschr* 91:851, 1966
18. AMMANN RW, TAGWERCHER E, KASHIWAGI H, ROSEMUND H: *Am J Dig Dis* 13:123, 1968
19. ANDERSCH MA, SZCZYPINSKI AJ: *Am J Clin Pathol* 17:571, 1947
20. ANDERSON EP, KALCKAR HM, KURAHASHI K, ISSELBACHER KJ: *J Lab Clin Med* 50:469, 1957
21. ANGELETTI P, MOORE B, SUNTZEFF V, GAYLE R: *Proc Soc Exp Biol Med* 108:53, 1961
22. ANNINO JS: *Am J Clin Pathol* 46:397, 1966
23. ANSON ML, *J Gen Physiol* 22:79, 1938
24. ANSON ML, MIRSKY AE: *J Gen Physiol* 16:59, 1932
25. ANSTALL HB, TRUJILLO JM: *Am J Clin Pathol* 47:296, 1967
26. APPELYARD J: *Biochem J* 42:596, 1948
27. APRISON MH, HANSON KM: *Proc Soc Exp Biol Med* 100:643, 1959

28. ARCHIBALD RM: *J Biol Chem 157*:507, 1945
29. ASHBY TM, SUGGS JE, JUE DL: *Clin Chem 16*:503, 1970
30. AUGUSTINSSON KB: *Methods Biochem Anal 5*:1, 1957
31. AUGUSTINSSON KB: *Methods Biochem Anal Vol (suppl)*:217, 1971
32. BABSON AL: *Clin Chem 11*:789, 1965
33. BABSON AL, ARNDT EG: *Clin Chem 16*:254, 1970
34. BABSON AL, ARNDT EG, SHARKEY LJ: *Clin Chim Acta 26*:419, 1969
35. BABSON AL, GREELEY SJ, COLEMAN CM, PHILLIPS GE: *Clin Chem 12*:482, 1966
36. BABSON AL, PHILLIPS GE: *Clin Chem 11*:533, 1965
37. BABSON AL, PHILLIPS GE: *Clin Chim Acta 13*:264, 1966
38. BABSON AL, READ PA: *Am J Clin Pathol 32*:88, 1959
39. BABSON AL, READ PA, PHILLIPS GE: *Am J Clin Pathol 32*:83, 1959
40. BABSON AL, SHAPIRO PO: *Clin Chim Acta 8*:327, 1963
41. BABSON AL, SHAPIRO PO, WILLIAMS PAR, PHILLIPS GE: *Clin Chim Acta 7*:199, 1962
42. BABSON AL, TENNEY SA, MEGRAW RE: *Clin Chem 16*:39, 1970
43. BABSON AL, WILLIAMS PAR, PHILLIPS GE: *Clin Chem 8*:62, 1962
44. BALINSKY D: *Biochim Biophys Acta 122*:537, 1966
45. BAMFORD KF, HARRIS H, LUFFMAN JE, ROBSON EB, CLEGHORN TE: *Lancet 1*:530, 1965
46. BARBERO GJ, SIBINGA MS, MARINO JM, SEIBEL R: *Am J Dis Child 112*:536, 1966
47. BARK CJ: *Am J Clin Pathol 52*:466, 1969
48. BARON DN: *Proc Assoc Clin Biochem 3*:215, 1965
49. BARTALOS M, GYÖRKEY F: *Clin Chim Acta 9*:273, 1964
50. BARTELS H: *Schweiz Med Wochenschr 96*:1304, 1966
51. BATSAKIS JG, BRIERE RO: *Clin Biochem 2*:171, 1969
52. BATSAKIS JG, PRESTON JA, BRIERE RO, GIESEN PC: *Clin Biochem 2*:125, 1968
53. BAUM G, WARD FB: *Anal Biochem 42*:487, 1971
54. BECK WS: *J Biol Chem 212*:847, 1955
55. BECKMANN J, LEYBOLD K, WEISBECKER L: *Z Klin Chem Klin Biochem 7*:18, 1969
56. BEHRENDT H: *Proc Soc Exp Biol Med 54*:268, 1943
57. BEISENHERZ G: *Methods in Enzymology.* Edited by SP Colowick, NO Kaplan. New York, Academic Press, 1955, Vol 1, p 387
58. BEISENHERZ G, BOLTZE HJ, BUCHER T, CZOK R, GARBADE KH, MEYER-ARENDT E, PFLEIDERER G: *Z Naturforsch 8b*:555, 1953
59. BÉLANGER A, PERREAULT J, COUTURE Y, DUNNIGAN J: *Can J Physiol Pharmacol 48*:758, 1970
60. BELFANTI S, CONTARDI A, ERCOLI A: *Biochem J 29*:1491, 1935
61. BELFIELD A, ELLIS G, GOLDBERG DM: *Clin Chem 16*:396, 1970
62. BELFIELD A, GOLDBERG DM: *Clin Chem 15*:931, 1969
63. BELFIELD A, GOLDBERG DM: *Clin Biochem 3*:105, 1970
64. BELFIELD A, GOLDBERG DM: *Z Klin Chem Klin Biochem 9*:197, 1971
65. BELL JL, BARON DN: *Clin Chim Acta 5*:740, 1960
66. BELL JL, BARON DN: *Enzymol Biol Clin 9*:393, 1968
67. BENSON PA, BENEDICT WH: *Am J Clin Pathol 45*:760, 1966
68. BERGER L, RUDOLPH GG: *Standard Methods in Clinical Chemistry.* Edited by S Meites, New York, Academic Press, 1965, Vol 5, p 211
69. BERGMEYER HU, BERNT E: *Methods of Enzymatic Analysis.* Edited by HU Bergmeyer. New York, Academic Press, 1963, p 846

70. BERGMEYER HU, KLOTZSCH H, MÖLLERING H, NELBÖCK-HOCHSTETTER M, BEAUCAMP K: *Methods of Enzymatic Analysis.* Edited by HU Bergmeyer. New York, Academic Press, 1963, p 997

71. BESSEY O, LOWRY OH, BROCK MJ: *J Biol Chem 164*:321, 1946

72. BEUTLER E: *J Lab Clin Med 49*:84, 1957

73. BEUTLER E: *Blood 28*:553, 1966

74. BEUTLER E: *Am J Clin Pathol 47*:303, 1967

75. BEUTLER E: *J Clin Invest 48*:1957, 1969

76. BEUTLER E: *Science 165*:613, 1969

77. BEUTLER E: *Red Cell Metabolism: A Manual of Biochemical Methods.* New York, Grune & Stratton, 1971

78. BEUTLER E, BALUDA MC: *Clin Chim Acta 13*:369, 1966

79. BEUTLER E, BALUDA MC: *J Lab Clin Med 68*:137, 1966

80. BEUTLER E, BALUDA MC, DONNELL GN: *J Lab Clin Med 64*:694, 1964

81. BEUTLER E, IRWIN HR, BLUMENFELD CM, GOLDENBURG EW, DAY RW: *JAMA 199*:501, 1967

82. BEUTLER E, MITCHELL M: *Blood 32*:816, 1968

83. BEUTLER E, MITCHELL M: *J Lab Clin Med 72*:527, 1968

84. BEUTLER E, YEH MKY: *Blood 21*:573, 1963

85. BEYER WF, LOCHER A: *Am J Med Technol 32*:363, 1966

86. BIETH J, METAIS P, WARTER J: *Clin Chim Acta 20*:69, 1968

87. BIGGS HG, CAREY S, MORRISON DB: *Am J Clin Pathol 30*:181, 1958

88. BIRKBECK JA, STEWART AG: *Can J Biochem Physiol 39*:257, 1961

89. BISERTE G, FARRIEUX JP, HOSTE A, GUINAMARD MP, FONTAINE G: *Ann Biol Clin (Paris) 24*:663, 1966

90. BISHOP C: *J Lab Clin Med 68*:149, 1966

91. BLOCK WD, CARMICHAEL R, JACKSON CE: *Proc Soc Exp Biol Med 115*:941, 1964

92. BLUMBERG WE, EISINGER J, AISEN P, MORELL AG, SCHEINBERG IH: *J Biol Chem 238*:1675, 1963

93. BLUME KG, RÜDIGER HW, LÖHR GW: *Biochim Biophys Acta 151*:686, 1968

94. BOASSON EH: *J Immunol 34*:281, 1938

95. BODANSKY O, KRUGMAN S, WARD R, SCHWARTZ MK, GILES JP, JACOBS AM: *Am J Dis Child 98*:166, 1959

96. BOSSCHE H VAN DEN: *Clin Chim Acta 12*:601, 1965

97. BOUTIN D, BRODEUR J: *Clin Biochem 2*:187, 1969

98. BOWERS GN Jr: *Clin Chem 5*:509, 1959

99. BOWERS GN Jr: *Clin Chem 7*:579, 1961

100. BOWERS GN Jr, KELLEY M, McCOMB RB: *Clin Chem 13*:608, 1967

101. BOWERS GN Jr, McCOMB RB: *Clin Chem 12*:70, 1966

102. BOWERS GN Jr, McCOMB RB: *Standard Methods of Clinical Chemistry.* Edited by RP MacDonald. New York, Academic Press, 1970, Vol 6, p 31

103. BOYD JW: *Biochem J 80*:18p, 1961

104. BOYD JW: *Biochem J 81*:434, 1961

105. BOYD JW: *Biochem J 81*:39p, 1961

106. BOYD JW: *Clin Chim Acta 7*:424, 1962

107. BOYDE TRC: *Z Klin Chem Klin Biochem 6*:431, 1968

108. BOYDE TRC, LATNER AL: *Biochem J 82*:51p, 1962

109. BRAY J, KING EJ: *J Pathol 55*:315, 1943

110. BRENNER BM, GILBERT VE: *Am J Med Sci 245*:31, 1963

111. BRIGGS RS, PERILLIE PE, FINCH SC: *J Histochem Cytochem 14*:167, 1966

112. BROHULT J: *Acta Med Scand 185*:357, 1969

113. BROMAN L: *Nature (Lond) 182*:1655, 1958
114. BROOKS L, OLKEN HG: *Clin Chem 11*:748, 1965
115. BROWN RW, GRISOLIA S: *J Lab Clin Med 54*:617, 1959
116. BRUNETTI P, NENCI G: *Enzymol Biol Clin 4*:51, 1964
117. BRUNETTI P, PUXEDDU A, NENCI G: *Haematologica (Pavia) 47*:505, 1962
118. BRUNS F: *Biochem Z 325*:156, 1954
119. BÜCHER T, PFLEIDERER A: *Methods in Enzymology.* Edited by SP Colowick, NO Kaplan. New York, Academic Press, 1955, Vol 1, p 435
120. BUNCH LD, EMERSON RL: *Clin Chem 2*:75, 1956
121. BURNEL D, GENETET F, NABET P, PAYSANT P: *Soc Biol (Paris) 163*:1651, 1969
122. BUSH D, HOFFBAUER RW, BLUME KG, LOEHR GW: *Red Cell Structure and Metabolism.* Edited by B Ramot. New York, Academic Press, 1971, p 193
123. CABAUD P, LEEPER R, WRÓBLEWSKI F: *Am J Clin Pathol 26*:1101, 1956
124. CAMPANINI RZ, TAPIA RA, SARNAT W, NATELSON S: *Clin Chem 16*:44, 1970
125. CAMPBELL DM: *Biochem J 82*:34p, 1962
126. CAMPBELL DM, MOSS DW: *Clin Chim Acta 6*:307, 1961
127. CAPPS RD, BATSAKIS JG, BRIERE RO, CALAM RRL: *Clin Chem 12*:406, 1966
128. CARAWAY WT: *Am J Clin Pathol 26*:945, 1956
129. CARAWAY WT: *Clin Chem 12*:187, 1966
130. CARMELI C, LIFSHITZ Y: *Anal Biochem 38*:309, 1970
131. CERIOTTI G, GAZZANIGA A: *Clin Chim Acta 14*:57, 1966
132. CERIOTTI G, GAZZANIGA A: *Clin Chim Acta 16*:436, 1967
133. CESKA M: *Clin Chim Acta 33*:135, 1971
134. CESKA M, BIRATH K, BROWN B: *Clin Chim Acta 26*:437, 1969
135. CHACKO CM, CHRISTIAN JC, NADLER HL: *J Pediat 78*:454, 1971
136. CHARNEY J, TOMARELLI RM: *J Biol Chem 171*:501, 1947
137. CHERRY IS, CRANDALL LA Jr: *Am J Physiol 100*:266, 1932
138. CHILSON OP, COSTELLO LA, KAPLAN NO: *Fed Proc 24*:S-55, 1965
139. CHINN A: *Gastroenterology 25*:14, 1953
140. CHINO H, GILBERT LI: *Anal Biochem 10*:395, 1965
141. CHIPMAN DM, SHARON N: *Science 165*:454, 1969
142. CHOW CM, ISLAM MF: *Clin Biochem 3*:295, 1970
143. CILLEY J, DALAL FR, WINSTEN S: *Clin Chem 17*:61, 1971
144. CIMASONI G: *Biochem J 99*:133, 1966
145. COHEN L, BLOCK J, DJORDJEVICH J: *Proc Soc Exp Biol Med 126*:55, 1967
146. COHEN L, DJORDJEVICH J, JACOBSEN S: *Med Clin North Am 50*:193, 1966
147. COHEN P, ROSEMEYER MA: *Eur J Biochem 8*:8, 1969
148. COLEMAN CM: *Clin Chim Acta 13*:401, 1966
149. COLENBRANDER HJ, GEUSKENS LM, VINK CLJ: *Clin Chim Acta 28*:329, 1970
150. COLLIER HB, ASHFORD DR, BELL RE: *Can Med Assoc J 95*:1188, 1966
151. COLMAN RF: *J Biol Chem 243*:2454, 1968
152. COLMAN RF: *Biochemistry 8*:888, 1969
153. COLMAN RF, SZETO RC, COHEN P: *Biochemistry 9*:4945, 1970
154. COLOMBO JP, RICHTERICH R: *Schweiz Med Wochenschr 94*:715, 1964
155. COLOMBO JP, RICHTERICH R, ROSSI E: *Klin Wochenschr 40*:37, 1962
156. COMFORT D, CAMPBELL DJ: *Clin Chim Acta 14*:263, 1966
157. COMFORT MW, OSTERBERG AE: *Gastroenterology 4*:85, 1945
158. CONCONI F, PEDRELLI M: *Acta Vitaminol Enzymol (Milano) 16*:205, 1962
159. CONN RB, ANIDO V: *Am J Clin Pathol 46*:177, 1966
160. COPENHAVER JH, BAUSCH LC, FITZGIBBONS JF: *Anal Biochem 30*:327, 1969

161. CORNELIUS CE, FREEDLAND R: *Cornell Vet 52*:344, 1962
162. CORNISH CJ, NEALE FC, POSEN S: *Am J Clin Pathol 53*:68, 1970
163. COULOMBE JJ, FAVREAU L: *Clin Chem 9*:102, 1963
164. COX DW: *J Lab Clin Med 68*:893, 1966
165. COYLE M, ARVAN DA: *Am J Clin Pathol 51*:777, 1969
166. CRITZ JB, MERRICK AW: *Proc Soc Exp Biol Med 109*:608, 1962
167. CURZON G: *Nature (Lond) 181*:115, 1958
168. CURZON G: *Biochem J 77*:66, 1960
169. CURZON G: *Biochem J 100*:295, 1966
170. CURZON G, SPEYER BE: *Biochem J 105*:243, 1967
171. DALE RA: *Clin Chim Acta 5*:652, 1960
172. DANIEL O, VAN ZYL JJW: *Lancet 1*:998, 1952
173. DANIELS JC, FUKUSHIMA M, LARSON DL, ABSTON S, RITZMANN SE: *Tex Rep Biol Med 29*:1, 1971
174. DARGEL R: *Z Klin Chem 4*:36, 1966
175. DAVIES DF: *Cryobiology 5*:87, 1968
176. DELLWEG H, BAERWALD G, GILDE W: *Monatsschr Brau 23*:152, 1970
177. DELORY GE, KING EJ: *Biochem J 39*:245, 1945
178. DEMETRIOU JA, BEATTIE JM: *Clin Chem 17*:290, 1971
179. DIETZ AA, HODGES LK, FOXWORTHY DT: *Clin Chem 13*:359, 1967
180. DIETZ AA, HODGES LK, RUBINSTEIN HM, BRINEY RR: *Clin Chem 13*:242, 1967
181. DIETZ AA, LUBRANO T, HODGES LK, RUBINSTEIN HM: *Clin Biochem 2*:431, 1969
182. DIGIORGIO J: *Clin Chem 17*:326, 1971
183. DIKOW AL: *Z Klin Chem Klin Biochem 6*:386, 1968
184. DIRSTINE PH, SOBEL C, HENRY RJ: *Clin Chem 14*:1097, 1968
185. DIXON TF, PURDOM M: *J Clin Pathol 7*:341, 1954
186. DOE RP, MELLINGER GT: *Metabolism 13*:445, 1964
187. DOE RP, MELLINGER GT, SEAL US: *Clin Chem 11*:943, 1965
188. DOLIN MI, JACOBSON KB: *J Biol Chem 239*:3007, 1964
189. DONAYRE J, PINCUS G: *J Clin Endocrinol Metab 25*:432, 1965
190. DONNELL GN, BERGREN WR, BRETTHAUER RK, HANSEN RG: *J Dis Child 98*:581, 1959
191. DORFMAN LE, AMADOR E, WACKER WEC: *JAMA 184*:1, 1963
192. DOUMAS B, BIGGS HG: *Clin Chim Acta 23*:75, 1969
193. DOUNCE AL, BARNETT SD, BEYER GT: *J Biol Chem 185*:769, 1950
194. DROR Y, SASSOON HF, WATSON JJ, JOHNSON BC: *Clin Chim Acta 28*:291, 1970
195. DUBACH UC: *Schweiz Med Wochenschr 87*:185, 1957
196. DUBE V, HUNTER DT, KNIGHT JA: *Am J Clin Pathol 50*:491, 1968
197. DYCK N, AMMANN R: *Am J Dig Dis 10*:530, 1965
198. EBASHI S, TOYOKURA Y, MOMOI H, SUGITA H: *J Biochem (Tokyo) 46*:103, 1959
199. EBELING J, MEIJERS CAM, SPIJKERS JBF: *Clin Chim Acta 27*:214, 1970
200. EBERHAGEN D, ISSMER A, BRANDMAIER B: *Z Klin Chem Klin Biochem 8*:284, 1970
201. ELLIOTT BA, JEPSON EM, WILKINSON JH: *Clin Sci 23*:305, 1962
202. ELLIOTT BA, WILKINSON JH: *Clin Sci 24*:343, 1963
203. ELLIS G, BELFIELD A, GOLDBERG DM: *Seventh International Congress Clinical Chemistry, Clinical Enzymology.* Edited by J Frei, M Jemelin. Basel, Karger, 1970, Vol 2, p 95

204. ELLIS G, BELFIELD A, GOLDBERG DM: *J Clin Pathol* 24:493, 1971
205. ELLIS G, GOLDBERG DM: *Am J Clin Pathol* 56:627, 1971
206. ELLIS G, GOLDBERG DM: *Clin Biochem* 4:175, 1971
207. ELLMAN GL, COURTNEY KD, ANDRES V Jr, FEATHERSTONE RM: *Biochem Pharmacol* 7:88, 1961
208. EMANUEL B, WEST M, ZIMMERMAN HJ: *Am J Dis Child* 105:261, 1963
209. EMMRICH R, ZIMMERMANN S: *Klin Wochenschr* 37:935, 1959
210. ENNOR AH, ROSENBERG H: *Biochem J* 57:203, 1954
211. Enzyme Nomenclature—Recommendation (1964) of the International Union of Biochemistry. New York, Elsevier, 1965
212. ESTBORN R: *Clin Chim Acta* 6:22, 1961
213. EVERED DF, STEENSON TI: *Nature (Lond)* 202:491, 1964
214. FAIRBANKS VF, FERNANDEZ MN: *JAMA* 208:316, 1969
215. FALKENBERG F, LEHMANN F-G, PFLEIDERER G: *Clin Chim Acta* 23:265, 1969
216. FASCE CF Jr, REJ R: *Clin Chem* 16:972, 1970
217. FAWCETT CP, CIOTTI MM, KAPLAN NO: *Biochim Biophys Acta* 54:210, 1961
218. FENTON MR, RICHARDSON KE: *Arch Biochem Biophys* 142:13, 1971
219. FISHER RA, HARRIS H: *Ann Hum Genet* 34:431, 1971
220. FISHMAN WH, GHOSH NK: *Adv Clin Chem* 10:255, 1967
221. FISHMAN WH, SIE H-G: *Enzymologia* 41:141, 1971
222. FLEISHER GA: *Standard Methods of Clinical Chemistry.* Edited by D Seligson. New York, Academic Press, 1961, Vol 3, p 14
223. FLEISHER GA: *Clin Chem* 13:233, 1967
224. FLEISHER GA, BUTT HR, HUIZENGA KA: *Ann NY Acad Sci* 75:363, 1958
225. FLEISHER GA, PANKOW M, WARMKA C: *Clin Chim Acta* 9:254, 1964
226. FLEISHER GA, POTTER CS, WAKIM KG: *Proc Soc Exp Biol Med* 103:229, 1960
227. FLEISHER M, SCHWARTZ MK: *Clin Chem* 17:417, 1971
228. FLEMING A: *Proc R Soc Lond [Biol]* 93:306, 1922
229. FLOCH MH, MONTALVO G, CROSBY P: *Am J Clin Pathol* 37:350, 1962
230. FORSELL OM, PALVA IP: *Scand J Clin Lab Invest* 13:131, 1961
231. FOULK WT, FLEISHER GA: *Mayo Clin Proc* 32:405, 1957
232. FOWLER WM Jr, CHOWDHURY SR, PEARSON CM, GARDNER G, BRATTON R: *J Appl Physiol* 17:943, 1962
233. FOY RB, KING RW: *Enzymol Biol Clin* 10:305, 1969
234. FRAJOLA WJ, WILLIAMS RD, AUSTAD RA: *Am J Clin Pathol* 43:261, 1965
235. FRASER GP, FENTON JCB: *J Clin Pathol* 21:764, 1968
236. FREE AH, MYERS VC: *J Lab Clin Med* 28:1387, 1943
237. FRIDHANDLER L, BERK JE: *Clin Chem* 16:911, 1970
238. FRIDHANDLER L, BERK JE, UEDA M: *Clin Chem* 17:423, 1971
239. FRIEDEL R, MATTENHEIMER H: *Z Anal Chem* 252:204, 1970
240. FRIEDMAN HS: *Anal Chem* 25:662, 1953
241. FRIEDMAN MM, TAYLOR TH: *Standard Methods of Clinical Chemistry.* Edited by D Seligson. New York, Academic Press, 1961, Vol 3, p 207
242. FÜBR J, STARY E: *Ärztl Lab* 15:55, 1969
243. FURUNO M, SHEENA A: *Clin Chem* 11:23, 1965
244. GAL EM, ROTH E: *Clin Chim Acta* 2:316, 1957
245. GAMBILL EE, MASON HL: *JAMA* 186:24, 1963
246. GARRY PJ: *Clin Chem* 17:183, 1971
247. GARRY PJ: *Clin Chem* 17:192, 1971
248. GAY RJ, McCOMB RB, BOWERS GN Jr: *Clin Chem* 14:740, 1968
249. GELDER BF VAN, VELDSEMA A: *Biochim Biophys Acta* 130:267, 1966

250. GEORGATSOS JG: *Arch Biochem Biophys 110*:354, 1965
251. GERLACH U, PAUL L, LATZEL H: *Enzymol Biol Clin 11*:251, 1970
252. GERRING EL, ALLEN EA: *Clin Chim Acta 24*:437, 1969
253. GHOSH BP, MITRA AK: *Med Exp 8*:28, 1963
254. GHOSH NK, FISHMAN WH: *Can J Biochem 47*:147, 1969
255. GILBOE DD, WILLIAMS JN: *Proc Soc Exp Biol Med 91*:535, 1956
256. GIRARD ML, ROUSSELET F, KOCH M: *Ann Biol Clin (Paris) 21*:557, 1963
257. GLATZLE D: *Seventh International Congress Clinical Chemistry, Clinical Enzymology*. Edited by J Frei, M Jemelin. Basel, Karger, 1970, Vol 2, p 89
258. GLENDENING MB, COHEN AM, PAGE EW: *Proc Soc Exp Biol Med 90*:25, 1955
259. GLOCK GE, McLEAN P: *Biochem J 55*:400, 1953
260. GOLDBARG JA, PINEDA EP, RUTENBURG AM: *Am J Clin Pathol 32*:571, 1959
261. GOLDBARG JA, PINEDA EP, SMITH EE, FRIEDMAN OM, RUTENBURG AM: *Gastroenterology 44*:127, 1963
262. GOLDBARG JA, RUTENBERG AM: *Cancer 11*:283, 1958
263. GOLDHAMMER H, WITTE I: *Ärztl Lab 10*:183, 1964
264. GOLDSTEIN NP, EPSTEIN JH, ROE JH: *J Lab Clin Med 33*:1047, 1948
265. GOMORI G: *Am J Clin Pathol 27*:170, 1957
266. GONNARD P, NGUYEN-PHILIPPON C: *Ann Biol Clin (Paris) 17*:206, 1959
267. GOODMAN RD, SANDOVAL E, HALSTED JA: *J Lab Clin Med 40*:872, 1952
268. GOTO I, KATSUKI S: *Clin Chim Acta 30*:795, 1970
269. GOTTS R, SKENDZEL L: *Clin Chim Acta 14*:505, 1966
270. GOWENLOCK AH: *Biochem J 53*:274, 1953
271. GRAIG FA, ROSS G: *Metabolism 12*:57, 1963
272. GRAIG FA, SMITH JC, FOLDES FF: *Clin Chim Acta 15*:107, 1967
273. GREEN JB, OLDEWURTEL HA, O'DOHERTY DS, FORSTER FM: *Arch Neurol Psychiat 80*:148, 1958
274. GREEN JB, OLDEWURTEL HA, O'DOHERTY DS, FORSTER FM, SANCHEZ-LONGO LP: *Neurology (Minneap) 7*:313, 1957
275. GREEN JB, PERRY M: *Neurology (Minneap) 13*:924, 1964
276. GREEN MN, TSOU K, BRESSLER R, SELIGMAN AM: *Arch Biochem Biophys 57*:458, 1955
277. GREEN RC, LITTLE C, O'BRIEN PJ: *Arch Biochem Biophys 142*:598, 1971
278. GREEN S, GIOVANNIELLO TJ, COTE RA, FISHMAN WH: *Automation in Analytical Chemistry*. Technicon Symposium. 1967, p 563
279. GREENBERG DM: *Methods in Enzymology*. Edited by CM Colowick, NO Kaplan. New York, Academic Press. 1955, p 368
280. GREENBERG WV, TEMPLE TE: *Am J Clin Pathol 48*:133, 1967
281. GRIFFITHS PD: *Clin Chim Acta 13*:413, 1966
282. GRIMES AJ, LEETS I, DACIE JV: *Br J Haematol 14*:309, 1968
283. GRUTTNER R, LODEN K, STORM H: *Z Kinderheilkd, 82*:548, 1959
284. GUILBAULT GG, HEISERMAN J: *Anal Chem 41*:2006, 1969
285. GUNNARSSON PO, PETTERSSON G, PETTERSSON I: *Eur J Biochem 17*:586, 1970
286. GÜRTLER B, LEUTHARDT F: *Helv Chim Acta 53*:654, 1970
287. GUTMAN AB, GUTMAN EB: *Proc Soc Exp Biol Med 38*:470, 1938
288. GUTMAN EB, GUTMAN AB: *J Biol Chem 136*:201, 1940
289. GUTMAN EB, GUTMAN AB: *Proc Soc Exp Biol Med 47*:513, 1941
290. HAGERMAN DD, WELLINGTON FM: *Am J Obstet Gynecol 77*:348, 1959
291. HALONEN PI, KONTTINEN A: *Nature (Lond) 193*:942, 1962
292. HAMFELT A: *Scand J Clin Lab Invest 18* (Suppl 92):181, 1967
293. HANES CS, HIRD FJR, ISHERWOOD FA: *Biochem J 51*:25, 1952

294. HAPPOLD FC, TURNER JM: *Nature (Lond) 179*:155, 1957
295. HARALAMBIE G: *Z Klin Chem Klin Biochem 7*:352, 1969
296. HARMS DR, CAMFIELD RN: *Am J Med Technol 32*:341, 1966
297. HARRIS JW, KELLERMEYER RW: *The Red Cell.* Cambridge, Harvard University Press, 1970, p 581
298. HARRISON GA: *Chemical Methods in Clinical Medicine.* Third Edition. New York, Grune & Stratton, 1947, p 506
299. HASCHEN RJ: *Clin Chim Acta 6*:322, 1961
300. HASCHEN RJ, FARR W, REICHELT D: *Z Klin Chem Klin Biochem 6*:11, 1968
301. HATHAWAY JA, HUNTER DT, BERRETT CR: *Clin Biochem 3*:217, 1970
302. HAVERBACK BJ, DYCE BJ, GUTENTAG PJ, MONTGOMERY DW: *Gastroenterology 44*:588, 1963
303. HAYWOOD BJ, STARKWEATHER WH: *Am J Clin Pathol 49*:275, 1968
304. HECHT G, STILLGER E: *Z Klin Chem Klin Biochem 5*:156, 1967
305. HEGESH E, AVRON M: *Biochim Biophys Acta 146*:91, 1967
306. HEGESH E, AVRON M: *Biochim Biophys Acta 146*:397, 1967
307. HEGESH E, CALMANOVICI N, AVRON M: *J Lab Clin Med 72*:339, 1968
308. HEGESH E, CALMANOVICI N, LUPO M, BOCHKOWSKY R: *Red Cell Structure and Metabolism.* Edited by B. Ramot. New York, Academic Press, 1971, p 113
309. HEIDEN CVD, DESPLANQUE J, STOOP JW, WADMAN SK: *Clin Chim Acta 22*:409, 1968
310. HELGER R, LANG H: *Ärztl Lab 11*:120, 1965
311. HENLEY KS, POLLARD HM: *J Lab Clin Med 46*:785, 1955
312. HENRY JB, BOWERS GN Jr: *An Introduction to Enzymology and Serum Enzyme Assay.*Chicago, American Society of Clinical Pathologists, 1960.
313. HENRY L: *J Clin Pathol 12*:131, 1959
314. HENRY RJ: *Clinical Chemistry, Principles and Technics.* Fourth edition. New York, Harper & Row, 1964, p 21
315. HENRY RJ: *Clinical Chemistry, Principles and Technics.* Fourth edition. New York, Harper & Row, 1964, p 469
316. HENRY RJ: *Clinical Chemistry, Principles and Technics.* Fourth edition. New York, Harper & Row, 1964, p 498
317. HENRY RJ: *Clinical Chemistry, Principles and Technics.* Fourth edition. New York, Harper & Row, 1964, p 533
318. HENRY RJ, CHIAMORI N: *Clin Chem 6*:434, 1960
319. HENRY RJ, CHIAMORI N, GOLUB OJ, BERKMAN S: *Am J Clin Pathol 34*:381, 1960
320. HENRY RJ, CHIAMORI N, JACOBS SL, SEGALOVE M: *Proc Soc Exp Biol Med 104*:620, 1960
321. HENRY RJ, SOBEL C, BERKMAN S: *Clin Chem 3*:77, 1957
322. HESS JW, MACDONALD RP: *J Michigan Med Soc 62*:1095, 1963
323. HESS JW, MACDONALD RP, FREDERICK RJ, JONES RN, NEELY J, GROSS D: *Ann Intern Med 61*:1015, 1964
324. HESS JW, MACDONALD RP, NATHO GJW, MURDOCK KJ: *Clin Chem 13*:994, 1967
325. HESS JW, MURDOCK KJ, NATHO GJW: *Am J Clin Pathol 50*:89, 1968
326. HESTRIN S: *J Biol Chem 180*:249, 1949
327. HICKS GP, UPDIKE SJ: *Anal Biochem 10*:290, 1965
328. HILL BR: *Cancer Res 16*:460, 1956
329. HILL PG, SAMMONS HG: *Clin Chim Acta 13*:739, 1966
330. HILL PG, SAMMONS HG: *Enzyme 12*:201, 1971
331. HILLIARD SD, O'DONNELL JG, SCHENKER S: *Clin Chem 11*:570, 1965

332. HOCHELLA NJ, WEINHOUSE S: *Anal Biochem 13*:322, 1965
333. HOFFMEISTER H, JUNGE B: *Z Klin Chem Klin Biochem 8*:613, 1970
334. HOLLAND P, LITTLE C: *Can J Biochem 49*:510, 1971
335. HOLMBERG CG, LAURELL CB: *Acta Chem Scand 2*:550, 1948
336. HOPKINSON DA, SPENCER N, HARRIS H: *Nature (Lond) 199*:969, 1963
337. HORN H-D: *Methods of Enzymatic Analysis.* Edited by H-U Bergmeyer. New York, Academic Press, 1963, p 875
338. HORNE M, CORNISH CJ, POSEN S: *J Lab Clin Med 72*:905, 1968
339. HOUCHIN OB: *Clin Chem 4*:519, 1958
340. HSIA DY-Y, HUANG J, DRISCOLL S: *Nature (Lond) 182*:1389, 1958
341. HSIEH H-S, JAFFÉ ER: *J Clin Invest 50*:196, 1971
342. HSIEH KM, BLUMENTHAL HT: *Proc Soc Exp Biol Med 91*:626, 1956
343. HUE AC, FREE AH: *Clin Chem 11*:708, 1965
344. HUGHES BP: *Clin Chim Acta 7*:597, 1962
345. HUMOLLER FL, HOLTHAUS JM, WALSH JR: *Clin Chem 3*:703, 1957
346. HUMOLLER FL, MAJKA FA, BARAK AJ, STEVENS JD, HOLTHAUS JM: *Clin Chem 4*:1, 1958
347. HUMMOLLER FL, MOCKLER MP, HOLTHAUS JM, MAHLER DJ: *J Lab Clin Med 56*:222, 1960
348. HUNTER A, DAUPHINEE JA: *J Biol Chem 85*:627, 1929
349. HUNTER A, DOWNS CE: *J Biol Chem 155*:173, 1944
350. HUNTER A, PETTIGREW JB: *Enzymologia 1*:341, 1936
351. HUNTER DT, KNIGHT JA, NOLAN JF: *Am J Med Technol 33*:366, 1967
352. HYLDGAARD-JENSEN J, VALENTA M, JENSEN SE, MOUSTGAARD J: *Clin Chim Acta 22*:497, 1968
353. IDEO G, DE FRANCHIS R: *Z Klin Chem Klin Biochem 8*:611, 1970
354. ITOH H, SRERE PA: *Clin Chem 17*:86, 1971
355. JACKS TJ, KIRCHER HW: *Anal Biochem 21*:279, 1967
356. JACOBS WLW: *Clin Chim Acta 31*:175, 1971
357. JACOBSSON K: *Scand J Clin Lab Invest 12*:367, 1960
358. JERGOVIC I, ZUZIC I, FISER-HERMAN M, STRAUS B: *Clin Chim Acta 30*:765, 1970
359. JOHANSSON BG, MALMQUIST J: *Scand J Clin Lab Invest 27*:255, 1971
360. JOHNSON DA, OSAKI S, FRIEDEN E: *Clin Chem 13*:142, 1967
361. JOHNSON RB Jr: *Clin Chem 15*:108, 1969
362. JOHNSTONE DE: *Am J Dis Child 84*:191, 1952
363. JOLLÉS P: *Angew Chem [Engl] 3*:28, 1964
364. JØRGENSEN K: *Scand J Clin Lab Invest 11*:282, 1959
365. JØRGENSEN MB: *Scand J Clin Lab Invest 6*:303, 1954
366. JÖSCH W, DUBACH UC: *Z Klin Chem Klin Biochem 5*:59, 1967
367. JUNGNER G: *Scand J Clin Lab Invest 10* (Suppl 31):280, 1957
368. JUUL P: *Clin Chim 13*:416, 1967
369. JUUL P: *Clin Chim Acta 19*:205, 1968
370. KACHMAR JF, BOYER PD: *J Biol Chem 200*:669, 1953
371. KALCKAR HM, ANDERSON EP, ISSELBACHER KJ: *Biochim Biophys Acta 20*:262, 1956
372. KALDOR J, SCHIAVONE DJ: *Clin Chem 14*:735, 1968
373. KALLOS J, KAHN D, RIZOK D: *Can J Biochem 42*:235, 1964
374. KALMANSOHN RB, KALMANSOHN RW: *Circulation 22*:769, 1960
375. KAPLAN J-C: *Nature (Lond) 217*:256, 1968
376. KAPLAN J-C: *Red Cell Structure and Metabolism.* Edited by B Ramot. New York, Academic Press, 1971, p 125

377. KAPLAN JC, KISSIN C: *Seventh International Congress, Clinical Chemistry, Clinical Enzymology*. Edited by J Frei, M Jemelin. Basel, Karger, 1970, Vol 2, p 1
378. KAPLAN J-C, NICOLAS A-M, HANZLICKOVA-LEROUX A, BEUTLER E: *Blood 36*:330, 1970
379. KAPLAN JC, SHORE N, BEUTLER E: *Am J Clin Pathol 50*:656, 1968
380. KAPLAN JC, TEEPLE L, SHORE N, BEUTLER E: *Biochem Biophys Res Commun 51*:768, 1968
381. KARMEN A: *J Clin Invest 34*:131, 1955
382. KARMEN A, WRÓBLEWSKI F, LADUE JS: *J Clin Invest 34*:126, 1955
383. KASPER CB, DEUTSCH HF, BEINERT H: *J Biol Chem 238*:2338, 1963
384. KATZ AM, KALOW W: *Can J Biochem 43*:1653, 1965
385. KAY HD: *J Biol Chem 89*:235, 1930
386. KELLEN J, ROMANCIK V: *Z Klin Chem Klin Biochem 4*:78, 1966
387. KELLER DF: *CRC Crit Rev Clin Lab Sci 1*:247, 1970
388. KELLY S, COPELAND W, SMITH RO: *Clin Chim Acta 21*:431, 1968
389. KIND PRN, KING EJ: *J Clin Path 7*:322, 1954
390. KING EJ, ARMSTRONG AR: *Can Med Assoc J 31*:376, 1934
391. KING EJ, CAMPBELL DM: *Clin Chim Acta 6*:301, 1961
392. KING EJ, DELORY GE: *Biochem J 33*:1185, 1939
393. KING J: *Practical Clinical Enzymology*. Princeton, NJ, Van Nostrand, 1965, p 75
394. KING J: *Practical Clinical Enzymology*. Princeton, NJ, Van Nostrand, 1965, p 108
395. KING J: *Practical Clinical Enzymology*. Princeton, NJ, Van Nostrand, 1965, p 208
396. KING J, MORGAN HG: *J Clin Pathol 23*:730, 1970
397. KING J, MORRIS MB: *Arch Dis Child 36*:604, 1961
398. KINGSLEY DPE, COOK J, VARTAN AE: *Clin Chim Acta 12*:489, 1965
399. KIRKMAN HN, BYNUM E: *Ann Hum Genet 23*:117, 1959
400. KLASSEN CHL, SIERTSEMA LH: *Ned Tijdschr Geneeskd 108*:1433, 1964
401. KLEIN B, AUERBACH J: *Clin Chem 12*:289, 1966
402. KLEIN B, AUERBACH J, MORGENSTERN S: *Clin Chem 11*:998, 1965
403. KLEIN B, FOREMAN JA, SEARCY RL: *Clin Chem 16*:32, 1970
404. KLOTZ AP, DUVALL MR: *J Lab Clin Med 50*:753, 1957
405. KNIGHT JA, HUNTER DT: *Clin Chem 14*:555, 1968
406. KNIGHT JA, HUNTER DT: *J Med Lab Technol 25*:106, 1968
407. KNIGHTS EM Jr, WHITEHOUSE JL, HUE AC, SANTOS CL: *J Lab Clin Med 65*:355, 1965
408. KOCH P, TONKS DB: *Clin Biochem 4*:186, 1971
409. KOEDAM JC: *Clin Chim Acta 23*:63, 1969
410. KOJ A, ZGLICZYNSKI JM, FRENDO J: *Clin Chim Acta 5*:339, 1960
411. KOKOT F, KUSKA J: *Clin Chim Acta 11*:118, 1965
412. KOLER RD, BIGLEY RH, STENZEL P: *Hereditary Disorders of Erythrocyte Metabolism*. Edited by E Beutler. New York, Grune & Stratton, 1968, p 249
413. KONTTINEN A: *Clin Chim Acta 18*:147, 1967
414. KONTTINEN A: *Clin Chim Acta 21*:29, 1968
415. KONTTINEN A, HALONEN PI: *Cardiology 43*:56, 1963
416. KONTTINEN A, PYORALA T: *Scand J Clin Lab Invest 15*:429, 1963
417. KOSSEL A, DAKIN HD: *Z Physiol Chem 41*:321, 1904
418. KOVE S, GOLDSTEIN S, WRÓBLEWSKI F: *JAMA 168*:860, 1958
419. KOWLESSAR D, HAEFFNER LJ, SLEISENGER MH: *J Clin Invest 39*:671, 1960
420. KRAMER HJ, GONICK HC: *Enzyme 12*:257, 1971
421. KRAMER HJ, WIGHT JE, PAUL WL, GONICK HC: *Enzymol Biol Clin 11*:435, 1970
422. KREUTZER HJH, KREUTZER HH: *Clin Chim Acta 11*:578, 1965

423. KUBY SA, NODA L, LARDY HA: *J Biol Chem 209*:191, 1963
424. KULHÁNEK V, DIMOV DM: *Clin Chim Acta 14*:619, 1966
425. KULHÁNEK V, MADĚROVÁ V, SINDELÁROVA K, VOKTIŠKOVA V: *Clin Chim Acta 8*:579, 1963
426. LADUE JS: *JAMA 165*:1776, 1957
427. LADUE JS, WRÓBLEWSKI F, KARMEN A: *Science 120*:497, 1954
428. LAGERLÖF HO: *Acta Med Scand [Suppl] 18*:1, 1942
429. LAMOTTA RV, WORONICK CL: *Clin Chem 17*:135, 1971
430. LANG N, MASSARRAT S: *Klin Wochenschr 43*:597, 1965
431. LARSEN EF: *Microelectrometric Method for Blood Cholinesterase Determinations.* Division of Laboratories, California State Department of Public Health, 1957
432. LATNER AL, SKILLEN AW: *Isoenzymes in Biology and Medicine.* New York, Academic Press, 1968, p 4
433. LAUBER K, RICHTERICH R: *Z Klin Chem Klin Biochem 4*:208, 1966
434. LAUDAHN G, HARTMANN E, ROSENFELD EM, WEYER H, MUTH HW: *Klin Wochenschr 48*:838, 1970
435. LAURSEN T: *Scand J Clin Lab Invest 10* (Suppl 31):285, 1957
436. LAURSEN T: *Scand J Clin Lab Invest 11*:134, 1959
437. LAURSEN T, ESPERSEN G: *Scand J Clin Lab Invest 11*:61, 1959
438. LAURSEN T, HANSEN PF: *Scand J Clin Lab Invest 10*:53, 1958
439. LAURSEN T, SCHMIDT A: *Scand J Clin Lab Invest 18* (Suppl 92):175, 1967
440. LAZAROFF N: *Z Med Labortech 11*:38, 1970
441. LEACH BE, COHEN M, HEATH RG: *Arch Neurol Psychiat 76*:635, 1956
442. LEBHERZ HG, RUTTER WJ: *Biochemistry 8*:109, 1969
443. LEE LW: *Am J Med Technol 32*:255, 1966
444. LELUAN G: *Ann Biol Clin (Paris) 22*:343, 1964
445. LENDING M, SLOBODY LB, STONE ML, MESTERN J: *J Dis Child 98*:487, 1959
446. LEVIN B, MACKAY HMM, OBERHOLZER VG: *Arch Dis Child 36*:622, 1961
447. LEVINE JB, HILL JB: *Automation in Analytical Chemistry.* Proceedings of the 1965 Technicon Symposium, New York, 1966, p 564
448. LEVINE JB, HILL JB: *Automation in Analytical Chemistry.* Proceedings of the 1965 Technicon Symposium, New York, 1966, p 569
449. LEYBOLD K, BECKMANN J, WEISBECKER L: *Z Klin Chem Klin Biochem 7*:25, 1969
450. LIEBERMAN J, DAIBER O, DULKIN SI, LOBSTEIN OE, KAPLAN MR: *N Engl J Med 257*:1201, 1957
451. LINDERSTRØM-LANG K, WEIL L, HOLTER H: *C R Trav Lab Carlsberg 21*:7, 1935
452. LIPPI U, GUIDI G: *Clin Chim Acta 28*:431, 1970
453. LIPPI U, GUIDI G, PAVAN R: *Clin Chim Acta 30*:845, 1970
454. LITWACK G: *Proc Soc Exp Biol Med 89*:401, 1955
455. LOCQUET JP, SAINT-BLANCARD J, JOLLÉS P: *Biochim Biophys Acta 167*:150, 1968
456. LOEB WF, STUHLMAN RA: *Clin Chem 15*:162, 1969
457. LÖHR GW, WALLER HD: *Methods of Enzymatic Analysis.* Edited by H-U Bergmeyer. New York, Academic Press, 1963, p 744
458. LOKEN MK, TERRILL KD, MARVIN JF, MOSSLER DG: *J Gen Physiol 42*:251, 1958
459. LONDON M, MARYMONT JH Jr, FULD J: *Pediatrics 33*:421, 1964
460. LONG WK: *Science 155*:712, 1967
461. LÓPEZ-GORGÉ J, VILLANUEVA E: *Rev Esp Fisiol 24*:63, 1968
462. LORENTZ K, NIEMANN E, *Z Gastroenterol 6*:204, 1968
463. LORENTZ K, OLTMANNS D: *Clin Chem 16*:300, 1970

464. LOUDERBACK AL, SHANBROM E: *JAMA 205*:294, 1968

465. LOWE ML, GIN JB: *Clin Chem 18*:1552, 1972

466. LOWE ML, STELLA AF, MOSHER BS, GIN JB, DEMETRIOU JA: *Clin Chem 18*:440, 1972

467. LOWRY OH, ROBERTS NR, WU M, HIXON WS, CRAWFORD EJ: *J Biol Chem 207*:19, 1954

468. LUBRAN M, JENSEN WE: *Clin Chim Acta 22*:125, 1968

469. LUDVIGSEN B: *J Lab Clin Med 61*:329, 1963

470. LUMENG J, BLICKENSTAFF L, MILLER J: *Am J Clin Pathol 55*:471, 1971

471. LUNDBLAD G, HULTIN E: *Scand J Clin Lab Invest 18*:201, 1966

472. MACDONALD RP, LEFAVE RO: *Clin Chem 8*:509, 1962

473. MAGDOFF-FAIRCHILD B, LOVELL FM, LOW BW: *J Biol Chem 244*:3497, 1969

474. MALACINSKI GM: *Am J Clin Pathol 56*:623, 1971

475. MANNUCCI PM, DIOGUARDI N: *Clin Chim Acta 14*:215, 1966

476. MANSO C, TARANTA A, NYDICK I: *Proc Soc Exp Biol Med 93*:84, 1956

477. MARON E, BONAVIDA B: *Biochim Biophys Acta 229*:273, 1971

478. MARSH ME, GREENBERG LD, RINEHART JF: *J Nutr 56*:115, 1955

479. MARSH WH, FINGERHUT B, KIRSCH E: *Clin Chem 5*:119, 1959

480. MARTINEK RG, BERGER L, BROIDA D: *Clin Chem 10*:1087, 1964

481. MARTINEK RG, JACOBS SL, HAMMER FE: *Clin Chim Acta 36*:75, 1971

482. MARUNA RFL: *Clin Chim Acta 25*:133, 1969

483. MARYMONT JH Jr, CAWLEY LP, HOFFMAN RG: *Am J Clin Pathol 49*:431, 1968

484. MASSION CG, SELIGSON D: *Am J Clin Pathol 48*:307, 1967

485. MASSOD MF, FRANEY RJ, THERRIEN ME, RIDEOUT PT, BABCOCK MT: *Am J Clin Pathol 42*:623, 1964

486. MATHAI CK, BEUTLER E: *Science 154*:1179, 1966

487. MATSUZAWA T, KATUNUMA N: *Anal Biochem 17*:143, 1966

488. MATTILA S: *Invest Urol 6*:337, 1969

489. MAXWELL ES, KALCKAR HM, BYNUM E: *J Lab Clin Med 50*:478, 1957

490. MCCOMB RB, BOWERS GN Jr: *Clin Chem 18*:97, 1972

491. MCCOMB RB, GAY RJ: *Clin Chem 14*:754, 1968

492. MEITES S, ROGOLS S: *Clin Chem 14*:1176, 1968

493. MEITES S, ROGOLS S: *CRC Crit Rev Clin Lab Sci 2*:103, 1971

494. MELLBYE OJ, SCOTT D: *Scand J Clin Lab Invest 16*:177, 1964

495. MELLERUP B: *Clin Chem 13*:900, 1967

496. MELLMAN WJ, TEDESCO TA: *J Lab Clin Med 66*:980, 1965

497. MENACHE R: *Ann Biol Clin (Paris) 17*:615, 1960

498. MENACHE R, RUBINSTEIN I, GAIST L, MARZIUK I: *Clin Chim Acta 19*:33, 1968

499. MERICAS G, ANAGNOSTOU E, HADZIYANNIS S, KAKARI S: *J Clin Pathol 17*:52, 1964

500. MEYER K, HAHNEL E: *J Biol Chem 163*:723, 1946

501. MEYER-BERTENRATH JG, KAFFARNIK H: *Hoppe Seylers Z Physiol Chem 349*:1071, 1968

502. MEZEY E, SLATER KC, HOLT PR: *Clin Chim Acta 25*:11, 1969

503. MICHEL HO: *J Lab Clin Med 34*:1564, 1949

504. MICHIE DD, BOOTH RW, CONLEY M, McGUIRE JJ: *Am J Clin Pathol 52*:329, 1969

505. MIRSKY IA, BLOCK S, OSHER S, BROH-KHAN RH: *J Clin Invest 27*:818, 1948

506. MIRSKY IA, FUTTERMAN P, KAPLAN S, BROH-KAHN RH: *J Lab Clin Med 40*:17, 1952

507. MIWA S, KANAI M, NOMOTO S: *Br J Haematol 13* (Suppl): 54, 1967

508. MOHUN AF, COOK IJY: *J Clin Pathol 10*:394, 1957
509. MONDORF AW, MACKENRODT G, HALBERSTADT E: *Klin Wochenschr 49*:61, 1971
510. MOODY JR, PURDY WC: *Anal Chim Acta 53*:31, 1971
511. MOORE FML: *Clin Chim Acta 15*:103, 1967
512. MORELL AG, AISEN P, SCHEINBERG IH: *J Biol Chem 237*:3455, 1962
513. MORETTI GF, MAHON R, STAEFFEN J, BALLAN P, CATANZANO G, VANDENDRIESSCHE M: *Ann Biol Clin (Paris) 21*:573, 1963
514. MORETTI G, STAEFFEN J, BALLAN P, FERRER J, ROQUES JC, BROUSTET A, BEYLOT J: *Ann Biol Clin (Paris) 28*:79, 1970
515. MORGENSTERN S, FLOR R, KESSLER G, KLEIN B: *Anal Biochem 13*:149, 1965
516. MORGENSTERN S, FLOR R, KESSLER G, KLEIN B: *Clin Chem 12*:274, 1966
517. MORGENSTERN S, KESSLER G, AUERBACH J, FLOR RV, KLEIN B: *Clin Chem 11*:876, 1965
518. MORGENSTERN S, OKLANDER M, AUERBACH J, KAUFMAN J, KLEIN B: *Clin Chem 12*:95, 1966
519. MOSLEY JW, GOODWIN RF: *Am J Clin Pathol 44*:591, 1965
520. MOSS DW, BARON DN, WALKER PG, WILKINSON JH: *J Clin Pathol 24*:740, 1971
521. MULLER G, HAUSLER M: *Z Klin Chem Klin Biochem 6*:78, 1968
522. NACHLAS MM, DAVIDSON MB, GOLDBERG JD, SELIGMAN AM: *J Lab Clin Med 62*:148, 1963
523. NACHLAS MM, MARGULIES SI, GOLDBERG JD, SELIGMAN AM: *Anal Biochem 11*:317, 1960
524. NAFTALIN L, SEXTON M, WHITAKER JF, TRACEY D: *Clin Chim Acta 26*:293, 1969
525. NAGEL W, WILLIG F, SCHMIDT FH: *Klin Wochenschr 42*:447, 1964
526. NARDI GL, LEES CW: *N Engl J Med 258*:797, 1958
527. NERDRUM HJ, NORDØY S: *Scand J Clin Lab Invest 16*:617, 1964
528. NEUMANN H: *Experientia 4*:74, 1948
529. NG WG, BERGREN WR, DONNELL GN: *Clin Chim Acta 10*:337, 1964
530. NG WG, BERGREN WR, DONNELL GN: *Nature (Lond) 203*:845, 1964
531. NG WG, BERGREN WR, DONNELL GN: *Clin Chim Acta 15*:489, 1967
532. NIGAM VN, DAVISON HM, FISHMAN WH: *J Biol Chem 234*:1550, 1959
533. NIGAM VN, FISHMAN WH: *J Biol Chem 234*:2394, 1959
534. NIHEI T, NODA L, MORALES MF: *J Biol Chem 236*:3203, 1961
535. NIXON JC, CLOUTIER LA: *Clin Biochem 2*:115, 1968
536. NORDIN JH, BRETTHAUER RK, HANSEN RG: *Clin Chim Acta 6*:578, 1961
537. NOTHMAN MM, CALLOW AD: *Gastroenterology 60*:82, 1971
538. NUTTALL FQ, JONES B: *J Lab Clin Med 71*:847, 1968
539. NUTTALL FQ, WEDIN, DS: *J Lab Clin Med 68*:324, 1966
540. OHMORI Y: *Enzymologia 4*:217, 1937
541. OLIVER IT: *Biochem J 61*:116, 1955
542. O'NEIL WR, GOCHMAN N: *Clin Chem 16*:985, 1970
543. ORLOWSKI M, MEISTER A: *Biochim Biophys Acta 73*:679, 1963
544. ORLOWSKI M, SZCZEKLIK A: *Clin Chim Acta 15*:387, 1967
545. ORLOWSKI M, SZEWCZUK A: *Clin Chim Acta 7*:755, 1962
546. OSAKI S: *J Biol Chem 241*:5053, 1966
547. OSAKI S, JOHNSON DA, FRIEDEN E: *J Biol Chem 241*:2746, 1966
548. OSAKI S, McDERMOTT JA, FRIEDEN E: *J Biol Chem 239*:3570, 1964
549. OSAKI S, McDERMOTT JA, FRIEDEN E: *J Biol Chem 239*:PC364, 1964

550. OSSERMAN EF, LAWLOR DP: *J Exp Med 124*:921, 1966
551. OVERBEEK GA: *Clin Chim Acta 2*:1, 1957
552. PARRY RM Jr, CHANDAN RC, SHAHANI KM: *Proc Soc Exp Biol Med 119*:384, 1965
553. PASSEN S, GENNARO W: *Am J Clin Pathol 46*:69, 1966
554. PAUNIER L, ROTTHAUWE HW: *Enzymol Biol Clin 3*:87, 1963
555. PEACOCK AC, BYRD LT, SULE AH: *Clin Chem 11*:505, 1965
556. PEACOCK AC, REED RA, HIGHSMITH EM: *Clin Chim Acta 8*:914, 1963
557. PEARSON JR, WALKER GF: *Arch Environ Health 16*:809, 1968
558. PEISACH J, LEVINE WG: *Biochim Biophys Acta 77*:615, 1963
559. PELIKAN V, KALAB M, TICHY J: *Clin Chim Acta 9*:141, 1964
560. PENG W-T, CHOU C-L: *Sheng Wu Hua Hsueh Yu Sheng Wu Wu Li Hsueh Pao 4*:493, 1965; *CA 62*:16557e, 1965
561. PERITI P: *Boll Soc Ital Biol Sper 37*:448, 1961
562. PERSIJN JP, VAN DER SLIK W, KRAMER K, deRUIJTER CA: *Z Klin Chem Klin Biochem 6*:441, 1968
563. PETTERSSON G: *Acta Chem Scand 23*:2317, 1969
564. PINTO PVC, KAPLAN A, VAN DREAL PA: *Clin Chem 15*:349, 1969
565. PINTO PVC, VAN DREAL PA, KAPLAN A: *Clin Chem 15*:339, 1969
566. PLOMTEUX G: *Ann Biol Clin (Paris) 28*:131, 1970
567. POILLON WN, BEARN AG: *Biochim Biophys Acta 127*:407, 1966
568. POJEROVA A, TOVAREK J: *Acta Paediatr Scand 49*:113, 1960
569. POLLOCK AL, LOEHR E, ROBERTS D: *J Lab Clin Med 73*:166, 1969
570. POSEN S, NEALE FC, CLUBB JS: *Ann Intern Med 62*:1234, 1965
571. POTTAGE J: *J Med Lab Technol 26*:397, 1969
572. POWELL MEA, SMITH MJH: *J Clin Pathol 7*:245, 1954
573. PRANKERD TAJ: *Seventh International Congress Clinical Chemistry, Clinical Enzymology*. Edited by J Frei, M Jemelin. Basel, Karger, 1970, Vol 2, p 19
574. PRASAD ALN, LITWACK G: *Anal Biochem 6*:328, 1963
575. PRICE CP, HILL PG, SAMMONS HG: *Clin Chim Acta 33*:260, 1971
576. PROCKOP DJ, DAVIDSON WD: *N Engl J Med 270*:269, 1964
577. RAABO E: *Scand J Clin Lab Invest 17*:265, 1965
578. RACKER E: *Methods in Enzymology*. Edited by SP Colowick, NO Kaplan. New York, Academic Press, 1957, Vol 2, p 722
579. RADERECHT HJ, SICKOR HJ, SCHULTZE M: *Z Med Labortech 5*:323, 1964
580. RAMOT B: *Seventh International Congress Clinical Chemistry, Clinical Enzymology*. Edited by J Frei, M Jemelin. Basel, Karger, 1970, Vol 2, p 23
581. RAMOT B, ASHKENAZI I, RIMON A, ADAM A, SHEBA C: *J Clin Invest 40*:611, 1961
582. RAPPAPORT F, FISCHL J, PINTO N: *Clin Chim Acta 4*:227, 1959
583. RAVIN HA: *Lancet 1*:726, 1956
584. RAVIN HA: *J Lab Clin Med 58*:161, 1961
585. Recommendations of the German Society for Clinical Chemistry: Standardization of methods for the estimation of enzyme activity in biological fluids. *Z Klin Chem Biochem 10*:281, 1972
585a. REED AH, CANNON DC, WINKLEMAN JW, BHASIN YP, HENRY RJ, PILEGGI VJ: *Clin Chem 18*:57, 1972
586. REED AH, HENRY RJ, MASON WB: *Clin Chem 17*:275, 1971
587. REICHARD H: *Enzymol Biol Clin 1*:47, 1961
588. REICHARD H: *Acta Med Scand 172* (Suppl):390, 1962
589. REICHARD H: *J Lab Clin Med 63*:1061, 1964
590. REICHARD H, REICHARD P: *J Lab Clin Med 52*:709, 1958

591. REIS JL: *Biochem J 48*:548, 1951
592. REITMAN S, FRANKEL S: *Am J Clin Pathol 28*:56, 1957
593. REMMERS AR, Jr, KALJOT V: *JAMA 185*:968, 1963
594. RENCHER JL, BEELER MF: *Am J Clin Pathol 47*:541, 1967
595. REVEL JP, BALL EG: *J Biol Chem 234*:577, 1959
596. REYNOLDS J, FOLLETTE JH, VALENTINE WN: *J Lab Clin Med 50*:78, 1957
597. RICE EW: *Anal Biochem 3*:452, 1962
598. RICHARDSON RW: *Clin Chim Acta 10*:152, 1964
599. RICK W: *Z Klin Chem Klin Biochem 7*:530, 1969
600. RIDER JA, HODGES JL Jr, SWADER J, WIGGINS AD: *J Lab Clin Med 50*:376, 1957
601. RINDERKNECHT H, MARBACH EP: *Clin Chim Acta 29*:107, 1970
602. RINDERKNECHT H, SILVERMAN P, GEOKAS MC, HAVERBACK BJ: *Clin Chim Acta 28*:239, 1970
603. RINDERKNECHT H, WILDING P, HAVERBACK BJ: *Experientia 23*:805, 1967
604. RINGOIR S, WIEME RJ: *Lancet 2*:906, 1965
605. ROCKERBIE RA, RASMUSSEN KL: *Clin Chim Acta 18*:183, 1967
606. ROE JH, BYLER RE: *Anal Biochem 6*:451, 1963
607. ROESER HP, LEE GR, NACHT S, CARTWRIGHT GE: *J Clin Invest 49*:2408, 1970
608. ROMAN W, FAVILLA I: *Enzymologia 26*:249, 1963
609. ROMEL WC, LaMANCUSA SJ: *Clin Chem 11*:131, 1965
610. RONWIN E: *Can J Biochem Physiol 40*:1725, 1962
611. ROSALKI SB: *J Clin Pathol 15*:566, 1962
612. ROSALKI SB: *J Lab Clin Med 69*:696, 1967
613. ROSALKI SB, WILKINSON JH: *Nature (Lond) 188*:1110, 1960
614. ROSALKI SB, WILKINSON JH: *JAMA 189*:61, 1964
615. ROSS JD: *Blood 21*:51, 1963
616. ROTH M: *Clin Chim Acta 9*:448, 1964
617. ROTTHAUWE HW, CERQUEIRO M: *Klin Wochenschr 41*:876, 1963
618. ROTTHAUWE HW, CERQUEIRO-RODRIGUEZ M: *Clin Chim Acta 10*:134, 1964
619. ROUSH A, NORRIS ER: *Arch Biochem 29*:124, 1950
620. ROY AV: *Clin Chem 16*:431, 1970
621. ROY AV, BROWER ME, HAYDEN JE: *Clin Chem 17*:1093, 1971
622. ROZACKY EE, SAWYER TH, BARTON RA, GRACY RW: *Arch Biochem Biophys 146*:312, 1971
623. ROZENSZAJN L, EPSTEIN Y, SHOHAM D, ARBER I: *J Lab Clin Med 72*:786, 1968
624. RUSSELL AS, STURGE RS, SMITH MA: *Br Med J 2*:428, 1971
625. RUTENBURG AM, SMITH EE, FISCHBEIN J: *J Lab Clin Med 69*:504, 1967
626. RUTSTEIN DD, INGENITO EF, REYNOLDS WE: *J Clin Invest 33*:211, 1954
627. SABATH LD, GERSTEIN DA, FINLAND M: *N Engl J Med 279*:1137, 1968
628. SALKIE ML, SIMPSON E: *J Clin Pathol 23*:708, 1970
629. SALL T, RICHARDS HK, HARRISON E, MYERSON RM: *J Lab Clin Med 50*:297, 1957
630. SALOMON LL, JAMES L, WEAVER PR: *Anal Chem 36*:1162, 1964
631. SALVATORE F, BOCCHINI V: *Clin Chim Acta 6*:109, 1961
632. SAMLOFF IM, KLEINMAN MS: *Gastroenterology 56*:30, 1969
633. SAMLOFF IM, TOWNES PL: *Gastroenterology 58*:462, 1970
634. SAMORAJSKI T, ROLSTEN C, ORDY JM: *Am J Clin Pathol 38*:645, 1962
635. SASS MD, CARUSO CJ, FARHANGI M: *J Lab Clin Med 70*:760, 1967
636. SASS-KORTSAK A: *Adv Clin Chem 8*:1, 1965

637. SAX SM, BRIDGEWATER AB, MOORE JJ: *Clin Chem 17*:311, 1971

638. SAX SM, MOORE JJ: *Clin Chem 11*:951, 1965

639. SAX SM, MOORE JJ: *Clin Chem 13*:175, 1967

640. SCHAFFERT RR, KINGSLEY GR, GETCHELL G: *Clin Chem 10*:519, 1964

641. SCHAPIRA F, DREYFUS J-C: *Ann Biol Clin (Paris) 22*:349, 1964

642. SCHEINBERG IH: *Neurochemistry*. Edited by SR Korey, JI Nurnberger. New York, Hoeber, 1956, p 52

643. SCHEINBERG IH, MORELL AG: *J Clin Invest 36*:1193, 1957

644. SCHIWARA HW: *Ärztl Lab 17*:340, 1971

645. SCHLAMOWITZ M, BODANSKY O: *J Biol Chem 234*:1433, 1959

646. SCHLANG HA, KIRKPATRICK CA: *Am J Med Sci 242*:338, 1961

647. SCHMIDT E, SCHMIDT FW: *Nature (Lond) 213*:1125, 1967

648. SCHMIDT H, KEHREL H: *Z Klin Chem Klin Biochem 5*:123, 1967

649. SCHMIDT JD: *Invest Urol 3*:405, 1966

650. SCHNEIDER AJ, WILLIS MJ: *Clin Chem 4*:392, 1958

651. SCHNEIDER AS: *Red Cell Genetics*. Edited by J Yunis. New York, Academic Press, 1969, p 189

652. SCHWARTZ JM, ROSS JM, PARESS PS, FAGELMAN K, FOGEL L: *Red Cell Structure and Metabolism*. Edited by B Ramot. New York, Academic Press, 1971, p 135

653. SCHWARTZ MK, KESSLER G, BODANSKY O: *J Biol Chem 236*:1207, 1961

654. SCHWARZ V, GOLDBERG L, KOMROWER GM, HOLZEL A: *Biochem J 59*:xxii, 1955

655. SCOTT EM: *J Clin Invest 39*:1176, 1960

656. SCOTT EM: *Clin Chim Acta 23*:495, 1969

657. SCOTT EM, DUNCAN IW, EKSTRAND V: *J Biol Chem 238*:3928, 1963

658. SCOTT EM, GRIFFITH IV: *Biochim Biophys Acta 34*:584, 1959

659. SEARCY RL, WILDING P, BERK JE: *Clin Chim Acta 15*:189, 1967

660. SEIFFERT UB: *Clin Chim Acta 28*:51, 1970

661. SEIJFFERS MJ, TURNER MD, MILLER LL, SEGAL HL: *Gastroenterology 48*:122, 1965

662. SELIGMAN AM, CHAUNCEY HH, NACHLAS MM, MANHEIMER LH, RAVIN HA: *J Biol Chem 190*:7, 1951

663. SELWYN JG, DACIE JV: *Blood 9*:414, 1954

664. *Seminar on Clinical Enzymology: Diagnostic Application of Lactic Dehydrogenase to Disease*. Edited by BM Wagner. New York, Columbia-Presbyterian Medical Center, 1969

665. SEVELA M: *Nature (Lond) 181*:915, 1958

666. SEWELL S: *Am J Med Sci 240*:593, 1960

667. SHAPIRA F: *Bull Soc Chim Biol 52*:1251, 1970

668. SHARMA AK, DATTA P: *Clin Chim Acta 32*:134, 1971

669. SHAW RF, PEARSON CM, CHOWDHURY SR: *Enzymol Biol Clin 6*:10, 1966

670. SHEPHERD HG Jr, McDONALD HJ: *Clin Chem 4*:13, 1958

671. SHERWIN AL, SIBER GR, ELHILALI MM: *Clin Chim Acta 17*:245, 1967

672. SHIGEMASA O, WALTER C, FRIEDEN E: *Biochem Biophys Res Comm 12*:1, 1963

673. SHINODA T: *J Biochem 64*:733, 1968

674. SHINOWARA GY, JONES LM, REINHART HL: *J Biol Chem 142*:921, 1942

675. SIAKOTOS AN, FILBERT M, HESTER R: *Biochem Med 3*:1, 1969

676. SIBLEY JA, FLEISHER GA: *Mayo Clin Proc 29*:591, 1954

677. SIDERS DB, BATSAKIS JG, STILES DE: *Am J Clin Pathol 50*:344, 1968

678. SIEGEL A, BING RJ: *Proc Soc Exp Biol Med 91*:604, 1956

679. SIEGEL AL, COHEN PS: *Clin Chem 12*:352, 1966
680. SIEKERT RG, FLEISHER GH: *Mayo Clinic Proc 31*:459, 1956
681. SINHA R, TREW JA: *Clin Biochem 3*:51, 1970
682. SMITH AF: *Lancet 2*:178, 1967
683. SMITH EE, PINEDA EP, RUTENBURG AM: *Proc Soc Exp Biol Med 110*:683, 1962
684. SMITH EL, SLONIM NB: *J Biol Chem 176*:835, 1948
685. SMITH JE, BEUTLER E: *Proc Soc Exp Biol Med 122*:671, 1966
686. SMITH JK, WHITBY LG: *Biochim Biophys Acta 151*:607, 1968
687. SMOLELIS AN, HARTSELL SW: *J Bacteriol 58*:731, 1949
688. SNODGRASS PJ, PARRY DJ: *J Lab Clin Med 73*:940, 1969
689. SOBEL C, BERKMAN S, SWABB N: *Am J Clin Pathol 26*:1477, 1956
690. SOMOGYI M: *Clin Chem 6*:23, 1960
691. SPIKESMAN AM, BROCK DJH: *Clin Chim Acta 26*:387, 1969
692. STAAL GEJ: *Bull Soc Chim Biol 52*:1297, 1970
693. STAAL GEJ, HELLEMAN PW, de WAEL J, VEEGER C: *Biochim Biophys Acta 185*:63, 1969
694. STAAL GEJ, HELLEMAN PW, de WAEL J, VEEGER C: *Seventh International Congress Clinical Chemistry, Clinical Enzymology.* Edited by J Frei, M Jemelin. Basel, Karger, 1970, Vol 2, p 129
695. STAAL GEJ, KOSTER JF, KAMP H, VAN MILLIGEN-BOERSMA L, VEEGER C: *Biochim Biophys Acta 227*:86, 1971
696. STAAL GEJ, KOSTER JF, VAN MILLIGEN-BOERSMA L: *Biochim Biophys Acta 220*:613, 1970
697. STAAL GEJ, VEEGER C: *Biochim Biophys Acta 185*:49, 1969
698. STAAL GEJ, VISSER J, VEEGER C: *Biochim Biophys Acta 185*:39, 1969
699. STANDEFER JC, MATUSIK EJ Jr, DAVIS JI, MATUSIK JE: *Anal Biochem 42*:72, 1971
700. STANKEWICZ MJ, CHENG S, MARTINEZ-CARRION M: *Biochemistry 10*:2877, 1971
701. STANTON RE, JOOS HA: *Pediatrics 24*:362, 1959
702. STARKWEATHER WH, SPENCER HH, SCHWARZ EL, SCHOCH HK: *J Lab Clin Med 67*:329, 1966
703. STEIN HH, LEWIS GJ: *Anal Biochem 15*:481, 1966
704. STEINBERG D, BALDWIN D, OSTROW BH: *J Lab Clin Med 48*:144, 1956
705. STEINBERG D, OSTROW BH: *Proc Soc Exp Biol Med 89*:31, 1955
706. STERLING RE, WILCOX AA, WARE AG, UMEHARA MK: *Clin Chem 10*:1112, 1964
707. STEWART AG, BIRKBECK JA: *J Pediatr 61*:395, 1962
708. STRANDJORD PE, CLAYSON KJ: *J Lab Clin Med 67*:131, 1966
709. STRANDJORD PE, CLAYSON KJ: *J Lab Clin Med 67,* 144, 1966
710. STRANDJORD PE, CLAYSON KJ: *J Lab Clin Med 67*:154, 1966
711. STRANDJORD PE, CLAYSON KJ, FREIER EF: *Fed Proc 21*:239, 1962
712. STREET HV, CLOSE JR: *Clin Chim Acta 1*:256, 1956
713. STREETO J: *Hartford Hosp Bull 16*:38, 1961
714. STRUMEYER DH: *Anal Biochem 19*:61, 1967
715. SUNDERMAN FW Jr, NOMOTO S: *Clin Chem 16*:903, 1970
716. SUSSMAN HH, SMALL PA Jr, COTLOVE E: *J Biol Chem 243*:160, 1968
717. SWINNEN J: *Clin Chim Acta 17*:255, 1967
718. SWINNEN J: *Z Klin Chem Klin Biochem 8*:557, 1970
719. SZASZ G: *Am J Clin Pathol 47*:607, 1967
720. SZASZ G: *Clin Chim Acta 19*:191, 1968
721. SZASZ G: *Clin Chem 15*:124, 1969

722. SZASZ G: *Z Klin Chem Klin Biochem* 8:1, 1970
723. SZCZEKLIK A: *Clin Chim Acta* 23:219, 1969
724. TAKAHARA K, NATELSON S: *Am J Clin Pathol* 47:693, 1967
724a. TAKAHARA K, NABANISHI S, NATELSON S, *Clin Chem* 15:397, 1969
725. TAKEMAKA T: *Bull Yamaguchi Med School* 11:57, 1964
726. TAN IK, WHITEHEAD TP: *Clin Chem* 15:467, 1969
727. TANAKA KR: *Biochemical Methods in Red Cell Genetics.* Edited by J Yunis. New York, Academic Press, 1969, p 167
728. TANZER ML, GILVARG C: *J Biol Chem* 234:3201, 1959
729. TAUBER H: *Proc Soc Exp Biol Med* 90:375, 1955
730. TAYLOR TH, FRIEDMAN ME: *Clin Chem* 6:208, 1960
731. TAYLOR WH: *Physiol Rev* 42:519, 1962
732. TEMLER RS, FELBER JP: *Biochim Biophys Acta* 236:78, 1971
733. *Temperature Conversion Tables.* Boehringer Mannheim Corporation, BMC 7037, 1970, p 10
734. THEISEN R, JACKSON CR, MORRISSEY J, PECKHAM B: *Obstet Gynecol* 17:183, 1961
735. THIERS RE, VALLEE BL: *Ann NY Acad Sci* 75:214, 1958
736. THOMPSON WHS: *Clin Chim Acta* 23:105, 1969
737. TIETZ NW, BORDEN T, STEPLETON JD: *Am J Clin Pathol* 31:148, 1959
738. TIETZ NW, GREEN A: *Clin Chim Acta* 14:566, 1966
739. TKESHELASHVILI MG: *Soobshch Adad Nauk Gruz SSR* 47:85, 1967, *CA* 67:114859a, 1967
740. TONHAZY NE, WHITE NG, UMBREIT WW: *Arch Biochem* 28:36, 1950
741. TRAINER DL: *Am J Med Technol* 33:167, 1967
742. TRAINER TD, GRUENIG D: *Clin Chim Acta* 21:151, 1968
743. TRINDER P, KIRKLAND JF: *Clin Chim Acta* 16:287, 1967
744. TRIP JAJ, QUE GS, van der HEM GK, MANDEMA E: *J Lab Clin Med* 75:403, 1970
745. TURNER MD, TUXILL JL, MILLER LL, SEGAL HL: *Anal Biochem* 16:487, 1966
746. UETE T, TSUCHIKURA H, HOSHIDA K: *Clin Biochem* 3:327, 1970
747. UETE T, TSUCHIKURA H, NINOMIYA K: *Clin Chem* 16:412, 1970
748. UETE T, WASA M, SHIMOGAMI A: *Clin Chem* 15:42, 1969
749. UGARTE G, PINO ME, VALENZUELA J: *J Lab Clin Med* 57:359, 1961
750. VALLEE BL: *Metabolism* 1:420, 1952
751. van der SLIK W, PERSIJN JP, ENGELSMAN E, RIETHORST A: *Clin Biochem* 3:59, 1970
752. Van Der VEEN KJ, WILLEBRANDS AF: *Clin Chim Acta* 13:312, 1966
753. van GOIDSENHOVEN GV, WILKOFF L, KIRSNER JB: *Gastroenterology* 34:421, 1958
754. van GORKOM WH: *Ned Tijdschr Geneesk* 106:297, 1962
755. Van SLYKE DD, ARCHIBALD RM: *J Biol Chem* 165:293, 1946
756. VAUGHAN A, GUILBAULT GG, HACKNEY D: *Anal Chem* 43:721, 1971
757. VERRI B, MACAGNO F, CORAZZA G: *Minerva Pediatr* 18:2256, 1966; *CA* 66:53465z, 1967
758. VESELL ES, FRITZ PJ, WHITE EL: *Biochim Biophys Acta* 159:236, 1968
759. VOGEL WC, ZIEVE L: *Clin Chem* 9:168, 1963
760. vom BRUCH CG: *Z Klin Chem* 3:41, 1965
761. WAGENKNECHT C, HASART E, ANDERS G: Ger. (EAST) 45,289 Jan 20, 1966, patent applied for Mar 30, 1965, 3pp
762. WALLER HD: *Hereditary Disorders of Erythrocyte Metabolism.* Edited by E Beutler. New York, Grune & Stratton, 1968, p 184
763. WANG CC, APPLHANZ I: *Clin Chem* 2:249, 1956

764. WARBURG O, CHRISTIAN W: *Biochem Z 314*:149, 1943
765. WEBER H: *Klin Wochenschr 41*:37, 1963
766. WEBER H: *Clin Chim Acta 10*:521, 1964
767. WEIL L, RUSSELL MA: *J Biol Chem 106*:505, 1934
768. WEINBERG AN: *Metabolism 10*:728, 1961
769. WENGER J, MUNRO M: *Clin Chem 16*:207, 1970
770. WEST M, ZIMMERMAN HJ: *Am J Med Sci 235*:443, 1958
771. WEST PM, ELLIS FW, SCOTT BL: *J Lab Clin Med 39*:159, 1952
772. WETSTONE HJ, BOWERS GN Jr: *Standard Methods in Clinical Chemistry*. Edited by D Seligson. New York, Academic Press, 1963, Vol 4, p 47
773. WETSTONE HJ, LaMOTTA RV: *Clin Chem 11*:653, 1965
774. WHETZEL LC: *Am J Clin Pathol 40*:345, 1963
775. WIEB DJ, FREIER EF: *Am J Med Technol 33*:87, 1967
776. WIESMAN U, COLOMBO JP, ADAM A, RICHTERICH R: *Enzymol Biol Clin 7*:266, 1966
777. WILDING P, COOKE WT, NICHOLSON GI: *Ann Intern Med 60*:1053, 1964
778. WILDING P, DAWSON HF: *Clin Biochem 1*:101, 1967
779. WILKINSON JH: *An Introduction to Diagnostic Enzymology*. Baltimore, Williams & Wilkins, 1962, p 106
780. WILKINSON JH: *Isoenzymes*. Philadelphia, Lippincott, 1965, p 43
781. WILKINSON JH, FUJIMOTO Y, SENESKY D, LUDWIG GD: *J Lab Clin Med 75*:109, 1970
782. WILKINSON JH, JENKINS FP, TUEY GAP: *Clin Chim Acta 19*:397, 1968
783. WILKINSON JH, STECIW B: *Clin Chem 16*:370, 1970
784. WILLIAMS AW: *J Clin Pathol 8*:85, 1955
785. WILLIS CE, NOSAL T, KING JW: *Automation in Analytical Chemistry: Technicon Symposia*, 1967, No I, p 579
786. WILSON AG: *NZ J Med Lab Technol 18*:119, 1964
787. WILSON SS, GUILLAN RA, HOCKER EV: *Am J Clin Pathol 48*:524, 1967
788. WINKELMAN J, NADLER S, DEMETRIOU J, PILEGGI VJ: *Am J Clin Pathol 57*:625, 1972
789. WINTERINGHAM FPW, DISNEY RW: *Bull (WHO) 30*:119, 1964
790. WITTER RF: *Arch Environ Health 6*:537, 1963
791. WITTER RF, GRUBBS LM, FARRIOR WL: *Clin Chim Acta 13*:76, 1966
792. WOLF M, DINWOODIE A, MORGAN HG: *Clin Chim Acta 24*:131, 1969
793. WOLFSON SK Jr, WILLIAMS-ASHMAN HG: *Proc Soc Exp Biol Med 96*:231, 1957
794. World Health Organization Commission: *Enzymologia 33*:59, 1967
795. WORTHY E, WHITEHEAD P, GOLDBERG DM: *Enzymol Biol Clin 11*:193, 1970
796. WRIGHT RK, ALEXANDER RL Jr: *Clin Chem 16*:294, 1970
797. WRÓBLEWSKI F: *Adv Clin Chem 1*:314, 1958
798. WRÓBLEWSKI F, CABAUD P: *Am J Clin Pathol 27*:235, 1957
799. WRÓBLEWSKI F, LaDUE JS: *Cancer 8*:1155, 1955
800. WRÓBLEWSKI F, LaDUE JS: *Proc Soc Exp Biol Med 90*:210, 1955
801. WRÓBLEWSKI F, LaDUE JS: *Proc Soc Exp Biol Med 91*:569, 1956
802. YAM LT, LI CY, LAM KW: *N Engl J Med 284*:357, 1971
803. YANG JS, BIGGS HG: *Clin Chem 17*:512, 1971
804. YATZIDIS H: *Nature (Lond) 186*:79, 1960
805. YONG JM: *J Clin Pathol 20*:647, 1967
806. ZAZVORKA Z: *Z Klin Chem 4*:79, 1966

807. ZETZEL L, KAPLAN H, DUSSIK KT: *Am J Dig Dis* 4:1027, 1959
808. ZIMMERMAN HJ, SILVERBERG IJ, WEST M: *Clin Chem* 6:216, 1960
809. ZINKHAM WH: *Pediatrics* 23:18, 1959
810. ZONDAG HA: *Determination and Diagnostic Significance of Lactate Dehydroge-nase.* Assen, Van Gorcum & Company, 1967
811. ZUCKER MB, BORELLI J: *Ann NY Acad Sci* 75:203, 1958
812. ZUCKER S, HANES DJ, VOGLER WR, EANES RZ: *J Lab Clin Med* 75:83, 1970

Chapter 22

Liver Function Tests, Including Bile Pigments

JAMES WINKELMAN, M.D.
DONALD C. CANNON, M.D., Ph.D.
S. LAWRENCE JACOBS, Ph.D.

The metabolic functions of the liver include important participation in so many synthetic, degradative, and detoxification pathways that disease of the liver is reflected by abnormalities in a tremendous array of substances found in blood, urine, and other body fluids. As further biochemical research reveals new metabolic activities, new liver function tests continue to be proposed. So many already exist that it is impossible to treat the subject exhaustively. The presentation that follows deals only with the most popularly employed tests in the United States today.

A number of tests are presented in other chapters which are conventionally thought of as measures of liver function. A partial directory of these includes total protein, protein fractionation, fibrinogen, and α_1-fetoprotein (Chapter 16); urea and ammonia (Chapter 17); virtually all the specific amino acids (Chapter 18); alkaline phosphatase, LDH, SGOT and SGPT, cholinesterase and ferroxidase, 5'-nucleotidase, guanase, LAP, OCT, and ASAL (Chapter 21); the porphyrins (Chapter 24); carbohydrates, their derivatives and metabolites (Chapters 25 and 26); and lipids and lipoproteins (Chapters 28 and 29).

All this notwithstanding, some important laboratory tests do not appear at all. For example, the Australia antigen test is emerging as a particularly useful one in the diagnosis of serum hepatitis, but since the antigen is probably a virus particle it is more properly covered in a treatise on immunology than clinical chemistry. Clotting tests that measure enzymatic activity dependent upon integrity of the liver also are not covered, nor is the entire field of *in vivo* radioisotopic scintillation scanning, although such procedures have been incorporated into the clinical laboratory in some settings.

HIPPURIC ACID TEST

The hippuric acid test was introduced as a liver function test by Quick (342) in 1933. As originally described, a dose of benzoic acid is administered orally and the urinary output of hippuric acid determined during the subsequent 4 hours. Since conjugation of benzoic acid with glycine to form hippuric acid takes place largely in the liver, the test is often considered to be a measure of the detoxifying ability of the liver.

The reaction requires the enzyme glucine-*N*-acylase and coenzyme A (377). The reaction rate is diminished in the presence of hepatic parenchymal damage, probably as a consequence of both decreased enzyme activity and diminished availability of glycine (338), but is not impaired by uncomplicated extrahepatic biliary obstruction (341). The test is obviously valid only in patients with normal renal function. About 98% of administered benzoic acid is normally excreted in conjugated form, chiefly as hippuric acid but also in minute quantities as benzoylglucuronic acid (479). Conjugation with glucuronic acid relative to glycine increases with large doses of benzoic acid or in the presence of liver disease.

Lack of a simple and accurate quantitative method for hippuric acid analysis has doubtlessly limited utilization of the hippuric acid test. Various methods of analysis have been described, but all have significant shortcomings (421). The major methods employed for determination of hippuric acid include: (a) *Gravimetric analysis.* Hippuric acid is precipitated from acidified urine to which a suitable salt such as ammonium sulfate or sodium chloride has been added, followed by drying and weighing (274). (b) *Acid-base titration.* Hippuric acid is hydrolyzed to benzoic acid, which is extracted into chloroform and titrated with sodium alcoholate (127). A later method involves precipitation of the hippuric acid with strong acid and sodium chloride followed by titration with standard alkali (454). (c) *Formol titration.* Ether extraction of hippuric acid is followed by acid hydrolysis and formol titration of the liberated glycine (340). (d) *Ultraviolet spectrophotometry.* Absorbance of hippuric acid is measured at 232 nm after removing creatinine and free amino acids with a cation exchange column (107) or following purification by gel filtration (398). (e) *Photometry.* Various methods include: conversion of hippuric acid to *N*-bromobenzamide (216); solution in pyridine and reaction with benzenesulfonyl chloride (427); bromine treatment followed by nitration, extraction of resultant mononitro compounds, reduction to the corresponding amino compounds, diazotization, and coupling with a dye-forming base (97). (f) *Fluorescence.* Hippuric acid fluoresces in 70% H_2SO_4 without any preliminary extraction or purification (110). (g) *Thin layer chromatography.* Following separation, spots are developed with *p*-dimethylaminobenzaldehyde in acetic acid anhydride acetone, removed with methanol, and quantified photometrically (421).

Two variations of the hippuric acid test are performed: (a) *Oral test* (274, 342). The test should be administered following an overnight fast or at least 2–3 hours after a light breakfast of toast and coffee. Dissolve 6.0 g sodium benzoate in 200 ml water. The patient drinks the sodium benzoate after voiding and discarding the urine. Urine is collected for the subsequent 4 hours as discrete 1 hour (341) or 2 hour (431) collections. Improved clinical discrimination has been reported when urine is collected in timed aliquots to reveal the pattern of excretion instead of pooling the entire 4 hour collection (431). (b) *Intravenous test* (341, 343). The intravenous test has the advantage that nausea and vomiting and the vagaries of gastrointestinal absorption are avoided. A solution of 1.77 g sodium benzoate in 20 ml water should be obtained from a pharmaceutical firm which can verify that the solution is sterile and pyrogen-free. The patient voids and discards the urine. The entire 20 ml sodium benzoate is then administered intravenously over a 5-min period by a licensed physician. Urine is collected 1 hour after injection. The patient should be cautioned to void as completely as possible.

DETERMINATION OF HIPPURIC ACID IN URINE

(method of Weichselbaum and Probstein, Ref 454, modified)

PRINCIPLE

Precipitation of hippuric acid from acidified urine is facilitated by the addition of sodium chloride and refrigeration. The precipitated hippuric acid is washed, dissolved, and titrated with standard alkali using phenolphthalein as an indicator. A standard correction is made for the solubility of hippuric acid under the conditions of precipitation.

REAGENTS

NaCl
H_2SO_4, conc
NaCl, 30% (w/v), cold. Store in refrigerator.
NaOH, about 0.1 N, standardized

PROCEDURE

1. Measure and record total volume of urine sample.
2. Transfer an aliquot of urine equivalent to one-tenth the total volume into a suitable test tube or centrifuge tube. Add 3.0 g NaCl for every 10 ml urine in the aliquot and dissolve (warm if necessary).
3. Add 0.1 ml conc H_2SO_4 for every 10 ml urine in the aliquot and mix. Place in refrigerator for 30 min. If precipitation of hippuric acid does not occur, scratch side of tube with glass stirring rod below fluid surface to initiate crystallization and return tube to refrigerator for another 30 min.
4. Centrifuge and discard the supernate (see *Note 2*).
5. Wash precipitate by adding 10 ml cold 30% NaCl, rinsing down the walls of the tube, mixing, and recentrifuging. Discard supernate. Wash again and discard supernate.
6. Dissolve precipitate in about 10 ml boiling water and transfer to small

flask. Titrate with standardized 0.1 N NaOH, using phenolphthalein as indicator.

Calculation:

The molecular weights of benzoic acid and hippuric acid are 122 and 179, respectively (ratio of 0.68). A correction must be made for the amount of hippuric acid remaining in solution following precipitation (*Note 2*) which comprises the second term of the following equation:

$$\text{g Hippuric acid} = 179 \times \frac{(\text{ml alkali})(N \text{ of alkali})}{1000} \times 10$$

$$+ 0.123 \times \frac{\text{ml urine in aliquot}}{100} \times 10$$

$$= (1.79 \times \text{ml alkali} \times N \text{ of alkali})$$
$$+ (0.0123 \times \text{ml urine in aliquot})$$

$$\begin{array}{l}\text{g Hippuric acid, expressed} \\ \text{as benzoic acid}\end{array} = 0.68 \times \text{g hippuric acid}$$

NOTES

1. Correction for endogenous hippuric acid. Hippuric acid is normally excreted in such small amounts, 123–739 mg ($\bar{x}$ = 392)/24 hours (431), that correction is not necessary.

2. Solubility of hippuric acid. The solubility of hippuric acid in strongly acidified urine is about 0.5%, but this is decreased to about 0.123% by 30% NaCl (454). Completeness of precipitation can be checked in step 4 by adding additional conc H_2SO_4 to the supernate (227).

3. Removal of bile. It is desirable to remove bilirubin if present since it may mask the endpoint of the indicator in titration. Kraus and Dulkin (227) recommend the following procedure: Measure urine volume and add 0.3 g Norit per 100 ml urine and boil for 1 min. More Norit may be added if excessive bilirubin is present. Filter while hot and acidify with acetic acid, monitoring with litmus. Boil to remove coagulable material and filter if necessary. Restore original volume with water; or if volume exceeds the original, make appropriate correction in final calculations.

ACCURACY AND PRECISION

Accuracy of the method is satisfactory for purposes of clinical testing. Small amounts of benzoic acid which may be excreted in the urine are precipitated along with hippuric acid and measured as such in this procedure as well as in various gravimetric methods.

The precision of the method is about ±5%.

NORMAL VALUES

The normal excretion of hippuric acid in the oral test is 3.0–3.5 g, expressed as benzoic acid in 4 hours (274, 342). Normally, 50%–85% of the administered benzoic acid is excreted during the first 2 hours and 70%–95% in 4 hours (431). In the presence of liver parenchymal disease, not only is the excretion of hippuric acid delayed but the total amount excreted during the 4 hour period is diminished. Normal excretion of hippuric acid in the intravenous test is 0.6–1.6 g, expressed as benzoic acid, in 1 hour (478). It is to be emphasized that in both the oral and intravenous tests the normal ranges represent minimal levels of excretion which individuals with normal hepatic and renal function may be epected to achieve; no known clinical significance can be attached to cases which exceed the upper limits of normal.

Repeat testing on successive days results in a significant increase in hippuric acid excretion in patients with liver disease but no increase in those with normal liver function, presumably as a result of a "priming" effect and glycine synthesis (479). It has been claimed that hippuric acid excretion increases with increasing urine volume (264). Other investigations, however, have revealed no correlation between hippuric acid excretion and urine volume (378) as would be expected from the fact that hippuric acid, as well as p-aminohippurate, is excreted by tubular transport (109). Hippuric acid excretion is presumably increased in free anxiety states (329).

It has been claimed that hippuric acid excretion increases with increasing body surface area and weight (388), but this has been disputed (264). Mathematical relationships correlating body surface area or weight with excretion have been developed (388).

GALACTOSE TOLERANCE TEST

Galactose is metabolically converted to glucose through a series of enzymatic reactions which occur largely in the liver. Galactokinase phosphorylates galactose in the presence of ATP to galactose-1-phosphate, which in turn is readily converted to glucose-1-phosphate by the action of phosphogalactose uridyl transferase and uridine diphosphate glucose. The reserve capacity of these enzymes is presumably not great since liver parenchymal damage may result in diminished ability to metabolize a loading dose of galactose. It is likely that a contributing factor is impaired ability to convert glucose to glycogen since glucose tolerance is also abnormal in the presence of hepatic parenchymal damage (344). Galactose which is not metabolized can be excreted in the urine. Bauer (25) took advantage of empirical observations related to these facts and introduced galactose tolerance as a liver function test in 1906. As originally described, the amount of galactose appearing in the urine during the 5 hours subsequent to a 40 g oral dose was determined. In early studies the amount of urinary reducing substances was measured by a technic such as the Benedict reaction applied directly to the urine or, in the case of concomitant glycosuria, after destruction of glucose by yeast fermentation. This approach was clinically imprecise as a consequence of the vagaries of intestinal absorption and variations in the renal threshold for galactose (202). The intravenous administration of galactose followed by timed measurements of blood galactose was introduced to circumvent the aforementioned difficulties (201).

Galactose tolerance is abnormal in a high percentage of cases of cirrhosis and acute hepatitis and frequently in the presence of carcinoma metastatic to the liver, but is usually normal in uncomplicated extrahepatic biliary obstruction (418). Unfortunately, until recently, technical methodology for the quantification of galactose was time-consuming, relatively imprecise, and not readily available. For these reasons galactose tolerance, although a reasonably discriminative test of liver function, has not achieved widespread popularity.

Two general methods were formerly used for quantification of galactose: (a) the orcinol reaction proposed by Brückner (54), and (b) measurement of reducing substances after the destruction of glucose either by yeast fermentation (346) or treatment with glucose oxidase (425, 436). Removal of glucose is not complete, however, and other reducing substances in plasma interfere to a greater or lesser extent depending upon the specificity of the particular method chosen for measuring reducing substances (425). A far more specific method became available in 1959 (84) with the isolation of galactose oxidase, which oxidizes galactose at the sixth carbon atom to yield D-galactohexodialdose (12). Galactose oxidase was soon applied to the determination of galactose in clinical specimens using the general reaction sequence (96):

$$\text{D-Galactose} + O_2 \xrightarrow{\text{galactose oxidase}} \text{D-galactohexodialdose} + H_2O_2$$

$$H_2O_2 + \text{oxygen acceptor} \xrightarrow{\text{peroxidase}} \text{chromogen} + H_2O$$
$$\text{(reduced leuco form)} \qquad\qquad \text{(oxidized form)}$$

Various redox chromogens have been used, including benzidine (364, 390), o-dianisidine (96, 128, 136), and o-cresol (118). In addition to its reaction with galactose, galactose oxidase catalyzes the oxidation of galactosamine and N-acetylgalactosamine (390). Xylose and ascorbic acid may also react in the enzyme system to a slight degree (128).

PROCEDURE WITH PATIENT

Oral Test (287). Following an overnight fast the patient drinks a solution of 40 g galactose in 250 ml water which may be flavored with lemon juice. Blood samples are drawn into heparin anticoagulant 30 and 60 min later.

Intravenous Test (22). The test should be conducted following an overnight fast or at least 2 hours after a light breakfast. A 50% (w/v) solution of galactose, 1 ml/kg body weight, is administered intravenously over a 4—5 min period. The galactose solution must be obtained from a pharmacy which can certify that it is sterile and pyrogen-free. Crystals of galactose may precipitate out of solution unless it is stored at 37°C. The intravenous injection must be performed by a licensed physician. A single heparinized blood sample is obtained 60 min after galactose administration.

DETERMINATION OF GALACTOSE

(method of Roth et al, Ref 364, modified)

PRINCIPLE

Galactose is oxidized to galactohexodialdose by the action of galactose oxidase. H_2O_2, which is formed stoichiometrically, oxidizes benzidine in the presence of catalase. The oxidized benzidine formed is measured by its absorbance at 310 nm, which is chosen because reduced benzidine also absorbs significantly at 288 nm, the peak absorbance of oxidized benzidine (Fig. 22-1).

REAGENTS

NaOH, 5 N. Dissolve 20 g NaOH in 50 ml water and dilute to 100 ml.

Stock Glycine Buffer, 0.5 M, pH 8.4. Dissolve 7.5 g glycine in about 150 ml water. Using a pH meter, add 5 N NaOH dropwise until pH reaches 8.4. Dilute to 200 ml and store in refrigerator.

Working Glycine Buffer, 0.05 M, pH 8.3. Dilute the Stock Glycine Buffer 1:10 with water. Stable 1 month in refrigerator.

Galactose Oxidase Solution, 35 units/ml. Dissolve galactose oxidase powder (Worthington Biochemical) in water to make a final solution of 35 units/ml based on the manufacturer's assay value. Mix well and dispense 1 ml aliquots into plastic vials. Stable at least 12 months frozen.

Peroxidase Solution, 1 mg/ml. Dissolve 10 mg horseradish peroxidase power (Calbiochem) in 10 ml water and mix well. Stable 1 month in refrigerator.

Recrystallized Benzidine. (**Caution**: Benzidine is extremely toxic and carcinogenic. Do not inhale the powder and avoid contact with the skin.) To 1 liter of boiling water, add 2 g benzidine powder and stir vigorously. Add 5 g activated charcoal and stir. While still hot, filter through Whatman 1 filter paper. Refrigerate the filtrate for 1 hour. Collect the recrystallized benzidine using a Büchner funnel with Whatman 54 filter paper. Vacuum desiccate the crystals overnight in the dark. Store in amber bottle.

Benzidine Solution, 10 mg/ml. Dissolve 100 mg recrystallized benzidine in 10 ml absolute ethanol. Mix thoroughly and dispense 0.5 ml aliquots into small plastic vials. Stable frozen for 1 month.

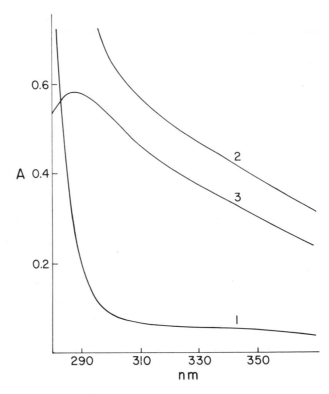

FIG. 22-1. Absorption curves in galactose determination: (1) reagent blank vs water, (2) galactose standard equivalent to 50 mg/100 ml vs water, (3) standard vs reagent blank. Cary Model 15 Spectrophotometer.

Working Enzyme Solution. Prepare within 15 min of use by mixing 10 ml working glycine buffer, 1 ml galactose oxidase solution, 0.5 ml peroxidase solution, and 0.25 ml benzidine solution.

Stock Galactose Standard, 50 mg/100 ml. Dissolve 50 mg D-galactose in water and dilute to 100 ml. Dispense 1 ml aliquots into plastic vials and store frozen. Stable indefinitely. Do not add preservatives (see *Note 3*).

Working Galactose Standard, 50 µg/ml. Dilute 1 ml stock galactose standard to 10 ml with water. Prepare fresh daily.

Zinc Sulfate, 5%. Dissolve 5 g $ZnSO_4 \cdot 7H_2O$ in water and dilute to 100 ml.

NaOH, 0.3 N. Dissolve 12 g NaOH in water and dilute to 1 liter. This is titrated against the 5% zinc sulfate using phenolphthalein as an indicator. Appropriately dilute the more concentrated solution to achieve perfect isoequivalence.

HCl, 5N

PROCEDURE

1. Add 0.5 ml blood to 2.5 ml water in a test tube and mix thoroughly. Allow 5 min for hemolysis and mix again.

2. To the hemolysate add 1.0 ml of 0.3 *N* NaOH dropwise while mixing.

3. Add 1.0 ml zinc sulfate solution and mix thoroughly. Allow to stand for 10 min and centrifuge.

4. In test tubes, set up the following:
   Blank: 0.50 ml water
   Standard: 0.50 ml working galactose standard
   Unknown: 0.50 ml clear supernate from step 3
   Incubate all tubes and the working enzyme solution at 37°C for 5 min.

5. At timed intervals add 0.5 ml working enzyme solution to each tube and mix thoroughly. Continue incubation at 37°C for 30 min.

6. After exactly 30 min add 1 drop of 5 *N* HCl to each tube at timed intervals. Mix thoroughly.

7. Using silica or quartz microcuvets, measure the absorbance of the solutions at 310 nm against a water blank.

Calculation:

$$\text{Blood galactose (mg/100 ml)} = \frac{A_x - A_b}{A_s - A_b} \times 50$$

Samples with calculated values greater than 50 mg/100 ml should be repeated with an appropriate dilution of the supernate from step 3.

NOTES

1. Beer's law. Beer's law is obeyed up to a concentration of 50 mg/100 ml. At higher concentrations moderate deviation from linearity occurs. Following acidification of the reaction, the UV absorbance is stable for at least 30 min.

2. Interfering substances. Oxidizing and reducing substances which can react with benzidine are capable of interfering, but in practice this is not a serious problem. Sulfate ions do not interfere with or precipitate benzidine under the conditions of the test. Theoretically, catalase does interfere, but the amounts present in blood are insignificant in relation to the peroxidase in the analytic system.

3. Sample stability. Blood samples are stable indefinitely frozen. Benzoic acid and cyanide inhibit galactase oxidase (364). Although peroxidase is reportedly inhibited by cyanide and fluoride (364), studies in our laboratory have demonstrated that sodium fluoride, 10 mg/ml, does not interfere and preserves galactose in blood samples for at least 2 weeks at 30°C. Our studies also indicate that oxalate anticoagulant does not interfere with the enzyme system.

ACCURACY AND PRECISION

The method is highly specific for galactose in blood since other substances with which galactose oxidase reacts such as galactosamine, *N*-acetylgalactosamine, and D-xylose are not present in blood. Glucose does not interfere.

The precision of the method is ±5%.

NORMAL VALUES

Following a 40 g dose of galactose in the oral test, the blood galactose ordinarily does not exceed about 40 (287) to 60 (237) mg/100 ml at 1 hour. Others have observed occasional values as high as 100 mg/100 ml in normal individuals (407).

In the intravenous test the blood galactose should not exceed 42 mg/100 ml at 60 min (478). Galactose clearance from the blood follows a straight line in a semilogarithmic plot with a half-life of 12.0 ± 2.6 min in adults (418). Clearance is even more rapid in infants and children (354).

CHOLEPHILIC DYE TESTS

The most sensitive and clinically useful tests for functional liver impairment are those which measure the ability of the liver to clear certain cholephilic dyes from the blood followed by excretion at concentrated levels in the bile. Such tests are thus dependent upon hepatic blood flow and biliary patency as well as hepatic parenchymal function. Detoxification, which implies metabolic degradation or biochemical conjugation, varies widely from one cholephilic dye to another and is not involved at all in the hepatic clearance and excretion of some dyes. Of the many chemicals which have been investigated, only three have had significant clinical acceptance—the tricarbocyamine dye, indocyanine green, and the phthalein dyes, rose bengal and bromsulfophthalein:

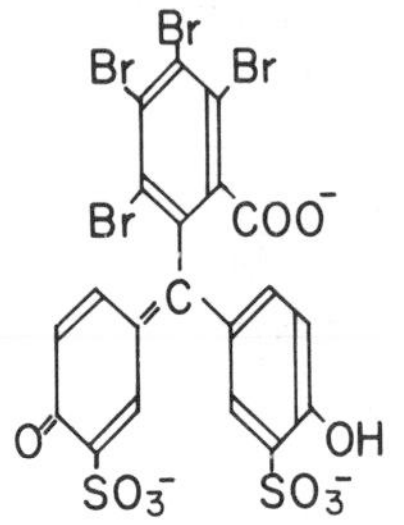

BROMSULFOPHTHALEIN

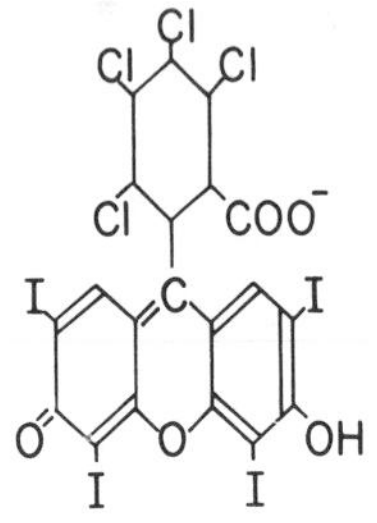

ROSE BENGAL

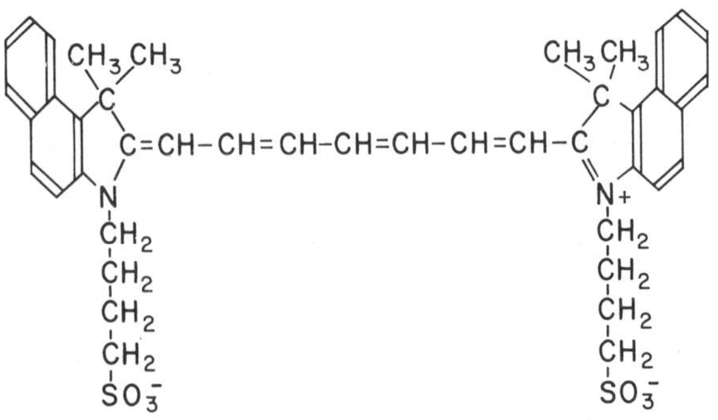

INDOCYANINE GREEN

INDOCYANINE GREEN

Indocyanine green was introduced in 1957 for the measurement of blood flow by the indicator-dilution method and shortly thereafter was recommended as a clinically useful test of liver function (188, 462). Following intravenous administration it is almost entirely bound to plasma proteins, chiefly α_1-lipoprotein (15). The dye can be readily measured spectrophotometrically at its absorbance peak of 805 nm in plasma. Indocyanine green is cleared solely by the liver at a constant percentage rate (188) resulting in a plasma decay curve with a half-life of 3.8 min in normal individuals (462). It is not chemically altered during hepatic excretion, and enterohepatic recirculation presumably does not occur (239). Unlike bromsulfophthalein, indocyanine green is nonirritating and nontoxic, and serious allergic reactions have not been reported (239). Earlier studies which indicated an inferior sensitivity in the detection of functional liver impairment may have been the result of inadequate dosages (239). In spite of its advantages, indocyanine green is not yet widely used.

ROSE BENGAL

Although rose bengal is usually considered to refer to the sodium or potassium salt of tetraiodotetrachlorofluorescein, confusion exists regarding its exact chemical identity since this name has also been applied to diiodotetrachlorofluorescein, tetraiodotetrabromofluorescein, tetraiododichlorofluorescein, and presumably various mixtures (194). Dyes called rose bengal have been used for clinical tests of liver function for almost 50 years. Acceptance has been limited, however, as a result of inferior clinical sensitivity, particularly in comparison with bromsulfophthalein. In recent years some interest has developed in the clinical use of ^{131}I-labeled rose bengal, but sensitivity in detecting impaired liver function has remained inadequate to justify its use in the nonjaundiced patient (295). Other studies have indicated limited usefulness for the labeled dye in distinguishing hepatocellular from obstructive jaundice (314).

BROMSULFOPHTHALEIN

The intravenous bromsulfophthalein (sulfobromophthalein, phenoltetra-bromphthalein sodium sulfonate, Bromsulphalein, commonly abbreviated BSP) test remains a clinically useful test of liver function. As originally described by Rosenthal and White in 1925 (362), the test involved the administration of 2 mg BSP/kg followed by the collection of blood samples 5 and 30 min later. This regimen was subsequently modified to include BSP dosages of 5, 7.5, 10, or even 20 mg/kg and collection of a single blood sample at times varying from 30–60 min later (194). The test has now become reasonably well standardized with a single dose of 5 mg/kg followed by a single blood sample 45 min later.

Following intravenous administration, BSP is largely bound to plasma proteins, chiefly albumin (16) but also in significant amounts to the α_1-lipoproteins (15). Equilibrium dialysis studies indicate that less than 0.1% of unconjugated BSP and approximately 1% of conjugated BSP remain unbound at plasma concentrations usually employed in clinical testing (16). The BSP-albumin complex is dissociated in the process of BSP uptake by the hepatic parenchymal cells. Although evidence has been presented which suggests that this is effected by the sinusoidal endothelium (9), the exact morphologic site has not been established with certainty.

Hepatic uptake of BSP is limited by two processes—hepatic storage and biliary excretion—and has been evaluated using intravenous infusions of BSP at rates sufficient to cause a continuous increase in the plasma concentrations (458). The concept of "liver mass" (Lm) was introduced in 1948 by Mason *et al* (276) and is defined as the amount of BSP taken up by the liver, in mg/min:

$$\text{Lm} = (Q - U) - \frac{V(P - P_1)}{T}$$

where

$$\begin{aligned}
Q &= \text{infusion rate (mg/min)} \\
U &= \text{urine loss of BSP (mg/min)} \\
V &= \text{plasma volume (ml)} \\
P \text{ and } P_1 &= \text{plasma concentrations (mg/ml) in samples separated} \\
&\quad \text{by time } T \text{ (min)}
\end{aligned}$$

Although recommended as being superior to the standard BSP test in terms of its ability to detect decreased functional capacity of the liver (434), the Lm or maximum capacity test has not been widely accepted. A theoretical objection to the concept of an Lm is that its value is not constant but rather is a variable which increases with the infusion dosage in the range usually employed clinically (466). At extremely high infusion rates—i.e., those exceeding 25 mg/min—a constant removal rate is obtained which is usually not greater than 20 mg/min and represents the true "maximal removal capacity" (322). This has been interpreted as an indication that hepatic uptake of BSP involves an active metabolic process rather than simple physicochemical diffusion.

Following hepatic uptake, BSP is largely (75%–85%) conjugated as a result of enzymic activity present in the soluble fraction of liver (79), which is probably S-arylglutathione transferase (194). A spectrophotometric method has been described for assaying this enzyme (149). The enzyme, which requires no cofactor, exhibits optimal activity in conjugating BSP with glutathione but also has limited activity with cysteine and cysteinylglycine (79). Studies with ^{35}S-labeled glutathione, however, have shown that ^{35}S is present in all BSP conjugates in the bile, suggesting that the cysteinylglycine and cysteinyl conjugates are at least in part formed as cleavage products of the fundamental BSP-glutathione conjugate (203). Bromine and hydrogen ion are released stoichiometrically during conjugation, thus indicating formation of a thio-ether linkage (78, 203). Conjugation does not appear to affect hepatic uptake of BSP directly but does facilitate its transport into the bile, perhaps as a consequence of reduced intracellular protein binding of the conjugated BSP (77). Conjugation is not absolutely essential for biliary excretion of BSP (458), however, and levels of the conjugating enzyme are not well correlated with the degree of BSP retention (45).

Biliary excretion is a rate-limited process, often designated *Tm* in analogy to the transport maxima for substances secreted by the renal tubules. The relatively small magnitude of the biliary excretion rate in relation to the hepatic storage capacity is the probable explanation for the fact that BSP continues to be excreted in the bile for several hours after its virtual disappearance from plasma (458). The biliary excretion rate becomes constant at plasma levels greater than about 3 mg/100 ml (457, 458). Although of physiologic importance, the calculation of *Tm* is fraught with systematic errors and is impractical for clinical diagnosis (8). It has been found, for instance, that the steady state required for the calculation is not always reached within 3 hours of infusion (466). Furthermore, the calculation of *Tm* is of no value in distinguishing obstructive from hepatocellular jaundice (385, 458).

Normally, less than 2% of administered BSP is excreted in the urine, probably by tubular secretion (464). In the presence of hepatic disease, up to 10% of the BSP may be excreted in the urine following a standard dose (464), although a much higher percentage has been reported using larger doses of BSP (139). An additional factor other than the elevated serum level of BSP that may contribute to the increased urinary excretion in liver disease is an alteration in serum protein binding (139). Urinary BSP is largely present as conjugates, both in the presence and absence of liver disease (62).

Extrahepatic removal of BSP from the plasma, other than renal excretion, is minimal. Nevertheless, other physiologic processes add to the complexity of the disappearance of BSP from the plasma. Conjugated BSP is regurgitated from the liver into the bloodstream in amounts that are normally small but which may increase greatly in liver disease (385). Furthermore, a limited enterohepatic recirculation undoubtedly occurs, as has been demonstrated by the fact that significant levels of BSP are found in the serum following intraduodenal administration (259).

In spite of the complex physiologic processes involved in BSP excretion, mathematical models have been developed which variously describe the plasma decay curve as biphasic (356) or triphasic (292). Others have concluded, however, that the decay curve is not a simple exponential function, and consequently an explicit mathematical solution cannot be given (465).

The concept of hepatic clearance was introduced in an attempt to increase the precision of the BSP test (252). Refinements have included the use of the clearance coefficient (K), defined as (238, 305):

$$K = \frac{\log R_1 - \log R_2}{t_2 - t_1}$$

where R_1 and R_2 are the percentages of BSP retention in plasma at times t_1 and t_2, respectively. The complexity of BSP disappearance from the blood prevents assignment of precise physiologic significance in hepatic clearance studies. It has also been shown that the clearance of BSP is not constant but decreases with increasing infusion dosage (466). Although most patients with abnormal liver function can be detected (151, 305), BSP clearance studies do not provide unique advantages over the simple one-sample retention test and have not been widely accepted for clinical use.

It is generally accepted that performance of the BSP test is superfluous in patients who are jaundiced as a result of liver disease. Nevertheless, determination of the relative amounts of free and conjugated BSP retained in the plasma has been suggested as a sensitive method for differentiating hepatic cellular damage from intra- and extrahepatic biliary obstruction (62, 385). In cases of hepatic cellular damage, BSP is poorly conjugated and that retained in the plasma is largely unconjugated. In contrast, biliary obstruction results in the regurgitation of large amounts of conjugated BSP.

An error is introduced into the standard one-sample BSP test by assuming that the excretory capacity of the liver is directly proportional to body weight. Some improvement in reliability of the test was demonstrated when the administered dose was based on lean body weight estimated by an anthropometric method (53). Others recommend a dosage based on body surface area (192). An error also results from the assumption that all individuals have the same plasma volume relative to body weight—i.e., 50 ml/kg. This error is particularly significant in individuals with expanded extracellular fluid volume as result of liver disease since false normal BSP results may occur (285).

That performance of the BSP test is accompanied by a certain risk of untoward effects on the patient deserves emphasis. Fatalities as a consequence of hypersensitivity reactions have been reported but fortunately are rare (439). For this reason a physician should be readily available during the performance of the test. Also occurring in a small percentage of patients are such adverse reactions as headache, nausea, generalized erythema, urticaria, chills, syncope, and thrombophlebitis at the site of injection (439). In addition to introducing a technical error in the test, subcutaneous infiltration of BSP during injection is painful and can result in skin sloughing.

Chemical determination of BSP is based on the fact that it is an indicator, with a pK of 8.8, which changes from colorless to purple as the pH is increased (389). The three most significant sources of error in the photometric measurement are bilirubin, hemoglobin, and turbidity usually resulting from increased lipids or lipoproteins. Gaebler (141) and Reinhold (353) attempted to correct for these interferences by reading absorbances at multiple wavelengths. A superior method was introduced by Seligson *et al* (389) who minimized errors from icterus, hemolysis, and lipemia by limiting the pH change between the colored sample and its acidified blank. This method has been adapted to automated analysis (430). A method of protein precipitation followed by acetone extraction of BSP was described by Henry *et al* (176) which gives accurate results in the presence of extreme icterus, lipemia, or hemolysis. As a result of binding of BSP to proteins, chiefly albumin, the maximum absorption of BSP occurs at 594 nm in the presence of serum and at 580 nm in its absence (389).

DETERMINATION OF BROMSULFOPHTHALEIN IN SERUM

(method of Seligson et al, Ref 389)

PRINCIPLE

An aliquot of serum is diluted with an alkaline buffer which contains *p*-toluenesulfonic acid to dissociate the BSP from serum proteins. The absorbance at 580 nm is then measured. Acid reagent is added to convert the BSP to its colorless form, and the absorbance is again measured. The difference in absorbances represents the absorbance of the BSP. The absorption curve of BSP at pH 10.3 is shown in Figure 22-2.

FIG. 22-2. Absorption curve of BSP, 0.008 mg/ml, in water at pH 10.6. Beckman DU spectrophotometer.

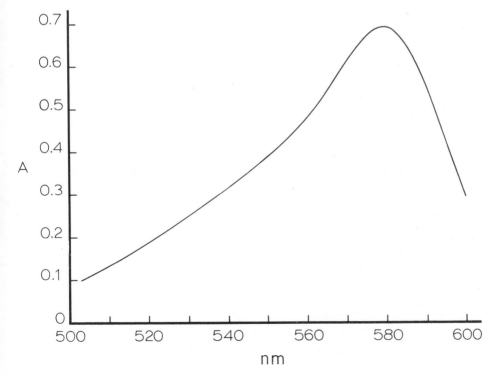

The test is usually performed with the patient in the fasting state. There is no justification for rigid adherence to this rule, however, except that fatty foods which promote lipemia must be avoided.

REAGENTS

Alkaline Buffer, pH 10.6–10.7. Dissolve 12.2 g $Na_2 HPO_4 \cdot 7H_2 O$ (or 6.46 g $Na_2 HPO_4$), 1.77 g $Na_3 PO_4 \cdot 12H_2 O$, and 3.2 g sodium *p*-toluenesulfonate in water to make 500 ml of solution. Check pH and adjust if necessary.

Acid Reagent, 2 M $NaH_2 PO_4$. Dissolve 27.6 g $NaH_2 PO_4 \cdot H_2 O$ in water to make 100 ml of solution.

BSP Standard, 5 mg/100 ml. The intravenous pharmaceutical solution, 50 mg/ml (available from Hynson, Westcott and Dunning, and Dade Reagents, distributed by Scientific Products) is diluted 1:1000. This solution is sufficiently accurate for standardization purposes. Crystals of BSP which may form upon exposure to cold must be redissolved by warming the vial. The concentration of 5 mg/100 ml is equivalent to the 50% retention standard for a test dose of 5 mg/kg assuming a plasma volume of 50 ml/kg. The diluted standard is stable for 1 week.

PROCEDURE

1. To 1.0 ml serum, add 7.0 ml alkaline buffer and mix.
2. Measure absorbance (A_x) at 580 nm or with a filter with nominal wavelength in this region (e.g., 540 or 560 nm) against water as a blank.
3. Add 0.2 ml acid reagent, mix, and again read absorbance (A_b).
4. To obtain absorbance (A_s) of the "50% retention standard", treat 1.0 ml of the standard as a serum sample, omitting step 3.

Calculation:

$$\% \text{ BSP retention} = \frac{A_x - A_b}{A_s} \times 50$$

The following corrections should be made for body weight (478):

110–149 lb, add 1% retention
150–169 lb, no correction
170–189 lb, subtract 1% retention
190–279 lb, subtract 3% retention

NOTES

1. Beer's law. The color in either the Seligson or the acetone method obeys Beer's law up to 100% retention.

2. Stability of color. The color in either the Seligson or the acetone method is stable for at least 24 hours.

3. Interferences.   Bilirubin up to a concentration of at least 20 mg/100 ml does not interfere with the BSP analysis. Hemoglobin in excess of about 50 mg/100 ml or moderate to severe lipemia causes significant errors in analysis. Such hemolyzed or lipemic serum samples can be analyzed by the method of acetone precipitation of serum proteins (176): To 2.0 ml serum, add 8.0 ml acidified acetone (0.5 ml glacial acetic acid added per liter of 88% (v/v) acetone), mix, stopper the tube, and centrifuge. Decant supernate and read at 585 nm or with a filter having a nominal wavelength in this region against a blank of water. Add 2 drops 10% NaOH, mix, and read again. If tubes are not optically clear after alkalinization (examine by Tyndall effect), they should be recentrifuged. The exception to this rule is the occasional lipemic serum which may have turbidity prior to alkalinization and therefore should not be clarified after alkalinization for the most accurate results. Quantities of serum and acetone can be scaled down if the cuvets employed require less supernate for readings. To obtain the absorbance of the "50% retention standard", treat 2.0 ml of the standard as a serum (the absorptivity of BSP in the acetone solution is lower than that in water alone). Calculations are the same as for the Seligson method.

4. Correction for BSP present prior to test.   In individuals free of hepatic disease, BSP is cleared from the blood at a sufficiently rapid rate so that the test can be performed on successive days without interference. In the presence of liver disease, particularly that of an obstructive etiology, BSP can be retained in the blood for extended periods of time. To correct for retained BSP, it is necessary to obtain a control serum sample prior to the BSP injection. This control sample is then analyzed for BSP and if a significant "% retention" is found, this value must be subtracted from the calculated "% retention" of the test sample.

ACCURACY AND PRECISION

The Seligson method is quite accurate for clear samples or those with minimal hemolysis or lipemia. In the acetone method there is a loss in the precipitation process of about 3% of the BSP present.

The precision of the methods is about ±3% (in terms of % of "% BSP retention") for values above 20% retention. Below 5% retention, the precision becomes progressively poorer.

NORMAL VALUES

A survey of the literature (174) gives the following normal values following a dosage of 5 mg/kg: <10% retention at 30 min, <6% retention at 45 min, and <3% retention at 60 min.

Although there is by no means unanimity of opinion, most relevant investigations have demonstrated a slower rate of BSP removal in infants as compared with adults (194). Newborns, both term and premature, have significant impairment in all phases of BSP metabolism—uptake, conjugation, and secretion—but this improves markedly during the first 20 days after birth (413). Increased BSP retention also occurs in higher age groups, over 50 years (194).

FLOCCULATION TESTS

The flocculation tests of liver function are designed to detect changes in plasma proteins, chiefly if not exclusively of a quantitative nature, which frequently accompany liver disease. The protein alterations result in turbidity or flocculation upon the addition of various reagents to serum or plasma under carefully prescribed conditions. In general the reactions are not "all or none" phenomena and must be quantified to distinguish normal from abnormal. Although abnormal reactions are typically characterized by augmented flocculation or turbidity, there are exceptions—e.g., the degree of abnormal protein alteration in the chloranilic acid test is inversely related to the degree of turbidity (74).

The serum protein changes which accompany liver disease vary considerably depending upon such diverse factors as the etiology, duration, and severity of the disease; specific host defense mechanisms; nutritional status; and the presence or absence of coexisting extrahepatic disease. Increased γ-globulin and decreased albumin are the most conspicuous protein alterations to accompany hepatic parenchymal damage. In contrast, uncomplicated extrahepatic biliary obstruction typically does not significantly affect albumin and globulin but may result in a marked increase in glycoproteins (159) and lipoproteins (332). γ-Globulin is by far the most common reactant in the various flocculation tests. Consequently, specificity is compromised by the fact that increases in γ-globulin which accompany a multitude of extrahepatic diseases, from infection to neoplasia, may cause positive reactions.

Macroglobulins (IgM) may have an important, perhaps essential, role in the initiation of flocculation (170). In many tests reactivity is augmented by a decrease in serum albumin, which normally acts as a stabilizing agent. A stabilizing role is also exhibited by certain glycoproteins, increases of which may inhibit reactivity (191). In some flocculation tests other protein alterations also influence the reactivity—e.g., lipids and lipoproteins which enhance the thymol turbidity reaction. It is important to emphasize that quantification of turbidity or flocculation is not reliably related to the severity of the underlying liver disease.

The flocculation tests have a diverse physicochemical basis that for the most part is not well defined; this is reflected in the name "empirical tests of liver function," which has been applied to this group. Reinhold (349) appropriately commented, "The complexity of a milieu that includes an array of proteins, lipids, and electrolytes is a formidable barrier to the development of a general hypothesis for the flocculation tests." Mechanisms which have been cited as being important in various tests include disruption of the hydration lattice of proteins, salt-like combinations between proteins and divalent cations, introduction of nonpolar residues on proteins, and electrostatic interactions (349).

During recent years the flocculation tests have become far less popular as more specific, precise, and informative laboratory methods have been implemented. Particularly important in this regard are the general availability of rapid and economical methods for electrophoresis of serum proteins and the ascendency of diagnostic enzymology to a position of paramount importance in clinical medicine. Methods are presented for the three tests which are still performed with any degree of frequency: cephalin-cholesterol flocculation, thymol turbidity, and zinc sulfate turbidity. Other tests that assumed some degree of general acceptance in the past but which are now largely obsolete are listed in Table 22-1. In addition, a few recently introduced tests are included. Table 22-1 is obviously not an exhaustive listing since at least 15 other flocculation tests have been proposed.

CEPHALIN-CHOLESTEROL FLOCCULATION TEST

In 1938 Hanger (166) observed that varying degrees of flocculation occur when sera from patients with liver parenchymal disease are added to an emulsion prepared from cholesterol and sheep brain cephalin. Flocculation, graded on an arbitrary scale from 0 to 4+, was estimated at 24 and 48 hours. The test gained rapid acceptance, particularly after suitable preparations of the cephalin-cholesterol reagent became commercially available. Many modifications have been introduced chiefly to decrease the time required for reading the result, to improve the quantitative basis for estimating the degree of flocculation, or to enhance the stability and reproducibility of the cephalin-cholesterol reagent.

One modification involves reading the degree of flocculation immediately after mixing and centrifugation (296). This rapid test reportedly achieves not only speed but also specificity with minimal loss of sensitivity. Other modifications have taken advantage of the fact that substantially increasing or decreasing the temperature from 25°C results in greater flocculation (371). Incubation at 37.5°C for 3 hours was found to yield flocculation comparable to that after incubation for 24 hours at room temperature (59). Still another application of temperature enhancement involves refrigeration of the test mixture for 4 hours followed by centrifugation and reading (361).

Initial attempts at improved quantification involved testing serial dilutions of serum, but this approach proved to be unacceptable when it was demonstrated that sufficient dilution of sera from normal individuals results in flocculation (294). Saifer (370) recommended quantitative determination of cholesterol in the centrifuged flocculum with results reported in cholesterol units. The supporting data, however, revealed considerable overlapping of cholesterol units for sera giving 1+ to 3+ reactions. Direct photometric measurement of the degree of clearing of the centrifuged supernatant fluid appears to give more discriminative results (59, 185, 220). Another approach is the direct photometric measurement of the degree of turbidity of the resuspended flocculum (361, 433).

Stability and reproducibility of the cephalin-cholesterol reagent have been improved by better methods of preparation, particularly of the cephalin moiety (349). One report claims improved stability from the substitution of desoxycholic acid for cephalin as an emulsifying agent (405). The practice of adding merthiolate to the working emulsion does not augment stability and may cause false-positive reactions (225).

TABLE 22-1. Principles of Some Flocculation Tests of Liver Function

Name of test	Principle
Acid-precipitable globulin (158)	Serum is added to dilute acetate buffer at pH 4.42 and turbidity is determined photometrically. Turbidity is the result of α_2- and β-globulins and is inhibited by γ-globulin. Turbidity is increased in obstructive biliary disease and usually normal in parenchymal liver disease.
Ammonium sulfate test (92)	Serum is added to a concentrated solution of $(NH_4)_2SO_4$ with added NaCl. About 90% of the γ-globulin is precipitated. Reaction is apparently not susceptible to the inhibitory action of albumin and mucoproteins (13).
Cadmium flocculation: Wunderly and Wuhrmann test (474)	Dilute cadmium sulfate is added to serum. Flocculation is the result of γ-globulin (94) and to a limited extent α- and β-globulins. Albumin appears to be a stabilizer (349).
Chloranilic acid test (74)	Chloranilic acid precipitates proteins of normal serum but not those in the serum of patients with cirrhosis.
Colloidal gold (24)	An adaptation to serum of the Lange method for cerebrospinal fluid proteins. A "paretic" curve occurs in liver disease (154). The major reactant is γ-globulin while α- and β-globulins and, to a limited extent, albumin are stabilizers (40).
Colloidal red: Scharlach red (267)	Similar to colloidal gold except that the reagent is composed of scarlet red dye. It has the advantages of lower reagent cost and purportedly more reproducible results (99).
Copper flocculation test: Kunkel test (231)	Serum is added to dilute copper sulfate in barbital buffer at pH 7.50. The flocculator is γ-globulin.
Distilled water turbidity (46); Sia test (395)	Addition of serum to distilled water results in flocculation in the presence of increased γ-globulin, especially IgM but also IgG. The test is nonspecific and is influenced by dissolved CO_2 in water as well as the pH of the serum (349).
Formol gel (147)	This test is not a flocculation reaction but rather is based on gelation which occurs when neutral formalin is added to serum. The reaction is nonspecific for liver disease and is best correlated with γ-globulin levels (349).

TABLE 22-1. Principles of Some Flocculation Tests of Liver Function (Continued)

Name of test	Principle
Gros test: Hayem test (163)	Mercuric chloride in Hayem's solution precipitates γ-globulin. Quantification is achieved by titrating serum with the reagent until permanent flocculation occurs.
Jirgl test (208)	Phosphotungstic acid is added to a filtrate obtained after the treatment of serum with dilute potassium hydroxide and sulfosalicylic acid. The resultant precipitate is dissolved in sodium carbonate and reacted with the Folin-Ciocalteu reagent. It is likely that phospholipids, not mucoproteins as initially thought, are the major reactants (332). Although recommended for distinguishing obstructive from hepatocellular jaundice, the test is frequently positive in cases of hepatic parenchymal damage (363).
Takata-Ara test (415)	Serum is diluted with a solution of sodium carbonate, mercuric chloride, and fuchsin (414). Flocculation is read at 24 hours. The major reactant is γ-globulin while albumin is inhibitory. Many modifications of the method exist including the Gros (Hayem) test.
Uranyl acetate test (219)	Serum is added to dilute uranyl acetate. Flocculation is said to be helpful in distinguishing cirrhosis and the icteric phase of viral hepatitis from obstructive jaundice.
Weltmann test (456)	Serum is serially diluted with increasing volumes of dilute calcium chloride. After heating flocculation occurs in tubes up to a critical dilution which defines the "band" or "zone." The major reactant is γ-globulin while α- and β-globulins (426) and mucoproteins (183) exert an inhibitory effect.

The physicochemical mechanism of the cephalin-cholesterol flocculation test appears to be fundamentally different in several respects from other empirical liver function tests, such as the thymol and zinc sulfate turbidity tests. The fact that cephalin-cholesterol flocculation is relatively insensitive to changes in pH and ionic strength supports the belief that the interaction between serum proteins and the emulsion involves nonpolar bonds. Reinhold (349) believes the reaction involves the introduction of large nonpolar residues on the protein molecules with consequent decrease in hydration and solubility. Also, in cephalin-cholesterol flocculation it is the sol which flocculates, along with coprecipitated γ-globulin, in contrast to most other tests in which serum lipids or proteins are the primary flocculators (349).

Some confusion has been engendered by the use of the term "antigen" to refer to the cephalin-cholesterol stock ether solution. The misnomer presumably originated from a false analogy to the cephalin-cholesterol reagent, which is mixed with cardiolipin from beef hearts and employed in various reagin tests for syphilis. In the hepatic function test no antigen-antibody reaction is involved.

The role of the various serum proteins in cephalin-cholesterol flocculation may be summarized as follows: (a) There is general agreement that γ-globulin is the principal flocculating protein (174, 349). Increases in serum γ-globulin, whether related to liver disease or of totally unrelated etiology, ordinarily result in increased flocculation (103). The more basic (migrating more slowly in an electrophoretic field) γ-globulin appears somewhat more potent in effecting flocculation (131). Although Hanger (167) initially suggested that flocculation depends on an altered globulin, qualitative changes in γ-globulin in liver disease have not been conclusively demonstrated. (b) Albumin is a potent stabilizer or inhibitor of flocculation (174, 349). Consequently, a decrease in albumin such as frequently occurs in liver disease, facilitates flocculation. The stabilizing role of albumin is more susceptible to dilution than is the flocculator role of γ-globulin, hence the fact that sufficient dilution of normal serum results in flocculation (349). Although it has been suggested that qualitative changes in albumin occur in liver disease (297), this has not been corroborated by chemical studies of electrophoretically isolated albumin (69). (c) The influence of α- and β-globulins and lipoproteins, a very minor role at best, is uncertain (174).

CEPHALIN-CHOLESTEROL FLOCCULATION TEST

(method of Hanger, Ref 166, modified by Bunch, Ref 59 and Knowlton, Ref 225)

PRINCIPLE

Serum is added to a cephalin-cholesterol emulsion. After incubation for 24 hours at 25°C, or 3 hours at 37°C followed by centrifugation, the degree of flocculation is read visually and reported on an arbitrary scale. The test is empiric, the physicochemical mechanism of flocculation not being completely understood. Minimal flocculation may occur with normal serum. Flocculation is augmented by an increase in serum γ-globulin or a decrease in serum albumin, both of which commonly occur in diseases affecting the liver parenchyma.

REAGENTS

Cephalin-Cholesterol Stock Ether Solution. The cephalin and cholesterol mixture is available commerically in the dried form from two sources: Difco (Bacto-Cephalin Cholesterol Antigen) and Wilson (Cephalin-Cholesterol Mixture). These are composed of 1 part specially prepared sheep-brain cephalin and 3 parts cholesterol. To prepare the stock ether solution, add 5 ml ether to the vial of Difco product or 8 ml ether to the vial of Wilson product. Stopper and shake well. If turbidity persists, add 1 drop water and reshake. Anesthesia grade ether usually contains sufficient moisture to effect solution. This solution is stable in the refrigerator for many months if kept tightly stoppered to prevent evaporation.

Cephalin-Cholesterol Emulsion. Warm 35 ml water to $65° - 70°C$ in a 50 ml graduated Erlenmeyer flask and add 1 ml stock ether solution slowly with stirring. Increase the temperature slowly to boiling and allow to simmer until final volume is approximately 30 ml. During heating, all coarse granular clumps should be dispersed, resulting in a stable, milky, translucent emulsion from which all traces of ether are eliminated. Cool to room temperature before use. After cooling to room temperature, examine the emulsion over a strong spotlight. If aggregations of a precipitate are present, centrifuge at high speed for about 20 min. A button of lipid may rise to the top. If so, push it aside and remove the smooth milky supernatant emulsion. The emulsion as originally prepared is sterile; it is stable for several months if sterility is maintained during refrigerator storage. If no attempt is made to maintain sterility, the emulsion, if stored in the refrigerator, is usable for 1–4 weeks. The appearance of microbial growth or curdled material in the emulsion results in false-positive reactions. Such an emulsion should be discarded and a new batch prepared.

NaCl, 0.85%

PROCEDURE

1. To 4.0 ml saline in a 12 ml tube (heavy-walled, conical-tipped centrifuge tubes are recommended), add 0.2 ml serum and mix.
2. With each series of tests an emulsion control consisting of 4 ml saline should be run. Negative and positive control sera should also be included when possible.
3. Add 1.0 ml cephalin-cholesterol emulsion to each tube, including the emulsion control, mix thoroughly, and stopper.
4. Proceed with one of the following two alternatives (see *Note 4* below):
   a. Place tubes in dark at $25° ± 3°C$ for 24 hours.
   b. Place tubes in dark at $37°C$ for 3 hours, then centrifuge at about 3000 rpm for 5 min.
5. Read the reaction as follows:

   Negative or 0, no flocculation or precipitation

   1+, minimal precipitation or flocculation

   2+, definite precipitation or flocculation

   3+, considerable precipitation or flocculation but definite residual turbidity in the supernate

   4+, precipitation complete and supernate clear.

ULTRAMICRO PROCEDURE

The test can be scaled down, keeping all proportions the same.

NOTES

1. Standardization. The original procedure of Hanger (167) for preparing partially oxidized cephalin from sheep brains resulted in considerable variation of reactivity of the cephalin-cholesterol emulsion. Commercial preparations utilizing cephalin aged under controlled conditions and subjected to thorough testing by independent laboratories to maintain uniformity have significantly diminished but not eliminated variations in sensitivity (174, 349). For example, it has been claimed that sensitivity varies from one vial to another within the same lot as well as between lots and manufacturers (225). There is no way to check an emulsion except by comparing it with another emulsion using serum samples which represent the entire spectrum from 0 to 4+. The saline-emulsion control is not a completely valid negative control but nevertheless should be included with each batch of tests. If more than minimal (1+ or greater) flocculation occurs, a new emulsion should be prepared. Ideally, known positive and negative serum controls should be run with each series of tests.

2. Effect of light. Neefe and Reinhold (312) demonstrated that exposure of the serum, emulsion, or reaction mixture to daylight or ultraviolet light markedly enhanced flocculation. Improvements in preparation of the cephalin-cholesterol reagent have diminished the photosensitivity (225). Sunlight and incandescent light have been reported to be more deleterious than fluorescent light (225). One report (302) indicated that sera from patients with liver disease are more photosensitive than those from normal individuals, but this was not confirmed in another study (23).

3. Cleanliness of glassware. It is necessary to use chemically clean glassware, since traces of heavy metals or strong acids result in false-positive tests (167). Running the test in duplicate usually rules out a false-positive result caused by a contaminated tube.

4. Effect of temperature. Flocculation increases as the temperature deviates either way from 25°C, although the effect is minimal within the range of ±3°C (371). Our laboratory has compared the results obtained at 25° and 37°C for 100 normal and abnormal sera and is in agreement with Bunch (59) that the results are the same in the vast majority of cases. Of the 100 comparisons, 81 were identical, 14 differed by 1+ in one direction or the other, and 5 differed by 2+.

5. Interference by lipemia. Evidence for such possible interference has been contradictory (58, 392).

6. Time of reading the test. Although Hanger (167) recommended reading the degree of flocculation at both 24 and 48 hours and this practice was widely accepted, there is no convincing evidence that any additional information of significance is obtained by the second reading (349). It is true that the degree of flocculation may increase slightly during the second 24 hour period but usually by no more than 1+. All things considered, the 3 hour reading after incubation at 37°C is preferable to the 24 hour reading at 25°C, particularly since, as discussed in *Note 4* above, the results agree very closely.

7. *Stability of samples.* Fresh serum should be used when possible. Opinions in the literature differ as to the stability of refrigerated serum (174). The stability of 31 normal and abnormal sera was studied in our laboratory. Usually there was no change within 2 days either at room (25°C) or refrigerator temperature, but occasionally a change of 1+ occurred, usually an increase. After 2 days the frequency and degree of change increased. Frozen serum may be used but should be thawed rapidly at 45°C to minimize protein alterations (225). False-positive results are obtained with sera inactivated for 30 min at 56°C.

ACCURACY AND PRECISION

Since the cephalin-cholesterol test is empiric, accuracy must be determined on a clinical rather than a chemical basis. The major difficulty continues to be the lack of any exact method for standardizing the cephalin-cholesterol reagent.

The test is reproducible within the limits of the visual reading of the test. For example, when a particular degree of flocculation lies between 2+ and 3+, some analysts may read it as 2+, others as 3+.

NORMAL VALUES

There is general agreement in the literature that at least 95% of normal individuals have values of 0 to 1+ when read after a 24 hour incubation at 25°C or after a 3 hour incubation at 37°C (174). Knowlton (225) found a 2+ or greater result in only 2% of healthy individuals. Since γ-globulin is the principal flocculating protein to react with the cephalin-cholesterol emulsion, the test is not reliable during the first year of life when serum γ-globulin is at a physiologically low concentration. Variations in serum γ-globulin and albumin concentration unrelated to liver disease are also primarily responsible for apparent racial and geographic differences in normal ranges. For example, in one study of 200 Nigerians who were free of liver disease, 29% had values of 2+ and 5% had values of 3+ (103).

THYMOL TURBIDITY TEST

While investigating the colloidal gold reaction, Maclagan noted that thymol added as preservative reacts strongly with sera from patients with liver disease to produce turbidity, and in 1944 introduced the thymol turbidity test as an indicator of hepatic parenchymal damage (265). As originally described, the test employed a saturated solution of thymol in a barbital buffer at pH 7.8 to which serum was added. The turbidity at 30 min was compared visually with gelatin standards and reported on an empirical scale. The test has subsequently undergone many technical modifications and improvements.

The reagent described by Maclagan proved to be imprecisely formulated particularly as a result of the rather wide variation in thymol concentration found in solutions saturated at "room temperature" (91). Preparation of the reagent as a 10% ethanolic solution was recommended to increase the solubility of thymol (91). This did not prove successful, however, because alcohol modifies the behavior of lipids and proteins so that turbidity is independent of the thymol (349). Furthermore, the reactivity of such a reagent is ultimately decreased as a result of thymol loss through crystal deposition on the walls of the container. Another difficulty with the thymol reagent as originally formulated is that the pH is 7.65 rather than the specified 7.80 (349). Mateer *et al* (277) recommended a thymol reagent in barbital buffer at pH 7.55 and claimed a sensitivity which was twice that of the original Maclagan reagent. Subsequent investigation indicated that measurements made at pH 7.55 are actually about 1.5 times as great as those at pH 7.80 (349). Glycylglycine buffer has been proposed as a substitute for barbital buffer (72) but apparently without wide acceptance. More recently tris-hydroxymethylaminomethane buffer was recommended for greater reagent stability, although decreased reactivity was noted toward serum triglycerides and chylomicrons (351).

Visual comparison of turbidity with egg albumin-gelatin-formazine standards as in the original Maclagan procedure proved imprecise. Shank and Hoagland (391) improved precision by using photometric measurement of the turbidity with barium sulfate suspensions as standards, a modification which is widely accepted today. The use of colored solutions, such as Evan's Blue or copper sulfate, as standards is unacceptable because of their dependence on the spectral band width of the photometer and failure to reproduce the light-scattering effects of thymol turbidity (349). Other standards which have been proposed include semipermanent colloidal glass suspensions (350) and stable latex particle suspensions (117).

A misprint in the original directions for preparing the barium sulfate standards engendered considerable confusion regarding the units of measurement of thymol turbidity. Directions in the Shank and Hoagland publication (391) called for an 0.0962 N BaCl$_2$ solution instead of the intended 0.0962 M solution. Use of 0.0962 M BaCl$_2$ would have given turbidities equivalent to the gelatin standards of Maclagan. The 0.0962 N solution, however, resulted in standards having a turbidity of one-half that of the Maclagan standards, so that 1 Maclagan unit = 2 Shank-Hoagland units. The situation was complicated by the fact that the error was corrected in reprints supplied by the authors. In 1955 the Commission on Liver Disease of the Armed Forces Epidemiological Board recommended adoption of the "incorrect" units of Shank and Hoagland (350). Although this recommendation has been generally adopted, at least in the United States, it is imperative that the unit of measurement, Maclagan or Shank-Hoagland, be designated.

The physicochemical mechanism of the thymol turbidity test has not been established with certainty. Other phenols react similarly with sera from patients with liver disease to produce turbidity, but thymol is the most active in this respect (265). Reinhold (349) advanced the theory that the nonpolar substituents of thymol lower the solubility of certain β- and γ-globulins by rupturing the hydration lattice of these proteins. Furthermore, the enhancement of γ-globulin reactivity by lipids, particularly phospholipids, may result from cleavage of the protein-lipid combination by thymol, which then reacts with the liberated lipid to produce turbidity. Two visibly different reactions do occur, a finely dispersed turbidity associated with sera rich in lipids and lipoproteins and a coarser turbidity associated with sera having increased γ-globulin (349). Attempts to define the role of the various serum protein fractions in thymol turbidity have resulted in considerable data, much of it conflicting (174, 349). Nevertheless, a few conclusions are warranted:

1. γ-Globulin, particularly the subfractions with highest isoelectric points, reacts with thymol reagent to produce turbidity and ultimately flocculation.
2. Lipoproteins can initiate or at least enhance turbidity. Reactivity is correlated with the entire S_f 0–400 group of lipoproteins but not with any particular subclass (333).
3. Triglycerides and chylomicrons are well correlated with positive thymol turbidity tests in postprandial hyperlipemia. This is a corollary of the effect of lipoproteins.
4. Albumin provides a stabilizing effect on serum and thus tends to inhibit thymol turbidity and flocculation. Albumin obtained from patients with liver disease reportedly has diminished stabilizing activity (11), although chemical studies have not demonstrated any qualitative change in albumin associated with liver disease (69).
5. Mucoproteins inhibit thymol turbidity.

Maclagan (265) observed that flocculation occurred in some tubes incubated overnight in the thymol turbidity test but did not feel that this phenomenon provided significant additional information. Neefe (311) subsequently recommended the "thymol flocculation test," concluding that it is a more sensitive indicator of liver disease than thymol turbidity. Flocculation is read visually after an 18 hour incubation on a 0 to 4+ scale using the same criteria as for the cephalin-cholesterol flocculation test. Values in excess of 1+ are considered abnormal (311). Flocculation occurs with sera having increased γ-globulin and is affected little if at all by lipoproteins and other lipids. Increasing temperature favors flocculation in contrast to its inhibitory effect on thymol turbidity (477). Although flocculation may occasionally occur subsequent to a normal thymol turbidity test (311), the two tests have a relatively high correlation coefficient of 0.78 (266). Opinions vary regarding the relative sensitivity of the 'two tests (215, 311). It is rather well established that thymol flocculation is less frequently positive in the acute stages of hepatic parenchymal disease (349).

DETERMINATION OF THE THYMOL TURBIDITY OF SERUM

(method of Maclagan, Ref 265, with modifications of Shank and Hoagland, Ref 391, and the thymol reagent of Reinhold and Yonan, Ref 352)

PRINCIPLE

Serum is added to a saturated solution of thymol in barbital buffer at a pH of 7.55. The resultant turbidity is measured photometrically at about 650 nm. The test is empiric, but increased turbidity is associated with quantitative increases in serum, γ-globulin, lipids, or lipoproteins, and a decrease in serum albumin.

REAGENTS

Thymol. Thymol of USP grade or better should be used. Thymol showing a yellow discoloration or yielding a turbid reagent should be recrystallized as follows: Dissolve a bulk amount of thymol in 95% ethanol and remove any insoluble matter by filtration. Precipitate the thymol by addition of cold water. Filter off the crystals and dry over calcium chloride for 2 or 3 days.

Thymol-Barbital Reagent, pH 7.55. Boil 1100 ml water for approximately 5 min to drive off CO_2. Cool to about 95°C. Add 300 ml to 6.0 g thymol crystals in a 2 liter Erlenmeyer flask. The thymol will melt and partially dissolve. Add 3.09 g barbital, 1.69 g sodium barbital, and 720 ml of the hot water. Stopper the flask and mix immediately by vigorous rotation for 5 min, periodically releasing the pressure by removing the stopper. Leave at room temperature until it has cooled to 25°C. Add 20 ml of previously boiled water to compensate for evaporation. Seed with approximately 1 g thymol crystals and shake flask vigorously until supernatant solution becomes clear. Leave at 25° ± 1°C until the following day. Mix and filter through Whatman 1 paper. The pH must be between 7.50 and 7.60 at 25°C. If outside these limits, adjustment must be made by altering the ratio of barbital to sodium barbital in a new preparation. Store reagent at 25° ± 1°C in a well stoppered container in order to protect against changes in pH from uptake of CO_2. Reagent that is slightly turbid can be used, but when distinct turbidity develops it should be discarded. Turbidity in fresh reagent is due to unduly prolonged heating, overheating, or impure thymol. The concentration of thymol in this reagent averages 106 mg/100 ml.

Barium Chloride Solution, 0.0962 N. Weigh 1.175 g $BaCl_2 \cdot 2H_2O$ into a 100 ml volumetric flask and add water to mark.

H_2SO_4, 0.2 N

STANDARDIZATION

Prepare a $BaSO_4$ suspension by diluting 3.0 ml 0.0962 *N* $BaCl_2$ to volume in a 100 ml volumetric flask by rapid addition of 0.2 *N* H_2SO_4 at

10°C. Temperature is important, and both solutions should be brought to 10° in a bath prior to mixing. A 10 unit turbidity standard is prepared by mixing 3.3 ml 0.2 N H_2SO_4 and 2.7 ml $BaSO_4$ suspension. Similarly, a 20 unit standard is prepared by mixing 0.6 ml 0.2 N H_2SO_4 and 5.4 ml $BaSO_4$ suspension. Absorbances (A) are read against water at 650 nm or with a filter of nominal wavelength in this region. At room temperature there is some tendency for the suspension to settle; suspensions should therefore be mixed immediately prior to reading. The constant K is calculated as follows:

$$K = \frac{\text{units of standard}}{A \text{ of standard}}$$

The 10 and 20 unit standards should yield the same value for K.

The absorbance of $BaSO_4$ suspensions is affected not only by the temperature and length of storage but also by the order of mixing the reagents and the temperature at the time of mixing (43, 135). Since it is practically impossible to reproduce the $BaSO_4$ suspensions exactly, it is strongly recommended that the value of K be averaged from three independently prepared batches of reagents.

PROCEDURE

1. Add 0.1 ml serum by TC pipet to 6.0 ml thymol-barbital buffer. Mix and place in a water bath at 25° ± 1°C (see *Note 1* below).
2. Read turbidity after 30 min against a blank of thymol-barbital buffer at 650 nm or with a filter having a nominal wavelength in this region.

Calculation:

Thymol turbidity units (Shank-Hoagland units) = $A \times K$

(for determination of constant K, see under *Standardization* above.)

N.B. It is recommended that results be rounded off to the nearest whole number.

NOTES

1. Temperature. Turbidity diminishes with increasing temperature probably as a result of increased solubility of the protein-reagent complex and also because the thymol solution is not saturated when used at temperatures greater than 25°C (477). This effect is partially counteracted by the decreased pH of the thymol reagent, and therefore enhanced reactivity which accompanies increasing temperature. The temperature coefficient is about 0.02 pH units per °C (351). A deviation of ±3° from 25° results in a maximal error of about 10% (477).

2. *Noninterference by bilirubin and hemoglobin.* Neither gross hemolysis nor bilirubin up to a concentration of at least 14 mg/100 ml interferes at the wavelength chosen for the test (391).

3. *Interference by lipemia.* Lipemia, including that occurring post-prandially, increases thymol turbidity readings (352). Corrections for lipemia are not possible, since the thymol reagent reacts directly with lipid components to produce turbidity.

4. *Variations caused by differences in photometers.* The absolute absorbance values obtained for a given turbidity and light path are a function of the arrangement of the optical system and light intensity of the photometer or spectrophotometer. Furthermore, different instruments can give different results even when properly standardized—e.g., the Evelyn and Coleman, Jr., photometers behave one way, whereas the Klett photometer and Beckman model DU spectrophotometer behave differently (350). This is ascribed to the fact that the turbidities produced by lipemic sera behave optically in a manner different from the turbidities resulting from the reaction with γ-globulin. The two groups of instruments apparently yield different turbidity measurements for the two types of suspensions.

5. *Coacervation.* Coacervation, in which the sol coalesces to form a separate liquid phase instead of precipitating, may rarely result in disappearance of turbidity once it has formed. In one study this occurred during an 18 hour observation period in only nine out of more than 2000 sera (50). All nine cases had elevated γ-globulin associated with various hematologic abnormalities including multiple myeloma, macroglobulinemia, anemia, and leukemia.

6. *Interference by heparin.* Heparin interferes with the development of turbidity and can reverse turbidity once it has formed (184).

7. *Stability of samples.* The stability of 31 normal and abnormal sera was studied in our laboratory. No critical change occurred in 1 day at room temperature, but at 2 days or more many of the values varied in an unpredictable fashion. Serum was more stable in the refrigerator, little change (up to about 2 units) occurring in 1 week. Serum inactivated at 56°C for 30 min is unacceptable for testing since false-negative results may occur (114).

ACCURACY AND PRECISION

Being an empiric test, thymol turbidity is measured in arbitrary units. It is therefore not possible to determine accuracy in a conventional sense as with quantitative chemical tests.

The precision of the test is about ±5%.

NORMAL VALUES

The normal range for the method presented is approximately 0–6.0 Shank-Hoagland units. The upper limit represents a compromise between the value for fasting serum, 5.0, and that for postcibal serum, 6.6 (349). Many other normal ranges are quoted (174, 349) which are in no small part a reflection of differences in reagent preparation, pH of the reagent, and standardization. The upper limit of normal is much lower in newborns, both premature and term, and may be the result of diminished blood lipids and β-globulins (95).

ZINC SULFATE TURBIDITY TEST

Kunkel (231) introduced the zinc sulfate turbidity test in 1947 as a method for estimating the amount of γ-globulin in serum. Only insofar as γ-globulin is frequently increased in liver disease is the test construed to be a liver function test. Although the zinc sulfate turbidity test is roughly correlated with the amount of γ-globulin in serum (349), the correlation does not justify the use of formulas devised to relate turbidity units to the γ-globulin content (232). In contrast to the thymol turbidity test in which the slowest migrating ("most basic") fraction of γ-globulin is most reactive, the γ-globulin of intermediate electrophoretic mobility has the greatest reactivity in the zinc sulfate test (131). Other serum proteins also influence the reaction—e.g., albumin (93) and mucoprotein (13) both depress the formation of turbidity.

The physicochemical mechanism of zinc sulfate turbidity is somewhat better understood than are other empirical tests of liver function. The concentration of zinc sulfate employed is too low for a salting-out effect (266). Turbidity results from the formation of an insoluble metallic complex with serum proteins, chiefly γ-globulin. This complex is chiefly a combination of zinc ions with the imidazole groups of histidine (164). Complexing zinc with carbamyl groups of the protein is probably also important in the production of turbidity and can explain the marked decrease in turbidity which results from loss of carbon dioxide from serum samples, independent of any change in pH (349):

DETERMINATION OF THE ZINC SULFATE TURBIDITY OF SERUM

(method of Kunkel, Ref 231)

PRINCIPLE

Serum is added to a zinc sulfate solution of low molarity buffered at pH 7.5 with barbital buffer. The resultant turbidity is measured photometrically at 650 nm and expressed in arbitrary units.

REAGENTS

Zinc Sulfate Reagent. Boil an adequate volume of water to expel dissolved CO_2 and allow to cool, protecting against CO_2 uptake by means of a soda lime tube. Dissolve 24 mg $ZnSO_4 \cdot 7H_2O$, 302 mg barbital, and 190 mg sodium barbital in the water and dilute to 1 liter. The pH must be 7.5 ± 0.05 and should be rechecked at intervals, the length of which depends on how often the bottle is opened to the air. This buffer composition is according to Reinhold (349); the directions given originally by Kunkel yield a reagent of higher pH.

STANDARDIZATION

Reagents and technic are identical to those given for the thymol turbidity test.

PROCEDURE

1. Add 0.1 ml serum by TC pipet to 6.0 ml zinc sulfate reagent. Mix and let sit at 25° ± 3°C.
2. After 30 min, mix thoroughly and read turbidity (absorbance, A) against a water blank at 650 nm or with a filter having a nominal wavelength in this region.

Calculation:

$$\text{Zinc sulfate turbidity units} = A \times K$$

(For determination of constant K, see *Determination of the Thymol Turbidity of Serum, Standardization.*)

NOTES

1. Standardization. Although Kunkel (231) claimed to have utilized the barium sulfate turbidity standards of Shank and Hoagland, the normality of barium in the 20 unit standard made according to Kunkel's directions is 0.00282 compared with 0.00259 in the method of Shank and Hoagland. It is unknown whether this represents a typographic error or a deliberate modification. The classic Shank-Hoagland standards have been generally accepted and are recommended in the method presented.

2. Temperature. Turbidity varies inversely with temperature, but the effect of temperature is less than in the thymol turbidity determination (477).

3. Noninterference by bilirubin and hemoglobin. As in the thymol turbidity test, bilirubin and hemoglobin do not interfere photometrically since the dilution factor and wavelength are the same in the two tests.

4. Interference by lipemia. Lipemia results in an increase in turbidity but to a lesser degree than in the thymol turbidity test (336).

5. *Interference by heparin.* Heparin interferes with the formation of turbidity in the zinc sulfate reaction (184).

6. *Stability of samples.* Unless the sample is protected from CO_2 loss, zinc sulfate turbidity decreases with length of storage of serum samples, even at room temperature. The reactivity of stored serum is almost completely restored if the serum is exposed to alveolar air or 5% CO_2 in N_2 for 1–2 min (349).

ACCURACY AND PRECISION

As discussed previously, the zinc sulfate turbidity test has only fair accuracy as a method for determining the amount of γ-globulin in serum. As with other empirical tests of liver function having arbitrary units of measurement, accuracy must be judged clinically.

The precision of the test is about ±5%.

NORMAL VALUES

Various normal ranges have been proposed (174, 349), reflecting at least in part variations in racial constitution of the population sample and geographic location. The normal adult range is 2–12 for a mixed Negroid and Caucasoid population and 2–9 for a Caucasoid population. Much higher values are found in some populations—e.g., Nigerians have nearly twice the average value of Europeans (103). The elevated values in Nigerians are a reflection of high serum γ-globulin levels unrelated to liver disease (103).

In normal infants zinc sulfate turbidity directly reflects the physiologic changes in serum γ-globulin (316). After birth there is a decrease which reaches a nadir within 15–24 weeks and then a rapid increase until about the age of 2 years. Values then slowly increase to adult levels by the 7th or 8th years. There is no significant sex difference or change with age during adulthood (258).

BILE PIGMENTS

The chemistry and biology of the bile pigments are quite complicated, and there are steps in the metabolic pathways still shrouded in mystery. An enormous volume of literature spanning more than a century has been considerably expanded during the past 15 years with important new insights into the origin, transport, conjugation, and excretion of bilirubin. These recent understandings can be treated in a cursory manner only. Excellent reviews are available for the reader requiring greater detail on these topics (33, 34, 86, 120, 146, 245, 358). Much of the older literature, including many contributions in foreign journals, is now obsolete and not of primary concern to the clinical chemist. No attempt is made to cover even all the literature pertinent to the laboratory analysis of bilirubin and its derivatives. The reader to whom this or the older background literature is of interest is referred to the exhaustive reviews of With (470, 471), Lemberg and Legge (241), and others (155, 156, 235, 335).

Bilirubin originates primarily from the breakdown of the heme moiety of hemoglobin in senescent erythrocytes by the reticuloendothelial system (383). This occurs primarily in the spleen, liver, and bone marrow. According to the most frequently proposed version of this catabolic pathway, the initial step is splitting of the iron-protoporphyrin moiety (hematin) from the globin. Hematin is not ordinarily detectable in peripheral blood, but in conditions of massive intravascular hemolysis it is bound to albumin as methemalbumin.

It is now believed that as much as 20% of bilirubin is derived from other sources—i.e., nonheme porphyrins, precursor pyrroles through a "shunt" pathway, and lysis of immature erythrocytes that occurs particularly in disease states characterized by ineffective erythropoiesis (358). This multiorigin component is sometimes referred to as "early labeled" bilirubin because it can be detected within hours or days from the introduction of radioisotopically labeled porphyrin precursors, in contrast to the major component of labeled bilirubin that appears after about 120 days, corresponding to the average survival of normal erythrocytes.

The enzymatic steps in the opening of the tetrapyrrol ring in the conversion of heme to bilirubin has been a controversial subject. The proposed heme α-methenyl oxygenase of Nakajima and co-workers (307) could not be substantiated by others (246, 463). Evidence for a microsomal heme oxygenase was recently presented (419).

Bilifuscin and other dipyrrolic compounds have been reported in stool and urine. Their origin, which was at one time thought to be from the further catabolism of bile pigments, is not established with certainty.

Measurement of the "early labeled" component of bilirubin has not thus far proved to be a clinically useful technic for routine diagnostic purposes. Its use in research has established, however, that this source contributes importantly to the elevated bilirubin found in disorders associated with ineffective erythropoiesis, such as thalassemia. In this disorder the "early labeled" peak is attributed to premature destruction of defective erythrocytes. The "early labeled" contribution to the elevated bilirubin found in pernicious anemia may also originate in the bone marrow from defective erythrocytes and erythrocyte precursors, but it has also been proposed that increased formation of bilirubin occurs in the liver (248, 476). Similarly, increased bilirubin formation in the liver has been implicated in congenital erythropoietic protoporphyria and other conditions (358).

Bilirubin formed from the breakdown of hemoglobin is transported in plasma bound to protein. Albumin is the major carrier, but small amounts of bilirubin have been identified with α-globulins and other serum proteins (20, 224). There appears to be sufficient albumin in normal serum to bind as much as 80 mg bilirubin/100 ml of serum (20). Coupling takes place between the phenolic hydroxyl groups of bilirubin and the primary amino groups in the protein, principally those of lysine (406). Anionic drugs such as salicylates and sulfa, or other anions such as free fatty acids, presumably can compete for binding sites and substantially reduce the bilirubin transport capacity of plasma. It is hypothesized that bilirubin dissociates from its carrier protein at the liver cell membrane and is transported intracellularly by some active process either unbound or attached to high-affinity, specific acceptor substance.

Conjugation of bilirubin with glucuronic acid, and to a lesser extent with sulfuric and possibly other acids, occurs in the liver. There is evidence that extrahepatic conjugation can also occur (455).

Glucuronic acid is conjugated with bilirubin in a transfer reaction from uridine diphosphate glucuronic acid (UDPGA) that is catalyzed by the microsomal enzyme glucuronyl transferase according to the reactions shown in Figure 22-3. For the sake of convenience, the reaction sequence is depicted as the consecutive conjugation of the carboxyl groups of the propionic acid moieties of bilirubin with glucuronic acid molecules. Conjugation occurs through an ester type linkage between the propionic side chain of bilirubin and the hydroxyl of the aldehyde group of glucuronic acid (33, 376). Although a bilirubin monoglucuronide must exist as an intermediary in the formation of the diglucuronide, it is apparently not found in the serum. As discussed subsequently, the "Pigment I" component of direct-reacting bilirubin in icteric serum, which was thought at one time to be bilirubin monoglucuronide, is probably an equimolar mixture of bilirubin and bilirubin-diglucuronide (161, 189, 199, 230, 453).

Conjugated bilirubin is excreted from the liver cell into the bile canaliculus by an active process, the exact mechanism of which is poorly understood. This process can be inhibited by administration of adrenergic steroids or by competition from anionic compounds. In the intestinal tract a small fraction of the conjugated bilirubin excreted in the bile is hydrolyzed and the unconjugated bilirubin reabsorbed. This is referred to as the enterohepatic circulation of bilirubin (244). The fate of most of the excreted bilirubin is conversion to urobilinogen by the flora of the large intestine. The steps in this further degradation are shown in Figure 22-4 (157, 260, 335). Bile pigments may be classified into bilanes, bilenes, biladienes, and bilatrienes, according to the number of double bonds attached to the bridge carbon atoms joining the pyrrole rings—e.g., bilirubin is a biladiene, urobilinogen is a bilane, etc.

Biliverdin is not detectable in fresh normal serum. In obstructive jaundice, however, a green pigment, presumably biliverdin, may be present in concentrations as high as 2.2 mg/100 ml of serum and can be determined spectrophotometrically (234). There is some question, however, whether this green pigment actually is biliverdin (75).

Classifications of the causes of hyperbilirubinemia should take into consideration disruptions that may occur in any of the various steps between production and excretion of bilirubin. The broad distinction between *retention* jaundice, signifying the increased bilirubin above the capacity for hepatic removal that occurs in hemolytic disease, and *regurgitation* jaundice, resulting from destruction of liver parenchyma or obstruction to the outflow of bile, has served generations of clinicians. Refinements are now possible,

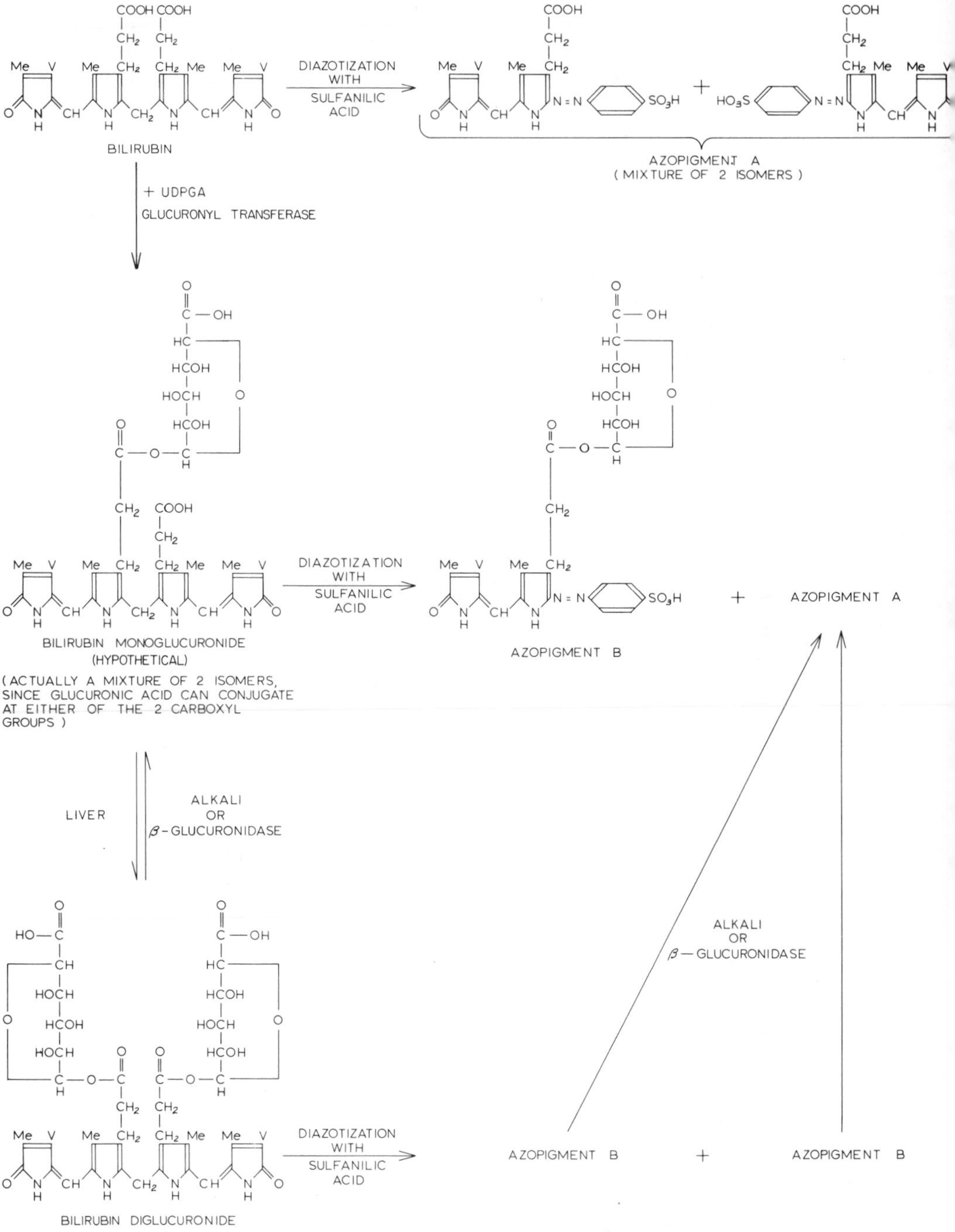

FIG. 22-3. Relationships among bilirubin, its glucuronides, and the azo pigments.

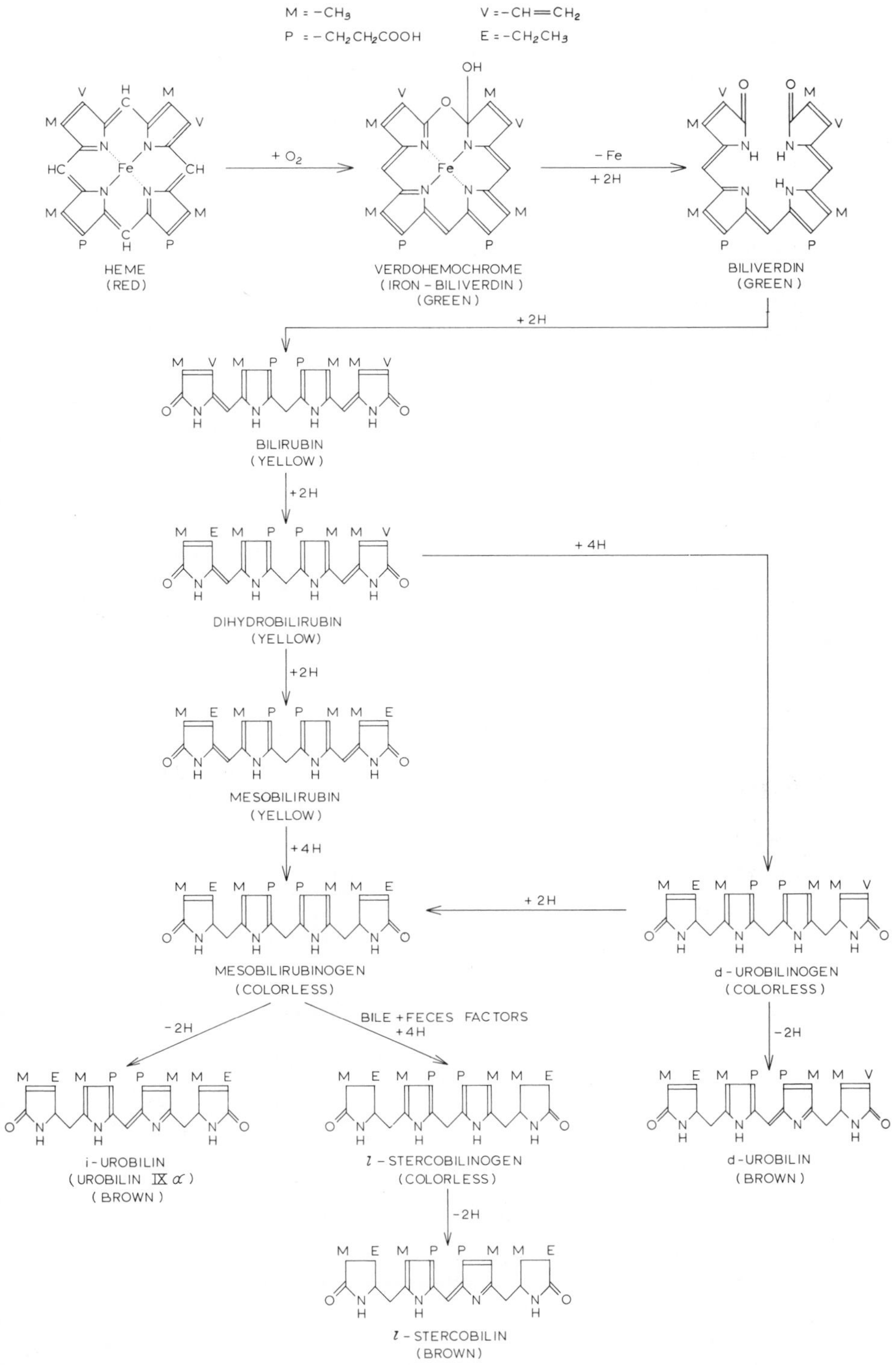

FIG. 22-4. Degradation of heme to bile pigments. Formulas of bile pigments shown in their bis-lactam form.

since congenital enzymatic or functional membrane defects for several steps in the uptake, conjugation, and excretion of bilirubin by the liver have been identified as additional causes for hyperbilirubinemia (381). A simple classification is given in Table 22-2.

TABLE 22-2. Causes of Hyperbilirubinemia (381)

Mechanism	Predominant form of bilirubin present in serum	Disease
"Retention"	Unconjugated	
Overproduction of pigment and/or inadequate hepatic uptake		Hemolytic disorders. Gilbert's disease (may also involve defective conjugation) (26, 37)
		Ineffective erythro- poiesis — e.g., thalassemia
Defective conjugation (Deficiency of glucuronyl transferase or other enzymes)		Neonatal jaundice Crigler-Najjar syndrome
"Regurgitation"	Conjugated	
Functional cholestasis (Defective excretory function from liver cells to biliary canaliculi)		Dubin-Johnson syndrome Rotor's syndrome
Biliary obstruction		Intrahepatic or extrahepatic obstruction
"Combined"	Conjugated and Unconjugated	
Disruption of hepatic architecture		Hepatitis Cirrhosis

BILIRUBIN

It is an interesting if somewhat chastening observation that despite the abundant new knowledge recently amassed about bilirubin, clinical labora- tories generally provide only the time honored determinations of total, direct, and indirect bilirubin. Examples of areas of increased understanding about bilirubin in health and disease include: the elucidation of the existence and sources of "early labeled" bilirubin; the molecular structure of the several bilirubin pigments that occur in serum; the inborn errors of metabolism and congenital enzymatic defects leading to hyperbilirubinemia; the enzymes involved in these pathways and the lessons learned about normal intracellular transport, conjugation, and excretion of bilirubin. The measures of serum bilirubin that were developed and used long before the underlying chemistry was as well understood as it is today continue to define to the satisfaction of medical practitioners virtually all clinical disorders in bilirubin metabolism. Such an extraordinary maintenance of the status quo can probably be accounted for by the requirement undoubtedly sensed by investigators in this field to relate their findings to these standard measures. It can probably also be concluded that the absence of substantial pressure to provide new tests related to bilirubin metabolism signifies that the differential diagnosis of the pertinent diseases is satisfactorily established by the currently available procedures. Therefore the following presentation, which finally emphasizes laboratory methods, presents only those for serum bilirubin, total and direct, and urinary bilirubin.

DETERMINATION OF SERUM BILIRUBIN BY THE DIAZO REACTION

In 1883 Ehrlich (104) introduced the diazo reaction for the detection of bilirubin in urine. The diazo reagent is also occasionally referred to as "Pauly's reagent" because Pauly (325) used it for histidine and tyrosine, although that work was not reported until 1904. Ehrlich showed that the diazo product, which henceforth is called azobilirubin, behaves as an indicator, appearing blue at strongly acidic and alkaline pH levels and red near neutrality. In 1913 Van den Bergh and Snapper (429) applied the diazo reaction to serum after removal of proteins with ethanol. A few years later Van den Bergh and Müller (428) discovered that two types of diazo reaction could be distinguished in sera: the *direct reaction* in which color develops within 30 seconds in the absence of alcohol, and the *indirect reaction* requiring the addition of alcohol for color development. By 1921 the serum diazo reaction, commonly referred to as the Van den Bergh test, was further broken down as follows (243): (a) *prompt direct,* color developing completely within several seconds without alcohol; (b) *delayed direct,* color developing only after 1 min without alcohol; (c) *biphasic,* color developing and increasing with time—i.e., a combination of (a) and (b); and (d) *indirect,* the color developing only after addition of alcohol. The plots shown in Figure 22-5 illustrate the common and practical important varieties of bilirubin color development encountered clinically. The curve labeled "erythroblastosis fetalis" typifies an *indirect* Van den Bergh reaction and the first part of the curve labeled "obstructive jaundice", that part developing before the addition of methanol, typifies the *direct* Van den Bergh reaction. It is important to clarify a point of occasional confusion in the literature. The total color existing after the addition of alcohol is the *total bilirubin*—i.e., the sum of the direct and indirect. The indirect fraction thus is the difference between the total and the direct. Some authors have incorrectly referred to the total as indirect.

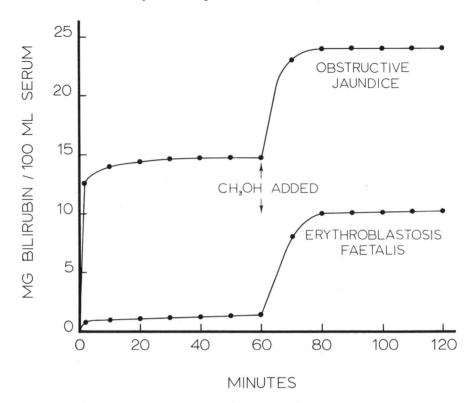

FIG. 22-5. Rate of color development in the Malloy-Evelyn technic for sera from cases of obstructive jaundice and erythroblastosis fetalis.

DIFFERENCE BETWEEN DIRECT AND INDIRECT BILIRUBIN

During at least four decades, through the 1950s, much investigative work was aimed at explaining the Van den Bergh test. Several chemical mechanisms were proposed for the difference between direct and indirect bilirubin, including: (a) the concentration of bilirubin *per se* is responsible for observed variations in rate; (b) the presence and absence of inhibitors and accelerations of the diazo reaction account for variation in the rate; (c) there is only one bilirubin which is bound to different substances; and (d) there are two different but closely related bilirubins.

The earlier edition of this book (174) reviewed in considerable detail the studies supporting and opposing these different proposals. The mass of currently available evidence has so clearly established the relation between conjugation of bilirubin and direct reaction in the Van den Bergh test that the controversies can be considered as laid to rest.

Terminology has occasionally been confused and various systems of nomenclature have evolved: bilirubin, bilirubin monoglucuronide, and bilirubin diglucuronide; Pigment I, Pigment II, and Pigment III; direct bilirubin and indirect bilirubin; or other classifications issuing from either the specific technics used for separation or the particular mixture that was analyzed, such as serum, urine, or bile. In the context of clinical chemistry, it makes the best sense to relate any particular designation to the Van den Bergh designations of direct and indirect bilirubin. It now appears that direct-reacting bilirubin includes both bilirubin diglucuronide (Pigment II or III, depending upon which study is referenced), and a 1:1 mixture of unconjugated bilirubin and bilirubin diglucuronide (Pigment I), whereas indirect bilirubin is unconjugated or free bilirubin. The evidence for this formulation is briefly presented.

The following properties of direct and indirect bilirubin had long been appreciated: (a) Direct bilirubin couples with the diazo reagent in the Van den Bergh reaction in aqueous solution, whereas indirect bilirubin does not (187). (b) Direct bilirubin is not extractable into organic solvents, whereas indirect bilirubin can be extracted by chloroform, benzene, ether, butyl alcohol, and toluene (83, 286, 432). In 1953 Cole and Lathe separated direct- and indirect-reacting bilirubin from icteric serum by reverse-phase chromatography on a column of siliconized Kieselguhr (75). Three fractions were resolved, and all were devoid of any protein moiety. Free bilirubin behaved in their system exactly like the indirect-reacting bilirubin fraction from serum. The direct-reacting fraction was divisible into two fractions by changing the solvent system. The authors named the most polar or water-soluble fraction Pigment II and the fraction with water solubility intermediate between this and free bilirubin as Pigment I. The same group also separated and identified free bilirubin, the mono- and the diglucuronide from icteric serum (76). Hydrolysis of the direct-reacting pigments with acid and with β-glucuronidase showed that they are, in fact, the glucuronide conjugates of bilirubin (33, 382, 416). It was originally thought that Pigment I was the monoglucuronide and Pigment II the diglucuronide (33). Later chromatographic, radioisotopic, and kinetic studies led to the conclusion that Pigment I is an approximately equimolar complex of free bilirubin and the diglucuronide (161, 230, 315, 453). Some observations had suggested that jaundice due to acute obstruction was associated with less of the Pigment I component of direct-acting bilirubin than was the case in chronic obstruction or hepatitis. Other reports appeared to the effect that the diglucuronide (Pigment II) predominates in biliary obstruction and that Pigment I predominates in hepatic parenchymal disease. In general, this further analysis of direct bilirubin has not found clinical usefulness because of significant overlap in the distribution of results from different disease states and the technical problems involved in the partition methods (14, 102, 162, 181, 280, 375, 409, 424).

METHODS FOR DETERMINATION OF TOTAL AND DIRECT BILIRUBIN

Most methods for the determination of bilirubin are based on the diazo reaction. The reaction products of the different serum bilirubin components in the diazo reaction are shown in Figure 22-3 (33, 35).

Fisher and Haberland (119) established in 1935 that the diazo reaction involves splitting a molecule of bilirubin at one side or the other of the central methene $(-CH_2-)$ group to form two dipyrroles, vinyl isoneo-xanthobilirubic acid, and vinyl neoxanthobilirubic acid. They believed that only the former combined with the diazonium salt to form azobilirubin. It is now known (321), however, that in the presence of sufficient coupling agent such as exists in the methods for serum bilirubin, formaldehyde is split off the vinyl neoxanthobilirubic acid yielding a second molecule of azobilirubin, this reaction proceeding at a rate slower than the first. The diazo reaction thus proceeds in two steps, each giving rise to equal amounts of azobilirubin. Brodersen (41) studied the course of reaction in sera and found that it could be described as a combination of three first-order processes, which he termed Phases I, II, and III. Phases I and II were related to conjugated bilirubin and Phase III to free bilirubin. The reaction velocity varied with the dilution of serum employed; this was ascribed to a protein effect and the fact that the amount of diazo reagent remaining after the initial decrease in reaction rate would be in inverse proportion to the amount of serum used. By reverse phase column chromatography (76) and paper chromatography (382), it has been shown that (a) free bilirubin forms azopigment A, (b) Pigment I forms equal amounts of azopigment A and azopigment B, and (c) Pigment II, the diglucuronide, forms azopigment B. Azopigment A is diazotized vinyl isoneoxanthobilirubic acid, and azopigment B is the diazo product of the glucuronide of vinyl isoneoxanthobilirubic acid. Although there is a difference in polarity between these pigments, permitting chromatographic separation, their absorption spectra are identical (76). Figure 22-3 shows the relationships among bilirubin, its glucuronides, and the two azopigments (33, 36). Azopigments A and B are the names given by Billings *et al* (33) to the products diazotized with aniline, not sulfanilic acid. Following the suggestion by Watson (448), the names are retained here for convenience.

Diazo methods that employ preliminary precipitation of protein have the advantage that problems with turbidity are eliminated and dilutions of serum of less than 1:10 can be used. Van den Bergh and Snapper (429) first precipitated the serum proteins with alcohol and then determined the bilirubin in the filtrate. It soon became recognized that some bilirubin, principally direct-reacting, coprecipitates with the protein (10). This of course resulted in low values for total bilirubin. In 1921 Thannhauser and Andersen (423) and later Laemmer and Beck (233) attempted to avoid loss of direct bilirubin in the precipitate by carrying out the diazo reaction first and then precipitating the proteins by alcohol and ammonium sulfate. This approach, however, did not completely succeed (205, 448). King and Coxon (221) revived the method, but it has been demonstrated (236) that results obtained are too low by 5%–15% because of incomplete coupling (alcohol concentration only 37%) and again it was claimed that there is loss on the precipitate. The further modification by Perryman *et al* (328), for which complete recovery of bilirubin was claimed, was stated by Watson and Rogers (451) to be low by up to 60%. Quantitative recovery was also claimed by Stoner and Weisberg (412) who added serum to the diazo reagent and then added concentrated hydrochloric acid before precipitating the protein with ammonium sulfate and ethanol.

Protein precipitation with its attendant problems has been completely avoided in some methods. One way in which this was accomplished was to decrease the alcohol concentration below that which causes the proteins to precipitate. In 1937 Malloy and Evelyn (270) introduced their technic using methanol in a final concentration of 50%. Concentrations of 33% (222), 40% (278), and 62% (153) have also been used, and the protein remains unprecipitated in dilute solution. The theoretical limitation to this general approach is that high dilution reduces the accuracy and precision of measurement at low serum concentrations. Even in such a dilute solution of serum and in spite of the use of blanks, problems with turbidity or hemolysis may be encountered (33).

Numerous modifications of the diazo reaction have been proposed which circumvent the problem of coprecipitation of bilirubin with protein simply by avoiding protein precipitation. An excellent review of details of these modifications and the use of catalysts and accelerators can be found in With's monograph (470).

Many substances promote the diazo coupling of free bilirubin in aqueous solution—e.g., caffeine, sodium benzoate, gum arabic, salicylate, pyridine, and bile salts (2, 145). Determination of total serum bilirubin employing caffeine and sodium benzoate was first introduced by Enriques and Sivo (133) in 1926, and modifications include those of Jendrassik and Cleghorn (204) and others (90, 186). Rappaport and Eichhorn (345) used caffeine, citrate, and urea. Brückner (55) used a combination of antipyrine or acetamide and methanol. Patterson *et al* (324) used urea, ethanol, and phenol; Powell (337) and others (317) employed urea and sodium benzoate. Sims and Horn (397) used sodium benzoate alone, finding the urea to be of no advantage.

An aqueous solution of free bilirubin does not react at acidic pH, presumably because it is insoluble in acid. The role of ethanol or methanol in bringing about the reaction in acidic or neutral solution is to put the free bilirubin into solution, a prerequisite for reaction. If the reaction is carried out at alkaline pH, free bilirubin reacts without alcohol since it is a soluble salt at this pH. As a matter of fact, a fairly sharp distinction between rates of reaction with free and conjugated bilirubins in aqueous solution can be made only at rather low pH levels (236, 315); for example, free bilirubin reacts almost completely in 30 min at pH 3.5, and even at pH 1.5, 4% of the free bilirubin reacts in 30 min (236). Billing (33) and Nosslin (315) expressed the opinion that the presence of conjugated bilirubin in normal sera is questionable and in most instances probably is an artifact of the coupling of free bilirubin in aqueous solution. Witmans *et al* (472) were unable to detect conjugated bilirubin in normal serum by reverse phase chromatography. Just how caffeine, etc., are able to bring about the reaction with free bilirubin at acidic or neutral pH in aqueous solution has not been elucidated.

The diazo methods referred to above develop and quantitate the diazo color at either acidic or neutral pH. Jendrassik and Grof (206) used caffeine and sodium benzoate in the coupling reaction and then added alkali, transforming the red azobilirubin into the alkaline blue form. Modifications of this method have appeared (122, 315, 469), including that of Michaëlsson (289) in whch dyphylline is used as accelerator instead of caffeine to avoid turbidities encountered with caffeine. It has been claimed (122) that the alkaline blue azobilirubin is more specific than the neutral red azobilirubin because of fewer interfering pigments. The alkaline pigment, however, is unstable (397). The use of sodium benzoate as a solubilizing agent at alkaline pH allows for quick determination of total bilirubin without high dilution. Free bilirubin, however, gives a direct reaction in alkaline solution so that the conventional "direct bilirubin" cannot be determined. Others have combined caffeine and sodium benzoate in phosphate buffer to stabilize azobilirubin, which ordinarily decomposes rapidly at alkaline pH (90).

Literally dozens of diazo procedures employing various combinations of solvents, promoters, and pH conditions have been proposed, including systems that incorporate caffeine-sodium benzoate, antipyrine-methanol, urea-ethanol-phenol, and urea-sodium benzoate. A systematic review of the coupling of bilirubin with different diazotized aromatic amines has been published (126). Since the diazo reaction is quite complex, not stoichiometric and affected by many variables, it is not surprising that the different diazo procedures yield different results. Henry published a figure showing widely variant sets of time curves for color development of the direct bilirubin from icteric serum using the most popular methods employed today (175). Equivalent results could not be obtained with these several methods regardless of the time selected for the reaction of "direct" bilirubin.

The original technic for determination of direct bilirubin was qualitative in nature, but it was only natural that attempts were made to put the test on a quantitative basis. The diazo color can be read photometrically before addition of the reagent used for coupling the indirect, free bilirubin. The difficulty is, however, that the direct color, after an initial rapid development, continues to increase at a slower pace so that it is difficult to know at what arbitrary point a reading should be taken to represent the direct-reacting fraction. Jendrassik and Cleghorn (204) arbitrarily chose 5 min as the time to read prior to addition of the caffeine and sodium benzoate. Malloy and Evelyn (270) recommended estimation of the direct fraction from a curve obtained by taking readings at 10, 30, 60, and 120 min after diazotization (without methanol). There appears to be a rather widespread mistaken impression in the literature (e.g., refs 32, 100, 155, 241) that Malloy and Evelyn recommended reading the direct at 30 min. In 1945 Ducci and Watson (100) recommended that a reading at 1 min in the Malloy-Evelyn technic be taken as an estimate of the prompt direct bilirubin and a second reading at 15 min to include the delayed direct bilirubin. Zieve *et al* (480) were in agreement with the proposals of Ducci and Watson. A 1 min reading has also been taken as a measure of direct bilirubin in a modification of the Thannhauser-Anderson method (412). The "1 min" direct bilirubin, although recognized to be an arbitrary measurement, has been widely accepted because of its clinical usefulness (383); the "15 min" delayed direct bilirubin has not proved to be of any particular additional value in diagnosis. Billing (32) examined the correlation between these estimates and the amounts of bilirubin glucuronides present and concluded:

(a) A reading at 30 min (Malloy-Evelyn method, no methanol) is a reliable estimate of the sum of the diglucuronides and what at that time was considered to be monoglucuronide but which now is thought to be the complex of unconjugated bilirubin and bilirubin diglucuronide (Pigment I). (b) The "1 min" bilirubin is not identifiable with either of the glucuronides but correlates well with the sum of the two, the correlation being:

$$1 \text{ min bilirubin} = (0.64 \times \Sigma \text{ glucuronides}) + 0.24$$

Lathe and Ruthven (236) recommended reading the direct at 5 min, and Nosslin (315) recommended 10 min. Brodersen (51), however, concluded from his studies of reaction rates that the "direct diazo reaction" cannot be taken as an exact measure of the concentration of conjugated bilirubin because a significant fraction of the conjugated bilirubin does not react in the direct reaction. Billing *et al* (33) reported that coupling of the monoglucuronide is incomplete in aqueous solution, the addition of ethanol resulting in an increase in color of as much as 40%. White and Duncan (459) studied the Malloy-Evelyn technic and concluded that the rate of the direct diazo reaction can be speeded up by employing threefold increases in sulfanilic acid and sodium nitrite concentrations. Lathe and Ruthven (236) observed a speedup of the indirect reaction after addition of methanol when an increased concentration of sulfanilic acid was employed but found no change in the rate of the direct reaction. Meites and Hogg (182, 283) reported an increase in rate of color development for total bilirubin when the sulfanilic acid, HCl, and sodium nitrite concentrations were increased five, four, and four times, respectively, over the concentrations used in the original Malloy-Evelyn technic. Our laboratory confirmed the findings of Meites and Hogg but not those of White and Duncan. Brodersen (51) concluded that only a part of the bilirubin present forms azobilirubin and that the yield increases with increasing concentration of diazo reagent.

Figure 22-5 shows the rates of color development in the Malloy-Evelyn technic for a serum sample from a case of obstructive jaundice and for a serum sample from a case of erythroblastosis fetalis. The color developing prior to the addition of methanol represents direct-reacting bilirubin principally, while the increase following methanol is due to reaction of the indirect bilirubin. The first observations made on each serum were at 1 min. Relatively speaking, there is much less direct-reacting bilirubin in the serum from the erythroblastotic patient than in that from the patient with obstructive jaundice.

Several completely different analytic approaches to the determination of total and direct bilirubin have been proposed, but none has found wide acceptance. One such approach was to determine the unconjugated bilirubin in serum by extraction into chloroform and benzene followed by evaporation and diazotization of the residue. The exact identity of such extracts has been questioned (100) but not yet reexamined with currently available technics.

Modified solvent extraction procedures or others that combine sequential extraction have also been proposed. Stevenson *et al* (409) devised a method in which total bilirubin is measured by direct photometric measurement of the yellow color after dilution of serum with acidified ethylene glycol, after which free bilirubin is extracted into chloroform. Schachter's (375) technic extracts azobilirubins and partitions them between *n*-butyl-CHCl₃ and aqueous acetate buffer, pH 3. Azopigment A goes into the organic and azopigment B into the aqueous phase. Technical problems (162, 315) and difficulty in the interpretation of results (14, 280) occurred with this technic. The Eberlein procedure (102), based on measuring the yellow color of serum partitioned between ethyl acetate and butanol, has also been criticized for similar reasons (162, 424).

The use of column chromatographic technics to quantitate the bilirubin pigments in serum not only presents too complicated and time-consuming a prospect for routine use, but has failed to provide important enough additional clinical information, as reviewed previously. The methods are inaccurate because of problems with recovery that occur even at modestly elevated levels of total serum bilirubin (181).

CHOICE OF METHOD

Selection of a method for determination of direct and total bilirubin presents a dilemma. The desire to have one method to determine both direct and total bilirubin immediately eliminates many of the proposed methods because they are concerned with estimation of total bilirubin alone. The technic of Malloy and Evelyn, which has enjoyed wide acceptance in the United States, suffers from the fact that the serum is greatly diluted in the test to avoid precipitation of the proteins with the methanol, thus rendering the precision very poor at normal bilirubin levels even if a cuvet with 2 cm light path is used, as in the original procedure. Methods using coupling agents such as caffeine-sodium benzoate are not saddled with this restriction and are definitely more precise at low bilirubin levels. There is some question, however, concerning the time required for complete reaction with such coupling agents (412, 451).

Numerous comparisons of diazo and other methods have appeared in the literature (289, 451). Diazo methods vary in such details as final sulfanilic acid, sodium nitrite, and acid concentrations; amount of sample used relative to final volume of test; time at which it is recommended that readings be taken; and final concentration of methanol in those methods using this means of effecting reaction of free bilirubin. Our laboratory compared seven methods for determination of total bilirubin and 11 methods for determination of direct-reacting bilirubin, using serum samples containing from about 3 to 15 mg total bilirubin per 100 ml (198). Values for total bilirubin agreed within about 10%, but values for direct bilirubin were quite variable. Furthermore, it was obvious from time curves of the development of diazo color of direct bilirubin that, for most diazo procedures, quite a range of values can be obtained depending on the time at which the color is read. Conditions apparently cannot be selected which allow for complete reaction of direct to the complete exclusion of indirect bilirubin. The reaction rates of the free and conjugated forms differ at differing pH values; some evidence exists that a significant fraction of conjugated bilirubin does not react in the direct reaction; the molar absorptivity of the diazo products of bilirubin conjugates in aqueous solution is different from their absorptivity after a miscible solvent such as methanol is added (209); the interference caused by hemolysis differs with different technics.

Unfortunately, no reference method for direct bilirubin has as yet been established. It would appear that a final judgment on the accuracy of any method for determination of conjugated or direct-reacting bilirubin must await the availability of standards of the conjugates in known purity.

Of the various methods studied, it is the authors' opinion that the method of Brückner (56) probably provides the most nearly correct results for direct bilirubin, although this cannot be proved for the reasons mentioned above. This method is also very precise for total and direct bilirubin in the normal and slightly elevated ranges. This great sensitivity, however, makes it necessary to dilute the final color in almost every instance of an elevated level. The Hogg-Meites method (182) gives absorbances only in the range 0.08–0.1 at bilirubin concentrations of 2 mg/100 ml, so great photometric accuracy with levels in this range cannot be expected. The Lathe and Ruthven method (236) requires a level of 7 mg/100 ml to provide an absorbance of 0.1. With Stoner and Weisberg's method (412), significant turbidity can be encountered (198). Conformance with Beer's law is observed over only a limited range with the method of Kingsley (222). The Jendrassik and Grof method for total bilirubin (206) or its commonly employed modifications including micro methods (142) have been favorably evaluated (289).

In the absence of any absolute measure of accuracy, selection of any method is somewhat arbitrary. We present the Malloy-Evelyn method because it is widely used in this country and we have had the greatest experience with it. A version of the Jendrassik and Grof method is also presented as the automated method for total bilirubin.

DETERMINATION OF DIRECT AND TOTAL BILIRUBIN BY THE DIAZO REACTION

(method of Malloy and Evelyn, Ref 270, modified)

PRINCIPLE

Bilirubin couples with *p*-benzenediazonium sulfonate to form azobilirubin as shown in Figure 22-3. Direct bilirubin is determined by photometric measurement of the purple color developed exactly 5 min after diazotization with sulfanilic acid in aqueous solution. This is the time suggested by Lathe and Ruthven (236). Methanol is then added in a concentration short of that causing precipitation of proteins but which releases indirect (free) bilirubin into solution, permitting it to undergo diazotization. The final color developed represents total bilirubin. For determination of direct bilirubin in the original procedures of Malloy and Evelyn, a separate tube was set up in which the methanol used in the determination of total bilirubin was replaced by water. In the modification presented here, the color is read without adding the water and then the methanol is added for development of total bilirubin color. This change was made because it simplifies the test and because results obtained for direct bilirubin are somewhat higher than are obtained in the original procedure. The present authors suspect that the higher values are more nearly correct and, in fact, may still be too low.

REAGENTS

Absolute Methanol, Reagent Grade. Michaëlsson (289) reports that the velocity and degree of color formation may be decreased ($\bar{x}$ 6%) with "old" methanol.

Diazo Reagent. Mix 0.3 ml 0.5% sodium nitrite with 10 ml sulfanilic acid stock solution shortly before use.

Sulfanilic Acid Stock Solution. Dissolve 100 mg sulfanilic acid in 1.5 ml conc HCl and add water to 100 ml.

Sodium Nitrite Stock Solution, 25%. Store in refrigerator.

Sodium Nitrite, 0.5%. Dilute stock solution of sodium nitrite 1:50 with water. Prepare just before use.

Diazo Blank Solution. Dilute 0.5 ml conc HCl to 100 ml with water.

MACRO PROCEDURE, DIRECT AND TOTAL

If only total bilirubin is to be determined, one-half of all volumes can be used in all steps. Step 3 is omitted and only the appropriate calculations are made.

1. Into each of two tubes, labeled B and X, place 0.40 ml serum or plasma and 3.6 ml distilled water.
2. Add 1.0 ml diazo blank solution to tube B and 1.0 ml of diazo reagent to tube X and mix immediately.
3. At *exactly* 5 min after addition of diazo reagent to tube X, read absorbance against B at 540 nm or with a filter of nominal wavelength in this region. This is $A_{5\ min}$.
4. Add 5.0 ml absolute methanol to each tube and mix by gentle inversion. Let stand at room temperature for 30 min and within the next 30 min read absorbance of X vs B. This is A_t. If solutions are transferred to cuvets before reading, wait for bubbles to rise before making readings; it may be necessary to tap cuvet to dislodge bubbles adhering to cuvet walls.

Calculation:

$$\text{mg 5 min direct bilirubin/} \atop \text{100 ml serum or plasma} = \frac{A_{5\ min}}{2A_s} \times 0.02 \times \frac{100}{0.4}$$

$$= \frac{A_{5\ min}}{A_s} \times 2.5$$

$$\text{mg total bilirubin/100 ml serum or plasma} = \frac{A_t}{A_s} \times 0.02 \times \frac{100}{0.4}$$

$$= \frac{A_t}{A_s} \times 5$$

$$\text{mg indirect bilirubin/100 ml serum or plasma} = \text{total bilirubin} - 5 \text{ min bilirubin}$$

A_s is the absorbance of the 2.0 μg/ml bilirubin standard defined in *Note 3.* The standard should be set up as a separate tube in the macro procedure. It is read at the same time as A_t. The effect on absorbance of the presence of methanol in the standard and its absence in tube X during the reading taken for direct bilirubin is accounted for as described in the same note.

NOTES

1. Beer's law. Beer's law is followed up to a serum bilirubin concentration of 15 mg/100 ml even with a filter photometer. The negative error at 20 mg/100 ml is about 5%–9%. At levels greater than 15 or 20 mg/100 ml, therefore, the test should be run on 0.2 ml of serum, adding 3.8 ml of water instead of 3.6 and multiplying the final results by 2.

2. Color stability. Maximal color develops faster with lower concentrations of bilirubin than with higher concentrations. Maximal color is attained and remains essentially constant at 25°C for 30–60 min after mixing at levels up to 20 mg/100 ml.

3. Standardization. The discussion in the previous edition (174) of the problems involved in standardization concluded that commercial availability of standards conforming to the specifications of the joint committee on bilirubin standardization, comprised of representatives of the American Academy of Pediatrics, College of American Pathologists, American Association of Clinical Chemists, and National Institutes of Health (347, 355), should lead to the abandonment of all artificial standards. Reference materials meeting these standards have been available for several years from sources including the US National Bureau of Standards (404). Our own laboratory and the kit method developed here still rely on artificial standards for daily check on photometric or spectrophotometric performance because of the instability of pure bilirubin standards prepared in serum or protein solutions. It is therefore recommended that the test be initially standardized with pure bilirubin and restandardized periodically in a similar manner.

Three alternatives for standardization are presented in order of decreasing preference:

(a) Standardization with bilirubin. Commercial bilirubin preparations vary considerably in purity (123, 284). Molar absorptivities of commercial bilirubins in $CHCl_3$ at 453 nm have ranged from 43 500 to 59 800 (180, 317). Other workers (73) reported the crystallization (140, 313, 379) or commercial availability of bilirubin from human and other species with absorptivities equal to or nearly equal to those specified by the joint

committee—viz, 1.0 cm molar absorptivity of 60 700 ± 1600 at 453 nm in chloroform. Attempts to prolong the utility of various preparations have involved storage in the deep freeze of dimethyl sulfoxide (DMSO) solutions of bilirubin to which albumin has been added (140), or the addition of ascorbic acid (401). Estimation of purity based on absorptivity in $CHCl_3$ is not an ideal situation, but in the absence of better criteria this can provide consistency between laboratories. Such a bilirubin preparation gives values of 64 100 and 59 800 for the absorptivities (based on an equivalent amount of bilirubin, molecular weight 584) of azobilirubin at 545 nm in the absence, and at 540 nm in the presence, of serum, respectively, when the diazo reaction is carried out by the technic presented below. The absorptivity of azobilirubin should be determined as follows: Exactly 20 mg bilirubin is weighed out and dissolved in and made up to 100 ml with 0.2% Na_2CO_3 in a volumetric flask in the dark at room temperature. This standard must be used as soon as possible, at least within 15 min. To one tube add 0.1 ml (20 µg) of bilirubin standard from a TC pipet to 4.9 ml of reagent grade methanol and mix. To a second tube add 4.0 ml distilled water, then 1.0 ml diazo reagent and mix. Mix the contents of the two tubes by pouring back and forth. Prepare a blank by mixing 1.0 ml diazo blank solution, 4.0 ml water, and 5.0 ml methanol. After 30 min read standard against blank at 545 nm. The absorbance should be 0.226. The absorptivity is then calculated as follows:

$$\text{Absorptivity} = A_{545} \times \frac{584}{0.00002 \times 100} \times 0.97$$

The factor 0.97 corrects for the 3% decrease in volume upon mixing equal volumes of water and methanol.

Bilirubin of high purity is rapidly and completely soluble in $CHCl_3$ and 0.2% Na_2CO_3. Impure preparations may not go into solution completely. Michaëlsson (289) states that some preparations contain sufficient calcium to render 1%—2% of the bilirubin insoluble in $CHCl_3$ as calcium bilirubinate.

Since the absorptivity of azobilirubin is significantly lower in the presence of serum than in its absence, the test must be standardized by the method of addition. A serum or serum pool low in bilirubin content is analyzed as given under the macro procedure above, with and without the addition of 0.1 ml of standard (20 mg/100 ml 0.2% Na_2CO_3, 0.1 ml containing 20 µg). The serum or serum pool used should have an absorbance of less than 0.100 at 414 nm and 0.040 at 460 nm at a dilution of 1:25 in 0.85% NaCl. The standard should be kept in the dark and must be used within 10 min of preparation. Thus one pair of tubes, B and X, is run without internal standard added and another pair is run with internal standard added. In the tubes for the internal standard, 0.1 ml standard is added to *both* the B and X tubes by a TC pipet after 0.40 ml serum and 3.5 ml (instead of 3.6 ml) water are added and mixed. Then:

A_s, equivalent to 5 mg/100 ml $= A_s$ with internal standard $- A_s$ without internal standard

There is no question but that the best approach to standardization is use of a bilirubin of very high purity. Less desirable is to determine the absorptivity of the preparation at hand and correct the standard readings to an assumed 100% pure standard. This can be done only if one is using a spectrophotometer (absorptivities shown are for cuvets with an effective light path of 1 cm).

$$\text{Corrected } A_s = \frac{64\ 100}{\text{absorptivity obtained without serum}}$$

$$\times\ A_s \text{ obtained with serum}$$

or

$$\text{Corrected } A_s = \frac{59\ 800}{\text{absorptivity obtained with serum}}$$

$$\times\ A_s \text{ obtained with serum}$$

Curve 1 of Figure 22-6 is the absorption of an azobilirubin standard, equivalent to 9.1 mg bilirubin per 100 ml of serum, devoid of protein, read against a water blank. The peak is at 545 nm. Curve 2 is a diazo blank of a serum to which bilirubin has been added to a concentration of 9.1 mg/100 ml, read against a diazo blank of the same serum without added bilirubin. Curve 3 is the azobilirubin of a serum containing 9.1 mg bilirubin per 100 ml read against a diazo blank of the same serum. Curve 3 drops to 0 absorbance at 480 nm and actually becomes negative at lower wavelengths, because below this wavelength the absorbance of the diazo blank containing unreacted bilirubin (see curve 2) becomes greater than the absorbance of the azobilirubin (see curve 1).

(b) Using an artificial standard. Because of the difficulties encountered in standardization, artificial standards have been used, even by Van den Bergh himself. Ferric thiocyanate, cobalt sulfate, potassium permanganate, phenolphthalein, and methyl red (186, 221) have been used. Some of these worked reasonably well for visual colorimetry but proved lacking in spectrophotometric methods because their absorption curves were not the

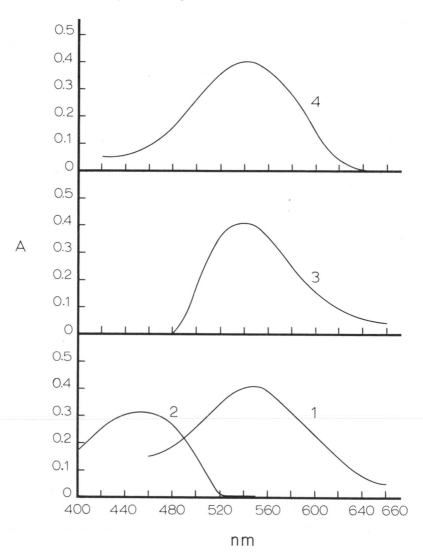

FIG. 22-6. Bilirubin standards. Standards equivalent to 9.1 mg bilirubin/100 ml serum. Curves obtained by Perkin-Elmer model 4000-A recording spectrophotometer. (1) Azobilirubin vs water, (2) bilirubin in serum vs serum blank, (3) azobilirubin in serum vs bilirubin in serum blank, (4) pontacyl violet 6R in 0.5% acetic acid vs water.

same as that of bilirubin. The methyl red standard is pink, having an absorption peak at 525 nm; it appears to be satisfactory for this technic because the azobilirubin color is developed and read at a higher pH, where it is pink. At the pH of the Malloy-Evelyn color (about 2.0), the color is purple so that the methyl red standard is unsatisfactory (236). Recently N-(1-naphthyl)ethylenediamine dihydrochloride was proposed (31), but it too has not found wide acceptance. Pontacyl violet 6R (Du Pont) in 0.5% acetic acid has an absorption curve similar to that of azobilirubin at pH 2.0. A stock dye solution of pontacyl violet 6R (Du Pont) is prepared by dissolving

100 mg in and making up to 1 liter with 0.5% acetic acid. A dye solution equivalent, at 540 nm on a spectrophotometer of high resolution, to the azobilirubin color formed from 5 mg bilirubin per 100 ml is prepared by diluting 7.5 ml of the stock dye solution to 100 ml with 0.5% acetic acid. These dye solutions are quite stable. The absorption curve of this dye at a concentration equivalent to 9.1 mg bilirubin per 100 ml is shown as curve 4 in Figure 22-6. When the artificial standard is used on a filter photometer, the equivalence of the dye color to that of azobilirubin must be checked with bilirubin standards.

(c) Assuming an absorptivity of azobilirubin in the presence of serum of 59 800. A standard equivalent to 5 mg bilirubin per 100 ml has an A_{540} of 0.211 (uncorrected for 3% decrease in volume). This can be used for A_s with a spectrophotometer such as a Beckman model DU or model B, using cuvets with a 1 cm light path, but the readings for filter photometers can vary too much to attempt to set up any similar constant for use with them.

4. Effect of pH. Azobilirubin has indicator properties. In the 50% methanol of the Malloy-Evelyn test it is blue below a pH of about 1.9 and becomes pink above pH 1.9. At alkaline pH it turns blue again. A change in pH from 1.25 to 3.0 is accompanied by a gradual shift in the absorption peak from 560 to 530 nm. Between the pH limits of 1.8 and 2.6 the absorbance at 540 nm with either a filter or a spectrophotometer changes, at the most, about 3%. The pH of a standard devoid of serum is about 1.9. The presence of serum raises the pH about 0.05—0.15 pH unit. At these pH values the azobilirubin is violet or purple. The pH of the solution when the 5 min bilirubin is read prior to addition of methanol is about 1.7.

The above pH values are those observed in our laboratory and are in substantial agreement with those reported by others (67, 236). The authors are, however, in disagreement with the claim that at 540 nm the absorbance is fairly constant in the pH range 1.2—1.6 and decreases above pH 1.6, as well as with the resultant recommendation that the acidity in the test should be doubled (67).

5. Effect of temperature. The reaction in the presence of methanol proceeds at a faster rate at 37° than at 25°C. In addition, increased absorbance of the blank occurs at 37° but not at 25°C (turbidity?). Temperature affects the rate of color development of the direct bilirubin as well as the indirect. Since the "5 min" bilirubin is an arbitrary test, it is imperative that a variable such as temperature, which can affect the results, be held in reasonable control. If room temperature exceeds about 30°C, it would seem advisable to place the tubes in a tap water or other bath in which the temperature is in the range 22—28°C.

6. Serum vs plasma. For reasons that are not apparent, one early report (393) indicated that plasma bilirubin was lower than serum bilirubin, while others (205, 365) claimed that bilirubin was significantly higher in plasma than in serum. No difference would be expected, and subsequent work (289) confirmed this expectation. Some workers, finding no difference between results with serum and plasma from infants by an adaptation of the Malloy-Evelyn method, recommended using plasma so that less sample is required (41). In general, prolonged contact with erythrocytes should be avoided with either sample since it has been shown that red blood cells can remove by adsorption up to 0.7 mg bilirubin per 10 ml cells (447).

7. *Interference by hemoglobin.* That hemoglobin causes low results for bilirubin determined by azobilirubin methods has been abundantly documented (174). That this effect is greater with direct than with indirect bilirubin was confirmed by our laboratory and others (279). Our studies indicated that the presence of 100, 250, and 1000 mg of hemoglobin per 100 ml of serum may cause direct bilirubin results to be low by 15, 40, and 75%, respectively, and total bilirubin results to be low by 10, 15, and 45%, respectively. The degree of interference is inversely related to the bilirubin level. Interference in total bilirubin determination by hemoglobin was assessed with five different and commonly used diazo methods by Michaëlsson *et al* (290). The Malloy-Evelyn method, which gave the highest results, and the Lathe and Ruthven method, which gave the lowest, as well as the Powell method and that of Ducci and Watson were all influenced by hemolysis. A modification of the Jendrassik and Grof method reported by Michaëlsson *et al* (290) showed not only the smallest methodologic error, but also that the results were insignificantly influenced by hemolysis in the sample. In this technic ascorbic acid is added to eliminate excess diazo reagent. Our laboratory was also unable to confirm the claim (445) that hemoglobin does not interfere in the method of Lathe and Ruthven (236).

One explanation offered (397) for the phenomenon of hemoglobin interference is that in the test the hemoglobin is partially converted to methemoglobin and that this conversion is more complete in the test (tube X) than in the blank because $NaNO_2$ is present in the former. At 540 nm, therefore, the contribution of hemoglobin to the absorbance is greater in the blank than in the test. It was claimed that this could be overcome by including nitrite in the blank and reading at 525 nm, the isobestic point for hemoglobin and methemoglobin. This work was performed by a technic employing sodium benzoate rather than methanol in the determination of total bilirubin. This explanation, however, does not hold for the interference observed in the Malloy-Evelyn technic since (a) inclusion of nitrite does not alter the absorption curves of blanks containing hemoglobin, and (b) hemoglobin lowers the absorptivity of azobilirubin read against water. It was recently observed (213) that methemalbumin and methemoglobin present in amniotic fluid do not affect the determination of bilirubin by a minor modification of the Malloy-Evelyn diazotization method, although hemoglobin itself does.

Engel (112) believed the interference to be due to the presence of oxyhemoglobin, which oxidizes the bilirubin, and claimed that the interference was eliminated by saturation with CO and addition of ascorbic acid (too much ascorbic acid prevented formation of diazo color). Michaëlsson (289) concluded that hemoglobin interference in the Malloy-Evelyn technic results from a combination of three effects: destruction of bilirubin after adding methanol, fading of the azobilirubin color, and increased blank absorbance.

8. *Stability of serum samples.* Both direct and indirect bilirubin undergo photooxidation or dehydrogenation to biliverdin or some intermediate

products upon exposure to white light, either artificial or sunlight. Wavelengths between 400 and 600 nm are more destructive than ultraviolet, and red light (600 nm) is ineffective. The effect *in vitro* has been shown to be dependent upon pH, the presence of albumin (319), metals, and chelating agents (125). There is contradictory evidence, however, as to which form of bilirubin is more susceptible. Thus, one group (87) claims that indirect bilirubin is about two or three times more sensitive to light than the direct form. Others (83, 171) find the reverse to be true. Unconjugated bilirubin exposed to weak light was altered so as to give an increased direct diazo reaction, while destruction of bilirubin occurred with intense illumination (38).

Serum bilirubin may drop as much as 50% upon a 1 hour exposure to sunlight (87, 317). Even storage in a shaded area leads to significant deterioration within 12 hours. For optimal stability, both darkness and low temperature are necessary. Frozen samples stored in the dark are stable at least 3 months. Samples are stable in the refrigerator for 4–7 days (178), but by 2 months they are virtually bilirubin-free (317). Even at room temperature samples may be stable for 2 days if kept in total darkness.

ACCURACY AND PRECISION

Of the known bilirubinoids, only bilirubin, dihydrobilirubin, mesobilirubin, and dihydromesobilirubin give the typical red diazo reaction; and of these, only bilirubin has been found in serum (315). The specificity of the determination for total bilirubin thus appears to be fairly good, although if the amount of hemoglobin present in serum is noticeable (at least 50 mg/100 ml) it may result in negative errors of up to about 5%. At elevated levels of bilirubin the hemoglobin level may be higher before becoming noticeable.

Brown color appearing in the diazo reaction of sera from uremic patients (10) is presumably indican or possibly indoxyl glucuronate (169). A correction can be made for this color by reading at two wavelengths (207). The contribution of this nonspecific color reaction to the diazo color in azotemic sera is insignificant with elevated bilirubin, but may be significant in azotemic serum with normal bilirubin concentration. A yellow diazo reaction has also been observed and attributed to the presence of urobilinogens (68) or to reaction with the histidine and tyrosine of serum proteins (315). Although urobilinogen gives an orange or yellow color with diazo reagent there is no detectable urobilinogen in normal sera, although it has been found in serum in association with increased urinary urobilinogen output (171). Biliverdin does not interfere because it does not react in the diazo reaction (104).

Other variables bearing on the accuracy of the diazo test include the complicated effect exerted by protein and the inhibition exhibited by NaCl at low pH (289, 315). An analytic artifact has been attributed to dextran in the sample, following treatment with that agent as a plasma expander (88). This effect was due to turbidity developing upon the addition of methanol.

The accuracy of the "5 min" bilirubin as a measure of the conjugated bilirubin is in considerable question, as previously discussed. The values obtained with normal sera may even be artifacts. Since at least the major portion of bilirubin reacting as "direct" bilirubin in normal serum is unconjugated, a "direct" bilirubin result of 0.2 mg/100 ml would be normal for a high normal total bilirubin but would be elevated if the total is low (315). At elevated levels, however, the results probably parallel fairly closely the concentration of conjugated bilirubin. There is evidence that the absorptivity of azobilirubin in aqueous solution from the direct-reacting bilirubin is significantly less than the absorptivity of the azobilirubin in water-methanol solution (209). The concentrations of sulfanilic acid, $NaNO_2$, and HCl are twice those in the original Malloy-Evelyn procedure for determination of the direct fraction because of omission of the water replacement for methanol. This results in somewhat higher values for direct bilirubin than are obtained by the original procedure.

It has been claimed (372) that the absorption curves resulting from the diazo reaction of the Malloy-Evelyn method carried out on normal sera show the typical maximum of azobilirubin in only about 30% of cases, throwing doubt on the formation of significant amounts of azobilirubin from normal sera. Curves on several dozen normal sera have been run in our laboratory, and in every case there was a detectable peak of azobilirubin.

There is a particularly large number of drugs that cause abnormal liver function. These have been reported to lead to elevated values for total bilirubin (106).

Reproducibility of the test in the normal range is so poor that no significance can be attached to any normal result other than it is normal (480). Besides the fact that the absorbance in this range is very low, and therefore in the range of very high relative analysis error, there is another complicating factor. Addition of methanol invariably results in some degree of turbidity, and it is difficult to reproduce the degree of turbidity exactly in the diazo blank and the serum test. Variations in turbidity may result from differences in rate of addition or mixing of the methanol. As a matter of fact, if the serum bilirubin is low, occasionally a negative serum bilirubin may be indicated. This is a spurious result caused by greater turbidity in the diazo blank than in the serum test.

The precision at levels greater than 3 mg/100 ml is about ±5%.

NORMAL VALUES

The upper normal limit for the "5 min" bilirubin is 0.35 mg/100 ml. The normal range for total bilirubin is up to 1.5 mg/100 ml (317, 471, 480). Small deviations from these limits have been reported in older literature under a variety of circumstances and with selected population subgroups: Levels for men appear to be slightly higher than those for women (317); bilirubin decreases after meals and increases during inanition and after muscular exercise (471); slight diurnal variation has been reported (19); fluctuations were attributed to weekend sunbathing in Scandinavia (52).

Fasting for 1–3 days purportedly causes increases of 22%–334% in total plasma bilirubin, attributed to reduced hepatic clearance (39). A recent large systematic study of normals failed to reveal statistically or medically important differences that can be attributed to sex, age, geography within the United States, body habitus, time of day the specimen was obtained, or other factors (348).

Total bilirubin in cerebrospinal fluid is normally less than 0.01 mg/100 ml but increases with plasma levels (6).

AUTOMATED DETERMINATION OF TOTAL AND DIRECT BILIRUBIN

(method of Jendrassik and Grof, Ref 206, adapted to AutoAnalyzer by Gambino and Schreiber, Ref 144)

PRINCIPLE

Total bilirubin is determined by diazotization in the presence of caffeine and sodium benzoate. Direct bilirubin is determined by diazotization at an acidic pH at which indirect bilirubin is insoluble and nonreacting. The absorbance of alkaline azobilirubin is measured at 600 nm. The absorbance of a blank, consisting of the sample and all reagents except the diazo reagent is subtracted.

REAGENTS

Caffeine Reagent. To approximately 400 ml water at 50°–60°C add 37.5 g caffeine, 57.0 g sodium benzoate, and 94.5 g sodium acetate. Mix, allow to cool, and dilute to 1 liter with water. Stable for at least 6 months.

Sulfanilic Acid Stock Solution. Add 10 g sulfanilic acid and 15.0 ml conc HCl to water; mix and dilute to 1 liter.

Sodium Nitrite, 0.5%. Dissolve 5 g sodium nitrite in water and dilute to 1 liter. Store in refrigerator in a stoppered container. Stable 1 month.

Diazo Reagent. Mix 2.5 ml 0.5% sodium nitrite with 100 ml 0.5% sodium nitrite with 100 ml sulfanilic acid stock solution. Make up fresh each day.

Ascorbic Acid, 4%. Dissolve 1 g ascorbic acid in water and dilute to 25 ml. Make up fresh each day.

Tartrate Buffer. Dissolve 50 g NaOH and 175 g potassium sodium tartrate in water and dilute to 1 liter. Stable at least 6 months.

HCl, 0.05 N.. Add 4.2 ml conc HCl to water and dilute to 1 liter.

Bilirubin Standards. Standardization should be performed by using standard bilirubin added to a normal serum (See *Determination of Direct and Total Bilirubin by the Diazo Reaction, Note 3*). Alternatively, freeze-dried serum standards containing elevated bilirubin levels are commercially available.

PROCEDURE

Set up manifold, other components, and operating conditions as shown in Figure 22-7. By following procedural guidelines as described in Chapter 10 steady-state interaction curves as shown in Figure 22-8 should be obtained and should be checked periodically.

The procedure is identical for total and direct bilirubin except that in the direct reaction 0.05 N HCl is substituted for the caffeine reagent. Blanks on fresh serum are usually low and relatively uniform; however, blanks should be run routinely and subtracted from the assay value. Results are calculated from the calibration curve in the standard manner for methods obeying Beer's law.

NOTES

1. See Notes 1, 2, 3, 6, 7, and 8 of manual method.

2. Blanks. Nonturbid specimens may be assumed to have a blank of about 0.1 mg/100 ml. Blanks for turbid specimens must be measured, since in some cases they may amount to over 1 mg/100 ml.

3. Light sensitivity. Red sampler tray covers are available to prevent specimen deterioration during the analysis.

4. Flow diagram. This method is modified from Technicon N-12b by increasing flow through the photometer cuvet to improve $W_{1/2}$ and by controlling bubble timing to decrease noise.

ACCURACY AND PRECISION

The precision for total bilirubin is ±5% at a level of 2 mg/100 ml. Precision for either total or direct bilirubin is correspondingly less at lower levels.

NORMAL VALUES

Same as for manual method.

ULTRAMICRO PROCEDURE

The macro procedure can be scaled down 20-fold without difficulty. Spectrophotometry is then carried out with Lowry-Bessey cuvets in a Beckman model DU. Several variations of this procedure were recently developed and favorably evaluated against reference methods (116, 190, 275, 369, 422, 438).

KIT APPROACH

A recent listing, which probably is already in need of expansion, showed 23 commercial concerns making kits available for determining direct only,

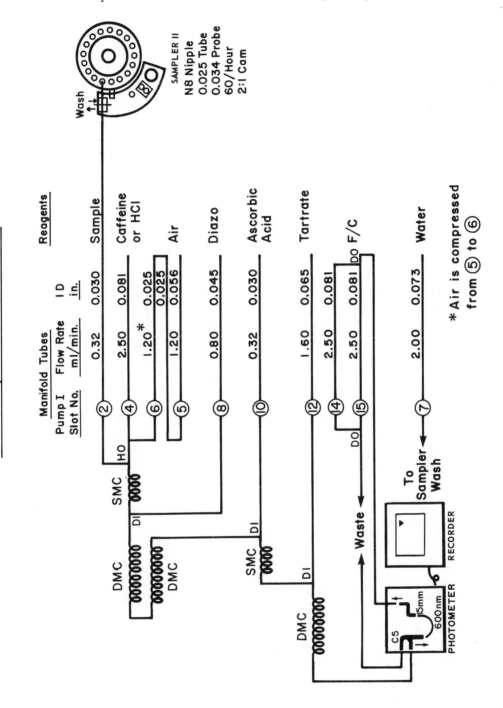

FIG. 22-7. Flow diagram for total or direct bilirubin in serum. Calibration range: 0–12 mg/100 ml. Conventions and symbols used in this diagram are explained in Chap. 10.

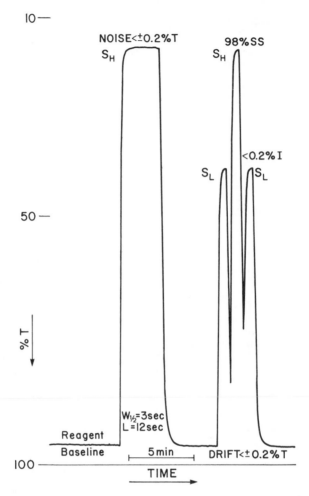

FIG. 22-8. Bilirubin strip-chart record. S_H is a 12.0 mg/100 ml standard, S_L a 6.0 mg/100 ml standard. Peaks reach 98% of steady state; interaction between samples is less than 0.2%. $W_{1/2}$ and L are explained in Chap. 10. Computer tolerance for noise is set at ±0.2% T, for base-line drift at ±0.2% T, and for standard drift at ±1% T.

total only, direct and total, or pediatric bilirubin (101). The majority of these procedures are photometric, but others are qualitative or spectrophotometric. Methodologies include several modifications of the Malloy-Evelyn and Jendrassik procedures. Standards include both bilirubin and bilirubin-albumin in different solvents, as well as artificial standards. Our laboratory developed and tested a modification of the Malloy-Evelyn method that is similar in all respects to the macro method given above except that an artificial standard of pontacyl violet 6R is employed routinely instead of a bilirubin standard (98). The reagent volumes are all scaled down because the sample consists of 0.2 ml serum in a micropipet. It is added directly into the reagent-containing vial that also serves as a cuvet for the specially designed photometer. A modification of this kit method for total bilirubin provides for the addition of sodium nitrite and methanol to the reaction mixture before the sample is added. The absorbance is determined at 5 min. A further modification provides a kit method suitable for total bilirubin determination in newborns by employing a 0.05-ml serum or plasma specimen and appropriately changing the calculations.

BILIRUBIN IN URINE

Over 70 tests have been proposed for the qualitative identification of bilirubin in urine; in general, they can be grouped into four categories according to the principle involved (134): (a) observation of the yellow or brown color of the urine; (b) addition of a dye—e.g., methylene blue—until the dye color dominates over the bilirubin color; (c) oxidation of bilirubin to characteristic colors; and (d) diazotization.

Table 22-3 lists some of the more commonly used qualitative tests and their reported sensitivities. The reader is referred to the first edition of this book (174) for documentation of these data. With (470) concluded from a review of the literature that the normal concentration of bilirubin in the urine of adults, children, and newborns is about 0.03 mg/100 ml. Ideally, a qualitative test is negative in the presence of normal amounts of a substance and positive when there is an abnormal increase. The sensitivity of a qualitative test for urine bilirubin therefore should be in the range of 0.05–0.1 mg/100 ml. Reference to Table 22-3 reveals that several tests possess sensitivity in this range, and some may be sensitive to normal concentrations of bilirubin. This is especially true for those involving a preliminary concentration of the bilirubin by coprecipitation onto the precipitate formed on the addition of $CaCl_2$ (precipitate composed of calcium sulfate and calcium phosphate; as pH decreases, less calcium phosphate precipitates). Certainly of equal importance to sensitivity is the specificity of the test. In general, isolation of bilirubin by coprecipitation increases specificity, and the diazo reaction is more specific than either oxidation or dye dilution tests.

The Gmelin and iodine ring tests, both of which date back to the 19th century, are regarded as insufficiently sensitive and difficult to interpret when pigments other than bilirubin are present. Some of the tests employing ferric chloride in acid as an oxidant have the proper degree of sensitivity, but ferric chloride yields a purple color with salicylates and the presence of increased levels of urobilinogen renders the test difficult to interpret.

The use of methylene blue for the semiquantitative determination of bilirubin in urine was introduced in 1897. The dye is added to urine until the green color which first appears becomes blue. There is evidence that the green color results predominantly from a blending of the blue with the yellow pigments (bilirubin, etc.) present, but that in addition there is a reaction forming a green compound composed of two equivalents of methylene blue per equivalent of bilirubin. In spite of reports as to its clinical usefulness, its sensitivity of about 2 mg/100 ml and its specificity are poor; urines from patients receiving penicillin or riboflavin yield false-positive tests.

TABLE 22-3. Qualitative Tests for Bilirubin in Urine

Name of test	Description	Sensitivity (mg/100 ml)
Oxidation		
Iodine ring test	Layer tincture of iodine over urine. Green ring (biliverdin) is positive for bilirubin.	0.3–1.0
Gmelin ring test	Layer fuming or conc HNO_3 over urine. Bilirubin yellow turns, in sequence, green, blue, violet, red-orange, and finally yellow again.	0.04–1.3
Fouchet	Oxidation to green color (biliverdin) by ferric chloride in trichloroacetic acid.	1
Kapsinow	Oxidation to green color (biliverdin) by ferric chloride in HCl (Obermayer's reagent). Shake with chloroform; indican is extracted, biliverdin not.	0.7
Huppert-Nakayama	Concentrate bilirubin with $BaCl_2$, add $FeCl_3$-ethanol-HCl reagent, heat. Green is positive, turning violet-red on addition of conc HNO_3.	0.1
Naumann	Concentrate bilirubin on talc. Add HNO_3 or Fouchet's reagent. Blue color is positive.	0.1
Zins	Concentrate bilirubin with $BaCl_2$. Oxidize to biliverdin with sodium nitrite and trichloroacetic acid.	
Harrison spot test	Concentrate bilirubin with $BaCl_2$, filter, add Fouchet's reagent. Blue to green color is positive.	0.005–0.1
Watson strip test	Strip modification of Harrison spot test. Thick filter paper strips are saturated with $BaCl_2$ and dried. Dip into urine, add Fouchet's reagent. Green color is positive.	0.1
Dye dilution		
Methylene blue	Add dye solution dropwise until urine becomes blue.	2

TABLE 22-3. Qualitative Tests for Bilirubin in Urine (Continued)

Name of test	Description	Sensitivity (mg/100 ml)
Diazotization		
Hunter	Concentrate bilirubin with $BaCl_2$. Apply diazo test to washed precipitate.	0.01
Müller	Apply diazo test to urine directly or first concentrate with $BaCl_2$.	
Godfried	Concentrate bilirubin with $BaCl_2$, filter, add diazo reagents to precipitate on filter paper.	0.005
Free and Free (Ictotest)	Drop urine onto an asbestos-cellulose mat. Place reagent tablet on mat and dampen with 2 drops water. Positive if blue or purple color develops within 30 sec.	0.05–0.1
Zinc complex		
Worth and Flitman	Color change from yellow to orange with NaOH and $ZnSO_4$.	0.5

A simple test for urine bilirubin is based on the ability of bilirubin to bind with zinc to give a color change from yellow to orange when the bilirubin concentration is over 0.5 mg/100 ml (473).

Free and Free (134) introduced in 1953 a tablet diazo test which is sold commercially (Ictotest, Ames Company). It has a sensitivity in the desired range, has a very high specificity, and is commonly regarded as a good qualitative test for bilirubin in urine (85, 148, 402, 417). This company also produces a dipstick modification (Ictostix).

The difficulties encountered in the qualitative tests for detection of abnormal levels of bilirubin in urine become even more bothersome when attempts are made at quantitative determination. The green color (biliverdin) resulting from oxidation with acidic ferric chloride has been measured photometrically, but as stated previously the oxidation reaction is not very specific. Diazotization of the urine directly has been proposed, but brown colors are obtained rather than the pure azobilirubin color. Approaches to increasing the specificity include the following: (a) correction for non-specific color by reading absorbance at two wavelengths; (b) extraction of azobilirubin into chloroform; (c) diazotization before and after oxidation with hypochlorite, the difference representing the bilirubin present; (d) isolation of the bilirubin by coprecipitation onto calcium phosphate followed by diazotization. The last technic has been criticized because of significant losses of bilirubin in the preliminary isolation step. There is little call for quantitative determination for clinical purposes.

The authors have chosen to present here, as the method of choice for qualitative detection, the diazo Ictotest. As a confirmatory or alternative test, the Harrison spot test is also presented.

ICTOTEST FOR BILIRUBIN IN URINE

(method of Free and Free, Ref 133, 134)

PRINCIPLE

Bilirubin is detected by diazotization, employing a tablet containing *p*-nitrobenzenediazonium *p*-toluenesulfonate, sodium bicarbonate, sulfosalicylic acid, and boric acid.

REAGENTS

Test mats and Ictotest reagent tablets are available in kit form from the Ames Co. The tablets should be protected from heat, moisture, and direct light.

PROCEDURE

1. Place 5 drops urine on one square of special test mat.
2. Place Ictotest reagent tablet in center of moistened area.
3. Flow 2 drops water onto tablet.
4. If mat around tablet turns blue or purple *within* 30 sec, test is positive. Ignore any color produced *after* 30 sec. The amount of bilirubin present is proportional to the speed and intensity of the color reaction. A slight pink or red color which may sometimes occur should be ignored.

HARRISON SPOT TEST FOR BILIRUBIN IN URINE

(method of Harrison, Ref 168)

PRINCIPLE

Bilirubin is concentrated by coprecipitation onto barium sulfate and phosphate and then oxidized by Fouchet's reagent to biliverdin, which is green, or to oxidation products of biliverdin that are blue.

REAGENTS

$BaCl_2$, 10% (w/v)

Fouchet's Reagent. Dissolve 25 g trichloroacetic acid in 100 ml distilled water and add 10 ml 10% (w/v) $FeCl_3 \cdot 6H_2O$.

PROCEDURE

1. To approximately 10 ml acidic urine (if alkaline, acidify with acetic acid), add approximately 5 ml 10% $BaCl_2$. If little or no precipitate forms, add 1 or 2 drops saturated aqueous solution of ammonium sulfate. Mix and filter.
2. Unfold filter paper and spread on another dry filter paper. Allow 1 drop Fouchet's reagent to fall on precipitate. A green or blue color is positive for bilirubin.

NOTES ON THE TWO QUALITATIVE TESTS

1. Semiquantitative application of the tests. A rough estimate of the urine bilirubin level can be made by applying either test to serial dilutions of a urine. The highest dilution giving a positive test can be assumed to possess a bilirubin concentration of 0.1 mg/100 ml, the approximate sensitivity of both tests.

2. Stability of urine samples. Bilirubin in urine is quite unstable and, as expected, degrades much faster at room temperature than in the refrigerator and much faster in sunlight than in darkness. Thus positive tests may become negative within 1 hour if the urine is exposed to sunlight. There is little decrease in the dark in the refrigerator after 24 hours. Since oxidation of bilirubin leads to biliverdin formation, the diazo tests are more likely to yield a false-negative test because of bilirubin degradation than are the oxidative tests.

Ascorbic acid in a concentration of 100 mg/100 ml has been used to preserve bilirubin in urine, even if alkaline, for 24 hours at room temperature. Ascorbic acid, however, interferes with diazo tests, causing them to be negative; it does not interfere with oxidative tests.

It is important to note that in chronic jaundice fresh urine may contain significant amounts of biliverdin (171).

ACCURACY

The diazo test (Ictotest) has a very high degree of specificity, but occasional unexplained false positives have been reported (85). Also, high levels of urobilin or indican give a red color, and salicylates react to give a color ranging from orange-yellow to red to red-purple. Phenazopyridine is reported to cause an elevated but atypical color reaction (106). The Harrison spot test is somewhat less specific. Urobilin and indican in high concentrations give a brown-purple color, and salicylates yield a purple color.

Increased levels of urobilinogen make interpretation difficult because in the reaction urobilinogen yields a rose color that masks the green of biliverdin by giving a final gray-brown color.

There are many drugs that cause changes in liver function. Just as has been mentioned for serum bilirubin, these are very likely to cause an increased or false-positive test for urine bilirubin (106).

DETERMINATION OF FECAL BILIRUBIN

Bilirubin is normally present in feces only until a few years before puberty, but occurs in adult feces as a result of very rapid transit through the gastrointestinal tract. Qualitative detection of bilirubin can be made by Schmidt's sublimate reaction (see *Schmidt Sublimate Test for Urobilin and Bilirubin in Feces* later in this chapter). Quantitative determination is rarely performed.

DETERMINATION OF SERUM BILIRUBIN FROM ITS COLOR

Since bilirubin in solution is yellow, it is only natural that many attempts have been made to quantitate the bilirubin in serum from its color. Some of these have been quite successful. In general, the approaches can be divided into two categories; (a) those in which correction is made for hemoglobin by reading absorbance in a spectrophotometer at multiple wavelengths; and (b) those in which the intensity of the yellow color is determined either by visual comparison or by reading absorbance at a single wavelength with a filter photometer or a spectrophotometer. Since the tests of the second category yield results inherently of relatively low accuracy, the results are expressed in arbitrary units and the test referred to as the *icterus index*.

Spectrophotometric Determination of Serum Bilirubin. Bilirubin in serum has its absorption peak at about 460 nm and an absorptivity of 52 300 at this wavelength (71). Bilirubin constitutes between 40% and 95% of the total yellow pigment in normal sera, with a mean of about 75% (171, 468). Carotene contributes about 8% and xanthophyll ester about 3%. Thus carotenoids (lipochromes) collectively have a normal concentration of about 0.1–0.2 mg/100 ml of serum. The remainder of the yellow color of normal sera (i.e., about 15%) is believed to be due to bilifuscin and mesobilifuscin derived from myoglobin or hemoglobin (468). These various conclusions have been derived from studies which did not specifically consider the effects of increased ingestion of vegetables or eggs (i.e., foods rich in carotenoids). Some early reports, not very quantitative in nature, indicated that the carotenoid share of the yellow color of normal serum can be materially increased by such a diet: Carrots increase the yellow color, even one or two carrots reportedly having a significant effect. Egg yolk causes a significant increase in lipochrome pigment. Up to 50% of the serum color may be due to lipochromes when on a high vegetable diet.

Human serum and plasma always contain some oxyhemoglobin. Various estimates of the top normal limit for plasma hemoglobin have ranged from about 0.6 to 5 mg/100 ml; the top limit for serum is about 20 mg/100 ml or even higher (see *Plasma and Serum Hemoglobin,* Chapter 23). Hemoglobin has three absorption peaks in the visible region: 414, 540, and 576 nm. The absorptivity at 414 nm is about tenfold that at the other two wavelengths. Since the concentrations normally present in serum are above the visibly detectable limit when present alone in water, oxyhemoglobin must contribute to some extent to the yellow color of serum. In the studies mentioned before on the various substances contributing to the yellow color, hemoglobin was not taken into consideration. Heilmeyer (171), however, believed that oxyhemoglobin plays a very minor role in the color of normal serum and plasma because of the shape of the visual perception curve.

The proportion of yellow color contributed by the various substances may be considerably different in certain abnormal situations. In diabetics the carotenoids may be as high as 0.7 mg/100 ml. When jaundice is waning after hepatitis or obstructive cholelithiasis the nonbilirubin yellow color may equal the bilirubin color. Icteric sera contain a greater amount of yellow nonbilirubin pigment than normal. Watson and Rogers (451) expressed the belief that some nonbilirubin pigments, perhaps dipyrroles such as bilifuscin, are present in serum containing much direct bilirubin. The present authors, however, regard the evidence on which this conclusion was based as very questionable. Watson (450) also presented evidence that plasma from infants with hepatitis may contain appreciable amounts of nondiazo coupling yellow pigments of unknown composition. The hemoglobin level of icteric sera is significantly higher than in normal sera. Hemolysis does not become visibly detectable in normal serum until the hemoglobin concentration exceeds about 50 mg/100 ml. Icterus raises the visual threshold of hemolysis—e.g., at a bilirubin concentration of 20 mg/100 ml, hemolysis is not detectable until the hemoglobin level is about 150 mg/100 ml. Drugs such as atabrine can also influence the color of serum and plasma. Table 22-4 lists the principal known pigments of serum and plasma with their absorption maxima.

There has been some confusion concerning the absorption peak of bilirubin in serum or plasma. Prior to the discovery of the chemical nature of direct bilirubin, direct and indirect bilirubin were believed to possess the same absorption spectrum. This conclusion was reached from examination of the absorption curves of sera from patients with high and low ratios of direct/indirect bilirubin. Photometric determination of the icterus index and spectrophotometric methods for the determination of serum bilirubin have assumed the presence of only one bilirubin peak. Some investigations, however, concluded that direct bilirubin has a different maximum than indirect, which if true would appear to invalidate photometric and spectrophotometric technics. The first edition of this book (174) offers a history of the evidence supporting this conclusion, and the observed spectral differences have been used to determine the conjugated and the unconjugated bilirubin in serum (124). However, the fact that spectrophotometric determination of total serum bilirubin (in which bilirubin is assumed to have one peak at 460 nm and correction for hemoglobin is made by measuring its

TABLE 22-4. Principal Pigments of Serum and Plasma and Their Absorption Peaks

Pigment	Absorption peaks (nm)
Bilirubin-protein complex	460
Bilirubin, unbound	420
Oxyhemoglobin	415
	541
	576
Methemalbumin	405
	500
	540
Carboxyhemoglobin	420
Neutral methemoglobin	405
Carotene	455
	490
Xanthophylls	445–451[a]
	475–483[a]
	508–517[a]

[a]Solution in CS_2.

absorbance at 415 or 575 nm) agrees so well with the total serum bilirubin as determined by the Malloy-Evelyn method is taken by the present authors as fairly conclusive proof that direct and indirect bilirubins have the same absorption curves in serum (71).

While it appears that essentially all of the bilirubin in serum has its peak absorption at about 460 nm, aqueous solutions of direct or indirect bilirubin, devoid of proteins, have a peak absorption at a significantly shorter wavelength. There is evidence that this peak is in the region of 410–430 nm. It is not unusual for the wavelength of peak absorption of a substance to shift as a result of combination with protein, but a shift of as much as 40 nm is quite unusual. It has been noted, however, that the peak in freshly prepared aqueous bilirubin is at 440 nm and gradually shifts to 420 nm on standing, possibly as a result of molecular aggregation.

Spectrophotometric methods for determination of serum bilirubin correct for heme pigments by using a formula for two component mixtures. They all measure absorbance in the region 450–460 nm, the absorption peak of bilirubin, and take a second absorption reading either at 415 nm (71, 178) or at the lesser peak of hemoglobin at 575 nm (82, 460). White *et al* (460) claimed that methods employing measurements at 415 nm—the Soret band of hemoglobin—are inferior because of changes that take place in the oxyhemoglobin spectrum after the blood is drawn. Two of the methods

make a correction of 0.44 mg/100 ml for the average nonbilirubin absorbance at 460 nm. In a comparison of results obtained by these various formulas with results yielded by the Malloy-Evelyn method on normal sera and icteric sera from cases of obstructive and hemolytic jaundice, the following observations and conclusions were made (71): (a) None of the methods is accurate for normal sera. (b) At bilirubin concentrations greater than about 2 mg/100 ml, the correlation was very good with all formulas, differences in results between formulas resulting primarily from differences in standardization with bilirubin. (c) Formulas correcting for hemoglobin at 415 nm appeared to be just as successful as those correcting at 575 nm. (d) The spectrophotometric methods referred to above are not applicable if significant turbidity is present. A correction for turbidity was proposed that utilized reading absorbance at a third wavelength. In the method of Chiamori *et al* (71) the serum is diluted with 0.85% saline 1:10–1:40, depending on the degree of icterus. Absorbances are read at 415 and 460 nm in a spectrophotometer of high resolution.
    Then:

$$\frac{\text{mg bilirubin/100 ml}}{\text{serum or plasma}} = \left[\text{dilution} \times (1.17\, A_{460} - 0.099\, A_{415}) \right] - 0.5$$

A procedure has been described in which the absorbance at 450 nm of serum diluted with acidified ethylene glycol is used to measure the total bilirubin content; free or unconjugated bilirubin is then extracted into chloroform and also measured at 450 nm (410).
    Several instruments have been designed for convenient direct spectrophotometric measurement of serum bilirubin (197, 250, 396). Nevertheless it is our opinion, borne out by others (47, 403), that diazo methods for serum bilirubin are still superior to direct spectrophotometric assays for routine use due to their greater sensitivity and hence greater accuracy in the range usually encountered. However, direct spectrophotometry can be useful for determination of total bilirubin in jaundiced neonates where the yellow color from lipochromes, carotenoids, and medications can be ignored and where differentiation between direct and indirect bilirubin is unimportant (1, 195, 394, 460). In a comparison of results obtained with jaundiced serum and bilirubin standards between direct spectrophotometry and the Malloy-Evelyn method, an advantage of the former was less of a decrease after exposure to light (249).
    Icterus Index. In 1919 Meulengracht (288) proposed an arbitrary method of quantitating the degree of icterus present in serum or plasma; 1 ml of sample was diluted with physiologic saline until the color intensity visually matched a 1:10 000 solution of potassium dichromate and the dilution factor was reported as units. Subsequently, a slight modification of this procedure came into use: Undiluted serum or plasma was compared with a series of graded concentrations of potassium dichromate, 1:10 000 being equivalent to 1 unit of icterus. Under ideal conditions, solutions of

potassium dichromate can be matched visually with an icteric sample, but anyone with experience with this method must admit that in many instances a color match is impossible. With the advent of photometers and spectrophotometers it was only natural that they be used in a more objective approach to measurement of the yellow color intensity. Unfortunately, in one of the earliest adaptions to a photometer, a filter having a nominal wavelength of 420 nm was employed. At this wavelength primarily oxyhemoglobin is measured, not bilirubin. Furthermore, potassium dichromate in acidic solution has an absorption curve completely different from hemoglobin. Modification of the icterus index determination for filter photometers and spectrophotometers was made by Henry *et al* (178). In this method serum or plasma is diluted with aqueous sodium citrate and the yellow color measured at 460 nm. Results are expressed in arbitrary units and standardized on acidic aqueous potassium dichromate. The reader is referred to the first edition of this book for details (174). Figure 22-9 shows the absorption curves of a typical icteric serum and acidic aqueous potassium dichromate.

This test is completely nonspecific in that any pigment absorbing at 460 nm interferes. In cases of established icterus, however, interfering pigments usually contribute insignificantly to the total yellow color, and the test is useful for following the course of the icterus. Even for diagnosis a normal result is reliable. A borderline or slightly elevated value may be subject to question. In a study of the correlation of the icterus index (determined by the original Meulengracht technic) with bilirubin concentration, the correlation coefficient was 0.69 for normal sera and 0.84 for sera from jaundice cases (212). The citrate dilution method yields results which are quite reproducible between different photometers and spectrophotometers (178). Normal values are 3–8 icterus units.

Frank hemolysis or turbidity render direct determination of the icterus index inaccurate. Both these interferences can be removed by precipitating the proteins, including hemoglobin, with acetone and measuring the bilirubin color in the filtrate. Adaptation of this method to a filter photometer and spectrophotometer was also made by Henry *et al* (178). Figure 22-9, Curve 2, shows the absorption curve of a typical acetone filtrate. This test too is completely nonspecific in that any pigment soluble in aqueous acetone solution and absorbing at the wavelength employed for measurement interferes. Lipochromes are soluble and appear in the filtrates. Apparently, direct bilirubin is coprecipitated with the proteins, and on the average about 35% of the total bilirubin in serum from cases of hepatic or obstructive jaundice does not appear in the filtrate. Sera from cases of hemolytic jaundice have a low proportion of direct bilirubin and thus yield a higher icterus index relative to the total bilirubin concentration. With the exception of cases of hemolytic jaundice, the icterus units obtained by the acetone extraction method are fairly close to results obtained by the citrate dilution method. As with the citrate dilution method, this test is reliable if the results are within normal limits, and it is useful for following the clinical course of jaundice. A correlation coefficient of 0.84 has been obtained in a

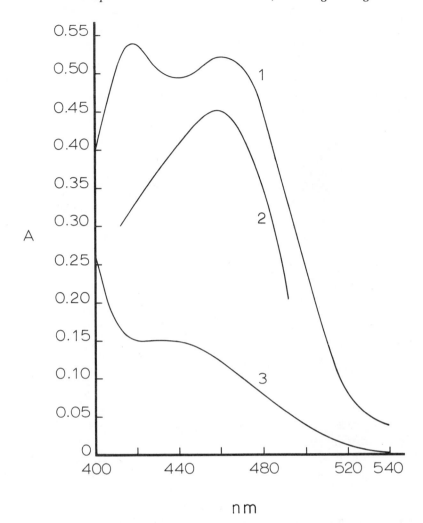

FIG. 22-9. Absorption curves in icterus index determination: (1) 1:40 dilution with 5% sodium citrate of serum containing 23 mg bilirubin per 100 ml vs citrate blank, (2) acetone extract of icteric serum vs 70% acetone blank, (3) $K_2Cr_2O_7$, aqueous 1:10 000, pH 2.8, vs water. Adapted from HENRY RJ, GOLUB OJ, BERKMAN S, SEGALOVE M: *Am J Clin Pathol* 23:841, 1953

comparison between the icterus index (acetone method) and the bilirubin concentration for bilirubin levels up to 3.8 mg/100 ml (408). The normal range is 2—8 icterus units.

BILIRUBIN IN AMNIOTIC FLUID

The treatment of this subject is included here because it naturally fits in with a review of bilirubin methodology, despite the fact that amniotic fluid bilirubin determinations are generally made to evaluate hemolytic disease in the fetus *in utero*.

The usefulness of estimating the amount of pigment in amniotic fluid as an indication of impending fetal distress from erythroblastosis fetalis was first suggested by Bevis in 1950 (27). In general, rising values of the pigment, regardless of the absolute level, indicate a severe condition requiring treatment or intervention, while diminishing values indicate a good prognosis (48, 254). Thus it is necessary to make repeated samplings of the amniotic fluid during the last trimester (48, 132, 143, 257, 282, 318, 334, 461). Virtually all schemes reported to date for assessing the severity of hemolytic disease in the fetus have given some erroneous predictions (48, 81, 281, 334, 357). Contamination of amniotic fluid by icteric fetal blood (132, 256), anencephaly (318, 411), atresia of the intestinal tract with regurgitation of bile by the fetus (254, 256, 318, 411), and infectious hepatitis (3) cause bilirubin to appear in the amniotic fluid. In order to correct for dilution of the amniotic fluid as gestation progresses, a more accurate evaluation has been proposed in which the pigment is reported as a ratio to the protein content of the fluid (70, 298). This approach, however, has not been widely used.

Bevis originally belived that bilirubin was not the pigment implicated (28, 29) but reversed his position on this in 1956 (30). Nevertheless, the nature of the pigment, although presumed to be bilirubin, was not firmly established. For several years various authors merely estimated the severity of the fetal condition by simple visual examination of the peak or hump at 450–460 nm in the spectrophotometric curve of the fluid (21, 251). Another estimate consisted of measuring the deviation of the absorbance in this region from that of a background presumed to be constant with wavelength (70, 282), or the deviation from an hypothetical background curve approximating amniotic fluid devoid of the pigment (61, 70, 132, 254, 257). On the supposition that the yellow color of the amniotic fluid from affected pregnancies was due to some form of bilirubin, diazo methods were employed (42, 143, 298, 411, 452) and very sensitive modifications were offered, such as that employing concentration of azobilirubin by adsorption and subsequent elution and formation of its cadmium complex (229). The major bile pigment of amniotic fluid from Rh-sensitized patients has now been identified with a reasonable degree of certainty as unconjugated bilirubin by careful examination of its spectrum and that of its azopigment, as well as by reverse phase chromatography (49) and paper chromatography (272). Consequently, more spectrophotometric technics have been presented whereby the bilirubin is quantitated by reading at several wavelengths so that corrections may be made for turbidity and the influence of other pigments whose absorptivities at these wavelengths are known (121). Absorbance differences between 490 and 520 nm (281, 374) and 480 and 500 nm (81, 320) are proportional to the bilirubin present. The authors using the latter range describe their approach as giving an estimate of $dA/d\lambda$ at 490 nm where this is maximal and proportional to bilirubin and is little influenced by other pigments or turbidity (81, 320). The usefulness of this method has been discussed and compared to that of direct spectrophotometry (7).

Although the spectrophotometric scanning methods are generally more reliable than the diazo methods, various heme pigments, due to their pronounced Soret peaks in the 400–415 nm region, can obscure the 450 nm peak of bilirubin to give clinically misleading results (214, 257). In addition, amniotic fluid normally exhibits a spectral curve which rises rapidly as the violet end of the spectrum is approached. It is of course this very fact that all authors acknowledge and have tried to cope with by the various approaches mentioned.

It is not surprising that what now appear to be the most accurate methods for the determination of unconjugated bilirubin in amniotic fluid are those which involve extraction of the solvent-soluble pigment with chloroform, resulting in an extract with an absorption spectrum identical to that of pure bilirubin (Figure 22-10) and leaving interfering pigments and other materials behind (48, 327). It has been recommended that aniline be added to the chloroform on the theory that a basic substance, soluble only in the nonaqueous phase, would facilitate liberation of the bilirubin from protein and its quantitative transfer to the solvent phase (269). The method of Brazie (48) is used in our laboratory and is presented here.

FIG. 22-10. Bilirubin in amniotic fluid. Curve a: spectral scan of a slightly turbid amniotic fluid Curve b: chloroform extract of same fluid. Curve c: pure bilirubin dissolved in chloroform.

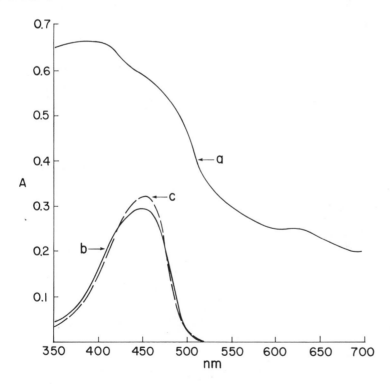

QUANTITATIVE DETERMINATION OF UNCONJUGATED BILIRUBIN IN AMNIOTIC FLUID

(method of Brazie, Bowes, and Ibbott, Ref 48, modified)

PRINCIPLE

Unconjugated bilirubin is extracted from amniotic fluid into chloroform and quantitated by reading the absorbance at 453 nm.

REAGENTS

Chloroform, Reagent Grade

PROCEDURE

Note: Protect reaction mixture from light throughout the whole procedure.

1. Pipet 3 ml amniotic fluid into a conical glass-stoppered centrifuge tube wrapped with aluminum foil.
2. Add 3 ml chloroform. Shake vigorously by hand for 30 sec.
3. Cool the tube in an ice bath for 5 min.
4. Centrifuge for 1 min. If protein mesh persists in the lower ($CHCl_3$) layer, break it up with a glass rod. Recentrifuge for 1 min.
5. Allow tube to return to room temperature.
6. Transfer $CHCl_3$ layer, which should be completely clear, into a cuvet.
7. Read absorbance (A) at 453 nm against $CHCl_3$ in a spectrophotometer of high resolution using a 1 cm light path cuvet. Photometers may also be used (but see *Note 1*).

Calculation:

$$\text{mg bilirubin}/100 \text{ ml amniotic fluid} = A$$

NOTES

1. Standardization of test. The rationale for the simple calculation given above is as follows: Since molar absorptivity of bilirubin in $CHCl_3$ at 453 nm in a 1 cm cuvet equals $60\,700 \pm 1600$ (80) and the molecular weight of bilirubin is 584.7, absorptivity of bilirubin in terms of

$$A/\text{mg}/100 \text{ ml } CHCl_3 = \frac{60\,700 \times 10}{584.7 \times 1000} = 1.04$$

It is satisfactory to use 1.0 as a close approximation for convenience. With a spectrophotometer in good calibration with respect to wavelength and absorbance scale, there is no need to prepare a bilirubin standard. Methods for ensuring this are described in Chapter 1. A photometer using a filter with maximal transmission at about 450 nm may be used, but it is then necessary to standardize the test with solutions of pure bilirubin (180) when the test is set up and periodically thereafter.

2. *Beer's law.* Solutions of bilirubin in $CHCl_3$ give a linear calibration curve to over 1 mg/100 ml on spectrophotometers as well as on instruments of broader bandwidth.

3. *Stability of samples.* It is important to *protect specimens from light* during storage and throughout the assay procedure. Although sterile specimens kept in the dark are reported to be stable for 9 months refrigerated and at least 30 days at room temperature (257), it is probably wiser to keep the sample frozen until it can be analyzed since it cannot be easily replaced if instability is suspected to have affected the result.

ACCURACY AND PRECISION

The accuracy of this test depends upon the absence of pigments other than the unconjugated bilirubin in the chloroform extract. Thus the conflicting reports on the presence of conjugated bilirubin (30, 173, 226, 452) are irrelevant, since the conjugated pigment is not soluble in chloroform. Likewise, heme pigments and vernix cannot interfere. While Brazie *et al* stated that meconium does not interfere in their method (48), consideration should be given to the fact that meconium does contain bilirubin (115, 254).

Errors in obtaining fluid for analysis have been reported (256) where maternal urine, fetal ascitic fluid, and amniotic cyst fluid were mistakenly collected as amniotic fluid. Recovery of bilirubin added to amniotic fluid is 90%–94%.

Precision of the analysis is about ±5%.

NORMAL VALUES

As already stated, it is important that a declining trend be observed in the bilirubin content of amniotic fluid in an isoimmunized pregnancy. A normal or only mildly affected infant results if the level is below about 0.06 mg/100 ml at or prior to 35–36 weeks (48). Additional data for normal pregnancies can be found in many of the references previously cited, but all are obtained by other methods (271) and generally reported in other terms.

UROBILINOGEN AND UROBILIN

Bilirubin is excreted in the bile predominantly in water-soluble, conjugated forms. In the intestinal tract the bilirubin conjugates are reduced by bacterial action. Part of the reduction products are excreted unchanged in the feces, and the remainder are reabsorbed into the portal circulation. Most of the reabsorbed reduction products are reexcreted by a normal liver, but a small amount normally is excreted into the urine. When there is liver damage, less is reexcreted by the liver and more appears in the urine. In complete hepatic obstruction the bilirubin does not reach the intestinal tract, but reduction can still occur if there is infection of the biliary passages and the gallbladder.

The interrelationships between the principal products of bilirubin metabolism are shown in Figure 22-4. Mesobilirubinogen (*i*-urobilinogen), *l*-stercobilinogen, and *d*-urobilinogen are colorless bilanes that are precursors of the colored urobilins, which are bilenes. The urobilins are called *pigments*, and their precursors *chromogens*. Collectively, they are called *urobilinoids*. The urobilinoids apparently are excreted both free and as glucuronides and other conjugates. The prefixes *d*-, *l*-, and *i*- applied to the urobilinoids refer to their optical activity—the designation is solely for convenience because they are not isomers. Upon exposure to air, reduction products begin to oxidize to urobilins.

Normal urine contains small amounts of both bilirubin and urobilinoid chromogens. After excretion, of course, the chromogens begin to oxidize to their respective pigments. Because of the observation of minor differences between fecal pigment and urine urobilin, the term *stercobilin* (L., *stercus* = dung) was introduced in 1871. It is now known, however, that stercobilinogen (*l*-urobilinogen) is usually the predominant chromogen in urine as well as feces, although mesobilirubinogen or *d*-urobilinogen sometimes predominates. Pathologic urines may also contain colorless dipyrroles called propentdyopents which, upon alkaline reduction, yield red pentdyopents (so called because the absorption maximum is at about 525 nm).

Normally in adults the feces contain no bilirubin, and it is probable that the urobilinoids of fresh feces occur only as the chromogens. Stercobilinogen predominates, but mesobilirubinogen also is present. Other pigments such as mesobiliviolin, mesobilifuscin, stercorubin, and stercofulvin may occur. The dipyrrole "fuscins," which in large measure account for the brown color of feces, appear to be anabolic in origin rather than products of hemoglobin breakdown. Some textbooks, however, still erroneously ascribe the brown color of feces to stercobilin (467). In newborns, where the intestinal microbial flora is incapable of transformation, no urobilinoids are formed and bilirubin itself is excreted. The urobilinoids show a gradual increase during childhood, and adult levels are reached a few years before puberty. Bilirubin may also occur in adult feces as a result of very rapid transit of feces through the gastrointestinal tract. Urobilinoids are also present in negligible amounts in feces and urine when the bacterial flora of the intestines is suppressed by antibiotics. *d*-Urobilin has been identified in infected fistula bile samples and in feces of patients treated with Aureomycin or Terramycin, following the withdrawal of antibiotic therapy, and occasionally in patients with no history of antibiotic treatment. *d*-Urobilinogen is formed by bacterial action, but whether it is only an abnormal or a transitory normal intermediary is not known. It has been postulated (444) that the finding of *d*-urobilinogen with no *i*-urobilin or stercobilin in the feces of patients receiving broad-spectrum antibiotics strongly indicates that the effect of the antibiotic is simply to halt bacterial reduction at the *d* stage, preventing normal further reduction to the *i* and *l* stages. With all this, there is still doubt as to the exact nature of the urobilinoids excreted in the feces (196).

Urobilinoids can be quantitated by two approaches: (a) Reduce any urobilins present to chromogens and determine chromogens by the benzaldehyde reaction, (b) Oxidize chromogens present to urobilins and determine urobilins by the zinc fluorescence reaction. Although the latter method is the more sensitive, the former method is preferred for quantitation of urobilinoids, at least in the present stage of development of the methods. All of the chromogens apparently react the same in the benzaldehyde reaction. Insofar as is known, they all have the same diagnostic significance. These facts simplify somewhat an analytic task which still proves difficult.

Separation of urobilinoids has been accomplished by chromatographic and electrophoretic technics (240, 262, 331) but thus far has been of no practical value (470). Likewise, no diagnostic importance is given to serum or plasma urobilinoid levels (240, 470).

For comprehensive reviews of the biochemistry of the urobilinoids, the papers by Watson are recommended (442, 443) as well as the discussions by With (470) and Schmid (381).

DETERMINATION OF UROBILINOGEN BY THE BENZALDEHYDE REACTION

For ease of presentation, the authors follow the custom of referring to the total chromogens as "urobilinogen." In 1901 Ehrlich (105) introduced the reaction with *p*-dimethylaminobenzaldehyde in concentrated HCl as a color test for urobilinogen and noted that the red color reaction was given by certain pathologic urines. This reaction is sometimes called the "Erhlich-Pröscher" reaction because Pröscher also studied it at about the same time (339). It was soon discovered that indole and skatole derivatives, which occur in urine and feces, also yield red color. In 1925 Terwen (420) introduced methods for urine and fecal urobilinogen which have formed the basis for nearly all the methods employed today. He found that the color due to urobilinogen develops with Ehrlich's reagent at once, and that replacement of the HCl with acetic acid by addition of sodium acetate both intensifies the urobilinogenaldehyde color and inhibits the color due to indole and skatole derivatives, thus making the test considerably more specific. Modifications of Terwen's methods have been presented by multiple authors (18, 172, 177, 309, 330, 387, 440, 441, 445, 446). The reaction has been used in a dipstick procedure (165), which is useful as a screening procedure in routine urinalysis although admittedly nonspecific.

In the test of Wallace and Diamond (437) introduced in 1925, Ehrlich's reagent is added directly to serial dilutions of urine. Since Ehrlich's color with urobilinogen is relatively weak unless sodium acetate is added, and since in most cases the color produced with pathologic urines can be obtained by the addition of HCl alone, the color produced in the Wallace-Diamond test is primarily not due to urobilinogen.

The quantitative technics that have been most widely used are those of Watson and co-workers and modifications thereof. Watson *et al* believed there was justification for two different procedures, the first a "semiquantitative method" (445, 446) in which Ehrlich's reaction is carried out directly on urine and feces but with the increased specificity afforded by the use of sodium acetate, and the second a "quantitative method" (387) which is more specific because the reaction is applied after a preliminary extraction that separates urobilinogen from various interfering substances. Urobilinogen in urine and feces is not very stable and in the "quantitative method" for urine and feces and in the "semiquantitative method" for feces the sample is first treated with ferrous hydroxide to reduce all urobilin back to urobilinogen, so that theoretically it does not matter how much oxidation has taken place. Watson and Hawkinson (445) pointed out the possibility that urobilinogen might be oxidized to a substance which cannot be reduced back to urobilinogen. Henry *et al* (177) studied the "quantitative method" for urine and arrived at the following conclusions: (a) Urobilinogen is oxidized partly to products other than urobilin. (b) There is very erratic recovery of urobilinogen or urobilin added to urine, being as low as 40% and apparently due to some interfering substance present in urine since recovery with aqueous solutions is satisfactory. (c) The interference changes with time of storage of urine samples. (d) Spurious absorption peaks may occur in the final color. (e) The reproducibility at the low levels in normal urine is very poor, which together with the highly variable recovery makes results in the subnormal range undependable. Attempts in our laboratory to solve these problems have not been fruitful. Although it may well be that erratic recovery also occurs in the "semiquantitative method", this method is being presented here as the method of choice simply because the authors know of no better method available. The more complicated "quantitative method" appears to possess no advantage over the "semiquantitative method" and therefore is not presented.

SEMIQUANTITATIVE DETERMINATION OF UROBILINOGEN IN URINE AND FECES

(method of Watson et al, Refs 445, 446, modified by Henry et al, Ref 177)

PRINCIPLE

Urobilinogen is determined photometrically by applying Ehrlich's aldehyde reaction with *p*-dimethylaminobenzaldehyde directly to urine or to an aqueous extract of feces following reduction of urobilin to urobilinogen by ferrous hydroxide. Ascorbic acid is also added as a reducing agent. The *p*-dimethylaminobenzaldehyde may add to the central active methylene bridge of urobilinogen, the condensation being accompanied by transformation of the benzene nucleus of *p*-dimethylaminobenzaldehyde into a *p*-quinone and by oxidation of the urobilinogen to urobilin. After formation of the "urobilinogenaldehyde," the acidity is decreased by addition of sodium acetate. This intensifies the urobilinogen-aldehyde color, at the same time inhibiting color formation by substances such as indole and skatole derivatives. A blank for correction of background absorption is prepared by adding sodium acetate at the same time as the Ehrlich's reagent, thus preventing development of the urobilinogen-aldehyde color. Apparently the effect of the sodium acetate is merely one of pH, although this point does not seem to have been investigated. As a matter of fact, some of the interfering colors can be produced merely by adding HCl to urine. In the original procedure the acetate was added prior to the Ehrlich's reagent for the blank. In the procedure presented here the two reagents are premixed, avoiding the production of local high acidity when the highly acidic Ehrlich's reagent is added as a separate reagent; thus somewhat lower blanks are obtained.

Phenolsulfonephthalein in alkaline solution is used as an artificial standard. Figure 22-11 shows the absorption curves of this standard and of urobilinogenaldehyde formed from urobilinogen extracted from a urine.

Since the direct method is only semiquantitative—i.e., not completely specific—results are expressed in Ehrlich units, where 1 Ehrlich unit = 1 mg urobilinogen.

REAGENTS

Ehrlich's Reagent. Dissolve 0.7 g *p*-dimethylaminobenzaldehyde in 150 ml conc HCl, reagent grade. Add 100 ml water and mix. Purity of the HCl is important. Reagent is stable.

Sodium Acetate, Saturated. Either the anhydrous acetate, reagent grade, can be used or the less expensive reagent grade triple hydrate; purity is important. To ensure saturation, some undissolved salt should be present.

FeSO$_4$, 20% (w/v). This solution is stable for 24 hours in the refrigerator.

NaOH, 10% (w/v), 2.5N

PSP Dye Standard. Dissolve 20.0 mg phenolsulfonephthalein (PSP, phenol red) in 100 ml 0.05% NaOH. Use the acidic form of the dye. Dilute 1:100 with 0.05% NaOH. This solution containing 0.20 mg/100 ml is equivalent in color to a solution of urobilinogenaldehyde containing 0.346 mg urobilinogen per 100 ml final colored solution read in the test.

One cannot assume that all batches of dye are of equal purity. The above working standard should have an absorbance of 0.384 at 562 nm on a spectrophotometer of high resolution. PSP salts contain about 4%–8% moisture, making their use in standardization more troublesome. The concentration of PSP intravenous injection solutions is not sufficiently controlled to permit their use for standardization.

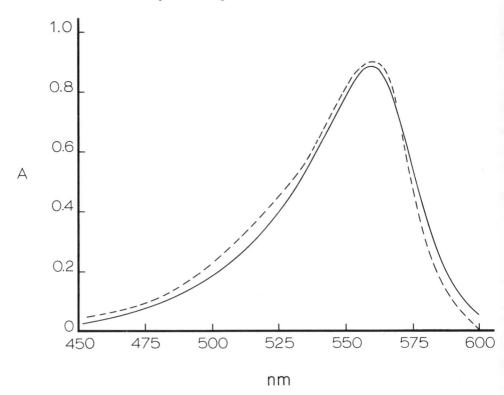

nm

FIG. 22-11. Absorption curves of urobilinogen-aldehyde (solid line) and phenolsulfoneph-thalein in 0.05% NaOH (broken line). Both solutions read against water. Concentration of phenolsulfonephthalein, 0.47 mg/100 ml 0.05% NaOH. Concentration of urobilinogen, 0.815 mg/100 ml final solution. Adapted from HENRY RJ, JACOBS SL, BERKMAN S: *Clin Chem* 7:231, 1961

PROCEDURE FOR URINE

Although the test can be applied to any urine, the test is customarily performed as follows: The bladder is emptied at 2 PM, and the patient then drinks a glass of water. The bladder is again emptied at 4 PM, this specimen being used for the analysis. The analysis must be carried out without delay, and the urine should not be exposed to sunlight or bright daylight prior to or during the analysis.

1. Measure volume of sample.
2. Test urine for bilirubin. If more than a faint trace is present, mix 2.0 ml 10% $BaCl_2$ with 8.0 ml urine and filter. The final result must then be multiplied by 1.25 to correct for this 4:5 dilution.
3. Dissolve 100 mg ascorbic acid in 10 ml clear urine (centrifuge if turbid) and place 1.5 ml aliquots in each of two test tubes or photometer cuvets, one labeled B for blank and the other X for unknown.
4. To tube B add 4.5 ml freshly prepared mixture of 1 vol. Ehrlich's reagent and 2 vol. saturated sodium acetate and mix.
5. To tube X add 1.5 ml Ehrlich's reagent, mix thoroughly, and immediately add 3.0 ml saturated sodium acetate (see *Note 4* below).

6. Measure absorbance of tubes X and B *within 5 min* against water at 562 nm or with a filter with nominal wavelength in this region (a 540 nm filter is satisfactory). The PSP standard is also read against water at the same wavelength, although the necessity for this to be done when using a spectrophotometer of high resolution *known* to give accurate absorbance readings at this wavelength is a moot point.

Calculation:

$$\text{Ehrlich units/100 ml urine} = \frac{A_x - A_b}{A_s} \times 0.346 \times \frac{6.0}{1.5}$$

$$= \frac{A_x - A_b}{A_s} \times 1.38$$

$$\text{Ehrlich units/2 hours} = \frac{A_x - A_b}{A_s} \times 0.0138 \times \text{vol in ml}$$

PROCEDURE FOR FECES

Any single *fresh* specimen may be used. *N.B.* Carry out analysis in the absence of sunlight or bright daylight.

1. Transfer 10 g feces to a large mortar. Fill a 250 ml graduated cylinder with water to the 190 ml mark. Pour about 20 ml water from graduate into mortar. Grind feces into a paste, add 80—90 ml more water, grind again, then set aside to settle. To a 500 ml Erlenmeyer flask, add 100 ml 20% $FeSO_4$ solution and the supernatant fine suspension of feces from mortar. Add more water from graduate to remaining fecal material, grind, and again transfer supernate to flask. Repeat, using last of the 190 ml of water. Slowly add 100 ml 10% NaOH to flask with shaking. Stopper flask and set aside in dark at room temperature for 1—3 hours.
2. Mix contents of flask well and filter part of contents. Dilute 5.0 ml of filtrate to 50 ml with water.
3. Dissolve 100 mg ascorbic acid in 10 ml dilute filtrate and place 1.5 ml aliquots in each of two test tubes or photometer cuvets, one labeled B for blank and the other X for unknown.
4. Proceed with steps 4 through 6 in method for urine.

Calculation:

$$\text{Ehrlich units/100 g wet feces} = \frac{A_x - A_b}{A_s} \times 0.346 \times \frac{6.0}{1.5} \times 400$$

$$= \frac{A_x - A_b}{A_s} \times 552$$

NOTES

1. Beer's law. The color obeys Beer's law in a filter photometer when a filter with a nominal wavelength of 565 nm is employed as well as in a spectrophotometer with a narrow band width.

2. Color stability. On standing, the urobilinogen-aldehyde color slowly decreases in intensity, especially that portion due to mesobilirubinogen-aldehyde. The decrease during the first 5 min, however, is negligible.

3. Interferences in the semiquantitative method. Nonurobilinogen substances which react with Ehrlich's reagent are discussed under *Accuracy and Precision,* below. Those substances which may be present in quantity sufficient to produce difficulty of one sort or another in the method as applied directly to urine, and which may be readily recognized, include the following:

Protein: The direct method cannot be employed in the presence of proteinuria because of turbidity formation (445).

Bilirubin: If the reaction is carried out in the presence of bilirubin, a green color may be produced—the so-called "green Ehrlich reaction"—due to oxidation of bilirubin to biliverdin by nitrous acid liberated by action of the acidic reagent on nitrites present in the urine. Bilirubin must be first removed by coprecipitation.

Bile acids: The strong acidity of Ehrlich's reagent converts bile salts to bile acids. If the bile salts are in high concentration, turbidity or even precipitation results. If the turbidity is slight and appears to be the same in the unknown and its blank, and if there is a strong color in the unknown, then no great error should be incurred. Unequal turbidity in the two tubes or a very low urobilinogen level would make results undependable.

Porphobilinogen: Porphobilinogen gives a reaction identical to urobilinogen but can be readily distinguished by adding several milliliters of chloroform and shaking. Porphobilinogen-aldehyde and the red color produced by urorosein, an indole derivative, are *not* chloroform-soluble. Urobilinogen-aldehyde and the red colors produced by indirubin (an indole derivative) and by phylloerythrinogen are extracted by chloroform.

4. Standardization. Urobilin is now available commercially, but it is expensive and the purity is sometimes suspect. Stercobilin or urobilin hydrochloride can be isolated from feces, but this procedure is time-consuming and not esthetically pleasant. *i*-Urobilin can be prepared by sodium amalgam reduction of bilirubin.

Lack of commercial availability of urobilin in the past led to adoption of artificial reference standards. Phenolphthalein, phenolsulfonephthalein, bromoauric acid, and a mixture of pontacyl dyes have been used for this purpose. Watson and co-workers (445, 446) introduced the mixture of pontacyl carmine 2B and pontacyl violet 6R, but Henry *et al* (179) showed that the absorption spectra of urobilinogen-aldehyde and the dye mixture differ to the extent that results of analyses in which the dye mixture is employed as standard vary considerably with change in spectral resolution of the photometer or spectrophotometer employed. On the other hand, PSP has an absorption curve identical to the urobilinogen-aldehyde between 530 and 570 nm (Figure 22-11) and results using this dye as a standard are independent of the instrument used in this region. The color equivalent of PSP presented is the mean of the four best preparations of urobilin available in the studies of Henry *et al* (179).

5. *Speed of acetate addition.* When Ehrlich's reagent is added to pure solutions of urobilinogen, full color develops promptly, but when added directly to urine the color increases with time due to slower reaction with substances other than urobilinogen. These slower reactions are stopped upon addition of saturated sodium acetate immediately or within 15 sec after the Ehrlich's reagent has been added and mixed.

6. *Two Hour Collection Period of the "Semiquantitative" Method.* Watson and co-workers (446) arbitrarily chose a 2 hour collection period for convenience and settled on 2–4 PM because they found the output to be higher during this period than in morning specimens. Other authors confirm or refute this observation (293, 326).

7. *Stability of samples.* Urobilinogen is not very stable, the rate of oxidation to urobilin being hastened by contact with air, increased temperature, and exposure to daylight. Urobilinogen is unstable in both feces and urine, about half being lost in 24 hours in the refrigerator. Oxidation can be effectively retarded in the case of urine by collection in a dark bottle under a layer of petroleum ether (387, 440). Sodium carbonate is added to prevent extraction of urobilinogen into the petroleum ether. Samples are stable for 1 month if frozen.

Analyses for urobilinogen must be carried out in the absence of sunlight or bright daylight. Decreases up to 87% have been reported merely as a result of performing the analysis in bright daylight (435).

ACCURACY AND PRECISION

In spite of the use of sodium acetate to increase the intensity of the urobilinogen-aldehyde color and to suppress the colors produced by various interfering substances, the semiquantitative method is not specific and regularly includes other Ehrlich-reacting substances. Substances known to react with Ehrlich's reagent and the colors produced are as follows: urobilinogen, porphobilinogen, opsopyrroledicarboxylic acid, opsopyrrole-carboxylic acid, melanogen (5,6-dihydroxyindole), melagin, excretion products of allylisopropylacetamide or Sedormid, indole-3-acetic acid, phylloerythrinogen, indole, and the indole derivatives urorosein and indirubin give a red color (see *Accuracy in Qualitative Determination of Porphobilinogen*, Chapter 24, for information concerning $CHCl_3$ solubility of these interferences); urea and indican give a yellow color; tryptophan orange; indoxyl sulfate orange-brown; bilirubin green; and skatole blue. Sulfonamides, phenothiazine and meprobamate metabolites, procaine, indoleacetic acid, and 5-hydroxyindoleacetic acid are also said to react with

Ehrlich's reagent. Ingestion of bananas leads to increased excretion of 5-hydroxyindoleacetic acid, and ingestion of coffee causes a large but temporary excretion of acetyltryptophan, another reactor. Paper chromatography has indicated that indole and skatole do not normally occur in urine, but that indican is always present in fresh urine. Traces of many other Ehrlich reactors have been detected in urine by paper chromatography. Phylloerythrinogen is a pyrrole pigment formed from chlorophyll by bacterial action in the large intestine and may occur in the urine of patients consuming large amounts of vegetables. Ingestion of the drug cascara sagrada leads to excretion of a substance giving a positive reaction, the color being extractable into $CHCl_3$.

Fortunately, from a diagnostic standpoint increases in the nonspecific, Ehrlich-reacting substances appear to be proportional to an increase in urobilinogen. Both false-positive and false-negative results can be obtained by the semiquantitative method.

Recovery of added urobilinogen varies between 40% and 68% with different urines. This indicates inhibition of color formation by substances in the urine thus far not identified. Accuracy of standardization is probably no better than ±10%. It is obvious that this test cannot be considered more than "semiquantitative".

The precision of the method is ±5%.

NORMAL VALUES

Urine. Values obtained in our laboratory on 11 normal adult females suggested 95% limits of 0—1.1 Ehrlich units/2 hours. For 12 normal adult males the 95% limits were 0.3—2.1 Ehrlich units. The sex difference was highly significant by the *t* test. This is in disagreement with a previous report claiming no sex difference (17). There is considerable day to day and hour to hour fluctuation in output, the hourly fluctuation apparently being related to the ingestion of food. The output is much lower at night and higher in the afternoon. Output tends to increase in alkaline urine (293). Normal daily excretion varies between 0 and 4 mg (470).

Feces. The normal range obtained by the original technic of Watson *et al* has been reported to be about 75—350 Ehrlich units/100 g feces (wet weight) (18, 445, 446). Normal values for feces have not been obtained in our laboratory using the new standardization. Comparison of the new normal values for urine with the old values—up to about 1.2 Ehrlich units/2 hours (18, 326, 446)—indicates that the new normal values for feces probably would be somewhat higher than the 75—350 range. Watson and Hawkinson (445) regard a level greater than 10 units as evidence of the passage of some bile into the intestinal tract and a level of less than 5 units as indicative of the absence of bile passage. The data of Balikov (17) indicate a significantly greater excretion by males than by females, a point apparently not examined by previous workers.

Urobilinogen is rarely found in the feces of newborns before the age of 3 weeks, and some infants fail to excrete urobilinogen up to 6 months.

DETERMINATION OF UROBILIN

Jaffe introduced the term *urobilin* in 1869 and was the first to show that the zinc salt of urobilin has a green fluorescence (200). The unmodified term "urobilin" today is used to refer collectively to *d*-urobilin, *i*-urobilin, and *l*-stercobilin. Schlesinger (380), who published in 1903, is usually but incorrectly credited with originating the zinc test.

Urobilinogen does not react in this test; the test is therefore made more sensitive by oxidation of urobilinogen to urobilin by I_2 or ammonium persulfate, otherwise only the urobilin present as a result of degradation of urobilinogen is available for reaction. The test has been applied directly to urine as well as to chloroform extracts of acidified urine. The zinc fluorescence test is more sensitive than Ehrlich's *p*-dimethylaminobenzaldehyde reaction for urobilinogen, and it is more specific than Ehrlich's test applied directly on urine or feces (308). So far, however, the zinc test has not been developed into a widely used quantitative procedure, although attempts at quantitation have been presented (218, 261).

Watson (444) has done a thorough study of normal and abnormal urines and feces in which urobilinogen was oxidized by iodine, and the *d*-urobilin, *i*-urobilin, and *l*-stercobilin were each quantitated from absorbance ratios at 492, 560, and 650 nm after further oxidation with ferric chloride.

In 1895 Schmidt (384) introduced the sublimate reaction with mercuric chloride as a simple qualitative test for urobilin and bilirubin in feces.

DETERMINATION OF UROBILIN IN URINE BY ZINC FLUORESCENCE REACTION

(method of Naumann, Ref 308, modified)

PRINCIPLE

Urobilinogen is oxidized to urobilin by iodine. Alcoholic zinc acetate solution is added, forming a zinc urobilin compound which fluoresces green. The sensitivity of the reaction is reported to be about 0.005 mg of urobilin per 100 ml. The following type of formula for the fluorescent compound has been proposed:

The zinc complex of *i*-urobilin has fluorescence maxima at 637, 601, 551, and 521, nm—the last being the most intense—whereas the complex of *l*-stercobilin has maxima at 638 and 520 nm only (470).

REAGENTS

Alcoholic Suspension of Zinc Acetate. Dissolve approximately 10 g zinc acetate in 100 ml ethanol. Ethanol denatured with methanol is satisfactory.
Conc HCl
Lugol's Solution. Dissolve 5 g iodine and 10 g KI in 100 ml water.

PROCEDURE

1. Mix 10 ml urine with 10 ml of well shaken alcoholic zinc acetate suspension. Filter. Divide into approximately equal portions in two test tubes of about 10 mm diameter.
2. Add 1 drop of conc HCl to one of the two tubes. Add 2 drops Lugol's solution to each tube and mix.
3. Observe tubes for fluorescence by observing against a black background with a strong light coming through the tubes from the side. An incandescent spot lamp, sunlight, or an ultraviolet light source can be used. Only a yellow-green fluorescence appearing in the unacidified tube which does not appear in the acidified tube constitutes a positive test for urobilin, since HCl destroys the fluorescence due to urobilin.

ACCURACY

Substances other than urobilin which give a fluorescence include the following: bilirubin, biliviolins, bilipurpurins, and porphyrins give a red fluorescence; phylloerythrinogen, the pyrrole pigment derived from chlorophyll which may occur in the urine of patients consuming large amounts of vegetables, gives a red fluorescence; riboflavin, acriflavin, and fluorescein yield a yellow-green fluorescence; and eosin, Mercurochrome, and erythrosin impart a pink color to the urine and yield a dull mauve fluorescence.

SCHMIDT SUBLIMATE TEST FOR UROBILIN AND BILIRUBIN IN FECES

(method of Schmidt, Ref 384)

Rub a pea-sized piece of feces into an evaporating dish and mix with several drops of a 5% aqueous solution of $HgCl_2$. Cover the dish and let stand at room temperature until the next day.

Urobilin yields a pink color, bilirubin a green color. No color indicates the absence of bile pigments. In the presence of an abnormal amount of fat in feces, the test should be applied to feces extracted with ether.

BILE ACIDS

The bile acids elaborated by the hepatic cells are C-24 steroids which are primarily conjugated at their carboxyl groups with the amino acids glycine and taurine. They are important excretory products of cholesterol. The principal bile acids found in the bile and serum are cholic and chenodesoxycholic acids:

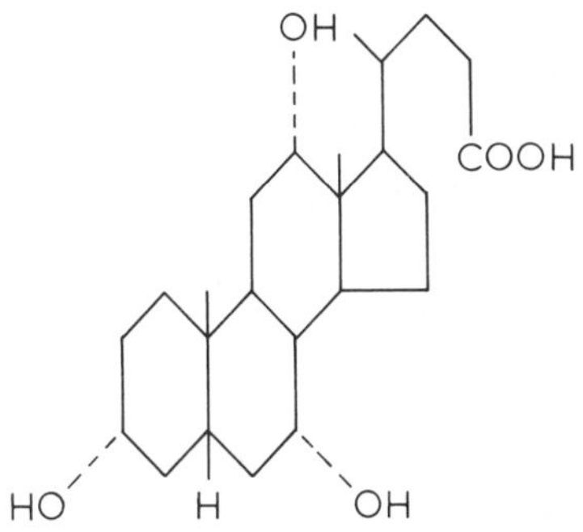

CHOLIC ACID

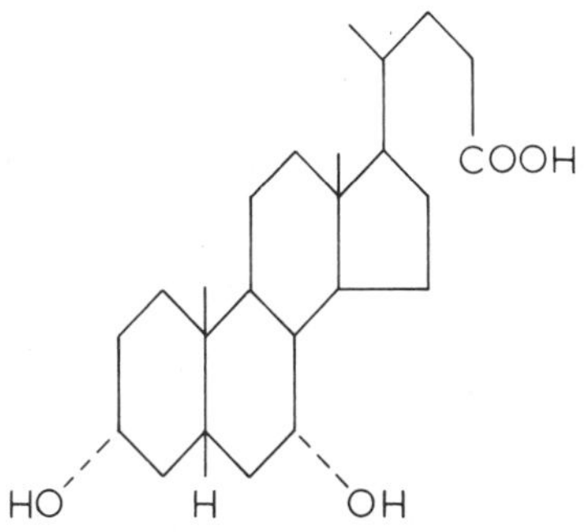

CHENODESOXYCHOLIC ACID

These primary bile acids are converted by deconjugation and dehydroxylation through bacterial action, mostly in the colon, to the free deoxycholic and lithocholic acids, respectively; a small amount of a complex mixture of other secondary acids is also formed (111). These secondary bile acids are excreted in the feces, but some of the more polar deoxycholic acid (two hydroxyls) is reabsorbed from the colon and is usually found in the bile and in small amounts in the serum. The primary bile acids, cholic and chenodesoxycholic, are usually absent from the feces of healthy adults (65, 386). There is about 12 g of bile salt in the total body pool. Two to three grams are delivered to the small intestine and recirculate through the enterohepatic circulation 2–3 times during a meal; and although 20–30 g pass through the enterohepatic circulation in 1 day, only 100–800 mg is lost in the stool (5, 130, 386).

Certain of the plasma globulins and especially plasma albumin readily bind bile acids (368). The free cholic acid is bound more strongly to albumin than is its taurine conjugate (60). This may be a factor contributing to the relatively high concentration of unconjugated bile acids in the plasma of patients with the stagnant loop syndrome in which there is excessive bacterial deconjugation in the intestine (253).

The determination of total conjugated bile acids in the serum has been said to be a more sensitive indicator of liver disease than are tests for bilirubin and SGPT (138). The ratio of the trihydroxy (or cholic) to the dihydroxy (or chenodesoxycholic) acids has been reported to be of diagnostic value when at least one of these fractions is elevated (64, 268, 310, 360, 367). Thus in cases of biliary obstruction and some types of hepatitis, the ratio is usually greater than 1.0. In the presence of severe damage to the liver cells, as in cirrhosis, there is a disproportionate increase in the level of dihydroxy acids, perhaps due to a deficiency of 12-α-hydroxylating enzyme, to produce a ratio of less than 1.0 (64, 360) or 1.5 (268). These observations, however, have been refuted by other workers (138, 323).

Bile acids are not normally found in the urine of healthy individuals. Elevated levels are reported in hepatocellular disease and obstructive jaundice, with the dihydroxy (chenodesoxycholic) acid rising proportionately higher than cholic acid in the former condition than in the latter (160, 366).

METHODS FOR THE DETERMINATION OF BILE ACIDS

Various methods have been used for the quantitation of bile acids. Most have been applied to bile, feces, and duodenal contents (111, 130, 210, 223, 301, 386, 400). (a) *Stalagmometric*. The qualitative Hay's test depends upon the fact that bile acids decrease the surface tension of fluids. In this test, which has been commonly used by the clinical laboratory for many years, the urine is cooled to 17°C or lower, and finely divided sulfur is sprinkled on the surface. If the sulfur sinks to the bottom, the test is positive. The test can be made more sensitive by acidification of the urine with HCl (242). The method has been used to obtain quantitative bile acid levels on bile and urine (300), but the finding of the unlikely level 350 mg/100 ml normal urine sheds doubt on its validity (299). (b) *Color reactions*. The Pettenkofer reaction in which cholic acid, furfural, and H_2SO_4 react to give a red color was applied in 1887 to detect this bile acid in urine (211, 304). This as well as the reactions of Szalkowski (using salicylaldehyde, H_2SO_4, and acetic acid) and of Minibeck (using acetic acid and H_2SO_4) have been used to determine the bile acids in serum (475). (c) *Hemolysis*. A method of quantitation has been based upon the ability of bile acids to cause hemolysis of erythrocytes (255). (d) *UV spectrophotometry*. The ultraviolet spectra of the bile acids in 65% H_2SO_4 have been used for quantitative determination subsequent to single solvent extraction of plasma or serum (217). This technic gives erroneously high results on normal samples. Spectrophotometry has also been applied after solvent extraction, solvent partitioning, hydrolysis and reextraction (64) or extraction, solvent partitioning, hydrolysis, and column partition chromatography (367). The latter two approaches give fairly specific results. (e) *Fluorescence*. Fluorometric and spectrophotofluorometric methods similar to those of UV spectrophotometry have given erroneously high results on normal serum (247). When combined with isolation of the bile acids by thin-layer chromotography results appear to be quite specific (323). (f) *Chromatography*. Other thin-layer chromatographic schemes using photometric and UV (89, 137, 138), densitometric (291), and enzymatic (57) quantitation have been reported, and a sensitive anisaldehyde-containing spray reagent has been suggested which gives different colors for the various bile acids (228). Paper chromatography has been applied to serum (66). Gas chromatography has been found useful in separating and quantitating the bile acids of serum and plasma (4, 268, 360, 373). In urine, where no bile acids are normally found, levels between 2 and 60 mg/24 hour excretion were observed in hepatocellular disease and obstructive jaundice (160). (g) *Enzymatic analysis.* The oxidation of α-hydroxysteroids by NAD is catalyzed by the enzyme α-hydroxysteroid dehydrogenase obtained from *Pseudomonas testosteroni* (273):

$$\text{HO} \quad + \text{ NAD}^+ \rightleftharpoons \quad \text{O} \quad + \text{ NADH} + \text{H}^+$$

The serum bile acids have been fractionated by thin-layer chromatography and then quantitated by this reaction, reading the NADH formed at 340 nm on a spectrophotometer (193). Subsequently, the reaction was used for total serum bile acids, purified by solvent partitioning, with fluorimetric quantitation of the NADH. This method appears to be both sensitive and specific (303).

The method presented here is that of Carey (64). It has been chosen from the many available since it seems to be reasonably specific for the conjugated bile acids, which are the important bile acids in serum. Furthermore, while demonstrating the relative concentration of the conjugated tri- and dihydroxy (or cholic and chenodesoxycholic) acids, this method avoids the taurine versus glycine conjugate approach which is not generally believed to be as useful clinically. The technic involves the interesting spectrophotometric manipulations necessary to deal with a two-component system as well as background absorbance. The authors feel that this type of technic should not be omitted from a compendium such as this. Typical ultraviolet absorption spectra for trihydroxy (cholic) and dihydroxy (chenodesoxycholic) acids in serum are illustrated in Figure 22-12.

A review on the separation and determination of bile acids was published in 1964 (399), and there have been several reviews on the biochemistry and physiology of the bile acids (44, 108, 129, 306).

DETERMINATION OF BILE ACIDS IN SERUM

(method of Carey, Ref 64)

PRINCIPLE

Serum proteins are precipitated with ethanol and barium; fats and neutral sterols are separated from the bile acids by solvent partitioning; and the

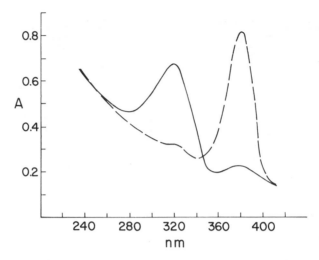

FIG. 22-12. Typical ultraviolet absorption spectra for trihydroxy (cholic: ———————)
and dihydroxy (chenodesoxycholic: – – – – – –) bile acids added to serum.

conjugated bile acids are hydrolyzed with alkali. The unconjugated bile acids
obtained are measured by their ultraviolet spectra after heating in strong
H_2SO_4. The trihydroxy bile acids exhibit a peak at 320 nm; the dihydroxy
acids at 380 nm.

REAGENTS

Ethanol. Ethanol denatured with methanol (formula 3A) is satisfactory.

Barium Reagent. Add 25 g anhydrous $Ba(OH)_2$ to 400 ml water. Heat
and filter. Add 4 g anhydrous barium acetate. This saturated solution is
stable for months.

Conc HCl

NaOH, 8%

H_2SO_4, *65%.* To 80 ml water, add 169 conc H_2SO_4 slowly while mixing
and cooling with tap water. When cool, dilute to 250 ml with water.

Diethyl Ether

Hexane, Peroxide-free

Ethanol-Water-Hexane-Ether, Equilibrated Solvent Mixture. Mix 25 ml
water, 25 ml ethanol, 125 ml hexane, and 125 ml diethyl ether. Shake
thoroughly and separate the phases.

Stock Cholic Acid Standard, 50 mg/100 ml. Dissolve 50 mg cholic acid in
100 ml ethanol.

Dilute Cholic Acid Standard. Dilute stock standard 1:10 with ethanol.
1 ml = 0.05 mg.

Stock Chenodesoxycholic Acid Standard, 50 mg/100 ml. Dissolve 50 mg
chenodesoxycholic acid in 100 ml ethanol.

Dilute Chenodesoxycholic Acid Standard. Dilute stock standard 1:10 with
ethanol. 1 ml = 0.05 mg.

PROCEDURE

1. To 25 ml ethanol and 1 ml water in a 25 × 150 mm test tube marked at 30 ml, add 1 ml barium reagent and 3 glass beads. Mix.

2. Add 3 ml serum dropwise while mixing thoroughly. It is convenient to use a mechanical mixing device. If less serum is to be used, reduce the volume of barium reagent used in step 1 proportionately, substituting water for the volume of serum plus barium reagent which was omitted.

3. Place in an 85°C water bath so that contents of the tube boil for 3 min, taking care to avoid bumping of sample due to erratic boiling. Then cool in ice water and adjust volume to 30 ml with ethanol; mix.

4. Centrifuge to remove the precipitate, approximately 20 min.

5. Transfer 25 ml supernate to a 250 ml round-bottom 24/40 flask.

6. Evaporate to dryness under reduced pressure on a rotary evaporator at 40°C or less.

7. Dissolve residue in 10 ml of the aqueous-ethanol phase of the equilibrated solvent mixture.

8. Adjust pH to less than 3.5 (pH paper) with conc HCl. One drop is usually required. Transfer quantitatively to a 125 ml separatory funnel (ungreased).

9. Wash solution with 25 ml of the ether phase of the equilibrated solvent mixture. Draw the lower layer off to a 250 ml round-bottom 24/40 flask.

10. Extract the ether wash with 5 ml of the aqueous-ethanol phase and combine both aqueous-ethanol extracts in the flask.

11. Adjust the extract to neutral pH with 8% NaOH. About 0.2 ml is required.

12. Evaporate to near dryness as in step 6. Do not take completely to dryness.

13. Add 5 ml 8% NaOH and stir to ensure complete solution.

14. Autoclave at 15 psi for 3 hours.

15. Add 5 ml water and adjust pH to less than 3.5 (pH paper) with conc HCl. Transfer quantitatively to a 125 ml separatory funnel (ungreased).

16. Extract 4 times with 20 ml portions of diethyl ether.

17. Wash the combined ether extract with 10 ml water.

18. Evaporate to dryness in a chromic acid-cleaned 18 × 150 mm test tube in a warm water bath at 37°C or less with a stream of N_2. The sample may be introduced in portions.

19. For standards, in chromic acid-cleaned 18 × 150 mm test tubes, dry down 1 ml of each standard.

20. To the dry residue add 5 ml 65% H_2SO_4 and heat immediately at 60°C for *exactly* 15 min.

21. Cool in ice water 10 min and then filter through a 2 cm, medium porosity, sintered glass funnel which is reserved for this test only. Filter with aid of suction into a side-arm test tube which has been chromic acid-cleaned.

22. Scan on a recording spectrophotometer in quartz cuvets versus reagent blank from 450 to 280 nm, or make a sufficient number of readings on a manual spectrophotometer of high resolution to describe the curves adequately.

Calculation: Corrected absorbance readings are obtained at 380 nm by measuring the difference between the reading at 380 nm and a straight line connecting the 360 and 400 nm readings or by using the formula

$$A(corr)_{380} = A_{380} - \tfrac{1}{2}(A_{360} + A_{400})$$

Corrected absorbance readings are obtained at 320 nm by measuring the difference between the reading at 320 nm and a straight line connecting the 300 and 340 nm readings or by using the formula

$$A(corr)_{320} = A_{320} - \tfrac{1}{2}(A_{300} + A_{340})$$

For *dihydroxy acids*:

$$\text{μg dihydroxy acids, as chenodesoxycholic acid/ml serum} = \frac{A_{380}\, A^T_{320} - A_{320}\, A^T_{380}}{A^D_{380}\, A^T_{320}} \times 20$$

For *trihydroxy acids*:

$$\text{μg trihydroxy acids, as cholic acid/ml serum} = \frac{A_{320}\, A^D_{380} - A_{380}\, A^D_{320}}{A^D_{380}\, A^T_{320}} \times 20$$

Where

A_{320} is the corrected absorbance of a sample at 320 nm

A_{380} is the corrected absorbance of a sample at 380 nm

A^D_{320} and A^D_{380} are the corresponding values for chenodesoxycholic acid standard (dihydroxy acid)

A^T_{320} and A^T_{380} are the corresponding values for cholic acid standard (trihydroxy acid)

See *Note 2*.

NOTES

1. Beer's law. The corrected absorbances for cholic and chenodesoxycholic acids follow Beer's law at 320 and 380 nm, respectively, to an absorbance of at least 1.5. The much lower readings for these bile acids at 380 and 320 nm, respectively, also follow Beer's law (64).

2. *Calculations.* The derivation of these equations is too lengthy to include here, but, in brief, they come from solving the two simultaneous equations:

$$A_{320} = A^D_{320} C^D + A^T_{320} C^T$$

$$A_{380} = A^D_{380} C^D + A^T_{380} C^T$$

where all the A terms are corrected absorbance readings as described above, and C^D and C^T are the concentrations of dihydroxy and trihydroxy acids, respectively. The denominator is the final equation actually turns out to be

$$A^D_{380} A^T_{320} - A^D_{320} A^T_{380}$$

The second term, however, is negligible in comparison to the first, and we have therefore dropped it from the equations. Carey's approach to these calculations (64) is a little different from that presented here, but the outcome is the same. The factor 20 accounts for sample and standard manipulations and dilutions.

3. *Stability of samples.* Sterile serum is stable for at least 1 week at room temperature. However, it might be predicted, on the basis of the action of intestinal flora, that bacterial growth in the serum might cause deconjugation, which would cause low recovery as well as changes in the bile acids that might alter the spectra (63). With no evidence of bacterial growth, sera have been found to be stable in the refrigerator for more than 6 months (63).

ACCURACY AND PRECISION

Recovery of conjugated bile acids by this procedure is approximately 80%. In checking recovery by putting bile acids through the entire serum procedure, it is important to remember that the method *removes unconjugated* compounds; thus *conjugated* bile acids must be used. If one wishes to make use of the *unconjugated* compounds, they must be added to the alcoholic extract after partitioning with ether-petroleum ether. Hemolysis causes low recoveries as well as altered UV spectra (63).

The precision of the test (95% limits) is about ±50% at the level of about 3 μg/ml and ±25% at the level of about 10 μg/ml.

NORMAL VALUES

Adult normal values have been reported to show no diurnal change and not to vary with age or sex or with eating (64); other reports indicate a rise in serum bile acids after a meal (247). There is not total agreement as to the nature or concentrations of the bile acids normally found in the serum (64, 66, 89, 138, 193, 247, 268, 303, 310, 323, 360, 367, 373). The values reported by Carey (64) are given here, but it is not unlikely that experience will suggest a downward revision:

Trihydroxy bile acids (as cholic acid): 0–3.4 μg/ml
Dihydroxy bile acids (as chenodesoxycholic acid): 0–1.9 μg/ml

REFERENCES

1. ABELSON NM, BOGGS TR Jr: *Pediatrics 17*:452, 1956
2. ADLER E, STRAUSS L: Z *Gesamte Exp Med 44*:43, 1924
3. AICKIN DR, CAMPBELL DG: *Obstet Gynecol 37*:687, 1971
4. ALI SS, JAVITT NB: *Can J Biochem 48*:1054, 1970
5. ALPERS D, WESSLER S, AVIOLI LV: *JAMA 215*:101, 1971
6. AMATUZIO DS, WEBER LJ, NESBITT S: *J Lab Clin Med 41*:615, 1953
7. AMSTEY MS, HOCHBERG CJ, CHOATE JW, WAX SH, LUND CJ: *Obstet Gynecol 39*:407, 1972
8. ANDERSSON E, NORBERG B, TEGER-NILSSON A-C: *Scand J Clin Lab Invest 15*:517, 1963
9. ANDREWS WHH, del RIO LOZANO I: *J Physiol (Lond) 163*:14P, 1962
10. ANDREWES CH: *Br J Exp Pathol 5*:213, 1924
11. ARMAS-CRUZ R, LOBO-PARA G, MADRID M, VELASCO C: *Gastroenterology 35*:298, 1958
12. AVIGAD G, ASENSIO C, AMARAL D, HORECKER BL: *Biochem Biophys Res Commun 4*:474, 1961
13. BADIN J, SCHMITT F: *Ann Biol Clin (Paris) 15*:313, 1957
14. BAIKIE AG: *Scott Med J 2*:359, 1957
15. BAKER KJ: *Proc Soc Exp Biol Med 122*:957, 1966
16. BAKER KJ, BRADLEY SE: *J Clin Invest 45*:281, 1966
17. BALIKOV B: *Clin Chem 3*:145, 1957
18. BALIKOV B: *Standard Methods of Clinical Chemistry*. Edited by D Seligson, New York, Academic Press, 1958, Vol 2, p 192
19. BALZER E: *Acta Med Scand [Suppl] 278*:67, 1953
20. BARAC G: *Arch Internatl Physiol 61*:129, 1953
21. BARTON DM, STANDER RW: *Am J Clin Pathol 42*:602, 1964
22. BASSETT AM, ALTHAUSEN TL, COLTRIN GC: *Am J Dig Dis 8*:432, 1941
23. BASSIR O, HALL J: *Biochem J 60*:XX, 1955
24. BAUER R: *Klin Wochenschr 16*:1570, 1937
25. BAUER R: *Wien Med Wochenschr 56*:2537, 1906
26. BERK PD, BLOOMER JR, HOWE RB, BERLIN NI: *Am J Med 49*:296, 1970
27. BEVIS DCA: *Lancet 2*:443, 1950
28. BEVIS DCA: *Lancet 1*:395, 1952
29. BEVIS DCA: *J Obstet Gynaecol Br Emp 60*:244, 1953
30. BEVIS DCA: *J Obstet Gynaecol Br Emp 63*:68, 1956
31. BILISSIS PK, SPEER RJ: *Clin Chem 9*:552, 1963
32. BILLING BH: *J Clin Pathol 8*:126, 1955
33. BILLING BH: *Adv Clin Chem 2*:267, 1959

34. BILLING BH: *Postgrad Med 39*:176, 1966
35. BILLING BH, LATHE GH: *Congr Intern Biochim Resumes Communs,* Third Congress, Brussels, 1955, p 123
36. BILLING BH, LATHE GH: *Biochem J 63*:6P, 1956
37. BLACK M, BILLING BH: *N Engl J Med 280*:1266, 1969
38. BLONDHEIM SH, KAUFMANN NA: *J Lab Clin Med 65*:659, 1965
39. BLOOMER JR, BARRETT PK, RODKEY FL, BERLIN NI: *Gastroenterology 61*:479, 1971
40. BLOOMFIELD N: *Am J Clin Pathol 41*:15, 1964
41. BOULANGER J-P, FIERRO L: *Am J Clin Pathol 42*:557, 1964
42. BOWER D, SWALE J: *Lancet 1*:1009, 1966
43. BOYD JF, SOMMERVILLE JW: *J Clin Pathol 13*:85, 1960
44. BOYD GS, PERCY-ROBB IW: *Am J Med 51*:580, 1971
45. BOYLAND E, GROVER PL: *Clin Chim Acta 16*:205, 1967
46. BRAHMACHARI UN: *Indian Med Gaz 52*:429, 1917
47. BRATLID D, WINSNES A: *Scand J Clin Lab Invest 28*:41, 1971
48. BRAZIE JV, BOWES WA Jr, IBBOTT FA: *Am J Obstet Gynecol 104*:80, 1969
49. BRAZIE JV, IBBOTT FA, BOWES WA Jr: *J Pediatr 69*:354, 1966
50. BREWS VAL: *J Clin Pathol 9*:390, 1956
51. BRODERSEN R: *Scand J Clin Lab Invest 12*:25, 1960
52. BRODERSEN R, FLODGAARD H, JACOBSEN J: *Scand J Clin Lab Invest 24*:227, 1969
53. BROHULT J: *Scand J Clin Lab Invest 19*:67, 1967
54. BRÜCKNER J: *Z Physiol Chem 268*:163, 1941
55. BRÜCKNER J: *Am J Clin Pathol 32*:513, 1959
56. BRÜCKNER J: *Clin Chim Acta 6*:370, 1961
57. BRUUSGAARD A: *Clin Chim Acta 28*:495, 1970
58. BUBB W, PEDRAZZINI A: *Schweiz Med Wochenschr 79*:167, 1949
59. BUNCH LD: *Am J Clin Pathol 28*:111, 1957
60. BURKE CW, LEWIS B, PANVELIWALLA D, TABAQCHALI S: *Clin Chim Acta 32*:207, 1971
61. BURNETT RW: *Clin Chem 18*:150, 1972
62. CARBONE JV, GRODSKY GM, HJELTE V: *J Clin Invest 38*:1989, 1959
63. CAREY JB Jr: *Personal Communication,* 1962
64. CAREY JB Jr: *J Clin Invest 37*:1494, 1958
65. CAREY JB Jr: *Gastroenterology 46*:490, 1964
66. CAREY JB Jr, FIGEN J, WATSON CJ: *J Lab Clin Med 46*:802, 1955
67. CARTER RE, McGANN CJ: *Clin Chem 5*:106, 1959
68. CASTEX MR, LÓPEZ GARCÍA A: *Arch Uruguay Med Cirurg Especial 18*:525, 1941
69. CHARLWOOD PA: *Biochem J 56*:480, 1954
70. CHERRY SH, KOCHWA S, ROSENFIELD RE: *Obstet Gynecol 26*:826, 1965
71. CHIAMORI N, HENRY RJ, GOLUB OJ: *Clin Chim Acta 6*:1, 1961
72. CHRISTENSEN HN, CHRISTENSEN AS: *Univ Michigan Med Bull 21*:417, 1955
73. CLARKE JT: *Clin Chem 11*:681, 1965
74. CLOSS K: *Lancet 1*:910, 1954
75. COLE PG, LATHE GH: *J Clin Pathol 6*:99, 1953
76. GOLE PG, LATHE GH, BILLING BH: *Biochem J 57*:514, 1954
77. COMBES B: *J Clin Invest 44*:1214, 1965
78. COMBES B, STAKELUM GS: *J Clin Invest 39*:1214, 1960
79. COMBES B, STAKELUM GS: *J Clin Invest 40*:981, 1961
80. Committee on Bilirubin Standardization: *Clin Chem 8*:405, 1962

81. CONNON AF: *Obstet Gynecol 33*:72, 1969
82. COOK VP, CAWLEY LP, FRITZ GE: *Clin Biochem 3*:261, 1970
83. COOLIDGE TB: *J Biol Chem 132*:119, 1940
84. COOPER JAD, SMITH W, BACILA M, MEDINA H: *J Biol Chem 234*:445, 1959
85. COUCH RD: *Am J Clin Pathol 53*:194, 1970
86. CRACCO JB, DOWER JC, HARRIS LE: *Mayo Clin Proc 40*:868, 1965
87. CREMER RJ, PERRYMAN PW, RICHARDS DH: *Lancet 1*:1094, 1958
88. CROWLEY LV: *Am J Clin Pathol 51*:425, 1969
89. CURTIUS H-CH: *Z Klin Chem 4*:27, 1966
90. DANGERFIELD WG, FINLAYSON R: *J Clin Pathol 6*:173, 1953
91. de la HUERGA J, POPPER H: *J Lab Clin Med 34*:877, 1949
92. de la HUERGA J, POPPER H: *J Lab Clin Med 35*:459, 1950
93. de la HUERGA J, POPPER H, FRANKLIN M, ROUTH JI: *J Lab Clin Med 35*:466, 1950
94. de LAVERGNE E, PEROT G, RACADOT A: *C R Soc Biol (Paris) 152*:987, 1958
95. DESMOND MM, ZIMMERMAN HJ, SWEET LK, THOMAS LJ: *Pediatrics 3*:49, 1949
96. de VERDIER C-H, HJELM M: *Clin Chim Acta 7*:742, 1962
97. DICKENS F, PEARSON J: *Biochem J 48*:216, 1951
98. *Dow Diagnostic Products Method Brochure*, Midland, Mich, The Dow Chemical Company, 1968
99. DUCCI H: *J Lab Clin Med 32*:1273, 1947
100. DUCCI H, WATSON CJ: *J Lab Clin Med 30*:293, 1945
101. EAVENSON E: *A List of Test Kits for Clinical Chemistry and Hematology.* Atlanta, Ga, US Department of Health, Education and Welfare, Center for Disease Control, November 1970
102. EBERLEIN WR: *Pediatrics 25*:878, 1960
103. EDOZIEN JC: *J Clin Pathol 11*:437, 1958
104. EHRLICH P: *Charité, Ann 8*:140, 1883
105. EHRLICH P: *Med Woche 1*:151, 1901
106. ELKING MP, KABAT HF: *Am J Hosp Pharm 25*:485, 1968
107. ELLIOTT HC Jr: *Anal Chem 29*:1712, 1957
108. ELLIOTT WH, HYDE PM: *Am J Med 51*:568, 1971
109. ELLIOTT HC, WALKER AA III: *Proc Soc Exp Biol Med 116*:268, 1964
110. ELLMAN GL, BURKHALTER A, LaDOU J: *J Lab Clin Med 57*:813, 1961
111. ENEROTH P, GORDON B, RYHAGE R, SJÖVALL J: *J Lipid Res 7*:511, 1966
112. ENGEL M: *Hoppe Seylers Z Physiol Chem 259*:75, 1939
113. ENRIQUES E, SIVO R: *Biochem Z 169*:152, 1926
114. ERNST RG, DOTTI LB: *Am J Med Sci 216*:316, 1948
115. FASHENA GJ: *Am J Dis Child 76*:196, 1948
116. FERRO PV, HAM AB: *Am J Clin Pathol 44*:111, 1965
117. FERRO PV, HAM AB: *Am J Clin Pathol 45*:166, 1966
118. FISCHER W, ZAPF J: *Z Physiol Chem 337*:186, 1964
119. FISCHER H, HABERLAND HW: *Hoppe Seylers Z Physiol Chem 232*:236, 1935
120. FLEISCHNER G, ARIAS IM: *Am J Med 49*:576, 1970
121. FLEMING AF, WOOLF AJ: *Clin Chim Acta 12*:67, 1965
122. FOG J: *Scand J Clin Lab Invest 10*:241, 1958
123. FOG J: *Scand J Clin Lab Invest 16*:49, 1964
124. FOG J, BAKKEN AF: *Scand J Clin Lab Invest 20*:88, 1967
125. FOG J, BUGGE-ASPERHEIM B: *Nature (Lond) 203*:756, 1964
126. FOG J, BUGGE-ASPERHEIM B, JELLUM E: *Scand J Clin Lab Invest 14*:567, 1962

127. FOLIN O, FLANDERS FF: *J Biol Chem 11*:257, 1912

128. FORD JD, HAWORTH JC: *Clin Chem 10*:1002, 1964

129. FORMAN DT: *Ann Clin Lab Sci 2*:137, 1972

130. FORMAN DT, PHILLIPS C, EISEMAN W, TAYLOR CB: *Clin Chem 14*:348, 1968

131. FRANKLIN EC: *Clin Chim Acta 4*:259, 1959

132. FREDA VJ: *Am J Obstet Gynecol 92*:341, 1965

133. FREE AH, FREE HM: *U.S. Patent* 2,854,317, 1958; *CA 53*:2343e, 1959

134. FREE AH, FREE HM: *Gastroenterology 24*:414, 1953

135. FRIEDMAN MH: *Gastroenterology 17*:57, 1951

136. FRINGS CS, PARDUE HL: *Anal Chem 36*:2477, 1964

137. FROSCH B: *Klin Wochenschr 43*:262, 1965

138. FROSCH B, WAGENER H: *Klin Wochenschr 46*:913, 1968

139. GABRIELI ER, RONCA PC, ORFANOS A, SULLIVAN M, SNELL FM: *Arch Intern Med 116*:894, 1965

140. GADD KG: *J Clin Pathol 19*:300, 1966

141. GAEBLER OH: *Am J Clin Pathol 15*:452, 1945

142. GAMBINO R: *Standard Methods in Clinical Chemistry.* New York, Academic Press, 1965, Vol 5, p 55

143. GAMBINO SR, FREDA VJ: *Am J Clin Pathol 46*:198, 1966

144. GAMBINO SR, SCHREIBER H: *Technicon Symposium Paper No. 54*, 1964

145. GARDIKAS C, KENCH JE, WILKINSON JF: *Nature (Lond) 159*:842, 1947

146. GARTNER LM, ARIAS IM: *N Engl J Med 280*:1339, 1969

147. GATÉ J, PAPACOSTAS G: *C R Soc Biol (Paris) 83*:1432, 1920

148. GIORDANO AS, WINSTEAD M: *Am J Clin Pathol 23*:610, 1953

149. GOLDSTEIN J, COMBES B: *J Lab Clin Med 67*:863, 1966

151. GOODMAN RD: *J Lab Clin Med 40*:531, 1952

153. GRAHAM JH: *Am J Med Sci 230*:633, 1955

154. GRAY SJ: *Proc Soc Exp Biol Med 41*:470, 1939

155. GRAY CH: *The Bile Pigments.* New York, Wiley, 1953

156. GRAY CH: *Biochemical Society Symposium No. 12*, 1954, p 46

157. GRAY CH, NICHOLSON DC: *Nature (Lond) 180*:336, 1957

158. GREENSPAN EM: *Mt Sinai J Med NY 21*:279, 1955

159. GREENSPAN EM, DREILING DA: *Gastroenterology 32*:500, 1957

160. GREGG JA: *Am J Clin Pathol 49*:404, 1968

161. GREGORY CH: *J Lab Clin Med 61*:917, 1963

162. GREGORY CH, WATSON CJ: *J Lab Clin Med 60*:1, 1962

163. GROS W: *Klin Wochenschr 18*:781, 1939

164. GURD FRN, GOODMAN DS: *J Am Chem Soc 74*:670, 1952

165. HAGER CB, FREE AH: *Am J Med Technol 36*:227, 1970

166. HANGER FM: *Trans Assoc Am Physicians 53*:148, 1938

167. HANGER FM: *J Clin Invest 18*:261, 1939

168. HARRISON GA: *Chemical Methods in Clinical Medicine.* Fourth edition. London, Churchill, 1957, p 77

169. HARRISON GA, BROMFIELD RJ: *Biochem J 22*:43, 1928

170. HARTMANN L, VIALLET A, FAUVERT R: *Clin Chim Acta 8*:872, 1963

171. HEILMEYER L: *Spectrophotometry in Medicine.* London, Adam Hilger Ltd, 1943

172. HEILMEYER L, KREBS W: *Biochem Z 231*:393, 1931

173. HEIRWEGH KPM, MEUWISSEN JA, JANSEN FH: *Biol Neonate 14*:74, 1969

174. HENRY RJ: *Clinical Chemistry, Principles and Technics.* New York, Harper & Row, 1964

175. HENRY RJ: *Hemoglobin, Its Precursors and Metabolites.* Edited by FW Sunderman. Philadelphia, Lippincott, 1964, p 225

176. HENRY RJ, CHIAMORI N, WARE AG: *Am J Clin Pathol 32*:201, 1959
177. HENRY RJ, FERNANDEZ AA, BERKMAN S: *Clin Chem 10*:440, 1964
178. HENRY RJ, GOLUB OJ, BERKMAN S, SEGALOVE M: *Am J Clin Pathol 23*:841, 1953
179. HENRY RJ, JACOBS SI, BERKMAN S: *Clin Chem 7*:231, 1961
180. HENRY RJ, JACOBS SI, CHIAMORI N: *Clin Chem 6*:529, 1960
181. HOFFMAN HN II, WHITCOMB FF Jr, BUTT HR, BOLLMAN JL: *J Clin Invest 39*:132, 1960
182. HOGG CK, MEITES S: *Am J Med Technol 25*:281, 1959
183. HOLOUBECK V: *Clin Chim Acta 1*:342, 1956
184. HORN Z, KOVACS E: *Acta Med Scand 164*:143, 1959
185. HOWARD RM, SCHMIDT CH, DER HOVANESIAN J Jr: *Am J Med Technol 18*:292, 1952
186. HSIA DY, HSIA H-H, GOLFSTEIN RM, WINTER A, GELLIS SS: *Pediatrics 18*:433, 1956
187. HUNTER G: *Br J Exp Pathol 11*:415, 1930
188. HUNTON DB, BOLLMAN JL, HOFFMAN HN: *Gastroenterology 39*:713, 1960
189. IBBOTT FA, O'BRIEN D: *Pediatrics 34*:418, 1964
190. ICHIDA T, NOBUOKA M: *Clin Chim Acta 19*:249, 1968
191. INFANTE F: *Clin Chim Acta 7*:77, 1962
192. INGELFINGER FJ, BRADLEY SE, MENDELOFF AI, KRAMER P: *Gastroenterology 11*:646, 1948
193. IWATA T, YAMASAKI K: *J Biochem 56*:424, 1964
194. JABLONSKI P, OWEN JA: *Adv Clin Chem 12*:309, 1969
195. JACKSON SH: *Clin Chem 11*:1051, 1965
196. JACKSON AH, SMITH KM, GRAY CH, NICHOLSON DC: *Nature (Lond) 209*:581, 1966
197. JACKSON SH, HERNANDEZ AH: *Clin Chem 16*:462, 1970
198. JACOBS SL, HENRY RJ, SEGALOVE M: *Clin Chem 10*:433, 1964
199. JACOBSEN J: *Scand J Clin Lab Invest 26*:395, 1970
200. JAFFE M: *Virchow Arch Pathol Anat 47*:405, 1869
201. JANKELSON IR, LERNER HH: *Am J Dig Dis Nutr 1*:310, 1934
202. JANKELSON IR, SEGAL M, AISNER M: *Am J Dig Dis Nutr 3*:889, 1937
203. JAVITT NB, WHEELER HO, BAKER KJ, RAMOS OL, BRADLEY SE: *J Clin Invest 39*:1570, 1960
204. JENDRASSIK L, CLEGHORN RA: *Biochem Z 289*:1, 1936
205. JENDRASSIK L, CZIKE A: *Z Ges Exp Med 60*:554, 1928
206. JENDRASSIK L, GRÓF P: *Biochem Z 297*:81, 1938
207. JENDRASSIK L, RÉBAY-SZABÓ M: *Biochem Z 294*:293, 1937
208. JIRGL V: *Klin Wochenschr 35*:938, 1957
209. JIRSA M, JIRSOVÁ M: *Clin Chem 5*:532, 1959
210. JONES DD: *Clin Chim Acta 19*:57, 1968
211. JOSEPHSON B: *Biochem J 29*:1519, 1935
212. JOSEPHSON B: *Acta Genet Statist Med 4*:231, 1953
213. KAPITULNIK J, KAUFMANN NA, BLONDHEIM SH: *Clin Chem 16*:756, 1970
214. KAPITULNIK J, SCHENKER JG, BLONDHEIM SH: *Am J Obstet Gynecol 110*:62, 1971
215. KATZ EJ, HASTERLIK RJ, SNAPP FE: *J Lab Clin Med 44*:353, 1954
216. KEHL H: *Clin Chem 13*:475, 1967
217. KIER LC: *J Lab Clin Med 40*:762, 1952
218. KERKHOFF JF, PETERS HJ: *Clin Chim Acta 21*:133, 1968
219. KHAYAT MH, AAS D, SCHLÜTZ GO: *Z Klin Chem 3*:60, 1965

220. KIBRICK AC, ROGERS HE, SKUPP SJ: *Am J Clin Pathol 22*:698, 1952
221. KING EJ, COXON RV: *J Clin Pathol 3*:248, 1950
222. KINGSLEY GR, GETCHELL G, SCHAFFERT RR: *Standard Methods of Clinical Chemistry.* Edited by M Reiner. New York, Academic Press, 1953, Vol 1, p 11
223. KLAASSEN CD: *Clin Chim Acta 35*:225, 1971
224. KLATSKIN G, BUNGARDS L: *J Clin Invest 35*:537, 1956
225. KNOWLTON M: *Standard Methods in Clinical Chemistry.* Edited by D Seligson. New York, Academic Press, 1958, Vol 2, p 12
226. KOPECKY P: *Geburtshilfe Frauenheilkd 29*:818, 1969
227. KRAUS I, DULKIN S: *J Lab Clin Med 26*:729, 1941
228. KRITCHEVSKY D, MARTAK DS, ROTHBLAT GH: *Anal Biochem 5*:388, 1963
229. KRUIJSWIJK H, KENNEDY JC, SCHAAP PAHM: *Clin Chim Acta 14*:561, 1966
230. KUENZLE EC, MAIER C, RÜTTNER JR: *J Lab Clin Med 67*:294, 1966
231. KUNKEL HG: *Proc Soc Exp Biol Med 66*:217, 1947
232. KUNKEL HG, AHRENS EH Jr, EISENMENGER WJ: *Gastroenterology 11*:499, 1948
233. LAEMMER M, BECK J: *C R Soc Biol (Paris) 116*:368, 1934
234. LARSON EA, EVANS GT, WATSON CJ: *J Lab Clin Med 32*:481, 1947
235. LATHE GH: *Biochemistry Society Symposium No. 12.* 1954 p 34
236. LATHE GH, RUTHVEN CRJ: *J Clin Pathol 11*:155, 1958
237. LAUCHENAUER C: *Gastroenterologia (Basel) 75*:193, 1949
238. LAVERS GD, COLE WH, KEETON RW, GEPHARDT MC, DYNIEWICZ JM: *J Lab Clin Med 34*:965, 1949
239. LEEVY CM, SMITH F, LONGUEVILLE J, PAUMGARTNER G, HOWARD MM: *JAMA 200*:236, 1967
240. LEHTONEN A, NÄNTÖ V, BRUMMER P: *Acta Med Scand 180*:235, 1966
241. LEMBERG R, LEGGE JW: *Hematin Compounds and Bile Pigments.* New York, Interscience, 1949
242. LEPEHNE GM: *N Engl J Med 241*:860, 1949
243. LEPEHNE G: *Dsch Arch Klin Med 135*:79, 1921
244. LESTER R, SCHMID R: *J Clin Invest 41*:1379, 1962
245. LESTER R, SCHMID R: *N Engl J Med 270*:779, 1964
246. LEVIN EY: *Biochim Biophys Acta 136*:155, 1967
247. LEVIN SJ, JOHNSTON CG: *J Lab Clin Med 59*:681, 1962
248. LEVITT M, SCHACTER BA, ZIPURSKY A, ISRAELS LG: *J Clin Invest 47*:1281, 1968
249. LEVKOFF AH, FINKLEA JF, WESTPHAL NC, PRIESTER LE Jr: *Israel J Med Sci 6*:432, 1970
250. LEVKOFF AH, WESTPHAL MC, FINKLEA JF: *Am J Clin Pathol 54*:562, 1970
251. LEWI S: *Ann Biol Clin (Paris) 22*:797, 1964
252. LEWIS AE: *Am J Clin Pathol 18*:789, 1948
253. LEWIS B, PANVELIWALLA D, TABAQCHALI S, WOOTTON IDP: *Lancet 1*:219, 1969
254. LEWIS F, SCHULMAN H, HAYASHI TT: *JAMA 190*:195, 1964
255. LICHTMANN SS: *J Biol Chem 107*:717, 1934
256. LILEY AW: *Am J Obstet Gynecol 86*:485, 1963
257. LILEY AW: *Am J Obstet Gynecol 82*:1359, 1961
258. LINDHOLM H: *Scand J Clin Lab Invest 8*:340, 1956
259. LORBER SH, SHAY H: *Gastroenterology 20*:262, 1952
260. LOWRY PT, ZIEGLER NR, CARDINAL R, WATSON CJ: *J Biol Chem 208*:543, 1954
261. LOZZIO BB, ROYER M: *Rev Soc Arg Biol 38*:8, 1962

262. LOZZIO BB, GORODISH S, ROYER M: *Clin Chim Acta 9*:78, 1964
263. MacDONALD RP: *Standard Methods in Clinical Chemistry.* Edited by S Meites. New York, Academic Press, 1965, Vol 5, p 65
264. MACHELLA TE, HELM JD, CHORNOCK FW: *J Clin Invest 21*:763, 1942
265. MACLAGAN NF: *Br J Exp Pathol 25*: 234, 1944
266. MACLAGAN NF, MARTIN NH, LUNNON JB: *J Clin Pathol 5*:1, 1952
267. MAIZELS M: *Lancet 2*:451, 1946
268. MAKINO I, NAKAGAWA S, MASHIMO K: *Gastroenterology 56*:1033, 1969
269. MALLIKARJUNESWARA VR, CLEMETSON CAB, CARR JJ: *Clin Chem 16*:180, 1970
270. MALLOY HT, EVELYN KA: *J Biol Chem 119*:481, 1937
271. MANDELBAUM B, LaCROIX GC, ROBINSON AR: *Obstet Gynecol 29*:471, 1967
272. MANDELBAUM B, ROBINSON AR: *Obstet Gynecol 28*:118, 1966
273. MARCUS PI, TALALAY P: *J Biol Chem 218*:661, 1956
274. MARRON TU: *J Lab Clin Med 27*:108, 1941
275. MARTINEK RG: *Clin Chim Acta 13*:161, 1966
276. MASON MF, HAWLEY G, SMITH A: *Am J Physiol 152*:42, 1948
277. MATEER JG, BALTZ JI, COMANDURAS PD, STEELE HH, BROUWER SW, YAGLE EM: *Gastroenterology 8*:52, 1947
278. MATHER A: *Pediatrics 26*:350, 1960
279. McGANN CJ, CARTER RE: *J Pediat 57*:199, 1960
280. McGILL DB, HOFFMAN HN II, BOLLMAN JL: *J Lab Clin Med 56*:925, 1960
281. McNAY A, OXLEY A, WALKER W: *Am J Clin Pathol 50*:122, 1968
282. MEISENHEIMER HR: *Obstet Gynecol 23*:485, 1964
283. MEITES S, HOGG CK: *Clin Chem 5*:470, 1959
284. MEITES S, TRAUBERT JW: *Clin Chem 11*:691, 1965
285. MENDENHALL CL, LEEVY CM: *N Engl J Med 264*:431, 1961
286. MENDIOROZ BA, CHARBONNIER A, BERNHARD R: *C R Soc Biol (Paris) 145*:1483, 1951
287. MERANZE DR, LIKOFF WB, SCHNEEBERG NG: *Am J Clin Pathol 12*:261, 1942
288. MEULENGRACHT E: *Ugeskrift Laeger 81*:1785, 1919
289. MICHAËLSSON M: *Scand J Clin Lab Invest 13* (Suppl. 56): 1, 1961
290. MICHAËLSSON M, NOSSLIN B, SJOLIN S: *Pediatrics 35*:925, 1965
291. MIHAESCO E, FAUVERT R: *Rev Fr Etud Clin Biol 13*:200, 1968
292. MIKULECKY M: *Clin Chim Acta 21*:43, 1968
293. MILNE MD: *Bilirubin Metabolism.* Edited by IAD Bouchier, BH Billing. Philadelphia, Davis, 1967, p 271
294. MIRSKY IM, von BRECHT R: *Science 98*:499, 1943
295. MOERTEL CG, OWEN CA: *J Lab Clin Med 52*:902, 1958
296. MOLONEY WC, DONOVAN AM, WHORISKEY FG: *Am J Clin Pathol 18*:568, 1948
297. MOORE DB, PIERSON PS, HANGER FM, MOORE DH: *J Clin Invest 24*:292, 1945
298. MORRIS ED, MURRAY J, RUTHVEN CRJ: *Br Med J 2*:352, 1967
299. MORRISON LM: *Am J Dig Dis 7*:527, 1940
300. MORRISON LM, SWALM WA: *J Lab Clin Med 25*:739, 1940
301. MOSBACH EH, KALINSKY HJ, HALPERN E, KENDALL FE: *Arch Biochem Biophys 51*:402, 1954
302. MOSES C: *J Lab Clin Med 30*:267, 1945
303. MURPHY GM, BILLING BH, BARON DN: *J Clin Pathol 23*:594, 1970
304. MYLIUS F: *Hoppe Seylers Z Physiol Chem 11*:492, 1887

305. NADEAU G: *Am J Clin Pathol* 24:740, 1954

306. NAIR PP, KRITCHEVSKY D, editors: *The Bile Acids; Chemistry, Physiology and Metabolism.* New York, Plenum Press, 1971, Vol 1

307. NAKAJIMA H, TAKEMURA T, NAKAJIMA O, YAMAOKA K: *J Biol Chem* 238:3784, 1963

308. NAUMANN HN: *Biochem J* 30:347, 1936

309. NAUMANN HN: *Biochem J* 30:1021, 1936

310. NEALE G, LEWIS B, WEAVER V, PANVELIWALLA D: *Gut* 12:145, 1971

311. NEEFE JR: *Gastroenterology* 7:1, 1946

312. NEEFE JR, REINHOLD JG: *Science* 100:83, 1944

313. NEWBOLD BT, LeBLANC G: *Can J Biochem* 42:1697, 1964

314. NORDYKE RA: *JAMA* 194:949, 1965

315. NOSSLIN B: *Scand J Clin Lab Invest* 12 (Suppl 49):1, 1960

316. OBERMAN HA, KULESH MH: *Am J Clin Pathol* 30:519, 1958

317. OHMORI Y: *Enzymologia* 4:217, 1937

318. OJALA A: *Acta Obstet Gynecol Scand 50 [Suppl]* 10:1, 1971

319. OSTROW JD, BRANHAM RV: *Gastroenterology* 58:15, 1970

320. OVENSTONE JA, CONNON AF: *Clin Chim Acta* 20:397, 1968

321. OVERBEEK JTG, VINK CLJ, DEENSTRA H: *Rec Trav Chim* 74:85, 1955

322. PAGLIARDI E, GIANGRANDI E, MOLINO G: *Am J Dig Dis* 8:251, 1963

323. PANVELIWALLA D, LEWIS B, WOOTON IDP, TABAQCHALI S: *J Clin Pathol* 23:309, 1970

324. PATTERSON J, SWALE J, MAGGS C: *Biochem J* 52:100, 1952

325. PAULY H: *Hoppe Seylers Z Physiol Chem* 42:508, 1904

326. PELLEGRINO E, PATEK AJ Jr, COLCHER A, DOMANSKI B: *J Lab Clin Med* 32:397, 1947

327. PENNINGTON GW, HALL R: *J Clin Pathol* 19:90, 1966

328. PERRYMAN PW, RICHARDS DH, HOLBROOK B: *Biochem J* 66:61P, 1957

329. PERSKY H, GRINKER RR, MIRSKY IA: *J Clin Invest* 29:110, 1950

330. PETERSON AC: *Am J Med Technol* 25:359, 1959

331. PETRYKA ZJ: *J Chromatogr* 50:447, 1970

332. PICARD J, SAMAIR P, AYRAULT-JARRIER M: *Clin Chim Acta* 13:514, 1966

333. PIERCE FT Jr, KIMMEL JR, BURNS TW: *Metabolism* 3:228, 1954

334. POLÁCEK K, ZWINGER A, VEDRA B: *J Obstet Gynaecol Br Commonw* 78:248, 1971

335. POPPER H, SCHAFFNER F: *Liver: Structure and Function.* New York, McGraw-Hill, 1957, p 70

336. POPPER H, STEIGMANN F, DYNIEWICZ H, DUBIN A: *J Lab Clin Med* 34:105, 1949

337. POWELL WN: *Am J Clin Pathol Techn Sec* 8:55, 1944

338. PROBSTEIN JG, LONDE S: *Ann Surg* 111:230, 1940

339. PRÖSCHER F: *Hoppe Seylers Z Physiol Chem* 31:520, 1900

340. QUICK AJ: *J Biol Chem* 67:477, 1926

341. QUICK AJ: *Am J Clin Pathol* 10:222, 1940

342. QUICK AJ: *Am J Med Sci* 185:630, 1933

343. QUICK AJ, OTTENSTEIN HN, WELTCHEK H: *Proc Soc Exp Biol Med* 38:77, 1938

344. RANKIN TJ, JENSON RL, DELP M: *Gastroenterology* 25:548, 1953

345. RAPPAPORT F, EICHHORN F: *Lancet* 244:62, 1943

346. RAYMOND AL, BLANCO JG: *J Biol Chem* 79:649, 1928

347. Recommendations of The College of American Pathologists Standards Committee: A uniform bilirubin standard. *Am J Clin Pathol* 39:90, 1963

348. REED AH, CANNON DC, WINKELMAN JW, BHASIN YP, HENRY RJ, PILEGGI VJ: *Clin Chem 18*:57, 1972
349. REINHOLD JG: *Adv Clin Chem 3*:83, 1960
350. REINHOLD JG: *Anal Chem 27*:239, 1955
351. REINHOLD JG: *Clin Chem 8*:475, 1962
352. REINHOLD JG, YONAN VL: *Am J Clin Pathol 26*:669, 1956
353. REINHOLD JG: *Medical and Public Health Laboratory Methods.* Edited by JS Simmons, DJ Gentzkow. Philadelphia, Lea & Febiger, 1955, p 102
354. RELANDER A: *Scand J Clin Lab Invest 22*:196, 1968
355. Representatives of the American Academy of Pediatrics, the College of American Pathologists, the American Association of Clinical Chemists, and the National Institutes of Health: Recommendation on a uniform bilirubin standard. *Standard Methods in Clinical Chemistry.* Edited by S. Meites. New York, Academic Press, 1965, Vol 5, p 75
356. RICHARDS TG, TINDALL VR, YOUNG A: *Clin Sci 18*:499, 1959
357. ROBERTSON JG: *Am J Obstet Gynecol 95*:120, 1966
358. ROBINSON SH: *N Engl J Med 279*:143, 1968
360. ROOVERS J, EVRARD E, VANDERHAEGHE H: *Clin Chim Acta 19*:449, 1968
361. ROSENBERG AA, EIMANN LG, O'LEARY J: *Clin Chem 11*:40, 1965
362. ROSENTHAL SM, WHITE EC: *JAMA 84*:1112, 1925
363. ROSENTHAL WS, DOUVERS PA: *Am J Dig Dis 10*:300, 1965
364. ROTH H, SEGAL S, BERTOLI D: *Anal Biochem 10*:32, 1965
365. ROTHE G: *Z Gesamte Exp Med 106*:338, 1939
366. RUDMAN D, KENDALL FE: *Fed Proc 15*:611, 1956
367. RUDMAN D, KENDALL FE: *J Clin Invest 36*:530, 1957
368. RUDMAN D, KENDALL FE: *J Clin Invest 36*:538, 1957
369. RUTOWSKI RB, deBAARE L: *Clin Chem 12*:432, 1966
370. SAIFER A: *J Clin Invest 27*:737, 1948
371. SAIFER A: *Am J Med Sci 219*:597, 1950
372. SALVINI M, GONZATO P: *Boll Soc Ital Biol Sper 23*:905, 1947
373. SANDBERG DH, SJOVALL J, SJOVALL K, TURNER DA: *J Lipid Res 6*:182, 1965
374. SAVAGE RD, WALKER W, FAIRWEATHER D VI, KNOX EG: *Lancet 2*:816, 1966
375. SCHACHTER D: *J Lab Clin Med 53*:557, 1959
376. SCHACHTER D: *Science 126*:507, 1957
377. SCHACHTER D, TAGGART JV: *J Biol Chem 208*:263, 1954
378. SCHEINBERG P, MYERS JD: *Proc Soc Exp Biol Med 68*:63, 1948
379. SCHELLONG G: *Klin Wochenschr 43*:814, 1965
380. SCHLESINGER W: *Dtsch Med Wochenschr 29*:561, 1903
381. SCHMID R: *The Metabolic Basis of Inherited Disease.* Third edition. Edited by JB Stanbury, JB Wyngaarden, DS Fredrickson. New York, McGraw-Hill, 1972, p 141
382. SCHMID R: *Biol Chem 229*:881, 1957
383. SCHMID R: *Arch Intern Med 101*:669, 1958
384. SCHMIDT A: *Verh Congr Inn Med 13*:320, 1895
385. SCHOENFIELD LJ, FOULK WT, BUTT HR: *J Clin Invest 43*:1409, 1964
386. SCHREIBER J, ERB W, BÖHLE E: *Verh Dtsch Ges Inn Med 75*:920, 1969
387. SCHWARTZ S, SBOROV V, WATSON CJ: *Am J Clin Pathol 14*:598, 1944
388. SCURRY MM, FIELD H Jr: *Am J Med Sci 206*:243, 1943
389. SELIGSON D, MARINO J, DODSON E: *Clin Chem 3*:638, 1957
390. SEMPERE JM, GANCEDO C, ASENSIO C: *Anal Biochem 12*:509, 1965
391. SHANK RE, HOAGLAND CL: *J Biol Chem 162*:133, 1946

392. SHAY H, BERK JE, SIPLET H: *Gastroenterology 9*:641, 1947
393. SHAY H, SCHLOSS EM: *J Lab Clin Med 15*:292, 1929
394. SHINOWARA GY: *Am J Clin Pathol 24*:696, 1954
395. SIA RHP: *China Med J 38*:35, 1924
396. SIGGAARD-ANDERSEN O, KOMARMY LE: *Am J Clin Pathol 49*:863, 1968
397. SIMS FH, HORN C: *Am J Clin Pathol 29*:412, 1958
398. SINHA SN, GABRIELI ER: *Clin Chim Acta 19*:313, 1968
399. SJÖVALL J: *Methods Biochem Anal 12*:97, 1964
400. SJÖVALL J: *Clin Chim Acta 5*:33, 1960
401. SMITH TB, RICHARDS JA: *Am J Med Technol 29*:291, 1963
402. SOBOTKA H, LUISADA-OPPER AV, REINER M: *Am J Clin Pathol 23*:607, 1953
403. SOINI R, DAUWALDER H, RICHTERICH R: *Schweiz Med Wochenschr 99*:1784, 1969
404. Standard Reference Material 916: *Bilirubin. US Department of Commerce, National Bureau of Standards Publication,* April 5, 1971
405. STEINBERG A: *J Lab Clin Med 34*:1049, 1949
406. STENHAGEN E, RIDEAL EK: *Biochem J 33*:1591, 1939
407. STENSTAM T: *Acta Med Scand [Suppl] 177*:61, 1946
408. STERNER JH, CUSACK M: *Am J Clin Pathol Tech Sect 9*:7, 1945
409. STEVENSON GW, JACOBS SL, HENRY RJ: *Clin Chem 8*:433, 1962
410. STEVENSON GW, JACOBS SL, HENRY RJ: *Clin Chem 10*:95, 1964
411. STEWART AG, TAYLOR WC: *J Obstet Gynaecol Br Commonw 71*:604, 1964
412. STONER RE, WEISBERG HF: *Clin Chem 3*:22, 1957
413. SUSSMAN S, CARBONE JV, GRODSKY G, HJELTE V, MILLER P: *Pediatrics 29*:899, 1962
414. TAKATA M: *Far East Association of Tropical Medicine. Transactions of The Sixth Congress, Tokyo,* 1925, Vol 1, p 693
415. TAKATA M, ARA K: *Far East Association of Tropical Medicine. Transactions of the Sixth Congress, Tokyo,* 1925, Vol 1, p 667
416. TALAFANT E: *Nature (Lond) 178*:312, 1956
417. TALLACK JA, SHERLOCK S: *Br Med J 2*:212, 1954
418. TENGSTRÖM B: *Acta Med Scand 183*:31, 1968
419. TENHUNEN R, MARVER HS, SCHMID R: *Proc Natl Acad Sci USA 61*:748, 1968
420. TERWEN AJL: *Dtsch Arch Klin Med 148*:72, 1925
421. TEUCHY H, VAN SUMERE CF: *Clin Chim Acta 25*:79, 1969
422. THALME B: *Acta Obstet Gynecol Scand 43*:78, 1964
423. THANNHAUSER SJ, ANDERSEN E: *Dtsch Arch Klin Med 137*:179, 1921
424. TISDALE WA, WELCH J: *J Lab Clin Med 59*:956, 1962
425. TYGSTRUP N, WINKLER K, LUND E, ENGELL HC: *Scand J Clin Lab Invest 6*:43, 1954
426. ULLMANN TD, KLEEBERG J, HEIMANN-HOLLAENDER E: *Clin Chim Acta 3*:531, 1958
427. UMBERGER CJ, FIORESE FF: *Clin Chem 9*:91, 1963
428. VAN den BERGH AAH, MULLER P: *Biochem Z 77*:90, 1916
429. VAN den BERGH AAH, SNAPPER J: *Dtsch Arch Klin Med 110*:540, 1913
430. VAN den BOSSCHE H: *Clin Chim Acta 9*:310, 1964
431. VAN SUMERE CF, TEUCHY H, PÉ H, VERBEKE R, BEKAERT J: *Clin Chim Acta 26*:85, 1969
432. VARELA FUENTES B: *Acta Med Scand 138*:65, 1950
433. VELU P, VELU M: *Ann Biol Clin (Paris) 17*:108, 1959
434. VERSCHURE JCM: *Acta Med Scand 142*:409, 1952
435. VOEGTLIN WL: *Am J Clin Pathol 18*:84, 1948

436. WALDSTEIN SS, DUBIN A, NEWCOMER A, McKENNA CH: *Clin Chem 10*:381, 1964
437. WALLACE GB, DIAMOND JS: *Arch Intern Med 35*:698, 1925
438. WALTERS MI, GERARDE HW: *Microchem J 15*:231, 1970
439. WANG RIH, JACOBSON J: *Am J Dig Dis 11*:973, 1966
440. WATSON CJ: *Am J Clin Pathol 6*:458, 1936
441. WATSON CJ: *Arch Intern Med 47*:698, 1931
442. WATSON CJ: *Ann Intern Med 70*:839, 1969
443. WATSON CJ: *J Clin Pathol 16*:1, 1963
444. WATSON CJ: *J Lab Clin Med 54*:1, 1959
445. WATSON CJ, HAWKINSON V: *Am J Clin Pathol 17*:108, 1947
446. WATSON CJ, SCHWARTZ S, SBOROV V, BERTIE E: *Am J Clin Pathol 14*:605, 1944
447. WATSON D: *Clin Chim Acta 5*:613, 1960
448. WATSON D: *Clin Chem 7*:603, 1961
449. WATSON D: *Clin Chim Acta 7*:733, 1962
450. WATSON D: *Clin Chim Acta 6*:737, 1961
451. WATSON D, ROGERS JA: *J Clin Pathol 14*:271, 1961
452. WATSON D, MacKAY EV, TREVELLA W: *Clin Chim Acta 12*:500, 1965
453. WEBER AP, SCHALM L, WITMANS J: *Acta Med Scand 173*:19, 1963
454. WEISHSELBAUM TE, PROBSTEIN JG: *J Lab Clin Med 24*:636, 1939
455. WEINBREN K, BILLING BH: Cited by BH Billing, GH Lathe: *Am J Med 24*:111, 1958
456. WELTMANN O: *Med Klin 26*:240, 1930
457. WHEELER HO, EPSTEIN RM, ROBINSON RR, SNELL ES: *J Clin Invest 39*:236, 1960
458. WHEELER HO, MELTZER JI, BRADLEY SE: *J Clin Invest 39*:1131, 1960
459. WHITE FD, DUNCAN D: *Can J Med Sci 30*:552, 1952
460. WHITE D, HAIDAR GA, REINHOLD JG: *Clin Chem 4*:211, 1958
461. WHITFIELD CR, NEELY RA, TELFORD ME: *J Obstet Gynaecol Br Commonw 75*:121, 1968
462. WIEGAND BD, KETTERER SG, RAPAPORT E: *Am J Dig Dis 5*:427, 1960
463. WINKELMAN J: *Experientia 23*:949, 1967
464. WINKLER K: *Scand J Clin Lab Invest 13*:44, 1961
465. WINKLER K, GRAM C: *Acta Med Scand 169*:263, 1961
466. WINKLER K, GRAM C: *Acta Med Scand 178*:439, 1965
467. WITH TK: *Bilirubin Metabolism.* Edited by IAD Bouchier, BH Billing. Philadelphia, Davis, 1967, p 135
468. WITH TK: *Nature (Lond) 158*:310, 1946
469. WITH TK: *Hoppe Seylers Z Physiol Chem 278*:130, 1943
470. WITH TK: *Bile Pigments; Chemical, Biological, and Clinical Aspects.* New York, Academic Press, 1968
471. WITH TK: *Biology of Bile Pigments.* Copenhagen, Arne Frost-Hanse, 1954
472. WITMANS J, SCHALM L, SCHULTE MJ: *Clin Chim Acta 6*:7, 1961
473. WORTH MH Jr, FLITMAN R: *J Lab Clin Med 70*:352, 1967
474. WUNDERLY C, WUHRMANN FH: *Schweiz Med Wochenschr 75*:1128, 1945
475. WYSOCKI AP, PORTMAN OW, MANN GV: *Arch Biochem Biophys 59*:213, 1955
476. YAMAMOTO T, SKANDERBEG J, ZIPURSKY A, ISRAELS LG: *J Clin Invest 44*:31, 1965
477. YONAN VL, REINHOLD JG: *Am J Clin Pathol 24*:232, 1954

478. ZIEVE L, HILL E: *Gastroenterology* 28:766, 1955
479. ZIEVE L, HANSON M: *J Lab Clin Med* 42:872, 1953
480. ZIEVE L, HILL E, HANSON M, FALCONE AB, WATSON CJ: *J Lab Clin Med* 38:446, 1951

Chapter 23

Hemoglobins, Myoglobins, and Haptoglobins

ANIS MAKAREM, Ph.D.

This chapter covers the detection of hemoglobin and its derivatives in urine, feces, and stomach contents; the determination of hemoglobin in whole blood and plasma; the determination of the hemoglobin derivatives methemoglobin, carboxyhemoglobin, and sulfhemoglobin in blood; the identification and determination of abnormal hemoglobins in blood; the detection and determination of myoglobin in urine or serum; and the determination of haptoglobins in serum.

Hemoglobin is the porphyrin-iron(II)-protein compound which gives blood its red color. The average man contains almost 1 kg, virtually all of it in the erythrocytes where it exists in practically saturated solution (over 30%). It enables the red blood cell to carry O_2, to assist in CO_2 transport, and to buffer the blood against sudden pH changes. The porphyrin portion of hemoglobin is protoporphyrin IX (type III), which combined with $Fe(II)$ forms heme. Heme plus the protein globin constitutes hemoglobin.

The hemoglobin molecule can be understood best in terms of its primary, secondary, tertiary, and quaternary structures. Its primary structure consists of four polypeptide chains each attached to a heme molecule. One amino acid sequence is common to two of the chains and another to the other two. Each of the two "alpha" chains has 141 amino acids in its sequence and each of the two "beta" chains 146. These polypeptide chains coil into alpha helices, which are considered to be the secondary structure of the molecule. This coiling tends to place the polar amino acid residues into the interior of the helix, leaving a relatively nonpolar surface. The folding of each coiled chain into a specific three-dimensional configuration forms the tertiary structure, which is the same for all normal globins regardless of length. One heme molecule is attached to each polypeptide chain. The attachment is at the same site in both types of chain if one counts from the carboxyl terminus of the molecule—i.e., the Fe atom of the heme is linked to a histidine residue, which is number 55 from the carboxyl terminus. The Fe is also very close to another histidine residue, which is number 84 from the carboxyl terminus, and the space between these two is the active site for O_2 transport (258).

It is the folding mentioned above which places the 55th and 84th residues on opposite sides of the planar heme molecule, as well as producing other important spatial relationships. These relationships include completely nonpolar environment for the heme, external areas polar enough to make hemoglobin very soluble, and highly specialized areas for intersubunit contact (see *Hemoglobin M Disease*, this chapter).

The four folded chains fit together closely and are held by noncovalent bonds. They occupy the four corners of a rough tetrahedron; this specific orientation is called the quaternary structure. This four-chain composite forms one molecule of hemoglobin, which has a gram molecular weight calculated to be 64 458 and Fe content of 0.347% by weight. Measured values agree closely with those calculated.

Varieties of hemoglobin containing five different types of globin subunit occur in normal blood at one or another stage of life. These are designated by the Greek letters α through ϵ. All consist of 146 amino acids except α, which has 141. The major constituent of normal adult hemoglobin has the globin formula $\alpha_2\beta_2$. The different subunits ordinarily occur in combinations which result in five types of normal hemoglobin, designated Gower-1, ϵ_4; Gower-2, $\alpha_2\epsilon_2$; fetal hemoglobin (Hb-F), $\alpha_2\gamma_2$; hemoglobin A_2 (Hb-A_2), $\alpha_2\delta_2$; and hemoglobin A (Hb-A), $\alpha_2\beta_2$. The proportion of these normal hemoglobins in blood changes rapidly from the earliest months of fetal existence to about 1 year, as shown in Table 23-1 and Figure 23-1 (254). The nomenclature of hemoglobins is discussed in greater detail below under *Hemoglobin Abnormalities*. Throughout this chapter the word hemoglobin is abbreviated Hb when it forms part of a name, as is the custom.

FIG. 23-1. Proportions of the various polypeptide chains of normal human hemoglobin through early life. Electrophoretic patterns at pH 8.6, typical for three periods, are shown along the top of the figure. (Refs. 164, 190, 254).

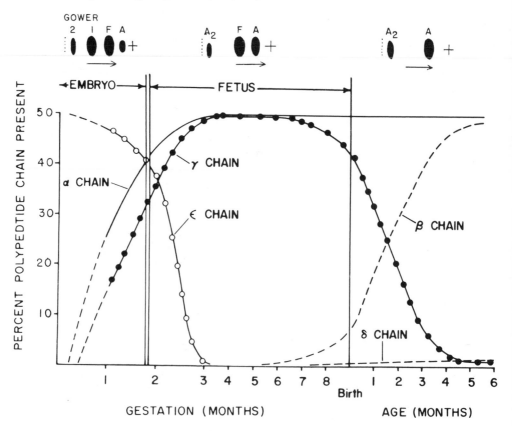

TABLE 23-1. Normal Human Hemoglobins (ref 254)

Class	Name	Subunit formula	Age present
Embryonic hemoglobins	Gower 1 Gower 2	ϵ_4 $\alpha_2\epsilon_2$	Predominates in first 2 months; disappears after third month of gestation
Fetal hemoglobin	Hb-F	$\alpha_2\gamma_2$	Predominates in fetal life; declines rapidly after birth; $< 2.0\%$ throughout life
Adult hemoglobins	Hb-A	$\alpha_2\beta_2$	Detectable even early in gestation; predominates post-partum
	Hb-A$_2$	$\alpha_2\delta_2$	Appears in late fetal life; exists throughout life in small amounts: $< 3.7\%$

Hemoglobin derivatives are compounds of medical and chemical importance which result from the reaction of hemoglobin with substances other than O_2, such as carbon monoxide, oxidizing agents, acids, alkalis, and cyanide. The chemical relationships between hemoglobin and its derivatives are shown in Figure 23-2. Different systems of nomenclature exist for these compounds. Table 23-2 lists the equivalents of the three commonly used systems. That presented in the first column is the oldest and is still the most widely used.

Hemoglobin and its derivatives are colored compounds with characteristic absorption properties at different wavelengths which are used for their identification and quantification. These properties, as shown by Pauling (250), result from the electrical and magnetic interactions of the Fe and the porphyrin ring. These interactions plus very complex stereochemical relationships (258) inhibit the Fe(II) from oxidizing or from forming further bonds with most substances other than O_2 and carbon monoxide, with which it bonds covalently. The derivatives of myoglobin are analogous to those of hemoglobin. Table 23-3 lists the wavelengths at which the hemoglobin and myoglobin derivatives have absorption maxima and minima.

Hemoglobin abnormalities are produced *in vivo* by two types of variations in hemoglobin synthesis. In one type the rate of synthesis of the protein or one of its constituent polypeptide chains is reduced, producing disorders called *thalassemias*. In the other type an abnormal protein is produced by substitution or deletion of one or more amino acids in the polypeptide chains. Hemoglobins with such abnormalities are known as *hemoglobin variants*. Also, abnormal hemoglobins may be produced by abnormal combinations of normal subunits at the quaternary level of structure.

TABLE 23-2. Nomenclature of Hemoglobin and Derivatives

Name in old system	Lemberg and Legge (207) and others	Pauling and Coryell (251)	Constitution
Hemoglobin	Hemoglobin	Ferrohemoglobin	Complex of ferroheme and native globin
Oxyhemoglobin	Oxyhemoglobin	Oxyferrohemoglobin	Complex of ferrohemoglobin and O_2
Methemoglobin	Hemiglobin	Ferrihemoglobin	Complex of ferriheme and native globin
Heme	Heme	Ferroheme	Ferrous porphyrin complex
Acid hematin	Acid hematin	Ferriheme	Ferric porphyrin complex
Hemochromogen	Hemochrome	Ferrohemochromogen	Complex of ferroheme and nitrogenous substance
Parahematin (kathematin)	Hemichrome	Ferrihemochromogen	Complex of ferriheme and nitrogenous substance
Alkaline hematin	Alkaline hematin	Ferriheme hydroxide	Ferriheme plus hydroxyl ions
Carboxyhemoglobin	Carboxyhemoglobin	Carbonmonoxohemoglobin	Complex of ferrohemoglobin and CO
Cyanmethemoglobin	Hemiglobin cyanide	Ferrihemoglobin cyanide	Complex of ferrihemoglobin and cyanide

FIG. 23-2. Interrelationships of hemoglobin and its derivatives.

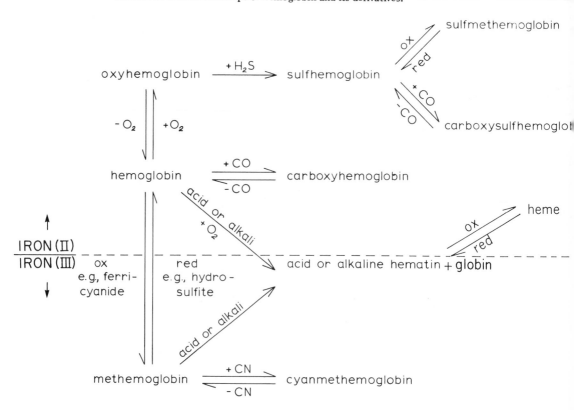

TABLE 23-3. Absorption Maxima and Minima of Hemoglobin, Myoglobin, and Their Derivatives

Compound	nm, maxima[a]	nm, minima	References
Hemoglobin	431, 554–556	478	338
Oxyhemoglobin	414, 531–543, 577	510, 560	338
Carboxyhemoglobin	420, 539, 568–569	496, 555	338
Sulfhemoglobin	540, 620		79, 83, 145, 186
Carboxysulfhemoglobin	615		186, 207
Methemoglobin			
pH 7.0–7.4	406, 500–502, 630	476, 610	338
pH 10–11	440, 540, 576	510, 560	145, 168
pH 5–6	404, 500, 630	470, 600	145
Cyanmethemoglobin	421, 540–542	504	338
Hematin			
Acid	376, 662	625	145
Alkaline	385, 540–580		145, 168
In serum (methemalbumin)	403, 623		1, 145
Hemochromogen (alkaline ferrohematin in serum)	425, 530, 560	495, 540	145, 168
Myoglobin	434, 560	484	31, 348
Oxymyoglobin	418, 542, 582	514, 564	26, 31, 38, 80, 144, 253, 348
Carboxymyoglobin	424, 540, 578	508, 563	26, 31, 38, 80, 144, 330, 348
Metmyoglobin			
pH 4.8–7.0	413, 500, 632	465, 607	31, 38, 348
pH 8.6–11.0	413, 540, 586	522, 565	103, 308, 309, 348, 362
Cyanmetmyoglobin	424, 543	514	348

[a]Values reported by different investigators may differ by up to about 2 nm.

Myoglobin is another porphyrin-iron(II)-globin compound, with function analogous to that of hemoglobin, which exists in tissue rather than red cells. Its molecular structure is that of one-fourth of a hemoglobin molecule—one heme attached to one polypeptide chain. The amino acid sequence of the globin is different from either the α or β chain of hemoglobin, although very close to the former. Myoglobin forms derivatives paralleling those of hemoglobin, and abnormal variants also exist.

Haptoglobins are a group of closely related plasma glycoproteins with α_2 mobility that migrate in the α_2 region in serum protein electrophoresis at pH 8.6 in barbital buffer. They combine avidly with free hemoglobin or globins. Thus hemoglobin released into plasma is promptly bound to haptoglobin, and the resulting complexes are eliminated from the plasma via the reticuloendothelial system without being excreted in the urine. Hemoglobinuria appears only when the amount of hemoglobin in the plasma exceeds the binding capacity of the haptoglobins (203).

HEMOGLOBIN

This section covers the qualitative detection and quantitative determination of hemoglobin. Detection is invariably applied to samples other than blood in a search for nonobvious bleeding. Determination is used to assist in detecting any of a variety of conditions which alter the normal hemoglobin concentration of blood—e.g., anemia or polycythemia.

DETECTION OF HEMOGLOBIN AND ITS DERIVATIVES

Hemoglobin, intact erythrocytes, and hemoglobin derivatives are designated *occult blood* when they exist in abnormal locations in such amounts as to be undetectable by eye. The clinical chemist must be able to detect occult blood in urine, feces, and stomach contents, and to distinguish abnormal from normal amounts.

Hemoglobin and its derivatives behave like the enzyme peroxidase and have been called *pseudoperoxidases.* These substances catalyze the redox reaction between hydrogen peroxide and benzidine, *o*-tolidine, diphenylamine, or similar compounds. In such reactions oxidation of the organic compound produces a color. Hematin has 50%–80% of the activity of hemoglobin, oxyhemoglobin, methemoglobin, and carboxyhemoglobin. The statement that alkaline hematin does not react (335) is incorrect. Myoglobin and its corresponding derivatives react in the same way, but porphyrins do not.

Adler and Adler in 1904 (3) were first to employ benzidine for detection of blood in urine and feces. The procedure was later modified by Gregerson (123) to increase color stability. *o*-Tolidine was first used in 1912 by Ruttan and Hardisty (287), and diphenylamine has been recognized as a sensitive redox indicator for many decades.

Benzidine is converted to the so-called "benzidine blue", which is a meriquinoid oxidation product (i.e., an addition compound) containing one molecule of the amine, one molecule of the imine, and two equivalents of a monovalent acid (85, 355). *o*-Tolidine, which is 3,3'-dimethylbenzidine, presumably reacts in the same manner.

BENZIDINE   BENZIDINE BLUE

The reaction of diphenylamine involves two steps and produces a chromogen chemically analogous to benzidine blue (85).

Benzidine and *o*-tolidine were for many years the most commonly used chromogens for determination of occult blood in feces. These compounds are now considered carcinogenic, and their continued use for routine analysis has been discouraged (378). Diphenylamine can easily replace them (366, 367). Other substrates which have been used in tests for occult blood include phenolphthalin (formed from phenolphthalein by boiling an alkaline solution of the indicator with zinc dust), *p*-phenylenediamine chlorhydrate, aminopyrine (amidopyrine, Pyramidon), 3-aminophthalic acid hydrazide, and leucomalachite green (148).

URINE

The term *hematuria* is applied to the presence of an abnormal number of erythrocytes in the urine and the term *hemoglobinuria* refers to the presence in urine of hemoglobin or its iron-containing derivatives in solution. Urine normally contains some erythrocytes. Addis (2) found up to 425 000 ($\bar{x}$ = 66 000) in 12 hour night urine specimens. Strictly speaking this is occult blood, since it is not visible to the naked eye. According to Cook *et al* (62) the usual microscopic examination of centrifuged sediment from 15 ml urine cannot distinguish between normal urine and a 1:1 000 000 dilution of blood in the same urine. This represents an addition of 5 million erythrocytes per liter or an increase above normal content of at least fivefold. Cook *et al* found the erythrocyte content of urine had to be about 50 million/liter before the presence of an abnormal content could be detected with certainty. Obviously, hemolysis is a factor in the sensitivity of the microscopic examination, hemolysis increasing with increasing pH and decreasing specific gravity (62). Factors contributing to false positive results in the microscopic examination of urine for erythrocytes include the presence of yeast cells, urates, oxalates, and fat droplets; factors contributing to false negative results include hemolysis and the presence of an obscuring sediment (92).

The desired sensitivity of chemical tests for occult blood in urine is about 0.0003 mg hemoglobin (or derivatives) per milliliter, which is equivalent to about 10 million erythrocytes per liter.

Specificity

Pus, iodides, and bromides may give false positive tests with guaiac, benzidine, and o-tolidine tests. Reactions to hemoglobin or derivatives are decreased after boiling the urine, whereas reactions due to iodides or bromides persist (148).

Sensitivity

Sensitivity in the literature is usually given as that dilution of blood with urine giving a positive reaction. If we assume that blood contains 15 g hemoglobin per 100 ml, a dilution of 1:1000 equals 15 mg/100 ml or 0.15 mg/ml. Tests on urine are usually performed on volumes that range from a few drops to 1 ml.

Sensitivities are much greater for aqueous solutions of hemoglobin than for hemoglobin in urine or feces, the difference being as much as 100-fold. This discrepancy is due to the presence of inhibitors—e.g., the sensitivity of the benzidine test decreases with increasing concentration of ascorbic acid, which is a reducing agent. Levels of ascorbic acid greater than 20 mg/100 ml definitely interfere, and addition of ascorbic acid actually reverses a positive test. Interference by ascorbic acid in urine can be avoided by acidification with acetic acid, extraction of resultant acid hematin into ether, evaporation, and performance of the test on the residue. Sensitivity may decrease with aging and decomposition of the urine sample. Erythrocytes may settle out on standing either by themselves or entrapped in phosphate precipitate, and not be included in the test solution.

Sensitivities of benzidine tests vary from 0.00015 to 0.025 mg/ml. Varying sensitivity is due to (a) variation in amount of sample used; (b) variation in proportion of reagents—e.g., sensitivity increases with benzidine concentration and H_2O_2 must not be in excess; (c) variation between lots of benzidine; and (d) variation between analysts—e.g., sensitivity of the same test procedure can vary fourfold between analysts simply because one analyst may think "peas" are smaller than does another analyst.

Sensitivities of o-tolidine tests vary from 0.000006 to 0.07 mg/ml. In fact the test can be made so sensitive that it gives a positive test in the absence of hemoglobin. As with other substrates, the sensitivity can be varied (148).

FECES

Normal fecal blood loss is up to about 2.5 ml/day as determined by radioactive chromium tracer studies (355). A good portion of the lost blood appears in the stool as protoporphyrin, deuteroporphyrin, and mesoporphyrin formed from hemoglobin by bacterial decomposition. Measured excretion equivalent to more than 3 ml of blood per day is indicative of clinically significant bleeding such as that from an ulcer or gastric carcinoma. Other sources of blood, however, such as mouth or nose bleeding, hemorrhoids, and menstrual flow must be excluded.

Clinically significant bleeding can be intermittent; a series of tests at regular intervals may be necessary for proper diagnosis. Severe gastric or duodenal bleeding (50–80 ml/day or more) results in the formation of glistening black stool ("tarry stool"). The presence of red streaks or spots in the stool suggests lower colon, rectal, or perianal bleeding.

Certain complications enter into detection of occult blood in feces which do not exist with urine. First, account must be taken of exogenous hemoglobin and myoglobin arising from ingestion of meat and fish. Second, hemoglobin and myoglobin on passing through the gastrointestinal tract are degraded to hematin and the corresponding pigment of myoglobin (which do react) and certain porphyrins (which do not react). The degree of degradation depends on the level in the tract where bleeding occurs and on transit time. It has been shown that the peroxidase activity of ingested blood decreases sixfold on passage through the alimentary canal. Pancreatin digestion of blood *in vitro* results in a 50% reduction of response to the benzidine test.

Specificity

Peroxidases of bacterial and plant origin are present in feces which may give false positive reactions. This has been solved in various ways: (a) Adjust the sensitivity of the test downward or ignore trace to 1+ reactions. This would appear to be a poor solution to the problem. (b) Boil a fecal suspension or heat it to 60°C for 5 min to denature the enzymes. Boiling apparently may also destroy some of the peroxidase activity of hemoglobin and derivatives since sensitivity of some tests may be decreased as much as 5- to 20-fold by boiling (149). (c) Acidify with acetic acid, extract the resultant acid hematin into ether or ethyl acetate, evaporate, and test the residue (148).

Sensitivity

Sensitivities of benzidine tests vary from 0.00015 to 0.0015 mg/g feces; 1–25 ml blood must be ingested to render the tests positive. Sensitivities of *o*-tolidine tests vary from 0.000015 to 0.0015 mg/g feces; 1–4.5 ml blood must be ingested to produce a positive test. If patients are on normal diets including meat, sensitivities are such that, although a test may be negative with some modifications, most benzidine or *o*-tolidine tests may be falsely positive. It is necessary, therefore, that patients be placed on a fish- and meat-free diet for 3 days prior to the test.

Mild gum bleeding from brushing teeth is not sufficient to produce a positive test even with benzidine or *o*-tolidine. False positive tests may occur from meat ingestion more readily in children and infants than in adults, possibly because of more rapid transit time and less degradation of hemoglobin (148). The use of the chromogen diphenylamine as reported by Woodman (366, 367) gives good agreement with the benzidine method. The sensitivity is similar to the Gregersen test as modified by Needham and Simpson (241), which is said to give a positive reaction when a minimum of 3–5 ml of whole blood is administered directly into the stomach of a healthy subject. With this sensitivity, a positive reaction is thought to provide conclusive evidence of significant bleeding. A negative reaction, however, is not conclusive. The advantage of this method is that significant bleeding should be detectable without the necessity of restricting the patient to a meat-free diet. Only blood-rich food such as liver and blood sausage need be restricted.

Quantitative Determination of Fecal Blood Loss

The fecal content of hemoglobin and derivatives in terms of hemoglobin has been determined quantitatively by photometric measurement of the color produced with benzidine. Loss of blood in the feces can also be determined by the radioactivity appearing in the feces following intravenous injection of ^{51}Cr-labeled erythrocytes, ^{59}Fe-labeled erythrocytes, or ^{59}Fe ferrous citrate, which appears in the circulating erythrocytes within 4 hours. For unknown reasons, results by the benzidine method are 12%–16% greater than results obtained with ^{51}Cr-labeled erythrocytes (148).

STOMACH CONTENTS

Normally blood should not be detectable in fasting stomach contents. Trauma from passage of the tube frequently produces some bleeding which of course appears as fresh blood. Blood which is brown (resembling "coffee grounds") represents hemoglobin degraded by stomach acid to hematin that was present prior to intubation. Tests for occult blood designed for urine or feces can of course be applied to stomach contents as well as to other body fluids such as sputum.

FILTER PAPER AND TABLET TESTS FOR OCCULT BLOOD

The use of filter paper strips impregnated with a buffered mixture of organic peroxide and various chromogens has been widely accepted as a routine method for the detection of occult blood in urine. *Hema-Combistix* strips (Ames) are impregnated with a buffered mixture of organic peroxide and *o*-tolidine. They are more sensitive to free hemoglobin and myoglobin than to intact erythrocytes. The sensitivity is of course reduced by high urinary concentrations of ascorbic acid. A positive reaction is produced by hemoglobin, myoglobin, and erythrocytes, and a false positive reaction is produced by certain oxidizing contaminants such as hypochlorites. The test is performed by dipping the Hema-Combistix in urine (148).

Watson-Williams (351) proposed a tablet containing citric acid, barium peroxide, sodium carbonate, and *o*-tolidine for detection of blood in urine. The sensitivity of this tablet was about 0.0015 mg/ml urine. In 1956 Free *et al* (97) introduced a tablet containing *o*-tolidine, strontium peroxide, calcium acetate, and tartaric acid for testing urine. This tablet is manufactured by Ames and sold as *Occultest*. The tablet today also contains sodium bicarbonate which upon solution with the tartaric acid effervesces to facilitate solution of the reactive ingredients. A red dye is also included to mask discoloration of the tablet. Tartaric acid and calcium acetate react with the strontium peroxide to form H_2O_2. One drop of uncentrifuged urine is placed in the center of a piece of filter paper provided, and a tablet then placed in the center of the moist area. Two drops of water are placed on the tablet. A diffuse blue color appearing on the filter paper around the tablet within 2 min constitutes a positive test. The amount of blood present is proportional to the intensity of the blue color, its area on the filter paper, and rapidity of color development. Sensitivity is reportedly 0.00015–0.0015 mg/ml.

Hematest is another tablet marketed by Ames which contains the same ingredients but the sensitivity is adjusted downward to 0.005–0.008 mg/ml. The manufacturer recommends its use with feces primarily. It is claimed that normal diet has no effect on results with feces, although iron preparations or ingestion of large amounts of meat or fresh, leafy green vegetables may cause a trace reaction. Ingestion of 8–10 ml blood results in 39% positive fecal samples. On the one hand it has been stated (67) that Hematest is not quite as good as Occultest for feces, whereas on the other it has been claimed (149, 170) that Hematest is too sensitive for feces (58% positive on normal diet, 35% positive on meat-free diet). Such claims of too great a sensitivity are rather difficult to understand unless the false positives are due to bacterial peroxidases (the test does not call for preliminary inactivation). Another criticism made of Hematest is that its sensitivity varies among lots of tablets (170). In our opinion this test is not sufficiently sensitive for testing urine (149).

Choice of Methods

Several studies have been made comparing various methods for detection of occult blood (62, 149, 170, 378). In general, the order of sensitivity is *o*-tolidine>benzidine>diphenylamine>guaiac. Sensitivities overlap, and within certain wide limits sensitivity can be controlled by test parameters such as reagent concentration. There does not seem to be any clear-cut difference in specificity. All three—benzidine, *o*-tolidine, and diphenylamine—have the required sensitivity for detection of occult blood in urine and feces. Guaiac is not sensitive enough. The diphenylamine test, as described by Woodman (366, 367), does not require the meat restriction necessary for the benzidine and o-tolidine tests. This plus the fact that the reagent is not carcinogenic makes it most suitable as a routine test for detection of occult blood in urine and feces.

DETECTION OF OCCULT BLOOD IN URINE, FECES, AND STOMACH CONTENTS

(method of Woodman, Refs 366, 367)

PRINCIPLE

O_2 removed from H_2O_2 by the peroxidase activity of hemoglobin, myoglobin, and some of their derivatives reacts with diphenylamine in acetic acid to produce a green chromophore. Fe also reacts, but the Na salt of N,N-di-(2-hydroxyethyl)glycine is added, which chelates Fe thus preventing interference from oral Fe therapy.

REAGENTS

Glacial Acetic Acid

Acetic Acid, 50%. Dilute 1 vol glacial acetic acid with 1 vol water.

Diphenylamine, 1%. Dissolve 1 g diphenylamine in 100 ml glacial acetic acid.

N,N-di-(2-hydroxyethyl)glycine, Na salt (NaDHEG). Other names: diethylolglycine; N,N-bis(2-hydroxyethyl)glycine, Na salt.

H_2O_2, 3%. Commercially available or make a 1:10 dilution of 30% H_2O_2 with water.

PROCEDURE

1. *Urine or stomach contents:* A few milliliters of mixed uncentrifuged sample is heated to boiling, then allowed to cool to room temperature. *Feces:* A portion of feces the size of a pea is added to 10 ml 50% acetic acid, heated to boiling, then allowed to cool to room temperature.
2. Transfer 0.5 ml diphenylamine reagent, 0.1 ml H_2O_2, and about 0.1 ml NaDHEG into a test tube. Wait 1 min before adding sample to ensure that no color appears due to contamination of tube or reagents.
3. Add 0.2 ml fecal suspension or 0.4 ml urine or stomach contents. Mix well and observe color at 90 and 120 sec.

INTERPRETATION

Blood produces a green color. The speed of color development is proportional to blood concentration. A distinct green color within 90 sec constitutes a positive reaction. A similar color between 90 and 120 sec is regarded as weakly positive. Any color developed after 120 sec or no color development indicates a negative reaction.

NOTES

1. False positives. Bile-contaminated feces yields a greenish color in the reaction. If this occurs at the end of 2 min add about 1.6 ml diethyl ether. Purple indicates a positive reaction. Color due to bile remains green. Ingestion of blood-rich foods—e.g., liver, blood sausage—causes false positive results and therefore should be avoided. Oxidizing contaminants in glassware or water cause a false positive reaction, which is checked in step 2 of the procedure.

2. Sample stability. Samples should be fairly stable at room temperature for at least several days and for much longer periods at refrigerator temperature. Peroxidase activity decreases, however, as the sample decomposes (108).

ACCURACY

The test appears to be reasonably specific for hemoglobin or myoglobin and their iron-containing derivatives. True peroxidases, if present, are destroyed by boiling. Iron salts do not produce false results. The sensitivity of the test, however, presumably may vary significantly between samples because of inhibitors present. Its average sensitivity is similar to that of the benzidine test as modified by Needham and Simpson (241). It is said to give a positive reaction when a minimum of 3–5 ml of blood is administered directly to the stomach of a healthy adult.

DETERMINATION OF HEMOGLOBIN

The quantitative determination of blood hemoglobin has been the most common clinical analysis for at least half a century. The technics used include iron determination, spectrophotometry, refractometry, specific gravity, pseudoperoxidase activity, and gasometric measurements. These methods are reviewed in detail by Henry in the first edition of this book (148). Because of recent agreements to standardize on particular methods, only the first two are of more than historic interest.

METHODS FOR DETERMINATION OF HEMOGLOBIN BASED ON SPECTRAL CHARACTERISTICS OF HEMOGLOBIN OR ITS DERIVATIVES

Acid Hematin

Hemoglobin is determined by measuring the brown color of acid hematin formed with dilute HCl. Sahli (288) in 1894 was the first to describe a simple method using this technic. The unknown was diluted to a color match with a standard. Many modifications of the technic were developed, one of the most popular being that of Haden (126–128) adapted to the Haden-Hauser hemoglobinometer: Blood was diluted 1:10 with acid in a white cell pipet, and after 30 min the color compared in a wedge chamber with a glass wedge. For many years some form of the acid hematin method was the most popular method for the routine determination of hemoglobin, at least in the United States. Although some studies of the acid hematin method have been favorable, most investigators found at least some of the modifications to be inaccurate: (a) The acid hematin color is affected by substances present other than hemoglobin. Wu (368) believed the difficulty to be that the acid hematin is in colloidal suspension rather than in true solution. (b) A long and variable time is required for full color development. (c) It has been reported (368) that the acid hematins from oxyhemoglobin, carboxyhemoglobin, methemoglobin, and cyanhemoglobin differ.

Cyanhematin

Dilute HCl is added, forming acid hematin, and then cyanide is added, forming cyanhematin. This technic is regarded favorably by a number of investigators.

Alkaline Hematin

Hemoglobin can be determined by conversion to alkaline hematin instead of acid hematin. Again, although some workers have found the technic to be accurate under some circumstances, others found that it suffers from the same sources of inaccuracy as the acid hematin method.

Pyridine Hemochromogen

Heme combines with a large variety of organic nitrogenous substances such as pyridine to form "hemochromogens." Two pyridine groups become attached to the iron in the reaction. In this method for determination of hemoglobin, blood is diluted with dilute alkali and pyridine, and hydrosulfite is added. Hemoglobin, carboxyhemoglobin, methemoglobin, sulfhemoglobin, and hematin are apparently measured satisfactorily by this technic.

Carboxyhemoglobin

Blood is diluted and saturated with CO (usually coal gas is used). An advantage of the technic is that any preformed carboxyhemoglobin is included in the measurement. A further advantage is the great stability of the color. Some workers have found the technic to be quite accurate. Others have found at least certain modifications to be inaccurate. The method is rather time-consuming, and although it has never become very popular in the United States it was regarded as the method of choice in Great Britain for many years.

Oxyhemoglobin

In oxyhemoglobin methods blood is diluted ($\sim$1:250) with a diluent that causes hemolysis and the resultant solution of hemoglobin is shaken with air to effect conversion of hemoglobin to oxyhemoglobin. Some of the earliest procedures for hemoglobin were based on determination of the intensity of the red color of the blood although it is not clear that the conditions were always such as to ensure complete oxygenation in every case. In 1878 Gowers (119) diluted blood to a color match with a solution of picrocarmine. Tallqvist in 1900 (327, 328) introduced his technic in which a drop of blood was placed on filter paper and the color compared with a series of lithographed color standards. Dare (71) placed a drop of blood between two small glass plates and compared the color with a graded colored glass disk which was rotated until the colors matched. In spite of the fact that it is well established that the Dare and the Tallqvist methods are liable to errors up to 3 g hemoglobin per 100 ml, the Tallqvist color scale is still being sold today.

In 1927 Davis and Sheard (72, 303, 304) presented the first photometric method for measuring oxyhemoglobin. They employed as diluent 0.1% Na_2CO_3, which has a pH of about 11.1. The instability of oxyhemoglobin in this diluent or in 0.4% ammonia (0.4 ml conc NH_4OH per 100 ml) is due to conversion to alkaline hematin, and some investigators believed that this conversion occurs stepwise through methemoglobin. Harboe (138), however, concluded that this conversion occurs directly without going through methemoglobin. The functions of alkali are to prevent turbidity from precipitation of proteins and to ensure complete destruction of erythrocyte ghosts and liberation of hemoglobin. Up to 10% of the hemoglobin may remain inside the ghosts following lysis with water, and few if any ghosts are destroyed, causing turbidity. The obvious solution to the problem of stability is to decrease the pH of the diluent. Oxyhemoglobin in 0.04% ammonia ($\sim$0.006 N) is stable for many hours, provided the reagent is devoid of cupric ions. Their effect can be negated by addition of EDTA. Phosphate or phosphate-tetraborate buffer, pH 8.0, or 0.01% Na_2CO_3 can be used, although the latter is not sufficiently alkaline to prevent turbidity completely.

Oxyhemoglobin has three absorption peaks: the Soret peak at 414 nm, and those at 540 and 576. The absorptivity at 414 nm is nearly 10 times that at the other two peaks, and although use of this peak has been proposed nearly all investigators have used a green filter with a photometer or the 540 nm peak with a spectrophotometer. Bell *et al* (12) stated that the differences in absorptivity are great between oxyhemoglobin and abnormal pigments at the blue end of the spectrum. Using a green filter with maximal transmission at about 520 nm, the error introduced by significant quantities of carboxyhemoglobin, methemoglobin, or sulfhemoglobin is no more than a few percent. Studies (148) have shown that the technic using a green filter is accurate and eminently suitable for routine analysis.

Cyanmethemoglobin

The total hemoglobin of blood is converted to cyanmethemoglobin, which is then determined from its absorbance at 540 nm. Oxyhemoglobin, hemoglobin, methemoglobin, and carboxyhemoglobin are all converted. In the original technic as proposed in 1920 by Stadie (315), two solutions were employed, a solution of potassium ferricyanide to convert hemoglobin to methemoglobin and a solution of potassium cyanide to convert the methemoglobin to cyanmethemoglobin. Later it was shown that the two reagents can be combined.

In 1958 the Panel on the Establishment of a Hemoglobin Standard of the Division of Medical Sciences, National Academy of Sciences, National Research Council (377), after reviewing several photometric methods in current use, concluded that the cyanmethemoglobin method was the best available for several reasons: (a) It involves dilution with a single reagent. (b) All forms of hemoglobin likely to occur in circulatory blood, with the exception of sulfhemoglobin, are determined. (c) The color is suitable for measurement in filter photometers as well as in narrow band spectrophotometers because its absorption band at 540 nm is broad and relatively flat. (d) Standards prepared from either crystalline hemoglobin or washed erythrocytes and stored in a brown glass container and in sterile condition are stable for at least 9 months (change <2%). The adopted absorptivity of cyanmethemoglobin per milligram-atom of iron per liter was 11.5. This has since been changed to 11.0 (338). Several modifications have been proposed (120, 339) designed to avoid precipitation of globulins which at times introduced difficulty.

In 1966 the International Congress of Hematology (82) recommended this method as the accepted routine procedure for measurement of hemoglobin in blood and an iron method as the standardization procedure.

Choice of Methods

A group of English workers, including Bartholomew, Donaldson, Sisson, King, Wootton, and MacFarlane (77), have been quite active in investigations of hemoglobin methodology. They found that the following technics were capable of giving results checking within about 3%: O_2 capacity, CO capacity, iron content, oxyhemoglobin, carboxyhemoglobin, acid-hematin, alkaline hematin, cyanhematin, and cyanmethemoglobin. Evidence has also been obtained that many of the older technics, including the Dare, Haden-Hauser, Hellige-wedge type, Haldane, Gowers, Sahli, Tallqvist, Cohen-Smith, and Newcomer, are liable to errors up to 3 g hemoglobin per 100 ml. Since it is inconceivable that any modern clinical laboratory would be without at least a photoelectric filter photometer, it must be recognized that although these methods served admirably in their day it is time to relegate them honorably to a state of obsolescence.

The determination of iron content of whole blood is firmly established as the most accurate method for quantitative determination of blood hemoglobin (130, 213) and therefore is recommended by the International Committee for Standardization in Hematology for standardization of the cyanmethemoglobin reference solution. Because the cyanmethemoglobin method has been widely accepted as the method of choice for routine hemoglobin determination, and the cyanmethemoglobin standard is used by many laboratories as the sole daily reference, it is indeed desirable to check the accuracy of these solutions periodically by the iron analysis method.

Various methods have been used for hemoglobin determination by iron analysis (7, 39, 61, 73, 193, 273, 373, 375). Obviously, hemoglobin and all its iron-containing derivatives are determined by this method. The nonhemoglobin iron of whole blood is infinitesimal in comparison to the hemoglobin iron and can be ignored.

The first step in most procedures for determination of iron in whole blood is ashing to eliminate organic material. Some early methods employed dry ashing but this technic may result in loss of iron by volatilization. The chloride is volatile at temperatures as low as 450°C, and the sulfate and phosphate as low as about 700°C. Although alkali can be added to prevent loss, most methods have used wet ashing. Reactions used for measurement of the iron in the digests have included the following: (a) *Titration of Fe^{3+} with Ti^{3+} ("titanium reduction method")*. This was one of the earliest technics but has been largely supplanted by easier, photometric methods. Among the disadvantages of the method are the facts that Cl^- interferes, O_2 must be excluded from the reagent and the titration itself, and the standard is unstable. (b) *Titration of Fe^{2+} with $K_2Cr_2O_7$*. (c) *Iodometric titration of Fe^{3+}*. Excess KI is added and the liberated I_2 titrated with thiosulfate. (d) *Titration of Fe^{2+} with $KMnO_4$*. Aside from the nuisance of having to carry out the titration in an atmosphere of H_2, the endpoint is difficult and unreliable. (e) *Titration of iron with EDTA*. (f) *Formation of complex with hydroxyquinoline, gravimetric*. (g) *Formation of complex with hydroxyquinoline, photometric*. (h) *Photometric measurement of color formed with thiocyanate*. This has been the most popular technic. In 1928 a modification using sulfuric acid and potassium persulfate was introduced, providing a digestion in which no heat is required. A recent innovation uses sodium hypochlorite, which, it is claimed, gives faster and smoother digestion than the H_2SO_4-persulfate mixture. (i) *Miscellaneous photometric methods*. These include color reactions with ferrocyanide (Prussian blue), thioglycolic acid (mercaptoacetic acid), phenanthroline dipyridyl, antipyrine, and ethylenediaminetetraacetic acid plus H_2O_2.

A survey of about 190 laboratories in 1955 showed that only 12% of these were using the cyanmethemoglobin technic for the quantitative measurement of blood hemoglobin. A similar survey in 1964, 5 years after the initiation of a program for the certification of cyanmethemoglobin solutions, revealed that 491 out of 531 laboratories (92.4%) were using the cyanmethemoglobin method; 89% of these depended solely on the cyanmethemoglobin standard for a reference (322).

At its 1964 Stockholm meeting the International Committee for Standardization in Hematology (ICSH) instituted an expert panel on hemoglobinometry to advise the ICSH board on proposals for an internationally acceptable standardized method for the determination of hemoglobin in blood. In 1966 the assembly of the ICSH gave final approval to the following recommendations (217): (a) Clinical laboratories should use the cyanmethemoglobin method exclusively for determination of blood hemoglobin. (b) The term gram (of hemoglobin) per 100 ml (of blood) should be used to express the hemoglobin concentration. (c) A sterile cyanmethemoglobin solution (60–85 mg/100 ml) prepared and tested to meet specifically defined conditions should be used as a standard for routine daily determinations. (d) Standard solutions should be used within 1 year. These recommendations met wide acceptance throughout the Western world.

The fact that cyanmethemoglobin (HbCN) is the only hemoglobin derivative with a stability of 1 year or longer was one factor in making it the preferred standard for the photometric determination of hemoglobin. Moreover, its spectral properties make the cyanmethemoglobin method accurate and reliable even by simpler filter photometry. The absorption spectrum of HbCN displays a rather flat maximum around 540 nm; thus it is easy to obtain a filter with a spectral band width within the region of maximum absorbance (339).

Cyanmethemoglobin solutions generally obey Beer's law at 540 nm. Since the diluent has negligible absorbance at this wavelength, it can be used as the zero standard and a calibration line can be obtained with only a single additional standard. Another stated advantage of the cyanmethemoglobin method is that all common hemoglobin derivatives can readily be converted to cyanmethemoglobin, thus eliminating unpredictable errors due to the presence of different hemoglobin derivatives (339). However, opinion on the last point is divided (148).

Although it is the accepted standard, the cyanmethemoglobin method has certain disadvantages which are procedural in nature. Several minutes are required for the formation of the cyanmethemoglobin under normal conditions, and if carboxyhemoglobin is present at least 90 min at room temperature may be necessary (282). This makes it less suitable for instrumentation or automation than the oxyhemoglobin method, for example. Also, red cell debris can cause varying degrees of turbidity. Finally, cyanide is poisonous and requires special care in handling.

An automatic device is available for determining hemoglobin in blood (102) employing a modification of the oxyhemoglobin technic. It employs a peristaltic proportioning pump and flow-through cuvet (see *Chapter 10*) to dilute and hemolyze whole blood samples, to measure them by photometry, and to give the result as a reading in g/100 ml, all in about 15 sec. The photometry is performed at 548.6 nm using an interference filter with a band pass of less than 2.5 nm and depends on the identical absorbance, at this wavelength, of hemoglobin, oxyhemoglobin, and carboxyhemoglobin.

DETERMINATION OF BLOOD HEMOGLOBIN, CYANMETHEMOGLOBIN METHOD

(method of Van Kampen and Zijlstra, Refs 79, 338, 339)

PRINCIPLE

The Fe(II) of heme in hemoglobin, oxyhemoglobin, and carboxyhemo-globin is oxidized to the ferric state by ferricyanide to form methemoglobin. Methemoglobin then combines with ionized cyanide to produce the stable, red cyanmethemoglobin which is measured photometrically at 540 nm (Fig. 23-3). The pH of this reagent is lower than earlier ones to increase the reaction rate, and a detergent is added to prevent the turbidity which would otherwise occur at the low pH.

REAGENTS

Hemoglobin Reagent (79, 339). Dissolve 200 mg $K_3Fe(CN)_6$, '50 mg KCN, and 140 mg KH_2PO_2 in about 900 ml water. Add 0.5 ml STEROX SE (Hartman-Leddon Co.) and mix gently until dissolved. Adjust the pH to 7.2

FIG. 23-3. Absorption curve of cyanmethemoglobin. Blood containing 14.8 g/100 ml hemoglobin vs reagent blank. Beckman DK-2A recording spectrophotometer.

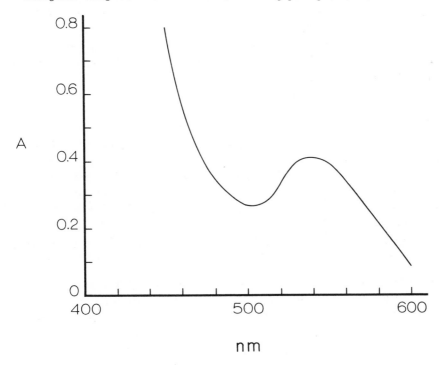

± 0.1 with 0.1 N H_3PO_4 or 0.1 N KOH and dilute to 1 liter with water. This solution is stable for several months if stored in a dark bottle in the refrigerator.

Standard, 60 mg/100 ml. Fresh human blood is collected and 2 ml 3.2% sodium citrate is added per 10 ml of blood. This is centrifuged and the plasma decanted off and discarded. The erythrocytes are washed twice with 0.9% saline and twice with 1.2% saline. To 1 vol of the washed cells, an equal vol of water is added, and then an equal vol of chloroform (reagent grade). The contents are thoroughly mixed, allowed to stand for 12 hours at 4°C, and then centrifuged at high speed for 30 min. The upper aqueous layer which contains the hemoglobin is transferred to a clean container, filtered through filter paper or sintered glass, and its concentration determined by the cyanmethemoglobin and the Fe methods. Hemoglobin reagent is then added in such a ratio to obtain the desired hemoglobin concentration of 60 mg/100 ml. The hemoglobin solution is then sterilized by filtering through a Seitz filter (EKS_2, 30 cm in diameter) and dispensed aseptically in 10 ml portions into brown glass ampules (type I, USP XVI). The ampules are sealed with sterilized rubber stoppers. Purity is checked by examining the absorption spectra between 450 and 700 nm, and by determining the ratio of absorbance at 540 and 504 nm. The ratio should be between 1.58 and 1.62. Turbidity is checked by observation of any Tyndall effect.

Standards containing 50% glycerol seem to be less susceptible to bacterial growth. If used, they should be read against blanks containing 50% glycerol. They are stable for at least 1 year in the dark at room temperature and should not be frozen. Commercial standards are available from various sources. The standard deviation of the absorbance of the standard cyanmethemoglobin solution, determined in 40 clinical laboratories, is about 1% of the established value (322).

PROCEDURE

1. Measure 5.0 ml hemoglobin reagent into cuvet. Transfer 20 μl well mixed blood (collected with anticoagulant) by a Sahli-type hemoglobin pipet or its equivalent. This is a TC pipet so that, after filling to the mark, the outside must be wiped clean with absorbent tissue or cotton and the contents rinsed into the diluent three times. Mix.
2. Allow cuvet to stand for at least 5 min. Measure absorbance against hemoglobin reagent blank at 540 nm or using a filter with nominal wavelength in this region.

Calculation (see *Notes 1* and *2* below): For spectrophotometer, 540 nm, narrow band width, cuvet with 1 cm light path, assuming absorptivity of 11.0/mg-atom Fe (55.85 mg) per liter (338):

$$\text{g hemoglobin/100 ml} = A_{540} \times 36.8$$

For filter photometer:

$$\text{g hemoglobin/100 ml} = A_{540} \times F$$

where

$$F = \frac{\text{g hemoglobin/100 ml (see } \textit{Note 2} \text{ below)}}{A_{540} \text{ on same sample by cyanmethemoglobin method}}$$

NOTES

1. Beer's law. Beer's law is obeyed in a spectrophotometer at 540 nm and in filter photometers using a filter with nominal wavelength at 540 nm.

2. Standardization. Although earlier studies gave the value 11.5 for the absorptivity of cyanmethemoglobin, overwhelming experimental evidence now places it at almost exactly 11.0, and this value has met with general acceptance (338).

The reliability of calculations of hemoglobin concentration from the absorptivity depends on the reliability of absorbance readings from any given spectrophotometer. As discussed in *Chapter 1, Photometry and Spectrophotometry*, absorbance readings on a spectrophotometer may be in error by at least several percent. Corrections may be applied by a technic analogous to that discussed in *Chapter 21* under *Spectrophotometric Determination of LDH, Note 13.*

Most hemoglobin analyses are run on a filter photometer rather than a spectrophotometer because of greater speed. The determination can be standardized on the filter photometer by analyzing several fresh blood samples on a spectrophotometer and correlating the values obtained with the photometer readings. From these a mean factor, *F*, can be calculated. For example, assume that we are standardizing the determination for use with a Klett photometer and a 54 filter, and we obtain values of 10.5, 16.4, and 12.8 g/100 ml for three blood samples using a Beckman DU spectrophotometer and the spectrophotometric equation for calculation. We obtain corresponding Klett readings (proportional to A_{540}) of 132, 204, and 162. These data provide *F* values of 0.0796, 0.0804, and 0.079. The mean is 0.080.

If the laboratory has no spectrophotometer, or if there are no means at the analyst's disposal for checking the accuracy of absorbance readings of a spectrophotometer, then standardization must be made by other means. This can best be done by determining the hemoglobin concentrations of a series of blood specimens by iron analysis and correlating the levels with photometer readings as outlined above. A simpler approach is to use cyanmethemoglobin standards which are available commercially: Ortho

(Acuglobulin Hemoglobin Standard); Hycel (Hycel Cyanmethemoglobin Standard); Armour Pharmaceutical (Metrix Hemoglobin Standard). The College of American Pathologists makes available as a service to commercial producers the certification of batches to within 2% of the stated value. Such standards are stable for at least 9–12 months if kept in the dark (338).

3. *Color stability.* At the pH of the hemoglobin reagent, less than 5 min is required for complete conversion of hemoglobin to cyanmethemoglobin (338). As stated in *Note 2*, above, the color is then stable for periods of 9–12 months.

4. *Lipemia.* Gross lipemia may cause a positive error up to 3 g hemoglobin per 100 ml (40).

5. *Arterial vs venous vs capillary blood.* Studies have detected no difference between arterial and venous blood (25) or between venous and capillary blood (122).

6. *Accuracy of Sahli pipets and their calibration.* Calibration of Sahli pipets obtained from various manufacturers has shown that only approximately two-thirds of them have an error less than 1%–3%, and as many as 6% have an error greater than 5%. It is best, therefore, that the calibrations on these pipets be checked.

7. *Stability of samples.* Stability studies in our laboratory indicate that samples are stable at 30°C for at least 1 week.

ACCURACY AND PRECISION

If we assume proper use of an accurate Sahli pipet, absence of interference by lipemia, and accurate standardization, the technic quite accurately determines the sum of oxyhemoglobin, hemoglobin, methemoglobin, and carboxyhemoglobin (339). Sulfhemoglobin is not converted to cyanmethemoglobin. Its presence, however, produces no serious error because A_{540} of the sulfmethemoglobin formed is 78% of A_{540} of a corresponding amount of cyanmethemoglobin (83).

The precision of the method is ±3% (300).

NORMAL VALUES

The normal adult ranges for blood hemoglobin are 13–18 and 11–16 g/100 ml for males and females, respectively (148). It is interesting to note that until 1944 the normal values in England were about 1 g less than in other countries; the discovery was not made until then that this difference was due to improper standardization by Haldane in 1901.

Among the variables of concern are the following.

Age. The normal range is about 14–23 g/100 ml for infants at birth, 12–16 at 1 month, and 10–14 at 1 year (148). After the age of 2 years there is a gradual increase to adult levels, which are reached during the teens. There is little or no change in values between the ages of 20 and 60 years. Observations on changes occurring after the age of 60 are contradictory.

Race. There is no difference among Caucasians, Negroes, and Orientals (122).

Exercise. There is a 3%—10% increase with exercise due to a decrease in blood volume. This is reversible, decreasing after the exercise is stopped. Athletic training results in a stable increase.

Season. A number of studies have indicated the existence of a seasonal variation amounting to a few percent. Usually the peak has been observed to occur in winter or spring (148).

Altitude. It is well established that individuals living at high altitudes have a higher normal level of hemoglobin (polycythemia) that is compensatory for the lower O_2 tension. At altitudes of 12 340 and 14 900 feet, for example, the ranges for females and males are 15.9—21.7 and 17.3—24.2 g/100 ml, respectively (148).

Posture. Lange (202) observed the hemoglobin to be about 7% lower after the individual has been lying down for an adequate period of time. Spealman *et al* (314) found no change with posture.

Temperature. Hemoglobin is affected through changes in blood volume. Thus concentration decreases in a warm environment and increases in a cold environment, but there is no change in the total circulating hemoglobin (314).

Within-day variation. There is a diurnal swing, being lower in the evening than in the morning by amounts up to about 1 g/100 ml (219, 224).

DETERMINATION OF BLOOD HEMOGLOBIN BY IRON ANALYSIS

(method of Sobel, Ref 148, modified)

PRINCIPLE

Blood is digested by wet digestion with a mixture of H_2SO_4, HNO_3, and $HClO_4$, Fe being split from its complex with any hemoglobin pigment and oxidized to Fe^{3+}. Much of the ferric iron is precipitated as $Fe_2(SO_4)_3$, so HCl is added to effect solution of the iron. The pH is adjusted to 4—5, hydroquinone is added to reduce Fe^{3+} to Fe^{2+}, and the color developed with *o*-phenanthroline is measured at 510 nm (Fig. 23-4).

REAGENTS

H_2SO_4, *Conc.* Reagent grade.
HNO_3, *Conc.* Reagent grade.
$HClO_4$, *70%.* Reagent grade.
HCl, Conc. Reagent grade.
$HClO_4$-HNO_3 *Mixture.* Mix 1 vol $HClO_4$ and 1 vol HNO_3 just before use.
Ammonium Acetate, 50%. Dissolve 50 g anhydrous ammonium acetate in 50 ml water. Dilute to 100 ml and mix thoroughly.

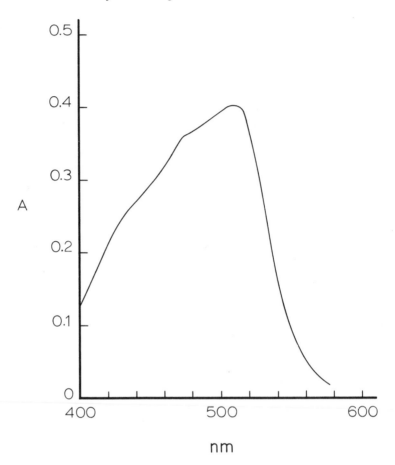

FIG. 23-4. Absorption curve of color formed in Fe determination. Standard (20 μg Fe) vs reagent blank. Beckman DK-2A recording spectrophotometer.

Hydroquinone, 1%. Dissolve 100 mg hydroquinone crystals in 5 ml water. Dilute to 10 ml and mix thoroughly. Prepare fresh before use.

o-*Phenanthroline, 0.2%.* Dissolve 0.2 g 1,10-phenanthroline monohydrate (G. Fredric Smith) in 50 ml water. Dilute to 100 ml and mix thoroughly. Stable in refrigerator for at least 1 year.

Stock Iron Standard, 100 mg/100 ml. Clean a piece of pure iron wire (at least 99.8% iron) with 6 N HCl and then with water and wipe dry. Place 100 mg of cleaned wire into a 250 ml Erlenmeyer flask then add 10 ml conc HNO_3 and about 40 ml water. Boil gently until iron is completely dissolved. Cool and transfer quantitatively to 100 ml volumetric flask. Dilute to 100 ml and mix thoroughly. Stable for at least 1 year.

Working Iron Standard, 1 mg/100 ml. Dilute 1 ml of stock iron standard with 99 ml water. Prepare fresh daily.

Glassware. All glassware used (reagent bottles, pipets, etc.) should be cleaned with hot HCl (conc acid diluted with an equal volume of water) and then rinsed with water.

Water. Single-distilled water is satisfactory for preparation of reagents and for use in the test itself.

PROCEDURE

1. Using a 1.0 ml TD Ostwald-Folin pipet, transfer 1.0 ml well mixed whole blood into the bottom of a 100 ml Kjeldahl digestion flask, allowing pipet to drain, then blow. Place 1.0 ml water in another 100 ml Kjeldahl digestion flask for the blank.

2. Carefully add to each flask 3.0 ml conc H_2SO_4 and 4 ml conc HNO_3. Digest until contents of sample flask are brown (*not black*).

3. Add 2 ml of $HClO_4$-HNO_3 mixture to each flask and digest to appearance of SO_3 fumes (adding these two acids as a mixture at this point, rather than separately avoids the possibility of a violent reaction). Digest will be at most pale yellow, and a precipitate of $Fe_2(SO_4)_3$ may be visible.

4. Cool, add 20 ml water, and heat again to the appearance of fumes.

5. Add 2 ml conc HCl and boil for 25 sec (longer boiling results in loss of HCl and the reprecipitation of $Fe_2(SO_4)_3$).

6. Add 20 ml water, heat to boiling, then cool to room temperature.

7. Dilute the digest to the 100 ml calibration mark with water.

8. Set up the following in tubes calibrated at 10 ml or in 10 ml volumetric flasks:

    Blank: 4.0 ml blank digest
    Standard: 4.0 ml blank digest + 2.0 ml working standard (20 μg)
    Unknown: 4.0 ml sample digest

9. Add 1.0 ml ammonium acetate to each tube, mix and check pH with a narrow-range pH paper. pH must be between 4 and 5; if not, adjust with more ammonium acetate or dilute HCl.

10. Add 1.0 ml 1% hydroquinone, mix, add 2.0 ml *o*-phenanthroline, mix, and then add water to a total volume of 10.0 ml and mix again.

11. Let stand 30 min and read absorbances against water at 510 nm.

Calculations:

$$\text{mg Fe/100 ml} = \frac{A_x - A_b}{A_s - A_b} \times 0.02 \times \frac{100}{4} \times 100$$

$$\text{g hemoglobin/100 ml blood} = \frac{\text{mg Fe/100 ml}}{347}$$

NOTES

1. Beer's law. Beer's law is obeyed on a Beckman DU spectrophotometer up to an absorbance of 2.0.

2. Color stability. The color is stable for at least 24 hours.

3. Stability of sample. As long as the iron is in a form that can be uniformly dispersed so that there is no sampling error, samples should be stable indefinitely.

ACCURACY AND PRECISION

This method determines "total hemoglobin"—i.e., hemoglobin and any of its iron-containing derivatives. No correction is made for the nonhemoglobin iron in the plasma, but even a level of 1000 μg Fe per 100 ml plasma introduces an error of only about 1% in the hemoglobin analysis.

The precision of the method is about ±3%.

PLASMA HEMOGLOBIN

Technics commonly used for determination of hemoglobin in whole blood cannot be used for plasma because they are not sufficiently sensitive. For plasma, accuracy and precision are required to a level of 0.1 mg hemoglobin per 100 ml. Technics that have been proposed fall into two groups: spectrophotometric methods and those utilizing the peroxidase activity of hemoglobin and certain of its derivatives.

SPECTROPHOTOMETRIC METHODS

Shinowara (305) and Childers and Barger (54) determined hemoglobin concentration from the difference between A_{575} (peak) and A_{560} (minimum) of diluted plasma. That this technic is not very sensitive is indicated by the fact that this difference for undiluted sample containing 10 mg hemoglobin per 100 ml is only about 0.031. Porter (268) employed the difference between A_{578} and A_{700}. According to Fleisch (93), methemoglobin and various other inactive hemoglobin pigments are not measured correctly in these procedures. If a sample contains 100 mg hemoglobin per 100 ml, methemalbumin is detectable in 3 hours, and in 24 hours up to half of the hemoglobin is in the form of methemalbumin (268). Chiamori *et al* (52) in our laboratory compared results obtained with these technics with results obtained by the procedure of Bing and Baker (22) and found that there was no significant correlation at levels below 50 mg/100 ml. They also determined hemoglobin concentration from measurement of absorbance at the Soret peak (415 nm) and corrected for bilirubin content by measurement at 460 nm. In spite of the much greater absorptivity of hemoglobin at the Soret peak compared to the peak at 575 nm, results at low levels did not appear to be much improved. Harboe (137) made measurements at 380, 415, and 450 nm and calculated concentration by the base line correction technic. No correction was made for bilirubin; hence the method could not be used with icteric sera.

Hunter *et al* (169) converted hemoglobin to alkaline ferrohematin with 10% NaOH and hydrosulfite and then calculated results from the ratio of A_{560} (peak) to A_{580}. Hemoglobin, methemoglobin, carboxyhemoglobin, and sulfhemoglobin are measured satisfactorily, but methemalbumin is overestimated (93). Flink and Watson (94) converted pigments to pyridine hemochromogen. Hemoglobin, methemoglobin, carboxyhemoglobin, sulfhemoglobin, and hematin are measured satisfactorily, but methemalbumin is overestimated (93). Heme combines with a large variety of organic nitrogenous substances such as pyridine to form "hemochromogens." Two pyridine groups become attached to the iron in the reaction. This technic has been criticized because of the varying turbidity produced (64). Other approaches include conversion of hemoglobin, methemoglobin, and sulfhemoglobin to carboxyhemoglobin, which is then determined by spectrophotometric measurement (methemalbumin is converted to similar pigment only after reduction with hydrosulfite) and conversion to cyanmethemoglobin followed by spectrophotometric measurement—methemalbumin gives different absorption curves than methemoglobin (93).

Absorption peaks may be shifted slightly when hemoglobin is complexed with various plasma proteins. It has also been reported that, although the 540 nm peaks of oxyhemoglobin do not shift on formation of the hemoglobin-haptoglobin complex, absorptivity increases at 540 nm and decreases at 575 nm. It is important in most spectrophotometric technics to standardize with hemoglobin added to plasma rather than using a solution of hemoglobin in water. Methods in which calculations are made from absorbance readings at a single wavelength are subject to positive error in the presence of turbidity. Theoretically, when calculations are made from the difference between absorbances at a maximum and a minimum, the effect of turbidity is canceled out. In practice, however, large errors can occur if the hemoglobin level is low and the turbidity very great.

It would appear that although spectrophotometric methods may be adequate for grossly elevated levels of hemoglobin in plasma they possess neither the accuracy nor precision required for normal or slightly elevated levels (148).

METHODS BASED ON THE PEROXIDASE ACTIVITY OF HEMOGLOBIN AND DERIVATIVES

Iron compounds of porphyrins exhibit peroxidase activity, and since they do not fulfill all the requirements of enzymes they are said to behave as "pseudoperoxidases"—i.e., they oxidize certain substrates in the presence of H_2O_2:

$$\text{Peroxidase} + H_2O_2 \rightarrow \text{peroxidase} \cdot H_2O_2$$

$$\text{Substrate} + \text{peroxidase} \cdot H_2O_2 \rightarrow \text{oxidized substrate} + H_2O + \text{peroxidase}$$

Among the substrates oxidized are pyrogallol, phenolphthalin and other leuco dyes, benzidine, *o*-tolidine, guaiac, hydriodic acid, and ascorbic acid. (The reactions with *o*-tolidine and benzidine are discussed in greater detail in the earlier section on *Detection of Hemoglobin and Its Derivatives*.) Oxyhemoglobin, reduced hemoglobin, methemoglobin, and carboxy-hemoglobin all reportedly react with the same activity; acid hematin has about 50%–80% of the activity of the foregoing, and methemalbumin has greater activity. One report (19) indicates, however, that methemoglobin has 30% greater activity than oxyhemoglobin. Obviously, any true peroxidase present in the test sample also reacts. The 10%–13% reduction in results obtained with whole blood and plasma effected by heating at 60°C for 30 min is presumably due to inactivation of true peroxidases. Such heat treatment has no effect on the activity of hemoglobin and derivatives. Wu (369) in his early studies, however, was unable to detect any true peroxidase in blood (148).

Wu (370) in 1923 was the first to introduce a method using benzidine and H_2O_2 for determination of hemoglobin. Bing and Baker (22) modified this method in 1931, and further modifications were made subsequently by others. Hanks *et al* (134, 135) succeeded in increasing the sensitivity of this method about tenfold. Vanzetti and Valente (179, 345) further improved this technic by introducing a double extraction procedure with ether and ethanol for removing inhibiting substances and interfering pigments. The sensitivity was also increased by reducing the final volume, thus making this method the most sensitive one available. The color is measured at 500 nm and is reasonably stable. These factors make this procedure the method of choice for the determination of plasma hemoglobin.

SPECIMEN COLLECTION AND NORMAL VALUES

Serum has been shown to be inappropriate for this determination because the clotting process itself may cause up to a 100-fold increase in the extracellular hemoglobin concentration (305). In obtaining plasma samples extreme care must be taken to prevent hemolysis during and after collection of the blood.

In early studies the upper normal limit for hemoglobin in plasma was believed to be about 6 mg/100 ml. Then in studies during the 1950s the upper normal limit started to decrease, until Hanks *et al* (134) in 1960 showed that the normal range was 0.16–0.58 ($\bar{x}$ 0.31) mg/100 ml with no assurance that it should not be still lower. They showed (134, 135) that the problem is strictly a matter of preventing hemolysis when obtaining the plasma sample. Their method of sample collection, which is used to obtain specimens for the determination in the following section, is as follows. Make a clean venipuncture with a 15 gauge needle attached to a 12 in. long polyvinyl tube (3 mm ID), and allow the blood to flow freely. Discard the first 4 ml, then collect 8 ml of freely flowing blood in each of two silicone-coated 15 ml test tubes whose interiors have been coated with a fine mist of 1% heparin, delivered from an atomizer. The volume of heparin solution in each tube should not exceed approximately 0.05 ml. Take care not to allow any blood to touch the wall of the tube above the final level of collected blood. (Tubes for collection of blood may be siliconized by placing dry, acid-washed tubes in a closed container overnight along with an open vessel of dichlorodimethylsilane. Then rinse them with distilled water and dry, taking care not to allow the tubes to be contaminated from the racks used for drying. Siliconized Vacutainers obtainable from Becton-Dickinson may also be used.) Rotate the tube gently to mix. Do not invert. Cap with Parafilm (American Can Co.) and centrifuge at 1100 g for 8 min. Remove the upper 3 ml of plasma from each tube into another clean tube, being careful not to disturb the erythrocytes, and centrifuge again.

DETERMINATION OF PLASMA HEMOGLOBIN

(method of Vanzetti and Valente, Refs 179, 345)

PRINCIPLE

Hemoglobin is converted to acid hematin with acetic acid and is extracted into ethyl ether. The polarity of the solvent is reduced by addition of petroleum ether, and the hematin reextracted into aqueous alcohol. The ethanolic extract is treated with benzidine and H_2O_2, and the stable purple secondary oxidation product of the benzidine is determined photometrically at 500 nm. The absorption spectrum is shown in Figure 23-5.

REAGENTS

Acetic Acid, 50% Aqueous. Mix equal volumes of glacial acetic acid and water.

Ethyl Ether, Peroxide-Free. Add 1 vol of an acid ferrous sulfate solution (prepared by dissolving 6.0 g $FeSO_4 \cdot 7H_2O$ in 110 ml water and 6 ml conc HCl) to 5 vol ethyl ether in a separatory funnel. Shake thoroughly for 3 min and discard the aqueous phase. Wash four times with water. It is convenient to prepare this reagent just before use. Redistillation is not required.

Petroleum Ether (b.p. 30°–60° C)

Ethanol, 30% (v/v) Aqueous. Dilute 3 vol absolute ethanol with 7 vol water. Ethanol which has been denatured with 5% methanol may be used in preparing this reagent.

Benzidine, 0.5 g/100 ml. Dissolve 500 mg of benzidine in 100 ml of 50% (v/v) acetic acid. This reagent is stable for 1 week in the refrigerator.

Hydrogen Peroxide, 30% Solution

Hemoglobin Standard Solutions. Determine total hemoglobin content in any sample of whole blood by an appropriate technic as by cyanmethemoglobin, oxyhemoglobin, or iron content. Dilute the blood with water so it contains 100 mg hemoglobin/100 ml and store in 2 ml portions in the freezer. Within 15 min of use, dilute 1.0 ml of this standard to 100 ml with water to obtain a 1 mg/100 ml standard.

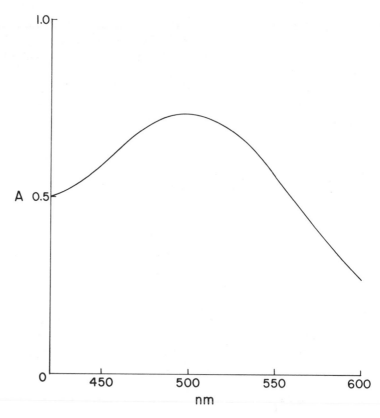

FIG. 23-5. Absorption curve of color formed in plasma hemoglobin determination. Standard (1 mg hemoglobin/100 ml) vs reagent blank.

Glassware. Soak all glassware in hot 6 N HCl for an hour and then rinse eight times with water.

PROCEDURE

1. To 0.6 ml of plasma in each of two (for duplicates) 15 ml glass-stoppered centrifuge tubes add 3 ml 50% acetic acid; mix. Do the same with duplicate 0.6 ml aliquots of the 1.0 mg/100 ml hemoglobin standard and 0.6 ml aliquots of water (blanks).
2. Let stand for 2 min and add 4.5 ml peroxide-free ether. Stopper the tubes and shake for 2 min. Loosen the stoppers and centrifuge for about 2 min to separate the phases.
3. Carefully remove and discard the aqueous layer with a capillary pipet, and add 4.5 ml petroleum ether and 1.5 ml 30% aqueous ethanol. Shake vigorously for 2 min.
4. Completely remove the upper layer by suction without losing any of the lower phase.

5. To each tube add 1.5 ml color reagent, which is prepared immediately before use by mixing 1 ml 30% H_2O_2 with 32 ml of the benzidine solution. Mix the contents of each tube as soon as the reagent is added and allow to stand at room temperature for 30 min to 1 hour.

6. Read the absorbance of the samples and standards within the next 15 min at 500 nm, using the reagent blank to zero the spectrophotometer. A photometer may be used with a filter having its maximum transmission at about 500 nm, but a series of standards should then be prepared to test the linearity of a calibration curve. If a sample has an absorbance greater than about 0.9, it should be diluted with 50% acetic acid and reread. Alternatively, the plasma may be diluted with water and the assay rerun.

Calculation:

$$\text{mg hemoglobin/100 ml plasma} = \frac{A_x}{A_s} \times 1.0 \times D$$

where D is a factor for dilution if required for reading high hemoglobin concentrations.

NOTE

1. Benzidine. Care should be exercised in handling this material, which is carcinogenic. Its application here is warranted by the need for extreme sensitivity.

ACCURACY AND PRECISION

As with other methods based on the oxidation of benzidine, etc., with H_2O_2 this method measures hemoglobin plus any other hematin pigments which may be present in plasma. The accuracy of the method is therefore unknown, since other methods of comparable sensitivity are based on the same reaction. In the first extraction with ether about 10% of the hematin is not recovered from the aqueous phase; however, this occurs with both standard and sample, and the error is thus canceled (345).

The precision (95% limits) is ±4%.

NORMAL VALUES

Hanks *et al* (134) found the mean hemoglobin concentration in the plasma of 15 ambulatory subjects to be 0.31 mg/100 ml, with a total range of 0.16–0.58; in 10 sleeping subjects it was 0.30 mg/100 ml with a total range of 0.20–0.57. Vanzetti and Valente (345), in 31 fasting healthy subjects, found a mean value of 0.41 mg/100 ml with a total range of 0.03–0.89. They also demonstrated that during intense physical exercise the concentration of hemoglobin in the plasma may increase up to 50-fold. Chaplin *et al* (45) found as much as a 20-fold increase in plasma hemoglobin after strenuous exercise. Sleep, diet, ordinary activity, induced lipemia, induced fever, menstruation, local circulatory stasis, and increased blood flow after injection of epinephrine or nicotinic acid had no effect on the concentration of plasma hemoglobin.

HEMOGLOBIN DERIVATIVES

The interrelationships of hemoglobin and the chemical compounds it forms are shown in Figure 23-2. Although only hemoglobin and oxyhemoglobin possess a useful function, carboxyhemoglobin, methemoglobin, and sulfhemoglobin are usually present in normal red cells in small amounts. In some circumstances they may, in sum, equal 10% or more of the total blood hemoglobin (109, 363). Because large changes in visible absorption spectra result from reactions at or near the Fe atoms of hemoglobin, the common hemoglobin derivatives differ markedly from each other in such spectra (Table 23-3).

Advantages of these differences can be taken to determine by spectrophotometric measurements the concentrations of individual derivatives in a mixture. Vierordt during the latter part of the 19th century proposed a mathematical treatment of this problem (reviewed by Heilmeyer, 145). Solution of equations for two-component systems requires measurement at two wavelengths, three-component systems at three wavelengths, etc. (see *Chapter 1*). A similar approach to analysis of two-component systems employs the use of Hüfner's quotients. These two approaches are used in spectrophotometric procedures for determination of carboxyhemoglobin, methemoglobin, and sulfhemoglobin in blood (323).

OXYHEMOGLOBIN

Hemoglobin combines with O_2 to form oxyhemoglobin. Each of the four subunits of the hemoglobin molecule can bind one oxygen molecule, which attaches itself to the Fe(II). The O_2 was thought to replace a water molecule previously held between the Fe and the adjacent distal histidine residue. However, recent studies show that the Fe in hemoglobin is attached only to four heme and one histidine nitrogens and is five-covalent (258). During oxygenation of each subunit the tertiary and quaternary structures of the hemoglobin molecule change in a complex fashion (258), and the Fe becomes six-covalent. The resulting alterations in a variety of intramolecular relationships interact to produce the well known "paradoxical properties" of hemoglobin — i.e., the sigmoid relationship between O_2 tension and degree of oxygenation (so-called heme-heme interaction) and the dependence of O_2 affinity on pH, pCO_2, and 2,3-diphosphoglycerate concentration. Only very recently have studies of composition, structure, and abnormal function of hemoglobin variants (see below) been combined to explain these phenomena (164, 258, 259, 261–263).

CARBOXYHEMOGLOBIN

Hemoglobin also binds carbon monoxide to form carboxyhemoglobin. The carbon monoxide molecule attaches at the Fe(II)-histidine site exactly as O_2 does, but the equilibrium constant for the reaction is such that carbon monoxide is bound over 200 times more strongly than oxygen. Therefore carboxyhemoglobin in red cells has lost the various functions of hemoglobin, accounting for the toxicity of carbon monoxide.

The carboxyhemoglobin content of normal blood varies considerably and depends on exposure. Thus individuals living in a city or in rural communities with minimal exposure have blood levels of 0.1–0.4 g/100 ml (109); smokers 0.26–1.2; street cleaners about 0.5; taxi drivers 1.1–3.2 (109). Levels increase two- or threefold following exercise, the increase persisting for several hours after the exercise is stopped.

The CO content of blood can be determined gasometrically by the Van Slyke apparatus, the most accurate technic being that in which the released CO is absorbed by cuprous chloride, or Winkler's solution (302, 343).

Various spectrophotometric procedures have been proposed for determination of carboxyhemoglobin (136, 139, 156, 157, 194, 218, 289, 320, 374). One method is based on the fact that hemoglobin is oxidized to methemoglobin rapidly by ferricyanide, whereas carboxyhemoglobin is oxidized slowly. In another method which is said to be simple, fast, and reliable for fresh specimens, the oxyhemoglobin of the blood is reduced by sodium hydrosulfite without significant reduction of any carboxyhemoglobin present (323). Spectrophotometry at 555 and 480 nm is used to quantify the two components. These methods work very well in synthetic mixtures, but for forensic purposes they must be used with care and understanding of their limitations (323). Infrared spectrophotometry has been used and is much more accurate, especially in the range below 10% saturation where the above methods lose precision and accuracy (101).

Carboxyhemoglobin decreases slightly in blood collected with oxalate as anticoagulant but is stable for at least 2 weeks at room temperature with fluoride (final concentration 0.3%) (109, 293). Blood is also stable for 2–4 days at room temperature if collected with sodium citrate as anticoagulant and placed under a layer of toluene (136).

METHEMOGLOBIN

Oxidizing agents react with hemoglobin to change the Fe(II) to Fe(III), producing methemoglobin. This oxidation occurs stepwise—i.e., one Fe at a time—and intermediate forms exist between hemoglobin and methemoglobin (175, 176), the ferroheme portions of which have increased affinity for O_2 compared to hemoglobin. O_2 itself oxidizes hemoglobin slowly because of the secondary and tertiary structure of the molecule. Other oxidants act quickly (e.g., ferricyanide, ozone, or permanganate) and some substances catalyze more rapid oxidation by O_2 (e.g., methylene blue). In the normal red cell, in spite of an abundance of O_2, the methemoglobin level is maintained at less than about 7% of the total hemoglobin (342). This reflects a balance between its slow formation and its reduction by enzymatic pathways involving NADH. Methemoglobin reacts with cyanide to form cyanmethemoglobin (Figure 23-2) and reacts with acids or bases. The last two split it into globin and acid or alkaline hematin, with the anion of the splitting reagent occupying the third valence of the iron. Reducing agents can reverse these valence changes.

Methemoglobin in the presence of oxyhemoglobin can be determined by spectrophotometric measurements. In alkaline solution (pH $\sim$ 10) methemoglobin has absorption peaks at 540 and 576 nm and minima at 510 and 560 nm. At neutral pH the main peaks are at 500 and 630 nm and the minima at 470 and 600 nm (Table 23-3). Among the spectrophotometric procedures proposed are the following: (a) Hüfner (163) employed the ratio A_{560}/A_{540} measured in alkaline solution. (b) Drabkin and Austin (78) found the ratio A_{560}/A_{575} to be superior to A_{560}/A_{540} because of a larger change in the ratio with change in the ratio of methemoglobin to oxyhemoglobin. Heilmeyer (145) used A_{560}/A_{576} or A_{576}/A_{590}. The ratio A_{560}/A_{540} is affected by pH, A_{560}/A_{575} and A_{575}/A_{590} being much less affected. (c) The ratios A_{526}/A_{564} and A_{556}/A_{564} have also been used but have been criticized (216). (d) Austin and Drabkin (5) made calculations from measurements at 540, 560, 575, and 630 nm. (e) Horecker and Brackett (157) took advantage of the high absorption of methemoglobin at 800 nm and made calculations from the difference in absorbance before and after conversion to cyanmethemoglobin. Their procedure also included calculation of carboxyhemoglobin conc. (f) Zijlstra and Muller (374) diluted blood 1:2 with water containing saponin and measured A_{558} and A_{523} in a cuvet with 0.13 mm light path. Carboxyhemoglobin content was calculated from A_{562} and A_{540}. (g) In another technic blood was diluted with water and measurements made at 520 and 620 nm. (h) Haurowitz (140) measured A_{623} before and after conversion of oxyhemoglobin to methemoglobin with ferricyanide. According to Haurowitz the absorption of methemoglobin at 623 nm is independent of pH. Michel and Harris (222), however, criticized the method because of interference by turbidity and the fact that the isobestic point may not be at 623 nm. Hamblin and Mangelsdorff (132) also developed a technic employing measurement of A_{623}. (i) Many workers believe that the best spectrophotometric method is based on the fact that methemoglobin has a peak at 630 nm, whereas cyanmethemoglobin has no peak at this wavelength. Calculations are made from ΔA_{630} before and after addition of cyanide. Absorption by oxyhemoglobin and cyanmethemoglobin is nearly independent of pH. Evelyn and Malloy (83) introduced the method in 1938, and modifications appeared subsequently. The technic also permits determination of oxyhemoglobin and sulfhemoglobin.

The validity of the spectrophotometric procedures is dependent on the presence of no more pigments than are assumed to be present by the particular method used. Furthermore, turbidity can introduce significant error in most of the methods.

Stadie (315, 344) in 1920 considered the difference between hemoglobin determined by O_2 capacity (gasometric) and total pigment as cyanmethemoglobin to represent methemoglobin content of the blood. Blood requires aeration for 15–60 min for 98%–100% oxygenation; however, if it is oxygenated and then deoxygenated, only 5 min of aeration is required. The possibility that this phenomenon is due to methemoglobin was rejected. Nicloux and Fontes (243) in 1924 introduced the principle of determining inactive hemoglobin with iron in the ferric state by reduction to active ferrohemoglobin with $Na_2S_2O_4$, followed by measurement of the resultant increase in CO or O_2 binding capacity. This method was further developed by Van Slyke (340–342) and others (159). It was shown (341), however, that CO also combines with various pigments formed from hydrosulfite reduction (e.g., hematin). Titanous tartrate (341) and sodium anthrahydroquinone β-sulfonate appeared to be superior as reducing agents when such pigments were present. The increase in CO capacity of normal blood following reduction is due to conversion of an "inactive" form of hemoglobin to an active form, but no more than a part of this inactive form appears to be methemoglobin (148).

Hereditary (Idiopathic) Methemoglobinemia

In hereditary or idiopathic methemoglobinemia, up to about 40% of the total hemoglobin pigment is in the form of methemoglobin because of failure of erythrocyte mechanisms normally present to maintain hemoglobin in the reduced (ferro) state. Ordinarily this defect is inherited as a recessive characteristic, although one case has been reported in which the condition was inherited as a dominant characteristic. Gibson (112) showed that methemoglobin in normal erythrocytes can be reduced by two pathways: in one, glucose is metabolized through triose phosphate to lactate, the enzyme system involved consisting of dehydrogenases, coenzyme I (NAD), and diaphorase I (coenzyme factor I); in the second, glucose is metabolized through hexose monophosphate to phosphogluconate, the enzyme system consisting of dehydrogenases, coenzyme II (NADP), and diaphorase II (coenzyme factor II). The second pathway is essentially dormant unless methylene blue is present. Some studies have indicated that erythrocyte diaphorase I is low in this disease, although another investigation found normal levels of flavin adenine dinucleotide, an essential component of diaphorase. Thus the first pathway is deficient in this disease and the availability of the efficient second pathway is taken advantage of in the treatment of patients with methylene blue. Townes and Lovell (334) expressed the opinion that the defect is not a deficiency of NADP or NAD methemoglobin reductase. Methemoglobin reductase is not a diaphorase but is a heme protein functioning as a NADPH oxidase, oxidizing NADPH and to a lesser extent NADH, with O_2, methemoglobin, or cytochrome C as terminal electron acceptor. Methylene blue is required as an electron carrier.

SULFHEMOGLOBIN

Like methemoglobin and carboxyhemoglobin this derivative is a cause of clinical cyanosis (363). However, its formation is not reversible by any known drugs or natural mechanisms, and it remains unchanged for the life of the red blood cell it occupies. The exact structure is still unknown, but recent work suggests that divalent sulfur breaks the carbon-carbon double bond of the pyrrole ring (see Chapter 24), and forms an episulfite bridge between the two-ring carbons (17, 18). This probably accounts for the paradoxical observation that in sulfhemoglobin the heme binds carbon monoxide but not O_2. An explanation for this paradox lies in the hypothesis that sulfhemoglobin can indeed bind O_2, exactly as hemoglobin does, but cannot do so at atmospheric pressure, because the sulfur bridge on the pyrrole ring adjacent to the Fe atom has drastically changed the equilibrium constant for the binding reaction. This has been demonstrated by an analogy with sulfmyoglobin (18). The equilibrium between O_2 and myoglobin is qualitatively identical to that between O_2 and sulfmyoglobin, but the equilibrium constant for the latter is such that 100% saturation with O_2 occurs only at about 100 atm.

Classically it is believed that the H_2S required for sulfhemoglobin formation originates in the intestine from bacterial action on sulfur compounds. In one study (220) of sulfhemoglobinemia, four of nine cases had an increased erythrocyte glutathione level which was believed to result in increased susceptibility to sulfhemoglobin formation. It was also believed probable that the source of the H_2S was in the erythrocyte, not the bowel.

Rarely do levels of sulfhemoglobin exceed 10% of the total hemoglobin pigment. It may be produced *in vitro* by adding phenylhydrazine and H_2S to hemoglobin. The absorption peak at 620 nm persists on adding cyanide, $Na_2S_2O_4$, or $Na_2S_2O_4$ plus CO but disappears on adding 3% H_2O_2 or $Na_2S_2O_4$ plus NaOH.

Sulfhemoglobin can be determined quantitatively by the spectrophotometric method of Evelyn and Malloy (83). The method is based on the fact that the absorption of sulfhemoglobin at 620 nm is unchanged by the addition of cyanide. Hence the concentration of sulfhemoglobin in a solution containing oxyhemoglobin, methemoglobin, and sulfhemoglobin is proportional to the residual absorbance after methemoglobin has been converted to cyanmethemoglobin. A correction is made for the small absorption of oxyhemoglobin and cyanmethemoglobin at 620 nm. Jope (186) also described a technic for determination of sulfhemoglobin. In the absence of methemoglobin, he diluted with saline and measured oxyhemoglobin at 540 nm and sulfhemoglobin at 620 nm. In the presence of methemoglobin he reduced the methemoglobin with $Na_2S_2O_4$ and then estimated sulfhemoglobin using its CO derivative.

DETERMINATION OF METHEMOGLOBIN AND SULFHEMOGLOBIN

(method of Evelyn and Malloy, Ref 83, modified)

PRINCIPLE

When NaCN is added to a solution of methemoglobin, its peak at 630 nm is almost completely abolished by its conversion to cyanmethemoglobin. ΔA_{630} is directly proportional to methemoglobin concentration. The absorption of sulfhemoglobin at its 620 nm peak is unchanged by the addition of the cyanide; hence the concentration of sulfhemoglobin in a solution containing oxyhemoglobin, methemoglobin, and sulfhemoglobin is proportional to the residual A_{620} after the methemoglobin has been converted to cyanmethemoglobin. A correction is made for the small absorption of oxyhemoglobin and cyanmethemoglobin at 620 nm. The concentration of oxyhemoglobin is obtained by subtracting the sum of methemoglobin and sulfhemoglobin from total hemoglobin determined as cyanmethemoglobin (conversion of all pigments to cyanmethemoglobin and measurement of A_{540}).

REAGENTS

Phosphate Buffer, M/15, pH 6.6. Dissolve 1.9 g anhydrous Na_2HPO_4 and 2.72 g anhydrous KH_2PO_4 in water in a 500 ml volumetric flask and dilute to mark.

Phosphate Buffer, M/60, pH 6.6. Dilute M/15 buffer 1:4 as required.

Potassium Ferricyanide, 20% Aqueous

NaCN, 10% Aqueous

Neutralized Cyanide Solution. A neutralized solution of NaCN is prepared within 1 hour of use by mixing equal volumes of 10% NaCN and 12% acetic acid.

NH_4OH, Conc

PROCEDURE

1. Transfer 0.2 ml blood (see *Note 1*, below) to 10.0 ml M/60 phosphate buffer by a TC pipet. Let stand 5 min, centrifuge, and measure A_{630} of supernate against water blank (A_1).

2. Add 1 drop of neutralized cyanide solution to the 10 ml from step 1, mix, let stand 2 min, and read again at 630 nm (A_2).

3. Add 1 drop conc NH_4OH, mix, and after 2 min measure A_{620} (A_3). Read also at 615, 618, 622, and 625 nm. A peak at 620 must be observed if sulfhemoglobin is to be reported as present.

4. Add 2.0 ml of solution from step 3 to 8.0 ml M/15 phosphate buffer containing 1 drop 20% potassium ferricyanide and mix. Let stand 2 min. Add 1 drop 10% NaCN, mix, let stand 2 min, and measure A_{540} (A_4).

Calculation (see Notes 2 and 3):

g total hemoglobin/100 ml $= 36.8 \times A_4$

g methemoglobin/100 ml $= 23.4 \times (A_1 - A_2)$

methemoglobin, % of total $= \dfrac{\text{g methemoglobin/100 ml}}{\text{g total hemoglobin/100 ml}} \times 100$

g sulfhemoglobin/100 ml $=$

$$6.5 \left(A_3 - \frac{\text{g total} - \text{methemoglobin/100 ml}}{205} - \frac{\text{g methemoglobin/100 ml}}{84} \right)$$

g oxyhemoglobin/100 ml $=$ total hemoglobin $-$ methemoglobin $-$ sulfhemoglobin

NOTES

1. Anticoagulant.   Heparin or oxalate can be used but the former appears to be preferable since oxalate reportedly favors formation of methemoglobin (171).

2. Photometer or spectrophotometer used.   This method has been employed in our laboratory only with a Beckman DU spectrophotometer or recording spectrophotometer of equivalent resolution. Leahy and Smith (204) adapted the method for use with the Coleman Jr. spectrophotometer. Evelyn and Malloy employed the Evelyn photometer using filters 635, 620, and 540. Presumably other filter photometers can be used with proper filters.

3. Standardization.   The constants used in the calculations were obtained in our laboratory using a Beckman DU spectrophotometer and apply only to a spectrophotometer of high resolution (bandwidth of 5 nm or less). Even when employing such an instrument it is advisable for each laboratory to check constants for total hemoglobin and methemoglobin with its own instrument. See *Note 2, Standardization,* in the *Determination of Hemoglobin Cyanmethemoglobin Method.*

Obtain a normal blood sample and determine its total hemoglobin content by iron analysis or by the standardized cyanmethemoglobin method. Run the blood through the procedure and obtain A_1, A_2, A_3, and A_4. Then substitute in the following equation and solve for K_1 (constant for calculation of total hemoglobin):

g total hemoglobin/100 ml $= K_1 \times A_4$

For obtaining the methemoglobin constant (K_2) and certain data needed in the sulfhemoglobin constants, proceed as follows: Transfer 0.2 ml of the blood to 10.0 ml $M/60$ phosphate buffer by a TC pipet. Add 1 drop 20% potassium ferricyanide, mix, let stand 2 min to effect oxidation to methemoglobin, and measure A_{630} (D_1). Add 1 drop neutralized cyanide, mix, wait 2 min, and read A_{630} again (D_2). Add 1 drop conc NH_4OH, mix, and read A_{620} (D_3). Substitute in the following equation and solve for K_2:

$$\text{g methemoglobin/100 ml} = K_2 (D_1 - D_2)$$

The factor 6.5 (K_3) used for calculation of sulfhemoglobin cannot be checked easily because sulfhemoglobin cannot be prepared quantitatively; in fact, it has never been prepared in pure form. The millimolar absorptivity at 620 nm, 1 cm light path, was estimated by Drabkin and Austin (79) to be about 11. Lemberg *et al* (206) pointed out that some choleglobin is formed simultaneously with sulfhemoglobin on treatment of hemoglobin with H_2S, and that the estimate of millimolar absorptivity by Drabkin and Austin was too low because of failure to take this into consideration. Lemberg *et al* found a value of 16 for the millimolar absorptivity of carboxysulfhemoglobin. Jope (186), by determining the relation between the spectral absorption of blood samples containing sulfhemoglobin before and after conversion to carboxyhemoglobin and carboxysulfhemoglobin, estimated the millimolar absorptivity of sulfhemoglobin at 617 nm to be 13. Using this value, the constant K_3 is obtained as follows:

$$\text{g sulfhemoglobin/100 ml} = A_3 \times \frac{d \times 1.65}{l \times 13}$$

$$= A_3 \times K_3$$

where

d = dilution of specimen (51.5 in method as presented)
l = light path (cm)
1.65 = g/100 ml of 1 mM solution, assuming minimal molecular weight to be 16 520

The factor 6.5 for K_3 thus applies to a spectrophotometer of high resolution and an effective light path of 1 cm. For filter photometers or round cuvets for which the effective light path is unknown, the factor may be calculated by assuming direct proportionality with another constant determined. For example, if an analyst obtains a value of 30.0 for K_1, then:

$$\frac{K_3}{6.5} = \frac{30.0}{36.8}$$

$$K_3 = 5.3$$

where 36.8 is the value for K_1, obtained in our laboratory.

A_3 must be corrected for the absorption of oxyhemoglobin and methemoglobin at 620 nm. Calculate K_4 and K_5 as follows:

$$K_4 = \frac{\text{g total hemoglobin (assumed to be all oxy)/100 ml}}{A_3}$$

Thus, g oxyhemoglobin/100 ml/K_4 = A_{620} due to oxyhemoglobin.

$$K_5 = \frac{\text{g total hemoglobin (assumed to be all met)/100 ml}}{D_3}$$

Thus, g methemoglobin/100 ml/K_5 = A_{620} due to methemoglobin. Then:

$$\text{g sulfhemoglobin/100 ml} = K_3 \left(A_3 - \frac{\text{g oxyhemoglobin/100 ml}}{K_4} - \frac{\text{g methemoglobin/100 ml}}{K_5} \right)$$

The value for "g oxyhemoglobin" is approximated by subtracting g methemoglobin per 100 ml from g total hemoglobin per 100 ml.

4. *Beer's law.* Beer's law is obeyed by oxyhemoglobin, methemoglobin, cyanmethemoglobin, and presumably sulfhemoglobin, in a spectrophotometer of high resolution. This may not be true with some filters.

5. *Interference by turbidity.* The chief difficulty encountered with this procedure is turbidity (186, 216). Before taking each measurement, the solution should be examined for turbidity by Tyndall effect. The purpose of adding NH_4OH prior to measurement of sulfhemoglobin is to effect clarification of the solution, but this is not always successful. A very slight degree of turbidity can be tolerated; if it is pronounced, however, results for total hemoglobin and sulfhemoglobin are not valid. Calculation of methemoglobin is made from ΔA_{630}; hence turbidity does not affect results unless the turbidity is affected by the addition of cyanide. Jope (186) obtained fair success by centrifuging at 20 000 rpm for 10 min (International centrifuge, high speed head chilled in acetone-Dry Ice mixture). Sometimes he was successful by using an angle head centrifuge for 20 min at 9000 rpm or 40–60 min at 6000 rpm. Heat must be avoided during centrifugation. If turbidity is due to severe lipemia, one can use washed erythrocytes in the determination instead of the whole blood (wash with 0.8 saline and add water back to volume of original blood sample).

6. *Stability.* Oxyhemoglobin in low concentration in water solution is very unstable, being oxidized to methemoglobin (53, 78, 83). However,

there is considerable contradictory evidence in the literature regarding the stability of methemoglobin in blood. In 1925 Neill and Hastings (242) studied the stability of hemoglobin in blood and reported that spontaneous conversion to methemoglobin occurred on standing *in vitro*. Blood was relatively stable when either completely oxygenated or completely deoxygenated, but unstable when a mixture of oxyhemoglobin and hemoglobin was present. They concluded that it was the deoxygenated hemoglobin which is oxidized, not oxyhemoglobin. When completely deoxygenated there is no oxygen available for oxidation. Other workers (63, 152, 356), however, reported that methemoglobin present in freshly drawn blood may rather quickly revert to hemoglobin on standing—e.g., as much as 50% conversion in 24 hours.

A sample containing 62% of its hemoglobin as methemoglobin has been observed in our laboratory to decrease to 30% on storage in the refrigerator for 36 hours. Under the circumstances, it is wise to perform the analysis for methemoglobin on fresh blood (no more than a few hours old).

The author is unaware of data on the stability of sulfhemoglobin, but it should be rather stable since its formation is irreversible.

ACCURACY AND PRECISION

The accuracy of the method has been assessed by Evelyn and Malloy as follows:

Total hemoglobin. Assuming correct standardization, the error is less than 0.2 g/100 ml when no sulfhemoglobin is present. Sulfhemoglobin is not converted to cyanmethemoglobin, but A_{540} of the sulfmethemoglobin produced by the ferricyanide is 78% of A_{540} of a corresponding amount of cyanmethemoglobin. For most routine work this can be ignored. If desired, a correction can be made:

$$\text{corrected g total hemoglobin/100 ml} = \text{g total hemoglobin/100 ml (calculated in routine method)} + (0.22 \times \text{g sulfhemoglobin/100 ml})$$

With this correction the error can be kept below 0.4 g/100 ml even in the presence of sulfhemoglobin. The error produced by turbidity can be serious—e.g., gross lipemia may cause a positive error up to 3 g hemoglobin per 100 ml.

Methemoglobin. The sensitivity of the method is about 0.2 g/100 ml and is accurate to about the same amount. The determination is based on ΔA and hence is not affected by the presence of other pigments or turbidity unless they are altered by cyanide. It should also be pointed out that radioactive tracer studies (177) have revealed six Fe(II) and two Fe(III) precursors of methemoglobin. The contribution of such precursors to various absorption measurements is unknown.

Sulfhemoglobin. The method is sensitive to about 0.1 g/100 ml. In the absence of significant turbidity the error of the method is probably less than 10%. This is difficult to assess exactly because pure sulfhemoglobin is not available. Any pigment absorbing at 620 nm is measured as sulfhemoglobin. For example, patients receiving sulfanilamide frequently exhibit cyanosis due to blue or purple derivatives formed from the drug. Such interference is almost entirely eliminated by the 50-fold dilution of the blood. A negative test is always reliable, but a false positive can occur. For this reason the presence of sulfhemoglobin should not be reported unless a peak is observed at 620 nm.

The precision of the results for total hemoglobin, methemoglobin, and sulfhemoglobin are about ±3%, ±5%, and ±5%, respectively.

NORMAL VALUES

Normal blood contains up to about 3% of its total hemoglobin as methemoglobin—i.e., up to about 0.5 g/100 ml. The average level is about 0.08 g/100 ml. Some workers (141) have found normal levels as high as 8% of the total hemoglobin.

Normally, sulfhemoglobin is not detectable.

HEMOGLOBIN ABNORMALITIES

Hemoglobin abnormalities may be classified into three main groups: thalassemias, hemoglobin variants, and abnormal hemoglobins (164). The term *thalassemia* covers a heterogeneous group of hereditary disorders characterized by a decreased rate of synthesis of one or more types of polypeptide chains in hemoglobin. The various thalassemias are classified according to the type of polypeptide chain involved—e.g., β-thalassemia is a disease caused by a decreased rate of synthesis of the β-polypeptide chain. The following thalassemias are known to exist: α, β, δ, and $\delta\beta$. These are subclassified according to the type of hemoglobin which predominates by default of the inadequately synthesized chain—e.g., β-thalassemia, A_2 type with increased Hb-A_2 ($\alpha_2\delta_2$), and F type with an excessive amount of Hb-F ($\alpha_2\gamma_2$).

Hemoglobin variants are hemoglobins which have some alteration in the primary structure (amino acid composition) of the polypeptide chains. These alterations may be as small as the substitution of one amino acid for another or as large as half or more of a given chain being replaced by that portion of a different type of chain. The latter is called a "fusion" variant because two different chain types have fused together. Such alterations are of course genetic expressions of gene variations. They lead to one or more of the following clinical conditions: (a) no observed clinical manifestations, (b) abnormal O_2 dissociation properties resulting in congenital or familial polycythemia or cyanosis, (c) unstable hemoglobin leading to hemolytic anemia.

Abnormal hemoglobins are hemoglobins made up of normal subunits but in abnormal combinations. Hemoglobin H (β_4) is one example. In it four β

subunits combine to form the hemoglobin H molecule. Hemoglobin Bart's (γ_4) is another example.

These three abnormalities may occur singly or in combination with each other, and they may interact, as illustrated in the following examples. β-Thalassemia has been found along with Hb-C and Hb-S diseases (164). The thalassemia causes elevation of the relative concentration of the hemoglobin variant, up to twice its level in the comparable nonthalassemic individual. An α-thalassemia has been observed with Hb-Lepore, a variant characterized by chains made up of fused parts of δ and β polypeptides. Another α-thalassemia is known to exist with Hb-H or Hb-Bart's.

THALASSEMIA

Thalassemias can be minor, intermediate, or major. An individual carrying one thalassemia gene and one normal gene (a heterozygote) generally has mild clinical manifestations and is said to be suffering from *thalassemia minor*, also known as *thalassemia trait*. The intermediate form, *thalassemia intermedia*, can be produced by a variety of genetic combinations. An individual who is homozygous for the milder forms of β-thalassemia such as Hb-A_2-F or Hb-F thalassemia (see below) may show this condition, as may one who is simultaneously a heterozygote for a thalassemia and a hemoglobin variant, or for two different forms of thalassemia. Thalassemia intermedia is fairly rare and follows a relatively benign course allowing affected individuals to live to adulthood. *Thalassemia major* is generally a life-threatening disease caused by the inheritance of two similar or identical thalassemia genes and characterized by an accelerated destruction of erythroid cells in the bone marrow and a shortened life span of peripheral erythrocytes in circulation. This is believed to be caused by the excessive accumulation in the cells of free polypeptide chains of the type synthesized at a normal rate, and by the adverse effect they exert on the cell membrane and its permeability. Chronic anemia, growth retardation, and various other severe clinical manifestations are observed in this condition. The various types of thalassemia and their clinically significant biochemical changes are summarized in Table 23-4.

HEMOGLOBIN VARIANTS

Although evidence existed for more than 50 years that there was more than one species of hemoglobin, it was the research of Pauling *et al* in 1949 (252) that demonstrated the presence of a hemoglobin variant in sickle cell anemia and established this disease as being molecular in nature. Since then about 140 hemoglobin variants have been discovered and new ones will probably continue to be reported until each amino acid in the hemoglobin molecule has been found to have the maximum number of mutations allowed by its genetic code. Despite the vast number of these mutations,

TABLE 23-4. Hemoglobin Abnormalities in Thalassemia

	Blood Hb (g/100 ml)	Hemoglobin[a] A2	Hemoglobin[a] F	Remarks	References
α-Thalassemias					
Minor	>10	N	N	Traces of Hb-H, 3.4%–14% Hb-Bart's	214, 239, 352, 354
Intermediate	7–10	L	H	10% Hb-H, 30% Hb-Bart's	167, 240, 277, 278
Major	<7	L	L	Incompatible with extrauterine life, 100% Hb-Bart's; positive sickle test	164, 208, 354
β-Thalassemias					
A2 thalassemia					
Minor	7–17	N-H	V		10, 57, 88, 164, 353
Intermediate	7–10	V	V		57
Major	<7	N-H	V		199, 307, 357, 360
F thalassemia	10–15	L-N	H		91, 96, 100, 214, 317, 353, 364, 376
A2-F thalassemia	5–13	H	H	Synthesis of δ chain decreased	301
δ-Thalassemia	—	L	N	Since δ chain is a minor one, no red cell changes occur	96, 332, 333
Thalassemia-like syndromes					
Hb-Lepore					
Minor	—	L-N	N-H	Hb-Lepore 5%–15%	57
Major	—	Absent	?	Hb-Lepore 8.6%–27%	57
Persistent fetal hemoglobin				Hb-F is homogeneous in RBC population	8, 59, 81, 166, 358
Negro type	—	L	15%–100%		—
Greek type	—	N	10%–19%		90
Swiss type	—	?	1%–3%		57
Georgia type	—	?	4%–5%		57

[a] N, normal; L, low; H, high; V, variable.

however, the relative rarity of congenital disease indicates that most hemoglobin mutants are benign or are pathogenic only in the homozygous form.

Knowledge of exact amino acid sequence and of precise steric relationships of the hemoglobin molecule, based on x-ray analysis, make variants meaningful and fruitful to study, even though some of them may be extremely rare. One can attempt to relate each amino acid substitution to its effects on one or more of secondary, tertiary, and quaternary structure, and thus gain insight into certain diseases at the molecular level (261). Before commenting further on this it is necessary to elaborate briefly on the structure and function of hemoglobin. It was pointed out earlier that the polypeptide chains of globin form α helices, with polar residues at the core and nonpolar ones at the surface of the helix. However, only about 80% of the length of the polypeptide is actually in helical configuration (165). In each chain there are eight helical sections, A through H (Figs 23-7 and 23-8, below). (The D helix is missing in α chains.) The helices are joined by short nonhelical strings of amino acid residues. These permit the chain to fold into a compact globular structure with a pocket in the center into which the heme fits. Amino acids having their side chains in the center of the globule are almost all nonpolar, as implied above, with the exception of functional side chains such as the histidines on either side of the heme iron.

The two α chains and the two β chains of hemoglobin fit together in a very specific fashion, such that almost no contact exists between one α chain and the other or between the two β chains and so that two types of contacts exist between α and β chains. Any given α chain contacts one β chain along 34 amino acid residues and contacts the other β chain along 19 residues. Each β chain also has the same two types of contacts. The larger intersubunit contact areas are firmly interlocked and rigid. The smaller ones seem to have two possible modes of contact which are shifted significantly from each other in hemoglobin and oxyhemoglobin. In fact the iron atoms of the β subunits are 7 Å. further apart in hemoglobin than in oxyhemoglobin (24). This might be represented diagrammatically as follows:

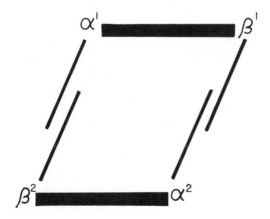

The heavy lines represent the larger rigid intersubunit contacts, generally called $\alpha^1\beta^1$ contacts, and the lighter, partially overlapping lines represent the shorter contacts with their bistable positional relationships, generally called $\alpha^1\beta^2$ contacts. The $\alpha^1\beta^2$ contacts break reversibly in hemoglobin but not in oxyhemoglobin, as follows:

$$\text{Hb } (\alpha^1\alpha^2\beta^1\beta^2) \rightleftharpoons \text{Hb } (\alpha^1\beta^1) + \text{Hb } (\alpha^2\beta^2)$$

$$\text{tetramer} \qquad\qquad \text{dimer} \qquad\quad \text{dimer}$$

The two half molecules on the right hand side of the equation are of course identical and are generally both called $\alpha^1\beta^1$ (197). The equilibrium is far to the left in normal hemoglobin *in vivo*. In the center of the tetrameric hemoglobin molecule there is a water-filled cavity which extends outward between the two α chains and between the two β chains.

Nomenclature in this field has become complex. Upon its discovery, the hemoglobin variant in sickle cell anemia was called Hb-S. Other mutants discovered after Hb-S were assigned other letters in the alphabet. When it became apparent that the number of possible hemoglobin variants was enormous, it was agreed that subsequent new hemoglobins would be named after a geographic location in which the specific hemoglobin was found—e.g., Hb-Mexico. Hemoglobins similar to those already characterized by an alphabet letter or name would have the geographic designation used as a subscript—e.g., Hb-O$_\text{Arab}$, Hb-Lepore$_\text{Boston}$. Hemoglobins which are completely characterized as to the amino acid substitution would be designated by a superscript after the polypeptide chain involved—e.g., Hb-G$_\text{Norfolk}$ α_2^{85} asp→asn β_2, or Hb-S $\alpha_2\beta_2^{6}$ glu→val (379).

Numbering amino acid residues in hemoglobin is conventionally started from the N terminus and is counted either from the very end or in the terms of each helix or interhelical section. Thus the difference between Hb-A and Hb-S mentioned above is given as β^6 glu→val or βA3 glu→val, indicating that the substitution is at the sixth residue from the N terminus and the third residue in helix A of the β chain. The interhelical sections are identified by the adjacent helices, so that amino acid residue α^{92} can also be called FG4 because it is the fourth residue in the nonhelical section between helices F and G. This chapter generally uses the number of the position in the whole chain rather than in each helix.

To the extent of our present knowledge of the locus of function in the intramolecular structure of hemoglobin, the pathology of hemoglobin variants can be explained quite rationally. For example, substitutions of amino acid residues which are involved in the $\alpha^1\beta^2$ contact might be expected to alter the so-called "heme-heme interaction" of hemoglobin, because this interaction appears to depend on a shifting of the subunits along the $\alpha^1\beta^2$ contact (259, 316). Such substitutions do indeed have this effect and can change the characteristic sigmoid shape of the O_2 dissociation curve of hemoglobin to a hyperbolic shape. This change greatly decreases the ability of hemoglobin to transport O_2.

The Bohr effect, which shifts the whole dissociation curve toward greater dissociation at the more acid pH and higher CO_2 tension characteristic of the capillaries, may also be affected by substitutions of specific amino acids. The enhancing effect of 2,3-diphosphoglycerate (2,3-DPG) in the red cell on O_2 dissociation may also be altered.

Amino acid substitutions can occur in various parts of the molecule: (a) on the outside surface, (b) on the inside of one of the four subunits, (c) in the heme pocket, (d) in the larger intersubunit contact, (e) in the smaller intersubunit contact, and (f) in the internal cavity. How great the pathologic effect of any substitution is depends upon the position of the amino acid involved and its type. Changes which involve the outside surface may not affect molecular stability even if they change the molecular charge and lead to a variant that is easily separated by electrophoresis. On the other hand substitution of a polar residue for a nonpolar one inside the molecule, particularly at some functional spot like a contact, is almost certain to cause gross spatial rearrangement and instability of the molecule.

Pathology arises not only from direct effects upon hemoglobin function, as exemplified above, but also from certain factors which cause hemoglobin molecules to bind together into structures larger than the functional quaternary one, and from factors which alter the stability of the molecule. Since intermolecular associations cause peculiar erythrocytic shapes called sickle cells, these hemoglobinopathies are grouped together as "sickle cell disease" and related syndromes, as distinct from hemoglobinopathies due to molecular instability.

SICKLE CELL DISEASE AND RELATED SYNDROMES

Sickle cell disease results from inheritance of genes for production of Hb-S. In this abnormal hemoglobin each β chain has one deviation from normal: β^6 glu→val. These two residues are at the surface of the molecule and are nonpolar. Their presence apparently causes molecules of hemoglobin, in the deoxygenated form, to adhere to complementary sites on similar molecules and to form linear aggregates, which have been demonstrated by electron microscopy (234, 260, 321). Hb-C$_{Georgetown}$, Hb-C $_{Harlem}$, and certain types of Hb-I (164) show similar properties. Figure 23-6 shows the structure of a tubular fiber of deoxyhemoglobin S that may extend through the entire erythrocyte or be wrapped around its inner surface. These fibers are packed together in parallel bundles forming either hexagonal or square lattices. Heterozygous inheritance of the trait for Hb-S results in blood which shows the sickling tendency to an extent which depends, in any one individual, upon the proportions of the sickling Hb-S to nonsickling hemoglobins A, A_2, and F. Homozygous inheritance of Hb-S frequently results in death during childhood.

The formation of aggregates lowers the solubility of deoxygenated Hb-S 50-fold compared to oxygenated Hb-S or normal oxyhemoglobin (14). The O_2 affinity of Hb-S is also drastically reduced (95), and the cells tend to become long and narrow or sickle shaped due to the aggregates (129). The dithionite-phosphate solubility test and the Murayama test, which are used as screening procedures for detection of Hb-S, utilize this insolubility (121, 237, 238).

FIG. 23-6. Structure of helical tube of deoxyhemoglobin S. The arrows and signs indicate the probable positions of the molecular dyads normal to the fiber axis. A length of eight rings is shown corresponding to the approximate repeat of the structure. The hand of the helix was investigated by tilting the specimen about the fiber axis and observing the shift of the superposition pattern. It was found to be right-handed. From FINCH JT, PERUTZ MF, BERTLES JR, DOBLER J: *Proc Nat Acad Sci USA 70*: 718, 1973

62Å

ELEVATION

PLAN

⊢50Å⊣

Testing for sickle cells merely involves lowering the O_2 tension to below 45 mm Hg by adding a reducing agent and then observing the cells under the microscope (see *Test for Sickling*, below). Electrophoresis of blood hemolysate of an individual with sickle cell anemia at pH 8.6 on cellulose acetate or starch gel shows Hb-S to migrate halfway between Hb-A and Hb-A$_2$. Variable amounts of Hb-F may be seen migrating slightly slower than Hb-A. As measured by denaturation tests, Hb-F may range from 2% to 30% of the total hemoglobin (58).

The blood of an individual heterozygous for Hb-S gives a positive sickling test, and electrophoresis shows that Hb-S constitutes less than 50% of the total hemoglobin. Blood from a homozygous individual also gives a positive test, of course, and electrophoresis shows 70%–90% of the hemoglobin to be Hb-S (58). A wide variety of other signs and symptoms accompany sickle cell disease.

Individuals other than those with active sickle cell disease may have most of their hemoglobin migrating as Hb-S by routine electrophoresis and their sickle test may be positive. Such findings have been reported in clinically asymptomatic individuals with the following conditions: (a) homozygosity for Hb-S and hereditary persistence of fetal hemoglobin (Hb-F); (b) heterozygosity for Hb-S and β-thalassemia; (c) heterozygosity for Hb-S and any of Hb-D, Hb-G, and Hb-Memphis. In such cases the following laboratory tests and family studies are useful for differential diagnosis. In condition (a) Hb-F is normally elevated and examination of a blood smear by the Kleihauer technic (see *Detection of Hb-F by Staining*, below; ref 191) demonstrates that Hb-F is homogeneously distributed among the erythrocytes. Condition (b) has two possibilities: heterozygosity for Hb-S and A$_2$ type or F type β-thalassemia (see Table 23-4). In condition (c) there may be heterozygosity for Hb-S and any of Hb-D, Hb-G, and Hb-Memphis; the ferrohemoglobin solubility in each case is higher than that in homozygous Hb-S (173, 174); and Hb-D, Hb-G, and Hb-Memphis can be separated from Hb-S by agar gel electrophoresis (311).

As mentioned earlier, hemoglobin variants other than Hb-S give positive sickling tests, including Hb-C$_{Georgetown}$ (β^6 glu→val and 83–120→?), and Hb-C$_{Harlem}$ (β^6 glu→val and β^{73} asp→asn). Blood cells with high concentrations of Hb-Bart's (γ_4) also sickle (58). Lowered susceptibility to malaria, characteristic of people with Hb-S, may have acted as a selective factor in its favor, a hypothesis supported by the world distribution of sickle cell disease (58).

Hemoglobin C Disease

The Hb-C variant is β^6 glu →lys, which migrates with Hb-A$_2$ on cellulose acetate (50). It is less soluble than Hb-A in dilute phosphate buffer and within the erythrocytes (58). In the homozygous individual the red blood cells are more rigid than normal and have a shortened mean life span of 30–55 days (331). The rate of hemoglobin production in this condition is 2.5–3 times normal (231).

Heterozygous Hb-C condition is asymptomatic, and the homozygous condition is characterized by moderate chronic hemolytic anemia and associated symptoms. It is prevalent in western Africa, mainly in the vicinity of northern Ghana where 17%–28% of the Negro population has been reported to carry Hb-C (50). An incidence of 3% is seen in the Negroes of the USA (297), and sporadic cases have been seen in Italy (76, 312).

Hemoglobin S-C Disease

This is caused by a heterozygous state in which the individual carries two abnormal β alleles—one for Hb-S and one for Hb-C—and has no Hb-A (58). The clinical manifestations are much less severe than in sickle cell disease because Hb-S and Hb-C are present in roughly equal proportions (230). Bizarre-shaped intraerythrocytic hemoglobin masses have been reported as a result of the combined tendencies of Hb-S to form aggregates and of Hb-C to form crystals (75). As with Hb-C disease the mean life span of the erythrocyte is decreased (18–56 days), and the rate of hemoglobin synthesis is increased up to fivefold (230).

Hemoglobin D Disease

This is caused by any of three β chain substitutions: Hb-D$_{Bushman}$, β^{16} gly→arg; Hb-D$_{Ibadan}$, β^{87} thr→lys; and Hb-D$_{Punjab}$, β^{121} glu→gln; and one substitution Hb-D$_{St\ Louis}$, α^{68} asn→lys. These all migrate with Hb-S on cellulose acetate and with Hb-A on agar gel at pH 6.2 (311). Individuals who are heterozygous for Hb-D are asymptomatic, even though Hb-D may form up to 50% of their total hemoglobin (49). Homozygous individuals are so rare that there is some question that one has been found (58).

Hemoglobin S-D Disease

This is caused by a heterozygous state in which Hb-S and Hb-D are formed, but no Hb-A. Traces of Hb-F may be present. It was first observed in a Caucasian family (173). The clinical symptoms are less severe than in pure Hb-S disease, and in one study even though from 25% to over 75% of the hemoglobin found in different affected individuals was Hb-S, no correlation between symptoms and amount could be found (44). The sickling test is positive, and the S and D hemoglobins migrate together on cellulose acetate at pH 8.6. Hence this condition may be confused with pure sickle cell anemia. Since the prognosis of the two conditions is so very different, correct identification is important. Hb-S and Hb-D can be separated on agar gel (see *Separation of Hemoglobin D and Hemoglobin E by Agar Gel Electrophoresis*, p.1188).

Hemoglobin E Disease

This results from substitution β^{26} glu→lys. Hb-E migrates on cellulose acetate at pH 8.6 slightly faster than Hb-C and at pH 6.5 slower than Hb-S.

It migrates with Hb-A on agar gel at pH 6.2. It has the same solubility as Hb-A. In heterozygotes it may constitute up to 45% of the hemoglobin, but such individuals are asymptomatic. Homozygotes have a mild anemia (51).

OTHER HEMOGLOBIN ABNORMALITIES

Sickle cell disease and related syndromes which in the main involve substitutions on the surface of the hemoglobin molecule are common and therefore clinically important. Mutations at the center of the molecule are less common but are of great interest because of the light they shed on the relationship between structure and function. Some substitutions alter the O_2 affinity of hemoglobin, others its Bohr effect, others the relative stability of its Fe(II) and Fe(III), and others the stability of the whole molecule.

For presentation of these substitutions to be most meaningful still further discussion of hemoglobin structure and function is necessary. There is no common pattern of ionized or polar residues on the surface of the molecules, but there is a total exclusion of polar residues from the interior. This expresses itself in a pattern of more than 30 sites where only nonpolar residues appear (261). The α and β chains of mammalian hemoglobins have a common pattern of sites which are invariably occupied by the same residues. These sites include all heme contacts and most intersubunit contacts. Mutations changing these patterns would be expected to have deleterious effects on function, and they almost invariably do. Discussion of all of the mutations covered in the figures and tables is beyond the scope of this chapter, but illustrative examples of various classes of mutation are given, along with the relationship between their altered structure and function.

Hemoglobin M Diseases

In 1948 Horlein and Weber (158) described the globin abnormality of hemoglobin leading to faulty oxygenation of hemoglobin in a German family with a history of hereditary cyanosis for four generations. Gerald *et al* (104, 105) were the first to isolate Hb-M$_{Boston}$ by starch block electrophoresis and demonstrate that it was chocolate in color and constituted about 30% of the total hemoglobin. There are five known types of Hb-M: Hb-M$_{Boston}$, Hb-M$_{Saskatoon}$, Hb-M$_{Iwate}$, Hb-M$_{Hyde Park}$, and Hb-M$_{Milwaukee}$. These hemoglobins have the following characteristics in common: (a) They are inherited in a dominant fashion and produce cyanosis secondary to methemoglobinemia. (b) Except for Hb-M$_{Milwaukee}$, they all involve the amino acid substitution of tyrosine for histidine, either proximal or distal to the heme. In Hb-M$_{Milwaukee}$, valine in position 67 of the β chain is replaced by glutamic acid.

Understanding of this molecular disease has been advanced by recent work of Perutz *et al* (258, 262) on the structure of hemoglobin. Using x-ray methods these workers constructed a model of the molecule which has a resolution of 2.8 Å. for both hemoglobin and oxyhemoglobin. The key item proves to be the steric relationship between the heme iron, the porphyrin ring, and the proximal histidine, which they can now measure to a fraction of 1 Å. Perutz *et al* find that the Fe(II) atom in hemoglobin is five-coordinate and is too large to fit into the space between the four porphyrin nitrogens. It therefore lies in a strained position 0.75 Å. out of the plane of the porphyrin toward the proximal histidine, whereas in oxyhemoglobin it is 6-coordinate and small enough to lie almost exactly in the plane of the porphyrin—a relaxed structure. This steric change is reciprocally related to changes in tertiary and quaternary structure which are involved in (a) stabilization of ferrous iron in hemoglobin and oxyhemoglobin (259), (b) the heme-heme interaction, (c) the Bohr effect, and (d) the 2,3-diphosphoglycerate effect. For example, in the β chains the space between the porphyrin iron and the distal histidine plus the adjacent valine β^{67} is too small to accommodate the O_2 molecule. However, structural changes upon oxygenation of either the α or β subunits, including the above-noted relaxation of the Fe-porphyrin ring structure, make the valine and histidine move away and thereby accommodate the O_2. Perutz demonstrated how the oxygenation of the α or β subunits can alter quaternary structure in such a way as to produce this relaxation in the other subunits, which then accept O_2 more readily than otherwise. However, any mutation which changes Fe(II) to Fe(III) and stabilizes it in the six-coordinate form produces a coplanar, relaxed, and nonfunctioning geometry.

All four types of Hb-M have a tyrosyl residue replacing either the proximal or distal histidine. When the tyrosine is distal to the heme, in place of the histidine, it provides an ionic bond via its phenol group, which occupies the sixth position of heme Fe(III), stabilizes it, and destroys its functionality. This is the case in Hb-M$_{Iwate}$ and Hb-M$_{Hyde\ Park}$. When the tyrosine is proximal to the heme, in place of the histidine which ordinarily occupies the fifth covalent position of heme Fe(II), it is believed that the distal histidine assumes the usual role of the proximal histidine; and the proximal tyrosine, via its phenol group provides the sixth stabilizing bond for Fe(III). This is the situation with Hb-M$_{Boston}$ and Hb-M$_{Saskatoon}$. The other Hb-M (Hb-M$_{Milwaukee}$) has a glutamate instead of a valine at β^{67}, which provides a carboxyl group distal to the Fe to form an ionic bond. In all of these cases Fe(III) is internally stabilized by the sixth bond, resulting in methemoglobinemia.

Cyanosis from early childhood characterizes these diseases, since blood O_2 affinity is grossly altered. In the β chain mutations the stability of the molecule is affected, and mild hemolytic anemia with low hemoglobin level may be seen. In the α chain mutations this does not occur, perhaps because of the larger available space distal to the heme Fe and consequently smaller effect of the substitution on quaternary structure.

The spectrum of Hb-M is different from that of Hb-A, as would be expected, due to the involvement of the porphyrin ring. The peak at 630 nm characteristic of methemoglobin A is not so pronounced, and the valley at 600 nm is replaced by a small peak which is different for different types of Hb-M (107). Also, ligands such as cyanide or carbon monoxide react with some types of Hb-M and not others, although Comings (58) says that α chain substitutions react and β chains do not, while Perutz and Lehman (261) state that distal substitutions react and proximal ones do not. Customary spectrophotemetric measurements are therefore invalid in the presence of Hb-M. Electrophoresis on agar gel at pH 7.0 or 7.6 is the best method of separating Hb-M from Hb-A.

One other mutation is known at the heme histidine, Hb-Zurich β^{63} his→arg. The distal histidine is replaced by the arginine residue, which is so large that it cannot remain in the heme pocket and must protrude at the surface leaving an empty space where the distal histidine belongs. Nevertheless combination with O_2 occurs, making one wonder about current ideas which emphasize the basic importance of the distal histidine (258). The molecule is less stable, however, and sulfa drugs cause hemolytic anemia (58).

Hemoglobins with Altered Oxygen Affinity

Hb-M has altered O_2 affinity, different in the various types, as would be expected from mutations so near the hemes. As Perutz and his co-workers discussed (258), the $\alpha^1\beta^2$ contacts are deeply involved in the changes in quaternary structure which produce heme-heme interaction. Correspondingly, these interactions are missing or diminished in virtually all known mutations involving these contacts. These include Hb-Chesapeake, Hb-Capetown, Hb-Yakima, Hb-Kempsey, Hb-Kansas, and Hb-Richmond (165)—see Table 23-5.

Those which increase O_2 affinity produce polycythemia in compensation for less efficient O_2 transport. Hb-Kansas is of particular interest because it shows decreased O_2 affinity (with little effect on transport) and decreased $\alpha_2\beta_2$ tetramer stability, as compared to the $\alpha\beta$ dimer (272). Table 23-5 summarizes variants which affect function.

Unstable Hemoglobins

Some hemoglobinopathies result from the fact that the hemoglobin variants which cause them are less stable than normal hemoglobin. The instability can be observed either in the presence of certain reagents or at elevated temperature. The variants in question are called "unstable hemoglobins." These hemoglobins are listed in Table 23-6. They are inherited as autosomal codominant mutations and can cause severe hemolytic anemia in the heterozygous state. They generally have the following characteristics in common: (a) instability of the hemoglobin molecule that causes hemolytic anemia which is increased in severity by sulfonamide or infections (15, 98, 276); (b) intracellular precipitation of unstable hemoglobins in peripheral erythrocytes (68, 172, 178, 295), which can be demonstrated by incubation of the blood with a redox dye such as brilliant cresyl blue or in certain cases by incubation at 37°C (68, 162, 270, 291, 295, 346); (c) the presence of free α or β chains which can be detected by starch gel electrophoresis and microscopic examination (42, 162, 198, 290, 295); (d) denaturation of the unstable hemoglobin at 50°C (68).

TABLE 23-5. Hemoglobin Variants with Abnormal Function

Hemoglobin variant	Amino acid substitution	Oxygen binding properties	% Hb Variant	Electrophoretic mobility on starch gel, pH 8.6	Blood Hb (g/100 ml)	Remarks
Hemoglobins with High O$_2$ Affinity						
Chesapeake	α^{92} Arg → Leu	Very low heme heme interaction	25–35	Like Hb-J	14.9–19.9 ⎫	Polycythemia; Hb-Rainier is alkali-resistant
Kempsey	β^{99} Asp → Asn		37	Like Hb-G	17.2–21.3 ⎬	
Rainier	β^{145} Tyr → His		30	Like Hb-A	15.9–22.9 ⎭	
Yakima	β^{99} Asp → His		38	Cathodal to Hb-A		
J Cape Town	α^{92} Arg → Gln	Very low heme heme interaction	35	Like Hb-J		Mild or no polycythemia
Hiroshima	β^{143} His → Asp	Low heme heme interaction; Bohr effect	50	Like Hb-J	14.4–17.0	Polycythemia
Ypsilanti	β^{99} Asp → Tyr	—	50	Cathodal to Hb-A	16.6–19.0	Polycythemia; abnormal polymerization
Hemoglobins with Low O$_2$ Affinity						
Kansas	β^{102} Asn → Thr	Absent heme heme interaction	50	Cathodal to Hb-A	—	Cyanosis due to reduced hemoglobin
M$_{Boston}$	α^{58} His → Tyr	Very low heme heme interaction; Bohr effect	22–42	Separates from Hb-A in methemoglobin form at pH 7.0	—	Cyanosis due to methemoglobinemia
M$_{Hyde\ Park}$	β^{92} His → Tyr	Low heme heme interaction	30–40		—	
M$_{Iwate}$	α^{87} His → Tyr	Absent heme heme interaction; Bohr effect	30	Separates from Hb-A in methemoglobin form at pH 7.0	—	Cyanosis due to methemoglobinemia
M$_{Milwaukee}$	β^{67} Val → Glu		26–30		—	
M$_{Saskatoon}$	β^{63} His → Tyr	Low heme heme	30–40		—	

Hemoglobinopathies associated with unstable hemoglobins are caused by various types of mutation. The instability is generally caused by failure of certain interactions in the molecule to occur either because of the size or the type of amino acid substitution involved—e.g., replacement of a hydrophobic amino acid residue by a hydrophilic one in the interior portion of a hemoglobin subunit invariably leads to severe instability of the molecule. This is illustrated by Hb-Wien (261).

TECHNICS FOR SEPARATION AND IDENTIFICATION OF HEMOGLOBINS

The first hemoglobin variant, Hb-S, was discovered in 1949 by Pauling *et al* (252) using electrophoresis on paper. Many hemoglobin variants are now known, and recent governmental interest in diseases associated with them has accelerated the already rapid pace of method development in this field of clinical chemistry. The many common chemical and physical properties of the different molecular species of hemoglobin make it necessary that a group of tests utilizing various technics be used for positive identification of hemoglobin variants. The following section discusses such technics in detail.

ELECTROPHORESIS

This technic utilizes the migration of charged molecules in an electrical field. This migration is toward either the anode or the cathode. Its rate is dependent on the size, shape, and net electrical charge of the molecule, and on the nature of the supporting medium used. The net charge can be altered by changing the pH of the solution, thus making it possible to select optimum conditions for the electrophoretic separation of different hemoglobin variants.

The two most common buffers for hemoglobin fractionation by electrophoresis are barbital and Tris, pH 8.4–8.8. The supporting media used are paper (116, 229), cellulose acetate (196), agar gel (221, 372), acrylamide gel (86), and starch gel (115, 117). Starch block electrophoresis (256) is useful for preparative separations. The resolution obtained increases roughly in the order listed, with paper far poorer than all of the others because of adsorption. However, gel preparation is too tedious and slow for routine laboratory work, and the superior simplicity, ease, and clarity of separation found with cellulose acetate has made it the most popular medium. Its introduction (196) coincided with recognition of hemolytic anemia due to abnormal hemoglobins as a major health problem in certain parts of the world, resulting in its rapid widespread adoption.

Of all the technics available for separation and identification of hemoglobin variants the electrophoretic technic using cellulose acetate or starch gel has been the most useful in screening human blood samples for the different molecular species of hemoglobin. This technic is especially helpful when the substitution or deletion of one or more amino acid residues in the globin portion of the hemoglobin molecule results in a significant change in the net electrical charge.

TABLE 23-6. Hemoglobin Variants Causing Molecular Instability

Hemoglobin	Amino acid substitution	% Hb Variant[a]	Blood Hb (g/100 ml)[a]	Heat denaturation	Electrophoretic mobility (ref 160)	Cause of molecular instability	Remarks	References
α Chain Abnormalities								
Ann Arbor	80 Leu → Arg	2–12	12	No	As Hb-S	Replacement of internal nonpolar residue by polar residue		164
Bibba	136 Leu → Pro	5–11	6.5–7.5	Yes	As Hb-S	Pro at heme contact breaks up H-helix	Inclusion bodies; increased tendency toward dissociation; anemia	192
Dakar	112 His → Glu	10	—	?	As Hb-J	Substituted surface residue stabilizes G- and B-helices	Mild hemolytic anemia	160
Etobicoke	84 Ser → Arg	15	Normal	?	As Hb-S	Substituted surface residue stabilizes F- and G-helices	No clinical abnormality; increased O_2 affinity	65
L-Ferrara	47 Asp → Glu	14–17	—	?	As Hb-S	Surface residue, cause of instability unknown		21,236,371
Hasharon	47 Asp → His	14–19	—	?	As Hb-S	Surface residue, cause of instability unknown	Hemolysis; decreased rate of synthesis	46,131,325

		8	11	?	As Hb-A	Val substitution at heme contact leaves empty gap	Sulfonamides cause hemolysis; about 5% methemoglobin; inclusion bodies; decreased O$_2$ affinity	15,160
Torino	43 Phe → Val	8	11	?	As Hb-A	Val substitution at heme contact leaves empty gap	Sulfonamides cause hemolysis; about 5% methemoglobin; inclusion bodies; decreased O$_2$ affinity	15,160
β Chain Abnormalities								
Borås	88 Leu → Arg	10	12.6	Yes	As Hb-S	Charged residue replacing nonpolar residue at heme contact	Hemolytic anemia; methemoglobin present; increased O$_2$ affinity	155,326
Bristol	67 Val → Asp	36	7–8	Yes	As Hb-A	Distorted heme pocket	Decreased O$_2$ affinity; hemolytic anemia; methemoglobin present	43,319
Bush Shepherd's	74 Gly → Asp	25	13	Yes	As Hb-J	Distorted heme pocket; too large and polar residue	Sulfonamides cause hemolytic anemia; increased O$_2$ affinity	361
Freiburg	23 Val deleted	30	—	Yes	Cathodal to Hb-A	Disruption of B-helix	Cyanosis; mild hemolysis; methemoglobin 10%	182
Genova	28 Leu → Pro	25	11.4	Yes	As Hb-S	Disruption of B-helix by pro	Hemolytic anemia	290,291
Gun Hill	93 through 97 deleted	30	13.5	Yes	As Hb-C	Disruption of heme contact	Increased rate of synthesis	32,33,274

TABLE 23-6. Hemoglobin Variants Causing Molecular Instability (Continued)

Hemoglobin	Amino acid substitution	% Hb Variant[a]	Blood Hb (g/100 ml)[a]	Heat denaturation	Electrophoretic mobility (ref 160)	Cause of molecular instability	Remarks	References
Hammersmith	42 Phe → Ser	30	6–8	Yes	As Hb-A	Gap in heme contact; smaller amino acid substituted	Decreased O_2 affinity; mild cyanosis	69,290
Köln	98 Val → Met	10–15	10–14 9–14	Yes	As Hb-S	Disruption of heme contact; larger amino acid substituted	Increased O_2 affinity; reduced heme-heme interaction; hemolytic anemia	13,41,172
Leiden	6 or 7 Glu deleted	30	—	?	Cathodal to Hb-A	Disruption of A-helix	Severe chronic hemolysis	74
Philly	35 Tyr → Phe	30–35	13.5	Yes	As Hb-A	Dissociates easily into monomers; weak $\alpha^1\beta^1$ contact bonds	Methemoglobin 3%; mild hemolytic anemia	275
Santa Ana	88 Leu → Pro	15	9–14	Yes	Cathodal to Hb-A	Pro breaks F-helix and disrupts heme contact	7% sulfhemoglobin + methemoglobin; hemolytic anemia; inclusion bodies	84,154,246
Sabine	91 Leu → Pro	7–10	8.6–10.6	Yes	As Hb-S	Disruption of F-helix and hydrophobic contact with heme	Acute hemolytic anemia; Hb-F about 12%	299

Seattle	76 Ala → Glu	43	10.4	Yes	As Hb-J	Glu-His interaction; polymerization	Decreased O_2 affinity; mild anemia	161,249
Sogn	14 Leu → Arg	30	Normal	?	As Hb-S	Charged substituted residue in internal position	No clinical abnormality	225
Sydney	67 Val → Ala	30	12	Yes	—	Disruption of heme contact; Val smaller than Ala	Hemolytic anemia; inclusion bodies	42,270
Toulouse	66 Lys → Glu	40	15.7	?	As Hb-J	Unknown	Increased methemoglobin; mild anemia	283
Wien	130 Tyr → Asp	—	—	Yes	As Hb-A	Internal nonpolar residue replaced with charged one	Mild hemolytic anemia	261
Zurich	63 His → Arg	25	14.7	Yes	As Hb-S	Disruption of heme pocket	Sulfonamides cause hemolytic anemia; increased $Hb\text{-}A_2$	6,98,226, 232,276

[a] Observed values.

CHROMATOGRAPHY

The main disadvantage of electrophoretic technics for separation of hemoglobin variants is the fact that mutants with the same net electrical charge cannot be separated by this method. Chromatography can separate some of these, and Amberlite IRC-50, DEAE cellulose, and carboxymethyl-cellulose have been used by many investigators for this purpose as well as for separation of the different free polypeptide chains (148, 153).

IMMUNOLOGY

Immunologic technics, including gel diffusion and immunoelectrophoresis (27, 28, 147, 153), have been applied to identification of hemoglobins. Hemoglobins A, S, and C either are nonantigenic (147) or at least do not differ enough immunologically to be separated by this technic (27). Hemoglobins A_2 and F are antigenic (27), and Hb-F has been quantitated by immunologic methods (48, 292).

ALKALI DENATURATION

In 1866 von Körber (349) observed that the hemoglobin in human placental blood (Hb-F) was more resistant than adult hemoglobin (Hb-A) to denaturation by alkali. Hb-F is the only hemoglobin known to be unquestionably alkali-resistant so far. Hemoglobin in the presence of alkali is denatured to alkaline hematin, and this conversion apparently proceeds as a monomolecular or first order reaction. A plot of log % unaltered hemo-globin versus time is a straight line, or nearly so. If one is dealing with a mixture of hemoglobins A and F where the velocity constant of Hb-A is large compared to that of Hb-F, extrapolation of the straight line of the semilog plot back to zero time gives the amount of the component with the lower denaturation velocity (i.e., Hb-F) in the mixture. Since the exact mechanism of denaturation is unknown and since it is possible that several reactions may be proceeding simultaneously, the reaction is referred to by some authors (269) as "pseudomonomolecular". The fact that different results are obtained if alkali denaturation is calculated from readings made at different wavelengths constitutes evidence that more than one step is involved in the denaturation process (148).

The amount of alkali-resistant hemoglobin present in a mixture has been determined by two different approaches. In the method of Brinkman and Jonxis (185) Hb-F is determined from the denaturation velocity measured spectrophotometrically by the difference in absorbance of oxyhemoglobin and the denaturation product at 576 nm. In the second approach, that of Singer *et al* (310) and others, Hb-A is completely denatured at pH 12.7 in 1 min and then precipitated with $(NH_4)_2SO_4$. The undenatured, unprecipi-tated Hb-F remaining in solution is measured photometrically at 540 nm. Studies have shown that these methods are not accurate when the mixture contains less than 10% Hb-F. The method of Singer *et al* also may be 10% too low at high levels (185). It is actually uncertain whether the two methods measure the same thing, it being possible that during denaturation with alkali the solubility of the protein is altered at a stage of the denaturation process different than the stage of alteration of the absorption spectrum. Other data bearing on the accuracy of the alkali denaturation method for determination of Hb-F are discussed later (see *Determination of Hemoglobin F by Alkali Denaturation, Accuracy and Precision*).

Kunzer (200, 201), using the precipitation method, converted hemoglobin to cyanmethemoglobin as a first step, claiming that this was superior to applying the test to oxyhemoglobin. Betke *et al* (20) agreed with this contention, but others have disagreed.

FERROHEMOGLOBIN SOLUBILITY

The appearance of the solubility curve of a protein is a classic criterion for heterogeneity and has been used to investigate hemoglobins. Two procedures are available: the variable solute and the variable solvent (salting-out) methods. In the "variable solvent" method of Roche *et al* (280, 281) a constant volume of hemoglobin solution is added to each of a series of dilutions of pH 6.5 phosphate buffer. After equilibration the mixtures are filtered and the absorbance at 542 nm measured and plotted against the phosphate concentrations. Results obtained by such solubility curves have led to many questionable conclusions. Many hemoglobin mixtures give rise to solid solutions, and in such a case the plot of protein concentration against concentration in the solid phase gives a smooth curve. This may give the false impression of heterogeneity. If successive experimental points are joined by straight lines, as many components are indicated as there are points. Even in the absence of solid solutions, interpretative errors are possible. For example, the conclusion of heterogeneity may be drawn from data, the precision of which does not warrant such a conclusion. As reviewed by Beaven and Gratzer (11), other possible explanations for observed discontinuities in solubility curves include interaction of hemoglobin with smaller molecules or ions, presence of impurities, and phase transitions at high salt concentration.

MISCELLANEOUS TECHNICS

A number of other tests are useful as laboratory aids in the identification of hemoglobinopathies. These include the following, which are presented in subsequent sections of this chapter: (a) test for sickling (118); (b) detection of Hb-F in red blood cells by staining (191); (c) heat denaturation test for detection of unstable hemoglobins (68); and (d) microscopic examination for inclusion bodies (68, 172, 178). Several rather sophisticated technics are very useful in research in the field of hemoglobin variants and occasionally are used as laboratory aids. These are mentioned here for reference: (a) The abnormal hemoglobin is separated from other hemoglobins by column chromatography. (b) The hemoglobin molecule is broken into its polypeptide chains with urea and mercaptoethanol, and the chains are separated by column chromatography (336). (c) One or more polypeptide chains is digested by trypsin and the resulting peptides are separated either by column chromatography or by a combination of chromatography and high voltage electrophoresis on 3MM Whatman paper ("finger printing"). (d) The abnormal peptides are then hydrolyzed and their amino acid composition is determined by column chromatography using ion exchange resin.

COLLECTION OF BLOOD SAMPLE AND PREPARATION OF HEMOLYSATE

Blood can be collected with oxalate as anticoagulant, but the hemoglobins are more stable if the blood is mixed with and kept in Alsever's solution. The latter is recommended, therefore, especially for samples sent through the mail. Blood is drawn in a syringe, and up to 10 ml is injected into vaccine bottles prepared in the following way: Dissolve 8.0 g sodium citrate dihydrate, 4.2 g NaCl, and 20.5 g glucose in approximately 800 ml water. Add 10% (w/v) citric acid until pH 6.1 is reached. Add water to a final volume of 1 liter. Filter through hard filter paper. Dispense 10 ml aliquots into 30 ml vaccine bottles and cap with rubber vaccine stoppers. Insert hypodermic needles through caps, autoclave ($118°-120°C$) for 10 min, cool, and remove needles.

The method of Friedman (99) is recommended for preparation of the hemolysate: Centrifuge up to 5 ml oxalated blood or 10 ml blood mixed with Alsever's solution. Pour off plasma and wash erythrocytes three times with 0.85% NaCl. To the packed cells add an equal volume of water and an equal volume of chloroform. Stopper, shake vigorously, store for 12–16 hours in refrigerator, shake vigorously again, centrifuge for 15 min, and take off supernatant hemoglobin solution. Freeze if there is to be a delay in analysis. Avoid repeated freezing and thawing because it leads to denaturation.

TEST FOR SICKLING

(method of Goldberg, Ref 118)

PRINCIPLE

In 1910 Herrick (151) recognized the capacity to sickle of the erythrocytes of patients with sickle cell anemia. Sickling depends on diminished O_2 tension. The early technic of sealing a drop of blood on a slide under a coverslip with Vaseline or petrolatum was not completely reliable because the development of low O_2 tension was dependent on O_2 consumption by the leukocytes. It is far more reliable to add a reducing agent (70). Sodium bisulfite is used in the method presented.

REAGENT

Sodium Bisulfite Solution, 1%. Dissolve 250 mg $NaHSO_3$ in freshly boiled and cooled water, and add water to 25 ml. Stable 1 week in refrigerator.

PROCEDURE

Very small drops of reagent and blood collected with anticoagulant are mixed on a microscope slide and covered with a coverslip. The edges of the coverslip are sealed with Vaseline. The preparation is examined microscopically for sickling immediately and every 15 min thereafter up to 1 hour.

INTERPRETATION

Sickling indicates one of the following: (a) homozygous Hb-SS (sickle cell anemia): (b) Hb-AS (sickle cell trait); (c) Hb-C $_{Harlem}$; (d) Hb-C $_{Georgetown}$; or (e) certain types of Hb-I.

NOTE

1. Stability. According to Daland and Castle (70) the ability to sickle may be lost if blood is kept for 24 hours in the presence of O_2. We find that sickling occurs in samples at least 1 week old.

DITHIONITE-PHOSPHATE SCREENING TEST FOR HEMOGLOBIN S

(method of Greenberg et al, Refs 121, 192, 512)

PRINCIPLE

Deoxygenated Hb-S is markedly less soluble in phosphate buffer than other hemoglobins, and forms insoluble linear aggregates which produce visible turbidity. The degree of turbidity in specimens from heterozygotes depends upon the Hb-S concentration and the type and concentration of other hemoglobins present. Hemoglobins C$_{Harlem}$, C$_{Georgetown}$, and high concentrations of Hb-Bart's also give a turbid solution. The mechanism for their insolubility is not wholly understood at present.

REAGENT

Phosphate Buffer. To a 1 liter volumetric flask containing about 300 ml water, add 216 g anhydrous K_2HPO_4. Dissolve completely and mix. Add 169 g KH_2PO_4. Dissolve completely and mix. Add 5 g sodium hydrosulfite.

Dissolve completely and mix. Add 1 g saponin. Dilute to 1 liter with water and mix again. Store in a plastic bottle at 4°C. This solution is stable for about 1 month.

PROCEDURE

1. Place 2.0 ml phosphate buffer into each of three test tubes
2. To one test tube add 20 μl sample (TC pipet). To a second test tube add 20 μl of a specimen known to contain Hb-S, and to a third add 20 μl of a specimen known to contain only Hb-A.
3. Mix and allow to stand at room temperature for 5 min.
4. Hold the tubes 1 in. in front of a white card on which have been ruled black lines 1 mm apart, and look through the tubes at the card.
5. The test is negative if the black lines are visible through the hemoglobin solution and positive if they are not.

INTERPRETATION

A positive test indicates the presence of Hb-S or, rarely, of Hb-C $_{Georgetown}$, Hb-C$_{Harlem}$, or Hb-Bart's. The degree of turbidity can be correlated with other information to assist in detecting homozygotes or heterozygotes for Hb-S.

NOTE

1. Stability. Blood samples for this test are stable for 4 days at room temperature and for 3 weeks at 4°C.

ACCURACY

Because this test is used for the detection of Hb-S, the presence of the hemoglobins mentioned in the interpretation section is considered to cause false positive results. Certain types of Hb-I may act similarly; however, such false positives provide valuable information.

Polycythemic samples may give false positive tests, as does any error in measurement in which too much blood is added. Blood from patients with multiple myeloma, cryoglobulinemia, or other dysglobulinemias may also show false positive results.

False negatives may occur in the presence of over 25% Hb-F, or in patients with severe anemia (10 g hemoglobin/100 ml or below). The test does not detect Hb-S at very low concentrations—e.g., immediately after tranfusion of normal blood to an individual with sickle cell trait.

NORMAL VALUES

The test is negative with normal blood samples.

FERROHEMOGLOBIN SOLUBILITY TEST

(method of Itano, Ref 174, modified by Goldberg, Refs 114, 118)

PRINCIPLE

Hemoglobins are converted to the reduced form by hydrosulfite, and then Hb-S is precipitated by a final phosphate concentration of 2.23 *M*. The unprecipitated hemoglobin in the filtrate is measured at 422 nm, and is expressed as percent of the total hemoglobin present in the control.

REAGENTS

Sodium Hydrosulfite, $Na_2 S_2 O_4$
Phosphate Buffer, 2.48 M. Dissolve 43.40 g anhydrous $K_2 HPO_4$ and 33.80 g anhydrous $KH_2 PO_4$ in CO_2-free water. Adjust volume to 200 ml. Store at room temperature in tightly stoppered polyethylene bottle.

PROCEDURE

1. Add 20 mg sodium hydrosulfite (can be approximated by a micro spatula) to 1.8 ml phosphate buffer in a small test tube. Add 0.2 ml hemolysate from a TC pipet, rinsing pipet out several times. Mix and let stand for 15 min.
2. Filter through Whatman 5 filter paper.
3. Add 0.2 ml filtrate by TC pipet to 4.8 ml phosphate buffer containing approximately 20 mg sodium hydrosulfite. Mix by inversion.
4. Measure absorbance against water blank at 422 nm in cuvet of same size as used for the alkali denaturation test (see *Determination of Hemoglobin F by Alkali Denaturation*). If A_{422} is too high, dilute with more buffer and make proper correction in calculations.
5. The "Control" is prepared as in step 5 of the alkali denaturation test. (If both tests are being run, use same control.) Measure absorbance at 422 nm against water.

Calculation:

$$\% \text{ Solubility} = \frac{A_{\text{unknown}}}{A_{\text{control}}} \times 100$$

NOTES

1. Wavelength and bandwidth. According to Goldberg (118) readings at 415 or 425 nm are satisfactory in broad bandwidth instruments such as the Coleman Jr. spectrophotometer. If a narrow bandwidth spectrophotometer is used, readings should be made at 422 nm, the isobestic point for oxy and reduced hemoglobin.

2. *Beer's law.* Beer's law is obeyed on a Beckman DU spectrophotometer but not on a Klett photometer with No. 42 filter. It is recommended, therefore, that conformance to Beer's law be checked before a "broad bandwidth instrument" (see *Note 1* above) is used for this test.

3. *Temperature.* Results are independent of temperature over a range of 20°–40°C.

4. *Stability.* See note under *Hemoglobin Fractionation by Electrophoresis on Cellulose Acetate.*

ACCURACY AND PRECISION

Precipitation of Hb-S is a function of pH and salt concentration; hence accuracy is quite dependent on the accuracy of the phosphate buffer preparation. Polosa *et al* (267) reported that this test does not give dependable data on the concentration of Hb-S in a natural mixture of hemoglobins.

According to Goldberg (114) the standard deviation of the test is 0.88% solubility, giving a precision of about ±2% at 100% solubility.

INTERPRETATION

Goldberg (114) obtained the following values for the indicated hemoglobin combinations:

Hemoglobin combination	% solubility (95% range)
AA	88–102
AS	35–68
SS	6–23
SC	36–44
AC	83–103

DETERMINATION OF HEMOGLOBIN F BY ALKALI DENATURATION

(method of Singer et al, Ref 310, modified)

PRINCIPLE

An aqueous solution of the hemoglobin is mixed with alkali. Denaturation is allowed to proceed for exactly 1 min; the mixture is then quickly neutralized and the denatured hemoglobin precipitated by ammonium sulfate. The precipitated hemoglobin is removed by filtration and the alkali-resistant hemoglobin remaining in solution is determined photometrically and expressed as percent of total hemoglobin present in the original hemolysate.

REAGENTS

NaOH, 0.083 N. Prepare from a concentrated carbonate-free stock solution and standardize by titrating against standard HCl. Stable in refrigerator stored in polyethylene bottle.

Precipitating Reagent. Dissolve 350 g $(NH_4)_2SO_4$ in 500 ml water at room temperature. A small amount remains undissolved. Filter and dilute 400 ml filtrate with 400 ml water (50% saturated solution). Add 2.0 ml 10 N HCl. This solution should be checked by mixing 3.4 ml with 1.6 ml 0.083 N alkali. Final pH should be 7. Reagent is stable indefinitely in a polyethylene bottle at room temperature.

EDTA-Na$_4$, 0.3%. Tetrasodium salt of ethylenediaminetetraacetic acid (obtainable from Dow Chemical). Stable at room temperature for at least several months. Filter if turbidity develops.

PROCEDURE

1. Measure 3.2 ml alkali into a test tube.
2. Add 0.2 ml hemoglobin solution (hemolysate) by a 0.2 ml TC pipet, rinsing pipet with the alkali solution several times. At moment of introduction start stopwatch and mix.
3. At *exactly 1 min,* add 6.6 ml precipitating solution rapidly. Stopper and invert tube immediately six times.
4. Filter through Whatman 5 filter paper.
5. Set up control by making a 1:251 dilution of hemolysate with 0.3% EDTA-Na$_4$ (i.e., 20 μl to 5 ml or 0.1 ml to 25 ml, etc.).
6. Read absorbances of control and filtrate from unknown against water blank at 540 nm or with a filter with nominal wavelength in this region.

Calculation:

$$\text{Alkali-resistant hemoglobin (Hb-F), \% of total} = \frac{A_{\text{unknown}}}{A_{\text{control}} \times 5} \times 100$$

$$= \frac{A_{\text{unknown}}}{A_{\text{control}}} \times 20$$

NOTES

1. Beer's law. Beer's law is obeyed with a spectrophotometer of narrow bandwidth, but there is slight deviation with glass filters. The error is negligible with a Klett photometer using a No. 54 filter.

2. Stability. See *Note* under *Hemoglobin Fractionation by Electrophoresis on Cellulose Acetate.*

ACCURACY AND PRECISION

It is generally conceded that the method is not very accurate when the hemoglobin mixture contains <10% Hb-F and certainly not at all accurate at levels as low as 2% or 3%. At such levels some Hb-A is being measured as well as Hb-F. Furthermore, results may be 10% too low at high levels. Beaven *et al* (9) concluded that the method was subject to 5%–10% error over the range of 10%–100% Hb-F (148).

Although Hb-F is the only common hemoglobin known to be alkali-resistant, one is not sure in every case that it is what is being measured: (a) Delayed denaturation has been observed in tests on burned patients. (b) Abnormal alkali denaturation curves have been observed in cases of pernicious anemia, the abnormality not being due to Hb-F. (c) Although Hb-F has been identified in normal adult blood by immunologic technics, there is evidence that at least part of the alkali-resistant hemoglobin in normal adults is not Hb-F (148). (d) Hemoglobins Alexandra, Bart's (89), and Cypriot (113) are reported to have increased alkali resistance.

Carboxyhemoglobin does not react the same as oxyhemoglobin; hence the presence of significant amounts of this derivative alters results (185).

The standard deviation of the procedure is about 0.4. The precision of the procedure is thus about ±40% at a level of "2% of total hemoglobin" and improves as the level increases (148).

NORMAL VALUES

At 20 weeks of pregnancy Hb-F comprises about 94% of the total hemoglobin. After 34 weeks of gestation its percent of the total may decrease at rates up to about 4% per week prenatally, which is about the same rate of decrease that occurs postnatally. The level at birth is about 64%–95%, and in postmature newborns the range is 54%–76%. At 2, 4, and 6 months, the values are about 50%, 15%, and 5%, respectively. Adult levels are attained at some time between 7 months and 2 years. The normal adult level by the test presented is <2%; however, because of the inaccuracy and imprecision inherent in the test at low levels, care must be exercised in interpretation of results <10% (148).

DETECTION OF FETAL HEMOGLOBIN IN RED BLOOD CELLS BY STAINING

(method of Kleihauer, Ref 191)

PRINCIPLE

Red blood cells are dried on a microscope slide and their protein denatured with ethanol. Hemoglobins other than Hb-F are then solubilized and washed off with citric acid-phosphate buffer. Hb-F, which is insoluble in this buffer, remains on the slide and is stained and identified microscopically.

REAGENTS

Ethanol, 80% (v/v) and 90% (v/v). Ethanol denatured with methanol is satisfactory.

Citric Acid, 0.1 M. Dissolve 21 g citric acid monohydrate in 500 ml water and dilute to 1 liter. This solution is stable for 6 months at 4°C.

Dibasic Sodium Phosphate, 0.2 M. Dissolve 35.6 g $Na_2HPO_4 \cdot 2H_2O$ in 500 ml water and dilute to 1 liter. This solution is stable for 6 months at 4°C.

Citric Acid-Phosphate Buffer, pH 3.3—3.5. Prepare fresh daily by mixing 72 ml of the citric acid solution with 28 ml of the dibasic sodium phosphate solution. Check pH.

Hematoxylin Stain. Dissolve 4 g hematoxylin crystals in 200 ml 90% ethanol. Add 8 ml 10% aqueous $NaIO_3$, 200 ml water, mix, and heat to boiling. Cool to about 37°C, then add 200 ml glycerine, 6 g $NH_4 Al(SO_4)_2$, and 200 ml glacial acetic acid. Mix thoroughly.

Erythrocin Stain. Dissolve 100 mg erythrocin B (eosin B) in 100 ml water.

PROCEDURE

1. Place 0.2 ml whole blood in a small test tube and dilute with 0.2 ml 0.9% saline. The blood should be less than 24 hours old, and EDTA or oxalate should be used as anticoagulant.

2. At least weekly prepare controls as in step 1 using (a) a normal adult blood for a negative control and (b) 1 ml of a fresh cord blood mixed with 2 ml fresh normal adult blood of the same ABO type for a positive control.

3. Prepare two very thin smears each for the patient, negative control, and positive control on clean microscope slides and allow to dry for 5 min.

4. Immerse slides in 80% ethanol for 5 min, then rinse gently with water.

5. Immerse slides in citric acid-phosphate buffer at 37°C for 6 min, swirling the buffer gently over the slides every 2 min.

6. Rinse gently with water, immerse in hematoxylin stain for 3 min, then rinse again with water.

7. Immerse in erythrocin stain for 3 min, rinse with water, and allow to dry.

8. Examine under microscope with high dry and oil immersion objectives.

INTERPRETATION

The negative control should show very little pink stain uptake, and red blood cells should appear as ghost cells. The positive control should show about two-thirds ghost cells and one-third pink cells. Two different types of positive results may be found: (a) In cases of persistent Hb-F, almost all cells are stained. (b) In cases of thalassemia only a minority of the cells are stained and the rest appear as ghost cells.

NOTES

1. Preparation of blood smears. Blood smears should be only one cell thick and should be denatured in ethanol within 1 hour after preparation.

2. Water pH. The pH of the water used to rinse slides should be between 6.6 and 7.1. Use boiled distilled water.

3. Stability. Samples up to 1 day old have been found adequate (36).

ACCURACY

The stain is quite specific for fetal hemoglobin. The distinction of thalassemia (where only certain cells are stained) and a condition of persistent Hb-F (where almost all cells are stained) is clear because in the former situation less than 50% of the cells are stained.

HEMOGLOBIN FRACTIONATION BY ELECTROPHORESIS ON CELLULOSE ACETATE

(method of Briere et al, Ref 35, modified by Schneider, Ref 296)

PRINCIPLE

Different molecular species of hemoglobin are separated from each other by electrophoresis at pH 8.4 on cellulose acetate. The separate hemoglobin bands are then stained with Ponceau S dye and identified by comparison with known hemoglobin standards separated and stained in the same manner.

REAGENTS AND EQUIPMENT

Tris-EDTA-Boric Acid (TEB) Buffer, pH 8.4. Dissolve 10.2 g tris(hydroxymethyl)aminomethane, 0.6 g ethylenediaminetetraacetic acid (EDTA), and 6.4 g boric acid in about 300 ml water. Dilute to 1 liter and store at room temperature. Check pH.

Ponceau S Dye, 0.5%, in 5% Trichloroacetic Acid (TCA). Dissolve 5 g TCA and 0.5 g Ponceau S dye (that available from Allied Chemical Co. has been found to be satisfactory) in 95 ml water. Store at room temperature. This solution is stable for about 1 month.

Acetic Acid, 5%

Absolute Methanol

Acetic Acid, 20% in Absolute Methanol

NaCl, 0.85%

Saponin Reagent, 0.1%. Dissolve 100 mg saponin (Coulter B3145-3 has been found to be satisfactory) in about 50 ml water. Dilute to 100 ml, mix, and store at room temperature.

Hemoglobin Standards. A standard containing equal concentrations of Hb-A, Hb-S, and Hb-C in the form of hemolysate can be purchased from commercial sources (Hyland Laboratories and Helena Laboratories, among others). Hemoglobin standards can also be prepared as described under *Collection of Blood Sample and Preparation of Hemolysate* if blood specimens with the desired hemoglobin concentrations are available. After hemolysate preparation the approximate concentration of the different hemoglobin variants in each hemolysate is determined by electrophoresis on cellulose acetate. Proper proportions of the different hemolysates are then mixed to give equal concentration of the hemoglobin standards desired. The hemolysate mixture is then aliquoted into small portions and stored at 4°C or frozen. The refrigerated solution is stable for at least 2 weeks and the frozen one for at least 1 month.

Cellulose Acetate Strips. A wide variety of supplies and equipment is available for this technic. Satisfactory results have been obtained in our laboratory with Titan III cellulose acetate plates, obtainable as No. 2560 from Photovolt or No. 3023 from Helena Laboratories.

Blotting Paper. Helena No. 5034 or 5098, Beckman paper wicks No. 319329, or equivalent.

Chamber Paper Wicks. Helena No. 5081 or equivalent.

Plastic Envelopes. Any envelope which holds one cellulose acetate strip neatly is satisfactory.

Electrophoresis Equipment. Equipment manufactured by Photovolt, Gelman, Beckman, Helena, and others has been used satisfactorily. This method is described as performed with Titan power supply No. 1500, Zip Zone chamber No. 1283, Zip Zone eight-sample application No. 4080, Zip Zone sample plate No. 4081, and Zip Zone aligning base No. 4082 from Helena Laboratories.

PROCEDURE

1. Pour 100 ml TEB buffer into each side of the chamber. Soak two chamber paper wicks and drape one over each support bridge making sure it makes buffer contact.

2. Immerse the cellulose acetate plate in the TEB buffer very slowly and allow to soak for at least 5 min or as specified by manufacturer.

3. Using capillary tubes, fill the troughs in the sample plate with hemolysate (see under *Collection of Blood Sample and Preparation of Hemolysate*), including a known hemoglobin standard containing Hb-A, Hb-S, and Hb-C, and prime the sample applicator.

4. Remove cellulose acetate plate from buffer and blot between two pieces of blotting paper quickly and evenly to remove excess buffer and position in aligning base.

5. Apply samples immediately to cellulose acetate side of the plate at a point 1 in. from the cathode and immediately position the plate, cellulose acetate side down, on the support bridges of the chamber. Place two glass microscope slides over plate to assure good contact.

6. Apply 450 volts for 15 min at room temperature.

7. Turn off power supply, remove plate from chamber, and stain for at least 10 min with the Ponceau S stain.

8. Remove plate from stain and wash in three successive dishes of 5% acetic acid, 2 min in each.

9. Fix plate in absolute methanol for 3–5 min.

10. Immerse plate in 20% glacial acetic acid in absolute methanol for 10 min to clear.

11. Dry plate in 65°C oven for 10 min, place in plastic envelope, and label.

INTERPRETATION

Hemoglobin bands are identified by comparison with known standards on the same plate. The rate of migration for the more common types is in the order Hb-H>Hb-A>Hb-F>Hb-S>Hb-C. If additional tests are needed proceed as is shown in Figure 23-8 under *Scheme of Laboratory Aids in Identification of Hemoglobinopathies*, below.

NOTE

1. Stability. Almost all samples of oxalated whole blood are stable for about 2 weeks at 4°C (255). However, insufficient information exists about the stability of many hemoglobin variants to permit a general statement on stability.

Bergren (16) found that blood collected in Alsever's solution is more stable than oxalated blood, and samples so collected are stable enough to permit mailing. Blood hemolysates have been found to be stable for months in our laboratory.

ACCURACY

It is possible for a given band in electrophoresis to result from the presence of more than one hemoglobin variant. For example Hb-D and Hb-S migrate at almost identical rates, as does the group Hb-C, Hb-E, and Hb-A_2. Therefore electrophoretic results cannot be interpreted unequivocally; they must be regarded as patterns to be correlated with clinical information and the results of other tests. A simple systematic approach to this overall problem is given under *Scheme of Laboratory Aids in the Identification of Hemoglobinopathies*, this chapter.

NORMAL VALUES

When normal blood is submitted to this procedure a very intense band is obtained at the location characteristic of Hb-A, with a light band at the location of Hb-A_2. The latter may be so light as to be missed, and the presence of Hb-A$_2$ may be demonstrable only by elution of an electrophoretic strip as described in the following section.

DETERMINATION OF HEMOGLOBIN A$_2$ BY ELECTROPHORESIS ON CELLULOSE ACETATE

(method of Bergren, Ref 16, modified)

PRINCIPLE

Hb-A$_2$ is separated from other hemoglobins by electrophoresis on cellulose acetate at pH 8.8. The Hb-A$_2$ band alone and all other hemoglobin bands as a group are then cut out of the strip, eluted separately with Tris-EDTA buffer, and quantified spectrophotometrically at 414 nm. Hb-A$_2$ is reported as percent of total hemoglobin.

REAGENTS AND EQUIPMENT

Tris-EDTA Buffer, pH 8.8. Dissolve 6.1 g tris hydroxymethyl-aminomethane (Schwarz/Mann buffer grade has been found satisfactory) and 0.4 g Na$_2$ EDTA in 700 ml water. Check pH. This solution is stable for about 1 month.

Na$_2$CO$_3$, 0.05 M. Dissolve 5.3 g Na$_2$CO$_3$ in 500 ml water. Dilute to 1 liter and store at room temperature.

Heathkit Power Supply. Heath Co. No. IP-17 or equivalent.

Electrophoresis Chamber. Gelman Instrument Co. No. 51101 with vinyl Magna-Grip tensioners or equivalent.

Cellulose Acetate Strips. Sepraphore III cellulose polyacetate strips, 2.5 x 12.0 in., manufactured by Gelman Instrument Co. and distributed by Scientific Products Co., or equivalent, have been found satisfactory. These strips are cut to 2.5 x 6.0 in. before use.

Blotting Paper. Helena No. 5034 or 5098, Beckman No. 319329, or equivalent has been found satisfactory.

Dixon Mile Master Pen. Joseph Dixon Crucible Co., or any other pen with nonmigrating ink may be used.

PROCEDURE

1. Prepare hemolysate as described under *Collection of Blood Sample and Preparation of Hemolysate.*
2. Using a Dixon Mile Master pen, number cellulose acetate strips on the anode (+) side, about 10 mm from the longitudinal end of the strip. Allow a 6−7 mm margin on each side of the strip and draw a 50 mm line along its width. Soak strips in Tris-EDTA buffer for at least 30 min by

floating the strips in the buffer until they are completely saturated and then sinking them in by moving the tray gently back and forth.

3. Fill the electrophoresis chamber to the required volume with 0.05 M Na_2CO_3.

4. Remove strips from buffer and blot them between two pieces of blotting paper, quickly and evenly, to remove excess buffer. Position them immediately in the electrophoresis chamber, in numerical order, using the Magna-Grip tensioner to hold strips in position and to make sure they are straight and flat. All positions in the chamber must be used to ensure uniform conditions of heat, evaporation, etc.

5. Immediately after strips are mounted, apply hemolysate as follows: Place a wooden block of proper height alongside the electrophoresis chamber to serve as a hand support for application of the sample. Using a Beckman-Spinco pipet No. 320016, or any other suitable pipet, apply about 10 μl hemolysate evenly along the 50 mm line. This is best done by resting the wrist on the wooden block and moving the tip of the pipet smoothly back and forth along the line until all of the hemolysate in the pipet is transferred to the strip. The task of sample application for all of the strips in the cell should take no more than 5 min.

6. Cover electrophoresis chamber and turn electrical power supply on, setting the voltage at 275 V.

7. Allow electrophoresis to proceed for exactly 45 min. Turn the power supply off and remove chamber cover.

8. Using a single-edge razor blade, cut the strips at the support points to prevent excess buffer flow. Place strips on Parafilm and, while moist, cut out the Hb-A$_2$ band, all other hemoglobin bands, and a blank strip equal in size to that of the Hb-A$_2$ band, as shown in Figure 23-7.

9. Cut each of the Hb-A$_2$ and Blank strips into four or more segments and place the segments of each strip separately in 0.75 x 6.0 in. test tubes. Cut the third strip, which has all other hemoglobin bands, into six or more segments and place into another 0.75 x 6.0 in. test tube.

10. Add 1.5 ml of Tris-EDTA buffer to the test tubes containing each of the Hb-A$_2$ and the Blank strips, and 15 ml to the tube containing the other hemoglobins. Shake gently, then allow to stand for 1 hour with gentle shaking every 15 min.

11. Pour the supernates containing the eluted hemoglobins and the Blank into separate cuvets and read their absorbances at 414 nm versus the Blank.

Calculation:

$$\text{Hb-}A_2\text{, as \% of total Hb} = \frac{A_{414} \text{ of Hb-}A_2 \times 100}{10(A_{414} \text{ of other Hbs}) + A_{414} \text{ of Hb-}A_2}$$

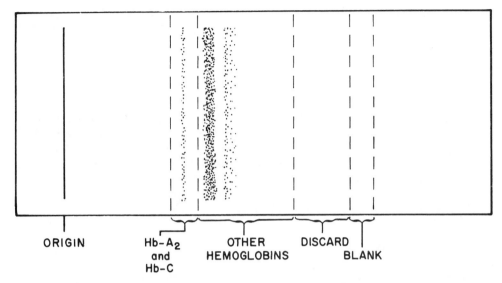

ORIGIN   Hb-A₂   OTHER   DISCARD
               and   HEMOGLOBINS   BLANK
               Hb-C

FIG. 23-7. Electrophoretic pattern obtained in method for determination of hemoglobin A^2.

NOTES

1. Distortion of pattern. If $Na_2 CO_3$ from the electrophoresis chamber splashes on the cellulose acetate strip a distorted migration pattern results.

2. Stability. See *Note 1* under *Hemoglobin Fractionation by Electrophoresis on Cellulose Acetate.*

ACCURACY AND PRECISION

Hb-A₂ cannot be quantitated by this technic when Hb-C is present, because their migration rates under these conditions are identical. In these rare cases starch gel electrophoresis must be used. Hb-A₂ values greater than 10% of the total hemoglobin are caused either by poor separation of Hb-A₂ from Hb-S, Hb-D, and other hemoglobins which migrate at a slightly different rate, or by the presence of Hb-C, Hb-E, and other hemoglobins that migrate at the same rate as Hb-A₂. In the former case, the problem can be identified by striping about 1 μl of Hb-S standard beside 1 μl of Hb-A₂ standard and checking for clear separation of the two hemoglobins. In the latter case the preceding method in this chapter, *Hemoglobin Electrophoresis on Cellulose Acetate,* should be performed to identify all h emoglobin bands present.

The precision (95% limits) of the method is ±6%.

NORMAL VALUES

In adults 1.8%–3.7% of total hemoglobin may be Hb-A₂. Less is present in children, as indicated in Figure 23-1.

SEPARATION OF HEMOGLOBIN D AND HEMOGLOBIN E BY AGAR GEL ELECTROPHORESIS

(method of Robinson et al, Ref 279, modified)

PRINCIPLE

Citrate agar and cellulose acetate electrophoresis in combination make it possible to identify Hb-D, Hb-S, Hb-E, or Hb-C. On agar gel Hb-D and Hb-E migrate together with Hb-A, Hb-S migrates faster than these, and Hb-C faster still. On cellulose acetate Hb-D and Hb-S migrate together, and so do Hb-C and Hb-E, but at a rate slower than the first two. These migration patterns are illustrated in Figure 23-8 under *Scheme of Laboratory Aids in the Identification of Hemoglobinopathies.*

REAGENTS AND EQUIPMENT

Sodium Citrate, 0.5 M. Dissolve 147 g sodium citrate in about 300 ml water and dilute to 1 liter.

Citric Acid, 30%. Dissolve 30 g citric acid crystals in about 60 ml water and dilute to 100 ml.

Citrate Buffer, 0.05 M, pH 6.2. Add 30 ml 0.5 *M* sodium citrate to 270 ml water. Adjust to pH 6.2 with 30% citric acid and dilute to 300 ml. Check pH.

Benzidine, 0.2% in 95% Ethanol. Dissolve 0.2 g benzidine powder in 100 ml 95% ethanol. Store in dark bottle at 4°C. Stable 1 month.

Acetic Acid, 3%

Sodium Nitroferricyanide, 1%. Dissolve 1 g sodium nitroferricyanide in about 60 ml water. Dilute to 100 ml. Store in dark bottle at 4°C. Stable 1 month.

Hydrogen Peroxide 3%. Prepare fresh daily from Superoxol (30% H_2O_2.)

Staining Solution. Mix immediately before use: 15 ml 3% acetic acid, 7.5 ml 0.2% benzidine, 1.5 ml 1% sodium nitroferricyanide, and 1.5 ml 3% hydrogen peroxide.

Agar, Purified. Bioquest, BBL Products No. 11853.

Heath Kit Power Supply. Heath Co. No. IP-17 or equivalent.

Electrophoresis Chamber. Helena No. 1283 or equivalent.

Microscope Slides, 25 × 75 mm.

Chamber Paper Wicks. Helena No. 5081 or equivalent.

Blotting Paper. Helena No. 5034 or 5098, or Beckman No. 319329 or equivalent.

Tooth Picks, Round, Wooden

PROCEDURE

1. Add 25 µl hemolysate (see under *Collection of Blood Sample and Preparation of Hemolysate*) to 0.5 ml water. Mix thoroughly.

2. Add 0.5 g agar to 50 ml pH 6.2 citrate buffer. Heat in boiling water bath until clear, then cool to 70°C.

3. Dispense 1.5 ml agar onto a microscope slide with a 5 ml pipet (surface must be level). Allow to solidify (5 min).

4. Add 100 ml cold buffer (4°C) to each side of electrophoresis chamber. Saturate two chamber paper wicks with buffer and drape one over each support bridge making sure it makes buffer contact.

5. Make sure agar covers complete slide surface and is uniform.

6. Dip the point of a round toothpick into the diluted hemolysate and remove excess by touching to inside of container. Apply unknown to the center of the slide about 5 mm from one edge. Apply reference standard to center of the slide 5 mm from the opposite edge.

7. Place slides in electrophoretic chamber, agar side in contact with wicks. Cover chamber and apply 90 V for 45 min. Place a tray of ice on chamber lid to prevent heating and denaturation.

8. Remove slides from chamber and place in a petri dish with agar layer up. Pour staining solution on slides to cover completely and allow to stand for 3 min.

9. Carefully pour off excess staining solution and wash slides in a gentle stream of tap water.

10. Place slides, agar side down, on blotting paper and allow to dry before moving (2 hours), or dry under heating lamp, agar side up.

INTERPRETATION

The order of migration toward the anode (+) is C>S>A=D=E with Hb-A remaining at the origin. Hb-F and Hb-A_2 migrate toward the cathode (−).

NOTE

1. Stability. Refer to *Note 1, Hemoglobin Fractionation by Electrophoresis on Cellulose Acetate.*

HEAT DENATURATION TEST FOR DETECTION OF UNSTABLE HEMOGLOBINS

(method of Dacie et al, Ref 68)

PRINCIPLE

Normal hemoglobin dissolved in phosphate buffer at pH 7.4 remains in solution when heated for 30 min at 60°C. Most unstable hemoglobins, however, when heated in this manner are almost completely denatured and form heavy flocculent precipitates.

REAGENTS

Monobasic Sodium Phosphate, 0.1 M. Dissolve 15.6 g $NaH_2PO_4 \cdot 2H_2O$ in about 500 ml water. Dilute to 1 liter.

Diabasic Sodium Phosphate, 0.1 M. Dissolve 14.2 g anhydrous Na_2HPO_4 crystals in about 500 ml water. Dilute to 1 liter.

Phosphate Buffer, 0.1 M, pH 7.4. Add 19.2 ml 0.1 M NaH_2PO_4 to 80.8 ml 0.1 M Na_2HPO_4. Mix and check pH.

NaCl, 0.85%

PROCEDURE

1. Centrifuge 1.0 ml anticoagulated blood (or 2.0 ml blood mixed with equal volume of Alsever's solution) in a 12 ml conical tipped test tube. Pour off plasma and wash erythrocytes three times with 0.85% NaCl.
2. To the packed cells add 5 ml water, mix, then add 5 ml phosphate buffer, mix, and centrifuge.
3. Transfer 2 ml of clear supernate into another test tube and place in a 60°C water bath for 30 min or in a 50°C water bath for 1 hour.
4. Set up a normal blood sample in the same manner as a control.
5. Examine visually for a flocculent precipitate.

INTERPRETATION

Normal hemoglobin should show little or no precipitate. Unstable hemoglobins such as Bibba, Bristol, Freiburg, Genova, Köln, Seattle, Sydney, Tacoma, Ube I, and Zurich produce a copious flocculent precipitate.

NOTE

1. Stability of samples. Blood is stable up to 3 days at 4°C (36).

ACCURACY

Hemolysis of the original specimen may cause false positive results (36) and therefore should be avoided.

INCLUSION BODIES TEST FOR HEMOGLOBIN H

(method of Lehman and Huntsman, Ref 205)

PRINCIPLE

Hb-H, Hb-I, and Hb-Bart's cannot be separated by electrophoresis on cellulose acetate at pH 8.4 because of their similar migration rates. However, one can detect the presence of Hb-H in erythrocytes because it denatures easily and forms inclusion bodies when incubated with brilliant cresyl blue dye. Hb-Bart's forms these bodies only occasionally, and Hb-I not at all. The test involves staining a blood smear and examining it under oil immersion lens.

REAGENTS

Sodium Citrate, 3%
NaCl, 0.85%
Sodium Citrate-NaCl. Mix 10 ml 3% sodium citrate with 40 ml 0.85% NaCl.
Brilliant Cresyl Blue Dye, 1%. Dissolve 1 g brilliant cresyl blue dye in 100 ml of sodium citrate-NaCl and store in refrigerator. Just before use mix thoroughly and filter the small amount required through Whatman 42 or 44 paper.

PROCEDURE

1. To 1 ml filtered brilliant cresyl blue dye in a small test tube add 0.1 ml (2 drops) fresh, anticoagulated blood. Set up another tube in the same manner, using normal adult blood as a control.
2. Incubate the tubes at 37°C for 20 min. Prepare two smears on clean microscope slides, dry, and examine under oil immersion lens for inclusion bodies.

INTERPRETATION

Inclusion bodies are numerous small intraerythrocytic masses which are blue in color and which can give the cell the general appearance of a golf ball, with its characteristic pitted pattern. They are found in mature red blood cells and are often peripherally arranged. Normal adult control blood or specimens of blood containing Hb-I do not show these bodies. Hb-Bart's is said to form them only very occasionally.

NOTE

1. Stability of sample. Fresh blood is required for this test.

ACCURACY

Reticulocytes are also stained by this test, but they look denser in color and the inclusions are coarser than in the case of Hb-H. In addition, when reticulocytes are present stained granules of reticulum appear both inside and outside of cells that have ruptured.

As noted above, occasional samples of blood containing Hb-Bart's may show a false positive test. If this is suspected, the two hemoglobins can be differentiated, when necessary, by separation of their polypeptide chains from the hemoglobin molecule and electrophoresis on cellulose acetate (336). Hb-H is β_4, while Hb-Bart's is γ_4.

SCHEME OF LABORATORY AIDS IN THE IDENTIFICATION OF HEMOGLOBINOPATHIES

It is advantageous to set up a systematic approach to the utilization of the tests for identification of hemoglobin abnormalities. Such a system is described in Figure 23-8. It is adapted from one given in reference 296, which is used in courses presented at the Center for Disease Control, US Department of Health, Education, and Welfare. In this system cellulose acetate electrophoresis is used as a primary screening technic, and other tests are used depending on the pattern obtained. The system is designed for screening a population—in contrast, for example, to the situation where a physician is dealing with the anemic sibling of a known heterozygote for Hb-S. In this situation he would be likely to request the Dithionite-Phosphate Screening Test for Hb-S. Electrophoresis has the obvious advantage that it detects a much broader spectrum of hemoglobin abnormalities than any other individual method.

MYOGLOBIN

In 1897 Mörner (228) was first to establish definitely that the muscle pigment is a special substance differing from hemoglobin in the globin fraction. This substance, myoglobin, is considered to serve mainly as a store for O_2 in the muscles.

Whereas a hemoglobin molecule is comprised of four iron atoms and four heme groups and has a molecular weight of 64 458, myoglobin has only one iron and one heme and a molecular weight of about 17 200. This and the fact that myoglobin is not bound to any significant extent by plasma protein account for its much faster urinary excretion rate and its low renal threshold. The mean maximal binding capacity of plasma proteins for myoglobin is about 21 mg/100 ml. The bound fraction migrates electrophoretically between α_2- and β-globulins. At concentrations below maximal binding capacity, hemoglobin is completely bound whereas 15%–50% of myoglobin is free. Hemoglobin interferes with myoglobin binding, suggesting that they are bound in part by the same protein. Spaet *et al* (313) pointed out that the combination of normal-appearing plasma and a benzidine-positive pigment in urine in the absence of hematuria indicates myoglobinuria; conversely, if the plasma is pink or red, hemoglobinuria is indicated because hemoglobin spills into the urine only when the plasma concentration exceeds the hemoglobin-binding capacity—i.e., about 125 mg/100 ml. Urine containing myoglobin or hemoglobin and/or their met derivatives is red to chocolate brown. Absence of fluorescence distinguishes these pigments from a porphyric urine.

Myoglobin separates into three components by paper and starch gel electrophoresis and into five components by chromatography on DEAE-cellulose (257, 284). However, all fractions have proved to be immuno-logically identical, and the addition of urea or 2-mercaptoethanol causes them to migrate electrophoretically as one. Hence molecular heterogeneity is not found in normal myoglobin (365), although abnormal variants exist (30).

The problem of the clinical chemist is to differentiate myoglobin from the hemoglobin and other red pigments in urine. Various procedures which have been used include the following: (a) *Spectrophotometric identification.* The absorption peaks of oxymyoglobin are a few nanometers different from the corresponding oxyhemoglobin peaks (see Table 21-2). The small magnitude of the differences, however, plus the presence of other pigments, plus the fact that the myoglobin in urine may be 80%–90% metmyoglobin and 10%–20% oxymyoglobin make direct spectrophotometric identification unreliable. Complete conversion of pigments to the met form can of course be effected with ferricyanide. Carboxyhemoglobin has a peak at 568 nm; the corresponding peak of carboxymyoglobin is at 578 nm, so identification from this peak is also somewhat better. de Duve (148) and Biorck (23) applied this approach to quantitation of hemoglobin and myoglobin in muscle extracts, reducing pigments with hydrosulfite and converting to the carboxy derivatives with CO. (b) *Ultracentrifugal identification.* Although the sedimentation constants for myoglobin and hemoglobin are about 2.2 and 4.4, respectively, reliance on this difference is difficult because of other urine proteins with similar sedimentation constants. (c) *Salt precipitation.* Based on the fact that myoglobin is soluble in 75%–80% $(NH_4)_2 SO_4$, whereas hemoglobin is not, Blondheim *et al* (26) proposed the following test: First, test for protein with sulfosalicylic acid. If the pigment is precipitated, it is protein: if the filtrate is the color of normal urine, there is no abnormal, nonprotein pigment. This establishes the fact that the pigment is myoglobin or hemoglobin. The heat and acetic acid test cannot be used since neither myoglobin nor hemoglobin is precipitated in this procedure. Differentiation is made by dissolving 2.8 g $(NH_4)_2 SO_4$ in 5.0 ml urine and filtering or centrifuging. If the filtrate is abnormal in color, myoglobin (or metmyoglobin) is present; if the filtrate is normal in color, hemoglobin is present. Duma *et al* (80) presented an instance of myoglo-binuria in which this test was falsely negative (d) *Electrophoresis.* Whisnant *et al* (359) differentiated myoglobin from hemoglobin by paper electrophoresis with pH 8.6 barbital buffer, μ of 0.05. They mixed 1 vol benzidine-positive urine with 2 vol human serum (to prevent absorption to paper) or used serum from the suspected patient directly. Following electrophoresis, they fixed the strips by heating at 125°C for 10 min and stained the strips with benzidine-$H_2 O_2$. Differentiation is made from the fact that myoglobin migrates half the distance of hemoglobin. Marti (215) converted the pigments to the cyanmet forms with ferricyanide and KCN, and used starch block electrophoresis at pH 8.6. Cyanmetmyoglobin and cyanmethemoglobin separate. If the urine pigment concentration was >20 mg/100 ml, the band derived from application of a 50–100 μl stripe was

FIG. 23-8. Scheme of laboratory aids in identification of hemoglobinopathies. Notes: (a) The small rectangles are diagrams of electrophoresis strips resulting from the method "hemoglobin fractionation by electrophoresis on cellulose acetate" (along left margin), or "detection of hemoglobin D and E by electrophoresis on citrate agar" (in body of figure). The anode (+) is on the right. The origin is indicated by the symbol resembling two colons. The bands are labeled below each strip. Hemoglobin variants which migrate together have a + between them; those in separate bands have a comma between them. As far as possible the identifying letters fall below

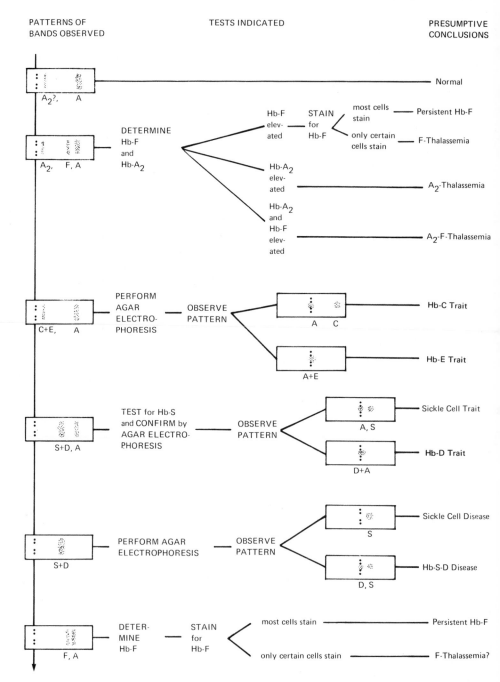

PATTERNS OF BANDS OBSERVED   TESTS INDICATED   PRESUMPTIVE CONCLUSIONS

the proper band. The symbol ? indicates that the variant may be present but the band is too light to observe. (b) The method "dithionite-phosphate screening test for hemoglobin S" may be used to differentiate between Hb-C$_{Harlem}$ or Hb-C$_{Georgetown}$ and other types of Hb-C. (c) "Test for Hb-S" means apply one or both of the methods "test for sickling" and "dithionite-phosphate screening test for Hb-S." (d) See section on accuracy in the method "inclusion bodies test for Hb-H."

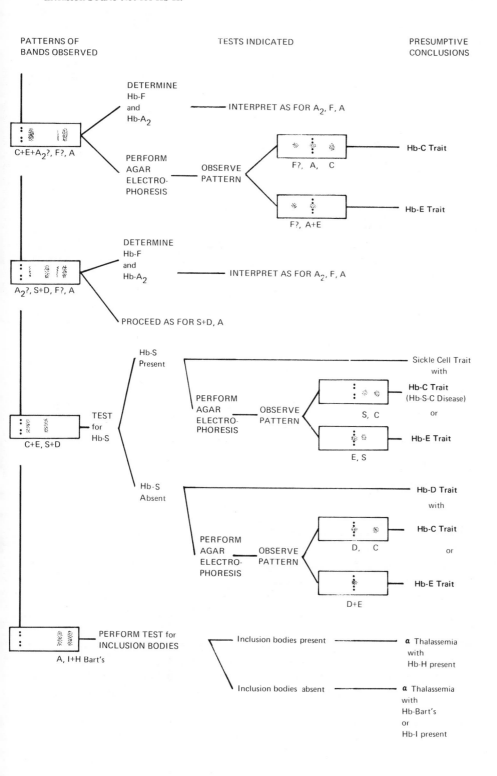

PATTERNS OF BANDS OBSERVED

TESTS INDICATED

PRESUMPTIVE CONCLUSIONS

visible to the eye and quantitation could be made by elution and measurement of absorbance at 540 nm. Benzidine-H_2O_2 was used for lower concentrations. (e) *Differentiation based on difference in molecular size.* A collodion membrane or Millipore filter (pore 9 ±2 nm) permits myoglobin but not hemoglobin to pass. (f) *Immunologic identification* (187). This technic has the advantage of higher sensitivity than any of the aforementioned methods. It can detect myoglobin levels as low as 5 mg/liter in contrast to the 10 mg/liter sensitivity of the spectrophotometric method. It also has the advantage of the specificity associated with antibody binding.

DETERMINATION OF MYOGLOBIN IN URINE OR SERUM

(method of Kagen et al, Ref 187)

PRINCIPLE

The double immunodiffusion technic in agar gel is used for quantitating myoglobin in urine or serum. Serial twofold dilutions of the patient's urine or serum are diffused against an antiserum prepared to react with purified myoglobin. Quantification is accomplished by comparing the specimen with a set of myoglobin standard solutions diffused against the same antiserum. All samples are first screened for pseudoperoxidase activity of hemoglobin or myoglobin using commercial test sticks. Samples with no activity need not be analyzed.

REAGENTS

Barbital Buffer, µ0.05, pH 8.2. Dissolve 7.93 g sodium barbital in about 500 ml water. Add 115 ml 0.1 N HCl and dilute to 1 liter. Check pH.

Anti-Human Myoglobin Serum. Obtainable from Cappel Laboratories Inc.

NaCl, 0.85%

Phosphate Buffer, 0.1 M, pH 6.6. Dissolve 6.53 g K_2HPO_4 and 0.37 g NaEDTA in about 700 ml water and dilute to 1 liter. Check pH.

Hemastix Strips. Ames Co.

Human Myoglobin Standard, 100–150 mg/liter. Two principal sources of human myoglobin are skeletal muscle, obtained from surgical or autopsy specimens, and urine from individuals with myoglobinuria. The product from muscle involves lengthy preparation and contains significant amounts of hemoglobin. Myoglobin from the urine of patients is very satisfactory and is prepared as follows. Select a urine sample 500 ml or more in volume which gives a strongly positive reaction with Hemastix and which comes from a patient with known myoglobinuria. A patient with normal-looking plasma, whose urine does not contain an abnormal number of erythrocytes but does give a positive test with Hemastix, probably is excreting myoglobin. Collect a positive urine sample. Centrifuge it. Concentrate the supernate by ultrafiltration on a Diaflo filter with UM-10 membrane (Amicon Corp.). Wash the retentate several times with 20 ml water and reduce to a volume of 5 ml.

Separate the myoglobin from salts and other proteins by gel filtration, using Sephadex G-75 (Pharmacia Inc.), in a column 2.5 x 100 cm. Place 1 ml of the concentrate on the column and elute with phosphate buffer, pH 6.6. Monitor the effluent by its ultraviolet absorption at 280 nm, and its reaction with Hemastix. Collect the myoglobin fraction. Repeat this step for successive 1 ml aliquots of the retentate until it is used up. Pool the collected myoglobin fractions. Determine the concentration of the myoglobin by direct measurement of its absorbance at 220 nm in a narrow band pass spectrophotometer such as the Beckman DU. At that wavelength, in a 1 cm cuvet, 1 mg of myoglobin per liter gives an absorbance of 0.0132. A check of the antigenic purity of the product can be made by immunologic methods. Adjust the concentration to 100–150 mg/liter with 0.85% NaCl. Divide into 5 ml aliquots and freeze until use. This solution is stable in the frozen state for at least 6 months.

Agar Gel Slides. Add 1 g agar (Ionagar from Colab Laboratories, Inc. was found to be satisfactory) to 100 ml barbital buffer, pH 8.2, in a 250 ml flask. Heat gently to boiling while constantly mixing by rotating the flask until the agar is completely dissolved, then add 100 mg merthiolate and mix. Do not boil longer than 2 min. Spread clean microscope slides on a level surface. Using an automatic dispenser, (Becton-Dickinson Cornwall automatic pipet is satisfactory) dispense immediately 4 ml hot agar onto the center of each slide. Allow to solidify for 30 min.

Cut a trough, 2 x 50 mm in the center of the slide, as shown in Figure 23-9 (Universal Agar Cutter, model 3-1052, from Buchler Instruments, is convenient for this purpose). On each side of the trough cut seven round holes, 2 mm in diameter, equidistant from each other as shown in Figure 23-9, using a perforator such as the one supplied with the agar cutter. This is easily done by placing the slide over a template or drawing (Figure 23-9). Store the slides in a moist chamber (such as a closed box with wet paper towels in it) at room temperature. Slides are stable for 5 days in a moist chamber. Extra agar gel can be aliquoted in 5 ml portions into screw-capped tubes and stored at room temperature for 3 months.

PROCEDURE

1. Centrifuge the specimen.
2. Test the supernate using a Hemastix strip. If the test is positive, proceed with following steps. If not, report as negative (< 4 mg/liter).

FIG. 23-9. Agar gel slide for determination of myoglobin.

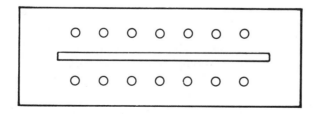

3. Fill the center trough with antimyoglobin antiserum and allow to diffuse for about 2 hours.

4. Prepare serial twofold dilutions of urine or serum with 0.85% NaCl up to 1:64.

5. Prepare serial dilutions of the standard in the same manner.

6. Using a single 50 μl capillary pipet and starting with the most dilute, fill wells 1 through 7 with the serial dilutions of standard. Using a different pipet, fill wells 8 through 14 on the opposite side with the urine or serum dilutions. Fill carefully to avoid spilling any sample over the edge of the holes.

7. Set the slides in a moisture chamber (closed box with wet paper towel) at room temperature and read at 24 and 48 hours.

Calculations:

Multiply the highest dilution of urine or serum which gives a precipitin line by the lowest concentration of human myoglobin standard which gives a precipitin line. Example: If 4 mg/liter is the smallest amount of myoglobin that gives a precipitin line and if the highest dilution of urine or serum that gives a precipitin line is 1:16, then the myoglobin concentration in the urine or serum is 4 x 16 = 64 mg/liter.

ACCURACY AND PRECISION

The accuracy of the method depends upon the specificity of the anti-human myoglobin antiserum. Acceptable material must have been checked (a) for absence of cross reactivity by immunoelectrophoresis against normal human serum, (b) for insensitivity to hemoglobin by double immunodiffusion against hemoglobin from human blood at the level of at least 1 g/100 ml, and (c) for sensitivity to myoglobin by quantitative double immunodiffusion against known standards of pure human myoglobin. Under these conditions no other substance is known to react except primate myoglobin, which is immunologically indistinguishable from human myoglobin.

The precision of the method is ± a twofold dilution.

NORMAL RANGE

Under 4 mg/liter.

HAPTOGLOBIN

Haptoglobins are a group of closely related glycoproteins with α_2 mobility at pH 8.6 in barbital buffer and with high affinity for globin whether free or in combination with heme compounds. They do not bind heme or myoglobin (148, 266). Haptoglobin is presumably synthesized in the liver (264), and the haptoglobin-hemoglobin complex is destroyed via the reticuloendothelial system as the first step in one mode of hemoglobin degradation (111). Haptoglobin prevents the buildup of free hemoglobin in plasma by binding that which is released from the red cells, thereby preventing loss of iron via the kidney. It also stabilizes hemoglobin by blocking the free exchange of heme between methemoglobin and albumin. Thus in any hemolytic situation one observes neither hematuria nor methemalbuminemia until the plasma hemoglobin level has exceeded the binding capacity of the plasma haptoglobin. Conversely, in cases of prolonged release of hemoglobin, plasma haptoglobin levels fall virtually to zero as its rate of binding and removal exceeds its rate of synthesis (37). The role of haptoglobin in hemoglobin catabolism has been questioned by Keene and Javell (188), and it has been calculated that at most only one-fifth of the total hemoglobin could be involved (111).

Haptoglobin contains about 20% carbohydrate (fucose, hexoses, glucosamine, sialic acid) and a variety of protein subunits which occur in three main types of genetically controlled combinations: types 1-1, 2-2, and 2-1. Type 1-1 is the simplest. It binds hemoglobin molecules $(\alpha_2\beta_2)$ and forms molecules, each of which contains one molecule of hemoglobin and one of haptoglobin. It also binds half-molecules of hemoglobin $(\alpha\beta)$ in a compound with one half-molecule of hemoglobin per molecule of haptoglobin. Type 2-2 forms haptoglobin polymers with two, three, or more units per molecule. It binds hemoglobin at various degrees of saturation, and in various ratios. Type 2-1 can be divided into five or six electrophoretic components which bind hemoglobin in various ratios. One of these has the same mobility as type 1-1 (211). These three types correspond to three genotypes produced by a single pair of allelic genes. There are multiple subtypes, and a third gene of type O has been postulated to explain the complete absence of haptoglobin in some population groups (111). Giblett (110) reports that in addition to the three common types of haptoglobin there are at least 11 variant phenotypes. Waks and Alfsen (350) studied the subunit structures of types 1-1 and 2-2 and found reversible dissociation as a function of pH or protein concentration.

The major role of haptoglobin may be to conserve hemoglobin and therefore Fe (111, 306). Haptoglobin levels are frequently elevated in inflammatory diseases and depressed following episodes of intravascular hemolysis (37, 245).

According to Javid *et al* (181) the finding of a low haptoglobin level after an episode of pigmenturia is almost diagnostic of hemoglobinuria, whereas a normal haptoglobin level within 3 days of the episode strongly suggests myoglobinuria or excretion of a nonheme pigment. Nyman (245) reviewed the clinical aspects of haptoglobin.

METHODS FOR THE DETERMINATION OF HAPTOGLOBIN

Methods for the routine determination of haptoglobin fall into several groups: (a) measurement of the peroxidase activity of the haptoglobin-hemoglobin complex, (b) electrophoresis of haptoglobin and the complex,

(c) gel filtration, (d) immunodiffusion, and (e) differential acid denaturation of hemoglobin and the complex, followed by spectrophotometry.

METHODS BASED ON PEROXIDASE ACTIVITY

In the method of Jayle (see ref 245) hemoglobin is added in excess of that required to saturate haptoglobin, the excess inactivated by adding a small amount of iodine, and the peroxidase activity of the haptoglobin-hemoglobin complex determined using ethyl hydroperoxide as substrate in the following reaction:

$$C_2 H_5 OOH + 2I^- + 2H^+ \rightarrow C_2 H_5 OH + I_2 + H_2 O$$

The I_2 released is determined by thiosulfate titration. Under the conditions of the test, the haptoglobin complex has about four times the peroxidase activity of free hemoglobin. The optimal pH for the peroxidase activity of hemoglobin is 5.6, whereas the optimum is 4.1–4.4 for the haptoglobin complex. The Jayle method has also been studied in detail by Nyman (245).

Connell and Smithies (60) and Owen *et al* (247) saturated haptoglobin by adding excess methemoglobin and determined the peroxidase activity of the haptoglobin-methemoglobin complex by photometric measurement of the tetraguaiacol formed from guaiacol and $H_2 O_2$. Under the conditions used, free methemoglobin has 10% or less of the peroxidase activity of the haptoglobin-methemoglobin complex.

Owen's method was criticized and made more reproducible by Veneziale and McCruckin (347), and the peroxidase method was modified for the AutoAnalyzer by Moretti *et al* (227) and by Lupovitch and Katase (212). Tarukoski (329) presents a method that employs *o*-dianisidine as chromogen and depends upon relative inactivation of free hemoglobin to avoid the necessity for a separation.

METHODS EMPLOYING ELECTROPHORESIS

The hemoglobin-binding capacity of haptoglobin separated by electrophoresis is determined. The supports used have included paper (245), cellulose acetate (55, 125, 337), acrylamide gel (87, 294), and starch blocks (66). Several approaches to this analysis have been made: (a) Laurell and Nyman (245) added hemoglobin to serum in a series of increasing quantities and then they employed a pH 7 phosphate buffer, under which conditions free hemoglobin migrates toward the cathode and the haptoglobin-hemoglobin complex toward the anode. The electrophoretic strips were stained with leucomalachite green in dilute acetic acid and $H_2 O_2$, hemoglobin staining green because of its conversion of the leuco dye. The concentration of haptoglobin, in terms of hemoglobin-binding capacity, lies between the hemoglobin concentration in the mixture just showing free hemoglobin and the hemoglobin concentration in the adjacent mixture producing no detectable free hemoglobin. Benzidine *o*-dianisidine, guaiacol, or *o*-tolidine can also be used for staining. It has been recommended (150) that hemoglobin C be used rather than hemoglobin A to obtain a more distinct separation of free and bound hemoglobin. (b) A known amount of hemoglobin (in excess of that required for saturation of haptoglobin) can be added, the strip stained following electrophoresis, and the relative amount of free hemoglobin and haptoglobin-hemoglobin complex determined by scanning (see *Determination of Haptoglobin in Serum,* below). (c) Following saturation of haptoglobin by an excess of hemoglobin, the haptoglobin-hemoglobin complex can be separated by paper electrophoresis, eluted, and quantitated with guaiacol (any other peroxidase reaction could be used). (d) The protein content of the haptoglobin area can be determined turbidimetrically with tannin, with and without hemoglobin added to the sample. (e) ^{59}Fe-labeled hemoglobin can be added. Then electrophoresis is carried out and the haptoglobin area cut out and counted.

Owen *et al* (248) compared various substrates of the peroxidase reaction and recommend *o*-dianisidine for the reaction on paper. The minimal concentration of hemoglobin detectable on paper was found to be 0.05, 0.10, 0.20, 0.40, >0.80, >0.80, and >0.80 $\mu g/cm^2$ for *o*-dianisidine, *o*-tolidine, benzidine, *p*-anisidine, amidopyrine, leucomalachite green, and guaiacol, respectively. Furthermore, the color with *o*-dianisidine was stable, the color with benzidine and *p*-anisidine faded in minutes, and *o*-tolidine gave a marked background color.

METHODS INVOLVING GEL FILTRATION

Lionetti *et al* (210) used a column of dextran gel to effect complete separation of bound and free hemoglobin in samples to which excess hemoglobin had been added. Absorbance measurements of both the free and bound fractions at 414 nm were used for quantitation.

METHODS USING IMMUNOLOGIC REACTIONS

Kluthe *et al* (195) used a method wherein the product of the reaction between haptoglobin and antibodies to it is measured by nephelometry, and pointed out its advantages over previous methods. However, Braun and Aly (34) showed that the existence of the various phenotypes of haptoglobin caused inaccuracies in immunologic methods, and specifically in radial immunodiffusion. In 1968 they showed that accuracy could be achieved for any sample provided the phenotypes it contained were known (4).

METHODS USING DIFFERENTIAL ACID DENATURATION

In dilute acid the hemoglobin-haptoglobin complex is more stable than hemoglobin alone, which forms acid hematin. This has been exploited in methods for determining haptoglobin. Roy *et al* (285) showed that at pH 3.7 the absorbance of free hemoglobin falls to half its original value in about 15 min, while that of hemoglobin complexed with haptoglobin does not, and they based a simple method on this fact. Tarukoski (329) used a similar technic but measured the peroxidase activity of the hemoglobin-haptoglobin complex.

COMPARISON OF METHODS

Direct comparison of the peroxidase methods has shown satisfactory agreement. Obviously, the electrophoretic methods must yield results lower than peroxidase methods by an amount equal to the plasma or serum hemoglobin concentration existing prior to the test. Hemolysis must therefore be avoided when using paper electrophoretic methods or else plasma or serum hemoglobin must be determined as part of the procedure. According to Nyman (245) the difference is no more than 5 mg/100 ml if the serum is nonhemolyzed to the naked eye, except in the case of an icteric serum for which an analysis for hemoglobin content must be made and the hemoglobin concentration added to the paper electrophoretic result for haptoglobin. Hemolysis in normal serum, however, may not become visible to the naked eye until the hemoglobin concentration exceeds 50 mg/100 ml and the degree of hemolysis occurring with collection of serum certainly may exceed 5 mg/100 ml. Use of properly collected serum or plasma (see earlier section, *Plasma Hemoglobin*) would seem to be a safer procedure to follow.

Comparison of the gel filtration and the spectrophotometric methods with peroxidase methods show good agreement (210, 329), but as mentioned above the immunologic methods are said to require knowledge of genotype for accuracy.

In our laboratory the most satisfactory method has proved to be a modification of the electrophoretic separation technic (180), using peroxidase activity to measure the bands.

DETERMINATION OF HAPTOGLOBIN IN SERUM

(method of Wybenga, Ibbott, and Pileggi, unpublished)

PRINCIPLE

A serum sample is spiked with a known amount of free hemoglobin in excess of the binding capacity of the haptoglobin present, and the mixture is allowed to incubate for 30 min at room temperature. The hemoglobin-haptoglobin complex and the free hemoglobin representing the excess added hemoglobin are separated electrophoretically on cellulose acetate strips, stained, and quantitated by densitometry. The hemoglobin-binding capacity is used as an indirect measure of haptoglobin concentration in serum.

REAGENTS AND ELECTROPHORESIS EQUIPMENT

o-Dianisidine Stock Stain. Add 1.43 g *o*-dianisidine to an Erlenmeyer flask containing 1 liter of ethanol (ethanol denatured with methanol is satisfactory). Wrap flask with aluminum foil to protect *o*-dianisidine from light and stir with a magnetic stirrer until stain is completely dissolved (about 30 min).

Phosphate Buffer, 0.05 M, pH 7.0. Dissolve 4.35 g Na_2HPO_4 and 2.64 g KH_2PO_4 in about 800 ml water. Dilute to 1 liter and check pH.

Acetate Buffer, 1.5 M, pH 4.7. Dissolve 102.2 g $NaC_2H_3O_2 \cdot 3H_2O$ in about 500 ml water. Add 43.5 ml glacial acetic acid and dilute to 1 liter. Check pH.

Hydrogen Peroxide, 3%. Dilute 1 vol of Superoxol (30% H_2O_2) to 10 vol with water. Prepare fresh just before use.

Methanol, Anhydrous

Clearing Solution. Mix 770 ml anhydrous methanol with 200 ml glacial acetic acid and 30 ml ethylene glycol. Prepare fresh daily.

o-Dianisidine Working Stain. Mix 140 ml *o*-dianisidine stock stain with 20 ml acetate buffer, 4 ml 3% H_2O_2, and 36 ml water. Pour immediately into a staining dish covered with aluminum foil. Prepare fresh after every 36 strips.

Hemolysate, 7.5%. Hemolysate is prepared as described under *Collection of Blood Sample and Preparation of Hemolysate,* this chapter. Adjust the hemoglobin concentration to 7.5% with water. Mix thoroughly and filter through Whatman 42 filter paper. Determine the exact hemoglobin concentration by running it in by cyanmethemoglobin method. Aliquot into small portions and freeze.

Electrophoresis Equipment. The following equipment or its equivalent is used: Heathkit power supply model IP-17; Photovolt integrating densitometer model 552 D with 445 nm filter; Photovolt model 55 electrophoresis chamber with strip rack and applicator guide; Photovolt No. 2555 sample applicator, No. 2562 strip basket, No. 2554 sample pick-up tray, and No. 2560 cellulose acetate strips.

PROCEDURE

1. To 0.2 ml of serum which shows no sign of hemolysis, in a small test tube, add 10 μl of the 7.5% hemolysate. Mix gently and allow to stand at room temperature for 30 min. If serum shows hemolysis, measure serum hemoglobin as in determination of plasma hemoglobin and incorporate this result into the calculation.

2. Fill electrophoresis chamber with 0.05 *M* phosphate buffer pH 7.0 (about 600 ml) and make sure it is level.

3. Place cellulose acetate strips (25 x 75 mm) in the strip basket and immerse slowly into dish containing phosphate buffer, pH 7.0, for 5–10 min. Remove strips from buffer and blot between filter paper wicks or paper towels to remove excess buffer.

4. Load soaked cellulose acetate strips on the chamber strip rack with the acetate side up. Cover electrophoresis chamber, turn power supply on, and allow strips to equilibrate for 3 min at 110 V.

5. Turn off power supply and insert the sample applicator guide into the strip rack. Prime sample applicator with 5 μl sample and insert into sample applicator guide and leave in place for about 2 min. Apply samples to the five remaining strips in the same manner. Sample application on the six strips should take not more than 5 min.

6. Cover electrophoresis chamber, turn power supply on, and allow electrophoresis to proceed for 20 min at 110 V.

7. Remove rack of strips from chamber, and immerse in anhydrous methanol for 2 min (replace methanol after every 36 strips).

8. Transfer strips from methanol to o-dianisidine working stain and allow to stain for 10 min (change working stain after every 36 strips).

9. Wash strips in three consecutive dishes of water by immersing for 5 min in each dish, gently agitating every 3 min. Change water when color becomes intense.

10. Immerse strips in clearing solution for exactly 1 min. Change the clearing solution when it shows a yellow tint.

11. Dry strips in dry heat oven at 70°C for about 8 min and scan on the densitometer using a 445 nm filter. Two major bands should be visible on the strip: the free hemoglobin band close to the origin and the hemoglobin-haptoglobin band migrating further away from the origin.

12. If on scanning the free hemoglobin band is not prominent, insufficient hemoglobin was added. Repeat the procedure from step 1 using 20 μl instead of 10 μl of hemoglobin and use 2 G instead of G in the calculation. If the band is again not prominent, repeat using 50 μl and 3 G.

13. The numbers obtained from the integrating densitometer represent the areas under the free hemoglobin and the hemoglobin-haptoglobin peaks, and they are used in the calculations.

Calculation:

mg haptoglobin/100 ml serum, as hemoglobin binding capacity

$$= \frac{area_{\text{Hb-bound}}}{area_{\text{Hb-free}} + area_{\text{HB-bound}}} \times F$$

where

$$F, \text{ in hemolysis-free serum} = G \times \frac{100}{0.2}$$

$$F, \text{ in serum with hemolysis} = G \times \frac{100}{0.2} + H$$

$$G = \text{mg Hb in 10 μl of the 7.5\% hemolysate}$$

$$H = \text{mg Hb/100 ml of sample showing hemolysis}$$

NOTES

1. Stability. Serum samples were found in our laboratory to be stable for 3 days at 30°C, 7 days at 4°C, and 1 year at −15°C (245).

2. Interference. On rare occasions abnormal serum proteins may be present which grossly change the appearance of the peak for either the bound or unbound fraction. In such cases this method cannot be used for determination of haptoglobin.

3. Methemalbumin. A peak due to methemalbumin is observed on the electrophoresis strip ahead of both the bound and free fractions. It is irrelevant to this determination.

ACCURACY AND PRECISION

Results by this method show excellent correlation with the peroxidase technic of Nyman (245) between 0 and 400 mg haptoglobin/100 ml serum. The precision (95% limits) of the method is ±10%.

NORMAL VALUES

Using nonparametric percentile estimation on data from 100 workers at our laboratory, the normal range (95% limits) was found to be 21−195 mg/100 ml. Other workers have found similar limits (245).

REFERENCES

1. ABER GM, ROWE DS: *Br J Haematol 6*:160, 1960
2. ADDIS T: *J Clin Invest 2*:409, 1926
3. ADLER O, ADLER R: *Hoppe Seylers Z Physiol Chem 41*:59, 1904
4. ALY FW, BRAUN HJ: *Klin Wochenschr 46*:385, 1968
5. AUSTIN JH, DRABKIN DL: *J Biol Chem 112*:67, 1935
6. BACHMANN F, MARTI HR: *Blood 20*:272, 1962
7. BAGINSKI ES, FOA PP, SUCHOCKA SM, ZAK B: *Microchem J 14*:293, 1969
8. BAGLIONI C: *Nature (Lond) 198*:1177, 1963
9. BEAVEN GH, ELLIS MJ, WHITE JC: *Br J Haematol 6*:1, 1960
10. BEAVEN GH, ELLIS MJ, WHITE JC: *Br J Haematol 7*:169, 1961
11. BEAVEN GH, GRATZER WB: *J Clin Pathol 12*:1, 1959
12. BELL GH, CHAMBERS JW, WADDELL MBR: *Biochem J 39*:60, 1945
13. BELLINGHAM AJ, HUEHNS ER: *Nature (Lond) 218*:924, 1968
14. BENESCH R, BENESCH RE, YU CI: *Proc Natl Acad Sci USA 59*:526, 1968
15. BERETTA A, PRATO V, GALLO E, LEHMANN H: *Nature (Lond) 217*:1016, 1968
16. BERGREN WR: *Children's Hospital*, Los Angeles, personal communication
17. BERZOFSKY JA, PEISACH J, BLUMBERG WE: *J Biol Chem 246*:3367, 1971
18. BERZOFSKY JA, PEISACH J, BLUMBERG WE: *J Biol Chem 246*:7366, 1971
19. BETKE K: *Biochem Z 321*:271, 1951
20. BETKE K, MARTI HR, SCHLICHT I: *Nature (Lond) 184*:1877, 1959
21. BIANCO I, MODIANO G, BOTTINI E, LUCCI R: *Nature (Lond) 198*:395, 1963
22. BING FC, BAKER RW: *J Biol Chem 92*:589, 1931

23. BIÖRCK G: *Acta Med Scand (Suppl) 226*:1, 1949
24. BIRD GWG: *Br Med J 1*:363, 1972
25. BLAKE TM, MERRILL JM, CALLAWAY JJ: *J Lab Clin Med 46*:900, 1955
26. BLONDHEIM SH, MARGOLIASH E, SHAFRIR E: *JAMA 167*:453, 1957
27. BOERMA FW, HUISMAN THJ, MANDEMA E: *Clin Chim Acta 5*:564, 1960
28. BOIVIN P, HUGOU MP, HARTMANN L: *Ann Biol Clin (Paris) 17*:193, 1959
29. BOTHA MC, BEALE D, ISAACS WA, LEHMANN H: *Nature (Lond) 212*:792, 1966
30. BOULTON FE, HUNTSMAN RG, LEHMANN H, LORKIN P, ROMERO-HERRERA AE: *Biochem J 119*:69P, 1970
31. BOWEN WJ: *J Biol Chem 179*:235, 1949
32. BRADLEY TB, RIEDER RF: *Blood 28*:975, 1966
33. BRADLEY TB, WOHL RC, RIEDER RF: *Science 157*:1581, 1967
34. BRAUN HJ, ALY FW: *Clin Chim Acta 26*:588, 1969
35. BRIERE RO, TIPTON G, BATSAKIS JG: *Am J Clin Pathol 44*:695, 1965
36. BROSIOUS E: Center for Disease Control, Atlanta, Ga, personal communication
37. BURS I, LEWIS SM: *Br J Haematol 5*:348, 1959
38. BYWATERS EGL, DELORY GE, RIMINGTON C, SMILES J: *Biochem J 35*:1164, 1941
39. CAMERON BF: *Anal Biochem 11*:164, 1965
40. CARAWAY WT: *Am J Clin Pathol 37*:445, 1962
41. CARRELL RW, LEHMANN H, HUTCHISON HE: *Nature (Lond) 210*:915, 1966
42. CARRELL RW, LEHMANN H, LORKIN PA, RAIK E, HUNTER E: *Nature (Lond) 215*:626, 1967
43. CATHIE IAB: *Gr Ormond St J 2*:43, 1952
44. CAWEIN MJ, LAPPET EJ, BRANGLE RW, FARLEY CH: *Ann Intern Med 64*:62, 1966
45. CHAPLIN H Jr, CASSELL M, HANKS GE: *J Lab Clin Med 57*:612, 1961
46. CHARACHE S, MONDZAC AM, GESSNER U, GAYLE EE: *J Clin Invest 48*:834, 1969
47. CHARACHE S, WEATHERALL DJ, CLEGG JB: *J Clin Invest 45*:813, 1966
48. CHERNOFF AI: *Blood 8*:399, 413, 1953
49. CHERNOFF AI: *Blood 13*:116, 1958
50. CHERNOFF AI: *Am J Hum Genet 13*:151, 1961
51. CHERNOFF AI, MINNICH V, NA-NAKORN S, TUCHINDA S, KASHEMSANT C, CHERNOFF RR: *J Lab Clin Med 47*:455, 1956
52. CHIAMORI N, HENRY RJ, GOLUB OJ: *Clin Chim Acta 6*:1, 1961
53. CHI CHIN LIANG: *Biochem J 66*:552, 1957
54. CHILDERS DM, BARGER JD: *Am J Clin Pathol 29*:546, 1958
55. COLFS B, VERHEYDEN J: *Clin Chim Acta 12*:470, 1965
56. COMINGS DE: The hemoglobinopathies and thalassemia. *Genetic Disorders in Man.* Edited by RM Goodman. Boston, Little, Brown, 1969
57. COMINGS DE: *Hematology.* Edited by WJ Williams, E Beutler, AJ Erslen, RW Rindles. New York, McGraw-Hill, 1972, p 328
58. COMINGS DE: *Hematology.* Edited by WJ Williams, E Beutler, AJ Erslen, RW Rindles. New York, McGraw-Hill, 1972, p 413
59. CONLEY CL, WEATHERALL DJ, RICHARDSON SN, SHEPHARD MK, CHARACHE C: *Blood 21*:261, 1963
60. CONNELL GE, SMITHIES O: *Biochem J 72*:115, 1959
61. CONNERTY HV, BRIGGS AR: *Clin Chem 8*:151, 1962
62. COOK MH, FREE HM, FREE AH: *Am J Med Technol 22*:218, 1956
63. COX WW, WENDELL WB: *J Biol Chem 143*:331, 1942

64. CREDITOR MC: *J Lab Clin Med 41*:307, 1953
65. CROOKSTON JH, FARQUHARSON HA, BEALE D, LEHMANN H: *Can J Biochem 47*:143, 1969
66. CULLIFORD BJ: *Nature (Lond) 198*:796, 1963
67. CULLIS JE: *J Clin Pathol 12*:486, 1959
68. DACIE JV, GRIMES AJ, MEISLER A, STEINGOLD L, HEMSTED EH, BEAVEN GH, WHITE JC: *Br J Haematol 10*:388, 1964
69. DACIE JV, SHINTON NK, GAFFNEY PJ, CARRELL RW, LEHMANN H: *Nature (Lond) 216*:603, 1967
70. DALAND GA, CASTLE WB: *J Lab Clin Med 33*:1082, 1948
71. DARE A: *Philadelphia Med J 6*:557, 1900
72. DAVIS GE, SHEARD C: *Arch Intern Med 40*:226, 1927
73. DAWES RLF, PARK C: *J Med Lab Technol 27*:55, 1970
74. DeJONG WWW, WENT LN, BERNINI LF: *Nature (Lond) 220*:788, 1968
75. DIGGS LW, BELL A: *Blood 25*:218, 1965
76. DIGGS LW, KRAUS AP, MORRISON DB, RUDNICKI RPT: *Blood 9*:1172, 1954
77. DONALDSON R, SISSON RB, KING EJ, WOOTTON IDP, MacFARLANE RG: *Lancet 1*:874, 1951
78. DRABKIN DL, AUSTIN JH: *J Biol Chem 98*:719, 1932
79. DRABKIN DL, AUSTIN JH: *J Biol Chem 112*:51, 1935
80. DUMA RJ, TRIGG JW, HAMMACK WJ: *Ann Intern Med 56*:97, 1962
81. EDINGTON GM, LEHMAN H: *Br Med J 1*:1308; *2*:1328, 1955
82. EILERS RJ: *Am J Clin Pathol 47*:212, 1967
83. EVELYN KA, MALLOY HT: *J Biol Chem 126*:655, 1938
84. FAIRBANKS VE, OPFELL RW, BURGERT EO: *Am J Med 46*:344, 1969
85. FEIGL F: *Qualitative Analysis by Spot Tests.* Amsterdam, Elsevier, 1946, p 55
86. FERRIS TG, EASTERLING RE, BUDD RE: *Blood 19*:479, 1962
87. FERRIS T. EASTERLING RE, NELSON K, BUDD R: *Am J Clin Pathol 46*:385, 1966
88. FESSAS P: *Abnormal Hemoglobins: A Symposium.* Edited by JHP Jonxis, JF Delafresnaye. Oxford, 1959, p 134
89. FESSAS P, MASTROXALOS N, FOSTIROPOULOS G: *Nature (Lond) 183*:30, 1959
90. FESSAS P, STAMATOYANNOPOULOS G: *Blood 24*:223, 1964
91. FESSAS P, STAMATOYANNOPOULOS G, KARALIS A: *Blood 19*:1, 1962
92. FETTER MC, FREE AH: *Am J Med Technol 28*:135, 1962
93. FLEISCH H: *Helv Med Acta 27*:383, 1960
94. FLINK EB, WATSON CJ: *J Biol Chem 146*:171, 1942
95. FRAIMOW W, RODMANN T, CLOSE HP, CATHCART R, PURCELL MK: *Am J Med Sci 236*:225, 1958
96. FRASER GR, STAMATOYANNOPOULOS G, KATTAMIS C, LOUKOPOULOS D, DEFARANAS B, KITSOS C, ZANNOS-MARIOLEA L, CHOREMIS C, FESSAS P, MOTULSKY AG: *Ann NY Acad Sci 119*:415, 1964
97. FREE HM, FREE AH, GIORDANO AS: *J Urol 75*:743, 1956. *CA 50*:9483, 1956
98. FRICK PG, HITZIG WH, BETKE K: *Blood 20*:261, 1962
99. FRIEDMAN HS: *Clin Chim Acta 7*:100, 1962
100. GABUZDA TG, NATHAN DG, GARDNER FH: *J Clin Invest 42*:1678, 1963
101. GAENSLER EA, CADIGAN JB Jr, ELLICOTT MF, JONES RH, MARKS A: *J Lab Clin Med 49*:945, 1957
102. GAMBINO SR, WARAKSA AJ: *Am J Clin Pathol 52*:557, 1969
103. GEORGE P, HANANIA G: *Biochem J 52*:517, 1952
104. GERALD PS: *Blood 13*:936, 1958

105. GERALD PS, COOK CD, DIAMOND LK: *Science 126*:300, 1957
106. GERALD PS, EFRON ML: *Proc Natl Acad Sci (USA)* 47:1758, 1961
107. GERALD PS, GEORGE P: *Science 129*:393, 1959
108. GETTLER AO, KAYE S: *Am J Clin Pathol Techn Sec 7*:77, 1943
109. GETTLER AO, MATTICE MR: *JAMA 100*:92, 1933
110. GIBLETT ER: *Cold Spring Harbor Symp Quant Biol 29*:321, 1965
111. GIBLETT ER: *Ser Haematol 1*:3, 1968
112. GIBSON QH: *Biochem J 42*:13, 1948
113. GILLESPIE JE, WHITE JC, ELLIS MJ, BEAVEN GH, GRATBER WB, SHOOTER EM, PARKHOUSE RME: *Nature (Lond) 184*:1876, 1959
114. GOLDBERG CAJ: *Clin Chem 4*:146, 1958
115. GOLDBERG CAJ: *Clin Chem 4*:484, 1958
116. GOLDBERG CAJ: *Clin Chem 5*:446, 1959
117. GOLDBERG CAJ: *Clin Chem 6*:254, 1960
118. GOLDBERG CAJ: *Standard Methods of Clinical Chemistry*. Edited by D Seligson. New York, Academic Press, 1961, Vol 3, p 131
119. GOWERS WR: *Lancet 2*:879, 1878
120. GREEN P, TEAL CFJ: *Am J Clin Pathol 32*:216, 1959
121. GREENBERG MS, HARVEY HA, MORGAN C: *N Engl J Med 286*:1143, 1972
122. GREENDYKE RM, MERIWETHER WA, THOMAS ET, FLINTJER JD, BAYLISS MW: *Am J Clin Pathol 37*:429, 1962
123. GREGERSON JP: *Arch Verdaungskr 25*:169, 1919
124. GRIMES AJ, MEISLER A, DACIE JV: *Br J Haematol 10*:281, 1964
125. GRUNBAUM BW, PACE N: *Microchem J 8*:317, 1964
126. HADEN RL: *J Lab Clin Med 16*:68, 1930
127. HADEN RL: *J Lab Clin Med 18*:1062, 1933
128. HADEN RL: *Am J Clin Pathol 3*:85, 1933
129. HAHN EV, GILLESPIE EB: *Arch Intern Med 39*:233, 1927
130. HAINLINE A: *Standard Methods of Clinical Chemistry*. Edited by D Seligson. New York, Academic Press, 1958, Vol 2, p 56
131. HALBRECHT I, ISAACS WA, LEHMANN H, BEN-PORAT F: *Israel J Med Sci 3*:827, 1967
132. HAMBLIN DO, MANGELSDORFF AF: *J Ind Hyg Toxicol 20*:523, 1938
133. HAMILTON HB, IUCHI I, MIYAJI T, SHIBATA S: *J Clin Invest 48*:525, 1969
134. HANKS GE, CASSELL M, RAY RN, CHAPLIN H Jr: *J Lab Clin Med 56*:486, 1960
135. HANKS GE, CHAPLIN H: *Fed Proc 18*:479, 1959
136. HARBOE M: *Scand J Clin Lab Invest 9*:317, 1957
137. HARBOE M: *Scand J Clin Lab Invest 11*:66, 1959
138. HARBOE M: *Scand J Clin Lab Invest 11*:138, 1959
139. HARTMANN H: *Ergeb Physiol 39*:413, 1937
140. HAUROWITZ F: *Hoppe Seylers Z Physiol Chem 138*:68, 1924
141. HAVEMANN R, JUNG F, VON ISSEKUTZ B Jr: *Biochem Z 301*:116, 1939
142. HAYASKI N, MOTOKAWA Y, KIKUCHI G: *J Biol Chem 241*:79, 1966
143. HAYASKI A, SUZUKI T, SHIMIZU A, IMAI K, MORIMOTO H, MIYAJI T, SHIBATA S: *Arch Biochem Biophys 125*:895, 1968
144. HED R: *Acta Med Scand 151* Suppl 303, 1955
145. HEILMEYER L: *Spectrophotometry in Medicine*. London, Adam Hilger Ltd, 1943
146. HELLER P, COLEMAN RD, YAKULIS VJ: *J Clin Invest 45*:1021, 1966
147. HELLER P, YAKULIS VJ, JOSEPHSON AM: *J Lab Clin Med 59*:401, 1962
148. HENRY RJ: *Clinical Chemistry: Principles and Technics*. New York, Harper & Row, 1964

149. HEPLER OE, WONG P, PIHL HD: *Am J Clin Pathol 23*:1263, 1953
150. HERMAN EC Jr: *J Lab Clin Med 57*:825, 1961
151. HERRICK JP: *Arch Intern Med 6*:517, 1910
152. HEUBNER W, STUHLMANN M: *Arch Exp Pathol Pharmakol 199*:1, 1942
153. HILL PJ, CRAIG LC: *J Am Chem Soc 81*:2272, 1959
154. HOLLÁN SR, SZELÉNYI JG, MILTÉNYIM, CHARLESWORTH D, LORKIN PA, LEHMANN H: *Haematologia (Budap) 4*:141, 1970
155. HOLLENDER A, LORKIN PA, LEHMANN H, SVENSSON B: *Nature (Lond) 222*:953, 1969
156. HORECKER BL: *J Biol Chem 148*:173, 1943
157. HORECKER BL, BRACKETT FS: *J Biol Chem 152*:669, 1944
158. HÖRLEIN A, WEBER G: *Dtsch Med Wochenschr 73*:476, 1948
159. HORVATH SM, ROUGHTON FJW: *J Biol Chem 144*:747, 1942
160. HUEHNS ER: *Ann Rev Med 21*:157, 1970
161. HUEHNS ER, HECHT F, YOSHIDA A, STAMATOYANNOPOULOS G, HARTMAN J, MOTULSKY AG: Cited by Huehns ER: *Ann Rev Med 21*:157, 1970
162. HUEHNS ER, SHOOTER EM: *J Med Genet 2*:48, 1965
163. HÜFNER G, *Arch Anat Physiol Abt* 1900, p 39
164. HUISMAN THJ: *Adv Clin Chem 15*:149, 1972
165. HUISMAN THJ, SCHROEDER WA: *Crit Rev Clin Lab Sci 1*:471, 1970
166. HUISMAN THJ, SCHROEDER WA, DOZY AM, SHELTON JR, SHELTON JB, BOYD EM, APELL G: *Ann NY Acad Sci 165*:320, 1969
167. HUNT JA, LEHMANN H: *Nature (Lond) 184*:872, 1959
168. HUNTER FT: *Quantitation of Mixtures of Hemoglobin Derivatives by Photoelectric Spectrophotometry.* Springfield, Ill, Charles C Thomas, 1951
169. HUNTER FT, GROVE-RASMUSSEN M, SOUTTER L: *Am J Clin Pathol 20*:429, 1950
170. HUNTSMAN RG, LIDDELL J: *J Clin Pathol 14*:436, 1961
171. HUTCHINSON EB: *Am J Med Technol 26*:75, 1960
172. HUTCHISON HE, PINKERTON PH, WATERS P, DOUGLAS AS, LEHMANN H, BEAGLE D: *Br Med J II*:1099, 1964
173. ITANO HA: *Proc Natl Acad Sci (USA) 37*:775, 1951
174. ITANO HA: *Arch Biochem 47*:148, 1953
175. ITANO HA, ROBINSON E: *J Am Chem Soc 78*:6415, 1956
176. ITANO HA, ROBINSON E: *Biochim Biophys Acta 29*:545, 1958
177. JACKSON H, THOMPSON R: *Biochem J 57*:619, 1954
178. JACKSON JM, WAY BJ, WOODLIFF HJ: *Br J Haematol 13*:474, 1967
179. JACOBS SL, FERNANDEZ AA: *Standard Methods of Clinical Chemistry.* Edited by RP MacDonald. New York, Academic Press, 1970, Vol VI, p 107
180. JAVID J, HOROWITZ HI: *Am J Clin Pathol 34*:35, 1960
181. JAVID J, HOROWITZ HI, SANDERS AR, SPAET TH: *Arch Intern Med 104*:628, 1959
182. JONES RT, BRIMHALL B, HUISMAN THJ, KLEIHAUER E, BETKE K: *Science 154*:1024, 1966
183. JONES RT, COLEMAN RD, HELLER P: *Fed Proc 23*:178, 1964
184. JONES RI, OSGOOD EE, BRIMHALL B, KOLER RD: *J Clin Invest 46*:1840, 1967
185. JONXIS JHP, HUISMAN THJ: *Blood 11*:1009, 1956
186. JOPE EM: *Br J Ind Med 3*:136, 1946
187. KAGEN LJ, CHRISTIAN CL: *Am J Physiol 211*:656, 1966
188. KEENE WR, JANDL JH: *Blood 26*:705, 1965
189. KIKUCHI G, HAYASHI N, TAMURA A: *Biochim Biophys Acta 90*:199, 1964

190. KLEIHAUER EF: *Physiology of the Perinatal Period*. Edited by U Stave. New York, Appleton-Century-Crofts, 1970, p 255

191. KLEIHAUER EF, BRAUN H, BETKE K: *Klin Wochenschr 35*:637, 1957

192. KLEIHAUER EF, REYNOLDS CA, DOZY AM, WILSON JB, MOORES RR, BERENSON MP, WRIGHT CS, HUISMAN THJ: *Biochim Biophys Acta 154*:220, 1968

193. KLEIN B, WEBER BK, LUCAS L, FORMAN JA, SEARCY RL: *Clin Chim Acta 26*:77, 1969

194. KLENDSHOJ NE, FELDSTEIN M, SPRAGUE AL: *J Biol Chem 183*:297, 1950

195. KLUTHE R, FAUL J, HEIMPEL H: *Nature (Lond) 205*:93, 1965

196. KOHN J: *Chromotographic and Electrophoretic Techniques*. Second edition. Edited by I Smith. New York, Interscience, 1960, Vol II, p 77

197. KOSHLAND DE: *Proc Natl Acad Sci (USA) 44*:98, 1958

198. KUNKEL HG, WALLENIUS G: *Science 122*:288, 1955

199. KUNKEL HG, CEPPELLINI R, MULLER-EBERHARD U, WOLF J: *J Clin Invest 36*:1615, 1957

200. KÜNZER W: *Z Kinderheilkd 76*:58, 1955

201. KÜNZER W: *Klin Wochenschr 35*:940, 1957

202. LANGE HF: *Acta Med Scand (Suppl) 176*, 1946

203. LATHEM W, WORLEY WE: *J Clin Invest 38*:474, 652, 1959

204. LEAHY T, SMITH R: *Clin Chem 6*:148, 1960

205. LEHMAN H, HUNTSMAN RG: *Man's Hemoglobins*. Amsterdam, North Holland Publishing Co, 1968, p 267

206. LEMBERG R, HOLDEN HF, LEGGE JW, LOCKWOOD WH: *Aust J Exp Biol Med Sci 20*:161, 1942

207. LEMBERG R, LEGGE JW: *Hematin Coumpounds and Bile Pigments*. New York, Interscience, 1949

208. LIE-INJO LE: *Blood 20*:581, 1962

209. LINES JC, McINTOSH R: *Nature (Lond) 215*:297, 1967

210. LIONETTI FJ, VALERI R, BOND JC, FORTIER NL: *J Lab Clin Med 64*:519, 1964

211. LOUDERBACK AL, SHANBROM E: *JAMA 206*:362, 1968

212. LUPOVITCH A, KATASE RY: *Clin Chim Acta 11*:566, 1965

213. MacFATE RP: *Hemoglobin Its Precursors and Metabolites*. Edited by FW Sunderman. Philadelphia, Lippincott, 1964, p 31

214. MALAMOS B, FESSAS P, STAMATOYANNOPOULOS G: *Br J Haematol 8*:5, 1962

215. MARTI HR: *Klin Wochenschr 39*:286, 1961

216. MARTIN GE, MUNN JI, BISKUP L: *J Assoc Off Agr Chemists 43*:743, 1960

217. MATSUBARA T, SHIBATA S: *Clin Chim Acta 23*:427, 1969

218. MAXFIELD ME, BAZETT HC, CHAMBERS CC: *Am J Physiol 133*:128, 1941

219. McCARTHY EF, VAN SLYKE DD: *J Biol Chem 128*:567, 1939

220. McCUTCHEON AD: *Lancet 2*:240, 1960

221. McDONALD CD Jr, HUISMAN THJ: *Clin Chim Acta 8*:639, 1963

222. MICHEL HO, HARRIS JS: *J Lab Clin Med 25*:445, 1940

223. MIHARA K, HAYASHI N, KIKUCHI G, SHIBATA S: *Biochem Biophys Res Commun 32*:763, 1968

224. MOLE RH: *J Physiol 104*:1, 1945

225. MONN E, GAFFNEY PJ, LEHMANN H: *Scand J Haematol 5*:353, 1968

226. MOORE WMO, BATTAGLIA FC, HELLEGERS AF: *Am J Obstet Gynecol 97*:63, 1967

227. MORETTI J, WAKS M, JAYLE MF: *Ann Biol Clin (Paris) 25*:149, 1967

228. MÖRNER KAH: *Nord Med Ark 30*:1, 1897
229. MOTULSKY AG, PAUL MH, DURRUM EL: *Blood 9*:897, 1954
230. MOVITT ER, MANGUM JF, PORTER WR: *Blood 21*:535, 1963
231. MOVITT ER, POLLYCOVE M, MANGUM JF, PORTER WR: *Am J Med Sci 247*:558, 1964
232. MULLER CJ, KINGMA S: *Biochim Biophys Acta 50*:595, 1961
233. MURAWSKI E, CARTA S, SORCINI M, TENTORI L, VIVALDI G, ANTONINI E, BRIMORE M, WYMAN J, BUCCI E, ROSSE-FANELLI A: *Arch Biochem Biophys 111*:197, 1965
234. MURAYAMA M: *J Cell Physiol 67 (Suppl 3)*:21, 1966
235. NAGEL RL, GIBSON QH, CHARACHE S: *Biochemistry 6*:2395, 1967
236. NAGEL RL, RANNEY HM, BRADLEY TB, JACOBS A, UDEM L: *Blood 34*:157, 1969
237. NALBANDIAN RM, HENRY RL, NICHOLS BM, CAMP FR Jr, WOLF PL: *Clin Chem 16*:945, 1970
238. NALBANDIAN RM, NICHOLS BM, CAMP FR Jr, LUSHER JM, CONTE NF, HENRY RL, WOLF PL: *Clin Chem 17*:1028, 1971
239. NA-NAKORN S, WASI P, PORNPATKUL M, POOTRAKUL S: *Nature (Lond) 223*:59, 1969
240. NECHELES TF, ALLEN DM, GERALD PS: *Ann NY Acad Sci 165*:5, 1969
241. NEEDHAM CD, SIMPSON RG: *Q J Med 21*:123, 1952
242. NEILL JM, HASTINGS AB: *J Biol Chem 63*:479, 1925
243. NICLOUX M, FONTES G: *Bull Soc Chim Biol 6*:728, 1924
244. NOVY MJ, EDWARDS MJ, METCALFE J: *J Clin Invest 46*:1848, 1967
245. NYMAN M: *Scand J Clin Lab Invest 11*:Suppl 39, 1959
246. OPFELL RW, LORKIN PA, LEHMANN H: *J Med Genet 5*:292, 1968
247. OWEN JA, BETTER FC, HOBAN J: *J Clin Pathol 13*:163, 1960
248. OWEN JA, SILBERMAN HJ, GOT C: *Nature (Lond) 182*:1373, 1958
249. PARER JT, STAMATOYANNOPUOLOS G, FINCH CA: *Clin Res 17*:153, 1969
250. PAULING L: *Stanford Med Bull 6*:215, 1948
251. PAULING L, CROYELL C: *Proc Natl Acad Sci (USA) 22*:210, 1936
252. PAULING L, ITANO HA, SINGER SJ, WELLS IC: *Science 110*:543, 1949
253. PEARSON CM, BECK WS, BLAHD WH: *Arch Intern Med 99*:376, 1957
254. PEARSON HA: *J Pediatr 69*:466, 1966
255. PEARSON HA, McCOO JV, LEIKIN SL: *Blood 17*:758, 1961
256. PEARSON HA, McFARLAND W: *US Armed Forces Med J 10*:693, 1959
257. PERKOFF GT, HILL RL, TYLER FH: *J Clin Invest 40*:1071, 1961
258. PERUTZ MF: *Nature (Lond) 228*:726, 1970
259. PERUTZ MF: *Nature (Lond) 237*:495, 1972
260. PERUTZ MF, LEHMANN H: *Nature (Lond) 219*:29, 1968
261. PERUTZ MF, LEHMANN H: *Nature (Lond) 219*:902, 1968
262. PERUTZ MF, MUIRHEAD H, COX JM, GOAMAW LCG: *Nature (Lond) 219*:131, 1968
263. PERUTZ MF, MUIRHEAD H, MAZZARELLA L, CROWTHER RA, GREEN J, KILMARTIN JV: *Nature (Lond) 222*:1240, 1969
264. PETERS JH, ALPER CA: *J Clin Invest 45*:314, 1966
265. PISCIOTTA AV, EBBE SN, HINZ JE: *J Lab Clin Med 54*:73, 1959
266. POLONOVSKI M, JAYLE MF: *C R Acad Sci (D) (Paris) 211*:517, 1940
267. POLOSA P, MOTTA L, FALSAPERLA A, NOLFO G: *Boll Soc Ital Biol Sper 35*:9, 1959
268. PORTER GA: *J Lab Clin Med 60*:339, 1962
269. PRINS HK: *J Chromatogr 2*:445, 1959

270. RAIKE, HUNTER EG, LINDSAY DA: *Med J Aust 1*:955, 1967
271. REED CS, HAMPSON R, GORDON S, JONES RT, NOVY MJ, BRIMHALL B, EDWARDS MJ, KOLER RD: *Blood 31*:623, 1968
272. REISSMANN KR, RUTH WE, NOMURA T: *J Clin Invest 40*:1826, 1961
273. RICE EW: *J Lab Clin Med 71*:319, 1968
274. RIEDER RF, BRADLEY TB Jr: *Blood 32*:355, 1968
275. RIEDER RF, OSKI FA, CLEGG JB: *J Clin Invest 48*:1627, 1969
276. RIEDER RF, ZINKHAM WH, HOLTZMAN NA: *Am J Med 39*:4, 1965
277. RIGAS DA, KOLER RD: *Blood 18*:1, 1961
278. RIGAS DA, KOLER RD, OSGOOD EE: *J Lab Clin Med 47*:51, 1956
279. ROBINSON AR, ROBSON M, HARRISON AP, ZUELZER WW: *J Lab Clin Med 50*:745, 1957
280. ROCHE J, DERRIEN Y: *Sang 24*:97, 1953
281. ROCHE J, DERRIEN Y, REYNAULD J, LAURENT G, ROQUES M: *Bull Soc Chim Biol 36*:51, 1954
282. RODKEY FL: *Clin Chem 13*:2, 1967
283. ROSA J, LABIE D, WAJCMAN H, BOIGNE JM, CABANNES R, BIERME R, RUFFIE J: *Nature (Lond) 223*:190, 191, 1969
284. ROSSI-FANELLI A, ANTONNI E: *Arch Biochem Biophys 65*:587, 1956
285. ROY RB, SHAW RW, CONNELL GE: *J Lab Clin Med 74*:698, 1969
286. RUCKNAGEL DL, GLYNN KP, SMITH JR: *Clin Res 15*:270, 1967
287. RUTTAN FF, HARDISTY RHM: *Can Med Assoc J 2*:995, 1912
288. SAHLI H: *Lehrbuch der Klinischen Untersuchungsmethoden.* Leipzig, Deuticke, 1894
289. SALT HB: *Analyst 76*:344, 1951
290. SANSONE G, CARRELL RW, LEHMANN H: *Nature (Lond) 214*:877, 1967
291. SANSONE G, PIK C: *Br J Haematol 11*:511, 1965
292. SATO K: *Iwate Igaku Zasshi 11*:193, 1959
293. SAYERS RR, O'BRIEN HR, JONES GW, OANT WP: *Public Health Rep 38*:2005, 1923
294. SCHLEYER F, SCHAIBLE P: *Z Klin Chem Klin Biochem 5*:32, 1967
295. SCHMID R, BRECHER G, CLEMENS T: *Blood 14*:991, 1959
296. SCHMIDT RM, BROSIUS EM: *Laboratory Methods of Hemoglobinopathy Detection.* US Department of Health, Education, and Welfare, Center for Disease Control, Atlanta, Ga.
297. SCHNEIDER RG: *J Lab Clin Med 44*:133, 1954
298. SCHNEIDER RG, UEDA S, ALPERIN JB, BRIMHALL G, JONES RT: *Am J Human Genet 20*:151, 1968
299. SCHNEIDER RG, UEDA S, ALPERIN JB, BRIMHALL B, JONES RT: *N Engl J Med 280*:739, 1969
300. SCHOEN I, SOLOMON M: *J Clin Pathol 15*:44, 1962
301. SCHOKKER RC, WENT LN, BOK J: *Nature (Lond) 209*:44, 1966
302. SENDROY J Jr, LIU SH: *J Biol Chem 89*:133, 1930
303. SHEARD C, SANFORD AH: *J Lab Clin Med 14*:558, 1929
304. SHEARD C, SANFORD AH: *JAMA 93*:151, 1929
305. SHINOWARA GY: *Am J Clin Pathol 24*:696, 1954
306. SHINTON NK, RICHARDSON RW, WILLIAMS JDF: *J Clin Pathol 18*:114, 1965
307. SILVESTRONI E, BIANCO I, GRAZIANI B: *Br J Haematol 14*:303, 1968
308. SINGER K, ANGELOPOULOS B, RAMOT B: *Blood 10*:979, 1955
309. SINGER K, ANGELOPOULOS B, RAMOT B: *Blood 10*:987, 1955
310. SINGER K, CHERNOFF AI, SINGER L: *Blood 6*:413, 1951
311. SMITH EW, CONLEY CL: *Ann Intern Med 50*:94, 1959

312. SMITH EW, KREVANS JR: *Bull Hopkins Hosp 104*:17, 1959
313. SPAET TH, ROSENTHAL MD, DAMESHEK W: *Blood 9*:881, 1954
314. SPEALMAN CR, NEWTON M, POST RL, MAXFIELD ME, BAZETT HC: *Am J Physiol 150*:628, 1947
315. STADIE WC: *J Biol Chem 41*:237, 1920
316. STAMATOYANNOUPULOS G, BELLINGHAM AJ, LENFAUT C, FINCH CA: *Ann Rev Med 22*:221, 1971
317. STAMATOYANNOPOULOS G, FESSAS P, PAPAYANNOPOULOU T: *Am J Med 47*:194, 1969
318. STAMATOYANNOPOULOS G, YOSHIDA A, ADAMSON J, HEINENBERG S: *Science 159*:741, 1968
319. STEADMAN JH, YATES A, HUEHNS ER: Cited by Huehns ER: *Ann Rev Med 21*:157, 1970
320. STEINMANN B: *Arch Exp Pathol Pharmakol 191*:237, 1939
321. STETSON CA Jr: *J Exp Med 123*:341, 1966
322. SUNDERMAN FW: *Am J Clin Pathol 43*:9, 1965
323. SUNDERMAN FW, SUNDERMAN FW Jr, editors: *Hemoglobin—Its Precursors and Metabolites*. Philadelphia, Lippincott, 1964
324. SUZUKI T, HAYASHI A, SHIMIZU A, YAMAMURA Y: *Biochim Biophys Acta 127*:280, 1966
325. SUZUKI T, HAYASHI A, YAMAMURA Y, ENOKI Y, TYUMA I: *Biochem Biophys Res Commun 19*:691, 1965
326. SVENSON B, STRAND L: *Scand J Haematol 4*:241, 1967
327. TALLQVIST TW: *Arch Gen Med 3*:421, 1900
328. TALLQVIST TW: *Z Klin Med 40*:137, 1900
329. TARUKOSKI PH: *Scand J Clin Lab Invest 18*:80, 1966
330. THEORELL H, DeDUVE C: *Arch Biochem 12*:113, 1947
331. THOMAS ED, MOTULSKY AG, WALTERS DH: *Am J Med 18*:832, 1955
332. THOMPSON RB, HEWITT B, ARD E, ODOM J, BELL WN: *Acta Haematol (Basel) 36*:412, 1966
333. THOMPSON RB, ODOM J, ARD E, BELL WN: *Acta Genet Statist Med 16*:340, 1966
334. TOWNES PL, LOVELL GR: *Blood 18*:18, 1961
335. TRAINER MT: *Proc Soc Exp Biol Med 52*:104, 1943
336. UEDA S, SCHNEIDER RG: *Blood 14*:230, 1969
337. VALERI CR, BOND JC, FOWLER K, SOBUCKI J: *Clin Chem 2*:581, 1965
338. VAN KAMPEN EJ, ZIJLSTRA WG: *Adv Clin Chem 8*:141, 1965
339. VAN KAMPEN EJ, ZIJLSTRA WG: *Clin Chim Acta 6*:538, 1961
340. VAN SLYKE DD: *J Biol Chem 66*:409, 1925
341. VAN SLYKE DD, HILLER A: *J Biol Chem 84*:205, 1929
342. VAN SLYKE DD, HILLER A, WEISIGER JR, CRUZ WO: *J Biol Chem 166*:121, 1946
343. VAN SLYKE DD, ROBSCHEIT-ROBBINS FS: *J Biol Chem 72*:39, 1927
344. VAN SLYKE DD, STADIE WC: *J Biol Chem 49*:1, 1921
345. VANZETTI G, VALENTE D: *Clin Chim Acta 11*:442, 1965
346. VAUGHN JONES R, GRIMES AJ, CARRELL RW, LEHMANN H: *Br J Haematol 13*:394, 1967
347. VENEZIALE CM, McGUCKIN WF: *Proc Mayo Clin 40*:751, 1965
348. VOGT H, GEISELER G: *Dtsch Arch Klin Med 189*:44, 1942
349. VON KÖRBER E: Cited by Bischoff H: *Z Ges Exp Med 48*:472, 1926
350. WAKS M, ALFSEN A: *Arch Biochem Biophys 123*:133, 1968
351. WATSON-WILLIAMS EJ: *Br Med J 1*:1511, 1955

352. WEATHERALL DJ: *Br J Haematol 9*:265, 1963

353. WEATHERALL DJ: *Ann NY Acad Sci 119*:450, 1964

354. WEATHERALL DJ: *Ann NY Acad Sci 119*:463, 1964

355. WEIS J: *Chem Ind London 16*:517, 1938

356. WENDEL WB: *J Clin Invest 18*:179, 1939

357. WENT LN, MacIVER JE: *Blood 17*:166, 1961

358. WHEELER JT, KREVANS JR: *Bull Hopkins Hosp 109*:217, 1961

359. WHISNANT CL Jr, OWINGS RH, CANTRELL CG, COOPER GR: *Ann Intern Med 51*:140, 1959

360. WHITE JC, BEAVEN GH: *Br Med Bull 15*:33, 1959

361. WHITE JM, BRAIN MC, LORKIN PA, LEHMANN H, SMITH M: *Nature (Lond) 225*:939, 1970

362. WHORTON CM, HUDGINS PC, CONNORS JJ: *N Engl J Med 265*:1242, 1961

363. WINTROBE MM: *Clinical Hematology.* Sixth edition. Philadelphia, Lea & Febiger, 1967, p 154

364. WOLFF JA, IGNATOV VG: *Am J Dis Child 105*:234, 1963

365. WOLFSON RY, YAKULIS V, COLEMAN RD, HELLER P: *J Lab Clin Med 69*:728, 1967

366. WOODMAN DD: *Clin Chim Acta 29*:249, 1970

367. WOODMAN DD: *Med Lab Technol 28*:121, 1971

368. WU H: *J Biochem 2*:173, 1922

369. WU H: *J Biochem 2*:181, 1922

370. WU H: *J Biochem 2*:189, 1922

371. YANASE T, HANADA M, SEITA M, OHYA I, OHTA Y, IMAMURA T, FUJIMURA T: *Jap J Hum Genet 13*:40, 1968

372. ZAK B, EGGERS EM, JARKOWSKI TL, WILLIAMS LA: *J Lab Clin Med 54*:288, 1959

373. ZETTNER A, MENSCH AH: *Am J Clin Pathol 48*:225, 1967

374. ZIJLSTRA WG, MULLER CJ: *Clin Chim Acta 2*:237, 1957

375. ZIJLSTRA WG, VAN KAMPEN EJ: *Am J Clin Pathol 50*:513, 1968

376. ZUELZER WW, ROBINSON AR, BOOKER CR: *Blood 17*:393, 1961

377. Division of Medical Sciences, National Academy of Sciences, National Research Council: *Science 127*:1376, 1958

378. Editorial: *Lancet 1*:819, 1970

379. Nomenclature of Abnormal Hemoglobins: *Am J Clin Pathol 43*:166, 1965

Chapter 24

Porphyrins and Their Precursors

S. LAWRENCE JACOBS, Ph.D.

PORPHYRINS AND THEIR PRECURSORS

To cover thoroughly the complex subject of porphyrin biochemistry is far beyond the scope of this book. The literature on this subject is prodigious and no attempt is made here to document the entire history of the subject. The reader is referred to excellent reviews by workers preeminent in the porphyrin field: Schwartz *et al* (172), With (223), Falk (59), Levere *et al* (117, 118), Waldenström (200), Goldberg and Rimington (76), Tschudy (190), Hebert (97), Marver and Schmid (129), and Pindyck *et al* (144).

NOMENCLATURE OF PORPHYRINS AND THEIR PRECURSORS

Porphyrins all possess a common basic structure called *porphin,* which is composed of four pyrrole rings joined in cyclic configuration by four methine bridges as follows:

PORPHIN

The constitutional formula of porphin above also indicates Fischer's system used for numbering the pyrrole rings, β-carbons (of the pyrrole rings), and the methine bridges. A system of shorthand has been devised in which the symbol ⊓ is used to represent each pyrrole ring. Porphin thus becomes:

PORPHIN

Porphyrins are derivatives of porphin in which hydrogens on the β-carbons are substituted by various combinations of the following:

$$
\begin{array}{lcl}
A & = & \text{acetic acid} \\
P & = & \text{propionic acid} \\
V & = & \text{vinyl} \\
M & = & \text{methyl} \\
E & = & \text{ethyl}
\end{array}
$$

Coproporphyrin, so named because it was originally identified in feces, has four methyl groups (M) and four propionic groups (P) substituted for the eight hydrogens. These can be positioned in four different ways, forming four isomers or "types" as follows:

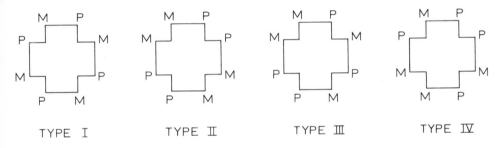

TYPE I TYPE II TYPE III TYPE IV

Roman numerals are used to identify the isomeric configuration. Thus, according to Fischer's nomenclature, coproporphyrin I is 1,3,5,7-tetramethyl porphintetrapropionic acid.

Uroporphyrin, so named because it was originally identified in urine, has four acetic and four propionic groups; thus there are four isomeric forms, uroporphyrin I being represented as:

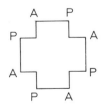

UROPORPHYRIN I

Deuteroporphyrin has the following constitutional formula:

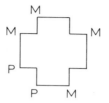

Protoporphyrin (heme of hemoglobin and myoglobin is iron-protoporphyrin) has four methyl, two vinyl, and two propionic groups; thus there are 15 possible isomers. The isomer designated protoporphyrin IX is the only one found in all known heme pigments. One can consider the methyl groups as one type of group and the other two (vinyl and propionic) together as a second type and visualize the correspondence of the 15 possible isomers as one of the four basic isomer types. Protoporphyrin IX corresponds to type III:

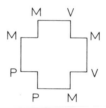

PROTOPORPHYRIN IX
(TYPE III)

In *hematoporphyrin* and *mesoporphyrin* the two vinyl groups of protoporphyrin are replaced by hydroxyethyl and ethyl groups, respectively. Thus these also correspond to type III porphyrins. Hematoporphyrin is formed only *in vitro*.

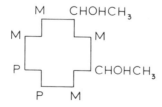

HEMATOPORPHYRIN
(TYPE III)

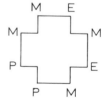

MESOPORPHYRIN
(TYPE III)

Etioporphyrins (aetioporphyrins) have four methyl and four ethyl groups.
Metalloporphyrins are formed by chelation of various metals to the centrally directed N atoms of the pyrrole rings. Zinc and Cu porphyrin

complexes are formed in this manner. Chlorophyll is a Mg-porphyrin complex and cytochrome and heme of hemoglobin are Fe-porphyrin complexes.

Porphyrinogens are the colorless cyclic tetrapyrrole compounds having methylene ($-CH_2-$) groups between the pyrrole rings as well as hydrogens on all four of the nitrogen atoms. The type III porphyrinogens are the important intermediates in the biosynthesis of heme. The corresponding porphyrins are formed by the stepwise oxidation of porphyrinogens.

The formulae for *δ-aminolevulinic acid* (ALA; has also been called δ-aminolevulic acid) and *porphobilinogen* (PBG) are given in the next section.

PORPHYRINS AND PRECURSORS IDENTIFIED
IN CLINICAL SAMPLES

δ-Aminolevulinic Acid (ALA). ALA, found so far in urine and serum, is the precursor of porphobilinogen (PBG). It was first identified by Shemin and Russel (178) in 1953. There is substantial evidence that it is synthesized from glycine and succinyl coenzyme A in the presence of the mitochondrial enzyme ALA synthetase and pyridoxal phosphate (73, 110, 129, 144, 190). CO_2 is released to form ALA; α-amino-β-ketoadipic acid may be an intermediate (178). This appears to be the rate-limiting step in the biosynthesis of heme (19), which along with closely related derivatives, can exert a negative feedback effect to control ALA synthetic activity. Two molecules of ALA are condensed, in the presence of δ-aminolevulinic acid dehydrase (also called PBG synthetase), by formation of a C-C bond at A and a C-N bond at B in the reaction shown below, with simultaneous removal of two molecules of water (74, 78, 178). A deficiency of this enzyme occurs in subclinical lead poisoning in children (213).

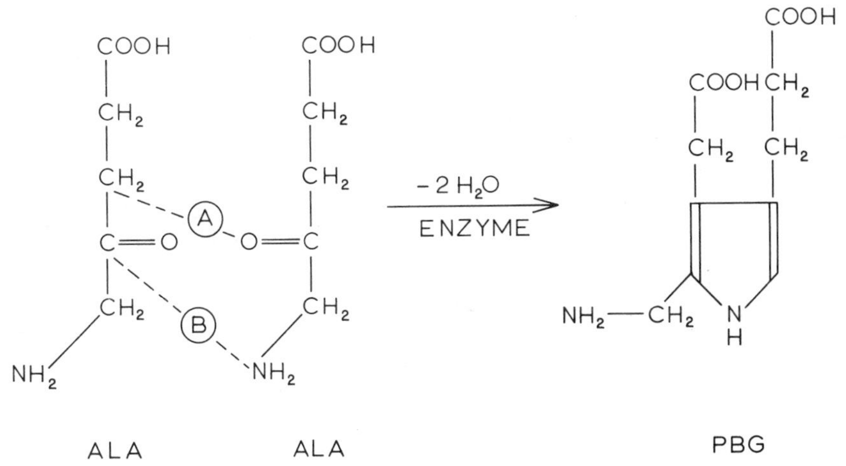

ALA   ALA   PBG

Porphobilinogen (PBG). Sachs (159) in 1931 apparently was first to distinguish the urine chromogen (PBG) in porphyria from urobilinogen, on the basis of the difference in chloroform solubility of the aldehyde derivative and the failure of PBG on oxidation and in the presence of zinc acetate to exhibit green fluorescence. Waldenström subsequently named this colorless precursor porphobilinogen and demonstrated that it was always excreted in acute porphyria (199). It was isolated by Westall (215) and its structure established by Cookson and Rimington (31, 32) as 2-aminomethyl-4,2-carboxyethyl-3-carboxymethylpyrrole.

Of what PBG is the precursor is not yet certain. Lockwood demonstrated (120) that PBG may go in at least two directions, depending on the availability of O_2:

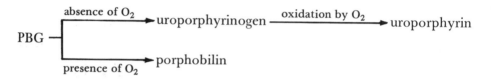

Conversion to porphyrin is spontaneous and is accelerated by heat, optimal conversion occurring in 30 min at $100°C$ and pH 5.0–5.5 (82). Uroporphyrinogens of types I, II, III, and IV may be obtained in such a nonenzymatic reaction (131). Porphobilin is a red, noncrystallizable chain compound which is ether-soluble, gives a positive biuret test and a negative Ehrlich aldehyde test, and is probably urobilene-(b), i.e., a bilene with the same side chains as uroporphyrin (201, 215).

Present thinking pictures the sequence in the biosynthetic pathway for heme as follows: Two or three molecules of PBG, under the influence of PBG deaminase, condense to a single pyrrylmethane and NH_3. Continued enzyme action yields uroporphyrinogen I, while under the influence of uroporphyrinogen isomerase uroporphyrinogen III is obtained. Mechanisms proposed for this reaction include condensation of two dimers, formation of a trimer which might split in two ways, and migration of side chains. Then condensation to the cyclic tetrapyrrole by action of these enzymes occurs (12, 13, 79).

Porphyrinogens in Urine. The true intermediates in the biosynthesis of heme are the colorless porphyrinogens. These are unstable and are easily oxidized to the corresponding porphyrins with the loss of six hydrogen atoms. The mechanism appears to be stepwise through porphomethenes and porphodimethenes (12, 79, 133). Once oxidation to the porphyrin has occurred, the molecule is no longer in the metabolic scheme but is only an excretion product, although it may produce certain pharmacologic effects such as skin photosensitivity. Stepwise enzymatic decarboxylation of the type III porphyrinogens occurs as follows:

$$\text{uro-porphyrinogen III} \xrightarrow[\text{decarboxylase}]{\text{uroporphyrinogen}} \text{copro-porphyrinogen III} \xrightarrow[\text{oxidase}]{\text{coproporphyrinogen}} \text{proto-porphyrinogen IX}$$

This is the important pathway in the synthesis of heme (14, 80, 133, 163). As stated above, uroporphyrinogen I is also produced, and this leads to uroporphyrin I by oxidation and by decarboxylation to coproporphyrinogen I, which is itself oxidized to coproporphyrin I. All of these type I chromogens and porphyrins are useless by-products, but as shall be seen later they have a significance in certain porphyrias. The oxidative decarboxylation of coproporphyrinogen I does not occur and thus no type I protoporphyrin is known (14, 76, 133, 163). No attempt is ordinarily made to determine porphyrinogens. In fact, it is common to convert the chromogens to the porphyrins by oxidation with I_2 prior to estimation. Protoporphyrinogen is not excreted in the urine.

Porphyrins in Urine. Only types I and III isomers of porphyrins have been positively identified in the body and its excreta (their configurations can be remembered most easily by the fact that type I has alternating side groups and type III is the only completely asymmetrical isomer). This does not mean, however, that types II and IV may not occur, at least in small amounts. As Cookson and Rimington have pointed out (32), for example, chromatographic methods usually can separate only types I and II from types III and IV. Eriksen *et al* (53, 56) obtained evidence from paper chromatographic studies of traces of coproporphyrin II even in normal urine.

The urinary porphyrins of greatest concern to clinical chemists are coproporphyrin (tetracarboxylic) and uroporphyrin (octacarboxylic). Porphyrins with 3, 5, 6, and 7 carboxyl groups, however, have been detected in pathologic and some normal urines (26, 30, 134). Presumably these result from progressive decarboxylation from uroporphyrinogen to protoporphyrinogen (dicarboxylic) with subsequent oxidation to the porphyrins. Eriksen *et al* (57) report that the porphyrin pattern in porphyria cutanea tarda differs from that of acute porphyria and the normal state in that the 5- to 7-carboxylic porphyrins are increased.

Although the isomeric distribution of coproporphyrin in normal urine is still somewhat controversial, it is increasingly certain that the major portion is type III (30, 53, 56, 221).

Waldenström (196) in 1934 reported the occurrence of ether-insoluble uroporphyrin in the urine of acute porphyria and believed it to be type III. Ever since, the term "Waldenström porphyrin" has been used to refer to the porphyrin extracted from urine by ethyl acetate, but not by ether, at pH 3.2–3.4. Considerable effort has been expended toward its identification, but a clear picture has not yet emerged. Cookson and Rimington (32) have expressed the opinion that Waldenström porphyrin is an artifact derived from PBG by chemical means and is a mixture of uroporphyrins, the composition varying depending on conditions under which it is formed.

There is general agreement that the uroporphyrin in normal urine is predominantly type I (172, 221).

Most workers have been unable to detect protoporphyrin IX (III) in urine (172, 199). Schlenker and Kitchell (167) reported finding it in normal urine, but they utilized a technic which should include coproporphyrin precursor in the "proto" fraction (extraction of copro from ether with 0.1 N HCl followed by extraction of proto from ether with 2.8 N HCl).

Porphyrins in Feces. Part of the porphyrins in feces is derived from endogenous metabolism and part is formed in the intestinal tract from hematin compounds or by microbial synthesis. Those from endogenous metabolism include protoporphyrin, coproporphyrins I and III, small amounts of coproporphyrin ester, uroporphyrin, and traces of penta- and heptacarboxyl porphyrins (40, 86, 134). Protoporphyrin and deuteroporphyrin are bacterial decomposition products of hemoglobin derived from gastrointestinal bleeding or meat digestion (40, 134, 172). There also may be a small amount of mesoporphyrin present, most likely formed by bacterial reduction of protoporphyrin (86, 134, 229).

Erythrocyte Porphyrins. Protoporphyrin IX (III), coproporphyrin, and uroporphyrin are all found in erythrocytes (169, 172, 174). The coproporphyrin which is found in normal erythrocytes is mainly type I which is a by-product not involved in heme biosynthesis (115).

Plasma Porphyrins. Normal plasma reportedly contains up to 0.7 µg porphyrin per 100 ml (129, 166), and significant elevations have been found in some porphyrias.

Clinical Aspects. A distinction must be made between *porphyrinurias* and *porphyrias*. The former is a symptom of some acute febrile states, lead or arsenic poisoning, pernicious anemia, cirrhosis, sprue, acute pancreatitis, blood dyscrasias, and malignancies, wherein the excretion of coproporphyrin may be increased. Porphyrias, on the other hand, are characteristic diseases of constitutional or induced types. A number of classifications of porphyrias have been proposed (76, 200, 205), and it has been difficult to correlate genetic type with biochemical features, particularly when attempting to formulate a classification which is generally applicable to various world areas (48). Much of the research of the past 20 years in the area of porphyrin biochemistry deals with the overproduction or deficiency of the various enzymes which catalyze the metabolic reactions of the porphyrin precursors, and is concerned with how these enzyme abnormalities occur and how they are reflected in the metabolic endproducts. In Figure 24-1 an attempt is made to show the major (less important intermediates are omitted) metabolic scheme, with indications of the enzymatic aberrations and the metabolites which are usually found to be significantly elevated in the various types of porphyria. Table 24-1 also summarizes the more common laboratory findings. According to With (223), "there is no agreement on the definition of porphyria itself" among the many investigators, but he presents as a practical definition "a marked disturbance of porphyrin synthesis which may be latent periodically or permanently."

The classification of porphyrias presented here is adapted from that of Levere and Kappas (117):

Classification of the Porphyrias

1. Erythropoietic
   a. Congenital erythropoietic porphyria (CEP)
   b. Congenital erythropoietic protoporphyria (EPP)

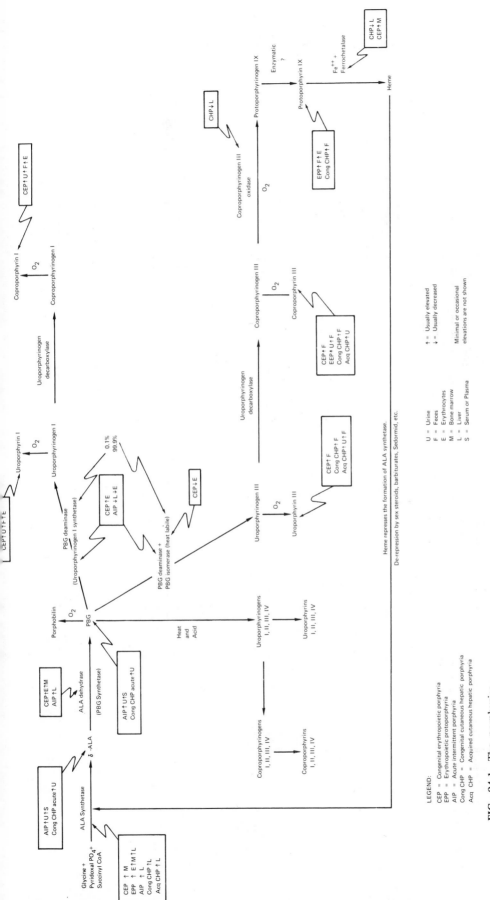

FIG. 24-1. The porphyrias.

TABLE 24-1. Summary of Findings in Lead Porphyrinuria and Porphyrias

	Lead Porphyrinuria	Porphyrias[a]				
		CEP	EPP	AIP	Congenital CHP	Acquired CHP
Urine						
ALA	Increased			Increased	Increased	
PBG				Increased	Increased	
Coproporphyrin	Increased	Increased	Increased		Increased	Increased
Uroporphyrin	Increased	Increased		Increased	Increased	Increased
Erythrocytes						
Protoporphyrin	Increased		Increased			
Coproporphyrin	Increased	Increased				
Uroporphyrin		Increased				
Feces						
Protoporphyrin			Increased		Increased	
Coproporphyrin		Increased	Increased		Increased	
Uroporphyrin		Increased			Increased	

[a]CEP, congenital erythropoietic porphyria; EPP, erythropoietic protoporphyria; AIP, acute intermittent porphyria; congenital CHP, congenital cutaneous hepatic porphyria; acquired CHP, acquired cutaneous hepatic porphyria.

2. Hepatic
   a. Hereditary
      (1) Acute intermittent porphyria (AIP) (Swedish porphyria, pyrroloporphyria)
      (2) Congenital cutaneous hepatic porphyria (congenital CHP) (South African porphyria, porphyria cutanea tarda, mixed porphyria, porphyria variegata, protocoproporphyria)
   b. Acquired cutaneous hepatic porphyria (acquired CHP) (porphyria cutanea tarda symptomatica, toxic, or drug-induced porphyrias)

In congenital erythropoietic porphyria (CEP) the defect is in the erythroblasts of the bone marrow where there is an overproduction of ALA synthetase resulting in large quantities of ALA and PBG, and an imbalance in the enzymes PBG deaminase and PBG isomerase, causing an unusual elevation of type I uro- and coproporphyrin in the urine, feces, and erythrocytes. Elevated fecal coproporphyrin III is also found. Urine may be red. The disease is recessive and the clinical onset is during early childhood. Skin photosensitivity and hemolytic anemia are manifestations of this disease.

Erythropoietic protoporphyria (EPP) (125) is unusual in that excessive urinary porphyrins or precursors are not always excreted although coproporphyrin III is likely to be found. It is necessary to analyze erythrocytes to identify the disease. Fecal protoporphyrin and coproporphyrin III are usually elevated. The enzymatic defect is in the marrow, erythrocytes, and liver and appears to be the overproduction of ALA synthetase with no deficiency in PBG isomerase as in CEP so that type III porphyrins are formed. It has been suggested that the major site of porphyrin production in EPP is the liver (43) and that the disease might be more properly called *erythrohepatic* protoporphyria (170). The urine is generally normal in color. The disease is inherited as a dominant characteristic and is usually manifested in childhood. There is mild skin photosensitivity.

Acute intermittent porphyria (AIP) is the hepatic porphyria most commonly found in northern Europe and North America. Acute abdominal pain, sometimes with constipation and vomiting, is the typical presenting symptom. Skin photosensitization is absent. It is inherited as a dominant characteristic. The overproduction of ALA synthetase in the liver causes large amounts of ALA and porphobilinogen to be produced and excreted in the urine. A deficiency of PBG deaminase and PBG isomerase rules out any increase in porphyrin excretion through the usual metabolic pathway, although nonenzymatic polymerization and oxidation converts some of the excreted PBG to the colored pigments uroporphyrin and porphobilin. Thus the urine may darken on standing. Patients with AIP are particularly susceptible to a number of drugs including barbiturates and sulfonamides, as well as to sex steroids. The effect of the latter is probably responsible for the onset of the disease, generally occurring after puberty. These compounds have the ability to induce production of ALA synthetase, probably by interfering with the negative feedback role of heme, the endproduct of porphyrin metabolism. Acute reduction in daily food intake has caused a rise in excretion of both ALA and PBG in asymptomatic patients with AIP (214) and has been reported to cause abnormal excretion of PBG and symptoms of AIP in an apparently normal but obese patient (111).

Hereditary cutaneous hepatic porphyria (congenital CHP) is in some ways similar to AIP, especially in its acute phase. ALA and PBG are excreted in abnormal amounts in the urine due to the increased hepatic production of ALA synthetase. However, since there is no deficiency of the enzymes PBG deaminase or PBG isomerase, uroporphyrin III, coproporphyrin III, and protoporphyrin are formed, and fecal excretion of these pigments is elevated; uro- and coproporphyrin III may also be found in the urine. Symptoms are similar to AIP with the addition of cutaneous photo-sensitivity. The effect of certain drugs is similar to that described for AIP. The disease is inherited as a dominant characteristic. Another form of hereditary hepatic porphyria is hereditary coproporphyria, having similarities to AIP and to congenital CHP. However, large amounts of coproporphyrin are found in the feces and urine; little protoporphyrin and only periodic elevations in the excretion of ALA and PBG occur.

Acquired cutaneous hepatic porphyria (acquired CHP) is manifested by photosensitivity. It is not a familial disease and affects apparently normal individuals. This is due to the ease with which chemicals can induce the synthesis of ALA synthetase. There are marked elevations in urinary uro- and coproporphyrins, primarily uroporphyrin.

Porphyrinurias found in alcoholic cirrhosis of the liver, certain anemias, and chemical poisoning are characterized by increased excretion of copro-porphyrin in the urine (up to several milligrams per day) but not in the feces, and with little or no change in uroporphyrin excretion. Stich (182) has concluded that lead interferes in porphyrin metabolism in three ways: (a) inhibition of enzymatic conversion of ALA to PBG; (b) inhibition of conversion of coproporphyrinogen III to protoporphyrin, leading to increased excretion of coproporphyrin III; and (c) interference with incorporation of iron into protoporphyrin, leading to increased blood protoporphyrin and coproporphyrin, and hypersideremia. The elevated levels of ALA often found in lead poisoning may be due to a blockage of ALA dehydrase or to induction of high levels of ALA synthetase (8). The usefulness of urinary ALA determinations for detection of lead poisoning is a matter of some controversy (11, 35, 194, 195). Certain drugs can cause disturbances of porphyrin metabolism. Elevated levels of urinary ALA, PBG, coproporphyrin, and uroporphyrin, and fecal coproporphyrin, protopor-phyrin, and uroporphyrin have been demonstrated in healthy subjects treated with 2-allyloxy-3-methylbenzamide (146).

DETERMINATION OF δ-AMINOLEVULINIC ACID AND PORPHO-BILINOGEN

Granick and coworkers (81, 132) introduced a technic for determination of ALA and PBG in urine in which these two compounds are first separated by adsorbing PBG onto Dowex 2, the ALA not being retained. The ALA is then adsorbed onto Dowex 50, eluted with sodium acetate, condensed with acetylacetone to form a pyrrole, 2-methyl-3-acetyl-4-(3-propionic acid)pyrrole, which reacts with Ehrlich's reagent to give a color that is read photometrically. Condensation with ethyl acetoacetate can also be made, but the color formed with Ehrlich's reagent is more susceptible to positive interference from high concentrations of amino acids, ammonia, and glucosamine (132, 179). Another photometric method consists of a modification of the Jaffe picrate reaction and depends on the reaction of an alkaline picrate complex of ALA with HCl to produce an orange-red color (179). Quantitative methods for ALA and PBG, using anion and cation exchange resins in tandem, have given results which appear to be satisfactory (45, 168). Simplified procedures for ALA only utilize a single cation exchange column (185, 216) or adsorption on charcoal (9). Electrophoresis and thin layer chromatography have been used in conjunction with Ehrlich's and other color reactions (69, 161, 162).

PBG reacts with Ehrlich's reagent (*p*-dimethylaminobenzaldehyde) in acid solution to form a red compound. The condensation product, which is not an aldehyde, is nevertheless commonly called "porphobilinogen aldehyde," referring to the aldehyde group of the *p*-dimethylaminobenzaldehyde through which combination occurs. The reaction is believed to proceed as follows (132, 155):

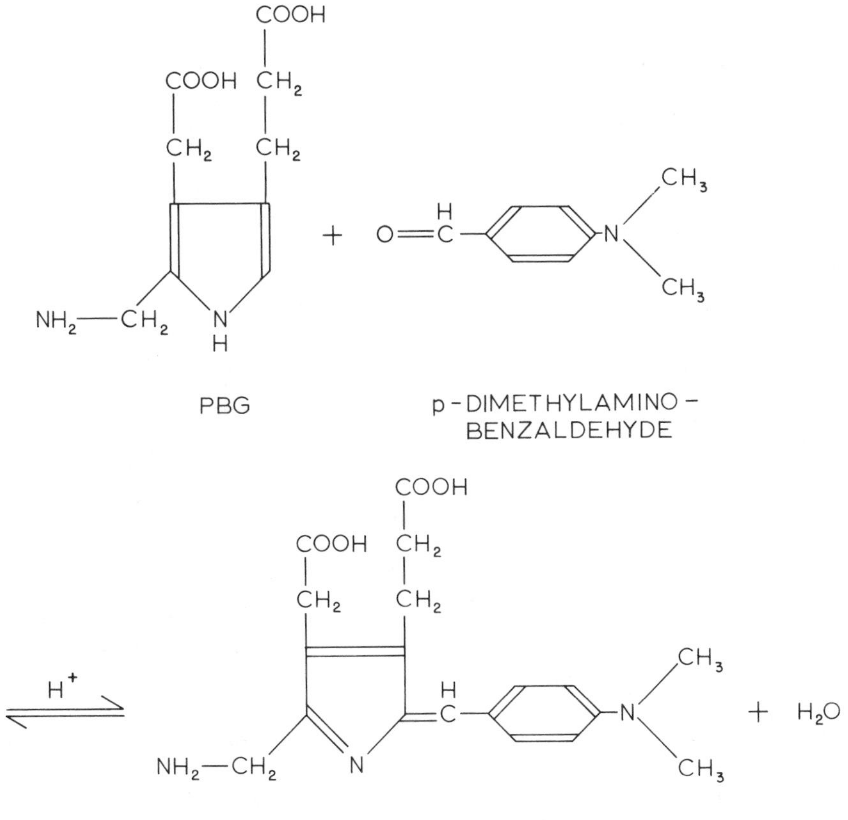

PBG

p-DIMETHYLAMINO-
BENZALDEHYDE

RED CONDENSATION PRODUCT
(RCP)

Color intensity increases rapidly and then diminishes slowly. This is believed to be the result of a second reaction in which the colored condensation product reacts with another molecule of PBG to form a colorless dipyrrylphenylmethane:

$$PBG + RCP \rightleftharpoons [\text{dipyrrylphenylmethane}] + H^+$$

The dipyrrylphenylmethane may in turn irreversibly lose dimethylaniline to form a dipyrrylmethene:

$$DIPYRRYLPHENYLMETHANE \rightarrow [\text{dimethylaniline}] + [\text{dipyrrylmethene}]$$

The color decay follows second order kinetics. Color is inhibited by thiols, probably by displacing the equilibrium normally attained, and possibly by competition in the strongly acid solution for the free α position of the pyrrole nucleus (155).

Vahlquist (191) in 1939 presented two technics for quantitative determination of PBG in urine: (a) application of the Ehrlich reaction directly to urine, and (b) application after prior isolation of the PBG on alumina. Direct determination appears to be satisfactory if the PBG level is high, results being only about 5% too high (155). Inhibition by urea and thiols is successfully minimized because of the great dilution involved when the level is high. Quantitation of normal or moderate levels is, however, a different story, and prior isolation of PBG from interferences is required. Rimington *et al* (155) and Mauzerall and Granick (132) presented methods in which the PBG is adsorbed onto a column of Dowex 2 ion exchange resin and then eluted with acetic acid. Aside from greater specificity and removal of color inhibitors, another advantage of prior isolation is that the color obeys Beer's law, a situation which does not exist when color is developed directly on urine.

Watson and Schwartz (211) in 1941 introduced a qualitative test for PBG in urine in which Ehrlich's reagent and sodium acetate are added to urine. In order to differentiate the color formed with PBG from urobilinogen or indole, extraction with $CHCl_3$ is made. PBG aldehyde color is not extracted whereas most others are. As discussed in detail later (see *Accuracy*, in the next section on *Qualitative Determination of Porphobilinogen*), there are a number of $CHCl_3$-insoluble reactants other than PBG but these nonspecific colors are soluble in butanol. Rimington (151) proposed extraction with an amyl-benzyl alcohol mixture (3:1) instead of butanol and without pre-liminary $CHCl_3$ extraction. Watson *et al* (206), though, believe it best to use $CHCl_3$ first, then butanol, or the procedure would fail to distinguish between urobilinogen and other nonporphobilinogen interferences. With (226) reported a screening test in which a drop of urine is placed on a talc plate and treated with Ehrlich's reagent. He also reported no false positive reactions in a study of 700 patients where 1 or 2 drops of urine were added directly to 1 ml of 2% *p*-dimethylaminobenzaldehyde in 6 *N* HCl. Doss (44) suggested a rapid screening test which involves isolation of the PBG on an ion exchange resin column, elution, and treatment with Ehrlich's reagent. A similar batch procedure has been used with readings taken on a spectro-photometer at two wavelengths to rule out interferences (138).

For routine diagnostic purposes it would appear that the qualitative test of Watson *et al* (206) for porphobilinogen is entirely adequate. Nevertheless, since a quantitative procedure is usually desirable for research purposes and since it can be incorporated into the quantitative procedure for ALA, a quantitative method is also presented later. False negative results have been observed when using the qualitative test on *random* urines, leading to the suggestion that 24 hour samples are preferable (107).

QUALITATIVE DETERMINATION OF PORPHOBILINOGEN

(method of Watson et al, Ref 206)

PRINCIPLE

PBG reacts with Ehrlich's *p*-dimethylaminobenzaldehyde reagent to form porphobilinogen aldehyde, which has a red color (this reaction was discussed in the previous section). Sodium acetate is next added to increase the pH so that maximal color occurs if urobilinogen is present and to render the urobilinogen aldehyde formed extractable with $CHCl_3$. If the reaction is positive $CHCl_3$ extraction is made, and if the aqueous phase is still red a butanol extraction is carried out. If appreciable red color remains in the aqueous phase, after butanol extraction a blank is also run on the urine with HCl alone. This is a modification of the original technic of Watson and Schwartz (211).

REAGENTS

Ehrlich's Reagent. Add 0.7 g reagent grade *p*-dimethylaminobenzaldehyde and 150 ml conc HCl to 100 ml water. The *p*-dimethylaminobenzaldehyde must be colorless or very light yellow, not brown or brown-red. Store reagent in an amber bottle.
Sodium Acetate, Saturated Aqueous
pH Test Paper. Must include pH range 4–5.
CHCl₃
n-Butanol
HCl, Conc

PROCEDURE

1. Mix 2.5 ml Ehrlich's reagent and 2.5 ml urine in a test tube. Add 5 ml saturated sodium acetate and mix. Check pH with pH paper (pH should be 4–5, assuring complete conversion of acid to acetic; if necessary, add more sodium acetate). If no color forms, the test is negative. If color forms, proceed with step 2.
2. Add 5 ml $CHCl_3$ and shake. If supernatant aqueous phase is colorless or is only faint pink, orange, or yellow, the test is negative. If significant color remains in the supernatant aqueous phase, proceed with step 3.
3. Transfer most of supernate and shake with 0.5 vol *n*-butanol. If aqueous phase is pink, red, or red-violet, proceed with step 4.
4. Add 1 ml conc HCl to 2.5 ml urine. If no pink, red, or red-violet color develops, and if aqueous phase at step 3 was pink, red, or red-violet, the test is considered positive for PBG.

For behavior of colors formed by substances other than PBG which are found in urine of patients with and without porphyria, see below under *Accuracy*.

NOTES

1. Stability of urine. The qualitative Ehrlich reaction may turn from positive to negative in a few hours if the urine is kept at room temperature (172). Apparently this is the result of two independent changes in the urine:

formation of an inhibitor of the reaction (206) and actual disappearance of PBG. PBG converts to porphyrin and to other substances as well. A small but significant decrease in PBG has been reported to occur at 0°C in 5 hours although no great change occurs in 100 hours; but after 10 hours at 20°–37°C as much at 70% may disappear (17). Cookson and Rimington (32) studied the stability of PBG in urine at 20°C and different pH's with the following results: at pH 6.5 and at pH >10 (Na_2CO_3 added), all PBG was gone in 17 days; in 0.5 N HCl, 55% disappeared in 17 days. It has also been reported that at least 2 weeks stability is achieved by adjusting urine to pH 6–7 and storing frozen or at 4°C (15). From a practical standpoint, however, it would be necessary for clinical diagnostic purposes only that sufficient PBG remains to give a positive qualitative test. So much PBG is usually found in urine of acute porphyria, e.g., up to about 10 mg/100 ml (81), that even after several days at room temperature enough would remain. Development of inhibitor, however, prevents the PBG remaining from being detected unless the quantitative procedure, which isolates PBG from the inhibitor, is used (cf *Stability* in the section on *Quantitative Determination of Porphobilinogen and* δ*-Aminolevulinic Acid in Urine*).

ACCURACY

In one study (92) in which the Watson-Schwartz test was applied to urines of 1000 miscellaneous patients, no false positive results were encountered; 16% were negative on addition of Ehrlich's reagent and sodium acetate and 84% were positive. All the latter were negative upon $CHCl_3$ extraction, i.e., the color went into the $CHCl_3$ (18%) or disappeared (yellow or slight orange tint in $CHCl_3$). Development of color after sodium acetate addition has been referred to as an "atypical porphobilinogen reaction" (75) or a "pseudo-porphobilinogen reaction" (155). Apparently, such a reaction may even occur with PBG, for it has been observed in some urines when PBG is added that color is suppressed until addition of sodium acetate unless large amounts of PBG are added (155). Two diseases in which false positive reactions were reported to occur are pellagra (203) and epilepsy (127). However, the latter could be due to treatment with promazine (107a); a variety of other drugs used do not interfere (177). It has been determined that extraction with butanol renders the test more specific than $CHCl_3$ extraction, about 10%–50% of the total urine aldehyde color being extractable in cases of acute porphyria (206). Extraction with $CHCl_3$ first, however, permits further differentiation of the interference. A false positive reaction, correctable by employing the butanol extraction after extraction with $CHCl_3$, has been reported in a case of partially obstructory carcinoma of the colon (66); 59 weakly positive reactions, obtained in a study of 1000 routine urines, were also rendered negative in this way (189). The following is a composite of published information on substances giving a red color in the test (20, 99, 123, 180, 203, 206, 212).

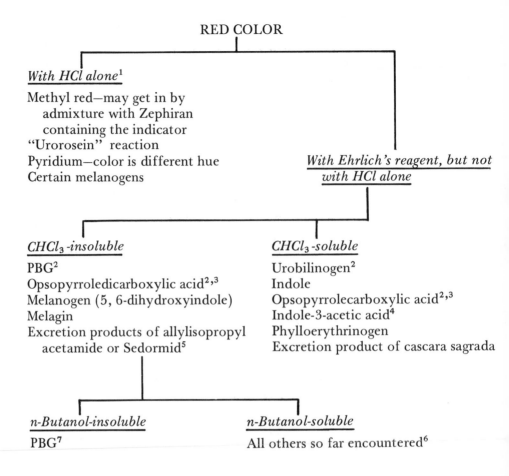

RED COLOR

With HCl alone[1]

Methyl red—may get in by
   admixture with Zephiran
   containing the indicator
"Urorosein" reaction
Pyridium—color is different hue
Certain melanogens

**With Ehrlich's reagent, but not
with HCl alone**

CHCl$_3$-insoluble

PBG[2]
Opsopyrroledicarboxylic acid[2,3]
Melanogen (5, 6-dihydroxyindole)
Melagin
Excretion products of allylisopropyl
   acetamide or Sedormid[5]

CHCl$_3$-soluble

Urobilinogen[2]
Indole
Opsopyrrolecarboxylic acid[2,3]
Indole-3-acetic acid[4]
Phylloerythrinogen
Excretion product of cascara sagrada

n-Butanol-insoluble

PBG[7]

n-Butanol-soluble

All others so far encountered[6]

NOTES

1. Urines containing large amounts of uro- or coproporphyrin are so deep red that the supernatant aqueous phase after CHCl$_3$ extraction is red and may be confused with a positive reaction. Comparison with a HCl control should be made.

2. With PBG, major color formation occurs before sodium acetate is added; with urobilinogen most color is formed after sodium acetate is added. Opsopyrrolecarboxylic and opsopyrroledicarboxylic acids are intermediate.

3. Found in some cases of acute porphyria (206).

4. Largely fades when sodium acetate is added.

5. From a study of rabbits intoxicated with these drugs (75). Intoxication of rabbits produces an hepatic form of porphyria with excretion of PBG, but prior to appearance of PBG in urine, an "atypical" or "pseudoporphobilinogen reaction" may be present. There may be no color until sodium acetate is added.

6. So stated by Watson *et al* (206), but no mention was made of the behavior of sulfonamides, procaine, 5-hydroxyindoleacetic acid, urea, indican, tryptophan, indoxyl sulfate, bilirubin, and skatole, which are all substances reported to react with Ehrlich's reagent. (These are discussed under *Semiquantitative Determination of Urobilinogen in Urine and Feces* in Chapter 22.) Urine contains traces of many other substances reacting with Ehrlich's reagent.

7. Tryptophanuria as seen in tryptophan loading tests is reported to be indistinguishable from PBG in this test (112).

Schwartz *et al* (172) state that final proof of the presence of PBG is its conversion to uroporphyrin on heating in dilute acid (i.e., 0.03 *N* HCl or 5% acetic acid). Rimington *et al* (155) stressed the importance of confirmation by paper chromatography.

Conversely, false negative results apparently can occur. Ludwig and Epstein (124) encountered several urines of porphyrics which yielded negative Watson-Schwartz tests in spite of elevated PBG levels. This was due to the presence of an inhibitor (not urea) which could be removed by adsorption onto Dowex 2 resin. "Inhibitors" of the reaction have been identified as certain indolic compounds (187, 212).

QUANTITATIVE DETERMINATION OF PORPHOBILINOGEN AND δ-AMINOLEVULINIC ACID IN URINE

(method of Mauzerall and Granick, Refs 64, 132)

PRINCIPLE

Porphobilinogen (PBG) and δ-aminolevulinic acid (ALA) are separated by adsorbing the PBG onto alumina and then eluting with dilute acetic acid. The ALA, which is not adsorbed on the alumina column, is then isolated on Dowex 50 and eluted with sodium acetate. The PBG is determined with Ehrlich's reagent, while the ALA is first condensed with acetylacetone to form a pyrrole, which then reacts with Ehrlich's reagent and is quantitated photometrically. Mauzerall and Granick used Dowex 2 ion exchange resin for separation of ALA from PBG. Alumina was found to be satisfactory for this separation and therefore is being substituted because it provides faster flow rate and avoids the necessity of preliminary conversion of Dowex 2 to the acetate form (64).

Figure 24-2 presents the absorption curves of the colors formed by PBG and the acetylacetone condensation product of ALA with Ehrlich's reagent.

REAGENTS

Alumina. Alcoa Activated, Grade F-20, Alcoa.

Dowex AG 50W X8, 200–400 Mesh, Hydrogen Form. Our laboratory obtains this from Bio-Rad. It was found necessary to recycle the resin since suspension in 4 *N* NaOH gave a dark brown supernate. One pound wet weight of resin is placed in 2 vol 4 *N* NaOH and warmed over a steam bath for about 8 hours with occasional stirring. The NaOH solution is changed twice during this time. Resin is filtered off with Whatman 42 paper in a Büchner funnel and rinsed with distilled water until free of alkali. Resin is suspended in 4 vol 2 *N* HCl, heated over a steam bath for about 8 hours, the HCl being changed once during this period. The resin is then filtered off with Whatman 42 paper and washed with distilled water until free of Cl$^-$ (AgNO$_3$ test). Air is sucked through the funnel to remove the bulk of the water and the resin then stored and used without further drying.

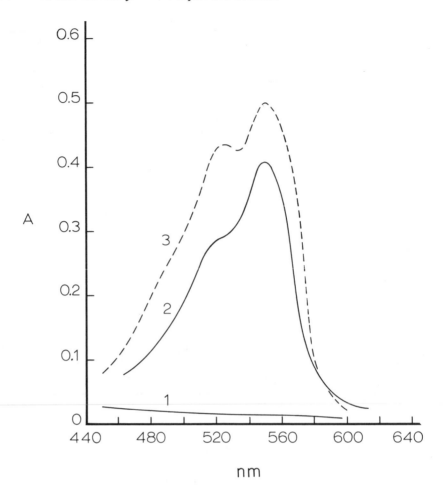

FIG. 24-2. Absorption curves of ALA and PBG. (1) ALA reagent blank vs water, (2) µg ALA standard vs water, (3) PBG from a urine vs reagent blank. Beckman DK-2A recording spectrophotometer.

Sodium Acetate, Saturated Aqueous
Sodium Acetate, 0.5 M. 6.8 g sodium acetate trihydrate per 100 ml.
Acetylacetone (2, 4-Pentanedione). Eastman Kodak, No. 1088.
Acetic Acid, 6%. 60 ml glacial acid per liter.
Acetate Buffer, pH 4.6. 5.7 ml glacial acetic acid and 13.6 g sodium acetate trihydrate per 100 ml.
Modified Ehrlich's Reagent. Dissolve 1.0 g *p*-dimethylaminobenzaldehyde in about 30 ml glacial acetic acid. Add 8.0 ml 70% perchloric acid. Dilute to 50 ml with glacial acetic acid. Prepare fresh daily; it is reportedly stable only about 6 hours (91).
Stock ALA Standard. 20.5 mg δ-aminolevulinic acid hydrochloride per 100 ml. A satisfactory grade is available from Calbiochem. This standard is stable for several weeks in the refrigerator. 1 ml = 0.16 mg ALA.

Working ALA Standard. Dilute stock 1:20 with pH 4.6 acetate buffer. 1 ml = 8 μg. The author does not know the stability of the working standard.

PROCEDURE

1. If urine is clear, adjust pH of 20 ml urine to 6—6.5 (wide range pH paper) using glacial acetic acid or a saturated solution of sodium acetate. If a precipitate is present in the original urine, acidify to about pH 4, mix to dissolve any phosphate present, then readjust an aliquot equivalent to 20 ml urine to pH 6—6.5.

2. Set up a small chromatographic column. Our laboratory uses a specially designed column which has a funnel top (28 mm ID), a 10 mm ID column which is 150 mm in length, and a 2 mm ID tip which is 20 mm in length; the overall length is 260 mm. These columns can be obtained from Cal-Glass for Research, Inc., catalog No. CG905. Insert a small pledget of glass wool and push down to capillary portion of column.

3. Weigh about 5 g alumina into a 100 ml beaker. Add about 50 ml water and swirl to stir up the alumina. Let stand 30 sec and decant supernate. Repeat this process six times or until supernate is clear. Transfer the alumina, suspended in water, to the chromatographic column with water, allowing the alumina to settle. *N.B. Alumina must at all times be covered with fluid.* Addition of alumina is continued until column is about 6 cm in height.

4. Transfer urine to the column. When urine level has nearly reached top of alumina, add 5 ml water to rinse column. Continue water wash until a total of 20 ml has been added. Collect eluate in a small flask after urine has been added. This fraction contains the ALA.

5. Elute PBG with 24 ml 6% acetic acid. Discard the first 4 ml and collect the next 20 ml in a small flask. See step 10 for PBG determination.

Determination of ALA

6. Set up column for Dowex 50 resin as follows: Use same type of column as in step 2. Prepare slurry of the resin in water and from this form a column 2.0 ± 0.1 cm in height. Place pledget of glass wool on top the column of resin.

7. Dilute eluate from step 4 to 80 ml with water. Transfer 4.0 ml to Dowex column. Wash column with 16 ml water (urea is washed through). Add 3.0 ml 0.5 *M* sodium acetate. After draining, elute ALA by adding 7 ml 0.5 *M* sodium acetate. Collect this eluate in a 10 ml volumetric flask. Add 0.2 ml acetylacetone, dilute to mark with pH 4.6 acetate buffer, and mix.

8. Set up in other 10 ml volumetric flasks:
   Blank: 7.0 ml 0.5 *M* sodium acetate, 1.0 ml distilled water, and 0.2 ml acetylacetone; dilute to mark with pH 4.6 acetate buffer. Mix.
   Standard: 7.0 ml 0.5 *M* sodium acetate, 1.00 ml working ALA standard, and 0.2 ml acetylacetone, diluted to mark with pH 4.6 acetate buffer. Mix.

9. Place stoppered flasks in boiling water for 10 min. Cool to room temperature. To 2.0 ml aliquots of blank, standard, and unknown add 2.0 ml modified Ehrlich's reagent, mix, and transfer to cuvets of 1 cm light path. After 15 min read absorbance of unknown (A_x), blank (A_b), and standard (A_s) at 553 nm against water.

Calculation:

$$\text{mg ALA/100 ml urine} = \frac{A_x - A_b}{A_s - A_b} \times 0.008 \times \frac{100}{20} \times \frac{80}{4}$$

$$= \frac{A_x - A_b}{A_s - A_b} \times 0.8$$

Determination of PBG

10. Add 2.0 ml modified Ehrlich's reagent to 2.0 ml acetic acid eluate from step 5. Mix.

11. Measure absorbance (1 cm light path) at 15 min at 553 nm against reagent blank composed of 2.0 ml water and 2.0 ml modified Ehrlich's reagent. If results are above normal, an absorption curve should be run in the region of 515–570 nm to make certain that the color obtained conforms to that given by PBG (Fig. 24-2).

Calculation:

$$\text{mg PBG/100 ml urine} = A_x \times 0.73$$

NOTES

1. Beer's law. Assuming a spectrophotometer is used with half-bandwidths of no more than a few nm:

ALA. It has been reported (91, 132) that linearity exists up to an absorbance of 1.2. Our laboratory, however, has observed deviation as low as about 0.8.

PBG. Haeger-Aronson (91) observed linearity up to only 0.20, but linearity was obtained in our laboratory up to an absorbance of 2.0 on a Beckman DU spectrophotometer.

2. Color stability.

ALA. It has been reported (10, 132) that the color reaches a maximum in 6–15 min and then remains stable for at least 15 min. This, however, has not been the experience in our laboratory where colors increase with time for long periods, the rates varying between standards and unknowns. Under such conditions the time at which color is read is entirely arbitrary, and it is obvious that this situation introduces an uncontrolled variable in the analysis.

PBG. Color reaches a maximum by 15 min and then remains stable for about 10 min (132).

3. Standardization.

ALA. According to Mauzerall and Granick (132) the molar absorptivity of the color formed is 7.2×10^4.

PBG. According to Mauzerall and Granick (132) the molar absorptivity of the color formed is 6.2×10^4. The factor 0.73 used in calculations was derived from this absorptivity. PBG is not available commercially but can be isolated from urine of patients with acute porphyria and purified by the method of Cookson and Rimington (32).

4. Stability of samples.

ALA. Urine appears to be stable as long as the pH remains <7 (15, 91). Thus it has been reported (91) that there was no change in refrigerated urine for at least 21 days if the pH did not become alkaline, and urine was stable at room temperature for 10 days with the ALA level decreasing after the pH rose above 8. About half of the ALA in unpreserved, light-exposed urines is lost in 24 hours at room temperature (195). Freezing or refrigeration is recommended (15), although stability for several days at 25°C has been reported in urines kept in the dark and preserved with tartaric acid (193).

PBG. This was discussed under *Stability* in the earlier section on *Qualitative Determination of Porphobilinogen.* For results to be valid from a quantitative standpoint it is best that fresh urine be analyzed, but if there is a delay the urine should be frozen. At room temperature it appears that PBG is most stable at pH <1.

5. Determination of ALA in serum and plasma. Haeger-Aronson (91) adapted the urine procedure of Mauzerall and Granick to serum, adding iodoacetamide to stabilize the Ehrlich color, presumably by removing interfering thiols. A similar adaption for plasma has been found useful in testing patients acutely ill with symptomatic lead intoxication and acute intermittent porphyria (22, 137).

ACCURACY

ALA. The method is not absolutely specific for ALA, since other aminoketones and glucosamine condense with acetylacetone to form colored products (89, 91, 130). Levels of 0.3 mg glucosamine, 2 mg ammonia, and 2 mg glycine per milliliter of sample each give an absorbance in the reaction of 0.01 (132), equivalent to about 0.5 µg ALA per milliliter. In the case of significantly elevated levels specificity appears to be adequate. For example, in a study of lead workers, most of the reacting material was identified as ALA (89, 91). Of the color formed with normal urine, however, only 10%–20% results from ALA, 20%–40% behaves as a pyrrole corresponding to an aminoketone without a carboxyl group (e.g., aminoacetone), and the remainder is organic solvent-insoluble material such as might be formed from glucosamine (132).

Conversion of ALA to 2-methyl-3-acetyl-4-(3-propionic acid)pyrrole is 95% of theoretical (132), but this is corrected for by calculating from ALA standards.

Recovery of ALA from aqueous solution or from urine using Dowex 2 resin followed by Dowex 50 resin is reported to be >91% (10, 91). Similar recoveries are obtained in our laboratory using alumina in place of the Dowex 2.

PBG. Schwartz et al (172) state that the level of PBG in normal urine is actually zero (at least it is undetectable). They were unable to detect any increase in uroporphyrin by fluorescence following treatment under conditions that should convert PBG to uroporphyrin. They also recommended determination of the difference between uro content of untreated urine and urine acidified and heated for confirmation of the presence of PBG.

Mauzerall and Granick (132) found their technic using Dowex 2 to be quite specific for PBG, although phenothiazine medications have been reported to give false positive results (149). There is no reason to believe that substitution of alumina should lessen the specificity. Recovery from Dowex 2 is quantitative within experimental error. Recovery of PBG from urine is >96% (91). Urea and ascorbic acid inhibit the Ehrlich reaction (147), but these are effectively removed in the procedure presented.

The use of a spectrophotometric curve for insuring the specificity of the PBG color reaction has been recommended (149) and is incorporated in the method presented here. Loriaux et al (122) used an Allen correction where significantly lower results for PBG were obtained by correcting the reading at 555 nm for those observed at 535 and 575 nm. They found this precaution unnecessary for ALA.

PRECISION (95% LIMITS)

ALA. ±0.04 mg/100 ml (91).
PBG. ±0.02 mg/100 ml (91).

NORMAL VALUES

ALA. The normal range for adults is about 0.1—0.6 mg/100 ml urine (45, 88, 90, 91, 126). Below the age of 15 years the top normal limit is 0.49 mg/100 ml (90). The normal daily output for adults is reported up to levels of about 7 mg (77, 126, 143, 182).

The normals in serum are variously reported with upper limits of 0.01—0.03 mg/100 ml (22, 62, 77, 91, 137). The normal content of feces is uncertain (172). That of erythrocytes has been reported up to about 52 μg/100 ml (77).

PBG. The normal range for adults is 0—0.2 mg/100 ml urine (45, 88, 90, 91, 126). The normal daily output is up to 2.0 mg (77, 126, 172). Peterka et al, however, report 0—4 mg/24 hours (143). The range for children below the age of 15 has been reported to be 0—0.11 mg/100 ml (90), while in a study of 339 children aged 0—12 years, the mean daily excretion was 1.61 mg/24 hours with 95% of the values under 0.08 mg/kg body weight/24 hours. This value did not vary with age, and higher results were found in winter than in summer (5).

The normal contents of feces and erythrocytes are uncertain (172). Normal serum PBG is reported to be 0–4.4 μg/100 ml (137).

DETERMINATION OF PORPHYRINS

Porphyrins are amphoteric substances with isoelectric points between about pH 3.0 and 4.5. Their weak base character depends on the two tertiary nitrogens in the pyrrole nuclei, and their acidic character depends on the carboxyl groups in the side chains. Depending on the pH, porphyrins exist in solution in free base, monocationic, or dicationic form. As the number of carboxyl groups increases, solubility decreases in organic solvents and increases in water. Porphyrins possessing four or less carboxyl groups (copro and proto) are soluble in ether and in ethyl acetate; uroporphyrin with eight carboxyl groups is not soluble in ether and soluble in ethyl acetate only under special conditions. Thus uroporphyrin III and Waldenström porphyrin are extractable from urine by ethyl acetate at pH 3.0–3.5, i.e., the region of the isoelectric point at which the solubility of porphyrin in water is minimal.

Porphyrins generally are extracted into organic solvents from dilute acetic acid (pH 4–6); the more polar the organic solvent, the more readily the porphyrin is extracted. There can, however, be complications. For example, two precautions are necessary in extracting coproporphyrin (and probably 5- to 7-carboxyl porphyrins) (172): (a) Even with proper pH, excess Na$^+$, which may result when a dilute HCl solution is neutralized with sodium acetate, may prevent extraction. (b) Excess acetic acid may prevent extraction. Both ether and ethyl acetate are used by various workers for extraction of coproporphyrin, there being arguments in favor of each (172): In favor of ethyl acetate are the facts that variation in Na$^+$ and acetic acid concentrations have less effect on copro extraction by aqueous washes, peroxides present in ether may cause destruction of the porphyrin upon subsequent extraction with HCl, and the fire hazard is less than with ether. On the other hand, ether gives a more rapid and clean separation of uroporphyrin from coproporphyrin.

Another approach to isolation of porphyrins from interfering impurities, as well as a means of concentrating porphyrins from a large volume of urine, is adsorption followed by elution. Among adsorbents used have been alumina (188, 197), calcium phosphate (49, 186), talc (30, 83, 220), kieselguhr (150, 156, 220), magnesium hydroxide (39), Florisil (165, 166), lead acetate (16, 34), and an anion exchange resin (94, 128).

Porphyrins are quantitated by spectrophotometric or fluorometric measurement. Their absorption characteristics are due to resonance of the molecule, and there is little difference in the spectra of their free or ester forms. Absorption curves differ somewhat in organic solvents and in acid aqueous and alkaline aqueous solutions. Ordinarily, absorbance measurements are made in analytic work with porphyrins in aqueous HCl solution, in which porphyrins have four absorption peaks in the following regions (listed in order of decreasing intensity): 400–410 nm (Soret band), 540–560 nm, 590–603 nm, 572–582 nm. Gamgee (72) in 1896 first described the Soret band (absorption peak) for porphyrins, and it is the wavelength used for quantitative measurement in spectrophotometric procedures. The Soret band is characteristic of all conjugated tetrapyrroles. To correct for nonspecific background absorbance many technics employ the Allen correction in which absorbance is measured at three wavelengths—the Soret peak and two others, one on each side of the Soret peak. The validity of this approach rests on the validity of the assumption of linearity of the background absorption, a situation that does not always exist (230). Table 24-2 lists the wavelength at the peak and the absorbance of a one percent solution in dilute HCl with a light path of one cm $(A^{1\%}_{1\,cm})$ for the three commonly determined porphyrins. The absorption peaks are narrow so that the $A^{1\%}_{1\,cm}$ values are valid only for measurements made with a spectrophotometer providing high resolution (narrow half-bandwidth). Peaks are shifted toward longer wavelength with increasing acid concentration, the shift being about 2–5 nm with a change from 0.1 to 7.5 N HCl (172, 232). Absorption spectra of I and III isomers do not differ significantly (108, 152). The spectra are somewhat different for the free base and monocationic and dicationic forms—state of ionization of the tertiary nitrogens of the pyrrole groups (230)—the position and intensity of the Soret band changing with pH, and the intensity being lowest at the isoelectric point (108, 139). For example, copro exists as the free base at pH 9.5 with the Soret peak at 391 nm, as the monocationic form at pH 5.6 with the peak at 372 nm, and as the dicationic form at pH 1 with the peak at 400 nm (139). Zinc porphyrins have, in addition to the Soret band, two equally intense peaks in the regions of 535–540 and 575–580 nm (172). Copper chelates of the porphyrins are reported to have more intense spectal absorption peaks than free porphyrins, and a spectrophotometric method exploiting this has been presented (37).

TABLE 24-2. Absorption Characteristics of Porphyrins in Dilute HCl[a]

Porphyrin	HCl (N)	Soret peak (nm)	$A^{1\%}_{1\,cm}$
Uroporphyrin	0.5	405	6520
Coproporphyrin	0.1	400	7470
Protoporphyrin	1.37	408	4890

[a]Values derived by Rimington, ref 152, as a compromise of his own data and the data of others.

When exposed to near-ultraviolet light, free porphyrins in acid solution exhibit orange fluorescence; in alkaline solution or in organic solvents they give a red fluorescence. They can be excited at other wavelengths, but the wavelength of the Soret peak is the most sensitive and specific. Fluorescence vanishes or becomes minimal at the isoelectric point (67).

Sensitivities for porphyrins by fluorometry and spectrophotometry (Soret band) are about 2×10^{-10} and 3×10^{-7} *M*, respectively; thus fluorometry is approximately 1000 times more sensitive. There is a difference of opinion as to which is the preferred technic, and both approaches have been presented as *Standard Methods* for porphyrins in urine (64, 188). Lemberg and Legge (116) and Rimington (152) prefer spectrophotometry when sufficiently sensitive because, in their opinion, fluorometry requires greater purification of the porphyrins prior to measurement, which is difficult to achieve without loss. Furthermore, fluorescence varies with pH and is affected by impurities, even inorganic ions, so that analytic errors are unpredictable and may be large. With (224) points out that fluorescence is proportional to porphyrin concentration over a comparatively narrow range, both positive and negative interferences are encountered, and fluorometry does not allow the critical observation of the position of the Soret peak, useful in identifying the porphyrin being measured. He therefore recommends the use of spectrophotometry rather than fluorometry unless the added sensitivity of the latter is needed.

Conversely, Schwartz *et al* (172) prefer fluorometry because, in their opinion, in addition to greater sensitivity it is more specific than spectrophotometry. Fluorometry is used for quantitation (128, 158, 172, 174, 188) of the extracts of biologic materials; but it is the use of the *spectrofluorometer* with its high resolution and sensitivity (128) which will no doubt provide the means to achieve more accurate and precise results than are obtainable by spectrophotometry (see Chapter 3). In this author's opinion, the major obstacle to the use of fluorescence in porphyrin quantitation is the need for materials of known purity for standardization.

The problem of the determination of porphyrins in biologic samples is rendered difficult by their multiplicity and their isomeric forms, the relative instability of some of them, and their low concentrations. As has been said (172), no procedure is known which gives an accurate analysis of all porphyrins in any biologic sample. The terms "coproporphyrin," "uroporphyrin," and "protoporphyrin" as used in conjunction with analytic procedures are actually arbitrary terms of convenience since copro and proto as usually isolated contain 5%–30% of other porphyrins, and uro usually contains up to 10% of others (172).

It is important to note that the spectrophotometric formulas (results calculated from readings at three wavelengths) given in some of the earlier methods referred to in the following sections are in error because of subsequent changes made in the accepted absorptivities (152, 217).

Determination of Coproporphyrin in Urine. Saillet (160) in 1896 introduced ether extraction of coproporphyrin from aqueous media acidified with acetic acid. Fisher (68) extended this by extracting the porphyrin in the ether into HCl. Among the approaches to the determination of coproporphyrin in urine are the following: (a) Urine is acidified with acetic acid to pH 4–6, extracted with ether, and the ether washed with water; copro is then extracted into 0.1 N HCl and determined spectrophotometrically (three wavelengths) (167, 232). A variation of this technic involves batch adsorption of PBG, uro, and copro onto Dowex 2 resin, elution of PBG and copro with 50% acetic acid, addition of sodium acetate to negative with Congo red, extraction of copro into ether, then back into 0.1 N HCl and spectrophotometry (94). As discussed previously, such procedures do not measure Zn coproporphyrin and/or coproporphyrin precursors since HCl stronger than 0.1 N HCl is necessary for extraction. (b) Urine is acidified with acetic acid, extracted with ether, and the ether washed with water; copro is extracted into 3 N HCl, taken to pH 5, extracted into 20:1 ether-acetic acid, and extracted into 0.1 N HCl, followed by spectrophotometry (50). It has been claimed that only about 50% of copro precursor is extracted by 3 N HCl (232). (c) Column chromatography on Hyflo Super-Cel follows esterification, then spectrophotometric measurement of the copro ester fraction (25). (d) Urine is acidified with acetate buffer, copro extracted with ethyl acetate (63, 64, 175) or ether (102), the extract washed with a solution of I_2 and sodium acetate (to convert chromogen), and copro extracted into 1.5 N HCl and measured fluorometrically or spectrophotometrically. The same general procedure using ether (65) or ethyl acetate (188) has been employed without I_2 treatment. (e) Urine is acidified with acetic acid, extracted with butanol-ethyl acetate, copro extracted into 10% NaOH, the alkaline solution acidified with acetic acid and extracted with ether, and the copro determined fluorometrically (18).

Determination of Uroporphyrin in Urine. Among the approaches to the determination of uroporphyrin in urine are the following: (a) Coproporphyrin is first removed by ether or ethyl acetate extraction (see above) and the uroporphyrin remaining in the aqueous phase is quantitated spectrophotometrically (three wavelengths) (167). This approach is satisfactory only with very high levels of uroporphyrin because of otherwise significant interference by other pigments present. (b) Coproporphyrin is removed, and the uroporphyrin in the aqueous phase is adsorbed onto calcium phosphate; the latter is separated and dissolved in 0.5 N HCl and spectrophotometry carried out (2, 157, 167, 186). It has been suggested (157) that the ratio of uro- to coproporphyrin usually is so high that removal of coproporphyrin may not always be required. This would beem to be a hazardous assumption to make. Adsorption on the precipitate from addition of lead acetate has also been used (16). (c) Coproporphyrin is removed, and the uroporphyrin in the aqueous phase is adsorbed onto alumina; it is then eluted with 1.5 N HCl and measured fluorometrically (188). (d) Coproporphyrin is removed and the uroporphyrin in the aqueous phase is extracted into cyclohexanone at pH 1.5 (46) or into butanol (38, 63, 64) and then quantitated. (e) Column chromatography on Hyflo Super-Cel after esterification followed by spectrophotometric measurement on the uro ester fraction (25). (f) PBG, uro, and copro are adsorbed onto Dowex 2 resin (batch treatment); PBG and copro are eluted, the resin dried, and uro is eluted with methanol saturated with HCl vapor; the solution is allowed to stand for 1 day at room temperature to effect esterification; the uro ester is extracted into $CHCl_3$, evaporated to dryness, taken up in 7 N HCl, and measured spectrophotometrically (three wavelengths) (94). (g) At pH 6.2 copro is in monocationic form and uro is in free base form. Copro interferes very little with spectrophotometric measurement (three wavelengths, Soret peak 399 nm) of uro at this pH (230).

Preformed and total uroporphyrin—i.e., preformed plus that which can be derived from PBG—can be determined from results obtained on samples with and without treatment either by heat (2) or by heat and acid (172, 188).

Screening Tests for Porphyrins in Urine. Although it is true that the presence of a purple-red color in urine or the darkening of a urine on standing is suggestive of the presence of an increased porphyrin concentration, such observations scarcely constitute proof. Conversely, the existence of a normal color does not rule out an elevated level. Even a porphyrin concentration of 500 μg/liter of urine is not visually detectable. Demonstration of an orange or red fluorescence upon activation with an ultraviolet lamp is certainly a better indication, but even this requires confirmation by a more exacting procedure.

A number of simplified procedures of a qualitative or quantitative nature have been suggested. These include the following: (a) Ehrlich's reagent (*p*-dimethylaminobenzaldehyde) is added to urine prior to qualitative detection of porphyrins to decrease nonspecific fluorescence (103). (b) There is a qualitative modification (7) of the fluorometric procedure of Schwartz *et al* (175) for coproporphyrin. (c) Porphyrins are adsorbed from urine onto lead acetate, eluted, chromatographed on filter paper in a test tube apparatus, and detected fluorometrically on the paper (34). (d) Urine is diluted with 0.5 N HCl and total porphyrins quantitated spectrophotometrically (three wavelengths) (157, 167). This procedure may not be very accurate because of a hyperbolic rather than linear background absorption (230). In any event it is applicable to urines with high porphyrin levels. (e) Urine is acidified to pH 2, porphyrins adsorbed onto a Florisil column, the column washed, porphyrins eluted with acetone, the acetone solution evaporated to dryness, the residue dissolved in 0.5 N HCl, and total porphyrins determined spectrophotometrically (three wavelengths) (165). (f) Porphyrins are isolated on an anion exchange resin, eluted with HCl, and fluorescence observed visually under a mercury lamp (44). (g) Porphyrins are adsorbed on magnesium hydroxide which is then examined for fluorescence (39).

It would appear to the present author that such shortcuts as listed above are of limited usefulness; either they are too nonspecific, only applicable to very high porphyrin levels, or, in the case of (e), the procedure for total porphyrin is not much less work than procedures for fractionation into uro and copro. It is true that some of these simplified methods may be adequate in certain circumstances, e.g., when it is a question of whether the urine is from a case of acute porphyria, but this is not the situation in most laboratories where urines arrive as unknowns.

Determination of Porphyrin Isomers and Other Means of Separation and Identification of Porphyrins. To date there is no clearcut indication from a clinical standpoint for quantitation of individual porphyrin isomers.

Schwartz *et al* (172, 173) introduced the "fluorescence quenching" method for coproporphyrin isomer analysis which depends on the difference in fluorescence stability of the two esters in 30%–35% aqueous acetone in the cold, III being much more stable than I. Methyl esters of a coproporphyrin mixture of I and III can be adsorbed onto alumina and eluted preferentially by pure acetone or by acetone-water (210). Still another technic involves esterification, evaporation to dryness, and extraction of residue with cold absolute methanol, which dissolves copro III ester but not copro I ester (41). Uroporphyrin I can be separated from III by differences in solubility characteristics. Thus the ethyl acetate-soluble fraction of uroporphyrins, after the uroporphyrins are adsorbed onto calcium phosphate and then dissolved in HCl, is the III isomer (186). Both uroporphyrins can be extracted into ethyl acetate at pH 3.0–3.2 and then the III isomer extracted into 1 N HCl (2).

Porphyrins or their esters can be separated quite successfully by paper chromatography (23, 24, 29, 51, 54, 58, 60, 135, 141, 148, 164, 219) and by thin layer chromatography (114, 176, 222, 225, 227). These technics are used for detection and rough estimation of the relative amounts of porphyrins with different numbers of carboxyl groups and of different isomers. It has been pointed out that prior esterification can give rise to artifacts (56). Fractionation has also been carried out by column chromatography on MgO and Hyflo Super-Cel (24), alumina (134), and cellulose powder (52). Porphyrins with four or more carboxyl groups can be separated by electrophoresis on paper (33, 93, 95, 106, 121, 181, 218) or in agar gel (55). Separation of porphyrins with fewer than four carboxyl groups is very limited. Detection and quantitation following electrophoresis have been made by fluorescence and spectrophotometry of the porphyrins themselves (181) or of the colored product formed in the benzidine reaction with iron salts of the porphyrins (106). Falk (58) reviewed the subjects of paper chromatography, column chromatography, and electrophoresis of porphyrins.

Schwartz *et al* (172) have emphasized that precise identification by these technics is generally possible at best only with extensive prior purification, which may result in differential porphyrin losses. Furthermore, authentic reference materials for all porphyrins are not available, and formation of molecular complexes may confuse the picture obtained.

Determination of Porphyrins in Serum or Plasma. Porphyrin in serum or plasma can be quantitated by adsorption onto Florisil and elution followed by spectrophotometric measurement (three wavelengths) (166). The porphyrin may be present in a combined form; the substance with which it is combined was called "x-substance," but it would be logical to assume that it is a protein (36). Normal levels of total porphyrins in plasma are generally reported up to 0.8 µg/100 ml (77, 166).

Determination of Porphyrins in Erythrocytes. The presence of free, nonhemoglobin porphyrin in human erythrocytes was reported by van den

Bergh and Hyman (100) in 1928. Their method involves extraction of the cells with ethyl acetate-acetic acid (3:1), washing the extract with water, extraction of the porphyrin into 5% HCl, and further purification by taking the porphyrin into ether after adding sodium acetate and then back into HCl. In the original procedure of van den Bergh and Hyman the porphyrin in the final extract was quantitated by fluorometry, whereas in a later modification by van den Bergh and Grotepass (101) photometric measurement was made. Other modifications using fluorometry or photometry have been presented, differing primarily in the method of purification of the final extract (84, 85, 87, 96, 136, 171, 174, 202, 228). An important step is removal of coproporphyrin, which can be achieved by extraction of the ethyl acetate solution with 0.1 N HCl, protoporphyrin remaining behind. Schwartz *et al* use iodine to convert porphyrinogens to porphyrins (172) although this has not generally been done in analyzing the erythrocytes. Rimington (154) has used fluorescence microscopy and an extraction procedure using 2 drops of blood with visual observation of fluorescence to detect abnormal levels of erythrocyte porphyrins.

Determination of Porphyrins in Feces. Analysis is complicated by the presence of relatively large amounts of red-fluorescing chlorophyll derivatives, the presence of large amounts of bile pigments, and the variable presence of protoporphyrin, deuteroporphyrin, and mesoporphyrin (172). Schwartz *et al* (172) presented a method for separation of the major porphyrin fractions (copro, uro, proto, and deutero) taking advantage of the solubility of the 2-carboxyl porphyrins (e.g., deuteroporphyrin) in $CHCl_3$ and the insolubility of chlorophyll derivatives in dilute HCl. They do not recommend the method of Dobriner (41, 42) because of incomplete recovery of the 2-carboxyl porphyrins, nor the method of Holti *et al* (102) and Barnes (6) because of poor separation of copro and proto and failure to include analysis of uroporphyrin. $CHCl_3$ extraction of dicarboxylic porphyrins from HCl solution was used by Herbert (98) who also provided for the ether-insoluble porphyrins. Rimington (153) extracts porphyrins and precursors into ether, converts coproporphyrinogen to copro by iodine treatment, then quantitates copro and proto spectrophotometrically after extraction into 0.1 N HCl and 5% HCl, respectively. The fecal porphyrins have also been separated, identified, and quantitated using countercurrent distribution (71).

QUANTITATIVE DETERMINATION OF PORPHYRINS IN URINE

(method of Fernandez et al, Refs. 63, 64)

PRINCIPLE

Urine is adjusted to pH 4.8 and the coproporphyrin extracted with ethyl acetate. After treating the washed extract with dilute iodine solution to convert any coproporphyrinogen into coproporphyrin, the porphyrin is reextracted into dilute HCl. Aqueous washes of the ethyl acetate extract are combined with the aqueous phase remaining after extraction with ethyl acetate and readjusted to pH 3. The uroporphyrin is then extracted with *n*-butanol. Petroleum ether is added to reduce the solvent polarity and to facilitate reextraction of the uroporphyrin into HCl. Quantitation is by spectrophotometric examination of the HCl extracts at the Soret peak (Fig. 24-3) and comparison with known absorptivities for the porphyrins.

REAGENTS

Ethyl Acetate

Sodium Acetate, Saturated Aqueous Solution. Dissolve 375 g sodium acetate, $CH_3COONa \cdot 3H_2O$, in 250 ml warm water, allow to cool, and store at room temperature (about 25°C). Sodium acetate crystals should be present at this temperature.

Acetate Buffer, pH 4.8. Mix 5 ml glacial acetic acid with 15 ml water and 20 ml saturated aqueous sodium acetate solution. Stable several weeks at room temperature.

Sodium Acetate, 1%. Dissolve 16.6 g $CH_3COONa \cdot 3H_2O$ in water and dilute to 1 liter. Stable several weeks at room temperature.

FIG. 24-3. Absorption curves of porphyrins in 1.5 *N* HCl: (1) uroporphyrin vs water, (2) coproporphyrin vs water. Beckman DK-2A recording spectrophotometer.

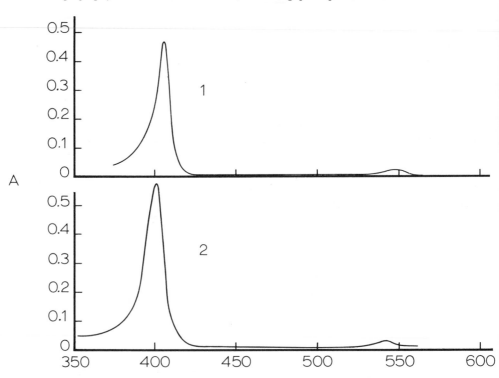

Iodine, 5 mg/100 ml, Aqueous. Prepare daily by diluting a stock 1% ethanolic iodine solution (1 g iodine crystals dissolved in 100 ml ethanol and kept under refrigeration) 1:200 with water. Ethanol denatured with methanol is satisfactory.

HCl, Conc

HCl, 1.5 N. Dilute 125 ml conc HCl to 1 liter with water.

n-Butanol

Acetic Acid, 0.5%. Dilute 5 ml glacial acetic acid to 1 liter with water.

Petroleum Ether, b.p. 30°–60°C

Copro- and *Uroporphyrin* may be purchased as the methyl esters or may be obtained as extracts from urine specimens containing large amounts of these compounds (188). These extracts may be diluted with water and purified by reprocessing the solutions as described below. The porphyrin concentrations are determined spectrophotometrically with the use of absorptivities and background correction formulas given under porphyrin *Calculations (vide infra).*

The methyl esters of copro- and uroporphyrin which several suppliers of biochemicals list in their catalogs are not always available and their purity is sometimes uncertain. They are useful in recovery studies but need not be used routinely as standards in spectrophotometric methods. It is only necessary to insure that the spectrophotometer is accurately adjusted for wavelength and that the accuracy of its absorbance scale be known (see Chapter 1).

PROCEDURE FOR PORPHYRINS

Coproporphyrin

1. Examine urines under 405 nm UV lamp (Ultra-Violet Products model UVL-56 long wave ultraviolet lamp is satisfactory) for red fluorescence. If fluorescence is observed, dilute 5 ml of the urine to 50 ml with water.

2. To 50 ml urine or diluted urine in a 250 ml separatory funnel, add 25 ml pH 4.8 acetate buffer and mix; then add 75 ml ethyl acetate.

3. Shake the mixture thoroughly for 1 min, then let stand until the two layers separate. Certain urines have a tendency to produce emulsions. This is apparent when the extraction is just begun. In such cases, *extract very gently,* since the very factors which cause the emulsification make vigorous shaking unnecessary to provide intimate contact between the organic and aqueous phase. Two or three drops of amyl alcohol are helpful in breaking an emulsion if it forms. Draw off the lower aqueous phase and set aside for the uroporphyrin determination.

4. Wash the ethyl acetate phase with 2.5 ml portions of 1% sodium acetate, repeating the washing until they show no red fluorescence in ultraviolet light. Combine the washings with the aqueous phase containing the uroporphyrin.

5. Shake the ethyl acetate solution gently with 5 ml freshly prepared aqueous iodine reagent and, after separating the phases, discard the aqueous layer. Contact with the iodine solution must be limited to 5 min.

6. Extract the coproporphyrin from the ethyl acetate with 2.5 ml portions of 1.5 N HCl until no more fluorescent material is removed. Three extractions are usually sufficient. Measure the volume of the pooled HCl extracts and centrifuge if not optically clear.

7. Determine absorbance in a spectrophotometer of fairly high resolution (1 cm light path) versus water at 380, 399, 400, 401, 402, 403, 404, and 430 nm. A recording spectrophotometer may be used to advantage. Maximum absorbance for coproporphyrin is obtained at about 401 nm on a carefully calibrated instrument. The readings at 380 and 430 nm are taken to correct for nonspecific background absorbance.

Uroporphyrin

1. Add conc HCl drop by drop to the aqueous uroporphyrin solution from step 3 above, until the pH drops to 3.0±0.2 (pH meter or narrow range pHydrion paper).

2. Transfer to 125 ml separatory funnel and add 10 ml *n*-butanol; shake the mixture thoroughly for about 1 min. As in the ethyl acetate extraction above, emulsions sometimes tend to form. Use the same technic described in step 3 for coproporphyrin to minimize emulsification. Allow phases to separate and remove the upper butanol layer to a 50 ml separatory funnel.

3. Repeat the extraction with two more 5 ml portions of butanol.

4. Combine the butanol extracts in the 50 ml separatory funnel and wash twice with 5 ml of 0.5% aqueous acetic acid; discard the lower aqueous layers.

5. Add 15 ml petroleum ether and 0.5 ml conc HCl to the butanol extract, shake well for 1 min. Draw off and save the lower aqueous layer (volume about 4 ml).

6. Reextract the organic layer using 1 ml 1.5 N HCl. Combine the acid extracts and measure the final volume (about 5 ml). Centrifuge if not optically clear.

7. Determine the absorbance (1 cm light path) at 380, 404, 405, 406, 407, 408 and 430 nm using a reference solution (blank) *prepared by extracting the remaining butanol once again with 3 ml of 1.5 N HCl.* Maximum absorbance is obtained at about 406 nm on a carefully calibrated instrument.

Calculations:

$$\mu g \text{ coproporphyrin/24 hours} = \frac{A_{corr}}{0.673} \times V \times \frac{TV}{50} \times D$$

where

$$A_{corr} = \frac{2A_{max} - (A_{380} + A_{430})}{1.835}$$

$$\mu g \text{ uroporphyrin/24 hours} = \frac{A_{corr}}{0.600} \times V \times \frac{TV}{50} \times D$$

where

$$A_{corr} = \frac{2A_{max} - (A_{380} + A_{430})}{1.844}$$

A_{max}	=	absorbance obtained at the peak
V	=	volume of acid extract
TV	=	24 hour urine volume in ml
D	=	number of times urine was diluted prior to extraction

NOTES

1. *Beer's law.* Assuming a spectrophotometer is used with half-bandwidths of no more than a few nm:
Copro. Linearity exists up to an absorbance of 1.5.
Uro. Linearity exists up to an absorbance of 1.3.
2. *Color stability. Copro and Uro.* Stable at least 3 days at room temperature.
3. *Standardization.*
Copro. The formula for calculation is derived as follows: If A_{corr} is absorbance of a pure standard at the absorption peak, then

$$A_{corr} = \frac{2A_{max} - (A_{380} + A_{430})}{K}$$

where $K = 1.835$ (152). The constant 0.673 in the formula is the A_{corr} for 1 $\mu g/ml$ with a 1 cm light path. Rimington (152) obtained an $A_{1cm}^{1\%}$ in 0.1 N HCl of 7470, giving an $A_{1cm}^{1\mu g/ml}$ of 0.747. According to Zondag and van Kampen (232) the absorbance in 1.5 N HCl is about 10% less; hence 0.9 x 0.747 = 0.673. Our laboratory also has obtained about an 8%– 10% lesser absorbance in 1.5 N HCl.
Uro. The formula for calculation is derived as follows: If A_{corr} is the absorbance of a pure standard at the absorption peak, then:

$$A_{corr} = \frac{2A_{max} - (A_{380} + A_{430})}{K}$$

where $K = 1.844$ (152). The constant 0.600 in the formula is the A_{corr} for 1 $\mu g/ml$ with a 1 cm light path. Rimington (152) obtained an $A_{1 cm}^{1\%}$ in 0.5 N HCl of 6520, giving an $A_{1cm}^{1\mu g/ml}$ of 0.6520. Our laboratory has obtained about an 8% lesser absorbance in 1.5 N HCl than in 0.5 N; hence 0.92 x 0.652 = 0.600.
4. *Stability of samples.*
Coproporphyrin. According to Schwartz *et al* (175) and Talman (188), copro is reasonably stable in urine between pH 6.0 and 9.5. Significant

conversion of copro precursor occurs in this pH range within 2 hours in the dark (50), and conversion is essentially complete in 24–48 hours (175, 232). Destruction of copro reportedly occurs at acid pH—e.g., at pH 4.0, 65% loss was observed in 24 hours at room temperature (175). Yet it has been observed in our laboratory that no decrease in the Soret peak of copro in 1.5 N HCl occurs in the dark at room temperature over a period of several days. Schwartz *et al* and Talman recommend using a final concentration of 0.1%–1% Na_2CO_3 as a preservative, stating that there is then no need for refrigeration of urine unless there is a delay of several days. Without Na_2CO_3 there is a loss of copro of 17%–39% in 1 day at room temperature in the dark (175), and a 31% loss in 1 day and 86% in 15 days if not protected from light (91). Erikson (50) found urine to be stable for 2 days in the light if alkaline; Haeger-Aronson (91) observed refrigerated alkaline urine to be stable 3 days and found a 32% decrease on the 15th day. Rogers (158) recommends testing a random urine specimen and relating the coproporphyrin excretion to that of creatinine so that the problem of stability during a 24 hour collection period may be avoided. Talman (188) also stated that toluene or thymol aids in stabilization (no data given) and recommended avoidance of metal caps or metal inserts in caps on bottles.

Our laboratory observed no change in coproporphyrin results obtained with several acute porphyria urines stored at 25°C for periods as long as 1 week in the dark.

To summarize, *urine should be collected in dark bottles or must at least be kept in the dark until analysis. A final concentration of 0.1% Na_2CO_3 is best for copro, but it must be remembered that ALA and PBG are not stable in alkalinized urine.*

Uroporphyrin. Conversion of PBG to uro is significant in 1 day at 37°C (82). Rate of conversion increases with increasing pH (32). Schwartz *et al* (172) point out that stabilization of porphyric urine has not been studied.

5. *Oxidation of coproporphyrinogen by I_2.* It has been shown (209) that significant amounts of coproporphyrin—e.g., approximately one-half in fresh normal urine—are present as a precursor, presumably coproporphyrinogen. Conversion can be effected by treatment with iodine. As discussed in *Note 4*, conversion can also occur by allowing the urine to stand at alkaline pH for at least 24 hours. Talman (188) points out, however, that the use of too much iodine or allowing the iodine to act for too long a time on fresh urine can result in loss of porphyrin. Furthermore, iodine treatment may not increase the amount of porphyrin detected in stored urine and may actually lead to a decrease in the amount found.

ACCURACY AND PRECISION

Coproporphyrin as isolated may contain 5%–30% of other porphyrins (172), and the uroporphyrin fraction probably somewhat less. If urine is contaminated with feces, the ethyl acetate-soluble fraction includes protoporphyrin as well as copro (188). There is also an ether-extractable chromogen, apparently an indole derivative, which forms a red pigment on acidification with HCl (50); presumably this is also extractable by ethyl acetate. Large amounts of nonporphyrin pigments exist in normal and porphyria urines, at least some of which are likely dipyrroles (fuscins), including urochrome B and porphobilin (221). At least some of these pigments may be carried along in the analysis to the final uroporphyrin fraction, which is read in the spectrophotometer. Most of the background which would otherwise complicate the spectrophotometric examination of the uroporphyrin fraction is effectively compensated for by zeroing the instrument on the third HCl extract of the butanol phase (63). Accuracy of the determination of copro in porphyric urines is also difficult to assess.

A factor which limits the accuracy of the test, particularly at low levels, is that the basic assumption of a background spectral absorption approximating a straight line is *not* always valid. Use of a recording spectrophotometer makes this problem obvious.

Maximum recovery of coproporphyrin is obtained when extraction is made with the pH of the urine between 2.5 and 4.5 (63). In order to minimize contamination by uroporphyrin, however, the extraction is made at pH 4.8, giving a recovery of uroporphyrin of about 90%. Quantitative recovery of uroporphyrin is obtained in this procedure (63). Sediment formation can cause loss of more than half of the urinary porphyrin, but this source of error can be overcome by thorough shaking of the urine prior to taking an aliquot for testing (47).

The precision (95% confidence limits) for this method at a level of about 200 μg/liter is approximately ±15% for coproporphyrin and for uroporphyrin. The fact that background absorption is not always linear and is not very reproducible materially affects the precision of the test at low levels.

NORMAL VALUES

It is the author's opinion that the normal range of 24 hour excretion of coproporphyrin is still in question. Fluorometric technics that include coproporphyrinogen in the results indicate a range of 8–250 μg/24 hours (3, 104, 109, 143, 172, 188). There were reports, especially prior to about 1952, of lower normal ranges obtained by fluorometry (183, 208), but according to Schwartz *et al* (172) the higher values obtained in later investigations were due to use of an improved fluorometric method. Nevertheless, a recent spectrofluorometric method suggests that *total* porphyrin excretion in the urine is about 22–140 μg/liter (128).

Early investigations employing spectrophotometric measurement indicated that the normal range for coproporphyrin was approximately 0–100 μg/24 hours. Fernandez *et al* (63) obtained a range of 0–161 μg/24 hours in the method presented here. Using a similar method, With (224) reports 0–150 μg/24 hours. Among the factors undoubtedly contributing to the disagreement in the data on normals are the following: (a) differences in sampling of normal subjects for normal range evaluation, (b) varying degrees of specificity of the methods used by different investigators, and (c) unknown recovery of copro complexes and precursors in the various methods, and hence the uncertain contribution of such substances to normal values obtained.

Coproporphyrin levels for normal males are reported to be somewhat higher than those for females (3, 172), but there is disagreement on this (63). Excretion correlates better with body weight (104, 183) than with age or surface area (104). The normals for children, therefore, are lower than adult normals (3, 4, 104, 140, 188). Day to day variation averages about ±6% (183). Output increases when there is an intake of 1 pint or more of liquor per day, this occurring without evidence of liver dysfunction (231). Fasting has been shown to increase urinary coproporphyrin levels (113).

The normal 24 hour output of uroporphyrin as determined by fluorometric measurements has been reported as up to 60 μg (109, 119, 143, 172, 188). Dresel *et al* (46), using spectrophotometric measurement, obtained a range of 2.7−14.6 μg/liter. Fernandez *et al* (63) found 0−26 μg/24 hours, and With (224) reported 0−50 μg/24 hours. The urine of normal children contains only traces of uro (1) and is directly proportional to the child's age and weight (4).

The urinary excretion of porphyrins is greater in daytime than at night, increases following heavy meals (192), and increases up to twofold following muscular exercise (70, 192).

The adult normal ranges used in our laboratory are 0−160 μg coproporphyrin and 0−26 μg uroporphyrin/24 hours.

QUANTITATIVE DETERMINATION OF PORPHYRINS IN FECES

(method of C. Sobel, unpublished)

PRINCIPLE

Feces are extracted with a mixture of ethyl acetate and acetic acid. The solvent mixture containing copro-, meso-, deutero-, proto-, and uroporphyrins is washed with sodium acetate solution, which removes the uroporphyrin. The uroporphyrin is precipitated on calcium phosphate, redissolved in dilute HCl, and determined spectrophotometrically. The copro-, meso-, and deuteroporphyrins in the solvent mixture are reextracted with 0.2 N HCl. This acid extract is partially neutralized and the meso- and deuteroporphyrins are removed by extraction with $CHCl_3$. The aqueous phase is adjusted to pH 5; the coproporphyrin is extracted with ethyl acetate and then back into 1.5 N HCl in which it is determined spectrophotometrically. The original solvent mixture (ethyl acetate-acetic acid) still contains the protoporphyrin, which is extracted with 2 N HCl and also determined spectrophotometrically. All readings are taken at the Soret peaks (401−408 nm) and corrected for background absorption from readings taken at 380 and 430 nm.

REAGENTS

Ethyl Acetate-Acetic Acid Mixture. Mix 3 vol ethyl acetate and 1 vol glacial acetic acid.

Sodium Acetate, 3%. Dissolve 49.8 g $CH_3COONa \cdot 3H_2O$ in water and dilute to 1 liter.

HCl, 0.2 N. Dilute 16.7 ml conc HCl to 1 liter with water.

NaOH, 0.2 N. Dissolve 8.0 g NaOH in water and dilute to 1 liter.

HCl, 1.5 N. Dilute 125 ml conc HCl to 1 liter with water.

HCl, 2 N. Dilute 167 ml conc HCl to 1 liter with water.

Sodium Phosphate, 0.4 M. Dissolve 76 g of $Na_3PO_4 \cdot 12H_2O$ in water and dilute to 500 ml.

Calcium Chloride Dihydrate, 30%. 150 g $CaCl_2 \cdot 2H_2O$ dissolved in 500 ml water.

NaOH, 40%. Dissolve 200 g NaOH in water, allow to cool, and dilute to 500 ml.

NaOH, 0.1 N. Dissolve 4.0 g NaOH in water and dilute to 1 liter.

PROCEDURE

1. Weigh out 2.5 g feces and transfer to a large mortar. With the help of a little water, grind to a paste. Then extract with ethyl acetate-acetic acid mixture by grinding with a pestle. Pour off fluid through a sintered glass filter (medium porosity), using suction. Repeat extracting with solvent mixture until supernatant fluid is colorless.

2. Wash combined extracts by shaking gently in a separatory funnel, with 1/5 of its volume of 3% sodium acetate. Make a total of 3 washes with the sodium acetate solution and one with water. *Combine all washes and reserve for uroporphyrin assay* (step 7).

3. Measure volume of solvent phase left in step 2 (graduated cylinder). Transfer 40% of this volume (equivalent to 1 g feces) to another separatory funnel. Extract with three 5 ml portions of 0.2 N HCl, combining all the washes *(the acid removes copro-, meso-, and deutero*porphyrin*)*. Save the solvent phase for step 6 in which *protoporphyrin* is extracted. Proceed to step 4 with the acid phase. Examine third acid extract under ultraviolet lamp for fluorescence. If appreciable red fluorescence is still present in third extract, make further extractions with 5 ml vol 0.2 N HCl until no more fluorescent material is extracted.

4. To the combined acid extracts from step 3, add 5 ml 0.2 N NaOH for each 1.5 ml acid extract. Extract two times with 5 ml vol chloroform and discard solvent extract (which may contain *deutero-* and *meso*-porphyrin). Neutralize the aqueous phase to pH 5 (narrow-range pH paper) with ammonia. Extract two times with 10 ml vol ethyl acetate. Extract the ethyl acetate solution two times with 5 ml vol 1.5 N HCl. Make a third acid extraction and examine under ultraviolet light for red fluorescence; if absent discard, if present add to previous two extractions and continue extraction with acid until no more red fluorescent material is extracted.

5. Combine the HCl extracts, measure volume in a graduated cylinder, and read absorbances in the Soret band (380, 399, 400, 401, 402, 403, 404, 405, and 430 nm). This solution contains the *coproporphyrin*, which gives a peak at 401–402 nm.

Calculation:

$$\mu\text{g coproporphyrin per 24 hour stool specimen} = \frac{2A_{max} - (A_{380} + A_{430})}{0.673 \times 1.835} \times V \times TW$$

V = volume of HCl extracts in ml

TW = weight in grams of 24 hour fecal specimen

6. The solvent phase left in step 3, after extraction with 0.2 N HCl, is now extracted two times with 5 ml vol of 2 N HCl, which removes the *protoporphyrin*. Read absorbances in Soret band (380, 404, 405, 406, 407, 408, 409, and 430 nm). Protoporphyrin shows a peak at 407–408 nm.

Calculation:

$$\mu\text{g protoporphyrin per 24 hour stool specimen} = \frac{2A_{max} - (A_{380} + A_{430})}{0.65 \times 1.84} \times V \times TW$$

7. Transfer the combined acetate washes from step 2 to a 50 ml glass-stoppered centrifuge tube. Add 1 ml phosphate and mix; then add 1 ml $CaCl_2$ reagent and mix. Add 40% NaOH, dropwise with mixing, until pH is 10. Centrifuge and discard supernate.
8. Stir up the precipitate with 10 ml 0.1 N NaOH, centrifuge, and discard supernate.
9. Stir up the precipitate with 10 ml water, centrifuge, and discard supernate.
10. Dissolve the precipitate in 10 ml 1.5 N HCl. Read absorbances in Soret band (380, 403, 404, 405, 406, 407, 408, 430 nm). *Uroporphyrin* gives peak at 406–407 nm.

Calculation:

$$\mu\text{g uroporphyrin per 24 hour stool specimen} = \frac{2A_{max} - (A_{380} + A_{430})}{0.600 \times 1.844} \times 0.4 \times V \times T$$

NOTES

1. Spectrophotometry and standardization. Refer to the notes on urinary porphyrins for pertinent information on copro- and uroporphyrins. The formula for the calculation of protoporphyrins is derived in the same way as those for copro and uro. The constant 0.65 is the A_{corr} for 1 μg/ml in 2 N HCl with 1 cm light path.

2. Stability of samples. Limited information suggests that fecal porphyrins are not very stable, and samples should therefore be frozen if the analysis cannot be performed on the same day.

PRECISION

The precision (95% limits) for each porphyrin fraction in this procedure is observed in our laboratory to be approximately ±30%. This may be due to difficulties in the accurate sampling of the feces for analysis.

NORMAL VALUES

Upper limits of normal for fecal coproporphyrin have been reported by several authors to lie between 13 and 40 $\mu g/g$ dry weight (6, 77, 102, 109, 143). Schwartz *et al* (172) give a normal range of 400−1100 $\mu g/$day, which, assuming a normal average daily excretion of 30 g feces dry weight, agrees with other data. Politzer and Kessel (145) found normal infants up to the age of 1 year to excrete 0−10 $\mu g/g$ dry weight. The coproporphyrin in normal feces is chiefly type I (41, 172).

The normal excretion of uroporphyrin in feces is reported to be about 10−40 $\mu g/$day (172), but the range of 0−4 $\mu g/g$ dry weight has also been observed (143).

Fecal protoporphyrin has been reported with upper limits of normal being about 60 (77, 143) and 100 (109) $\mu g/g$ dry weight.

QUANTITATIVE DETERMINATION OF PROTOPORPHYRIN AND COPROPORPHYRIN IN ERYTHROCYTES

(method of Schwartz and Wikoff, Ref 174, modified)

PRINCIPLE

Cells are extracted with a mixture of ethyl acetate and acetic acid. The porphyrins are then reextracted with 3 N HCl and then back into ethyl acetate after adjusting solution to pH 5.5−6.0. Coproporphyrin is then reextracted with 0.1 N HCl and the protoporphyrin with 3 N HCl. The porphyrins are then estimated from their fluorescence compared to that of standards.

REAGENTS

Extraction Mixture.   400 ml ethyl acetate + 100 ml glacial acetic acid.
Saturated Sodium Acetate
Sodium Acetate Trihydrate, 3%.   Dissolve 30 g $CH_3 COONa \cdot 3H_2O$ in water and dilute to 1 liter.
HCl, Conc
HCl, 3 N.   Dilute 250 ml conc HCl to 1 liter with water.
HCl, 0.1 N.   Dilute 8.3 ml conc HCl to 1 liter with water.
Saline, 0.85%.   Dissolve 8.5 g NaCl in water and dilute to 1 liter.
Ethyl Acetate
Chloroform

Protoporphyrin Standard, approx 100–140 µg/100 ml in 3 N HCl. Standardized by spectrophotometric reading of this concentration of standard compared to $A_{1\,cm}^{1\%} = 4890$ at 405 nm. Weigh about 3 mg protoporphyrin IX into a 250 ml volumetric flask. Wash in with 150 ml water and add 2 ml conc $NH_4 OH$. Mix to dissolve. Add 63 ml conc HCl, dilute to 250 ml with water and mix. Dilute 1 ml of this solution with 4 ml 3 N HCl to get a standard for reading on spectrophotometer. The purity of some preparations of protoporphyrin IX may be only 50%; preparations of >95% purity are available from Calbiochem and Schwarz/Mann.

$$\frac{A_s}{0.489} \times 100 = \mu g \text{ protoporphyrin/100 ml standard solution}$$

Dilute 1 ml of the standard solution read on the spectrophotometer with 19 ml 3 N HCl to obtain a working standard of approx 5–7 µg/100 ml. Then, 0.05 x conc of standard read on spectrophotometer = µg protoporphyrin/ 100 ml standard solution to be used for fluorescence measurements.

Coproporphyrin Standard. Dissolve about 1 mg coproporphyrin III tetramethyl ester (may be obtained from Calbiochem) in 10 ml 7.5 N HCl in a 50 ml volumetric flask and allow to stand at room temperature overnight to effect hydrolysis of the ester. Dilute to 50 ml with water. Stable several months when refrigerated. Dilute 1 ml of this to 10 ml with water. Determine concentration from absorbance at 400 nm.

$$\mu g/100 \text{ ml} = \frac{A_s}{0.673} \times 100$$

Dilute 1 ml of this further with 19 ml 1.5 N HCl. Then, 0.05 x conc of standard solution read on spectrophotometer = µg coproporphyrin/100 ml standard solution to be used for fluorescence measurements.

PROCEDURE

1. Pipet 4.0 ml heparinized blood into a 12–15 ml calibrated, glass-stoppered centrifuge tube.
2. Centrifuge to remove and discard plasma. Then wash the cells twice with 5 ml portions of 0.85% saline, mixing gently and centrifuging each time. Finally centrifuge in any standard centrifuge which allows the tubes to spin in a horizontal position at 2000 rpm for 10 min. Remove the saline and note volume of packed cells.
3. Add 10 ml of extraction mixture, mix with a glass rod, and transfer to a medium porosity sintered glass filter. Grind cells with a footed rod. Pull solution through filter into a small flask with slight suction.
4. Wash original tube with 10 ml of extraction mixture. Add to funnel, grind again, and combine filtrates.

5. Repeat step 4 three times more.
6. Transfer the combined extracts to a 100 ml separatory funnel (no grease) and wash with 10 ml 3% sodium acetate which has been mixed with 20 ml 2 *N* NaOH. pH must be>5.5; if not, add saturated sodium acetate until it is.
7. Extract the porphyrins with 5 ml portions of 3 *N* HCl until no more red fluorescence is observed in extract. Then extract once more. Observe fluorescence in dark with 405 nm ultraviolet lamp.
8. To the combined extracts, add saturated sodium acetate to get pH 5.5–6.0 (approx 2 ml is required for each ml 3 *N* HCl).
9. Extract three times with 30 ml portions ethyl acetate.
10. Wash the combined ethyl acetate extract twice with 5 ml portions of water.
11. Extract the ethyl acetate three times with 5 ml portions of 0.1 *N* HCl. Check for fluorescence in last extraction; if positive, extract twice more. Combine these HCl extracts. This combined 0.1 *N* HCl extract contains the coproporphyrin. Wash it twice with 3 ml portions of chloroform and add the chloroform washes to the ethyl acetate solution from step 10.
12. Bring the acid extract from step 11 to 1.5 *N* by the addition of 1.3 ml conc HCl per 10 ml of 0.1 *N* HCl extract. Note the total volume of this adjusted extract.
13. Extract the ethyl acetate solution remaining from step 11 with 5 ml portions of 3 *N* HCl and then extract three times with 2 ml portions. Combine all these HCl extracts which contain the protoporphyrin. Mix and measure volume.
14. Centrifuge the solution and transfer some of the clear solution to a clean tube with a pipet.
15. Read the acid solutions obtained in steps 12 and 14 together with the coproporphyrin and protoporphyrin standards. For coproporphyrin the optimum wavelength for activation should be about 398 nm, and that for fluorescence should be about 595 nm. For protoporphyrin these wavelengths are approximately 400 nm and 600 nm, respectively.

Calculations:

$$\mu g/100 \text{ ml cells } = \frac{F_x}{F_s} \times \mu g/100 \text{ ml in dilute std} \times \frac{V_1}{V_2}$$

where

F_x = fluorescence of sample
F_s = fluorescence of dilute copro- or protoporphyrin standard
V_1 = volume of combined extract in steps 12 or 13
V_2 = volume of cells measured in step 2

NOTES

1. Linearity. Using a spectrofluorometer with exciting radiation of 400 nm and observing fluorescence at 600 nm, the author obtained linearity of fluorescence versus protoporphyrin concentration up to 100 μg/100 ml 3 N HCl. This would correspond to about 800 μg/100 ml erythrocytes.

2. Fluorescence stability. With ordinary care, the acid solution of the porphyrins is quite stable for fluorescence measurements. Grinstein and Watson (84) reported that after 24 hours in the dark an acid solution of protoporphyrin emits only about half of the fluorescence as when fresh. Its absorbance at the Soret peak, however, is not affected. Schwartz *et al* (172) attribute deterioration of fluorescence to peroxides in the ether used in the analysis.

3. Stability of samples. Limited experience in our laboratory indicates that protoporphyrin is stable in the blood for 4 days at room temperature. Coproporphyrin probably is as stable.

ACCURACY AND PRECISION

It is recommended by the author that fluorescence measurements be made on a spectrofluorometer so that the specificity of the determination can be assessed by examination of spectra. The wavelengths given here for maximum excitation for fluorescence of the porphyrins are uncorrected for lamp output and photomultiplier tube response. For best results a very sensitive instrument should be used with a photomultiplier tube having a high response in the 600–700 nm region. Wavelengths should be carefully checked, using standard solutions in the individual instrument to be used for the determinations.

The recovery of proto- and coproporphyrin is about 70%–80% (228).

The precision (95% limits) of the method for protoporphyrin is about ±16 μg/100 ml erythrocytes in the normal range. The precision for coproporphyrin is probably about the same at equivalent concentrations.

NORMAL VALUES

The normal protoporphyrin content of erythrocytes of children and adults is about 15–100 μg/100 ml erythrocytes. This is a compromise of studies which are not in very good agreement on the normal range, the upper normal limit varying from about 46 to 155 μg/100 ml (21, 77, 87, 105, 172, 184, 202, 207, 228). Values for females are reportedly somewhat higher than those for males (172).

The normal coproporphyrin content of erythrocytes in adults and children is generally reported in the range 0.5–2.0 μg/100 ml (105, 143, 172, 204, 228), although some authors have found values up to 6 μg/100 ml (77).

Uroporphyrin normally is not detectable in erythrocytes (172), but traces have been reported (77).

REFERENCES

1. ALDRICH RA, LABBE RF, TALMAN EL: *Am J Med Sci 230*:675, 1955
2. ASKEVOLD R: *Scand J Clin Lab Invest 3*:318, 1951
3. AZIZ MA, SCHWARTZ S, WATSON CJ: *J Lab Clin Med 63*:585, 1964
4. BAKHMET'EV P: *Pediatriya (Sofia) 69*:623, 1969; *CA 74*:62265b, 1971
5. BARLTROP D: *Acta Paediat Scand 56*:265, 1967
6. BARNES HD: *S African Med J 32*:680, 1958
7. BENSON PF, CHISOLM JJ, Jr: *J Pediat 56*:759, 1960
8. BERK PD, TSCHUDY DP, SHEPLEY LA, WAGGONER JG, BERLIN NI: *Am J Med 48*:137, 1970
9. BERKÓ GY, DURKÓ I: *Clin Chim Acta 37*:443, 1972
10. BILITCH M, STERLING RE, REDEKER AG: *Univ S Calif Med Bull 10*:20, 1958
11. BLANKSMA LA, SACHS HK, MURRAY EF, O'CONNELL MJ: *Am J Clin Pathol 53*:956, 1970
12. BOGORAD L: *J Biol Chem 233*:501, 1958
13. BOGORAD L: *J Biol Chem 233*:510, 1958
14. BOGORAD L: *J Biol Chem 233*:516, 1958
15. BOSSENMAIER I, CARDINAL R: *Clin Chem 14*:610, 1968
16. BRUGSCH J: *Z Ges Inn Med Ihre Grenzgebiete 4*:253, 1949
17. BRUGSCH J, MAASSEN J: *Z Ges Inn Med Ihre Grenzgebiete 15*:1023, 1960
18. BRUNSTING LA, MASON HL, ALDRICH RA: *J. Am Med Assoc 146*:1207, 1951
19. BURNHAM BF, LASCELLES J: *Biochem J 87*:462, 1963
20. CARMICHAEL R, NEILL DW: *Clin Chim Acta 6*:590, 1961
21. CARTWRIGHT GE, HUGULEY CM Jr, ASHENBRUCKER H, FAY J, WINTROBE MM: *Blood 3*:501, 1948
22. CHISOLM JJ, Jr: *Anal Biochem 22*:54, 1968
23. CHU TC, CHU EJ: *J Biol Chem 208*:537, 1954
24. CHU TC, CHU EJ: *J Biol Chem 227*:505, 1957
25. CHU TC, CHU EJ: *Anal Chem 30*:1678, 1958
26. CHU TC, CHU EJ: *J Biol Chem 234*:2741, 1959
27. CHU TC, CHU EJ: *Clin Chem 12*:647, 1966
28. CHU TC, CHU EJ: *Clin Chem 13*:371, 1967
29. CHU TC, GREEN AA, CHU EJ: *J Biol Chem 190*:643, 1951
30. COMFORT A, MOORE H, WEATHERALL M: *Biochem J 58*:177, 1954
31. COOKSON GH: *Nature 172*:457, 1953
32. COOKSON GH, RIMINGTON C: *Biochem J 57*:476, 1954
33. COQUELET ML, DE TRAVERSE PM: *Ann Biol Clin (Paris) 17*:498, 1959
34. CORWIN LM, ORTEN JM: *Anal Chem 26*:608, 1954
35. DAVIS JR: *Am J Clin Pathol 53*:967, 1970
36. DE LANGEN CD: *Acta Med Scand 133*:73, 1949
37. DEŽELIĆ M, CETINIĆ F: *Clin Chim Acta 13*:652, 1966
38. DILLAHA CJ, HICKLIN W: *J Invest Dermatol 19*:489, 1952
39. DJURIC D: *Arch Environ Health 9*:742, 1964
40. DOBRINER K: *Proc Soc Exp Biol Med 35*:175, 1936
41. DOBRINER K: *J Biol Chem 120*:115, 1937
42. DOBRINER K, RHOADS CP: *Physiol Rev 20*:416, 1940
43. DONALDSON EM, McCALL AJ, MAGNUS IA, SIMPSON JR, CALDWELL RA, HARGREAVES T: *Brit J Dermatol 84*:14, 1971
44. DOSS MO: *Lancet 2*:983, 1971
45. DOSS M, SCHMIDT A: *Z Klin Chem Klin Biochem 9*:99, 1971

46. DRESEL EIB, RIMINGTON C, TOOTH BE: *Scand J Clin Lab Invest 8*:73, 1956
47. DURST J, PÁSZTOR G, NAGY I: *Ärztl Lab 16*:378, 1970
48. EDITORIAL: *Lancet 1*:926, 1961
49. ERIKSEN L: *Scand J Clin Lab Invest 3*:121, 1951
50. ERIKSEN L: *Scand J Clin Lab Invest 4*:55, 1952
51. ERIKSEN L: *Scand J Clin Lab Invest 5*:155, 1953
52. ERIKSEN L: *Scand J Clin Lab Invest 9*:97, 1957
53. ERIKSEN L: *IV Intern Congr Biochem*, Vienna, 1958, Vol 15, Great Britain, Pergamon Press, 1960, p 173
54. ERIKSEN L: *Scand J Clin Lab Invest 10*:319, 1958
55. ERIKSEN L: *Scand J Clin Lab Invest 10*:39, 1958
56. ERIKSEN L, ERIKSEN N, HAAVALDSEN S: *Scand J Clin Lab Invest 14*:1, 1962
57. ERIKSEN L, ERIKSEN N, HAAVALDSEN S: *Scand J Clin Lab Invest 14*:6, 1962
58. FALK JE: *J Chromatog 5*:277, 1961
59. FALK JE: *Porphyrins and Metalloporphyrins*. New York, Elsevier, 1964
60. FALK JE, BENSON A: *Biochem J 55*:101, 1953
62. FELDMAN F, LICHTMAN HC, ORANSKY S, ANA ES, REISER L: *J Pediat 74*:917, 1969
63. FERNANDEZ AA, HENRY RJ, GOLDENBERG H: *Clin Chem 12*:463, 1966
64. FERNANDEZ AA, JACOBS SL: *Standard Methods of Clinical Chemistry*. Edited by R.P. MacDonald. New York, Academic Press, 1970, Vol 6, p 57
65. FIKENTSCHER R: *Biochem Z 249*:257, 1932
66. FINE MH, KAPLAN AI: *J Am Med Assoc 197*:584, 1966
67. FINK H, HOERBURGER W: *Hoppe Seylers Z Physiol Chem 218*:181, 1933
68. FISCHER H: *Abderhalden's Handbuch der biologischen Arbeitsmethoden*. Berlin, Urban und Schwarzenberg, 1936, Section 1, Part 11, p 169
69. FISCHL J, EICHHORN F, RUTTENBERG A, MAJOR Ch: *Clin Chem 16*:331, 1970
70. FRANKE K, FIKENTSCHER R: *Muench Med Wochschr 82*:171, 1935
71. FRENCH JM, ENGLAND MT, LINES J, THONGER E: *Arch Biochem Biophys 107*:404, 1964
72. GAMGEE A: *Z Biol 34*:505, 1896
73. GIBSON KD, LAVER WG, NEUBERGER A: *Biochem J 70*:71, 1958
74. GIBSON KD, NEUBERGER A, SCOTT JJ: *Biochem J 61*:618, 1955
75. GOLDBERG A, RIMINGTON C: *Proc Roy Soc London B 143*:257, 1955
76. GOLDBERG A, RIMINGTON C: *Diseases of Porphyrin Metabolism*. Springfield, Ill., Charles C Thomas, 1962
77. GORECZKY L, RÓTH I, BRECKNER M: *Z Klin Chem Klin Biochem 6*:489, 1968
78. GRANICK S: *Science 120*:1105, 1954
79. GRANICK S, MAUZERALL D: *J Biol Chem 232*:1119, 1958
80. GRANICK S, MAUZERALL D: *Fed Proc 17*:233, 1958
81. GRANICK S, VANDEN SCHRIECK HG: *Proc Soc Exp Biol Med 88*:270, 1955
82. GRIEG A, ASKEVOLD R, SVEINSSON SL: *Scand J Clin Lab Invest 2*:1, 1950
83. GRINSTEIN M, SCHWARTZ S, WATSON CJ: *J Biol Chem 157*:323, 1945
84. GRINSTEIN M, WATSON CJ: *J Biol Chem 147*:675, 1943
85. GRINSTEIN M, WINTROBE MM: *J Biol Chem 172*:459, 1948
86. GROTEPASS W, DELFALQUE A: *Hoppe Seylers Z Physiol Chem 252*:155, 1938
87. GUTNIAK O: *Med Pracy 10*:407, 1959; *CA 54*:13235a, 1960
88. HAEGER B: *Scand J Clin Lab Invest 9*:211, 1957
89. HAEGER B: *Scand J Clin Lab Invest 10*:229, 1958
90. HAEGER B: *Lancet 2*:606, 1958
91. HAEGER-ARONSON B: *Scand J Clin Lab Invest 12 (Suppl 47)*, 1960
92. HAMMOND RL, WELCKER ML: *J Lab Clin Med 33*:1254, 1948

93. HEIKEL T: *Scand J Clin Lab Invest 7*:347, 1955
94. HEIKEL T: *Scand J Clin Lab Invest 10*:193, 1958
95. HEILMEYER L, CLOTTEN R, WEHINGER H: *Deut Med Wochschr 87*:131, 1962
96. HELLER SR, LABBE RF, NUTTER J: *Clin Chem 17*:525, 1971
97. HERBERT FK: *Clin Chim Acta 13*:19, 1966
98. HERBERT FK: *Clin Chim Acta 13*:38, 1966
99. HERTER CA: *J Biol Chem 4*:253, 1908
100. HIJMANS VAN DEN BERGH AA, HYMAN AJ: *Deut Med Wochschr 54*:1492, 1928
101. HIJMANS VAN DEN BERGH AA, GROTEPASS W: *Klin Wochschr 12*:586, 1933
102. HOLTI G, RIMINGTON C, TATE BC, THOMAS G: *Quart J Med 27*:1, 1958
103. HOSCHEK R: *Zentr Arbeitsmed Arbeitsschutz 10*:134, 1960
104. HSIA DY, PAGE M: *Proc Soc Exp Biol Med 85*:86, 1954
105. HSIA DY, PAGE M: *Proc Soc Exp Biol Med 86*:89, 1954
106. IPPEN H, SCHLINK H: *Arzneimittel-Forsch 10*:1034, 1960
107. JACKSON CE, BLOCK WD: *J Lab Clin Med 62*:887, 1963
107a. JANCAR J, PHILPOT GR: *Brit Med J 1*:1498, 1965
108. JOPE EM, O'BRIEN JRP: *Biochem J 39*:239, 1945
109. KAUFMAN L, MARVER HS: *New Engl J Med 283*:954, 1970
110. KIKUCHI G, KUMAR A, TALMAGE P, SHEMIN D: *J Biol Chem 233*:1214, 1958
111. KNUDSEN KB, SPARBERG M, LECOCQ F: *New Engl J Med 277*:350, 1967
112. KOHN R, WEDEL Jv: *Klin Wochschr 46*:210, 1968
113. KOSKELO P, PELKONEN R: *New Engl J Med 278*:856, 1968
114. KOSKELO P, TOIVONEN I: *Scand J Clin Lab Invest 18*:543, 1966
115. KOSKELO P, TOIVONEN I: *Clin Chim Acta 21*:291, 1968
116. LEMBERG R, LEGGE JW: *Hematin Compounds and Bile Pigments*, New York, Interscience, 1949
117. LEVERE RD, KAPPAS A: *Advances Clin Chem 11*:133, 1968
118. LEVERE RD, PINDYCK J: *Ann Clin Lab Sci 1*:101, 1971
119. LOCKWOOD WH: *Austral J Exp Biol Med Sci 31*:453, 1953
120. LOCKWOOD WH: *Austral J Exp Biol Med Sci 31*:457, 1953
121. LOCKWOOD WH, DAVIES JL: *Clin Chim Acta 7*:301, 1962
122. LORIAUX L, DELENA S, BROWN H: *Clin Chem 15*:292, 1969
123. LUDWIG GD: *J Clin Invest 37*:914, 1958
124. LUDWIG GD, EPSTEIN IS: *Ann Internal Med 55*:81, 1961
125. MAGNUS IA, JARRETT A, PRANKERD TAJ, RIMINGTON C: *Lancet 2*:448, 1961
126. MALOOLY DA, HIGHTOWER NC Jr: *J Lab Clin Med 59*:568, 1962
127. MARKOVITZ M: *J Lab Clin Med 50*:367, 1957
128. MARTINEZ CA, MILLS GC: *Clin Chem 17*:199, 1971
129. MARVER HS, SCHMID R: *The Metabolic Basis of Inherited Disease*, 3rd ed. Edited by J.B. Stanbury, J.B. Wyngaarden and D.S. Fredrickson. New York, McGraw-Hill, 1972, p 1087
130. MARVER HS, TSCHUDY DP, PERLROTH MG, COLLINS A, HUNTER G, Jr: *Anal Biochem 14*:53, 1966
131. MATHEWSON JH, CORWIN AH: *J Am Chem Soc 83*:135, 1961
132. MAUZERALL D, GRANICK S: *J Biol Chem 219*:435, 1956
133. MAUZERALL D, GRANICK S: *J Biol Chem 232*:1141, 1958
134. McSWINEY RR, NICHOLAS REH, PRUNTY FTG: *Biochem J 46*:147, 1950
135. MICHALEC Č, KOMÁRKOVÁ A: *Naturwissenschaften 43*:19, 1956
136. MINGIOLI ES: *Anal Biochem 22*:47, 1968
137. MIYAGI K, CARDINAL R, BOSSENMAIER I, WATSON CJ: *J Lab Clin Med 78*:683, 1971

138. MOORE DJ, LABBE RF: *Clin Chem 10*:1105, 1964
139. NEUBERGER A, SCOTT JJ: *Proc Roy Soc (London) A 213*:307, 1952
140. NEVÉ RA, ALDRICH RA: *Pediatrics 15*:553, 1955
141. NICHOLAS REH, RIMINGTON C: *Scand J Clin Lab Invest 1*:12, 1949
142. NICHOLAS REH, RIMINGTON C: *Biochem J 48*:306, 1951
143. PETERKA ES, FUSARO RM, RUNGE WJ, JAFFE MO, WATSON CJ: *J Am Med Assoc 193*:1036, 1965
144. PINDYCK J, KAPPAS A, LEVERE RD: *CRC Crit Rev Clin Lab Sci 2*:639, 1971
145. POLITZER WM, KESSEL I: *J Clin Pathol 11*:183, 1958
146. PRATO V, MAZZA U, RAMELLO A: *Lancet 2*:1276, 1964
147. PRUNTY FTG: *Biochem J 39*:446, 1945
148. RAPPOPORT DA, CALVERT CR, LOEFFLER RK, GAST JH: *Anal Chem 27*:820, 1955
149. REIO L, WETTERBERG L: *J Am Med Assoc 207*:148, 1969
150. RIMINGTON C: *Biochem J 37*:443, 1943
151. RIMINGTON C: *Assoc Clin Pathol Broadsheet No. 20 (new series)*, November 1958
152. RIMINGTON C: *Biochem J 75*:620, 1960
153. RIMINGTON C: *Assoc Clin Pathol Broadsheet No. 36 (new series)*, 1961
154. RIMINGTON C, CRIPPS DJ: *Lancet 1*:624, 1965
155. RIMINGTON C, KROL S, TOOTH B: *Scand J Clin Lab Invest 8*:251, 1956
156. RIMINGTON C, MILES PA: *Biochem J 50*:202, 1951
157. RIMINGTON C, SVEINSSON SL: *Scand J Clin Lab Invest 2*:209, 1950
158. ROGERS CJ: *Clin Chem 10*:678, 1964
159. SACHS P: *Klin Wochschr 10*:1123, 1931
160. SAILLET: *Rev Med 16*:542, 1896
161. SAMUELS S, FISHER C: *Arch Environ Health 21*:728, 1970
162. SAMUELS S, VELIZ G: *Clin Chem 17*:51, 1971
163. SANO S, GRANICK S: *J Biol Chem 236*:1173, 1961
164. SAWADA T, SHIBATA Y, KAJIWARA T: *Kyushu J Med Sci 7*:219, 1956
165. SCHLENKER FS, DAVIS CL, KITCHELL CL: *Am J Clin Pathol 32*:103, 1959
166. SCHLENKER FS, DAVIS CL, KITCHELL CL: *Am J Clin Pathol 36*:31, 1961
167. SCHLENKER FS, KITCHELL CL: *Am J Clin Pathol 29*:593, 1958
168. SCHLENKER FS, TAYLOR NA, KIEHN BP: *Am J Clin Pathol 42*:349, 1964
169. SCHMID R, SCHWARTZ S, SUNDBERG D: *Blood 10*:416, 1955
170. SCHOLNICK P, MARVER H, SCHMID R: *Clin Research 17*:278, 1969
171. SCHUMM O: *Z Ges Exp Med 106*:252, 1939
172. SCHWARTZ S, BERG MH, BOSSENMAIER I, DINSMORE H: *Methods of Biochemical Analysis*, edited by Glick D. New York, Interscience, 1960, Vol 8, p 221
173. SCHWARTZ S, HAWKINSON V, COHEN J, WATSON CJ: *J Biol Chem 168*:133, 1947
174. SCHWARTZ S, WIKOFF HM: *J Biol Chem 194*:563, 1952
175. SCHWARTZ S, ZIEVE L, WATSON CJ: *J Lab Clin Med 37*:843, 1951
176. SCOTT CR, LABBE RF, NUTTER J: *Clin Chem 13*:493, 1967
177. SCOTT DF: *Brit Med J 1*:997, 1965
178. SHEMIN D, RUSSELL CS: *J Am Chem Soc 75*:4873, 1953
179. SHUSTER L: *Biochem J 64*:101, 1956
180. STANLEY TE, SCHOOLS PE, FIELD H Jr: *Virginia Med Month 79*:371, 1952
181. STERLING RE, REDEKER AG: *Scand J Clin Lab Invest 9*:407, 1957
182. STICH W: *Klin Wochschr 39*:338, 1961
183. STRAIT LA, BIERMAN HR, EDDY B, HRENOFF M, EILER JJ: *J Appl Physiol 4*:699, 1952
184. STURGEON P: *Pediatrics 13*:107, 1954

185. SUN M, STEIN E, GRUEN FW: *Clin Chem 15*:183, 1969
186. SVEINSSON SL, RIMINGTON C, BARNES HD: *Scand J Clin Lab Invest 1*:2, 1949
187. TADDEINI L, KAY IT, WATSON CJ: *Clin Chim Acta 7*:890, 1962
188. TALMAN EL: *Standard Methods of Clinical Chemistry*. Edited by D. Seligson. New York, Academic Press, 1958, Vol 2, p 137
189. TOWNSEND JD: *Ann Internal Med 60*:306, 1964
190. TSCHUDY DP: *J Am Med Assoc 191*:718, 1965
191. VAHLQUIST B: *Hoppe Seylers Z Physiol Chem 259*:213, 1939
192. VANNOTTI A: *Porphyrins,* translated by Rimington C. London, Hilger and Watts Ltd., 1954
193. VINCENT WF, ULLMAN WW: *Clin Chem 16*:612, 1970
194. VINCENT WF, ULLMAN WW: *Ann Clin Lab Sci 2*:31, 1972
195. VINCENT WF, ULLMANN WW, WEIDNER GL: *Am J Clin Pathol 53*:963, 1970
196. WALDENSTRÖM J: *Acta Med Scand 83*:281, 1934
197. WALDENSTRÖM J: *Deut Arch Klin Med 178*:38, 1935
199. WALDENSTRÖM J: *Acta Med Scand Suppl 82*:1, 1937
200. WALDENSTRÖM J: *Am J Med 22*:758, 1957
201. WALDENSTRÖM J, VAHLQUIST B: *Hoppe Seylers Z Physiol Chem 260*:189, 1939
202. WARD E, MASON HL: *J Clin Invest 29*:905, 1950
203. WATSON CJ: *Proc Soc Exp Biol Med 41*:591, 1939
204. WATSON CJ: *Arch Internal Med 86*:797, 1950
205. WATSON CJ: *New Engl J Med 263,* 1205, 1960
206. WATSON CJ, BOSSENMAIER I, CARDINAL R: *J Am Med Assoc 175*:1087, 1961
207. WATSON CJ, GRINSTEIN M, HAWKINSON V: *J Clin Invest 23*:69, 1944
208. WATSON CJ, HAWKINSON V, SCHWARTZ S, SUTHERLAND D: *J Clin Invest 28*:447, 1949
209. WATSON CJ, PIMENTA DE MELLO R, SCHWARTZ S, HAWKINSON V, BOSSENMAIER I: *J Lab Clin Med 37*:831, 1951
210. WATSON CJ, SCHWARTZ S: *Proc Soc Exp Biol Med 44*:7, 1940
211. WATSON CJ, SCHWARTZ: *Proc Soc Exp Biol Med 47*:393, 1941
212. WATSON CJ, TADDEINI L, BOSSENMAIER I: *J Am Med Assoc 190*:501, 1964
213. WEISSBERG JB, LIPSCHUTZ F, OSKI FA: *New Engl J Med 284*:565, 1971
214. WELLAND FH, HELLMAN ES, GADDIS EM, COLLINS A, HUNTER GW, Jr, TSCHUDY DP: *Metab Clin Exptl 13*:232, 1964
215. WESTALL KG: *Nature 170*:614, 1952
216. WILLIAMS MK, FEW JD: *Brit J Indus Med 24*:294, 1967
217. WITH TK: *Scand J Clin Lab Invest 7*:193, 1955
218. WITH TK: *Scand J Clin Lab Invest 8*:113, 1956
219. WITH TK: *Scand J Clin Lab Invest 9*:395, 1957
220. WITH TK: *Biochem J 68*:717, 1958
221. WITH TK: *Scand J Clin Lab Invest 10*:297, 1958
222. WITH TK: *Clin Biochem 1*:30, 1967
223. WITH TK: *Clin Biochem 1*:224, 1968
224. WITH TK: *Clin Biochem 2*:97, 1968
225. WITH TK: *J Chromatog 42*:389, 1969
226. WITH TK: *Lancet 2*:1187, 1970
227. WITH TK: *Lancet 1*:240, 1971
228. WRANNE L: *Acta Paediat 49 (Suppl 124),* 1960
229. ZEILE K, RAU B: *Hoppe Seylers Z Physiol Chem 250*:197, 1937
230. ZENKER N: *Anal Biochem 2*:89, 1961
231. ZIEVE L, HILL E: *J Lab Clin Med 42*:705, 1953
232. ZONDAG HA, VAN KAMPEN EJ: *Clin Chim Acta 1*:127, 1956

Chapter 25

Carbohydrates

VINCENT J. PILEGGI, Ph.D.,
CHRISTOPHER P. SZUSTKIEWICZ, Ph.D.

The detection, identification, and quantification of carbohydrates in blood, urine, and other biologic fluids have challenged clinical chemists and their predecessors for over a century. Interest in carbohydrates has largely been stimulated by the high incidence of diabetes mellitus and its serious clinical consequences. Other clinical conditions, however, including pentosuria, lactosuria, and galactosemia, dictate the need to detect and identify carbohydrates other than glucose in biologic fluids.

Table 25-1 gives the reported mean concentrations of various carbohydrates found in normal serum or plasma, whole blood, and urine of adults. Normal urine also contains small amounts of the following (242, 382): ribose, desoxyribose, glucuronolactone, fucose, fructose, sucrose, mannoheptulose, sedoheptulose, inositol, mannose, rhamnose, glucodifructose, and other unidentified carbohydrates. Traces of mannose and inositol have also been identified in normal blood (242).

IDENTIFICATION OF CARBOHYDRATES

Technics that have been used for the detection and identification of clinically important carbohydrates in body fluids include: (a) *chemical tests,* e.g., color tests for pentoses, tests for reducing sugars, and formation of osazones; (b) *enzymatic tests,* e.g., fermentation test and glucose oxidase test; (c) *paper electrophoresis;* (d) *paper chromatography;* (e) *gas-liquid chromatography;* (f) *thin layer chromatography (TLC).*

The fermentation test (133) in conjunction with the reducing test has been widely used in the past, and a detailed description of the method appears in the first edition of this book. The usefulness of this test is based on the fact that not all of the sugars of physiologic interest are fermentable by baker's or brewer's yeast, nor are they all reducing sugars. When one combines these two bits of information, the identity of the carbohydrate in question can be narrowed down to a few possibilities.

Until recently, paper chromatography has been the chief procedure for identification of sugars in the clinical laboratory. A major disadvantage of this technic, however, is the long development time, which makes it unsuitable when rapid diagnosis is required.

The first attempts to separate sugar mixtures by TLC were made during the early 1960s. Since then various support media have been used for the separation of carbohydrates, e.g., silica gel G, Kieselguhr, cellulose, and

TABLE 25-1. Carbohydrates in Normal Body Fluids (Figures Are Reported Average Concentrations)

Carbohydrate	Serum or Plasma mg/100 ml	Refs	Whole Blood mg/100 ml	Refs	Urine mg/100 ml	Refs	mg/24 hours	Refs
Glucose	83	40	74	39	5.2	168	52	6
							29	27
Lactose	<0.5	158			9.5	168	23	65
Fructose	0.87	168			2.1	168	60	6
Galactose	0.70	168			4	168	14	65
Glycogen	0	372	5	372				
			50	241				
Polysaccharides	110	311,306						
Glucosamine	70	311						
Hexuronates, as glucuronic acid	0.8	69	6	93			370	179
	2.5	290	7	274				
Pentoses, total	3.7	122	1.4	219			225[a]	355
Xylose					2.8	168	49	65
Arabinose					6.9	168	38	65
L-Xyloketose	<0.3	109,35					4	109
D-Ribulose							1	115
DL-Arabitol							50	355
Sucrose	0.06	168			2.2	168		
Raffinose	0.24	168			0.42	168		
Mannose	1.15	168						

[a]Subjects on fruit-free diet.

TABLE 25-2. Behavior of Carbohydrates in Reduction and Fermentation Tests

Carbohydrate	Reducing	Fermentable[a]
Glucose	+	+
Lactose	+	−
Fructose	+	+
Galactose	+	−
Maltose	+	+
Mannose	+	+
Sucrose	−	+
Trehalose	−	+
Pentoses	+	−

[a]Refers to fermentability by *Saccharomyces cerevisiae* (baker's yeast).

gypsum. The application of various support media, solvents, and detection reagents to the identification of mono- and disaccharides has been the subject of several comprehensive reviews (232, 301, 394). Numerous TLC methods utilizing precoated silica gel G and cellulose plates were examined in our laboratory, but none of them fully met the requirements of good separation, speed, and simplicity. A simplified procedure was subsequently developed (345) which is convenient and rapid for both plasma and urine specimens. A test tube desalting method is used for preparation of urine samples. The small volume (75 μl) of blood or plasma that is required makes the method practical for pediatric purposes.

IDENTIFICATION OF SUGARS BY THIN LAYER CHROMATOGRAPHY

(method of Szustkiewicz and Demetriou, Ref 345)

PRINCIPLE

The separation of some clinically important mono- and disaccharides occurring in plasma and urine is accomplished by means of double-pass TLC on silica gel G glass plates using a solvent of pyridine, ethyl acetate, and water (26:66:8). Substances interfering with the application and migration of the saccharides are removed by resin treatment of urine and ultrafiltration of plasma samples. Visualization of the sugars is made with aniline-phthalic acid reagent. Identification is made by comparing mobility of the unknown with that of the standard on the same chromatogram.

REAGENTS

TLC Glass Plates. 20 x 20 cm plates precoated with a 250 μ silica gel layer (Brinkmann, Cat. No. 68 10200-6).

Chromatogram Developing Solvent. Mix pyridine, ethyl acetate, and water in proportions of 26:66:8.

Standard Solutions. Dissolve 100 mg each of glucose, galactose, ribose, lactose, arabinose, and 200 mg each of fructose and xylulose in 100 ml of 50% isopropanol.

Resin. Mixed bed resin AG 501 x 8D (Bio-Rad Laboratories).

Ultrafiltration Cones. Ultrafiltration membrane cone CF_{50} (Amicon Corp.).

Benedict's Qualitative Reagent. Dissolve 17.3 g $CuSO_4 \cdot 5H_2O$ in about 100 ml hot water. With the aid of heat dissolve 173 g sodium citrate and 100 g anhydrous Na_2CO_3 in about 800 ml water. When cool, pour the second solution into the first while stirring and dilute to 1 liter. Stable at room temperature.

Detection Reagent. Dissolve 1.66 g *o*-phthalic acid and 0.95 ml aniline in a solution consisting of 48 ml isobutanol, 48 ml ethyl ether, and 4 ml water.

PROCEDURE

1. *Preparation of samples:*

   Plasma. Centrifuge 200 μl plasma in an Amicon ultrafiltration membrane cone CF_{50} for 5 min. An aliquot of the centrifugate is applied to the chromatogram in step 3.

   Urine. Urine is applied directly to the chromatogram in step 3. Prior to chromatography all urine samples are tested with Benedict's reagent to eliminate unnecessary analyses of negative samples (*cf Qualitative Test for Reducing Substances in Urine*). When the presence of lactose is suspected, the urine sample is treated with resin AG 501 x 8D as follows: An equal volume of resin is added to 1 ml urine in a 12 ml calibrated centrifuge tube. Stopper the tubes and mix for 4 min on a rotary mixer. Add 2 ml 5% acetic acid and mix until the resin turns to a gold color. Aliquot 1 ml to a calibrated test tube, and evaporate to approximately one-third of its volume with a stream of N_2. To facilitate rapid evaporation the sample may be placed in a 45°C water bath.

2. Prior to application of the unknowns, nine sample application spaces, each 1.5 cm wide and separated by 0.5 cm, are demarcated 1 cm above the bottom edge of the plate. Score the film horizontally 14 cm above the origin. Write sample identification numbers above this finish line. Subsequently, the chromatogram is sprayed lightly with the chromatogram developing solvent and dried at 105°C for 2 hours.

3. Using micropipets apply 10 μl samples as 1.5 cm stripes at 0.5 cm intervals. Apply a single stripe of the standard solution in the middle lane of the plate. Dry applications with a stream of warm air to minimize diffusion of the samples.

4. Carry out ascending chromatography in closed glass tanks lined with Whatman 1 paper saturated with the developing reagent. Allow solvent to ascend to a height of approximately 14 cm.

5. Dry chromatogram in a 50°C oven for 10 min and then return to the tank for a second identical development in the same solvent mixture. Dry plate with hot air (80°C for 5 min).

6. Spray with aniline reagent and heat for 10 min at 105°C. Locate sugars with the aid of long wave UV light.

INTERPRETATION

The approximate R_f values for carbohydrates of clinical interest are shown in Table 25-3. Since it is not possible to reproduce them from run to run, interpretation is more accurately made by comparing unknown spots with those of the standard mixture.

The seven-sugar mixture is not complete from a clinical point of view (maltose, sucrose, xylose, and raffinose are absent), but it contains the saccharides which are important for screening for carbohydrate disorders. Specifically, this method can be used as a diagnostic aid in conditions involving abnormal carbohydrate metabolism; e.g., diabetes mellitus, essential and alimentary pentosuria, ribosuria, lactose intolerance, and galactosemia.

TABLE 25-3. Relative Mobilities[a] of Carbohydrates on Silica Gel G Solvent: Pyridine-Ethyl Acetate-Water (26:66:8); Detection: Aniline (0.95 ml)-phthalic acid (1.66 g)-ethyl ether (48 ml)-isobutanol (48 ml)-water (4 ml).[b]

Carbohydrate	Approximate R_f
Xylulose	0.70
Ribose	0.64
Arabinose	0.44
Fructose	0.39
Glucose	0.34
Galactose	0.27
Lactose	0.15

[a] Double-pass chromatography.
[b] From Szustkiewicz C, Demetriou J: *Clin Chim Acta 32:*355, 1971.

NOTES

1. Prespraying of plates. Prespraying of the chromatogram with the solvent mixture followed by drying for 2 hours at 105°C prior to application reduces the variation in the mobility of carbohydrates and improves their resolution.

2. Development. Single-pass development provides fair separation of galactose, glucose, fructose, and arabinose, and may be employed to save time if only those sugars with well separated migration values are suspected. The time required to complete first development is about 75 min, and the subsequent development takes approximately 60 min.

3. Stability of samples. Whole blood, plasma, serum, and urine specimens containing 10 mg NaF/ml are stable for at least 7 days at 30°C. If fluoride is not added, samples should be analyzed within 4 hours of collection.

GLUCOSE

Although methods for the determination of glucose in blood and urine were described over 100 years ago, the search for simpler and more specific methods continues. Some of these represent automated versions of manual methods described earlier, while others involve different approaches or modifications. Basically, glucose methods fall into two categories: chemical methods and enzymatic methods. These two groups can in turn be broken down into many different categories based on (a) the type of measurement finally employed, e.g., photometric, fluorometric, or coulometric; (b) the enzyme system(s) used; etc. Before proceeding into a discussion of the various methodologic approaches, it is helpful to review briefly the structure of glucose and some of its unique properties.

Glucose exists in several different forms in solution:

```
      OH                    O                   OH
    HC                    HC                   HC
    |                     ||                   ||
    HCOH                  HCOH                 COH
    |                     |                    |
    HOCH   O     ⇌        HOCH      ⇌          HOCH
    |                     |                    |
    HCOH                  HCOH                 HCOH
    |                     |                    |
    HC                    HCOH                 HCOH
    |                     |                    |
    CH₂OH                 CH₂OH                CH₂OH

  RING FORM
  (PYRANOSE)              ALDEHYDE            ENEDIOL
```

The predominance of a given form is a function of the pH and solvent. Although glucose is referred to as an aldohexose, the aldehyde form of glucose is highly unstable, and only small amounts are present in aqueous solution. The enediol form is favored by alkaline conditions and is the form which gives glucose its familiar reducing properties. It is important to note that since glucose, fructose, and mannose differ only in the second carbon, they all form the same enediol:

```
      OH                    OH                   O
    HC                    HC                   HC
    |                     ||                   ||
    HCOH                  COH                  HOCH
    |                     |                    |
    HOCH     ⇌            HOCH      ⇌          HOCH
    |                     |                    |
    HCOH                  HCOH                 HCOH
    |                     |                    |
    HCOH                  HCOH                 HCOH
    |                     |                    |
    CH₂OH                 CH₂OH                CH₂OH

  D-GLUCOSE            COMMON ENEDIOL        D-MANNOSE
                           ⇅
                        CH₂OH
                        |
                        C=O
                        |
                        HOCH
                        |
                        HCOH
                        |
                        HCOH
                        |
                        CH₂OH

                     D-FRUCTOSE
```

Interconversion is catalyzed by dilute alkali and is commonly referred to as the Lobry de Bruyn transformation. The consequence of this is that any reaction involving the reactivity of the enediol form does not differentiate between glucose, fructose, and mannose. This applies to all of the methods based on the reducing properties of glucose.

In neutral or weakly acidic solution, the predominant forms are the hemiacetal structures in which the first and fifth carbon atoms are joined via an oxygen bridge, to yield a stable six membered ring referred to as a pyranose. The formation of the ring structure imparts asymmetry to carbon atom 1, and so two forms are possible. These are referred to as α and β-D-glucopyranose. Trace amounts (0.5%–1.0%) of another ring form also exist in solution in which the first and fourth carbon atoms are involved.

```
        OH                      O                     H
     HC⁄                     HC⁄                   HOC⁄
      |                       |                      |
     HCOH       |            HCOH                   HCOH       |
      |                       |                      |
     HOCH       O   ⇌        HOCH      ⇌           HOCH       O
      |                       |                      |
     HCOH       |            HCOH                   HCOH       |
      |                       |                      |
     HC ────┘               HCOH                   HC ────┘
      |                       |                      |
     CH₂OH                   CH₂OH                  CH₂OH

 α-D-GLUCOPYRANOSE                              β-D-GLUCOPYRANOSE
```

These five member ring structures are called α- and β-D-glucofuranose.

Pure glucose crystallized from aqueous solution at room temperature is chiefly in the α-D-glucopyranose form. Because of the asymmetrical carbon atoms, glucose exhibits the property of rotating polarized light. Freshly dissolved α-D-glucose has a specific rotary power of +112° which gradually decreases on standing to a constant value of +52.5°. This phenomenon, known as mutarotation, is due to the attainment of equilibrium between the α-D-glucopyranose and β-D-glucopyranose forms. At equilibrium, 36% of glucose is in the α form and 64% in the β form. These optical isomers are formed via the intermediate formation of the aldehyde structure. β-D-Glucopyranose, which can be prepared by crystallization from water at temperatures above 98°C, has a specific rotary power of +18.7°. The importance of these two forms will be made clear when methods employing glucose oxidase, which acts only on the β form, are discussed.

The spatial arrangement of the ring structures for glucose is more accurately represented by the three-dimensional structures first proposed by Haworth. In this structure the functional groups are shown as below and above the plane of the ring which is perpendicular to the plane of the paper:

α – D-GLUCOPYRANOSE

The β form would have the OH and H groups on carbon 1 reversed.

QUANTITATIVE METHODS FOR BLOOD GLUCOSE

REDUCING METHODS

The enediol form of glucose is highly reactive and easily oxidized, even by gaseous oxygen. Upon oxidation, the carbon chain is ruptured with the formation of acids of shorter chains. The reaction is not stoichiometric. Many different products are formed varying with the extent of oxidation, which is dependent on pH, temperature, salt concentration, and concentration of oxidant. Rigid standardization of conditions, however, permits the use of the reducing property as a quantitative tool.

Picric Acid as Oxidant.

Of historic interest only is the use of picric acid in hot alkaline solution. This method, which involves the reduction of picric acid to the red salt of picramic acid, was introduced in 1931 by Lewis and Benedict (205) but fell into disuse rapidly because of nonspecificity (60).

Ferricyanide as Oxidant.

Yellow ferricyanide ions, $Fe(CN)_6^{3-}$, are reduced in alkaline solution to colorless ferrocyanide ions $Fe(CN)_6^{4-}$. One of the advantages of this reaction is that ferrocyanide is not readily reoxidized by air. Numerous modifications have been applied to glucose determinations:

Direct Redox Titration. Protein-free filtrates of blood are added dropwise to hot alkaline ferricyanide until the yellow color disappears and the solution becomes colorless (318).

Timed Decolorization of Ferricyanide. An amount of ferricyanide is used which is less than the equivalent of glucose to be determined. Under such conditions the time for decolorization varies inversely with the log of glucose concentration (137).

Measurement of Excess Ferricyanide. Glucose is determined from the difference in ferricyanide before and after reaction. *(a) Iodometric titration.* In 1923 Hagedorn and Jensen (126) introduced the technic in which ferricyanide is reacted with KI in acid medium, releasing an equivalent amount of I_2, which is titrated with thiosulfate and starch:

$$2\ Fe(CN)_6^{3-}\ +\ 2\ I^-\ \rightleftharpoons\ 2\ Fe(CN)_6^{4-}\ +\ I_2$$

$$2\ S_2O_3^{2-}\ +\ I_2\ \longrightarrow\ 2\ I^-\ +\ S_4O_6^{2-}$$

The first reaction is forced to completion by precipitation of the ferrocyanide with Zn^{2+}. Although the method yields results which are higher than the true glucose values by 20–30 mg/100 ml (193), it proved to be a useful method which was widely used for many years. One screening test based on this method permitted classification of blood glucose levels as being above 180 or below 170 mg/100 ml (136, 384). *(b) Gasometric determination.* N_2 liberated by the action of ferricyanide on hydrazine can be measured (367). *(c) Photometric measurement of ferricyanide.* Introduced by Hoffman in 1937 (153), this is the most convenient of the ferricyanide glucose methods. The decrease in yellow color is equivalent to the ferricyanide consumed since ferrocyanide is colorless. This method was adapted for the AutoAnalyzer (167) but was subsequently superseded by direct measurement of the ferrocyanide (90, 161). *(d) Titration with sodium indigo sulfonate.* Ferricyanide oxidizes the blue dye to isatin, which is colorless at the dilution used (273).

Measurement of Ferrocyanide. Basically this is a more precise approach than that of determining ferricyanide consumed in the reaction, because calculation based on a difference between two measurements is avoided by determining ferrocyanide formation directly. *(a) Photometric measurement by Prussian blue color reaction.* The reaction is as follows:

$$3 \ Fe(CN)_6{}^{4-} + 4 \ Fe^{3+} \longrightarrow Fe_4 [Fe(CN)_6]_3$$
$$\text{ferric ferrocyanide}$$
$$\text{(Prussian blue)}$$

The blue color must be measured in the presence of the excess unreduced ferricyanide and excess Fe^{3+}, both of which are yellow. This technic was introduced by Folin (97) in 1928. The fact that Prussian blue is an unstable, hydrophobic colloidal sol has prompted many modifications. Hydrophilic colloids such as gum ghatti, gum arabic, or Duponol are added in most modifications as protective agents to prevent flocculation of the Prussian blue. An alternative procedure is to peptize the sol with oxalic acid (220). Ceriotti (45) eliminated the color instability of Prussian blue by using a $Cd(OH)_2$ deproteinization of the blood followed by removal of Cd^{2+} with borate and color development with a ferric ammonium sulfate solution made up in 50% acetic acid. *(b) Photometric measurement of the yellow-brown molybdenum ferrocyanide* (332). *(c) Photometric measurement of the brown uranyl ferrocyanide* (238). *(d) Titration with ceric ions.* When first introduced (237), Setopaline C was used as an indicator for the endpoint (appearance of excess ceric ions). Subsequently an *o*-phenathroline-ferrous complex was substituted for Setopaline C (139). *(e) Manganimetric titration* (91). *(f) Potentiometric titration with Zn^{2+}* (365).

Potentiometric Measurement with a Ferri-Ferrocyanide Electrode (309).

Copper As Oxidant

Copper (Cu^{2+}) in hot alkaline solution is reduced to Cu^+, which combines with OH^- to form yellow CuOH. The heat converts CuOH to red

Cu_2O, both cuprous compounds being insoluble. To prevent precipitation of $Cu(OH)_2$ or $CuCO_3$ in the reagent, the Cu^{2+} is complexed by citrate or tartrate (Rochelle salt). These soluble complexes dissociate sufficiently to provide a continuous supply of Cu^{2+} during the redox reaction. Citrate reagents are less sensitive than tartrate reagents (96, 320). Copper methods are basically more specific than ferricyanide methods because of the lower oxidation-reduction potential of the former; i.e., ferricyanide more easily oxidizes substances other than sugars. The extent of oxidation of sugars by Cu^{2+}—i.e., the sensitivity of the method (320)—varies inversely with the alkalinity of the copper reagent. The stability of the reagent also varies inversely with alkalinity. Numerous modifications of the copper reagent appeared during the 1920s and 1930s, foremost among them being those of Folin, Benedict, and Somogyi, all stemming from a search for a reagent with greater stability, sensitivity, and specificity.

Under carefully controlled conditions the Cu_2O produced is directly proportional to the glucose present. Among the approaches to determination of the Cu_2O are the following:

Measurement of Color Produced by the Folin-Denis Phenol Reagent (100). This reagent contained phosphotungstic and phosphomolybdic acids. The method was rapidly abandoned because of interference by phenols.

Measurement of Color Produced by Phosphomolybdate. Under suitable conditions, orthophosphoric and molybdic acids condense to form heteropoly complex compounds, the molybdiphosphoric or phosphomolybdic acids (340).

$$H_3PO_4 + H_2MoO_4 \longrightarrow H_4[P(Mo_3O_{10})_4]$$

Reduction of phosphomolybdic acid by Cu_2O yields "molybdenum blue," a compound of uncertain composition. The ratio of orthophosphoric/molybdic acid in the color reagent is important. If the ratio is too low, some molybdate is reduced as well as the heteropoly complex, thus resulting in a high reagent blank; if the ratio is too high, there is a decrease in the blue color formed. This reagent does not react with phenols. Folin and Wu (101) introduced this reagent in 1920, and their method using tungstic acid precipitation of proteins or a zinc hydroxide (Somogyi) filtrate was probably the most widely used method in the United States for the ensuing four decades. Because of the ease of reoxidation by air of the Cu_2O, Folin and Wu designed a reaction tube with a construction so that the liquid-air interface is limited to a diameter no greater than 11 mm, which allows only negligible reoxidation. Best known of the modifications of the technic and reagents were those of Folin (98) and Benedict (20), the latter achieving greater specificity by adding bisulfite and alanine to the copper reagent. One of the problems encountered with the method is instability of the molybdenum blue color, which increases for a time before gradually decreasing. There is actually a change in the absorption spectrum with time, the least change occurring in the green region (127). Fading, which increases as the color deepens and thus is not proportional to the glucose concentration, is a serious obstacle to accuracy. A number of different approaches have been used to stabilize the color. One of the more convenient and effective methods involves placing the final colored solutions in boiling water, after addition of phosphomolybdate but prior to dilution with water (92).

Phosphotungstate Color Reaction. Benedict (19), prior to switching to phosphomolybdate, used the tungstic-arsenic-phosphoric acid reagent from his uric acid method.

Arsenomolybdate Color Reaction. Nelson (252) in 1944 reported the use of arsenomolybdate in conjunction with Somogyi's copper reagent (commonly referred to as the Nelson-Somogyi method). The color with arsenomolybdate is more stable and more sensitive than that obtained with phosphomolybdate, and consequently this method achieved widespread use. A critical examination of the Somogyi modification (323) of Nelson's method was made by Marais *et al* (217).

Neocuproine Color Reaction. Neocuproine (2,9-dimethyl-1,10-phenanthroline hydrochloride) is specific for Cu^+, and the stable color formed is 30 times more intense than the color formed with phosphomolybdic acid. Substitution of neocuproine for phosphomolybdic acid in the Folin-Wu glucose method decreases the blood sample requirement to $10-30$ μl (38). A modification in which serum is deproteinized with sodium tungstate and copper sulfate has been described (41).

Iodometric Titration. The best known application of this technic was introduced in 1921 by Shaffer and Hartmann (308). An alkaline-citrate-oxalate-copper sulfate reagent containing KIO_3 and excess KI is heated with the solution to be analyzed and then immediately acidified. The IO_3^- sets free an equivalent amount of I_2 by the following reaction:

$$5I^- + IO_3^- + 6\,H^+ \longrightarrow 3H_2O + 3I_2$$

The Cu_2O formed in the oxidation of glucose reacts at once with part of the I_2 formed in the previous reaction:

$$2\,Cu^+ + I_2 \rightleftharpoons 2\,Cu^{2+} + 2I^-$$

The oxalate in the reagent forces the reaction to completion to the right by formation of nonionized cupric oxalate. The difference in titration of a blank and the unknown is equivalent to the reducing sugar present. Somogyi's modification (322), one of many, is the most popular and became known as the Somogyi-Shaffer-Hartmann method. A study of the variables involved in the reactions was subsequently reported by Shaffer and Somogyi (307).

METHODS DEPENDING ON FORMATION OF COLOR WITH
PHENOLS IN CONCENTRATED H_2SO_4

These methods are based on the formation of hydroxymethylfurfural
when glucose is heated with strong acids:

$$D\text{-GLUCOSE} \xrightarrow[\text{HEAT}]{H_2SO_4} HOCH_2 \cdot \text{—} \cdot CHO + 3H_2O$$

HYDROXYMETHYLFURFURAL

$$\longrightarrow HCOOH + CH_3COCH_2CH_2COOH$$

FORMIC   LEVULINIC
ACID   ACID

Although the reaction proceeds further, the amount of hydroxymethylfurfural formed under standard conditions is directly proportional to glucose
concentration. Photometric measurement of the color produced by condensing the aldehyde group of hydroxymethylfurfural with a phenolic compound
is the basis for a number of methods which have been used for detecting and
quantitating blood glucose. Phenolic compounds used include a-naphthol
(Molisch test for sugars) (192), resorcinol (10), and anthrone (73). A
solution of anthrone in concentrated H_2SO_4, when heated with glucose,
yields a green color. The use of anthrone for quantitative determination of
glucose in blood was described in 1948 (244). The method is not specific for
glucose, since a color is given by any carbohydrate yielding furfural (e.g.,
from pentoses) or hydroxymethylfurfural (e.g., from hexoses, disaccharides,
polysaccharides, glycosides, carbohydrate ethers, and esters) (31). The
specificity for glucose in blood is about the same as the copper reduction
methods when applied to zinc hydroxide filtrates (81), but the reaction can
be used to determine fructose in the presence of glucose by omitting the
heating period (31). A serious drawback to the anthrone methods is that the
corrosiveness of the reagent creates serious problems of safety and disposal.

ALDOSE-AROMATIC AMINE CONDENSATION METHODS

Glucose and other aldoses condense with aromatic amines in hot acetic
acid solutions to yield glycosylamines. The use of benzidine in glacial acetic
acid for the photometric determination of various aldoses, including glucose,
was described in 1954 (169), but the technic was not applied to blood or
urine. Among various arylamines tested, *o*-aminodiphenyl in glacial acetic
acid has been found to be useful for detecting aldoses on paper chromatograms as well as for quantitation of aldoses (350). The products of the
reaction may be equilibrium mixture of the cyclic N-glycoside (I) and the
corresponding acyclic Schiff base (II):

There is, however, much evidence (327) to support the cyclic N-glycoside form (I) as the probable structure of glycosylamines.

This reaction was initially applied to glucose in deproteinized blood using *p*-aminosalicylic acid and glacial acetic acid (76). Subsequently, *p*-amino-benzoic acid and *m*-aminophenol were used to determine glucose and fructose in blood, serum, and urine (75). *p*-Aminobenzoic acid in glacial acetic acid reacts with aldosugars and only to a small extent with ketosugars, whereas *m*-aminophenol reacts with both aldo- and ketosugars. The use of *o*-aminobiphenyl in glacial acetic acid has also been proposed (8). The 45 min heating period in boiling water was later shortened by heating under pressure (102). The determination was applied to both Folin-Wu and Somogyi-Nelson (Ba(OH)$_2$-ZnSO$_4$) deproteinized blood filtrates, and the results were found to be in good agreement with the Somogyi-Nelson copper method.

In 1959 Hultman (160) reported the use of *o*-toluidine in glacial acetic acid in a procedure that requires only an 8 min heating period and which can be used with serum or plasma without protein precipitation. High sensitivity allows concentrations as low as 20 mg/100 ml to be measured on 20 μl samples. Numerous modifications have been proposed including the use of thiourea to stabilize the reagent and reduce the blank (162), the use of borate to stabilize the color (47, 120), and the elimination of glacial acetic acid (46, 135, 386). A complete study of Hultman's *o*-toluidine method was carried out by Dubowski (74) who proposed use of a filtrate. This procedure has been the basis of many minor modifications which are widely used and adapted to the AutoAnalyzer (110, 243, 336, 380, 397).

MISCELLANEOUS CHEMICAL METHODS

Photometric measurement of the color produced with the following reagents has been used for the determination of glucose in biologic fluids: dinitrosalicylic acid (240, 339), phenylhydrazine (335, 373), chromatropic acid (185), *p*-anisyltetrazolium blue (49), triphenyltetrazolium chloride (211), 3,4-dinitrobenzoic acid (189), hot H$_2$SO$_4$ (formation of hydroxy-methylfurfural which undergoes further changes to a pink color product) (229, 230), and *p*-bromoaniline in acetic acid (67). A fluorescence technic with extremely high sensitivity (a deproteinized aliquot equivalent to 0.5 μl of blood is used) involves the reaction of glucose with 5-hydroxytetralone to yield benzonaphthenone (33).

ENZYMATIC METHODS

None of the methods discussed previously is absolutely specific for glucose. The use of enzymes as a means of achieving ultimate specificity initially involved the use of yeast—viz, obtaining the difference between reducing sugars before and after yeast fermentation. The inadequacy of this approach was recognized as early as 1925 by Hiller *et al* (148) who observed that yeast fermentation of normal blood left a residue which would reduce ferricyanide and Cu^{2+}. Other complicating factors which limited the specificity of yeast methods include: (a) other sugars present in normal blood and urine are fermented by yeast: and (b) sugar phosphates are fermented although at a rate less than 5% of the rate of sugar (287). Measurement of the CO_2 produced by yeast fermentation, volumetrically (277), or after microdiffusion (256) has been used as a method for glucose estimation, but CO_2 produced from amino acids and α-keto acids (366) is a source of error.

Greater specificity would be expected by use of a single purified enzyme which acts on glucose but not other physiologically occurring carbohydrates or derivatives. Glucose oxidase, discovered by Müller in 1928 (247), is specific for β-D-glucose, the only notable exception being its action on 2-desoxy-D-glucose which is oxidized at about 12% of the rate of glucose (222). Glucose oxidase oxidizes glucose according to the following reaction:

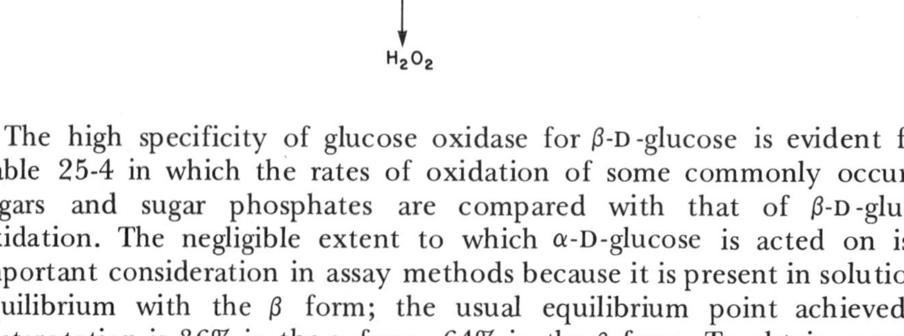

The high specificity of glucose oxidase for β-D-glucose is evident from Table 25-4 in which the rates of oxidation of some commonly occurring sugars and sugar phosphates are compared with that of β-D-glucose oxidation. The negligible extent to which α-D-glucose is acted on is an important consideration in assay methods because it is present in solution in equilibrium with the β form; the usual equilibrium point achieved via mutarotation is 36% in the α form, 64% in the β form. To obtain complete oxidation of glucose with glucose oxidase, all glucose must be converted into the β form. The speed at which mutarotation occurs is affected by pH, increased temperature, and more specifically by an enzyme mutarotase

(113). Some but not all commercial preparations of glucose oxidase contain glucomutarotase (2, 113).

TABLE 25-4. Rates of Oxidation by Glucose Oxidase, Taking Glucose as 100

Sugar	Rate	Refs
β-D-Glucose	100	
α-D-Glucose	0.64	176
Galactose	0.14	176
	0.06	159
	0.03	2
Mannose	0.98	176
	0.8	159
	0.03	2
	1	174
D-Xylose	0.03	2
	1	174
D- and L-Arabinose	0	151
D-Fructose	0	151
2-Desoxy-D-glucose	12	222
Glucose-1-phosphate	0	159
Glucose-6-phosphate	0	159

In 1948, Keilin and Hartree (175) demonstrated the applicability of glucose oxidase for determining glucose by a manometric technic in which the O_2 consumed in the reaction was measured. Initial applications to glucose in biologic fluids involved the determination of reducing substances before and after treatment with glucose oxidase (113). Direct photometric measurement of glucose was made possible through use of a coupled enzyme system in which the H_2O_2 formed is coupled via peroxidase to a chromogenic O_2 acceptor (181, 347). Compounds used as O_2 acceptors include o-dianisidine (181, 347), o-anisidine (376), o-tolidine (234), indophenol (70), diethyl-p-phenylenediamine (349), and adrenaline (357). Free (106) wrote an excellent review of the history and properties of glucose oxidase and its application to the measurement of glucose in biologic fluids. The chromogens are colloidal, and means of stabilization have been suggested (123, 235). The H_2O_2 produced has also been reacted, as formed, with iodide in the presence of molybdate catalyst to form triiodide, which is measured by its strong absorption at 360 nm (215) or by formation of the starch-iodine chromogen (9).

Deproteinization is required for whole blood to remove numerous interfering substances primarily present in the red blood cell. When the methods are applied directly to serum or plasma the problem of inhibitors is still present. Negative errors are caused by ascorbic acid concentrations greater than 5 mg/100 ml (44, 196), bilirubin (271), and hemolysis (44). Uric acid causes a negative error of about 1 mg/100 ml for each mg uric acid/100 ml (223, 271). These inhibitors, as well as catechols, glutathione, and cysteine, presumably act by competing with the chromogen as H^+ donors (81). The interferences by bilirubin, hemolysis, and uric acid can be avoided by deproteinization with the Somogyi zinc hydroxide method. It has been claimed that preincubation of serum with iodine avoids interference from hemolysis, lipemia, bilirubin, and reducing substances oxidized by iodine, thereby permitting accurate measurement of glucose in serum and cerebrospinal fluid without deproteinization (375). A fluorometric glucose oxidase method utilizes peroxidase and homovanillic acid, the latter being converted to a highly fluorescent product (266). The procedure is carried out directly on plasma and is free from interference by uric acid, bilirubin up to levels of 20 mg/100 ml, or mild hemolysis (100 mg/100 ml). Direct measurement of H_2O_2 formed by chelation is the basis of an automated method which can be used directly on serum (346).

Other approaches include the use of a polarographic oxygen electrode (171, 214) which permits direct determination of glucose in blood or plasma within 20 seconds. This principle is the basis of the Beckman Glucose Analyzer. A coulometric method has also been described for application to deproteinized serum specimens (313). Updike and Hicks (363) reported on the use of an "enzyme electrode" which consists of an immobilized enzyme membrane over a polarographic oxygen electrode. The O_2 which diffuses through the membrane is reduced in the presence of glucose oxidase and glucose.

Automated glucose oxidase methods have been described for the AutoAnalyzer (84, 147, 282) and for an automatic potentiometric system (216).

Dextrostix (Ames), a paper strip impregnated with a glucose oxidase-peroxidase-chromogen system, is used for semiquantitative estimates of blood glucose levels. The indicator is oxidized in the presence of glucose, and the color produced as well as its intensity are related to the glucose content of the blood. There appears to be general agreement that Dextrostix are useful for screening but not for quantitative purposes (48, 262, 297, 360). Some studies (154, 283) have revealed high coefficients of variation when comparing results performed by different technicians. Use of a reflectance meter capable of quantitating the color change purportedly increases considerably the quantitative precision of the Dextrostix blood glucose estimation (166).

Another enzymatic method for glucose uses hexokinase and glucose 6-phosphate dehydrogenase to convert glucose ultimately to 6-phosphogluconate (302):

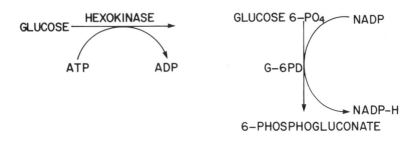

The NADP·H formed can be measured by the increase in absorbance at 340 or 366 nm. This approach, commonly referred to as the hexokinase method, has been used for determination of glucose in blood, urine, and spinal fluid (177, 333) and yields results in excellent agreement with the glucose oxidase methods (149). A fully automated method utilizing the AutoAnalyzer has also been described (130).

Another enzymatic method proposed for glucose estimation involves the enzyme acylphosphate: D-glucose-6-phosphotransferase (23).

SPECIFICITY OF GLUCOSE METHODS–CHOICE OF A QUANTITATIVE METHOD

The specificity of glucose methods is usually described in terms of how closely the values agree with those obtained by glucose oxidase or hexokinase methods. Values obtained by enzymatic methods are frequently referred to as "true glucose," but as previously discussed the presence of inhibitors may decrease the accuracy of results.

Prior to the development of quantitative enzyme methods or the relatively specific chemical method utilizing o-toluidine, most widely used methods were based on the reducing properties of glucose. They were therefore susceptible to positive errors caused by the presence of nonglucose-reducing substances present in normal blood. The term "saccharoid" was applied to these substances by Benedict in 1931 (20) who defined them as nonglucose, nonfermentable reactants. The substances identified as belonging to the saccharoid fraction include glutathione (82, 142), ergothioneine (321), creatinine, uric acid, ascorbic acid (83), certain amino acids (353), homogentisic acid, creatine, phenols (369), glucuronic acid (83, 369), and sugar phosphates (81).

Much of the effort devoted to improving early glucose methods focused on elimination of the error caused by inclusion of all or part of the saccharoids. This was accomplished with varying degrees of success both by improvements in the copper reagents and by the use of various protein precipitants to remove the interfering substances by coprecipitation. One of the most successful combinations of these improvements was the use of a zinc sulfate-barium hydroxide protein precipitant and an arsenomolybdate copper reagent, commonly referred to as the Nelson-Somogyi glucose method (252). Thus, whereas glutathione constitutes some 37% of the saccharoid fraction (82) in the Folin-Wu method (tungstic acid filtrate), the use of $Zn(OH)_2$ almost completely eliminates glutathione as a source of error. The reader is referred to the first edition of this book for a complete review of the saccharoid levels found by the various glucose methods applied to whole blood. These data are not included here since many such methods are of historic interest only.

The Nelson-Somogyi method is still used today and, when applied to deproteinized serum or plasma, yields results which are only slightly higher (2–3 mg/100 ml) than values obtained by the glucose oxidase and hexokinase technics (213). One of the glucose methods commonly used with the AutoAnalyzer employs the ferricyanide reaction and gives values which average about 7% higher than glucose oxidase results (342).

Of the nonenzymatic technics, the *o*-toluidine method is rapidly becoming the most widely used because of its high specificity, simplicity, and applicability to serum without deproteinization. Since the reaction is not based on the reducing properties of glucose, only a few physiologically occurring compounds react with *o*-toluidine to yield substances absorbing at the wavelength used to measure the colored compound formed with glucose. Galactose and mannose, and to a lesser degree lactose and xylose, are potential sources of error but are not normally present in serum or plasma in significant amounts. Dubowski (74) studied the specificity of the *o*-toluidine method applied to urine, serum, and cerebrospinal fluid deproteinized with trichloroacetic acid by making determinations before and after incubation with baker's yeast. Residual or blank levels were only 4 mg/100 ml for serum, 1 mg/100 ml for CSF, and 5 mg/100 ml for urine. In another study no residual "glucose" was found with an *o*-toluidine method in blood specimens incubated with glucose oxidase for 24 hours (162). Azotemia, which causes erroneously high glucose values with copper reduction methods, does not interfere with *o*-toluidine methods (243).

It is the opinion of many investigators that when all factors are considered—e.g., susceptibility of enzymatic methods to inhibitors, expense of reagents, stability of reagents, and ease of performance—the *o*-toluidine method, either automated or manual, is the choice for the routine clinical chemistry laboratory (279, 336, 380). It is the authors' opinion that there is much merit in the above recommendation, and we therefore chose to present both a manual and an automated *o*-toluidine method. Recognizing the preference of some for enzymatic technics, a glucose oxidase method is also included. The reader desirous of utilizing a copper reduction method with a zinc hydroxide protein precipitation step should refer to the first edition of this book.

Some comments concerning the type of specimen to use, i.e., whole blood or serum, are pertinent since the decision can influence results significantly, depending upon the method of glucose analysis. The bulk of the saccharoids resides in the erythrocytes. Glutathione and ergothionine, for example, occur only in erythrocytes. Thus for glucose methods based on the reducing properties of glucose, the method of protein precipitation is an important consideration, as discussed previously. Likewise, the inhibitory action of glutathione and other substances concentrated in red cells makes whole blood less desirable than serum or plasma for enzymatic methods. Although glucose is freely diffusable between plasma and erythrocytes and is in the same concentration in erythrocyte water as in plasma water (187), the different water content of erythrocytes and plasma (about 72 g/100 ml for the former and 94 g/100 ml for the latter) makes true glucose levels in whole blood a function of the hematocrit. It is not possible, therefore, to calculate precisely whole blood glucose from plasma or serum glucose or vice versa without a hematocrit. It is desirable from several viewpoints to use only serum or plasma for glucose analyses. Although the trend is slowly moving in this direction, whole blood is still commonly used, and the problems associated with it must be considered in both the reporting and interpretation of results.

DETERMINATION OF GLUCOSE IN BLOOD

(method of Hultman, Ref 160, modified by Dubowski, Ref 74, and Hyravinen and Nikila, Ref 162)

PRINCIPLE

Glucose reacts with *o*-toluidine in glacial acetic acid in the presence of heat to yield a blue-green N-glucosylamine, the absorbance of which is measured at 625 nm (Fig. 25-1). The method can be applied directly to serum, plasma, cerebrospinal fluid, and urine. For whole blood or moderately hemolyzed serum, deproteinization is required.

REAGENTS

Tungstic Acid Reagent (Stabilized). Dissolve 1.0 g polyvinyl alcohol (Elvanol 70-05, DuPont) in about 100 ml water with gentle warming *(caution*: do not boil). Cool and transfer to a 1 liter volumetric flask

FIG. 25-1. Absorption curves for glucose determination by the *o*-toluidine-glacial acetic acid method: (1) reagent blank vs water. (2) glucose standard equivalent to 100 mg/100 ml vs water. Beckman DBG recording spectrophotometer.

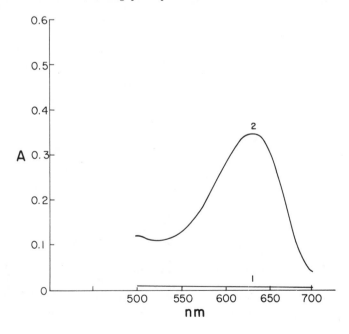

containing 11.1 g sodium tungstate ($Na_2WO_4 \cdot 2H_2O$), reagent grade, previously dissolved in about 100 ml water. Mix by swirling. In a separate vessel add 2.10 ml conc H_2SO_4, reagent grade, to about 300 ml water, mix, and then add to the tungstate solution in the volumetric flask. Mix and dilute to 1 liter with water. This reagent is stable for up to 1 year at room temperature.

o-Toluidine Reagent (Stabilized). To 5.0 g thiourea, reagent grade, add 90.0 ml o-toluidine (Reliable Chem. Co.) and dilute to 1 liter with glacial acetic acid. Stored in an amber bottle, the reagent is stable for up to 2 years at refrigerator temperatures. As the reagent ages, it yields less color for a given amount of glucose, but proportionality between standards and unknowns is maintained.

Glucose Standard, 100 mg/100 ml. Dissolve 1.00 g pure anhydrous glucose (dextrose) in 1 liter of water containing 1.5 g benzoic acid. This solution is stable up to 2 years in the refrigerator.

PROCEDURE FOR SERUM OR PLASMA–DIRECT

1. Transfer 5.0 ml o-toluidine reagent to test tubes or vials of about 10 ml capacity. These tubes may be matched photometer or spectrophotometer tubes if desired, since final absorbance readings can be made without further transfer. Label tubes "blank", "standard", and "unknown."
2. Add 0.1 ml glucose standard (100 mg/100 ml) to the "standard" tube (TC pipet). Mix thoroughly.
3. Add 0.1 ml serum or plasma (TC pipet) to the "unknown" tube. Mix thoroughly.
4. Loosen the stoppers or caps and place all tubes in a boiling water bath or a heating block preset at 100°C. Remove all tubes after 10 min and place in a cold tap water bath.
5. When tubes have cooled, read absorbances of standard and unknowns against the blank at 625 nm or with a filter having a nominal wavelength in this region.

Calculation:

$$\text{mg glucose/100 ml} = \frac{A_x}{A_s} \times 0.1 \times \frac{100}{0.1}$$

$$= \frac{A_x}{A_s} \times 100$$

PROCEDURE FOR CEREBROSPINAL FLUID–DIRECT

Same as for serum or plasma except use 0.2 ml aliquot of spinal fluid and add 0.1 ml water to "standard" tube in addition to 0.1 ml glucose standard.

Calculation:

$$\text{mg glucose/100 ml} = \frac{A_x}{A_s} \times 0.1 \times \frac{100}{0.2}$$

$$= \frac{A_x}{A_s} \times 50$$

PROCEDURE FOR WHOLE BLOOD

1. Prepare a deproteinized sample by transferring 0.2 ml blood sample into a test tube containing 1.8 ml tungstic acid reagent (stabilized). Mix, let stand 5 min, and centrifuge.
2. Add a 1.0 ml aliquot of the centrifugate to 5.0 ml *o*-toluidine reagent, mix, and label "unknown." Set up the blank and standard as follows:

Blank: 5.0 ml *o*-toluidine reagent + 1.0 ml water
Standard: 5.0 ml *o*-toluidine reagent + 0.9 ml water + 0.1 ml glucose standard (TC pipet).

3. Proceed, starting with step 4 of the "direct" method.

NOTES

1. Beer's law. The relationship between absorbance and concentration at 625 nm and with most filter instruments is linear up to absorbance readings of at least 0.9.

2. Color stability. The color of the cooled reaction mixture is stable up to 30 min and then the color slowly decreases.

3. Color formation. The color intensity produced for a given amount of glucose is a function of the reagent composition (different lots of *o*-toluidine may yield more or less sensitive reagents), age of the reagent (older reagents give less color), and heating time and temperature. It is therefore important to adhere closely to the recommended time of 10 min and to the 100°C heating temperature. It is of far greater importance, however, for the blank, standard, and unknowns to be exposed to the same time and temperature. Batch sizes therefore should be restricted to numbers which allow for good control of reaction conditions.

4. Interfering substances. Hemoglobin up to 350 mg/100 ml and bilirubin up to 20 mg/100 ml do not interfere in the direct method (no deproteinization). Greater concentration of hemoglobin results in positive errors, and bilirubin in negative errors. Both sources of interference are completely avoided by first deproteinizing with tungstic acid (see *Procedure for Whole Blood*). Galactose and mannose react and are "positive" sources of error. This is not a serious problem since these two sugars are normally present at noninterfering levels. In suspected galactosemics, this possible source of error should be considered. Plasma expanders (dextrans) cause a positive error since they are insoluble in the *o*-toluidine reagent and cause turbidity (111). Some recent work (135) indicates that if the glacial acetic acid is removed from the reagent and replaced with a mixture of glycollic acid and ethyleneglycolmonomethyl ether, the interference caused by plasma expanders can be reduced or eliminated.

5. *Stability of samples.* Glucose disappears from whole blood on standing as a result of enzymic action (glycolysis). The rate of loss decreases with decreasing temperature. The loss is about 10–20 mg/100 ml/hour at 37°C (352), but only 5–10 mg/100 ml at room temperature (341). The rate of glycolysis is significantly increased in the presence of leukocytosis (341). Since erythrocytes and leukocytes are responsible for glycolysis, rapid separation of the plasma or serum (within 10 min of the time the blood is taken) is required to avoid significant loss of glucose. If this is not practical, it is necessary to use inhibitors of the glycolytic enzymes. Alternatively, dilution of the blood with water at 1:80 (v/v) effectively inhibits glycolysis for 2 hours (227).

The most widely used preservative is NaF alone or in combination with other compounds. Since the problem of preserving blood or serum involves not only the inhibition of glycolysis but also the prevention of bacterial growth, thymol is frequently used in combination with NaF, the usual formulation being 10 mg NaF plus 1 mg thymol per ml blood. In one study comparing the efficacy of KF, NaF, and combinations with thymol, all showed about the same effectiveness (172), with 98% of samples having glucose levels within 10% of the initial value for up to 144 hours. NaF is rather insoluble and is best used as the dry powder. After adding blood, one must shake well to prevent clotting. $KF \cdot 2H_2O$ in contrast is very soluble, and 0.1 ml aliquots (2 drops) of an 80% solution can be conveniently dispensed into tubes or plastic vials and the water removed by evaporation at 50°C overnight. This provides enough KF to preserve 5 ml whole blood satisfactorily.

The stability of glucose in serum or plasma which has been promptly separated and removed from the cells (clot) is extremely variable. Our own experience reveals that most unpreserved samples show no change for up to 4 days at room temperature, but some show almost complete loss of glucose. It is almost certain that this represents variable bacterial contamination. The use of NaF or KF (10 mg/ml serum or plasma) is recommended if any significant delay in analysis is anticipated. This gives stability for 7 days at 30°C.

It is to be emphasized that the method of analysis employed must be considered when using glucose preservatives. Thus thymol yields a false positive error in the AutoAnalyzer method based on Hoffman's ferricyanide method, and high levels of fluoride (10 mg/ml or higher) interfere in the glucose oxidase methods (218, 377).

ACCURACY AND PRECISION

The specificity of the *o*-toluidine method has been discussed previously. The values obtained are quite close to "true" glucose values due to the specificity of the reaction for aldohexoses.

The precision (95% limits) of the method is ±5% for glucose values between 50 and 300 mg/100 ml.

NORMAL VALUES

The normal adult fasting ranges for whole blood, and serum or plasma, by the method presented are 60–100 and 70–110 mg/100 ml, respectively. These values are based on published values for whole blood (4, 162, 201) and the addition of 10 mg/100 ml to the whole blood values to obtain serum or plasma ranges. There is no sex difference (271, 276), and the ranges cited are applicable to children after the first few weeks of life (58).

Glucose levels of cerebrospinal fluid parallel venous blood levels and are about 40%–80% of blood values (95, 331). Levels in ventricular fluid are higher than those in cisternal or lumbar fluid (331). The normal fasting range for lumbar fluid of infants at birth is 75–150 mg/100 ml (182). The glucose concentration drops within 3–6 hours (182) to the range 60–90 mg/100 ml, which holds up to the age of about 15 years (61, 182, 289, 331). The adult normal fasting range is 40–80 mg/100 ml (61, 95, 264).

AUTOMATED DETERMINATION OF GLUCOSE

(method of Hultman, Ref 160, adapted to AutoAnalyzer by Frings et al, Ref 110, modified)

PRINCIPLE

Glucose reacts with *o*-toluidine in glacial acetic acid in the presence of heat, to yield blue-green N-glucosylamine, the absorbance of which is measured at 660 nm. The method is applicable to serum or plasma, and does not require a dialyzer.

REAGENTS

Same as manual reagents except prepare additional glucose standards to yield concentrations of 50, 100, 200, and 300 mg/100 ml.

PROCEDURE

Set up manifold, other components, and operating conditions as shown in Figure 25-2. By following procedural guidelines as described in Chapter 10, steady state interaction curves as shown in Figure 25-3 should be obtained and checked periodically. Results are calculated from the calibration curve in the standard manner for methods obeying Beer's law.

NOTES

1. *See Notes 3, 4, and 5 of manual method.*
2. *Operating procedure.* The pump should not be run at high speed to

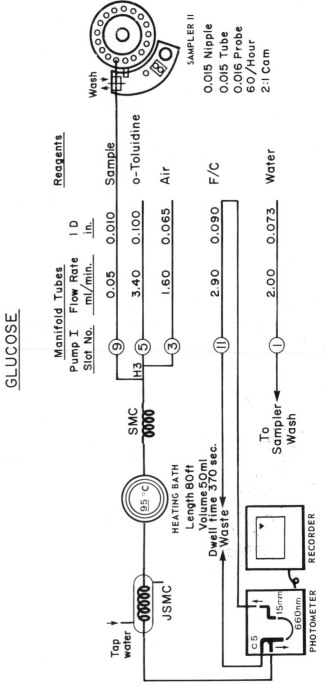

FIG. 25-2. Flow diagram for glucose in serum. Calibration range is 0–300 mg/100 ml in serum. Conventions and symbols used in this diagram are explained in Chapter 10.

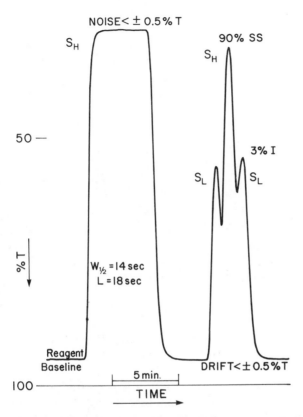

FIG. 25-3. Glucose strip chart record. S_H is a 200 mg/100 ml standard, S_L a 100 mg/100 ml standard. Peaks reach 90% of steady state; interaction between samples is 6%. $W_{1/2}$ and L are explained in Chapter 10. Computer tolerance for noise is set at ±0.5% T, for baseline drift at ±1% T and for standard drift at ±1% T.

establish a reagent base line because of the danger of spurting the highly acid *o*-toluidine reagent from ruptured or disconnected lines. It is recommended that the manifold reagent line be changed frequently. All samples must be free of red blood cells and fibrin clots, which can cause the microprobe and sample tubing to become occluded.

3. *Shutdown*. The *o*-toluidine reagent is strongly corrosive to the tubing and should not be left in the system when inoperative for extended periods of time. To shut down, first pump all reagent out of the system and then pump water through to clean the lines.

ACCURACY AND PRECISION

The accuracy of the *o*-toluidine method has been discussed previously under the section *Specificity of Methods for Glucose* and in *Note 5* of the manual procedure. The precision (95% limits) is ±5% for values between 50 and 300 mg/100 ml.

NORMAL VALUES

Same as for manual method.

DETERMINATION OF SERUM GLUCOSE BY GLUCOSE OXIDASE

(method of Ware and Marbach, Ref 375)

PRINCIPLE

Serum is preincubated with a buffered iodide-iodine complex. Glucose oxidase is added, and the H_2O_2 produced by its action on glucose reacts with iodide in the presence of a catalyst (ammonium molybdate) to form molecular iodine. The amount of iodine formed is proportional to the glucose concentration in the serum and is measured photometrically at 420 nm in the presence of polyvinyl pyrrolidone (PVP). PVP shifts the maximum absorption of iodine from the near ultraviolet toward the blue portion of the visible spectrum which results in a two- to threefold increase in the absorption of iodine in the 400–470 nm region (Fig. 25-4).

REAGENTS

Buffered Iodine Color Reagent. Dissolve 25 g KH_2PO_4, 12.8 g K_2HPO_4, 4 g ammonium molybdate, $(NH_4)_6Mo_7O_{24} \cdot H_2O$, 25 g KI, 0.4 g polyvinyl pyrrolidone, and 0.2 g I_2 in water. Make up to 1 liter and adjust to pH 6.3 ± 0.1 at 25°C if necessary. Bring to a boil and add additional I_2 or boil away I_2 until an absorbance of approximately 0.3 is obtained when read at 420 nm (1 cm light path) against water. Stable refrigerated but gradually loses I_2 when left for long periods of time at higher temperatures.

Glucose Oxidase. Dissolve purified glucose oxidase (Calbiochem purified enzyme, approximately 16 IU/mg or Fermco Fermcozyme No. 653 AM, 750 IU/ml have been found to be satisfactory) in the color reagent to produce a concentration of 190 IU/ml. Stable for at least 1 week refrigerated.

Stock Glucose Standard, 1 g/100 ml. Dissolve 1 g glucose, reagent grade, in water containing 0.2% benzoic acid and dilute to 100 ml.

Working Glucose Standard, 100 mg/100 ml. Dilute the stock glucose standard 1:10 with water containing 0.2% benzoic acid.

PROCEDURE

1. Transfer 5.0 ml color reagent to cuvets and add 0.02 ml serum or working glucose standard (TC pipet). Mix and place tubes in a 37°C heating bath or block.

2. Allow to stand in the heating block for at least 5 min and then remove and immediately read absorbances against water at 420 nm or with a filter having a nominal wavelength in this region.

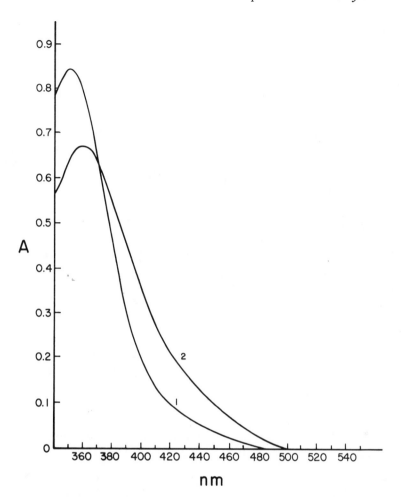

FIG. 25-4. Absorption curves for buffered iodine color reagent: (1) reagent without polyvinylpyrrolidone (PVP) vs water. (2) reagent with PVP (0.4 g per liter) vs water. Carey model 15 recording spectrophotometer.

3. Return to the 37°C heating block, add 0.2 ml glucose oxidase reagent to each tube, and mix.
4. Read the absorbance again after 15 min.

Calculation:

$$\text{mg glucose/100 ml} = \frac{A_x\ (\text{step 4}) - A_x\ (\text{step 2})}{A_s\ (\text{step 4}) - A_s\ (\text{step 2})} \times 0.02 \times \frac{100}{0.02}$$

$$= \frac{A_x\ (\text{step 4}) - A_x\ (\text{step 2})}{A_s\ (\text{step 4}) - A_s\ (\text{step 2})} \times 100$$

NOTES

1. Beer's law. Absorbance change is linear with concentration up to 350 mg/100 ml with most spectrophotometers. Linearity should be checked with the instrument used.

2. Color stability. The reaction reaches completion in 10—15 min and can be verified by a constant absorbance. After 25 min the absorbance tends to decrease gradually. Readings therefore should be made between 15 and 25 min.

3. Temperature control. Absorbance of the iodine PVP complex varies inversely with temperature by approximately 1%/1°C. It is important, therefore, to make absorbance readings immediately after removing the cuvet from the 37°C bath or heating block.

4. Application to other biologic fluids. Cerebrospinal fluid can be substituted for serum. For increased precision at low levels of glucose, use a 0.04 ml sample with appropriate correction in the calculation. The method cannot be applied directly to urine due to unpredictable amounts of reducing substances which react with the iodine reagent.

ACCURACY AND PRECISION

The method is specific for glucose and yields results which average about 8 mg/100 ml lower than those obtained with the ferricyanide method by continuous flow analysis (167). Bilirubin, hemoglobin, azotemia, and moderate lipemia do not interfere. Recoveries of glucose added to normal serum average 98%.

The precision (95% limits) of the test at levels of 70 and 150 mg/100 ml is about ±6% and ±3%, respectively.

NORMAL VALUES

The normal adult fasting range for venous serum is 70—100 mg/100 ml (375).

QUALITATIVE SCREENING TESTS FOR URINE GLUCOSE

Trommer (358) in 1841 introduced the metallic oxide reduction method for detection of reducing sugars. For over a century the routine qualitative testing for glucose in urine was performed almost exclusively by copper or bismuth reduction methods. Because of the recognized nonspecificity of such tests, other procedures were employed when it was wished to rule out a nonglucose reactant. Examination of the crystal structure and melting point of the insoluble osazone formed with phenylhydrazine (glucose, fructose, and mannose form the same osazone), polarimetry, and yeast fermentation have also been used with variable success. Accurate identification has become very easy today with the advent of qualitative glucose oxidase tests. The desired sensitivity of a qualitative test for glucose has never been accurately defined. Since the top normal glucose concentration in urine appears to be approximately 10 mg/100 ml, the desired sensitivity must be somewhat above this concentration.

COPPER REDUCTION TESTS

In 1848 Fehling (86) introduced a modification of Trommer's test in which tartrate was incorporated into the alkaline copper reagent to prevent precipitation of $Cu(OH)_2$ and black CuO, which obscures the color of the Cu_2O precipitate formed by sugar reduction. Fehling used a two reagent system: one containing KOH and tartrate, the other cupric sulfate. This was required because of the instability of Cu^{2+} in highly alkaline solution. By 1932 it was reported that some 90 modifications of Fehling's reagent had been described in the literature (68). By far the most significant of these was made in 1909 by Benedict (17), who used a stable reagent of cupric sulfate with citrate (replaces tartrate) and Na_2CO_3 (replaces KOH). The reagent was tenfold more sensitive than Fehling's reagent for glucose in urine, although this was not the case for pure solutions of glucose. The increased sensitivity of Benedict's modification for urine glucose was explained by the fact that cuprous creatinine, which is formed in urine, is soluble in the high alkalinity of Fehling's reagent but is insoluble (white precipitate) in the carbonate of Benedict's reagent (292). Thus there is an enormous magnification of the volume of Cu_2O precipitate. Creatinine is effective only in the concentration range 30–150 mg/100 ml, i.e., at levels usually found. These levels sensitize Benedict's test in the "total reducing substances" range of 50–300 mg/100 ml. Higher creatinine concentrations inhibit or delay precipitate formation. At lower concentrations of creatinine, urine behaves as a pure solution, glucose levels of 100–300 mg/100 ml yielding a barely visible red haze which would ordinarily be read as a negative test. Uric acid, which forms a yellow flocculent precipitate with cuprous ion, may also contribute to the high sensitivity of Benedict's reagent (245).

In 1944 Kasper and Jeffrey (173) introduced a tablet version of Benedict's copper reduction test which is marketed by Ames Co. under the name Clinitest. The tablet contains anhydrous copper sulfate, NaOH, citric acid, and $NaHCO_3$. Either 5 or 2 drops of urine are mixed with 10 drops of water in a test tube. One tablet is added, and the mixture is allowed to stand undisturbed for 15 seconds, remixed, and observed for color by comparison with a color chart. The heat generated by the interaction of excess NaOH and water is sufficient for the redox reaction. The sensitivity of Clinitest is about 150 mg/100 ml (108) with a semiquantitative range up to 2 g/100 ml using 5 drops of urine. With over 4 g/100 ml there is a reversal of color which can be mistaken for 0.75–1.0 g/100 ml (15). This phenomenon is referred to as "pass through." When 2 drops of urine are used, the pass through is delayed until the sugar concentration reaches 10 g/100 ml (15, 55). In one study the pass through phenomenon with 5 drops of urine occurred 70 times in 191 urines from diabetic children, whereas it did not occur using 2 drops of urine (15). Some have recommended that the 2 drop method be adopted for routine use (15, 71). The range of the 5 drop method may also be extended by diluting the urine with a negative urine (118).

MISCELLANEOUS CHEMICAL TESTS

Antimony oxychloride can be used instead of the bismuth salt in a reduction test (64). A blue color is obtained in a reaction of glucose with benzidine and perborate (295). A test has been devised for the blind diabetic in which yeast and urine are placed in a test tube and a rubber finger cot fastened over the mouth with a rubber band. Distention of the rubber from CO_2 formed can be felt after 30–60 min (59).

GLUCOSE OXIDASE TESTS

A number of commercially available filter paper strip tests employing the glucose oxidase reaction for detecting glucose have been designed. Among the more popular and widely used are *Clinistix* (Ames) and *Tes-Tape* (Eli Lilly). Clinistix is filter paper impregnated with glucose oxidase, peroxidase, and *o*-tolidine (3, 107). Combination strip tests are also available—e.g., Combistix (Ames)—for glucose, protein, and pH. The test end of these paper strips is dipped into the urine, removed, and the color compared with a color chart after a fixed time. Clinistix yields a blue color if the "glucose" content of the urine exceeds about 100 mg/100 ml. The sensitivity and speed of reaction is affected by urine pH, temperature, and concentrations of inhibitors if present. For this reason the test is read as either positive or negative with no attempt at quantification. The following substances have been demonstrated to interfere (false negatives): ascorbic acid (249, 250), homogentisic acid (87, 249), bilirubin, hydroquinone, adrenaline (249), dypyrone, sodium meralluride (250), and 5-hydroxyindole acetic acid (117). The test is negative with galactose, fructose, lactose, mannose, maltose, sucrose, xylose, D-ribose, D-arabinose, L-arabinose, and L-xylulose at concentrations of 20 g/100 ml (108).

Tes-Tape is impregnated with glucose oxidase, peroxidase, catalase (impurity in peroxidase), F.D.C. yellow No. 5 dye, and *o*-tolidine (52). The strip is moistened with urine, and at 1 min the color is compared with a chart. The color scale, which goes from yellow through shades of green, gives colors corresponding to 0, 0.1, 0.75 and 2 or more g/100 ml. Since the same variables affect the validity of this test as described for Clinistix, the accuracy of the semiquantitative estimation is questionable.

Numerous studies have been made in which the sensitivity and accuracy of the various screening tests for urine glucose have been studied and compared. Earlier reports indicated that the enzyme paper tests are about equal in sensitivity to Clinitest (copper reduction) (210, 259), but there is now considerable evidence which indicates a greater sensitivity for the enzyme screening tests (50, 104, 183, 250). The problem of false negatives with the enzyme tests, however, is real and must be considered (250). Furthermore, as a result of high specificity the enzyme tests do not detect pentosuria and lactosuria (210).

It has been reported that tapes with continuous impregnation of glucose oxidase are more suitable for testing in the presence of interfering substances than dipsticks with narrow bands of impregnation (117). The explanation for this is that, in some cases, as the urine ascends the paper by capillary action the inhibitors are adsorbed to the paper (chromatography) and a positive reaction occurs at the solvent front.

False positives, although not expected, have been reported as a result of contamination with H_2O_2 or hypochlorite (42). Some detergents contain hypochlorite, and catheters are sometimes rinsed in H_2O_2.

Glucose test papers deteriorate upon extended storage and from exposure to direct sunlight (14, 52), heat (14), and air (344).

CHOICE OF QUALITATIVE TESTS FOR URINE GLUCOSE

Because of specificity the immediate tendency is to choose a glucose oxidase test for routine analysis. By using such a specific test, however, instances of increased levels of homogentisic acid, pentose, fructose, and galactose are missed. They would not be missed by a copper reduction test. Admittedly, such cases occur only in about 40 per million population, but this appears to constitute sufficient reason to recommend Benedict's qualitative copper reduction test (or Clinitest) as the routine test; if positive, the presence of glucose can be confirmed by a glucose oxidase paper strip test (303). Presence of an inhibitor can be checked by adding 300 mg glucose to 10 ml urine, waiting 10 min, and then retesting.

QUALITATIVE TEST FOR REDUCING SUBSTANCES IN URINE

(method of Benedict, Ref 16)

PRINCIPLE

The Cu^{2+} of Benedict's qualitative glucose reagent is reduced by glucose and other reducing substances present, and precipitated as yellow or red Cu_2O.

REAGENT

Benedict's Qualitative Reagent. Dissolve 17.3 g $CuSO_4 \cdot 5H_2O$ in about 100 ml hot water. With the aid of heat, dissolve 173 g sodium citrate and 100 g anhydrous Na_2CO_3 in about 800 ml water. When cool, pour the second solution into the first while stirring, and dilute to 1 liter with water. Stable at room temperature.

PROCEDURE

Place 5.0 ml Benedict's reagent and 8 drops urine in a test tube and mix. Boil over a flame for 2 min or place in boiling water bath for 3 min. Read immediately after heating.

INTERPRETATION

The sensitivity of the test is, under optimal conditions, approximately 50–80 mg glucose per 100 ml (6, 27, 108, 391). The sensitivity for any given urine, however, is quite variable (27, 292, 391). Many investigators feel that opaque green reactions should be ignored (391), even though an occasional urine giving such a reaction may contain as much as 700 mg fermentable sugar per 100 ml (391). Glucose concentrations of urines yielding a green reaction plus a yellow precipitate fall in the range of 100–500 mg/100 ml (27, 52).

It is recommended that the reaction be interpreted and reported as follows:

Reaction	Report as	Approximate average glucose concentration (mg/100 ml)
Clear blue or green opacity, no precipitate	0	<100
Green with yellow precipitate	1+	250
Yellow to olive	2+	800
Brown	3+	1400
Orange to red	4+	2000 or more

There have been several variations in the technic of running and interpreting Benedict's test. Benedict originally observed that at low glucose concentrations a precipitate formed only on cooling, and his original directions were to read the test following spontaneous cooling to room temperature. Since the present authors recommend reading trace reactions as negative, we are following the suggestion of reading the test before cooling (291). Benedict also concluded that the bulk of the precipitate, not its color, is the criterion of the degree of positivity of the reaction. Subsequent workers appear to rely as much or more on the color, and some have attached even more significance to the rapidity with which the precipitate forms. Although some workers have stated that reductions by substances other than glucose may sometimes be detected by an unusual appearance of precipitate or its manner of development, these do not appear to be dependable criteria (391).

NOTES

1. Time of boiling. Benedict's original directions were to boil for 1–2 min and then let cool (16). Folin and McEllroy (99) recommended boiling 1 min or placing in a boiling water bath for 3–5 min. This apparently led many workers to use, and many texts to recommend, 5 min in a boiling water bath. This leads to many doubtful or false positive tests because of reaction with nonglucose reductants (292). Three minutes in boiling water is equivalent to boiling for 2 min (291).

2. *Stability.* See *Note 2* under *Quantitative Method for Urine Glucose.*

ACCURACY

The reduction of Cu by substances other than glucose has already been discussed in the earlier section *Specificity of Glucose Methods.* Among the substances occurring in urine reported as giving a positive test with Benedict's reagent are fructose, lactose, galactose, maltose, arabinose, xylose, ribose, uric acid, creatinine, ketone bodies, certain amino acids, oxalic acid, phenolic substances, hippuric acid, homogentisic acid, glucuronic acid, conjugated glucuronates, caronamide, chloral, formaldehyde, isoniazid, *p*-aminosalicylic acid, salicylates, cinchophen, uronates (following hyaluroni-dase), and salicyluric acid (54, 163). As discussed later in the section on *Pentoses*, L-xyluloketose also gives a positive test. Ascorbic acid at a concentration of 50 mg/100 ml gives a positive reaction (52). Terramycin, streptomycin, chlortetracycline, and chloromycetin cause false positive tests at high concentrations (10 mg/ml) (207). Neither penicillin (100,000 units/ml) (207), nor sulfanilamide (351) interferes.

NORMAL VALUES

The urine tests of normal children and adults are negative. Normal infants during the first 5 days of life may excrete sufficient galactose to give a positive reaction (28).

QUANTITATIVE METHODS FOR URINE GLUCOSE

Most of the approaches to determination of glucose in blood have also been applied to urine. The problem of specificity is, however, more challenging for urine measurements. The methods are reviewed briefly. For a discussion of the reactions involved, refer to the section *Quantitative Methods for Blood.*

Of historic interest is the use of picramic acid (18). As with blood, it was soon abandoned because of nonspecificity. A number of methods employing reduction of ferricyanide were used based on time required for decoloriza-tion (138), gasometric (366), titrimetric (275), potentiometric (393), and photometric (251) measurements.

The earliest methods employing Cu reduction were titrimetric. In the technic introduced by Benedict (16) in 1911 and which is still used in some laboratories today, the hot copper reagent containing citrate and KCNS is titrated with urine until the blue color of the cupric ion disappears. A white precipitate of CuCNS forms in the reduction instead of red Cu_2O. Procedures involving titration of residual unreduced Cu^{2+} have been described, using either an iodometric method (132) or EDTA with murexide as indicator (334). Other titrimetric methods include oxidation with I_2 and titration of the excess I_2 with thiosulfate (254), and oxidimetric titration with potassium cupri-3-periodate or potassium cupri-3-tellurate (12).

A photometric method which involves measuring the residual blue color in the supernatant solution after applying Benedict's qualitative copper test was introduced for semiquantitative measurement (89). Photometric measurement of the yellow color formed when glucose is heated with alkali has been used as a quantitative procedure (319). In 1924 Sumner (338) introduced a method in which the mahogany red color formed with dinitrosalicylic acid is measured. Specificity of the method was subsequently improved (37, 337). The color produced by condensation with *p*-aminobenzoic acid or *m*-aminophenol in strong acid has also been measured (75). The use of *o*-toluidine in glacial acetic acid, first described by Hultman (160) in 1959 proved to be the simplest and most specific (43) of the nonenzymatic methods for urine glucose. Glucose oxidase methods have been applied to urine. Removal of interfering substances, e.g., uric acid, is required for valid results. This has been accomplished by adsorption with charcoal and Lloyd's reagent (184), or by oxidation with potassium biiodate (170). Another approach involves heating urine with 45% $HClO_4$ for 10 min and using a centrifugate for analysis (202). Ion exchange resins have also been used with good results (209, 378). A number of semiautomated glucose oxidase methods for urine glucose have evolved utilizing the AutoAnalyzer (190, 209). The hexokinase method has also been adapted for urine glucose utilizing spectrophotometric (265) or fluorometric (299) technics, and a semiautomated version has been described (298).

QUANTITATIVE DETERMINATION OF URINE GLUCOSE

(method of Hultman, Ref 160, modified)

PRINCIPLE

The glucose in a diluted urine specimen is reacted with *o*-toluidine in glacial acetic acid with heat to yield a colored glucosylamine which is measured photometrically.

REAGENTS

All reagents are the same as those described under the method for blood. In addition, Benedict's qualitative sugar reagent is used (see under *Qualitative Test for Reducing Substances in Urine*).

PROCEDURE

1. First carry out a qualitative Benedict's test for reducing substances. Make a dilution of the urine specimen with water according to the following:

Benedict's screening result	Dilution
4 +	1:25
3 +	1:10
2 +	1:5
1 +	None
Neg	None

2. Transfer 5.0 ml *o*-toluidine reagent to test tubes or vials of about 10 ml capacity. These tubes may be matched photometer or spectrophotometer tubes if desired, since final absorbance readings can be made without further transfer. Label tubes "reagent blank," "standard," and "unknown." A urine blank is also required for each unknown and is set up by adding 0.2 ml urine (or diluted urine) to 5.0 ml glacial acetic acid.

3. Add 0.1 ml glucose standard (100 mg/100 ml) plus 0.1 ml water to the "standard" and 0.2 ml of urine, diluted according to step 1 above, to the "unknown" (use TC pipets). Mix thoroughly.

4. Loosen stoppers or caps if used and place all tubes in a boiling water bath or heating block at 100°C. Remove all tubes after 10 min and place in a cold tap water bath.

5. When tubes have cooled, read absorbances of standard, unknowns, and urine blanks against the reagent blank at 625 nm or with a filter having a nominal wavelength in this region.

Calculation:

$$\text{g glucose/100 ml} = \frac{A_x - A_{b \text{ (urine)}}}{A_s} \times \frac{0.1}{1000} \times \frac{100}{0.2}$$

$$= \frac{A_x - A_{b \text{ (urine)}}}{A_s} \times 0.050 \times D$$

where D = dilution (1 for undiluted urine, 10 for 1:10 dilution, etc.).

$$\text{g glucose/24 hours} = \text{g glucose/100 ml} \times \frac{24 \text{ hour vol in ml}}{100}$$

NOTES

1. *See Notes 1, 2, and 3 of manual o-toluidine method.*

2. *Stability.* Glucose in urine is stable if microbial growth is prevented. Without added preservative 4 hours at 30°C or 24 hours at refrigerator temperature should not be exceeded prior to analysis. Useful preservatives

which do not interfere with the o-toluidine method include NaF at a concentration of 10 mg/ml, formaldehyde, chloroform, toluene, and thymol.

ACCURACY AND PRECISION

The specificity for glucose is good, as has been discussed previously. Galactose and mannose would be included in the result if present.

Precision (95% limits) is about ±5%.

NORMAL VALUES

The normal 24 hour output of glucose as determined by the o-toluidine method is 0 to 0.25 g/24 hours. This is based on cited ranges for enzymatic technics (300) and the equivalence of results by the o-toluidine and enzymatic methods.

GLUCOSE TOLERANCE TESTS

The most important use of glucose tolerance tests is in the diagnosis of diabetes mellitus, although abnormal responses occur in other diseases such as nephritis, hyperthyroidism, and certain other endocrine disorders. The test measures the ability of the individual to remove an added glucose load from the circulation. Normally, this is accomplished at such a rate that the blood glucose level does not significantly exceed the renal threshold, and little or no glucose appears in the urine. Because of the observation that carbohydrate restriction results in decreased tolerance for glucose, some authorities have recommended that the test subject be placed on a standard preparatory diet prior to the test—e.g., 300 g carbohydrate and 3000 calories per day for 3 days (53). The more common practice is to have the patient eat a normal diet but advise him to eat pastries, potatoes, or sugar. Full meals with 55%–60% carbohydrates assure an intake of more than 250 g carbohydrates per day (389). Tests are performed in the morning after an overnight fast since the recent ingestion of a meal affects the tolerance curve (131).

The most widely used glucose tolerance test is the oral one dose test. It is common practice with adults to give 100 g glucose dissolved in cold water flavored with lemon juice or one of a number of flavored commercial products, although a 50 g glucose challenge is also commonly used (314, 364). An alternative is to give glucose, 1 g/kg body weight, after the age of 12 (152). The dosage commonly used for children is 1.75 g/kg with a minimum of 10 g and a maximum of 50 g (13). Blood and urine samples are taken for analysis at 0, 0.5, 1, 2, and 3 hours, and occasionally at 4, 5, and 6 hours if postprandial hypoglycemia is suspected. Interpretation of the curve is made from the appearance of glucose in the urine, the peak glucose concentration attained in the blood, and the rate at which the blood level returns to a normal level. The peak blood level is usually reached between 30 and 60 min with a return to normal at the end of 2 hours.

Exton and Rose (78) in 1934 introduced the 1 hour two dose oral test, applying Allen's paradoxical law (in normals, the more sugar given, the more sugar utilized, but the reverse being true in diabetics). The test enjoyed popularity for some time, being considered by some to be more sensitive and reliable than the one dose tests (78, 121). Others, however, have taken the position that the test is too sensitive (246), and that it is based on false presumptions and should be discarded (198).

A practical variation of the one dose test is to obtain a single blood sample and urine sample after a breakfast containing approximately 100 g carbohydrates, 30 g protein, and 30 g fat. It has been recommended that the samples be taken 1 (24), 1.5 (36), and 2 hours (246) after breakfast. A 1 hour period should be used if the test is to be evaluated from the level reached near the peak. Mitchell and Strauss (239) compared the 2 hour postprandial test with the oral glucose tolerance test and found the former ineffective in screening for diabetes. They concluded that the most effective practical method of screening is to assay blood glucose exactly 2 hours after an oral dose of 50 g of glucose. This abbreviated 2 hour tolerance test is generally viewed as the best individual test for diabetes screening (164, 362), although there is some disagreement (314).

The use of intravenous glucose administration has been proposed (359) because it eliminates some of the variables which can affect an oral glucose tolerance test, e.g., rate of absorption from the intestine and evacuation time from the stomach. Numerous mathematical treatments of the data obtained from IV glucose tolerance tests have been described. These involve determination of removal rate constants by plotting some function of the blood glucose level versus time (150).

There is a relatively high incidence (about 10%) of unpleasant side reactions following intravenous administration of glucose which include pyrexia, malaise, headache, and phlebitis (359). Furthermore, some investigators (225, 246) have failed to demonstrate any significant advantage of the IV test over the oral test, or have in fact found the latter to be more useful for diabetes detection.

Glucose tolerance tests are not without error in the detection of diabetes. In addition to certain disease states other than diabetes which result in abnormal curves, a number of other variables have been shown to affect results (78, 248). Intraindividual variability between repeat tests has been recognized for many years (1, 362). The conclusion reached by three different groups of investigators (191, 224, 314) who have made extensive studies on the reproducibility of the glucose tolerance test is that it is unwise to make or exclude a diagnosis of diabetes on the basis of one tolerance test.

Another potential source of variability is the type of blood sample used for glucose analysis (381). Venous blood is ordinarily used, but the use of capillary blood has also been proposed, not only for small children but for general screening (51, 197). Whichelow *et al* (381) found that capillary blood taken from the warmed ear lobe gives an accurate reflection of the arterial blood sugar level which they consider better than venous blood in the glucose tolerance test.

ORAL GLUCOSE TOLERANCE TEST

(Based on Report of the Committee on Statistics of the American Diabetes Association—Standardization of the Oral Glucose Tolerance Test, Ref 281)

PATIENT PREPARATION

Diet: The patient should be on a normal diet containing at least 150 g carbohydrate per day for at least 3 days preceding the test.

Food and other restrictions: There should be an interval of fasting before the test, of at least 8 hours and no more than 16 hours. No food should be consumed after midnight prior to the morning of the test; water is allowed.

PROCEDURE

Blood and urine samples are taken prior to administration of the glucose (0 time samples) and at 0.5, 1, 2, and 3 hours after ingestion.

Size of glucose load: for all adults, 1 g/kg body weight. In most adults the glucose load ranges from 50 to 100 g. For children below the age of 12, give 0.8 g glucose per pound body weight, with a minimum of 10 g and a maximum of 50 g. Commercial preparations (e.g., Glu-co-tol, which is a carbonated lemon-flavored preparation sold by Steri-Kem) are suitable, but avoid the use of preparations which are not pure glucose (i.e., those labeled "glucose equivalent"). These may contain disaccharides and/or polysaccharides and could influence results due to their influence on intestinal absorption rates. If a commercial product is not used, a practical way of preparing the required glucose solution is to dissolve the required amount in 1–2 glasses of ice water flavored with lemon juice.

The glucose content of the blood samples (serum or plasma is preferable) is determined quantitatively, and the urine samples are tested qualitatively.

INTERPRETATION

All urines should be negative for glucose. Three methods of interpretation of the blood levels (80, 186, 385) have been recommended for use by the Committee on Statistics of the American Diabetes Association (281). Kobberling and Creutzfeldt (188) compared these three methods of evaluation with their own and pointed out differences in sensitivity and lack of agreement. They reported further that the different methods of evaluation did not always indicate the same subject as being diabetic. The criterion recommended by these investigators for diagnosis of diabetes is that the sum of the 1 hour and the 2 hour values is 300 mg/100 ml or more.

It is clear that no single method of choice exists for interpretation of the glucose tolerance curve. The Wilkerson Point System (385), one of the more widely used interpretative schemes, is an example of one commonly used approach:

| | Glucose (mg/100 ml) | | |
Time	Blood	Plasma	Points
Fasting	≥ 110	≥ 130	1
1 hour	≥ 170	≥ 195	½
2 hours	≥ 120	≥ 140	½
3 hours	≥ 110	≥ 130	1

Two or more points are judged diagnostic of diabetes. This chart is based on a 100 g oral dose of glucose with determinations on venous blood by the Somogyi-Nelson method. If plasma is used, the critical values are changed by multiplying the blood value by 1.15 and adding 6 mg/100 ml. All values are rounded off to the nearest 5 mg/100 ml. The chart can also be used if glucose is determined by the o-toluidine or glucose oxidase method, since Somogyi-Nelson values were only slightly higher (2–3 mg/100 ml).

Decreased glucose tolerance has been observed with increasing age (21, 34, 233), but there is disagreement as to whether this represents true diabetes mellitus (21, 233). Pickens *et al* (267) reported results on normal children ranging in age from 1 to 13 years, and propose the use of their norms for interpretation of tolerance tests performed on children.

Other factors which have been reported to influence the glucose tolerance test include pregnancy (11, 316), phase of the menstrual cycle (212), physical inactivity (29), illness and trauma (310, 324), and certain drugs including oral contraceptives (325), salicylates (88), nicotinic acid (124), and diuretic agents. The intraindividual variation between tests run on different days is significant as discussed previously.

FRUCTOSE

Fructose accounts for one-sixth to one-third of the total carbohydrate intake and is present at low concentrations in normal blood and urine. Elevated blood levels and increased excretion occur in *essential fructosuria*, a relatively rare condition characterized by inability to utilize dietary fructose completely. Lasker (199) estimated the incidence of this asymptomatic metabolic disorder in the general population at approximately 1:130,000. An autosomal recessive inheritance is suspected by most investigators. A recent study showed the pathogenetic role of a primary lack of hepatic fructokinase in this disorder (296). A familial condition called *fructose intolerance* has also been described in which there is a strong aversion to fruits and sweets with severe gastrointestinal, neurologic, and mental symptoms and signs (72). The primary enzyme defect in hereditary fructose intolerance is a lack of fructose-1-phosphate-aldolase, which causes intracellular accumulation of fructose-1-phosphate (263). Most of the reported cases are compatible with an autosomal recessive inheritance (112). A form of fructose intolerance has been described in association with galactose intolerance and hyperinsulinism (72). The pathogenesis of this syndrome is not yet completely understood. Increased fructose excretion may also occur in hepatic failure (194). Aside from the apparently rare case of fructose intolerance, fructosuria is important only because of confusion with glucosuria since fructose is a reducing sugar.

Among the qualitative tests for fructose applied to urine are the following: (a) *Borchardt's test* (32). To a few milliliters urine add an equal volume 25% HCl and mix. Add a few granules of resorcinol and bring just to boiling. If a red color appears, cool quickly in running tap water, make alkaline with solid Na_2CO_3, and shake with ethyl acetate. If fructose is present the ethyl acetate becomes yellow. Interference occurs when nitrites and indican are simultaneously present. (b) *Seliwanoffs' resorcinol test* (312). Boil urine with an equal volume of 25% HCl. Add some resorcinol and boil again for 10 sec. A heavy red precipitate which is soluble in ethanol forms with keto sugars.

Sarma (294) developed a diagnostic test for fructosuria which detects as little as 8 μg of fructose. Identification is made by adding 2 drops of a fructose solution to approximately 0.2 g of KOH on a spot plate. It is claimed that this test can be used for a rapid diagnosis of fructosuria if at least 0.5 g/100 ml is excreted in the urine. Glucose does not interfere when more than 0.1 mg of fructose and less than 0.9 mg of glucose is present in the mixture. In order to prove that a reducing substance is fructose, one should resort to identification by thin layer chromatography (344).

Quantitative photometric determination of fructose in blood and urine includes methods based on color reactions with the following reagents: resorcinol (66, 284), diphenylamine (141, 286), thymol (261), skatole (268), anthrone (253), thiobarbituric acid (6), and bile salts (305). Most of these reactions have also been used for determination of inulin, the polysaccharide of fructose, and are discussed in somewhat greater detail under *Inulin*. The method presented for determination of inulin, reaction with indole-3-acetic acid, is recommended by the authors for quantitative determination of fructose.

Schmidt (302) has introduced an enzymatic method for determination of fructose. The fructose and glucose in a Ba^{2+}-Zn^{2+} filtrate are converted to fructose-6-phosphate and glucose-6-phosphate, respectively, by reacting with adenosine triphosphate (ATP) in the presence of hexokinase. Glucose-6-phosphate dehydrogenase and TPN are added, converting glucose-6-phosphate to 6-phosphogluconate with an equivalent amount of TPNH being formed. The increase in absorbance at 366 nm due to TPNH is a measure of the glucose originally present. Phosphohexose isomerase is next added, converting fructose-6-phosphate to glucose-6-phosphate, which reacts with TPN to form more TPNH. This second increment in A_{366} is equivalent to the fructose originally present.

A positive test for fructose in an alkaline urine must be viewed with suspicion unless the urine is fresh, since fructose may be formed from glucose under such conditions (chemical inversion) (199).

GALACTOSE

Galactose is present in normal blood and urine in very low concentrations (cf Table 25-1). Increased levels occur occasionally, especially in nursing infants with disorders of digestive function, in rare cases of the familial disease called *galactose intolerance*, and in the congenital disease *galacto-semia* (cf *Galactose-1-Phosphate Uridyl Transferase, Chapter 17*). Increased urinary excretion of galactose frequently occurs in newborn infants, especially during the second to sixth days after birth (62). Galactosemia and various methods available for investigating specific enzyme deficiencies have been reviewed by Hsia (157).

Rapid screening tests for galactosemia include a fluorescence assay of blood developed by Beutler and Baluda (25) which can differentiate between homozygous and heterozygous states, and a test paper assay (63) designed to detect increased urinary excretion of galactose. It should be noted that a positive urine test for reducing substances (e.g., Clinitest) combined with a negative glucose oxidase test can be useful in the recognition of galacto-semia. In this case, TLC can be used to confirm the diagnosis (345). Qualitative tests for galactose include the following: (a) *mucic acid test* (cf under *Lactose*)—does not differentiate from lactose; (b) *phloroglucinol test* (cf under *Lactose*)—does not differentiate from lactose or glucose; (c) *formation of hydrazone* (103). Reaction with o-tolylhydrazine forms an insoluble hydrazone which differs markedly from osazone crystals yielded by glucose or fructose.

Methods for quantitative determination of galactose are considered under *Galactose Tolerance* in chapter 22.

Epimerization and transformation of galactose occur with alkalinization of aging urine (356).

PENTOSES

Adults normally excrete pentoses at about 2–5 mg/kg/24 hours if on a fruit-free diet, while children excrete somewhat more (356). There are at least three types of pentosuria, i.e., increased urinary excretion of pentoses (326): (a) *Alimentary*. This is a temporary situation following ingestion of large amounts of pentose-rich fruits such as prunes, cherries, grapes, or plums. Arabinose and/or xylose may well be the pentoses involved, although this apparently has not been well documented by definitive procedures. (b) *Toxic* or *drug-induced*. This is caused by fever, allergy, and drugs, including morphine, antipyretics, cortisone, and thyroid hormone. (c) *Essential* or *chronic*. This is a harmless congenital abnormality, governed by a recessive gene and occurring in approximately one in every 50,000 persons (392). It occurs almost exclusively in people of Jewish ancestry from Russia, although there have been cases among others (269). This disease is characterized by increased output of L-xyloketose—also called L-xylulose and L-threopentulose (203). The metabolic defect in pentosurics is an abnormal red cell xylotol dehydrogenase which has a decreased affinity for NADP compared with the enzyme in nonpentosuric subjects (374). Pentosurics excrete about 2–4 g/day. The fasting serum level of L-xyloketose may be somewhat higher in pentosurics than in normals (35), but the difference becomes considerably more striking following oral administration of glucuronolactone, which results in no increase in normals but a significant increase in pentosurics. An increase also occurs in heterozygotes (109, 374). Elevated levels of L-xylulose have also been observed in the serum of fasting diabetic patients (387), but the physiologic significance is still unknown.

Actually, there may be a fourth type of pentosuria which is associated with one or more specific diseases, e.g., increased excretion of ribose in pseudohypertrophic muscular dystrophy (356).

The qualitative tests for urine pentose can be divided into two groups: (a) *Color reactions with furfural*. The furfural is produced from the sugar by heat and acid. Reagents employed to produce color include phloroglucinol (Tollens' test; see also under *Lactose*), benzidine (221), and orcinol (Bial's test) (26, 236). These tests are not specific. In addition to free and combined pentoses, a positive test is also given by certain uronic acids (glucuronic, galacturonic, etc.) which decompose with heat and acid to form pentoses— e.g., the entire color obtained by the orcinol test on a urine may be due to glucuronides present (356). (b) *Reduction tests*. Lasker and Enklewitz (200) introduced a procedure by which Benedict's qualitative reagent may be used to aid in identification of urine sugars. 1 ml urine is added to 5 ml reagent and placed in a water bath at 50°C; the time for reduction (definite yellow precipitate) is noted. Xyloketose, fructose, xylose, arabinose, glucose, and lactose in 0.3% concentration require 4–8, 20, 43, 58, 68, and 88 min, respectively. The time required for visible reduction, however, is inversely related to the concentration of the particular sugar. This has led to some confusion in various texts describing this test. It has been found in our laboratory that at 1% concentration fructose gives a positive test within 10 min, whereas glucose is negative; at >4% concentration glucose is positive. Ribose causes reduction in the following modification (257): 8–10 drops of urine are added to 5 ml Benedict's qualitative reagent and the tube placed in boiling water for 45 min. Pentoses also reduce in the dinitrosalicylic acid method at a lower temperature than other sugars (79). Enklewitz reported (77) that addition of 3% H_2O_2 to urine at room temperature results in rapid loss of reducing properties of pentoses but not of other sugars.

Roe and Rice (285) presented a quantitative method wherein furfural is formed by action of heat and acetic acid in the presence of thiourea as antioxidant. The furfural is then reacted with *p*-bromoaniline acetate to form a pink complex. Kerstell (180) proposed a simplified modification. Combined and keto and desoxy forms of pentose are not determined by these procedures (356). An enzymatic method for determination of D- and L-xyloketose, alone or in mixtures, has been introduced (145). The assay depends on the specificities of TPN-xylitol (L-xyloketose) dehydrogenase and DPN-xylitol (D-xyloketose) dehydrogenase. McKay (226) investigated the orcinol reaction of Bial (26) and applied it to quantitative estimation of pentoses in plasma. He showed that the result is influenced not only by the glucose concentration but also by the small differences in concentration of reagents, especially orcinol. A simple micromethod for measuring pentoses in body fluids has been described (119).

The clinical chemist ordinarily faces only the problems of qualitative identification of a urinary substance giving a positive copper reduction test. A negative glucose oxidase test (see under *Qualitative Tests for Urine Glucose*) rules out glucose. Identification of a pentose should be made by TLC (see earlier section, *Identification of Sugars by Thin Layer Chromatography*).

XYLOSE ABSORPTION TEST

In 1937 Helmer and Fouts (140) introduced a test of carbohydrate absorption using xylose. This sugar has the following advantages: absorption does not entail phosphorylation, does not occur against a concentration gradient, is not normally present *in vivo*, and is unaltered by the liver. Renal rate of excretion is dependent on plasma concentration and not on rate of urinary flow, although kidney impairment can influence urinary excretion.

The test in its original form is run by giving 25 g D-xylose orally in water to a fasting subject and collecting urine for 5 hours. Normals excrete 4.1–9.0 g during the 5 hour period (278). There is no sex difference, and values are highest in the 35–44 year age group, decreasing thereafter (368). Collection for 24 hours has been recommended (368) because excretion may continue past the 5 hour period. The normal excretion in 24 hours is 5.0–10.4 g (348). Because some subjects experience abdominal cramps and diarrhea with the 25 g dose, a modification has been suggested using 5 g. With this dose normals excrete 1.2–2.4 g in 5 hours (293). Serial blood levels generally parallel urinary excretion, but they are too variable and show too much overlap between normals and abnormals to be used for diagnostic purposes (22). Maximal blood levels occur between 1 and 3 hours, and normally are 33–67 mg/100 ml (348).

Our laboratory has employed the method of Roe and Rice (285) for determination of urinary pentose on occasions when knowledge was not readily available as to the glucose content of the patient's urine. Certainly if the urine is normal in this respect—and this ordinarily can be ascertained by performing a qualitative reduction test (e.g., Benedict's test) on a sample of urine obtained prior to commencement of the absorption test—any standard quantitative glucose procedure based on reduction can be used. The Folin-Wu blood method has been used in our laboratory without difficulty, since if the 25 g test dose is given the urine must be diluted approximately 1:20 for analysis. Xylose and glucose have nearly identical reducing equivalents in the Folin-Wu method. Xylose standards should be used instead of glucose, however, since they do not necessarily have identical reducing equivalents in all methods for reducing substances.

Several methods for determining xylose in serum and urine have recently been reported (57, 119, 316), including an automated urine method (316). The latter procedure permits undiluted urine to be sampled, and blanks are not required.

DISACCHARIDES

The chief clinical importance of disaccharides is physiologic intolerance which may result from a congenital enzyme defect or may be secondary to other diseases of the small intestine. The condition is relatively common, especially in infants; however, it may be mistaken for cystic fibrosis, irritable bowel syndrome, oral administration of neomycin or kanamycin, and *Giardia lambia* infestation (280). Over a dozen fatalities resulting from the syndrome have now been described (156).

Sugar malabsorption may be due either to the defective absorption of the monosaccharides or to deficiency of the disaccharidases in the small intestine. Three different forms have been recognized: sucrose-isomaltose malabsorption, lactose malabsorption, and glucose-galactose malabsorption. It has been suggested that all these conditions are transmitted as autosomal recessive traits (156). Sucrose-isomaltose malabsorption results from deficiency of sucrase and isomaltase in the intestinal mucosa (379). A deficiency of lactase results in lactose malabsorption (155). These conditions are characterized by diarrhea with watery acid feces and no significant rise in blood glucose levels after an oral dose of the particular disaccharide. Following ingestion of the disaccharide, usually in a dose of 2 g/kg body weight or 50 g/sq meter body surface, an increase in blood glucose of less than 20 mg/100 ml is considered abnormal (116). Direct confirmation of deficient intestinal disaccharidases may be accomplished by measuring the enzyme activity in mucosal biopsy specimens.

In lactose intolerance mild lactosuria may be present, usually after prolonged diarrhea (343). An increased output of lactose may also be encountered during the nursing period or soon after weaning, as well as following a long-term, exclusive milk diet. Lactosuria is of no significance except that it may be mistaken for glycosuria when employing a test for reducing sugars. The presence of lactose in physiologic fluids can be easily confirmed by TLC (345). One should note that normal adult urine contains approximately 12–40 mg lactose per 24 hours (65), and a normal infant's urine contains up to about 1.5 mg/100 ml (388).

Chemical tests for lactose include the following: (a) *Mucic acid test* (*Tollens test*) (260). Add 12 ml conc HNO_3 to 50 ml urine in an evaporating dish and reduce the volume to about 10 ml on a steam bath. Cool, add 10 ml water, and let stand overnight. Presence of white precipitate of mucic acid crystals, seen under the microscope, indicates presence of galactose or lactose. (b) *Phloroglucinol test* (Oshima and Tollens, 258). Phloroglucinol reacts with methylfurfural in the cold, yielding a color. Substances which yield methylfurfural with heat plus HCl give a positive test. Many textbooks perpetuate misinformation about this test, e.g., that pentoses and galactose yield a red color. It has even been indicated that lactose does not react (260). Fowweather (103) has shown, and it has been confirmed in our laboratory, that xylose gives a red color, and galactose and glucose a red-brown or yellow-brown (depending on concentration). Lactose behaves as galactose. This test is of no value in differentiating glucose, galactose, and lactose in urine. (c) *Rubner's test* (288). Add 3 g lead acetate to 15 ml urine, shake mixture well, and filter. Bring filtrate to a boil, add 2 ml conc NH_4 OH, and boil mixture again. When lactose is present the solution turns brick red followed by formation of a red precipitate, leaving the supernate colorless. Glucose yields a yellow solution and a precipitate. (d) *Fearon's methylamine test* (7, 85). To 5 ml urine, add 1 ml 0.2% aqueous solution of methylamine hydrochloride ($CH_3 NH_2 \cdot HCl$) and 0.2 ml 10% NaOH. Mix, cover tube with glass bulb or marble, and heat in water bath at 56°C for 30 min. An intense red color develops by about 20 min if lactose concentration is 0.5%. If no red color appears by end of heating period, let stand at room temperature. At 0.05% lactose a slight but definite red color appears after 30 min. Weak colors are best detected by comparing with a blank. The only sugars giving a red color are lactose, maltose, and reducing disaccharides. Glucose, fructose, xylose, galactose, and sucrose in large amounts give a yellow color. When the concentration of glucose is one-half that of lactose, the red reaction with lactose is less definite; when the concentrations are equal, negative results are obtained (7).

Recently a specific method was described for the determination of plasma and urine lactose (354). The technic depends upon the quantitation of NADPH, which is directly proportional to the increase in the amount of glucose liberated from lactose by hydrolysis with β-galactosidase. The absorbance is read at 340 nm. A similar method has been reported for the determination of lactose in urine (370). The procedure calls for the enzymatic hydrolysis of lactose to glucose and galactose, which then are measured by the *o*-toluidine method.

Glucose-galactose malabsorption was first described by Lindquist and Meeuwisse (206). This condition is characterized by the inability of mucosa to absorb glucose normally. Glucose and galactose loading tests readily substantiate the diagnosis. To eliminate the possibility of nonspecific malabsorption, a fructose loading test should result in a significant increase of blood glucose.

INULIN

Inulin is a polysaccharide having a molecular weight of approximately 5000 and composed of D-fructose units joined as follows:

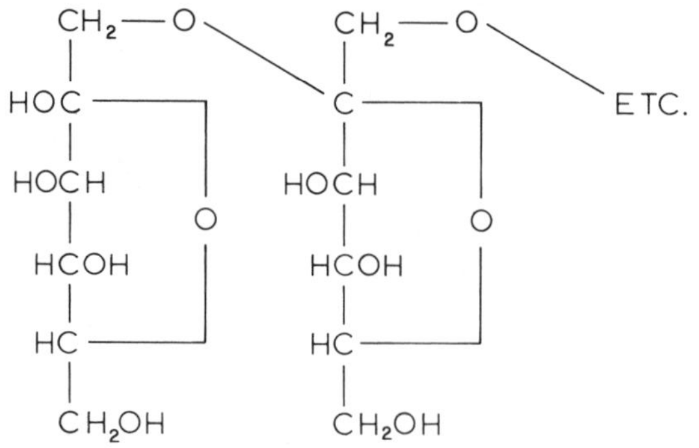

It is of concern to the clinical chemist because of its use in determining the glomerular filtration rate—inulin is excreted by glomular filtration alone and is not subject to tubular reabsorption or excretion. This test is called the *inulin clearance test*.

The various methods for determination of inulin are all based on reactions with the fructose yielded upon hydrolysis. In one approach reducing substances are determined following hydrolysis (105). The glucose can be removed by yeast fermentation (315) or glucose oxidase (113), or a correction can be made by running a blank on a sample obtained prior to inulin administration (105). Menne *et al* (231) obtained good agreement between reduction and resorcinol methods. Steinitz (329), however, criticized methods employing blank corrections for glucose since such corrections may be rather high.

Hexoses heated with strong acid form hydroxymethylfurfural, which reacts to form a colored condensation product with either diphenylamine (blue color) or resorcinol (red color). Alving *et al* (5) and Corcoran and Page (57) simultaneously introduced photometric methods for inulin employing diphenylamine, which were subsequently modified by others (30, 134, 178). Glucose is the only other substance normally present which gives some color in this reaction. This interference has been handled in several ways: (a) Removal of glucose by yeast fermentation (5, 134). Alving *et al* (5) state that commercial inulin contains appreciable amounts of noninulin-fermentable material which gives a blue color in the diphenylamine reaction; they maintain, therefore, that it is essential that it be removed from both blood and urine samples by fermentation. (b) Removal of glucose by autooxidation in alkaline medium plus heat (208). (c) Isolation of inulin by precipitation with $Ca(OH)_2$ (143). (d) Modification of the reaction conditions so that glucose color is so low it can be ignored if the inulin/glucose ratio is sufficiently high (30).

Photometric measurement of the red color formed in the reaction between fructose and resorcinol (Seliwanoff reaction) has also been used (114, 304). The resorcinol reaction is less sensitive to glucose than the diphenylamine reaction. In one technic 200 mg glucose per 100 ml is equivalent to 1 mg inulin (146). Most workers correct for the slight interference normally encountered by running blanks on blood and urine obtained prior to administration of inulin and subtracting the values obtained from subsequent samples analyzed (114). Such "inulinoid" blanks may average the equivalent of about 1.6 mg/100 ml (304). Several approaches have been employed to cope with the situation of a high glucose level: (a) Removal of glucose by yeast fermentation (195); (b) removal of glucose by glucose oxidase (5); (c) determination of glucose level and adding equivalent amount of glucose to the fructose standard (146).

Color formed by reaction of fructose with anthrone has also formed the basis of determination of inulin (128, 395). Interference by glucose has been avoided in a number of ways: (a) Removal by digestion with alkali (396), an approach that has been criticized (129) because of an irregular distribution of the alkali-labile fraction of inulin between blood and urine; (b) carrying out the reaction at a temperature at which fructose reacts but glucose does not (128); (c) destruction of glucose with invertase (394); (d) destruction by yeast (383).

Inulin can be separated from glucose and other small molecular weight substances by passing the mixture through a column of Sephadex G-25 (165).

Several other color reactions have been employed for determination of inulin: (a) A color is formed with skatole and Reinecke's ethanolic HCl solution (272). (b) A purple-violet color is formed with indole-3-acetic acid in conc HCl (144). (c) Heating with vanillin in acid solution produces a red color (204). Glucose interference is apparently negligible in these three color reactions.

Steinitz and Blasbalg (330) compared several methods for inulin and concluded that Heyrovsky's method (144), using the color formed with indole-3-acetic acid, is the best because of (a) the strong, stable color obeying Beer's law and (b) the small interference by glucose.

Commercial inulin is not a homogeneous substance. Cotlove in a personal communication to Young and Raisz (396) reported that commercial inulin contains three fractions: (a) Approximately 1% is a reducing material diffusing like fructose; (b) 15%–40% is nonreducing but alkali-labile and diffuses like a much larger molecule; and (c) the remainder is nonreducing and alkali-stable, and diffuses even more slowly than the second fraction. Young and Raisz themselves found the alkali-labile fraction to be about 10%–25% of the total and observed no difference in renal clearance and distribution in muscle between total nonreducing and alkali-stable fractions. Commercial inulin contains appreciable amounts of noninulin, fermentable material (228). These facts must be taken into account in considering specificity of analytic results and validity of inulin clearance values. As previously stated, Alving *et al* (5) pointed out that the noninulin-fermentable material reacts with diphenylamine, and stated that its removal from samples by fermentation is essential. Ullmann and Ullmann (361) used the resorcinol-thiourea reaction after removal of glucose by alkaline autooxidation but pointed out that valid inulin clearance results could be obtained by this method only if the alkali-labile fraction was less than 13% because the presence of glucose protected the alkali-labile fraction of inulin under certain conditions.

Several automated methods (94, 125, 390) for determination of inulin in both plasma and urine have been reported. A fluorometric procedure has been described for the determination of nanogram quantities (371). According to the author, the presence of five times more glucose than inulin does not affect the assay.

DETERMINATION OF INULIN

(method of Heyrovsky, Ref 144)

PRINCIPLE

Inulin is hydrolyzed to fructose and the fructose determined photometrically by the purple-violet color formed upon reaction with indole-3-acetic acid in HCl. Figure 25-5 shows the absorption curve of color formed.

REAGENTS

Indole-3-Acetic Acid Reagent, 0.5% in Ethanol. Ethanol denatured with methanol is satisfactory. Store in refrigerator. If indole-3-acetic acid is not pure white, recrystallize from hot dilute ethanol after treatment with charcoal.

HCl, Conc

Trichloroacetic Acid, 10%, Aqueous

Inulin Standard, 0.05 mg/ml. Dry inulin in vacuum desiccator over anhydrous $CaCl_2$ overnight. Rub 12.5 mg with 1—2 drops water in 100 ml breaker until no lumps remain. Add rapidly 50—75 ml distilled water which is nearly at its boiling point. Rinse solution into 250 ml volumetric flask with water and dilute to volume. This standard can be preserved by saturation with benzoic acid (329), which does not interfere with the color reaction.

PROCEDURE

1. For serum or plasma, add 4.0 ml 10% trichloroacetic acid slowly to 1.0 ml sample with mixing. Let stand 10 min, centrifuge and, if necessary, filter. Urines should usually be diluted 1:100 with water. Samples should now contain 0.01—0.10 mg inulin per milliliter.
2. To 1.0 ml aliquots of filtrate, diluted urine, standard, and water (reagent blank), add 0.2 ml indole-3-acetic acid reagent and 8.0 ml conc HCl and mix.

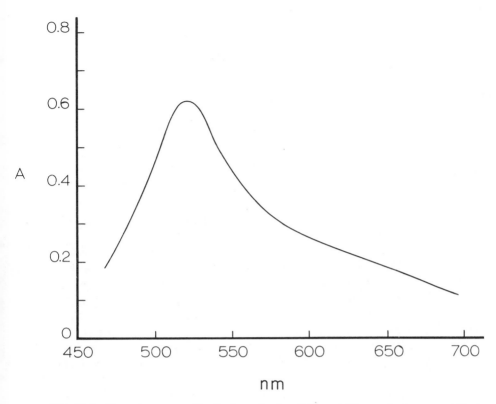

FIG. 25-5. Absorption curve of inulin determination; 0.08 mg inulin standard vs reagent blank. Reagent blank had no absorption in this wavelength range when read vs water. Perkin-Elmer model 4000-A recording spectrophotometer.

3. Place in 37°C water bath for 75 min. It is important that this temperature is not exceeded to avoid interference from other sugars and instability of color.

4. Cool to room temperature and read absorbances of unknowns, standard, and reagent blank versus water at 520 nm or with filter with nominal wavelength in this region. The reagent blank should have negligible absorbance. All tubes must be read rapidly one after the other since the color intensity increases approximately 0.5%/min.

Calculation:

Serum or plasma:

$$\text{mg inulin/100 ml} = \frac{A_x - A_b}{A_s - A_b} \times 0.05 \times \frac{100}{0.2}$$

$$= \frac{A_x - A_b}{A_s - A_b} \times 25$$

Urine:

$$\text{mg inulin/100 ml} = \frac{A_x - A_b}{A_s - A_b} \times 0.05 \times \frac{100}{0.01}$$

$$= \frac{A_x - A_b}{A_s - A_b} \times 500$$

NOTES

1. Beer's law. Beer's law was obeyed up to absorbance of 1.1 on a Beckman DU spectrophotometer and up to 0.1 mg inulin in final color with Klett filters 52 and 54.

2. Serum vs plasma vs whole blood. Identical results are obtained on serum and plasma. Values for serum or plasma, however, are less than those calculated from whole blood values and the hematocrit, because approximately 8%–10% of the inulin is "trapped" by the erythrocytes (128).

3. Interference by dextran (270). The presence of dextran in the sample causes a positive interference in the indole-3-acetic acid method presented as well as in the diphenylamine and resorcinol methods. For example, addition of dextran to serum in final concentrations of 2.5 and 5.0 g/100 ml gave inulin values of 6.2 and 14.3 mg/100 ml, respectively. Dextran can be removed from the sample by precipitation with ethanol.

4. Serum blanks. Normal serum devoid of inulin gives color equivalent to approximately 3.5 mg fructose per 100 ml, of which approximately 0.4 mg/100 ml is due to glucose.

ACCURACY AND PRECISION

The chief possible source of inaccuracy is glucose. Heyrovsky (144), however, showed that glucose levels of 10 and 1 g/100 ml in urine and plasma, respectively, caused errors of only about 2%–4%. Steinitz and Blasbalg (330) also found negligible interference by glucose. It was found in our laboratory that 100 mg glucose per 100 ml sample gave color equivalent to 0.4 mg inulin per 100 ml.

REFERENCES

1. ACLAND JD, CLAYTON H, MITCHELL B: *J Appl Physiol 17*:119, 1962
2. ADAMS EC Jr, MAST RL, FREE AH: *Arch Biochem Biophys 91*:230, 1960
3. ADAMS EC Jr. BURKHART CE, FREE AH: *Science 125*:1082, 1957
4. AHLERT G, HOFER E, HOFFMAN W, BESTVATER G: *Deut Gesund Heitsw 19*:2256, 1964; *Chem Abstr 62*:15057, 1965
5. ALVING AS, FLOX J, PITESKI I, MILLER FB: *J Lab Clin Med 27*:115, 1941
6. APTHORP GH: *J Clin Pathol 10*:84, 1957

7. ARCHER H, HARAM B: *Lancet 1*:558, 1948
8. ATHANAIL G, CABAUD PG: *J Lab Clin Med 51*:321, 1958
9. AW SE: *Clin Chim Acta 26*:235, 1969
10. BARAC G, DELVENNE J: *Bull Soc Chim Biol 29*:1094, 1947
11. BEASER SB: *J Am Med Assoc 199*:990, 1967
12. BECK G: *Mikrochemia ver Mikrochim Acta 35*:169, 150; *C A 44*:7709b, 1950
13. BEHRENDT H: *Diagnostic Tests for Infants and Children.* New York, Interscience, 1949, p103
14. BELL WN, JUMPER E: *J Am Med Assoc 166*:2145, 1958
15. BELMONTE MM, SARKOZY E, HARPUR ER: *Diabetes 16*:557, 1967
16. BENEDICT SR: *J Am Med Assoc 57*:1194, 1911
17. BENEDICT SR: *J Biol Chem 5*:485, 1909
18. BENEDICT SR, OSTERBERG E: *J Biol Chem 34*:195, 1918
19. BENEDICT SR: *J Biol Chem 64*:207, 1925
20. BENEDICT SR: *J Biol Chem 92*:141, 1931
21. BENNETT PH, STEINBERG AG, MILLER M, BURCH TA: *J Lab Clin Med 66*:852, 1965
22. BENSON JA Jr, CULVER PJ, RAGLAND S, JONES CM, DRUMMEY GD, BOUGAS E: *New Engl J Med 256*:335, 1957
23. BERGMEYER HU, MOELLERING H: *Clin Chim Acta 14*:74, 1966
24. BERNARDY T: *J Am Med Assoc 174*:2001, 1960
25. BEUTLER E, BALUDA M: *J Lab Clin Med 68*:137, 1966
26. BIAL M: *Deut Med Wochschr 28*:253, 1902
27. BICKEL H: *J Pediat 59*:641, 1961
28. BICKEL H: *Mod Probl Pediat 4*:136, 1959
29. BLOTNER H: *Arch Internal Med 75*:39, 1945
30. BOJESEN E: *Acta Med Scand 142 (Suppl 266)*:275, 1952
31. BONTING SL: *Arch Biochem Biophys 52*:272, 1954
32. BORCHARDT L: *Hoppe Seylers Z Physiol Chem 55*:241, 1908
33. BOURNE BB: *Clin Chem 10*:1121, 1964
34. BOYNS DR, CORSSLEY JN, ABRAMS ME, JARRETT RJ, KEEN H: *Brit Med J 1*:595, 1969
35. BOZIAN R, TOUSTER O: *Nature 184*:463, 1959
36. BRILL IC: *J Lab Clin Med 8*:727, 1923
37. BRODERSEN R, RICKETTS HT: *J Lab Clin Med 34*:1447, 1949
38. BROWN ME: *Diabetes 10*:60, 1961
39. BUCKLEY I, DRURY MI: *Irish J Med Sci Series 6*:272, 1960
40. BUTLER TJ: *Am J Med Technol 27*:205, 1961
41. CAMPBELL DM, KING EJ: *J Clin Pathol 16*:173, 1963
42. CARAWAY WT: *Am J Clin Pathol 37*:445, 1962
43. CARAWAY WT: *Fundamentals of Clinical Chemistry*, edited by Tietz NW. Philadelphia, Saunders, 1970, p 169
44. CAWLEY LP, SPEAR FE, KENDALL R: *Am J Clin Pathol 32*:195, 1959
45. CERIOTTI G: *Clin Chim Acta 8*:157, 1963
46. CERIOTTI G: *Clin Chem 17*:440, 1971
47. CERIOTTI G, DE NADAI FRANK A: *Clin Chim Acta 24*:311, 1969
48. CHANTLER C, BAUM JD, NORMAN DA: *Lancet II, 7531*:1395, 1967
49. CHERONIS ND, ZYMARIS MC: *Mikrochim Acta 6*:769, 1957; *C A 52*:15627, 1958
50. CHERTACK MM: *J Am Med Assoc 166*:48, 1957
51. COLE HS, BILDER JH: *Diabetes 19*:176, 1970
52. COMER JP: *Anal Chem 28*:1748, 1956

53. CONN JW: *Am J Med Sci 199*:555, 1940
54. COOK MH, FREE AH, GIORDANO AS: *Am J Med Technol 19*:283, 1953
55. COOK MH, FREE AH: *Am J Med Technol 24*:305, 1958
56. COOPER JDH: *Med Lab Technol 29*:51, 1972
57. CORCORAN AC, PAGE IH: *J Biol Chem 127*:601, 1939
58. CORNBLATH M, REISNER SH: *New Engl J Med 273*:378, 1965
59. COVEY EP: *Am J Clin Pathol 19*:500, 1949
60. COWIE DM, PARSONS JP: *Arch Internal Med 26*:333, 1920
61. CRAIG RG, BERGQUIST LM, SEARCY RL: *Anal Biochem 1*:433, 1960
62. DAHLQVIST A, SVENNINGSEN N: *J Pediat 75*:454, 1969
63. DAHLQVIST A, ROSENTHAL I: *PAP Int Conf,* 1967. 1969, p287
64. DANNENBERG, E: *Munch Med Wochschr 96*:410, 1954
65. DATE JW: *Scand J Clin Lab Invest 10*:155, 1958
66. DAVIS J, GANDER J: *Anal Biochem 19*:72, 1967
67. DECKERT T: *Scand J Clin Lab Invest 20*:217, 1967
68. DEHN WM, JACKSON KE, BALLARD DA: *Anal Chem 4*:413, 1932
69. DEICHMAN WB, DIERKER M: *J Biol Chem 163*:753, 1946
70. DOBRICK LA, HARRER HA: *J Biol Chem 231*:403, 1958
71. DOBSON HL, SHAFFER R, BURNS R: *Diabetes 17*:281, 1968
72. DORMANDY T, PORTER R: *Lancet 1*:1189, 1961
73. DREYWOOD R: *Ind Eng Chem (Anal Ed) 18*:499, 1946
74. DUBOWSKI KM: *Clin Chem 8*:215, 1962
75. EK J, HULTMAN E: *Nature 181*:780, 1958
76. EK J, HULTMAN E: *Scand J Clin Lab Invest 9*:315, 1957
77. ENKLEWITZ M: *J Biol Chem 116*:47, 1936
78. EXTON WG, ROSE AR: *Am J Clin Pathol 4*:381, 1934
79. EXTON WG, ROSE A, ROEHL E: *Proc Soc Exp Biol Med 32*:417, 1934
80. FAJANS SS, CONN JW: *On the Nature and Treatment of Diabetes.* New York, Excerpta Medica International Congress Series 84, 1965, Chap. 46, p 641
81. FALES FW, RUSSELL JA, FAIN JN: *Clin Chem 7*:289, 1961
82. FASHENA GJ: *J Biol Chem 100*:357, 1933
83. FASHENA GJ, STIFF HA: *J Biol Chem 137*:21, 1941
84. FAULKNER DE: *Analyst 90*:736, 1965
85. FEARON WR: *Introduction to Biochemistry.* London, Heinemann, 1947
86. FEHLING HV: *Arch Physiol Heilk 7*:64, 1848
87. FELDMAN JM, KELLEY WN, LEBOVITZ HE: *Diabetes 19*:337, 1970
88. FIELD JB, BOYLE C, REMER A: *Lancet I*:1191, 1967
89. FINE J: *Brit Med J 2*:167, 1934
90. FINGERHUT B, FERZOLA R, MARSH WH: *Clin Chim Acta 8*:953, 1963
91. FIORENTINO M, BONI P: *Diagnost e Tech Lab (Napoli) Riv Mens 11*:449, 1940; *C A 41*:3155f, 1947
92. FIORENTINO M, GIANNETTASIO G: *J Lab Clin Med 25*:866, 1939–40
93. FISHMAN WH, SMITH M, THOMPSON DB, BONNER CD, KASDON SC, HOMBURGER F: *J Clin Invest 30*:685, 1951
94. FJELDBO W, STAMEY TA: *J Lab Clin Med 72*:353, 1968
95. FLEXNER LB: *Physiol Rev 14*:161, 1934
96. FOLIN O: *J Biol Chem 67*:357, 1926
97. FOLIN O: *J Biol Chem 77*:421, 1928
98. FOLIN O: *J Biol Chem 82*:83, 1929
99. FOLIN O, McELLROY WS: *J Biol Chem 33*:513, 1918
100. FOLIN O, WU H: *J Biol Chem 38*:81, 1919
101. FOLIN O, WU H: *J Biol Chem 41*:367, 1920

102. FORSELL OM, PALVA IP: *Scand J Clin Lab Invest 11*:409, 1959
103. FOWWEATHER FS: *Biochem J 55*:718, 1953
104. FOX RE, ROBERTS HK, OPPENHEIMER HE, GOLDENBERG S, BETTON-VILLE PJ, MAHE GA: *J Am Med Assoc 182*:622, 1962
105. FRANK H, SCHEIFFARTH F: *Klin Wochschr 34*:914, 1956
106. FREE AH: *Advan Clin Chem 6*:67, 1963
107. FREE AH: *U S Patent No 2,848,308*, August 19, 1958, *C A 53*:11503, 1959
108. FREE AH, ADAMS EC, KERCHER ML, FREE HM, COOK MH: *Clin Chem 3*:163, 1957
109. FREEDBERG I, FEINGOLD D, HIATT M: *Biochem Biophys Res Commun 1*:328, 1959
110. FRINGS CS, RATLIFF CR, DUNN RT: *Clin Chem 16*:282, 1970
111. FRINGS CS: *Clin Chem 16*:618, 1970
112. FROESCH E, WOLF H, BAITSCH H, PRADER A, LABHART A: *Am J Med 34*:151, 1963
113. FROESCH ER, RENOLD AE: *Diabetes 5*:1, 1956
114. FROESCH ER, REARDON B, RENOLD AE: *J Lab Clin Med 50*:918, 1957
115. FUTTERMAN S, ROE JH: *J Biol Chem 215*:257, 1955
116. GARDNER L: *Endocrine and Genetic Diseases of Childhood*, edited by Launiala K, Perheentupa J, Hallman N. Philadelphia, Saunders, 1969, p 837
117. GIFFORD H, BERGERMAN J: *J Am Med Assoc 178*:423, 1961
118. GIORDANO AS, POPE JL, HOGAN B: *Am J Med Technol 22*:29, 1956
119. GOODWIN J: *Clin Chem 17*:397, 1971
120. GOODWIN JF: *Clin Chem 16*:85, 1970
121. GOULD SE, ALTSHULER SS, MELLEN HS: *Am J Med Sci 193*:611, 1937
122. GREEN HN, STONER HB, WHITELEY HJ, EGLIN D: *Clin Sci 8*:65, 1949
123. GUIDOTTI G, COLOMBO J-P, FOA PO: *Anal Chem 33*:151, 1961
124. GURIAN H, ADLERSBERG D: *Am J Med Sci 237*:12, 1959
125. HAFF AC: *Biochem Med 2*:190, 1968
126. HAGEDORN HC, JENSEN BN: *Biochem Z 135*:46, 1923
127. HAM AB: *Am J Med Technol 22*:339, 1956
128. HANDELSMAN MB, SASS M: *J Lab Clin Med 48*:759, 1956
129. HANDELSMAN MB, DRABKIN J: *Proc Soc Exp Biol Med 86*:356, 1954
130. HARDING U, HEINZEL G: *Z Klin Chem Klin Biochem 7*:640, 1969
131. HARDING VJ, SELBY DL: *Biochem J 25*:1815, 1931
132. HARDING VJ, DOWNS CE: *J Biol Chem 101*:487, 1933
133. HARRISON GA: *Chemical Methods in Clinical Medicine*, 3rd ed. New York, Grune & Stratton, 1947, p 506
134. HARRISON HE: *Proc Soc Exp Biol Med 49*:111, 1942
135. HARTEL AV, HELGER R, LANG H: *Z Klin Chem Klin Biochem 7*:14, 1969
136. HAUNZ EA, KERANEN MT: *Proc Am Diabetes Assoc 10*:200, 1950
137. HAWKINS JA, VAN SLYKE DD: *J Biol Chem 81*:459, 1929
138. HAWKINS JA: *J Biol Chem 84*:69, 1929
139. HECK K, BROWN WH, KIRK PL: *Mikrochemie 22*:306, 1937
140. HELMER O, FOUTS P: *J Clin Invest 16*:343, 1937
141. HERBERT F: *Biochem J 32*:815, 1938
142. HERBERT FK, BOURNE MC, GROEN J: *Biochem J 24*:291, 1930
143. HERZ N, SHAPIRO B: *J Lab Clin Med 32*:1159, 1947
144. HEYROVSKY A: *Clin Chim Acta 1*:470, 1956
145. HICKMAN J, ASHWELL G: *J Biol Chem 234*:758, 1959
146. HIGASHI A, PETERS J: *J Lab Clin Med 35*:475, 1950
147. HILL JB, KESSLER G: *J Lab Clin Med 57*:907, 1961

148. HILLER A, LINDER GC, VAN SLYKE DD: *J Biol Chem 64*:625, 1925
149. HJELM M: *Scand J Clin Lab Invest 18 (Suppl 92)*:85, 1966
150. HLAD CJ JR, ELRICK H: *J Clin Endocrinol Metab 19*:1258, 1959
151. HLAING TT, HUMMEL JP, MONTGOMERY R: *Arch Biochem Biophys 93*:321, 1961
152. HOFFMAN WS: *Biochemistry of Clinical Medicine*. Chicago, Year Book Publishers, 1954
153. HOFFMAN WS: *J Biol Chem 120*:51, 1937
154. HOLLISTER L, HELMKE E, WRIGHT A: *Diabetes 15*:691, 1966
155. HOLZEL A, SCHWARZ V, STUCLIFFE K: *Lancet 1*:1126,1959
156. HSIA DY: *Inborn Errors of Metabolism*. Chicago, Year Book Publishers, 1966, pp 202, 204
157. HSIA DY: *Metab Clin Exp 35*:419, 1967
158. HUBBARD RS, BROCK HJ: *J Biol Chem 110*:411, 1935
159. HUGGETT A, NIXON DA: *Lancet 2*:368, 1957
160. HULTMAN E: *Nature 183*:108, 1959
161. HUNTER WH: *J Med Lab Tech 24*:293, 1967
162. HYVÄRINEN A, NIKKILÄ EA: *Clin Chim Acta 7*:140, 1962
163. IUCHI I, SHIBATA S: *Clin Chim Acta 5*:42, 1960
164. JACKSON WPU, GOLDBERG MD, MARINE N, VINIK AI: *Lancet II, 7578*:1101, 1968
165. JACOBSON L: *Clin Chim Acta 7*:180, 1962
166. JARRETT RJ, KEEN H, HARDWICK C: *Diabetes 19*:724, 1970
167. JOHNSON J: *Am J Med Technol 24*:271, 1958
168. JOLLEY R, WARREN K, SCOTT C, JAINCHILL J, FREEMAN M: *Am J Clin Pathol 53*:793, 1970
169. JONES JKN, PRIDHAM JB: *Biochem J 58*:288, 1954
170. JONNARD R, NEWMAN H, RUSSOMANNO A, SCHATTNER F: American Chemical Society, Atlantic City Meeting, Div. Biol. Chem., Abstract 84c, 1959
171. KADISH AH, LITLE RL, STERNBERG JC: *Clin Chem 14*:116, 1968
172. KANTOR N, WILKERSON HLC: *Diabetes 6*:176, 1957
173. KASPER JA, JEFFREY IA: *Am J Clin Pathol Tech Sect 8*:117, 1944
174. KEILIN D, HARTREE EF: *Biochem J 42*:221, 1948
175. KEILIN D, HARTREE EF: *Biochem J 42*:230, 1948
176. KEILIN D, HARTREE EF: *Biochem J 50*:331, 1952
177. KELLER DM: *Clin Chem 11*:471, 1965
178. KENDRICK AB: *Furman Univ Bull 3*:89, 1956
179. KERBY GC: *J Clin Invest 33*:1168, 1954
180. KERSTELL J: *Scand J Clin Lab Invest 13*:637, 1961
181. KESTON AS: Abstract of Papers, 129th Meeting, Am. Chem. Soc., Dallas, Texas, 1956, p 31
182. KETTERINGHAM A, AUSTIN BR: *Am J Med Sci 195*:328, 1938
183. KING JW, HAINLINE A: *Cleveland Clinic Quart 23*:215, 1956
184. KINGSLEY GR, GETCHELL G: *Clin Chem 6*:466, 1960
185. KLEIN B, WEISSMAN M: *Anal Chem 25*:771, 1953
186. KLIMT CR, MEINERT CL, MILLER M, KNOWLES HC Jr: University Group Diabetes Program, Brook Lodge Symposium, August 1967. New York, Excerpta Med. Int. Congress Series No. 149, 1967.
187. KLINGHOFFER KA: *Am J Physiol 111*:231, 1935
188. KOBBERLING J, CREUTZFELDT W: *Diabetes 19*:870, 1970; *C A* Biochem. Sections 74 No. 5, 1971, p 193 (40468t)
189. KOIKE H, KURODA Y, AMINO S: *Nisshin Igaku 43*:575, 1956; *CA 51*:10631g, 1957

190. KONDON C, O'SULLIVAN JB: *Public Health Rept* (US) *81*:743, 1966; *C A 65*, No. 13, p 20502b, 1966
191. KOSAKA K, MIZUNO Y, KUZUYA T: *Diabetes 15*:901, 1966
192. KRAINICK HG: *Mikrochemia Mikrochim Acta 29*:45, 1941
193. KRAMER H, STEINER A: *Biochem J 25*:161, 1931
194. KRANE S: *Metabolic Basis of Inherited Disease*, edited by Stanbury JB, Wyngaarden JB, Fredrickson DS, New York, McGraw-Hill, 1960, p 144
195. KRUHOFER P: *Acta Physiol Scand 11*:1, 1946
196. KUTTER D: *Das Arztliche Lab 7*:175, 1961
197. LAMBERT TH, JOHNSON RB: *J Lab Clin Med 60*:976, 1962
198. LANGNER PH, ROMANSKY MJ, ROBIN ED: *Am J Med Sci 212*:466, 1946
199. LASKER M: *Human Biol 13*:51, 1941
200. LASKER M, ENKLEWITZ M: *J Biol Chem 101*:289, 1933
201. LEHMANN J: *Scand J Clin Lab Invest 16 (Suppl 76)*, 1964
202. LESKE R, MAYERSBACH H: *Biochim Biol Sper 1*:193, 1961; *C A 62*:4317b, 1965
203. LEVENE P, LA FORGE F: *J Biol Chem 18*:319, 1914
204. LEVINE VE, BECKER WW: *Clin Chem 5*:142, 1959
205. LEWIS RC, BENEDICT SR: *Proc Soc Exp Biol Med 11*:57, 1913–14
206. LINDQUIST B, MEEUWISSE G: *Acta Pediat 51*:674, 1962
207. LIPPMAN RW: *Am J Clin Pathol 22*:1186, 1952
208. LITTLE JM: *J Biol Chem 180*:747, 1949
209. LOGAN JE, HAIGHT DE: *Clin Chem 11*:367, 1965
210. LONGFIELD GM, HOLLAND DE, LAKE AJ, KNIGHTS EM Jr: *Am J Clin Pathol 33*:550, 1960
211. LORENTZ K: *Clin Chim Acta 13*:660, 1966
212. MacDONALD I, CROSSLEY JN: *Diabetes 19*:450, 1970
213. MAGER M, FARESE G: *Am J Clin Pathol 44*:104, 1965
214. MAKIN HLJ, WARREN PJ: *Clin Chim Acta 29*:493, 1970
215. MALMSTADT HV, HADJIIOANNOU SI: *Anal Chem 34*:452, 1962
216. MALMSTADT HV, PARDUE HL: *Clin Chem 8*:606, 1962
217. MARAIS JP, de WIT JL, QUICKE GV: *Anal Biochem 15*:373, 1966
218. MARKS V: *Clin Chim Acta 4*:395, 1959
219. MASTURZO M, NEGRO L: *Boll Soc Ital Biol Sper 29*:255, 1953
220. MATELES RI: *Nature 187*:241, 1960
221. McCANCE R: *Biochem J 20*:1111, 1926
222. McCOMB RB, YUSHOK WD, BATT DB: *J Franklin Inst 263*:161, 1957
223. McCOMB RB, YUSHOK WD: *J Franklin Inst 265*:417, 1958
224. McDONALD GW, FISHER GF, BURNHAM C: *Diabetes 14*:473, 1965
225. McINTYRE N, HODSWORTH CD, TURNER DS: *Lancet II, 7349*:20, 1964
226. McKAY E: *Clin Chim Acta 10*:320, 1964
227. MEITES S, BOHMAN N: *Clin Chem 9*:594, 1963
228. MELOCHE V: *Am Soc Testing Mater, Spec Tech Pub 116*:3, 1952
229. MENDEL B, BAUCH M: *Klin Wochschr 5*:1329, 1926
230. MENDEL B, HOOGLAND PL: *Lancet 259*:16, 1950
231. MENNE F, WETTER O, CRAMER L: *Z Ges Exp Med 123*:523, 1954
232. MES J, KAMM L: *J Chromatog 38*:120, 1968
233. METZ R, SURMACZYNSKA B, BERGER S, SOBEL G: *Ann Internal Med 64*:1042, 1966
234. MIDDLETON JE, GRIFFITHS WJ: *Brit Med J 2*:1525, 1957
235. MIDDLETON JE: *Clin Chim Acta 22*:433, 1968
236. MILITZER WE: *Arch Biochem 9*:85, 1946
237. MILLER BF, VAN SLYKE DD: *J Biol Chem 114*:583, 1936
238. MILTON RF: *Analyst 67*:183, 1942

239. MITCHELL FL, STRUASS WT: *Lancet 1*:1185, 1964
240. MOHUN AF, COOK IJY: *J Clin Pathol 15*:169, 1962
241. MONTGOMERY R: *Arch Biochem Biophys 67*:378, 1957
242. MONTREUIL J, BOULANGER P: *Compt Rend Soc Biol 236*:337, 1953
243. MOOREHEAD WR, SASSE EA: *Clin Chem 16*:285, 1970
244. MORRIS DL: *Science 107*:254, 1948
245. MOYA F, DEWAR J: *J Lab Clin Med 47*:314, 1956
246. MOYER JH, WOMACK CR: *Am J Med Sci 219*:161, 1950
247. MULLER D: *Biochem Ztschr 199*:136, 1928
248. MYERS GB, McKEAN RM: *Am J Clin Pathol 5*:299, 1935
249. NAGANNA B, RAJAMMA M, RAO KV: *Clin Chim Acta 17*:219, 1967
250. NAKAMURA RM, REILLY EB, FUJITA K, BROWN J, KUNITAKE GM: *Diabetes 14*:224, 1965
251. NAZARIO M: *Rev Asoc Bioquim Arg 21*:6, 1956
252. NELSON N: *J Biol Chem 153*:375, 1944
253. NIXON D: *Clin Chim Acta 26*:167, 1969
254. NOYONS EC: *Rec Trav Chim 65*:485, 1946; *C A 41*:171d, 1947
255. NYLANDER E: *Hoppe-Seylers Z Physiol Chem 8*:175, 1883
256. O'MALLEY E, CONWAY EJ, FITZGERALD O: *Biochem J 37*:278, 1943
257. ORR W, MINOT A: *Arch Neurol Psychiat 67*:483, 1952
258. OSHIMA K, TOLLENS B: *Ber 34*:1425, 1901
259. PACKER H, ACKERMAN RF: *Diabetes 7*:312, 1958
260. PAGE L, CULVER P: *Syllabus of Laboratory Examinations in Clinical Diagnosis*, revised ed. Cambridge, Mass., Harvard Univ. Press, 1960
261. PAKIANATHAN S: *Enzymologia 26*:155, 1963
262. PARROTT LH: *Am J Clin Pathol 49*:877, 1968
263. PERHEENTUPA J, PITKANEN E, NIKKILA E, SOMERSALO O, HAKOSALO J: *Ann Paediat Fenniae 8*:221, 1962
264. PERSSON L: *Scand J Clin Lab Invest 7*:279, 1955
265. PETERSON JI: *Clin Chem 14*:513, 1968
266. PHILLIPS RE, ELEVITCH FR: *Am J Clin Pathol 49*:622, 1968
267. PICKENS JM, BURKEHOLDER JN, WOMACK WN: *Diabetes 16*:11, 1967
268. POGELL B: *J Biol Chem 211*:143, 1954
269. POLITZER WM, FLEISCHMANN H: *Am J Human Genet 14*:256, 1962
270. PREEDY JRK: *J Biol Chem 210*:651, 1954
271. RAABO E, TERKILDSEN TC: *Scand J Clin Lab Invest 12*:402, 1960
272. RANNEY H, McCUNE DJ: *J Biol Chem 150*:311, 1943
273. RAPPAPORT F, EICHHORN E: *Am J Clin Pathol 20*:834, 1950
274. RATISH HD, BULLOWA JGM: *Arch Biochem Biophys 2*:381, 1943
275. RECANT L: *J Lab Clin Med 48*:165, 1956
276. REED AH, CANNON DC, WINKELMAN JW, BHASIN YP, HENRY RJ, PILEGGI VJ: *Clin Chem 18*:57, 1972
277. REINER M, FENICHEL RL, STERN KG: *Acta Hematol 3*:202, 1950
278. REINER M, CHEUNG H: *Standard Methods in Clinical Chemistry*, edited by Meites S. New York, Academic Press, 1965, Vol 5, p 264
279. RELANDER A, RAIHA CE: *Scand J Clin Lab Invest 15*:221, 1963
280. RENEHER L, BEELER M: *Clinical Diagnosis by Laboratory Methods*, edited by Davidsohn I, Henry J. Philadelphia, Saunders, 1969, p 791
281. Report of the Committee on Statistics of the Amer. Diabetes Assoc. June 14, 1968: *Diabetes 18*:299, 1969
282. ROBIN M, SAIFER A: *Clin Chem 11*:840, 1965
283. ROCK J, GERENDE L: *J Am Med Assoc 198*:321, 1966

284. ROE J: *J Biol Chem 107*:15, 1934
285. ROE J, RICE E: *J Biol Chem 173*:507, 1948
286. ROLF D, SURTSHIN A, WHITE H: *Proc Soc Exp Biol Med 72*:351, 1949
287. ROTHSTEIN A, MEIER R: *J Cellular Comp Physiol 34*:97, 1949
288. RUBNER M: *Z Biol 20*:397, 1884
289. RUDESILL CL, HENDERSON RA: *Am J Disease Children 61*:108, 1941
290. SALTZMAN A, CARAWAY WT, BECK IA: *Metab 3*:11, 1954
291. SAMSON M: *Am J Clin Pathol 22*:1106, 1952
292. SAMSON M: *J Am Chem Soc 61*:2389, 1939
293. SANTINI R, Jr, SHEEHY TW, MARTINEZ DE JESUS J: *Gastroenterology 40*:772, 1961
294. SARMA P: *Clin Chem 10*:224, 1964
295. SAWICKI E: *Chemist-Analyst 45*:45, 1956
296. SCHAPIRA F, SCHAPIRA G, DREYFUS J: *Enzymol Biol Clin 1*:170, 1962
297. SCHERSTEN B: *Acta Med Scand 178*:583, 1965
298. SCHERSTEN B,, TIBBLING G: *Clin Chem 14*:243, 1968
299. SCHERSTEN B, TIBBLING G: *Clin Chim Acta 18*:383, 1967
300. SCHERSTEN B, FRITZ H: *J Am Med Assoc 201*:949, 1967
301. SCHERZ H, STEHLIK G, BANCHER E, KAINDL K: *Chromatog Rev 10*:1, 1968
302. SCHMIDT FH: *Klin Wochschr 39*:1244, 1961
303. SCHOEN I: *Am J Clin Pathol 34*:494, 1960
304. SCHREINER GE: *Proc Soc Exp Biol Med 74*:117, 1950
305. SCOTT LD: *Biochem J 29*:1012, 1935
306. SEIBERT FB, ATNO J: *J Biol Chem 163*:511, 1946
307. SHAFFER PA, SOMOGYI M: *J Biol Chem 100*:695, 1933
308. SHAFFER PA, HARTMANN AF: *J Biol Chem 45*:365, 1921
309. SHAFFER PA, WILLIAMS RD: *J Biol Chem 111*:707, 1935
310. SHAMBAUGH GE, BEISEL WR: *Diabetes 16*:369, 1967
311. SHETLAR MR, FOSTER JV, KELLY KH, EVERETT MR: *Proc Soc Exp Biol Med 69*:507, 1948
312. SILBER S, REINER M: *Arch Internal Med 54*:412, 1934
313. SIMON RK, CHRISTIAN GD, PURDY WC: *Clin Chem 14*:463, 1968
314. SISK CW, BURNHAM CE, STEWART J, McDONALD GW: *Diabetes 19*:852, 1970
315. SMITH HW, GOLDRING W, CHASIS H: *J Clin Invest 17*:263, 1938
316. SMITH MB, BRAIDWOOD JL: *Clin Biochem 4*:118, 1971
317. SOLER NG, MALINS JM: *Lancet II*:724, 1971
318. SOLOMOS GI: *Bull Soc Chim Biol 17*:1465, 1935
319. SOMOGYI M: *J Lab Clin Med 26*:1220, 1941
320. SOMOGYI M: *J Biol Chem 70*:599, 1926
321. SOMOGYI M: *J Biol Chem 75*:33, 1927
322. SOMOGYI M: *J Biol Chem 86*:655, 1930
323. SOMOGYI M: *J Biol Chem 195*:19, 1952
324. SOWTON E: *Brit Med J 1*:84, 1962
325. SPELLACY WN, CARLSON KL, BIRK SA, SCHADE SL: *Metab Clin Exptl 17*:496, 1968
326. STANBURY J, WYNGAARDEN J, FREDRICKSON D: *The Metabolic Basis of Inherited Disease*. New York, McGraw-Hill, 1966
327. STANEK J, CERNY M, KOCOUREK J, PACAK J: *The Monosaccharides*, edited by Ernest I, Hebky J. New York, Academic Press, 1963, p 453
328. STANLEY P: *Am J Med Technol 6*:263, 1940
329. STEINITZ K: *J Biol Chem 126*:589, 1938
330. STEINITZ K, BLASBALG H: *Scand J Clin Lab Invest 10 (Suppl 31)*:310, 1957

331. STEWART D: *Arch Disease Childhood 3*:96, 1928
332. ST. LORANT I: *J Clin Pathol 10*:136, 1957
333. STORK H, SCHMIDT FH: *Klin Wochschr 46*:789, 1968
334. STREET HV: *Analyst 83*:628, 1958
335. STROES JAP, ZONDAG HA, CORNELISSEN PJH CHR: *Clin Chim Acta 8*:152, 1963
336. SUDDUTH NC, WISISH JR, MOORE JL: *Am J Clin Pathol 53*:181, 1970
337. SUMNER JB: *J Biol Chem 65*:393, 1925
338. SUMNER JB: *J Biol Chem 62*:287, 1924
339. SUMNER JB, SISLER EB: *Arch Biochem 4*:333, 1944
340. SUNDERMAN FW, MacFATE RP, EVANS GT, FULLER JB: *Am J Clin Pathol 21*:901, 1951
341. SUNDERMAN FW, COPELAND BE, MacFATE RP, MARTENS VE, NAUMANN HN, STEVENSON GF: *Am J Clin Pathol 26*:1355, 1956
342. SUNDERMAN FW Jr, SUNDERMAN FW: *Am J Clin Pathol 36*:75, 1961
343. SUNSHINE P, KRETCHMER N: *Pediat 34*:38, 1964
344. SUTHERLAND HW, STOWERS JM, CHRISTIE RJ: *Lancet 1*:1071, 1970
345. SZUSTKIEWICZ C, DEMETRIOU J: *Clin Chim Acta 32*:355, 1971
346. TAMMES AR, NORDSCHOW CD: *Am J Clin Pathol 49*:613, 1968
347. TELLER JD: Abstr. 130th Meeting Am. Chem. Soc., Atlantic City, 1956, p 69c
348. THAYSEN EH, MULLERTZ S: *Acta Med Scand 171*:521, 1962
349. THOMPSON RH: *Clin Chim Acta 25*:475, 1969
350. TIMELL TE, GLAUDEMANS CPJ, CURRIE AL: *Anal Chem 28*:1916, 1956
351. TODD WR, DODSON MC, TRAINER JB, McKEE J: *Arch Biochem 4*:337, 1944
352. TOLSTOI E: *J Biol Chem 60*:69, 1924
353. TOMPSETT SL: *Biochem J 24*:1148, 1930
354. TOSELAND P: *J Clin Pathol 21*:112, 1968
355. TOUSTER O, HARWELL S: *J Biol Chem 230*:1031, 1958
356. TOWER DB, PETERS EL, POGORELSKIN MA: *Neurology 6*:37, 125, 1956
357. TRINDER P: *J Clin Pathol 22*:158, 1969
358. TROMMER CA, quoted by MITSCHERLICH E: *Ann Chem Pharm 39*:360, 1841
359. TUNBRIDGE RE, ALLIBONE EC: *Quart J Med 33*:11, 1940
360. TURPIN B, DOWNEY P: *Am J Med Technol 32*:327, 1966
361. ULLMANN TD, ULLMANN L: *J Lab Clin Med 49*:793, 1957
362. UNGER RH: *Ann Internal Med 47*:1138, 1957
363. UPDIKE SJ, HICKS GP: *Nature 214*:986, 1967
364. US Public Health Service: *Glucose Tolerance of Adults.* Publication No. 1000, Series 11, No. 2, Vital and Health Statistics, 1964
365. VAN PINXTEREN JAC: *Pharm Weekblad 93*:753, 1958; *C A 52*:19716b, 1958
366. VAN SLYKE DD, HAWKINS JA: *J Biol Chem 83*:51, 1929
367. VAN SLYKE DD, HAWKINS JA: *J Biol Chem 79*:739, 1928
368. VARTIO T: *Scand J Clin Lab Invest 14*:36, 1962
369. VOLK BW, SAIFER A, LAZARUS SS: *J Lab Clin Med 57*:367, 1961
370. VON HUMBEL R, LUDWIG S: *Z Klin Chem Klin Biochem 8*:318, 1970
371. VUREK G, PEGRAM S: *Anal Biochem 16*:409, 1966
372. WAGNER R: *Arch Biochem Biophys 11*:249, 1946
373. WAHBA N, HANNA S, EL-SADR MM: *Analyst 81*:430, 1956
374. WANG Y, van EYS J: *New Engl J Med 282*:892, 1970
375. WARE AG, MARBACH EP: *Clin Chem 14*:548, 1968
376. WASHKO ME: *Federation Proc 19*:81x, 1960
377. WATSON D: *Anal Biochem 3*:131, 1962
378. WEATHERBURN MW, LOGAN JE: *Diabetes 15*:127, 1966

379. WEIJERS H, van de KAMER J: *Lancet 2*:296, 1960
380. WENK RE, CRENO RJ, LOOCK V, HENRY JB: *Clin Chem 15*:1162, 1969
381. WHICHELOW MJ, WIGGLESWORTH A, COX BD, BUTTERFIELD WJH, ABRAMS ME: *Diabetes 16*:219, 1967
382. WHITE AA, HESS WC: *Arch Biochem Biophys 64*:57, 1956
383. WHITE RP, SAMSON FE: *J Lab Clin Med 43*:475, 1954
384. WILKERSON HLC, HEFTMANN E: *J Lab Clin Med 33*:236, 1948
385. WILKERSON HLC, HYMAN H, KAUFMAN M, McCUISTION AC, FRANCES JO: *New Engl J Med 262*:1047, 1960
386. WINCKERS PLM, JACOBS PH: *Clin Chim Acta 34*:401, 1971
387. WINEGRAD A, BURDEN C: *New Engl J Med 274*:298, 1966
388. WOLF L, NORMAN J: *Pediat 50*:271, 1957
389. *World Health Organ, Tech Rep Ser 310*:6, 1965
390. WRIGHT HK, CANN DS: *J Lab Clin Med 67*:689, 1966
391. WRIGHT WT: *New Engl J Med 254*:570, 1956
392. WRIGHT WT: *New Engl J Med 265*:1154, 1961
393. WUILLEY E: *J Med Lab Technol 13*:158, 1955
394. YOUNG DS, JACKSON A: *Clin Chem 16*:954, 1970
395. YOUNG MK Jr, PRUDEN JF: *J Lab Clin Med 44*:160, 1954
396. YOUNG MK Jr, RAISZ LG: *Proc Soc Exp Biol Med 80*:771, 1952
397. ZENDER R: *Clin Chim Acta 8*:351, 1963

Chapter 26

Carbohydrate Derivatives and Metabolites

PATRICIA A. DREWES, Ph.D.

LACTIC ACID

The classic method for determination of lactic acid involves its separation and conversion to insoluble zinc lactate by boiling with ZnO or $ZnCO_3$. The zinc lactate is then weighed. This method was applied to blood by Wolf (249) in 1914. It is time-consuming and requires large samples, and therefore has been little used in clinical chemical analysis.

Most analytic methods have been based on determination of one of the reaction products in either of the following reactions:

$$CH_3CHOHCOOH \xrightarrow[+\,\Delta]{\overset{\text{conc}}{\underset{}{H_2SO_4}}} CH_3CHO + CO + H_2O \qquad (1)$$

$$2\,CH_3CHOHCOOH + O_2 \rightarrow 2\,CH_3CHO + 2\,CO_2 + 2\,H_2O \qquad (2)$$

Technics employed for estimation of the acetaldehyde have included the following: (a) *Iodometric titration.* In the earliest method, that of Boas (23) in 1893, the aldehyde was distilled into alkaline iodine solution and the reduced iodine was determined. Von Fürth and Charnass (70) found this direct titration inaccurate; they distilled into bisulfite and used the Ripper iodometric titration to determine the amount of aldehyde bound by bisulfite. The bisulfite complexed with aldehyde does not react with iodine, so that the aldehyde is equivalent to the difference in titrations before and after the distillation. Clausen (37) distilled the aldehyde into acid bisulfite solution, titrated the free bisulfite with iodine, then split the aldehyde-bisulfite addition compound by alkaline pH and continued titration of the released bisulfite directly. The Clausen titration was modified by others with a view to increasing specificity, sensitivity, and yield (51, 62, 63, 67, 80, 137). For instance, $MnSO_4$ was found to increase the rate and yield of acetaldehyde (62); tungstic acid or $CuSO_4$-$Ca(OH)_2$ precipitation removed interfering materials (67). The method can be adapted to urine as well as blood (63). However, the variable binding of aldehyde to bisulfite in the presence of other bisulfite-binding compounds remains (172). (b) *Photometric determination.* A red color is formed by reaction of acetaldehyde with rosaniline · HCl bleached with SO_2 (194), guaiacol (89), or 1,2-dimethoxybenzene (157). When tyrosine is present during reaction 1, an orange-red color appears (43). Barker and Summerson (14) applied the p-hydroxydiphenyl (Eegriwe) color reaction (161) to a protein-free filtrate after removal of interferences by the $CuSO_4$-$Ca(OH)_2$ treatment of Van Slyke, thus eliminating distillation. The method is more sensitive than the foregoing methods. In addition to precipitating proteins, the $CuSO_4$-$Ca(OH)_2$ treatment also removes carbohydrates, tartaric, citric and glyceric acids, and many saccharinic or polyatomic alcohols (67). The technic was subsequently improved by others (25, 102, 182). According to one study ascorbic, glucuronic, and oxalacetic acids are completely removed and α-ketoglutaric, pyruvic, and malic acids, and choline partially removed (25). Acetaldehyde can also be trapped with semicarbazide in a Conway diffusion vessel and the semicarbazone read at 261.5 nm (217) or 224 nm (193).

The CO formed in reaction 1 has been measured by means of a nitrometer (202) and titrimetrically by the iodine pentoxide method (188), but values obtained for both urine and blood were spuriously high by this approach (152, 188). The CO_2 formed in reaction 2 has been measured in the Van Slyke manometric apparatus (12).

Lactate has been determined in tissue extracts by fluorescent reaction with α-naphthol in the presence of H_2SO_4 (247). A simple sensitive method for blood employs gas chromatography after oxidation of lactate to acetaldehyde with periodic acid (97). Savory and Kaplan (200), however, found contamination problems with periodic acid and trichloroacetic acid, as well as with conversion of sugars to formaldehyde at the 100°C reaction temperature. They oxidized the lactate in a protein-free filtrate with Long's $Ce(SO_4)_2$ procedure, then injected the vapor as in Natelson and Stellate's blood alcohol method. A micro modification has also been described (227).

Lactic acid can be determined directly by the use of lactic acid dehydrogenase (LDH) and nicotinamide-adenine dinucleotide (NAD). The amount of NAD reduced to NADH, as lactate is enzymatically oxidized to pyruvate, is measured by the increase in absorbance at wavelengths of 334–366 nm (92, 98, 100, 141, 171, 177, 240). In these methods the original pyruvate is either destroyed by H_2O_2 (240) or trapped by semicarbazide or hydrazine (100, 141, 171) to drive the reaction to completion. The method of Hohorst (98) has proved to be reliable. The availability of this procedure in commercial kit form has added to its popularity. The specificity inherent in the enzymatic approach as well as its simplicity make it the method of choice. It also adapts to fecal (36) and cerebrospinal fluid (169) determinations. Micro determination (228), fluorometric measurement of the NADH produced (138, 196), and use of 3-acetyl-pyridine-NAD analog (99) are variations of the basic technic. With the NAD-independent yeast LDH enzyme, ferric compounds have been used as electron acceptors, leading to visible color changes at 405–510 nm (130, 192, 203). Alternatively, the enzyme reaction has been coupled to diaphorase in the presence of a diazo salt, 3-*p*-nitrophenyl-2-*p*-indophenyl tetrazolium chloride (INT), to allow measurement in the visible spectrum (68).

Automated methods are based on the 340 nm absorbance measurement of the LDH-NAD system (85), the ferricyanide color reaction with the LDH-NAD enzyme system (144), the Barker-Summerson color reaction (93), the "INT" diazo photometric LDH-NAD method (93), and the fluorometric enzyme technic (7, 40). These would appear to be promising where large numbers of tests are run. The "INT" method cannot be used for urine since ascorbic acid reduces the dye (93); however, the LDH-NAD method has been applied to urine, tissues, milk, and sugar factory juice (85).

DETERMINATION OF LACTIC ACID IN BLOOD, PLASMA, OR SERUM

(method of Hohorst, Ref 98, modified)

PRINCIPLE

Lactic dehydrogenase (LDH) catalyzes the oxidation of lactic acid by NAD:

$$\text{L-lactate} + \text{NAD} \overset{\text{LDH}}{\rightleftharpoons} \text{pyruvate} + \text{NADH} + \text{H}^+$$

The equilibrium of the reaction, which would normally lie far to the left, is shifted by removal of the reaction products so that the reaction proceeds completely to the right as written. Thus hydrazine traps the pyruvate and the hydrogen ion is low because of the high pH. Increase in absorbance due to the formation of NADH is a measure of enzymatically oxidized lactate.

REAGENTS

Perchloric Acid, 7% (w/v). Dilute 6.0 ml of 71% (w/w) $HClO_4$ (sp. gr. 1.68) to 100 ml with water.

Hydrazine-Glycine Buffer, 0.4 M Hydrazine and 0.5 M Glycine. Dissolve 3.76 g glycine and 5.2 g hydrazine sulfate in about 80 ml of water. Adjust pH to 9.0 with NaOH (5 N is convenient) and bring total volume to 100 ml with water. Stable for 1 week in the refrigerator.

Lactic Dehydrogenase (LDH) Solution. Dilute the rabbit muscle enzyme suspension (that from Boehringer, suspended in ammonium sulfate solution, has been found to be satisfactory) with water to a concentration of 2 mg (approx 720 U) per ml. Store in refrigerator. Stable for at least 1 year.

NAD Solution. Dissolve 100 mg NAD (that from Boehringer has been found to be satisfactory) in 5 ml water. Stable for several weeks in the refrigerator.

Lactic Acid Standard, 10 mg/100 ml. Dissolve 21.3 mg lithium lactate in water in a 200 ml volumetric flask. Add 0.05 ml conc H_2SO_4 and bring to volume with water. Stable indefinitely in the refrigerator.

COLLECTION OF BLOOD SPECIMEN

Blood should be obtained without constriction of the vein (see *Note 2*) and immediately deproteinized as in step 1 below. For determination in plasma or serum, blood should be centrifuged as soon as possible in the cold (use refrigerated centrifuge if available; otherwise place centrifuge in refrigerator for this procedure).

PROCEDURE

1. Transfer 1.0 ml 7% perchloric acid to a test tube.
2. Add 1.0 ml blood, plasma, serum, or standard (see *Note 3*) slowly with shaking. Mix well, particularly in the case of blood samples.
3. Let stand for 10 min, then centrifuge.
4. Set up a series of small test tubes for specimens, blank, and standard to contain 2.0 ml hydrazine-glycine buffer, 0.1 ml supernate (from step 3), 3.5% perchloric acid (blank), or standard, 0.03 ml LDH solution, and 0.2 ml NAD solution.
5. Mix and incubate for exactly 1 hour in a 25°C water bath.
6. Transfer incubation mixture to a 2 ml cuvet (1 cm light path) and read absorbance at 340 nm vs blank.

Calculation:

$$\text{mg lactic acid}/100\text{ ml} = \frac{A \times \text{dilution factor} \times 10^3 \times 90 \times 100 \times 2.33}{6.22 \times 10^6 \times \text{ml supernate}}$$

$$= A \times \text{dilution factor} \times 33.7$$

where the dilution factor is 2.0 for serum, plasma, and standard, and 1.85 for blood; 6.22×10^6 is the molar absorptivity of NADH; 90 is the molecular weight of lactic acid; and 2.33 ml is the reaction volume. If the result is greater than 35 mg/100 ml, the specimen should be rerun at an appropriate dilution of supernate with water.

NOTES

1. Skin contamination. Barker and Summerson (14) reported that contamination of the sample with lactic acid from the skin during venipuncture is a major source of error. The skin surface therefore should be well cleansed. The analyst should be aware of this source of contamination and avoid handling any glassware surface which could come in contact with specimen or reagents.

2. Effect of stasis during venipuncture. Venous stasis may cause large positive errors in the determination and therefore must be avoided. If a tourniquet is necessary for venipuncture, the constriction time should be minimized, and after releasing the tourniquet there should be a wait of about 30−60 seconds before drawing the blood.

3. Standardization. Zinc lactate is available but contains a variable amount of water. Lithium lactate is preferable because it has no water of crystallization and is not hygroscopic. Lithium lactate is available commercially but, if desired, can be prepared from USP 85% lactic acid as follows: Dilute with equal volume of water. Add few drops phenol red solution and titrate to slight alkalinity with 20% lithium hydroxide or solid lithium carbonate. Heat to boiling and add lithium hydroxide or carbonate to slight excess. After solution has been cooled, stir in 4 vol ethanol. Filter crystals in Büchner funnel and wash several times with cold 80% ethanol. Recrystallize by dissolving in small volume of boiling water, filter while hot, and reprecipitate when cool by adding cold ethanol. Dry over $CaCl_2$. While the measurement of lactic acid is directly related mol for mol to the amount of NAD reduced to NADH, a lactate standard is useful in monitoring the overall system, including the absorbance scale of the spectrophotometer. An absorbance of about 0.148 is expected in the above procedure with a 1 cm cuvet in a Beckman DU spectrophotometer.

Standardization of the spectrophotometer is necessary when absolute absorbance units for NADH are the basis for standardization of the method. Dry $K_2Cr_2O_7$ (National Bureau of Standards) at $100°-110°C$ for at least 4 hours and cool in a desiccator. Prepare a 40.0 mg/liter solution of $K_2Cr_2O_7$ by dissolving 160 mg in 4 liters of $0.01\ N\ H_2SO_4$. The absorbance at 350 nm (absorption peak) in a 1 cm cuvet, read against a water blank, should be 0.428. If it is not, electrical adjustment of the instrument can be made to obtain a reading of 0.428, or the absorbance of NADH at 340 nm can be corrected by the following equation (151):

$$\text{corrected } A_{340} = \text{observed } A_{340} \times \frac{0.428}{\text{observed } A_{350} \text{ for } K_2Cr_2O_7 \text{ solution}}$$

4. Preparation of protein-free filtrate. Perchloric acid is usually the protein precipitant of choice in this type of method, although trichloroacetic acid may also be used (86). Metaphosphoric acid (7, 145), $ZnSO_4$-NaOH (144) and $ZnSO_4$-$Ba(OH)_2$ (85) have also been used. Because of the instability of lactic acid (see *Note 5*) there is agreement that it is best to deproteinize without delay. It is permissible, however, to store heparinized blood at 4°C for 24 hours before separating plasma (7). Deproteinization has been avoided in some enzyme methods (93, 138, 192).

5. Stability of samples. As soon as blood is withdrawn from the body the lactic acid level begins to increase as a result of breakdown of glucose (glycolysis) (29, 102). For example, at 25°C increases of 19% and 70% have been found by 3 and 30 min, respectively (102). At $-10°C$ glycolysis begins at a slower rate; after about 10 min the lactic acid level begins to decrease, and eventually the level is lower than initially (102). If untreated blood is immediately chilled and spun within 15 min, the plasma is stable at $-20°C$ for 38 days but decreases about 36% at 5°C for the same period of time (169). Samples rapidly hemolyzed after sampling maintain stable lactic acid levels for 4 days at 24°C before falling (51). It has been reported (29) that neither sodium fluoride nor sodium iodoacetate alone is completely effective in stopping glycolysis, but that a mixture of the two does succeed at concentrations of 10 mg of each per ml blood.

Others (200) have noted incomplete cessation of glycolysis under these conditions but have found glycolysis to be completely arrested for 24 hours at 4°C with 10 mg sodium fluoride and 2 mg potassium oxalate per ml blood. Sodium fluoride and sodium iodoacetate, however, may inhibit enzymatic methods for determination of lactic acid (200). Cetyltrimethylammonium bromide, acid citrate buffer, and sodium fluoride can stabilize lactic acid levels in blood (137). Heparin does not appear to affect stability (7, 144). Lactic acid is stable in tungstic filtrates for 3 days refrigerated (188), and Ba-Zn filtrates are stable for several weeks at room temperature (144). Trichloroacetic acid filtrates are stable at least 3 days in the refrigerator (64). It is quite probable that trichloroacetic acid filtrates are stable at room temperature and that the blood-trichloroacetic acid mixture is also stable. Blood can be measured directly from the syringe into the acid and mixed. In our laboratory perchloric filtrates of oxalated blood have been kept for 1 week at 30°C before being analyzed by the procedure described without loss of lactic acid.

ACCURACY AND PRECISION

Rabbit muscle LDH reacts only with the L(+) form but not the D(−) form of lactic acid (98, 140). Interferences from the following list of substances at equal concentrations only amount to 12.5% for DL-glyceric acid; 3% for DL-malic acid, methylglyoxal, and methanol; 2% for α-keto-*n*-valeric acid, DL-glycericaldehyde, and citric acid; 1% for α-ketobutyric acid, formaldehyde, and propyleneglycol; and 0% for α-hydroxyisobutyric acid, acetaldehyde, acrolein, and cystine (140). Recovery of added lactic acid is essentially 100% (228).

Precision of the test (95% limits) is approximately ±6%.

NORMAL VALUES

The normal range for fasting venous blood withdrawn with the patient at rest can be taken tentatively as 5–20 mg/100 ml. This range is a compromise of results obtained by iodometric titration methods (29, 119), the semicarbazide method (193), the *p*-hydroxydiphenyl method (104, 105, 119), the gas chromatography method (227), and the enzyme method (145). The level is similar in heel blood of infants (228). There is general agreement on the lower limit, but the normal upper limit has varied from 10 to 45 mg/100 ml. The arterial blood normal range is reported as 3.1–7 mg/100 ml (103, 104). That severe muscular work causes an increase in blood lactate is well known, but mild exercise such as a brisk walk reportedly has no effect (66). Although short periods of exertion such as the last minute spurt of a 35 min ski race may increase the blood lactic acid to 139 mg/100 ml, a prolonged effort such as a 3 hour race results in only 39 mg/100 ml (10). Bed rest results in a mean decrease in the arterial blood level of about 3 mg/100 ml (103). While lying down, however, contracting the muscles elevates blood lactate twofold, and exercising the legs threefold (104). After breakfast the venous blood level increases to a peak occurring at 1−2 hours, followed by a return to the base line in 3−4 hours. The magnitude of the increase is variable, being 20%−50% of the fasting value (66).

The plasma level is 7% higher than the whole blood level, and the ratio plasma concentration/erythrocyte concentration is 1.37 (102).

The normal 24 hour urine output of lactic acid is 50−200 mg (222). Normal fecal levels are 3−120 mg/100 g (36). Cerebrospinal fluid lactic acid concentrations normally are close to the range observed in blood (182).

α-KETO ACIDS

Carbonyl compounds present in blood and urine have been identified by paper chromatography (4, 21, 34, 158, 170, 226). The following have been detected in normal blood and urine (a few are not always detectable): pyruvic, α-ketoglutaric, α-ketoisocaproic, oxalacetic, α-ketoisovaleric, α-keto-β-methyl-*n*-valeric, and glyoxylic acids, and succinic semialdehyde (49, 234). α-Ketoisocaproic, α-ketoisovaleric, and α-keto-β-methyl-*n*-valeric acids are excreted in large amounts in "maple syrup disease" (see *Amino Acids and their Metabolites, Chapter 18*). The α-keto acids are detected as the 2,4-dinitrophenylhydrazones in a reaction used by Dakin and Dudley (45) in establishing the relationship between amino acids and α-keto acids. In one approach dinitrophenylhydrazones are formed on the paper following chromatography. In another, the hydrazones are formed and then chromatographed. The latter technic has the unavoidable drawback of isomerization of the parent compound, leading to the appearance of two spots (146). The colored spots may be eluted from the paper for quantitation. The concentration of total carbonyl compounds of normal blood other than pyruvic acid has been found to be approximately 1−2 mg/100 ml (49). It is difficult, however, to assess much of the published chromatographic data because results have not always been corrected for losses (112). The methods are far too elaborate for routine use (125). Keto acids in blood, urine, and cerebrospinal fluid have also been successfully separated by paper electrophoresis (254).

The predominant α-keto acid normally is pyruvic acid. Many methods for its determination have been developed and are given in the following section. Oxalacetic acid can be determined in blood by converting it to malic acid in the presence of malic dehydrogenase, and measuring the decrease in absorbance at 340 nm as the coenzyme NADH is concomitantly oxidized to NAD (112). By this method the average oxalacetic acid content of blood is 0.38 mg/100 ml. By a similar NADH-linked enzymatic reaction, α-ketoglutaric acid can be determined. Glutamic dehydrogenase converts α-ketoglutaric acid and ammonium ion in the presence of NADH to glutamic acid and NAD. This simple spectrophotometric method has been applied to blood, urine, plasma, and tissue (17, 24, 147, 208). Blood requires prior precipitation of protein by means of perchloric acid (147); this step is omitted for protein-free urine. The mean normal value for blood is 0.12 mg/100 ml (24, 147). The 24 hour urinary excretion has been reported as 7–29 mg (147) and 10–50 mg (24).

PYRUVIC ACID

Pyruvic acid of tissues may be determined by forming a compound with bisulfite, removing the excess bisulfite with iodine, adding bicarbonate, and titrating the liberated bisulfite with iodine (38). This is the Clausen titration (see above under *Lactic Acid, Iodometric Titration*) and determines total aldehydes and ketones. Most aldehydes and other ketones may be destroyed by heat in alkaline solution prior to the reaction, pyruvate being stable under these conditions (38). It has been claimed that this method yields low recoveries (135), and certainly other technics appear to be more specific and suitable for blood and urine.

Another approach applied to blood and urine which has not found much favor involves zinc reduction of pyruvic acid to lactic acid, followed by determination of the lactic acid (113, 124).

In 1936 Straub (220) reacted pyruvic acid with salicylaldehyde in alkaline solution and measured the orange color formed. Although modified by others (19), the method has not been widely used. It cannot be highly specific (cf *Reaction with Salicylaldehyde* under *Ketone Bodies, Quantitative Methods,* later in this chapter).

Since its introduction in 1931 by Simon and Neuberg (213), the most widely used approach to the determination of pyruvic acid in blood and urine has been based on the formation of the hydrazone with 2,4-dinitrophenylhydrazine. Many modifications have appeared (31, 65, 77, 102, 118, 121, 125, 135, 139). The general procedure involves formation of the hydrazone in acid, extraction into an organic solvent such as ethyl acetate, extraction into a solution of Na_2CO_3, addition of NaOH, and photometric measurement of the resultant brown color. The color reaction is by no means specific for pyruvic acid, being a general one for keto groups. Other keto acids present which give color include acetoacetic, levulinic, α-ketoglutaric, and oxalacetic (26, 79, 121, 139). Friedemann and Haugen (65) called the color produced directly without extraction the "total hydrazones," and found the difference between the total and those extractable by xylene, toluene, or benzene to be 0.12–0.69 mg/100 ml blood. Xylene has been used by others (84, 102), and it has been believed that this solvent is more specific than ethyl acetate for the hydrazone of pyruvic acid. In one study (125), however, it was shown that the hydrazone of α-ketoglutaric acid also was extracted and that pyruvic acid accounted for only approximately 70% of the final color. Interference by acetoacetic acid can be avoided by permitting it to degrade prior to analysis (26, 118). Spectrophotometric measurement of the yellow color of the hydrazone in $NaCO_3$ solution increases sensitivity threefold (26), and reading the color at multiple wavelengths increases specificity, since the absorption curves of the hydrazones of various keto acids differ somewhat (26).

The chromatographic approach to determination of pyruvate was discussed earlier (see *Keto Acids*). By this method pyruvate normally accounts for 42%–75% ($\bar{x} = 60$) of the total keto acids of blood (52). It is difficult, however, to assess quantitatively much of the chromatographic data published because results have not always been corrected for losses (125).

Methods far easier to perform and greatly superior by reason of their specificity are those based on the ultraviolet spectrophotometric determination of NADH, which is oxidized to NAD as pyruvate is reduced to lactate in the presence of lactic dehydrogenase (LDH). Segal *et al* (207) and Henley *et al* (91) originally performed the enzymatic reaction on perchloric acid filtrates of blood and measured the change in absorbance at 340 nm. Modifications of the method have included use of a NADH blank for better accuracy (128), substitution of trichloroacetic for perchloric acid for purported better recovery (75, 86) in spite of other findings of 100% recovery with perchloric acid (189, 228), and introduction of 5% metaphosphoric acid as precipitant to counteract nonenzymatic losses of NADH (leading to positive errors in pyruvate results) observed with trichloroacetic acid (145). A micro modification of the enzyme technic has been published (228). Marks (147) combined consecutively the enzymatic methods for pyruvate and α-ketoglutarate. Urine specimens analyzed by the spectrophotometric method require no precipitation if protein-free. Hadjivassiliou and Rieder (86), using trichloroacetic filtrates (96%–101% recovery), have simplified the method by establishing conditions under which the running of standards and NADH blanks is unnecessary.

Semiautomated procedures for fluorometric determination of pyruvate in perchloric filtrates by the LDH-NADH system have been introduced (7, 40).

DETERMINATION OF PYRUVIC ACID IN BLOOD AND URINE

(method of Segal et al, Ref 207, modified)

PRINCIPLE

Catalytic reduction of pyruvic acid in a perchloric acid filtrate by lactic acid dehydrogenase (LDH) results in oxidation of an equivalent amount of NADH. The amount of NADH oxidized is determined from the decrease in absorbance at 340 nm (ΔA_{340}).

REAGENTS

Perchloric Acid, 7% (w/v). Dilute 6 ml 71% (w/w) perchloric acid (sp. gr. 1.68) to 100 ml with water.

Tropaeolin 00 Indicator Solution. Dissolve 10 mg Tropaeolin 00 (Eastman Kodak is satisfactory) in 100 ml water.

KOH, 5 N. Dissolve 280 g KOH pellets in water and bring to total volume of 1 liter.

Tris Buffer, 0.1 M. Dissolve 12.1 g tris(hydroxymethyl)aminomethane in about 800 ml water. Adjust pH to 7.4±0.1 with conc HCl and bring to total volume of 1 liter with water.

NADH Solution. Dissolve 2.0 mg NADH in 1 ml water. Prepare fresh for use or store in the freezer.

Lactic Dehydrogenase (LDH) Solution. Dilute 0.1 ml of twice crystallized suspension of muscle LDH (that from Worthington Biochemical was found to be satisfactory) to 10 ml with cold (refrigerator temperature) 0.01 M NaCl (0.59 g per liter).

Pyruvic Acid, Stock Standard. Dissolve 125 mg sodium pyruvate (A grade, over 95% pure) in 100 ml 0.1 N HCl (8.3 ml conc HCl per liter of water). Store in refrigerator.

Pyruvic Acid, Working Standard. Dilute the stock standard 1:50 with water. This contains 2 mg pyruvic acid per 100 ml. Pyruvic acid polymerizes in solution after a time (243). We have not checked to see whether the polymer behaves differently in the enzymatic reaction, so it is recommended that the standard be prepared fresh.

SPECIMEN COLLECTION

Blood. Draw slightly more than 5 ml of blood into a syringe (see *Note 2*, below) and immediately add exactly 5.0 ml of the blood from the syringe to 5.0 ml of 7% perchloric acid. Mix vigorously. The sample is stable in this condition for 3 days at room temperature.

Urine. Urine can be preserved by adding H_2SO_4 to a final concentration of 0.05−0.1 N (see *Note 3*, below). Use 0.5 ml of 20 N H_2SO_4 (slowly add 55.5 ml conc H_2SO_4 to 44.5 ml water with stirring, while keeping container cold) for a single 100 ml specimen, 5 ml for an entire 24 hour collection.

PROCEDURE

1. Centrifuge blood-perchloric acid mixture and filter supernate through Whatman 1 paper. For urine specimens, add 5.0 ml of 7% perchloric acid to 5.0 ml urine, mix, and filter.

2. Transfer 2 ml filtrate to a test tube. Cool in an ice bath. Add 1 drop indicator. Using a fine-tipped dropper, add 5 N KOH until a pH of 3−4 is reached (about 0.2 ml required for blood filtrate). A slight change in color from pink to yellow signals the approaching desired pH, which is checked by narrow-range pH paper. It is very easy to overshoot the endpoint; consequently, additions of KOH should be made in minute increments once the color change occurs.

3. Allow to stand in ice bath for 5 min, then centrifuge to remove potassium perchlorate precipitate.

4. To 1.0 ml of supernate from step 3, add 2.0 ml Tris buffer and 0.10 ml of

NADH solution. Mix. Set up blank by substituting water for 1.0 ml supernate.

5. Read absorbance (*A*) of unknowns and blank versus water at 340 nm in a spectrophotometer with half-bandwidth no greater than 5 nm.

6. Add 0.10 ml LDH solution to each tube. Mix.

7. Allow to stand at room temperature for 5–7 min and read again at 340 nm versus water. A_{340} (step 5) − A_{340} (step 7) = ΔA. If ΔA is greater than 0.20, repeat test from step 4 using 1.0 ml of an appropriate dilution of supernate from step 3. ΔA for blank is usually only 0.001–0.002.

Calculation:

$$\text{mg pyruvic acid}/100 \text{ ml} = \frac{\Delta A_x - \Delta A_b}{6.22 \times 10^6} \times 88 \times \frac{10^3 \times 2.22 \times 3.2 \times 100}{\text{ml supernate used in step 4}}$$

$$= \Delta A_x - \Delta A_b \times \frac{10}{\text{ml supernate used in step 4}}$$

where 6.22×10^6 reflects the molar absorptivity of NADH, 88 is the molecular weight of pyruvic acid, 2.22 is a dilution factor, and 3.2 ml is the reaction volume.

NOTES

1. Standardization. Since calculations are made from the absorbance change in NADH in the reaction, it is not necessary to run standards each time unknowns are analyzed. Nevertheless, at least when setting up the test originally, it is recommended that standards be run to make certain the test is working (it must also be remembered that the absorbance scale of a spectrophotometer may have up to a 10% error). Add 1.0 ml 7% perchloric acid to 1.0 ml working standard and proceed, beginning at step 2. Theoretically $\Delta A_x - \Delta A_b$ is 0.200, but must be corrected for the purity of the sodium pyruvate used. For direct standardization of spectrophotometer and correction for absorbance scale errors, see *Note 3* under *Determination of Lactic Acid.*

2. Effect of stasis during venipuncture. There is no alteration in blood pyruvate level if the tourniquet is left on the arm for a period of 2 min, even if the fist is repeatedly clenched during venipuncture (64, 77, 207).

3. Stability of samples.

Blood. Pyruvate is extremely unstable in blood, a significant decrease being observable as early as 1 min after withdrawing the blood (31) and reaching 15%–50% loss within 30 min (31, 75, 102). After the initial decrease the pyruvate concentration in whole, heparinized, or defibrinated blood subsequently increases to levels above the original value (30, 91). There is a difference in opinion as to whether this increase takes place in oxalated blood (30, 136). The changes occurring in pyruvate concentration are believed to be the net of the following two sets of reactions (136):

$$\text{pyruvate} + \text{triose phosphate} \rightleftharpoons \text{lactate} + \text{3-phosphoglycerate} \tag{1}$$

$$\text{3-phosphoglycerate} \rightleftharpoons \text{2-phosphoglycerate} \rightleftharpoons \text{phosphopyruvate} \rightarrow \text{pyruvate} \tag{2}$$

The result of reaction 1 proceeding to the right is loss of pyruvate; in reaction 2, pyruvate is formed. It is probable that some of the strange effects of preservatives on blood pyruvate can be explained by their activator or inhibitor effect on one or the other of these reactions. Oxalate inhibits reaction 2 (136), thus causing loss of pyruvate (64). Cyanide causes an increased loss of pyruvate by activation of cocarboxylase (31). Fluoride also causes increased loss (31, 64, 103), approximately 90% being removed as lactic acid (30). Iodoacetate interferes positively in some way with methods employing the *p*-hydroxydiphenyl color reaction (31, 64, 77, 103, 136). Citrate tends to stabilize blood pyruvate, there being only about a 10% decrease in 30 min (136). The apparent stabilizing effect of fluoride plus iodoacetate added to oxalated blood is likely the result of the balance between many reactions of gain and loss of pyruvate (64). Hemolyzed blood buffered to pH 4–5 and containing fluoride is stable for 3 weeks (136). Pyruvate in trichloroacetic or perchloric acid filtrates of blood (or in the mixture without separation) is stable for periods between 2 days and 1 month at refrigerator temperature (65, 207). It has been reported (65) that the level may increase by 0.05–0.20 mg/100 ml in such filtrates held at 35°–37°C, but no change has been observed in our laboratory within 3 days at 30°C employing perchloric acid as in the method presented. After 35 days at room temperature a 27% loss has been noted (147). When metaphosphoric acid is used for protein precipitation, samples are stable for 42 days frozen, 8 days refrigerated, and 6 days at room temperature; the slight turbidity which sometimes occurs in the last situation is caused by incompletely precipitated protein (145). However, see section on metaphosphoric acid in *Chapter 15, Preparation of Protein-free Filtrate.*

Urine. Urine can be preserved for 24 hours in the refrigerator by adding H_2SO_4 to a final concentration of 0.05–0.1 N (65).

ACCURACY AND PRECISION

Accuracy should be limited only by the specificity of the LDH reaction. α-Ketoglutaric, phenylpyruvic, oxalacetic, acetoacetic, and β-hydroxybutyric acids and acetone do not react (207). Gloster and Harris (75) have claimed that only 76% of added pyruvate is recovered in perchloric acid filtrates of whole blood, whereas 97% is recovered in trichloroacetic acid filtrates. They believed that this was due to incomplete inactivation of lactic dehydrogenase in the erythrocytes by perchloric acid. Recoveries of 85%–97% have been obtained in our laboratory and, as stated previously, the filtrates are quite stable, indicating complete inactivation of any lactic dehydrogenase present. In a more accurate calculation (86), the change in volume before and after addition of LDH is taken into account, but this can be ignored for routine work.

The precision (95% limits) of the test at levels of 2 and 1 mg/100 ml is about ±5% and ±10%, respectively, and the precision decreases further as levels approach zero.

NORMAL VALUES

Based on values obtained by the enzymatic method, it is suggested that the normal limits be as follows: 0.3–0.9 mg/100 ml for fasting whole blood (5, 128, 147, 207) and less than 9 mg/24 hours for urine (5). Older chromatographic and hydrazone methods tend to give higher values (75). Apparently there is no sex difference (77). Based on chromatographic results which agree with enzymatic results for blood, normal cerebrospinal fluid would be expected to contain 0.5–1.7 mg/100 ml (156).

Levels in the newborn during the first few days are somewhat higher than those of adults (78). Adult levels apply to children (250). Values increase 20%–50% after eating, reaching a peak at 1–2 hours and returning to fasting base line at 3–4 hours (61, 147). Values increase following exercise (65, 77), returning to the base line by 2 hours (77). Blood levels are lower after bed rest (103). Usual activities of office or laboratory work cause no increase (66). The within-day variation is about 30% (207). Urine excretion of pyruvate is lower at night than during the day (156) and is higher during ambulation than when bedridden (254).

The ratio plasma pyruvate/erythrocyte pyruvate is 2.3, so that plasma pyruvate is about 21% greater than whole blood pyruvate (102).

CITRIC ACID

Nearly all methods for determination of citric acid involve its oxidation in the presence of bromine to pentabromoacetone, followed by some means of determination of the pentabromoacetone. Potassium permanganate, manganese dioxide, and vanadic acid have been used as oxidizing agents. Cahours in 1847 presented the first description of pentabromoacetone (32), and Stahre in 1895 employed the reaction as a qualitative test for citric acid (216). Kunz (126) in 1914 was the first to use it in a quantitative way, filtering the pentabromoacetone precipitate and weighing it. He determined the citric acid content of milk, marmalade, and fruit syrups. The reaction is believed to proceed as follows (76):

$$
\begin{array}{c}
\text{COOH} \\
|\\
\text{CH}_2\text{ COOH} \\
|\ \diagup \\
\text{C} \\
|\ \diagdown \\
\text{CH}_2\quad\text{OH} \\
|\\
\text{COOH}
\end{array}
\xrightarrow[\text{+ DECARBOXYLATION}]{\text{OXIDATION}}
\begin{array}{c}
\text{COOH} \\
|\\
\text{CH}_2 \\
|\\
\text{C}=\text{O} \\
|\\
\text{CH}_2 \\
|\\
\text{COOH}
\end{array}
\xrightarrow[\text{+ BROMINATION}]{\text{DECARBOXYLATION}}
\begin{array}{c}
\text{CHBr}_2 \\
|\\
\text{C}=\text{O} \\
|\\
\text{CBr}_3
\end{array}
$$

CITRIC ACID   ACETONE –   PENTABROMOACETONE
DICARBOXYLIC ACID

Tetra- and hexabromoacetone are also produced, the relative amounts of each bromacetone being governed by acidity, citrate concentration, and temperature (76).

The following approaches have been employed for determining the pentabromoacetone: (a) *Titration of Br⁻*. Pentabromoacetone is decomposed, and the Br^- liberated is titrated with $AgNO_3$ (76, 184). (b) *Photometric measurement of yellow color formed in solution with $Na_2 S$.* This method was introduced by Pucher and coworkers (183) and was subsequently modified by others (28, 109, 122, 242). (c) *Photometric measurement of color formed with thiourea in alkaline solution* (20, 57, 167). The color is more stable, of somewhat greater intensity, and of approximately the same specificity as the color with $Na_2 S$. (d) *Photometric measurement of color formed with $Na_2 S$ + thiourea* (155). (e) *Photometric measurement of yellow iodine complex.* In this method of Taussky and Schorr (224, 225) and Cartier and Pin (33) the reaction of Kometiani is employed. In the reaction between pentabromoacetone and NaI in acid solution the bromine is substituted by iodine with concomitant liberation of a yellow iodine complex. The color is more stable and intense than that formed with $Na_2 S$ or thiourea. (f) *Photometric determination of bromide by molybdate and phosphate* (248). Sulfite and phosphate are added, followed by heat, hypochlorite, formate, molybdate, and KI. A violet color is formed. (g) *Photometric measurement of color formed with alkaline pyridine* (53, 195). The color is pink, but in the presence of excess acetic anhydride it is yellow (195). Sensitivity is about the same as that in the Kometiani reaction. (h) *Radioisotopic measurement of pentabromoacetone* (106). ^{82}Br is incorporated into the brominating agent and the labeled pentabromoacetone is counted directly.

Salant and Wise (196) oxidized citric acid to acetone with permanganate and measured the Denigès mercury precipitate gravimetrically. Krog (123) converted citric acid to acetone in a similar manner and determined the acetone after distillation by photometric measurement of the color formed with salicylaldehyde. Values obtained by the latter method were about 40% lower than those yielded by a pentabromoacetone procedure, although complete recoveries of added citrate were obtained.

Thunberg in 1929 introduced an enzymatic method in which citric acid dehydrogenase catalyzes the dehydrogenation (oxidation) of citric acid in the presence of methylene blue as hydrogen acceptor. The time required for decolorization (reduction) of methylene blue in anerobic conditions is measured (231). Sjöström (214) modified the method, and Grönvall (83) used indigo trisulfonate as the acceptor dye. Pucher *et al* (183) stated that the enzymatic method has many difficulties and is far from satisfactory.

An equilibrium exists between citric acid, *cis*-aconitic acid, and isocitric acid as follows:

$$
\begin{array}{ccc}
\text{COOH} & \text{COOH} & \text{COOH} \\
| & | & | \\
\text{CH}_2 & \text{CH} & \text{CHOH} \\
| & \| & | \\
\text{HOC}-\text{COOH} \underset{+\text{H}_2\text{O}}{\overset{-\text{H}_2\text{O}}{\rightleftharpoons}} & \text{C}-\text{COOH} \underset{-\text{H}_2\text{O}}{\overset{+\text{H}_2\text{O}}{\rightleftharpoons}} & \text{HC}-\text{COOH} \\
| & | & | \\
\text{CH}_2 & \text{CH}_2 & \text{CH}_2 \\
| & | & | \\
\text{COOH} & \text{COOH} & \text{COOH}
\end{array}
$$

CITRIC ACID	CIS–ACONITIC	ISOCITRIC ACID
80 %	ACID	16 %
	4 %	

The enzyme aconitase catalyzes these equilibrium reactions. Martensson (150) states that the citric acid dehydrogenase method measures preformed *cis*-aconitic and isocitric acids as well as citric acid, whereas pentabromoacetone methods are specific for citric acid; for this reason, lower results are obtained by pentabromoacetone methods. An enzymatic method with absolute specificity for citrate was introduced by Moellering and Gruber (162). Citrate in a protein-free filtrate is cleaved by citrate lyase to acetate and oxaloacetate, and the oxaloacetate is converted to malate by malic dehydrogenase in the presence of NADH. Decrease in absorbance of NADH at 340 nm is a measure of the citrate. Trichloroacetic and perchloric acids were found to be the best for removing protein without loss of citrate by adsorption onto the precipitate (108). Some investigators have omitted prior deproteinization of plasma entirely (255).

Automated methods for citrate are based on the formation of a pentabromoacetone-thiourea-borax complex (6) or conversion to citraconic anhydride and condensation with pyridine to form a fluorescent product in alkaline solution (175).

Choosing a method presents a dilemma. Our laboratory has had experience with the ethanolic NaI method of Taussky and Schorr (225), which is very sensitive but the most complicated; with the alkaline pyridine method of Ettinger *et al* (53), which is equally sensitive and much simpler but too odiferous for many workers; and with the thiourea-borax method of Beutler and Yeh (20), which is relatively simple to perform but is too insensitive to obtain good results on normal serum. The method of Taussky and Schorr therefore won by default.

DETERMINATION OF CITRIC ACID

(*method of Taussky and Schorr, Refs 224, 225, modified*)

PRINCIPLE

The ketone bodies in a protein-free filtrate are brominated by treatment with a strong acid solution of bromide and bromate. Excess free bromine is reduced by $FeSO_4$. Hydrazine was used in the original method (225), but many workers get skin reactions with this substance. The brominated ketone bodies are removed by extraction with heptane (only low levels of ketone bodies are successfully removed by this preliminary treatment). The citric acid in the aqueous phase is then oxidized to pentabromoacetone by a mixture of manganese sulfate, bromide-bromate, and permanganate. The oxidation reaction is autocatalytic and it is probable that MnO_2 is the ultimate oxidant. Excess Br_2 and MnO_2 are reduced with $FeSO_4$. Pentabromoacetone is extracted into heptane, which is then washed with water. Alcoholic NaI is added to the extract. A yellow iodine complex is formed as a result of substitution of iodine for bromine in the penta-bromoacetone, and this colored complex appears in the alcohol phase. This substitution reaction is believed to take place as follows:

Thus one molecule of citric acid consumes 6 I^- in this reaction with the formation of 3 I_2. Figure 26-1 shows the absorption curve of the color formed. It is identical to that of free iodine in an ethanolic solution of NaI.

REAGENTS

Trichloroacetic Acid, 10%

Trichloroacetic Acid, 9.3%

H_2SO_4, *27 N.* Add 750 ml conc acid slowly to 250 ml water; after cooling, dilute to 1 liter.

Bromide-Bromate Solution, 2 N. Dissolve 42.88 g NaBr and 12.58 g $NaBrO_3$ and dilute to 250 ml with water.

Ferrous Sulfate, Aqueous Saturated. About 40 g $FeSO_4 \cdot 7H_2O$ is diluted to 100 ml with water and 1 ml 2 N H_2SO_4 added. Shake for about 5 min to ensure saturation.

$KMnO_4$, 5%. Dissolve 50 g and dilute with water to 1 liter. Boil for 10 min, let cool to room temperature, filter through glass wool, and adjust back to volume.

FIG. 26-1. Absorption curves in citric acid determination: (1) reagent blank vs water, (2) standard equivalent to 0.5 mg/100 ml (6 µg at step 2) vs water, (3) standard vs blank. Perkin-Elmer model 4000-A recording spectrophotometer.

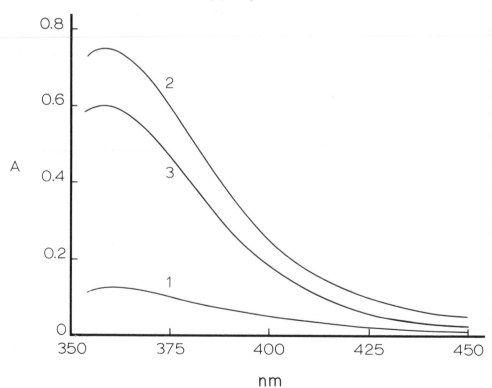

MnSO₄·H₂O, 40%. Requires heating to 50°C to effect solution. After cooling to room temperature, filter.

Normal Heptane, b.p. 98°C. Purification is sometimes required. Hexane (British Drug House) has been found in our laboratory to be satisfactory without purification.

Ethanolic NaI Solution, 10%. Make up in 95% ethanol, using an amber colored volumetric flask to minimize liberation of iodine by light. Ethanol denatured with methanol is satisfactory. Prepare fresh before use.

Bromcresol Green Indicator Solution, 0.04%. Dissolve 400 mg in 6 ml 0.1 *N* NaOH and dilute to 1 liter with water.

Citric Acid Standards. Stock. Dissolve 1.69 g potassium salt ($K_3 C_6 H_5 O_7 \cdot H_2 O$) and dilute to 1 liter with 1 *N* $H_2 SO_4$ (equivalent to 1 mg citric acid/ml). Stable for 1 year. *Working.* 0.96 ml stock standard + 9.04 ml water. 1 ml = 0.096 mg. Prepare fresh.

PROCEDURE

1. For serum or plasma (see *Note 4*), add 1.5 ml to 11 ml 10% trichloroacetic acid. Mix, let stand about 10 min, and centrifuge. Filter through Whatman 42 filter paper. For urine, add 1.0 ml 1:5 aqueous dilution of urine (negative for glucose, see *Note 7*) to 19 ml 9.3% trichloroacetic acid; if a protein precipitate occurs, filter.

2. Set up the following in a fume hood in 30 ml glass stoppered Pyrex tubes:
   Blank. 9.5 ml 9.3% trichloroacetic acid + 0.5 ml water
   Standard. 9.5 ml 9.3% trichloroacetic acid + 0.5 ml working standard (0.048 mg)
   Unknown. 10 ml of filtrate from step 1
   Add to each tube 5 ml 27 *N* $H_2 SO_4$ and 1 ml bromide-bromate solution.

3. Let stand at room temperature for about 30 min. Remove Br_2 fumes from above liquid surface by suction (water aspirator pump). Add 2.0 ml ferrous sulfate solution and shake.

4. Add 6 ml heptane, stopper, and shake for 3 min (shaking machine is convenient). After separation of phases (centrifugation may be necessary), transfer 15 ml lower aqueous phase to 60 ml glass-stoppered Pyrex bottle or tube and cool in water bath at about 18°C (running tap water may be at this temperature).

5. Add 1.0 ml $MnSO_4$ solution and 1.0 ml bromide-bromate solution. Then slowly add 2.5 ml $KMnO_4$ solution while mixing. Let stand for about 30 min in 18°C bath. Aspirate Br_2 fumes as before.

6. Reduce excess Br_2 and MnO_2 by addition of 6.0 ml ferrous sulfate solution. Shake and let stand for about 3 min. Precautions must be taken to ensure complete reduction of free Br_2, which may adhere to sides of stopper. Br_2 fumes are removed again by aspiration. Solutions

are orange-yellow at this stage due to Fe^{3+}. This color does not enter heptane phase at next step.

7. Add 11.0 ml heptane, stopper, and shake 6 min. Transfer contents to 60 ml separatory funnel (transfer need not be quantitative). After separation of phases, discard lower aqueous phase.

8. Wash heptane extract and neck and stopper twice with jet of water from wash bottle. Wash water should contain a few drops of bromcresol green indicator. Then add about 10 ml of wash water, stopper the funnel, and shake for about 20 sec. Discard wash water phase. This washing procedure is repeated until there is no change in color of the wash water. Two or three washes usually suffice.

9. Let mixture sit for about 30 min to ensure complete drainage of water to bottom of funnel. Discard lower aqueous phase. The procedure can be interrupted at this stage, if desired, since the heptane extract is stable in the refrigerator for at least 2 weeks (225).

10. Transfer 10.0 ml heptane extract into small amber colored, glass-stoppered flask or bottle. Add 3.0 ml ethanolic NaI reagent, stopper, and shake. Let sit in dark for about 70 min.

11. Read absorbances of alcohol phases of blank, standard, and unknown versus water at 420 nm or with filter with nominal wavelength in this region (see *Note 1*). The volume of the alcohol phase is 2.8 ml.

Calculation:

$$\text{mg citric acid/100 ml serum or plasma} = \frac{A_x - A_b}{A_s - A_b} \times \frac{0.048 \times 100 \times 12.5}{1.5 \times 10}$$

$$= \frac{A_x - A_b}{A_s - A_b} \times 4$$

$$\text{mg citric acid/ml urine} = \frac{A_x - A_b}{A_s - A_b} \times \frac{0.048 \times 20}{0.2 \times 10}$$

$$= \frac{A_x - A_b}{A_s - A_b} \times 0.48$$

$$\text{mg citric acid/24 hr} = \text{mg/ml} \times 24 \text{ hr volume in ml}$$

NOTES

1. Choice of wavelength and Beer's law. Taussky and Schorr (225) used a Klett 42 filter, although as seen in Figure 26-1 sensitivity is about sixfold greater when colors are read at 360 nm, the wavelength of peak absorption.

Our laboratory, however, has not compared results obtained on unknowns at the two wavelengths. Taussky and Schorr also reported that the color obeyed Beer's law up to about 0.06 mg citric acid in the final colored solution on a Beckman DU or with a Klett 42 filter. They stated, however, that not all Klett 42 filters gave uniform performance with respect to Beer's law. In our experience, Beer's law is obeyed up to only about 0.036 mg in the final colored solution (equivalent to 4 mg/100 ml) even with a Beckman DU at 420 nm and up to one-half this concentration with the Klett 42 filter we used. Obviously, a calibration curve must be run by the analyst when first setting up the method.

The volume of the final alcohol phase is 2.8 ml. For a Klett photometer, either volumes must be doubled or the instrument itself must be altered to permit reading as little as 2.8 ml (225).

2. Color stability. Approximately 90% of full color develops immediately on mixing the heptane extract with the ethanolic NaI. Full color is reached in about 1 hour. Thereafter there is very slow uniform increase throughout a set of determinations due to a slow oxidation of NaI to I_2.

3. Reagent blank. Taussky and Schorr (225) avoided running a reagent blank by running two standards of different concentrations and calculating the "apparent blank" as in the following example: Standards equivalent to 2 and 4 mg/100 ml gave absorbances of 0.395 and 0.760, respectively, when read against water; then:

$$\text{calculated blank} = (2 \times 0.395) - 0.760 = 0.030$$

In our laboratory, however, running a reagent blank in the conventional way is preferred.

4. Anticoagulants. Either serum or oxalated plasma should be used; heparin interferes in the analysis (225).

5. Reproducibility of standard and blanks. Standards and blanks fluctuate somewhat from run to run unexplainably. They must therefore be run with each set of analyses.

6. Elevated levels of ketone bodies. Such levels are not removed in the preliminary bromination and heptane extraction step, and therefore must be removed prior to analysis. The treatment employed removes acetone and acetoacetic acid but not β-hydroxybutyric acid (225).

Procedure for serum or plasma. Add 3.0 ml serum or plasma to 22 ml 10% trichloroacetic acid. Transfer 20 ml protein-free filtrate to a 50 ml Erlenmeyer flask and reduce volume to about 6 ml by boiling on a hot plate. Filter into a 25 ml graduated cylinder or volumetric flask and dilute to 20 ml with water (use part of water to rinse through the filter paper). Mix and use 10.0 ml at step 2 in the analysis.

Procedure for urine. Add 1.0 ml 27 N H_2SO_4 to 25 ml urine and boil on hot plate until volume is reduced to about 10 ml. Cool and dilute to 25 ml with water.

7. Volume of heptane extract used. In Taussky and Schorr's (225) original technic, the aliquots of heptane extract processed for urine unknowns were different from those for standards. They also recommended that if the final color of a urine unknown was too high or too low the final color development step could be repeated taking a smaller or larger aliquot of the extract. In our experience this approach is likely to introduce significant error because changing the volume of extract used may alter the reagent blank absorbance.

8. *Stability of samples.* Whole blood is stable for 12 hours in the refrigerator, and there is slight increase after 24 hours (225). Plasma is stable for at least 16 days frozen but only 8 hours at 20°C (255). Serum rapidly separated from the clot was found in our laboratory to be stable for a few hours at room temperature. Without preservatives, significant bacterial contamination in urine samples may lead to appreciable diminution in levels in a few hours at room temperature (76). Urine preserved by acidification (25 ml 18 N H_2SO_4 for a 24 hour collection) is stable for at least a week in the refrigerator (225). In our laboratory preserved urine was found to be stable at room temperature for 1 week and for several weeks frozen.

ACCURACY AND PRECISION

Taussky and Schorr (224, 225) obtained 92%—108% recovery of citric acid added to urine or plasma. Values obtained by the original photometric method of Taussky and Schorr ranged from 93%—107% of those obtained by the titrimetric procedure of Goldberg and Bernheim—titration of Br^- released from pentabromoacetone with $AgNO_3$ (76). Taussky's modification (224) yielded results averaging about 7%—8% lower, and she concluded that the modification was more specific. β-Hydroxybutyric acid interferes and is not destroyed by the procedure recommended for samples with elevated levels of ketone bodies but is partially eliminated by the bromination step (225). Aromatic compounds such as salicylates are similarly removed at this step (242). Taussky and Schorr found no interference by the following substances when 1 mg amounts were added to 20 μg citric acid: pyruvic, succinic, lactic, hippuric, fumaric, malonic, glutaric, α-ketoglutaric, malic, uric, tartaric, oxalic, and ascorbic acids, urea, glucose, thymol, toluene, creatine, and creatinine. Glucose levels up to 100 g/liter urine did not interfere in the technic as presented. *cis*-Aconitic and isocitric acid do not form pentabromoacetone (53).

The precision of the method (95% limits) is about ±5% to ±8%.

NORMAL VALUES

The adult normal range for serum or plasma obtained by pentabromoace-tone methods is about 1.7—3.0 mg/100 ml (168, 248). Since approximately 76% (range 64%—82%) of the citric acid is in the plasma and but 18%—36% in the erythrocytes, the normal range for whole blood is lower than that for plasma, i.e., about 1.3—2.3 mg/100 ml (248). The ranges for serum obtained by the enzymatic methods are about 1.9—2.9 mg/100 ml with citric dehydrogenase (2, 214) and 0.9—2.5 mg/100 ml with the more specific citric lyase (162, 255) methods. Exercise causes an increase (2). Values in children are higher than adult normals (57, 88, 168), and levels in newborns are higher still (168).

The normal 24 hour urine output is about 0.3–0.9 g (212, 221). In females the excretion is low during menses and high during midcycle, the swing being as much as 225–500 mg/24 hour output (212). Excretion of citric acid is reported to be higher in women than in men (95, 164) but is unrelated to output, age, weight, or body type (164).

OXALIC ACID

In the earliest methods for the determination of oxalic acid in urine, oxalate was directly precipitated as the calcium salt, followed by classic gravimetric procedures for ignition and weighing of CaO (129, 204). Since there was difficulty in obtaining quantitative precipitation of the calcium oxalate, later attempts were made, either to isolate the oxalic acid from urine by extracting the acidified urine with ether before precipitation (44, 197), or in the reverse order–to precipitate, then extract with ether, and finally ignite to CaO (11).

Dakin (44) titrated the final precipitate with potassium permanganate, a shorter procedure which has many modifications. Unlike simple aqueous solutions, urine contains many pigments and salts which either inhibit precipitation of calcium oxalate or are carried down with the precipitate and react with permanganate. To remove these interferences, one method (50) employs esterification and distillation *in vacuo* from acid-alcohol solution, 6 hour extraction, and gasometric instead of titrimetric determination. In another approach (180) the acidified urine is extracted for 6 hours, precipitated from 60% acid-alcohol and titrated; 10% of the oxalate is lost in the extraction, but this negative error is compensated for by the contribution of calcium citrate which precipitates with calcium oxalate and yields a titration value equivalent to 10% of the oxalate. Improved recovery can be obtained with peroxide-free ether and a more efficient extraction apparatus (251). Unfortunately, all of these modifications are time-consuming. The procedure described by Archer *et al* (8) simplified previous methods so that large numbers of samples could be determined simultaneously. Samples were adjusted to pH 5.0–5.2, $CaCl_2$ added, and the oxalate allowed to precipitate for 16 hours; the precipitate was then separated and washed with dilute ammonia, dissolved in 1 N H_2SO_4, and titrated with permanganate at 60°–70°C. Good reproducibility and approximately 92% recovery have made this method widely accepted. Some investigators, however, have reported difficulty in exceeding 40% recovery with this direct method and, after finding separation of oxalate from interfering salts and pigments by Dowex 1X8 too time-consuming, have reverted to the use of ether extraction in a somewhat easier to use extraction apparatus with inner and outer compartments (178). A simpler direct method involves precipitation with

CaCl$_2$ followed by titration of surplus Ca^{2+} in the supernate with murexide and complexon while stirring the solution in the cuvet of a photometer (74). Atomic absorption instead of titration of the calcium has given greater sensitivity to the precipitation method and has made handling of large numbers of samples easier (58).

Europium has also been employed for direct precipitation of oxalate from urine (237). After removal of inorganic phosphate the europium in the washed precipitate was measured by polarography. The oxalate precipitated by calcium has been determined in a photometric reaction with indole, with greater specificity than permanaganate titration and 95%–110% recovery (90). Extraction by tri-n-butyl phosphate instead of ether can shorten the time for quantitative recovery of microgram amounts of oxalate to 5 min at room temperature (253). A further innovation is the reduction of oxalate to glyoxylic acid and reaction with resorcinol (253). In another reexamination of conditions for quantitative precipitation of oxalic acid from urine, the addition of an internal standard and saturation of the washing solution with calcium oxalate was suggested (120). Specificity may be increased by replacing permanganate titration by reduction of oxalate to glycolate and forming a color with 2,7-dihydroxynaphthalene-3,6-disulfonic acid (47). A method for the direct determination of oxalic acid in acidified urine by reaction with 2,7-dioxynaphthalin to form a color has been reported (176).

Probably the most promising of all are the enzymatic methods for determination of oxalic acid. The enzyme employed, first described in a wood rot fungus, can specifically decarboxylate oxalic acid to stoichiometric amounts of formaldehyde and CO$_2$ (210). It was used to prove the Archer method capable of a mean recovery of 95% (153). As a method for oxalate, the precipitation with calcium is carried out for 3 hours at room temperature, then left at 4°C overnight. After centrifugation at 0°C the unwashed precipitate is dissolved in citrate buffer and incubated in the presence of the enzyme suspension and EDTA (to suppress enzyme inhibitors precipitated with the calcium oxalate). The CO$_2$ evolved is measured by the Warburg manometric technic. A similar enzymatic determination can be performed on an ultrafiltrate of urine concentrated under pressure (186). The commercial availability of oxalic decarboxylase should stimulate additional investigation into this type of assay.

The use of ^{14}C-labeled oxalic acid avoids the problems of quantitative extraction and precipitation. Hockaday et al (94) added ^{14}C-oxalic acid to urine, precipitated the calcium salt, reduced the oxalate to glycolate, and isolated the glycolate by Dowex 1X8. Counting the ^{14}C and determining glycolate by the 2,7-dihydroxynaphthalene reaction showed recovery to be 99%–103%. In another technic ^{14}C-oxalic acid was isolated from urine by continuous extraction with ether, and the extracted oxalate was precipitated to constant activity as the calcium salt (46). Good agreement with these values was found (73) when tri-n-butyl phosphate was used for extraction instead of ether.

Many early methods for determination of oxalic acid in blood or plasma were not reliable. In general, a discrepancy exists between results of the early methods and concentrations predicted from isotope dilution studies (96).

The sensitivity of the enzymatic method (42) lies near the upper limit of the normal range. Methods based on the esterification of oxalic acid appear to be more sensitive (96). Methodology for oxalate in cerebrospinal fluid, soft tissues, bone, and feces has been briefly reviewed by Hodgkinson (96).

The method chosen for presentation is a modification of Fraser and Campbell's precipitation method (58). It is similar in principle to the widely accepted Archer method but possesses the advantages of even greater speed and ease of handling large numbers of samples, and in our hands has not been prone to recovery losses.

DETERMINATION OF OXALATE IN URINE

(method of Fraser and Campbell, Ref 58, modified)

PRINCIPLE

Oxalate is precipitated from acidified urine, in the presence of a known amount of added oxalate, by an excess of calcium chloride at pH 4.5; the washed calcium oxalate precipitate is then dissolved in H_2SO_4 and the calcium determined by atomic absorption.

REAGENTS

Sodium Oxalate Solution, 2 mg/ml. Dissolve 0.2 g $Na_2C_2O_4$ in water and bring to a total volume of 100 ml. Store at 4°C and discard if microbial growth is suspected.

Calcium Chloride Solution, 10%. Dissolve 10 g anhydrous $CaCl_2$ in water and bring to a total volume of 100 ml.

Pentachlorophenol Solution, 2 mg/ml. Dissolve 0.2 g pentachlorophenol (Eastman Kodak) in about 75 ml ethanol and bring to a total volume of 100 ml with ethanol. Ethanol denatured with methanol is satisfactory.

Saturated Calcium Oxalate Wash Solution. Add 1 g powdered CaC_2O_4 to 2 liters of water. Add 1 ml pentachlorophenol solution. Mix by stirring on magnetic stirrer for 24 hours, then allow to settle. Filter through asbestos or Whatman 1 filter paper disc in a sintered glass funnel with the aid of suction.

H_2SO_4, 1 N, 0.2 N, 0.1 N, 0.01 N. Carefully add 28 ml conc H_2SO_4 to water with stirring. Bring to a total volume of 1 liter. This 1 N solution is diluted further to 0.2 N, 0.1 N and 0.01 N for convenient concentrations of acid for standards and for adjusting pH.

NH_4OH, approx 6 N. Add 75 ml of 28%–30% NH_4OH to 25 ml water. Further dilution 10- or 100-fold makes convenient solutions for adjusting pH.

Calcium Standards

Primary stock solution, 1 mg/ml. Transfer 249.7 mg $CaCO_3$ (primary standard grade, available from G. Frederick Smith Chemical Company, has been found suitable) to about 50 ml water. Add 6 ml 1 N H_2SO_4. After dissolution is complete, bring total volume to 100 ml with water.

Intermediate stock solutions. Pipet 2.5, 5.0, 10.0, and 15.0 ml, respectively, of primary stock solution into 100 ml volumetric flasks and dilute with water (2.5, 5, 10, and 15 mg Ca/100 ml).

PROCEDURE

1. Clarify about 25 ml of acidified urine by centrifugation (see *Note 1*).
2. Pipet 20 ml of clear urine into a 35 ml glass-stoppered centrifuge tube.
3. Adjust the pH to 4.5±0.1 with 6 N NH_4OH, using a pH meter. Back-titrate with 1 N H_2SO_4 if necessary.
4. Add 0.4 ml $Na_2C_2O_4$ solution and mix.
5. Add 0.4 ml $CaCl_2$ solution and mix. Stopper tubes.
6. Place tubes in a boiling water bath for 20 min, then cool to room temperature in running tap water.
7. Readjust pH to 4.5 with 1 N H_2SO_4 and 6 N NH_4OH, using a pH meter.
8. Centrifuge to pack the precipitate firmly. Remove as much supernate as possible by aspiration with capillary tip pipet connected to water aspirator. Do not disturb precipitate.
9. Add 3 ml saturated CaC_2O_4 wash solution and resuspend precipitate. Wash down walls with another 3 ml of wash solution.
10. Repeat step 8.
11. Repeat the entire washing procedure (steps 9 and 10). The test may be interrupted at this point; stopper tubes and store refrigerated.
12. Add 1.0 ml 1 N H_2SO_4 and disperse the precipitate. Stopper tube and heat in 60°C water bath for 5 min to dissolve the calcium oxalate precipitate.
13. Add 4.0 ml water and mix.
14. Transfer 1.0 ml to a 25 ml volumetric flask and dilute to mark with water.
15. Transfer 1.0 ml aliquots of 0.2 N H_2SO_4 into five 25 ml volumetric flasks. To four of them add 1.0 ml aliquots of the 2.5, 5, 10 and 15 mg Ca/100 ml standards, respectively. The fifth flask serves as the zero standard. Dilute all five with water to volume (final concentrations 0, 0.1, 0.2, 0.4, and 0.6 mg Ca/100 ml).
16. Analyze standards and unknowns in atomic absorption spectrophotometer, first setting null meter to zero with the zero standard (% absorption zero). For treatment of general technic, see *Chapter 2, Flame Photometry and Atomic Absorption.*

Calculation:

On one-cycle, semilog graph paper, plot % absorption of the standards on the arithmetic scale versus final concentrations of standards in mg/100 ml on the logarithmic scale and read unknowns from the graph.

$$\text{mg oxalate, as oxalic acid dihydrate/24 hours} = \left[\left(\text{mg Ca/100 ml}\right.\right.$$

$$\left.\left. \times \frac{5}{20} \times \frac{126}{40} \times \frac{25}{1} \times \frac{1000}{100}\right) - 38\right] \times 24 \text{ hour vol in liters}$$

$$= \left[\left(\text{mg Ca/100 ml}\right.\right.$$

$$\left.\left. \times 197\right) - 38\right] \times 24 \text{ hour vol in liters}$$

where 126 is the molecular weight of oxalic acid dihydrate, 40 is the atomic weight of calcium, 38 is the equivalent in oxalic acid dihydrate of the 40 mg/liter of $Na_2C_2O_4$ added to the sample, and the remaining numbers denote volume conversion factors. Results are expressed as "mg oxalate, as oxalic acid dihydrate" because for some strange reason workers in the field have customarily given results (normals, etc.) in this way.

NOTES

1. *Collection of specimen.* The urine specimen may be acidified throughout 24 hour collection period by the preaddition of 10 ml conc HCl to the collection container. Alternatively, it may be collected first and the conc HCl added later to bring pH to below 3. Heat at $60°C$ for 15 min to dissolve the oxalate, then cool.

2. *Instrument settings.* For the Perkin-Elmer model 303 atomic absorption spectrophotometer the settings are as follows; burner position 0; wavelength 211.8 nm; visible range; slit 4; meter response 1; scale 1; 10 ma hollow cathode light source; acetylene fuel; fuel flow 9; oxidizer air; oxidizer flow 7.5.

3. *Stability of samples.* In our laboratory oxalate in urine has been found to be essentially stable for 7 days at $30°C$. Slow chemical changes are not prevented by storing urine in refrigerator or freezer, or by addition of HCl (96).

4. *Standard diluent.* Lanthanum (or strontium) should not be used in the diluent for standards, as is customary for calcium, since it inhibits the atomic absorption of calcium standards containing oxalate (58).

5. *Wash solution.* Use of a wash solution saturated with calcium oxalate prevents the losses normally encountered in precipitation methods using simple ammonia washes. The presence of pentachlorophenol prevents bacterial decomposition of oxalate, which would render the wash solution no longer saturated and thus lead to losses.

Positive errors due to the use of a wash solution saturated with CaC_2O_4 (solubility: 0.00067 g/100 ml) are negligible. Assuming that 0.2 ml of wash solution could be trapped in the precipitate, this contribution to a specimen of 1400 ml 24 hour volume would increase the output for the 24 hours by less than 0.1 mg.

6. Standard curve. The standard curve should be linear between 5 and 15 mg/100 ml, and samples may be read off this portion. The addition of a known amount of oxalate at step 4 is for the purpose of bringing the specimen up to this linear portion of the standard curve. For elevated samples the final 25 ml solution may be further diluted so as to fall on-scale when it is re-analyzed. Make the dilution in water containing 4 ml 0.2 N H_2SO_4/100 ml.

7. Effect of pH on calcium oxalate precipitation. Experiments have shown that oxalate results are not affected by variations in pH from 4.0 to 5.2 in the calcium precipitation step (120). In our laboratory, however, we have encountered problems with a copious precipitate at pH 4.9–5.2. Since after boiling, the pH shifts up to 0.5 pH points higher, we therefore prefer to operate near the midrange point of 4.5.

ACCURACY AND PRECISION

Fraser and Campbell (58) obtained 80%–115% recoveries in urine samples to which known increments of oxalate were added. In our laboratory recovery studies yielded a range of 90%–115% for samples analyzed by the modified method. Aqueous standards carried through the procedure yield unsatisfactory recovery, e.g., 75%–90% at the 50 mg/liter level (58).

Precision (95% limits) of the procedure is about ±15%.

NORMAL VALUES

The adult normal range for oxalic acid dihydrate in urine is generally reported to be up to 40 mg/24 hours under normal dietary conditions (8, 58, 96, 180). One study extends the lower normal limit to zero (74), while another extends the upper normal limit to 64 mg/24 hours (127). Studies in our laboratory have verified the 10–40 mg/24 hour range for most normals with occasional outliers from 0–65 mg/24 hours. A temporary rise in urinary oxalate may persist for 3 hours following a meal (96). Excretion of oxalate may be higher during the day than during the night (96). Levels are slightly higher in males than in females (96), and while the same author reports children's values to be lower than those of adults, another study of 25 healthy children showed 10–45 mg/liter (178). Ingestion of large doses of ascorbic acid increases urinary oxalate levels. The effect is negligible with daily doses below 4 g for a week; but daily doses of 4, 8, and 9 g for a week have been found to increase urinary oxalate levels by 12, 45, and 68 mg, respectively (127). In view of dietary effects on oxalate and variations in dietary preferences, it is understandable that the normal range is less clearly defined than for most tests.

KETONE BODIES

Acetoacetic acid (diacetic acid) is a normal endproduct of fatty acid oxidation in the liver. It is also produced to a very limited extent by oxidative breakdown of leucine, phenylalanine, and tyrosine. β-Hydroxybutyric acid is formed from acetoacetic by reversible reduction, and acetone is formed spontaneously by nonreversible decarboxylation:

$$CH_3\!-\!\overset{\overset{\textstyle O}{\|}}{C}\!-\!CH_2\!-\!COOH \xrightarrow{\;-CO_2\;} CH_3\!-\!\overset{\overset{\textstyle O}{\|}}{C}\!-\!CH_3$$

ACETOACETIC ACID   ACETONE

$$-2H \Big\Updownarrow +2H$$

$$CH_3\!-\!\overset{\overset{\textstyle OH}{|}}{\underset{\underset{\textstyle H}{|}}{C}}\!-\!CH_2\!-\!COOH$$

β − HYDROXYBUTYRIC ACID

These three substances are called *ketone bodies* or *acetone bodies.* Any condition in the body (diabetes mellitus, starvation, etc.) which limits carbohydrate utilization with resultant increase in fat utilization leads to production of ketone bodies by the liver at a rate faster than peripheral tissues can oxidize them. This condition is *ketosis,* and the abnormally high levels of ketone bodies in the blood and urine are referred to as *ketonemia* and *ketonuria,* respectively.

QUANTITATIVE METHODS

The various methods proposed for determination of ketone bodies employ reactions more or less specific for either acetone or acetoacetic acid. Until recently, no specific reaction for β-hydroxybutyric acid was available. For many years it was believed advantageous to determine the level of each ketone body individually, since it was thought that mild acidosis produced only acetonuria whereas more severe acidosis also produced acetoacetic acid. The basic approach to fractionation has been as follows: Preformed acetone is first determined. Preformed acetone plus that formed from acetoacetic acid by heat and acid is next measured. Finally, total ketone bodies are determined by conversion of acetoacetic and β-hydroxybutyric acids to acetone by treatment with acid dichromate. The acetone in each case is isolated by distillation. The acetoacetic and β-hydroxybutyric acid concentrations are calculated by difference.

Although the proportions are variable, the total ketone bodies are comprised approximately of 2% acetone, 20% acetoacetic acid, and 78% β-hydroxybutyric acid (187, 218). Fortunately, it now appears that one ketone body has as much but no more significance than the others (187), so that there is no need for fractionation in routine practice.

The reactions employed for quantitation of ketone bodies can be outlined as follows: (a) *Bisulfite binding.* In one method (117) blood filtrate is freed of glucose and then added to bisulfite. Aldehydes and ketones form complexes with bisulfite. The excess bisulfite is destroyed by iodine, alkali is added, and the bisulfite released is titrated with iodine (Clausen titration: cf *Lactic Acid,* section on *Iodometric titration*). This method is not specific, other aldehydes and ketones (pyruvic and α-ketoglutaric acids) being included. Considerably greater specificity is achieved if the acetone is first isolated by isothermal distillation (246). (b) *Iodometric titration.* Messinger (159) in 1888 distilled acetone into alkaline iodine solution, forming iodoform with utilization of six atoms of iodine per acetone molecule, and titrated the excess iodine with thiosulfate. Modifications of this technic applied to blood and urine were proposed by Folin (55), Shaffer (209), Hubbard (101), and Rappaport and Baner (185). The reaction is not very specific, and multiple distillations are required to free acetone of interferences. (c) *Precipitation with mercuric cyanide.* In 1911 Scott-Wilson (206) precipitated acetone from alkaline solution with mercuric cyanide, redissolved it in HNO_3, and determined the mercury by titration with sulfocyanate. Nephelometric measurement of the precipitate was employed by several workers (56, 148, 149, 211), but this technic is difficult to perform accurately. (d) *Precipitation with mercuric sulfate.* In 1898 Denigès (48) boiled acetone with mercuric sulfate and H_2SO_4, forming a crystalline precipitate weighing 20 times as much as acetone. Van Slyke developed this approach for urine (235) and blood (236), either weighing the precipitate or dissolving it in HCl and titrating the mercury with KI. The mercury has also been determined by its specific interference with the color formed when thiocyanate is added to excess ferric nitrate (41). Precipitation of acetone in these original methods was carried out directly in blood or urine filtrates, but Weichselbaum and Somogyi (241) subsequently showed that lactic acid causes significant positive interference, necessitating preliminary isolation of acetone by distillation. Barnes and Wick (15) precipitated acetone as the mercury complex first, dissociated the complex in HCl, and then distilled the acetone and performed a Messinger titration. In a somewhat different approach, the acetone of the sample is distilled isothermally either into bisulfite, nesslerized, and the amount of precipitate compared visually with a series of standards (1); or directly into Nessler's reagent and the time measured for appearance of precipitate, which is proportional to the acetone concentration (244). This latter method is said to lack reproducibility (246). (e). *Reaction with salicylaldehyde.* Fabinyi (54) in 1900 showed that warming acetone with salicylaldehyde in alkaline solution produced a red color. The mechanism of this reaction has been a matter of some dispute. Braunstein (27) believed that it occurs with all compounds containing a CH_3CO group linked directly to a C or H atom. The first stage involves condensation with the carbonyl compound to form an oxybenzylidene derivative (Perkin's reaction). If a methyl group is attached to the carbonyl of the original compound, enolization ensues in strong alkali, leading to formation of enolate, the intense yellow color of which is due to a system of cumulated double bonds. Thomson (230) disagreed, believing that the reaction is less specific than postulated by Braunstein and that the mechanism is condensation with a methylene group in the α position to an unsaturated group such as carbonyl and formation of a simple alkali salt of the resulting compound. Frommer (69) applied the reaction as a qualitative test for acetone in urine. This reaction has been the basis of numerous methods for quantitative determination of ketone bodies in blood and urine (16, 22, 165, 173, 174, 181, 223, 229). (f) *Reaction with dinitrophenyl-hydrazine.* Greenberg and Lester (82) introduced a technic in which the colored hydrazone is formed and extracted into CCl_4 for photometric measurement. This method has been modified by others (81, 87, 154, 160, 233). Acetaldehyde (154, 233) interferes; ethanol interferes when total ketone bodies are determined (154). (g) *Enzymatic method.* Williamson *et al* (245) have introduced a method using partially purified $D(-)-\beta$-hydroxybutyric dehydrogenase, which catalyzes the following reaction:

$$\beta\text{-hydroxybutyrate} + NAD \rightleftharpoons acetoacetate + NADH + H^+$$

At pH 8.5, β-hydroxybutyrate is determined by the increase in absorbance at 340 nm of NADH in the presence of hydrazine; at pH 7.0 acetoacetate is measured by the decrease in absorbance of NADH at 340 nm. Using a crystalline enzyme the pH optimum of the forward reaction was shifted to 9.5 (18). Measurement of NADH by fluorometry has been employed in micro methods (72, 252). This technic has greater specificity than most chemical methods, and the availability of the enzyme commercially makes it relatively simple. (h) *Miscellaneous reactions.* Vanillin in alkaline solution reacts with acetone to form red vanillalacetone or divanillalacetone (132). Vanillin is an aromatic aldehyde with a phenol group similar to salicyl-aldehyde, but opinion on the relative merits of each of these reactions is divided (13, 132). Acetone condenses with furfural in the presence of strong base to form difurfurylidene-acetone, which develops red to violet colors in strong acid (142, 205). A red color forms when acetone and acetic acid in deproteinized blood are treated with trinitrobenzene (163). Ketone bodies may be quantitated in serum or urine by the formation of a purple color with nitroprusside in buffered alkaline medium (143, 201). Condensation of acetoacetic acid with resorcinol and HCl yields a fluorescent product (39). Acetoacetic acid couples with 4-nitrobenzenediazonium salt in buffered carbonate solution to form a product which is green upon alkalinization and which can be extracted into a mixture of benzene and butanol (190). This method has been applied to blood but cannot be used for urine. In a modification of this approach, acetoacetic acid reacts with diazotized *p*-nitroaniline (238). Acetone can be measured rapidly by gas chroma-tography (232). (i) *Automated methods.* Acetoacetic acid is reacted with nitroferricyanide (116) or coupled with 2,5-dichlorobenzene di-azonium chloride (198) to form products that are measured photo-metrically.

Lack of specificity and poor or unpredictable recovery are attested to by the normal values observed in utilizing various chemical methods. In this regard the availability of enzymatic methods with a greater potential for specificity, accuracy, and speed is a promising development. The approximate top normal limits for total ketone bodies in urine obtained by various methods are as follows (expressed as mg acetone/100 ml): 28 by direct Denigès precipitation (235), 1.7 by Denigès precipitation after preliminary distillation of acetone (218), 4.7 by the salicylaldehyde reaction (16), and 42 by formation of the 2,4-dinitrophenylhydrazone (107). The situation is no better for blood: Various studies using the Denigès precipitation after preliminary distillation of acetone give the top limit as 0.52 (241) and 1.1 (111) mg (as acetone)/100 ml; using hydrazone methods the top limit has been reported to be 0.6 (154), 3.7 (219), and 8.1 (107).

Our laboratory studied the diffusion method of Nadeau (165), in which acetone is isothermally distilled into salicylaldehyde, and we were unable to select conditions giving anywhere near quantitative recovery. This was due apparently to the fact that by the time distillation was complete the color formed by the acetone distilled earlier was deteriorating. Accordingly the method was modified by diffusing the acetone into an alkaline vanillin reagent. Vanillin has the further advantage over salicylaldehyde that it is a more reproducible reagent; different brands of salicylaldehyde give different color intensities (132).

DETERMINATION OF ACETONE PLUS ACETOACETIC ACID IN BLOOD OR SERUM

(method of Nadeau, Ref 165, modified)

PRINCIPLE

Preformed acetone and acetone derived from acetoacetic acid, which is decarboxylated at the temperature of the test, are isothermally distilled at $50°-55°C$ into alkaline vanillin. Vanillalacetone or divanillalacetone is formed as follows:

VANILLIN   VANILLALACETONE

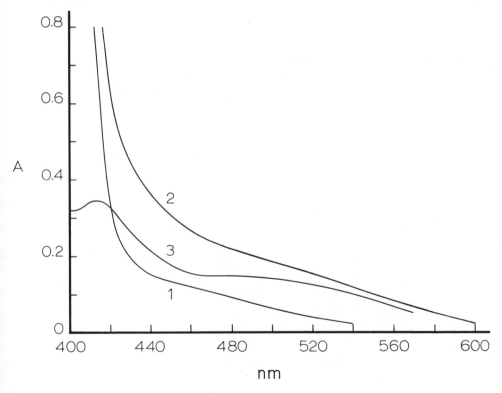

DIVANILLALACETONE

Figure 26-2 shows the absorption curve of the red product.

APPARATUS

Micro Diffusion Flask, 50 ml. The following are satisfactory: flask, alcohol, Cavett, Curtin 258-780; flask, diffusion, micro, Scientific Glass Apparatus JM 5505.

FIG. 26-2. Absorption curves with acetone: (1) reagent blank vs water, (2) 10 μg acetone standard vs water, (3) standard vs reagent blank, Perkin-Elmer model 4000-A recording spectrophotometer.

REAGENTS

Vanillin Reagent, 2% in 4 N KOH. Prepare daily.

Stock Acetone Standard, 1.0 g/liter. Dilute 1.26 ml reagent grade acetone (sp. gr. 0.791) to 1 liter with water. 1 ml = 1 mg acetone.

Dilute Acetone Standard. Dilute stock 1:20 with water. 1 ml = 0.05 mg acetone.

PROCEDURE

1. Place 2.0 ml aliquots of vanillin reagent into bottoms of micro diffusion flasks for standards, unknowns, and blank.
2. Place 0.2 ml blood, plasma, or serum in the cup of the apparatus attached to flask stopper. In a similar manner set up standard using 0.2 ml dilute acetone standard, and reagent blank using 0.2 ml water.
3. Insert stoppers into flasks, taking care not to touch the sides, and use springs or rubber bands to hold stoppers in place tightly. Place in $50°-55°C$ water bath or incubator for 60 min.
4. Cool to room temperature, add 3.0 ml water to each flask, mix, and read absorbance of standard (A_s) and unknown (A_x) against reagent blank at 415 nm (see *Notes 1* and *2*, below).

Calculation:

$$\text{mg acetone/100 ml} = \frac{A_x}{A_s} \times 0.01 \times \frac{100}{0.2}$$

$$= \frac{A_x}{A_s} \times 5$$

$$\text{mmol acetone/liter} = \frac{A_x}{A_s} \times 0.86$$

NOTES

1. Beer's law. Beer's law is followed at 415 and 500 nm on a Beckman DU up to an absorbance of 1.3. There is slight deviation with a Klett 42 filter.

2. Color stability. Color is stable for at least 1 hour.

3. Stability. It has been reported (233) that blood kept in a syringe is stable for 24 hours in the refrigerator. Our laboratory has found serum or whole blood to be stable for 2 days in the refrigerator or in the frozen state. Levels decrease about 30% in 1 day at room temperature.

ACCURACY AND PRECISION

Our laboratory obtained about 95% recoveries of acetone added to blood. Recovery of acetoacetic acid was not studied. Nadeau (165) claimed that acetoacetic acid is also measured because of decomposition to acetone at the temperature and pressure of incubation but presented no experimental data on this point. Pawan (173) employed aniline to decarboxylate acetoacetic acid catalytically to acetone in a modification of Nadeau's method, but again no experimental data were presented to document its efficacy or, in fact, to show that it was necessary. It is quite possible that the method is fairly specific for the preformed acetone plus whatever acetone is derived from acetoacetic acid, but quite frankly proof of this is lacking.

The precision of the method in the normal range is ±15%.

NORMAL VALUES

A range for whole blood of 0.5–3.0 mg/100 ml for nine normal adult males and nine normal adult females was obtained in our laboratory, there being no evidence of a sex difference. Values for serum averaged 10% less than those for whole blood. The concentration of ketone bodies in serum or plasma has been reported to be about twice that in erythrocytes (218). It may be that the distribution between plasma and erythrocytes is not the same for the three ketone bodies. Postprandial values for total ketone bodies in infants and children are roughly comparable to adult nonfasting levels, except for the 2–5 day group whose levels are slightly higher (174). Differences in the 2–7 day values have been explained by functional immaturity of the kidneys (3).

QUALITATIVE TESTS

For proper design and assessment of qualitative tests it is necessary first to decide on the desired sensitivity. In the case of ketone bodies this must be done rather empirically in view of the confusion regarding normal ranges previously discussed.

In 1865 Gerhardt (71) described the reaction with ferric chloride. Ferric chloride solution is added to urine drop by drop until no more precipitate of ferric phosphate is formed. A positive test is the appearance of a Bordeaux red color. Ferric chloride reacts with acetoacetic acid, not with acetone (60) or β-hydroxybutyric acid (215). This test is usually regarded as insufficiently sensitive, being sensitive to approximately 25–50 mg acetoacetic acid per 100 ml (166, 215), and not very specific. Salicylates interfere (187).

Wallhauser (239) applied the Scott-Wilson reaction to the detection of acetone in urine. In his technic 1 drop of mercuric cyanide reagent is placed on a microscope slide which is then inverted over the mouth of the urine container. After a 2 min exposure as a hanging drop, the reagent is inspected for white clouding or precipitate. Faint clouding occurs with 0.005% acetone and a heavy precipitate with 0.1%. Samson (199) has regarded this test as one of the simplest and best for acetone.

Two tests employing reaction with iodine have been proposed. In Lindemann's test (134) the urine is acidified with acetic acid, Lugol's solution is added, and the mixture is extracted with $CHCl_3$. If acetoacetic acid is present, the $CHCl_3$ phase remains colorless. This method is said to be quite insensitive and nonspecific (61). Lieben's test (133) depends on formation of a crystalline yellow precipitate of iodoform with its characteristic odor upon addition of alkaline Lugol's solution.

Frommer's (69) salicylaldehyde reaction for acetone and Arnold's (9) *p*-amidoacetophenone reaction for acetoacetic acid have also been used as qualitative tests. The color reaction of acetone with vanillin, an aromatic aldehyde with a phenol group similar to salicylaldehyde, has also been proposed as a qualitative test for urine and blood (132). A semiquantitative test for acetoacetic acid is based on reaction with *p*-benzoquinone and ammonia to form blue colors (115).

Le Nobel (131) in 1884 described the formation of a red color when nitroprusside reacts with acetoacetic acid, followed by alkalinization. Acidification with acetic acid turns the color to magenta. Rothera (191) found that creatinine reacts when NaOH is used as alkali but that use of NH_4OH obviates this interference. Essentially all the color produced in this test (over 96%) is due to acetoacetic acid because of its relative insensitivity to acetone and because ordinarily the acetone concentration is about one-tenth that of acetoacetic acid (35, 60). The sensitivity of the original Rothera test (qualitative nitroprusside tests are frequently so called) is about 1–5 mg acetoacetic acid or 10–25 mg acetone per 100 ml (60, 110, 166, 215), regarded by some as too sensitive (110, 166). For example, it is reported (215) that it may be positive for urine obtained after an overnight fast. In one modification (110) the sensitivity has been adjusted to 25 mg/100 ml.

Ames markets two forms of the Rothera test. *Ketostix* is a paper strip coated with sodium nitroprusside, glycine, and Na_2HPO_4. *Acetest* is a tablet containing the same mixture plus lactose. The sensitivities of the strip and tablet are claimed by the company to be 20 and 10 mg acetoacetic acid per 100 ml, respectively. Reports (59, 114, 166) have given the sensitivity as about 5 mg/100 ml. Both forms of the test have color charts with three colors representing different ranges of positivity. Studies with clinical samples have shown that the tablet and strip are essentially equivalent (35, 61, 114) and are superior to the Gerhardt ferric chloride test (61, 166) and the original Rothera nitroprusside test (166). False positives were not encountered for a group of 68 drugs (187); however, levodopa can interfere with Ketostix while having no effect on Acetest (179). The two tests can be used with serum or plasma, although for Acetest it is necessary to add 2 drops of water to the drop of serum on the tablet so that the tablet can absorb the fluid (35).

For routine use, Ketostix, Acetest, or the following modification of Rothera's test is recommended.

QUALITATIVE DETECTION OF KETONE BODIES IN URINE

(method of Rothera, Ref 191, modification of Free and Free, Ref 60)

PRINCIPLE

Nitroprusside in alkaline solution reacts principally with acetoacetic acid to form a purple color. The test is carried out as a ring test.

REAGENTS

$(NH_4)_2 SO_4$
$NH_4 OH$, Conc
Sodium Nitroprusside, 5% Aqueous. Prepare fresh.

PROCEDURE

Approximately 200 mg solid $(NH_4)_2 SO_4$, 5 drops urine, and 2 drops fresh nitroprusside reagent are mixed thoroughly in a small test tube (7 x 70 mm). Conc $NH_4 OH$, about 10–15 drops, is layered on top. After 10 min the interface is observed for purple color.

INTERPRETATION

Trace of purple	= 5 mg acetoacetic acid (or 20 mg acetone) per 100 ml
Moderate purple	= 30 mg acetoacetic acid (or 250 mg acetone) per 100 ml
Strong purple	= 80 mg acetoacetic acid (or 800 mg acetone) per 100 ml

NOTE

1. Stability. Acetoacetic acid is quite stable in some urines at room temperature but disappears rapidly in others, the disappearance appearing to be related to microbial action. In the presence of bacteria or yeast it may disappear completely in less than 24 hours, whereas if sterile it is stable for 8–10 days (60). Acetone is not formed from acetoacetic acid by microbial action. About 20% of the acetone present disappears in 24 hours at room temperature (1), but none is lost in the same period if the urine is kept in a closed container in the refrigerator (233). Toluene has some protective action against the loss of acetone, but the amount used must be limited since acetone is half as soluble in toluene as in water (1). Ketone bodies may also disappear from urine *in vivo* in the presence of a urinary tract infection (60).

REFERENCES

1. ABELS JC: *J Biol Chem 119:*663, 1937
2. AGRELL IG: *Acta Physiol Scand 12:*372, 1946
3. AKERBLOM H, AHOLA T, SOMERSALO O: *Ann Paediat Fenniae 11:*108, 1965
4. ALTMANN SM, CROOK EM, DATTA SP: *Biochem J 49:* lxiii, 1951
5. ANDERSON J, COMTY CM, MAZZA R: *Lancet 2:*1093, 1964
6. ANTENER I, VUATAZ L, BRUSCHI A, KAESER M: *Z Klin Chem 4:*296, 1966
7. ANTONIS A, CLARK M, PILKINGTON TRE: *J Lab Clin Med 68:*340, 1966
8. ARCHER HF, DORMER AE, SCOWEN EF, WATTS RWE: *Clin Sci 16:*405, 1957
9. ARNOLD V: *Wien Klin Wochschr 12:*541, 1899
10. ASTRAND PO, HALLBACK I, HEDMAN R, SALTIN B: *J Appl Physiol 18:*619, 1963
11. AUTENRIETH W, BARTH H: *Hoppe-Seyler's Z Physiol Chem 35:*327, 1902
12. AVERY BF, HASTINGS AB: *J Biol Chem 94:*273, 1931
13. BAHNER F, SCHULZE G: *Z Klin Chem 3:*10, 1965
14. BARKER SB, SUMMERSON WH: *J Biol Chem 138:*535, 1941
15. BARNES RH, WICK AW: *J Biol Chem 131:*413, 1939
16. BEHRE JA, BENEDICT SR: *J Biol Chem 70:*487, 1926
17. BERGMEYER HU, BERNT E: *Methods of Enzyme Analysis,* Edited by Bergmeyer HU. New York, Academic Press, 1963, p 324
18. BERGMEYER HU, BERNT E: *Enzymol Biol Clin 5:*65, 1965
19. BERNTSSON S: *Anal Chem 27:*1659, 1955
20. BEUTLER E, YEH MKY: *J Lab Clin Med 54:*125, 1959
21. BISERTE G, DASSONVILLE B: *Clin Chim Acta 1:*49, 1956
22. BLOOM WL: *J Lab Clin Med 51:*824, 1958
23. BOAS I: *Deut Med Woch 19:*940, 1893
24. BOEHM M, HENSE D: *Nutr Dieta 5:*92, 1963
25. BONTING SL: *Arch Biochem Biophys 56:*307, 1955
26. BONTING SL: *Arch Biochem Biophys 58:*100, 1955
27. BRAUNSTEIN AE: *Nature (London) 140:*427, 1937
28. BREUSCH FL, TULUS R: *Biochim Biophys Acta 1:*77, 1947
29. BUEDING E, GOLDFARB W: *J Biol Chem 141:*539, 1941
30. BUEDING E, GOODHART R: *J Biol Chem 141:*931, 1941
31. BUEDING E, WORTIS H: *J Biol Chem 133:*585, 1940
32. CAHOURS A: *Ann Chim 19:*484, 1847
33. CARTIER P, PIN P: *Bull Soc Chim Biol 31:*1176, 1949
34. CAVALLINI D, FRANTALI N, TOSCHI G: *Nature (London) 164:*792, 1949
35. CHERTACK MM, SHERRICK JC: *J Am Med Assoc 167:*1621, 1958
36. CLARKE AD, PODMORE DA: *Clin Chim Acta 13:*725, 1966
37. CLAUSEN SW: *J Biol Chem 52:*263, 1922
38. CLIFT FP, COOK RP: *Biochem J 26:*1788, 1932
39. COQUOIN R, LUPU R: *Compt Rend Soc Biol 109:*801, 1932
40. CRAMP DG: *J Clin Pathol 21:*171, 1968
41. CRANDALL LA Jr: *J Biol Chem 133:*539, 1940
42. CRAWHALL JC, WATTS RWE: *Clin Sci 20:*357, 1961
43. CRESTA LA: *Rev Asoc Bioquim Arg 30:*57, 1965

44. DAKIN HD: *J Biol Chem 3*:57, 1907
45. DAKIN HD, DUDLEY HW: *J Biol Chem 15*:127, 1913
46. DEAN BM, GRIFFIN WJ: *Nature (London) 205*:598, 1965
47. DEMPSEY EF, FORBES AP, MELICK RA, HENNEMAN PH: *Metabolism 9*:52, 1960
48. DENIGÈS G: *Compt Rend Acad Sci 127*:963, 1898
49. de SCHEPPER P, PARMENTIER G, VANDERHAEGHE H: *Biochim Biophys Acta 28*:507, 1958
50. DODDS EC, GALLIMORE EJ: *Biochem J 26*:1242, 1932
51. EDWARDS HT: *J Biol Chem 125*:571, 1938
52. EL HAWARY MFS, THOMPSON RHS: *Biochem J 53*:340, 1953
53. ETTINGER RH, GOLDBAUM LR, SMITH LH Jr: *J Biol Chem 199*:531, 1952
54. FABINYI R: *Chem Centralblatt 2*:302, 1900
55. FOLIN O: *J Biol Chem 3*:177, 1907
56. FOLIN O, DENIS W: *J Biol Chem 18*:263, 1914
57. FORFAR JO, TOMPSETT SL, FORSHALL W: *Arch Disease Childhood 34*:525, 1959
58. FRASER J, CAMPBELL DJ: *Clin Biochem 5*:99, 1972
59. FRASER J, FETTER MC, MAST RL, FREE AH: *Clin Chim Acta 11*:372, 1965
60. FREE AH, FREE HM: *Am J Clin Pathol 30*:7, 1958
61. FREE HM, SMEBY RR, COOK MH, FREE AH: *Clin Chem 4*:323, 1958
62. FRIEDEMANN TE, COTONIO M, SHAFFER PA: *J Biol Chem 73*:335, 1927
63. FRIEDEMANN TE, GRAESER JB: *J Biol Chem 100*:291, 1933
64. FRIEDEMANN TE, HAUGEN GE: *J Biol Chem 144*:67, 1942
65. FRIEDEMANN TE, HAUGEN GE: *J Biol Chem 147*:415, 1943
66. FRIEDEMANN TE, HAUGEN GE, KMIECIAK TC: *J Biol Chem 157*:673, 1945
67. FRIEDEMANN TE, KENDALL AI: *J Biol Chem 82*:23, 1929
68. FRIEDLAND IM, DIETRICH LS: *Anal Biochem 2*:390, 1961
69. FROMMER V: *Ber Klin Wochschr 42*:1008, 1905
70. FÜRTH O von, CHARNASS D: *Biochem Z 26*:199, 1910
71. GERHARDT C: *Wein Med Pr 6*:672, 1865
72. GIBBARD S, WATKINS PJ: *Clin Chim Acta 19*:511, 1968
73. GIBBS DA, WATTS RWE: *J Lab Clin Med 73*:901, 1969
74. GITERSON AL, SLOOFF PAM, SCHOUTEN H: *Clin Chim Acta 29*:342, 1970
75. GLOSTER JA, HARRIS P: *Clin Chim Acta 7*:206, 1962
76. GOLDBERG AS, BERNHEIM AR: *J Biol Chem 156*:33, 1944
77. GOLDBERG L, GILMAN T: *S African J Med Sci 8*:117, 1943
78. GONZALES RF, GARDNER LI: *Pediatrics 19*:844, 1957
79. GOODWIN TW, WILLIAMS GR: *Biochem J 51*:708, 1952
80. GORDON JJ, QUASTEL JH: *Biochem J 33*:1332, 1939
81. GÖSCHKE H: *Clin Chim Acta 28*:359, 1970
82. GREENBERG LA, LESTER D: *J Biol Chem 154*:177, 1944
83. GRÖNVALL H: *Acta Ophthalmol Suppl 14*:1, 1937
84. GRÜNDIG E: *Clin Chim Acta 6*:331, 1961
85. HADJIIOANNOU TP, SISKOS PA, VALKANA CG: *Clin Chem 15*:940, 1969
86. HADJIVASSILIOU AG, RIEDER SV: *Clin Chim Acta 19*:357, 1968
87. HANSEN O: *Scand J Clin Lab Invest 11*:259, 1959
88. HARRISON HE, HARRISON HC: *Yale J Biol Med 24*:273, 1952
89. HARROP GA Jr: *Proc Soc Exp Biol Med 17*:162, 1919
90. HAUSMAN ER, McANALLY JS, LEWIS GT: *Clin Chem 2*:439, 1956
91. HENLEY KS, WIGGINS HS, POLLARD HM: *J Lab Clin Med 47*:978, 1956
92. HESS B: *Biochem Z 328*:110, 1956

93. HOCHELLA NJ, WEINHOUSE S: *Anal Biochem 10*:304, 1965

94. HOCKADAY TDR, FREDERICK EW, CLAYTON JE, SMITH LH: *J Lab Clin Med 65*:677, 1965

95. HODGKINSON A: *Clin Sci 23*:203, 1962

96. HODGKINSON A: *Clin Chem 16*:547, 1970

97. HOFFMAN NE, BARBORIAK JJ, HARDMAN HF: *Anal Biochem 9*:175, 1964

98. HOHORST HJ: *Methods of Enzymatic Analysis*, edited by Bergmeyer HU. New York, Academic Press, 1963, p 266

99. HOLZER H, SÖLING HD: *Biochem Z 336*:201, 1962

100. HORN HD, BRUNS FH: *Biochim Biophys Acta 21*:378, 1956

101. HUBBARD RS: *J Biol Chem 49*:357, 1921

102. HUCKABEE WE: *J Appl Physiol 9*:163, 1956

103. HUCKABEE WE: *J Clin Invest 37*:244, 1958

104. HUCKABEE WE: *Am J Med 30*:833, 1961

105. HUMMEL JP: *J Biol Chem 180*:1225, 1949

106. JACOBS SL, LEE ND: *J Nucl Med 5*:297, 1964

107. JOHNSON RE, SARGENT F II, PASSMORE R: *Quart J Exptl Physiol 43*:339, 1958

108. JONES GB, BELLING GB: *Anal Biochem 37*:105, 1970

109. JOSEPHSON B, FORSSBERG V: *J Lab Clin Med 27*:267, 1941

110. JUHASZ B: *Kiserletes Orvostudomany 8*:215, 1956; Excerpta Med 10; Section II, Abstr. No. 4708, 1957

111. KARTIN BL, MAN EB, WINKLER AW, PETERS JP: *J Clin Invest 23*:824, 1944

112. KELLER W, DENZ L: *Z Physiol Chem 314*:153, 1959

113. KENDALL AI, FRIEDEMANN TE: *J Infect Diseases 47*:176, 1930

114. KILLANDER J, SJÖLIN S, ZAAR B: *Scand J Clin Lab Invest 14*:311, 1962

115. KLEEBERG J: *Clin Chim Acta 13*:779, 1966

116. KLEIN B, OKLANDER M: *Clin Chem 12*:606, 1966

117. KLEIN D: *J Biol Chem 135*:143, 1940

118. KLEIN D: *J Biol Chem 137*:311, 1941

119. KLEIN D: *J Biol Chem 145*:35, 1942

120. KOCH GH, STRONG FM: *Anal Biochem 27*:162, 1969

121. KOEPSELL HJ, SHARPE ES: *Arch Biochem Biophys 38*:443, 1952

122. KRAUS P: *Clin Chim Acta 12*:462, 1965

123. KROG PW: *Acta Physiol Scand 9*:68, 1945

124. KRUSIUS FE: *Acta Physiol Scand 2 (Suppl 3)*, 1940

125. KULONEN E, CARPEN E, RUOKOLAINEN T: *Scand J Clin Lab Invest 4*:189, 1952

126. KUNZ R: *Arch Chem Mikr 7*:285, 1914

127. LAMDEN MP, CHRYSTOWSKI GA: *Proc Soc Exp Biol NY 85*:190, 1954

128. LANDON J, FAWCETT JK, WYNN V: *J Clin Pathol 15*:579, 1962

129. LEHMAN H: *Chemie Physiologique*, 1st ed. Paris, W. Engelmann, 1850

130. LEHMANN J: *Scand Arch Physiol 80*:237, 1938

131. LE NOBEL C: *Naunyn-Schmiedebergs Arch Exptl Pathol Pharmakol 18*:6, 1884

132. LEVINE VE, TATERKA M: *Clin Chem 3*:646, 1957

133. LIEBEN A: *Ann Chemie Pharmacie Suppl 7*:218, 1870

134. LINDEMANN L: *Muench Med Wochschr 52*:1386, 1905

135. LONG C: *Biochem J 36*:807, 1942

136. LONG C: *Biochem J 38*:447, 1944

137. LONG C: *Biochem J 40*:27, 1946

138. LOOMIS ME: *J Lab Clin Med 57*:966, 1961

139. LU GD: *Biochem J 33*:249, 1939

140. LUNDHOLM L, MOHME-LUNDHOLM E, SVEDMYR N: *Scand J Clin Lab Invest* 15:311, 1963
141. LUNDHOLM L, MOHME-LUNDHOLM E, VAMOS N: *Acta Physiol Scand 58*:243, 1963
142. LYON JB, BLOOM WL: *Can J Biochem Physiol 36*:1047, 1958
143. MADONIA JP: *Am J Clin Pathol 39*:206, 1963
144. MANN GV, SHUTE E: *Clin Chem 16*:849, 1970
145. MARBACH EP, WEIL MH: *Clin Chem 13*:314, 1967
146. MARKEES S: *Biochem J 56*:703, 1954
147. MARKS V: *Clin Chim Acta 6*:724, 1961
148. MARRIOTT WM: *J Biol Chem 16*:289, 1913
149. MARRIOTT WM: *J. Biol Chem 16*:293, 1913
150. MARTENSSON J: *Acta Physiol Scand Suppl 2*:1, 1940
151. MARTINEK RG, JACOBS SL, HAMMER FE: *Clin Chim Acta 36*:75, 1972
152. MAVER MC: *J Biol Chem 32*:71, 1917
153. MAYER GG, MARKOW D, KARP F: *Clin Chem 9*:334, 1963
154. MAYES PA, ROBSON W: *Biochem J 67*:11, 1957
155. McARDLE B: *Biochem J 60*:647, 1955
156. McARDLE B: *Biochem J 66*:144, 1957
157. MENDEL B, GOLDSCHEIDER I: *Biochem Z 164*:163, 1925
158. MENKES JH: *Am J Diseases Children 99*:500, 1960
159. MESSINGER O: *Ber Deut Chem Ges 21*:366, 1888
160. MICHAELS GD, MORGEN S, LIEBERT G, KINSELL LW: *J Clin Invest 30*:1483, 1951
161. MILLER BF, MUNTZ JA: *J Biol Chem 126*:413, 1938
162. MOELLERING H, GRUBER W: *Anal Biochem 17*:369, 1966
163. MOMOSE T, OHKURA Y, KOHASHI K, NAGATA R: *Chem Pharm Bull (Tokyo) 11*:973, 1963; *C A 60*:837c, 1964
164. MURATA T: *Kumamoto Daigaku Taishitsu Igaku Kenkyusho Hokoku 13*:481, 1963; *C A 63*:18782h, 1965
165. NADEAU G: *Can Med Assoc J 67*:158, 1952
166. NASH J, LISTER J, VOBES DH: *Lancet 1*:801, 1954
167. NATELSON S, PINCUS JB, LUGOVOY JK: *J Biol Chem 175*:745, 1948
168. NATELSON S, PINCUS JB, LUGOVOY JK: *J Clin Invest 27*:446, 1948
169. NELSON SR, KUGLER KK: *Biochem Med 2*:325, 1969
170. NORDMANN R, GAUCHERY O, du RUISSEAU J, THOMAS Y, NORDMANN J: *Bull Soc Chim Biol 36*:1461, 1954
171. OLSON GF: *Clin Chem 8*:1, 1962
172. PARKINSON E, WAGNER EC: *Anal Chem 6*:433, 1934
173. PAWAN GLS: *Biochem J 68*:33p, 1958
174. PEDEN VH: *J Lab Clin Med 63*:332, 1964
175. PELLET MV, SEIGNER C, COHEN H: *Pathol Biol 17*:909, 1969
176. PEREIRA RS: *Mikrochemie 36/37*:398, 1951
177. PFLEIDERER G, DOSE K: *Biochem Z 326*:436, 1955
178. PIK C, KERCKHOFFS HPM: *Clin Chim Acta 8*:300, 1963
179. POCELINKO R, SOLOMON HM, GANT ZN: *New Engl J Med 281*:1075, 1969
180. POWERS HH, LEVATIN P: *J Biol Chem 154*:207, 1944
181. PROCOS J: *Clin Chem 7*:97, 1961
182. PRYCE JD: *Analyst 94*:1151, 1969
183. PUCHER GW, SHERMAN CC, VICKERY HB: *J Biol Chem 113*:235, 1936
184. PUCHER GW, VICKERY HB, LEAVENWORTH CS: *Anal Chem 6*:190, 1934
185. RAPPAPORT F, BANER B: *J Lab Clin Med 28*:1770, 1943

186. RIBEIRO ME, ELLIOTT JS: *Invest Urol* 2:78, 1964
187. RIEKERS H, MIALE JB: *Am J Clin Pathol* 30:530, 1958
188. RONZONI E, WALLEN-LAWRENCE Z: *J Biol Chem* 74:363, 1927
189. ROSENBERG JC, RUSH BF: *Clin Chem* 12:299, 1966
190. ROSENTHAL SM: *J Biol Chem* 179:1235, 1949
191. ROTHERA ACH: *J Physiol* 37:491, 1908
192. RUTKOWSKI RB, DeBARRE L: *Am J Clin Pathol* 46:405, 1966
193. RYAN H: *Analyst* 83:528, 1958
194. RYFFEL JH: *J Physiol* 39:v, 1909-10
195. SAFFRAN M, DENSTEDT OF: *J Biol Chem* 175:849, 1948
196. SALANT W, WISE LE: *J Biol Chem* 28:27, 1916
197. SALKOWSKI E: *Z Physiol Chem* 29:437, 1900
198. SALWAY JG: *Clin Chim Acta* 25:109, 1969
199. SAMSON M: *Am J Clin Pathol* 22:1106, 1952
200. SAVORY J, KAPLAN A: *Clin Chem* 12:559, 1966
201. SCHILKE RE, JOHNSON RE: *Am J Clin Pathol* 43:539, 1965
202. SCHNEYER J: *Biochem Z* 70:294, 1915
203. SCHON R: *Anal Biochem* 12:413, 1965
204. SCHULTZEN O: *Arch Anat Physiol* 6:719, 1868
205. SCHUSTER HG, BAASCH G: *Z Med Labortech* 12:312, 1971
206. SCOTT-WILSON H: *J Physiol* 42:444, 1911
207. SEGAL S, BLAIR AE, WYNGAARDEN JB: *J Lab Clin Med* 48:137, 1956
208. SELLECK B, COHEN JJ, RANDALL HM Jr: *Anal Biochem* 7:178, 1964
209. SHAFFER PA: *J Biol Chem* 5:211, 1908
210. SHIMAZONO H, HAYAISHI O: *J Biol Chem* 227:151, 1957
211. SHIPLEY RA, LONG CNH: *Biochem J* 32:2242, 1938
212. SHORR E, BERNHEIM A, TAUSSKY H: *Science* 95:606, 1942
213. SIMON E, NEUBERG C: *Biochem Z* 232:479, 1931
214. SJÖSTRÖM P: *Acta Chir Scand Suppl* 49:1, 1937
215. SMITH MJH: *J Clin Pathol* 10:101, 1957
216. STAHRE L: *Nord Tidskr Pharm* 2:141, 1895
217. STARCK HJ: *Klin Wochschr* 34:153, 1956
218. STARK IE, SOMOGYI M: *J Biol Chem* 147:319, 1943
219. STORMONT JM, MACKIE JE, DAVIDSON CS: *Proc Soc Exptl Biol Med* 106:642, 1961
220. STRAUB FB: *Hoppe-Seyler's Z Physiol Chem* 244:117, 1936
221. SÜLLMANN H, SCHAERER E: *Schweiz Med Wochschr* 62:619, 1932
222. SUNDERMAN FW, BOERNER F: *Normal Values in Clinical Medicine.* Philadelphia, Saunders, 1949, p 368
223. TANAYAMA S, UI M: *Chem Pharm Bull* 11:835, 1963
224. TAUSSKY HH: *J Biol Chem* 181:195, 1949
225. TAUSSKY HH, SCHORR E: *J Biol Chem* 169:103, 1947
226. TAYLOR KW, SMITH MJH: *Analyst* 80:607, 1955
227. TEAFORD ME, KAPLAN A: *Clin Chim Acta* 15:133, 1967
228. TFELT-HANSEN P, SIGGAARD-ANDERSEN O: *Scand J Clin Lab Invest* 27:15, 1971
229. THIN C, ROBERTSON A: *Biochem J* 51:218, 1952
230. THOMSON T: *Nature (London)* 141:917, 1938
231. THUNBERG T: *Biochem Z* 206:109, 1929
232. TROTTER MD, SULWAY MJ, TROTTER E: *Clin Chim Acta* 35:137, 1971
233. TSAO MU, LOWREY GH, GRAHAM EJ: *Anal Chem* 31:311, 1959
234. VAN DER HORST CJG: *Nature (London)* 187:146, 1960

235. VAN SLYKE DD: *J Biol Chem 32*:455, 1917
236. VAN SLYKE DD, FITZ R: *J Biol Chem 32*:495, 1917
237. VITTU CH, LEMAHIEU JC: *Ann Biol Clin (Paris) 23*:913, 1965
238. WALKER PG: *Biochem J 58*:699, 1954
239. WALLHAUSER A: *J Am Med Assoc 91*:21, 1928
240. WARBURG O, GAWEHN K, GEISSLER AW: *Hoppe-Seyler's Z Physiol Chem 320*:227, 1960
241. WEICHSELBAUM TE, SOMOGYI M: *J Biol Chem 140*:5, 1941
242. WEIL-MALHERBE H, BONE AD: *Biochem J 45*:377, 1949
243. WENDEL WB: *J Biol Chem 94*:717, 1932
244. WERCH SC: *J Lab Clin Med 25*:414, 1940
245. WILLIAMSON DH, MELLANBY J, KREBS HA: *Biochem J 82*:90, 1962
246. WINNICK T: *J Biol Chem 141*:115, 1941
247. WOHNLICH J: *Bull Soc Chim Biol 48*:736, 1966
248. WOLCOTT GH, BOYER PD: *J Biol Chem 172*:729, 1948
249. WOLF CGL: *J Physiol 48*:341, 1914
250. WORTIS H, GOODHART RS, BUEDING E: *Am J Diseases Children 61*:226, 1941
251. YARBRO CL, SIMPSON RE: *J Lab Clin Med 48*:304, 1956
252. YOUNG DAB, RENOLD AE: *Clin Chim Acta 13*:791, 1966
253. ZAREMBSKI PM, HODGKINSON A: *Biochem J 96*:717, 1965
254. ZELNICEK E: *Nature (London) 184*:727, 1959
255. ZENDER R: *Clin Chim Acta 24*:335, 1969

Chapter 27

Vitamins

JAMES A. DEMETRIOU , Ph.D.

The term vitamin is used to designate an organic compound that is required from exogenous sources in small amounts and is essential for growth and normal health. Two major classes of vitamins, based on solubility properties, are the water-soluble and fat-soluble vitamins. This difference in chemical properties is used in the methods of quantification for the individual vitamins.

The principle functions of vitamins are as biologic catalysts in combination with specific types of enzymes. This type of enzyme, comprised of the vitamin *(coenzyme)* and a protein moiety *(apoenzyme)*, is referred to as a *holoenzyme*. In their biochemical forms as coenzymes, the vitamins may be incorporated either directly as an integral part of the enzyme, as a simple phosphorylated derivative (e.g., pyridoxal phosphate, thiamine pyrophosphate), or in the form of a complex derivative (e.g., flavin adenine dinucleotide, nicotinamide adenine dinucleotide).

In the performance of vitamin assays, the choice of the appropriate biologic fluid for analysis that reflects the vitamin status of the individual still remains a problem. Pearson (164) reviewed the relationships of blood and urinary levels to the body tissue stores. Tissue, blood, and urine levels of thiamine, riboflavin, ascorbic acid, vitamin A, and carotene in vitamin depletion studies are presented. An equally important study of the distribution of 11 vitamins in red blood cells and plasma before and during a 24 hour period following intravenous administration of a multivitamin loading dose was reported by Baker *et al* (14). Of considerable importance from a nutritional standpoint was the variation in urinary excretion of metabolically active vitamins from the loading dose, from less than 1% for vitamins A, E, and B_6 to greater than 50% for folic acid, thiamine, ascorbic acid, riboflavin, and biotin.

In 1971 the International Congress on Nutrition (158) presented its recommendations on tentative rules for generic descriptors and trivial names for vitamins and related compounds. The terms vitamin A, D, E, K, B_6, B_{12}, and C have been retained. Nicotinamide and folic acid are the preferred terms instead of niacin and folacin. For those vitamins that have different forms, the summation term "equivalents" should be used (e.g., retinol equivalents, α-tocopherol equivalents, etc).

Some of the types of assays used for the quantification of vitamins in biologic fluids are listed in the following outline:

    I. Biologic assays
        A. Animal tests
        B. Microbiologic

II. Chemical assays
   A. Photometric
   B. Fluorometric

III. Biochemical assays (so-called because specific proteins are used as assay tools)
   A. Enzymatic
   B. Competitive protein binding
   C. Radioimmunologic

IV. Gas chromatography assays

Difficulties encountered with sensitivity and specificity with some of the vitamin assays may in the near future be resolved by application of the competitive protein binding (155) and radioimmunoassay technics (25). The synthesis of hapten-protein antigens and their use to produce hapten-specific antibodies in the endocrine field will undoubtedly be used as the models for the development of vitamin radioimmunoassay methods (23).

It is important that one be aware that most vitamin assays described not only measure the principal vitamin, but also many of its related analogs. For example, the assay for vitamin B_{12} does not distinguish it from the hydroxo, aquo, or desoxyadenosyl derivatives. Similarly, tocopherol determination measures the α, β, and γ forms. This type of nonspecificity, however, does not detract from the utility of the determination in the assessment of vitamin status of individuals.

VITAMIN A AND CAROTENE

Vitamin A is the term used as the generic descriptor for the β-ionone derivatives. Retinol, an unsaturated primary alcohol, has also been named vitamin A alcohol, vitamin A_1, and axerophthol. A number of naturally occurring pigments, called carotenoids, are precursors to vitamin A. Carotenoids are pigments with colors ranging from yellow to purple, and are characterized by an aliphatic chain with attached methyl groups and conjugated double bonds. Since β-carotene is the principal provitamin of retinol, both compounds are discussed and the technics for their assay are presented in this section. The structures of these two compounds are:

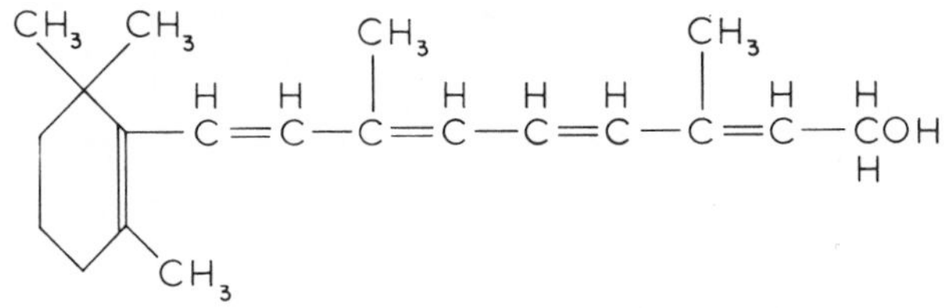

RETINOL

β – CAROTENE

Vitamin A is predominantly present as the alcohol in fasting blood samples (129). β-Carotene, lycopene, and xanthophyll are the major carotenoids in serum, and are exclusively located in the lipoprotein fractions. Vitamin A esters following absorption are also present in the low-density lipoproteins. A retinol-binding protein was recently characterized in the prealbumin fraction which is firmly bound to thyroxine-binding prealbumin (121). Kahan (120) and others (202) found that the protein complexes of retinol and the carotenes can easily be dissociated by the addition of varying amounts of alcohol to serum. Less than 1% of a loading dose of vitamin A appears in the red cells (120).

An isoprenoid polyene, phytofluene, has been detected in serum and is of dietary origin, being especially high in tomatoes (202). This compound has an absorption maximum at 349 nm, a fluorescence maximum at 475 nm, and interferes with the fluorescence assay of vitamin A (79, 202).

Reviews dealing with methods of bioassay, physicochemical determinations, and metabolic roles of vitamin A (67, 98, 144) and carotene (27, 144) appeared in 1957, 1960, and 1967, respectively. Therefore many of the approaches and method modifications to the measurement of retinol and carotene are not presented in detail.

The reaction of vitamin A with antimony trichloride to yield a brilliant blue color was described by Carr and Price in 1926 (43). Modifications of this reaction have been made by substitution of 1,3-dichloro-2-propanol (glycerol dichlorohydrin) (192), trifluoroacetic acid (64), and trichloroacetic acid (21) for the antimony trichloride. These compounds all produce a blue or violet chromogen that lacks stability and is sensitive to traces of moisture. Carotene also yields a similar chromogen with this reaction, and a correction factor is necessary in the estimation of vitamin A (21, 191).

Bessey *et al* (26) experienced difficulties with the Carr-Price reaction in their micro sample method (0.1 ml or less), and resorted to a previously described method of measuring the absorption maximum of retinol at 328 nm before and after ultraviolet irradiation (49). Carotene was quantitated by a measurement at 460 nm. These investigators incorporated a saponification step to hydrolyze the vitamin esters, but reinvestigation using fluorescence measurement (120) or reaction with 1,3-dichloro-2-propanol (191) showed no increase in yield of retinol by this step. However, saponification was found to be necessary when measuring abnormally high values (e.g., in vitamin A tolerance tests) by the dichloropropanol method (191).

The initial use of fluorescence measurements as a technic for quantification was attempted in 1943 (193). Later a survey of the fluorescence properties of biologic compounds described an activation maximum of 325 nm and a fluorescence maximum at 470 nm for retinol in ethanol (66). Kahan (120) reevaluated the fluorescence approach and devised a sensitive method (0.1–0.2 ml serum) for the measurement of retinol fluorescence in cyclohexane at 490 nm with excitation at 345 nm. Two other methods described the use of photochemically stable acetate (96) or palmitate (186) esters as the reference standards. A study of various solvents for extraction efficiency (120) and fluorescence yield has resulted in the recommendation of xylene as the best (186).

In view of the presence of variable amounts of interfering phytofluene in serum (202), Garry *et al* (79) used a small silicic acid column to separate this polyene from retinol. Alumina and silica gel thin layer chromatography have also been used to fractionate the carotenoids, esters, and vitamin A (63, 206).

DETERMINATION OF VITAMIN A AND "CAROTENE"

(method of Sobel and Snow, Ref 191, modified)

PRINCIPLE

Vitamin A and "carotene" (includes other carotenoids) are split from their protein complexes, and saponification is effected by warming with ethanolic KOH. The sample is extracted with isooctane, and one aliquot is transferred to a microcuvet for measurement of "carotene" content at 450 nm (Fig. 27-1). Another aliquot is evaporated to dryness, the residue is dissolved in chloroform, and then reacted with dichloropropanol. Vitamin A forms a blue color which changes to violet and its absorbance is measured at 550 nm (Fig. 27-2).

REAGENTS

Ethanolic KOH. Add 1 vol of 1 N KOH (56 g/liter) to 10 vol absolute ethanol. This reagent must be prepared fresh. The stock solution of KOH is stable several months. Denatured ethanol, formula 3A (95% ethanol, 5% methanol), is satisfactory.

Isooctane, Reagent Grade

1,3-Dichloro-2-propanol. See *Note 3.*

Stock β-Carotene Standard. Dissolve 2.5 mg in 250 ml isooctane. The product from Eastman Organic Chemicals is satisfactory.

Working β-Carotene Standard, 100 µg/100 ml. Dilute stock 1:10 with isooctane.

Stock Vitamin A Standard. Weigh out 200 mg USP Vitamin A Reference Standard (6 mg vitamin A in 200 mg oil) or 6 mg of crystalline vitamin A (Eastman Organic Chemicals). Dissolve in chloroform, add 1 g butylated hydroxytoluene, and dilute to 100 ml. Store in light-protected container at −20°C.

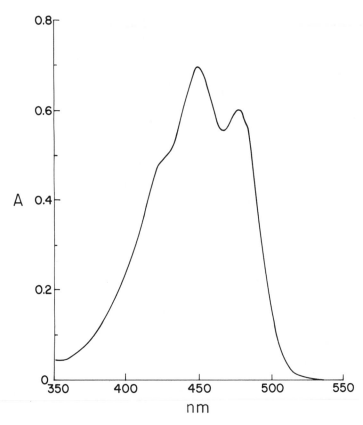

FIG. 27-1. Absorption spectrum of β-carotene (2.7 μg/ml in isooctane). Spectrum recorded with a Cary model 15 spectrophotometer and a 1 cm light path.

Working Vitamin A Standard, 60 μg/100 ml. Transfer 1 ml of stock to a volumetric flask, add 200 mg butylated hydroxytoluene, and dilute with chloroform to 100 ml. This standard is stable 1 month when stored in a light-protected container at 4°C.

PROCEDURE

Procedure should be carried out in the absence of direct light, preferably in dim light.
1. Place 1.5 ml serum into a glass-stoppered, 15 ml centrifuge tube and add 1.5 ml ethanolic KOH solution. Mix, stopper lightly, and heat at 60°C for 20 min.
2. Cool tube, add 4.5 ml isooctane, stopper tightly, and shake for 10 min. A shaking machine is convenient.
3. Centrifuge 3 min at 2000 rpm.
4. For "carotene," transfer 1.0 ml supernate to a Lowry-Bessey cuvet (1 cm light path, capacity 1 ml or less). Place 1.0 ml of working standard

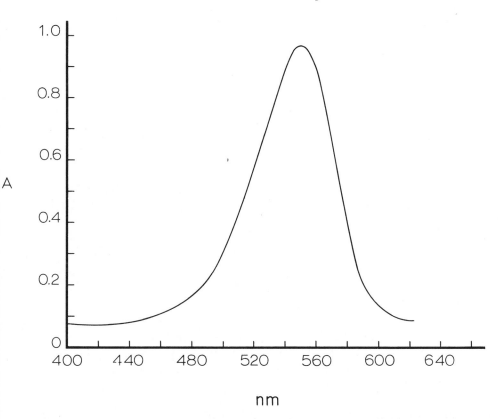

FIG. 27-2. Absorption curve of vitamin A in dichloropropanol: 1 ml chloroform solution of vitamin A (6.43 mg USP/100 ml) plus 4 ml dichloropropanol vs reagent blank at 2 min. Spectrum recorded with a Beckman DK-2A spectrophotometer, 1 cm light path.

solution in a cuvet. Measure absorbances of sample (A_x) and standard (A_s) against a blank of isooctane at 450 nm.

5. For vitamin A, transfer 3.0 ml supernate to a test tube (13 x 100 mm) and evaporate to dryness with stream of N_2 on a 37°C water bath. Rinse down tube walls with 0.5 ml $CHCl_3$; reevaporate.

6. Transfer 1.0 ml working standard solution to a test tube and evaporate to dryness. Rinse tube walls with 0.5 ml $CHCl_3$; reevaporate.

7. Prepare a reagent blank of 0.2 ml $CHCl_3$ plus 0.8 ml dichloropropanol. Mix, being careful not to entrap air bubbles in the solution, and transfer to a Lowry-Bessey cuvet.

8. To the specimen (only one specimen is measured at a time) add 0.2 ml $CHCl_3$ to dissolve residue, and 0.8 ml dicholoropropanol. Mix carefully, allow to stand 1 min, and transfer to the cuvet. The working standard residue is treated in the same way.

9. Measure the absorbances of the standard (A_s) and the sample (A_x) at 550 nm against the reagent blank. Use the maximum reading for calculations.

Calculation (see *Note* 5):

$$\mu g \text{ "carotene"}/100 \text{ ml} = \frac{A_x}{A_s} \times 1.0 \times \frac{4.5}{1.0} \times \frac{100}{1.5}$$

$$= \frac{A_x}{A_s} \times 300$$

$$\mu g \text{ vitamin A}/100 \text{ ml} = \frac{A_x}{A_s} \times 0.6 \times \frac{4.5}{3.0} \times \frac{100}{1.5} - (\mu g \text{ carotene}/100 \text{ ml} \times 0.11)$$

$$= \frac{A_x}{A_s} \times 60 - (\mu g \text{ carotene}/100 \text{ ml} \times 0.11)$$

NOTES

1. Standardization. The weighing out of the β-carotene standard is checked by comparing the calculated absorbance using the absorptivity at 450 nm of 2500 for a 1% solution in isooctane. Small adjustments in the volume of the standard can be made to correct for any difference in absorbances of the standard.

The amount of vitamin A in the chloroform solution may be checked by preparing an aliquot of the stock standard in isooctane and comparing the absorbance with the $A_{1\text{ cm}}^{1\%}$ value of 1830 at 325 nm.

One International Unit (IU) of vitamin A is equal to 0.3 μg of the free alcohol, and 1 unit of β-carotene is equivalent to 0.6 μg. In some countries molar concentrations are being used for reporting clinical laboratory results (120). The conversion factor for μg/ml to μM for retinol is 3.491.

2. Beer's law. The dichloropropanol reaction with vitamin A follows Beer's law in a spectrophotometer to an absorbance of 1.0. Similarly the absorbance of β-carotene is linear to 1.0.

3. Color stability. The color produced with vitamin A is stable 4 min, commencing 1 min after addition of the reagent. Although the color was initially reported to be stable for 2–10 min (192), we found a slow loss after 5 min. On occasion, various lots of dichloropropanol from different suppliers give either a rapidly fading color or a low yield of color. Our studies have shown that addition of 1 g antimony trichloride to 100 g dichloropropanol stabilizes the color. For those reagent lots that give a low but stable color, vacuum distillation at 10–40 mm pressure, discarding the first and last fractions, restores the color yield to an acceptable value.

4. Photodecomposition. All sera and standards should be protected from light. The simplest approach is to work in a room with low light intensity, but low actinic glassware can be used or the containers can be covered with aluminum foil. Stock solutions are stable several months at −20°C in the dark. Working solutions are stable several weeks at 4°C in the dark (120).

The addition of butylated hydroxytoluene to the chloroform improved the stability of vitamin A standard solutions in our laboratory.

5. *Mutual interference.* According to Sobel and Snow (191), 100 μg β-carotene per 100 ml serum contributes about 11 μg equivalents of vitamin A per 100 ml at 550 nm. Similar results have been obtained in our laboratory; therefore vitamin A values are corrected by 11% of the amount of carotene present. Vitamin A does not contribute significantly to the measurement of carotene at 450 nm.

6. *Stability of serum.* Vials of serum that are kept in the dark show no loss of vitamin A or carotene at 30°C for 4 days and for 2 weeks at 4°C (120). Our laboratory checked the stability at 30°C of vitamin A and carotene in sera of six individuals taken before and after oral ingestion of vitamin A (vitamin A absorption test) and no significant change was observed in 4 days.

ACCURACY AND PRECISION

Vitamin A. The accuracy of the dichloropropanol method is indirectly supported by the agreement of the normal range values by this method with the ranges reported with antimony trichloride (43), trichloroacetic acid (21), and fluorometric methods (79, 120).

Precision of the method for vitamin A performed in duplicate is ±15%.

"Carotene." The result obtained is actually "total carotenoids, in terms of β-carotene." The accuracy of the method is impossible to assess with the data available, since the contribution of xanthophyll and lycopene, the other major pigments, to the results has not been determined. Nevertheless, "carotene" values obtained from the dichloropropanol reaction agree quite well with values calculated from the absorbance of petroleum ether extracts at 440 nm (191).

The precision of the method is ±10% at normal levels.

NORMAL VALUES

Vitamin A. The adult normal range of serum vitamin A is about 20–80 μg/100 ml (65–276 IU/100 ml) (63, 120, 166). Values for adult men may be slightly higher than those for women (166). The normals for children over 2 years of age appear to be the same as for adults (9, 79, 175, 199). The normals for infants appear to be somewhat lower, about 10–60 μg/100 ml (199).

Variation between days is about ±20% (175). Some studies (175), but by no means all, have indicated higher levels in autumn and winter than in the spring. Levels are not affected by individual meals but are affected by general dietary intake (175). Seasonal variation, where observed, may be a reflection of change in dietary intake.

"Carotene." The normal adult range for "carotene" in serum is about 50–250 μg/100 ml (88, 213). Phillips *et al* (166) reported a lower normal range in fasting blood samples and slightly higher values for women than for

men. The normals for children over 2 years of age are about the same as for adults (175). Normal values for infants are quite low, e.g., up to about 50 $\mu g/100$ ml in infants less than 1 week old (199). Infants on artificial diets have lower carotene levels than those who are breast-fed. Levels are reportedly higher in autumn and winter than in the spring (175). Day to day variation may be up to ±50% (175). Such seasonal and between-day variations presumably are reflections of variations in dietary intake.

VITAMIN A ABSORPTION TEST

Absorption of vitamin A from the gastrointestinal tract following an oral dose, as signaled by an increase in serum level, is used as a method for evaluating fat absorption. The test is also referred to as "vitamin A tolerance test." As an index of fat absorption it can only be regarded as a screening test since it is open to criticism on a number of points, including the facts that the absorption curve is not always an accurate measure of vitamin A absorption from the intestine, and that there is no close correlation between fat absorption and vitamin A absorption (20, 160). Accuracy can be improved by preliminary priming of the patient with vitamin A to ensure adequate tissue storage when the test is performed and by simultaneous measurement of fecal loss (20).

One of the principal factors in absorption is the degree of "wetting" or dispersion of the vitamin A administered. Chesney and McCoord, who introduced the test in 1934 (48), used vitamin A in oil. Absorption is greater and more rapid, fecal loss is decreased, and serum levels are increased, however, when the vitamin A is given as an aqueous dispersion made with a Tween compound (20, 167). A number of modifications of the test have been proposed (160, 163, 167), but most studies used approximately 7500 IU/kg body weight and found the peak serum level to occur between 4 and 5 hours.

RECOMMENDED PROCEDURE FOR PERFORMANCE OF TEST

(Taken mostly from Paterson and Wiggins, Ref 163)

Regular meals are not withheld, but no vitamin A concentrates are given for 24 hours preceding the test. A control sample of serum is taken for analysis of carotene and vitamin A. The test dose is 7500 IU/kg body weight with a maximum of 350 000. Oleum percomorph (60 000 IU/g) can be used, but this preparation causes nausea in some individuals. Other oily preparations (e.g., arachis oil, peanut oil) can be used. The vitamin A can be in the form of either the free alcohol or an ester such as the palmitate. Another serum sample is taken at 4 hours.

INTERPRETATION

The degree of increase in vitamin A concentration in serum in normal individuals is not completely agreed upon. Paterson and Wiggins regarded an increase to 500 or more IU (150 μg)/100 ml as normal, whereas others (48, 160) believe that normals should increase to at least 1000 IU (300 μg)/100 ml. The serum carotene level is unchanged (167).

THIAMINE (Vitamin B_1)

Thiamine is present in plasma as the free form and in red cells as the phosphorylated derivative. This derivative, thiamine diphosphate, is the coenzyme for many enzymatic reactions.

THIAMINE

To a certain extent, thiamine is excreted in the urine along with the metabolites of either the pyrimidine or thiazole moieties. The ratio of thiamine in red cells to that in plasma is 5:1 (14), with almost 90% of the red cell content as the phosphorylated derivative and, conversely, with negligible amounts of the diphosphate in plasma (40, 84).

The preferred method for determination of thiamine in body fluids and tissues is by an oxidative conversion with ferricyanide in an alkaline solution to thiochrome. This product is highly fluorescent and is extracted from the aqueous phase with isobutanol for subsequent fluorometric measurement (210). Oxidation of thiamine diphosphate with ferricyanide yields a thiochrome derivative that is not extractable with isobutanol (123). Kinnersley and Peters (124) were the first to discover that diastase is suitable for hydrolysis of the phosphate derivative to thiamine. By pretreatment of blood or tissues with any of several phosphatase preparations, total thiamine content can be determined by the thiochrome-isobutanol extraction technic. Mickelson and Yamamoto (148) reviewed the many modifications of the thiochrome method from both the chemical aspects and applications to natural products. They concluded that none of the reviewed methods is completely suitable for assay of thiamine in natural products, and therefore they recommended a procedure that is a composite of several methods. Haugen (100) presented a method that eliminates the blank in urine samples by incubation at pH 10 to destroy the thiamine prior to the ferricyanide oxidation step. A simplified thiochrome method employing acid hydrolysis in an autoclave followed by enzyme hydrolysis afforded the elimination of sample deproteinization, dilution, and purification by an adsorption column (156). Another method applies a deproteinized solution to a gel filtration column and determines both the free and phosphoric acid esters by the thiochrome method without hydrolysis (215).

Other chemical methods have been reported that utilize diazotized *p*-aminoacetophenone (146) or ethyl-*p*-aminobenzoate (125). The red products can be measured photometrically, but these methods lack the sensitivity of the thiochrome method.

Microbiologic assays for thiamine and its urinary metabolites have been reviewed by Baker and Frank (11). Their conclusions were that *Ochromonas danica* requires intact thiamine and is the recommended test organism for this assay. Other microorganisms can utilize the intact thiamine, the pyrimidine moiety, the thiazole moiety, or combinations of these products.

An enzymatic method for assessment of thiamine levels has been developed by Brin *et al* (36, 37) that utilizes the measurement of red cell transketolase activity before and after the addition of thiamine diphosphate. The percent stimulation of enzyme activity by the added thiamine diphosphate is related to thiamine adequacy; thus normals show 0%–15% stimulation, marginally deficient 15%–25%, and severely deficient ≥ 25%. Dreyfus (62) concluded that as little as a 10% elevation is compatible with thiamine deficiency.

Normal ranges of thiamine differ significantly when the levels are determined by the thiochrome method and by microbiologic assay. By the thiochrome method (156) the normal ranges for whole blood and serum are 11–48 and 0–20 ng/ml, respectively. Normal ranges by *Ochromonas danica* assay (11) for whole blood and serum are 25–75 and 15–42 ng/ml, respectively.

RIBOFLAVIN (VITAMIN B$_2$)

The chemical forms of vitamin B$_2$ in blood and tissues are free riboflavin, riboflavin monophosphate (FMN), and flavin adenine dinucleotide (FAD). Riboflavin is excreted in urine along with a compound called uroflavin (72). Thin layer chromatography of the riboflavin fraction of urine indicates a mixture of compounds (101).

RIBOFLAVIN

Fluorometric (72) and photometric (68) methods requiring pretreatment with potassium permanganate have been described for serum and urine. The fluorometric method has been extended to the determination of free riboflavin, FMN, and FAD in serum, and total riboflavin in red and white blood cells (39). Attempts at eliminating the interfering compounds present in urine include adsorption onto lead sulfite (68) or Florisil (153). Morell and Slater (153) reported the presence of "apparent" riboflavin and precursors of "apparent" riboflavin in urine; they also reported that these products could be affected by the reducing agents, oxidizing agents, or Florisil that had been used in earlier methods. A simple method utilizing a talc column to adsorb urinary riboflavin followed by stepwise development to remove extraneous material and elution of a riboflavin fraction was described by Wahba (208). The sensitivity and specificity of this latter method was subsequently increased by direct fluorometry of a thin layer chromatogram of the riboflavin fraction from the talc column (101).

Microbiologic assays for riboflavin in blood and urine have been reported using either *Lactobacillus casei* (190, 198) or *Tetrahymena pyriformis* (16). Urea inhibits *Lactobacillus casei* assays of urine, especially at low levels (117).

The normal range of serum riboflavin in adults is 2.6–3.7 $\mu g/100$ ml as determined by a fluorometric method (39), or 4–24 $\mu g/100$ ml by microbiologic assay (16). Urine riboflavin levels vary with diet, and the range is dependent on the specificity of the method (101). Haworth *et al (101)* demonstrated the nonspecificity of the fluorometric method when applied to urine.

VITAMIN B$_6$

Vitamin B$_6$ is the designation used to refer to either one or all three of the following compounds:

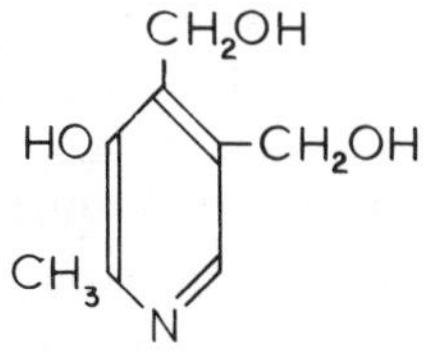

PYRIDOXINE
(2-METHYL-3-HYDROXY-4,5-
  DIHYDROXYMETHYLPYRIDINE)

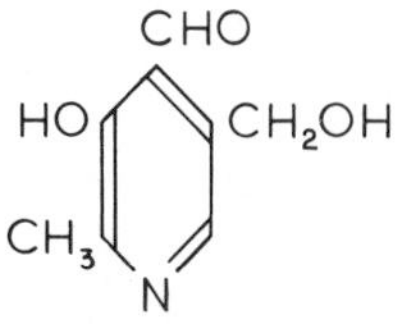

PYRIDOXAL

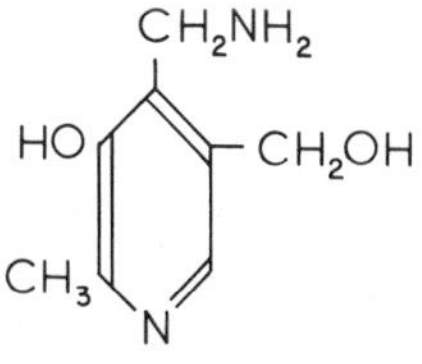

PYRIDOXAMINE

The phosphorylated derivatives are the active coenzyme forms of this vitamin. Of the three coenzymes, pyridoxal phosphate is probably the principal form in blood and tissue cells. Studies with radioactively labeled pyridoxine have shown rapid incorporation into the red cells and conversion to pyridoxal phosphate (4). Pyridoxamine was less efficiently converted to the active forms by the red cells. Although protein binding of the coenzymes for many enzymes is well established, the binding of this vitamin by plasma and tissue proteins is still undefined. The nature of the protein binding is an important factor in the establishment of a valid method for a total vitamin B_6 determination. The major urinary product of vitamin B_6 is 4-pyridoxic acid (3-hydroxy-4-carboxy-5-hydroxymethyl pyridine) and accounts for a major portion of an ingested dose (113, 172).

Storvick and Peters (196) reviewed the microbiologic assays of vitamin B_6. Six different microorganisms and 14 hydrolysis procedures have been applied to the measurement of vitamin B_6 in blood. Although hydrolysis increases the amount of assayable vitamin, the response of the different test organisms may differ markedly when protein-free filtrate is tested before and after hydrolysis. Other unanswered problems related to hydrolysis are: (a) generation of an inhibitor, (b) rebinding of the vitamin to a component in the hydrolysate, and (c) destruction of the vitamin in different biologic samples by the hydrolysis technic.

Several fluorometric methods have been described for estimating vitamin B_6 compounds in blood. Fukita *et al* (76) oxidized pyridoxine and pyridoxal to 4-pyridoxic acid and converted this acid to the fluorescent lactone by heating in an acid medium. Another fluorometric method employs peroxide to destroy the fluorescence of pyridoxal and ultraviolet irradiation for pyridoxamine to measure each compound selectively (52). Reaction of pyridoxal and pyridoxal phosphate with cyanide to yield fluorescent derivatives of these compounds has been reported (33). Contractor and Shane (51) utilized the fluorescence of the cyanohydrin and lactone derivatives in the quantification of the vitamin B_6 compounds and coenzymes. The fluorometric determination of 4-pyridoxic acid in urine (113) was modified to a more specific test by the use of ion exchange columns to eliminate interfering substances (51, 172, 217).

Since pyridoxal phosphate is the principal coenzyme for many enzymatic reactions, application of these reactions to the evaluation of the vitamin status of human subjects has been attempted. Boxer *et al* (35) described a manometric method for pyridoxal phosphate that involves the decarboxylation of tyrosine. Activation of a yeast apotransaminase by the addition of pyridoxal phosphate has also been described (109). Another approach to the enzymatic method for evaluation of the vitamin B_6 nutritional status is measurement of the *in vitro* stimulation of erythrocyte glutamic oxaloacetic transaminase activity by pyridoxal phosphate (170). Other investigators studied glutamic-pyruvic acid transaminase activation by the addition of coenzyme as a criterion of vitamin B_6 deficiency (47, 50).

Pyridoxine deficiency has also been assessed by the tryptophan challenge test of Greenberg *et al* (90). This test is based on ingestion of 10 g tryptophan and the excretion of increased amounts of 3-hydroxykynurenine and xanthurenic acid if a vitamin B_6 deficiency exists. Reinvestigation of this test has shown an inverse relationship of urinary vitamin B_6 with xanthurenic acid excretion and deficiency status (8). Excretion of 3-hydroxykynurenine is of limited use in evaluation of vitamin B_6 requirements.

The range of blood pyridoxal levels is 20—90 ng/ml (51, 196) and that of urinary pyridoxic acid excretion levels is 0.65—4.8 mg/24 hours (51, 217). A composite range for total vitamin B_6 in blood by microbiologic assay is 30—255 ng/ml (196).

VITAMIN B_{12}

This vitamin is a complex molecule consisting of a porphyrin-like structure, called a corrin nucleus, with a central cobalt atom. A nucleotide, 5,6-dimethylbenziminazole-1'α-ribosyl-3'-phosphate, linked with 1-amino-2-propanol and a propionic acid group, are also part of the total molecule. *Cobalamin*, vitamin B_{12} without the cyanide group, is the natural form of the molecule. Two other forms are *hydroxocobalamin* (HO linked to the cobalt ion) and *aquocobalamin* (water linked to the cobalt ion). Cyano, hydroxo, and other forms of vitamin B_{12} are biologically converted to the two types of coenzymes, 5'-desoxyadenosyl-cobalamin or methylcobalamin. Both of these substituent groups are linked to the cobalt atom. The stable cyanocobalamin is not naturally occurring but is formed in the isolation of the vitamin. These coenzymes are mainly involved in the metabolism of methyl malonic acid, the transfer of methyl groups, and the conversion of ribonucleotides to desoxyribonucleotides in mammalian systems (138, 194).

There is a slightly higher level of B_{12} in plasma than in red cells. In plasma the vitamin is bound by three glycoproteins that have been designated as transcobalamin I, II, and III (31, 95). Type I migrates electrophoretically in the α-globulin region, whereas II and III migrate as β-globulins. Fasting increases the binding capacity of type I and II but not that of type III (31). Cobalamin-binding proteins are also found in gastric juice (intrinsic factor), saliva, tears, milk, leukocytes, erythrocytes, and other cellular fluids (87).

Before 1960 assays of vitamin B_{12} were amenable only to microbiologic technics (86). Many of these assays were suitable, but *Euglena gracilis* was the most sensitive, and *Ochromonas malhanesis* the most specific but insensitive. In a study of the activities of the different cobalamins with *E. gracilis*, the methyl and 5'-desoxyadenosyl cobalamins were found to be biologically less active than equivalent concentrations of cyano- and hydroxocobalamins (3).

In 1961 Barakat and Ekins (19) published a simple radioisotope method that utilized the vitamin B_{12}-binding proteins in plasma and separated the free and bound fractions by dialysis. This technic of "saturation analysis" involves saturation of the binding sites by exogenous vitamin and cobalamin labeled with radioactive cobalt; excess vitamins remain in the free form (18). The ratio of free to bound fractions depends on the amount of exogenous vitamin B_{12}. Other methods for separating the free and bound fractions include an immunosorbent technic (44, 214), precipitation of the bound fraction with barium hydroxide-zinc sulfate (181), and adsorption of the free fraction onto albumin- or hemoglobin-coated charcoal (132) or DEAE cellulose (75, 203). The binding protein used in these assays has generally been either intrinsic factor (44, 132, 214) or plasma proteins (18, 75, 181). In most methods binding by intrinsic factor was at a pH of less than 6.0. Two studies of the pH binding curve of intrinsic factor shows a considerable drop in binding below pH 6.0 (83, 188). This pH effect may be the reason for some of the inconsistent results that have been reported in those vitamin B_{12} assays using intrinsic factor. Another problem with intrinsic factor reported by other investigators (157), was the adsorption of the factor to glass at pH 4.0. Addition of albumin or use of siliconized glass tubes or polystyrene tubes prevents this effect.

Although boiling releases a considerable amount of the protein-bound vitamin B_{12}, the addition of cyanide to the serum sample increases the recovery from 15% to 50% (44, 74, 203). Others (44, 137) incorporated the cyanide in the buffer system of the assay. Ceska and Lundkvist (44) used a glutamic acid-KCN buffer that after boiling gave no protein precipitate at a final pH of 4.1.

Vitamin B_{12} binding capacity of serum was investigated by Miller (151) using a simple charcoal adsorption technic. Later, radioactive cobalt provided a more rapid method (92) than the microbiologic assays. Albumin-coated charcoal was introduced to eliminate charcoal adsorption of both free and protein-bound vitamin B_{12} (85). Another approach was the use of gel filtration to separate the free and bound forms of the vitamin for the estimation of binding capacity (93).

An expression of vitamin B_{12} deficiency is also seen by increased excretion of methyl malonic acid (53). Photometric (81) and thin layer chromatographic technics (94) have been described to measure and detect this compound.

DETERMINATION OF VITAMIN B_{12} IN SERUM

(method of Lau et al, Ref 84, and Ceska and Lundkvist, Ref 44, modified)

PRINCIPLE

The protein-bound forms of vitamin B_{12} are dissociated by heating at an acid pH in the presence of cyanide ions to form stable cyanocobalamin. After the addition of ^{57}Co-vitamin B_{12}, a measured amount of binding protein (intrinsic factor) is added and the mixture is allowed to equilibrate. The free ^{57}co-vitamin B_{12} fraction is separated from the protein-bound fraction by adsorption onto hemoglobin-coated charcoal and subsequent centrifugation. The protein-bound fraction in the supernate is decanted into another tube and counted. Concentration of vitamin B_{12} in serum is calculated by comparing the counts or observed with results obtained from vitamin B_{12} standard solutions treated in the same manner.

REAGENTS

All glassware, after washing, must be soaked in 6 N HCl and rinsed with deionized or distilled water.

NaCl, 0.85%

KCN 0.4%

L-Glutamic Acid Solution, 0.6%. Dissolve 6 g of L-glutamic acid in 1 liter of water by heating at $50°-60°C$. This solution is stable.

L-Glutamic Acid–KCN Solution, pH 3.3. Add 2.0 ml 0.4% KCN to 200 ml L-glutamic acid solution, mix, and check pH on a pH meter. Prepare this solution fresh.

L-Glutamic Acid–KCN Solution, pH 4.1. Adjust 100 ml of the above solution to pH 4.1 with 1 N NaOH on a pH meter. Prepare fresh.

Hemoglobin Solution, 10 g/100 ml. Prepare hemoglobin solution from discarded human red cells. Wash the cells 3 times with 0.85% NaCl and hemolyze with an equal volume of water. Add 1/2 vol $CHCl_3$ and vigorously shake the mixture for 5 min. Centrifuge at 3000 rpm for 15 min and pass the upper hemoglobin layer through Whatman 1 filter paper. The hemoglobin content is determined and the filtrate diluted with water to a final concentration of 10 g/100 ml. Frozen aliquots (15 ml) are stable for 6 months.

Hemoglobin-Coated Charcoal. Add 25 g charcoal (Norit A Charcoal, Neutral, Pharmaceutical Grade, Amend Drug and Chemical Co. has been found to be satisfactory) to 500 ml water in a 1 liter flask. Add 12.5 ml hemoglobin solution to a 500 ml volumetric flask and dilute to volume with water. Pour the hemoglobin solution into the charcoal suspension and mix. This mixture containing 0.125% hemoglobin and 2.5% charcoal is stable for 1 month at $4°C$.

Human Albumin Solution, 5%. Dissolve 0.75 g human albumin in 15 ml 0.85% NaCl. Prepare this solution fresh.

Human Serum Pool. Dispense 2.5 ml aliquots of a serum pool into vials for storage at $-20°C$. Thaw one vial for each run and mix by gentle inversion before use. This pool serves as the *serum blank* to correct for the nonspecific binding of radioactive vitamin B_{12}.

Stock Cyanocobalamin Solution, 16 ng/ml. Dilute exactly 0.20 ml of Cyanocobalamin Injection USP solution (Squibb Rubramin p.c., 100 μg/ml) to 10 ml with 0.85% NaCl in a volumetric flask. Pipet 1.60 ml of this dilution into a 200 ml volumetric flask and dilute to volume with glutamic acid-KCN pH 4.1 solution. This solution is stable 1 month in the refrigerator.

Working Cyanocobalamin Solutions. Make the following serial dilutions into plastic vials with plastic disposable pipets. These dilutions are stable 1 week in the refrigerator. All dilutions are made with the glutamic acid-KCN pH 4.1 solution.

pg/ml	Dilution factors
1600	1.0 ml of 16 ng/ml standard + 9.0 ml diluent
800	5.0 ml of 1600 pg/ml standard + 5.0 ml diluent
400	5.0 ml of 800 pg/ml standard + 5.0 ml diluent
200	5.0 ml of 400 pg/ml standard + 5.0 ml diluent
100	5.0 ml of 200 pg/ml standard + 5.0 ml diluent
50	5.0 ml of 100 pg/ml standard + 5.0 ml diluent

^{57}*Cobalt-Cyanocobalamin, 400 pg/ml.* Transfer a volume of ^{57}Co-cyanocobalamin to a flask and add a calculated amount of 0.85% NaCl to give a final concentration of 400 pg/ml; 0.1 ml of this solution should contain a minimum of 10 000 counts per minute (CPM)—hence the specific activity of the initial product should be about 200 μCi/μg or greater. The product from Amersham Searle has been found to be suitable. Store in light-protected container in refrigerator.

Stock Intrinsic Factor Solution, 1 mg/ml. Dissolve 20 mg Intrinsic Factor Concentrate (Squibb 0690, 1 NF-Xl unit per capsule) in 20 ml 0.85% NaCl in a polyethylene bottle. Mix well and store 1 ml aliquots at $-20°$C.

Working Intrinsic Factor Solution. Add 50 ml 0.85% NaCl to the 1 ml stock solution and mix well. This solution is prepared fresh. *N.B.* The vitamin B_{12} binding capacity of intrinsic factor may be different for individual lot numbers. Therefore serial dilutions of each new lot of intrinsic factor are prepared to contain between 10–40 μg factor/ml. Test each concentration of intrinsic factor with 0 and 400 pg/ml of vitamin B_{12} according to the procedure. Calculate for each concentration:

$$\text{ratio of } {}^{57}\text{Co-}B_{12} \text{ bound at 0 pg/ml} = \frac{\text{CPM}_{0 \text{ pg supernate}}}{\text{CPM}_{\text{total counts}}}$$

$$\text{ratio of } {}^{57}\text{Co-}B_{12} \text{ bound at 400 pg/ml} = \frac{\text{CPM}_{400 \text{ pg supernate}}}{\text{CPM}_{0 \text{ pg supernate}}}$$

The concentration of intrinsic factor that yields ratios of 0.5±0.1 is optimal for use in the assay.

PROCEDURE

1. Set up the following 15 x 125 mm polypropylene test tubes. Only the standard curve tubes are in duplicate. Add the appropriate volumes of each solution to the respective tubes as designated.

	Standard aliquot	5% Albumin	pH 4.1 Buffer	pH 3.3 Buffer	Serum aliquot
Total counts			2.50		
Standard blank			1.50		
0 pg/ml	0.20	0.10	0.70		
50 pg/ml	0.20	0.10	0.70		
100 pg/ml	0.20	0.10	0.70		
200 pg/ml	0.20	0.10	0.70		
400 pg/ml	0.20	0.10	0.70		
800 pg/ml	0.20	0.10	0.70		
1600 pg/ml	0.20	0.10	0.70		
Serum blank			0.50	0.80	0.20
Serum sample				0.80	0.20

2. Mix the contents and place a marble or boiling cap on each tube.
3. Immerse tubes in a boiling water bath for 15 min. Be sure the level of water bath is above the level of the liquid contents of the tubes.
4. Cool tubes in cold running tap water.
5. Add 0.10 ml radioactive vitamin B_{12} to all tubes and mix well.
6. Add 0.50 ml of intrinsic factor to standard and serum-containing tubes *only*, mix, and allow to stand 30 min. (Those not receiving intrinsic factor are the Total Counts, Standard Blank, and Serum Blank tubes.)
7. Mix the hemoglobin-charcoal solution by swirling to insure a uniform suspension, and dispense 1.0 ml into each tube with the exception of the "total counts" tube. The suspension must be well mixed *between* each addition to insure reproducibility in the method.
8. Mix each tube thoroughly for 5 sec.
9. Immediately centrifuge all tubes for 15 min at 2500 rpm (see *Note 4*).
10. Without delay, carefully decant the supernates to counting tubes, cap, and count 1 min in a gamma spectrometer.

Calculation:

Subtract the counts per minute (CPM) of the standard blank from the CPM of each standard. Plot the net CPM for each standard on linear graph paper and the picograms of each standard on the arithmetic scale (see example in Figure 27-3). Subtract the CPM of the serum blank from each serum sample and read the pg/ml from the curve.

NOTES

1. Serum vs plasma. Either may be used with the exception of heparinized plasma. Heparin binds vitamin B_{12}, and this complex is fairly resistant to the hydrolysis step; hence falsely low results are obtained.

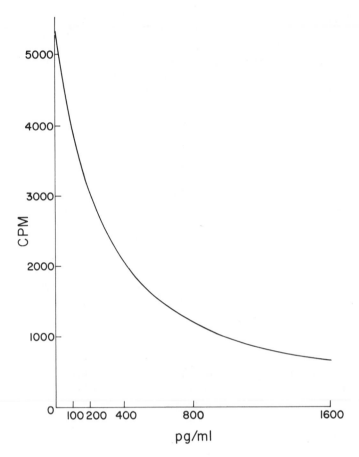

FIG. 27-3. A representative standard curve for the determination of serum vitamin B_{12}.

2. *Intrinsic factor.* Care must be exercised in the preparation and standardization of the binding protein solution, otherwise considerable variation can occur in the sensitivity and precision of the test. The limited stability of the working solution of intrinsic factor is due to the nonspecific adsorption of the protein by glass (157).

3. *Hydrolysis step.* Heating the boiling water bath evenly and a level of water at least 1 in. above the liquid content of the tubes are essential for uniform and consistent hydrolysis between runs. The presence of opalescence due to denatured protein does not affect the assay, but in the event of excessive turbidity the solutions may be filtered through filter paper without a measureable loss of the vitamin (137).

4. *Centrifugation.* It is important to obtain a supernate clear of charcoal particles for counting, otherwise charcoal fines in suspension result in a low answer and variable results. Periodically, the supernate should be examined to insure that a clear solution is obtained with the specified speed and time. A relative centrifugal force of 1250 *g* has been found to be adequate.

5. *Stability.* Serum is stable at room temperature or 4°C for 3 weeks provided no bacterial growth occurs.

ACCURACY AND PRECISION

Good agreement of values by the radioisotopic and microbiologic technics has been reported by numerous investigators (18, 83, 132, 181). Recoveries of added vitamin B_{12} range from 95% to 105%.

Precision varies from ±15% over a large part of the normal range up to 20% at the low and high ends of the range.

NORMAL VALUES

Normal serum vitamin B_{12} values by the method presented range from 300–1000 pg/ml. There is no sex difference in the normal range. This range is similar to those reported by other investigators (80, 214). Serum vitamin B_{12} levels in pregnant women were reported by Craft *et al* (54) to be slightly lower. Vitamin B_{12} levels in erythrocytes range from 85 to 225 pg/ml packed cells (99).

VITAMIN B_{12} UNSATURATED BINDING CAPACITY OF SERUM

(method of Gottlieb et al, Ref 85, modified)

PRINCIPLE

To a buffered serum sample is added an excess of [57] Co-vitamin B_{12} to saturate the transcobalamins. The excess radioactive vitamin not bound to the transcobalamins is removed by adsorption to hemoglobin-coated charcoal and centrifugation. A fraction of the supernate containing the vitamin B_{12} saturated proteins is removed and counted.

REAGENTS

Phosphate Buffer, 0.1 M, pH 7.0. Dissolve 4.55 g KH_2PO_4 and 9.35 g Na_2HPO_4 in water and dilute to 1 liter.

HCl, 0.0001 N

NaCl, 0.85%

Stock Cyanocobalamin Solution, 100 ng/ml. Dilute exactly 1.00 ml of Cyanocobalamin Injection USP solution (Squibb Rubramin p.c., 100 μg/ml) to 1 liter with 0.0001 N HCl. This solution is stable 1 month in the refrigerator.

Working Cyanocobalamin Solution, 1000 pg/ml. Dilute 1.00 ml of stock solution to 100 ml with 0.85% NaCl. Prepare this solution fresh.

[57]*Cobalt-Cyanocobalamin Solution 12.5 μCi/liter, 1000 ng/liter.* Transfer a volume of the radioactive cyanocobalamin solution containing 12.5 μCi to a 1 liter volumetric flask. Calculate the nanograms (N) of vitamin B_{12} in that volume. Since the total amount of vitamin B_{12} required (radioactive plus nonradioactive) is 1000 ng/liter, 1000 − N is equal to the volume in milliliters of the 1000 pg/ml nonradioactive cyanocobalamin solution to be added to the flask. Add 0.0001 N HCl to 1 liter.

Example: If the ^{57}Co-Vitamin B_{12} obtained has 10 μCi/ml and 100 ng/ml, then 1.25 ml of the radioactive solution is required to prepare the 12.5 μCi/liter solution. This aliquot contains 125 ng; therefore the amount of nonradioactive cyanocobalamin to be added is $1000 - 125$ ng/liter = 875 ng. The volume of 1000 pg/ml of cyanocobalamin to be added is 875 ml.

Hemoglobin-Coated Charcoal. Prepared as described in the vitamin B_{12} method.

PROCEDURE

1. Set up the following tubes in duplicate (it is recommended that the test be run in duplicate; see under *Accuracy and Precision*):
   Standard: 2.5 ml buffer
   Blank: 2.5 ml buffer
   Unknown: 0.3 ml serum + 2.2 ml buffer
2. Add 1.0 ml of ^{57}Co-cyanocobalamin to the Blank and Unknown. Add 2.5 ml of radioactive vitamin to the Standard.
3. Mix the contents of all tubes for 10 sec and allow to stand for 10 min.
4. Mix hemoglobin-coated charcoal reagent well and add 2.0 ml to Blank and Unknown.
5. Mix tube contents for 10 sec and centrifuge at 2500 rpm for 15 min.
6. Immediately after centrifugation transfer 2.0 ml of the supernate from each tube to counting tubes.
7. Count for 1 min the Blank (CPM_B), Standard (CPM_S), and Unknown (CPM_X) tubes.

Calculation:

pg unsaturated vitamin B_{12} binding capacity/ml serum =

$$\frac{CPM_X - CPM_B}{CPM_S - CPM_B} \times \text{pg vit } B_{12} \times \frac{\text{ml final volume reaction mixture}}{\text{ml vol supernate} \times \text{ml vol of serum}}$$

$$= \frac{CPM_X - CPM_B}{CPM_S - CPM_B} \times 1000 \times \frac{5.5}{2 \times 0.3}$$

$$= \frac{CPM_X - CPM_B}{CPM_S - CPM_B} \times 9167$$

NOTES

1. Serum vs plasma. Serum is preferred because artifacts can be introduced by the use of heparinized or EDTA-treated plasma (95, 132).

2. Vitamin B$_{12}$ solution. Protection from undue light exposure is recommended because exposed solutions have been reported to yield higher binding values (93).

3. Hemoglobin-coated charcoal supernate. It is essential that the supernate be removed soon after centrifugation, otherwise the sedimented charcoal tends to become more easily dislodged from the bottom of the tube by the pipeting step.

4. Nutritional factors. At least 72 hours from the last injection of vitamin is necessary for the binding capacity to return to its baseline level (93); therefore it is recommended that the subject be withdrawn from the vitamin supplements and an overnight fasting sample be collected for testing. A 12-hour fast increases the total unsaturated binding capacity (31).

5. Stability. Serum samples are stable for 4 days at room temperature and for 7 days at −15°C.

ACCURACY AND PRECISION

Unsaturated vitamin B$_{12}$ binding capacity is a reflection of binding by transcobalamin I, II, and III, and therefore the value determined is the result of the combined effect of these three proteins. According to Gilbert *et al* (80), approximately 20% of the binding capacity is due to the α-globulin fraction and 80% to the β-globulins. Distribution of the binding capacities of the three transcobalamins has been reported to be 6%, 78%, and 15% for transcobalamin I, II, and III, respectively (31).

No definitive statement can be made about the accuracy of this test. Relative to the binding of vitamin B$_{12}$ by plasma proteins, Hall (95) enumerated the factors of nonspecific binders, the difference in binding of hydroxycobalamin and cyanocobalamin, and the effects of anticoagulants, freezing, and thawing, and low ionic strength buffers.

The precision of this test when run in single is about ±7%–8%. It is recommended, therefore, that the test be run in duplicate.

NORMAL VALUES

In our laboratory the normal range for vitamin B$_{12}$ unsaturated binding capacity in 40 subjects has been found to be between 1000–2000 pg/ml. Our normal range is in agreement with Gilbert *et al* (80) who used a similar technic. Measurement of the serum vitamin B$_{12}$ binding capacity by the gel filtration technic has yielded a lower range (93).

VITAMIN C (ASCORBIC ACID)

This vitamin exists in the body mainly as ascorbic acid and to a very minor extent as dehydroascorbic acid.

L-ASCORBIC
ACID

DEHYDROASCORBIC
ACID

Studies on the catabolic rate of carbon-14-labeled ascorbic acid have shown the principle excretory products to be ascorbic acid, dehydroascorbic acid, and oxalate (7). There is some uncertainty about the origin of dehydro-ascorbic acid since it can be formed readily by simple oxidation of ascorbic acid in an aqueous solution. Furthermore, chromatographic analysis of some of the radioactive labeled ascorbic acid preparations have shown the presence of both dehydroascorbic acid and diketogulonic acid in varying proportions. The metabolic pathway in the degradation of ascorbic acid to oxalate via diketogulonic acid is not considered a major pathway, other-wise appreciable amounts of radioactive carbon dioxide would have been formed by the spontaneous decarboxylation of the unstable diketo acid (7). Approximately 23% of an injected dose of ^{14}C-labeled ascorbic acid is excreted as urinary oxalate. Further metabolic studies with radioactive labeled ascorbic acid have described the isolation of ascorbate-3-sulfate in the urine of human subjects with scurvy (6). Interestingly, it was reported that under conditions of high intake of ascorbic acid, most of the vitamin is excreted unchanged and that rapid decomposition of the free acid occurs in urine. Mumma *et al* (154) have synthesized and given chromatographic systems along with various chemical sprays for the detection of this ascorbic acid conjugate.

Although Lowry *et al* (136) clearly showed the relationship of ascorbic acid in white blood cells to total body concentration in 1946, many reports have since appeared describing methods for blood (103). Conclusions reached on reexamination of leukocyte and plasma ascorbic acid concentra-tions are: (a) Plasma levels are indicative of ascorbic acid metabolic turnover rate. (b) Leukocyte concentrations provide a measure of availability of the vitamin for storage. (c) The concentration in leukocytes alone is not indicative of the tissue status of ascorbic acid (133).

Methods and technics employed in the assay of ascorbic acid before 1960 are comprehensively reviewed by Henry (103) and Roe (178). The 2,4-dinitrophenylhydrazine method of Roe and Keuther (179) has been reexamined in regard to its specificity several times (177, 218). One of the critical factors in preparing the derivative is the temperature of the reaction and the incubation time (218). The most suitable conditions defined were 37°C for 3–4 hours or 60°C for 1.5 hours (218).

Some of the more recent adaptations and approaches to the quantification of ascorbic acid are the quantitative reduction of mercuric to mercurous chloride by urinary ascorbic acid (130), resin extraction of urinary ascorbic acid and measurement by 2,6-dichloroindophenol (115), reduction of ferric to ferrous ion and photometric measurement with bathophenanthroline (205), reaction with diazotized 4-nitroaniline-2,5-dimethoxyaniline (147), oxidation of ascorbic acid with 2,6-dichloroindophenol followed by formation of hydrazones (165), and the reduction of 1,2-naphthoquinone by ascorbic acid to the highly fluorescent dihydro derivative (112). Two automated methods have been described, one that utilizes the dichloroindophenol dye (78), and the other the 4-methoxy-2-nitroaniline reaction (216).

DETERMINATION OF ASCORBIC ACID

(method of Roe, Ref 176, modified)

PRINCIPLE

A protein-free filtrate of serum, plasma, or urine is prepared with trichloroacetic acid. Charcoal is added and the ascorbic acid in the filtrate is oxidized to dehydroascorbic acid. Dehydroascorbic acid is coupled with 2,4-dinitrophenylhydrazine to form the 2,4-dinitrophenylosazone. Treatment of the osazone with strong sulfuric acid causes rearrangement to yield a reddish complex which is measured at 515 nm (Fig. 27-4).

FIG. 27-4. Absorption spectrum of 2,4-dinitrophenylhydrazine derivative of dehydroascorbic acid (5 μg/ml). Curve recorded with a Cary model 15 spectrophotometer and a 1 cm light path.

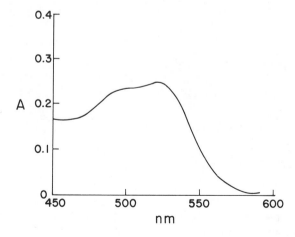

REAGENTS

Trichloroacetic Acid, 7.5%. Dissolve 75 g acid in water and dilute to 1 liter.

Trichloroacetic Acid, 4%. To 400 ml of the 7.5% solution add 350 ml water and mix.

H_2SO_4, *9 N.* Carefully add 25 ml conc H_2SO_4 to 75 ml water with mixing and cooling. Dilute to 100 ml.

2,4-Dinitrophenylhydrazine Reagent. Dissolve 2.0 g 2,4-dinitrophenyl-hydrazine (2,4-DNP) in 100 ml 9 N H_2SO_4. Add 4.0 g thiourea and, when completely dissolved, filter the reagent through Whatman 42 filter paper. This reagent is stable 1 month in the refrigerator and may be refiltered when necessary.

Acid-Washed Charcoal. May be purchased as acid-washed Norit (Fisher Scientific Co. or Pfanstiehl Labs Inc.).

H_2SO_4, *85%.* To 100 ml water carefully add 900 ml conc H_2SO_4 with mixing and cooling.

Oxalic Acid, 0.5%

Stock Ascorbic Acid Standard, 25 mg/100 ml. Dissolve 25 mg ascorbic acid in 100 ml 0.5% oxalic acid. This standard is stable for 1 month at 4°C.

Working Ascorbic Acid Standard, 0.5 mg/100 ml. Dilute 1.00 ml of the stock standard to 50 ml with 4% trichloroacetic acid. Prepare fresh for each run.

PROCEDURE

1. Place 7.0 ml 7.5% trichloroacetic acid in a tube and then add dropwise with mixing 3.0 ml serum or plasma. For urine, mix 0.5 ml with 9.5 ml of 4% trichloroacetic acid. Mix vigorously, allow to stand 5 min, and centrifuge.

2. Transfer the supernate into 60 ml plastic centrifuge tubes containing about 0.25 g Norit.

3. Dispense 10 ml 4% trichloroacetic acid (Blank) and 10 ml working standard into plastic centrifuge tubes containing about 0.25 g Norit.

4. Stopper tubes and shake for 5 min, centrifuge, then filter the supernate through Whatman 42 paper. These filtrates are stable several days at 4°C.

5 Place 2.0 ml aliquots of each filtrate into two test tubes. Mark the tubes as Blank Control (A_{bb}), Blank Test (A_b), Standard Control (A_{sb}), Standard Test (A_s), Sample Control (A_{xb}), and Sample Test (A_x).

6. Add 0.50 ml 2,4-DNP reagent to Blank Test, Standard Test, and Sample Test tubes, and mix.

7. Place tubes in a 37°C water bath for 4.5 hours.

8. Transfer tubes to an ice water bath and allow to stand at least 5 min.

9. With the tubes in the ice water bath, add dropwise with mixing 2.5 ml 85% H_2SO_4 to all tubes. It is important not to allow the temperature to rise appreciably, otherwise charring of organic matter can occur.

10. To the tubes marked Control add 0.50 ml 2,4-DNP reagent and mix while still in the ice water bath.

11. Remove all tubes from the bath and after 40 min wipe each tube and read the adsorbance of each tube at 515 nm versus a water blank.

Calculations:

$$\text{mg ascorbic acid/100 ml serum} = \frac{(A_x - A_{xb}) - (A_b - A_{bb})}{(A_s - A_{sb}) - (A_b - A_{bb})} \times 0.01 \times \frac{100}{0.6}$$

$$= \frac{(A_x - A_{xb}) - (A_b - A_{bb})}{(A_s - A_{xb}) - (A_b - A_{bb})} \times 1.66$$

$$\text{mg ascorbic acid/liter urine} = \frac{(A_s - A_{xb}) - (A_b - A_{bb})}{(A_s - A_{sb}) - (A_b - A_{bb})} \times 0.01 \times \frac{1000}{0.2}$$

$$= \frac{(A_x - A_{xb}) - (A_b - A_{bb})}{(A_s - A_{sb}) - (A_b - A_{bb})} \times 50$$

NOTES

1. Beer's law. The color obeys Beer's law on a spectrophotometer at 515 nm and on a photometer with a Klett 52 filter.

2. Serum vs plasma. Due to the instability of ascorbic acid in both serum and plasma, plasma is recommended for analysis because of the greater speed with which it can be obtained from blood. One study (195) has shown the rate of loss of ascorbic acid is slightly greater when plasma is left in contact with the red cells. Serum with hemolyzed blood added resulted in a 38% loss of the vitamin when incubated at 38°C for 1 hour (135).

3. Stability. Our laboratory found that the addition of 20 mg oxalic acid per milliliter of serum or plasma stabilizes the ascorbic acid for 1 week at room temperature as determined by the method presented here. In a study of the effect of storage on the ascorbic acid values of serum, Lowry *et al* (135) reported no losses with acidified samples stored up to 13 days at room temperature, 4°C, or −20°C. Metaphosphoric acid or trichloroacetic acid have both been reported to be useful (78, 135). Nonacidified serum samples stored 6 days at −20°C showed a 10% loss. A final concentration of approximately 0.5% oxalic acid or metaphosphoric acid is sufficient to stabilize vitamin C in urine (176).

Variability in the stability of ascorbic acid in serum, plasma, whole blood, with added KCN, at different temperatures or acidified samples as enumerated by Henry (103) may be due to the method used to measure the losses in ascorbic acid levels. Steward *et al* (195) found greater losses by the indophenol method than with the dinitrophenylhydrazine assay. The indophenol method measures ascorbic acid; hence conversion of ascorbic acid to dehydroascorbic acid by oxidative processes during storage would result in losses of measureable vitamin. The dinitrophenylhydrazine method measures the dehydroascorbic acid, which is stable under acid conditions, and the oxidative formation of this product would not affect quantitation by this reagent.

4. Formation of osazone. The osazone formation of dehydroascorbic acid is both temperature- and time-dependent (218). Although greater sensitivity can be obtained by using higher incubation temperatures, the possibilities exist that decomposition of oxidized ascorbic acid and formation of hydrazones and osazones of other substances could occur at such temperatures. In view of these potential disadvantages, a 37°C incubation temperature is recommended for routine assays.

5. Stability of color. No change is observed in 40 min and only a slight decrease after standing 18 hours (179).

ACCURACY AND PRECISION

Recovery studies by this method range from 96% to 104% (179). This by itself, of course, does not prove accuracy. Agreement of normal range values by other methods, however, supports the accuracy of the dinitrophenylhydrazine method (147, 165). According to Zloch *et al* (218) only 49% of the ascorbic acid reacts with the phenylhydrazine, but as long as the standards are treated in the same fashion this shortcoming is not serious. The final standing time at room temperature improves the specificity of the reaction because the dehydroascorbic acid derivative is highly stable at the final acidity of the reaction while carbohydrate derivatives are not (176). Agreement of values obtained with tissue extracts at incubation temperatures of 15°C or 37°C was proposed as an indication of the specificity of the method (177).

The precision (95% limits) of the method in the normal range is ±5%.

NORMAL VALUES

The normal range for serum or plasma ascorbic acid is about 0.2–2.0 mg/100 ml (176). Similar ranges have been reported for preschool children (78) and elderly subjects (133).

Urinary excretion of ascorbic acid varies according to the individual's daily diet; hence urinary levels are only meaningful in nutritional studies in which the intake is controlled.

VITAMIN D

The two types of vitamin D are the animal type, *cholecalciferol (D₃)*, and the plant type, *ergocalciferol (D₂)*.

CHOLECALCIFEROL (VITAMIN D₃)

The precursor of cholecalciferol is 7-dehydrocholesterol while ergocalciferol is formed from the plant sterol, ergosterol. The principal physiologic effects of vitamin D and the biologically active hydroxylated derivatives are to increase intestinal calcium adsorption and to mediate bone mineral mobilization and calcification. Within the past few years, four new derivatives of cholecalciferol have been isolated and characterized. These are the 25-hydroxy (32), the 1,25-dihydroxy (107), 21,25-dihydroxy, and the 25,26-dihydroxy (106) derivatives. The 25-hydroxy and 21,25-dihydroxy cholecalciferols have been isolated and identified in human plasma (106). According to DeLuca, vitamin D itself must be considered as the storage or inert form of the vitamin, and the hydroxylated derivatives are the physiologically active compounds (60).

The transport protein of vitamin D in human serum has been identified as an α-globulin in one paper (201), or albumin and α_2-globulin in another (59). Instead of bioassaying the protein fractions obtained by electrophoresis, Smith and Goodman (189) injected radioactive vitamin D₃, fractionated the plasma proteins by gel filtration and acrylamide gel electrophoresis, and found 94% of the radioactive products in the protein fractions with a density greater than 1.21. The electrophoretic position of the bound fraction was slightly in front of the albumin peak, yet not in the prealbumin fraction.

Methods for the measurement of vitamin D have been reviewed by DeLuca and Blunt (61). Antimony trichloride, UV adsorption at 264 nm, and gas-liquid chromatographic analyses are discussed from the standpoints of applicability and sensitivity. The practical bioassays with rats or chicks that are used in their laboratory are described. Conversion of vitamin D to fluorescent products with acetic anhydride-sulfuric acid (46) or ethanol-sulfuric acid (162) have been reported, but other vitamins and related sterols were noted to cause marked interference.

A competitive protein binding technic has been described for the assay of vitamin D and 25-hydroxy vitamin D (24). Silicic acid column chromatography was used to separate the two vitamins. A specific vitamin D-binding protein in the sera of rats on a vitamin D-deficient diet was employed in the assay. Specific binding proteins in rat intestine cytoplasm for vitamin D_3, 25-hydroxycholecalciferol, and 1,25-dihydroxycholecalciferol have been reported (111). Intestinal cytoplasm could provide another source of vitamin D binding proteins for a competitive protein binding assay.

One international unit of vitamin D is equivalent to 25 ng; therefore the reported adult range for serum vitamin D of 0.5–1.35 IU/ml is equivalent to 12–30 ng/ml (30). Belsey *et al* (24) reported ranges of 24–40 ng/ml for vitamin D_3, and 18–36 ng/ml for 25-hydroxy-vitamin D_3.

VITAMIN E

The vitamin E complex, known as *tocopherols*, is composed of at least eight compounds. Of the eight compounds, the α, β, and γ derivatives are the predominant forms found in man. α-Tocopherol has the greatest biologic activity.

α-Tocopherol . . .(5, 7, 8,- Trimethyltocol)
β-Tocopherol . .(5, 8, - Dimethyltocol)
γ-Tocopherol . .(7, 8, - Dimethyltocol)

Analysis of serum vitamin E by thin layer chromatography has shown the presence of the α, β, and γ forms (28). No δ-isomer was detectable. Approximately 87% of the total vitamin content is in the α form. Studies have also revealed α-tocopherol to be the major constituent present in human tissues (169). Hansen and Warwick (97) reported 80% of the vitamin E in serum was in the free form and 20% appeared as the acetate. Serum tocopherol is found primarily in the low density lipoproteins (142).

The original method of Emmerie and Engel (69), which utilized oxidation of tocopherol with ferric chloride and interaction of the ferrous ion formed with α, α'-dipyridyl, has been modified by using tripyridyl triazine (141) and bathophenanthroline (71) to complex the ferrous ion. One of the problems with this reaction has been its lack of specificity due to interference by carotene (71) and ubichromenol (149). These interfering compounds have been removed by chromatography (5, 149) or by incorporating phosphoric acid to decrease the oxidation of the ferric ions by carotene (5, 71).

Other methods for the quantification of tocopherols have been summarized in a review by Bunnell (38). Photometric, fluorometric, and chromatographic (thin layer, gas, column) methods are discussed.

Duggan (65) first described a spectrophotofluorometric method for the quantification of vitamin E in 2 ml plasma. This technic was subsequently modified to a micro method that required only 0.1 ml (97). Cholesterol, carotene, and vitamin A were found not to interfere with the fluorometric method.

SERUM FREE VITAMIN E

(method of Hansen and Warwick, Ref 97)

PRINCIPLE

After precipitation of serum proteins, the vitamin is extracted into hexane and then quantitated by measuring relative fluorescence at specific activation and emission wavelengths (Fig. 27-5).

FIG. 27-5. Activation and fluorescence spectra of vitamin E (1.2 μg/ml) in hexane. Fluorescence spectrum (peak at 330 nm) obtained upon activation at 295 nm; activation spectrum obtained with fluorescence monochrometer set at 330 nm. The spectra were recorded with an Aminco-Bowman spectrophotofluorometer.

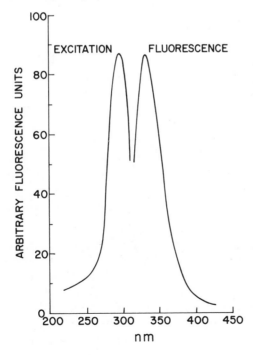

REAGENTS

Hexane, Distilled. Distill 1 liter of reagent grade solvent, discarding the first 25 and last 100 ml.

Ethanol, Distilled. Distill 1 liter over 20 g of KOH, discarding first 25 ml and last 100 ml.

Water, Redistilled. Distill deionized water or redistill distilled water in an all glass distillation apparatus.

Stock Vitamin E, Standard, 2000 µg/ml. Dissolve 500 mg *dl*-α-tocopherol in 100 ml of the distilled ethanol and dilute to 250 ml. The product from Merck was found to be suitable. Stable for 6 months.

Working Vitamin E Standard, 20 µg/ml. Dilute 1.0 ml stock standard to 100 ml with the distilled ethanol. This standard is stable for 6 months.

PROCEDURE

1. Set up the following in glass-stoppered 15 ml centrifuge tubes:
   Blank: 0.2 ml water
   Standard: 0.2 ml working standard
   Unknown: 0.2 ml serum
2. Add 1 ml water to the Blank and Unknown tubes. To the Standard tube add 1.2 ml water. Mix all tubes 30 sec.
3. Add 2 ml distilled ethanol, with mixing, to the Blank and Unknown tubes. Add 1.8 ml to the Standard tube. Mix all tubes again for 30 sec.
4. Add 5 ml hexane to all tubes, stopper, and shake for 5 min by hand or mechanical mixer.
5. Centrifuge the tubes 5 min at 2000 rpm.
6. Transfer hexane layer to a quartz cuvet.
7. Set the spectrophotofluorometer with the activation wavelength at 295 nm and the emission wavelength at 330 nm. Read the relative fluorescence of the Blank (F_b), the Standard (F_s), and the Unknown (F_x).

Calculation:

$$\mu\text{g free vitamin E/ml} = \frac{F_x - F_b}{F_s - F_b} \times 20$$

NOTES

1. Glassware. All glassware used for the test must be soaked at least 1 hour in 8 *N* nitric acid, rinsed thoroughly with water, and dried.

2. Interfering factors. Glass distilled or spectrograde solvents should be checked for fluorescence before use in the procedure. Solutions that give less than 5 fluorescence units are acceptable.

3. Photodecomposition. Maximum fluorescence is reported to be attained within 1 min and then a 22% decrease occurs after 70 min of continual irradiation at 295 nm (97). Our experience shows a more rapid decomposition, perhaps attributable to the greater transmission of the ultraviolet light by quartz cuvets than by the Pyrex tubes used by Hansen and Warwick (97).

4. *Linearity of measurement.* We find linearity up to 40 µg/ml. Occasionally, however, a check should be made for linearity by running 10, 20, 30, and 40 µg/ml standards. If nonlinearity is observed, calculations must be made from a calibration curve.

5. *Stability.* Vitamin levels are stable in sera stored 2 weeks at room temperature or 5°C, and 2 months at −20°C (141).

ACCURACY AND PRECISION

Recovery studies of added vitamins range from 90% to 104%. Indirect evidence of the accuracy of the measurement of free serum vitamin E is provided by the good agreement of the normal ranges obtained by the fluorometric (97), ferric chloride-tripyridyltriazine, and bathophenanthroline technics (71, 141).

The precision (95% limits) of the method is ±10%.

NORMAL VALUES

The range for free vitamin E levels is 5−20 µg/ml serum (29). Adolescents have a range of 3−15 µg/ml. There is no sex difference in the levels of the vitamin, but there is a significant correlation with serum carotene levels.

BIOTIN

Methods for the assay of biotin have been limited to microbiologic methods because of the low levels in biologic fluids and tissues (10). Both biotin and the ε-N-biotinyl-L-lysine derivative promote the growth of biotin-dependent microorganisms.

BIOTIN (β-biotin)

Biotin is mainly in a protein-bound form with only small amounts in the unbound state. Liver, fibrin, and casein have to be hydrolyzed with either proteolytic enzymes or acid in order to release biotin (34). Another study of animal livers showed that 92% of the activity in liver homogenates was bound to protein (152). Similar findings were obtained with the particulate fractions of liver tissue with the exception of the microsomal fraction (23% protein-bound). Serum and blood required pretreatment with the proteolytic enzyme papain to liberate biotin activity (17).

According to Baker *et al* (10) the flagellate *Ochromonas danica* is the preferred microbiologic system for biotin assay. *Lactobacillus casei* or *L. arabinosis* and *Saccharomyces cervisiae* require other growth factors and do not have the specific biotin dependency of *O. danica*.

The stoichiometric interaction of biotin and avidin has been applied to the measurement of either component. Green (89) described a competitive protein binding approach that utilized the dye 4-hydroxyazo-benzene-2'-carboxylic acid (HABA), avidin, and biotin. Measurement of the decrease in absorbance of the avidin-HABA dye complex after the addition of biotin was proportional to the amount of the vitamin. A sensitivity of ΔA_{500} of 0.069 per nanogram of biotin in 2 ml was reported for this method. A radioligand binding assay has been described for the measurement of the egg white protein avidin based on the binding of ^{14}C-biotin (128). This same approach could be used for the development of a simple and sensitive method for the measurement of biotin instead of avidin.

Ranges of normal serum biotin levels for adults and adolescent children are 200–500 pg/ml and 220–680 pg/ml, respectively. Human urine contains 6–50 μg biotin per 24 hour sample (10).

FOLIC ACID

Folic acid or *pteroylglutamic acid* is the parent compound of the folate complex. The molecule consists of three constituents: pteridine, *p*-aminobenzoic acid, and L-glutamic acid.

FOLIC ACID

This vitamin has a multitude of analogs or derivatives that are found in animal and plant tissue. Some of the derivatives are dihydrofolic acid, tetrahydrofolic acid, five types of one carbon analogs, and polyglutamates (2–10 glutamic acid residues). The coenzyme functions of folic acid involve one carbon metabolism for simple compounds and purine and pyrimidine synthesis.

Folate activity, based on microbiologic assays, is approximately 95% in the red cells (91, 105). Of the activity in serum, about 40% is protein-bound as determined by dialysis or hemoglobin-charcoal equilibrium experiments (174). There also appears to be labile fraction of folate activity in serum that varies from 65% to 94% of the total activity (211). This fraction is destroyed by autoclaving but is protected by the addition of ascorbic acid. According to Herbert *et al* (104) the principal compound in serum is N^5-methyl tetrahydrofolate. Other investigators report the presence of at least six components with folate activity in whole blood (204).

Assays for folate activity have been made mainly by the microbiologic technic because of its sensitivity and convenience. Reviews dealing with the advantages, disadvantages, and interferences have covered all aspects of this method (12, 82, 122). *Lactobacillus casei* is the principal organism used for the assay of serum and red cell folate activity, but *Streptococcus faecalis* and *Pediococcus cerevisiae* have also been utilized (91). The greater growth response with *L. casei* is attributed to the utilization of polyglutamates and other folates by this organism.

The demonstration that ascorbic acid and phosphate buffer protect folate activity (204) has resulted in their use in establishing methods for the assay of serum and red cell folates (105, 211). However, normal ranges for folic acid as determined with the *L. casei* bioassay have varied widely. Incubation time, variation in the number of viable inoculated organisms, and the difference in growth patterns of serum or folic acid-enriched cultures are critically examined by Streeter and O'Neill (197). Another problem is that the presence of antibiotics can negate bioassay results (173). However, a chloroamphenicol-resistant strain of *L. casei* has afforded the measurement of folate concentrations without previous deproteinization (150) and the elimination of sterilization or aseptic addition (58).

Recently, the technics of competitive protein-binding (212) and radio-immunoassay (56) have been applied to the measurement of serum folate levels. The competitive protein-binding method is based on powdered milk protein as the folate binder and the use of ^{3}H-methyl-tetrahydrofolic acid or ^{3}H-pteroylglutamic acid and hemoglobin-coated charcoal to separate the free and protein-bound fractions. Test results are available in four hours and the values obtained are identical to those by *L. casei* bioassay (212). Da Costa and Rothenberg (56), utilizing antibodies which bind folic acid, ^{3}H-folic acid, and albumin-coated charcoal, developed an assay with a range of 25–400 pg of folic acid per milliliter reaction mixture. Interestingly, the addition of ascorbic acid to serum resulted in low values, whereas merely boiling the serum gave significantly larger amounts of folate. Reduced folates did not inhibit the binding of ^{3}H-folic acid; therefore the immunoreactive folate in serum and serum extracts by this method is an unreduced folate.

DETERMINATION OF SERUM FOLIC ACID ACTIVITY

(method of Waters and Molin, Ref 211, modified)

PRINCIPLE

Lactobacillus casei, in a culture medium that is nutritionally complete with the exception of folic acid, grows on the addition of folic acid. The extent of growth is proportional to the amount of folic acid present in serum. Growth densities are measured turbidimetrically.

REAGENTS

Phosphate buffer, 0.05 M, pH 6.1. Dissolve 6.9 g $NaH_2PO_4 \cdot H_2O$ and 17.9 g $Na_2HPO_4 \cdot 12H_2O$ in water and dilute to 1 liter. Check pH.

Ascorbic-Phosphate Buffer. Dissolve 100 mg ascorbic acid in 100 ml phosphate buffer. Prepare fresh.

Stock Folic Acid Standard, 2000 ng/ml. Dissolve 20 mg in a solution of 20 ml absolute ethanol and 1 ml 1 N NaOH, then dilute to 100 ml with double distilled water (this solution is stable 3 months at $-20°C$). From this solution prepare a 1:100 dilution with double distilled water and store at $-20°C$. Solution is stable 1 month.

Working Folic Acid Standard, 0.8 ng/ml. This standard is prepared by making a 1:100 dilution with double distilled water and then diluting 2 ml to 50 ml with ascorbic acid-phosphate buffer. Prepare fresh.

Lactobacillus casei. A pure culture can be obtained from the American Type Culture Collection. Subspecies *rhamnosus* (*rogosa*) is also suitable for folate and other vitamin assays. The culture is maintained by periodic transfer (subculture) to Lactobacilli Agar, AOAC.

Lactobacilli Agar, AOAC. Dissolve 48 g agar in 1 liter water by heating just to boiling temperature. Agar stabs are prepared according to the directions on the container. The product from Difco was found to be suitable.

Lactobacilli Broth, AOAC. Suspend 38 g dehydrated medium in 1 liter water. Ten ml amounts are distributed in culture tubes, capped, and sterilized for 15 min at 121°C. The product from Difco was found to be suitable and is a nutritionally complete medium (broth).

L. casei Broth (Folic Acid-Free Preparation). Dissolve 7.8 g medium per 100 ml double distilled water. Folic Acid Assay PGA Broth from Baltimore Biological Laboratories was found to be a suitable medium. Heat to boiling for 2–3 min. Cool rapidly in running water.

Folic Acid-Free Wash Medium (FAFW). Prepare 6 screw-capped tubes containing 6 ml of *L. casei* broth and add 4 ml water. Autoclave these tubes along with the assay tubes. This FAFW is used to wash the *L. casei* culture and to make the final suspension for inoculation.

L. casei Inoculum. Inoculum is maintained in AOAC agar (stab culture) at 4°C and a transfer is made once a month along with a check for culture purity. On the day before the test, make a transfer from the stock AOAC agar to 10 ml of AOAC broth. During the morning of the test, transfer 0.5 ml of the 18 hour culture to 10 ml of AOAC broth. After 6–8 hours centrifuge and wash culture 4 times with 10 ml vol FAFW. After final wash, resuspend cells in 10 ml FAFW and add 0.5 ml to 10 ml FAFW.

Formalin, 4%. Dilute 10 ml of 40% formalin to 100 ml with water.

PROCEDURE

1. Mix 0.5 ml serum and 4.5 ml ascorbic-phosphate buffer.
2. Autoclave all assay tubes at 121°C for 1 min, then place in cold tap water bath.
3. Shake all tubes and centrifuge at approximately 2200 rpm for 15 min.
4. Transfer 0.5 ml of clear supernate, in duplicate, to 15 x 125 mm tubes containing 1.5 ml double distilled water. Decant the rest of the supernate and freeze the contents in the event a sample needs to be diluted or repeated.
5. Set aside one of the duplicate tubes as the serum blank for each specimen and two tubes with 2.0 ml water to serve as Media Blank tubes.
6. Prepare a standard curve in duplicate with working standard:

Folic acid ng/tube	Standard ml/tube	Water ml/tube
0.0 (Blank)	0.00	2.00
0.2	0.25	1.75
0.4	0.50	1.50
0.8	1.00	1.00

7. Add 3.00 ml *L. casei* broth to all tubes.
8. Place all serum blank tubes and two Media Blank tubes into separate racks, for these tubes are not inoculated.
9. Cover tubes with aluminum foil and autoclave all tubes at 121°C for 2.5 min. Avoid overheating.
10. After cooling, add with a sterile capillary pipet 0.025 ml of *L. casei* inoculum to all tubes except the Blanks.
11. Incubate at 37°C for 20 hours.
12. Remove tubes from incubator and place in an ice bath until read. Alternatively, if there is to be a delay before the tubes are read, then add 0.5 ml 4% formalin to each tube and mix well.
13. Mix the tube contents well, allow tubes to stand until bubbles disappear, and then read turbidity with a photometer using a 540 nm filter. Set photometer at zero with Media Blank tubes and record the test reading (see *Note 2* on serum blanks). Read the Standards against the Blank set at zero.

Alternatively, the samples may be read by using an AutoAnalyzer with the manifold, components, and operating condition set up as shown in Figure 27-6. The recorder is set at 95% transmission with distilled water and the standard curve tubes are positioned to be sampled in an order of increasing concentration. By following procedural guidelines as described in Chapter 10, steady state interaction curves as shown in Figure 27-7 should be obtained and checked periodically.

Calculation:

Plot the percent transmission and the concentration of folic acid from the standard curve as ng/ml on one cycle log graph paper as shown in Figure 27-8. With this graph, convert the sample reading to ng/ml and use this value for the calculation:

$$\text{ng folic acid activity/ml} = \text{ng/ml} \times \frac{0.5}{1.0} \times \frac{5.0}{0.5} \times \frac{2.0}{0.5}$$

$$= \text{ng/ml} \times 20$$

DETERMINATION OF RED BLOOD CELL FOLIC ACID ACTIVITY
(method of Hoffbrand et al, Ref 105)

PRINCIPLE

Whole blood folate and serum folic acid activities are performed with the *L. casei* bioassay. Red cell folate is calculated by the difference of the two assays.

REAGENTS

Reagents of Method for Folic Acid Activity of Serum.
Ascorbic Acid, 1%. Prepare this reagent fresh with each assay.

PROCEDURE

1. Collect whole blood with EDTA (4 mg/ml blood) as the anticoagulant.
2. Determine hematocrit of the whole blood to obtain the packed cell volume (PCV).
3. Prepare the hemolysate by slowly pipeting 0.5 ml whole blood into 4.5 ml 1% ascorbic acid solution. Also prepare a 1:10 dilution of this hemolysate with the ascorbic acid-phosphate buffer. These hemolysates can be stored in the frozen state if desired.
4. Centrifuge the whole blood, then transfer and dilute 0.5 ml serum with 4.5 ml ascorbic acid-phosphate buffer.

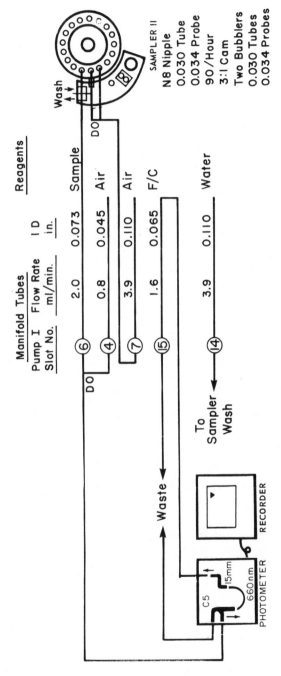

FIG. 27-6. Flow diagram for folic acid in serum. Turbidity of bacterial growth medium is measured. Calibration range: 0–30 ng/ml. Conventions and symbols used in this diagram are explained in Chapter 10. Air bubblers on the sampler are used to mix the samples before aspiration.

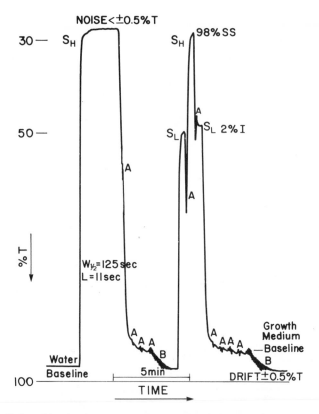

FIG. 27-7. Folic acid strip chart record. S_{II} is a 16 ng/ml pooled sample, S_L a 4 ng/ml pooled sample. Peaks reach 98% of steady state; interaction between samples is 2%. $W_{1/2}$ and L are explained in Chapter 10. Computer tolerance for noise is set at ±0.5% T, for baseline drift at ±1% T, and for standard drift at ±1.4% T. Noise at points marked A is caused by schlieren effects at interfaces between water and growth medium. Such interfaces wash out of the photometer cuvet very slowly when water follows medium, as at points marked B.

5. Process both dilutions of the hemolysate and the serum dilution according to the procedure for *Folic Acid Activity in Serum*.

Calculation:

Using a standard curve graph as in Figure 27-8, convert the sample readings to ng/ml. With these values the calculations are:

$$\text{ng folic acid activity/ml}^a \text{ sample} = \text{ng/ml} \times \frac{0.5}{5.0} \times \frac{5.0^b}{0.5} \times \frac{2.0}{0.5}$$

$$= \text{ng/ml} \times 20$$

[a]The same factors apply for whole blood or serum.

[b]This factor becomes 5.00/0.05 for the 1:10 dilution of the whole blood hemolysate.

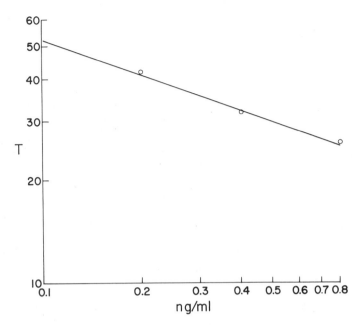

FIG. 27-8. A representative standard curve for the calculation of serum folic acid with *Lactobacillus casei*.

$$\text{ng folic acid activity/ml packed cells} = \frac{\text{ng/ml blood} - \text{ng/ml serum}\left(\frac{1 - \text{PCV}}{100}\right)}{\text{PCV}/100}$$

NOTES

1. Sample preparation. Serum must be carefully separated from the clot to avoid hemolysis because of the high folic acid content of red cells. Two hemolysate dilutions are tested because of the wide range of folic acid content found in red cells.

2. Serum blank. Blanks are uniformly low and are not measured. Only those sera that have a definite color or opalescence after autoclaving and incubation are read against the respective blank tube.

3. Antibiotic interferences. Erythromycin, lincomycin, novobiocin, and other antibiotics suppress bacterial growth (187, 211). Subject must not have taken antibiotics for at least 1 week prior to the assay.

4. Stability of specimens. Serum is stable 4–5 days at room temperature with 15 mg ascorbic acid added per milliliter. For storage at −20°C for up to 3 months add 5 mg ascorbic acid per ml of serum (211). Nonhemolyzed blood samples show no loss in activity for 7–10 days when stored at 4°C. Hemolysates prepared with a 1% ascorbic acid solution are stable 3–5 months at −20°C.

5. Glassware. All glassware and pipets to be used for reagent preparation for the assay should be filled or covered with water, autoclaved, rinsed with water, and oven-dried. This procedure is essential in preventing folic acid contamination of the glassware.

ACCURACY AND PRECISION

The accuracy of the folate assay is problematic because of the multiple forms of the vitamins in blood and the relative response of the bioassay system to the vitamin complex. Studies with added folic acid have shown 99%–113% recovery for serum and 97%–106% for red blood cells (211).

The precision (95% limits) for serum folic acid activity is ±20%. Precision for the red cell assay is ±15%.

NORMAL VALUES

The folic acid activity range is 5–21 ng/ml serum and 160–640 ng/ml packed red cells (105, 211). Both serum and red cell folate values are reduced during pregnancy (182).

NICOTINIC ACID AND NICOTINAMIDE

Nicotinic acid is readily converted to nicotinamide, and this latter compound is incorporated into two important coenzymes, nicotinamide adenine dinucleotide (NAD) and its phosphate derivative (NADP). In plasma, nicotinic acid (41) and two metabolites (N'-methyl-2-pyridone-5-carboxamide and N'-methyl-4-pyridone-3-carboxamide) are normally present (2). After oral ingestion of nicotinamide the plasma concentrations of both metabolites increase, but nicotinamide is not detectable.

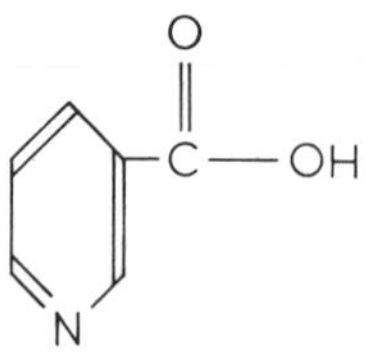

NICOTINIC ACID
(PYRIDINE–3–CARBOXYLIC
ACID)

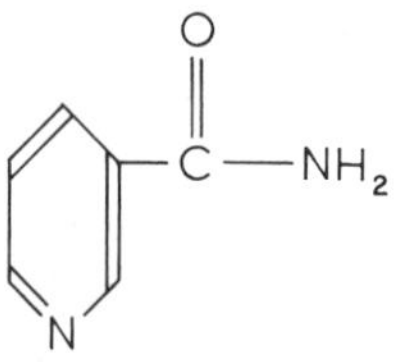

NICOTINAMIDE
(PYRIDINE-3-CARBOXYLIC
ACID AMIDE)

Study of the excretory products in urine has revealed that only 1% of the administered dose of the vitamin is excreted as nicotinic acid and 4% as nicotinamide (118). Holman and de Lange (108) identified the principle urinary metabolites as N'-methyl-2-pyridone-5-carboxamide and N'-methylnicotinamide. There are considerable differences in the amounts of each metabolite reported by different investigators, but Walters *et al* (209) found that the proportions of each metabolite varied according to the amount of vitamin ingested. Following the ingestion of 500 mg nicotinamide, the urinary product N'-methyl-6-pyridone-3-carboxamide was identified and measured by a fluorometric method (180).

A photometric method for nicotinic acid in blood, plasma, and urine was described by Melnick and Field (145) using cyanogen bromide. The resultant yellow color of this reaction was found to be intensified by the addition of aniline (145) or p-methylaminophenol sulfate (Elon) (126). Free nicotinic acid in plasma has also been measured by the chloramine T reaction (114). The urinary N'-methylnicotinamide is converted to a fluorescent derivative by condensing it with acetone (114) or methyl ethyl ketone (42) in strong alkali. The initial charcoal treatment of the urine to remove the interfering substances from this determination was later replaced by ion exchangers (207). A method for the measurement of the other major metabolite, N-methyl-2-pyridone-5-carboxamide, required nitration followed by alkalinization to form a product with an absorption maximum at 450 nm (108). Ion exchange resins were later substituted for Fuller's earth for the removal of extraneous materials in the procedure for the determination of the metabolite in urine (168) and blood (119).

Various microbiologic assays have been used to assay nicotinic acid and nicotinamide, but according to Baker and Frank (13) the protozoan *Tetrahymena pyriformis* is the preferred test organism because of its specificity for the vitamins *per se*, rather than the metabolites.

The following normal ranges are from either individual reports or composites of different methods that are in reasonable agreement. The ranges for whole blood and plasma levels of nicotinic acid are $3.0-8.3$ μg/ml (13, 73) and $0.2-2.4$ μg/ml (73, 126), respectively. The range for the metabolite N'-methyl-2-pyridone-5-carboxamide in serum is $0.2-0.6$ μg/ml (2), and for urine $7-26.9$ mg/24 hour (168, 209). The range of N'-methyl-6-pyridone-3-carboxamide levels is $2.5-12$ mg/24 hour urine sample. The range for N'-methylnicotinamide in urine is $3-17$ mg/24 hours (114, 207).

In view of the variety of nicotinic acid metabolites in blood and urine (108, 180), the effect of the level of ingested vitamin on the proportion of metabolites, the qualified statements on measureable nicotinic acid activity by microbiologic assays (13), and the different chemical reactions required to measure the different metabolites (108, 114, 145), it is difficult to assess whether one is really measuring the nicotinic acid status of an individual. The most recent report on the measurement of the pyridone metabolites in blood indicates the absence of detectable products in fasting blood samples (1).

MISCELLANEOUS VITAMINS

CHOLINE

In biologic fluids and tissue extracts, choline is found in the free form, as acetylcholine or cytidine diphosphocholine and as a constituent of phospholipids. Choline has been measured gravimetrically, by oxidation to yield trimethylamine, or as reineckate or periodate complexes (70). The most commonly used method has been the reineckate complex technic. Beattie (22) reported the solubilization of the choline-reineckate precipitate with acetone and established a photometric technic for the vitamin. An improvement in the detection limits was described based on oxidation of the chromium of the reineckate followed by formation of the red-violet chromium-diphenylcarbazide complex (140).

Neurospora crassa (cholineless mutant) has also been used to measure choline in natural products (110). This microbiologic assay was used to quantify choline levels in plasma and urine (139). The ranges of free and total choline in plasma reported were 0.45–0.75 and 2.6–3.5 μg/100 ml, respectively. Choline in urine is entirely in the free form and its range is 5.6–9.0 mg/24 hours.

COENZYME Q_{10} OR UBIQUINONE

This vitamin is present in microbial, plant, and animal cells and is found in high concentrations in the mitochondrial fraction. The various types of coenzyme Q are designated by the number of isoprenoid units in the side chain that are attached to the tetrasubstituted benzoquinone nucleus. A succinct description of the four methods to measure coenzyme Q have been reviewed, and the reaction with ethyl cyanoacetate under alkaline conditions to form a blue complex was considered to be the most specific method (55).

The normal range of blood levels of coenzyme Q_{10} is 40–180 μg/100 ml (171). The normal range of urinary levels in women is 0–40 μg/24 hours, while adult males excrete slightly larger amounts with the range being 0–100 μg/100 ml (127).

INOSITOL

This vitamin has also been referred to as myoinositol and mesoinositol. In nature it is found as the free form, mono- or polyphosphate and lipid inositol (102). All of the assay methods require free inositol; therefore strong acid hydrolysis methods are used to liberate the phosphate esters or lipid complexes. Various yeasts and molds have been used to effect the microbiologic assay of inositol (143). An enzymic assay has been reported with inositol dehydrogenase from *Aerobacter suboxydans* (131). This enzyme method was modified by adding pig heart diaphorase and measuring the decolorization of 2,6-dichlorophenol-indophenol (45). A photometric method was described that utilized the periodate oxidation with measurement of liberated iodine at 352 nm (77). Another chemical method employed sulfuric acid, chromium ion, and hydroquinone to form a complex with an absorption maximum at 450 nm (134). Both of the chemical methods required ion exchange fractionation to remove interfering sugars and polysaccharides. The total inositol content of human plasma samples, by microbiologic assay, is 0.9–1.8 mg/100 ml (200).

VITAMIN K$_2$

The predominant form of vitamin K$_2$ in mammals is the naphthoquinone nucleus with four isoprenoid units in the side chain. A method for the estimation of vitamin K was first described by Dam *et al* (57); it employed sodium ethylate and yielded a red complex. Another reaction with diethyl dithiocarbamate in alcoholic alkali, giving a cobalt-blue color, was reported to be stable for only 5 min (116). Scudi and Buhs described a method with the sensitivity to measure vitamin K levels in biologic materials that involved the catalytic reduction of the naphthoquinone nucleus to vitamin K hydroquinone (185). The decrease in color of the 2,6-dichlorophenol-indophenol is a measure of the hydroquinone. The formation of a complex of vitamin K with 5-imino-3-thione-1,2,4-dithazolidine with an absorption maximum at 440 nm has been described as a specific method (183). The level of vitamin K in urine is reported to be less than 0.075 μg/ml (185).

PANTOTHENIC ACID

Little is known about the form of pantothenic acid in blood, but in cells it is present as coenzyme A. The bound form is liberated by incubation with a multienzyme preparation (Clarase) (15), intestinal phosphatase, or fresh pigeon liver extract (159). *Lactobacillus plantarum* and *Tetrahymena pyriformis* have been used to measure pantothenic acid levels in biologic fluids (15). No free pantothenic acid is demonstrable in blood, whereas it is entirely in the free form in urine.

Photometric (184) and fluorometric (161) methods have been used to assay the vitamin in pharmaceutical products, but neither method has been applied to biologic materials.

The normal range of pantothenic acid levels as determined by microbiologic assay have been reported as 11–190 μg/100 ml blood, 10–170 μg/100 ml cerebrospinal fluid, and 60–480 μg/100 ml urine (15).

REFERENCES

1. ABELSON DM, BOYLE A:*Clin Chim Acta 14*:483, 1966
2. ABELSON D, BOYLE A, DEPATIE C, SELIGSON H: *Clin Chim Acta 8*:603, 1963
3. ADAMS JF, McEWAN F: *J Clin Pathol 24*:15, 1971
4. ANDERSON BB, FULFORD-JONES CE, CHILD JA, BEARD MEJ, BATEMAN CJT: *J Clin Invest 50*:1901, 1971
5. ATLAS SJ, PINTER KG: *Anal Biochem 17*:258, 1966
6. BAKER EM III, HAMMER DC, MARSH SC, TOLBERT BM, CANHAM JE: *Science 173*:826, 1971
7. BAKER EM, SAARI JC, TOLBERT BM: *Am J Clin Nutr 19*:371, 1966
8. BAKER EM, CANHAM JE, NUNES WT, SAUBERLICH HE, McDOWELL ME: *Am J Clin Nutr 15*:59, 1964

9. BAKER H: *Am J Clin Nutr 20*:850, 1967

10. BAKER H, FRANK O: *Clinical Vitaminology, Methods and Interpretation.* New York, Interscience, 1968, p 22

11. BAKER H, FRANK O: *Clinical Vitaminology; Methods and Interpretation.* New York, Interscience, 1968, p 7

12. BAKER H, FRANK O: *Clinical Vitaminology, Methods and Interpretation.* New York, Interscience, 1968, p 87

13. BAKER H, FRANK O: *Clinical Vitaminology, Methods and Interpretation.* New York, Interscience, 1968, p 31

14. BAKER H, FRANK O, THOMSON AD, FEINGOLD S: *Am J Clin Nutr 22*:1469, 1969

15. BAKER H, FRANK O, PASHER I, DINNERSTEIN A, SOBOTKA H: *Clin Chem 6*:36, 1960

16. BAKER H, FRANK O, FEINGOLD S, GELLENE RA, LEEVY CM, HUTNER SH: *Am J Clin Nutr 19*:17, 1966

17. BAKER H, FRANK O, MATOVITCH VB, PASHER I, AARONSON S, HUTNER SH, SOBOTKA H: *Anal Biochem 3*:31, 1962

18. BARAKAT RM, EKINS RP: *Blood 21*:70, 1963

19. BARAKAT RM, EKINS RP: *Lancet 2*:25, 1961

20. BARNES BC, WOLLAEGER EC, MASON HL: *J Clin Invest 29*:982, 1950

21. BAYFIELD RF: *Anal Biochem 39*:282, 1971

22. BEATTIE FJR: *Biochem J 30*:1554, 1936

23. BEISER SM, BUTLER VP Jr, ERLANGER BF: *Textbook of Immunopathology,* edited by Miescher PA, Muller-Eberhard HJ. New York, Grune & Stratton, 1968, p 15

24. BELSEY R, DeLUCA HF, POTTS JT Jr: *J Clin Endocrinol Metab 33*:554, 1971

25. BERSON SA, YALOW RS: *Clin Chim Acta 22*:51, 1968

26. BESSEY OA, LOWRY OH, BROCK MJ, LOPEZ JA: *J Biol Chem 166*:177, 1946

27. BICKOFF EM: *Methods Biochem Analy 4*:1, 1957

28. BIERI JG, PRIVAL EL: *Proc Soc Exp Biol Med 120*:554, 1965

29. BIERI JG, TEETS L, BELAVADY B, ANDREWS EL: *Proc Soc Exp Biol Med 117*:131, 1964

30. BILLS CE: *The Vitamins; Chemistry Physiology and Pathology,* edited by Sebrell WH, Harris RS. New York, Academic Press, 1964, Vol II, p 132

31. BLOOMFIELD FJ, SCOTT JM: *Brit J Haematol 22*:33, 1972

32. BLUNT JW, DeLUCA HF, SCHNOES HK: *Biochemistry 7*:3317, 1968

33. BONAVITA V: *Arch Biochem Biophys 88*:366, 1960

34. BOWDEN JP, PETERSON WH: *J Biol Chem 178*:533, 1949

35. BOXER GE, PRUSS MP, GOODHART RS: *J Nutr 63*:623, 1957

36. BRIN M: *Methods in Enzymology.* Vol 18: Vitamins and Coenzymes, edited by McCormick DB, Wright LD. New York, Academic Press, 1971, Part A, p 125

37. BRIN M, TAI M, OSTASHEVER AS, KALINSKY H: *J Nutr 71*:273, 1960

38. BUNNELL RA: *Lipids 6*:245, 1971

39. BURCH HB, BESSEY OA, LOWRY OH: *J Biol Chem 175*:457, 1948

40. BURCH HB, BESSEY OA, LOVE RH, LOWRY OH: *J Biol Chem 198*:477, 1952

41. CARLSON LA: *Clin Chim Acta 13*:349, 1966

42. CARPENTER KJ, KODICEK E: *Biochem J 46*:421, 1950

43. CARR FH, PRICE EA: *Biochem J 20*:497, 1926

44. CESKA M, LUNDKVIST U: *Clin Chim Acta 32*:339, 1971

45. CHARALAMPOUS FC, ABRAHAMS P: *J Biol Chem 225*:575, 1957

46. CHEN PS Jr, TEREPKA AR, LANE K: *Anal Biochem 8*:34, 1964

47. CHENEY M, SABRY ZI, BEATON GH: *Am J Clin Nutr 16*:337, 1965
48. CHESNEY J, McCOORD AB: *Proc Soc Exp Biol Med 31*:887, 1934
49. CHEVALLIER A, CHORON Y, MATHERSON R: *Compt Rend Soc Biol 127*:541, 1938
50. CINNAMON AD, BEATON JR: *Am J Clin Nutr 23*:696, 1970
51. CONTRACTOR SF, SHANE B: *Clin Chim Acta 21*:71, 1968
52. COURSIN DB, BROWN VC: *Proc Soc Exper Biol Med 98*:315, 1958
53. COX EV, WHITE AM: *Lancet 2*:853, 1962
54. CRAFT IL, MATTHEWS DM, LINNELL JC: *J Clin Pathol 24*:449, 1971
55. CRANE FL, DILLEY RA: *Methods Biochem Analy 11*:279, 1963
56. Da COSTA M, ROTHENBERG SP: *Brit J Haematol 21*:121, 1971
57. DAM H, GEIGER A, GLAVIND S, KARRER P, KARRER W, ROTHSCHILD E, SALOMON H: *Helv Chim Acta 22*:310, 1939
58. DAVIS RE, NICOL DJ, KELLY A: *J Clin Pathol 23*:47, 1970
59. deCROUSAZ P, BLANC B, ANTENER I: *Helv Odontol Acta 9*:151, 1965
60. DeLUCA HF: *Nutr Rev 29*:179, 1971
61. DeLUCA HF, BLUNT JW: *Methods in Enzymology. Vol 18: Vitamins and Coenzymes*, edited by McCormick DM, Wright LD. New York, Academic Press, 1971, Part C, p 709
62. DREYFUS PM: *New Engl J Med 267*:596, 1962
63. DRUJAN BD, CASTILLON R, GUERRERO E: *Anal Biochem 23*:44, 1968
64. DUGAN RE, FRIGERIO NA, SIEBERT JM: *Anal Chem 36*:114, 1964
65. DUGGAN DE: *Arch Biochem Biophys 84*:116, 1959
66. DUGGAN DE, BOWMAN RL, BRODIE BB, UDENFRIEND S: *Arch Biochem Biophys 68*:1, 1957
67. EMBREE ND, AMES SR, LEHMAN RW, HARRIS PL: *Methods Biochem Analy 4*:43, 1957
68. EMMERIE A: *Nature 138*:164, 1936
69. EMMERIE A, ENGEL C: *Rec Trav Chim 57*:1351, 1938
70. ENGEL RW, SALMON WD, ACKERMAN CJ: *Methods Biochem Analy 1*:265, 1954
71. FABIANEK J, DeFILIPPI J, RICKARDS T, HERP A: *Clin Chem 14*:456, 1968
72. FERREBEE JW: *J Clin Invest 19*:251, 1940
73. FIELD H Jr, MELNICK D, ROBINSON WD, WILKINSON CF Jr: *J Clin Invest 20*:379, 1941
74. FRENKEL EP, McCALL MS, WHITE JD: *Am J Clin Pathol 53*:891, 1970
75. FRENKEL E, KELLER S, McCALL MS: *J Lab Clin Med 68*:510, 1966
76. FUKITA A, FUJITA D, FUJINO K: *J Vitaminol 1*:279, 1955
77. GAITONDE MK, GRIFFITHS M: *Anal Biochem 15*:532, 1966
78. GARRY PJ, OWEN GM: *Technicon Symp 1*:507, 1967
79. GARRY PJ, POLLACK JD, OWEN GM: *Clin Chem 16*:766, 1970
80. GILBERT HS, KRAUSS S, PASTERNACK B, HERBERT V, WASSERMAN LR: *Ann Internal Med 71*:719, 1969
81. GIORGIO AJ, PLAUT GWE: *J Lab Clin Med 66*:667, 1965
82. GIRDWOOD RH: *Advan Clin Chem 3*:235, 1960
83. GOLDBERG LS, FUDENBERG HH: *J Lab Clin Med 73*:469, 1969
84. GOODHART RS, SINCLAIR HM: *Biochem J 33*:1099, 1939
85. GOTTLIEB C, LAU KS, WASSERMAN LR, HERBERT V: *Blood 25*:875, 1965
86. GRÅSBECK R: *Advan Clin Chem 3*:299, 1960
87. GRÅSBECK R: *Progress in Hematology*, edited by Brown EB, Moore CV. New York, Grune & Stratton, 1969, Vol 6, p 233

88. GRAVESEN KJ: *Scand J Clin Lab Invest 20*:57, 1967
89. GREEN NM: *Methods in Enzymology.* Vol 18: *Vitamins and Coenzymes*, edited by McCormick DB, Wright LD. New York, Academic Press, 1971, Part A, p 418
90. GREENBERG LD, BOHR D, McGRATH H, RINEHART JF: *Arch Biochem Biophys 21*:237, 1949
91. GROSSOWICZ N, MANDELBAUM-SHAVIT F, DAVIDOFF R, ARONOVITCH J: *Blood 20*:609, 1962
92. GROSSOWICZ N, SULITZEANU D, MERZBACH D: *Proc Soc Exp Biol Med 109*:604, 1962
93. GULLBERG R: *Clin Chim Acta 27*:251, 1970
94. GUTTERIDGE JMC, WRIGHT EB: *Clin Chim Acta 27*:289, 1970
95. HALL CA: *Ann Internal Med 75*:297, 1971
96. HANSEN LG, WARWICK WJ: *Am J Clin Pathol 50*:525, 1968
97. HANSEN LG, WARWICK WJ: *Am J Clin Pathol 46*:133, 1966
98. HARRIS RS, INGLE DJ: *Vitamins Hormones 18*:291, 1960
99. HARRISON RJ: *J Clin Pathol 23*:219, 1970
100. HAUGEN HN: *Scand J Clin Lab Invest Suppl 84*:252, 1965
101. HAWORTH C, OLIVER RWA, SWAILE RA: *Analyst 96*:432, 1971
102. HAWTHORNE JN: *J Lipid Res 1*:255, 1960
103. HENRY RJ: *Clinical Chemistry, Principles and Technics*, 4th edition. New York, Harper & Row, 1964, p 711
104. HERBERT N, LARRABEE AR, BUCHANAN JM: *J Clin Invest 41*:1134, 1962
105. HOFFBRAND AV, NEWCOMBE BFA, MOLLIN DL: *J Clin Pathol 19*:17, 1966
106. HOLICK MF, DeLUCA HF, AVIOLI LV: *Arch Internal Med 129*:56, 1972
107. HOLICK MF, SCHNOES HK, DeLUCA HF: *Proc Nat Acad Sci 68*:803, 1971
108. HOLMAN WIM, DeLANGE DJ: *Biochem J 45*:559, 1949
109. HOLZER H, GERLACH U: *Methods of Enzymatic Analysis*, edited by Bergmeyer HV. New York, Academic Press, 1963, p 606
110. HOROWITZ NH, BEADLE GW: *J Biol Chem 150*:325, 1943
111. HOSOYA N, OKU T: *J Vitaminol 17*:119, 1971
112. HUBMANN B, MONNIER D, ROTH M: *Clin Chim Acta 25*:161, 1969
113. HUFF JW, PERLZWEIG WA: *J Biol Chem 155*:345, 1944
114. HUFF JW, PERLZWEIG WA: *J Biol Chem 167*:157, 1947
115. HUGHES RE: *Analyst 89*:618, 1964
116. IRREVERE F, SULLIVAN MX: *Science 94*:497, 1941
117. ISBELL H, FRASER HF: *Public Health Rept (US) 56*:282, 1941
118. JOHNSON BC, HAMILTON TS, MITCHELL HH: *J Biol Chem 159*:231, 1945
119. JOUBERT CP, de LANGE DJ: *Proc Nutr Soc S Africa 3*:60, 1962
120. KAHAN J: *Scand J Clin Lab Invest 18*:679, 1966
121. KANAI M, RAZ A, GOODMAN DS: *J Clin Invest 47*:2025, 1968
122. KAUFMAN BT, BAKERMAN HA: *Hemoglobin, Its Precursors and Metabolites*, edited by Sunderman FW. Philadelphia, Lippincott, 1964, p 187
123. KINNERSLEY HW, PETERS RA: *Biochem J 32*:697, 1938
124. KINNERSLEY HW, PETERS RA: *Biochem J 32*:1516, 1938
125. KIRCH ER, BERGEIM O: *J Biol Chem 143*:575, 1942
126. KLEIN JR, PERLZWEIG WA, HANDLER P: *J Biol Chem 145*:27, 1942
127. KONIUSZY FR, GALE PH, PAGE AC Jr, FOLKERS K: *Arch Biochem Biophys 87*:298, 1960
128. KORENMAN SG, O'MALLEY BW: *Methods in Enzymology.* Vol 18: *Vitamins and Coenzymes*, edited by McCormick DB, Wright LD. New York, Academic Press, 1971, Part A, p 427
129. KRINSKY NI, CORNWELL DG, ONCLEY JL: *Arch Biochem Biophys 73*:233, 1958

130. KUM-TATT L, LEONG PC: *Clin Chem 10*:575, 1964
131. LARNER J, JACKSON WT, GRAVES DJ, STAMER JR: *Arch Biochem Biophys 60*:352, 1956
132. LAU KS, GOTTLIEB C, WASSERMAN LR, HERBERT V: *Blood 26*:202, 1965
133. LOH HS, WILSON CWM: *Brit Med J 3*:733, 1971
134. LORNITZO FA: *Anal Biochem 25*:396, 1968
135. LOWRY O, LOPEZ JA, BESSEY OA: *J Biol Chem 160*:609, 1945
136. LOWRY OH, BESSEY OA, BROCK MJ, LOPEZ JA: *J Biol Chem 166*:111, 1946
137. LUBRAN MM: *Ann Clin Lab Sci 1*:245, 1971
138. LUBRAN MM: *Ann Clin Lab Sci 1*:236, 1971
139. LUECKE RW, PEARSON PB: *J Biol Chem 153*:259, 1944
140. MARENZI AD, CARDINI CE: *J Biol Chem 147*:363, 1943
141. MARTINEK RG: *Clin Chem 10*:1078, 1964
142. McCORMICK EC, CORNWELL DG, BROWN JB: *J Lipid Res 1*:221, 1960
143. McKIBBIN JM: *Meth Biochem Analy 7*:111, 1959
144. McLAREN DS, READ WWC, AWDEH ZL, TCHALIAN M: *Methods Biochem Analy 15*:1, 1967
145. MELNICK D, FIELD H Jr: *J Biol Chem 134*:1 1940
146. MELNICK D, FIELD H Jr: *J Biol Chem 130*:97, 1939
147. MICHAËLSSON G, MICHAËLSSON M: *Scand J Clin Lab Invest 20*:97, 1967
148. MICKELSEN O, YAMAMOTO RS: *Methods Biochem Analy 6*:191, 1958
149. MILLAR KR, CARAVAGGI C: *New Zealand J Sci 13*:329, 1970
150. MILLBANK L, DAVIS RE, RAWLINS M, WATERS AH: *J Clin Pathol 23*:54, 1970
151. MILLER ON: *Arch Biochem Biophys 68*:255, 1957
152. MISTRY SP, DAKSINAMURTI K: *Vitamins Hormones 22*:1, 1964
153. MORELL DB, SLATER EC: *Biochem J 40*:652, 1946
154. MUMMA RO, VERLANGIERI AJ, WEBER WW II: *Carbohyd Res 19*:127, 1971
155. MURPHY BEP: *Nature 201*:679, 1964
156. MYINT T, HOUSER HB: *Clin Chem 11*:617, 1965
157. NEWMARK P, PATEL N: *Blood 38*:524, 1971
158. Nomenclature, tentative rules for generic descriptors and trivial names for vitamins and related compounds. *J Nutr 101*:133, 1971
159. NOVELLI GD, KAPLAN NO, LIPMANN F: *J Biol Chem 177*:97, 1949
160. OLLENDORFF P: *Nord Med 61*:683, 1958; *C A 53*:12380, 1959
161. PANIER RG, CLOSE JA: *J Pharm Sci 53*:108, 1964
162. PASSANNANTE AJ, AVIOLI LV: *Anal Biochem 15*:287, 1966
163. PATERSON JCS, WIGGINS HS: *J Clin Pathol 7*:56, 1954
164. PEARSON WN: *Am J Clin Nutr 20*:514, 1967
165. PELLETIER O: *J Lab Clin Med 72*:674, 1968
166. PHILLIPS WEJ, MURRAY TK, CAMPBELL JS: *Can Med Assoc J 102*:1085, 1970
167. PRATT EL, FAHEY KR: *Am J Diseases Children 68*:83, 1944
168. PRICE JM: *J Biol Chem 211*:117, 1954
169. QUAIFE ML, DJU MY: *J Biol Chem 180*:263, 1949
170. RAICA N, SAUBERLICH HE: *Am J Clin Nutr 15*:67, 1964
171. REDALIEU E, NILSSON IM, FARLEY TM, FOLKERS K: *Anal Biochem 23*:132, 1968
172. REDDY SK, REYNOLDS MS, PRICE JM: *J Biol Chem 233*:691, 1958
173. REIZENSTEIN P: *Acta Med Scand 178*:133, 1965
174. RETIEF FP, HUSKISSON YJ: *J Clin Pathol 23*:703, 1970
175. ROBINSON A, LESHER M, HARRISON AP, MOYER EZ, GRESOCH MC, SAUNDERS C: *J Am Dietet Assoc 24*:410, 1948
176. ROE JH: *Standard Methods of Clinical Chemistry*, edited by Seligson D. New York, Academic Press, 1961, Vol 3, p 35

177. ROE JH: *J Biol Chem 236*:1611, 1961
178. ROE JH: *Meth Biochem Analy 1*:115, 1960
179. ROE JH, KUETHER CA: *J Biol Chem 147*:399, 1943
180. ROSEN F, PERLZWEIG WA, LEDER IG: *J Biol Chem 179*:157, 1949
181. ROTHENBERG SP: *J Clin Invest 42*:1391, 1963
182. ROTHMAN D: *Am J Obstet Gynecol 108*:149, 1970
183. SCHILLING K, DAM H: *Acta Chem Scand 12*:347, 1958
184. SCHMALL M, WOLLISH EG: *Anal Chem 29*:1509, 1957
185. SCUDI JV, BUHS RP: *J Biol Chem 141*:451, 1941
186. SELVARAJ RJ, SUSHEELA TP: *Clin Chim Acta 27*:165, 1970
187. SHOJANIA AM, HORNADY G: *Am J Clin Pathol 52*:454, 1969
188. SHUM HY, O'NEILL BJ, STREETER AM: *J Clin Pathol 24*:239, 1971
189. SMITH JE, GOODMAN DS: *J Clin Invest 50*:2159, 1971
190. SNELL EE, STRONG FM: *Anal Chem 11*:346, 1939
191. SOBEL AE, SNOW SD: *J Biol Chem 171*:617, 1947
192. SOBEL AE, WERBIN H: *J Biol Chem 159*:681, 1945
193. SOBOTKA H, KANN S, LOEWENSTEIN E: *J Am Chem Soc 65*:1959, 1943
194. STADTMAN TC: *Science 171*:859, 1971
195. STEWARD CP, HORN DB, ROBSON JS: *Biochem J 53*:254, 1953
196. STORVICK CA, PETERS JM: *Vitamins Hormones 22*:833, 1964
197. STREETER AM, O'NEILL BJ: *Blood 34*:216, 1969
198. STRONG FM, FEENEY RE, MOORE B, PARSONS HT: *J Biol Chem 137*:363, 1941
199. SZYMANSKI BB, LONGWELL BB: *J Nutr 45*:431, 1951
200. TAYLOR WE, McKIBBIN JM: *J Biol Chem 201*:609, 1953
201. THOMAS WC Jr, MORGAN HG, CONNOR TB, HADDOCK L, BILLS CE, HOWARD JE: *J Clin Invest 38*:1078, 1959
202. THOMPSON JN, ERDODY P, BRIEN R, MURRAY TK: *Biochem Med 5*:67, 1971
203. TIBBLING G: *Clin Chim Acta 23*:209, 1969
204. USDIN E, PHILLIPS PM, TOENNIES G: *J Biol Chem 221*:865, 1956
205. VANN LS: *Clin Chem 11*:979, 1965
206. VARMA TNR, PANALAKS T, MURRAY TK: *Anal Chem 36*:1864, 1964
207. VIVIAN VM, REYNOLDS MS, PRICE JM: *Anal Biochem 10*:274, 1965
208. WAHBA N: *Analyst 94*:904, 1969
209. WALTERS CJ, BROWN RR, KAIHARA M, PRICE JM: *J Biol Chem 217*:489, 1955
210. WANG YL, HARRIS LJ: *Biochem J 33*:1356, 1939
211. WATERS AH, MOLLIN DL: *J Clin Pathol 14*:335, 1961
212. WAXMAN S, SCHREIBER C, HERBERT V: *Blood 38*:219, 1971
213. WENGER J, KIRSNER JB, PALMER WL: *Am J Med 22*:373, 1957
214. WIDE L, KILLANDER A: *Scand J Clin Lab Invest 27*:151, 1971
215. WILDEMAN L: *Z Klin Chem Klin Biochem 7*:509, 1969
216. WILSON SS, GUILLAN RA: *Clin Chem 15*:282, 1969
217. WOODRING MJ, FISHER DH, STORVICK CA: *Clin Chem 10*:479, 1964
218. ZLOCH Z, ČERVEŇ J, GINTER E: *Anal Biochem 43*:99, 1971

Chapter 28

Lipids

DONALD R. WYBENGA, B.M.,
JAMES A. INKPEN, Ph.D.

The classic definition of lipids imposes the following requirements: (a) The substance must be insoluble in water but soluble in lipid solvents such as diethyl ether, chloroform, and benzene; (b) the substance must be actually or potentially an ester of fatty acids; and (c) the substance must be utilized by living organisms. Requirement (c) is essential to exclude mineral oil derivatives. These conditions must not be interpreted too literally. For a thorough review of the nomenclature and classification of lipids, the reader is referred to the works of Deuel (97) and Lovern (251).

The clinical chemist is chiefly concerned with the lipids of serum or plasma and feces. The components of major interest are the triglycerides, phospholipids, cholesterol, cholesterol esters, and nonesterified fatty acids. In blood at least 95% of the lipids exist in combination with proteins as the lipoproteins (91)–cf *Lipoproteins* in Chapter 29. The subject of this chapter is the determination of the various serum lipid components after disruption of lipoprotein complexes, as well as the determination of fecal lipids.

Solvents immiscible with water, such as pentane or benzene, can extract only a small fraction of the blood lipids because they are unable to cause more than a minimal disruption of the lipoprotein complexes (270). To effect simultaneous disruption of the complexes and extraction of the lipid from serum or plasma it is necessary to combine a water-soluble solvent with the water-immiscible solvent. The most commonly used solvent mixtures for manual determination of lipids are ethanol-ether–Bloor's solvent mixture (28, 35, 71)–and chloroform-methanol (274, 360). With the advent of specific semiautomated procedures for the analysis of various lipid classes which require a solvent with low volatility, isopropanol has been used extensively to extract serum lipids (71, 212).

Before proceeding with discussion of individual lipids, mention should be made of general technics for simultaneous isolation of all classes of lipids. This has been accomplished by column chromatography on silicic acid (183, 274, 411), alumina (376), and florisil (53); by chromatography on paper impregnated with silicic acid (4, 75); and by thin layer chromatography (159, 403).

TOTAL SERUM LIPIDS

The total serum lipids include triglycerides, phospholipids, cholesterol, cholesterol esters, nonesterified fatty acids, glycolipids (cerebrosides), acetal phosphatides (plasmalogens), phosphatidic acids, higher alcohols, carotenoids, steroid hormones, and vitamins A, D, and E.

There have been several approaches to the determination of total serum lipids, but unfortunately no method which is both simple and reliable has been forthcoming. Indirect methods—e.g., oxidimetric technics (156, 181, 196) or photometric technics (181) for the determination of the total fatty acids derived from triglycerides, phospholipids, cholesterol esters, and nonesterified fatty acids—are inaccurate (181, 359).

In the strictest meaning of the term, gravimetric procedures are the only direct quantitative technics available for the determination of total lipids. In addition to gravimetric procedures, turbidimetric and summation procedures are used extensively and are reviewed. The sulfophosphovanillin reaction has been introduced as a procedure for total lipids, and since it appears to be promising it is also discussed.

Gravimetric Technics

In this technic the lipids are extracted, dried, and weighed. In one method (195, 407) the lipids in serum are first extracted with Bloor's solvent mixture (3 vol ethanol + 1 vol ethyl ether), dried, reextracted with petroleum ether, dried, and weighed. The ethanol-ether mixture extracts the serum lipids and also significant amounts of nonlipid substances such as urea, glucose, glutathione, uric acid, creatinine, and inorganic salts. Reextraction into petroleum ether eliminates the bulk of these nonlipid components, although the petroleum ether extracts do contain small amounts of nonlipid materials, e.g., urea and amino acids (132, 388). Such nonlipid material by itself is not soluble in petroleum ether but is soluble in the presence of lipid.

In the Folch method as modified by Sperry and Brand (359, 360), lipids in serum are extracted into chloroform-methanol (3:1, v/v). The nonlipid material and methanol are removed by diffusion into water, and an aliquot of the chloroform phase is evaporated to dryness and weighed to determine the total lipids. It is claimed (359) that this procedure eliminates the inclusion of most nonlipid material in the extract. In a comparison of the Bloor extraction with the Sperry and Brand method (195), the latter was more accurate but had the disadvantages of requiring more time and being a more elaborate method.

Van Slyke and Plazin (388) proposed a method in which the lipids and proteins in plasma or serum are coprecipitated. The water-soluble material is removed from the precipitate by washing with water. The lipids are then extracted from the precipitate with ethanol-ether or chloroform-methanol. Tungstic acid, colloidal iron, and zinc hydroxide were compared as precipitants. The combination of zinc hydroxide as the precipitant and chloroform-methanol as the extracting solvent was preferred. This procedure purportedly gave less contamination with nonlipid materials as demonstrated by an average nitrogen/phosphorus ratio of 1.3 in the chloroform-methanol extract as compared with 4.2 for extracts obtained with the Sperry-Brand method. Friedmann (144, 145) compared the precipitation method with the

Sperry-Brand procedure (359). Using an acid zinc reagent as precipitant and chloroform-methanol as the solvent the precipitation procedure gave results which averaged 6% higher than the Sperry-Brand method. The C.V. of 2.4% was greater than that obtained with the reference method, but the precipitation procedure was simpler and easier to perform.

If lipid extracts are to be analyzed for individual components, they should be dried with N_2 or CO_2 or with air after adding hydroquinone to the extract.

Turbidimetric Technics

In spite of their drawbacks turbidimetric procedures are popular and are presented by many authors (54, 136, 149, 224, 238, 265, 268, 302, 338, 350), both for quantitative determination of serum lipids and as screening procedures. In one approach serum lipids are extracted with Bloor's solvent. The extract is dried and redissolved in dioxane. After addition of H_2SO_4 the turbidity is measured (90). In another approach serum is diluted with 1% phenol at high salt concentration, causing precipitation of lipid without protein precipitation (234). The disadvantage inherent in turbidimetric technics stems from the difficulty in controlling particle size. Temperature fluctuations have a considerable influence on the results (121), and various lipid components also give different turbidities when other factors are constant (29). Turbidimetric procedures generally do not correlate well with gravimetric procedures (59, 76).

Summation Technic

In this technic the major lipid classes—e.g., triglycerides, phospholipids, cholesterol, and cholesterol esters—are determined separately by direct methods and the total serum lipids calculated by summation. This technic obviously yields much more information than the single value of total serum lipids but excludes about 3%–5% of the total lipids in normal serum from the analysis (71). The lipid classes which are excluded are the nonesterified fatty acids, higher alcohols, steroid hormones, and fat-soluble vitamins.

Either a serum sample is extracted manually with a suitable solvent (ethanol-ether, chloroform-methanol, or isopropanol) and individual assays performed on the extract, or separate extracts of the serum are made for each assay. Alternatively, the lipid classes of an extract are isolated by column chromatography (193, 274) or thin layer chromatography (159) prior to assay. A comparison of the results obtained with and without column chromatography (274) showed that after chromatography average cholesterol values were higher by about 3 mg/100 ml and triglyceride values were lower. This was attributed to loss of mono- and diglycerides in the chromatography. The phospholipid levels did not change after chromatography. From a laboratory standpoint, the use of column chromatography to purify and separate lipid classes adds very little and introduces extra work. In a comparison study (59) excellent correlation was obtained between the summation technic and gravimetric procedures.

Thin layer chromatography has been proposed as a method for purifying and separating the lipid classes in a serum extract. The separate classes can be determined chemically after elution from the adsorbent that has been scraped from the chromatogram (159). This method suffers from the drawbacks already mentioned for column chromatography.

Alternately, thin layer chromatography has been proposed as a method for separation of the lipid classes and quantitation by densitometric measurements on the plate (403). The major advantage of thin layer chromatography of serum lipids lies in the screening capabilities of this technic (4).

Sulfophosphovanillin (SPV) Reaction

The total lipids in serum have been assayed using the SPV reaction, in which serum is added to conc H_2SO_4 and heated, after which a portion of the heated acid-serum solution is added to concentrated phosphoric acid. A vanillin solution is added and the color measured at 530 nm after 10 min. This method was first published by Chabrol and Charonnat (56, 57), and reinvestigated by Zoeliner and Kirsch (427). The postulated reaction is that lipids containing double bonds form ketones or ketols with the hot H_2SO_4, which then react with vanillin in phosphoric acid to yield color. Molecules such as linoleic acid, which contain two adjacent double bonds, contribute the same amount of color as oleic acid, which contains one double bond. There is apparently no additive effect with increasing unsaturation within the molecule (119). The SPV reaction has also been automated (155, 208, 350), and correlation with gravimetric (107, 147, 350) and summation (155) procedures is very good.

The SPV procedure appears to be an excellent single procedure for approximating (119) total serum lipids. The drawback is that the results depend on the amount of unsaturated compounds present in the lipid material.

Miscellaneous Technics

Various other approaches which have been used for the determination of total lipids, such as acidimetric, oxidimetric, and volumetric methods, are discussed in the first edition. These methods are little used today in clinical laboratories and are only of historical interest.

CHOICE OF METHOD

The authors believe the gravimetric method of Sperry and Brand (359, 360) to be the most accurate and precise available for determination of total lipids in serum or plasma, and this method is presented as a reference procedure.

DETERMINATION OF TOTAL LIPIDS IN SERUM

(method of Sperry and Brand, Refs 359, 360, modified by Jacobs and Henry, Ref 195)

PRINCIPLE

Lipids are determined gravimetrically after extraction into chloroform-methanol and further purification by diffusion of water-soluble substances into water.

REAGENTS

Methanol, Absolute
Chloroform, Reagent Grade
Chloroform-Methanol Mixture (2:1, v/v)

PROCEDURE

1. Transfer 2.0 ml serum (use 1.0 ml if lipemic) to a 50 ml volumetric flask; add 16.0 ml of methanol rapidly with mixing.
2. Add 16.0 ml of chloroform and bring to a boil on a steam bath. Cool.
3. Dilute to 50 ml with chloroform; mix by inversion and filter rapidly through Whatman 1 filter paper.
4. Transfer 40 ml filtrate to a 100 ml beaker. Slowly add water to within 1 cm of the top of the beaker. Carefully transfer the beaker to the bottom of a 1 liter beaker and fill the 1 liter beaker to about 1 inch from the top with water. The 100 ml beaker will be completely immersed in water so that the methanol and other water-soluble materials diffuse into the aqueous phase, leaving the lipids behind in the chloroform and in the interface. Let stand overnight.
5. Carefully aspirate the water from the large beaker. Then aspirate the water from the 100 ml beaker with extreme caution so as not to disturb the interface. A small amount of water may be left in the beaker.
6. Evaporate the solvent to dryness on a steam bath with the aid of a stream of dry air or N_2. Methanol may be added to aid in the evaporation of water. The beaker should be set in a water bath so that the temperature of the sample does not exceed 40°C.
7. The residue is dissolved in 2 ml of warm chloroform-methanol (2:1) and filtered through a sintered-glass filter under vacuum into a tared, clean, dry, 10 ml volumetric flask. The beaker is rinsed with 4 successive 2 ml portions of the solvent, and the washings are filtered into the tared flask.
8. Evaporate the solvent to dryness on a steam bath with the aid of a stream of N_2 or dry air. The external surface of the flask is rinsed with water and thoroughly wiped with a clean dry towel.
9. Transfer the flask to a vacuum desiccator containing anhydrous $CaCl_2$,

dry overnight under vacuum, and weigh to the nearest 0.1 mg. Subtract the weight of the flask from the weight of the flask plus lipid to obtain weight of lipid. The weight of the dried lipid represents the lipid in 1.6 ml of serum.

Calculation:

$$\text{mg lipid/100 ml serum} = \text{mg lipid} \times \frac{100}{1.6}$$

$$= \text{mg lipid} \times 62.5$$

NOTES

1. *Stability of samples.* In a study of eight serum samples kept at 25°C our laboratory found average decreases of 5% and 10% at 3 and 6 days, respectively. The losses are believed to be due to loss of glycerol and volatile fatty acids after hydrolysis of esters.

2. *Drying of extracts.* If the dried lipid samples are to be used for further analysis, oxidation can be prevented by carrying out the drying procedures under a stream of CO_2 or N_2, or if dry air is to be used hydroquinone can be added at levels of 0.01 mg or less.

ACCURACY AND PRECISION

The final lipid residue contains nitrogen equivalent to about 3 mg/100 ml serum, and chloride (as NaCl) equivalent to about 17 mg/100 ml (195). Among the nitrogen compounds present are amino acids, urea, and uric acid (388).

The precision of the method is reported to be ±5% (145, 195, 359).

NORMAL VALUES

A compromise of the published adult fasting normal ranges for total lipid in serum or plasma is 450–1000 mg/100 ml (195). There is evidence that the total lipids increase steadily with age (104, 149). There is no sex difference (195) except during pregnancy (10) when total lipid levels increase. At birth the levels are about 100–200 mg/100 ml and there is nearly a two-fold increase within the first few days of life, with the adult level reached and maintained after the age of 1 year (195). The total lipid level increases after a meal containing fat, principally because of an increase in the neutral fat fraction, although there may also be an increase in phospholipids (181).

CHOLESTEROL AND CHOLESTEROL ESTERS

Cook (73) and Kritchevsky (233) have thoroughly covered the chemistry, physiology, and pathology of cholesterol and cholesterol esters in their publications. Because of its long recognized importance in clinical medicine, no other lipid of the human body has been so intensely studied. It is therefore understandable that a great number of methods for its determination has been published. A search is still being made for faster technics, especially automated technics, which will provide greater accuracy and require smaller volumes of sample.

CHOLESTEROL IN THE BLOOD

The formula for cholesterol (Δ^5-cholesterol, cholest-5-en-3β-ol) is (233):

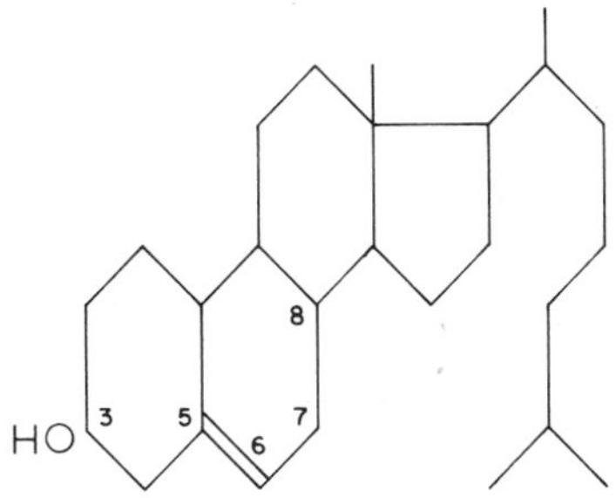

Most of the reactions of cholesterol occur at the hydroxyl group, at the double bond or at carbon 7. The hydroxyl group is equatorial and thus is freely reactive. This explains why ester formation takes place at the hydroxyl group. Hydrogenation of the double bond leads to stereoisomers: *coprostanol* (coprosterol, 5β-cholestan-3β-ol) and *cholestanol* (dihydrocholesterol, 5α-cholestan-3β-ol). Coprostanol is formed from cholesterol in the intestine but is not found in serum (320). Cholestanol, however, appears in normal serum in a concentration of about 1%–3% of total cholesterol (137, 346). Δ^7-Cholestanol, an isomer of cholesterol with the double bond at 7-8 instead of 5-6, is also found in normal serum, representing about 0.4%–1.4% of total cholesterol (127, 283). 7-Dehydrocholesterol (double bonds at 5-6 and 7-8) has been identified in human serum and is present in concentrations ranging from 5 to 40/mg/100 ml ($\bar{x}$ = 20), about 80% of which is esterified (137, 226). Small amounts of other derivatives may also be present. The remainder of the total "cholesterol" of serum or plasma, i.e., about 85%, is comprised of cholesterol and its esters. Approximately 25% of the total cholesterol exists free, and 75% as esters.

The total cholesterol of erythrocytes is some 10%–30% less than that of serum (133). One study (319) found the normal total cholesterol content to range from 1.3 to 1.44 mg/ml of red cells. Erythrocytes, however, contain considerably less cholesterol esters than serum; findings have ranged from 0 (42) to about 50 mg/100 ml (36, 133). The ratio of free to ester cholesterol is about 4:1 in red cells and about 1:3 in serum or plasma (73). The serum cholesterol level is altered by certain factors which do not affect the red cell level (73, 125, 257), and serum or plasma is therefore the specimen of choice for the determination of cholesterol and cholesterol esters.

EXTRACTION OF CHOLESTEROL

The "reference" cholesterol methods all employ an extraction step. Recently "direct" methods (serum added directly to color reagents) have been developed, but they all are compared with extraction procedures to establish accuracy. Complete extraction of cholesterol requires dissociation of the lipid-protein bond. Cold chloroform by itself extracts only 25 mg of cholesterol per 100 ml of serum (135). Equipment was devised (188) which dries serum on filter paper and extracts cholesterol in a high yield with chloroform in one continuous operation. Ether by itself extracts only a small fraction of cholesterol from normal serum (106). Petroleum ether on the other hand completely extracts free and esterified cholesterol (397). Ethanol-ether, which was introduced by Bloor in 1914 (28), effects complete extraction of cholesterol and has been employed in many methods including the most commonly used reference method, that of Abell *et al* (1, 2). Other solvents have also been used successfully: Acetic anhydride-dioxane (333, 334), acetic anhydride (52, 250), acetone (92), acetone-ethanol (3, 25, 80, 278, 361), acetic acid (419), chloroform-H_2SO_4 (431), chloroform-methanol (131, 278), carbon tetrachloride (385), methanol (315), methylal-methanol (93), ethanol (137), isopropanol (212, 240, 278), and ethyl acetate-ethanol (337). Frequently the extraction process is aided by heat. Extractions with ethanol-ether are complete only if the serum is greatly diluted with solvent. A minimal dilution of serum with solvent of 1:25 to 1:50 is required for complete extraction with this solvent system (5). This dilution is required to reduce the water content in the final mixture, since otherwise the partition coefficient for cholesterol does not favor the solvent.

ISOLATION OF CHOLESTEROL AS THE DIGITONIDE

The organic solvent extracts contain cholesterol and its esters plus many other substances—urea, glucose, etc. Many cholesterol methods include a step in which cholesterol is isolated by precipitation. Digitonin has been the most commonly employed precipitating reagent for cholesterol. This naturally occurring glucoside forms a 1:1 complex with cholesterol. The cholesterol-digitonide complex is quite insoluble in most solvents: 0.0006% in boiling water, 0.0007% in ether (279), 0.02% in 96% ethanol, 0.09% in absolute ethanol, and 0.5% in absolute methanol (344). The complex can be split and the cholesterol recovered by treatment with pyridine or acetic anhydride. Naturally occurring sterols having either a 3α-hydroxyl or no hydroxyl group form digitonides. Cholesterol esters do not form digitonides, since they do not have free hydroxyl groups. The precipitation of cholesterol digitonide has been facilitated by $Al(OH)_3$ (38, 293) and Al^{3+} in acid solution (38, 383). Digitonin itself reacts in the Liebermann-Burchard reaction (374, 417), causing a positive error unless the digitonin is removed prior to color formation. Furthermore, digitonin precipitation is no guarantee of specificity (203, 348) since stanols, which comprise about 5% of the serum sterols (168, 275), also precipitate.

HYDROLYSIS (SAPONIFICATION) OF CHOLESTEROL ESTERS

There are three reasons why, in some methods, hydrolysis of esters is performed: first, to permit digitonide precipitation of all the cholesterol; second, to obtain more complete extraction into certain solvents; and third, to allow a color reaction to be made with one form of cholesterol only. The latter is important in those instances in which free cholesterol and its esters yield different color equivalents. The most widely used technic of hydrolysis is treatment with warm alcoholic KOH. Cholesterol may decompose during this treatment. The alkali may also produce a substance from the ethanol of ethanolic KOH which yields a color in the cholesterol color reaction (154). Addition of benzyl trimethylammonium hydroxide to ethanol-ether extracts causes hydrolysis during evaporation of the extract (2, 69).

METHODS FOR DETERMINATION OF CHOLESTEROL

Several early methods might be mentioned briefly primarily because of historic interest: (a) *Gravimetry*. Cholesterol is precipitated as the digitonide after saponification and the digitonide weighed (358). (b) *Nephelometry*. Cholesterol digitonide can be suspended and quantitated by nephelometric measurements (308). (c) *Gas chromatrography* (55, 86, 229). (d) *Thin layer chromatography*. Free and esterified cholesterol are first separated by thin layer chromatography and are then determined separately with the ferric ammonium chloride method (23). (e) *Coulometry*. Cholesterol is separated by thin layer chromatography and quantitated by coulometric titration (378). (f) *Fluorometry*. These methods are based on the Liebermann-Burchard reaction (51, 326, 355) or the Tschugaeff reaction (373), and the resultant fluorescence is determined. (g) *Turbidimetry*. The turbidity produced upon addition of sodium alcoholate is measured (207, 221).

The photometric methods, however, are the ones most frequently employed. Before discussing in some detail the various photometric methods, the relationship between the principal color reactions should be considered together (73, 233): Cholesterol, as well as other steroids, gives intense colors when treated with acid reagents (Lewis acids). The actual color produced depends on the acid used, its concentration, and other variables. In the Liebermann-Burchard reaction cholesterol in chloroform is treated with acetic anhydride and concentrated H_2SO_4 resulting in a green color. It is believed that the acetic anhydride serves only as a diluent for the H_2SO_4 since it can be replaced by acetic acid, ethyl acetate, or butanol. The Salkowski reaction eliminates the acetic anhydride and a red-purple color is formed. Steroids having two double bonds, or those capable of forming them by dehydrogenation, give color when treated with acid alone. There is evidence that the steroid is ultimately converted to polymeric unsaturated hydrocarbons by way of dehydration, sulfonation, and polymerization. Other color reactions, such as the Tschugaeff (reaction with acetyl chloride and $ZnCl_2$ in glacial acetic acid) and $FeCl_3$-H_2SO_4 reactions, probably have a similar basis and either employ weaker acids or include an oxidizing agent which can increase the length of the conjugated diene chain. The reaction with $FeCl_3$-H_2SO_4 appears to involve oxidation because a definite amount of Fe^{3+} is reduced to Fe^{2+}, the amount being proportional to the cholesterol present (418).

Methods Employing the Liebermann-Burchard Reaction

Liebermann in 1885 described the reaction of cholesterol in acetic anhydride with H_2SO_4, in which colors are produced going from red through violet to blue-green. Burchard (44) applied this reaction to cholesterol in $CHCl_3$. Numerous technics employing the Liebermann-Burchard color reaction have been introduced since then. The technics most widely used are as follows: Bloor (30) and its modification (262), of Schoenheimer and Sperry (345) and modifications (25, 39, 153, 361), of Carr and Drekter (52), and that of Abell *et al* (1, 2) and modifications (20, 72, 201, 385). Methods in which this color reaction is carried out directly on serum without prior extraction have become very popular (126, 191, 219, 286, 331, 431). Tonks (374) reported in his critical review of methods that these direct methods give spuriously high results and that inaccuracies occur if bilirubin levels are elevated above normal. As we shall see later, however, this criticism does not apply to the direct method presented in this chapter.

The Liebermann-Burchard color reaction is complex and is affected by many variables. These variables include the concentration of H_2SO_4 (95, 262) and water (262), acetic acid content of the acetic anhydride (95, 262), solvent employed (262), time of reaction (95, 262), effects of light (202, 262), and temperature (95, 209, 262). The absorption peaks of this color are at 430 and 630 nm. The yellow component of the green Liebermann-Burchard color is more stable than the blue component but is affected by light (374). Some workers have recommended absorbance readings at 430 nm (209, 262) while others prefer 540 nm (189, 205). The 630 nm peak is susceptible to time and temperature (262). The Liebermann-Burchard reaction carried out at room temperature in chloroform solution yields 10%–30% more color with cholesterol esters than with cholesterol and therefore gives falsely high values of total cholesterol if a saponification step is not included (52, 95, 96, 211). The same color intensity is given by free cholesterol and its esters if the reaction is performed in acetic acid (52). In the method of Schoenheimer and Sperry (345) and its modifications, the esters are saponified prior to color development and the inaccuracy caused by differences in color equivalence is eliminated. The method of Abell *et al*, however, which compares well with the Schoenheimer-Sperry procedure (22, 216, 413) and is easier to perform, is now generally accepted as the reference method. Tonks (374) in his review of cholesterol methods also came to this conclusion and recommended the Schoenheimer-Sperry method as the reference method for cholesterol esters.

Methods Employing the p-Toluenesulfonic Acid Reaction

In 1952 Pearson, Stern, and McGavack (299) introduced a method in which cholesterol in serum is reacted directly with p-toluenesulfonic acid. The color is read at 505 nm against a serum blank to correct for interferences. Pearson *et al* (298) claimed that results agreed within 3.5% of those given by the method of Schoenheimer and Sperry and that free cholesterol and cholesterol acetate yielded the same amount of color. Kenny *et al* (210) and Jamieson (198), employing this reaction, confirmed the good agreement with an extraction method in which the extract is subjected to the Liebermann-Burchard reaction. The latter author stated that under his conditions cholesterol and its esters showed equimolar color production, which has also been claimed by others (85, 242). One report (52) stated that Pearson's method gave results 10% too low, while other evidence indicated that results are too high by about 10% (242, 413). Subsequent to a reported explosion with p-toluenesulfonic acid (200), sulfosalicylic acid (317, 389, 413), or 2,5-dimethylbenzenesulfonic acid has been employed (395). Bilirubin interference is significant with p-toluenesulfonic acid, and corrections have been made by subtracting 5 mg cholesterol/100 ml for each mg bilirubin/100 ml of serum (198, 210). Watson (395) demonstrated that the blank used by Pearson *et al* is not valid in the case of an icteric or hemolyzed sample since it did not include an oxidizing agent. The procedure of Watson (395), which uses 2,5-dimethylbenzenesulfonic acid was recently improved by Richardson *et al* (323), who corrected for pigments by replacing acetic acid in their blanks with a solution of 30% phosphoric acid and using H_2O_2 and peroxidase to oxidize the pigments. They found that their blank readings represented an apparent absorbance equivalent to 5 mg cholesterol/100 ml for every 1 mg bilirubin/100 ml serum. The results of this method are said to agree with the method of Abell *et al*.

Methods Employing the Tschugaeff Reaction

The reaction of cholesterol with acetyl chloride and $ZnCl_2$ in glacial acetic acid was described by Tschugaeff (379, 380) in 1900. The red color formed in this reaction is read at 528 nm. A number of technics employing this reaction have been published (176, 178, 347), and the use of o-nitrobenzoyl chloride was recommended in one (292). The stringent requirements for anhydrous $ZnCl_2$ (233) reduced the popularity of this method despite the fact that it is 15 times more sensitive than the Liebermann-Burchard reaction (328).

Methods Employing the $FeCl_3$-H_2SO_4 Reaction

This reaction, originally described by Lipschitz in 1907 and published as a method in 1953 by Zlatkis, Zak, and Boyle (426) is four or five times more sensitive than the Liebermann-Burchard reaction (422). The absorption peak of the color produced is at 560 nm and is the same for free cholesterol and its esters (60, 422, 426). The majority of investigators (60, 422, 426) have found the absorptivities of free cholesterol and its esters to be within 5% of each other. Webster (396) showed conclusively that cholesterol and its esters give the same intensity of color when reacted with $FeCl_3$ reagent in acetic acid followed by H_2SO_4 separately. Franey and Amador (137) confirmed again that free cholesterol and its esters give equivalent absorbances when the $FeCl_3$-H_2SO_4 reagent is added to an ethanol extract. Badzio (12) and others (337) demonstrated that marked differences may occur in the absorptivity of free and esterified cholesterol due to incomplete color development of the esters which is influenced by temperature, H_2SO_4 concentration, and amounts of ester determined.

The method introduced by Zlatkis *et al* (426) was applied directly to serum without any preliminary extraction of the cholesterol; and they, as well as others (254), claimed accuracy for the procedure. Many other investigators (40, 70, 186, 230, 400) found the direct method to be completely unsatisfactory. The method has been modified (a) by preliminary extraction with methanol (315), ethanol (33, 137), ethanol-acetone (89, 419, 421), ethanol-ether (70, 82), ethanol-ethyl acetate (11, 199, 225, 337), ethanol-petroleum ether (87, 175, 398), ethanol-chloroform followed by an alkali wash (420), ethanol-acetone-trichloroethylene (157), petroleum ether after saponification (182, 258); (b) by precipitation of protein with acetic acid (128) or the $FeCl_3$-acetic acid reagent (60, 180, 263, 418); and (c) by treatment with ion exchange resin followed by extraction with chloroform-methanol (393). Results obtained by such modifications reportedly agree with values given by the methods of Schoenheimer-Sperry, Sperry-Webb, and Abell *et al* (27, 79, 82, 137, 175, 199, 398, 421).

A direct method (414) for determination of cholesterol was published in 1970 in which a single stable reagent comprised of H_2SO_4, ethyl acetate, and ferric perchlorate is used. This method, which requires no extraction of the serum, uses 50 μl of sample. Serum cholesterol concentrations determined by this method did not differ statistically from those obtained by the reference method of Abel *et al* (1, 2), and for the first time a direct method that is relatively free of interference by bilirubin or hemoglobin became available.

SEPARATION OF FREE CHOLESTEROL FROM ITS ESTERS AND ESTER FRACTIONATION

The most commonly employed method for separation of free cholesterol from its esters has been precipitation of the former with digitonin (11, 241, 263, 337, 398, 421). Formation of this complex was discussed in an earlier section. Other means of separating free cholesterol from its esters are as follows: precipitation of free cholesterol as pyridium cholesteryl sulfate (353); precipitation of free cholesterol by the glycoside tomatin (115, 204,

324); oxidation of free cholesterol to cholest-4-ene-3,6-dione with CrO_3 and H_2SO_4 in acetone (288); reaction of free cholesterol with $ZnSO_4$ in acetic acid and H_2SO_4 (239); reaction of free cholesterol at a specific temperature (9, 157); Craig countercurrent distribution (309); paper chromatography (148, 310); column chromatography on silicic acid (84, 416), silicic acid-Celite (271), or alumina (180, 229, 262, 397); thin layer chromatography (13, 23, 229, 295, 382, 423); gas chromatography (86, 229); infrared spectrophotometry (139); oscillopolarography (277); partitioning with a binary solvent system (151) and petroleum ether-ethanol-water system after incubation (61).

It has always been assumed that "free" cholesterol is cholesterol by itself, not esterified. There is some evidence, however, indicating that cholesterol, so defined, may not be equivalent to that fraction precipitated by digitonin. Thus cholesterol possessing a free hydroxyl group is completely precipitable as pyridinium cholesteryl sulfate, yet the values for "free" cholesterol obtained by this precipitation method are about one-third those obtained by digitonin (353). Cholesterol therefore must exist in serum in more fractions than just "ester" and "free." In addition to "free," there must be "loosely bound" cholesterol, not of fatty acid ester type, which can be precipitated by digitonin but not as pyridinium cholesteryl sulfate. The nature of these cholesterol complexes, which are either split in the process of digitonin precipitation or precipitated in their entirety, is not known.

The approximate distribution of the fatty acids of cholesterol esters of normal serum, expressed as percent of total, is as follows (224, 244, 272, 412): (a) *saturated fatty acids*—10% palmitic, 3% stearic, 1% myristic, 1% arachidic; (b) *monoenoic acids* (one double bond)—24% oleic, 6% palmitoleic; (c) *dienoic acid*—43% linoleic; (d) *trienoic acid*—4% linolenic; (e) *tetranoic acid*—6% arachidonic; (f) *pentenoic acids*—1%; (g) *hexenoic acids*—1%.

CHOICE OF A CHOLESTEROL METHOD

As stated previously, the method of Abell *et al* (1, 2) or its modification by Anderson and Keys (286a) is generally accepted today as the reference method. This technic is presented. The method, however, does not lend itself well to large numbers of determinations. Two studies have compared the relative accuracy and precision of a number of methods; one (186) regarded the Carr-Drekter (52) as the method of choice, the other (277a) concluded that the method of Anderson and Keys or that of Trinder (376a), which can be performed more quickly, is the most suitable for routine analyses. Tonks (374), who recently critically reviewed the various classes of cholesterol methods, recommended Abell's reference method for total cholesterol and the method of Sperry and Webb (361) for cholesterol esters. Our laboratory has had considerable experience with the ferric perchlorate-ethyl acetate-H_2SO_4 reaction applied to serum directly. Good precision as well as the excellent agreement obtained with the procedure of Abell *et al* (1, 2), justifies its recommendation as a routine method.

DETERMINATION OF TOTAL CHOLESTEROL

(method of Abell et al, Refs 1, 2)

PRINCIPLE

The cholesterol esters of serum or plasma are saponified by incubation with alcoholic KOH. Free cholesterol is extracted into petroleum ether. An aliquot of the extract is dried and the cholesterol determined photometrically by a modified Liebermann-Burchard reagent (acetic anhydride-H_2SO_4-acetic acid). Figure 28-1 shows the absorption curve of the color formed.

REAGENTS

Absolute Ethanol. Ethanol denatured with methanol is satisfactory.
Petroleum Ether (b.p. approx 68° C) or hexane
Glacial Acetic Acid
H_2SO_4, *Conc*
Acetic Anhydride, Free of HCl
KOH, 33% (w/w). 10 g reagent grade KOH in 20 ml water.
Alcoholic KOH. 6 ml 33% KOH + 94 ml absolute ethanol. Ethanol denatured with methanol is satisfactory. Prepare just before use.
Modified Liebermann-Burchard Reagent. Chill 20 vol acetic anhydride to $<10°C$ in a glass-stoppered bottle. Add 1 vol conc H_2SO_4, mix well, and keep cold for 9 min. Add 10 vol acetic acid, mix, and warm to room temperature. Use within 1 hour.
Cholesterol Standard, 0.4 mg/ml. Dissolve 100 mg cholesterol in 250 ml absolute ethanol (ethanol denatured with methanol is satisfactory). Stable at least several months in refrigerator. The products of Pfanstiehl and Pfizer are pure as indicated by infrared analysis (186). If there is any question about the purity of the preparation at hand, it should be recrystallized four times from absolute ethanol (ethanol denatured with methanol is satisfactory) and dried to constant weight. The purity of cholesterol standards is discussed in more detail under *Note 6.*

PROCEDURE

1. *Standards.* Mix duplicate 5.0 ml aliquots of cholesterol standard with 0.3 ml 33% KOH in 25 ml glass-stoppered centrifuge tubes.

    Unknown. Transfer 0.50 ml serum or plasma to a 25 ml glass stoppered centrifuge tube. Add 5 ml alcoholic KOH, mix, and stopper.
2. Incubate at $37° - 40°C$ in water bath for 55 min.
3. Cool to room temperature, add 10 ml petroleum ether, and mix.
4. Add 5 ml water, stopper, and shake vigorously for 1 min. Centrifuge at about 1000 rpm for 5 min.
5. *Standards.* Transfer 3 ml aliquots of petroleum ether layers, equivalent to 0.6 mg of cholesterol, to test tubes (about 6 x 3/4 inch). See *Note 1* below.

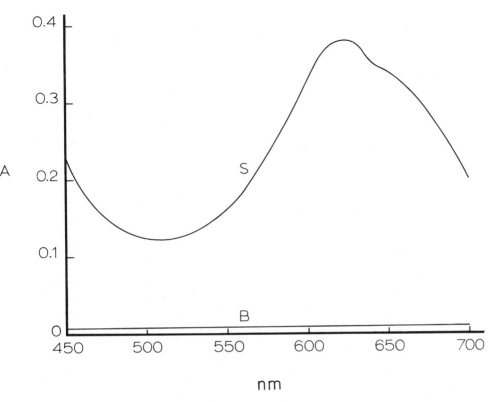

FIG. 28-1. Absorption curves in method of Abell *et al*: (B) reagent blank vs water, (S) 0.7 mg cholesterol standard vs water. Perkin-Elmer model 4000-A recording spectrophotometer.

Unknown. Transfer duplicate 4.0 ml aliquots of petroleum ether layer to test tubes.

6. Evaporate petroleum ether at 60°C with aid of a stream of air or N_2 (see *Note 2* below). Cool to room temperature and stopper with clean dry corks.

7. Set up empty tube as blank.

8. Place all tubes in 25°C water bath. Add 6 ml Liebermann-Burchard reagent to each tube at staggered intervals, mix, cork tightly, and return to bath. Incubate and read absorbances against reagent blank at 620 nm in a spectrophotometer such as a Beckman DU or B (see *Note 1*) exactly 30 min after addition of reagent. Keep tubes out of intense light.

Calculation:

$$\text{mg cholesterol}/100 \text{ ml} = \frac{A_x}{A_s} \times 0.6 \times \frac{100}{0.2}$$

$$= \frac{A_x}{A_s} \times 300$$

NOTES

1. Beer's law. Beer's law is obeyed with a spectrophotometer of high resolution such as a Beckman DU or B. According to Abell *et al* deviation from linearity may occur with an occasional batch of reagents, and they recommend running three standards, 0.2, 0.4, and 0.6 mg at the beginning and end of each set of unknowns. For routine work, however, it would seem that an initial check with each batch of reagents with 0.2, 0.4, 0.6, and 1.2 mg standards is sufficient. The color shows deviation from linearity with Klett 62 and 66 filters.

2. Evaporation of petroleum ether: air vs N_2. Evaporation at step 6 may be hastened by a stream of air instead of N_2, but the air used must be checked by comparing standards dried with the air and with N_2. Air-dried standards have been found in our laboratories to yield as much as 50% less color than N_2-dried standards. Filtering the air through a column of anhydrous $CaCl_2$ yielded dried standards giving color only a few percent less than that obtained with N_2-dried standards.

3. Determination of cholesterol esters. Determination of cholesterol esters was attempted in our laboratory by a modification of Abell *et al*, but certain difficulties were encountered. Free cholesterol was precipitated by digitonin from an ethanol-ether (Bloor) extract of serum. Esters were extracted from the dried residue with petroleum ether and the extract dried. The esters were saponified with ethanolic KOH and the analysis continued as for total cholesterol. Recovery of added cholesterol esters was low, apparently due to lack of solubility of cholesterol in the final color solution (a small amount of white particulate matter was observed). Addition of 2 ml $CHCl_3$ at the final step resulted in complete recovery, apparently being effective by keeping the cholesterol in solution.

ACCURACY AND PRECISION

Discussed after the automated method.

DETERMINATION OF TOTAL CHOLESTEROL

(method of Wybenga et al, Ref 414)

PRINCIPLE

Cholesterol, free and esterified, reacts with a combined reagent composed of ethyl acetate, sulfuric acid, and ferric perchlorate. The resultant purple color is measured photometrically at 610 nm. Figure 28-2 shows the absorption curve of the chromogen formed.

REAGENTS

H_2SO_4 Conc. Reagent grade.
Ethyl Acetate. "Spectroquality" solvent (Matheson, Coleman and Bell), used without redistillation.

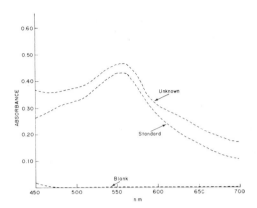

FIG. 28-2. Absorption curves recorded for a 200 mg/100 ml cholesterol standard and a sample of serum. A 50 μl aliquot of standard or serum was added to 5.0 ml reagent and the color developed in a heating block, set at 100°C, for 1.5 min.

Ferric Perchlorate. "Nonyellow, containing excess $HClO_4$" (catalog No. 40, G. Frederick Smith Chemical Co.). This reagent usually contains 59%–75% ferric perchlorate. Keep desiccated after reagent bottle has been opened.

Cholesterol Reagent. Dissolve 520 mg ferric perchlorate in 600 ml ethyl acetate contained in a 2 liter Erlenmeyer flask. Place the flask in an ice bath and cool the contents to 4°C. Add gradually, in small portions, 400 ml cold conc H_2SO_4. Mix after each portion is added but do not allow the temperature to exceed 45°C. The reagent is stable for at least a year when stored in an amber bottle at 25°C, or two years when refrigerated.

Cholesterol Standard. 200 mg cholesterol/100 ml glacial acetic acid (Pfanstiehl Labs, Inc.) The National Bureau of Standards' cholesterol standard could be used, but this is generally not used routinely because of its high cost (see *Note 6*).

PROCEDURE

1. To each of three screw capped vials (13 x 100 mm) with Teflon lining labeled "blank," "standard," and "unknown," add 5.0 ml cholesterol reagent.

2. Add 50 μl cholesterol standard and 50 μl serum or plasma (do not use oxalated plasma) to the "standard" and "unknown" vials, respectively, and mix the contents of each vial thoroughly for at least 10 sec.

3. Insert all vials into a heating block (such as the Dow Diagnostest Heating Block, catalog No. 58770) set at 100°C. "Standard" and "Unknown" vials should be heated for the exact same period.

4. Remove all vials exactly 1.5 min later and immerse them in tap water (20°C or cooler) for 5 min. Remove, dry the exterior of the vials, screw on the caps tightly, and mix their contents by inversion (if screw-capped vials are not available, inversion can be accomplished with regular test tubes by placing Parafilm on the top and holding it on tightly with the thumb).

5. Read A_x and A_s versus the blank at 610 nm or with a filter with nominal wavelength in this region.

Calculation:

$$\text{mg cholesterol/100 ml serum} = \frac{A_x}{A_s} \times 0.1 \times \frac{100}{0.05}$$

$$= \frac{A_x}{A_s} \times 200$$

NOTES

1. Beer's law. The color obeys Beer's law with many filters with a nominal wavelength in the region of 610 nm up to a cholesterol concentration of 400 mg/100 ml. For higher concentrations the serum should be diluted prior to analysis.

2. Color development and stability. Sufficient color develops when the reaction mixture is heated for 1.5 minutes in a heating block set at 100°C, during which time the temperature of the mixture reaches 70°C. Sample and standard develop color proportionally as long as the heating time is between 1 and 2 minutes under conditions described in the method. Heating for less than 1 min or more than 2 min gives low and high results, respectively. Color development for 30 min at room temperature gives unsatisfactory results, since the increase of color is not the same in unknown and standard.

3. Interferences by hemoglobin and bilirubin. Hemoglobin and bilirubin have a negligible effect on the cholesterol results at concentrations of 600 mg/100 ml and 5 mg/100 ml, respectively.

4. Serum vs plasma. As would be predicted, anticoagulants such as oxalate, citrate, etc. cause lower cholesterol values in plasma than are found in serum because of the shift in water from erythrocytes (166, 276).

5. Stability of samples. Total cholesterol has been reported (165, 378) to be stable at room temperature for at least 7 days, and this has been confirmed in our laboratory. At −20°C it is stable for 5 years (20). One report (83) stated that the total cholesterol is stable for 6 months frozen, provided there is only one freezing and thawing, while another states that it is stable even if frozen and thawed three times.

6. Purity of cholesterol primary standards. Radin (312) described several cholesterol recrystallization methods and concluded that the dibromide derivative method is the method of choice because the cholesterol preparation so prepared gives the greatest absorptivity in the Liebermann-Burchard reaction. Williams *et al* (404) purified cholesterol preparations from 14 commercial sources by the dibromide method and demonstrated that the purified samples had higher contents of cholesterol, higher melting points, narrower melting point ranges, and exhibited fewer contaminants than the original preparations. A certification program has evolved (280) which ensures the reproducibility and stability of crystalline cholesterol standards. The National Bureau of Standards issued a 99.4% pure cholesterol as Standard Reference Material No. 1911 for clinical and pathology laboratories in 1967. Its purpose is for calibration of methods and to assist in manufacturing clinical products to specifications. Cholesterol preparations recommended for use as primary standards were studied, three samples in 1968 and seven samples in 1969 (405). Two lots of one preparation contained only 97% cholesterol while three other products studied in 1969 contained oxidized products. It is therefore recommended that commercial preparations be checked against the standard reference material of the National Bureau of Standards by the individual user.

ACCURACY AND PRECISION

This topic is discussed under *Automated Determination of Total Cholesterol.*

DETERMINATION OF CHOLESTEROL ESTERS

(method of Creech and Sewell, Ref 84, modified)

PRINCIPLE

Free and esterified cholesterol are separated by column chromatography on silicic acid. Free cholesterol is eluted with chloroform and the eluate evaporated to dryness under N_2. The concentration of the free cholesterol is determined in the dried down eluate by the total cholesterol method described previously. The concentration of the cholesterol esters is then calculated by subtracting free cholesterol from total cholesterol.

REAGENTS

Chloroform, Redistilled. Use an all-glass still.
Chloroform: Methanol (Anhydrous), 2:1 (v/v)
Benzene: Hexane, 1:1 (v/v)
"Unisil" Activated Silicic Acid, 100–200 Mesh. (Clarkson Chemical Co. Inc.)
Cholesterol Standard, 200 mg/100 ml. Dissolve 200 mg cholesterol (Pfanstiehl Labs, Inc.) in approximately 80 ml glacial acetic acid and make up to 100 ml with glacial acetic acid.
Cholesterol Reagent. See under *Reagents* of *Total Cholesterol Method.*

PROCEDURE

1. Transfer 0.2 ml serum to a 15 x 125 mm test tube.
2. Add 3.8 ml $CHCl_3$:CH_3OH reagent dropwise while mixing on a

mechanical mixer. Filter through a glass wool funnel into a 10 ml volumetric flask. Wash tube and funnel with an additional 4 ml of the $CHCl_3:CH_3OH$ reagent and dilute to 10 ml with the $CHCl_3:CH_3OH$.

3. Transfer 2.5 ml to a 10 x 100 mm test tube and evaporate to dryness under N_2 in a 60°C water bath.

4. Redissolve in 0.5 ml benzene:hexane.

Column Chromatography

5. Weigh 1 g activated silicic acid and transfer to a 50 ml beaker.

6. Prepare a slurry of the resin with approximately 3 ml benzene:hexane.

7. Quantitatively transfer to a microchromatography column containing a small glass wool plug at the top of the narrow part of the column.

8. Wash column with 20 ml benzene:hexane.

9. Apply the benzene:hexane extract obtained in step 4 to the column.

10. Wash out the remainder of the extract with 5 ml benzene:hexane and apply also to the column. Repeat this twice, adding each washing to the column.

11. Elute free cholesterol with 20 ml $CHCl_3$ and collect in an 18 x 150 mm test tube.

12. Evaporate eluate to dryness under N_2 in a 60° H_2O bath.

Color Development

13. Mark the tube containing the evaporated eluate "unknown" and mark two empty tubes "standard" and "blank", respectively. Transfer 0.05 ml glacial acetic acid, 0.05 ml cholesterol standard, and 0.05 ml glacial acetic acid to the respective tubes.

14. Add 5 ml cholesterol reagent to each tube and mix thoroughly using a mechanical mixer.

15. Place all tubes in a heating block set at 100°C.

16. Remove all tubes at the same time 90 sec later and cool them in cold tap water (not higher than 20°C) for 5 min. Mix the contents of the tubes after cooling.

17. Read A_x and A_s versus the blank at 610 nm.

Calculation:

$$\text{mg free cholesterol/100 ml serum} \quad = \frac{A_x}{A_s} \times 0.1 \times \frac{100}{0.05}$$

$$= \frac{A_x}{A_s} \times 200$$

$$\text{mg cholesterol esters/100 ml serum} = \text{mg total cholesterol/100 ml} - \text{mg free cholesterol/100 ml}$$

NOTES

1. Beer's law and color stability. See *Note 1* under *Determination of Total Cholesterol.*

2. Silicic acid. The difficulties with preparation and activation of silicic acid can be avoided by the use of "Unisil." This material is prepared and activated specifically for use with a benzene-hexane solvent, and it can be used for this method without further treatment.

3. Cholesterol esters. The esters can be determined directly with this method by collecting the benzene-hexane wash. We determine the free cholesterol since the molar absorptivities for cholesterol and its esters employing the described cholesterol reagent are not identical for all esters.

4. Stability of samples. On standing, free cholesterol decreases and cholesterol esters increase as a result of enzymatic action by cholesterol esterase (122, 357, 392). The optimal pH for the esterase activity is at 8.0, and heat inactivation at $55°-60°C$ for 30 min destroys the enzyme (357, 381). At room temperature the rate of esterification is rather slow. Our laboratory studied eight normal sera at $25°C$, and the decrease in free cholesterol was about 15% in 6 days with a concomitant increase of about 8% in cholesterol esters. In the frozen state there is no change over a period of 6 months (45, 83).

ACCURACY AND PRECISION

This topic is discussed under *Automated Determination of Total Cholesterol.*

AUTOMATED DETERMINATION OF TOTAL CHOLESTEROL.

(method of Wybenga et al, Ref 414)

PRINCIPLE

Free and esterified cholesterol reacts with a combined reagent composed of ethyl acetate, H_2SO_4, and ferric perchlorate. The resultant purple color is measured photometrically at 560 nm.

REAGENTS

Saline with BRIJ. (For sampler wash.) Add 8.5 g NaCl to a 1 liter volumetric flask containing approximately 600 ml water. Add 0.6 ml of a 35% BRIJ solution, bring to mark with water, and mix.

Cholesterol Reagent. See under *Reagents* for *Determination of Total Cholesterol* by the direct method of Wybenga *et al.*

Cholesterol Standard. A serum pool is used as the standard. The value for the pool must be accurately determined by the manual direct method. It is recommended that the cholesterol concentration of this pool be established by analyzing in triplicate on several days and taking the average value. Several liters of this serum standard are aliquoted in 1 ml portions and kept frozen. The established cholesterol value of this serum is rechecked monthly.

PROCEDURE

Set up manifold, other components, and operating conditions as shown in Figure 28-3. By following procedural guidelines as described in Chapter 10, steady state interaction curves as shown in Figure 28-4 should be obtained and checked periodically. Serum blanks do not have to be run. Results are calculated from the calibration curve in the standard manner for methods obeying Beer's law.

NOTES

1. See Notes 1, 2, and 3 of manual direct method.

2. Lipemic samples. Moderately lipemic specimens should be extracted with isopropanol as follows. Add 0.5 ml serum dropwise to 4.5 ml isopropanol, mixing after each drop on a mechanical mixer. Continue mixing for at least 30 sec after the last drop of serum has been added. Mix the contents by using a mechanical mixer at high speed for 15 min and then centrifuge. Run the isopropanol extracts at the end of a tray preceded by three isopropanol washes and multiply the results by 10 to correct for the dilution.

3. Serum standard. Standards in isopropanol or in glacial acetic acid cannot be used because the narrow sample line accentuates the difference in viscosity between isopropanol standards and serum samples. Results obtained with isopropanol standards do not compare as well with those obtained with the reference method as do results calculated with serum standards.

ACCURACY AND PRECISION

The method of Abell *et al* is generally regarded as the most accurate one available today for total cholesterol. Countercurrent distribution studies by these authors indicated that the method was more than 99% specific. Results obtained by the direct manual method presented were compared with those by the procedure of Abell *et al*. A paired test revealed no statistically significant difference between the two means. The automated method was compared with the manual method and no statistically significant difference could be demonstrated. Moderate lipemia does not interfere with the manual procedure but does affect the automated procedure. Specimens demonstrating moderate lipemia therefore must first be extracted with isopropanol before they can be analyzed by the automated method.

CHOLESTEROL

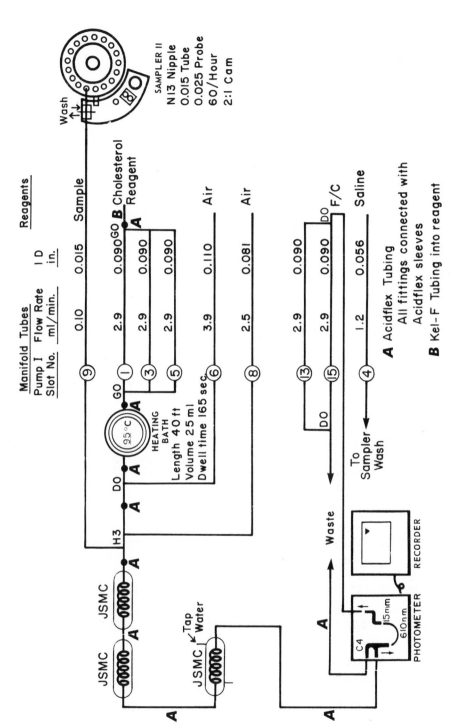

FIG. 28-3. Flow diagram for cholesterol in serum. Calibration range: 0–400 mg/100 ml. Conventions and symbols used in this diagram are explained in Chapter 10.

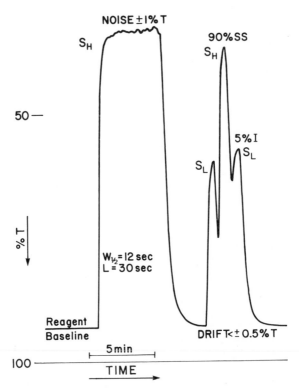

FIG. 28-4. Cholesterol strip chart record. S_H is a 244 mg/100 ml control serum, S_L a 122 mg/100 ml control serum. Peaks reach 90% of steady state; interaction between samples is 5%. $W_{1/2}$ and L are explained in Chapter 10. Computer tolerance for noise is set at $\pm 1.0\%$ T, for baseline drift at $\pm 1.5\%$ T, and standard drift at $\pm 1.0\%$ T.

Results obtained with the procedure for cholesterol esters were compared with those obtained by the reference method of Sperry and Webb, and there was very good agreement. Recovery data on cholesterol palmitate and cholesterol demonstrated recoveries of 99%–100.7%.

Precision (95% limits) of the manual and automated total cholesterol methods are ±5% for serum cholesterol concentrations between 100 and 400 mg per 100 ml. The precision for the cholesterol ester method is ±8%.

NORMAL VALUES

Normal Values for Total Cholesterol. Reed *et al* (17) examined the cholesterol results on approximately 1419 sera and, applying nonparametric statistical analysis, reported the normal ranges for total cholesterol shown in Table 28-1.

TABLE 28-1. Normal Range Estimates for Total Cholesterol[a]

Age (years)	Total cholesterol (mg/100 ml)
20–29	144–275
30–39	165–295
40–49	170–315
50+	177–340[b]

[a]From Reed AH, Cannon DC, Winkelman JW, Bhasin YP, Henry RJ, Pileggi VJ: *Clin Chem 18*:57, 1972.
[b]The range here may be higher for females than for males.

Although all subjects included in this study were carefully screened by history and physical examination and adjudged to be normal, three important points must be kept in mind: (a) The groups may well have included some with subclinical atherosclerosis or hypothyroidism. (b) The so-called "normal values" may not necessarily be desirable values. For example, one worker (311) has proposed a top "desirable" limit as one standard deviation below the mean of the coronary group of the corresponding age groups (for males in the age group of 40–50 this would be 212 mg/100 ml). (c) There is a great overlap in the normal and atherosclerotic distributions so that an individual with actual or incipient atherosclerosis need not have a cholesterol level above the upper limit of normal. For example, in one study (237) the normal range for "normal" men at age 45 was found to be 140–340 mg/100 ml, and the range in men of the same age, post infarct, was 147–403 mg/100 ml.

Normal ranges reported by others (46, 217, 237, 245, 252, 294, 300, 349) are in good agreement with those in Table 28-1.

Age and sex. The normal range for total cholesterol at birth is about 45–100 mg/100 ml (232, 337) with no difference between premature and full-term infants (401). During the first few days of life the level approximately doubles (206, 314). The normal range reached at 1 month apparently holds until the age of 20 years (184, 228). Serial cholesterol determinations in 327 premature infants on various dietary regimens for the first 6 months of life showed consistently high values for cholesterol in those infants fed a prepared cows' milk formula containing 2.7% protein and 2.8% butter fat (170). Consistently low values were registered in the infants fed a prepared cows' milk formula with 1.5% protein and 3.5% of a fat mixture consisting mainly of vegetable oils with a high proportion of polyunsaturated fatty acids.

After the age of 20 there is a steady increase in normals with age (46, 172, 201, 252, 294, 318, 339, 363, 375). No sex difference up to 30 or 35 was stated in one report (363), another (318) claimed no difference up to 50 years, and Reed *et al* (17) did not find a sex difference at any age.

Diet. The mean cholesterol level is 25%–50% higher when the diet is high in fat—about 40% of caloric intake (216). A study of total serum cholesterol performed on 476 healthy men between the ages of 17 and 44 years revealed a direct relation between serum cholesterol and obesity, which was defined in terms of weight in excess of the predicted standard body weight (129). The response of serum cholesterol to changes in dietary lipids was studied (215), and it was found that saturated acids with 12–17 carbon atoms and dietary cholesterol, expressed as the square root of the mg cholesterol per 1000 kilocalories of diet, raise the serum level and that the two effects are addictive.

Menses. Some workers have observed a postovulatory increase, a premenstrual increase, a decrease during the menses, and the highest level at ovulation, the total swing being up to about 25% of the total cholesterol level (241). Others (111), however, observed lowest values at ovulation.

Stress and exercise. Severe emotional stress (e.g., final exams in medical school) causes increases of 3–50 mg/100 ml (108, 143). Increases as high as 100 mg/100 ml occurring within 4 hours after stress have been reported (306). A highly significant inverse relationship between exercise and cholesterol levels was reported in one study (366).

Within-day variation. The serum cholesterol level in an individual varies during a 24 hour period about ±10% from the average for the day (15, 20, 24). Even wider variations have been reported in some cases (16, 369), e.g., shifts as great as 100 mg/100 ml within 1 hour (304, 305). Hourly variations of 2.0%–8.6% were reported in a study in which total cholesterol levels were determined at hourly intervals in 22 human subjects (295). The fluctuations were not related to meals, normal activity, or the time of day.

Day to day variation. There is a wide difference between individuals in their day to day variation, ranging in most cases from ±5 to ±50 mg/100 ml (325, 369, 394). One report (295) stated day to day variation of fasting samples as much as 14.5%. Hollister *et al* (187) determined serum cholesterol in 28 men twice weekly over a 12 week period and found coefficients of variation between 2.8% and 14.9%. In some individuals the day to day difference may be as much as 200 mg/100 ml (304).

Seasonal variations. Thomas *et al* (370) reported that cholesterol levels of healthy young men averaged 35 mg/100 ml higher in December and January than in May and June. The mean serum cholesterol concentration of older men was highest during the months of February and March (19). Another report (105) also stated that the highest cholesterol concentrations occurred in February. A study in our laboratory (204) confirmed that cholesterol levels in fall and winter are higher than in spring and summer.

Normal Values for Free and Ester Cholesterol. A normal range for free cholesterol of 22%–30% of total cholesterol was chosen from the studies of a number of workers using digitonin precipitation (67, 303, 354, 356). Consequently the normal range for cholesterol esters as percent of total cholesterol is 70%–78%. There is no variation in the ratio of free to ester cholesterol with sex or age (228, 246).

NONESTERIFIED FATTY ACIDS

The terms "nonesterified fatty acids" (NEFA), "unesterified fatty acids" (UFA), and "free fatty acids" are all used in the literature synonymously, but since almost all of these fatty acids are bound to proteins the last term probably should be reserved for the unbound fraction (21).

Fluorometric technics for NEFA use β-methylumbelliferone as indicator for the endpoint while titrimetric methods for NEFA employ thymol blue, phenolphthalein, or a pH meter for determination of the endpoint after preliminary extraction into ethyl ether (163), petroleum ether (169, 352), isopropanol-heptane (58, 101, 103, 142, 377), methanol-chloroform (367), acetic acid-isooctane (164), chloroform (8), and heptane (371). It has been claimed (367) that both ethyl ether and petroleum ether do not completely extract the fraction of NEFA bound to albumin. Two substances, lactic acid and phospholipids, which are extracted by some of these solvents, cause spuriously high results unless removed (142, 164, 367, 377). For example, in the method of Dole (101), each milligram of lactic acid per 100 ml causes an apparent increase in NEFA of about 0.002 meq/liter (142, 377). Exceptional levels of acetic, acetoacetic, or β-hydroxybutyric acids may also interfere (102). Removal of phospholipids by adsorption on silicic acid (48, 236, 367) has been accomplished. A photometric method has been introduced (48) in which NEFA are isolated on an ion exchange resin or florisil (269), then methylated and converted to hydroxamic acids. Other photometric methods (249, 264) based on change of color of a buffered acid-base indicator have been described. The photometric autotitration method (214) is a modification of the earlier two-phase titration procedure and is performed in a one-phase medium by means of an autoburet using tetrabutylammonium hydroxide. Quantitative determinations of NEFA using gas chromatography (171, 301) or thin layer chromatography have also been performed. Photometric methods (8, 112, 194, 236) based on the formation of NEFA-Cu soaps have been modified (117, 291) by extracting cobalt rather than copper soaps with a solvent lighter than water instead of chloroform.

The distribution of NEFA of one normal plasma has been determined by column chromatography with the following results (100): 6% stearic, 34% palmitic, 54% oleic, and 6% unidentified. Gas chromatographic analysis (63, 351) has yielded somewhat different results and also identified small amounts of lauric, myristic, myristoleic, palmitoleic, linoleic, and arachidonic acids. Perry *et al* (301), using gas chromatography, reported the following serum concentrations in mg/ml: acetic acid, 1.85–8.6; propionic acid, trace–1.80; isobutyric acid, 0–0.22; butyric acid, 0–0.29; isovaleric (may also include α-methylbutyric acid), 0–0.42; caproic acid, 0–0.32.

DETERMINATION OF NONESTERIFIED FATTY ACIDS

(method of Trout et al, Ref 377)

PRINCIPLE

Serum or plasma is extracted with an isopropanol-heptane mixture containing $H_2 SO_4$. The extract is then washed with 0.05% $H_2 SO_4$ to remove lactic acid and an acetone-insoluble material that interferes. The fatty acids are titrated with alkali and thymolphthalein as indicator, using a stream of N_2 for mixing.

REAGENTS

Extraction Mixture. 40 vol redistilled isopropanol, plus 10 vol redistilled heptane, plus 1 vol 1 N $H_2 SO_4$.
Heptane (Redistilled)
$H_2 SO_4$, 0.05%. Dilute 0.5 ml conc $H_2 SO_4$ to 1 liter with water.
Thymolphthalein. Make up as 0.01% in 90% ethanol.
NaOH, Approx 0.02 N. Make up with boiled water.
Stock Standard. Dissolve 32.0 mg palmitic acid in 25 ml heptane: 1 ml equivalent to 1.28 mg or 5 μeq.
Working Standard. Dilute stock standard 1:10 with heptane: 1 ml equivalent to 0.5 μeq.

PROCEDURE

1. Add 10 ml extraction mixture to 2.0 ml sample in glass stoppered tube and shake vigorously for a few seconds. Let stand 10 min or longer. While waiting set up the following in glass stoppered tubes:

   Blanks: 10 ml extraction mixture, 6.0 ml heptane, 6.0 ml water

   Standard: 2.0 ml working standard, 10 ml extraction mixture, 4.0 ml heptane, 6.0 ml water

2. Add 6.0 ml heptane and 4.0 ml water to tube containing sample (unknown). Gently invert all tubes 10–15 times. The two phases should separate without centrifugation.

3. Transfer 4–5 ml of each upper phase to glass stoppered centrifuge tubes and shake vigorously with an equal volume of 0.05% $H_2 SO_4$. Centrifuge.

4. Transfer 3.0 ml aliquots of upper phases to 15 ml conical centrifuge tubes containing 1 ml indicator and titrate with alkali from ultramicro buret calibrated in 0.1 or 0.2 μl, bubbling N_2 (alkali-washed) through the mixture for mixing. Endpoint is greenish yellow and is best seen by interrupting flow of N_2 and observing color of alcohol phase by a fluorescent lamp placed just above and in front of tube placed against a white background. Blank (T_b) and standard (T_s) should be titrated first, and the unknown (T_x) then titrated to the color endpoint matching them.

Calculation:

$$\text{meq NEFA/liter} = \frac{T_x - T_b}{T_s - T_b} \times 0.5$$

NOTES

1. Standardization of alkali. Since calculations are based on titration of a known standard, it is not necessary to standardize the alkali unless absolute values are desired for both unknown and standard. If this is done, the volumes of the upper phases must also be measured.

2. Serum vs plasma. It has been customary to employ plasma with heparin as anticoagulant. There appears to be no reason, however, why serum cannot be used if the serum is removed from the clot and analyzed as soon as possible.

3. Stability of samples. At 25°–30°C there is an increase in NEFA of about 16% in 6 hours and 38%–50% in 24 hours (134, 281). At 0°–10°C there is an increase of about 2% in 6 hours and 12%–25% in 24 hours. One report (164) states that frozen samples are stable several weeks; another claims 12%–50% increase in 24 hours (134). Refrigerated heptane extracts, however, are stable at least several days (134). Broechoven (37) reported no change in plasma concentration for 17 days if stored at +4°C and at −20°C. He also confirmed the increase in NEFA concentration at 20°C even after 24 hours. The heptane extracts were found to be stable for 7 days at −20°, +4°, and +20°C. Increase in NEFA apparently occurs as a result of release of fatty acids, by conversion of lecithin to lysolecithin catalyzed by lecithinase A (425).

ACCURACY AND PRECISION

The method presented yields results averaging 0.012 meq/liter greater than values obtained by the method of Gordon *et al*, which is accepted at present as the reference method (164, 377). This latter method, however, does not lend itself to routine use because lyophilization is involved. Recoveries of fatty acids in the present method are about 93%, but correction for this is presumably automatic since a standard is taken through the entire procedure (377).

The precision (95% limits) of the technic is about ±5%.

NORMAL VALUES

The normal adult range is about 0.45–0.90 meq/liter (48, 163, 367). Day to day variation in fasting values is 10%–30% (281). Ingestion of food produces a decrease in NEFA (54). One group (50) reported that the percentage of stearic acid in the free fatty acid fraction was markedly decreased during work. Values for children and obese adults are higher (78).

TRIGLYCERIDES

Simple triglycerides are triesters of a fatty acid with glycerol and carry the name of the component acid, with the prefix "tri" and the suffix "in" replacing the terminal "ic"—e.g., a triglyceride containing only oleic acid is called triolein. Mixed triglycerides contain more than one type of fatty acid per molecule. The triglycerides are held in solution in serum complexed with protein, principally as β-lipoproteins and pre-β-lipoproteins (cf Chapter 29, *Lipoproteins*). Mono- and diglycerides are also present in small amounts (183, 430). The appearance of a transient lipemia after a high fat meal is due primarily to triglycerides in the form of chylomicrons in the serum. The entire subject of the physicochemical form of lipoproteins in normal and disease states is covered fully in Chapter 29, *Lipoproteins*.

Triglycerides in plasma are higher than those in serum from the same subject (316) due to enmeshment during clot formation; nevertheless, nearly all studies of triglycerides appear to have been performed on serum.

The main approaches to the determination of triglycerides are (a) "by difference," and (b) determination of triglyceride glycerol. These approaches are reviewed here.

DETERMINATION OF TRIGLYCERIDES "BY DIFFERENCE"

If total lipid, phospholipids, cholesterol, and cholesterol esters are determined gravimetrically or by a method permitting calculation of results in gravimetric terms, then:

triglycerides = total lipid − phospholipid − cholesterol − cholesterol esters

Lipid P can be expressed as phospholipid by multiplying by 25, and cholesterol esters expressed as free cholesterol can be converted to cholesterol esters by multiplying by 1.67 (181):

$$\frac{\bar{x}\ \text{mol wt fatty acid esters of cholesterol}}{\text{mol wt cholesterol}} = \frac{646}{387} = 1.67$$

Hence:

mg triglyceride/100 ml = [mg total lipid/100 ml]

$\qquad$ − [25 × mg lipid P/100 ml]

$\qquad$ − [mg free cholesterol/100 ml]

$\qquad$ − [1.67 × mg cholesterol esters (expressed as free cholesterol)/100 ml]

The disadvantages of this approach are as follows: (a) The factors for phospholipid and cholesterol esters (25 and 1.67, respectively) are averages which do not apply exactly to any one sample. (b) Gravimetric determinations of total lipids include lipids other than triglycerides, phospholipid, cholesterol, and cholesterol esters; thus in some conditions spuriously high values for triglycerides can be obtained. (c) Normally, the triglycerides are only about one-tenth to one-fifth of the total lipids, and with lower levels of triglycerides the precision of this method of their estimation is extremely poor.

In another approach to the determination of triglycerides "by difference" the triglycerides are calculated from determinations of total fatty acids, cholesterol esters, and phospholipids. After the total lipids are extracted and saponified to release the esterified fatty acids, the total fatty acids can be determined by titration and corrections made for fatty acids obtained from cholesterol esters, phospholipids, and the nonesterified fatty acids (321). In another technic the total lipids are extracted and the total esterified fatty acids (from triglycerides, phospholipid, and cholesterol esters) are converted to their hydroxamates which, with ferric salts in a weak acid solution, produce a red to violet color (71, 321). Results are corrected for cholesterol esters and phospholipid fatty acids that have been determined separately. A variation on the hydroxamic acid technic involves removal of phospholipids from the total lipid extract by adsorption onto Doucil (a zeolite) (321) before determination of the fatty acids. The cholesterol esters containing higher fatty acids (18 carbons or more) have been reported as not yielding hydroxamate derivatives. High levels of cholesterol and cholesterol esters cause turbidity and interfere with color measurements in the hydroxamic acid technic (300). The color developed is unstable, limiting the number of standards and specimens which can be assayed (321).

Total esterified fatty acids are usually expressed as milliequivalents of triglyceride fatty acid per liter. The corrections for the fatty acid content in phospholipid and/or cholesterol esters are based on average milliequivalents of fatty acids in phospholipid and cholesterol esters of normal subjects. When the results are expressed as mg triglyceride per 100 ml serum (71) a further error is thus introduced by the use of a factor which is not precise.

In still another approach to determination of triglycerides "by difference" the total lipid extract is saponified and the total lipid glycerol determined. The glycerol from phospholipid, calculated from a determination of lipid phosphorus, is subtracted and a value for triglyceride glycerol obtained. Glycerol from mono- and diglycerides is also included as triglyceride glycerol (26). The determination of glycerol is reviewed in greater detail in the next section.

DETERMINATION OF TRIGLYCERIDE GLYCEROL

Serum or plasma triglycerides are most frequently determined by analysis of the glycerol released from triglycerides. The factors involved in the determination of triglyceride glycerol are as follows: (a) extraction, (b) removal of phospholipids and other chromogens, (c) saponification or transesterification to free the glycerol, (d) determination of the glycerol. The extraction of lipid material from serum has been discussed earlier in this chapter (cf section on *Total Lipids*). A number of materials have been used for the removal of phospholipids from serum extracts, including silicic acid, zeolite, Lloyd's reagent, alumina, Florisil, and diatomaceous earth (47, 49, 212, 232, 248, 330, 336, 386). In some cases mixtures of zeolite, Lloyd's reagent, $Ca(OH)_2$, and $CuSO_4$ have been used, in which the $Ca(OH)_2$ in conjunction with $CuSO_4$ is used to remove glucose (212, 213). The most common method for the hydrolysis of triglycerides to release glycerol is saponification with alkali. Transesterification, which is the reaction between a triglyceride (an ester) and an alcohol in the presence of an acid or base, is also used to release glycerol from triglycerides. Glycerol released can be determined by either enzymatic or chemical methods.

ENZYMATIC DETERMINATION OF GLYCEROL

The enzymatic determination of glycerol is based on the following series of reactions:

$$\text{glycerol} + \text{ATP} \xrightarrow{\text{glycerol kinase}} \text{glycerol phosphate} + \text{ADP}$$

$$\text{ADP} + \text{phosphoenol pyruvate} \xrightarrow[\text{kinase}]{\text{pyruvate}} \text{ATP} + \text{pyruvate}$$

$$\text{pyruvate} + \text{NADH} \xrightarrow[\text{dehydrogenase}]{\text{lactate}} \text{lactate} + \text{NAD}$$

The amount of glycerol present is proportional to the decrease in absorbance of NADH (116).

To determine glycerol enzymatically, the total serum lipids are extracted, phospholipids removed by adsorption onto silicic acid, the extract saponified with KOH, neutralized with perchloric acid, and the glycerol then measured (116, 118, 197, 266, 342). Factors which must be controlled to obtain optimal results (197, 342) include: (a) use of glycerol-free KOH; (b) removal of the turbidity caused by fatty acids by precipitation with a magnesium salt; (c) use of enzymes as reagents within their shelf life. The number of samples which can be run in one batch is limited and a high level of technical competence is required.

DETERMINATION OF GLYCEROL BY CHEMICAL PROCEDURES

Blix (26) introduced a method for determining the glycerol of triglycerides and phospholipids in which hydroiodic acid hydrolyzes the lipids and converts the liberated glycerol to isopropyl iodide. The iodine is determined by titration and the amount of glycerol calculated. This approach has been modified by several workers for use in the determination of triglycerides in which, after removal of the phospholipids, the lipid extract is saponified with alcoholic KOH (49, 322, 386) or by transesterification of glycerides with sodium methylate (330) to release free glycerol. The glycerol derived from mono-, di-, and triglycerides is oxidized by periodic acid to formaldehyde according to the procedure of Lambert and Neish (274). The excess periodic acid can be titrated (391) and the amount of glycerol calculated; or, after removal of excess periodic acid, the formaldehyde can be determined spectrophotometrically (a) by reaction with phenylhydrazine and potassium ferricyanide in which a reddish compound, 1,5-diphenylformazone, is formed (43, 152, 247); (b) by reaction with diacetylacetone and ammonia to give a dihydrolutidine derivative which can be measured photometrically or fluorometrically (113, 196, 336, 365); (c) by condensation with chromotropic acid to give a violet compound (47, 49, 64, 66, 94, 192, 196, 218, 386).

Manual Determination of Glycerol

Van Handel and Zilversmit (386) presented a method in which serum lipids were extracted with chloroform-methanol and the phospholipids removed by adsorption onto Doucil (a zeolite). The Doucil was later replaced by a different zeolite in order to obtain more consistent results (218). Triglycerides were hydrolyzed, glycerol oxidized with periodic acid, and the resulting formaldehyde determined using chromotropic acid. Carlson and Wadström (49) used column chromatography on silicic acid to remove the phospholipids, and later simplified the procedure by removal of the phospholipids by batch process with the silicic acid (47). The remainder of the procedure is essentially as described by Van Handel and Zilversmit.

The Van Handel and Zilversmit procedure using zeolite gave higher results than the technic using silicic acid column chromatography (274). The difference was attributed to inclusion of mono- and diglycerides with the triglycerides when zeolite was used. In the procedure of Van Handel and Zilversmit lower triglyceride results were obtained when a serum blank (sample run without saponification) was used rather than a reagent blank. Van Handel and Zilversmit (386) indicated that a reagent blank gave less than 5% differences compared with a serum blank, so that inclusion of a serum blank was an unnecessary refinement. Most workers (212, 322) have modified the Van Handel and Zilversmit procedure by replacing the zeolite with silicic acid.

When acetone-ethanol is the solvent used for extraction of triglycerides, and florisil column chromatography is used for removal of phospholipids, higher results are found with samples than those obtained with the procedure of Van Handel and Zilversmit (62).

Semiautomated Determination of Glycerol

In 1964 Lofland (248) described a procedure for determination of triglyceride glycerol in which the final steps were performed on an AutoAnalyzer. Serum is extracted with isopropanol, phospholipids removed, and the extract saponified manually. The saponified extracts in 0.2 N H_2SO_4 are placed in sample cups, and the oxidation of glycerol to formaldehyde, addition of chromotropic acid, and measurement and recording of absorbance are done automatically. There were no differences in results by this method and the manual procedure of Van Handel and Zilversmit.

In the modification of Kessler and Lederer (212, 213) the serum is extracted with isopropanol, phospholipid removed with zeolite, and the isopropanol extracts placed in the sample cups. The saponification to liberate glycerol, the oxidation, the color development by condensation of the formaldehyde with diacetylacetone and ammonia, and the measurement and recording of fluorescence is done automatically in a continuous flow system (AutoAnalyzer).

Claude *et al* (68) compared Kessler and Lederer's semiautomated procedure with the manual procedure of Van Handel and Zilversmit and found good correlation except that hypertriglyceridemic sera gave higher values by the semiautomated procedure. These workers also suggested a modification in the semiautomated procedure consisting of preparation of a standard curve and utilization of unsaponified blanks to obtain better reproducibility. Others (289) have recommended further minor procedural changes to improve the accuracy of the Kessler and Lederer procedure. It has been suggested that unsaponified serum blanks are unnecessary if reagent volumes and concentrations are reduced and the flow pattern thus improved (81). Automated procedures other than Lofland's that use chromotropic acid for color development have been proposed (232), but this reagent has the disadvantage of requiring high concentrations of H_2SO_4.

In both the manual and semiautomated procedures the glyceride glycerol is calculated from a triglyceride standard, usually triolein, and is reported as mg/100 ml (322); other workers (322) recommend that results be reported as millimoles of glyceride glycerol per liter. Carlson and Wadström (49) have pointed out that triolein, but not tripalmitin or tristearin, gives higher results than the equivalent weight of glycerol upon reaction with chromotropic acid. Since serum triglycerides contain a mixture of saturated and unsaturated fatty acids, the fatty acids should be removed after saponification. This problem has not been reported to occur when diacetylacetone and ammonia are used for the color reaction.

MANUAL DETERMINATION OF SERUM TRIGLYCERIDES

(method of Van Handel and Zilversmit, Refs 322, 386, modified)

PRINCIPLE

Triglycerides are extracted from serum with chloroform-methanol. Phospholipids are removed by adsorption on silicic acid. Triglycerides are then saponified to give glycerol, which is oxidized by periodic acid to formaldehyde. The formaldehyde is determined photometrically by reaction with chromotropic acid. The major absorption peak of the colored product is at 564 nm (Fig. 28-5).

REAGENTS

Chloroform.   Reagent grade.
Methanol.   Anhydrous (absolute, acetone-free).
Chloroform-Methanol Mixture.   Mix 64 ml chloroform and 34 ml methanol. Store in dark glass.
NaCl, Saturated, Aqueous.   Add about 400 g reagent grade NaCl to 1 liter

FIG. 28-5. Absorption curves for triglyceride glycerol with chromotropic acid: (a) reagent blank vs water, (b) 20 µg glycerol vs water, (c) 20 µg glycerol vs reagent blank. Beckman DK-2A recording spectrophotometer.

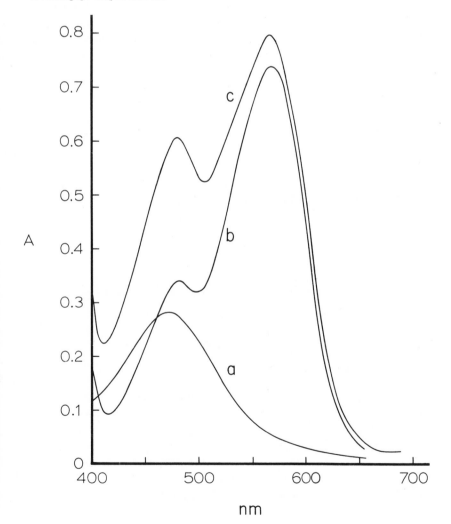

water. Mix thoroughly, allow excess solid to remain in contact with the solution, and use the clear supernatant saturated solution.

Silicic Acid. 100 mesh, suitable for chromatographic analysis. Activate the material by heating at 120°C for 3 hours. Store in a tightly stoppered container.

Alcoholic KOH Solution. Dissolve 0.1 g reagent grade KOH in 95% ethanol and dilute to 100 ml with 95% ethanol (95% ethanol denatured with methanol is satisfactory).

H_2SO_4, 0.2 N. Add 3 ml reagent grade conc H_2SO_4 to 450 ml water and dilute to 500 ml.

Sodium Metaperiodate Solution, 0.5%. Reagent grade.

Sodium Bisulfite Solution, 5%. Reagent grade. Stable 3—4 weeks at room temperature.

Chromotropic Acid Reagent. Dissolve 1.14 g disodium salt of chromotropic acid (disodium salt of 4,5-hydroxy-2,7-naphthalene disulfonic acid) in 1 ml of water. Add slowly 300 ml reagent grade H_2SO_4 to 150 ml water in a separate container. After the acid solution has cooled to room temperature add it cautiously, with cooling under tap water, to the chromotropic acid solution. Store the reagent in a dark glass bottle in a freezer and do not expose unduly to light. Stable for at least a month.

Thiourea Solution, 7.0% Solution is stable indefinitely at room temperature.

Stock Triolein Standard. Dissolve 0.100 g triolein in chloroform contained in a 100 ml volumetric flask and dilute to volume with chloroform. Tightly stoppered, this stock standard is stable for several months in the dark.

Working Triolein Standard. Prepare daily before use by diluting 1 vol stock standard with 9 vol chloroform. 1 ml = 0.10 mg.

PROCEDURE

1. Into a 15 ml glass stoppered or screw-capped centrifuge tube containing 9.8 ml chloroform-methanol mixture, add 0.2 ml serum or plasma. Stopper the tube, shake vigorously, and let stand at least 30 min with intermittent shaking, then centrifuge for several minutes.

2. Transfer 4.0 ml of the lipid extract to a 15 ml glass stoppered or screw-capped centrifuge tube containing 8.0 ml saturated NaCl solution. Stopper the tube, shake vigorously, let stand for 1 hour or longer, and centrifuge for several minutes. The methanol diffuses quantitatively into the aqueous phase. This yields a two-phase system consisting of a solution of serum lipids in essentially pure chloroform and a supernatant layer of aqueous methanol.

3. Using a capillary pipet and a water aspirator, carefully draw off and discard the supernatant aqueous-methanol phase leaving the washed chloroform layer containing the lipids.

4. Wet a folded 7 cm Whatman 541 filter paper disc with chloroform, shake

off the excess fluid, place the damp disc in a short-stem filtering funnel of 5 cm diameter, and immediately filter the chloroform extract of the serum lipids into a 10 ml glass stoppered or screw-capped conical centrifuge tube. The extract filters rapidly.

5. If only triglycerides are to be determined, add 0.2 g of activated silicic acid to the filtered lipid extract in the centrifuge tube. Shake the mixture gently to obtain a suspension, allow the stoppered tube to stand for at least 30 min with occasional shaking, and then centrifuge for about 5 min. A spatula of appropriate size may be used to measure the silicic acid.

6. Carefully transfer 0.50 ml of the supernatant phospholipid-free extract into a 18 x 150 mm test tube and evaporate to dryness in a water bath at 70°C. Use a stream of N_2 or CO_2 to accelerate removal of the last traces of solvent.

7. At this point, likewise evaporate 0.50 ml triolein working standard solution contained in another test tube.

8. To all tubes and to an empty tube (reagent blank) add 0.5 g alcoholic KOH solution, mix, and saponify the mixtures in a 60°–70°C water bath for 20 min. Stopper each tube to minimize evaporation, and immerse only the bottoms of the tubes in the water in order to avoid possible loss of glycerol by volatilization.

9. Add 0.5 ml 0.2 N H_2SO_4 to all tubes and place them immersed to half their length in a gently boiling water bath until the odor of ethanol has disappeared (usually takes 9–12 min).

10. Cool the tubes, add 0.1 ml sodium metaperiodate solution, mix, and allow to stand for 10 min.

11. Reduce the excess metaperiodate by adding 0.1 ml sodium bisulfite solution, mix, and allow to stand for 10 min.

12. Add 5.0 ml chromotropic acid reagent to each tube, mix thoroughly, preferably with a mechanical mixer, and stopper. Place the tubes in a boiling water bath or in a thermostatically controlled aluminum block heater at 100°C for 30 min away from direct strong daylight or artificial light and then cool to room temperature in tap water.

13. Add 0.5 ml thiourea solution to each tube, mix thoroughly, preferably with a mechanical mixer, and measure A_x and A_s at 570 nm versus the reagent blank. The color is stable for at least 2 hours under normal laboratory conditions.

Calculation:

$$\text{mg triglycerides/100 ml} = \frac{A_x}{A_s} \times 0.05 \times \frac{100}{0.0156}$$

$$= \frac{A_x}{A_s} \times 320$$

$$\text{mmol triglycerides/liter} = \frac{A_x}{A_s} \times \frac{0.05}{885} \times \frac{1000}{0.0156}$$

$$= \frac{A_x}{A_s} \times 3.62$$

The figure 885 is the molecular weight of triolein.

NOTES

1. Beer's law. Beer's law is obeyed in a Beckman Du spectrophotometer and with a Klett 54 but not with a Klett 56 filter.

2. Color stability. The color is stable for at least 2 hours.

3. Recovery. In our experience 100% recoveries were obtained when tripalmitin or triolein were compared with glycerol standards subjected to oxidation and the color reaction. It is recommended that recovery studies be done when the method is first set up.

4. Stability of serum samples. Serum triglycerides are reported to be stable for 3 days at 37° (181) or 5 days at 25° (322). Our laboratory has observed a 0%—10% decrease in 6 days at 25°.

ACCURACY AND PRECISION

The accuracy of this procedure, insofar as it is known, is quite good. Separation from phospholipids which would give positive interference is complete and, as previously stated, recovery of triglycerides is complete.

The precision (95% limits) of the method is about ±5% (322).

NORMAL VALUES

The normal range for triglycerides in serum from fasting individuals determined in our laboratory is 30—135 mg/100 ml, expressed as triolein. This range agrees very well with the data of Van Handel and Zilversmit (386) and that of Carlson and Wadström (49) using essentially the same procedure.

Serum triglyceride levels in males are reported to be higher than in females of the same age (130). Triglyceride levels are not affected by the menstrual cycle but are increased during pregnancy (243). With use of oral contraceptives the triglyceride levels are increased by about 45% (179, 416). Fredrickson (138a) observed the upper limit of the normal range increasing with age.

Levels of triglycerides increase following a meal containing fat (181). The daily intraindividual variation is about ±44% from the daily mean (181).

SEMIAUTOMATED DETERMINATION OF TRIGLYCERIDES IN SERUM

(method of Kessler and Lederer, Ref 213, modified)

PRINCIPLE

Isopropanol extracts of serum are treated with a zeolite-Lloyd reagent for removal of phospholipids, glucose, and other chromogenic material. The isopropanol, phospholipid-free extract is mixed with KOH to saponify triglycerides. The liberated glycerol is oxidized with periodic acid, and the formaldehyde formed is mixed with diacetylacetone and ammonia to give a product which is measured fluorometrically.

REAGENTS

Isopropanol. Reagent grade, aldehyde-free.
Zeolite. W.A. Taylor Co. Grind in a Waring Blendor and dry overnight at 100°C.
Lloyd's Reagent. Available from Harleco.
$Ca(OH)_2$. Reagent grade, 50–100 mesh.

FIG. 28-6. Excitation-emission spectra for triglyceride glycerol with diacetyl acetone, on an Aminco Bowman spectrophotofluorometer: (a) excitation spectra; (b) and (c) emission spectra of glycerol from triolein (1 mg/ml) and of blank, unsaponified triolein (1 mg/ml), respectively. For the excitation scan emission wavelength was set at 480 nm, and for the emission scan excitation wavelength was set at 405 nm.

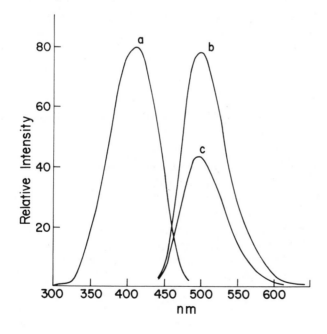

CuSO₄·5H₂O. Reagent grade, 50–100 mesh.

Zeolite Mixture. Mix thoroughly 200 g Zeolite, 20 g Lloyd's reagent, 20 g Ca(OH)₂, and 10 g CuSO₄·5H₂O. Store in a tightly sealed container.

KOH, 0.2%

Base Reagent. Mix 250 ml 0.2% KOH with 750 ml isopropanol.

Sodium Metaperiodate, 0.025 M, in 2 M Acetic Acid. Dissolve 5.35 g reagent grade sodium metaperiodate in 115 ml glacial acetic acid and dilute to 1 liter with water.

Ammonium Acetate, 2 M, pH 6.0. Dissolve 154 g reagent grade ammonium acetate in approx 900 ml water. Adjust to pH 6.0±0.1 with glacial acetic acid (approx 7 ml) and then dilute to 1 liter with water.

Diacetylacetone Reagent. Add 7.5 ml diacetylacetone (2,4-pentane-dione) to 25 ml isopropanol and mix. Then add one liter 2 M ammonium acetate (pH 6.0) and mix well. *Prepare fresh daily.*

NaC1, 0.85%.

Stock Triolein Standard, 10 mg/ml. Dissolve 1.00 g triolein in 100 ml isopropanol.

Working Triolein Standard. Dilute 5, 10, 20 and 30 ml aliquots of stock standard with isopropanol to 100 ml. The resulting working standards contain 50, 100, 200, and 300 mg triolein per 100 ml, respectively.

PROCEDURE

Preparation of isopropanol extract

1. Set up the following in 16 x 150 mm screw capped tubes:

    Standards. To separate tubes containing 9.0 ml isopropanol, add with mixing 0.5 ml of each working standard and 0.5 ml of 0.85% NaC1.

    Unknown. To tubes containing 9.5 ml isopropanol, add with mixing 0.5 ml serum and continue mixing for 15 sec. If only a smaller amount of serum is available, the volume of the reagents can be reduced proportion-ately. If serum is lipemic take a 0.1 or 0.2 ml aliquot and make up the volume difference with 0.85% NaCl.

2. Add 2 g Zeolite mixture to all the extraction tubes, cap with a Teflon-lined screw cap, and mix vigorously on a mechanical mixer for 15 min. A scoop calibrated to contain 2 g may be used here.

3. Centrifuge all tubes and decant supernates into clean tubes and cap. If extracts are to stand overnight, store capped tubes in a refrigerator.

Automated technic

Set up manifold, other components, and operating conditions as shown in Figure 28-7. By following procedural guidelines as described in Chapter 10 steady state interaction curves as shown in Figure 28-8 should be obtained and checked periodically. Each sample tray should contain a set of extracted working standards followed by an isopropanol wash. Extracts of a control pool and the specimens are then placed in the tray. As indicated in Figure 28-6 serum blanks are run routinely and subtracted from the assay value. Results are calculated from the calibration curve in the standard manner for linear fluorescence methods.

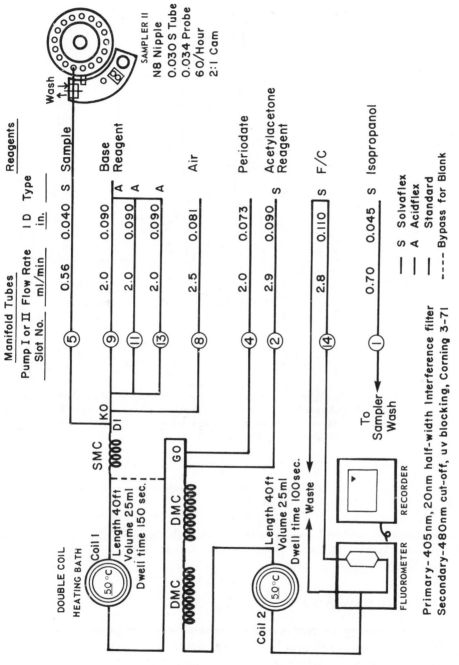

FIG 28-7. Flow diagram for triglycerides in serum. Calibration range: 0–800 mg/100 ml. Conventions and symbols used in this diagram are explained in Chapter 10.

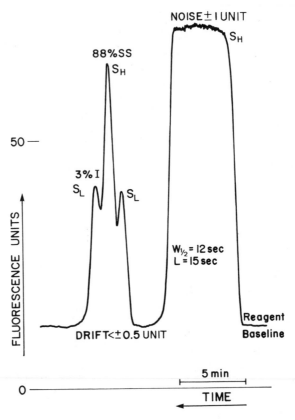

FIG. 28-8. Triglycerides strip chart record. S_H is a 400 mg/100 ml standard, S_L a 200 mg/100 ml standard. Peaks reach 88% of steady state; interaction between samples is 3%. $W_{1/2}$ and L are explained in Chapter 10. Computer tolerance for noise is set at ±0.7% T, for baseline drift at ±1.5% T, and standard drift at ±0.8% T.

NOTES

1. See Notes 1, 3, and 4 of manual method.

2. Serum blanks. The use of a blank is recommended to correct for the presence of glycerol from sources other than triglycerides and phospholipids. In our experience fluorescence from the drug Aldomet gives falsely elevated triglyceride levels.

ACCURACY AND PRECISION

The recovery of triglycerides added to extracts or serum is 95%–104%. Separation from phospholipids is quantitative. The method compares well with the manual procedure of Carlson and Wadström (49, 311) and that of Van Handel and Zilversmit.

The precision (95% limits) of the method in our laboratory is ±10%.

NORMAL RANGE

Same as for manual method.

PHOSPHOLIPIDS

Phospholipids (phosphatides) are conjugated lipids containing a phosphate group. The chief members of the group also contain nitrogenous components. Structural formulas of the principle phospholipids are shown in Figure 28-9.

FIG. 28-9. Structural formulas of phospholipids.

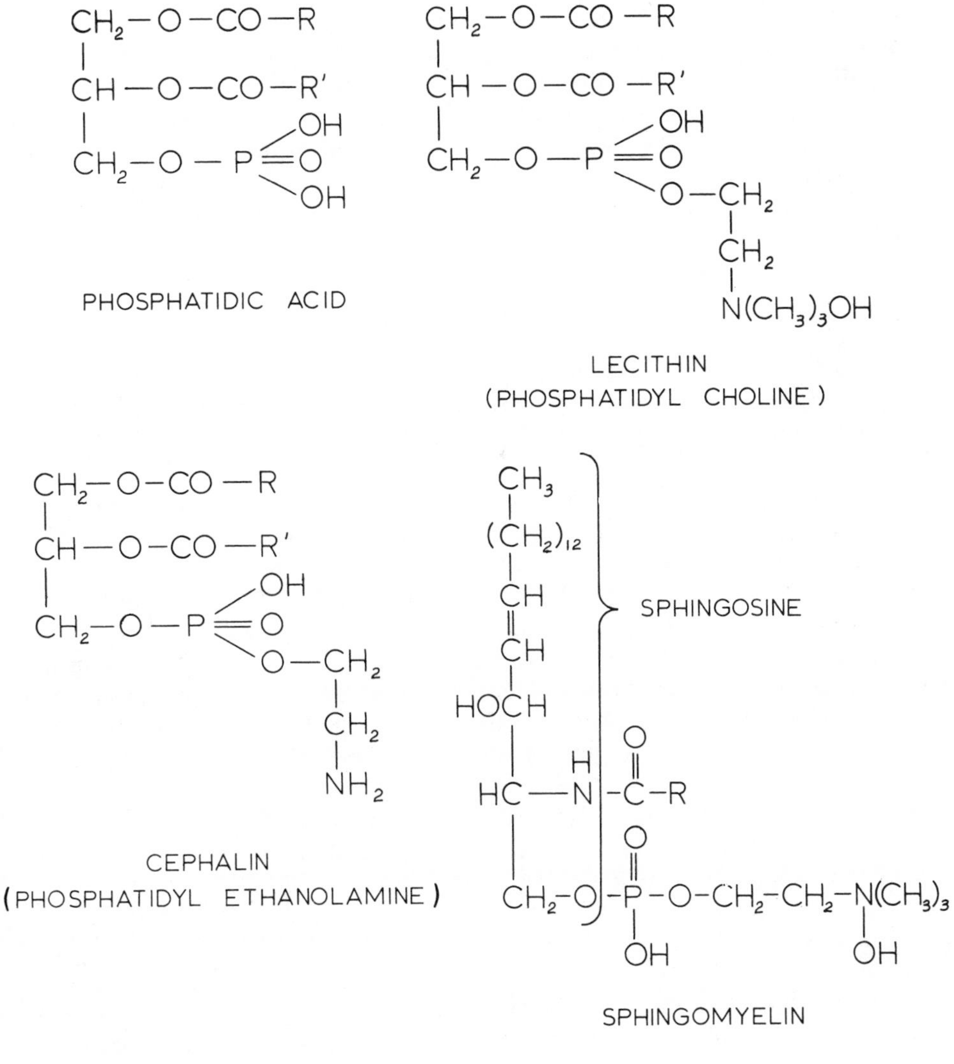

PHOSPHATIDIC ACID

LECITHIN
(PHOSPHATIDYL CHOLINE)

CEPHALIN
(PHOSPHATIDYL ETHANOLAMINE)

SPHINGOSINE

SPHINGOMYELIN

The fractionation of phospholipids has been subjected to considerable study due to their involvement in transport of material across cell membranes. The older physicochemical fractionation methods have been largely replaced by chromatographic procedures (260, 329). Column chromatography, usually with silicic acid, is used mainly to separate phospholipids as a class from other lipids (274). The separation of phospholipids into lecithins, sphingomyelins, etc. has been done by chromatography on paper (329) or paper impregnated with silicic acid (260, 390), or by thin layer chromatography (158, 327, 406, 409). Thin layer chromatography and phosphate analysis have been used for the quantitative determination of individual phospholipids in human serum (158, 327, 406). The phospholipids usually determined are lecithins, sphingomyelins, phosphatidyl ethanolamines, and lysolecithins. The approximate distribution of phospholipids in normal serum or plasma expressed as percent of toal lipid P is as follows: lecithin 68%–73%, lysolecithin 5%–8%, and sphingomyelin 17%–18%, the remainder being made up mainly of cephalin or ethanolamine (158, 161, 406). After taking into account the variations of the results and the small number of normals examined, there appears to be essentially no difference in distribution of phospholipids between serum and plasma.

The approaches to the determination of total phospholipids in serum can be grouped into two main categories: determination of phosphorus in an extract of serum, and determination of phosphorus in a protein precipitate of serum.

Determination of Phosphorus in Organic Solvent Extract

In this technic the lipids of serum are extracted, the extract digested to break down organic material and yield orthophosphate, and the phosphate content determined by a photometric method.

Extraction. The most commonly used materials for extraction are: ethanol-ether, petroleum ether (14, 406), and chloroform-methanol (274) for manual procedures and isopropanol (212) for semiautomated procedures. Studies with radioactive inorganic phosphate indicate that ethanol-ether extracts of serum contain less than 0.1% of the added radioactivity (123), indicating that virtually no inorganic phosphate is extracted. The phospholipids in a dried ether-ethanol extract of serum have been found to be 95%–100% soluble in petroleum ether (31, 256), but care must be taken (256, 364) to obtain complete solution of the ethanol-ether extract in petroleum ether.

The phospholipids in ethanol-ether extracts are heat-stable but easily oxidized, and should be evaporated to dryness with N_2 but not with air (256). This is important when phospholipid P is to be determined in the lipid extract after a gravimetric determination of total lipids. When chloroform-methanol was used as the extracting solvent, 80%–98% of the P in these serum extracts was found to be from phospholipids (261). This variation was not observed by Moline and Barron (274). These workers found that chloroform-methanol, the solvent that they recommended, extracted less P than ethanol-ether. Washing decreased the amount of P in the ethanol-ether extract but not in the chloroform-methanol extract. They demonstrated that the P in the chloroform-methanol was entirely from phospholipids by performing column chromatography with silicic acid on the extract, and obtaining similar results with and without column chromatography and washing. They did not, however, duplicate these steps with the ethanol-ether extract and did not determine if the washing removed phospholipid which had not been extracted by chloroform-methanol.

Isopropanol has been shown to extract phospholipids quantitatively, but not inorganic P, from serum (212). Extraction with isopropanol has been used because it is less volatile than other solvents and thus is better for use in semiautomated procedures.

Digestion. Serum extracts containing phospholipids are digested to destroy organic matter and to yield orthophosphate. The digestion can be carried out with H_2SO_4-H_2O_2 (18, 181, 212, 255), $HClO_4$ (41, 162, 231), or with HNO_3 (14), and is usually done manually although automated digestion procedures are available (402).

Color Development. The phosphate in the digestion mixture is usually reacted with ammonium molybdate, and the ammonium phosphomolybdate formed is reduced under conditions which do not reduce the excess molybdate. The molybdenum blue that is formed is measured photometrically. A second method is called the molybdivanadophosphoric acid method. These methods are discussed in detail in Chapter 19, *Determination of Inorganic Phosphorus.*

The photometric analysis of phosphate has been automated by several workers (212, 223, 402) using AutoAnalyzer equipment. Kraml (231), using either ethanol-ether extraction or the trichloroacetic acid precipitation technic of Zilversmit and Davis (425)—cf the next section, *Determination of Phosphorus in Protein Precipitates*—and digestion with $HClO_4$, automated the photometric assay using molybdic acid and $SnCl_2$-hydrazine as a reducing agent. Color development was rapid at room temperature, and the sensitivity was equal to that obtained with procedures requiring heating. Kessler (212), in a similar procedure using isopropanol for extraction of the total lipid and H_2SO_4-H_2O_2 for digestion, automated the photometric analysis with molybdic acid and $SnCl_2$ as the reducing agent.

Determination of Phosphorus in Protein Precipitates

Zilversmit and Davis (425) in 1950 introduced a technic in which phosphate is determined in a trichloroacetic acid precipitate of serum after digestion with $HClO_4$. Results obtained with trichloroacetic acid precipitates agree with determinations of P in ethanol-ether extracts (220, 231). The trichloroacetic acid precipitation technic has been adapted to a semiautomated procedure (402), in which the precipitates are prepared manually, dissolved in aqueous NaOH, and transferred to plastic sample cups. The digestion step and photometric analysis are automated. The automation of the digestion step has been criticized (231) because cross contamination occurs unless samples are separated by washes. With the usual reducing reagents the method is less sensitive and slower than those methods which use a manual digestion step and automated photometric analysis. Kraml (231) used manual precipitation and digestion and automated photometric analysis, and found that the trichloroacetic acid precipitation and ethanol-ether extractions gave comparable results.

DETERMINATION OF SERUM PHOSPHOLIPID

(method of Baumann, Ref 18, using method of Dryer et al, Ref 110, for phosphate determination)

PRINCIPLE

Lipids are extracted and proteins precipitated by ethanol-ether. Total P in the extract is determined photometrically by the *p*-semidine method, following oxidation of phospholipid P to inorganic phosphate by digestion with H_2SO_4 and H_2O_2. The P can be expressed as phospholipid by multiplying by a factor of 25 (174).

FIG. 28-10. Absorption curves in phosphorus determination: (a) reagent blank vs water, (b) 5 μg phosphorus standard vs reagent blank. Beckman model DU spectrophotometer.

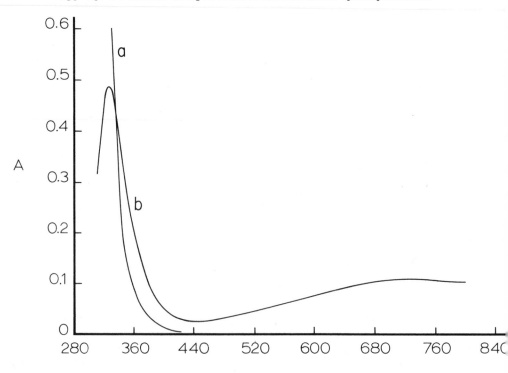

A

nm

REAGENTS

Ethanol-Ether Solvent Mixture. 3 vol ethanol + 1 vol ether. Ethanol denatured with methanol is satisfactory.

H_2SO_4, *30% (v/v)*

H_2O_2 *(Superoxol), 30%.* Store in refrigerator.

Ammonium Molybdate, 0.0202 M. Dissolve 25 g $(NH_4)_6Mo_7O_{24} \cdot 4H_2O$ in approximately 700 ml water. Cautiously add, with mixing, 84 ml conc H_2SO_4. Cool, dilute to 1 liter with water, and mix. Store in a polyethylene bottle.

p-Semidine Reagent. p-Semidine is obtainable from Eastman Organic Chemicals as N-phenyl-*p*-phenylenediamine hydrochloride, No. 2043. Place 50 mg in a flask, wet the solid salt with a few drops of 95% ethanol, and add 100 ml 1% $NaHSO_3$ with shaking. Filter off insoluble residue. Store in refrigerator. Stable about 1 month. A slight discoloration may appear which does not impair the effectiveness of the reagent but does increase the absorbance of the reagent blank. After 1 month in the refrigerator, the absorbance of standards slowly decreases.

Stock Phosphorus Standard, 1 mg P/ml. Place 439 mg KH_2PO_4 in a 100 ml volumetric flask and add water to volume. Add a few drops of choloroform as preservative.

Working Phosphorus Standard, 40 μg P/ml. Dilute 1 ml stock standard to 25 ml with water.

PROCEDURE

1. Transfer 1.0 ml serum or plasma to a dry 25 ml volumetric flask. With shaking, add about 15 ml ethanol-ether; heat on steam bath until solvent boils. Cool to room temperature and dilute to mark with solvent. Mix and filter.

2. Set up in tubes about 20 mm in diameter and calibrated at 12.5 ml:

   Blank: 1 ml 30% H_2SO_4

   Standard: 1 ml 30% H_2SO_4 + 1.0 ml standard $(40 \mu g\ P)$

   Unknown: 10 ml filtrate

3. Add 2 glass beads to each tube. Place unknown in metal test tube rack on hot plate and heat to dryness. Add 1.0 ml 30% H_2SO_4 to unknown.

4. Heat over an open flame or in a heating block at $250°-275°C$ until water is driven off and dense white fumes of H_2SO_4 begin to fill the tube.

5. Remove tubes from heat for about 30 sec and add 3 drops 30% H_2O_2 to each tube while holding the tubes vertically, letting the drops fall directly into the bottom of the tubes. Resume digestion until H_2SO_4 fumes fill the tubes and continue digestion for 2 min. Discontinue heating (if solution is not colorless, repeat addition of H_2O_2) and allow the tubes to cool. Use same number of drops of 30% H_2O_2 for blank and standard as for unknown.

6. Dilute cooled digest to 12.5 ml with water and mix. Transfer 2.0 ml from each tube to labeled tubes; add 0.4 ml molybdate reagent and 4.0 ml *p*-semidine reagent. Mix.

7. After 10 min read all tubes against water at 770 nm or with a filter having a nominal wavelength in this region (a 660 nm filter is satisfactory). The color is stable at least 1.5 hours. If A_x is too high (> 0.8), set up color again with a smaller aliquot of diluted digest, make up difference with water, and apply proper correction in calculation.

Calculation:

$$\text{mg phospholipid P/100 ml} = \frac{A_x - A_b}{A_s - A_b} \times 0.04 \times \frac{100}{0.4}$$

$$= \frac{A_x - A_b}{A_s - A_b} \times 10$$

$$\text{mg phospholipid, as lecithin/100 ml} = \frac{A_x - A_b}{A_s - A_b} \times 250$$

NOTES

1. Beer's law. Beer's law is obeyed with a spectrophotometer or a Corning glass filter with nominal wavelength of 660 nm.

2. Stability of samples. No change was observed in our laboratory in eight samples held at 25°C for 6 days. Apparently, however, lecithinase B is present in human serum and eventually causes decomposition of the phospholipids (368). In one study (123) during incubation of serum at 37°C only the lecithins were attacked and degradation reached as high as 40%.

ACCURACY AND PRECISION

The method appears to possess a fairly high degree of accuracy. This conclusion is based on radioisotope studies (387) and on the finding that 95%–100% of the P present in the ethanol-ether extracts is soluble in petroleum ether (256).

The precision (95% limits) is about ±9%.

NORMAL VALUES

The fasting adult normal range for phospholipids in serum or plasma is 5–12 mg/100 ml expressed as P, or 125–300 mg/100 ml, expressed as lecithin (181). Both the upper and lower limits of the reported ranges have varied somewhat (104, 114, 160, 222, 246, 274, 340, 343, 428). The normal range in cord serum is about 80–170 mg/100 ml, as lecithin (181), with a 20%–80% increase during the first days of life. During adult life there appears to be general agreement that there is an increase with age (46, 246, 340). In one large survey of about 1000 individuals (340) there was no change in males up to the age of 20, but there was then an increase until the age of 32, when the levels became constant again; in females the levels were constant up to the age of 32, when a steady increase began.

Day to day variation has been reported to be about ±15% from the individual's average (34), there being no pattern to the change (190). There is no pattern identifiable with the menstrual cycle (173). There is no identifiable sex difference (343) except during pregnancy, when there is an increase in the levels (10, 185). With oral contraceptive therapy the phospholipid levels are increased (10).

SEMIAUTOMATED DETERMINATION OF SERUM PHOSPHOLIPIDS

(method of Kessler, Ref 212)

PRINCIPLE

Lipids are extracted from serum with isopropanol, and an aliquot of the extract is subjected to hot acid digestion with H_2SO_4 and H_2O_2 to convert organic phosphate to inorganic phosphate. The photometric determination of phosphate using molybdic acid-$SnCl_2$ is performed on an AutoAnalyzer.

REAGENTS

Isopropanol. Reagent grade.

30% H_2O_2 (Superoxol). Store in refrigerator.

H_2SO_4, *2.5 N.* Carefully add 137.5 ml conc H_2SO_4 to approx 1500 ml water. Mix well and dilute to 2 liters.

Molybdic Acid Reagent. Dissolve 15 g ammonium molybdate

FIG. 28-11. Absorption curves in phospholipid phosphorus determination: (a) reagent blank vs water, (b) 40 µg phosphorus standard vs water, (c) 40 µg phosphorus standard vs reagent blank.

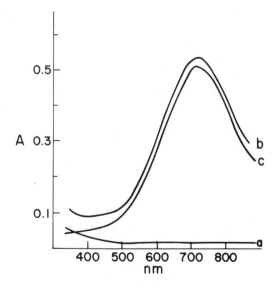

$(NH_4)_6 Mo_7 O_{24} \cdot 4H_2 O$ in approx 600 ml water and add, with mixing, 55 ml conc $H_2 SO_4$. Dilute to 1 liter with water.

Stock SnCl₂ Reagent. 10 g $SnCl_2$ per 25 ml conc HCl. Store in refrigerator.

Working SnCl₂ Reagent. Dilute 1 ml stock reagent to 1 liter with water.

Blank Acid Reagent. To 100 ml water add 7.0 ml conc $H_2 SO_4$.

Stock Phosphorus Standard, 50 µg/ml. Dissolve 0.219 g $KH_2 PO_4$ (dried overnight at 110°C) in 1 liter water.

Working Phosphorus Standards. Dilute 5, 7.5, 10, 12.5, and 15 ml of stock standard to 100 ml with water. Add 7 ml conc $H_2 SO_4$. The resulting working standards are equivalent to 5, 7.5, 10, 12.5, and 15 mg P/100 ml when carried through the automated part of the procedure.

PROCEDURE

Preparation of extracts

1. To a 16x150 mm screw-capped test tube add 9.5 ml isopropanol.
2. While mixing the contents of the tube on a mechanical mixer, rapidly add 0.5 ml serum or blank (0.85% NaCl). Keep mixing for 15 sec to produce a finely divided precipitate. Cap the tube with a Teflon-lined screw cap and allow to stand for 5 min.
3. Centrifuge the capped tubes at high speed for 5 min. Decant supernate into clean tubes and cap. The supernate should not be in prolonged contact with the precipitated protein. Separated supernates can be stored in refrigerator at this point.

Digestion

1. Transfer 1.0 ml aliquots of the isopropanolic lipid extract to 18x150 mm Pyrex ignition tubes (Corning No. 9860).
2. Add 1.0 ml 2.5 N $H_2 SO_4$ to all tubes and mix. All steps from this point on should be carried out in a fume hood.
3. Place tubes in a heating block preheated to 125°C to evaporate solvent and to initiate digestion (isopropanol volatilizes at this temperature without bumping). Continue heating until temperature reaches 250° – 275°C. The digestion mixture should be charred at this point.
4. Remove tubes from heating block and allow to cool slightly. Carefully add 3 drops 30% $H_2 O_2$ at the side of the tube near the bottom. Replace tubes in heating block and continue digestion until mixture is colorless, usually about 5 min. Additional $H_2 O_2$ may be added if necessary.
5. Remove tubes from heating block and allow to cool. The volume of digest is approx 0.07 ml. If digests are to stand overnight, store capped tubes in refrigerator.
6. Add 1.0 ml water to each tube and mix thoroughly on a mechanical mixer.

Automated technic

Set up manifold, other components, and operating conditions as shown in Figure 28-12. By following procedural guidelines as described in Chapter 10, steady state interaction curves as shown in Figure 28-13 should be obtained

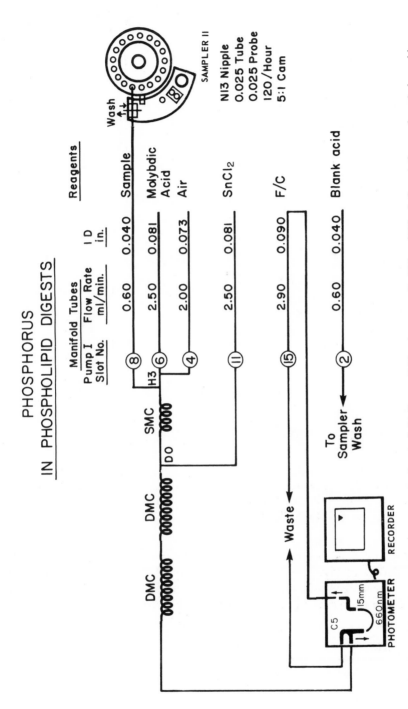

FIG. 28-12. Flow diagram for phosphorus in phospholipid-digests. Calibration range: 0–900 mg/100 ml. Conventions and symbols used in this diagram are explained in Chapter 10.

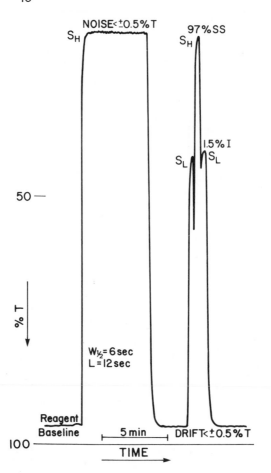

FIG. 28-13. Phosphorus in phospholipid digests strip chart record. S_H is a 450 mg/100 ml standard, S_L a 225 mg/100 ml standard. Peaks reach 97% of steady state; interaction between samples is 2%. $W_{1/2}$ and L are explained in Chapter 10. Computer tolerance for noise is set at ±0.7% T, for baseline drift at ±0.7% T, and for standard drift at ±0.7% T.

and checked periodically. Each sample tray should contain a set of working standards followed by a blank. An unmeasured aliquot of the digested extracts of specimens is transferred to sample cups. Results are calculated from the calibration curve in the standard manner for methods obeying Beer's law.

NOTES

1. See Notes 1 and 2 of manual method.

ACCURACY AND PRECISION

The accuracy is the same as described for the manual method. The precision (95% limits) is about ±8%.

Same as for manual method.

FECAL LIPIDS

Fecal lipids are derived from four sources: unabsorbed dietary residue, cellular debris from the intestinal wall, bacteria or other gastrointestinal organisms, and excretions of the body into the intestinal lumen (167). The sum of the last three is called the "metabolic lipid." On a lipid-free diet, fecal output falls to a basal level of 1–4 g/day, which must represent the sum of the nondietary components (141). In normal individuals a dietary intake of lipid up to 150 g/day makes relatively little change in total lipid excretion (141). In cases of defective absorption, the bulk of fecal lipid is unabsorbed ingested lipid.

Table 28-2 gives the approximate distribution of lipids in normal feces (32, 181). Various procedures for the fractionation of fecal lipids have

TABLE 28-2. Approximate Distribution of Lipids in Normal Feces

Component	Percent of total lipid	Comments
Unsaponifiable matter (other than free fatty acids)	30	Coprostanol, cholestanol, epicoprostanol, β-sitosterol,
Sterols	10	cholestan-3-one, coprostan-3-one, cholesterol, phytosterols from vegetables, other sterols; 70–90% in free form
Other than sterols	20	Higher alcohols, paraffins, carotenoids from vegetables
Acids	60	
Volatile	10	Acetic, propionic, butyric, valeric, hexanoic, higher acids
Nonvolatile	40	
Solid	25	Myristic, palmitic, stearic, higher acids
Liquid	15	Oleic
Insoluble in petroleum ether	10	Includes colored N-containing compounds
Triglycerides	10–15	Also mono- and diglycerides
Cholesterol esters	<10	
Phospholipids	<10	

been presented (32, 120, 273, 372), but the primary diagnosis of the malabsorption syndrome is still considered to rest most satisfactorily upon determination of the daily total fecal lipid output (138).

The determination of the split fat—i.e., the sum of the free fatty acids and fatty acid esters (soaps) or of triglycerides (unsplit fat) in fecal lipid—is considered to be of no value as a method for assessment of the degree of hydrolysis of fat in the intestinal tract (138, 167, 181, 384). Rather than use split versus unsplit fat as an index of pancreatic function, it has been recommended that direct examination of samples of small intestinal content for pancreatic enzymes be used for this purpose (138).

DETERMINATION OF TOTAL FECAL LIPIDS

At the present time there are two primary groups of procedures for quantitative determination of fecal lipids—gravimetric and titrimetric.

Gravimetric Technics

These procedures for total fecal lipids involve acidification of the fecal material, extraction of this material with an organic solvent, removal of the organic solvent, and weighing of the dried extract. Gravimetric procedures as usually performed include the nonsaponifiable lipids.

Acidification of the fecal material with mineral acids such as HCl or H_2SO_4 converts Ca or Mg salts of the fatty acids to the free fatty acid. The Ca and Mg salts are difficult to extract and also cause problems by forming emulsions (335). The mineral acids can cause hydrolysis of fatty acid esters, which can be largely avoided by carrying out the extraction step within a reasonable period of time. Ethanol, which may be used in the collection and extraction of fecal lipids, in the presence of mineral acids can react with fatty acids to form ethyl esters (177). The choice of organic solvent used in lipid extraction is influenced by two factors: the ability to extract lipids quantitatively, and the specificity of extraction for lipid. Since no solvent completely meets these requirements, the solvent chosen must of necessity be a compromise. Chloroform and ether-ethanol are excellent solvents for extraction of fecal lipid, but they also extract large amounts of nonlipid material (146, 181). The lipid extracted with these solvents is usually dried and reextracted into petroleum ether. Nonlipid materials may still be dissolved in petroleum ether because of the presence of lipid, contributing to erroneously high results.

Toluene has been used extensively as a solvent for extraction of hydroxystearic acid from feces. However, toluene extracts nonlipid material which should not be included in the fecal lipids (146, 181).

Feces collected under ethanol, then acidified, emulsified, and extracted with petroleum ether yield larger amounts of fecal lipid. The collection procedure has the advantages of eliminating some of the loss of volatile materials and inhibiting the bacterial hydrolysis of triglycerides (146, 167).

Our laboratory compared results obtained gravimetrically with the following extraction procedures: (a) acidify, extract with ethanol-petroleum ether; (b) acidify, boil with ethanol, extract with petroleum ether; (c) acidify, extract with ethanol-ether, dry, extract with petroleum ether; (d) acidify, extract with ispropyl ether, dry, extract with petroleum ether. Methods (a), (c), and (d) yielded comparable results and method (b) gave results about 12% lower (181).

Titrimetric Technics

The most popular titrimetric procedure is that of Van de Kamer *et al* (384), in which the lipids are saponified, the mixture acidified, then extracted with petroleum ether. The petroleum ether is transferred and evaporated to dryness. The dried extract is redissolved in ethanol and the fatty acids titrated with aqueous NaOH or with ethanolic KOH. The weight of the fatty acids is calculated, using a mean molecular weight of 265 for fatty acid. Anderson (6) criticized the titrimetric procedure because high losses of volatile fatty acids occur and because of a 15% variation in the titration factor based upon molecular weight.

The titrimetric procedure excludes the nonsaponifiable lipid and measures only free fatty acids; after saponification and acidification this includes fatty acids hydrolyzed from mono-, di-, and triglycerides, cholesterol esters, and phospholipids.

Comparison of Gravimetric and Titrimetric Methods

Assuming quantitative extraction and absolute specificity for lipids, gravimetric methods determine all lipid components. Titrimetric procedures measure only fatty acids. In normal individuals fatty acids in various chemical forms comprise 60%–70% of the total fecal lipids. In our laboratory, results obtained by the titrimetric method averaged 78% of the gravimetric results. A drawback of the titrimetric procedure is that further study of the fecal lipids from the same sample is impossible because of the saponification step. If information is needed on the distribution of lipid classes other than fatty acids in the total lipid, the gravimetric procedure is the method of choice.

In cases where the free fatty acids, fatty acid salts, and total fecal lipid are to be determined, the gravimetric procedure can be used to determine the total lipid and the free fatty acids can then be determined by titration of the same extract. In patients with steatorrhea the amount of free fatty acid present in fecal lipid was formerly taken as an index of hydrolysis of ingested triglycerides by pancreatic lipase. Assessment of the amount of free fatty acids is considered to be quite unreliable as a guide to pancreatic enzyme activity because of the presence of lipase activity of fecal bacteria (32, 416).

MISCELLANEOUS METHODS

Microscopy

Various dyes are capable of staining neutral lipids and free fatty acids. Slides of fecal material can be stained with Sudan III (109) and examined microscopically for yellow or orange refractile globules of neutral fat. Deep-orange globules of fatty acids can be brought out by adding acetic acid and heating. This dye stains the background and fat a similar color. Nile blue sulfate (235) has the advantage of staining fat red or orange against a blue background.

While this technic is claimed by some (109) to be satisfactory for use as a screening procedure, others (74, 399) feel it is unreliable due to poor sampling of the feces.

Specific Gravity

The decrease in specific gravity of chloroform caused by the fecal lipids extracted into chloroform can be used as a semiquantitative measure of fecal lipids (285). If the screening indicates that the fecal liquids are above an acceptable maximum, an aliquot of the chloroform extract can then be assayed gravimetrically. Changes in specific gravity, however, can be due to extraction of nonlipid material.

Chromatography

Column and thin layer chromatography have been used to separate and quantitate various lipid fractions (32, 120, 273, 372). The feces must be extracted as soon as possible after collection in order to minimize hydrolysis of triglycerides, phospholipids, and sterol esters.

FAT BALANCE STUDIES

Normally, nearly all of the fat ingested is absorbed. The clinician is interested in fecal lipids to learn whether there is malabsorption of fat. Theoretically in such a case greater accuracy should be obtained if the exact amount ingested is known along with the amount excreted. The percent *coefficient of fat retention* can then be calculated: (fat ingested—fat excreted) x 100/fat ingested. Before proceeding further, a point in terminology must be clarified. It was stated in the introduction to this chapter that the term "fat" would be restricted to mixtures of triglycerides. The "fat content" of foods usually refers to the triglyceride content (includes free fatty acids) since the method of analysis employed involves extraction, saponification, and titration. As seen previously, however, the methods employed for fecal analysis may or may not measure additional lipid components. An exception is being made here, therefore, and the term "fat" is used regardless of the method used for fecal analysis.

Several approaches have been used to determine the coefficient of fat retention, the main approach being fat balance studies in which a diet of known fat content is fed, the feces collected, and the fat excreted in the feces determined. This approach is claimed to be valid, provided more than 50 g of fat is ingested per day; when the intake falls below 50 g, the proportion of the total excreted fat made up of nondietary fat becomes large, making the coefficient meaningless (384).

The absorption of specific compounds such as vitamin A, carotene, and ^{131}I-labeled triolein or oleic acid has been used to detect malabsorption of fat (77, 138, 150, 161, 284). These compounds can be used to distinguish between unabsorbed dietary fat and that derived from other sources (138). The measurement of vitamin A or carotene absorption has been criticized as being an unreliable index of the intestinal handling of lipid compared with measurement of ^{131}I triolein absorption or fat balance studies (161). Studies using ^{131}I give the best correlation with fat balance studies (77, 161). The use of ^{131}I-labeled lipid has been criticized because the stability of the iodine label in the intestinal tract is uncertain (138).

To determine the coefficient of fat retention with a fat balance study, the individual must be placed on a diet of known fat content. Diets for use in lipid balance studies have been described by Nothman (290), Polak and Pontes (307), and by Annegers *et al* (7). The Nothman diet as described in the next section contains about 135 g fat per day. It is interesting that the average urban diet in the United States contains 3250 cal and 140 g fat—40% of total calories derived from fat (216, 259). Thus an "average" diet in the United States is equivalent to the Nothman diet in fat content. The normal coefficient of fat retention on these diets for individuals over the age of 1 year is 95% or higher (384, 408, 410). In normal infants less than 1 year of age the coefficient is normally greater than 83% if formula-fed and greater than 93% if breast-fed (384). In premature infants the coefficient may be much lower (average of 38%–73% depending on formula)(88).

Instructions for Lipid Balance Study with the Nothman Diet (290)

The diet contains 105 g protein, 135 g lipid, and 180 g carbohydrate per day. The day as written out below is only slightly modified from Nothman, and the quantities have been put in units familiar to laymen. As presented, the assumption is made that the diet begins to be excreted within 48 hours of its initiation. If one does not desire to make this assumption, markers may be used as mentioned in the following section. The following are instructions which may be given to a patient:

Nothman Diet

7:30 AM	1	egg fried with 1 teaspoonful butter
	2	oz orange juice
	2	oz cream (in coffee or tea if desired)
	1	level teaspoonful sugar in the orange juice, coffee, or tea
	1	slice toast spread with 1 level teaspoonful butter

10:30 AM	1	slice toast spread with 1 level teaspoonful butter
	1	glass whole milk
12:30 PM	2	oz raw lean beef pattie, broiled
	1	medium sized serving mashed potato
	1	level teaspoonful butter
	1	glass whole milk
3:00 PM	1	slice bread spread with 1 level teaspoonful butter
	1	egg custard
6:00 PM	2	oz raw lean beef pattie, broiled
	1	slice of bread spread with 1½ level teaspoonsful butter
	2	oz orange juice
	1	glass whole milk
8:30–12:00 PM	1	egg, hard boiled, soft boiled, or coddled
	1	cup coffee, containing 1 level teaspoonful sugar and 2 oz cream
	1	glass whole milk
	1	slice toast spread with 1 level teaspoonful butter

Note: If the midmorning and midafternoon snacks are inconvenient, the 10:30 AM snack can be added to the lunch and the 3:00 PM snack can be added to the 6:00 PM meal.

Instructions to patient. The test takes 5 days, keeping to the above diet for the entire 5-day period. Water or black coffee can be drunk as freely as desired.

1st day: Begin special diet. Do *not* save feces.

2nd day: Do *not* save feces.

3rd, 4th, and 5th day: Save all feces passed, collecting them in plastic or cardboard ice cream containers. If possible, deliver each day's collection to laboratory on the day following collection.

Analysis of Wet vs Dry Feces

It has been reported (167) that analysis of dried feces gives less consistent results and lower values than analysis of wet feces because of a less efficient extraction and loss of volatile fatty acids during the drying process. Expressing the results as grams of fecal fat per 24 hours has been recommended (138, 181) as this eliminates the variable due to water content if results are expressed as grams per gram of wet feces. Feces normally contain 60%–95% water by weight (6). The weight and fecal solids (dry weight) of feces excreted per 24 hours varies with age and diet (6, 88). On low residue diets the net weight is 45–62 g/day (181); on vegetable diets (high residue) the range may go up to 350 g; on an average mixed diet the range is up to 170 g with a mean of about 100 g/day (181).

FECES COLLECTION

Because of the variability of fecal output from day to day, this variation should be minimized by pooling a 3–5 day collection (138, 313). In lipid balance studies there is the additional problem of collecting only those feces representative of an ingested diet. This can be partially overcome by the use of inert unabsorbed materials as markers for fecal collection (99, 140, 253, 290). An alternate is to begin the diet two days before fecal collection begins, as the transit time for the bulk of ingested material in normal individuals is between 26–37 hours (98). Results obtained on single random fecal samples are generally considered to be useless (181). Certainly such an analysis is not likely to be definitive unless, perhaps, the results are markedly abnormal.

Fecal samples can be kept for up to 4 days at refrigerator temperatures without a major alteration of the amount of total lipid, even though considerable hydrolysis of the triglyceride fraction occurs. Since the triglyceride fraction makes up only 10%–20% of the total fecal lipid, the loss of the glycerol moiety is considered insignificant. If the free fatty acids are to be determined or if lipid classes are to be quantitated, the fecal samples should be frozen as soon as possible after collection.

DETERMINATION OF FECAL LIPIDS
(method of Sobel, Ref 181)

PRINCIPLE

For determination of total lipids, a sample of wet feces is emulsified and acidified, ethanol is added, and the mixture is extracted with petroleum ether. The petroleum ether is evaporated and the residue weighed. To determine lipid other than free fatty acids and soaps, the emulsified feces are made alkaline and this fraction extracted. The remaining emulsion is then acidified and the free fatty acids are extracted. The extracts are dried and weighed.

REAGENTS

HCl, conc
NaOH, 10%
Petroleum Ether
Ethanol. Denatured is satisfactory.

PROCEDURE FOR TOTAL LIPIDS

1. Mix stool specimen thoroughly. Emulsify 10 g feces and 20 ml water in micro cup of Waring Blendor.
2. Transfer 3.0 g emulsion into a 50 ml glass-stoppered centrifuge tube. Add 2 drops conc HCl and mix (pH must be less than 3). Add 5 ml ethanol and 20 ml petroleum ether.
3. Stopper and shake tube for about 10 min in a shaking machine. Centrifuge.

4. By means of pipet transfer all supernatant petroleum ether extract to weighed 50 ml beaker. Beakers for this purpose should be stored in a desiccator prior to use. (Aluminum muffin pans about 2.25 inches in diameter can be used for this purpose and discarded after a single usage.) Evaporate solvent on steam bath.

5. Make two more extractions with 20 ml portions of petroleum ether, transferring extract each time to the weighed container and evaporating to dryness.

6. Place the container with dried lipid extract in a vacuum desiccator over anhydrous $CaCl_2$ overnight. The weight of the residue represents total lipid in 1 g feces.

NOTES

1. Stability. Total lipids are stable in feces for at least 48 hours in the refrigerator, but hydrolysis of fatty acid esters proceeds even at this temperature (6, 167). Stability studies in our laboratory have given the following results: (a) The fatty acids of normal feces increases significantly within 24 hours at room or refrigerator temperature. There is no significant change in total lipid even in 5 days at room temperature, but this is presumbly because triglycerides normally make up only about 10% of the total lipid. (b) Glycerol trioleate (not glycerol tripalmitate, however) is hydrolyzed by normal feces even in the refrigerator at a rate indicating that feces high in neutral fat would not be stable at such a temperature for even 1 day.

2. Mineral oil. This medication is included in the total lipid.

ACCURACY AND PRECISION

Completeness and specificity of extraction have not been adequately evaluated in gravimetric procedures. The final residue in the method presented contains some nonlipid material. Washing the petroleum ether extract with water lowers total lipid results by 7%. The accuracy, however, appears adequate for most purposes, and although greater accuracy presumably could be obtained by a second extraction, the added labor would seem to be a poor investment.

The precision (95% limits) of the technic is about ±10%.

NORMAL VALUES

The adult normal range for total lipids in feces is 1–7 g/24 hours (7, 120, 408). Free fatty acids and fatty acid salts are usually reported to make up more than 40% of the total lipid (181), but average levels for normals of about 9% as determined with thin layer chromatography have been reported (120). Children from the age of 2 months to 6 years normally excrete about 0.3–2 g lipid per 24 hours, of which more than 40% is free fatty acids and fatty acid salts (6, 297). Bottle-fed infants excrete significantly more lipid than breast-fed infants: 27%–52% of dry weight compared with 9%–38% (287, 384).

The total lipid concentration in feces is also sometimes reported as percent of dry weight. In earlier studies a normal range of 7%–27% total lipid as percentage dry weight is reported; later studies indicated that this range may be too high, and according to Gravesen (88) 95% of healthy subjects between the ages of 10 and 70 should have levels below 20%.

REFERENCES

1. ABELL LL, LEVY BB, BRODIE BB, KENDALL FE: *Standard Methods of Clinical Chemistry*, edited by Seligson D, New York, Academic Press, 1958, Vol 2, p 26
2. ABELL LL, LEVY BB, BRODIE BB, KENDALL FE: *J Biol Chem 195*:357, 1952
3. ADAMSON LF: *J Nutr 71*:27, 1960
4. ADRIAENSSENS K, VANHEULE R, KARCHER D, MARDENS Y: *Clin Chim Acta 23*:449, 1969
5. ALEXANDER B, LEVI JE: *J Biol Chem 146*:399, 1942
6. ANDERSEN DH: *Am J Diseases Children 69*:141, 1945
6a. ANDERSON JT, KEYS A: *Clin Chem 2*:145, 1956
7. ANNEVERS JH, BOUTWELL JH, IVY AC: *Gastroenterology 10*:486, 1948
8. ANSTALL HB, TRUJILLO JM: *Clin Chem 11*:741, 1965
9. ASSOUS EF, GIRARD ML: *Midicamenta 16*:274, 1964
10. AURELL M, CRAMER K: *Clin Chim Acta 13*:278, 1966
11. BABSON AL, SHAPIRO PO, PHILLIPS GE: *Clin Chim Acta 7*:800, 1962
12. BADZIO T: *Clin Chim Acta 11*:53, 1965
13. BADZIO T, BOCZON H: *Clin Chim Acta 13*:794, 1966
14. BAGINSKI E, ZAK B: *Clin Chim Acta 5*:834, 1960
15. BAILEY JM, WHELAN WJ: *J Biol Chem 236*:969, 1961
16. BAIRD HW III: *J Pediat 52*:715, 1958
17. BARAKAT MZ, SHEHAB SK, EL-SADR MM: *Arch Biochem Biophys 66*:444, 1957
18. BAUMANN EJ: *J Biol Chem 59*:667, 1924
19. BEILER RE, YEARICK ES, SCHNUR SS, SINGSON IL, OHLSON MA: *Am J Clin Nutr 12*:12, 1963
20. BENSON ES, STRANDJORD PE: *Multiple Laboratory Screening*. New York, Academic Press, 1949, p 147
21. BERNARD H, TISSIER M, POLONOVSKI M, CAJDOS A: *Compt Rend Soc Biol 143*:1520, 1949
22. BEST M, VAN LOON EJ, WATHEM J, SEGER AJ: *Am J Med 16*:601, 1954
23. BEUKERS H, VELTKAMP WA, HOOGHWINKEL GJM: *Clin Chim Acta 25*:403, 1969
24. BEVERIDGE JMR, CONNELL WF, HAUST HL, MAYER GA: *Can J Biochem Physiol 37*:575, 1959
25. BILLIMORIA JD, JAMES DCO: *Clin Chim Acta 15*:644, 1960
26. BLIX G: *Biochem Z 305*:145, 1940
27. BLOCK WD, JARRETT K, LEVINE J: *Clin Chem 12*:681, 1966
28. BLOOR WR: *J Biol Chem 17*:377, 1914
29. BLOOR WR: *J Biol Chem 77*:53, 1928
30. BLOOR WR: *J Biol Chem 24*:227, 1916

31. BLOOR WR: *J Biol Chem 82*:273, 1929
32. BOWERS MA, LUND PK, MATHIES JC: *Clin Chim Acta 9*:344, 1964
33. BOWMAN RE, WOLF RC: *Clin Chem 8*:302, 1962
34. BOYD EM: *J Biol Chem 110*:6, 1935
35. BOYD EM: *J Biol Chem 114*:223, 1936
36. BOYD EM: *J Lab Clin Med 22*:237, 1936
37. BROECHOVEN C, PARIJS J: *Clin Chim Acta 20*:530, 1968
38. BROWN HH, ZLATKIS A, ZAK B, BOYLE AJ: *Anal Chem 26*:397, 1954
39. BROWN WD: *Australian J Exp Biol Med Sci 39*:223, 1961
40. BROWN WD: *Australian J Exp Biol Med Sci 39*:209, 1961
41. BROWN WD: *Australian J Exp Biol Med Sci 32*:677, 1954
42. BRUN GC: *Acta Med Scand Suppl 99*:1, 1939
43. BUCKLEY CE, CUTLER JM, LITTLE JA: *Can Med Assoc J 94*:886, 1966
44. BURCHARD H: *Chem Zentr 61*:25, 1890
45. BUTLER LC, CHILDS MT, FORSYTHE AJ: *J Nutr 59*:469, 1956
46. CARLSON LA: *Acta Med Scand 167*:377, 1960
47. CARLSON LA: *J Atherosclerosis Res 3*:334, 1963
48. CARLSON LA, WADSTRÖM LB: *Scand J Clin Lab Invest 10*:407, 1958
49. CARLSON LA, WADSTRÖM LB: *Clin Chim Acta 4*:197, 1959
50. CARLSTEN A, HALLGREN B, JAGENBURG R, SVANBORG A, WERKO L: *Scand J Clin Lab Invest 14*:185, 1962
51. CARPENTER KJ, GOTSIS A, HEGSTED DM: *Clin Chem 3*:233, 1957
52. CARR JJ, DREKTER IJ: *Clin Chem 2*:353, 1956
53. CARROLL KK: *J Lipid Res 2*:135, 1961
54. CASTALDO A, PETTI L, RAGNO I: *Rass Med Sper 11*:201, 1964
55. CAWLEY LP, MUSSER BO, CAMPBELL S, FAUCETTE W: *Am J Clin Pathol 39*:450, 1963
56. CHABROL E, CASTELLANO A: *J Lab Clin Med 57*:300, 1961
57. CHABROL E, CHARRONAT R: *Presse Med 45*:1713, 1937
58. CHAKRABARTI B: *Clin Chim Acta 7*:451, 1962
59. CHEEK CS, WEASE DF: *Clin Chem 15*:102, 1969
60. CHIAMORI N, HENRY RJ: *Am J Clin Pathol 31*:305, 1959
61. CHIN HP, BLANKENHORN DH, CHIN TJ: *Lipids 1*:285, 1966
62. CHIN HP, ABD EL-MEGUID SS, BLANKENHORN DH: *Clin Chim Acta 31*:381, 1971
63. CHLOUVERAKIS C, HARRIS P: *Nature 188*:1111, 1960
64. CHLOUVERAKIS C, HANLEY T, BUTTERFIELD WJH: *Guy's Hosp Rept 112*:193, 1963
65. CHRISTENSEN NC, HORDER M, SIMONSEN EE, TOFT H: *Scand J Clin Lab Invest 23*:355, 1969
66. CHRISTIAN JC, JAKOVCIC S, HSIA DYY: *J Lab Clin Med 64*:756, 1964
67. CLARKE DH, MARNEY AF: *J Lab Clin Med 30*:615, 1945
68. CLAUDE JR, CORRE F, LEVALLOIS C: *Clin Chim Acta 19*:231, 1968
69. CLAYTON MM, ADAMS PA, MAHONEY GB, RANDALL SW, SCHWARTZ ET: *Clin Chem 5*:426, 1959
70. COHEN L, JONES RJ, BATRA KV: *Clin Chim Acta 6*:613, 1961
71. CONNERTY HV, BRIGGS AR, EATON EH Jr: *Clin Chem 7*:37, 1961
72. CONRAD SM: *Can J Med Technol 27*:205, 1965
73. COOK RP: *Cholesterol.* New York, Academic Press, 1958
74. COOKE WT, ELKES JJ, FRAZER AC, PARKES J, PEENEY ALP, SAMMONS HG, THOMAS G: *Quart J Med 15*:141, 1946
75. CORMIER M, JOUAN P, GIRRE L: *Bull Soc Chim Biol 41*:1037, 1959

76. CORNWELL DG, KRUGER FA: *J Lipid Res* 2:110, 1961
77. CORREIA JP, COELHO CS, GODINHO F, BARROS F, MAGALHAES EM: *Am J Digest Diseases* 8:649, 1963
78. CORVILAIN J, LOEB H, CHAMPENOIS A, ABRAMOW M: *Lancet 1*:534, 1961
79. COURCHAINE AJ, MILLER WH, STEIN DB Jr: *Clin Chem* 5:609, 1959
80. CRAMER K, ISAKSSON B: *Scand J Clin Lab Invest* 11:213, 1959
81. CRAMP DG, ROBERTSON G: *Anal Biochem* 25:246, 1968
82. CRAWFORD N: *Clin Chim Acta* 3:357, 1958
83. CRAWFORD N: *Clin Chim Acta* 4:494, 1959
84. CREECH BG, SEWELL BW: *Anal Biochem* 3:119, 1962
85. CROCKSON RA: *J Med Lab Technol* 19:243, 1962
86. CURTIUS H-CH, BURGI W: *Z Klin Chem* 4:38, 1966
87. DAIKOS GK, MATTHEOU P, ATHANASIADOU M: *Lancet* 2:488, 1959
88. DAVIDSON M, BAUER CH: *Pediatrics* 25:375, 1960
89. DEARCY RL, BERGQUIST LM: *Am J Med Technol* 25:237, 1959
90. DE LA HUERGA J, YESINICK C, POPPER H: *Am J Clin Pathol* 23:1163, 1953
91. DE LALLA OF, GOFMAN JW: *Methods of Biochemical Analysis*, edited by Glick D. New York, Interscience, 1954, Vol 1, p 459
92. DELSAL JL: *Compt Rend Soc Biol* 144:66, 1950
93. DELSAL JL: *Bull Soc Chim Biol* 26:99, 1944
94. DE METS M: *Bull Soc Chim Belges* 72:508, 1963
95. DE TRAVERSE PM, LAVERGNE GH, DEPRAITERE R: *Ann Biol Clin (Paris)* 14:236, 1956
96. DE TRAVERSE PM, DEPRAITERE R, THIROUX G: *Ann Biol Clin (Paris)* 17:384, 1959
97. DEUEL HJ Jr: *The Lipids, Their Chemistry and Biochemistry*. New York Interscience, 1951, Vol 1
98. DIAO EK, JOHNSTON FA: *Gastroenterology* 33:605, 1957
99. DICK M: *J Clin Pathol* 22:378, 1969
100. DOLE VP: *Proc Soc Exp Biol Med* 93:532, 1956
101. DOLE VP: *J Clin Invest* 35:150, 1956
102. DOLE VP, MEINERTZ H: *J Biol Chem* 235:2595, 1960
103. DOMANSKI J, KONIECZNY L: *Clin Chim Acta* 6:360, 1961
104. DONEDA G, ZANETTI A, BOBBIO-PALLAVICINI E, ZANPAGLIONE G: *Haematologica Suppl* 12:1421, 1965
105. DOYLE JT, KINCH SH, BROWN DF: *J Chronic Diseases* 18:657, 1965
106. DREKTER IJ, BERNARD A, LEOPOLD JS: *J Biol Chem* 110:541, 1935
107. DREVON B, SCHMIT JM: *Bull Trav Soc Pharm* 8:173, 1964
108. DREYFUSS F, CZACZKES JW: *AMA Arch Internal Med* 103:708, 1959
109. DRUMMEY GD, BENSON JA Jr, JONES CM: *New Engl J Med* 264:85, 1961
110. DRYER RL, TAMMES AR, ROUTH JI: *J Biol Chem* 225:177, 1957
111. DUBOFF GS, STEVENSON WW: *Clin Chem* 8:105, 1962
112. DUNCOMBE WB: *Clin Chim Acta* 9:122, 1964
113. DUNSBACH VF: *Z Klin Chim* 4:262, 1966
114. EDER HA, RUSS EM, PRITCHETT RAR, WILBER MM, BARR DP: *J Clin Invest* 34:1149, 1955
115. EDWARDS CH, EDWARDS GA, GADSDEN EL: *Anal Chem* 36:420, 1964
116. EGGSTEIN M, KRUETZ FH: *Klin Wochschr* 44:262, 1966
117. ELPHICK MC: *J Clin Pathol* 21:567, 1968
118. ENTENMAN C: *Methods in Enzymology*, edited by Colwick SP, Kaplan NO. New York, Academic Press, 1957, Vol 3, p 323
119. EPSTEIN E, BAGINSKI ES, ZAK B: *Ann Clin Lab Sci* 2:244, 1972

120. ERB W, BOEHLE E: *Z Klin Chim Klin Biochem 6*:379, 1968
121. ERIKSEN N: *J Lab Clin Med 51*:521, 1958
122. ETIENNE J, POLONOVSKI J: *Bull Soc Chim Biol 61*:805, 1959
123. ETIENNE J, POLONOVSKI J: *Bull Soc Chim Biol 62*:857, 1960
124. FARSTAD M: *Clin Chim Acta 14*:341, 1966
125. FELS G, KANABROCKI E, KAPLAN E: *Clin Chem 7*:16, 1961
126. FERRO PV, HAM AB: *Am J Clin Pathol 33*:545, 1960
127. FIESER LF: *J Am Chem Soc 73*:5007, 1951
128. FIGUEROA IS, PIKE RL: *Anal Biochem 2*:97, 1960
129. FISCHER RA: *J Am Dietet Assoc 42*:511, 1963
130. FLETCHER MJ: *Clin Chim Acta 22*:393, 1968
131. FOLCH J, ASCOLI I, LEES M, MEATH JA, LEBARON FN: *J Biol Chem 191*:833, 1951
132. FOLCH J, VAN SLYKE DD: *Proc Soc Exp Biol Med 41*:514, 1939
133. FOLDES FF, MURPHY AJ: *Proc Soc Exp Biol Med 62*:215, 1946
134. FORBES AL, CAMLIN JA: *Proc Soc Exp Biol Med 102*:709, 1959
135. FORBES JC, DILLARD GHL, PORTER WB, PETTERSON O: *Proc Soc Exp Biol Med 68*:240, 1948
136. FOSBROOKE AS, RUDD BT: *Clin Chim Acta 13*:251, 1966
137. FRANEY RJ, AMADOR E: *Clin Chim Acta 21*:255, 1968
138. FRAZER AC: *Advances in Clinical Chemistry*, edited by Sobotka H, Stewart CP. New York, Academic Press, 1962, Vol 5, p 70
138a. FREDRICKSON DS: *Hosp Pract 3*:54, 1968
139. FREEMAN NK: *J Lipid Res 5*:236, 1964
140. FRENCH AB, BROWN IF, GOOD CJ, McLEOD GM: *Am J Digest Diseases 13*:558, 1968
141. FRENCH JM: *Lipid Metabolism*, Biochemical Society Symposium, No. 9. London, Cambridge University Press, 1952, p 30
142. FRIEDBERG SJ, HARLAN WR Jr, TROUT DL, ESTES EH Jr: *J Clin Invest 39*:215, 1960
143. FRIEDMAN HS: *Clin Chim Acta 7*:100, 1962
144. FRIEDMAN HS: *Clin Chim Acta 25*:173, 1969
145. FRIEDMAN HS: *Clin Chim Acta 19*:291, 1968
146. FRIEDNER S, MOBERG S: *Clin Chim Acta 18*:345, 1967
147. FRINGS CS, DUNN RT: *Am J Clin Pathol 53*:89, 1970
148. GABBAY KH, WATERHOUSE C: *J Chromatog 11*:241, 1963
149. GABL F, CIRESA M: *Arztl Lab 14*:37, 1968
150. GABRIEL JB, SOLOMON NA, FIERST SM, SASS M: *Am J Digest Diseases 8*:280, 1963
151. GALANOS DS, AIVAZIS GAM, KAPOULAS VM: *J Lipid Res 5*:242, 1964
152. GALLETTI F: *Clin Chim Acta 15*:184, 1967
153. GALLOWAY LS, NIELSON PW, ETHELWYN B, LANTZ EM: *Clin Chem 3*:226, 1957
154. GARDNER JA, FOX FW: *Biochem J 15*:376, 1921
155. GENETET F, NABET P, PAYSANT P: *Ann Biol Clin (Paris) 26*:1129, 1968
156. GIARNIERI D: *Lab Diagnosi Med 8*:1, 1963
157. GIRARD M, ASSOUS E: *Ann Biol Clin (Paris) 20*:335, 1962
158. GJONE E, ORNING OM: *Scand J Clin Lab Invest 18*:209, 1966
159. GLOSTER J, FLETCHER RF: *Clin Chim Acta 13*:235, 1966
160. GOLDBLOOM AA: *Am J Digest Diseases 19*:9, 1952
161. GOLDBLOOM RB, BLAKE RM, CAMERON D: *Pediatrics 34*:814, 1964
162. GOODWIN JF, THIBERT R, McCANN D, BOYLE AJ: *Anal Chem 30*:1097, 1958

163. GORDON RS Jr, CHERKES A: *J Clin Invest 35*:206, 1956
164. GORDON RS Jr, CHERKES A, GATES H: *J Clin Invest 36*:810, 1957
165. GRAFNETTER D, FODOR J, TEPLY V, ZACEK K: *Clin Chim Acta 16*:33, 1967
166. GRANDE F, AMATUZIO DS, WADA S: *Clin Chem 10*:619, 1964
167. GRAVESEN KJ: *Acta Med Scand 175*:257, 1964
168. GREIG M, COOK RP: *Biochem J 77*:16, 1960
169. GROSSMAN MI, STADLER J, CUSHING A, PALM L: *Proc Soc Exp Biol Med 88*:132, 1955
170. GYORGY P, ROSE CS, CHU EH: *Am J Diseases Children 106*:165, 1963
171. HAGENFELDT L: *Clin Chim Acta 13*:266, 1966
172. HALLBERG L, HÖGDAHL AM, SVANBORG A, VIKROT O: *Acta Med Scand 180*:697, 1966
173. HALLBERG L, HÖGDAHL AM, SVANBORG A, VIKROT O: *Acta Med Scand 181*:143, 1967
174. HALLGREN B, STENHAGEN S, SVANBORG A, SVENNERHOLM L: *J Clin Invest 39*:1424, 1960
175. HAM AB: *Am J Med Technol 32*:182, 1966
176. HANEL HK, DAM H: *Acta Chem Scand 9*:677, 1955
177. HARRISON GA: *Brit J Exp Pathol 6*:139, 1925
178. HAUGE JG, NICOLAYSON R: *Acta Physiol Scand 43*:359, 1958
179. HAZZARD WR, SPIGER MJ, BAGDADE JD, BIERMAN EL: *New Engl J Med 280*:471, 1969
180. HENLY AA: *Analyst 82*:286, 1957
181. HENRY RJ: *Clinical Chemistry, Principles and Technics.* New York, Harper & Row, 1964
182. HERMAN RG: *Proc Soc Exp Biol Med 94*:503, 1957
183. HIRSCH J, AHRENS EJ Jr: *J Biol Chem 233*:311, 1958
184. HODGES RG, SPERRY WM, ANDERSON DH: *Am J Diseases Children 65*:858, 1943
185. HÖGDAHL AM, VIKROT O: *Acta Med Scand 178*:637, 1965
186. HOLLINGER NF, AUSTIN E, CHANDLER D, LANSING RK: *Clin Chem 5*:458, 1959
187. HOLLISTER LE, BECKMAN WG, BAKER M: *Am J Med Sci 248*:329, 1964
188. HOLMES FE: *Clin Chem 11*:44, 1965
189. HOLMGARD A: *Scand J Clin Lab Invest 9*:99, 1957
190. HUANG NN, KUO PT: *Am J Diseases Children 98*:598, 1959
191. HUANG TC, CHEN CP, WEFLER V, RAFTERY A: *Anal Chem 33*:1405, 1961
192. IGNATOWSKA H: *Polskie Arch Med Wewnetrzneg 34*:273, 1964
193. IKLES KG: US Department of Commerce, Office Technical Service AD, volume 266: p 166, 1961
194. ITAYA K, KADOWAKI T: *Clin Chim Acta 26*:401, 1969
195. JACOBS SL, HENRY RJ: *Clin Chim Acta 7*:270, 1962
196. JAGANNATHAN SJ: *Can J Biochem 42*:566, 1964
197. JAHNKE K, HERBERG M: *Arztl Lab 15*:201, 1969
198. JAMISON A: *Clin Chim Acta 10*:530, 1964
199. JEROME H, FONTY P, VERNIN H, LEPLAIDEUR F: *Ann Biol Clin (Paris) 20*:117, 1962
200. JONES BJ, MORELAND FB: *Clin Chem 1*:345, 1955
201. JURAND J, ALBERT-RECHT F: *Clin Chim Acta 7*:522, 1962
202. KABARA JJ: *J Lab Clin Med 44*:246, 1954
203. KABARA JJ: *J Lab Clin Med 50*:146, 1957
204. KABARA JJ, McLAUGHLIN JT, RIEGEL CA: *Anal Chem 33*:305, 1961

205. KANTER SL, GOODMAN JR, YARBOROUGH J: *J Lab Clin Med 40*:303, 1952
206. KARKKAINEN VJ, HARTEL G: *Naturwissenschaften 43*:373, 1956
207. KAWADE M, SAIKI K: *Mie Med J 12*:241, 1963
208. KCHOUK M, HORAK MM: *Arch Inst Pasteur Tunis 46*:339, 1969
209. KENNY AP: *Biochem J 52*:611, 1952
210. KENNY AP, JAMIESON A: *Clin Chim Acta 10*:536, 1964
211. KERR LMH, BAULD WS: *Biochem J 55*:872, 1953
212. KESSLER G: *Advances in Clinical Chemistry*, edited by Bodansky O, Stewart CP. New York, Academic Press, 1967, Vol 10, p 45
213. KESSLER G, LEDERER H: *Automation in Analytical Chemistry*, edited by Skeggs LT Jr. New Jersey, Mediad, 1965, p 341
214. KEUL J, LINNET N, ESCHENBRUCH E: *Z Klin Chem Klin Biochem 6*:394, 1968
215. KEYS A, PARLIN RW: *Am J Clin Nutr 19*:175, 1966
216. KEYS A, ANDERSON JT, FIDANZA F, KEYS MH, SWAHN B: *Clin Chem 1*:34, 1955
217. KEYS A, MICKELSEN O, MILLER EVO, HAYES ER, TODD RL: *J Clin Invest 29*:1347, 1950
218. KIBRICK AC, SKUPP SJ: *Clin Chem 1*:317, 1955
219. KIM E, GOLDBERG M: *Clin Chem 15*:1171, 1969
220. KING EJ, WEEDON W: *Biochem J 48*:1, 1951
221. KINGSLEY GR, ROBNETT O: *Anal Chem 33*:561, 1961
222. KIRK EJ: *J Biol Chem 123*:637, 1938
223. KIRK E, PAGE IH, VAN SLYKE DD: *J Biol Chem 106*:203, 1934
224. KLEIN PD, JANSSEN ET: *J Biol Chem 234*:1417, 1959
225. KLUNGSÖYR L, HAUKENES E, CLOSS K: *Clin Chim Acta 3*:514, 1958
226. KOEHLER AE, HILL E: *Federation Proc 12*:232, 1953
227. KONRODE M: *Mie Med J 11*:399, 1962
228. KORNERUP V: *Arch Internal Med 85*:398, 1950
229. KOTODA K, OKAZAKI N, MURAKOSHI M, ARAKI E: *Seibutsu Butsuri Kagaku 11*:335, 1966
230. KOVAL GJ: *J Lipid Res 2*:419, 1961
231. KRAML M: *Clin Chim Acta 13*:442, 1966
232. KRAML M, COSYNS L: *Clin Biochem 2*:373, 1969
233. KRITCHEVSKY D: *Cholesterol*. New York, Wiley, 1958
234. KUNKEL HG, AHRENS EH Jr, EISENMENGER WJ: *Gastroenterology 11*:499, 1948
235. LANE RF: *J Med Lab Technol 18*:106, 1961
236. LAURELL S, TIBBLING G: *Clin Chim Acta 16*:57, 1967
237. LAWRY EY, MANN GV, PETERSON A, SYSOCKI AP, O'CONNELL R, STARE FJ: *Am J Med 22*:605, 1957
238. LAZAROFF N: *Z Med Labortech 7*:242, 1966
239. LAZAROV N: *Z Med Labortech 5*:277, 1964
240. LEFFLER HH: *Am J Clin Pathol 31*:310, 1959
241. LEFFLER HH, McDOUGALD CH: *Am J Clin Pathol 39*:311, 1963
242. LEPPANEN V: *Scand J Clin Lab Invest 8*:201, 1956
243. LEREN P, HAABREKKE O: *Acta Med Scand 189*:501, 1971
244. LEWIS B: *Lancet 2*:71, 1958
245. LEWIS LA, OLMSTED F, PAGE IH, LAWRY EY, MANN GV, STARE FJ, HANIG M, LAUFFER MA, GORDON T, MOORE FE: *Circulation 16*:227, 1957
246. LINDHOLM H: *Scand J Clin Lab Invest (Suppl 23) 8*, 1956
247. LLOYD MR, GOLDRICK RB: *Med J Australia 2*:493, 1968
248. LOFLAND HB Jr: *Anal Biochem 9*:393, 1964

249. LORCH E, GEY KF: *Anal Biochem 16*:244, 1966
250. LOVE EB: *Brit Med J 1*:701, 1957
251. LOVERN JA: *Chemistry of Lipids of Biochemical Significance.* New York, Wiley, 1955
252. LUND JC, SILVERTSSEN E, GODAL HC: *Acta Med Scand 169*:623, 1961
253. LUTWAK L, BURTON BT: *Am J Clin Nutr 14*:109, 1964
254. MacINTYRE I, RALSTON M: *Biochem J 56*: xliii, 1954
255. MACLAY E: *Ann J Med Technol 17*:265, 1951
256. MAN EB: *J Biol Chem 117*:183, 1937
257. MANCINI M, KEYS A: *Proc Soc Exp Biol Med 104*:371, 1960
258. MANN GV: *Clin Chem 7*:275, 1961
259. MANN GV, MUNOZ JA, SCRIMSHAW NS: *Am J Med 19*:25, 1955
260. MARINETTI GV: *New Biochemical Separations*, edited by James AT, Morris LJ. Princeton N.J., Van Nostrand, 1964, p 339
261. MARINETTI GV, ALBRECHT M, FORD T, STOTZ E: *Biochim Biophys Acta 36*:4, 1959
262. MARTENSSON EH: *Scand J Clin Lab Invest 15 (Suppl 69)*:164, 1963
263. MARTINEK RG: *Clin Chem 11*:495, 1965
264. MASSION CG, SELIGSON D: *Am J Clin Pathol 48*:301, 1967
265. MASSMANN W: *Arztl Lab 13*:351, 1967
266. MATSUMIYA K, OKISHIO T, OMORI K: *Rinsho Byori 18*:55, 1970
267. McMAHON A, ALLEN HN, WEBER CJ, MISSEY WC: *Am West Med Surg 6*:629, 1952
268. MEDVEDLEV VP: *Federation Proc 23*:T142, 1964
269. MICHAELS G: *Metab Clin Exptl 11*:833, 1962
270. MICHAELS GD, WHEELER P, BARCELLINI A, KINSELL LW: *Am J Clin Nutr 8*:38, 1960
271. MICHAELS GD, FUKAYAMA G, CHIN HP, WHEELER P: *Proc Soc Exp Biol Med 98*:826, 1958
272. MICHAELS GD, WHEELER P, FUKAYAMA G, KINSELL LW: *Ann NY Acad Sci 72*:633, 1959
273. MITCHELL WD, DIVER MJ: *Lipids 2*:467, 1967
274. MOLINE C, BARRON EJ: *Clin Biochem 2*:321, 1969
275. MONASTERIO G, BERTI G: *Klin Wochschr 30*:111, 1952
276. MOORE RV, BOYLE E Jr: *Clin Chem 9*:156, 1963
277. MORAVEK V: *Arch Biochem Biophys 110*:403, 1965
277a. MORRIS TG: *J Clin Pathol 12*:518, 1959
278. MUELLER G: *Z Med Labortech 7*:332, 1967
279. MUELLER JH: *J Biol Chem 30*:39, 1917
280. MUELLING RJ, COPELAND BE: *Am J Clin Pathol 47*:654, 1967
281. MUNKNER C: *Scand J Clin Lab Invest 11*:388, 1959
282. MUNKNER C: *Scand J Clin Lab Invest 11*:394, 1959
283. NAKANISHI K, BHATTACHARYYA BK, FIESER LF: *J Am Chem Soc 75*:4415, 1953
284. NEERHOUT RC, LANZKOWSKY P, KIMNEL JR, LLOYD EA, WILSON JF, LAHEY ME: *J Pediat 65*:701, 1964
285. NELSON WR, O'HOPP SJ: *Clin Chem 15*:1062, 1969
286. NESS AT, PASTEWKA JV, PEACOCK AC: *Clin Chim Acta 10*:229, 1964
287. NIELSEN G: *Acta Paediat 31*:225, 1944
288. NISHINA T, KIMURA M: *Chem Pharm Bull 12*:521, 1964
289. NOBLE RP, CAMPBELL FM: *Clin Chem 16*:166, 1970
290. NOTHMAN MM: *Ann Internal Med 34*:1358, 1951

291. NOVAK M: *J Lipid Res* 6:431, 1965
292. OBERMER E, MILTON R: *J Lab Clin Med* 22:943, 1937
293. OBERMER E, MILTON R: *Biochem J* 27:345, 1933
294. ORVIS HH, THOMAS RE, FAWAL IA, EVANS JM: *Am J Med Sci* 241:167, 1961
295. PAGE IH, MOINUDDIN M: *J Atherosclerosis Res* 2:181, 1962
296. PANDE SV, KHAN RP, VENKITASUBRAMANIAN TA: *Anal Biochem* 6:415, 1963
297. PARSONS LG: *Am J Diseases Children* 43:1293, 1932
298. PEARSON S, STERN S, McGAVACK TH: *Anal Chem* 25:813, 1953
299. PEARSON S, STERN S, McGAVACK TH: *J Clin Endocrinol* 12:1245, 1952
300. PELL S, KONIECKI WB: *Delaware Med J* 32:400, 1960
301. PERRY TL, HANSEN S, DIAMOND S, BULLIS B, MOK C, MELANCON SB: *Clin Chim Acta* 29:369, 1970
302. PERTSOVSKII AI, KROMSKAYA TV: *Lab Delo No* 7:427, 1967
303. PETERS JP, MAN EB: *J Clin Invest* 22:707, 1943
304. PETERSON JE, WILCOX AA, HALEY MI, KEITH RA: *Circulation* 22:247, 1960
305. PETERSON JE, KEITH RA, WILCOX AA: *Circulation* 25:798, 1962
306. PFLEIDERER T, OTTO P, HARDEGG W: *Klin Wochschr* 37:39, 1959
307. POLAK M, PONTES JF: *Gastroenterology* 83:224, 1955
308. POLLAK OJ, WADLER B: *J Lab Clin Med* 39:791, 1952
309. POLONOVSKI J, JARRIER M: *Ann Biol Clin (Paris)* 13:564, 1955
310. QUAIFE ML, GEYER RP, BOLLINGER HR: *Anal Chem* 31:950, 1959
311. Questions and Answers: *J Am Med Assoc* 166:310, 1958
312. RADIN N: *Standard Methods in Clinical Chemistry*, edited by Meites S. New York, Academic Press, 1965, Vol 5, p 91
313. RAFFENSPERGER EC, D'AGOSTINO F, MANFREDO H, RAMIREZ M, BROOKS FP, O'NEILL F: *Arch Internal Med* 119:573, 1967
314. RAFSTEDT S, SWAHN B: *Acta Paediat* 43:221, 1954
315. RAMANATHA AN, SAVITHRI TS: *Current Sci (India)* 32:407, 1963
316. RANDRUP A: *Scand J Clin Lab Invest* 12:1, 1960
317. RAPPAPORT F, EICHHORN F: *Clin Chim Acta* 5:161, 1960
318. REED AH, CANNON DC, WINKELMAN JW, BHASIN YP, HENRY RJ, PILEGGI VJ: *Clin Chem* 18:57, 1972
319. REED CF, SWISHER SN, MARINETTI GV, EDEN EG: *J Lab Clin Med* 56:281, 1960
320. REINHOLD JG: *Am J Med Sci* 189:302, 1935
321. REINHOLD JG, YONAN VL, GERSHMAN ER: *Standard Methods of Clinical Chemistry*, edited by Seligson, D. New York, Academic Press, 1963, Vol 4, p 85
322. RICE EW: *Standard Methods in Clinical Chemistry*, edited by MacDonald RP. New York, Academic Press, 1970, Vol 6, p 215
323. RICHARDSON RW, SETCHELL KD, WOODMAN DD: *Clin Chim Acta* 31:403, 1971
324. RINEHART RK, DELANEY SE, SHEPPARD H: *J Lipid Res* 3:383, 1962
325. RIVIN AU, YOSHINO J, SHICKMAN M, SCHJEIDE OA: *J Am Med Assoc* 166:2108, 1958
326. ROBERTSON G, CRAMP DG: *J Clin Pathol* 23:243, 1970
327. ROBINSON N, PHILLIPS BM: *Clin Chim Acta* 8:385, 1963
328. ROSE AR, SCHATTNER F, EXTON WG: *Am J Clin Pathol* 5:19, 1941
329. ROUSER G, KRITCHEVSKY G, YAMAMOTO A: *Lipid Chromatographic Analysis*, edited by Marinetti GV. London, Edward Arnold, Ltd., 1967, Vol 1, p 99
330. ROYER ME, KO H: *Anal Biochem* 29:405, 1969
331. RUNDE I: *Scand J Clin Lab Invest* 18:461, 1966

332. RUSS EM, EDER HA, BARR DP: *J Clin Invest 33*:1662, 1954
333. SAIFER A: *Am J Clin Pathol 21*:24, 1951
334. SAIFER A, KAMMERER OF: *J Biol Chem 164*:657, 1946
335. SAMMONS HG, WIGGS SW: *Clin Chim Acta 5*:141, 1960
336. SARDESAI VM, MANNING JA: *Clin Chem 14*:156, 1968
337. SAX SM, DAUGHETEE LA: *Clin Chim Acta 13*:79, 1966
338. SCALA E, MEROLLA R, CASTALDO A: *Boll Soc Ital Biol Sper 40*:60, 1964
339. SCHAEFER LE: *Am J Med 36*:262, 1964
340. SCHAEFER LE, ALDERSBERG D, STEINBERG AG: *Circulation 18*:341, 1958
341. SCHILLING FJ, CHRISTAKIS GJ, BENNETT NJ, COYLE JF: *Am J Pub Health 54*:461, 1964
342. SCHMIDT FH, VAN DAHL K: *Z Klin Chem Klin Biochem 6*:156, 1968
343. SCHOENBECK M, JAKAB T, TISCH D, WERNING C: *Med Klin 65*:644, 1970
344. SCHOENHEIMER R, DAM H: *Hoppe-Seylers Z Physiol Chem 215*:59, 1933
345. SCHOENHEIMER R, SPERRY WM: *J Biol Chem 106*:745, 1934
346. SCHOENHEIMER R, BEHRJNG RH, HUMMEL R: *Hoppe-Seylers Z Physiol Chem 192*:93, 1930
347. SCHÖN H, GEY F: *Hoppe-Seylers Z Physiol Chem 303*:81, 1956
348. SCHWENK E, WERTHESSEN NT: *Arch Biochem Biophys 40*:334, 1952
349. SEARCY RL, CARROLL VP Jr, BERGQUIST LM: *Am J Med Technol 28*:161, 1962
350. SEARCY RL, KOROTZER JL, BERGQUIST LM: *Clin Chim Acta 8*:376, 1963
351. SEIBERT FB, PFAFF ML, SEIBERT MV: *Arch Biochem 18*:279, 1948
352. SHORE B, NICHOLS AV, FREEMAN NK: *Proc Soc Exp Biol Med 83*:216, 1953
353. SOBEL AE, EICHEN S: *Proc Soc Exp Biol Med 79*:629, 1952
354. SOBEL AE, MAYER AM: *J Biol Chem 157*:255, 1945
355. SOLOW EB, FREEMAN LW: *Clin Chem 16*:472, 1970
356. SPERRY WM: *J Biol Chem 114*:125, 1936
357. SPERRY WM: *J Biol Chem 111*:467, 1935
358. SPERRY WM: *J Lipid Res 4*:221, 1963
359. SPERRY WM: *Standard Methods of Clinical Chemistry*, edited by Seligson D. New York, Academic Press, 1963, Vol 4, p 173
360. SPERRY WM, BRAND FC: *J Biol Chem 213*:69, 1955
361. SPERRY WM, WEBB M: *J Biol Chem 187*:97, 1950
362. SPERRY WM, WEBB M: *J Biol Chem 187*:107, 1950
363. SPIGAI C, CARACRISTI R, LOMBARDI V: *Giorn Gerontol 12*:1300, 1964
364. STEVEN MA, LYMAN RL: *Proc Soc Exp Biol Med 114*:16, 1963
365. STOLZ P, ROST G, HONIGMANN G: *Z Med Labortech 9*:215, 1968
366. STULB S, McDONOUGH JR, GREENBERG BG, HAMES CG: *Am J Clin Nutr 16*:238, 1965
367. SVANBORG A, SVENNERHOLM L: *Clin Chim Acta 3*:443, 1958
368. TAYEAU F, NIVET R, DUMAS M: *Bull Soc Pharm Bordeaux 95*:202, 1956
369. THOMAS CB, EISENBERG FF: *J Chronic Diseases 6*:1, 1957
370. THOMAS CB, HOLLJES HWD, EISENBERG FF: *Ann Internal Med 54*:413, 1961
371. THOMITZEK WD, *Z Med Labortech 11*:227, 1970
372. THOMPSON JB, LANGLEY RL, HESS DR, WELSH JD: *J Lab Clin Med 73*:512, 1969
373. TISHLER F, BATHISH JN: *J Pharm Sci 54*:1786, 1965
374. TONKS DB: *Clin Biochem 1*:12, 1967
375. TOURTELLOTTE WW, SKRENTNY BA, DeJONG RN: *J Lab Clin Med 54*:197, 1959
376. TRAPPE W: *Biochem Z 306*:316, 1940

376a. TRINDER P: *Analyst* 77:321, 1952

377. TROUT DL, ESTES EH, FRIEDBERG SJ: *J Lipid Res* 1:199, 1960

378. TROY RH, PURDY WC: *Clin Chim Acta* 26:155, 1969

379. TSCHUGAEFF LA: *Chemiker Ztg* 24:542, 1900

380. TSCHUGAEFF L, GASTEFF A: *Ber Deut Chem Ges* 42:4631, 1909

381. TURNER KB, PRATT V: *Proc Soc Exp Biol Med* 71:633, 1949

382. VAHOUNY GV, BORJA CR, WEERSING S: *Anal Biochem* 6:555, 1963

383. VAHOUNY GV, BORJA CR, MAYER RM, TREADWELL CR: *Anal Biochem* 1:371, 1960

384. VAN DE KAMER JH: *Standard Methods of Clinical Chemistry*, edited by Seligson D. New York, Academic Press, 1958, Vol 2, p 34

385. VAN DER HONING J, SAARLOOS CC, STIP J: *Clin Chem* 14:960, 1968

386. VAN HANDEL E, ZILVERSMIT DB: *J Lab Clin Med* 50:152, 1957

387. VAN SLYKE DD, SACKS J: *J Biol Chem* 200:525, 1953

388. VAN SLYKE DD, PLAZIN J: *Clin Chim Acta* 12:46, 1965

389. VINCENT D, LAFFITTE P, LACROIX R: *Ann Biol Clin (Paris)* 19:165, 1961

390. VOGEL WC, ZIEVE L, CARLETON RO: *J Lab Clin Med* 59:335, 1962

391. VORIS L, ELLIS G, MAYNARD LA: *J Biol Chem* 133:491, 1940

392. WADSWORTH JV, AARON AH: *Am J Surg* 7:480, 1929

393. WARKANY J: *Biochem Z* 293:415, 1937

394. WATKIN DM, LAWRY EY, MANN GV, HALPERIN M: *J Clin Invest* 13:874, 1954

395. WATSON D: *Clin Chim Acta* 15:637, 1960

396. WEBSTER D: *Clin Chim Acta* 8:731, 1963

397. WEBSTER D: *Clin Chim Acta* 8:19, 1963

398. WEBSTER D: *Clin Chim Acta* 7:277, 1962

399. WEIJERS HA, VAN DE KAMER JH: *Acta Paediat* 42:24, 1953

400. WELLER H: *Arztl Lab* 7:215, 1961

401. WHITELAW MJ: *J Clin Invest* 27:260, 1948

402. WHITLEY RW, ALBURN HE: *Technicon International Symposium*, New York, 1964

403. WILDGRUBE J, ERB W, BOHLE E: *Z Klin Chem Klin Biochem* 7:514, 1969

404. WILLIAMS JH, KUCHMAK M, WITTER RF: *J Lipid Res* 6:461, 1965

405. WILLIAMS JH, KUCHMAK M, WITTER RF: *Clin Chem* 16:423, 1970

406. WILLIAMS JH, KUCHMAK M, WITTER RF: *Lipids* 1:89, 1966

407. WILSON WR, HANNER JP: *J Biol Chem* 106:323, 1934

408. WOLLAEGER EE, COMFORT MW, OSTERBERG AE: *Gastroenterology* 9:272, 1947

409. WOOD PDS, HOLTON S: *Proc Soc Exp Biol Med* 115:990, 1964

410. WOODMAN D, YEOMAN WD: *J Clin Pathol* 8:79, 1955

411. WREN JJ, MITCHELL HK: *Proc Soc Exp Biol Med* 99:431, 1958

412. WRIGHT AS, PITT GA, MORTON RA: *Lancet* 2:594, 1959

413. WRIGHT LA, TONKS DB, ALLEN RH: *Clin Chem* 6:243, 1960

414. WYBENGA DR, PILEGGI VJ, DIRSTINE PH, DI GIORGIO J: *Clin Chem* 16:980, 1970

415. WYCOFF HD, PARSONS J: *Science* 125:347, 1957

416. WYNN V, MILLS GL, DOAR JWH, STOKES T: *Lancet* 7624:756, 1969

417. YASUDA M: *J Biochem* 24:429, 1936

418. ZAK B: *Am J Clin Pathol* 27:583, 1957

419. ZAK B: *Standard Methods in Clinical Chemistry*, edited by Meites S: New York, Academic Press, 1965, Vol 5, p 79

420. ZAK B, EPSTEIN E: *Clin Chim Acta* 6:72, 1961

421. ZAK B, LUZ DA, FISHER M: *Am J Med Technol 23*:283, 1957
422. ZAK B, DICKENMAN RL, WHITE EG, BURNET H, CHERNEY PG: *Am J Clin Pathol 24*:1307, 1954
423. ZEITMAN BB: *J Lipid Res 6*:578, 1965
424. ZIEVE L, VOGEL WC: *J Lab Clin Med 57*:586, 1961
425. ZILVERSMIT DB: *Standard Methods of Clinical Chemistry*, edited by Seligson D. New York, Academic Press, 1958, Vol 2, p 132
426. ZLATKIS A, ZAK B, BOYLE AJ: *J Lab Clin Med 41*:486, 1953
427. ZOELINER N, KIRSCH K: *Z Ges Exp Med 135*:545, 1962
428. ZÖLLNER N: *Deut Med Wochschr 84*:386, 1959
429. ZÖLLNER N, WOLFRAM H, WOLFRAM G: *Z Klin Chem Klin Biochem 7*:339, 1969
430. ZUCKERMAN JL, NATELSON S: *J Lab Clin Med 33*:1322, 1948
431. ZURKOWSKI P: *Clin Chem 10*:451, 1964

Chapter 29

Lipoproteins

JAMES W. WINKELMAN, M.D.,
DONALD R. WYBENGA, B.M.

The diagnostic usefulness of various determinations of serum lipoproteins (LP) depends on the thoroughness of the clinical correlations established for the particular test. Recent advances in this field have provided a strong impetus to the reevaluation of older tests and the introduction of newer ones. A fairly complete treatment of the background chemistry and a more extensive consideration of the clinical application than that given for most other topics covered in the book is therefore necessary.

CHARACTERIZATION, CLASSIFICATION, AND PROPERTIES OF LIPOPROTEINS

Several different characterizations of serum lipoproteins have developed from investigations of their chemical, physical, and immunologic properties. In the course of these investigations various nomenclatures have been derived that arose directly from the different methods employed. Clinical chemistry has increasingly accepted for the most medically important LP classes designations based on sedimentation characteristics: very low density lipoproteins (VLDL), low density lipoproteins (LDL), and high density lipoproteins (HDL). Terminology based on electrophoretic mobilities can conveniently be equated with that based on sedimentation, e.g., VLDL with pre-β-, LDL with β-, and HDL with α-LPs. The more specific and quantitative designations that are developed by analytic ultracentrifugation are also shown in Table 29-1, which summarizes several LP classification systems. Included in this table is a listing of the N-terminal amino acids found for each class which has served to divide LPs into three groups on the basis of the apoprotein component; apo A (Asp), apo B (Gln), and apo C (Ser + Thr). Further subgroupings have already been made and are likely to be further expanded as immunochemical and other methods and reagents are further developed (2, 3, 8, 9, 52, 79, 113).

The major forces holding lipoproteins together are almost certainly noncovalent, although covalent bonding has been demonstrated for a small amount of lipid (8). It is thought that the chylomicrons, VLDLs, and LDLs, which are composed of less than 30% protein, are micellar. According to this model, a hydrophobic center of triglycerides and cholesterol esters is believed to be surrounded by a hydrophilic coat of protein, phospholipid, and free cholesterol. A small amount of carbohydrate is associated with the protein fraction derived from β-LPs of normal and hypercholesterolemic

TABLE 29-1. Comparison of Lipoprotein Classification Systems

	LP class[a]						
	Chylomicrons	VLDL (pre-β)	LDL (β)	HDL (α)	HDL$_2$	HDL$_3$	VHDL
Density range (g/ml)	<0.95	0.95–1.006	1.006–1.063	1.063–1.21	1.063–1.125	1.125–1.21	>1.21
Size (diameter)	1 200–11 000 Å	300–700 Å	170–260 Å	74–98 Å			<100 Å
Electrophoretic behavior							
Paper	Origin	Pre-β	β	α			
Cellulose acetate	Origin	Pre-β	β	α			
Agarose	Origin	Pre-β	β	α			
Starch block	α$_2$, β	α$_2$	β$_1$	α$_1$			
Disc	Origin (loading gel)	α$_2$	β$_1$	α$_1$			
Flotation characteristics[b]							
S_f	>400	20–400	0–20	0–9			
F_{1-21}	—	—	—	—			
N-Terminal amino acids[c]							
Major	Ser, Thr	Ser, Ala, Val	Ser	Gln, Thr			
Minor	Gln, Asp	Gln, Thr	Ala	Ala, Val			
ApoProtein	Apo B	Apo A	Apo B	Apo A			Apo A
	Apo A (secondary)	Apo B	Apo C (?)				
		Apo C (?)					
Approximate molecular weight (in millions)	10^3–10^4	5–100	2–3	0.25			

[a]These abbreviations are as follows: VLDL, very low density lipoproteins; LDL, low density lipoproteins; HDL, high density lipoprotein; VHDL, very high density lipoprotein.

[b]S_f designates the negative sedimentation coefficient in Svedbergs in density 1.063 g/ml NaCl solution at 26°C. F_{1-21} is also a negative sedimentation coefficient at density 1.21 g/ml and at 26°C.

[c]Major, > 30% of detectable end groups; minor, 10%–30% of detectable end groups.

individuals (82). Electron micrographs of chylomicrons (107), VLDL, and LDL show them to be spherical, as predicted on chemical and physical grounds. A chemical model, then, for micellar LPs is a sphere with a protein surface, the polar side of which is exposed to the surrounding aqueous medium and the inner surface of which interacts with a 1:1 phospholipid-cholesterol complex, interior to which are the hydrophobic triglycerides and cholesterol esters.

HDL and very high density lipoproteins (VHDL), in which protein constitutes more than 30%, are referred to as pseudomolecular, i.e., as lipid-protein subunits associated in some definite quaternary structure. Electron micrographs of HDL show what appear to be particles with a quaternary structure of 2–6 subunits. Both the micellar and pseudomolecular forms of LPs allow for transport of the predominantly hydrophobic lipids found in serum through the aqueous medium.

Only recently apoproteins from micellar LPs were isolated in lipid-free, water-soluble form (52). As yet, none has been prepared in high yield and with sufficient purity that definitive characterization has been possible (109). Better apoprotein purification has been achieved from LDL, which is purportedly apo B (54). Other data have appeared which are not completely consistent (109), and clarification of this problem requires further study. Recently the major apoproteins of LDL were compared with a predominant fraction of the VLDL apoprotein. They were indistinguishable in amino acid composition, in circular dichroic spectra, in their influence on radioimmunoassay displacement curves, and in the formation of immunoprecipitation arcs (53). The VLDL fraction was similar in all these properties in sera from the various phenotypes. Other detailed information about LP synthesis and metabolism also recently appeared from the same laboratory (73). The structural specificity of the LPs and perhaps the key to some of their behavior is believed to lie in the apoproteins. At this time there is still no theoretical basis for any diagnostic significance of the apoproteins, and a practical procedure for determining apoproteins is not yet available (75).

The micellar LPs show a continuous tendency toward decreasing size with increasing particle density and protein content. As size decreases, the triglyceride content decreases and the phospholipid and cholesterol content increases. The pseudomolecular LPs show increases in the ratios of esterified cholesterol to free cholesterol, and lecithin to sphingomyelin, with increasing density (47). In the VHDL most of the lipid is phospholipid, and the bulk of this is lysolecithin. However, there is not believed to be a continuous spectrum in the composition or properties among either the LPs in general or within any one particular LP class. The β-LPs, for example, have been adequately described by a model composed of two components of fixed composition (84). A recent publication described five separate LDLs in the β-LPs of a type II hyperlipoproteinemic patient (55). The protein per mole of LP remained constant and only the lipid content varied, in specific increments consisting of multiples of 6.25×10^5 g/mole of LP. The resultant five discrete LDLs presumably constitute thermodynamically favored associations of protein and lipid. Differences in size, density, and chemical composition that do exist within a class of LPs have been attributed either to reciprocal changes in triglycerides and total cholesterol solely, as claimed by one group that reported no difference in the amount of protein or phospholipid in human LDLs of different densities (74), or to removal of protein-lipid units together (109).

Immunologic properties of lipoproteins have been investigated by using isoimmune antibodies, usually produced in humans who were immunized in the course of numerous transfusions, and by using heteroimmune antibodies, produced experimentally in rabbits or other animals by repeated injection of LDL isolated from single individuals. Anti-β-LP (anti-LDL) sera precipitates LDL and VLDL, but not HDL. Anti-α-LP (anti-HDL) sera precipitates native and lipid-free HDL, but not LDL or VLDL. However, partially extracted or sonically disrupted VLDL reacts with anti-HDL. This and other evidence indicates that whereas apo A and apo B are the sole protein moieties of α- and β-LPs, respectively, both are present in the pre-β-LPs.

MODEL OF LIPOPROTEIN METABOLISM

According to the most current model of LP metabolism, chylomicrons are synthesized in the intestine, released into the lymphatics, and eventually reach the systemic circulation, at which time they acquire additional protein (38–42). This represents the major means of transport of exogenous triglycerides. VLDL are synthesized in the liver in response to excess dietary carbohydrate or in response to fatty acids mobilized from adipose tissue. This is the main form of transport of endogenous triglycerides and cholesterol. As chylomicrons are cleared from the circulation, "secondary particles" appear in increasing amounts. These are micellar LPs of "endogenous" origin with pre-β-LP (VLDL) sedimentation characteristics, but they are of a particle size that extends into the range of the "primary particles"—i.e., chylomicrons—of exogenous origin (104). Their electrophoretic behavior is believed to be responsible for the trailing from the pre-β band to the origin that had previously been ascribed solely to "endogenous particles" (76). HDL and VHDL are synthesized in the liver.

It is postulated that apo A, originating from HDL or VHDL, becomes bound to chylomicrons and VLDL when the latter classes enter the circulation (109). Differences in the uptake of apo A by subclasses within the VLDL may account for so-called primary and secondary VLDL particles, which migrate with α_2 and β mobilities, respectively, on starch block electrophoresis, but which apparently have the same size distribution as determined by electron microscopy. Lipoprotein lipase activity (LPase) in the serum, which may be due to a single enzyme or isoenzymes from various sources, catalyzes the removal of triglycerides from chylomicrons and VLDL. Schumaker and Adams (109) proposed that a concomitant reduction in the hydrophilic surface of the micellar LPs occurs from removal of lecithin and free cholesterol, in a 1:1 ratio, by the serum enzyme lecithin-cholesterol acyltransferase (LCAT). The apo A bound to chylomicrons and VLDL may function to activate both LPase and LCAT. This mechanism would provide the basis by which micellar LPs retain their spherical shape during their reduction in size (diameter) by a coordinated reduction of the lipid portion of the lipid-protein membrane forming the hydrophilic surface and the triglycerides of the hydrophobic interior. LCAT catalyzes the transfer of an acyl group from the two position of lecithin to the 3-hydroxyl of cholesterol. Its substrates are free cholesterol and lecithin, and its products are lysolecithin and cholesterol ester. The lysolecithin, which has appreciable aqueous solubility, is presumably transferred to VHDL. The cholesterol ester either enters the interior of the micelle or is transferred to LDL or HDL. According to this proposed model, LDL and HDL are generated during the removal of triglycerides from, and breakdown of, chylomicrons and VLDLs. The eventual fate and physiologic or pathophysiologic roles of LDL are uncertain. There is evidence that these particles may be phagocytized by the reticuloendothelial system. HDL is believed to be catabolized eventually by the liver.

RELATION OF BLOOD LIPIDS AND LIPOPROTEINS TO ATHEROSCLEROSIS

An association between blood lipids and atherosclerosis has been established by vast numbers of both experimental and epidemiologic studies. The pathophysiologic mechanisms involved in atherosclerosis are still very largely unknown, and controversy exists concerning virtually every conclusion drawn about the role of serum lipid measurements in prognosis, diagnosis, or monitoring of therapy. Historically, emphasis has been placed at one time or another on measurement of serum total lipid, cholesterol, triglyceride, or phospholipid, on ratios of these, and on various lipoprotein analyses (13). The best known studies have related cholesterol levels and cholesterol metabolism to atherosclerosis. Cholesterol has often been demonstrated as a prominent constituent of atherosclerotic plaques (126). Cholesterol feeding in certain species has induced lesions very similar to human atherosclerosis. Epidemiologic studies have related dietary intake of cholesterol and blood levels with the incidence of arteriosclerosis and several of its major manifestations, including cardiovascular disease and stroke (67, 70). Long-term prospective studies have also established a relationship between hypercholesterolemia and ischemic heart disease (98). As early as 1949, however, it was suggested that the physicochemical state of cholesterol in blood was more important than simply its concentration (27). Gofman and his co-workers shortly thereafter initiated studies that delineated the composition of different classes of lipoproteins (49, 64). These measures, particularly ultracentrifugal analysis, were applied to the LPs of subjects in the Framingham study to determine whether certain classes of LPs were elevated in those who subsequently developed ischemic heart disease (119). The numbers of patients were insufficient during the period included in that prospective study to permit firm conclusions to be drawn. However, the basis for a calculated "Atherogenic Index" which had been developed assigned weighting factors to two LP classes, S_f 0–12 and S_f 12–400, that were apparently prominent (46, 50). In 1954 another large cooperative study compared serum cholesterol and lipoprotein measurements as predictors of diseases which are manifestations of atherosclerosis (120). Most workers in that study concluded that the cholesterol measurement, which was simple, rapid, and inexpensive, and which could be performed in every clinical chemistry laboratory, was as useful as LP analysis by ultracentrifugation, which required costly equipment that was found in very few laboratories.

In 1959 attention was directed to the serum triglycerides, rather than cholesterol, by Albrink and Man (4). They noted an association of hyperglyceridemia with ischemic heart disease. During the ensuing several years studies compared the diagnostic value of triglyceride and cholesterol measurements, although the rationale or desirability of measuring them exclusively or independent of one another was undercut by the already available information that both circulated in the serum as components of lipoproteins.

The formation of atheromatous plaques was originally understood as a consequence of inert filtration of fat from the blood. Although it was subsequently shown that the aortic wall has the capability of fatty acid synthesis, it is now known that the cholesterol in atheromas comes from the blood and not from *in situ* synthesis. An attempt to combine the synthetic and filtration theories suggests that cholesterol increases the fatty acid synthesis in the intima (127). These fatty acids are hypothesized to esterify cholesterol, thereby fixing them and inhibiting their diffusion. Whole LDL lipoprotein enters the intima by passive filtration (135). The concentration of LDL in the intima increases with serum cholesterol (116). The higher ratio of lipoprotein to albumin in the atheroma as compared to the serum suggests that reduction in the egress of lipoproteins either by binding or destruction is also an important factor.

IMPORTANCE OF LIPOPROTEIN ELECTROPHORESIS

Electrophoresis of lipoproteins has been performed by the relatively simple technic of Jencks and Durrum since 1955 (63). The modification by Strauss and Wurm (117, 136), which included the convention of expressing results in terms of the ratios of several bands, achieved limited popularity among clinicians who sought a helpful laboratory test that related LPs to atherosclerosis. In this test lipalbumin and β-lipoprotein were reported as a percent of total, and the ratios of β:α+lipalbumin and β:lipalbumin were also reported.

A major improvement in technic was developed by Lees and Hatch (78), the importance of which became established when Fredrickson and co-workers (38–42, 77) classified clinically distinguishable disorders of lipid metabolism according to the patterns formed by the classes of LP separated by this technic. The major technologic improvement embodied in this technic, compared with previous electrophoretic methods, is that pre-β-LP and β-LP are separated. This relatively simple electrophoretic technic therefore was able to give much the same information as had previously been available only with analytic ultracentrifugation (37). For example, it became possible to specify from an electrophoretic strip that a patient has familial hypercholesterolemia, a conclusion establishable earlier only by the ultra-centrifugal finding of an elevated S_f 0–20 class of LPs. The clinical correlations developed with thoroughly studied patients at the Clinical Center of the NIH stand today as the best available basis for the use of laboratory testing of LP for the evaluation of susceptibility to atherosclerosis and for selection and monitoring of therapy (38–42). By 1970 an international panel of experts was able to conclude that the most rational basis for describing hyperlipidemia at present is in terms of the lipoproteins, and that the determination of lipoprotein concentration by this electrophoretic technic offers more information than lipid analyses alone (10).

Basically, five different electrophoretic patterns of hyperlipoproteinemia were found (38–42). They were designated as phenotypes on the basis of family studies that demonstrated inheritance patterns. It is believed that the phenotypes reflect genetically determined features in the transport and metabolism of lipids of affected individuals. The greatest significance of this classification from the point of view of the clinician is that some phenotypes are and others are not associated with the early and severe development of atherosclerosis and its consequent disease state manifestations. Furthermore, of those associated with atherosclerosis, phenotyping separates hyper-lipidemic individuals into categories for which different drug and dietary regimens are recommended. With the combination of lipoprotein phenotyp-ing and other clinical information, a physician is able to institute an ap-propriate therapeutic program. Despite the fact that the effectiveness of the programs currently available for each phenotype are not tantamount to cure, and the therapy is still more empiric than rational, today the determination of serum LP phenotypes is considered by many authorities a virtual requirement for proper medical practice in these cases. By no means, however, is the role of serum LPs in atherosclerosis completely understood. There is much still unknown about the basic chemistry of the absorption, transport, clearing, synthesis, and utilization of LPs (89). The pathogenesis of actual lesions certainly involves genetic and constitutional factors yet to be discovered. The relationship between the familial hyperlipoproteinemias and nonfamilial or acquired hyperlipoproteinemias giving the same labora-tory findings is uncertain. Discoveries in any of these areas could signifi-cantly change the selection and utilization of tests for serum LPs.

CORRELATION OF LP ELECTROPHORETIC PATTERNS WITH DISORDERS OF LIPID METABOLISM

The reporting and interpretation of LP test results is so dependent upon the current conception of the disorders of lipid metabolism that a brief clinical exposition of the phenotypes is required. The accompanying Table 29-2 summarizes current information about both clinical and laboratory features of the LP phenotypes.

A practical definition of hyperlipidemia is an elevation in total lipids or of any one of the major constituents above the age- and sex-adjusted normals. The phenotypes are numbered according to the system of Fredrickson and colleagues (38–42). That classification was developed for familial or genetic hyperlipoproteinemias. Definitive identification of a genetic basis in any particular case with the laboratory findings that are typical of a primary hyperlipoproteinemia can be established by proving that the same findings occur in other close family members. Among the primary hyperlipoproteinemias some electrophoretic patterns—e.g., that seen in type I—are more indicative than others that the phenotype reflects the genotype. Nonfamilial or "sporadic" occurrence caused by environmental or unknown factors, with or without constitutional predispositions, produces essentially equivalent laboratory and clinical findings and carries the same prognostic implications. Estimates vary about the proportion of "familial" versus "sporadic" cases of hyperlipidemia found in the general population. It appears probable that the majority of patients with hyperlipoproteinemia phenotypes do not have sufficient data available about family members to warrant the conclusion that they have the familial variety. In a third and separate category of causes for the appearance of certain of the laboratory findings of hyperlipoproteinemias, including the typical electrophoretic pattern, are the "secondary" hyperlipoproteinemias. Table 29-3 lists known diseases, medications, or dietary regimens that cause such patterns (19, 38–42, 48, 96, 108).

CLINICAL SIGNIFICANCE OF LIPOPROTEIN DISORDERS

HYPERLIPOPROTEINEMIAS

Type I patients are characterized by massive chylomicronemia while on a normal diet, presumably due to a deficiency in lipoprotein lipase. They have creamy plasma, elevated cholesterol and triglycerides, and may develop splenomegaly, eruptive xanthomas, and acute abdominal pain, but they do not experience the accelerated development of atherosclerosis. The response to moderate restriction of dietary fat is dramatic. The type I pattern alone is virtually sufficient to establish the phenotype. Nonfamilial type I patterns are extraordinarily rare.

The type II syndrome consists of elevated β-lipoproteins with consequent hypercholesterolemia without marked hypertriglyceridemia. It has been shown that the elevated β-LPs found in type II are indistinguishable from the β-LPs found in normal controls with respect to lipid composition and protein content (115). The cholesterol-carrying capacity of the β-LPs in type II is not increased; rather, it is the total amount that is elevated. Current opinion holds that the common mutation(s) giving rise to familial type II hyper-β-lipoproteinemia probably causes a rate-limiting defect in the catabolism of either cholesterol or LDL (35). It may represent expression of a single gene of relatively high frequency. Very early age of onset and rapid rate of progression of the associated xanthomatosis and atherosclerosis are

TABLE 29-2. Characteristics of the Hyperlipoproteinemia Phenotypes

	Type				
	I	II	III	IV	V
Description	Hyperchylomicronemia	Hyper-β-lipoproteinemia	Hyper-β and pre-β lipoproteinemia	Hypertriglyceridemia	Combined types I and IV
Synonyms	Exogenous hyperlipidemia; Burger-Grutz disease	Hypercholesterolemia; essential or familial hypercholesterolemia	Broad beta "floating beta;" idiopathic hyperlipidemia	Endogenous hyperlipidemia; idiopathic hyperlipidemia	Mixed hyperlipidemia
Major electrophoretic abnormality	Chylomicrons present	Increased LDL (β-LP)	Abnormal (broad) β-LP	Increased VLDL (pre-β-LP)	Chylomicrons present and increased VLDL (pre-β-LP)
β-LP (LDL)	Normal or slight ↓	↑	Abnormal β with electrophoretic mobility between β and pre-β, or showing both bands	Normal	Normal or ↓
Pre-β (VLDL)	Normal or slight ↓	Normal or ↑		↑	↑
α-LP (HDL)	Normal	Normal	Normal	Normal	Normal or ↓
Cholesterol	↑	↑	↑	Normal or ↑	↑
Triglycerides	↑	Normal	↑	↑	↑
Cholesterol/triglycerides	<0.2	>1.6	0.6–1.6	<0.6	<0.6
Gross appearance of plasma on standing	Creamy layer over clear infranatant	Clear or slightly turbid	Turbid	Turbid	Creamy layer over turbid infranatant

	Early childhood	Any age; early childhood in severe cases	Adulthood	Adulthood	Early adulthood
Age of detection of familial form	Early childhood	Any age; early childhood in severe cases	Adulthood	Adulthood	Early adulthood
Clinical features	Eruptive xanthomas; hepatosplenomegaly; lipemia retinalis; abdominal pain	Xanthomatosis: tendonous, tuberous; xanthelasma, arcus senilus	Xanthomatosis: tuberous, palmar; tubereruptive	Xanthomas; xanthelasma; hepatosplenomegaly; hyperuricemia	Same as type I hyperuricemia
Accelerated development of atherosclerotic diseases	No	Yes	Yes	Yes	Uncertain
Carbohydrate-inducible	No	No	Yes	Yes	Uncertain
Treatment					
Diet	Low fat	Low saturated fat and low cholesterol	Low carbohydrate (40%) and low fat	Low carbohydrate; low fat and high protein	Low fat
Drug	—	Cholestyramine; thyroxine; nicotinic acid; clofibrate	Clofibrate; thyroxine; nicotinic acid	Clofibrate; nicotinic acid	—

TABLE 29-3. Causes of Inaccuracies in Phenotype Reporting Requiring Exclusion in Interpretation of Results as Primary Hyperlipidemias[a]

Type	Drugs or diet	Primary diseases
I	—	Dysgammaglobulinemia; diabetes; systemic lupus erythematosus
II	Very high cholesterol diet; triglyceride-lowering drugs in types III and IV	Nephrosis; obstructive liver disease; portal cirrhosis; viral hepatitis, acute phase; myxedema; porphyria; stress; anorexia nervosa; idiopathic hypercalcemia
III	Triglyceride-lowering drugs in type IV	Myxedema; dysgammaglobulinemia; primary biliary cirrhosis
IV	Caffeine prior to testing; alcohol; oral contraceptives; chlorthiazide; cholesterol-lowering drugs	Diabetes; pancreatitis; nephrotic syndrome; glycogen-storage disease; hypothyroidism; dysglobulinemia
V	—	Myeloma; macroglobulinemia; nephrosis; pancreatitis; diabetes (insulin-independent)

[a]Failure to adhere to strict 12–14 hour fasting of patient prior to specimen collection, unbalanced diet in the period preceding collection (40), delay beyond the 4 day stability period of the serum without refrigeration or freezing (59) can generally cause patterns to appear that are erroneously interpreted.

thought to indicate a double dose of the abnormal gene. The genetic basis may, however, be considerably more complex. A genetic variant called "double β-lipoprotein" has been described (110). In three generations of a family with this variant, a β-LP was found that differed from normal β-lipoprotein with respect to molecular size, density, and electrophoretic mobility. A subgrouping of type II into types IIa and IIb was recently proposed (10). Type IIa has the electrophoretic pattern described above— i.e., an abnormally heavy β (LDL) band—but type IIb shows, in addition, an increase in the pre-β (VLDL) band. Both patterns have been found in the same kindreds affected with familial hyper-β-lipoproteinemia. Some authors (10) pointed out that the importance of identifying type IIb is that treatment additional to that required for "pure" hypercholesterolemia is recommended; hence this type may be equivalent to one of the two electrophoretic varieties described under type III below. We have found numerous cases with cellulose acetate electrophoresis (CAE) patterns showing abnormally heavily stained, completely separated β and pre-β bands which were confirmed as type III by analytic ultracentrifugation and by the direct and indirect confirmatory tests for type III described later in the chapter.

The requirements for definitive phenotyping of type II are still in a state of development. The type II pattern can occur secondarily to myxedema or in patients with a highly cholesterogenic diet. It is currently believed that familial type II phenotyping is established if the patient with a typical pattern has the characteristic xanthomatosis and one first degree relative with a similar electrophoretic pattern. Alternatively it is also sufficient to demonstrate that the β-lipoproteins contain greater than 250 mg cholesterol per 100 ml serum. The indirect confirmatory test described later in this chapter appears to be very helpful in cases where uncertainty remains after more routine testing has been performed. The same comment applies to questionable cases of types III and IV.

The type III syndrome is characterized by the presence of a heavily stained broad abnormal β band or completely separated heavily stained β and pre-β bands on cellulose acetate electrophoretic strips. At present there is still uncertainty over whether both patterns definitely represent type III (36), and additional supportive evidence is necessary for greater accuracy. Typical of type III is the presence in the serum of a β-migrating VLDL—an abnormal complex of apo-LDL and triglycerides. It has been suggested that this alteration is caused by a block in the breakdown of normal VLDL having pre-β mobility. β-Lipoproteins containing an unusually high concentration of triglyceride then accumulate and as a result of their slow turnover give rise to an elevated serum triglyceride level (101).

A subfractionation of the VLDL from a patient with type III hyperlipidemia into S_f 20–60, 60–100, and 100–400 showed that only the S_f 20–60 fraction had β mobility on paper and agarose electrophoresis. All these subfractions from a patient with type IV hyperlipidemia had the expected β mobility. On the other hand the cholesterol ester/triglyceride ratio of all subfractions was higher in the type III than in the type IV (60).

Further recent work has identified two subspecies of the VLDL (S_f 20–400) found in patients with type III hyperlipoproteinemia. The properties of one of these, called α_2-VLDL, were essentially identical to those of the VLDL found in normal subjects. These include the lipid and protein content, physical characteristics, apoprotein immunoprecipitation reactions (positive for apo B and apo A), and electrophoretic mobility faster than that of β-LPs. The other subspecies, called β-VLDL, had properties more akin to the β-LPs (LDL)—viz., relatively more cholesterol and less triglyceride relative to protein, only apo B immunochemically detectable, and electrophoretic mobility similar to LPs (100). Both the α_2- and β-VLDL were found throughout the S_f 20–400 flotation range. Apolipoprotein different from apo A and apo B has also been reported in the VLDL of type III patients (111), so the situation is probably even more complex. The relative amounts of each of these may account for the occurrence of the two different patterns that we have found to be present in type III cases that are substantiated by both direct and indirect confirmatory testing.

Both serum cholesterol and triglycerides are usually elevated in type III.

Associated clinical findings often include ischemic heart disease, mild diabetes, and xanthomatosis. The hyperlipemia is said to be very responsive to restriction of dietary cholesterol and saturated fat, and possibly also to carbohydrate restriction, along with attainment of ideal body weight. Pharmacologic agents are also efficacious. These are shown in Table 29-3. The benefits of proper management are so great that the establishment of this phenotype is particularly important.

Patients with the type IV syndrome have increased pre-β-lipoproteins and elevated plasma triglycerides without an impressive hypercholesterolemia. Very severe examples have elevations in both triglycerides and cholesterol, but with the former still disproportionately increased. Abnormal glucose tolerance curves are usually present. The presence of hypertension, hyperuricemia, and gout has also been reported in type IV (104). Those with type IV patterns tend to have early development of symptomatic ischemic heart disease. The high serum levels of triglyceride are usually very responsive to dietary restriction of carbohydrates.

Perhaps 50% of patients with primary type IV patterns are without familial basis. Many of these individuals do, however, respond to carbohydrate restriction, but the degree of response and its determinants are unclear at this time.

The type V syndrome patients present a mixture of findings characteristic of types I and IV. They usually show a diabetic tendency, and their tendency to ischemic heart disease is not yet known. They, like type IV, may prove to be responsive to carbohydrate restriction, high protein diet, and maintenance of ideal body weight in a clinically important way.

HYPOLIPOPROTEINEMIAS

Abeta-lipoproteinemia. This rare, genetically determined disorder is characterized clinically by malabsorption of fat, retarded growth, and degenerative changes in the retina and in the central nervous system with consequent neuromuscular symptoms. Laboratory findings include acanthocytosis, severe hypocholesterolemia, and the diagnostic lipoprotein electrophoretic pattern in which there is absence of the β- and pre-β-lipoproteins (121).

Hypo-β-lipoproteinemia. The familial occurrence of reductions to 10%–50% of normal of the β-lipoproteins, with comparable decreases in plasma cholesterol, phospholipids, and triglycerides, has been reported. Reduction of β-lipoproteins is also seen in cases of severe debilitating illnesses or malabsorption of various etiologies.

Familial α-lipoprotein deficiency (Tangier disease). This rare genetic disorder, consisting of the absence of α-lipoprotein with very low serum cholesterol and gross deposition of cholesterol esters in reticuloendothelial tissues, was discovered in the highly inbred population of Tangier Island in the Chesapeake Bay of Maryland. Lipoprotein electrophoresis shows a diagnostic pattern of absent α and pre-β staining and a broad β band.

DETERMINATION OF LOW DENSITY LIPOPROTEINS AND VERY LOW DENSITY LIPOPROTEINS BY ULTRACENTRIFUGATION

(method of DeLalla and Gofman, Ref 24)

PRINCIPLE

The low density lipoproteins (LDL) and the very low density lipoproteins (VLDL) are separated from the more dense lipoproteins and proteins by preparative ultracentrifugation in a solution of density 1.063 g/ml. The LDL and VLDL are then subjected to analytic ultracentrifugation, 52 640 rpm at 26°C, yielding a film record from which the concentrations of the lipoproteins are calculated.

APPARATUS

Beckman model "L" preparative ultracentrifuge with type 40.3 rotor, and Beckman model "E" analytic ultracentrifuge with analytical-F Rotor.

REAGENTS

NaCl Solution, D_4^{20} = 1.1168±0.0005 g/ml. Prepare by dissolving 192.32 g NaCl in 1 liter distilled water. The density of the solution is then adjusted by pycnometry to a final density of 1.1168±0.0005 g/ml at 20°C.

NaCl Solution, D_4^{20} = 1.0073±0.0005 g/ml ("mock" serum). Prepare by dissolving 12.88 g NaCl in 1 liter distilled water. Then proceed as for NaCl Solution D_4^{20} = 1.1168 g/ml to adjust the density to the stated value by pycnometry.

NaCl Solution, D_4^{26} = 1.0630±0.0005 g/ml. Prepare by dissolving 96.50 g NaCl in 1 liter distilled water. Then proceed to adjust density to specified value by pycnometry as for "NaCl Solution D_4^{20} = 1.1168 g/ml," with the exception of the temperature being 26°C.

PROCEDURE

Preparative Run

1. Centrifuge 4 ml serum at 2500 rpm for 15 min.
2. Transfer 3 ml clear serum to a cellulose nitrate tube (0.5 x 2.5 in.) and add 3 ml of NaCl solution, D_4^{20} = 1.1168.
3. Cap tube tightly and mix by inverting several times. Check for leaks.
4. Transfer 0.15 ml H_2O to each rotor's hole and load each tube carefully.
5. Ultracentrifuge in Beckman model L for 18 hours at 39 000 rpm and 14°C using type 40.3 rotor (approximately 110 000 *g*).
6. Carefully remove cap and place tube in tube slicer. Slice immediately below top lipid layer. This must be done while tube is still cool.

7. Transfer upper layer to a 2.5 ml centrifuge tube graduated in 0.1 ml divisions, and record the volume.

Analytic Run

1. Assemble analytic ultracentrifuge cells and tighten them until torque meter indicates 110 in.-lb.
2. Transfer 0.42 ml of NaCl solution, $D_4^{26} = 1.0630$, to left side of analytic cell with 1.0 ml Bristol syringe and B-D 24-gauge needle.
3. Transfer 0.42 ml of supernate obtained in step 7 of preparative run to right side of analytic cell.
4. Place the analytic cell in the rotor and balance the rotor with a dummy cell or a second analytic cell.
5. Align the cells properly in the rotor by matching scribed lines on cell and rotor.
6. Place the rotor carefully in position and tighten it with the coupling wrench. Handle rotor at all times with extreme care.
7. Bring the rotor assembly to $26.0°±0.5°C$ and ultracentrifuge at 52 640 rpm.
8. Set camera for automatic film exposures at 0, 6, 22 and 30 min after attainment of the full speed of 52 640 rpm.

Quantitative Analysis

The template as described by de Lalla *et al* (25) is employed (Fig. 29-1).
1. Develop the ultracentrifuge film and project the film patterns on the template using a photographic enlarger. Then trace the projected patterns.
2. Determine the boundary limits of the lipoprotein classes from the template and measure the areas by use of a polar planimeter.
3. Convert the template areas to concentration by using a conversion table. See *Note 4* for detailed example of this conversion.
4. Apply the Ogsten-Johnston correction from Table 29-4. See the example of calculation given below for use of this table.
5. Apply the correction for effects of inhomogeneous force field and radial concentration (Table 29-5).

Calculation:

$$\text{Atherogenic index (A.I.)} = 0.1 \ (S_f 0{-}12) \ \text{mg/100 ml}$$
$$+ \ 0.175 \ (S_f 12{-}400) \ \text{mg/100 ml}$$

Example of calculation:

1. Align the template so that the cell base line of 0-time frame is superimposed on template line AA'. See Figure 29-1 for this and all subsequent references to projected lines and areas on the template. The position of the reference hole line for the 0-time frame is indicated by pp'.

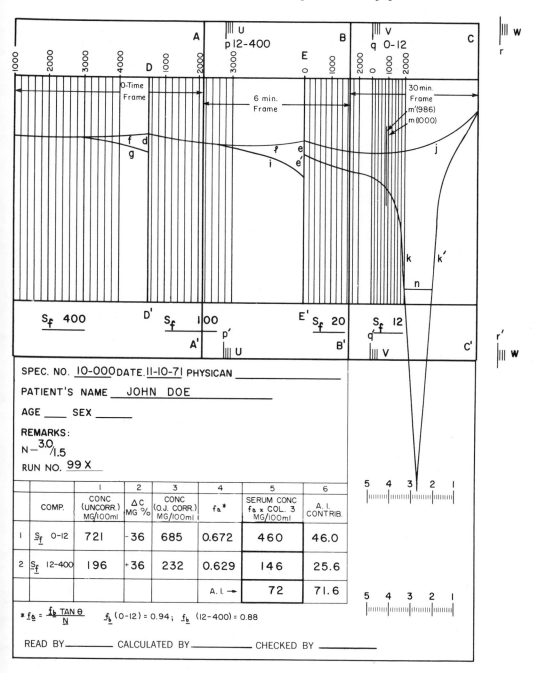

Ultracentrifuge Template For Low Density Lipoproteins
(52,640 rpm AT 26°C; Solution Density = 1.063 G/ml, Enlarged 5X; O=55°)

FIG. 29-1. Ultracentrifuge template for low density lipoproteins (52 640 rpm at 26°C; solution density = 1.063 g/ml, enlarged 5 X; 0 = 55°).
From DE LALLA OF, TANDY K, LOEB HG: *Clin Chem* 13:85, 1967

TABLE 29-4. Ogston-Johnson Corrections (ΔC)

$S_f\,0{-}12$ (mg/100 ml)		$S_f\,12{-}400$ (mg/100 ml)	Correction (ΔC)
>3000	*or*	>3000	0.2 × $S_f\,0{-}12$
<300	*and*	<300	0
		300–1500	0.05 × $S_f\,0{-}12$
		1500–3000	0.10 × $S_f\,0{-}12$
		>3000	0.20 × $S_f\,0{-}12$
300–1500	*and*	<300	0.05 × $S_f\,0{-}12$
		300–1500	0.10 × $S_f\,0{-}12$
		1500–3000	0.15 × $S_f\,0{-}12$
		>3000	0.20 × $S_f\,0{-}12$
1500–3000	*and*	<300	0.10 × $S_f\,0{-}12$
		300–1500	0.10 × $S_f\,0{-}12$
		1500–3000	0.15 × $S_f\,0{-}12$
		>3000	0.20 × $S_f\,0{-}12$

TABLE 29-5. Corrections for Effects of Inhomogeneous Force Field and Radial Concentration

N	$S_f\,0{-}12$ f_a *factor*	$S_f\,12{-}400$
3.0/2.9	0.896	0.839
3.0/1.5	0.672	0.629
3.0/1.0	0.447	0.418
2.5/3.0	1.613	1.510
2.5/2.0	1.075	1.006
2.5/1.5	0.806	0.755
2.5/1.0	0.538	0.503
2.0/2.0	1.344	1.258
2.0/1.5	1.008	0.944
2.0/1.0	0.672	0.629
2.0/3.0	2.016	1.888
1.5/2.0	1.792	1.677
1.5/1.0	0.896	0.839

$$N = \frac{\text{vol serum taken for preparative run}}{\text{vol of top fraction taken for analytic run}}$$

$$f_a = \frac{\text{tan of instrument phase angle}}{N} \times \left[\frac{R_a}{R_o}\right]^2$$

R_a = distance (cm) from center of rotation to peak position of lipoprotein fraction

R_o = distance (cm) from center of rotation to base of cell

The positions of the reference hole lines for the 6 min and the 30 min frames are then correspondingly indicated by gg' and rr', respectively.

2. Project and align the film image for the 0-time frame so that the line for the outer reference hole of the rotor is exactly superimposed on line pp'; then align the cell base line on AA'. The film contains lines both for reference and sample solutions, since "double cell" analytic cells are used. Trace on the template that portion of the reference solution and sample solution patterns left on line DD' (indicated as f and g, respectively, on Figure 29-1).

3. Align the 6 min frame similarly, with the line representing the outer reference hole of the rotor exactly superimposed on line qq'. Trace the reference h and sample i patterns of 6 min frame between DD' and EE'.

4. Align the 30 min frame similarly on line rr'. Trace the reference j and sample k patterns between lines EE' and CC'.

5. *Sf 0–12 (30 min frame).* In the example, the pattern extends down beyond the border of the template. The area down to the line n is measured by planimetry, and the remaining area calculated as the area of a triangle.

   a. Measure the area enclosed by $jeknk'$ by planimetry, giving a result of 14.0 cm^2.

   b. Calculate the area of the triangle with base n, giving a result of 4.6 cm^2.

   c. Sum the two areas: $14.0 + 4.6 = 18.6$ cm^2.

   d. In this system in our laboratory 1 cm^2 is equivalent to 53 mg/100 ml; therefore the estimate is $18.6 \times 53 = 986$ mg/100 ml.

   e. Draw line m' representing 986 parallel to line m representing 1000.

   f. Measure the area enclosed by $jm'knk'$ giving a result of 9.0 cm^2.

   g. Add the area calculated in step b, above, to the area calculated in step f: $9.0 + 4.6 = 13.6$ cm^2.

   h. Calculate again the concentration: $13.6 \times 53 = 721$ mg/100 ml.

   i. Enter this value on the template form in the Sf 0–12 row under the heading "uncorrected concentration."

6. *Sf 12–400 (0 time and 30 min frames).*

   a. Measure the area enclosed by fdg using planimetry giving a result of 1.0 cm^2.

   b. Measure the area enclosed by $h(ee')i$ in the same way, giving a result of 2.7 cm^2.

   c. Sum the two areas: $1.0 + 2.7 = 3.7$ cm^2.

   d. Calculate the concentration: $3.7 \times 53 = 196$ mg/100 ml.

   e. Enter this value in the Sf 12–400 row under the heading "uncorrected concentration."

7. *Ogston-Johnson correction*

   a. Find the appropriate correction factor in Table 29-4, which in this case is 0.05.

b. Multiply the Sf 0–12 concentration with this factor: $0.05 \times 721 = 36$ mg/100 ml.

c. Enter this value with the positive sign in the Sf 12–400 row, column 2, and with the negative sign in the Sf 0–12 row, column 2.

d. Enter in column 3 of each row the corrected concentration:

$$Sf\, 0\text{–}12: \quad 721 - 36 = 685 \text{ mg/100 ml}$$

$$Sf\, 12\text{–}400: \quad 196 + 36 = 232 \text{ mg/100 ml}$$

8. *Correction for effects of inhomogeneous force field and radial concentration*

a. Find the appropriate f_a factor in Table 29-5. Knowing that $N = 3.0/1.5$, the f_a factors for Sf 0–12 and Sf 12–400 are 0.672 and 0.629, respectively.

b. Enter these factors in column 4 and calculate the serum concentrations:

$$Sf\, 0\text{–}12 = 0.672 \times 685 = 460 \text{ mg/100 ml}$$

$$Sf\, 12\text{–}400 = 0.629 \times 232 = 146 \text{ mg/100 ml}$$

Then

$$\text{A.I.} = (0.1 \times 460) + (0.175 \times 146) = 72$$

NOTES

1. Use of "mock" serum. If the specimen provided for analysis is less than the 3.0 ml called for in step 1 in the preparative run of the procedure, use NaCl Solution $D^{20}_4 = 1.1073$ g/ml to bring to 3.0 ml. This maneuver can be employed only if the sample is not less than 2 ml.

2. Operational details. Before starting and during a run always check the following:

a. The outline of the optical procedure.

b. The rotor chamber and mercury pool for spillage of mercury.

c. The diffusion pump; check that it is actually working after turning it on by observing the pressure, which should rise to $20\ \mu$ in approximately 10 sec.

d. Possible leakage of cell, especially when reaching full speed.

3. Film analysis. The line representing the outer reference hole of the rotor is always sharp and is therefore used as the reference point for locating the cell base for each frame. To establish the distance from the cell base line to the outer reference hole, any of the four frames of the film having a sharply delineated cell base line can be used. The measured planimeter areas are expressed in cm² and can be converted to mg lipoprotein per 100 ml by an appropriate factor. It is more convenient, however, to calibrate the planimeter to read in mg/100 ml, and in our system 1 cm² is equivalent to 53 mg/100 ml.

The factors for f_b (see *Note 4*) remain constant and are provided on the template. The f_a values given on the template can be used if 3 ml serum is employed for the preparative run, if 1.5 ml is removed from the top fraction for the analytic run, and if the analytic ultracentrifuge is operated at a phase angle of 55°.

4. *Conversion of template areas to concentration.* In case conversion tables have not been prepared the lipoprotein concentrations can be calculated using the following formula:

$$C = A (f_a) K$$

where

A = template area in cm²

$$f_a = \frac{\tan \theta}{N} (f_b)$$

$$f_b = \left[\frac{R_a}{R_o}\right]^2$$

R_a = distance (cm) from center of rotation to peak position of lipoprotein fraction

R_o = distance (cm) from center of rotation to base of cell

$\tan \theta$ = tangent of instrument phase angle

$$N = \frac{\text{vol of serum taken for preparative run (ml)}}{\text{vol of top fraction taken for analytic run (ml)}}$$

$$K = \frac{10^3}{L \times M \times m \times T \times \Delta n \times E^2}$$

L = optical lever arm of instrument (cm)

M = vertical magnification of instrument optical system

m = horizontal magnification of instrument optical system

$$T \quad = \quad \text{cell thickness (cm)}$$

$$\Delta n \quad = \quad \text{specific refractive increment for serum LDL}$$
$$(1.54 \ \times \ 10^{-3}/\text{g}/100 \ \text{ml})$$

$$E \quad = \quad \text{template enlargement factor}$$

5. *Stability of samples.* DeLalla and Gofman (24) stored sera at $0°-4°C$ and demonstrated no significant alteration in lipoprotein composition for at least 28 days. The effect of storage on lipoprotein patterns as determined by ultracentrifugation has not been studied in our laboratory. It is probably safe to assume, however, that the stability data presented under *Phenotyping of Hyperlipoproteinemias* also applies to the lipoprotein determination by ultracentrifugation.

ACCURACY

The determination of low density and very low density lipoproteins by ultracentrifugation utilizing the Gofman differential-density separation method, is generally regarded as the most accurate and precise available today. The specific lipoprotein classes isolated by this technic have been characterized by the determination of sedimentation and diffusion coefficients, hydrated density, molecular weight, peptide pattern, and lipid and protein analyses. Electron micrographs (59) of the various lipoprotein molecules show the lipoproteins to be discrete entities and are employed to obtain confirmatory results. Greater accuracy may be achieved with the aid of a computer and a modified procedure for reading the Schlieren films (30).

NORMAL VALUES

In 1954 Glazier *et al* (46), using the method of deLalla and Gofman (24), reported normal "standard" values for the four major low density lipoprotein classes. The originally calculated normal limits assumed a normal distribution of the raw data. Since the range was more than twofold we thought it reasonable to assume that a better range could be obtained by assuming a log distribution. Raw data were not available for direct application of our currently preferred nonparametric method of normal range calculation. Therefore the log transformation of Glazier's data on healthy subjects for the classes $Sf\,0-12$ and $Sf\,12-400$ is presented in Table 29-6. Hatch (57), using the ultracentrifugation method of Ewing *et al* (30), found normal levels for males (35–55 years) and females (35–55 years) to be for $Sf\,0-12$, 288–450 mg/100 ml and 267–339 mg/100 ml, respectively, and for $Sf\,12-400$, 43–275 mg/100 ml and 0–116 mg/100 ml, respectively. Nichols (89), who also used Ewing's method, studied 16 males and 16 females and found normal values for VLDL of 7–251 mg/100 ml and 0–100 mg/100 ml, respectively; for LDL, ranges of 340–538 mg/100 ml and 301–459 mg/100 ml, respectively. These normal values agree well with those given in Table 29.6.

TABLE 29-6. Normal Ranges for the S_f 0–12 and S_f 12–400 Low Density Lipoprotein Classes

Mean age	Standard values, nonfasting serum (mg/100 ml)					Mean A.I.
	Std S_f 0–12		Std S_f 12–400			
	Mean	Range[a]	Mean	Range[a]		
Males						
25	292	172–494	161	66–395		57
35	329	198–548	205	71–595		69
44	355	225–559	233	92–592		76
54	357	221–576	206	77–553		72
61	352	228–542	220	99–492		74
Females						
24	274	166–453	97	40–236		44
35	295	180–483	126	45–348		52
·44	337	212–535	166	63–441		63
54	354	231–542	216	85–550		73
62	361	235–553	270	118–682		83

[a]Range equals mean ±2s.

Age and Sex. Walton and Scott (124) found a significant sex difference after the age of 12 years. These authors stated that lipoprotein levels in males leveled off during the third decade, gradually increased until the sixth decade, and then declined. The increase in the levels of females on the other hand seemed to resemble that reported by Glazier *et al* (Table 29-6, ref 46).

Diet. It is well established that dietary habits affect serum lipid and lipoprotein concentrations. High fat and low carbohydrate diets increase LDL. These lipoproteins are also affected by the chain length of the fatty acids, the degree of unsaturation, and the amount of cholesterol in the diet. VLDL concentrations are increased after a low fat and high carbohydrate diet.

CHOICE OF METHOD FOR PHENOTYPING HYPERLIPOPROTEINEMIAS

None of the methods commonly used has been generally accepted as the reference method. Our laboratory has several years' experience with the method presented in this chapter, and we believe that its recommendation as a routine method is justified by the discussion below.

Supporting Medium

Lees and Hatch (78) developed a technic for lipoprotein separation using paper electrophoresis. The major innovation in this method was the inclusion of albumin in the buffer, which facilitated separation of the β- and pre-β-lipoproteins. This technic was used by Fredrickson *et al* (38) when they introduced phenotyping of hyperlipoproteinemias in 1967. Moinuddin and Taylor (86) reported a paper electrophoresis method without albumin in the buffer. They claimed that the advantages over the procedure with albumin were low background staining, and sharper a bands and a 4 hour electrophoresis (run with sharper and denser lipoprotein bands than with the 16 hour run).

Winkelman *et al* (128) evaluated the Lees and Hatch (78) paper electrophoresis technic and found that only approximately one-third of a large series of specimens could be assigned a definitive type. Furthermore, only 58% of the strip readings were consistent with lipid fractionation data. These authors stated that the inconsistencies occurred largely because of poor definition or separation of β and pre-β bands. Lane (72) observed in his evaluation of paper electrophoresis that it has no diagnostic value if the blood lipids are not above the normal levels. Buckley *et al* (14) studied serum lipoprotein patterns using paper electrophoresis in healthy subjects and established that in 21% of 240 subjects the lipoprotein pattern was slightly abnormal, while cholesterol and triglyceride results were well within the normal range. Winkelman *et al* (131) compared paper electrophoresis (78) with the electrophoretic technic using cellulose acetate. These authors concluded that sharper delineation and more complete separation of the β- and pre-β-lipoprotein bands was achieved with cellulose acetate. They claimed that the improved readability resulted in an improvement in consistency with lipid fractionation data and that the error rate was lower and less critical clinically. Results were compared with cellulose membranes supplied by Shandon Scientific Company, Titan III cellulose acetate plate from Helena Laboratories, Sepraphore III strips from Gelman, and Sartorius cellulose acetate strips from Brinkmann Instruments. Best results were obtained with those from Shandon. The Sepraphore III strips, however, appear to be the most commonly used cellulose acetate strips. Farber and associates (31) also compared cellulose acetate and paper technics and demonstrated an excellent qualitative correlation. The major advantage of cellulose acetate presented by this group was the significant reduction in time. Many investigators (11, 17, 18, 21, 34, 68, 90, 99) performing lipoprotein electrophoresis have used cellulose acetate as support medium, and some (12, 20, 122) employed it in the gel form. All confirmed the findings of Winkelman *et al* (131) with respect to resolution of the lipoprotein bands and the relatively short time required for the test. A support medium gaining popularity has been agarose gel (21, 29, 66, 91, 92, 95, 96). Agarose gel electrophoresis is slightly more difficult than cellulose acetate electrophoresis, but resolution of the lipoprotein bands is as good as obtained with cellulose acetate, and the time required for staining is shorter. Other supporting media used are agar gel (62), starch gel (80), starch granules (102), and polyacrylamide gel (22, 44, 87, 88, 132–134) with disc electrophoresis.

Visualization of Lipoprotein Bands

The four most commonly used stains for visualization of the lipoprotein fractions are oil red O, Sudan black B, amido black, and Schiff's reagent after oxidation with ozone. Narayan *et al* (87) compared the first three and found that amido black was superior to the other two for isolated lipoprotein fractions. Since oil red O and Sudan black B stain only the lipoproteins and not the proteins, these two stains were preferred when working with serum or plasma. Gabl *et al* (45) compared oil red O with Sudan black B and concluded that the two stains were of equal merit for staining lipoprotein fractions. Noble (91) demonstrated that there was no difference between these two stains in uptake by lipoprotein. Although the preparation of oil red O requires a little bit more care, and the solution is not as stable as Sudan black B, it appeared to be the preferred stain. The oil red O stain can be prepared in ethanol (58, 91, 131) or in methanol (16, 18, 34). In both cases it should be prepared and used at temperatures between 35° and 40°C. In our experience the staining solution prepared in ethanol has always given satisfactory results. Sudan black B is commonly employed since it can be used at room temperature and requires only a 30 min (112) or 2 hour (18, 22, 26, 44, 62, 66, 123, 114) staining period, versus 6—18 hours with oil red O. Beckering and Ellefson (11), using oil red O for staining, managed to reduce the staining time to 50 min by adding water to the oil red O solution immediately before staining. The staining of lipoprotein bands with Schiff's reagent (31, 68, 90, 99, 103) has not been very popular because of the requirement for oxidation with ozone prior to staining. Elevitch and Austin (29) employed fat red 7B and stained the lipoproteins for only 15 min. This stain, however, requires bleaching the background. Bertrand and associates (12), describing a micromethod for the identification of lipoproteins, stained with Ciba red T 192.

Quantitation

Semiquantitative determinations of lipoproteins using densitometric scans of paper electrophoretic strips were compared with analytic ultracentrifugation results by Hatch *et al* (58). These authors claimed that the electrophoretic results were within ±30% of those obtained with the ultracentrifuge. They felt that this degree of accuracy justified the semiquantitation of lipoproteins by paper electrophoresis, since significant lipoprotein disorders involve larger changes. Fredrickson and co-workers (43), however, discussing paper and agarose-gel electrophoresis of lipoproteins, stated that comparisons between quantitative electrophoretic results and those obtained by ultracentrifugation were needed in a large number of hyperlipoproteinemic patients with all the different phenotypes before quantitative densitometry of electrophoretic strips could improve "phenotyping." Iammarino *et al* (62), employing agar gel lipoprotein electrophoresis, demonstrated that a statistically significant correlation existed between the % total β (pre-β + β fraction) calculated from the densitometer scan of the electrophoretic strip and the % LDL obtained by ultracentrifugation. Noble and associates (92) reported highly significant correlation coefficients between electrophoretic methods (paper and agarose gel) and analytic ultracentrifugation for fractions β and Sf 0—20, pre-β and Sf 20—400, and α and total HDL. They indicated that both electrophoretic methods could be used semiquantitatively. Dyerberg and Hjorne (28) calculated factors to correct for the different uptake of the dye Sudan black by the various lipoproteins. These factors were based on the average lipid composition of the different lipoproteins and on uptake by pure lipids equivalent to those composing the lipid moiety. These correction factors, applicable to the extinction values obtained after elution of a stained lipoprotein, were tested on the lipoprotein electrophoretic results of healthy persons. The calculated lipoprotein values agreed well with those reported for the ultracentrifugation method. Sirtori *et al* (112) performed statistical analyses on data obtained by densitometric measurements of agarose gel electrophoretic scans from 97 patients with type II or type IV hyperlipoproteinemia. The statistical results placed a patient with a pre-β of less than 30% of the total lipoprotein in the type II category, and the ones with a pre-β larger than 35% in the type IV group. For those with pre-β between 30% and 35%, the percent β was used; if the latter was less than 40% the patient was designated as a phenotype IV, and when greater than 40% the type was II. The validity of this screening approach has not yet been confirmed by other investigators. Chin and Blankenhorn (18) determined the relative peak areas of the lipoprotein bands by scanning the cellulose acetate strips in a densitometer and found the percentages of the fractions to be helpful in studying lipoproteins in normals and patients with hyperlipoproteinemias. Postma and Stroes (99), also using cellulose acetate, established normal values for the relative peak areas of β-, pre-β-, and α-lipoprotein bands by densitometry. These relative peak areas were used to calculate the lipoproteins in mg of total lipids/100 ml. The values for α-, pre-β-, and β-lipoprotein bands obtained in this manner from patients with hyperlipoproteinemias were helpful in determining the phenotype. Chin and Blankenhorn (17) demonstrated that the precision of lipoprotein electrophoresis on cellulose acetate was good enough to monitor day to day lipoprotein changes induced by a liquid diet in which 87% of total calories were from carbohydrate. The average differences between 50 duplicate determinations with their technic were: α 2.5%, pre-β 3.6%, and β 3.5%. They indicated that this precision made cellulose acetate very useful in the study of changes within individual patients with hyperlipoproteinemia phenotypes II and IV. Ross and Brown (105) quantitated the lipoproteins on cellulose acetate by scanning the lipoprotein bands and obtained the following mean values on serum samples of 25 normal individuals: β 48.6%, pre-β 36.7%, and α-lipoproteins 16.7%. These authors successfully used the numerical density values of the lipoprotein fractions for classification of the electrophoretic patterns into Fredrickson's phenotypes. Winkelman *et al* (130) separated the β-, pre-β-, and α-lipoprotein bands by electrophoresis on cellulose acetate in each of the five recognized hyperlipoproteinemia phenotypes, and quantitated the lipoprotein bands by microdensitometry. Their scanning procedure had the following precision: β ±3.3%, pre-β ±0.8%, α ±2.2%. Specific types of errors of strip readings that occurred using visual evaluation alone were eliminated by the quantitative densitometric values. These values, however, did not help to eliminate errors in strip readings of type III. Furthermore, the improvement in accuracy obtained with the densitometer scans required a virtual doubling of technician time.

PHENOTYPING HYPERLIPOPROTEINEMIAS

(method of Winkelman et al, Ref 131)

PRINCIPLE

The lipoprotein fractions of serum or plasma are separated by electrophoresis using cellulose acetate strips. The lipoproteins are stained with oil red O, and the pattern is interpreted according to the system of Fredrickson *et al* (39) by visual comparison with known normal and abnormal patterns.

REAGENTS

Barbital Buffer, (pH 8.6, μ 0.050). Dissolve 41.2 g sodium barbital, 7.36 g barbital, and 0.1488 g disodium salt of ethylenediaminetetraacetic acid in 3.5 liters distilled water, and make up to 4 liters with water.

Oil Red O Stain. Add 0.8 g dry oil red O dye to 1200 ml of 95% ethanol (ethanol denatured with methanol is satisfactory) in a 3—4 liter flask. Allow the mixture to stand for about 15 min, occasionally swirling the flask. Then add 800 ml of boiling water, cover mouth of flask with aluminum foil, and heat to boiling on an electric hot plate. Let mixture stand overnight in a water bath or on a hotplate at 37°C with stirring. On the following day heat to 41°±1°C and filter through a prewarmed funnel into a prewarmed vessel using large Whatman 1 filter paper. *Do not allow the temperature of the dye solution to fall below 37°C at any time.* Store in 37°C incubator and use at this temperature to stain strips. Prepare *fresh* every 14 days.

Glycerol

PROCEDURE

The following method is used in our laboratory with the Photovolt or Gelman electrophoresis chamber employing Shandon cellulose acetate strips (25 x 130 mm).

1. Preparation of electrophoresis chamber: All four compartments of the chamber are filled with buffer (approximately 750 ml). Be sure that the liquid level is uniform. Paper wicks (Beckman) are impregnated with buffer solution and placed over the shoulders with both edges dipping in the buffer. Change the paper wicks daily and the buffer after every 40 strips.
2. Preparation of cellulose acetate strips:
   a. Mark the starting line (1 in. from end of strip) with a nonsmearing ballpoint pen.
   b. Impregnate the strips with buffer in the following way: Let the strip float on buffer and allow it to be saturated slowly with buffer, then submerge. Do not use the strips if white spots are present, which indicates trapped air.
   c. Blot strips gently with filter paper. Position strips in the chamber (eight strips per chamber) and keep them taut using small magnets.

    d. Equilibrate the strips for 5 min employing 150 V (7 to 8 mA) per chamber.

3. Apply 10 μl serum to the strip in the form of a fine uniform streak along the previously marked line using a Beckman micropipet as applicator. The streak of sample should finish at least 5 mm from each edge of the strip to avoid irregularities in the lipoprotein pattern.

4. Run the electrophoresis for 90 min using a constant voltage of 150 V.

5. Take the strips out, blot the ends with filter paper, and stain overnight at 37°C with oil red O.

6. Rinse the stained strips twice with distilled water. Blot the strips gently with filter paper, cut to a length of 10 cm, and place in glycerol for 5 min.

7. Mount the strips in plastic envelopes (8.5 x 11.0 in.).

8. Visually compare the pattern with those of known pools run in same chamber.

INTERPRETATION OF ELECTROPHORETIC STRIPS AND LIPID RESULTS

The electrophoretic strips are examined and the findings are recorded by marking the appropriate box of a reporting scheme (Tables 29-7 and 29-8). After the electrophoretic result is determined, it is checked for compatibility with cholesterol (138) and triglyceride (69) values obtained on the same specimen. The criteria used in determining whether lipid results are consistent with the electrophoretic pattern are given in Table 29-9. If the cholesterol and triglyceride results are consistent with the electrophoretic pattern, the phenotype is reported. If the lipid results are not consistent with the electrophoretic pattern, all determinations are repeated. If results are still not compatible after repeating, the phenotype as established by electrophoresis is reported with a note stating the inconsistency and possible causes.

NOTES

1. Stability of sample. Winkelman and co-workers (129) showed that samples kept at room temperature for 3 days are suitable for cellulose acetate electrophoresis. They also found satisfactory stability after storage at refrigerator temperature for at least 28 days, and after storage at freezer temperatures for 14 days if only one freeze-thaw cycle occurs. Other reports on stability of samples for lipoprotein determinations differ in some details from the above quoted results. Fredrickson *et al* (38–42) stated that freezing irreversibly altered the paper electrophoretic lipoprotein pattern, although Adlersberg *et al* (1), Dangerfield and Smith (23), and Mills and Wilkenson (85) reported that paper electrophoretic patterns were not affected by frozen storage for periods of several months. Another report by Gottfried *et al* (51) claimed that β-lipoprotein decreased 10%, 20%, and 25% after 1, 2, and 3 days, respectively, at room or refrigerator temperatures

Some of these differences may well be attributable to different analytic technics, e.g., paper rather than cellulose acetate used as the electrophoretic medium.

TABLE 29-7. Definitive Patterns

Result	Observation	Pattern appears consistent with (ref 40)
☐	Heavy staining at origin	Type I hyperlipoproteinemia
☐	Heavy staining β; some pre-β may be present	Type IIa hyperlipoproteinemia
☐	Heavy staining broad β *or* heavy staining discrete β and pre-β	Type III or type IIb hyperlipoproteinemia
☐	Heavy staining pre-β; some β may be present	Type IV hyperlipoproteinemia
☐	Heavy staining origin and pre-β, some β present	Type V hyperlipoproteinemia
☐	No staining except α	Abeta-lipoproteinemia
☐	Markedly decreased β-lipoprotein	Hypo-β-lipoproteinemia
☐	No α stain; broad β band	Tangier disease

TABLE 29-8. Nondefinitive Patterns

Result	Observation	Possible interpretations
☐	Normal staining β (ND_1)	Normal, nonhyperlipidemic patient Type IIa, borderline or treated
☐	Normal staining β and normal staining discrete pre-β (ND_2)	Normal, nonhyperlipidemic patient Type IIb, borderline or treated Type III, borderline or treated Type IV, borderline or treated
☐	Normal staining β with incompletely separated not very heavy staining, broad advancing margin; possible abnormal β (ND_3)	Normal, nonhyperlipidemic patient Type III, borderline or treated Type IV, borderline or treated
☐	Other	Unknown

TABLE 29-9. Criteria for Phenotyping of Hyperlipoproteinemias

Type	Cholesterol (mg/100 ml)		Triglycerides (mg/100 ml)		Chol/trig
ND_1, ND_2, ND_3[a]	<315	and	<200		—
II	>315	and	<200	and	>1.6
III	>315	and	>200	and	>0.6 and <1.6
IV	<315		>200	and	<0.6

[a]For definition of the normal patterns, designated types ND_1, ND_2, and ND_3, see Table 29-8.

DIRECT CONFIRMATION OF HYPERLIPOPROTEINEMIA PHENOTYPE III BY ULTRACENTRIFUGATION

(method of Fredrickson et al, Ref 37)

PRINCIPLE

Serum is diluted with a solution of NaCl and ultracentrifuged. The fractions with densities less than and greater than 1.006 are electrophoresed. In cases of phenotype III, the abnormal floating β-lipoprotein can be demonstrated in the fraction with density less than 1.006.

REAGENTS

NaCl, 0.15 M. Dissolve 7.66 g NaCl in 1 liter water.

PROCEDURE

1. Transfer 1.5—3.0 ml serum to a cellulose nitrate tube (0.5 × 2.5 in.)
2. Fill the tube with 0.15 *M* NaCl and cap.
3. Ultracentrifuge in Beckman model L for 18 hours at 39 000 rpm and at 14°C using type 40.3 rotor (approximately 110 000 *g*).
4. Carefully remove cap and place tube in tube slicer. Slice immediately below the top lipid layer. This must be done while the tube is still cool. Leave tubes in rotor prior to slicing.
5. Transfer upper and lower layers to separate tubes and mix contents. Place tubes in ice until electrophoresed.
6. Stripe aliquots from upper and lower layers (densities below and above 1.006, respectively) and electrophorese the lipoproteins by the method given previously.
7. Examine the electrophoretic strip on which the upper layer was applied and determine if lipoprotein with β mobility is present. If this is the case, type III is confirmed, otherwise type III is ruled out (see Fig. 29-2). The

PHENOTYPE III

ABNORMAL VLDL
β-LIPOPROTEIN IN
ULTRACENTRIFUGAL FRACTION
WITH d<1.006

β-LIPOPROTEIN IN
ULTRACENTRIFUGAL FRACTION
WITH d>1.006

FIG. 29-2. Cellulose acetate electrophoretic strips demonstrating a phenotype III, the presence of the abnormal VLDL β-lipoprotein in ultracentrifugal fraction with d < 1.006 and the normal β-lipoprotein in ultracentrifugal fraction with d > 1.006.

electrophoretic strip on which the bottom layer was applied will show a β-lipoprotein band representing the nonfloating β-lipoproteins. This band aids in establishing whether the band appearing on the strip on which the upper layer was applied is in the β or pre-β position.

NOTES

1. Electrophoretic results. Specimens from patients with hyperlipoproteinemia types I, II, IV, and V, as well as those of normal individuals, do not show a lipoprotein band when the fraction with density less than 1.006 is subjected to electrophoresis. Type IIb, which resembles type II in electrophoretic appearance and in most of the lipid parameters, also does not demonstrate the abnormal floating β-lipoprotein. Two serum electrophoretic patterns are accepted as typical of type III. In one, a single broad abnormal band smears from the β to pre-β region. In the other, two discrete abnormally intense bands occur, β and pre-β. Both patterns are found in sera that have cholesterol and triglyceride values consistent with type III. In our experience both patterns give a positive confirmatory test by ultracentrifugation except when the effects of diet or drug therapy cause misleading results (137). Demonstration of the abnormal β-lipoprotein with density less than 1.006 in the confirmatory test, however, definitely establishes type III.

2. Stability of samples. See under *Notes* of *Phenotyping of Hyperlipoproteinemias.*

INDIRECT CONFIRMATION OF HYPERLIPOPROTEINEMIA PHENOTYPES II, III, AND IV BY ULTRACENTRIFUGATION

(method of Wybenga et al, Ref 137)

PRINCIPLE

Serum is diluted with a solution of NaCl and ultracentrifuged. An analysis for cholesterol is performed on the supernate which contains the very low density lipoproteins (VLDL, density less than 1.006). The VLDL cholesterol concentration is expressed as percent of serum total cholesterol.

REAGENTS

NaCl, 0.15 M. Dissolve 7.66 g NaCl in 1 liter water.

Isopropyl Alcohol (Aldehyde-Free). A.C.S. reagent grade.

Acetic Acid, Glacial. Reagent grade.

Sulfuric Acid. Reagent grade.

Ethyl Acetate. "Spectroquality" solvent (Matheson, Coleman and Bell), used without redistillation.

Ferric Perchlorate. "Nonyellow," containing excess $HClO_4$ (catalog No. 40, G. Frederick Smith Chemical Co.). This reagent usually contains 68%–75% ferric perchlorate. Keep in desiccator after reagent bottle has been opened.

Cholesterol Reagent. Dissolve 520 mg ferric perchlorate reagent in 600 ml ethyl acetate contained in a 2 liter Erlenmeyer flask. Place the flask in an ice bath and cool the contents to 4°C. Add gradually, in small portions, 400 ml cold conc H_2SO_4. Mix after each portion is added but do not allow the temperature to exceed 45°C (use thermometer). The reagent is stable for at least 1 year when stored in amber bottle at 25°C, or for 2 years when refrigerated.

Stock Cholesterol Standard, 200 mg/100 ml. Dissolve 200 mg cholesterol in approximately 60 ml chloroform and make up to 100 ml with chloroform. Store in refrigerator.

Working Cholesterol Standard, 2 mg/100 ml. Dilute stock standard 1:100 with chloroform. Dispense 5.0 ml volumes into vials. Evaporate to dryness under N_2 in a 60°C waterbath. Store dried standards in refrigerator.

PROCEDURE

1. Transfer 3.0 ml serum to a cellulose nitrate tube (0.5 x 2.5 in.). If less than 3.0 ml is available, any amount between 1.5 and 3.0 ml can be used. In this case the serum volume used should be recorded for substitution in the calculation.

2. Add sufficient 0.15 *M* NaCl to bring volume to 6.5 ml and cap tube.

3. Ultracentrifuge in Beckman model L for 18 hours at 39 000 rpm and at 14°C using type 40.3 rotor (approximately 110 000 *g*).

4. Carefully remove cap and place tube in tube slicer. Slice immediately below the top lipid layer. This must be done while the tube is still cool. Leave tubes in rotor prior to slicing.

5. Transfer upper layer to a 2.5 ml graduated centrifuge tube, graduated in 0.1 ml divisions. Record the volume of the supernate.

6. Add 0.1 ml supernate to 1.9 ml isopropanol and mix using a mechanical mixer.

7. Centrifuge 10 min at 2500 rpm.

8. Transfer 1 ml supernate to a vial marked Unknown.
9. Evaporate to dryness under N_2 in a 60 °C waterbath.
10. Mark an empty vial Blank and take a standard vial containing previously dried-down working standard (see *Reagents*) and mark this vial Standard.
11. Add 100 µl glacial acetic acid to each vial and cap. Shake each vial for 10 sec and let them stand for 15 min with a brief shake after every 5 min.
12. Add 5 ml cholesterol reagent to each vial and shake each vial vigorously for 30 sec.
13. Simultaneously insert all vials in a heating block set at 100°C.
14. Simultaneously remove all vials 1.5 min later and immerse them in tap water (20°C or cooler) for 5 min. Remove, dry the exterior of the vials, and mix their contents by inversion.
15. Read absorbances against the blank at 610 nm in a spectrophotometer such as a Beckman DU. Other photometers can be used with appropriate settings and standardization.

Calculation:

$$\text{mg VLDL cholesterol/100 ml serum} = \frac{A_x}{A_s} \times 200 \times \frac{\text{ml supernate volume (step 5)}}{\text{ml serum volume (step 1)}}$$

$$\text{LDL cholesterol, as \% of total cholesterol} = \frac{\text{mg VLDL cholesterol/100 ml}}{\text{mg total cholesterol/100 ml}} \times 100$$

INTERPRETATION OF VLDL CHOLESTEROL RANGES

The ranges of VLDL cholesterol as percentage of total cholesterol are:

0% − 24%   consistent with type II
25% − 50%   consistent with type III
51% − 100% consistent with type IV

The range for normal individuals is similar to that of type II. The cholesterol levels of a phenotype II, however, are markedly higher than those of "normals" and can therefore be used to distinguish one group from the other.

NOTES

1. VLDL cholesterol. Fredrickson (37) determined the VLDL cholesterol as the difference between total cholesterol and cholesterol in the fraction with density greater than 1.006; he found in his four type III patients values for VLDL cholesterol, expressed as percentage of total cholesterol, which were slightly higher than reported here. In the same publication (37), however, it was reported that in the type III pool, which was comprised of two male and two female type III patients, the VLDL cholesterol was 36% of total cholesterol. This value falls well within the range for type III given above.

2. Stability of samples. See under "Notes" of *Phenotyping of Hyperlipoproteinemias.*

β-LIPOPROTEIN SCREENING

(method of Hartmann, Ref 56)

PRINCIPLE

β-Lipoproteins form a precipitate with heparin in the presence of Ca^{2+}, and the resultant turbidity is a measure of β-lipoprotein concentration.

REAGENTS

Heparin, 0.1%. 40 000 USP units/ml = 330 mg/ml. Dilute 0.303 ml of heparin solution to 100 ml with distilled water.

$CaCl_2$, 0.025 M. Dissolve 3.6755 g $CACl_2 \cdot 2H_2O$ in 1 liter distilled water.

PROCEDURE

1. Pipet 0.4 ml 0.1% heparin into a 125 x 15 mm test tube, using a 1 ml serologic pipet.
2. Add 25 μl serum and mix.
3. Add 5 ml $CaCl_2$. Mix and allow to stand for 5 min.
4. Mix immediately before transferring to 1 cm cuvet. Read absorbance at 655 nm versus water on a spectrophotometer such as the Beckman DU or model B (see *Note 1*).

Calculation:

$$units = A \times 1000$$

NOTES

1. Standardization. A procedure using turbidimetric measurements should be carefully standardized with each spectrophotometer employed. In our laboratory this was accomplished by using a normal serum pool of which the β-lipoprotein concentration was determined by ultracentrifugation. This pool was then used to standardize the Beckman DU and model B. Walton and Scott (124), using high molecular weight dextran sulfate for the estimation of β-lipoproteins, prepared low density lipoprotein standard solutions from lipoprotein freeze dried in sucrose and isolated by the method of Oncley *et al* (93). Their method was then calibrated against a series of standard solutions of purified low density lipoprotein of known concentration. Lopez and associates (81) determined turbidimetrically the cholesterol level in β-lipoproteins using a cholesterol calibration curve, and then calculated the β-lipoprotein concentration by assuming β-lipoprotein to contain 46.9% cholesterol. The standardization of each method, however, should always be checked against a reference method.

2. Stability of samples. See under *Notes* of *Phenotyping of Hyperlipoproteinemias.*

ACCURACY AND PRECISION

The screening method for β-lipoproteins presented is a modification of the method of Burstein and Samaille (15). The high degree of specificity of the latter method was several times proved by ultracentrifugation (5, 37, 71, 106); it is therefore recommended both with and without preparative ultracentrifugation as one of the preferred methods for determination of β-lipoproteins (10).

Comparison of results of the test "phenotyping of hyperlipoproteinemias" with those obtained by the lipoprotein screening procedure have shown good correlation (56). Samples with normal cholesterol, triglyceride, and electrophoretic values as well as those of phenotype IV had normal β-lipoprotein levels, while samples of types II and III showed an increase of β-lipoproteins. The method of Burstein and Samaille compares very well with the accurate ultracentrifugation method (5, 37, 71, 106). The method presented here (Hartmann) is only slightly modified from that of Burstein and Samaille.

The precision (95% limits) of the method is ±5%.

NORMAL VALUES

The 95% normal range calculated in our laboratory from 60 normal individuals using a nonparametric method was 18—54 units and 8—40 units employing the Beckman model B and DU, respectively. A sex difference for β-lipoprotein concentrations employing Hartmann's method was not found.

REFERENCES

1. ALDERSBERG D, BOSSAK ET, SHER IH, SOBOTKA H: *Clin Chem 1*:18, 1955
2. ALAUPOVIC P, SANBAR SS, FURMAN RH, SULLIVAN ML, WALRAVEN SL: *Biochemistry 5*:4044, 1966
3. ALBERS JJ, ALBERS LV, ALADJEM F: *Biochem Med 5*:48, 1971
4. ALBRINK MJ, MAN EB: *Arch Internal Med 103*:4, 1959
5. ANTONIADES HN, TULLIS JL, SARGEANT LH, PENNELL RB, ONCLEY JL: *J Lab Clin Med 51*:630, 1958
6. ARMSTRONG SJ: *Am J Med Technol 30*:47, 1964
7. AUBRY F, LAPIERRE Y, NOËL C, DAVIGNON J: *Ann Internal Med 75*:231, 1971
8. AYRAULT-JARRIER M, LÉVY G, WALD R, POLONOVSKI J: *Bull Soc Chim Biol 45*:349, 1963

9. BARCLAY M, BARCLAY RK, TEREBUS-KEKISH O, SHAH EB, SKIPSKI VP: *Clin Chim Acta 8*:721, 1963

10. BEAUMONT JL, CARLSON LA, COOPER GR, FEJFAR Z, FREDRICKSON DS, STRASSER T: *Bull World Health Org 43*:891, 1970

11. BECKERING RE Jr, ELLEFSON RD: *Am J Clin Pathol 53*:84, 1970

12. BERTRAND F, WATELET M, GENETET F, NABET P, PAYSANT P: *Ann Biol Clin (Paris) 27*:735, 1969

13. BROWN DF: *Am J Med 46*:691, 1969

14. BUCKLEY GC, LITTLE JA, CSIMA A, KOENIG E, YANO R, SULLIVAN K: *Can Med Ass J 102*:943, 1970

15. BURSTEIN M, SAMAILLE J: *Clin Chim Acta 5*:609, 1960

16. CHARMAN RC, LANDOWNE RA: *Anal Biochem 19*:177, 1967

17. CHIN HP, BLANKENHORN DH: *Clin Chim Acta 23*:239, 1969

18. CHIN HP, BLANKENHORN DH: *Clin Chim Acta 20*:305, 1968

19. CIVIN WH, FREDRICKSON DS, POWELL JB, WENDT JE: *Patient Care* Feb 28, 1971, p 28

20. COLFS B, VERHEYDEN J: *Clin Chim Acta 18*:325, 1967

21. CORNELISSEN PJHC, ROELOFS-FIZAAN LEAM: *Clin Chim Acta 29*:344, 1970

22. DANGERFIELD WG, PRATT JJ: *Clin Chim Acta 30*:273, 1970

23. DANGERFIELD WG, SMITH EB: *J Clin Pathol 8*:132, 1955

24. DE LALLA OF, GOFMAN JW: *Methods of Biochemical Analysis*, edited by GLICK D. New York, Interscience, 1954, Vol 1, p 459

25. DE LALLA OF, TANDY RK, LOEB HG: *Clin Chem 13*:85, 1967

26. DiLEO FP: *SIPS Sci Tech 4*:110, 1960

27. DUFF GL, McMILLAN GC: *J Exper Med 89*:611, 1949

28. DYERBERG J, HJORNE N: *Clin Chim Acta 30*:407, 1970

29. ELEVITCH FR, AUSTIN DG: *Lipoprotein Electrophoresis in Thin Agarose Gel.* Palo Alto, Calif., Analytical Chemists, Inc.

30. EWING AM, FREEMAN NK, LINDGREN FT: *Advan Lipid Res 3*:25, 1965

31. FARBER ER, BATSAKIS JG, GIESEN PC, THIESSEN M: *Am J Clin Pathol 51*:523, 1969

32. FISHER WR, GURIN S: *Science 143*:362, 1964

33. FLEISCHMAJER R: *Arch Derm 100*:401, 1969

34. FLETCHER MJ, STYLIOU MH: *Clin Chem 16*:362, 1970

35. FREDRICKSON DS: *Proc Nat Acad Sci 64*:1138, 1969

36. FREDRICKSON DS, LEVY RI: *Lancet 1*:191, 1970

37. FREDRICKSON DS, LEVY RI, LINDGREN FT: *J Clin Invest 47*:2446, 1968

38. FREDRICKSON DS, LEVY RI, LEES RS: *New Engl J Med 276*:34, 1967

39. FREDRICKSON DS, LEVY RI, LEES RS: *New Engl J Med 276*:94, 1967

40. FREDRICKSON DS, LEVY RI, LEES RS: *New Engl J Med 276*:148, 1967

41. FREDRICKSON DS, LEVY RI, LEES RS: *New Engl J Med 276*:215, 1967

42. FREDRICKSON DS, LEVY RI, LEES RS: *New Engl J Med 276*:273, 1967

43. FREDRICKSON DS, LEVY RI, KWITEROVICH PO Jr, JOVER A: *C A Biochem Sections 72*:183, 1970

44. FRINGS CS, FOSTER LB, COHEN PS: *Clin Chem 17*:111, 1971

45. GABL F, SAILER S, SANDHOFER F, WACHTER H: *Arztl Lab 12*:1, 1966

46. GLAZIER FW, TAMPLIN AR, STRISOWER B, de LALLA OF, GOFMAN JW, DAWBERT R, PHILLIPS E: *J Gerontol 9*:395, 1954

47. GLOMSET JA, JANSSEN ET, KENNEDY R, DOBBINS J: *J Lipid Res 7*:639, 1966

48. GLUECK CJ, LEVY RI, GLUECK HI, GRALNICK HR, GRETEN H, FREDRICKSON D: *Am J Med 47*:318, 1969

49. GOFMAN JW, LINDGREN FT, ELLIOTT H: *J Biol Chem 179*:973, 1949
50. GOFMAN JW, STRISOWER B, de LALLA OF, TAMPLIN A, JONES H, LINDGREN FT: *Mod Med 21*:119, 1953
51. GOTTFRIED SP, POPE RH, FRIEDMAN NH, AKERSON IB, DiMAURO S: *Clin Chem 1*:253, 1955
52. GOTTO AM Jr: *Proc Nat Acad Sci 64*:1119, 1969
53. GOTTO AM, BROWN WV, LEVY RI, BIRNBAUMER ME, FREDRICKSON DS: *J Clin Invest 51*:1486, 1972
54. GUSTAFSON A, ALAUPOVIC P, FURMAN RH: *Biochemistry 5*:632, 1966
55. HAMMOND MG, FISHER WR: *J Biol Chem 246*:5454, 1971
56. HARTMANN G: *Arztl Lab 12*:65, 1966
57. HATCH FT, LEES RS: *Advan Lipid Res 6*:1, 1968
58. HATCH FT, MAZRIMAS JA, MOORE JL, LINDGREN FT, JENSEN LC, WILLIS RD, ADAMSON GL: *Clin Biochem 3*:115, 1970
59. HAYES TE, HEWITT JE: *Univ California Lawrence Radiation Lab Report* 3681, February 1957
60. HAZZARD WR, LINDGREN FT, BIERMAN EL: *Biochim Biophys Acta 202*:517, 1970
61. HEISKELL CL, FISK RT, FLORSHEIM WH, TACHI A, GOODMAN JR, CARPENTER CM: *Am J Clin Pathol 35*:222, 1961
62. IAMMARINO RM, HUMPHREY M, ANTOLIK P: *Clin Chem 15*:1218, 1969
63. JENCKS WP, DURRUM EL: *J Clin Invest 34*:1437, 1955
64. JONES H, GOFMAN JW, LINDGREN FT, LYON T, GRAHAM D, STRISOWER B: *Am J Med 11*:358, 1951
65. KAHAN J, SUNDBLAD L: *Scand J Clin Lab Invest 24*:61, 1969
66. KAHLKE W, SCHLIERF G: *Klin Wschr 46*:330, 1968
67. KANNEL WB, KAGAN A, DAWBER TR, REVOTSKIE N: *Geriatrics 17*:675, 1962
68. KANNO T, SAKURADA T, MARUYAMA Y: *C A Biochem Sections 71*:80 (78030v), 1969
69. KESSLER G, LEDERER H: *Technicon Symposia.* New York, Mediad, Inc., 1965, p 341
70. KEYS A: *J Chron Dis 6*:552, 1967
71. KRITCHEVSKY D, TEPPER SA, ALAUPOVIC P, FURMAN RH: *Proc Soc Exp Biol Med 112*:259, 1963
72. LANE RF: *J Med Lab Technol 26*:212, 1969
73. LANGER T, STROBER W, LEVY RI: *J Clin Invest 51*:1528, 1972
74. LEE D, ALAUPOVIC P, FURMAN RH: *Circulation 34*:III—18, 1966
75. LEES RS: *Science 169*:493, 1970
76. LEES RS, FREDRICKSON DS: *J Clin Invest 44*:1968, 1965
77. LEES RS, FREDRICKSON DS: *Circulation 30*:20, 1964
78. LEES RS, HATCH FT: *J Lab Clin Med 61*:518, 1963
79. LEVY RI, FREDRICKSON DS: *J Clin Invest 44*:426, 1965
80. LEWIS LA: *Lipids 4*:60, 1969
81. LOPEZ SA, SRINIVASAN SR, DUGAN FA, RADHAKRISHNAMURTHY B, BERENSON GS: *Clin Chim Acta 31*:123, 1971
82. MARSHALL WE, KUMMEROW FA: *Arch Biochem Biophys 98*:271, 1962
83. McGLASHAN DAK, PILKINGTON TRE: *Clin Chim Acta 22*:646, 1968
84. MILLS GL, TAYLAUR CE: *Clin Chim Acta 22*:251, 1968
85. MILLS GL, WILKINSON PA: *Clin Chim Acta 7*:685, 1962
86. MOINUDDIN M, TAYLOR L: *Lipids 4*:186, 1969
87. NARAYAN KA, KUMMEROW FA: *Clin Chim Acta 13*:532, 1966
88. NARAYAN KA, NARAYAN S, KUMMEROW FA: *Nature 205*:246, 1965

89. NICHOLS AV: *Proc Nat Acad Sci 64*:1128, 1969
90. NIKKARI T, VIIKARI J: *Clin Chim Acta 24*:473, 1969
91. NOBLE RP: *J Lipid Res 9*:693, 1968
92. NOBLE RP, HATCH FE, MAZRIMAS JA, LINDGREN FT, JENSEN LC, ADAMSON GL: *Lipids 4*:55, 1969
93. ONCLEY JL, WALTON KW, CORNWELL DG: *J Am Chem Soc 79*:4666, 1957
94. ORVIS HH, BURGER D: *Med Ann District Columbia 32*:44, 1963
95. PAPADOPOULOS NM, KINTZIOS JA: *Clin Chem 17*:427, 1971
96. PAPADOPOULOS NM, KINTZIOS JA: *Anal Biochem 30*:421, 1969
97. PAPADOPOULOS NM, CHARLES MA: *Proc Soc Exp Biol Med 134*:797, 1970
98. PAUL O, LEPPER MH, PHELAN WH, DUPERTUIS GW, MacMILLAN A, McKEAN H, PARK H: *Circulation 28*:20, 1963
99. POSTMA T, STROES JP: *Clin Chim Acta 22*:569, 1968
100. QUARFORDT SH, LEVY RI, FREDRICKSON DS: *J Clin Invest 50*:754, 1971
101. QUARFORDT SH, LEVY RI, FRANK A, FREDRICKSON DS: *J Clin Invest 47*:81a, 1968
102. RIESSEL PK, HAGOPIAN LM, HATCH FT: *J Lipid Res 7*:551, 1966
103. ROSATI LA, NAPIER ER Jr: *Clin Biochem 3*:171, 1970
104. ROSE HG: *J Lab Clin Med 76*:92, 1970
105. ROSS DL, BROWN K: *Am J Med Technol 35*:540, 1969
106. SAKAGAMI T, ZILVERSMIT DB: *J Lipid Res 3*:111, 1962
107. SALPETER MM, ZILVERSMIT DB: *J Lipid Res 9*:187, 1968
108. SANBAR SS: *Hyperlipidemia and Hyperlipoproteinemia*. Boston, Little, Brown, 1969
109. SCHUMAKER VN, ADAMS GH: *Ann Rev Biochem 38*:113, 1969
110. SEEGERS W, HIRSCHHORN K, BURNETT L, ROBSON E, HARRIS H: *Science 149*:303, 1965
111. SEIDEL D, GRETEN H: *Clin Chim Acta 30*:31, 1970
112. SIRTORI C, HASSANEIN KM, HASSANEIN R, BOULOS BM: *Clin Chim Acta 31*:305, 1971
113. SKIPSKI VP, BARCLAY M, BARCLAY RK, FETZER VA, GOOD JJ, ARCHIBALD FM: *Biochem J 104*:340, 1967
114. SLACK J, MILLS GL: *Clin Chim Acta 29*:15, 1970
115. SMITH EB: *Lancet 2*:530, 1962
116. SMITH EB, SLATER RS: *Lancet 1*:463, 1972
117. STRAUSS R, WURM M: *Am J Clin Pathol 29*:581, 1958
118. SWAHN B: *Scand J Clin Lab Invest 5*:5, 1953
119. TAMPLIN AR, STRISOWER B, DE LALLA OF, GOFMAN J, GLAZIER F: *J Gerontol 9*:403, 1954
120. Technical Group of the Committee on Lipoproteins and Atherosclerosis. *Circulation 14*:725, 1954
121. VAN BUCHEM FSP, POL G, DE GIER J, BÖTTCHER CJF, PRIES C: *Am J Med 40*:794, 1966
122. VERGANI C, RHO GL, VANNOTTI M, VECCHI G: *Clin Chem 17*:551, 1971
123. WACHTER H, ZELGER J: *Arztl Lab 11*:263, 1965
124. WALTON KW, SCOTT PJ: *J Clin Path 17*:627, 1964
125. WARBURTON FG, NIXON JV: *Clin Chim Acta 18*:75, 1967
126. WATTS HF: *Ann NY Acad Sci 149*:725, 1968
127. WHEREAT AF: *Adv Lipid Res 9*:119, 1971
128. WINKELMAN JW, IBBOTT FA: *Clin Chim Acta 26*:25, 1969
129. WINKELMAN JW, WYBENGA DR, IBBOTT FA: *Clin Chim Acta 16*:507, 1970
130. WINKELMAN JW, WYBENGA DR, IBBOTT FA: *Clin Chim Acta 27*:181, 1970

131. WINKELMAN JW, IBBOTT FA, SOBEL C, WYBENGA DR: *Clin Chim Acta* 26:33, 1969
132. WOLFMAN L, SACHS BA: *Clin Chem* 16:620, 1970
133. WOLLENWEBER J, KAHLKE W: *Clin Chim Acta* 29:411, 1970
134. WOLLENWEBER J, KAHLKE W, SCHLIERF G: *Verh Deut Ges Inn Med* 74:254, 1968
135. WOOLF N, PILKINGTON TRE: *J Pathol Bacteriol* 90:459, 1965
136. WURM M, KISITCHEK R, STRAUSS R: *Circulation* 21:526, 1960
137. WYBENGA DR, IBBOTT FA, WINKELMAN JW: *Clin Chim Acta* 40:121, 1972
138. WYBENGA DR, PILEGGI VJ, DIRSTINE PH, DI GIORGIO J: *Clin Chem* 16:980, 1970

Chapter 30

Kidney Function Tests

DONALD C. CANNON, M.D., Ph.D.

Renal function tests in general are influenced by both renal and extrarenal factors. Methods presented in this chapter are those that are reasonably specific for renal physiologic function and are commonly performed in the clinical laboratory. Other tests occasionally indicate altered renal physiology although they ordinarily reflect extrarenal aspects of metabolism or abnormal function of some other organ system. Urine pH, for example, is influenced primarily by diet and general metabolism but may reveal abnormal renal function in cases of renal tubular acidosis. It must be emphasized that the myriad metabolic and excretory activities performed by the kidneys may require multiple tests to describe adequately the status of renal function.

RENAL CLEARANCE

The concept of renal clearance was introduced in 1921 by Austin, Stillman, and Van Slyke (3) in relation to their investigations of the excretion rate of urea. By definition renal clearance of a substance is that volume of plasma that contains an amount of the substance equivalent to the amount excreted in the urine per unit time. Although any units of volume and time may be chosen, clearances are usually expressed in ml/min. The mathematical equation is thus:

$$C = UV/P$$

where

C = clearance in ml/min
U = urine concentration of the substance in mg/ml
V = urine flow rate in ml/min
P = plasma concentration of substance in mg/ml

Renal clearance is more closely related to body surface area than to body weight (51, 57) and is classically corrected to 1.73 m^2, which is considered to be the standard body surface area of an adult (51). Body surface area is readily calculated to an accuracy of ±5% from weight and height by the formula of DuBois and DuBois (19):

$$A = W^{0.425} \, H^{0.725} \times 71.84$$

where

$$A = \text{body surface area in cm}^2$$
$$W = \text{weight in kg}$$
$$H = \text{height in cm}$$

This formula is more conveniently expressed logarithmically as

$$\log S = 0.425 \ \log W + 725 \ \log H - 2.144$$

where

$$S = \text{body surface area in m}^2$$

Various nomograms have also been constructed for determining the body surface area (7).

It is to be emphasized that clearance refers to a virtual rather than to a real plasma volume and is the net result of both glomerular filtration and renal tubular excretion or reabsorption. A substance that is freely filtered from the plasma into the glomerular ultrafiltrate and that is neither excreted nor reabsorbed by the renal tubules will have a clearance equal to the glomerular filtration rate. In addition, the ideal exogenous substance for the measurement of glomerular filtration rate will be nontoxic, readily soluble in aqueous medium at body temperature, subject to accurate analysis, and will not be metabolized or otherwise eliminated extrarenally (54). These criteria are satisfied by inulin, which is the classical clearance method for determining the glomerular filtration rate. Inulin clearance, in common with other exogenous clearance methods, has never achieved widespread diagnostic application largely as a consequence of the clinical manipulations required, including carefully controlled constant intravenous infusions and, according to some protocols, bladder catheterizations and washouts. Furthermore, the chemical methods for measurement of inulin are not readily available in clinical laboratories. Analytical difficulties have been alleviated through the use of other substances with clearances approximating the glomerular filtration rate, particularly such radiopharmaceuticals as carboxyl-[14] C-inulin, ^{131}I-allyl inulin, and [57]Co-cyanocobalamin (54).

More commonly used as estimates of glomerular filtration rate are the clearances of endogenous creatinine and urea. The clearances of both substances have significant shortcomings, however, chief among which is their susceptibility to renal tubular modifications of the glomerular filtrate.

The concentrations of creatinine or urea in serum and nonprotein nitrogen in blood, which are discussed in Chapter 17, are commonly used as indirect indices of glomerular filtration. Concentrations of these substances are inversely related to their respective clearances, as is evident from the fundamental clearance equation (67):

$$\text{clearance} = \text{excretion rate/plasma concentration}$$

Thus at a constant excretion rate, a 50% reduction in clearance theoretically results in a doubling of the plasma (serum or blood) concentration. In practice, however, plasma concentrations are not a sensitive indicator of deterioration in renal function. It is commonly stated that increases in plasma creatinine or urea are not reliably evident until clearances have decreased to less than 50% of normal (64, 67). This discrepancy between theory and practice is the result of several factors: (a) A doubling of the plasma concentration may occur while the upper limits of the rather broad normal ranges are minimally exceeded if at all. (b) Plasma concentrations are the net result of both production and renal excretion. Plasma urea, for example, is significantly influenced by such diverse factors as diet, endogenous protein catabolism, state of hydration, gastrointestinal bleeding, liver disease, and tetracycline therapy (64). The practically obsolete nonprotein nitrogen determination is even more variable since it includes not only urea and creatinine but also uric acid, amino acids, and other minor components each of which is metabolically unique and may be treated differently by the renal excretory mechanisms. Plasma creatinine is relatively constant since it is related to muscle mass but is not greatly influenced by other factors (18). Attempts have been made to improve the predictive value of the plasma creatinine concentration through linear transformations of the hyperbolic relationships between plasma creatinine and creatinine clearance. Linear transformations have been obtained by plotting the log of plasma creatinine concentration versus the log of the 24 hour clearance (21) or by plotting the reciprocal of plasma creatinine versus clearance (20, 22).

CREATININE CLEARANCE

Creatinine clearance was first proposed as an estimate of the glomerular filtration rate in 1926 by Rehberg (66), who augmented the plasma levels with exogenous creatinine. It was not until 1938 that the endogenous creatinine test was introduced by Miller and Winkler (56). Although not an ideal method for estimating glomerular filtration, the endogenous creatinine clearance test has many practical advantages. Creatinine is a normal end product of metabolism with a relatively constant plasma concentration and daily urinary excretion that are not greatly influenced by diet, urine flow rate, or exercise (18). In addition the analytic method for creatinine is readily available in clinical chemistry laboratories.

The major disadvantage of the test is that although creatinine is freely filtered by the glomeruli, significant amounts may be added to the urine filtrate by active tubular secretion. Consequently most studies indicate a creatinine clearance equal to or somewhat exceeding the inulin clearance (67, 76, 77). Whether tubular reabsorption of creatinine occurs to any extent under physiologic conditions has not been established with certainty although there is evidence suggesting its occurrence in certain renal circulatory disorders (6). Creatinine is reported to reach a maximum concentration in the urine at flow rates below about 0.35 to 0.5 ml/min so that the endogenous plasma creatinine clearance then assumes a linear dependence upon the urine flow rate (13). Consequently, the patient should be adequately hydrated prior to the test. It is likely that poor correlations between creatinine and inulin clearances that have been reported (6, 16, 55) are in part the result of variations in tubular transport of creatinine under the conditions of study (6).

A factor of particular importance in the ratio of creatinine clearance: inulin clearance is the specificity of the chemical method used for measuring creatinine (37). As discussed in Chap. 17, varying amounts of noncreatinine chromogens are included in the less specific methods for creatinine analysis. For example, one study of the Jaffé alkaline picrate reaction indicated that noncreatinine chromogens constitute approximately 20% and 5% of the plasma and urine creatinine measurements, respectively, when Lloyd's reagent is not used in the reaction (60). Consequently the calculated creatinine clearance, C_{creat}, is less when nonspecific chemical methods are used (18, 53) as would be evident from the formula

$$C_{creat} = \frac{U_{creat}V}{P_{creat}}$$

where U_{creat} and P_{creat} are the urine and plasma creatinine concentrations, respectively. The importance of noncreatinine chromogens in the calculation of clearance is inversely related to the plasma creatinine concentration since the increase in plasma creatinine associated with renal impairment is not accompanied by a proportional increase in noncreatinine chromogen (18). The fact that the less specific methods for creatinine measurement give clearance results that approximate the inulin clearance quite closely, at least in the presence of normal renal function, is largely fortuitous (67): Compared with inulin clearance, the inclusion of nonspecific chromogens increases the denominator, P_{creat}, of the clearance formula by about the same proportion as the numerator, $U_{creat}V$, is increased by tubular secretion of creatinine.

Tubular secretion of creatinine increases significantly as plasma levels increase, so that creatinine clearance becomes a less reliable estimate of the glomerular filtration rate (35, 55). This fact does not, however, invalidate the usefulness of creatinine clearance for monitoring the patient with renal impairment (18, 76). Clearance calculations based on true rather than nonspecific creatinine measurements may be somewhat more satisfactory for following the functional progress of renal disease because of the discrepancy in clearance rates between creatinine and the nonspecific chromogens (67). Augmented tubular secretion of creatinine also explains the fact that clearances are some 10%–40% higher in the exogenous test than in the endogenous test (67).

Procedure for Creatinine Clearance. The creatinine clearance test may be performed over any accurately timed period (76), but 1, 4, 12, or 24 hour intervals are most commonly selected. Reproducibility of the 1-hour test is reportedly comparable to that of the 24-hour collection (18, 77). The 24-hour test does have the purported advantage that it allows better representation of the full range of the patient's activities (67), but this would not seem to outweigh the greater difficulty involved in prolonged urine collection (54). In order to assure an adequate urine flow and thereby minimize the error resulting from incomplete bladder emptying, especially when short collection periods are employed, the patient must be adequately hydrated prior to the test. This can be achieved if the patient drinks about 500 ml of water over a 10–15 min period immediately prior to the test. Although diurnal variation in creatinine clearance is somewhat enhanced by a high meat diet (1), dietary restrictions are not necessary for the test (18). Furthermore, although the 1- and 4-hour tests are commonly performed in the morning, neither time of day nor extent of physical activity has any significant effect on the test (18).

At the start of the test, the patient voids as completely as possible and discards the urine. All urine is then collected through the conclusion of the timed period when the patient again voids as completely as possible. It is essential that urine be appropriately preserved, usually by refrigeration (see Chap. 17), particularly when 24-hour collections are involved. A suitable blood sample is obtained at any convenient time immediately before or after the test or while it is in progress. The calculated clearance is usually corrected to standard body surface of 1.73 m^2 by multiplying by 1.73/S, where S is the body surface area in m^2.

Normal Values. As previously indicated the clearance values are dependent upon the specificity of the analytic method for creatinine. Clearances based on less specific methods may be as much as 25% lower than those based on true creatinine values, but this difference decreases significantly as the plasma creatinine increases (53). For 24-hour tests, Doolan *et al* (18) found normal ranges for true creatinine (Lloyd's reagent with alkaline picrate reaction) clearances of 84–162 ml/min/1.73 m^2 for males and 82–146 ml/min/1.73 m^2 for females. In contrast, when Lloyd's reagent was omitted the ranges were 71–135 ml/min/1.73 m^2 for males and 78–116 ml/min/1.73 m^2 for females. These ranges are similar to those reported by other investigators (20, 39, 76). The significant difference in ranges between males and females can be eliminated by correcting for lean body weight (18). Creatinine clearance is diminished in very young infants but increases to reach adult levels, when corrected for body surface area, by the age of 1–3 years (79). Clearance decreases progressively after 60 years of age (77). A marked increase in creatinine clearance which occurs in pregnancy parallels the increase in glomerular filtration rate (73).

In one study clearances for the 1-hour test were some 7%–11% higher than those of the 24-hour test (18) probably as a result of the augmented hydration and diuresis in the 1-hour test. Coefficients of variation of about 15% and 11% have been reported for the 24-hour and 1-hour clearance tests, respectively, repeated on successive days in the same individuals (18).

UREA CLEARANCE

Urea clearance was the first clearance method to be used clinically (3), and it continues to enjoy widespread, largely unwarranted, popularity. Urea is freely filtered by the glomeruli, but varying amounts are passively absorbed with water as the filtrate becomes concentrated in the renal tubules. Consequently urea clearance underestimates the actual glomerular filtration rate. The fraction of filtered urea that is reabsorbed is inversely related to the urine flow rate (12). In their initial studies, however, Austin, Stillman, and Van Slyke (3) concluded that at urine flow rates greater than 2 ml/min, a rate which they termed the "augmentation limit," the clearance of urea is relatively constant. At urine flow rates less than the augmentation limit, the clearance of urea is approximately proportional to the square root of the rate of urine flow.

In 1928, Möller, McIntosh, and Van Slyke (57) introduced the terms *maximum clearance* (C_m) and *standard clearance* (C_s) for instances in which the urine flow is greater than or less than 2 ml/min, respectively. Formulas which they introduced for calculation of clearances are still used:

$$C_m = \frac{U_{urea} V}{B_{urea}}$$

$$C_s = \frac{U_{urea} \sqrt{V}}{B_{urea}}$$

where U_{urea} and B_{urea} are the concentrations of urea in urine and blood, respectively, and V is the urine flow rate in ml/min.

It is important to emphasize that the standard clearance is not strictly speaking a clearance since it does not signify a volume of blood cleared of urea. Rather it represents an empiric correction of a clearance to make the formula fit observed data. Other empiric formulas have been proposed (10, 17).

Classically the urine flow rates in the above formulas are corrected for variation from the assumed average size of an adult male by multiplying the observed V by 1.73/surface area in m². The corrected volume flow rate, V_{corr}, may then be introduced into the appropriate formula depending upon whether it is greater or less than 2 ml/min. One investigation, however, indicated an augmentation limit of 1.7 ml/min/1.73 m² in children above the age of 3 years (51).

The expressions of maximum and standard clearances are frequently expressed as percentages of normal by comparing them with mean normal values of 75 and 54 ml/min/1.73 m², respectively (57). The formulas thus become

$$\% \text{ of mean normal } C_m = \frac{100\, U_{urea} V_{corr}}{75\, B_{urea}} = \frac{1.33\, U_{urea} V_{corr}}{B_{urea}}$$

$$\% \text{ of mean normal } C_s = \frac{100\, U_{urea} \sqrt{(V_{corr})}}{54\, B_{urea}} = \frac{1.85\, U_{urea} \sqrt{(V_{corr})}}{B_{urea}}$$

The use of percentages of normal is obviously only an approximation. The practice has been criticized on the basis of mathematical considerations of the curvilinear relationship between urea clearance and urine flow rate (17). Also it has been demonstrated that the relationship between urea clearance and urine flow rate is such that clearance values far exceeding 75 ml/min and even approaching the glomerular filtration rate can be achieved at very high urine flow rates (10).

Procedure for Urea Clearance. The test is customarily performed in the morning either with the subject fasting or following a light breakfast. Avoidance of high protein foods and diuretic beverages, such as coffee, has been recommended (57), but others (49, 52) claim that meals have no effect on the urea clearance. Posture has no effect on the test (49), but vigorous physical activity should be avoided (57). Clearance may be higher in the afternoon (52), but this has been disputed (49).

The maximum clearance test is generally conceded to be more precise than the standard clearance (67), particularly since any error introduced by incomplete emptying of the bladder increases as the urine volume decreases. To achieve a urine flow greater than 2 ml/min, which is somewhat more than twice the usual flow rate, the patient must be deliberately hydrated prior to the test. Excessive water intake must be carefully avoided, however, in patients with seriously impaired renal function. The patient should drink about 500 ml of water over a 10–15 min period prior to the test. At the beginning of the test the patient is instructed to void as completely as possible, and the urine is discarded. Urine is then collected in two successive periods of 1 hour each, with the patient voiding as completely as possible at the conclusion of each period. Toward the end of the first collection period an appropriate blood sample is obtained for urea nitrogen determination, and the patient is given another 500 ml of water to drink. The urine flow rate for each of the two collection periods is then calculated, corrected to 1.73 m^2, and substituted in the appropriate formula for urea clearance. Calculation of clearances for each of the two collection periods provides a check on such factors as accuracy of timing and completeness of urine collection. It has been stated that correction to 1.73 m^2 may be ignored in the calculation of C_m if the patient is between 65 and 69 inches in height and for C_s if between 62 and 71 inches, since in such cases the corrections will ordinarily alter the calculated clearance by no more than 5% (51).

It is usually preferable to determine clearances using serum or plasma samples rather than whole blood. In the case of the urea clearance, however, most investigational work has been done on whole blood clearances. Fortunately, the urea concentration in serum or plasma is only slightly greater than that of whole blood so that for practical purposes the use of any of the three samples yields approximately the same result. It is to be emphasized that the formulas for calculating urea clearance are intended for use with analytic methods that are reasonably specific for urea. For example, whereas blood contains negligible amounts of ammonia, the preformed ammonia in urine is about 5% of the total present after conversion of urea to ammonia in the urease method. With this method, preformed ammonia may be determined and subtracted from the total after urease treatment as discussed in Chap. 17.

Longer urine collection periods may be employed if they are accurately timed. If prolonged collections are chosen, such as 12 or 24 hours, it is desirable to obtain two blood samples, e.g., early morning and evening, and use the average of the two urea nitrogen determinations for the calculation of clearance (46). Proper preservation of the urine pool, usually by refrigeration or freezing, becomes increasingly more important as the collection period is extended since otherwise urea may be destroyed by the growth of urease-producing bacteria.

Normal Values. The normal ranges reported by Möller *et al* (57) were 59–95 ml/min/1.73 m^2 for C_m and 41–65 ml/min/1.73 m^2 for C_s. These values are about 75%–125% of the average normal clearances of 75 and 54 ml/min/1.73 m^2 for C_m and C_s, respectively. Hayman *et al* (36) reported approximately the same mean values but a broader range for normals, from 51%–150% for C_m and 59%–131% for C_s. The upper limits of normal do not have clinical significance in view of the fact that the urea clearance may be greatly enhanced by increased urine flow in individuals with normal renal function (10). The reproducibility of urea clearances in the same individual is such that C_s varies over a range of about ±20% from the average while C_m varies even more (57).

Urea clearance corrected to standard body surface area is decreased in newborn infants in relation to adult ranges (32, 50) and is depressed even further in premature newborns (32). Urea clearances for babies aged 2–12 months do not differ significantly from adult ranges (72). Urea clearance decreases with increasing age from about 100% of normal at age 40 to 55% at age 89 (49).

CONCENTRATION TESTS

Concentration of fluid in the nephron occurs chiefly in the distal tubules and is dependent upon various factors, particularly the active tubular transport of Na, an adequate secretion of antidiuretic hormone by the posterior pituitary, and ability of the renal tubules to respond to antidiuretic hormone with augmented permeability to water. Being both informative and technically simple, urine concentration tests have become firmly established as the primary tests of renal tubular function. It is to be emphasized, however, that the renal tubules perform so many different transport and other metabolic functions related to the formation of urine that no single test can completely measure renal tubular function (41, 67). For example, reabsorption of glucose as well as the secretion of para-aminohippurate and iodopyracet (Diodrast) takes place in the proximal tubules. Calculation of the tubular transport maximum, T_m, for these and other substances has been the basis for tests of renal function. Such tests are seldom used for diagnosis because they are clinically cumbersome and difficult to control.

Urine concentration is measured by three methods: specific gravity, total solids or indirect estimation of specific gravity by refractometry, and osmolality.

Specific Gravity

Specific gravity of a solution refers to the ratio of its weight to that of an equal volume of water at the same temperature. For a solution such as urine, the specific gravity is a function of the number, density, and weight of the various species of dissolved solute particles (65). Price *et al* (65) have calculated that the percentage of specific gravity accounted for by the major constituents of urine is a value almost identical with the percentage of total solids which these constituents comprise.

The specific gravity of urine is usually determined by a hydrometer, commonly called a "urinometer." There are five major difficulties with urinometers:

(a) There may be difficulty in reading the meniscus. Froth or bubbles on the urine surface should be removed (touch with a piece of filter paper) and the reading taken at the bottom of the meniscus. On some models the scale is compressed into such a small segment of the stem that accurate readings are not possible.

(b) There is a tendency for the urinometer to cling to the side of the cylinder containing the urine, thus leading to an error in position when it comes to rest. This problem can be alleviated somewhat by gently spinning the urinometer when it is placed in the cylinder. The urinometer is particularly apt to cling to the side when the cylinder is not exactly vertical, i.e., is not resting on a level surface. Another cause is lateral displacement of the ballast in the bottom of the urinometer.

(c) Specific gravity varies with temperature. Most urinometers are calibrated at 15.6°C (60°F) although other temperatures may be chosen. Measurements should be corrected by adding 0.001 for each 3°C that the urine temperature is above the reference temperature and subtracting 0.001 for each 3°C that it is below the temperature.

(d) Urinometers may be inaccurately calibrated. It is essential that each urinometer be standardized after purchase. The entire range of values must be checked and for this purpose solutions of sucrose can be used: 1.6, 2.8, 4.1, 5.3, 6.6, 7.8, and 9.0% (w/w) solutions have specific gravities at 20°/15°C of 1.005, 1.010, 1.015, 1.020, 1.025, 1.030, and 1.035, respectively. These values have been calculated from specific gravities given at 20°/20° (81) by multiplying by 0.9991, the ratio of density of water at 20°C to the density at 15°C. The specific gravities listed above are the values which should be obtained on a urinometer, calibrated at 15°C, when the sucrose solutions are read at 20°C.

(e) The quantity of urine sample is frequently not adequate for floating a urinometer. In such cases the specific gravity can often be measured by diluting the urine:

$$\text{actual specific gravity} = 1.000 + \left[\left(\text{specific gravity of diluted sample} - 1.000\right) \times \text{dilution factor}\right]$$

Values so obtained are less accurate than measurements on undiluted urine since the reading error is also multiplied by the dilution factor.

Urine specific gravity should ordinarily be corrected for protein and glucose if either is present. Protein increases the specific gravity by 0.001 for each 0.4 g/100 ml and glucose increases it by 0.001 for each 0.27 g/100 ml (61). When specific gravities in excess of 1.040 are encountered, particularly in the absence of glycosuria, the presence of abnormal substances in the urine, particularly radiographic contrast media, should be suspected.

Specific gravity of urine can also be measured by the falling drop technic whereby a drop of urine is allowed to fall into a gradient column of organic solvents such as bromobenzene and kerosene (9). The point at which the drop of urine comes to rest indicates the specific gravity on a scale that has been calibrated by solutions of known density. A modification of the falling drop method involves measurement of the time required for the drop of urine to traverse a fixed distance in a column of organic solution of uniform density (80). The specific gravity is read from a graph of values for solutions of known density. The various drop methods can be quite precise and have the further advantage that minimal urine volumes are required. They are, however, awkward, time consuming to prepare and standardize, and consequently seldom used.

Refractometry

Although it provides only an indirect estimate of specific gravity, refractometry has several important advantages over direct measurement of specific gravity: (a) The method is more rapid and precise. (b) Only a single drop of urine is required thus eliminating the problem of insufficient sample volume. (c) Refractometers are available with automatic temperature compensation over extended ranges, e.g., 60°–100° F. (d) Refractometers are relatively rugged and infrequently require adjustment or recalibration. Several models of refractometers are commercially available, but the cost is high.

The refractive index of a solution is linearly related to the total solids or weight of solute per unit volume of solution. The physical basis for this relationship has been derived by Glover and Goulden (30). The linear relationship between total solids and refractive index holds true for a complex solution such as urine as well as for simple solutions (69). Slight deviations from linearity occur when relating total solids, and therefore refractive index, to specific gravity (69). For practical purposes these deviations can be ignored within the range of urine specific gravities commonly encountered, i.e., up to 1.035.

Osmolality

In contrast to specific gravity, osmolality is a colligative property of a solution, which means that it is a function only of the number of particles effectively in solution. The concept of osmolality has been summarized as follows (42): In the case of a solution of an ideal nonelectrolyte (non-dissociation), osmolality equals the molal concentration of the substance. In the case of a substance that dissociates completely, behaving like an ideal electrolyte, osmolality = NC, where N is the number of ions into which the substance dissociates and C is the molal concentration of the substance. Deviations from ideal behavior of an electrolyte solution result from incomplete dissociation of the salt, which is a function of the concentration of the salt and its activity. Activity of a substance may in turn depend on pH, temperature, and concentration.

Although osmolality refers to the osmotic pressure in osmols or milliosmols per kilogram of water, it is usually measured through the effect of the colligative properties on the freezing point of the solution. The freezing point is depressed 1.86°C by a solute concentration of 1000 mOsm/kg of water. Various osmometers are commercially available which vary from manual to fully automated operation. Technical advantages of osmolality include the following: (a) Temperature correction is obviously not necessary. (b) Small volumes of specimen (as little as 0.2 ml) are required. (c) Correction for the presence of glucose, protein, or other large molecules in urine is not necessary since their contribution to osmolality is not significant (40).

Although a close correlation exists between osmolality and specific gravity in a solution containing a single solute, the correlation is not good for a complex solution such as urine (27, 40, 67, 70). For example, urine specific gravities of 1.016 corresponded to osmolalities ranging from 550 to 910 mOsm/kg in one study (40). The correlation between specific gravity and osmolality is somewhat better in dilute or concentrated urines, specific gravities less than 1.005 or greater than 1.020, respectively, than in the range of 1.005–1.020 (70).

In addition to greater accuracy and precision, osmolality has other distinct advantages over specific gravity or total solids measurements for clinical testing of renal function. Of paramount importance is the fact that the physiology of the renal concentrating process is best described in terms of solute particles per unit of solvent instead of total solids per unit volume of solution (67). Furthermore, osmolality measurement allows urine to be compared precisely with plasma and other high protein body fluids in evaluating water and electrolyte metabolism. In this regard the concepts of osmolal clearance, C_{osm}, and free water clearance, C_{H_2O}, were introduced by Wesson and Anslow (78). These are represented mathematically by the formulas:

$$C_{osm} = \frac{U_{osm}}{P_{osm}} V$$

and

$$C_{H_2O} = V - C_{osm} = \left(1 - \frac{U_{osm}}{P_{osm}}\right) V$$

where

$$C_{\text{osm}} = \text{serum cleared of solute in ml/min}$$
$$C_{\text{H}_2\text{O}} = \text{water removed from plasma in ml/min}$$
$$V = \text{urine flow rate in ml/min}$$
$$U_{\text{osm}} = \text{urine osmolality}$$
$$P_{\text{osm}} = \text{plasma osmolality.}$$

Negative values of $C_{\text{H}_2\text{O}}$ are conceptually equivalent to the tubular reabsorption rate for water, sometimes represented as $T^c_{\text{H}_2\text{O}}$. The importance of these concepts to renal physiology and clinical diagnosis have been discussed by various investigators (8, 23, 24, 43, 45, 67).

Concentration and Dilution Tests

If the corrected specific gravity of a random urine sample equals or exceeds 1.023 (67) to 1.025 (34), the renal concentrating mechanism may be considered to be intact, and further testing is ordinarily not required. Similarly urine osmolality in normal dehydrated subjects should be in the range of 800–1400 mOsm/liter (67), and values in excess of 850 mOsm/liter indicate normal concentration (40). Renal concentrating ability decreases slightly with advanced age (49). Diet has an important bearing on urine concentration, and maximum concentration may not be achieved if solute intake and excretion are inadequate (24). An adequate protein intake prior to testing is therefore indicated. Maximum urine concentration may also not be achieved in compulsive water drinkers (23). Urine concentration may also be compromised by an inadequate glomerular filtration rate (48). In the presence of severe renal impairment the urine specific gravity tends to be fixed at a low level, commonly quoted as 1.010 (25, 67) although levels as low as 1.004 may be encountered (34).

Various test protocols have been described. Addis and Shevky (2) recommended a 24-hour period of fluid deprivation beginning after breakfast. A regular diet is allowed except that liquid foods are prohibited. The specific gravity of an overnight urine, collected during the final 12 hours of the test, averaged 1.032 (2).

Although the designation "Fishberg concentration test" is often loosely applied to any period of fluid deprivation, the test as originally described involves rigorous preparation (25). The patient has the usual breakfast including fluids. Fluids are then proscribed for the remainder of the test. A dry lunch and supper with ample protein are ingested. Prior to retiring the patient voids and discards this urine as well as any voided during the night. Upon awakening, three urine specimens are collected: immediately upon awakening, 1 hour after awakening but prior to arising, and 1 hour after resumption of normal activity. With normal concentrating ability the specific gravity of at least one specimen should be equal to or greater than 1.022. For reasons which are not clear, maximum concentration frequently occurs in the first specimen. On the other hand, mobilization of edema fluid during sleep may result in a lowered concentration of the first specimen (25).

For most purposes an overnight period of fluid deprivation, such as the 14-hour protocol of Jacobson *et al* (40), is satisfactory: After a light dry supper no fluid is allowed until completion of the test. Urine specimens are collected 13 and 14 hours later, and a sample of venous blood is obtained at 14 hours. Urine osmolality should equal or exceed 850 mOsm/liter and the urine/plasma osmolality ratio should equal or exceed 3.0.

Urine dilution tests are also a measure of renal tubular function but are infrequently performed. Dilution tests are generally less sensitive than concentration tests as indicators of imparied renal function, but the two tests are independently influenced by several renal and extrarenal conditions (67). For example, maximal diluting ability is impaired by emotion or nausea but is not dependent upon antidiuretic hormone.

Schoen (71) recommended the following protocol for testing diluting ability: In the morning the patient drinks 1000–1500 ml of water. Urine samples are collected at 15–60 min intervals for several hours. The osmolality of at least one sample should fall below 80 mOsm/liter and may decrease to 37 mOsm/liter. The specific gravity of at least one sample should be less than 1.003 (67).

PHENOLSULFONEPHTHALEIN TEST

The phenolsulfonephthalein (PSP, phenol red) test for renal function was introduced in 1912 by Rowntree and Geraghty (68), who concluded that the appearance of this dye in urine following parenteral administration is the result of its being "excreted mostly by the tubules but probably also to a slight extent by the glomeruli." It was subsequently determined that PSP excretion is about 94% by tubular secretion and only 6% by glomerular filtration (31). The renal clearance of PSP is 394 ± 45 ml plasma/1.73 m^2/min, which is approximately 3.2 times the clearance of inulin and thus consistent with the predominant tubular secretion (74). Saturation of the renal tubular mechanism for secreting PSP normally occurs at plasma levels of about 1 mg/100 ml, which is roughly five times the plasma level achieved by the usual dose of 6 mg (67). Consequently renal tubular secretion *per se* does not become the limiting factor in PSP excretion until tubular function is reduced by disease to about 20% of normal. As usually performed the PSP excretion test is properly considered to be an index of effective renal plasma flow (67). The renal clearance of PSP is only about one-half the effective renal plasma flow as classically measured by iodopyracet or paraaminohippurate (PAH), however (74). Following intravenous administration PSP is 80% bound to albumin (31), but this is not the major reason for the decreased clearance of PSP relative to iodopyracet (59). One investigation failed to prove a close correlation between PSP and PAH clearances in renal disease (44). In renal disease the glomerular filtration rate tends to deteriorate as the effective renal blood flow decreases, hence the 15-min PSP excretion test has been correlated with clearances of both creatinine (47) and inulin (44). The 15-min PSP excretion test may be more sensitive than the endogenous creatinine clearance test in detecting early renal disease (47), while the creatinine clearance test is a more reliable index in severe renal insufficiency (76).

The discriminative value of the PSP test was enhanced by the fractional method of Chapman and Halstead (11) that involves urine collections at 15, 30, 60, and 120 min after intravenous PSP administration. The 15-min excretion is of the greatest significance, however, since even a damaged kidney may remove a normal amount of PSP from the plasma as a result of continuous renal recirculation of the dye over longer time intervals (11, 67). Rowntree and Geraghty (68), for example, found that experimental removal of one-half the kidney mass failed to alter the 120-min excretion. Chapman and Halstead (11) found that one-third of patients with renal disease had decreased 15-min excretions but normal 120-min excretions. Difficulties are sometimes encountered in attempting to obtain urine samples precisely at 15 min. Pierce *et al* (63) recommended correcting overtime collections by subtracting 1.25% from the excretion value for each minute in excess of 15 min, up to a maximum of 15 min.

In addition to intrinsic renal disease, PSP excretion is altered by various other factors. Excretion is decreased by penicillin, PAH, iodopyracet, probenecid, sulfinpyrazone, and other uricosuric agents as a result of competitive inhibition at the level of the renal tubular transport mechanism (58, 75). Extrarenal factors which result in diminished renal blood flow, such as renal arterial disease, oligemia, and congestive heart failure, will reduce PSP excretion (67). Excretion of PSP is 10%–13% greater in the supine than in the standing position (14). Small amounts of PSP, estimated to be as much as 10%–15% of an injected dose, are normally removed by the liver and excreted in the bile (33). In liver disease, hepatic removal of PSP is impaired and renal excretion is correspondingly augmented (33). A decrease in plasma albumin may result in increased PSP excretion as a result of diminished protein binding (59, 74). Edema and ascites do not significantly alter the effective PSP extracellular compartment (28). PSP excretion is decreased in the third trimester of pregnancy (26).

Modifications of the PSP excretion test have been developed for specific purposes. A diaper test has been described for detecting silent urogenital anomalies in newborns (5). Gault and Fox (29) described a plasma PSP test based on the concentration of PSP in plasma 60 min after an intravenous injection of 1 mg/kg. This approach was recommended for renal function testing when accurate urine collection is not possible, e.g., cases of urine retention, urinary fistulae, or ureteral implantation into the bowel (29). Modifications of the PSP test have also been used to measure residual urine volume (4, 15).

Performance of the Test

Although toxic reactions to PSP have not been reported (67), the test should be conducted by a licensed physician. Care must be taken to avoid overhydration of the patient with significantly reduced renal function. The patient is instructed to drink about 400 ml of water approximately 30 min before injection of the dye and not to void thereafter until the test is begun. Additional water may be taken during the test. 1 ml (6 mg) PSP, commercially available in sterile, sealed ampules, is injected intravenously by the physician. Subcutaneous or intramuscular injection is proscribed because of the vagaries of absorption by these routes.

The exact time of injection is noted, and urine is collected at 15, 30, 60, and 120 min. The test is not valid in the presence of significant urinary retention unless catheterization is performed—a decision that must be made by the physician.

DETERMINATION OF PHENOLSULFONEPHTHALEIN IN URINE

PRINCIPLE

PSP is a weak organic acid that changes color from yellow to red in the pH range 6.8—8.4. The red color at alkaline pH is measured photometrically.

REAGENTS

NaOH, approx 4 N
PSP Standard, 1.8 mg/liter. Transfer 0.30 ml intravenous solution containing 6 mg/ml to a 1 liter volumetric flask. Add 3 ml 4 *N* NaOH, dilute to volume, and mix (See Note 1).

PROCEDURE

1. Transfer each timed urine collection to a 1 liter volumetric flask, add 3 ml 4 *N* NaOH, add water to volume, and mix.
2. Read absorbances of unknown and standard against a water blank at 562 nm or with a filter with nominal wavelength in this region.

Calculation:

$$\% \text{ PSP excreted in sample } = \frac{A_x}{A_s} \times \frac{1.8}{6} \times 100$$

$$= \frac{A_x}{A_s} \times 30$$

NOTES

1. Standardization. The actual concentration of PSP in commercial ampules may differ from the stated concentration by as much as 17% (38). These deviations presumably will be constant within lots. It is imperative, therefore, that standardization of the test be performed at least with each new lot of ampules.

2. Beer's law. Beer's law is obeyed with a Beckman DU Spectrophotometer and with Klett filter Nos. 54 and 56.

3. Color stability. The color is stable.

4. Interference by bile and hemoglobin (62). These substances give positive interference. They can be removed by adding an equal volume of zinc acetate reagent (add 50 g zinc acetate to 100 ml absolute methanol), shake vigorously, let stand 10 min, and filter. Analyze clear filtrate for PSP.

ACCURACY AND PRECISION

The accuracy of the determination of dye concentration, assuming standardization with the same dye solution as that injected, is inversely related to the amount of other pigments present with significant absorption at 562 nm at alkaline pH. Ordinarily, other pigments would be negligible in relation to the PSP present. If there is any question, however, a baseline analysis can be performed on urine obtained prior to injection of dye. Even in the presence of normal liver function, the PSP test should not be performed within 24 hours of a bromosulfophthalein test.

The precision of the analysis is about ±3%.

NORMAL VALUES

Normal ranges for excretion during various time periods after intravenous injection are as follows (11): 0–15 min, 28%–52%; 15–30 min, 13%–25%; 30–60 min, 9%–17%; 60–120 min, 3%–10%.

REFERENCES

1. ADDIS T, BARRETT E, POO LJ, UREEN HJ, LIPPMAN RW: *J Clin Invest 30*:206, 1951
2. ADDIS T, SHEVKY MC: *Arch Internal Med 30*:558, 1922
3. AUSTIN JH, STILLMAN E, VAN SLYKE DD: *J Biol Chem 46*:91, 1921
4. AXELROD DR: *Arch Internal Med 117*:74, 1966
5. BARSOCCHINI LM, SMITH LM: *J Urol 91*:195, 1964
6. BERLYNE GM, NILWARANGKUR S, VARLEY H, HOERNI M: *Lancet 2*:874, 1964
7. BOOTHBY WM, SANDIFORD RB: *Boston Med Surg J 185*:337, 1921
8. BOYARSKY S, SMITH HW: *J Urol 78*:511, 1957
9. BROWN ME: *Am J Clin Pathol 29*:188, 1958
10. BRULLES A, GRAS J, MAGRIÑA N, TORRES N, CARALPS A: *Clin Chim Acta 24*:261, 1969
11. CHAPMAN EM, HALSTED JA: *Am J Med Sci 186*:223, 1933
12. CHASIS H, SMITH HW: *J Clin Invest 17*:347, 1938
13. CHESLEY LC: *J Clin Invest 17*:591, 1938
14. CORDERO N, FRIEDMAN MH: *Arch Internal Med 41*:279, 1928
15. COTRAN RS, KASS EH: *N Engl J Med 259*:337, 1958
16. DODGE WF, TRAVIS LB, DAESCHNER CW: *Am J Diseases Children 113*:683, 1967

17. DOMINGUEZ R, POMERENE E: *J Clin Invest 22*:1, 1943
18. DOOLAN PD, ALPEN EL, THEIL GB: *Am J Med 32*:65, 1962
19. DUBOIS D, DUBOIS EF: *Arch Internal Med 17*:863, 1916
20. EDWARDS KDG, WHYTE HM: *Australasian Ann Med 8*:218, 1959
21. EFFERSØE P: *Acta Med Scand 156*:429, 1957
22. ENGER E, BLEGEN EM: *Scand J Clin Lab Invest 16*:273, 1964
23. EPSTEIN FH, KLEEMAN CR, HENDRIKX A: *J Clin Invest 36*:629, 1957
24. EPSTEIN FH, KLEEMAN CR, PURSEL S, HENDRIKX A: *J Clin Invest 36*:635, 1957
25. FISHBERG AM: *Hypertension and Nephritis,* 5th edition. Philadelphia, Lea and Febiger, 1954, p. 97
26. FREYBERG RH: *J Am Med Assoc 105*:1575, 1935
27. GALAMBOS JT, HERNDON EG Jr, REYNOLDS GH: *N Engl J Med 270*:506, 1964
28. GAULT MH: *Can Med Assoc J 94*:61, 1966
29. GAULT MH, FOX I: *Am J Clin Pathol 52*:345, 1969
30. GLOVER FA, GOULDEN JDS: *Nature (London) 200*:1165, 1963
31. GOLDRING W, CLARKE RW, SMITH HW: *J Clin Invest 15*:221, 1936
32. GORDON HH, HARRISON HE, McNAMARA H: *J Clin Invest 21*:499, 1942
33. HANNER JP, WHIPPLE GH: *Arch Internal Med 48*:598, 1931
34. HARROW BR, SLOANE JA: *Postgrad Med 37*:A-48, 1965
35. HAUGEN HN, BLEGEN EM: *Scand J Clin Lab Invest 5*:67, 1953
36. HAYMAN JM Jr, HALSTED JA, SEYLER LE: *J Clin Invest 12*:861, 1933
37. HEALY JK, GRAEME ER: *Am J Med 44*:348, 1968
38. HENRY RJ, JACOBS SL, BERKMAN S: *Clin Chem 7*:231, 1961
39. HOPPER J Jr: *Bull Univ Calif M Center 2*:315, 1951
40. JACOBSON MH, LEVY SE, KAUFMAN RM, GALLINEK WE, DONNELLY OW: *Arch Internal Med 110*:83, 1962
41. JOSEPHSON B, EK J: *Advan Clin Chem 1*:41, 1958
42. KARHAUSEN LR: *Nature (London) 194*:1234, 1962
43. KLEEMAN CR, EPSTEIN FH, WHITE C: *J Clin Invest 35*:749, 1956
44. KRETCHMAR LH, McDONALD DF: *J Urol 89*:753, 1963
45. LADD M: *J Appl Physiol 4*:602, 1952
46. LANDIS EM, ELSOM KA, BOTT PA, SHIELS E: *J Clin Invest 14*:525, 1935
47. LAPIDES J, BOBBITT JM: *J Am Med Assoc 166*:866, 1958
48. LEVINSKY NG, DAVIDSON DG, BERLINER RW: *J Clin Invest 38*:730, 1959
49. LEWIS WH Jr, ALVING AS: *Am J Physiol 123*:500, 1938
50. McCANCE RA, YOUNG WF: *J Physiol (London) 99*:265, 1941
51. McINTOSH JF, MÖLLER E, VAN SLYKE DD: *J Clin Invest 6*:467, 1928
52. MacKAY EM: *J Clin Invest 6*:505, 1928
53. MANDEL EE, JONES FL, WILLIS MJ, CARGILL WH: *J Lab Clin Med 42*:621, 1953
54. MATERSON BJ: *CRC Critical Reviews in Clinical Laboratory Sciences 2*:1, 1971
55. MILLER BF, LEAF A, MAMBY AR, MILLER Z: *J Clin Invest 31*:309,1952
56. MILLER BF, WINKLER AW: *J Clin Invest 17*:31, 1938
57. MÖLLER E, McINTOSH JF, VAN SLYKE DD: *J Clin Invest 6*:427, 1929
58. NEWCOMBE DS, COHEN AS: *Arch Internal Med 112*:738, 1963
59. OCHWADT BK, PITTS RF: *Am J Physiol 187*:318, 1956
60. OWEN JA, IGGO B, SCANDRETT FJ, STEWART CP: *Biochem J 58*:426, 1954
61. PAGE LB, CULVER PJ: *Syllabus of Laboratory Examinations in Clinical Diagnosis,* revised edition. Cambridge, Mass., Harvard U.P., 1960, p. 297
62. PAGE LB, CULVER PJ: Ref. 61, p. 352

63. PIERCE JM Jr, RUZUMNA R, SEGAR R: *J Am Med Assoc 175*:711, 1961
64. PILLAY VKG: *Med Clin N Am 55*:231, 1971
65. PRICE JW, MILLER M, HAYMAN JM: *J Clin Invest 19*:537, 1940
66. REHBERG PB: *Biochem J 20*:447, 1926
67. RELMAN AS, LEVINSKY NG: *Diseases of the Kidney,* edited by Strauss MB and Welt LG. Boston, Little, Brown, 1963, p. 80
68. ROWNTREE LG, GERAGHTY JT: *Arch Internal Med 9*:284, 1912
69. RUBINI ME, WOLF AV: *J Biol Chem 225*:869, 1957
70. SCHOEN EJ, YOUNG G, WEISSMAN A: *J Lab Clin Med 54*:277, 1959
71. SCHOEN EJ: *J Appl Physiol 10*:267, 1957
72. SCHOENTHAL L, LURIE D, KELLY M: *Am J Diseases Children 45*:41, 1933
73. SIMS EAH, KRANTZ KE: *J Clin Invest 37*:1764, 1958
74. SMITH HW, GOLDRING W, CHASIS H: *J Clin Invest 17*:263, 1938
75. TAGGART JV: *Am J Med 24*:774, 1958
76 TOBIAS GJ, McLAUGHLIN RF Jr, HOPPER J: *N Engl J Med 266*:317, 1962
77. VAN PILSUM JF, SELJESKOG EL: *Proc Soc Exp Biol Med 97*:270, 1958
78. WESSON LG Jr, ANSLOW WP Jr: *Am J Physiol 170*:255, 1952
79. WINBERG J: *Acta Paediat 48*:443, 1959
80. WINSTEAD M, SISSON ND, MARKEY RL: *Am J Clin Pathol 39*:235, 1963
81. *Official Methods of Analysis,* Assoc. of Official Agricultural Chemists, P.O. Box 540, Benjamin Franklin Station, Washington, D.C., 1960

Chapter 31

Gastric Analysis

DONALD C. CANNON, M.D., Ph.D.

Gastric secretion is a complex mixture containing as major components hydrochloric acid, pepsin, mucus, and interstitial fluid. Minor components include the digestive enzymes, rennin and gastric lipase (both of which, paradoxically, are inactivated at the low pH of stimulated gastric fluid), albumin and γ-globulin from plasma, various nondigestive enzymes (lactate dehydrogenase, isocitric dehydrogenase, ribonuclease, and numerous others) that probably only reflect the active metabolism of the stomach, and miscellaneous substances such as intrinsic factor. By complexing with vitamin B-12 and thereby facilitating its absorption in the ileum, intrinsic factor is the most essential ingredient of gastric secretion from the standpoint of general physiology, although it is a very minor component in terms of volume. In approximately 80% of individuals, those whose genetic constitution includes the dominant secretor gene, the water soluble ABH blood group substances will be present in gastric secretion as well as in other body fluids.

The composition of gastric fluid varies considerably depending upon the particular physiologic state of the individual. Hollander (19) has proposed a two-component hypothesis in an attempt to explain the chemical composition of gastric fluid. According to this hypothesis the electrolyte composition is a result of the dilution and partial neutralization of parietal cell secretion, principally HCl at a concentration of about 160 meq/liter, by a nonparietal alkaline component derived from multiple sources. Quantitative considerations indicate that the nonparietal component is derived principally from interstitial fluid with minimal contributions from mucus and peptic secretions (30). The factors that stimulate gastric secretion are similarly complex and include neurogenic stimuli transmitted by the vagus nerves; hormonal stimuli, principally gastrin secreted into the blood stream by cells of the gastric antrum; and chemical stimuli, including the breakdown products of ingested protein, which can stimulate parietal and peptic secretions by direct contact with the gastric mucosa and indirectly via the gastrin mechanism.

Various tests have been proposed for analyzing gastric secretion, but the measurement of acid has been generally accepted by virtue of its technical simplicity and superior correlation with gastroduodenal disease. Measurement of the pepsin and rennin activities in gastric secretion, formerly common tests, do not add significant additional information and are not indicated for routine use. In recent years the use of gastric analysis has diminished as more precise and informative diagnostic procedures have become generally available, especially gastroscopy, improved roentgenographic technics, and exfoliative cytology. It is likely that the use of gastric analysis will be further curtailed as newer laboratory procedures, such as the radioimmunoassay for gastrin and various methods for measuring intrinsic factor, become better established in clinical diagnosis.

It is important to emphasize that the results of gastric analysis, perhaps more so than any other clinical laboratory test, must be interpreted in proper clinical perspective since considerable overlap exists between normal individuals and those with most types of gastrointestinal disease. Such an example is gastric carcinoma, which is associated with a mean acid output that does not differ greatly from that of normal individuals (23). Formerly, screening tests relied upon anacidity for detecting gastric carcinoma although it is now known that anacidity occurs in a minority of cases and then usually only in advanced stages of the disease. Anacidity is helpful in the diagnosis of pernicious anemia but is obviously not diagnostic since in addition to gastric carcinoma it may be associated with hypochromic anemia, rheumatoid arthritis, steatorrhea, aplastic anemia, and may be found in healthy relatives of pernicious anemia patients (7). At the other extreme of acid secretion, patients with duodenal ulcer tend to have increased acid output, but only in about 40% of cases is the upper limit of the normal range exceeded (32). Hypersecretion is not pathognomonic of duodenal ulcer, however, and frequently occurs with postoperative stomal ulcer and invariably with the Zollinger-Ellison syndrome consisting of severe peptic ulceration associated with gastrin-secreting adenomas of the pancreas. Gastric analysis is thus indicated in the diagnosis of diseases associated with the extremes of gastric secretion, either anacidity or marked hypersecretion. In addition the test is sometimes utilized to determine the completeness of vagotomy (20), to determine the efficacy of other surgical, medical, or roentgen therapy, or as a guide in selecting the particular surgical procedure to be performed.

GASTRIC ACID

MEASUREMENT

For many years, measurements of gastric secretion emphasized the importance of "free," "combined," and "total" acid. This was based on the belief that a large amount of hydrochloric acid is combined with various proteins and peptones as organic salts. Accordingly "free" acid could exist in solution only when the buffering capacity was exceeded, which supposedly occurred when the pH fell below 3.0–3.5. Consequently "free" acid was measured by titration with sodium hydroxide to an endpoint with p-dimethyl-aminoazobenzene (Töpfer's reagent), which has a color change in the pH range of 2.8–3.5. "Total" acid was determined by titration to a phenolphthalein endpoint (pH 8.2–10.0). "Combined" acid was calculated as the difference between "total" and "free" acid. Results were commonly expressed as "clinical units" or "degrees of acidity," which equaled the number of milliliters of 0.1 N NaOH required to titrate 100 ml of gastric secretion. The studies of Michaelis (36) purported to render a scientific basis for this concept of gastric acid by demonstrating that below pH 2.8 the titration curve of gastric secretion was identical to that of pure hydrochloric acid while at higher pH the titration curve resembled that of a weak acid and its salt. However, such studies are invalidated by the fact that allowance was not made for the buffering action of various test meals used as gastric stimulants prior to specimen collection (5).

Studies of gastric secretion, collected without adulteration by ingested food, indicate a very close resemblance of the titration curve to that of pure hydrochloric acid (5), although slight deviations, indicating minor buffering effects are evident at pH values above about 4.0 (37). Consequently it is desirable to replace the older terminology related to gastric acid with titrable acidity measured in milliequivalents per liter (1). Titratable acidity is readily determined by titrating an appropriate aliquot of gastric secretion with 0.1 N NaOH to an endpoint of pH 7.0 (or 7.4, the physiologically "neutral" pH preferred by some). In addition, for each specimen the volume and initial pH should be reported (1). The acid output can be calculated as:

$$\text{acid output} = \frac{\text{specimen volume (ml)} \times \text{titratable acidity}}{1000}$$

The measurement of pH is obviously much more discriminative than titratable acidity in detecting small amounts of acid secretion. For example, a pH of 5.0, which effectively excludes anacidity, represents a titratable acidity of about 0.01 meq/liter, which is not readily measurable in the titration methods usually employed.

Moore and Scarlata (37) have recently recommended the estimation of hydrogen ion concentration by measuring hydrogen ion activity with a glass electrode and dividing by the hydrogen ion activity coefficient. At any given pH, the hydrogen ion activity coefficient is a function of total ionic strength of the solution, which in the case of gastric secretion is determined largely by the concentration of sodium and potassium cations. By measuring these two cations the ionic strength of the solution, and therefore the activity coefficient of hydrogen ion, can be approximated for a given pH from a table of experimentally determined values. This more rigorous physicochemical approach does not appear to be justified for routine clinical use.

Terms used to describe gastric acidity must be used with caution since the contributions of various investigators have often resulted in multiple definitions for the same term. *Anacidity* has been variously defined as the absence of "free" acid, evidenced by a pH greater than 3.5, or failure of the pH to fall below either 7.0 or 6.0 with maximal histamine stimulation. The most useful definition of anacidity for clinical diagnosis, although not strictly correct from a physicochemical standpoint, would be failure of the pH to fall below 6.0 with maximal histamine stimulation. With this definition, all adults with pernicious anemia will have anacidity, including the few whose gastric pH will fall between 6.0 and 7.0 following histamine stimulation (7). *Achlorhydria* is considered to be synonymous with anacidity by some investigators (9). Another definition of achlorhydria which has gained some acceptance is a pH always greater than 3.5 with a decrease of less than 1 pH unit after maximal histamine stimulation (7). Achlorhydria is distinguished from *hypochlorhydria*, which is defined as a gastric pH always greater than 3.5 but which falls more than 1 pH unit after maximal histamine stimulation (7). It would seem desirable to avoid completely these definitions of achlorhydria and hypochlorhydria since they have neither physiological validity nor any particular clinical usefulness.

GASTRIC STIMULANTS

Various substances have been recommended for stimulating gastric secretion. Of historical importance only are various test meals such as that described by Ewald (13) consisting of two dry rolls without butter and a cup of tea without cream or sugar. Food ingestion is a submaximal secretory stimulus, however, and its use frequently results in incorrect diagnoses of anacidity. The same is true for ethanol and caffeine, which also were commonly used in the past.

Histamine has been used as a gastric stimulant for about 50 years. Initially only small doses were used, such as 0.01 mg histamine acid phosphate in the "standard histamine test." In 1953 Kay (22) introduced the augmented histamine test with a dose of 0.04 mg/kg, which is administered 30 min after the injection of a suitable antihistamine. This dosage, which has now been generally accepted, results in maximal gastric stimulation of somewhat limited duration. The antihistamine prevents many of the extragastric effects of histamine, but untoward side effects such as drowsiness, headache, tachycardia, erythema, and hypertension occur rather frequently (7). Serious side effects are rare but syncope and profound shock have been reported (4). Continuous intravenous infusion of histamine has also been described and has the advantage that a steady state of maximal acid output can be maintained (27).

Various terms have been introduced that relate to the augmented histamine test. The *maximal histamine response* was defined by Kay (22) as the acid output in the period 15–45 min after histamine injection. The *maximal acid output* is defined as the acid output during the 60 min subsequent to histamine (8). The *peak acid output*, which is the highest output of acid in any two consecutive 15-min posthistamine samples, is reportedly the most discriminating and reproducible measurement (2).

Betazole (Histalog), a histamine analog, was introduced in 1951 as a potent stimulant to gastric secretion without the severe side effects of histamine (42). The betazole stimulus is more prolonged than histamine but the achievement of maximal acid output is somewhat delayed (53). A maximal betazole test has been described, using 1.7 mg/kg, which is analagous to the maximal or augmented histamine test (53). Serious side effects resulting from betazole are rare but syncope and shock have been known to occur (4).

Insulin induced hypoglycemia is a potent stimulus to gastric secretion, but its only unique advantage is in testing for completeness of vagotomy

(20). Results are often difficult to interpret and prolonged specimen collection is indicated (50). Excessive hypoglycemia is a significant risk, and close medical supervision is required throughout the test.

Purified gastrin of swine origin provides a physiologic stimulus to acid secretion which is 30 times as potent as histamine on a weight basis (31). After subcutaneous administration the response is somewhat slower than with histamine, but the peak response is approximately 19% greater and more prolonged. The hormone is not, however, approved for routine clinical testing. A variety of active analogs have been produced. Pentagastrin or pentapeptide, a synthetic peptide related to the active nucleus of gastrin, has been shown to provide a safe, reproducible stimulus to gastric secretion that is more potent than histamine and virtually free of side effects (25).

SPECIMEN COLLECTION

Gastric analysis should be performed after an overnight fast. Better recovery of gastric contents is achieved when intubation, by either the oral or nasal route, is performed with the patient in the sitting rather than the supine position (21). It is important that the tube be directed to the most dependent portion of the stomach as confirmed by fluoroscopic examination; consequently intubation should be performed by a physician. Intubation solely by "clinical judgment" results in incorrect placement of the tube in about 50% of cases (7). Although mechanical means, such as the Stedman pump, have been recommended for gastric aspiration, superior results are achieved by applying continuous tension on a common hypodermic syringe (21). The patient should be instructed to expectorate saliva and not to drink water while the test is in progress. Smoking is proscribed immediately prior to or during the test in order to avoid spurious stimulation of gastric secretion (38). Although usually a benign (albeit uncomfortable) procedure, gastric intubation is ordinarily contraindicated for patients with aortic aneurysm, congestive heart failure, recent severe gastric hemorrhage, varices, stenosis, diverticula, or malignant neoplasm of the esophagus, and during the last half of pregnancy.

Gastric contents are collected either under basal physiologic conditions or following pharmacologic stimulation. For basal collection it is imperative that the patient be fasting, quiet, and free of influential medications. Formerly the basal test was performed as a 12-hour overnight collection, but this had the disadvantage that hospitalization with continual supervision was required, and the patient was subjected to unnecessarily prolonged discomfort. The 1-hour basal test performed after an overnight fast has been found to compare favorably with the 12-hour test in terms of representative sampling and is the preferred method (28). The 1-hour test is usually collected in four 15-min samples after discarding one 15-min sample to allow for adjustment to the procedure. In spite of careful performance the basal test has poor reproducibility from day to day, and for that matter, from hour to hour in the same patient in terms of both secretory volume and total acid output (51).

The stimulated tests involve collection for a designated time period after administration of the particular pharmacologic agent and are characteristically performed immediately following the basal test. The collection period is usually 60 min following histamine and 90–120 min after betazole. The specimen is collected in 15-min aliquots in either case.

MEASUREMENT OF GASTRIC ACID SECRETION

(method of Baron, Ref. 1, modified)

PRINCIPLE

Gastric secretion is collected either under basal physiologic conditions or following stimulation with a pharmacologic agent such as histamine or betazole. The titratable acidity of each specimen is determined by titration to pH 7.0 with sodium hydroxide, and the acid output is calculated.

REAGENTS

NaOH, 0.1 N

PROCEDURE

1. Measure the volume of each specimen and determine the pH with a suitable pH meter.
2. Titrate a 5 ml aliquot of each specimen to pH 7.0 with 0.1 N NaOH (see Note 1).

Calculation:

$$\text{titratable acidity (meq/liter)} = \text{ml NaOH} \times 20$$

$$\text{acid output (meq } H^+\text{)} = \frac{\text{titratable acidity} \times \text{specimen vol in ml}}{1000}$$

NOTES

1. If a suitable pH meter is not available for monitoring the titration, titratable acidity may be determined using phenol red indicator, 0.1% aqueous (dissolve 0.1 g acid form in 5.7 ml 0.05 N NaOH and dilute to 100 ml with water). Add 1–2 drops of indicator to the aliquot of gastric secretion. Titration should be continued until the color of the indicator matches that of a suitable control buffered at pH 7.0 (e.g., phosphate buffer, see Appendix).

2. If the specimen of gastric secretion is of inadequate volume for titration, water may be added and appropriate correction made in the calculation for titratable acidity.

ACCURACY AND PRECISION

Technical precision of the titration is well within ±5%. This is adequate for clinical purposes particularly since this is considerably less than the physiologic variation inherent in basal secretion and the error in specimen collection in either basal or stimulated tests.

NORMAL VALUES

In the basal secretion test, mean acid output for normal males is approximately 3 meq/hour but a broad range exists. Anacidity is common, and, at the other extreme, values up to at least 10 meq/hour may be found (32). Values greater than 10 meq/hour are found in only about 4% of normal males and in 13%–19% of duodenal ulcer patients (32). In 25 patients with the Zollinger-Ellison syndrome, values ranged from 11 to greater than 80 meq/hour in the basal test (12). In the augmented histamine test, mean acid output for normal males is approximately 23 meq/hour with a range from less than 1 to about 40 meq/hour (32). In one study, about 40% of males with duodenal ulcer excreted more than 40 meq/hour in the augmented histamine test (32). Values for both basal and augmented histamine tests tend to be somewhat less for females (33) and with aging (2).

As with titratable acidity, measurement of pH is not of great discriminative value in the basal test. Following histamine stimulation, pH will usually fall to 2 or below in normal individuals. Failure of the pH to fall below about 3.5–4.0 is indicative of diminished acid secretion (7).

TUBELESS GASTRIC ANALYSIS

The investment of time required to perform proper gastric intubation as well as the attendant discomfort to the patient stimulated efforts to find a simpler method for quantifying gastric acid. In 1950, Segal *et al* (45) introduced the use of carboxylic cationic resins for this purpose. Initially either Amberlite IRC or XE-96 was combined with quinine as the indicator cation. Quinine is released by hydrogen ion in the gastric secretion at or below the pH range 3.01–3.20. The liberated quinine is subsequently absorbed in the small intestine and excreted in the urine in which it may be measured fluorometrically. Various fluorescent medications such as quinidine, atabrine, riboflavin, and nicotinic acid derivatives interfere with the measurement (44). In order to circumvent the difficulties of fluorometry, various dye cations were substituted for quinine, the most advantageous of which is azure A (3-amino-7-dimethylaminophenazathionium chloride) commercially available in combination with resin as Diagnex Blue (46). The quantity of azure A excreted may be estimated by visual comparison with color standards as supplied in the commercially available Diagnex kit or with dye standards (6) or the dye may be measured photometrically (24). Preliminary complexing with picric acid followed by chloroform extraction has been recommended (29) but offers no advantage. As a result of reducing substances present in the urine, particularly ascorbic acid, some azure A may be excreted in the colorless reduced form. This may be reoxidized by heating in the presence of copper sulfate as a catalyst (41).

The clinical reliability of the tubeless gastric analysis with azure A is questionable. Out of all tests performed perhaps 1%–3% false positive results and 10%–15% false negative results can be anticipated. Correia *et al* (11), for example, reviewed 1462 cases of tubeless gastric analysis with azure A reported in the literature and found 28 false positives (1.9%) and 144 false negatives (9.8%). In these cases the tubeless tests were performed with betazole or caffeine stimulation and compared with intubated gastric analysis using caffeine, standard histamine, or maximal histamine stimulation as a reference procedure. Consequently the percentage of false negatives must represent a minimum number since many of the reference tests utilized submaximal stimuli. In a selected group of patients Marks and Shay (34) compared the azure A tubeless test using 500 mg caffeine sodium benzoate with intubation and augmented histamine as a reference test and found 1 incorrect result out of 55 positives (i.e., those testing as acid secretors) and 9 incorrect results out of 22 negatives.

False positive results may be caused by the following: (a) Presence of cations in the gastric contents that have strong affinity for the resin and are capable of displacing azure A, e.g., aluminum, magnesium, barium, calcium, iron, or kaolin (18). Such medication should be discontinued at least 48 hours prior to testing. Sodium and potassium, which have relatively little affinity for the resin, may also displace significant amounts of the dye as a consequence of their relatively high concentration in gastrointestinal secretions (18). (b) Partial gastrectomy, which facilitates rapid passage of the resin into the small intestine where it may react with cations in the succus entericus (39). (c) Gastric hypermotility with rapid emptying (34). (d) Diverticulosis of the intestine which presumably causes alteration of the bacterial flora (15).

False negative results may be caused by: (a) Insufficient stimulation of gastric secretion by caffeine sodium benzoate. This can be circumvented by substituting betazole (16, 48) or histamine (39, 40, 44) as a stimulant. Approximately 50% of persons who test anacid with the caffeine stimulus will produce acid if betazole is used (48). Extending the urine collection from 2–4 hours in combination with an appropriately increased normal value has also been recommended as a means of decreasing false negative results (10). (b) Pyloric obstruction (44) and malabsorption syndromes (46) that interfere with intestinal absorption of the liberated azure A cations. (c) Renal disease or urinary tract obstruction (40, 46). Measurement of azure A in serum has been recommended as a means of circumventing these difficulties (47). (d) Advanced liver disease may result in delayed excretion (14, 44), perhaps as a result of expanded extracellular fluid volume. (e) Subtotal gastrectomy that may result in such rapid gastric emptying that hydrogen ions in the gastric secretion are unable to displace the azure cations prior to neutralization by intestinal secretion (49).

Various other approaches to tubeless gastric analysis have been introduced: (a) Determination of isotopically labeled calcium (^{45}Ca) in blood after oral administration of labeled calcium carbonate (35). (b) Determination of urinary excretion of quinine following oral administration of quinine carbonate (17). This approach has the purported advantage that intestinal secretions and food and drug cations, exclusive of fluorescent materials, do not interfere. (c) Determination of 2,6-diamino-3-phenylazopyridine in urine following oral administration of this dye in combination with protein (3). This reagent is commercially available in Europe as Gastrotest. (d) Intragastric measurement of pH with various electrodes. This technic has been used for several decades as a research method, but special glass electrodes have now been developed which are feasible for routine clinical diagnosis (26). Recent interest has focused on the Heidelberg pH capsule (52).

TUBELESS GASTRIC ANALYSIS WITH AZURE A

(method of Rosenthal and Buscaglia, Ref. 41, modified)

PRINCIPLE

Following an overnight fast, gastric secretion is stimulated with caffeine sodium benzoate, betazole, or histamine. Approximately 2 g of azure A resin are then ingested with ample water. (In the commercially available Diagnex Blue test, stimulation is achieved by oral administration of 500 mg caffeine sodium benzoate followed in 1 hour by ingestion of azure A resin.) If the gastric secretion develops a pH of 3.0 or lower, some azure A cations will be liberated from the resin by hydrogen ions and will subsequently be absorbed in the small intestine. Maximum liberation of azure A occurs at pH 1.5 (46). Urine is collected for 2 hours subsequent to resin ingestion during which time about 20% of the dye cations freed in the stomach will be excreted (43). Some of the azure A may be excreted in the colorless reduced form, thus requiring reoxidation by boiling in the presence of copper ion to achieve maximum color development. The color is measured photometrically at 630 nm against a urine blank prepared by reduction of the azure A content to the colorless form with ascorbic acid.

REAGENTS

Acid Copper Sulfate. Dissolve 0.2 g $CuSO_4 \cdot 5H_2O$ in 50 ml water and dilute to 100 ml with concentrated HCl.

Stock Azure A Standard. Dissolve 64 mg azure A (Azure I, Methylene Azure A, with dye content of approximately 77%, obtainable from Matheson, Coleman and Bell) in about 50 ml water. Add 0.5 ml concentrated HCl, and dilute to 100 ml with water. 1 ml = 0.5 mg.

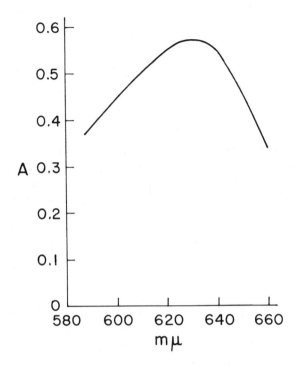

FIG. 31-1. Absorption curve of azure A. Standard containing 0.0025 mg/ml 0.006 *N* HCl vs water blank. Beckman DU Spectrophotometer.

Dilute Azure A Standards. Dilute stock standard 1:10 and 1:20 with 0.06 *N* HC1 (0.5 ml concentrated HC1 per 100 ml). 1 ml = 0.05 and 0.025 mg, respectively.
Ascorbic Acid, U. S. P.

PROCEDURE

1. Measure volume of 2-hour specimen.
2. To 10 ml of urine in a graduated centrifuge tube add 0.1 ml acid copper sulfate reagent.
3. Place in boiling water for 10 min. Cool at room temperature for 2 hours. Add water to restore volume to 10 ml.
4. Centrifuge. Set up the following in test tubes:
   Test. 1.0 ml urine supernate +9.0 ml water.
   Blank. 1.0 ml urine supernate +9.0 ml water + approximately 300 mg ascorbic acid.
   Standards. 1.0 ml dilute azure A Standard +9.0 ml water.
5. Read absorbance of test (A_x) vs blank and standards (A_s) vs water at 630 nm or with filter with nominal wavelength in this region (see *Note 1*).

Calculation (see Note 1):

$$\frac{\text{mg azure A excreted}}{\text{in 2 hours}} = \frac{A_x}{A_s} \times \frac{\text{mg dye in}}{\text{standard}} \times \frac{\text{ml urine excreted}}{\text{in 2 hours}}$$

NOTES

1. Beer's law. Slight deviation from linearity (about 50% was observed in our laboratory with a Beckman DU Spectrophotometer. Considerable deviation occurred, however, with a Klett No. 62 filter. The analyst should run a calibration curve when first setting up the test. If the deviation is small, as it was found to be in our laboratory with a Beckman DU, then it is satisfactory to run two standards and use that which is closer to the unknown. In the case of marked deviation, one must resort to a calibration curve.

2. Stability of urine. Considerable color is lost in 1–2 weeks at refrigerator temperature without acidification. The loss is about 20% in this period of time if the urine is acidified to pH 1–2 (46).

3. Temperature. It is imperative that the specimen be allowed to cool to room temperature for a full 2 hours following heating in order to obtain maximum color development (46).

4. Delayed excretion. Repeat testing, if necessary, should be performed only after an interval of at least 4–5 days to allow for delayed excretion from the first test (43).

ACCURACY AND PRECISION

Limitations to accuracy of the test are chiefly of a clinical, as opposed to a technical, nature and include such factors as adequacy of gastric acid stimulation, integrity of intestinal absorption and renal function, and the completeness of urine collection. False low or negative results can be significantly diminished by using maximal histamine or betazole stimulation instead of caffeine sodium benzoate. The tubeless test does not reliably compare on a quantitative basis with acid output values determined by intubated gastric analysis (16, 46).

Use of the urine specimen as its own photometric blank by reduction with ascorbic acid obviates objections to the older procedure of measuring absorbance against a baseline urine obtained immediately prior to administration of the azure A resin.

Precision of the chemical method is about ±4%.

NORMAL VALUES

Segal *et al* (46) found 30 normal adults to excrete 0.95–3.91 mg/2 hour with an average of 1.84 mg. Most physicians interpret results according to the guidelines of Segal (46): 0–0.3 mg as anacidity, 0.3–0.6 mg as equivocal, and >0.6 mg as normal.

REFERENCES

1. BARON JH: *Gastroenterology 45*:118, 1963
2. BARON JH: *Gut 4*:136, 1963
3. BIANCHETTI E, GERBER T: *Schweiz Med Wochschr 88*:736, 1958
4. BLUM NI, MAYORAL LG, KALSER MH: *J Am Med Assoc 191*:339, 1965
5. BOCK OAA: *Lancet 2*:1101, 1962
6. BOLT RJ, OSSIUS TG, POLLARD HM: *Gastroenterology 32*:34, 1957
7. CALLENDER ST, RETIEF FP, WITTS LJ: *Gut 1*:326, 1960
8. CARD WI, MARKS IN: *Clin Sci 19*:147, 1960
9. CARD WI, SIRCUS W: *Modern Trends in Gastroenterology*, Series 2, edited by Jones FA. New York, Hoeber, 1958, p. 177
10. CHANG P, DUNNER DL, HARDING J, SUN DCH: *Am J Gastroenterol 42*:165, 1964
11. CORREIA JP, DE MOURA MC, DA COSTA CG: *Brit J Clin Pract 21*:227, 1967
12. ELLISON EH, WILSON SD: *Ann Surg 160*:512, 1964
13. EWALD CA: *Berlin Klin Wochschr 23*:33, 1886
14. FENTRESS V, SANDWEISS DJ: *J Am Med Assoc 165*:21, 1957
15. FORSTER GM: *J Am Med Assoc 176*:619, 1961
16. GALAMBOS JT, KIRSNER JB: *Arch Internal Med 96*:752, 1955
17. HARKNESS J: *Scand J Clin Lab Invest 10*, suppl. 31:276, 1957
18. HARKNESS J, DURANT JA: *J Clin Pathol 6*:178, 1953
19. HOLLANDER F: *Federation Proc 11*:706, 1952
20. HOLLANDER F: *Gastroenterology 7*:607, 1946
21. JOHNSTON DH, McCRAW BH: *Gastroenterology 35*:512, 1958
22. KAY AW: *Brit Med J 2*:77, 1953
23. KAY AW: *Gastroenterology 53*:834, 1967
24. KLEIN B, WEISSMAN M: *Clin Chem 5*:115, 1959
25. KONTUREK SJ, LANKOSZ J: *Scand J Gastroent 2*:112, 1967
26. KRISTENSEN M: *Acta Med Scand 177*:415, 1965
27. LAWRIE JH, SMITH GMR, FORREST APM: *Lancet 2*:270, 1964
28. LEVIN E, KIRSNER JB, PALMER WL: *Gastroenterology 19*:88, 1951
29. LUBRAN M: *Clin Chim Acta 6*:582, 1961
30. MAKHLOUF GM, McMANUS JPA, CARD WI: *Gastroenterology 51*:149, 1966
31. MAKHLOUF GM, McMANUS JPA, CARD WI: *Lancet 2*:485, 1961
32. MARKS IN: *Gastroenterology 41*:599, 1961
33. MARKS IN, BANK S, MOSHAL MG, LOUW JH: *S Afr J Surg 1*:53, 1963
34. MARKS IN, SHAY H: *Am J Digest Diseases 5*:1, 1960
35. MAURER W, BASTEN H, BECKER W, NIKLAS A, PUCHTLER H: *Klin Wochschr 29*:89, 1951
36. MICHAELIS L: *Harvey Lectures, Ser. 22*:59, 1928
37. MOORE EW, SCARLATA RW: *Gastroenterology 49*:178, 1965
38. PIPER DW, RAINE JM: *Lancet 1*:696, 1959
39. POLINER IJ, HAYES MA, SPIRO HM: *N Engl J Med 256*:1051, 1957
40. RODMAN T, GUTMAN A, MYERSON RM: *J Am Med Assoc 167*:172, 1958

41. ROSENTHAL HL, BUSCAGLIA S: *J Am Med Assoc 168*:409, 1958
42. ROSIERE CE, GROSSMAN MI: *Science 113*:651, 1951
43. SEGAL HL: *Ann NY Acad Sci 140*:896, 1967
44. SEGAL HL: *Ann NY Acad Sci 57*:308, 1953
45. SEGAL HL, MILLER LL, MORTON JJ: *Proc Soc Exp Biol Med 74*:218, 1950
46. SEGAL HL, MILLER LL, PLUMB EJ: *Gastroenterology 28*:402, 1955
47. SEGAL HL, PLOSSCOWE RP, GERLE RD, RUSSELL PK: *J Lab Clin Med 55*:815, 1960
48. SEGAL HL, RUMBOLD JC, FRIEDMAN BL, FINIGAN MM: *N Engl J Med 261*:544, 1959
49. SHAY H, OSTROVE R, SIPLET H: *J Am Med Assoc 156*:224, 1954
50. STEMPIEN SJ: *Am J Digest Diseases 7*:138, 1962
51. SUN DCH, SHAY H, CIMINERA JL: *J Am Med Assoc 158*:713, 1955
52. YARBROUGH DR III, McALHANY JC, COOPER N, WEIDNER MG Jr: *Am J Surg 117*:185, 1969
53. ZATERKA S, NEVES DP: *Gastroenterology 47*:251, 1964

Chapter 32

Calculi

THOMAS S. LA GANGA, PH.D.

Calculi occur in the urinary tract, gall bladder, prostate, and occasionally in other sites such as tonsils and salivary glands. They are formed through several incompletely understood physicochemical processes.

URINARY TRACT CALCULI

Urinary tract calculi can be found in the renal pelvis, ureter, bladder, or urethra. Calculi in the renal pelvis or ureter are of particular clinical significance because of their frequent association with serious renal disease or as the etiologic agent of renal colic. Considerable study has been devoted to all aspects of urolithiasis, but Keyser's (24) statement that "in spite of the volume of work, we are much bewildered and confused," still retains a certain pertinency.

Urinary calculi are among the oldest afflictions of man. Today they are common in all parts of the world, and the occurrence in some areas is so alarming that they have become known as "Stone belts" (13). The incidence of renal calculi from approximately 25 000 autopsies was reported as 1.12% with 0.38% of the deaths directly related to their presence (1). The incidence of ureteral calculi was reported as 0.29% of 1000 hospital admissions (12). Children in the United States have less than 1% of all calculi found (41) whereas calculous disease is a major health problem in small boys in various parts of the world (13). Distinct seasonal evidence of calculous disease was reported for the southeastern States, with the highest incidence during the hot summer months (41). The sex incidence of ureteral calculi among 309 hospitalized patients was found to be 76% males and 24% females (12) while the sex ratio found in hospitalized patients with urinary tract calculi in general has been reported as 54% males and 46% females (2). The disease is rare among Negroes (2) and Eskimos and unknown among the Bantu (34). Investigations of the origin of calcareous renal calculi suggest that the initial focus may develop within a renal tubule through the sequence of tissue injury, calcium phosphate aggregation, desquamation of damaged and partially calcified tissue, and finally crystal growth (49). An attractive theory for the origin of primary renal calculi was proposed by Randall (42). Randall's plaque theory is based upon a submucosal lesion of unknown nature in the renal papilla that is exposed to urine by erosion of the overlying epithelium. Urinary salts are deposited on the resultant plaque and augmented by accretion. The deposit eventually breaks away as a free calculus. Such a calculus has a definite indentation where it has grown, as if molded over the papilla. A small, shallow, rough depression near the center

1570

of the indentation is indeed often seen and is the point of attachment according to Randall's theory. In a study (43) of human autopsies, Randall described minute "milk patches" on the papilla that were sometimes associated with tiny adherent concretions. On microscopic examination, the milk patches appeared to be subepithelial foci of calcification. Randall's attempts to produce papillary plaques in laboratory animals, as confirmation of his concept, were unsuccessful (43). Others have also reported a relationship between calculus formation and the renal papillae (15, 49). Although Randall's plaques are important in the formation of some renal calculi, their presence does not necessarily result in calculus formation (45).

Experimental animal studies in calculus formation have highlighted the production of renal and vesical phosphate calculi in vitamin A deficient rats (37) and hyperoxaluria and oxalate nephrocalcinocis in vitamin B_6 deficient rats (14). Evidence is controversial as to whether these vitamins have a role in the etiology of urinary calculus formation in man (13).

In patients who develop calculi during long periods of immobilization, the cause may be damage to the urinary tract epithelium similar to that of plaque formation, or disturbed calcium metabolism (25). Winer (51) listed seven factors that may contribute to the formation of calculi: (a) metabolic disturbances such as cystinuria and gout; (b) endocrinopathies such as hyperparathyroidism; (c) urinary obstruction; (d) urinary tract infection; (e) mucosal metaplasia such as occurs in vitamin A deficiency; (f) systemic conditions such as dehydration, dietary excess, drug excess, or chemotherapy; and (g) isohydruria, i.e., fixation of pH over a narrow range throughout most of a 24-hour cycle with a loss of normal acid-alkaline tides. These tides normally fluctuate widely over a range from pH 4.5 to 7.5. Three types of isohydruria are described: (a) Acid type (pH 4.5–5.5) is associated with precipitation of uric acid; (b) mild acid to neutral type (pH 5.5–6) is associated with precipitation of calcium oxalate; (c) alkaline type (pH 7–7.5) is associated with precipitation of phosphates. Increased urinary excretion of high molecular weight polyelectrolytes and urea are also believed to be factors promoting calculus formation (53).

There is little question that some calculi are formed about a foreign body, e.g., dead epithelium, or a blood clot, which acts as a nidus for crystal deposition. Less common items reported as a nidus include a piece of catheter (9), a rubber band, and a hairpin (13).

Most urinary calculi do in fact contain a nucleus demonstrable by careful dissection or by grinding and polishing (40). If the nucleus is of the same chemical composition as the remainder of the calculus, it is a "pseudo" nucleus, in contrast to a "true" nucleus, which is of different composition from the remainder (10). Calculi are of two basic types (6): (a) concretionary, which have concentric laminations about a discrete center (nucleus), and many have radial striations (6); (b) sedimentary, where crystallization occurs from multiple centers and is arrested while very small by adsorption of organic material. Subsequent consolidation binds the centers into a calculus of random orientation.

Although usually round or oval, the nucleus may be irregular and varies up to several millimeters (40). A concentric matrix arranged in compact parallel fibrils between successive laminations is present in every human concretion from center to surface. The radial striations that occur in many calculi, especially those of predominantly uric acid content, are composed of a layer of fibrous matrix running at right angles to the concentric laminations. This disposition suggests the "cracks" that cross the annular rings of drying soft wood logs (3). Matrix composition is remarkably constant for calculi of any type of crystalline content. Formation of matrix derives from the macronucleus suggesting that matrix deposition unquestionably precedes crystal formation (3). The shortest time interval necessary for the formation of calculi has been reported as 74 days with 1200 days as the longest (25).

Organic material is incorporated during crystallization and makes up about 3% of the calculus weight (39). This material has histochemical properties similar to bone matrix (19) and is a combination of mucoprotein and mucopolysaccharide (5) analytically similar to urinary mucoprotein (26). This crystallizable matrix is regarded as an essential ingredient in concrement formation (4).

Calcium is a major constituent in 90% of all renal calculi in the United States (47). Prien and Frondel (40) in an extensive mineralogic and x-ray diffraction study of 1000 renal calculi reported 327 to be pure calcium oxalate. The chemical composition for approximately 4000 calculi from adults is listed in Table 32-1 (21, 22, 36, 40).

TABLE 32-1. Distribution of Urinary Calculi by Chemical Analysis

Analysis	% of total
Pure calcium oxalate	35.8
Calcium oxalate + apatite (carbonate, phosphate)	40.8
Magnesium ammonium phosphate	17
Uric acid	5.2
Cystine	1.2
Xanthine	0.03

Table 32-2 gives the results of this study (40) together with pertinent data on constituents of stones including the physical description of various types of calculi. Calculi composed of tricalcium phosphate (whitlockite), indigo, or xanthine are not included in the table because they did not occur in the study.

Herring (17) in a crystallographic and x-ray diffraction study of 10 000 calculi reported that, with rare exceptions, urinary tract calculi in humans are composed of calcium oxalate (as the mono- or dihydrate), calcium phosphate (as hydroxyapatite), uric acid, magenesium ammonium phosphate, and (or) cystine. Any one of these crystallites may comprise almost all of a calculus, but 9 of every 10 calculi contain more than 10% of one or more additional crystalline components. Urinary calculi may contain trace amounts of many other urinary substances presumably mechanically incorporated in the growing calculus. The contributions of the various substances to calculi varies geographically. While apatite exceeds all other components in urinary calculi of Americans, ammonium acid urate and calcium monohydrate were found to be the most common constituent of 200 calculi from Thailand.

Several approaches exist for the analysis of calculi: (a) *Thin section technic.* Sections of the calculi are ground and polished. The structure (nucleus, laminations) can be examined by reflected light or by polarized light if sections are made thin enough to transmit light (44). This difficult technic is useful chiefly for photographic reproductions and has very limited success in identification of components. (b) *Infrared absorption spectroscopy* (50). Powdered sample is incorporated in a KBr disk. This technic permits precise qualitative identification of both single components and mixtures and some semiquantitative estimations. (c) *X-ray diffraction* (17). This technic is applied to powdered sample, requires only 0.1 mg or less, and offers a rigorous means of identification. (d) *Optical method.* The calculus is fractured, dissected, and individual crystals identified by determination of the refractive index. This is accomplished under a chemical microscope by immersion in a series of fluids of known refractive index. Optical sign, axial angle, dispersion, extinction angle, sign of elongation, etc., may be measured at the same time and aid in identification. By making grain counts in a random ground sample of the calculus, it is possible to estimate percent composition to within 5%–10%. It has been claimed that opaque impurities are likely to render this technic useless (22). Prien and Frondel (40), however, encountered only four that could not be identified by optical methods because of foreign matter (usually blood intimately mixed with the crystalline material). The optical method, however, is by no means an easily acquired technic. The analyst must be well versed in chemical microscopy, and even then, it is a rather slow and tedious procedure. (e) *Chemical analysis.* Qualitative chemical analysis has been and still is by far the most widely used technic. In 1860 Heller (16) published a comprehensive scheme of analysis. Many schemes have been presented, differing primarily in the choice of test made for each component (10, 33, 52). Tests may be made for the following substances (see Table 32-3): calcium, magnesium, ammonium, oxalate, phosphate, carbonate, uric acid (or urate), cystine, xanthine, indigo, sulfonamide, and urostealith. Schemes for quantitative chemical analysis of calculi have also been recommended (31, 35). Chemical microscopy has been employed as a useful method of calculus analysis for all the common constituents (29). The success of this technic is dependent upon a reasonably good mechanical separation of individual crystals. Chemical reactions are used as confirmatory evidence to identify substances identified by crystal morphology.

The specificity of the optical, infrared, and x-ray diffraction methods is unquestioned. Prien and Frondel (40) are of the opinion that the chemical methods leave much to be desired: (a) There is confusion regarding the exact nature of reactions in qualitative tests; (b) interfering organic substances of unknown composition may invalidate the reactions; and (c) complex mixtures that occur are not amenable to resolution by chemical methods.

TABLE 32-2. Data on Urinary Calculi

Constituent	Relative incidence % of total	Mineralogic name	Characteristics of urine in which found	Description
Pure calcium oxalate		Whewellite	Usually acid and sterile	Monohydrate: very hard, porous, smooth, lustrous surface; light honey-brown, red-brown, or black-brown. Three types: (a) small, smooth, ovoid "hemp seed," usually multiple with concentric laminations; (b) "mulberry," variable size and shape with irregular rounded mammillary processes; (c) "jack stone," dense central mass with radiating spicules.
Monohydrate	13.7 ⎫	Wheddellite		
Dihydrate	0.4 ⎬ 32.7			Dihydrate: pale yellow-white to honey-brown, octahedral crystals, loose or compactly interlocking aggregates.
Mixed	18.6 ⎭			
Apatite plus			Usually acid and sterile	Porous, granular. Apatite exists in interstices or cavities.
Calcium oxalate · H_2O	7.2 ⎫			
Calcium oxalate · $2H_2O$	4.7 ⎬ 34.3			
Calcium oxalate, mixed	22.4 ⎭			
Apatite, pure	3.4	Carbonate apatite[a] Hydroxyl apatite[b]	Acid or alkaline	Fine grained, soft, and compact, usually laminated; chalk white, cream white, yellow-brown.

Composition	%	Reaction	Name	Description
$MgNH_4PO_4 \cdot 6H_2O$	0.3	Usually alkaline and infected	Struvite	Densely granular, creamy white, resembling lump sugar.
$MgNH_4PO_4 \cdot 6H_2O$ plus apatite	15.5	Usually alkaline and infected		Dirty to cream white, resembling cancellous bone in structure. Forms the "staghorn" calculi.
$MgNH_4PO_4 \cdot 6H_2O$ plus apatite plus mixed calcium oxalate	3.2	Usually alkaline and infected		
$CaHPO_4 \cdot 2H_2O$ Pure Mixed	$\left.\begin{array}{c}0.2\\1.7\end{array}\right\}$ 1.9	Acid	Brushite	Cream white or light brown, moderately soft, fracture surfaces pearly or silky luster, radially fibrous or bladed structure.
Uric acid Pure Mixed	$\left.\begin{array}{c}4.7\\1.1\end{array}\right\}$ 5.8			Dense, fine grain, typical oblate or flattened pebblelike with smooth but not polished surface (somewhat warty); usually laminated brown, relatively soft, and easily crushed.
Cystine Pure Mixed	$\left.\begin{array}{c}2.2\\0.7\end{array}\right\}$ 2.9			Porous aggregates of relatively well formed short hexagonal prisms or tablets, sometimes compact granular center. Honey-yellow to yellow-white. Waxy luster, very soft unless densely granular.
Urostealith	0.1			Consistency of beeswax.

[a] $Ca_{10}(PO_4 \cdot CO_3 OH)_6(OH)_2$
[b] $Ca_{10}(PO_4)_6(OH)_2$

TABLE 32-3. Qualitative Tests Used for Analysis of Calculi

Substance	Method	References
Carbonate	1. Add HCl $\longrightarrow$ effervescence of CO_2	52
Calcium oxalate	1. Resorcinol + H_2SO_4 $\longrightarrow$ blue green	52
	2. HCl + MnO_2 $\longrightarrow$ gas bubbles	52
	3. Make HCl solution of stone alkaline with NH_4OH, add acetic acid, examine crystals; to confirm, add H_2SO_4, heat, add $KMnO_4$	10
	4. Diphenylamine + H_3PO_4 + heat + alcohol $\longrightarrow$ blue	18
	5. HCl solution + saturated sodium acetate to pH 5 $\longrightarrow$ white precipitate	33
	6. Convert to $CaCO_3$ by heat, add HCl $\longrightarrow$ CO_2	27
Calcium	1. Add oxalic acid after removing calcium oxalate $\longrightarrow$ white precipitate	33
	2. Picrolinate	18
Magnesium	1. HCl extract + alkali + p-nitrobenzene-azoresorcinol $\longrightarrow$ blue color	23
	2. Remove calcium oxalate and other Ca, add NH_4OH to pH 8, add Na_2HPO_4 $\longrightarrow$ $MgNH_4PO_4$ precipitate	33
	3. Acid extract + quinalizarin + NH_3 $\longrightarrow$ blue	18
	4. Titan yellow	30
Ammonium	1. Nessler's reaction	52
	2. Add NaOH $\longrightarrow$ odor of NH_3	27
Phosphate	1. HNO_3 extract + ammonium molybdate + heat $\longrightarrow$ yellow precipitate	52
	2. Acid molybdate + $SnCl_2$ $\longrightarrow$ blue	23
	3. Acid extract + molybdate + aminonaphtholsulfonic acid $\longrightarrow$ blue	27
Uric acid	1. Alkaline extract + phosphotungstate $\longrightarrow$ blue	52
	2. Alkaline extract + arsenophosphotungstate $\longrightarrow$ blue	23
	3. Murexide reaction: add HNO_3, evaporate to dryness, add NH_4OH $\longrightarrow$ brilliant purple	52
Cystine	1. Add NH_4OH + NaCN + sodium nitroprusside $\longrightarrow$ red	52
	2. Evaporate NH_4OH extract on micro slide $\longrightarrow$ hexagonal plates	10
	3. Add HCl + NaCN + 1, 2-naphthoquinone-4-sulfonate + Na_2SO_3 + NaOH $\longrightarrow$ red brown, turning deep red on adding $Na_2S_2O_4$ in NaOH	23
	4. Add NaOH, heat, lead acetate, heat $\longrightarrow$ PbS precipitate (black)	27
Xanthine	1. Murexide reaction: Add HNO_3, evaporate, add NaOH $\longrightarrow$ orange, turning red with heat	23
	2. Ehrlich's diazo reaction	33
	3. Add NaOH to residue from murexide test $\longrightarrow$ red; add H_2O $\longrightarrow$ yellow, which when evaporated $\longrightarrow$ red-violet	10
Indigo	1. Reduce to leuco form, reoxidize to indigo	18
	2. $CHCl_3$ extract is blue	23
Urostealith	1. Evaporate ether or $CHCl_3$ extract, stain with Sudan III	23
Sulfonamides	1. Add HCl + $NaNO_2$ + N-(1-naphthyl)-ethylenediamine 2 HCl $\longrightarrow$ pink or magenta	52
Cholesterol	1. Liebermann-Burchard color reaction	23
Fibrin	1. Millon's reagent + heat $\longrightarrow$ red precipitate	23

Lonsdale (32) has stated, however, that physical methods are not beyond reproach as regards qualitative accuracy and may need to be supplemented by chemical tests for certain individual components, e.g., distinguishing between carbonate apatite and hydroxy apatite, which is not easily done by other methods. In one study (21) x-ray and petrographic analyses of certain primary stones revealed no phosphates, yet chemical analysis yielded phosphate contents 0.75%–5% of dry weight. Qualitative chemical analysis appears to be completely adequate for routine use.

QUALITATIVE IDENTIFICATION OF URINARY TRACT CALCULI

(method of Simmons and Gentzkow, Ref. 46, except magnesium, Ref. 31, and ammonium, Ref. 7, modified)

PRINCIPLE

Urates are detected by reduction of alkaline phosphotungstate to tungsten blue; cystine by the red color formed in the cyanide-nitroprusside test; carbonate by the release of CO_2 upon acidification; calcium oxalate by precipitation at pH 3–4; other calcium by precipitation with oxalate after removal of intrinsic calcium oxalate; phosphate by precipitation as ammonium phosphomolybdate; magnesium by the Titan-yellow reaction, and xanthine by the murexide reaction and by its uv absorption spectrum. The presence of ammonium ion is indicated by the blue color formed in the alkaline-hypochlorite reaction.

REAGENTS

KH_2PO_4, 7%
HCl, conc
Potassium acetate, 160 g/100 ml
Potassium oxalate, 3%
HNO_3 conc
Phosphate test reagent. Add 30 ml conc HNO_3 to 70 ml water in which have been dissolved 5 g ammonium molybdate and 7.5 g ammonium nitrate. Then add 15 g tartaric acid and let stand overnight in 37°C incubator before use; then transfer to a glass-stoppered glass bottle.
Na_2CO_3, 14%
NaCN, 5%. Store in refrigerator.
Titan Yellow, 0.05%. Store in dark in amber colored bottle. Stable about 1 week.
NaOH, 20%
NaOH, 0.1 N. 4 g/liter
Phosphotungstic acid reagent (see Chap. 17, section on uric acid).
Sodium nitroprusside. Prepare fresh each time by dissolving a few crystals in a few milliliters of water.
Phenol color reagent. 50.0 g reagent grade phenol and 0.25 g reagent grade sodium nitroprusside per liter. Stable at least 2 months if kept cool and in amber bottle protected from light.

Alkali-hypochlorite reagent. 25.0 g reagent grade NaOH and 2.1 g sodium hypochlorite per liter. Commercial bleach (Clorox), which contains 5.25 g NaOCl/100 ml can be used: 25.0 g NaOH, 40 ml Clorox, dilute to 1 liter. Stable at least 2 months if kept cool and in amber bottle protected from light.

Ammonia, conc

KOH, 1 N. 5.6 g/liter

PROCEDURE A (used when more than 2 mg of calculus available)

1. Describe, where possible, the size, shape, color (zones), number, type of surface, hardness, and brittleness of the calculi.

2. Pulverize the calculus in a test tube with the help of a glass rod. Transfer approximately 2 mg of the powered calculus to another test tube for ammonium analysis. Always keep some crushed calculus in reserve. *If only a small fragment is available, proceed immediately (after step 1) to procedure B.*

3. Transfer 5—10 mg powdered calculus to a 13 × 100 mm Pyrex test tube. Add 1.0 ml of 0.1 N NaOH and warm in 60° C water bath for 5 min. Shake tube 2—3 times during this period. Then centrifuge and decant supernate into another test tube of same size; reserve supernate for urate and cystine tests (step 4). Rinse walls of tube containing specimen with 1 ml water, mix, centrifuge, and discard supernate. Repeat this washing procedure once more. Save residue for further tests (beginning with step 5).

4. (a) Transfer a drop of NaOH extract from step 3 to a spot plate. Add 1 drop of sodium carbonate, mix with glass rod. Add one drop of phosphotungstic acid and mix. A *deep blue* color indicates *uric acid* and/or *urates*. (Faint blue color is not significant.)

   (b) Transfer a drop of NaOH extract from step 3 to a spot plate, add a drop of ammonia and a drop of sodium cyanide reagent. Let stand 5 min and then add 1 drop of sodium nitroprusside reagent. A deep red color indicates *cystine*.

5. Tilt the tube containing residue from NaOH extraction (step 3), and allow a drop of conc HCl to run down the side. Watch for the moment of contact with the solid residue. A momentary but *copious* evolution of gas bubbles indicates presence of *carbonate*. Now add an additional 0.2 ml HCl and 0.5 ml distilled water. Boil gently for a few seconds, cool, and centrifuge.

6. Transfer 0.3 ml supernate from step 5 to a 13 × 100 mm test tube. Add potassium acetate reagent to pH 3.0—4.0 (use narrow range pH paper). Let stand 10 min. Appearance of a precipitate or strong turbidity indicates presence of *calcium oxalate* (cystine may give false positive). Centrifuge.

7. Transfer supernate from step 6 to another 13 × 100 test tube. Add 2 drops of potassium oxalate solution and sufficient water to approximately double the volume of the contents of the tube. Let stand 10 min.

Appearance of a precipitate or strong turbidity indicates presence of *calcium other than oxalate*. Centrifuge.

8. Transfer 1 drop of supernate from step 5 to a 13 × 100 mm test tube. Add 1 drop of conc HNO_3 and boil gently for 5 sec. Add 2 drops of phosphomolybdate reagent and boil gently for 3–4 sec. Appearance of a yellow precipitate indicates *phosphate*.

9. Transfer remainder of supernate from step 7 to a 13 × 100 mm test tube. Add 2 drops of Titan yellow reagent and 0.5 ml of 20% NaOH solution. Appearance of a red color, changing in about a minute or two to a red precipitate, indicates *magnesium*.

10. If a considerable portion of the calculus dissolves in NaOH (step 3 of procedure above) and the tests for cystine and uric acid are negative, then test for presence of xanthine in the NaOH extract as follows:

    (a) Transfer 4 drops of NaOH extract to a clean porcelain crucible and add 2 drops of conc HNO_3. Evaporate to dryness on steam bath. Add another drop of HNO_3 and again evaporate to dryness. Add 2 drops of 1 N KOH. If a purple-red color appears, urates are present. Warm the crucible on the steam bath for about 1 min. If only urates are present, the purple-red color will disappear; *if xanthine is present, a yellow color is present before heating, and a purple-red color appears only after heating.*

    (b) To 0.2 ml NaOH extract add approximately 3 ml water. Add 7% KH_2PO_4, dropwise, to pH 8 (narrow range pH paper). Run an absorption curve. *Xanthine* shows a major peak at about 270 nm and a minor peak at about 240 nm. *Uric acid* peaks at about 293 nm and 235 nm.

11. To pulverized calculus in second test tube add 0.1 ml conc HCl carefully. Add 0.3 ml water, mix, and boil. Transfer 2 drops of this acid extract to another test tube and add 0.5 ml water, 3 drops 20% NaOH, and 0.1 ml Phenol Color Reagent and mix. Add 0.1 ml alkaline hypochlorite solution and mix. Incubate at 37°C for 20 min. Also run a blank omitting the acid extract and the 20% NaOH.

    A distinct blue color indicates *ammonium*. A light blue color, even if a little bluer than the blank, should be disregarded. A positive control can be run by running the test on 4 ml water to which has been added one small crystal of $(NH_4)_2SO_4$.

PROCEDURE B (used when 2 mg or less of calculus is available)

1. Transfer fragment to a micro centrifuge tube. Add 0.2 ml of 0.1 N NaOH, warm for 5 min in 60°C water bath and centrifuge. Decant supernate and use for uric acid and cystine tests (step 4 above).

2. Wash tube two times with 0.2 ml vol of water, centrifuge, and discard supernate.

3. Allow 1 drop of conc HCl to run down side of tube and watch for effervescence when acid makes contact with residue. Then add 0.4 ml water, heat gently for 5 sec over micro burner, and centrifuge.

4. Decant supernate into a 13 x 100 mm test tube and proceed to tests described in steps 6–9 of procedure A.

Report findings as in the following example:

Gross description:	One calculus, about 2 X 3 mm. Irregular surface, dark brown, hard, and brittle.
Chemical analysis:	Carbonate—negative
	Calcium oxalate—positive
	Calcium other than oxalate—positive
	Magnesium—negative
	Phosphate—positive
	Ammonium—negative
	Uric acid—negative
	Cystine—negative
Conclusion:	Calculus contains calcium oxalate and calcium phosphate.

NOTES

1. An anion must be found if a cation is present and vice versa.

2. The analyst must be alert to the possibility of submission of "calculi" from malingerers that are not of biologic origin. On several occasions our laboratory has received stones that could not be crushed with a hammer, sometimes much too large to pass the urethra, presumably picked up by the patient in his gravel driveway. Herring (17) reports that artifactual materials encountered in his study were of considerable diversity, the most common being smooth, rounded, quartzite pebbles.

3. Occasionally the calculus submitted is extremely small. The whole scheme must then be scaled down appropriately and reactions read with the aid of a magnifying glass.

PROSTATIC CALCULI

Prostatic calculi are the most common calculi occurring in man and their incidence increases with aging. They are, however, usually of little or no clinical significance. Prostatic calculi vary from microscopic size to about 1 cm (20). They contain approximately 20% organic material (protein, cholesterol, and citrate) (20), and the inorganic component is principally carbonate-apatite, $C_{10}(PO_4,CO_3OH)_6(OH)_2$ (20, 40), with smaller amounts of $MgNH_4PO_4 \cdot 6H_2O$ (36). Infrared analysis has been applied to prostatic calculi (28).

BILIARY CALCULI

Edwards *et al* (11) obtained the following distribution in a study of thirty gallstones: fourteen "pure" stones, including eight of cholesterol, five of calcium bilirubinate, and one of calcium carbonate; twelve "mixed" stones, composed chiefly of two or three of the components found in the pure stones; four "combined" stones (stones made up of a nucleus and shell of different compositions), three of which had nuclei presumed to be calcium bilirubinate enclosed in a shell of calcium carbonate plus cholesterol. Cholesterol and calcium salts of bile pigments, carbonate, and phosphate are probably present, at least in minute amounts, in all gallstones. Small amounts of fats, soaps, lecithin, mucus, Cu, Fe, and Mn are also found. X-ray diffraction and infrared technics have been applied to the analysis of gallstones (8). A qualitative chemical analytical scheme is available for protein, silica, fat, bile pigment, cholesterol, calcium, iron, and phosphate (18).

QUALITATIVE IDENTIFICATION OF BILIARY CALCULI

(scheme of Oser, Ref. 38, modified)

PRINCIPLE

Cholesterol is detected by the Liebermann-Burchard reaction. Calcium is precipitated by addition of oxalate. Phosphate is precipitated as ammonium phosphomolybdate. Bilirubin is detected by the diazo color reaction.

REAGENTS

Ether
HCl, 0.2N
Chloroform
Ethanol. Formula 3A denatured ethanol is satisfactory.
Acetic anhydride
H_2SO_4, *conc*
Sodium acetate, saturated
Potassium oxalate, 10%
HNO_3, *conc*
Phosphate test reagent. Same as for urinary calculi.
Methanol
Diazo reagent (see Chap. 22 on liver function tests)
$NaNO_2$, *0.5%*
Sulfanilic acid reagent. 100 mg sulfanilic acid + 1.5 ml conc HCl/100 ml water.

PROCEDURE

1. Describe, where possible, the size, shape, color, number, hardness, and brittleness of calculi.
2. Grind dry calculus to powder in a test tube. Always try to keep some crushed calculus in reserve.
3. Extract several times with 3 ml portions of ether. Filter and combine the extracts.

4. Evaporate ether filtrate to dryness and test for cholesterol by Liebermann-Burchard reaction: Add 5 ml $CHCl_3$ to dried extract. Add 2 ml of acetic anhydride and 2 drops of conc H_2SO_4. Place in dark for 40 min. Green color indicates *cholesterol*.

5. To ether-extracted residue in tube from step 2 add 3 ml 0.2 N HCl. Mix and pour through filter used for ether extraction. Repeat treatment with acid.

6. Test 0.5 ml of acid filtrate for calcium: Add saturated sodium acetate to pH of about 4.0 (pH paper). Add 2 drops 10% potassium oxalate. Let stand 10 min. A precipitate indicates presence of *calcium*.

7. Test 0.5 ml of acid filtrate for phosphate: Add 0.5 ml conc HNO_3 and boil gently for 5 sec. Add 1 ml phosphate test reagent (urinary calculi procedure) and boil gently for 3–4 sec. A yellow precipitate indicates the presence of *phosphate*.

8. Wash filter paper (from step 5) with water and dry. Extract several times with hot chloroform. Save filter paper for later step.

9. Chloroform filtrate will be golden yellow if bilirubin is present. Dry down extract under N_2 and test for bilirubin by diazo reaction. Dissolve in 5 ml methanol. Add 1 ml diazo reagent (made by adding 0.3 ml $NaNO_2$ reagent to 10 ml sulfanilic acid reagent). A violet color indicates *bilirubin*.

10. Extract filter paper from step 8 with hot ethanol. Filtrate will be green if *biliverdin* is present.

NOTES

1. An anion must be found if a cation is present and vice versa.

2. Occasionally the calculus submitted is extremely small. The whole scheme must then be scaled down appropriately and reactions read with the aid of a magnifying glass.

REFERENCES

1. BELL ET: *Renal Disease*. Philadelphia, Lea and Febiger, 1946, p 395
2. BOYCE WH, GARVEY FK, STRAWCUTTER HE: *J Am Med Assoc 161*:1437, 1956
3. BOYCE WH: *Am J Med 45*:673, 1968
4. BOYCE WH, KING JS Jr: *J Urol 81*:351, 1959
5. BOYCE WH, SULKIN NM: *J Clin Invest 35*:1067, 1956
6. CARR JA: *Brit J Urol 28*:240, 1956
7. CHANEY AL, MARBACH EP: *Clin Chem 8*:131, 1962
8. CHIHARA G, YAMAMOTO S, KAMEDA H: *Chem Pharm Bull 6*:50, 1958
9. CHUTE R: *J Urol 87*:355, 1962
10. DOMANSKI TJ: *J Urol 37*:399, 1937
11. EDWARDS JD Jr, ADAMS WD, HALPERT B: *Am J Clin Pathol 29*:236, 1958
12. FETTER TR, ZIMSKIND PD, GRAHAM RH, BRODIE DE: *J Am Med Assoc 186*:21, 1963

13. GERSHOFF SN: *Metabolism 13*:875, 1964
14. GERSHOFF SN, FARAGALLA FF, NELSON DA, ANDRUS SB: *Am J Med 27*:72, 1959
15. HARRISON HE, HARRISON HC: *J Clin Invest 34*:1662, 1955
16. HELLER JF: *Die Harnconcretionen, ihre Entstehung, Erkennung, und Analyse mit besonderer Rücksicht auf Diagnose und Therapie der Nieren-und Blasenerkrankung. Wein,* Berlin, Tendler u Comp, 1860
17. HERRING LC: *J Urol 88*:545, 1962
18. HOLT PF, TARNOKY *AL: J Clin Pathol 6*:114, 1953
19. HOWARD JE: *J Urol 72*:999, 1954
20. HUGGINS C, BEAR RS: *J Urol 51*:37, 1944
21. HUGHES J, COPPRIDGE WM, ROBERTS LC, MANN VI: *J Am Med Assoc 172*:774, 1960
22. JENSEN AT: *Acta Chir Scand 84*:207, 1940
23. KAMLET J: *J Lab Clin Med 23*:321, 1937
24. KEYSER LD: *J Urol 50*:169, 1943
25. KIMBROUGH JC, DENSLOW JC: *J Urol 61*:837, 1949
26. KING JS Jr, BOYCE WH: *Arch Biochem Biophys 82*:455, 1959
27. KIRBY JK, PELPHREY CF, RAINEY JR Jr: *Am J Clin Pathol 27*:360, 1957
28. KLEIN B, WEISSMAN M, BERKOWITZ J: *Clin Chem 6*:453, 1960
29. LASKOWSKI DE: *Anal Chem 37*:1399, 1965
30. LEONARD RH: *Clin Chem 7*:546, 1961
31. LEONARD RH, BUTT AJ: *Clin Chem 1*:241, 1955
32. LONSDALE K: *Science 159*:1199, 1968
33. McINTOSH JF, SALTER RW: *J Clin Invest 21*:751, 1942
34. MODLIN M: *Ann Roy Coll Surg Engl 40*:155, 1967
35. NICHOLAS HO: *Clin Chem 4*:261, 1958
36. NICHOLAS HO, LEIFESTE HF: *Clin Chem 4*:267, 1958
37. OSBORNE TB, MENDEL LB, FERRY EL: *J Am Med Assoc 69*:32, 1917
38. OSER BL: ed., Hawk's Physiological Chemistry, 14th ed, New York, McGraw-Hill, 1965, p 497
39. PHILIPSBORN H: *Urol Intern 7*:28, 1958
40. PRIEN EL, FRONDEL C: *J Urol 57*:949, 1947
41. PRINCE CL, SCARDINO PL: *J Urol 83*:561, 1960
42. RANDALL A: *New Engl J Med 214*:234, 1936
43. RANDALL A: *Internat Abst Surg 71*:209, 1940
44. RANDALL A: *J Urol 48*:642, 1942
45. ROSENOW EC Jr: *J Urol 44*:19, 1940
46. SIMMONS JS, GENTZKOW CJ: Laboratory Methods of the U.S. Army, 5th ed. Philadelphia, Lea and Febiger, 1944, p 182
47. SMITH LH: *Am J Med 45*:649, 1968
48. THOMAS WC Jr: *Ann Intern Med 69*:165, 1968
49. VERMOTTEN V: *J Urol 48*:27, 1942
50. WEISSMAN M, KLEIN B, BERKOWITZ J: *Anal Chem 31*:1334, 1959
51. WINER JH: *J Am Med Assoc 169*:1715, 1959
52. WINER JH, MATTICE MR: *J Lab Clin Med 28*:898, 1943
53. ZINSSER HH: *J Am Med Assoc 174*:2062, 1960

Appendix

TABLE A—1. pH indicators

Common name	Chemical name	Working range and colors			Preparation[a]
Methyl violet	Pentamethylbenzyl-p-rosaniline · HCl	Yellow	0.4–1.0	Blue	0.25% aqueous
Thymol blue (acid range)	Thymolsulfonephthalein	Red	1.2–2.8	Yellow	b
Metacresol purple (acid range)	m-Cresolsulfonephthalein	Red	1.2–2.8	Yellow	b
Töpfer's reagent	Dimethylaminoazobenzene	Red	2.9–4.0	Yellow	0.5% in 95% ethanol
Bromphenol blue	Tetrabromophenolsulfonephthalein	Yellow	3.0–4.6	Blue	b
Congo red	Na tetrazodiphenylnaphthionate	Blue	3.0–5.0	Red	0.1% aqueous
Methyl orange	Na p-dimethylaminobenzenesulfonate	Red	3.1–4.4	Yellow	0.1% aqueous
Bromcresol green	Tetrabromo-m-cresolsulfonephthalein	Yellow	4.0–5.6	Blue	b
Methyl red	Dimethylaminoazobenzene-o-carbonic acid	Red	4.2–6.3	Yellow	b
Litmus	Obtained from lichens	Red	4.5–8.3	Blue	0.5% aqueous
Alizarin	Dihydroxyanthraquinone	Yellow	5.0–6.8	Red	0.1% in 95% ethanol
Chlorphenol red	Dichlorophenolsulfonephthalein	Yellow	5.1–6.7	Red	b
Bromphenol red	Dibromophenolsulfonephthalein	Yellow	5.2–6.8	Red	b
Bromcresol purple	Dibromo-o-cresolsulfonephthalein	Yellow	5.4–7.0	Red	b
Bromthymol blue	Dibromothymolsulfonephthalein	Yellow	6.0–7.6	Blue	b
Phenol red	Phenolsulfonephthalein	Yellow	6.8–8.4	Red	b
Neutral red	Aminodimethylaminotoluphenazine · HCl	Red	6.8–8.0	Yellow	0.1% in 95% ethanol
Cresol red	o-Cresolsulfonephthalein	Yellow	7.2–8.8	Red	b
Metacresol purple (alkaline range)	m-Cresolsulfonephthalein	Yellow	7.4–9.0	Purple	b
Thymol blue (alkaline range)	Thymolsulfonephthalein	Yellow	8.0–9.6	Blue	b
Phenolphthalein		Colorless	7.8–10.0	Red	0.1% in 95% ethanol
Thymolphthalein		Colorless	9.3–10.5	Blue	0.1% in 95% ethanol
Alizarin yellow GG	Na m-nitrobenzeneazosalicylate	Yellow	10.0–12.0	Lilac	0.1% in 50% ethanol
1,3,5-Trinitrobenzene		Colorless	12.0–14.3	Orange	0.25% in 50% ethanol

[a] Water used for preparation should be boiled to free it from dissolved CO_2. Alcohol (ethanol denatured with methanol is satisfactory) should be neutral when used as solvent.

[b] Sulfonephthalein dyes are available either in the acid or salt form. For volumetric titrations a 0.04% solution in 95% ethanol is satisfactory. For pH determination a 0.04% aqueous solution of the salt form is used. If the salt form is not available the acid form can be triturated with the proper amount of alkali for conversion to the salt form (amounts given in The Merck Index and Handbook of Chemistry and Physics). The aqueous solutions are equally satisfactory for volumetric titrations.

TABLE A–2. pH Reference Standards for pH Meter

Buffer	pH				Composition	Natl. Bur. Std. Catalog No.	Refs	Remarks
	0°C	10°C	25°C	40°C				
HCl, 0.1 N	1.10	1.10	1.10	1.10	Standardized by titration		1	pH constant to 95°C; uncertainty of about ±0.02 pH unit
Potassium tetraoxalate, 0.05 M	1.67	1.67	1.68	1.70	12.7 g $KH_3 (C_2O_4)_2 \cdot 2H_2O$/liter		1,3,7	Must not heat salt above 60°C
Potassium acid tartrate, saturated at 25° ± 3°C			3.56	3.55	Saturated at room temperature, about 0.034 M; absolute saturation not critical; preferable to remove excess	188	1,3,7	Supports mold growth; accompanied by pH increase up to 0.1 pH unit
Potassium acid phthalate, 0.05 M	4.01	4.00	4.01	4.03[a]	10.21 g $KHC_8H_4O_4$/liter	185	1,7	Dry salt 1 hour at 105°C; has low buffering capacity
CH_3COOH, 0.1 N/ CH_3COONa, 0.1 N			4.64	4.635	Equal vol 0.2 N CH_3COOH (standardized by titration) and 0.2 N CH_3COONa (27.22 g/liter)			
Sodium acid succinate, 0.025 M/ sodium succinate, 0.025 M	5.46	5.42	5.40	5.41	For preparation of salts, see ref 8		3	Supports mold growth

Buffer	pH	pH	pH	pH	Preparation		References	Remarks
KH_2PO_4, 0.025 M/ Na_2HPO_4, 0.025 M	6.98	6.92	6.86	6.84[a]	3.40 g KH_2PO_4 + 3.55 g Na_2HPO_4/liter	186 Ib, 186 IIb	1,2	If any question about anhydrous state of salts, dry 2 hours at 110°–130°C; supports mold
KH_2PO_4/ Na_2HPO_4	7.531	7.474	7.413	7.379	1.179 g KH_2PO_4 + 4.303 g Na_2HPO_4/liter	186 Ib, 186 IIb	4	Recommended for blood pH standard; $\mu = 0.1$
Borax, 0.01 M	9.46	9.33	9.18	9.07[a]	3.81 g $Na_2B_4O_7 \cdot H_2O$/liter; exact concentration not critical	187 a	1,6	Na error negligible; must not heat salt above room temperature
$NaHCO_3$, 0.025 M/ Na_2CO_3, 0.025 M	10.32	10.18	10.02	9.91	2.10 g $NaHCO_3$ + 2.65 g Na_2CO_3/liter		3	Na_2CO_3 should be ignited at 250°–300°C for 2 hours
Na_3PO_4, 0.01 M			11.72	11.38	Mix equal vol 0.02 M NaOH and 0.02 M Na_2HPO_4		3,7	
NaOH, 0.1 M	13.83	13.43	12.88	12.42	Prepare from carbonate-free solution of NaOH; standardize		1,3	Uncertainty of about 0.03 pH unit
$BaCl_2$/ $Ba(OH)_2$		13.81		13.74	Add excess $Ba(OH)_2$ to 0.1–0.2 N HCl; let stand overnight before use		1,3,5	Stable at least 30 days

[a]pH values at temperatures to 95°C given in ref 7.

References: 1. Bates RG: *Electrometric pH Determinations.* New York, Wiley, 1954. 2. Bates RG, Acree SF: *J Res Natl Bur Std (US)* 34:373, 1945. 3. Bates RG, Pinching GD, Smith ER: *J Res Natl Bur Std (US)* 45:418, 1950. 4. Bower VE, Paabo M, Bates RG: *Clin Chem* 7:292, 1961. 5. Brooke M: *Chemist-Analyst* 42:28, 1953. 6. Manov GG, DeLollis NJ, Lindvall PW, Acree SF: *J Res Natl Bur Std (US)* 36:543, 1946. 7. National Bureau of Standards, Letter Circ, LC 993, Washington, DC, 1950. 8. Pinching GD, Bates RG: *J Res Natl Bur Std (US)* 45:1950.

TABLE A—3a. HCl-KCl Buffer of Clark and Lubs (no pK_a'; constant ionic strength, $\mu = 0.1$)

KCl, 0.2 M: 14.91 of reagent grade in water to 1 liter.

HCl, 0.2 M: 20 ml conc reagent grade HCl + 1 liter water.

Determine normality by titration versus Na_2CO_3 or a standardized solution of NaOH. Then adjust to 0.2 M by proper dilution (it always titrates higher than 0.2 M) or you can convert figures below to an equivalent volume of the M found.

Each mixture is made up to 100 ml with water. For practical purposes these buffer mixtures have the same pH up to 60°C.

pH (20°C)	HCl, 0.2 M (ml)	KCl, 0.2 M (ml)	pH (20°C)	HCl, 0.2 M (ml)	KCl, 0.2 M (ml)
1.1	47.3	2.7	1.7	11.9	38.1
1.2	37.6	12.5	1.8	9.4	40.6
1.3	29.8	20.2	1.9	7.5	42.5
1.4	23.7	26.3	2.0	6.0	44.1
1.5	18.8	31.2	2.1	4.7	45.3
1.6	15.0	35.0	2.2	3.8	46.2

Reference: Clark WM: *The Determination of Hydrogen Ions*, 3rd ed. Baltimore, Williams & Wilkins, 1928.

TABLE A—3b. Potassium Biphthalate-HCl Buffer of Clark and Lubs ($pK_a' = 2.8$)

Potassium biphthalate ($KHC_8H_4O_4$), 0.2 M: 40.83 g reagent grade to 1 liter with water.

HCl, 0.1 N: see *Michaelis "Universal Buffer,"* Table A—3m.

Each mixture is made up to 100 ml with water.

There is no appreciable change (about 0.02 pH unit) in the pH of these buffer mixtures at 37°C.

pH (20°C)	HCl, 0.1 N (ml)	$KHC_8H_4O_4$, 0.2 M (ml)	pH (20°C)	HCl, 0.1 N (ml)	$KHC_8H_4O_4$, 0.2 M (ml)
2.2	46.60	25	3.2	14.80	25
2.4	39.60	25	3.4	9.95	25
2.6	33.00	25	3.6	6.00	25
2.8	26.50	25	3.8	2.65	25
3.0	20.40	25			

References: Clark WM, Lubs HA: *J Biol Chem* 25:479, 1916. Hamer WJ, Pinching GD, Acree SF: *J Res Natl Bur Std (US)* 35:539, 1945.

TABLE A–3c. Acetic Acid-Sodium Acetate Buffer ($pK_a' = 4.64$)

CH_3COONa, 0.2 N: 16.41 g reagent grade CH_3COONa or 27.22 g reagent grade $CH_3COONa \cdot 3H_2O$ diluted to 1 liter with water.

CH_3COOH, 0.2 N: dilute 11.5 ml reagent grade glacial CH_3COOH to 1 liter with water. Standardize against 0.1 N NaOH with phenolphthalein as indicator. Dilute to exactly 0.2 N or add an equivalent amount and make up the difference with water.

At 38°C these buffer mixtures are about 0.05 pH unit lower.

pH (25°C)	CH_3COOH, 0.2 N (ml)	CH_3COONa, 0.2 N (ml)	pH (25°C)	CH_3COOH, 0.2 N (ml)	CH_3COONa, 0.2 N (ml)
3.6	92.5	7.5	4.8	41.0	59.0
3.8	88.0	12.0	5.0	30.0	70.0
4.0	82.0	18.0	5.2	21.0	79.0
4.2	73.5	26.5	5.4	14.0	86.0
4.4	63.0	37.0	5.6	9.0	91.0
4.6	52.0	48.0	5.8	6.0	94.0

References: Bates RG, Siegel GL, Acree SF: *J Res Natl Bur Std (US)* 30:347, 1943. Walpole GS: *J Chem Soc 105*:2501, 1914.

TABLE A–3d. Potassium Biphthalate-NaOH Buffer of Clark and Lubs ($pK_a' = 5.0$)

Potassium biphthalate ($KHC_8H_4O_4$); 0.2 M: 40.83 reagent grade to 1 liter with water.

NaOH; 0.1 N: If not exactly 0.1 N, can add equivalent amount.

Each mixture is made up to 100 ml with water.

There is no appreciable change (about 0.02 pH unit) in the pH of these buffer mixtures at 37°C.

pH (20°C)	NaOH, 0.1 N (ml)	$KHC_8H_4O_4$, 0.2 M (ml)	pH (20°C)	NaOH, 0.1 N (ml)	$KHC_8H_4O_4$, 0.2 M (ml)
4.0	0.40	25	5.2	29.75	25
4.2	3.65	25	5.4	35.25	25
4.4	7.35	25	5.6	39.70	25
4.6	12.00	25	5.8	43.10	25
4.8	17.50	25	6.0	45.40	25
5.0	23.65	25	6.2	47.00	25

References: Clark WM, Lubs, HA: *J Biol Chem 25*:479, 1916.
Hamer WJ, Acree, SF: *J Res Natl Bur Std (US) 35*:381, 1945.

TABLE A–3e. Sørensen's Phosphate Buffer (pK_a' = 6.7)

Na_2HPO_4 ; $M/15$: 9.47 g anhydrous, reagent grade to 1 liter with water.
Reagent grade contains no more than 0.3% moisture but takes up water on exposure to air when the relative humidity is greater than 41%. If there is any question about its condition it should be dried at 110°C for 2 hours or in a desiccator over $CaCl_2$.

KH_2PO_4 ; $M/15$: 9.08 g reagent grade to 1 liter with water.
The pH values in the table below are for 20°C; at 37°C the pH values are only about 0.03 pH unit less.

pH (20°C)	Na_2HPO_4, $M/15$ (ml)	KH_2PO_4, $M/15$ (ml)	pH (20°C)	Na_2HPO_4, $M/15$ (ml)	KH_2PO_4, $M/15$ (ml)
5.4	3.0	97.0	7.0	61.1	38.9
5.6	5.0	95.0	7.1	66.6	33.4
5.8	7.8	92.2	7.2	71.5	28.5
5.9	9.9	90.1	7.3	76.8	23.2
6.0	12.0	88.0	7.4	80.4	19.6
6.1	15.3	84.7	7.5	84.1	15.9
6.2	18.5	81.5	7.6	86.8	13.2
6.3	22.4	77.6	7.7	89.4	10.6
6.4	26.5	73.5	7.8	91.4	8.6
6.5	31.8	68.2	7.9	93.2	6.8
6.6	37.5	62.5	8.0	94.5	5.5
6.7	43.5	56.5	8.1	95.8	4.2
6.8	50.0	50.0	8.2	97.0	3.0
6.9	55.4	44.6			

References: Bates RG: *J Res Natl Bur Std (US)* 39:411, 1947. Bates RG, Acree SF: *J Res Natl Bur Std (US) 34*:373, 1945. Sørensen SPL: *Biochem Z 21*:131, 1909; *22*:352, 1909.

TABLE A–3f. Imidazole Buffer (pK_a' = approximately 7)

Imidazole; 0.2 M: 1.362 g imidazole (glyoxaline) to 100 ml with water. Obtainable from Eastman Kodak; should be dried in a desiccator before use.

HCl; 0.1 N: See under *Michaelis "Universal Buffer,"* Table A–3m.
To prepare 0.05 M buffer mixtures of the pHs listed in the table below, add the corresponding quantity of 0.1 N HCl, or its equivalent, to 25 ml 0.2 M imidazole and dilute to 100 ml with water. Buffers are stable at room temperature for at least several months.

pH (25°C)	HCl, 0.1 N (ml)	pH (25°C)	HCl, 0.1 N (ml)	pH (25°C)	HCl, 0.1 N (ml)
6.2	42.9	6.8	30.4	7.4	13.6
6.4	39.8	7.0	24.3	7.6	9.3
6.6	35.5	7.2	18.6	7.8	6.0

Reference: Mertz ET, Owen CA: *Proc Soc Exp Biol Med 43*:204, 1940.

TABLE A—3g. Buffers of Gomori

Collidine, 0.2 M: 2.64 ml 2,4,6-collidine (s-collidine) to 100 ml with water. Obtainable from Eastman Kodak.

Tris(hydroxymethyl)aminomethane (known as Tris), 0.2 M: 2.43 g to 100 ml with water. Obtainable from G.F. Smith and Commercial Solvents.

2-Amino-2-methyl-1,3-propanediol, 0.2 M: 2.1 g to 100 ml with water. Obtainable from Commercial Solvents. This material is slightly hygroscopic and should therefore be dried in a desiccator.

HCl, 0.1 N: See under *Michaelis "Universal Buffer,"* Table A—3m.

To prepare 0.05 M buffer mixtures of the pHs listed in the table below, add the corresponding quantity of 0.1 N HCl, or its equivalent, to 25 ml of the respective 0.2 M solutions above and dilute to 100 ml with water. The quantities were calculated graphically from the data of Gomori. Buffers are stable at room temperature for at least several months. The pHs in the table are for 23°C; at 37°C the pHs are approximately 0.1 pH unit lower.

	ml 0.1 N HCl to be added to 25 ml of following 0.2 M solutions		
pH (23°C)	Collidine $pK_b' = 6.6$	Tris(hydroxymethyl)-aminomethane $pK_b' = 5.76$	2-Amino-2-methyl-1,3-propanediol $pK_b' = 5.22$
6.5	44.2	llll	
6.6	42.8		
6.7	41.4		
6.8	40.0		
6.9	38.0		
7.0	35.7		
7.1	33.1		
7.2	30.5	45.0	
7.3	27.5	43.5	
7.4	25.0	42.0	
7.5	22.5[a]	40.5	
7.6	19.0	38.9	
7.7	16.5	36.6	
7.8	14.0	34.0	
7.9	12.0	31.8	43.9
8.0	10.0	29.0	42.5
8.1	8.4	26.2	41.2
8.2	7.0	23.3[a]	39.6
8.3	5.8	20.5	37.5
8.4		17.5	35.0
8.5		15.0	32.5
8.6		12.8	30.0
8.7		10.8	27.5
8.8		9.0	24.5[a]
8.9		7.8	21.6
9.0		6.3	18.8
9.1		5.0	16.2
9.2			13.6
9.3			11.5
9.4			9.6
9.5			8.2
9.6			6.7
9.7			5.3

[a] Approximate pH at half-neutralization of base.

Reference: Gomori G: *Proc Soc Exp Biol Med* 62:33, 1946.

TABLE A–3h. Veronal Buffer (pK_a' = 8.0)

Sodium Veronal, 0.1 M: 20.60 g sodium diethylbarbiturate (sodium Veronal, sodium barbital) in water to 1 liter. USP grade satisfactory without recrystallization or drying. 10 ml of this solution should require 10 ml N HCl to reach endpoint with methyl red indicator.

HCl, 0.1 N: See under *Michaelis "Universal Buffer,"* Table A-3m.

At 37°C the buffer mixtures are about 0.1 pH unit lower.

pH (25°C)	Na Veronal, 0.1 N (ml)	HCl, 0.1 N (ml)	pH (25°C)	Na Veronal, 0.1 M (ml)	HCl, 0.1 N (ml)
6.8	52.2	47.8	8.4	82.3	17.7
7.0	53.6	46.4	8.6	87.1	12.9
7.2	55.4	44.6	8.8	90.8	9.2
7.4	58.1	41.9	9.0	93.6	6.4
7.6	61.5	38.5	9.2	95.2	4.8
7.8	66.2	33.8	9.4	97.4	2.6
8.0	71.6	28.4	9.6	98.5	1.5
8.2	76.9	23.1			

References: Manov GG, Schuette KE, Kirk FS: *J Res Natl Bur Std (US)* 48:84, 1952. Michaelis L: *J Biol Chem* 87:33, 1930.

TABLE A–3i. Veronal-Sodium Veronal Buffer (pK_a' = 8.0)

Sodium Veronal, 0.04 M: 8.24 g sodium diethylbarbiturate (sodium Veronal, sodium barbital) in water to 1 liter. USP grade satisfactory without recrystallization or drying.

Veronal, 0.04 M: 7.36 g diethylbarbituric acid (Veronal, barbital, barbitone) in water to 1 liter. USP grade satisfactory without recrystallization or drying. May be necessary to heat to effect complete solution.

The table below was obtained by graphic interpolation of unpublished data obtained on solutions made as above and using a Beckman model G pH meter. At 37°C the buffer mixtures read about 0.1 pH unit lower.

pH (25°C)	Na Veronal, 0.04 M (ml)	Veronal 0.04 M (ml)	pH (25°C)	Na Veronal 0.04 M (ml)	Veronal 0.04 M (ml)
7.0	10	90	8.0	50	50
7.1	12.5	87.5	8.1	57.5	42.5
7.2	15.5	84.5	8.2	65	35
7.3	19	81	8.3	72	28
7.4	22.5	77.5	8.4	76.5	23.5
7.5	26	74	8.5	80	20
7.6	30	70	8.6	83	17
7.7	34.5	65.5	8.7	85.5	14.5
7.8	39.5	60.5	8.8	88	12
7.9	44.5	55.5	8.9	90	10

Reference: Manov GG, Schuette KE, Kirk FS: *J Res Natl Bur Std (US)* 48:84, 1952.

TABLE A–3j. H_3BO_3 –KCl–NaOH Buffer of Clark and Lubs (pK_a' = 9.2)

H_3BO_3, 0.2 M/KCl, 0.2 M: 12.37 g reagent grade H_3BO_3 + 14.91 g reagent grade KCl to 1 liter with water.

NaOH, 0.1 N: If not exactly 0.1 N, can add equivalent amount. Each mixture listed in the table is made up to 100 ml with water.

The pH of these buffer mixtures is about 0.1 pH unit lower at 37°C.

pH (20°C)	NaOH, 0.1 N (ml)	H_3BO_3/KCl, 0.2 M (ml)	pH (20°C)	NaOH, 0.1 N (ml)	H_3BO_3/KCl, 0.2 M (ml)
7.8	2.65	25	9.0	21.40	25
8.0	4.00	25	9.2	26.70	25
8.2	5.90	25	9.4	32.00	25
8.4	8.55	25	9.6	36.85	25
8.6	12.00	25	9.8	40.80	25
8.8	16.40	25	10.0	43.90	25

References: Clark WM, Lubs HA: *J Biol Chem* 25:479, 1916. Manov GG, DeLollis NJ, Lindvall PW, Acree SF: *J Res Natl Bur Std (US)* 36:543, 1946.

TABLE A–3k. Sodium Carbonate-Bicarbonate Buffer (pK_a' = 9.9)

Na_2CO_3, 0.1 M: 10.60 g reagent grade anhydrous salt to 1 liter with water. Reagent grade contains a maximum moisture content of 1% and is hygroscopic; if there is any question of the moisture being greater than 1% the substance should be dried in a desiccator or at 250°–300°C.

$NaHCO_3$, 0.1 M: 8.40 g reagent grade to 1 liter with water.

These buffer mixtures are stable up to 6 months in well stoppered polyethylene containers. The table below was obtained by graphic interpolation of unpublished data obtained on solutions made as above and using a Beckman model G pH meter. At 37°C the buffer mixtures read about 0.1 pH unit lower.

pH (25°C)	Na_2CO_3, 0.1 M	$NaHCO_3$, 0.1 M	pH (25°C)	Na_2CO_3, 0.1 M	$NaHCO_3$, 0.1 M
9.1	11.3	88.7	9.9	51.5	48.5
9.2	14	86	10.0	58	42
9.3	18	82	10.1	64	36
9.4	22	78	10.2	69.5	30.5
9.5	27	73	10.3	74.5	25.5
9.6	32.5	67.5	10.4	79	21
9.7	38.5	61.5	10.5	83.5	16.5
9.8	45	55	10.6	88	12

TABLE A–3l. Sørensen's Glycine-Sodium Hydroxide Buffer.

Glycine, 0.1 M/NaCl, 0.1 M: 7.505 g reagent grade glycine (glycocoll, aminoacetic acid) + 5.85 g reagent grade NaCl to 1 liter with water.

NaOH, 0.1 N.

The values below are interpolated from Sørensen's data for $25°$C; pH values at $37°$C would be about 0.3 pH unit lower.

pH (25°C)	Glycine, 0.1 M/ NaCl, 0.1 M (ml)	NaOH, 0.1 N (ml)	pH (25°C)	Glycine, 0.1 M/ NaCl, 0.1 M (ml)	NaOH, 0.1 N (ml)
8.4	95	5	11.1	50	50
8.7	90	10	11.4	49	51
9.1	80	20	11.8	45	55
9.5	70	30	12.2	40	60
9.9	60	40	12.4	30	70
10.3	55	45	12.6	20	80
10.8	51	49	12.8	10	90

Reference: Sørensen SPL: *Ergeb Physiol 12*:393, 1912.

TABLE A–3m. Michaelis "Universal Buffer" ($pK_a' = 4.7, 8.0$)

All pH values are at constant ionic strength ($\mu = 0.16$) and isotonic with whole blood. This system has no insoluble Ca salt. Prepare 9.714 g $CH_3COONa·3H_2O$ + 14.714 g USP grade sodium diethylbarbiturate (sodium Veronal, sodium barbital) in water to 500 ml. To each 5 ml of this solution add 2 ml 8.5% NaCl + a ml 0.1 N HCl + $(18 - a)$ ml water.

HCl 0.1 N is prepared by adding 10 ml conc reagent grade HCl to 1 liter water. Determine normality by titration versus Na_2CO_3 or a standardized solution of NaOH. Then adjust to 0.1 N by proper dilution (it always titrates higher than 0.1 N).

pH	a	pH	a	pH	a	pH	a
2.62	15.0	4.66	9.0	7.25	5.5	8.68	0.75
3.20	14.0	4.93	8.0	7.42	5.0	8.90	0.50
3.62	13.0	5.32	7.0	7.66	4.0	9.16	0.25
3.88	12.0	6.12	6.5	7.90	3.0	9.64	0.0
4.13	11.0	6.99	6.0	8.18	2.0		
4.33	10.0			8.55	1.0		

Reference: Michaelis L: *Biochem Z 234*:139, 1931.

TABLE A—3n. McIlvaine Standard Buffer ($pK_a' = 2.1, 3.1, 4.7, 6.4, 6.7$)

McIlvaine employed a mixture of $0.2\,M$ Na_2HPO_4 and $0.1\,M$ citric acid. The citrate functions as a buffer in the pH region between that buffered by the H_3PO_4/NaH_2PO_4 system (see Bates) and the NaH_2PO_4/Na_2HPO_4 system. The ionic strength can be maintained constant by addition of proper amounts of KCl (see Elving *et al*).

Na_2HPO_4, $0.2\,M$: 28.41 g anhydrous reagent grade to 1 liter with water. Reagent grade contains no more than 0.3% moisture but takes up water on exposure to air. If there is any question about its condition it should be dried at $105°C$ or in a desiccator over $CaCl_2$.

Citric acid, $0.1\,M$: Reagent grade available as the monohydrate, which is efflorescent and must therefore be kept in a tightly closed container. Can convert to anhydrous by heating to constant weight at $110°C$. To 19.21 g of anhydrous acid or 21.01 g of the monohydrate, add water to 1 liter.

pH	Na_2HPO_4, $0.2\,M$ (ml)	Citric acid, $0.1\,M$ (ml)	pH	Na_2HPO_4, $0.2\,M$ (ml)	Citric acid, $0.1\,M$ (ml)
2.2	0.40	19.60	5.2	10.72	9.28
2.4	1.24	18.76	5.4	11.15	8.85
2.6	2.18	17.82	5.6	11.60	8.40
2.8	3.17	16.83	5.8	12.09	7.91
3.0	4.11	15.89	6.0	12.63	7.37
3.2	4.94	15.06	6.2	13.22	6.78
3.4	5.70	14.30	6.4	13.85	6.15
3.6	6.44	13.56	6.6	14.55	5.45
3.8	7.10	12.90	6.8	15.45	4.55
4.0	7.71	12.29	7.0	16.47	3.53
4.2	8.28	11.72	7.2	17.39	2.61
4.4	8.82	11.18	7.4	18.17	1.83
4.6	9.35	10.65	7.6	18.73	1.27
4.8	9.86	10.14	7.8	19.15	0.85
5.0	10.30	9.70	8.0	19.45	0.55

References: Bates RG: *Electrometric pH Determinations*. New York, Wiley, 1954. Elving PJ, Markowitz JM, Rosenthal I: *Anal Chem* 28:1179, 1956. McIlvaine TC: *J Biol Chem* 49:183, 1921.

TABLE A—4. Primary Standards for Acid-Base and Redox Titrations

Primary standards	% purity of reagent grade, ACS specs.	Natl. Bur. Std. catalog No.	Properties	mEq weight (grams)
Acids				
Potassium biphthalate	>99.8	84d	Nonhygroscopic, stable	0.2042
Benzoic acid	99.9—100.2	39g		0.1221
Tartaric acid	99.8—100.2			0.07505
Oxalic acid	99.8—100.2		Effloresces in warm, dry air	0.06304
Base				
Na_2CO_3, anhydrous	Max. 1% moisture		Hygroscopic; heat to $250°—300°C$ to constant weight	0.0530
Oxidants				
$K_2Cr_2O_7$	99.7—100.2	136		0.04904
KIO_3	99.8—100.3			0.03567
Reductants				
Sodium oxalate	99.9—100.2	40e		0.06700
Oxalic acid	99.8—100.2		Effloresces in warm, dry air	0.06304

TABLE A—5. Preparation of Working Standards for Acid-Base and Redox Titration

Solution	Preparation of approx 0.1 N	Standardization
HCl	Dilute 9.0 ml to 1 liter.	Titrate vs primary standard Na_2CO_3 or standardized NaOH, using methyl orange, methyl red, or alizarin red. Phenolphthalein satisfactory if N is 0.1 or greater.
NaOH	Add approx. 110 g NaOH to 100 ml water in 300 ml flask, slowly with stirring. Stopper and let stand several days. Clear supernate contains ca 75 g/100 ml. Dilute 5.33 ml to 1 liter. Store in polyethylene bottle. Freshly boiled (CO_2-free) water must be used for preparation of solutions <0.01 N.	Titrate vs primary standard acid or standardized HCl using phenolphthalein.
$KMnO_4$	Add 3.2 g to 1 liter in glass-stoppered bottle or flask. Let stand in dark several days. Filter through glass wool. Store in amber glass-stoppered bottle in refrigerator.	Titrate vs primary standard sodium oxalate or oxalic acid. Weigh out 100 mg, add 25 ml 1 N H_2SO_4, and heat to ca $70°C$. Titrate to pink that persists at least 30 sec.
$Na_2S_2O_3$	Add 25 g to 1 liter. Store in refrigerator in amber glass-stoppered bottle.	Weigh out 100 mg KIO_3 or $K_2Cr_2O_7$. Add 25 ml water, approximately 1 g KI, and several drops conc HCl. Titrate to pale yellow. Add 1 ml 1% starch (1 g soluble starch in 10 ml boiling water, add 90 ml saturated solution of NaCl) and continue titration until blue color is discharged and solution is pale green.

TABLE A—6. Information on Drying Agents

Chemical	Proprietary name	Manufacturer	Drying intensity, approx mm Hg partial pressure or mg H_2O left in 1 liter gas dried at room temp	Capacity for H_2O,[a] % of initial weight	Method of regeneration	Comments
Na_2SO_4			12	127	Not reused	For drying liquids only
$CaCl_2$ "anhydrous"			1.25–1.5	144	Not reused	Average composition $CaCl_2 \cdot 0.25H_2O$ to $CaCl_2 \cdot H_2O$
$CaCl_2$, fused			0.36	162	Heat $CaCl_2$ to fusion in open Pt dish; break up while hot	Must be prepared by analyst; not available commercially Also absorbs alcohol
H_2SO_4, conc			0.3	Good to ca 15	Not reused	Also absorbs alcohol
Silica gel (SiO_2)		Davison Chemical	0.01	45	175°C several hours	"Tel-Tale" Grade 42, the indicator grade, goes from blue to pink on exhaustion
$CaSO_4$, anhydrous	Drierite	Hammond Drierite	0.005	11	235°–250°C 1–2 hours	"Indicating Drierite" contains $CoCl_2$, going from blue to red on exhaustion
Alumina (Al_2O_3)	Activated Alumina, Grade F-1	Alcoa	0.0009	19	175–315°C overnight	Grade F-6, the indicator grade, contains $CoCl_2$, going from blue to red on exhaustion
BaO	Barco Porous Ba	Barium and Chemicals, Inc.	0.0006			
$Mg(ClO_4)_2$, anhydrous	Anhydrone Baker Chemical; G.F. Smith Dehydrite[b] A.H. Thomas		< 0.00002	48	Heat at high vacuum in metal vacuum oven[c]	Also absorbs NH_3
P_2O_5			< 0.00002	13	Not reused	Scrape surface before use; channels badly

[a] In the case of compounds forming hydrates the capacity given is that to the highest hydrate.

[b] Trade name formerly applied to the trihydrate.

[c] If not hydrated beyond diaquo form, heat at 200°–250°C for 6–10 hours; if hexaquo form, heat to 130°–135° for 6–10 hours then at 200°–250° for 6–10 hours.

References: Bower JH: *J Res Nat Bur Std (US) 12*:241, 1934. Robertson GR: *Laboratory Practice of Organic Chemistry.* New York, Macmillan, 1943. *Activated*

TABLE A—7. Adsorbents Acting by Physical Adsorption

Adsorbent	Trade name	Manufacturer	Comments
Sucrose			
Talc			A natural magnesium silicate
Magnesium silicate	Florisil	Floridin	A synthetic product
Magnesium trisilicate		Baker Chemical	
Slaked lime: $Ca(OH)_2$			
$CaCO_3$			
$Ca_3(PO_4)_2$			
Magnesia (MgO)			
$MgCO_3$			
Silicic acid (H_4SiO_4)			
Alumina (Al_2O_3)	Activated Alumina, Grade F-20	Alcoa	80–200 mesh
	"Chromatography Grade"	Merck	
	Florite	Floridin	
	Woelm alumina	M. Woelm,	Each form available
	Basic form	distributed	in five grades (vary in
	Neutral form	by Waters	H_2O content); basic
	Acid form	Associates,	form acts as cation
		Inc.	exchanger in aqueous
			solution; acid form in
			aqueous and alcoholic
			solution serves as an
			anion exchanger
	Fluka alumina	Fluka AG	Activity grade I
	Type 5016 A, basic		Mean grain size = 0.13
	pH 9.5±0.2		mm
	Type 507 C, neutral pH 7.0±0.5		
	Type 506 C, weakly acid, pH 6.0±0.5		
	Type 504 C, acid, pH 4.5±0.3		
	Activated alumina	Bio-Rad	Each form available
	AG 10, basic form	Laboratories	in five grades (vary in
	AG 7, neutral form		H_2O content) and in
	AG 4, acid form		3 meshes: 50–100,
			100–200, and minus
			200
Activated charcoal	Norit (from wood)	Greef	Norit N.F. is in accordance
	Washed grades		with National Formulary
	Norit N.F.		specs; Norit L.I. has only
	Norit L.I.		traces of iron
	Nuchar T	Westvaco	Nuchar T is neutral and
			conforms with USP
			requirements for
			activated carbons with
			varying properties
			available
	Carbo-Dur	Permutit	

(Continued)

TABLE A–7 (continued)

Adsorbent	Trade name	Manufacturer	Comments
Aluminum silicates	Kaolin (china clay)		Properties of clays vary
	Bentonite		with origin; unless used
	Super Filtrol	Filtrol	with aqueous solution,
	Special Filtrol		activate at 650°C for
			1–2 hours before use
	Fuller's earths		
	Lloyd's reagent	A purified fuller's earth sold through pharmaceutical supply houses	
	Florex	Floridin	
	Florigel		
	Diluex		
	Florite (a bauxite)	Floridin	

TABLE A–8. Adsorbents Acting by Partitioning

Adsorbent	Trade name	Manufacturer	Comments
Starch Cellulose powder	"For Chromatography"	Whatman, distrib. by H. Reeve Angel	Two qualities available 1. Ashless—acid washed 2. "B" quality—not acid washed Both available in two grades 1. Standard—close pack (200 mesh) 2. Coarse—loose pack
Silica gel ($SiO_2 \cdot xH_2O$)		Davison Chemical	Regular grades: 912, 922, 963 Purified grades (minimal metal oxides): 923, 950 Special grades: 70, 35
Diatomaceous earth (kieselguhr)	Celite filter-aids Analytical Filter-Aid Others in order of flow rate: Filter-Cel Celite 505 Standard Super-Cel Celite 512 Hyflo Super-Cel Celite 501 Celite 503 Celite 535 Celite 545	Johns-Manville	Low in metallic oxides

TABLE A–9. Adsorbents for Thin Layer Chromatography

Adsorbent	Trade name	Manufacturer	Comments
Silicic acid (H_4SiO_4)	Silica Gel Type 60 (with $CaSO_4$)	E. Merck	Mean particle size $10-40\ \mu$
	Silica Gel H (without binder)	E. Merck	Mean particle size $10-40\ \mu$
	Silica Gel HF (with inorganic fluorescent indicator, 254 A; no binder)	E. Merck	Mean particle size $10-40\ \mu$
Alumina (Al_2O_3)	Aluminum Oxide G (neutral) (with $CaSO_4$)	E. Merck	
	Fluka Type DO (50512)	Fluka AG	Mean grain size approximately $2\ \mu$
	Fluka Type D5 (50512a)	Fluka AG	Contains 5% $CaSO_4$
	Fluka Type D5F (50512b)	Fluka AG	Contains 5% $CaSO_4$ plus fluorescent indicator
Aluminum silicates	Kieselguhr G (with $CaSO_4$)	E. Merck	
Cellulose powders			
Normal	MN 300 (without binder)	Macherey, Nagel	Obtained through Brinkmann
Acetylated powders (10%)	MN 300/AC (without binder)	Macherey, Nagel	Obtained through Brinkmann
Ion exchange powders	MN 300 CM (without binder)	Macherey, Nagel	Carboxymethylcellulose, obtained through Brinkmann
	MN 300 P (without binder)	Macherey, Nagel	Cellulose phosphate (Brinkmann)
	MN 300 DEAE	Macherey, Nagel	Diethylaminoethyl-cellulose (Brinkmann)
Anion exchange powders	MN 300 ECTEOLA (without binder)	Macherey, Nagel	Cellulose prepared with ECTEOLA (Brinkmann)

TABLE A—10. Ion Exchange Crystals

Characteristics, name	Type	Form in which sold	Operating pH range
Cation exchanger, very strong			
Bio-Rad ZP-1	Zirconium phosphate micro crystalline gel	20—50 mesh 50—100 mesh 100—200 mesh	<1—13
Bio-Rad AMP-1	Ammonium molybdo-phosphate micro crystals	Microcrystalline	<1—6
Bio-Rad HZO-1 (in basic media)	Hydrous zirconium oxide micro crystalline gel	20—50 mesh 50—100 mesh 100—200 mesh	7—14
Anion exchanger, very strong			
Bio-Rad HZO-1 (in acidic media)	Hydrous zirconium oxide micro crystalline gel	20—50 mesh 50—100 mesh 100—200 mesh	1—7

Products in this table are available from Bio-Rad.

TABLE A–11. Ion Exchange Resins

Characteristic, name	Manufacturer[a]	Type	Form in which sold	Operating pH range
Cation exchangers				
Strong				
Amberlite IR-120[c]	A	Styrene-divinylbenzene copolymer ($-SO_3H$)	Na	1–14
Dowex 50[b]	C	Styrene-divinylbenzene copolymer ($-SO_3H$)	20–50 mesh Na; other meshes H	1–14
Permutit Q[h]	B	Sulfonated polystyrene ($-SO_3H$)	Na and H	0–13
Zeo-Karb[h]	B	Sulfonated coal ($-SO_3H$, $-COOH$, and $-OH$)	Na and H	0–11
Zeo-Dur[g, h]	B	A processed glauconite (naturally occurring greensand)	Na	6.2–8.7
Decalso[g, h, i]	B	A precipated gel-type sodium aluminosilicate	Na	6.9–7.9
Weak				
Amberlite IRC-50[c]	A	Styrene-divinylbenzene copolymer ($-COOH$)	H	7–14
Permutit H-70[h]	B	Styrene-divinylbenzene copolymer ($-COOH$)	H	3.5–12
Anion exchangers				
Strong				
Amberlite IRA-400[c]	A	Styrene-divinylbenzene copolymer ($-NR_3{}^+OH^-$)	Cl	0–12
Amberlite IRA-401[c, d]	A	Styrene-divinylbenzene copolymer ($-NR_3{}^+OH^-$)	Cl	0–12
Dowex 1[b, e]	C	Styrene-divinylbenzene copolymer (all three R groups are methyl groups: $-NR_3{}^+OH^-$)	Cl	0–12

		copolymer (one of the three methyl groups replaced by an ethanol group: $-NR_3{}^+OH^-$)		
Permutit S^h	B	Styrene-divinylbenzene copolymer ($-NR_3{}^+OH^-$)	SO_4	0–13.9
Permutit A^{f,h}	B	Styrene-divinylbenzene copolymer ($-NR_2H^+OH^-$ and $-NR_3{}^+OH^-$)	SO_2	0–13.9
Weak				
Amberlite IR-4B^c	A	Phenol-formaldehyde resin, polyamine	OH	0–7
Amberlite IR-45^c	A	Styrene-divinylbenzene copolymer, polyamine	OH	0–7
Dowex^{b,h}	C	Styrene-divinylbenzene copolymer ($-NH_3{}^+OH^-$, $-NRH_2{}^+OH^-$, and $-NR_2H^+OH^-$)	OH	0–7
Permutit W^h	B	Styrene-divinylbenzene copolymer ($-NR_2H^+OH$)	SO_4	0–13.9
DeAcidite^h	B	Styrene-divinylbenzene copolymer ($-NRH_2{}^+OH^-$ and $-NR_2H^+OH$)	SO_4	0–12
DEAE-Sephadex A-25 A-50 (for mol. wt. > 10 000)	D	Cross linked dextran with added diethylaminoethyl ether	Cl$^-$; coarse (50–100 mesh), medium (100–250 mesh), fine (250–400 mesh)	1–14

^{a}Code to manufacturers: A, Rohm and Haas; B, Permutit; C, Dow Chemical; D, Pharmacia.

^{b}Dowex resins are available in different degrees of cross linkage: X1, X2, X4, X8, X10, and X12, the numbers corresponding to percent divinylbenzene. The following meshes are available: 20–50, 50–100, 100–200, 200–400, and colloidal. Analytic grade Dowex resins are available from Bio-Rad.

^{c}Amberlite resins are available in about 20–50 mesh. Most resins also available in analytic grade. Amberlite resins can be purchased in 1 and 5 pound quantities from Fisher Scientific.

^{d}Amberlite IRA-401 is a more porous form of IRA-400.

^{e}Dowex 2 is slightly less basic than Dowex 1.

^{f}Permutit A is actually a "medium strength anionic resin."

^{g}Usual operating pH range is 6–8. Regenerate with salts, not acids.

^{h}Exchangers available in about 10–50 mesh.

^{i}"Permutit, Folin" is similar to Decalso but of finer mesh.

TABLE A–12. Approximate Composition of Various Types of Water

| Composition | ACS specs., no more than | Typical US water supplies | Distilled water | | Deionized water |
			Single	Double (from glass)	
Specific conductance (micromhos)			2–6		0.1
pH		7.3–8.3	5.8–6.6		ca 7
ppm					
Total solids	1	31–1119	1.5–4		27
SiO_2		2.1–25	0.00		22.8
Fe		0–13	0.000–0.006		0.001
Ca		4.4–143	0.15		0.01
Mg		0.5–21	0.004		0.5
Na		1.7–?	0.33		0.0
K		0.7–?			0.0
Heavy metals (as Pb)	0.01		0.000–0.055	0.0025	0.0015
Cu			0.002–0.074	0.0016	0.0035
Zn			0.00–0.02	0.00	0.00
NH_3 (as N)	0.1		0.01		
HCO_3		14–451	7		
SO_4		4.9–382	0.00		
Cl	0.1	1.4–276	0.00		0.00
O_2			6		

References: Kunin R, Myers RJ: *Ion Exchange Resins.* New York, Wiley, 1950. Liebig GF Jr, Vanselow AP, Chapman HD: *Soil Sci 55:371, 1943.*

Gas	Grade	Purity (%)	Maximum cylinder pressure (psi)	State of gas in cylinder	b.p (°C)	Flammable limits in air (vol %)	Cu ft/lb at 21°C	Special precautions
Acetylene	Research	99.6	250	Dissolved in acetone	−84.0	2.5–81	14.5	Never use free gas outside cylinder at pressure greater than 15 psi. Never use with Cu tubing.
Air			2200–2400	Gas	−191.5		ca 15	
Ammonia	Research	99.5	114	Liquid	−33.4	15–28	22.6	Do not use with brass.
Butane	C.P.	99.9	17	Liquid	−0.5	1.9–8.5	6.4	Extremely flammable.
	Technical	95.0						
Carbon dioxide	"Bone dry"	99.956	830	Liquid	−78.5		8.76	
	"Beverage"	99.5						
Carbon monoxide		96.8	1500	Gas	−191.5	12–75	13.8	Odorless poison.
Hydrogen		99.9[a]	2000	Gas	−252.9	4–75	191.7	If valve is opened wide, may ignite by friction; close valve. Extremely flammable.
Hydrogen sulfide		99.9	250	Liquid	−60.3	4.3–45	11.2	
Nitrogen			2200	Gas	−195.8		13.8	
Oxygen		99.6	2200	Gas	−183		12.1	Never use oil or grease on valves or fittings.
Propane	Instrument grade	99.9[a]	109	Liquid	−42.7	2.4–9.5	8.5	Extremely flammable.
	C.P.	99						
	Natural	96						
	Commercial	65						
Sulfur dioxide	Anhydrous	99.988	35	Liquid	−10.0		5.9	Always use a trap when bubbling into liquid to prevent suck-back into cylinder. Use only in well ventilated area.
	Commercial	99.90						

[a]Research grade also available with an accompanying analysis.

Reference: Braker W, Mossman AL: *Matheson Gas Data Handbook*, 5th ed. East Rutherford, NJ, Matheson Gas Products, 1971.

TABLE A—14 Source List

Arthur H. Thomas Co., P.O. Box 779, Philadelphia, Pa. 19105

Abbott Laboratories, Abbott Scientific Products Div., 820 Mission St. South Pasadena, Ca. 91030

Abbott Laboratories, Radio-Pharmaceutical Products Division, North Chicago, Il. 60064.

AGA Corporation, 467 Forbes Blvd., South San Francisco, Ca. 94080

Aluminum Company of America, Chemical Div., 1501 Alcoa Building, Pittsburg, Pa. 15219

Alfa American Corporation, 5420 Walker Ave., Rockford, Il. 61111

Allied Chemical Corporation, Specialty Chemicals Div., P.O. Box 1087R, Morristown, N.J. 07960

Amend Drug and Chemical Co., 83 Cordier St., Irvington, N. J. 07111

American Can Co., Chemical Products Dept., American Lane, Greenwich, Ct. 06830

American Cyanamid Co., Berdan Ave., Wayne, N. J. 07470

American Instrument Co., Inc., 8030 Georgia Ave., Silver Springs, Md 20910

American Monitor Corporation, P.O. Box 68505, Indianapolis, In. 46268

American Type Culture Collection 12301 Parklawn Dr., Rockville, Md. 20852

Amersham Searle, 2636 S. Clearbrook Drive, Arlington Heights, Il., 60005

Ames Co., Div. of Miles Laboratories, 810 McNaughton Ave., Elkhart In. 46514

Amicon Corporation, 21 Hartwell Ave., Lexington, Ma. 02173

Aminco, see American Instrument Co.

Armour Pharmaceutical Co., Metrix Clinical and Diagnostic Div., 530 E. 31st St., Chicago, Il. 60616

AutoAnalyzer, See Technicon.

Baltimore Biological Laboratories, Div. of Bioquest, Cockeysville, Md. 21030

Baird-Atomic, Inc., 125 Middlesex Tpke., Bedford, Ma. 01730

Baker. See J. T. Baker Chemical Co.

Barium and Chemicals, Inc., P.O. Box 218, Steubenville, Oh. 43952

Bausch and Lomb, Inc., Scientific Instrument Div., 77471 Bausch St., Rochester, N.Y. 14602

Bausch and Lomb, Inc., Analytical Systems Div., 820 Linden Ave., Rochester, N.Y. 14625

Beckman Instruments, Inc., 2500 Harbor Blvd., Fullerton, Ca. 92634

Beckman Instruments, Inc., Spinco Div., 1117 California Ave., Palo Alto, Ca. 94304

Becton-Dickinson, Div. of Becton, Dickinson and Co., Rutherford, N.J. 07070

Bio-Logics, Inc., 1 Research Rd., Salt Lake City, Ut, 84112

Bioquest, P.O. Box 243, Cockeysville, Md. 21030

Bio-Rad Laboratories, 32nd and Griffin Ave., Richmond, Ca. 94804

Bio-Science Laboratories, 7600 Tyrone Ave., Van Nuys, Ca. 91405

C. F. Boehringer Mannheim Corp., 219 E. 44th St., New York, N.Y. 10017

Brinkmann Instruments, Inc., Cantiague St., Westbury, N.Y. 11590

The British Drug House, Ltd., Barclay Ave., Quennsway, Toronto 550, Ontario, Canada

Buchler Instruments Div., Nuclear Chicago Corp., 1327 16th St., Fort Lee, N.J. 07024

Calbiochem, P.O. Box 12087, San Diego, Ca. 92112

Cal-Glass for Research, Inc., 3012 Enterprise Ave., Costa Mesa, Ca. 92626

California Laboratory Equipment Co., 1399 64th St., Emeryville, Ca. 94608

Cambridge Instrument Co., 73 Spring St., Ossining, N.Y. 10562

Cappel Laboratories, Inc., Box 156, Downington, Pa. 19335

Cargille, see R. P. Cargille.

Carl Zeiss, Inc., 444 Fifth Ave., New York, N.Y. 10018

Table continued

TABLE A–14 (continued)

Cary Instruments, 2724 So. Peck Rd., Monrovia, Ca. 91016
Clarkson Chemical Co., Inc., P.O. Box 97, Williamsport, Pa. 17701
Colab Laboratories, Inc., 3 Science Rd., Glenwood, Il. 60425
Coleman Instruments, Inc., Div. of Perkin Elmer, 42 Madison St., Maywood, Il. 60153
Commercial Solvents Corporation, 245 Park Ave., New York, N.Y. 10017
Cordis Laboratories, P.O. Box 428, Miami, Fl. 33137
Corning Glass Works, Corning, N.Y. 14830
Corning Scientific Instruments, Medfield Industrial Park, Medfield, Ma. 02052
Coulter Electronics, Inc., 590 W. 20th St., Hialeah, Fl. 33010
Curtin Scientific Co., P.O. Box 1546, Houston, Tx. 77001.
Dade Division, American Hospital Supply Corporation, P.O. Box 672, Miami, Fl. 33152
Davison Chemical Div., W. R. Grace and Co., 101 N. Charles St., Baltimore, Md. 21203
Diagnostic Aids, 5192 Washington St., West Roxbury, Ma. 02132
Difco Laboratories, P.O. Box 1058A, Detroit, Mi. 48232
Digital Equipment Corporation, 146 Main St., Maynard, Ma. 01754
Dow Chemical Co., Midland, Mi. 48640
Dow Diagnostics, P.O. Box 1656, Indianapolis, In. 46206
Drummond Scientific Co., 500 Parkway, Broomall, Pa. 19008.
E. I. DuPont de Nemours and Co., Inc., Organic Chemicals Dept., Dyes and Chemicals Div., Nemours Bldg., Wilmington, De. 19898
E. I. DuPont de Nemours and Co., Instrument Products Div. 1007 Market St., Wilmington, De. 19898
Eastman Kodak Co., 343 State St., Rochester, N.Y. 14650
Eastman Organic Chemicals, Div. of Eastman Kodak, 343 State St., Rochester, N.Y. 14650
Electro-Nucleonics, Inc., 368 Passaic Ave., Fairfield, N.J. 07006
E. Leitz, Inc., Link Dr., Rockleigh, N.J. 07647
Eli Lilly and Co., 740 S. Alabama St., Indianapolis, In. 46206
E. Merck, Darmstat, see EM Laboratories, Inc.
EM Laboratories, Inc., 500 Executive Blvd., Elmsford, N.Y. 10523
Eppendorf Div., Brinkmann Instruments, Inc., Cantiague Rd., Westbury, N.Y. 11590
Farrand Optical Co., Inc., 117 Wall St., Valhalla, N.Y. 10595
Fermco Laboratories, Div. of G. D. Searle and Co., P.O. Box 5110, Chicago, Ill. 60680
Filtrol Corporation, 3250 E. Washington Blvd., Los Angeles, Ca. 90023
Fish-Schurman Corporation, 70 Portman Rd., New Rochelle, N.Y. 10802
Fisher Scientific Co., 711 Forbes Ave., Pittsburg, Pa. 15219
Floridin Co., 3 Penn Center, Pittsburg, Pa. 15235
Fluka AG, Buchs SG, Switzerland
G. Frederick Smith Chemical Co., 867 McKinley Ave., Columbus, Oh. 43223
G. K. Turner Associates, 2524 Pulgas Ave., Palo Alto, Ca. 94303
Gelman Instrument Co., 600 So Wagner Rd., Ann Arbor, Mi. 48106
General Diagnostics Div., Warner-Chilcott Laboratories, 201 Tabor Rd., Morris Plains, N.J. 07950
Gilford Instrument Laboratories, Inc., 132 Artino St., Oberlin, Oh. 44074
Roger Gilmont Instruments, Inc., 161 Great Neck Rd., Great Neck, N.Y. 11021
R. W. Greef and Co., Inc., 1 Rockefeller Plaza, New York, N.Y. 10020
George T. Gurr, Ltd. U.S. Distr: Bio/medical Specialties, P.O. Box 48641, Los Angeles, Ca., 90048

Table continued

TABLE A–14 (continued)

H. Reeve Angel and Co., 9 Bridewell Pl., Clifton, N.J. 07014

Haake Instrument, Inc., 244 Saddle River Rd., Saddlebrook, N.J. 07662

Hamilton Co., 4960 Energy Way, P.O. Box 7500, Reno, Nv. 89502

W. A. Hammond Drierite Co., P.O. Box 460, Xenia, Oh. 45385

Harleco, Div. of American Hospital Supply, 60th St. and Woodland Ave., Philadelphia, Pa. 19143

Hartman-Leddon Co., see Harleco.

The Heath Co., Scientific Instrument Div., Hilltop Rd., Benton Harbor, Mi. 49022

Helena Laboratories, P.O. Box 752, Beaumont, Tx. 77704

Hoffman-La Roche, Inc., Kingsland Rd., Nutley, N.J. 07110

Hycel, Inc., 7920 Westpark Dr., Houston, Tx. 77042

Hyland Laboratories, Div. of Travenol Laboratories, Inc., 3300 Hyland Ave., Costa Mesa, Ca. 92626

Hynson, Westcott and Dunning, Inc., Charles and Chase Sts., Baltimore, Md. 21201

Instrumentation Laboratory/Harleco, see Instrumentation Laboratory, Inc.

Instrumentation Laboratory, Inc., 113 Hartwell Ave., Lexington, Ma. 02173

International Equipment Co., 300 Second Ave., Needham Heights, Ma. 02194

Ionac Chemical Co., Div. of Sybron Corporation, Birmingham, N.J. 08011

Jarrell Ash, Div. of Fisher Scientific Co., 590 Lincoln St., Waltham, Ma. 02154

Johns Manville Products Corporation, 22 E. 40th St., New York, N.Y. 10016

The Joseph Dixon Crucible Co., Writing Products Div., 167 Wayne St., Jersey City, N.J. 07302

J. T. Baker Chemical Co., 222 Red School Lane, Phillipsburg, N.J. 08865

Kimble Products, Div. Owens-Illinois, Inc., P.O. Box 1035, Toledo, Oh. 43601

Klett Manufacturing Co., Inc., 179 E. 87th St., New York, N.Y. 10028

Knox Gelatin, Inc., Johnstown, N.Y. 12095

Laboratory Computing, Inc., 4915 Monona Dr., Madison, Wi. 53716

Leitz, see E. Leitz.

Libecap Enterprises, 204 S. Haskell Ave., P.O. Box 26406, Dallas, Tx. 75226

LKB Instruments, Inc., 12221 Parklawn Dr., Rockville, Md. 20852

The London Company, 811 Sharon Dr., Cleveland, Oh. 44145

Mallinckrodt Chemical Works, P.O. Box 5439, St. Louis, Mo. 63160

Manostat, 20 No. Moore St., New York, N.Y. 10013

Materials Research Corporation, Route 303, Orangeburg, N.Y. 10962

Matheson, Coleman and Bell, see MC/B Manufacturing Chemists.

Matheson Gas Products Div., Will Ross, Inc., P.O. Box 85, East Rutherford, N.J. 07073

MC/B Manufacturing Chemists, 2909 Highland Ave., Norwood, Oh. 45212

McKesson Chemical Co., 155 E. 44th St., New York, N.Y. 10017

Medical Laboratory Automation, Inc., 520 Nuber Ave., Mt. Vernon, N.Y. 10550

Merck and Co., Inc., 126 E. Lincoln Ave., Rahway, N.J. 07065

Microchemical Specialties Co., 1825 Eastshore Hwy., Berkeley, Ca. 94710

Micromedic Systems, Inc., Independence Mall West, Philadelphia, Pa. 19105

Micrometric Instrument Co., 7777 Exchange, Cleveland, Oh. 44125

Miles Laboratories, Inc., Research Products, Research Division, P.O. Box 272, Kankakee, Il. 60901

Millipore Corporation, Ashby Rd., Bedford, Ma. 01730

Monsanto Chemical Co., 800 N. Lindbergh Blvd., St. Louis, Mo. 63166

Table continued

TABLE A–14 (continued)

National Bureau of Standards, Connecticut Ave. and Van Ness St., N.W., Washington, D.C. 20008

National Instrument Laboratories, Inc., 12300 Parklawn Dr., Rockville, Md. 20852

New England Nuclear, 575 Albany St., Boston, Ma. 02118

Nuclear-Chicago Corporation, 2000 Nuclear Dr., Des Plaines, Il. 60018

Old Monk Co., 718 N. Aberdeen, Chicago, Il. 60607

Olin Chemicals, Olin Corporation, 120 Long Ridge Rd., Stamford, Ct. 06904

Orion Research, Inc., 11 Blackstone St., Cambridge, Ma. 02139

Ortho Pharmaceutical Corporation, U.S. Hiway 202, Raritan, N.J. 08869

Oxford Laboratories, 107 N. Bayshore Blvd., San Mateo, Ca. 94401

Oxo, Ltd., London E. C. 4, England

Packard Instrument Co., Inc., 2200 Warrenville Rd., Downers Grove, Il. 60515

Parke-Davis, and Co., P.O. Box 118GPO, Detroit, Mi. 48232

Perkin-Elmer Corporation, Instrument Div., 800 Main Ave., Norwalk, Ct. 06852

The Permutit Co., Div. of Pfaudler Permutit, Inc., E. 49 Midland Ave., Paramus, N.J. 07652

Pfanstiehl Laboratories, Inc., 1219 Glen Rock Ave., Waukegan, Il. 60085

Pfizer, Inc., Milwaukee Operations, 4215 N. Port Washington Ave., Milwaukee, Wi. 53212

Pharmacia Laboratories, Inc., 800 Centennial Ave., Piscataway, N.J. 08854

Photovolt Corporation, 1115 Broadway, New York, N.Y. 10010

Picker Corporation, Nuclear Division, 333 State St., No. Haven, Ct. 06473

P-L Biochemicals, Inc., 1037 W. McKinley Ave., Milwaukee, Wi. 53205

Precision Sampling Corporation, P.O. Box 15119, Baton Rouge, La. 70815

Precision Scientific Co., 3737 W. Cortland St., Chicago, Il. 60647

Pyrocell Manufacturing Co., Inc., 91 Carver Ave., Westwood, N.J. 07675

Quantitest Chemical Corporation, 61 Moulton St., Cambridge, Ma. 02138

R. P. Cargille Laboratories, Inc., 55 Commerce Rd., Cedar Grove, N.J. 07009

Radiometer-Copenhagen, see The London Company

Reliable Chemical Co., P.O. Box 1099, Maryland Heights, Mo. 63043

Research Products International Corporation, 2692 Delta Lane, Elk Grove Village, Il. 60007

Rohm and Haas Co., Independence Mall West, Philadelphia, Pa. 19105

Schleicher and Schnell, Inc., 543 Washington St., Keene, N.H. 03431

Schoeffel Instrument Corporation, 24 Booker St., Westwood, N.J. 07675

Schwarz/Mann Div., Becton, Dickinson and Co., Mountain View Ave., Orangeburg, N.Y. 10962

Scientific Industries, Inc., 15 Park St., Springfield, Ma. 01103

Scientific Products, Div. of American Hospital Supply Corporation, 1430 Waukegan Rd., McGraw-Park, Il. 60085

Scientific Glass Apparatus Co., Inc., 735 Broad St., Bloomfield, N.J. 07003

Shandon Scientific Co., Inc., 515 Broad St., Sewickley, Pa. 15143

Society of Nuclear Medical Technologists, 1201 Waukegan Rd., Glenview, Il. 60025

E. R. Squibb and Sons, Inc., Georges Rd., New Brunswick, N.J. 08840

Standard Reagents Co., 127 N. 15th St., Philadelphia, Pa. 19102

Steri-Kem, Inc., 671 W. Putnam Dr., Whittier, Ca. 90602

Technicon Instruments Corporation, 511 Benedict Ave., Tarrytown, N.Y. 10591

Table continued

TABLE A–14 (continued)

Turner, see G. K. Turner Associates
Ultra-Violet Products, Inc., 5114 Walnut Grove Ave., San Gabriel, Ca. 91778
Unimetrics Corporation, 1853 Raymond Ave., Anaheim, Ca. 92801
Union Carbide Corporation, Food Products Div., 6733 W. 65th St., Chicago, Il. 60638
Union Carbide Corporation, Research Instruments Div., 270 Park Ave., New York, N.Y. 10017
Vickers Medi Computer Corporation, 581 W. Putnam Ave., Greenwich, Ct. 06830
W. A. Taylor Co., 7300 York Rd., Baltimore, Md. 21204
Waters Associates, Inc., 61 Fountain Ave., Framingham, Ma. 01701
West Co., Inc., West Bridge St., Phoenixville, Pa. 19460
Westvaco, Chemical Div., Carbon Dept., Covington, Va. 24426
Wilson Laboratories, 4221 S. Western Ave., Chicago, Il. 60609
Winthrop Laboratories, 90 Park Ave., New York, N.Y. 10017
Worthington Biochemical Corporation, Halls Mills Rd., Freehold, N.J. 07728

Table A-15. Atomic Weights
Abridged Table

(Values as of 1971, based on $^{12}C = 12$, rounded off to 4 significant figures)

Aluminum	Al	26.98	Magnesium	Mg	24.31
Antimony	Sb	121.8	Manganese	Mn	54.94
Arsenic	As	74.92	Mercury	Hg	200.6
Barium	Ba	137.3	Molybdenum	Mo	95.94
Beryllium	Be	9.012	Nickel	Ni	58.71
Bismuth	Bi	209.0	Nitrogen	N	14.01
Boron	B	10.81	Oxygen	O	16.00
Bromine	Br	79.90	Palladium	Pd	106.4
Cadmium	Cd	112.4	Phosphorus	P	30.97
Calcium	Ca	40.08	Platinum	Pt	195.1
Carbon	C	12.01	Potassium	K	39.10
Cerium	Ce	140.1	Selenium	Se	78.96
Chlorine	Cl	35.45	Silicon	Si	28.09
Chromium	Cr	52.00	Silver	Ag	107.9
Cobalt	Co	58.93	Sodium	Na	22.99
Copper	Cu	63.55	Strontium	Sr	87.62
Fluorine	F	19.00	Sulfur	S	32.06
Gold	Au	197.0	Tellurium	Te	127.6
Helium	He	4.003	Thallium	Tl	204.4
Hydrogen	H	1.008	Tin	Sn	118.7
Iodine	I	126.9	Titanium	Ti	47.90
Iron	Fe	55.85	Tungsten	W	183.8
Lanthanum	La	138.9	Uranium	U	238.0
Lead	Pb	207.2	Vanadium	V	50.94
Lithium	Li	6.941	Zinc	Zn	65.37

Table A-16. Useful Information about Concentrated Acids and Bases

Chemical, reagent grade	Sp gr at room temperature	Percent wt in wt (w/w)	Reagent (g/ml)	Molecular weight	Approximate N in acid-base reactions
H_2SO_4, conc	1.84	95–98	1.76	98.08	36
HNO_3, conc	1.42	69–71	1.0	63.02	16
HCl, conc	1.18	36.5–38	0.44	36.46	12
CH_3COOH, glacial	1.05	99–100	1.04	60.05	17.5
H_3PO_4, ortho	1.7	85	1.45	98.00	15, 30, 45 (depending on reaction)
Lactic acid	1.2	85	1.02	90.08	11
$HClO_4$, 60%	1.53	60–62	0.92	100.46	9
71%	1.68	70–72	1.19	100.46	12
NH_4OH, conc	0.90	28–30 (as NH_3)	0.25 (as NH_3)	35.05 (NH_4OH) 17.03 (NH_3)	14.8
NaOH, saturated solution	1.5	50	0.75	40.00	19

INDEX

75 76 77 78   9 8 7 6 5 4 3